ANNUAL REVIEW
OF NEUROSCIENCE

VOLUME 24, 2001

W. MAXWELL COWAN, *Editor*
Howard Hughes Medical Institute

STEVEN E. HYMAN, *Associate Editor*
National Institute of Mental Health

ERIC M. SHOOTER, *Associate Editor*
Stanford University School of Medicine

CHARLES F. STEVENS, *Associate Editor*
Salk Institute for Biological Studies

www.AnnualReviews.org   science@AnnualReviews.org   650-493-4400

ANNUAL REVIEWS
4139 El Camino Way • P.O. BOX 10139 • Palo Alto, California 94303-0139

ANNUAL REVIEWS
Palo Alto, California, USA

International Standard Serial Number: 0147-006X
International Standard Book Number: 0-8243-2424-2

Annual Review and publication titles are registered trademarks of Annual Reviews.

⊗ The paper used in this publication meets the minimum requirements of American
National Standards for Information Sciences—Permanence of Paper for Printed Library
Materials, ANSI Z39.48-1992.

Annual Reviews and the Editors of its publications assume no responsibility for the statements expressed by the contributors to this *Annual Review*.

Typeset by TechBooks, Fairfax, VA
Printed and Bound in the United States of America

Annual Review of Neuroscience
Volume 24, 2001

CONTENTS

ERRATA
An online log of corrections to *Annual Review of Neuroscience* chapters
(if any have yet been occasioned, 1997 to the present) may be found
at http://neuro.AnnualReviews.org/errata.shtml

RELATED ARTICLES

From the *Annual Review of Biochemistry*, Volume 69 (2000)

Cryptochrome: The Second Photoactive Pigment in the Eye and its Role in Circadian Photoreception, Aziz Sancar

From the *Annual Review of Biomedical Engineering*, Volume 2 (2000)

Bioengineering Models of Cell Signaling, Anand R. Asthagiri and Douglas A. Lauffenburger

New Currents in Electrical Stimulation of Excitable Tissues, Peter J. Basser and Bradley J. Roth

Two-Photon Excitation Fluorescence Microscopy, Peter T. C. So, Chen Y. Dong, Barry R. Masters, and Keith M. Berland

Magnetic Resonance Studies of Brain Function and Neurochemistry, Kâmil Ugurbil, Gregor Adriany, Peter Andersen, Wei Chen, Rolf Gruetter, Xiaoping Hu, Hellmut Merkle, Dae-Shik Kim, Seong-Gi Kim, John Strupp, Xiao Hong Zhu, and Seiji Ogawa

From the *Annual Review of Cell and Developmental Biology*, Volume 16 (2000)

Mechanisms of Synaptic Vesicle Exocytosis, Richard C. Lin and Richard H. Scheller

Structural Basis for the Interaction of Tubulin with Proteins and Drugs that Affect Microtubule Dynamics, Kenneth H. Downing

Cajal Bodies: The First 100 Years, Joseph G. Gall

Dynamin and its Role in Membrane Fission, J. E. Hinshaw

Structure and Regulation of Voltage-Gated Ca^{2+} Channels, William A. Catterall

From the *Annual Review of Genetics*, Volume 34 (2000)

Genetics of the Mammalian Circadian System: Photic Entrainment, Circadian Pacemaker Mechanisms, and Posttranslational Regulation, Phillip L. Lowrey and Joseph S. Takahashi

From the *Annual Review of Medicine*, Volume 52 (2001)

Effects of Neuropeptides and Leptin on Nutrient Partitioning: Dysregulations in Obesity, Bernard Jeanrenaud and Françoise Rohner-Jeanrenaud

Annu. Rev. Neurosci. 2001. 24:1–29

PDZ Domains and the Organization of Supramolecular Complexes

Morgan Sheng and Carlo Sala

Howard Hughes Medical Institute and Department of Neurobiology, Massachusetts General Hospital and Harvard Medical School, Boston, Massachusetts 02114; e-mail: sheng@helix.mgh.harvard.edu, sala@helix.mgh.harvard.edu

Key Words   protein targeting, scaffold protein, postsynaptic density, INAD, PSD-95

■ **Abstract**   PDZ domains are modular protein interaction domains that bind in a sequence-specific fashion to short C-terminal peptides or internal peptides that fold in a β-finger. The diversity of PDZ binding specificities can be explained by variable amino acids lining the peptide-binding groove of the PDZ domain. Abundantly represented in *Caenorhabditis elegans, Drosophila melanogaster*, and mammalian genomes, PDZ domains are frequently found in multiple copies or are associated with other protein-binding motifs in multidomain scaffold proteins. PDZ-containing proteins are typically involved in the assembly of supramolecular complexes that perform localized signaling functions at particular subcellular locations. Organization around a PDZ-based scaffold allows the stable localization of interacting proteins and enhances the rate and fidelity of signal transduction within the complex. Some PDZ-containing proteins are more dynamically regulated in distribution and may also be involved in the trafficking of interacting proteins within the cell.

INTRODUCTION

First recognized as sequence repeats approximately 10 years ago, PDZ domains have emerged in the past few years as important modular protein interaction domains found in a wide variety of eukaryotic proteins. The sequence-specificity of PDZ domain interactions has been a source of confusion and misunderstanding, so this area is covered in some detail in the first half of this review. In the second half, we focus on the function of PDZ domains in the assembly of large protein complexes and in subcellular targeting of proteins, as illustrated by several examples that are particularly pertinent to neuroscience. Indeed, the word complex seems inadequate to reflect the scope and complexity of PDZ-based structures, hence the use of "supramolecular complex" in the title of this review.

0147-006X/01/0621-0001$14.00

PDZ DOMAINS

Prevalence of PDZ Domains

PDZ domains were originally recognized as ~90 amino acid–long repeated sequences in the synaptic protein PSD-95/SAP90 (PSD for postsynaptic density the Drosophila septate junction protein Discs-large, and the epithelial tight junction protein ZO-1 (hence the acronym PDZ). They have also been termed Discs-large homology regions or GLGF repeats (after a signature Gly-Leu-Gly-Phe sequence found in the PDZ domain). PSD-95, Discs-large, and ZO-1 belong to a superfamily of proteins dubbed membrane-associated guanylate kinases (MAGUKs), which contain an SH3 domain and a guanylate kinase-like (GK) domain in addition to one or more PDZ domains (Figure 1, see color insert). Although first noted in MAGUKs, PDZ domains are present in a wide variety of proteins in diverse organisms, including yeast (Ponting 1997) (Figure 1). They are among the most common protein domains: 60 and 69 PDZ-containing proteins are encoded in the *Caenorhabditis elegans* and *Drosophila melanogaster* genomes, respectively. In addition to being widespread, PDZ domains show considerable sequence variation, which presumably reflects their diversity of binding specificities and functional roles.

PDZ Domains as Protein Interaction Modules

A key function of PDZ domains came to light when the N-terminal two PDZ domains of PSD-95 were shown to bind the specific peptide motif (-E-T/S-D/E-V) found at the very C terminus of Shaker-type K^+channels (Kim et al 1995) and NMDA receptor NR2 subunits (Kornau et al 1995, Niethammer et al 1996). Around the same time, a PDZ domain of the protein tyrosine phosphatase PTP1E/FAP1 was found to bind the C terminus of the cell surface receptor Fas (Sato et al 1995). These discoveries defined the primary role of PDZ domains as protein interaction domains that bind in a sequence-specific manner to C-terminal peptides. Moreover, they established that PDZ domains were modular, in that they could be transplanted into heterologous proteins while retaining their activity of binding to C-terminal peptides.

Although C-terminal peptides are the typical ligands, PDZ domains can participate in other modes of protein interaction. In particular PDZ domains can bind to internal peptide sequences that bend in a β-hairpin structure (Hillier et al 1999) and to other PDZ domains.

STRUCTURE AND SPECIFICITY OF PDZ DOMAINS

The three-dimensional structure of the PDZ domain, in some cases complexed with a specific ligand, has been solved for several PDZ domains: the third PDZ domain (PDZ3) of PSD-95 (Doyle et al 1996, Morais Cabral et al 1996); PDZ2 of PSD-95 (Tochio et al 2000); the PDZ domain of CASK (Daniels et al 1998);

the PDZ domain of syntrophin (Schultz et al 1998); the PDZ domain of neuronal nitric oxide synthase (nNOS) (Hillier et al 1999, Tochio et al 1999); and PDZ2 from tyrosine phosphatase PTP1E (Kozlov et al 2000).

PDZ domains have similar architectures consisting of six antiparallel β-strands and two α-helices; their overall fold approximates a six-strand β-sandwich flanked by two α-helices (Figure 2, see color insert). Peptide binding occurs in a groove between the βB strand and the αB helix. The N and C termini of the PDZ domain lie close to each other on the opposite side of the domain relative to the peptide-binding groove, an arrangement common to protein interaction modules.

Peptide-Binding Groove

Each PDZ domain binds a single peptide ligand. The peptide binds in a groove between βB and αB on the surface of the domain; but the side chain of the C-terminal residue (defined here as position 0) dips into a hydrophobic cavity formed by conserved amino acid residues distributed throughout the primary sequence of the domain (Figure 2). In the first cocrystal of a PDZ domain (PDZ3 of PSD-95) complexed with its ligand (CRIPT), the main chain of the last four amino acids of the peptide ligand were highly ordered (Doyle et al 1996). This was important evidence that PDZs recognize more than just the terminal three residues, as is often suggested by sequence comparisons of PDZ ligands. The main chain of the βB strand runs the length of the peptide-binding groove and provides important interactions with the main chain of the peptide. Via these main-chain interactions, the peptide ligand is stitched as an additional strand into the antiparallel β-sheet on the surface of the PDZ domain. Although contributing to the affinity of binding, the main-chain interactions between βB and the peptide ligand cannot be responsible for sequence-specific recognition by PDZs.

Why Do PDZs Bind Selectively to C-Terminal Peptides?

The architecture of the PDZ domain is designed for binding to a free carboxylate group at the end of the peptide. The carboxylate-binding loop lies between the βA and βB strands, extending from a highly conserved arginine or lysine residue (Arg 318 in PDZ3 of PSD-95) to the signature Gly-Leu-Gly-Phe (GLGF) motif. Three main-chain amide protons of the GLGF motif form hydrogen bonds with the terminal carboxylate of the peptide. In addition, the guanidinium group of Arg-318 interacts with the carboxylate anion via a water molecule (Doyle et al 1996). Since a free carboxylate group occurs only at the very C terminus of the peptide main chain, the interactions between the carboxylate-binding loop and the carboxylate oxygens form the structural basis for PDZ recognition of C-terminal peptides. The carboxylate-binding loop (R/K-XXX-GLGF) is highly conserved among PDZ domains. The second and fourth residues of the GLGF motif are invariably hydrophobic. The second of the two glycines is absolutely conserved, but a serine, threonine, or proline replaces the first glycine in a minority of PDZs.

Specificity for a Hydrophobic Residue at the C Terminus   Binding of the carboxylate-binding loop to the terminal carboxylate group orients the 0 position residue such that its side chain projects into the hydrophobic pocket of the PDZ domain. The amino acids of the PDZ domain lining the pocket determine the nature of the C-terminal residue selected. In PDZ3 of PSD-95, valine is the preferred 0 position residue (Niethammer et al 1998). The residues lining the pocket in diverse PDZ domains are united in their hydrophobic character; this presumably accounts for the generalization that most (if not all) PDZ domains select for peptides with a hydrophobic C-terminal residue. Variations in the size and geometry of the hydrophobic pocket presumably account for the differential preference of various PDZ domains for valine, leucine, isoleucine, phenylalanine, or alanine at the very end of peptide ligands (Songyang et al 1997). However, selectivity by hydrophobic interactions is usually not highly stringent, and this could account for the variation of C-terminal residue specified by individual PDZs. For instance, the PDZ domains of PSD-95 can also bind to sequences ending with isoleucine or leucine, although they prefer valine-terminating peptides (Cohen et al 1996, Songyang et al 1997).

Recognition of the −2 Position

PDZ domains generally show greater specificity for amino acids upstream of the 0 position in peptide ligands. The −2 residue of the ligand is a particularly important determinant for PDZ binding and forms the basis for classification of PDZ specificity (Table 1). The PDZ domains of PSD-95 and of many other proteins select for a threonine or serine at the −2 position (these are defined as class I PDZ domains) (Table 1). In the crystal structure of PDZ3 of PSD-95 complexed with CRIPT (last four residues −QTSV), the hydroxyl group of the −2 threonine of the ligand forms a strong hydrogen bond with the N-3 nitrogen of His-372, the first residue of helix αB (position αB1) of the PDZ domain (Figure 2). This interaction would also accommodate the hydroxyl group of serine, thereby rationalizing the specific recognition of threonine or serine in the −2 position. In keeping with this, histidine is conserved at the αB1 position in class I PDZ domains. Significantly, the nature of the αB1 residue differs in PDZ domains with distinct specificities for the −2 position and is a good predictor of the −2 amino acid in preferred peptide ligands. For instance, the nNOS PDZ, a class III domain that selects for negatively charged amino acids at the −2 position of C-terminal peptide ligands (Table 1) (Stricker et al 1997), has tyrosine at the αB1 position. The hydroxyl group of the αB1 tyrosine makes a hydrogen bond with the carboxyl group of the −2 aspartate in the peptide ligand (Tochio et al 1999). On the other hand, hydrophobic residues at the αB1 position determine the preference of class II PDZs for a hydrophobic residue in the −2 position (Table 1). Examples among many include the LIN-2/CASK and p55 PDZ domains, which contain valine at αB1 and which select for phenylalanine or tyrosine at the −2 position of peptide ligands (Songyang et al 1997).

TABLE 1   Classification of PDZ domains according to specificity for C-terminal peptides[a]

Class	C-term sequence	Interacting protein	PDZ domain	References
Class I				
X-S/T-X-V	E-S-D-V	NMDAR2A, B	PSD-95 (PDZ2)	Kornau et al 1995
	E-T-D-V	Shaker channel		Kim et al 1995
	Q-S-S-V	Citron	PSD-95 (PDZ3)	Zhang et al 1999
	Q-T-S-V	CRIPT		Niethammer et al 1998
	T-T-R-V	Neuroligin		Irie et al 1997
	E-T-S-V	PMCA4b	PSD-95 (PDZ1/2/3)	E Kim et al 1998
	E-S-L-V	Voltage-gated sodium channel	Syntrophin	Gee et al 1998
X-S/T-X-L	Q-T-R-L	GKAP	Shank	Naisbitt et al 1999
	S-S-T-L	mGluR5		Tu et al 1999
	D-S-S-L	β2-adrenergic receptor	NHERF (PDZ1)	Hall et al 1998
	P-T-R-L	GRK6A		Hall et al 1999
	D-T-R-L	CFTR		Wang et al 1998
Class II				
X-ϕ-X-ϕ	E-Y-Y-V	Neurexin	CASK	Hata et al 1996
	E-F-Y-A	Syndecan	CASK, syntenin	Hsueh et al 1998
	E-Y-F-I	Glycophorin C	p55	Grootjans et al 1997; Marfatia et al 1997
	S-V-K-I	GluR2	GRIP (PDZ5), PICK-1	Dong et al 1997 Xia et al 1998
	S-V-E-V	EphB2	GRIP (PDZ6), PICK-1	Torres et al 1998
	G-I-Q-V	EphA7	GRIP (PDZ6), PICK-1 syntenin	
	Y-Y-K-V	EphrinB1	GRIP (PDZ6), PICK-1 Syntenin	
Class III				
X-D-X-V	V-D-S-V	Melatonin receptor	nNOS	Stricker et al 1997

[a]This table illustrates the range of PDZ binding specificities but is not comprehensive. Left column shows consensus sequence recognized by each class of PDZ domain. Only the last four amino acids of the interacting proteins are shown, although PDZ specificity may sometimes involve more than these residues (see text). ϕ, Hydrophobic amino acid; X, unspecified amino acid.

Recognition of the −3 Position

An important conclusion of structural studies was that the amino acid at the −3 position of the peptide ligand also makes specific contact with the PDZ domain. The −3 glutamine of CRIPT forms hydrogen bonds via its side chain oxygen with PDZ residues Asn-326 (βB2) and Ser-339 (βC4) in PSD-95 (Doyle et al 1996)

(Figure 2). A glutamate residue could participate in the same interactions, and indeed, a glutamate at the -3 position is specifically enriched from peptide libraries screened with PDZ domains of PSD-95 (Songyang et al 1997). A -3 glutamate is found commonly in natural ligands of PSD-95's PDZ1 and PDZ2 domains, e.g. NR2 subunits of the NMDA receptor ($-$ESDV) and Shaker potassium channels ($-$ETDV). Although it does not directly contact the -3 residue in the PDZ3 crystal, the βC5 residue of the PDZ domain (which is phenylalanine in PDZ3 and lysine in PDZ1 and PDZ2 of PSD-95) may be involved in selectivity for the -3 position (Songyang et al 1997). In general, the -3 amino acid in the peptide ligand is less stringently specified by individual PDZ domains than is the residue at the -2 position (Songyang et al 1997).

Recognition of the -1 Position

The side chain of the -1 residue of CRIPT (serine) showed no contacts with the PDZ3 domain of PSD-95 (Doyle et al 1996), which suggests that the -1 residue plays no role in PDZ specificity. This structural conclusion fails to explain why the -1 position is influential in PDZ binding in other studies. For instance, substitution of the -1 position of CRIPT from $-$QT$\underline{S}$V to $-$QT$\underline{D}$V converted the CRIPT peptide from a PDZ3-preferring to a PDZ2-preferring ligand (Niethammer et al 1998). Similarly, using a peptide library screening approach, most PDZ domains selected for specific amino acids at the -1 position, albeit not as strongly as at the -2 (Songyang et al 1997, Stricker et al 1997). Thus there is little doubt that many PDZ domains can discriminate between amino acids at the -1 position of the C-terminal peptide, but the structural basis of this specificity is uncertain. Mutational analysis of PDZ3 of PSD-95 implicates residue βB2 of the PDZ domain in the specification of the -1 position of the peptide ligand (Niethammer et al 1998). The βC5 residue may also play a role in selection of the -1 side chain (Songyang et al 1997).

Four or More C-Terminal Residues Important for PDZ Binding

Consensus sequences for PDZ binding are often written as a C-terminal tripeptide, for instance the "T/SXV" motif for binding to PSD-95's PDZ domains (where X is any amino acid) (Kornau et al 1995). Based on structural studies, however, it is clear that PDZ domains interact specifically with at least a C-terminal tetrapeptide. It would therefore be more accurate to describe PDZ-binding consensus sequences in terms of four amino acids, such as E/Q-T/S-X-V for PSD-95 (Doyle et al 1996, Songyang et al 1997). However, even a tetrapeptide consensus sequence probably underestimates the specificity of PDZ binding. CRIPT requires more than six C-terminal residues to bind optimally to PDZ3 of PSD-95 (Niethammer et al 1998). Nuclear magnetic resonance studies suggest that as many as six residues of the Fas C terminus interact with PDZ2 of PTP1E (Kozlov et al 2000). Peptide library screens reveal differential selection by some PDZ domains for specific amino acids as far back as the -8 position (Songyang et al 1997). The interaction between muscle sodium channel and the syntrophin PDZ domain also involves residues N-terminal of the C-terminal tetrapeptide (Gee et al 1998, Schultz et al 1998). How PDZ domains recognize amino acids upstream of the -3 position is not clear, but it probably involves the loop between βB and βC, the αB helix, and the loop between αB and βE (Kozlov et al 2000, Tochio et al 2000).

PDZ domains vary in the degree of their specificity of peptide binding, i.e. some PDZs have more stringent sequence requirements than others. An extreme example of this seems to be the PDZ domain of PICK-1, which was originally identified as a binding site for protein kinase C (PKC) (C-terminal sequence $-$QSAV) (Staudinger et al 1997). Subsequently, the PICK-1 PDZ domain was also shown to bind specifically the C terminus of AMPA receptor GluR2 subunit (C-terminal sequence $-$SVKI) (Xia et al 1999). Thus PICK-1's PDZ (which has a lysine residue at the $\alpha B1$) falls into both class I and class II in terms of its preference for the -2 amino acid.

Affinity of PDZ Binding

A wide range of binding affinities (K_ds from 10^{-8} M to 10^{-6} M) have been measured for various PDZ domain–C-terminal peptide interactions. Part of this variation undoubtedly reflects real differences in optimal binding affinity between various PDZs; in addition, different peptide ligands probably have different affinities for a given PDZ domain. Another source of variation is the diversity of in vitro binding assays that have been used to measure K_ds of PDZ domains. Using surface plasmon resonance (Biacore), the affinity of PDZ2 of mDlg (a close relative of PSD-95) for its optimal peptide ligand was 42 nM, whereas that of PDZ3 of PTP1E was 154 nM (Songyang et al 1997). By comparison, an apparent affinity of 10^{-8} M was obtained by an ELISA-based assay for the interaction of PDZ2 of SAP102 (another relative of PSD-95) with the last nine amino acids of its natural ligand NR2B (Müller et al 1996). Using the in-solution method of fluorescence polarization, the affinity of CRIPT for PDZ3 of PSD-95 was determined to be $\sim$1 μM (Niethammer et al 1998).

PDZ Binding to Internal Sequences

A less common mode of PDZ domain binding is with internal rather than C-terminal peptide sequences. This has been best characterized for the interaction between neuronal nitric oxide synthase (nNOS) and the PDZ domain of PSD-95 or syntrophin (Brenman et al 1996a,b). The region of nNOS required for binding to the PDZ domain of PSD-95 includes the canonical PDZ domain of nNOS plus an additional $\sim$30 amino acids flanking the C-terminal side. Because the tertiary structures of both PDZ domains are required, this interaction has been considered a PDZ-PDZ interaction. Crystallographic studies of the nNOS-syntrophin complex revealed that the interaction is mediated primarily by contacts between the PDZ domain of syntrophin and a nonterminal hairpin turn ("β-finger") in the

PDZ-flanking region of nNOS (Hillier et al 1999). The first strand of the nNOS β-finger (sequence $-$ETTF$-$) mimics a canonical C-terminal peptide in its sequence-specific interactions with the peptide-binding groove of syntrophin's PDZ (which can also bind to conventional C-terminal peptides) (Gee et al 1998). The C-terminal free carboxylate group normally required for PDZ binding is replaced by the sharp β-turn at the tip of the β-finger. The second strand of the β-finger adds another antiparallel strand to the β-sheet on the surface of the PDZ domain. The PDZ domain of nNOS is important because it makes extensive contacts with the β-finger and appears to be required for its stabilization and presentation (Hillier et al 1999, Tochio et al 1999). In addition, the βA strand of the nNOS PDZ domain packs against the syntrophin αB helix, thereby increasing the buried surface area on the syntrophin PDZ domain ($\sim$800 A^2 compared with $\sim$400 A^2 for a conventional C-terminal peptide interaction) (Hillier et al 1999). It is interesting that the β-finger projects from the "backside" of the nNOS PDZ, so that binding of the β-finger to syntrophin's PDZ domain still leaves the peptide-binding groove of the nNOS PDZ free to engage its own ligands (Hillier et al 1999, Tochio et al 1999).

Other examples of PDZ interactions with internal peptide sequences have been reported (see Hillier et al 1999 for references), but these internal peptide sequences are not typically adjacent to PDZ domains (as is the nNOS β-finger). Thus the prevalence of PDZ–β-finger or PDZ-PDZ interactions remains to be determined.

REGULATION OF PDZ-PEPTIDE INTERACTION BY PHOSPHORYLATION

It is probably not a coincidence that the -2 residue of PDZ-binding C-terminal peptides (arguably the most critical residue for PDZ recognition) is frequently a phosphorylable amino acid such as threonine, serine, or tyrosine (Table 1). The -2 serine of inward rectifier K$^+$ channel Kir2.3 falls within a consensus sequence for protein kinase A (PKA); phosphorylation of this site by PKA abolishes Kir2.3 interaction with PSD-95's PDZ domains (Cohen et al 1996). Phosphorylation of the -2 serine of the β2-adrenergic receptor (C-terminal sequence $-$DSSL) by G-protein–coupled receptor kinase GRK5 disrupts receptor binding to the PDZ domain of NHERF (Cao et al 1999, Hall et al 1998).

Phosphorylatable residues need not be at the -2 position to affect PDZ binding. For instance, the -3 serine of the AMPA receptor subunit GluR2 C terminus ($-$SVKI) can be phosphorylated by PKC, and this modification prevents GluR2 binding to the PDZ domain protein GRIP (Matsuda et al 1999) (RL Huganir, personal communication). Phosphorylation of residues near the C terminus is likely to be a common mechanism for regulating PDZ interaction; these modifications may be catalyzed by known broad-spectrum protein kinases (such as PKA or PKC) or by yet-to-be-characterized kinases that specifically phosphorylate C-terminal sequences.

MULTIMERIZATION AND LINKAGE OF PDZ DOMAINS

PDZ-containing proteins often self-associate to form multimers, which is of obvious utility in assembly of macromolecular complexes. PDZ proteins can multimerize via PDZ-independent mechanisms, e.g. PSD-95 multimerizes by its N-terminal region (Hsueh et al 1997). However, it appears that some PDZ domains themselves have the propensity to undergo multimerization. For instance, GRIP and ABP (containing six or seven PDZs and no other recognizable domain) form homomultimers and heteromultimers via association of their PDZ4-6 domains (Dong et al 1999, Srivastava et al 1998). INAD (for inactivation-no-afterpotential-D) (a protein containing five PDZs) multimerizes via PDZ3 and PDZ4, apparently without interfering with PDZ-ligand binding (Xu et al 1998). Thus distinct surfaces of the PDZ domain may be used for multimerization and peptide binding. This mode of interaction differs from the PDZ–β-finger–PDZ interaction involved in nNOS association with syntrophin or PSD-95, which is competitively inhibited by the peptide ligand (Brenman et al 1996a).

Although modular in design, PDZ domains are often linked together in groups within a multi-PDZ protein. For instance, the first two PDZ domains in the PSD-95 subfamily of MAGUKs are immediately adjacent to each other and form a protease-resistant unit (Lue et al 1996). In the GRIP family of proteins, PDZ1-3 and PDZ4-6 are consistently grouped together. The functional significance of such linkages of PDZ domains is uncertain. We have found that the combined PDZ1-2 of PSD-95 shows significantly different binding specificity for C-terminal sequences than do the individual PDZ1 and PDZ2 domains (M Niethammer, M Sheng, unpublished observations).

PDZ DOMAINS IN THE ORGANIZATION OF SIGNALING COMPLEXES AND MEMBRANE MICRODOMAINS

In recent years, it has become clear that a major function of PDZ-containing proteins is to act as scaffolds for the assembly of large protein complexes at specific subcellular locations, particularly at the cell surface. Modular PDZ domains are well suited to such a function because they easily can be strung together to form multi-PDZ proteins (e.g. INAD, GRIP) or combined with other modular protein interaction domains to generate more complex scaffolds (e.g. CASK/LIN-2, PSD-95) (Figure 1, see color insert). Because they recognize just a few amino acids at the C-termini of proteins, PDZ domains are able to interact with the great majority of transmembrane proteins that have their C termini facing the cytoplasm, even when these cytoplasmic tails are short. Hence PDZ proteins are well suited for functions at the membrane. Because individual PDZ domains have different binding specificities, the combination of PDZs within a scaffold protein would determine the composition of the protein complex assembled around the scaffold.

The propensity of PDZ-containing proteins to multimerize would increase the size and potentially the heterogeneity of the PDZ-based complex.

Here, we focus on some of the best-characterized supramolecular complexes that are organized around PDZ proteins in neurons: INAD, PSD-95, LIN-2/LIN-7/LIN-10, and GRIP. Several other PDZ-proteins implicated in the organization of cell junctions, signal transduction, and protein sorting have been recently reviewed, including NHERF/EBP50 (Minkoff et al 1999), the erythrocyte membrane MAGUK p55 (Chishti 1998), and the tight junction MAGUKs ZO-1/ZO-2/ZO-3 (Mitic & Anderson 1998).

ORGANIZATION OF A PHOTOTRANSDUCTION SIGNALING CASCADE BY INAD

The Phototransduction Cascade in Rhabdomeres

The phototransduction cascade in the Drosophila eye provides a compelling example of a signaling complex organized around a PDZ scaffold at a specific subcellular location (reviewed in Montell 1998, 1999; Scott & Zuker 1998a; Tsunoda & Zuker 1999).

Light-induced isomerization of rhodopsin activates a heterotrimeric G protein of the Gq family whose effector is phospholipase C (PLC) (encoded by the *norpA* gene). PLC catalyzes the conversion of phosphatidylinositol-4,5-bisphosphate to inositol-1,3,5-triphosphate and diacylglycerol and is required for the photoresponse. Activation of PLC leads to opening of light-responsive cation channels termed transient receptor potential (TRP) and TRP-like (TRPL). The mechanism of TRP gating by PLC is controversial, but the end result is depolarization of the photoreceptor cell. The deactivation of the light response is a calcium-dependent process involving an eye-specific PKC, calmodulin (CaM), and NINAC (an unconventional myosin III). Both activation ($\sim$20 ms) and deactivation of the light response (less than 100 ms) are extremely rapid, which suggests that the components of the transduction machinery are closely linked. Indeed, phototransduction in the Drosophila eye occurs in rhabdomeres, specialized microvillar structures that project from the photoreceptor cell surface (Figure 3). The signaling proteins required for phototransduction (rhodopsin, PLC, PKC, TRP, TRPL), are spatially confined to the rhabdomere. Within the rhabdomere, phototransduction proteins are further organized into a protein complex by INAD, a protein containing five PDZ domains (Figure 1).

Protein Interactions of INAD

All five PDZ domains of INAD have been shown to interact with various phototransduction proteins, thus showcasing the scaffolding function of INAD in this signaling pathway (Figure 3). The first protein demonstrated to bind to INAD was the TRP channel (Huber et al 1996a, Shieh & Zhu 1996). TRP binding is mediated by PDZ3 of INAD and apparently an internal sequence in the C-terminal cytoplasmic tail of TRP; this interaction is lost in the original *inaD* mutant (*inaD^{P215}*), which contains a mutation in PDZ3 (Chesevich et al 1997, Shieh & Zhu 1996, Tsunoda et al 1997). PLC binds to two separate PDZ domains of INAD using different mechanisms: to PDZ1 with its C terminus (last four amino acids −EFCA), and to PDZ5 via an internal sequence (Tsunoda et al 1997, van Huizen et al 1998, Xu et al 1998). The region of PLC that binds to PDZ5 overlaps with the region that interacts with the G protein, which suggests a possible competition between INAD and the activated G protein (van Huizen et al 1998). Eye-PKC can bind to both PDZ2 and PDZ4 of INAD; the C terminus of PKC (-ITII, a type I PDZ binding motif) is critical for the interaction (Adamski et al 1998, Huber et al 1996b, Tsunoda et al 1997, Xu et al 1998). NINAC binds to PDZ1 via its C terminus (Wes et al 1999), whereas CaM shows affinity for a region between PDZ1 and PDZ2 (Xu et al 1998) (Figure 3).

Rhodopsin and TRPL are also reported to interact with INAD, directly or indirectly, via PDZ3 and/or PDZ4 (Xu et al 1998). However, these results are disputed by others (see Tsunoda & Zuker 1999). It is possible that rhodopsin and TRPL interactions with INAD are low affinity or dynamically regulated (Montell 1998). An example of a transient interaction with the INAD complex is the recent finding that activated visual Gαq subunit associates with INAD, apparently via binding to PLC (Bahner et al 2000).

Thus molecular analysis has identified multiple binding partners for INAD among the proteins of the phototransduction pathway (Figure 3, see color insert). TRP, PKC, and PLC are generally considered to be stoichiometric core components of the INAD complex, whereas the association of rhodopsin, TRPL, CaM, NINAC, and Gαq with INAD is more dynamic and/or controversial. All these interactions (except CaM) occur at PDZ domains and are mediated by classical C-terminal as well as internal peptide interactions. The seductive idea that each PDZ domain of INAD might bind to one specific phototransduction protein was not borne out. Nevertheless, the notion of a comprehensive signaling pathway built around INAD remains feasible because INAD multimerizes via its PDZ3 and PDZ4 domains (Xu et al 1998), thereby increasing the potential binding capacity of the INAD scaffold.

Function of INAD

What is the physiological role of INAD and its protein-protein interactions? At the cell biological level, INAD functions to localize its major interacting proteins to the rhabdomere. This is established for the three core proteins of the INAD complex: TRP, PLC, and PKC. Mutations that prevent the binding of any of these proteins to INAD result in the loss of that protein from rhabdomeres. In the null mutant *inaD1*, TRP, PLC, and PKC are all mislocalized (Tsunoda et al 1997). In *inaD^{P215}* mutants, which contain a PDZ3 mutation, TRP is diffusely distributed throughout the plasma membrane and intracellular compartments (Chesevich et al

1997, Tsunoda et al 1997). In the *inaD2* mutant, which affects PDZ5, just PLC is specifically mislocalized (Tsunoda et al 1997). Truncating the C terminus of PLC (which binds PDZ1 of INAD) also results in abnormal distribution of PLC (van Huizen et al 1998). Together, these studies provide compelling in vivo evidence for the subcellular targeting function of INAD, at the same time highlighting the independent modular nature of its PDZ-mediated interactions. By contrast, the rhabdomere targeting of TRPL and NINAC does not depend on INAD (Tsunoda et al 1997, Wes et al 1999). Loss of INAD binding and rhabdomere localization is also correlated with destabilization of TRP, PLC, and PKC (Tsunoda et al 1997).

The targeting function of INAD, however, probably reflects a more fundamental role, which is to organize PLC, TRP, PKC, NINAC, and perhaps other molecules into a tightly coupled signaling unit (Figure 3). In a general sense, binding to a common INAD scaffold would bring the visual signaling proteins into close mutual proximity, isolate them from other influences, and increase the rates of interaction between each component. Indeed, the original *inaD* mutant ($inaD^{P215}$) was isolated by virtue of an abnormal electroretinogram reflecting aberrant kinetics of light-induced currents (Pak 1979, Shieh & Niemeyer 1995). The electrophysiological phenotype appears to be due to increased latency of activation of the light response (Tsunoda et al 1997). Subsequent genetic analysis has confirmed that INAD is important for the efficiency and rapid kinetics of activation of the photoresponse, as evidenced by the diminished amplitude and increased latency of response in *inaD* null mutants and in mutants that prevent the INAD-PLC interaction (Scott & Zuker 1998b, Shieh et al 1997, Tsunoda et al 1997, van Huizen et al 1998). Although the mechanism of gating of TRP channels remains unclear, it is tempting to speculate that INAD contributes to the rapidity of the activation process by allowing the second messengers inositol-1,3,5-triphosphate/diacylglycerol to act in the immediate vicinity of their production by PLC. Alternatively, because INAD binds directly to TRP, it may play a more direct role in the gating of the channel.

INAD also plays a role in the termination of the photoresponse, a Ca^{2+}-dependent feedback mechanism involving CaM and PKC. Mutations that are null for INAD or that interfere with its interactions with TRP, PLC, and PKC show abnormal deactivation of the light response (Adamski et al 1998, Shieh & Zhu 1996, Shieh et al 1997, Tsunoda et al 1997). One way that INAD can ensure a rapid, sensitive, and localized negative feedback response is by assembling the Ca^{2+}-regulated molecules CaM and PKC in the proximity of Ca^{2+}-permeant TRP channels. In essence then, INAD serves as a scaffold to couple TRP channels physically and functionally to both activation and deactivation mechanisms. Speed, specificity, sensitivity, and feedback regulation are optimized through the creation of supramolecular microdomains constructed around INAD. Such an organized protein assembly, however, sacrifices some of the amplification that occurs when components of the pathway are able to diffuse and interact with multiple downstream effectors (Scott & Zuker 1998b).

In INAD mutants, the altered kinetics of the photoresponse are usually associated with the loss of rhabdomere localization of specific INAD-binding proteins.

The subcellular targeting and signal coupling functions of INAD seem to be dissociated in the case of NINAC. A C-terminal mutation in NINAC that disrupts its interaction with INAD results in a delayed termination of the light response but does not affect the rhabdomere localization of NINAC (Wes et al 1999). In this instance then, INAD's primary function appears to be the coupling of NINAC to the phototransduction mechanism rather than the concentration of NINAC in the rhabdomere. As an atypical myosin, NINAC is the only putative cytoskeleton-associated protein thought to interact with INAD, and its molecular function in phototransduction is unclear. It remains to be determined how, or whether, the INAD complex is anchored to the actin cytoskeleton in the microvilli of rhabdomeres.

INAD represents the best example to date of a PDZ protein whose interactions mediate the subcellular localization and functional coupling of a signaling cascade. This gave rise to the influential "transducisome" model of the INAD protein complex (Tsunoda et al 1997, Tsunoda & Zuker 1999). Genetic evidence indicates that the INAD transducisome represents the signaling unit that underlies quantum bumps, the visual response activated by single photons (Scott & Zuker 1998b). An alternative model, based on the ability of INAD to polymerize and to associate with perhaps all major components of the phototransduction process, posits that INAD organizes a massive signaling web dubbed a signalplex (Montell 1998, Xu et al 1998). The signalplex metaphor implies something more extensive in space and more complex in composition than a transducisome, as well as something stoichiometrically less well-defined. The verisimilitude of the transducisome and signalplex concepts should become clearer as we gain a more detailed understanding of the structure, composition, and dynamics of the INAD complex.

ORGANIZATION OF THE POSTSYNAPTIC DENSITY BY PSD-95/SAP90

In excitatory synapses of the brain, the postsynaptic membrane is specialized for responding rapidly to glutamate release from the presynaptic terminal. Associated with the postsynaptic membrane of excitatory synapses is the postsynaptic density (PSD), an electron-dense structure ~40–50 nm thick and up to ~500 nm wide containing filamentous and particulate elements (Figure 4, see color insert). Functionally, the PSD can be regarded as a postsynaptic organelle specialized for glutamatergic signal transduction. Structurally, it consists of a high concentration of glutamate receptors integrated into a complex network of regulatory, scaffolding, and cytoskeletal proteins (reviewed in Kennedy 1997, Kim & Huganir 1999, Sheng & Pak 2000, Ziff 1997). The molecular architecture of the PSD is largely built around PDZ-containing scaffold proteins, in particular the PSD-95 (also known as SAP90) family of MAGUK proteins.

MAGUKs

The PSD-95/SAP90 family of MAGUK proteins (which in mammals comprises PSD-95/SAP90, PSD93/chapsyn-110, SAP97/hDlg, and SAP102) contain three PDZs in addition to SH3 and GK domains (Figure 1). MAGUKs are typically associated with cell junctions or other specialized sites on the plasma membrane; for instance, the ZO-1 family of MAGUK proteins are localized in the tight junctions of epithelial cells (Mitic & Anderson 1998). MAGUKs assemble a specific protein complex by binding to a set of membrane and cytoplasmic proteins; in this way, they are believed to organize specialized membrane microdomains (Fanning & Anderson 1999). The PSD-95 subfamily of MAGUKs are specifically localized to the PSD of excitatory synapses and play an important role in the organization of the postsynaptic specialization.

Protein Interactions of PSD-95/SAP90

PSD-95 was first identified as an abundant polypeptide highly enriched in PSD preparations (Cho et al 1992, Kistner et al 1993). The central role of PSD-95 in the organization of the PSD was first uncovered through the discovery that NMDA receptors bind to PSD-95. NMDA receptors (the major glutamate receptor subtype in the PSD) are composed of NR1 and NR2 subunits, the latter having long cytoplasmic tails with a conserved C-terminal sequence (−ESDV or −ESEV). This C-terminal motif mediates binding to the first two PDZ domains of the PSD-95 family of proteins (Kornau et al 1995, Müller et al 1996, Niethammer et al 1996) (Figure 4).

Membrane Proteins In addition to NMDA receptors, other less abundant transmembrane proteins are also enriched in the postsynaptic membrane; several of these bind to the PDZ domains of PSD-95 via their cytoplasmic C termini. Neuroligin, a postsynaptic transmembrane ligand for neurexin, binds to PDZ3 of PSD-95 (Irie et al 1997, Song et al 1999). ErbB4, a receptor tyrosine kinase for neuregulin, binds to PDZ1/2 of PSD-95 (Garcia et al 2000, Huang et al 2000). Additional membrane proteins that can bind to PSD-95 via C terminus–PDZ interactions include voltage-gated K^+ channels (Kim et al 1995), inward rectifying K^+ channels (Cohen et al 1996, Hibino et al 2000, Nehring et al 2000), and plasma membrane calcium pumps (E Kim et al 1998). However, the localization of these proteins in the PSD is unproven.

Ionotropic glutamate receptors other than NMDA receptors are also concentrated in the postsynaptic membrane. GluR6, a subunit of kainate-type glutamate receptors, binds via its C terminus (−ETMA) to PDZ1 of PSD-95 family MAGUKs, whereas another kainate receptor subunit, KA2, binds to the SH3 and GK domains (Garcia et al 1998). The AMPA-type glutamate receptor is generally not associated with PSD-95; it interacts instead with GRIP and ABP, scaffold proteins containing six or seven PDZs (Dong et al 1997, Srivastava et al 1998). A curiosity is that one subunit of AMPA receptors (GluR1) is specifically associated with SAP97, a member of the PSD-95 family (Leonard et al 1998).

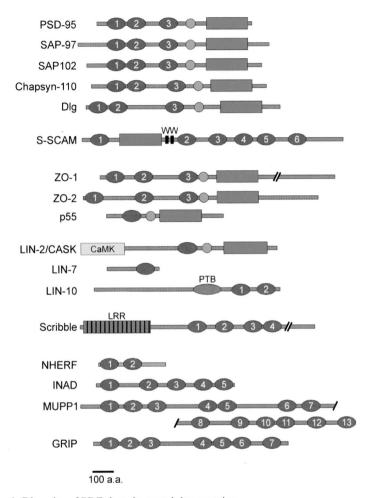

Figure 1 Diversity of PDZ domain-containing proteins.
Domain structures are shown for a variety of PDZ-containing proteins. PSD-95/SAP90, SAP97, PSD-93/chapsyn-110 and SAP102 comprise the mammalian PSD-95 subfamily of MAGUKs. Discs large (Dlg) is the Drosophila homologue of the PSD-95 family. S-SCAM (also known as WWP3/MAGI-1) contains two WW domains in addition to GK and PDZ domains (Dobrosotskaya et al 1997, Hirao et al 1998). ZO-1 and ZO-2 are MAGUKs found at tight junctions of epithelial cells. p55 is a MAGUK localized at the cell membrane of erythrocytes. LIN-2/CASK is a MAGUK protein with a calmodulin-dependent protein kinase (CaMK)-like domain. LIN-2 forms a ternary complex with LIN-7 and LIN-10. Scribble contains leucine-rich repeats (LRR) and is required for epithelial cell polarity (Bilder & Perrimon 2000). NHERF, INAD, GRIP and MUPP1 (Ullmer et al 1998) are examples of multi-PDZ proteins. See text for references to other proteins. PDZ domains are depicted as red ovals (numbered from N-terminus); SH3 domains as green circles; GK domains as blue rectangles. PTB, phosphotyrosine binding domain.

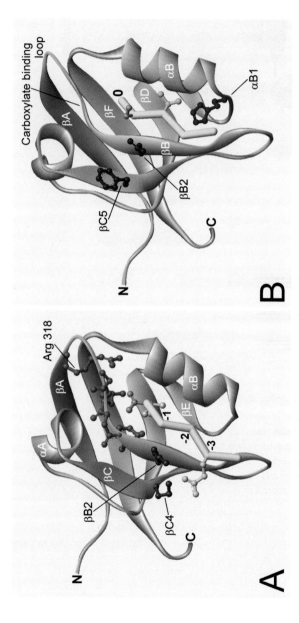

Figure 2 Structure of a PDZ domain complexed with a C-terminal peptide ligand, based on PDZ3 of PSD-95 complexed with CRIPT (Doyle, 1996; Niethammer, 1998). The ribbon diagram of the PDZ domain (gray) is shown bound to the peptide ligand (main chain represented in yellow). The structures in A and B are slightly rotated relative to each other to better show particular sets of interactions. A. The free carboxylate group (orange) of the C-terminal residue (0 position) of the peptide interacts with the conserved amino acids (Arg-318 and Gly-Leu-Gly-Phe) of the carboxylate binding loop (red). The side chain of the -3 residue (glutamine; light green) interacts with βB2 (asparagine) and βC4 (serine; dark green). B. The hydroxyl group of the -2 residue (threonine; light blue) interacts with the sidechain of αB1 (histidine; dark blue). The side chain of the -1 residue (serine; light purple) of the CRIPT peptide shows no interactions with the PDZ domain. However, βB2 and βC5 residues (dark purple) are likely to influence selectivity at the -1 position of the peptide ligand.

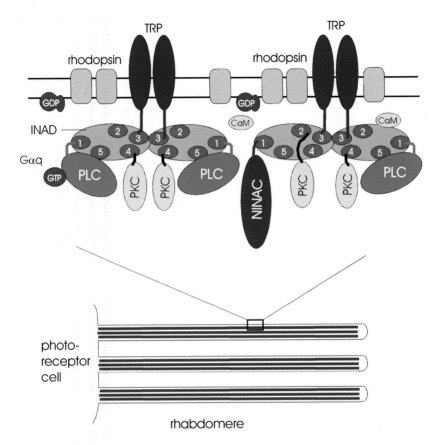

Figure 3 Organization of the phototransduction complex by INAD in Drosophila photoreceptors. A schematic of actin-filled microvilli of rhabdomeres is shown at bottom. INAD is depicted beneath the plasma membrane multimerized via its PDZ3/PDZ4 domains. The major interacting proteins are shown binding to specific PDZ domains of INAD. G protein is depicted in dark gray, associated with the membrane in its GDP-bound form. Activated (GTP-bound) Gαq subunit interacts with phospholipase C (PLC). PDZ domains (numbered) are represented by red ovals, F-actin by purple lines.

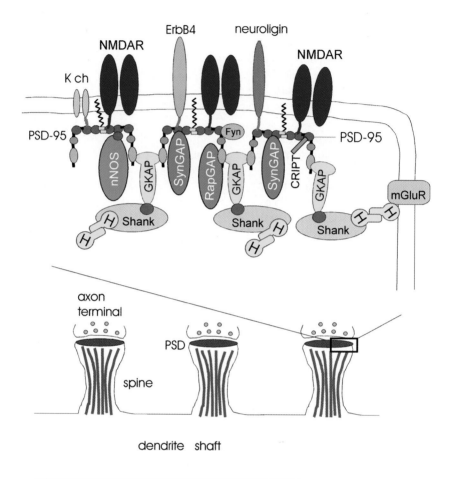

Figure 4 Organization of the PSD of excitatory synapses by PSD-95.
The PSD is usually located on dendritic spines and is directly aligned with the active zone of the presynaptic terminal (schematized at bottom). PSD-95 is shown beneath the post-synaptic membrane multimerized via its N-terminal region, which is palmitoylated (represented by jagged line). The figure depicts an incomplete list of proteins binding to PSD-95, either directly or indirectly. Metabotropic glutamate receptors (mGluR) are shown on the edge of the PSD. PDZ domains are represented by red ovals, F-actin by purple lines. NMDAR, NMDA receptor, H, Homer; K ch, K+ channel.

One current view is that PSD-95 forms a two-dimensional lattice directly beneath the postsynaptic membrane (Figure 4). This lattice, which arises at least in part from head-to-head multimerization of PSD-95 (Hsueh et al 1997), offers an array of PDZ domains as docking sites for transmembrane proteins with the appropriate cytoplasmic C-terminal binding sequence (consensus -E/Q-T/S-X-V). In this way PSD-95 clusters in the postsynaptic membrane with a heterogeneous set of membrane proteins, of which NMDA receptors are a major representative.

Cytoplasmic Signaling Proteins   As with INAD, the binding partners of PSD-95 are not limited to integral membrane proteins such as receptors and ion channels. PSD-95 also binds to a variety of cytoplasmic proteins that are probably involved in signal transduction by NMDA receptors and/or other postsynaptic receptors (Figure 4). Thus, PSD-95 can assemble a complex of specific signaling molecules that is physically and functionally coupled to postsynaptic receptors. For example, PSD-95 binds to nNOS via a PDZ–β-finger interaction (Brenman et al 1996a), thereby bringing nNOS (a Ca^{2+}-calmodulin regulated enzyme) within the sphere of influence of the NMDA receptor (a Ca^{2+}-permeant channel). Antisense suppression of PSD-95 inhibits nNOS activation by NMDA receptors (Sattler et al 1999), consistent with the idea that PSD-95 acts as an adaptor to couple NMDA receptors and nNOS in the PSD.

Other signaling molecules whose postsynaptic roles are less well understood also interact with PSD-95. Regulators or effectors of small GTPases Ras, Rho, and Rap have been found to bind to PSD-95 and to be associated with the PSD (Sheng & Pak 2000). The most prominent of these is SynGAP, a GTPase activating protein for Ras that interacts via its C terminus ($-$QTRV) with all three PDZ domains of PSD-95 (Chen et al 1998, JH Kim et al 1998). SynGAP would be well positioned to act on Ras that has been activated by NMDA receptors or by postsynaptic receptor tyrosine kinases. Although there is no known case of a GTPase of the Ras superfamily binding directly to PSD-95, it is plausible that they interact with the PSD-95 complex in a regulated fashion, akin to the dynamic association of Gαq with the INAD complex.

Fyn and other Src-family tyrosine kinases are associated with NMDA receptors in vivo and involved in NMDA receptor modulation (Lu et al 1998, Tezuka et al 1999). PSD-95 interacts with Fyn, apparently via binding of the Fyn SH2 domain to PDZ3 of PSD-95. In heterologous cells, PSD-95 enhances the phosphorylation of NMDA receptors by Fyn, presumably by bringing together NMDA receptor and Fyn in a ternary complex with PSD-95 (Tezuka et al 1999). Finally, PSD-95 family proteins can bind via a PDZ–C terminus interaction to APC (Matsumine et al 1996), a protein well-known in the context of tumorigenesis and Wingless/Wnt signaling (Bienz 1999) but whose role in neuronal synapses is unclear.

PSD-95 family proteins also interact with cytoplasmic proteins via their GK domain. The GK domain binds to GKAP (also known as SAPAP or DAP) (Kim et al 1997, Naisbitt et al 1997, Satoh et al 1997, Takeuchi et al 1997) and BEGAIN (Deguchi et al 1998), but the functions of these proteins remain to be determined. GKAP in turn binds to the Shank family of PDZ-containing scaffold proteins, thereby greatly increasing the extent and complexity of the protein network in which PSD-95 is integrated (Naisbitt et al 1999, Sheng & Kim 2000, Tu et al 1999). Because Shank interacts with Homer, a metabotropic glutamate receptor-binding protein, the PSD-95–GKAP–Shank chain of interactions could link NMDA receptors to metabotropic glutamate receptors in the postsynaptic membrane (Sheng & Kim 2000). Although the significance of these protein interactions emanating from the GK domain of PSD-95 is largely unknown, they emphasize the complexity and size of the supramolecular architecture constructed around the PSD-95 lattice. The complex web-like nature of the PSD-95 complex in excitatory synapses is reminiscent of the signalplex concept proposed for INAD in rhabdomeres. Indeed, a reasonable argument can be made that the extended PSD-95–based protein network is the molecular basis of the PSD, a morphologically definable and biochemically purifiable entity.

Cytoskeletal Proteins   Several sets of protein-protein interaction have been identified that might anchor PSD-95 to the postsynaptic cytoskeleton. Band 4.1, an actin/spectrin-binding protein of the ezrin-radixin-moesin family, binds in vitro to SAP97 and perhaps to other members of the PSD-95 family (Lue et al 1994, 1996; Marfatia et al 1996, Wu et al 1998). Such an interaction has the potential to link the PSD-95 complex to F-actin, which is the predominant cytoskeleton of dendritic spines (where the vast majority of PSDs are located).

PSD-95 also interacts with microtubule-associated proteins. PDZ3 of PSD-95 binds to CRIPT, a small polypeptide that associates with microtubules (Niethammer et al 1998, Passafaro et al 1999). PSD-95 family proteins can also bind directly to the microtubule-associated protein MAP1A via the GK domain (Brenman et al 1998). The interaction of PSD-95 family proteins with microtubule-binding proteins such as CRIPT or MAP1A may link the PSD to a tubulin-based cytoskeleton, which is somewhat surprising because microtubules are generally thought to be sparse or absent from dendritic spines. However, microtubule anchoring may be relevant for the minority of excitatory synapses that exist on dendritic shafts (where microtubules are abundant). Alternatively, these putative microtubule interactions may be involved in trafficking of the PSD-95 complex. Despite these potential cytoskeletal anchoring mechanisms, the actual synaptic clustering of PSD-95 and NMDA receptors is largely unaffected by depolymerization of microtubules or F-actin, at least over a time course of many hours (Allison et al 2000).

Function of PSD-95

Our understanding of the physiological functions of PSD-95 lags behind that of INAD. One activity ascribed to PSD-95 is that of clustering of its binding partners at the cell surface. When PSD-95 is coexpressed with NMDA receptors or Shaker-type K^+ channels in heterologous cells, the receptor/channel proteins associate with PSD-95 to form macroscopic coclusters (Kim et al 1995, 1996). The clustering activity of PSD-95 in heterologous cells depends on its N-terminal region (upstream of PDZ1), which contains a pair of cysteines, and on PDZ1 or PDZ2, which bind to NMDA receptors and K^+ channels (Hsueh et al 1997, Hsueh & Sheng 1999b). The N-terminal region of PSD-95 and its pair of cysteines are also essential for multimerization (Hsueh et al 1997, Hsueh & Sheng 1999b), palmitoylation, and synaptic targeting of PSD-95 (Craven et al 1999, El-Husseini et al 2000, Topinka & Bredt 1998). It remains unclear how receptor/channel clustering by PSD-95 in heterologous cells relates to in vivo clustering of receptors and ion channels in the postsynaptic membrane.

By comparison with INAD, genetic analysis of PSD-95 function is at a primitive stage; this is compounded by the existence of four genes in the PSD-95 family in mammals. Most progress has been achieved in Drosophila, which seems to express only a single homologue of the PSD-95 family: Discs-large (Dlg) (Woods & Bryant 1991). Dlg is concentrated in the glutamatergic neuromuscular junction of Drosophila (Budnik et al 1996), where it colocalizes with the Shaker K^+ channel and the fasciclin II cell adhesion molecule, two transmembrane proteins whose C termini bind to the PDZ domains of Dlg. *dlg* mutants show loss of the normal synaptic localization of Shaker and fasciclin II (Tejedor et al 1997, Thomas et al 1997, Zito et al 1997). Moreover, the C terminus of Shaker is sufficient to confer synaptic targeting on a heterologous protein in wild-type but not in *dlg* mutant flies (Zito et al 1997). Taken together, these studies indicate that Dlg is important in vivo for synaptic localization of its binding partners. Dlg is also localized to the septate junctions (akin to the tight junctions in vertebrates) of epithelial cells in Drosophila and is important for the organization of these cell junctions and in maintenance of cell polarity (Bryant 1997).

By analogy with Dlg in Drosophila, the PSD-95 family of proteins in mammals may be involved in the targeting of its PDZ ligands (such as NMDA receptors) to the PSD. However, a targeted mutation of the PSD-95 gene in mice (a severe hypomorph) caused no obvious defect in synaptic localization of NMDA receptors (Migaud et al 1998). The lack of effect could be due to functional redundancy of PSD-95 and its close relatives. Alternatively, other independent mechanisms may exist for proper localization of NMDA receptors, e.g. via interactions with the NR1 subunit (reviewed in Sheng & Pak 2000). The importance of the NR2 cytoplasmic tail (which binds to PSD-95) for the synaptic targeting of NMDA receptors has also been studied in knockout mice. These studies detected either partial loss of synaptic localization (Mori et al 1998) or no obvious defect in synaptic localization of the mutant NMDA receptors (Sprengel et al 1998). Passafaro et al (1999) found that dispersing PSD-95 family proteins with a PDZ3-binding peptide had no effect on synaptic localization of NMDA receptors. The overall conclusion from these studies seems to be that PSD-95 interaction with the C terminus of NMDA receptors is not essential for postsynaptic targeting of NMDA receptors; however, other postsynaptic binding partners of PSD-95 need to be examined in this regard.

The evidence that PSD-95 is important for postsynaptic signaling (as opposed to postsynaptic localization) of NMDA receptors is much more compelling. The

PSD-95 knockout had dramatic effects on NMDA receptor-dependent long-term potentiation in hippocampus (Migaud et al 1998). The magnitude of long-term potentiation in PSD-95–deficient mice was enhanced at the expense of long-term depression. Thus PSD-95 somehow regulates the threshold between induction of long-term potentiation and long-term depression. The precise molecular explanation for this phenotype is not clear. Targeted mutations that removed the cytoplasmic tails of NR2 subunits of the NMDA receptor also profoundly impaired NMDA receptor signaling, without dramatically affecting NMDA receptor channel activity (Sprengel et al 1998).

Dynamic Regulation of the PSD-95 Complex

Although NMDA receptors are relatively stably incorporated into the PSD, their subcellular distribution can be regulated by synaptic activity over a timescale of days (reviewed in Craig 1998). In keeping with a dynamic regulation of the PSD-95–based complex, time-lapse studies of GFP-tagged PSD-95 in cultured neurons reveal an activity-dependent turnover of >20% of PSD-95 clusters within 24 h (Okabe et al 1999).

The time course of appearance of components of the PSD during synapse development suggests that the PSD-95 complex is not transported to the synapse in a fully preassembled state (Lee & Sheng 2000, Rao et al 1998). Construction of the complex occurs at least in part at the destination site. Thus the structure and composition of the PSD-95 supramolecular complex is subject to both local assembly and turnover, which makes sense in terms of activity-dependent plasticity of synapses.

PDZ DOMAINS IN PROTEIN SORTING/TARGETING

INAD and PSD-95 are examples of PDZ scaffold proteins that are concentrated in specific subcellular sites and that organize large signaling complexes at those locations. We now discuss additional examples of PDZ proteins that are important for the subcellular targeting of their partner proteins: the LIN-2/LIN-7/LIN-10 complex and the GRIP proteins. In these cases, however, the PDZ proteins may act not only as localized scaffolds, but also as mediators of the trafficking of their binding partners.

The LIN-2/LIN-7/LIN-10 Complex and Basolateral/Postsynaptic Sorting

lin-2, *lin-7* and *lin-10* were initially identified as genes important for vulval differentiation in *C. elegans*. All three genes encode PDZ-containing proteins (Figure 1), and each is required in *C. elegans* for the normal basolateral localization of LET-23 (an EGF receptor homolog) in vulval precursor epithelial cells (Bredt 1998; Kim 1997, Simske et al 1996, Whitfield et al 1999). *lin-10*, but not *lin-2* or *lin-7*, is also required for postsynaptic localization of the glutamate receptor GLR-1 in *C. elegans* neurons (Rongo et al 1998), lending partial support to the idea that the dendritic compartment of neurons is analogous to the basolateral domain of polarized epithelial cells. Homologs of the LIN-2, LIN-7, and LIN-10 proteins are present in other organisms, including mammals, where they have alternate names (mLIN-2 or CASK; mLIN-7, Veli, or MALS; mLIN-10, Mint, or X11). LIN-2/CASK belongs to the MAGUK superfamily of proteins. In both nematode and mammals, LIN-2 binds directly to LIN-7 and LIN-10 to form a ternary complex (Borg et al 1998, Butz et al 1998, Kaech et al 1998). These interactions do not involve the PDZ domains, thus leaving them available for engagement of C-terminal motifs in transmembrane proteins.

In *C. elegans*, the PDZ domain of LIN-7 binds directly to the C terminus (−ETCL) of the LET-23 receptor, a typical class I PDZ interaction (Kaech et al 1998, Simske et al 1996). In mammals, LIN-7's PDZ has been shown to interact with the C terminus of the epithelial GABA transporter (−ETHL) (Perego et al 1999) and NMDA receptor NR2 subunits (−ESDV) (Jo et al 1999). LIN-7 is enriched at neuronal synapses and epithelial cell junctions (Irie et al 1999, Jo et al 1999, Simske et al 1996).

Specific binding partners have also been identified for the PDZ domain of LIN-2/CASK (a class II PDZ): These include the putative cell adhesion receptor neurexin (C-terminal sequence-EYYV) (Hata et al 1996) and the cell surface heparan sulfate proteoglycan syndecan (−EFYA) (Cohen et al 1998, Hsueh et al 1998). Mammalian syndecan and LIN-2/CASK colocalize at basolateral membranes of epithelial cells (Cohen et al 1998) and in synaptic junctions and along axons of neurons (Hsueh & Sheng 1999a, Hsueh et al 1998). CASK/LIN-2 can bind to protein 4.1, thereby attaching to the actin cytoskeleton (Cohen et al 1998). LIN-10 colocalizes with GLR-1 in synaptic puncta in *C. elegans* (Rongo et al 1998), although there is no evidence that these proteins bind each other directly.

The above findings are consistent with the idea that the LIN-2/LIN-7/LIN-10 complex functions analogously to INAD and PSD-95 in directly anchoring or retaining their binding partners at specific subcellular sites (such as the basolateral membrane or the postsynaptic membrane). However, certain features of LIN-2, LIN-7, and LIN-10 suggest that this conclusion is too simplistic. Unlike PSD-95 and INAD, which are highly localized in the PSD and in the rhabdomere, respectively, LIN-2, LIN-7, and LIN-10 show relatively wide subcellular distributions. More important, LIN-2, LIN-7, and LIN-10 are not tightly and specifically colocalized with each other or with their cell surface protein partners. For instance, in both worm and mammalian cells, the main concentration of LIN-10 seems to be in the Golgi or *trans*-Golgi network (Borg et al 1998; Rongo et al 1998, Whitfield et al 1999). On the other hand, LIN-7 is most enriched at the adherens junctions, whereas its binding partner LET-23 is enriched throughout the basolateral membrane (Simske et al 1996). CASK/LIN-2 is widely distributed in mammalian neurons both on the cell surface and in intracellular compartments, and it can even translocate into the nucleus to regulate gene transcription (Hsueh et al 2000). The discrepancies between the subcellular distributions of LIN-2, LIN-7, LIN-10, and their ligands has raised the possibility that this complex is involved more in trafficking of its membrane protein partners than in retention at their final destination. Consistent with a role in protein trafficking is that LIN-10/Mint1 also interacts with Munc18, a protein required for exocytosis (Okamoto & Südhof 1997). The issue is complicated by the fact that members of the mammalian LIN-2, LIN-7, and LIN-10 protein families show differential interactions among themselves and additional interactions with other PDZ-containing proteins (Butz et al 1998, Kamberov et al 2000). Protein trafficking to a particular subcellular destination and protein anchoring at that location are not necessarily mutually exclusive functions; future research should clarify the relative cell biological roles of the LIN-2/LIN-7/LIN-10 complex.

GRIP

GRIP1 and ABP (also known as GRIP2) comprise a family of proteins (here generically termed GRIP) containing six or seven PDZ domains (Figure 1). They were originally identified by their interaction with postsynaptic AMPA-type glutamate receptors, which is mediated by the fifth PDZ domain of GRIP binding to the C-terminal −SVKI sequence of GluR2 and GluR3 subunits of AMPA receptors (Dong et al 1997, Srivastava et al 1998, Wyszynski et al 1998). However, GRIP is only modestly enriched in brain synapses when compared with PSD-95, and GRIP distribution overall shows only weak overlap with AMPA receptors by immunostaining (Dong et al 1999, Wyszynski et al 1999). GRIP and AMPA receptors differ from PSD-95 and NMDA receptors in being relatively abundant in intracellular compartments (including putative transport vesicles) of neurons (Dong et al 1999, Petralia et al 1999, Wyszynski et al 1999). Taken together, these findings raise the possibility that GRIP is involved in the trafficking of AMPA receptors rather than, or in addition to, the anchoring of AMPA receptors at synapses. Overexpression of the C-terminal tail of GluR2 in neurons inhibits synaptic localization of AMPA receptors (Dong et al 1997), a result consistent with either an anchoring or a trafficking role for GRIP. Peptide interference with the GRIP PDZ-GluR2/3 interaction also prevents dynamic recruitment of AMPA receptors to the synapse, as measured in electrophysiological experiments (Li et al 1999).

In addition to AMPA receptors, GRIP binds to EphB2 and EphA7, members of the large family of Eph receptor tyrosine kinases, and to the EphrinB ligands for Eph receptors (Bruckner et al 1999, Torres et al 1998). Liprins, proteins that bind to the LAR family of receptor tyrosine phosphatases (Serra-Pagès et al 1998), also bind to GRIP, utilizing PDZ6 (M Wyszynski, M Sheng, unpublished observations). Liprin has been implicated in presynaptic differentiation (Zhen & Jin 1999) and LAR tyrosine phosphatase in axon guidance (Van Vactor 1998). Consistent with its diverse interactions and a possible role in protein trafficking, GRIP shows a wide distribution within neurons; it is localized on both presynaptic and postsynaptic sides of the synapse, present in axons and dendrites, and associated with intracellular membrane compartments like ER (Dong et al 1999, Wyszynski et al 1999).

In summary, GRIP resembles LIN-2/LIN-7/LIN-10 in the sense that it is more widely distributed within cells than are INAD and PSD-95, and its localization correlates less tightly with its cell surface binding partners. GRIP is clearly involved in targeting of proteins to specific locations (such as synapses) in neurons, but it might do so via protein trafficking rather than (or in addition to) protein anchoring mechanisms. A precedent for the involvement of a PDZ interaction in a protein sorting decision is provided by NHERF/EBP50, a two-PDZ protein that binds to the C terminus of β2-adrenergic receptors (Cao et al 1999, Hall et al 1998). Interaction with NHERF/EBP50 is required for normal recycling of the endocytosed β2-adrenergic receptors; disrupting this PDZ–C terminus interaction (for instance by GRK5 phosphorylation of the -2 serine residue) results in missorting of the endocytosed receptor to lysosomes (Cao et al 1999).

It should be emphasized that the difference between LIN-2/LIN-7/LIN-10 and GRIP on the one hand and PSD-95 and INAD on the other may be quantitative rather than qualitative. Because PSD-95 and INAD also must be sorted from their sites of synthesis in the cell body to their destinations in the PSD and rhabdomere, it is possible that some components of these respective complexes will be cotransported in a preassembled state with the PDZ scaffold. In this sense, PSD-95 and INAD would be involved in the trafficking of such binding partners. The key difference seems to be that at steady state, the vast majority of PSD-95 and INAD are localized in a specific subcellular microdomain, whereas LIN-2/LIN-7/LIN-10 and GRIP populate many more compartments, including those that are probably involved in intracellular protein transport. This difference may reflect the more dynamic regulation of the LIN-2/LIN-7/LIN-10– and GRIP-based macromolecular complexes compared with the more stable architectures organized by PSD-95 and INAD.

CONCLUSIONS

PDZ interactions are involved in the sorting, targeting, and assembly of supramolecular complexes. The pleiomorphic functions of PDZ domains are not surprising given their differential binding specificities and the structural diversity of PDZ-containing proteins. This young field will continue to evolve. Future conceptual advances will come from understanding the molecular stoichiometry and geometry of PDZ-based macromolecular complexes and from studying the dynamics of PDZ-based interactions on both short (activity-dependent) and long (developmental) timescales.

ACKNOWLEDGMENTS

MS is Assistant Investigator of Howard Hughes Medical Institute. This work was supported by grants from NIH [NS35050 (MS)] and the Harvard Armenise Foundation (CS). We thank Daniel Pak, Craig Blackstone, Mike Wyszynski, Yi-Ping Hsueh, Heike Hering, and Mingjie Zhang for critical reading of this manuscript.

Visit the Annual Reviews home page at www.AnnualReviews.org

LITERATURE CITED

Adamski FM, Zhu MY, Bahiraei F, Shieh BH. 1998. Interaction of eye protein kinase C and INAD in Drosophila. Localization of binding domains and electrophysiological characterization of a loss of association in transgenic flies. *J. Biol. Chem.* 273:17713–19

Allison DW, Chervin AS, Gelfand VI, Craig AM. 2000. Postsynaptic scaffolds of excitatory and inhibitory synapses in hippocampal neurons: maintenance components independent of actin filaments and microtubules. *J. Neurosci.* 20:4545–54

Bahner M, Sander P, Paulsen R, Huber A. 2000. The visual G protein of fly photoreceptors interacts with the PDZ domain assembled INAD signaling complex via direct binding of activated Galpha(q) to phospholipase cbeta. *J. Biol. Chem.* 275:2901–4

Bienz M. 1999. APC: the plot thickens. *Curr. Opin. Genet. Dev.* 9:595–603

Bilder D, Perrimon N. 2000. Localization of apical epithelial determinants by the basolateral PDZ protein Scribble. *Nature* 403: 676–80

Borg JP, Straight SW, Kaech SM, de Taddeo-Borg M, Kroon DE, et al. 1998. Identification of an evolutionarily conserved heterotrimeric protein complex involved in protein targeting. *J. Biol. Chem.* 273:31633–36

Bredt DS. 1998. Sorting out genes that regulate epithelial and neuronal polarity. *Cell* 94: 691–94

Brenman JE, Chao DS, Gee SH, McGee AW, Craven SE, et al. 1996a. Interaction of nitric oxide synthase with the postsynaptic density protein PSD-95 and α1-syntrophin mediated by PDZ domains. *Cell* 84:757–67

Brenman JE, Christopherson KS, Craven SE, McGee AW, Bredt DS. 1996b. Cloning and characterization of postsynaptic density 93, a nitric oxide synthase interacting protein. *J. Neurosci.* 16:7407–15

Brenman JE, Topinka RJ, Cooper EC, McGee AW, Rosen J, et al. 1998. Localization of postsynaptic density-93 to dendritic microtubules and interaction with microtubule-associated protein 1A. *J. Neurosci.* 18: 8805–13

Bruckner K, Pablo Labrador J, Scheiffele P, Herb A, Seeburg PH, Klein R. 1999. EphrinB ligands recruit GRIP family PDZ adaptor proteins into raft membrane microdomains. *Neuron* 22:511–24

Bryant PJ. 1997. Junction genetics. *Dev. Genet.* 20:75–90

Budnik V, Koh Y-H, Guan B, Hartmann B, Hough C, et al. 1996. Regulation of synapse structure and function by the Drosophila tumor suppressor gene dlg. *Neuron* 17:627–40

Butz S, Okamoto M, Südhof TC. 1998. A tripartite protein complex with the potential to couple synaptic vesicle exocytosis to cell adhesion in brain. *Cell* 94:773–82

Cao TT, Deacon HW, Reczek D, Bretscher A, van Zastrow M. 1999. A kinase-regulated PDZ-domain interaction controls endocytic sorting of the beta2-adrenergic receptor. *Nature* 401:286–90

Chen HJ, Rojas-Sota M, Oguni A, Kennedy MB. 1998. A synaptic RasGTPase activating protein inhibited by CaM kinase II. *Neuron* 20:895–904

Chesevich J, Kreuz AJ, Montell C. 1997. Requirement for the PDZ domain protein, INAD, for localization of the TRP store-operated channel to a signaling complex. *Neuron* 18:95–105

Chishti AH. 1998. Function of p55 and its non-erythroid homologues. *Curr. Opin. Hematol.* 5:116–21

Cho K-O, Hunt CA, Kennedy MB. 1992. The rat brain postsynaptic density fraction contains a homolog of the Drosophila discs-large tumor suppressor protein. *Neuron* 9: 929–42

Cohen AR, Woods DF, Marfatia SM, Walther Z, Chishti AH, Anderson JM. 1998. Human Cask/Lin-2 binds syndecan-2 and protein 4.1 and localizes to the basolateral membrane of epithelial cells. *J. Cell. Biol.* 142:129–38

Cohen NA, Brenman JE, Snyder S, Bredt DS. 1996. Binding of the inward rectifier K^+ channel Kir 2.3 to PSD-95 is regulated by protein kinase A phosphorylation. *Neuron* 17:759–67

Craig AM. 1998. Activity and synaptic receptor targeting: the long view. *Neuron* 21:459–62

Craven SE, El-Husseini AE, Bredt DS. 1999. Synaptic targeting of the postsynaptic density protein PSD-95 mediated by lipid and protein motifs. *Neuron* 22:497–509

Daniels DL, Cohen AR, Anderson JM, Brunger AT. 1998. Crystal structure of the hCASK PDZ domain reveals the structural basis of class II PDZ domain target recognition. *Nat. Struct. Biol.* 5:317–25

Deguchi M, Hata Y, Takeuchi M, Ide N, Hirao K, et al. 1998. BEGAIN (brain-enriched guanylate kinase-associated protein), a novel neuronal PSD-95/SAP90-binding protein. *J. Biol. Chem.* 273:26269–72

Dobrosotskaya I, Guy RK, James GL. 1997. MAGI-1, a membrane-associated guanylate kinase with a unique arrangement of protein-protein interaction domains. *J. Biol. Chem.* 272:31589–97

Dong H, O'Brien RJ, Fung ET, Lanahan AA, Worley PF, Huganir RL. 1997. GRIP: a synaptic PDZ domain-containing protein that interacts with AMPA receptors. *Nature* 386:279–84

Dong H, Zhang P, Song I, Petralia RS, Liao D, Huganir RL. 1999. Characterization of the glutamate receptor-interacting proteins GRIP1 and GRIP2. *J. Neurosci.* 19:-6930–41

Doyle DA, Lee A, Lewis J, Kim E, Sheng M, MacKinnon R. 1996. Crystal structures of a complexed and peptide-free membrane protein-binding domain: molecular basis of peptide recognition by PDZ. *Cell* 85: 1067–76

El-Husseini AE, Craven SE, Chetkovich DM, Firestein BL, Schnell E, et al. 2000. Dual palmitoylation of PSD-95 mediates its vesiculotubular sorting, postsynaptic targeting, and ion channel clustering. *J. Cell. Biol.* 148:159–72

Fanning AS, Anderson JM. 1999. Protein modules as organizers of membrane structure. *Curr. Opin. Cell. Biol.* 11:432–39

Garcia EP, Mehta S, Blair LA, Wells DG, Shang J, et al. 1998. SAP90 binds and clusters kainate receptors causing incomplete desensitization. *Neuron* 21:727–39

Garcia RA, Vasudevan K, Buonanno A. 2000. The neuregulin receptor ErbB-4 interacts with PDZ-containing proteins at neuronal synapses. *Proc. Natl. Acad. Sci. USA* 97:3596–601

Gee SH, Madhavan R, Levinson SR, Caldwell JH, Sealock R, Froehner SC. 1998. Interaction of muscle and brain sodium channels with multiple members of the syntrophin family of dystrophin-associated protein. *J. Neurosci.* 18:128–37

Grootjans JJ, Zimmermann P, Reekmans G, Smets A, Degeest G, et al. 1997. Syntenin, a PDZ protein that binds syndecan cytoplasmic domains. *Proc. Natl. Acad. Sci. USA* 94:13683–88

Hall RA, Premont RT, Chow C-W, Blitzer JT, Pitcher JA, et al. 1998. The β2-adrenergic receptor interacts with the Na^+/H^+-exchanger regulatory factor to control Na^+/H^+ exchange. *Nature* 392:626–30

Hall RA, Spurney RF, Premont RT, Rahman N, Blitzer JT, et al. 1999. G protein-coupled receptor kinase 6A phosphorylates the Na(+)/H(+) exchanger regulatory factor via a PDZ domain-mediated interaction. *J. Biol. Chem.* 274:24328–34

Hata Y, Butz S, Südhof TC. 1996. CASK: a novel *dlg*/PSD95 homolog with an N-terminal calmodulin-dependent protein kinase domain identified by interaction with neurexins. *J. Neurosci.* 16:2488–94

Hibino H, Inanobe A, Tanemoto M, Fujita A, Doi K, et al. 2000. Anchoring proteins confer G protein sensitivity to an inward-rectifier K(+) channel through the GK domain. *EMBO J.* 19:78–83

Hillier BJ, Christopherson KS, Prehoda KE, Bredt DS, Lim WA. 1999. Unexpected modes of PDZ domain scaffolding revealed by structure of nNOS-syntrophin complex. *Science* 284:812–15

Hirao K, Hata Y, Ide N, Takeuchi M, Irie M, et al. 1998. A novel multiple PDZ domain-containing molecule interacting with N-methyl-D-aspartate receptors and neuronal cell adhesion proteins. *J. Biol. Chem.* 273:21105–10

Hsueh Y-P, Kim E, Sheng M. 1997. Disulfide-linked head-to-head multimerization in the mechanism of ion channel clustering by PSD-95. *Neuron* 18:803–14

Hsueh Y-P, Sheng M. 1999a. Regulated expression and subcellular localization of syndecan heparan sulfate proteoglycans and the syndecan-binding protein CASK/LIN-2 during rat brain development. *J. Neurosci.* 19:7415–25

Hsueh Y-P, Sheng M. 1999b. Requirement of N-terminal cysteines of PSD-95 for PSD-95 multimerization and ternary complex formation, but not for binding to potassium channel Kv1.4. *J Biol Chem* 174:532–36

Hsueh Y-P, Wang TF, Yang FC, Sheng M. 2000. Nuclear translocation and transcription regulation by the membrane-associated guanylate kinase CASK/LIN-2. *Nature* 404:298–302

Hsueh Y-P, Yang F-C, Kharazia V, Naisbitt S, Cohen AR, et al. 1998. Direct interaction of CASK/LIN-2 and syndecan heparan sulfate proteoglycan and their overlapping distribution in neuronal synapses. *J. Cell. Biol.* 142:139–51

Huang YZ, Won S, Ali DW, Wang Q, Tanowitz M, et al. 2000. Regulation of Neuregulin signaling by PSD-95 interacting with Erb4 at CNS synapses. *Neuron* 26:443–55

Huber A, Sander P, Gobert A, Bähner M, Hermann R, Paulsen R. 1996a. The transient receptor potential protein (Trp), a putative store-operated Ca^{2+} channel essential for phosphoinositide-mediated photoreception, forms a signaling complex with NorpA, InaC and InaD. *EMBO J.* 15:7036–45

Huber A, Sander P, Paulsen R. 1996b. Phosphorylation of the InaD gene product, a photoreceptor membrane protein required for recovery of visual excitation. *J. Biol. Chem.* 271:11710–17

Irie M, Hata Y, Deguchi M, Ide N, Hirao K, et al. 1999. Isolation and characterization of mammalian homologues of *Caenorhabditis elegans* lin-7: localization at cell-cell junctions. *Oncogene* 18:2811–17

Irie M, Hata Y, Takeuchi M, Ichtchenko K, Toyoda A, et al. 1997. Binding of neuroligins to PSD-95. *Science* 277:1511–15E

Jo K, Derin R, Li M, Bredt DS. 1999. Characterization of MALS/Velis-1, -2, and -3: a family of mammalian LIN-7 homologs enriched at brain synapses in association with the postsynaptic density-95/NMDA receptor postsynaptic complex. *J. Neurosci.* 19:4189–99

Kaech SM, Whitfield CW, Kim SK. 1998. The LIN-2/LIN-7/LIN-10 complex mediates basolateral membrane localization of the *C. elegans* EGF receptor LET-23 in vulval epithelial cells. *Cell* 94:761–71

Kamberov E, Makarova O, Roh M, Liu A, Karnak D, et al. 2000. Molecular cloning and characterization of pals, proteins associated with mLin-7. *J. Biol. Chem.* 275:11425–31

Kennedy MB. 1997. The postsynaptic density at glutamatergic synapses. *Trends Neurosci.* 20:264–68

Kim E, Cho K-O, Rothschild A, Sheng M. 1996. Heteromultimerization and NMDA receptor-clustering activity of chapsyn-110, a member of the PSD-95 family of proteins. *Neuron* 17:103–13

Kim E, DeMarco SJ, Marfatia SM, Chishti AH, Sheng M, Strehler EE. 1998. Plasma membrane Ca2+ ATPase isoform 4b binds to membrane-associated guanylate kinase (MAGUK) proteins via their PDZ (PSD-95/Dlg/ZO-1) domains. *J. Biol. Chem.* 273(3):1591–95

Kim E, Naisbitt S, Hsueh Y-P, Rao A, Rothschild A, et al. 1997. GKAP, a novel synaptic protein that interacts with the guanylate kinase-like domain of the PSD-95/SAP90 family of channel clustering molecules. *J. Cell. Biol.* 136:669–78

Kim E, Niethammer M, Rothschild A, Jan YN, Sheng M. 1995. Clustering of shaker-type K^+ channels by interaction with a family of membrane- associated guanylate kinases. *Nature* 378:85–88

Kim JH, Huganir RL. 1999. Organization and regulation of proteins at synapses. *Curr. Opin. Cell. Biol.* 11:248–54

Kim JH, Liao D, Lau LF, Huganir RL. 1998. SynGAP: a synaptic RasGAP that associates with the PSD-95/SAP90 protein family. *Neuron* 20:683–91

Kim SK. 1997. Polarized signaling: basolateral receptor localization in epithelial cells by PDZ-containing proteins. *Curr. Opin. Cell. Biol.* 9:853–59

Kistner U, Wenzel BM, Veh RW, Cases-Langhoff C, Garner AM, et al. 1993. SAP90, a rat presynaptic protein related to the product of the *Drosophila* tumor suppressor gene *dlg*-A. *J. Biol. Chem.* 268:4580–83

Kornau H-C, Schenker LT, Kennedy MB, Seeburg PH. 1995. Domain interaction between NMDA receptor subunits and the postsynaptic density protein PSD-95. *Science* 269:1737–40

Kozlov G, Gehring K, Ekiel I. 2000. Solution structure of the PDZ2 domain from human phosphatase hPTP1E and its interac-tions with C-terminal peptides from the fas receptor. *Biochemistry* 39:2572–80

Lee SH, Sheng M. 2000. Development of neuron-neuron synapses. *Curr. Opin. Neurobiol.* 10:125–31

Leonard AS, Davare MA, Horne MC, Garner CC, Hell JW. 1998. SAP97 is associated with the alpha-amino-3-hydroxy-5-methylisoxazole-4-propionic acid receptor GluR1 subunit. *J. Biol. Chem.* 273:19518–24

Li P, Kerchner GA, Sala C, Wei F, Huettner JE, et al. 1999. AMPA receptor-PDZ interactions in facilitation of spinal sensory synapses. *Nat. Neurosci.* 2:972–77

Lu YM, Roder JC, Davidow J, Salter MW. 1998. Src activation in the induction of long-term potentiation in CA1 hippocampal neurons. *Science* 279:1363–67

Lue RA, Brandin E, Chan EP, Branton D. 1996. Two independent domains of hDlg are sufficient for subcellular targeting: the PDZ1–2 conformational unit and an alternatively spliced domain. *J. Cell. Biol.* 135:1125–37

Lue RA, Marfatia SM, Branton D, Chishti AH. 1994. Cloning and characterization of hdlg: the human homologue of the Drosophila discs large tumor suppressor binds to protein 4.1. *Proc. Natl. Acad. Sci. USA* 91:9818–22

Marfatia SM, Cabral JH, Lin L, Hough C, Bryant PJ, et al. 1996. Modular organization of the PDZ domains in the human discs-large protein suggests a mechanism for coupling PDZ domain-binding proteins to ATP and the membrane cytoskeleton. *J. Cell. Biol.* 135:753–66

Marfatia SM, Morais-Cabral JH, Kim AC, Byron O, Chishti AH. 1997. The PDZ domain of human erythrocyte p55 mediates its binding to the cytoplasmic carboxyl terminus of glycophorin C. Analysis of the binding interface by in vitro mutagenesis. *J. Biol. Chem.* 272:24191–97

Matsuda S, Mikawa S, Hirai H. 1999. Phosphorylation of serine-880 in GluR2 by protein kinase C prevents its C terminus from binding with glutamate receptor-interacting protein. *J. Neurochem.* 73:1765–68

Matsumine A, Ogai A, Senda T, Okumura N, Satoh K, et al. 1996. Binding of APC to the human homolog of the *Drosophila* discs large tumor suppressor protein. *Science* 272: 1020–23

Migaud M, Charlesworth P, Dempster M, Webster LC, Watabe AM, et al. 1998. Enhanced long-term potentiation and impaired learning in mice with mutant postsynaptic density-95 protein. *Nature* 396:433–39

Minkoff C, Shenolikar S, Weinman EJ. 1999. Assembly of signaling complexes by the sodium-hydrogen exchanger regulatory factor family of PDZ-containing proteins. *Curr. Opin. Nephrol. Hypertens.* 8:603–8

Mitic LL, Anderson JM. 1998. Molecular architecture of tight junctions. *Annu. Rev. Physiol.* 60:121–42

Montell C. 1998. TRP trapped in fly signaling web. *Curr. Opin. Neurobiol.* 8:389–97

Montell C. 1999. Visual transduction in Drosophila. *Annu. Rev. Cell. Dev. Biol.* 15:231–68

Morais Cabral JH, Petosa C, Sutcliffe MJ, Raza S, Byron O, et al. 1996. Crystal structure of a PDZ domain. *Nature* 382:649–52

Mori H, Manabe T, Watanabe M, Sath Y, Suzuki N, et al. 1998. Role of the carboxy-terminal region of the GluR epsilon2 subunit in synaptic localization of the NMDA receptor channel. *Neuron* 21:571–80

Müller BM, Kistner U, Kindler S, Chung WJ, Kuhlendahl S, et al. 1996. SAP102, a novel postsynaptic protein that interacts with the cytoplasmic tail of the NMDA receptor subunit NR2B. *Neuron* 17:255–65

Naisbitt S, Kim E, Tu JC, Xiao B, Sala C, et al. 1999. Shank, a novel family of postsynaptic density proteins that binds to the NMDA receptor/PSD-95/GKAP complex and cortactin. *Neuron* 23:569–82

Naisbitt S, Kim E, Weinberg RJ, Rao A, Yang F-C, et al. 1997. Characterization of guanylate kinase-associated protein, a postsynaptic density protein at excitatory synapses that interacts directly with postsynaptic density-95/synapse-associated protein 90. *J. Neurosci.* 17:5687–96

Nehring RB, Wischmeyer E, Doring F, Veh RW, Sheng M, Karschin A. 2000. Neuronal inwardly rectifying K(+) channels differentially couple to PDZ proteins of the PSD-95/SAP90 family. *J. Neurosci.* 20:156–62

Niethammer M, Kim E, Sheng M. 1996. Interaction between the C terminus of NMDA receptor subunits and multiple members of the PSD-95 family of membrane-associated guanylate kinases. *J. Neurosci.* 16:2157–63

Niethammer M, Valtschanoff JG, Kapoor TM, Allison DW, Weinberg RJ, et al. 1998. CRIPT, a novel postsynaptic protein that binds to the third PDZ domain of PSD-95/SAP90. *Neuron* 20:693–707

Okabe S, Kim HD, Miwa A, Kuriu T, Okado H. 1999. Continual remodeling of postsynaptic density and its regulation by synaptic activity. *Nat. Neurosci.* 2:804–11

Okamoto M, Südhof TC. 1997. Mints, munc18-interacting proteins in synaptic vesicle exocytosis. *J. Biol. Chem.* 272:31459–64

Pak WL. 1979. Study of photoreceptor function using *Drosophila* mutants. *In Neurogenetics, Genetic Approaches to the Nervous System,* ed. XO Breakfield, pp. 67–99. New York: Elsevier

Passafaro M, Sala C, Niethammer M, Sheng M. 1999. Microtubule binding by CRIPT and its potential role in the synaptic clustering of PSD-95. *Nat. Neurosci.* 2:1063–69

Perego C, Vanoni C, Villa A, Longhi R, Kaech SM, et al. 1999. PDZ-mediated interactions retain the epithelial GABA transporter on the basolateral surface of polarized epithelial cells. *EMBO J.* 18:2384–93

Petralia RS, Esteban JA, Wang Y-X, Partridge JG, Zhao H-M, et al. 1999. Selective acquisition of AMPA receptors over postnatal development suggests a molecular basis for silent synapses. *Nat. Neurosci.* 2:31–36

Ponting CP. 1997. Evidence for PDZ domains in bacteria, yeast, and plants. *Protein. Sci.* 6:464–68

Rao A, Kim E, Sheng M, Craig AM. 1998. Heterogeneity in the molecular composition of excitatory postsynaptic sites during development of hippocampal neurons in culture. *J. Neurosci.* 18:1217–29

Rongo C, Whitfield CW, Rodal A, Kim SK, Kaplan JM. 1998. LIN-10 is a shared component of the polarized protein localization pathways in neurons and epithelia. *Cell* 94:751–59

Sato T, Irie S, Kitada S, Reed JC. 1995. FAP-1: a protein tyrosine phosphatase that associates with Fas. *Science* 268:411–15

Satoh K, Yanai H, Senda T, Kohu K, Nakamura T, et al. 1997. DAP-1, a novel protein that interacts with the guanylate kinase-like domains of hDLG and PSD-95. *Genes Cells* 2:415–24

Sattler R, Xiong Z, Lu WY, Hafner M, MacDonald JF, Tymianski M.1999. Specific coupling of NMDA receptor activation to nitric oxide neurotoxicity by PSD-95 protein. *Science* 284:1845–48

Schultz J, Hoffmüller U, Krause G, Ashurst J, Macias MJ, et al. 1998. Specific interactions between the syntrophin PDZ domain and voltage-gated sodium channels. *Nat. Struct. Biol.* 5:19–24

Scott K, Zuker C. 1998a. TRP, TRPL and trouble in photoreceptor cells. *Curr. Opin. Neurobiol.* 8:383–88

Scott K, Zuker CS. 1998b. Assembly of the Drosophila phototransduction cascade into a signalling complex shapes elementary responses. *Nature* 395:805–8

Serra-Pagès C, Medley QG, Tang M, Hart A, Streuli M. 1998. Liprins, a family of LAR transmembrane protein-tyrosine phosphatase-interacting proteins. *J. Biol. Chem.* 273:15611–20

Sheng M, Kim E. 2000. The Shank family of scaffold proteins. *J. Cell. Sci.* 113:1851–56

Sheng M, Pak DTS. 2000. Ligand-gated ion channel interactions with cytoskeletal and signaling proteins. *Annu. Rev. Physiol.* 62:755–78

Shieh B-H, Niemeyer B. 1995. A novel protein encoded by the InaD gene regulates recovery of visual transduction in Drosophila. *Neuron* 14:201–10

Shieh B-H, Zhu M-Y. 1996. Regulation of the TRP Ca2+ channel by INAD in drosophila photoreceptors. *Neuron* 16:991–98

Shieh B-H, Zhu MY, Lee JK, Kelly IM, Bahiraei F. 1997. Association of INAD with NORPA is essential for controlled activation and deactivation of Drosophila phototransduction in vivo. *Proc. Natl. Acad. Sci. USA* 94:12682–87

Simske JS, Kaech SM, Harp SA, Kim SK. 1996. LET-23 receptor localization by the cell junction protein LIN-7 during C. elegans vulval induction. *Cell* 85:195–204

Song JY, Ichtchenko K, Sudhof TC, Brose N. 1999. Neuroligin 1 is a postsynaptic cell-adhesion molecule of excitatory synapse. *Proc. Natl. Acad. Sci. USA* 96:1100–5

Songyang Z, Fanning AS, Fu C, Xu J, Marfatia SM, et al. 1997. Recognition of unique carboxyl-terminal motifs by distinct PDZ domains. *Science* 275:73–77

Sprengel R, Suchanek B, Amico C, Brusa R, Burnasheve N, et al. 1998. Importance of the intracellular domain of NR2 subunits for NMDA receptor function in vivo. *Cell* 92:279–89

Srivastava S, Osten P, Vilim F, Khatri L, Inman G, et al. 1998. Novel anchorage of GluR2/3 to the postsynaptic density by the AMPA receptor- binding protein ABP. *Neuron* 21:581–91

Staudinger J, Lu J, Olson EN. 1997. Specific interaction of the PDZ domain protein PICK1 with the COOH terminus of protein kinase C-alpha. *J. Biol. Chem.* 272:32019–24

Stricker NL, Christopherson KS, Yi BA, Schatz PJ, Raab RW, et al. 1997. PDZ domain of neuronal nitric oxide synthase recognizes novel C-terminal peptide sequences. *Nat. Biotech.* 15:336–42

Takeuchi M, Hata Y, Hirao K, Toyoda A, Irie M, Takai Y. 1997. SAPAPs a family of PSD-95/SAP90-associated proteins localized at postsynaptic density. *J. Biol. Chem.* 272:11943–51

Tejedor FJ, Bokhari A, Rogero O, Gorczyca M, Zhang J, et al. 1997. Essential role for *dlg* in synaptic clustering of Shaker K^+ channels *in vivo*. *J. Neurosci.* 17:152–59

Tezuka T, Umemori H, Akiyama T, Nakanishi S, Yamamoto T. 1999. PSD-95 promotes fyn-mediated tyrosine phosphorylation of the N-methyl-D-aspartate receptor subunit NR2A. *Proc. Natl. Acad. Sci. USA* 96:435–40

Thomas U, Kim E, Kuhlendahl S, Ho Koh Y, Gundelfinger ED, et al. 1997. Synaptic clustering of the cell adhesion molecule fasciclin II by discs-large and its role in the regulation of presynaptic structure. *Neuron* 19:787–99

Tochio H, Hung F, Li M, Bredt DS, Zhang M. 2000. Solution structure and backbone dynamics of the second PDZ domain of postsynaptic density-95. *J. Mol. Biol.* 295:225–37

Tochio H, Zhang Q, Mandal P, Li M, Zhang M. 1999. Solution structure of the extended neuronal nitric oxide synthase PDZ domain complexed with an associated peptide. *Nat. Struct. Biol.* 6:417–21

Topinka JR, Bredt DS. 1998. N-terminal palmitoylation of PSD-95 regulates association with cell membranes and interaction with K^+ channel Kv1.4. *Neuron* 20:125–34

Torres R, Firestein BL, Dong H, Staudinger J, Olson EN, et al. 1998. PDZ proteins bind cluster and synaptically colocalize with Eph receptors and their ephrin ligands. *Neuron* 21:1453–63

Tsunoda S, Sierralta J, Sun Y, Bodner R, Suzuki E, et al. 1997. A multivalent PDZ-domain protein assembles signalling complexes in a G- protein-coupled cascade. *Nature* 388:243–49

Tsunoda S, Zuker CS. 1999. The organization of INAD-signaling complexes by a multivalent PDZ domain protein in Drosophila photoreceptor cells ensures sensitivity and speed of signaling. *Cell Calcium* 26:165–71

Tu JC, Xiao B, Naisbitt S, Yuan JP, Petralia RS, et al. 1999. Coupling of mGluR/Homer and PSD-95 complexes by the Shank family of postsynaptic density proteins. *Neuron* 23:583–92

Ullmer C, Schmuck K, Figge A, Lubbert H. 1998. Cloning and characterization of MUPP1 a novel PDZ domain protein. *FEBS Lett.* 424:63–68

van Huizen R, Miller K, Chen DM, Li Y, Lai ZC, et al. 1998. Two distantly positioned PDZ domains mediate multivalent INAD-phospholipase C interactions essential for G protein-coupled signaling. *EMBO J.* 17:2285–97

Van Vactor D. 1998. Protein tyrosine phosphatases in the developing nervous system. *Curr. Opin. Cell. Biol.* 10:174–81

Wang S, Raab RW, Schatz PJ, Guggino WB, Li M. 1998. Peptide binding consensus of the NHE-RF-PDZ1 domain matches the C-terminal sequence of cystic fibrosis transmembrane conductance regulator (CFTR). *FEBS Lett.* 427:103–8

Wes PD, Xu XZ, Li HS, Chien F, Doberstein SK, Montell C. 1999. Termination of phototransduction requires binding of the NINAC myosin III and the PDZ protein INAD. *Nat. Neurosci.* 2:447–53

Whitfield CW, Benard C, Barnes T, Hekimi S, Kim SK. 1999. Basolateral localization of the *Caenorhabditis elegans* epidermal growth factor receptor in epithelial cells by the PDZ protein LIN-10. *Mol. Biol. Cell.* 10:2087–100

Woods DF, Bryant PJ. 1991. The discs-large tumor suppressor gene of Drosophila encodes a guanylate kinase homolog localized at septate junctions. *Cell* 66:451–64

Wu H, Reuver SM, Kuhlendahl S, Chung WJ, Garner CC. 1998. Subcellular targeting and cytoskeletal attachment of SAP97 to the epithelial lateral membrane. *J. Cell Sci.* 111:2365–76

Wyszynski M, Kim E, Yang F-C, Sheng M. 1998. Biochemical and immunocytochemical characterization of GRIP a putative AMPA receptor anchoring protein in rat brain. *Neuropharmacology* 37:1335–44

Wyszynski M, Valtschanoff JG, Naisbitt S, Dunah AW, Kim E, et al. 1999. Association of AMPA receptors with a subset of glutamate receptor-interacting protein in vivo. *J. Neurosci.* 19:6528–37

Xia J, Zhang X, Staudinger J, Huganir RL. 1999. Clustering of AMPA receptors by the synaptic PDZ domain-containing protein PICK1. *Neuron* 22:179–87

Xu X-ZS, Choudhury A, Li X, Montell C. 1998. Coordination of an array of signaling proteins through homo-and heteromeric interactions between PDZ domains and target proteins. *J. Cell Biol.* 142:545–55

Zhang W, Vazquez L, Apperson M, Kennedy MB. 1999. Citron binds to PSD-95 at gluta-matergic synapses on inhibitory neurons in the hippocampus. *J. Neurosci.* 19:96–108

Zhen M, Jin Y. 1999. The liprin protein SYD-2 regulates the differentiation of presynaptic termini in *C. elegans. Nature* 401:371–75

Ziff EB. 1997. Enlightening the postsynaptic density. *Neuron* 19:1163–74

Zito K, Fetter RD, Goodman CS, Isacoff EY. 1997. Synaptic clustering of fasciclin II and shaker: essential targeting sequences and role of dlg. *Neuron* 19:1007–16

Annu. Rev. Neurosci. 2001. 24:31–55

THE ROLE AND REGULATION OF ADENOSINE IN THE CENTRAL NERVOUS SYSTEM

Thomas V. Dunwiddie[1,2] and Susan A. Masino[1]

[1]*Department of Pharmacology and Program in Neuroscience, University of Colorado Health Sciences Center, Denver, Colorado 80262; e-mail: susan.masino@uchsc.edu*
[2]*Denver Veterans Administration Medical Center, Denver, Colorado 80220:*
e-mail: tom.dunwiddie@uchsc.edu

Key Words neuromodulation, synaptic transmission, epilepsy, neuroprotection, sleep

■ **Abstract** Adenosine is a modulator that has a pervasive and generally inhibitory effect on neuronal activity. Tonic activation of adenosine receptors by adenosine that is normally present in the extracellular space in brain tissue leads to inhibitory effects that appear to be mediated by both adenosine A_1 and A_{2A} receptors. Relief from this tonic inhibition by receptor antagonists such as caffeine accounts for the excitatory actions of these agents. Characterization of the effects of adenosine receptor agonists and antagonists has led to numerous hypotheses concerning the role of this nucleoside. Previous work has established a role for adenosine in a diverse array of neural phenomena, which include regulation of sleep and the level of arousal, neuroprotection, regulation of seizure susceptibility, locomotor effects, analgesia, mediation of the effects of ethanol, and chronic drug use.

INTRODUCTION

Purines and purine nucleotides are essential constituents of all living cells. ATP is used as an energy source for nearly all cellular activity, whereas adenine is a component of nucleic acids. Perhaps as a result of their ubiquitous nature, purines have also evolved as important molecules for both intracellular and extracellular signaling, roles that are distinct from their activity related to energy metabolism and the genetic transmission of information. ATP itself interacts with two general classes of extracellular receptors, the ionotropic P2X receptors and the metabotropic P2Y receptors (for reviews, see Ralevic & Burnstock 1998, Harden et al 1995), and cAMP is an intracellular messenger that plays a key role in regulating intracellular activity. Adenosine is a third "purinergic messenger" that regulates many physiological processes, particularly in excitable tissues such as heart and brain. Many of the actions of adenosine either reduce the activity of excitable tissues (e.g. by

0147-006X/01/0621-0031$14.00

slowing the heart rate) or increase the delivery of metabolic substrates (e.g. by inducing vasodilation) and, thus, help to couple the rate of energy expenditure to the energy supply. However, this type of unitary role for adenosine is not sufficient to explain many of its actions, and it is clear that adenosine plays a variety of different roles as an intercellular messenger. This is particularly the case in the brain, which expresses high concentrations of adenosine receptors, and where adenosine has been shown to be involved in both normal and pathophysiological processes, including regulation of sleep, arousal, neuroprotection, and epilepsy. The pharmacological actions of caffeine, which is the most widely used psychoactive drug in the world, are largely attributable to its activity as an adenosine receptor antagonist (Fredholm et al 1999). A challenging issue with respect to the functional role(s) played by adenosine in the brain is to understand why antagonizing the effects of endogenous adenosine produce what are generally considered to be improvements in mental function and performance, whereas antagonism of most other neurotransmitter receptors produce either deficits or pathological effects. The primary intent of this review is to explore the functional role of adenosine in the nervous system and to discuss the mechanisms by which extracellular concentrations of adenosine are regulated.

ADENOSINE RECEPTORS AND TRANSDUCTION MECHANISMS

Adenosine Receptor Subtypes

Adenosine receptors have been intensively studied, and to date four different adenosine receptors have been cloned in a variety of species, including man (Table 1) (for a review, see Olah & Stiles 1995). Because exhaustive efforts to identify other adenosine receptors have been unsuccessful, it appears unlikely that additional receptors will be identified. All of the adenosine receptors are seven transmembrane domain, G-protein–coupled receptors, and they are linked to a variety of transduction mechanisms. The A_1 receptor has the highest abundance in the brain and is coupled to activation of K^+ channels (Trussell & Jackson 1985) and inhibition of Ca^{2+} channels (Macdonald et al 1986), both of which would inhibit neuronal activity. The A_{2A} receptor is expressed at high levels in only a few regions of the brain and is primarily linked to activation of adenylyl cyclase. Antagonism of both A_1 and A_{2A} receptors appears to be responsible for the stimulant effects of adenosine receptor antagonists, at least in rodents (Marston et al 1998), although stimulation of locomotor activity may be primarily an A_{2A} effect (Ongini 1997, El Yacoubi et al 2000). The A_{2B} receptor, which also activates adenylyl cyclase, is thought to be fairly ubiquitous in the brain, but it has been difficult to link this receptor to specific physiological or behavioral responses because of the paucity of A_{2B}-specific agonists or antagonists (for a review, see Feoktistov & Biaggioni 1997). The A_3 receptor is also somewhat poorly characterized, but it has been

TABLE 1 Adenosine receptors in the brain

Receptor	Adenosine affinity	G-protein	Transduction mechanisms[a]	Physiological actions in brain	Distribution in brain
A_1	~70 nM	G_i and G_o	Inhibits adenylyl cyclase Activates GIRKs Inhibits Ca^{2+} channels Activates PLC	Inhibits synaptic transmission Hyperpolarizes neurons	Widespread[b]
A_{2A}	~150 nM	G_s^{c}, G_{olf}	Activates adenylyl cyclase Inhibits Ca^{2+} channels Activates Ca^{2+} channels (?)	Facilitates transmitter release[d] Inhibition of transmitter release[d]	Primarily striatum, olfactory tubercle, nucleus accumbens[e]
A_{2B}	~5100 nM	G_s^{c}	Activates adenylyl cyclase Activates PLC	Increases in cAMP in brain slices Modulation of Ca^{2+} channel function (?)	Widespread[f]
A_3	~6500 nM	G_{i3}, G_q	Activates PLC Inhibits adenylyl cyclase Increases intracellular Ca^{2+}	Uncouples A_1, mGlu receptors (?)	Widespread[f]

[a]GIRKs, G-protein–dependent inwardly rectifying K^+ channels; PLC, phospholipase C.

[b]Based upon numerous ligand binding studies and in situ hybridization studies.

[c]Primary mechanism of coupling.

[d]For reviews, see Latini et al 1996, Edwards & Robertson 1999.

[e]Ligand binding and in situ hybridization generally shows high levels in these regions, very low levels elsewhere; reverse transcriptase–polymerase chain reaction (RT-PCR) shows more widespread expression.

[f]Relatively low levels, not detectable with in situ hybridization but apparent with RT-PCR (Dixon et al 1996).

reported to uncouple A_1 and metabotropic glutamate receptors via a protein kinase C–dependent mechanism (Dunwiddie et al 1997a, Macek et al 1998), and thus, one of its functions may be to modulate the activity of other receptors.

From a pharmacological standpoint, it has been extremely difficult to develop tissue-specific drugs that interact with adenosine receptors, primarily because of their ubiquitous nature. For example, although there are highly A_1-selective agonists and antagonists, the A_1 receptor that slows the heart rate appears to be identical to the A_1 receptor that depresses neural activity. Although there may be tissue differences in spare receptors (Shryock et al 1998), G-protein coupling, and transduction mechanisms (Linden et al 1998), there are few differences that can be exploited pharmacologically.

ACTIONS OF ADENOSINE AT THE CELLULAR LEVEL

Actions of Adenosine Mediated by Effects on K^+ and Ca^{2+} Channels

In terms of cellular physiology, adenosine has a number of actions that would be considered neuromodulatory but not neurotransmission per se. Adenosine does not appear to be released in a classical Ca^{2+}-dependent fashion, nor is it stored in vesicles, and there is no evidence for synapses where the primary transmitter is adenosine. However, A_1 receptors are linked to inhibition of the release of virtually every classical neurotransmitter (including glutamate, gamma-aminobutyric acid (GABA), acetylcholine, norepinephrine, 5-hydroxytryptamine (5-HT), dopamine, and other transmitters as well). The most prominent inhibitory actions are generally on excitatory glutamatergic systems (e.g. Dunwiddie & Hoffer 1980, Kocsis et al 1984), where synaptic transmission can often be completely blocked by adenosine. Inhibitory modulation of inhibitory (e.g. GABA) systems is less frequently observed, so that the net effect of adenosine receptor activation in nearly all regions of the brain is to reduce excitability. The mechanism of inhibitory modulation of transmitter release has been extensively studied, and it appears to reflect a G-protein–coupled inhibition of Ca^{2+} channels in nerve endings, although this is still the subject of debate. Other mechanisms may contribute to this effect as well, because adenosine also inhibits the spontaneous Ca^{2+}-independent release of neurotransmitter (Scanziani et al 1992), but under normal physiological conditions the inhibition of Ca^{2+} influx appears to be the primary inhibitory mechanism (Fredholm & Dunwiddie 1988, Wu & Saggau 1997). Adenosine receptors may also enhance neurotransmitter release (Cunha et al 1994), but these actions are less common than the inhibition of neurotransmitter release. Another major action of A_1 receptors is a hyperpolarization of the resting membrane potential mediated via a G-protein–dependent activation of inwardly rectifying K^+ channels (GIRKs). GIRKs are activated by many other receptors as well (e.g. in hippocampal pyramidal neurons by A_1, $GABA_B$, $5HT_{1A}$, and somatostatin receptors), and the effects of these agents typically occlude, which suggests that they act on a common population of G-proteins and/or K^+ channels.

Interactions Between Adenosine Receptors and Other Receptor Systems

One interesting aspect of adenosine receptors pertains to interactions between adenosine receptors and other types of G-protein–coupled receptors. Synergistic interactions have been reported between low concentrations of A_1 and $GABA_B$ agonists on GIRKs (Sodickson & Bean 1998), which suggests that the tonic, low level of occupation of A_1 receptors might regulate the strength of $GABA_B$ synapses. There is also extensive evidence from studies primarily in the striatum for direct interactions between A_{2A} receptors and D1 receptors, and between A_1 and D2 receptors (for a review, see Fuxe et al 1998).

REGULATION OF EXTRACELLULAR ADENOSINE

In many systems, basal extracellular adenosine concentrations are sufficient to tonically activate a substantial fraction of high-affinity (A_1 and A_{2A}) adenosine receptors. Estimates of this basal concentration span a wide range, but most estimates using pharmacological approaches (Dunwiddie & Diao 1994) or microdialysis of the brain (Ballarin et al 1991) are in the range of 25–250 nM. Given the affinity of adenosine for its receptors (Table 1), this would suggest that interactions with A_1 and A_{2A} receptors are primarily responsible for the basal purinergic "tone" that is seen in most systems. The stimulatory effects of such drugs as caffeine stem from their ability to antagonize the actions of endogenous adenosine and, hence, reverse this tonic inhibition. Little is known about this basal tone, which in the brain may differ markedly from region to region (Delaney & Geiger 1996). Basal concentrations of adenosine probably reflect an equilibrium between the multiple mechanisms that increase extracellular adenosine and its uptake and metabolism. The recent observation that A_1 receptors appear to play a role in the regulation of adenosine concentrations in neuronal cultures (Andresen et al 1999) suggests that there may be some interesting unknown aspects to the regulation of extracellular adenosine concentrations.

Extracellular Conversion of Adenine Nucleotides as a Source for Adenosine

There are two primary mechanisms by which adenosine can reach the extracellular space of the brain, and these are via dephosphorylation of adenine nucleotides by ecto-nucleotidases and release of adenosine from cells via transporters. The first of these depends on ecto-nucleotidases, ecto-phosphodiesterases, and apyrases that can dephosphorylate virtually any adenine nucleotide to 5′-AMP, which is subsequently dephosphorylated by 5'-nucleotidase to adenosine. There are a wide variety of such ectonucleotidases, which have been the subject of a recent review (Zimmermann & Braun 1999). These ecto-enzymes are highly expressed in the brain, have rather broad specificity, and are generally rapid in their action. Recent studies have suggested that most nucleotides (with the exception of cAMP) are converted to adenosine in less than a second (Dunwiddie et al 1997b). Even "stable" ATP analogs can be substrates for these nucleotidases (Cunha et al 1998), and there is evidence that the nucleotidases may be present in close physical proximity to presynaptic inhibitory A_1 receptors.

There are multiple mechanisms by which adenine nucleotides are known to be released into the extracellular space. ATP is colocalized with such neurotransmitters as acetylcholine, dopamine, 5-HT, and norepinephrine, and is coreleased on electrical stimulation (e.g. White 1977, Fredholm et al 1982), where it is subsequently hydrolyzed to adenosine. In many systems, cAMP is released into the extracellular space by a probenecid-sensitive transporter (Rosenberg & Li 1995). The amounts of cAMP released in this fashion are sufficient to produce large increases in extracellular adenosine. This can be observed with forskolin stimulation of adenylyl cyclase (Dunwiddie et al 1992, Brundege et al 1997) and following receptor-mediated activation of adenylyl cyclase (Gereau & Conn 1994).

There may be yet other mechanisms for nucleotide release in the brain as well. Proteins that are members of the ATP-binding cassette family of proteins, such as P-glycoprotein (Abraham et al 1993) and the cystic fibrosis transmembrane conductance regulator (Prat et al 1996), appear to be able to function as ATP-conducting ion channels, although this has not been demonstrated in the brain. ATP can also be released by activation of stretch-activated receptors (Hazama et al 1999).

Release of Adenosine Via Facilitated Diffusion Transporters

Another mechanism by which adenosine levels in the extracellular space are regulated is by facilitated diffusion nucleoside transporters. There are two known forms of this transporter, which have been distinguished by their sensitivity to the transport inhibitor nitrobenzylthioinosine (for a review, see Cass et al 1998). These transporters are passive, in that they do not depend on ATP or ionic gradients to transport adenosine, and they equilibrate the concentration of adenosine across cellular membranes. Because of the relatively high activity of intracellular adenosine kinase, adenosine concentrations inside cells are normally low, so the net flux through these transporters is inwardly directed. However, under conditions where intracellular adenosine concentrations rise, these transporters can release adenosine. There are also active transport mechanisms for adenosine, which depend on the Na^+ gradient to provide the energy for transport. Some of these transporters have been cloned (Cass et al 1998), but their relative importance in the regulation of extracellular adenosine concentrations is unclear, largely because of a lack of selective pharmacological tools for these transporters. It is also possible that these transporters could be driven in reverse when intracellular adenosine is high and the Na^+ gradient is reduced, such as during hypoxia, ischemia, and seizures, and thus could become mechanisms for adenosine release as well.

Regulation of Intracellular Adenosine

Because of the presence of equilibrative transporters, regulation of intracellular adenosine concentrations is critical to the regulation of extracellular adenosine, and this control is exerted in two ways. First, if intracellular concentrations of adenosine rise, the ability of these transporters to take up adenosine formed extracellularly from nucleotides is lost, as the adenosine gradient is reduced. Second, if intracellular adenosine concentrations rise even further, direct efflux of adenosine will occur when the intracellular concentration of adenosine exceeds the extracellular.

Although the basic metabolic pathways for intracellular nucleotides in the brain are known, the precise regulation of adenosine kinase and cytosolic 5'-nucleotidase is not well understood (Figure 1). In other tissues, such as heart (Kroll et al 1993), and in hepatocytes (Bontemps et al 1983), there is a high rate of flux in a futile cycle involving these two enzymes, and partly as a consequence, inhibition of adenosine kinase in the heart (and in the brain) leads to very large increases in adenosine.

Physiological Stimuli that Release Adenosine in the Brain

Many physiological manipulations can increase extracellular adenosine, often by cellular mechanisms that are not well understood (Table 2). It would appear likely that the regulation of the activity of key enzymes in intracellular adenine nucleotide/adenosine metabolism (cytosolic 5' nucleotidase-I and adenosine kinase, but also possibly S-adenosylhomocysteine hydrolase and adenosine deaminase) is central to the mechanisms by which diverse stimuli elevate extracellular adenosine in the brain. The diversion of adenine nucleotides into the AMP-adenosine cycle (e.g. by the breakdown of ATP to ADP and AMP during ischemia) would also be expected to contribute to increases in adenosine by mass action, independently of any regulation of enzyme activity.

Although there is wide diversity in the stimuli that will release adenosine, there seem to be some common elements. Manipulations that cause the energy requirements of brain to outstrip its ability to synthesize ATP profoundly increase adenosine release. This can occur either through a large increase in energy requirements (e.g. during seizures) or because of a loss of metabolic substrates (e.g. ischemia). Under these conditions, ATP levels are reduced, and the levels of other adenine nucleotides and adenosine are increased. Because intracellular ATP concentrations are high (typically estimated to be in the range of 3 mM), even a 1% conversion of ATP to adenosine would result in an approximate 100-fold increase in intracellular adenosine and a corresponding increase in the extracellular concentration. However, there is also evidence that adenosine can be released

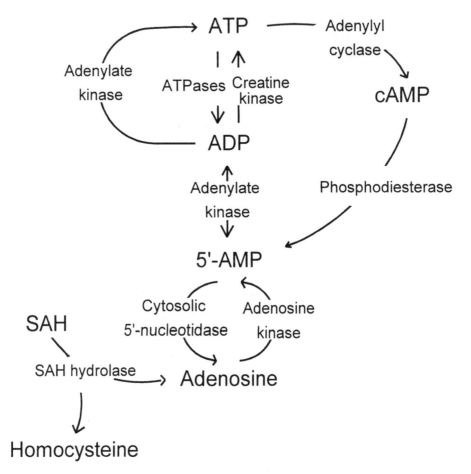

Figure 1 The primary intracellular pathways for the formation of adenosine. Adenosine is formed from 5′-AMP by the cytosolic 5′-nucleotidase and is converted back to 5′-AMP by adenosine kinase (which requires ATP as a phosphate donor). Under resting conditions, there is often a substantial flux through this futile cycle. Adenosine may also be formed by the action of S-adenosylhomocysteine (SAH) hydrolase.

under conditions that should preserve ATP levels (Doolette 1997). Similarly, inhibition of adenosine kinase probably has little effect on ATP levels, but it profoundly increases adenosine release (Pak et al 1994, Lloyd & Fredholm 1995, Brundege & Dunwiddie 1998). In the heart, hypoxia produces a profound inhibition of adenosine kinase activity (to as low as 6% of normal activity) (Decking et al 1997), whereas 5′-nucleotidase activity does not appear to be greatly affected. This generates large amounts of adenosine, and a similar mechanism might underlie adenosine release in the brain as well.

TABLE 2   Experimental manipulations that stimulate adenosine release in the brain[a]

Stimulus[b]	References
Physiological	
Hypoxia, anoxia	Fowler 1989,1993a,b; Gribkoff et al 1990; Lloyd et al 1993; Zetterström et al 1982; Zhu & Krnjevic 1994
Ischemia	Fowler 1993a,b; Pedata et al 1993; Phillis et al 1987
Hypoglycemia	Fowler 1993a,b
Seizures	During & Spencer 1993, Lewin & Bleck 1981, Schrader et al 1980, Winn et al 1980
Increases in temperature	Gabriel et al 1998, Masino & Dunwiddie 1999
Free radicals	Delaney et al 1998, Masino et al 1999
Electrical stimulation	Lloyd et al 1993, Pull & McIlwain 1972, Schrader et al 1980, Yawo & Chuhma 1993
K^+ depolarization	Hoehn & White 1990
Synaptic stimulation	Grover & Teyler 1993, Manzoni et al 1994, Mitchell et al 1993
Pharmacological	
Adenosine kinase inhibitors	Brundege & Dunwiddie 1998, Doolette 1997, Lloyd & Fredholm 1995, Pak et al 1994
Lipopolysaccharides, interleukin-1β	Luk et al 1999, Wang & White 1999
Intracellular acidification	SA Masino, unpublished data
Metabolic inhibitors	
Cyanide	Doolette 1997
Dinitrophenol	Doolette 1997
Na^+ replacement	Fowler 1995
Opiate receptor activation	Stone et al 1989, Sweeney et al 1991
AMPA receptor activation	
Increase	Craig & White 1993
No change	Delaney et al 1998
Kainate receptor activation	Craig & White 1993, Delaney et al 1998
NMDA receptor activation	Chen et al 1992, Craig & White 1993, Delaney et al 1998, Manzoni et al 1994
5-HT receptor activation	Sweeney et al 1990
Forskolin (via cAMP)	Brundege & Dunwiddie 1998, Dunwiddie et al 1992

[a]The preceding list includes not only agents that are thought to directly lead to the efflux of adenosine, but also ones that may release an adenine nucleotide that is subsequently converted to adenosine [e.g. N-methyl-D-aspartate (NMDA) receptor activation]. In addition, some of these stimuli may act indirectly via other mechanisms (e.g. K^+ stimulated release is at least partially the result of glutamate release) (Hoehn & White 1990).

[b]AMPA, α-amino-3-hydroxy-5-methyl-4-isoxazolepropionic acid; 5-HT, 5-hydroxytryptamine.

There are yet other kinds of stimuli [e.g. N-methyl-D-aspartate (NMDA) receptor activation], where the cellular mechanism underlying the release is largely unknown, and where increased ATP breakdown seems unlikely to account for release. Biochemical experiments have suggested that NMDA receptor activation releases an unknown nucleotide, which is then converted to adenosine (Craig & White 1993), whereas electrophysiological experiments have suggested that activation of NMDA receptors releases adenosine per se (Manzoni et al 1994). It seems unlikely that a common mechanism could account for release under all of these conditions, although Doolette (1997) has suggested that intracellular acidification, which can be induced by nearly all the stimuli listed in Table 2, might be a common factor. Further experiments will be required to evaluate the merit of this hypothesis.

Removal of Adenosine from the Extracellular Space

The mechanisms that are responsible for clearing adenosine from the extracellular space are not completely understood, but the transport of adenosine into cells, either by facilitated diffusion or by active transport, appears to be the primary mechanism. Inhibition of facilitated diffusion transport leads to a slowly developing but substantial increase in extracellular adenosine (Dunwiddie & Diao 1994, Zhu & Krnjevic 1994). These increases would probably occur more rapidly except for the fact that inhibitors of transport inhibit both efflux as well as uptake. Thus, the adenosine that builds up extracellularly must come from other sources, such as metabolism of nucleotides by ecto-nucleotidases.

An alternative pathway for the inactivation of extracellular adenosine is its metabolic transformation to inosine by adenosine deaminase. Under either basal conditions or during stimulated release of adenosine from slices of brain tissue, adenosine usually comprises <10% of the total purine efflux, whereas the remainder appears as the adenosine metabolites inosine, hypoxanthine, or xanthine (Pedata et al 1990, Lloyd et al 1993). Although this might imply that adenosine deaminase is relatively important in clearing the extracellular space of adenosine, this is not the case. Adenosine deaminase inhibitors have little or no influence on the concentration of extracellular adenosine (Pak et al 1994, Zhu & Krnjevic 1994; TV Dunwiddie, unpublished data), whereas uptake inhibitors substantially increase adenosine concentrations (Dunwiddie & Diao 1994). The resolution of these seemingly paradoxical observations is that the majority of adenosine in the extracellular space is cleared via reuptake; however, any metabolites that are formed are much more likely than adenosine to diffuse out of the slice without being recaptured and, hence, make a disproportionate contribution to purine efflux. Nevertheless, during hypoxia and ischemia, adenosine deaminase assumes a prominent role in regulating extracellular adenosine concentrations (Lloyd & Fredholm 1995, Barankiewicz et al 1997, Dupere et al 1999). Under these conditions, the adenosine transporters probably are largely inactive, so adenosine deaminase becomes important in the absence of any other mechanisms for adenosine removal.

Abnormalities in Adenosine Regulation

Little is known about the possibility that levels of extracellular adenosine in the brain may differ between individuals, and possibly in certain disease states. However, it has been known for some time that in Down syndrome, purine levels are generally elevated by approximately 50% (Pant et al 1968). Most notable, AMP and ADP levels are significantly elevated, whereas ATP is not (Stocchi et al 1985). A number of the known abnormalities in Down syndrome, such as daytime sedation/sleepiness, reduced pain sensitivity (Martinez-Cue et al 1999), learning disorders (Siarey et al 1999), and central sleep apneas (Ferri et al 1997), are consistent with increased adenosine concentrations, because in experimental models adenosine can produce all these effects. One prediction based on these observations would be that there might be an elevated sensitivity to adenosine receptor antagonists in Down syndrome, but this has apparently never been tested.

PHYSIOLOGICAL ROLES OF ADENOSINE

Role of Adenosine in Normal Physiology

Adenosine appears to subserve a number of diverse roles in normal physiology, which include promoting and/or maintaining sleep, regulating the general state of arousal as well as local neuronal excitability, and coupling cerebral blood flow to energy demand. Selective adenosine receptor antagonists have been used frequently in the past to provide evidence concerning these proposed roles for adenosine. The more recent development of knockout mice for the A_{2A} receptor (Ledent et al 1997, Chen et al 1999), A_3 receptor (Zhao et al 2000), and A_1 receptor (BB Fredholm, personal communication) have provided additional tools with which to characterize the functions of these receptors.

Sleep and Regulation of Arousal   The idea that adenosine plays a role in sleep is a natural outgrowth of the observation that adenosine receptor antagonists such as caffeine promote wakefulness and disrupt normal sleep. Evidence to support this hypothesis generally has fallen into two categories. First, direct measurement of endogenous adenosine in the basal forebrain of cats using microdialysis has shown that adenosine levels progressively increase during prolonged wakefulness and decrease during subsequent recovery sleep (Porkka-Heiskanen et al 1997, Porkka-Heiskanen 1999). A similar relationship between behavioral state and endogenous adenosine appears to exist in the hippocampus but not in the thalamus (Huston et al 1996). Second, pharmacological manipulations involving adenosine receptors have shown that agonists generally promote sleep (Portas et al 1997), whereas antagonists reduce sleep (Lin et al 1997). Some of the most compelling evidence along these lines comes from studies showing that adenosine inhibits neuronal activity in cholinergic nuclei that are thought to regulate arousal (Rainnie et al 1994), and that adenosine dialysis into these regions in vivo promotes sleep and reduces the level of arousal as measured by EEG activity (Portas et al 1997). Parallel noncholinergic systems that contribute to sleep regulation may exist as well; for example, infusion of adenosine or a selective A_1 agonist into the preoptic area has been shown to reduce sleep latency, increase total sleep time, and increase slow-wave sleep (Ticho & Radulovacki 1991, Mendelson 2000). Although many studies have implicated A_1 receptors in both sleep and decreased arousal (Dunwiddie & Worth 1982, Fulga & Stone 1998), there is also evidence that A_{2A} receptors may be involved, particularly in the rostral basal forebrain, where the A_{2A} agonist CGS21680 promotes both REM (rapid eye movement) and non-REM sleep (Satoh et al 1999).

Adenosine as a Retrograde Synaptic Messenger   Although adenosine does not appear to be a classical neurotransmitter, there is some evidence that adenosine could serve as a retrograde synaptic messenger. If an individual neuron is loaded with adenosine via patch pipette, adenosine efflux from that cell is sufficient to significantly inhibit its synaptic inputs, whereas synaptic communication to other nearby cells is unaffected (Brundege & Dunwiddie 1996). The precise subcellular localization of these transporters would be important with respect to this kind of speculative mechanism. However, previous localization studies regarding these transporters are not definitive, or they have lacked the resolution necessary to evaluate this possibility. However, now that these transporters have been cloned, the distribution of transporters should be characterized with more precision and resolution.

Adenosine as a Mechanism for Coupling Energy Demand to Cerebral Blood Flow   Adenosine has long been recognized to be involved in the autoregulation of cerebral blood flow (Berne et al 1974, Winn et al 1981, Wahl & Schilling 1993), where it modulates vascular resistance via A_{2A} receptors (Phillis 1989, Coney & Marshall 1998). Adenosine applied externally to cerebral blood vessels induces vasodilation (Hylland et al 1994), and there is evidence that endogenous adenosine is a tonic regulator of vascular smooth muscle tone. Thus, application of adenosine antagonists causes vasoconstriction and reverses adenosine-mediated vasodilation (Ko et al 1990, Dirnagl et al 1994, Hylland et al 1994). Accordingly, any stimulus that promotes release of additional adenosine from neurons or glia will induce vasodilation.

It has been suggested that this relationship between adenosine and cerebral blood flow is a mechanism that couples increased cell energy expenditure (seen as increased ATP utilization and demand) with increased oxygen and glucose delivery via the cerebral vasculature. The increased adenosine released during such conditions as ischemia would serve to increase cerebral blood flow and could ameliorate the effects of ischemia. However, nonpathological changes in energy requirements have a similar effect; for example, increased activity in somatosensory cortex due to peripheral sensory stimulation is sufficient to induce vasodilation mediated via adenosine, indicating that adenosine is a component of the autoregulatory mechanisms that act on the cerebral vasculature (Ko et al 1990, Dirnagl et al 1994).

Role of Adenosine in Pathological Conditions

Extracellular brain concentrations of adenosine are markedly elevated by a diverse array of pathological stimuli (Table 2). Many of the effects of adenosine that are observed to a minor extent under normal conditions (e.g. presynaptic inhibition of glutamate release) are greatly augmented during pathological events and are neuroprotective in that context. In addition to having acute protective effects, transient activation of adenosine receptors offers protection against damage induced by a subsequent hypoxic or ischemic event. This phenomenon, which is referred to as preconditioning, occurs not only in brain but also in other excitable tissues, such as heart (Miura & Tsuchida 1999).

Neuroprotective Effects of Adenosine in Hypoxia and Ischemia *Acute protective effects* Endogenous adenosine released by hypoxia (Gribkoff & Bauman 1992, Fowler 1993a,b), ischemia (Lloyd et al 1993, Latini et al 1999), electrical activity (Arvin et al 1989, Lloyd et al 1993), and hypo- or aglycemia (Fowler 1993b, Hsu et al 1994, Calabresi et al 1997) reduces the subsequent damage to neuronal tissue. This neuroprotection offered by adenosine is also effective against other kinds of damage that are not as directly related to energy metabolism, such as mechanical cell injury (Mitchell et al 1995) and methamphetamine-induced neurotoxicity (Delle Donne & Sonsalla 1994). Conversely, applying adenosine receptor antagonists in conjunction with any of these conditions exacerbates the consequent damage (Arvin et al 1989, Hsu et al 1994, Mitchell et al 1995).

The neuroprotective actions of adenosine are mediated primarily via A_1 receptor activation, and at least three cellular mechanisms may be involved. Adenosine strongly inhibits transmitter release (and glutamate in particular), hyperpolarizes neurons, and directly inhibits certain kinds of Ca^{2+} channels. All these actions could reduce excitotoxicity by limiting Ca^{2+} entry, which is thought to be a key step in excitotoxic damage, and by reducing metabolic demand, which would help to preserve ATP stores that are essential for pumping Ca^{2+} out of the cell. Experiments with cardiac tissue suggest that the number of A_1 receptors may be a limiting factor in acute protection because overexpression of A_1 receptors provides additional protection against ischemia-reperfusion injury (Matherne et al 1997, Headrick et al 1998). A similar protective effect may be possible in neuronal tissue because an allosteric enhancer of A_1 receptor binding has been shown to offer neuroprotection in neonates (Halle et al 1997). The utility of an alternative strategy, i.e. enhancing the local release of adenosine (e.g. by inhibiting adenosine kinase), is not clear, although positive effects have been reported (Jiang et al 1997). The concentrations of adenosine in the extracellular space during ischemia probably saturate A_1 receptors, so the primary effect of enhancing adenosine release would be expected to be in marginally affected regions, where adenosine concentrations are not as high.

Alternatively, some of the protective effects could be mediated by other receptors (e.g. the A_3 receptor), which has a substantially lower affinity for adenosine and thus, would require higher concentrations for maximal activation. The A_{2A} receptor, on the other hand, may actually contribute to ischemic tissue damage, because mice lacking A_{2A} receptors show reduced brain damage following focal ischemia (Chen et al 1999). The neuroprotective role of adenosine has been reviewed recently (Deckert & Gleiter 1994, Schubert et al 1997, Fredholm 1997) and continues to be an area of rapid development.

Preconditioning A brief episode of mild hypoxia or ischemia that produces little or no damage has been shown to afford protection against a subsequent challenge of greater severity presented hours or even days later. This effect, which is observed in both cardiac and neuronal tissues, has been termed preconditioning and seems to involve A_3 as well as A_1 receptors (Stambaugh et al 1997, Liang & Jacobson 1998). In the brain, adenosine release, A_1 receptor activation, and the opening of ATP-dependent K^+ channels appear to play a central role in preconditioning (Heurteaux et al 1995). Recently, it has been observed that cross-tolerance exists between potentially damaging stimuli, and many of these interactions involve adenosine receptors. For example, a sublethal kainate seizure will protect against subsequent ischemia, and vice versa (Plamondon et al 1999). Chemical inhibition of oxidative phosphorylation provides protection against hypoxia within an hour and lasts for 24 h (Riepe et al 1997), and it may protect against other insults as well. Much clinical interest is focused on determining how to maximize acute neuroprotection, and how to take advantage of the preconditioning phenomenon in both the brain and the heart to improve patient outcome (Liang & Jacobson 1999, Schwarz et al 1999).

Epilepsy   Consistent with its role as an inhibitory neuromodulator, adenosine exhibits anticonvulsant effects in experimental models of epilepsy (for a recent review, see Dunwiddie 1999b). Exogenously administered adenosine receptor agonists reduce seizure activity (Dunwiddie & Worth 1982, Barraco et al 1984, Zhang et al 1990), whereas adenosine receptor antagonists have proconvulsant effects (Dunwiddie 1980, Ault et al 1987), which in hippocampus are mediated by A_1 receptors (Alzheimer et al 1989). Because endogenous levels of adenosine rise markedly during seizure activity (Table 2), it has been proposed that adenosine functions as an "endogenous anticonvulsant" (Dragunow 1988). However, neither the loss of A_1 receptors in knockout mice (BB Fredholm, personal communication) nor the antagonism of adenosine receptors by such antagonists as caffeine lead directly to seizures. Very high concentrations of caffeine can induce convulsions, but this occurs in concentrations where actions other than adenosine receptor antagonism are probably involved. The anticonvulsant effects of adenosine appear to be mediated primarily by A_1 receptors (Murray et al 1992, Zhang et al 1994), although there may be A_{2A} involvement in some regions of the brain. Audiogenic seizures in DBA/2 mice are inhibited by both A_1 and A_{2A} receptor agonists, and selective antagonists for each subtype promote seizures (De Sarro et al 1999).

Beyond the acute anticonvulsant effects of adenosine acting at A_1 receptors, a chronic reduction of A_1 receptors has been found in epileptic tissue, in both humans (Glass et al 1996) and rats (Ochiishi et al 1999). A loss of the tonic inhibitory effects of adenosine may contribute to the hyperexcitability and recurrent seizures that characterize epilepsy.

Despite its profound anticonvulsant effects, adenosine agonists have not proved clinically useful in the treatment of epilepsy because of the peripheral effects of adenosine, which include decreased heart rate, blood pressure, and body temperature (Dunwiddie 1999b). Effective strategies that enhance the protective effects of adenosine near a seizure focus may require a novel approach, such as a method for local release of adenosine. Using this type of technique, Boison et al (1999) have produced a profound reduction in seizure activity in kindled animals by implanting an adenosine-releasing polymer into the cerebral ventricle. Alternatively, a pharmacological strategy that potentiates the effect of endogenous adenosine, such as inhibiting adenosine kinase (Kowaluk & Jarvis 2000) may have clinical potential.

Adenosine and the Actions of Drugs of Abuse

As discussed above, adenosine is centrally involved in the actions of caffeine, which is a relatively nonselective adenosine receptor antagonist, and this pharmacological action is largely responsible for the effects of caffeine on the central nervous system (Fredholm et al 1999). However, there is also evidence that the effects of drugs of abuse may be linked in some manner to adenosine as well.

Ethanol Among the various drugs of abuse, ethanol is perhaps most closely linked mechanistically to adenosine. Three general mechanisms have been proposed to account for this interaction, involving changes in adenosine formation, adenosine uptake, and effects on adenosine receptor coupling. One potential interaction relates to the fact that substantial concentrations (1–2 mM) of acetate are formed as a result of the metabolism of ethanol (Carmichael et al 1991), which is then incorporated into acetyl-coenzyme A with the concomitant formation of AMP. The increase in AMP could then lead directly to increased adenosine formation (Figure 1). Ethanol has also been reported to inhibit facilitated diffusion transporters (Diamond et al 1991, Krauss et al 1993), which would increase extracellular brain concentrations of adenosine by inhibiting uptake. Finally, ethanol can facilitate the receptor-mediated activation of adenylyl cyclase by various hormones and neurotransmitters (Rabin & Molinoff 1981, Hoffman & Tabakoff 1990). Because all the known adenosine receptors can interact with adenylyl cyclase, this provides a third mechanism by which ethanol could modulate effects mediated via adenosine receptors. The general subject of ethanol-adenosine interactions has been discussed extensively in a recent review (Dunwiddie 1999a).

Opiates A number of studies have suggested that opioids in particular, and possibly psychomotor stimulants such as cocaine as well, can interact with adenosine systems. As far as the opioids are concerned, agonists such as morphine have been shown to release adenosine in the brain, spinal cord, and peripheral nervous system (Fredholm & Vernet 1978, Stone 1981, Cahill et al 1996), and this release occurs via a facilitated diffusion transporter (Sweeney et al 1993). The early observation that opioid analgesia can be at least partially antagonized by adenosine receptor antagonists (Ho et al 1973) has been confirmed in more recent studies as well (for a review, see Sawynok et al 1989) and has led to the hypothesis that some opiate actions are mediated indirectly via release of adenosine. The well-established analgesic properties of adenosine receptor agonists (Herrick-Davis et al 1989, Sosnowski et al 1989, Sawynok 1998) provide further support for this hypothesis. In behavioral studies, adenosine receptor antagonists elicit a response termed the quasi-morphine withdrawal syndrome (Francis et al 1975), which in many respects is similar to the response to naloxone in opiate-tolerant animals. Thus, there are strong parallels between the pharmacological effects of opiate and adenosine agonists, and also between opiate and adenosine antagonists. Finally, an interesting purinergic role has also emerged in terms of the effects of chronic opioids as well as cocaine; in animals withdrawn from chronic treatment with either morphine or cocaine, there are persistent increases in extracellular adenosine in the ventral tegmental region, a brain region intimately involved in the rewarding effects of these drugs (Bonci & Williams 1996, Shoji et al 1999, Fiorillo & Williams 2000). The source of this adenosine appears to be from the release of cAMP and subsequent extracellular catabolism to adenosine.

SUMMARY AND CONCLUSIONS

Adenosine is involved in a diverse array of functions in the central nervous system. Although in a general sense many of its effects are inhibitory, consistent with its proposed roles as an endogenous anticonvulsant, neuroprotectant, and sleep-inducing factor, this differs depending on the brain system and the complement of adenosine receptors that are present. There is little evidence that adenosine is a neurotransmitter; rather, it appears to be a neuromodulator that is released in some unconventional ways to regulate and modulate neuronal activity. A current challenge in this field is to better define the mechanisms underlying the release of adenosine evoked by pathological and nonpathological stimuli. These kinds of studies should help to clarify the role of adenosine as a signaling agent in the brain and to relate this function to its other actions such as neuroprotection.

Visit the Annual Reviews home page at www.AnnualReviews.org

LITERATURE CITED

Abraham EH, Prat AG, Gerweck L, Seneveratne T, Arceci RJ, et al. 1993. The multidrug resistance (mdr1) gene product functions as an ATP channel. *Proc. Natl. Acad. Sci. USA* 90:312–16

Alzheimer C, Sutor B, ten Bruggencate G. 1989. Transient and selective blockade of adenosine A_1-receptors by 8-cyclopentyl-1, 3-dipropylxanthine (DPCPX) causes sustained epileptiform activity in hippocampal

CA3 neurons of guinea pigs. *Neurosci. Lett.* 99:107–12

Andresen BT, Gillespie DG, Mi ZC, Dubey RK, Jackson EK. 1999. Role of adenosine A_1 receptors in modulating extracellular adenosine levels. *J. Pharmacol. Exp. Ther.* 291: 76–80

Arvin B, Neville LF, Pan J, Roberts PJ. 1989. 2-Chloroadenosine attenuates kainic acid-induced toxicity within the rat striatum: relationship to release of glutamate and Ca^{2+} influx. *Br. J. Pharmacol.* 98:225–35

Ault B, Olney MA, Joyner JL, Boyer CE, Notrica MA, et al. 1987. Pro-convulsant actions of theophylline and caffeine in the hippocampus: implications for the management of temporal lobe epilepsy. *Brain Res.* 426:93–102

Ballarin M, Fredholm BB, Ambrosio S, Mahy N. 1991. Extracellular levels of adenosine and its metabolites in the striatum of awake rats: inhibition of uptake and metabolism. *Acta Physiol. Scand.* 142:97–103

Barankiewicz J, Danks AM, Abushanab E, Makings L, Wiemann T, et al. 1997. Regulation of adenosine concentration and cytoprotective effects of novel reversible adenosine deaminase inhibitors. *J. Pharmacol. Exp. Ther.* 283:1230–38

Barraco RA, Swanson TH, Phillis JW, Berman RF. 1984. Anticonvulsant effects of adenosine analogues on amygdaloid-kindled seizures in rats. *Neurosci. Lett.* 46:317–22

Berne RM, Rubio R, Curnish RR. 1974. Release of adenosine from ischemic brain. Effect on cerebral vascular resistance and incorporation into cerebral adenine nucleotides. *Circ. Res.* 35:262–71

Boison D, Scheurer L, Tseng JL, Aebischer P, Mohler H. 1999. Seizure suppression in kindled rats by intraventricular grafting of an adenosine releasing synthetic polymer. *Exp. Neurol.* 160:164–74

Bonci A, Williams JT. 1996. A common mechanism mediates long-term changes in synaptic transmission after chronic cocaine and morphine. *Neuron* 16:631–39

Bontemps F, Van den Berghe G, Hers HG. 1983. Evidence for a substrate cycle between AMP and adenosine in isolated hepatocytes. *Proc. Natl. Acad. Sci. USA* 80:2829–33

Brundege JM, Diao LH, Proctor WR, Dunwiddie TV. 1997. The role of cyclic AMP as a precursor of extracellular adenosine in the rat hippocampus. *Neuropharmacology* 36:1201–10

Brundege JM, Dunwiddie TV. 1996. Modulation of excitatory synaptic transmission by adenosine released from single hippocampal pyramidal neurons. *J. Neurosci.* 16: 5603–12

Brundege JM, Dunwiddie TV. 1998. Metabolic regulation of endogenous adenosine release from single neurons. *NeuroReport* 9: 3007–11

Cahill CM, White TD, Sawynok J. 1996. Synergy between μ/δ-opioid receptors mediates adenosine release from spinal cord synaptosomes. *Eur. J. Pharmacol.* 298: 45–49

Calabresi P, Centonze D, Pisani A, Bernardi G. 1997. Endogenous adenosine mediates the presynaptic inhibition induced by aglycemia at corticostriatal synapses. *J. Neurosci.* 17: 4509–16

Carmichael FJ, Israel Y, Crawford M, Minhas K, Saldivia V, et al. 1991. Central nervous system effects of acetate: contribution to the central effects of ethanol. *J. Pharmacol. Exp. Ther.* 259:403–8

Cass CE, Young JD, Baldwin SA. 1998. Recent advances in the molecular biology of nucleoside transporters of mammalian cells. *Biochem. Cell Biol.* 76:761–70

Chen JF, Huang ZH, Ma JY, Zhu JM, Moratalla R, et al. 1999. A_{2A} adenosine receptor deficiency attenuates brain injury induced by transient focal ischemia in mice. *J. Neurosci.* 19:9192–200

Chen Y, Graham DI, Stone TW. 1992. Release of endogenous adenosine and its metabolites by the activation of NMDA receptors in the rat hippocampus in vivo. *Br. J. Pharmacol.* 106:632–38

Coney AM, Marshall JM. 1998. Role of adenosine and its receptors in the vasodilatation induced in the cerebral cortex of the rat by systemic hypoxia. *J. Physiol.* 509:507–18

Craig CG, White TD. 1993. N-methyl-D-aspartate- and non-N-methyl-D-aspartate-evoked adenosine release from rat cortical slices: distinct purinergic sources and mechanisms of release. *J. Neurochem.* 60: 1073–80

Cunha RA, Milusheva E, Vizi ES, Ribeiro JA, Sebastiao AM. 1994. Excitatory and inhibitory effects of A_1 and A_{2A} adenosine receptor activation on the electrically evoked [^{3}H] acetylcholine release from different areas of the rat hippocampus. *J. Neurochem.* 63:207–14

Cunha RA, Sebastiao AM, Ribeiro JA. 1998. Inhibition by ATP of hippocampal synaptic transmission requires localized extracellular catabolism by ecto-nucleotidases into adenosine and channeling to adenosine A_1 receptors. *J. Neurosci.* 18:1987–95

Deckert J, Gleiter CH. 1994. Adenosine–an endogenous neuroprotective metabolite and neuromodulator. *J. Neural Transm. Suppl.* 43:23–31

Decking UKM, Schlieper G, Kroll K, Schrader J. 1997. Hypoxia-induced inhibition of adenosine kinase potentiates cardiac adenosine release. *Circ. Res.* 81:154–64

Delaney SM, Geiger JD. 1996. Brain regional levels of adenosine and adenosine nucleotides in rats killed by high-energy focused microwave irradiation. *J. Neurosci. Methods* 64:151–56

Delaney SM, Shepel PN, Geiger JD. 1998. Levels of endogenous adenosine in rat striatum. I. Regulation by ionotropic glutamate receptors, nitric oxide and free radicals. *J. Pharmacol. Exp. Ther.* 285:561–67

Delle Donne KT, Sonsalla PK. 1994. Protection against methamphetamine-induced neurotoxicity to neostriatal dopaminergic neurons by adenosine receptor activation. *J. Pharmacol. Exp. Ther.* 271:1320–26

De Sarro G, De Sarro A, di Paola ED, Bertorelli R. 1999. Effects of adenosine receptor agonists and antagonists on audiogenic seizure-sensible DBA/2 mice. *Eur. J. Pharmacol.* 371:137–45

Diamond I, Nagy L, Mochly-Rosen D, Gordon A. 1991. The role of adenosine and adenosine transport in ethanol-induced cellular tolerance and dependence. Possible biologic and genetic markers of alcoholism. *Ann. NY Acad. Sci.* 625:473–87

Dirnagl U, Niwa K, Lindauer U, Villringer A. 1994. Coupling of cerebral blood flow to neuronal activation: role of adenosine and nitric oxide. *Am. J. Physiol.* 267: H296–301

Dixon AK, Gubitz AK, Sirinathsinghji DJ, Richardson PJ, Freeman TC. 1996. Tissue distribution of adenosine receptor mRNAs in the rat. *Br. J. Pharmacol.* 118:1461–68

Doolette DJ. 1997. Mechanism of adenosine accumulation in the hippocampal slice during energy deprivation. *Neurochem. Int.* 30:211–23

Dragunow M. 1988. Purinergic mechanisms in epilepsy. *Prog. Neurobiol.* 31:85–108

Dunwiddie TV. 1980. Endogenously released adenosine regulates excitability in the *in vitro* hippocampus. *Epilepsia* 21:541–48

Dunwiddie TV. 1999a. Adenosine and ethanol: is there a caffeine connection in the actions of ethanol? In *The "Drunken" Synapse: Studies of Alcohol-Related Disorders*, ed. Y Liu, WA Hunt, pp. 119–33. New York: Kluwer/Plenum

Dunwiddie TV. 1999b. Adenosine and suppression of seizures. In *Jasper's Basic Mechanisms of the Epilepsies*, ed. A Delgado-Escueta, WA Wilson, RW Olsen, RJ Porter, pp. 1001–10. Philadelphia: Lippincott Williams & Wilkins

Dunwiddie TV, Diao LH. 1994. Extracellular adenosine concentrations in hippocampal brain slices and the tonic inhibitory modulation of evoked excitatory responses. *J. Pharmacol. Exp. Ther.* 268:537–45

Dunwiddie TV, Diao LH, Kim HO, Jiang JL,

Jacobson KA. 1997a. Activation of hippocampal adenosine A_3 receptors produces a desensitization of A_1 receptor-mediated responses in rat hippocampus. *J. Neurosci.* 17:607–14

Dunwiddie TV, Diao LH, Proctor WR. 1997b. Adenine nucleotides undergo rapid, quantitative conversion to adenosine in the extracellular space in rat hippocampus. *J. Neurosci.* 17:7673–82

Dunwiddie TV, Hoffer BJ. 1980. Adenine nucleotides and synaptic transmission in the *in vitro* rat hippocampus. *Br. J. Pharmacol.* 69:59–68

Dunwiddie TV, Taylor M, Heginbotham LR, Proctor WR. 1992. Long-term increases in excitability in the CA1 region of rat hippocampus induced by beta-adrenergic stimulation: possible mediation by cAMP. *J. Neurosci.* 12:506–17

Dunwiddie TV, Worth TS. 1982. Sedative and anticonvulsant effects of adenosine analogs in mouse and rat. *J. Pharmacol. Exp. Ther.* 220:70–76

Dupere JRB, Dale TJ, Starkey SJ, Xie XM. 1999. The anticonvulsant BW534U87 depresses epileptiform activity in rat hippocampal slices by an adenosine-dependent mechanism and through inhibition of voltage-gated Na^+ channels. *Br. J. Pharmacol.* 128:1011–20

During MJ, Spencer DD. 1993. Adenosine: a potential mediator of seizure arrest and postictal refractoriness. *Ann. Neurol.* 32: 618–24

Edwards FA, Robertson SJ. 1999. The function of A_2 adenosine receptors in the mammalian brain: evidence for inhibition vs. enhancement of voltage gated calcium channels and neurotransmitter release. *Prog. Brain Res.* 120:265–73

El Yacoubi M, Ledent C, Ménard JF, Parmentier M, Costentin J, et al. 2000. The stimulant effects of caffeine on locomotor behaviour in mice are mediated through its blockade of adenosine A_{2A} receptors. *Br. J. Pharmacol.* 129:1465–73

Feoktistov I, Biaggioni I. 1997. Adenosine A_{2B} receptors. *Pharm. Rev.* 49:381–402

Ferri R, Curzi-Dascalova L, Del Gracco S, Elia M, Musumeci SA, et al. 1997. Respiratory patterns during sleep in Down's syndrome: importance of central apnoeas. *J. Sleep Res.* 6:134–41

Fiorillo CD, Williams JT. 2000. Selective inhibition by adenosine of mGluR IPSPs in dopamine neurons after cocaine treatment. *J. Neurophysiol.* 83:1307–14

Fowler JC. 1989. Adenosine antagonists delay hypoxia-induced depression of neuronal activity in hippocampal brain slice. *Brain Res.* 490:378–84

Fowler JC. 1993a. Changes in extracellular adenosine levels and population spike amplitude during graded hypoxia in the rat hippocampal slice. *Naunyn Schmiedebergs Arch. Pharmacol.* 347:73–78

Fowler JC. 1993b. Purine release and inhibition of synaptic transmission during hypoxia and hypoglycemia in rat hippocampal slices. *Neurosci. Lett.* 157:83–86

Fowler JC. 1995. Choline substitution for sodium triggers glutamate and adenosine release from rat hippocampal slices. *Neurosci. Lett.* 197:97–100

Francis DL, Roy AC, Collier HOJ. 1975. Morphine abstinence and quasi-abstinence effects after phosphodiesterase inhibitors and naloxone. *Life Sci.* 16:1901–6

Fredholm BB. 1997. Adenosine and neuroprotection. *Int. Rev. Neurobiol.* 40:259–80

Fredholm BB, Battig K, Holmen J, Nehlig A, Zvartau EE. 1999. Actions of caffeine in the brain with special reference to factors that contribute to its widespread use. *Pharm. Rev.* 51:83–133

Fredholm BB, Dunwiddie TV. 1988. How does adenosine inhibit transmitter release? *Trends Pharmacol.* 9:130–34

Fredholm BB, Fried G, Hedqvist P. 1982. Origin of adenosine released from rat vas deferens by nerve stimulation. *Eur. J. Pharmacol.* 79:233–43

Fredholm BB, Vernet L. 1978. Morphine increases depolarization induced purine release from rat cortical slices. *Acta Physiol. Scand.* 104:502–4

Fulga I, Stone TW. 1998. Comparison of an adenosine A_1 receptor agonist and antagonist on the rat EEG. *Neurosci. Lett.* 244:55–59

Fuxe K, Ferre S, Zoli M, Agnati LF. 1998. Integrated events in central dopamine transmission as analyzed at multiple levels. Evidence for intramembrane adenosine A_{2A}/dopamine D_2 and adenosine A_1/dopamine D_1 receptor interactions in the basal ganglia. *Brain Res. Rev.* 26:258–73

Gabriel A, Klussmann FW, Igelmund P. 1998. Rapid temperature changes induce adenosine-mediated depression of synaptic transmission in hippocampal slices from rats (non-hibernators) but not in slices from golden hamsters (hibernators). *Neuroscience* 86:67–77

Gereau RW, Conn PJ. 1994. Potentiation of cAMP responses by metabotropic glutamate receptors depresses excitatory synaptic transmission by a kinase-independent mechanism. *Neuron* 12:1121–29

Glass M, Faull RL, Bullock JY, Jansen K, Mee EW, et al. 1996. Loss of A_1 adenosine receptors in human temporal lobe epilepsy. *Brain Res.* 710:56–68

Gribkoff VK, Bauman LA. 1992. Endogenous adenosine contributes to hypoxic synaptic depression in hippocampus from young and aged rats. *J. Neurophysiol.* 68:620–28

Gribkoff VK, Bauman LA, VanderMaelen CP. 1990. The adenosine antagonist 8-cyclopentyltheophylline reduces the depression of hippocampal neuronal responses during hypoxia. *Brain Res.* 512:353–57

Grover LM, Teyler TJ. 1993. Role of adenosine in heterosynaptic, posttetanic depression in area CA1 of hippocampus. *Neurosci. Lett.* 154:39–42

Halle JN, Kasper CE, Gidday JM, Koos BJ. 1997. Enhancing adenosine A_1 receptor binding reduces hypoxic-ischemic brain injury in newborn rats. *Brain Res.* 759:309–12

Harden TK, Boyer JL, Nicholas RA. 1995. P_2-purinergic receptors: subtype-associated signaling responses and structure. *Annu. Rev. Pharmacol. Toxicol.* 35:541–79

Hazama A, Shimizu T, Ando-Akatsuka Y, Hayashi S, Tanaka S, et al. 1999. Swelling-induced, CFTR-independent ATP release from a human epithelial cell line: lack of correlation with volume-sensitive cl(−) channels. *J. Gen. Physiol.* 114:525–33

Headrick JP, Gauthier NS, Berr SS, Morrison RR, Matherne GP. 1998. Transgenic A_1 adenosine receptor overexpression markedly improves myocardial energy state during ischemia-reperfusion. *J. Mol. Cell. Cardiol.* 30:1059–64

Herrick-Davis K, Chippari S, Luttinger D, Ward SJ. 1989. Evaluation of adenosine agonists as potential analgesics. *Eur. J. Pharmacol.* 162:365–69

Heurteaux C, Lauritzen I, Widmann C, Lazdunski M. 1995. Essential role of adenosine, adenosine A_1 receptors, and ATP-sensitive K^+ channels in cerebral ischemic preconditioning. *Proc. Natl. Acad. Sci. USA* 92:4666–70

Ho IK, Loh HH, Way EL. 1973. Cyclic adenosine monophosphate antagonism of morphine analgesia. *J. Pharmacol. Exp. Ther.* 185:336–46

Hoehn K, White TD. 1990. Role of excitatory amino acid receptors in K^+- and glutamate-evoked release of endogenous adenosine from rat cortical slices. *J. Neurochem.* 54:256–65

Hoffman PL, Tabakoff B. 1990. Ethanol and guanine nucleotide binding proteins: a selective interaction. *FASEB J.* 4:2612–22

Hsu SS, Newell DW, Tucker A, Malouf AT, Winn HR. 1994. Adenosinergic modulation of CA1 neuronal tolerance to glucose deprivation in organotypic hippocampal cultures. *Neurosci. Lett.* 178:189–92

Huston JP, Haas HL, Boix F, Pfister M, Decking U, et al. 1996. Extracellular adenosine levels in neostriatum and hippocampus during rest and activity periods of rats. *Neuroscience* 73:99–107

Hylland P, Nilsson GE, Lutz PL. 1994. Time course of anoxia-induced increase in cerebral blood flow rate in turtles: evidence for a role of adenosine. *J. Cereb. Blood Flow Metab.* 14:877–81

Jiang N, Kowaluk EA, Lee CH, Mazdiyasni H, Chopp M. 1997. Adenosine kinase inhibition protects brain against transient focal ischemia in rats. *Eur. J. Pharmacol.* 320:131–37

Ko KR, Ngai AC, Winn HR. 1990. Role of adenosine in regulation of regional cerebral blood flow in sensory cortex. *Am. J. Physiol.* 259:H1703–8

Kocsis JD, Eng DL, Bhisitkul RB. 1984. Adenosine selectively blocks parallel-fiber-mediated synaptic potentials in rat cerebellar cortex. *Proc. Natl. Acad. Sci. USA* 81:6531–34

Kowaluk EA, Jarvis MF. 2000. Therapeutic potential of adenosine kinase inhibitors. *Exp. Opin. Invest. Drugs* 9:1–14

Krauss SW, Ghirnikar RB, Diamond I, Gordon AS. 1993. Inhibition of adenosine uptake by ethanol is specific for one class of nucleoside transporters. *Mol. Pharmacol.* 44:1021–26

Kroll K, Decking UK, Dreikorn K, Schrader J. 1993. Rapid turnover of the AMP-adenosine metabolic cycle in the guinea pig heart. *Circ. Res.* 73:846–56

Latini S, Bordoni F, Pedata F, Corradetti R. 1999. Extracellular adenosine concentrations during *in vitro* ischaemia in rat hippocampal slices. *Br. J. Pharmacol.* 127:729–39

Latini S, Pazzagli M, Pepeu G, Pedata F. 1996. A_2 adenosine receptors: their presence and neuromodulatory role in the central nervous system. *Gen. Pharmacol.* 27:925–33

Ledent C, Vaugeois J-M, Schiffmann SN, Pedrazzini T, El Yacoubi M, et al. 1997. Aggressiveness, hypoalgesia and high blood pressure in mice lacking the adenosine A2a receptor. *Nature* 388:674–78

Lewin E, Bleck V. 1981. Electroshock seizures in mice: effect on brain adenosine and its metabolites. *Epilepsia* 22:577–81

Liang BT, Jacobson KA. 1998. A physiological role of the adenosine A_3 receptor: sustained cardioprotection. *Proc. Natl. Acad. Sci. USA* 95:6995–99

Liang BT, Jacobson KA. 1999. Adenosine and ischemic preconditioning. *Curr. Pharm. Design* 5:1029–41

Lin AS, Uhde TW, Slate SO, McCann UD. 1997. Effects of intravenous caffeine administered to healthy males during sleep. *Depress. Anxiety* 5:21–28

Linden J, Auchampach JA, Jin XW, Figler RA. 1998. The structure and function of A_1 and A_{2B} adenosine receptors. *Life Sci.* 62: 1519–24

Lloyd HGE, Fredholm BB. 1995. Involvement of adenosine deaminase and adenosine kinase in regulating extracellular adenosine concentration in rat hippocampal slices. *Neurochem. Int.* 26:387–95

Lloyd HGE, Lindström K, Fredholm BB. 1993. Intracellular formation and release of adenosine from rat hippocampal slices evoked by electrical stimulation or energy depletion. *Neurochem. Int.* 23:173–85

Luk WP, Zhang Y, White TD, Lue FA, Wu CP, et al. 1999. Adenosine: a mediator of interleukin-1β-induced hippocampal synaptic inhibition. *J. Neurosci.* 19:4238–44

Macdonald RL, Skerritt JH, Werz MA. 1986. Adenosine agonists reduce voltage-dependent calcium conductance of mouse sensory neurones in cell culture. *J. Physiol.* 370:75–90

Macek TA, Schaffhauser H, Conn PJ. 1998. Protein kinase C and A_3 adenosine receptor activation inhibit presynaptic metabotropic glutamate receptor (mGluR) function and uncouple mGluRs from GTP-binding proteins. *J. Neurosci.* 18:6138–46

Manzoni OJ, Manabe T, Nicoll RA. 1994. Release of adenosine by activation of NMDA receptors in hippocampus. *Science* 265: 2098–101

Marston HM, Finlayson K, Maemoto T, Olverman HJ, Akahane A, et al. 1998. Pharmacological characterization of a simple behavioral response mediated selectively by central adenosine A_1 receptors, using in vivo and in vitro techniques. *J. Pharmacol. Exp. Ther.* 285:1023–30

Martinez-Cue C, Baamonde C, Lumbreras MA, Vallina IF, Dierssen M, et al. 1999. A murine model for Down syndrome shows reduced responsiveness to pain. *NeuroReport* 10:1119–22

Masino SA, Dunwiddie TV. 1999. Temperature-dependent modulation of excitatory transmission in hippocampal slices is mediated by extracellular adenosine. *J. Neurosci.* 19:1932–39

Masino SA, Mesches MH, Bickford PC, Dunwiddie TV. 1999. Acute peroxide treatment of rat hippocampal slices induced adenosine-mediated inhibition of excitatory transmission in area CA1. *Neurosci. Lett.* 274:91–94

Matherne GP, Linden J, Byford AM, Gauthier NS, Headrick JP. 1997. Transgenic A_1 adenosine receptor overexpression increases myocardial resistance to ischemia. *Proc. Natl. Acad. Sci. USA* 94:6541–46

Mendelson WB. 2000. Sleep-inducing effects of adenosine microinjections into the medial preoptic area are blocked by flumazenil. *Brain Res.* 852:479–81

Mitchell HL, Frisella WA, Brooker RW, Yoon KW. 1995. Attenuation of traumatic cell death by an adenosine A_1 agonist in rat hippocampal cells. *Neurosurgery* 36:1003–7

Mitchell JB, Lupica CR, Dunwiddie TV. 1993. Activity-dependent release of endogenous adenosine modulates synaptic responses in the rat hippocampus. *J. Neurosci.* 13:3439–47

Miura T, Tsuchida A. 1999. Adenosine and preconditioning revisited. *Clin. Exp. Pharmacol. Physiol.* 26:92–99

Murray TF, Franklin PH, Zhang G, Tripp E. 1992. A_1 adenosine receptors express seizure-suppressant activity in the rat prepiriform cortex. *Epilepsy Res. Suppl.* 8:255–61

Ochiishi T, Takita M, Ikemoto M, Nakata H, Suzuki SS. 1999. Immunohistochemical analysis on the role of adenosine A_1 receptors in epilepsy. *NeuroReport* 10:3535–41

Olah ME, Stiles GL. 1995. Adenosine receptor subtypes: characterization and therapeutic regulation. *Annu. Rev. Pharmacol. Toxicol.* 35:581–606

Ongini E. 1997. SCH 58261: a selective A_{2A} adenosine receptor antagonist. *Drug. Dev. Res.* 42:63–70

Pak MA, Haas HL, Decking UK, Schrader J. 1994. Inhibition of adenosine kinase increases endogenous adenosine and depresses neuronal activity in hippocampal slices. *Neuropharmacology* 33:1049–53

Pant SS, Moser HW, Krane SM. 1968. Hyperuricemia in Down's syndrome. *J. Clin. Endocrinol. Metab.* 28:472–78

Pedata F, Latini S, Pugliese AM, Pepeu G. 1993. Investigations into the adenosine outflow from hippocampal slices evoked by ischemia-like conditions. *J. Neurochem.* 61:284–89

Pedata F, Pazzagli M, Tilli S, Pepeu G. 1990. Regional differences in the electrically stimulated release of endogenous and radioactive adenosine and purine derivatives from rat brain slices. *Naunyn Schmiedebergs Arch. Pharmacol.* 342:447–53

Phillis JW. 1989. Adenosine in the control of the cerebral circulation. *Cerebrovasc. Brain Metab. Rev.* 1:26–54

Phillis JW, Walter GA, O'Regan MH, Stair RE. 1987. Increases in cerebral cortical perfusate adenosine and inosine concentrations during hypoxia and ischemia. *J. Cereb. Blood Flow Metab.* 7:679–86

Plamondon H, Blondeau N, Heurteaux C, Lazdunski M. 1999. Mutually protective actions of kainic acid epileptic preconditioning and sublethal global ischemia on hippocampal neuronal death: involvement of adenosine A_1 receptors and $K_{(ATP)}$ channels. *J. Cereb. Blood Flow Metab.* 19:1296–308

Porkka-Heiskanen T. 1999. Adenosine in sleep and wakefulness. *Ann. Med.* 31:125–29

Porkka-Heiskanen T, Strecker RE, Thakkar M, Bjorkum AA, Greene RW, et al. 1997. Adenosine: a mediator of the sleep-inducing effects of prolonged wakefulness. *Science* 276:1265–68

Portas CM, Thakkar M, Rainnie DG, Greene RW, McCarley RW. 1997. Role of adenosine in behavioral state modulation: a microdialysis study in the freely moving cat. *Neuroscience* 79:225–35

Prat AG, Reisin IL, Ausiello DA, Cantiello HF. 1996. Cellular ATP release by the cystic fibrosis transmembrane conductance regulator. *Am. J. Physiol.* 270:C538–45

Pull I, McIlwain H. 1972. Adenine derivatives as neurohumoral agents in the brain. The quantities liberated on excitation of superfused cerebral tissues. *Biochem. J.* 130:975–81

Rabin RA, Molinoff PB. 1981. Activation of adenylate cyclase by ethanol in mouse striatal tissue. *J. Pharmacol. Exp. Ther.* 216:129–34

Rainnie DG, Grunze HC, McCarley RW, Greene RW. 1994. Adenosine inhibition of mesopontine cholinergic neurons: implications for EEG arousal. *Science* 263:689–92

Ralevic V, Burnstock G. 1998. Receptors for purines and pyrimidines. *Pharmacol. Rev.* 50:413–92

Riepe MW, Esclaire F, Kasischke K, Schreiber S, Nakase H, et al. 1997. Increased hypoxic tolerance by chemical inhibition of oxidative phosphorylation: "chemical preconditioning." *J. Cereb. Blood Flow Metab.* 17:257–64

Rosenberg PA, Li Y. 1995. Adenylyl cyclase activation underlies intracellular cyclic AMP accumulation, cyclic AMP transport, and extracellular adenosine accumulation evoked by β-adrenergic receptor stimulation in mixed cultures of neurons and astrocytes derived from rat cerebral cortex. *Brain Res.* 692:227–32

Satoh S, Matsumura H, Koike N, Tokunaga Y, Maeda T, et al. 1999. Region-dependent difference in the sleep-promoting potency of an adenosine A_{2A} receptor agonist. *Eur. J. Neurosci.* 11:1587–97

Sawynok J. 1998. Adenosine receptor activation and nociception. *Eur. J. Pharmacol.* 347:1–11

Sawynok J, Sweeney MI, White TD. 1989. Adenosine release may mediate spinal analgesia by morphine. *Trends Pharmacol.* 10:186–89

Scanziani M, Capogna M, Gahwiler BH, Thompson SM. 1992. Presynaptic inhibition of miniature excitatory synaptic currents by baclofen and adenosine in the hippocampus. *Neuron* 9:919–27

Schrader J, Wahl M, Kuschinsky W, Kreutzberg GW. 1980. Increase of adenosine content in cerebral cortex of the cat during bicuculline-induced seizure. *Pflugers Arch.* 387:245–51

Schubert P, Ogata T, Marchini C, Ferroni S, Rudolphi K. 1997. Protective mechanisms of adenosine in neurons and glial cells. *Ann. NY Acad. Sci.* 825:1–10

Schwarz ER, Reffelmann T, Kloner RA. 1999. Clinical effects of ischemic preconditioning. *Curr. Opin. Cardiol.* 14:340–48

Shoji Y, Delfs J, Williams JT. 1999. Presynaptic inhibition of $GABA_B$-mediated synaptic potentials in the ventral tegmental area during morphine withdrawal. *J. Neurosci.* 19:2347–55

Shryock JC, Snowdy S, Baraldi PG, Cacciari B, Spalluto G, et al. 1998. A_{2A}-adenosine receptor reserve for coronary vasodilation. *Circulation* 98:711–18

Siarey RJ, Carlson EJ, Epstein CJ, Balbo A, Rapoport SI, et al. 1999. Increased synaptic depression in the Ts65Dn mouse, a model for mental retardation in Down syndrome. *Neuropharmacology* 38:1917–20

Sodickson DL, Bean BP. 1998. Neurotransmitter activation of inwardly rectifying potassium current in dissociated hippocampal CA3 neurons: interactions among multiple receptors. *J. Neurosci.* 18:8153–62

Sosnowski M, Stevens CW, Yaksh TL. 1989. Assessment of the role of A_1/A_2 adenosine receptors mediating the purine antinociception, motor and autonomic function in the rat spinal cord. *J. Pharmacol. Exp. Ther.* 250:915–22

Stambaugh K, Jacobson KA, Jiang JL, Liang BT. 1997. A novel cardioprotective function of adenosine A_1 and A_3 receptors during prolonged simulated ischemia. *Am. J. Physiol. Heart Circ. Physiol.* 273:H501–5

Stocchi V, Magnani M, Cucchiarini L, Novelli G, Dallapiccola B. 1985. Red blood cell adenine nucleotides abnormalities in Down syndrome. *Am. J. Med. Genet.* 20:131–35

Stone TW. 1981. The effects of 4-aminopyridine on the isolated vas deferens and its effects on the inhibitory properties of adenosine, morphine, noradrenaline and gamma-aminobutyric acid. *Br. J. Pharmacol.* 73: 791–96

Stone TW, Fredholm BB, Phillis JW. 1989. Adenosine and morphine. *Trends Pharmacol.* 10:316–16

Sweeney MI, White TD, Sawynok J. 1990. 5-Hydroxytryptamine releases adenosine and cyclic AMP from primary afferent nerve terminals in the spinal cord in vivo. *Brain Res.* 528:55–61

Sweeney MI, White TD, Sawynok J. 1991. Intracerebroventricular morphine releases adenosine and adenosine 3′, 5′-cyclic monophosphate from the spinal cord via a serotonergic mechanism. *J. Pharmacol. Exp. Ther.* 259:1013–18

Sweeney MI, White TD, Sawynok J. 1993. Morphine-evoked release of adenosine from the spinal cord occurs via a nucleoside carrier with differential sensitivity to dipyridamole and nitrobenzylthioinosine. *Brain Res.* 614:301–7

Ticho SR, Radulovacki M. 1991. Role of adenosine in sleep and temperature regulation in the preoptic area of rats. *Pharmacol. Biochem. Behav.* 40:33–40

Trussell LO, Jackson MB. 1985. Adenosine-activated potassium conductance in cultured striatal neurons. *Proc. Natl. Acad. Sci. USA* 82:4857–61

Wahl M, Schilling L. 1993. Regulation of cerebral blood flow—a brief review. *Acta Neurochir. Suppl.* 59:3–10

Wang YSS, White TD. 1999. The bacterial endotoxin lipopolysaccharide causes rapid inappropriate excitation in rat cortex. *J. Neurochem.* 72:652–60

White TD. 1977. Direct detection of depolarisation-induced release of ATP from a synaptosomal preparation. *Nature* 267:67–68

Winn HR, Rubio GR, Berne RM. 1981. The role of adenosine in the regulation of cerebral blood flow. *J. Cereb. Blood Flow Metab.* 1:239–44

Winn HR, Welsh JE, Rubio R, Berne RM. 1980. Changes in brain adenosine during bicuculline-induced seizures in rats. Effects of hypoxia and altered systemic blood pressure. *Circ. Res.* 47:868–77

Wu LG, Saggau P. 1997. Presynaptic inhibition of elicited neurotransmitter release. *Trends Neurosci.* 20:204–12

Yawo H, Chuhma N. 1993. Preferential inhibition of omega-conotoxin-sensitive presynaptic Ca^{2+} channels by adenosine autoreceptors. *Nature* 365:256–58

Zetterström TS, Vernet L, Ungerstedt U, Tossman U, Jonzon B. 1982. Purine levels in the intact rat brain. Studies with an implanted perfused hollow fibre. *Neurosci. Lett.* 29:111–15

Zhang G, Franklin PH, Murray TF. 1990. Anticonvulsant effect of N-ethylcarboxamidoadenosine against kainic acid-induced behavioral seizures in the rat prepiriform cortex. *Neurosci. Lett.* 114:345–50

Zhang G, Franklin PH, Murray TF. 1994. Activation of adenosine A_1 receptors underlies anticonvulsant effect of CGS21680. *Eur. J. Pharmacol.* 255:239–43

Zhao ZH, Makaritsis K, Francis CE, Gavras H, Ravid K. 2000. A role for the A_3 adenosine receptor in determining tissue levels of cAMP and blood pressure: studies in knockout mice. *Biochim. Biophys. Acta Mol. Basis Dis.* 1500:280–90

Zhu PJ, Krnjevic K. 1994. Endogenous adenosine deaminase does not modulate synaptic transmission in rat hippocampal slices under normoxic or hypoxic conditions. *Neuroscience* 63:489–97

Zimmermann H, Braun N. 1999. Ecto-nucleotidases–molecular structures, catalytic properties, and functional roles in the nervous system. *Prog. Brain Res.* 120:371–85

Annu. Rev. Neurosci. 2001. 24:57–86

LOCALIZATION AND GLOBALIZATION
IN CONSCIOUS VISION

S. Zeki

*Wellcome Department of Cognitive Neurology, University College London,
London WC1E 6BT, United Kingdom; e-mail: zeki.pa@ucl.ac.uk*

Key Words   motion, color, chronoarchitectonic maps, space, time

■ **Abstract**   The primate visual brain consists of many separate, functionally specialized processing systems, each consisting of several apparently hierarchical stages
or nodes. The evidence reviewed here leads me to speculate (*a*) that the processing
systems are autonomous with respect to one another, (*b*) that activity at each node
reaches a perceptual end point at a different time, resulting in a perceptual asynchrony
in vision, and (*c*) that, consequently, activity at each node generates a microconsciousness. Visual consciousness is therefore distributed in space and time, with the universal
organizing principle of abstraction applied separately within each processing system.
The consequence of spatially and temporally distributed microconsciousnesses is that
their integration is a multistage, nonhierarchical process that may involve a neural
"glue."

INTRODUCTION

This article is not a summary of past achievements but a speculative and conceptual gaze into the future through the achievements of the past. It is written with a
philosophical ideology, namely that the visual brain is an epistemic system whose
function is to acquire knowledge about the world, which it does not so much by
"representing" the visual world as by constructing it according to its own laws as
well as to laws inherent in the physical world (Mountcastle 1998, Zeki 1993). It is
a modern, neurobiological version of a view expressed by Immanuel Kant (1787).
A believer in the unity of knowledge, Kant sought to understand the formal contribution that the mind (in our case, the brain) makes, the organizing principle that
it applies, as well as the limitations that it imposes in acquiring all knowledge.
Explicit knowledge is of course intimately linked to consciousness, for it is difficult to acquire knowledge except in the conscious state. To question the unity of
knowledge in terms of the underlying organizing principles, as I do here, becomes
therefore a questioning, too, of the concept of the unity of consciousness. The
term "unity of consciousness" is widely used and susceptible to more than one
interpretation. I use it in the sense intended by Kant. He wrote that all perceptions

0147-006X/01/0621-0057$14.00   **57**

"have a necessary reference to a *possible* empirical consciousness.... But all empirical consciousness has a necessary reference to a transcendental consciousness (a consciousness that precedes all particular experience), viz. the consciousness of myself," thus leading to "[t]he synthetic proposition that all the varied *empirical consciousness* must be combined in one single self-consciousness" (Kant 1787, original emphasis). The emphasis on "possible" amounts to an admission that there may be many empirical consciousnesses, but that any understanding depends ultimately on the one "transcendental" consciousness. The evidence reviewed here leads me to the view that there are many functionally specialized microconsciousnesses (perhaps equivalent to Kant's "empirical consciousnesses"), each one tied to a different knowledge-acquiring system of the visual brain; hence both the knowledge-acquiring system and the visual consciousness are distributed in space. Kant's conjecture that there is a synthetic, transcendental "one single self-consciousness" (myself) into which all the "possible empirical consciousness" (microconsciousnesses) are synthesized raises the question of whether there is a seat of consciousness in the cortex, and hence a "unity of consciousness." I am of the view that this "one single self-consciousness," just like "awareness of being aware," is possible through the linguistic system alone and therefore do not consider it further here. Instead, using color and motion as examples, I explore the proposition that the knowledge-acquiring systems are fairly autonomous and that each has its own organizing principle and conscious correlate. I also consider what neural "glue" may be used to bring the results of activity in them together, to give us our coherent and conscious view of the visual world.

THE UNIFYING THEME IN VISION

The supposition that there is a universal organizing principle in acquiring all knowledge is plausible because there is, in fact, a unifying theme in the brain's quest for knowledge, including visual knowledge. This is to be found in the doctrine that Plato attributed to Heraclitus, and which is therefore known as the Heraclitan doctrine of flux. It refers to the fact that things are never the same from moment to moment, making it difficult to obtain knowledge of them. Translated into neurobiological terms, this becomes the problem of constancy: In vision, we speak of color constancy, of object constancy, of size constancy, and so on. The problem of flux led both Plato and Kant (with differences) to suppose that we can never obtain knowledge about what is around us through sensation but only through ideals (the Platonic Ideal and the Kantian Thing-in-Itself), constructed by a thought process that alone, they believed, can give us real knowledge. I suggest that the "thought-process" is an unconscious neurological process applied in different areas to different incoming visual signals, the common aim of which is to make the brain independent of the particular by giving it knowledge of general properties. There are consequently many "thought processes," each one tailored for a different attribute and applied in a different area or group of areas, or even transient areas (see below). The argument rests on the demonstration that there are separate cortical areas that undertake separate tasks autonomously, and that activity in each can result in a separate conscious correlate.

Some Basic General Facts

Certain general and indisputable facts, which are nevertheless open to more than one interpretation, have led me to my view.

1. The visual cortex consists of many different areas, each one part of a chain or system that consists of several stations or nodes (Zeki & Bartels 1999). A node refers not only to an area such as V5 (middle temporal—MT) but also to a specialized compartment of an area, such as the blobs of V1 or the thin stripes of V2 (Figure 1). There are therefore several parallel, distributed systems in the visual brain. The presence of several nodes within each processing system raises the question of whether activity at each is always implicit and not perceived until a "terminal" stage of processing, where perception is enshrined, is reached.

2. Apart from V1, all these areas reside within an expanse of cytoarchitectonically uniform cortex consisting of the basic six layers. This cytoarchitectonic uniformity naturally prompts speculation about whether, in addition to the specialized functions imputed to each, there is any common operation that all areas perform. The notion of a uniform operation, repetitively applied in all cortical areas, has been especially championed by Mountcastle (1998).

3. There is compelling evidence that the different parallel systems, and the nodes comprising them, are specialized for different visual functions (Zeki 1978, DeYoe & van Essen 1988, Livingstone & Hubel 1988, Zeki & Shipp 1988). The contrary view, which I do not agree with, states that there is no specialization within visual areas (Lennie et al 1990, Leventhal et al 1995, Gegenfurtner et al 1996) or in the visual brain at large (Schiller 1997). I have traced this specialization to the brain's need to undertake different operations to acquire knowledge about different attributes and believe that it has found it more efficient to separate the different machineries for these operations into separate areas or systems (see below) (Zeki 1993). The knowledge-acquiring system of the visual brain is therefore distributed throughout much of the cerebral cortex. Implicit in this view is the supposition that if there is a universal organizing principle in the acquisition of all knowledge, it must be superimposed upon microorganizing principles, tailored to the demands of acquiring knowledge about particular attributes.

4. Clinical evidence shows that damage to one processing system need not affect the other systems and, conversely, that a spared system can still function when much of the other systems are damaged or inactive. I

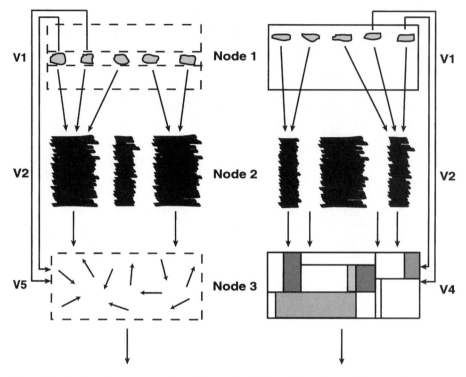

Figure 1   A schematic representation of the motion (*left*) and the color (*right*) processing systems of the primate visual cortex. Each system consists of more than three nodes. In the motion system, the cells of layer 4B that project directly, or through the thick stripes of V2, to V5 constitute the first node; the thick stripes of V2 constitute the second node and V5 the third node. In the color systems, the wavelength-selective cells in the blobs of Vl that project directly, or through the thin stripes of V2, to V4 constitute the first node whereas the second and third nodes are constituted by the thin stripes of V2 and V4, respectively. Both systems have further cortical projections that would thus constitute further nodes, but these are not considered in this article.

interpret this to mean that the different systems have fair autonomy in their operations (Zeki & Bartels 1999).

5. Recent psychophysical evidence shows that some attributes, e.g. color, are perceived before others, e.g. motion (Moutoussis & Zeki 1997a). I interpret this to mean that different systems reach a perceptual end point at different times, and independently of each other, thus supporting further the notion of autonomy.

6. Each node, or area, of each processing system has multiple inputs and outputs, both cortical and subcortical. Thus, activity in each is of consequence to several others. That each cortical node has multiple outputs (a rule applicable to the cortex at large) implies that there is no

terminal station in the cortex, at least in anatomical terms (Zeki 1993). This leads me to consider the possibility that activity at each node has a conscious correlate, a possibility that would confer substantial perceptual advantages. If so, then the conclusion seems inescapable that activity at each node of one specialized system must be capable of being integrated with activity at another node of another specialized processing system, no matter where in the hierarchy each node may be. This, the basis of the theory of multistage integration, raises interesting questions about integration and binding in the cortex. Multistage integration is made potentially possible by a universal rule of cortical connectivity, with no known exceptions: An area A that projects to B also has return projections from B (Rockland & Pandya 1979, Felleman & van Essen 1991).

THE APPLICATION OF "MICROALGORITHMS" TO VISUAL SIGNALS

Because there is a unifying theme in the brain's quest for knowledge, any or all algorithms applied to the incoming visual must ultimately share a basic property, namely the capacity to discount all changes that are unnecessary for acquiring knowledge about the permanent, essential, and nonchanging properties of all that is around us. But it is more than likely that even if they share certain characteristics, different rather than the same algorithms are applied in different subsystems of the visual brain. Here, I go beyond and suggest that the algorithms applied in the visual brain consist of microalgorithms, which are applied at different nodes or areas of a system. The microalgorithms belonging to any one subsystem, for example that of color, are also distributed in space. If deprived of the relevant microalgorithm, the incoming visual signals cannot be rendered meaningful. Color vision provides a good example.

The Microalgorithms of Color and Motion

Although no one knows the detailed neural mechanisms that generate colors in the brain, it is likely that the algorithm applied to color signals is similar to the one postulated by Land (1974) and Land & McCann (1971). This consists of two steps that may well reflect the application of two distinct microalgorithms cooperatively at two or more sites of the color pathways. (*a*) One is a lightness microalgorithm to determine the lightness record of a complex scene in different wavebands (Figure 2, see color insert). This stage, which, as conceptually formulated, consists also of several steps, is capable of "discounting" changes in lighting conditions. The lightness (relative brightness) of a surface that is part of a complex scene does not change with changes in the intensity of the light coming from any individual area because a change in the illuminant in which a scene is viewed will entail a change in the amount of light coming from every part, keeping unaltered the ratios between what one area and what the surrounds reflect. (*b*) The second is a comparator microalgorithm, which compares the lightness records of a complex scene generated by the three (or more) wavebands, leading to the construction of color. A change in the wavelength composition (by changing the illuminant) will mean an overall change in the wavelength composition of the light reflected from every point, again keeping the ratios the same. Thus, the stage that compares the lightnesses generated by the three or more wavebands (second microalgorithm) actually compares stable ratios at each waveband, where changes have already been discounted. Together, the two microalgorithms constitute the formal contribution of the brain to the acquisition of knowledge about color, the limitations being imposed by other factors, including the capacities of the photoreceptors for absorbing light of different wavelengths.

The visual motion subsystem is the other one most intensively studied. Of the algorithms proposed for it (see Ullman 1982), that of Movshon et al (1985) is as neat as any. Here again, what I interpret to be two microalgorithms are involved, and like the color microalgorithms, they are distributed in space. The first one, based in V1, computes the local direction of small parts of a moving object, whereas the second, based in V5, extracts the overall direction of motion of the object. (There may be more than two microalgorithms, as in the color system, for no one has provided an adequate formulation of what happens to the motion or color signals within V2.) A further similarity with the color algorithm is that changes that are not essential for detecting the overall direction of motion must be discounted.

TIME AND SPACE IN ALGORITHMS

Kant believed there are two innate intuitions, time and space, which antedate all experience, but into which all experience is read. Space "must already lie at the basis in order for certain sensations to be referred to something outside me ... [and] ... as being in different locations" whereas "simultaneity or succession would not even enter our perception if the presentation of time did not underlie them a priori" (Kant 1787).

The color and motion processing systems share a common requirement, in that the motion or the color must be referred to a position outside the observer. But though space and time are important for calculations in both domains, there are substantial differences between the requirements of the two systems, making it not implausible to suppose that this is another reason for the physical separation of the two processing systems. With color vision, the brain has to determine simultaneously the amount of light of a given waveband composition reflected (*a*) from a surface located in one region of space and (*b*) from its surrounds (which can be arbitrarily arranged) located in a different region. With motion, the requirements are different. Now the direction, and therefore the relation of at least two spatial points, successively in time, becomes important.

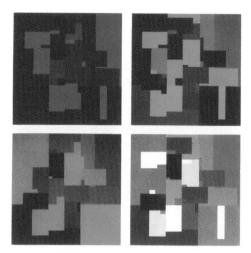

Figure 2 A multicolored display as it appears when illuminated with light of all wavebands (*lower right*) and when illuminated by long-wave (red), middle-wave (green), and short-wave (blue) light. The latter constitute th0e lightness records of the multicolored display in the three wavebands. See text for details.

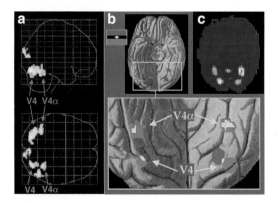

Figure 3 The human color center (the V4 complex) is located in the fusiform gyrus and consists of two subdivisions. (*a*) Here it is activated when subjects view a multicolored display that is illuminated by a light source whose wavelength composition changes continually—the subjects however perceive constant colors. (*b*) Retinotopic stimulation shows that area V4 is retinotopically organized whereas V4α is not. (*c*) An independent component analysis separates independent maps of brain activity without a priori knowledge about the stimulus conditions. The isolation of the complete V4 complex from all other brain activity (shown here in the glass-brain view of a single subject's brain) indicates that V4 and V4α act independently of other areas as a functional unit. Because it is damage to this very area that leads to the syndrome of cerebral achromatopsia, when subjects are no longer able to see the world in color but only in shades of grey, we are led to equate the processing system with the perceptual system. [From Bartels & Zeki (2000).]

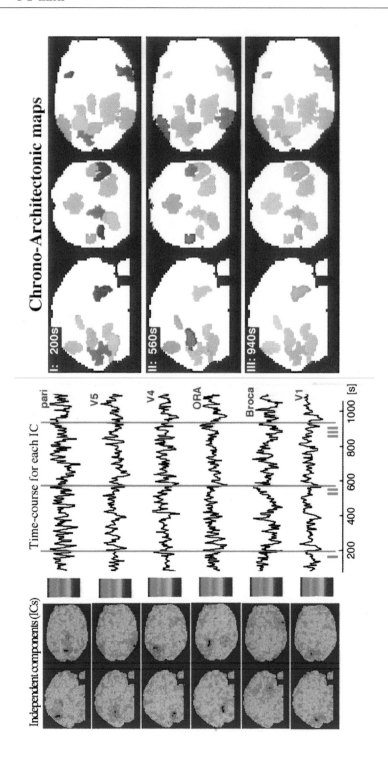

Chrono-Architectonic maps

I: 200s

II: 560s

III: 940s

Time-course for each IC

pari

V5

V4

ORA

Broca

V1

200 400 600 800 1000 [s]

Independent components (ICs)

Figure 7 (*Left*) Independent component analysis (ICA) applied to whole-brain functional magnetic resonance imaging (fMRI) data. ICA was applied to the 320 whole-brain images obtained when a single subject watched 20 min of the opening sequence of the James Bond movie *Tomorrow Never Dies*. ICA isolated functionally specialized areas in separate components together with the time courses of their activations. It did so without a priori knowledge, utilizing the fact that each functionally specialized area has a unique spatial extent and a time course that differs from all the others. From the many areas isolated we show here only six (shown from top down) 1. the middle parietal cortex along with a part of the posterior cingulate cortex; 2. the visual motion area V5 (middle temporal); 3. the left visual color area V4; 4. the visual face/object recognition area; 5. Broca's speech generation area (presumed); 6. left primary visual cortex (V1). The color code next to the time courses is used in Figure 7 (*right*) to demonstrate the relative activity that each of the six maps has at different points in time (indicated by vertical lines I—III across the time courses), thus revealing the chronoarchitecture of the human brain when so stimulated (from Bartels & Zeki 1999). (*Right*) The chronoarchitecture of the human brain revealed by application of ICA to fMRI data. The independent components shown in Figure 7 (*left*) were thresholded and superimposed onto a single glass-brain view after being color-coded according to their relative activity at a given time. This procedure was applied three times to the same six independent components from the previous figure, creating three chronoarchitectonic maps, each corresponding to one point in time [see Figure 7 (*left*): lines I—III intersecting the time courses at 200 s (I), 560 s (II), and 940 s (III)]. The three maps reveal how the pattern of activity of the functionally specialized areas changes across time, with different areas changing their activity in individual ways across time. We refer to these types of maps as to chronoarchitectonic maps, in this case of a single human brain watching a James Bond movie. For example the activity of the visual motion area V5, isolated in component number 2 in Figure 7 (*left*), is medium active (green) at time (I), low at time (II), and medium high at time (III), whereas Broca s area is low at (I), high at (II), and medium at (III). [From Bartels & Zeki 1999).]

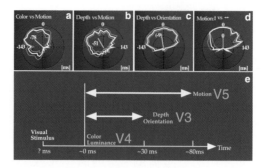

Figure 6 A schematization of the theory of perceptual sites, which supposes that attributes processed in the same node are perceived simultaneously, whereas attributes processed in different nodes are perceived asynchronously. The diagrams indicate the latencies with which one attribute is perceived with respect to another, each averaged for seven or more subjects. If two attributes (e.g. left-right motion and up-down motion) are perceived at the same time, the vector will show no displacement from 0 ms. Any displacement of the vector will indicate that one of two attributes is perceived before the other. In the examples given above, color (*a*) and depth (*b*) are perceived before motion whereas depth and orientation (*c*) as well as left-right motion and up-down motion (*d*) are perceived simultaneously. The relative perceptual times for different attributes are summarized in (*e*). [Modified from Zeki & Bartels (1999).]

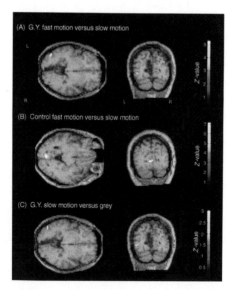

Figure 9 Imaging experiments. Statistically significant increases in BOLD signal (shown in color) superimposed on transverse and coronal sections of GY's brain. (*A*) The increases in cerebral activity comparing fast with slow motion in GY. (*B*) The increases in cerebral activity comparing fast with slow motion in a control subject. (*C*). The increases in cerebral activity comparing slow motion with an isoluminant gray control.

The application of the microalgorithm to the incoming signals is a process we are not normally aware of. Leibnitz (1714) thought that we cannot be conscious of all the automatic (thought) processes that are necessary for us to perceive the reality behind all natural phenomenon, and that therefore there must be an unconscious mind. I translate the unconscious mind of Leibnitz to mean the microalgorithm, from which I conclude that a different unconscious process underlies each of the (micro)conscious events that we experience. The clinical evidence shows that each microalgorithm can give a certain limited knowledge about the visual world without necessarily involving other microalgorithms. In the color system for example, application of the first microalgorithm gives knowledge about the reflectance of surfaces at a given waveband (reflectance being the amount of light reflected from a surface as a percentage of the light incident on it), whereas the second constructs color and thus gives knowledge about the reflectance of surface for light of all wavebands, color being strictly an interpretation that the brain gives to the reflectance of a surface for light of different wavebands. The second microalgorithm is contingent upon the first, but the reverse is not the case. In the motion system, the second microalgorithm can, it seems, be applied even in the absence of the first.

THE CORTICAL SITES OF THE COLOR AND MOTION ALGORITHMS

In the color system, the first microalgorithm is applied at a prebinocular stage and may involve the entire visual pathway up to the monocular cells of V1, whereas the second microalgorithm is at a postbinocular stage (Moutoussis & Zeki 2000), and probably in the V4 complex (Bartels & Zeki 2000).

The Color Center in the Human Brain

The human color center is located in the fusiform gyrus and consists of two subdivisions, V4 and V4α (Bartels & Zeki 2000) (Figure 3, see color insert), just as in monkeys (Zeki 1977, Shipp & Zeki 1995). In both, the two subdivisions together constitute the V4 complex and represent the entire contralateral hemifield, retinotopically in V4 (McKeefry & Zeki 1997, Hadjikhani et al 1998) and nonretinotopically within V4α (Bartels & Zeki 2000). The presence of two subdivisions, both apparently cooperatively active and involved in color, raises the question of whether there is more than one microalgorithm operating within the V4 complex itself. The V4 complex is a pivotal center in the color system and receives input from V1 and V2 (Zeki & Shipp 1989, DeYoe & van Essen 1988, Nakamura et al 1993). It is not a terminal station; it sends outputs to the inferior temporal cortex (Desimone et al 1980) where, in monkeys, cells selective for color have been found (Komatsu et al 1992) [note that the posterior part at least of the area called temporo-occipital (TEO) is almost certainly a ventral extension of V4 (Zeki 1996)]. In humans, imaging evidence also implicates the inferior temporal cortex in color when it is a property of objects (Zeki & Marini 1998). The projection fields of the V4 complex are not considered further here.

A recent terminological confusion introduced by Hadjikhani et al (1998) makes it necessary to digress briefly. The human V4 complex is located ventrally in the fusiform gyrus. Hadjikhani et al (1998) have confirmed the presence of a retinotopically organized color-selective area here but have called it area "V8," thus causing unnecessary confusion and highlighting a problem with their area "V4v," also located in the fusiform gyrus but posterior to our V4. Apparently V4v represents upper fields alone, making it into one of Kaas' (1993) "improbable areas." Hadjikhani et al (1998) described "V8," lying anterior to their "V4v," as a new, "previously undifferentiated cortical area," a conclusion accepted uncritically by Heywood & Cowey (1998), who thought that it is damage to "area V8, not the favorite candidate V4" that produces cortical color blindness when damaged, making "V8" "a ready candidate for a region responsible for our conscious perception of a colored world." In fact, their "new" area (V3) has the identical Talairach coordinates as our previously defined V4 (see Hadjikhani et al 1998; see also Figure 4) and is therefore the same area, which they have tried to rename. Their reason for calling it "V8," ostensibly at any rate, is that it lies in front of putative area "V4v" defined earlier by Tootell et al (1996; see also Sereno et al 1995). But the existence of "V4v" as an area separated from V4, in which the upper visual field alone is mapped, and in which there is no selectivity for color, as claimed by Hadjikhani et al (1998), is in doubt. Others (DeYoe et al 1996) have found that the area called "V4v" corresponds at least partially to the color-selective region (V4) defined earlier by us (Lueck et al 1989, Zeki et al 1991). Moreover, we (Bartels & Zeki 2000) and others (Kastner et al 1998) have not been able to

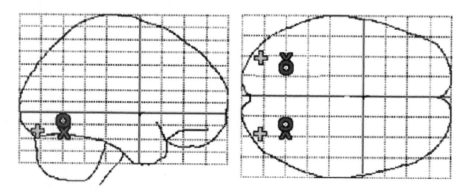

Figure 4 The locations of the three areas that are discussed in the text, in a glass-brain projection. The areas were located by using the Talairach coordinates of the three areas given in the paper by Hadjikhani et al (1998). O corresponds to area V4, as defined elsewhere (Lueck et al 1989, Zeki et al 1991, McKeefry & Zeki 1997); X corresponds to the "new" area "V8" of Hadjikhani et al (1998); + corresponds to area V4v, as defined by Sereno et al (1995). [From Bartels & Zeki (2000).]

confirm the existence of an area "V4v" separate from V4. The reasons for this difficulty in confirming the result of Tootell et al have been discussed elsewhere (Bartels & Zeki 2000) and may be traceable, at least in part, to the way in which the phase-encoded retinal stimulation method of Engel et al (1994) has been used by these authors. In sum it is evident that "V8" is identical to V4, but it remains doubtful whether area "V4v" extsts.

The Motion Center of the Human Brain

The motion center of the human brain is centered around area V5 and is too well known to be described in detail here. Suffice it to say that V5 is also part of a chain that, like the color system, includes different compartments of V1 and V2 (see Figure 1). There are thus more than two cortical nodes involved up to the level of V5, and clinical evidence also suggests that the microalgorithms of motion are distributed between the nodes. V5 is surrounded by other areas that are functionally specialized for different kinds of motion and are anatomically linked to V5, thus constituting the V5 complex (Wurtz et al 1990, Howard et al 1996, Orban 1997). But clinical evidence shows that not all the parts of the V5 complex depend on the integrity of V5 itself, which suggests the relative autonomy of nodes.

CONSEQUENCES OF DAMAGE TO THE COLOR AND THE MOTION SYSTEMS

In general terms, the clinical evidence shows that damage to the V4 complex causes cerebral achromatopsia, or an incapacity to see the world in color (Meadows 1974, Zeki 1990a), whereas damage to V5 causes akinetopsia, or an inability to see objects when they are in motion (Zihl et al 1983, 1991; Zeki 1991). Lesions in the V4 complex do not cause akinetopsia or defects of depth or form vision, whereas those in V5 do not cause achromatopsia or defects in depth vision or form vision for stationary objects. The two processing systems are therefore, at this level, substantially autonomous, a conclusion that is reinforced by the clinical evidence for other kinds of specific deficit, produced by other specific lesions (for a review, see Zeki & Bartels 1999). But clinical evidence also shows that activity at individual nodes of a processing system can be perceptually explicit, that is they require no further processing, and have a conscious correlate. Moreover, when distal parts of a system are damaged, the activity in the remaining parts is all that becomes explicit.

In classical cases of achromatopsia, the ocular media, the retina, and the optic pathways up to V1 are intact (e.g. Mollon et al 1980), as if the signals can only be rendered meaningful if a microalgorithm is applied to them (within the V4 complex). My examination of achromatopsic patients showed that they can experience lightnesses at any waveband consciously, though without being able to see colors (Zeki 1990a; S Zeki, unpublished results). This implies that it is the second microalgorithm that is applied within the V4 complex, and that the microalgorithms applied to the incoming signals at each of these two nodes of the color pathway can result in a conscious percept. It is of course possible that in patients with V4 lesions, the lightnesses are generated in another system, as argued elsewhere (Zeki 1990a). But because patients rendered achromatopsic by damage to the V4 complex can distinguish between light of different wavebands, though with elevated thresholds and without being able to assign colors to them (Vaina 1994), we are still left with the conclusion that activity at an earlier node of the color system may be sufficient to elicit a conscious correlate, even if this activity is only implicit (that is, requires further processing) in people with normal vision. Moreover, some patients with (presumably subtotal) V4 complex lesions have a color vision that is wavelength dominated (Kennard et al 1995), resulting in an incapacity to achieve the color constancy that is the hallmark of a normal brain. This may be the result of the activity of cells in V1 and perhaps V2, which are known to be wavelength but not color selective (Zeki 1983). Normally, the activity of the wavelength-selective cells at this node is something we are not aware of, and it remains implicit until the next node, the V4 complex. But in brain-damaged patients, the responses at the earlier node can become perceptually explicit. This conclusion contains within it the germ of an idea that is worth entertaining: that cells whose activity is only implicit can, in the right circumstances, become explicit. Put more boldly, cells can have double duties, rendering the incoming signals explicit or not, depending on the activity at the next node, an idea explored more below. In fact, one may be led to an even more outrageous conclusion: that in the normal brain, what can become perceptually explicit and therefore conscious at an earlier node may actually be suppressed by activity at subsequent nodes.

That, in a compromised system, the activity at nodes that are left intact by the damage not only leads to a percept but is the only one to do so receives support from recent studies of patient PB (Zeki et al 1999), suffering from a syndrome first described by Wechsler (1933), in which blindness resulting from carbon monoxide poisoning can spare color vision. I have tried to explain this by postulating that the richer vasculature of the blobs in V1 (Zheng et al 1991) protects them and the wavelength-selective cells in them from the effects of hypoxia (Zeki 1993). The suggestion raises questions about whether activity in V1 can become perceptually explicit, which Crick & Koch (1995) have supposed is unlikely.

PB's vascular insufficiency was provoked by a cardiac arrest. After recovery he was found to be blind although he could see colors (Humphrey et al 1995). His color-constancy mechanisms are impaired and his color vision is very much wavelength based. The activity in his brain when he viewed, on a TV monitor, colors that he could name accurately (red, green, blue—all generated with maximum phosphor purity) was restricted to the calcarine sulcus (area V1), which suggests that his restricted but conscious color vision is mediated through V1 (Figure 5). Based on their work in macaques, Walsh et al (1993) had in fact suggested that mechanisms within V1 are capable of mediating perception of the basic color categories. Unfortunately, evidence derived from functional magnetic resonance

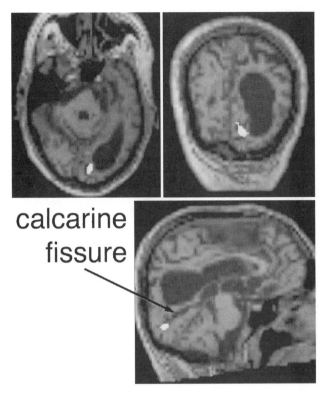

calcarine
fissure

Figure 5 Coregistered structural and functional magnetic resonance imaging data from patient PB showing, in coronal, sagittal, and transverse sections, the activation in the calcarine fissure (area V1) when PB viewed colored stimuli. The white area shows the cluster of voxels exceeding an uncorrected significance level of <0.001. [From Zeki et al (1999)]

imaging (fMRI) cannot exclude any residual activity in other areas that may have contributed to PB's conscious color vision but that was too weak to attain significance in the activity maps. The clinical evidence thus makes it plausible that we may become aware of activity in V1, without providing conclusive proof for it. What seems certain is that activity in V1 is not always a prerequisite for conscious visual experience, and that activity at visual nodes disconnected from it can have a conscious correlate. (Note that, for the sake of simplicity, I have written about the prebinocular stage, including the monocular level of V1, as if it constituted a single node. Except in the functional sense used here, that of generating lightnesses, this is of course not true because the prebinocular stage involves the cells of the retina and the lateral geniculate nucleus. It is possible to consider the entire prebinocular stage as cooperatively involved in generating lightnesses, which in any case probably involves at least three stages.)

The Riddoch Syndrome

In the motion system, unlike the color system, damage to the first (or earlier) part does not impair the second part completely. This shows that under certain pathological conditions, we can become aware of activity in visual cortical areas that are not fed by V1. In other words, activity in a visual area without antecedent, parallel, or subsequent activity in V1 can mediate conscious vision, even if this is extremely crude, which of course raises the question of whether "preprocessing" by V1 is always necessary.

If V5 is intact in a patient with a damaged V1, signals reaching it from subcortical centers without passing through V1 (see below) can still mediate a conscious awareness of fast motion (Barbur et al 1993, Weiskrantz 1995, Zeki & ffytche 1998), thus showing that activity in a node fed with an appropriate visual input can have a conscious correlate. Long before such experiments were undertaken with modern techniques, the notion that conscious vision without V1 is possible should have become apparent from the studies of Riddoch (1917). He had shown that some of his soldiers blinded by gunshot wounds that had damaged their V1 were nevertheless "conscious" of having seen motion (and only motion) in their blind fields. In the days before mathematical pitchforks and statistics invested scientific findings with a respectability they do not always deserve, language was used with greater care. Riddoch's use of the word "conscious" was deliberate—his subjects reported their sensations verbally. Riddoch nevertheless interpreted his results very conservatively, within the context of what was known at the time, and that did not include any knowledge about V5. Presumably, it seemed inconceivable to him that anyone should consciously see a visual attribute without the participation of V1. He supposed therefore that his patients were seeing motion in their blind fields because the gunshot wounds had spared those subdivisions of V1 that mediate motion. This conclusion was easy to dismiss and was quickly done so by Sir Gordon Holmes (1918). He wrote emphatically: "Occipital lesions do not produce true dissociations of function with intact retinal sensibility."

The examination of patient GY who has a long-standing lesion in V1 has given interesting insights into what may be called the Riddoch Syndrome (Zeki & ffytche 1998). Although described as a "blindsight" patient (Weiskrantz 1986), he is commonly conscious of what he discriminates and unconscious of what he cannot discriminate. His visual capacity is crude: The motion that he can see and discriminate consciously is from high-contrast, fast-moving ($>8°$ s^{-1}) stimuli, signals that reach V5 without passing through V1. He is not apparently able to either discriminate or see consciously signals from slowly moving ($<2°$ s^{-1}) stimuli, which are relayed to V5 through V1 (ffytche et al 1995). His descriptions of what he was conscious of when presented with such stimuli are compellingly similar to what Riddoch had described earlier, consisting of "shadows." Imaging studies show that the activity in his brain when he was viewing fast-moving stimuli was restricted to V5. These imaging experiments and previous clinical evidence (Ceccaldi et al 1992, Mestre et al 1992) thus show that activity in V5, without parallel or antecedent activity in V1, can have a conscious correlate. This probably reflects the anatomical and physiological picture that there is a direct input from the pulvinar to V5, and that directional selectivity, which is a hallmark of V5, is not abolished when macaque V5 is disconnected from V1 (Rodman et al 1989, Girard et al 1992). It is thus not true to say that "conscious vision is not possible without V1" (Stoerig & Cowey 1995, Stoerig 1996). The same picture could well turn out to be true of V4, which, in monkeys, also receives a direct input from the pulvinar (Cragg 1969, Standage & Benevento 1983). There is a report, not yet confirmed by other studies, of one blind patient who could report colors verbally and correctly, constituting a sort of residual vision for color (Blythe et al 1987).

Similar experiments have not yet been undertaken for other systems, but the results from this system alone are compelling enough to raise the question of whether activity in one area or an isolated group of areas is not sufficient to elicit a microconsciousness for the relevant attribute. The reverse is also interesting, in that a patient with an intact V1 but a damaged V5, though unable to experience fast motion, can nevertheless discriminate and experience slow motion (Hess et al 1989, Baker et al 1991, Shipp et al 1994), presumably because fast- and slow-motion signals take different routes to reach the cortex (see below).

The evidence from the color and motion systems thus suggests that nodes can act autonomously of one another, even if there are reverse, transverse, and forward connections between them. That activity at individual nodes within each system can have a conscious correlate provides further support for the proposition that the knowledge-acquiring system of the brain is distributed. Moreover, it is not at all clear that the microconsciousness generated by activity at a given node must necessarily be reported to a "center" for consciousness or a "Cartesian Theater" (see Dennett 1991). Hemiachromatopsic patients, for example, are often unaware of their color loss (e.g. Paulson et al 1994; S Zeki, A Bartels, personal observations), which suggests that there is no putative consciousness center to record the gap in the patients' visual capacity.

The above evidence and discussion makes it plausible to suppose that there is no single organizing principle in acquiring all visual knowledge. Yet the idea of a universal principle, repetitively applied, is appealing. What could that grander principle be? I have suggested elsewhere (Zeki 1999) that it may lie in two linked factors that must be characteristic of any efficient knowledge-acquiring system. The first is abstraction and the second the formation of ideals. The former is a selective and eliminative process, allowing the brain to find some property or relation that is common to many particulars, and thus making itself independent of the particular to which it would otherwise be enslaved. This capacity is probably also imposed on the brain by the limitations of its memory system. As Descartes saw, memory could not be trusted in an unqualified way, even in the certain world of mathematics, where one has to rely on earlier steps in a deductive process. And there is no guarantee that the memory process itself may not be at fault, as it often is. The process of abstraction leads naturally to the formation of ideals, used in the Platonic sense of a universal in contrast to a particular. Again, I suppose that this is the product of an unconscious neurological process, a "thought process." An interesting example of a neurological "thought process" is provided by the experiments of Logothetis et al (1995), who found a small proportion of cells in the inferior temporal cortex of monkeys that are able to respond in a view-invariant manner to nonsense objects that the monkeys had been previously exposed to. None of the cells, however, responded to views that the monkeys had not been exposed to. Thus the responses of this small proportion of cells is, in a sense, a synthesis of all the views, making a particular view irrelevant. This evidence comes tantalizingly close to suggesting that underlying the Platonic Ideal, which gives real knowledge about objects, is an unconscious "thought process" whose neural implementation is currently opaque, though it may become clear soon. Here, I put forward the suggestion that each of the many areas of the visual brain may be endowed with the capacity to abstract, though not necessarily on its own. On the other hand, abstraction in color is not possible without the V4 complex and is not dependent on V5, whereas abstraction in motion depends on V5 and is independent of V4. Hence abstraction, though it may well be a general organizing principle applied repetitively in cortical areas, is also likely to be tailored to the requirements of specific processing systems and the attributes they are specialized to abstract.

TEMPORAL HIERARCHIES IN CORTICAL ACTIVATION, PERCEPTUAL HIERARCHIES IN VISION, AND A CHRONOLOGY OF CONSCIOUSNESS

The application of an algorithm to the incoming signals must take a finite time, the processing time. By that I mean the time taken by a nervous structure to bring the activity within it to a perceptually explicit end point, which requires no further processing and of which the subject is conscious. In color vision, we may say that the processing time can be defined by the time it takes signals to reach the V4 complex and the time it takes for the subject to perceive a color. The issue is more complicated than that, however, if we accept that the overall algorithm is constituted by microalgorithms distributed in space and time. If activity at each node of, say, the color system is capable of becoming perceptually explicit, then we must think of microprocessing times underlying each perceptually explicit activity. Microprocessing time can be defined as the time taken by a node to render the activity within it perceptually explicit. We may postulate, for example, that there is (*a*) a finite microprocessing time required to bring the lightness calculations to a perceptually explicit stage, perhaps within V1, perhaps elsewhere, and (*b*) a finite microprocessing time to generate colors from lightness comparisons, probably within the V4 complex. This leads to the supposition that there are many separate, spatially and temporally distributed, unconscious processes (microprocesses) whose results, the microconsciousnesses, are also spatially and temporally distributed.

There is, in fact, direct psychophysical evidence to suggest that there is a temporal hierarchy in visual perception, from which we surmise that the (micro) processing times required to bring different attributes to a perceptual end point are different. The observed hierarchy could not have been predicted from either the anatomical arrangement or the physiological evidence. V5 receives the most strongly myelinated, and therefore fast-conducting, fibers (Cragg 1969, Allman & Kaas 1971). The shortest latencies recorded have been from V5, not V1 (Raiguel et al 1989), with signals from fast-moving stimuli reaching it in about 30 ms, compared with ~70 ms, which signals related to color take to reach V4 (Buchner et al 1994). It takes about 30 ms for the signals to reach V1 from V5 (ffytche et al 1995, Beckers & Zeki 1995). This reverse V5-to-V1 temporal hierarchy can be contrasted with the V1-to-V5 hierarchy, related to signals from slowly moving ($<2°$ s^{-1}) stimuli carried by the classical retino-geniculo-cortical pathway, with signals taking about 60 ms to travel from V1 to V5. In fact, the chronology is more complex than this. V5 has an ipsilateral input, in addition to a contralateral one (Tootell et al 1995, 1998). The ipsilateral input is callosally mediated and takes about 11 ms to reach it from the contralateral side (ffytche et al 2000). The consequence is that both V5s are activated by fast-moving stimuli before the contralateral V1 is activated.

But it is color, not motion, that is perceived first, by about 80 ms (Moutoussis & Zeki 1997a,b), prompting the supposition that the accelerated arrival of motion signals in the cortex is an evolutionary compensation for the slower processing time for motion compared with color. Further studies have shown that different attributes, such as color, orientation, motion, depth, faces, facial expressions, natural scenes, and so on, are perceived at different times (Zeki & Bartels 1999; S Zeki, A Bartels, M Self, L Dell'Acqua-Bellavitis, unpublished results). It is interesting that we are not aware of this perceptual asynchrony, which is only revealed by relatively sophisticated psychophysical experiments. By contrast, some pairs of attributes are perceived at the same time. Depth, for example, is perceived at the same time as orientation, and upward-downward motion at the same time as left-right motion (Figure 6, see color insert). The theory of perceptual sites (Zeki & Bartels 1999) supposes that attributes perceived at the same time are processed at the same place whereas those perceived at different times are processed at different places. Though all that we have so far observed is consistent with this theory, it may yet turn out to be wrong or only partially correct. The temporal hierarchy in perception thus obeys neither the anatomical nor the physiological picture; instead, it cuts across both, raising the suspicion that it depends not so much on speed of input as on (micro)processing time in the cortex, and giving us hints about how the visual brain operates. If the activity in different visual areas becomes perceptually explicit at different times, leading to a temporal hierarchy in visual perception, and if perceiving something is being conscious of it, then it follows that we become conscious of different attributes at different times. This in turn implies that consciousness is distributed in time. And because the temporally distributed activity occurs in different areas, consciousness is distributed in space as well.

That a temporal asynchrony can be demonstrated at all implies two things: that over very brief time windows, the activity in different brain nodes is autonomous, and that therefore the brain is able to bind not what happens in real time but only the end points of its own processing systems. Thus the brain "mis-binds" in terms of real time, for example ascribing the "wrong" direction of motion to the "right" color (Moutoussis & Zeki 1997a). The brain, in other words, does not wait for all the processing systems to terminate their tasks—it has no standard, or zero, time. Over very brief periods of time the brain has not found it necessary to adopt a standard, presumably because no integration occurs, given the autonomy of the separate knowledge-acquiring systems. This raises the question of how integration occurs over a longer time course.

REVERSE HIERARCHIES AND CHRONOARCHITECTONIC MAPS IN THE VISUAL BRAIN

If temporally distributed, do these microconsciousnesses reflect the overall hierarchical organization of the visual brain and that of its subsystems? The traditional view that the brain is hierarchical in connections and function has much merit. For over half a century anatomists have been rightly impressed by the successive nature of the relays, from say V1 to V2 to V4 to the inferior temporal cortex. Many electroencephalographic studies have interpreted V1 to be the first visual area to be activated (Barrett et al 1976, Halliday 1993). The demonstration of parallelism within the visual brain does little, in itself, to compromise this view because each one of the parallel systems is itself hierarchical in nature. Physiologically, the cells of V1 are, broadly, functionally less complex than those of, say, V4 or V5, rendering V1 into what has been commonly regarded as a "preprocessing" stage, whatever that may mean. The major input to the prestriate areas comes from area V1 through the retino-geniculo-cortical pathway. Anatomical studies have schematized the connections of cortical areas into a broad hierarchy, not always very satisfactory but consistent enough to be compelling (Felleman & van Essen 1991). But this impressive evidence in favor of hierarchies has important exceptions, which may give hints about how the brain operates.

The most compelling evidence in favor of a reverse hierarchy comes from studies of area V5. Several lines of evidence show that signals from fast-moving objects reach V5 before they reach V1 (Beckers & Zeki 1995, ffytche et al 1995, Zeki & ffytche 1998). A reverse hierarchy therefore operates in this instance, with important consequences for understanding the functioning of the visual brain. V5 can, in fact, be active without activity in V1. The activity in V5 in such instances is potent enough to result in a conscious experience of fast motion, though not of much else (see above). It is therefore obvious that, in this instance at least, signals from fast-moving objects do not have to be "preprocessed" in V1 and that if in the normal brain any "preprocessing" is executed on fast-moving signals, V5 is the more likely site for it, thus reversing the traditional hierarchy.

This could of course be an exception, but it could equally well be an exception that shows the difficulty of applying the principle of hierarchy derived from anatomical studies to the following:

1. the latency with which signals arrive in the cortex: V5 is not unique in receiving a visual input that bypasses V1. Other areas, including V4, are in the same league; much of the prestriate cortex receives a direct input from the pulvinar (Cragg 1969, Standage & Benevento 1983), and the pulvinar itself has been shown to contain a variety of physiological categories of cells, including ones that are apparently selective for colors (Bender 1981). It remains for the future to determine whether certain categories of signals reach these prestriate areas before they reach V1, just as fast-motion signals reach V5 before V1.

2. processing and perceptual times: Psychophysical evidence suggests that processing time is longer for motion than for color, and that color is perceived before motion, even though motion signals arrive in the cortex first (Moutoussis & Zeki 1997a, Zeki & Moutoussis 1997).

The task is to learn what role such reverse hierarchies have and to exploit them to study the functioning of the visual brain. Reverse hierarchies are, in a sense, the product of a system in which different nodes have fair autonomy, and in which there is no terminal stage at which perception is enshrined. The presence of reverse hierarchies, if it is indeed based even partially on autonomy of processing, also necessitates feedback and transverse connections because the processes rendered explicit at one stage of hierarchy must then be communicated to the processes rendered explicit at earlier hierarchical stages of the same, and other, systems. In such a system of internode communication, the concept of a reverse hierarchy may itself be an oversimplification. There may indeed be many crisscrossing hierarchies (Zeki 1998).

If the knowledge-acquiring system of the visual brain and the attendant micro-consciousnesses are distributed in time and space, then the sequence with which different nodes reach a perceptual end point can perhaps in the future be represented by what I call chronoarchitectonic maps of the cerebral cortex (Figure 7, see color insert). To date, the major emphasis of cerebral studies has been to associate distinct functions with histologically or anatomically distinct parts of the cerebral cortex, an emphasis that has reached new heights with the development of brain imaging techniques. Whatever their shortcomings, the common jibe that these are expressions of a "modern phrenology" does little justice to the great advances in schematization that these maps have helped bring about, and the continued use of the Brodmann maps in the interpretation of human imaging studies even today attests to their value. Yet the chronology of activation sketched out above, though for only a small part of the brain, makes one wonder whether the time is not now ripe for locating activity in time and thus developing a new map of the cortex, the chronoarchitectonic map. Such a map would reveal the time course of activity for each area or coactive groups of areas, thus reconstituting in time the relationship of activity between different cortical areas. The spatial map so produced would be different at any given time and with different visual stimuli. Ideally, the data used to create such a map should have a spatial resolution high enough to differentiate between cortical layers and a temporal resolution that would allow it to reveal, for example, the activity at different cortical processing stages. Though crude, a first approximation to this is the fMRI data, which have a typical spatial resolution of 3 mm^3 and about 4000 ms in time. The algorithm necessary is in fact available even now, in the form of the independent component analysis (ICA) method of Bell & Sejnowski (1995). To date, the full potential of the ICA method has not been exploited, partly because the most widely used method of spatial localization, that of fMRI, has, at 4000 ms, a very poor temporal resolution in terms of cerebral chronology. Even in spite of these limitations, the ICA method has shown that when humans view a colored Mondrian display in which the wavelength composition of the light coming from every point changes continually, only two areas in the ventral occipital lobe are simultaneously active, V4 and V4α, which together constitute the human V4 complex (Bartels & Zeki 2000). This gives hope that the method will isolate areas that are simultaneously active at any one time and distinguishes them from areas that are simultaneously active at a subsequent or antecedent time.

Chronoarchitectonic maps would be predicted, from the known facts, to present a very different picture from the current maps, and Figure 7 shows one very early, and therefore crude, such map. In fact, chronoarchitectonic maps would be expected to vary with the stimuli to which a viewer is exposed and with the motor action that may be involved. They would therefore be context dependent. With fast-moving stimuli, the chronology would be V5, then V1 and V2, then V5 again, followed by other visual areas in the parietal and inferior temporal regions. If the stimulus contains both color and motion, an even more complex and unconventional map would be predicted to result. In fact, there may be a number of different chronoarchitectonic maps. One could imagine, for example, a chronoarchitectonic map that displays the chronology of arrival of signals in the brain. Within the visual cortex, V5 would of course take precedence over both V1 and V4. A chronoarchitectonic map depicting conscious experience might well have a different configuration because we now presume that V4 will take precedence over V5. It is possible to envisage a situation, in the remote future, when a chronoarchitectonic map depicting the hierarchy of consciousnesses will become available. Such a chronoarchitectonic map would be different from the one that maps temporally the binding of attributes, as discussed below.

BINDING AND INTEGRATION

Given that in our daily life all the different attributes of vision are seen in temporal and spatial registration, it seems natural to suppose that the activity in the different processing systems, and in the nodes within them, is integrated or bound. What is generally implied by the term binding is the bringing together of what is processed by different systems or, more commonly, the binding of the responses of cells within a single processing system. In the latter instance, binding is thought to distinguish the firing of the "bound" cells from that of all others, thus constituting the neural basis for the kind of perceptual salience that is evident in, for example, figure-ground segregation (Engel et al 1999). There have been previous classifications of binding (Crick & Koch 1990, Zeki & Bartels 1999). Here, I restrict myself to a discussion of parallel binding between nodes or areas. This refers to the coupling of the activity of cells—e.g. through synchronous or oscillatory firing or any other form of communication—within a single area or across different areas. It is postconscious because it is the microconsciousness generated at a given node of one processing system that is bound to the microconsciousness generated at a given node of another (or the same) processing system.

A Theory of Multistage Integration

In the absence of an exclusive hierarchy and a terminal perceptual station, it becomes natural to suppose that the perceptually explicit activity at each node of one subsystem should potentially be capable of being integrated with activity at each node of another subsystem, in other words that the binding is that of microconsciousnesses generated at the different nodes. In the mesh of connections linking the different nodes, the forward ones obey the like-with-like principle, whereas the reverse and transverse ones do not. Thus, transverse connections can be of the like-with-like variety, e.g. between the blobs of V1 (Livingstone & Hubel 1984), or of the like-with-unlike variety, e.g. the connections between blobs and nonblobs (Yoshioka et al 1996, Hubener & Bolz 1992) or the lateral connections that link the thick and thin stripes of V2 (Rockland & Lund 1983, Rockland 1985, Lund et al 1993, Levitt et al 1994). Reverse connections are often of the like-with-unlike variety (Shipp & Zeki 1989, Rockland et al 1994, Rockland & van Hoesen 1994). For example, unlike the forward connections from V1 to V5, the reverse connections from V5 to V1 or V2 are not restricted to the territory of the thick stripes in V2 or the territory of layer 4B cells of V1, which project to V5. Because the like-with-like pathway is hierarchical whereas the other two are not, it follows that integration itself can be hierarchical or not, or it may even form a hierarchy of its own. This constitutes the basis of the theory of multistage integration (Zeki 1990b, Bartels & Zeki 1998).

Assuming that activity at each node has a perceptual correlate, multistage integration leads to an increased number of different possible combinations and thus of perceptual repertoires, by bringing about integration between different perceptual correlates. The repertoires would be reduced if the processing systems had to report to a "terminal" station—either a common one or individual ones—for integration to occur. Such a hypothetical integrator area would have to code the results of processing at each node in a perceptually explicit way, separately and in the required combinations. The number of pairwise connections between N nodes $= N \times (N - 1)/2$. Even given the constraints of cortical connectivity, this would still create a vast repertoire that would not be possible if integration could only occur between hypothetical "terminal" stages. Moreover, if the result of activity at a given stage is not made perceptually explicit, it would be lost in subsequent perceptual stages and would therefore no longer be perceptually accessible. Multistage integration makes the activity at different nodes accessible to each other.

Third Area Involvement in Parallel Binding of the Activity in Two or More Nodes

What decides that activity at a particular node should be bound to activity at another? One inevitably thinks of the strength of activity at different nodes and their timing. What actual strength is required is unclear, but if parallel binding is indeed postconscious, then one would postulate a strength that is sufficient to have a conscious correlate. Timing creates a problem, if different nodes reach a perceptual end point at different times. One is therefore led to ask whether there is any third area, besides the general "enabling" system in the reticular formation (see below), that is systematically involved in supervising, overseeing, or simply determining that binding and integration occurs, by perhaps synchronizing the activity of the different nodes in space and time.

In theory, there is no mandatory need for third-area involvement in binding two simple attributes, such as may be found in, say, moving green dots, when the wavelength composition of the dots is continually changing. One could postulate a direct anatomical connection between the centers involved, V4 and V5, for which there is evidence (though in our experience these connections are not strong). Given that color is perceived before motion, one would also need to postulate some timing adjustment or monitoring. But the situation may be more complex when, say, a form is generated from color and motion. Form vision correlates with activity in the fusiform gyrus (Bly & Kosslyn 1997), distinct from both V4 and V5. Hence, any stimulus that has a form generated from form and motion should, in theory, activate areas V4 and V5 and the fusiform gyrus. Evidence from a study in which subjects were asked to view shapes that can only be recognized by binding motion from one part of the shape with color from another part (Perry & Zeki 2000) suggests that a third area, located in the superior parietal lobule, is also specifically engaged in the binding, even when allowance has been made for the spatial distribution of the stimulus that emerges from color and motion. We note that object recognition does not correlate with activity in the parietal cortex (Bly & Kosslyn 1997) whereas lesions in it lead to problems of binding (Treisman 1996).

"Transient" Cortical Areas

The involvement of third areas in parallel binding has important implications for future studies of binding. It is possible that only when the consequence of parallel binding is the emergence of a new category, not encoded in the two (or more) areas involved, that the binding is monitored and determined by a third area. It is still not clear whether the superior parietal lobule is specifically involved in the binding of color and motion, or whether it is involved in all types of binding; nor is it known whether other kinds of binding involve other cortical areas, distinct from the superior parietal cortex, either in the parietal lobe or elsewhere. But the involvement of third areas in binding, even though presently demonstrated only for the superior parietal lobule and for binding between color and motion, raises the interesting possibility of what I shall call dynamic or "transient" areas. When subjects are stimulated with motion alone or color alone, the parietal lobe is either weakly activated or not activated at all; when they view objects, the activation is restricted to the fusiform gyrus and does not involve the parietal cortex. However when they view objects constructed from motion and color, the parietal cortex suddenly becomes active (Perry & Zeki 2000), as if it has established a temporary functional connection with the fusiform gyrus and with areas V4 and V5. I refer to this as a transient area because I envisage it to be a dynamic entity, cutting across areas. Transient areas would be established by anatomical connections that, though permanently present, are not continuously active. Anatomical studies show that there are strong connections between the parietal and inferior temporal areas (Morel & Bullier 1990) and weak ones between V4 and V5 (Shipp & Zeki 1995). The areas are thus envisaged to be temporary only in terms of time; the very same "area" should become active whenever the same, or a similar, stimulus is viewed. A "thought process," leading to abstraction, might involve such temporary areas, at least in its initial stages. If there are indeed such temporary areas, one would expect an even more complex chronoarchitectonic map with stimuli that bring these areas momentarily into play. Note, again, that we are not conscious of the activity within such putative temporary areas but only of the final result, which in the example given above is the activity within the fusiform gyrus alone.

GLOBALIZATION OF CONSCIOUS VISION

So far, I have adopted what many would regard as an extreme position: fractionating consciousness and the unconscious processes underlying it, and supposing that there are many distributed mechanisms for creating ideals and for abstraction. This view may turn out to be partially or completely wrong. Currently, I find it disturbingly seductive: disturbing because this is not how I imagined the brain functions, and seductive because the evidence that has impressed me most points in that direction. Taken to further extremes, it would imply that activity in an area like V5, disconnected from the rest of the cerebral cortex but able to receive signals from a healthy retina, should have a conscious correlate. This is unlikely to be true, and it raises the question of what the minimum cortical machinery might be to allow an area like V5 to function in the way that it does. There is no ready answer to that question today but there are certain guides. Among these are the following:

1. the presence of an enabling system. Although patient GY is usually able to discriminate fast-moving stimuli and is unaware that he cannot discriminate slow-moving stimuli, in our experiments he was not aware during every fast-motion trial nor was he unaware during every slow-motion one (Zeki & ffytche 1998). We designed our experiments in such a way that we could compare the activity produced in his brain when "aware" trials were compared with "unaware" trials. The comparison showed that the only significant activity in his brain was located inferior to the ponto-medullary junction, which we interpret to be the reticular formation (Figure 8). This suggests that there must be an "enabling" system located in the reticular formation (Zeki & ffytche 1998), a zone long ago implicated in consciousness (Moruzzi & Magoun 1949), and that V5 is not acting in isolation.

2. a minimum strength of activity. Although GY is, in general, not aware of and cannot discriminate slowly moving ($<2°$ s^{-1}) stimuli whereas he is both aware of and can discriminate fast-moving ($>10°$ s^{-1}) stimuli, GY's brain activity when he views slowly moving stimuli compared with when he views a blank screen shows that there is activity within V5 (Zeki & ffytche 1998) (Figure 9, see color insert). That he is not aware of these stimuli while being aware of fast-moving stimuli, which produce a more pronounced activity in V5, makes it plausible to suppose that a conscious correlate depends on a minimum strength of activity at a node. This of course raises the critical question of whether there are any particular cells within, say, V5 whose activity is critical for generating consciousness, although the question itself makes a questionable presumption that the activity of one set of cells can be separated, in this context, from the activity of another. We have no ready answer to this question, but the studies of Logothetis (1998) have shown that there are many cells in each cortical area whose activity correlates with perception, though their distribution according to layers remains uncharted. The proportion of such cells increases as one proceeds from what are traditionally regarded as lower areas to higher ones. Conversely, there are also many cells in each area whose activity does not correlate with perception, which is not to say that they are not intimately involved, in that their activity is crucial to the emergence of cells whose activity does correlate with perception.

3. the fact that V5 is really part of a motion complex (Wurtz et al 1990, Howard et al 1995, Orban 1997) and it is possible to postulate that at least the rest of the V5 complex must be functioning normally for the activity in V5 to be expressed at all. There is no definite proof of this, however. If anything, proof testifies to the contrary, in that a patient with a damaged V5

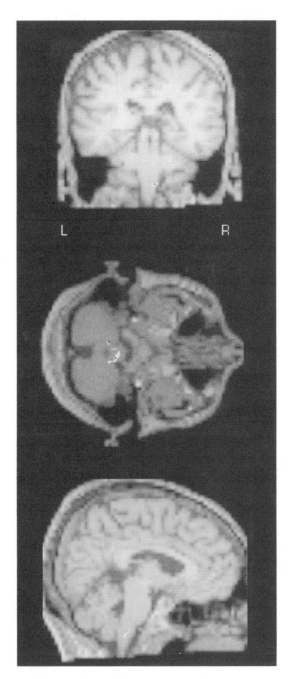

Figure 8 Imaging of aware versus unaware trials. A factorial analysis has been performed testing for the main effect of awareness (fast aware and slow aware versus fast unaware and slow unaware). The gray control stimulus is included in the analysis as a covariate of no interest. The statistically significant increases in BOLD signal are superimposed on coronal, transverse, and sagittal sections of GY's brain.

can apparently still experience biological motion (Vaina et al 1990). This implies that the healthy functioning of the whole of the motion complex is dispensable, at least for the conscious experience of biological motion.

4. reciprocal activation-deactivation. The intimate link between V5 and other areas, not obviously concerned with motion, comes from imaging evidence that shows that activation of area V4 with color stimuli is accompanied by the inactivation of area V5 (Zeki 1997) (Figure 10), and possibly other areas as well, whereas activation of parts of the limbic system with faces of loved persons is accompanied by widespread deactivation of other parts of the limbic system (Bartels & Zeki 2000). Such reciprocal activation-deactivation processes suggest that many areas may be involved in an unspecified way in the generation of a perceptually explicit, conscious, correlate to activity within an area, but what their precise role may be is anyone's guess. In emphasizing the importance of involvement of large parts of the cortex in generating consciousness, the notions of dynamic cores (Tononi et al 1994) and of complexity (Tononi & Edelman 1998) have much to recommend them and deserve careful pursuit. They may constitute further organizing principles in the Kantian sense, applicable to the generation of consciousness at large. But as currently formulated they do not address the question of what, in terms of cortical areas and processes involved, would constitute a dynamic core for generating a conscious correlate of activity at a single node, for example of fast motion through the activity of a V5 that is disconnected from V1.

CONCLUSION

In trying to account for conscious vision, we thus have two competing sets of facts that have somehow to be reconciled. On the one hand are the facts of anatomy, physiology, pathology, and psychophysics which tell that activity in the specialized processing systems and the nodes within them can have a conscious correlate, even in a vastly impoverished cortex. On the other, we have the knowledge that a greatly enhanced and sophisticated repertoire is the preserve of a hugely expanded and complexly interconnected cerebral cortex. On the one hand, we have to account for the microorganizing principles that underlie the activity at individual nodes and result in a conscious correlate, and on the other we have to try to understand whether there is an overall general organizing principle that not only controls the microorganizing principles but also enhances their capacity. If biology is anything to go by, the solution, when found, will probably turn out to be dazzlingly simple and elegant. But so far it has been elusive and beyond the reach of all.

ACKNOWLEDGMENT

The work of this laboratory is supported by the Wellcome Trust, London.

Increases in rCBF

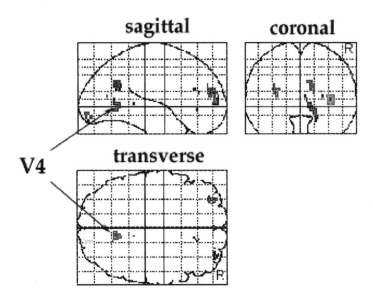

sagittal   coronal

V4   transverse

Decreases in rCBF

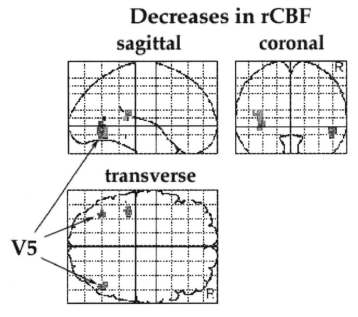

sagittal   coronal

transverse

V5

Figure 10   Measurement of change in regional cerebral blood flow (rCBF) by the technique of positron emission tomography when subjects view a multicolored Mondrian display shows that the increase in activity within V4 is coupled to a decrease of activity in V5 (average of six subjects).

LITERATURE CITED

Allman JM, Kaas JH. 1971. A representation of the visual field in the caudal third of the middle temporal gyrus of the owl monkey (*Aotus trivirgatus*). *Brain Res.* 31:85–105

Baker CL, Hess RH, Zihl J. 1991. Residual motion perception in a "motion-blind" patient, assessed with limited lifetime random dots. *J. Neurosci.* 112:454–61

Barbur JL, Watson JDG, Frackowiak RSJ, Zeki S. 1993. Conscious visual perception without V1. *Brain* 116:1293–302

Barrett G, Blumhardt L, Halliday AM, Halliday E, Kriss A. 1976. A paradox in the lateralisation of the visual evoked response. *Nature* 261:253–55

Bartels A, Zeki S. 1998. The theory of multistage integration in the visual brain. *Proc. R. Soc. London Ser. B* 265:2327–32

Bartels A, Zeki S. 1999. Can independent component analysis (ICA) isolate cortical areas from fMRI data? *Soc. Neurosci. Abstr.* 766:5

Bartels A, Zeki S. 2000. The architecture of the colour centre in the human visual brain: new results and a review. *Eur. J. Neurosci.* 12:172–93

Beckers G, Zeki S. 1995. The consequences of inactivating areas V1 and V5 on visual-motion perception. *Brain* 118:49–60

Bell AJ, Sejnowski TJ. 1995. An information maximization approach to blind separation and blind deconvolution. *Neural Comput.* 7:1129–59

Bender DB. 1981. Retinotopic organization of macaque pulvinar. *J. Neurophysiol.* 46:672–93

Bly BM, Kosslyn SM. 1997. Functional anatomy of object recognition in humans: evidence from positron emission tomography and functional magnetic resonance imaging. *Curr. Opin. Neurol.* 10:5–9

Blythe IM, Kennard C, Ruddock KH. 1987. Residual vision in patients with retrogenic-ulate lesions of the visual pathways. *Brain* 110:887–905

Buchner H, Weyen U, Frackowiak RSJ, Romaya J, Zeki S. 1994. The timing of visual-evoked potential activity in human area V4. *Proc. R. Soc. London Ser. B* 257:99–104

Ceccaldi M, Mestre D, Brouchon M, Balzamo M, Poncet M. 1992. Autonomie déambulatoire et perception visuelle du mouvement dans un cas de cécit' corticale quasi totale. *Rev. Neurol. Paris* 148:343–49

Cragg BG. 1969. The topography of the afferent projections in circumstriate visual cortex studied by the Nauta method. *Vis. Res.* 9:733–47

Crick F, Koch C. 1990. Towards a neurobiological theory of consciousness. *Semin. Neurosci.* 2:263–75

Crick F, Koch C. 1995. Are we aware of neural activity in primary visual cortex? *Nature* 375:121–23

Dennett D. 1991. *Consciousness Explained.* Boston: Little Brown

Desimone R, Fleming J, Gross CG. 1980. Prestriate afferents to inferior temporal cortex: an HRP study. *Brain Res.* 184:41–55

DeYoe EA, Carman GJ, Bandettini P, Glickman S, Wieser J, et al. 1996. Mapping striate and extrastriate visual areas in human cerebral cortex. *Proc. Natl. Acad. Sci. USA* 93:2382–86

DeYoe EA, van Essen DC. 1988. Concurrent processing streams in monkey visual cortex. *Trends Neurosci.* 11:219–26

Engel AK, Fries P, Roelfsema PR, König P, Singer W. 1999. Temporal binding, binocular rivalry, and consciousness. *Conscious. Cogn.* 8:155–58

Engel SA, Rummelhart DE, Wandell BA, Lee AH, Glover GH, et al. 1994. fMRI of human visual cortex. *Nature* 369:525

Felleman DJ, van Essen DC. 1991. Distributed

hierarchical processing in the primate cerebral cortex. *Cerebr. Cortex* 1:1–47

ffytche DH, Guy CN, Zeki S. 1995. The parallel visual motion inputs into areas V1 and V5 of human cerebral cortex. *Brain* 118:1375–94

ffytche DH, Howseman A, Edwards R, Sandeman DR, Zeki S. 2000. Human area V5 and motion in the ipsilateral visual field. *Eur. J. Neurosci.* In press

Gegenfurtner KR, Kiper DC, Fenstemaker SB. 1996. Processing of color, form, and motion in macaque area V2. *Vis. Neurosci.* 13:161–72

Girard P, Salin PA, Bullier J. 1992. Response selectivity of neurons in area MT of the macaque monkey during reversible inactivation of area V1. *J. Neurophysiol.* 67:1437–46

Hadjikhani N, Liu AK, Dale A, Cavanagh P, Tootell RBH. 1998. Retinotopy and color sensitivity in human visual cortical area V8. *Nat. Neurosci.* 1:235–41

Halliday AM. 1993. *Evoked Potentials in Clinical Testing.* London: Churchill Livingstone

Hess RH, Baker CL, Zihl J. 1989. The "motion-blind" patient: low level spatial and temporal filters. *J. Neurosci.* 9:1628–40

Heywood C, Cowey A. 1998. With color in mind. *Nat. Neurosci.* 1:171–73

Holmes G. 1918. Disturbances of vision caused by cerebral lesions. *Br. J. Ophthalmol.* 2:353–84

Howard RJ, Brammer M, Wright I, Woodruff PW, Bullmore ET, Zeki S. 1996. A direct demonstration of functional specialization within motion-related visual and auditory-cortex of the human brain. *Curr. Biol.* 6:1015–19

Howard RJ, Bullmore E, Brammer M, Williams SCR, Mellers J, et al. 1995. Activation of area V5 by visual-perception of motion demonstrated with echoplanar Mr-imaging. *Magn. Res. Imaging* 13:907–9

Hubener M, Bolz J. 1992. Relationships between dendritic morphology and cytochrome oxidase compartments in monkey striate cortex. *J. Comp. Neurol.* 324:67–80

Humphrey GK, Goodale MA, Corbetta M, Aglioti S. 1995. The McCollough effect reveals orientation discrimination in a case of cortical blindness. *Curr. Biol.* 5:545–51

Kaas JH. 1993. The organization of the visual cortex in primates: problems, conclusions and the use of comparative studies in understanding the human brain. In *The Functional Organization of the Human Visual Cortex,* ed. B Gulyas, D Ottoson, PE Roland, pp. 1–11. Oxford, UK: Pergamon

Kant I. 1787. *Critique of Pure Reason.* Transl. WS Pluhar, 1996. Indianapolis, IN: Hackett (From German)

Kastner S, DeWeerd P, Desimone R, Ungerleider LC. 1998. Mechanisms of directed attention in the human extrastriate cortex as revealed by functional MRI. *Science* 282:108–11

Kennard C, Lawden M, Morland AB, Ruddock KH. 1995. Colour identification and colour constancy are impaired in a patient with incomplete achromatopsia associated with pre-striate cortical lesions. *Proc. R. Soc. London Ser. B* 260:169–75

Komatsu H, Ideura Y, Kaji S, Yamane S. 1992. Color selectivity of neurons in inferior temporal cortex of the awake macaque monkey. *J. Neurosci.* 12:408–24

Land E. 1974. The retinex theory of colour vision. *Proc. R. Inst. Great Br.* 47:23–58

Land EH, McCann JJ. 1971. Lightness and retinex theory. *J. Opt. Soc. Am.* 61:1–11

Leibnitz GW. 1714. *Monadolagie.* Transl. R Latta, 1898. Oxford, UK: Clarendon (From French)

Lennie P, Krauskopf J, Sclar G. 1990. Chromatic mechanisms in striate cortex of macaque. *J. Neurosci.* 10:649–69

Leventhal AG, Thompson KG, Liu D, Zhou Y, Ault SJ. 1995. Concomitant sensitivity to orientation, direction, and color of cells in layers 2, 3, and 4 of monkey striate cortex. *J. Neurosci.* 153(1):1808–18

Levitt JB, Yoshioka T, Lund JS. 1994. Intrinsic cortical connections in macaque visual area V2: evidence for interaction between different functional streams. *J. Comp. Neurol.* 342:551–70

Livingstone MS, Hubel DH. 1984. Specificity of intrinsic connections in primate primary visual cortex. *J. Neurosci.* 4:2830–35

Livingstone MS, Hubel DH. 1988. Segregation of form, color, movement, and depth: anatomy, physiology, and perception. *Science* 240:740–49

Logothetis NK. 1998. Single units and conscious vision. *Philos. Trans. R. Soc. London Ser. B* 353:1801–18

Logothetis NK, Pauls J, Poggio T. 1995. Shape representation in the inferior temporal cortex of monkeys. *Curr. Biol.* 5:552–63

Lueck CJ, Zeki S, Friston KJ, Deiber MP, Cope P, et al. 1989. The colour centre in the cerebral cortex of man. *Nature* 340:386–89

Lund JS, Yoshioka T, Levitt JB. 1993. Comparison of intrinsic connectivity in different areas of macaque monkey cerebral cortex. *Cerebr. Cortex* 3:148–62

McKeefry D, Zeki S. 1997. The position and topography of the human colour centre as revealed by functional magnetic resonance imaging. *Brain* 120:2229–42

Meadows JC. 1974. Disturbed perception of colours associated with localized cerebral lesions. *Brain* 97:615–32

Mestre DR, Brouchon M, Ceccaldi M, Poncet M. 1992. Perception of optical flow in cortical blindness: a case report. *Neuropsychologia* 30:783–95

Mollon JD, Newcombe F, Polden PG, Ratcliff G. 1980. On the presence of three cone mechanisms in a case of total achromatopsia. In *Colour Vision Deficiencies,* ed. VG Verriest, 5:130–35. Bristol, UK: Hilger

Morel A, Bullier J. 1990. Anatomical segregation of two cortical visual pathways. *Vis. Neurosci.* 4:555–78

Moruzzi G, Magoun HW. 1949. Brain stem reticular formation and activation of the EEG. *EEG Clin. Neurophysiol.* 1:455–73

Mountcastle VB. 1998. *The Cerebral Cortex.* Cambridge, MA: Harvard Univ. Press. 486 pp.

Moutoussis K, Zeki S. 1997a. A direct demonstration of perceptual asynchrony in vision. *Proc. R. Soc. London Ser. B* 264:393–99

Moutoussis K, Zeki S. 1997b. Functional segregation and temporal hierarchy of the visual perceptive systems. *Proc. R. Soc. London Ser. B* 264:1407–14

Moutoussis K, Zeki S. 2000. A psychophysical dissection of the brain sites involved in colour generating comparisons. *Proc. Natl. Acad. Sci. USA.* In press

Movshon JA, Adelson EH, Gizzi MS, Newsome WT. 1985. The analysis of moving visual patterns. In *Pattern Recognition Mechanisms,* ed. C Chagas, R Gattass, CG Gross, pp. 117–51. Vatican City: Pontifical Acad. Sci.

Nakamura M, Gattass R, Desimone R, Ungerleider LG. 1993. The modular organization of projections from areas V1 and V2 to areas V4 and TEO in macaques. *J. Neurosci.* 13:3681–91

Orban GA. 1997. Visual processing in macaque area MT/V5 and its satellites (MSTd and MSTv). See Rockland et al. 1997, pp. 359–434

Paulson HL, Galetta SL, Grossman M, Alavi A. 1994. Hemiachromatopsia of unilateral occipitotemporal infarcts. *Am. J. Ophthalmol.* 118:518–23

Perry RJ, Zeki S. 2000. Integrating motion and colour within the visual brain: an fMRI approach to the binding problem. *Soc. Neurosci.* 25:547. Abstr. 221.1

Raiguel SE, Lagae L, Gulyas B, Orban GA. 1989. Response latencies of visual cells in macaque areas V1, V2 and V5. *Brain Res.* 493:155–59

Riddoch G. 1917. Dissociations of visual perception due to occipital injuries, with especial reference to appreciation of movement. *Brain* 40:15–57

Rockland KS. 1985. A reticular pattern of intrinsic connections in primate area V2 (area 18). *J. Comp. Neurol.* 235:467–78

Rockland KS, Kaas JH, Peters A, eds. 1997. *Cerebral Cortex: Extrastriate Cortex in Primates,* Vol. 12. New York/London: Plenum

Rockland KS, Lund JS. 1983. Intrinsic laminar lattice connections in primate visual cortex. *J. Comp. Neurol.* 216:303–18

Rockland KS, Pandya DN. 1979. Laminar origins and terminations of cortical connections of the occipital lobe in the rhesus monkey. *Brain Res.* 179:3–20

Rockland KS, Saleem KS, Tanaka K. 1994. Divergent feedback connections from area V4 and TEO in the macaque. *Vis. Neurosci.* 11:579–600

Rockland KS, van Hoesen GW. 1994. Direct temporal-occipital feedback connections to striate cortex (V1) in the macaque monkey. *Cerebr. Cortex* 4:300–13

Rodman R, Gross CG, Albright TD. 1989. Afferent basis of visual response properties in area MT of the macaque. I. Effects of striate cortex removal. *J. Neurosci.* 9:2033–50

Schiller PH. 1997. Past and present ideas about how the visual scene is analyzed by the brain. See Rockland et al. 1997, pp. 59–90

Sereno MI, Dale AM, Reppas JB, Kwong KK, Belliveau JW, et al. 1995. Borders of multiple visual areas in humans revealed by functional magnetic resonance imaging. *Science* 268:889–93

Shipp S, de Jong BM, Zihl J, Frackowiak RSJ, Zeki S. 1994. The brain activity related to residual motion vision in a patient with bilateral lesions of V5. *Brain* 117:1023–38

Shipp S, Zeki S. 1989. The organization of connections between areas V5 and V1 in macaque monkey visual cortex. *Eur. J. Neurosci.* 14:309–32

Shipp S, Zeki S. 1995. Segregation and integration of specialised pathways in macaque monkey visual cortex. *J. Anat.* 187:547–62

Standage GP, Benevento LA. 1983. The organization of connections between the pulvinar and visual area MT in the macaque monkey. *Brain Res.* 262:288–94

Stoerig P. 1996. Varieties of vision—from blind responses to conscious recognition. *Trends Neurosci.* 199:401–6

Stoerig P, Cowey A. 1995. Visual-perception and phenomenal consciousness. *Behav. Brain Res.* 711(2):147–56

Tononi G, Edelman GM. 1998. Consciousness and complexity. *Science* 282:1846–51

Tononi G, Sporns O, Edelman GM. 1994. A measure for brain complexity: relating functional segregation and integration in the nervous system. *Proc. Natl. Acad. Sci. USA* 91:5033–37

Tootell RB, Dale AM, Sereno MI, Malach R. 1996. New images from human visual cortex. *Trends Neurosci.* 19:481–89

Tootell RBH, Mendola JD, Hadjikhani NK, Liu AK, Dale AM. 1998. The representation of the ipsilateral visual field in human cerebral cortex. *Proc. Natl. Acad. Sci. USA* 95:818–24

Tootell RBH, Reppas JB, Kwong KK, Malach R, Born RT, et al. 1995. Functional analysis of human MT and related visual cortical areas using magnetic resonance imaging. *J. Neurosci.* 15:3215–30

Treisman A. 1996. The binding problem. *Curr. Opin. Neurobiol.* 6:171–78

Ullman S. 1982. *The Interpretation of Visual Motion.* Cambridge, MA: MIT Press. 229 pp.

Vaina LM. 1994. Functional segregation of color and motion processing in the human visual cortex: clinical evidence. *Cerebr. Cortex* 4:555–72

Vaina LM, Lemay M, Bienfang DC, Choi AY, Nakayama K. 1990. Intact biological motion and structure from motion perception in a patient with impaired motion mechanisms—a case-study. *Vis. Neurosci.* 5:353–69

Walsh V, Carden D, Butler SR, Kulikowski JJ. 1993. The effects of V4 lesions on the visual abilities of macaques: Hue discrimination and colour constancy. *Behav. Brain Res.* 531(2):51–62

Wechsler IS. 1933. Partial cortical blindness with preservation of color vision. *Arch. Ophthalmol.* 9:957–65

Weiskrantz L. 1986. *Blindsight.* Oxford, UK: Oxford Univ. Press

Weiskrantz L. 1995. Blindsight—not an island unto itself. *Curr. Dir. Psychol. Sci.* 4:146–51

Wurtz RH, Yamasaki DS, Duffy CJ, Roy JP.

1990. Functional specialization for visual motion processing in primate cerebral cortex. *Cold Spring Harbor Symp. Quant. Biol.* 55:717–27

Yoshioka T, Blasdel GG, Levitt JB, Lund JS. 1996. Relations between patterns of intrinsic lateral connectivity, ocular dominance, and cytochrome oxidase-reactive regions in macaque monkey striate cortex. *Cerebr. Cortex* 6:297–310

Zeki S. 1977. Colour coding in the superior temporal sulcus of rhesus monkey visual cortex. *Proc. R. Soc. London Ser. B* 197:195–223

Zeki S. 1978. Functional specialization in the visual cortex of the monkey. *Nature* 274:423–28

Zeki S. 1983. Colour coding in the cerebral cortex: the reaction of cells in monkey visual cortex to wavelengths and colours. *Neuroscience* 9:741–65

Zeki S. 1990a. A century of cerebral achromatopsia. *Brain* 113(6):1721–77

Zeki S. 1990b. A theory of multi-stage integration in the visual cortex. In *The Principles of Design and Operation of the Brain*, ed. JC Eccles, O Creutzfeldt, pp. 137–54. Vatican City: Pontifical Acad. Sci.

Zeki S. 1991. Cerebral akinetopsia (visual motion blindness). A review. *Brain* 114(2):811–24

Zeki S. 1993. *A Vision of the Brain*. Oxford, UK: Blackwell. 366 pp.

Zeki S. 1996. Are areas teo and pit of monkey visual-cortex wholly distinct from the 4th visual complex (V4 complex)? *Proc. R. Soc. London Ser. B* 263:1539–44

Zeki S. 1997. The Woodhull Lecture. 1995: Visual art and the visual brain. *Proc. R. Inst. Great Br.* 68:29–63

Zeki S. 1998. Parallel processing, asynchronous perception and a distributed system of consciousness in vision. *Neuroscientist* 4:365–72

Zeki S. 1999. Splendours and miseries of the brain. *Philos. Trans. R. Soc. London Ser. B* 354:2053–65

Zeki S, Aglioti S, McKeefry D, Berlucchi G. 1999. The neurobiological basis of conscious color perception in a blind patient. *Proc. Natl. Acad. Sci. USA* 96:14124–29

Zeki S, Bartels A. 1999. Towards a theory of visual consciousness. *Conscious. Cogn.* 8:225–59

Zeki S, ffytche D. 1998. The Riddoch Syndrome: insights into the neurobiology of conscious vision. *Brain* 121:25–45

Zeki S, Marini L. 1998. Three cortical stages of colour processing in the human brain. *Brain* 121:1669–85

Zeki S, Moutoussis K. 1997. Temporal hierarchy of the visual perceptive systems in the Mondrian world. *Proc. R. Soc. London Ser. B* 264:1415–19

Zeki S, Shipp S. 1988. The functional logic of cortical connections. *Nature* 335:311–17

Zeki S, Shipp S. 1989. Modular connections between areas V2 and V4 of macaque monkey visual cortex. *Eur. J. Neurosci.* 1:494–506

Zeki S, Watson JDG, Lueck CJ, Friston KJ, Kennard C, Frackowiak RSJ. 1991. A direct demonstration of functional specialization in human visual cortex. *J. Neurosci.* 11:641–49

Zheng D, LaMantia AS, Purves D. 1991. Specialized vascularization of the primate visual cortex. *J. Neurosci.* 118:2622–29

Zihl J, Von Cramon D, Mai N. 1983. Selective disturbance of movement vision after bilateral brain damage. *Brain* 106:313–40

Zihl J, Von Cramon D, Mai N, Schmid CH. 1991. Disturbance of movement vision after bilateral posterior brain damage. Further evidence and follow up observations. *Brain* 114:2235–52

Annu. Rev. Neurosci. 2001. 24:87–105

GLIAL CONTROL OF NEURONAL DEVELOPMENT

Greg Lemke

Molecular Neurobiology Laboratory, The Salk Institute, La Jolla, California 92037;
e-mail: lemke@salk.edu

Key Words   glia, axon guidance, neuronal survival, midline

■ **Abstract**   Reciprocal interactions between differentiating glial cells and neurons define the course of nervous system development even before the point at which these two cell types become definitively recognizable. Glial cells control the survival of associated neurons in both *Drosophila* and mammals, but this control is dependent on the prior neuronal triggering of glial cell fate commitment and trophic factor expression. In mammals, the growth factor neuregulin-1 and its receptors of the ErbB family play crucial roles in both events. Similarly, early differentiating neurons and their associated glia rely on reciprocal signaling to establish the basic axon scaffolds from which neuronal connections evolve. The importance of this interactive signaling is illustrated by the action of glial transcription factors and of glial axon guidance cues such as netrin and slit, which together regulate the commissural crossing of pioneer axons at the neural midline. In these and related events, the defining principle is one of mutually reinforced and mutually dependent signaling that occurs in a network of developing neurons and glia.

INTRODUCTION

Conceptualizations in neural development have tended to be neuronocentric. That is, developmental events have typically been viewed from the standpoint of differentiating neurons: how cells adopt specific neuronal fates, how the proliferation of neuronal precursors and the survival of differentiated neurons may be regulated, or how neuronal migration, axon guidance, and synapse formation may be coordinated. Recent observations, however, have highlighted the extent to which these aspects of nervous system development are often events over which neurons exercise little or no control, in that they are largely or entirely regulated by glia and their precursors.

In higher mammals, glial cells—most broadly defined as nonneuronal cells that arise from the neuroepithelium or from secondary neural anlagen—vastly outnumber neurons, and even in the nervous systems of invertebrates, glia are numerous. It is becoming increasingly clear that neurons and glia interact with each other from the earliest times of development, and that these interactions, which are generally reciprocal in nature, are crucial for the organization and operation of

0147-006X/01/0621-0087$14.00

the nervous system. In this review, I consider selected examples, in *Drosophila* and mice, of studies that illustrate the central role played by glia and their precursors in the specification, survival, differentiation, and interconnection of neurons.

LIFE AND DEATH

The "neurotrophic hypothesis" is a by-now-classical explanation for the observed matching of the number of neurons within a developing ganglion or nucleus to the size of its innervation target (Purves 1986, Oppenheim 1991). The central tenet of this hypothesis is that the survival of developing neurons (prominent examples in vertebrates are motor neurons in the ventral spinal cord and sensory and sympathetic neurons in peripheral ganglia) depends on one or more target-derived neurotrophic factors, which are present in limiting amounts. It is the job of the target cells to synthesize and provide these factors—for example, nerve growth factor—to the neurons, which through their axons and terminals then compete for the limiting supply. Competition is fierce because losing neurons die (Davies 1996, Henderson 1996). In contrast to this neuron-centered perspective, genetic studies in the mouse have now suggested that the survival of many developing neurons is more dependent on their ability to provide trophic factors to their targets, including their associated glia, than on the ability of these targets to provide trophic factors to the neurons.

Neuregulin-1 and Neuregulin Receptors

One of the most important trophins that central and peripheral neurons provide to Schwann cells, the glial cells of peripheral nerves and ganglia, is neuregulin-1 (NRG1). This protein, formerly known as glial growth factor, is structurally related to what is perhaps the best studied trophic factor—epidermal growth factor (EGF) (Lemke & Brockes 1984, Marchionni et al 1993, Falls et al 1993, Lemke 1996). Multiple isoforms of NRG1, some transmembrane and some secreted, are generated by alternative splicing; these are produced by many neurons, including motor, sensory, and sympathetic neurons, very soon after they are born (Marchionni et al 1993, Falls et al 1993, Corfas et al 1995). NRG1 proteins are then transported down the length of axons, where they are presented to axon-associated Schwann cells and to peripheral targets, including muscles (Yang et al 1998). A large number of cell culture and in vivo perturbation experiments indicate that NRG1 plays important roles at multiple stages in Schwann cell development. At the very first stage, for example, it appears to drive multipotent neural crest stem cells toward a Schwann cell fate (Shah et al 1994, Dong et al 1995). As glial growth factor, NRG1 was originally identified, named, and purified based on its ability to markedly stimulate Schwann cell proliferation in vitro (Lemke & Brockes 1984, Porter et al 1986, Levi et al 1995), and several lines of evidence indicate that it plays this same mitogenic role in the expansion of Schwann cell populations in vivo (e.g. Garratt et al 2000). Finally, recent studies demonstrate that neuronally produced NRG1 protects against programmed Schwann cell death, both in vitro and in vivo (Trachtenberg & Thompson 1996, Grinspan et al 1996, Syroid et al 1996). The results of these cell survival experiments have generally been interpreted as providing a mechanism for matching the number of NRG1-requiring Schwann cells in a given peripheral nerve to the number of NRG1-supplying axons—a "gliotrophic" formulation that turns the neurotrophic hypothesis on its head.

NRG1 signaling is transduced through the cell surface receptors ErbB2, ErbB3, and ErbB4, receptor-configured protein-tyrosine kinases that are structurally and functionally related to the EGF receptor, with which they interact (Carraway & Cantley 1994, Lemke 1996, Burden & Yarden 1997). ErbB proteins partner with one another to form dimeric NRG receptors, and an ErbB2 + ErbB3 heterodimer functions as the principal neuregulin receptor in most developing Schwann cells (Levi et al 1995, Syroid et al 1996). The *ErbB2* and *ErbB3* genes, as well as the *NRG1* gene, have been conventionally and conditionally inactivated in mice (Lee et al 1995, Meyer & Birchmeier 1995, Erickson et al 1997, Riethmacher et al 1997, Woldeyesus et al 1999, Garratt et al 2000). Schwann cell defects are not readily studied in conventional *ErbB2* and *NRG1* knockouts, which like *ErbB4* mouse mutants (Gassmann et al 1995) die at embryonic day (E) 10.5, from a failure of cardiac myocyte differentiation (Lemke 1996). (This is a point at which significant Schwann cell differentiation has yet to occur in most areas of the developing mouse embryo.) ErbB3 does not function in cardiac muscle differentiation, however, and *ErbB3* knockouts survive well past the mid to late embryonic period in which Schwann cells differentiate from multipotent neural crest cells and take up their positions in association with peripheral axons and their targets (Erickson et al 1997, Riethmacher et al 1997). Survival throughout the course of embryogenesis is also seen for two variants of the *ErbB2* mutants. In one of these, the regulatory elements of the cardiac-specific α-myosin heavy chain (α-*MHC*) gene were used to transgenically drive wild-type ErbB2 receptor expression in the heart muscle of ErbB2 mutants, leaving most of the remaining tissues in the embryo, including all neural tissues, mutant for the receptor (Morris et al 1999). In a related approach, Woldeyesus et al (1999) used homologous recombination methods to "knock-in" an *ErbB2* cDNA into the cardiac-specific *Nkx2.5* gene and then crossed the resulting knock-in mice with the *ErbB2* null lines.

At E12.5, a time at which wild-type peripheral nerves in mice are normally well populated with Schwann cells and their precursors, the same nerves in the *ErbB3* nulls, and in both of the *ErbB2*-rescued mutants, are devoid of Schwann cells (Riethmacher et al 1997, Morris et al 1999, Woldeyesus et al 1999). This has important consequences for fasciculation of the naked axons that make up the mutant nerves, and for the ability of sensory, sympathetic, and motor neuron populations to maintain synapses at their targets. Even more dramatic, however, are the effects seen on the survival of these neuronal populations from E12.5 through E18.5 in the mutants. In both *Nkx2.5-ErbB2* and *ErbB3*$^{-/-}$ mice, for example, nearly 90% of dorsal root ganglion neurons are subject to apoptotic death by E18.5, and similarly, the number of cervical motor neurons in these same mice is reduced by ~80% (Riethmacher et al 1997, Woldeyesus et al 1999). These neuronal losses, which occur after the loss of Schwann cells, are much more extensive than any of the neuronal losses seen in knockouts of the *TrkA, TrkB,* or *TrkC* neurotrophin receptor genes, or of the neurotrophin genes themselves (Klein 1994). It is important that these effects on neuronal survival are cell nonautonomous: The neurons that die do not express ErbB2 or ErbB3. Rather, it is the cells with which these neurons interact—Schwann cells in nerves and muscles in the periphery—that are ErbB2$^+$ and ErbB3$^+$. Thus, the survival of neurons is dependent on the survival of Schwann cells and presumably on a set of neurotrophic factors—including CNTF, GDNF, BDNF, LIF, PDGF, FGF, and NT-3—that Schwann cells are known to produce (Davies 1998).

These observations demonstrate the extent to which trophic signaling and support between developing neurons and glia may be reciprocal and interactive: Neurons provide trophic factors, such as NRG1, which promote the survival and differentiation of Schwann cells, and these glia in turn provide a distinct set of trophins that promotes the survival and differentiation of neurons (see Figure 1). They are in keeping with several similar demonstrations of the extent to which the survival and functioning of mammalian neurons in general is dependent on the survival and functioning of associated glia. Loss of enteric glia in a GFAP-thymidine kinase transgenic mouse ablation line, for example, leads to secondary cell nonautonomous loss of enteric neurons, which in turn results in severe jejuno-ileitis (Bush et al 1998). Similarly, a host of mouse mutations in glial-specific myelination-related genes, both engineered and naturally occurring, lead to primary defects in myelinating glia, but in addition, to secondary axonal degeneration and neuronal loss in neurons whose axons are enwrapped by the defective myelin (e.g. Giese et al 1992, Griffiths et al 1998).

--→

Figure 1 A mammalian example of the dependence of neuronal survival on neuron-mediated trophic support of glia. In a wild-type mouse embryo (+/+, top panel), motor neurons in the central nervous system (CNS) and sensory neurons in the dorsal root ganglia (DRG) produce neuregulin-1 (NRG1). This protein acts as a trophic factor for Schwann cells and for targets cells (such as muscles), which express the neuregulin receptors ErbB2 and ErbB3. Motor and sensory neurons in the caudal CNS do not express these receptors. In mice in which the *ErbB2* and *ErbB3* genes are selectively inactivated in all noncardiac tissues (see text), Schwann cells are found to be absent from peripheral nerves at midembryogenesis (middle panel). This is followed over succeeding days by progressive loss of motor neurons, particularly in the cervical and lumbar spinal cord, and of sensory neurons at all axial levels (lower panel). Innervation of target muscles by the axons of cervical motor neurons—e.g. of the diaphragm by the phrenic nerve—is also lost. Less-pronounced effects on motor neuron survival are seen at thoracic levels of the cord (Woldeyesus et al 1999). See text for details.

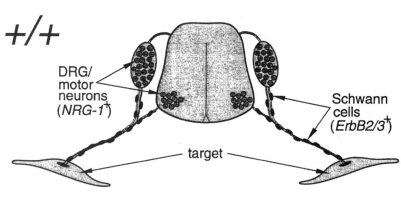

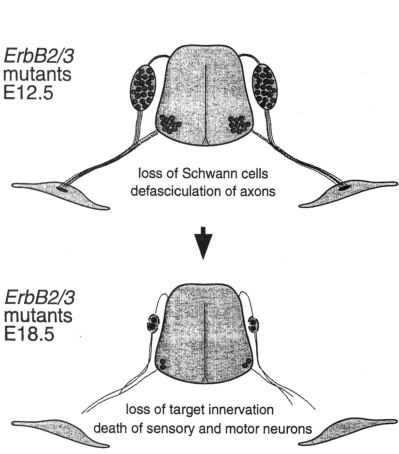

Glial Control of Neuronal Survival in *Drosophila*

Recent studies on the phenotypes seen in flies mutant for the *glial-cells-missing* (*gcm*) gene (Hosoya et al 1995, Jones et al 1995) provide one example, among several, of a similar interdependence in the developing central nervous system (CNS) of *Drosophila* (Booth et al 2000). The ventral nerve cord of *Drosophila* larvae is formed by an axon scaffold composed of two longitudinal connectives, which run the length of the embryo, together with reiterated sets of anterior and posterior commissures that connect the longitudinal connectives in each segment. When "interface glia," which are associated with and enwrap the axons of the longitudinal pathways, are killed using glial-specific expression of the toxin Ricin, the neurons whose axons are associated with these glia subsequently die through apoptotis (Booth et al 2000). When the same interface glia are transfated to neurons, in flies that lack the glial-specific transcription factor gcm, a similar secondary neuronal loss, also apoptotic, is observed (Booth et al 2000). As for the neuronal losses in the *ErbB2* and *ErbB3* mutants, the death of the longitudinally projecting neurons is, in both the Ricin ablations and the *gcm* mutants, cell nonautonomous, i.e. it is due to the loss of glia rather than to any direct effect on the neurons that die (Booth et al 2000).

These findings parallel similar earlier observations of neuronal loss in *Drosophila* mutants in other glial-expressed genes, including *reversed-polarity* (*repo*) (Xiong et al 1994, Halter et al 1995) and *drop-dead* (Buchanan & Benzer 1993). *Repo* is a homeobox gene that is expressed in most glia in the developing CNS, with the exception of the glial cells that populate the CNS midline (see below). Among *repo*[+] cells are laminar glia in the optic lobes of the fly brain, which receive visual input from photoreceptors in the retina. Loss of repo function in these laminar glia leads to the death of *repo* laminar neurons, which suggests that laminar glia supply factors essential for neuronal survival (Xiong & Montell 1995). Subsequent to this neuronal loss in the brain, neurons in the *repo* mutant retina also degenerate, demonstrating a further, retrograde neuronal dependence on *repo*[+] glia.

NEURONAL MIGRATION

Radial Glia, Bergmann Glia, and CNS Morphogenesis

Embryonic neurons are typically born at some distance from their sites of residence and action in the mature nervous system, and this is particularly true for neurons in the midbrains and forebrains of mammalian central nervous systems (Hatten 1999). The pathways taken by newborn neurons are specific and stereotyped and are dependent on substrate, cell surface, and diffusible guidance cues. Many of these cues are provided by glia and their precursors.

Perhaps the most widely appreciated neuronal migration is the radial migration of newborn neurons during the development and lamination of the mammalian neocortex. This process has long been known to involve a set of specialized neuroepithelial cells termed radial glia (Rakic 1971, 1972), which provide the migration scaffold used by 80%–90% of cortical neurons. Several lines of evidence indicate that these glia, which early on elaborate processes that span the full extent of the neural tube (Ramon y Cajal 1995), are precursors to astrocytes in the mature brain (Culican et al 1990). Newborn neurons that emerge from the ventricular zone attach themselves to radial glial processes and then migrate, in a saltatory fashion, along these processes toward the pial surface of the developing brain. An analogous process occurs during the migration of newborn granule cells, along Bergmann glial processes, from the external to the internal granule cell layer of the developing postnatal cerebellum (Hatten 1999).

These migrations involve several reciprocal signaling interactions between newborn neurons and glia, and again, neuronally produced NRG1 appears to be an important mediator of these interactions. In cell culture, NRG1 promotes the adoption of a radial glial phenotype by both cortical and cerebellar astrocytes, as monitored by morphology (number, orientation, and length of processes) and by the up-regulation of the radial glial markers BLBP (brain lipid binding protein) and RC2 (Anton et al 1997, Rio et al 1997). In the developing cerebellum, NRG1 is expressed by the granule cells that migrate along the processes of Bergmann glia, both before the onset of and during granule cell migration. Bergmann glia express the ErbB4 NRG1 receptor, rather than ErbB2 or ErbB3 (Anton et al 1997, Rio et al 1997). When a dominant-negative antagonist of ErbB4 is introduced into cultured cerebellar astroglia, it blocks their NRG1-induced in vitro transformation into radial glia and severely inhibits the ability of granule cell neurons to migrate on these glia (Rio et al 1997). Similarly, in the developing neocortex, NRG1 is expressed by migrating neurons, its receptors ErbB2 and ErbB4 are expressed by radial glia, and exogenous NRG1 stimulates cortical neuron migration and the adoption of a radial glial phenotype in cortical imprint assays (Anton et al 1997). There is therefore good evidence that in neuronal migration as well, NRG1-mediated neuronal signaling of glia is an important component of the glial-facilitated migration of neurons. Given that many of these migratory events occur from midembryogenesis through early postnatal life in mice, a definitive genetic test of these models must await the development of conditional mouse mutants because, as noted above, the conventional knockouts of *NRG1*, *ErbB2*, and *ErbB4* genes are lethal mutations at E10 (Lemke 1996).

AXON GUIDANCE

Guidance at the *Drosophila* Midline

Most metazoans are bilaterally symmetric, and many features of mature neural function—including the interpretation of sensory information and the coordination of locomotion—are dependent on coherent communication between the two halves of the nervous system. This communication is established during embryonic development, when axons of neurons located on one side of the nervous system must recognize, grow toward, and then choose between crossing or not crossing the embryonic midline. The presentation of navigational cues at the midline is therefore of fundamental importance to neural organization during embryogenesis. Work over the past several years has demonstrated that many, if not most, of these navigational cues are provided by glia and their precursors.

In the ventral nerve cord of *Drosophila*, a set of glial cells pioneer and then populate the midline of fly CNS (Granderath & Klambt 1999). These midline glia and their processes comprise the cellular environment that commissural axons encounter as they make their way from one side of the ventral nerve cord to the other. Genetic and biochemical experiments suggest that two dual-function secreted cues are especially important to this process—netrin (Serafini et al 1994, Mitchell et al 1996) and slit (Rothberg et al 1990, Kidd et al 1999). Both of these cues are the products of midline glia (Mitchell et al 1996, Hummel et al 1999, Rothberg et al 1990).

Netrins were originally purified, characterized, and cloned not in flies but in vertebrates, as long-range, secreted chemoattractants expressed by the floor plate. The floor plate is composed of a set of specialized nonneuronal cells located at the ventral midline of the developing spinal cord (Tessier-Lavigne et al 1988, Placzek et al 1990, Serafini et al 1994, Kennedy et al 1994). These cells express several radial glial markers. Netrin expression by the floor plate plays a chemotropic role in directing the growth of commissural axons—those that arise from dorsal projection neurons, grow toward and cross the ventral midline, and then innervate targets on the opposite side of the spinal cord. (Netrin plays the same role in commissural axon growth in more anterior regions of the vertebrate CNS—see below.) Netrins are secreted, modular proteins in which a γ-laminin domain is followed by three EGF-like domains and a C-terminal basic domain (Serafini et al 1994). In *Drosophila*, two closely related netrin genes, *netrin-A* and *-B*, are expressed by midline glia during the period in which commissural axons are first negotiating their way toward and across the midline (Harris et al 1996, Mitchell et al 1996). The localized glial expression of netrins at the midline is important because ectopic misexpression of netrins throughout the fly CNS leads to defects in commissural axon crossing that are in several respects similar to those seen in mutants that lack both netrin-A and -B (Mitchell et al 1996).

There are several receptors for the netrins, but these fall into two broad groups— those related to the DCC (deleted in colorectal cancer) protein and those related to the *Caenorhabditis elegans* protein unc-5 (Chan et al 1996, Keino-Masu et al 1996, Leonardo et al 1997). Combinatorial pairing of these receptor proteins in part determines how, in terms of both magnitude and sign, neurons and their axons respond to netrin signaling (Seeger & Beattie 1999). For guidance at the midline in *Drosophila*, the netrins function as chemoattractants, and netrin signaling is transduced by a *Drosophila* DCC-related receptor designated frazzled, which is expressed by commissural axons and their growth cones (Kolodziej et al 1996). The restricted expression of the *frazzled* and *netrin* genes means that the cells that are "calling the shots" with respect to midline crossing in *Drosophila* are glial. Commissural axons, rather than making an informed decision, faithfully respond to instructions that orginate in their glial partners.

Crossing or not crossing the midline appears to reflect a balance in expression and activity of axon attractants and repellents. Prominent among the latter, in both invertebrates and vertebrates, are proteins related to *Drosophila* slit (Rothberg et al 1990). Structurally, these proteins are also large, secreted molecules, which like the netrins bind to components of the extracellular matrix. They are composed of four leucine-rich repeats, followed by nine EGF-like repeats, and a cysteine knot at the *C*-terminus. The single *Drosophila* slit protein is also a product of midline glia (Rothberg et al 1990, Brose & Tessier-Lavigne 2000), where it acts as a strong repulsive cue for commissural axons. *Slit* loss-of-function mutants exhibit a "collapsed midline" axon guidance phenotype, in which commissural axons extend toward the midline but then never leave it, forming one large axon fasicle. It was initially thought that many midline glia were missing in the *Drosophila slit* mutants, and that the observed axon crossing defects were secondary to this loss, but it is now clear the glia are only displaced at the midline (Sonnenfeld & Jacobs 1994).

The axonal receptors for *Drosophila* slit are the roundabout (robo) proteins (Brose et al 1999, Kidd et al 1999), so-named for the axon guidance phenotype exhibited by *robo* mutants. In a wild-type ventral nerve cord, commissural axons typically cross the midline only once, and then turn anteriorly or posteriorly to join other axons in the longitudinal connectives, never again crossing the midline. In *robo* mutants, commissural axons cross the midline repeatedly, typically turning anteriorly or posteriorly after crossing once, and then crossing again at the next available anterior or posterior commissure, and so on (Seeger et al 1993). Slit binds to robo in cell-based interaction assays, and the slit and robo mutations interact genetically. The difference between the collapsed single commissure of the slit mutants, in which all axons remain at the midline, and the hyper-crossing phenotype of the *robo* mutants is thought to be due to the continued presence of robo-2 and robo-3 in the commissural neurons of the latter, which prevents a pile-up of axons at the midline.

If slit is a strong repellent for the commissural axons in the *Drosophila* ventral nerve cord, why do they ever cross the midline? One answer to this question involves the protein commissureless (Comm), which was also named based on a loss-of-function phenotype. In a *comm* mutant, most axons never cross the midline of the ventral nerve cord and instead run almost exclusively in two unconnected longitudinal bundles (Seeger et al 1993). Comm is a type II transmembrane protein, which, by a largely unknown mechanism, appears to down-regulate robo expression in axons as they approach and then transit the midline. Comm appears to act only at the midline, and in axons that have already crossed over to the other side of the ventral nerve cord, robo expression is markedly up-regulated. It is this up-regulation that is thought to normally prevent recrossing of the midline. Flies in which comm is transgenically elevated by modest amounts at the midline exhibit a *robo*-like axon guidance phenotype, and those in which Comm is elevated to high levels exhibit a *slit*-like phenotype (Kidd et al 1998, 1999). As for the other important guidance cues, Comm is expressed not by the commissural neurons but by the midline glia. Thus, all three of the central regulators of axon guidance at the *Drosophila* midline—as defined by genetics, biochemistry, and cell biology—are the products of midline glia, and the localization of these cues to these midline cells is essential to their mode of action. By these criteria, the glia at the center of the *Drosophila* ventral nerve cord are the masters of its midline (see Figure 2).

Although the midline glia of flies have received the most attention with respect to axon guidance, they are but one *Drosophila* example. Others include the recent demonstration of essential glial roles in axon fasciculation, defasciculation, and growth cone guidance in the formation of the longitudinal connectives of ventral nerve cord (Hidalgo & Booth 2000), and of an important role for retinal basal glia in the guidance of axons from eye into a developing optic stalk (Rangarajan et al 1999).

Axon Guidance at the Vertebrate Midline

There are clear parallels to these *Drosophila* guidance phenomena that are seen at the midline of the developing vertebrate nervous system. As noted above, netrins were originally identified and purified through their action in the vertebrate floor plate, a triangularly shaped group of specialized, nonneuronal cells at the base of the caudal neural tube, i.e. at the ventral midline of the developing spinal cord. These nonneuronal cells occupy a position that is topologically equivalent to that of midline glia of *Drosophila*. In addition to netrin-1 and netrin-2, they express slit1, 2, and 3, three vertebrate homologs of *Drosophila* slit (Brose et al 1999, Li et al 1999, Yuan et al 1999). Similarly, commissural axons of the developing spinal cord express vertebrate robo homologs (Robo-1 and Rig-1) and DCC receptors (Keino-Masu et al 1996). As in *Drosophila* netrin mutants, mouse *netrin-1* gene knockouts exhibit defects in the guidance of commissural axons in the embryonic spinal cord that are consistent with netrin-1 acting as a chemotropic factor for guiding these axons toward the ventral midline (Serafini et al 1996). Mouse netrin-1 mutants also exhibit pronounced defects in commissural axon crossing in forebrain midline loci, such as the corpus callosum, where the floor plate is not evident, but where other netrin-1[+] ventral midline cells, including what are thought to be guidance glia, are located (Serafini et al 1996).

Indeed, a striking demonstration of glial control of axon guidance at the vertebrate midline in these more-anterior regions of the nervous system has recently been provided by the analysis of mice carrying mutations in the gene encoding a transcription factor expressed by ventral midline cells. This protein—designated Vax1, for ventral anterior homeobox 1—is a member of a new subfamily of homeodomain proteins that appear to be specific to vertebrates (Hallonet et al 1998, Bertuzzi et al 1999). It is expressed in a relatively small group of ventral midline cells specifically in the forebrain, including the preoptic area and optic chiasm

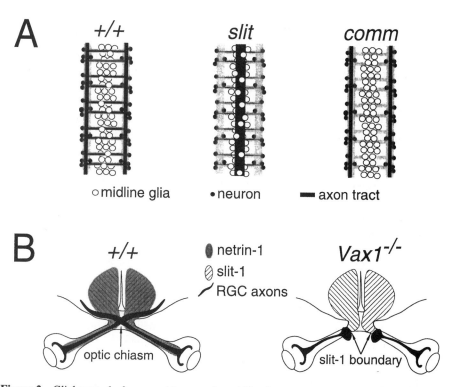

Figure 2 Glial control of axon guidance at the midline in *Drosophila* (A) and in mice (B). The ventral nerve cord of a wild-type *Drosophila* embryo (A, *left*) is composed of two longitudinal connectives that run anterior-posterior and, within each segment, two commissures that link these connectives. Glial cells are interposed between the longitudinal connectives, at the midline. In flies homozygous for a loss-of-function mutation in the gene encoding slit, a chemorepellent produced by midline glia, commissural axons extend to the midline, but then never leave it, forming one large connective that runs the length of the embryo (A, *middle*). In flies homozygous for a loss-of-function mutation in the gene encoding commissureless, a glial membrane protein required for the p precrossing down-regulation of the receptors for slit (the roundabout proteins), axons are blocked from crossing the midline (A, *right*). In wild-type mice (B, *left*), retinal ganglion cell (RGC) axons from each eye extend through the two optic nerves, meet at the ventral midline of the brain at the ventral diencephalon, and then cross at the optic chiasm and proceed on to the superior colliculus and the lateral geniculate nucleus. In mice homozygous for a loss-of-function mutation in the gene encoding Vax1, RGC axons are blocked from crossing at the optic chiasm or even from entering the brain; they pile up at its base (B, *right*). Vax1 is a homeodomain trancription factor, expressed by midline cells, which is required for the expression of midline chemoattractants, such as netrin-1, but not for midline chemorepellents, such as slit-1 (Bertuzzi et al 1999). See text for details.

of the diencephalon, cells in the septum, and cells of the optic stalk and disk (Hallonet et al 1998, 1999; Bertuzzi et al 1999). Although the identity of many of these Vax1$^+$ midline cells is not yet certain, those of the optic stalk are clearly astrocyte precursors; in culture, Vax1 is largely restricted to cells of glial lineage (Bertuzzi et al 1999). Glia at the vertebrate midline are widely thought, based on position and intimate early axonal association, to play essential roles in pioneering and directing commissural growth at forebrain structures, such as the corpus callosum and optic chiasm (Silver 1993; Marcus et al 1995, 1999; Wang et al 1995), and Vax1 is expressed in each of these forebrain structures.

Mice that lack Vax1 exhibit failures in midline axonal crossing in these regions (Bertuzzi et al 1999, Hallonet et al 1999): They lack a corpus callosum, anterior commissure, and optic chiasm. Indeed, retinal ganglion cell (RGC) axons fail to even penetrate the brain at the ventral diencephalon. In the optic stalk, glia that are normally Vax1$^+$ and normally intimately associated with RGC axons are completely displaced from the axons, which are left naked (Bertuzzi et al 1999). These failures in axon recognition and guidance across the midline are cell nonautonomous because the neurons whose axons are mistargeted do not express Vax1. Instead, it is the normally Vax1$^+$ glia that are defective. These cells are not missing (dead) in knockouts, but their guidance properties are altered (Bertuzzi et al 1999). One of these properties is expression of netrin-1, which appears to be an important downstream target, direct or indirect, of Vax1 action. Netrin-1 and Vax1 are expressed by many of the same cells at the ventral midline of the forebrain, and netrin-1 expression in these cells is lost in Vax1 mutants (Bertuzzi et al 1999). In addition, many of the commissural crossing defects seen in Vax1 mutants are also present in mouse netrin-1 knockouts (Serafini et al 1996). Although netrin-1 expression is clearly dependent on Vax1 activity, that of slit is not. Slit-1 is also normally expressed by the Vax1$^+$netrin-1$^+$ cells at the ventral anterior midline, and this expression is maintained in Vax1 knockouts (Bertuzzi et al 1999). Commissural axons are therefore confronted by a midline environment in which the normally balanced presentation of chemoattractants and chemorepellents is shifted in favor of the latter (see Figure 2). In vitro collagen gel explant studies support this interpretation (Bertuzzi et al 1999).

In addition to netrin-1, Vax1 is required for the expression of other midline glial guidance cues that have been shown by mutational analyses to regulate commissural axon crossing in the forebrain. Two of these are the closely related receptor protein-tyrosine kinases EphB2 and EphB3 (Eph Nomenclature Committee 1997). Mouse double knockouts of the *EphB2* and *EphB3* genes, which are frequently coexpressed in neural cells, also exhibit midline crossing defects in the forebrain (Orioli et al 1996, Henkemeyer et al 1996). These double knockouts are typically acallosal, lack the posterior branch of the anterior commissure, and exhibit highly abnormal habenular commissures. Milder, less penetrant manifestations of these phenotypes are also seen in the single mutants. When these commissural defects were first observed, it was assumed that the neurons whose axons fail to cross the midline were the cells that normally express EphB2/B3 because most receptor protein-tyrosine kinase mutations act cell autonomously. However, it was subsequently discovered that in fact these receptors are expressed not by the commissurally projecting neurons but by the midline glia and their precursors with which the axons of these neurons must interact (Orioli et al 1996, Henkemeyer et al 1996). One interpretation of these and related data is that the EphB/ephrin-B signaling system operates bidirectionally (Holland et al 1996, Bruckner et al 1997), and that the $EphB2^+/B3^+$ midline cells in the forebrain correspond to the signal-sending, rather than the signal-receiving, population. These same $EphB^+$ cells are $Vax1^+$, and in Vax1 mutants, the expression of *EphB3* is lost with a time course that parallels both the loss of netrin-1 and the appearance of axon guidance defects (Bertuzzi et al 1999).

REGENERATION AND REPAIR

Regeneration in the Spinal Cord

There are a number of instances in which damage to the nervous system—to the CNS, to peripheral nerves, or to innervated targets—is followed by a repair response. The extent to which this response is successful is frequently a function of the presence, nature, and activity of glia. Following traumatic injury to the adult spinal cord, for example, axonal regeneration into and through the lesion site is almost invariably initiated and then aborted. Historically, several causes for this failure in CNS regeneration have been laid at the door of glia. These include (*a*) the formation of an astrocytic "glial scar," which is widely believed to act as a physical and chemical barrier to axon growth (Fawcett & Asher 1999), (*b*) an inflammatory response mediated by microglia, a CNS macrophage population (Kreutzberg 1996), and (*c*) inhibitors to axon growth, such as the myelin-associated glycoprotein and the IN-1 antigens, which are produced by oligodendrocytes and are present in CNS myelin (Filbin 1996, Caroni & Schwab 1988). The IN-1 antigens, composed predominantly of two polypeptides designated NI35 and NI250, were originally identified through the activity of the IN-1 monoclonal antibody. This monoclonal inhibits a myelin-associated neurite extension antagonist in in vitro assays; it also modestly potentiates axon regrowth through a spinal cord lesion when infused into animals (Caroni & Schwab 1988, Schnell & Schwab 1990, Bregman et al 1995). NI35 and NIF250 are prominently expressed in CNS myelin, but not in PNS myelin, which is consistent with the observation that peripheral nerve grafts support, rather than inhibit, the regrowth of CNS axons. The IN-1 antigens, now named Nogo, have recently been cloned; they turn out members of the Reticulon family of membrane proteins and appear to normally be prominently expressed in endoplasmic reticulum membranes (Chen et al 2000, GrandPre et al 2000). Nogo is made by oligodendrocytes, but not by Schwann cells, and is present in CNS but not PNS myelin. When tested in in vitro assays, both full-length Nogo and a lumenal/extracellular fragment of the protein inhibited the extension of neurites from DRG neurons (Chen et al 2000, GrandPre et al 2000), and antisera against the cloned Nogo protein neutralized much of the axon-extension inhibitory activity of CNS white matter in explant assays (Chen et al 2000). Together with a large body of earlier work on myelin-associated glycoprotein, these data suggest that much of the inhibition of axon elongation and outgrowth in the mature CNS is due to the action of myelinating glia. Thus, the regulatory activities that glia exert over the growth and development of neurons, particularly in the context of injury repair, are not always beneficial.

Sprouting at the Neuromuscular Junction

When the end organs of peripheral nerves are partially denervated because of injury to a portion of the nerve, axons from remaining synapses on the organ sprout to occupy synaptic sites whose innervation has been lost. This phenomenon has been most thoroughly studied at the neuromuscular junction, where branches from axons associated with an intact endplate extend toward and then form new synapses at nearby endplates whose axons have been lost. Although the phenomenology of these events has been well studied (Gordon et al 1988), it is only recently that the cellular and molecular mechanisms underlying sprouting have begun to be elucidated. Again, glial cells have been found to play a central role. At the neuromuscular junction, the growth of axons from an intact to a denervated endplate occurs via lamellapodial Schwann cell processes or "bridges," which pioneer a pathway that axons then follow (Son & Thompson 1995, Son et al 1996). Each motor endplate onto a muscle fiber is encapsulated by a large, specialized peripheral glial cell termed the terminal Schwann cell. On retraction of an axon following partial denervation, these terminal Schwann cells become reactive, probably as a result of signals sent by the denervated muscle (Love & Thompson 1999), and begin to extend finger-like projections across the surface of the muscle. When one of these membrane projections contacts a neighboring denervated endplate, it is stabilized in the form of a bridge between the two endplates. Axons are only then seen to grow along these Schwann cell bridges (Son et al 1996).

It is interesting that in the early postnatal period in rodents, synaptic sprouting following denervation at the neuromuscular junction is markedly less robust than it is in adults (Trachtenberg & Thompson 1996). This difference has been attributed to a difference in the dependence of terminal Schwann cell survival on the presence of axons. In neonates, but not in adults, loss of axon terminals is followed quickly by the apoptotic death of terminal Schwann cells. The dependence of these neonatal Schwann cells on axons again appears to reflect a "gliotrophic" influence that is accounted for by NRG1. NRG1 is expressed by motor and sensory axons at their presynaptic terminals, the terminal Schwann cells express the requisite NRG1 receptors, and most important, exogenously applied NRG1 completely rescues neonatal terminal Schwann cells from programmed cell death following denervation (Trachtenberg & Thompson 1996). Very similar phenomena have been observed for the Schwann cells associated with the terminals of axons innervating sensory end organs—the Pacian corpuscles and the Golgi tendon organs—which are also subject to denervation-induced apoptosis (Kopp et al 1997). NRG1, which is normally neuronally supplied, rescues these glial cells as well.

CONCLUSIONS

Just as it is probably old-fashioned to view glia as "support cells" whose role is to ensure the contentedness of neurons, so it is probably equally inaccurate, the title of this review notwithstanding, to view developing neurons as mere puppets whose strings are pulled by glia. The observations summarized above highlight the extent to which cellular ensembles in the developing nervous system represent interactive networks whose differentiation as functional units is dependent on reciprocal signaling. In such networks—which may take the form of peripheral ganglia, CNS nuclei, axon scaffolds, or neurons linked to their end organs—distinct sets of signals controlling cell fate commitment, proliferation, differentiation, and survival are conveyed both from neurons to glia (e.g. NRG1) and from glia to neurons (e.g. netrin-1). This signaling reciprocity reinforces and temporally coordinates differentiation of the component cells of the network and thereby assures that it contains the requisite number of specialized neuronal and nonneuronal cell types. This in turn assures that neurons and glia are appropriately configured with respect to each other to enable optimal operation of the network when the nervous system is mature.

Visit the Annual Reviews home page at www.AnnualReviews.org

LITERATURE CITED

Anton ES, Marchionni MA, Lee KF, Rakic P. 1997. Role of GGF/neuregulin signaling in interactions between migrating neurons and radial glia in the developing cerebral cortex. *Development* 124:3501–10

Bertuzzi S, Hindges R, Mui SH, O'Leary DD, Lemke G. 1999. The homeodomain protein vax1 is required for axon guidance and major tract formation in the developing forebrain. *Genes Dev.* 13:3092–105

Booth GE, Kinrade EF, Hidalgo A. 2000. Glia maintain follower neuron survival during Drosophila CNS development. *Development* 127(2):237–44

Brose K, Bland KS, Wang KH, Arnott D, Henzel W, et al. 1999. Slit proteins bind Robo receptors and have an evolutionarily conserved role in repulsive axon guidance. *Cell* 96:795–806

Brose K, Tessier-Lavigne M. 2000. Slit proteins: key regulators of axon guidance, ax-onal branching, and cell migration. *Curr. Opin. Neurobiol.* 10:95–102

Bruckner K, Pasquale EB, Klein R. 1997. Tyrosine phosphorylation of transmembrane ligands for Eph receptors. *Science* 275:1640–43

Buchanan RL, Benzer S. 1993. Defective glia in the Drosophila brain degeneration mutant drop-dead. *Neuron* 10:839–50

Burden S, Yarden Y. 1997. Neuregulins and their receptors: a versatile signaling module in organogenesis and oncogenesis. *Neuron* 18:847–55

Bush TG, Savidge TC, Freeman TC, Cox HJ, Campbell EA, et al. 1998. Fulminant jejuno-ileitis following ablation of enteric glia in adult transgenic mice. *Cell* 93:189–201

Caroni P, Schwab ME. 1988. Antibody against myelin-associated inhibitor of neurite growth neutralizes nonpermissive substrate properties of CNS white matter. *Neuron* 1:85–96

Carraway KL III, Cantley LC. 1994. A neu acquaintance for erbB3 and erbB4: a role for receptor heterodimerization in growth signaling. *Cell* 78:5–8

Chan SS, Zheng H, Su MW, Wilk R, Killeen MT, et al. 1996. UNC-40, a *C. elegans* homolog of DCC (deleted in colorectal cancer), is required in motile cells responding to UNC-6 netrin cues. *Cell* 87:187–95

Chen MS, Huber AB, van der Haar ME, Frank M, Schnell L, et al. 2000. Nogo-A is a myelin-associated neurite outgrowth inhibitor and an antigen for monoclonal antibody IN-1. *Nature* 403:434–39

Corfas G, Rosen KM, Aratake H, Krauss R, Fischbach GD. 1995. Differential expression of ARIA isoforms in the rat brain. *Neuron* 14:103–15

Culican SM, Baumrind NL, Yamamoto M, Pearlman AL. 1990. Cortical radial glia: identification in tissue culture and evidence for their transformation to astrocytes. *J. Neurosci.* 10:684–92

Davies AM. 1996. The neurotrophic hypothesis: Where does it stand? *Philos. Trans. R. Soc. London. Ser. B* 351:389–94

Davies AM. 1998. Neuronal survival: early dependence on Schwann cells. *Curr. Biol.* 8:R15–18

Dong Z, Brennan A, Liu N, Yarden Y, Lefkowitz G, et al. 1995. Neu differentiation factor is a neuron-glia signal and regulates survival, proliferation, and maturation of rat Schwann cell precursors. *Neuron* 15:585–96

Eph Nomenclature Committee. 1997. Unified nomenclature for Eph family receptors and their ligands, the ephrins. *Cell* 90:403–4

Erickson SL, O'Shea KS, Ghaboosi N, Loverro L, Frantz G, et al. 1997. ErbB3 is required for normal cerebellar and cardiac development: a comparison with ErbB2- and heregulin-deficient mice. *Development* 124:4999–5011

Falls DL, Rosen KM, Corfas G, Lane WS, Fischbach GD. 1993. ARIA, a protein that stimulates acetylcholine receptor synthesis, is a member of the neu ligand family. *Cell* 72:801–15

Fawcett JW, Asher RA. 1999. The glial scar and central nervous system repair. *Brain Res. Bull.* 49:377–91

Filbin MT. 1996. The muddle with MAG. *Mol. Cell. Neurosci.* 8:84–92

Garratt AN, Voiculescu O, Topilko P, Charnay P, Birchmeier C. 2000. A dual role of erbB2 in myelination and in expansion of the Schwann cell precursor pool. *J. Cell. Biol.* 148:1035–46

Gassmann M, Casagranda F, Orioli D, Simon H, Lai C, et al. 1995. Aberrant neural and cardiac development in mice lacking the ErbB4 neuregulin receptor. *Nature* 378:390–94

Giese KP, Martini R, Lemke G, Soriano P, Schachner M. 1992. Mouse P0 gene disruption leads to hypomyelination, abnormal expression of recognition molecules, and degeneration of myelin and axons. *Cell* 71:565–76

Gordon T, Bambrick L, Orozco R. 1988. Comparison of injury and development in the neuromuscular system. *Ciba Found. Symp.* 138:210–26

Granderath S, Klambt C. 1999. Glia development in the embryonic CNS of *Drosophila*. *Curr. Opin. Neurobiol.* 9:531–36

GrandPre T, Nakamura F, Vartanian T, Strittmatter SM. 2000. Identification of the Nogo inhibitor of axon regeneration as a reticulon protein. *Nature* 403:439–44

Griffiths I, Klugmann M, Anderson T, Yool D, Thomson C, et al. 1998. Axonal swellings and degeneration in mice lacking the major proteolipid of myelin. *Science* 280:1610–13

Grinspan JB, Marchionni MA, Reeves M, Coulaloglou M, Scherer SS. 1996. Axonal interactions regulate Schwann cell apoptosis in developing peripheral nerve: neuregulin receptors and the role of neuregulins. *J. Neurosci.* 16:6107–18

Hallonet M, Hollemann T, Pieler T, Gruss P. 1999. Vax1, a novel homeobox-containing gene, directs development of the basal forebrain and visual system. *Genes. Dev.* 13:3106–14

Hallonet M, Hollemann T, Wehr R, Jenkins NA, Copeland NG, et al. 1998. Vax1 is a novel homeobox-containing gene expressed in the developing anterior ventral forebrain. *Development* 125:2599–610

Halter DA, Urban J, Rickert C, Ner SS, Ito K, et al. 1995. The homeobox gene repo is required for the differentiation and maintenance of glia function in the embryonic nervous system of *Drosophila melanogaster*. *Development* 121:317–32

Harris R, Sabatelli LM, Seeger MA. 1996. Guidance cues at the *Drosophila* CNS midline: identification and characterization of two *Drosophila* Netrin/UNC-6 homologs. *Neuron* 17:217–28

Hatten ME. 1999. Central nervous system neuronal migration. *Annu. Rev. Neurosci.* 22:511–39

Henderson CE. 1996. Role of neurotrophic factors in neuronal development. *Curr. Opin. Neurobiol.* 6:64–70

Henkemeyer M, Orioli D, Henderson JT, Saxton TM, Roder J, et al. 1996. Nuk controls pathfinding of commissural axons in the mammalian central nervous system. *Cell* 86:35–46

Hidalgo A, Booth GE. 2000. Glia dictate pioneer axon trajectories in the Drosophila embryonic CNS. *Development* 127:393–40

Holland SJ, Gale NW, Mbamalu G, Yancopoulos GD, Henkemeyer M, Pawson T. 1996. Bidirectional signalling through the EPH-family receptor Nuk and its transmembrane ligands. *Nature* 383:722–25

Hosoya T, Takizawa K, Nitta K, Hotta Y. 1995. glial cells missing: a binary switch between neuronal and glial determination in Drosophila. *Cell* 82:1025–36

Hummel T, Schimmelpfeng K, Klambt C. 1999. Commissure formation in the embryonic CNS of *Drosophila*. *Development* 126:771–79

Jones BW, Fetter RD, Tear G, Goodman CS. 1995. Glial cells missing: a genetic switch that controls glial versus neuronal fate. *Cell* 82:1013–23

Keino-Masu K, Masu M, Hinck L, Leonardo

ED, Chan SS, et al. 1996. Deleted in colorectal cancer (DCC) encodes a netrin receptor. *Cell* 87:175–85

Kennedy TE, Serafini T, de la Torre JR, Tessier-Lavigne M. 1994. Netrins are diffusible chemotropic factors for commissural axons in the embryonic spinal cord. *Cell* 78:425–35

Kidd T, Bland KS, Goodman CS. 1999. Slit is the midline repellent for the robo receptor in Drosophila. *Cell* 96:785–94

Kidd T, Russell C, Goodman CS, Tear G. 1998. Dosage-sensitive and complementary functions of roundabout and commissureless control axon crossing of the CNS midline. *Neuron* 20:25–33

Klein R. 1994. Role of neurotrophins in mouse neuronal development. *FASEB J.* 8:738–44

Kolodziej PA, Timpe LC, Mitchell KJ, Fried SR, Goodman CS, et al. 1996. frazzled encodes a Drosophila member of the DCC immunoglobulin subfamily and is required for CNS and motor axon guidance. *Cell* 87:197–204

Kopp DM, Trachtenberg JT, Thompson WJ. 1997. Glial growth factor rescues Schwann cells of mechanoreceptors from denervation-induced apoptosis. *J. Neurosci.* 17:6697–706

Kreutzberg GW. 1996. Microglia: a sensor for pathological events in the CNS. *Trends Neurosci.* 19:312–18

Lee KF, Simon H, Chen H, Bates B, Hung MC, Hauser C. 1995. Requirement for neuregulin receptor erbB2 in neural and cardiac development. *Nature* 378:394–98

Lemke G. 1996. Neuregulins in development. *Mol. Cell. Neurosci.* 7:247–62

Lemke GE, Brockes JP. 1984. Identification and purification of glial growth factor. *J. Neurosci.* 4:75–83

Leonardo ED, Hinck L, Masu M, Keino-Masu K, Ackerman SL, Tessier-Lavigne M. 1997. Vertebrate homologues of *C. elegans* UNC-5 are candidate netrin receptors. *Nature* 386:833–38

Levi AD, Bunge RP, Lofgren JA, Meima L,

Hefti F, et al. 1995. The influence of heregulins on human Schwann cell proliferation. *J. Neurosci.* 15:1329–40

Li HS, Chen JH, Wu W, Fagaly T, Zhou L, et al. 1999. Vertebrate slit, a secreted ligand for the transmembrane protein roundabout, is a repellent for olfactory bulb axons. *Cell* 96: 807–18

Love FM, Thompson WJ. 1999. Glial cells promote muscle reinnervation by responding to activity-dependent postsynaptic signals. *J. Neurosci.* 19:10390–96

Marchionni MA, Goodearl AD, Chen MS, Bermingham-McDonogh O, Kirk C, et al. 1993. Glial growth factors are alternatively spliced erbB2 ligands expressed in the nervous system. *Nature* 362:312–18

Marcus RC, Blazeski R, Godement P, Mason CA. 1995. Retinal axon divergence in the optic chiasm: uncrossed axons diverge from crossed axons within a midline glial specialization. *J. Neurosci.* 15:3716–29

Marcus RC, Shimamura K, Sretavan D, Lai E, Rubenstein JL, Mason CA. 1999. Domains of regulatory gene expression and the developing optic chiasm: correspondence with retinal axon paths and candidate signaling cells. *J. Comp. Neurol.* 403:346–58

Meyer D, Birchmeier C. 1995. Multiple essential functions of neuregulin in development. *Nature* 378:386–90

Mitchell KJ, Doyle JL, Serafini T, Kennedy TE, Tessier-Lavigne M, et al. 1996. Genetic analysis of Netrin genes in *Drosophila*: Netrins guide CNS commissural axons and peripheral motor axons. *Neuron* 17:203–15

Morris JK, Lin W, Hauser C, Marchuk Y, Getman D, Lee KF. 1999. Rescue of the cardiac defect in ErbB2 mutant mice reveals essential roles of ErbB2 in peripheral nervous system development. *Neuron* 23:273–83

Oppenheim RW. 1991. Cell death during development of the nervous system. *Annu. Rev. Neurosci.* 14:453–501

Orioli D, Henkemeyer M, Lemke G, Klein R, Pawson T. 1996. Sek4 and Nuk receptors cooperate in guidance of commissural axons

and in palate formation. *EMBO J.* 15:6035–49

Placzek M, Tessier-Lavigne M, Yamada T, Jessell T, Dodd J. 1990. Mesodermal control of neural cell identity: floor plate induction by the notochord. *Science* 250:985–88

Porter S, Clark MB, Glaser L, Bunge RP. 1986. Schwann cells stimulated to proliferate in the absence of neurons retain full functional capability. *J. Neurosci.* 6:3070–78

Purves D. 1986. The trophic theory of neural connections. *Trends Neurosci.* 9:496–99

Rakic P. 1971. Neuron-glia relationship during granule cell migration in developing cerebellar cortex. *J. Comp. Neurol.* 141:283–312

Rakic P. 1972. Mode of cell migration to the superficial layers of fetal monkey cortex. *J. Comp. Neurol.* 145:61–84

Ramon y Cajal S. 1911. *Histology of the Nervous System of Man and Vertebrates*, Vol. 1. Transl. L Swanson, N Swanson, 1995. Oxford, UK: Oxford Univ. Press (From Spanish)

Rangarajan R, Gong Q, Gaul U. 1999. Migration and function of glia in the developing Drosophila eye. *Development* 126:3285–92

Riethmacher D, Sonnenberg-Riethmacher E, Brinkmann V, Yamaai T, Lewin GR, Birchmeier C. 1997. Severe neuropathies in mice with targeted mutations in the ErbB3 receptor. *Nature* 389:725–30

Rio C, Rieff HI, Qi P, Khurana TS, Corfas G. 1997. Neuregulin and erbB receptors play a critical role in neuronal migration. *Neuron* 19:39–50

Rothberg JM, Jacobs JR, Goodman CS, Artavanis-Tsakonas S. 1990. slit: an extracellular protein necessary for development of midline glia and commissural axon pathways contains both EGF and LRR domains. *Genes Dev.* 4:2169–87

Schnell L, Schwab ME. 1990. Axonal regeneration in the rat spinal cord produced by an antibody against myelin-associated neurite growth inhibitors. *Nature* 343:269–72

Seeger M, Tear G, Ferres-Marco D, Goodman CS. 1993. Mutations affecting growth cone guidance in *Drosophila*: genes necessary for guidance toward or away from the midline. *Neuron* 10:409–26

Seeger MA, Beattie CE. 1999. Attraction versus repulsion: modular receptors make the difference in axon guidance. *Cell* 97:821–24

Serafini T, Colamarino SA, Leonardo ED, Wang H, Beddington R, et al. 1996. Netrin-1 is required for commissural axon guidance in the developing vertebrate nervous system. *Cell* 87:1001–14

Serafini T, Kennedy TE, Galko MJ, Mirzayan C, Jessell TM, Tessier-Lavigne M. 1994. The netrins define a family of axon outgrowth-promoting proteins homologous to *C. elegans* UNC-6. *Cell* 78:409–24

Shah NM, Marchionni MA, Isaacs I, Stroobant P, Anderson DJ. 1994. Glial growth factor restricts mammalian neural crest stem cells to a glial fate. *Cell* 77:349–60

Silver J. 1993. Glia-neuron interactions at the midline of the developing mammalian brain and spinal cord. *Perspect. Dev. Neurobiol.* 1:227–36

Son YJ, Thompson WJ. 1995. Nerve sprouting in muscle is induced and guided by processes extended by Schwann cells. *Neuron* 14:133–41

Son YJ, Trachtenberg JT, Thompson WJ. 1996. Schwann cells induce and guide sprouting and reinnervation of neuromuscular junctions. *Trends Neurosci.* 19:280–85

Sonnenfeld MJ, Jacobs JR. 1994. Mesectodermal cell fate analysis in *Drosophila* midline mutants. *Mech. Dev.* 46:3–13

Syroid DE, Maycox PR, Burrola PG, Liu N, Wen D, et al. 1996. Cell death in the Schwann cell lineage and its regulation by neuregulin.

Proc. Natl. Acad. Sci. USA 93:9229–34

Tessier-Lavigne M, Placzek M, Lumsden AG, Dodd J, Jessell TM. 1988. Chemotropic guidance of developing axons in the mammalian central nervous system. *Nature* 336:775–78

Trachtenberg JT, Thompson WJ. 1996. Schwann cell apoptosis at developing neuromuscular junctions is regulated by glial growth factor. *Nature* 379:174–77

Wang LC, Dani J, Godement P, Marcus RC, Mason CA. 1995. Crossed and uncrossed retinal axons respond differently to cells of the optic chiasm midline in vitro. *Neuron* 15:1349–64

Woldeyesus MT, Britsch S, Riethmacher D, Xu L, Sonnenberg-Riethmacher E, et al. 1999. Peripheral nervous system defects in erbB2 mutants following genetic rescue of heart development. *Genes Dev.* 13:2538–48

Xiong WC, Montell C. 1995. Defective glia induce neuronal apoptosis in the repo visual system of Drosophila. *Neuron* 14:581–90

Xiong WC, Okano H, Patel NH, Blendy JA, Montell C. 1994. repo encodes a glial-specific homeo domain protein required in the Drosophila nervous system. *Genes Dev.* 8:981–94

Yang X, Kuo Y, Devay P, Yu C, Role L. 1998. A cysteine-rich isoform of neuregulin controls the level of expression of neuronal nicotinic receptor channels during synaptogenesis. *Neuron* 20:255–70

Yuan W, Zhou L, Chen JH, Wu JY, Rao Y, Ornitz DM. 1999. The mouse SLIT family: secreted ligands for ROBO expressed in patterns that suggest a role in morphogenesis and axon guidance. *Dev. Biol.* 212:290–306

Annu. Rev. Neurosci. 2001. 24:107–37

TOUCH AND GO: Decision-Making Mechanisms in Somatosensation

Ranulfo Romo

*Instituto de Fisiología Celular, Universidad Nacional Autónoma de México,
04510 México, D.F. México; e-mail: rromo@ifisiol.unam.mx*

Emilio Salinas

*Computational Neurobiology Laboratory, Howard Hughes Medical Institute,
The Salk Institute, La Jolla, California 92037; e-mail: emilio@salk.edu*

Key Words   somatosensory cortex, discrimination, categorization, working
memory, cortical microstimulation, attention

■ **Abstract**   A complex sequence of neural events unfolds between sensory receptor
activation and motor activity. To understand the underlying decision-making mechanisms linking somatic sensation and action, we ask what components of the neural
activity evoked by a stimulus are directly related to psychophysical performance, and
how are they related. We find that single-neuron responses in primary and secondary
somatosensory cortices account for the observed performance of monkeys in vibrotactile discrimination tasks, and that neuronal and behavioral responses covary in single
trials. This sensory activity, which provides input to memory and decision-making
mechanisms, is modulated by attention and behavioral context, and microstimulation
experiments indicate that it may trigger normal perceptual experiences. Responses
recorded in motor areas seem to reflect the output of decision-making operations, which
suggests that the ability to make decisions occurs at the sensory-motor interface.

INTRODUCTION

Decision making involves weighing evidence in favor of several possible courses
of action. This process may be conceived as a chain of neural operations: encoding recent sensory stimuli, extracting from them currently relevant features,
maintaining these features in working memory, comparing them to prior information (which might include the conditions of the task), and communicating the
result to the motor apparatus so that an action is produced accordingly. The divisions between these steps may be artificial, but breaking the problem into pieces is
helpful. Here we discuss several such pieces, although, in the long run, we aim for
an integrated understanding of the decision-making process, at least to the extent
possible within the minimalist environment of a laboratory task.

0147-006X/01/0621-0107$14.00

In many paradigms, the strictly sensory and motor components seem relatively well understood. But in the past decade, neurophysiologists have unearthed a wealth of complexity in the intermediate decision-making steps (Salzman et al 1990, 1992; Leon & Shadlen 1998; Kim & Shadlen 1999; Schall & Thompson 1999; Gold & Shadlen 2000). Physiologists in the motor system would call this the sensory-motor interface. However, the problem should not be understated. To explore decision-making, two key elements are required: first, comprehending the neural codes of the cells activated during the task (DeCharms & Zador 2000) and second, experimental preparations in which behavioral and neuronal activity are quantified simultaneously. Thus, psychophysical techniques are essential (Werner 1980, Newsome et al 1989, Hernández et al 2000). The rationale underlying these experiments is that the responses of neurons involved in the decision-making process should correlate with the animal's behavior, and the specifics of these correlations should provide an indication of functional significance.

In the somatosensory system, combined psychophysical/neurophysiological experiments have a long tradition that goes back to Mountcastle and colleagues, who pioneered this approach almost four decades ago (Werner & Mountcastle 1965, Talbot et al 1968). Here we focus on recent studies that take this approach using controlled behavioral tasks based on tactile stimuli. We also mention some important studies in vision that are germane to our subject. After a brief section on the general organization of the somatosensory system, we discuss the cortical representation of somatosensory stimuli, its relation to behavior and perception, its dependence on behavioral context, and its persistence in working memory, all crucial ingredients in decision making. The final set of results describes neural responses found in areas traditionally implicated in motor control that seem to reflect the output of a categorical decision-making process.

OVERVIEW OF THE SOMATOSENSORY SYSTEM

Cutaneous Primary Afferents

Four types of cutaneous afferent fibers can be found in humans (Darian-Smith 1984, Vallbo & Johansson 1984, Vallbo 1995). Two of them adapt rapidly to constant skin indentations and two adapt slowly. The rapidly adapting fibers are anatomically associated with Meissner (RA) and Pacini (PC) mechanoreceptors under the skin, and the slowly adapting fibers are linked to Merkel (SA) and Ruffini receptors. These primary afferents transmit cutaneous information to the central nervous system; other fibers exist that transmit proprioceptive information, but these are not discussed here. Monkeys have the first three kinds of afferents but lack those linked to Ruffini organs. Although all these fibers respond to stimulation on the skin surface, they are selective for different spatiotemporal features (Talbot et al 1968; Phillips & Johnson 1981a,b; Phillips et al 1988, 1992). Signals from the mechanoreceptive transduction process are conveyed with exquisite fidelity to the cortex: Even a single spike from a single RA primary afferent may be reliably perceived (Vallbo & Johansson 1984, Vallbo 1995).

Anatomical Pathways: From Sensation to Action in the Neocortex

After a relay in the nuclei of the dorsal column and the basal complex of the thalamus, somatosensory information reaches primary somatosensory cortex (S1). In primates, S1 is composed of four areas, areas 3a, 3b, 1, and 2. In each of them, the receptive fields of the neurons form an orderly map of the body (Kaas et al 1979, Nelson et al 1980). Areas 3b, 1, and 2 are interconnected (Shanks et al 1985) and are the most heavily involved in processing tactile information. Neurons in S1 are traditionally classified as RA, SA, or PC because their firing is typically similar to that of the corresponding afferent fibers (Mountcastle et al 1969, Powell & Mountcastle 1959, Talbot et al 1968). These three subtypes are organized in columns (Mountcastle 1957, Powell & Mountcastle 1959, Sur et al 1984).

In the visual system, information originating in primary visual cortex flows along two pathways, the dorsal and ventral streams, which show some degree of functional specialization (Mishkin 1979, Ungerleider & Mishkin 1982, Goodale & Milner 1992). It is interesting that somatosensory information from S1 also proceeds along dorsal and ventral routes that might also be specialized (Mishkin 1979, Murray & Mishkin 1984). The dorsal stream flows through the posterior parietal cortex, via areas 5 and 7b (Cavada & Goldman-Rakic 1989, Pearson & Powell 1985, Shanks et al 1985), and the ventral stream flows through the lateral somatosensory areas (Burton et al 1995; Krubitzer et al 1995; Pons et al 1987, 1992).

The dorsal stream projects to premotor and primary motor (M1) cortices (Cavada & Goldman-Rakic 1989, Godschalk et al 1984, Leichnetz 1986, Tokuno & Tanji 1993) and could be important for self-initiated or stimulus-triggered voluntary movements involving somatic processing. The ventral stream is more likely associated with fine discrimination and recognition of stimulus patterns. This pathway reaches the lateral premotor area (Cavada & Goldman-Rakic 1989, Godschalk et al 1984, Leichnetz 1986) and the prefrontal cortex (Preuss & Goldman-Rakic 1989, Carmichael & Price 1995) and might be associated with fine discrimination or identification of objects. As in vision, receptive-field size and complexity increase progressively (Sinclair & Burton 1993, Iwamura et al 1994). Both streams reach M1 (Leichnetz 1986, Tokuno & Tanji 1993), so both should influence motor activity during sensory tasks that require limb movements. The functional specializations of these streams still need to be elucidated, but it would be very interesting if a clear distinction could be drawn.

Peripheral Neural Codes for Tactile Stimuli

Coding of Flutter-Vibration　　In the somatosensory system there is a nice correspondence between perceptual and anatomical submodalities. When mechanical vibrations are applied to the skin, two sensations can be evoked. With stimuli of low frequency, between approximately 5 and 50 Hz, the sensation of flutter is produced; with stimuli of higher frequency, between 60 and 300 Hz, the distinct sensation of vibration is felt. Talbot et al (1968) demonstrated that flutter is mediated by RA afferents, whereas vibration is mediated by PC afferents. They also found a close, quantitative correspondence between perceptual and neuronal detection thresholds, which was later confirmed by recording and microstimulation of afferent fibers in awake human subjects (Macefield et al 1990, Ochoa & Torebjörk 1983, Vallbo & Johansson 1984, Vallbo 1995). This work established the experimental paradigm in which psychophysical and neurophysiological responses are directly compared.

Two additional observations were made in the same study (Talbot et al 1968); we comment on them because they have had a strong influence in the field. First, when periodic, sinusoidal stimuli were used, the evoked spike trains from RA and PC afferents were also periodic, with individual action potentials being strongly phase locked to the mechanical inputs. Second, the firing rates of PC afferents increased with increasing stimulus frequency (within their range of 60–300 Hz), but the rates of RA afferents seemed to change little inside their 5- to 50-Hz range. This posed a problem: High frequencies could be encoded by firing rate but low frequencies could not, because the same number of RA action potentials would be seen in any given time window regardless of frequency. Talbot and colleagues (1968) concluded that some mechanism downstream had to identify or distinguish the intervals between incoming RA spikes (see Ahissar & Vaadia 1990, Ahissar 1998) in order to generate different percepts for different frequencies. In other words, a temporal neural code (Shadlen & Newsome 1994, Singer & Gray 1995) had to be used. Subsequent studies reported similar findings in S1 (Mountcastle et al 1969).

Although some contradicting evidence appeared (Ochoa & Torebjörk 1983), the suggestion that a temporal code is used to represent flutter frequency in the cortex remained unchallenged until recently (Hernández et al 2000, Salinas et al 2000a). The new results do not eliminate this possibility, but they show, first, that the firing rate of S1 neurons is substantially modulated by flutter frequency and, second, that the observed rate modulations are highly consistent with psychophysical performance in flutter discrimination (see below).

Coding of Spatiotemporal Patterns   Johnson & Hsiao (1992) and Johnson et al (1996) have studied the neural representations of texture and form in primary afferents. First they characterized the responses of SA, RA, and PC fibers to embossed letters of the alphabet scanned across afferent receptive fields. Experiments using anesthetized monkeys (Phillips et al 1981a,b) and attending humans (Phillips et al 1992) showed that the spikes of SA and RA afferents provided high-quality spatiotemporal images of the letters, which could account for human psychophysical performance in pattern recognition. Another study (Phillips et al 1988) suggested that neurons in area 3b with SA properties could also account for psychophysical performance, but that neurons with RA properties could not.

Johnson and coworkers also investigated the neural basis of texture perception (Johnson & Hsiao 1992, Johnson et al 1996). In this case a more rigorous comparison between neural and behavioral responses could be made because the stimuli consisted of arrays of raised dots of different textures, and texture could be parameterized. Subjective roughness estimates were made by human subjects, where roughness was manipulated by varying dot diameter, dot height, and inter-dot spacing, and recordings were again obtained from single SA, RA, and PC afferents in anesthetized monkeys. The question they asked was, what measures of neural activity may provide a consistent code for texture, where consistency simply referred to the condition that any given value of the response measure had to correspond to a single value of subjective roughness. Their approach to this problem is a textbook example of the scientific method: A large number of neural codes were methodically eliminated, until only one was left (Connor et al 1990, Connor & Johnson 1992, Johnson et al 1996, Blake et al 1997). Their conclusion was that information about texture had to be conveyed mostly by the differences in activity between SA fibers with receptive fields at different locations, which they referred to as SA spatial variations (Johnson et al 1996, Blake et al 1997). This led to the prediction that cortical somatosensory neurons should have receptive fields with excitatory and inhibitory subregions, which is what was indeed found later (DiCarlo et al 1998).

The results reviewed in this section highlight the roles that different cutaneous afferents play in representing somatosensory stimuli. The SA system transmits information regarding spatial features, whereas the RA and PC systems encode temporal features. The response properties of primary afferents must limit the psychophysical capacity to detect, recognize, and discriminate tactile stimuli, and indeed, a quantitative agreement has been observed repeatedly.

NEURAL ACTIVITY IN SOMATOSENSORY CORTEX DURING BEHAVIOR

The first step in studying central decision-making mechanisms is to understand which neurons encode the quantities being decided on and how they do it. Important developments concerning this issue are reviewed in this section.

Stimulus Representations in S1

The flutter submodality provides a convenient channel to study decision-making mechanisms and cognitive tasks based on temporal patterns (Talbot et al 1968; Mountcastle et al 1969, 1990, 1992; Recanzone et al 1992; Hernández et al 1997, 2000; Romo et al 1998, 1999, 2000). First, the same set of afferent fibers are activated regardless of stimulus frequency, because the stimulator moves perpendicularly to the skin. Second, this sensation is conveyed by a highly specific set of primary afferents and primary somatosensory neurons: those with RA properties.

Third, humans and monkeys have similar detection thresholds and discrimination capacities (Mountcastle et al 1990, Hernández et al 1997). However, for a long time the neural code for flutter frequency remained a subject for speculation because it seemed to depend on the temporal configuration of fixed numbers of spikes.

Mountcastle and colleagues investigated not only the responses of cutaneous afferents (Talbot et al 1968), but also those of S1 neurons, both in anesthetized (Mountcastle et al 1969) and in behaving (Mountcastle et al 1990) monkeys. They concluded that RA cortical neurons were similar to RA afferents: They responded briskly to flutter, firing in phase with applied mechanical oscillations, and their firing rates changed little as functions of frequency. A later study by Recanzone and coworkers (1992) agreed with these conclusions, which were further supported by comparisons between psychophysical performance and variations in periodic spike firing (Mountcastle et al 1969, Recanzone et al 1992). It indeed appeared that "frequency discrimination is made by a central neural mechanism capable of measuring the lengths of the dominant periodic intervals in the [evoked] trains of impulses" (Mountcastle et al 1967).

However, the key piece of evidence here, that the firing rate of S1 neurons is insensitive to flutter frequency, was based on a small sample (17 neurons) (Mountcastle et al 1990), and on tests within a narrow range of frequencies using anesthetized animals (Recanzone et al 1992). Recent studies focusing on this particular issue have reached other conclusions (Hernández et al 2000, Salinas et al 2000a).

Hernández et al (1997) trained monkeys to discriminate between the frequencies of two flutter stimuli presented sequentially on a fingertip; a schematic diagram of the task is shown in Figure 1a. They then recorded neurons with RA properties in areas 3b and 1 of S1 while the monkeys performed the task (Hernández et al 2000, Salinas et al 2000a). They confirmed that the evoked periodic spikes of S1 neurons were indeed extremely regular, but they also found that the firing rates of more than 50% of the neurons were significantly modulated by stimulus frequency (Hernández et al 2000, Salinas et al 2000a). This showed that the neural code for flutter frequency could be based on firing rate, but how could such a code be compared with one based on periodicity?

As a first approach to this problem, Shannon's information (Cover & Thomas 1991) was used to quantify how well stimulus frequency was encoded (a) by the firing rates of the neurons and (b) by a response measure based on the periodicity of the interspike intervals (Salinas et al 2000a). On average, the periodicity-based response provided 1.71 bits of information about frequency, which roughly means that, by measuring the periodic intervals evoked in any single trial, stimulus frequency could be perfectly localized within a range of about 6.1 Hz (i.e. the full frequency range used, of approximately 20 Hz, could be divided into $2^{1.71} = 3.3$ perfectly distinguishable bins). For the neurons with the highest periodicity, the accuracy was about 3.5 Hz. In contrast, firing rate modulations carried much less information about stimulus frequency, giving accuracies of 16.6 and 10 Hz for average and best neurons, respectively.

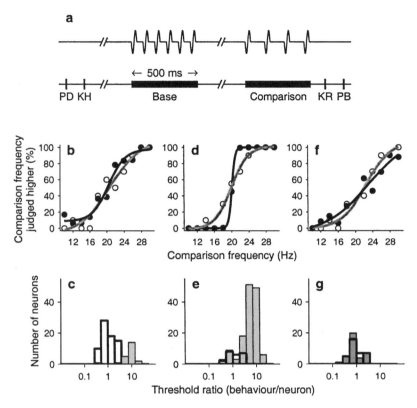

Figure 1 Comparison between neuronal and psychophysical responses during vibrotactile discrimination. (*a*) Sequence of events in the task. The probe moves down (PD) and indents a fingertip. The monkey reacts by holding a key (KH). After a delay, the probe oscillates vertically at a base frequency. After another delay, the probe oscillates again, now at a comparison frequency. The monkey releases the key (KR) after the end of the comparison period and presses one of two push buttons (PB) to indicate whether the comparison frequency was higher or lower than the base. (*b*) Percentage of trials in which the comparison was called higher than the base. Continuous lines are sigmoidal fits to the data; for each curve, the threshold is proportional to its maximum steepness (see text). Open circles and gray curves indicate the monkey's performance during one discrimination run using periodic stimuli. Filled circles and black curves indicate performance of an ideal observer that based his decisions on the evoked firing rates of a single neuron recorded while the monkey discriminated. (*c*) Numbers of S1 neurons with the indicated threshold ratios. Thick lines correspond to neurometric thresholds based on evoked firing rate; gray bars correspond to neurometric thresholds based on periodicity of evoked spike trains. Data are from all neurons with significant rate modulation. (*d*) As in panel *a*, but the observer discriminated based on the periodicity of the evoked interspike intervals of another neuron recorded in a different session. (*e*) As in panel *c*, but data are from all neurons with significant periodic entrainment. (*f*) As in panel *a*, but aperiodic stimuli were applied. Data are from another neuron. (*g*) Threshold ratios from all neurons tested with both periodic and aperiodic stimuli. Neurometric thresholds were computed from the firing rates in periodic (thick lines) and aperiodic (gray bars) conditions. (Modified from Hernández et al 2000.)

Thus, a neural code for flutter frequency based on firing rate was entirely possible, but in principle, a code based on spike periodicity could be up to three times more accurate.

Neuron Versus Monkey

Discrimination Based on Periodic Stimuli   Given the large discrepancy in information between neural codes based on firing rate and periodicity, Hernández and collaborators (2000) used the methods of signal detection theory (Green & Swets 1966) to find out which code, if any, was in agreement with the observed psychophysical behavior of the monkeys. In the flutter discrimination task, two stimuli, called base and comparison, are presented one after the other, with a pause in between, and the subject has to indicate whether the frequency of the comparison was higher or lower than the frequency of the base. Hernández et al (2000) first computed the expected performance of an ideal observer who solved the task using the following simple (but optimal) rule: If the number of spikes evoked during the comparison period is higher than the number evoked during the base, then the comparison frequency was higher than the base. This rule is applied in each trial. An example of the resulting neurometric curve for a single S1 neuron is shown in Figure 1*b*. The threshold of the curve is defined as the stimulus frequency identified as higher than the standard in 75% of trials, minus the frequency identified as higher in 25% of the trials, divided by 2. Lower thresholds correspond to steeper curves and better performance. For the neurometric data shown in Figure 1*b*, the threshold was 2.5 Hz. The sigmoidal fit for the psychophysical performance of the monkey in the same trials had a threshold of 3.2 Hz, so the threshold ratio (behavior/neuron) was 1.3, close to 1. In this example the neurometric and psychometric curves were in excellent agreement. This was the case for a large number of S1 neurons whose firing rates varied significantly as functions of frequency, as shown in Figure 1*c*. Therefore, the recorded rate modulations accounted for the observed phsychophysical performance.

The strategy of the ideal observer can be used with other response measures (Green & Swets 1966), so Hernández et al (2000) also applied it to the periodicity-based response used before (Salinas et al 2000a). Again, the procedure simply consisted of determining whether the response was larger during the base or during the comparison periods. A neurometric function constructed from the periodicity-based responses of a single S1 neuron is shown in Figure 1*d*, along with the monkey's performance in the recording trials. In this case, neurometric and psychometric thresholds differed by a factor of 10; using the periodicity-based response, the ideal observer could perform the task much more accurately than the monkey. Similar results were obtained with other neurons that fired in phase with the stimuli, as shown in Figure 1*e*. On average over all neurons analyzed, discrimination thresholds based on periodicity were about four times smaller than those based on firing rate, in agreement with the previous information analysis (Salinas et al 2000a). Therefore, according to the psychophysics, either a firing rate code was being used efficiently or a periodicity-based code was being used very inefficiently.

Discrimination Based on Aperiodic Stimuli   Having practiced the discrimination task for months using periodic stimuli, the monkeys were presented with aperiodic stimulation patterns. These consisted of trains of short, mechanical pulses separated by random times. The numbers of pulses, however, still depended on the chosen frequency; for instance, if the stimulation period lasted 0.5 s, 20 Hz corresponded to 10 pulses. These stimuli were designed so that even highly stimulus-entrained neurons could not carry information about stimulus frequency in their periodicity; in this case the periodicity-based response was essentially random (Salinas et al 2000a). Experiments with these patterns produced three results. First, the monkeys' performance was, from the initial sessions, practically the same as with periodic stimuli; they effortlessly extracted a mean frequency from the new patterns (Romo et al 1998, Salinas et al 2000a). Second, firing rate modulations in S1 were, on average, equally strong with periodic and aperiodic stimuli (Salinas et al 2000a). And third, consistent with the latter result, neurometric and psychometric curves were again in good agreement (Figure 1*f,g*) (Hernández et al 2000).

In conclusion, when periodicity was broken, both monkeys and firing rates behaved essentially the same as with periodic stimuli.

Single-Trial Covariations in Activity   A further test of association between neural activity and behavior is to search for changes in response properties between correct and incorrect discriminations (hits and errors) in trials with identical sensory stimuli (Britten et al 1996). This analysis can also be applied to different kinds of responses. Salinas et al (2000a) found no significant covariations between periodicity and behavior; the chances of the monkey making a correct discrimination were the same whether the spike trains were more periodic or less periodic than average. In contrast, a few single cells were found whose firing rates fluctuated along with the monkeys' decisions; the numbers of cells that did this were small but highly significant. In addition, this effect was evident at the population level.

Stimulus Representations in S2

What is the neural representation of flutter stimuli in structures central to S1? An interesting area to explore is the secondary somatosensory cortex (S2), which belongs to the ventral somatosensory stream (Mishkin 1979, Murray & Mishkin al 1984). S2 is strongly connected to S1 (Burton et al 1995, Krubitzer et al 1995), and at least in primates, information seems to be processed mostly serially from S1 to S2 (Pons et al 1987, 1992).

When S2 neurons were recorded during the flutter discrimination task, significant firing rate modulations were observed, as in S1, but several differences were found between the two areas (Salinas et al 2000a). First, very few neurons in S2 were significantly entrained by flutter stimuli (less than 8%, versus 84% in S1). Second, although in both areas the firing rates were approximately monotonic functions of frequency, negative slopes were uncommon in S1 (8%), whereas in S2 they were almost as abundant as positive slopes (40% were negative). It is interesting that a similar change in representation from S1 to S2 has been reported using textured surfaces (Sinclair & Burton 1993; see also Jiang et al 1997). Third, several neurons in S2 (13%), but none in S1, sustained their responses for a few hundred milliseconds after stimulus offset, carrying information about the base frequency into the early component of the delay period. Fourth, attentional or task-dependent changes in firing rate were observed in both areas but were much stronger in S2 (see Hsiao et al 1993). Finally, single-trial covariations between behavioral and neuronal responses were detected in both areas, but these were again much more evident in S2. Here about 9% of the neurons showed significant differences in responses in correct versus incorrect discriminations, and the effect at the population level was strong (Salinas et al 2000a).

Thus, although periodicity is a prominent feature of the activity evoked by flutter in S1, comparisons between neuronal and behavioral responses suggest that the neural code for flutter frequency in this area probably relies on firing rate. In addition, neurons in S2 show a minimal trace of periodicity but still modulate their firing rates as functions of frequency, and most important, they seem to be more strongly correlated with the animal's decision on a single-trial basis.

INSERTING ARTIFICIAL PERCEPTS THROUGH MICROSTIMULATION

Lesion effects, electrophysiological recordings, and imaging techniques can re-veal correlations between neuronal activity and observable behavior, but a causal relationship is usually hard to prove. Intracortical microstimulation can produce results that are much closer to such proof (Salzman et al 1990, 1992; Britten & van Wezel 1998; Romo et al 1998, 2000). For this technique to work, the coactivated neurons should have similar functional properties. Flutter is thus well suited for microstimulation experiments because S1 is known to be organized in modules of specific submodalities. In particular, neurons with RA properties are found in columns (Mountcastle 1957, Powell & Mountcastle 1959, Jones et al 1975, Sur et al 1984). Another advantage of flutter is that humans and monkeys have comparable discrimination capacities (Mountcastle et al 1990, Hernández et al 1997).

Romo and colleagues (1998, 2000) injected microstimulation current in area 3b of monkeys trained in the flutter discrimination task (Figure 1a). In this paradigm subjects pay attention to the frequency of the first (base) stimulus, store a trace of it during a delay period, and compare it with the frequency of a second (comparison) stimulus (Hernández et al 1997). Microstimulation sites in S1 were required to have RA neurons with receptive fields on the fingertip at the location of the me-chanical stimulating probe. In the initial experiment (Romo et al 1998), the second stimulus was substituted in half of the trials with a train of current bursts delivered at the comparison frequency. Regular discrimination trials included natural stimuli only and were intermixed with microstimulation trials. Figure 2a shows a diagram of this paradigm and a plot with the results. The monkeys were able to discriminate between mechanical (base) and electrical (comparison) signals with performance profiles that were indistinguishable from those obtained with natural stimuli only (Romo et al 1998). In view of the high discrimination accuracy with electrical signals, the artificially induced sensation probably resembled natural flutter quite closely (Romo et al 1998, Wickersham & Groh 1998). Further experiments supported this conclusion.

The microstimulation experiment was repeated using aperiodic stimuli during the comparison period (Romo et al 1998) (see above). During aperiodic stimulation, the same average frequencies were used, and the monkeys had to judge the differences between base and comparison frequencies as before. Figure 2b illustrates the two kinds of trials that were compared in this experiment. The graph shows that performance in microstimulation trials was again practically identical to performance with natural stimuli. Monkeys had no difficulty working with aperiodic vibrations, not even in the first sessions, so spike periodicity was not critical for flutter discrimination (see also Salinas et al 2000a).

By design, in this task the frequency of the second stimulus is compared against a memory trace of the first stimulus (Hernández et al 1997). To test whether the artificially induced percept could be stored in memory, the base stimulus was substituted with electrical microstimulation patterns while the comparison remained natural, mechanical (Romo et al 2000). This paradigm is schematized in Figure 2c. The graph shows that, once more, performance with artificial sensations was indistinguishable from that observed with mechanical stimuli only. This experiment showed that monkeys can memorize the frequency of an artificial stimulus just like that of natural flutter.

Going one step further, both base and comparison stimuli were substituted with electrical pulses in the same trials (Romo et al 2000). As shown in Figure 2d, in this condition the monkeys did not reach the same discrimination levels as with natural stimuli only, but their performance was only slightly worse, still well above chance levels. However, in 2 out of 11 sessions, this experiment with two artificial stimuli did not work; average performance dropped way below 75% correct. The reason for this failure is uncertain, but it may have to do with the location of the stimulating microelectrode relative to clusters of RA and SA neurons. The flutter sensation is mediated specifically by RA primary afferents (Mountcastle et al 1969, Ochoa & Torebjörk 1983) and previous studies suggested that the specificity of the RA circuit for flutter is maintained in S1 (Mountcastle et al 1990, Recanzone et al 1992). This was also tested.

Figure 3 shows the results of a single experimental session in which the stimulating microelectrode encountered clusters of SA and RA neurons at different depths along the same penetration (Romo et al 2000). Separate discrimination runs were performed with the electrode positioned at three locations. The same paradigm was used in the three runs: In microstimulation trials, the base stimulus was electrical and the comparison was mechanical; in interspersed trials, both stimuli were natural (as in Figure 2c). When the electrode tip was placed near neurons with SA properties, the monkey performed at chance levels in microstimulation trials (Figure 3b), although it had no problem discriminating with natural stimuli. On the other hand, performance in microstimulation trials was just as good as with natural stimuli when the electrode tip was placed around RA neurons. When microstimulation was delivered near the border between RA and SA clusters, performance was better than chance but significantly below the levels seen with mechanical stimuli. The results of Figure 3 were confirmed in four other sessions.

In summary, neuronal activity elicited by direct electrical microstimulation in area 3b is localized (at least in a functional sense) and can trigger perceptual experiences. Furthermore, as far as can be measured psychophysically, this activity can be stored in memory and used to make quantitative comparisons and decisions in practically the same ways as naturally evoked activity.

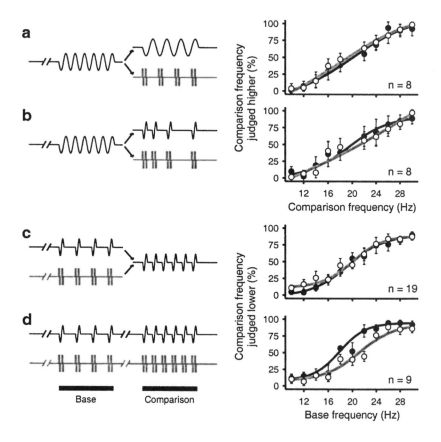

ATTENTIONAL AND CONTEXT-DEPENDENT
MODULATION OF ACTIVITY IN S1 AND S2

The responses of neurons to sensory stimuli may be affected by their behavioral context; holding someone's hand may lead to completely opposite reactions, depending on whose hand is held. Interactions between stimulus-driven and contextual information should become stronger as activity reaches stages where decision-making becomes imminent.

Attentional effects have been thoroughly investigated in vision (Desimone & Duncan 1995). In several experimental paradigms and visual areas, attention sets a gain factor by which whole stimulus-response curves are multiplied (Connor et al 1997, McAdams & Maunsell 1999a, Treue & Martínez-Trujillo 1999). Whether attention acts similarly on somatosensory responses is unknown, but it should be investigated. Other interactions related to spatial or feature selection have been described as well (Reynolds & Desimone 1999). In somatosensory areas, modulations in firing rate have been documented, as in vision, but in addition, effects on spike timing have also been observed. This is discussed below.

Hyvarinen et al (1980) and Poranen & Hyvarinen (1982) first attempted to record attentional effects in S1 neurons. Neural responses to flutter stimuli were compared

---

←

Figure 2   Psychophysical performance in frequency discrimination with natural, mechanical stimuli and with artificial, electrical stimuli. Monkeys were trained in the flutter discrimination task (Figure 1a). (*Left*) The protocols used in four experiments. In all cases, two kinds of trials were interleaved: In natural discrimination trials, the monkeys compared two mechanical vibrations (black traces) delivered on a fingertip; in microstimulation trials, one or both stimuli were replaced by trains of current bursts (gray traces) microinjected into area 3b at the corresponding frequencies. Base and comparison frequencies varied across trials. (*Right*) The animals' average performance in the situations schematized on the left. Filled circles and black lines indicate discrimination with natural stimuli; open circles and gray lines indicate discrimination in microstimulation trials. (*a*) The base stimulus consisted of periodic, mechanical vibrations (sinusoidal); the comparison stimulus was periodic and could be either mechanical (sinusoidal) or electrical. (*b*) The base stimulus consisted of periodic, mechanical vibrations (sinusoidal); the comparison stimulus was aperiodic and could be either mechanical (a series of short, single-sinusoid pulses) or electrical. (*c*) The base stimulus was periodic and could be either mechanical (single-sinusoid pulses) or electrical; the comparison stimulus was always periodic and mechanical (single-sinusoid pulses). (*d*) In half of the trials, both base and comparison stimuli were periodic and mechanical (single-sinusoid pulses). In the other half, both stimuli were periodic and electrical. In all experiments, natural and microstimulation trials were interleaved and were equally frequent. Scale bars equal 500 ms. Data points shown are means (± 1 standard deviation) over the indicated numbers of data collection runs. In panels *a* and *b*, only a subset of the collected data for which the base frequency was 20 Hz is shown. In panels *c* and *d*, only the subset of points for which the comparison frequency was 20 Hz is shown. (Modified from Romo et al 1998, 2000.)

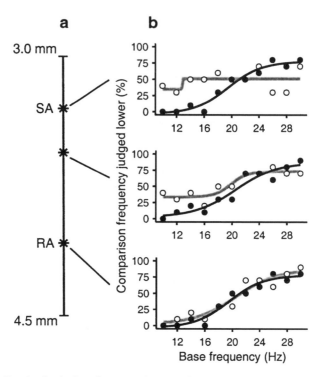

Figure 3 Psychophysical performance in natural and microstimulation trials for three electrode locations. Neurons with slowly adapting (SA) and rapidly adapting (RA) properties were encountered along the same penetration. (*a*) Electrode depths. Asterisks mark microstimulation sites along the track. (*b*) Psychophysical performance measured when microstimulation was applied at the sites indicated in panel a. (*Top*) Microstimulation delivered around SA neurons; (*center*) microstimulation delivered near the border between RA and SA clusters; (*bottom*) microstimulation delivered around RA neurons. Symbols have the same meaning as in Figure 2; the protocol was the same as in Figure 2*c*. Each data point represents 10 trials collected during successive runs on the same day. (Modified from Romo et al 2000.)

in two conditions: when stimuli were delivered passively to the fingertips, without any behavioral reaction, and when the monkeys performed a simple detection task. A few neurons did fire at higher rates during the task compared with the no-task condition, but the effects were small. In retrospect this is not surprising, for two reasons. First, no effort was made to divert attention away from the stimulus in the passive condition, and second, attentional effects may increase substantially with increasing task difficulty (Posner et al 1978, Whang et al 1991). Much larger effects on firing rate were found later, both in S1 and S2, using more challenging paradigms (Hsiao et al 1993, Burton et al 1997, Steinmetz et al 2000). Studies using noninvasive techniques are consistent with these fingings (Mima et al 1998,

Burton et al 1999). As in other modalities, typically stronger activity is seen when attention is focused on a stimulus that is relevant to the recorded neurons.

Attentional Effects on Spike Timing

Motivated by models of cortical interactions in which spike timing plays a key role (Niebur & Koch 1994), Steinmetz et al (2000) recorded pairs of neighboring S2 neurons to investigate whether attention is related to spike synchrony. In their experiments, tactile and visual stimuli were always presented simultaneously. Their monkeys were trained to switch between visual and somatosensory discrimination tasks, depending on a cue at the beginning of each block of trials. Across conditions, stimulus sets and arm motions used to indicate behavioral choices were identical. To perform the somatosensory task accurately, monkeys had to pay attention to the tactile stimuli and ignore the visual display, and vice versa during the visual task; mistakes indicated that attention was misdirected.

As in earlier work (Hsiao et al 1993), these investigators observed that the firing rates of a large proportion of S2 neurons (80%) changed significantly between conditions (Steinmetz et al 2000). They also found that 66% of all neuron pairs were significantly synchronized, regardless of the task. This was simply a consequence of receptive field overlap because neurons sharing common inputs should exhibit some degree of synchrony. The most novel finding was that, on average, synchrony was stronger during tactile discrimination than during visual discrimination: About 9% of all S2 neuron pairs fired more synchronously when the monkeys attended the tactile stimuli. Changes in synchrony also increased with task difficulty and could not be explained in terms of variations in firing rate. The implication is that shifting the attentional focus has an impact on spike synchrony. Two facts cannot be overlooked, however. First, we do not know whether changes in synchrony in S2 translate into functional changes downstream (Niebur & Koch 1994, Shadlen & Newsome 1994, Salinas & Sejnowski 2000). And second, the attentional modulation of firing rate seems much more prominent, at least judging from the fractions of neurons with significant effects (Hsiao et al 1993, Steinmetz et al 2000).

Other experiments using flutter stimuli confirmed that attention may influence the timing of evoked spikes. Salinas et al (2000a) compared neural responses in S1 recorded during active discrimination (Figure 1*a*) versus responses obtained during passive stimulation. In blocks of passive trials, the same stimuli used for discrimination were applied, but the responding arm was restrained, no behavioral reaction was required, and no reward was delivered. A measure of periodicity was constructed based on the Fourier spectrum of the evoked spike trains; this was just the fraction of total power found around the frequency of stimulation. Periodicity was significantly higher in active than in passive tests (Salinas et al 2000a). In other words, the timing of evoked S1 spikes relative to the mechanical stimulation pulses was more regular during active discrimination; tighter phase locking occurred in this condition.

As with the results of Steinmetz et al (2000), the functional implications of these timing effects are uncertain, but they are extremely interesting, if nothing else, because they constrain the microcircuitry that may underlie attentional modulation of stimulus-evoked activity.

Attentional Effects on Signal Quality

It is not surprising that Salinas et al (2000a) also observed that firing rates changed between active and passive conditions in S1 and S2. However, they noticed that the rates depended more strongly on stimulus frequency during active discrimination. Several measures were computed to evaluate the association between firing rate and frequency in the two conditions. These measures included the mean trial-to-trial variability in spike count, a signal-to-noise ratio, and Shannon's information (Cover & Thomas 1991). The same conclusion was reached with all quantities tested: When the animals performed the task and presumably paid attention to the stimuli, the evoked firing rates were less variable across trials and were better correlated with frequency (Salinas et al 2000a). For this analysis, the random cofluctuations in firing rate between pairs of neurons, which could, in principle, limit the quality of the overall population signal (Zohary et al 1994, Abbott & Dayan 1999), were taken into account. Thus, the quality of the rate-based representation of flutter frequency was higher during active discrimination than during passive stimulation. These effects were seen in both S1 and S2, although they were stronger in S2. Attention is commonly thought of as a mechanism that amplifies or selects relevant signals, filtering out irrelevant ones (Desimone & Duncan 1995, Connor et al 1997, McAdams & Maunsell 1999a, Reynolds & Desimone 1999), but it may also produce clearer signals (McAdams & Maunsell 1999b).

Thus, context or attention typically modulate the intensity of stimulus-evoked neural responses, but they can also influence the times at which spikes are fired and the effective signal-to-noise ratio, even in a primary sensory area.

NEURAL CORRELATE OF SOMATOSENSORY WORKING MEMORY IN THE PREFRONTAL CORTEX

In the flutter discrimination task described above (Figure 1a), a neural representation of the base frequency needs to be remembered during the delay period following the base stimulus. Where in the brain is such mnemonic trace kept and what is the stimulus representation?

Salinas et al (2000a) found neurons in S2 that prolonged their frequency-specific responses into the early component of the delay period. Their rates typically varied with base frequency in the same direction as during stimulation. This sustained activity typically lasted a few hundred milliseconds (one second at most) and was not seen in S1 (but see Zhou & Fuster 1996). Its functional role is unknown but could be related to working memory.

Romo and colleagues (1999) recorded in the prefrontal cortex, a structure known to be involved in memory processes (Funahashi et al 1989, Fuster 1989). They found delay responses whose dependence on stimulus frequency was monotonic, like that observed in S2 (Salinas et al 2000a): Some prefrontal neurons increased their firing rates steadily with increasing frequency, whereas others had firing rates that varied in the opposite direction, firing most intensely at low base frequencies (Figure 4). Most delay neurons fell into one of three groups: early, late, or

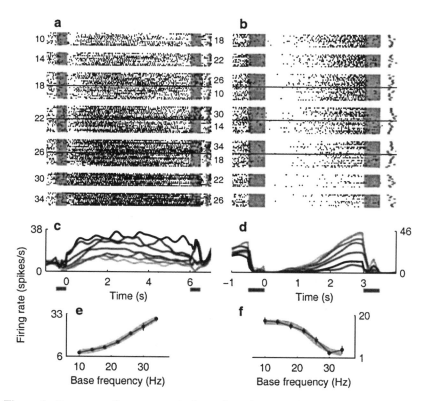

Figure 4 Responses of two neurons in the prefrontal cortex with delay activity during the flutter discrimination task (Figure 1a). (a) Spike raster plot. Each row of dots corresponds to one trial, and each dot represents one action potential. Trials are shown sorted into blocks of equal base frequency (*left*), in hertz. Long horizontal lines divide groups with equal base but different comparison frequency (*right*), in hertz. Intervals in gray indicate base and comparison stimulation periods (500 ms). The time axis is shown on the plot below. The firing rate of this neuron increased as a function of base frequency. (b) As in panel a, but for a neuron whose firing rate decreased as a function of base frequency. (c, d) Spike densities (firing rate as a function of time) for the data shown above. Base frequency is indicated by the level of gray: The lightest line corresponds to 10 Hz and the darkest to 34 Hz. (e, f) Mean firing rate(±1 standard error of the mean) averaged over the entire delay period as a function of base frequency. Gray lines are sigmoidal fits to the data. (Modified from Romo et al 1999.)

persistent. Persistent neurons carried a significant signal related to base frequency during the full delay period, early neurons carried it during the first second of the delay but not during the last one, and late neurons carried such signal during the last second of the delay but not during the first. Thus, early neurons resembled those units in S2 with prolonged activity, which is consistent with the anatomical projections from S2 to prefrontal cortex (Preuss & Goldman-Rakic 1989, Carmichael & Price 1995).

The response characteristics of these prefrontal neurons were not static. When the delay was extended from three to six seconds, most late neurons shifted the onset of their response, again developing a significant signal about 1 s before the end of the delay period (Romo et al 1999; see also Kojima & Goldman-Rakic 1982). These shifts took place in the course of a few trials and probably reflect temporal expectations: With a fixed delay, the monkey can anticipate the onset time of the comparison stimulus.

Anticipation is a crucial factor in memory tasks because it blurs the distinction between sensory- and motor-related activity (Romo et al 1999). When monkeys use different arm movements to indicate their choices, delay responses seen to vary across sensory conditions may actually be associated with the timing (Fuster 1989) or identity (Kojima & Goldman-Rakic 1982, Funahashi et al 1989, Hoshi et al 1998) of the various motor acts. Therefore, the observed delay activity (Romo et al 1999) could have been related to motor preparation or motor intention. To test this, a special stimulus set was designed in which approximately equal numbers of arm motions were made toward the two push-buttons for any given base frequency (only those two arm motions were possible). If the sustained responses had been related to motor anticipation only, no variation in firing rate should have been obtained as a function of base frequency, but this is not what happened. With this set the same kinds of monotonic responses were observed. Thus, it is unlikely that these delay neurons are involved in motor preparation. The observed monotonic encoding might be a general, basic kind of sensory representation used in working memory because, in principle, it may be used for any scalar, analog quantity.

NEURONAL REPRESENTATIONS OF A DECISION

A Categorization Task for Exploring Decision-Making Mechanisms

Romo and colleagues (1993, 1996) designed a task in which the speed of a moving tactile stimulus could be categorized as either low or high. Figure 5a shows the sequence of events. In each trial, a mechanical probe traversed a fixed distance across a fingertip at either of 10 speeds: Five of them were considered low (12, 14, 16, 18, and 20 mm/s), and five were considered high (22, 24, 26, 28, and 30 mm/s). The monkey had to decide whether the stimulus belonged to the low or high category, and the choice was indicated by pressing one of two push buttons.

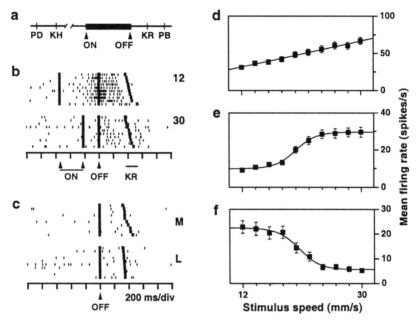

Figure 5   Responses related to categorical decision making. (*a*) Sequence of events in the categorization paradigm. The tip of a probe was lowered (PD) and touched a fingertip. The monkey reacted by holding a key (KH). Following a delay, the probe started moving (ON) at one of ten speeds. After traversing a constant distance, the probe stopped moving (OFF). The monkey released the key (KR) and pressed one of two push buttons (PB) to indicate whether the speed was low (<20 mm/s) or high (>20 mm/s). (*b*) Spike rasters from an M1 neuron that fired more intensely at low speeds. Each line corresponds to one trial, and each small tick represents an action potential. Large squares indicate behavioral events: onset (ON) and offset (OFF) of probe movement, or key release (KR), which marked initiation of the hand-arm movement toward the buttons. Responses are shown for only two speeds, 12 and 30 mm/s. (*c*) Spike rasters from the same neuron in panel *b* recorded while the monkey made identical arm movements to the medial (M) or lateral (L) buttons but was cued by the dimming of an LED (light emitting diode) light, with no tactile stimulation. In each trial, the first square indicates the visual go signal (OFF), and the second square indicates key release. (*d*) Mean firing rate (±1 standard error of the mean) as a function of stimulus speed averaged over a population of 27 S1 neurons. Continuous line indicates best linear fit. (*e*) As in panel *d*, but for a population of 40 supplementary motor area (SMA) neurons that fired significantly more strongly at high speeds than at low speeds. Continuous line indicates sigmoidal fit. Similar responses were also recorded in M1. (*f*) As in panel *e*, but for a population of 20 M1 neurons selective for low speeds. Similar responses were also recorded in the SMA. (Modified from Salinas & Romo 1998a,b.)

The animal had to discover which speeds belonged to which category by trial and error.

Three quantities are essential in this paradigm: stimulus speed, speed category, and arm motion (which could take two values). Here, computing-speed category is equivalent to reaching a decision, and understanding the decision-making process amounts to answering three key questions. (*a*) Where are the neural representations of the three quantities, and what are the neural codes used? (*b*) How are these representations constructed? (*c*) How do they correlate with a subject's observable behavior? The second question is the hardest (but see Gold & Shadlen 2000), but neurophysiological experiments have provided some insight into the other two.

Representation of Tactile Motion

Neurons in S1 respond to motion on the skin. DiCarlo & Johnson (1999, 2000) characterized this dependence in detail using random-dot patterns. When a single moving probe is used, the evoked S1 activity can be represented by a population vector (Georgopoulos et al 1988) whose direction and magnitude correspond to the direction and speed, respectively, of the moving tactile stimulus (Ruiz et al 1995).

Recordings from S1 were obtained from monkeys performing the categorization task (Romo et al 1996). Two kinds of neurons were found. One kind increased their firing rates during stimulation regardless of stimulus speed; these neurons signaled the presence of a stimulus but not its properties. However, other neurons fired at rates that increased monotonically with increasing speed, and the relationships were, to a good approximation, linear (Romo et al 1996, Salinas & Romo 1998b; see also DiCarlo & Johnson 1999) (Figure 5*d*). These neural responses did not covary with behavior: For any given speed, the firing rates for correct and incorrect categorizations were indistinguishable (Salinas & Romo 1998b). In addition, when the same stimuli were delivered passively, the responses were unchanged (in passive trials the monkey's responding arm was restrained, no movements were performed, and no reward was given).

These results differ from those of Salinas et al (2000a), who found attentional effects and single-trial covariations between neuronal and behavioral responses in S1. Two factors (other than differences in stimuli) might account for the discrepancy. First, fewer data were available in the categorization experiments, so small effects could have been missed. Second, categorization is easier than discrimination (Hernández et al 1997), and attentional effects may depend on task difficulty (Posner et al 1978, Whang et al 1991).

Thus, S1 neurons responded in a purely sensory fashion: First, the firing rate curves reflected stimulus speed, not speed category; second, passive and active responses were indistinguishable; and third, no cofluctuations with behavior were detected. However, lesions in S1 (areas 3b, 1, and 2) drastically impair the capacity to categorize, with deficits persisting for months (Zainos et al 1997). This suggests that S1 is not directly involved in the categorization process, although it provides a neural representation of tactile motion that is essential for downstream decision-making mechanisms, and probably for perception as well (Romo et al 1998, 2000).

Categorical Decision-Making in Motor Networks

Where should one search for the decision-making mechanisms that operate during somatosensation? The motor areas of the frontal lobe are good candidates, first, because they must be kept informed about any decisions in order to generate the appropriate motor actions (Shadlen & Newsome 1996, Leon & Shadlen 1998, Schall & Thompson 1999) and second, because they receive strong projections from somatosensory areas. For instance, the supplementary motor area (SMA) (also known as medial premotor cortex) receives direct inputs from the posterior parietal cortex and the lateral somatosensory areas (Jurgens 1984, Cavada & Goldman-Rakic 1989, Luppino et al 1993). Neurophysiologists have characterized the SMA as related to motor preparation or execution (Alexander & Crutcher 1990a,b; Kurata & Tanji 1985; Tanji & Kurata 1985), but strong responses to visual, auditory, and somatic stimuli are also seen when animals use them to initiate arm movements (Kurata & Tanji 1985, Tanji & Kurata 1985, Romo & Schultz 1987). Therefore, the role of motor areas in sensory-motor integration (and possibly decision-making processes) may be more important than generally accepted (Shen & Alexander 1997, Zhang et al 1997).

This possibility was investigated by recording the activity of single neurons in the SMA (Romo et al 1993, 1997) and in primary motor cortex (M1) (Salinas & Romo 1998a) of awake monkeys performing the tactile categorization task. Many neurons in these areas displayed activity related to hand-arm motions, as expected from previous reports (Schwartz et al 1988; Alexander & Crutcher 1990a,b; Georgopoulos 1994). These responses did not vary as functions of stimulus speed. However, a small but significant fraction of the neurons in these areas (191/745 in SMA, 71/477 in M1) responded differentially to speed category. These neurons were recorded along the same microelectrode penetrations as the classical motor-related units (Salinas & Romo 1998a). Their firing rates did not vary linearly with stimulus speed, as in S1. Instead, they typically traced sharp sigmoids that either increased or decreased (Figures $5e, f$). These neurons showed an almost binary specificity for one or another category. In both areas, about equal proportions of differential neurons preferred low and high speeds (Romo et al 1993, 1997; Salinas & Romo 1998a,b).

Motor and premotor neurons are typically tuned to the direction of arm motion (Georgopoulos et al 1988, Schwartz et al 1988, Georgopoulos 1994). Because low and high speeds always corresponded to the same two push buttons, did the observed differences in rate arise simply from differences between movement trajectories? This is unlikely because tuning curves for movement direction are much broader than the 11 degrees of separation used between push buttons. Based on this distance, the maximum change in firing rate expected for a typical M1 neuron was about 1 spike/s (Salinas & Romo 1998a; see also Schwartz et al 1988).

The observed rate modulation was, on average, much larger than expected from classic directional tuning (Figures 5e,f).

Distinguishing Responses Related to Sensory, Motor, and Decision-Making Processes

The plots of firing rate versus speed suggested that the differential responses could represent the output of a decision-making mechanism (speed category), but other possibilities had to be considered (see Horwitz & Newsome 1998). For instance, they could still have been related to preparation or execution of arm movements, but with an exceptionally sharp tuning. What were these neurons coding? Additional control experiments and analyses were performed to address this.

First, none of the differential neurons responded to passive stimulation (0/30 in SMA, 0/5 in M1) (Romo et al 1997, Salinas & Romo 1998a). These cells were not just driven by the sensory stimulus, as in S1.

Second, the animals were also trained to make arm movements toward the push buttons in response to visual cues only, in the absence of tactile input. In this condition, the category-specific neurons either stopped responding altogether or did fire at higher rates compared with baseline, but with the same intensity for both arm movements (Romo et al 1997, Salinas & Romo 1998a) (Figure 5b,c). Overall, about three quarters of the differential neurons tested during visually guided reaching stopped responding differentially (57/71 in SMA, 29/42 in M1). The category-specific responses did not correlate unconditionally with motor behavior.

Third, the psychophysical performance of the monkeys was compared with the performance of an ideal observer who based his judgements on the responses of a population of neurons (Salinas & Romo 1998a,b; see also Britten et al 1992, Shadlen et al 1996). The neurometric curves constructed from S1 responses and the measured psychometric curves had different shapes, which suggests that behavioral performance is based on a different representation of speed. However, the neurometric curves derived from category-specific neurons in the SMA and M1 were almost identical to the monkeys' performance curves (Salinas & Romo 1998a,b), indicating that the output of the categorical decision-making process is likely to be coded by units that have sharp, increasing and decreasing sigmoidal dependencies on speed.

Fourth, the category-specific responses were compared in correct (hits) versus incorrect categorizations (errors) to determine whether they covaried with sensory input, motor behavior, or both (Salinas & Romo 1998b; see also Britten et al 1996). For example, at any given speed, an ideal motor neuron should produce different responses in hit trials and error trials because, having fixed the sensory input, hits and errors correspond to different arm movements. The resulting patterns of hit-and-error responses exhibited both sensory and motor characteristics in a variety of combinations. This was consistent with an analysis of response latencies, which were not consistently time locked to either sensory (stimulus onset and offset) or motor (key release) events (Salinas & Romo 1998a). These neurons seemed part of a sensory-motor interface.

Widely Distributed Category Signals

In additional experiments, category-specific neurons were also found in the neostriatum, and these failed to respond differentially during visually guided reaching and during passive stimulation, as in the SMA and M1 (Romo et al 1995, Merchant et al 1997). This is not entirely surprising, considering the bilateral inputs that originate in the SMA (Kunzle 1975, Jurgens 1984, McGuire et al 1991) and the similarities between striatal and motor cortical responses (Alexander & Crutcher 1990a,b; Schultz & Romo 1992; Romo et al 1992; Romo & Schultz 1992; Salinas et al 2000b). Whatever the message the category-specific neurons are sending, it is widely broadcast throughout vast motor networks.

The results in this section can be summarized as follows. In S1, firing rate varies linearly with the speed of tactile motion (Romo et al 1996). This representation is essential for categorization (Zainos et al 1997), but the animal's choice, low or high, is probably encoded by category-specific neurons or by units with similar responses (Salinas & Romo 1998a,b). These neurons are found in a large number of motor structures, but to activate them, both sensory and motor components of the task are required (Romo et al 1997, Merchant et al 1997, Salinas & Romo 1998a). On the other hand, their activity covaries with both sensory and motor variables in single trials (Salinas & Romo 1998a,b). Therefore, category-specific neurons might represent an interface between the output of the sensory categorization process and the motor command used to indicate the monkey's choice. Studies in vision have also revealed a tight association between sensory and motor activity during decision making (Leon & Shadlen 1998, Kim & Shadlen 1999, Schall & Thompson 1999, Gold & Shadlen 2000). Whether other areas contain dedicated neurons that encode independently of motor activity the categorical decision being made has yet to be investigated (Horwitz & Newsome 1998).

CONCLUDING REMARKS

Decision making is present even in the simplest tasks. When the afferent input needs to be heavily transformed or analyzed, sensory areas become more involved in the process, and at least in some cases, their activity matches psychophysical performance (Newsome et al 1989, Vogels & Orban 1990, Britten et al 1992, Hernández et al 2000). Thus, the relevant sensory properties—receptive field, feature selectivity, response sensitivity, etc—should be thoroughly investigated, as has been done, for instance, for visual motion or for flutter. This is critical in studying decision making because we need to understand the neural codes for the sensory quantities being decided on (DeCharms & Zador 2000, Salinas et al 2000a), as well as their dependence on factors such as attention or memory

(Romo et al 1999, Salinas et al 2000a). Intuitively, the circuits that generate motor commands should stand at the other end of the decision-making processes because their output needs to be expressed physically. But is this intuition correct?

First, decision-making might involve neural mechanisms and interactions not yet understood or maybe not even identified so far. The problem of timing may be one of greatest importance. Consider, for example, the following result. When S1 is lesioned, monkeys can no longer categorize the speed of a moving tactile stimulus (Zainos et al 1997). However, they can still detect its presence, so they keep performing the task. A crucial component of the decision-making process is clearly missing, but a decision to press one switch or the other is made anyway; in fact, it is made with the right timing because the reaction times of lesioned animals remain normal (Zainos et al 1997). This result is the converse of the experiment by Seidemann et al (1999), who found that neural activity that would normally have a strong influence on a perceptual decision had absolutely no effect when it was out of synch relative to the normal sequence of events in the task (although this could be modified through training). Therefore, the decision-making process seems to proceed as if it were part of an encompassing, established plan. Maybe it is better to think of making a decision as the creation of a highly flexible motor plan that can be delayed (if the action requires a go signal to be triggered) or rapidly reconfigured (if the motor output is not specified ahead of time), for example. As illustrated in the previous section, the signals that come closest to encoding the output of a decision-making process have been found in areas involved in motor control: the SMA (Romo et al 1997, Salinas & Romo 1998b), M1 (Zhang et al 1997, Shen & Alexander 1997, Salinas & Romo 1998a), and the lateral intraparietal area (Shadlen & Newsome 1996). Is there any difference between making a decision and planning a motor action that can be postponed indefinitely? The recent study by Gold & Shadlen (2000) is particularly revealing. They recorded neural activity and injected microstimulation current in the frontal eye field of monkeys performing a two-alternative motion-discrimination task. The frontal eye field is involved in generating eye movements. Gold & Shadlen found that this area gradually accumulates evidence for motion in one or another direction, such that the process of forming a decision and motor preparation seem to be indistinguishable.

Another intriguing consideration is this. Making decisions may involve information coming from various sources and gathered at widely different times, such as information arriving from sensory areas, working memory, or long-term memory. Thus, the circuits toward which multiple pieces of information converge would seem to require a common format for the incoming signals, a common code. McAdams & Maunsell (1999b) pointed out that sensory events and internal behavioral state may affect cortical responses in equivalent ways, through similar circuits and synaptic mechanisms that differ only in the information they deliver. It is interesting that in several instances, the firing of somatosensory neurons has been shown to be linearly related (approximately) to stimulus parameters and to subjective intensity (Mountcastle et al 1963, Sinclair & Burton 1993, Jiang et al

1997, Johnson et al 1996, Blake et al 1997, Romo et al 1996, Salinas et al 2000a). This includes the prefrontal responses that encode flutter frequency during a delay period (Romo et al 1999). Is there a common neural substrate, independent of modality or of stimulus parameters, shared by most discrimination tasks or decision-making processes?

These and related questions will probably be central to future neurophysiological studies of decision making in all modalities.

ACKNOWLEDGMENTS

The research of RR was supported by an International Research Scholars award from the Howard Hughes Medical Institute and by grants from CONACYT and DGAPA-UNAM. ES also thanks Terry Sejnowski and the Howard Hughes Medical Institute for their support. We appreciate the assistance of Adrián Hernández in preparing the figures.

Visit the Annual Reviews home page at www.AnnualReviews.org

LITERATURE CITED

Abbott LF, Dayan P. 1999. The effect of correlated activity on the accuracy of a population code. *Neural Comput.* 11:91–101

Ahissar E. 1998. Temporal-code to rate-code conversion by neuronal phase-locked loops. *Neural Comput.* 10:597–650

Ahissar E, Vaadia E. 1990. Oscillatory activity of single units in a somatosensory cortex of an awake monkey and their possible role in texture analysis. *Proc. Natl. Acad. Sci. USA* 87:8935–39

Alexander GE, Crutcher MD. 1990a. Preparation for movement: neural representations of intended direction in three motor areas of the monkey. *J. Neurophysiol.* 64:133–50

Alexander GE, Crutcher MD. 1990b. Neural representation of the target (goal) of visually guided arm movements in three motor areas of the monkey. *J. Neurophysiol.* 64:164–78

Blake DT, Hsiao SS, Johnson KO. 1997. Neural coding mechanisms in tactile pattern recognition: the relative contributions of slowly and rapidly adapting mechanoreceptors to perceived roughness. *J. Neurosci.* 17:7480–89

Britten KH, Newsome WT, Shadlen MN, Cele-brini S, Movshon JA. 1996. A relationship between behavioral choice and the visual responses of neurons in macaque MT. *Vis. Neurosci.* 13:87–100

Britten KH, Shadlen MN, Newsome WT, Movshon JA. 1992. The analysis of visual motion: a comparison of neuronal and psychophysical performance. *J. Neurosci.* 12:4745–65

Britten KH, van Wezel RJ. 1998. Electrical microstimulation of cortical area MST biases heading perception in monkeys. *Nat. Neurosci.* 1:59–63

Burton H, Abend NS, MacLeod AM, Sinclair RJ, Snyder AZ, Raichle ME. 1999. Tactile attention tasks enhance activation in somatosensory regions of parietal cortex: a positron emission tomography study. *Cereb. Cortex* 9:662–74

Burton H, Fabri M, Alloway K. 1995. Cortical areas within the lateral sulcus connected to cutaneous representations in areas 3b and 1: a revisited interpretation of the second somatosensory area in macaque monkeys. *J. Comp. Neurol.* 355:539–62

Burton H, Sinclair RJ, Hong SY, Pruett JR,

Whang KC. 1997. Tactile-spatial and cross-modal attention effects in the second somatosensory and 7b cortical areas of rhesus monkeys. *Somat. Motil. Res.* 14:237–67

Carmichael ST, Price JL. 1995. Sensory and premotor connections of orbital and medial prefrontal cortex of macaque monkeys. *J. Comp. Neurol.* 363:642–64

Cavada C, Goldman-Rakic PS. 1989. The posterior parietal cortex in rhesus monkeys: I. Parcellation of areas based on distinctive limbic and sensory corticocortical connections. *J. Comp. Neurol.* 287:393–421

Connor CE, Hsiao SS, Phillips JR, Johnson KO. 1990. Tactile roughness: neural codes that account for psychophysical magnitude estimates. *J. Neurosci.* 10:3823–36

Connor CE, Johnson KO. 1992. Neural coding of tactile texture: comparisons of spatial and temporal mechanisms for roughness perception. *J. Neurosci.* 12:3414–26

Connor CE, Preddie DC, Gallant JL, Van Essen DC. 1997. Spatial attention effects in macaque area V4. *J. Neurosci.* 17:3201–14

Cover TM, Thomas JA. 1991. *Elements of Information Theory.* New York: Wiley

Darian-Smith I. 1984. The sense of touch: performance and peripheral neural processes. In *Handbook of Physiology. I. The Nervous System.* Vol. 3. *Sensory Processes, Part 2,* ed. JM Brookhart, VB Mountcastle, pp. 739–88. Bethesda, MD: Am. Physiol. Soc.

DeCharms CR, Zador A. 2000. Neural representation and the cortical code. *Annu. Rev. Neurosci.* 23:613–47

Desimone R, Duncan J. 1995. Neural mechanisms of selective visual attention. *Annu. Rev. Neurosci.* 18:193–222

DiCarlo JJ, Johnson KO. 1999. Velocity invariance of receptive field structure in somatosensory cortical area 3b of the alert monkey. *J. Neurosci.* 19:401–19

DiCarlo JJ, Johnson KO. 2000. Spatial and temporal structure of receptive fields in primate somatosensory area 3b: effects of stimulus scanning direction and orientation. *J. Neurosci.* 20:495–510

DiCarlo JJ, Johnson KO, Hsiao SS. 1998. Structure of receptive fields in area 3b of primary somatosensory cortex in the alert monkey. *J. Neurosci.* 18:2626–45

Funahashi S, Bruce CJ, Goldman-Rakic PS. 1989. Mnemonic coding of visual space in the monkey's dorsolateral prefrontal cortex. *J. Neurophysiol.* 61:331–49

Fuster J. 1989. *The Prefrontal Cortex.* New York: Raven

Georgopoulos AP. 1994. New concepts in generation of movement. *Neuron* 13:257–68

Georgopoulos AP, Kettener RE, Schwartz AB. 1988. Primate motor cortex and free arm movements to visual targets in three-dimensional space. II. Coding of the direction of movement by a neuronal population. *J. Neurosci.* 8:2928–37

Godshalk M, Lemon RB, Kuypers HG, Ronday HK. 1984. Cortical afferents and efferents of monkey postarcuate area: an anatomical and electrophysiological study. *Exp. Brain Res.* 56:410–24

Gold JI, Shadlen MN. 2000. Representation of a perceptual decision in developing oculomotor commands. *Nature* 404:390–94

Goodale MA, Milner AD. 1992. Separate visual pathways for perception and action. *Trends Neurosci.* 15:20–25

Green DM, Swets JA. 1966. *Signal Detection Theory and Psychophysics.* New York: Wiley

Hernández A, Salinas E, García R, Romo R. 1997. Discrimination in the sense of flutter: new psychophysical measurements in monkeys. *J. Neurosci.* 17:6391–400

Hernández A, Zainos A, Romo R. 2000. Neuronal correlates of sensory discrimination in the somatosensory cortex. *Proc. Natl. Acad. Sci. USA* 97:6191–96

Horwitz GD, Newsome WT. 1998. Sensing and categorizing. *Curr. Biol.* 8:R376–78

Hoshi E, Shima K, Tanji J. 1998. Task-dependent selectivity of movement-related neuronal activity in the primate prefronal cortex. *J. Neurophysiol.* 80:3392–97

Hsiao SS, Johnson KO, O'Shaughnessy DM.

1993. Effects of selective attention of spatial form processing in monkey primary and secondary somatosensory cortex. *J. Neurophysiol.* 70:444–47

Hyvarinen J, Poranen A, Jokinen Y. 1980. Influence of attentive behavior on neuronal responses to vibration in primary somatosensory cortex of the monkey. *J. Neurophysiol.* 43:870–82

Iwamura Y, Iriki A, Tanaka M. 1994. Bilateral hand representation in the postcentral somatosensory cortex. *Nature* 369:554–56

Jiang W, Tremblay F, Chapman CE. 1997. Neuronal encoding of texture changes in the primary and the secondary somatosensory cortical areas of monkeys during passive texture discrimination. *J. Neurophysiol.* 77:1656–62

Johnson KO, Hsiao SS. 1992. Neural mechanisms of tactual form and texture perception. *Annu. Rev. Neurosci.* 15:227–50

Johnson KO, Hsiao SS, Blake DT. 1996. Linearity as the basic law of psychophysics: evidence from studies of the neural mechanisms of roughness magnitude estimation. In *Somesthesis and Neurobiology of the Somatosensory System*, ed. O Franzen, RS Johansson, L Terenius, pp. 213–28. Basel, Switzerland: Birkhaüser

Jones EG, Burton H, Porter R. 1975. Commissural and cortico-cortical 'columns' in the somatic sensory cortex of primates. *Science* 190:572–74

Jurgens U. 1984. The efferent and afferent connections of the supplementary motor area. *Brain Res.* 300:63–81

Kaas JH, Nelson RJ, Sur M, Lin CS, Merzenich MM. 1979. Multiple representations of the body within the primary somatosensory cortex of primates. *Science* 204:521–23

Kim JN, Shadlen MN. 1999. Neural correlates of a decision in the dorsolateral prefrontal cortex of the macaque. *Nat. Neurosci.* 2:176–85

Kojima S, Goldman-Rakic PS. 1982. Delay-related activity of prefrontal neurons in rhesus-monkeys performing delayed-response. *Brain Res.* 248:43–49

Krubitzer L, Clarey J, Tweendale R, Elston G, Calford MA. 1995. A redefinition of somatosensory areas in the lateral sulcus of macaque monkeys. *J. Neurosci.* 15:3821–39

Kunzle H. 1975. Bilateral projections from the precentral motor cortex to the putamen and other parts of the basal ganglia. An autoradiographic study in *Macaca fascicularis. Brain Res.* 88:195–209

Kurata K, Tanji J. 1985. Contrasting neuronal activity in supplementary motor area and precentral motor cortex of monkeys. II. Responses to movement triggering versus nontriggering sensory signals. *J. Neurophysiol.* 53:142–52

Leichnetz GR. 1986. Afferent and efferent connections of the dorsolateral precentral gyrus (area 4, hand/arm region) in the macaque monkey, with comparisons to area 8. *J. Comp. Neurol.* 254:460–92

Leon MI, Shadlen MN. 1998. Exploring the neurophysiology of decisions. *Neuron* 669–72

Luppino G, Mattelli M, Camarda RM, Rizzolatti GM. 1993. Cortico-cortical connections of area F3 (SMA-proper) and area F6 (Pre-SMA) in the macaque monkey. *J. Comp. Neurol.* 338:114–40

Macefield G, Gandevia SC, Burke D. 1990. Perceptual responses to microstimulation of single afferents innervating joints, muscles and skin of the human hand. *J. Physiol.* 429:113–29

McAdams CJ, Maunsell JHR. 1999a. Effects of attention on orientation tuning functions of single neurons in macaque cortical area V4. *J. Neurosci.* 19:431–41

McAdams CJ, Maunsell JHR. 1999b. Effects of attention on the reliability of individual neurons in monkey visual cortex. *Neuron* 23:765–73

McGuire KP, Bates JF, Goldman-Rakic PS. 1991. Interhemispheric integration. II. Symmetry and convergence of the corticostriatal projections of the left and the right principal sulcus (PS) and the left and the right supplementary motor area (SMA) of the rhesus monkey. *Cereb. Cortex* 1:409–17

Merchant H, Zainos A, Hernández A, Salinas E, Romo R. 1997. Functional properties of primate putamen neurons during the categorization of tactile stimuli. *J. Neurophysiol.* 77:1132–54

Mima T, Nagamine T, Nakamura K, Shibasaki H. 1998. Attention modulates both primary and second somatosensory cortical activities in humans: a magnetoencephalographic study. *J. Neurophysiol.* 80:2215–21

Mishkin M. 1979. Analogous neural models for tactual and visual learning. *Neurophychologia* 17:139–51

Mountcastle VB. 1957. Modality and topographic properties of single neurons of cat's somatic sensory cortex. *J. Neurophysiol.* 20:408–34

Mountcastle VB, Atluri P, Romo R. 1992. Selective output-discriminative signals in the motor cortex of waking monkeys. *Cereb. Cortex* 2:277–94

Mountcastle VB, Poggio GF, Werner G. 1963. The relation of thalamic cell response to peripheral stimuli varied over an intensive continuum. *J. Neurophysiol.* 26:807–34

Mountcastle VB, Steinmetz MA, Romo R. 1990. Frequency discrimination in the sense of flutter: psychophysical measurements correlated with postcentral events in behaving monkeys. *J. Neurosci.* 10:3032–44

Mountcastle VB, Talbot WH, Darian-Smith I, Kornhuber HH. 1967. Neural basis of the sense of flutter-vibration. *Science* 155:597–600

Mountcastle VB, Talbot WH, Sakata H, Hyvarinen J. 1969. Cortical neuronal mechanisms in flutter-vibration studied in unanesthetized monkeys. Neuronal periodicity and frequency discrimination. *J. Neurophysiol.* 32:452–84

Murray EA, Mishkin M. 1984. Relative contributions of SII and area 5 to tactile discrimination in monkeys. *Behav. Brain Res.* 11:67–83

Nelson RJ, Sur M, Felleman DJ, Kaas JH. 1980. Representations of the body surface in the postcentral parietal cortex of *Macaca fascicularis. J. Comp. Neurol.* 192:611–43

Newsome WT, Britten KH, Movshon JA. 1989. Neuronal correlates of a perceptual decision. *Nature* 341:52–54

Niebur E, Koch C. 1994. A model for the neuronal implementation of selective visual attention based on temporal correlation among neurons. *J. Comput. Neurosci.* 1:141–58

Ochoa J, Torebjörk E. 1983. Sensations evoked by intraneural microstimulation of single mechanoreceptor units innervating the human hand. *J. Physiol.* 42:633–54

Pearson RC, Powell TPS. 1985. The projection of primary somatic sensory cortex upon area 5 in the monkey. *Brain Res.* 356:89–107

Phillips JR, Johnson KO. 1981a. Tactile spatial resolution. I. Two-point discrimination, gap detection, grating resolution, and letter recognition. *J. Neurophysiol.* 46:1177–91

Phillips JR, Johnson KO. 1981b. Tactile spatial resolution. II. Neural representations of bars, edges, and gratings in monkey primary afferents. *J. Neurophysiol.* 46:1192–203

Phillips JR, Johnson KO, Hsiao SS. 1988. Spatial pattern representation and transformation in monkey somatosensory cortex. *Proc. Natl. Acad. Sci. USA* 85:1317–21

Phillips JR, Johansson RS, Johnson KO. 1992. Responses of human mechanoreceptive afferents to embossed dot arrays scanned across fingerpad skin. *J. Neurosci.* 12:827–39

Pons TP, Garraghty PE, Friedman DP, Mishkin M. 1987. Physiological evidence for serial processing in somatosensory cortex. *Science* 237:417–20

Pons TP, Garraghty PE, Mishkin M. 1992. Serial and parallel processing of tactual information in somatosensory cortex of rhesus monkeys. *J. Neurophysiol.* 68:518–27

Poranen A, Hyvarinen J. 1982. Effects of attention on multiunit responses to vibration in the somatosensory regions of the monkey's brain. *Electroencephalogr. Clin. Neurophysiol.* 53:525–37

Posner MI, Nissen MJ, Ogden WC. 1978.

Attended and unattended processing modes: the role of set of spatial locations. In *Models of Processing and Perceiving Information*, ed. HL Pick, IJ Salzmann, pp. 288–321. Hillsdale, NJ: Erlbaum

Powell TPS, Mountcastle VB. 1959. Some aspects of the functional organization of the cortex of the postcentral gyrus of the monkey: a correlation of findings obtained in a single unit analysis with cytoarchitecture. *Bull. Johns Hopkins Hosp.* 105:133–62

Preuss TM, Goldman-Rakic PS. 1989. Connections of the ventral granular frontal cortex of macaques with perisylvian premotor and somatosensory areas: anatomical evidence for somatic representation in primate frontal association cortex. *J. Comp. Neurol.* 282:293–316

Recanzone GH, Merzenich MM, Schreiner CE. 1992. Changes in the distributed temporal response properties of SI cortical neurons reflect improvements in performance on a temporally based tactile discrimination task. *J. Neurophysiol.* 67:1071–91

Reynolds J, Desimone R. 1999. Competitive mechanisms subserve attention in macaque areas V2 and V4. *J. Neurosci.* 19:1736–53

Romo R, Brody CD, Hernández A, Lemus L. 1999. Neuronal correlates of parametric working memory in the prefrontal cortex. *Nature* 339:470–73

Romo R, Hernández A, Zainos A, Brody CD, Lemus L. 2000. Sensing without touching: psychophysical performance based on cortical microstimulation. *Neuron* 26:273–78

Romo R, Hernández A, Zainos A, Salinas E. 1998. Somatosensory discrimination based on cortical microstimulation. *Nature* 392:387–90

Romo R, Merchant H, Ruiz S, Crespo P, Zainos A. 1995. Neuronal activity of primate putamen during categorical perception of somaesthetic stimuli. *NeuroReport* 6:1013–17

Romo R, Merchant H, Zainos A, Hernández A. 1996. Categorization of somaesthetic stimuli: sensorimotor performance and neuronal activity in primary somatic sensory cortex of awake monkeys. *NeuroReport* 7:1273–79

Romo R, Merchant H, Zainos A, Hernández A. 1997. Categorical perception of somesthetic stimuli: psychophysical measurements correlated with neuronal events in primate medial premotor cortex. *Cereb. Cortex* 7:317–26

Romo R, Ruiz S, Crespo P, Zainos A, Merchant H. 1993. Representation of tactile signals in primate supplementary motor area. *J. Neurophysiol.* 70:2690–94

Romo R, Scarnati E, Schultz W. 1992. Role of primate basal ganglia and frontal cortex in the internal generation of movements. II. Movement related activity in the striatum. *Exp. Brain Res.* 91:385–95

Romo R, Schultz W. 1987. Neuronal activity preceding self-initiated or externally timed arm movements in area 6 of monkey cortex. *Exp. Brain Res.* 67:656–62

Romo R, Schultz W. 1992. Role of primate basal ganglia and frontal cortex in the internal generation of movements. III. Neuronal activity in the supplementary motor area. *Exp. Brain Res.* 91:396–407

Ruiz S, Crespo P, Romo R. 1995. Representation of moving tactile stimuli in the somatic cortex of awake monkeys. *J. Neurophysiol.* 73:525–37

Salinas E, Hernández H, Zainos A, Romo R. 2000a. Periodicity and firing rate as candidate neural codes for the frequency of vibrotactile stimuli. *J. Neurosci.* 20:5503–15

Salinas E, Opris I, Zainos A, Hernández A, Romo R. 2000b. Motor and non-motor roles of the cortico-basal ganglia circuitry. In *Brain Dynamics and the Striatal Complex*, ed. R Miller, JR Wickens, pp. 237–55. Amsterdam: Harwood Acad.

Salinas E, Romo R. 1998a. Conversion of sensory signals into motor commands in primary motor cortex. *J. Neurosci.* 18:499–511

Salinas E, Romo R. 1998b. Neuronal representations in categorization task: sensory to motor transformation. In *Computational*

Neuroscience: Trends in Research 98, ed. J Bower, pp. 599–604. New York: Plenum

Salinas E, Sejnowski TJ. 2000. Impact of correlated synaptic input on output firing rate and variability in simple neuronal models. *J. Neurosci.* In press

Salzman CD, Murasugi CM, Britten KH, Newsome WT. 1992. Microstimulation in visual area MT: effects on direction discrimination performance. *J. Neurosci.* 12:2331–55

Salzman D, Britten K, Newsome WT. 1990. Cortical microstimulation influences perceptual judgements of motion direction. *Nature* 346:174–77

Schall JD, Thompson KG. 1999. Neural selection and control of visually guided eye movements. *Annu. Rev. Neurosci.* 22:241–59

Schultz W, Romo R. 1992. Role of primate basal ganglia and frontal cortex in the internal generation of movements. I. Preparatory activity in the anterior striatum. *Exp. Brain Res.* 91:363–84

Schwartz A, Kettner RE, Georgopoulos AP. 1988. Primate motor cortex and free arm movements to visual targets in three-dimensional space. I. Relations between single cell discharge and direction of movement. *J. Neurosci.* 8:2913–27

Seidemann E, Zohary U, Newsome WT. 1999. Temporal gating of neural signals during performance of a visual discrimination task. *Nature* 394:72–75

Shadlen MN, Britten KH, Newsome WT, Movshon JA. 1996. A computational analysis of the relationship between neuronal and behavioral responses to visual motion. *J. Neurosci.* 16:1486–510

Shadlen MN, Newsome WT. 1994. Noise, neural codes and cortical organization. *Curr. Opin. Neurobiol.* 4:569–79

Shadlen MN, Newsome WT. 1996. Motion perception: seeing and deciding. *Proc. Natl. Acad. Sci. USA* 93:628–33

Shanks MF, Person RC, Powell TPS. 1985. The ipsilateral cortico-cortical connexions between the cytoarchitectonic subdivisions of the primary somatic sensory cortex in the monkey. *Brain Res.* 356:67–88

Shen L, Alexander GE. 1997. Neural correlates of a spatial sensory-to-motor transformation in primary motor cortex. *J. Neurophysiol.* 77:1171–94

Sinclair RJ, Burton H. 1993. Neuronal activity in the second somatosensory cortex of monkeys (*Macaca mulatta*) during active touch of gratings. *J. Neurophysiol.* 70:331–50

Singer W, Gray CM. 1995. Visual feature integration and the temporal correlation hypothesis. *Annu. Rev. Neurosci.* 18:555–86

Steinmetz PN, Roy A, Fitzgerald PJ, Hsiao SS, Johnson KO, Niebur E. 2000. Attention modulates synchronized neuronal firing in primate somatosensory cortex. *Nature* 404:187–90

Sur M, Wall JT, Kaas JH. 1984. Modular distribution of neurons with slowly adapting and rapidly adapting responses in area 3b of somatosensory cortex in monkeys. *J. Neurophysiol.* 51:724–44

Talbot WH, Darian-Smith I, Kornhuber HH, Mountcastle VB. 1968. The sense of flutter-vibration: comparison of the human capacity response patterns of mechanoreceptive afferents from the monkey hand. *J. Neurophysiol.* 31:301–34

Tanji J, Kurata K. 1985. Contrasting neuronal activity in supplementary motor area and precentral motor cortex of monkeys. I. Responses to instructions determining motor responses to forthcoming signals of different modalities. *J. Neurophysiol.* 53:129–41

Tokuno H, Tanji J. 1993. Input organization of distal and proximal forelimb areas in the monkey primary motor cortex: retrograde double labeling study. *J. Comp. Neurol.* 333:199–209

Treue S, Martínez-Trujillo JC. 1999. Feature-based attention influences motion processing gain in macaque visual cortex. *Nature* 399:575–79

Ungerleider LG, Mishkin M. 1982. The two cortical visual systems. In *Analysis of Visual Behavior*, ed. DJ Ingle, MA Goodale,

RJW Mansfield, pp. 549–86. Cambridge, MA: MIT Press

Vallbo AB. 1995. Single-afferent neurons and somatic sensation in humans. In *The Cognitive Neurosciences*, ed. MS Gazzaniga, pp. 237–52. Cambridge, MA: MIT Press

Vallbo AB, Johansson RS. 1984. Properties of cutaneous mechanoreceptors in the human hand related to touch sensations. *Hum. Neurobiol.* 3:3–14

Vogels R, Orban GA. 1990. How well do response changes of striate neurons signal differences in orientation: a study in the discriminating monkey. *J. Neurosci.* 10:3543–58

Werner G. 1980. The study of sensation in physiology. In *Medical Physiology*, ed. VB Mountcastle, 1:605–28. St. Louis, MO: Mosby

Werner G, Mountcastle VB. 1965. Neural activity in mechanoreceptive cutaneous afferents: stimulus-response relations, Weber functions, and information transmission. *J. Neurophysiol.* 28:359–97

Whang KC, Burton H, Shulman GL. 1991. Selective attention in vobrotactile tasks: detecting the presence and absence of amplitude change. *Percept. Psychophysiol.* 50:157–65

Wickersham I, Groh JM. 1998. Electrically evoking sensory experience. *Curr. Biol.* 8:R412–14

Zainos A, Merchant H, Hernández A, Salinas E, Romo R. 1997. Role of primary somatic sensory cortex in the categorization of tactile stimuli: effects of lesions. *Exp. Brain Res.* 115:357–60

Zhang J, Riehle A, Requin J, Kornblum S. 1997. Dynamics of single neuron activity in monkey primary motor cortex related to sensorimotor transformation. *J. Neurosci.* 17:2227–46

Zhou YD, Fuster JM. 1996. Mnemonic neuronal activity in somatosensory cortex. *Proc. Natl. Acad. Sci. USA* 93:10533–37

Zohary E, Shadlen MN, Newsome WT. 1994. Correlated neuronal discharge rate and its implications for psychophysical performance. *Nature* 370:140–43

Annu. Rev. Neurosci. 2001. 24:139–66

SYNAPTIC MODIFICATION BY CORRELATED ACTIVITY: Hebb's Postulate Revisited

Guo-qiang Bi[1] and Mu-ming Poo

Department of Molecular & Cell Biology, University of California at Berkeley, Berkeley, CA 94720-3200; email: mpoo@uclink4.berkeley.edu

Key Words   Hebbian synapse, LTP, LTD, spike timing, input specificity

■ **Abstract**   Correlated spiking of pre- and postsynaptic neurons can result in strengthening or weakening of synapses, depending on the temporal order of spiking. Recent findings indicate that there are narrow and cell type–specific temporal windows for such synaptic modification and that the generally accepted input- (or synapse-) specific rule for modification appears not to be strictly adhered to. Spike timing–dependent modifications, together with selective spread of synaptic changes, provide a set of cellular mechanisms that are likely to be important for the development and functioning of neural networks.

> When an axon of cell A is near enough to excite cell B or repeatedly or consistently takes part in firing it, some growth or metabolic change takes place in one or both cells such that A's efficiency, as one of the cells firing B, is increased.
>
> Donald Hebb (1949)

INTRODUCTION

Half a century since the publication of his famous treatise (Hebb 1949), Hebb's postulate of synaptic modification by correlated activity has become a cornerstone in our understanding of activity-dependent neural development and the cellular basis of learning and memory. This postulate was originally proposed by Hebb as a mechanism for the growth of "cell assembly," a hypothetical group of neurons that act briefly as a closed system after stimulation has ceased and that serve for the first stage of perception. Over the past several decades, Hebb's idea has been extended into various forms of correlation-based rules for synaptic modification and successfully used in many learning networks and in the analysis of activity-driven refinement of developing circuits (Stent 1973, Sejnowski & Tesauro 1989, Brown et al 1990, Fregnac & Bienenstock 1998, Sejnowski 1999). Here, we review

[1]Present address: Department of Neurobiology, University of Pittsburgh School of Medicine, Pittsburgh, PA 15261; e-mail: gqbi@pitt.edu

0147-006X/01/0621-0139$14.00

139

recent findings that shed new light on two aspects of Hebb's postulate: (*a*) the temporal specificity in the correlated activity required for the induction of synaptic modification—the importance of temporal order in the pre- and postsynaptic spiking; and (*b*) the spatial specificity in the induced synaptic changes—the notion that only the activated synapse becomes modified. Implications of these findings for the development and functioning of the nervous system are also addressed.

TEMPORAL SPECIFICITY IN ACTIVITY-INDUCED SYNAPTIC MODIFICATION

A central feature of Hebb's postulate is temporal specificity: The synaptic connection is strengthened only if cell A "takes part in firing" cell B, i.e. cell A fires before cell B. Such temporal specificity of activity-induced synaptic modification may be relevant for physiological functions, such as learning and memory, which are known to be temporally specific. Early studies have demonstrated a requirement for temporal contiguity in associative synaptic modification in rat hippocampus (Levy & Steward 1983) and *Aplysia* ganglia (Hawkins et al 1983, Walters & Byrne 1983). Recent experiments have revealed the importance of the temporal order of pre- and postsynaptic spiking for synaptic modification and have further defined the "critical windows" of spike timing, with precision on the order of milliseconds (Markram et al 1997, Magee & Johnston 1997, Bell et al 1997, Debanne et al 1998, Zhang et al 1998, Bi & Poo 1998, Egger et al 1999). The precise profile of critical windows appears to depend on the synapse type, and the underlying molecular mechanisms remain to be fully understood.

Temporal Requirements for Associative Synaptic Modification

Since the first description of long-term potentiation (LTP) in rabbit hippocampal formation (Bliss & Lømo 1973, Bliss & Gardner-Medwin 1973), similar synaptic potentiation has been found in many areas of the central and peripheral nervous systems of both vertebrates and invertebrates (Teyler & DiScenna 1987, Brown et al 1990, Bliss & Collingridge 1993, Buonomano & Merzenich 1998, Milner et al 1998, Malenka & Nicoll 1999). Certain forms of LTP in hippocampus have been shown to be cooperative and associative, properties relating directly to the theoretical construct of the Hebbian synapse. Cooperativity results from the existence of an intensity threshold for inducing LTP by tetanic stimulation (Bliss & Lømo 1973, McNaughton et al 1978), whereas associativity usually refers to the induction of LTP in a "weak" input when it is coactive with a "strong" convergent input (Levy & Steward 1979, Barrionuevo & Brown 1983). In a sense, cooperatively induced LTP is a special case of associative LTP: both originate from the requirement in the synchrony of inputs for postsynaptic activation (Teyler & DiScenna 1987).

Levy & Steward (1983) studied in more detail the temporal specificity in associative synaptic modification. By stimulating a weak and a strong input from the entorhinal cortex to the denate gyrus of hippocampus, they found that associative induction of LTP did not require perfectly synchronous activation of the two pathways. Instead, the temporal order of the activation was crucial. LTP of the weak input could be induced when the strong input was activated concurrently with, or following the activation of, the weak input by as much as 20 ms. When the temporal order was reversed, long-term depression (LTD) was induced. This and other early studies (Kelso & Brown 1986, Gustafsson & Wigström 1986) have clearly indicated the existence of a stringent temporal specificity in the activity-induced synaptic modification.

In the marine mollusk *Aplysia*, temporally specific synaptic modifications have been shown to account for classical conditioning in several reflex systems (Carew et al 1981, Hawkins et al 1983, Walters & Byrne 1983). For example, in the siphon-withdrawal reflex, paired stimulation of the siphon sensory neuron [conditioned stimulus (CS)] and the tail nerve [unconditioned stimulus (US)] produces greater associative synaptic facilitation at sensorimotor connections when the CS precedes the US than if the temporal order is reversed (Hawkins et al 1983, Clark et al 1994). Such synaptic enhancement was considered non-Hebbian because the underlying cellular mechanism appeared to be purely presynaptic (Carew et al 1984). More recent studies, however, have shown postsynaptic involvement in associative synaptic facilitation of the same *Aplysia* system (Lin & Glanzman 1994, Murphy & Glanzman 1997). Temporally specific synaptic modification in invertebrates may thus be mediated by both presynaptic non-Hebbian and postsynaptic Hebbian mechanisms (Lechner & Byrne 1998).

The Role of Postsynaptic Spiking

Low-frequency presynaptic stimulation coupled with concurrent postsynaptic depolarization (the "pairing protocol") can induce LTP in the hippocampus and different cortical areas (Kelso et al 1986, Sastry et al 1986, Wigström et al 1986, Malenka & Nicoll 1999), demonstrating the importance of coincident activity in synaptic modification, as suggested by Hebb. However, this pairing protocol, with postsynaptic depolarization on the order of seconds to minutes, does not address the precise temporal specificity in the induction of LTP/LTD. Under natural conditions, postsynaptic neurons fire action potentials as their normal functional output, but whether these spikes are involved in inducing synaptic changes has been debated (for review, see Linden 1999). Active properties of dendrites (Regehr et al 1992, Lasser-Ross & Ross 1992, Johnston et al 1996) allow the spike initiated at the axon hillock to back-propagate into dendrites (Stuart & Sakmann 1994, Buzsáki et al 1996, Hoffman et al 1997). The back-propagating spike provides a precise signal capable of informing the synapses whether and precisely when the postsynaptic cell has fired, thus may play an active role in associative synaptic modification. The first definitive support for this notion came from recent studies in cortical and hippocampal slices (Markram et al 1997, Magee & Johnston 1997). Using dual whole-cell recording from two interconnected layer 5 pyramidal neurons, Markram et al (1997) found that when the spikes were triggered 10 ms after the onset of excitatory postsynaptic potentials (EPSPs), LTP was induced by repetitive pairing of postsynaptic spiking (induced by current injection) with the EPSP, whereas when the temporal order of the spikes and EPSPs was reversed, LTD was induced. Magee & Johnston (1997) found that back- propagating spikes in CA1 pyramidal neurons of hippocampus, when coupled immediately after the onset of subthreshold EPSPs, evoked a significantly higher Ca^{2+} influx at the synaptic site, resulting in LTP. In both studies, neither EPSPs nor postsynaptic spiking alone was sufficient to induce any synaptic modification. Thus, back-propagating postsynaptic spikes indeed can function as an associative signal for synaptic modification, and relative timing of pre- and postsynaptic activity is critical. Similar results have also been obtained in slice cultures (Debanne et al 1998), in developing *Xenopus* retinotectal projections in vivo (Zhang et al 1998), and in cultured hippocampal neurons (Bi & Poo 1998).

Critical Windows of Spike Timing for Synaptic Modification

In cell culture, a critical window (Figure 1) for the induction of LTP/LTD has been characterized by systematically varying the spike timing (defined as the time interval between the onset of EPSP and postsynaptic spike) during repetitive correlated stimulation at a low frequency of 1 Hz (Bi & Poo 1998). The window for modification is about 40 ms in width and is temporally asymmetric. Postsynaptic spiking within about 20 ms after presynaptic activation (positive intervals) results in LTP, whereas that within about 20 ms before presynaptic activation (negative intervals) results in LTD. The transition from LTD to LTP occurs within a few milliseconds of change in spike timing. It is interesting that an essentially identical window has been found in developing retinotectal system of *Xenopus* tadpoles (Zhang et al 1998). At Schaffer collateral synapses in the CA1 region of rat hippocampal slices, LTP/LTD can also be induced by similar correlated pre- and postsynaptic activation at a low frequency (5 Hz), with a characteristic critical window of spike timing that differs slightly from that found in culture. An additional LTD window was observed at positive intervals of 15–20 ms (Nishiyama et al 2000).

In hippocampal cultures, GABAergic transmission can also be modified by correlated pre- and postsynaptic spiking. Such modification exhibits a symmetric spiking-timing window that lasts about 50 ms and is independent of N-methyl-D-aspartate receptors (NMDARs) (K Ganguly, M-m Poo, unpublished results). In the cerebellum-like structure of electric fish, synapses formed by parallel fiber onto Purkinje-like cells have a window opposite to that associated with the excitatory synapses described above: Spikes of positive intervals induce LTD and those of negative intervals induce LTP (Bell et al 1997). This opposite polarity in temporal specificity may be appropriate for the function this particular circuit, given that the Purkinje cell is an inhibitory projection neuron. In somatosensory cortex, excitatory neurons at different layers have markedly different windows. For synapses between layer 5 pyramidal neurons in somatosensory cortex, the

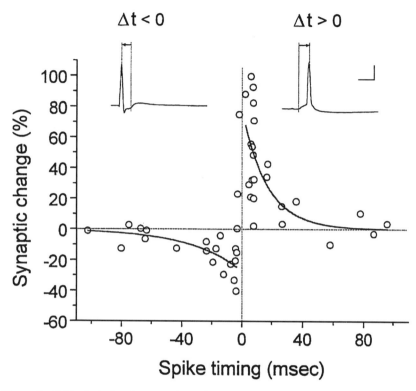

Figure 1 Critical window for synaptic modifications. Long-term potentiation (LTP)/long-term depression (LTD) were induced by correlated pre- and postsynaptic spiking at synapses between hippocampal glutamatergic neurons in culture. The percentage change in the excitatory postsynaptic current (EPSC) amplitude at 20–30 min after repetitive correlated spiking (pulses at 1 Hz) was plotted against spike timing, which is defined as the time interval (Δt) between the onset of the EPSP and the peak of the postsynaptic action potential during each pair of correlated spiking, as illustrated by the traces above. LTP and LTD windows are each fitted with an exponential function: $\Delta W = A \times e^{(-\Delta t/\tau)}$. For LTP and LTD, respectively, $A = 0.777$ and -0.273; $\tau = 16.8$ and -33.7 ms. Note that ΔW represents the total amount of change in synaptic strength after 60 pairs of correlated spiking. Assuming effective unitary change Δw (due to a single pair) has similar time constants as those of ΔW, i.e. $\Delta w = a \times e^{(\Delta t/\tau)}$, thus $(1 + A) = (1 + a)^{60}$, yielding $a = 0.0096$ and -0.0053 and the total areas under the unitary LTP and LTD curves $a \times \tau = 0.16$ and 0.18 for LTP and LTD, respectively. Scales: 50 mV, 10 ms. [Data from Bi & Poo (1998).]

window is presumably of similar profile as that in Figure 1 (Markram et al 1997). However, synapses between layer 4 spiny stellate neurons appear to have a symmetric depression window (Egger et al 1999). Various spike-timing windows for synaptic modification reflect the diversity of synaptic machinery and may serve for specific functions of information processing at different stages of neural pathways.

Mechanisms for the Detection of Coincidence and Temporal Order

Previous studies on cellular mechanisms of LTP have addressed the role of the coincident pre- and postsynaptic activity in the induction of LTP (Bourne & Nicoll 1993). In some brain regions, including the CA1 area of hippocampus, synaptic activation of postsynaptic NMDAR, a subtype of glutamate receptor channel, concurrently with membrane depolarization underlies the coincidence requirement for LTP (for reviews, see Collingridge et al 1983, Bourne & Nicoll 1993). The depolarization-dependent removal of the Mg^{2+} block of the NMDAR (Mayer et al 1984, Nowak et al 1984, Jahr & Stevens 1987) allows the channel to serve as a molecular detector for the coincidence of presynaptic activation (glutamate release) and postsynaptic depolarization. The resultant Ca^{2+} influx through the NMDAR triggers downstream kinases that lead to synaptic potentiation. Induction of LTD also requires coincident pre- and postsynaptic activation and, in some cases, NMDAR activation, although lower-level postsynaptic depolarization is needed (Linden & Connor 1995, Bear & Abraham 1996, Goda & Stevens 1996). It is now generally believed that postsynaptic Ca^{2+} elevation is crucial for both LTP and LTD: Transient high-level Ca^{2+} elevation may lead to activation of certain protein kinases and LTP, whereas sustained low-level Ca^{2+} elevation may activate phosphatases and result in LTD (Lisman 1989, Malenka et al 1989, Malinow et al 1989, Silva et al 1992, Mulkey et al 1994, Yang et al 1999).

In principle, coincidence detection can occur at any converging point of signaling pathways where integration is nonlinear, as is the case for many biological processes. Thus, many components of signal transduction pathways may function as molecular coincidence detectors (Bourne & Nicoll 1993). In *Aplysia* sensorimotor synapse, presynaptic adenylyl cyclase that is modulated by G-protein–coupled serotinin receptors and Ca^{2+}/calmodulin detects the coincidence of Ca^{2+} elevation (due to sensory neuron activity, or CS) and serotonin release (from the facilitating interneuron due to US). The resultant rise in cAMP is at least partly responsible for the associative synaptic facilitation at this synapse (Abrams et al 1991, Yovell et al 1992, Lechner & Byrne 1998). In vertebrate brain, adenylyl cyclase is inhibited via G-protein by group II metobotropic receptors (mGluRs), which are known to be involved in several forms of LTP/LTD (Tzounopoulos et al 1998, Egger et al 1999, Yeckel et al 1999, Bortolotto et al 1999). Interaction between group II mGluR pathway and depolarization-induced Ca^{2+} signaling pathway is implicated in the coincidence detection at these forms of synaptic changes. Similarly, LTD in cerebellum and other forms of LTP/LTD in hippocampus require group I mGluRs that initiate Ca^{2+} release from intracellular stores via the activating phospholipase C (Ito et al 1982, Linden & Connor 1995, Bear & Abraham 1996, Bortolotto et al 1999). A synergistic release of Ca^{2+} from inositol-1,4,5-triphosphate (IP3)-sensitive stores, evoked by synaptic activation of mGluRs, paired with back-propagating action potentials has been observed in the apical dendrites of CA1 pyramidal neurons (Nakamura et al 1999).

What cellular mechanisms are responsible for detecting the temporal order of pre- and postsynaptic spiking? Both LTP and LTD induced by correlated spiking depend on the activation of NMDARs (Markram et al 1997, Magee & Johnston 1997, Debanne et al 1998, Zhang et al 1998, Bi & Poo 1998), and LTD also requires functional L-type calcium channels (Bi & Poo 1998). These requirements are similar to LTP and LTD induced by the conventional tetanic stimulation or pairing protocols. For correlated spiking with positive intervals, spiking following synaptic inputs apparently helps to open NMDARs, allowing high-level Ca^{2+} influx, thus LTP. With negative intervals, spiking allows low-level Ca^{2+} influx through voltage-gated Ca^{2+} channels, which was followed immediately by additional low-level Ca^{2+} influx due to NMDARs, resulting in LTD. Indeed, imaging studies have shown that Ca^{2+} elevation was supralinear when presynaptic stimulation immediately preceded the spike and was sublinear when the temporal order was reversed (Koester & Sakmann 1998). Because the time constant of glutamate binding to NMDARs is much longer than 20 ms (Mayer et al 1984, Hollmann & Heinemann 1994), the narrow window observed in different systems must be accounted for by additional mechanisms, e.g. nonlinearity in the activation of downstream effectors (e.g. calmodulin), or EPSP-induced, dendritic A-type K^+ channel inactivation (time constant $\sim$10 ms) that permits synaptic invasion of back-propagating spikes (Hoffman et al 1997). Additional LTD at positive intervals found in hippocampal slices (Nishiyama et al 2000) may be due to differences in NMDAR properties, and K^+ channel activity, as well as the presence of local inhibitory inputs.

IMPLICATIONS OF TEMPORAL SPECIFICITY

The observation of stringent spike-timing windows highlights the potential role of individual spikes in synaptic modification (Linden 1999, Sejnowski 1999). The capability of synapses to detect precise timing of individual spikes is consonant with the idea that neural information may be encoded in spike timing (Perkel & Bullock 1968, Rieke et al 1997). Temporally asymmetric windows, such as that shown in Figure 1, appear to be the most predominant form in different systems (see Table 1). These windows provide a basis for the formulation of a spike-based, temporally asymmetric Hebbian learning rule (Gerstner et al 1996, Gerstner & Abbott 1997, Kempter et al 1999, Abbott & Song 1999, Rao & Sejnowski 2000, Mehta et al 2000, Senn et al 2000, Paulsen & Sejnowski 2000). Here, we focus on several properties of this new learning rule, namely the spike- rather than rate-based computation, self-normalization, and temporal asymmetry. Its functional implications for classical conditioning and temporal-to-spatial conversion of information in neural networks are also discussed.

TABLE 1   Spike timing (Δt) for the induction of LTP/LTD in different systems[a]

System	Δt LTD (ms)	Δt LTP (ms)	References
Hippocampal dentate gyrus	<0[b]	0 ~ +20	Levy & Steward 1983
Hippocampal slice CA1	None[c]	0 ~ +40 0 ~ +15	Gustafsson & Wigström 1986, Gustafsson et al 1987, Magee & Johnston 1997
	−30 ~ 10; +15 ~ +20	0 ~ +15	Nishiyama et al 2000
Cortical slice layer 5	−10[d]	+10[d]	Markram et al 1997
Cerebellum-like structure	0 ~ +50	−50 ~ 0	Bell et al 1997
Hippocampal slice culture	−100 ~ −15[d]	+15[d]	Debanne et al 1998
Retino-tectal projection	−20 ~ 0	0 ~ +20	Zhang et al 1998
Dissociated cell culture	−30 ~ 0	0 ~ +20	Bi & Poo 1998
Cortical slice layer 4	−10 ~ +10	None[e]	Egger et al 1999

[a]LTP, long-term potentiation; LTD, long-term depression.

[b]Heterosynaptic depression may contribute to LTD observed in this study.

[c]LTD was not observed or tested in the experiments.

[d]Single time points were tested for LTP/LTD in the experiments.

[e]No significant LTP was observed in the experiment.

Spike-Based, Temporally Asymmetric Learning Rule

A "Microscopic" Learning Rule   One distinct feature of the spike-based learning rule is that, in the computation of synaptic modification, it emphasizes the precise timing of each individual spike rather than the average rate of a population of spikes over a defined interval. Thus, this rule can be considered "microscopic," as opposed to earlier statistical or "macroscopic" rate-based formulations, e.g. the covariance and the Bienenstock-Cooper-Munro (BCM) rules (Sejnowski 1977b, Bienenstock et al 1982), in both of which only the average spike rate is important. Using the spike-based rule and assuming particular firing characteristics of pre- and postsynaptic neurons, one can derive different statistical rules (Kempter et al 1999, Senn et al 2000) and readily incorporate them into more conventional rate-based models. An exciting application of spike-based learning rules is in a new type of neural network models, the "pulsed neural networks" (Maass & Bishop 1999), that use precise timing of individual spikes to encode information. Spike-timing or temporal coding may enable construction of more versatile and powerful networks because it provides larger coding capacity and easier handling of temporal information. However, whether the brain actually uses spike timing for coding information has been a long-standing issue in neuroscience (Perkel & Bullock 1968, Rieke et al 1997, Singer 1999). The critical question is, timewise, how precisely can a neuron (or a group of neurons) detect synaptic inputs and generate as well as transmit spikes? The issues of precision and reliability of synaptic transmission, spike initiation, and propagation have been addressed by many theoretical and experimental studies (Calvin & Stevens 1968, Bullock 1970, Bekkers et al 1990, Mainen & Sejnowski 1995, Marsálek et al 1997, Stevens & Zador 1998, Diesmann et al 1999). It is interesting that theoretical analysis suggests that in multilayer networks, synchronous spiking activity may propagate from one ensemble to the next while maintaining high precision of spike timing (Marsálek et al 1997, Diesmann et al 1999). Such "mode-locked" spike propagation in the brain is consistent with the highly synchronous activity found in many areas of the brain (Abeles et al 1993, deCharms & Merzenich 1996, Riehle et al 1997, Roelfsema et al 1997, Dan et al 1998, Singer 1999). The existence of narrow windows of spike timing for synaptic modification indicates that synapses can indeed "read" information coded in the timing of pre- and postsynaptic spikes, with a precision on the order of a few milliseconds. Therefore, spike timing–based learning rules and the pulsed network models may provide powerful tools as well as insights into computational principles of the nervous system.

A "Self-Normalizing" Rule   Hebb's original postulate described only activity-dependent strengthening of a synapse. However, a complete learning rule must include both strengthening (LTP) and weakening (LTD) of synapses in order to avoid saturation, and weakening must be as specific as strengthening (rather than nonspecific decay) for serving long-term memories (Sejnowski 1977a, Stevens 1996). The new asymmetric Hebbian rule provides an immediate mechanism for self-normalization of synaptic weights and the output firing rate of a network (Kempter et al 1999, Abbott & Song 1999, Senn et al 2000). Under normal conditions, such a balance is achieved if, in the spike-timing window of unitary modification (the effect of a single pairing event), the total integral of the LTD curve is equal to or slightly larger than that of the LTP curve. The data shown in Figure 1 indeed largely satisfy this requirement of self-normalization. In some synapses, the spike-timing windows for correlated spiking-induced modification appear not self-normalized (Egger et al 1999). It is possible that there is a low-level (undetectable) timing-independent LTD/LTP or activity-independent LTD/LTP that serves for nonspecific normalization of these synapses.

A "Temporally Asymmetric" Rule   Perhaps the most attractive feature of the new learning rule is its sharp asymmetry in the window of spike timing. Association between two events of a specific temporal order can be established by asymmetric synaptic modification. This is fundamentally different from most formulations of Hebbian rules, in which only coincidence of activity (or associated between two concomitant events) was considered (Brown et al 1990). Hebb's original statement was remarkably accurate in implying the role of temporal order: A synaptic input is strengthened only when it "takes part in firing" the postsynaptic cell. Thus, a "true" Hebbian synapse behaves as a causality detector rather than a coincidence detector. Networks of neurons with such synapses can learn and predict sequences (Minai & Levy 1993, Abbott & Blum 1996, Roberts 1999) and form navigational maps that have been observed in hippocampus (Mehta et al 1997, Gerstner & Abbott 1997, Mehta et al 2000). More generally, the predictive nature of the temporally asymmetric learning rule is directly related to the implementation of classical conditioning (see below).

Classical Conditioning

A hallmark of classical conditioning is the requirement of temporal contiguity: the conditioned stimulus (CS) must precede the unconditioned stimulus (US) during conditioning in order to elicit conditioned response. This ability to detect and remember causal relationships between events, to predict the about-to-happen, and to respond in advance are obviously advantageous to all animal species. Classical conditioning has been modeled successfully in the past by temporal-difference learning algorithm in reinforcement learning (Sutton & Barto 1981, Tesauro 1986, Montague & Sejnowski 1994, Moore et al 1998). It has been shown recently that the asymmetric Hebbian rule is mathematically equivalent to the temporal-difference learning algorithm (Roberts 1999, Rao & Sejnowski 2000). Therefore, the asymmetric synaptic modification by correlated spiking provides a natural cellular mechanism of temporal-difference learning at the synaptic level.

Classical conditioning normally operates over a timescale on the order of seconds. In *Aplysia* siphon-withdrawal reflex, temporally specific synaptic enhancement at sensorimotor connections can be induced when stimulation of the tail (US) follows stimulation of the siphon sensory neuron (CS) within seconds (Hawkins et al 1983, Walters & Byrne 1983, Clark et al 1994). In this case, cellular process with slow kinetics at the presynaptic terminal—the interaction between Ca^{2+} and adenylyl cyclase—provides a detection mechanism for a temporal order of sensory events (Yovell & Abrams 1992). In vertebrate brain, processes such as synergistic release of postsynaptic Ca^{2+} stores by mGluR activation paired with back-propagating spikes (Nakamura et al 1999) may also provide temporally specific detection with a longer timescale than that mediated by NMDARs. Does correlated-spiking–induced synaptic modification found for hippocampal and cortical synapses, with the spike-timing window lasting tens of milliseconds, play a role in more elaborated forms of classical conditioning in vertebrates known to have temporal windows that last for seconds (Squire 1987, Thompson & Krupa 1994)? Many different areas are likely to be involved in classical conditioning within the vertebrate brain, with polysynaptic and recurrent circuitry that can effectively extend the timescale of computation. For example, polysynaptic pathways may introduce long transmission delays so that a "remote" synapse may detect and become modified by two temporally distant stimuli that arrive at the site of modification within a narrow window of spike timing (Bi & Poo 1999).

Temporal-to-Spatial Conversion: Delay-Line Mechanism

An essential process in learning and memory is the transformation of temporal information into spatially distributed information in the brain. Temporally varying sensory information, e.g. that associated with vision and audition, are processed and stored by the brain with remarkable precision (Carr 1993, Singer & Gray 1995). The essential temporal information may be coded in the timing of transmitted spike trains (Rieke et al 1997, Singer 1999). In addition, inputs that are not explicitly temporal, e.g. odorant stimuli, may also be coded in spike timing (Hopfield 1995, Laurent 1997). Relative timing of spikes among neurons within functional ensembles may be used directly to represent internal states of functioning circuits (Riehle et al 1997). For synaptic modification to be a mechanism for long-term storage of memory, such information coded in spike timing must be converted into and stored as spatially distributed synaptic modifications in the brain. This requires that each synapse be sensitive to spike timing, as discussed above, and that a mechanism "assign" the temporal information specifically to different synaptic locations. One strategy for accomplishing this conversion relies on intrinsic synaptic properties (Buonomano & Merzenich 1995), such as short-term plasticity and slow inhibitory currents that provide differential neuronal sensitivity to different spikes in an input train (Buonomano et al 1997). Another strategy is to take advantage of network architectures of "delay-lines" (Jeffress 1948, Braitenberg 1967, Tank & Hopfield 1987) that have been demonstrated in several systems where the delay is due to axonal conduction of action potentials (Carr 1993). Recently, we have shown that in networks of cultured neurons, polysynaptic transmission pathways may form functional delay-lines (Bi & Poo 1999). Repetitive paired-pulse stimulation of one neuron results in correlated spiking and LTP/LTD at remote synaptic sites, when the interpulse interval matches the delay difference between convergent pathways leading to the remote sites (Bi & Poo 1999). Because neurons in such cultures tend to hyper-innervate each other, these polysynaptic pathways in culture may be analogous to chains of synchronously firing cell ensembles in vivo that have been proposed for cortical processing (Abeles 1991). With such delay-lines and spike-timing dependent LTP/LTD, information coded in the spike train may be stored at selective connections between cell ensembles in the form of long-term synaptic modification.

Activity-Driven Refinement of Developing Networks

Since the early works of Hubel and Wiesel on the effects of monocular deprivation (Wiesel & Hubel 1963) and artificial strabismus (Hubel & Wiesel 1965) on the developing visual system, it has been generally recognized that neuronal activity is critical for refining the connections in the brain (for reviews, see Katz & Shatz 1996, Constantine-Paton et al 1990). There is also evidence that activity may provide an instructive (rather than permissive) role. For example, ocular dominance columns in the cortex were abolished if artificially synchronized activity was imposed to both eyes, whereas imposing asychonous activity between the eyes of the same frequency was not effective (Stryker & Strickland 1984). Development of orientation selectivity and responsiveness of visual cortical neurons can also be reduced by artificial stimulation of optic nerves (Weliky & Katz 1997). Although patterned visual experience may not be necessary for the initial development of either ocular dominance or orientation maps (Horton & Hocking 1996, Crair et al 1998, Crowley & Katz 1999), the endogenous patterns of spontaneous activity (Wong et al 1995, Penn et al 1998), in the retina as well as in higher visual areas, could still play a role.

In addressing the cellular mechanism by which asynchronous activity between the two eyes drives the formation of ocular dominance columns, Stent (1973) proposed an extension of Hebb's postulate: Synchronous activity strengthens the synapse, whereas asynchronous activity weakens the synapse. This simple rule of "neurons that fire together wire together" can explain activity-dependent competitive interactions between converging inputs on the same postsynaptic neurons. Synchronous spiking is more likely to occur for inputs from the same eye; thus, there is eye-specific segregation of inputs in lateral geniculate nucleus and in the visual cortex. Correlation of spiking is stronger for inputs from neighboring neurons in the same retina; thus, a crude retinotopic map (presumably formed initially by chemical cues) can be refined into a more fine-grained map (Constantine-Paton et al 1990). Simple synchrony- or correlation-based rules have also been successfully used in modeling activity-driven development of simple receptive fields and the ordered arrangement of ocular dominance and orientation columns (Miller et al 1989, Miller 1994), as well as ocularly matched orientation and ocular dominance maps (Erwin & Miller 1998).

The finding of spike-timing–dependent modifications raises several questions. Is a simple correlation-based Hebbian rule a realistic or complete description of activity-driven processes? Are there activity-dependent processes other than the modification of synaptic strength that are correlation based but spike-timing independent? Does introduction of spike-timing dependency into correlation-based rule offer new features for the more complex aspects of circuit developments? It is interesting to note that spike-timing–dependent rules have been used successfully in modeling the development of fine temporal discrimination in the auditory system (Gerstner et al 1996) and direction-sensitive receptive field organization of mammalian visual cortical neurons (Rao & Sejnowski 2000).

SPATIAL SPECIFICITY IN ACTIVITY-INDUCED SYNAPTIC MODIFICATION

Although not explicitly stated in Hebb's postulate, it is generally assumed that Hebbian synaptic modifications are synapse (or input) specific—only synapses experiencing correlated activity become modified. Input specificity was indeed observed in many studies of LTP/LTD (Andersen et al 1977, Lynch et al 1977, Dudek & Bear 1992, Mulkey & Malenka 1992, Nicoll & Malenka 1997). Thus, Hebbian rule is regarded as a "local" rule, in which individual synapses are independent of one another. Although the simplicity of this local rule has been

TABLE 2 Spread of LTP/LTD observed in different systems[a]

Induction	Spread[b]	Systems (References)
LTP	Presynaptic lateral (diverging outputs)	Hippocampal slice CA1 (Schuman & Madison 1994)
		Hippocampal slice culture (Bonhoeffer et al 1989)
		Visual cortex slice (Kossel et al 1990)
		Hippocampal culture (Tao et al 2000)
	Postsynaptic lateral (converging inputs)	Hippocampal slice CA3 (Bradler & Barrioneuvo 1989)
		Hippocampal slice CA1 (Muller et al 1995)
		Hippocampal slice culture (Engert & Bonhoeffer 1997)
		Hippocampal aspinous interneurons (Cowan et al 1998)
	Back propagation	Hippocampal culture (Tao et al 2000)
LTD	Presynaptic lateral	Hippocampal culture (Fitzsimonds et al 1997)
	Postsynaptic lateral	Hippocampal slice CA1 (Muller et al 1995,
		Staubli & Ji 1996, Nishiyama et al 2000)
		Hippocampal aspinous interneurons (Cowan et al 1998)
		Hippocampal interneurons (McMahon & Kauer 1997)
		Hippocampal culture (Fitzsimonds et al 1997)
	Back propagation	Hippocampal culture (Fitzsimonds et al 1997)

[a]LTP, long-term potentiation; LTD, long-term depression.
[b]See text and Figure 2 for definitions.

beneficial for many neural network models (Brown et al 1990, Rolls & Treves 1998), there is now increasing evidence that activity-induced synaptic modification may be accompanied by changes in some other synapses within a neural network (see Table 2).

Heterosynaptic LTD

Early studies of LTP in the CA1 region of hippocampus have shown a depression of unstimulated commissural-CA1 synapses following LTP-inducing tetanic stimulation of Schaffer collaterals (Lynch et al 1977, Dunwiddie & Lynch 1978). This phenomenon, termed heterosynaptic LTD, has since been observed in the cortex (Tsumoto & Suda 1979, Hirsch et al 1992), dentate gyrus (Levy & Steward 1979, Abraham & Goddard 1983), and the CA3 area (Bradler & Barrioneuvo 1989) of hippocampus, as well as at embryonic neuromuscular junctions in culture (Lo & Poo 1991). The conditions for the induction and expression of heterosynaptic LTD vary between different systems, but appear to be closely related to that for homosynaptic LTD (for reviews, see Linden & Connor 1995, Bear & Abraham 1996). A moderate rise in postsynaptic Ca^{2+}, through influx mediated by NMDARs and other Ca^{2+} channels as well as through Ca^{2+} release from internal stores, has been implicated in the induction mechanism, although the signaling pathways downstream of Ca^{2+} elevation and the mechanism of expression remain largely unknown. Heterosynaptic LTD induced by the induction of LTP represents perhaps a special case of more general non-local effects of synaptic modification (see below).

Spread of LTP/LTD in Slices

Although heterosynaptic LTD indicates the existence of interaction between distant synapses, it is not considered a violation of input specificity because changes at nonstimulated sites are opposite in polarity from those at induction sites. However, exceptions to this rule have been observed since the early days of LTP studies (for review, see Teyler & DiScenna 1987). For example, in area CA3 of hippocampus, mossy fiber tetanization leads to heterosynaptic LTP, as reflected in an increased field potential evoked by other converging but independent inputs, in addition to the conventional homosynaptic LTP (Yamamoto & Chujo 1978, Misgeld et al 1979). This heterosynaptic LTP may be caused by an increase in the polysynaptic component of CA3 field potentials, resulting from enhanced transmission among CA3 neurons (Higashima & Yamamoto 1985). It is interesting that Bradler & Barrioneuvo (1989) found that among three inputs to the CA3 area, tetanus-induced LTP in the mossy fiber input was accompanied by heterosynaptic LTP of Schaffer collateral and fimbrial responses, whereas LTP in the Schaffer pathway was associated with the LTP of fimbrial responses and LTD of mossy fiber responses. In contrast, LTP induced in the fimbrial response was input specific.

More recent works have revealed different forms of "breakdown" of synapse specificity of LTP in the CA1 region of hippocampus. In slice cultures, Bonhoeffer et al (1989) found that LTP induced by pairing presynaptic stimulation with postsynaptic depolarization is accompanied by potentiation at synapses made by the stimulated fibers on neighboring postsynaptic cells. Similar "spread" of potentiation was also observed in the visual cortex, where the spread of potentiation was shown to occur only for synapses made by the stimulated, but not unstimulated, inputs (Kossel et al 1990). A closer examination using local perfusion indicated that, in hippocampal slice cultures, LTP can spread to other inputs on the same postsynaptic cell as long as the unstimulated synapses are within ~70 μm of the site of LTP induction (Engert & Bonhoeffer 1997). Also observed at excitatory synapses was postsynaptic spread of LTP onto aspinous dendrites of interneurons whose somata were located in the CA1 pyramidal cell layer (Cowan et al 1998). It is interesting that pairing-induced LTP in CA1 pyramidal cells of acute hippocampal slices also spread to synapses made by the same set of afferent fibers onto neighboring postsynaptic cells (Schuman & Madison 1994). Taken together, these studies revealed two forms of spread of LTP. In one form, synaptic sites of induction and spread are apparently made by the same (group of) presynaptic axons on different postsynaptic cells (Bonhoeffer et al 1989, Kossel et al 1990, Schuman & Madison 1994). In the other form, the synapses of induction and spread are on the same postsynaptic cell but made by different presynaptic axons (Engert & Bonhoeffer 1997, Cowan et al 1998). For convenience, we define these two forms presynaptic spread and postsynaptic spread, respectively (Figure 2). Note that presynaptic spread can only be defined unequivocally when the stimulation is applied to a single presynaptic neuron. Thus, the cases using extracellular stimulation of afferent fibers (Bonhoeffer et al 1989, Kossel et al 1990, Schuman & Madison 1994) can only be regarded as consistent with the definition of presynaptic spread.

Several studies of conventional homosynaptic LTD have also suggested a breakdown of synapse specificity under different conditions. In CA1, homosynaptically induced LTD was found to be associated with a heterosynaptic reversal of the LTP previously induced in a separate pathway, and the effect was absent for naïve inputs and dependent on both NMDAR and calcineurin (Muller et al 1995). Another study, however, suggested that LTD at one of two independent Schaffer-collateral/commissural inputs may spread to the other (presumably naïve) input when the level of synaptic activation during low-frequency stimulation was strong enough to induce postsynaptic spiking (Staubli & Ji 1996). Spread of a different form of LTD has been observed in excitatory synapses onto GABAergic interneurons in the CA1 area of hippocampus: Tetanic stimulation of the input fiber resulted in LTD that spread to neighboring excitatory synapses onto the same postsynaptic interneurons (McMahon & Kauer 1997). In all these cases, only postsynaptic spread has been examined. These results differ from conventional heterosynaptic LTD in that the change at the induction pathway is also LTD rather than LTP.

The issue of input specificity of LTP/LTD induced by correlated pre- and postsynaptic spiking has been examined recently for Schaffer collateral-CA1 synapses in hippocampal slices (Nishiyama et al 2000). It was found that correlated activity of positive intervals induces input-specific homosynaptic LTP, whereas that of negative intervals induces LTD that spreads to heterosynaptic unstimulated inputs. Reduction of postsynaptic Ca^{2+} influx by partial blockade of NMDA receptors resulted in a conversion of LTP to LTD and a "breakdown" of input specificity, with LTD appearing at heterosynaptic inputs. The induction of LTD at homo- and heterosynaptic sites requires functional ryanodine receptors and IP_3 receptors (IP_3R), respectively. Functional blockade or genetic deletion of type 1 IP_3R led to a conversion of LTD to LTP and elimination of heterosynaptic LTD, whereas blocking ryanodine receptors eliminates only homosynaptic LTD. These results indicate that both the polarity and input specificity of activity-induced synaptic modifications are tightly regulated by postsynaptic Ca^{2+}, derived from Ca^{2+} influx and differential release of Ca^{2+} from internal stores. Therefore, input specificity should be viewed not as an intrinsic property associated with LTP/LTD but as a dynamic variable linked to the spread of dendritic Ca^{2+} elevation.

Selective Spread of LTP/LTD in Cell Cultures

Recent studies on synaptic plasticity in cell cultures have revealed extensive but selective spread of both LTP and LTD from the site of induction to other synapses

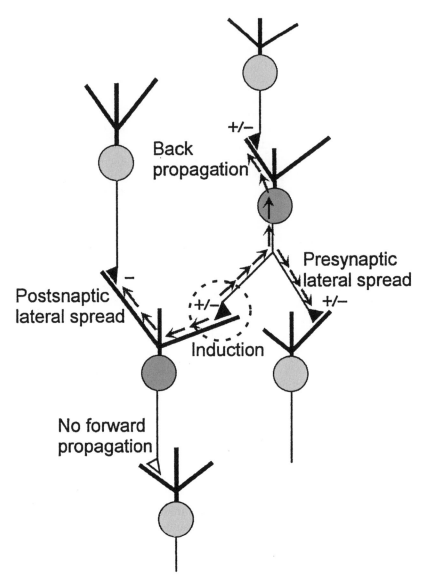

Figure 2   Different types of spread of long-term potentiation (LTP) and long-term depression (LTD). Induction of LTP/LTD at a synapse between glutamatergic neurons (black triangle in dashed circle) leads to the spread of LTP/LTD (marked as +/−) to other synapses in the network. Arrows indicate direction of signaling underlying the spread of LTP/LTD to divergent outputs of the presynaptic neuron (presynaptic lateral spread) and synapses on the dendrites of presynaptic neuron (back propagation), as well as the spread of LTD to convergent inputs on the postsynaptic neuron (postsynaptic lateral spread). Spread of LTP/LTD to the output of postsynaptic neurons (forward propagation) has not been observed.

in a network. In *Xenopus* nerve-muscle cultures, LTD induced by postsynaptic elevation of Ca^{2+} can spread to synapses made by the same presynaptic neuron on other myocytes, apparently by signaling within the cytoplasm of the presynaptic neuron (Cash et al 1996). In small networks of cultured hippocampal neurons, LTD induced at synapses between two glutamatergic neurons can spread to other synapses made by divergent outputs of the same presynaptic neuron (presynaptic lateral propagation) and to synapses made by other convergent inputs on the same postsynaptic cell (postsynaptic lateral propagation). Furthermore, LTD can also spread in a retrograde direction to depress synapses on the dendrite of the presynaptic neuron (back propagation) but not in anterograde direction (forward propagation) to depress output synapses of the postsynaptic neuron (Fitzsimonds et al 1997) (see Figure 2). It is interesting that in the same culture system, LTP induced at synapses between two glutamatergic neurons by correlated spiking exhibits a selective spread only to those synapses associated with the presynaptic neuron (i.e. both lateral and back propagation), not to those associated with the postsynaptic neuron (no lateral or forward propagation) (Tao et al 2000). In addition, the presynaptic lateral propagation of LTP is target cell specific: Divergent outputs to GABAergic postsynaptic neurons are not modified. This specificity is similar to the target specificity found in the induction of LTP in both cultures and slices (McMahon & Kauer 1997, Maccaferri et al 1998, Bi & Poo 1998, Reyes et al 1998, Schinder et al 2000). Finally, neither LTP nor LTD was found to spread further up- or downstream beyond the immediate neighboring synapses (Fitzsimonds et al 1997, Tao et al 2000). Although findings in cell culture may represent an exaggeration of phenomena occurring in vivo, the simplicity of the culture system enables investigation of cellular processes and network principles that may be present to varying degrees in more complex neural systems.

Development of Input Specificity

Recent studies of activity-induced synaptic modification in the developing *Xenopus* retinotectal system have suggested a cellular basis for the input specificity associated with LTP (H-z Tao, L Zhang, F Engert, M-m Poo, unpublished results). The induction of LTP at retinotectal synapses by theta burst stimulation of the retinal ganglion cells was not input-specific: LTP induced in one retinal ganglion cell pathway spreads to other converging retinal ganglion cell inputs on the same tectal neuron. Inputs on other adjacent tectal neurons were not affected, which suggests that the spread of potentiation was due to signaling within the postsynaptic cytoplasm. As the animal matures, LTP of retinotectal synapses induced by the same theta burst activity becomes input specific, a change that correlates with increased complexity of the dendritic arbor of tectal neurons and more restricted distribution of dendritic Ca^{2+} elevation evoked by retinal inputs. In contrast to that found for theta burst activity, LTP induced by low-frequency correlated pre- and postsynaptic spiking was input specific throughout the development. These results showed that a "break-down" of input specificity of activity-induced LTP can occur during in vivo development. The extent of the spread of synaptic potentiation, which correlates with spatial distribution of dendritic Ca^{2+} elevation, depends on the developmental stage as well as the pattern of synaptic activity.

IMPLICATIONS OF SPATIAL SPECIFICITY (OR LACK OF IT)

Long-Range Signaling in and Across Neurons

The spread of LTP/LTD and heterosynaptic LTD indicates that there are interactions among different synapses by long-range signaling either within the neuronal cytoplasm or across the neurons through the extracellular space. It has been suggested that membrane-permeable diffusible factors are responsible for retrograde signaling during the induction of LTP and LTD (for review, see Bliss & Collingridge 1993, Fitzsimonds & Poo 1998). These diffusible factors may affect other synapses close to the sites of induction. Schuman & Madison (1994) have suggested that nitric oxide may mediate the spread of LTP in hippocampus because inhibition of nitric oxide synthase in the postsynaptic cell blocked induction of LTP as well as its spread. However, a different signal triggered by the retrograde signal associated with LTP (e.g. nitric oxide) could be responsible for the spread of potentiation, and the signal could be confined intracellularly.

The extensive but highly selective spread of LTP/LTD found in culture argues for the existence of a rapid long-range intracellular signaling within the cytoplasm, with a speed of at least a few micrometers per second. This requires cytoskeleton-based axonal transport or regenerative waves as a means of signaling. Induction of LTP/LTD may generate Ca^{2+} waves (Berridge 1998) in either pre- or postsynaptic neuron and, through its diverse downstream effectors, results in either heterosynaptic LTD or spread of LTP/LTD. At Schaffer collateral-CA1 synapses, LTD induced by correlated pre- and postsynaptic activation (of negative intervals) spreads to unstimulated inputs to the same CA1 neuron (Nishiyama et al 2000). This spread was abolished by postsynaptic loading of a specific function-blocking antibody of type 1 IP_3R and was absent in hippocampal slices obtained from IP_3R-deficient mice, consistent with the notion that an IP_3R-dependent wave may be responsible for postsynaptic spread of LTD. Another second-messenger cAMP is known to be involved in different forms of synaptic plasticity and can diffuse rapidly in neuronal cytoplasm (Hempel et al 1996). Local elevation of cAMP/PKA activity in a developing neuron has been shown to exert long-range actions on distant parts of the neuron (Zheng et al 1994). In *Aplysia*, restricted application of serotonin to the cell bodies of sensory neurons can cause long-term enhancement at synaptic connections of the same sensory neuron to distant motoneurons

(Clark & Kandel 1993, Emptage & Carew 1993). Thus, cAMP or its downstream effectors can serve for cytoplasmic signaling involved in the spread of synaptic modification.

Long-range cytoplasmic signaling may also involve motor protein-based transport of vesicular membranes (Vale et al 1985, Vallee & Bloom 1991, Kuznetsov et al 1992, Bi et al 1997). Fast axonal transport of vesicles carrying internalized retrograde factors (e.g. neurotrophins) or vesicle-associated cytoplasmic effectors triggered by the induction of LTP/LTD may spread the signal for potentiation/depression within the cytoplasm. Restricted application of glutamate to somata of identified presynaptic neurons led to LTP of its output synapses, and this LTP can be blocked by pretreatment of microtubule-depolymerizing reagent colchicine (Lux & Veselovsky 1994). However, depolymerization of microtubule may also disrupt the normal structure of endoplasmic reticulum that supports the propagation of Ca^{2+} waves. Axotomy of postganglionic fibers leads to regression of synaptic inputs from preganglionic axon, a phenomenon that can be simulated by local chochicine block of the postganglionic fiber and can be prevented by supplying an exogenous neurotrophin nerve growth factor (Purves 1975). In hippocampal cell cultures, brain-derived neurotrophic factor induces synaptic potentiation by elevating transmitter secretion in a target cell–specific manner (Schinder et al 2000), i.e. only for those terminals innervating glutamatergic but not GABAergic neurons. This specificity has also been seen in the induction and spread of LTP in the same cultures (Bi & Poo 1998, Tao et al 2000). Thus, brain-derived neurotrophic factor may also be a good candidate for the cytoplasmic signal for presynaptic spread of LTP.

Non-Local Rules in Neural Networks

Back-Propagation   Although synapse specificity has been a "central dogma" in the concept of synaptic modification, a popular network model of supervised learning adopted a non-local rule, with which signals for synaptic modification can propagate backward through connections in multilayer networks (Rumelhart et al 1986a,b). This "back-propagation" algorithm, which uses the "error" of the output of a neuron to proportionally adjust the synaptic strength of its inputs, has been extremely powerful in training simple artificial networks to perform a wide range of tasks. The back-propagation algorithm has been considered nonbiological because it directly violates the rule of synapse specificity and is inconsistent with the known direction of signaling in neurons, i.e. from the input to the output (cf Rolls & Treves 1998). It is interesting that back propagation of LTP/LTD observed in cell cultures appears to fit qualitatively the requirement for back propagation of errors: If the output synapses of a neuron undergo LTP/LTD, then its input synapses undergo similar changes. However, it is not clear whether the spreading LTP/LTD can be integrated by the neuron, analogous to the linear summation of error signals in the back-propagation algorithm. In addition, the spread of LTP/LTD in culture does not appear to go beyond the immediate layer of neurons involved in the induction of LTP/LTD.

The Unit of Synaptic Modification   The exclusively "presynaptic" pattern of LTP spread in culture (Tao et al 2000) suggests that many output synapses of a single excitatory neuron may be modified as an integral unit in activity-dependent synaptic plasticity, with changes at each synapse affecting the others. Such coordinated changes among specific sets of synapses may help to recruit new cells into an existing ensemble and contribute to the development of synchronous firing in the cortex. Furthermore, both the induction and the presynaptic spread of LTP are target cell specific (do not occur at glutamatergic synapses made onto GABAergic postsynaptic neurons), whereas GABAergic synapses may be modified directly by correlated activity and indirectly by spread of LTP/LTD from neighboring sites (Fitzsimonds et al 1997, Aizenman et al 1998, Tao et al 2000; K Ganguly, M-m Poo, unpublished results). Target cell–specific induction and spread of LTP/LTD indicate that the unit of synaptic modification is the subset of presynaptic nerve terminals innervating the same postsynaptic cell type. The ability of GABAergic synapses to be modified suggests an active role for these inhibitory connections in the learning and memory function of neural circuits. Finally, we note that for converging inputs in networks of dissociated hippocampal neurons in culture and in CA1 pyramidal neurons in hippocampal slices, input specificity appears to be more strictly followed for LTP than for LTD. The implication of such asymmetry in input specificity in the development and functioning of neural network remains to be investigated.

CONCLUDING REMARKS

The induction and expression of LTP/LTD represent only the initial consequence of correlated activity. A myriad of subsequent events, including local activation of enzymes, posttranslational modification of proteins, and changes in protein synthesis and gene expression, occur in both pre- and postsynaptic cells, leading to long-lasting structural and functional alterations of the synapses as well as the entire neuron (Milner et al 1998). The existence of long-range and selective signaling within the neuron, as examplified by the selective spread of LTP/LTD, argues that activity-induced local synaptic modification must be considered as an integral part of global neuronal changes. The existence of reciprocal and selective transsynaptic signaling between neurons further suggests that synaptic and neuronal plasticity must be considered within the context of the interacting neurons in the neural network.

Synaptic modification by correlated activity was proposed by Hebb as a mechanism for the formation of "cell assembly," a network consolidated by patterned activities, which serves as a fundamental unit for perceptual functions of the brain. Although Hebb's original notion that development of the network is intimately linked to the perceptual function of the network remains to be validated, his postulate on synaptic modification has become a recurrent theme for understanding the synaptic basis underlying both activity-driven refinement of developing networks and learning and memory functions of mature networks. Developmental refinement may indeed use synaptic mechanisms similar to that used for learning and memory. In the spirit of Hebb, one may further ask whether the development of new networks is a basis of learning and memory in mature brain and, conversely, whether the structure of mature brain is but a reflection of perceptual experience during its early development.

Visit the Annual Reviews home page at www.AnnualReviews.org

LITERATURE CITED

Abbott LF, Blum KI. 1996. Functional significance of long-term potentiation for sequence learning and prediction, *Cereb. Cortex* 6:406–16

Abbott LF, Song S. 1999. Temporally asymmetric Hebbian learning, spike timing and neuronal response variability. In *Advances in Neural Information Processing Systems*, ed. MS Kearns, SA Solla, DA Cohn, 11:69–75. Cambridge, MA: MIT Press

Abeles M. 1991. *Corticonics*. Cambridge, UK: Cambridge Univ. Press

Abeles M, Bergman H, Margalit E, Vaadia E. 1993. Spatiotemporal firing patterns in the frontal cortex of behaving monkeys. *J. Neurophysiol* 70:1629–38

Abraham WC, Goddard GV. 1983. Asymmetric relationships between homosynaptic long-term potentiation and heterosynaptic long-term depression. *Nature* 305:717–19

Abrams TW, Karl KA, Kandel ER. 1991. Biochemical studies of stimulus convergence during classical conditioning in Aplysia: dual regulation of adenylate cyclase by Ca^{2+}/calmodulin and transmitter. *J. Neurosci.* 11:2655–65

Aizenman CD, Manis PB, Linden DJ. 1998. Polarity of long-term synaptic gain change is related to postsynaptic spike firing at a cerebellar inhibitory synapse. *Neuron* 21:827–35

Andersen P, Sundberg SH, Sveen O, Wigström H. 1977. Specific long-lasting potentiation of synaptic transmission in hippocampal slices. *Nature* 266:736–37

Barrionuevo G, Brown TH. 1983. Associative long-term potentiation in hippocampal slices. *Proc. Natl. Acad. Sci. USA* 80:7347–51

Bear MF, Abraham WC. 1996. Long-term depression in hippocampus. *Annu. Rev. Neurosci.* 19:437–62

Bekkers JM, Richerson GB, Stevens CF. 1990. Origin of variability in quantal size in cultured hippocampal neurons and hippocampal slices. *Proc. Natl. Acad. Sci. USA* 87:5359–62

Bell CC, Han VZ, Sugawara Y, Grant K. 1997. Synaptic plasticity in a cerebellum-like structure depends on temporal order. *Nature* 387:278–81

Berridge MJ. 1998. Neuronal calcium signaling. *Neuron* 21:13–26

Bi G-q, Morris RL, Liao GC, Alderton JM, Scholey JM, Steinhardt RA. 1997. Kinesin- and myosin-driven steps of vesicle recruitment for Ca^{2+}-regulated exocytosis. *J. Cell Biol.* 138:999–1008

Bi G-q, Poo M-m. 1998. Synaptic modifications in cultured hippocampal neurons: Dependence on spike timing, synaptic strength, and postsynaptic cell type. *J. Neurosci.* 18:10464–72

Bi G-q, Poo M-m. 1999. Distributed synaptic modification in neural networks induced by patterned stimulation. *Nature* 401:792–96

Bienenstock EL, Cooper LN, Munro PW. 1982. Theory for the development of neuron selectivity: orientation specificity and binocular interaction in visual cortex. *J. Neurosci.* 2:32–48

Bliss TV, Collingridge GL. 1993. A synaptic model of memory: long-term potentiation in the hippocampus. *Nature* 361:31–39

Bliss TV, Gardner-Medwin AR. 1973. Long-lasting potentiation of synaptic transmission in the dentate area of the unanaestetized rabbit following stimulation of the perforant path. *J. Physiol.* 232:357–74

Bliss TV, Lømo T. 1973. Long-lasting potentiation of synaptic transmission in the dentate area of the anaesthetized rabbit following stimulation of the perforant path. *J. Physiol.* 232:331–56

Bonhoeffer T, Staiger V, Aertsen A. 1989. Synaptic plasticity in rat hippocampal slice cultures: local "Hebbian" conjunction of pre- and postsynaptic stimulation leads to distributed synaptic enhancement. *Proc. Natl. Acad. Sci. USA* 86:8113–17

Bortolotto ZA, Fitzjohn SM, Collingridge GL. 1999. Roles of metabotropic glutamate receptors in LTP and LTD in the hippocampus. *Curr. Opin. Neurobiol.* 9:299–304

Bourne HR, Nicoll R. 1993. Molecular machines integrate coincident synaptic signals. *Cell* 72(Suppl.):65–75

Bradler JE, Barrioneuvo G. 1989. Long-term potentiation in hippocampal CA3 neurons: tetanized input regulates heterosynaptic efficacy. *Synapse* 4:132–42

Braitenberg V. 1967. Is the cerebellar cortex a biological clock in the millisecond range? *Prog. Brain Res.* 25:334–46

Brown TH, Kairiss EW, Keenan CL. 1990. Hebbian synapses: biophysical mechanisms and algorithms *Annu. Rev. Neurosci.* 13:475–511

Bullock TH. 1970. The reliability of neurons. *J. Gen. Physiol.* 55:565–84

Buonomano DV, Hickmott PW, Merzenich MM. 1997. Context-sensitive synaptic plasticity and temporal-to-spatial transformations in hippocampal slices. *Proc. Natl. Acad. Sci. USA* 94:10403–8

Buonomano DV, Merzenich MM. 1995. Temporal information transformed into a spatial code by a neural network with realistic properties. *Science* 267:1028–30

Buonomano DV, Merzenich MM. 1998. Cortical plasticity: from synapses to maps. *Annu. Rev. Neurosci.* 21:149–86

Buzsáki G, Penttonen M, Nádasdy Z, Bragin A. 1996. Pattern and inhibition-dependent invasion of pyramidal cell dendrites by fast spikes in the hippocampus in vivo. *Proc. Natl. Acad. Sci. USA* 93:9921–25

Calvin WH, Stevens CF. 1968. Synaptic noise and other sources of randomness in motoneuron interspike intervals. *J. Neurophysiol.* 31:574–87

Carew TJ, Hawkins RD, Abrams TW, Kandel ER. 1984. A test of Hebb's postulate at identified synapses which mediate classical conditioning in Aplysia. *J. Neurosci.* 4:1217–24

Carew TJ, Walters ET, Kandel ER. 1981. Classical conditioning in a simple withdrawal reflex in *Aplysia californica*. *J. Neurosci.* 1:1426–37

Carr CE. 1993. Processing of temporal information in the brain. *Annu. Rev. Neurosci.* 16:223–43

Cash S, Zucker RS, Poo M-m. 1996. Spread of synaptic depression mediated by presynaptic cytoplasmic signaling. *Science* 272:998–1001

Clark GA, Hawkins RD, Kandel ER. 1994. Activity-dependent enhancement of presynaptic facilitation provides a cellular mechanism for the temporal specificity of classical conditioning in Aplysia. *Learn. Mem.* 1:243–57

Clark GA, Kandel ER. 1993. Induction of long-term facilitation in Aplysia sensory neurons by local application of serotonin to remote synapses. *Proc. Natl. Acad. Sci. USA* 90:11411–15

Collingridge GL, Kehl SJ, McLennan H. 1983. Excitatory amino acids in synaptic transmission in the Schaffer collateral-commissural pathway of the rat hippocampus. *J. Physiol.* 334:33–46

Constantine-Paton M, Cline HT, Debski E. 1990. Patterned activity, synaptic convergence, and the NMDA receptor in developing visual pathways. *Annu. Rev. Neurosci.* 13:129–54

Cowan AI, Stricker C, Reece LJ, Redman SJ. 1998. Long-term plasticity at excitatory synapses on aspinous interneurons in area CA1 lacks synaptic specificity. *J. Neurophysiol.* 79:13–20

Crair MC, Gillespie DC, Stryker MP. 1998. The role of visual experience in the development of columns in cat visual cortex. *Science* 279:566–70

Crowley JC, Katz LC. 1999. Development of ocular dominance columns in the absence of retinal input. *Nat. Neurosci.* 2:1125–30

Dan Y, Alonso JM, Usrey WM, Reid RC. 1998. Coding of visual information by precisely correlated spikes in the lateral geniculate nucleus. *Nat. Neurosci.* 1:501–7

Debanne D, Gahwiler BH, Thompson SM. 1998. Long-term synaptic plasticity between pairs of individual CA3 pyramidal cells in rat hippocampal slice cultures. *J. Physiol.* 507:237–47

deCharms RC, Merzenich MM. 1996. Primary cortical representation of sounds by the coordination of action-potential timing. *Nature* 381:610–13

Diesmann M, Gewaltig MO, Aertsen A. 1999. Stable propagation of synchronous spiking in cortical neural networks. *Nature* 402:529–33

Dudek SM, Bear MF. 1992. Homosynaptic long-term depression in area CA1 of hippocampus and effects of N-methyl-D-aspartate receptor blockade. *Proc. Natl. Acad. Sci. USA* 89:4363–67

Dunwiddie T, Lynch G. 1978. Long-term potentiation and depression of synaptic responses in the rat hippocampus: localization and frequency dependency. *J. Physiol.* 276:353–67

Egger V, Feldmeyer D, Sakmann B. 1999. Coincidence detection and changes of synaptic efficacy in spiny stellate neurons in rat barrel cortex. *Nat. Neurosci.* 2:1098–105

Emptage NJ, Carew TJ. 1993. Long-term synaptic facilitation in the absence of short-term facilitation in Aplysia neurons. *Science* 262:253–56

Engert F, Bonhoeffer T. 1997. Synapse specificity of long-term potentiation breaks down at short distances. *Nature* 388:279–84

Erwin E, Miller KD. 1998. Correlation-based development of ocularly matched orientation and ocular dominance maps: determination of required input activities. *J. Neurosci.* 18:9870–95

Fitzsimonds RM, Poo M-m. 1998. Retrograde signaling in the development and modification of synapses. *Physiol. Rev.* 78:143–70

Fitzsimonds RM, Song H-j, Poo M-m. 1997. Propagation of activity-dependent synaptic depression in simple neural networks. *Nature* 388:439–48

Fregnac Y, Bienenstock E. 1998. Correlational models of synaptic plasticity: development, learning, and cortical dynamics of mental representations. In *Mechanistic Relationships Between Development and Learning*, ed. T Carew, R Menzel, C Shatz. pp. 113–48. New York: Wiley

Gerstner W, Abbott LF. 1997. Learning navigational maps through potentiation and modulation of hippocampal place cells. *J. Comput. Neurosci.* 4:79–94

Gerstner W, Kempter R, van Hemmen JL, Wagner H. 1996. A neuronal learning rule for sub-millisecond temporal coding. *Nature* 383:76–81

Goda Y, Stevens CF. 1996. Long-term depression properties in a simple system. *Neuron* 16:103–11

Gustafsson B, Wigström H. 1986. Hippocampal long-lasting potentiation produced by pairing single volleys and brief conditioning tetani evoked in separate afferents. *J. Neurosci.* 6:1575–82

Gustafsson B, Wigström H, Abraham WC, Huang YY. 1987. Long-term potentiation in the hippocampus using depolarizing current pulses as the conditioning stimulus to single volley synaptic potentials. *J. Neurosci.* 7:774–80

Hawkins RD, Abrams TW, Carew TJ, Kandel ER. 1983. A cellular mechanism of classical conditioning in Aplysia: activity-dependent amplification of presynaptic facilitation. *Science* 219:400–5

Hebb D. 1949. The *Organization of Behavior*. New York: Wiley

Hempel CM, Vincent P, Adams SR, Tsien RY, Selverston AI. 1996. Spatio-temporal dynamics of cyclic AMP signals in an intact neural circuit. *Nature* 384:166–69

Higashima M, Yamamoto C. 1985. Two components of long-term potentiation in mossy fiber-induced excitation in hippocampus. *Exp. Neurol.* 90:529–39

Hirsch JC, Barrionuevo G, Crepel F. 1992. Homo- and heterosynaptic changes in efficacy are expressed in prefrontal neurons: an in vitro study in the rat. *Synapse* 12:82–85

Hoffman DA, Magee JC, Colbert CM, Johnston D. 1997. K+ channel regulation of signal propagation in dendrites of hippocampal pyramidal neurons. *Nature* 387:869–75

Hollmann M, Heinemann S. 1994. Cloned glutamate receptors. *Annu. Rev. Neurosci.* 17:31–108

Hopfield JJ. 1995. Pattern recognition computation using action potential timing for stimulus representation. *Nature* 376:33–36

Horton JC, Hocking DR. 1996. An adult-like pattern of ocular dominance columns in striate cortex of newborn monkeys prior to visual experience. *J. Neurosci.* 16:1791–807

Hubel DH, Wiesel TN. 1965. Binocular interaction in striate cortex of kittens reared with artificial squint. *J. Neurophysiol.* 28:1041–59

Ito M, Sakurai M, Tongroach P. 1982. Climbing fibre induced depression of both mossy fibre responsiveness and glutamate sensitivity of cerebellar Purkinje cells. *J. Physiol.* 324:113–34

Jahr CE, Stevens CF. 1987. Glutamate activates multiple single channel conductances in hippocampal neurons. *Nature* 325:522–5

Jeffress LA. 1948. A place theory of sound localization. *J. Comp. Physiol. Psychol.* 41:35–39

Johnston D, Magee JC, Colbert CM, Cristie BR. 1996. Active properties of neuronal dendrites. *Annu. Rev. Neurosci.* 19:165–86

Katz LC, Shatz CJ. 1996. Synaptic activity and the construction of cortical circuits. *Science* 274:1133–38

Kelso SR, Brown TH. 1986. Differential conditioning of associative synaptic enhancement in hippocampal brain slices. *Science* 232:85–87

Kelso SR, Ganong AH, Brown TH. 1986. Hebbian synapses in hippocampus. *Proc. Natl. Acad. Sci. USA* 83:5326–30

Kempter R, Gerstner W, van Hemmen JL. 2000. Hebbian learning and spiking neurons. *Phys. Rev. E* 59. In press

Koester HJ, Sakmann B. 1998. Calcium dynamics in single spines during coincident pre- and postsynaptic activity depend on relative timing of back-propagating action potentials and subthreshold excitatory postsynaptic potentials *Proc. Natl. Acad. Sci. USA* 95:9596–601

Kossel A, Bonhoeffer T, Bolz J. 1990. Non-Hebbian synapses in rat visual cortex. *NeuroReport* 1:115–18

Kuznetsov SA, Langford GM, Weiss DG. 1992. Actin-dependent organelle movement in squid axoplasm. *Nature* 356:722–25

Lasser-Ross N, Ross WN. 1992. Imaging voltage and synaptically activated sodium transients in cerebellar Purkinje cells. *Proc. R. Soc. London Ser. B* 247:35–39

Laurent G. 1997. Olfactory processing: maps, time and codes. *Curr. Opin. Neurobiol.* 7:547–53

Lechner HA, Byrne JH. 1998. New perspectives on classical conditioning: a synthesis of Hebbian and non-Hebbian mechanisms. *Neuron* 20:355–58

Levy WB, Steward O. 1979. Synapses as associative memory elements in the hippocampal formation. *Brain Res.* 175:233–45

Levy WB, Steward O. 1983. Temporal contiguity requirements for long-term associative potentiation/depression in the hippocampus. *Neuroscience* 8:791–97

Lin XY, Glanzman DL. 1994. Hebbian induction of long-term potentiation of Aplysia sensorimotor synapses: partial requirement for activation of an NMDA-related receptor. *Proc. R. Soc. London Ser. B* 255:215–21

Linden DJ. 1999. The return of the spike: postsynaptic action potentials and the induction of LTP and LTD. *Neuron* 22:661–66

Linden DJ, Connor JA. 1995. Long-term synaptic depression. *Annu. Rev. Neurosci.* 18:319–57

Lisman J. 1989. A mechanism for the Hebb and the anti-Hebb processes underlying learning and memory. *Proc. Natl. Acad. Sci. USA* 86:9574–78

Lo YJ, Poo MM. 1991. Activity-dependent synaptic competition in vitro: heterosynaptic suppression of developing synapses. *Science* 254:1019–22

Lux HD, Veselovsky NS. 1994. Glutamate-produced long-term potentiation by selective challenge of presynaptic neurons in rat hippocampal cultures. *Neurosci. Lett.* 178:231–34

Lynch GS, Dunwiddie T, Gribkoff V. 1977. Heterosynaptic depression: a postsynaptic correlate of long-term potentiation. *Nature* 266:737–39

Maass W, Bishop CM, eds. 1999. *Pulsed Neural Network.* Cambridge, MA: MIT Press. 377 pp.

Maccaferri G, Toth K, McBain CJ. 1998. Target-specific expression of presynaptic mossy fiber plasticity. *Science* 279:1368–70

Magee JC, Johnston D. 1997. A synaptically controlled, associative signal for Hebbian plasticity in hippocampal neurons. *Science* 275:209–13

Mainen ZF, Sejnowski TJ. 1995. Reliability of spike timing in neocortical neurons. *Science* 268:1503–6

Malenka RC, Kauer JA, Perkel DJ, Mauk MD, Kelly PT, et al. 1989. An essential role for postsynaptic calmodulin and protein kinase activity in long-term potentiation. *Nature* 340:554–57

Malenka RC, Nicoll RA. 1999. Long-term potentiation—a decade of progress. *Science* 285:1870–74

Malinow R, Schulman H, Tsien RW. 1989. Inhibition of postsynaptic PKC or CaMKII blocks induction but not expression of LTP. *Science* 245:862–66

Markram H, Lubke J, Frotscher M, Sakmann B. 1997. Regulation of synaptic efficacy by coincidence of postsynaptic APs and EPSPs. *Science* 275:213–15

Marsálek P, Koch C, Maunsell J. 1997. On the relationship between synaptic input and spike output jitter in individual neurons. *Proc. Natl. Acad. Sci. USA* 94:735–40

Mayer ML, Westbrook GL, Guthrie PB. 1984. Voltage-dependent block by Mg^{2+} of NMDA responses in spinal cord neurones. *Nature* 309:261–63

McMahon LL, Kauer JA. 1997. Hippocampal interneurons express a novel form of synaptic plasticity. *Neuron* 18:295–305

McNaughton BL, Douglas RM, Goddard GV. 1978. Synaptic enhancement in fascia dentata: cooperativity among coactive afferents. *Brain Res.* 157:277–93

Mehta MR, Barnes CA, McNaughton BL. 1997. Experience-dependent, asymmetric expansion of hippocampal place fields. *Proc. Natl. Acad. Sci. USA* 94:8918–21

Mehta MR, Quirk MC, Wilson MA. 2000. Experience-dependent asymmetric shape of hippocampal receptive fields. *Neuron* 25:707–15

Miller KD. 1994. A model for the development of simple cell receptive fields and the ordered arrangement of orientation columns through activity-dependent competition between ON- and OFF-center inputs. *J. Neurosci.* 14:409–41

Miller KD, Keller JB, Stryker MP. 1989. Ocular dominance column development: analysis and simulation. *Science* 245:605–15

Milner B, Squire LR, Kandel ER. 1998. Cognitive neuroscience and the study of memory. *Neuron* 20:445–68

Minai AA, Levy WB. 1993. Sequence learning in a single trial. *INNS World Congr. Neural Netw.* 2:505–8

Misgeld U, Sarvey JM, Klee MR. 1979. Heterosynaptic postactivation potentiation in hippocampal CA 3 neurons: long-term changes of the postsynaptic potentials. *Exp. Brain Res.* 37:217–29

Montague PR, Sejnowski TJ. 1994. The predictive brain: temporal coincidence and temporal order in synaptic learning mechanisms *Learn. Mem.* 1:1–33

Moore JW, Choi J-S, Brunzell DH. 1998. Predictive timing under temporal uncertainty: the time derivative model of the conditioned response. In *Timing of Behavior*, ed. DA Risenbaum, CE Collyer, pp. 3–34. Cambridge, MA: MIT Press

Mulkey RM, Endo S, Shenolikar S, Malenka RC. 1994. Involvement of a calcineurin/inhibitor-1 phosphatase cascade in hippocampal long-term depression. *Nature* 369:486–88

Mulkey RM, Malenka RC. 1992. Mechanisms underlying induction of homosynaptic long-term depression in area CA1 of the hippocampus. *Neuron* 9:967–75

Muller D, Hefft S, Figurov A. 1995. Heterosynaptic interactions between LTP and LTD in CA1 hippocampal slices. *Neuron* 14:599–605

Murphy GG, Glanzman DL. 1997. Mediation of classical conditioning in *Aplysia californica* by long-term potentiation of sensorimotor synapses. *Science* 278:467–71

Nakamura T, Barbara JG, Nakamura K, Ross WN. 1999. Synergistic release of Ca^{2+} from IP3-sensitive stores evoked by synaptic activation of mGluRs paired with backpropagating action potentials. *Neuron* 24:727–37

Nicoll RA, Malenka RC. 1997. Neurobiology.

Long-distance long-term depression. *Nature* 388:427–28

Nishiyama M, Hong K, Mikoshiba K, Poo M-m, Kato K. 2000. Calcium release from internal stores regulates polarity and input specificity of synaptic modification. *Nature.* In press

Nowak L, Bregestovski P, Ascher P, Herbet A, Prochiantz A. 1984. Magnesium gates glutamate-activated channels in mouse central neurones. *Nature* 307:462–65

Paulsen O, Sejnowski TJ. 2000. Natural patterns of activity and long-term synaptic plasticity. *Curr. Opin. Neurobiol.* 10:172–79

Penn AA, Riquelme PA, Feller MB, Shatz CJ. 1998. Competition in retinogeniculate patterning driven by spontaneous activity. *Science* 279:2108–12

Perkel DH, Bullock TH. 1968. Neural coding. *Neurosci. Res. Prog. Sum.* 3:405–527

Purves D. 1975. Functional and structural changes in mammalian sympathetic neurones following interruption of their axons. *J. Physiol.* 252:429–63

Rao RPN, Sejnowski TJ. 2000. Predictive learning of temporal sequences in recurrent neocortical circuits In *Advances in Neural Information Processing Systems*, Vol. 12, ed. SA Solla, TK Leen, K-R Muller. Cambridge, MA: MIT Press. In press

Regehr WG, Konnerth A, Armstrong CM. 1992. Sodium action potentials in the dendrites of cerebellar Purkinje cells. *Proc. Natl. Acad. Sci. USA* 89:5492–96

Reyes A, Lujan R, Burnashev N, Somogyi P, Sakmann B. 1998. Target-cell-specific facilitation and depression in neocortical circuits. *Nat. Neurosci.* 1:279–85

Riehle A, Grün S, Diesmann M, Aertsen A. 1997. Spike synchronization and rate modulation differentially involved in motor cortical function. *Science* 278:1950–53

Rieke F, Warland D, de Ruyter van Steveninck R, Bialek W. 1997. *Spikes: Exploring the Neural Code.* Cambridge, MA: MIT Press. 395 pp.

Roberts PD. 1999. Computational consequences of temporally asymmetric learning rules. I. Differential hebbian learning. *J. Comput. Neurosci.* 7:235–46

Roelfsema PR, Engel AK, König P, Singer W. 1997. Visuomotor integration is associated with zero time-lag synchronization among cortical areas. *Nature* 385:157–61

Rolls ET, Treves A. 1998. *Neural Networks and Brain Function.* New York: Oxford Univ. Press

Rumelhart DE, Hinton GE, Williams RJ. 1986a. Learning internal representations by error propagation. In *Parallel Distributed Processing,* ed. JA Feldman, PJ Hayes, DE Rumelhart. 1:318–62. Cambridge, MA: MIT Press

Rumelhart DE, Hinton GE, Williams RJ. 1986b. Learning representation by back-propagating errors. *Nature* 323:533–36

Sastry BR, Goh JW, Auyeung A. 1986. Associative induction of posttetanic and long-term potentiation in CA1 neurons of rat hippocampus. *Science* 232:988–90

Schinder AF, Berninger B, Poo M. 2000. Postsynaptic target specificity of neurotrophin-induced presynaptic potentiation. *Neuron* 25:151–63

Schuman EM, Madison DV. 1994. Locally distributed synaptic potentiation in the hippocampus. *Science* 263:532–36

Sejnowski TJ. 1977a. Statistical constraints on synaptic plasticity. *J. Theor. Biol.* 69:385–89

Sejnowski TJ. 1977b. Storing covariance with nonlinearly interacting neurons. *J. Math. Biol.* 4:303–21

Sejnowski TJ. 1999. The book of Hebb. *Neuron* 24:773–76

Sejnowski TJ, Tesauro G. 1989. The Hebb rule for synaptic plasticity: algorithms and implementations. In *Neural Models of Plasticity: Experimental and Theoretical Approaches,* ed. John H. Byrne. pp. 94–103. San Diego, CA: Academic

Senn W, Markram H, Tsodyks M. 2000. An algorithm for modifying neurotransmitter release probability based on pre- and post-synaptic spike timing. *Neural Comput.* In press

Silva AJ, Stevens CF, Tonegawa S, Wang Y. 1992. Deficient hippocampal long-term potentiation in alpha-calcium-calmodulin kinase II mutant mice. *Science* 257:201–6

Singer W. 1999. Time as coding space? *Curr. Opin. Neurobiol.* 9:189–94

Singer W, Gray CM. 1995. Visual feature integration and the temporal correlation hypothesis *Annu. Rev. Neurosci.* 18:555–86

Squire LR. 1987. *Memory and Brain.* New York: Oxford Univ. Press

Staubli UV, Ji ZX. 1996. The induction of homo- vs. heterosynaptic LTD in area CA1 of hippocampal slices from adult rats. *Brain Res.* 714:169–76

Stent GS. 1973. A physiological mechanism for Hebb's postulate of learning. *Proc. Natl. Acad. Sci. USA* 70:997–1001

Stevens CF. 1996. Strengths and weaknesses in memory. *Nature* 381:471–72

Stevens CF, Zador AM. 1998. Input synchrony and the irregular firing of cortical neurons. *Nat. Neurosci.* 1:210–17

Stryker MP, Strickland SL. 1984. Physiological segregation of ocular dominance columns depends on patterns of afferent activity. *Invest. Ophthalmol. Suppl.* 25:278

Stuart GJ, Sakmann B. 1994. Active propagation of somatic action potentials into neocortical pyramidal cell dendrites. *Nature* 367:69–72

Sutton RS, Barto AG. 1981. Toward a modern theory of adaptive networks: expectation and prediction. *Psychol. Rev.* 88:135–70

Tank DW, Hopfield JJ. 1987. Neural computation by concentrating information in time. *Proc. Natl. Acad. Sci. USA* 84:1896–900

Tao HW, Zhang LI, Bi G-q, Poo M-m. 2000. Selective presynaptic propagation of long-term potentiation in defined neural networks. *J. Neurosci.* 20:3233–43

Tesauro G. 1986. Simple neural models of classical conditioning. *Biol. Cybern.* 55:187–200

Teyler TJ, DiScenna P. 1987. Long-term potentiation. *Annu. Rev. Neurosci.* 10:131–61

Thompson RF, Krupa DJ. 1994. Organization of memory traces in the mammalian brain. *Annu. Rev. Neurosci.* 17:519–49

Tsumoto T, Suda K. 1979. Cross-depression: an electrophysiological manifestation of binocular competition in the developing visual cortex. *Brain Res.* 168:190–94

Tzounopoulos T, Janz R, Südhof TC, Nicoll RA, Malenka RC. 1998. A role for cAMP in long-term depression at hippocampal mossy fiber synapses. *Neuron* 21:837–45

Vale RD, Reese TS, Sheetz MP. 1985. Identification of a novel force-generating protein, kinesin, involved in microtubule-based motility. *Cell* 42:39–50

Vallee RB, Bloom GS. 1991. Mechanisms of fast and slow axonal transport. *Annu. Rev. Neurosci.* 14:59–92

Walters ET, Byrne JH. 1983. Associative conditioning of single sensory neurons suggests a cellular mechanism for learning. *Science* 219:405–8

Weliky M, Katz LC. 1997. Disruption of orientation tuning in visual cortex by artificially correlated neuronal activity. *Nature* 386:680–85

Wiesel TN, Hubel DH. 1963. Single cell responses in striate cortex of kittens deprived of vision in one eye. *J. Neurophysiol.* 26:1003–7

Wigström H, Gustafsson B, Huang YY, Abraham WC. 1986. Hippocampal long-term potentiation is induced by pairing single afferent volleys with intracellularly injected depolarizing current pulses. *Acta Physiol. Scand.* 126:317–19

Wong RO, Chernjavsky A, Smith SJ, Shatz CJ. 1995. Early functional neural networks in the developing retina. *Nature* 374:716–18

Yamamoto C, Chujo T. 1978. Long-term potentiation in thin hippocampal sections studied by intracellular and extracellular recordings. *Exp. Neurol.* 58:242–50

Yang SN, Tang YG, Zucker RS. 1999. Selective induction of LTP and LTD by postsynaptic $[Ca^{2+}]$ elevation. *J. Neurophysiol.* 81:781–87

Yeckel MF, Kapur A, Johnston D. 1999. Multiple forms of LTP in hippocampal CA3 neurons use a common postsynaptic mechanism. *Nat. Neurosci.* 2:625–33

Yovell Y, Abrams TW. 1992. Temporal asymmetry in activation of Aplysia adenylyl cyclase by calcium and transmitter may explain temporal requirements of conditioning. *Proc. Natl. Acad. Sci. USA* 89:6526–30

Yovell Y, Kandel ER, Dudai Y, Abrams TW. 1992. A quantitative study of the Ca^{2+}/calmodulin sensitivity of adenylyl cyclase in Aplysia, Drosophila, and rat. *J. Neurochem.* 59:1736–44

Zhang LI, Tao HW, Holt CE, Harris WA, Poo Mm. 1998. A critical window for cooperation and competition among developing retinotectal synapses. *Nature* 395:37–44

Zheng JQ, Zheng Z, Poo M. 1994. Long-range signaling in growing neurons after local elevation of cyclic AMP-dependent activity. *J. Cell Biol.* 127:1693–701

Annu. Rev. Neurosci. 2001. 24:167–202

AN INTEGRATIVE THEORY OF PREFRONTAL CORTEX FUNCTION

Earl K. Miller

Center for Learning and Memory, RIKEN-MIT Neuroscience Research Center and Department of Brain and Cognitive Sciences, Massachusetts Institute of Technology, Cambridge, Massachusetts 02139; e-mail: ekm@ai.mit.edu

Jonathan D. Cohen

Center for the Study of Brain, Mind, and Behavior and Department of Psychology, Princeton University, Princeton, New Jersey 08544; e-mail: jdc@princeton.edu

Key Words frontal lobes, cognition, executive control, working memory, attention

■ **Abstract** The prefrontal cortex has long been suspected to play an important role in cognitive control, in the ability to orchestrate thought and action in accordance with internal goals. Its neural basis, however, has remained a mystery. Here, we propose that cognitive control stems from the active maintenance of patterns of activity in the prefrontal cortex that represent goals and the means to achieve them. They provide bias signals to other brain structures whose net effect is to guide the flow of activity along neural pathways that establish the proper mappings between inputs, internal states, and outputs needed to perform a given task. We review neurophysiological, neurobiological, neuroimaging, and computational studies that support this theory and discuss its implications as well as further issues to be addressed.

INTRODUCTION

One of the fundamental mysteries of neuroscience is how coordinated, purposeful behavior arises from the distributed activity of billions of neurons in the brain. Simple behaviors can rely on relatively straightforward interactions between the brain's input and output systems. Animals with fewer than a hundred thousand neurons (in the human brain there are 100 billion or more neurons) can approach food and avoid predators. For animals with larger brains, behavior is more flexible. But flexibility carries a cost: Although our elaborate sensory and motor systems provide detailed information about the external world and make available a large repertoire of actions, this introduces greater potential for interference and confusion. The richer information we have about the world and the greater number of options for behavior require appropriate attentional, decision-making, and coordinative functions, lest uncertainty prevail. To deal with this multitude of

0147-006X/01/0621-0167$14.00

possibilities and to curtail confusion, we have evolved mechanisms that coordinate lower-level sensory and motor processes along a common theme, an internal goal. This ability for cognitive control no doubt involves neural circuitry that extends over much of the brain, but it is commonly held that the prefrontal cortex (PFC) is particularly important.

The PFC is the neocortical region that is most elaborated in primates, animals known for their diverse and flexible behavioral repertoire. It is well positioned to coordinate a wide range of neural processes: The PFC is a collection of interconnected neocortical areas that sends and receives projections from virtually all cortical sensory systems, motor systems, and many subcortical structures (Figure 1). Neurophysiological studies in nonhuman primates have begun to define many of the detailed properties of PFC, and human neuropsychology and neuroimaging studies have begun to provide a broad view of the task conditions under which it is engaged. However, an understanding of the mechanisms by which the PFC executes control has remained elusive. The aim of this article is to describe a theory of PFC function that integrates these diverse findings, and more precisely defines its role in cognitive control.

The Role of the PFC in Top-Down Control of Behavior

The PFC is not critical for performing simple, automatic behaviors, such as our tendency to automatically orient to an unexpected sound or movement. These behaviors can be innate or they can develop gradually with experience as learning mechanisms potentiate existing pathways or form new ones. These "hardwired" pathways are advantageous because they allow highly familiar behaviors to be executed quickly and automatically (i.e. without demanding attention). However, these behaviors are inflexible, stereotyped reactions elicited by just the right stimulus. They do not generalize well to novel situations, and they take extensive time and experience to develop. These sorts of automatic behaviors can be thought of as relying primarily on "bottom-up" processing; that is, they are determined largely by the nature of the sensory stimuli and well-established neural pathways that connect these with corresponding responses.

By contrast, the PFC is important when "top-down" processing is needed; that is, when behavior must be guided by internal states or intentions. The PFC is critical in situations when the mappings between sensory inputs, thoughts, and actions either are weakly established relative to other existing ones or are rapidly changing. This is when we need to use the "rules of the game," internal representations of goals and the means to achieve them. Several investigators have argued that this is a cardinal function of the PFC (Cohen & Servan-Schreiber 1992, Passingham 1993, Grafman 1994, Wise et al 1996, Miller 1999). Two classic tasks illustrate this point: the Stroop task and the Wisconsin card sort task (WCST).

In the Stroop task (Stroop 1935, MacLeod 1991), subjects either read words or name the color in which they are written. To perform this task, subjects must selectively attend to one attribute. This is especially so when naming the color

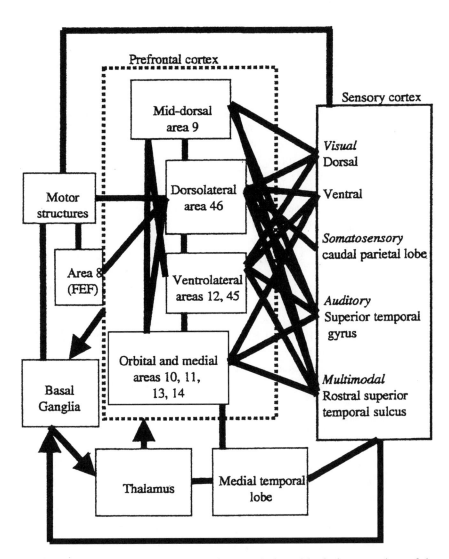

Figure 1 Schematic diagram of some of the extrinsic and intrinsic connections of the prefrontal cortex. The partial convergence of inputs from many brain systems and internal connections of the prefrontal cortex (PFC) may allow it to play a central role in the synthesis of diverse information needed for complex behavior. Most connections are reciprocal; the exceptions are indicated by arrows. The frontal eye field (FEF) has variously been considered either adjacent to, or part of, the PFC. Here, we compromise by depicting it as adjacent to, yet touching, the PFC.

of a conflict stimulus (e.g. the word GREEN displayed in red), because there is a strong prepotent tendency to read the word ("green"), which competes with the response to the color ("red"). This illustrates one of the most fundamental aspects of cognitive control and goal-directed behavior: the ability to select a weaker, task-relevant response (or source of information) in the face of competition from an otherwise stronger, but task-irrelevant one. Patients with frontal impairment have difficulty with this task (e.g. Perrett 1974, Cohen & Servan-Schreiber 1992, Vendrell et al 1995), especially when the instructions vary frequently (Dunbar & Sussman 1995, Cohen et al 1999), which suggests that they have difficulty adhering to the goal of the task or its rules in the face of a competing stronger (i.e. more salient or habitual) response.

Similar findings are evident in the WCST. Subjects are instructed to sort cards according to the shape, color, or number of symbols appearing on them and the sorting rule varies periodically. Thus, any given card can be associated with several possible actions, no single stimulus-response mapping will work, and the correct one changes and is dictated by whichever rule is currently in effect. Humans with PFC damage show stereotyped deficits in the WCST. They are able to acquire the initial mapping without much difficulty but are unable to adapt their behavior when the rule varies (Milner 1963). Monkeys with PFC lesions are impaired in an analog of this task (Dias et al 1996b, 1997) and in others when they must switch between different rules (Rossi et al 1999).

The Stroop task and WCST are variously described as tapping the cognitive functions of either selective attention, behavioral inhibition, working memory, or rule-based or goal-directed behavior. In this article, we argue that all these functions depend on the representation of goals and rules in the form of patterns of activity in the PFC, which configure processing in other parts of the brain in accordance with current task demands. These top-down signals favor weak (but task-relevant) stimulus-response mappings when they are in competition with more habitual, stronger ones (such as in the Stroop task), especially when flexibility is needed (such as in the WCST). We believe that this can account for the wide range of other tasks found to be sensitive to PFC damage, such as A-not-B (Piaget 1954, Diamond & Goldman-Rakic 1989), Tower of London (Shallice 1982, 1988; Owen et al 1990), and others (Duncan 1986, Duncan et al 1996), Stuss & Benson 1986).

We build on the fundamental principle that processing in the brain is competitive: Different pathways, carrying different sources of information, compete for expression in behavior, and the winners are those with the strongest sources of support. Desimone & Duncan (1995) have proposed a model that clearly articulates such a view with regard to visual attention. These authors assume that visual cortical neurons processing different aspects of a scene compete with each other via mutually inhibitory interactions. The neurons that "win" the competition and remain active reach higher levels of activity than those with which they share inhibitory interactions. Voluntary shifts of attention result from the influence of excitatory top-down signals representing the to-be-attended features of the scene. These bias the competition among neurons representing the scene, increasing the activity of neurons representing the to-be-attended features and, by virtue of mutual inhibition, suppressing activity of neurons processing other features. Desimone & Duncan suggest that the PFC is an important source of such top-down biasing. However, they left unspecified the mechanisms by which this occurs. That is the focus of this article.

We begin by outlining a theory that extends the notion of biased competition and proposes that it provides a fundamental mechanism by which the PFC exerts control over a wide range of processes in the service of goal-directed behavior. We describe the minimal set of functional properties that such a system must exhibit if it can serve as a mechanism of cognitive control. We then review the existing literature that provides support for this set of properties, followed by a discussion of recent computational modeling efforts that illustrate how a system with these properties can support elementary forms of control. Finally, we consider unresolved issues that provide a challenge for future empirical and theoretical research.

Overview of the Theory

We assume that the PFC serves a specific function in cognitive control: the active maintenance of patterns of activity that represent goals and the means to achieve them. They provide bias signals throughout much of the rest of the brain, affecting not only visual processes but also other sensory modalities, as well as systems responsible for response execution, memory retrieval, emotional evaluation, etc. The aggregate effect of these bias signals is to guide the flow of neural activity along pathways that establish the proper mappings between inputs, internal states, and outputs needed to perform a given task. This is especially important whenever stimuli are ambiguous (i.e. they activate more than one input representation), or when multiple responses are possible and the task-appropriate response must compete with stronger alternatives. From this perspective, the constellation of PFC biases—which resolves competition, guides activity along appropriate pathways, and establishes the mappings needed to perform the task—can be viewed as the neural implementation of attentional templates, rules, or goals, depending on the target of their biasing influence.

To help understand how this might work, consider the schematic shown in Figure 2 (see color insert). Processing units are shown that correspond to cues (C1, C2, C3). They can be thought of as neural representations of sensory events, internal states (e.g. stored memories, emotions, etc), or combinations of these. Also shown are units corresponding to the motor circuits mediating two responses (R1 and R2), as well as intervening or "hidden" units that define processing pathways between cue and response units. We have set up the type of situation for which the PFC is thought to be important. Namely, one cue (C1) can lead to either of two responses (R1 or R2) depending on the situation (C2 or C3), and appropriate behavior depends on establishing the correct mapping from C1 to R1 or R2. For example, imagine you are standing at the corner of a street (cue C1). Your natural reaction is to look left before crossing (R1), and this is the correct thing to do in most of the world (C2). However, if you are in England (C3), you should look right (R2). This is a classic example of a circumstance requiring cognitive control, which we assume depends on the PFC. How does the PFC mediate the correct behavior?

We assume that cues in the environment activate internal representations within the PFC that can select the appropriate action. This is important when the course of action is uncertain, and especially if one of the alternatives is stronger (i.e. more habitual or more salient) but produces the incorrect behavior. Thus, standing at the corner (C1), your "automatic" response would be to look left (R1). However, other cues in the environment "remind" you that you are in England (C3). That is, the cues activate the corresponding PFC representation, which includes information about the appropriate action. This produces excitatory bias signals that guide neural activity along the pathway leading you to look right (e.g. C1 $\rightarrow \cdots \rightarrow$ R2). Note that activation of this PFC representation is necessary for you to perform the correct behavior. That is, you had to keep "in mind" the knowledge that you were in England. You might even be able to cross a few streets correctly while keeping this knowledge in mind, that is, while activity of the appropriate representation is maintained in the PFC. However, if this activity subsides—that is, if you "forget" you are in England—you are likely to revert to the more habitual response and look left. Repeated selection can strengthen the pathway from C1 to R2 and allow it to become independent of the PFC. As this happens, the behavior becomes more automatic, so you can look right without having to keep in mind that you are in England. An important question is how the PFC develops the representations needed to produce the contextually appropriate response.

In an unfamiliar situation you may try various behaviors to achieve a desired goal, perhaps starting with some that have been useful in a similar circumstance (looking to the left for oncoming traffic) and, if these fail, trying others until you meet with success (e.g. by looking right). We assume that each of these is associated with some pattern of activity within the PFC (as in Figure 2). When a behavior meets with success, reinforcement signals augment the corresponding pattern of activity by strengthening connections between the PFC neurons activated by that behavior. This process also strengthens connections between these neurons and those whose activity represents the situation in which the behavior was useful, establishing an association between these circumstances and the PFC pattern that supports the correct behavior. With time (and repeated iterations of this process), the PFC representation can be further elaborated as subtler combinations of events and contingencies between them and the requisite actions are learned. As is discussed below, brainstem neuromodulatory systems may provide the relevant reinforcement signals, allowing the system to "bootstrap" in this way.

Obviously, many details need to be added before we fully understand the complexity of cognitive control. But we believe that this general notion can explain many of the posited functions of the PFC. The biasing influence of PFC feedback signals on sensory systems may mediate its role in directing attention (Stuss & Benson 1986; Knight 1984, 1997; Banich et al 2000), signals to the motor system may be responsible for response selection and inhibitory control (Fuster 1980, Diamond 1988), and signals to intermediate systems may support short-term (or working) memory (Goldman-Rakic 1987) and guide retrieval from long-term memory (Schachter 1997, Janowsky et al 1989, Gershberg & Shimamura 1995). Without the PFC, the most frequently used (and thus best established) neural pathways would predominate or, where these don't exist, behavior would be haphazard. Such impulsive, inappropriate, or disorganized behavior is a hallmark of PFC dysfunction in humans (e.g. Bianchi 1922, Duncan 1986, Luria 1969, Lhermitte 1983, Shallice & Burgess 1996, Stuss & Benson 1986).

Minimal Requirements for a Mechanism of Top-Down Control

There are several critical features of our theory. First, the PFC must provide a source of activity that can exert the required pattern of biasing signals to other structures. We can thus think of PFC function as "active memory in the service of control." It follows, therefore, that the PFC must maintain its activity robustly against distractions until a goal is achieved, yet also be flexible enough to update its representations when needed. It must also house the appropriate representations, those that can select the neural pathways needed for the task. Insofar as primates are capable of tasks that involve diverse combinations of stimuli, internal states, and responses, representations in the PFC must have access to and be able to influence a similarly wide range of information in other brain regions. That is, PFC representations must have a high capacity for multimodality and integration. Finally, as we can acquire new goals and means, the PFC must also exhibit a high degree of plasticity. Of course, it must be possible to exhibit all these properties without the need to invoke some other mechanism of control to explain them, lest our theory be subject to perennial concerns of a hidden "homunculus."

The rapidly accumulating body of findings regarding the PFC suggests that it meets these requirements. Fuster (1971, 1973, 1995), Goldman-Rakic (1987, 1996), and others have extensively explored the ability of PFC neurons to maintain task-relevant information. Miller et al (1996) have shown that this is robust to interference from distraction. Fuster has long advocated the role of the PFC in integrating diverse information (Fuster 1985, 1995). The earliest descriptions of the effects of frontal lobe damage suggested its role in attention and the control of behavior (Ferrier 1876, Bianchi 1922), and investigators since have interpreted the pattern of deficits following PFC damage as a loss of the ability to acquire and use behavior-guiding rules (Shallice 1982, Duncan 1986, Passingham 1993, Grafman 1994, Wise et al 1996). Recent empirical studies have begun to identify neural correlates of plasticity in the PFC (Asaad et al 1998, Bichot et al 1996, Schultz & Dickinson 2000), and recent computational studies suggest how these may operate as mechanisms for self-organization (Braver & Cohen 2000, Egelman et al 1998). Our purpose in this article is to bring these various observations and arguments together, and to illustrate that a reasonably coherent, and mechanistically explicit, theory of PFC function is beginning to emerge. The view presented here draws on previous work that has begun to outline such a theory (e.g. Cohen & Servan-Schreiber 1992; Cohen et al 1996; O'Reilly et al 1999; Miller 1999, 2000). In the sections that follow, we review neurobiological, neuropsychological, and neuroimaging findings that support this theory, and computational modeling studies that have begun to make explicit the processing mechanisms involved.

PROPERTIES OF THE PFC

Convergence of Diverse Information

One of the critical features for a system of cognitive control is the requirement that it have access to diverse information about both the internal state of the system and the external state of the world. The PFC is anatomically well situated to meet this requirement. The cytoarchitectonic areas that comprise the monkey PFC are often grouped into regional subdivisions, the orbital and medial, the lateral, and the mid-dorsal (see Figure 1). Collectively, these areas have interconnections with virtually all sensory systems, with cortical and subcortical motor system structures, and with limbic and midbrain structures involved in affect, memory, and reward. The subdivisions have partly unique, but overlapping, patterns of connections with the rest of the brain, which suggests some regional specialization. However, as in much of the neocortex, many PFC connections are local; there are extensive connections between different PFC areas that are likely to support an intermixing of disparate information. Such intermixing provides a basis for synthesizing results from, and coordinating the regulation of, a wide variety of brain processes, as would be required of a brain area responsible for the orchestration of complex behavior.

Sensory Inputs   The lateral and mid-dorsal PFC is more closely associated with sensory neocortex than is the ventromedial PFC (see Figure 1). It receives visual, somatosensory, and auditory information from the occipital, temporal, and parietal cortices (Barbas & Pandya 1989, 1991; Goldman-Rakic & Schwartz 1982; Pandya & Barnes 1987; Pandya & Yeterian 1990; Petrides & Pandya 1984, 1999; Seltzer & Pandya 1989). Many PFC areas receive converging inputs from at least two sensory modalities (Chavis & Pandya 1976; Jones & Powell 1970). For example, the dorsolateral (DL) (areas 8, 9, and 46) and ventrolateral (12 and 45) PFC both receive projections from visual, auditory, and somatosensory cortex. Furthermore, the PFC is connected with other cortical regions that are themselves sites of multimodal convergence. Many PFC areas (9, 12, 46, and 45) receive inputs from the rostral superior temporal sulcus, which has neurons with bimodal or trimodal (visual, auditory, and somatosensory) responses (Bruce et al 1981, Pandya & Barnes 1987). The arcuate sulcus region (areas 8 and 45) and area 12 seem to be particularly multimodal. They contain zones that receive overlapping inputs from three sensory modalities (Pandya & Barnes 1987). In all these cases, the PFC is directly connected with secondary or "association" but not primary sensory cortex.

Motor Outputs   The dorsal PFC, particularly DL area 46, has preferential connections with motor system structures that may be central to how the PFC exerts control over behavior. The DL area 46 is interconnected (*a*) with motor areas in the medial frontal lobe such as the supplementary motor area, the pre–supplementary motor area, and the rostral cingulate, (*b*) with the premotor cortex on the lateral frontal lobe, and (*c*) with cerebellum and superior colliculus (Bates & Goldman-Rakic 1993, Goldman & Nauta 1976, Lu et al 1994, Schmahmann & Pandya 1997). The DL area 46 also sends projections to area 8, which contains the frontal eye fields, a region important for voluntary shifts of gaze. There are no direct connections between the PFC and primary motor cortex, but they are extensive with premotor areas that, in turn, send projections to primary motor cortex and the spinal cord. Also important are the dense interconnections between the PFC and basal ganglia (Alexander et al 1986), a structure that is likely to be crucial for automating behavior. The basal ganglia receives inputs from much of the cerebral cortex, but its major output (via the thalamus) is frontal cortex (see Figure 1).

Limbic Connections   The orbital and medial PFC are closely associated with medial temporal limbic structures critical for long-term memory and the processing of internal states, such as affect and motivation. This includes direct and indirect (via the medial dorsal thalamus) connections with the hippocampus and associated neocortex, the amygdala, and the hypothalamus (Amaral & Price 1984, Barbas & De Olmos 1990, Barbas & Pandya 1989, Goldman-Rakic et al 1984, Porrino et al 1981, Van Hoesen et al 1972). Other PFC regions have access to these systems both through connections with the orbital and medial PFC and through other intervening structures.

Intrinsic Connections   Most PFC regions are interconnected with most other PFC regions. There are not only interconnections between all three major subdivisions (ventromedial, lateral, and mid-dorsal) but also between their constituent areas (Barbas & Pandya 1991, Pandya & Barnes 1987). The lateral PFC is particularly well connected. Ventrolateral areas 12 and 45 are interconnected with DL areas 46 and 8, with dorsal area 9, as well as with ventromedial areas 11 and 13. Intrinsic connections within the PFC allow information from regional afferents and processes to be distributed to other parts of the PFC. Thus, the PFC provides a venue by which information from wide-ranging brain systems can interact through relatively local circuitry.

Convergence and Plasticity

Given that goal-directed behavior depends on our ability to piece together relationships between a wide range of external and internal information, it stands to reason that top-down control must come from PFC representations that reflect a wide range of learned associations. There is mounting neurophysiological evidence that this is the case. Asaad et al (1998) trained monkeys to associate, on different blocks of trials, each of two cue objects with a saccade to the right or a saccade to the left. They found relatively few lateral PF neurons whose activity simply reflected a cue or response. Instead, the modal group of neurons (44% of the population) showed activity that reflected the current association between a visual cue and a directional saccade it instructed. For example, a given cell might only be strongly activated when object "A" instructed "saccade left" and not when object "B" instructed the same saccade or when object "A" instructed another saccade (Figure 3A). Lateral PFC neurons can also convey the degree of association between a cue and a response (Quintana & Fuster 1992).

Other studies indicate that PFC neurons acquire selectivity for features to which they are initially insensitive but are behaviorally relevant. For example, Bichot et al (1996) observed that neurons in the frontal eye fields (in the bow of the arcuate sulcus)—ordinarily not selective to the form and color of stimuli—became so as the animal learned eye movements that were contingent on these features. Similarly, Watanabe (1990, 1992) has trained monkeys to recognize that certain visual and auditory stimuli signaled whether or not, on different trials, a reward (a drop of juice) would be delivered. He found that neurons in lateral PFC (around the arcuate sulcus and posterior end of the principal sulcus) came to reflect specific cue-reward associations. For example, a given neuron could show strong activation to one of the two auditory (and none of the visual) cues, but only when it signaled reward. Other neurons were bimodal, activated by both visual and auditory cues but also strongly modulated by their reward status.

More complicated behaviors depend not on simple contingencies between cues and responses or rewards but on general principles or rules that may involve more-complex mapping. PFC activity also seems to represent this information. Barone & Joseph (1989) observed cells near the arcuate sulcus that were responsive to specific light stimuli, but only when they occurred at a particular point in a particular

---

Figure 3 (A) Shown is the activity of four single prefrontal (PF) neurons when each of two objects, on different trials, instructed either a saccade to the right or a saccade to the left. The lines connect the average values obtained when a given object cued one or the other saccade. The error bars show the standard error of the mean. Note that in each case, the neuron's activity depends on both the cue object and the saccade direction and that the tuning is nonlinear or conjunctive. That is, the level of activity to a given combination of object and saccade cannot be predicted from the neuron's response to the other combinations. [Adapted from Asaad et al (1998).] (B) A PF neuron whose neural response to a cue object was highly dependent on task context. The bottom half shows an example of a single PF neuron's response to the same cue object during an object task (delayed matching to sample) and during an associative task (conditional visual motor). Note that the neuron is responsive to the cue during one task but not during the other, even though sensory stimulation is identical across the tasks. [Adapted from Asaad et al (2000).]

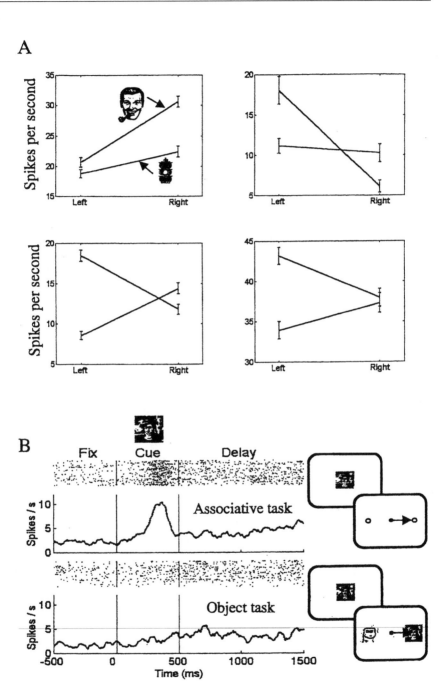

sequence that the monkey had to imitate. White & Wise (1999) trained a monkey to orient to a visual target according to two different rules. One of four cue patterns briefly appeared at one of four locations. The cue indicated where a target would eventually appear. It did so by one of two rules, a spatial rule (the cue appeared at the same location that the target would appear) or an associative rule (the identity of the cue instructed the location, e.g. cue A indicated the right, cue B the left, etc). They found that up to half of lateral PFC neurons showed activity that varied with the rule. Another example was provided by Asaad et al (2000), who trained monkeys to alternate between tasks that employed the same cues and responses but three different rules: matching (delayed matching to sample), associative (conditional visuomotor), and spatial (spatial delayed response). Over half of lateral PFC neurons were rule dependent. Neural responses to a given cue or forthcoming saccade often depended on which rule was current in the task (Figure 3B). Plus, the baseline activity of many neurons (54%) varied with the rule. Hoshi et al (1998) have also observed PFC neurons that were modulated by whether the monkey was using a shape-matching or location-matching rule. Recently, Wallis et al (2000) have shown that lateral and orbitofrontal PFC neurons reflect whether the monkey is currently using a "matching" or "nonmatching" rule to select a test object.

Studies of monkeys and humans with PFC damage also suggest that the PFC is critical for learning rules. For example, Petrides found that following PFC damage, patients could no longer learn arbitrary associations between visual patterns and hand gestures (Petrides 1985, 1990). In monkeys, damage to ventrolateral area 12 or to the arcuate sulcus region also impairs the ability to learn arbitrary cue-response associations (Halsband & Passingham 1985, Murray et al 2000, Petrides 1982, 1985). Learning of visual stimulus-response conditional associations is also impaired by damage to PFC inputs from the temporal cortex (Eacott & Gaffan 1992, Gaffan & Harrison 1988, Parker & Gaffan 1998). Passingham (1993) argues that most, if not all, tasks that are disrupted following PFC damage depend on acquiring conditional associations (if-then rules).

In sum, these results indicate that PFC neural activity represents the rules, or mappings required to perform a particular task, and not just single stimuli or forthcoming actions. We assume that this activity within the PFC establishes these mappings by biasing competition in other parts of the brain responsible for actually performing the task. These signals favor task-relevant sensory inputs (attention), memories (recall), and motor outputs (response selection) and thus guide activity along the pathways that connect them (conditional association).

Feedback to Other Brain Areas

Our model of PFC function requires feedback signals from the PFC to reach widespread targets throughout the brain. The PFC has the neural machinery to provide these feedback signals; it sends projections to much of the neocortex (Pandya & Barnes 1987, Pandya & Yeterian 1990). Physiological studies have yielded results consistent with this notion.

Fuster et al (1985) and Chafee & Goldman-Rakić (2000) have found that deactivating the lateral PFC cortex attenuates the activity of visual cortical (inferior temporal and posterior parietal) neurons to a behaviorally relevant cue. Tomita et al (1999) directly explored the role of top-down PFC signals in the recall of visual memories stored in the inferior temporal (IT) cortex. Appearance of a cue object instructed monkeys to recall and then choose another object that was associated with the cue during training. In the intact brain, information is shared between IT cortices in the two cerebral hemispheres. By severing the connecting fibers, each IT cortex could only "see" (receive bottom-up inputs from) visual stimuli in the contralateral visual field. The fibers connecting the PFC in each hemisphere were left intact. When Tomita et al examined activity of single neurons in an IT cortex that could not "see" the cue, it nonetheless reflected the recalled object, albeit with a long latency. It appeared that visual information took a circuitous route, traveling from the opposite IT cortex (which could "see" the cue) to the still-connected PFC in each hemisphere and then down to the "blind" IT cortex. This was confirmed by severing the PFC in the two hemispheres and eliminating the feedback, which abolished the IT activity and disrupted task performance.

Other evidence suggestive of PFC-IT interactions also comes from investigations, by Miller & Desimone (1994) and Miller et al (1996), of the respective roles of the PFC and IT cortex in working memory. During each trial, monkeys were shown first a sample stimulus. Then, one to four test stimuli appeared in sequence. If a test stimulus matched the sample, the monkey indicated so by releasing a lever. Sometimes, one of the intervening nonmatch stimuli could be repeated. For example, the sample stimulus "A" might be followed by "B ... B ... C ... A." The monkey was only rewarded for responding to the final match ("A") and thus had to maintain a specific representation of the sample rather than respond to any repetition of any stimulus. As noted in the next section, neurons were found in the PFC that exhibited sustained sample-specific activity that survived the presentation of intervening distractors. This was not so for IT cortex. However, neurons in both areas showed a selective enhancement of responses to a match of the sample. The fact that IT neurons had not maintained a representation of this stimulus suggests that their enhanced response to the match might have resulted from interactions with the representation maintained in the PFC. This is consistent with the recent finding indicating that, in a target detection task, target-specific activity appears simultaneously within the PFC and the visual cortex (Anderson et al 1999). Together, these findings suggest that identification of an intended stimulus relies on interactions between the PFC and the posterior cortex.

Active Maintenance

If the PFC represents the rules of a task in its pattern of neural activity, it must maintain this activity as long as the rule is required. Usually this extends beyond the eliciting event and must span other intervening, irrelevant, and potentially interfering events. The capacity to support sustained activity in the face of interference is one of the distinguishing characteristics of the PFC.

Sustained neural activity within the PFC was first reported by Fuster (1971) and Kubota & Niki (1971) and has subsequently been reported in a large number of studies. These have demonstrated that neurons within the PFC remain active during the delay between a transiently presented cue and the later execution of a contingent response. Such delay period activity is often specific to a particular type of information, such as the location and/or identity of a stimulus (di Pellegrino & Wise 1991; Funahashi et al 1989; Fuster 1973; Fuster & Alexander 1971; Kubota & Niki 1971; Rainer et al 1998a,b, 1999; Rao et al 1997; Romo et al 1999), forthcoming actions (Asaad et al 1998, Ferrera et al 1999, Quintana & Fuster 1992), expected rewards (Leon & Shadlen 1999, Tremblay et al 1998, Watanabe 1996), and more-complex properties such as the sequential position of a stimulus within an ordered series (Barone & Joseph 1989) or a particular association between a stimulus and its corresponding response (Asaad et al 1998). Functional neuroimaging studies have begun to yield similar results with humans (Cohen et al 1997, Courtney et al 1997, Prabhakaran et al 2000).

Other areas of the brain exhibit a simple form of sustained activity. For example, in many cortical visual areas, a brief visual stimulus will evoke activity that persists from several hundred milliseconds to several seconds (Fuster & Jervey 1981; Gnadt & Andersen 1988, Miller et al 1993, Miyashita & Chang 1988). What appears to distinguish the PFC is the ability to sustain such activity in the face of intervening distractions. When monkeys must sustain the memory of a sample object over a delay filled with visual distractors, each of which must be attended and processed, sustained activity in the PFC can maintain the sample memory across the distractors (Miller et al 1996). By contrast, sustained activity in extrastriate visual areas (such as the IT and posterior parietal cortex) is easily disrupted by distractors (Constantinidis & Steinmetz 1996; Miller et al 1993, 1996). Thus, posterior cortical neurons seem to reflect the most recent input regardless of its relevance, whereas the PFC selectively maintains task-relevant information.

Learning "Across Time" Within the PFC

Typically, the internal representation of goals and associated rules must be activated in anticipation of the behavior they govern. Furthermore, as we have seen, rules often involve learning associations between stimuli and behaviors that are separated in time. How can associations be learned between a rule or event that occurs at one point in time and contingent behaviors or rewards that occur later? The capacity of the PFC for active maintenance, coupled with its innervation by brainstem dopaminergic systems, suggests one way in which this might occur.

The capacity to actively maintain representations over time is fundamental to associative learning, as it allows information about fleeting events and actions to comingle that would otherwise be separated in time (Fuster 1985). For example, consider the Asaad et al (1998) study discussed above, in which the monkey needed to associate a cue object with the direction of a saccade that had to be made after the cue was no longer present. Presumably, this was made possible by sustained activity within the PFC that insured that a representation of the cue persisted until a saccade was made. Furthermore, as learning progressed, activity related to the forthcoming saccade direction was triggered progressively earlier. Thus, even though the cue and action were separated in time, information about each was simultaneously present in the PFC, permitting an association to be formed between them. In addition to associating temporally separate events that were needed to form task rules, the PFC must be able to associate those rules, and the conditions that elicited them, with subsequent reward. This is necessary if patterns of PFC activity responsible for achieving a goal are to be reinforced, so that they are likely to recur under the appropriate conditions in the future. This function may be served by dopaminergic projections from the midbrain ventral tegmental area.

Midbrain dopamine (DA) neurons exhibit relatively low levels of spontaneous firing but give bursts of activity to behaviorally salient events, especially the delivery of unpredicted, desirable stimuli, such as food or juice rewards (Mirenowicz & Schultz 1994, 1996). As learning progresses, however, DA neurons become activated progressively earlier in time, by events that predict reward, and cease their activation to the now-expected reward (Schultz et al 1993). If the predicted reward fails to appear, activity is inhibited at the expected time of its delivery (Hollerman & Schultz 1998), and if the reward (or an event that has come to predict it) appears earlier than expected, it will again elicit DA neural responses. Thus, midbrain DA neurons seem to be coding "prediction error," the degree to which a reward, or a cue associated with reward, is surprising (Montague et al 1996, Schultz 1998). There is growing evidence that this mechanism operates within PFC. Neurons throughout the PFC (e.g. in lateral as well as ventromedial areas) convey information about expected rewards and show enhanced activity as the size and desirability of an expected reward increases (Leon & Shadlen 1999; Tremblay & Schultz 1999; Watanabe 1990, 1992, 1996). Similar observations have been made in human neuroimaging studies (London et al 2000, O'Doherty et al 2000).

The aim of the cognitive system is not only to predict reward but to pursue the actions that will ensure its procurement. The prediction error signal could help mediate this learning by selectively strengthening not only connections among neurons that provide information about the prediction of reward (Schultz et al 1997), but also their connections with representations in the PFC that guide the behavior needed to achieve it. The role of the PFC in mediating this relationship is suggested by the observation that frontally damaged patients exhibit disturbances in learning and decision-making tasks that involve the evaluation of reward (Bechara et al 1994, 1997; Rolls 2000). Evidence that a predictive reinforcement-learning mechanism may operate within the PFC was also observed by Asaad et al (1998). In this experiment, monkeys learned to associate visual cues with one of two saccadic responses. Initially, the monkeys chose their responses at random, but learned the correct cue-response pairing over a few trials. As they learned the association, neural activity representing the forthcoming saccadic response appeared progressively earlier on successive trials. In other words, the initiation of response-related delay activity gradually shifted, with learning, from a point in time just before the execution of the response and reward delivery to progressively earlier points in time, until it was nearly coincident with the presentation of the cue. This evolution with learning closely resembles that of the reward-prediction signal thought to be mediated by the DA system (Montague et al 1996). In the section that follows, we discuss computational modeling work that provides further support for the plausibility of a DA-based learning mechanism that can establish an association between anticipation of reward and activation of representations in the PFC needed to achieve it.

Summary

Our review of studies in monkeys and humans shows that the PFC exhibits the properties required to support a role in cognitive control: sustained activity that is robust to interference; multimodal convergence and integration of behaviorally relevant information; feedback pathways that can exert biasing influences on other structures throughout the brain; and ongoing plasticity that is adaptive to the demands of new tasks. Of course, these properties are not unique to the PFC. They can be found elsewhere in the brain, to varying degrees and in various combinations. However, we argue that the PFC represents a specialization along this particular combination of dimensions that is optimal for a role in the brain-wide control and coordination of processing. In the section that follows, we consider the theoretical implications of this model, with reference to computational analyses that illustrate these in explicit and concrete form.

A GUIDED ACTIVATION THEORY OF PFC FUNCTION

Explorations of how properties of the PFC might be implemented to mediate control have been conducted using neural network models. Such models attempt to simulate the behavioral performance of human subjects (or animals) in cognitive tasks using neurobiologically plausible mechanisms (e.g. the spread of activity among simple processing units along weighted connections) in order to identify the principles that are most relevant to behavior. Using this approach, Dehaene & Changeux (1989, 1992), Levine & Prueitt (1989), Cohen & Servan-Schreiber (1992), and Braver et al (1995) have all described models of PFC function and have used these to simulate the performance of normal and frontally damaged patients in tasks that are sensitive to PFC damage, such as the Stroop task, WCST, and others. These models capture many of the PFC properties reviewed in the first part of this article and suggest how they might interact to engender cognitive control.

A Simple Model of PFC Function

Most neural network models that address the function of the PFC simulate it as the activation of a set of "rule" units whose activation leads to the production of a

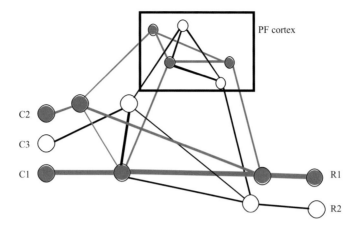

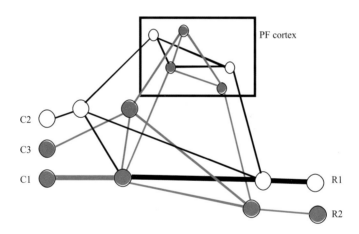

Figure 2 Schematic diagram illustrating our suggested role for the PF cortex in cognitive control. Shown are processing units representing cues such as sensory inputs, current motivational state, memories, etc. (C1, C2, and C3), and those representing two voluntary actions (e.g., "responses", R1 and R2). Also shown are internal or "hidden" units that represent more central stages of processing. The PF cortex is not heavily connected with primary sensory or motor cortices but instead connected with higher-level "association" and premotor cortices. Hence, we illustrate connections between the PFC and the hidden units. Reward signals foster the formation of a task model, a neural representation that reflects the learned associations between task-relevant information. A subset of the information (e.g., C1 and C2) can then evoke the entire model, including information about the appropriate response (e.g., R1). Excitatory signals from the PF cortex feeds back to other brain systems to enable task-relevant neural pathways. Thick lines indicate well-established pathways mediating a prepotent behavior. Red indicates active units or pathways.

A. No Control

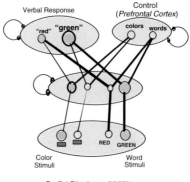

B. Color Naming

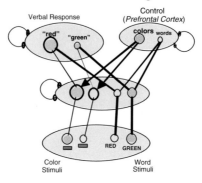

Figure 4 Schematic of the Stroop model. Circles represent processing units, corresponding to a population of neurons assumed to code a given piece of information. Lines represent connections between units, with heavier ones indicating stronger connections. Looped connections with small black circles indicate mutual inhibition among units within that layer (e.g., between the "red" and "green" output units). Adapted from Cohen, Dunbar & McClelland (1990).

A. No control. Activation of conflicting inputs in the two pathways produces a response associated with the word, due to the stronger connections in the word reading pathway.

B. Presentation of a conflict stimulus. The color unit is activated (indicated by the orange fill), representing the current intent to name the color. This passes activation to the intermediate units in the color naming pathway (indicated by arrows), which primes those units (indicated by larger size), and biases processing in favor of activity flowing along this pathway. This biasing effect favors activation of the response unit corresponding to the color input, even though the connection weights in this pathway are weaker than in the word pathway.

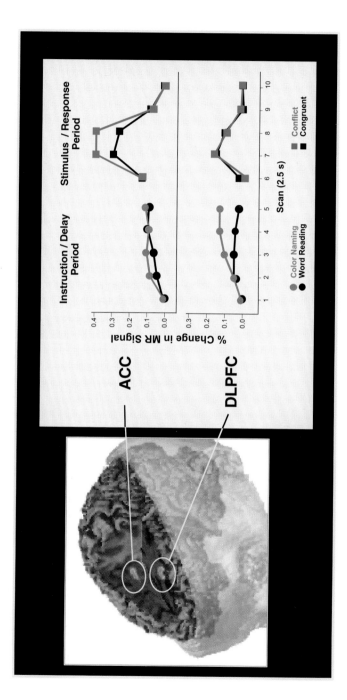

Figure 5 Time course of fMRI activity in dorsolateral prefrontal cortex (DLPFC) and anterior cingulate cortex (ACC) during two phases of a trial in the instructed Stroop task. During the instruction and preparatory period, there is significantly greater activation of DLPFC for color naming than word reading, but no difference in ACC. During the stimulus and response phase, there is greater activation of ACC for conflict than congruent stimuli, but no difference between these for DLPFC. Adapted from MacDonald et al. (2000).

response other than the one most strongly associated with a given input. In most models, the PFC units themselves are not responsible for carrying out input-output mappings needed for performance. Rather, they influence the activity of other units whose responsibility is making the needed mappings. This principle is illustrated in its simplest form by a model of the Stroop task developed by Cohen et al (1990).

This model (Figure 4, see color insert) is made up of five sets of units required for carrying out each of the two possible tasks in a Stroop experiment, color naming and word reading: two sets of input units representing each of the two types of stimulus features (e.g. the colors red and green, and the orthographic features associated with the words RED and GREEN); a set of output units representing each potential response (e.g. the articulatory codes for "red" and "green"); and two sets of intermediate units that provide a pathway between each set of input units and the output units. Connections along the word-reading pathway are stronger as a consequence of more extensive and consistent use. The result is that when a Stroop conflict stimulus is presented (such as the word GREEN printed in red ink), information flowing along the word pathway dominates competition at the response level, and the model responds to the word (see Figure 4*A*). This captures the fact that in the absence of instructions, subjects routinely read the word (i.e. say "green"). However, when they are instructed to do so, subjects can instead name the color (i.e. say "red").

The ability to engage the weaker pathway requires the addition of a set of units (labeled "control" in Figure 4), which in this case, represent the two dimensions of the stimulus (color and word). Each of these control units is connected to intermediate units in the corresponding processing pathway. Activating one of these units biases processing in favor of that pathway by providing additional input to (i.e. "priming") the intermediate units along that pathway. In the case of the color pathway, this allows them to more effectively compete with and prevail over activity flowing along the stronger word pathway (see Figure 4*B*). This biasing effect corresponds to the role of top-down attentional control in the biased competition model proposed by Desimone & Duncan (1995). We assume that the control units in the Stroop model represent the function of neurons within the PFC; they establish the mapping between stimuli and responses required to perform the task. This model provides a concrete implementation of the scheme diagrammed in Figure 2. It and closely related models have been used to simulate quantitative features of the performance of both normal subjects and patients with frontal damage in a wide range of tasks that rely on cognitive control (Braver et al 1995; Cohen & Servan-Schreiber 1992; Cohen et al 1992, 1994a,b, 1996; Dehaene & Changeux 1992; Mozer 1991; Phaf et al 1990).

Guided Activation as a Mechanism of Cognitive Control

The Stroop model brings several features of our theory into focus. First, it emphasizes that the role of the PFC is modulatory rather than transmissive. That is, the pathway from input to output does not "run through" the PFC. Instead, the PFC guides activity flow along task-relevant pathways in more posterior and/or subcortical areas. In this respect, the function of the PFC can be likened to that of a switch operator in a system of railroad tracks. We can think of the brain as a set of tracks (pathways) connecting various origins (e.g. stimuli) to destinations (responses). The goal is to get the trains (activity carrying information) at each origin to their proper destination as efficiently as possible, avoiding any collisions. When the track is clear (i.e. a train can get from its origin to destination without risk of running into any others), then no intervention is needed (i.e. the behavior can be carried out automatically and will not rely on the PFC). However, if two trains must cross the same bit of track, then some coordination is needed to guide them safely to their destinations. Patterns of PFC activity can be thought of as a map that specifies which pattern of "tracks" is needed to solve the task. In the brain, this is achieved by the biasing influence that patterns of PFC activity have on the flow of activity in other parts of the brain, guiding it along pathways responsible for task performance, just as activation of the color-control unit in the Stroop model biased processing in favor of the color-naming pathway. Note that this function need not be restricted to mappings from stimuli to responses but applies equally well to mappings involving internal states (e.g. thoughts, memories, emotions, etc), either as "origins" or "destinations," or both. Thus, depending on their target of influence, we can think of representations in the PFC as attentional templates, retrieval cues, rules, or goals, depending on whether the biasing influences target sensory processes, internal processes, particular courses of action, or their intended outcomes.

This distinction between modulation vs transmission is consistent with the classic pattern of neuropsychological deficits associated with frontal lobe damage. The components of a complex behavior are usually left intact, but the subject is not able to coordinate them in a task-appropriate way (for example, a patient who, when preparing coffee, first stirred and then added cream) (Shallice 1982, Levine et al 1998, Duncan et al 1996). The notion that the function of the PFC is primarily modulatory also makes some interesting and testable predictions. For example, in neuroimaging studies, it should be possible to find circumstances that activate more posterior cortical areas without activation of the PFC whereas it should be much less common to activate the PFC without associated posterior structures. In other words, although there should be circumstances under which transmission can occur without the need for modulation (e.g. word reading in the Stroop task), it does not make sense to have modulation in the absence of transmission.

Active Maintenance in the Service of Control

The Stroop model also illustrates another critical feature of our theory: the importance of sustained activity as a mechanism of control. For a representation to have a biasing influence, it must be activated over the course of performing a task. This feature of the model brings theories of PFC function into direct contact with cognitive psychological constructs, such as the relationship between controlled (PFC-mediated) and automatic processing. The model suggests that this is a continuum, defined by the relative strength of the pathway supporting a task-relevant process compared with those carrying competing information (Cohen et al 1990). Thus, weaker pathways (such as for color naming) rely more on top-down support (i.e. the activity of control units), especially when they face competition from a stronger pathway (e.g. word reading).

This suggests that an increase in the demand for control requires greater or more-enduring PFC activation, which concurs with accumulating evidence from the neuroimaging literature that tasks thought to rely more heavily on controlled processing consistently engage the PFC (Baker et al 1996; Cohen et al 1994a,b, 1997; Frith et al 1991; MacDonald et al 2000; Smith & Jonides 1999; Banich et al 2000). Furthermore, it provides a mechanistic account of the long-standing observation that as a task becomes more practiced, its reliance on control (and the PFC) is reduced. This happens because practice strengthens the connections along the task-relevant pathway in other brain structures. Simulations using the Stroop model (Cohen et al 1990) capture detailed quantitative effects of practice both on measures of performance (e.g. power law improvements in speed of response) and concurrent changes in the reliance on control (e.g. Stroop interference). From a neural perspective, as a pathway is repeatedly selected by PFC bias signals, activity-dependent plasticity mechanisms can strengthen them. Over time, these circuits can function independently of the PFC, and performance of the task becomes more automatic. This concurs with studies showing that PFC damage impairs new learning while sparing well-practiced tasks (Rushworth et al 1997) and neuroimaging and neurophysiological studies that demonstrate greater PFC activation during initial learning and weaker activity as a task becomes more practiced (Knight 1984, 1997; Yamaguchi & Knight 1991; Asaad et al 1998; Shadmehr & Holcomb 1997, Petersen et al 1998).

Our view also provides an interpretation of the relationship between PFC function and working memory. Traditional theories of working memory have distinguished between storage and executive components (Baddeley 1986), with the former responsible for maintaining information online (i.e. in an activated state), and the latter responsible for its manipulation (i.e. the execution of control). Neuropsychological interpretations have placed the storage component in more posterior sensory and motor systems (e.g. Gathercole 1994), whereas the executive control component has been assigned to the PFC. By contrast, as reviewed above, early monkey neurophysiological studies have emphasized the role of the PFC in maintenance. Our theory offers a possible resolution of this dilemma. It suggests that executive control involves the active maintenance of a particular type of information: the goals and rules of a task. This view concurs with cognitive psychological theories based on production system architectures (e.g. ACT*) (Anderson 1983), which posit that executive control relies on the activation of representations that correspond to the goals of a behavior and the rules for achieving it.

This perspective also provides a unifying view of the role of the PFC in other cognitive functions with which it has been associated, most commonly attention and inhibition. Both can be seen as varying reflections, in behavior, of the operation of a single underlying mechanism of cognitive control: the biasing effects of PFC activity on processing in pathways responsible for task performance. As suggested by the biased competition model of Desimone & Duncan (1995), selective attention and behavioral inhibition are two sides of the same coin: Attention is the effect of biasing competition in favor of task-relevant information, and inhibition is the consequence that this has for the irrelevant information. Note that according to this view, inhibition occurs because of local competition among conflicting representations (e.g. between the two responses in the Stroop model) rather than centrally by the PFC. The "binding" function of selective attention (e.g. Treisman & Gelade 1980) can also be explained by such a mechanism if, in this case, we think of PFC representations as selecting the desired combination of stimulus features to be mapped onto the response over other competing combinations.

Finally, it is important to distinguish the form of activity-dependent control that we have ascribed to PFC from other forms of control that may occur in the brain. In particular, we believe that PFC-mediated control is complemented by another form of control dependent on the hippocampal system. The hippocampus is important for binding together information into a memory of a specific episode (Eichenbaum et al 1999, McClelland et al 1995, Squire 1992, Zola-Morgan & Squire 1993). By contrast, we suggest that the PFC, like other neocortical areas, is more important for extracting the regularities across episodes—in the case of the PFC, those corresponding to goals and task rules, rather than episodic memories of actually performing the task. We further posit that the PFC uses "activity-based" control; that is, its ongoing activity specifies the pattern of neural pathways that are currently needed. If PFC activity changes, so does the selected pattern of pathways. By contrast, the hippocampus may provide a form of "weighted-based" control; it helps consolidate permanent associative links between the pieces of information that define a long-term memory (Cohen & O'Reilly 1996, O'Reilly et al 1999, O'Reilly 2000). To use the railroad metaphor, the hippocampus is responsible for laying down new tracks and the PFC is responsible for flexibly switching between them. As noted below, interactions between the PFC and the hippocampus may provide a basis for understanding prospective forms of control, such as planning.

Updating of PFC Representations

In the real world, cognitive control is highly dynamic. People move from one task to the next, and new goals replace old ones. A major benefit of the activity-based mechanism of control that we have proposed is that it is highly flexible. So long as suitable representations exist within the PFC, activating them can quickly invoke a goal or rule, which can be flexibly switched to others as circumstances demand. That is, it is easier and faster (and perhaps less costly) to switch between existing tracks than it is to lay new ones down. This is clearly illustrated by models of PFC function in the WCST (Dehaene & Changeux 1992) and recent variants of this task (RC O'Reilly, DC Noelle, TS Braver, JD Cohen, submitted for publication), and it is supported by the fact that damage to the PFC impairs such flexibility

(Milner 1963, Dias et al 1996b, Rossi et al 1999). It should be noted, however, that the mechanisms responsible for updating representations within the PFC must be able to satisfy two conflicting demands: On the one hand, they must be responsive to relevant changes in the environment (adaptive); on the other, they must be resistant to updating by irrelevant changes (robust). As described above, neurophysiological studies suggest that PFC representations are selectively responsive to task-relevant stimuli (Rainer et al 1998b), yet they are robust to interference from distractors (Miller et al 1996). Conversely, two hallmarks of damage to the PFC are perseveration (inadequate updating) and increased distractibility (inappropriate updating) (e.g. Mishkin 1964, Chao & Knight 1997). These observations suggest the operation of mechanisms that ensure the appropriate updating of PFC activity in response to behavioral demands.

Cohen et al (1996) and Braver & Cohen (2000) have proposed that DA may play an important role in this function. They hypothesize that DA release may "gate" access to the PFC by modulating the influence of its afferent connections. A similar role for DA in the PFC has been suggested by Durstewitz et al (1999, 2000). Timing is a critical feature of such a gating mechanism: The signal-to-gate input must be rapid and coincide with the conditions under which an update is needed. This is consistent with recent studies indicating that DA release (once thought to be slow and nonspecific) has a phasic component with timing characteristics consistent with its proposed role in gating (Schultz 1998, Schultz & Dickinson 2000). As discussed above, midbrain DA neurons give bursts of activity to stimuli that are not predicted but that predict a later reward. This is precisely the timing required for a gating signal responsible for updating goal representations. For example, imagine that you are walking to work and out of the corner of your eye you notice a $20 bill lying on the ground. This unpredicted stimulus predicts reward, but only if you update your current goal and bend down to pick up the bill.

It is intriguing that the properties of midbrain DA neurons that may drive PFC associative learning mechanisms (discussed above) are formally equivalent to those used in models that simulate PFC updating mechanisms. These dual and concurrent influences of DA on gating and learning suggest that if the system learns while it gates, then perhaps it can learn on its own when to gate. That is, if an exploratory DA-mediated gating signal leads to a successful behavior, its coincident reinforcing effects will strengthen the association of this signal with cues representing the current context and the pattern of activity within the PFC that produced the behavior. This will increase the probability that in the future, the same context will elicit a gating signal and reactivation of the PFC activity that led to the rewarded behavior, which in turn will produce further reinforcement of these associations, etc. Recent computational modeling studies establish the plausibility of this bootstrapping mechanism (Braver & Cohen 2000). This ability to self-organize averts the problem of a theoretical regress regarding control (i.e. the invocation of a "homunculus") by allowing the system to learn on its own what signals should produce an updating of the contents of the PFC and when this should occur. However, an important issue that requires further exploration is whether this system can support "subgoaling," or hierarchical updating—that is, the updating of some representations (e.g. the next chess move to make) while preserving others (e.g. the strategy being pursued). A closely related issue concerns the proper sequencing of actions. These abilities are fundamental to virtually all higher cognitive faculties, such as reasoning and problem solving (Newell & Simon 1972), and are known to involve the PFC (e.g. Duncan 1986; Duncan et al 1996; Baker et al 1996; Koechlin 1999; Nichelli et al 1994; Shallice 1982, 1988). Recent modeling work suggests that hierarchical updating and the sequencing of actions may rely on interactions between the PFC and the basal ganglia (Gobbel 1995, Houck 1995); however, a full elaboration of the mechanisms involved remains a challenge for future work.

REMAINING ISSUES

A detailed consideration of all the issues relevant to cognitive control is beyond the scope of this article. However, some additional issues are important for a more complete theory of PFC function and cognitive control. Here, we briefly review several important remaining issues.

Representational Power of the PFC

The tremendous range of tasks of which people are capable raises important questions about the ability of the PFC to support the necessary scope of representations. The large size of the PFC (over 30% of the cortical mass), coupled with its anatomic connectivity discussed earlier, suggests that it can support a wide number and range of mappings. However, there are a nearly limitless number of tasks that a person can be asked to perform (e.g. "wink whenever I say bumblydoodle"), and it seems unlikely that all possible mappings are represented within the PFC. It is possible that some as-yet-undiscovered representational scheme supports a wide enough range of mappings to account for the flexibility of human behavior. However, more likely, it seems that plasticity may also play an important role in PFC function, establishing new representations as they are needed. This need is accentuated by virtue of the fact that the PFC must be able to modulate processes in other parts of the brain that are themselves plastic. Conversely, the circuitry within the PFC that supports older, well-established behaviors is likely to be "reclaimed" as PFC-independent pathways become responsible for them. Above, we reviewed evidence that the PFC exhibits a high degree of plasticity. As yet, however, the mechanisms that govern this plasticity are largely unknown, at either the neurobiological or the computational levels. At the neural level, this could involve the modification of existing synapses, the formation of new ones (perhaps with the assistance of rapid-learning mechanisms in the hippocampus), or even the recruitment of entirely new neurons (Gould et al 1999). At the computational level, we have suggested how predictive reinforcement-learning mechanisms can (*a*) strengthen patterns of PFC activity that appropriately guide behavior and (*b*) associate them with the circumstances in which they are useful. However, the ability of such mechanisms to account for the wide range and extraordinary flexibility of human behavior remain to be established.

PFC Functional Organization

One important question concerns how PFC representations are functionally organized. Understanding the principles of PFC organization is likely to provide insights into how its representations develop and function. Various schemes have been proposed. For example, one possibility is that the PFC is organized by function, with different regions carrying out qualitatively different operations. One long-standing view is that orbital and medial areas are associated with behavioral inhibition, whereas ventrolateral and dorsal regions are associated with memory or attentional functions (Fuster 1989, Goldman-Rakic 1987). Another recent suggestion is that ventral regions support maintenance of information (memory), whereas dorsal regions are responsible for the manipulation of such information (Owen et al 1996, Petrides 1996). Such distinctions have heuristic appeal. However, our theory suggests an intriguing alternative.

If different regions of the PFC emphasize different types of information, then perhaps variations in the biasing signals that they provide can account for apparent dissociations of function. For example, both activity and deficits of orbital PFC are most frequently associated with tasks involving social, emotional, and appetitive stimuli (Hecaen & Albert 1978, O'Doherty et al 2000, Price 1999, Stuss & Benson 1986, Swedo et al 1989), whereas more-dorsal regions are activated in tasks involving more-"cognitive" dimensions of stimuli (form, location, sequential order, etc). Social and appetitive stimuli are "hot," meaning they are more likely to elicit reflexive (and often inappropriate) reactions. Thus, the impression that the orbital PFC subserves an inhibitory function may be explained by the fact that it is more involved in biasing task-relevant processes against strong competing alternatives. In contrast, more-cognitive stimuli (e.g. shapes, locations, etc) that engage more-dorsal regions are "cold," meaning they are less likely to engage responses with such asymmetries of strength. Thus, their competition is likely to be less fierce. Neuropsychological studies of monkeys support this notion (Dias et al 1996a, Roberts & Wallis 2000), and recent computational modeling has shown that dissociations in performance interpreted as evidence of a distinction between inhibitory and memory processes within the PFC can, alternatively, be explained in terms of a single processing mechanism operating over different types of representations (RC O'Reilly, DC Boelle, TS Braver, JD Cohen, submitted for publication).

Other organizational schemes have also been proposed for the PFC, including those based on stimulus dimensions, sensory vs motor, and sequential order (e.g. Barone & Joseph 1989, Wilson et al 1993, Wagner 1999, Casey et al 2000). Although our theory does not provide deep insights into which, if any, is most likely to be correct, it does make strong claims that are related to this issue. First, although it allows for the possibility of broad categories or gradients of organization, it suggests that it is unlikely that different classes of information will be represented in a modular, or discretely localized, form. Complex behavior requires that we recognize and respond to relationships across diverse dimensions, and the role we have ascribed to the PFC involves representing these relationships. Both neurophysiological (Asaad et al 1998, Bichot et al 1996, Quintana & Fuster 1992, Rainer et al 1998a, Rao et al 1997, Watanabe 1990, White & Wise 1999) and neuorimaging (Cohen 2000, Nystrom et al 2000, Prabhakaran et al 2000) findings support this view, which suggests that most regions of the PFC can respond to a variety of different types of information.

Second, our theory suggests that learning will play an important role in the formation of representations in the PFC and, thus, may have an important influence on representational organization. This has been illustrated in a computational model of the PFC. Braver et al (1996) trained a network on a "spatial" memory task and an "object" memory task. The PFC module included some units with projections from both spatial and object pathways, and some units with projections from only one or the other. When the network was trained on each task separately, the PFC module relied primarily on the segregated location units and object units to perform the task. When training on the tasks was intermixed, the activity and number of the multimodal units were increased. This result has found support in empirical studies. Evidence of segregation in the PFC by stimulus domain has been reported by studies that separate training of different stimulus attributes or in monkeys passively viewing stimuli (i.e. in tasks that do not engage PFC function) (O Scalaidhe et al 1997, Wilson et al 1993). Evidence for PFC integration has come from tasks in which stimulus domains are intermixed or their integration is relevant (Asaad et al 1998, Bichot et al 1996, Fuster et al 1982, Prabhakaran et al 2000, Rainer et al 1998a, Rao et al 1997, White & Wise 1999).

Monitoring and the Allocation of Control

Previous discussion has focused on the need to appropriately update representations in the PFC as new goals arise and new rules are applied. However, as noted earlier, people also show a facility for adapting the degree of control they allocate to a task. For example, you pay closer attention to the road on a dark and rainy night than on a bright, sunny day. Such adjustments are adaptive, in view of the well-recognized capacity limits on cognitive control (discussed below). In our model, such adjustments would correspond to strength of the PFC pattern of activity (e.g. the strength of the color unit in the Stroop model). A stronger pattern of activity within the PFC produces stronger biasing effects for a particular pathway, but possibly at the expense of other ones (e.g. through competition among PFC representations). Recent studies have suggested that the allocation of control may depend on signals from the anterior cingulate cortex (ACC) that detect conflict in processing (e.g. Carter et al 1998, 2000; Botvinick et al 1999). Drawing on our train track analogy, conflict occurs when two trains are destined to cross tracks at the same time. In neural terms, this corresponds to the coactivation of competing (i.e. mutually inhibitory) sets of units (e.g. for the responses "red" and "green" in the Stroop task). Such conflict produces uncertainly in processing, and an increased probability of errors. Thus, conflict signals the need for the allocation of additional control. Modeling work (M Botvinick, TS Brauer, CS Carter, DM Barch, JD Cohen, submitted for publication) has shown that coupling the conflict signal (detected by ACC) to adjustments in the allocation of control (amplification of PFC pattern of activity) can accurately simulate trial-based adjustments that subjects make in their behavior in experimental tasks (e.g. Botvinick et al 1999, Gratton et al 1992, Laming 1968, Logan et al 1983, Tzelgov et al 1992).

The tight coupling of conflict detection and allocation of control may explain the pervasive finding of coactivation of the PFC and the ACC in most neuroimaging studies (Owen & Duncan 2000). However, the distinct roles of the PFC and ACC are illustrated by a dissociation of their activity in a recent functional magnetic resonance imaging study by MacDonald et al (2000), using an instructed version of the Stroop task (Figure 5, see color insert). In each trial, subjects were given a cue indicating whether they were to name the color or read the word in the subsequent display. The cue was followed by a delay of several seconds, and then either a congruent or a conflict stimulus was displayed. Figure 5 shows that during the delay, increasing activity was observed within a region of the DL PFC, greater for color naming (the more conrol demanding task) than word reading. There was no differential activation observed within ACC during this period. In contrast, strong activation was observed in ACC during the period of stimulus presentation and responding. This activity was greater for conflict than congruent stimuli. There was no differential response for these trial types within PFC during this period. These findings provide strong support for several of the hypotheses we have discussed: The demands for control are associated with an increase in PFC activity; tasks demanding greater control elicit stronger activity within the PFC; and the ACC responds selectively to conflict in processing. However, further work is needed to establish the causal relationship between detection of conflict within the ACC and the augmentation of control by the PFC.

Mechanisms of Active Maintenance

Our theory of the PFC, like many others, emphasizes its capacity for active maintenance. However, there has been relatively little empirical research on the mechanisms responsible for sustained activity. There are a number of theoretical possibilities, which can be roughly divided into two classes: cellular and circuit based. Cellular models propose neuron bistability as the basis of sustained activity, which is dependent on the biophysical properties of individual cells. The transitions between states are triggered by inputs to the PFC but maintained via the activation of specific voltage-dependent conductances (Wang 1999). Circuit-based models, on the other hand, propose that the recirculation of activity through closed (or "recurrent") loops of interconnected neurons, or attractor networks (Hopfield 1982), support self-sustained activity (Zipser et al 1993). These loops could be intrinsic to the PFC (Pucak et al 1996, Melchitzky et al 1998), or they might involve other structures, such as the cortex-striatal-globus pallidus-thalamus-cortex loops (Alexander et al 1996). In either case, it should be noted that a mechanism is needed for regulating the updating of activity within the PFC, as discussed above.

Capacity Limits of Control

A better understanding of the mechanisms underlying active maintenance may provide insight into one of the most perplexing properties of cognitive control: its severely limited capacity. This has long been recognized in cognitive psychology (Broadbent 1958, Posner & Snyder 1975, Shiffrin & Schneider 1977) and is painfully apparent to anyone who has tried to talk on the phone and read e-mail at the same time. It is important to distinguish between this form of capacity, which has to do with how many representations can be actively maintained at the same time, and the issue of representational power discussed above, which has to do with the range of representations that are available to draw from in the PFC. It may also be important to distinguish the capacity limits of cognitive control from those of short-term storage of item information (e.g. verbal or visual short term memory) (Miller 1956, Baddeley 1986). Limits of control presumably reflect properties of PFC function. The limited capacity of short-term memory may involve mechanisms (e.g. articulatory rehearsal) and structures (e.g. sustained activity in posterior cortical areas) that are not central to cognitive control, and that may or may not rely on PFC function. The capacity limits of cognitive control have been used to explain many features of human cognition (e.g. Cowen 1988, Engle et al 1999, Just & Carpenter 1992, Posner & Snyder 1975, Shiffrin & Schneider 1977). However, to date, no theory has provided an explanation of the capacity limitation itself. This could reflect an inherent physiological constraint, such as the energetic requirements of actively maintaining representations in the PFC. More likely, it reflects fundamental computational properties of the system, such as an inherent limit on the number of representations that can be actively maintained and kept independent of one another within an attractor network (e.g. Usher & Cohen 1999). In any event, capacity constraints are a sine qua non of cognitive control and, therefore, provide an important benchmark for theories that seek to explain its underlying mechanisms.

Prospective Control and Planning

Perhaps the most impressive feature of human cognition is its ability to plan for the future. We often forgo pursuing a given goal until a more appropriate time. Active maintenance cannot account for this. When we plan in the morning to go to the grocery story on the way home from work, it seems unlikely that we actively maintain this information in the PFC throughout the day. More likely, this information is stored elsewhere and then activated at the appropriate time. This may involve interactions between the PFC and other brain systems capable of rapid learning, such as the hippocampus (cf Cohen & O'Reilly 1996, O'Reilly & McClelland 1994). Thus, it is possible that the hippocampus rapidly encodes an association between the desired goal representation(s) within the PFC (e.g. go to the grocery store) and features of the circumstance under which the goal should be evoked (e.g. commuting home). Then, as these circumstances arise, the appropriate representation within the PFC is associatively activated, guiding performance in accord with the goal and its associated rules (e.g. turning right at the light toward the store rather than the habitual left toward home). Neurophysiological studies suggest that the PFC is important for the ability to prospectively activate long-term memories (Rainer et al 1999, Tomita et al 1999). However, the detailed nature of such interactions, and their relationship to the dopaminergic gating and learning mechanisms described above, remain to be fully specified.

CONCLUSIONS

One of the great mysteries of the brain is cognitive control. How can interactions between millions of neurons result in behavior that is coordinated and appears willful and voluntary? There is consensus that it depends on the PFC, but there has been little understanding of the neural mechanisms that endow it with the properties needed for executive control. Here, we have suggested that this stems from several critical features of the PFC: the ability of experience to modify its distinctive anatomy; its wide-ranging inputs and intrinsic connections that provide a substrate suitable for synthesizing and representing diverse forms of information needed to guide performance in complex tasks; its capacity for actively maintaining such representations; and its regulation by brainstem neuromodulatory systems that provide a means for appropriately updating these representations and learning when to do so. We have noted that depending on their target of influence, representations in the PFC can function variously as attentional templates, rules, or goals by providing top-down bias signals to other parts of the brain that guide the flow of activity along the pathways needed to perform a task. We have pointed to a rapidly accumulating and diverse body of evidence that supports this view, including findings from neurophysiological, neuroanatomical, human behavioral and neuroimaging, and computational modeling studies.

The theory we have described provides a framework within which to formulate hypotheses about the specific mechanisms underlying the role of the PFC in cognitive control. We have reviewed a number of these, some of which have begun to take explicit form in computational models. We have also provided a sampling of the many questions that remain about these mechanisms and the functioning of the PFC. Regardless of whether the particular hypotheses we have outlined accurately describe PFC function, they offer an example of how neurally plausible mechanisms can exhibit the properties of self-organization and self-regulation required to account for cognitive control without recourse to a "homunculus." At the very least, we hope that they provide some useful examples of how the use of a computational and empirical framework, in an effort to be mechanistically explicit, can provide valuable leads in this conceptually demanding pursuit. We believe that future efforts to address the vexing, but important, questions surrounding PFC function and cognitive control will benefit by ever tighter coupling of neurobiological experiments and detailed computational analysis and modeling.

ACKNOWLEDGMENTS

The order of authorship was arbitrarily determined; both authors made equal contributions to the research and preparation of this manuscript. Our work has been supported by grants from the NINDS, the NIMH, the RIKEN-MIT Neuroscience Research Center, the Pew Charitable Trusts, the McKnight Foundation, the Whitehall Foundation, the John Merck Fund, the Alfred P. Sloan Foundation, and NARSAD. We thank Wael Asaad, Mark Histed, Jonathan Wallis, Richard Wehby, and Marlene Wicherski for valuable comments on this manuscript. EKM would also like to thank Wael Asaad for valuable conversations. JDC would like to thank BJ Casey and Leigh Nystrom for thoughtful discussions and comments regarding this manuscript, as well as Todd Braver, David Noelle, and Randy O'Reilly, with whom many of the theoretical ideas presented in this article were developed.

Visit the Annual Reviews home page at www.AnnualReviews.org

LITERATURE CITED

Alexander GE, Delong MR, Strick PL. 1986. Parallel organization of functionally segregated circuits linking basal ganglia and cortex. *Annu. Rev. Neurosci.* 9:357–81

Amaral DG, Price JL. 1984. Amygdalo-cortical projections in the monkey (Macaca fascicularis). *J. Comp. Neurol.* 230:465–96

Anderson JR. 1983. *The Architecture of Cognition.* Cambridge, MA: Harvard Univ. Press

Anderson KC, Asaad WF, Wallis JD, Miller EK. 1999. Simultaneous recordings from monkey prefrontal (PF) and posterior parietal (PP) cortices during visual search. *Soc. Neurosci. Abstr.* 25:885

Asaad WF, Rainer G, Miller EK. 1998. Neural activity in the primate prefrontal cortex during associative learning. *Neuron* 21:1399–407

Asaad WF, Rainer G, Miller EK. 2000. Task-specific neural activity in the primate prefrontal cortex. *J. Neurophysiol.* 84:451–59

Baddeley A. 1986. *Working Memory.* Oxford: Clarendon

Baker SC, Rogers RD, Owen AM, Frith CD, Dolan RJ, et al. 1996. Neural systems engaged by planning: a PET study of the Tower of London Task. *Neuropsychologia* 34:515–26

Banich MT, Milham MP, Atchley R, Cohen NJ, Webb A, et al. 2000. Prefrontal regions play a predominant role in imposing an attentional "set": evidence from fMRI. *Cogn. Brain Res.* In press

Barbas H, De Olmos J. 1990. Projections from the amygdala to basoventral and mediodorsal prefrontal regions in the rhesus monkey. *J. Comp. Neurol.* 300:549–71

Barbas H, Pandya D. 1991. Patterns of connections of the prefrontal cortex in the rhesus monkey associated with cortical architecture. In *Frontal Lobe Function and Dysfunction,* ed. HS Levin, HM Eisenberg, AL Benton, pp. 35–58. New York: Oxford Univ. Press

Barbas H, Pandya DN. 1989. Architecture and intrinsic connections of the prefrontal cortex in the rhesus monkey. *J. Comp. Neurol.* 286:353–75

Barone P, Joseph JP. 1989. Prefrontal cortex and spatial sequencing in macaque monkey. *Exp. Brain Res.* 78:447–64

Bates JF, Goldman-Rakic PS. 1993. Prefrontal connections of medial motor areas in the rhesus monkey. *J. Comp. Neurol.* 336:211–28

Bechara A, Damasio AR, Damasio H, Anderson SW. 1994. Insensitivity to future consequences following damage to human prefrontal cortex. *Cognition* 50:7–15

Bechara A, Damasio H, Tranel D, Damasio AR. 1997. Deciding advantageously before knowing the advantageous strategy. *Science* 275:1293–95

Bianchi L. 1922. *The Mechanism of the Brain and the Function of the Frontal Lobes.* Edinburgh: Livingstone

Bichot NP, Schall JD, Thompson KG. 1996. Visual feature selectivity in frontal eye fields induced by experience in mature macaques. *Nature* 381:697–99

Botvinick M, Nystrom LE, Fissell K, Carter CS, Cohen JD. 1999. Conflict monitoring versus selection-for-action in anterior cingulate cortex. *Nature* 402:179–81

Braver TS, Cohen JD, eds. 2000. *On the Control of Control: The Role of Dopamine in Regulating Prefrontal Function and Working Memory.* Cambridge, MA: MIT Press. In press

Braver TS, Cohen JD, Servan-Schreiber D. 1995. A computational model of prefrontal cortex function. In *Advances in Neural Information Processing Systems*, ed. DS Touretzky, G Tesauro, TK Leen, pp. 141–48. Cambridge, MA: MIT Press

Broadbent DE. 1958. *Perception and Communication.* London: Pergamon

Bruce C, Desimone R, Gross CG. 1981. Visual properties of neurons in a polysensory area in superior temporal sulcus of the macaque. *J. Neurophysiol.* 46:369–84

Carter CS, Braver TS, Barch DM, Botvinick MM, Noll D, Cohen JD. 1998. Anterior cingulate cortex, error detection, and the online monitoring of performance. *Science* 280:747–49

Carter CS, Macdonald AM, Botvinick M, Ross LL, Stenger VA, et al. 2000. Parsing executive processes: strategic vs. evaluative functions of the anterior cingulate cortex. *Proc. Natl. Acad. Sci. USA* 97:1944–48

Casey BJ, Forman SD, Franzen P, Berkowitz A, Braver TS, et al. 2000. Sensitivity of prefrontal cortex to changes in target probability: a functional MRI study. *Hum. Brain Mapp.* In press

Chafee MV, Goldman-Rakic PS. 2000. Inactivation of parietal and prefrontal cortex reveals interdependence of neural activity during memory-guided saccades. *J. Neurophysiol.* 83:1550–66

Chao LL, Knight RT. 1997. Prefrontal deficits in attention and inhibitory control with aging. *Cereb. Cortex* 7:63–9

Chavis DA, Pandya DN. 1976. Further observations on cortico-frontal connections in the rhesus monkey. *Brain Res.* 117:369–86

Cohen JD. 2000. Special issue: functional topography of prefrontal cortex. *Neuroimage* 11:378–79

Cohen JD, Barch DM, Carter CS, Servan-Schreiber D. 1999. Schizophrenic deficits in the processing of context: converging evidence from three theoretically motivated cognitive tasks. *J. Abnorm. Psychol.* 108:120–33

Cohen JD, Braver TS, O'Reilly RC. 1996. A computational approach to prefrontal cortex, cognitive control, and schizophrenia: Recent developments and current challenges. *Philos. Trans. Roy. Soc. London B.* 351:1515–1527

Cohen JD, Dunbar K, McClelland JL. 1990. On the control of automatic processes: a parallel distributed processing account of the Stroop effect. *Psychol. Rev.* 97:332–61

Cohen JD, Forman SD, Braver TS, Casey BJ, Servan-Schreiber D, Noll DC. 1994a. Activation of prefrontal cortex in a nonspatial working memory task with functional MRI. *Hum. Brain Mapp.* 1:293–304

Cohen JD, O'Reilly RC. 1996. A preliminary theory of the interactions between the prefrontal cortex and hippocampus that contribute to planning and prospective memory. In *Prospective Memory: Theory and Applications*, ed. M Brandimonte, G Einstein, M McDaniel. Hillsdale, New Jersey: Erlbaum

Cohen JD, Perlstein WM, Braver TS, Nystrom LE, Noll DC, et al. 1997. Temporal dynamics of brain activation during a working memory task. *Nature* 386:604–8

Cohen JD, Romero RD, Farah MJ, Servan-Schreiber D. 1994b. Mechanisms of spatial attention: the relation of macrostructure to microstructure in parietal neglect. *J. Cogn. Neurosci.* 6:377–87

Cohen JD, Servan-Schreiber D. 1992. Context, cortex and dopamine: a connectionist approach to behavior and biology in schizophrenia. *Psychol. Rev.* 99:45–77

Cohen JD, Servan-Schreiber D, McClelland JL. 1992. A parallel distributed processing approach to automaticity. *Am. J. Psychol.* 105:239–69

Constantinidis C, Steinmetz MA. 1996. Neuronal activity in posterior parietal area 7a during the delay periods of a spatial memory task. *J. Neurophysiol.* 76:1352–55

Courtney SM, Ungerleider LG, Keil K, Haxby JV. 1997. Transient and sustained activity in a distributed neural system for human working memory. *Nature* 386:608–12

Cowen N. 1998. Evolving conceptions of memory storage, selective attention, and their mutual constraints within the human information processing system. *Psychol. Bull.* 104:163–91

Dehaene S, Changeux JP. 1989. A simple model of prefrontal cortex function in delayed-response tasks. *J. Cogn. Neurosci.* 1:244–61

Dehaene S, Changeux JP. 1992. The Wisconsin card sorting test: theoretical analysis and modeling in a neuronal network. *Cerebr. Cortex* 1:62–79

Desimone R, Duncan J. 1995. Neural mecha-

nisms of selective visual attention. *Annu. Rev. Neurosci.* 18:193–222

Diamond A. 1988. Abilities and neural mechanisms underlying AB performance. *Child Dev.* 59:523–27

Diamond A, Goldman-Rakic PS. 1989. Comparison of human infants and rhesus monkeys on Piaget's A-not-B task: evidence for dependence on dorsolateral prefrontal cortex. *Exp. Brain Res.* 74:24–40

Dias R, Robbins TW, Roberts AC. 1996a. Dissociation in prefrontal cortex of affective and attentional shifts. *Nature* 380:69–72

Dias R, Robbins TW, Roberts AC. 1996b. Primate analogue of the Wisconsin Card Sorting Test: effects of excitotoxic lesions of the prefrontal cortex in the marmoset. *Behav. Neurosci.* 110:872–86

Dias R, Robbins TW, Roberts AC. 1997. Dissociable forms of inhibitory control within prefrontal cortex with an analog of the Wisconsin Card Sort Test: restriction to novel situations and independence from "on-line" processing. *J. Neurosci* 17:9285–97

di Pellegrino G, Wise SP. 1991. A neurophysiological comparison of three distinct regions of the primate frontal lobe. *Brain* 114:951–78

Dunbar K, Sussman D. 1995. Toward a cognitive account of frontal lobe function: simulating frontal lobe deficits in normal subjects. *Ann. NY Acad. Sci.* 769:289–304

Duncan J. 1986. Disorganization of behaviour after frontal lobe damage. *Cogn. Neuropsychol.* 3:271–90

Duncan J, Emslie H, Williams P, Johnson R, Freer C. 1996. Intelligence and the frontal lobe: the organization of goal-directed behavior. *Cogn. Psychol.* 30:257–303

Duncan J, Owen AM. 2000. Common regions of the human frontal lobe recruited by diverse cognitive demands. *Trends Neurosci.* In press

Durstewitz D, Kelc M, Gunturkun O. 1999. A neurocomputational theory of the dopaminergic modulation of working memory functions. *J. Neurosci.* 19:2807–22

Durstewitz D, Seamans JK, Sejnowski TJ. 2000. Dopamine-mediated stabilization of delay-period activity in a network model of the prefrontal cortex. *J. Neurophysiol.* 83:1733–50

Eacott MJ, Gaffan D. 1992. Inferotemporal-frontal disconnection—the uncinate fascicle and visual associative learning in monkeys. *Eur. J. Neurosci.* 4:1320–32

Eichenbaum H, Dudchenko P, Wood E, Shapiro M, Tanila H. 1999. The hippocampus, memory, and place cells: Is it spatial memory or a memory space? *Neuron* 23:209–26

Engel RW, Kane M, Tuholski S. 1999a. Individual differences in working memory capacity and what they tell us about controlled attention, general fluid intelligence, and functions of the prefrontal cortex. In *Mechanisms of Active Maintenance and Executive Control*, ed. A Miyake, P Shah. New York: Cambridge Univ. Press

Engel RW, Tuholski SW, Laughlin JE, Conway AR. 1999b. Working memory, short-term memory, and general fluid intelligence: a latent-variable approach. *J. Exp. Psychol. Gen.* 128:309–31

Ferrera VP, Cohen J, Lee BB. 1999. Activity of prefrontal neurons during location and color delayed matching tasks. *NeuroReport* 10:1315–22

Ferrier D. 1876. *The Functions of the Brain.* London: Smith, Elder

Frith CD, Friston K, Liddle PF, Frackowiak RSJ. 1991. Willed action and the prefrontal cortex in man: a study with PET. *Proc. R. Soc. London Ser. B* 244:241–46

Funahashi S, Bruce CJ, Goldman-Rakic PS. 1989. Mnemonic coding of visual space in the monkey's dorsolateral prefrontal cortex. *J. Neurophysiol.* 61:331–49

Fuster JM. 1973. Unit activity in prefrontal cortex during delayed-response performance: neuronal correlates of transient memory. *J. Neurophysiol.* 36:61–78

Fuster JM. 1980. *The Prefrontal Cortex.* New York: Raven

Fuster JM. 1985. The prefrontal cortex, mediator of cross-temporal contingencies. *Hum. Neurobiol.* 4:169–79

Fuster JM. 1989. *The Prefrontal Cortex*, Vol. 2. New York: Raven

Fuster JM. 1995. *Memory in the Cerebral Cortex.* Cambridge, MA: MIT Press

Fuster JM, Alexander GE. 1971. Neuron activity related to short-term memory. *Science* 173:652–54

Fuster JM, Bauer RH, Jervey JP. 1982. Cellular discharge in the dorsolateral prefrontal cortex of the monkey in cognitive tasks. *Exp. Neurol.* 77:679–94

Fuster JM, Bauer RH, Jervey JP. 1985. Functional interactions between inferotemporal and prefrontal cortex in a cognitive task. *Brain Res.* 330:299–307

Fuster JM, Jervey JP. 1981. Inferotemporal neurons distinguish and retain behaviorally relevant features of visual stimuli. *Science* 212:952–55

Gaffan D, Harrison S. 1988. Inferotemporal-frontal disconnection and fornix transection in visuomotor conditional learning by monkeys. *Behav. Brain Res.* 31:149–63

Gathercole SE. 1994. Neuropsychology and working memory: a review. *Neuropsychology* 8:494–505

Gershberg FB, Shimamura AP. 1995. Impaired use of organizational strategies in free recall following frontal lobe damage. *Neuropsychologia* 13:1305–33

Gnadt JW, Andersen RA. 1988. Memory related motor planning activity in posterior parietal cortex of macaque. *Exp. Brain Res.* 70:216–20

Gobbel JR. 1995. A biophysically-based model of the neostriatum as a reconfigurable network. *Proc. Swed. Conf. Connectionism, 2nd.* Hillsdale, NJ: Erlbaum

Goldman PS, Nauta WJ. 1976. Autoradiographic demonstration of a projection from prefrontal association cortex to the superior colliculus in the rhesus monkey. *Brain Res.* 116:145–49

Goldman-Rakic PS. 1987. Circuitry of primate prefrontal cortex and regulation of behavior by representational memory. In *Handbook of Physiology: The Nervous System*, ed. F Plum, pp. 373–417. Bethesda, MD: Am. Physiol. Soc.

Goldman-Rakic PS. 1996. The prefrontal landscape: implications of functional architecture for understanding human mentation and the central executive. *Philos. Trans. R. Soc. London Ser. B* 351:1445–53

Goldman-Rakic PS, Schwartz ML. 1982. Interdigitation of contralateral and ipsilateral columnar projections to frontal association cortex in primates. *Science* 216:755–57

Goldman-Rakic PS, Selemon LD, Schwartz ML. 1984. Dual pathways connecting the dorsolateral prefrontal cortex with the hippocampal formation and parahippocampal cortex in the rhesus monkey. *Neuroscience* 12:719–43

Gould E, Reeves AJ, Graziano MS, Gross CG. 1999. Neurogenesis in the neocortex of adult primates. *Science* 286:548–52

Grafman J. 1994. Alternative frameworks for the conceptualization of prefrontal functions. In *Handbook of Neuropsychology*, ed. F Boller, J Grafman, pp. 187. Amsterdam: Elsevier

Gratton G, Coles MGH, Donchin E. 1992. Optimizing the use of information: strategic control of activation of responses. *J. Exp. Psychol.* 121:480–506

Halsband U, Passingham RE. 1985. Premotor cortex and the conditions for movement in monkeys. *Behav. Brain Res.* 18:269–76

Hecaen H, Albert ML. 1978. *Human Neuropsychology.* New York: Wiley

Hollerman JR, Schultz W. 1998. Dopamine neurons report an error in the temporal prediction of reward during learning. *Nat. Neurosci. USA* 1:304–9

Hopfield JJ. 1982. Neural networks and physical systems with emergent collective computational abilities. *Proc. Natl. Acad. Sci. USA* 79:2554–58

Hoshi E, Shima K, Tanji J. 1998. Task-dependent selectivity of movement-related neuronal activity in the primate prefrontal cortex. *J. Neurophysiol.* 80:3392–97

Houck JC. 1995. *Models of Information in the Basal Ganglia.* Cambridge, MA: MIT Press

Janowsky JS, Shimamura AP, Kritchevsky M, Squire LR. 1989. Cognitive impairment following frontal lobe damage and its relevance to human amnesia. *Behav. Neurosci.* 103:548–60

Jones EG, Powell TPS. 1970. An anatomical study of converging sensory pathways within the cerebral cortex of the monkey. *Brain* 93:793–820

Just MA, Carpenter PA. 1992. A capacity theory of comprehension: individual differences in working memory. *Psychol. Rev.* 99:122–49

Knight RT. 1984. Decreased response to novel stimuli after prefrontal lesions in man. *Clin. Neurophysiol.* 59:9–20

Knight RT. 1997. Distributed cortical network for visual attention. *J. Cogn. Neurosci.* 9:75–91

Koechlin E, Basso G, Pietrini P, Panzer S, Grafman J. 1999. The role of the anterior prefrontal cortex in human cognition. *Nature* 399:148–51

Kubota K, Niki H. 1971. Prefrontal cortical unit activity and delayed alternation performance in monkeys. *J. Neurophysiol.* 34:337–47

Laming DRJ. 1968. *Information Theory of Choice-Reaction Times.* London: Academic

Leon MI, Shadlen MN. 1999. Effect of expected reward magnitude on the response of neurons in the dorsolateral prefrontal cortex of the macaque. *Neuron* 24:415–25

Levine B, Stuss DT, Milberg WP, Alexander MP, Schwartz M, Macdonald R. 1998. The effects of focal and diffuse brain damage on strategy application: evidence from focal lesions, traumatic brain injury and normal aging. *J. Int. Neuropsychol. Soc.* 4:247–64

Levine DS, Prueitt PS. 1989. Modeling some effects of frontal lobe damage-novelty and perseveration. *Neural Networks* 2:103–16

Lhermitte F. 1983. "Utilization behaviour" and its relation to lesions of the frontal lobes. *Brain* 106:237–55

Logan GD, Zbrodoff NJ, Fostey ARW. 1983.

Costs and benefits of strategy construction in a speeded discrimination task. *Mem. Cogn.* 11:485–93

London ED, Ernst M, Grant S, Bonson K, Weinstein A. 2000. Orbitofrontal cortex and human drug abuse: functional imaging. *Cereb. Cortex* 10:334–42

Lu MT, Preston JB, Strick PL. 1994. Interconnections between the prefrontal cortex and the premotor areas in the frontal lobe. *J. Comp. Neurol.* 341:375–92

Luria AR. 1969. Frontal lobe syndromes. In *Handbook of Clinical Neurology*, ed. PJ Vinken, GW Bruyn, pp. 725–57. New York: Elsevier

MacDonald AW, Cohen JD, Stenger VA, Carter CS. 2000. Dissociating the role of dorsolateral prefrontal cortex and anterior cingulate cortex in cognitive control. *Science* 288:1835–38

MacLeod CM. 1991. Half a century of research on the Stroop effect: an integrative review. *Psychol. Bull.* 109:163–203

McClelland JL, McNaughton BL, O'Reilly RC. 1995. Why there are complementary learning systems in the hippocampus and neocortex: insights from the successes and failures of connectionist models of learning and memory. *Psychol. Rev.* 102:419–57

Melchitzky DS, Sesack SR, Pucak ML, Lewis DA. 1998. Synaptic targets of pyramidal neurons providing intrinsic horizontal connections in monkey prefrontal cortex. *J. Comp. Neurol.* 390:211–24

Miller EK. 1999. The prefrontal cortex: complex neural properties for complex behavior. *Neuron* 22:15–17

Miller EK. 2000. The neural basis of top-down control of visual attention in the prefrontal cortex. In *Attention and Performance*, ed. S Monsell, J Driver, 18:In press. Cambridge, MA: MIT Press

Miller EK, Desimone R. 1994. Parallel neuronal mechanisms for short-term memory. *Science* 263:520–22

Miller EK, Erickson CA, Desimone R. 1996. Neural mechanisms of visual working memory in prefrontal cortex of the macaque. *J. Neurosci.* 16:5154–67

Miller EK, Li L, Desimone R. 1993. Activity of neurons in anterior inferior temporal cortex during a short-term memory task. *J. Neurosci.* 13:1460–78

Miller GA. 1956. The magical number seven plus or minus two: some limits on our capacity for processing information. *Psychol. Rev.* 63:81–97

Milner B. 1963. Effects of different brain lesions on card sorting. *Arch. Neurol.* 9:90

Mirenowicz J, Schultz W. 1994. Importance of unpredictability for reward responses in primate dopamine neurons. *J. Neurophysiol.* 72:1024–27

Mirenowicz J, Schultz W. 1996. Preferential activation of midbrain dopamine neurons by appetitive rather than aversive stimuli. *Nature* 379:449–51

Mishkin M. 1964. Perseveration of central sets after frontal lesions in monkeys. In *The Frontal Granular Cortex and Behavior*, ed. JM Warren, K Abert, pp. 219–41. New York: McGraw-Hill

Miyashita Y, Chang HS. 1988. Neuronal correlate of pictorial short-term memory in the primate temporal cortex. *Nature* 331:68–70

Montague PR, Dayan P, Sejnowski TJ. 1996. A framework for mesencephalic dopamine systems based on predictive Hebbian learning. *J. Neurosci.* 16:1936–47

Mozer MC. 1991. *The Perception of Multiple Objects: A Connectionist Approach.* Cambridge, MA: MIT Press

Murray EA, Bussey TJ, Wise SP. 2000. Role of prefrontal cortex in a network for arbitrary visuomotor mapping. *Exp. Brain Res.* In press

Newell A, Simon HA. 1972. *Human Problem Solving.* Englewood Cliffs, NJ: Prentice Hall

Nichelli P, Grafman J, Pietrini P, Alway D, Carton JC, Miletich R. 1994. Brain activity in chess playing. *Nature* 369:191

Nystrom LE, Braver TS, Sabb FW, Delgado MR, Noll DC, Cohen JD. 2000. Working memory for letters, shapes and locations: fMRI evidence against stimulus-based regional organization in human prefrontal cortex. *Neuroimage.* In press

O'Doherty J, Rolls ET, Francis S, Bowtell R, McGlone F, et al. 2000. Sensory-specific satiety-related olfactory activation of the human orbitofrontal cortex. *NeuroReport* 11:893–97

O'Reilly RC, Braver TS, Cohen JD. 1999. A biologically-based computational model of working memory. In *Models of Working Memory: Mechanisms of Active Maintenance and Executive Control,* ed. A Miyake, P Shah. New York: Cambridge Univ. Press

O'Reilly RC, McClelland JL. 1994. Hippocampal conjunctive coding, storage, and recall: avoiding a tradeoff. *Hippocampus* 4:661–82

O'Reilly RC, Munakata Y. 2000. *Computational Explorations in Cognitive Neuroscience: Understanding the Mind.* Cambridge: MIT Press

O Scalaidhe SP, Wilson FA, Goldman-Rakic PS. 1997. Areal segregation of face-processing neurons in prefrontal cortex. *Science* 278:1135–38

Owen AM, Downes JJ, Sahakian BJ, Polkey CE, Robbins TW. 1990. Planning and spatial working memory following frontal lobe lesions in man. *Neuropsychologia* 28:1021–34

Owen AM, Evans AC, Petrides M. 1996. Evidence for a two-stage model of spatial working memory processing within the lateral frontal cortex: a positron emission tomography study. *Cerebr. Cortex* 6:31–38

Pandya DN, Barnes CL. 1987. Architecture and connections of the frontal lobe. In *The Frontal Lobes Revisited,* ed. E Perecman, pp. 41–72. New York: IRBN

Pandya DN, Yeterian EH. 1990. Prefrontal cortex in relation to other cortical areas in rhesus monkey—architecture and connections. *Prog. Brain Res.* 85:63–94

Parker A, Gaffan D. 1998. Memory after frontal/temporal disconnection in monkeys: conditional and non-conditional tasks, unilateral and bilateral frontal lesions. *Neuropsychologia* 36:259–71

Passingham R. 1993. *The Frontal Lobes and Voluntary Action.* Oxford, UK: Oxford Univ. Press

Perret E. 1974. The left frontal lobe of man and the suppression of habitual responses in verbal categorical behaviour. *Neuropsychologia* 12:323–30

Petersen SE, van Mier H, Fiez JA, Raichle ME. 1998. The effects of practice on the functional anatomy of task performance. *Proc. Natl. Acad. Sci. USA* 95:853–60

Petrides M. 1982. Motor conditional associative-learning after selective prefrontal lesions in the monkey. *Behav. Brain Res.* 5:407–13

Petrides M. 1985. Deficits in non-spatial conditional associative learning after periarcuate lesions in the monkey. *Behav. Brain Res.* 16:95–101

Petrides M. 1990. Nonspatial conditional learning impaired in patients with unilateral frontal but not unilateral temporal lobe excisions. *Neuropsychologia* 28:137–49

Petrides M. 1996. Specialized systems for the processing of mnemonic information within the primate frontal cortex. *Philos. Trans. R. Soc. London Ser. B* 351:1455–61

Petrides M, Pandya DN. 1984. Projections to the frontal cortex from the posterior parietal region in the rhesus monkey. *J. Comp. Neurol.* 228:105–16

Petrides M, Pandya DN. 1999. Dorsolateral prefrontal cortex: comparative cytoarchitectonic analysis in the human and the macaque brain and corticocortical connection patterns. *Eur. J. Neurosci.* 11:1011–36

Phaf RH, Van der Heiden AHC, Hudson PTW. 1990. SLAM: a connectionist model for attention in visual selection tasks. *Cogn. Psychol.* 22:273–341

Piaget J. 1954 (1937). *The Origins of Intelligence in Children.* New York: Basic Books

Porrino LJ, Crane AM, Goldman-Rakic PS. 1981. Direct and indirect pathways from the amygdala to the frontal lobe in rhesus monkeys. *J. Comp. Neurol.* 198:121–36

Posner MI, Snyder CRR. 1975. Attention and cognitive control. In *Information Processing and Cognition*, ed. RL Solso. Hillsdale, NJ: Erlbaum

Prabhakaran V, Narayanan K, Zhao Z, Gabrieli JD. 2000. Integration of diverse information in working memory within the frontal lobe. *Nat. Neurosci.* 3:85–90

Price JL. 1999. Prefrontal cortical networks related to visceral function and mood. *Ann. NY Acad. Sci.* 877:383–96

Pucak ML, Levitt JB, Lund JS, Lewis DA. 1996. Patterns of intrinsic and associational circuitry in monkey prefrontal cortex. *J. Comp. Neurol.* 376:614–30

Quintana J, Fuster JM. 1992. Mnemonic and predictive functions of cortical neurons in a memory task. *NeuroReport* 3:721–24

Rainer G, Asaad WF, Miller EK. 1998a. Memory fields of neurons in the primate prefrontal cortex. *Proc. Natl. Acad. Sci. USA* 95:15008–13

Rainer G, Asaad WF, Miller EK. 1998b. Selective representation of relevant information by neurons in the primate prefrontal cortex. *Nature* 393:577–79

Rainer G, Rao SC, Miller EK. 1999. Prospective coding for objects in the primate prefrontal cortex. *J. Neurosci.* 19:5493–505

Rao SC, Rainer G, Miller EK. 1997. Integration of what and where in the primate prefrontal cortex. *Science* 276:821–24

Roberts AC, Wallis JD. 2000. Inhibitory control and affective processing in the prefrontal cortex: neuropsychological studies in the common marmoset. *Cerebr. Cortex* 10:252–62

Rolls ET. 2000. The orbitofrontal cortex and reward. *Cereb. Cortex* 10:284–94

Romo R, Brody CD, Hernandez A, Lemus L. 1999. Neuronal correlates of parametric working memory in the prefrontal cortex. *Nature* 399:470–73

Rossi AF, Rotter PS, Desimone R, Ungerleider LG. 1999. Prefrontal lesions produce impairments in feature-cued attention. *Soc. Neurosci. Abstr.* 25:3

Rushworth MF, Nixon PD, Eacott MJ, Passing-ham RE. 1997. Ventral prefrontal cortex is not essential for working memory. *J. Neurosci.* 17:4829–38

Schacter DL. 1997. The cognitive neuroscience of memory: perspectives from neuroimaging research. *Philos. Trans. R. Soc. London Ser. B* 352:1689–95

Schmahmann JD, Pandya DN. 1997. Anatomic organization of the basilar pontine projections from prefrontal cortices in rhesus monkey. *J. Neurosci.* 17:438–58

Schultz W. 1998. Predictive reward signal of dopamine neurons. *J. Neurophysiol.* 80:1–27

Schultz W, Apicella P, Ljungberg T. 1993. Responses of monkey dopamine neurons to reward and conditioned stimuli during successive steps of learning a delayed response task. *J. Neurosci.* 13:900–13

Schultz W, Dickinson A. 2000. Neuronal coding of prediction errors. *Annu. Rev. Neurosci.* 23:473–500

Schultz W, Dayan P, Montague PR. 1997. A neural substrate of prediction and reward. *Science* 275:1593–99

Seltzer B, Pandya DN. 1989. Frontal lobe connections of the superior temporal sulcus in the rhesus monkey. *J. Comp. Neurol.* 281:97–113

Shadmehr R, Holcomb H. 1997. Neural correlates of motor memory consolidation. *Science* 277:821–24

Shallice T. 1982. Specific impairments of planning. *Philos. Trans. R. Soc. London Ser. B* 298:199–209

Shallice T. 1988. *From Neuropsychology to Mental Structure*. Cambridge, UK: Cambridge Univ. Press

Shallic T, Burgess P. 1996. The domain of supervisory processes and temporal organization of behaviour. *Philos. Trans. R. Soc. London Ser. B* 351:1405–11

Shiffrin RM, Schneider W. 1977. Controlled and automatic information processing: II. Perceptual learning, automatic attending, and a general theory. *Psychol. Rev.* 84:127–90

Smith EE, Jonides J. 1999. Storage and executive processes in the frontal lobes. *Science* 283:1657–61

Squire LR. 1992. Memory and the hippocampus: a synthesis from findings with rats, monkeys, and humans. *Psychol. Rev.* 99:195–231

Stroop JR. 1935. Studies of interference in serial verbal reactions. *J. Exp. Psychol.* 18:643–62

Stuss DT, Benson DF. 1986. *The Frontal Lobes.* New York: Raven

Swedo SE, Shapiro MB, Grady CL, Cheslow DL, Leonard HL, et al. 1989. Cerebral glucose metabolism in childhood-onset OCD. *Arch. Gen. Psychiatr.* 46:518–23

Tomita H, Ohbayashi M, Nakahara K, Hasegawa I, Miyashita Y. 1999. Top-down signal from prefrontal cortex in executive control of memory retrieval. *Nature* 401:699–703

Treisman A, Gelade G. 1980. A feature integration theory of attention. *Cogn. Psychol.* 12:97–136

Tremblay L, Hollerman JR, Schultz W. 1998. Modifications of reward expectation-related neuronal activity during learning in primate striatum. *J. Neurophysiol.* 80:964–77

Tremblay L, Schultz W. 1999. Relative reward preference in primate orbitofrontal cortex. *Nature* 398:704–8

Tzelgov J, Henik A, Berger J. 1992. Controlling Stroop effects by manipulating expectations for color words. *Mem. Cogn.* 20:727–35

Usher M, Cohen JD. 1997. *Interference-based capacity limitations in active memory.* Presented at Abstr. Psychonom. Soc., Philadelphia

Van Hoesen GW, Pandya DN, Butters N. 1972. Cortical afferents to the entorhinal cortex of the rhesus monkey. *Science* 175:1471–73

Vendrell P, Junque C, Pujol J, Jurado MA, Molet J, Grafman J. 1995. The role of prefrontal regions in the Stroop task. *Neuropsychologia* 33:341–52

Wagner AD. 1999. Working memory contributions to human learning and remembering. *Neuron* 22:19–22

Wallis JD, Anderson KC, Miller EK. 2000. Neuronal representation of abstract rules in the orbital and lateral prefrontal cortices (PFC). *Soc. Neurosci. Abstr.* In press

Wang XJ. 1999. Synaptic basis of cortical persistent activity: the importance of NMDA receptors to working memory. *J. Neurosci.* 19:9587–603

Watanabe M. 1990. Prefrontal unit activity during associative learning in the monkey. *Exp. Brain Res.* 80:296–309

Watanabe M. 1992. Frontal units of the monkey coding the associative significance of visual and auditory stimuli. *Exp. Brain Res.* 89:233–47

Watanabe M. 1996. Reward expectancy in primate prefrontal neurons. *Nature* 382:629–32

White IM, Wise SP. 1999. Rule-dependent neuronal activity in the prefrontal cortex. *Exp. Brain Res.* 126:315–35

Wilson FAW, O Scalaidhe SP, Goldman-Rakic PS. 1993. Dissociation of object and spatial processing domains in primate prefrontal cortex. *Science* 260:1955–58

Wise SP, Murray EA, Gerfen CR. 1996. The frontal-basal ganglia system in primates. *Crit. Rev. Neurobiol.* 10:317–56

Yamaguchi S, Knight RT. 1991. Anterior and posterior association cortex contributions to the somatosensory P300. *J. Neurosci.* 11:2039–54

Zipser D, Kehoe B, Littlewort G, Fuster J. 1993. A spiking network model of short-term active memory.13:3406–20

Zola-Morgan S, Squire LR. 1993. Neuroanatomy of memory. *Annu. Rev. Neurosci.* 16:547–63

Annu. Rev. Neurosci. 2001. 24:203–38

THE PHYSIOLOGY OF STEREOPSIS

B. G. Cumming

*University Laboratory of Physiology, Oxford, OX1 3PT,[1]
United Kingdom; e-mail: bgc@lsr.nei.nih.gov*

G. C. DeAngelis

*Department of Anatomy and Neurobiology, Washington University School of Medicine,
St. Louis, Missouri 63110-1093; e-mail: gregd@thalamus.wustl.edu*

Key Words   binocular vision, disparity, striate cortex, extrastriate cortex,
depth perception

■ **Abstract**   Binocular disparity provides the visual system with information concerning the three-dimensional layout of the environment. Recent physiological studies in the primary visual cortex provide a successful account of the mechanisms by which single neurons are able to signal disparity. This work also reveals that additional processing is required to make explicit the types of signal required for depth perception (such as the ability to match features correctly between the two monocular images). Some of these signals, such as those encoding relative disparity, are found in extrastriate cortex. Several other lines of evidence also suggest that the link between perception and neuronal activity is stronger in extrastriate cortex (especially MT) than in the primary visual cortex.

INTRODUCTION

A central problem faced by the visual system is providing information about a three-dimensional environment from two-dimensional retinal images. In many animals, one of the most precise sources of information arises from the fact that the two eyes have different vantage points. This means that the images on the two retinae are not identical (see Figure 1). The differences between the locations of matching features on the retinae are termed binocular disparities, and the ability to perceive depth from these disparities is stereopsis. This review focuses on the neuronal basis for such depth judgements and so does not discuss all published studies of disparity selectivity. A more encyclopedic review of much of this material has appeared recently (Gonzalez & Perez 1998b).

Before it is possible to determine the disparity of an image feature, it is essential to match features in the left eye with appropriate features in the right eye (see

---

[1]Current address: Laboratory of Sensorimotor Research, National Eye Institute, National Institutes of Health, Bethesda, Maryland, 20892

0147-006X/01/0621-0203$14.00

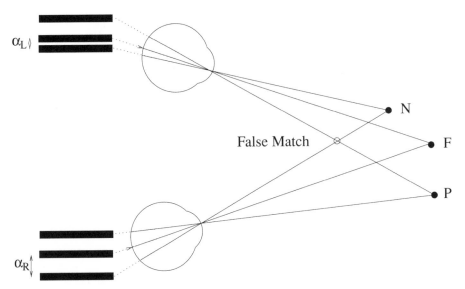

Figure 1   Geometry of binocular vision. Both eyes fixate bar F, so the image of F falls on the fovea in each eye. The images of a nearer bar, N, fall on noncorresponding retinal locations. The angular distances from the fovea (a convenient reference, defining corresponding locations) are marked by α_L and α_R, and the difference between these angles is the binocular disparity of N. This also illustrates the correspondence problem: The image of N in the right eye combined with the image of P in the left eye forms a binocular image with a disparity corresponding to the open circle labeled "false match." No object is perceived at this depth because the brain matches only correctly corresponding features on the two retinae.

Figure 1). This "stereo correspondence problem" was highlighted by the random dot stereogram (RDS) (Julesz 1971). Here a set of random dots is shown to each eye. The dots within a region of one eye's image are displaced a small distance horizontally, thus introducing a binocular disparity. When fused, this gives rise to a vivid depth sensation, even though the two monocular images look homogeneous, with no distinctive features. Although it was the work of Julesz that led to the modern use of the RDS for studying stereopsis, the phenomenon had been noticed 100 years earlier by Cajal (Bergua & Skrandies 2000).

In the primary visual cortex (V1), a good understanding of the mechanism of disparity selectivity has been achieved in recent years, so the first half of the review focuses on this. The second half describes those properties of extrastriate areas that suggest a greater involvement in depth perception.

Measuring Disparity Selectivity

Before discussing the possible roles of disparity-selective neurons in stereopsis, it is important to recognize some of the difficulties in establishing that individual neurons signal disparity. This is usually assessed by presenting some stimulus at a range of disparities, and neurons are classified as disparity-selective if they fire more action potentials in response to some disparities than in response to others. There are two potential pitfalls in this approach. First, in the absence of any change in the visual scene, changes in the animals' fixation distance (convergence) alter the disparity of the retinal stimulus. Second, changes in the disparity of a stimulus are inevitably associated with changes to at least one of the monocular images presented, so it is vital to dissociate binocular and monocular effects of manipulating disparity.

Eye Movements   In preparations that involve a paralyzed, anesthetized animal, the vergence state is probably stable over short periods of time (long enough to characterize disparity selectivity in one cell), but is likely to drift slowly over the course of a long experiment. This problem has been circumvented in some studies by use of a reference cell technique—a second neuron is recorded from a different electrode and held as long as possible. Repeatedly plotting the receptive field (RF) locations of the reference cell allows compensation for drifts in eye position (Hubel & Wiesel 1970, Ferster 1981). Although this technique compensates for changes in vergence, it still does not permit absolute calibration of vergence state. Thus, it is not possible to say with certainty that a neuron's preferred disparity is crossed (nearer than the fixation point), uncrossed (farther than the fixation point), or zero. In some studies, visual identification of retinal landmarks has been used to determine corresponding locations and hence to determine eye position. However, this is fairly imprecise compared with the precision of disparity tuning in many cells.

In an awake animal, the vergence state may change within the course of a single trial; thus, knowing the position of both eyes is essential for the interpretation of disparity tuning data. If the animal is converging correctly, then the absolute value of stimulus disparities is known. Early studies of awake monkeys that recorded the positions of both eyes clearly demonstrated the existence of neurons selective for nonzero disparities in primate V1 (Poggio & Talbot 1981). Other studies have recorded the position of only one eye (e.g. Poggio & Fisher 1977, Trotter et al 1996, Janssen et al 1999), under the assumption that if the animal is fixating with one eye, it is probably also converging correctly. This assumption is not always secure, since small changes in vergence (and therefore disparity) can have a significant effect on firing rate in sharply tuned cells.

Monocular and Binocular Effects of Disparity   Changing the disparity of a stimulus inevitably changes at least one of the monocular half-images. Consider a bar stimulus flashed at different disparities. As the disparity is changed, the monocular position of the bar changes. With a sufficiently large disparity the bar may fall completely off the RF in one eye (or even both eyes). Obviously the failure to respond to such a stimulus need not indicate disparity selectivity. Careful use of a sweeping bar can avoid its falling off the RF altogether, but changes in the monocular stimuli alone may still elicit changes in firing rate.

Some studies have applied criteria to the neural responses to reduce the chance of obtaining a misleading appearance of disparity selectivity. Hubel & Wiesel (1970) and Hubel & Livingstone (1987) required that the disparity tuning width be much narrower than the RF width. But this criterion might exclude cells that are genuinely disparity selective. It may be for this reason that Hubel & Wiesel (1970, p. 42) "studied hundreds of cells in area 17" and "found no convincing examples of binocular depth cells."

Two different solutions to the problem of monocular artifacts have been effective. The first is to present a dichoptic bar stimulus at all possible combinations of positions in the two eyes. In this way the effects of disparity can be separated from the effects of monocular position. This is the approach taken in the reverse-correlation methods (Ohzawa et al 1990). The second approach is to use RDS stimuli (Julesz 1971), first applied to physiological recording in the pioneering work of Poggio et al (1985, 1988). Here, changes in disparity are not associated with any discernible changes in the monocular images. Disparity selectivity in response to such stimuli identifies a specific response related to binocular correlation.

Fortunately, most of the disparity-selective phenomena reported in early studies have been replicated with stimuli that eliminate monocular artifacts. However, there are certain observations that have only been reported using simple bar stimuli that should therefore be treated with caution. These include observations on the properties of near/far cells (see section on Classes of Disparity Tuning) and the combination of vertical and horizontal disparities (section on Horizontal and Vertical Disparities).

PRIMARY VISUAL CORTEX

Although the responses of many cells in the lateral geniculate nucleus can be modulated by stimuli in the nondominant eye (Suzuki & Kato 1966, Singer 1970, Marocco & McClurkin 1979, Rodieck & Dreher 1979), this does not produce disparity-selective responses (Xue et al 1987). V1 is the first site at which single neurons can be activated by stimuli in both eyes. The first studies to document disparity selectivity in V1 (Pettigrew et al 1968, Barlow et al 1967) used sweeping bar stimuli in anesthetized cats. These studies demonstrated that some V1 neurons encode information specifically about the relationship between the images in the two eyes. The data are compatible with a variety of different mechanisms. At one extreme is the possibility that the monocular processing is complicated: A distinctive "trigger feature" such as an oriented edge is identified (Barlow et al 1967), and the neuron responds maximally when this feature appears at the preferred disparity. At the other extreme is a trivial possibility that these neurons are activated whenever any excitatory stimulus is present in each monocular RF. Although such neurons could carry some information about disparity, it would be confounded by effects of monocular stimulus location.

Between these extremes is the possibility that each monocular RF performs a relatively simple operation on the image, and the cell fires maximally when this calculation produces a large result in both eyes. In this scheme it does not matter whether the visual stimulus is the same in both eyes, only that the stimulus in each eye produces a strong output from the monocular filter. An important consequence of such a scheme is that the monocular RF shape largely determines the shape of the disparity response, whereas a scheme based on trigger features requires no special relationship between the structure of the monocular RF and the shape of the disparity tuning function. We now examine these relationships for the two main physiological cell types in V1: simple cells and complex cells.

Simple Cells

A defining characteristic of simple cells is that they show linear spatial summation (Hubel & Wiesel 1962, Movshon et al 1978b). Thus, their responses to monocular stimuli can be summarized by a RF map that describes the response to small bright and dark spots presented at different locations in space (Jones & Palmer 1987, DeAngelis et al 1993a). These RF maps are well described by Gabor functions (a sinewave multiplied by a Gaussian envelope), and the response of a simple cell can be reasonably well predicted by convolving a visual pattern with the RF map (Jones & Palmer 1987, Field & Tolhurst 1986, DeAngelis et al 1993b). (Convolving here means multiplying the image brightness at each point with the value of the RF at that point and summing all the products.) Disparity selectivity in simple cells might then be understood as follows. A convolution is performed in each eye, and the results are added (so a negative result in the left eye can cancel excitation from the right eye). After this binocular summation, the output is half-wave rectified (negative values are discarded). The cell will fire roughly in proportion to the result of this binocular summation. In such a scheme, the key to understanding disparity selectivity would be to understand the differences between the two monocular RFs.

Ohzawa & Freeman (1986b) performed the first quantitative comparison of monocular and binocular responses at different disparities in simple cells. They presented sinusoidal luminance gratings and used the monocular responses to drifting gratings to predict binocular responses to a range of interocular phase differences. The majority of responses were well described by this linear model, and nearly all could be explained by a linear interaction followed by a threshold. [An earlier study, using bars, suggested the same conclusion (Ferster 1981).]

Up until this time, it had generally been thought that neurons had closely matched RF profiles in the two eyes (e.g. Hubel & Wiesel 1973, Maske et al 1984). Disparity selectivity was thought to result from these RFs being placed in different positions on the two retinae. Ohzawa & Freeman (1986b) pointed out that similar disparity selectivity could be produced by cells that have RFs in corresponding retinal locations but that have different RF shapes in the two eyes (see Figure 2). Indeed Bishop et al (1971) had noted such differences in some cells in the cat. To investigate this possibility explicitly, Freeman and colleagues (DeAngelis et al 1991, Ohzawa et al 1996) fitted Gabor functions to RF profiles

a)   b)

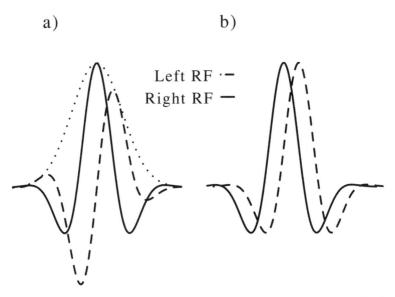

Left RF ·−
Right RF −

Figure 2   Phase and position disparity mechanisms. Receptive field (RF) profiles in the left eye (dashed lines) and right eye (solid lines) are shown for two possible binocular neurons. The profiles are Gabor functions, the product of a sinewave and a Gaussian envelope (dotted line). In (*a*), the location of this envelope is the same in both eyes, but the phase of the sinusoidal component is different (by $\pi/2$ here). In (*b*), the RF has the same shape (determined by the phase of the sinewave relative to the envelope) but different positions in the two eyes.

measured separately for each eye. Differences in the internal structure of the RF were quantified as a difference in the phase of the sinusoidal component relative to the center of the Gaussian envelope (see Figure 2). This revealed a wide range of interocular phase differences, called phase disparities, in binocular simple cells.

Anzai et al (1999b) went on to compare monocular and binocular responses of simple cells by showing uncorrelated one-dimensional noise patterns to the two eyes. Monocular RF profiles were constructed by computing the average effect of black or white lines at different locations in each eye separately. Binocular RF profiles were constructed independently from looking at the average effect of lines of the same contrast polarity in the two eyes (black-black or white-white pairs) compared with the effect of lines of opposite polarity in the two eyes. They found a good agreement between the monocular RF structure and the shape of the binocular disparity response. Taken together, these studies clearly showed that phase differences between monocular RFs do occur in simple cells, and that these differences account for the shape of the binocular interaction profile. However, this conclusion does not imply that position disparities are not also used, as discussed below.

Complex Cells

For complex cells, it is much less straightforward to understand disparity selectivity in terms of monocular RF structure because these neurons are spatially nonlinear. They respond to oriented contours over a range of positions, but are nonetheless quite selective for the luminance structure of the stimulus (Hubel & Wiesel 1962). With monocular sinusoidal gratings, complex cells are insensitive to the spatial phase of the grating yet remain selective for the spatial frequency. For disparity-selective complex cells, this gives rise to an interesting property: They are insensitive to the phase of the grating when tested in either eye alone, yet they are sensitive to the phase difference between the eyes (Ohzawa & Freeman 1986a).

An extension of the earliest model of complex cells (Movshon et al 1978a) to the binocular case provides a possible explanation of this phenomenon. This disparity "energy" model (Ohzawa et al 1990) simply proposes that a complex cell is constructed from a set of simple cells. As shown in Figure 3, all of the constituent simple cells have the same disparity tuning, but their monocular RFs are in quadrature (meaning that all spatial frequency components are shifted by $\pi/2$, so that the responses are orthogonal). If a stimulus is at the complex cell's preferred disparity, then at least one of the simple cells is activated, no matter where in the RF a stimulus falls. However, if a stimulus is at the null disparity, none of the simple cells is active, so the complex cell does not fire either. This model produces a complex cell that is sensitive to the correlation between images in the two eyes (Qian 1994, Fleet et al 1996).

This model explains many properties of disparity-selective neurons in V1. First, it explains the results obtained with sinusoidal gratings by Ohzawa & Freeman (1986a). Second, it explains the shape of disparity tuning functions measured with broadband stimuli: Because the RF profiles of the constituent simple cells are well described as Gabor functions, the shape of the disparity tuning curve is as well (Ohzawa et al 1990, 1997; SJD Prince, AD Pointon, BG Cumming, AJ Parker, submitted for publication). In the energy model, the spatial period of the Gabor function describing the disparity response is closely related to the monocular spatial frequency tuning. In practice, however, only a weak correlation has been observed between these two measures (Ohzawa et al 1997; SJD Prince, AD Pointon, BG Cumming, AJ Parker, submitted for publication).

The energy model also explains the responses of complex cells to stimuli of opposite polarity in the two eyes. Most complex cells can be activated monocularly both by dark bars (against a grey background) and by bright bars. But when a dark bar is shown to one eye, while a bright bar is shown to the other eye at the preferred disparity, disparity-selective cells are generally suppressed. Similarly, they show activation for those disparities where same-polarity bars produce inhibition (Ohzawa et al 1990). This occurs because in each simple cell subunit, the maximum binocular response occurs when the bar in both eyes causes maximum excitation. If the polarity of a bar in only one eye is reversed, then that bar

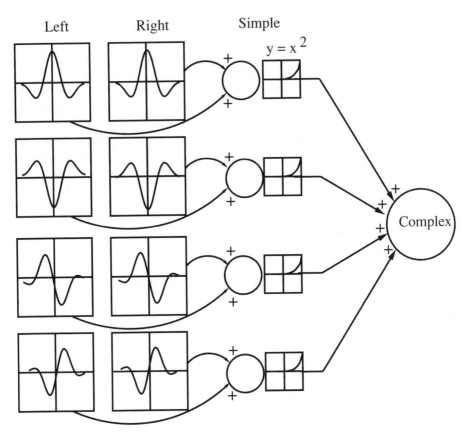

Figure 3   A model for disparity selectivity in complex cells. Each complex cell receives inputs from a minimum of four simple cells. These consist of two pairs that are in quadrature, so that the sum of outputs is invariant to monocular spatial phase. At least one simple cell is activated by a stimulus in any phase. Those simple cells that are inhibited contribute nothing to the response (because the output of each simple cell is rectified). Thus all stimulus phases are excitatory. In this example, the receptive field (RF) profiles are identical in both eyes, so the complex cell is maximally activated by stimuli at zero disparity. If a stimulus is presented with a disparity equal to one half cycle of the RF, then the monocular responses cancel one another in each simple cell. Because responses from the two eyes are added before rectification, there is no response to this disparity. Hence this model explains the preservation of sensitivity to interocular phase differences, despite an insensitivity to spatial phase in each eye. If the same interocular phase shift or position shift is added to each of the simple cells, then the complex cell is maximally activated by nonzero disparities. If vairables L and R are the results of convolving the stimulus in each eye with the corresponding RF profile, then the output of each simple cell is $(L + R)^2 = L^2 + R^2 + 2LR$. Thus, by virtue of the last term in this expression, the half-squaring nonlinearity makes the cell sensitive to binocular correlation. Adapted from Ohzawa et al (1990).

becomes a suppressive stimulus. The position of this monocular stimulus must then be altered to produce excitation. Since this position change is required for only one eye, it results in a change in disparity.

In summary, the responses of disparity tuned cells can be explained with the energy model or some similar model in which there is no substantial nonlinearity in monocular processing prior to binocular combination. There is no need to postulate any complex feature detection prior to the representation of disparity in V1.

Horizontal and Vertical Disparities

Although we have discussed disparity encoding as if it were one-dimensional, RFs are two-dimensional. Thus, responses of the energy model depend on both horizontal and vertical disparities. A plot of responses to all combinations of vertical and horizontal disparities will reflect the structure and orientation of the monocular RFs (see Figure 4). A binocular neuron with perfectly matched RFs in the two eyes will be maximally activated by an RDS stimulus at zero disparity, and applying either a horizontal or a vertical disparity will reduce the response. The response should change more rapidly when disparities are applied orthogonal to the RF orientation because the structure of the monocular RFs changes most rapidly in that direction. However, if the visual stimulus is one-dimensional, such as a bar or a grating, then disparities applied parallel to the stimulus orientation have no effect (Figure 4f). Therefore, when evaluating responses to vertical and horizontal disparities, it is important to use stimuli that are orientation broadband, like RDS. With such stimuli, the energy model predicts that disparity tuning depends on RF orientation, phase disparity, and position disparity (both horizontal and vertical components).

This prediction of the energy model remains largely untested: The only study using combinations of vertical and horizontal disparities used bar stimuli (Maske et al 1986). It is an important prediction to test because stereopsis does not require equally precise information about all types of disparity. In most viewing situations, disparities in the central part of the retina will be larger horizontally than vertically. If the primary function of such cells is stereopsis, this should be reflected in the direction of neuronal disparity preferences. Alternatively, V1 neurons may measure binocular correlation for disparities in all directions. Such measurements would be useful for many binocular functions, including stereopsis and the control of vergence eye movements (which maintain vertical and horizontal alignment of the eyes). In this view, one would expect V1 neurons to represent horizontal and vertical disparities equally (isotropic).

Although data in the format of Figure 4 have not been obtained with RDS, two other experimental approaches have been used to determine whether disparity encoding is isotropic. The first has been to examine whether there is a relationship between orientation preference and the strength of disparity tuning. All the studies that have examined this quantitatively have found no correlation (Ohzawa &

Zero phase and zero position disparity RDS stimulus

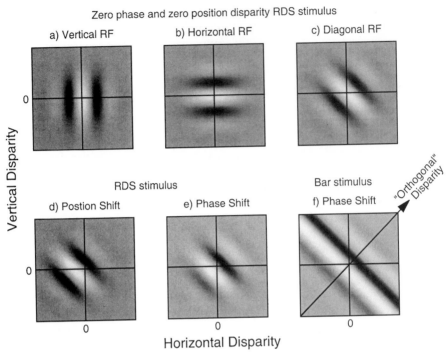

Horizontal Disparity

Figure 4   Responses of the energy model to vertical and horizontal disparities. The brightness of each point represents the response to a combination of horizontal and vertical disparity. Bright areas are strong responses; dark areas are weak responses. (*a–e*) Responses to orientation broadband stimuli [like random dot stereogram (RDS)]; (*f*) Responses to a one-dimensional (e.g. oriented bar) stimulus. (*a*) Responses of a neuron with identical RFs in the two eyes and a vertical preferred orientation are shown. Such a neuron is most sensitive to small changes in horizontal disparity. (*b*) Responses of a cell with matched RFs and a horizontal orientation. Because of its Gaussian envelope, this cell can also signal horizontal disparity for broadband stimuli. Note, however, that the most rapid changes in response result from vertical disparities (orthogonal to the RF orientation). (*c*) Responses of a cell with matched RFs and a diagonal RF orientation. (*d*) The effect of adding a horizontal position disparity to the neuron in (*c*). Note that the response profile has a diagonal axis of mirror symmetry because there is no phase disparity. The disparity that produces the greatest response is a horizontal disparity because the position shift is horizontal, but disparities orthogonal to the RF orientation produce the steepest change in response. (*e*) The effect of adding a phase disparity to the neuron in (*c*). Now, there is no axis of mirror symmetry parallell to the RF, and the neuron's largest response is produced by a combination of horizontal and vertical disparities along the direction orthogonal to the RF orientation. (*f*) When a long bar stimulus is used, only disparity changes orthogonal to the stimulus orientation elicit changes in response. Displacements parallel to the bar produce no change in the stimulus within the RFs. For this reason, many studies using oriented bars or gratings have only applied disparities orthogonal to the stimulus orientation.

Freeman 1986a,b; Smith et al 1997; SJD Prince, AD Pointon, BG Cumming, AJ Parker, submitted for publication).

The second approach has been to look for a relationship between orientation preference and the range of disparities encoded. This approach is hazardous in anesthetized animals because the measured range can be influenced by drifts in eye position. Perhaps this explains why several studies of anesthetized cats have obtained conflicting results (e.g. Barlow et al 1967, Nikara et al 1968, von der Heydt et al 1978, Maske et al 1986). The only quantitative study of awake animals found no relationship between orientation preference and the range of disparities encoded (SJD Prince, AD Pointon, BG Cumming, AJ Parker, submitted for publication). One measurement that is not influenced by slow drifts in eye position is the interocular phase difference between the two monocular RFs. In simple cells from cat V1, DeAngelis et al (1991) found that neurons preferring near-vertical orientations exhibited a larger range of phase differences than those preferring horizontal orientations. A similar, but less clear, correlation was observed by Anzai et al (1999b). However, in complex cells Anzai et al (1999c) found a correlation in the opposite direction (vertically oriented cells showed smaller phase differences). In awake monkeys, Prince et al (SJD Prince, BG Cumming, AJ Parker, submitted for publication) found no correlation between orientation preference and either phase shift or horizontal position shift.

Overall, then, there is limited evidence to support the view that V1 preferentially represents the directions of disparity that are most useful for stereopsis. However, no single study has gathered all the data needed to test this hypothesis conclusively.

Phase and Position Mechanisms

Many complex cells have disparity tuning curves the shape of which indicates an interocular phase difference (Ohzawa et al 1990, 1997; Anzai et al 1999c; SJD Prince, BG Cumming, AJ Parker, submitted for publication). A tuning curve that is even-symmetric (like that labeled T0 in Figure 5) suggests that the cell has similar RF structures in the two eyes. A curve that has odd-symmetry (like those labeled FA and NE in Figure 5) indicates a 90° phase shift between the subunits in the two eyes. Another way to distinguish phase and position mechanisms is to measure disparity selectivity with sinewave gratings at different spatial frequencies. If only a position shift is present, then the peaks of the disparity tuning curves should coincide at a value equal to that shift. If only a phase shift is present, then the modulation in the disparity tuning should show a consistent phase of modulation (for discussion see Fleet et al 1996, Zhu & Qian 1996). Wagner & Frost (1993, 1994) reported that in the Wulst of the barn owl, multi-unit activity showed consistent peak positions, whereas using this method with single-cell recording in awake monkeys indicates both position and phase shifts (SJD Prince, BG Cumming, AJ Parker, submitted for publication). Furthermore, Prince et al found that the phase shift estimated by this method agreed with the estimate derived from analysis of tuning curve shape. Nieder & Wagner (2000) recently analyzed the shape of

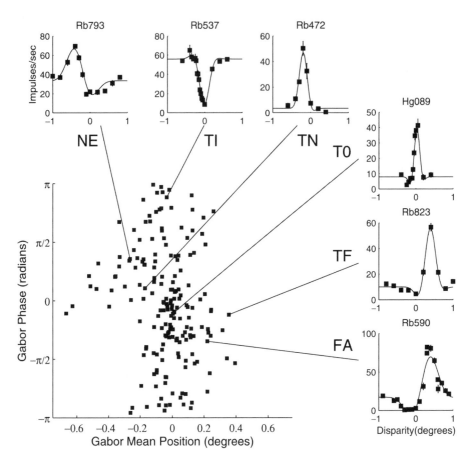

Figure 5 Distribution of phase and position disparities in a population of disparity-selective neurons (SJD Prince, BG Cumming, AJ Parker, submitted for publication). Tuning curves for horizontal disparity in random dot stereograms were fitted with Gabor functions. (Such curves are equivalent to horizontal cross sections through the surfaces shown in Figure 4.) For each neuron, the fitted phase is plotted against the fitted position of the Gaussian envelope. Examples of each of the classes identified by Poggio and collaborators are shown: NE, near; TI, tuned inhibitory, TN, tuned near; TO, tuned zero; TF, tuned far; FA, far. However, there is no tendency for a grouping around any of these shapes. Rather, the shapes of disparity tuning curves for V1 seem to form a continuum.

tuning curves recorded from the Wulst of the barn owl, and reported a range of phase shifts similar to that found in the cat and monkey. Taken together, these observations suggest that the study by Wagner and Frost probably underestimated that contribution of phase shifts.

Two studies have compared the relative contributions of phase and position mechanisms. Anzai et al (1997) used a reference cell method, recording from simple cells in anesthetized cats. In awake monkeys, Prince et al (SJD Prince,

BG Cumming, AJ Parker, submitted for publication) looked at responses of simple and complex cells to horizontal disparities in RDS. The phase and position of fitted Gabor functions were used to estimate underlying phase and position disparities. The data from both studies is shown in Figure 6: When converted into equivalent position shifts, phase shifts encode a slightly larger range of disparities than do position shifts (by 60% in cats and 25% in monkeys). However, this comparison is somewhat difficult to interpret:

1. Position shifts are measured in terms of visual angle; phase shifts are expressed in units of phase angle. This can be converted numerically into units of visual angle by scaling with the spatial period of the RF, but care is required in interpreting these numbers. Phase shifts outside the range $\pm\pi/2$ are simply inverted versions of phase shifts within the range $\pm\pi/2$, so it is not clear how much additional information they convey about disparity (SJD Prince, BG Cumming, AJ Parker, submitted for publication).

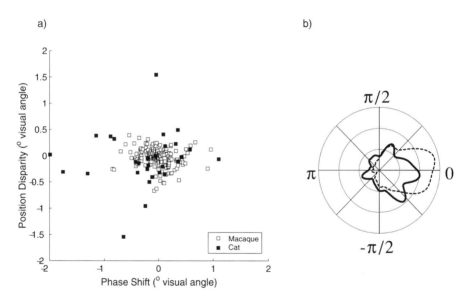

Figure 6 (*a*) Relative magnitudes of phase and position shifts in simple cells from cats (solid symbols, data from Anzai et al 1999a) and all cell types from monkeys (open symbols, data from SJD Prince, BG Cumming, AJ Parker, submitted for publication). Phase disparities have been converted into equivalent position disparities. The range of phase disparities is larger than position disparities. The pattern of results is broadly similar in the two species. (*b*) Compares the probability distributions for phase differences in monkeys (solid line Prince et al 2000a) and cats (dashed line, data combined from DeAngelis et al 1991, Anzai et al 1999a,c). These are plotted as polar probability density functions: The distance of each point from the origin indicates the probability of finding a fitted phase equal to the point's polar angle. The distribution is similar in the two animals, both showing a bias towards even-symmetric tuning (phase shifts near zero).

2. Position shifts are inherently two-dimensional (Anzai et al 1997), whereas phase shifts are one-dimensional. As shown in Figure 4, position shifts can encode useful information about disparities parallel to the RF orientation, but phase shifts cannot.

Given these difficulties and the modest difference in reported magnitudes, it seems likely that both phase and position mechanisms contribute importantly to disparity encoding, and this is similar in monkeys and cats. It is unclear what advantage is derived from employing both coding mechanisms, although Erwin & Miller (1999) offer one possible explanation.

The very existence of significant phase differences between the RF structure in the two eyes (in both cats and monkeys) has important implications. It argues strongly against the view that disparity selectivity depends on monocular responses to distinctive "trigger features," since cells with phase differences are responding to different features in the two eyes. Also, phase disparities enforce a "size-disparity correlation"—neurons can only encode disparities up to $\pm 1/2$ of the preferred spatial period. This limitation is computationally useful, since it restricts the number of false matches (e.g. Marr & Poggio 1979). It is interesting that even position shifts tend not to exceed this half cycle limit (SJD Prince, BG Cumming, AJ Parker, submitted for publication). Some aspects of psychophysical performance also show a size-disparity correlation (discussed in Prince & Eagle 2000, Smallman & Macleod 1994): This may be a reflection of the underlying physiological substrate (DeAngelis et al 1995).

Classes of Disparity Tuning

Phase disparities also provide a rationale for understanding the different shapes of disparity tuning curves that are observed (Nomura et al 1990). Poggio et al (1988) and Poggio (1995) distinguish three classes of disparity tuning curve (Poggio 1995):

1. "Tuned excitatory" (TE) neurons respond maximally to zero or near-zero disparities and show a roughly symmetrical response profile. These are subdivided into tuned zero (T0, maximal response to zero disparity), tuned near (TN) (maximal response to small crossed disparities), and tuned far (TF) (maximal response to small uncrossed disparities). The shape of these disparity tuning curves can be explained by supposing that the phase disparity is near zero.

2. Tuned inhibitory (TI) neurons are similar to TE cells, but inverted, showing maximal suppression for near zero disparities. This can be explained by a phase disparity near π.

3. Near and far neurons have asymmetrical response profiles (or more correctly, odd-symmetric), responding only to crossed (near cells) or uncrossed (far cells) disparities. The typical description of these cells also includes that their responses are "extended rather than tuned" (Poggio 1995)—there is a broad range of disparities over which the responses

change little. This particular feature is less easily reconciled with a simple phase disparity. Phase disparities near $\pi/2$ or $-\pi/2$ produce odd-symmetric curves that are just as tuned as even-symmetric curves. However, all the published examples of near/far cells showing these extended responses have used bar stimuli. With such stimuli, it is hard to exclude a contribution from monocular changes in the stimulus (see section on Measuring Disparity Selectivity). Our experience is that when a sufficiently large range of disparities is explored using random dot stimuli, no clear "plateau" is observed in the tuning of near/far cells, and they are well described by odd-symmetric Gabor functions.

Viewed from the perspective of phase disparities, it seems more natural to view these tuning curves as points on a continuum (as suggested by LeVay & Voigt 1988; Freeman & Ohzawa 1990) rather than as distinct classes. Prince et al (SJD Prince, BG Cumming, AJ Parker, submitted for publication) examined the distributions of both phase and position disparities in a large population of disparity-selective cells from monkeys and found no evidence of distinct classes (see Figure 5).

Depth Perception and Disparity-Selective Neurons in V1

It appears that we have a good understanding of the mechanism by which V1 neurons signal disparity. Here we consider how well these neuronal properties can account for the perceptual properties of stereopsis. In this context, we consider (*a*) the stereo correspondence problem, (*b*) the distinction between relative and absolute disparities, (*c*) the statistical reliability of neuronal signals and psychophysical judgements, and (*d*) the relationship between disparity and depth.

The Correspondence Problem   If a single random dot pattern is convolved with monocular filters in each eye, there will usually be several disparities that elicit similar responses in both eyes. The pattern will activate binocular filters tuned to different disparities, but only one of these is perceived. In order to discard the "false" matches, the correspondence problem must be solved. (This need not entail considering matches dot by dot. The number of false matches depends on the monocular filters that are applied.)

It is of course possible that V1 neurons are more sophisticated than the energy model and distinguish false matches from correct matches. In order to test this possibility, it is necessary to place false matches in the neuronal receptive field. Most experiments with RDS have used dynamic RDS—each frame of the display contains a fresh pattern of dots, but the disparity relationships remain constant. Unfortunately, the disparities at which false matches occur depend on both the monocular filters and on the particular dot pattern used. Therefore, if one dot pattern contains a false match at some disparity, the dot pattern displayed on the next frame will in general not contain a false match at the same disparity. For this reason, averaged across many RDS frames, even the energy model responds maximally to the correct matches (Qian 1994, Fleet et al 1996, Cumming & Parker 1997).

One stimulus manipulation that clearly differentiates the properties of the energy model from those of visual perception is to reverse the contrast of the image in one eye. Each bright feature on one retina is then paired geometrically with a dark feature on the other retina, and vice versa. Such stereograms are called anticorrelated because the correlation coefficient between luminance values in the two images is a negative one. Ohzawa et al (1990) and Livingstone & Tsao (1999) examined the responses with bar stimuli and (as described above) found an inversion of the disparity tuning. This is exactly what one would expect from the energy model, since its response reflects binocular correlation (see legend to Figure 3). Cumming & Parker (1997) found very similar results using anticorrelated RDS in awake monkeys (for a comparison, see Ohzawa 1998). In both cases, the effects of anticorrelation on neuronal activity are quite different from the perceptual effects. With bar stimuli, human observers perceive depth in the geometrically correct direction (Helmholtz 1909, Cogan et al 1995, Cumming et al 1998). In anticorrelated RDS of the type used by Cumming & Parker (1997), no depth is perceived (Julesz 1971, Cogan et al 1993, Cumming et al 1998). In this case observers appear unable to access the information about disparity contained in the firing rate of single V1 neurons. Both types of anticorrelated stimulus activate disparity-selective neurons without observers perceiving a stimulus at the equivalent depth. Thus, the psychophysical matching process appears to discard these responses as false matches. The neural responses are not associated only with psychophysically matched disparities. This dissociation between neuronal firing and perceived depth does not imply that the perception of depth is completely independent of activity in V1 neurons. They may perform an initial analysis of binocular correlation that extrastriate areas use to solve the correspondence problem.

In one quantitative respect, the neuronal responses to anticorrelation deviate from the predictions of the energy model: Although the disparity tuning curves are generally inverted by anticorrelation, the magnitude of the disparity-induced modulation is often smaller for anticorrelated stimuli than for their correlated counterparts (Cumming & Parker 1997, Ohzawa et al 1997). The energy model predicts that these magnitudes will be the same. It remains to be seen whether major changes in the model are required to accommodate this observation.

A different examination of the role of V1 in stereo correspondence was presented by Cumming & Parker (2000). They used circular patches of sinusoidal luminance gratings. For two disparities differing by the spatial period of the grating, the stimulus within the RF is identical, although the perceived depth is different. Consider a grating at a crossed disparity equal to one grating period. Within the RF, each bar of one monocular grating superimposes on the next bar of the other monocular grating. Thus, within the RF it is identical to a grating at zero disparity. Nonetheless, what is perceived is a patch of grating standing in front of the fixation point. The perceptual effects were demonstrated psychophysically in the animals from whom neurons were recorded, and a similar psychophysical result had been reported in humans using rows of dots (Mitchison 1988, McKee & Mitchison 1988). Cumming & Parker (2000) found that for the vast majority of V1 neurons, the response was determined by the local disparities within the RF, regardless of the perceived depth. This reinforces the view that additional processing is required beyond striate cortex to account for how depth is perceived in stereograms.

Relative and Absolute Disparities   The mechanisms discussed so far signal the disparity of a feature in retinal coordinates (how far the two images fall from corresponding retinal locations). This is called the absolute disparity. The difference between the absolute disparities of two features is called their relative disparity. A major advantage of relative disparities is that they are unaffected by vergence eye movements, whereas changes in vergence alter the values of absolute disparities. Cumming & Parker (1999) controlled vergence movements in a feedback loop to manipulate absolute disparities independent of relative disparities. The results showed clearly that neurons in monkey V1 signal absolute, not relative, disparity. In contrast, a number of psychophysical studies have suggested that stereopsis relies primarily on relative disparities. Stereoacuity, when measured using a single isolated feature (absolute disparity threshold), is fivefold poorer than when relative disparities are provided by a simultaneously visible reference stimulus (Westheimer 1979). When an absolute disparity is applied uniformly to a large display, substantial changes in disparity are not detected (Erkelens & Collewijn 1985a,b; Regan et al 1986).

Neuronal and Psychophysical Sensitivity   Both the experiments on stereo correspondence and those on relative disparity indicate that signals that determine depth perception are different from those carried by single V1 neurons. Further elaboration of stereo signals probably occurs outside V1, and this may produce signals that could be used more directly for depth perception. If these signals were derived from V1 neurons, then the precision with which V1 neurons are able to signal disparity imposes limits on the precision of subsequent processing and psychophysical performance. This was examined explicitly by Prince et al (2000), who measured the smallest disparity change that single V1 neurons could detect with a given reliability (the neurometric threshold). The performance of the animals was measured with the same stimuli (psychometric thresholds). Many neuronal thresholds were as low as the psychometric thresholds, which indicates that a modest degree of pooling from V1 responses is sufficient to account for observed stereoacuity. Note that this result applied when the animals' task was a relative disparity judgment. When the animals were forced to rely on absolute disparities alone, the psychometric thresholds were generally larger (poorer performance) than the neurometric thresholds. Under these circumstances, the animals were not able to discriminate between stimuli even when information available in single V1 neurons made the discrimination possible.

Disparity and Depth   There are many unresolved questions concerning how a map of angular disparities might be converted into a representation of the three-dimensional world. Are all possible relative disparities encoded? Are they converted into a depth map with some fixed coordinate frame? Most of these complex questions have not been addressed at the neurophysiological level. One exception is the effect of viewing distance. The disparity produced by a fixed depth difference depends on viewing distance. A few studies (Trotter et al 1992, 1996; Gonzalez & Perez 1998a) have reported that disparity-selective neurons in V1 alter their response to a fixed stimulus disparity when the viewing distance is changed. The change in viewing distance requires a change in vergence angle, but as these studies did not measure vergence, it is possible that inaccuracies in vergence resulted in changes in horizontal disparity. Also, changes in viewing distance induce changes in vertical disparity (Mayhew & Longuet-Higgins 1982), which should also affect response rates. (Disparity tuning curves performed at different viewing distances correspond to different cross sections through the surfaces shown in Figure 4.) At present it is not possible to be sure that effects like those observed by Trotter et al are not the result of changes in the vertical and horizontal disparities of the retinal stimulus.

Conclusion

The major features of disparity-selective responses in cat and monkey V1 are captured by a relatively simple model (the energy model of Ohzawa et al 1990). The model is certainly a simplification—it is likely that real complex cells receive input from more than four subunits. Nonetheless, this simple model has been very successful—only two failures have been noted to date: (a) a poor correlation betwen monocular spatial frequency tuning and the spatial scale of the disparity tuning function, and (b) a failure to explain the reduced amplitude of responses to anticorrelated stimuli.

Although the mechanism of disparity-selectivity in V1 seems to be well understood, there are several substantial differences between the properties of stereopsis and the properties of V1 neurons. Conversely, disparity-selective V1 neurons seem well suited to the control of vergence eye movements: Anticorrelated RDS elicit reversed vergence movements (Masson et al 1997); vergence depends on absolute rather than relative disparity; and maintaining alignment of the eyes requires signals about horizontal and vertical disparities. It is even possible that vergence control is the primary role of these V1 neurons—there is no definitive evidence that stereopsis is mediated by disparity-selective V1 neurons—but it seems more likely that V1 serves as an initial stage in stereo processing. Extrastriate areas may then make explicit the signals that support depth perception (discarding false matches and representing relative disparity). We now turn our attention to the role of these areas.

EXTRASTRIATE CORTEX

Considerably less is known about stereoscopic processing outside V1 than within V1. Disparity-selective neurons can be found in many different areas of the brain (for a review see Gonzalez & Perez 1998b). In cats, disparity-selective neurons have been reported to occur in extrastriate areas 18, 19, and 21, in the superior colliculus, and in the accessory optic system (Ferster 1981; LeVay & Voigt 1988; Guillemot et al 1993a,b; Pettigrew & Dreher 1987; Wang & Dreher 1996; Vickery & Morley 1999; Berman et al 1975; Bacon et al 1998; Grasse 1994). In monkeys, disparity-selective neurons can be found in extrastriate areas V2, V3, V3A, VP, MT, MST (both subdivisions), and IT (Hubel & Wiesel 1970, Poggio & Fisher 1977, Poggio et al 1988, Burkhalter & van Essen 1986, Felleman & van Essen 1987, Maunsell & van Essen 1983, Roy et al 1992, Janssen et al 1999, Uka et al 2000), as well as in some visuomotor regions of parietal and frontal cortex (Gnadt & Mays 1995, Ferraina et al 2000).

To understand why binocular disparity is repeatedly represented in different areas, it is useful to identify ways in which disparity processing differs from V1. This requires more than just measurements of disparity tuning to simple stimuli: It is essential to record and/or manipulate the activity of disparity-selective neurons in cortical areas during a variety of stereoscopic tasks. In this section, we consider how the representation of binocular disparity in extrastriate cortex differs from that in V1, and we review emerging evidence for more direct links between neuronal activity and perception.

Columnar Architecture for Disparity

Columnar architecture is a common feature of the organization of cerebral cortex. In many cortical areas, neurons within a column normal to the cortical surface have similar functional properties, and these properties usually vary systematically from column to column, thus forming a topographic map (Mountcastle 1997). Thus, one might expect to find a map of binocular disparity in areas that are important for stereopsis. DeAngelis & Newsome (1999) provide compelling evidence for a map of binocular disparity in visual area MT. Using RDS, they showed that disparity selectivity often occurred in discrete patches (typically 0.5–1 mm in extent) that were interspersed among similar-sized patches of cortex with weak disparity tuning. Within the disparity tuned patches, preferred disparities changed smoothly across the surface of MT, but there was little change in disparity selectivity along penetrations normal to the cortical surface. This suggests strongly that there are disparity columns in MT, in addition to the well-known columns for direction of motion (Albright et al 1984). A similar methodology applied in monkey V1 by Prince et al (SJD Prince, AD Pointon, BG Cumming, AJ Parker, submitted for publication) also found evidence for a clustering of disparity selectivity, although this was much weaker than in MT (see Figure 7). Earlier investigations in V1 using bar stimuli yielded conflicting results: Blakemore (1970) qualitatively described "constant depth" columns, whereas LeVay & Voigt (1988) reported a weak clustering of disparity preference.

There is also a clustering of disparity-selective neurons in the thick stripes of V2 (Hubel & Livingstone 1987, Peterhans & von der Heydt 1993, Roe & Ts'o

a)   b)

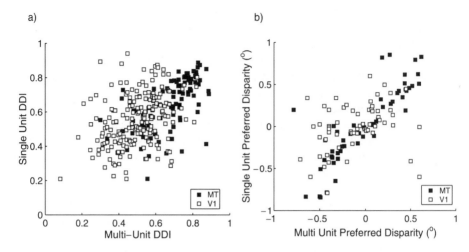

Figure 7 Clustering of disparity preference in areas V1 (open symbols) (SJD Prince, AD Pointon, BG Cumming, AJ Parker, submitted for publication) and MT (solid symbols, DeAngelis & Newsome 1999), assessed by comparing properties of isolated single units (SU) with multi-unit (MU) recordings at the same site. (*a*) Plots the modulation of firing rate induced by disparity for MU and SU data. This is measured using the disparity discrimination index (DDI) (SJD Prince, AD Pointon, BG Cumming, AJ Parker, submitted for publication). DDI = (Max − Min)/(Max − Min + 2SD), where SD is an estimate of the standard deviation of firing calculated across all disparities. Although there is a significant correlation in both V1 and MT, the latter is stronger. It is also clear that both MU and SU responses are generally more strongly tuned for disparity in MT than in V1. (*b*) Plots the disparity that produces maximal activation for MU and SU (some data points from MT fall outside the range plotted here). Again there is a significant correlation for both areas, but the correlation is much stronger in MT ($r = 0.91$) than in V1 ($r = 0.30$).

1995). No quantitative electrophysiological studies have demonstrated an orderly map of disparity across adjacent columns, although this was reported in a recent optical imaging study (Burkitt et al 1998). Although all of these studies have used bar stimuli in anesthetized animals, the results are sufficient to suggest that there is a topographic map of disparity within the thick stripes of V2. Less is known about columnar architecture for disparity in ventral stream areas; however, Uka et al (2000) have recently reported modest clustering for disparity in inferotemporal cortex.

Can Extrastriate Responses to Disparity be Derived from V1?

Two differences between the shapes of disparity tuning curves in striate and extrastriate cortex have frequently been noted (see for example Poggio 1995). First, neurons in extrastriate cortex tend to be more coarsely tuned to disparity than neurons in V1 and have peak responses at larger disparities. Second, while the majority of V1 neurons show symmetrical tuning (like the T0 cell in Figure 5), in extrastriate areas odd-symmetric tuning (near and far cells) predominates. Both of these observations suggest that the outputs of disparity-selective neurons in V1 must be combined in specific ways to generate extrastriate neuronal responses.

If neurons in extrastriate cortex have coarser disparity tuning than V1 neurons, it indicates that there is a range of large disparities that have no effect on the firing of V1 neurons, but do alter the firing of neurons in extrastriate cortex. This implies that the extrastriate responses are not derived from disparity-selective neurons in V1, but are constructed de novo. However, it is important to consider the effect of stimulus eccentricity. Most disparity-selective neurons studied in V1 have had parafoveal RFs, whereas studies in extrastriate cortex typically involve more eccentric stimulation. No study has compared disparity tuning to the same stimuli at matched eccentricities across brain areas. Figure 8 therefore compares

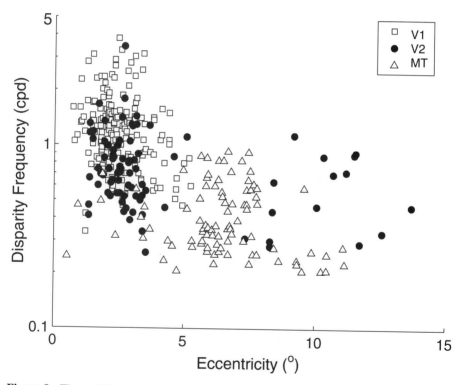

Figure 8 The spatial scale of disparity tuning, as a function of eccentricity, compared across cortical areas. "Disparity frequency" plots the peak frequency in the continuous Fourier transform of the disparity tuning curve. Narrow tuning curves have high peak frequencies, broad tuning curves have low frequencies. Although disparity tuning curves recorded in MT are generally coarser than those in V1, this may largely reflect the eccentricity at which they were recorded. Data taken from Prince et al (SJD Prince, AD Pointon, BG Cumming, AJ Parker, submitted for publication), DeAngelis & Newsome (1999), and Thomas et al (OM Thomas, BG Cumming, AJ Parker, submitted for publication).

the responses of V1,V2, and MT neurons to disparity in RDS. The spatial scale of each disparity tuning curve is estimated from the dominant spatial frequency in the Fourier transform of the tuning curve (SJD Prince, AD Pointon, BG Cumming, AJ Parker, submitted for publication), and this is plotted as a function of stimulus eccentricity. At matched eccentricities, there is a sustantial overlap between the data of different areas, although there is a tendency for the extrastriate neurons to show coarser tunning.

The claim that the symmetry of disparity selectivity differs between cortical areas rests largely on the results of classifying neurons manually into the categories proposed by Poggio & Fisher (1977). The few studies that have attempted to measure this property quantitatively have used different measures and different stimuli (LeVay & Voigt 1988, Roy et al 1992) and so are hard to compare. Figure 9 therefore applies the same metric (the phase of a fitted Gabor) to data gathered with RDS from different brain areas. The data used were from area V1 (SJD Prince, BG Cumming, AJ Parker, submitted for publication), V2 (OM Thomas, BG Cumming, AJ Parker, submitted for publiction), MT (DeAngelis & Newsome 1999), and the dorsal part of MST (MSTd) (Takemura et al 1999). The fitted phase of the Gabor measures the symmetry of the tuning curve (see Figure 5).

In accord with earlier claims, V1 shows a preponderance of even symmetry, while other areas do not. V2 and MT contain many neurons with phases intermediate between even and odd symmetry, and MSTd shows a preponderance of odd symmetry. This suggests that the shape of tuning curves for extrastriate neurons is not simply inherited from V1 neurons. It might be that an appropriate combination of even-symmetric inputs (e.g. inhibition from cells with peaks at crossed disparities, excitation from cells with peaks at uncrossed disparities) is used to construct odd-symmetric responses outside V1, but this too requires more than a simple pooling of inputs from V1.

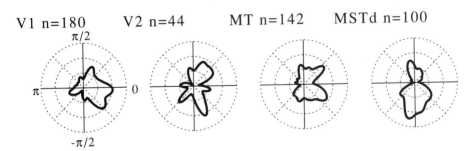

Figure 9 The distribution of phases for Gabor functions fitted to disparity tuning data in different areas of the macaque brain. Fitted phases near zero indicate symmetrical tuning, phases near $\pm\pi/2$ indicate odd symmetry (near and far cell types). There seems to be a systematic progression toward increasing odd symmetry from V1 to MSTd.

Representation of Relative Disparity

As discussed above, stereopsis is strongly dependent on relative disparities between different locations in the visual field, and yet V1 neurons signal only absolute disparities (Cumming & Parker 1999). One possibility is that relative disparity might be explicitly represented at the level of single neurons somewhere in extrastriate cortex. This could be achieved by spatial interactions between the classical RF and the nonclassical surround, which are prevalent in many visual cortical areas (Allman et al 1985). Recent studies have demonstrated center-surround interactions that depend on binocular disparity in area MT (Bradley & Andersen 1998) and in the lateral portion of area MST (Eifuku & Wurtz 1999).

To examine relative disparity encoding more directly, OM Thomas, BG Cumming, AJ Parker (submitted for publication) presented RDS consisting of a center and a surround while recording from V2 neurons. The horizontal disparity of both regions (Figure 10A) was varied independently. The center patch was sized to match the classical RF. Figure 10B shows the type of interaction that yields relative disparity encoding, with a strong diagonal structure in the response map. As the surround disparity changes, the preferred center disparity changes proportionally so that response remains roughly invariant along diagonal lines of constant relative disparity. Thomas et al measured the response to a range of center disparities at different surround disparities (e.g. horizontal cross sections in Figure 10B). If a neuron encodes relative disparity, then its preferred center disparity should shift by an amount equal to the surround disparity.

Figure 11 shows example tuning curves and summarizes the shifts in tuning for populations of neurons from areas V1 and V2. For a handful of V2 neurons, the shift is consistent with relative disparity coding, whereas other V2 neurons show a partial but significant shift in the direction of relative disparity encoding. The remaining V2 neurons, as well as virtually all neurons tested in V1, do not show any significant shift in their disparity preference with changing surround disparity. These latter neurons appear to encode only absolute disparities. These results strongly suggest that some V2 neurons encode relative disparity.

Eifuku & Wurtz (1999) have also suggested that neurons in the lateral portion of area MST encode relative disparity. In this study, the authors measured responses to variable center disparities at a surround disparity of zero and responses to variable surround disparities at a center disparity of zero. This corresponds to horizontal and vertical cross-sections, respectively, through the center of the two-dimensional map (black lines in Figure 10B,C). For a number of cells, the tuning curve for surround disparities was roughly the inverse of the tuning curve for center disparities. Although this pattern of results might reflect encoding of relative disparities (Figure 10B), these data could also have arisen from a separable interaction between center and surround disparities, as depicted in Figure 10C. By a separable interaction, we mean that the response to combinations of center and surround disparities is proportional to the product of the responses to center and surround disparities alone. A separable interaction does not indicate selectivity for

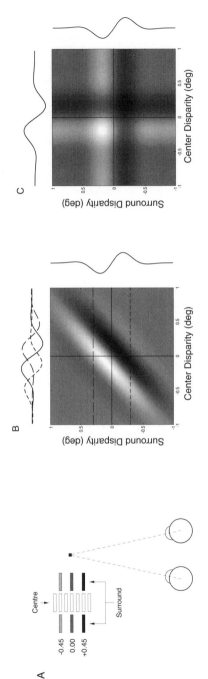

Figure 10 Measurement of relative disparity selectivity. (*a*) A bipartite field of random dots allows the disparity of a central region to be manipulated independently of the disparity of the surround. (*b*) Idealized response pattern for a neuron selective to the relative disparity between center and surround regions. Dark regions denote weak responses; bright regions indicate strong responses. Configurations with a constant disparity difference fall along diagonal lines, hence the strong diagonal structure. The lines at the top of panel (*b*) show horizontal cross sections (disparity tuning curves for the center) taken at different heights (surround disparity). The curves all have the same shape but are translated along the disparity axis relative to one another. Note also that the tuning for surround disparity (right) is the opposite of the tuning to center disparity. (*c*) Response pattern for a neuron that is not selective for relative disparity but has a separable interaction between center and surround disparities. Note that the horizontal and vertical cross sections through zero disparity are identical to those in panel *b*. Thus, these cross sections alone are insufficient to demonstrate relative disparity selectivity.

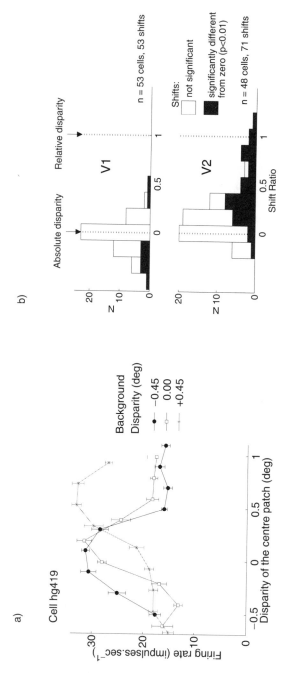

Figure 11 (*a*) The responses of one neuron in V2 to the disparity of a random dot stereogram covering the minimum response field. Surrounding this was a background of random dots, whose disparity was also altered (see key). Change in the background disparity produced systematic shifts in the preferred central disparity, so that the neuron appears to encode the relative disparity between center and surround. (*b*) The magnitude of the shift in preferred disparity, as a fraction of the change in background disparity, measures the extent to which the neuron signals relative disparity (shift ratio 1.0) or absolute disparity (shift 0.0). Unlike neurons in V1 (data from Cumming & Parker 1999), a fraction of neurons in V2 shows some selectivity to relative disparity.

relative disparity, so the results of Eifuku & Wurtz are are not conclusive. Further studies, in which the center-surround disparity space is mapped more finely, will be valuable for understanding the encoding of relative disparities.

The Correspondence Problem

Neurons in V1 respond to binocular matches that are not perceived (i.e. "false" stereo matches). If extrastriate areas combine the outputs of V1 neurons appropriately, they might produce responses more similar to the psychophysical sensations. This might be achieved by combining responses of V1 neurons with different spatial scales (Fleet et al 1996), which could also eliminate the modulation of responses to anticorrelated RDS. Two preliminary reports have examined this, in areas MT (Krug et al 1999) and MSTd (Takemura et al 1999). Both found disparity induced modulations in response to anticorrelated RDS, similar to those already reported in V1 (Cumming & Parker 1997). In this respect at least, disparity-selective responses in MT and MSTd are no closer to psychophysical stereo matching than V1.

Links Between Disparity-Selective Neurons and Perception

Neurons that signal binocular disparity do not necessarily contribute to stereopsis. To establish that a candidate set of neurons contributes to performance of a specific stereoscopic task, additional criteria must be met (Parker & Newsome 1998). First, neuronal activity should be recorded during performance of the task, and it should be shown that the candidate neurons are sufficiently sensitive to mediate task performance (so far only demonstrated for V1 neurons; Prince et al 2000). Second, neuronal activity should be shown to covary with perceptual judgements near psychophysical threshold. Third, artificial manipulation of neuronal activity (either activation or suppression) should be shown to alter performance of the task. Below, we review experiments that begin to address these requirements.

Covariation of Neuronal Firing and Depth Perception If a group of neurons contributes strongly to a three-dimensional percept, then the activity of those neurons should covary with perceptual reports under circumstances in which the visual stimulus is near threshold or ambiguous. Two groups have recently probed for this type of covariation (Bradley et al 1998, Parker et al 2000). Monkeys were trained to report the direction of rotation of a three-dimensional cylinder defined by random dots (Figure 12). When the depth of the cylinder is defined by binocular disparity, direction of rotation is unambiguously perceived. In contrast, when the disparity cues are removed, the percept becomes bistable: For the same visual stimulus, clockwise rotation is seen in some trials and counterclockwise rotation is seen in other trials (Wallach & O'Connell 1953). Bradley et al (1998) recorded from neurons in area MT that are selective for conjunctions of motion and disparity (e.g. rightward and near), and showed that these neurons can encode the direction of rotation of unambiguous cylinders defined with disparity. More importantly, they showed that the average responses of

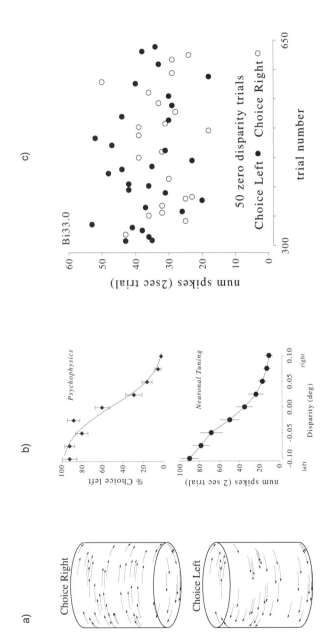

Figure 12 Moving dots portraying a transparent rotating cylinder give rise to an ambiguous percept (*a*). The perceived direction of rotation, which is bistable, depends on whether the leftward moving dots are perceived in the front plane (top) or back plane (bottom). The stimulus can be rendered unambiguous by the addition of disparities defining the depth relationships, and animals report one direction of rotation unambiguously for small disparities (*b*, top). The activity of an example MT neuron, selective for the direction of rotation defined by disparity, is shown in the bottom half of panel *b*. (*c*) The response rates for each trial of the zero disparity stimulus only: Filled symbols indicate those trials where the animal reported leftward rotation at the end of the trial; open symbols indicate rightward choices. Trials associated with "left" choices are associated with higher firing rates. Since this neuron is also selective for leftward rotation defined by disparity, the choice probability is >0.5. For this neuron the choice probability was 0.66, close to the population mean (0.68).

some MT neurons covary with perceived direction of rotation, separate from any disparity-induced modulation. The activity of these neurons appears to reflect the percept and not just the physical stimulus.

Parker et al (2000) extended this observation, analyzing only responses to the ambiguous, zero-disparity stimulus (see Figure 12) and quantifying the covariation with choice probabilities (Britten et al 1996). These define the probability that the behavioral outcome of a trial can be predicted from the firing rate of a single neuron (choice probabilities >0.5 indicate a positive correlation). It is interesting that the average choice probability for the cylinder task (0.68) is substantially higher than the average choice probability (0.56) exhibited by MT neurons during a direction discrimination task (Britten et al 1996). This means that fluctuations in activity of MT neurons are more tightly linked to fluctuations in perceptual reports for the three-dimensional cylinder task. Further investigation of what aspects of the stimulus or task influence the magnitude of the choice probability is required in order to interpret this difference in choice probabilities.

Effects of Lesions   Strong trial by trial covariation between firing rate and perceptual reports engenders confidence that the neurons under study contribute to depth perception. However, even these measures are only correlative in nature; thus, choice probabilities do not establish a causal linkage between neuronal activity and perception. Such a link could be established if localized brain lesions produce deficits in stereoscopic vision without degrading other visual capabilities. It is important to ensure that eye movements are unaffected: If a lesion disrupts vergence control then this will be detrimental to stereo tasks. Unfortunately none of the following studies measured or controlled vergence, so the results are inconclusive.

Human Studies.   Several studies have examined the effects of cerebral lesions on stereopsis in human patients (Carmon & Bechtoldt 1969, Benton & Hecaen 1970, Rothstein & Sacks 1972, Lehmann & Wächli 1975, Danta et al 1978, Hamsher 1978, Ross 1983, Vaina 1989, Ptito et al 1991). In general, because the lesions are poorly localized, these studies reveal little about the contributions of specific cortical areas to stereopsis. Moreover, equivalent, nonstereoscopic control tasks were generally not performed; thus, it is difficult to be sure that the observed deficits are specific to stereopsis. Nonetheless, a few observations are worth noting.

A few studies have reported that depth perception in RDS ("global" stereopsis) is selectively impaired by lesions to the right cerebral hemisphere (Carmon & Bechtoldt 1969, Benton & Hecaen 1970, Hamsher 1978, Ross 1983, Vaina 1989). In contrast, local stereopsis (stereoacuity measured with isolated stimuli) seems to be equally impaired by left and right hemisphere lesions (Rothstein & Sacks 1972, Lehmann & Wächli 1975, Danta et al 1978, Hamsher 1978). However, only Hamsher (1978) studied global stereopsis and stereoacuity in the same group of patients. The range of disparities used for the two tasks was nearly nonoverlapping, so any interhemispheric difference may be in the range of disparities processed.

Vaina (1989) has reported that subjects with right occipital/temporal lesions were able to see depth in RDSs but were unable to identify the shape of regions defined by disparity. In contrast, patients with occipital/parietal lesions failed to see any depth at all in the same stimuli. Thus, Vaina posits that occipital/parietal areas may be necessary for establishing binocular correspondence, whereas occipital/temporal areas are needed for extracting cyclopean form after the correspondence problem is solved.

Animal Studies.   A major advantage of animal studies is that surgically induced lesions can be fairly well localized. Cowey & Porter (1979) trained monkeys to discriminate depth in RDSs. They found no deficits following lesions of the central visual field representation in V1 or V2. In contrast, they report substantial deficits following temporal lobe lesions, which appear to include most of inferotemporal cortex as well as substantial portions of prestriate cortex. Although the authors conclude that "global stereopsis is mediated in temporal lobe areas," this conclusion must be treated with care. Lesions of V1 and V2 were restricted to a central region of the visual field smaller than the center portion of the stereograms. Moreover, monkeys were not trained to maintain fixation. Thus, the animals may have simply fixated eccentrically and used portions of V1 and V2 unaffected by the lesions.

Cowey & Wilkinson (1991) tested the stereoacuity of monkeys following similar lesions. Following V1 and V2 lesions, monkeys could still perform the task, but their thresholds were elevated roughly 2- to 10-fold. Because fixation was not enforced, it is unclear whether this residual capacity should be attributed to other brain structures, or whether it resulted from animals fixating eccentrically to perform the task. Lesions of inferotemporal cortex produced only a mild increase in thresholds (1.5- to 2-fold), which suggests that these areas may not be critical for fine stereo judgments.

Schiller (1993) evaluated the effects of V4 and MT lesions on stereopsis in monkeys. Fixation (but not vergence) was tightly controlled, and performance was compared between lesioned and intact portions of the visual field. Neither V4 nor MT lesions, nor the combination of the two, produced any discernible effects on performance in the detection or discrimination tasks used in this study. This is surprising, given that MT and V4 are central stages along the dorsal and ventral processing streams, respectively. This finding might indicate that lower visual areas (e.g. V1, V2, and V3) are sufficient to mediate performance on these tasks, or it might reflect the fact that some of these areas have alternative projections to the temporal and parietal lobes (Felleman & van Essen 1991). However, there are two reasons for interpreting these results cautiously. First, the monkeys were working well above psychophysical threshold; thus, the task conditions did not force the animals to rely on the most sensitive neurons. The effects of lesions might have been much larger near threshold. Second, data are reported only from sessions in which performance had stabilized after the lesions: Transient deficits in stereopsis may have gone unnoticed. As the tasks were not performed at threshold, disparity-selective neurons in other visual areas may have been able to compensate for the loss of MT and V4.

In cats, unlike monkeys, lesions of V1 spare many visual functions, including grating and vernier acuity (e.g. Berkley & Sprague 1979, Ptito et al 1992). This presumably reflects a more important role of retino-tectal pathways in felines. In this light, it is interesting to note that combined lesions of areas 17 and 18 are reporetd to completely abolish stereopsis in cats (Kaye et al 1981, Ptito et al 1992).

Microstimulation   Many of the difficulties with lesion studies (effect of recovery, control of vergence) can be avoided in microstimulation studies, where stimulated and nonstimulated trials are interleaved. Microstimulation studies have previously established that areas MT and MST play a central role in motion perception (Salzman et al 1992, Salzman & Newsome 1994, Celebrini & Newsome 1995).

DeAngelis et al (1998) used microstimulation to probe the role that area MT plays in stereopsis. Monkeys were trained to discriminate between two suprathreshold disparities (e.g. one near, one far) in the presence of disparity noise. The relative proportions of signal and noise dots were varied around psychophysical threshold, and microstimulation was applied during half of the trials. Because of the columnar organization for disparity in MT (DeAngelis & Newsome 1999), electrical stimulation could be applied to a cluster of neurons with similar disparity selectivity. At locations in MT with strong disparity tuning, microstimulation biased the monkey's judgments in favor of the preferred disparity of the stimulated neurons, with no decrement in psychophysical sensitivity. That is, microstimulation of a cluster of far-preferring neurons shifted the monkeys' psychometric function, resulting in more far choices. In contrast, there was generally little or no effect at locations in MT with poor disparity tuning. Thus, injecting an artificial signal into the disparity map within area MT caused a predictable bias in depth judgments. This result establishes the first causal linkage between a population of disparity-selective neurons and stereopsis.

FUTURE DIRECTIONS

The mechanisms by which responses to disparity are produced in striate cortex are now well characterized. These result in signals that differ in many ways from the perception of depth: V1 neurons respond to false matches in the RF and do not signal relative disparities. Rather, V1 seems to measure binocular correlation over a range of vertical and horizontal disparities. These preliminary computations may be exploited in extrastriate cortex for a number of different tasks: stereopsis, vergence control, scene segmentation, and three-dimensional heading judgments.

How each of these more sophisticated judgments is derived from the activity of neurons in striate and extrastriate cortex is largely unknown. Functional imaging in human subjects is likely to provide valuable insights into which brain areas are specialized for each of these tasks.

It is clear that certain extrastriate areas contain signals that are more closely related and causally linked to the perception of depth. V2 is able to signal relative disparities, and center-surround interactions in MT and MST may achieve similar results. Microstimulation in MT systematically biases depth judgments. These observations raise many further questions, such as how all the possible combinations of relative disparities are represented and how this information is maintained as the eyes move. A few studies have examined the neurophysiological representation of surfaces (Shikata et al 1996, Janssen et al 1999, Taira et al 2000), but much remains to be done in this field.

The quantitative study of responses to disparity in extrastriate cortex, combined with matching psychophysical studies, promises to clarify the ways in which a number of binocular tasks are carried out by the brain. Combining this with the study of how extrastriate responses are derived from those in V1 offers the prospect of a system in which the gap between neuronal mechanisms and perceptual phenomena is significantly narrowed.

ACKNOWLEDGMENTS

We are especially grateful to Simon Prince, who prepared many of the figures (including the data analysis) for us, as well as providing penetrating criticisms of the manuscript. We also thank Andrew Parker, Aki Anzai, Holly Bridge, Ralph Freeman, Alex Foulkes, Ichiro Fujita, Alice Gardner, Andrew Glennerster, Kristine Krug, Katrina Pearce, Izumi Ohzawa, Owen Thomas and Takanori Uka for helpful discussions and comments.

Visit the Annual Reviews home page at www.AnnualReviews.org

LITERATURE CITED

Albright TD, Desimone R, Gross CG. 1984. Columnar organization of directionally selective cells in visual area MT of the macaque. *J. Neurophysiol.* 51:16–31

Allman J, Miezin F, McGuinness E. 1985. Stimulus specific responses from beyond the classical receptive field: neurophysiological mechanisms for local-global comparisons in visual neurons. *Annu. Rev. Neurosci.* 8: 407–30

Anzai A, Ohzawa I, Freeman RD. 1997. Neural mechanisms underlying binocular fusion and stereopsis: position vs. phase. *Proc. Natl. Acad. Sci. USA* 94:5438–43

Anzai A, Ohzawa I, Freeman RD. 1999a.

Neural mechanisms for encoding binocular disparity: receptive field position versus phase. *J. Neurophysiol.* 82:874–90

Anzai A, Ohzawa I, Freeman RD. 1999b. Neural mechanisms for processing binocular information. I. Simple cells. *J. Neurophysiol.* 82:891–908

Anzai A, Ohzawa I, Freeman RD. 1999c. Neural mechanisms for processing binocular information. II. Complex cells. *J. Neurophysiol.* 82:909–24

Bacon BA, Lepore F, Guillemot JP. 1998. Striate, extrastriate and collicular processing of spatial disparity cues. *Arch. Physiol. Biochem.* 106:236–44

Barlow H, Blakemore C, Pettigrew J. 1967. The neural mechanisms of binocular depth discrimination. *J. Physiol. London* 193: 327–42

Benton AL, Hecaen H. 1970. Stereoscopic vision in patients with unilateral cerebral disease. *Neurology* 20:1084–88

Bergua A, Skrandies W. 2000. An early antecedent to modern random dot stereograms—'The Secret Stereoscopic Writings' of Ramón y Cajal *Int J. Psychophysiol.* 36:69–72

Berkley MA, Sprague JM. 1979. Striate cortex and visual acuity functions in the cat. *J. Comp. Neurol.* 187:679–702

Berman N, Blakemore C, Cynader M. 1975. Binocular interaction in the cat's superior colliculus. *J. Physiol. London* 246:595–615

Bishop PO, Henry GH, Smith CJ. 1971. Binocular interaction fields of single units in the cat striate cortex. *J. Physiol. London* 216:39–68

Blakemore C. 1970. The representation of three-dimensional visual space in the cat's striate cortex. *J. Physiol. London* 209: 155–78

Bradley DC, Andersen RA. 1998. Center-surround antagonism based on disparity in primate area MT. *J. Neurosci.* 18: 7552–65

Bradley DC, Chang GC, Andersen RA. 1998. Encoding of three-dimensional structure-from-motion by primate area MT neurons. *Nature* 392:714–17

Britten KH, Shadlen MN, Celebrini S, Newsome WT, Movshon JA. 1996. A relationship between behavioural choice and the visual responses of neurons in macaque MT. *Vis. Neurosci.* 13:87–100

Burkhalter A, Van Essen DC. 1986. Processing of color, form and disparity information in visual areas VP and V2 of ventral extrastriate cortex in the macaque monkey. *J. Neurosci.* 6:2327–51

Burkitt GR, Lee J, Ts'o DY. 1998. Functional organization of disparity in visual area V2 of the macaque monkey. *Soc. Neurosci. Abstr.* 24:1978 (Abstr.)

Carmon A, Bechtoldt HP. 1969. Dominance of the right cerebral hemisphere for stereopsis. *Neuropsychologia* 7:29–39

Celebrini S, Newsome WT. 1995. Microstimulation of extrastriate area MST influences performance on a direction discrimination task. *J. Neurophysiol.* 73:437–48

Cogan A, Kontsevich L, Lomakin A, Halpern D, Blake R. 1995. Binocular disparity processing with opposite contrast stimuli. *Perception* 24:33–47

Cogan A, Lomakin A, Rossi A. 1993. Depth in anticorrelated stereograms. *Vis. Res.* 33: 1959–75

Cowey A, Porter J. 1979. Brain damage and global stereopsis. *Proc. R. Soc. London Ser. B* 204(1157):399–407

Cowey A, Wilkinson F. 1991. The role of the corpus callosum and extrastriate visual areas in stereoacuity in macaque monkeys. *Neuropsychologia* 29(6):465–79

Cumming BG, Parker AJ. 1997. Responses of primary visual cortical neurons to binocular disparity without the perception of depth. *Nature* 389:280–83

Cumming BG, Parker AJ. 1999. Binocular neurons in V1 of awake monkeys are selective for absolute, not relative, disparity. *J. Neurosci.* 19:5602–18

Cumming BG, Parker AJ. 2000. Local disparity, not perceived depth, is signaled by binocular neurons in cortical area V1 of the macaque. *J. Neurosci.* 20:4758–67

Cumming BG, Shapiro SE, Parker AJ. 1998. Disparity detection in anticorrelated stereograms. *Perception* 27:1367–77

Danta G, Hilton RC, O'Boyle DJ. 1978. Hemisphere function and binocular depth perception. *Brain* 101:569–89

DeAngelis GK, Ohzawa I, Freeman RD. 1991. Depth is encoded in the visual cortex by a specialized receptive field structure. *Nature* 352:156–59

DeAngelis GC, Cumming BG, Newsome WT. 1998. Cortical area MT and the perception of stereoscopic depth. *Nature* 394:677–80

DeAngelis GC, Newsome WT. 1999. Organization of disparity-selective neurons in macaque area MT. *J. Neurosci.* 19:1398–415

DeAngelis GC, Ohzawa I, Freeman RD. 1993a. Spatiotemporal organization of simple-cell receptive fields in the cat's striate cortex. I. General characteristics and postnatal development. *J. Neurophysiol.* 75:1091–117

DeAngelis GC, Ohzawa I, Freeman RD. 1993b. Spatiotemporal organization of simple-cell receptive fields in the cat's striate cortex. II. Linearity of temporal and spatial summation. *J. Neurophysiol.* 75:1118–35

DeAngelis GC, Ohzawa I, Freeman RD. 1995. Neuronal mechanisms underlying stereopsis: How do simple cells in the visual cortex encode binocular disparity? *Perception* 24: 3–31

Eifuku S, Wurtz RH. 1999. Response to motion in extrastriate area MSTl: disparity sensitivity. *J. Neurophysiol.* 82:2462–75

Erkelens CJ, Collewijn H. 1985a. Eye movements and stereopsis during dichoptic viewing of moving random-dot stereograms. *Vis. Res.* 25:1689–700

Erkelens CJ, Collewijn H. 1985b. Motion perception during dichoptic viewing of moving random-dot stereograms. *Vis. Res.* 25: 583–88

Erwin E, Miller K. 1999. The subregion correspondence model of binocular simple cells. *J. Neurosci.* 19:7212–29

Felleman DJ, van Essen DC. 1987. Receptive field properties of neurons in area V3 of macaque monkey extrastriate cortex. *J. Neurophysiol.* 57:889–920

Felleman DJ, van Essen DC. 1991. Distributed hierarchical processing in the primate cerebral cortex. *Cereb Cortex* 1:1–47

Ferraina S, Pare M, Wurtz RH. 2000. Disparity sensitivity of frontal eye field neurons. *J. Neurophysiol.* 83:625–29

Ferster D. 1981. A comparison of binocular depth mechanisms in areas 17 and 18 of cat visual cortex. *J. Physiol. London* 315:623–55

Field DJ, Tolhurst DJ. 1986. The structure and symmetry of simple-cell receptive-field profiles in the cat's visual cortex. *Proc. R. Soc. London Ser. B* 228:379–400

Fleet DJ, Wagner H, Heeger DJ. 1996. Neural encoding of bincoular disparity: energy models, position shifts and phase shifts. *Vis. Res.* 36:1839–57

Freeman R, Ohzawa I. 1990. On the neurophysiological organization of binocular vision *Vis. Res.* 30:1661–76

Gnadt JW, Mays LE. 1995. Neurons in monkey parietal area LIP are tuned for eye-movement parameters in three-dimensional space. *J. Neurophysiol.* 73:280–97

Gonzalez F, Perez R. 1998a. Modulation of cell responses to horizontal disparities by ocular vergence in the visual cortex of the awake *macaca mulatta* monkey. *Neurosci. Lett.* 245:101–4

Gonzalez F, Perez R. 1998b. Neural mechanisms underlying stereoscopic vision. *Prog. Neurobiol.* 55:191–224

Grasse KL. 1994. Positional disparity sensitivity of neurons in the cat accessory optic system. *Vis. Res.* 34:1673–89

Guillemot JP, Paradis MC, Samson A, Ptito M, Richer L, Lepore F. 1993a. Binocular interaction and disparity coding in area 19 of visual cortex in normal and split-chiasm cats. *Exp. Brain Res.* 94:405–17

Guillemot JP, Richer L, Ptito M, Lepore F. 1993b. Disparity coding in the cat: a comparison between areas 17-18 and area 19. *Prog. Brain Res.* 95:179–87

Hamsher KD. 1978. Stereopsis and unilateral brain disease. *Invest. Ophthalmol. Vis. Sci.* 17:453–59

Helmholtz HL. 1909. *Treatise on Physiological Optics.* New York: Dover

Hubel DH, Wiesel TN. 1962. Receptive fields, binocular interaction and functional architecture in the cat's visual cortex. *J. Physiol. London* 160:106–54

Hubel DH, Livingstone MS. 1987. Segregation of form, color, and stereopsis in primate area 18. *J. Neurosci.* 7:3378–415

Hubel DH, Wiesel TN. 1970. Stereoscopic vision in macaque monkey cells sensitive to binocular depth in area 18 of the macaque monkey cortex. *Nature* 225:41–42

Hubel DH, Wiesel TN. 1973. A re-examination of stereoscopic mechanisms in area 17 of the cat. *J. Physiol. London* 232:29–30P (Abstr.).

Janssen P, Vogels R, Orban GA. 1999. Macaque inferior temporal neurons are selective for disparity-defined three-dimensional shapes. *Proc. Natl. Acad. Sci. USA* 96:8217–22

Jones JP, Palmer LA. 1987. The two-dimensional spatial structure of simple cell receptive fields in cat striate cortex. *J. Neurophysiol.* 58:1187–211

Julesz B. 1971. *Foundations of Cyclopean Perception.* Chicago: Univ. Chicago

Kaye M, Mitchell DE, Cynader M. 1981. Selective loss of binocular depth perception after ablation of cat visual cortex. *Nature* 293:60–62

Krug K, Cumming BG, Parker AJ. 1999. Responses of single MT neurons to anticorrelated stereograms in the awake macaque. *Soc. Neurosci. Abstr.* 25:275

Lehmann D, Wächli P. 1975. Depth perception and location of brain lesions. *J. Neurol.* 209:157–64

LeVay S, Voigt T. 1988. Ocular dominance and disparity coding in the cat visual cortex. *Vis. Neurosci.* 1:395–414

Livingstone MS, Hubel DH. 1987. Connections between layer 4b of area 17 and the thick cytochrome oxidase stripes of area 18 in the squirrel monkey. *J. Neurosci.* 7:3371–77

Livingstone MS, Tsao DY. 1999. Receptive fields of disparity-selective neurons in macaque striate cortex. *Nat. Neurosci.* 2:825–32

Marocco RT, McClurkin JW. 1979. Binocular interaction in the lateral geniculate nucleus of the monkey. *Brain Res.* 168:633–37

Marr D, Poggio T. 1979. A computational theory of human stereo vision. *Proc. R. Soc. London Ser. B* 204:301–28

Maske R, Yamane S, Bishop PO. 1984. Binocular simple cells for local stereopsis: comparison of receptive field organizations for the two eyes. *Vis. Res.* 24:1921–29

Maske R, Yamane S, Bishop PO. 1986. End-stopped cells and binocular depth discrimination in the striate cortex of cats. *Proc. R. Soc. London Ser. B* 229:257–76

Masson G, Bussettini C, Miles F. 1997. Vergence eye movements in response to binocular disparity without the perception of depth. *Nature* 389:283–86

Maunsell JHR, van Essen DC. 1983. Functional properties of neurons in middle temporal visual area of the macaque monkey. II. Binocular interactions and sensitivity to binocular disparity. *J. Neurophysiol.* 49:1148–66

Mayhew J, Longuet-Higgins HC. 1982. A computational model of binocular depth perception. *Nature* 297:376–78

McKee SP, Mitchison GJ. 1988. The role of retinal correspondence in stereoscopic matching. *Vis. Res.* 28:1001–12

Mitchison GJ. 1988. Planarity and segmentation in stereoscopic matching. *Perception* 17:753–82

Mountcastle V. 1997. The columnar organization of the neocortex. *Brain* 120:701–22

Movshon JA, Thompson ID, Tolhurst DJ. 1978a. Receptive field organization of complex cells in the cat's striate cortex. *J. Physiol. London* 283:79–99

Movshon JA, Thompson ID, Tolhurst DJ. 1978b. Spatial summation in the receptive fields of simple cells in the cat's striate cortex. *J. Physiol. London* 283:53–77

Nieder A, Wagner H. 2000. Horizontal-disparity tuning of neurons in the visual forebrain of the behaving Barn Owl. *J. Neurophysiol.* 83:2967–79

Nikara T, Bishop PO, Pettigrew JD. 1968. Analysis of retinal correspondence by studying receptive fields of binocular single units in cat striate cortex. *Exp. Brain Res.* 6:353–72

Nomura M, Matsumoto G, Fujiwara S. 1990. A binocular model for the simple cell. *Biol. Cybern.* 63:237–42

Ohzawa I. 1998. Mechanisms of stereoscopic vision: the disparity energy model. *Curr. Opin. Biol.* 8:509–15

Ohzawa I, DeAngelis GC, Freeman RD. 1990. Stereoscopic depth discrimination in the visual cortex: neurons ideally suited as disparity detectors. *Science* 249:1037–41

Ohzawa I, DeAngelis GC, Freeman RD. 1996. Encoding of binocular disparity by simple cells in the cat's visual cortex. *J. Neurophysiol.* 75:1779–805

Ohzawa I, DeAngelis GC, Freeman RD. 1997. Encoding of binocular disparity by complex cells in the cat's visual cortex. *J. Neurophysiol.* 77:2879–909

Ohzawa I, Freeman RD. 1986a. The binocular organization of complex cells in the cat's visual cortex. *J. Neurophysiol.* 56:243–59

Ohzawa I, Freeman RD. 1986b. The binocular organization of simple cells in the cat's visual cortex. *J. Neurophysiol.* 56:221–42

Parker AJ, Cumming BG, Dodd JV. 2000. Binocular neurons and the perception of depth. In *The Cognitive Neurosciences*, ed. M Gazzaniga, pp. 263–78. Cambridge, MA: MIT Press

Parker AJ, Newsome WT. 1998. Sense and the single neuron: probing the physiology of perception. *Annu. Rev. Neurosci.* 21:227–77

Peterhans E, von der Heydt R. 1993. Functional organization of area V2 in the alert macaque. *Eur. J. Neurosci.* 5:509–24

Pettigrew JD, Dreher B. 1987. Parallel parallel processing of binocular disparity in the cat's retinogeniculocortical pathways. *Proc. R. Soc. London Ser. B* 232:297–321

Pettigrew JD, Nikara T, Bishop PO. 1968. Binocular interaction on single units in cat striate cortex: simultaneous stimulation by moving slit with receptive fields in correspondence. *Exp. Brain Res.* 6:353–72

Poggio G, Gonzalez F, Krause F. 1988. Stereoscopic mechanisms in monkey visual cortex: binocular correlation and disparity selectivity. *J. Neurosci.* 8:4531–50

Poggio GF. 1995. Mechanisms of stereopsis in monkey visual cortex. *Cereb. Cortex* 3:193–204

Poggio GF, Fisher B. 1977. Binocular interactions and depth sensitivity in striate and prestriate cortex of behaving rhesus monkey. *J. Neurophysiol.* 40:1392–405

Poggio GF, Motter BC, Squatrito S, Trotter Y. 1985. Responses of neurons in visual cortex (V1 and V2) of the alert macaque to dynamic random dot stereograms. *Vis. Res.* 25:397–406

Poggio GF, Talbot WH. 1981. Mechanisms of static and dynamic stereopsis in foveal cortex of the rhesus monkey. *J. Physiol. London* 315:469–92

Prince SJD, Eagle RA. 2000. Stereo correspondence in one-dimensional Gabor stimuli. *Vis. Res.* 40:913–24

Prince SJD, Pointon AD, Cumming BG, Parker AJ. 2000. The precision of single neuron responses in cortical area V1 during stereoscopic depth judgements. *J. Neurosci.* 20:3387–400

Ptito A, Zatorre R, Larson W, Tosoni C. 1991. Stereopsis after unilateral anterior temporal lobectomy. *Brain* 114:1323–33

Ptito M, Lepore F, Guillemot JP. 1992. Loss of stereopsis following lesions of cortical areas 17-18 in the cat. *Exp. Brain Res.* 89:521–30

Qian N. 1994. Computing stereo disparity and motion with known binocular properties. *Neural Comput.* 6:390–404

Regan D, Erkelens CJ, Collewijn H. 1986. Necessary conditions for the perception of motion-in-depth. *Invest. Ophthalmol. Vis. Sci.* 27:584–97

Rodieck RW, Dreher B. 1979. Visual suppression from nondominant eye in the lateral geniculate nucleus: a comparison of cat and monkey. *Exp. Brain Res.* 35:465–77

Roe A, Ts'o D. 1995. Visual topography in primate V2: multiple representation across functional stripes. *J. Neurosci.* 15:3689–715

Ross JE. 1983. Disturbance of stereoscopic vision in patients with unilateral stroke. *Behav. Brain Res.* 7:99–112

Rothstein TB, Sacks JG. 1972. Defective stereopsis in lesions of the parietal lobe. *Am. J. Ophthalmol.* 73:281–84

Roy JP, Komatsu H, Wurtz RH. 1992. Disparity

sensitivity of neurons in monkey extrastriate area MST. *J. Neurosci.* 12:2478–92

Salzman CD, Murasugi CM, Britten KH, Newsome WT. 1992. Microstimulation in visual area MT: effects on direction discrimination performance. *J. Neurosci.* 12:2331–55

Salzman CD, Newsome WT. 1994. Neural mechanisms for forming a perceptual decision. *Science* 264:231–37

Schiller PH. 1993. The effects of V4 and middle temporal (MT) area lesions on visual performance in the rhesus monkey. *Vis. Neurosci.* 10:717–46

Shikata E, Tanaka Y, Nakamura H, Taira M, Sakata H. 1996. Selectivity of the parietal visual neurones in 3D orientation of surface of stereoscopic stimuli. *NeuroReport* 7: 2389–94

Singer W. 1970. Inhibitory binocular interaction in the lateral geniculate body of the cat. *Brain Res.* 18:165–70

Smallman HS, Macleod DIA. 1994. Size-disparity correlation in stereopsis at contrast threshold. *J. Opt. Soc. Am. A* 11:2169–83

Smith EL, Chino YM, Ni J, Ridder WH, Crawford MLJ. 1997. Binocular spatial phase tuning characteristics of neurons in the macaque striate cortex. *J. Neurophysiol.* 78:351–65

Suzuki H, Kato E. 1966. Binocular interaction at cat's lateral geniculate body. *J. Neurophysiol.* 29:909–20

Taira M, Tsutsui K, Jiang M, Yara K, Sakata H. 2000. Parietal neurons represent surface orientation.

Takemura A, Inoue Y, Kawano K, Quaia C, Miles FA. 1999. Evidence that disparity-sensitive cells in medial superior temporal area contribute to short-latency vergence eye movements. *Soc. Neurosci. Abstr.* 25: 1400

Trotter Y, Celebrini S, Stricanne B, Thorpe S, Imbert M. 1992. Modulation of neural stereoscopic processing in primate area-V1 by the viewing distance. *Science* 257(5074): 1279–81

Trotter Y, Celebrini S, Stricanne B, Thorpe S, Imbert M. 1997. Neural processing of stereopsis as a function of viewing distance in primate visual cortical area V1. *J. Neurophysiol.* 76:2872–85

Uka T, Tanaka H, Yoshiyama K, Kato M, Fujita I. 2000. Disparity selectivity of neurons in monkey inferior temporal cortex. *J. Neurophysiol.* 84:120–32

Vaina LM. 1989. Selective impairment of visual motion interpretation following lesions of the right occipito-parietal area in humans. *Biol. Cybern.* 61:347–59

Vickery RM, Morley JW. 1999. Binocular phase interactions in area 21a of the cat. *J. Physiol. London* 514:541–49

von der Heydt R, Adorjani C, Hänny P, Baumgartner G. 1978. Disparity sensitivity and receptive field incongruity of units in the cat striate cortex. *Exp. Brain Res.* 31:523–45

Wagner H, Frost B. 1993. Disparity-sensitive cells in the owl have a characteristic disparity. *Nature* 364:796–98

Wagner H, Frost B. 1994. Binocular responses of neurons in the barn owl's visual Wulst. *J. Comp. Physiol.* 174:661–70

Wallach H, O'Connell DN. 1953. The kinetic depth effect. *J. Exp. Psychol.* 45:205–17

Wang C, Dreher B. 1996. Binocular interactions and disparity coding in area 21a of cat extrastriate visual cortex. *Exp. Brain Res.* 108: 257–72

Westheimer G. 1979. Cooperative neural processes involved in stereoscopic acuity. *Exp. Brain Res.* 36:585–97

Xue JT, Ramoa AS, Carney T, Freeman RD. 1987. Binocular interaction in the dorsal lateral geniculate nucleus of the cat. *Exp. Brain Res.* 68:305–10

Zhu YD, Qian N. 1996. Binocular receptive field models, disparity tuning, and characteristic disparity. *Neural Comput.* 8: 1611–41

Annu. Rev. Neurosci. 2001. 24:239–62

PARANEOPLASTIC NEUROLOGIC DISEASE ANTIGENS: RNA-Binding Proteins and Signaling Proteins in Neuronal Degeneration

Kiran Musunuru and Robert B. Darnell

Laboratory of Molecular Neuro-Oncology, The Rockefeller University, New York, New York 10021; e-mail: darnelr@rockvax.rockefeller.edu, musunuk@rockvax.rockefeller.edu

Key Words onconeural antigens, neuronal survival, neuron-specific proteins, autoimmune disease, tumor immunity

■ **Abstract** Studies of the disorders known as paraneoplastic neurologic degenerations exemplify the successful application of modern molecular biological techniques to diseases, yielding, even for these extremely rare disorders, wide-ranging insight into basic neurobiology, tumor immunity, and autoimmune neurologic disease. Immune responses to paraneoplastic neurologic degeneration antigens, also called onconeural antigens, have been exploited to clone and characterize a number of neuron-specific proteins, including several RNA-binding proteins and new kinds of signaling molecules. The biology and functions of these proteins are reviewed, and a model in which their functions are related to the pathogenesis of autoimmune neurologic disease is discussed.

INTRODUCTION

Neurologists have recognized a relationship between specific degenerations of the nervous system and the presence of cancer in the body for over a century and have termed such disorders paraneoplastic neurologic degenerations (PNDs) (see Posner & Furneaux 1990 for a detailed historical review). In the 1980s, Posner and colleagues established an immunologic link between the neuronal degeneration and cancer seen in PND patients by recognizing that these patients harbor high-titer antibodies circulating in their sera and cerebrospinal fluid (CSF). These antibodies recognize the same proteins that are expressed in the patients' tumor cells and in neurons undergoing degeneration, and we have termed them onconeural antigens (Darnell et al 1991). In this review we focus on the characterization of onconeural antigens whose functions have been well studied and appear to relate to their roles in disease pathogenesis.

0147-006X/01/0621-0239$14.00   **239**

Our emerging model for the pathogenesis of the PNDs has three essential features. First, the onconeural antigens are ordinarily expressed solely in immune-privileged sites, such as neurons, and for this reason are recognized by the immune system as foreign proteins when they are expressed in systemic tumors. Second, this immune response against onconeural antigens ectopically expressed in cancer cells provides effective tumor immunity. Third, a subset of patients with onconeural-antigen-directed tumor immunity exhibits a breakdown in their immune tolerance to neurons, caused by an unknown mechanism, and thereby develop autoimmune neurologic disease.

In this review we explore a fourth feature of a model for PNDs that relates the biology of the target antigens to their roles in disease. This aspect of our model derives from the understanding of the neuronal functions of these proteins that has emerged over the past 5 years with the use of onconeural antibodies to clone complementary DNAs (cDNAs) encoding PND target antigens. These studies have demonstrated that a number of different onconeural antigens play critical roles in neuron survival. After reviewing these functional studies, we discuss how converging observations of the functions of these proteins in neurons suggest ways in which they may participate in neuronal degeneration.

FROM BEDSIDE TO BENCH: Identification
of the Onconeural Antigens

Historically, the diagnosis of PNDs has been a clinical challenge, complicated by the fact that neuronal degeneration typically emerges before an underlying malignancy is discovered. Posner and colleagues have suggested several criteria to assist in identifying a specific PND (Anderson et al 1987). The neurologic signs should bear similarity to a syndrome that has been observed to be frequently associated with cancer. PNDs generally have a subacute evolution over a period of weeks to months. PNDs of the central nervous system (CNS) are associated with perturbations in the CSF, including an early mild pleocytosis and a prolonged elevation of protein and immunoglobulin G levels.

Perhaps most helpful in clarifying the diagnosis of the PNDs has been the identification in patient serum and CSF of high-titer antibodies that recognize onconeural antigens. The presence of such antibodies correlates with effective tumor immunity, is considered pathognomonic for a specific PND, and predicts the presence of a particular type of underlying cancer. In the past decade, high-titer antibodies from PND patients have been used to clone and characterize a host of onconeural antigens, which has established the feasibility of using antisera from patients with such disorders to screen expression vector cDNA libraries to identify autoimmunity-related target antigens.

PND antibodies provide important biological reagents. They have been used to identify a set of proteins whose normal biology is of special interest to neuroscientists because they are normally expressed only in neurons. A significant body of data reviewed here suggests that many of these proteins have critical physiologic functions whose disruption by antibodies may trigger neuronal death and, thereby, neurologic disease. A relatively unexplored question is the significance of the dysregulation of gene expression and cellular function that the ectopic expression of onconeural antigens confers on tumor cells. The proteins are consistently and selectively expressed in certain tumor types, underscoring their importance in tumor biology. Lastly, the proteins are intriguing from the point of view of immunology. Determining how the antigens elicit successful antitumor responses from the immune system may shed light on the poorly understood phenomenon of tumor immunity and lead the way to novel therapeutic approaches for combating cancer.

CLASSIFICATION OF THE ONCONEURAL ANTIGENS

The onconeural antigens have traditionally been grouped according to the clinical syndromes in which they are implicated, proceeding naturally from their original characterization as immune system targets. For example, at least six different antigens [Ma1 (Dalmau et al 1999), Tr (Graus et al 1997b), Nb (Darnell et al 1991), Yo/cdr2 (Peterson et al 1992), Nova (Buckanovich et al 1993, Luque et al 1991), and Hu (Dalmau et al 1991)] have been implicated in the pathogenesis of paraneoplastic cerebellar dysfunction and as such have been classified clinically as a distinct family of antigens. Although this scheme has been useful from a medical standpoint, it is less adequate in capturing the biological parameters of each of the antigens. The six antigens mentioned are expressed in different subsets of neurons, are ectopically expressed in different tumor types, and have widely divergent cellular functions, being involved in processes ranging from RNA metabolism to neuronal vesicle formation.

Because our foremost concern here is reviewing the biology of the onconeural antigens, we use an alternative classification scheme in which the proteins are grouped by their biological characteristics into four categories (Darnell 1996): neuromuscular junction proteins, vesicle-associated nerve-terminal proteins, neuron-specific RNA-binding proteins (RBPs), and neuronal signaling proteins. This classification is particularly useful in elucidation of common features of the onconeural antigens that lead to their specific expression in tumor cells and susceptibility to neuronal immunity. Such features may include protein function and interaction, cellular location, antigenicity, and vulnerability to functional disruption by antibody.

Rather than surveying the dozen or so onconeural antigens characterized to date and giving a brief treatment for each, we discuss antigens from two of the four categories—neuron-specific RBPs and neuronal signaling proteins— whose functions appear to be critical for neuronal survival and therefore are proteins that may have a common pathophysiology. Several recent reviews discuss PND-associated nerve-terminal vesicle-associated proteins (Darnell 1996,

Floyd & De Camilli 1998), ion channels present at the neuromuscular junction (Whitney & McNamara 1999), and general aspects of the PNDs (Dalmau & Posner 1999).

NEURON-SPECIFIC RNA-BINDING PROTEINS:
The Nova and Hu Antigens

Two groups of onconeural antigens, Nova and Hu, are RBPs of particular interest to neurobiologists because they are uniquely expressed in neurons. RBPs are critical regulators of gene expression at every level, including transcription, processing, transport, localization, stability, and translation. These functions are likely to be particularly important in neurons, which constitute a system of extraordinary complexity. Regulation of neuronal differentiation and function is likely to depend on a large number of neuron-specific genes; it has been estimated that 30%–50% of all genes are specifically transcribed within the brain (He & Rosenfeld 1991). Expression of these genes is almost certainly regulated, at least in part, by posttranscriptional controls. Many brain pre-mRNAs undergo alternative splicing, allowing one primary transcript to encode a diverse set of proteins; in some cases, brain-specific alternative splicing is restricted to neurons [e.g. the splicing of calcitonin/CGRP (calcitonin gene-related peptide) (Amara et al 1982, Rosenfeld et al 1983), n-src (Martinez et al 1987), and the rat clathrin light chain B (Stamm et al 1992)]. The 5' and 3' untranslated regions of mRNAs are involved in translational efficiency (Stripecke et al 1994) and mRNA stability (Sachs 1993), thereby controlling protein levels. Several neuronal RNAs are thought to be transported to postsynaptic densities within neurons, thereby permitting spatial regulation of mRNA translation (Steward & Banker 1992). A functionally critical position of the glutamate receptor is controlled by RNA editing that occurs in neuronal nuclei (Higuchi et al 1993, Simpson & Emeson 1996). These and other posttranscriptional controls occurring in neurons are likely to be mediated by neuron-specific RBPs such as the Nova and Hu onconeural antigens.

Nova: The Paraneoplastic Opsoclonus-Myoclonus Ataxia Antigen

Paraneoplastic opsoclonus-myoclonus ataxia (POMA), as the name suggests, is a syndrome marked by a loss of inhibitory motor control in the eyes, limbs, and trunk (Anderson et al 1988b, Budde-Steffen et al 1988, Digre 1986). A subset of POMA patients further develop encephalopathy and cortical deficits, with features ranging from mild emotional lability to dementia, coma, and death (Pranzatelli 1992). POMA arises primarily in patients with gynecological or small-cell lung cancer who harbor a high-titer antibody termed Ri (Luque et al 1991). Ri antisera cross-react with tumor tissue obtained from POMA patients (Luque et al 1991) and with subsets of neurons found in the CNS (Buckanovich et al 1993, Buckanovich et al 1996, Graus et al 1993).

Ri antisera recognize antigens of 50–55 kDa and 70–80 kDa (Buckanovich et al 1993, Luque et al 1991) and have been used to identify two distinct POMA antigens, designated Nova-1 and Nova-2 (Buckanovich et al 1993, Yang et al 1998), by expression vector cloning. Immunohistochemical and in situ hybridization studies revealed that Nova-1 and Nova-2 are strictly neuron specific throughout development and have distinct regional patterns within the CNS. Nova-1 expression is restricted to the hindbrain and spinal cord, whereas Nova-2 is preferentially expressed in areas where Nova-1 is not, such as the forebrain and thalamus (Buckanovich et al 1996, Yang et al 1998). Nova proteins are most abundantly expressed in the neuronal nucleus but are also seen in the somatodendritic compartment. Sequencing of cDNAs encoding the two Nova proteins placed them in an expanding family of RBPs, implying that they may function to regulate RNA metabolism in neurons.

Structure of the Nova Antigens Several conserved motifs with divergent functions have been identified in RBPs (Burd & Dreyfuss 1994). The two most common motifs are the K homology (KH) motif, present in the Nova antigens, and the RNA recognition motif (RRM), present in the Hu antigens (see below). The KH motif was originally described as a conserved element repeated three times in the heterogeneous nuclear ribonucleoprotein K protein. The KH motif has since been identified in more than 50 proteins from a wide variety of organisms, mediating functions that range from regulation of alternative splicing of RNA transcripts to messenger RNA (mRNA) localization. The canonical KH motif spans about 70 residues, with a characteristic α/β fold consisting of a three-stranded β-sheet backed by three α-helices, as well an invariant Gly-X-X-Gly loop and a second loop of variable length (Lewis et al 1999, Musco et al 1996).

KH motifs have been discerned in eubacterial and archaeal genomes, as well as those of various eukaryotes. Some prokaryotic examples are the ribonuclease plynucleotide phosphorylase (PNP), the transcription elongation factor NusA, and the ribosomal protein S3; eukaryotic examples include the yeast Mer1, *Drosophila* P-element somatic inhibitor protein (PSI), and mammalian SF1 and KH-type splicing regulatory protein (KSRP) splicing factors, the mammalian α-globin messenger ribonucleoprotein (mRNP) complex proteins heterogeneous nuclear ribonucleoprotein E1 and E2, and the fragile X mental retardation syndrome protein (FMRP), in addition to Nova (see Lewis et al 1999). The high degree of conservation of the KH motif implicates it as an ancient and important structural motif.

The single common feature of the numerous KH-containing proteins is that each has been shown to be associated with RNA in some fashion, and it is generally accepted that the KH motif can function in vivo as an RNA-interacting module. Each of the Nova antigens harbors three repeats of the KH motif. The KH domains in the mouse Nova-1 and -2 isoforms and in the single Nova-related protein identified in the *Drosophila* genome are all highly homologous. The Nova KH domains may recognize similar RNA sequences across species as a component of performing highly conserved functions in neurons.

Mechanism of Nova Sequence-Specific RNA Binding   In vitro RNA selection experiments demonstrated that the Nova-1 and Nova-2 proteins bind with nanomolar affinity to RNAs in which one or more repeats of the tetranucleotide RNA sequence UCAY [where Y is a pyrimidine (either uracil or cytosine)] are present in the context of a stem-loop (Buckanovich & Darnell 1997, Yang et al 1998). Mutational analysis revealed that this element is preferentially recognized by the Nova-1 KH3 domain (Buckanovich & Darnell 1997), whereas KH1 and KH2 appear to prefer distinct tetranucleotide sequences (RB Darnell, unpublished observations). The specificity of Nova KH3 domains was confirmed by additional RNA selection experiments using isolated Nova KH3 domains (Jensen et al 2000b). In these experiments, the single KH3 domain selected stem-loop RNAs harboring a single UCAY loop sequence. These observations suggest that the KH3 domain plays a major role in determining the sequence specificity of the full-length Nova protein.

Structural studies in which the Nova KH3 domain was cocrystallized (Lewis et al 2000) with a stem-loop RNA optimized for binding to this isolated domain were performed (Jensen et al 2000b). The X-ray structure of the complex, determined at 2.4-Å resolution, revealed the mechanism by which Nova KH3 recognizes the UCAY tetranucleotide. Contrary to expectations, this cocrystal structure shows that the RNA is supported on neither the β-sheet nor the α-helical face. Rather, the RNA is gripped in a "molecular vise" whose jaws are formed by the two conserved loops emerging from the side of the sandwich, the invariant Gly-X-X-Gly loop and the loop of variable length. The RNA bases then stack atop a hydrophobic platform formed by aliphatic residues contributed by α-helices and β-strands on either side. Polar and charged residues surrounding the hydrophobic platform form Watson-Crick-like hydrogen bonds with the RNA bases, thereby conferring specificity for the RNA ligand. For example, a glutamate and an arginine residue coordinate to form a molecular mimic of a guanine base, which base pairs with the cytosine in the second position of the UCAY tetranucleotide. No base besides cytosine can be readily accommodated in this position.

How does the full-length Nova-1 protein bind and recognize UCAY repeat elements? The cocrystal structure suggests that of the Nova KH domains, only KH3 recognizes the UCAY tetranucleotide sequence. The particular residues of KH3 involved in UCAY RNA recognition differ in Nova KH1 and KH2. Moreover, the cocrystal structure indicates that the individual Nova KH domains are incapable of recognizing anything more than four or five consecutive nucleotides in length. However, several of the in vitro-selected Nova RNA targets, as well as confirmed in vivo targets (see below), contain repeats of the UCAY tetranucleotide, more than one of which is required for optimal Nova binding (Buckanovich & Darnell 1997). The simplest model consistent with all available data is that a physiologic dimer of the Nova-1 molecule simultaneously binds to distinct UCAY repeats

(Lewis et al 1999, 2000). This dimer model for Nova binding is reminiscent of some dimeric transcription factors, including the E2 and the steroid/nuclear receptors (reviewed in Patikoglou & Burley 1997), and would allow for increased sequence complexity in the RNA targets that Nova proteins recognize.

The significance of Nova function in sequence-specific RNA binding was underscored by the discovery that in all studied cases of POMA, the paraneoplastic antibodies specifically recognize an epitope in the Nova KH3 domain (Buckanovich et al 1996). We are thus presented with the intriguing possibility that the autoimmune antibodies in POMA contribute to neurologic disease by interfering with the ability of Nova to bind specific RNA targets. This premise has experimental support from in vitro RNA-binding studies in which affinity-purified POMA antisera abrogated RNA binding by recombinant Nova protein (Buckanovich & Darnell 1997, Buckanovich et al 1996). Because the Nova antigens are exclusively intracellular proteins, these findings suggest a model for the pathogenesis of POMA in which antibodies are taken up by neurons and, thereafter, bind to Nova and disrupt its activity (see below).

Nova Functions as a Splicing Regulator and Is Essential for Neuronal Survival   Knowledge of the UCAY repeat consensus sequence led to the detection of one candidate Nova-1 RNA target (Buckanovich & Darnell 1997) present in a sequence library of ~350 neuronal pre-mRNAs known to undergo alternative splicing (Stamm et al 1994). The UCAY repeat motif in the inhibitory glycine receptor α2 (GlyRα2) pre-mRNA is located in an intron 85 nucleotides upstream of exon 3A, which is alternatively spliced in a mutually exclusive fashion with the downstream exon 3B. A similar UCAY repeat motif within the pre-mRNA of Nova-1 itself is located in an intron just downstream of the alternatively spliced exon H. Coimmunoprecipitation and cross-linking experiments have demonstrated that Nova-1 specifically binds these two pre-mRNAs in the mouse brain (Buckanovich & Darnell 1997, Jensen et al 2000a).

To establish a system to assess the function of Nova-1 in vivo, particularly its putative role in regulation of GlyRα2 splicing, Nova-1-null mice were generated by homologous recombination (Jensen et al 2000a). Nova-1-null mice are phenotypically indistinguishable from their littermates at birth, yet by the second or third postnatal day they are noticeably smaller than their wild-type littermates. Thereafter, they demonstrate progressive motor dysfunction, including tremulousness and overt weakness, and die 7–10 days after birth. These motor defects in Nova-1-null mice phenocopy many of the neurologic features of POMA, although they are more severe than those seen in the human disease.

A pathologic examination of Nova-1-null mice revealed normal gross structure, neuronal development, and morphology. However, histologic analysis revealed a marked increase in pyknotic cells (relative to wild-type mice) in the brain stem and ventral, but not dorsal, spinal cord, precisely those regions that express Nova-1 (Jensen et al 2000a). TUNEL (TdT-dUTP terminal nick-end labeling) staining revealed that these cells underwent an apoptotic death. Thus, the hallmark change in postnatal Nova-1-null mice is an increase in apoptosis in hindbrain and ventral spinal cord neurons in the postnatal mouse.

Nova-1-null mice were used to evaluate a role for Nova protein in neuronal premRNA splicing. GlyRα2 splicing was assessed in Nova-1-null mice by several complementary methods (Jensen et al 2000a), in all cases revealing decreased utilization of exon 3A relative to exon 3B in the null mice. This result was consistent with cell transfection data in which excess Nova-1 enhanced the utilization of GlyRα2 exon 3A, assayed from a cotransfected GlyRα2 minigene. These effects could be abrogated by altering the intronic UCAU repeat element within the GlyRα2 minigene to a nonbinding UAAU element, which demonstrates that a direct action of Nova on the UCAU repeat element mediates alternative splice site selection.

In a survey of an additional six alternatively spliced neuronal transcripts, only one, that encoding the GABA$_A$ receptor γ2 subunit, was also altered in Nova-1-null mice (Jensen et al 2000a). It is striking that the altered splicing patterns for GlyRα2 and GABA$_A$ γ2 occur exclusively in the portions of the CNS that normally express Nova-1; no aberrations are seen in the forebrain tissues of Nova-1-null mice. Thus, Nova-1 regulates the alternative splicing of GlyRα2 and GABA$_A$ γ2 receptor pre-mRNAs in neurons in a cell-autonomous manner and is the tissue-specific splicing regulator identified in mammals (Grabowski 2000).

Mechanism and Significance of Splicing Regulation by Nova   Yeast two-hybrid screens using Nova protein were undertaken to identify protein interactors that might modulate Nova function. These studies uncovered a novel isoform of the polypyrimidine tract-binding protein (PTB) that is specifically enriched in brain, termed brPTB (Polydorides et al 2000). A number of studies implicate PTB in the negative regulation of alternative splicing (i.e. the suppression of exon inclusion) in neurons and other tissues (Chan & Black 1997, Grabowski 1998, Lin & Patton 1995, Lou et al 1996, Singh et al 1995, Valcarcel & Gebauer 1997). Consistent with this premise, brPTB was found to bind Nova-1 in vivo and to antagonize the Nova-dependent increase of GlyRα2 exon 3A utilization in the minigene cotransfection assay described above (Polydorides et al 2000). Furthermore, a putative pyrimidine-rich brPTB binding site was identified on the GlyRα2 pre-mRNA just upstream of the UCAY repeat Nova binding site. Mutagenesis of this pyrimidine-rich sequence abolished the ability of brPTB to antagonize Nova's regulatory activity in the cotransfection splicing assay. Together, these observations suggest a model in which brPTB binds the GlyRα2 pre-mRNA at a site adjacent to Nova-1 and mediates an inhibitory effect on Nova-dependent exon inclusion, presumably by preventing assembly of a multiprotein complex necessary to activate splicing.

Hu: The Paraneoplastic Encephalomyelitis and Sensory Neuropathy Antigens

The members of the Hu family of onconeural antigens are targeted in a diverse set of neurodegenerative disorders, collectively termed paraneoplastic encephalomyelitis and sensory neuropathy (PEM/SN), most commonly affecting the dorsal root ganglia and variably involving the cerebellum, brain stem, limbic system, motor neurons, or autonomic nervous system (Dalmau et al 1991). PEM/SN is most commonly associated with small-cell lung cancer (SCLC), in addition to most neuroblastomas and some types of sarcoma and prostate cancer (Ball & King 1997, Dalmau et al 1992, Manley et al 1995). All SCLCs appear to express Hu antigens, but the syndrome is rare, probably complicating considerably less than 1/1000 of SCLC cases (Anderson et al 1987).

Patients with the Hu syndrome harbor high titers of the Hu antibody in their sera and CSF, and this antibody was used to clone cDNAs encoding the HuC and HuD target antigens (Sakai et al 1993a, Szabo et al 1991), which are expressed in neurons and small-cell lung tumors (Manley et al 1995, Szabo et al 1991). In addition, a third neuronal homologue, termed Hel-N1 or HuB, and a nonneuronal homologue, termed HuR or HuA, have been cloned from human and mouse cDNA libraries (Levine et al 1993, Ma et al 1996, Okano & Darnell 1997).

It is interesting that a significant subset of SCLC patients without neurologic disease (∼17%) nonetheless harbor low but detectable titers of the Hu antibody, and, like patients with PEM/SN, these patients have limited-stage disease and an improved clinical prognosis (Dalmau et al 1990, Graus et al 1997a). This suggests that effective tumor immune responses to the Hu antigens may be dissociated from autoimmune disease, underscoring the clinical significance of the study of tumor immunity in PNDs (Darnell 1999).

Function of the Hu Family of Onconeuronal Antigens   Hu proteins bear many similarities to Nova proteins. Each of the four Hu family members has the same overall architecture (although no sequence homology) as the Nova proteins, harboring three RNA-binding (RRM) motifs with a spacer between the second and third RRMs. Hu proteins are highly homologous within their RRM domains to the *Drosophila* ELAV and sex lethal (sxl) proteins; like the latter in fruit flies, the Hu antigens are believed to be critical for the early development of vertebrate neurons and for neuronal survival in the adult. Both the Hu and Nova proteins are largely localized in the neuronal nucleus but can also be detected in the soma/dendrite. Hu proteins shuttle between the nucleus and cytoplasm (Keene 1999), using a conserved element termed the Hu nucleocytoplasmic shuttling (HNS) motif (Fan & Steitz 1998a), and Nova proteins harbor a distinct nuclear export signal (NES)-like element that may be involved in shuttling (RB Darnell, unpublished observations). The *Drosophila* ELAV protein has been implicated in the regulation of alternative splicing of the neuroglian and erect wing (EWG) transcripts (Koushika et al 1996, 2000), although to date there are no data indicating a direct role for sequence-specific RNA binding for Hu proteins in the regulation of splicing, as demonstrated for Nova. Within the soma-dendritic compartment, Hu proteins are present in RNA granules in association with microtubules, where they are presumed to be involved in RNA localization in neurons (Antic & Keene 1998), and Nova protein colocalizes with GlyRα2 mRNA in the dendrites of spinal motor neurons (A Triller & RB Darnell, unpublished observations).

Perhaps the best-characterized biochemical function for the Hu proteins relates to their ability to bind to 3′ untranslated-region AU-rich elements (AREs) (Keene 1999) that are believed to be involved in mRNA stability. This activity was initially revealed by the use of in vitro RNA selection analysis to determine the sequence-specific RNA-binding preferences of the HuB protein (Gao et al 1994; reviewed in Antic & Keene 1997). The RNA selection studies using Hu (Gao et al 1994) and Nova (Buckanovich & Darnell 1997) proteins are the only instances in which this technique has been used to elucidate previously unknown cellular RNA targets and functions of RBPs.

In nonneuronal cells, Hu proteins are able to bind AREs in a manner that correlates with the ability of these elements to stabilize ARE-containing mRNAs that are overexpressed in transiently transfected cells (Fan & Steitz 1998b, Myer et al 1997, Peng et al 1998). Hu proteins can also be cross-linked to poly(A)$^+$ mRNA in tissue culture (Gallouzi et al 2000) and can be coimmunoprecipitated with mRNAs (neurofilament-medium–NF-M) harboring AREs (Antic et al 1999). The HuA protein accumulates in the cytoplasm after cytokine stimulation of T cells (Atasoy et al 1998) or heat shock of HeLa cells (Gallouzi et al 2000), treatments associated with stabilization of ARE-containing mRNAs. In vitro Hu binding to AREs is able to stabilize deadenylated mRNAs (Ford et al 1999), suggesting one mechanism by which the protein may mediate mRNA stability.

mRNA translation and ARE-mediated stability may be linked (Jacobson & Peltz 1996). Within the cytoplasm, Hu proteins are associated with polysomes (Gallouzi et al 2000, Jain et al 1997) and may participate in translational control via polysome recruitment and translation initiation (Antic et al 1999, Jain et al 1997).

Hu binding to AREs may also be related to its localization in the nucleus. After heat shock, the majority of Hu protein cross-linked to poly(A)$^+$ mRNA is not associated with cytoplasmic mRNA, as expected, but rather is present in the nucleus (Gallouzi et al 2000). It has been suggested that heat shock treatment may dissociate Hu binding of nuclear mRNA from the former's HNS-dependent export. Additional factors may be necessary for export of Hu-containing RNA-protein complexes (Gallouzi et al 2000), suggesting a role for AREs in the assembly of specific mRNAs for nuclear export (see also Keene 1999).

How these activities relate to the role of Hu proteins in neurons remains to be determined. Hu immunoreactivity has been detected in the cell bodies and dendrites of cortical neurons (Gao & Keene 1996). This has led to the suggestion that the binding of Hu proteins to 3′ untranslated-region elements might mediate mRNA stability, translation, and localization in neurons (Antic & Keene 1997).

Hu Proteins in Neuronal Development and Survival   A number of lines of evidence suggest that Hu proteins function in neuronal development and survival. Hu proteins are expressed in early postmitotic neurons in a cascading fashion as these cells leave the periventricular zone and migrate to the cortical plate in the cerebral cortex or granule layer of the cerebellum (Okano & Darnell 1997, Wakamatsu & Weston 1997). In vitro, the upregulation of Hu proteins correlates with neuronal differentiation in embryonic P19 cells (Gao & Keene 1996). Hu proteins are able to induce a neuronal phenotype when overexpressed in transfected tissue culture cells (Akamatsu et al 1999, Antic et al 1999) and, ectopically, in transfected E9.5 mouse embryos (Akamatsu et al 1999).

As in the case of Nova, disruption of RNA binding by the Hu proteins may contribute to disease. Hu antibodies are directed against an epitope within the first two RRM domains (Manley et al 1995), and these domains are essential for RNA binding of the HuD antigen. Antibody recognition of a single Hu antigen, with subsequent targeting of the other Hu proteins, may account for the complex evolution of and heterogeneity in the PEM/SN syndrome. The observation that Hu antibodies target a domain essential for RNA binding and may thereby abrogate function has consequences for the model of pathogenesis of PEM/SN and suggests that, as in the case of Nova, identification of the in vivo RNA targets of Hu may be of particular neurobiological interest.

NEURONAL SIGNALING PROTEINS

cdr2: The Paraneoplastic Cerebellar Degeneration Antigen

Paraneoplastic cerebellar degeneration (PCD) is associated with a high-titer antibody termed Yo (Anderson et al 1988a) and with breast or ovarian cancer (Peterson et al 1992). PCD antisera recognize target antigens expressed in discrete populations of neurons, particularly the Purkinje cells of the cerebellum (Cunningham et al 1986, Peterson et al 1992), and those present in PCD tumor extracts (Corradi et al 1997, Furneaux et al 1990). Expression cloning studies with PCD sera led to the identification of three cerebellar-degeneration–related (cdr) antigens (Dropcho et al 1987, Fathallah-Shaykh et al 1991, Sakai et al 1990). RNA expression analysis revealed that only one of the three *cdr* genes, *cdr2*, was transcribed in human PCD tumors (Corradi et al 1997).

The cdr2 protein has a predicted size of 52 kDa and contains a helix-leucine zipper (HLZ) motif at its N terminus. Although identification of the HLZ motif initially suggested that cdr2 might act as a nuclear transcriptional regulator (Fathallah-Shaykh et al 1991), the protein resides in the cytoplasm (Cunningham et al 1986, Greenlee & Brashear 1983, Hida et al 1994, Jaeckle et al 1985) and has no basic DNA-binding motif.

cdr2 is unique among the onconeural antigens in that its tissue-specific expression is restricted at a post-transcriptional level (Corradi et al 1997). Western blot and immunohistochemistry analyses using PCD antisera demonstrate that antigen expression is tightly restricted to the brain and testis, and this pattern is consistent with the paradigm proposed for PND pathogenesis. Unexpectedly, however, the transcript encoding cdr2 is detected in all tissues, with cellular specificity evident within some organs (e.g. the spleen), as revealed by in situ hybridization

(Corradi et al 1997). The mechanism by which this strict control is maintained is uncertain, although at least one other protein, the transcription factor BTEB (basic transcription element binding protein), with apparently similar posttranscriptional restriction of expression to brain and testis has been described (Imataka et al 1994).

Function of the cdr2 Antigen   Because HLZ domains are often found to facilitate protein dimerization, a yeast two-hybrid screen using the cdr2 HLZ domain as bait was performed (Okano et al 1999). A specifically interacting clone containing the HLZ domain of the c-myc transcription factor was identified. In vitro and immunoprecipitation experiments confirmed that cdr2 and c-myc are bona fide protein interactors; moreover, they colocalize in the cytoplasm of Purkinje neurons. Functionally, cdr2 acts to inhibit c-myc transcriptional activity. Overexpression of cdr2 in cell culture lines altered c-myc distribution from a predominantly nuclear to a cytoplasmic profile, and in separate experiments cdr2 reversed the ability of c-myc to induce transcription from a heterologous reporter (Okano et al 1999). Taken together, these experiments indicate that cdr2 functions to inhibit c-myc activity, most likely by sequestering the protein in the cytoplasm.

The observation that cdr2 interacts with an important regulatory protein through the HLZ domain, together with the finding that the cdr2 target epitope recognized by PCD sera is the leucine zipper domain (Sakai et al 1993b; RB Darnell, unpublished observations), suggests that the antisera obtained from patients with this disease might have important biological activity. Confirming this idea and the importance of the HLZ domains in the cdr2/c-myc interaction, six of six PCD patients' sera specifically interrupted the protein-protein complex (Okano et al 1999).

These findings lead to a model for cdr2 function that bears on its role as a disease-related antigen (Okano et al 1999). In several of the PNDs, it has been suggested that neuronal antibodies may be taken up by neurons (see below). In PCD, entry of Yo antibodies in Purkinje neurons is predicted to disrupt cdr2/c-myc binding, leading to excess c-myc activity. c-myc activity can stimulate cell cycle pathways or induce apoptosis in dividing cells (Evan & Littlewood 1998). In neurons, excess c-myc activity is proposed to induce apoptosis (Okano et al 1999). This suggestion is based in part on an analogous example of apoptotic death in Purkinje neurons which express the simian virus 40 T antigen (Feddersen et al 1992). A consequence of disruption of cdr2 protein activity by PCD antisera may be neuronal apoptosis.

Recoverin: The Cancer-Associated Retinopathy Antigen

Cancer-associated retinopathy (CAR) was originally defined as a paraneoplastic disease of the retina that causes blindness in association with SCLC, owing to the specific destruction of photoreceptor cells (Buchanan et al 1984, Sawyer et al 1976). An autoimmune etiology was implicated in the retinal degeneration after the detection of high-titer autoantibodies in patient sera (Keltner et al 1983; Thirkill et al 1987, 1989). In many cases the antibodies were found to be targeted against a 23-kDa photoreceptor antigen. Two independent techniques, purification of the antigen by chromatography and subsequent sequencing (Polans et al 1991) and expression vector cloning (Thirkill et al 1992) identified the protein as a calcium-binding protein harboring EF-hand motifs, termed recoverin (Dizhoor et al 1991). Consistent with the general model of PND pathogenesis, aberrant expression of recoverin is detectable in the tumors of CAR patients (Matsubara et al 1996, Polans et al 1995).

Recoverin plays a critical role in the modulation of the retinal phototransduction cascade by which the eye adapts to either darkness or light (Molday 1998). Light-activated rhodopsin triggers a signaling pathway that culminates in the cleavage of cyclic GMP by phosphodiesterase, the closure of cyclic GMP-gated membrane ion channels, and a net efflux of Ca^{2+}. In adaptation to light, low intracellular Ca^{2+} levels permit rhodopsin kinase (RK) to be activated to rapidly phosphorylate rhodopsin, thereby desensitizing the photoreceptor and shutting off the signal cascade. In adaptation to darkness, high intracellular Ca^{2+} levels inhibit RK activity, allowing the photosignal to persist and thereby rendering the photoreceptor more sensitive to low light levels. Recoverin acts as a molecular Ca^{2+} sensor to relay information to RK: binding of Ca^{2+} to the second and third EF hands of recoverin induces a conformational change in the protein that enables it to bind and inhibit RK (Dizhoor et al 1993, Flaherty et al 1993, Gorodovikova et al 1994, Kawamura 1993, Matsuda et al 1998, Zozulya & Stryer 1992). Epitope mapping of recoverin antibodies from several CAR patients showed a preference for an alpha-helical portion of the second EF hand (Adamus et al 1993, Polans et al 1995). An attractive idea with some supporting biochemical evidence is that CAR antibodies can interfere in vivo with Ca^{2+}-stimulated binding of RK by recoverin (Gorodovikova et al 1994). One consequence of autoimmunity might be the dysregulation of cyclic GMP and Ca^{2+} in the photoreceptor, followed by cell death.

Three lines of evidence establish that antibody-mediated disruption of recoverin activity is a reasonable mechanism for neuronal degeneration and, conversely, that recoverin is important in neuronal survival. First, perturbation of various components of the phototransduction signaling machinery with which recoverin interacts led to photoreceptor destruction (Lem et al 1992, Olsson et al 1992, Smith et al 1991). Second, immunization of Lewis rats with either the full-length recoverin protein or the peptide epitope recognized by CAR antisera elicited high-titer recoverin antibodies and retinal degeneration (Adamus et al 1994, Gery et al 1994, Polans et al 1995). Third, injection of recoverin antibodies from CAR patients (in contrast to control antibodies) directly into the vitreous humor of the eyes of Lewis rats resulted in significant apoptotic death in the retinal layers with attendant retinal dysfunction (Adamus et al 1998, Ohguro et al 1999).

A GENERAL MODEL OF PARANEOPLASTIC NEUROLOGIC DEGENERATION PATHOGENESIS

Onconeural Antigens and Neuronal Survival

The functions of each of the four onconeural antigens discussed here suggest that each protein has a specific role in neuronal survival. For example, the loss of Nova-1 function in mice leads to neuronal apoptosis; the Hu homologue ELAV is essential for neuronal survival in flies; cdr2 inhibits the function of c-myc, whose inappropriate activity may in turn lead to Purkinje neuronal apoptosis; and inhibition of recoverin activity is linked to apoptotic death in the retina. How might these observations relate to PND disease pathogenesis?

In a review of autoantibodies and autoimmunity, Tan (1991) noted that the target epitopes of a number of autoimmune antigens were critical functional domains of the proteins. For the onconeural antigens discussed here, in cases in which epitopes have been mapped, the immunogenic protein domains can be tied directly to neuronal survival pathways. POMA antisera all target Nova-1 KH3, and these antibodies disrupt protein interactions with specific target RNAs (Buckanovich et al 1996); in addition, the loss of Nova-1 function in mice leads to neuronal apoptosis. Hu antisera target RRM-1 and -2, although it has not yet been determined if the antisera disrupt protein interactions with RNA targets. All PCD antisera bind the cdr2 leucine zipper domain and abrogate the ability of cdr2 to interact with c-myc (Okano et al 1999). Disruption of this interaction may trigger neuronal apoptosis (Okano et al 1999). Finally, CAR antisera recognize the Ca^{2+}-binding sensor of the recoverin antigen, and disruption of this activity can be plausibly tied to neuronal death.

These observations suggest a model in which the entry of PND antibodies into neurons would disrupt critical functional epitopes of critical neuronal proteins, leading to apoptotic neuronal death. As discussed below, such PND antibody-mediated neuronal apoptosis may be necessary but is not sufficient to trigger the neurologic degeneration that is clinically evident in persons with PNDs. A corollary of this suggestion is that the infrequency with which the PNDs are seen clinically may result from a confluence of uncommon pathogenic events—the expression of a tissue-restricted (neuronal) antigen in tumors whose function can be disrupted by an internalized autoantibody and that can thereby trigger apoptotic neuronal death.

Immune Mechanisms in Paraneoplastic Neurologic Degeneration Pathogenesis

Although PND antibodies may participate in disease pathogenesis, a number of observations suggest that they are not likely to be sufficient to cause disease. In general, treatments that reduce the antibody titer in PND patients fail to ameliorate the progression of disease, and passive transfer of antibody or immunization of animals with recombinant fusion proteins fails to reproduce the disease (reviewed in Darnell 1996). Although exceptions may exist—CAR antibodies may be able to reproduce some aspects of retinal degeneration, anti-GAD (glutamate decarboxylase) antibodies in stiff-man syndrome may be sufficient to cause reversible neuronal dysfunction (see Darnell 1996 for a discussion), and there are anecdotal reports of neurologic improvement following immunosuppression or tumor removal in some PND patients (Posner 1995)—the severe neuronal degeneration refractory to treatments that reduce antibodies evident in most PNDs and the intracellular location of the target antigens suggest that additional factors play important roles in the disorders.

T Cells in Disease Pathogenesis   For the reasons described in the previous section, a role for cellular immune responses in PND pathogenesis has been examined. PCD patients harbor cdr2-specific killer T cells capable of killing targets expressing cdr2 on major histocompatibility complex (MHC) class I molecules (Albert et al 1998a). Intracellular proteins are presented, via small peptide fragments, on MHC class I molecules on the cell surface. $CD8^+$-T-cell recognition and killing of target cells such as that evident in PCD provides a ready mechanism for the immunity to PND-expressing tumor cells evident clinically in the PND cases. The significance of identifying the immune mechanism for tumor recognition in PND patients is underscored by the observation that the tumors in these patients are typically indolent and in some cases spontaneously regress. Moreover, several of the onconeural antigens have been found to be widely expressed in the general population of cancer patients [Hu antigens are expressed in all SCLCs, and cdr2 is expressed in a high percentage of gynecologic cancers (Darnell et al 2000; see Darnell 1999 for a discussion of these issues)].

Some data suggest that the presence of activated T cells in the body may not be sufficient to generate autoimmune neurologic disease, although these cells can be detected in the CSF of PND patients (Albert et al 2000) and are therefore likely to be necessary for the widespread destruction of neurons evident pathologically in most PNDs. It has been noted that a high percentage of SCLC patients have antibody immune responses to the Hu antigen present in their blood in the absence of neurologic disease (Dalmau et al 1990, Graus et al 1997a) and that such patients have evidence of tumor immunity (limited-stage disease, improved response to chemotherapy, and improved prognosis). We hypothesize that these patients harbor activated Hu-specific killer T cells in their blood in the absence of neurologic disease.

Several phenomena may preclude onconeural–antigen-specific T cells in the blood from being sufficient to generate autoimmune neurologic disease. Activated T cells routinely cross into the CSF, albeit transiently and in small numbers (Couraud 1994, Williams & Hickey 1995), and presumably PND-specific T cells do so in some neurologically normal cancer patients, such as those with immunity to the Hu antigen. Additional events are likely to be necessary for activated T cells to remain in the CSF and instigate autoimmune neurologic disease.

One event that may be necessary for stimulation of T cells that have crossed into the CSF is antigen presentation within the CNS. Hickey and colleagues have shown that phagocytic perivascular cells populated from bone-marrow derived monocytes, perhaps including cells that are the equivalent of the potent antigen presenting cells (APCs) termed dendritic cells (DCs), are important for the homing of myelin-basic protein (MBP)-specific T cells into the CNS (Hickey & Kimura 1988). Like these perivascular cells, microglial cells within the brain parenchyma have features suggesting that they too may be the equivalent of DCs, such as the high level expression of MHC molecules and co-stimulatory molecules (Becher et al 2000). Peripheral T cells may be attracted into the CNS, perhaps sequentially, by perivascular and parenchymal APCs presenting neuronal antigens.

Another phenomenon necessary for the development of T–cell–mediated autoimmunity within the CNS is antigen presentation on MHC molecules on the surfaces of target cells. Neurons have long been thought to lack expression of MHC class I molecules, which are necessary for recognition by cytotoxic CD8[+] T cells; although this dogma has recently been challenged (Corriveau et al 1998), levels of MHC I expression may be low in some populations of neurons, and expression may require induction by such factors as cytokines (Neumann et al 1995, Neumann et al 1997) to be able to effectively engage PND-specific T cells (Darnell 1998).

Antibodies in Disease Pathogenesis   We suggest that PND antibodies may be part of the trigger that allows a peripheral tumor immune response to develop into neurologic disease (Figure 1; see color insert). Just as activated T cells present in the blood are able to penetrate the CNS, so too do peripheral antibodies enter the CNS. IgG is believed to gain access to the CSF by passive diffusion at a molar ratio (serum:CSF) of about 800:1 (Felgenhauer 1974, Fishman et al 1990, 1991). Although no experimental evidence conclusively addresses whether antibodies may enter neurons once in the CNS, a large number of reports suggest that such a phenomenon may be possible (Fishman et al 1990, 1991; Tezel & Wax 2000; reviewed in Alarcon-Segovia et al 1996, Darnell 1996). What role might such antibodies in the CSF play in PND pathogenesis if they are not sufficient to cause disease?

There are several features common to the PND antibodies discussed herein— Nova, Hu, cdr2, and recoverin—which suggest that they have a potential role in triggering autoimmune neurologic disease. Each antibody targets critical functional epitopes that disrupt the function of the target neuronal antigen, and disruption of the activity of these antigens is linked to apoptotic neuronal death. Although there is evidence that neurons may imbibe large macromolecules, even antibodies (Borges et al 1985, Graus et al 1991, Greenlee et al 1993; see discussion in Darnell 1996), this proposed mechanism does not seem likely to be sufficiently effective to lead to the complete loss of neurons evident in such PNDs as PCD, but it is more

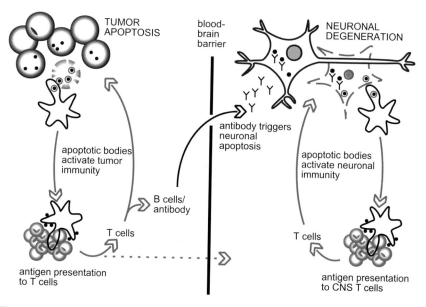

Figure 1 A general model for the pathogenesis of PNDs. Tumor cells (*blue*) aberrantly express an onconeural antigen, represented as a *black spot*. As tumor cells undergo apoptosis (*red*), dendritic cells (DCs) take up apoptotic fragments containing the onconeural antigen. DCs present the onconeural antigen on MHC class I, along with costimulatory signals, to T cells (*green*), thereby activating antigen-specific CD8$^+$ T cells. These T cells are competent to mediate effective tumor immunity but not to induce neurologic disease. DCs also activate CD4$^+$ T cells, which in turn help activate B cells that produce onconeural antibodies (represented as symbol Y). Some of this antibody crosses the blood-brain barrier by passive diffusion. Neurons (*blue*) expressing the onconeural antigen take up the antibody (inefficiently), which binds to the antigen and thereby impairs onconeural activities essential for neuronal survival and triggers rare events of neuronal apoptosis. Apoptosis of even a small number of neurons will generate apoptotic material (*red*) that is very efficiently scavenged by DCs. DCs resident in the CNS stimulate extant antigen-specific T cells (*green*) transiently crossing the blood-brain barrier to enter the CNS and mediate widespread neuron killing, giving rise to clinical neurologic dysfunction.

likely, we propose, to act as an immunologic trigger for neuronal degeneration in the setting of an extant T-cell-mediated tumor immune response.

Apoptotic Neuronal Death as a Trigger for Autoimmune Central Nervous System Disease   Inefficient antibody entry into the CNS and into neurons may account for uncommon events of neuronal apoptosis (Figure 1, see color insert). Potent APCs such as dendritic cells (Banchereau & Steinman 1998) effectively scavenge apoptotic material for presentation to and stimulation of both CD4$^+$ and CD8$^+$ T cells (Albert et al 1998b, Inaba et al 1998) in regional lymph nodes.

The mechanism for antigen presentation to T cells in the brain is unknown, although it appears that DCs do reside within the brain (as discussed above; see also McMenamin 1999). Antigen presentation to T cells in the CSF, perhaps initially by APCs at the endothelial lining of the blood–brain barrier (Becher et al 2000) and subsequently by parenchymal APCs, may provide a stimulus that generates an intrathecal T-cell-mediated immune response to onconeural antigens. Such stimulation of T cells within the CNS is likely to be accompanied by an inflammatory cascade, including the production of cytokines that promote permeability of the blood–brain barrier, upregulate neuronal MHC class I expression and promote further entry and activation of T cells in the CNS (Becher et al 2000).

Thus, inefficient killing of neurons by antibody-mediated interruption of specific onconeural antigen functions and subsequent apoptotic death may trigger widespread and efficient CD8$^+$ T–cell killing of the majority of susceptible antigen-expressing neurons. Such a model is consistent with both the remarkable functional specificity of PND antibodies and the evidence of activated T cells in the CSF of PND patients with clinically evident neuronal destruction (Albert et al 2000). Further, this model suggests that approaches to preventing tumor immunity from developing into CNS autoimmunity must consider initiating events for the latter, including antibody-mediated neuronal death and the activation of peripheral immune cells within the CNS.

CONCLUSIONS

The paraneoplastic neurologic disorders provide an exciting window into neuroscience—the disease-related antibodies identify target antigens that are neuron-specific proteins, specifically expressed in certain tumor types, whose functions are critical for neuronal survival. The immune response to onconeural antigens is of importance to tumor biology because the rare phenomena that occur in PNDs appear to hold important clues for understanding cancer immunology. A complete understanding of the means by which the immune response to onconeural antigens in tumors evolves into autoimmune neurologic disease remains elusive and will require the establishment of animal models of the PNDs. A complementary approach to understanding the disease pathogenesis lies in further elucidating the biology of the onconeural antigens.

Visit the Annual Reviews home page at www.AnnualReviews.org

LITERATURE CITED

Adamus G, Guy J, Schmied JL, Arendt A, Hargrave PA. 1993. Role of anti-recoverin autoantibodies in cancer-associated retinopathy. *Invest. Ophthalmol. Vis. Sci.* 34:2626–33

Adamus G, Machnicki M, Elerding H, Sugden B, Blocker YS, Fox DA. 1998. Antibodies to recoverin induce apoptosis of photoreceptor and bipolar cells in vivo. *J. Autoimmun.* 11:523–33

Adamus G, Ortega H, Witkowska D, Polans A. 1994. Recoverin: a potent uveitogen for the induction of photoreceptor degeneration in Lewis rats. *Exp. Eye Res.* 59:447–55

Akamatsu W, Okano HJ, Osumi N, Inoue T, Nakamura S, et al. 1999. Mammalian ELAV-like neuronal RNA-binding proteins HuB and HuC promote neuronal development in both the central and the peripheral nervous systems. *Proc. Natl. Acad. Sci. USA* 96:9885–90

Alarcon-Segovia D, Ruiz-Argüelles A, Llorente L. 1996. Broken dogma: penetration of autoantibodies into living cells. *Immunol. Today* 17:163–64

Albert ML, Austin LM, Darnell RB. 2000. Detection and treatment of activated T cells in the cerebrospinal fluid of patients with paraneoplastic cerebellar degeneration. *Ann. Neurol.* 47:9–17

Albert ML, Darnell JC, Bender A, Francisco LM, Bhardwaj N, Darnell RB. 1998a. Tumor-specific killer cells in paraneoplastic cerebellar degeneration. *Nat. Med.* 4:1321–24

Albert ML, Sauter B, Bhardwaj N. 1998b. Dendritic cells acquire antigen from apoptotic cells and induce class I-restricted CTLs. *Nature* 392:86–89

Amara SG, Jonas V, Rosenfeld MG, Ong ES, Evans R. 1982. Alternative RNA processing in calcitonin gene expression generates mRNAs encoding different polypeptide products. *Nature* 298:240–44

Anderson N, Rosenblum M, Graus F, Wiley R, Posner J. 1988a. Autoantibodies in paraneoplastic syndromes associated with small-cell lung cancer. *Neurology* 38:1391–98

Anderson NE, Budde-Steffen C, Rosenblum MK, Graus F, Ford D, et al. 1988b. Opsoclonus, myoclonus, ataxia, and encephalopathy in adults with cancer: a distinct paraneoplastic syndrome. *Medicine* 67:100–9

Anderson NE, Cunningham JM, Posner JB. 1987. Autoimmune pathogenesis of paraneoplastic neurological syndromes. *Crit. Rev. Neurobiol.* 3:245–99

Antic D, Keene JD. 1997. Embryonic lethal abnormal visual RNA-binding proteins involved in growth, differentiation, and posttranscriptional gene expression. *Am. J. Hum. Genet.* 61:273–78

Antic D, Keene JD. 1998. Messenger ribonucleoprotein complexes containing human ELAV proteins: interactions with cytoskeleton and translational apparatus. *J. Cell Sci.* 111:183–97

Antic D, Lu N, Keene JD. 1999. ELAV tumor antigen, Hel-N1, increases translation of neurofilament M mRNA and induces formation of neurites in human teratocarcinoma cells. *Genes Dev.* 13:449–61

Atasoy U, Watson J, Patel D, Keene JD. 1998. ELAV protein HuA (HuR) can redistribute between nucleus and cytoplasm and is upregulated during serum stimulation and T cell activation. *J. Cell Sci.* 111:3145–56

Ball NS, King PH. 1997. Neuron-specific hel-N1 and HuD as novel molecular markers of neuroblastoma: a correlation of HuD messenger RNA levels with favorable prognostic features. *Clin. Cancer Res.* 3:1859–65

Banchereau J, Steinman RM. 1998. Dendritic cells and the control of immunity. *Nature* 392:245–52

Becher B, Prat A, Antel JP. 2000. Brain-immune connection: immuno-regulatory properties of CNS-resident cells. *Glia* 29: 293–304

Borges LF, Elliott PJ, Gill R, Iversen SD, Iversen LL. 1985. Selective extraction of small and large molecules from the cerebrospinal fluid by Purkinje neurons. *Science* 228:346–48

Buchanan T, Gardiner T, Archer D. 1984. An ultrastructural study of retinal photoreceptor degeneration associated with bronchial carcinoma. *Am. J. Ophthalmol.* 97:277–87

Buckanovich RJ, Darnell RB. 1997. The neuronal RNA binding protein Nova-1 recognizes specific RNA targets in vitro and in vivo. *Mol. Cell. Biol.* 17:3194–201

Buckanovich RJ, Posner JB, Darnell RB. 1993. Nova, the paraneoplastic Ri antigen, is homologous to an RNA-binding protein and is specifically expressed in the developing motor system. *Neuron* 11:657–72

Buckanovich RJ, Yang YY, Darnell RB. 1996. The onconeural antigen Nova-1 is a neuron-specific RNA-binding protein, the activity of which is inhibited by paraneoplastic antibodies. *J. Neurosci.* 16:1114–22

Budde-Steffen C, Anderson N, Rosenblum M, Graus F, Ford D, et al. 1988. An antineuronal autoantibody in paraneoplastic opsoclonus. *Ann. Neurol.* 23:528–31

Burd CG, Dreyfuss G. 1994. Conserved structures and diversity of functions of RNA-binding proteins. *Science* 265:615–21

Chan RC, Black DL. 1997. The polypyrimidine tract binding protein binds upstream of neural cell-specific c-src exon N1 to repress the splicing of the intron downstream. *Mol. Cell. Biol.* 17:4667–76

Corradi JP, Yang CW, Darnell JC, Dalmau J, Darnell RB. 1997. A post-transcriptional regulatory mechanism restricts expression of the paraneoplastic cerebellar degeneration antigen cdr2 to immune privileged tissues. *J. Neurosci.* 17:1406–15

Corriveau RA, Huh GS, Shatz CJ. 1998. Regulation of class I MHC gene expression in the developing and mature CNS by neural activity. *Neuron* 21:505–20

Couraud PO. 1994. Interactions between lymphocytes, macrophages, and central nervous system cells. *J. Leukocyte Biol.* 56:407–15

Cunningham J, Graus F, Anderson N, Posner JB. 1986. Partial characterization of the Purkinje cell antigens in paraneoplastic cerebellar degeneration. *Neurology* 36:1163–68

Dalmau J, Furneaux HM, Cordon-Cardo C, Posner JB. 1992. The expression of the Hu (paraneoplastic encephalomyelitis/sensory neuronopathy) antigen in human normal and tumor tissues. *Am. J. Pathol.* 141:881–86

Dalmau J, Furneaux HM, Gralla RJ, Kris MG, Posner JB. 1990. Detection of the anti-Hu antibody in the serum of patients with small cell lung cancer—a quantitative Western blot analysis. *Ann. Neurol.* 27:544–52

Dalmau J, Graus F, Rosenblum MK, Posner JB. 1991. Anti-Hu associated paraneoplastic encephalomyelitis/sensory neuropathy: a clinical study of 71 patients. *Medicine* 71:59–72

Dalmau J, Gultekin SH, Voltz R, Hoard R, DesChamps T, et al. 1999. Ma1, a novel neuron- and testis-specific protein, is recognized by the serum of patients with paraneoplastic neurological disorders. *Brain* 122:27–39

Dalmau JO, Posner JB. 1999. Paraneoplastic syndromes. *Arch. Neurol.* 56:405–8

Darnell JC, Albert ML, Darnell RB. 2000. Cdr2, a target antigen of naturally occurring tumor immunity, is widely expressed in gynecologic tumors. *Cancer Res.* 60:2136–39

Darnell RB. 1996. Onconeural antigens and the paraneoplastic neurologic disorders: at the intersection of cancer, immunity and the brain. *Proc. Natl. Acad. Sci. USA* 93:4529–36

Darnell RB. 1998. Immunologic complexity in neurons. *Neuron* 21:947–50

Darnell RB. 1999. The importance of defining the paraneoplastic neurologic disorders. *N. Engl. J. Med.* 340:1831–33

Darnell RB, Furneaux HM, Posner JR. 1991. Antiserum from a patient with cerebellar degeneration identifies a novel protein in Purkinje cells, cortical neurons, and neuroectodermal tumors. *J. Neurosci.* 11:1224–30

Digre K. 1986. Opsoclonus in adults: report of three cases and review of the literature. *Arch. Neurol.* 43:1165–75

Dizhoor AM, Chen CK, Olshevskaya E, Sinelnikova VV, Phillipov P, Hurley JB. 1993. Role of the acylated amino terminus of recoverin in Ca^{2+}-dependent membrane interaction. *Science* 259:829–32

Dizhoor AM, Ray S, Kumar S, Niemi G, Spencer M, et al. 1991. Recoverin: a calcium sensitive activator of retinal rod guanylate cyclase. *Science* 251:915–18

Dropcho E, Chen Y, Posner J, Old L. 1987. Cloning of a brain protein identified by autoantibodies from a patient with paraneoplastic cerebellar degeneration. *Proc. Natl. Acad. Sci. USA* 84:4552–56

Evan G, Littlewood T. 1998. A matter of life and cell death. *Science* 281:1317–21

Fan XC, Steitz JA. 1998a. HNS, a nuclear-cytoplasmic shuttling sequence in HuR. *Proc. Natl. Acad. Sci. USA* 95:15293–98

Fan XC, Steitz JA. 1998b. Overexpression of HuR, a nuclear-cytoplasmic shuttling protein, increases the in vivo stability of ARE-containing mRNAs. *EMBO J.* 17:3448–60

Fathallah-Shaykh H, Wolf S, Wong E, Posner J, Furneaux H. 1991. Cloning of a leucine-zipper protein recognized by the sera of patients with antibody-associated paraneoplastic cerebellar degeneration. *Proc. Natl. Acad. Sci. USA* 88:3451–54

Feddersen RM, Ehlenfeldt R, Yunis WS, Clark HB, Orr HT. 1992. Disrupted cerebellar cortical development and progressive degeneration of Purkinje cells in SV40 T antigen transgenic mice. *Neuron* 9:955–66

Felgenhauer K. 1974. Protein size and cerebrospinal fluid composition. *Klin. Wochenschr.* 52:1158–64

Fishman PS, Farrand DA, Kristt DA. 1990. Internalization of plasma proteins by cerebellar Purkinje cells. *J. Neurol. Sci.* 100:43–49

Fishman PS, Farrand DA, Kristt DA. 1991. Penetration and internalization of plasma proteins in the human spinal cord. *J. Neurol. Sci.* 104:166–75

Flaherty KM, Zozulya S, Stryer L, McKay DB. 1993. Three-dimensional structure of recoverin, a calcium sensor in vision. *Cell* 75:709–16

Floyd S, De Camilli P. 1998. Endocytosis proteins and cancer: a potential link? *Trends Cell Biol.* 8:299–301

Ford LP, Watson J, Keene JD, Wilusz J. 1999. ELAV proteins stabilize deadenylated intermediates in a novel in vitro mRNA deadenylation/degradation system. *Genes Dev.* 13:188–201

Furneaux HM, Rosenblum MK, Dalmau J, Wong E, Woodruff P, et al. 1990. Selective expression of Purkinje-cell antigens in tumor tissue from patients with paraneoplastic cerebellar degeneration. *N. Engl. J. Med.* 322:1844–51

Gallouzi IE, Brennan CM, Stenberg MG, Swanson MS, Eversole A, et al. 2000. HuR binding to cytoplasmic mRNA is perturbed by heat shock. *Proc. Natl. Acad. Sci. USA* 97:3073–78

Gao FB, Carson CC, Levine T, Keene JD. 1994. Selection of a subset of mRNAs from combinatorial 3' untranslated region libraries using neuronal RNA-binding protein Hel-N1. *Proc. Natl. Acad. Sci. USA* 91:11207–11

Gao FB, Keene JD. 1996. Hel-N1/Hel-N2 proteins are bound to poly(A)$^+$ mRNA in granular RNP structures and are implicated in neuronal differentiation. *J. Cell Sci.* 109:579–89

Gery I, Chanaud NP 3rd, Anglade E. 1994. Recoverin is highly uveitogenic in Lewis rats. *Invest. Ophthalmol. Vis. Sci.* 35:3342–45

Gorodovikova EN, Gimelbrant AA, Senin II, Philippov PP. 1994. Recoverin mediates the calcium effect upon rhodopsin phosphorylation and cGMP hydrolysis in bovine retina rod cells. *FEBS Lett.* 349:187–90

Grabowski PJ. 1998. Splicing regulation in neurons: tinkering with cell-specific control. *Cell* 92:709–12

Grabowski PJ. 2000. Genetic evidence for a Nova regulator of alternative splicing in the brain. *Neuron* 25:254–56

Graus F, Dalmau J, Rene R, Tora M, Malats N, et al. 1997a. Anti-Hu antibodies in patients with small-cell lung cancer: association with complete response to therapy and improved survival. *J. Clin. Oncol.* 15:2866–72

Graus F, Dalmau J, Valldeoriola F, Ferrer I, Rene R, et al. 1997b. Immunological characterization of a neuronal antibody (anti-Tr) associated with paraneoplastic cerebellar degeneration and Hodgkin's disease. *J. Neuroimmunol.* 74:55–61

Graus F, Illa I, Agusti M, Ribalta T, Cruz-Sanchez F, Juarez C. 1991. Effect of intraventricular injection of an anti-Purkinje cell antibody (anti-Yo) in a guinea pig model. *J. Neurol. Sci.* 106:82–87

Graus F, Rowe G, Fueyo J, Darnell RB, Dalmau J. 1993. The neuronal nuclear antigen recognized by the human anti-Ri autoantibody is expressed in central but not peripheral nervous system neurons. *Neurosci. Lett.* 150:212–14

Greenlee J, Brashear H. 1983. Antibodies to cerebellar Purkinje cells in patients with paraneoplastic cerebellar degeneration and ovarian carcinoma. *Ann. Neurol.* 14:609–13

Greenlee JE, Parks TN, Jaeckle KA. 1993. Type IIa ('anti-Hu') antineuronal antibodies produce destruction of rat cerebellar granule neurons in vitro. *Neurology* 43:2049–54

He X, Rosenfeld M. 1991. Mechanisms of complex transcriptional regulation: implications for brain development. *Neuron* 7:183–96

Hickey WF, Kimura H. 1988. Perivascular microglial cells of the CNS are bone marrow-derived and present antigen in vivo. *Science* 239:290–92

Hida C, Tsukamoto T, Awano H, Yamamoto T. 1994. Ultrastructural localization of anti-Purkinje cell antibody-binding sites in paraneoplastic cerebellar degeneration. *Arch. Neurol.* 51:555–58

Higuchi M, Single F, Kohler M, Sommer B, Sprengel R, Seeburg P. 1993. RNA editing of AMPA receptor subunit GluR-B: A base-paired intron-exon structure determines position and efficiency. *Cell* 75:1361–70

Imataka H, Nakayama K, Yasumoto K, Mizuno A, Fujii-Kuriyama Y, Hayami M. 1994. Cell-specific translational control of transcription factor BTEB expression. *J. Biol. Chem.* 269:20668–73

Inaba K, Turley S, Yamaide F, Iyoda T, Mahnke K, et al. 1998. Efficient presentation of phagocytosed cellular fragments on the major histocompatibility complex class II products of dendritic cells. *J. Exp. Med.* 188:2163–73

Jacobson A, Peltz SW. 1996. Interrelationships of the pathways of mRNA decay and translation in eukaryotic cells. *Annu. Rev. Biochem.* 65:693–739

Jaeckle K, Graus F, Houghton A, Cordon-Cardo C, Nielsen SL, Posner JB. 1985. Autoimmune response of patients with paraneoplastic cerebellar degeneration to a Purkinje cell cytoplasmic protein antigen. *Ann. Neurol.* 18:592–600

Jain RG, Andrews LG, McGowan KM, Pekala PH, Keene JD. 1997. Ectopic expression of Hel-N1, an RNA-binding protein, increases glucose transporter (GluT1) expression in 3T3-L1 adipocytes. *Mol. Cell. Biol.* 17:954–62

Jensen KB, Dredge BK, Stefani G, Zhong R, Buckanovich RJ, et al. 2000a. Nova-1 regulates neuron-specific alternative splicing and is essential for neuronal viability. *Neuron* 25:359–71

Jensen KB, Musunuru K, Lewis HA, Burley SK, Darnell RB. 2000b. The tetranucleotide UCAY directs the specific recognition of RNA by the Nova KH3 domain. *Proc. Natl. Acad. Sci. USA* 97:5740–45

Kawamura S. 1993. Rhodopsin phosphorylation as a mechanism of cyclic GMP phosphodiesterase regulation by S-modulin. *Nature* 362:855–57

Keene JD. 1999. Why is Hu where? Shuttling of early-response-gene messenger RNA subsets. *Proc. Natl. Acad. Sci. USA* 96:5–7

Keltner J, Roth A, Chang R. 1983. Photoreceptor degeneration. Possible autoimmune disorder. *Arch. Ophthalmol.* 101:564–69

Koushika SP, Lisbin MJ, White K. 1996. ELAV, a *Drosophila* neuron-specific protein, mediates the generation of an alternatively spliced neural protein isoform. *Curr. Biol.* 6:1634–41

Koushika SP, Soller M, White K. 2000. The neuron-enriched splicing pattern of *Drosophila* erect wing is dependent on the presence of ELAV protein. *Mol. Cell. Biol.* 20:1836–45

Lem J, Flannery JG, Li T, Applebury ML, Farber DB, Simon MI. 1992. Retinal degeneration is rescued in transgenic *rd* mice by expression of the cGMP phosphodiesterase β subunit. *Proc. Natl. Acad. Sci. USA* 89:4422–26

Levine TD, Gao F, King PH, Andrews LG, Keene JD. 1993. Hel-N1: an autoimmune RNA-binding protein with specificity for 3′ uridylate-rich untranslated regions of growth factor mRNA's. *Mol. Cell. Biol.* 13:3494–504

Lewis HA, Chen H, Edo C, Buckanovich RJ, Yang YY, et al. 1999. Crystal structures of Nova-1 and Nova-2 K-homology RNA-binding domains. *Structure* 7:191–203

Lewis HA, Musunuru K, Jensen KB, Edo C, Chen H, et al. 2000. Sequence-specific RNA binding by a Nova KH domain: implications for paraneoplastic disease and the fragile X syndrome. *Cell* 100:323–32

Lin C-H, Patton J. 1995. Regulation of alternative 3′ splice site selection by constitutive splicing factors. *RNA* 1:234–45

Lou H, Gagel RF, Berget SM. 1996. An intron enhancer recognized by splicing factors activates polyadenylation. *Genes Dev.* 10:208–19

Luque F, Furneaux H, Ferziger R, Rosenblum M, Wray S, et al. 1991. Anti-Ri: an antibody associated with paraneoplastic opsoclonus and breast cancer. *Ann. Neurol.* 29:241–51

Ma W, Cheng S, Campbell C, Wright A, Furneaux H. 1996. Cloning and characterization of HuR, a ubiquitously expressed Elav-like protein. *J. Biol. Chem.* 271:8144–51

Manley GT, Smitt PS, Dalmau J, Posner JB. 1995. Hu antigens: reactivity with Hu antibodies, tumor expression, and major immunogenic sites. *Ann. Neurol.* 38:102–10

Martinez R, Mathey-Prevot B, Bernards A, Baltimore D. 1987. Neuronal pp60^{c-src} contains a six-amino acid insertion relative to its nonneuronal counterpart. *Science* 237:411–15

Matsubara S, Yamaji Y, Sato M, Fujita J, Takahara J. 1996. Expression of a photoreceptor protein, recoverin, as a cancer-associated retinopathy autoantigen in human lung cancer cell lines. *Br. J. Cancer* 74:1419–22

Matsuda S, Hisatomi O, Ishino T, Kobayashi Y, Tokunaga F. 1998. The role of calcium-binding sites in S-modulin function. *J. Biol. Chem.* 273:20223–27

McMenamin PG. 1999. Distribution and phenotype of dendritic cells and resident tissue macrophages in the dura mater, leptomeninges, and choroid plexus of the rat brain as demonstrated in wholemount preparations. *J. Comp. Neurol.* 405:553–62

Molday RS. 1998. Photoreceptor membrane proteins, phototransduction, and retinal degenerative diseases. The Friedenwald Lecture. *Invest. Ophthalmol. Vis. Sci.* 39:2491–513

Musco G, Stier G, Joseph C, Morelli MAC, Nilges M, et al. 1996. Three-dimensional structure and stability of the KH domain: molecular insights into the fragile X syndrome. *Cell* 85:237–45

Myer VE, Fan XHC, Steitz JA. 1997. Identification of HuR as a protein implicated in AUUUA-mediated mRNA decay. *EMBO J.* 16:2130–39

Neumann H, Cavalie A, Jenne DE, Wekerle H. 1995. Induction of MHC class I genes in neurons. *Science* 269:549–52

Neumann H, Schmidt H, Cavalie A, Jenne D, Wekerle H. 1997. Major histocompatibility complex (MHC) class I gene expression in single neurons of the central nervous system: differential regulation by interferon (IFN)-gamma and tumor necrosis factor (TNF)-alpha. *J. Exp. Med.* 185:305–16

Ohguro H, Ogawa K, Maeda T, Maeda A, Maruyama I. 1999. Cancer-associated retinopathy induced by both anti-recoverin and anti-hsc70 antibodies in vivo. *Invest. Ophthalmol. Vis. Sci.* 40:3160–67

Okano HJ, Darnell RB. 1997. A hierarchy of Hu RNA binding proteins in developing and adult neurons. *J. Neurosci.* 17:3024–37

Okano HJ, Park WY, Corradi JP, Darnell RB. 1999. The cytoplasmic Purkinje antigen cdr2 downregulates Myc function: implications for neuronal and tumor cell survival. *Genes Dev.* 13:2087–98

Olsson JE, Gordon JW, Pawlyk BS, Roof D, Hayes A, et al. 1992. Transgenic mice with a rhodopsin mutation (Pro23His): a mouse model of autosomal dominant retinitis pigmentosa. *Neuron* 9:815–30

Patikoglou G, Burley SK. 1997. Eukaryotic transcription factor-DNA complexes. *Annu. Rev. Biophys. Biomol. Struct.* 26:289–325

Peng SS, Chen CY, Xu N, Shyu AB. 1998. RNA stabilization by the AU-rich element binding protein, HuR, an ELAV protein. *EMBO J.* 17:3461–70

Peterson K, Rosenblum MK, Kotanides H, Posner JB. 1992. Paraneoplastic cerebellar degeneration. I. A clinical analysis of 55 anti-Yo antibody-positive patients. *Neurology* 42:1931–37

Polans AS, Buczylko J, Crabb J, Palczewski K. 1991. A photoreceptor calcium binding protein is recognized by autoantibodies obtained from patients with cancer-associated retinopathy. *J. Cell Biol.* 112:981–89

Polans AS, Witkowska D, Haley TL, Amundson D, Baizer L, Adamus G. 1995. Recoverin, a photoreceptor-specific calcium-binding protein, is expressed by the tumor of a patient with cancer-associated retinopathy. *Proc. Natl. Acad. Sci. USA* 92:9176–80

Polydorides AD, Okano HJ, Yang YY, Stefani G, Darnell RB. 2000. A brain-enriched polypyrimidine tract-binding protein antagonizes the ability of Nova to regulate neuron-specific alternative splicing. *Proc. Natl. Acad. Sci. USA* 97:6350–55

Posner JB. 1995. *Neurologic Complications of Cancer.* Philadelphia: F A Davis

Posner JB, Furneaux HM. 1990. Paraneoplastic syndromes. In *Immunologic Mechanisms in Neurologic and Psychiatric Disease*, ed. BH Waksman, pp. 187–219. New York: Raven

Pranzatelli MR. 1992. The neurobiology of the opsoclonus-myoclonus syndrome. *Clin. Neuropharmacol.* 15:186–228

Rosenfeld MG, Mermod JJ, Amara SG, Swanson LW, Sawchenko PE, et al. 1983. Production of a novel neuropeptide encoded by the calcitonin gene via tissue-specific RNA processing. *Nature* 304:129–35

Sachs A. 1993. Messenger RNA degradation in eukaryotes. *Cell* 74:413–21

Sakai K, Gofuku M, Kitagawa Y, Ogasawara T, Hirose G, et al. 1993a. A hippocampal protein associated with paraneoplastic neurologic syndrome and small cell lung carcinoma. *Biochem. Biophys. Res. Commun.* 199:1200–8

Sakai K, Mitchell DJ, Tsukamoto T, Steinman L. 1990. Isolation of a complementary DNA clone encoding an autoantigen recognized by an anti-neuronal antibody from a patient with paraneoplastic cerebellar degeneration. *Ann. Neurol.* 28:692–98

Sakai K, Ogasawara T, Hirose G, Jaeckle KA, Greenlee JE. 1993b. Analysis of autoantibody binding to 52-kd paraneoplastic cerebellar degeneration-associated antigen expressed in recombinant proteins. *Ann. Neurol.* 33:373–80

Sawyer R, Selhorst J, Zimmerman L, Hoyt W. 1976. Blindness caused by photoreceptor degeneration as a remote effect of cancer. *Am. J. Ophthalmol.* 81:606–13

Simpson L, Emeson RB. 1996. RNA editing. *Annu. Rev. Neurosci.* 19:27–52

Singh R, Valcarcel J, Green MR. 1995. Distinct binding specificities and functions of higher eukaryotic polypyrimidine tract-binding proteins. *Science* 268:1173–76

Smith DP, Ranganathan R, Hardy RW, Marx J, Tsuchida T, Zuker CS. 1991. Photoreceptor deactivation and retinal degeneration mediated by a photoreceptor-specific protein kinase C. *Science* 254:1478–84

Stamm S, Casper D, Dinsmore J, Kaufmann C, Brosius J, Helfman D. 1992. Clathrin light chain B: gene structure and neuron-specific splicing. *Nucleic Acids Res.* 20:5097–103

Stamm S, Zhang MQ, Marr TG, Helfman DM. 1994. A sequence compilation and comparison of exons that are alternatively spliced in neurons. *Nucleic Acids Res.* 9:1515–26

Steward O, Banker G. 1992. Getting the message from the gene to the synapse: sorting and intracellular transport of RNA in neurons. *Trends Neurosci.* 15:180–86

Stripecke R, Oliveira C, McCarthy J, Hentze M. 1994. Proteins binding to 5' untranslated region sites: a general mechanism for translation regulation of mRNAs in human and yeast cells. *Mol. Cell. Biol.* 14:5898–909

Szabo A, Dalmau J, Manley G, Rosenfeld M, Wong E, et al. 1991. HuD, a paraneoplastic encephalomyelitis antigen, contains RNA-binding domains and is homologous to Elav Sex-lethal. *Cell* 67:325–33

Tan E. 1991. Autoantibodies in pathology and cell biology. *Cell* 67:841–42

Tezel G, Wax MB. 2000. The mechanisms of hsp27 antibody-mediated apoptosis in retinal neuronal cells. *J. Neurosci.* 20:3552–62

Thirkill C, Roth A, Kelner J. 1987. Cancer-associated retinopathy. *Arch. Ophthalmol.* 105:372–75

Thirkill CE, Fitzgerald RC, Sergott AM, Roth NK, Tyler NK, Keltner JL. 1989. Cancer-associated retinopathy (CAR syndrome) with antibodies reacting with retina, optic-nerve and cancer cells. *N. Engl. J. Med.* 321:1589–94

Thirkill CE, Tait RC, Tyler NK, Roth AM, Keltner JL. 1992. The cancer-associated retinopathy antigen is a recoverin-like protein. *Invest. Ophthalmol. Vis. Sci.* 33:2768–72

Valcarcel J, Gebauer F. 1997. Post-transcriptional regulation: the dawn of PTB. *Curr. Biol.* 7:R705–8

Wakamatsu Y, Weston JA. 1997. Sequential expression and role of Hu RNA-binding proteins during neurogenesis. *Development* 124:3449–60

Whitney KD, McNamara JO. 1999. Autoimmunity and neurological disease: antibody modulation of synaptic transmission. *Annu. Rev. Neurosci.* 22:175–95

Williams KC, Hickey WF. 1995. Traffic of hematogenous cells through the central nervous system. *Curr. Top. Microbiol. Immunol.* 202:221–46

Yang YYL, Yin GL, Darnell RB. 1998. The neuronal RNA binding protein Nova-2 is implicated as the autoantigen targeted in POMA patients with dementia. *Proc. Natl. Acad. Sci. USA* 95:13254–59

Zozulya S, Stryer L. 1992. Calcium-myristoyl protein switch. *Proc. Natl. Acad. Sci. USA* 89:11569–73

Annu. Rev. Neurosci. 2001. 24:263–97

ODOR ENCODING AS AN ACTIVE, DYNAMICAL PROCESS: Experiments, Computation, and Theory

Gilles Laurent,[1] Mark Stopfer,[1] Rainer W Friedrich,[1]
Misha I Rabinovich,[2] Alexander Volkovskii,[2]
and Henry DI Abarbanel,[2,3]

[1]*California Institute of Technology, Division of Biology 139-74, Pasadena, California
91125; e-mail: laurentg@caltech.edu, stopfer@cns.caltech.edu, friedric@cns.caltech.edu*
[2]*University of California San Diego, Institute for Nonlinear Science, La Jolla, California
92093-0402; e-mail: mrabinovich@ucsd.edu, avolkovskii@ucsd.edu,
hdia@hamilton.ucsd.edu*
[3]*Department of Physics and Marine Physical Laboratory, Scripps Institution of
Oceanography, University of California San Diego, La Jolla, California 92093*

Key Words   olfaction, olfactory bulb, antennal lobe, learning, dynamical systems,
oscillations

■ **Abstract**   We examine early olfactory processing in the vertebrate and insect
olfactory systems, using a computational perspective. What transformations occur
between the first and second olfactory processing stages? What are the causes and
consequences of these transformations? To answer these questions, we focus on the
functions of olfactory circuit structure and on the role of time in odor-evoked integrative
processes. We argue that early olfactory relays are active and dynamical networks,
whose actions change the format of odor-related information in very specific ways, so
as to refine stimulus identification. Finally, we introduce a new theoretical framework
("winnerless competition") for the interpretation of these data.

INTRODUCTION

The olfactory brain converts generally complex air- or water-borne chemical mixtures into singular signatures, experienced as vivid percepts. Such transformations are achieved by way of only a few brain stations—olfactory circuits are
shallower than their visual and auditory counterparts—and are strongly tied to
emotions and to memories acquired through other modalities. Understanding olfactory coding is thus an ambitious enterprise whose scope goes much beyond
that of this review. We focus here on the sensory transformations accomplished

0147-006X/01/0621-0263$14.00

by the first two processing stages, that is, olfactory receptors and postsynaptic structures.

As with any sense, understanding olfaction first requires defining the problems it has evolved to solve (Attneave 1954, Barlow 1969): segmenting an odor into its various constituents, as a chemist might do, does not appear to be one of these functions (Lawless 1997, Cain & Potts 1996). Rather, olfaction is a synthetic sense par excellence. Olfaction enables pattern learning, storage, recognition, tracking, or localization and attaches "meaning" to these patterns. By meaning we imply the richer set of associations acquired through other senses as well as hedonic (pleasant/unpleasant) and emotional valence—both of which have no physical reality outside the brain. Each one of these tasks needs to be better defined; recognition, for example, encompasses at least categorization, identification, and separation. The abilities to categorize and to identify a priori each imply very different kinds of processing; for example, categorization disregards small differences, whereas identification emphasizes them. We show how a single circuit can in fact accomplish both, through the use of dynamics.

We should also exploit our understanding of the physics of odors. In vision, much attention has been given to the statistics of natural images (Field 1987, 1994; Olshausen & Field 1996; Rudderman 1994). Correlations across space and time make natural images highly nonstochastic. Also, the spatial-frequency (f) content of a natural image, be it a face or a landscape, obeys a $1/f^{\alpha}$ distribution (at any scale), implying large areas of low contrast cut by edges whose frequency of occurrence decreases as sharpness increases. Knowing these rules helps us understand or predict the existence of specific spatial filters in early vision. Have equivalent rules been extracted from our knowledge of natural odors? It seems not. Many natural odors (such as families of flower fragrances) consist of similar combinations of compounds, implying the existence of natural correlations across volatile molecules; however, it seems as well that many unnatural combinations concocted in the laboratory evoke odors that are just as real or learnable. This impression remains anecdotal; we do not know of systematic investigations of odor statistics or of an animal's ability to learn and recognize composite odors. Help may eventually come from studies of olfactory receptor functional "specificity" (Buck 1995), although the complexity of chemical space may preclude such simple analysis. Some physical features of odors, however, are well known. Odor plumes, for instance, discretize stimulus delivery (Murlis et al 1992). This temporal structure is reinforced by odor-sampling behavior, which is usually repetitive (e.g. sniffing, appendage flicking, and casting behavior [Mellon 1997, Murlis et al 1992]). We examine the consequences of such intermittency on odor representations and their implications for olfactory codes. Finally, whereas sounds and images are typically updated rapidly, thus conveying information at potentially high rates that are exploitable for high-throughput communication, odors change more slowly. Olfaction is generally a low-temporal (and often spatial)-bandwidth sense. This means that time can be used as a coding dimension for the representation of nontemporal features of odors (Laurent 1999).

GOALS AND CONCEPTUAL FRAMEWORK

The cloning of odorant receptors and the structure and mapping of early olfactory circuits have been comprehensively reviewed over the last decade (Axel 1995; Boeckh & Tolbert 1993; Buck 1996; Hildebrand & Shepherd 1997; Mombaerts et al 1996b; Satou 1990; Scott et al 1993; Shepherd 1993, 1994; Shipley & Ennis 1996). We here focus on a computational view of odor encoding within the vertebrate main olfactory bulb (OB) and the analogous circuit in insects, the main antennal lobe (AL), excluding macroglomerular or specialized pheromonal centers (Dulac 1997, Keverne 1999). The OB and AL are organized according to similar anatomical principles (Hildebrand & Shepherd 1997); these similarities are only reinforced by physiological observations. Our approach thus tends to emphasize potential principles rather than peculiarities.

The foundation of this review is that early olfactory circuits should be studied as a system. The OB and AL form richly interconnected circuits whose global mode of action is captured accurately by neither static nor isolated samples (e.g. neither anatomical studies nor single-cell recordings); these circuits cannot be viewed as passive relays of odor-related information. Because olfactory-neuron responses are complex and correlated across space, a study of odor encoding must take into account these interactions and their distributed consequences. These consequences unfold over time; the early olfactory system behaves as an active nonstationary system; its dynamics occur on multiple time scales. Hence, the transfer to olfaction of functional models developed with other sensory systems is often inappropriate.

For example, analyses of visual, auditory, or somatosensory physiology rely heavily on the concept of the tuning curve, applied to single-neuron studies (e.g. Hubel & Wiesel 1968). This approach implies a scalar description of the stimulus (e.g. position or frequency; i.e. in a space with one or a few independent dimensions) and a one-dimensional scoring of a neuron's response (usually mean firing rate). While empirically powerful, both descriptions are ill adapted to much of olfactory physiology. One reason is that the relevant olfactory stimulus dimensions are likely to be numerous and difficult to order in relation to one another; a second reason is that neuronal response profiles are usually temporally complex, and the information they carry is not conveyed fully by firing rates. Moreover, because most "classical" studies focused, for technical reasons, on unitary recordings, their significance is always examined in a framework in which the single neuron is the relevant informational entity: the unspoken implication is that "whoever" reads activity within the brain does it either one neuron at a time or by simple spatial or temporal averaging of a population's activity (Shadlen & Newsome 1995), disregarding interneuronal correlations. Our results indicate that this is not appropriate for olfactory codes (Laurent 1996, Laurent et al 1996, MacLeod et al 1998, Stopfer et al 1997, Wehr & Laurent 1996). We argue that understanding olfactory coding requires a shift to a different framework, in which relevance is measured globally and over time.

Walter Freeman's many seminal contributions to the development of a dynamical perspective on olfaction (Freeman 1978, 1992, 2000; Freeman & Skarda 1985) must be recognized here. Our approach and interpretations, however, differ from Freeman's in at least three important ways. The first lies in the nature of the data. While we recognize the importance of macro- or mesoscopic signals (e.g. EEGs and field potentials) as experimental tools, we believe that they are not of the appropriate scale for analysis. The spatiotemporal phenomena that cause recognizable features in field potentials (e.g. local synchronization and nonstationary behavior) are indeed functionally relevant; but field potentials are only "shadows" of underlying distributed but precise neural-activity patterns, which need to be deciphered. The second difference lies in our theoretical model of population behavior. "Winnerless competition," introduced later, depends to a significant extent on a neuron-resolution-mechanistic understanding of odor signal processing. The third difference is that our experimental approach, using small olfactory systems (insects and fish), tries to separate stimulus-evoked activity from centrifugal "higher" influences providing contextual information. Our goal, illustrated here, is to understand the "unsupervised" sensory formatting of odor representations by early olfactory circuits first, although we agree that expectation influences odor-evoked neural activity (Pager 1983, Kay & Freeman 1998, Kay & Laurent 1999).

DISTRIBUTED, CLUSTERED REPRESENTATIONS BY AFFERENT ARRAYS

Olfactory-Receptor-Neuron Tuning and Combinatorial Codes

Odor encoding is a spatially distributed process. Since the classical work of Adrian (1942, 1950, 1953), it has been shown repeatedly and with increasing clarity that single odorants activate neural activity over a wide area of the nasal epithelium and its postsynaptic target neuropils (Cinelli & Kauer 1992, Duchamp 1982, Duchamp-Viret & Duchamp 1997, Friedrich & Korsching 1997, Joerges et al 1997, Kauer 1987, Leveteau & MacLeod 1966, Moulton 1967, Rubin & Katz 1999, Stewart et al 1979). Such broad activation is now, through remarkable advances in the molecular biology of odorant receptors (Buck & Axel 1991, Buck 1996), understandable in terms of olfactory receptor neuron (ORN) tuning and axon projections; single ORNs express a limited number of odorant receptor types (probably a single one in mammals), and all ORNs expressing the same receptor type converge to one or a few glomeruli in their target areas (Bozza & Kauer 1997, Malnic et al 1999, Mombaerts et al 1996a, Ressler et al 1993, Vassar et al 1994, Wang et al 1998). Because single odors can activate broad overlapping regions, it is clear that individual neurons—and thus possibly the receptor proteins themselves—can be activated by many odorants, including ones that belong to different chemical families, underlying different odor qualities. Calcium imaging on dissociated ORNs (Bozza & Kauer 1997, Malnic et al 1999) and older studies in situ (Duchamp et al 1974, Getchell 1986, Kang & Caprio 1995) are consistent with such results. The molecular/structural mechanisms underlying receptor-ligand recognition remain poorly understood. In conclusion, odors, be they mono- or multimolecular, generally activate an array of receptor types and, hence, a distributed population of glomeruli. Different odors activate different combinations of glomeruli; the odor identity code across ORNs thus has a critical combinatorial component.

Clustering Within Odor Representations by Olfactory Receptor Neurons

Many studies have illustrated the distributed nature of odor-evoked activity (Buck 1996); the clearest demonstration of distributed representations across ORNs, however, comes from recent calcium- and voltage-sensitive imaging experiments in zebrafish (Friedrich & Korsching 1997, 1998). Indeed, while glomeruli are generally described as the functional units of the early olfactory system, often overlooked is that they contain projections from afferent fibers (ORNs), intrinsic neurons (periglomerular cells), and output neurons (mitral and tufted cells [M/TCs]). Describing glomerular activity without assessing the respective contributions of pre- and postsynaptic elements precludes any understanding of the computations carried out there. Friedrich & Korsching's (1997) studies are significant in that only ORNs were labeled, ensuring that patterns of glomerular activation reported only ORN activity (although the possible contribution of presynaptic inhibition and thus of intrinsic neurons cannot be excluded). They showed that natural amino acid odorants activate overlapping combinations of glomeruli (i.e. ORN populations) and that odors represented by similar combinations share chemical features (e.g. acidic, basic, or long-chain neutral). Hence, the molecular underpinnings of olfactory transduction result in redundant (overlapping) odor representations (representation clusters) by afferent arrays (see also Duchamp-Viret & Duchamp 1997). Less direct evidence suggests that this is true also for other species (Bozza & Kauer 1997, Imamura et al 1992, Joerges et al 1997, Katoh et al 1993). Note that such clustering might provide a physical substrate for perceptual odor categorization. We see later that dynamical processing within the OB alters this initial clustered format.

OLFACTORY-RECEPTOR-NEURON SCATTERING, AXONAL CONVERGENCE, AND NOISE REDUCTION

ORNs that express the same receptor are, at least in fish and rodents (Ngai et al 1993, Ressler et al 1993, Weth et al 1996), distributed randomly within wide and overlapping zones across the nasal epithelium. Much interest is now focused on how the axons of these ORNs converge on the same glomeruli in the OB (or AL) during development (Lin et al 2000, Zheng et al 2000). This distributed structure itself, however, raises an important question: Are there computational gains in scattering rather than grouping idiotypic ORNs? One advantage might be noise reduction. The reasoning is twofold. First, scattering ORNs across the receptive sheet ensures that global activity across the ORN array is minimally affected by local fluctuations; because air (or water) flow along turbinates is turbulent, noisy fluctuations in odor concentration affecting some receptors are unlikely to be correlated across the entire ORN array. Second, ORN-to-OB/AL output neuron convergence ratios are generally very large ($\sim$1000:1 in mammals and 100:1 in many insects). A mitral cell or its insect analog (projection neuron [PN]) thus forms its responses to odors from very large numbers of converging inputs. These two conditions (uncorrelated noise and convergence) should ensure that postsynaptic averaging increases signal-to-noise ratios. Amphibian ORNs can display periodic synchronization, independent of downstream activity (Dorries & Kauer 2000, Ottoson 1959). Although indicating correlated modulation of ORN input, this does not contradict the above hypothesis, provided that noise from external sources remains uncorrelated across the ORN population. We shall see below that stimulus-noise reduction is highly desirable if further processing seeks to amplify the differences between overlapping representations (within-cluster patterns) for fine odor discrimination.

LATERAL CONNECTIONS WITHIN THE OLFACTORY BULB/ANTENNAL LOBE

Connections

Olfactory-Receptor-Neuron Output   Once odors activate groups of ORNs and their corresponding glomerular targets, information does not simply flow through the OB/AL, channeled directly by projection neurons to downstream areas. Rather, glomerular target sites and projection neurons form widespread lateral connections within the OB/AL. In the vertebrate OB, for example, M/TCs sometimes possess dendrites in several glomeruli (e.g. rabbit and turtle [Mori et al 1981, 1983]) and thus combine inputs with different chemical sources. Such patterns are common in lower vertebrates (Dryer & Graziadei 1994, Nieuwenhuys 1967) and in the AL of many insects (e.g. locusts and wasps) (Laurent 1996, Masson & Mustaparta 1990).

Local Inhibition   Local inhibitory neurons (granule cells [GCs] and periglomerular cells in vertebrates and local neurons [LNs] in insects) interconnect excitatory projection cells via excitatory inhibitory pathways. Resulting interactions can be very local, such as those mediated by periglomerular cells between neighboring glomeruli in the mammalian OB, or widespread, such as those mediated by GCs between M/TCs with deep secondary dendrites in the external plexiform layer (Dryer & Graziadei 1994). GC-mediated pathways may indeed link M/TCs whose somata and input glomeruli lie millimeters apart (Shipley & Ennis 1996).

The strength of this functional linkage, however, may vary with distance, owing to biophysical properties of M/TC secondary dendrites (Chen et al 1997). Note also that, in rodents at least, ORN projections bifurcate to two hemi-OBs; it is not clear yet whether the glomerular topology is identical in both. If local glomerular neighborhoods differ between the two halves, yet more combinations of lateral interactions between different input types can be created. If hemi-OBs are identical, GC-mediated interactions across the border between the hemi-OBs should allow interglomerular interactions that are different from those between the glomeruli within each hemi-OB. Insect inhibitory LNs often (and in certain species, always) have widespread arborizations, giving them access, in principle, to almost any neuron in the system, e.g. in *Drosophila melanogaster* (Heisenberg et al 1985) and in locusts (Laurent 1996).

Other Pathways   Interactions between projection cells or between LNs allow for more complex polysynaptic pathways. In the turtle and mammalian OB for example, evidence exists for lateral long-lasting excitatory interactions between M/TCs (Aroniadou-Anderjaska et al 1999, Isaacson 1999, Nicoll 1971a, Nicoll & Jahr 1982). In insects, PN-PN synaptic connections have been directly observed (Leitch & Laurent 1996; Malun 1991a, 1991b). Similarly, GCs or LNs can inhibit each other (Leitch & Laurent 1996, Shipley & Ennis 1996), enabling disinhibition, that is, context-dependent excitation. Finally, output neurons may send axonal collaterals within the OB/AL, thus influencing other neurons directly or polysynaptically (Nicoll 1971b, Nicoll & Jahr 1982, Gray & Skinner 1988). In short, ORNs affect neurons other than their immediate targets in glomeruli, even if the underlying lateral circuits vary across species (Dryer & Graziadei 1994). The OB (like the AL) is not a simple point-to-point relay (Kauer 1991). This observation underlies the dynamical framework introduced below.

Functions

The existence of GC-mediated lateral inhibition has been known for many decades; inhibitory neurons unquestionably make lateral contacts. We question, however, the common functional interpretation that these contacts serve to sharpen single-M/TC tuning through a process akin to retinal "lateral inhibition" (DeVries & Baylor 1993, Mori & Shepherd 1994). We find unconvincing the physiological evidence for such single-cell "sharpening" and the rationale for these connections' role (Imamura et al 1992, Katoh et al 1993, Mori et al 1992, Yokoi et al 1995) (see Laurent 1999). A "retinal" understanding of lateral inhibition, for example, predicts that cells "best tuned" to a given odor should be selected at the expense of suboptimally activated neurons (winner take all) over the duration of a stimulus. We found no evidence for such trends (RW Friedrich & G Laurent, submitted manuscript). We also observe that an individual neuron often responds to many odors (hindering the fusion of anatomical and functional lateral inhibitions) and that responses are not stationary (see below). We thus propose a different interpretation—that contacts mediated by GCs and other neurons contribute together to a global reformatting of odor representations, in the form of a stimulus-dependent, temporal redistribution of activity across the OB/AL. One can thus think of lateral interactions as a means to sharpen odor representations by the population in a manner often not assessable from single-neuron data. We now review the supporting evidence.

TEMPORAL FEATURES OF OLFACTORY RESPONSES

Historical Background

Oscillations   Adrian (1942, 1950) first reported the existence of oscillatory activity in the olfactory systems of anaesthetized mammals. Later studies (Freeman 2000, Freeman & Skarda 1985, Gray & Skinner 1988) showed that spontaneous or odor-driven oscillatory activity could contain complex frequency spectra. Recent physiological and imaging studies also revealed that the spatial coherence of oscillatory activity across or between olfactory areas can be complex (Lam et al 1999) and, in behaving animals, context dependent (Kay & Freeman 1998). Odor-evoked oscillatory activity has since been observed in most animal classes, including mollusks (Gelperin & Tank 1990), insects (Laurent & Naraghi 1994), fish (Satou 1990), amphibians (Ottoson 1959), reptiles (Beuerman 1975), and mammals, including primates (Hughes & Mazurowski 1962). Early computational studies indicated that lateral and reciprocal synapses between mitral cells (MCs) and GCs could generate the oscillatory patterning observed experimentally (Rall & Shepherd 1968). Other studies predicted also that such connections should impose a 90° average phase shift between MC and GC activity (Freeman 1975). The precise synaptic and biophysical mechanisms underlying or participating in oscillatory synchronization vary, and yet, odor-driven oscillations are ubiquitous in olfactory systems, including downstream areas, such as piriform and entorhinal cortices in mammals (Haberly 1990). As with other brain areas where stimulus-evoked oscillations occur (Gray 1994, Singer & Gray 1995), a major question emerges: Is oscillatory synchrony functionally relevant?

Slow Single-Unit Patterning   Following Adrian's pioneering observations, many single-unit studies on mammals, fish, and amphibians revealed that MC responses to odors are not fully described by firing rates (Hamilton & Kauer 1989, Kauer & Moulton 1974, Macrides & Chorover 1972, Meredith 1986, Wellis et al 1989). Extracellular recordings of OB units in anaesthetized rodents, for example, showed that neurons exhibit odor-dependent temporal discharge patterns relative to an imposed inhalation rhythm (Macrides & Chorover 1972). The structure of the patterns evoked by one odor appeared to be more stable over the duration of an experiment than firing rates and was largely independent of the imposed inhalation schedule or the concentration. Intracellular recordings from salamander MCs documented the deep sculpting of their responses by phasic inhibition (Kauer & Moulton 1974). Later intracellular studies on the insect pheromonal system (Burrows et al 1982) described similar features. Finally, voltage-sensitive dye imaging in salamanders provided macroscopic evidence for spatiotemporal patterns evoked by odors over large areas of the OB (Cinelli & Kauer 1992, Cinelli et al 1995). As anticipated (Kauer 1991), "presumably, ... this spatio-temporal array [of MCs] now carries re-encoded information about the stimulus to the next level of integration, the olfactory cortices"... Our recent work in insects and fish builds on these seminal studies. It links together many past and some new observations, attempts to provide a functional understanding of this re-encoding, and tests its relevance.

The Locust Model System

Insects provide an accessible model for forms of odor processing observed throughout a broad range of animals (Hildebrand & Shepherd 1997).

Circuits   The insect main AL contains the processes of three main neuron populations: ORN axon terminals, LNs, and PNs. LNs are inhibitory, although their population is, in some species, heterogeneous. PNs are excitatory and project to two areas: the mushroom body, an area involved in multimodal processing (Heisenberg 1998, Strausfeld et al 1998) and associative memory (including olfactory; Heisenberg et al 1985, Menzel 1987), and the lateral protocerebral lobe. Details about these pathways vary somewhat with species (Masson & Mustaparta 1990). In addition, the AL contains the terminals of aminergic or peptidergic neuromodulatory neurons, whose projections are generally widespread within the brain (Hammer 1993, Sun et al 1993) and whose effects on associative learning can be critical (Hammer & Menzel 1995). The AL has a glomerular architecture, but the number of glomeruli varies (e.g. ~50 in *D. melanogaster* and ~1000 in locusts or wasps). Species with few large glomeruli are described as macroglomerular; the others are microglomerular. Accordingly, PN projections can be uniglomerular (as in macroglomerular species, e.g. *D. melanogaster*, bees, and cockroaches) or multiglomerular (e.g. wasps and locusts). These morphological subtypes are not correlated with categories of physiological output: PN oscillatory synchronization and slow response patterning (see below) are observed equally in micro- and macroglomerular species (Heinbockel et al 1998, Laurent & Naraghi 1994, Stopfer et al 1997). In locusts, the AL contains the terminals of ~90,000 ORNs, ~300 LNs, and 830 PNs. LN projections are axonless and extend over the entire AL. Each PN has a planar dendritic tree with 10–20 radial dendrites each ending in one or two glomeruli (Laurent et al 1996). Neither LNs nor PNs appear to express intrinsic oscillatory properties (Laurent & Davidowitz 1994). Upon natural stimulation, odor-specific subgroups of LNs and PNs become activated. Their responses contain several interlocked features, reviewed below.

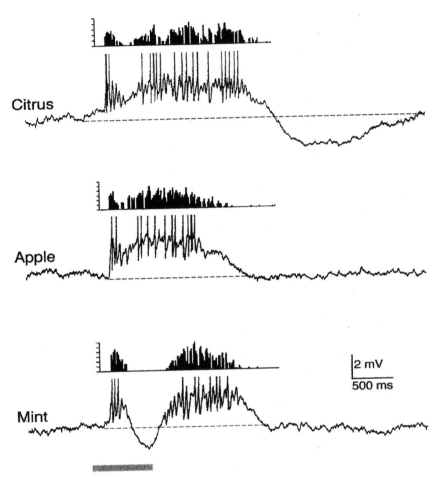

Figure 1   Intracellular recording from one locust antennal lobe projection neuron illustrating the subthreshold oscillatory activity (50-ms cycles) giving rise to periodic spiking superimposed on slower, odor-specific temporal-response patterns. All three recordings are from the same PN (see Laurent & Davidowitz 1994, Laurent et al 1996).

Oscillations   A typical LN or PN intracellular record shows odor-evoked subthreshold oscillations in the 20- to 30-Hz band (Figure 1). Paired intracellular recordings show that coactive neurons within either population are phase-locked with 0° mean phase, whereas LNs and PNs are phase shifted relative to each other by 90° (Laurent & Davidowitz 1994). Oscillation frequency is independent of odor identity, and air alone evokes no oscillatory activity. Because PNs project to a layered structure (the mushroom body calyx), PN coherence can be measured from local field potential (LFP) oscillations recorded there (Laurent & Naraghi 1994).

Whereas insect ORNs do not show odor-evoked oscillatory synchronization, sustained electrical stimulation of ORN axons causes oscillatory activity within and across PNs (Wehr & Laurent 1999). Oscillatory synchronization of LNs and PNs is therefore the result of AL circuit dynamics, driven by ORN-evoked excitation. At rest, low-level spontaneous oscillatory activity can be detected from spectral analysis of mushroom body LFPs. Upon transsection of ORN axons, however, this spectral peak disappears. Baseline ORN activity thus sets AL dynamics on the threshold of coherent oscillatory behavior, even though basal PN activity is generally low (1–5 Hz).

Slow Patterning   The odor-evoked responses of LNs and PNs also contain prolonged and successive periods of increased and decreased activity (Laurent 1996, Laurent & Davidowitz 1994) shaped, in part at least, by slow, picrotoxin-insensitive synaptic inhibition (MacLeod & Laurent 1996). These patterns are cell and odor specific (Figure 1) and are stable from trial to trial; a given odor thus evokes activity in a PN assembly whose composition changes reliably throughout the response. In addition, the action potentials produced by a PN during its odor-specific phases of activity are not necessarily all phase-locked to the LFP (Laurent et al 1996). For each odor-PN combination, however, precise and consistent epochs of phase-locked or non-phase-locked activity can usually be identified. Hence, pairwise oscillatory synchronization of PNs is generally transient (Laurent 1996, Laurent & Davidowitz 1994), while the LFP, which reflects a larger fraction of the population's coherent activity, usually shows continuous oscillatory activity throughout the stimulus. When phase-locked, PN spikes occur within a ± 5-ms window around the ensemble mean. Hence, oscillatory synchronization and slow patterning together shape a complex, distributed representation in which odor-specific information appears both in the identity and in the time of recruitment and phase-locking of PNs. None of these features can be deciphered from LFPs or single-cell recordings alone.

Substrate for Odor Encoding   Upon odor stimulation, large-amplitude oscillations appear in the LFP. The time scale of PN update during a response is the oscillation cycle (Wehr & Laurent 1996). Using multiple trials with the same odor, one can assign a firing probability to each responding PN for each one of the successive cycles of the population oscillation. For a given odor, PN, and cycle, this probability can be close to 1; that is, each cycle of the population response contains one spike from each of several reliable PNs and less reliable spikes from many PNs with intermediate firing probabilities. The latter group is interesting because it allows one to estimate the functional coupling of PNs by measuring conditional firing probabilities. Take the example of two simultaneously recorded PNs of which one (PN_1) fired one action potential during cycle 5 in only 16 of 25 trials. Given this, what was the output of PN_2 during or around cycle 5 over the nine trials when PN_1 did not fire? Firing probabilities were often significantly correlated (positively or negatively) across PNs, so that the firing of a given PN in one trial would predict an increased or decreased firing probability in the other PN, during the same or even a different cycle of that trial (Wehr & Laurent 1996). This indicates that each PN's firing is correlated with the present and past behavior of other PNs in the network. It does not, however, identify the causes of this correlation.

Over the few seconds that an odor response can be sustained (before ORN adaptation sets in), these patterns rarely settle or repeat themselves. Using the terminology of dynamical systems, an odor representation by PNs can be thought of as a complex trajectory through phase space (with an uncertainty given by the limited reliability of many participating PNs) with no fixed-point or limit-cycle attractor (see theory below). When the stimulus stops, PNs return to baseline firing levels either immediately or, in some instances, after a short period of deep inhibition or rapid desynchronized firing. In conclusion, specific odors at a given concentration activate odor-specific PN assemblies whose components are recruited during precise and sometimes multiple epochs of the response. Odor-specific differences can be resolved at the temporal scale of one oscillation cycle ($\sim$50 ms). No evidence was ever obtained for a within-cycle phase code, suggested for this or other sensory systems (Hopfield 1995, von der Malsburg & Schneider 1986).

FUNCTIONAL RELEVANCE OF OSCILLATORY SYNCHRONIZATION

The functional relevance of these patterns was recently tested, exploiting the knowledge that oscillatory synchronization results from LN-mediated fast inhibition (MacLeod & Laurent 1996).

Selective Projection Neuron De-synchronization by Disruption of Fast Local Inhibition

LNs make γ-aminobutyric acid-immunoreactive contacts onto both LNs and PNs (Leitch & Laurent 1996, Malun 1991b). Among several Cl-channel or γ-aminobutyric acid receptor antagonists tested, picrotoxin was found to block selectively the oscillatory synchronization of AL neurons while sparing the slow inhibition responsible for slow patterning (MacLeod et al 1998, MacLeod & Laurent 1996); the slow odor response patterns of individual PNs remained unchanged, except for the introduced temporal jitter of their spikes. Odor identification using temporal information contained in individual PN spike trains, for example, was not affected by desynchronization (MacLeod et al 1998). Similarly, picrotoxin never caused the appearance of PN responses to new odors (neither in locusts [MacLeod et al 1998, MacLeod & Laurent 1996] nor in honey bees [Stopfer et al 1997]).

Behavioral Assay: Projection Neuron Synchronization and Odor Discrimination

Evaluating the functional significance of oscillatory synchronization requires a behavioral measure of perception. Honeybees can be trained to recognize odors in a well-defined proboscis extension behavioral paradigm; after associative pairing(s) of an odor with a reward (sucrose solution or nectar), bees can predict the delivery of the reward upon presentation of the conditioned odor alone (Kuwabara 1957, Bitterman et al 1983, Smith & Menzel 1989, Hammer & Menzel 1995). Conditioning strength is assayed by the probability of proboscis extension during single extinction trials, carried out over a population of individuals. Because the physiological responses of bee AL neurons closely resemble those described in locusts (odor-activated oscillating neural assemblies and picrotoxin-sensitive oscillatory coherence [Stopfer et al 1997]), we could assess the putative role of PN synchronization on odor perception.

One group of bees (controls) received an application of saline into (or at the surface of) each AL; the second (picrotoxin group) received similar treatments with picrotoxin. Odor-sucrose pairing started 10 min after treatment, using an aliphatic alcohol as the conditioning stimulus (C). Both groups learned to respond to C equally well. Each bee's discrimination was then tested using three odorants: C, the alcohol used for training; S, a similar aliphatic alcohol; and D, a chemically dissimilar odorant (a terpene). These odors were chosen because bees generalize partly across the alcohols but very little from either alcohol to the terpene; this triple test thus enabled us to better assess the effects of desynchronization on odor discrimination. The control animals responded vigorously to C, but significantly less to S and D, indicating that they could readily discriminate each odor. The picrotoxin group responded vigorously to C but hardly at all to D, indicating that picrotoxin had not affected the animals' ability to memorize the association of C with a reward; picrotoxin did not cause widespread generalization. This group, however, could not perform the more difficult discrimination task between C and S. In a further study, honeybees were given two successive injections of saline or picrotoxin, one before each of the conditioning and testing periods; there were thus four groups: saline-saline, saline-picrotoxin, picrotoxin-saline, and picrotoxin-picrotoxin groups. All groups were tested with odors C, S, and D. The results were identical to those obtained previously (Stopfer et al 1997); all groups discriminated C and D, but only the saline-saline group could distinguish C from S (Hosler et al 2000). We conclude that oscillatory synchronization of AL neurons is functionally relevant for tasks that require fine, but not coarse, odor discrimination (Hosler et al 2000, Stopfer et al 1997). Synchronization thus seems to enable the use of an additional coding dimension, time, which becomes important when recognition requires fine discrimination between overlapping assemblies.

Recent results obtained in mollusks support these conclusions. In *Limax* odors modulate the coherence of ongoing, slow ($\sim$1 Hz) LFP oscillations in its procerebral lobe (Gelperin & Tank 1990). These oscillations can be reversibly suppressed by the NO-synthase inhibitor L-NAME (Gelperin 1994). By scoring the intensity of the neural correlate of an odor-elicited behavior (tentacle positioning), odor discrimination could be assessed upon pharmacological desynchronization of procerebral-lobe networks. Saline-treated preparations were able to discriminate between similar odorants; L-NAME-treated preparations could not. The ability to recognize conditioned odorants per se, however, was not impaired by desynchronization (Teyke & Gelperin 1999). Thus, in honeybees, AL neurons were desynchronized by blocking fast γ-aminobutyric acid-mediated inhibition; in *Limax*, procerebral-lobe neurons were desynchronized by blocking an NO-mediated pathway. In both species, network desynchronization led to the same specific deficit—a loss of precise olfactory discrimination.

Physiological Assay: Projection Neuron Synchronization and Tuning of Downstream Decoders

The above experiments suggested that some individual neurons or groups of neurons downstream of the AL PNs are sensitive to the presence (or absence) of synchronized inputs. We identified in locusts a population of odor-sensitive neurons downstream of the AL. Rather than focus on neurons postsynaptic to PNs (e.g. the Kenyon cells of the mushroom body), we chose neurons postsynaptic to these—a subgroup of mushroom body extrinsic neurons called beta-lobe neurons (β-LNs [MacLeod et al 1998]). The rationale was that PNs are greatly outnumbered by Kenyon cells, whereas Kenyon cells converge to many fewer β-LNs, implying the existence of a bottleneck, potentially useful as a physiological read-out of distributed PN activity.

β-Lobe Neuron Detuning by Input Desynchronization   We compared β-LN odor responses recorded before and after picrotoxin injection into the AL (MacLeod et al 1998). PN de-synchronization caused, in β-LNs, the appearance of responses to odors to which they had been unresponsive before treatment. Conversely, however, desynchronization never led to the disappearance of β-LN odor responses observed before treatment. PN desynchronization also caused a loss of odor specificity in the temporal response patterns observed in controls. Hence, β-LN odor selectivity depends on PN synchronization. This shows that high-order neurons decode temporal correlations between their inputs and use them to fine-tune their sensory properties; information is contained in interneuronal temporal correlations, which cannot be deciphered from serial sampling of neurons within a population. These results also imply that some aspects at least of the information distributed across PNs eventually converge to single neurons (although it does not exclude the possibility that still more information could be retrieved from correlations across β-LNs—such tests have yet to be carried out). Finally, these data provide physiological support for the behavioral experiments in bees (Stopfer et al 1997).

Possible Mechanisms of Coincidence Detection   Because optimal odor tuning in β-LNs depends on PN synchronization, β-LNs or the neurons interposed between them and the PNs—the Kenyon cells—should possess biophysical mechanisms that favor coincident inputs (in the 20-Hz range) over noncoincident ones. While no information exists on putative mechanisms in β-LNs, in vivo intracellular Kenyon cell recordings revealed subthreshold, voltage-dependent, tetrodotoxin-resistant properties, able to amplify PN-evoked excitatory postsynaptic potentials (EPSPs; Laurent & Naraghi 1994). The time constant of the active EPSPs was considerably shorter than the oscillation period (due to as-yet-uncharacterized active repolarizing conductances). A Kenyon cell intracellular response to an appropriate odor thus typically contains, over its consecutive oscillation cycles, some passive (low-amplitude) EPSPs, some active (sharp, high-amplitude) EPSPs, and some action potentials, at an instantaneous frequency generally lower than the oscillation's 20 Hz (Laurent & Naraghi 1994). The dendrites of Kenyon cells, by virtue of this boosting nonlinearity, could thus act as coincidence detectors of synchronized PN inputs.

SHARPENING OF TEMPORAL STRUCTURE THROUGH EXPERIENCE

Response Strength and Information

Much of sensory neurobiology depends on assuming a positive correlation between firing rate and stimulus "preference." The following study in locusts indicated that this need not be the case. Under most environmental conditions, turbulence (Murlis et al 1992) and olfactory behavior (Mellon 1997) discretize odor sampling. We found that the neural representations of odors in the AL change reliably when an animal so experiences a stimulus. By delivering odors using discrete pulses and starting from a naive state, we found that, while the response to the first pulse was the most intense, it lacked the fine temporal definition described above; neither PNs nor LNs showed periodic subthreshold activity; the LFP contained very little power in the 20- to 30-Hz band; and oscillatory coherence between PNs or LNs was low. In summary, the synchronized, evolving ensemble response could not be identified. As odor exposures ensued, however, the ensemble response changed (Figure 2A). PN firing rates declined by half, and strong subthreshold oscillatory ripples appeared in the membrane potentials of PNs and LNs; a sharp 20- to 30-Hz spectral peak emerged in the LFP; the odor-elicited PN spikes, although sparser, became increasingly locked to the LFP; the temporal patterns of relative PN firing (Wehr & Laurent 1996) appeared and stabilized (Figure 2B). Within the delivery of 5 to 10 stimuli, the ensemble responses ceased to evolve further. These changes did not depend on any associative pairing of the odor with a reinforcer. They could also result from a variety of stimulation regimes

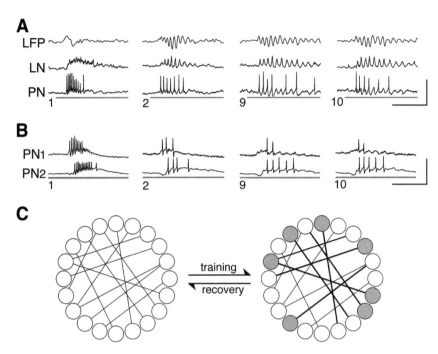

Figure 2 Nonstationarity of network dynamics. Repeated exposure to an odor causes a decrease in response intensity but an increase in oscillatory coherence and spike time precision. (*A*) Simultaneous local field potential (LFP) and intracellular recordings from a local (LN) and projection (PN) neurons during early (1–2) and later (9–10) trials. Horizontal bar indicates odor delivery. Calibration: horizontal, 300 ms; vertical (mV), .8 (LFP), 10 (LN), and 40 (PN). (*B*) From a separate experiment, odor-elicited responses in two simultaneously recorded PNs illustrate increasing spike time precision over successive stimulus trials. Calibration: horizontal, 200 ms; vertical: top trace, 70 mV; bottom trace, 40 mV. (*C*) Putative mechanisms for use-dependent changes in network dynamics; when the naïve AL receives repeated stimulations, only the activated neurons and/or their interconnections undergo (as yet uncharacterized) modifications (training) that endure for several minutes in the absence of further odor stimulation and spontaneously returns to the naïve state (recovery) once stimulation ends (see Stopfer & Laurent 1999).

(e.g. interstimulus intervals of ≤20 s, variable consecutive interstimulus intervals, or fewer but longer individual stimuli). Once established, the state change persisted for several minutes in the absence of further stimulation or even in the presence of interposed stimulation with other odors (Stopfer & Laurent 1999). Thus, this evolution of olfactory network dynamics is the expression of a form of short-term memory, with a time constant of ∼5 to 10 min. A similar phenomenon was observed also in zebrafish (RW Friedrich & G Laurent, manuscript in preparation).

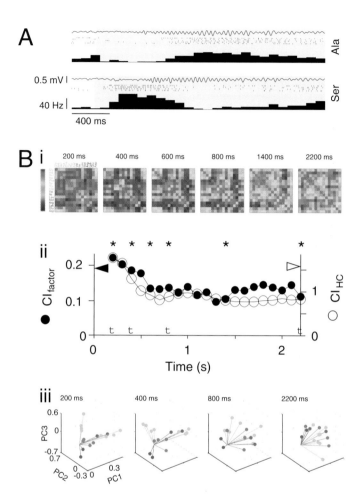

Figure 3 Temporal declustering of odor representations by MC assemblies in the zebrafish olfactory bulb. (*A*) Responses of one MC to 2 of 16 amino acids tested. *Top*, LFP; *middle*, spike raster; *bottom*, peristimulus time histogram. *Gray shadow*, odor stimulation. Note the odor-specific temporal-firing patterns and the late onset of LFP oscillations. (*B*) Quantification of declustering over time. *i*, correlation matrix plotting similarity between odor representations, as measured from a 50-mitral-cell assembly (see RW Friedrich & G Laurent, submitted manuscript). Each matrix is constructed from firing rates measured over a 400-ms-long epoch, starting at odor onset. Times above matrices indicate middle of each epoch (corresponding to * in *ii*). Note clear odor clusters along diagonal in leftmost matrix and their progressive dissolution. *ii*, clustering indices (see above reference for details) as functions of time throughout odor response; declustering reaches steady-state after ~800 ms. *iii*, principal component analysis of same data. Projection of odor representations in the space defined by first three principal components for four epochs (*t* in part *ii*). Note dissolution of vector groups.

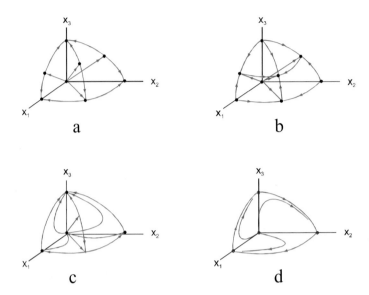

Figure 4 Competitive systems dynamics. Schematic illustrations of four types of behavior. (*a*) Multistability, (*b*) weak competition, (*c*) winner take all, (*d*) winnerless competition.

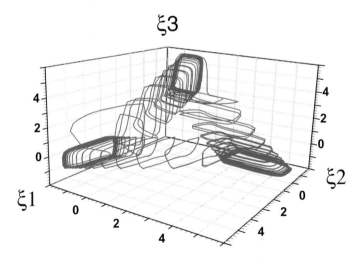

Figure 5 Heteroclinic loop for a simple network of three oscillating projection neurons (PNs), activated by a given input pattern. Each axis maps the activity of one PN through a state variable ξ. The sequence connecting saddle limit cycles and the intervals between them are functions of the stimulus. A different stimulus would thus be represented by a different orbit and thus, a different temporal pattern of activation of the PNs.

Specificity

This memory was odor specific, carrying over to chemically related, but not to chemically distinct odorants. If, for example, oscillatory synchrony had been established by presentations of 1-hexanol, then 1-octanol would immediately elicit a synchronized (although not identical) ensemble response. If, however, the new odorant were geraniol (a terpene), then a naive (strong, unsynchronized) response immediately appeared. If the new odorant were a blend containing the familiar odor, the oscillatory power fell between those of the naïve and established states. This specificity indicates that the cellular or synaptic modifications underlying the phenomenon occur only within the subset of olfactory neurons repeatedly activated by the odor used during training (Figure 2C).

Sites of Changes

In locusts, ORNs are distributed along the lengths of the antennae; we could therefore stimulate separate populations of receptors by selective odor delivery. Once strong network oscillations had been established by repeatedly stimulating one set of receptors with one odor, the same stimulus delivered to naive ORNs gave rise to a trained (sparse and coherent) response pattern (Stopfer & Laurent 1999). Thus, the neural modifications underlying the evolution to a coherent population state must reside, at least in part, within the AL. Finally, this evolution of circuit dynamics from naïve to coherent state itself defined a consistent trajectory: two training sessions with the same odor separated by a long enough interval both produced similar "trained" PN response patterns. This self-organizing network evolution is thus deterministic and likely depends on the network's "basal" connectivity matrix (Figure 2C). The mechanisms underlying this evolution are so far unknown.

Functions

Is this sharpening of distributed odor representations relevant? Our previous work showed that the relative timing of individual PN action potentials contains information about odor identity and that oscillatory synchronization is required for refined olfactory discrimination. Because the network dynamics necessary to gain access to this relational information emerge only after some initial exposure, we predict that repeated sampling should improve behavioral odor discrimination; repeated (naturalistic) experience provides downstream decoders with more informative patterns (individual or corelational) of PN activity. Then, one might ask, why not ensure coherent and temporally precise activity at all times? We suspect that there is a trade-off between quantitative and qualitative sensitivity; unsupervised learning allows a system to have a low detectability threshold for all odors while naïve (at the expense of precision) and to have high discriminability once self-trained. When classification has been made (i.e. once the subcircuits likely to be challenged again by the next sample have been primed), all of the computational resources can be used to refine identification within this limited region of coding space. In the vocabulary of dynamical systems, the orbits representing individual odors progressively increase their relative distances from one another. Finally, because the effects last only a few minutes in the absence of continued stimulation, this short-term memory can be seen as a "working" feature, useful only while the animal either forms a memory or attempts to match an ongoing experience to a stored memory. What representations (early and coarse, late and refined, or possibly all) a brain ends up storing for future use remain unknown. This question probably depends primarily on associative mechanisms (Hammer & Menzel 1995, Heisenberg et al 1985, Smith 1998) and, thus, on the precise time of pairing between odor-evoked activity and reinforcer. It would also be interesting to see whether the time constant of this memory is in any way related to the statistics of odor pulsing experienced by the animal.

Multiple Time Scales of Circuit Dynamics and Consequences

We have identified at least three time scales that are relevant for our understanding of olfactory dynamics in insects and fish (see below): a fast, periodic one (40- to 50-ms time scale) for 20- to 30-Hz oscillations; an intermediate, aperiodic one (hundreds of milliseconds) for slow response patterning; and a slow one (seconds to tens of seconds) for repeated or prolonged odor sampling. The first two are useful to characterize any individual odor response (although oscillatory power will be absent from naïve responses). The third one is useful to describe response sharpening during familiarization. This slow familiarization has several practical consequences for the experimenter. First, it suggests that multiple-trial averaging may often be inappropriate. Second, because of the nature of in vivo physiology, an experimenter usually spends much time searching for a match between neuron and stimulus; this inevitably leads to a modification of the system under study. The degree to which this search time biases the results towards a "familiarized state," not necessarily representative of a natural stimulation sequence, may thus need to be assessed. Third, it is also traditional to interleave trials while studying a neuron's response to a family of stimuli. Given that the refinement of a representation depends on frequent enough samples, some experiments may, contrary to point two above, fail to evoke the natural response evolution.

DYNAMIC DECOMPOSITION OF REPRESENTATION CLUSTERS

Our analyses of odor representations by AL PNs in insects showed that useful information is contained in individual neuron's temporal response patterns as well as in temporal correlations among them. Assume that downstream targets decode this distributed information over some limited time. Could early and late portions of a single response contain different information about the stimulus? Would decoders profit from reading an early or a late segment of the population response? We addressed these possibilities in zebrafish, by first comparing the global structure of amino acid odor representations by MCs, the OB's output, to those carried by their inputs, the ORNs (Friedrich & Korsching 1997, 1998). (Given the nonstationary behavior described above and observed in fish also, we focus here on the changes that occur during single responses to familiar odors.)

Dynamic Declustering of Odor Representations by Mitral-Cell Assemblies

Amino acid representations by zebrafish ORNs are clustered: related odors evoke overlapping patterns of glomerular activity (Friedrich & Korsching 1997), imposing that fine discrimination rely on small differences between within-cluster representations. We first examined whether clustering was found also among amino acid representations by MCs. We characterized the responses of 50 MCs to 16 amino acids and represented each odor by a 50-dimensional vector, constructed using mean response rates. Whereas small clusters could be found, they were fewer and much less distinct than those observed among the afferent representations (RW Friedrich & G Laurent, submitted manuscript). We then examined whether time played a role in the observed cluster reduction.

MC responses to amino acids are temporally structured (Figure 3A, see color insert); while each neuron usually responds to several odors, it often responds to many with different patterns. As in the insect AL, therefore, odor-encoding neuron assemblies in the fish OB evolve over the stimulus duration. Clustering of odor representations by MCs was thus now analyzed over the response duration; each odor was represented by a sequence of 50-dimensional vectors, with each vector in the sequence constructed from MC firing rates calculated over a short window (e.g. 400 ms), scrolled in 50- or 100-ms steps. Potential changes in odor representations by MCs could thus be examined by following the grouping of those vectors over the duration of a response (Figure 3B, see color insert). This procedure revealed a dramatic evolution; initial MC responses were markedly clustered, with odor groupings identical to those observed using afferent data. Over the first 800 ms of the response, however, clusters disappeared; representations that were initially similar became decorrelated while dissimilar ones became slightly more correlated. Odor representations by MC assemblies are therefore more distinct from one another than those across ORNs, but only if the response patterns are given enough time to evolve (several hundred milliseconds); time thus seems to underlie a self-organized transformation of odor representations in the OB (RW Friedrich & G Laurent, submitted manuscript).

Functions

Declustering did not result simply from an increase in MC response variability over time; the standard deviation of MC responses over trials decreased, if anything, from beginning to end of a response. In addition, odor identification by matching single-trial odor vectors to templates constructed from other single trials became more reliable as the response progressed. This improvement matched the temporal progress of declustering. Hence, representation declustering facilitates fine odor discrimination, given the uncertainties introduced by noise, by making each distributed pattern less similar to any of those evoked by other odors, including chemically related ones. In other words, redundancy is reduced by better occupying coding space. Concurrently, however, useful information about odor classes (clusters) is lost as a response progresses. The OB thus appears to solve two antagonistic problems of pattern recognition, using circuit dynamics: early responses, based (passively) on the afferents' responses, provide information relevant for perceptual grouping; later responses, resulting from afferent input and active reformatting by the OB, provide information relevant for fine identification (RW Friedrich & G Laurent, submitted manuscript). These results reinforce three key points introduced earlier: first, stimulus representations should be studied over cell assemblies rather than single cells; second, the resolution needed to uncover these phenomena in the OB/AL is the single neuron; third, responses are not stationary and their odor-driven evolution is not random; rather, it appears to be optimized to facilitate recognition, without complete loss of grouping information.

Interesting studies in the frog OB (Duchamp & Sicard 1984) seem, at first glance, to contradict our results. In these accounts, clustering of odor representations by ORNs was less than that observed across MCs. This study, however, did not use natural odor stimuli (for which each system may have been optimized) and did not analyze representations as a function of time. In addition, the clusters found each corresponded to very different chemical classes; the clusters found by us, by contrast, might be seen as subgroups within one of their clusters (amino acids). Declustering might thus be confined to representations within such larger classes, which, in fish at least, are processed within specialized subregions of the OB (Friedrich & Korsching 1998). Indeed, in the frog, representations within a cluster appear to be less similar across MCs than across ORNs, whereas the clusters to which they belong are better defined across MCs than ORNs (Duchamp-Viret & Duchamp 1997). It would be interesting to see these data reanalyzed over response time.

Dynamic transformations of stimulus representations occur also in other sensory systems. In monkey inferotemporal cortex, the information content of responses of face-selective neurons changes over time (Sugase et al 1999). Early response phases contain information mainly about stimulus categories (e.g. distinguishing monkey faces from human faces or abstract objects), whereas later phases contain more information about details (e.g. individual monkey faces and facial expressions). Although responses were analyzed in single neurons only, these results support the contention that circuit dynamics underlies a refinement in stimulus identification.

Mechanisms

Noise Reduction   Because this temporal declustering process was not observed among the afferents, it must result from OB circuit dynamics. Because the evolution of a representation is stimulus specific, it must rely also specifically on the stimulus. Overall, this phenomenon appears to amplify, through OB-driven processes, the small differences between odor representations first formed by afferent arrays. This suggests that stimulus noise must be efficiently controlled so as to amplify the signal preferentially. Hence, efficient noise-reducing mechanisms, possibly including the spatial averaging hypothesized earlier and chemical low-pass filters (Pelosi 1996), must exist at the earliest levels of odor processing.

Oscillatory Synchronization   Zebrafish MC responses to odors also caused 20-Hz LFP oscillations. Oscillatory synchronization, however, generally lagged the odor-induced population firing rate increase by ~500 ms. Hence, the earliest phase of odor-evoked MC activity, that for which odor clusters could be defined, was desynchronized (RW Friedrich & G Laurent, submitted manuscript). The process of cluster dissolution, by contrast, was accompanied by the progressive development of oscillatory synchronization. Because oscillatory synchronization results from local feedback within the OB (Gray & Skinner 1988), we conclude that the reformatting of odor representations is correlated with the development of OB circuit dynamics shaped by lateral interactions. Whether this correlation is causal remains to be determined.

Lateral Inhibition vs Redistribution   One form of lateral interaction that might underlie both synchronization and declustering is GC-mediated lateral inhibition, in its classical sense (i.e. as a means to sharpen MC tuning, DeVries & Baylor 1993, Mori & Shepherd 1994). This hypothesis was tested in two ways. First, we assessed the sparseness (Rolls & Tovee 1995, Vinje & Gallant 2000) of odor representations as a function of time, over single responses. If classical lateral inhibition were responsible for declustering, one would predict (*a*) that the average representation sparseness across MCs would be more than across ORNs (MC representations should be "sharper" than across ORNs) and (*b*) that the sparseness of MC representations should increase over response duration (as the better-tuned MCs inhibit those less well tuned). Neither of the two predictions was met (RW Friedrich & G Laurent, submitted manuscript). Second, we examined the representation of all amino acids across five MCs, selected because of their similar early responses to three related amino acids; these five MCs, therefore, defined an early cluster. By simply following the "tuning" of these five cells to all 16 amino acids over the response duration, we observed no systematic sharpening or strengthening of initial responses. Rather, we noted a clear redistribution of activity of all five MCs across all 16 odors, in ways that could not be predicted from the odors' structures. The only clear trend was for a disappearance of the initial cluster, contradicting a prediction of the "lateral-inhibition" hypothesis. We conclude that classical lateral inhibition, the evidence for which is lacking, does not underlie declustering. To be very explicit, we do not hereby deny the existence of GC-mediated inhibition; rather we mean that such inhibition does not, as far as we can assess, mediate the sharpening of single-MC odor tuning, as is usually proposed.

In short, the declustering of odor representations involves odor-induced OB activity, including lateral interactions between MCs and other neurons. The consequences of this activity are not to sharpen MC response profiles but rather to redistribute activity within the OB. This redistribution is odor specific and such that late phases of the response contain the activity of more unique (i.e. more specific) MC assemblies. Response sharpness indeed increases, but only when measured over the cell population. The OB's internal circuits thus exploit stimulus-driven dynamics so as to optimize odor encoding during a response.

THEORETICAL FRAMEWORK: WINNERLESS COMPETITION

Physiological and imaging data show that odor encoding in the OB/AL is a complex, distributed dynamical process occurring over several time scales. We introduce here a theoretical framework for these observations but limit our description, for brevity's sake, to the slow evolution of OB/AL networks upon stimulation with a familiar odor (i.e. addressing deterministic, stimulus-specific slow-firing patterns in PNs and MCs, odor-specific intercell firing correlations, sustained evolution of the network during an odor, spontaneous return to baseline upon stimulus termination, and pattern reliability over multiple trials, but excluding cycle-by-cycle network update and unsupervised learning over repeated presentations of a new odor).

Introduction to the Language of Nonlinear Dynamics

Nonlinear dynamics (NLD) is useful to describe neuronal-network function because it can often explain much of the observed complexity with simple models and few assumptions. It also provides a qualitative, geometric view of the behavior of a system of interdependent parts, reducing it to only a few dimensions, in which one's intuition can be trained and applied (Figure 4, see color insert). The critical aspect of NLD models is that they present a theoretical foundation in which instabilities, such as those in which small changes in state are amplified rapidly, are permitted. Critically, dissipation and saturation (see below) constrain a system's motions such that a return to the starting regions of state space is possible. Consequently, a system's motion can remain within compact regions of state space and yet retain an unstable flavor. This aspect, characteristic of nonlinear systems only and well suited for olfactory processing, is central to the formal description proposed below.

The language of dynamical systems is that of differential equations or discrete time maps. The idea of instability is easily explained with a one-dimensional differential equation in a variable $x_i(t) > 0$,

$$dx_i/dt = aS^n x_i(t) - x_i^2(t) + S_i^s,$$

1

where $x_i(t)$, S_i^s, and S^n represent, respectively, the firing rate of neuron i, the stimulus current provided to neuron i, and a trigger function for neuron i. Consider the simple case where $S_i^s = 0$ (no external stimulus, i.e. an "autonomous" system) and $S^n = 1$. If $a < 0$, the time derivative of x_i is negative, and $x_i(t)$ rapidly goes to 0, whatever its initial value. If $a > 0$, however, the state $x_i(t) = 0$ is unstable; any state starting near $x_i = 0$ moves away from it, grows exponentially rapidly towards $x_i = a$, and stops there. $x^2(t)$, a very simple nonlinearity, shows that instabilities in nonlinear systems can lead to change without disaster. This example illustrates a second important point: If $a > 0$ and the starting state is $x(0) > a$, the system also moves to $x(t) = a$. This is the simplest example of a dynamical system with an attractor which depends on one parameter: If $a < 0$, the attractor is the origin; if $a > 0$, the attractor is $x = a$. In both cases, the attractors are fixed points.

The single dynamic variable $x(t)$ is not adequate to describe or understand how complex neuron assemblies might work. With more dynamical variables or degrees of freedom—that is, higher dimensional differential equations—richer behaviors may arise. For example, with three or more dynamical variables, $[x(t), y(t), z(t), \ldots]$, in addition to time-independent states ($x = 0$ or $x = a$ in the above example) and periodic oscillations $\{[x(t + T), y(t + T), z(t + T), \ldots] = [x(t), y(t), z(t), \ldots]\}$, nonperiodic behaviors become possible; these characterize chaotic motions of a system. In each of these cases, instabilities in some region of state space (e.g. $x = 0$ when $a > 0$) play a central role. With chaotic oscillations, for example, there is an infinite set of unstable points, and the system evolves from one unstable region of state space to another. We can classify these instabilities by an exponent, indicative of how rapidly two nearby states move away from each other. In our simple example, near $x = 0$ and when $a > 0$, the solution to the state evolution equation is $x(t) \approx x(0)e^{at}$. The growth index a is called a Lyapunov or instability exponent. In a multidimensional system, there is an index for each dimension of state space. In chaotic, dissipative, and long-term motions, one or more of these exponents is positive, some are negative, and the sum of all is always negative.

Olfactory Networks as Dissipative and Active Systems

Plasticity notwithstanding, we observe that, to be useful to the animal, odor representation should be reproducible, that is, by and large insensitive to the network's initial state (internal noise). In the language of NLD, this is possible only if the system is strongly dissipative, in other words, if it can rapidly forget its initial state. In dissipative systems, the initial phase volume is rapidly compressed and all trajectories converge to attractors (fixed points, closed trajectories or limit cycles, strange attractors, or other specific trajectories such as homoclinic/heteroclinic trajectories). On the other hand, a useful olfactory system should be sensitive to small variations in the inputs, so that fine discriminations between similar but not identical stimuli are possible. How can both conditions coexist? This is only possible if a system is active; an active system uses external sources of energy to increase a small distance between initial states (representations) caused by similar stimuli, independent of the initial state of the network. Because the system is active, small initial differences between representations can grow rapidly over time. Time thus plays a critical role in separating representations. The OB/AL should therefore behave as active, nonlinear, dissipative systems. Although many existing models of early olfactory systems satisfy these conditions, these models present limitations, briefly examined below.

Coding with Attractors

The most common nonlinear dynamical models of early olfactory processing (representation and recognition) lead to the idea of "coding with attractors" ("Hopfield nets" [Cohen & Grossberg 1982, Hopfield 1982]). In these models, each stimulus or odor is represented by a specific behavior (attractor) of the neural network. The behavior expressed depends on the connections among the network's elements. The number of different attractors determines the number of different stimuli that the system can represent or recognize. Because many stable states must coexist, the system is called multistable (Figure 4a, see color insert). Each attractor possesses its own basin of attraction, and different basins are separated from each other by boundaries whose shapes can be complex (usually fractal). The main idea behind a Hopfield network is that, during a learning stage, a long-acting stimulus is used to modify specific sets of connections until a steady-state behavior (attractor) specific for that stimulus is obtained. Through training, each stimulus thus generates its own attractor. After learning, this trained, autonomous system becomes multistable (the unstimulated system possesses several basins of attraction); for recognition, future incoming stimuli will simply play the role of initial conditions. Hence, although a Hopfield network is a dynamical system, it is static (or stationary) after convergence. Time does not play an intrinsic role in the encoding or decoding of the input. This appears to conflict with experimental results on early olfactory processing, as reported above.

A very important feature of such a network is that, during recognition, the evolution of the system within a basin of attraction is resistant to corruption of the input (noise or missing features). Conversely, however, the number m of stimuli (attractors) allowed in a system of N neurons is relatively limited: $m < 0.14 N$ (Hertz et al 1991). This limit is due to increasingly complex boundaries between basins of attraction as the number of desired stimuli rises; small mistakes, such as ones resulting from noise in the initial conditions, lead to increasing error rates. Such systems therefore have hard capacity limits. For these reasons, we think that such models, although very important from a conceptual point of view, do not capture some essential features of biological olfactory systems.

Population Models and Chaos

Walter Freeman and collaborators have, over the years, followed a different approach, merging experiments and NLD theory. They developed dynamical models that simulate the global behavior of neuron population (Freeman 1975, 1992, 2000) as manifested in EEG recordings from the mammalian OB. These models of population dynamics (called "mesoscopic" models because of their intermediate scale) reproduce several important aspects of EEG data. For example, at rest, the models produce irregular oscillations whose origin lies in dynamical chaos. The mathematical signature of this behavior is called a strange attractor. There is now much experimental evidence that EEG time series can indeed sometimes be interpreted as expressions of dynamical chaos (e.g. Elbert et al 1994). In addition, Freeman's models generate transient and reproducible waveforms that uniquely depend on the stimulus; they can thus be used for recognition. These models therefore have a true temporal component. On the other hand, they do not explicitly model individual neurons, groups of synchronized neurons, or network topology (Freeman 1987, Yao & Freeman 1990). They propose to explain macroscopic signals (as indicated by EEG) whose functional relevance per se is debatable and, in any case, difficult to assess without explicit description of their underlying causes. Hence, we find it difficult to transpose these spatiotemporal models into the biological realm.

A New Theoretical Framework: "Winnerless Competition"

The simple model in Equation 1 embodies part of what is needed for a dynamical system with input-specific behavior. If $x_i(t)$ and a, respectively, represent neuron i's firing rate and the state of the input ($a \leq 0$, stimulus off; $a \geq 0$, stimulus on), then the neuron can move between 0 (stimulus off) and a (stimulus on). This system, however, has a very limited capacity. To encode many stimuli, the dynamics should allow stimulus-specific trajectories $x_i(t)$ in state space. We introduce the conditions required for such dynamics.

First, many neurons are required. For the sake of simplicity, $x_i(t)$ will represent the firing rate of the ith neuron or, collectively, that of the ith group of synchronized neurons. The functional unit of the network is thus either the neuron or a small group of coactive, synchronized neurons. The identity of the elements of a group may change over time, in which case the identity of the group will similarly evolve. The mechanisms underlying this evolution will become clear as we proceed.

Second, these neurons (or groups) must interact at least in part through inhibitory connections. If $x_i(t)$ and $x_j(t)$ characterize the activities of groups i and j, respectively, the system's behavior can be described by

$$\frac{dx_i}{dt} = x_i(t)\left[1 - \sum_{j=1}^{N} \rho_{ij} x_j(t)\right] + S_i^s, \qquad 2$$

where $\rho_{ij} > 0$ characterizes the strength of inhibition of i by j and S_i^s is the input current contributed by stimulus s to i. The simplest dynamics arise if these inhibitory connections are symmetrical ($\rho_{ij} = \rho_{ji}$). In this case, the autonomous system ($S_i^s = 0$) is "potential" or gradient, and Equation 2 can be rewritten as

$$\frac{dx_i}{dt} = -\frac{\partial F}{\partial x_i} \qquad 3$$

and

$$\frac{dF}{dt} \leq 0, \qquad 4$$

where F is the free energy or Lyapunov function. Equation 4 implies that the system cannot generate complex temporal patterns; rather, free energy must decrease monotonically along any trajectory in state space or remain constant; the system can only move to a local minimum and stay there. If $\rho_{ij} > \rho_{ii}$, the free-energy landscape for a system with N neurons (or groups) has N or fewer minima and behaves as a multistable Hopfield network. If $\rho_{ij} < \rho_{ii}$, the system has only one global attractor, corresponding to simultaneous activity of all N elements ("weak competition") (Figure 4b, see color insert).

If connections are not symmetrical, two cases arise, depending on the nature of the asymmetries. If the asymmetry is only partial (e.g. for $N = 3$, $\rho_{12} \sim \rho_{21} > 1$; $\rho_{13}, \rho_{23} > 1$; $\rho_{31}, \rho_{32} < 1$), only one attractor, corresponding to the activity of one neuron, will result (Figure 4c, see color insert). This is a "winner-take-all" circuit (e.g. Yuille & Grzywacs 1989). These and Hopfield-like networks are well studied in dynamical systems theory (Morse-Smale systems [e.g. see Guckenheimer & Holmes 1983]) and generally display behaviors too simple to account for experimental data. If the asymmetries are cyclic ($\rho_{ij} > \rho_{ii}$, but $\rho_{ji} < \rho_{ii}$; e.g. for $N = 3$, ρ_{12}, ρ_{23}, and $\rho_{31} > 1$, and ρ_{21}, ρ_{32}, and $\rho_{13} < 1$) however, more interesting behaviors arise in which activity "bounces off" between neurons (groups), in a stimulus-dependent manner (Figure 4d, see color insert). We call this behavior "winnerless competition". Its geometrical description is a heteroclinic orbit, that is, a trajectory linking quasi-stationary states within phase space (Figure 5, see color insert). One critical point, developed further below, is that heteroclinic orbits are very sensitive to the stimulus. Consequently, each orbit can be associated with a stimulus and thus "encode" it. A second (related) critical point is that the state space of the system is enlarged upon reception of the input. We propose that this form of nonlinear dynamical behavior best describes odor-evoked activity in the early olfactory system. How and why does it work?

Closed Topology

An important condition for this form of dynamical behavior is that the network topology (connectivity) be closed. This condition is fulfilled by AL and OB circuits, whose synaptology is well known (Masson & Mustaparta 1990, Scott 1986, Shipley & Ennis 1996); a mitral cell, for example, will excite other mitral or granule cells which in turn will feed back onto it through mono- or polysynaptic pathways. In addition, if reciprocal connections exist between local inhibitory neurons and the principal neurons that excite them, those inhibitory connections should not be so powerful as to cause, on their own, an immediate silencing of the principal neurons. This condition appears to be fulfilled in the locust AL, where no evidence for functional reciprocal inhibitory connections has been found either morphologically (Leitch & Laurent 1996) or physiologically from recordings of connected pairs (MacLeod & Laurent 1996; G Laurent, personal observation). In the vertebrate OB, well-documented reciprocal inhibition via granule cells never prevents prolonged MC firing caused either by an odor stimulus or by direct depolarizing current injection (Isaacson 1999, Isaacson & Strowbridge 1998, Nicoll & Jahr 1982). These features are important because they act to disperse activity across the neural assembly once a few elements have been activated by a stimulus.

Activity thus proceeds across parts of the network by sequentially activating and deactivating subgroups of neurons. Individual neurons can belong to several subgroups recruited at different times during the stimulation, and different subgroups join and depart over successive, sometimes overlapping epochs. Hence, at any time during the stimulation, each neuron may receive inputs from the afferents or from other neurons in the network and thereby contribute, by its own activity combined with that of the afferents, to the excitation and inhibition of other neurons. This activation path within the network (or, correspondingly, this heteroclinic orbit in the state space that describes it) is determined by and thus "represents" the stimulus. Once the stimulus ceases (either because it was withdrawn or because of receptor adaptation), each active neuron returns to its baseline activity, controlled by intrinsic properties, basal-connection strengths, and noise; correspondingly, the system returns to the neighborhood of the origin in state space. A stimulus can thus be thought of as a perturbational or informational signal that reorganizes the global attractor in a stimulus-specific manner, forcing the system to evolve through state space along a complex, but deterministic path joining unstable "saddle states." Once the stimulus disappears, the system resets itself autonomously, because no stable attractor other than the resting state could be found. Hence, the system's only attractors are the stimulus-evoked heteroclinic orbits (one for each stimulus) or, when no stimulus is present, the origin.

Advantages of Winnerless Competition

This dynamical-systems description of early olfactory processing took shape as a means to explain experimental data (Laurent et al 1996, Wehr & Laurent 1996; RW Friedrich & G Laurent, submitted manuscript). Although its global features appear to match the data, our description should also provide insights as to what is gained by such dynamical behavior.

Global Stability Winnerless competition provides global stability to the representation. Although each neuron or subgroup participating in the representation might, left to its own devices, wander during a response, the effect of the population behavior is to provide local and temporary stability to each, by attracting them to the neighborhood of a succession of quasi-stable saddle states. Each neuron's local activity thus inherits the global stability of the heteroclinic orbit defined by the system. This stability can be verified by examining the reproducibility of stimulus-evoked trajectories in model networks of simplified LNs and PNs (MI Rabinovich, R Huerta, A Volkovskii, HDI Abarbanel, G Laurent, in preparation) or in experimental data (Laurent et al 1996, Wehr & Laurent 1996).

Sensitivity   Such dynamics are very sensitive to the forcing stimulus. This is because the heteroclinic linking of a specific set of saddle points is always unique. Two like stimuli, activating greatly overlapping subsets of a network, may thus become easily separated because small initial differences will become amplified in time. This feature is central to our experimental observations (RW Friedrich & G Laurent, submitted manuscript; Stopfer & Laurent 1999) and is characteristic of an active, dissipative, nonlinear dynamical system. We do not know yet, however, whether divergence can be optimized by fine-tuning of the internal connectivity. Intermittent-stimulation experiments indeed suggest that short-term changes can increase the separability of odor representations (Stopfer & Laurent 1999). The rules for such unsupervised improvements are so far unknown.

Capacity   A heteroclinic (spatiotemporal) representation provides greatly increased capacity to the system. Because sequences of activity are combinatorial across neurons and time, overlap between representations can be reduced, and the distance in phase space between orbits can be increased, thus reducing the effects of noise. This feature, given by a dynamic representation, is absent from a Hopfield-like network, where the intricacy of the boundaries between neighboring basins of attraction can easily lead to classification errors.

Implementation Details   Finally, does the expression of such dynamics depend on the details of the network or neuron implementation? Indeed, to be general, this theory should accommodate the many idiosyncrasies of OBs, ALs, and other equivalent circuits and rely on broad organizational principles (Hildebrand & Shepherd 1997). Preliminary tests indicate that this is indeed the case; winnerless competition arises from a few simple and biologically relevant rules.

GENERAL CONCLUSIONS

The thrust of this review is that the transfer of odor-evoked signals from receptors to OB/AL circuits is accompanied by a fundamental reformatting of odor representations. This reshaping of odor codes results from the internal connectivity of early olfactory circuits and from the global dynamics that these connections produce. This reshaping is useful in that it removes ambiguities about stimulus identity, exploiting time as an additional coding dimension. We do not mean to suggest, however, that time is the only coding dimension for olfaction. Rather we tried to explain that neural identity and temporal recruitment of neurons are, in these circuits at least, two sides of the same coin; because the OB/ALs are networks of interacting neurons and because odor stimuli are usually sustained, the structure of odor-encoding neural assemblies is shaped both by the stimulus and by the distributed patterns forced by the stimulus. We propose that odor-encoding neural assemblies in the OB/AL are dynamical, by virtue of the physical laws that govern such systems' behavior (winnerless competition).

This perspective has several practical correlates. First, it suggests that, because information is distributed across neural assemblies, traditional measures such as single-neuron tuning should be appropriately weighted. By analogy, take a symphonic piece in which the piccolo plays one measure only, but one whose role is critical to the melodic line that emerges from the orchestra. The relevance of the piccolo to the global output is accurately conveyed not by the strength of its contribution but by its relation to the other instruments' output. Melodic representation is, in this context, a property of the ensemble. Second, this perspective (and our results) casts doubts on the traditional ("single-neuron") functional interpretation of lateral inhibitory contacts. Third, our results demonstrate the importance of system nonstationarity at several time scales. The critical observation is that, over time and without supervision, olfactory networks become increasingly attuned to a stimulus's precise identity.

Finally, we must emphasize that many important aspects (e.g. intensity coding, decoding mechanisms, and the role of high-level features such as expectation) remain to be studied in detail, incorporated, and tested within this framework. Similarly, physiological studies on larger vertebrates such as reptiles and mammals show a greater phenomenological complexity. These features will eventually also need to be accounted for. Our hope is that a computational/theoretical perspective combined with experiments on small and tractable systems can help both in framing questions of general relevance and in designing more direct experimental tests.

ACKNOWLEDGMENTS

The work described here was supported by the NIDCD, the Sloan Center for Theoretical Neuroscience at Caltech, the Keck Foundation, the McKnight Foundation (GL), and the US Department of Energy, Office of Basic Energy Sciences, Division of Engineering and Geosciences (HDIA, MIR).

Visit the Annual Reviews home page at www.AnnualReviews.org

LITERATURE CITED

Adrian ED. 1942. Olfactory reactions in the brain of the hedgehog. *J. Physiol.* 100:459–73

Adrian ED. 1950. The electrical activity of the olfactory bulb. *Electroencephalogr. Clin. Neurophysiol.* 2:377–88

Adrian ED. 1953. Sensory messages and sensation. The responses of the olfactory organ to different smells. *Acta Physiol. Scand.* 29:5–14

Aroniadou-Anderjaska V, Ennis M, Shipley M.

1999. Dendrodendritic recurrent excitation in mitral cells of the rat olfactory bulb. *J. Neurophysiol.* 82:489–94

Attneave F. 1954. Informational aspects of visual perception. *Psychol. Rev.* 61:183–93

Axel R. 1995. The molecular logic of smell. *Sci. Am.* 273:130–37

Barlow H. 1969. Pattern recognition and the responses of sensory neurones. *Ann. NY Acad. Sci.* 156:872–81

Beuerman R. 1975. Slow potentials of the turtle olfactory bulb in response to odor stimulation of the nose. *Brain Res.* 97:61–78

Bitterman M, Menzel R, Fietz A, Schaefer S. 1983. Classical conditioning of proboscis extension in honeybees (*Apis mellifera*). *J. Comp. Psychol.* 97:107–19

Boeckh J, Tolbert LP. 1993. Synaptic organization and development of the antennal lobe in insects. *Microsc. Res. Tech.* 24:260–80

Bozza TC, Kauer JS. 1997. Odorant response properties of convergent olfactory receptor neurons. *J. Neurosci.* 18:4560–69

Buck LB. 1995. Unraveling chemosensory diversity. *Cell* 83:349–52

Buck LB. 1996. Information coding in the vertebrate olfactory system. *Annu. Rev. Neurosci.* 19:517–44

Buck LB, Axel R. 1991. A novel multigene family may encode odorant receptors: a molecular basis for odor recognition. *Cell* 65:175–87

Burrows M, Boeckh J, Esslen J. 1982. Physiological and morphological properties of interneurons in the deutocerebrum of male cockroaches with responses to female pheromones. *J. Comp. Physiol. A* 145:447–57

Cain WS, Potts BC. 1996. Switch and bait: probing the discriminative basis of odor identification via recognition memory. *Chem. Senses* 21:35–44

Chen WR, Midtgaard J, Shepherd GM. 1997. Forward and backward propagation of dendritic impulses and their synaptic control in mitral cells. *Science* 278:463–67

Cinelli AR, Hamilton KA, Kauer JS. 1995. Salamander olfactory bulb neuronal activity observed by video rate, voltage-sensitive dye imaging. III. Spatial and temporal properties of responses evoked by odorant stimulation. *J. Neurophysiol.* 73:2053–71

Cinelli AR, Kauer JS. 1992. Voltage-sensitive dyes and functional activity in the olfactory pathway. *Annu. Rev. Neurosci.* 15:321–51

Cohen M, Grossberg S. 1982. Neural networks and physical systems with emergent computational abilities. *Proc. Natl. Acad. Sci. USA* 79:2554–58

DeVries SH, Baylor DA. 1993. Synaptic circuitry of the retina and olfactory bulb. *Cell* 10:139–49

Dorries KM, Kauer JS. 2000. Relationships between odor-elicited oscillations in the salamander olfactory epithelium and olfactory bulb. *J. Neurophysiol.* 83:754–65

Dryer L, Graziadei PPC. 1994. Mitral cell dendrites: a comparative approach. *Anat. Embryol.* 189:91–106

Duchamp A. 1982. Electrophysiological responses of olfactory bulb neurons to odour stimuli in the frog: a comparison with receptor cells. *Chem. Senses* 7:191–210

Duchamp A, Revial M, Holley A, MacLeod P. 1974. Odor discrimination by frog olfactory receptors. *Chem. Senses* 1:213–33

Duchamp A, Sicard G. 1984. Odour discrimination by olfactory bulb neurons: statistical analysis of electrophysiological responses and comparison with odour discrimination by receptor cells. *Chem. Senses* 9:1–14

Duchamp-Viret P, Duchamp A. 1997. Odor processing in the frog olfactory system. *Prog. Neurobiol.* 53:561–602

Dulac C. 1997. Molecular biology of pheromone perception in mammals. *Semin. Cell Dev. Biol.* 8:197–205

Elbert T, Ray W, Kowalik A, Skinner J, Graf K, Birbaumer N. 1994. Chaos and physiology: deterministic chaos in excitable cell assemblies. *Physiol. Rev.* 74:1–47

Field D. 1987. Reactions between the statistics of natural images and the response properties of cortical cells. *J. Opt. Soc. Am. A* 4:2379–94

Field D. 1994. What is the goal of sensory coding? *Neural Comput.* 6:559–601

Freeman WJ. 1975. *Mass Action in the Nervous System.* New York: Academic

Freeman WJ. 1978. Spatial properties of an EEG event in the olfactory bulb and cortex. *Electroencephalogr. Clin. Neurophysiol.* 44:586–605

Freeman WJ. 1987. Simulation of chaotic EEG patterns with a dynamic model of the olfactory system. *Biol. Cybern.* 56:139–50

Freeman WJ. 1992. Tutorial on neurobiology: from single neurons to brain chaos. *Int. J. Bifurcation Chaos* 2:451–82

Freeman WJ. 2000. *Neurodynamics: An Exploration in Mesoscopic Brain Dynamics.* London: Springer. 397 pp.

Freeman WJ, Skarda CA. 1985. Spatial EEG patterns, non-linear dynamics and perception: the neo-Sherringtonian view. *Brain Res.* 357:147–75

Friedrich RW, Korsching SI. 1997. Combinatorial and chemotopic odorant coding in the zebrafish olfactory bulb visualized by optical imaging. *Neuron* 18:737–52

Friedrich RW, Korsching SI. 1998. Chemotopic, combinatorial and noncombinatorial odorant representations in the olfactory bulb revealed using a voltage-sensitive axon tracer. *J. Neurosci.* 18:9977–88

Gelperin A. 1994. Nitric oxide mediates network oscillations of olfactory interneurons in a terrestrial mollusc. *Nature* 369:61–63

Gelperin A, Tank DW. 1990. Odour-modulated collective network oscillations of olfactory interneurons in a terrestrial mollusc. *Nature* 345:437–40

Getchell TV. 1986. Functional properties of vertebrate olfactory receptor neurons. *Physiol. Rev.* 66:772–818

Gray C. 1994. Synchronous oscillations in neuronal systems: mechanisms and function. *J. Comput. Neurosci.* 1:11–38

Gray C, Skinner J. 1988. Centrifugal regulation of neuronal activity in the olfactory bulb of the waking rabbit as revealed by reversible cryogenic blockade. *Exp. Brain Res.* 69:378–86

Guckenheimer J, Holmes P. 1983. *Nonlinear Oscillations, Dynamical Systems and Bifurcations of Vector Fields,* pp. 64–453. New York: Springer-Verlag

Haberly LB. 1990. Comparative aspects of olfactory cortex. In *Cerebral Cortex,* ed. EG Jones, A Peters, pp. 137–66. New York: Plenum

Hamilton KA, Kauer JS. 1989. Patterns of intracellular potentials in salamander mitral/tufted cells in response to odor stimulation. *J. Neurophysiol.* 62:609–25

Hammer M. 1993. An identified neuron mediates the unconditioned stimulus in associative olfactory learning in honeybees. *Nature* 366:59–63

Hammer M, Menzel R. 1995. Learning and memory in the honeybee. *J. Neurosci.* 15:1617–30

Heinbockel T, Kloppenburg P, Hildebrand J. 1998. Pheromone-evoked potentials and oscillations in the antennal lobes of the sphinx moth *Manduca sexta. J. Comp. Physiol. A* 182:702–14

Heisenberg M. 1998. What do the mushroom bodies do for the insect brain? An introduction. *Learn. Mem.* 5:1–10

Heisenberg M, Borst A, Wagner S, Byers D. 1985. Drosophila mushroom body mutants are deficient in olfactory learning. *J. Neurogenet.* 2:1–30

Hertz J, Krogh A, Palmer R. 1991. *Introduction to the Theory of Neural Computation.* Santa Fe: Addison-Wesley

Hildebrand JG, Shepherd GM. 1997. Mechanisms of olfactory discrimination: converging evidence for common principles across phyla. *Annu. Rev. Neurosci.* 20:595–631

Hopfield JJ. 1982. Neural networks and physical systems with emergent computational abilities. *Proc. Natl. Acad. Sci. USA* 79:2554–58

Hopfield JJ. 1995. Pattern recognition computation using action potential timing for stimulus representation. *Nature* 376:33–36

Hosler J, Buxton K, Smith B. 2000. Impairment of olfactory discrimination by blockade of GABA and nitric oxide activity in the honey bee antennal lobes. *Behav. Neurosci.* In press

Hubel D, Wiesel T. 1968. Receptive fields and functional architecture of monkey striate cortex. *J. Physiol.* 195:215–43

Hughes J, Mazurowski J. 1962. Studies on the supracallosial mesial cortex of unanesthetized, conscious mammals. II. Monkeys. B. Responses from the olfactory bulb. *Electroencephalogr. Clin. Neurophysiol.* 14:635–45

Imamura K, Mataga N, Mori K. 1992. Coding of odor molecules by mitral/tufted cells in rabbit olfactory bulb. I. Aliphatic compounds. *J. Neurophysiol.* 68:1986–2002

Isaacson JS. 1999. Glutamate spillover mediates excitatory transmission in the rat olfactory bulb. *Neuron* 23:377–84

Isaacson JS, Strowbridge BW. 1998. Olfactory reciprocal synapses: dendritic signaling in the CNS. *Neuron* 20:749–61

Joerges J, Küttner A, Galizia CG, Menzel R. 1997. Representations of odours and odour mixtures visualized in the honeybee brain. *Nature* 387:285–88

Kang J, Caprio J. 1995. In vivo responses of single olfactory receptor neurons in the channel catfish, *Ictalurus punctatus. J. Neurophysiol.* 73:172–77

Katoh K, Koshimoto H, Tani A, Mori K. 1993. Coding of odor molecules by mitral/tufted cells in rabbit olfactory bulb. II. Aromatic compounds. *J. Neurophysiol.* 70:2161–75

Kauer JS. 1987. Coding in the olfactory system. In *Neurobiology of Taste and Smell*, ed. TE Finger, WL Silver, pp. 205–22. New York: Wiley

Kauer JS. 1991. Contributions of topography and parallel processing to odor coding in the vertebrate olfactory pathway. *Trends Neurosci.* 14:79–85

Kauer JS, Moulton D. 1974. Responses of olfactory bulb neurones to odour stimulation of small nasal areas in the salamander. *J. Physiol.* 243:717–37

Kay L, Freeman W. 1998. Bidirectional processing in the olfactory-limbic axis during olfactory behavior. *Behav. Neurosci.* 112:541–53

Kay L, Laurent G. 1999. Odor- and context-dependent modulation of mitral cell activity in behaving rats. *Nat. Neurosci.* 2:1003–9

Keverne E. 1999. The vomeronasal organ. *Science* 286:716–20

Kuwabara M. 1957. Bildung des bedingten Reflexes von Pavlovs Typus bei der Honigbiene, *Apis mellifica. J. Fac. Sci. Hokkaido Univ. Ser. VI Zool.* 13:458–64

Lam Y-W, Cohen L, Wachowiak M, Zochowski M. 1999. Odors elicit three different oscillations in the turtle olfactory bulb. *J. Neurosci.* 19:749–62

Laurent G. 1996. Dynamical representation of odors by oscillating and evolving neural assemblies. *Trends Neurosci.* 19:489–96

Laurent G. 1999. A systems perspective on early olfactory coding. *Science* 286:723–28

Laurent G, Davidowitz H. 1994. Encoding of olfactory information with oscillating neural assemblies. *Science* 265:1872–75

Laurent G, Naraghi M. 1994. Odorant-induced oscillations in the mushroom bodies of the locust. *J. Neurosci.* 14:2993–3004

Laurent G, Wehr M, Davidowitz H. 1996. Temporal representations of odors in an olfactory network. *J. Neurosci.* 16:3837–47

Lawless H. 1997. *Olfactory Psychophysics*. San Diego, CA: Academic

Leitch B, Laurent G. 1996. GABAergic synapses in the antennal lobe and mushroom body of the locust olfactory system. *J. Comp. Neurol.* 372:487–514

Leveteau J, MacLeod P. 1966. Olfactory discrimination in the rabbit olfactory glomerulus. *Science* 153:175–76

Lin D, Wang F, Lowe G, Gold G, Axel R, et al. 2000. Formation of precise connections in the olfactory bulb occurs in the absence of odorant-evoked neuronal activity. *Neuron* 26:69–80

MacLeod K, Bäcker A, Laurent G. 1998. Who reads temporal information contained across synchronized and oscillatory spike trains? *Nature* 395:693–98

MacLeod K, Laurent G. 1996. Distinct mechanisms for synchronization and temporal patterning of odor-encoding neural assemblies. *Science* 274:976–79

Macrides F, Chorover SL. 1972. Olfactory bulb units: activity correlated with inhalation cycles and odor quality. *Science* 185:84–87

Malnic B, Hirono J, Sato T, Buck LB. 1999. Combinatorial receptor codes for odors. *Cell* 96:713–23

Malun D. 1991a. Inventory and distribution of synapses of identified uniglomerular projection neurons in the antennal lobe of *Periplaneta americana*. *J. Comp. Neurol.* 305:348–60

Malun D. 1991b. Synaptic relationships between GABA-immunoreactive neurons and an identified uniglomerular projection neuron in the antennal lobe of *Periplaneta americana*: a double labeling electron microscopic study. *Histochemistry* 96:197–207

Masson C, Mustaparta H. 1990. Chemical information processing in the olfactory system of insects. *Physiol. Rev.* 70:199–245

Mellon D. 1997. Physiological characterization of antennular flicking reflexes in the crayfish. *J. Comp. Physiol. A.* 180:553–65

Menzel R. 1987. Memory traces in honeybees. In *Neurobiology and Behaviour of Honeybees*, ed. R Menzel, A Mercer, pp. 310–25. Berlin: Springer-Verlag

Meredith M. 1986. Patterned response to odor in mammalian olfactory bulb: the influence of intensity. *J. Neurophysiol.* 56:572–97

Mombaerts P, Wang F, Dulac C, Chao SK, Nemes A, et al. 1996a. Visualizing an olfactory sensory map. *Cell* 87:675–86

Mombaerts P, Wang F, Dulac C, Vassar R, Chao SK, et al. 1996b. The molecular biology of olfactory perception. *Cold Spring Harb. Symp. Quant. Biol.* 61:135–45

Mori K, Kishi K, Ojima H. 1983. Distribution of dendrites of mitral, displaced mitral, tufted, and granule cells in the rabbit olfactory bulb. *J. Comp. Neurol.* 219:339–55

Mori K, Mataga N, Imamura K. 1992. Differential specificities of single mitral cells in rabbit olfactory bulb for a homologous series of fatty acid odor molecules. *J. Neurophysiol.* 67:786–89

Mori K, Nowycky MC, Shepherd GM. 1981. Analysis of synaptic potentials in mitral cells in the isolated turtle olfactory bulb. *J. Physiol.* 314:295–309

Mori K, Shepherd GM. 1994. Emerging principles of molecular signal processing by mitral/tufted cells in the olfactory bulb. *Semin. Cell Biol.* 5:65–74

Moulton DG. 1967. Spatio-temporal patterning of response in the olfactory system. In *Olfaction and Taste*, vol. II, ed. T Hayashi, pp. 109–16. Oxford/New York: Pergamon

Murlis J, Elkington J, Carde R. 1992. Odor plumes and how insects use them. *Annu. Rev. Entomol.* 37:505–32

Ngai J, Chess A, Dowling MM, Necles N, Macagno ER, Axel R. 1993. Coding of olfactory information: topography of odorant receptor expression in the catfish olfactory epithelium. *Cell* 72:667–80

Nicoll R. 1971a. Pharmacological evidence for GABA as the transmitter in granule cell inhibition in the olfactory bulb. *Brain Res.* 35:137–49

Nicoll R. 1971b. Recurrent excitation of secondary olfactory neurons: a possible mechanism for signal amplification. *Science* 171:824–25

Nicoll RA, Jahr CE. 1982. Self-excitation of olfactory bulb neurones. *Nature* 296:441–44

Nieuwenhuys R. 1967. Comparative anatomy of olfactory centers and tracts. *Prog. Brain Res.* 23:1–63

Olshausen BA, Field DJ. 1996. Emergence of simple-cell receptive field properties by learning a sparse code for natural images. *Nature* 381:607–9

Ottoson D. 1959. Comparison of slow potentials evoked in the frog's nasal mucosa and olfactory bulb by natural stimulation. *Acta Physiol. Scand.* 47:149–59

Pager J. 1983. Unit responses changing with behavioral outcome in the olfactory bulb of unrestrained rats. *Brain Res.* 289:87–98

Pelosi P. 1996. Perireceptor events in olfaction. *J. Neurobiol.* 30:3–19

Rall W, Shepherd G. 1968. Theoretical reconstruction of field potentials and dendrodendritic synaptic interactions in olfactory bulb. *J. Neurophysiol.* 31:884–915

Ressler KJ, Sullivan SL, Buck LB. 1993. A zonal organization of odorant receptor gene expression in the olfactory epithelium. *Cell* 73:597–609

Rolls ET, Tovee MJ. 1995. Sparseness of the neuronal representation of stimuli in the primate temporal visual-cortex. *J. Neurophysiol.* 73:713–26

Rubin BD, Katz LC. 1999. Optical imaging of odorant representations in the mammalian olfactory bulb. *Neuron* 23:499–511

Rudderman D. 1994. Statistics of natural images. *Network* 5:517–48

Satou M. 1990. Synaptic organization, local neuronal circuitry, and functional segregation of the teleost olfactory bulb. *Prog. Neurobiol.* 34:115–42

Scott J. 1986. The olfactory bulb and central pathways. *Experientia* 42:223–32

Scott JW, Wellis DP, Riggott MJ, Bounviso N. 1993. Functional organization of the main olfactory bulb. *Microsc. Res. Tech.* 24:142–56

Shadlen MN, Newsome WT. 1995. Is there signal in the noise? *Curr. Opin. Neurobiol.* 5:248–50

Shepherd GM. 1993. Principles of specificity and redundancy underlying the organization of the olfactory system. *Microsc. Res. Tech.* 24:106–12

Shepherd GM. 1994. Discrimination of molecular signals by the olfactory receptor neuron. *Neuron* 13:771–90

Shipley MT, Ennis M. 1996. Functional organization of olfactory system. *J. Neurobiol.* 30:123–76

Singer W, Gray C. 1995. Visual feature integration and the temporal correlation hypothesis. *Annu. Rev. Neurosci.* 18:555–86

Smith BH. 1998. Analysis of interactions in binary odorant mixtures. *Physiol. Behav.* 65:397–407

Smith BH, Menzel R. 1989. The use of electromyogram recordings to quantify odorant discrimination in the honeybee, *Apis mellifera. J. Insect Physiol.* 35:369–75

Stewart WB, Kauer JS, Shepherd GM. 1979. Functional organization of rat olfactory bulb analysed by the 2-deoxyglucose method. *J. Comp. Neurol.* 185:715–34

Stopfer M, Bhagavan S, Smith BH, Laurent G. 1997. Impaired odour discrimination on desynchronization of odour-encoding neural assemblies. *Nature* 390:70–74

Stopfer M, Laurent G. 1999. Short-term memory in olfactory network dynamics. *Nature* 402:664–68

Strausfeld N, Hansen L, Li Y, Gomez R, Ito K. 1998. Evolution, discovery and interpretations of arthropod mushroom bodies. *Learn. Mem.* 5:11–37

Sugase Y, Yamane S, Ueno S, Kawano K. 1999. Global and fine information coded by single neurons in the temporal visual cortex. *Nature* 400:869–73

Sun X, Tolbert L, Hildebrand J. 1993. Ramification pattern and ultrastructural characteristics of the serotonin-immunoreactive neuron in the antennal lobe of the moth *Manduca sexta. J. Comp. Neurol.* 338:5–16

Teyke T, Gelperin A. 1999. Olfactory oscillations augment odor discrimination, not odor identification by Limax CNS. *NeuroReport* 10:1–8

Vassar R, Chao SK, Sitcheran R, Nunez JM, Vosshall LB, Axel R. 1994. Topographic organization of sensory projections to the olfactory bulb. *Cell* 79:981–91

Vinje WE, Gallant JL. 2000. Sparse coding and decorrelation in primary visual cortex during natural vision. *Science* 287:1273–76

von der Malsburg C, Schneider W. 1986. A neural cocktail-party processor. *Biol. Cybern.* 54:29–40

Wang F, Nemes A, Mendelsohn M, Axel R. 1998. Odorant receptors govern the formation of a precise topographic map. *Cell* 93:47–60

Wehr M, Laurent G. 1996 Odor encoding by temporal sequences of firing in oscillating neural assemblies. *Nature* 384:162–66

Wehr M, Laurent G. 1999. Relationship between afferent and central temporal patterns in the locust olfacory system. *J. Neurosci.* 19:381–90

Wellis DP, Scott JW Harrison TA. 1989 Discrimination among odorants by single neurons of the rat olfactory bulb. *J. Neurophysiol.* 61:1161–77

Weth F, Nadler W, Korsching S. 1996. Nested expression domains for odorant receptors in zebrafish olfactory epithelium. *Proc. Natl. Acad. Sci. USA* 93:13321–26

Yao Y, Freeman WJ. 1990. Model of biological pattern recognition with spatially chaotic dynamics. *Neural Netw.* 3:153–70

Yokoi M, Mori K, Nakanishi S. 1995. Refinement of odor molecule tuning by dendrodendritic synaptic inhibition in the olfactory bulb. *Proc. Natl. Acad. Sci. USA* 92:3371–75

Yuilli A, Grzywacs N. 1989. A winner-take-all mechanism based on presynaptic inhibition. *Neural Comput.* 1:334–47

Zheng C, Feinstein P, Bozza T, Rodriguez I, Mombaerts P. 2000. Peripheral olfactory projections are differentially affected in mice deficient in a cyclic nucleotide-gated channel subunit. *Neuron* 26:81–91

Annu. Rev. Neurosci. 2001. 24:299–325

PROTEIN SYNTHESIS AT SYNAPTIC SITES ON DENDRITES

Oswald Steward[1] and Erin M. Schuman[2]

[1]*Reeve-Irvine Research Center and Departments of Anatomy/Neurobiology and Neurobiology and Behavior, College of Medicine, University of California at Irvine, Irvine, California 92697; e-mail: osteward@uci.edu*
[2]*Division of Biology, 216-76 Howard Hughes Medical Institute, California Institute of Technology, Pasadena, California 91125; e-mail: schumane@its.caltech.edu*

Key Words   synaptic plasticity, long-term potentiation, gene expression, hippocampus

■ **Abstract**   Studies over the past 20 years have revealed that gene expression in neurons is carried out by a distributed network of translational machinery. One component of this network is localized in dendrites, where polyribosomes and associated membranous elements are positioned beneath synapses and translate a particular population of dendritic mRNAs. The localization of translation machinery and mRNAs at synapses endows individual synapses with the capability to independently control synaptic strength through the local synthesis of proteins. The present review discusses recent studies linking synaptic plasticity to dendritic protein synthesis and mRNA trafficking and considers how these processes are regulated. We summarize recent information about how synaptic signaling is coupled to local translation and to the delivery of newly transcribed mRNAs to activated synaptic sites and how local translation may play a role in activity-dependent synaptic modification.

INTRODUCTION

Synapses can undergo long-lasting changes in strength that likely contribute to learning and memory. Many studies that have used protein synthesis inhibitors have shown that long-lasting forms of behavioral and synaptic plasticity require protein synthesis (Davis & Squire 1984, Bailey et al 1996, Kang & Schuman 1996, Mayford et al 1996, Nguyen & Kandel 1996, Schuman 1999). According to this idea, behavioral experience or electrical activity at synapses induces the synthesis of particular proteins that are critical for establishing enduring modifications. Possible roles for newly synthesized proteins include replacing degraded proteins, increasing the levels of existing proteins, or expressing novel or alternatively spliced forms of proteins.

Newly synthesized proteins must be made from messenger RNAs (mRNAs) that are present constitutively or from mRNAs that are synthesized as a consequence of transcriptional activation. If the required proteins are made using existing mRNAs, then there must be a mechanism for regulating translation via synaptic activity. As discussed below, there exist signal transduction pathways that are activated by synaptic events, resulting in the stimulation of translation. If new gene expression is required, there must be signaling from the synapse to the transcriptional machinery of the neuron.

Recently synthesized proteins also need to be made available, selectively, to those synapses undergoing changes in strength. The observation that individual neurons are endowed with so many synapses suggests the capability of independently processing and storing many bits of information. Electrophysiological experiments have confirmed this to some extent; anatomically isolated groups of synapses on the same cell can be modified independently. That is, given two sets of afferents that converge on the same postsynaptic cell, the enhancement of synaptic strength at one set of afferents does not spread to enhance synaptic strength at the other set of afferents (e.g. Andersen et al 1977). This property of synaptic plasticity is often referred to as "input specificity."

In the CA1 region of the hippocampus, input specificity holds only if the two sets of afferents are spatially remote. If they are close ($< \sim 50\ \mu$m), there may be some heterosynaptic enhancement (Bonhoeffer et al 1989, Engert & Bonhoeffer 1996, Schuman & Madison 1994). Hence, it may be more accurate to describe long-term potentiation (LTP) in the CA1 region as "site" rather than "input specific." In the dentate gyrus, the situation is somewhat different in that activation of one set of afferents to a particular dendritic segment produces LTD at inactive synapses that terminate nearby (a form of heterosynaptic depression; see Levy & Steward 1979, 1983). Again, this heterosynaptic interaction does not occur when inputs are spatially separate (White et al 1988, White et al 1990).

Given these issues of specificity and protein synthesis dependence, there must exist cell-biological mechanisms that allow synapses to independently control their strength. If the source of new proteins is the soma, neurons could utilize known protein transport and targeting mechanisms coupled with a "tag" or "marker" generated by synaptic activity (Martin et al 1997, Frey & Morris 1997). According to this idea, newly synthesized proteins are selectively deposited or stabilized at sites containing the tag. Another solution involves the local, rather than somatic, synthesis of proteins at synaptic sites. This idea circumvents the need for a protein trafficking/capture mechanism, but does potentially require a mechanism for targeting the mRNA to activated synapses.

The notion of dendritic protein synthesis had its roots in the discovery of synapse-associated polyribosome complexes (SPRCs)—polyribosomes and associated membranous cisterns that are selectively localized beneath postsynaptic sites on the dendrites of central nervous system (CNS) neurons (Steward 1983, Steward & Fass 1983, Steward & Levy 1982). Based on the assumption that form implies function, the highly selective localization of SPRCs beneath synapses suggested the following: (*a*) the translation machinery might synthesize key molecular constituents of the synapse, and (*b*) translation might be regulated by activity at the individual postsynaptic sites. As discussed below, there is now substantial evidence to support these ideas. In addition, many of the mRNAs that are present in dendrites have been identified; there is now considerable evidence that these mRNAs enable a local synthesis of the encoded proteins (Steward et al 1996, Steward & Singer 1997).

What has been missing until recently is a link between synaptic activation and either the transport of mRNAs into dendrites or the local translation of these mRNAs on site. In this review, we summarize some of the recent data that establish this link and also begin to address the mechanisms underlying mRNA trafficking and local translation. We begin with a brief review of the nature of the protein-synthetic machinery that is present at synaptic sites and provide an update regarding mRNAs that are localized in dendrites (and thus potentially are present at SPRCs). We summarize recent evidence that local translation of mRNA at the synapse may be regulated by synaptic activity and then summarize new information regarding how afferent activity regulates the translation and trafficking of dendritic mRNAs. We review the data suggesting that dendritic protein synthesis plays a key role in long-term synaptic modifications. Finally, we consider how these pieces might fit together to suggest a mechanism by which protein synthesis at the synapse mediates long-term synaptic modifications induced by activity.

THE MACHINERY FOR PROTEIN SYNTHESIS AT SYNAPTIC SITES

Localization

Synapse-associated polyribosome complexes are precisely localized in the post-synaptic cytoplasm. One very important feature of SPRCs is the selectivity of their localization. Quantitative electron-microscopic analyses have revealed that the vast majority of the polyribosomes that are present in dendrites are precisely positioned beneath postsynaptic sites and are absent from other parts of the dendrite (Steward & Levy 1982). SPRCs are most often localized at the base of the spine, in the small moundlike structures from which the neck of the spine emerges. Thus, SPRCs are located within or near the portal between the spine neck and the shaft of the dendrite—the route through which current must flow when spine synapses are activated. In this location, SPRCs are ideally situated to be influenced by electrical and/or chemical signals from the synapse as well as by events within the dendrite proper. An important implication of this selective localization is that there be some mechanism that causes ribosomes, mRNA, and other components of the translational machinery to dock selectively in the postsynaptic cytoplasm. The mechanisms underlying this highly selective localization remain to be established.

Although most dendritic polyribosomes are localized beneath synapses, a few clusters of ribosomes are localized within the core of the dendritic shaft (Steward & Levy 1982). It is not yet known whether these represent a different population than the synapse-associated polyribosomes. One possibility is that the polyribosomes in the dendritic core are associated with mRNAs that encode proteins that are not destined for synaptic sites but play some other role in dendritic function. This speculation is of particular interest given the functional diversity of the mRNAs that have been identified in dendrites (see below). Alternatively, the clusters of polyribosomes in the core of the dendrite may represent packets of mRNAs and ribosomes that are in transit from the cell body.

SPRCs are often associated with membranous organelles in a rough endoplasmic reticulum (RER)-like configuration. Serial section reconstructions of mid-proximodistal dendrites of dentate granule cells and hippocampal pyramidal cells revealed that ~50% of the polyribosomes are found in association with tubular cisterns (Steward & Reeves 1988). A common configuration is one in which the ribosomes seem to surround a blind end of a cistern. Thus, the SPRC/cisternal complex may be a form of RER that could allow the synthesis of integral membrane proteins or soluble proteins destined for release. Evidence in support of this hypothesis is considered in more detail below.

Interestingly, the cisterns with which SPRCs are associated are sometimes connected with a spine apparatus (Steward & Reeves 1988). The significance of these connections is not known. One interesting hypothesis is that the spine apparatus may be involved in some aspect of posttranslational processing of proteins that are synthesized at the SPRCs (more on this below).

Synapse-Associated Polyribosome Complexes at Different Types of Synapses

SPRCs are present at spine synapses on different neuron types. Quantitative analyses of polyribosomes have been carried out on dentate granule cells, hippocampal pyramidal cells, cortical neurons, and cerebellar Purkinje cells. These analyses reveal that SPRCs are present in a roughly similar configuration in all of the spine-bearing neurons that have been evaluated.

Estimates of the incidence of polyribosomes at spine synapses vary depending on the quantitative methods used. In evaluations of single sections, ~11%–15% of the identified spines have underlying polyribosomes (Steward & Levy 1982). However, this is clearly an underestimate because not all of the area under a spine is contained within a single section. Serial section reconstructions of dendrites in the dentate gyrus reveal that the actual incidence of polyribosomes in spines on mid-proximodistal dendrites is ~25% (Steward & Levy 1982). The estimates of incidence also depend on the counting criteria. Studies that have used serial section reconstruction techniques to evaluate the distribution of individual ribosomes (not polyribosomes) yield higher estimates of incidence (Spacek & Hartmann 1983). For example, in pyramidal neurons in the cerebral cortex, 82% of the reconstructed spines had ribosomes in the head, 42% had ribosomes in the neck, and 62% had ribosomes at the base. In cerebellar Purkinje cells, 13% of the spines had ribosomes in the head, and 22% had ribosomes at the base. It is likely that an important reason for the higher incidence values in this study is that single ribosomes were counted rather than polyribosomes. In any case, it is clear that polyribosomes are a ubiquitous component of the postsynaptic cytoplasm in a variety of neuron types.

SPRCs are also present at nonspine synapses. There have been no detailed quantitative evaluations of polyribosome distribution in the dendrites of nonspiny neurons, but it is clear that the same basic relationships exist as in spiny dendrites. For example, polyribosomes are often present beneath shaft synapses in association with submembranous cisterns and are found beneath both asymmetric (presumed excitatory) and symmetric (presumed inhibitory) synapses. Polyribosomes are also localized beneath synapses on axon initial segments (Steward & Ribak 1986). This localization is noteworthy for two reasons: first, it extends the generality of SPRCs to yet another type of postsynaptic location; second, most (perhaps all) synapses on axon initial segments use γ-aminobutyric acid as their neurotransmitter and are thus inhibitory. Axon initial segments also contain organelles that appear identical to spine apparatuses, known cisternal organelles. Based on the localization of subsynaptic cisternal organelles beneath both excitatory and inhibitory synapses on axon initial segments, it may be worthwhile to reconsider the possible functions of these enigmatic organelles. This is especially true because previous hypotheses have focused on functions that would be especially important at excitatory synapses and perhaps of minimal importance at inhibitory synapses (e.g. Ca^{2+} sequestration).

If protein synthetic machinery is localized at synapses in order to synthesize some of the components of the synaptic junction, one would expect SPRCs to be especially prominent at synapses during periods of synapse growth. This is the case. Polyribosomes are very abundant in the dendrites of developing neurons and again appear to be preferentially localized beneath postsynaptic sites, although the degree of selectivity has not been evaluated quantitatively (Steward & Falk 1986).

PROTEINS SYNTHESIZED AT SYNAPSES

The discovery of polyribosomes beneath synapses focused attention on the question of what proteins were synthesized in the postsynaptic cytoplasm. The approaches that have been used to address this question include the following: (*a*) biochemical studies of proteins synthesized by subcellular fractions enriched in pinched-off dendrites, (*b*) in situ hybridization analyses of the subcellular distribution of mRNAs in neurons, and (*c*) molecular biological analyses of the complement of mRNAs in isolated dendrites, most often from immature neurons grown in culture.

Proteins Synthesized in Subcellular Fractions

Biochemical approaches take advantage of subcellular fractionation techniques that allow the isolation of synaptosomes with attached fragments of dendrites that retain their cytoplasmic constituents, including polyribosomes and associated mRNAs. We have called these "synaptodendrosomes" (Rao & Steward 1991a). Others have used similar fractions prepared by filtration, which are termed "synaptoneurosomes" (Weiler & Greenough 1991, 1993; Weiler et al 1997).

The major limitation in using synaptodendrosomes or synaptoneurosomes to study dendritic protein synthesis is that the fractions are contaminated with fragments of neuronal and glial cell bodies. For example, high levels of the mRNA encoding glial fibrillary acidic protein are present (Chicurel et al 1990, Rao & Steward 1991b), and it is likely that there are also fragments of neuronal cell bodies containing mRNAs that are normally not present in dendrites. The problem of contamination can be partially circumvented by focusing on proteins that are synthesized in synaptosomes and then assembled into synaptic structures. For example, pulse-labeling techniques have been used to label the proteins synthesized within synaptosomes and then subcellular fractionation and detergent extraction techniques have been used to prepare synaptic plasma membranes and fractions enriched in synaptic junctional complexes (the postsynaptic membrane specialization and associated membrane). The proteins that were synthesized within the synaptodendrosomes and assembled into the synaptic plasma membrane and synaptic junctional complex were then characterized using polyacrylamide gel electrophoresis combined with fluorography (Leski & Steward 1996, Rao & Steward 1991a). This strategy has revealed characteristics of the labeled bands, but so far, the approach has not provided definitive identification of the labeled proteins. This combined strategy also has the limitation that it is useful only for proteins that are assembled into the synaptic membrane or synaptic junctional complex. Thus, proteins that are not assembled into the synapse are not detected.

Studies using synaptoneurosome fractions without the secondary purification step of subcellular fractionation have provided evidence for the dendritic synthesis of one novel protein that had not previously been identified—fragile X mental retardation protein (FMRP) (Weiler et al 1997). FMRP is encoded by the *fmr1* gene, which is affected in human fragile X syndrome. Treatment of synaptoneurosomes with agonists for metabotropic glutamate receptors caused a rapid increase in the amount of FMRP in the synaptoneurosome fractions, as determined by Western blot analysis. These data suggested that FMRP was being synthesized within the fractions and that the synthesis was enhanced by metabotropic glutamate receptor (mGluR) activation. This evidence has led to the idea that the neuronal dysfunction that is part of fragile X syndrome may result from a disruption of local synthesis of protein at synapses (Comery et al 1997, Weiler et al 1997).

There are some inconsistencies, however, in the story regarding fragile X mental retardation protein (FMRP). In the first place, in situ hybridization analyses indicate that FMRP mRNA is not evident in dendrites of neurons in vivo (Hinds et al 1993,

Valentine et al 2000). One can conceive of reasons why an mRNA might not be detected in dendrites by in situ hybridization. For example, the mRNA could be present at levels that are below the threshold for detection by standard in situ hybridization techniques. This possibility must be reconciled with the biochemical data, however, which indicate an almost twofold increase in the amount of FMRP detectable by Western blots within 5 min after treatment with mGluR agonists (Weiler et al 1997). Presumably such a large change in protein concentration could be achieved only if the levels of the mRNA were substantial.

It has also been suggested that FMRP plays a role in the regulation of translation of mRNAs at synapses (Feng et al 1997). This hypothesis is based on two facts: (*a*) FMRP is an RNA-binding protein, and (*b*) immunocytochemical studies reveal that the protein is localized at polyribosome clusters in neuronal dendrites. Localization at dendritic polyribosomes is also consistent with the hypothesis that FMRP is synthesized within dendrites. It is clear that the story regarding FMRP is an evolving one.

The only other identified protein that has been shown to be synthesized in synaptoneurosomes is the alpha subunit of calcium-/calmodulin-dependent protein kinase II (CAMKII) (Sheetz et al 2000). These studies involved metabolic labeling with ^{35}S-methionine followed by two-dimensional gel electrophoresis and fluorography or by immunoprecipitation of metabolically labeled protein using an antibody against CAMKII. Synthesis by fragments of neuronal-cell bodies or glia cannot be excluded in these experiments, but dendritic synthesis is supported by the observation that the labeling was regulated by treatment with neurotransmitters. Although it was already known that the CAMKII message is present in dendrites, this study documents that this mRNA is in actuality translated in synaptoneurosomes.

There have been continuing efforts to refine subcellular fractionation approaches so as to yield fractions of greater purity. In this regard, one recent study reported a fractionation approach that yields synaptosomes in which glial-fibrillary-acidic-protein mRNA is not detected by reverse transcription-polymerase chain reaction (Bagni et al 2000), suggesting a lack of contamination by glial fragments. Previously identified dendritic mRNAs such as those encoding CAMKII, Arc, and an inositol 1,4,5 triphosphate (InsP3) receptor InsP3R1were detected in the same fractions. These studies also demonstrated that the mRNA for CAMKII was associated with polysomes and that the fraction associated with polysomes was increased by neurotransmitter activation. Interestingly, the mRNA for FMRP was also detected in these fractions by reverse transcription-polymerase chain reaction. Thus, these fractions may represent a purer population of synaptodendrosomes than has been available previously, which could provide a means to identify novel dendritic mRNAs.

mRNAs in Dendrites in Vivo

Important clues about the identity of the proteins that may be synthesized at synapses have been obtained from in situ hybridization analyses that document the presence of particular mRNAs in dendrites. In most studies, dendritic localization has been inferred by the pattern of labeling in brain regions where neuronal cell bodies are concentrated in discrete layers and where there are distinct neuropil layers that contain dendrites and axons but few neuronal cell bodies (e.g. cortical regions including the hippocampus and the cerebellar cortex). Definitive evidence that the mRNA is in fact present in dendrites (and not in glial cells) can be obtained using nonisotopic in situ hybridization techniques. Dendritic localization can also be confirmed by studies of neurons in culture, although one must consider the possibility that neurons in culture may express an unusual complement of mRNAs or sort mRNAs in different ways than neurons in vivo.

One important caveat is that the presence of a particular mRNA in dendrites does not establish that the mRNA is translated at synapses. There are polyribosomes in dendrites that are not localized beneath synapses, and these polyribosomes could be associated with a different set of mRNAs than are translated in the postsynaptic cytoplasm (Steward & Reeves 1988). Nevertheless, identification of mRNAs that are present in dendrites provides candidates for synapse-associated mRNAs that can be further evaluated in other ways.

Table 1 lists the mRNAs for which the evidence for dendritic localization in vivo is strong. All of the RNAs listed extend for several hundred micrometers from the cell body. Certain other mRNAs that are localized primarily in cell bodies may extend slightly into proximal dendrites. For example, it has been reported that the mRNAs for two protein kinase C substrates (F1/GAP43 and RC3) extend somewhat further into the proximal dendrites of forebrain neurons than other "cell body" mRNAs. (Laudry et al 1993). The differences in the distribution of mRNAs encoding F1/GAP43 and RC3 vs other cell body mRNAs are slight and indeed were not evident in studies using nonisotopic in situ hybridization techniques that produced heavy labeling over cell bodies (Paradies & Steward 1997).

Taken together, the information on mRNAs in dendrites allows several generalizations:

1. A number of different mRNAs that encode unrelated proteins are present in dendrites. The proteins encoded by mRNAs that are present in dendrites include a variety of different classes of protein (Table 1), including cytoplasmic, cytoskeletal, integral-membrane, and membrane-associated proteins. The proteins also have very different functions. Thus, it is likely that the translation of these mRNAs subserves different aspects of cellular and synaptic function [for additional discussion of this point, see Steward (1997)].

2. All of the dendritic mRNAs that have been identified so far are expressed differentially by different types of neurons. This is especially evident when considering the mRNAs that are present in the dendrites of forebrain neurons vs cerebellar Purkinje cells. For example, the mRNAs for MAP2, CAMKII, dendrin, and Arc are found in forebrain neurons, but are not expressed at high levels by Purkinje cells. Purkinje cells, on the other

hand, express a different complement of mRNAs including the InsP3 receptor and other Ca^{2+}-interacting proteins, for example L7 and PEP19 (Bian et al 1996). These mRNAs are expressed by Purkinje cells and a few other neuron types. The fact that different mixtures of mRNAs are present in the dendrites of different cell types suggests that dendritic protein synthesis may have different purposes in different cell types.

3. Although a number of mRNAs are present in the dendrites of forebrain neurons, the patterns of expression and subcellular distributions of the mRNAs are different. The mRNAs encoding CAMKII, dendrin, and Arc (when induced) are localized throughout the dendrites. In contrast, the mRNA for microtubule-associated protein 2 (MAP2) is found at high levels in the proximal one third to one half of the total dendritic length, but it is not detectable in distal dendrites of most neurons. The mRNAs that are present in the dendrites of Purkinje cells also exhibit different localization patterns. For example, the mRNA for the InsP3 receptor is present throughout dendrites but is concentrated in the proximal one third of the total dendritic length. The mRNA for L7 appears to be more uniformly distributed throughout the dendrites (Bian et al 1996). Studies of the trafficking of the immediate early gene *Arc* also reveal that the complement of dendritic mRNAs can vary over time in an activity-dependent fashion (see below). These findings indicate that the capability exists for a different mixture of proteins to be synthesized locally at different times in different neuron types and different dendritic domains, providing a considerable complexity in the mechanisms underlying protein synthesis at synapses. Also, the variety of subcellular distributions implies that there must be multiple signals mediating mRNA localization within dendrites.

4. The presence of mRNAs encoding different classes of proteins implies the existence of different types of translational machinery in dendrites. MAP2, CAMKII, and Arc are nonmembrane proteins and thus would presumably be synthesized by free polysomes. In contrast, the InsP3 receptor is an integral membrane protein that presumably must be synthesized by membrane-bound ribosomes (RER). As noted above, electron-microscopic studies have revealed subsynaptic polyribosomes closely associated with membranous cisterns that may represent a form of RER (Steward & Reeves 1988). However, the InsP3 receptor is also a glycoprotein; thus, there is the question of how the newly synthesized protein is glycosylated (see below).

5. The dendritic localization of some mRNAs may be developmentally regulated. For example, the mRNA for calmodulin can be detected by in situ hybridization in dendritic laminae of developing, but not mature animals (Berry & Brown 1996). This is consistent with the idea that local dendritic protein synthesis is especially important during periods of synaptogenesis (Palacios-Pru et al 1981, 1988; Steward & Falk 1986). Several other mRNAs have been shown to be present in the dendrites

TABLE 1 mRNAs that have been shown to be localized within dendrites of neurons in vivo by in situ hybridization

mRNA	Cell type	Localization in dendrites	Class of protein	Protein function
MAP2[a]	Cortex, hippocampus, dentate gyrus	Proximal 1/3 – 1/2	Cytoskeletal	Microtubule associated
CAMII kinase[b] alpha subunit	Cortex, hippocampus, dentate gyrus	Throughout	Membrane-associated postsynaptic density	Multifunctional kinase Ca^{2+} signaling
Arc/Arg 3.1[c]	Cortex, hippocampus, dentate gyrus depending on inducing stimulus	Throughout (when induced)	Cytoskeleton-associated	Actin-binding synaptic junctional protein
Dendrin[d]	Hippocampus, dentate gyrus, cerebral cortex	Throughout	Putative membrane	Unknown
G-protein[e] gamma subunit	Cortex, hippocampus, dentate gyrus, striatum	Throughout	Membrane-associated	Metabotrophic receptor signaling
Calmodulin[f]	Cortex, hippocampus, Purkinje cells	Proximal-middle (during synaptogenesis)	Cytoplasm- and membrane-associated	Ca^{2+} signaling in conjunction with CAMII kinase
NMDAR1[g]	Dentate gyrus	Proximal-middle?	Integral membrane	Receptor
Glycine receptor[h] alpha subunit	Motoneurons	Proximal	Integral membrane	Receptor

Gene/Protein	Cell type	Location	Type	Function
Vasopressin[i]	Hypothalamohypophyseal	Proximal-middle	Soluble	Neuropeptide
Neurofilament protein 68[j]	Vestibular neurons	Proximal-middle	Cytoskeletal	Neurofilament
InsP3 receptor[k]	Purkinje cells	Throughout (concentrated proximally)	Integral membrane (endoplasmic reticulum)	Ca^{2+} signaling
L7[l]	Purkinje cells	Throughout	Cytoplasmic?	Homology to PDGF oncogene signaling?
PEP19[l]	Purkinje cells	Proximal one third	Cytoplasmic	Ca^{2+} binding

Not shown are mRNAs that are localized only in the most proximal segments.

[a] Garner et al 1988.
[b] Burgin et al 1990.
[c] Link et al 1995, Lyford et al 1995.
[d] Herb et al 1997.
[e] Watson et al 1994.
[f] Berry & Brown 1996.
[g] Gazzaley et al 1997.
[h] Racca et al 1997.
[i] Prakash et al 1997.
[j] Paradies & Steward 1997.
[k] Furuichi et al 1993.
[l] Bian et al 1996.

of young neurons developing in vitro. For example, the mRNAs for brain-derived neurotrophic factor (BDNF) and tyrosine kinase (trkB) receptors extend into the proximal 30% of the total dendritic length of hippocampal neurons in culture (Tongiorgi et al 1997). Potassium-induced depolarization (Tongiorgi et al 1997) or BDNF treatment (Righi et al 2000) increases the extent of dendritic labeling, so that the mRNAs extend to an average of 60%–70% of the total dendritic length. Despite the easily detectable dendritic labeling in neurons in vivo, the mRNAs for BDNF and trkB receptors appear to be largely restricted to the region of the cell body in young neurons in vivo (Dugich et al 1992). It remains to be seen whether a dendritic localization can be induced in neurons in vivo by manipulating neuronal activity.

Another mRNA that has been demonstrated in the dendrites of hippocampal neurons in vitro encodes the fatty acylated membrane-bound protein ligatin. Fluorescent in situ hybridization analyses indicate that the mRNA for ligatin extends for >100 μm into the dendrites of hippocampal neurons in culture; indeed, the dendritic labeling produced by ligatin probes is nearly as extensive as that produced by probes for CAMKII (Severt et al 2000). Nevertheless, previous studies of ligatin mRNA distribution in neurons in vivo reveal that the mRNA is largely restricted to the cell body region of hippocampal neurons in vivo (Perlin et al 1993). It is possible that the radioisotopic in situ hybridization techniques used in the earlier study in vivo were not sufficiently sensitive to detect mRNA in dendritic laminae.

mRNAs in Dendrites of Neurons in Vitro

One approach to identifying dendritic mRNAs has been to use patch pipettes to aspirate the cytoplasmic contents of individual dendrites of neurons grown in culture and then use RNA amplification techniques to clone the mRNAs (Miyashiro et al 1994). This study provided intriguing evidence that there may be a substantial number of mRNAs in dendrites, many of which remain to be characterized. However, there are certain inconsistencies between these and other findings. For example, the mRNAs for GluR1 receptors were detected, although the mRNAs for these receptors have not been detected by in situ hybridization in dendrites of neurons in vivo or in vitro. The reason for the disparity of results is not clear.

There are several possible explanations for the inconsistencies. One possibility is that Miyashiro et al analyzed cytoplasm from dendrites of very young neurons developing in culture. These might contain a different complement of mRNAs than the dendrites of mature neurons. Another possibility is that the amplification techniques detect mRNAs that are present in such low abundance that they are not easily detected using routine in situ hybridization techniques. It is also conceivable that some of the mRNAs in dendrites are in a form that somehow interferes with hybridization by complementary probes.

New Dendritic mRNAs

It is almost certain that there are more dendritic mRNAs yet to be found. For example, biochemical studies of proteins synthesized within synaptodendrosomes suggest that several as yet unidentified constituents of synaptic junctions are locally synthesized within dendrites (Rao & Steward 1991a, Steward et al 1991). The most prominent of these are not in a molecular-weight range that would be consistent with their being the translation products of known dendritic mRNAs. Systematic searches for new members of the family of dendritic mRNAs must deal with the problem of how to obtain sufficient quantities of mRNA from dendrites that are not contaminated by mRNA from neuronal cell bodies or supporting cells. So far, systematic searches have not yet identified new members of the family of dendritic mRNAs whose presence in dendrites in vivo was later confirmed by in situ hybridization analyses.

Posttranslational Processing Within Dendrites

The presence of mRNAs encoding integral membrane proteins raises the question of whether dendrites contain the machinery for posttranslational processing of recently synthesized proteins (specifically, components of the RER and Golgi apparatus [GA]). This question has been evaluated by assessing the distribution of protein markers and enzyme activities characteristic of the RER and GA (especially glycosyltransferase activities).

Initial studies evaluated the distribution of glycosyltransferase activities by pulse-labeling neurons with the sugar precursors that are the substrates of various glycosyltransferases that are present in the RER and GA. For example, mannose is added to nascent glycoproteins in the RER. Thus, when neurons are pulse-labeled with [^{3}H]mannose under conditions in which transport of recently synthesized proteins is blocked, the sites of mannose incorporation can be revealed autoradiographically. The same strategy can be applied to higher-order glycosyltransferases that are characteristic of the GA, for example fucosyltransferase and galactosyltransferase. Studies of this type provided the initial evidence for the presence of both mannosyltransferase and higher-order (Golgi-like) glycosyltransferase activity in the dendrites of hippocampal neurons grown in culture (Torre & Steward 1996).

Additional evidence regarding the localization of the RER and GA in dendrites has come from immunocytochemical studies of the subcellular distribution of proteins that are considered markers of the two endomembrane systems (Torre & Steward 1996). Immunocytochemical studies of the distribution of ribophorin I (an RER marker) reveal staining that extends well into dendrites. In general, however, immunostaining for Golgi markers extended only into proximal dendrites. The immunocytochemical data were generally consistent with the autoradiographic evidence regarding the intracellular distribution of glycosyltransferase activity characteristic of the RER and GA.

These data raise the question of what membranous organelles are actually responsible for the activities characteristic of the RER and GA. Recent electron-microscopic immunocytochemical studies indicate that the membranous cisterns present near spine synapses stain for Sec6Ialpha protein complex, which is part of the machinery for translocation of proteins through the RER during their synthesis (Pierce et al 2000). The cisterns exhibiting labeling have the same appearance as the cisterns that represent the membranous component of SPRCs. Moreover, immunostaining for ribosomal protein S3 revealed labeling over the same membranous cisterns that were labeled for Sec6Ialpha.

The organelle responsible for higher-order glycosyltransferase (Golgi-like) activity remains to be identified. The studies of RER and GA distribution in dendrites imply that RER is present throughout dendrites whereas machinery for higher-order glycosylation may be present only in proximal dendrites, at least in hippocampal neurons in culture (Torre & Steward 1996). Thus, if any integral membrane proteins are synthesized in distal dendrites of forebrain neurons, their glycosylation may be incomplete. In this regard, nonglycosylated membrane receptors can still function, although they may have different properties than fully glycosylated versions of the receptor (Giovannelli et al 1991). It will be of interest to determine whether the story is different in Purkinje cells, where the mRNA for at least one integral membrane protein is present throughout dendrites.

REGULATION OF mRNA TRAFFICKING IN DENDRITES

Evidence that synaptic activation triggers the transport of new mRNA transcripts to the synapse has come from studies of the immediate early gene *Arc*. *Arc* was discovered in screens for novel immediate early genes, defined as genes that are induced by activity in a protein synthesis-independent fashion (Link et al 1995, Lyford et al 1995). In both studies, the inducing stimulus was a single electroconvulsive seizure, and the protein synthesis independence was ensured by treating animals with cycloheximide to block protein synthesis (and hence the synthesis of secondary response genes). *Arc* was one of a number of novel immediate early genes that were identified using this paradigm.

Recent studies have revealed an interesting feature of the trafficking of *Arc* mRNA (Steward et al 1998). Patterned synaptic activation both induces *Arc* and causes the newly synthesized mRNA to localize selectively to activated dendritic domains. This was demonstrated in studies in which *Arc* expression was induced by stimulating the entorhinal cortical projections to the dentate gyrus in anesthetized rats. The projection from the entorhinal cortex to the dentate gyrus (the perforant path) terminates in a topographically organized fashion along the dendrites of dentate granule cells. Projections from the medial entorhinal cortex terminate in the middle molecular layer of the dentate gyrus, whereas projections from the lateral entorhinal cortex terminate in the outer molecular layer. Hence, by positioning a stimulating electrode in the medial entorhinal cortex, it is possible to selectively activate a band of synapses that terminate on mid-proximodistal dendrites.

When the medial perforant path was activated using a stimulation paradigm that is commonly used to induce LTP (400-Hz trains, 8 pulses per train, delivered at a rate of 1 pulse/10 s), *Arc* expression was strongly induced. The newly synthesized mRNA migrated into dendrites and accumulated selectively in the middle molecular layer in exactly the location of the band of synapses that had been activated. Similarly, activation of other afferent systems that terminate at different locations in the molecular layer caused the newly synthesized mRNA to localize selectively in other dendritic laminae. For example, simulation of the lateral entorhinal cortex produced a band of labeling for *Arc* mRNA in the outer molecular layer; stimulation of the commissural projection produced a band of labeling in the inner molecular layer (Steward et al 1998). Thus, these experiments revealed that synaptic activation generated a signal that caused *Arc* mRNA to localize near the active synapses. The nature of this docking signal remains to be defined.

Localization of *Arc* mRNA in activated dendritic laminae is associated with a local accumulation of Arc protein. Immunostaining of tissue sections from stimulated animals using an Arc-specific antibody revealed a band of newly synthesized protein in the same dendritic laminae in which *Arc* mRNA was concentrated (Steward et al 1998). The fact that synaptic activation leads to the selective targeting of both recently synthesized mRNA and protein suggests that the targeting of the mRNA underlies a local synthesis of the protein. Given that Arc protein is usually highly localized in postsynaptic junctions, it is likely that the newly synthesized Arc protein is targeted to the postsynaptic sites in the activated region. This remains to be established, however.

A number of pieces of the puzzle are still missing. First, it remains to be established whether *Arc* is directly involved in activity-induced synaptic modification. Additional clues to *Arc*'s function will likely come from studies of the protein itself and its interactions with other functional molecules of the synaptic junctional region. But even if *Arc* does turn out to be a red herring, these studies have delineated RNA trafficking mechanisms that could be used for sorting other mRNAs that participate in activity-dependent modifications.

In addition to the induced *Arc* mRNA there are other mRNAs present in dendrites constitutively. These include the mRNAs for molecules that have already been strongly implicated in activity-dependent synaptic modification (e.g. the alpha subunit of CAMII kinase). The mRNAs that are present constitutively provide an opportunity for local regulation of the synthesis of key signaling molecules via translational regulation (see below). Hence, protein synthesis at individual synapses may be regulated in a complex fashion, first through the regulation of the mRNAs available for translation (i.e. Arc) and then by regulation of the translation of the mix of mRNAs that are in place, including those present constitutively. How this is coordinated and how all of these molecules actually fit in to the molecular consolidation process remain to be established.

mRNA Binding Proteins

It is very likely that mRNAs are transported into dendrites in RNA-protein complexes, which contain *trans*-acting factors responsible for RNA transportation as well as components of translational machinery (Knowles et al 1996). In the future, identification of the minimal *cis*-elements sufficient for dendritic RNA targeting should facilitate the characterization of the *trans*-acting factors that are involved in this process. Several proteins that are involved in RNA targeting and asymmetric distribution have been described in *Drosophila* embryos and *Xenopus* oocytes. One such molecule is staufen, a protein required for the asymmetric distribution of specific mRNAs in *Drosophila* oocytes and neuroblasts. Recently multiple mammalian staufen-related genes have been identified (Kiebler et al 1999, Marion et al 1999, Tang et al 1999, Wickham et al 1999). At least two of them are present in hippocampal neurons. Immunostaining experiments indicate somatodendritic distribution patterns of the staufen protein in hippocampal neurons, with higher signals in the proximal dendrites compared with distal dendrites in adult hippocampal slices. The staufen distribution pattern overlaps with that of dendritic RNA in cultured neurons (Tang et al 1999). In addition, RNA-containing staufen protein particles move along microtubules in the dendrites (Kohrmann et al 1999, Tang et al 1999). Expression of a truncated staufen, lacking the microtubule-binding domain, decreases the amount of RNA detected in dendrites (Tang et al 1999). In addition, overexpression of staufen can increase the amount of clustering of RNA in cultured hippocampal dendrites (Tang et al 1999). These observations are consistent with a role for staufen in dendritic RNA targeting. The specific mRNAs that staufen binds in vivo remain to be determined, as does the potential regulation of staufen trafficking during synaptic plasticity.

REGULATION OF mRNA TRANSLATION AT SYNAPSES

The selective localization of ribosomes at synapses provides a mechanism for locally regulating the production of proteins; this is an ideal mechanism to produce proteins that are necessary for synaptic modification. Local protein synthesis could involve mRNAs already in place and/or mRNAs that are induced by synaptic activity and delivered into dendrites. Until recently, there was no experimental link between synaptic activity and either the transport of mRNAs into dendrites or the local translation of these mRNAs at synapses. Recent studies, however, have shown that synaptic activity can trigger the transport of new mRNA transcripts to synaptic sites, as discussed above, and modulate the translation of mRNAs already in place, as discussed below.

Initial evidence for synaptic modulation of protein synthesis within dendrites came from studies of protein precursor incorporation by hippocampal slices in vitro (Feig & Lipton 1993). When the Schaffer collaterals were activated in the presence of the cholinergic agonist carbachol, there was an increase in [^{3}H]leucine incorporation in dendritic laminae as revealed by high-resolution autoradiography.

Interestingly, neither Schaffer collateral stimulation nor carbachol alone produced a similar increase in labeling. While the combination of stimulation and carbachol caused a clear increase in protein synthesis, there was no detectable change in synaptic strength.

Given the abundance of the mRNAs for the alpha-subunit of CAMKII and MAP2 in the dendrites of hippocampal neurons (Burgin et al 1990), one obvious experiment is to evaluate whether the dendritic synthesis of CAMKII or MAP2 can be modulated by synaptic activity. Two studies have addressed this question by stimulating afferent projections to the hippocampus and dentate gyrus and evaluating CAMKII levels using immunocytochemistry. Ouyang et al (1997, 1999) approached the question by delivering high-frequency stimulation to the Schaffer collateral system in hippocampal slices in vitro. They observed increased immunostaining for both phosphorylated and nonphosphorylated CAMKII in the dendritic laminae. The increase in nonphosphorylated CAMKII occurred within 5 min of high-frequency stimulation and was blocked by anisomycin, strongly suggesting that the kinase was synthesized in the dendrites. A fast transport of CAMKII from the cell bodies seemed unlikely given the distance of the CAMKII increases from the cell bodies (e.g. $\sim$150–200 μm) as well as the fact that these experiments were conducted at room temperature. In a related study, Steward and colleagues demonstrated that high-frequency (400-Hz) stimulation of the perforant path projections to the dentate gyrus in vivo also causes increases in immunostaining for CAMKII in the activated dendritic laminae (Steward & Halpain 1999). There were also alterations in the pattern of immunostaining for MAP2, but the nature of the changes was different than was the case for CAMKII. Specifically, the increases in immunostaining for MAP2 occurred in the laminae on each side of the activated lamina. The changes in immunostaining for MAP2 were diminished but not eliminated by systemic or local application of protein synthesis inhibitors. Surprisingly, however, the increases in immunostaining for CAMKII were not affected by inhibiting protein synthesis. Thus, high-frequency synaptic activity can cause domain-specific alterations in the molecular composition of dendrites, but only a portion of the change may be attributable to local protein synthesis. In comparing these results to those of Ouyang et al (1997, 1999), it is possible that different stimulation frequencies (e.g. 100 vs 400 Hz) differentially invoke local synthesis vs trafficking of CAMKII. In addition, these studies were conducted in different areas of the hippocampus.

Although these combined results suggest synaptic regulation of dendritic protein synthesis, the possibility that newly synthesized proteins were actually produced in the neuronal cell body and rapidly transported into the dendrites cannot be ruled out. In this regard, parallel studies in subcellular fractions provide important complementary evidence.

Studies in Subcellular Fractions

In addition to the FMRP studies discussed above, recent studies of synapto-neurosomes prepared from frog tectum have revealed a novel mechanism for synaptic regulation of the translation of the alpha subunit of CAMKII. There is evidence that the formation of eye-specific projections in the tectum is regulated by activity and that the *N*-methyl-D-aspartate (NMDA) receptors play an important role in this process (Sheetz et al 1997). Based on these findings and the fact that CAMKII mRNA is present in dendrites, Sheetz et al (2000) evaluated how NMDA receptor activation of synaptoneurosomes from the tectum modulates CAMKII synthesis. Their results revealed a surprising and complex translation regulation mechanism. NMDA receptor activation did enhance CAMKII synthesis within the subcellular fractions (presumably within the synaptoneurosomes). At the same time, however, there was an increase in the phosphorylation of the initiation factor IF2, which would be expected to decrease the rate of polypeptide elongation (slowing the overall protein synthesis rate). This apparent paradox can be explained by the fact that studies of mRNA competition in translation assays reveal that decreases in elongation favor the translation of weakly initiated mRNAs. CAMKII is one of the mRNAs for which initiation is inefficient, and so general decreases in elongation consequent to IF2 phosphorylation could lead to increases in CAMKII synthesis. Sheetz et al (2000) also provided evidence in support of this idea by showing that low to moderate concentrations of cycloheximide (which partially inhibited overall protein synthesis) caused increases in the synthesis of CAMKII at the same time that overall levels of protein synthesis were diminished. These results thus provided evidence for regulation of CAMKII mRNA translation via NMDA receptor activation.

Other recent studies using synaptoneurosomes have revealed another novel mechanism for the control of the translation of CAMKII mRNA at synapses (Wu et al 1998). An unusual feature of CAMKII mRNA is that it contains a sequence in its 3'-untranslated region that is a consensus sequence for the binding of a cytoplasmic polyadenylation element (CPE). CPEs are known to play a key role in regulating the translation of maternal mRNAs in oocytes. These mRNAs, which are inherited from the mother, have a selective localization in the oocyte cytoplasm. The proteins encoded by these maternal mRNAs are often transcription factors that regulate the development of polarity in the embryo (i.e. defining dorsal vs ventral, anterior vs posterior, and particular body regions).

Maternal mRNAs are translationally repressed until fertilization. Upon fertilization, translation repression is relieved through the action of the CPE. Translationally repressed maternal mRNAs have very short poly-A tails. At fertilization, CPE is activated and triggers an elongation of the poly-A tail, resulting in translation induction. In an interesting experiment, Wu et al (1998) demonstrated that the translation of CAMKII kinase was regulated in a similar way in brain. In particular, NMDA receptor activation triggered polyadenylation of the mRNA for CAMII kinase, which in turn increased the synthesis of CAMKII protein. They further showed that this activation could be triggered by behavioral experience (light exposure for animals raised in the dark). This study thus revealed a second mechanism through which CAMKII protein synthesis could be regulated at the synapse. It remains to be seen how this mechanism interacts with the mechanism suggested in the experiments by Sheetz et al (2000).

Taken together, this evidence suggests a complex regulation of the translation of mRNAs at synapses. Activation of particular neurotransmitter receptors appears to play a role in regulating translation, and both metabotropic and NMDA-type receptors have been implicated in this process. It remains to be determined whether different mRNAs are controlled in different ways or whether different control mechanisms are present at different types of synapses.

Coupling Synaptic Activity to Translation

The signal transduction events that couple synaptic events to the protein translation machinery are not yet well understood. One signaling pathway that is stimulated by growth factors and results in the translation of several mRNAs includes the rapamycin-sensitive kinase mammalian target of rapamycin [mTOR, also known as FRAP kinase and RAFT-1; see Brown & Schreiber (1996) for a review]. Several components of this translational signaling pathway, including mTOR, 4E-BP1, 4E-BP-2, and eIF-4E, are present in the rat hippocampus, as shown by Western blot analysis and immunostaining studies (Tang et al 1998). In cultured hippocampal neurons, the distribution of these factors overlaps substantially with a synaptic protein, synapsin-I, suggesting synaptic localization. Disruption of mTOR activity by rapamycin results in deficits in late-phase LTP expression induced by high-frequency stimulation, while the early phase of LTP is unaffected (Tang et al 1998). Rapamycin also blocks the synaptic enhancement induced by BDNF in hippocampal slices (Tang et al 1998) as well as the long-term synaptic facilitation induced by repeated presentations of 5-hydroxytryptamine (5-HT) at *Aplysia* synapses in culture (Casadio et al 1999). These results imply an essential role for the rapamycin-sensitive signaling in three different forms of synaptic plasticity that require new protein synthesis. The localization of this translational signaling pathway at synaptic sites may provide a mechanism that controls local protein synthesis at potentiated synapses.

ROLE OF LOCAL SYNTHESIS IN SYNAPTIC PLASTICITY

Long-Term Potentiation

It is now well accepted that the late phase of LTP requires both transcription and translation. Indeed, the term late-phase LTP is operationally defined as the temporal phase of potentiation that can be blocked by inhibitors of protein synthesis. Initial evidence came from studies of hippocampal LTP in vitro, which reported that three different translation inhibitors, emetine, cycloheximide, and puromycin, reduced the proportion of slices exhibiting LTP (Stanton & Sarvey 1984). A fourth inhibitor, anisomycin, reduced only the magnitude of LTP of the population spike. In subsequent studies of LTP in the dentate gyrus in vivo, it was found that pretreatment with anisomycin did not block LTP induction, but did result in the decay of LTP over time (Frey et al 1988, Otani & Abraham 1989, Otani et al 1989). Subsequent studies of LTP in hippocampal slices yielded similar results (protein synthesis inhibitors block the late phase of LTP but not the early phase). (Nguyen & Kandel 1997, Osten et al 1996). These studies and others provide the basis for the widely held conclusion that the synthesis of new proteins is required for long lasting (e.g. $\geq \sim 1$ h) potentiation at both the Schaffer collateral-CA1 synapses in the hippocampus and perforant path synapses in the dentate gyrus.

A recent study (Raymond et al 2000) has suggested that the "priming" of LTP by mGluRs is also mediated by stimulation of local protein synthesis. Priming refers to the transformation of a small, decaying potentiation into a long-lasting enhancement by either an agonist or a synaptic stimulation protocol. Activation of mGluRs (by a selective agonist) 20 min before a tetanus promotes LTP induction and persistence (Cohen & Abraham 1996, Cohen et al 1998). The application of a protein synthesis inhibitor during mGluR activation prevents the priming effect (Raymond et al 2000). Given the rapidity of the priming effect (<20 min), these data are consistent with a mechanism involving mGluR stimulation of translation in the dendrites. This suggestion is further supported by the observation (discussed earlier) that mGluR agonists can stimulate protein synthesis in synaptoneurosomes (Weiler et al 1997).

Although the above studies support the conclusion that the late phase of LTP requires transcription and translation, they do not indicate whether the new proteins are synthesized in the cell body or in the dendrites or, for that matter, whether the required protein synthesis occurs pre- or postsynaptically.

Dissecting Sites of the Protein Synthesis Required for LTP

One of the earliest studies linking dendritic protein synthesis to synaptic plasticity examined the protein synthesis dependence of potentiation induced by the growth factors BDNF and neurotrophin-3 (Kang & Schuman 1995, 1996). Application of either BDNF or neurotrophin-3 to CA1 synapses caused a large and long lasting (2- to 3-h) enhancement of synaptic transmission (Kang & Schuman 1995). Surprisingly, pretreatment with a protein translation inhibitor blocked both the early and late phases of the enhancement (Kang & Schuman 1996). This early (e.g. within 10-min) requirement for protein synthesis is not consistent with a somatic origin, given the distance between the recording site (distal dendrites) and the cell bodies. Microlesion experiments, in which the synaptic neuropil was isolated from the cell bodies, showed directly that the protein synthesis source was not the cell bodies. Dendrites isolated from their cell bodies still exhibited growth-factor-induced synaptic potentiation that was sensitive to translation inhibitors. More recent experiments have shown that BDNF can stimulate the translation of a dendritically localized green fluorescent protein mRNA in isolated dendrites (Smith et al 1999).

Long-Term Depression

Proteins synthesized in dendrites may also contribute to long-term decreases in synaptic strength or long-term depression (LTD). There are two forms of LTD, one that requires the activation of NMDA receptors and the other that requires the activation of mGluRs. Capitalizing on the link between mGluR activity and stimulation of dendritic protein synthesis, Bear and colleagues examined whether mGluR-dependent LTD requires protein synthesis that is local (Huber et al 2000). The induction of LTD by an mGluR agonist was inhibited by bath application of anisomycin or the postsynaptic infusion of mRNA cap analog m^7GpppG (Huber et al 2000). In addition, dendrites dissociated from their cell bodies still showed mGluR agonist-induced LTD, although the anisomycin sensitivity of this plasticity was not examined. Taken together with the results above, these data indicate that locally synthesized proteins can contribute to bidirectional changes in synaptic strength.

Long-Term Facilitation in *Aplysia* Species

There is abundant evidence that long-term synaptic modifications in invertebrates also require local protein synthesis at or near synapses. An entire review could be written on this topic, so here we summarize the evidence only briefly. In *Aplysia* the molecular bases of behavioral sensitization have been studied in a variety of reduced preparations, including the synapses formed between isolated sensory and motoneurons in culture (Montarolo et al 1986). In this system, repeated applications of 5-HT result in a protein synthesis-dependent long-term facilitation of synaptic transmission between the sensory and motoneurons (Clark & Kandel 1993, Martin et al 1997). When a single sensory neuron with a bifurcating axon contacts two different motoneurons, five pulses of 5-HT to one synapse, but not the other, results in synapse-specific long-term facilitation at the 5-HT-treated synapse (Martin et al 1997). This long-term facilitation requires local protein synthesis: restricted application of a translation inhibitor to the 5-HT-treated synapse blocks the facilitation (Martin et al 1997). The long-term facilitation can be "captured" by the neighboring "naive" synapse if it is treated with single application of 5-HT within a few hours (Casadio et al 1999, Martin et al 1997). The long-term (e.g. 72-h) expression of synaptic capture also requires local protein synthesis. If a translation inhibitor is coapplied with the single 5-HT pulse to the "naive," this synapse will no longer exhibit long-term facilitation (Casadio et al 1999).

SUMMARY

The presence of polyribosomes, translation machinery, and mRNAs in dendrites endows individual synapses with the capability to independently control synaptic strength through the local synthesis of proteins. Studies in the past few years have provided strong evidence linking synaptic plasticity to dendritic protein synthesis and mRNA trafficking. Relatively little is know about how these processes are regulated including the coupling of synaptic signaling to translation machinery, the selective translation of specific mRNAs present at synapses, and the specific delivery of newly transcribed mRNAs to activated synaptic sites. Future studies will no doubt continue to strengthen the idea that local control of synaptic mRNAs and proteins is essential for maintaining the complexity of synaptic connections in the nervous system.

ACKNOWLEDGMENTS

This work was supported by NIH NS12333 (OS), NIH NS37292, NIMH MH49176, and HHMI (EMS).

Visit the Annual Reviews home page at www.AnnualReviews.org

LITERATURE CITED

Andersen P, Sundberg SH, Sveen O, Wigstrom H. 1977. Specific long-lasting potentiation of synaptic transmission in hippocampal slices. *Nature* 266:736–37

Bagni C, Mannucci L, Dotti CG, Amaldi F. 2000. Chemical stimulation of synaptosomes modulates alpha-Ca^{2+}/calmodulin-dependent protein kinase II mRNA association to polysomes. *J. Neurosci.* 20(RC76): 1–6

Bailey CH, Bartsch D, Kandel ER. 1996. Toward a molecular definition of long-term memory storage. *Proc. Natl. Acad. Sci. USA* 93:13445–52

Berry FB, Brown IR. 1996. CaM I mRNA is localized to apical dendrites during postnatal development of neurons in the rat brain. *J. Neurosci. Res.* 43:565–75

Bian F, Chu T, Schilling K, Oberdick J. 1996. Differential mRNA transport and the regulation of protein synthesis: selective sensitivity of Purkinje cell dendritic mRNAs to translational inhibition. *Mol. Cell. Neurosci.* 7: 116–33

Bonhoeffer T, Staiger V, Aertsen A. 1989. Synaptic plasticity in rat hippocampal slice cultures: Local "Hebbian" conjunction of pre- and postsynaptic stimulation leads to distributed synaptic enhancement. *Proc. Natl. Acad. Sci. USA* 86:8113–17

Brown EJ, Schreiber SL. 1996. A signaling pathway to translational control. *Cell* 86: 517–20

Burgin KE, Waxham MN, Rickling S, Westgate SA, Mobley WC, Kelly PT. 1990. In situ hybridization histochemisty of Ca/calmodulin dependent protein kinase in developing rat brain. *J. Neurosci.* 10:1788–89

Casadio A, Martin KC, Giustetto M, Zhu H, Chen M, et al. 1999. A transient, neuron-wide form of CREB-mediated long-term facilitation can be stabilized at specific synapses by local protein synthesis. *Cell* 99:221–37

Chicurel ME, Terrian DM, Potter H. 1990. Subcellular localization of mRNA: isolation and characterization of mRNA from an enriched preparation of hippocampal dendritic spines. *Soc. Neurosci. Abstr.* 16:353

Clark GA, Kandel ER. 1993. Induction of long-term facilitation in Aplysia sensory neurons by local application of serotonin to remote synapses. *Proc. Natl. Acad. Sci. USA* 90:11411–15

Cohen AS, Abraham WC. 1996. Facilitation of long-term potentiation by prior activation of metabotropic glutamate receptors. *J. Neurophysiol.* 76:953–62

Cohen AS, Raymond CR, Abraham WC. 1998. Priming of long-term potentiation induced by activation of metabotropic glutamate receptors coupled to phospholipase C. *Hippocampus* 8:160–70

Comery TA, Harris JB, Willems PJ, Oostra BA, Irwin SA, et al. 1997. Abnormal dendritic spines in fragile X knockout mice: maturation and pruning deficits. *Proc. Natl. Acad. Sci. USA* 94:5401–4

Davis HP, Squire LR. 1984. Protein synthesis and memory: a review. *Psychol. Bull.* 96:518–59

Dugich MM, Tocco G, Willoughby DA, Najm I, Pasinetti G, et al. 1992. BDNF mRNA expression in the developing rat brain following kainic acid-induced seizure activity. *Neuron* 8:1127–38

Engert F, Bonhoeffer T. 1996. Pairing-induced LTP in hippocampal slice cultures is not strictly input-specific. *Soc. Neurosci. Abstr.* 22:516

Feig S, Lipton P. 1993. Pairing the cholinergic agonist carbachol with patterned scaffer collateral stimulation initiates protein synthesis in hippocampal CA1 pyramidal cell dendrites via a muscarinic, NMDA-dependent mechanism. *J. Neurosci.* 13:1010–21

Feng Y, Gutekunst C-A, Eberhart DE, Yi H, Warren ST, Hersch SM. 1997. Fragile X mental retardation protein: nucleocytoplasmic shuttling and association with somatodendritic ribosomes. *J. Neurosci.* 17: 1539–47

Frey U, Krug M, Reymann KG, Matthies H. 1988. Anisomycin, an inhibitor of protein synthesis, blocks late phases of LTP phenomena in the hippocampal CA1 region *in vitro*. *Brain Res.* 452:57–65

Frey U, Morris RGM. 1997. Synaptic tagging and long-term potentiation. *Nature* 385:533–36

Furuichi T, Simon-Chazottes D, Fujino I, Yamada N, Hasegawa M, et al. 1993. Widespread expression of inositol 1,4,5-triphosphate receptor type 1 gene (Insp3r1) in the mouse central nervous system. *Recept. Channels* 1:11–24

Garner CC, Tucker RP, Matus A. 1988. Selective localization of messenger RNA for cytoskeletal protein MAP2 in dendrites. *Nature* 336:674–77

Gazzaley AH, Benson DL, Huntley GW, Morrison JH. 1997. Differential subcellular regulation of NMDAR1 protein and mRNA in dendrites of dentate gyrus granule cells after perforant path transection. *J. Neursci.* 17:2006–17

Giovannelli A, Grassi F, Eusebi F, Miledi R. 1991. Tunicamycin increases desensitization of acetylcholine receptors in cultured mouse muscle cells. *Proc. Natl. Acad. Sci. USA* 88:1808–11

Herb A, Wisden W, Catania DMV, Marechal D, Dresse A, Seeberg PH. 1997. Prominent dendritic localization in forebrain neurons of a novel mRNA and its product, dendrin. *Mol. Cell Neurosci.* 8:367–74

Hinds HL, Ashley CT, Sutcliffe JS, Nelson DL, Warren ST, et al. 1993. Tissue specific expression of FMR-1 provides evidence for a functional role in fragile X syndrome. *Nat. Genet.* 3:36–43

Huber KM, Kayser MS, Bear MF. 2000. Role for rapid dendritic protein synthesis in hippocampal mGluR-dependent long-term depression. *Science* 288:1254–56

Kang H, Schuman EM. 1995. Long-lasting neurotrophin-induced enhancement of synaptic transmission in the adult hippocampus. *Science* 267:1658–62

Kang H, Schuman EM. 1996. A requirement for local protein synthesis in neurotrophin-induced synaptic plasticity. *Science* 273: 1402–6

Kiebler MA, Hemraj I, Verkade P, Kohrmann M, Fortes P, et al. 1999. The mammalian Staufen protein localizes to the somatodendritic domain of cultured hippocampal neurons: implications for its involvement in mRNA transport. *J. Neurosci.* 19:288–97

Knowles RB, Sabry JH, Martone ME, Deerinck TJ, Ellisman MH, et al. 1996. Translocation of RNA granules in living neurons. *J. Neurosci.* 16:7812–20

Kohrmann M, Luo M, Kaether C, DesGroseiller L, Dotti CG, Kiebler MA. 1999. Microtubule-dependent recruitment of Staufen-green fluorecent protein into

RNA-containing granules and subsequent dendritic transport in living hippocampal neurons. *Mol. Biol. Cell* 10:2945–53

Laudry CF, Watson JB, Handley VW, Campagnoni AT. 1993. Distribution of neuronal and glial mRNAs within neuronal cell bodies and processes. *Soc. Neurosci. Abstr.* 19:1745

Leski ML, Steward O. 1996. Synthesis of proteins within dendrites: ionic and neurotransmitter modulation of synthesis of particular polypeptides characterized by gel electrophoresis. *Neurochem. Res.* 21:681–90

Levy WB, Steward O. 1979. Synapses as associative memory elements in the hippocampal formation. *Brain Res.* 175:233–45

Levy WB, Steward O. 1983. Temporal contiguity requirements for long-term associative potentiation/depression in the hippocampus. *Neuroscience* 7:791–97

Link W, Konietzko G, Kauselmann G, Krug M, Schwanke B, et al. 1995. Somatodendritic expression of an immediate early gene is regulated by synaptic activity. *Proc. Natl. Acad. Sci. USA* 92:5734–38

Lyford G, Yamagata K, Kaufmann W, Barnes C, Sanders L, et al. 1995. Arc, a growth factor and activity-regulated gene, encodes a novel cytoskeleton-associated protein that is enriched in neuronal dendrites. *Neuron* 14:433–45

Marion RM, Fortes P, Beloso A, Dotti C, Ortin J. 1999. A human sequence homology of staufen is an RNA-binding protein that is associated with polysomes and localizes to the endoplasmic reticulum. *Mol. Cell. Biol.* 19:2212–19

Martin KC, Casadio A, Zhu H, E Y, Rose JC, et al. 1997. Synapse-specific, long-term facilitation of Aplysia sensory to motor synapses: a function for local protein synthesis in memory storage. *Cell* 91:927–38

Mayford M, Bach ME, Huang Y-Y, Wang L, Hawkins RD, Kandel ER. 1996. Control of memory formation through regulated expression of a CaMKII transgene. *Science* 274:1678–83

Miyashiro K, Dichter M, Eberwine J. 1994.

On the nature and differential distribution of mRNAs in hippocampal neurites: implications for neuronal functioning. *Proc. Natl. Acad. Sci. USA* 91:10800–4

Montarolo PG, Goelet P, Castellucci V, Morgan J, Kandel ER, Schacher S. 1986. A critical period for macromolecular synthesis in long-term heterosynaptic facilitation in aplysia. *Science* 234:1249–54

Nguyen PV, Kandel ER. 1996. A macromolecular synthesis-dependent late phase of long-term potentiation requiring cAMP in the medial perforant pathway of rat hippocampal slices. *J. Neurosci.* 16:3189–98

Nguyen PV, Kandel ER. 1997. Brief theta burst stimulation induced a transcription-dependent late phase of LTP requiring cAMP in area CA1 of the mouse hippocampus. *Learn. Mem.* 4:230–43

Osten P, Valsamis L, Harris A. 1996. Protein synthesis-dependent formation of protein kinase M zeta in long-term potentiation. *J. Neurosci.* 16:2444–51

Otani S, Abraham WC. 1989. Inhibition of protein synthesis in the dentate gyrus, but not the entorhinal cortex, blocks maintenance of long-term potentiation in rats. *Neurosci. Lett.* 106:175–80

Otani S, Marshall CJ, Tate WP, Goddard GV, Abraham WC. 1989. Maintenance of long-term potentiation in rat dentate gyrus requires protein synthesis but not messenger RNA synthesis immediate post-tetanization. *Neuroscience* 28:519–26

Ouyang Y, Kantor DB, Harris KM, Schuman EM, Kennedy MB. 1997. Visualization of the distribution of autophosphorylated calcium/calmodulin-dependent protein kinase II after tetanic stimulation in the CA1 area of the hippocampus. *J. Neurosci.* 17:5416–27

Ouyang Y, Rosenstein A, Kreiman G, Schuman EM, Kennedy MB. 1999. Tetanic stimulation leads to increased accumulation of Ca(2+)/calmodulin-dependent protein kinase II via dendritic protein synthesis in hippocampal neurons. *J. Neurosci.* 19:7823–33

Palacios-Pru EL, Miranda-Contreras L, Mendoza RV, Zambrano E. 1988. Dendritic RNA and postsynaptic density formation in chick cerebellar synaptogenesis. *Neuroscience* 24:111–18

Palacios-Pru EL, Palacios L, Mendoza RV. 1981. Synaptogenetic mechanisms during chick cerebellar cortex development. *J. Submicrosc. Cytol.* 13:145–67

Paradies MA, Steward O. 1997. Multiple subcellular mRNA distribution patterns in neurons: a nonisotopic *in situ* hybridization analysis. *J. Neurobiol.* 33:473–93

Perlin JB, Gerwin CM, Panchision DM, Vick RS, Jakoi ER, DeLorenzo RJ. 1993. Kindling produces long-lasting and selective changes in gene expression in hippocampal neurons. *Proc. Natl. Acad. Sci. USA* 90:1741–45

Pierce JP, van Leyen K, McCarthy JB. 2000. Translocation machinery for synthesis of integral membrane and secretory proteins in dendritic spines. *Nat. Neurosci.* 3:311–13

Prakash N, Fehr S. Mohr E. Richter D. 1997. Dendritic localization of rat vasopressin mRNA: ultrastructural analysis and mapping of targeting elements. *Eur. J. Neurosci.* 9:523–32

Racca C, Gardiol A, Triller A. 1997. Dendritic and postsynaptic localizations of glycine receptor alpha subunit mRNAs. *J. Neurosci.* 17:1691–1700

Rao A, Steward O. 1991a. Evidence that protein constituents of postsynaptic membrane specializations are locally synthesized: analysis of proteins synthesized within synaptosomes. *J. Neurosci.* 11:2881–95

Rao A, Steward O. 1991b. Synaptosomal RNA: assessment of contamination by glia and comparison with total RNA. *Soc. Neurosci. Abstr.* 17:379

Raymond CR, Thompson VA, Tate WR, Abraham WC. 2000. Metabotropic glutamate receptors trigger homosynaptic protein synthesis to prolong long-term potentiation. *J. Neurosci.* 20:969–76

Righi M, Tongiorgi E, Cattaneo A. 2000. Brain-derived neurotrophic factor (BDNF) induces dendritic targeting of BDNF and tyrosine kinase B mRNAs in hippocampal neurons through a phosphatidylinositol-3 kinase-dependent pathway. *J. Neurosci.* 20:3165–74

Schuman EM. 1999. mRNA trafficking and local protein synthesis at the synapse. *Neuron* 23:645–8

Schuman EM, Madison DV. 1994. Locally distributed synaptic potentiation in the hippocampus. *Science* 263:532–36

Severt WL, Biber TUL, Wu X-Q, Hecht NB, DeLorenzo RJ, Jakoi ER. 2000. The suppression of testis-brain RNA binding protein and kinesin heavy chain disrupts mRNA sorting in dendrites. *J. Cell. Sci.* 112:3691–702

Sheetz AJ, Nairn AC, Constantine-Paton M. 1997. N-methyl-D-aspartate receptor activation and visual activity induce elongation factor-2 phosphorylation in amphibian tecta: a role for N-methyl-D-aspartate receptors in controlling protein synthesis. *Proc. Natl. Acad. Sci. USA* 94:14770–75

Sheetz AJ, Nairn AC, Constantine-Paton M. 2000. NMDA receptor-mediated control of protein synthesis at developing synapses. *Nat. Neurosci.* 3:211–16

Smith WB, Aakalu GN, Tsung MI., Reis GF, Schuman EM. 1999. Neurotrophin-induced increase in local protein synthesis in cultured hippocampal neurons. *Soc. Neurosci. Abstr.* 25:467

Spacek J, Hartmann M. 1983. Three-dimensional analysis of dendritic spines. I. Quantitative observations related to dendritic spines and synaptic morphology in cerebral and cerebellar cortices. *Anat. Embryol.* 167:289–310

Stanton PK, Sarvey JM. 1984. Blockade of long-term potentiation in rat hippocampal CA1 region by inhibitors of protein synthesis. *J. Neurosci.* 4:3080–84

Steward O. 1983. Polyribosomes at the base of dendritic spines of CNS neurons: their possible role in synapse construction and modification. *Cold Spring Harbor Symp. Quant. Biol.* 48:745–59

Steward O. 1997. mRNA localization in neurons: a multipurpose mechanism. *Neuron* 18: 9–12

Steward O, Falk PM. 1986. Protein synthetic machinery at postsynaptic sites during synaptogenesis; a quantitative study of the association between polyribosomes and developing synapses. *J. Neurosci.* 6: 412–23

Steward O, Falk PM, Torre ER. 1996. Ultrastructural basis for gene expression at the synapse: synapse-associated polyribosome complexes. *J. Neurocytol.* 25:717–34

Steward O, Fass B. 1983. Polyribosomes associated with dendritic spines in the denervated dentate gyrus: evidence for local regulation of protein synthesis during reinnervation. *Prog. Brain Res.* 58:131–36

Steward O, Halpain S. 1999. Lamina-specific synaptic activation causes domain-specific alterations in dendritic immunostaining for MAP2 and CAM kinase II. *J. Neurosci.* 15:7834–45

Steward O, Levy WB. 1982. Preferential localization of polyribosomes under the base of dendritic spines in granule cells of the dentate gyrus. *J. Neurosci.* 2:284–91

Steward O, Pollack A, Rao A. 1991. Evidence that protein constituents of postsynaptic membrane specializations are locally synthesized: time course of appearance of recently synthesized proteins in synaptic junctions. *J. Neurosci. Res.* 30:649–60

Steward O, Reeves TM. 1988. Protein synthetic machinery beneath postsynaptic sites on CNS neurons: association between polyribosomes and other organelles at the synaptic site. *J. Neurosci.* 8:176–84

Steward O, Ribak CE. 1986. Polyribosomes associated with synaptic sites on axon initial segments: localization of protein synthetic machinery at inhibitory synapses. *J. Neurosci.* 6:3079–85

Steward O, Singer RH. 1997. The intracellular mRNA sorting system: postal zones, zip codes, mail bags and mail boxes. In *mRNA Metabolism and Post-Transcriptional Gene Regulation*, ed. JB Hartford, DR Morris, pp. 127–46. New York: Wiley-Liss

Steward O, Wallace CS, Lyford GL, Worley PF. 1998. Synaptic activation causes the mRNA for the IEG Arc to localize selectively near activated postsynaptic sites on dendrites. *Neuron* 21:741–51

Tang SJ, Meulemans D, Vasquez L, Schuman EM. 1999. The role of a rat staufen-like protein in dendritic RNA targeting of hippocampal neurons. *Proc. Annu. Meet. Soc. Neurosci.* 25:468

Tang SJ, Smith WB, Schuman ME. 1998. Identification of components of a translational signaling pathway at synaptic sites in the hippocampus. *Proc. Annu. Meet. Soc. Neurosci.* 24:328

Tongiorgi E, Righi M, Cattaneo A. 1997. Activity-dependent dendritic targeting of BDNF and TrkB mRNAs in hippocampal neurons. *J. Neurosci.* 17:9492–505

Torre ER, Steward O. 1996. Protein synthesis within dendrites: distribution of the endoplasmic reticulum and the Golgi apparatus in dendrites of hippocampal neurons in culture. *J. Neurosci.* 16:5967–78

Valentine G, Chakravarty S, Sarvey J, Bramham C, Herkenham M. 2000. Fragile X (fmr1) mRNA expression is differentially regulated in two adult models of activity-dependent gene expression. *Mol. Brain Res.* 75:337–41

Watson JB, Coulter PM, Margulies JE, de Lecea L, Danielson PE, et al. 1994. G-protein gamma7 subunit is selectively expressed in medium-sized neurons and dendrites of the rat neostriatum. *J. Neurosci. Res.* 39:108–16

Weiler IJ, Greenough WT. 1991. Potassium ion stimulation triggers protein translation in synaptoneuronsomal polyribosomes. *Mol. Cell. Neurosci.* 2:305–14

Weiler IJ, Greenough WT. 1993. Metabotropic glutamate receptors trigger postsynaptic protein synthesis. *Proc. Natl. Acad. Sci. USA* 90:7168–71

Weiler IJ, Irwin SA, Klintsova AY, Spencer CM, Brazelton AD, et al. 1997. Fragile X mental retardation protein is translated near synapses in response to neurotransmitter activation. *Proc. Natl. Acad. Sci. USA* 94:5395–400

White G, Levy WB, Steward O. 1988. Evidence that associative interactions between afferents during the induction of long-term potentiation occur within local dendritic domains. *Proc. Natl. Acad. Sci. USA* 85: 2368–72

White G, Levy WB, Steward O. 1990. Spatial overlap between populations of synapses determines the extent of their associative interaction during the induction of long term potentiation and depression. *J. Neurophysiol.* 64:1186–98

Wickham L, Duchaine T, Luo M, Nabi IR, DesGroseillers L. 1999. Mammalian staufen is a double-stranded RNA- and tubulin-binding protein which localizes to the rough endoplasmic reticulum. *Mol. Cell. Biol.* 19:2220–30

Wu L, Wells D, Tay J, Mendis D, Abborr M-A, et al. 1998. CPEB-mediated cytoplasmic polyadenylation and the regulation of experience-dependent translation of alpha-CaMKII at synapses. *Neuron* 21:1129–39

Annu. Rev. Neurosci. 2001. 24:327–55

Signaling and Transcriptional Mechanisms in Pituitary Development

Jeremy S. Dasen and Michael G. Rosenfeld

*Howard Hughes Medical Institute, Cellular and Molecular Medicine,
University of California, San Diego, La Jolla, California 92093-0648;
e-mail: jdasen@ucsd.edu; mrosenfeld@ucsd.edu*

Key Words transcription factor, morphogenesis, synergy, BMP, FGF

■ **Abstract** During the development of the pituitary gland, distinct hormone-producing cell types arise from a common population of ectodermal progenitors, providing an instructive model system for elucidating the molecular mechanisms of patterning and cell type specification in mammalian organogenesis. Recent studies have established that the development of the pituitary occurs through multiple sequential steps, allowing the coordinate control of the commitment, early patterning, proliferation, and positional determination of pituitary cell lineages in response to extrinsic and intrinsic signals. The early phases of pituitary development appear to be mediated through the activities of multiple signaling gradients emanating from key organizing centers that give rise to temporally and spatially distinct patterns of transcription factor expression. The induction of these transcriptional mediators in turn acts to positionally organize specific pituitary cell lineages within an apparently uniform field of ectodermal progenitors. Ultimately, pituitary cell types have proven to be both specified and maintained through the combinatorial interactions of a series of cell-type-restricted transcription factors that dictate the cell autonomous programs of differentiation in response to the transient signaling events.

INTRODUCTION

Understanding the molecular mechanisms by which diverse and specialized cell types emerge during the development of multicellular organisms remains a critical question in biology. Based initially on genetic studies in *Drosophila melanogaster*, early patterning events in vertebrate organogenesis have been found to be coordinated through the interplay of highly organized signaling cues (Edlund & Jessell 1999). Extrinsic signals, provided in the form of secreted morphogens or through transmembrane signaling receptors, create local environments for the positional determination of progenitor cell types and orchestrate the patterning of organs through the induction of specific proliferative or apoptotic events. These signals are subsequently translated into the intrinsic or cell-autonomous determination

0147-006X/01/0621-0327$14.00

327

programs, in part through modulation in the activity or expression of cell type-restricted transcriptional regulators. Extrinsic signaling events have been demonstrated to be mediated through the activities of multiple members of a relatively small family of different classes of molecules that include members of the transforming-growth-factor beta superfamily, Wnts, hedgehogs, fibroblast growth factors (FGFs), epidermal growth factors and retinoids (reviewed by Hogan 1999). How such an apparently small repertoire of signaling molecules can lead to the patterning of diverse organs across multiple species remains to be fully understood.

The development of the pituitary gland has provided a particularly useful model system in which to study these complex processes because the cell types are derived from a common ectodermal primordium and arise in a distinct spatial and temporal fashion in response to intrinsic and extrinsic signals. The pituitary gland is of dual embryonic origin and arises through intimate association of neural ectoderm and oral roof ectoderm. The mature pituitary gland is ultimately composed of three lobes; the anterior and intermediate lobes contain the six hormone-secreting cell types and are derived from oral ectoderm, whereas the posterior lobe, containing the axonal projections emanating from the hypothalamus, is derived from neural ectoderm.

Each cell type of the anterior pituitary gland is characterized by the secretion of one or more trophic hormones that regulate a diverse range of important biological processes in response to signals from the hypothalamus and peripheral organs (Figure 1, see color insert). Two cell types synthesize proopiomelanocortin (POMC), which is cleaved by proteolytic processing to generate melanocyte-stimulating hormone in melanotropes and adrenocorticotropin (ACTH) in corticotropes. ACTH regulates metabolic function through stimulation of glucocorticoid synthesis in the adrenal cortex whereas melanocyte-stimulating hormone regulates melatonin synthesis in the epidermis in some vertebrate species. Somatotropes produce growth hormone (GH) and regulate linear growth and metabolism, whereas lactotropes secrete prolactin (Prl) which regulates milk production in females. Thyrotropes produce thyroid-stimulating hormone (TSH), which controls the secretion of thyroid hormone from the thyroid gland. Gonadotropes produce luteinizing hormone (LH) and follicle-stimulating hormone (FSH) which regulate reproductive development and function. TSH, LH, and FSH are heterodimeric glycoproteins consisting of a common α subunit (α-GSU) and a specific β subunit.

In recent years considerable progress has been made in identifying the critical transcription factors responsible for the appearance of these distinct cell phenotypes (Watkins-Chow & Camper 1998, Dasen & Rosenfeld 1999b, Sheng & Westphal 1999). An unresolved issue has been the mechanisms by which these cell types arise from a common population of progenitor cells in response to morphogen cues. In this review, we describe recent studies that have investigated the roles of signaling gradients that govern the organ commitment, patterning, and positional determination of cell types in the developing pituitary gland. We illustrate the mechanisms by which multiple signaling centers can act to positionally establish cell types through the induction of overlapping patterns of transcription factor expression. We also focus on the aspects of pituitary cell type specification that are ultimately mediated through the combinatorial interactions of a series of cell type-specific or -restricted transcription factors which serve to establish the cell type-specific molecular memory of the transient signaling events.

SIGNALING MECHANISMS IN PITUITARY ORGANOGENESIS

The cell types of the pituitary gland are initially derived from the most anterior midline portion of the embryo in a region contiguous with the anterior neural ridge (Figure 2, see color insert) (Couly & Le Douarin 1988, Eagleson & Harris 1990, Rubenstein et al 1998). The anterior neural ridge is displaced ventrally upon embryonic head turn fold to form the oral epithelium, which gives rise to the roof of the mouth and its derived structures. The onset of pituitary organogenesis coincides with a thickening of this initially uniform oral roof ectoderm on embryonic-day (e) 8.5 (e8.5) in mouse ontogeny, which invaginates to form the structure referred to as Rathke's pouch on e9.0. The pouch epithelium continues to proliferate as it closes and separates from the underlying oral ectoderm with the progenitors of the hormone-secreting cell types arising from the ventral proliferation of cells that will populate the anterior lobe of the pituitary gland.

Coincident with the invagination of oral ectoderm, a portion of the ventral diencephalon (the infundibulum) evaginates and makes direct contact with the dorsal portion of Rathke's pouch. Classical embryological experiments have demonstrated that this contact of neuroectoderm with Rathke's pouch is required for the determination of pituitary cell types, providing early evidence that the infundibulum acts as a key organizing center (Ferrand 1972; Daikoku et al 1982; Watanabe 1982a,b; Fedtsova & Barabanov 1990; Kawamura & Kikuyama 1992, 1995). This observation received genetic confirmation initially from analysis of mice disrupted for the homeodomain protein T/ebp (Nkx2.1) which is expressed in discrete regions of the brain, including the ventral diencephalon, but is excluded from expression within Rathke's pouch. In the *T/ebp* gene-deleted mice, ventral regions of the brain including the infundibulum are absent, providing a genetic model with which to study the influence of the infundibular signals (Kimura et al 1996). Consistent with a critical extrinsic signaling role for the infundibulum in pituitary organogenesis, all three lobes of the pituitary gland are absent in *T/ebp*-knockout mice. A central objective has thus been to define and characterize the putative infundibular-signaling molecules involved in the early inductive processes.

BMP4 Signaling and Pituitary Organ Commitment

The transforming-growth-factor beta superfamily of secreted signaling molecules, which includes several bone-morphogenetic proteins (BMPs), has been demonstrated to play critical roles in patterning and cell type specification in several species [reviewed by Hogan (1996)]. *BMP4* is expressed in the ventral diencephalon as the infundibulum makes direct contact with Rathke's pouch at e8.5–e9.0 and may be one of the early signaling factors required for the initial commitment of a subpopulation of oral ectodermal cells to form the pituitary gland (Figure 2, see color insert) (Ericson et al 1998, Treier et al 1998). The role of BMP4 signaling in pituitary organogenesis has been investigated in vivo by targeted expression of the BMP2/4 antagonist Noggin, using the regulatory sequences of the *Pitx1* gene to target expression throughout the oral ectoderm and within Rathke's pouch. In Pitx1/Noggin-transgenic mice, pituitary development is arrested at e10, and there is a complete absence of all pituitary cell types and a failure of the characteristic ventral proliferation of cells from the pouch beginning at e11.5 (Treier et al 1998). Based on the similarity of this phenotype with that of mice with a targeted disruption of the *Lhx3* (*P-Lim/mLim3*) gene (Seidah et al 1994, Bach et al 1995, Zhadanov et al 1995), a LIM homeodomain protein critical for the determination of most pituitary cell types (Sheng et al 1996), BMP4 signaling appears to be required in the initial phases of organ commitment.

The role of BMP signaling in early pituitary development has also been investigated in *BMP4* gene-deleted animals, in which the initial invagination of Rathke's pouch fails to occur (Takuma et al 1998). Because most *BMP4* knockout animals die at or near the time of pituitary organ commitment (Winnier et al 1995), definitive proof of a requirement for BMP4 signaling in the initial invagination event will await the generation of a neural-specific knockout animal. Thus the signals required for the initial invagination of Rathke's pouch are still unclear, although a recent study has suggested a potential role for the notochord in this process (Gleiberman et al 1999).

Infundibular Fibroblast Growth Factor Signaling Provides Proliferative and Positional Cues to Rathke's Pouch

In addition to BMP4, multiple members of the FGF family are expressed in the infundibulum and have been demonstrated to play critical roles in both the organ morphogenesis and positionally restricted determination of pituitary cell lineages. Subsequent to *BMP4* induction, *FGF8* and *FGF10* are expressed in a temporally and spatially overlapping manner within the infundibulum throughout the early phases of pituitary development. The role of FGF8 has been investigated both in vitro using pituitary explant cocultures and in vivo through generation and analysis of transgenic and gene-disrupted animals. Cultivation of Rathke's pouch with ventral diencephalon has provided strong evidence for an instructive role for FGF signaling in the early phases of pituitary development (Treier et al 1998, Ericson et al 1998). In culture, the infundibulum is both required and sufficient for the induction of *Lhx3/P-Lim* gene expression in pouch explants. In the absence of the infundibulum, *Lhx3* expression can be induced by culture of explants with FGF8 or FGF2, suggesting that FGFs are required for the maintenance or initial induction of *Lhx3* gene expression. Additionally, expression of *Lhx3* early in development has been observed to be highest in the dorsal aspect of the developing gland (Ericson et al 1998), consistent with its expression being regulated, in part, by dorsal signaling molecules.

The role of dorsal FGF signaling has also been studied in vivo through generation of transgenic animals misexpressing FGF8 in the ventral regions of the pituitary under control of the regulatory sequences for the αGSU gene, which targets expression to Rathke's pouch and later to ventral cell types. In αGSU/FGF8-transgenic animals, most ventral and intermediate cell types are absent, and the pituitary is characterized by a hyperplasia of corticotropes and melanotropes, consistent with a patterning role for FGFs in the positional determination of dorsally arising pituitary cell types (Treier et al 1998). This phenotype is also associated with an expanded population of pouch ectoderm, similar to the dysmorphogenesis of Rathke's pouch observed in Ames dwarf mice (Sornson et al 1996, Gage et al 1996a), suggesting that FGFs also contribute to the proliferation of pituitary progenitor cells. The requirement for FGF8 has been further suggested based on analysis of *T/ebp* gene-deleted mice, in which *FGF8* fails to be expressed in the ventral diencephalon, and this absence of FGF8 signaling has been suggested to be directly linked to a loss of *Lhx3* expression (Takuma et al 1998).

Analysis of mice deleted for FGF receptor type 2 (FGFR2) has provided further genetic evidence for an essential role for FGF signaling. The *FGFR2* gene is differentially spliced into two receptor isoforms, each of which displays different ligand specificities. By gene-targeting strategies, mice lacking the IIIb isoform of FGFR2 were generated, which presumably would abolish signaling through FGFs 1, 3, 7, and 10 (De Moerlooze et al 2000). In FGFR2 (IIIb)-null mice, Rathke's pouch forms but rapidly undergoes apoptosis with the pituitary becoming completely absent by e14.5. These studies suggest a critical role for FGF10 signaling for the continued proliferation of the pouch ectoderm, although pituitary defects in *FGF10*-deleted mice (Min et al 1998, Sekine et al 1999) have not yet been reported. Thus, similar to the well-established roles of FGFs in limb bud and lung morphogenesis (Martin 1998), FGF8 and FGF10 appear to play critical roles in the early patterning and proliferation events in pituitary organogenesis.

Bone-Morphogenetic Protein 2 Signaling Positionally Specifies Ventral Pituitary Cell Lineages

In addition to the dorsal BMP4 and FGF8/10 signals emanating from the infundibulum, early patterning events in pituitary organogenesis are also governed through the activities of ventral and pouch-intrinsic signals, which contribute to the establishment of the positional identity of ventral pituitary cell types. Expression of *BMP2* is initially detected in the most ventral aspect of the invaginating gland at e9.5, at the ventral boundary between Rathke's pouch and *Sonic hedgehog* (*Shh*) expression throughout the oral ectoderm (Treier et al 1998). *BMP2* expression later expands throughout the pouch by e12.5 with expression of BMP2/4 antagonist *chordin* in the caudal mesenchyme, potentially serving to maintain a ventrodorsal BMP2 gradient (Figure 2, see color insert). Subsequent to the closure of Rathke's pouch and its separation from the oral ectoderm, *BMP2* expression is also detected in the ventral juxtapituitary mesenchyme, in a region adjacent to the pituitary cell types characterized by expression of ventrally induced transcription factors *GATA-2, Isl-1*, and *P-Frk* as well as the hormone subunit αGSU (Treier et al 1998, Ericson et al 1998).

The role of BMP2 signaling has been investigated in vivo through ventralized overexpression of BMP2/4 under the control of *αGSU* regulatory information, which leads to a dorsal expansion in the expression domains of ventral lineage markers *Isl-1* and *Msx-1* and the direct transcriptional induction of *GATA-2* gene expression (Treier et al 1998). Similarly, cultivation of Rathke's pouch in the presence of BMP2 is sufficient for the induction of *Isl-1* and αGSU expression (Ericson et al 1998). Based on the ability of BMP2 to induce αGSU gene expression, an early marker for the thyrotrope and gonadotrope cell populations, these studies suggest that BMP2 signaling specifies the progenitors that will later give rise to ventral pituitary cell types. The attenuation of BMP signaling, however, is also required for the developmental progression of pituitary cell types because overexpression of BMP in vivo prevents terminal differentiation, possibly due to maintained expression of *Msx-1* (Treier et al 1998), a repressor homeodomain factor known to inhibit terminal differentiation in myogenesis (Song et al 1992).

Opposing Bone-Morphogenetic Protein- and Fibroblast Growth Factor-Signaling Gradients

A critical component by which physically opposing signaling gradients can act to positionally determine cell types has emerged from studies on dorsal-FGF and ventral-BMP2 signaling in pituitary development. Acting on a uniform cell population, multiple signaling pathways can be hypothesized to exert either cooperative or antagonistic effects on cellular proliferation and/or differentiation. In pituitary explant cultures, the ability of the infundibulum or FGFs to induce *Lhx3* gene expression coincides with the restricted expression of the BMP2-induced genes *Isl-1* and αGSU away from the source of the FGF signal (Ericson et al 1998). Similarly, the ability of ventralized expression of FGF8 to prevent the appearance of ventral and intermediate cell types in vivo can be attributed to the inhibition of ventral BMP2 signaling (Treier et al 1998). Conversely, while cultivation of Rathke's pouch with BMP2/4 initiates the expression of the ventral markers *Isl-1* and αGSU, it inhibits the expression of more dorsal cell type markers such as ACTH in vitro (Ericson et al 1998) and *Pit-1* in vivo (Treier et al 1998). Thus antagonistic and opposing dorsal → ventral FGF8 and ventral → dorsal BMP2 gradients appear to be associated with the positional determination of dorsal- and ventral-cell types, respectively. These opposing gradients parallel events in tooth development, in which opposing BMP and FGF gradients establish differential domains of *Pax9* expression and define specific regions of tooth morphogenesis (Neubuser et al 1997, Peters & Balling 1999).

Ventral Sonic Hedgehog Signaling and Compartmentalization of the Pouch

In the development of *Drosophila* appendages, the secreted protein Hedgehog (Hh) plays a crucial role in defining the border between the anterior and posterior compartments in the imaginal disk (Dahmann & Basler 1999). A vertebrate homolog of Hh, sonic hedgehog (Shh), is expressed in several organizing centers during embryogenesis, where it exerts crucial patterning roles (Hammerschmidt et al 1997). During the early development of the pituitary gland, *Shh* is expressed throughout the oral ectoderm but is excluded from the region that forms Rathke's pouch, creating a potential molecular compartmental boundary within the continuous ectoderm (Figure 2, see color insert).

Studies in several species have provided suggestive evidence for a role of Hedgehog signaling in early phases of pituitary morphogenesis. In *Xenopus laevis* it has been demonstrated that, in animal cap explants cultured with *banded hedgehog*, the homolog of the mammalian *Indian hedgehog* (*ihh*) gene, the expression domains of pituitary-restricted factors such as *XANF-2* [a homolog of the mammalian homeodomain factor *rpx/Hesx1* (Thomas & Rathjen 1992, Hermesz et al 1996)] are expanded, consistent with a role for Hedgehog signaling in control of proliferative events in pituitary development (Ekker et al 1995). Similarly, in vivo ventral overexpression of *Shh* can expand the population of ventral pituitary cell types and modify levels of *Lhx3* gene expression (MT Treier & MG Rosenfeld, unpublished observation).

Evidence for a patterning role for hedgehog signaling in pituitary-cell-type specification has emerged from genetic studies in zebrafish. The *you-too* (*yot*) mutant was initially characterized by defects in optic chiasm and somite development and more recently has been shown to display defects in pituitary development. Positional cloning of the *yot* locus identified the defective gene as the zebrafish homolog of the mammalian Gli2 zinc finger protein (Karlstrom et al 1999). The Gli family of transcription factors are well established as downstream mediators of Hedgehog signaling and they act as positive or negative transcriptional regulators in multiple-organ systems (reviewed by Ruiz & Altaba [1999]). In *yot* mutants, the rostral expression domains (analogous to the ventral domains in mice) of pituitary-specific transcription factors such as *lim3* (*Lhx3*) and *six3* are lost, with other pituitary-restricted factors such as *nk 2.2* (*Nkx2.2*) completely absent. Given the sequential and cooperative roles that BMPs and Hedgehogs exert in limb and neural-tube development (Laufer et al 1994, Dale et al 1997), where Shh acts to induce expression of BMPs, Shh may act in a signaling cascade with BMP2 in the determination of ventral pituitary cell lineages.

Other Potential Morphogenetic Factors

Several other classes of signaling molecules have been implicated to exert roles in pituitary-cell-type specification. The Wnt proteins are a family of secreted proteins involved in a variety of early embryonic events and functions through transcriptional mediators such β-catenin and T-cell factor (reviewed by Eastman & Grosschedl 1999). In the pituitary, Wnt4 and Wnt5a are expressed in the ventral diencephalon and within the cells of Rathke's pouch, respectively. In Wnt4-mutant mice, the pituitary is mildly hypocellular, with the ventral-cell types showing normal differentiation but incomplete expansion. Additionally, cultivation of Rathke's pouch with Wnt5a and BMP4 can induce expression of the early cell type marker αGSU (Treier et al 1998). Thus Wnts and BMPs may act in synergy to expand pituitary cell lineages and induce cell determination programs.

Very little is currently known about the contribution of cytokines to development of mammalian organs outside the hematopoietic system, despite the potent ability of cytokines such as leukemia inhibitory factor (LIF) to maintain mouse embryonic stem cells in an undifferentiated state. A potential role for LIF in pituitary development was initially investigated in pituitary-derived cell lines where it was demonstrated that LIF can activate synthesis of the corticotrope-specific gene *ACTH* in combination with the hypothalamic peptide corticotropin-releasing hormone (Bousquet et al 1997). The role of LIF was further investigated in vivo through generation of transgenic mice expressing LIF in differentiated cell types and during early pituitary ontogeny. In transgenic animals expressing LIF under control of αGSU regulatory information, LIF has a potent morphogenetic effect on pituitary development, because most cell types fail to properly differentiate, and the pituitary is characterized by the formation of ciliated cysts of Rathke's pouch and corticotrope hyperplasia (Yano et al 1998). The formation of ciliated cells similar in morphology to cells present in the nasal epithelium suggests that many of the progenitor cells have been directed down a different, although related, developmental pathway. LIF therefore may act in concert with other signaling molecules to establish the identity of dorsal-cell phenotypes.

Retinoids are well established as morphogens in vertebrate organogenesis, and are involved in the patterning of cell types in many organs including the hindbrain, limbs, and neural tube [reviewed by Eichele (1997)]. Investigations into the role of retinoic acid (RA) signaling in pituitary development have thus far focused primarily on the direct regulation of pituitary-specific genes, such as the genes for growth hormone (*GH*) and *Pit-1*, where RA can activate gene expression through ligand-dependent synergy of the RA receptor (RAR) with the POU homeodomain protein Pit-1 (Rhodes et al 1993, Sanchez-Pacheco et al 1998, Palomino et al 1998). Because the activation of these genes occurs subsequently to the initial patterning events, the contribution of retinoids to the earlier phases of pituitary development, if any, is unresolved. The expression profile of the *RALDH2* gene, which encodes the enzyme that catalyzes the final reaction step in RA synthesis, has provided a useful marker for identifying the sites of RA activity during development. In addition to many other discrete regions of the embryo, *RALDH2* is highly expressed in the developing pituitary as early as e10.5 (Niederreither et al 1997), suggestive of a role for retinoid signaling in earlier phases of pituitary development.

TRANSCRIPTION FACTOR HIERARCHIES GOVERNING THE EARLY PHASES OF PITUITARY DEVELOPMENT

LIM Homeodomain Factors as Selector Genes

Multiple members of the LIM homeodomain family of transcription factors are expressed in Rathke's pouch, including *Lhx3* (*P-Lim/mLim3*), *Lhx4* (*Gsh4*), and *Isl-1*. *Lhx3* gene expression initiates coincident with the formation of Rathke's pouch at e9.5 and is one of the first transcription factors that is specifically expressed in the pouch and not throughout the oral ectoderm (Bach et al 1995, Zhadanov et al 1995). Targeted disruption of the genes for *Lhx3* and *Lhx4* have established that these factors play crucial roles in the earliest phases of pituitary organogenesis. In *Lhx3*-mutant mice, the rudiment of Rathke's pouch forms normally, but the pouch ectoderm fails to continue to proliferate and virtually all pituitary cell types are absent, with only a few corticotropes remaining (Sheng et al 1996). A highly related factor, Lhx4, serves a redundant role to Lhx3 as *Lhx4-/-* mice also show severe defects in pituitary development but only within the context of *Lhx3* heterozygote mutants (Sheng et al 1997). In addition to the established roles of Lhx3 and Lhx4 in pituitary development, Lhx3 and Lhx4 are also required for the appearance of specific motor neuron types within the neural tube (Sharma et al 1998).

Whereas *Lhx3* and *Lhx4* are expressed broadly throughout Rathke's pouch in early pituitary ontogeny, *Isl-1* is initially expressed throughout the pouch, but it later becomes restricted to the ventral-cell populations by e10.5. In *Isl-1*-mutant mice (Pfaff et al 1996), the invagination of Rathke's pouch occurs normally, but the epithelial cells forming Rathke's pouch fail to proliferate (Takuma et al 1998). Because *Isl-1* expression can be induced by BMPs both in vivo and in vitro, its expression in the pituitary gland has been suggested to be under the dual regulation of both the dorsal-BMP4 and ventral-BMP2 signals (Ericson et al 1998). Consistent with this hypothesis, the ventral restriction of *Isl-1* expression correlates with down-regulation of the BMP4 signal in the infundibulum and the appearance of BMP2 within the pouch and the ventral juxtapituitary mesenchyme.

Pitx Homeodomain Factors

Two *bicoid*-related Pitx homeodomain factors are expressed throughout pituitary ontogeny, demonstrating distinct and overlapping patterns of expression. Pitx1 (formerly referred to as P-Otx and Ptx1) was identified in screens for factors interacting with the N terminus of the pituitary-specific POU homeodomain protein

Pit-1 (Szeto et al 1996) and in a screen for factors regulating expression of the *POMC* gene promoter (Lamonerie et al 1996). The *Pitx1* gene is expressed in the earliest stages of pituitary organogenesis, initially at the anterior region of the neural plate in early mouse development, later becoming expressed throughout the oral ectoderm and subsequently in all pituitary cell types. *Pitx1* is also expressed in several other discrete regions of the embryo, including the first branchial arch and hind limbs (Szeto et al 1996, Lanctot et al 1997). Targeted disruption of the *Pitx1* gene leads to diminished expression of the terminal-differentiation markers for the ventral-gonadotrope and -thyrotrope pituitary cell types (Szeto et al 1999) as well as defects in craniofacial and hind-limb morphogenesis (Szeto et al 1999, Lanctot et al 1999b). The defects in two pituitary cell types are consistent with the observation that the *Pitx1* gene is preferentially expressed in the most ventral cell lineages (Lanctot et al 1999a), suggesting that Pitx1 is either negatively regulated by dorsal-signaling or positively regulated by ventral-signaling molecules. The reduced cell numbers in only two lineages, despite the pan-pituitary expression of Pitx1, also suggests that the second Pitx gene, *Pitx2*, can compensate for loss of *Pitx1* in the more dorsal cell types.

In the development of limbs, Pitx1 has also been shown to be critical for the components of transcriptional pathways that distinguish hind-limb from forelimb identity (Szeto et al 1999, Lanctot et al 1999b). In *Pitx1*-null mice, the hind limb assumes morphological features of the forelimb, and hind-limb-specific structures such as the patella are absent. Expression of the hind-limb-specific transcription factor *Tbx4* is also markedly decreased in *Pitx1*-mutant mice, and Pitx1 is sufficient to induce expression of *Tbx4* when misexpressed in the forelimb (Szeto et al 1999, Logan & Tabin 1999). Thus Pitx1, in addition to its role in ventral-pituitary-cell types, is critical in developmental decisions in other organs. The *Pitx2* gene was identified through the positional cloning of the gene responsible for the human disease Rieger syndrome, characterized by defects in eye, tooth, and umbilical cord development (Semina et al 1996). Like *Pitx1, Pitx2* gene expression is detectable in most cell types throughout early and late pituitary ontogeny but is also expressed in several other developing organs including the primordia of the heart, tooth, lung, gut, and eye. *Pitx2* gene-deleted mice are characterized by multiple severe developmental defects including a failure of ventral-body-wall closure, altered cardiac positioning, lung isomerisms, and defects in tooth and pituitary organogenesis (Gage et al 1999, Kitamura et al 1999, Lin et al 1999, Lu et al 1999). In the pituitary, organogenesis is arrested in the early stages of development, subsequent to the contact of the infundibulum with Rathke's pouch and establishment of early signaling gradients. Although the initial invagination of the pouch and induction of *Lhx3* gene expression occur normally, in *Pitx2* mutant mice, the gland fails to progress beyond e10.5, and it is characterized by a hypoplasia of the pouch ectoderm and a failure of the ectodermal cells to proliferate ventrally and populate the pituitary gland.

Much attention has also been given to the role of Pitx2 in events governing the establishment of asymmetry in the early embryo [reviewed by Capdevila et al (2000)], providing a potential model to understand *Pitx2* gene regulation in the developing pituitary in response to early signaling events. In vertebrate embryogenesis, the earliest asymmetric event is manifested in the looping of the heart, which is governed by inductive events occurring through the node, notochord, and the lateral-plate mesoderm (LPM). *Pitx2* is asymmetrically expressed in the left LPM, acts as a key determinant of "leftness" in multiple organ systems, and is regulated by the integrated activities of Shh, FGF8, and the transforming-growth-factor beta family member nodal. In mice, nodal and FGF8 appear to be required for the induction of *Pitx2* expression on the left side of the LPM, whereas Shh prevents its expression on the right side because *FGF8* mutants fail to express *Pitx2* and *Shh* mutants show bilateral expression of *Pitx2* on the LPM (Meyers & Martin 1999). It is intriguing that a similar cross-regulatory cascade of signaling molecules modulates expression of *Pitx2* within the pituitary.

Pitx and Lhx Factors Synergistically Regulate Pituitary Organogenesis and Pituitary-Specific Gene Expression

The striking similarity of the phenotypes of *Pitx2* and *Lhx3* gene-deleted mice suggests that these two classes of homeodomain factors collaborate to regulate the same or overlapping cohorts of pituitary-specific target genes. Indeed, LIM homeodomain factors can act synergistically with Pitx factors to activate expression of pituitary-specific genes such as α*GSU* (Bach et al 1997). The synergy between Lhx and Pitx factors has been further shown to be mediated by the broadly expressed cofactors CLIM1/Lbd-2 and CLIM2/Lbd-1/NLI, which were identified in screens for proteins interacting with the LIM domain zinc finger (Agulnick et al 1996, Jurata et al 1996, Bach et al 1997). A *Drosophila* homolog of CLIM/Lbd factors, called Chip, also modulates the activity of the LIM domain containing homeodomain Apterous (Morcillo et al 1997), suggesting that this interaction represents a conserved aspect of the function of LIM homeodomain factors. In summary, the induction of *Lhx3* gene expression in response to infundibular FGF signaling appears to be the critical step in the selection of oral ectoderm to assume the pituitary fate, allowing synergistic activation of pituitary-specific gene programs by Lhx3, Pitx2, and CLIM cofactors.

Ventral Progression and the Paired-Like Homeodomain Factors Prop-1 and Rpx

While the commitment and early proliferation of Rathke's pouch require the activities of the *Lhx* and *Pitx* genes, other factors are needed for the asymmetric progression of the ectoderm to generate the precursors of the hormone-secreting cell types. The proliferation of cell types from Rathke's pouch to populate the anterior lobe of the pituitary gland requires the activities of another homeodomain factor identified through positional cloning of the Ames dwarf (*df*) locus. Genetic and phenotypic analysis of *df* mice that were deficient in at least three pituitary cell types ultimately led to the isolation of the pituitary-specific *paired-like* homeodomain factor *Prophet of Pit-1* (*Prop-1*) (Buckwalter et al 1991, Andersen et al

1995, Gage et al 1996b, Sornson et al 1996, Watkins-Chow et al 1997). *Prop-1* expression is detected in the pituitary coincident with closure of Rathke's pouch at e10.5, and it is expressed throughout the pouch, becoming down-regulated coincident with the appearance of pituitary-cell-type terminal-differentiation markers between e15.5 and e16.5. Although the induction of *Pitx* and *Lhx* gene family members occurs and is maintained normally, *Prop-1*-defective mice fail to activate *Pit-1* gene expression, and subsequently fail to generate somatotropes, lactotropes, and thyrotropes. In Ames dwarf mice, the epithelial cells surrounding the lumen of Rathke's pouch fail to populate the anterior pituitary, generating expanded lumen and dysmorphogenesis of the pouch ectoderm (Sornson et al 1996, Gage et al 1999a). Prop-1 may therefore function to allow nascent pituitary cell types to delaminate or asymmetrically progress from the epithelium of the pouch (Figure 3, see color insert).

In addition to its role in the determination of the three Pit-1-dependent cell lineages (somatotropes, lactotropes, and thyrotropes), Prop-1 also appears to be required for the generation of a fourth anterior pituitary cell type, the gonadotrope, because patients with combined pituitary hormone deficiency-bearing mutations in the human *PROP1* gene are defective in gonadotropin hormone synthesis (Wu et al 1998, Deladoey et al 1999, Rosenbloom et al 1999). This difference in the severity of phenotype between the mouse and human mutations can be attributed to differences in the functional consequences of various Prop-1 mutations. Whereas the mouse Prop-1 point mutation in the first helix of the homeodomain (S83P) still retains some ability to bind DNA, a subset of human *PROP1* mutations in which the DNA-binding third helix of the homeodomain is deleted consistently are presented with severe defects in gonadotropin synthesis (Figure 3, see color insert). These more severe human *PROP1* mutations have also been reported to be present with defects in ACTH synthesis (Pernasetti et al 2000), suggesting that Prop-1 functions in the appearance or expansion of all anterior pituitary cell lineages, including the non–Pit-1-dependent gonadotrope and corticotrope.

A second *paired*-like homeodomain factor expressed in the pituitary, rpx/Hesx1, belongs to diverse family of homeodomain proteins containing a conserved N terminal "eh-1" repressor domain originally identified in the *Drosophila* engrailed homeodomain (Smith & Jaynes 1996). Like the *Pitx* genes, *rpx* expression initiates at the anterior neural plate and is later restricted to the oral ectoderm and Rathke's pouch. Attenuation of *rpx* expression coincides with the appearance of terminal differentiation markers for anterior pituitary cell types (Hermesz et al 1996, Sornson et al 1996), suggesting that *rpx* down-regulation is required for their developmental progression.

In Prop-1-defective mice, there is an extension of *rpx* expression beyond e13.5 (Gage et al 1996a, Sornson et al 1996), and *rpx/Hesx-1*–deleted mice often show pituitary dysmorphogenesis, although quite distinct from the defect in Prop-1-defective Ames mice (Dattani et al 1998). Rpx is capable of heterodimerization with Prop-1 on consensus-binding sites and can inhibit Prop-1 activity, suggesting that Rpx acts to antagonize Prop-1 function in vivo (Sornson et al 1996). A similar type of antagonism based on heterodimerization between two *paired*-like homeodomain factors has also been demonstrated in *Xenopus* development, in which Mix1 can block the axis duplication induced by the siamois homeodomain protein (Mead et al 1996). The functional consequences of Prop-1/Rpx interactions may therefore define the proper spatial and temporal proliferation of the pouch ectoderm into the nascent anterior pituitary gland.

Other Homeodomain Factors in Pituitary Development

The expression of several other classes of homeodomain factors in early and late pituitary ontogeny has been described, and their roles are currently under investigation. In addition to its well-documented role in the development of the eye, the paired homeobox factor Pax-6 is expressed in the oral ectoderm and becomes restricted to Rathke's pouch in early development. In the developing pituitary, Pax-6 is transiently expressed in the dorsal region of Rathke's pouch becoming down-regulated coincident with cell type differentiation. In the pituitaries of Pax-6 mutant small eye (*sey*) mice, there is an expansion of the ventral αGSU-expressing cell types, predominantly thyrotropes, with a reciprocal loss of the more dorsal somatotrope lineage (Bentley et al 1999, Kioussi et al 1999). Pax-6 therefore appears to be required for refining the dorsal/ventral boundaries between the thyrotrope/gonadotrope and somatotrope/lactotrope progenitor fields. This phenotype is reminiscent of the role that Pax6 exerts in the dorsal/ventral patterning of cell types in the neural tube, where Pax-6 acts as a negative regulator of ventral Shh signaling (Ericson et al 1997).

In the development of the *Drosophila* eye, the Pax-6 homolog *eyeless* has been implicated to play a role in collaboration with several other nuclear factors including the *Drosophila* proteins sine oculus (so) and eyes absent (eya). Expression of the mammalian homologs of these proteins has been reported in the developing pituitary, including the *so*-related homeodomain factors Six-3 and Six-6 (Sornson et al 1996, Jean et al 1999). Mutations in the human *SIX6* gene have been reported to be responsible for pituitary anomalies (Gallardo et al 1999). The mammalian homologs of these *Drosophila* factors may also therefore contribute to the early development of the pituitary.

TRANSCRIPTIONAL CONTROL OF PITUITARY-CELL-TYPE SPECIFICATION AND HOMEOSTASIS

Pit-1 Is Required for the Generation of Three Pituitary Cell Lineages

Pit-1 is prototypic of a large family of POU domain-containing transcription factors, which contains a bipartite DNA-binding domain. The POU domain consists of an amino-terminal POU-specific domain separated by a short linker to a carboxy-terminal POU homeodomain. Pit-1 (also referred to as GHF-1) was originally identified through analysis of the proteins regulating the transcription of the growth hormone and prolactin genes (Ingraham et al 1988, Bodner et al 1988). Genetic analysis of the Snell and Jackson dwarf mice further established that Pit-1 is required for generation of three pituitary cell lineages; somatotropes, lactotropes, and thyrotropes (Camper et al 1990, Li et al 1990). Because Pit-1 is directly involved in the transcriptional regulation of the hormonal markers for terminal differentiation in these cell types (e.g. *GH* in somatotropes, *Prl* in lactotropes, and *TSHβ* in thyrotropes), a central goal has been to address how Pit-1, in response to diverse signaling cascades and in collaboration with other factors, directs cell-specific expression of its multiple targets.

Control of Somatotrope-Specific Gene Expression

Extensive studies on the *cis*-active elements within the genes encoding the terminal differentiation markers of pituitary cell types have provided important insights into the mechanisms by which Pit-1 can direct cell-specific gene expression. Pit-1 binding is required to direct pituitary-specific expression in the proximal promoter region of the growth hormone gene in somatotropes. As little as 320 bp of rat *GH* promoter sequence is sufficient to target expression to somatotropes in vivo and contains binding sites for Pit-1, a novel zinc finger protein, and nuclear receptors (Lipkin et al 1993). Several groups have reported cooperatively between Pit-1 and the RAR and thyroid hormone nuclear receptor in the control of *GH* gene expression (Schaufele et al 1992, Palomino et al 1998), and recent genetic studies have supported the proposed roles for one of these nuclear receptors in the regulation *GH* gene expression. Mice deleted for all known isoforms of the thyroid hormone receptor show significant decreases in growth hormone expression and somatotrope cell numbers, with a reciprocal pronounced increase in *TSHβ* expression and thyrotropes (Gothe et al 1999).

Further insight into the nature of somatotrope-specific gene expression has emerged from extensive promoter/enhancer mapping analyses of the human growth hormone gene. The human growth hormone locus is a cluster of five related genes, in which the most 5′ gene (*hGH-N*) is expressed exclusively in the pituitary gland. High levels of cell-specific expression of the *hGH-N* require a locus control region located 14–16 kb upstream of the promoter and contain multiple DNase I hypersensitivity sites (Bennani-Baiti et al 1998), a characteristic feature of an open chromatin conformation and gene activity. In vivo and in vitro mapping studies have revealed that putative Pit-1 binding sites within the *hGH-N* locus control region are required for cell-specific expression (Jin et al 1999, Shewchuk et al 1999), further establishing the requirement of Pit-1 in the direct regulation of growth hormone gene expression.

Control of Lactotrope-Specific Gene Expression

Unlike most cell types of the pituitary gland, in which hormone gene expression can be both positively and negatively regulated by releasing hormones secreted from the hypothalamus, *prolactin* gene expression is controlled primarily through the negative effects of dopamine on lactotropes. *Prl* gene transcription is controlled through a series of distal and proximal enhancer elements containing multiple binding sites for Pit-1, which are both required and sufficient to direct cell-specific expression in vivo (Crenshaw et al 1989). Early evidence for a cell type-specific Pit-1 synergy partner derived from studies on the *Prl* gene in which it was demonstrated that the estrogen nuclear receptor (ER) can activate *Prl* expression in cooperation with Pit-1 at a distal enhancer site (Simmons et al 1990). Consistent with a requirement for ER-mediated synergy, it has been demonstrated that mice deleted for the α isoform of the *ER* gene show a dramatic decrease in *Prl* gene expression and lactotrope cell numbers, with other pituitary cell types unchanged or even expanded (Scully et al 1997).

Other factors can cooperatively regulate *Prl* gene expression with Pit-1 at its proximal enhancer elements. This regulation appears to involve the interplay between two ETS-domain-containing factors, Ets-1 and ERF (Ets-2 repressor factor). A pituitary-specific role for Ets-1 was initially identified through the characterization of a composite Pit-1/Ets-1 binding site in the *Prl* gene, which confers synergy between these proteins (Howard & Maurer 1995, Bradford et al 1997) and is regulated through a mitogen-activated protein kinase pathway. The interaction of Pit-1 with Ets-1 is mediated by the POU homeodomain, and the functional consequences of this interaction may be modulated through use of alternative Pit-1 isoforms (Bradford et al 2000). Furthermore, the synergy between Ets-1 and Pit-1 can be abrogated by an Ets repressor factor, ERF, apparently by inhibition of Pit-1 binding to this site (Day et al 1998). The interplay between Ets-1, ERF, and Pit-1 thus may provide a component of the molecular mechanism for the inhibitory effects of dopamine on *Prl* gene expression.

A third class of factors that have been implicated in the control of *Prl* expression is the *Pitx* genes, which can also act as synergy partners for Pit-1. Both Pitx1 and Pitx2 have been demonstrated to physically associate with Pit-1, leading to synergistic activation of several pituitary-restricted genes (Szeto et al 1996, Amendt et al 1998, Tremblay et al 1998). Mapping of functional domains revealed that a 39-amino-acid carboxy-terminal tail in Pitx2 acts as an autorepressive domain for DNA binding to a canonical *bicoid* site. Interactions between Pit-1 and Pitx2 apparently relieve this auto-repression by the carboxy-terminal tail, allowing for enhanced Pitx2 DNA binding and synergistic activation of *Prl* gene transcription (Amendt et al 1999). Thus Pitx2/Pit-1 interactions appear to be another component of the complex regulation of *Prl* gene transcription.

Control of Thyrotrope and Gonadotrope Cell Type Specification

The third Pit-1-dependent cell type, the caudomedial thyrotrope, shares features and molecular markers in common with the non-Pit-1-dependent gonadotrope lineage. The conserved structure of the β subunits of FSH, LH (expressed in gonadotropes), and TSH (expressed in thyrotropes) and the shared common sub-unit αGSU suggests that these two most ventrally arising pituitary cell types evolved from a common ancestral origin. Both cell types share expression of a series of ventrally induced transcription factors including *GATA-2*, *Isl-1*, *Brn4*, and *P-Frk*, and they are determined in part by the ventral → dorsal BMP2 gradient (Treier et al 1998, Ericson et al 1998). Thus, a central question has been to address the mechanisms which govern the appearance of these two similar but distinct cell types.

An important distinction between these two lineages is the presence of Pit-1 in the thyrotrope and its absence in the more ventrally arising gonadotrope. As ventralized expression of Pit-1 proves sufficient to convert gonadotropes to thyrotropes in vivo (Dasen et al 1999a), it is likely that factor(s) controlling thyrotrope development in collaboration with Pit-1 are present in gonadotropes. One of these factors has proven to be the zinc finger protein GATA-2, originally identified to be critical for the development of the hematopoietic system (Tsai et al 1994) and was later shown to be expressed in permanent pituitary cell lines derived from the αGSU lineage (Steger et al 1994). GATA-2 has been demonstrated to play an important role in the determination of both gonadotropes and thyrotropes and is a direct transcriptional target of the ventral BMP2 signal (Figure 4). In transgenic animals, dorsal misexpression of GATA-2 under control of *Pit-1* regulatory in-formation is alone sufficient to convert all of the Pit-1-dependent lineages to the gonadotrope fate in vivo. High levels of expression of GATA-2 are also associated with the inhibition of endogenous *Pit-1* gene expression (Treier et al 1998, Dasen et al 1999a), suggesting that high levels of GATA-2 in the ventral presumptive gonadotrope delineate this cell type from the more dorsal thyrotrope.

In presumptive caudomedial thyrotropes, the levels of GATA-2 are proposed to be insufficient to inhibit the *Pit-1* gene, allowing emergence of a cell type that expresses both Pit-1 and GATA-2. Within thyrotropes, Pit-1 and GATA-2 can phys-ically interact, leading to either synergistic activation of thyrotrope-specific genes, such as *TSHβ*, which contains adjacent Pit-1- and GATA-2-binding sites (Lin et al 1994, Haugen et al 1996, Gordon et al 1997), or inhibition of gonadotrope-specific genes because Pit-1 can inhibit GATA-2 binding to promoters not containing an adjacent Pit-1 site (Dasen et al 1999a). The inhibition of GATA-2 binding ac-counts for the ability of Pit-1, independently of its own ability to bind DNA, to inhibit expression of gonadotrope-specific genes in thyrotropes. This observa-tion has also received genetic confirmation based on analysis of Pit-1-defective Snell dwarf mice, in which the *W48C* mutation in the Pit-1 POU homeodomain disrupts Pit-1/GATA-2 interactions and causes the thyrotrope to assume a go-nadotrope fate. Similar context-dependent consequences of GATA/homeodomain interactions have been also observed in cardiac development, in which the inter-action between Nkx-2.5 and GATA-4 can have either synergistic (Durocher et al 1997) or inhibitory (Shiojima et al 1999) effects on GATA-dependent transcription, depending on whether there is an adjacent homeodomain-binding site, suggest-ing that promoter-dependent functional consequences of GATA-homeodomain or

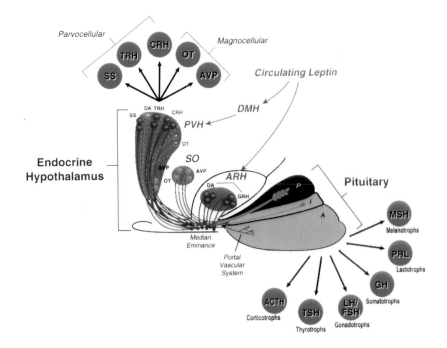

Figure 1 The hypothalamic-pituitary axis. Hormone synthesis and secretion from the pituitary gland are regulated by a series of peptide hormones released from hypothalamic neurons. The magnocellular neurosecretory system includes neurons in the paraventricular hypothalamus (PVH) and supraoptic (SO) nuclei that synthesize the peptide hormones oxytocin (OT) and arginine vasopressin (AVP) and release them in an activity-dependent manner from axonal terminals in the posterior lobe (P) of the pituitary gland. In addition, the PVH harbors separate populations of parvocellular cells, which synthesize corticotropin-releasing hormone (CRH) and thyrotropin-releasing hormone (TRH). These neuropeptides are delivered to the median eminence for conveyance via the hypophyseal-portal vascular system to modulate the synthesis and release of ACTH and TSH in the anterior pituitary gland (A). Centered in ventrally contiguous cell groups in the anterior periventricular (AVP) or the arcuate nuclei (ARH) of the hypothalamus are hypophysiotrophic neurons that provide both the dopaminergic (DA) control of prolactin (PRL) secretion and somatostatin (SS) or growth hormone-releasing hormone (GHRH) which impart the principal inhibitory and stimulatory regulation of growth hormone (GH), respectively. Cells of the intermediate lobe (I) of the pituitary gland produce melanocyte-stimulating hormone (MSH) by proteolytic processing of POMC (not shown).

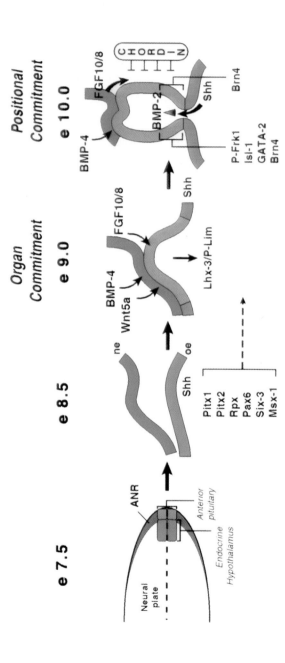

Figure 2 Signaling molecules and transcription factors expressed in the early phases of pituitary organogenesis. The most anterior midline portion of the neural ridge gives rise to the primordia of the pituitary and endocrine hypothalamus. An initially uniform oral ectoderm (oe) makes contact with the overlying neural epithelium (ne) by e9.0 in the mouse with Sonic hedgehog (Shh) becoming excluded from the region which forms Rathke's pouch. The neural epithelium of the ventral diencephalon, which makes contact with the pouch (the infundibulum), expresses BMP4, FGF8, and Wnt5a, while BMP2 is expressed within Rathke's pouch at the Shh restriction boundary and within the ventral mesenchyme. Multiple transcription factors are expressed throughout the oral ectoderm, with the expression of the transcription factor Lhx3 restricted to Rathke's pouch. Opposing dorsal FGF8 and ventral BMP2 gradients are associated with the spatially restricted expression patterns of transcription factors by e10—e11, including a series of dorsally or ventrally induced factors, and they allow for the positional commitment of cell lineages.

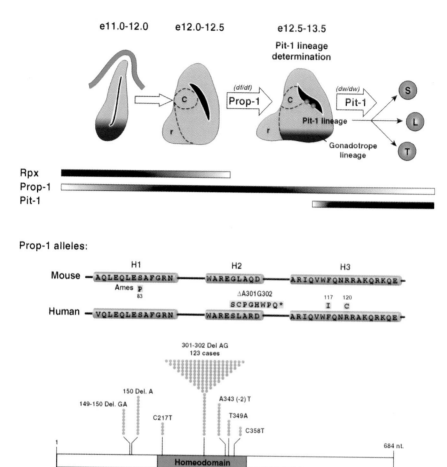

Figure 3 Role of Prop-1 and Pit-1 in murine and human pituitary development. Based on the molecular analysis of Ames (*df*) mice, the homeodomain factor Prop-1 has been shown to be required for the ventral progression of anterior pituitary cell types and extinction of Rpx expression between days e12 and e13. Analysis of Snell (*dw*) mice has revealed a requirement for the POU homeodomain factor Pit-1 in generation of three cell lineages. Multiple human alleles for mutations in the human *PROP1* gene have revealed a requirement for Prop-1 in the generation of a non-Pit-1-dependent cell lineage, the gonadotrope.

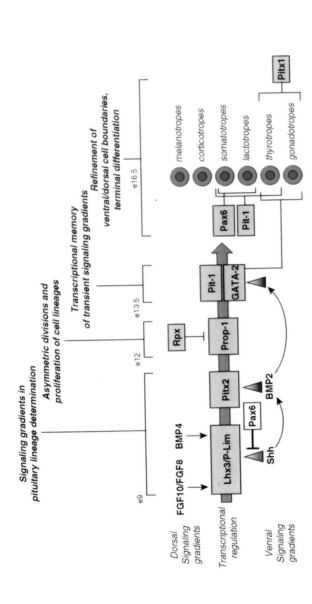

Figure 5 Model for the progression of pituitary cell lineages in response to extrinsic and intrinsic signals. Primordial-cell types derived from the anterior neural ridge are committed to the pituitary fate through the induction of *Lhx3/P-LIM* expression, which may require the combinatorial actions of FGF8, BMP4, and Shh. Pitx2 is required for the expansion of these precursors within Rathke's pouch, with Prop-1 required for the asymmetric ventral proliferation and determination of at least four cell types. The transcriptional activities of Prop-1 are hypothesized to be antagonized by the homeodomain repressor factor Rpx/Hesx1. Pit-1 is subsequently required for the cell fate determination of three cell types (somatotropes, lactotropes, and thyrotropes), whereas GATA-2 is hypothesized to be required for the thyrotrope and gonadotrope cell lineages, based on the presence or absence of Pit-1, respectively.

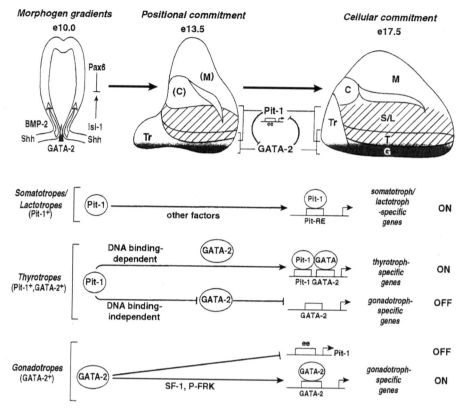

Figure 4   Cell lineage determination in the early phases of pituitary development. Pax6 acts to refine dorsal/ventral cell boundaries in the early phases of lineage determination. The reciprocal interaction of Pit-1 and GATA-2 are hypothesized to mediate the effects of the early signaling events and are sufficient to determine multiple ventral cell lineages.

similar types of interactions between DNA-binding proteins are critical for the development of cell types in other organ systems.

Molecular Memory in the Enhancer Switching of the *Pit-1* Gene

The regulation of *Pit-1* gene expression has also presented an interesting model gene for understanding the pathways leading to the serial activation of transcription factors in pituitary development because its expression is initiated prior to the appearance of terminal differentiation markers for pituitary cell types and is temporally regulated by two distinct enhancers. *Pit-1* gene expression is initiated at e13.5 by an early enhancer and subsequently switches to a late, autoregulatory enhancer between e16.5 and birth (Rhodes et al 1993, DiMattia et al 1997). Although the factors that govern the initiation of *Pit-1* expression are yet to be identified and may include the homeodomain factor Prop-1 (Sornson et al 1996), a component of the regulation of *Pit-1* gene expression appears to involve its restriction from the gonadotrope lineage by the ventral BMP2 signal (Treier et al 1998, Dasen et al 1999a).

As in the examples of the regulation of *GH* and *Prl* genes, a nuclear receptor/Pit-1 interaction has been demonstrated to be involved in the autoregulation of the *Pit-1* gene, where RAR can act in a ligand-dependent manner to synergistically activate *Pit-1* gene expression (Rhodes et al 1993). It has been demonstrated that human combined-pituitary-hormone-deficiency patients with a mutation in the homeodomain of Pit-1 that fails to affect DNA binding show attenuated synergy of Pit-1 and RAR, providing suggestive genetic evidence that this type of nuclear receptor-homeodomain interaction is critical for determination of the Pit-1-dependent cell lineages (Cohen et al 1999b). Thus, the switch from an early to a late acting enhancer in the *Pit-1* gene may reflect the transition from a signaling factor-induced gene activation event to a permanent molecular memory loop through *Pit-1* autoregulation within the three Pit-1-dependent pituitary cell lineages.

Integration of Intracellular Signaling Events by Pit-1

Within the three Pit-1-dependent pituitary cell types, a component to the regulation of cell type-specific gene expression has proved to require combinatorial interactions of Pit-1 with other cell-type-restricted factors. Because different intracellular-signaling cascades have been demonstrated to predominate within these cell types, a second level of regulation in Pit-1 activity may involve its regulated interactions with different coactivators and corepressors in response to the hypothalamic signals that activate these cascades. For example, in somatotropes *GH* gene transcription is regulated by the hypothalamic peptide growth hormone releasing hormone (GHRH), which binds to the GHRH receptor to activate a cAMP-mediated protein kinase A response. The importance of signaling through the GHRH receptor in somatotropes has been revealed through the analysis of *little* dwarf mice, in which the receptor is mutated and leads to severely reduced *GH* expression and somatotrope cell numbers (Godfrey et al 1993, Lin et al 1993). Other signaling pathways appear to predominate within other Pit-1-dependent cell types, such as the mitogen-activated-protein-kinase pathway in lactotropes, and phospholipase C pathway in thyrotropes. Thus it has been speculated that these different cascades might be involved in the differential activities of Pit-1 in different cell types or in response to the activation of different signaling pathways.

Insight into this issue has emerged from studies on the interactions of Pit-1 with transcriptional-coregulator complexes, which for activation include CREB-binding protein (CBP) and histone acetyltransferases (HATs) and, for transcriptional repression, the nuclear receptor corepressor and other histone deactylase-associated factors. In quiescent cells, Pit-1 is a very weak transcriptional activator as it associates with members of the repressor machinery including nuclear receptor corepressor and histone deacytylases (Xu et al 1998). Upon activation of cultured cells with cAMP or growth factors (such as insulin and epidermal growth factor), Pit-1 becomes an effective transcriptional activator, and this activity is mediated in part through interactions with coactivator CBP (Xu et al 1998, Cohen et al 1999a).

Interestingly, activation of Pit-1 by cAMP or growth factors requires a different domain of CBP, with the amino-terminal region of CBP required for Pit-1 activation by cAMP and a carboxy-terminal domain required for activation by growth factors. These required domains also correspond to the interaction domains of CBP on Pit-1 because the Pit-1 POU domain interacts with a cystine–histidine-rich domain and a novel amino-terminal domain in CBP (Xu et al 1998, Zanger et al 1999). Furthermore, the stimulation of Pit-1 activity by cAMP and growth factors also requires different HAT-containing coregulators because cAMP-stimulated cells require the HAT domain of CBP whereas cells stimulated with growth factors require the HAT domain of another transcriptional coregulator, p/CAF. Thus, the activity of Pit-1 in response to different signals can be regulated through modulation of its interactions with components of the transcriptional-activation machinery, likely leading to cell-type-specific effects on the chromatin associated with the regulatory sequences of specific subsets of genes.

Control of Gonadotrope-Specific Gene Expression

Of the non–Pit-1-dependent pituitary cell lineages, the best characterized at the molecular level are the gonadotropes, which arise in the most ventral region of the pituitary gland. The regulated control of the genes for the terminal-differentiation markers for gonadotrope development, which include *LHβ* and *FSHβ*, require the activities of multiple transcription factors, including SF-1, Egr-1, and Pitx1 (Ingraham et al 1994, Lee et al 1996, Topilko et al 1998, Szeto et al 1999). Egr-1 (also referred to as NGFI-A, Krox-24, and Zif 268) is a zinc finger containing protein homologous to the Wilms tumor gene product, which synergizes with the orphan nuclear receptor SF-1 on the *LHβ* promoter (Lee et al 1996) and is rapidly induced in gonadotrope-derived pituitary cell lines treated with gonadotropin-releasing hormone (GnRH) (Tremblay & Drouin 1999). Pitx1 and SF-1 also cooperatively regulate *LHβ*, through association of the carboxyl terminus of Pitx1 with the amino terminus of SF-1. The interaction of Pitx1 and SF-1 is reminiscent of interactions of Pitx factors and Pit-1, and this transcriptional synergy does not appear to require Pitx1 DNA binding.

Analysis of mice bearing targeted mutations in the genes for *Egr-1, SF-1*, and *Pitx1* have revealed distinctive roles for each of these factors in gonadotropin synthesis. Mice disrupted in *Egr-1* show specific defects in *LHβ* expression, with expression of the *FSHβ* gene unchanged (Lee et al 1996, Topilko et al 1998). Surprisingly, Egr-1-deficient mice are also characterized by reduced somatotrope numbers, although this defect appears to be background strain specific (Topilko et al 1998). Defects in gonadotrope proliferation and hormone synthesis are also observed in *SF-1* gene-deleted mice, although it is unclear whether this defect is due to defective GnRH synthesis because SF-1 is also required in the development of hypothalamic GnRH-producing neurons. Similarly, *Pitx1* gene-deleted animals are also characterized by reduced gonadotrope cell numbers, although specification of the lineage appears normal.

Understanding the mechanisms controlling expression of the αGSU gene regulation has presented an interesting problem in pituitary development because its expression can be controlled by distinct mechanisms within gonadotropes and thyrotropes, and it is regulated by series of distinct *cis*-acting elements directing expression in a cell-restricted manner. As little as 313 bp of the proximal bovine promoter sequences, containing sites for Lhx/Pitx/CLIM synergy as well as GATA-2/3 and SF-1, are required to target αGSU expression to gonadotropes in transgenic mice (Kendall et al 1991, Steger et al 1994, Bach et al 1997). Full activity and cell type-specific expression of the murine αGSU gene require a more distal enhancer element (Brinkmeier et al 1998, Kendall et al 1994), and this element is also apparently required for the restriction of αGSU expression from other pituitary cell types because it can confer repression of basal αGSU promoter activity in somatotrope-representative pituitary cell lines (Wood et al 1999). The regulatory region of the αGSU gene therefore presumably contains elements required for both its cell-specific activation in thyrotropes and gonadotropes and restriction from other pituitary cell types. This regulatory region contains potential binding sites for several factors including Ets, GATA, and HLH factors (Wood et al 1999), although the contribution of these elements to the activation and restriction of αGSU expression in different pituitary cell types is yet to be determined.

Control of Melanotrope- and Corticotrope-Specific Gene Expression

Considerably less is known about the cell-autonomous factors governing the appearance of the two pituitary cell types that share expression of the *POMC* gene, corticotropes and melanotropes. Of the currently existing mouse dwarf strains and gene-deleted animals for pituitary-restricted transcription factors, none is characterized by the complete absence of corticotropes and melanotropes; thus genetic mouse models to address the development of these cell types are lacking. Although the signaling factors that positionally determine the POMC lineages appear to include LIF and FGFs, the specific transcription factors that specify these cell types remain elusive. Analysis of regulatory elements in the *POMC* promoter have revealed binding sites for several factors including a *bicoid*-binding site and the orphan nuclear receptor Nur77 (Philips et al 1997). The relevance of these sites in the regulation of *POMC* in vivo remains to be determined.

The best characterized component of the transcriptional regulation of corticotrope- and melanotrope-specific gene expression has been the negative-feedback regulation of the *POMC* gene by the glucocorticoid receptor (GR) (Therrien & Drouin 1993). The *POMC* gene contains a negative DNA response element in which GR acts in a ligand-dependent manner to inhibit *POMC* expression in corticotropes in response to elevated levels of glucocorticoids. Consistent with this observation, in GR gene-deleted animals there is dramatic increase in *POMC* expression in corticotropes within the anterior lobe of the pituitary (Reichardt & Schutz 1996). Interestingly, the inhibition of *POMC* expression is not dependent on GR DNA binding because mice carrying a targeted mutation in the GR DNA-binding domain are also characterized by increased *POMC* expression (Reichardt et al 1998). This observation is consistent with the hypothesis that nuclear receptors can act to *trans*-repress certain target genes off DNA. Furthermore, there appear to be significant differences between the regulation of *POMC* in the anterior lobe (corticotropes) and the intermediate lobe (melanotropes) in mice. In the intermediate lobe, GR appears to be required to activate *POMC* because *POMC* expression is dramatically reduced in GR-knockout animals. The molecular mechanisms behind this cell-type-specific functional difference remain to be elucidated.

CONCLUSIONS AND FUTURE DIRECTIONS

Taken together, the studies described here begin to define a model encompassing the extrinsic and intrinsic signaling mechanisms governing the early and late aspects of pituitary development (Figure 5, see color insert). A series of signaling molecules secreted from multiple organizing centers appears to coordinate the commitment, early patterning, proliferation, and positional determination of six hormone-producing pituitary cell types. Secretion of BMP4 from the ventral diencephalon appears to be required for the initial phases of pituitary organ commitment, with subsequent opposing and antagonistic dorsal FGF8 and ventral BMP2 gradients governing the patterning and positional determination of cell types through the induction of overlapping patterns of transcription factor gene expression. The establishment of these distinct expression patterns allows the positional determination of pituitary cell types to occur long before the cell-type-specific terminal differentiation markers appear. Future studies will attempt to encompass the still undefined roles for other signaling molecules such as Shh, RA, and LIF, which may be involved in the induction or modulation of other signaling gradients and transcription factor function.

The earliest phases of pituitary development have been established to be mediated through the activities of the transcription factors Pitx2 and Lhx3/4, which are required for the early patterning and proliferation events within Rathke's pouch. The critical step in defining the activities of these early factors is the induction of *Lhx3* gene expression in response to dorsal FGF gradients originating from the infundibulum. Later proliferation and cellular-determination events require the activities of several induced factors including Prop-1, Pit-1, and GATA-2. Given the established roles of these factors in the generation of specific cell phenotypes, a complete understanding of how broadly expressed signaling molecules can generate an organ-specific determination program will require knowledge of how these and other pituitary-restricted transcription factors are induced. This will undoubtedly require further investigations into the multiple downstream targets of the signaling pathways and the elucidation of the factors whose activities are directly modulated by the early morphogen gradients.

ACKNOWLEDGMENTS

We thank P Meyer for her expertise and assistance in preparation of illustrations. J Dasen is supported by a postdoctoral fellowship from the Nation Institutes of Health. M Rosenfeld is an Investigator with the Howard Hughes Medical Institute.

Visit the Annual Reviews home page at www.AnnualReviews.org

LITERATURE CITED

Agulnick AD, Taira M, Breen JJ, Tanaka T, Dawid IB, Westphal H. 1996. Interactions of the LIM-domain-binding factor Ldb1 with LIM homeodomain proteins. *Nature* 384:270–72

Amendt BA, Sutherland LB, Russo AF. 1999. Multifunctional role of the Pitx2 homeodomain protein C-terminal tail. *Mol. Cell. Biol.* 19:7001–10

Amendt BA, Sutherland LB, Semina EV, Russo AF. 1998. The molecular basis of Rieger syndrome: analysis of Pitx2 homeodomain protein activities. *J. Biol. Chem.* 273:20066–72

Andersen B, Pearse RV II, Jenne K, Sornson M, Lin SC, et al. 1995. The Ames dwarf gene is required for Pit-1 gene activation. *Dev. Biol.* 172:495–503

Bach I, Carriere C, Ostendorff HP, Andersen B, Rosenfeld MG. 1997. A family of LIM domain-associated cofactors confer transcriptional synergism between LIM and Otx homeodomain proteins. *Genes Dev.* 11:1370–80

Bach I, Rhodes SJ, Pearse RV II, Heinzel T, Gloss B, et al. 1995. P-Lim, a LIM homeodomain factor, is expressed during pituitary organ and cell commitment and synergizes with Pit-1. *Proc. Natl. Acad. Sci. USA* 92:2720–24

Bennani-Baiti IM, Asa SL, Song D, Iratni R, Liebhaber SA, Cooke NE. 1998. DNase I-hypersensitive sites I and II of the human growth hormone locus control region are a major developmental activator of somatotrope gene expression. *Proc. Natl. Acad. Sci. USA* 95:10655–60

Bentley CA, Zidehsarai MP, Grindley JC, Parlow AF, Barth-Hall S, Roberts VJ. 1999. Pax6 is implicated in murine pituitary endocrine function. *Endocrine* 10:171–77

Bodner M, Castrillo JL, Theill LE, Deerinck T, Ellisman M, Karin M. 1988. The pituitary-specific transcription factor GHF-1 is a homeobox-containing protein. *Cell* 55:505–18

Bousquet C, Ray DW, Melmed S. 1997. A common pro-opiomelanocortin-binding element mediates leukemia inhibitory factor and corticotropin-releasing hormone transcriptional synergy. *J. Biol. Chem.* 272:10551–57

Bradford AP, Brodsky KS, Diamond SE, Kuhn LC, Liu Y, Gutierrez-Hartmann A. 2000. The Pit-1 homeodomain and beta-domain interact with Ets-1 and modulate synergistic activation of the rat prolactin promoter. *J. Biol. Chem.* 275:3100–6

Bradford AP, Wasylyk C, Wasylyk B, Gutierrez-Hartmann A. 1997. Interaction of Ets-1 and the POU-homeodomain protein GHF-1/Pit-1 reconstitutes pituitary-specific gene expression. *Mol. Cell. Biol.* 17:1065–74

Brinkmeier ML, Gordon DF, Dowding JM, Saunders TL, Kendall SK, et al. 1998. Cell-specific expression of the mouse glycoprotein hormone alpha-subunit gene requires multiple interacting DNA elements in transgenic mice and cultured cells. *Mol. Endocrinol.* 12:622–33

Buckwalter MS, Katz RW, Camper SA. 1991. Localization of the panhypopituitary dwarf mutation (df) on mouse chromosome 11 in an intersubspecific backcross. *Genomics* 10:515–26

Camper SA, Saunders TL, Katz RW, Reeves RH. 1990. The Pit-1 transcription factor gene is a candidate for the murine Snell dwarf mutation. *Genomics* 8:586–90

Capdevila J, Vogan KJ, Tabin CJ, Izpisua Belmonte JC. 2000. Mechanisms of left-right determination in vertebrates. *Cell* 101:9–21

Cohen LE, Hashimoto Y, Zanger K, Wondisford F, Radovick S. 1999a. CREB-independent regulation by CBP is a novel mechanism of human growth hormone gene expression. *J. Clin. Invest.* 104:1123–30

Cohen LE, Zanger K, Brue T, Wondisford FE, Radovick S. 1999b. Defective retinoic acid regulation of the Pit-1 gene enhancer: a novel mechanism of combined pituitary hormone deficiency. *Mol. Endocrinol.* 13:476–84

Couly G, Le Douarin NM. 1988. The fate map of the cephalic neural primordium at the presomitic to the 3-somite stage in the avian embryo. *Development* 103(Suppl.):101–13

Crenshaw EB III, Kalla K, Simmons DM, Swanson LW, Rosenfeld MG. 1989. Cell-specific expression of the prolactin gene in transgenic mice is controlled by synergistic interactions between promoter and enhancer elements. *Genes Dev.* 3:959–72

Dahmann C, Basler K. 1999. Compartment boundaries: at the edge of development. *Trends Genet.* 15:320–26

Daikoku S, Chikamori M, Adachi T, Maki Y. 1982. Effect of the basal diencephalon on the development of Rathke's pouch in rats: a study in combined organ cultures. *Dev. Biol.* 90:198–202

Dale JK, Vesque C, Lints TJ, Sampath TK, Furley A, et al. 1997. Cooperation of BMP7 and SHH in the induction of forebrain ventral

midline cells by prechordal mesoderm. *Cell* 90:257–69

Dasen JS, O'Connell SM, Flynn SE, Treier M, Gleiberman AS, et al. 1999a. Reciprocal interactions of Pit1 and GATA2 mediate signaling gradient induced determination of pituitary cell types. *Cell* 97: 587–98

Dasen JS, Rosenfeld MG. 1999b. Combinatorial codes in signaling and synergy: lessons from pituitary development. *Curr. Opin. Genet. Dev.* 9:566–74

Dattani MT, Martinez-Barbera JP, Thomas PQ, Brickman JM, Gupta R, et al. 1998. Mutations in the homeobox gene HESX1/Hesx1 associated with septo-optic dysplasia in human and mouse. *Nat. Genet.* 19:125–33

Day RN, Liu J, Sundmark V, Kawecki M, Berry D, Elsholtz HP. 1998. Selective inhibition of prolactin gene transcription by the ETS-2 repressor factor. *J. Biol. Chem.* 273:31909–15

Deladoey J, Fluck C, Buyukgebiz A, Kuhlmann BV, Eble A, et al. 1999. "Hot spot" in the PROP1 gene responsible for combined pituitary hormone deficiency. *J. Clin. Endocrinol. Metab.* 84:1645–50

De Moerlooze L, Spencer-Dene B, Revest J, Hajihosseini M, Rosewell I, Dickson C. 2000. An important role for the IIIb isoform of fibroblast growth factor receptor 2 (FGFR2) in mesenchymal-epithelial signaling during mouse organogenesis. *Development* 127:483–92

DiMattia GE, Rhodes SJ, Krones A, Carriere C, O'Connell S, et al. 1997. The Pit-1 gene is regulated by distinct early and late pituitary-specific enhancers. *Dev. Biol.* 182:180–90

Durocher D, Charron F, Warren R, Schwartz RJ, Nemer M. 1997. The cardiac transcription factors Nkx2-5 and GATA-4 are mutual cofactors. *EMBO J.* 16:5687–96

Eagleson GW, Harris WA. 1990. Mapping of the presumptive brain regions in the neural plate of *Xenopus laevis*. *J. Neurobiol.* 21:427–40

Eastman Q, Grosschedl R. 1999. Regulation of LEF-1/TCF transcription factors by Wnt and other signals. *Curr. Opin. Cell Biol.* 11:233–40

Edlund T, Jessell TM. 1999. Progression from extrinsic to intrinsic signaling in cell fate specification: a view from the nervous system. *Cell* 96:211–24

Eichele G. 1997. Retinoids: from hindbrain patterning to Parkinson disease. *Trends Genet.* 13:343–45

Ekker SC, McGrew LL, Lai CJ, Lee JJ, von Kessler DP, et al. 1995. Distinct expression and shared activities of members of the hedgehog gene family of *Xenopus laevis*. *Development* 121:2337–47

Ericson J, Norlin S, Jessell T, Edlund T. 1998. Integrated FGF and BMP signaling controls the progression of progenitor cell differentiation and the emergence of pattern in the embryonic anterior pituitary. *Development* 125:1005–15

Ericson J, Rashbass P, Schedl A, Brenner-Morton S, Kawakami A, et al. 1997. Pax6 controls progenitor cell identity and neuronal fate in response to graded Shh signaling. *Cell* 90:169–80

Fedtsova NG, Barabanov VM. 1990. The distribution of competence for adenohypophysis development in the ectoderm of chick embryos. *Ontogenez* 21:254–60

Ferrand. 1972. Experimental study of the factors in cytological differentiation of the adenohypophysis in the chick embryo. *Arch. Biol.* 83:297–371

Gage PJ, Brinkmeier ML, Scarlett LM, Knapp LT, Camper SA, Mahon KA. 1996a. The Ames dwarf gene, df, is required early in pituitary ontogeny for the extinction of Rpx transcription and initiation of lineage-specific cell proliferation. *Mol. Endocrinol.* 10:1570–81

Gage PJ, Roller ML, Saunders TL, Scarlett LM, Camper SA. 1996b. Anterior pituitary cells defective in the cell-autonomous factor, df, undergo cell lineage specification but not expansion. *Development* 122:151–60

Gage PJ, Suh H, Camper SA. 1999. Dosage requirement of Pitx2 for development of multiple organs. *Development* 126:4643–51

Gallardo ME, Lopez-Rios J, Fernaud-Espinosa I, Granadino B, Sanz R, et al. 1999. Genomic cloning and characterization of the human homeobox gene SIX6 reveals a cluster of SIX genes in chromosome 14 and associates SIX6 hemizygosity with bilateral anophthalmia and pituitary anomalies. *Genomics* 61:82–91

Gleiberman AS, Fedtsova NG, Rosenfeld MG. 1999. Tissue interactions in the induction of anterior pituitary: role of the ventral diencephalon, mesenchyme, and notochord. *Dev. Biol.* 213:340–53

Godfrey P, Rahal JO, Beamer WG, Copeland NG, Jenkins NA, Mayo KE. 1993. GHRH receptor of little mice contains a missense mutation in the extracellular domain that disrupts receptor function. *Nat. Genet.* 4:227–32

Gordon DF, Lewis SR, Haugen BR, James RA, McDermott MT, et al. 1997. Pit-1 and GATA-2 interact and functionally cooperate to activate the thyrotropin beta-subunit promoter. *J. Biol. Chem.* 272:24339–47

Gothe S, Wang Z, Ng L, Kindblom JM, Barros AC, et al. 1999. Mice devoid of all known thyroid hormone receptors are viable but exhibit disorders of the pituitary-thyroid axis, growth, and bone maturation. *Genes Dev.* 13:1329–41

Hammerschmidt M, Brook A, McMahon AP. 1997. The world according to hedgehog. *Trends Genet.* 13:14–21

Haugen BR, McDermott MT, Gordon DF, Rupp CL, Wood WM, Ridgway EC. 1996. Determinants of thyrotrope-specific thyrotropin beta promoter activation: cooperation of Pit-1 with another factor. *J. Biol. Chem.* 271:385–89

Hermesz E, Mackem S, Mahon KA. 1996. Rpx: a novel anterior-restricted homeobox gene progressively activated in the prechordal plate, anterior neural plate and Rathke's pouch of the mouse embryo. *Development* 122:41–52

Hogan BL. 1996. Bone morphogenetic proteins in development. *Curr. Opin. Genet. Dev.* 6:432–38

Hogan BL. 1999. Morphogenesis. *Cell* 96:225–33

Howard PW, Maurer RA. 1995. A composite

Ets/Pit-1 binding site in the prolactin gene can mediate transcriptional responses to multiple signal transduction pathways. *J. Biol. Chem.* 270:20930–36

Ingraham HA, Chen RP, Mangalam HJ, Elsholtz HP, Flynn SE, et al. 1988. A tissue-specific transcription factor containing a homeodomain specifies a pituitary phenotype. *Cell* 55:519–29

Ingraham HA, Lala DS, Ikeda Y, Luo X, Shen WH, et al. 1994. The nuclear receptor steroidogenic factor 1 acts at multiple levels of the reproductive axis. *Genes Dev.* 8:2302–12

Jean D, Bernier G, Gruss P. 1999. Six6 (Optx2) is a novel murine Six3-related homeobox gene that demarcates the presumptive pituitary/hypothalamic axis and the ventral optic stalk. *Mech. Dev.* 84:31–40

Jin Y, Surabhi RM, Fresnoza A, Lytras A, Cattini PA. 1999. A role for A/T-rich sequences and Pit-1/GHF-1 in a distal enhancer located in the human growth hormone locus control region with preferential pituitary activity in culture and transgenic mice. *Mol. Endocrinol.* 13:1249–66

Jurata LW, Kenny DA, Gill GN. 1996. Nuclear LIM interactor, a rhombotin and LIM homeodomain interacting protein, is expressed early in neuronal development. *Proc. Natl. Acad. Sci. USA* 93:11693–98

Karlstrom RO, Talbot WS, Schier AF. 1999. Comparative synteny cloning of zebrafish *you-too*: mutations in the Hedgehog target gli2 affect ventral forebrain patterning. *Genes Dev.* 13:388–93

Kawamura K, Kikuyama S. 1992. Evidence that hypophysis and hypothalamus constitute a single entity from the primary stage of histogenesis. *Development* 115:1–9

Kawamura K, Kikuyama S. 1995. Induction from posterior hypothalamus is essential for the development of the pituitary proopiomelacortin (POMC) cells of the toad (*Bufo japonicus*). *Cell Tissue Res.* 279:233–39

Kendall SK, Gordon DF, Birkmeier TS, Petrey D, Sarapura VD, et al. 1994. Enhancer-mediated high level expression of mouse pituitary glycoprotein hormone alpha-subunit transgene in thyrotropes, gonadotropes, and developing pituitary gland. *Mol. Endocrinol.* 8:1420–33

Kendall SK, Saunders TL, Jin L, Lloyd RV, Glode LM, et al. 1991. Targeted ablation of pituitary gonadotropes in transgenic mice. *Mol. Endocrinol.* 5:2025–36

Kimura S, Hara Y, Pineau T, Fernandez-Salguero P, Fox CH, et al. 1996. The T/ebp null mouse thyroid-specific enhancer-binding protein is essential for the organogenesis of the thyroid, lung, ventral forebrain, and pituitary. *Genes Dev.* 10:60–69

Kioussi C, O'Connell S, St-Onge L, Treier M, Gleiberman AS, et al. 1999. Pax6 is essential for establishing ventral-dorsal cell boundaries in pituitary gland development. *Proc. Natl. Acad. Sci. USA* 96:14378–82

Kitamura K, Miura H, Miyagawa-Tomita S, Yanazawa M, Katoh-Fukui Y, et al. 1999. Mouse Pitx2 deficiency leads to anomalies of the ventral body wall, heart, extra and periocular mesoderm and right pulmonary isomerism. *Development* 126:5749–58

Lamonerie T, Tremblay JJ, Lanctot C, Therrien M, Gauthier Y, Drouin J. 1996. Ptx1, a bicoid-related homeobox transcription factor involved in transcription of the proopiomelanocortin gene. *Genes Dev.* 10:1284–95

Lanctot C, Gauthier Y, Drouin J. 1999a. Pituitary homeobox 1 (Ptx1) is differentially expressed during pituitary development. *Endocrinology* 140:1416–22

Lanctot C, Lamolet B, Drouin J. 1997. The bicoid-related homeoprotein Ptx1 defines the most anterior domain of the embryo and differentiates posterior from anterior lateral mesoderm. *Development* 124:2807–17

Lanctot C, Moreau A, Chamberland M, Tremblay ML, Drouin J. 1999b. Hindlimb patterning and mandible development require the Ptx1 gene. *Development* 129:1805–10

Laufer E, Nelson CE, Johnson RL, Morgan BA, Tabin C. 1994. Sonic hedgehog and Fgf-4 act through a signaling cascade and feedback loop to integrate growth and patterning of the developing limb bud. *Cell* 79:993–1003

Lee SL, Sadovsky Y, Swirnoff AH, Polish JA, Goda P, et al. 1996. Luteinizing hormone deficiency and female infertility in mice lacking the transcription factor NGFI-A (Egr-1). *Science* 273:1219–21

Li S, Crenshaw EB III, Rawson EJ, Simmons DM, Swanson LW, Rosenfeld MG. 1990. Dwarf locus mutants lacking three pituitary cell-types result from mutations in the POU-domain gene pit-1. *Nature* 347:528–33

Lin CR, Koussi C, O'Connell S, Briata P, Szeto D, et al. 1999. Pitx2 regulates lung asymmetry, cardiac positioning and pituitary and tooth morphogenesis. *Nature* 401:279–82

Lin SC, Li S, Drolet DW, Rosenfeld MG. 1994. Pituitary ontogeny of the Snell dwarf mouse reveals Pit-1-independent and Pit-1-dependent origins of the thyrotrope. *Development* 120:515–22

Lin SC, Lin CR, Gukovsky I, Lusis AJ, Sawchenko PE, Rosenfeld MG. 1993. Molecular basis of the little mouse phenotype and implications for cell type-specific growth. *Nature* 364:208–13

Lipkin SM, Naar AM, Kalla KA, Sack RA, Rosenfeld MG. 1993. Identification of a novel zinc finger protein binding a conserved element critical for Pit-1-dependent growth hormone gene expression. *Genes Dev.* 7:1674–87

Logan M, Tabin CJ. 1999. Role of Pitx1 upstream of Tbx4 in specification of hindlimb identity. *Science* 283:1736–39

Lu MF, Pressman C, Dyer R, Johnson RL, Martin JF. 1999. Function of Rieger syndrome gene in left-right asymmetry and craniofacial development. *Nature* 401:276–78

Martin GR. 1998. The roles of FGFs in the early development of vertebrate limbs. *Genes Dev.* 12:1571–86

Mead PE, Brivanlou IH, Kelley CM, Zon LI. 1996. BMP-4-responsive regulation of dorsal-ventral patterning by the homeobox protein Mix.1. *Nature* 382:357–60

Meyers EN, Martin GR. 1999. Differences in left-right axis pathways in mouse and chick: functions of FGF8 and SHH. *Science* 285:403–6

Min H, Danilenko DM, Scully SA, Bolon B, Ring BD, et al. 1998. Fgf-10 is required for both limb and lung development and exhibits striking functional similarity to Drosophila branchless. *Genes Dev.* 12:3156–61

Morcillo P, Rosen C, Baylies MK, Dorsett D. 1997. Chip, a widely expressed chromosomal protein required for segmentation and activity of a remote wing margin enhancer in Drosophila. *Genes Dev.* 11:2729–40

Neubuser A, Peters H, Balling R, Martin GR. 1997. Antagonistic interactions between FGF and BMP signaling pathways: a mechanism for positioning the sites of tooth formation. *Cell* 90:247–55

Niederreither K, McCaffery P, Drager UC, Chambon P, Dolle P. 1997. Restricted expression and retinoic acid-induced downregulation of the retinaldehyde dehydrogenase type 2 (RALDH-2) gene during mouse development. *Mech. Dev.* 62:67–78

Palomino T, Barettino D, Aranda A. 1998. Role of GHF-1 in the regulation of the rat growth hormone gene promoter by thyroid hormone and retinoic acid receptors. *J. Biol. Chem.* 273:27541–47

Pernasetti F, Toledo SP, Vasilyev VV, Hayashida CY, Cogan JD, et al. 2000. Impaired adrenocorticotropin-adrenal axis in combined pituitary hormone deficiency caused by a two-base pair deletion (301-302delAG) in the prophet of Pit-1 gene. *J. Clin. Endocrinol. Metab.* 85:390–97

Peters H, Balling R. 1999. Teeth: where and how to make them. *Trends Genet.* 15:59–65

Pfaff SL, Mendelsohn M, Stewart CL, Edlund T, Jessell TM. 1996. Requirement for LIM homeobox gene Isl1 in motor neuron generation reveals a motor neuron-dependent step in interneuron differentiation. *Cell* 84:309–20

Philips A, Lesage S, Gingras R, Maira MH, Gauthier Y, et al. 1997. Novel dimeric Nur77 signaling mechanism in endocrine and lymphoid cells. *Mol. Cell. Biol.* 17:5946–51

Reichardt HM, Kaestner KH, Tuckermann J, Kretz O, Wessely O, et al. 1998. DNA binding of the glucocorticoid receptor is not essential for survival. *Cell* 93:531–41

Reichardt HM, Schutz G. 1996. Feedback control of glucocorticoid production is established during fetal development. *Mol. Med.* 2:735–44

Rhodes SJ, Chen R, DiMattia GE, Scully KM, Kalla KA, et al. 1993. A tissue-specific enhancer confers Pit-1-dependent morphogen inducibility and autoregulation on the pit-1 gene. *Genes Dev.* 7:913–32

Rosenbloom AL, Almonte AS, Brown MR, Fisher DA, Baumbach L, Parks JS. 1999. Clinical and biochemical phenotype of familial anterior hypopituitarism from mutation of the PROP1 gene. *J. Clin. Endocrinol. Metab.* 84:50–57

Rubenstein JL, Shimamura K, Martinez S, Puelles L. 1998. Regionalization of the prosencephalic neural plate. *Annu. Rev. Neurosci.* 21:445–77

Ruiz I, Altaba A. 1999. Gli proteins and Hedgehog signaling: development and cancer. *Trends Genet.* 15:418–25

Sanchez-Pacheco A, Pena P, Palomino T, Guell A, Castrillo JL, Aranda A. 1998. The transcription factor GHF-1, but not the splice variant GHF-2, cooperates with thyroid hormone and retinoic acid receptors to stimulate rat growth hormone gene expression. *FEBS Lett.* 422:103–7

Schaufele F, West BL, Baxter JD. 1992. Synergistic activation of the rat growth hormone promoter by Pit-1 and the thyroid hormone receptor. *Mol. Endocrinol.* 6:656–65

Scully KM, Gleiberman AS, Lindzey J, Lubahn DB, Korach KS, Rosenfeld MG. 1997. Role of estrogen receptor-alpha in the anterior pituitary gland. *Mol. Endocrinol.* 11:674–81

Seidah NG, Barale JC, Marcinkiewicz M, Mattei MG, Day R, Chretien M. 1994. The mouse homeoprotein mLIM-3 is expressed early in cells derived from the neuroepithelium and persists in adult pituitary. *DNA Cell Biol.* 13:1163–80

Sekine K, Ohuchi H, Fujiwara M, Yamasaki M, Yoshizawa T, et al. 1999. Fgf10 is essential for limb and lung formation. *Nat. Genet.* 21:138–41

Semina EV, Reiter R, Leysens NJ, Alward WL, Small KW, et al. 1996. Cloning and characterization of a novel bicoid-related homeobox transcription factor gene, RIEG, involved in Rieger syndrome. *Nat. Genet.* 14:392–99

Sharma K, Sheng HZ, Lettieri K, Li H, Karavanov A, et al. 1998. LIM homeodomain factors Lhx3 and Lhx4 assign subtype identities for motor neurons. *Cell* 95:817–28

Sheng HZ, Moriyama K, Yamashita T, Li H, Potter SS, et al. 1997. Multistep control of pituitary organogenesis. *Science* 278:1809–12

Sheng HZ, Westphal H. 1999. Early steps in pituitary organogenesis. *Trends Genet.* 15:236–40

Sheng HZ, Zhadanov AB, Mosinger B Jr, Fujii T, Bertuzzi S, et al. 1996. Specification of pituitary cell lineages by the LIM homeobox gene Lhx3. *Science* 272:1004–7

Shewchuk BM, Asa SL, Cooke NE, Liebhaber SA. 1999. Pit-1 binding sites at the somatotrope-specific DNase I hypersensitive sites I, II of the human growth hormone locus control region are essential for in vivo hGH-N gene activation. *J. Biol. Chem.* 274:35725–33

Shiojima I, Komuro I, Oka T, Hiroi Y, Mizuno T, et al. 1999. Context-dependent transcriptional cooperation mediated by cardiac transcription factors Csx/Nkx-2.5 and GATA-4. *J. Biol. Chem.* 274:8231–39

Simmons DM, Voss JW, Ingraham HA, Holloway JM, Broide RS, et al. 1990. Pituitary cell phenotypes involve cell-specific Pit-1 mRNA translation and synergistic interactions with other classes of transcription factors. *Genes Dev.* 4:695–711

Smith ST, Jaynes JB. 1996. A conserved region of engrailed, shared among all en-, gsc-, Nk1-, Nk2- and msh-class homeoproteins, mediates active transcriptional repression in vivo. *Development* 122:3141–50

Song K, Wang Y, Sassoon D. 1992. Expression of Hox-7.1 in myoblasts inhibits terminal differentiation and induces cell transformation. *Nature* 360:477–81

Sornson MW, Wu W, Dasen JS, Flynn SE, Norman DJ, et al. 1996. Pituitary lineage determination by the prophet of Pit-1 homeodomain factor defective in Ames dwarfism. *Nature* 384:327–33

Steger DJ, Hecht JH, Mellon PL. 1994. GATA-binding proteins regulate the human gonadotropin alpha-subunit gene in the placenta and pituitary gland. *Mol. Cell. Biol.* 8:5592–602

Szeto DP, Rodriguez-Esteban C, Ryan AK, O'Connell SM, Liu F, et al. 1999. Role of the Bicoid-related homeodomain factor Pitx1 in specifying hindlimb morphogenesis and pituitary development. *Genes Dev.* 15:484–94

Szeto DP, Ryan AK, O'Connell SM, Rosenfeld MG. 1996. P-OTX: a PIT-1-interacting homeodomain factor expressed during anterior pituitary gland development. *Proc. Natl. Acad. Sci. USA* 5:7706–10

Takuma N, Sheng HZ, Furuta Y, Ward JM, Sharma K, et al. 1998. Formation of Rathke's pouch requires dual induction from the diencephalon. *Development* 125:4835–40

Therrien M, Drouin J. 1993. Molecular determinants for cell specificity and glucocorticoid repression of the proopiomelanocortin gene. *Ann. NY Acad. Sci.* 680:663–71

Thomas PQ, Rathjen PD. 1992. HES-1, a novel homeobox gene expressed by murine embryonic stem cells, identifies a new class of homeobox genes. *Nucleic Acids Res.* 20:5840

Topilko P, Schneider-Maunoury S, Levi G, Trembleau A, Gourdji D, et al. 1998. Multiple pituitary and ovarian defects in Krox-24 (NGFI-A, Egr-1)-targeted mice. *Mol. Endocrinol.* 12:107–122

Treier M, Gleiberman AS, O'Connell SM, Szeto DP, McMahon JA, et al. 1998.

Multistep signaling requirements for pituitary organogenesis in vivo. *Genes Dev.* 12:1691–704

Tremblay JJ, Drouin J. 1999. Egr-1 is a downstream effector of GnRH and synergizes by direct interaction with Ptx1 and SF-1 to enhance luteinizing hormone beta gene transcription. *Mol. Cell. Biol.* 19:2567–76

Tremblay JJ, Lanctot C, Drouin J. 1998. The pan-pituitary activator of transcription, Ptx1 (pituitary homeobox 1), acts in synergy with SF-1 and Pit1 and is an upstream regulator of the Lim-homeodomain gene Lim3/Lhx3. *Mol. Endocrinol.* 12:428–41

Tsai FY, Keller G, Kuo FC, Weiss M, Chen J, et al. 1994. An early haematopoietic defect in mice lacking the transcription factor GATA-2. *Nature* 371:221–26

Watanabe YG. 1982a. An organ culture study on the site of determination of ACTH and LH cells in the rat adenohypophysis. *Cell Tissue Res.* 227:267–75

Watanabe YG. 1982b. Effects of brain and mesenchyme upon the cytogenesis of rat adenohypophysis in vitro. I. Differentiation of adrenocorticotropes. *Cell Tissue Res.* 227:257–66

Watkins-Chow DE, Camper SA. 1998. How many homeobox genes does it take to make a pituitary gland? *Trends Genet.* 14:284–90

Watkins-Chow DE, Douglas KR, Buckwalter MS, Probst FJ, Camper SA. 1997. Construction of a 3-Mb contig and partial transcript map of the central region of mouse chromosome 11. *Genomics* 45:147–57

Winnier G, Blessing M, Labosky PA, Hogan BL. 1995. Bone morphogenetic protein-4 is required for mesoderm formation and patterning in the mouse. *Genes Dev.* 9:2105–16

Wood WM, Dowding JM, Gordon DF, Ridgway EC. 1999. An upstream regulator of the glycoprotein hormone alpha-subunit gene mediates pituitary cell type activation and repression by different mechanisms. *J. Biol. Chem.* 274:15526–32

Wu W, Cogan JD, Pfaffle RW, Dasen JS, Frisch H, et al. 1998. Mutations in PROP1 cause familial combined pituitary hormone deficiency. *Nat. Genet.* 18:147–49

Xu L, Lavinsky RM, Dasen JS, Flynn SE, McInerney EM, et al. 1998. Signal-specific coactivator domain requirements for Pit-1 activation. *Nature* 395:301–6

Yano H, Readhead C, Nakashima M, Ren SG, Melmed S. 1998. Pituitary-directed leukemia inhibitory factor transgene causes Cushing's syndrome: neuro-immune-endocrine modulation of pituitary development. *Mol. Endocrinol.* 12:1708–20

Zanger K, Cohen LE, Hashimoto K, Radovick S, Wondisford FE. 1999. A novel mechanism for cyclic adenosine $3',5'$-monophosphate regulation of gene expression by CREB-binding protein. *Mol. Endocrinol.* 13:268–75

Zhadanov AB, Bertuzzi S, Taira M, Dawid IB, Westphal H. 1995. Expression pattern of the murine LIM class homeobox gene Lhx3 in subsets of neural and neuroendocrine tissues. *Dev. Dyn.* 202:354–64

Annu. Rev. Neurosci. 2001. 24:357–84

Neuropeptides and the Integration of Motor Responses to Dehydration

Alan G. Watts

The Neuroscience Program and the Department of Biological Sciences,
University of Southern California, Los Angeles, California 90089-2520;
e-mail: watts@rcf.usc.edu

Key Words anorexia, hypothalamus, parabrachial nucleus, amygdala, hormones

■ **Abstract** Drinking and eating are critically important motivated behaviors whose expression is usually tightly linked; under conditions of spontaneous intake, disruption of one usually disturbs the other. This characteristic is exemplified by dehydration-induced anorexia in which increasing plasma osmolality leads to a centrally generated reduction in food intake, which is then rapidly reversed as water is again made available. This review discusses, at a systems level, how the brain is organized to generate these behaviors and how dehydration affects the expression of neuropeptides in sets of anatomically defined forebrain circuits that contribute to the integration of motor outputs. These findings are then used to consider how altered neuropeptidergic signaling operates within motor drive networks and how these changes may impact the way neuroendocrine, autonomic, and behavioral motor systems respond to this fundamental homeostatic challenge.

INTRODUCTION

When and how animals locate and ingest the substances essential for maintaining metabolism, growth, and overall bodily integrity are controlled by a triad of motivated behaviors. Thus, feeding controls the intake of nutrients (fat, protein, and carbohydrates), vitamins, and cofactors; targeted appetites control the intake of specific ions (principally sodium, but also, to a lesser extent, other essential ions such as calcium); and drinking regulates water intake. Which particular behavior an animal chooses to express at a particular time is mediated by complex neural systems that have components distributed throughout the brain. The need to replenish or anticipate deficits, the desire to find suitable goal objects, and knowledge of their whereabouts are transformed into appropriate sets of behavioral and visceral motor outputs by the actions of these systems. Defining their architecture is a fundamental neuroscientific problem with a long history of research that has endeavored to explain the neural bases of motivated behaviors.

0147-006X/01/0621-0357$14.00

Given the complexity of the systems involved, many experimental strategies have understandably concentrated on the simpler aspects of ingestive behaviors, particularly those generated by homeostatic challenges. Measuring the amount of water or food consumed during a fixed period after administration of agents into the brain exemplifies this strategy. However, despite revealing a great deal about the basic organization of the circuits underlying the sensory control of behaviors, this type of approach often addresses only the simpler reflex aspects of ingestive behavior that have few anticipatory or complex foraging components. In this manner, many details remain unclear as to which processes account for how animals learn to anticipate homeostatic deficits by structuring drinking and eating in regular bouts, how one ingestive behavior interacts with another, or how ingestive behaviors might interact with other types of motivated behavior.

Although we are still some way from providing detailed explanations of these problems, recent physiological and functional neuroanatomical studies are beginning to reveal some of the underlying processes. These include which circuits stimulate, inhibit, or switch particular ingestive behaviors; how behavior is coordinated with visceromotor function; and how neuropeptides might act to facilitate these complex interactions.

This review focuses on recent work that addresses these issues. The first part describes the general organization of neural systems concerned with regulating the motor responses to homeostatic challenges, of which dehydration is a classic example. The second part describes how the dynamic expression of neuropeptides contained in a variety of forebrain cell groups can modulate the neural networks concerned with controlling neuroendocrine, autonomic, and behavioral motor function. Endocrine, behavioral, and functional neuroanatomical data are also described, and recent work is highlighted that uses a physiological model of anorexia in which animals actively reduce their food intake as they become chronically dehydrated (DE). Under these circumstances, energy balance is sacrificed at the expense of protecting the fluid compartment, as indicated by the classic signs of negative energy balance, including the endocrine and hypothalamic neuropeptidergic signatures usually associated with increased hunger. However, the effects of these signals on feeding behavior are suppressed until the reinstatement of drinking water, when they are dramatically released and animals engage in a brisk compensatory-feeding episode. These experiments have not only disclosed potentially novel neuropeptidergic substrates for behavioral integration and switching, they have also emphasized the value of correlating neural events with highly reproducible and well-characterized behavioral sequences.

Neural Systems

The idea of behaviorally specific drive networks embedded within motor control systems (Watts 2001) evolved from the idea of discretely localized hypothalamic satiety and hunger centers developed during the 1950s and 1960s. During the past 30 years studies on behaviorally specific drive networks have replaced the notion of isolated centers with a scheme where sets of more widely distributed but highly interconnected drive networks direct the motor responses for particular behaviors. Each drive network contains sets of circuits that either stimulate, inhibit, or disinhibit a particular motor event. The exact nature of the expressed behavior, or whether it is expressed at all, is determined by the integrated output of these circuits.

Rather than being confined to a single locus, results from a variety of studies have shown that each drive network has component cell groups distributed in the hindbrain, midbrain, telencephalon, and hypothalamus (for reviews, see Elmquist et al 1999, Grill & Kaplan 1990, Kelley 1999, Risold et al 1997, Sawchenko 1998, Stricker 1990, Swanson 2000, Watts 1991, 2000). However, added complexity arises because the cellular components of these different control networks may be found within a single traditional cytoarchitectonically defined region. This is emphasized by the fact that hypothalamic cell groups often contain elements that can stimulate or inhibit particular behaviors, as evidenced by the expression in the same cell group of neuropeptides with opposing behavioral actions (Table 1). This feature makes delineating drive networks particularly difficult and has likely contributed to the continuing confusion concerning the precise function of some individual hypothalamic nuclei.

Based on these data, Figure 1 illustrates, without reference to anatomical loci, an arrangement of neural elements that accounts for a wide range of motor actions activated by an array of sensory inputs. These motor functions vary from those anticipatory behavioral, autonomic, and neuroendocrine motor events that require circadian timing and complex telenecephalic neural processing (inputs 1 and 2, Figure 1) through motor actions produced by deficit signals (e.g. the thirst arising from dehydration or hunger arising from starvation) requiring the hypothalamus and forebrain (input 3, Figure 1), to the simple relex actions (e.g. vasopressin release in resonse to increased osmolality, reflex rejection of unpalatable foods or

TABLE 1 Neuropeptides that stimulate or inhibit eating are expressed by neurons located in the same hypothalamic regions

	Arcuate nucleus	Lateral hypothalamic area	Paraventricular nucleus
Orexigenic peptides	Neuropeptide Y Agouti-related peptide galanin	Melanin-concentrating hormone Dynorphin Galanin Hypocretin/orexin	Dynorphin
Anorexigenic peptides	Cocaine- and amphet-amine-regulated transcript α-Melanocyte-stimulating hormone Neurotensin	Cocaine- and amphet-amine-regulated transcript Neurotensin CRH	Cocaine- and amphet-amine-regulated transcript Oxytocin Neurotensin CRH

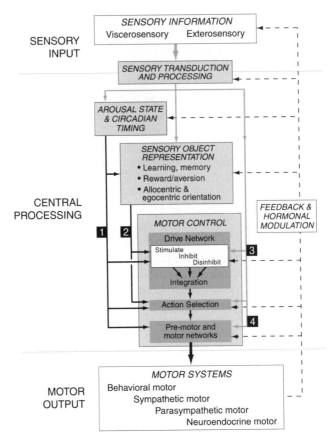

Figure 1   Motor control networks are organized at three levels: drive networks that consist of components that can either stimulate, inhibit, or disinhibit specific behaviors; action selection networks that integrate the outputs of multiple drive networks; and executive pre-motor and motor neuron networks. The generation of motivated behavioral actions by motor control networks can be initiated from four different sets of inputs: (1) systems controlling arousal state and circadian timing; (2) systems generating representations of sensory objects; (3) direct modulatory hormone and the sensory signals encoding physiological deficits; and (4) sensory signals that generate reflex actions directly from pre-motor and motor neuron networks. Sensory inputs are shown as grey lines, central neural connections as black lines, and hormonal and feedback signals as dashed lines.

fluids) generated by the more or less direct sensory inputs to the pre-motor and motor networks (input 4, Figure 1).

The scheme described in Figure 1 is consistent with the fact that some aspects of motor function still operate when one part of a drive network is removed or is disconnected from others located elsewhere in the brain. For example, animals that have their hindbrain separated from the forebrain still reject unpalatable food and show responses involving short-term comparisons of gustatory stimuli, but they fail to generate compensatory responses to challenges requiring longer-term processing (Grill & Kaplan 1990, Kaplan et al 1993, Grigson et al 1997). Moreover, large excitotoxic lesions of the lateral hypothalamic area (LHA) are consistent with the relatively normal maintenance of body weight and with normal compensatory eating and drinking responses to deprivation (which has definite exterosensory components), but not to signals that are primarily internal in origin such as glucoprivation or hypovolemia (Winn 1995).

Sensory Input and Motor Output

Neurons in the motor drive networks are regulated by sensory inputs that fall into two categories (Figure 1): viscerosensory signals that encode internal states and exterosensory inputs that encode features of fluids and food such as smell, taste, temperature, tactile properties, and appearance. Some of these sensory signals directly access drive networks, as typified by the drinking initiated by the neurally targeted effects of increasing plasma osmolality or angiotensin II (A-II; Fitzsimons 1998); in feeding behavior, plasma leptin and insulin act as adiposity signals to affect feeding by actions in the hypothalamus (Baskin et al 1999, Elmquist et al 1999, Schwartz et al 2000). Sensory information also includes feedback signals encoding the magnitude and consequences of generated motor actions and can control the length of an ingestive episode. For example, postabsorptive humoral feedback (e.g. decreasing plasma osmolality in some species) and viscerosensory signals (e.g. gastric distension and oropharyngeal metering; Grossman 1990) lead to drink termination and subsequent ingestive behavioral refractoriness. Evidence suggests that these feedback signals may act to increase the activity of inhibitory components in the drive networks (Stricker 1990).

Sensory Object Representation and Behavioral State Control

In addition to the motor control networks that target specific motor events, there are separate neural structures whose functions are behaviorally nonspecific and influence a wide range of motor functions. These regions contribute to two complex and widely distributed systems that control either arousal state or sensory object representation.

Parts of the brain provide timing information and control arousal state, so enabling motor control networks to generate the types of actions that anticipate deficits. This system includes the suprachiasmatic nucleus, which generates the circadian timing signal that entrains virtually all neural activity within limits determined by the prevailing photoperiod (Watts 1991, Moore 1997). Catecholamine cell groups in the hindbrain (e.g. the locus coeruleus), histaminergic neurons in the tuberomammillary nucleus, the ventrolateral preoptic nucleus, and the recently identified hypocretin/orexin neurons in the LHA (Sherin et al 1998, Willie et al 2001) also supply information that is likely to be important for controlling the arousal state.

Structures that generate the neural representations of sensory objects that are important for motor control include learning and memory mechanisms in the telencephalon and cerebellum as well as the reward/aversion systems that most likely involve the midbrain ventral tegmentum (Schultz 1998), parts of the basal forebrain (particularly the nucleus accumbens; Kelley 1999), the amygdala, and parts of the cortex, particularly the prefrontal (regions (Gallagher et al 1999). Areas in the telencephalon are also important for navigation and orientation within the environment. The majority of exterosensory information is processed through these networks, which collectively assign "incentive value" to a particular goal object (Toates 1986). The neural pathways mediating the interactions between the object representation and motor control systems are not fully understood, but sets of bidirectional connections between the hypothalamus and cortical structures such as the prefrontal cortex and hippocampus, as well as subcortical regions such as the amygdala, septal nuclei, bed nuclei of the stria terminalis (BST), and basal ganglia are all likely to be critical in the integrative operations that designate and coordinate the full spectrum of motor functions associated with dehydration (Denton et al 1999a,b; Saper 1985; Risold & Swanson 1996, 1997; Risold et al 1997; Swanson & Petrovich 1998).

PHYSIOLOGICAL CHALLENGES TO THE FLUID COMPARTMENTS

The composition of the fluid compartment is defended at the expense of virtually all other functions. Two parameters are used to signal changes in fluid compartment composition to the brain: plasma osmolality (with sodium ions being particularly important) and extracellular fluid volume (ECV). Increases in osmolality are detected directly by osmoreceptors located primarily at the rostral end of the third ventricle. Decreases in ECV are transduced by two mechanisms (Grossman 1990). The first involves renal juxtaglomerular cells and elevates circulating A-II that interacts with receptors in the subfornical organ (SFO), whereas in the second mechanism, low-pressure baroreceptors in the venous circulation alter afferent sensory input to the nucleus of the solitary tract in the hindbrain.

Motor Responses Directly Generated by Dehydration

Engaging these sensory mechanisms leads to well-characterized neuroendocrine, autonomic, and behavioral motor responses that can be usefully categorized as protective, adaptive, or restorative (Dicker & Nunn 1957, Flanagan et al 1989, Grossman 1990).

The principal protective responses involve the neuroendocrine and autonomic motor systems. They include increased release of vasopressin and oxytocin from the posterior pituitary, which are very rapidly initiated by neuroendocrine reflexes after dehydration. These hormones help maintain blood pressure and increase renal water retention and natriuresis (Grossman 1990, Huang et al 1996). Other protective responses occur after hemorrhage or hypovolemia and include modifications to cardiovascular function and the rapid activation of all levels of the hypothalamo-pituitary-adrenal axis (Gann et al 1978, Tanimura et al 1998). In contrast, hyperosmolemia tends to depress the activity of the hypothalamo-pituitary-adrenal axis (Watts 1996, 2000).

Hyperosmolemia generates a set of adaptive motor responses that target gastrointestinal function to conserve water. They include reduced salivation, altered gastric and intestinal motility, and anorexia (Flanagan et al 1989). Finally, restorative mechanisms modify behavior to return water and sodium (in the case of ECV challenges) back into the body. The selection, timing, and magnitude of all of these responses ultimately depend on the intensity and kinetics of the stimulus. For example, mild hyperosmolemia will stimulate arginine vasopressin (AVP) secretion but not anorexia (Grossman 1990, Watts 1999).

Dehydration-Generated Anorexia

The adaptive responses triggered by increased plasma osmolality reduce how much food enters the gut and retard digestion. These actions help maintain fluid compartmental integrity because digestion, particularly of dry chow, requires significant amounts of water and because absorption of food adds additional osmoles to an already burdened system.

Two types of dehydration-generated anorexia are distinguishable depending on the rate at which hyperosmolality increases. The first quickly develops after large and rapid increases in plasma osmolality of the type produced by injections of hypertonic saline. It is most easily seen as an inhibition either of compensatory feeding after deprivation or of eating expressed when food is presented on a restricted schedule. This rapidly initiated anorexia has a prominent reflex component, because acute hyperosmolemia in animals with surgically isolated hindbrains still reduces the volume of sucrose solution accepted during periodic feeding schedules (Flynn et al 1995). However, if the more effective parts of the inhibitory control network for this type of anorexia are in the hindbrain, the fact that applying oxytocin antagonist partially blunts its expression (Olson et al 1991) suggests that forebrain components are still required for full development, because preautonomic neurons in the hypothalamic paraventricular nucleus (PVH) are the only source of afferent fibers to the hindbrain (Swanson 1987).

A second type of dehydration-generated anorexia is evident in animals on a continuous feeding schedule in which drinking and eating have a strong circadian-driven component requiring the forebrain (Grill & Kaplan 1990). In this case, spontaneous nocturnal feeding is strongly inhibited by the chronic dehydration produced by drinking hypertonic saline. During a 5-day period of dehydration, nocturnal food intake gradually falls to ~20% of normal, whereas the small amount of food eaten during the day remains unaffected (Watts 1999). The development of dehydration anorexia is relatively slow, presumably because the gradually increasing burden placed on plasma osmolality is well buffered at this time (Watts et al 1995).

Motor Responses Generated by the Consequences of Dehydration-Generated Anorexia

A variety of other responses to dehydration are less concerned with regulating fluid balance but instead originate primarily as a consequence of the negative energy balance generated by steadily diminishing food intake. Thus, plasma leptin, insulin, and corticosterone concentrations are virtually identical in DE animals to those seen in pair-fed food-restricted animals; leptin and insulin decrease while corticosterone increases compared with euhydrated animals (Watts et al 1999). The status of these hormones during dehydration is critical because they have major effects on the expression of neuropeptides that are intimately concerned with neuroendocrine, autonomic, and behavioral motor functions. Collectively, these feedback signals in DE animals reflect the increasingly negative energy balance that, under normal circumstances, should stimulate eating. However, following dehydration, eating is increasingly suppressed, presumably by the actions of inhibitory control networks. The involvement of a third network that helps mediate behavioral switching is revealed when DE rats are again allowed to drink water.

The Rapid Release of Eating Behavior in Dehydrated Rats

As chronically DE rats drink water, their behavior during the subsequent hour is manifested as a temporally ordered and highly reproducible sequence. The first action, not unexpectedly, is a bout of drinking that lasts about 8 min, at which point drinking abruptly terminates, to be followed almost immediately by eating. Although the sensory signals responsible for this behavioral switch are unknown, reductions in peripheral plasma osmolality after water absorption do not seem to be responsible (Watts 2000). The eating episode continues for about 10 min, followed by a further 10 min in which eating and drinking alternate. About 35 min from the beginning of the sequence, both ingestive behaviors are reduced and are replaced by a brisk increase in general activity that continues for a further 20 min. During this time, short bouts of intense activity are particularly striking, suggesting that rehydration also interacts with the central mechanisms that activate general arousal (Watts & Sanchez-Watts 2000). Collectively, these observations suggest that rehydration has rapid, powerful, and temporally ordered effects on the central processes regulating ingestive behaviors and general arousal. These behaviors are clearly the ones most useful to the animal for correcting the deficits incurred during dehydration.

THE STRUCTURE OF CONTROL NETWORKS

Determining the anatomical substrates of these networks has proved quite difficult. However, the wealth of data cataloging the stimulatory or inhibitory effects of hypothalamic neuropeptides on particular motor functions clearly concurs with earlier lesion and stimulation experiments that the hypothalamus is a key locus for components of the drive networks. Many groups have identified which hypothalamic neurons synthesize these neuropeptides and have determined where they project. One critical point that has emerged from these studies is that placing individual hypothalamic cell groups within one or another of these networks is rather difficult. This is because traditionally defined cell groups such as the arcuate nucleus (ARH), LHA, and PVH appear to contain elements that, in terms of function, belong to more than one type of drive network, and it seems unlikely that there is a tight "one cell group to one network" relationship in the hypothalamus (Table 1).

The most familiar example of a neuropeptide that increases drinking is the stimulation of the SFO by circulating A-II (Fitzsimons 1998); results from many studies place this structure centrally in a circuit that directly and specifically stimulates water intake. The SFO provides efferents, most of which also contain A-II (Swanson 1987), to a relatively limited set of structures, including parts of the prefrontal cortex, substantia innominata, medial preoptic area, BST, zona incerta, PVH, supraoptic nucleus, and LHA (Swanson & Lind 1986), and presumably these regions constitute part of the stimulatory network that controls the motor aspects of drinking.

The best documented example of a stimulatory eating mechanism involves neuropeptide Y (NPY) neurons in the ARH; results from many studies show that NPY contributes to a circuit that directly stimulates food intake (Elmquist et al 1999, Kalra et al 1999, Schwatrz et al 2000). However, the recent report that NPY knock-out mice have no impairment of eating behavior shows that this peptide is not an absolute requirement for either anticipatory or deficit-controlled eating (Palmiter et al 1998).

Inhibition of drinking is for the most part effected by sensory signals derived from postabsorptive mechanisms (Grossman 1990), but the role of neuropeptides in this function is unclear. The role of neuropeptides as components of inhibitory networks involved with feeding is better documented. For example, α-melanocyte-stimulating hormone synthesized in ARH neurons from the gene encoding proopiomelanocortin (*POMC*) provides an inhibitory signal to feeding that acts at melanocortin (MC) 4 receptors expressed by neurons in the LHA and PVH (Cowley et al 1999, Elmquist et al 1999). POMC neurons in the ARH and retrochiasmatic area also express cocaine- and amphetamine-regulated transcript. This widely expressed neuropeptide will inhibit eating without affecting spontaneous drinking when injected intracerebroventricularly (Kask et al 2000, Kuhar & Dall Vechia 1999).

Because it is unclear precisely how the neural components that comprise the motor control systems are organized at each of the functional levels illustrated in Figure 1 or exactly how they interact, we do not know where the divergence occurs that accounts for the distinct types of motor output. That there must be divergence is exemplified by the different motor effects of two humoral viscerosensory inputs; A-II can stimulate drinking behavior and autonomic and neuroendocrine output (Herbert 1993), while leptin suppresses eating behavior but increases thyroid function and sympathetic output (Elias et al 1998a, Haynes et al 1997, Kim et al 2000). Are there separate sets of drive networks associated with each of the three types of motor output, or are divergent outputs from "core" networks directed towards the specific premotor networks associated with the different types of motor output?

FOREBRAIN NEUROPEPTIDE GENE EXPRESSION DURING DEHYDRATION-GENERATED ANOREXIA OR PAIRED FOOD RESTRICTION

The question now arises as to the nature of the changes occurring within control networks that can account for the overall behavior of animals as dehydration challenges their fluid balance. As shown above, neuropeptides can control specific motor functions, and it seems likely that they are critical elements in generating the modified motor events after dehydration. To address this question, we recently compared the effects of dehydration and food restriction on the expression of those forebrain neuropeptide genes implicated in regulating ingestive behaviors (Watts et al 1999). We hypothesized that we would see at least two patterns of neuropeptide gene expression; changes seen during both dehydration and paired food restriction were most likely the consequences of negative energy balance, whereas changes seen in dehydration but not paired food restriction would potentially be causative agents for anorexia.

Hypothalamus

Both groups of animals have the characteristic patterns of gene expression generally seen during negative energy balance: increased NPY, decreased POMC, and neurotensin mRNAs in the ARH (Table 2). The response of the corticotropin-releasing hormone (CRH)-R2 receptor mRNA in the hypothalamic ventromedial nucleus—which may be the target of urocortin (Ohata et al 2000)—also fits this pattern (Kay-Nishiyama & Watts 1998). However, DE animals are strongly anorexic despite these characteristics, demonstrating that the output of hormone-sensitive ARH neurons that ordinarily stimulate eating must be inhibited. The neuropeptide mRNA pattern seen only in DE-anorexic animals has two components, both of which could potentially generate anorexia (Table 2). The first is evident in the PVH where CRH mRNA is reduced only in DE animals (Watts et al 1995, 1999). Although it is difficult to conceive how these neuroendocrine neurons can directly affect behavior, it has been postulated that altered CRH mRNA seen there in some situations (particularly after glucocorticoid manipulation) is associated with reduced food intake (see Porte et al 1998 for review). For the anorexia after dehydration, however, neither increased corticosterone nor the reduced CRH mRNA is required because PVH CRH mRNA in DE adrenalectomized animals (which also develop anorexia; Watts et al 1999) is unchanged from

TABLE 2   Comparison of effects of food restriction (hunger) and dehydration (anorexia) on the levels of neuropeptide mRNAs in hypothalamic and telencephalic regions

Condition	Brain region	Changes in		
		mRNA[a]	FR[b]	DE[b]
Hunger and anorexia	Arcuate nucleus	NPY	+	+
		POMC	−	−
		NT	−	−
	Ventromedial nucleus	CRH-R2	−	−
	Bed nuclei of the stria terminalis–fusiform	CRH	−	+
	Paraventricular nucleus[c] (adrenalectomized animals)	CRH	+	+
Anorexia only	Paraventricular nucleus[b] (intact animals)	CRH	0	−
	Retrochiasmatic area	CRH	nd	+
		NT	0	+
	Lateral hypothalamic area	CRH	nd	+
		NT	0	+
Hunger only	Central nucleus of the amygdala–lateral part	CRH	−	0
		NT	−	0

[a]NPY, neuropeptide Y; POMC, proopiomelanocortin; NT, neurotensin; CRH, corticotropin-releasing hormone; CRH-R2, CRH type 2 receptor.

[b]+, increased vs control; −, decreased vs control; 0, unchanged vs control; nd, not detectable, FR, food restricted; DE, dehydration.

[c]Parvicellular neuroendocrine part.

euhydrated adrenalectomized controls (Watts & Sanchez-Watts 1995a, Watts et al 1999).

The second configuration occurs in the LHA and retrochiasmatic areas of DE animals (Table 2) where CRH and neurotensin gene expression and the numbers of CRH-immunoreactive (ir) neurons increase (Kay-Nishiyama & Watts 1999, Watts et al 1995). Both of these regions are implicated in regulating feeding and autonomic function (Elias et al 1998a, Elmquist et al 1999). CRH and neurotensin are extensively colocalized in these neurons, and the majority of these CRH neurons are also GABAergic (Sanchez-Watts et al 2000). A possible inhibitory role for these CRH neurons during anorexia is supported by the fact that CRH mRNA in the LHA is strongly correlated with the intensity of anorexia; increased LHA CRH gene expression precedes the onset of anorexia; and DE-adrenalectomized animals also have elevated LHA CRH and neurotensin gene expression, showing a distribution pattern similar to that of DE intact animals (Watts 1999, Watts et al 1995, 1999). Neurotensin is also implicated as an anorexigenic peptide when injected introcerebroventricularly (Levine et al 1983), possibly through actions on melanin-concentrating-hormone neurons (Tritos et al 1998).

Telencephalon

CRH and neurotensin mRNAs in the lateral part of the central nucleus of the amygdala (CEAl) and the oval (BSTov) and fusiform (BSTfus) parts of the BST also respond to dehydration or food restriction (Table 2). The importance of these telencephalic regions is that their connections form part of the link between the cortex and hippocampus and the hypothalamus and hindbrain (Risold & Swanson 1997, Swanson & Petrovich 1998). CRH and neurotensin mRNAs are colocalized in the CEAl and BSTov (Shimada et al 1989) and show similar responses to dehydration; levels initially decrease but then return to control values (Watts et al 1995). In the BSTfus, CRH mRNA continues to increase as dehydration persists (Watts et al 1995). However, a different profile occurs in paired food-restricted animals in which CRH and neurotensin mRNA levels in the CEAl but not the BSTov are significantly reduced when compared with levels after 5 days of dehydration or with control euhydrated animals. Furthermore, food restriction significantly reduces CRH mRNA in the BSTfus (C Kay-Nishiyama & AG Watts, unpublished observations). The importance of these findings with regard to integrative processes is illustrated later when the efferent connections of these nuclei are examined (Figure 2).

Mechanisms Mediating the Changes in Neuropeptide Gene Expression Following Dehydration

Arcuate Nucleus   The patterns of gene expression seen in the ARH after dehydration and food restriction (Table 2) are driven, at least in part, by altered hormone profiles, particularly the markedly reduced leptin levels seen as a consequence of the negative energy balance in both sets of animals (Elmquist et al 1999, Nishiyama et al 1999, Sahu 1998).

Paraventricular Nucleus   The marked reduction of CRH mRNA in the neuroendocrine parvicellular part of the PVH of DE animals is determined by at least two factors: the presence of corticosterone and neural afferents from the ventral part of the lamina terminalis (Kovács & Sawchenko 1993, Watts & Sanchez-Watts 1995a). In the absence of either of these factors, reductions in CRH mRNA in DE animals do not occur. However, corticosterone has no impact on the increase in CRH gene expression in magnocellular neuroendocrine neurons, which is driven by afferent input from the ventral lamina terminalis.

Lateral Hypothalamic Area   Unilateral knife cuts placed between the lamina terminalis and the LHA significantly impair the response to dehydration of CRH and neurotensin gene expression on the lesioned side of the LHA (Kelly & Watts 1996) suggesting that fibers passing through this region are critical for activating gene

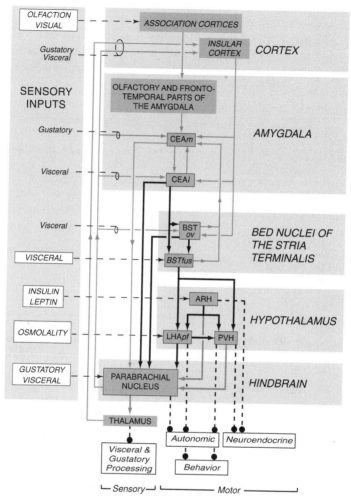

Figure 2 A schematic representation of some of the major connections between the cortex, amygdala, bed nuclei of the stria terminalis, hypothalamus, and parabrachial nucleus. Sensory inputs are shown by long dashed lines. Projections involved with motor functions are shown by short dashed lines. Projections where alterations in neuropeptide mRNAs have been reported following dehydration or food restriction are shown in black; others are shown in grey. Abbreviations: ARH, arcuate nucleus; BSTfus, fusiform nucleus of the bed nuclei of the stria terminalis; BSTov, oval nucleus of the bed nuclei of the stria terminalis; CEAl, lateral part of the central nucleus of the amygdala; CEAm, medial part of the central nucleus of the amygdala; LHApf, lateral hypothalamic area, perifornical zone; PVH, paraventricular nucleus of the hypothalamus.

expression. Because osmosensitive neurons in the SFO and the median preoptic nucleus (both of which are critical for the development of behavioral and neuroendocrine responses to dehydration) project directly to the LHA, these inputs may well play a critical role in regulating gene expression (Kelly & Watts 1996). However, unlike the DE-dependent reduction in CRH gene expression in the PVH, corticosterone may not mediate changes in gene expression in the LHA (Watts & Sanchez-Watts 1995b).

Telencephalon   The mechanisms underlying the changes in CRH and neurotensin gene expression in DE or food-restricted animals are not known. The mRNAs for both peptides are increased in the CEAl and, to a lesser extent, in the BSTov after chronic corticosterone administration, but CRH gene expression in the BSTfus is virtually insensitive to circulating glucocorticoids (Swanson & Simmons 1989, Makino et al 1994, Watts & Sanchez-Watts 1995b). However, because changes in neuropeptide gene expression are more complex after altered ingestive behaviors than can be accounted for by a simple response to elevated corticosterone (Watts et al 1995), other factors must contribute to these effects.

INTEGRATION

Coordinating patterns of behavioral, autonomic, and neuroendocrine output with homeostatic demands requires that the outputs from the drive networks, sensory object representation, and arousal-state control systems be integrated to generate the appropriate responses (Figure 1). Although this is probably a relatively simple process for reflex events in which sensory input is processed directly by premotor and motor networks, more complete behaviors require forebrain, midbrain, and cortical structures (Grillner et al 1997). Undoubtedly, the integrative process is complex, but understanding its nature will ultimately reveal how animals generate appropriate sets of responses to a given challenge.

With regard to dehydration anorexia, how and where might this integration occur? We know that the changes in neuropeptide gene expression in the ARH are common to both food restriction and dehydration and conform to a pattern indicative of negative energy balance. Under these circumstances, the integrated output from those parts of the drive network to which ARH neurons contribute should stimulate eating. On the other hand, neurons in which patterns of gene expression are seen only in dehydration are likely candidates for generating the inhibition that leads to anorexia. Are there regions that receive inputs from both of these control networks that can mediate this integration? Identifying these regions requires considering both their function in controlling ingestive behaviors and detailed information about the organization of their afferent and efferent connections, which are summarized in Figure 2 (see also Watts 2000).

Afferent and Efferent Connections of Those Regions Exhibiting Changes in Neuropeptide Gene Expression during Dehydration

Arcuate Nucleus ARH neurons project both to the PVH and to melanin-concentrating-hormone-containing neurons in the LHA (Broberger et al 1998a, Elias et al 1998b, Li et al 2000). ARH projections to the LHA may provide a link between neurons directly engaged by hormones that signal changes in energy balance and neurons in the LHA that project to those parts of the brain involved with planning and executing motivated behaviors (Elmquist et al 1999, Risold et al 1997, Sawchenko 1998). Based on the presence of agouti-related protein (AGRP) ir fibers (which is synthesized only in the ARH), ARH efferent connections also target other parts of the brain implicated in regulating autonomic and behavioral aspects of feeding: the lateral septal complex, parts of the BST and amygdala, the parabrachial nucleus, and the medulla (Bagnol et al 1999; Broberger et al 1998b).

Lateral Hypothalamic Area The LHA is a large and heterogeneous collection of neurons that have diverse and complex connections throughout the brain. Critically, these neurons project to and receive extensive inputs from parts of the cortex and hippocampus, nucleus accumbens and substantia innominata, nuclei of the septal complex, amygdala, and BST (Kelly & Watts 1996, Risold & Swanson 1997, Risold et al 1997, Saper 1985, Swanson 1987). Furthermore, the LHA receives projections from both the SFO and the median preoptic nucleus, which are both critical for regulating drinking behavior (Swanson & Lind 1986, Kelly & Watts 1996). The LHA also has strong projections to the periaqueductal gray, parabrachial nucleus, dorsal medulla (Moga et al 1990a; Kelly & Watts, 1998; Swanson 1987), and to some parts of the periventricular hypothalamus that control neuroendocrine output (Larsen et al 1994, Swanson 1987, Watts et al 1999).

Considering that the ARH, PVH, and LHA contain neurons that either stimulate or inhibit ingestive behavior (Table 1), it is worth noting that most telencephalic regions provide stronger connections to the LHA than to the ARH or PVH, suggesting that the LHA is better positioned to mediate the more volitional aspects of ingestive behavior. Similarly, the LHA provides the most extensive hypothalamic connections back to the telencephalon. On the other hand, the ARH—because of its hormonal inputs and its central position in neuroendocrine function—may be more important for initiating those feeding-associated motor events mediated by alterations in internal state. Collectively, these connections place the LHA in a pivotal position for incorporating volitional aspects of ingestive behaviors into the motor patterns organized by the hypothalamus.

Central Nucleus of the Amygdala The CEA is part of the striatal amygdala, and consists of three parts: a large medial part (CEAm), which provides extensive projections to the hindbrain and quantitatively constitutes the main output from the CEA; a smaller, more compact CEAl that contains a significant population of neuropeptides and has a restricted set of efferents; and a capsular part implicated in nociception (Bernard et al 1993, Swanson & Petrovich 1998). Most CEA neurons are GABAergic and control a wide variety of autonomic functions (Swanson & Petrovich 1998). Peptidergic neurons in the CEAl project heavily to restricted parts of the BST, as well to the parabrachial nucleus, but they provide only sparse projections to the hypothalamus (Moga et al 1989, 1990b; Petrovich & Swanson 1997, Prewitt & Herman 1998). The CEAm provides a major input to the CEAl and indirectly receives exterosensory inputs from a variety of cortical areas by way of the olfactory and frontotemporal parts of the amygdala. In addition, the insular cortex—an area critical for processing gustatory and vagal sensory information—projects directly to both parts of the CEA (Cechetto & Saper 1990). In this way, the CEA is well-positioned to modulate activity within downstream BST, hypothalamic, and hindbrain regions using information derived from a variety of cortical sources (Figure 2).

Bed Nuclei of the Stria Terminalis The BST comprises ~15 variously sized and often poorly differentiated cell groups located immediately rostral and dorsal to the preoptic area of the hypothalamus. Their functions are collectively influenced by sets of topographically organized projections from the amygdala and hippocampus and range from the autonomic and behavioral responses to various stressors to sexually dimorphic functions (Herman & Cullinan 1997, Simerly 1999).

Based on dynamic patterns of neuropeptide gene expression, two parts of the BST have particularly prominent responses to dehydration: the BSTov and BSTfus (Watts et al 1995). The BSTov is located dorsally to the anterior commissure and contains a number of neuropeptides, including CRH, neurotensin, substance P, and enkephalin (Ju et al 1989). Although it has strong projections to the parabrachial nucleus (Moga et al 1989) and bidirectional connections with the CEA, particularly the CEAl, its hypothalamic projections are sparse (Dong et al 2000, Petrovich & Swanson 1997, Prewitt & Herman 1998). The BSTfus is a small nucleus located ventrally to the anterior commissure (Ju & Swanson 1989). Although often considered together with the BSTov as forming the lateral part of the BST, the BSTfus is a distinct nucleus. Unlike the BSTov, BSTfus neurons contain CRH but not neurotensin (Ju et al 1989, Moga et al 1989) and have few parabrachial projections. Its hypothalamic projections are very different from those of the BSTov and CEAl in that the BSTfus projects strongly to the PVH and LHA (Moga & Saper 1994, Kelly & Watts 1996, Petrovich & Swanson 1997, Dong et al 2000). Finally, corticosterone-dependent regulation of CRH gene expression in the BSTfus differs significantly from that in the BSTov or CEAl (Watts & Sanchez-Watts 1995b).

Regarding function during dehydration, Fos activation implicates both the CEAl and the lateral parts of the BST in responses to A-II (Herbert 1993) and to treatments that alter ingestive behaviors, for example the complex reward-encoding features of opioid systems and anorexic responses to dexfenfluramine or cytokines (Carr et al 1999, Day et al 1999, Li & Rowland 1996, Pomonis et al 1997). Although these studies clearly show that some aspect of neuronal physiology is altered after specific experimental manipulations, we currently do not know whether their role is directly related to behavioral or autonomic aspects of feeding or it is subordinated to other functions more indirectly related to ingestive behaviors.

Integrative Regions

If those parts of the forebrain showing changes in neuropeptide gene expression during dehydration contribute to different components of the drive networks, which regions might be important for the integrative process during dehydration? If we assume that dehydration-sensitive CRH neurons in the LHA are involved in some parts of the anorexic process, one approach is to look where CRH and neurotensin-ir correlate with the efferent projections from this part of the LHA. This reveals two regions as candidates for integration—the PVH and the parabrachial nucleus in the hindbrain.

Hypothalamic Paraventricular Nucleus   The PVH is a critical hypothalamic cell group that regulates many of the motor functions precipitated by dehydration. Its neurons control the AVP and oxytocin neuroendocrine response to dehydration and a wide variety of motor functions associated with energy balance. Neuroendocrine responses are mediated by CRH motor neurons that ultimately control glucocorticoid secretion (Watts 1996), along with groups of thyrotropin-releasing hormones and somatostatin neuroendocrine neurons that are pivotally placed to regulate endocrine control of metabolism (Swanson 1987). In addition, the PVH contains caudally projecting neurons to the periaqueductal gray, parabrachial nucleus, dorsal vagal complex, and preganglionic neurons in the dorsal and ventrolateral medulla and intermediolateral column of the spinal cord. These PVH neurons contain CRH, oxytocin, dynorphin, or AVP, and they control a wide range of autonomic functions associated with fluid homeostasis including gastric motility and cardiovascular function.

To regulate these diverse motor functions, PVH neurons receive an array of afferents. These include ascending, predominantly catecholaminergic inputs that relay vagally mediated information from the viscera critical for coordinating feeding responses with peripheral requirements and leptin- and insulin-related viscerosensory information from the ARH and dorsomedial nucleus (Baskin et al 1999, Elmquist et al 1998, Swanson & Sawchenko 1983).

ARH neuropeptide gene expression patterns in DE animals show that these neurons respond to negative energy balance in a manner indistinguishable from that of food-restricted animals (Table 2). Under these circumstances, how might the neuropeptidergic signals received from the ARH impact PVH motor control? One group of ARH neurons expresses NPY and AGRP and regulates a variety of ingestive motor functions. AGRP injected into the PVH reduces plasma thyroid hormone concentrations in a manner consistent with reduced energy expenditure (Kim et al 2000). Injections of NPY into the PVH stimulate feeding (Stanley

& Leibowitz 1985), whereas food restriction increases NPY release in the PVH (Kalra et al 1991). NPY injections also increase corticosterone secretion, decrease brown fat thermogenesis, and increase lipolysis in white fat, suggesting that this peptide can control a range of important autonomic and neuroendocrine motor functions important for energy balance (Billington et al 1994, Wahlestedt et al 1987). However, as a caveat, it should be noted that, although part of the NPY innervation in the PVH originates from the ARH—some of which terminates on CRH and thyrotropin-releasing hormones neurons (Broberger et al 1999; Li et al 2000)—the remainder is colocalized in noradrenergic afferents from the hindbrain (Swanson 1987). The ARH also provides a peptidergic input to the PVH from POMC/cocaine- and amphetamine-regulated transcript neurons that are thought to reduce ingestive behaviors (Schwartz et al 2000). The PVH contains both POMC-ir fibers and cognate MC4 receptors (Cowley et al 1999). During dehydration, however, ARH *POMC* gene expression is significantly reduced, suggesting that this type of anorexia is not generated by increased endogenous agonist activity at MC receptors.

Collectively these data show that the PVH receives both anorexic and orexigenic signals from hormone-sensitive neurons in the ARH and that some PVH neurons can integrate these neuropeptidergic inputs (Kalra et al 1999). At this point it is also worth noting that virtually all ARH neurons are GABAergic and that alterations of neuropeptide signaling most likely occur on a background of GABAergic neurotransmission (Kalra et al 1999). How these effects are mediated is unknown, although individual PVH neurons can electrophysiologically integrate orexigenic and anorexigenic stimuli encoded by neuropeptide signals originating from the ARH (Cowley et al 1999).

Based on the neuropeptide profile present in the ARH, both anorexic and orexigenic neuropeptide signals in DE animals are set to control neuroendocrine, autonomic, and behavioral motor output in a manner compatible with negative energy balance. However, at least in terms of behavior, part of this signaling is inhibited until water is again made available. Dehydration-sensitive CRH and neurotensin neurons in the LHA may partly mediate this inhibition because of their direct projections to the PVH. Injections of anterograde tracer into that part of the LHA containing these neurons preferentially labels fibers in PVH regions that contain preautonomic neurons (Watts et al 1999), whereas CRH neurons in the LHA are retrogradely labeled from the PVH (Champagne et al 1998).

Parabrachial Nucleus   The parabrachial nucleus is a large complex cell group located in the dorsal part of the pons and mesencephalon. It is divided by the superior cerebellar peduncle into lateral (predominantly concerned with autonomic and behavioral functions) and medial (predominantly concerned with gustatory function) parts. The parabrachial nucleus appears to function as an integrative center where inputs from a variety of hindbrain and forebrain pathways converge to modulate autonomic motor output, gustatory and vesceral sensory information flow to the thalamus and insular cortex, and information flow to the limbic forebrain (Figure 3).

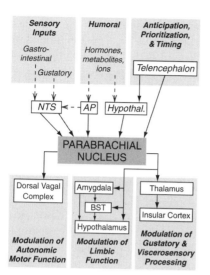

Figure 3 A schematic representation of the afferent and efferent connections of the parabrachial nucleus. Abbreviations: AP, area postrema; NTS, nucleus of the solitary tract.

Thus, the parabrachial receives cholecystokinin (CCK)-associated satiety signals (Fulwiler & Saper 1985, Takaki et al 1990) is critical for taste-aversion learning (Reilly 1999), feeding responses stimulated by mercaptoacetate (Calingasan & Ritter 1993), serotonin-mediated anorexia (Lee et al 1998, Li et al 1994), and some aspects of sodium appetite and fluid balance (Edwards & Johnson 1991). It is also important for nociceptive processing and may function as part of a network that incorporates pain information into autonomic processes (Bernard & Bandler 1998). The parabrachial nucleus has wide-ranging efferent and afferent connections with the medulla, hypothalamus (including the PVH, LHA, and ARH), amygdala, BST, parts of the thalamus, and cortex (particularly the insular cortex) (Alden et al 1994, Bernard et al 1993, Bester et al 1999, Broberger et al 1998b, Herbert et al 1990, Karimnamazi & Travers 1998, Kelly & Watts 1998, Moga et al 1989, Moga et al 1990a,b, Petrovich & Swanson 1997, Touzani et al 1993).

THE OPERATION OF NEUROPEPTIDERGIC CIRCUITS DURING DEHYDRATION

Over the past 30 years, a wealth of data has accumulated from injecting neuropeptides into the brain to show that they can regulate individual behaviors and their attendant neuroendocrine and autonomic motor functions in quite specific ways. Presumably, neuropeptides act within the different regulatory networks to alter neuronal function. But how neuropeptides act at the cellular level to achieve these effects is largely unknown. One possibility is that they modulate the premotor and motor networks that constitute pattern generators. Although this action has been extensively studied in invertebrates, in which neuropeptides switch the functions of motor pattern generators (e.g. Meyrand et al 1991), such models are less well characterized in vertebrates (Barthe & Clarac 1997).

The actions of neuropeptides at higher levels of control (Figure 1) are not clear. However, some indication of how they might act in the parabrachial nucleus is suggested by the work of Saleh and coworkers. They have identified that parabrachial neurons form a mandatory link in the transfer of vagal sensory information to the thalamus and infralimbic cortex (Saleh & Cechetto 1994) and that this function is modulated by neuropeptides including substance P, cholecystokinin, and neurotensin (Saleh & Cechetto 1993). In turn, stimulation of the vagus nerve releases neuropeptides in the parabrachial nucleus where they have differential effects on excitatory synaptic transmission (Saleh 1997, Saleh et al 1997). The fact that the parabrachial nucleus is a key component of gustatory processing raises the possibility that this type of modulation might also affect the processing of information for this sensory modality.

In the context of dehydration, we have seen that this homeostatic challenge modifies the expression of genes encoding many of the neuropeptides implicated in regulating ingestive behaviors (Table 2). In turn, the diverse populations of neurons in which these changes occur project to a variety of hypothalamic and hindbrain sites critical for controlling functions associated with ingestive behaviors (Figure 2). How might neuropeptides act within these circuits to alter neural function in dehydrated animals?

If we consider the PVH and parabrachial nucleus as sites of integration, we see that their neuropeptidergic inputs are differentially modified (as implied from changes in gene expression) by the three different physiological states—euhydration, food restriction, and dehydration (Figure 4). In the PVH, neuropeptidergic inputs from the ARH, BSTfus, and LHA each encode very different types of information; those from the ARH mediate the actions of leptin and insulin, whereas inputs containing CRH and possible neurotensin from the BSTfus and LHA may encode information from the telencephalon that coordinates neuroendocrine, autonomic, and behavioral motor patterns. Figure 4 shows that, for each of the three challenges, there may be quite different patterns of neuropeptide release in the PVH that generate the different patterns of motor functions required by these three states. In the parabrachial nucleus, a feature common to forebrain cell groups that contain CRH and neurotensin (the LHA, BSTov, and CEAl) is that they all project rather heavily to the pontine part of the parabrachial nucleus (Figure 2), although each innervates somewhat different subdivisions (Dong et al 2000, Kelly & Watts 1998, Petrovich & Swanson 1997). In turn, because neuropeptidergic gene expression in each of these forebrain cell groups responds differently to food restriction and dehydration, the neuropeptidergic "read out" within the parabrachial nucleus may, like that in the PVH, change for each of these homeostatic challenges (Figure 4). This signaling complexity is increased still further as we consider the accompanying role of conventional neurotransmitters coexpressed in these pathways (Swanson 2000).

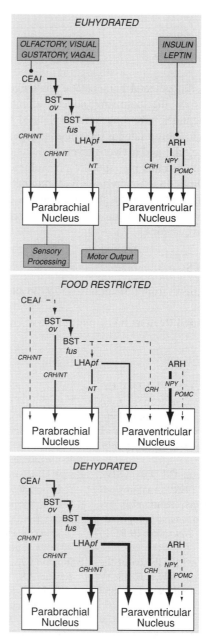

Figure 4 A representation of how the neuropeptidergic inputs from forebrain components to the parabrachial and paraventricular nuclei are differentially altered by food restriction and dehydration. These alterations can then modify the output functions of these two integrative regions. Decreased or increased activity compared with euhydrated animals is indicated by *dashed* or *thick* lines, respectively. Abbreviations: CRH/NT, cortiotropin releasing hormone/neurotensin; NPY, neuropeptide Y; POMC, proopiomelanocortin; see legend to Figure 2 for other abbreviations.

CONCLUSION

Dehydration is a fundamental homeostatic challenge that initiates sets of coordinated motor actions that both limit adverse effects and restore fluid balance. Although neuropeptides have long been implicated as critical regulators for the sets of motor events appropriate for specific challenges, how they realize this function at both the cellular and network levels remains unclear. The work reviewed in this article demonstrates that specific physiological challenges differentially alter neuropeptidergic function in those forebrain networks heavily implicated in controlling specific motor actions. One of the challenges now facing neuroscience is to determine precisely how neuropeptidergic systems can achieve these highly complex integrative actions.

ACKNOWLEDGMENTS

I acknowledge support from NIH grants NS29728 and NS KO-4 01833.

Visit the Annual Reviews home page at www.AnnualReviews.org

LITERATURE CITED

Alden M, Besson JM, Bernard JF. 1994. Organization of the efferent projections from the pontine parabrachial area to the bed nucleus of the stria terminalis and neighboring regions: a PHA-L study in the rat. *J. Comp. Neurol.* 341:289–314

Bagnol D, Lu X-Y, Kaelin CB, Day HEW, Ollmann M, et al. 1999. Anatomy of an endogenous antagonist: relationship between agouti-related protein and proopiomelanocortin in brain. *J. Neurosci.* 19:RC26 (1–7)

Barthe JY, Clarac F. 1997. Modulation of the spinal network for locomotion by substance P in the neonatal rat. *Exp. Brain Res.* 115:485–92

Baskin DG, Figlewicz Lattemann D, Seeley RJ, Woods SC, Porte D Jr, Schwartz MW. 1999. Insulin and leptin: dual adiposity signals to the brain for the regulation of food intake and body weight. *Brain Res.* 848: 114–23

Bernard JF, Alden M, Besson JM. 1993. The organization of the efferent projections from the pontine parabrachial area to the amyg-

daloid complex: a Phaseolus vulgaris leucoagglutinin (PHA-L) study in the rat. *J. Comp. Neurol.* 329:201–29

Bernard JF, Bandler R. 1998. Parallel circuits for emotional coping behaviour: new pieces in the puzzle. *J. Comp. Neurol.* 401:429–36

Bester H, Bourgeais L, Villanueva L, Besson JM, Bernard JF. 1999. Differential projections to the intralaminar and gustatory thalamus from the parabrachial area: a PHA-L study in the rat. *J. Comp. Neurol.* 405: 421–49

Billington CJ, Briggs JE, Harker S, Grace M, Levine AS. 1994. Neuropeptide Y in hypothalamic paraventricular nucleus: a center coordinating energy metabolism. *Am. J. Physiol. Regul. Integr. Comp. Physiol.* 266:R1765–R1770

Broberger C, De Lecea L, Sutcliffe JG, Hokfelt T. 1998a. Hypocretin/orexin- and melanin-concentrating hormone-expressing cells form distinct populations in the rodent lateral hypothalamus: relationship to the neuropeptide Y and agouti gene-related protein systems. *J. Comp. Neurol.* 402:460–74

Broberger C, Johansen J, Johansson C, Schalling M, Hökfelt T. 1998b. The neuropeptide Y/agouti gene-related protein (AGRP) brain circuitry in normal, anorectic, and monosodium glutamate-treated mice. *Proc. Natl. Acad. Sci. USA* 95:15043–48

Broberger C, Visser TJ, Kuhar MJ, Hokfelt T. 1999. Neuropeptide Y innervation and neuropeptide-Y-Y1-receptor-expressing neurons in the paraventricular hypothalamic nucleus of the mouse. *Neuroendocrinology* 70:295–305

Calingasan NY, Ritter S. 1993. Lateral parabrachial subnucleus lesions abolish feeding induced by mercaptoacetate but not by 2-deoxy-D-glucose. *Am. J. Physiol.* 265:R1168–78

Carr KD, Kutchukhidze N, Park TH. 1999. Differential effects of mu and kappa opioid antagonists on Fos-like immunoreactivity in extended amygdala. *Brain Res.* 822:34–42

Cechetto D, Saper CB. 1990. Role of the cerebral cortex in autonomic function. In *Central Regulation of Autonomic Function.* ed. AD Loewy, KM Spyer, pp. 208–23. New York: Oxford Univ. Press

Champagne D, Beaulieu J, Drolet G. 1998. CRFergic innervation of the paraventricular nucleus of the rat hypothalamus—a tract-tracing study. *J. Neuroendocr.* 10:119–31

Cowley MA, Pronchuk N, Fan W, Dinulescu DM, Colmers WF, et al. 1999. Integration of NPY, AGRP, and melanocortin signals in the hypothalamic paraventricular nucleus: evidence of a cellular basis for the adipostat. *Neuron* 24:155–63

Day HE, Curran EJ, Watson SJ Jr, Akil H. 1999. Distinct neurochemical populations in the rat central nucleus of the amygdala and bed nucleus of the stria terminalis: evidence for their selective activation by interleukin-1beta. *J. Comp. Neurol.* 413:113–28

Denton D, Shade R, Zamarippa F, Egan G, Blair-West J, et al. 1999a. Correlation of regional cerebral blood flow and change of plasma sodium concentration during genesis and satiation of thirst. *Proc. Natl. Acad. Sci. USA* 96:2532–37

Denton D, Shade R, Zamarippa F, Egan G, Blair-West J, et al. 1999b. Neuroimaging of genesis and satiation of thirst and an interoceptor-driven theory of origins of primary consciousness. *Proc. Natl. Acad. Sci. USA* 96:5304–9

Dicker SE, Nunn J. 1957. The role of antidiuretic hormone during water deprivation in rats. *J. Physiol.* 136:235–48

Dong H-W, Petrovich GD, Watts AG, Swanson LW. 2000. Organization of projections from the oval and fusiform nuclei of the BST. *Soc. Neurosci. Abstr.* 26:2219

Edwards GL, Johnson AK. 1991. Enhanced drinking after excitotoxic lesions of the parabrachial nucleus in the rat. *Am. J. Physiol. Regul. Integr. Comp. Physiol.* 261:R1039–R1044

Elias CF, Lee C, Kelly J, Aschkenasi C, Ahima RS, et al. 1998a. Leptin activates hypothalamic CART neurons projecting to the spinal cord. *Neuron* 21:1375–85

Elias CF, Saper CB, Maratos-Flier E, Tritos NA, Lee C, et al. 1998b. Chemically defined projections linking the mediobasal hypothalamus and the lateral hypothalamic area. *J. Comp. Neurol.* 402:442–59

Elmquist JK, Ahima RS, Elias CF, Flier JS, Saper CB. 1998. Leptin activates distinct projections from the dorsomedial and ventromedial hypothalamic nuclei. *Proc. Natl. Acad. Sci. USA* 95:741–46

Elmquist JK, Elias CF, Saper CB. 1999. From lesions to leptin: hypothalamic control of food intake and body weight. *Neuron* 22:221–32

Fitzsimons JT. 1998. Angiotensin, thirst, and sodium appetite. *Physiol. Rev.* 78:583–686

Flanagan LM, Verbalis JG, Stricker EM. 1989. Effects of anorexigenic treatments on gastric motility in rats. *Am. J. Physiol. Regul. Integr. Comp. Physiol.* 256:R955–R961

Flynn FW, Curtis KS, Verbalis JG, Stricker EM. 1995. Dehydration anorexia in decerebrate rats. *Behav. Neurosci.* 109:1009–1012

Fulwiler CE, Saper CB. 1985. Cholecystokinin-immunoreactive innervation of the ventromedial hypothalamus in the rat: possible substrate for autonomic regulation of feeding. *Neurosci. Lett.* 53:289–96

Gallagher M, McMahan RW, Schoenbaum G. 1999. Orbitofrontal cortex and representation of incentive value in associative learning. *J. Neurosci.* 19:6610–6614

Gann DS, Ward DG, Carlson DE. 1978. Neural control of ACTH: a homeostatic reflex. *Recent Prog. Horm. Res.* 34:357–400

Grigson PS, Kaplan JM, Roitman MF, Norgren R, Grill HJ. 1997. Reward comparison in chronic decerebrate rats. *Am. J. Physiol. Regul. Integr. Comp. Physiol.* 273: R479–R486

Grill HJ, Kaplan JM. 1990. Caudal brainstem participates in the distributed neural control of feeding. In *Neurobiology of Food and Fluid Intake. Handbook of Behavioral Neurobiology*, ed. EM Stricker, 10:125–50. New York: Plenum. 553 pp.

Grillner S, Georgopoulos AP, Jordan LM. 1997. Selection and intiation of motor behavior. In *Neurons, Networks, and Motor Behavior*, ed. PSG Stein, S Grillner, A Selveston, DG Stuart, pp. 3–19. Cambridge, MA: MIT Press. 305 pp.

Grossman SP. 1990. *Thirst and sodium appetite.* San Diego, CA: Academic. 289 pp.

Haynes WG, Sivitz WI, Morgan DA, Walsh SA, Mark AL. 1997. Sympathetic and cardiorenal actions of leptin. *Hypertension* 30: 619–23

Herbert H, Moga MM, Saper CB. 1990. Connections of the parabrachial nucleus with the nucleus of the solitary tract and the medullary reticular formation in the rat. *J. Comp. Neurol.* 293:540–80

Herbert J. 1993. Peptides in the limbic system: neurochemical codes for co-ordinated adaptive responses to behavioural and physiological demand. *Prog. Neurobiol.* 41: 723–91

Herman JP, Cullinan W. 1997. Neurocircuitry of stress—central control of the hypothalamo-pituitary-adrenocortical axis. *Trends Neurosci.* 20:78–84

Huang W, Lee SL, Arnason SS, Sjoquist M. 1996. Dehydration natriuresis in male rats is mediated by oxytocin. *Am. J. Physiol. Regul. Integr. Comp. Physiol.* 270:R427–R433

Ju G, Swanson LW. 1989. Studies on the cellular architecture of the bed nuclei of the stria terminalis in the rat: I. Cytoarchitecture. *J. Comp. Neurol.* 280:587–602

Ju G, Swanson LW, Simerly RB. 1989. Studies on the cellular architecture of the bed nuclei of the stria terminalis in the rat. II. Chemoarchitecture. *J. Comp. Neurol.* 280: 603–21

Kalra SP, Dube MG, Pu S, Xu B, Horvath TL, Kalra PS. 1999. Interacting appetite-regulating pathways in the hypothalamic regulation of body weight. *Endocr. Rev.* 20:68–100

Kalra SP, Dube MG, Sahu A, Phelps CP, Kalra PS. 1991. Neuropeptide Y secretion increases in the paraventricular nucleus in association with increased appetite for food. *Proc. Natl. Acad. Sci. USA* 88: 10931–35

Kaplan JM, Seeley RJ, Grill HJ. 1993. Daily caloric intake in intact and chronic decerebrate rats. *Behav. Neurosci.* 107:876–81

Karimnamazi H, Travers JB. 1998. Differential projections from gustatory responsive regions of the parabrachial nucleus to the medulla and forebrain. *Brain Res.* 813:283–302

Kask A, Schioth HB, Mutulis F, Wikberg JE, Rago L. 2000. Anorexigenic cocaine- and amphetamine-regulated transcript peptide intensifies fear reactions in rats. *Brain Res.* 857:283–85

Kay-Nishiyama C, Watts AG. 1998. CRH in dehydration-induced anorexia: CRH immunoreactivity in non-colchicine treated rats, and CRH R2 receptor mRNA levels in the ventromedial hypothalamic nucleus. *Soc. Neurosci. Abstr.* 24:449

Kay-Nishiyama C, Watts AG. 1999. Dehydration modifies somal CRH immunoreactivity in the rat hypothalamus: an immunocyto-chemical study in the absence of colchicine. *Brain Res.* 822:251–55

Kelley AE. 1999. Functional specificity of ventral striatal compartments in appetitive behaviors. *Ann. NY Acad. Sci.* 877:71–90

Kelly AB, Watts AG. 1996. The mediation of dehydration-induced peptidergic gene expression in the rat lateral hypothalamic area by forebrain afferent projections. *J. Comp. Neurol.* 370:231–46

Kelly AB, Watts AG. 1998. The region of the pontine parabrachial nucleus is a major target of dehydration-sensitive CRH neurons in the rat lateral hypothalamic area. *J. Comp. Neurol.* 394:48–63

Kim MS, Small CJ, Stanley SA, Morgan DG, Seal LJ, et al. 2000. The central melanocortin system affects the hypothalamo-pituitary thyroid axis and may mediate the effect of leptin. *J. Clin. Invest.* 105:1005–11

Kovács KJ, Sawchenko PE. 1993. Mediation of osmoregulatory influences on neuroendocrine corticotropin-releasing factor expression by the ventral lamina terminalis. *Proc. Natl. Acad. Sci. USA* 90:7681–85

Kuhar MJ, Dall Vechia SE. 1999. CART peptides: novel addiction- and feeding-related neuropeptides. *Trends Neurosci.* 22:316–20

Larsen PJ, Hay-Schmidt A, Mikkelsen JD. 1994. Efferent connections from the lateral hypothalamic region and the lateral preoptic area to the hypothalamic paraventricular nucleus of the rat. *J. Comp. Neurol.* 342:299–319

Lee MD, Aloyo VJ, Fluharty SJ, Simansky KJ. 1998. Infusion of the serotonin1B (5-HT1B) agonist CP-93,129 into the parabrachial nucleus potently and selectively reduces food intake in rats. *Psychopharmacology* 136:304–7

Levine AS, Kneip J, Grace M, Morley JE. 1983. Effect of centrally administered neurotensin on multiple feeding paradigms. *Pharmacol. Biochem. Behav.* 18:19–23

Li BH, Rowland NE. 1996. Effect of chronic dexfenfluramine on Fos in rat brain. *Brain Res.* 728:188–92

Li BH, Spector AC, Rowland NE. 1994. Reversal of dexfenfluramine-induced anorexia and c-Fos/c-Jun expression by lesion in the lateral parabrachial nucleus. *Brain Res.* 640:255–67

Li C, Chen PL, Smith MS. 2000. Corticotropin releasing hormone neurons in the paraventricular nucleus are direct targets for neuropeptide Y neurons in the arcuate nucleus: an anterograde tracing study. *Brain Res.* 854:122–29

Makino S, Gold PW, Schulkin J. 1994. Effects of corticosterone on CRH mRNA and content in the bed nucleus of the stria terminalis: comparison with the effects in the central nucleus of the amygdala and the paraventricular nucleus of the hypothalamus. *Brain Res.* 657:141–49

Meyrand P, Simmers J, Moulins M. 1991. Construction of a pattern-generating circuit with neurons of different networks. *Nature* 351:60–63

Moga MM, Herbert H, Hurley KM, Yasui Y, Gray TS, et al. 1990a. Organization of cortical, basal forebrain, and hypothalamic afferents to the parabrachial nucleus in the rat. *J. Comp. Neurol.* 295:624–61

Moga MM, Saper CB. 1994. Neuropeptide-immunoreactive neurons projecting to the paraventricular hypothalamic nucleus in the rat. *J. Comp. Neurol.* 346:137–50

Moga MM, Saper CB, Gray TS. 1989. Bed nucleus of the stria terminalis: cytoarchitecture, immunohistochemistry, and projection to the parabrachial nucleus in the rat. *J. Comp. Neurol.* 283:315–32

Moga MM, Saper CB, Gray TS. 1990b. Neuropeptide organization of the hypothalamic projection to the parabrachial nucleus in the rat. *J. Comp. Neurol.* 295:662–82

Moore RY. 1997. Circadian rhythms: basic neurobiology and clinical applications. *Annu. Rev. Med.* 48:253–66

Nishiyama M, Makino S, Asaba K, Hashimoto K. 1999. Leptin effects on the expression of type-2 CRH receptor mRNA in the ventromedial hypothalamus in the rat. *J. Neuroendocr.* 11:307–14

Ohata H, Suzuki K, Oki Y, Shibasaki T. 2000. Urocortin in the ventromedial hypothalamic nucleus acts as an inhibitor of feeding behavior in rats. *Brain Res.* 861:1–7

Olson BR, Drutarosky MD, Stricker EM, Verbalis JG. 1991. Brain oxytocin receptor antagonism blunts the effects of anorexigenic treatments in rats: evidence for central oxytocin inhibition of food intake. *Endocrinology* 129:785–91

Palmiter RD, Erickson JC, Hollopeter G, Baraban SC, Schwartz MW. 1998. Life without neuropeptide Y. *Recent Prog. Horm. Res.* 53:163–99

Petrovich GD, Swanson LW. 1997. Projections from the lateral part of the central amygdalar nucleus to the postulated fear conditioning circuit. *Brain Res.* 763:247–54

Pomonis JD, Levine AS, Billington CJ. 1997. Interaction of the hypothalamic paraventricular nucleus and central nucleus of the amygdala in naloxone blockade of neuropeptide Y-induced feeding revealed by c-fos expression. *J. Neurosci.* 17:5175–82

Porte D Jr, Seeley RJ, Woods SC, Baskin DG, Figlewicz DP, et al. 1998. Obesity, diabetes and the central nervous system. *Diabetologia* 41:863–81

Prewitt CM, Herman JP. 1998. Anatomical interactions between the central amygdaloid nucleus and the hypothalamic paraventricular nucleus of the rat: a dual tract-tracing analysis. *J. Chem. Neuroanat.* 15:173–85

Reilly S. 1999. The parabrachial nucleus and conditioned taste aversion. *Brain Res. Bull.* 48:239–54

Risold PY, Swanson LW. 1996. Structural evidence for functional domains in the rat hippocampus. *Science* 272:1484–86

Risold PY, Swanson LW. 1997. Connections of the rat lateral septal complex. *Brain Res. Rev.* 24:115–95

Risold PY, Thompson RH, Swanson LW. 1997. The structural organization of connections between hypothalamus and cerebral-cortex. *Brain Res. Rev.* 24:197–254

Sahu A. 1998. Evidence suggesting that galanin (GAL), melanin-concentrating hormone (MCH), neurotensin (NT), proopiomelanocortin (POMC) and neuropeptide Y (NPY) are targets of leptin signaling in the hypothalamus. *Endocrinology* 139:795–98

Saleh TM. 1997. Visceral afferent stimulation-evoked changes in the release of peptides into the parabrachial nucleus in vivo. *Brain Res.* 778:56–63

Saleh TM, Cechetto DF. 1993. Peptides in the parabrachial nucleus modulate visceral input to the thalamus. *Am. J. Physiol. Regul. Integr. Comp. Physiol.* 264:R668–R675

Saleh TM, Cechetto DF. 1994. Neurotransmitters in the parabrachial nucleus mediating visceral input to the thalamus in rats. *Am. J. Physiol. Regul. Integr. Comp. Physiol.* 266:R1287–R1296

Saleh TM, Kombian SB, Zidichouski JA, Pittman QJ. 1997. Cholecystokinin and neurotensin inversely modulate excitatory synaptic transmission in the parabrachial nucleus in vitro. *Neuroscience* 77:23–35

Sanchez-Watts G, Kay-Nishiyama CA, Watts AG. 2000. Neuropeptide colocalization in the lateral hypothalamus of dehydrated-anorexic rats. *Soc. Neurosci. Abstr.* 26:2040

Saper CB. 1985. Organization of cerebral cortical afferent systems in the rat. II. Hypothalamocortical projections. *J. Comp. Neurol.* 237:21–46

Sawchenko PE. 1998. Toward a new neurobiology of energy balance, appetite, and obesity: the anatomists weigh in. *J. Comp. Neurol.* 402:435–41

Schultz W. 1998. Predictive reward signal of dopamine neurons. *J. Neurophysiol.* 80:1–27

Schwartz MW, Woods SC, Porte D Jr, Seeley RJ, Baskin DG. 2000. Central nervous system control of food intake. *Nature* 404:661–71

Sherin JE, Elmquist JK, Torrealba F, Saper CB. 1998. Innervation of histaminergic tubero-mammillary neurons by GABAergic and galaninergic neurons in the ventrolateral pre-optic nucleus of the rat. *J. Neurosci.* 18:4705–21

Shimada S, Inagaki S, Kubota Y, Ogawa N, Shibasaki T, et al. 1989. Coexistence of peptides (CRF/neurotensin and sub-stance P/somatostatin) in the bed nucleus of the stria terminalis and central amyg-daloid nucleus of the rat. *Neuroscience* 30: 377–83

Simerly RB. 1999. Development of sexually dimorphic forebrain pathways. In *Sexual-break Differentiation of the Brain*, ed. A Mat-sumoto, pp. 175–202. Boca Raton, FL: CRC Press. 344 pp.

Stanley BG, Leibowitz SF. 1985. Neuropep-tide Y injected in the paraventricular hy-pothalamus: a powerful stimulant of feed-ing behavior. *Proc. Natl. Acad. Sci. USA* 82: 3940–43

Stricker EM. 1990. Homeostatic origins of in-gestive behavior. In *Neurobiology of Food and Fluid Intake. Handbook of Behavioral Neurobiology*, ed. EM Stricker, 10:45–60 New York:Plenum. 553 pp.

Swanson LW. 1987. The hypothalamus. In *Handbook of Chemical Neuroanatomy*, ed. A Bjorklund, T Hökfelt, LW Swanson, 5: 1–124. Amsterdam: Elsevier. 459 pp.

Swanson LW. 2000. Cerebral hemispheric reg-ulation of motivated behavior. *Brain Res.* In press

Swanson LW, Lind RW. 1986. Neural projec-tions subserving the initiation of a specific motivated behavior in the rat: new projec-tions from the subfornical organ. *Brain Res.* 379:399–403

Swanson LW, Petrovich GD. 1998. What is the amygdala? *Trends Neurosci.* 21:323–331

Swanson LW, Sawchenko PE. 1983. Hypothala-mic integration: organization of the paraven-tricular and supraoptic nuclei. *Annu. Rev. Neurosci.* 6:269–324

Swanson LW, Simmons DM. 1989. Differen-tial steroid hormone and neural influences on peptide mRNA levels in CRH cells of the paraventricular nucleus: a hybridization his-tochemical study in the rat. *J. Comp. Neurol.* 285:413–35

Takaki A, Nagai K, Takaki S, Yanaihara N, Nakagawa H. 1990. Satiety function of neu-rons containing a CCK-like substance in the dorsal parabrachial nucleus. *Physiol. Behav.* 48:865–71

Tanimura SM, Sanchez-Watts G, Watts AG. 1998. Peptide gene activation, secretion, and steriod feedback during stimulation of rat neuroendocrine CRH neurons. *Endocrinol-ogy* 139:3822–29

Toates F. 1986. *Motivational Systems.* Cam-bridge, UK: Cambridge Univ. Press

Touzani K. Tramu G, Nahon J-L, Velley L. 1993. Hypothalamic melanin-concentrating hormone and alpha-neoendorphin-immuno-reactive neurons project to the medial part of the rat parabrachial area. *Neuroscience* 53:865–67

Tritos NA, Vicent D, Gillette J, Ludwig DS, Flier ES, Maratos-Flier E. 1998. Functional interactions between melanin-concentrating hormone, neuropeptide Y, and anorectic neu-ropeptides in the rat hypothalamus. *Diabetes* 47:1687–92

Wahlestedt C, Skagerberg G, Ekman R, Heilig M, Sundler F, Hakanson R. 1987. Neuropep-tide Y (NPY) in the area of the hypotha-lamic paraventricular nucleus activates the pituitary-adrenocortical axis in the rat. *Brain Res.* 417:33–38

Watts AG. 1991. The efferent projections of the suprachiasmatic nucleus: anatom-ical insights into the control of circa-dian rhythms. In *The Suprachiasmatic Nu-cleus: The Mind's Clock*, ed. D Klein, RY Moore, SM Reppert, pp. 75–104. New York: Oxford Univ. Press

Watts AG. 1996. The impact of physio-logical stimulation on the expression of corticotropin-releasing hormone and other neuropeptide genes. *Front. Neuroendocr.* 17: 281–326

Watts AG. 1999. Dehydration-associated anorexia: development and rapid reversal. *Physiol. Behav.* 65:871–78

Watts AG. 2000. Understanding the neural control of ingestive behaviors: helping to seperate cause from effect with dehydration-associated anorexia. *Horm. Behav.* 37:261–83

Watts AG. 2001. Motivation, neural substrates. In *The Handbook of Brain Theory and Neural Networks*, ed. M Arbib. Cambridge: MIT Press. 2nd ed. In press

Watts AG, Kelly AB, Sanchez-Watts G. 1995. Neuropeptides and thirst: the temporal response of CRH and neurotensin/neuromedin N gene expression in rat limbic forebrain neurons to drinking hypertonic saline. *Behav. Neurosci.* 109:1146–57

Watts AG, Sanchez-Watts G. 1995a. A cell-specific role for the adrenal gland in regulating CRH mRNA levels in rat hypothalamic neurosecretory neurons after cellular dehydration. *Brain Res.* 687:63–70

Watts AG, Sanchez-Watts G. 1995b. Region-specific regulation of neuropeptide mRNAs in rat limbic forebrain neurons by aldosterone and corticosterone. *J. Physiol.* 484:721–36

Watts AG, Sanchez-Watts G. 2000. Lateral hypothalamic fos expression in hypocretin/orexin, but not CRH or MCH neurons is activated after reversal of dehydration-anorexia. *Soc. Neurosci. Abstr.* 26:2041

Watts AG, Sanchez-Watts G, Kelly AB. 1999. Distinct and similar patterns of neuropeptide gene expression are present in rat hypothalamus following dehydration-induced anorexia or paired food restriction. *J. Neurosci.* 19:6111–21

Willie JT, Chemelli RM, Sinton CM, Yanagisawa, M. 2001. To eat or to sleep? Orexin in the regulation of feeding and wakefulness. *Annu. Rev. Neurosci.* 24:429–58

Winn P. 1995. The lateral hypothalamus and motivated behavior: and old syndrome reassessed and a new perspective gained. *Curr. Dir. Psychol. Sci.* 4:182–87

Annu. Rev. Neurosci. 2001. 24:385–428

THE DEVELOPMENTAL BIOLOGY
OF BRAIN TUMORS

Robert Wechsler-Reya and Matthew P. Scott

Departments of Developmental Biology and Genetics and Howard Hughes Medical Institute, Stanford University School of Medicine, Stanford, California, 94305-5329; e-mail: rwreya@cmgm.stanford.edu; scott@cmgm.stanford.edu

Key Words medulloblastoma, glioblastoma, retinoblastoma, development, cancer

■ **Abstract** Tumors of the central nervous system (CNS) can be devastating because they often affect children, are difficult to treat, and frequently cause mental impairment or death. New insights into the causes and potential treatment of CNS tumors have come from discovering connections with genes that control cell growth, differentiation, and death during normal development. Links between tumorigenesis and normal development are illustrated by three common CNS tumors: retinoblastoma, glioblastoma, and medulloblastoma. For example, the retinoblastoma (Rb) tumor suppressor protein is crucial for control of normal neuronal differentiation and apoptosis. Excessive activity of the epidermal growth factor receptor and loss of the phosphatase PTEN are associated with glioblastoma, and both genes are required for normal growth and development. The membrane protein Patched1 (Ptc1), which controls cell fate in many tissues, regulates cell growth in the cerebellum, and reduced Ptc1 function contributes to medulloblastoma. Just as elucidating the mechanisms that control normal development can lead to the identification of new cancer-related genes and signaling pathways, studies of tumor biology can increase our understanding of normal development. Learning that Ptc1 is a medulloblastoma tumor suppressor led directly to the identification of the Ptc1 ligand, Sonic hedgehog, as a powerful mitogen for cerebellar granule cell precursors. Much remains to be learned about the genetic events that lead to brain tumors and how each event regulates cell cycle progression, apoptosis, and differentiation. The prospects for beneficial work at the boundary between oncology and developmental biology are great.

INTRODUCTION

More than a century ago, in trying to understand the origin of cancer, the German pathologist Julius Cohnheim noted the similarities between tumors and embryonic cells (Rather 1978). Both types of cells are morphologically simple, both can differentiate into cells of various shapes and sizes, and, most importantly, both have the capacity for extensive growth. Based on these observations, Cohnheim proposed that cancer might originate from embryonic cells. During development,

0147-006X/01/0621-0385$14.00

he suggested, more cells might be produced than are necessary for the construction of a particular tissue. These excess cells (which he called "embryonic rests") would persist in that tissue until later in life and, because they were embryonic in origin, would retain the capacity for growth. Tumorigenesis, then, would result from abnormal activation of a growth program in these cells.

Although some of the details of Cohnheim's theory differ from our current understanding of tumorigenesis, his recognition of the relationship between development and cancer was remarkably prescient. In the last few years, striking parallels between cancer and normal development have begun to emerge. Molecules originally discovered based on their role in cancer—oncogenes and tumor suppressors—have now been shown to function as fundamental regulators of cell growth and differentiation during development. Similarly, genes identified as regulators of pattern formation in invertebrates and vertebrates have been implicated in a variety of human cancers. In light of these findings, the fields of developmental biology and tumor biology, which were distinct for many years, have begun to converge and inform each other in fascinating ways.

In the context of the nervous system, the interface between development and cancer is the study of brain tumors. Although brain tumors are rare compared with many other types of cancer, they are not uncommon; about 20,000 new primary brain tumors are diagnosed in the United States each year. That they often affect children and young adults and frequently lead to mental impairment or death makes them particularly devastating. This area of investigation offers a unique opportunity for synergistic interactions between basic scientists and clinicians. For developmental neurobiologists, brain tumors represent a kind of natural genetic screen, which can provide valuable information about genes that regulate proliferation, differentiation, and death in the nervous system. For neurooncologists, exploring the mechanisms that control cell fate in the developing nervous system can yield important insights into the mechanisms of tumorigenesis and potentially yield new targets for therapy.

We focus here on three of the most common and widely studied central nervous system (CNS) tumors: retinoblastoma, glioblastoma, and medulloblastoma. Each of these tumors arises from a distinct cell type, and each has important implications for our understanding of how neurons and glia grow, differentiate, and die. For each tumor, we (*a*) briefly review what is known about normal development of the cell type from which the tumor is thought to originate, (*b*) highlight the molecules implicated in transformation of this cell type, and (*c*) discuss the insights gained and the questions raised by comparing the mechanisms of development and tumorigenesis.

RETINAL DEVELOPMENT AND RETINOBLASTOMA

Overview of Retinal Development

The retina, a highly ordered array of neurons and glia, is optimized for sensing, transducing, and transmitting visual information. It is derived from the rostral neural tube, from a region that evaginates early in embryonic development to form a pouch called the optic vesicle (Robinson 1991). As the optic vesicle grows and comes into contact with the overlying ectoderm, which will give rise to the lens, it forms a concave structure called the optic cup. Neuroepithelial cells in the optic cup initially undergo symmetric divisions to generate a large pool of retinoblasts. Retinoblasts then begin to divide asymmetrically, producing a variety of neurons and glia that migrate outward to form distinct layers. The mature retina consists of an outer nuclear layer containing photoreceptors (rods and cones), an inner nuclear layer made up of interneurons (horizontal, bipolar, and amacrine cells) and Müller glial cells, and a ganglion layer, containing the retinal ganglion cells whose axons form the optic nerve.

Retroviral lineage-tracing experiments have demonstrated that retinoblasts are multipotent; they each have the capacity to give rise to all of the different neurons and glia in the retina (Holt et al 1988, Turner et al 1990, Wetts et al 1989). The cells that retinoblasts actually generate change during the course of development (Reh 1992, Reh & Kljavin 1989). Initially they produce primarily ganglion cells, cones, and horizontal cells; later they give rise to amacrine cells and rods. During the last phase of retinal development, retinoblasts generate rods, bipolar cells, and Müller glia. This sequential pattern of cell generation is believed to result from changes in cell-cell interactions and soluble factors in the retinoblast microenvironment (Cepko 1999, Reh 1992, Watanabe & Raff 1990).

A variety of soluble and cell-bound factors influence retinoblast growth and differentiation (reviewed in Reh & Levine 1998, Cepko 1999, Levine et al 2000). For example, epidermal growth factor (EGF), basic fibroblast growth factor (bFGF), transforming growth factors alpha (TGF-α) and beta-3 (TGF-β3), and Sonic hedgehog (Shh) all induce retinal precursors to enter mitosis. Many of these factors have optimal activity at particular stages of retinal development and regulate not only cell growth but also the types of neurons that retinoblasts generate. Thus, bFGF is a potent mitogen early in embryogenesis and favors production of ganglion cells, TGF-α promotes proliferation of later retinoblasts and causes them to produce more amacrine cells, and TGF-β3 is most active on postnatal retinoblasts and promotes proliferation of Müller glia. In addition to soluble factors, integral membrane proteins of the Notch and Delta families have been shown to promote growth of retinoblasts and inhibit their differentiation into various cell types (Austin et al 1995, Perron & Harris 2000). The timing and direction of differentiation of each retinoblast is determined by the sum of the signals to which it is exposed at a particular stage of development.

Transcription factors that integrate these signals and control retinal growth and development have also been identified [reviewed by Cepko (1999), Mathers & Jamrich (2000), and Perron & Harris (2000)]. The homeobox gene *pax6* is mutated in the mouse mutant "*small-eye*" and in humans with aniridia (Glaser et al 1992, Quinn et al 1996). *Chx10* is disrupted in the "ocular retardation" mouse (Burmeister et al 1996). *Rx* plays important roles in early eye development, controlling specification of the retinal primordium and growth of retinoblasts in the optic cup (Mathers et al 1997). Loss of any of these homeobox genes leads to severe defects in formation of the retina and surrounding eye structures.

Retinoblastoma and Identification of the *Rb* Tumor Suppressor Gene

Retinoblastoma is one of the best-studied tumors of the CNS and perhaps the best example of a tumor that has taught us about the molecular mechanisms of development. Retinoblastoma occurs most commonly in children. Among 200 to 300 new cases of retinoblastoma each year, 90% occur before the age of 5 (Brodeur 1995). Untreated retinoblastoma is almost always fatal, but with early detection and treatment most cases can be cured. Unfortunately, treatment often results in loss of vision, and survivors have a relatively high incidence of other tumors.

The cell type from which retinoblastoma originates has been debated. Analysis of primary tumors indicates that the majority of cells have a morphology and antigenic profile reminiscent of photoreceptors (Nork et al 1995, Tajima et al 1994), suggesting transformation of a cell with relatively restricted potential. On the other hand, retinoblastoma cells can differentiate in vitro into cells resembling conventional neurons, photoreceptors, glia, and pigment epithelial cells (Kyritsis et al 1984, 1986; Tsokos et al 1986), consistent with transformation of a more primitive neuroectodermal precursor.

The discovery of the genetic basis of retinoblastoma began with the observation that 30%–40% of cases of retinoblastoma are hereditary. These cases tend to be more severe; often both eyes are affected (bilateral disease) and multiple tumor foci occur in each eye. In the early 1970s, Knudson (Knudson 1971, Knudson et al 1975) compared the age of onset of bilateral (hereditary) vs unilateral (sporadic) retinoblastoma and noted that bilateral cases arise significantly earlier than unilateral ones. Based on the age of onset and estimates of mutation rates, he proposed that bilateral cases result from one genetic event, whereas unilateral cases result from two. His two-mutation (or two-hit) hypothesis suggested that retinoblastomas result from loss or inactivation of both alleles of a gene; in hereditary cases, one mutant allele is inherited and the other is disrupted by somatic mutation, whereas in sporadic cases both alleles are inactivated by somatic events.

The implication of the Knudson model was that complete inactivation of a single gene might result in a tumor. The gene, later called a tumor suppressor, would normally act as a brake on cell growth, and loss of the gene would therefore lead to uncontrolled growth. The concept of a tumor suppressor was based on theoretical data, at a time when limited molecular techniques were available. Cytogenetic studies indicated that many retinoblastoma patients had abnormalities of chromosome 13, and improved molecular techniques made it possible to compare chromosomal changes among patients and narrow down the region that was most commonly affected (Dryja et al 1986). In 1986, the first tumor suppressor gene was isolated and named *Rb-1* (Friend et al 1986). *Rb-1* mutations were found in all patients with hereditary retinoblastoma and in the majority of retinoblastoma patients with nonhereditary disease (Friend et al 1986, Fung et al 1987, Lee et al 1987).

Although *Rb* was first identified for its function in the retina, it is expressed in most tissues. *Rb* encodes a phosphoprotein that plays a critical role in cell cycle regulation (reviewed in Lipinski & Jacks 1999). Rb regulates the earliest checkpoint in the cell cycle, progression into late G_1 phase (Figure 1). Members of the E2F family of transcription factors, whose targets include many genes required for cell cycle progression, normally mediate passage through this checkpoint (DeGregori et al 1997, Nevins et al 1997). In nondividing cells, E2Fs form complexes with Rb, which renders the E2Fs inactive. In response to mitogens, cells increase expression of D- and E-type cyclins, which activate cyclin-dependent kinases (Cdk4 and Cdk2, respectively). Cdks phosphorylate Rb, causing it to dissociate from E2Fs and freeing E2Fs to activate transcription and cell cycle progression. Rb can also associate with a number of viral proteins, including simian virus 40 (SV40), large-T antigen (TAg), adenovirus E1a protein, and papillomavirus E7 protein (DeCaprio et al 1988, Dyson et al 1989, Nevins 1994, Whyte et al 1988). Binding of any of these oncogenic viral proteins dissociates Rb from E2Fs, thus promoting cell cycle progression and tumor formation. TAg and E1A can also associate with the p53 tumor suppressor and with p300/CBP (CREB binding protein) transcriptional coactivators, and this contributes to their transforming activity as well (Giordano & Avantaggiati 1999, Ludlow & Skuse 1995, Snowden & Perkins 1998).

Based on its regulation of cell cycle progression, one function of Rb in tumorigenesis is clear: loss or inactivation of both copies of the Rb gene unleashes E2F proteins to initiate cell cycle progression even in the absence of mitogens. However, the Rb story has become much more complex. First, proteins with structures and functions similar to Rb are made in many cell types. The proteins p107 and p130 resemble Rb, especially in a "pocket domain" that is crucial for binding E2Fs and viral proteins such as TAg, E1A, and E7 (Lipinski & Jacks 1999). The levels of Rb, p107, and p130 differ among cell types and at different stages in the cell cycle, but their binding of E2F proteins and their growth-regulating activities overlap substantially. In addition, a structurally unrelated protein called Necdin, which is expressed at high levels in neurons, shares many properties of the pocket proteins including the ability to bind E2Fs and viral proteins and to suppress cell growth (Aizawa et al 1992, Hayashi et al 1995, Taniura et al 1998, Yoshikawa 2000). Second, Rb regulates cell survival and differentiation in addition to cell cycle progression (Lipinski & Jacks 1999). Much of this information has come from studies of the expression and function of Rb and other pocket proteins in development. Here we focus on roles of the Rb family in neural development.

Functions of Rb in Neural Development

Rb and the other pocket proteins are widely produced in the developing nervous system (Jiang et al 1997). Rb is found in the ventricular zone, where neuroblasts divide, as well as in regions that contain only postmitotic cells. Expression of *p107* overlaps with that of *Rb*, but it is restricted to proliferating cells in the ventricular zone. *p130* is expressed at low levels in the nervous system throughout embryogenesis. Members of the E2F family are also present in developing neurons,

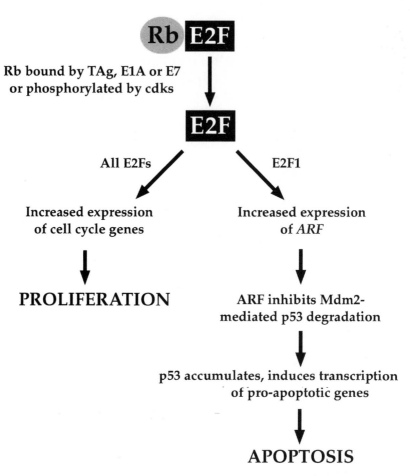

Figure 1   Regulation of proliferation and apoptosis by the Rb/E2F pathway. In non-dividing cells, E2F transcription factors are bound to the Retinoblastoma (Rb) protein and are thereby rendered inactive. Phosphorylation of Rb by cyclin dependent kinases (cdks) or binding of Rb by viral oncoproteins such as SV40 T antigen (TAg), adenovirus E1A, or papillomavirus E7 leads to dissociation of Rb from E2Fs. E2Fs can then activate transcription of genes necessary for cell cycle progression, allowing cells to proliferate. In some cells, inactivation of Rb can also lead to apoptosis. This may be mediated by E2F1, which can promote transcription of the ARF tumor suppressor. ARF binds to a complex of Mdm2 and p53, and in so doing, prevents the Mdm2-mediated degradation of p53. The resulting accumulation of p53 allows transcription of pro-apoptotic genes, causing cells to undergo apoptosis. The fact that Rb can inhibit apoptosis as well as proliferation means that loss of Rb may lead to cell death rather than tumorigenesis; however, loss of other apoptotic regulators (such as p53) can synergize with loss of Rb to promote tumor formation.

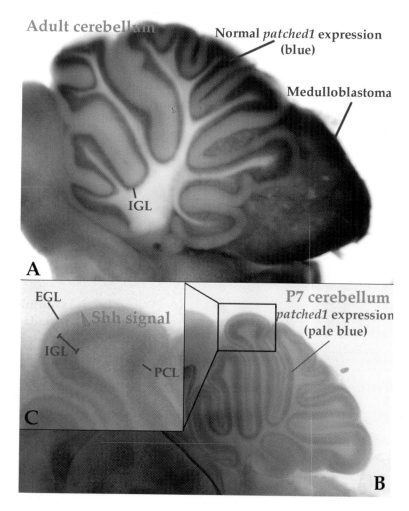

Figure 4 Mouse model of medulloblastoma. (*A*) Saggital section of adult cerebellum from heterozygous *patched1/+* mouse, showing normal cerebellum adjacent to a tumor. (*B*) Normal cerebellum at seven days after birth. (*C*) Enlargement of box in C, showing the layers of the cerebellum and the normal signaling process. Shh, originating from the Purkinje cell layer (PCL), signals to the external germinal layer (EGL) where granule cell precursors (GCPs) proliferate in response (red arrowhead). The blue stain in all three panels shows the pattern of patched1 expression, primarily in the internal granule cell layer (IGL) to which differentiating granule cells migrate from the EGL.

with maximal expression in the ventricular zone (Dagnino et al 1997). This expression pattern suggests that the Rb family might play an important role in regulating proliferation of neuronal precursors.

Studies of mice lacking *Rb* function ("knockout mice") support this notion (Clarke et al 1992, Jacks et al 1992, Lee et al 1992). These mice die between embryonic days 13 (E13) and 15, and they have defects in hematopoietic, lens, and neural development. Abnormal patterns of cell division occur in the central and peripheral nervous systems of homozygous *Rb*$^{-/-}$ mice (Lee et al 1992, Lee et al 1994). Dividing CNS cells are normally restricted to the ventricular zone, but in *Rb* mutants, cells well outside this region divide. Extra cell division was expected, given the role of Rb in cell cycle regulation. What was not expected was the dramatic neuronal cell death seen in the knockout mice. This death, most prominent in the hindbrain, spinal cord, and sensory ganglia, often affected the ectopic proliferating cells. The surviving neurons in the mutant mice were also not normal; many had abnormal morphology and failed to express differentiation markers such as neuronal βII tubulin. Similar effects have been seen in some strains of *p130* knockout mice (LeCouter et al 1998). *p107* knockouts do not have significant neuronal defects, but *Rb/p107* double homozygotes die earlier than the *Rb* knockout mice and have even more severe apoptosis in the CNS, which suggests that *p107* and *Rb* may be partially redundant (Lee et al 1996). These results indicate that Rb and other pocket proteins are essential not only for proliferation of neuronal precursors but also for survival and differentiation of postmitotic neurons.

Rb as a Regulator of Cell Cycle Exit and Survival

The antiapoptotic function of Rb family proteins is most apparent as neurons exit the cell cycle and begin to differentiate. In many cases, increased Rb levels accompany cell cycle exit. For example, Rb and p130 levels increase dramatically in embryonal carcinoma cells that have been induced to differentiate into neurons (Gill et al 1998, Kranenburg et al 1995a, Slack et al 1993). Similarly, in the quail retina, cell cycle exit at E6-E7 is accompanied by a sudden rise in the amount of E2F-1/Rb complexes (Kastner et al 1998). Pocket proteins appear to be necessary for both cell cycle exit and survival. The pocket protein inhibitor E1A both prevents cell cycle exit and increases cell death in cultured retinal precursors (Kastner et al 1998). Similarly, introduction of E1A into embryonal carcinoma cells, cortical neuronal precursors, or striatal stem cells inhibits cell cycle exit and promotes apoptosis (Callaghan et al 1999; Slack et al 1995, 1998).

The contributions of Rb, p107, and p130 to cell cycle exit and survival may differ. Cultured neuronal precursors from *Rb* knockout mice have delayed terminal mitosis, but they survive (Callaghan et al 1999, Slack et al 1998). However, these cells have increased levels of p107 compared with wild-type cells, suggesting that p107 may compensate for Rb in protection from apoptosis. In support of this notion, inactivation of all pocket proteins by expression of E1A does result in increased apoptosis of *Rb*$^{-/-}$ cells (Callaghan et al 1999). Similar compensation is presumably behind the increased apoptosis seen in *Rb/p107* double knockouts compared with the *Rb* single knockouts (Lee et al 1996). The molecular basis of apoptosis in $Rb^{-/-}$ mice has been elucidated by crossing these mice to other knockout strains. Apoptosis in the central—but not in the peripheral—nervous system of $Rb^{-/-}$ mice is dependent on E2F1 and p53, because it is dramatically reduced in mice lacking either of these genes (Macleod et al 1996, Tsai et al 1998). It is interesting to note that the ectopic cell proliferation in the *Rb* knockout is not suppressed by p53 deficiency, indicating that apoptosis and cell cycle exit are separately regulated. The involvement of E2F1 and p53 suggests how apoptosis is controlled (Figure 1). Loss of Rb leads to increased E2F1 transcriptional activity. Although E2F proteins can induce cell cycle progression, E2F1 is unique in that it can also promote apoptosis (DeGregori et al 1997, Field et al 1996, Hsieh et al 1997, Kowalik et al 1998, Phillips et al 1997, Qin et al 1994, Wu & Levine 1994). It does this, in part, by inducing expression of the tumor suppressor gene *ARF* (encoded by an alternative reading frame in the *CDKN2A/p16* locus; Bates et al 1998, Zhu et al 1999). ARF binds to the oncoprotein Mdm2 (murine double minute 2) and prevents Mdm2-mediated p53 degradation (Chin et al 1998, Kamijo et al 1998, Pomerantz et al 1998, Zhang et al 1998). The resulting increase in p53 then promotes apoptosis by activating transcription of proapoptotic genes such as *bax, fas,* and *killer/dr5* (el-Deiry 1998).

Another potential mediator of apoptosis in *Rb*-deficient mice is a nuclear protein called N5 (Doostzadeh-Cizeron et al 1999). N5 contains a region of sequence similarity to death domain proteins involved in apoptosis. Overexpression of N5 can promote apoptosis in certain cells. Rb can associate with N5 and inhibit N5-induced apoptosis, so loss of Rb might lead to increased apoptosis via N5.

Rb and Prevention of Cell Cycle Reentry

In addition to ensuring proper cell cycle exit and survival, Rb and other pocket proteins may prevent postmitotic neurons from resuming cell division. The most striking evidence for this comes from studies of transgenic mice in which SV40 TAg is targeted to cerebellar Purkinje cells (Feddersen et al 1992). These mice develop ataxia as a result of Purkinje cell degeneration. The mechanisms of Purkinje cell death in the TAg mice have been studied in detail (Athanasiou et al 1998; Feddersen et al 1995, 1997). TAg expression induces inappropriate cell cycle entry, followed by apoptosis. The apoptosis is dependent on binding of TAg to Rb, because mutant forms of TAg that cannot bind Rb are unable to induce apoptosis. Purkinje cells expressing TAg have elevated E2F in their nuclei, suggesting that E2F is involved in the apoptosis of these cells. Although transgenic mice overexpressing E2F1 in their Purkinje cells do not have Purkinje cell degeneration or ataxia, crossing these mice to the TAg mice results in accelerated Purkinje cell loss and ataxia. Together, these studies suggest that Rb/E2F function may prevent postmitotic neurons from reentering the cell cycle.

Rb as a Regulator of Neuronal Differentiation

Rb knockout mice, in addition to ectopic proliferation and apoptosis, have defects in expression of differentiation markers such as βII tubulin and the neurotrophin receptors TrkA, TrkB, and p75 (Lee et al 1994). Primary cultures of sensory ganglion cells from *Rb* mutant embryos have reduced neurite outgrowth, even in the presence of appropriate neurotrophins (Lee et al 1994). A role for Rb in differentiation is also suggested by studies of some neuronal cell lines. In these cells, decreased Cdk activity and dephosphorylation of Rb accompany differentiation, and overexpression of Rb or Cdk inhibitors can promote differentiation and neurite outgrowth (Dobashi et al 1995, Kranenburg et al 1995b).

The mechanisms by which Rb promotes differentiation are unknown. One potential mediator of Rb's effects on neurite outgrowth is Cdk5, a kinase that has structural homology to cyclin-dependent kinases but does not control the cell cycle (Lee et al 1997b, Tang et al 1996). Cdk5, in conjunction with its activators p35 and p39, plays a critical role in neuron migration and axon growth. In cultured cortical and hippocampal neurons, overexpression of Cdk5, p35, or p39 stimulates growth of neurites, whereas dominant-negative mutants of Cdk5 inhibit neurite growth (Nikolic et al 1996, Tang & Wang 1996, Xiong et al 1997). Cdk5 can bind and phosphorylate Rb (Lee et al 1997a), so Rb may affect neurite growth by binding to Cdk5.

Another protein that may be involved in Rb-mediated neurite outgrowth is NRP/B (nuclear restricted protein/brain) (Kim et al 1998). NRP/B is a nuclear matrix protein whose expression increases during neuronal differentiation. NRP/B overexpressed in neuroblastoma cells can promote neurite growth. Antisense NRP/B oligonucleotides can inhibit neurite development in primary hippocampal neurons and in PC12 cells. During differentiation of neuroblastoma cells induced by retinoic acid, NRP/B associates with hypophosphorylated Rb. Thus, NRP/B interaction with Rb may be necessary for neuronal differentiation.

The Limits to Growth: Why Do Rb-Related Tumors Usually Form in the Retina?

It has been almost 15 years since Rb was first isolated. In light of what we have learned about Rb function in neurons and other cells, do we now understand the mechanisms of tumorigenesis in retinoblastoma? One great mystery is why *Rb* mutations in humans cause tumors in the retina but not in the many other cell types in which *Rb* is expressed. *Rb*-heterozygous mice, in contrast to people, do not develop retinoblastoma (Clarke et al 1992, Jacks et al 1992, Lee et al 1992). However, retinal tumors do occur in transgenic mice that overexpress SV40 TAg (which inactivates all pocket proteins) in the retina and in chimeric mice derived from *Rb/p107* knockout embryonic stem cells (al-Ubaidi et al 1992, Robanus-Maandag et al 1998). These studies suggest that in the mouse, loss of Rb does not cause retinoblastoma because other pocket proteins in the retina can compensate.

In people, similar compensation may be absent from the retina but present in other tissues, thus accounting for the focus of human tumors in the retina.

However, pocket protein redundancy may not be the whole story. SV40 TAg can bind and inactivate pocket proteins, but it can also inactivate the p53 tumor suppressor. p53 is necessary for apoptosis, so tumors may arise in TAg transgenic mice because both p53 and pocket protein functions are reduced. To address this, mice expressing the human papillomavirus E7 protein in photoreceptors were generated (Howes et al 1994). E7 shares with TAg the ability to bind pocket proteins, but it cannot bind p53. Strikingly, the E7 mice do not develop retinal tumors but instead exhibit retinal apoptosis. If the E7 mice are crossed to *p53*-knockout mice, they develop tumors. Thus, the status of p53 may determine whether retinoblastoma develops. *p53* mutations have been observed in human retinoblastoma, although they are not common (Emre et al 1996, Kato et al 1996). This suggests that other components of the apoptotic machinery may need to be inactivated for retinoblastoma to develop.

In fact, genes other than *Rb* are likely to play a role in the development of retinoblastoma. The frequency of mouse retinoblastoma induced by overexpression of papilloma virus *E6* and *E7* genes (which together inactivate both Rb and p53) is strain dependent (Griep et al 1998). In families that carry *Rb* mutations, the penetrance of retinoblastoma varies from 20% to 95% (Griep et al 1998, Hamel et al 1993). The majority of retinoblastomas contain at least one other genetic change besides inactivation of *Rb* (Benedict et al 1983, Hamel et al 1993, Kusnetsova et al 1982, Potluri et al 1986, Squire et al 1985). By far the most common changes are extra copies of the long arm of chromosome 1, also seen in other tumors, and extra copies of the short arm of chromosome 6, which are unique to retinoblastoma. Cloning the genes at these loci is likely to provide important insights into the basis of retinoblastoma and the regulation of normal neural development.

ASTROCYTE DEVELOPMENT AND GLIOBLASTOMA

Overview of Astrocyte Development

Astrocytes perform diverse functions in the CNS, including regulating neuronal growth and survival (Arenander & de Vellis 1992, Kornblum et al 1998, Richardson 1994), guiding cell migration and axon growth during development (Bentivoglio & Mazzarello 1999, Hatten & Mason 1990, Komuro & Rakic 1998, Mason & Sretavan 1997, Powell et al 1997, van den Pol & Spencer 2000), promoting synapse formation and modulating synaptic transmission (Araque et al 1999, Bacci et al 1999, Pfrieger & Barres 1996, Vesce et al 1999), and orchestrating inflammatory and immune responses during brain infection and injury (Aschner 1998, Montgomery 1994). In most regions of the brain, astrocytes begin to develop later than neurons and are still being generated long after neurons have stopped being produced (Altman 1966; Schubert & Rudolphi 1998; Sturrock 1982, 1987). Many astrocytes or their precursors retain the capacity for division throughout life. This makes them uniquely susceptible to transformation and is presumably one reason that astrocytic tumors are the most common brain tumors (Collins 1998, Rasheed et al 1999, Salcman 1995).

Astrocytes arise from multipotent neural stem cells (NSCs) (Pringle et al 1998) that have the capacity to self-renew and to produce a variety of classes of neurons, astrocytes, and oligodendrocytes (Davis & Temple 1994, Kalyani et al 1997, Quinn et al 1999, Rao 1999, Reynolds et al 1992). Differentiation along each of these lineages involves generation of progressively more restricted precursor cells (Lee et al 2000). Thus, NSCs give rise to neuron-restricted precursors that produce only neurons (Luskin et al 1993, Mayer-Proschel et al 1997) and glial-restricted precursors, which cannot generate neurons but can produce astrocytes and oligodendrocytes (Rao & Mayer-Proschel 1997, Rao et al 1998). Glial-restricted precursors, in turn, generate even more restricted precursors, which produce either astrocytes or oligodendrocytes but not both. Oligodendrocyte precursor cells and astrocyte precursor cells (APCs) (Mi & Barres 1999, Noble et al 1995, Seidman et al 1997, Tang et al 2000) are presumed to be direct progenitors of glia in the CNS.

Progenitor cell proliferation and astrocytic differentiation are controlled, in part, by extracellular signals (Lee et al 2000). NSCs isolated from various parts of the CNS at various stages of development undergo self-renewal in response to fibroblast growth factors (FGFs) and epidermal growth factors (EGFs) (Ben-Hur et al 1998, Kalyani et al 1997, Quinn et al 1999, Reynolds et al 1992, Weiss et al 1996). EGF may also promote astrocyte differentiation, since infection of progenitors with a retrovirus that increases expression of EGF receptor (EGFR) results in increased generation of astrocytes in vitro and in vivo (Burrows et al 2000, Lillien 1995). Morever, EGFR-knockout mice exhibit delayed astrocyte differentiation and reduced numbers of astrocytes in certain parts of the brain (Kornblum et al 1998, Sibilia et al 1998).

Other signals that regulate differentiation of NSCs into astrocytes include cytokines of the ciliary neurotrophic factor (CNTF)/leukemia inhibitory factor (LIF) family, whose signals are transduced through the gp130 receptor chain and the Jak/Stat signaling pathway (Segal & Greenberg 1996, Touw et al 2000). CNTF and LIF can promote astrocytic differentiation of NSCs and more restricted progenitors in culture (Bonni et al 1997, Hughes et al 1988, Mi & Barres 1999, Park et al 1999, Rajan & McKay 1998), and knockout mice lacking either the LIF receptor or gp130 have severe defects in astrocyte generation (Koblar et al 1998, Nakashima et al 1999a). Bone morphogenetic proteins (BMPs) can also promote astrocyte development (Kawabata et al 1998, Mehler et al 1997). BMPs activate a signaling pathway distinct from that triggered by LIF and CNTF and can promote astrocyte generation on their own or in synergy with LIF (D'Alessandro et al 1994; Gross et al 1996; Mabie et al 1997; Mehler et al 2000; Nakashima et al 1999b, 1999c). Whether CNTF/LIF, BMPs, and EGF act on the same precursors, and whether they generate the same types of astrocytes is not yet clear.

Glioblastoma

Glioblastoma multiforme is the most common brain tumor in adults (Collins 1998, Salcman 1995). It is a highly malignant tumor that is thought to arise from astrocytes or astrocyte precursors, but the heterogeneity of tumor morphology and behavior (indicated by the term "multiforme") makes conclusions about its origin extremely difficult (Lopes et al 1995). Although retinoblastoma and medulloblastoma can often be treated successfully, glioblastoma is almost invariably fatal. Glioblastomas are often divided into two subtypes: progressive, which arises from lower-grade astrocytic tumors, and de novo, which does not (Collins 1998, Rasheed et al 1999).

The etiology of glioblastoma is complex, and almost always involves mutation or overexpression of multiple genes. Cytogenetically, glioblastomas exhibit losses of portions of chromosomes 6, 9, 10, 13, 17, 22, and Y and amplification or gain of material on chromosomes 7, 12, and 19 (Collins 1998, Rasheed et al 1999). Some of the relevant genes have been identified (Table 1), and they include growth factor receptors (e.g. EGFR), components of the cell cycle machinery (Rb, cdk4, and the cdk inhibitor p16), and regulators of apoptosis (p53, mdm2, ARF, and PTEN).

Epidermal Growth Factor Receptor Gene Amplification

Amplification of the *EGFR* gene was among the first genetic abnormalities to be linked to glioblastoma (Libermann et al 1985a, 1985b). *EGFR* amplification

TABLE 1   Genetic events that correlate with glioblastoma

Chromosome	Amplified (+) or lost (−)	Candidate genes
6	−	?
7	+	*EGFR*
9	−	*CDKN2A/p16, CDKN2B/p15, ARF*
10	−	*PTEN, DMBT1, LGI1, MxiI, h-neu*
12	+	*CDK4, SAS, MDM2, GLI, GAS41*
13	−	*Rb*
17	−	*p53*
19	+	?
22	−	*NF2?*
Y	−	?

[1]Abbreviations: EGFR, epidermal growth factor receptor; CDKN2A/p16, cyclin dependent kinase inhibitor 2A; CDKN2B/p15, cyclin-dependent kinase inhibitor 2B; ARF, tumor suppressor encoded by alternative reading frame of CDKN2A/p16 exon 2; PTEN, phosphatase/tensin homolog on chromosome 10; DMBT1, deleted in malignant brain tumors 1; LGI1, Leucine-rich gene-Glioma Inactivated 1; MxiI, Max interactor 1; h-neu, human homolog of Drosophila *neuralized* gene; CDK4, cyclin dependent kinase 4; SAS, sarcoma amplified sequence; MDM2, murine double minute 2; GLI, glioblastoma amplified gene; GAS41, glioma amplified sequence 41; Rb, retinoblastoma; p53, 53-kilodalton tumor suppressor protein; NF2, neurofibromatosis 2. "?" refers to chromosomal deletions or amplifications for which no strong candidates have been identified.

occurs in 40%–50% of glioblastomas and usually results in elevated levels of EGFR expression. The fact that many glioblastomas contain at least one ligand capable of activating the receptor (Ekstrand et al 1991, Mishima et al 1998) suggests that in some cases, tumor growth might be promoted by autocrine or paracrine stimulation of the EGFR-signaling pathway. However, another view of the significance of *EGFR* amplification came from sequencing of the amplified EFGR genes, which revealed that they are often mutated or rearranged to generate a protein that lacks part of the extracellular domain (Ekstrand et al 1992, Sugawa et al 1990, Wong et al 1992). The mutant EGFRs cannot bind ligand, but they have constitutive tyrosine kinase activity, promote increased growth of tumor cells in vitro, and enhance tumorigenicity of glioblastoma cells that are transferred into mice (Ekstrand et al 1994, Hoi Sang et al 1995a, Nagane et al 1998, Nishikawa et al 1994). Furthermore, overexpression of mutant EGFRs in astrocytes (or their precursors) can promote development of glioblastoma (Holland et al 1998). Thus, in glioblastoma cells EGFR signaling may be activated in a ligand-dependent or -independent manner.

The activation of EGFR in a large percentage of glioblastomas suggests an important role for EGFR in astrocyte development. Recent studies of mice lacking EGFR function support this idea. The mice often die before birth, but in some genetic backgrounds, they are able to survive for ≤ 3 weeks after birth (Sibilia & Wagner 1995, Threadgill et al 1995). In the surviving mice the effects on brain development are striking: a significant reduction in the number of glial fibrillary acidic protein-expressing astrocytes in many brain regions, and massive degeneration of neurons in the frontal cortex, olfactory bulb, and thalamus (Kornblum et al 1998, Sibilia et al 1998). Because the glial defects appear before the neuronal degeneration and because many of the affected neurons do not normally express EGFR, some of the neuronal death observed may result from the loss of trophic support from glia. In any case, EGFR signaling appears to be critical for normal development of astrocytes.

EGFR can be activated by a number of ligands, including EGF, TGF-α, amphiregulin/schwannoma-derived growth factor, betacellulin, epiregulin, and heparin-binding EGF (Khazaie et al 1993, Kimura et al 1990, Komurasaki et al 1997, Prigent & Lemoine 1992, Riese et al 1996, Shoyab et al 1989). EGFR ligands have different affinities for EGFR and other receptors, yet the consequences of their binding to EGFR, at least in astrocytes, appear to be similar. At least four ligands— EGF, TGF-α, amphiregulin, and heparin-binding EGF—promote proliferation of astrocytes in culture (Kimura et al 1990; Kornblum et al 1998, 1999; Leutz & Schachner 1981; Simpson et al 1982). Although glioblastomas are often presumed to arise from cells committed to the astrocyte lineage, some tumors may originate from less differentiated progenitors (Noble et al 1995, Shoshan et al 1999). In this regard, it is worth noting that EGFR ligands can also promote growth of astrocyte precursors (Ben-Hur et al 1998, Seidman et al 1997) and of multipotent CNS stem cells that give rise to neurons and glia (Kornblum et al 1999, Reynolds et al 1992). Thus, EGFR activation may contribute to glioblastoma by promoting proliferation of astrocytes or their precursors.

In addition to their growth-inducing effects, EGFR ligands promote process extension and migration in both normal astrocytes and astrocytic tumor cells (Faber-Elman et al 1996, Hoi Sang et al 1995b, Rabchevsky et al 1998, Westermark et al 1982). EGFR activation may therefore also contribute to tumor malignancy by promoting invasion and metastasis (Nishikawa et al 1994). Perhaps this explains why EGFR amplification and rearrangement are most commonly observed in high-grade, malignant astrocytic tumors (Collins 1998, Rasheed et al 1999). Whatever its mechanisms of action, EGFR activation in glioblastoma is extremely potent and has therefore become the target of a variety of pharmacologic, immunologic, and gene-transfer-based therapies (Halatsch et al 2000, Pfosser et al 1999, Pu et al 2000, Tian et al 1998).

Impaired Control of Cell Cycle and Apoptosis

Disruption of the cell cycle machinery also contributes to glioblastoma. Changes in the genes encoding cyclins, cdks, cdk inhibitors, and Rb have been observed at high frequency in glioblastomas (Biernat et al 1998, Ichimura et al 1996, Schmidt et al 1994, Simon et al 1999). Altogether, two-thirds of glioblastomas have abnormalities in one of these genes (Collins 1998, Rasheed et al 1999). Homozygous deletions or mutations of chromosome 9, in a region that encodes the cdk inhibitors CDKN2A/p16 and CDKN2B/p15, are found in 30%–40% of tumors. At least 14% of tumors have inactivated both copies of the *Rb* gene on chromosome 13. About 15% of glioblastomas have amplifications of a segment of chromosome 12 that includes the *cdk4* gene. A small number of tumors have amplification and overexpression of *cyclin D1* or *cyclin D3* (Buschges et al 1999). The different cell cycle machinery changes are rarely seen in the same tumor, so any one of them may be sufficient for loss of cell cycle control.

Many cells that enter the cell cycle inappropriately, for example as a result of loss of Rb, subsequently undergo apoptosis (Lipinski & Jacks 1999). Thus, tumor progression may require not only cell cycle dysregulation but also loss of the ability to die. In glioblastoma this is often achieved by disruption of the p53 pathway (Biernat et al 1998, Ichimura et al 2000, Rasheed et al 1999, Simon et al 1999). At least 25% of glioblastomas have mutations or deletions of the p53 gene itself, and another 5%–12% have amplifications of the locus on chromosome 12 that encodes Mdm2, which promotes degradation of p53. ARF, which inhibits Mdm2-mediated p53 degradation, is generated by alternative splicing of a transcript from the *CDKN2A/p16* locus on chromosome 9. This locus is deleted or mutated in 30%–40% of glioblastomas, resulting in loss of functional ARF and decreased levels of p53. Overall, about 75% of glioblastomas have reduced ability to undergo apoptosis due to p53-related lesions (Ichimura et al 2000). The critical importance of *Rb* and *p53* defects in glioblastoma has raised hopes of treating tumors using gene therapy directed at components of the Rb and p53 pathways (Fueyo et al 1998, 1999; Gomez-Manzano et al 1999; Lang et al 1999).

PTEN: A Multifunctional Tumor Suppressor

A significant advance in the understanding of glioblastoma etiology came with the discovery of the tumor suppressor gene *PTEN* (Besson et al 1999, Cantley & Neel 1999, Di Cristofano & Pandolfi 2000). A large percentage of glioblastomas have part of the long arm of chromosome 10 deleted (Rasheed et al 1992). This region is also frequently deleted in many other tumors, including breast, prostate, and endometrial carcinoma and melanoma (Cantley & Neel 1999, Di Cristofano & Pandolfi 2000). In 1997, three groups independently cloned a gene that was eliminated by the deletion (Li & Sun 1997, Li et al 1997a, Steck et al 1997). The gene was called *PTEN* (phosphatase/tensin homolog on chromosome 10), MMAC (mutated in multiple advanced cancers), or TEP-1 (TGFβ- regulated and epithelial cell-enriched phosphatase 1). Both alleles of *PTEN* are inactivated in $\sim$30% of glioblastomas. *PTEN* is also the target of mutation in three rare inherited disorders: Cowden disease (CD), Lhermitte-Duclos disease, and the Bannayan-Zonana syndrome (Liaw et al 1997; Marsh et al 1997, 1998). Patients with these disorders develop widespread benign tumors (harmartomas) in the skin, mouth, thyroid, breast, intestine, and other tissues. They also have an increased risk of breast and thyroid cancer, cerebellar tumors, and mental retardation. The association of PTEN mutations with sporadic as well as inherited tumor syndromes suggested that PTEN might act as a regulator of cell growth, differentiation, or survival, but its mechanism of action was initially unclear.

PTEN Protein: Structure and Function   The PTEN protein has sequences similar to protein tyrosine phosphatases (Cantley & Neel 1999, Li & Sun 1997, Li et al 1997a, Steck et al 1997), specifically dual-specificity phosphatases which can dephosphorylate both tyrosine and threonine/serine residues (Neel & Tonks 1997, Tonks & Neel 1996). The significance of the sequence similarity was reinforced by mutations found in most tumors, which cluster in the putative phosphatase domain (Ali et al 1999, Marsh et al 1998). Surprisingly, in vitro assays indicated that PTEN was unable to dephosphorylate many conventional protein tyrosine phosphatase substrates (Li et al 1997b, Myers et al 1997). The proteins it could dephosphorylate did not appear to be critical for PTEN function because some disease-associated PTEN mutant proteins could still dephosphorylate them (Furnari et al 1998; Myers et al 1997, 1998). These observations suggested that PTEN might have a biological function that did not depend on protein phosphatase activity.

That function became clear with the discovery that PTEN is a more effective phosphatase for lipids than it is for proteins (Maehama & Dixon 1998, Myers et al 1998). One important reaction catalyzed by PTEN is the conversion of phosphatidylinositol (3,4,5)-trisphosphate (PIP3) to phosphatidylinositol (4,5)-bisphosphate (PIP2). This ability placed PTEN in a well-characterized signaling pathway linking growth factor and extracellular matrix signaling to cell survival (Downward 1998, Marte & Downward 1997, Stambolic et al 1999) (Figure 2).

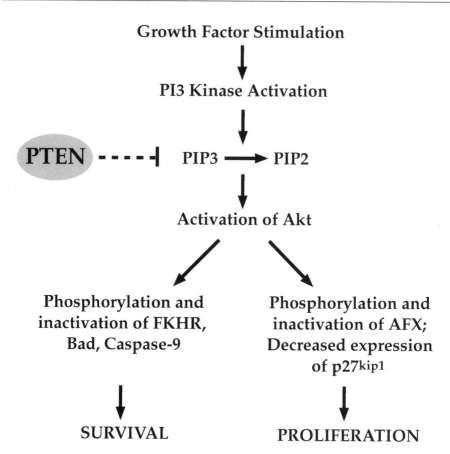

Figure 2   Loss of PTEN promotes growth and survival by activating the PI3 kinase/Akt signaling pathway. In normal cells, growth factor stimulation causes activation of phosphatidylinositol-3′ kinase (PI3 Kinase), which catalyzes with conversion of phosphatidylinositol (3,4,5)-trisphosphate (PIP3) into phosphatidylinositol (4,5)-bisphosphate (PIP2). Membrane-associated PIP2 attracts and activates the Akt kinase, which promotes survival by phosphorylating and inactivating pro-apoptotic proteins such as the transcription factor FKHR, the Bcl2-like protein Bad, and the cysteine protease caspase-9. Akt activation also promotes cell growth by phosphorylating and inactivating the transcription factor AFX, which would otherwise induce expression of the cdk inhibitor p27[kip1]. PTEN is a lipid phosphatase that blocks the activation of Akt by converting PIP3 to PIP2 and thereby promotes cell cycle arrest and apoptosis. Loss of PTEN function increases cell growth and survival and can contribute to glioblastoma.

Growth factors activate phosphatidylinositol-3′ kinase (PI3K), which increases cellular levels of PIP3. PIP3 serves as a second messenger that attracts proteins containing pleckstrin homology domains to the plasma membrane. One such protein is the serine-threonine kinase Akt (also called protein kinase B), which binds to PIP3 and becomes activated. Akt then phosphorylates and inactivates a number of proteins necessary for apoptosis (e.g. the Bcl2-like protein Bad, the apoptotic effector caspase 9, and the forkhead rhabdomyosarcoma transcription factor (FKHR), thereby promoting cell survival (Brunet et al 1999, Cardone et al 1998, Datta et al 1997, del Peso et al 1997). In this scheme, reduction of cellular PIP3 levels by PTEN would reduce Akt activation and make cells more likely to undergo apoptosis. Loss of PTEN would have the opposite effect: cells lacking wild-type PTEN would be resistant to apoptosis and would be much more likely to form a tumor.

PTEN Regulates Cell Survival, Growth, and Adhesion   The function of PTEN in the PIP3 pathway has now been validated by many studies. For example, fibroblasts from PTEN-deficient mice have elevated levels of PIP3, constitutively activated Akt and resistance to induction of apoptosis by a number of different stimuli (Stambolic et al 1998). Introduction of wild-type PTEN into these cells restores normal PIP3 levels and Akt activity and promotes apoptosis. Similarly, epithelial cells transfected with *PTEN* undergo apoptosis, and this can be inhibited by cotransfection of *Akt* (Li et al 1998). These findings indicate that PTEN is a regulator of the PI3K-Akt signaling pathway and is necessary to maintain the sensitivity of cells to apoptosis.

The loss of PTEN observed in many glioblastomas is consistent with this model in most respects. Glioblastoma cells containing mutant PTEN have abnormally high levels of PIP3 and elevated Akt activity, and expression of wild-type PTEN in these cells normalizes both of these parameters (Davies et al 1998, Haas-Kogan et al 1998). However, there is an important discrepancy between predictions of the model and the observed effects of PTEN in glioblastoma cells. Rather than inducing apoptosis, introduced PTEN causes most cells to arrest in the G1 phase of the cell cycle (Davies et al 1998, Furnari et al 1997, Tian et al 1999). Cells expressing exogenous PTEN have increased glial fibrillary acidic protein and develop long cytoplasmic processes, characteristics of astrocyte differentiation (Adachi et al 1999, Tian et al 1999). Glioblastoma cells do undergo apoptosis if they are prevented from binding to a substrate—a phenomenon termed "anoikis" (Davies et al 1998, Tamura et al 1999a).

The mechanisms of PTEN-mediated growth arrest are still not fully understood, but some potential mediators of the effect have recently been identified (Figure 2). The process requires the phospholipid phosphatase activity of PTEN (Furnari et al 1998) and appears to involve regulation of Akt because a constitutively active form of Akt can override PTEN-induced growth arrest (Ramaswamy et al 1999). An important substrate of Akt in this pathway may be the forkhead transcription factor AFX (acute lymphocytic leukemia fusion, X chromosome) (Kops et al 1999, Medema et al 2000, Takaishi et al 1999). AFX can induce expression of the cdk inhibitor p27[kip1], which promotes cell cycle arrest by inactivating cyclin E/cdk2 complexes. Phosphorylation by Akt prevents AFX from entering the nucleus and inducing p27[kip1] and thereby allows cells to proliferate. Since PTEN inhibits Akt activation, it is expected to increase AFX function and p27[kip1] expression. Indeed, introduction of exogenous PTEN causes increased production of p27^{kip1} in various cell types (Cheney et al 1999, Lu et al 1999, Sun et al 1999, Wu et al 2000a). This results in reduced phosphorylation of Rb, which in turn prevents cell cycle progression. In other words, PTEN inhibits growth by promoting activation of Rb, and loss of PTEN contributes to tumorigenesis, at least in part, by reducing Rb function (Paramio et al 1999).

In addition to its function in cell cycle control, PTEN can regulate cell adhesion and migration. PTEN contains extensive homology to tensin, a protein that interacts with actin filaments at focal adhesions (Li et al 1997a, Steck et al 1997). PTEN also contains a sequence that binds to PDZ domains, which are frequently involved in assembling multiprotein complexes at membrane/cytoskeletal interfaces (Wu et al 2000b). These features raise the possibility that PTEN associates with and regulates phosphorylation of cytoskeleton-associated molecules.

A number of studies indicate that PTEN can regulate cell shape and movement, and that this function depends primarily on its ability to dephosphorylate proteins rather than lipids. Although PTEN is inefficient at dephosphorylating most proteins, among the substrates it can act on are focal adhesion kinase (Fak) and the SH2-containing adaptor protein Shc. Fak and Shc mediate signal transduction by integrins and regulate cell adhesion and migration (Gu et al 1999; Tamura et al 1998, 1999b). Overexpression of PTEN inhibits cell spreading and cell migration induced by integrins, whereas reduction of PTEN levels by using antisense oligonucleotides has the opposite effect (Tamura et al 1998). The inhibitory effects of PTEN can be blocked by overexpression of Fak or Shc, suggesting that these substrates may be important mediators of PTEN's effects. Interestingly, the types of cell movement mediated by Fak and Shc appear to be different; Fak induces extensive reorganization of the actin cytoskeleton, formation of focal adhesions, and directional migration, whereas Shc causes less actin reorganization and more random cell movement (Gu et al 1999). Both forms of movement are inhibited by PTEN. Thus, cells lacking functional PTEN would be expected to display increased migration and an increased tendency to undergo metastasis. In keeping with this notion, PTEN mutations are found in glioblastoma but not in most lower-grade (less metastatic) astrocytic tumors.

The interaction of PTEN with Fak also provides a rationale for the ability of PTEN to induce growth arrest under some circumstances and apoptosis or anoikis under others (Davies et al 1998, Lu et al 1999, Tamura et al 1999b). Fak is normally phosphorylated in response to integrin ligation at sites of cell adhesion. This phosphorylation activates Fak, which in turn promotes survival by activating the PI3K/Akt-signaling pathway. When normal cells are prevented from attaching to the extracellular matrix, Fak becomes dephosphorylated, survival signaling is reduced, and cells undergo programmed cell death (anoikis). However, in cancer cells lacking wild-type PTEN, Fak cannot be effectively dephosphorylated upon cell detachment. Persistent phosphorylated Fak results in sustained PI3-K and Akt activity, and prolonged cell survival. Thus, loss of PTEN helps make tumor cells anchorage-independent. Overexpression of PTEN inhibits cell growth by inducing $p27^{kip1}$, but it also makes cells anchorage dependent by inhibiting the PI3K/Akt pathway. If cells are allowed to attach to the extracellular matrix, the first pathway predominates, and they undergo cell cycle arrest. If they are unable to attach, the second pathway takes over, and they undergo anoikis. In this regard, PTEN can be seen as a critical integrator of cell adhesion, growth, and survival.

Role of PTEN in Normal Development and Tumorigenesis   PTEN is expressed throughout the embryo, and has critically important functions in early embryonic development (Luukko et al 1999, Podsypanina et al 1999). Three different lines of *PTEN*-knockout mice have been established, and all exhibit early (6.5–9.5 days of gestation) embryonic lethality (Di Cristofano et al 1998, Podsypanina et al 1999, Suzuki et al 1998). Homozygous *PTEN*-knockout mice apparently die because they fail to establish normal connections with the maternal circulation (Suzuki et al 1998). At the time of death they also have disorganized germ layers and severe overgrowth of cephalic and caudal regions. Embryonic stem cells derived from these mice are impaired in their ability to differentiate into endoderm, ectoderm, and mesoderm in vitro and after implantation into normal blastocysts (Di Cristofano et al 1998).

PTEN also functions at later stages of development, as an important regulator of proliferation and survival. These functions have been investigated using heterozygous ($PTEN^{+/-}$) mice and using chimeras made with $PTEN^{+/-}$ embryonic stem cells (Di Cristofano et al 1998, Podsypanina et al 1999, Suzuki et al 1998). Both types of mice are viable and fertile, but they have marked hyperplasia in multiple tissues including the colon, skin, testes, and prostate. Full-blown tumors of the colon, testes, prostate, thyroid, liver, and immune system (leukemias and lymphomas) occur in 12%–15% of the mice. These tumors occur at an early age, often by 3.5 months in heterozygotes (Di Cristofano et al 1998), and their appearance is accelerated by exposure to gamma irradiation (Suzuki et al 1998). Although the tumors occur in heterozygous mice, in many cases the wild-type allele of the *PTEN* locus has been deleted or mutated in the tumor cells (Podsypanina et al 1999, Suzuki et al 1998). *PTEN* heterozygotes also have a high incidence of autoimmune disease (Di Cristofano et al 1999). This may reflect an inability of self-reactive lymphocytes to undergo apoptosis through a Fas-mediated pathway.

Given the frequency of *PTEN* mutations in human glioblastoma, it is surprising that *PTEN* heterozygotes have no defects in the nervous system and that none develop brain tumors. One possible explanation is that PTEN has a different function in human vs mouse astrocytes. Alternatively, defects in glial cells might require inactivation of the second *PTEN* allele, and this event may be less common in murine brain than in other tissues. Finally, the loss of *PTEN*, on its own, may not be sufficient to promote glial hyperplasia or tumor formation. Most human glioblastomas have mutations or amplifications of multiple genes that regulate proliferation, differentiation, and cell death, so it may not be surprising that loss of *PTEN* alone does not cause glioblastoma in mice.

Other Genes on Chromosome 10 May Also Contribute to Glioblastoma

Chromosome 10 deletions in glioblastoma frequently disrupt genes other than *PTEN*. A gene called *DMBT* (deleted in malignant brain tumors), located at chromosome 10q25, is lost from 20%–40% of glioblastomas and from many lower-grade astrocytic tumors (Lin et al 1998, Mollenhauer et al 1997, Somerville et al 1998). The *DMBT* gene product (also called hensin or collectin-binding protein/gp-340) is an extracellular matrix protein of the scavenger receptor cysteine-rich superfamily, and it has been implicated in epithelial-cell differentiation and immune responses (Mollenhauer et al 2000). However, the protein has no known function in the nervous system. Although it is frequently lost in glioblastoma, this loss does not correlate with changes in tumor cell growth in vitro or with increased severity of tumors in vivo (Lin et al 1998, Steck et al 1999).

Mxi1 (Max interactor 1), also located on chromosome 10q25, is a more promising candidate for a tumor suppressor gene deleted in glioblastoma (Fults et al 1998). *Mxi1* encodes an antagonist of the Myc oncoprotein, and suppresses growth in a number of tissues (Foley & Eisenman 1999, Schreiber-Agus & DePinho 1998). In glioblastoma cells that lack endogenous *Mxi1* expression, introduction of an exogenous gene inhibits proliferation and induces cell cycle arrest (Wechsler et al 1997). Thus loss of *Mxi1* could contribute to excessive growth of tumor cells. Two other genes in this region—*LGI1* (leucine-rich gene-glioma inactivated), which encodes a putative adhesion molecule (Chernova et al 1998), and h-*neu*, a human homolog of the *Drosophila* neuralized gene which encodes a transcription factor (Nakamura et al 1998) – have also been suggested to play a role in glioblastoma.

CEREBELLAR GRANULE CELL DEVELOPMENT AND MEDULLOBLASTOMA

Granule Cell Development

The cerebellum plays an important role in motor coordination and learning, and it has been implicated in a variety of cognitive and affective functions (Altman & Bayer 1997, Leiner et al 1993). The cerebellar cortex contains intricate circuitry composed of five different types of neurons. Granule cells are by far the most abundant. They regulate the activity of Purkinje cells and thereby control the output from the cerebellum to other parts of the brain. The murine cerebellum contains $\sim 10^8$ granule cells, more than the number of all neurons in the rest of the brain. The critical importance of granule cells is evident from mutant mice in which the loss of granule cells causes severe ataxia (Mullen et al 1997) and from patients with congenital granule cell degeneration, who have severe deficits in motor coordination, language use, and cognitive function (Pascual-Castroviejo et al 1994).

Granule cell development has extraordinary features that set it apart from other kinds of neurogenesis. Most neurons are born around the ventricles and then migrate outward toward the surface of the brain. In contrast, granule cells are generated on the outside of the cerebellum and then migrate inward. They arise from a dorsal hindbrain structure called the rhombic lip. During late embryogenesis, neural precursor cells leave the rhombic lip and stream across the surface of the early cerebellum to form the external germinal layer (EGL). At birth (in rodents), the EGL consists of a single layer of undifferentiated cells. During the next few days, these cells undergo extensive proliferation to generate a large pool of granule cell precursors (GCPs). As new GCPs are generated, older cells begin to exit the cell cycle and differentiate. The differentiating cells extend axons that contact Purkinje cell dendrites, and then their cell bodies migrate inward past the Purkinje cell bodies to their final destination, the internal granule layer. The waves of GCP proliferation and differentiation continue until ~3 weeks of age, at which time the EGL dwindles, and all GCPs mature into granule cells.

The molecular mechanisms that regulate the proliferation, differentiation, and migration of GCPs are not well understood. However, some insight into these mechanisms has come from recent studies of the cerebellar tumor medulloblastoma.

Medulloblastoma

Medulloblastoma is the most common malignant brain tumor in children. There are ~350 new cases in the United States each year, most occurring at between 5 and 10 years of age (Novakovic 1994). Most medulloblastoma cases are treated with surgery followed by radiation and/or chemotherapy. Although these approaches are often effective at shrinking the primary tumor, recurrence and metastasis are common, and only ~50% of patients survive for 5 years after diagnosis (Packer et al 1994).

Medulloblastoma cells are small and round with little cytoplasm. The cells usually appear to be undifferentiated, although they may express markers of neurons and glia. The tumor cells are frequently found near the surface of the cerebellum. These properties are not completely consistent with any normal cerebellar cell, but they are most similar to GCPs. For this reason, many investigators believe that medulloblastomas originate from GCPs that become transformed and fail to undergo normal differentiation.

Genetic factors clearly affect the rate of medulloblastoma formation, because a number of chromosomal abnormalities have been associated with the disease (Rasheed & Bigner 1991, Thapar et al 1995). The most common of these abnormalities is loss or mutation of a portion of chromosome 17. The medulloblastoma gene on chromosome 17 appears to be distinct from p53 (which is also on that chromosome) and has not yet been identified. In addition to chromosome loss, medulloblastomas overexpress certain genes, including c-*myc*, *pax5*, and *zic*, all of which encode transcription factors (Kozmik et al 1995, Rasheed & Bigner 1991, Yokota et al 1996). Whether these genes contribute to the initiation or progression of medulloblastoma or whether their expression is a consequence of the transformed cell type remains to be determined.

Patched1 Mutations in Medulloblastoma

New insight into the molecular basis of medulloblastoma has come from recent studies of the Hedgehog-Patched signaling pathway (Figure 3), which was first discovered for its role in *Drosophila* embryo segmentation. Sonic hedgehog (Shh), a vertebrate homolog of the *Drosophila* segment polarity gene product Hedgehog, is a secreted protein that plays a critical role in patterning of the nervous system, the limbs, the skin, and other tissues (Goodrich & Scott 1998, Hammerschmidt et al 1997). Patched1 (Ptc1) is a transmembrane protein that can bind Shh and appears to serve as a receptor. In the absence of Shh, Ptc1 actively represses target gene expression. In the presence of Shh, Ptc1-mediated repression is relieved, and the transcription of target genes is induced. The signaling cascade involves the transmembrane protein Smoothened (Smo) and a family of zinc finger transcription factors, Gli1, Gli2, and Gli3. Gli proteins have activating and repressing effects on transcription, and their effects may depend on proteolytic processing (Dai et al 1999, Ruiz i Altaba 1997, Sasaki et al 1999). Current research is directed at learning how proteolysis, subcellular location, and functional activation of Gli proteins are controlled. The net effect of Shh is to influence Gli proteins to induce target gene transcription. One common target gene is *ptc1* itself, so Shh induces production of its antagonist. The induced build-up of Ptc1 may limit the duration of the Shh effect on a cell. In flies, at least, Ptc also slows the movement of Hedgehog across cells, perhaps by binding and sequestering it (Chen & Struhl 1998).

The first evidence that Shh signaling might be involved in medulloblastoma came from studies of the human *PATCHED1* (*PTCH1*) gene (Hahn et al 1996, Johnson et al 1996). The gene is mutated in basal cell nevus syndrome (or Gorlin's syndrome), a disease characterized by widespread skin tumors, craniofacial and skeletal abnormalities, and an increased incidence of medulloblastoma (Gorlin 1987). Of sporadic medulloblastomas, 10%–20% contain mutations at the *PTCH* locus (Pietsch et al 1997, Raffel et al 1997, Xie et al 1997), suggesting that errors in Shh/Ptc signaling contribute to the etiology of cerebellar tumors.

A causal link between *PTCH1* mutations and medulloblastoma was demonstrated using *ptc1*-knockout mice. Homozygous *ptc1*$^{-/-}$ mice have multiple defects in the neural tube, heart, and other tissues, and die at 9–10 days of gestation. Heterozygotes are viable and initially the only phenotypes are increased body size and a low incidence of polydactyly (Goodrich et al 1997). Between 4 and 6 months of age, however, 15%–25% of the heterozygotes develop aggressive cerebellar

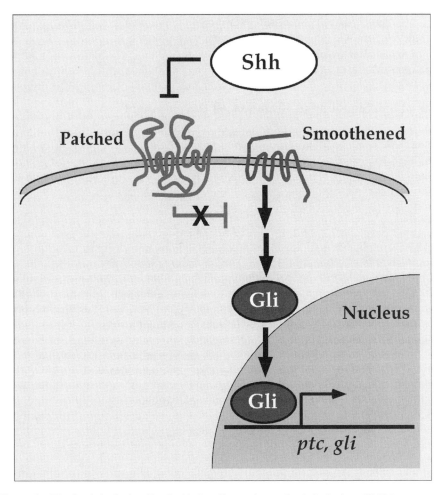

Figure 3 The Sonic hedgehog/Patched1 signaling pathway. Sonic hedgehog (Shh) is a secreted signaling protein, Patched1 is a receptor for Shh (predicted to have 12-transmembrane-domains), and Smoothened is another membrane protein (predicted to have 7-transmembrane-domains). The bulk of Patched and Smoothened are found in intracellular organelles. The Gli proteins are transcription factors with activating or repressing activities for Shh target genes. In the absence of Shh signal, Patched1 inhibits Smoothened and Gli proteins remain in the transcriptional repression mode. The Shh signal inactivates Patched1, allowing Smoothened to become active. Active Smoothened transduces signals that convert Gli proteins into transcriptional activators, which turn on target genes. Among the targets commonly induced are those encoding components of the Shh pathway, *patched1* and *gli1*. Inactivating mutations in *patched1*, or activating mutations in *smoothened*, can lead to medulloblastoma because target genes are inappropriately active.

tumors (Goodrich et al 1997, Wetmore et al 2000; Figure 4, see color insert). The tumors consist of small, round cells on the surface of the cerebellum and are morphologically similar to human medulloblastoma. *ptc1* heterozygotes also develop muscle tumors (rhabdomyosarcomas) (Hahn et al 1998, 1999), and they have an increased sensitivity to radiation-induced basal cell carcinomas of the skin (Aszterbaum et al 1999).

Role of Shh Signaling in Granule Cell Development

The association of *ptc1* mutations with cerebellar tumors suggested that Shh signaling might play a role in normal cerebellar development. Shh had been studied for its role in patterning the ventral neural tube—the floor plate, motor neurons, and ventral interneurons—but its role in the dorsal neural tube, from which the cerebellum forms, was unknown. Initial studies indicated that *shh* was expressed by Purkinje cells (Millen et al 1995, Traiffort et al 1998). Components of the Shh signaling pathway—the membrane proteins Ptc1, Ptc2, and Smo and the transcription factors Gli1 and Gli2—were found to be produced in GCPs in the EGL (Dahmane & Ruiz i Altaba 1999, Wechsler-Reya & Scott 1999). Finally, ablating Purkinje cells leads to local failure of GCP proliferation, suggesting that a Purkinje cell-derived signal is necessary for GCP proliferation (Smeyne et al 1995). Together, these findings raised the possibility that Shh from Pukinje cells might be required for GCP proliferation (Figure 4, see color insert).

The mitogenic function of Shh signaling was tested by adding Shh to dissociated cerebellar cells or to slices of cerebellum in culture (Dahmane & Ruiz i Altaba 1999, Wallace 1999, Wechsler-Reya & Scott 1999). Soluble Shh protein induced proliferation of GCPs in culture and could inhibit their differentiation and migration into the internal granule layer in cerebellar slices. Shh also promoted the differentiation of glial cells in the postnatal cerebellum (Dahmane & Ruiz i Altaba 1999). The mitogenic effects of Shh were found to be critical for granule cell development, because Shh-blocking antibodies produced in the brain during postnatal development reduced GCP proliferation and caused a thinning of the EGL (Wechsler-Reya & Scott 1999).

The fact that Shh could promote proliferation of GCPs raised another important question: How do the progeny of GCPs stop proliferating and differentiate into mature granule cells? GCPs exit the cell cycle in the middle of the EGL just as they approach the source of Shh, Purkinje cells. Thus, the signal that stops GCPs from dividing must overcome the effects of an increasing concentration of Shh. Two stimuli—bFGF and forskolin, an activator of protein kinase A— have this capacity. Each of them, added to GCPs together with Shh, can prevent the Shh-induced proliferative response (Wechsler-Reya & Scott 1999). In the developing cerebellum, bFGF is made by astrocytes and Purkinje cells (Hatten et al 1988, Matsuda et al 1994) and can stimulate granule cell differentiation and neurite extension (Hatten et al 1988, Saffell et al 1997). Pituitary adenylate cyclase-activating polypeptide (PACAP), a potent physiologic activator of protein kinase A, is made by Purkinje cells (Nielsen et al 1998, Skoglosa et al 1999). PACAP promotes granule cell survival and neurite outgrowth (Basille et al 1993, Gonzalez et al 1997). Thus, FGF and PACAP are good candidates for regulators of GCP cell cycle exit and differentiation.

Contributions of Shh/Ptc Signaling to Tumorigenesis

The mitogenic effect of Shh on GCPs is consistent with a role for *ptc1* mutations in medulloblastoma. Shh and Ptc1 are antagonists, so loss of Ptc1 is equivalent to overstimulation with Shh. The tumors can arise long after the EGL has disappeared, so either EGL-like cells persist when Ptc1 function is insufficient or a new type of rapidly dividing cell forms later. Loss of *ptc1* alone is unlikely to be sufficient to cause tumors. Among mouse *ptc1* heterozygotes, only 15%–25% develop medulloblastomas. Tumorigenesis could require loss of the second copy of *ptc1*, and that might occur only in a subset of the animals. However, recent studies of the *ptc1* heterozygous mice indicate that a wild-type (nonmutated) *ptc1* allele is expressed in the majority of tumors. Therefore, complete loss of *ptc1* function is not required for tumor formation (Wetmore et al 2000, Zurawel et al 1998). *ptc1* haploinsufficiency predisposes to medulloblastoma, but additional genetic alterations are probably necessary to produce a full-blown tumor.

Wnt Signaling in Medulloblastoma and in Cerebellar Development

Just as studies of basal cell nevus syndrome led to an appreciation of the role of Shh signaling in medulloblastoma, studies of another hereditary disease—Turcot's syndrome—suggest a role for the Wnt signaling pathway. Patients with Turcot's syndrome have a high incidence of both colorectal cancers and brain tumors, especially medulloblastomas (Hamilton et al 1995). The genetic lesion in many cases of Turcot's syndrome is a mutation in the adenomatous polyposis coli (*APC*) gene (Kadin et al 1970), which was originally identified for its role in familial and sporadic colon cancers. Subsequent studies have revealed that APC is an important element of the Wnt signaling pathway (Peifer & Polakis 2000, Wodarz & Nusse 1998).

Wnt proteins, like Hedgehogs, are secreted molecules that play a critical role in pattern formation in invertebrates as well as vertebrates. In vertebrates, they are required for cell growth and cell fate determination in the nervous system, limbs, and other organs and tissues. Wnts bind to receptors called Frizzleds and modulate the cellular levels of β-catenin (Figure 5). In the absence of Wnt signaling, β-catenin is phosphorylated and degraded by a complex of proteins that includes the serine-threonine kinase GSK3β (glycogen synthase kinase 3β), a scaffolding protein called Axin, and APC. Binding of Wnts to Frizzleds causes inactivation of GSK3β, which in turn results in reduced phosphorylation and degradation of β-catenin. The consequent release of β-catenin from the destruction complex leads to β-catenin accumulation and its translocation to the nucleus. There it interacts with transcription factors of the LEF/TCF family, and converts them from

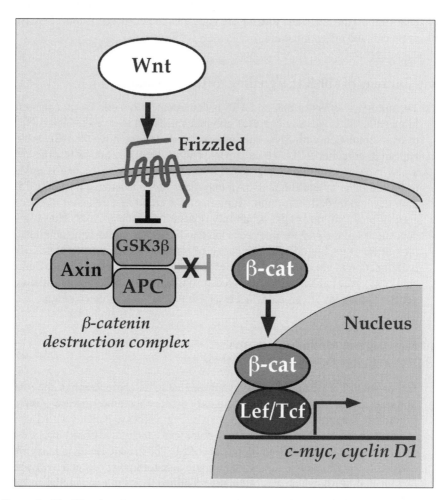

Figure 5   The Wnt signaling pathway. Secreted proteins of the Wnt family act by binding to 7-transmembrane-receptors called Frizzleds and by regulating the levels of intracellular beta-catenin (β-cat). In the absence of Wnt signal, β-cat is phosphorylated and targeted for degradation by a complex of proteins (the "destruction complex") that includes the serine-threonine kinase Glycogen synthase kinase 3β (GSK3β), the scaffolding protein Axin, and the tumor suppressor protein adenomatous polypsis coli (APC). Binding of Wnts to Frizzleds inactivates GSK3β, reducing the phosphorylation and degradation of β-cat and allowing it to accumulate in the cell. As β-cat builds up, it translocates to the nucleus, where it binds to transcription factors of the Lef/Tcf family and converts them from transcriptional repressors into activators. β-cat-Lef/TCF complexes induce expression of target genes such as c-myc and cyclin D1, which promote cell cycle progression. Mutations that inactivate APC or Axin, or render β-cat resistant to degradation, are associated with a wide variety of different tumor types.

transcriptional repressors to transcriptional activators. Activated LEF/TCF transcription factors induce transcription of *c-myc* and *cyclin D1*, among other genes. C-myc and cyclin D1 promote cell cycle progression and thereby contribute to tumorigenesis.

Mutations in different Wnt pathway components have been implicated in a wide variety of cancers. The founding member of the Wnt family, Wnt-1, was originally identified as an oncogene activated by the mouse mammary tumor virus in murine breast cancer (Nusse & Varmus 1982). Deletion or inactivation of APC, which normally promotes degradation of β-catenin, is associated with colon cancer. Inactivation of Axin, another component of the β-catenin destruction complex, has been observed in hepatocellular carcinoma (de La Coste et al 1998). β-catenin mutations that render the protein resistant to phosphorylation or degradation have been identified in colon, prostate, and ovarian cancers (Polakis 1999).

The association of Wnt signaling with medulloblastoma was first suggested by the identification of APC mutations in Turcot's syndrome. Since then, a number of investigators have looked for APC mutations in sporadic medulloblastomas and in most cases have not found any (Yong et al 1995). However, a recent study identified APC mutations in 4% of the sporadic medulloblastomas examined (Huang et al 2000). Activating β-catenin mutations have been found in 8%–15% of sporadic medulloblastomas (Eberhart et al 2000, Huang et al 2000, Zurawel et al 1998).

By analogy with the Shh/Ptc signaling pathway, Wnt signaling might well have important functions in cerebellar growth or differentiation. Wnt-1 is required during early embryogenesis for specification of the midbrain/hindbrain boundary (isthmus) from which the cerebellum arises (McMahon & Bradley 1990, Thomas et al 1991). Wnt-1 mutant mice have serious defects in the midbrain and little or no discernable cerebellum. Wnt-7a has been implicated in axon branching and synapse formation in granule cells and mossy fibers (Lucas et al 1998). These early events may or may not be related to the formation of tumors much later. On the other hand, lithium chloride, which can inhibit GSK3β and thereby mimic Wnt signaling, promotes proliferation of cultured granule cell precursors (Cui et al 1998). If this effect is indicative of a physiologic function of Wnt signaling, it might provide an explanation for the association of APC and β-catenin mutations with medulloblastoma.

Chromosome 17 Deletions

One of the most important genes involved in medulloblastoma has yet to be identified. Of human medulloblastomas, 30%–50% have a deletion or rearrangement of part of chromosome 17 (Bigner et al 1997, Burnett et al 1997, Cogen & McDonald 1996). In most cases the short arm (17p) is lost, and head-to-head apposition of the long arms (17q) occurs, which is referred to as isochromosome 17q [i(17q)]. This rearrangement is frequently detected in leukemias, lymphomas, and cancers of the stomach, colon, and cervix. The loss of 17p in a number of types of cancer suggests that at least one potent tumor suppressor gene is located there. Fine mapping of deletions from different tumors has narrowed the region of interest considerably. Most investigators now believe the putative tumor suppressor is located at 17p13.3, a region of ~20 known genes including those encoding the lissencephaly-associated protein Lis1, the breakpoint cluster region (BCR)-related gene *ABR*, and the Max-binding protein Mnt. Despite considerable effort, none of these genes has been clearly implicated in the etiology of medulloblastoma. When the chromosome 17 tumor suppressor is identified, it is likely to provide important insights into the basis of cerebellar tumors as well as into normal cerebellar development.

CONCLUSIONS

Consistent themes emerge from the developmental biology and genetics of brain tumors.

1. Brain tumor cells bear some resemblance to certain normal cell types, often rapidly dividing cell types, but there is little reason to think that the tumor cells are exactly the same as any normal cell type. Ideas about "cell type of origin" are especially relevant to defining when and where the tumors arise. The properties of the normal cells that are most similar to the tumor cells may also be informative for devising therapeutic interventions. Much more needs to be learned about the extent to which the earliest tumor cells resemble particular types of normal cells. The normal function of Rb may be to control proliferation of neuronal precursors and promote differentiation. In the human retina, Rb family members do not adequately replace lost Rb function, and some type of neuronal precursor divides when it should not. EGF signaling is critical for normal astrocyte development. Glioblastomas often have constitutively active EGF receptors, so astrocyte precursor cells that are potentially responsive to EGFR ligands are inappropriately entering or remaining in the cell cycle. Ptc1 normally restricts growth of cerebellar granule cells; for them to grow, an Shh signal from Purkinje cells blocks the action of Ptc1. Reduced Ptc1 function allows excess growth. In medulloblastoma as in glioblastoma and retinoblastoma, the normal developmental regulation systems are informative about tumorigenesis.

2. Multiple genetic events are usually necessary for tumorigenesis. Prospective tumor cells must remain in the cell cycle beyond the appropriate time (or reenter the cycle). Differentiation into a nondividing cell type must be blocked, and apoptosis must be avoided, either by specific inactivation of the apoptotic machinery or by attaining a differentiation state that is resistant to programmed cell death. In retinoblastoma, loss of Rb function promotes cell cycle progression, but may also predispose cells to apoptosis; other genetic changes may be necessary for retinal tumors to occur. In glioblastoma, mutations that cause aberrant proliferation (e.g. CDK4 amplication, CDKN2A/B deletion) are invariably accompanied by genetic events that prevent differentiation and block apoptosis (e.g. *EGFR*

amplification, *PTEN* mutation, loss of *p53*). Although medulloblastoma occurs in people and mice that are heterozygous for mutations in *patched1*, other factors clearly contribute. For example, the chromosome 17 gene that is a common correlate with the tumor has yet to be identified.

3. The effect of a particular regulator (or its absence) is highly dependent on the cell type or stage of development. During normal development, cells that are powerfully affected by a regulator at one moment may ignore it at another time. That particular mutations transform only cells with certain properties seems likely to reflect normal limitations on responses to regulators. Rb deficiency may cause inappropriate growth of cells that are in a differentiation state with limited ability to die, even in the absence of Rb. Understanding what makes cells receptive to signals or mutations, such as the history of the cell or a convergence of regulators, will be critical for understanding how particular genetic changes lead to tumors.

Viewing brain tumors as an aberration of normal development has helped to put the events of tumorigenesis in proper context. Studies of developmental biology have enlarged our knowledge of the pathways that are affected by tumorigenic mutations and have led to the recognition of new oncogenes and tumor suppressors. The construction of better mouse models of human disease, guided by human genetics and knowledge of developmental pathways, has led to promising new ways to search for and test therapies. Deeper knowledge of the relationship between tumor cells and normal cells will help guide us to still hidden susceptibilities of tumor cells.

ACKNOWLEDGMENTS

We thank Drs. S McConnell and B Barres for their helpful criticism. We thank L Milenkovic for the images in Figure 4. R Wechsler-Reya is an Associate, and MPS an Investigator, of the Howard Hughes Medical Institute.

Visit the Annual Reviews home page at www.AnnualReviews.org

LITERATURE CITED

Adachi J, Ohbayashi K, Suzuki T, Sasaki T. 1999. Cell cycle arrest and astrocytic differentiation resulting from PTEN expression in glioma cells. *J. Neurosurg.* 91:822–30

Aizawa T, Maruyama K, Kondo H, Yoshikawa K. 1992. Expression of necdin, an embryonal carcinoma-derived nuclear protein, in developing mouse brain. *Brain Res. Dev. Brain Res.* 68:265–74

Ali IU, Schriml LM, Dean M. 1999. Mutational spectra of PETEN/MMAC1 gene: a tumor suppressor with lipid phosphatase activity. *J. Natl. Cancer Inst.* 91:1922–32

Altman J. 1966. Proliferation and migration of undifferentiated precursor cells in the rat during postnatal gliogenesis. *Exp. Neurol.* 16:263–78

Altman J, Bayer SA. 1997. *Development of the Cerebellar System: In Relation to Its*

Evolution, Structure and Functions. Boca Raton, FL: CRC Press. 783 pp.

al-Ubaidi MR, Font RL, Quiambao AB, Keener MJ, Liou GI, et al. 1992. Bilateral retinal and brain tumors in transgenic mice expressing simian virus 40 large T antigen under control of the human interphotoreceptor retinoid-binding protein promoter. *J. Cell Biol.* 119:1681–87

Araque A, Sanzgiri RP, Parpura V, Haydon PG. 1999. Astrocyte-induced modulation of synaptic transmission. *Can. J. Physiol. Pharmacol.* 77:699–706

Arenander A, de Vellis J. 1992. Early response gene induction in astrocytes as a mechanism for encoding and integrating neuronal signals. *Prog. Brain Res.* 94:177–88

Aschner M. 1998. Astrocytes as mediators of immune and inflammatory responses in the CNS. *Neurotoxicology* 19:269–81

Aszterbaum M, Epstein J, Oro A, Douglas V, LeBoit PE, et al. 1999. Ultraviolet and ionizing radiation enhance the growth of BCCs and trichoblastomas in patched heterozygous knockout mice. *Nat. Med.* 5:1285–91

Athanasiou MC, Yunis W, Coleman N, Ehlenfeldt R, Clark HB, et al. 1998. The transcription factor E2F-1 in SV40 T antigen-induced cerebellar Purkinje cell degeneration. *Mol. Cell. Neurosci.* 12:16–28

Austin CP, Feldman DE, Ida JA Jr, Cepko CL. 1995. Vertebrate retinal ganglion cells are selected from competent progenitors by the action of Notch. *Development* 121:3637–50

Bacci A, Verderio C, Pravettoni E, Matteoli M. 1999. The role of glial cells in synaptic function. *Philos. Trans. R. Soc. London Ser. B Biol. Sci.* 354:403–9

Basille M, Gonzalez BJ, Leroux P, Jeandel L, Fournier A, Vaudry H. 1993. Localization and characterization of PACAP receptors in the rat cerebellum during development: evidence for a stimulatory effect of PACAP on immature cerebellar granule cells. *Neuroscience* 57:329–38

Bates S, Phillips AC, Clark PA, Stott F, Peters G, et al. 1998. p14ARF links the tumour suppressors RB and p53. *Nature* 395:124–25

Benedict WF, Banerjee A, Mark C, Murphree AL. 1983. Nonrandom chromosomal changes in untreated retinoblastomas. *Cancer Genet. Cytogenet.* 10:311–33

Ben-Hur T, Rogister B, Murray K, Rougon G, Dubois-Dalcq M. 1998. Growth and fate of PSA-NCAM+ precursors of the postnatal brain. *J. Neurosci.* 18:5777–88

Bentivoglio M, Mazzarello P. 1999. The history of radial glia. *Brain Res. Bull.* 49:305–15

Besson A, Robbins SM, Yong VW. 1999. PTEN/MMAC1/TEP1 in signal transduction and tumorigenesis. *Eur. J. Biochem.* 263:605–11

Biernat W, Debiec-Rychter M, Liberski PP. 1998. Mutations of TP53, amplification of EGFR, MDM2 and CDK4, and deletions of CDKN2A in malignant astrocytomas. *Pol. J. Pathol.* 49:267–71

Bigner SH, McLendon RE, Fuchs H, McKeever PE, Friedman HS. 1997. Chromosomal characteristics of childhood brain tumors. *Cancer Genet. Cytogenet.* 97:125–34

Bonni A, Sun Y, Nadal-Vicens M, Bhatt A, Frank DA, et al. 1997. Regulation of gliogenesis in the central nervous system by the JAK-STAT signaling pathway. *Science* 278:477–83

Brodeur GM. 1995. Genetics of embryonal tumours of childhood: retinoblastoma, Wilms' tumour and neuroblastoma. *Cancer Surv.* 25:67–99

Brunet A, Bonni A, Zigmond MJ, Lin MZ, Juo P, et al. 1999. Akt promotes cell survival by phosphorylating and inhibiting a Forkhead transcription factor. *Cell* 96:857–68

Burmeister M, Novak J, Liang MY, Basu S, Ploder L, et al. 1996. Ocular retardation mouse caused by Chx10 homeobox null allele: impaired retinal progenitor proliferation and bipolar cell differentiation. *Nat. Genet.* 12:376–84

Burnett ME, White EC, Sih S, von Haken MS, Cogen PH. 1997. Chromosome arm 17p deletion analysis reveals molecular genetic heterogeneity in supratentorial and infratentorial primitive neuroectodermal tumors of the central nervous system. *Cancer Genet. Cytogenet.* 97:25–31

Burrows RC, Lillien L, Levitt P. 2000. Mechanisms of progenitor maturation are conserved in the striatum and cortex. *Dev. Neurosci.* 22:7–15

Buschges R, Weber RG, Actor B, Lichter P, Collins VP, Reifenberger G. 1999. Amplification and expression of cyclin D genes (CCND1, CCND2 and CCND3) in human malignant gliomas. *Brain Pathol.* 9:435–42

Callaghan DA, Dong L, Callaghan SM, Hou YX, Dagnino L, Slack RS. 1999. Neural precursor cells differentiating in the absence of Rb exhibit delayed terminal mitosis and deregulated E2F 1 and 3 activity. *Dev. Biol.* 207:257–70

Cantley LC, Neel BG. 1999. New insights into tumor suppression: PTEN suppresses tumor formation by restraining the phosphoinositide 3-kinase/AKT pathway. *Proc. Natl. Acad. Sci. USA* 96:4240–45

Cardone MH, Roy N, Stennicke HR, Salvesen GS, Franke TF, et al. 1998. Regulation of cell death protease caspase-9 by phosphorylation. *Science* 282:1318–21

Cepko CL. 1999. The roles of intrinsic and extrinsic cues and bHLH genes in the determination of retinal cell fates. *Curr. Opin. Neurobiol.* 9:37–46

Chen Y, Struhl G. 1998. In vivo evidence that patched and smoothened constitute distinct binding and transducing components of a Hedgehog receptor complex. *Development* 125:4943–48

Cheney IW, Neuteboom ST, Vaillancourt MT Ramachandra M, Bookstein R. 1999. Adenovirus-mediated gene transfer of MMAC1/PTEN to glioblastoma cells inhibits S phase entry by the recruitment of p27Kip1 into cyclin E/CDK2 complexes. *Cancer Res.* 59:2318–23

Chernova OB, Somerville RP, Cowell JK. 1998. A novel gene, LGI1, from 10q24 is rearranged and downregulated in malignant brain tumors. *Oncogene* 17:2873–81

Chin L, Pomerantz J, DePinho RA. 1998. The INK4a/ARF tumor suppressor: one gene—two products—two pathways. *Trends Biochem. Sci.* 23:291–96

Clarke AR, Maandag ER, van Roon M, van der Lugt NM, van der Valk M, et al. 1992. Requirement for a functional Rb-1 gene in murine development. *Nature* 359:328–30

Cogen PH, McDonald JD. 1996. Tumor suppressor genes and medulloblastoma. *J. Neuro-Oncol.* 29:103–12

Collins VP. 1998. Gliomas. *Cancer Surv.* 32:37–51

Cui H, Meng Y, Bulleit RF. 1998. Inhibition of glycogen synthase kinase 3-beta activity regulates proliferation of cultured cerebellar granule cells. *Brain Res. Dev. Brain Res.* 111:177–88.

D'Alessandro JS, Yetz-Aldape J, Wang EA. 1994. Bone morphogenetic proteins induce differentiation in astrocyte lineage cells. *Growth Factors* 11:53–69

Dagnino L, Fry CJ, Bartley SM, Farnham P, Gallie BL, Phillips RA. 1997. Expression patterns of the E2F family of transcription factors during mouse nervous system development. *Mech. Dev.* 66:13–25

Dahmane N, Ruiz i Altaba A. 1999. Sonic hedgehog regulates the growth and patterning of the cerebellum. *Development* 126:3089–100

Dai P, Akimaru H, Tanaka Y, Maekawa T, Nakafuku M, Ishii S. 1999. Sonic hedgehog-induced activation of the Gli1 promoter is mediated by GLI3. *J. Biol. Chem.* 274:8143–52

Datta SR, Dudek H, Tao X, Masters S, Fu H, et al. 1997. Akt phosphorylation of BAD couples survival signals to the cell-intrinsic death machinery. *Cell* 91:231–41

Davies MA, Lu Y, Sano T, Fang X, Tang P, et al. 1998. Adenoviral transgene expression of MMAC/PTEN in human glioma cells inhibits Akt activation and induces anoikis. *Cancer Res.* 58:5285–90. Erratum. 1999. *Cancer Res.* 59(5):1167

Davis AA, Temple S. 1994. A self-renewing

multipotential stem cell in embryonic rat cerebral cortex. *Nature* 372:263–66

DeCaprio JA, Ludlow JW, Figge J, Shew JY, Huang CM, et al. 1988. SV40 large tumor antigen forms a specific complex with the product of the retinoblastoma susceptibility gene. *Cell* 54:275–83

DeGregori J, Leone G, Miron A, Jakoi L, Nevins JR. 1997. Distinct roles for E2F proteins in cell growth control and apoptosis. *Proc. Natl. Acad. Sci. USA* 94:7245–50

de La Coste A, Romagnolo B, Billuart P, Renard CA, Buendia MA, et al. 1998. Somatic mutations of the beta-catenin gene are frequent in mouse and human hepatocellular carcinomas. *Proc. Natl. Acad. Sci. USA* 95:8847–51

del Peso L, Gonzalez-Garcia M, Page C, Herrera R, Nunez G. 1997. Interleukin-3-induced phosphorylation of BAD through the protein kinase Akt. *Science* 278:687–89

Di Cristofano A, Kotsi P, Peng YF, Cordon-Cardo C, Elkon KB, Pandolfi PP. 1999. Impaired Fas response and autoimmunity in Pten+/− mice. *Science* 285:2122–25

Di Cristofano A, Pandolfi PP. 2000. The multiple roles of PTEN in tumor suppression. *Cell* 100:387–90

Di Cristofano A, Pesce B, Cordon-Cardo C, Pandolfi PP. 1998. Pten is essential for embryonic development and tumour suppression. *Nat. Genet.* 19:348–55

Dobashi Y, Kudoh T, Matsumine A, Toyoshima K, Akiyama T. 1995. Constitutive overexpression of CDK2 inhibits neuronal differentiation of rat pheochromocytoma PC12 cells. *J. Biol. Chem.* 270:23031–37

Doostzadeh-Cizeron J, Evans R, Yin S, Goodrich DW. 1999. Apoptosis induced by the nuclear death domain protein p84N5 is inhibited by association with Rb protein. *Mol. Biol. Cell* 10:3251–61

Downward J. 1998. Mechanisms and consequences of activation of protein kinase B/Akt. *Curr. Opin. Cell Biol.* 10:262–67

Dryja TP, Friend S, Weinberg RA. 1986. Genetic sequences that predispose to retinoblastoma and osteosarcoma.

Symp. Fundam. Cancer Res. 39:115–19

Dyson N, Howley PM, Munger K, Harlow E. 1989. The human papilloma virus-16 E7 oncoprotein is able to bind to the retinoblastoma gene product. *Science* 243:934–47

Eberhart CG, Tihan T, Burger PC. 2000. Nuclear localization and mutation of beta-catenin in medulloblastomas. *J. Neuropathol. Exp. Neurol.* 59:333–37

Ekstrand AJ, James CD, Cevenee WK, Seliger B, Pettersson RF, Collins VP. 1991. Genes for epidermal growth factor receptor, transforming growth factor alpha, and epidermal growth factor and their expression in human gliomas in vivo. *Cancer Res.* 51:2164–72

Ekstrand AJ, Longo N, Hamid ML, Olson JJ, Liu L, et al. 1994. Functional characterization of an EGF receptor with a truncated extracellular domain expressed in glioblastomas with EGFR gene amplification. *Oncogene* 9:2313–20

Ekstrand AJ, Sugawa N, James CD, Collins VP. 1992. Amplified and rearranged epidermal growth factor receptor genes in human glioblastomas reveal deletions of sequences encoding portions of the N- and/or C-terminal tails. *Proc. Natl. Acad. Sci. USA* 89:4309–13

el-Deiry WS. 1998. Regulation of p53 downstream genes. *Semin. Cancer Biol.* 8:345–57

Emre S, Sungur A, Bilgic S, Buyukpamukcu M, Gunalp I, Ozguc M. 1996. Loss of heterozygosity in the VNTR region of intron 1 of P53 in two retinoblastoma cases. *Pediatr. Hematol. Oncol.* 13:253–56

Faber-Elman A, Solomon A, Abraham JA, Marikovsky M, Schwartz M. 1996. Involvement of wound-associated factors in rat brain astrocyte migratory response to axonal injury: in vitro simulation. *J. Clin. Invest.* 97:162–71

Feddersen RM, Clark HB, Yunis WS, Orr HT. 1995. In vivo viability of postmitotic Purkinje neurons requires pRb family member function. *Mol. Cell. Neurosci.* 6:153–67

Feddersen RM, Ehlenfeldt R, Yunis WS, Clark HB, Orr HT. 1992. Disrupted cerebellar cortical development and progressive degeneration of Purkinje cells in SV40 T antigen transgenic mice. *Neuron* 9:955–66

Feddersen RM, Yunis WS, O'Donnell MA, Ebner TJ, Shen L, et al. 1997. Susceptibility to cell death induced by mutant SV40 T-antigen correlates with Purkinje neuron functional development. *Mol. Cell. Neurosci.* 9:42–62

Field SJ, Tsai FY, Kuo F, Zubiaga AM, Kaelin WG Jr, et al. 1996. E2F-1 functions in mice to promote apoptosis and suppress proliferation. *Cell* 85:549–61

Foley KP, Eisenman RN. 1999. Two MAD tails: what the recent knockouts of Mad1 and Mxi1 tell us about the MYC/MAX/MAD network. *Biochim. Biophys. Acta* 1423:M37–47

Friend SH, Bernards R, Rogelj S, Weinberg RA, Rapaport JM, et al. 1986. A human DNA segment with properties of the gene that predisposes to retinoblastoma and osteosarcoma. *Nature* 323:643–46

Fueyo J, Gomez-Manzano C, Yung WK, Kyritsis AP. 1999. Targeting in gene therapy for gliomas. *Arch. Neurol.* 56:445–48

Fueyo J, Gomez-Manzano C, Yung WK, Liu TJ, Alemany R, et al. 1998. Suppression of human glioma growth by adenovirus-mediated Rb gene transfer. *Neurology* 50:1307–15

Fults D, Pedone CA, Thompson GE, Uchiyama CM, Gumpper KL, et al. 1998. Microsatellite deletion mapping on chromosome 10q and mutation analysis of MMAC1, FAS, and MXI1 in human glioblastoma multiforme. *Int. J. Oncol.* 12:905–10

Fung YK, Murphree AL, T'Ang A, Qian J, Hinrichs SH, Benedict WF. 1987. Structural evidence for the authenticity of the human retinoblastoma gene. *Science* 236:1657–61

Furnari FB, Huang HJ, Cavenee WK. 1998. The phosphoinositol phosphatase activity of PTEN mediates a serum-sensitive G1 growth arrest in glioma cells. *Cancer Res.* 58:5002–8

Furnari FB, Lin H, Huang HS, Cavenee WK. 1997. Growth suppression of glioma cells by PTEN requires a functional phosphatase catalytic domain.

Proc. Natl. Acad. Sci. USA 94:12479–84

Gill RM, Slack R, Kiess M, Hamel PA. 1998. Regulation of expression and activity of distinct pRB, E2F, D-type cyclin, and CKI family members during terminal differentiation of P19 cells. *Exp. Cell Res.* 244:157–70

Giordano A, Avantaggiati ML. 1999. p300 and CBP: partners for life and death. *J. Cell. Physiol.* 181:218–30

Glaser T, Walton DS, Maas RL. 1992. Genomic structure, evolutionary conservation and aniridia mutations in the human PAX6 gene. *Nat. Genet.* 2:232–39

Gomez-Manzano C, Fueyo J, Alameda F, Kyritsis AP, Yung WK. 1999. Gene therapy for gliomas: p53 and E2F-1 proteins and the target of apoptosis. *Int. J. Mol. Med.* 3:81–85

Gonzalez BJ, Basille M, Vaudry D, Fournier A, Vaudry H. 1997. Pituitary adenylate cyclase-activating polypeptide promotes cell survival and neurite outgrowth in rat cerebellar neuroblasts. *Neuroscience* 78:419–30

Goodrich LV, Milenkovic L, Higgins KM, Scott MP. 1997. Altered neural cell fates and medulloblastoma in mouse patched mutants. *Science* 277:1109–13

Goodrich LV, Scott MP. 1998. Hedgehog and patched in neural development and disease. *Neuron* 21:1243–57

Gorlin RJ. 1987. Nevoid basal-cell carcinoma syndrome. *Medicine* 66:98–113

Griep AE, Krawcek J, Lee D, Liem A, Albert DM, et al. 1998. Multiple genetic loci modify risk for retinoblastoma in transgenic mice. *Invest. Ophthalmol. Vis. Sci.* 39:2723–32

Gross RE, Mehler MF, Mabie PC, Zang Z, Santschi L, Kessler JA. 1996. Bone morphogenetic proteins promote astroglial lineage commitment by mammalian subventricular zone progenitor cells. *Neuron* 17:595–606

Gu J, Tamura M, Pankov R, Danen EH, Takino T, et al. 1999. Shc and FAK differentially regulate cell motility and directionality modulated by PTEN. *J. Cell Biol.* 146:389–403

Haas-Kogan D, Shalev N, Wong M, Mills G, Yount G, Stokoe D. 1998. Protein kinase B (PKB/Akt) activity is elevated in glioblastoma cells due to mutation of the tumor suppressor PTEN/MMAC. *Curr. Biol.* 8:1195–98

Hahn H, Wicking C, Zaphiropoulous PG, Gailani MR, Shanley S, et al. 1996. Mutations of the human homolog of Drosophila patched in the nevoid basal cell carcinoma syndrome. *Cell* 85:841–51

Hahn H, Wojnowski L, Miller G, Zimmer A. 1999. The patched signaling pathway in tumorigenesis and development: lessons from animal models. *J. Mol. Med.* 77:459–68

Hahn H, Wojnowski L, Zimmer AM, Hall J, Miller G, Zimmer A. 1998. Rhabdomyosarcomas and radiation hypersensitivity in a mouse model of Gorlin syndrome. *Nat. Med.* 4:619–22

Halatsch ME, Schmidt U, Botefur IC, Holland JF, Ohnuma T. 2000. Marked inhibition of glioblastoma target cell tumorigenicity in vitro by retrovirus-mediated transfer of a hairpin ribozyme against deletion-mutant epidermal growth factor receptor messenger RNA. *J. Neurosurg.* 92:297–305

Hamel PA, Phillips RA, Muncaster M, Gallie BL. 1993. Speculations on the roles of RB1 in tissue-specific differentiation, tumor initiation, and tumor progression. *FASEB J.* 7:846–54

Hamilton SR, Liu B, Parsons RE, Papadopoulos N, Jen J, et al. 1995. The molecular basis of Turcot's syndrome. *N Engl. J. Med.* 332:839–47

Hammerschmidt M, Brook A, McMohan AP. 1997. The world according to Hedgehog. *Trends Genet.* 13:14–21

Hatten ME, Lynch M, Rydel RE, Sanchez J, Joseph- Silverstein J, et al. 1988. In vitro neurite extension by granule neurons is dependent upon astroglial-derived fibroblast growth factor. *Dev. Biol.* 125:280–89

Hatten ME, Mason CA. 1990. Mechanisms of glial-guided neuronal migration in vitro and in vivo. *Experientia* 46:907–16

Hayashi Y, Matsuyama K, Takagi K, Sugiura H, Yoshikawa K. 1995. Arrest of cell growth by necdin, a nuclear protein expressed in postmitotic neurons. *Biochem. Biophys. Res. Commun.* 213:317–24

Hoi Sang U, Espiritu OD, Kelley PY, Klauber MR, Hatton JD. 1995a. The role of the epidermal growth factor receptor in human gliomas: I. The control of cell growth. *J. Neurosurg.* 82:841–46

Hoi Sang U, Espiritu OD, Kelley PY, Klauber MR, Hatton JD. 1995b. The role of the epidermal growth factor receptor in human gliomas: II. The control of glial process extension and the expression of glial fibrillary acidic protein. *J. Neurosurg.* 82:847–57

Holland EC, Hively WP, DePinho RA, Varmus HE. 1998. A constitutively active epidermal growth factor receptor cooperates with disruption of G1 cell-cycle arrest pathways to induce glioma-like lesions in mice. *Genes Dev.* 12:3675–85

Holt CE, Bertsch TW, Ellis HM, Harris WA. 1988. Cellular determination in the Xenopus retina is independent of lineage and birth date. *Neuron* 1:15–26

Howes KA, Ransom N, Papermaster DS, Lasudry JG, Albert DM, Windle JJ. 1994. Apoptosis or retinoblastoma: alternative fates of photoreceptors expressing the HPV-16 E7 gene in the presence or absence of p53. *Genes Dev.* 8:1300–10. Erratum. 1994. *Genes Dev.* 8(14):1738

Hsieh JK, Fredersdorf S, Kouzarides T, Martin K, Lu X. 1997. E2F1-induced apoptosis requires DNA binding but not transactivation and is inhibited by the retinoblastoma protein through direct interaction. *Genes Dev.* 11:1840–52

Huang H, Mahler-Araujo BM, Sankila A, Chimelli L, Yonekawa Y, et al. 2000. APC mutations in sporadic medulloblastomas. *Am. J. Pathol.* 156:433–37

Hughes SM, Lillien LE, Raff MC, Rohrer H, Sendtner M. 1988. Ciliary neurotrophic factor induces type-2 astrocyte differentiation in culture. *Nature* 335:70–73

Ichimura K, Bolin MB, Goike HM, Schmidt EE, Moshref A, Collins VP. 2000. Deregulation of the p14ARF/MDM2/p53 pathway is a prerequisite for human astrocytic gliomas with G1-S transition control gene abnormalities. *Cancer Res.* 60:417–24

Ichimura K, Schmidt EE, Goike HM, Collins VP. 1996. Human glioblastomas with no alterations of the CDKN2A (p16INK4A, MTS1) and CDK4 genes have frequent mutations of the retinoblastoma gene. *Oncogene* 13:1065–72

Jacks T, Fazeli A, Schmitt EM, Bronson RT, Goodell MA, Weinberg RA. 1992. Effects of an Rb mutation in the mouse. *Nature* 359:295–300

Jiang Z, Zacksenhaus E, Gallie BL, Phillips RA. 1997. The retinoblastoma gene family is differentially expressed during embryogenesis. *Oncogene* 14:1789–97

Johnson RL, Rothman AL, Xie J, Goodrich LV, Bare JW, et al. 1996. Human homolog of patched, a candidate gene for the basal cell nevus syndrome. *Science* 272:1668–71

Kadin ME, Rubinstein LJ, Nelson JS. 1970. Neonatal cerebellar medulloblastoma originating from the fetal external granular layer. *J. Neuropathol. Exp. Neurol.* 29:583–600

Kalyani A, Hobson K, Rao MS. 1997. Neuroepithelial stem cells from the embryonic spinal cord: isolation, characterization, and clonal analysis. *Dev. Biol.* 186:202–23

Kamijo T, Weber JD, Zambetti G, Zindy F, Roussel MF, Sherr CJ. 1998. Functional and physical interactions of the ARF tumor suppressor with p53 and Mdm2. *Proc. Natl. Acad. Sci. USA* 95:8292–97

Kastner A, Espanel X, Brun G. 1998. Transient accumulation of retinoblastoma/E2F-1 protein complexes correlates with the onset of neuronal differentiation in the developing quail neural retina. *Cell Growth Differ.* 9:857–67

Kato MV, Shimizu T, Ishizaki K, Kaneko A, Yandell DW, et al. 1996. Loss of heterozygosity on chromosome 17 and mutation of the p53 gene in retinoblastoma. *Cancer Lett.* 106:75–82

Kawabata M, Imamura T, Miyazono K. 1998. Signal transduction by bone morphogenetic proteins. *Cytokine Growth Factor Rev.* 9:49–61

Khazaie K, Schirrmacher V, Lichtner RB. 1993. EGF receptor in neoplasia and metastasis. *Cancer Metastasis Rev.* 12:255–74

Kim RA, Lim J, Ota S, Raja S, Rogers R, et al. 1998. NRP/B, a novel nuclear matrix protein, associates with p110 (RB) and is involved in neuronal differentiation. *J. Cell Biol.* 141:553–66

Kimura H, Fischer WH, Schubert D. 1990. Structure, expression and function of a schwannoma-derived growth factor. *Nature* 348:257–60

Knudson AG Jr. 1971. Mutation and cancer: statistical study of retinoblastoma. *Proc. Natl. Acad. Sci. USA* 68:820–23

Knudson AG Jr, Hethcote HW, Brown BW. 1975. Mutation and childhood cancer: a probabilistic model for the incidence of retinoblastoma. *Proc. Natl. Acad. Sci. USA* 72:5116–20

Koblar SA, Turnley AM, Classon BJ, Reid KL, Ware CB, et al. 1998. Neural precursor differentiation into astrocytes requires signaling through the leukemia inhibitory factor receptor. *Proc. Natl. Acad. Sci. USA* 95:3178–81

Komurasaki T, Toyoda H, Uchida D, Morimoto S. 1997. Epiregulin binds to epidermal growth factor receptor and ErbB-4 and induces tyrosine phosphorylation of epidermal growth factor receptor, ErbB-2, ErbB-3 and ErbB-4. *Oncogene* 15:2841–48

Komuro H, Rakic P. 1998. Distinct modes of neuronal migration in different domains of developing cerebellar cortex. *J. Neurosci.* 18:1478–90

Kops GJ, de Ruiter ND, De Vries-Smits AM, Powell DR, Bos JL, Burgering BM. 1999. Direct control of the Forkhead transcription factor AFX by protein kinase B. *Nature* 398:630–4.

Kornblum HI, Hussain R, Wiesen J, Miettinen P, Zurcher SD, et al. 1998. Abnormal astrocyte development and neuronal death in mice lacking the epidermal growth factor receptor. *J. Neurosci. Res.* 53:697–717

Kornblum HI, Zurcher SD, Werb Z, Derynck R, Seroogy KB. 1999. Multiple trophic actions of heparin-binding epidermal growth factor (HB-EGF) in the central nervous system. *Eur. J. Neurosci.* 11:3236–46

Kowalik TF, DeGregori J, Leone G, Jakoi L, Nevins JR. 1998. E2F1-specific induction of apoptosis and p53 accumulation, which is blocked by Mdm2. *Cell Growth Differ.* 9:113–18

Kozmik Z, Sure U, Ruedi D, Busslinger M, Aguzzi A. 1995. Deregulated expression of PAX5 in medulloblastoma. *Proc. Natl. Acad. Sci. USA* 92:5709–13

Kranenburg O, de Groot RP, Van der Eb AJ, Zantema A. 1995a. Differentiation of P19 EC cells leads to differential modulation of cyclin-dependent kinase activities and to changes in the cell cycle profile. *Oncogene* 10:87–95

Kranenburg O, Scharnhorst V, Van der Eb AJ, Zantema A. 1995b. Inhibition of cyclin-dependent kinase activity triggers neuronal differentiation of mouse neuroblastoma cells. *J. Cell Biol.* 131:227–34

Kusnetsova LE, Prigogina EL, Pogosianz HE, Belkina BM. 1982. Similar chromosomal abnormalities in several retinoblastomas. *Hum. Genet.* 61:201–4

Kyritsis AP, Tsokos M, Triche TJ, Chader GJ. 1984. Retinoblastoma—origin from a primitive neuroectodermal cell? *Nature* 307:471–73

Kyritsis AP, Tsokos M, Triche TJ, Chader GJ. 1986. Retinoblastoma: a primitive tumor with multipotential characteristics. *Invest. Ophthalmol. Vis. Sci.* 27:1760–64

Lang FF, Yung WK, Sawaya R, Tofilon PJ. 1999. Adenovirus-mediated p53 gene therapy for human gliomas. *Neurosurgery* 45:1093–104

LeCouter JE, Kablar B, Whyte PF, Ying C, Rudnicki MA. 1998. Strain-dependent embryonic lethality in mice lacking the retinoblastoma-related p130 gene. *Development* 125:4669–79

Lee EY, Chang CY, Hu N, Wang YC, Lai CC, et al. 1992. Mice deficient for Rb are non-viable and show defects in neurogenesis and haematopoiesis. *Nature* 359:288–94

Lee EY, Hu N, Yuan SS, Cox LA, Bradley A, et al. 1994. Dual roles of the retinoblastoma protein in cell cycle regulation and neuron differentiation. *Genes Dev.* 8:2008–21

Lee JC, Mayer-Proschel M, Rao MS. 2000. Gliogenesis in the central nervous system. *Glia* 30:105–21

Lee KY, Helbing CC, Choi KS, Johnston RN, Wang JH. 1997a. Neuronal Cdc2-like kinase (Nclk) binds and phosphorylates the retinoblastoma protein. *J. Biol. Chem.* 272:5622–26

Lee KY, Qi Z, Yu YP, Wang JH. 1997b. Neuronal Cdc2-like kinases: neuron-specific forms of Cdk5. *Int. J. Biochem. Cell Biol.* 29:951–58

Lee MH, Williams BO, Mulligan G, Mukai S, Bronson RT, et al. 1996. Targeted disruption of p107: functional overlap between p107 and Rb. *Genes Dev.* 10:1621–32

Lee WH, Bookstein R, Hong F, Young LJ, Shew JY, Lee EY. 1987. Human retinoblastoma susceptibility gene: cloning, identification, and sequence. *Science* 235:1394–99

Leiner HC, Leiner AL, Dow RS. 1993. Cognitive and language functions of the human cerebellum. *Trends Neurosci.* 16:444–47

Leutz A, Schachner M. 1981. Epidermal growth factor stimulates DNA-synthesis of astrocytes in primary cerebellar cultures. *Cell Tissue Res.* 220:393–404

Levine E-M, Fuhrmann S, Reh TA. 2000. Soluble factors and the development of rod photoreceptors. *Cell. Mol. Life Sci.* 57:224–34

Li DM, Sun H. 1997. TEP1, encoded by a candidate tumor suppressor locus, is a novel protein tyrosine phosphatase regulated by transforming growth factor beta. *Cancer Res.* 57:2124–29

Li J, Simpson L, Takahashi M, Miliaresis C, Myers MP, et al. 1998. The PTEN/MMAC1 tumor suppressor induces cell death that is rescued by the AKT/protein kinase B oncogene. *Cancer Res.* 58:5667–72

Li J, Yen C, Liaw D, Podsypanina K, Bose S, et al. 1997a. PTEN, a putative protein tyrosine phosphatase gene mutated in human brain, breast, and prostate cancer. *Science* 275:1943–47

Li L, Ernsting BR, Wishart MJ, Lohse DL, Dixon JE. 1997b. A family of putative tumor suppressors is structurally and functionally conserved in humans and yeast. *J. Biol. Chem.* 272:29403–6

Liaw D, Marsh DJ, Li J, Dahia PL, Wang SI, et al. 1997. Germline mutations of the PTEN gene in Cowden disease, an inherited breast and thyroid cancer syndrome. *Nat. Genet.* 16:64–67

Libermann TA, Nusbaum HR, Razon N, Kris R, Lax I, et al. 1985a. Amplification and overexpression of the EGF receptor gene in primary human glioblastomas. *J. Cell Sci. Suppl.* 3:161–72

Libermann TA, Nusbaum HR, Razon N, Kris R, Lax I, et al. 1985b. Amplification, enhanced expression and possible rearrangement of EGF receptor gene in primary human brain tumours of glial origin. *Nature* 313:144–47

Lillien L. 1995. Changes in retinal cell fate induced by overexpression of EGF receptor. *Nature* 377:158–62

Lin H, Bondy ML, Langford LA, Hess KR, Delclos GL, et al. 1998. Allelic deletion analyses of MMAC/PTEN and DMBT1 loci in gliomas: relationship to prognostic significance. *Clin. Cancer Res.* 4:2447–54

Lipinski MM, Jacks T. 1999. The retinoblastoma gene family in differentiation and development. *Oncogene* 18:7873–82

Lopes MBS, VandenBerg SR, Scheithauer BW. 1995. Histopathology, immunochemistry and ultrastructure of brain tumors. In *Brain Tumors: An Encyclopedic Approach,* ed. AH Kaye, ER Laws. Edinburgh, UK: Churchill Livingstone, pp. 125–162

Lu Y, Lin YZ, LaPushin R, Cuevas B, Fang X, et al. 1999. The PTEN/MMAC1/TEP tumor suppressor gene decreases cell growth and induces apoptosis and anoikis in breast cancer cells. *Oncogene* 18:7034–45

Lucas FR, Goold RG, Gordon-Weeks PR, Salinas PC. 1998. Inhibition of GSK-3beta leading to the loss of phosphorylated MAP-1B is an early event in axonal remodelling induced by WNT-7a or lithium. *J. Cell Sci.* 111:1351–61

Ludlow JW, Skuse GR. 1995. Viral oncoprotein binding to pRB, p107, p130, and p300. *Virus Res.* 35:113–21

Luskin MB, Parnavelas JG, Barfield JA. 1993. Neurons, astrocytes, and oligodendrocytes of the rate cerebral cortex originate from separate progenitor cells: an ultrastructural analysis of clonally related cells. *J. Neurosci.* 13:1730–50

Luukko K, Ylikorkala A, Tiainen M, Makela TP. 1999. Expression of LKB1 and PTEN tumor suppressor genes during mouse embryonic development. *Mech. Dev.* 83:187–90

Mabie PC, Mehler MF, Marmur R, Papavasiliou A, Song Q, Kessler JA. 1997. Bone morphogenetic proteins induce astroglial differentiation of oligodendroglial-astroglial progenitor cells. *J. Neurosci.* 17:4112–20

Macleod KF, Hu Y, Jacks T. 1996. Loss of Rb activates both p53-dependent and independent cell death pathways in the developing mouse nervous system. *EMBO J.* 15:6178–88

Maehama T, Dixon JE. 1998. The tumor suppressor, PTEN/MMAC1, dephosphorylates the lipid second messenger, phosphatidylinositol 3,4,5-trisphosphate. *J. Biol. Chem.* 273:13375–78

Marsh DJ, Coulon V, Lunetta KL, Rocca-Serra P, Dahia PL, et al. 1998. Mutation spectrum and genotype-phenotype analyses in Cowden disease and Bannayan-Zonana syndrome, two hamartoma syndromes with germline PTEN mutation. *Hum. Mol. Genet.* 7:507–15

Marsh DJ, Dahia PL, Zheng Z, Liaw D, Parsons R, et al. 1997. Germline mutations in PTEN are present in Bannayan-Zonana syndrome. *Nat. Genet.* 16:333–34

Marte BM, Downward J. 1997. PKB/Akt:

connecting phosphoinositide 3-kinase to cell survival and beyond. *Trends Biochem. Sci.* 22:355–58

Mason CA, Sretavan DW. 1997. Glia, neurons, and axon pathfinding during optic chiasm development. *Curr. Opin. Neurobiol.* 7:647–53

Mathers PH, Grinberg A, Mahon KA, Jamrich M. 1997. The Rx homeobox gene is essential for vertebrate eye development. *Nature* 387:603–37

Mathers PH, Jamrich M. 2000. Regulation of eye formation by the Rx and pax6 homeobox genes. *Cell. Mol. Life Sci.* 57:186–94

Matsuda S, Ii Y, Desaki J, Yoshimura H, Okumura N, Sakanaka M. 1994. Development of Purkinje cell bodies and processes with basic fibroblast growth factor-like immunoreactivity in the rat cerebellum. *Neuroscience* 59:651–62

Mayer-Proschel M, Kalyani AJ, Mujtaba T, Rao MS. 1997. Isolation of lineage-restricted neuronal precursors from multipotent neuroepithelial stem cells. *Neuron* 19:773–85

McMahon AP, Bradley A. 1990. The Wnt-1 (int-1) proto-oncogene is required for development of a large region of the mouse brain. *Cell* 62:1073–85

Medema RH, Kops GJ, Bos JL, Burgering BM. 2000. AFX-like Forkhead transcription factors mediate cell-cycle regulation by Ras and PKB through p27kip1. *Nature* 404:782–87.

Mehler MF, Mabie PC, Zhang D, Kessler JA. 1997. Bone morphogenetic proteins in the nervous system. *Trends Neurosci.* 20:309–17

Mehler MF, Mabie PC, Zhu G, Gokhan S, Kessler JA. 2000. Developmental changes in progenitor cell responsiveness to bone morphogenetic proteins differentially modulate progressive CNS lineage fate. *Dev. Neurosci.* 22:74–85

Mi H, Barres BA. 1999. Purification and characterization of astrocyte precursor cells in the developing rat optic nerve. *J. Neurosci.* 19:1049–61

Millen KJ, Hui CC, Joyner AL. 1995. A role for En-2 and other murine homologues of Drosophila segment polarity genes in regulating positional information in the developing cerebellum. *Development* 121:3935–45

Mishima K, Higashiyama S, Asai A, Yamaoka K, Nagashima Y, et al. 1998. Heparin-binding epidermal growth factor-like growth factor stimulates mitogenic signaling and is highly expressed in human malignant gliomas. *Acta Neuropathol.* 96:322–28

Mollenhauer J, Herbertz S, Holmskov U, Tolnay M, Krebs I, et al. 2000. DMBT1 encodes a protein involved in the immune defense and in epithelial differentiation and is highly unstable in cancer. *Cancer Res.* 60:1704–10

Mollenhauer J, Wiemann S, Scheurlen W, Korn B, Hayashi Y, et al. 1997. DMBT1, a new member of the SRCR superfamily, on chromosome 10q25.3-26.1 is deleted in malignant brain tumours. *Nat. Genet.* 17:32–39

Montgomery DL. 1994. Astrocytes: form, functions, and roles in disease. *Vet. Pathol.* 31:145–67

Mullen RJ, Hamre KM, Goldowitz D. 1997. Cerebellar mutant mice and chimeras revisited. *Perspect. Dev. Neurobiol.* 5:43–55

Myers MP, Pass I, Batty IH, Van der Kaay J, Stolarov JP, et al. 1998. The lipid phosphatase activity of PTEN is critical for its tumor suppressor function. *Proc. Natl. Acad. Sci. USA* 95:13513–18

Myers MP, Stolarov JP, Eng C, Li J, Wang SI, et al. 1997. P-TEN, the tumor suppressor from human chromosome 10q23, is a dual-specificity phosphatase. *Proc. Natl. Acad. Sci. USA* 94:9052–57

Nagane M, Levitzki A, Gazit A, Cavenee WK, Huang HJ. 1998. Drug resistance of human glioblastoma cells conferred by a tumor-specific mutant epidermal growth factor receptor through modulation of Bcl-XL and caspase-3-like proteases. *Proc. Natl. Acad. Sci. USA* 95:5724–29

Nakamura H, Yoshida M, Tsuiki H, Ito K, Ueno M, et al. 1998. Identification of a human homolog of the Drosophila neuralized gene within the 10q25.1 malignant astrocytoma deletion region. *Oncogene* 16:1009–19

Nakashima K, Wiese S, Yanagisawa M,

Arakawa H, Kimura N, et al. 1999a. Developmental requirement of gp130 signaling in neuronal survival and astrocyte differentiation. *J. Neurosci.* 19:5429–34

Nakashima K, Yanagisawa M, Arakawa H, Kimura N, Hisatsune T, et al. 1999b. Synergistic signaling in fetal brain by STAT3-Smad1 complex bridged by p300. *Science* 284:479–82

Nakashima K, Yanagisawa M, Arakawa H, Taga T. 1999c. Astrocyte differentiation mediated by LIF in cooperation with BMP2. *FEBS Lett.* 457:43–46

Neel BG, Tonks NK. 1997. Protein tyrosine phosphatases in signal transduction. *Curr. Opin. Cell Biol.* 9:193–204

Nevins JR. 1994. Cell cycle targets of the DNA tumor viruses. *Curr. Opin. Genet. Dev.* 4:130–34

Nevins JR, Leone G, DeGregori J, Jakoi L. 1997. Role of the Rb/E2F pathway in cell growth control. *J. Cell. Physiol.* 173:233–36

Nielsen HS, Hannibal J, Fahrenkrug J. 1998. Expression of pituitary adenylate cyclase activating polypeptide (PACAP) in the postnatal and adult rat cerebellar cortex. *NeuroReport* 9:2639–42

Nikolic M, Dudek H, Kwon YT, Ramos YF, Tsai LH. 1996. The cdk5/p35 kinase is essential for neurite outgrowth during neuronal differentiation. *Genes Dev.* 10:816–25

Nishikawa R, Ji XD, Harmon RC, Lazar CS, Gill GN, et al. 1994. A mutant epidermal growth factor receptor common in human glioma confers enhanced tumorigenicity. *Proc. Natl. Acad. Sci. USA* 91:7727–31

Noble M, Gutowski N, Bevan K, Engel U, Linskey M, et al. 1995. From rodent glial precursor cell to human glial neoplasia in the oligodendrocyte-type-2 astrocyte lineage. *Glia* 15:222–30

Nork TM, Schwartz, TL, Doshi HM, Millecchia LL. 1995. Retinoblastoma. Cell of origin. *Arch. Ophthalmol.* 113:791–802

Novakovic B. 1994. U.S. childhood cancer survival, 1973–1987. *Med. Pediatr. Oncol.* 23:480–86

Nusse R, Varmus HE. 1982. Many tumors induced by the mouse mammary tumor virus contain a provirus integrated in the same region of the host genome. *Cell* 31:99–109

Packer RJ, Sutton LN, Elterman R, Lange B, Goldwein J, et al. 1994. Outcome for children with medulloblastoma treated with radiation and cisplatin, CCNU, and vincristine chemotherapy. *J. Neurosurg.* 81:690–98

Paramio JM, Navarro M, Segrelles C, Gomez-Casero E, Jorcano JL. 1999. PTEN tumour suppressor is linked to the cell cycle control through the retinoblastoma protein. *Oncogene* 18:7462–68

Park JK, Williams BP, Alberta JA, Stiles CD. 1999. Bipotent cortical progenitor cells process conflicting cues for neurons and glia in a hierarchical manner. *J. Neurosci.* 19:10383–89

Pascual-Castroviejo I, Gutierrez M, Morales C, Gonzalez-Mediero I, Martinez-Bermejo A, Pascual-Pascual SI. 1994. Primary degeneration of the granular layer of the cerebellum. A study of 14 patients and review of the literature. *Neuropediatrics* 25:183–90

Peifer M, Polakis P. 2000. Wnt signaling in oncogenesis and embryogenesis—a look outside the nucleus. *Science* 287:1606–9

Perron M, Harris WA. 2000. Determination of vertebrate retinal progenitor cell fate by the Notch pathway and basic helix-loop-helix transcription factors. *Cell. Mol. Life Sci.* 57:215–23

Pfosser A, Brandl M, Salih H, Grosse-Hovest L, Jung G. 1999. Role of target antigen in bispecific-antibody-mediated killing of human glioblastoma cells: a preclinical study. *Int. J. Cancer* 80:612–16

Pfrieger FW, Barres BA. 1996. New views on synapse-glia interactions. *Curr. Opin. Neurobiol.* 6:615–21

Phillips AC, Bates S, Ryan KM, Helin K, Vousden KH. 1997. Induction of DNA synthesis and apoptosis are separable functions of E2F-1. *Genes Dev.* 11:1853–63

Pietsch T, Waha A, Koch A, Kraus J, Albrecht S, et al. 1997. Medulloblastomas of the desmoplastic variant carry mutations of the human homologue of Drosophila patched. *Cancer Res.* 57:2085–88

Podsypanina K, Ellenson LH, Nemes A, Gu J, Tamura M, et al. 1999. Mutation of Pten/Mmac1 in mice causes neoplasia in multiple organ system. *Proc. Natl. Acad. Sci. USA* 96:1563–68

Polakis P. 1999. The oncogenic activation of beta-catenin. *Curr. Opin. Genet. Dev.* 9:15–21

Pomerantz J, Schreiber-Agus N, Liegeois NJ, Silverman A, Alland L, et al. 1998. The Ink4a tumor suppressor gene product, p19Arf, interacts with MDM2 and neutralizes MDM2's inhibition of p53. *Cell* 92:713–23

Potluri VR, Helson L, Ellsworth RM, Reid T, Gilbert F. 1986. Chromosomal abnormalities in human retinoblastoma: a review. *Cancer* 58:663–71

Powell EM, Meiners S, DiProspero NA, Geller HM. 1997. Mechanisms of astrocyte-directed neurite guidance. *Cell Tissue Res.* 290:385–93

Prigent SA, Lemoine NR. 1992. The type 1 (EGFR-related) family of growth factor receptors and their ligands. *Prog. Growth Factor Res.* 4:1–24

Pringle NP, Guthrie S, Lumsden A, Richardson WD. 1998. Dorsal spinal cord neuroepithelium generates astrocytes but not oligodendrocytes. *Neuron* 20:883–93

Pu P, Liu X, Liu A, Cui J, Zhang Y. 2000. Inhibitory effect of antisense epidermal growth factor receptor RNA on the proliferation of rat C6 glioma cells in vitro and in vivo. *J. Neurosurg.* 92:132–39

Qin XQ, Livingston DM, Kaelin WG Jr, Adams PD. 1994. Deregulated transcription factor E2F-1 expression leads to S-phase entry and p53-mediated apoptosis. *Proc. Natl. Acad. Sci. USA* 91:10918–22

Quinn JC, West JD, Hill RE. 1996. Multiple functions for Pax6 in mouse eye and nasal development. *Genes Dev.* 10:435–46

Quinn SM, Walters WM, Vescovi AL, Whittemore SR. 1999. Lineage restriction of neuroepithelial precursor cells from fetal human spinal cord. *J. Neurosci. Res.* 57:590–602

Rabchevsky AG, Weinitz JM, Coulpier M, Fages C, Tinel M, Junier MP. 1998. A role for transforming growth factor alpha as an inducer of astrogliosis. *J. Neurosci.* 18:10541–52

Raffel C, Jenkins RB, Frederick L, Hebrink D, Alderete B, et al. 1997. Sporadic medulloblastomas contain PTCH mutations. *Cancer Res.* 57:842–45

Rajan P, McKay RD. 1998. Multiple routes to astrocytic differentiation in the CNS. *J. Neurosci.* 18:3620–29

Ramaswamy S, Nakamura N, Vazquez F, Batt DB, Perera S, et al. 1999. Regulation of G1 progression by the PTEN tumor suppressor protein in linked to inhibition of the phosphatidylinositol 3-kinase/Akt pathway. *Proc. Natl. Acad. Sci. USA* 96:2110–15

Rao MS. 1999. Multipotent and restricted precursors in the central nervous system. *Anat. Rec.* 257:137–48

Rao MS, Mayer-Proschel M. 1997. Glial-restricted precursors are derived from multipotent neuroepithelial stem cells. *Dev. Biol.* 188:48–63

Rao MS, Noble M, Mayer-Proschel M. 1998. A tripotential glial precursor cell is present in the developing spinal cord. *Proc. Natl. Acad. Sci. USA* 95:3996–4001

Rasheed BK, Bigner SH. 1991. Genetic alterations in glioma and medulloblastoma. *Cancer Metastasis Rev.* 10:289–99

Rasheed BK, Fuller GN, Friedman AH, Bigner DD, Bigner SH. 1992. Loss of heterozygosity for 10q loci in human gliomas. *Genes Chromosomes Cancer* 5:75–82

Rasheed BK, Wiltshire RN, Bigner SH, Bigner DD. 1999. Molecular pathogenesis of malignant gliomas. *Curr. Opin. Oncol.* 11:162–67

Rather LJ. 1978. *The Genesis of Cancer: A Study in the History of Ideas.* Baltimore, MD: Johns Hopkins Univ. Press

Reh TA. 1992. Cellular interactions determine neuronal phenotypes in rodent retinal cultures. *J. Neurobiol.* 23:1067–83

Reh TA, Kljavin IJ. 1989. Age of differentiation determines rat retinal germinal cell phenotype: induction of differentiation by dissociation. *J. Neurosci.* 9:4179–89

Reh TA, Levine EM. 1998. Multipotential stem cells and progenitors in the vertebrate retina. *J. Neurobiol.* 36:206–20

Reynolds BA, Tetzlaff W, Weiss S. 1992. A multipotent EGF-responsive striatal embryonic progenitor cell produces neurons and astrocytes. *J. Neurosci.* 12:4565–74

Richardson PM. 1994. Ciliary neurotrophic factor: a review. *Pharmacol. Ther.* 63:187–98

Riese DJ 2nd, Bermingham Y, van Raaij TM, Buckley S, Plowman GD, Stern DF. 1996. Betacellulin activates the epidermal growth factor receptor and erbB-4, and induces cellular response patterns distinct from those stimulated by epidermal growth factor or neuregulin-beta. *Oncogene* 12:345–53

Robanus-Maandag E, Dekker M, van der Valk M, Carrozza ML, Jeanny JC, et al. 1998. p107 is a suppressor of retinoblastoma development in pRb-deficient mice. *Genes Dev.* 12:1599–60

Robinson SR. 1991. Development of the mammalian retina. In *Neuroanatomy of the Visual Pathways and Their Development, Vol. 3, Vision and Visual Dysfunction*, ed. B. Dreher and SR. Robinson. pp. 69–128. Boca Raton, FL: CRC Press

Ruiz i Altaba A. 1997. Catching a Gli-mpse of Hedgehog. *Cell* 90:193–96

Saffell JL, Williams EJ, Mason IJ, Walsh FS, Doherty P. 1997. Expression of a dominant negative FGF receptor inhibits axonal growth and FGF receptor phosphorylation stimulated by CAMs. *Neuron* 18:231–42. Erratum. 1998. *Neuron* 20:619

Salcman M. 1995. Glioblastoma and malignant astrocytoma. In *Brain Tumors: An Encyclopedic Approach*, ed. AH Kaye, ER Laws. Edinburgh, UK: Churchill Livingstone

Sasaki H, Nishizaki Y, Hui C, Nakafuku M, Kondoh H. 1999. Regulation of Gli2 and Gli3 activities by an amino-terminal repression domain: implication of Gli2 and Gli3 as

primary mediators of Shh signaling. *Development* 126:3915–24

Schmidt EE, Ichimura K, Reifenberger G, Collins VP. 1994. CDKN2 (p16/MTS1) gene deletion or CDK4 amplification occurs in the majority of glioblastomas. *Cancer Res.* 54:6321–24

Schreiber-Agus N, DePinho RA. 1998. Repression by the Mad(Mxi1)-Sin3 complex. *BioEssays* 20:808–18

Schubert P, Rudolphi K. 1998. Interfering with the pathologic activation of microglial cells and astrocytes in dementia. *Alzheimer Dis. Assoc. Disord.* 12:S21–28

Segal RA, Greenberg ME. 1996. Intracellular signaling pathways activated by neurotrophic factors. *Annu. Rev. Neurosci.* 19:463–89

Seidman KJ, Teng AL, Rosenkopf R, Spilotro P, Weyhenmeyer JA. 1997. Isolation, cloning and characterization of a putative type-1 astrocyte cell line. *Brain Res.* 753:18–26

Shoshan Y, Nishiyama A, Chang A, Mork S, Barnett GH, et al. 1999. Expression of oligodendrocyte progenitor cell antigens by gliomas: implications for the histogenesis of brain tumors. *Proc. Natl. Acad. Sci. USA* 96:10361–66

Shoyab M, Plowman GD, McDonald VL, Bradley JG, Todaro GJ. 1989. Structure and function of human amphiregulin: a member of the epidermal growth factor family. *Science* 243:1074–76

Sibilia M, Steinbach JP, Stingl L, Aguzzi A, Wagner EF. 1998. A strain-independent postnatal neurodegeneration in mice lacking the EGF receptor. *EMBO J.* 17:719–31

Sibilia M, Wagner E-F. 1995. Strain-dependent epithelial defects in mice lacking the EGF receptor. *Science* 269:234–38. Erratum. 1995. *Science* (5226)269:909

Simon M, Koster G, Menon AG, Schramm J. 1999. Functional evidence for a role of combined CDKN2A (p16-p14(ARF))/CDKN2B (p15) gene inactivation in malignant gliomas. *Acta Neuropathol.* 98:444–52

Simpson DL, Morrison R, de Vellis J, Herschman HR. 1982. Epidermal growth factor binding and mitogenic activity on purified populations of cells from the central nervous system. *J. Neurosci. Res.* 8:453–62

Skoglosa Y, Patrone C, Lindholm D. 1999. Pituitary adenylate cyclase activating polypepetide is expressed by developing rat Purkinje cells and decreases the number of cerebellar gamma-amino butyric acid positive neurons in culture. *Neurosci. Lett.* 265:207–10

Slack RS, El-Bizri H, Wong J, Belliveau DJ, Miller FD. 1998. A critical temporal requirement for the retinoblastoma protein family during neuronal determination. *J. Cell Biol.* 140:1497–509

Slack RS, Hamel PA, Bladon TS, Gill RM, McBurney MW. 1993. Regulated expression of the retinoblastoma gene in differentiating embryonal carcinoma cells. *Oncogene* 8:1585–91

Slack RS, Skerjanc IS, Lach B, Craig J, Jardine K, McBurney MW. 1995. Cells differentiating into neuroectoderm undergo apoptosis in the absence of functional retinoblastoma family proteins. *J. Cell Biol.* 129:779–88

Smeyne RJ, Chu T, Lewin A, Bian FS, Crisman S, et al. 1995. Local control of granule cell generation by cerebellar Purkinje cells. *Mol. Cell. Neurosci.* 6:230–51

Snowden AW, Perkins ND. 1998. Cell cycle regulation of the transcriptional coactivators p300 and CREB binding protein. *Biochem. Pharmacol.* 55:1947–54

Somerville RP, Shoshan Y, Eng C, Barnett G, Miller D, Cowell JK. 1998. Molecular analysis of two putative tumour suppressor genes, PTEN and DMBT, which have been implicated in glioblastoma multiforme disease progression. *Oncogene* 17:1755–57

Squire J, Gallie BL, Phillips RA. 1985. A detailed analysis of chromosomal changes in heritable and non-heritable retinoblastoma. *Hum. Genet.* 70:291–301

Stambolic V, Mak TW, Woodgett JR. 1999. Modulation of cellular apoptotic potential: contributions to oncogenesis. *Oncogene* 18:6094–103

Stambolic V, Suzuki A, de la Pompa JL, Broth-

ers GM, Mirtsos C, et al. 1998. Negative regulation of PKB/Akt-dependent cell survival by the tumor suppressor PTEN. *Cell* 95:29

Steck PA, Lin H, Langford LA, Jasser SA, Koul D, et al. 1999. Functional and molecular analyses of 10q deletions in human gliomas. *Genes Chromosomes Cancer* 24:135–43

Steck PA, Pershouse MA, Jasser SA, Yung WK, Lin H, et al. 1997. Identification of a candidate tumour suppressor gene, MMAC1, at chromosome 10q23.3 that is mutated in multiple advanced cancers. *Nat. Genet.* 15:356

Sturrock RR. 1982. Gliogenesis in the prenatal rabbit spinal cord. *J. Anat.* 134:771–93

Sturrock RR. 1987. A quantitative histological study of cell division and changes in cell number in the meningeal sheath of the embryonic human optic nerve. *J. Anat.* 155:133

Sugawa N, Ekstrand AJ, James CD, Collins VP. 1990. Identical splicing of aberrant epidermal growth factor receptor transcripts from amplified rearranged genes in human glioblastomas. *Proc. Natl. Acad. Sci. USA* 87:8602–6

Sun H, Lesche R, Li DM, Liliental J, Zhang H, et al. 1999. PTEN modulates cell cycle progression and cell survival by regulating phosphatidylinositol 3,4,5-trisphosphate and Akt/protein kinase B signaling pathway. *Proc. Natl. Acad. Sci. USA* 96:6199–204

Suzuki A, de la Pompa JL, Stambolic V, Elia AJ, Sasaki T, et al. 1998. High cancer susceptibility and embryonic lethality associated with mutation of the PTEN tumor suppressor gene in mice. *Curr. Biol.* 8:1169–78

Tajima Y, Munakata S, Ishida Y, Nakajima T, Sugano I, et al. 1994. Photoreceptor differentiation of retinoblastoma: an electron microscopic study of 29 retinoblastomas *Pathol. Int.* 44:837–43

Takaishi H, Konishi H, Matsuzaki H, Ono Y, Shirai Y, et al. 1999. Regulation of nuclear translocation of Forkhead transcription factor AFX by protin kinase B. *Proc. Natl Acad Sci USA* 96:11836–41

Tamura M, Gu J, Danen EH, Takino T, Miyamoto S, Yamada KM. 1999a. PTEN interactions with focal adhesion kinase and suppression of the extracellular matrix-dependent phosphatidylinositol 3-kinase/Akt cell survival pathway. *J. Biol. Chem.* 274:20693–703

Tamura M, Gu J, Matsumoto K, Aota S, Parsons R, Yamada KM. 1998. Inhibition of cell migration, spreading, and focal adhesions by tumor suppressor PTEN. *Science* 280:1614

Tamura M, Gu J, Tran H, Yamada KM. 1999b. PTEN gene and integrin signaling in cancer. *J. Natl. Cancer Inst.* 91:1820–28

Tang D, Lee KY, Qi Z, Matsuura I, Wang JH. 1996. Neuronal Cdc2-like kinase: from cell cycle to neuronal function. *Biochem. Cell Biol.* 74:419–29

Tang D, Wang JH. 1996. Cyclin-dependent kinase 5 (Cdk5) and neuron-specific Cdk5 activators. *Prog. Cell Cycle Res.* 2:205–16

Tang DG, Tokumoto YM, Raff MC. 2000. Long-term culture of purified postnatal oligodendrocyte precursor cells: evidence for an intrinsic maturation program that plays out over months. *J. Cell Biol.* 148:971–84

Taniura H, Taniguchi N, Hara M, Yoshikawa K. 1998. Necdin, a postmitotic neuron-specific growth suppressor, interacts with viral transforming proteins and cellular transcription factor E2F1. *J. Biol. Chem.* 273:720–28

Thapar K, Fukuyama K, Rutka JT. 1995. Neurogenetics and the molecular biology of human brain tumors. In *Brain Tumors: An Encyclopedic Approach*, ed. AH Kaye, ER Laws. Edinburgh, UK: Churchill Livingstone

Thomas KR, Musci TS, Neumann PE, Capecchi MR. 1991. Swaying is a mutant allele of the proto-oncogene Wnt-1. *Cell* 67:969–76

Threadgill DW, Dlugosz AA, Hansen LA, Tennenbaum T, Lichti U, et al. 1995. Targeted disruption of mouse EGF receptor: effect of genetic background on mutant phenotype. *Science* 269:230–34

Tian XX, Lam PY, Chen J, Pang JC, To SS, et al. 1998. Antisense epidermal growth factor receptor RNA transfection in human malignant glioma cells leads to inhibition of proliferation and induction of differentiation.

Neuropathol. Appl. Neurobiol. 24:389–96

Tian XX, Pang JC, To SS, Ng HK. 1999. Restoration of wild-type PTEN expression leads to apoptosis, induces differentiation, and reduces telomerase activity in human glioma cells. *J. Neuropathol. Exp. Neurol.* 58:472–79

Tonks NK, Neel BG. 1996. From form to function: signaling by protein tyrosine phosphatases. *Cell* 87:365–68

Touw IP, De Koning JP, Ward AC, Hermans MH. 2000. Signaling mechanisms of cytokine receptors and their perturbances in disease. *Mol. Cell. Endocrinol.* 160:1–9

Traiffort E, Charytoniuk DA, Faure H, Ruat M. 1998. Regional distribution of Sonic hedgehog, patched, and smoothened mRNA in the adult rat brain. *J. Neurochem.* 70:1327–30

Tsai KY, Hu Y, Macleod KF, Crowley D, Yamasaki L, Jacks T. 1998. Mutation of E2f-1 suppresses apoptosis and inappropriate S phase entry and extends survival of Rb-deficient mouse embryos. *Mol. Cell* 2:293–304

Tsokos M, Kyritsis AP, Chader GJ, Triche TJ. 1986. Differentiation of human retinoblastoma in vitro into cell types with characteristics observed in embryonal or mature retina. *Am. J. Pathol.* 123:542–52

Turner DL, Snyder EY, Cepko CL. 1990. Lineage-independent determination of cell type in the embryonic mouse retina. *Neuron* 4:833–45

van den Pol AN, Spencer DD. 2000. Differential neurite growth on astrocyte substrates: interspecies facilitation in green fluorescent protein-transfected rat and human neurons. *Neuroscience* 95:603–16

Vesce S, Bezzi P, Volterra A. 1999. The highly integrated dialogue between neurons and astrocytes in brain function. *Sci. Prog.* 82:251

Wallace VA. 1999. Purkinje-cell-derived Sonic hedgehog regulates granule neuron precursor cell proliferation in the developing mouse cerebellum. *Curr. Biol.* 9:445–48

Watanabe T, Raff MC. 1990. Rod photoreceptor development in vitro: intrinsic properties of proliferating neuroepithelial cells change as development proceeds in the rat retina. *Neuron* 4:461–67

Wechsler DS, Shelly CA, Petroff CA, Dang CV. 1997. MXI1, a putative tumor suppressor gene, suppresses growth of human glioblastoma cells. *Cancer Res.* 57:4905–12

Wechsler-Reya RJ, Scott MP. 1999. Control of neuronal precursor proliferation in the cerebellum by Sonic hedgehog. *Neuron* 22:103

Weiss S, Dunne C, Hewson J, Wohl C, Wheatley M, et al. 1996. Multipotent CNS stem cells are present in the adult mammalian spinal cord and ventricular neuroaxis. *J. Neurosci.* 16:7599–609

Westermark B, Magnusson A, Heldin CH. 1982. Effect of epidermal growth factor on membrane motility and cell locomotion in cultures of human clonal glioma cells. *J. Neurosci. Res.* 8:491–507

Wetmore C, Eberhart DE, Curran T. 2000. The normal patched allele is expressed in medulloblastomas from mice with heterozygous germ-line mutation of patched. *Cancer Res.* 60:2239–46

Wetts R, Serbedzija GN, Fraser SE. 1989. Cell lineage analysis reveals multipotent precursors in the ciliary margin of the frog retina. *Dev. Biol.* 136:254–63

Whyte P, Buchkovich KJ, Horowitz JM, Friend SH, Raybuck M, et al. 1988. Association between an oncogene and an anti-oncogene: the adenovirus E1A proteins bind to the retinoblastoma gene product. *Nature* 334:124–29

Wodarz A, Nusse R. 1998. Mechanisms of Wnt signaling in development. *Annu. Rev. Cell Dev. Biol.* 14:59–88

Wong AJ, Ruppert JM, Bigner SH, Grzeschik CH, Humphrey PA, et al. 1992. Structural alterations of the epidermal growth factor receptor gene in human gliomas. *Proc. Natl. Acad. Sci. USA* 89:2965–69

Wu RC, Li X, Schonthal AH. 2000a. Transcriptional activation of p21WAF1 by PTEN/MMAC1 tumor suppressor. *Mol. Cell. Biochem.* 203:59–71

Wu X, Hepner K, Castelino-Prabhu S, Do D, Kaye MB, et al. 2000b. Evidence for regulation of the PTEN tumor suppressor by a membrane–localized multi-PDZ domain containing scaffold protein MAGI-2. *Proc. Natl. Acad. Sci. USA* 97:4233–38

Wu X, Levine AJ. 1994. p53 and E2F-1 cooperate to mediate apoptosis. *Proc. Natl. Acad. Sci. USA* 91:3602–6

Xie J, Johnson RL, Zhang X, Bare JW, Waldman FM, et al. 1997. Mutations of the PATCHED gene in several types of sporadic extracutaneous tumors. *Cancer Res.* 57:2369–72

Xiong W, Pestell R, Rosner MR. 1997. Role of cyclins in neuronal differentiation of immortalized hippocampal cells. *Mol. Cell. Biol.* 17:6585–97

Yokota N, Aruga J, Takai S, Yamada K, Hamazaki M, et al. 1996. Predominant expression of human zic in cerebellar granule cell lineage and medulloblastoma. *Cancer Res.* 56:377–83

Yong WH, Raffel C, von Deimling A, Louis DN. 1995. The APC gene in Turcot's syndrome. *N. Engl. J. Med.* 333:524

Yoshikawa K. 2000. Cell cycle regulators in neural stem cells and postmitotic neurons. *Neurosci. Res.* 37:1–14

Zhang Y, Xiong Y, Yarbrough WG. 1998. ARF promotes MDM2 degradation and stabilizes p53: ARF-INK4a locus deletion impairs both the Rb and p53 tumor suppression pathways. *Cell* 92:725–34

Zhu JW, DeRyckere D, Li FX, Wan YY, De-Gregori J. 1999. A role for E2F1 in the induction of ARF, p53, and apoptosis during thymic negative selection. *Cell Growth Differ.* 10:829–38

Zurawel RH, Chiappa SA, Allen C, Raffel C. 1998. Sporadic medulloblastomas contain oncogenic beta-catenin mutations. *Cancer Res.* 58:896–99

Annu. Rev. Neurosci. 2001. 24:429–58

To Eat or to Sleep? Orexin in the Regulation of Feeding and Wakefulness

Jon T. Willie,[1,2] Richard M. Chemelli,[1,2,3]
Christopher M. Sinton,[4] and Masashi Yanagisawa[1,2]

[1]Howard Hughes Medical Institute
[2]Department of Molecular Genetics
[3]Department of Pediatrics
[4]Department of Psychiatry
University of Texas Southwestern Medical Center at Dallas
Dallas, Texas 75390-9050; e-mail: willie.jon@tumora.swmed.edu, docvette@aol.com,
christopher.sinton@utsouthwestern.edu, myanagisawa@aol.com

Key Words appetite, metabolism, arousal, narcolepsy, lateral hypothalamus

■ **Abstract** Orexin-A and orexin-B are neuropeptides originally identified as endogenous ligands for two orphan G-protein–coupled receptors. Orexin neuropeptides (also known as hypocretins) are produced by a small group of neurons in the lateral hypothalamic and perifornical areas, a region classically implicated in the control of mammalian feeding behavior. Orexin neurons project throughout the central nervous system (CNS) to nuclei known to be important in the control of feeding, sleep-wakefulness, neuroendocrine homeostasis, and autonomic regulation. orexin mRNA expression is upregulated by fasting and insulin-induced hypoglycemia. C-fos expression in orexin neurons, an indicator of neuronal activation, is positively correlated with wakefulness and negatively correlated with rapid eye movement (REM) and non-REM sleep states. Intracerebroventricular administration of orexins has been shown to significantly increase food consumption, wakefulness, and locomotor activity in rodent models. Conversely, an orexin receptor antagonist inhibits food consumption. Targeted disruption of the orexin gene in mice produces a syndrome remarkably similar to human and canine narcolepsy, a sleep disorder characterized by excessive daytime sleepiness, cataplexy, and other pathological manifestations of the intrusion of REM sleep-related features into wakefulness. Furthermore, orexin knockout mice are hypophagic compared with weight and age-matched littermates, suggesting a role in modulating energy metabolism. These findings suggest that the orexin neuropeptide system plays a significant role in feeding and sleep-wakefulness regulation, possibly by coordinating the complex behavioral and physiologic responses of these complementary homeostatic functions.

0147-006X/01/0621-0429$14.00

INTRODUCTION

Feeding behavior is dependent upon the integration of metabolic, autonomic, endocrine, and environmental factors coordinated with an appropriate state of cortical arousal (wakefulness). Historically, the hypothalamus has been recognized to play a critical role in maintaining energy homeostasis by integrating these factors and coordinating the behavioral, metabolic, and neuroendocrine responses (Oomura 1980; Bernardis & Bellinger 1993, 1996). In mammals, the neurons of the lateral hypothalamic area (LHA) are particularly important for feeding and behavioral arousal. Animal models with lesions of the LHA exhibit hypophagia, an increased metabolic rate and decreased arousal that frequently leads to death by starvation. Furthermore, they consistently fail to respond to homeostatic challenges such as fasting with appropriate adaptive behavioral and physiologic responses (Bernardis & Bellinger 1996). Therefore, the LHA has classically been regarded as the hypothalamic "feeding center" and as an important component of the autonomic nervous system with extensive projections within the hypothalamus and throughout the entire neuroaxis. With the ability to influence nuclei throughout the CNS, the LHA appears to be anatomically well placed to coordinate the metabolic, motivational, motor, autonomic, and arousal processes necessary to elicit environmentally appropriate feeding-related behaviors (Bernardis & Bellinger 1993, 1996).

The past 10 years have witnessed a dramatic increase in our knowledge regarding the number of central and peripheral mediators of energy homeostasis and their complex physiologic and neuro-anatomic inter-relationships (for a detailed review of peripheral and central mechanisms of feeding, see Woods et al 1998, Elmquist et al 1999, Kalra et al 1999, and Salton et al 2000). An understanding of the complex neural network that controls feeding behavior is emerging from these studies. The regulation of the known hypothalamic neuropeptide mediators are highly influenced by the peripheral lipostat leptin. Principal components of this leptin-sensitive network include antagonistic and complementary appetite-stimulating (orexigenic) and appetite-suppressing (anorectic) pathways. Several authors have suggested that the central orexigenic pathways involving neuropeptide Y (NPY), agouti-related peptide (AgRP) and melanin concentrating hormone (MCH) are redundant, or that some of these factors lack physiologic relevance in vivo (Flier & Maratos-Flier 1998, Salton et al 2000). For example, NPY potently increases food consumption when given centrally but *Npy*-null mutant mice fail to demonstrate a significant feeding phenotype. Anorectic pathways, such as those involving alpha-melanocyte stimulating hormone (α-MSH) and leptin appear to be less redundant since null mouse mutants for these factors and their receptors are well described models of obesity (Flier & Maratos-Flier 1998, Salton et al 2000). Still, it is likely that many other central and peripheral mediators of energy homeostasis remain unidentified, and our understanding of the complex interactions among the many feeding-related pathways is still limited.

Until recently, melanin-concentrating hormone (MCH) was the only neuropeptide implicated in feeding regulation known to be produced solely within the LHA.

MCH dose-dependently increases food consumption when administered centrally in rodents. Genetic disruption of the MCH gene in mice results in hypophagia and reduced body weight compared with wild-type littermates (Salton et al 2000, Shimada et al 1998). Now, after the discovery of orexin neuropeptides A and B (Greek: *Orexis* = appetite), two more neuropeptides from the LHA have been identified that mediate feeding behavior and are likely to play a significant role in energy homeostasis.

OREXIN SIGNALING PATHWAY

Orexin Neuropeptides

Our group isolated orexin-A and orexin-B while screening high-resolution high-performance liquid chromatography fractions from brain extracts for stimulation of signal transduction in cell lines expressing orphan G-protein-coupled receptors (GPCRs) (Sakurai et al 1998). Mammalian orexin-A is a 33 amino-acid peptide of 3562 Da with an N-terminal pyroglutamyl residue, C-terminal amidation (both typical of neuropeptides), and two sets of intrachain disulfide bonds (Figure 1, see color insert). The primary structure of orexin-A is completely conserved among human, rat, mouse, cow, and pig genera (Sakurai et al 1999, Dyer et al 1999). Mammalian orexin-B is a 28-amino acid, C-terminally amidated peptide of 2937 Da with 46% (13/28) amino acid identity to the orexin-A sequence (Figure 1). Mouse and rat orexin-B peptides are identical, but human orexin-B has two amino acid substitutions compared with the rodent sequences (Sakurai et al 1998). Orexins have also been cloned in the amphibian *Xenopus laevis* and found to have a high amino acid identity with their mammalian counterparts, especially at the carboxyl terminus (Figure 1; Shibahara et al 1999).

A single gene composed of two exons and an intervening intron encodes the orexin neuropeptides. This structure is conserved within rodent and human genomes (Sakurai et al 1998, Sakurai et al 1999). *prepro-orexin* cDNAs encode 130-residue and 131-residue polypeptides in rat and human neurons, respectively. These polypeptides have typical secretory signal sequences and are cleaved to form mature orexin-A and orexin-B peptides that are post-translationally modified as neuropeptides.

A messenger RNA encoding the same neuropeptide precursor was independently isolated from a hypothalamus-enriched cDNA library by using a differential cloning approach, and the putative encoded peptides were named hypocretins by de Lecea et al (1998). Nucleotide sequence alignment shows that the base-pair sequences of hypocretins-1 and -2 are the same as orexins-A and -B, but that the mature peptides predicted in this report had additional amino acids not found in native orexins. The hypocretin peptides, synthesized according to sequences predicted by de Lecea et al (1998), are markedly less potent agonists compared with orexins on transfected cells expressing human orexin receptors (Smart et al 2000). The hypocretins were named for the limited identity of hypocretin-2 (orexin-B) with the gut hormone secretin, but native orexin peptides are, in fact, distantly similar to the bombesin neuropeptide family and not the secretin family (Figure 1, see color insert). Nevertheless, both names are used interchangeably in the literature.

Orexin Receptors

Sakurai et al (1998) identified two orexin receptor subtypes, named orexin-1 receptor (OX_1R) and orexin-2 receptor (OX_2R), that are structurally similar to other G-protein-coupled neuropeptide receptors. The OX_1R is the orphan G-protein-coupled receptor used during ligand hunting to first identify and then purify the orexins. A search of the GenBank dbEST database with the OX_1R amino acid sequence revealed two candidate ESTs. Using PCR, with primers designed from these ESTs, the OX_2R receptor was discovered and found to have a 64% amino-acid identity with OX_1R. Competitive radio-ligand binding assays reveal that the orexin receptors have different binding profiles for the respective orexin peptides. The OX_1R has a 1-order-of-magnitude greater affinity for orexin-A [50% Inhibitory Concentration (IC_{50} = 20 nM)] compared with orexin-B (IC_{50} = 250 nM). In contrast, orexins-A and -B bind the OX_2R with equal affinity (IC_{50} = 20 nM). Therefore, it appears that the OX_1R is moderately selective for orexin-A, whereas OX_2R is a nonselective receptor for both neuropeptide agonists. Evidence from receptor-transfected cell lines and isolated receptor-expressing hypothalamic neurons suggest that the OX_1R is coupled exclusively to the G_q subclass of heterotrimeric G proteins, whereas OX_2R may couple to $G_{i/o}$, and/or G_q (Sakurai et al 1998; van den Pol et al 1998; T Yada, S Muroya, H Funahashi, S Shioda, A Yamanaka, et al, submitted for publication).

Neuroanatomy of the Orexin System

In the rodent CNS, orexin-producing cells (Figure 2) are a small group of neurons restricted to the lateral and posterior hypothalamus and perifornical areas (Chemelli et al 1999, Date et al 1999, de Lecea et al 1998, Peyron et al 1998, Sakurai et al 1998). Despite their highly restricted origin, immunohistochemistry studies using orexin antibodies have shown that orexin neurons project widely throughout the entire neuroaxis. Particularly abundant projections are those found in the cerebral cortex, olfactory bulb, hippocampus, amygdala, septum, diagonal band of Broca, bed nucleus of the stria terminalis, thalamus, anterior and posterior hypothalamus, midbrain, brainstem, and spinal cord (Peyron et al 1998, Date et al 1999, Nambu et al 1999, van den Pol 1999). Orexin immunoreactivity is also reported in the enteric nervous system and pancreas (Kirchgessner & Liu 1999), and *orexin* mRNA expression has been found in the testes (Sakurai et al 1998).

In situ hybridization studies with orexin receptor riboprobes demonstrate that orexin receptors are expressed in a pattern consistent with orexin projections, but that they have a marked differential distribution (Marcus et al 2000, Trivedi et al 1998). *ox1r* mRNA is highly expressed in the prefrontal cortex, hippocampus,

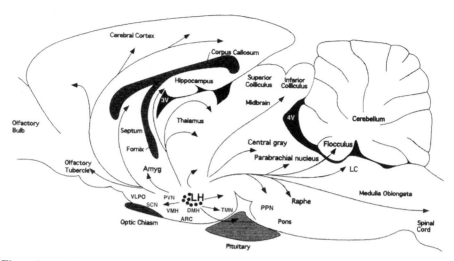

Figure 2   Orexin neurons are found only in the lateral hypothalamic area and project to the entire central nervous system. Schematic drawing of a sagittal section through the rat brain summarizes the organization of the orexin neuronal system. Abbreviations: 3V, third ventricle; 4V, fourth ventricle; Amyg, amygdala; VLPO, ventrolateral preoptic area; SCN, suprachiasmatic nucleus; PVN, paraventricular nucleus; VMH, ventromedial hypothalamus; ARC, arcuate nucleus; DMH, dorsomedial hypothalamus; LH, lateral hypothalamus; TMN, tuberomamillary nucleus; PPN, pedunculopontine nucleus; LC, locus coeruleus. Modified with permission from T Sakurai (1999).

paraventricular thalamus, ventromedial hypothalamus (VMH), arcuate nucleus (ARC), dorsal raphe nucleus, and locus coeruleus (LC). *ox2r* mRNA is found in the cerebral cortex, septal nuclei, hippocampus, medial thalamic groups, dorsal and median raphe nuclei, and many hypothalamic nuclei including the tubero-mammillary nucleus (TMN), dorsomedial hypothalamus (DMH), paraventricular hypothalamic nucleus (PVN), and ventral premammillary nucleus. Orexin receptor mRNA expression has also been reported in the adrenal gland (Malendowicz et al 1999), enteric nervous system, and pancreas (Kirchgessner & Liu 1999).

Intracerebroventricular (i.c.v.) injection studies of orexin neuropeptide using Fos as an immunohistochemical marker of neuronal activation demonstrated that the distribution of neurons activated by either orexin-A or orexin-B is similar (Date et al 1999). The pattern of Fos immunoreactivity is consistent with orexin immunohistochemistry studies and orexin receptor in situ hybridization studies. Areas with the strongest activation include the arcuate nucleus (ARC), PVN, supraoptic nucleus, paraventricular thalamic nucleus, LC, central gray, dorsal raphe, nucleus of the solitary tract, dorsal motor nucleus of the vagus, and suprachiasmatic nucleus. However, no conclusions regarding direct receptor-specific activation can be derived from this study. Furthermore, these studies should be interpreted cautiously since nuclei in close proximity to the ventricular space would be preferentially activated from ventricular delivery, and no conclusions regarding orexin-mediated inhibition of neuronal pathways were obtained. The potential importance of orexin-mediated inhibition is confirmed by the work of van den Pol et al (1998). These investigators found that orexins increase the release of the inhibitory neurotransmitter γ-aminobutyric acid (GABA), as well as the excitatory neurotransmitter glutamate, by acting directly on axon terminals of neuroendocrine cells in the ARC nucleus.

Orexin neurons also express mRNAs for the orexigenic opioid dynorphin and the secretory marker secretogranin II (Risold et al 1999), and the biosynthesis of these three peptides may be similarly regulated (Griffond et al 1999). Immunoreactivity for the appetite stimulating neuropeptide galanin has also been identified in orexin neurons (Hakansson et al 1999). MCH neurons, like orexin neurons, are found in the LHA and project diffusely throughout the entire neuroaxis (Elmquist et al 1999). However, definitive immunohistochemistry studies have demonstrated that orexin and MCH neurons are distinct and independent neuronal populations within the LHA (Broberger et al 1998, Elias et al 1998).

OREXINS MAINTAIN WAKEFULNESS

The regulation of sleep-wakefulness cycling within the context of circadian and environmental influences is critical for the efficient maintenance of energy homeostasis. In mammals, the sleep-wakefulness cycle can be divided into periods of waking, non-rapid-eye-movement (non-REM), and rapid-eye-movement (REM) sleep on the basis of distinct behavioral and electroencephalographic (EEG) criteria. A role for LHA in sleep regulation was suggested by early animal experiments that showed electrical stimulation of the LHA promoted wakefulness while destruction of the LHA caused somnolence and inattentiveness (Bernardis & Bellinger 1996, Levitt & Teitelbaum 1975, Danguir & Nicolaidis 1980a). More recent studies have identified posterior lateral hypothalamic neurons whose firing rates vary with the sleep-wakefulness cycle (Steininger et al 1999). Sleep studies in rats found that Fos immunoreactivity in orexin neurons is positively correlated with wakefulness and negatively correlated with the amount of non-REM and REM sleep, suggesting that these neurons are "waking-active" (Estabrooke et al 2000). They further demonstrated that orexin neuronal activity appears to be under strong circadian control since the temporal relationship with wakefulness and the usual onset of the dark phase, is preserved even under conditions of constant darkness. Fos is also increased after mild sleep deprivation produced by gentle handling, and after administration of wake-promoting stimulants such as modafinil and amphetamine (Chemelli et al 1999, Estabrooke et al 2000). I.c.v. administration of orexin-A in rodents dose-dependently increases wakefulness and suppresses non-REM and REM sleep providing further evidence that orexins play a causative role in sleep-wakefulness regulation (Hagan et al 1999). Orexin i.c.v. injection in rats is also associated with behavioral changes indicative of an aroused state including increased locomotor activity, rearing, grooming, burrowing, searching behaviors, and food consumption (discussed below) (Hagan et al 1999, Ida et al 1999, Sakurai et al 1998).

Two recent reports highlight the importance of orexin signaling in the promotion and consolidation of wakefulness. Our group discovered that orexin neuropeptide knockout mice have a phenotype remarkably similar to the human sleep disorder narcolepsy (Chemelli et al 1999). And, in a complementary but independent study, Mignot and colleagues found that the genetic defect in a narcoleptic dog model is in the *ox2r* gene (Lin et al 1999). Idiopathic narcolepsy is a debilitating, lifelong neurologic disease characterized by excessive daytime sleepiness, cataplexy, and other pathologic intrusions of REM-associated phenomena into wakefulness. The ability of narcoleptic patients to lead normal lives is compromised by frequent involuntary or irresistible daytime "sleep attacks" and sporadic episodes of sudden bilateral skeletal muscle weakness (cataplexy). Cataplexy is often provoked by strong emotion and can be partial or complete (causing collapse to the floor) while consciousness is maintained (Honda 1988). "Sleep-onset REM periods," REM sleep occurring at or near sleep onset, are pathognomonic for narcolepsy and confirm the clinical diagnosis. The uncontrollable intrusion of REM sleep into wakefulness is thought to form the physiologic basis of narcoleptic symptomatology (Bassetti & Aldrich 1996). In contrast to animal narcolepsy models, human narcolepsy is rarely familial and is thought to involve environmental factors acting on a genetically susceptible background (Mignot 1998). The strong association between human leukocyte antigen class II haplotypes DR2 and DQB1*0602 with sporadic human narcolepsy has suggested that an autoimmune process may play a role in the etiology of this disorder (Mignot 1998).

Several recent studies confirm the contribution of disrupted orexin signaling to the etiology of human narcolepsy. Nishino et al (2000) found that orexin-A was undetected in the cerebrospinal fluid (CSF) of 7 of 9 narcoleptic patients, but readily and consistently detected in normal controls. Peyron et al (2000) studied human brains and reported that *orexin* mRNA and peptides are completely absent from narcoleptics but consistently detected in control brains. Futhermore, an unusually severe, early onset case of human narcolepsy is associated with a mutation in the secretory signaling sequence of the *orexin* locus (Peyron et al 2000).

Based on behavioral and polysomnographic criteria, *orexin* knockout mice exhibit a phenotype strikingly similar to human narcoleptics (Chemelli et al 1999). Homozygous knockout mice have frequent narcoleptic/cataplectic attacks during the dark phase, when mice spend the most time awake and active. EEG/EMG recordings reveal significant disruptions of sleep-wakefulness cycling that are also primarily restricted to the dark phase. Graphs illustrating typical vigilance state cycling (hypnograms) for wild-type and *orexin* knockout mice are presented in Figure 3. The *orexin*-null mouse hypnogram is characterized by frequent direct transitions into REM sleep from wakefulness (arrowheads in Figure 3*B*) and marked fragmentation of waking. Quantitative sleep state parameters for *orexin*

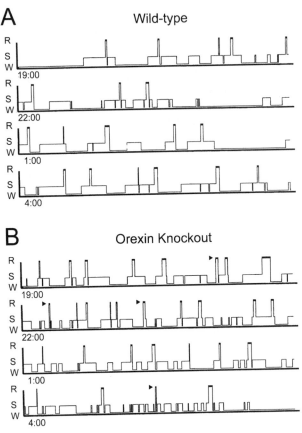

Figure 3 Representative 12-h dark-phase hypnogram of an *orexin* knockout mouse (B) illustrates sleep-onset REM periods (arrowheads), marked sleep fragmentation, and reduced wakefulness compared to a wild-type littermate (A). Hypnograms were obtained by concatenating 20-s epoch stage scores from concurrent EEG/EMG recordings of individually housed male C57Bl/6J-129/SvEv F2 mice. The height of the horizontal line indicates the vigilance state score that is plotted against the time (min) from the beginning of the recording period. Baseline, W, represents a period of wakefulness; S, non-REM sleep; R, REM sleep. Modified with permission from Chemelli et al (1999).

knockout mice reveal significantly decreased waking time, increased non-REM and REM sleep time, decreased REM sleep latency, and, perhaps most importantly, a decreased duration of waking episodes during the dark phase.

A ten-year positional cloning effort to find the canine narcolepsy gene recently identified defects in the *ox2r* gene as the cause of well-characterized autosomal recessive model of narcolepsy in dogs (Lin et al 1999). Mignot's team found intron deletions in the canine gene that cause defective mRNA splicing and consequent production of non-functional OX_2R receptors. Our lab has also created mice with targeted disruptions of the *ox1r* and *ox2r* receptor genes. Preliminary studies in the receptor knockouts show that, as expected, the *ox2r* knockout mice have characteristics of narcolepsy. However, the behavioral and electroencephalographic phenotype of the *ox2r*-null mice is less severe than that found in the *orexin* neuropeptide knockout mice (Chemelli et al 1999, Chemelli et al 2000). Interestingly, the *ox1r* knockouts do not have any overt behavioral abnormalities and exhibit only increased fragmentation of sleep wakefulness cycles (Kisanuki et al 2000). Double receptor knockouts (*ox1r*- and *ox2r*-null mice) appear to be a phenocopy of the ligand knockout mice (Kisanuki et al 2000). This suggests that despite the lack of an overt *ox1r* phenotype, loss of signaling through both receptor pathways is necessary for the severe narcoleptic characteristics of the *orexin*-null mice.

Cataplexy is frequently observed in dogs and mice with disrupted orexin signaling. Strong, generally positive emotional stimuli such as laughter are known to trigger cataplexy in humans with narcolepsy. This implies that orexin neurons may play a role in the physiologic responses associated with emotions. Orexin neuronal projections to the limbic system, dopaminergic ventral tegmental area, and the basal forebrain cholinergic centers are consistent with this suggestion (Date et al 1999, Peyron et al 1998, Sakurai et al 1998, Nakamura et al 2000). Furthermore, neurons controlling cardiovascular responses to emotion map to the perifornical nucleus (Smith et al 1990), an area of the LHA rich in orexin neurons. Centrally administered orexin-A potently stimulates grooming behavior in rats, a behavior that is often associated with a stress response. Therefore, orexin-A may also promote emotional arousal. Orexin-induced grooming can be partially inhibited by a corticotropin-releasing-factor (CRF) antagonist further strengthening the assertion that central stress and orexin signaling pathways may be related (Ida et al 2000). Conversely, orexin signaling is unlikely to increase anxiety because rats exhibit normal exploratory behavior after orexin-A administration (Hagan et al 1999).

The ascending cortical activating system (ACAS) is a diffuse collection of brainstem nuclei grouped together based on their ability to stimulate cortical arousal and wakefulness. Dense projections from orexin neurons to the ACAS including the histaminergic TMN, noradrenergic LC, serotonergic dorsal raphe, and cholinergic pedunculopontine nucleus provide further anatomic evidence of orexin's important role in sleep-wakefulness regulation (Chemelli et al 1999, Date et al 1999, Peyron et al 1998). Slice electrophysiology studies of the LC, an important vigilance-promoting nucleus, found that application of orexin-A dose-dependently increases its intrinsic neuronal firing rate (Hagan et al 1999, Horvath et al 1999b). Taheri et al (2000) found that orexin-A immunoreactivity varies diurnally in the pons, the location of LC, and peaks during the dark phase in rats. This finding appears consistent with a significant role in promoting wakefulness since orexin is most abundant in the region of the LC during the normal waking period in rats. Orexin neurons also innervate the "sleep-active" ventrolateral preoptic area (VLPO). The VLPO is thought to regulate the transition between non-REM and REM sleep by inhibiting the TMN, LC, and median raphe nuclei (Peyron et al 1998, Sherin e 1996, Sherin et al 1998). Orexin-A immunoreactivity in the preoptic/ante hypothalamus, the location of VLPO, also exhibits diurnal variation with the p occurring during the light phase when VLPO neurons are most active (Taheri e 2000). Direct innervation of ventral tegmental dopaminergic neurons (Nakam et al 2000) and orexin receptor expression in the substantia nigra (Marcus e 2000) suggest that orexin may modulate dopaminergic involvement in corti emotional, and motor arousal. Interestingly, dopamine D1 and D2 antagon can dose-dependently suppress orexin-induced hyperlocomotor and groon behaviors (Nakamura et al 2000).

OREXINS INFLUENCE INGESTIVE BEHAVIORS AND METABOLISM

Feeding

Early lesioning experiments of the LHA consistently caused a syndrome decreased food and water intake that lowered body weight set-point to a 75%–80% that of sham-operated controls in several species (Teitelbaum & Eps 1962, Bernardis & Bellinger 1996). Complementary electrical stimulation stu of the LHA found that acute stimulation causes hyperphagia and that chronic sti lation can cause obesity. Electrical self-stimulation studies of the LHA demonst that the LHA participates in a dopaminergic transmission dependent reward sys that is markedly facilitated by food deprivation (Bauco et al 1993, Fouriezc Francis 1992, Carr & Wolinsky 1993, Goldstein et al 1970, Bernardis & Belli 1996). This suggests that energy balance may significantly influence the excit ity of the LHA feeding circuitry.

Acute injection of orexin-A into the lateral ventricles of fed rats, during early light phase, significantly and dose-dependently stimulates food consu tion (Sakurai et al 1998, Edwards et al 1999, Haynes et al 1999, Yamanaka 2000). Similar experiments with orexin-B were inconsistent, but positive stu found the feeding effects to be shorter-lived than those of orexin-A. This sugg that orexin-A may be more resistant to inactivating peptidases due to its disu bonds or that only a subset of the orexin-A feeding pathways (presumably O and OX_2R mediated) may be activated by orexin-B (presumably OX_2R media It should also be noted that orexin-A is significantly less potent at stimulating consumption than NPY under the same conditions. However, its duration of ac is apparently longer than that of NPY (Sakurai et al 1998), and the magni of the maximum effect of orexins is similar to that of other appetite stimula peptides such as MCH and galanin (Edwards et al 1999). The physiologic vance of feeding effects of orexin is supported by the finding that central ad istration of a neutralizing anti-orexin antibody significantly and dose-depende

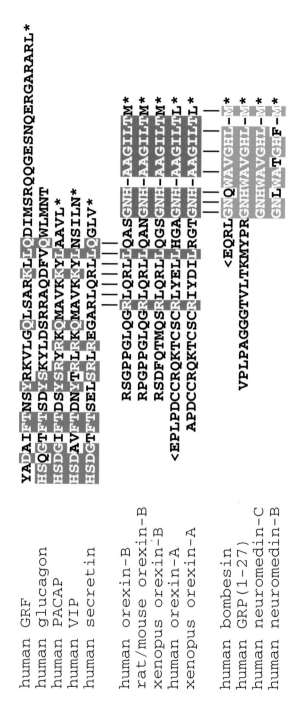

Figure 1 Comparison of orexin sequences with secretin and the bombesin families. "Signature" sequences of peptides related to secretin (green highlights) are found primarily at the amino-terminus. Orexin-B is similar to the carboxy-terminus of secretin, but neither orexin peptide shares significant identity with the other members of the secretin family. In contrast, characteristic sequences of the bombesin family (blue highlights) reside in the carboxy-terminus. In this region, both orexin peptides share significant identity with all bombesin family members. Red highlights depict absolute interspecific and interisopeptide identity among orexins. Note that there are two intrachain disulfide bonds in orexin-A (Cys6-Cys12 and Cys7-Cys14) but not in orexin-B. Abbreviations and symbols: GRF, growth hormone-releasing factor; PACAP, pituitary adenylyl cyclase-activating peptide; VIP, vasoactive intestinal peptide; GRP, gastrin-releasing peptide; *, C-terminal amide; <E, pyroglutamyl residue.

suppresses spontaneous feeding in fasted rats (Yamada et al 2000). In addition, a selective OX_1R antagonist can inhibit natural feeding over several days as well as feeding stimulated by fasting or i.c.v. injection of orexin-A (Arch 2000). Furthermore, orexin-A dose-dependently increases gastric acid secretion only when given centrally and with an intact vagus nerve, suggesting a role for orexin in the cephalic phase of digestion and the brain-gut axis (Takahashi et al 1999). A confounding factor in orexin-feeding studies to date is the potential that these effects may be partly or completely secondary to changes in other unmeasured variables, such as wakefulness.

Circadian processes have a marked influence on feeding behavior and disruptions of normal circadian feeding patterns are well-described effects of LHA lesions (Bernardis & Bellinger 1996). Interestingly, Haynes et al (1999) found that the feeding response to acute i.c.v. injection of orexin-A is highly dependent on the time of day. They found the largest increase in orexin-stimulated food intake occurred in the early light phase and 6 hr into the dark phase when normal food intake is at a nadir or slowing. Injection of orexin-A at the beginning of the dark phase, when the normal feeding rate is at its highest, had no effect. Similarly, orexin-A was ineffective at increasing intake in the first hour of refeeding after a fast. It seems reasonable to speculate that at the beginning of the dark cycle and immediately after a fast, orexin-stimulated feeding pathways are already maximally activated and therefore unresponsive to additional pharmacologic stimulation by orexin-A.

Chronic infusion of orexin-A over several days disrupts the normal circadian feeding pattern in rats by increasing daytime and decreasing nighttime food intake (Haynes et al 1999, Yamanaka et al 1999). These studies showed no effect on total daily food intake, adiposity, or body weight. Interestingly, while MCH and galanin yield similar findings, continuous infusion of NPY is the only known neuropeptide that can induce obesity from over-eating (Flier & Maratos-Flier 1998). The inability of chronic orexin-A infusions to increase overall food consumption or bodyweight may be due to a circadian variation in the relative responsiveness to exogenous orexin. Food intake is increased during the day when orexin responsiveness is high (and endogenous orexin activity is low), but this daytime hyperphagia may result in counter-regulatory measures that later reduce the drive to eat during periods of relative orexin-insensitivity (when endogenous orexin activity is high) (Haynes et al 1999). Increased wakefulness is another possible confounding factor: in continuously infused rats, Yamanaka et al (1999) observed increased daytime wakefulness that per se may have resulted in increased feeding.

Nuclei-specific studies performed by Dube et al (1999) found that orexin-A stimulates feeding when microinjected into the PVN, DMN, LHA, and perifornical area of rats. Parallel experiments using orexin-B showed no effect on feeding. Micro-injection into the ARC, VMH, preoptic area, central nucleus of the amygdala, and NTS with either orexin-A or -B did not stimulate feeding (Dube et al 1999). Others, however, have found that orexin-A stimulates feeding in the ARC nucleus (T Yada, S Muroya, H Funahashi, S Shioda, A Yamanaka, et al, submitted for publication). The reasons for this discrepancy are unclear.

There are several examples of spontaneous single-gene mutations and targeted deletions of anorectic signals, satiety factors, and metabolic enzymes that can cause obesity in animal models (for review see Salton et al 2000, Flier & Maratos-Flier 1998). Conversely, mice with null disruptions of orexigenic factors such as NPY, galanin, and β-endorphin typically exhibit normal feeding behavior, body weight, and metabolic homeostasis. Based on these observations, several authors have suggested that there is a robust redundancy among appetite-generating pathways. Disruption of the MCH signaling pathway appears to be the only published exception. Mice with targeted deletions of the MCH gene are hypophagic with decreased body weight and increased oxygen consumption suggesting a hypermetabolic state (Shimada et al 1998). This pattern of findings is remarkably similar to the LHA lesion syndrome. *orexin* knockout mice are also significantly hypophagic (Figure 4), but have normal body weight suggesting differences in energy homeostasis and metabolic rate (RM Chemelli, unpublished observations). Whether decreased feeding is caused by disruptions in appetite pathways, arousal pathways, or both will need to be examined more closely.

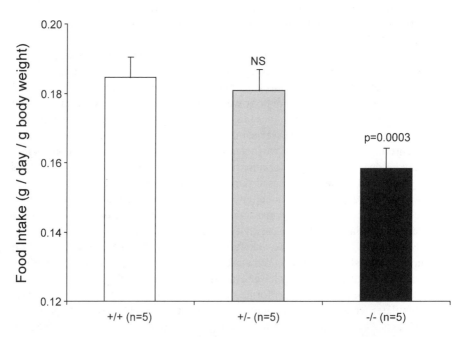

Figure 4   *Orexin* knockout mice are hypophagic. Data represent 7-day mean food intake adjusted for body weight. Age- and weight-matched group-housed male C57B1/6J-129/SvEv F2 mice were fed freely throughout the study. Despite hypophagia, these mice maintain normal growth curves, suggesting a reduced metabolic rate (RM Chemelli, unpublished observations).

Drinking

Early electrical stimulation studies also suggested that the LHA is a "primary drinking center" (Greer 1955, Rowland 1976). LHA lesioning studies cause adipsia that always outlasts aphagia as these animals approach a lower body weight set-point. Interestingly, animals with lesions become "prandial drinkers;" they only drink when they eat and drink just enough to enable mastication (Bernardis & Bellinger 1996). Centrally injected orexins increase drinking at doses similar to those that elicit increased feeding (Kunii et al 1999). This effect is similar in potency to angiotensin II but orexin-stimulated water intake is longer–lasting. These effects are consistent with orexin innervation of the subfornical organ and the area postrema, regions known to be important in fluid homeostasis. In addition, *prepro-orexin* mRNA levels are upregulated during fluid deprivation (Kunii et al 1999).

Metabolic, Autonomic, and Endocrine Effects

The LHA is also important in controlling metabolic rate. Animal models with lesions in the LHA consistently become hypercatabolic and remain so, even after they reach a lower body weight set-point (Bernardis & Bellinger 1993, Powley & Keesey 1970). The apparent hypermetabolic phenotype of MCH knockout mice is consistent with these findings (Shimada et al 1998). Conversely, the finding of decreased food intake and normal body weight suggests that *orexin* knockout mice are likely to be hypometabolic. This is supported by experiments performed by Lubkin & Stricker-Kongrad (1998) who studied the metabolic effects of i.c.v. orexin injection using indirect calorimetry. They found that orexin-A injection during the light phase increases oxygen consumption and the respiratory quotient by an apparent increase in carbohydrate metabolism. Interestingly, the effect on respiratory quotient is dependent on circadian phase because, although oxygen consumption increases, the respiratory quotient decreases when orexin-A was given in the dark phase.

Several studies suggest that orexin neurons may also influence neuroendocrine homeostasis and modulate pituitary hormone release. Orexin neurons project densely to other nuclei within the hypothalamus and have lesser projections to the pituitary (de Lecea et al 1998, Peyron et al 1998, van den Pol et al 1998, Date et al 1999). Neuroendocrine cells from the arcuate nucleus, in a slice electrophysiology preparation, demonstrate a dose-dependent increase in inhibitory post-synaptic potentials in response to orexin neuropeptide application (van den Pol et al 1998). Several studies show changes in pituitary hormone secretion in response to orexin administration. Hagan et al (1999) found that i.c.v. orexin-A significantly decreases prolactin and growth hormone and increases corticosterone plasma levels. Pulsatile luteinizing hormone secretion is also suppressed suggesting a role in the coordination of metabolic and reproductive functions (Pu et al 1998, Tamura et al 1999).

Orexin projections to the nucleus of the solitary tract, dorsal motor nucleus of the vagus, and sympathetic neurons in the intermediolateral column of the spinal cord suggest that orexins may modulate autonomic function and participate in the stress response (Date et al 1999, van den Pol et al 1999). Both central and peripheral administration of orexin increase plasma levels of corticosterone (Hagan et al 1999, Malendowicz et al 1999), and orexins can stimulate corticosterone release directly from adrenocortical cells in vitro (Malendowicz et al 1999). Furthermore, *ox1r* and *ox2r* mRNAs are expressed in the adrenal medulla, suggesting that orexins may modulate systemic epinephrine release and therefore influence vascular tone (Lopez et al 1999). This proposal is consistent with studies by several investigators who found that injection of orexin dose-dependently increases heart rate and blood pressure (Chen et al 2000, Samson et al 1999, Shirasaka et al 1999). The sympathomimetic effects of orexin may also indirectly influence orexin-mediated changes in oxygen consumption and substrate utilization.

INTERACTION WITH HYPOTHALAMIC FEEDING PATHWAYS

Leptin is an anorectic protein produced and secreted in proportion to adipocyte fat stores and therefore it is commonly referred to as the "adipostat." Exogenously administered leptin can inhibit the increased feeding stimulated by several orexigenic neuropeptides. In experiments to test whether orexin-mediated feeding is sensitive to leptin, Yamanaka et al (A Yamanaka, K Kunii, N Tsujino, I Matsuzaki, K Goto, T Sakurai, submitted for publication) pre-treated rats with leptin at doses that completely inhibit NPY- and galanin-stimulated feeding. They found that leptin only partially inhibited orexin-mediated feeding. The effect of anorectic peptides, thought to mediate leptin effects on food intake, were then examined. Pre-treatment with either CART (cocaine- and amphetamine-related transcript), αMSH (proopiomelanocortin-derived α-melanocyte-stimulating hormone), or GLP-1 (glucagon-like peptide-1), at doses that abolished NPY-induced feeding, again only partially inhibited orexin-induced feeding. These studies suggest that appetite-stimulating pathways involving NPY and galanin are leptin-sensitive, while orexin-mediated feeding involves both leptin-sensitive and -insensitive pathways. Because leptin has recently been shown to promote slow wave (delta) at the expense of REM sleep (Sinton et al 1999), the effect of leptin on orexin-mediated arousal should be studied.

The ARC nucleus is a major site of leptin-responsive neurons and is regarded as an important "satiety center" on the basis of classic lesioning studies (Kalra et al 1999, Elmquist et al 1999). This nucleus is a complex collection of neurons that express most of the well-described orexigenic and anorectic neuropeptides. Leptin-mediated inhibition of orexigenic NPY/agouti-related protein (AgRP)-coexpressing neurons and excitation of anorectic proopiomelanocortin (POMC)/CART-coexpressing neurons are thought to underlie the suppression of appetite by leptin (Hakansson et al 1998, Elmquist et al 1999). Orexin neurons project to the ARC nucleus (Peyron et al 1998, Date et al 1999) and specifically innervate NPY neurons (Horvath et al 1999a). Reciprocal connections from NPY/AgRP neurons to orexin and MCH neurons in LHA have also been identified (Horvath et al 1999a, Elias et al 1998). Furthermore, orexin and MCH neurons are also innervated by αMSH-immunoreactive fibers from the ARC (Elias et al 1998). Orexin neurons may also be directly inhibited by leptin because leptin receptor immunoreactivity and STAT-3, a transcription factor activated by leptin, are found in orexin neurons (Horvath et al 1999a, Hakansson et al 1999). It seems likely that orexin neuronal activity is influenced directly by leptin, indirectly through down-stream leptin sensitive pathways, and can provide feedback to these same afferent feeding pathways through reciprocal projections.

Fos expression in NPY neurons of the ARC nucleus is induced by i.c.v. injection of orexin, suggesting that orexin-stimulated feeding may occur through NPY pathways (Date et al 1999, Yamanaka et al 2000). To test this hypothesis, Yamanaka et al (2000) studied orexin-induced feeding during complete pharmacologic inhibition of NPY-stimulated feeding. They found that orexin-stimulated feeding was only partially inhibited by pre-treatment with a selective NPY Y1 receptor antagonist. This suggests that activation of NPY-mediated feeding pathways is only partially responsible for orexin-stimulated feeding. It is unlikely that these results were confounded by decreased arousal because NPY inhibits arousal-related behaviors (Ida et al 1999) and has been shown to promote non-REM sleep (Steiger et al 1998). NPY receptor antagonists would therefore be expected to increase arousal and wakefulness. Furthermore, microinjection of orexin-A or -B directly into the ARC nucleus selectively increases feeding without increasing locomotor and stereotypic behaviors associated with i.c.v. orexin (T Yada, S Muroya, H Funahashi, S Shioda, A Yamanaka, et al, submitted for publication). These experiments suggest that orexin-stimulated food intake is not simply due to increased arousal or prolonged wakefulness.

Isolated ARC neurons were studied to determine their responses to orexin and leptin using an in vitro fura-2 microfluorimetry system (T Yada, S Muroya, H Funahashi, S Shioda, A Yamanaka, et al, submitted for publication). Application of orexin-A or -B to POMC neurons decreased cytosolic calcium ($[Ca^{2+}]_i$) and inhibited $[Ca^{2+}]_i$ oscillations with a picomolar sensitivity. This effect had an equimolar sensitivity to orexin-A and orexin-B and was inhibited by pertussis toxin suggesting that it is mediated through an OX_2R pathway that is coupled to G_i and/or G_o. Interestingly, many of the POMC neurons inhibited by orexin were excited by picomolar concentrations of leptin. In contrast, arcuate NPY neurons are inhibited by leptin and excited by orexin-A and -B, with orexin-B 100-fold less potent than orexin-A. Selective inhibitors of phospholipase C, protein kinase C, and inositol tri-phosphate-dependent Ca^{2+} channels block these excitatory effects suggesting the involvement of OX_1R signaling through a G_q-coupled pathway.

The VMH is another important component of the "satiety center" and is characterized by a high density of glucose and insulin-responsive neurons (Oomura & Kita

1981). Although orexin neurons project to the VMH (Peyron et al 1998), no feeding response was found by Dube et al (1999) when either orexin-A or –B was microinjected into this site. VMH neurons are, however, inhibited by orexin and excited by leptin in vitro (T Yada, S Muroya, H Funahashi, S Shioda, A Yamanaka et al, submitted for publication). The molecular antagonism between orexin and leptin within the satiety-promoting neurons of the VMH appear to mimic the opposing functions of the LHA and VMH in feeding regulation (Bernardis & Bellinger 1996).

There is considerable evidence to support a role for orexin in directly stimulating food intake and in coordinating the orexigenic and anorectic feeding pathways of the ARC and VMH nuclei to increase feeding. The antagonism between orexin and leptin-mediated effects in these nuclei is consistent with a model in which orexin-stimulated feeding occurs partially within the context of pathways inhibited by leptin. The potential executive function of orexin in coordinating the known hypothalamic feeding pathways is exemplified by its unique ability to provide simultaneous stimulation of orexigenic (OX_1R-mediated) and inhibition of anorectic (OX_2R-mediated) feeding circuits. It is unknown whether orexin signaling mechanisms identified in the ARC and VMH nuclei, characterized by inputs that are both excitatory (OX_1R) and inhibitory (OX_2R), are generalizable. However, these mechanisms may provide a model of orexin action at other sites.

REGULATION OF OREXIN NEURONS

The LHA receives innervation from much of the neuroaxis and can also be influenced by actively transported peripheral factors, such as leptin, insulin, and other hormones, as well as diffusible factors including glucose, electrolytes, amino acids, and peptides (Bernardis & Bellinger 1996). Therefore the regulation of orexin neurons is potentially complex. Because of orexin's restriction to the LHA, and its early association with feeding (Sakurai et al 1998), most of the work on defining molecular regulatory mechanisms has focused predominantly on feeding and metabolism rather than on pathways involving sleep and wakefulness.

Circadian Influences

Orexin neurons are significantly influenced by circadian processes because Fos studies indicate that orexin neurons are primarily "waking active" (Estabrooke et al 2000), and circadian feeding patterns are disrupted by chronic central orexin-A injection (Haynes et al 1999). The suprachiasmatic nucleus (SCN), the CNS circadian oscillator, appears to play a prominent role in controlling diurnal feeding patterns (Zucker & Stephan 1973). The SCN projects to most of the hypothalamus including the LHA, and SCN lesions eliminate the circadian patterns of sleep-wakefulness and food intake as well as the circadian electrical activity of many LHA neurons (Bernardis & Bellinger 1996, Edgar et al 1993). The retino-geniculo-hypothalamic tract, a source of photic information to the SCN, projects to the LHA and may influence circadian activity in the LHA (Mikkelsen 1990). Interestingly, Kurumiya & Kawamura (1991) found that the LHA has a weak oscillatory capacity of its own when the SCN is lesioned. Although orexin activates SCN neurons (Date et al 1999), direct SCN and/or retino-geniculo-hypothalamic efferents to orexin neurons have not yet been reported. The SCN may also indirectly influence orexin neurons through other hypothalamic and thalamic nuclei that in turn project to the LHA. The paraventricular thalamic nucleus, for example, receives input from the SCN, and the paraventricular thalamic nucleus projects to the LHA (Watts et al 1987, Moga et al 1995). Spontaneous activity of glucose-sensitive LHA neurons, that probably include orexin neurons (discussed below), exhibit a distinct circadian firing pattern (Schmitt 1973), suggesting that metabolic and circadian factors influence these neurons. It is unclear whether orexin signaling pathways secondarily mediate circadian feeding and sleep patterns based on SCN signals, or if circadian orexin activity is just an epi-phenomenon of these processes. The latter seems more likely though, because daytime sleep patterns are unaffected (Chemelli et al 1999) and diurnal variation in food intake is intact (RM Chemelli, unpublished observations) in *orexin* knockout mice.

Nutritional and Visceral Satiety Signals

The LHA reward system is modulated by inhibitory signals from the VMH (Margules & Olds 1962) and it is highly dependent on nutritional state, demonstrating marked potentiation following food deprivation (Carr & Wolinsky 1993, Goldstein et al 1970). LHA neuronal activity is suppressed by elevated blood glucose suggesting short-term nutrient sensing. In contrast, VMH neurons are excited by glucose, consistent with the antagonistic roles of the LHA and VMH in feeding regulation (Bernardis & Bellinger 1996). LHA neurons respond to the presence of food in the gut partly as a result of vagally transmitted afferent intestinal signals that are relayed through the nucleus of the solitary tract (Bernardis & Bellinger 1996).

Orexin neurons are activated by acute hypoglycemia. Bahjaoui-Bouhaddi et al (1994a) found that insulin, given to acutely lower blood glucose levels in rats, caused a marked increase in Fos-immunoreactivity in LHA neurons now known to express orexin (recently confirmed in Moriguchi et al 1999). When insulin and glucose were given together so that euglycemia was maintained, no increase above baseline Fos levels was found. This, in combination with the report that insulin receptor mRNA is not expressed in the LHA (Marks et al 1990), suggest that insulin is unlikely to act directly on these neurons. Also, it is unlikely that reduced plasma osmolarity is a confounding factor in these experiments because there were no significant differences in plasma osmolality between the control and experimental groups. Insulin-induced hypoglycemia does not increase Fos in MCH neurons (Bahjaoui-Bouhaddi et al 1994b), suggesting that the mechanisms regulating orexin and MCH neurons are different.

Cai et al (1999) performed a detailed study of *orexin* mRNA expression in rats after physiologic and pharmacologic manipulations designed to stimulate food intake through different mechanisms. These authors fasted rats for 48 hours and found that *orexin* mRNA levels significantly increased over controls, accompanied by decreases in serum leptin and glucose. No significant difference in *orexin* mRNA levels were found when rats were restricted to 50% of their usual daily intake for six days. Serum leptin and glucose levels at the end of this experiment were 38% and 83% of control values, respectively. These starvation studies suggest that *orexin* expression increase markedly in response to acute metabolic challenges, but approach normal after sub-acute stresses when compensatory metabolic processes are likely to intervene (glucose and leptin returning toward baseline values). These authors then studied rats after 6 hours of acute hypoglycemia and after six days of chronic hypoglycemia induced by subcutaneous insulin injections. A significant increase in *orexin* mRNA expression was found after acute hypoglycemia when rats were food-deprived after insulin injection, but no change was seen when they were allowed to feed freely. Leptin and glucose levels were 228% and 49% vs. 228% and 17% of control values in the freely fed vs. food-deprived conditions, respectively. An increase in *orexin* mRNA expression that approached significance was observed after chronic hypoglycemia and freely-fed conditions, with serum leptin and glucose levels 288% and 47% of controls. These authors identified subnormal plasma glucose levels and absence of food intake as key factors associated with increased *orexin* mRNA expression. They further suggested that orexin neurons may belong to a glucose-sensitive sub-population of LHA neurons that are stimulated by falls in serum glucose and inhibited by vagally-mediated prandial signals, such as gastric distension and/or rising portal venous glucose concentrations (Bernardis & Bellinger 1996). These hypoglycemia studies also suggest that orexin neurons are more responsive to nutrient depletion (hypoglycemia) than nutrient excess signals (hyperleptinemia) when faced with antagonistic influences.

orexin expression may also be regulated by changes in serum osmolarity. *orexin* mRNA is significantly upregulated after 48 hours of water deprivation and Fos immunoreactivity is increased under hyperosmotic conditions (Kunii et al 1999). However, no studies of whether orexin neurons are directly osmosensitive are available.

Leptin and Hypothalamic Feeding Peptides

The LHA receives afferent projections from many CNS areas known to be important in energy homeostasis (Bernardis & Bellinger 1996). Broberger et al (1998) and Elias et al (1998) have presented convincing neuroanatomic evidence of extensive reciprocal innervation between the leptin-sensitive feeding pathways of the ARC nucleus and orexin neurons of the LHA. Leptin receptor immunoreactivity has also been identified in orexin neurons, raising the possibility of leptin directly influencing these neurons (Horvath et al 1999a, Hakansson et al 1999). Interestingly, leptin was recently demonstrated to inhibit LHA reward mechanisms in a self-stimulation paradigm (Fulton et al 2000).

Leptin regulates the expression of many orexigenic and anorectic signals (Kalra et al 1999). Lopez et al (2000) found that leptin injection suppresses the well-described fasting-induced increase in *orexin* mRNA expression in rats. They also found increased *ox1r* receptor mRNA expression with fasting that was suppressed by leptin treatment. No change in *ox2r* receptor mRNA expression was detected in either fasted or leptin-treated conditions. These results suggest that leptin influences *orexin* gene expression in food-restricted rats and that the OX_1R signaling pathway is leptin-sensitive and involved in the efferent response to fasting. Interestingly, no effect on *orexin*, *ox1r*, or *ox2r* mRNA expression was found after leptin-treatment under freely fed conditions.

Genetic mouse models of obesity have been instrumental in identifying and characterizing many hypothalamic feeding pathways. Deficient leptin signaling causes hyperphagia, abnormal glucose utilization, infertility, and early-onset obesity in *ob/ob* and *db/db* mice. The leptin gene has a nonsense mutation in the *ob/ob* mouse whereas the *db/db* mouse has an inactivating mutation in the leptin receptor gene (Zhang et al 1994, Chen et al 1996). In lethal yellow (A^y/a) mice, a genetic rearrangement causes ubiquitous over-expression of the coat color signaling protein agouti (Miller et al 1993). Peripheral over-expression causes the characteristic yellow coat color while central over-expression antagonizes α-MSH at the melanocortin-4 receptor (MC4-R), mimicking the native actions of hypothalamic AgRP (Barsh 1999, Salton et al 2000). Unregulated chronic antagonism of the anorectic neuropeptide α-MSH causes a dominantly inherited maturity onset obesity syndrome associated with hyperphagia, hyperinsulinemia, hyperglycemia, and hyperleptinemia (Salton et al 2000, Maffei et al 1995). Orexin and MCH neurons are strongly innervated by leptin-sensitive AgRP, α-MSH, and NPY neurons (Elmquist et al 1999), and *mcr-4* mRNA is highly expressed within the lateral hypothalamus (Elias et al 1998).

Studies of *Mch*, *Npy*, and *orexin* gene expression in these mouse models of obesity suggest that orexin is regulated in a manner opposite to that of MCH and NPY. In freely-fed *ob/ob*, *db/db*, and A^y/a mice, *Mch* and *Npy* expression are upregulated compared with normal controls (Qu et al 1996, Hanada et al 2000, Stephens et al 1995, Kesterson et al 1997), whereas *orexin* expression is decreased (Yamamoto et al 1999, Hanada et al 2000). This suggests that the MCH and NPY pathways are normally inhibited by leptin and that increased *Mch* and *Npy* expression may contribute to hyperphagia in these mice. This is supported by decreased food intake and partial amelioration of obesity in *Npy-null*, *ob/ob* double mutants (Erickson et al 1996, Salton et al 2000). The finding of decreased *orexin* expression in these obese mutants is contrary to a model in which *orexin* is predicted to be upregulated if it is directly inhibited by leptin. It is however, consistent with upregulation of leptin-sensitive NPY and MCH efferents that cause hyperphagia and contradictory signals of nutritional excess, such as hyperglycemia, increased free fatty acids, and visceral satiety afferents that consequently override the lack of leptin signaling and decrease *orexin* expression. Regardless of the specific mechanisms involved, it is clear that orexin is unlikely to contribute to the characteristic hyperphagia in these mouse models of obesity. It is intriguing to speculate, however, that decreased *orexin* expression contributes to the decreased metabolic rate that is believed necessary to explain the extreme obesity of these genetic models since it cannot be explained on grounds of increased food intake alone.

Apparently contradictory results have been obtained from studies examining *orexin* expression when *ob/ob* and A^y/a mice are fasted. Yamamoto et al (1999) found that *orexin* expression is increased in *ob/ob* mice food restricted for 2 weeks compared with freely fed controls. Whereas Hanada et al (2000) found that *orexin* mRNA expression was unchanged in A^y/a mice fasted for 48 hours despite a marked improvement in hyperglycemia. Although different experimental methods (food restriction versus fasting) may partially explain these results, these models are not directly comparable because *ob/ob* mice lack leptin while A^y/a mice are hyperleptinemic. In the absence of leptin (ob/ob mice) and with concurrent signals of nutritional depletion, such as falling glucose and decreased visceral satiety afferents, upregulation of *orexin* in food-restricted *ob/ob* mice makes physiologic sense. Conversely, persistent marked hyperleptinemia accompanied by normalization of plasma glucose might not lead to changes in *orexin* expression in the fasted A^y/a mouse. It is also possible that *orexin* expression in mouse models of obesity might be a reflection of obesity-induced changes in wakefulness that can be influenced by fasting, but these studies have not been performed.

DISCUSSION

Feeding behavior is critically dependent on appropriate sleep-wakefulness cycling at environmentally advantageous times and in response to homeostatic needs. While feeding and sleep are mutually exclusive behaviors, wakefulness, with increased sensory awareness and motor activity, is required for successful feeding. The relationship among waking, feeding, and environment are highly species dependent. While humans consolidate waking and feeding cycles during the daytime, rodents are awake and feed primarily at night. It can be argued that the balance between food availability and the relative risks of predation were the evolutionary driving forces behind these divergent behavioral patterns. Similarly, during periods of nutritional depletion, central mechanisms capable of augmenting wakefulness and therefore feeding opportunities within the proper circadian phase would confer an advantage. Because of the co-dependency between sleep-wakefulness and feeding, stressors that primarily affect one very often affect the other. Indeed, comorbid disturbances of sleep, appetite, and metabolism are well-described symptoms of obesity, anorexia nervosa, depression, and Cushings disease (Crisp et al 1970, Crisp et al 1971, Fleming 1989, Wurtman & Wurtman 1995, Shipley et al 1992).

Substantial evidence is accumulating to support the hypothesis that pathologic disruption of orexin signaling causes narcolepsy in humans, mice, and dogs. Several studies suggest that altered energy homeostasis may accompany narcolepsy in these species. Honda et al (1986) found an increased incidence of non-insulin-dependent diabetes mellitus in an adult Japanese population with narcolepsy. Another study found lower daily caloric intake in narcoleptics compared with matched controls, and that the difference was due primarily to reduced carbohydrate intake (Lammers et al 1996). Schuld et al (2000) found that narcoleptic patients have an increased body mass index suggesting a tendency toward obesity. Further support for the proposed link between altered metabolism and narcolepsy include many case reports of obesity in pediatric narcolepsy (Dahl et al 1994, Allsopp et al 1992, Kotagal et al 1990). The finding of decreased caloric intake combined with an increased body mass index suggest that narcoleptics have reduced energy expenditure or metabolic rate. Interestingly, *orexin* knockout mice consume significantly less food (Figure 4) but maintain the same body weight as young age-matched littermates suggesting differences in energy homeostasis and metabolic rate (RM Chemelli unpublished observations).

No feeding or metabolic abnormalities have been reported in the *ox2r*-null dog model of narcolepsy. While this may be explained by a lack of surveillance for a subtle phenotype, it may indicate that the OX_1R-signaling pathway is more important in orexin-mediated energy homeostasis. Several lines of evidence support this hypothesis: There is a marked differential expression pattern of orexin receptors in the CNS, and importantly, a high density of OX_1Rs are found in the ARC and VMH nuclei of the hypothalamus (Marcus et al 2000, Trivedi et al 1998). An OX_1R-selective antagonist can inhibit orexin-mediated feeding and natural feeding for several days (Arch 2000). And, Lopez et al (2000) found increased *ox1r* expression in rats fasted for 48 hours and that this response could be ameliorated by leptin, but there was no change in *ox2r* expression. It is clear that further study of narcoleptic patients and animal models will be important to uncover the physiologic mechanisms underlying these metabolic changes.

Researchers have long recognized that fasting alters sleep-wakefulness patterns in several species (Borbely 1977, Danguir & Nicolaidas 1979, Faradji et al 1979, Rashotte et al 1998). These investigators found that food deprivation increases waking, decreases non-REM, and markedly reduces REM sleep times. Interestingly, these changes occur only during the normal circadian feeding period of the respective species, except in profound starvation when increased wakefulness extends into the normal sleep period. The magnitude of the changes in wakefulness and non-REM sleep diminish as food deprivation is prolonged, but the striking REM suppression continues. The magnitudes of the fasting-induced changes in vigilance states are also highly dependent on prior nutritional state. Danguir and Nicolaidis (1979) found highly significant increases in waking and reductions in non-REM and REM sleep time in fasted lean rats (240–250 g) but minimal changes in fasted obese rats (300–380 g). The fasting-induced changes found in the lean rats could be reversed by glucose infusion, suggesting that homeostatic sensing of nutritional depletion rather than absence of food in the gut was the critical factor. Conversely, these authors also found that meal size is positively correlated with the subsequent duration of REM and non-REM sleep. In summary, these results suggest that acute reductions in energy substrate availability (e.g., glucose, free fatty acids, and amino acids) stimulate increased waking time, thus presumably increasing the chance of feeding opportunities. The magnitude of this effect is highly dependent upon endogenous energy reserves (possibly through leptin as an indicator of fat stores and metabolic state). And, as food deprivation is prolonged, compensatory mechanisms that conserve energy are engaged and energetically costly increases in wakefulness diminish. It is interesting to speculate that the persistent suppression of REM may also increase feeding opportunities by shifting sleep-rest periods into vigilance states with increased sensory arousal and motor tone.

The regulation of both sleeping and feeding behaviors in mammals is thought to result from the integration of homeostatic and circadian influences (Borbely 1998, Kalra et al 1999). Sleep is promoted homeostatically by prior wakefulness (Borbely 1998), and biologically active "sleep factors" such as adenosine (Porkka-Heiskanen 1997) have been identified that accumulate in the brain during wakefulness and dissipate during sleep. Similarly, feeding behavior and central metabolic control are regulated homeostatically by prior nutritional state. Acute systemic state is mirrored by circulating metabolites, including glucose, free fatty acids, and amino acids, whereas long-term energy reserves are typically reflected in circulating leptin levels (Kalra et al 1999). These homeostatic mechanisms may indicate that the central regulation of both sleep and feeding behaviors is "ischymetric," defined as depending on the rate and degree of utilization of circulating metabolites at the cellular level (Danguir & Nicolaidis 1980b, Nicolaidis & Even 1992). Indeed, there is substantial evidence to suggest that the LHA is an important site for the ischymetric regulation of feeding-related behaviors (Bernardis & Bellinger 1996). Furthermore, the LHA is increasingly recognized as an important center for sleep-wakefulness regulation based on neuroanatomic (Date et al 1999, Peyron et al 1998), physiologic (Bernardis & Bellinger 1996, Levitt & Teitelbaum 1975, Danguir & Nicolaidis 1980a, Steininger et al 1999), and recent molecular genetic findings (Chemelli et al 1999, Lin et al 1999, Peyron et al 2000). It seems likely that the LHA plays a significant role in coordinating these processes because fasting-induced changes in sleep are highly dependent on endogenous energy stores (Danguir & Nicolaidis, 1979), and non-REM sleep is promoted by feeding neuropeptides such as leptin, galanin, and NPY (Sinton et al 1999, Steiger et al 1998). In addition to homeostatic influences, the LHA is also anatomically positioned to integrate circadian rhythms from the SCN (Bernardis & Bellinger 1996, Edgar et al 1993) and photic information via the retino-geniculo-hypothalamic tracts (Mikkelsen 1990).

Orexin neurons have the requisite anatomic connections, physiologic effects, interactions with hypothalamic feeding pathways, and regulation by circadian and nutritional factors to suggest that they may be an important cellular and molecular link in the integration of sleep and energy homeostasis (Figure 5). Efferent and afferent innervation of orexin cells suggest interactions between these cells and important feeding centers in the ARC and VMH nuclei, arousal and sleep-wakefulness centers in the ACAS and VLPO, sympathetic and parasympathetic nuclei, dopaminergic motor centers, and the limbic system. Numerous studies

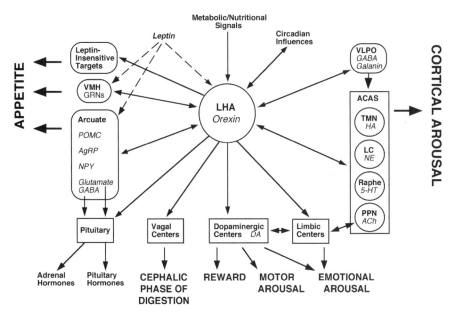

Figure 5 Model of orexin function in the coordination of energy and sleep homeostasis. Peripheral metabolic signals, leptin, and circadian rhythms influence orexin neuronal activity. Orexin neurons stimulate leptin-sensitive and -insensitive targets to promote appetite and the cephalic phase of digestion. Orexins increase cortical arousal and promote wakefulness through the ascending cortical activating system (ACAS) and other sleep-related nuclei. Stimulation of dopaminergic, limbic, and cholinergic centers by orexins can modulate reward systems, motor activity, and emotional arousal. Endocrine function is influenced by orexin efferents within hypothalamic and pituitary sites. See text for details. Abbreviations and symbols: VMH, ventromedial hypothalamus; GRNs, glucose-responsive neurons; POMC, pro-opiomelanocortin; AgRP, agouti-related protein; NPY, neuropeptide Y; LHA, lateral hypothalamic area; VLPO, ventrolateral preoptic area; GABA, γ-aminobutyric acid; TMN, tuberomamillary nucleus; HA, histamine; LC, locus coeruleus; NE, norepinephrine; 5-HT, serotonin; ACh, acetylcholine; DA, dopamine.

have documented that acute orexin injection dose-dependently increases arousal-associated behavioral and physiologic process including food-intake, waking time, motor activity, and metabolic rate, as well as heart rate and blood pressure. It is unlikely, however, that orexin-signaling plays a significant role in long-term "lipostat" mechanisms since chronic administration fails to increase body weight, total daily food intake, or adiposity. Furthermore, it is likely that orexin-mediated food intake results partly from stimulation of ARC feeding pathways such as those involving NPY. The increased expression of *Npy* and *Mch* and decreased expression of *orexin* found in hyperphagic mouse models of obesity are consistent with this hypothesis. Fasting studies in normal and fat-mutant mice suggest that orexin neurons are more responsive to acute changes in energy substrate availability than endogenous energy stores. The potential role of orexin as a barometer of acute energy homeostasis is further exemplified by its upregulation when integrating contradictory metabolic signals such as hypoglycemia in the face of hyperleptinemia. The strong "waking active" character of orexin neurons also reinforces the hypothesis that orexin signaling is important in the acute regulation of energy homeostasis: short-term changes in substrate availability likely lead to increased orexin neuronal firing that in turn increases sensory arousal, wakefulness, motor activity, and substrate mobilization from energy stores, while stimulating and reinforcing hypothalamic feeding pathways.

The orexin neuropeptide signaling system is an important central pathway that promotes adaptive behavioral and physiologic responses in response to metabolic and environmental signals. During periods of nutritional depletion, orexin-induced increases in arousal and reinforcement of appetite/feeding pathways may be an evolutionarily conserved mechanism that helps to ensure survival. In summary, the orexin neuropeptide system plays a significant role in coordination of feeding and sleep-wakefulness regulation, probably by coordinating the complex behavioral and physiologic responses of these complementary homeostatic functions.

ACKNOWLEDGMENTS

We thank M Brown, S Tokita, Y Kisanuki, C Beuckmann, and H Gershenfeld for helpful discussions. We also thank T Sakurai, T Yada, J Elmquist, and T Scammell for communication of results prior to publication. JT Willie is a joint fellow of the Medical Scientist Training Program and the Department of Cell and Molecular Biology of UT Southwestern Medical Center at Dallas. M Yanagisawa is an Investigator of the Howard Hughes Medical Institute. This work is supported in part by research funds from the WM Keck Foundation and the Perot Family Foundation.

Visit the Annual Reviews home page at www.AnnualReviews.org

LITERATURE CITED

Allsopp MR, Zaiwalla Z. 1992. Narcolepsy. *Arch. Dis. Child.* 67:302–306

Arch JRS. 2000. The role of orexins in the regulation of feeding: a perspective. *Regul. Pept.* 89:51 (Abstr.)

Bahjaoui-Bouhaddi M, Fellmann D, Bugnon C. 1994a. Induction of Fos immunoreactivity in prolactin-like containing neurons of the rat lateral hypothalamus after insulin treatment. *Neurosci. Lett.* 168(1–2):11–15

Bahjaoui-Bouhaddi M, Fellmann D, Griffond B, Bugnon C. 1994b. Insulin treatment stimulates the rat melanin-concentrating hormone-producing neurons. *Neuropeptides* 27(4):251–58

Barsh G. 1999. From Agouti to Pomc—100 years of fat blonde mice. *Nat. Med.* 5:984–85

Bassetti C, Aldrich MS. 1996. Narcolepsy. *Neurol. Clin.* 14:545–71

Bauco P, Wang Y, Wise RA. 1993. Lack of sensitization of tolerance to the facilitating effect of ventral tegmental area morphine on lateral hypothalamic brain stimulation reward. *Brain Res.* 617:303–308

Bernardis LL, Bellinger LL. 1993. The lateral

hypothalamic area revisited: neuroanatomy, body weight regulation, neuroendocrinology and metabolism. *Neurosci. Biobehav. Rev.* 17:141–93

Bernardis LL, Bellinger LL. 1996. The lateral hypothalamic area revisited: ingestive behavior. *Neurosci. Biobehav. Rev.* 20:189–287

Borbely AA. 1977. Sleep in the rat during food deprivation and subsequent restitution of food. *Brain Res.* 124:457–71

Borbely AA. 1998. Processes underlying sleep regulation. *Horm. Res.* 49:114–17

Broberger C, De Lecea L, Sutcliffe JG, Hokfelt T. 1998. Hypocretin/orexin- and melanin-concentrating hormone-expressing cells form distinct populations in the rodent lateral hypothalamus: relationship to the neuropeptide Y and agouti gene-related protein systems. *J. Comp. Neurol.* 402:460–74

Cai XJ, Lister CA, Buckinghan RE, Pickavance L, Wilding J, et al. 2000. Down-regulation of orexin gene expression by severe obesity in the rats: studies in Zucker fatty and Zucker diabetic fatty rats and effects of rosiglitazone. *Mol. Brain Res.* 77:131–37

Cai XJ, Widdowson PS, Harrold J, Wilson S, Buckingham RE, et al. 1999. Hypothalamic orexin expression: modulation by blood glucose and feeding. *Diabetes* 48:2132–37

Carr KD, Wolinsky TD. 1993. Chronic food restriction and weight loss produce opioid facilitation of perifornical hypothalamic self-stimulation. *Brain Res.* 607:141–48

Chemelli RM, Sinton CM, Yanagisawa M. 2000. Polysomnographic characterization of orexin-2 receptor knockout mice. *Sleep* 23:A296-A97. (Abstr.)

Chemelli RM, Willie JT, Sinton CM, Elmquist JK, Scammell T, et al. 1999. Narcolepsy in orexin knockout mice: molecular genetics of sleep regulation. *Cell* 98:437–51

Chen CT, Hwang LL, Chang JK, Dun NJ. 2000. Pressor effects of orexins injected intracisternally and to rostral ventrolateral medulla of anesthetized rats. *Am. J. Physiol.* 278: R692–97

Chen H, Charlat O, Tartaglia LA, Woolf EA, Weng X, et al. 1996. Evidence that the diabetes gene encodes the leptin receptor: identification of a mutation in the leptin receptor gene in db/db mice. *Cell* 84:491–95

Crisp AH, Stonehill E, Fenton GW. 1970. An aspect of the biological basis of the mind-body apparatus: the relationship between sleep, nutritional state and mood in disorders of weight. *Psychother. Psychosom.* 18:161–75

Crisp AH, Stonehill E, Fenton GW. 1971. The relationship between sleep, nutrition and mood: a study of patients with anorexia nervosa. *Postgrad. Med. J.* 47:207–13

Dahl RE, Holttum J, Trubnick L. 1994. A clinical picture of child and adolescent narcolepsy. *J. Am. Acad. Child Adolesc. Psychol.* 33:834–41

Danguir J, Nicolaidis S. 1979. Dependence of sleep on nutrients' availability. *Physiol. Behav.* 22:735–40

Danguir J, Nicolaidis S. 1980a. Cortical activity and sleep in the rat lateral hypothalmic syndrome. *Brain Res.* 185:305–21

Danguir J, Nicolaidis S. 1980b. Intravenous infusions of nutrients and sleep in the rat: an ischymetric sleep regulation hypothesis. *Am. J. Physiol. Endocrinol. Metab.* 238:E307–E12

Date Y, Ueta Y, Yamashita H, Yamaguchi H, Matsukura S, et al. 1999. Orexins, orexigenic hypothalamic peptides, interact with autonomic, neuroendocrine and neuroregulatory systems. *Proc. Natl. Acad. Sci. USA* 96: 748–53

de Lecea L, Kilduff TS, Peyron C, Gao X, Foye PE, et al. 1998. The hypocretins: hypothalamus-specific peptides with neuroexcitatory activity. *Proc. Natl. Acad. Sci. USA* 95:322–27

Dube MG, Kalra SP, Kalra PS. 1999. Food intake elicited by central administration of orexins/hypocretins: identification of hypothalamic sites of action. *Brain Res.* 842:473–77

Dyer CJ, Touchette KJ, Carroll JA, Allee GL,

Matteri RL. 1999. Cloning of porcine prepro-orexin cDNA and effects of an intramuscular injection of synthetic porcine orexin-B on feed intake of young pigs. *Domest. Anim. Endocrinol.* 16(3):145–48

Edgar DM, Dement WC, Fuller CA. 1993. Effect of SCN lesions on sleep in squirrel monkeys: evidence for opponent processes in sleep-wake regulation. *J. Neurosci.* 13: 1065–79

Edwards CM, Abusnana S, Sunter D, Murphy KG, Ghatei MA, Bloom SR. 1999. The effect of the orexins on food intake: comparison with neuropeptide Y, melanin-concentrating hormone and galanin. *J. Endocrinol.* 160:R7–12

Elias CF, Saper CB, Maratos-Flier E, Tritos NA, Lee C, et al. 1998. Chemically defined projections linking the mediobasal hypothalamus and the lateral hypothalamic area. *J. Comp. Neurol.* 402:442–59

Elmquist JK, Elias CF, Saper CB. 1999. From lesions to leptin: hypothalamic control of food intake and body weight. *Neuron* 22:221–32

Erickson JC, Hollopeter G, Palmiter RD. 1996. Attenuation of the obesity syndrome of ob/ob mice by the loss of neuropeptide Y. *Science* 274:1704–7

Estabrooke IV, McCarthy MT, Ko E, Chou T, Chemelli RM, et al. 2000. Orexin neuron activity varies with behavioral state. *J. Neurosci.* In press

Faradji H, Cespuglio R, Valatx J, Jouvet M. 1979. Effets du jeûne sur les phénomènes phasiques du sommeil paradoxal de la souris. *Physiol. Behav.* 23:539–46

Fleming J. 1989. Sleep architecture changes in depression: interesting finding or clinically useful. *Prog. Neuro-Psychopharmacol. Biol. Psychol.* 13:419–29

Flier JS, Maratos-Flier E. 1998. Obesity and the hypothalamus: novel peptides for new pathways. *Cell* 92:437–40

Fouriezos G, Francis S. 1992. Apomorphine and electrical self-stimulation of rat brain. *Behav. Brain Res.* 52:73–80

Fulton S, Woodside B, Shizgal P. 2000. Modulation of brain reward circuitry by leptin. *Science* 287:125–28

Goldstein R, Hill SY, Templer DL. 1970. Effect of food deprivation on hypothalamic self-stimulation in stimulus-bound eaters and non-eaters. *Physiol. Behav.* 5:915–18

Greer M. 1955. Suggestive evidence of a primary drinking center in hypothalamus of the rat. *Proc. Soc. Exp. Biol. Med.* 89:59–62

Griffond B, Risold PY, Jacquemard C, Colard C, Fellmann D. 1999. Insulin-induced hypoglycemia increases preprohypocretin (orexin) mRNA in the rat lateral hypothalamic area. *Neurosci. Lett.* 262:77–80

Hagan JJ, Leslie RA, Patel S, Evans ML, Wattam TA, et al. 1999. Orexin A activates locus coeruleus cell firing and increases arousal in the rat. *Proc. Natl. Acad. Sci. USA* 96:10911–16

Hakansson M, de Lecea L, Sutcliffe JG, Yanagisawa M, Meister B. 1999. Leptin receptor- and STAT3-immunoreactivities in hypocretin/orexin neurons of the lateral hypothalamus. *J. Neuroendocrinol.* 11:653–63

Hakansson ML, Brown H, Ghilardi N, Skoda RC, Meister B. 1998. Leptin receptor immunoreactivity in chemically defined target neurons of the hypothalamus. *J. Neurosci.* 18:559–72

Hanada R, Nakazato M, Matsukura S, Murakami N, Yoshimatsu H, Sakata T. 2000. Differential regulation of melanin-concentrating hormone and orexin genes in the agouti-related protein/melanocortin-4 receptor system. *Biochem. Biophys. Res. Commun.* 268:88–91

Haynes AC, Jackson B, Overend P, Buckingham RE, Wilson S, et al. 1999. Effects of single and chronic intracerebroventricular administration of the orexins on feeding in the rat. *Peptides* 20:1099–105

Honda Y. 1988. Clinical features of narcolepsy: Japanese experiences. In *HLA in Narcolepsy*, ed. T Honda, T Juji, pp. 24–27. Berlin: Springer-Verlag. 208 pp.

Honda Y, Doi Y, Ninomiya R, Ninomiya C.

1986. Increased frequency of non-insulin-dependent diabetes mellitus among narcoleptic patients. *Sleep* 9:254–59

Horvath TL, Diano S, van den Pol AN. 1999a. Synaptic interaction between hypocretin (orexin) and neuropeptide Y cells in the rodent and primate hypothalamus: a novel circuit implicated in metabolic and endocrine regulations. *J. Neurosci.* 19:1072–87

Horvath TL, Peyron C, Diano S, Ivanov A, Aston-Jones G, et al. 1999b. Hypocretin (orexin) activation and synaptic innervation of the locus coeruleus noradrenergic system. *J. Comp. Neurol.* 415:145–59

Ida T, Nakahara K, Katayama T, Murakami N, Nakazato M. 1999. Effect of lateral cerebroventricular injection of the appetite-stimulating neuropeptide, orexin and neuropeptide Y, on the various behavioral activities of rats. *Brain Res.* 821:526–29

Ida T, Nakahara K, Murakami T, Hanada R, Nakazato M, Murakami N. 2000. Possible involvement of orexin in the stress reaction in rats. *Biochem. Biophys. Res. Commun.* 270:318–23

Kalra SP, Dube MG, Pu S, Xu B, Horvath TL, Kalra PS. 1999. Interacting appetite-regulating pathways in the hypothalamic regulation of body weight. *Endocr. Rev.* 20: 68–100

Kesterson RA, Huszar D, Lynch CA, Simerly RB, Cone RD. 1997. Induction of neuropeptide Y gene expression in the dorsal medial hypothalamic nucleus in two models of the agouti obesity syndrome. *Mol. Endocrinol.* 11:630–37

Kirchgessner AL, Liu M. 1999. Orexin synthesis and response in the gut. *Neuron* 24: 941–51

Kisanuki YY, Chemelli RM, Sinton CM, Williams SC, Richardson JA, et al. 2000. The role of orexin receptor type-1 (OX1R) in the regulation of sleep. *Sleep* 23: A91 (Abstr.)

Kotagal S, Hartse KM, Walsh JK. 1990. Characteristics of narcolepsy in preteenaged children. *Pediatrics* 85:205–9

Kunii K, Yamanaka A, Nambu T, Matsuzaki I, Goto K, Sakurai T. 1999. Orexins/hypocretins regulate drinking behaviour. *Brain Res.* 842:256–61

Kurumiya S, Kawamura H. 1991. Damped oscillation of the lateral hypothalamic multineuronal activity synchronized to daily feeding schedules in rats with suprachiasmatic lesions. *J. Biol. Rhythms* 6:115–27

Lammers GJ, Pijl H, Iestra J, Langius JA, Buunk G, Meinders AE. 1996. Spontaneous food choice in narcolepsy. *Sleep* 19:75–76

Levitt DR, Teitelbaum P. 1975. Somnolence, akinesia, and sensory activation of motivated behavior in the lateral hypothalamic syndrome. *Proc. Natl. Acad. Sci. USA* 72(7):2819–23

Lin L, Faraco J, Li R, Kadotani H, Rogers W, et al. 1999. The sleep disorder canine narcolepsy is caused by a mutation in the hypocretin (orexin) receptor 2 gene. *Cell* 98:365–76

Lopez M, Senaris R, Gallego R, Garcia-Caballero T, Lago F, et al. 1999. Orexin receptors are expressed in the adrenal medulla of the rat. *Endocrinology* 140:5991–94

Lopez M, Seoane L, Garcia MC, Lago F, Casanueva FF, et al. 2000. Leptin regulation of prepro-orexin and orexin receptor mRNA levels in the hypothalamus. *Biochem. Biophys. Res. Commun.* 269:41–45

Lubkin M, Stricker-Krongrad A. 1998. Independent feeding and metabolic actions of orexins in mice. *Biochem. Biophys. Res. Commun.* 253:241–45

Maffei M, Halaas J, Ravussin E, Pratley RE, Lee GH, et al. 1995. Leptin levels in human and rodent: Measurement of plasma leptin and *ob* RNA in obese and weight-reduced subjects. *Nat. Med.* 1:1155–61

Malendowicz LK, Tortorella C, Nussdorfer GG. 1999. Orexins stimulate corticosterone secretion of rat adrenocortical cells, through the activation of the adenylate cyclase-dependent signaling cascade. *J. Steroid Biochem. Mol. Biol.* 70:185–88

Marcus JN, Aschkenasi CJ, Chemelli RM,

Saper CB, Yanagisawa M, Elmquist JK. 2000. Differential expression of orexin receptors 1 and 2 in the rat brain. *J. Comp. Neurol.* In press

Margules DL, Olds J. 1962. Identical "feeding" and "rewarding" systems in the lateral hypothalamus of rats. *Science* 135:374–75

Marks JL, Porte D Jr, Stahl WL, Baskin DG. 1990. Localization of insulin mRNA in rat brain by in situ hybridization. *Endocrinology* 127:3234–36

Mignot E. 1998. Genetic and familial aspects of narcolepsy. *Neurology* 50:S16–22 (Suppl.)

Mikkelsen JD. 1990. A neuronal projection from the lateral geniculate nucleus to the lateral hypothalamus of the rat demonstrated with Phaseolus vulgaris leucoagglutinin tracing. *Neurosci. Lett.* 116:58–63

Miller MW, Duhl DMJ, Vrieling H, Cordes SP, Ollmann MM, et al. 1993. Cloning of the mouse *agouti* gene predicts a secreted protein ubiquitously expressed in mice carrying the *lethal yellow* mutation. *Genes and Devel.* 7:454–67

Moga MM, Weis RP, Moore RY. 1995. Efferent projections of the paraventricular thalamic nucleus in the rat. *J. Comp. Neurol.* 359: 221–38

Moriguchi T, Sakurai T, Nambu T, Yanagisawa M, Goto K. 1999. Neurons containing orexin in the lateral hypothalamic area of the adult rat brain are activated by insulin-induced acute hypoglycemia. *Neurosci. Lett.* 264:101–4

Nakamura T, Uramura K, Nambu T, Yada T, Goto K, et al. 2000. Orexin-induced hyperlocomotion and stereotypy are mediated by the dopaminergic system. *Brain Res.* 873:181–87

Nambu T, Sakurai T, Mizukami K, Hosoya Y, Yanagisawa M, Goto K. 1999. Distribution of orexin neurons in the adult rat brain. *Brain Res.* 827:243–60

Nicolaidis S, Even P. 1992. The metabolic signal of hunger and satiety, and its pharmacological manipulation. *Int. J. Obes. Rel. Metab. Dis.* 16:S31–41

Nishino S, Ripley B, Overeem S, Lammers GJ, Mignot E. 2000. Hypocretin (orexin) deficiency in human narcolepsy. *Lancet* 355: 39–40

Oomura Y. 1980. Input-output organization in the hypothalamus relating to food intake behavior. In *Handbook of the Hypothalamus: Physiology of the Hypothalamus*, ed. PJ Morgane, J Panksepp, 2:577–620. New York: Marcel Dekker

Oomura Y, Kita H. 1981. Insulin acting as modulators of feeding through the hypothalamus. *Diabetologia* 20:290–98 (Suppl.)

Peyron C, Faraco J, Rogers W, Ripley B, Overeem S, et al. 2000. A mutation in a case of early onset narcolepsy and a generalized absence of hypocretin peptides in human narcoleptic brains. *Nat. Med.* 6:991–97

Peyron C, Tighe DK, van den Pol AN, de Lecea L, Heller HC, et al. 1998. Neurons containing hypocretin (orexin) project to multiple neuronal systems. *J. Neurosci.* 18:9996–10015

Porkka-Heiskanen T. 1997. Adenosine: a mediator of the sleep-inducing effects of prolonged wakefulness. *Science* 276:1265–68

Powley TL, Keesey RE. 1970. Relationship of body weight to the lateral hypothalamic syndrome. *J. Comp. Physiol. Psychol.* 70: 25–36

Pu S, Jain MR, Kalra PS, Kalra SP. 1998. Orexins, a novel family of hypothalamic neuropeptides, modulate pituitary luteinizing hormone secretion in an ovarian steroid-dependent manner. *Regul. Pept.* 78:133–36

Qu D, Ludwig DS, Gammeltoft S, Piper M, Pelleymounter MA, et al. 1996. A role for melanin-concentrating hormone in the central regulation of feeding behaviour. *Nature* 380:243–47

Rashotte ME, Pastukhov IF, Poliakov EL, Henderson RP. 1998. Vigilance states and body temperature during the circadian cycle in fed and fasted pigeons *(Columbia livia)*. *Am. J. Physiol.* 275:R1690–702

Risold PY, Griffond B, Kilduff TS, Sutcliffe JG, Fellmann D. 1999. Preprohypocretin (orexin) and prolactin-like immunoreactivities are coexpressed by neurons of the rat lateral hypothalamic area. *Neurosci. Lett.* 259(3):153–56

Rowland N. 1976. Recovery of regulatory drinking following lateral hypothalamic lesions: nature of residual deficits analyzed by NaCl and water infusions. *Exp. Neurol.* 53:488–507

Sakurai T. 1999. Orexins and orexin receptors: implication in feeding behavior. *Regul. Pept.* 85(1):25–30

Sakurai T, Amemiya A, Ishii M, Matsuzaki I, Chemelli RM, et al. 1998. Orexins and orexin receptors: a family of hypothalamic neuropeptides and G protein-coupled receptors that regulate feeding behavior. *Cell* 92: 573–85

Sakurai T, Moriguchi T, Furuya K, Kajiwara N, Nakamura T, et al. 1999. Structure and function of human prepro-orexin gene. *J. Biol. Chem.* 274:17771–76

Salton SR, Hahm S, Mizuno TM. 2000. Of mice and MEN: what transgenic models tell us about hypothalamic control of energy balance. *Neuron* 25:265–68

Samson WK, Gosnell B, Chang JK, Resch ZT, Murphy TC. 1999. Cardiovascular regulatory actions of the hypocretins in brain. *Brain Res.* 831:248–53

Schmitt M. 1973. Circadian rhythmicity in responses of cells in the lateral hypothalamus. *Am. J. Physiol.* 225:1096–101

Schuld A, Hebebrand J, Geller F, Pollmacher T. 2000. Increased body-mass index in patients with narcolepsy. *Lancet* 355:1274–5

Sherin JE, Elmquist JK, Torrealba F, Saper CB. 1998. Innervation of histaminergic tubero-mammillary neurons by GABAergic and galaninergic neurons in the ventrolateral preoptic nucleus of the rat. *J. Neurosci.* 18:4705–21

Sherin JE, Shiromani PJ, McCarley RW, Saper CB. 1996. Activation of ventrolateral preoptic neurons during sleep. *Science* 271:216–19

Shibahara M, Sakurai T, Nambu T, Takenouchi

T, Iwaasa H, et al. 1999. Structure, tissue distribution, and pharmacological characterization of Xenopus orexins. *Peptides* 20:1169–76

Shimada M, Tritos NA, Lowell BB, Flier JS, Maratos-Flier E. 1998. Mice lacking melanin-concentrating hormone are hypophagic and lean. *Nature* 396:670–74

Shipley JE, Schteingart DE, Tandon R, Pande AC, Grunhaus L, et al. 1992. EEG sleep in Cushing's disease and Cushing's syndrome: comparison with patients with major depressive disorder. *Biol. Psychiatry* 32:146–55

Shirasaka T, Nakazato M, Matsukura S, Takasaki M, Kannan H. 1999. Sympathetic and cardiovascular actions of orexins in conscious rats. *Am. J. Physiol. Regulatory Integrating Comp. Physiol.* 277:R1780–R85

Sinton CM, Fitch TE, Gershenfeld HK. 1999. The effects of leptin on REM sleep and slow wave delta in rats are reversed by food deprivation. *J. Sleep Res.* 8:197–203

Smart D, Jerman JC, Brough SJ, Neville WA, Jewitt F, Porter RA. 2000. The hypocretins are weak agonists at recombinant human orexin-1 and orexin-2 receptors. *Br. J. Pharmacol.* 129:1289–91

Smith OA, DeVito JL, Astley CA. 1990. Neurons controlling cardiovascular responses to emotion are located in lateral hypothalamus-perifornical region. *Am. J. Physiol.* 259:R943–R54

Steiger A, Antonijevic IA, Bohlhalter S, Frieboes RM, Friess E, Murck H. 1998. Effects of hormones on sleep. *Horm. Res.* 49:125–30

Steininger TL, Alam MN, Gong H, Szymusiak R, McGinty D. 1999. Sleep-waking discharge of neurons in the posterior lateral hypothalamus of the albino rat. *Brain Res.* 840:138–47

Stephens TW, Basinski M, Bristow PK, Bue-Valleskey JM, Burgett SG, et al. 1995. The role of neuropeptide Y in the antiobesity action of the obese gene product. *Nature* 377:530–32

Taheri S, Sunter D, Dakin C, Moyes S, Seal L, et al. 2000. Diurnal variation in orexin A immunoreactivity and prepro-orexin mRNA in the rat central nervous system. *Neurosci. Lett.* 279:109–12

Takahashi N, Okumura T, Yamada H, Kohgo Y. 1999. Stimulation of gastric acid secretion by centrally administered orexin-A in conscious rats. *Biochem. Biophys. Res. Commun.* 254:623–27

Tamura T, Irahara M, Tezuka M, Kiyokawa M, Aono T. 1999. Orexins, orexigenic hypothalamic neuropeptides, suppress the pulsatile secretion of luteinizing hormone in ovariectomized female rats. *Biochem. Biophys. Res. Commun.* 264:759–62

Teitelbaum P, Epstein AN. 1962. The lateral hypothalamic syndrome. *Psychol. Rev.* 69:74–90

Trivedi P, Yu H, MacNeil DJ, Van der Ploeg LH, Guan XM. 1998. Distribution of orexin receptor mRNA in the rat brain. *FEBS Lett.* 438:71–75

van den Pol AN. 1999. Hypothalamic hypocretin (orexin): robust innervation of the spinal cord. *J. Neurosci.* 19:3171–82

van den Pol AN, Gao XB, Obrietan K, Kilduff TS, Belousov AB. 1998. Presynaptic and postsynaptic actions and modulation of neuroendocrine neurons by a new hypothalamic peptide, hypocretin/orexin. *J. Neurosci.* 18:7962–71

Volkoff H, Bjorklund JM, Peter RE. 1999. Stimulation of feeding behavior and food consumption in the goldfish, Carassius auratus, by orexin-A and orexin-B. *Brain Res.* 846:204–209

Watts AG, Swanson LW, Sanchez-Watts G. 1987. Efferent projections of the suprachi-asmatic nucleus. I. Studies using anterograde transport of *Phaseolus vulgaris* leucoagglutinin in the rat. *J. Comp. Neurol.* 258:204–29

Woods SC, Seeley RJ, Porte D, Schwartz MW. 1998. Signals that regulate food intake and energy homeostasis. *Science* 280:1378–83

Wurtman RJ, Wurtman JJ. 1995. Brain serotonin, carbohydrate-craving, obesity and depression. *Obes. Res.* 3:477S–80S

Yamada H, Okumura T, Motomura W, Kobayashi Y, Kohgo Y. 2000. Inhibition of food intake by central injection of anti-orexin antibody in fasted rats. *Biochem. Biophys. Res. Commun.* 267:527–31

Yamamoto Y, Ueta Y, Date Y, Nakazato M, Hara Y, et al. 1999. Down regulation of the prepro-orexin gene expression in genetically obese mice. *Brain Res. Mol. Brain Res.* 65:14–22

Yamanaka A, Kunii K, Nambu T, Tsujino N, Sakai A, et al. 2000. Orexin-induced food intake involves neuropeptide Y pathway. *Brain Res.* 859:404–9

Yamanaka A, Sakurai T, Katsumoto T, Yanagisawa M, Goto K. 1999. Chronic intracerebroventricular administration of orexin-A to rats increases food intake in daytime, but has no effect on body weight. *Brain Res.* 849:248–52

Zhang Y, Proenca R, Maffei M, Barone M, Leopold L, Friedman JM. 1994. Positional cloning of the mouse obese gene and its human homologue. *Nature* 372:425–32. Erratum. 1995. *Nature* 30:479

Zucker I, Stephan FK. 1973. Light-dark rhythms in hamster eating, drinking and locomotor behaviors. *Physiol. Behav.* 11:239–50

Annu. Rev. Neurosci. 2001. 24:459–86

SPATIAL PROCESSING IN THE BRAIN:
The Activity of Hippocampal Place Cells

Phillip J. Best,[1] Aaron M. White,[2] and Ali Minai[3]

[1]*Department of Psychology and Center for Neuroscience, Miami University, Oxford, Ohio 45056; e-mail: bestpj@muohio.edu*
[2]*Department of Psychiatry, Duke University Medical Center, Durham, North Carolina 27705; e-mail: amwhite@duke.edu*
[3]*Department of Electrical and Computer Engineering, University of Cincinnati, Cincinnati, Ohio 45221-0030; e-mail: aminai@eces.uc.edu*

Key Words   unit recording, computational models

■ **Abstract**   The startling discovery by O'Keefe & Dostrovsky (*Brain Res.* 1971; 34: 171–75) that hippocampal neurons fire selectively in different regions or "place fields" of an environment and the subsequent development of the comprehensive theory by O'Keefe & Nadel (*The Hippocampus as a Cognitive Map.* Oxford, UK: Clarendon, 1978) that the hippocampus serves as a cognitive map have stimulated a substantial body of literature on the characteristics of hippocampal "place cells" and their relevance for our understanding of the mechanisms by which the brain processes spatial information. This paper reviews the major dimensions of the empirical research on place-cell activity and the development of computational models to explain various characteristics of place fields.

INTRODUCTION

When humans and other complex animals move about the environment, they naturally form representations of their experiences. These representations permit them to recall what happened, when it happened, and where it happened. This review addresses the study of brain mechanisms for processing the "where" of experience. Because space appears to exist as a mental construct, how then is it constructed in the brain? How would a brain be organized such that representations of space would emerge from its operation? What forms of brain activity could we observe that would provide evidence for brain mechanisms of spatial processing? The current review focuses on the study of place-field activity of hippocampal neurons in freely behaving animals and the implications of this research for the nature of spatial processing in the brain. The first part of the review describes the empirical study of place-field activity of hippocampal neurons. The second part describes the use of computational models to understand the mechanism of place-field activity in particular and spatial processing in general. For additional information about hippocampal place cells and current theories of hippocampal function, the reader is referred to a recent special issue of the journal *Hippocampus* (Vol. 9, number 4, 1999) and to a recent book by Redish (1999).

The internal representation of the "where" of experiences could take many forms at the formal conceptual, cognitive level, as well as at the basic cellular neurobiological level. The nature of that representation is of critical importance for developing navigational strategies. A number of formal strategies can be used for effective navigation, such as dead reckoning, path integration, or the formation of cognitive maps (see Redish 1999). The consequent ability of the organism to navigate effectively depends on the nature of those internal representations. For example, birds with hippocampal lesions can migrate accurately many miles by using dead-reckoning processes, but they get lost in their local neighborhood because they lack the ability to form cognitive maps (Bingman & Yates 1992).

In the 1930s and 1940s, when American behavioral psychology was dominated by stimulus-response or stimulus-stimulus explanations of behavior, Tolman proposed that animals do not learn to complete maze tasks solely on the basis of stimulus-response relationships. He believed that "In the course of learning, something like a field map of the environment gets established in the rat's brain" (Tolman 1948). Such maps represent the environment as a configuration that allows the subject to navigate flexibly from any given location. They free organisms from the confines of fixed stimulus-response interactions with the environment and allow them to anticipate the existence of important stimuli from a distance. For instance, they provide organisms with the capacity to take various novel paths to avoid locations in the world where predators have been encountered and to approach distant locations known to contain food, water, shelter, or mates. Although the idea of cognitive maps was dismissed by the behavioral mainstream in the 1950s (see Restle 1957), some investigators continued to see in animal behavior strong evidence for the existence of internal representations of space or cognitive maps (Hebb 1961).

The Discovery of Place Cells

The first demonstration of an internal representation of space in the brain was presented by O'Keefe & Dostrovsky (1971), who reported that the activity of some cells recorded in the hippocampi of freely behaving rats was closely related to the locations of these animals in an open field. These so-called "place cells" fired maximally when an animal was in a rather small, well-defined region of the environment, the "place field," and were virtually silent when the animal was elsewhere. Even though the behavior of the place cells was quite remarkable, only 10.5% (8 of 76) of the cells studied showed place field activity.

In 1973, Ranck reported a comprehensive study of the relationship between hippocampal cellular activity and various behaviors. He found that there are two distinct classes of cells in the hippocampus; cells with complex spikes, which were known to be pyramidal cells (the primary output cells of the hippocampus) and the more rapidly firing theta cells, which were later shown to be interneurons (Fox & Ranck 1975, 1981). Ranck's (1973) study found that the activity of individual hippocampal complex-spike cells fell into a small number of discrete classes, each with a very specific relationship to various aspects of an animal's behavior. He also found that theta cells fired faster during locomotion, when the hippocampal EEG was in the theta frequency band. However, his study found no close relationship between cellular activity and an animal's location in the environment. In a later study, originally planned to further analyze the relationship between cellular activity and various behavioral categories, 16 of 18 complex-spike cells showed place-field activity. Five theta cells fired faster during locomotion but showed no obvious relationship to a rat's location (Best & Ranck 1975, 1982). The nature of the place-field activity was so compelling that, when naïve observers examined videotapes containing images of the behaving rat and the oscilloscope record, they had no difficulty identifying the place-field activity of the cells.

O'Keefe (1976) next published a study reporting on the activity of 50 hippocampal neurons, 26 complex-spike cells that exhibited place-field activity, and 16 cells that had the electrophysiological and behavioral characteristics of Ranck's theta cells. A subsequent study by O'Keefe & Conway (1978) demonstrated that the place fields of hippocampal complex-spike cells were affected in predictable and logical ways by various environmental manipulations. They recorded units as rats performed in an open T-maze in a room containing four salient extramaze stimuli: a light, a fan, a white cue card, and a noise generator. For all place cells, rotation of the distal cues was accompanied by rotation of the place fields. Furthermore, although removal of any two distal stimuli had a negligible effect on place-cell activity, removal of any three stimuli typically caused the place fields to disintegrate. These early reports began to stimulate the study of place-field activity in other laboratories and provided compelling evidence that the hippocampus plays a central role in spatial processing. An example of the place field activity of a hippocampal neuron is illustrated in Figure 1.

The Cognitive-Map Theory

The discovery of place cells, coupled with the results of several hippocampal-lesion studies, led O'Keefe & Nadel (1978) to propose a theory that the hippocampus serves as the neural substrate for cognitive maps. They further proposed that hippocampal place cells are the basic units of the map. According to this theory, an environment is represented by a collection of place cells, each of which represents a specific region of space. The specific configuration of place cells provides an internal representation of the environment that affords an animal knowledge of its position relative to important locations. Whereas the classical sensory systems are designed to process egocentric space, that is, the location

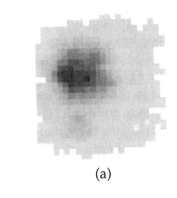

(a)

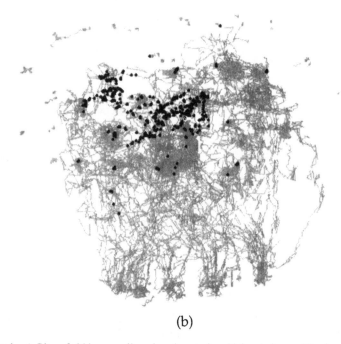

(b)

Figure 1   *A*. Place field in a small enclosed arena in which rats foraged for food that was randomly dropped in the arena. *Darker shading* indicates areas in which a higher rate of firing was seen. *B*. Place field in a large open arena. *Dots* indicate location of the animal when a spike was fired. *Gray lines* indicate trajectory of the animal. [From Figure B.6 of Redish 1999, with permission of author and publisher (copyright 1997, MIT Press). Data courtesy of D. Nitz, K. Gothard, B. Skaggs, K. Moore, C. Barnes, and B. McNaughton].

of stimuli with respect to the receptors and the organism, the hippocampus, according to the cognitive-mapping theory, encodes allocentric space, the location of the organism with respect to important places in the environment. Thus, the nature of the information that causes place cells to fire in their place fields must affect the organism's awareness of its location in the environment such that it can engage in the appropriate locomotor behavior to navigate effectively to another location. This information comes not only from the configuration of exteroceptive stimuli, but also from the vestibular system and other proprioceptive systems, as well as from already existing representations of space in the organism's memory.

This theory also predicts that lesions of the hippocampus and related structures should cause learning, memory, and performance deficits on behavioral tasks that require cognitive-mapping strategies, while having little or no effect on behavioral tasks that do not require spatial cognitive processing. A discussion of the hundreds of studies on the effects of hippocampal lesions on spatial behavior goes beyond the scope of this review. Suffice it to say that the overwhelming majority of these studies support the cognitive mapping theory, but several do lend support to other interpretations of hippocampal function. For a thorough review of the lesion literature, see Redish (1999).

WHAT INFORMATION DETERMINES WHERE PLACE CELLS FIRE?

After the initial characterization of place cells and the development of the cognitive-mapping theory, researchers began to address questions regarding the nature of the information that drives place-cell firing. For instance, does a particular sensory modality dominate in determining place-cell activity, or are a variety of types of sensory information equally weighted? As might be expected, research addressing this issue has revealed that the answer to the question is complicated. Distal visual cues, when present, appear to provide the preferred source of information used to support place-cell activity. Other forms of external sensory information, such as olfactory, auditory, and local visual cues, can influence place-cell activity. Additionally internal proprioceptive and vestibular systems can provide self-motion or "idiothetic" cues that influence place-cells (Knierim et al 1998). Furthermore, under some circumstances, distal, proximal, and idiothetic cues interact to determine place-cell activity.

Distal Visual Information Can Influence Place-Cell Firing

A number of studies have shown that distal sensory information can exert a strong influence on place-cell activity. O'Keefe & Conway (1978) addressed this problem by rotating the distal cues around a stationary open maze. As discussed above, under such circumstances, place fields rotate with the distal cues. Another way

A.

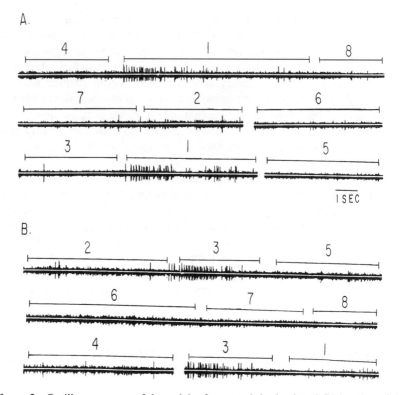

Figure 2   Oscilloscope trace of the activity from a unit in the dorsal CA1 region of the hippocampus as the animal traversed an eight-arm radial arm maze. The numbers represent the arms of the maze. *A.* The maze is in the initial position. Place field is on arm 1. *B.* The maze has been rotated 90°, such that arm 3 is now in the position originally occupied by arm 1. The place field is now on arm 3. (From Figure 8 of Miller & Best 1980, with permission of authors and Elsevier Scientific. Copyright 1980.)

to investigate the relative influence of distal and proximal cues on place-field activity is to rotate a symmetric maze in the fixed environment. When place fields are located on an elevated, open radial-arm maze and the maze is subsequently rotated, the place fields remain in the same positions relative to the room and not on the same physical arm of the maze (Olton et al 1978, Miller & Best 1980; see Figure 2).

Subsequent work has demonstrated that, of the various types of distal information, stable distal visual cues exert a particularly powerful influence over place-cell firing. Muller & Kubie (1987) demonstrated that a single visual cue, a white card attached to the inside wall of a cylindrical arena, could exert almost total control over the location of place fields. Rotations of the card were accompanied by equal rotations of the place fields. Removal of the card did not alter the size or shape of the fields, but did cause the fields to move to new positions (Muller & Kubie 1987).

Thus, once established, the location of place fields can drift in the absence of stable distal cues to anchor them.

Hetherington & Shapiro (1997) recorded place-field activity while subjects moved about in a square recording chamber with cue cards on three of the four walls. Rotation of the cues led to concomitant rotation of the place fields. Removal of individual visual cues altered both the size of the place fields and the within-field firing rates of the cells. Place-field size and within-field firing rates were decreased by removal of a cue located near the place field, but they were increased by removal of a cue located far from the field. Thus, individual visual cues influence place fields, and the degree of influence depends on their proximity to the place field.

Distal visual cues can lose their control over place-cell activity if the rat learns that the cues are unstable. Jeffery & O'Keefe (1999) recorded activity from place cells as rats foraged for food on a platform located in a curtained enclosure containing a single, prominent visual cue. After the locations of place fields were determined, the cue was moved to a new location. When rats were not able to see the experimenter move the cue, the place fields rotated with the cue. However, when rats were able to see the experimenter move the cue, place fields did not reliably rotate with the cue. Similar findings have been reported in other studies (e.g. Knierim et al 1995).

Proximal Sensory Information Can Influence Place-Cell Firing

Although stable distal cues, when present, exert a powerful influence over place-cell firing, a number of studies suggest that under certain conditions place-cell activity can be influenced by proximal cues. Muller et al (1987) investigated the impact of introducing novel, salient proximal cues to an environment on place-cell firing. Introduction of a barrier into the place field typically caused the field to disappear. This occurred even if the barrier placed in the field was transparent. Introduction of the barrier to a portion of the environment outside the place field exerted little impact on activity of the cell.

Cressant et al (1997, 1999) also found that proximal cues located outside a place field exert little influence over place-cell firing. The authors placed three objects near the center of a cylindrical chamber that contained a single distal visual cue. When the positions of the objects were rotated, firing fields remained fixed relative to the distal cue. When the objects were moved against the walls of the cylinder, they exerted powerful control over place-cell firing; that is, they then acted as distal cues.

Shapiro et al (1997) recorded activity from hippocampal place cells while rats traversed a four-arm radial arm maze located in the center of a four-sided curtained enclosure. A prominent visual cue was located on each curtain, and each arm of the maze contained a unique combination of tactile, visual, and olfactory information. After the place fields had been established, a trial ensued in which distal cues were rotated 90° in one direction and proximal maze cues were rotated 90° in the opposite direction. The fields of some cells rotated with the distal cues, while many other fields rotated with the proximal cues. In another condition, the relative positions of distal and proximal cues were scrambled. Some fields moved with particular proximal stimuli and some with particular distal stimuli, and some cells fired maximally only when the proximal and distal cues were oriented in a particular way relative to one another. Thus, place-cell activity can be controlled by distal or proximal cues or by a combination of the two.

Place-Cell Activity in the Absence of Visual Information

A number of studies suggest that visual information is often neither necessary nor sufficient to govern place-cell firing. Hill & Best (1980) found well-defined reliable place fields in rats that were both deafened and blindfolded as the subjects traversed a six-armed radial maze. After maze rotation, the place fields of the majority of the cells rotated with the physical apparatus. In the absence of visual and auditory information, place fields for those cells were determined by proximal cues on the maze. In one-third of the cells, place fields remained stationary with respect to the real world after maze rotation. Because these subjects were receiving no visual or auditory information, it appears that they were relying on idiothetic information to track changes in location in an internal map of the environment. The subjects were later removed from the maze and spun to disrupt their vestibular system. After spinning, the place fields rotated with the physical apparatus, indicating that, in the absence of reliable idiothetic information, the place-field locations were now determined by proximal cues on the maze.

Save et al (1998) also examined place-cell activity in animals that did not have access to visual information. Place cells were recorded from both normal, sighted rats and rats that were blinded shortly after birth. Recording took place in a cylindrical arena containing three objects (a wooden cone, a plastic cylinder, and a bottle of wine) near the wall of the apparatus. When the objects in the arena were rotated, place fields from both groups rotated with them. However, in blind rats, place-field activity did not begin until at least one of the three objects was approached and explored by the animals. In contrast, for sighted animals, place-field activity typically began during the subject's first pass through the field. The place fields recorded from animals bereft of visual information were similar to those recorded from sighted animals, but were defined by interactions with proximal stimuli.

The Influence of "Intrahead" Variables on Place-Cell Firing

It has become clear that place-cell activity is influenced by a variety of intra-head variables, besides pure idiothetic information, such as the animal's memory and motivational state. Under certain circumstances, these intrahead variables can override the influence of external sensory information on place-cell firing. For instance, when rats are placed in an environment in the presence of room lights, turning the lights off does not disrupt the locations of place fields. However, when the rats are placed in the environment in the dark, many cells develop new place fields, the majority of which persist when the room lights are turned on (Quirk et al 1990). Such evidence strongly suggests that, in some cases, an animal's recent experiences in an environment can exert a more powerful influence on place-field firing than presently available stable distal cues.

Additional studies suggest that place-cell activity can be influenced by an animal's experiences in an environment. Breese et al (1989) recorded unit activity as rats navigated on a platform that contained five watering cups, only one of which contained water. When the position of the cup containing the water was shifted, the place fields of some cells shifted in the direction of the rewarded cup. Similarly, Markus et al (1995) observed that changing the nature of the search strategy that subjects needed to retrieve food reward on an open platform, from random to directed searching, caused a relocation of place fields in approximately one-third of the cells. The dramatic effect of behavioral task and/or location of reward on the location of place-cell firing stands in stark contrast to the stability of place fields in constant environments. In unchanging environments, place fields have been found to remain stable for ≤ 153 days (Thompson & Best 1989).

A recent study by Skaggs & McNaughton (1998) also demonstrates that intra-head cues influence place-cell firing. The authors recorded place-cell activity while rats shuttled back and forth between two virtually identical boxes that were connected by a corridor. The subjects first foraged for food in one box (box A) and then moved via the corridor to the other box (box B). If external sensory information alone controlled the location of place fields, then cells with place fields in box A would have place fields in the same relative locations in box B. Generally, this was not the case. Cells with fields in box A often had firing fields in box B, but the fields were typically in different locations.

It is possible that the two environments were different enough that subjects could distinguish between them. However, in a second daily recording session, the relative positions of boxes were switched. The subjects were placed in box B, which was now located in box A's position. If subjects recognized the two environments as distinct, then the place fields should have moved with the particular boxes. In no case did the place fields move with the boxes. Similar findings were recently reported by Tanila (1999) using two identical environments connected by a hidden door.

The obvious question that arises from the findings described above is how were subjects able to distinguish between the two environments if they were virtually identical? The most likely explanation is that subjects were aware that they were leaving one environment and entering another based on self-generated motion information or idiothetic cues, and they were therefore aware that the environments differed. Thus, the findings clearly indicate that (*a*) current external sensory information alone does not control place-field locations and (*b*) idiothetic information plays a role in governing place-cell activity.

Further evidence that hippocampal place cells are not merely sensory neurons but are highly influenced by various intrahead variables was provided by O'Keefe & Speakman (1987). The authors trained rats on a radial-arm maze to select a goal-arm that was defined by its location with respect to a set of salient distal cues. The set of cues was rotated to a different position before each trial. Subjects rapidly learned to choose the correct goal-arm, and the activity of most of the place cells rotated with the cues. On some (memory) trials, the cues were removed before subjects were allowed to choose the goal arm. Most of the time, the rats chose correctly, and the place fields remained constant relative to the now absent cues. So, even in the absence of the cues, the subject's memory of the room layout was sufficient to maintain both the accuracy of choice behavior and the location of the place fields. On some memory trials, the animal chose the wrong arm. On these trials, the locations of the place fields were far from random. Their locations were highly predictable based on the location of the arm that the subject chose as the goal arm. In other words, the cells fired when animals evidently thought they were in the place fields. These results indicate that the place cells are not merely responding to exteroceptive sensory input, but are instead important components of a coherent allocentric cognitive map.

Interaction Between "Intrahead" and Distal Visual Cues

As detailed above, there is mounting evidence that information from idiothetic cues influences place-cell activity. A few recent studies have recorded place-cell activity in instances in which information from idiothetic cues was incongruent with information from distal visual cues. For instance, in the study by Jeffrey & O'Keefe (1999) described above, place-cell activity was examined when a prominent distal visual cue was moved either in clear sight of subjects or out of the subjects' views. When subjects were unable to see the visual cue being moved, place-field locations rotated with the cue. When subjects were able to see the cue being moved, place fields did not consistently rotate with the cue. Such evidence suggests that, when the subjects were unaware that the cue was moved, the visual cues dominated over idiothetic information in determining place-field locations, but when the subjects were aware that the cue was unstable, idiothetic cues dominated.

Similar findings were reported by Knierim et al (1998), who recorded place cell activity as rats foraged in a cylindrical apparatus located in the center of a curtained enclosure bearing a single, prominent visual cue. After place fields were initially characterized, the apparatus was rotated abruptly by either 45° or 180°. Place-field locations remained under the control of the distal visual stimulus when the apparatus was rotated 45°. Responses to the 180° rotation were variable. In some instances, distal visual cues continued to define place-field locations, whereas in other cases, complete remapping of hippocampal place fields occurred (i.e. place fields disappeared, and new firing fields emerged). Such findings suggest that, when the mismatch between idiothetic information and distal visual information is small, external visual cues govern place-field locations. However, for many place cells, when the mismatch between idiothetic information and distal visual information is large, neither source of information completely controls place-field activity, and the cells form new fields.

A similar experiment was conducted by Rotenberg & Muller (1997), except that the distal visual cue, rather than the apparatus, was moved by various amounts in the animal's presence. If the cue was rotated in increments of 45°, place fields rotated with the cue, but if it was rotated by 180° in one move, the place fields retained their original positions. If the 180° rotation was followed by four 45° rotations, the place fields did rotate with these smaller movements, so that a return to the original cue configuration left the fields 180° out of phase with their original positions. The results again support the hypothesis of competitive control of place fields by external and idiothetic cues.

In another experiment, Gothard et al (1996b) studied the effect of mismatch between visual cues and idiothetic estimates of position via path integration. The animals were trained to run from a movable start box to a fixed food box at the other end of a linear track and then return to the start box. While a rat was running to the fixed box, the start box was moved up the track by various amounts so that the return journey was shortened. Recordings showed that place fields fired at fixed distances from the start box on the outward journey and return journeys. During shortened return journeys, if the box had been moved a small distance, the place fields moved gradually up the track to compensate. However, if the box was moved a great distance, the place fields jumped abruptly to correct the mismatch between the perceived and expected positions. The authors saw this as evidence that place fields are controlled competitively by idiothetic and external cues. Under these circumstances, the idiothetic cues actually appeared to dominate unless there was a large mismatch between idiothetic and external cues, which triggered a correction based on external cues.

Normally, external information and idiothetic information are congruent as an animal navigates in an environment. The above studies reveal the relative importance of different types of information by dissociating external and internal information. Although both external and vestibular information are easily disrupted or dissociated from each other and external information can be easily eliminated, it has been difficult to eliminate vestibular information. To determine whether the absence of vestibular information would reduce the consistency of hippocampal place coding, Knierim et al (2000) recorded place-cell activity in rats during flight on the space shuttle. They found that the place fields on a three-dimensional track were as robust in the weightless environment as in earth gravity.

Ensemble Characteristics

Recent technical advances in recording and analysis techniques have permitted the development of procedures to record unit activity from many electrodes and from many cells on the same electrode. Thus, one can now record from an ensemble of many neurons simultaneously in an individual behaving animal and can examine the relationship among the activity of individual units in the ensemble. An example of simultaneous recording from five individual neurons is illustrated in Figure 3, see color insert. Using this approach, Wilson & McNaughton (1994) recorded the activity of up to 89 place cells simultaneously from individual animals during spatial tasks and slow-wave sleep. Cells that had overlapping place fields in the awake state also tended to fire together during slow-wave sleep, indicating enhanced synaptic connectivity between them and providing indirect evidence for Hebbian plasticity. This plasticity can lead to the emergence of cell assemblies that code for specific environments and capture the topology of these environments because the connection strength between place cells becomes inversely related to the distance between their place fields. Such assemblies can provide a means for localization (Samsonovich & McNaughton 1997, Redish & Touretzky 1998), prediction (Blum & Abbott 1996, Gerstner & Abbott 1996), and efficient navigation (Muller & Stead 1996). One consequence of associative, temporally asymmetric Hebbian learning (Levy & Steward 1983) between place fields would be an experience-dependent, asymmetric expansion of place fields as the animal explores an environment. The existence of this phenomenon was verified experimentally by Mehta et al (1997). Wilson & McNaughton's work on place-cell reactivation during sleep was refined and extended by Skaggs & McNaughton (1996) and Kudrimoti et al (1999), who verified the dependence of the effect on experience and showed that it does not occur during "REM" (rapid-eye-movement) sleep.

Remapping Due to Dissonant Cue Configurations

A very noticeable feature of hippocampal place codes is that, when sensory cues are dramatically disrupted or the animal is moved to a new environment, the result is an almost complete reconfiguration of place codes (Bostock et al 1991). Wilson & McNaughton (1993) recorded from ensembles of hippocampal place cells while animals experienced a dramatic change in their environment. Animals foraged for food in an environment consisting of two identical boxes, A and B, separated by a partition. Cellular activity was first recorded as the animals were familiarized with box A. Then the partition was removed, and the animals were allowed to move around in both boxes. Eventually the partition was replaced, and the animals were confined again to box A. Between 73 and 148 neurons were recorded in each animal. The ensemble place code for the novel part of the environment (box B) was initially quite noisy but stabilized after 10 min of exploration. The opening of box B and experience in it did not alter place fields in box A. Only a fraction of the cells in the recorded population showed place fields in each box, and, for those cells, fields in box B were distinct from fields in box A. The authors concluded that the activity of ∼140 place cells provided sufficient information to localize an animal's position to within 1 cm, indicating that the place code is extremely redundant. These results suggest that the hippocampal place cells provide a reliable, robust, and plastic distributed code for the animals' positions in familiar and novel environments.

Tanila et al (Tanila et al 1997a, Shapiro et al 1997) used the ensemble approach to examine the impact of a dramatic reconfiguration of distal and proximal cues on place-cell firing. In these experiments, distal and local cues were rotated in opposite

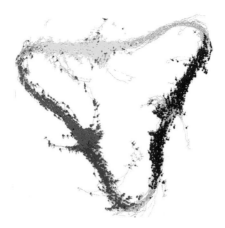

Figure 3 Map of the place fields of five simultaneously recorded units on an elevated triangular maze. The light gray trace represents the rat's path as it traversed the maze. The dots in each color indicate the locations in which each of the five cells fired. (Data courtesy of C Barnes & B McNaughton.)

directions (double rotations), scrambled, or removed, and ensembles of place cells were recorded. In the double-rotation experiment, place cells responded in one of many ways: Some remained consistent with distal cues while others went with the local cues, and others lost their fields or acquired new fields. However, discordant responses were more common in older animals than younger ones, which tended to remap completely (Tanila et al 1997a,b). Repeated double rotations increased the incidence of complete remapping and reduced that of partial remapping. When cues were scrambled or removed, place cells tended to remap or remain consistent with a subset of cues. These results led the authors to conclude that place cells encode a hierarchical representation of the environment with different cells tied to a broader or narrower subset of cues.

Results like those discussed above suggest the presence of global cognitive maps rather than collections of unconnected place cells. Kubie & Muller (1991) speculated that CA3/CA1 place cells might be configured into self-supporting cell assemblies, each corresponding to a cognitive map. This idea was later developed into the notions of reference frames (Gothard et al 1996a, McNaughton et al 1996, Redish & Touretzky 1996, 1998), charts (Samsonovich & McNaughton 1997), and latent attractors (Doboli et al 2000; see below). Several researchers have recently performed experiments to study reference frames in more detail.

Gothard et al (1996a) studied behavioral correlates of CA1 pyramidal cells in a spatial task in which reward locations in an open cylindrical environment were consistently indicated by a pair of local landmarks. These landmarks were moved relative to distal cues from trial to trial, but they retained their position relative to each other. Numerous cells were recorded simultaneously from each animal. The authors discovered four distinct classes of cells: (*a*) place cells that depended on location with respect to the distal cues; (*b*) goal/landmark cells that fired close to the reward location or the landmarks, independently of their position in the environment; (*c*) cells that fired only upon entering or leaving the start box; and (*d*) cells that coded conjunctively or disjunctively over more than one of these dimensions. The authors concluded that the hippocampus had constructed multiple representations of the environment, each with its own frame of reference, and the appropriate ones were activated in different situations.

THE RELATIONSHIP OF THE HIPPOCAMPUS TO THE REST OF THE BRAIN

If the hippocampus serves as the substrate for the cognitive map and place cells represent the basic units of the map, then not only should disruption of the hippocampus selectively disrupt spatial behaviors that rely on a functioning cognitive map, but disruption of inputs to the hippocampus should disrupt both spatial behavior and place-field activity. Furthermore, there should be some evidence of spatially related neuronal activity in some of the areas of the brain that are connected to the hippocampus. As mentioned above, a large number of studies have demonstrated that lesions of the hippocampus or its connections with the rest of the brain produce deficits in a wide variety of spatial tasks, while leaving performance on a variety of nonspatial tasks unaffected.

The Effects of Lesions of Hippocampal Afferents on Place-Cell Activity

A number of lesion studies have been conducted to determine the influence of afferent inputs on hippocampal place-cell activity. Of the many structures providing input to the hippocampus, the two regions most critical to hippocampal function appear to be the medial septum and entorhinal cortex. The medial septum projects rhythmically bursting GABAergic and acetylcholinergic fibers to the hippocampus via the fimbria-fornix. These fibers synchronize the activity of hippocampal neurons, giving rise to the hippocampal theta rhythm (Ylinen et al 1995). The hippocampus receives its major cortical input from the entorhinal cortex (Jones 1993). The entorhinal cortex projects to the hippocampus via the perforant path and provides the hippocampus with highly processed sensory information (Lopes da Silva et al 1990).

Lesions of the medial septum or entorhinal cortex or their projections to the hippocampus disrupt the hippocampal EEG and impair spatial memory (Partlo & Sainsbury 1996, Marighetto et al 1998). It therefore seems reasonable to expect that such lesions should also disrupt hippocampal place-cell firing. Interestingly, lesion studies have produced mixed results. Entorhinal lesions virtually abolish place-field activity in hippocampal neurons (Miller & Best 1980). Fimbria-fornix lesions reduce the precision of place-field activity by increasing the rate of activity outside the field and increasing the field size. They also change the nature of the external stimuli that influence field locations. As in blindfolded and deafened rats, the locations of place fields in lesioned animals are more influenced by the local intramaze cues than by the distal room cues (Miller & Best 1980, Shapiro et al 1989).

In contrast to the dramatic effects described above, a number of other studies have revealed much more subtle effects of damage to afferent systems on place-cell activity. Mizumori et al (1989) recorded place-cell activity during reversible septal inactivation via tetracaine. Whereas septal inactivation disrupted the activity of CA3 place cells, it did not alter the activity of CA1 place cells. These findings indicate that place-cell activity in CA1 occurs independently of normal input from either the septum or CA3. A study by McNaughton et al (1989) indicates that place-cell activity in CA3 and CA1 is not dependent on information flow along the classic trisynaptic loop. Colchicine lesions were made in the dentate gyrus, destroying roughly 75% of dentate granule cells. Dentate lesions impaired performance in a variety of spatial tasks, but produced relatively minor changes in the firing properties of place cells recorded from both CA3 and CA1. A recent study by Leutgeb & Mizumori (1999) suggests that the impact of septal lesions on place-cell activity correlates with the impact of septal lesions on spatial memory. Place-cell activity was recorded after subjects received excitotoxic lesions of the septum. Damage to the septum impaired spatial working memory but produced only mild effects on place-cell firing. Lesions did not disrupt spatial specificity (i.e. ratio of in-field firing to out-of-field firing), but they did increase the variability in place-field activity from trial to trial. The magnitude of the lesion effect on the variability was related to both the size of the lesion and the impairment in working memory. Interestingly, whereas the place fields of intact animals remapped or new fields were developed when a subject was placed in a new environment, place fields of animals with lesions were much less likely to remap in the new environment. Such findings suggest that septal lesions reduce the flexibility of place-cell activity.

The Role of the NMDA Receptor in the Formation of Stable Place Fields

Recent studies have demonstrated that NMDA receptor-mediated synaptic plasticity plays a role in the formation of normal place fields. McHugh et al (1996) examined place-cell activity in a line of genetically engineered mice known as CA1 knockout (CA1-KO) mice. In these animals, the gene encoding the NMDA1 receptor is missing exclusively in CA1. Long-term potentiation (LTP) at Schaffer collateral-CA1 synapses does not occur in hippocampal slices from these animals, and they show significant acquisition deficits in the water maze (Tsien et al 1996[P1], Wilson & Tonegawa 1997). Interestingly, despite the fact that CA1-KO mice exhibit deficient LTP and impaired spatial learning, CA1 pyramidal cells in these animals continued to exhibit salient location-specific increases in activity. The size of the place fields for these cells are roughly one-third larger than fields in wild controls, but the fields remain stable for periods of ≥ 1 h. The firing rates of cells with overlapping place fields are also far less correlated than in normal controls, suggesting that the quality of the spatial information leaving the hippocampus might be diminished in these animals. Thus, deletion of the NMDA1 receptor from CA1 leads to subtle changes in the general characteristics of place cells, but does not block the formation or maintenance of place fields over a ≥ 1-h period.

Additional studies have investigated the effects of widespread NMDA receptor blockade on place-cell activity in normal lab rats. Kentros et al (1998) found that NMDA receptor blockade by 3-[($\pm$)-2-carboypiperazin-4-yl]propyl-1-phosphonate, a competitive NMDA receptor antagonist, did not disrupt activity in place fields that were established before drug administration, indicating that NMDA receptor activation is not needed for the maintenance of previously established place fields. When drug-treated subjects were placed in a novel environment, new place fields emerged and persisted for ≥ 1.5 h. However, fields formed under 3-[($\pm$)-2-carboypiperazin-4-yl]propyl-1-phosphonate were not present when subjects were returned to the environment the following day, indicating that NMDA receptor blockade prevented the long-term stabilization of the fields. Further research (Shapiro & Eichenbaum 1999) revealed that NMDA receptor blockade can prevent even the short-term stabilization of place fields. As before, when drug-treated animals were placed in a novel environment, new place fields were formed. However, when the room lights were turned off for 5 min and then turned back on, place fields in drug-treated animals moved to new locations.

The Location of Cells with Spatial Properties

Most of the studies of place-field activity discussed above have concentrated on pyramidal cells in the CA1 and CA3 regions of the dorsal hippocampus. Place cells are also found in ventral hippocampus (Poucet et al 1994). However, fewer cells in ventral hippocampus have place fields, and the fields tend to be less spatially selective than dorsal cells (Jung et al 1994).

Although theta cells are generally assumed not to have spatial properties, they have been found to fire at slightly differential rates in different locations in the environment. However, their fields are very large, sometimes encompassing >50% of the test environment, and the modulation in rate by location is not at all as great as their modulation by movement (Kubie et al 1990). Granule cells in the dentate gyrus, which provide inputs to CA3 and CA1 pyramidal cells, also have been found to have spatial properties. Their fields show slightly more spatial specificity than theta cells, but nowhere near the spatial specificity of pyramidal cells (Jung & McNaughton 1993).

Cells exhibiting location-sensitive activity have been recorded in other areas of the brain that are connected to the hippocampus. In no studies have cells been found which have spatial specificity approaching that of hippocampal pyramidal cells. Cells with spatial properties have been found in entorhinal cortex (Quirk et al 1992, Barnes et al 1990), parasubiculum (Taube 1995b), and subiculum (Phillips & Eichenbaum 1998, Sharp & Green 1994, Sharp 1997). Whereas the locations of the place fields of hippocampal pyramidal cells are unrelated in different environments, the relative locations of firing fields recorded from subicular cells are maintained when the subject is moved from one environment to another; that is, if a subicular cell fired near the north wall of one arena, it will also fire near the north wall of a second arena. Sharp (1999) suggests that cells in the subiculum, in concert with cells in the entorhinal cortex, form generic or universal maps that are used to represent multiple environments. Novel combinations of input from the generic map and input regarding sensory information from the environment give rise to the location-specific firing of hippocampal pyramidal cells.

Many cells in postsubiculum, called "head direction cells," are very active when the rat's head is oriented in one direction in the horizontal plane and fire more slowly as the head moves away from the preferred direction. This directional specificity is found in all parts of the environment and is independent of the nature of the environment, the animal's location, or the orientation of the animal's head relative to its body (Ranck 1985, Taube et al 1990a,b). Typically head direction cells are under the control of distal stimuli, like place cells, and have different preferred directions in different environments. Head direction cells have also been found in anterior thalamus (Taube 1995a, Blair & Sharp 1995). Under conditions that disorient the rat's sense of direction, the stability of hippocampal place cells and thalamic head direction cells are strongly coupled (Knierem et al 1995). It appears that the strength of external stimulus control over both place cells and head direction cells depends on the rat's learned perception of the stability of the stimuli.

Cells in the medial and lateral septum do not show spatial specificity nor do cells in medial prefrontal cortex, a target of CA1 cells of the temporal hippocampus (Poucet 1997).

COMPUTATIONAL MODELING AND SYSTEM LEVEL THEORIES

As the sensory, perceptual, and contextual dependencies of place activity have become clearer, so has the need for a more subtle and complex systemic understanding of the hippocampus's role in spatial processing. Computational models have been developed to explore and refine various theories of spatial processing in the hippocampus. Because of their computational nature, these models allow researchers to explore ensemble dynamics under a variety of simulated circumstances. The remainder of this review discusses computational-modeling studies of hippocampal place cells.

Place-Field Models

The first detailed computational model of the hippocampal place-cell system was developed by Zipser (1985, 1986). Zipser (1985) used a two-layer, feed-forward neural-network model to show how place fields could emerge from information about a small set of visual landmarks/cues. Sharp (1991) presented a particularly simple and insightful model for landmark-based place fields. The model uses a feed-forward neural network with a sensory input layer and two processing layers of competitive units. The model is able to produce very realistic place cells and can even account for the different degrees of directionality seen in place fields in open environments (Muller et al 1987) and arm mazes (McNaughton et al 1983). Another model for the emergence of place fields was reported by Shapiro & Hetherington (1993), who use the back-propagation algorithm to train a three-layer neural network with input neurons tuned to visual landmarks. More recently, O'Keefe & Burgess have developed a detailed computational model for the dependence of place fields on visual cues (O'Keefe & Burgess 1996, Burgess & O'Keefe 1996). This model, which constructs place fields from Gaussian firing fields tuned to individual cues, is able to replicate a variety of results from cue manipulation experiments.

The Role of Learning in the Place-Cell System

Several authors have considered the possible role of LTP in place representations. Using a model of CA3 with detailed, multicompartment neurons, Wallenstein &

Hasselmo (1997) have shown that LTP in the recurrent excitatory connections and selective modulation of GABA B inhibition by the theta rhythm spontaneously produce place tuning in the CA3 cells. Kubie & Muller (1991) suggest that, due to the associative and temporal characteristics of LTP, synapses between pairs of CA3 cells should, over time, come to reflect the distance between the cells' place fields. This effectively encodes the topology of the environment in the system architecture. Muller & Stead (1996) demonstrated quantitatively that the information in such a "cognitive graph" can be used by a conventional shortest-path algorithm to discover short cuts between arbitrary locations in a familiar environment.

LTP between CA3 place cells can also help in learning routes by strengthening associations between consecutively activated cells along frequent routes. Experimental studies (Wilson & McNaughton 1994, Mehta et al 1997, Kudrimoti et al 1999) have shown strong evidence that sequences of hippocampal-activity patterns elicited during spatial navigation are indeed learned by the system. Such learning can be useful in goal-directed navigation, and this has been considered in several computational models of the hippocampus (Blum & Abbott 1996, Gerstner & Abbott 1996, Samsonovich & McNaughton 1997, Redish & Touretzky 1998).

Models of Navigation

The first model of hippocampally guided navigation was developed by Zipser (1986), who used his place-field model (Zipser 1985) as the basis of two computational models for landmark-based navigation such as that seen in the Morris water maze (Morris et al 1982). The first of these models associates goal directions to directional place fields by using Hebbian learning. The goal is first encountered through random search, and its direction is tracked thereafter by idiothetic update. The second navigation model by Zipser (1986) is much more abstract, building a transformation matrix that maps any landmark conjuration to a goal location.

An early model by O'Keefe (1991) uses an approach similar to Zipser's, that is, casting the navigation problem into a coordinate transformation framework. The model addresses how a global, allocentric cognitive map can be used to guide navigation via transformation to egocentric coordinates. Similar models by McNaughton et al (1989) and McNaughton & Nadel (1990) investigate whether a CA3-like recurrent network could use LTP to learn associations between current location, motor actions, and the resulting locations, thus performing path integration (Mittelstaedt & Mittelstaedt 1980). Hetherington & Shapiro (1993) also proposed a recurrent neural-network model that learns to produce a sequence of location codes from a starting location to the indicated goal location.

Burgess et al (1993, 1994) presented a connectionist model of the hippocampal system comprising place cells, subicular cells, and goal cells. In the model, the subicular cells develop large place fields via LTP on the synapses from the place cells. In the course of exploration, the goal cells become estimators of the animal's distance and direction from the goal(s) and are used to determine motor commands during navigation. The model makes use of the phase precession effect—firing at successively earlier phases of the theta cycle as the animal moves through the cell's field (O'Keefe & Recce 1993; see below).

Wan et al (1994a,b) hypothesized that the place code provides a mechanism for maintaining consistency between an animal's sensory and idiothetic perceptions of location and direction in a familiar environment. In their model, the place code, the visual input, and the head direction signal are associated in a local-view system, which can reset the head direction based on the place and visual signals if the animal becomes disoriented. The visual input can also reset the path integrator if the latter becomes incorrect due to disorientation or drift. Elaborations of these models have been developed by Redish & Touretzky (1997a,b, 1998) using an explicit connectionist implementation with attractor networks for localization, path integration, and head direction coding. A detailed account of the theory underlying this model is presented by Touretzky & Redish (1996).

Several other models of navigation based on hippocampal place codes have been reported in the literature. Recce & Harris (1996) proposed a model that uses the hippocampus as an auto-associative memory, which stores cortical cognitive maps by using place cells and is used to complete cortical representations during navigation. Blum & Abbott (1996) modeled a CA3-like network of place cells that uses temporally asymmetric LTP (Levy & Steward 1983, Gustafsson et al 1987) to learn routes towards a specific goal. Work by Gerstner & Abbott (1996) provided theoretical analysis of this model and also extended it by allowing the possibility of multiple goals by introducing a goal-dependent modulation of the place fields. Sharp et al (1996; see also Brown & Sharp 1995) hypothesized that directed motion is controlled by cells in the nucleus accumbens that receive input from hippocampal place cells and from the head direction cells in the postsubiculum and the anterior thalamic nucleus. The paper also presented a comprehensive model of head direction cells in the postsubiculum and the anterior thalamic nucleus. Recently, Foster et al (2000) proposed a model of navigation in the water maze by using the principle of temporal difference learning (Sutton 1988). Another model developed by Arleo & Gerstner (2000) views CA3/CA1 place cells as providing basis functions for the construction of goal-dependent reward surfaces for guiding navigation. As in the model by Foster et al (2000), a temporal-difference procedure is used to learn directed navigation. The authors have implemented their model using an actual robot, and they demonstrate that the model works effectively for a number of realistic situations.

Reference Frames

The issue of reference frames has been of great interest recently, motivated by experiments showing that place codes can remap in certain situations (Quirk et al 1990, Bostock et al 1991, Gothard et al 1996a,b; Barnes et al 1997). Redish & Touretzky (1997a,b, 1998) proposed that the reference frame for an environment is instantiated in a preconfigured attractor network of head direction cells whose activity becomes associated with the place code. Different attractors in the head direction network correspond to different reference frames. A comprehensive model by Samsonovich & McNaughton (1997) also hypothesized that reference frames are implemented via continuous, stimulus-dependent attractors called charts. The CA3 is proposed as the location of these charts. The function of the place code, according to this model, is to maintain consistency between exteroceptive and idiothetic place estimates. A detailed mathematical analysis of the chart model was presented by Battaglia & Treves (1998). Another recent attempt to model reference frames is the latent attractor model of Doboli et al (2000). This model hypothesizes that the dentate gyrus-hilus system functions as an attractor-based biasing system that imposes a contextual element on CA3/CA1 place-cell activity. An accessible overview of attractor-based models of hippocampal place coding was provided by Tsodyks (1999).

Other Issues Addressed by Modeling

Several other aspects of hippocampal place representations have been explored through computational modeling. Brunel & Trullier (1998) used LTP of synapses between CA3 pyramidal cells to explain the observation that place cells are non-directional in open environments but show directionality in constrained ones (Markus et al 1995). A similar approach was used by Arleo & Gerstner (2000).

The precession of place-cell firing relative to theta phase as an animal moves through the cell's field (O'Keefe & Recce 1993, Skaggs et al 1996) has been modeled variously as arising from asymmetric interactions between place fields (Tsodyks et al 1996), attentional dynamics during each theta cycle (Burgess et al 1994, Burgess & O'Keefe 1996), interplay between attractor dynamics and path integration (Samsonovich & McNaughton 1997), and partial replay of previously learned routes (Wallenstein & Hasselmo 1997). However, the phenomenon is still not fully understood.

Finally, modeling has also been used to quantify the information content of place firing. Skaggs et al (1993) used information theoretic arguments to measure the information in single spikes. Subsequently, several researchers have studied whether a moving animal's position can be determined accurately from the ensemble activity of a place-cell subpopulation by using template matching (Wilson & McNaughton 1993), Bayesian estimation (Zhang et al 1998), and reconstruction based on inhomogeneous point processes (Brown et al 1998). All of these studies have demonstrated that place cells provide a very accurate and redundant distributed coding of an animal's position.

Modeling has contributed significantly to the current understanding of spatial processing in the hippocampus. However, the models have been most useful in evaluating system-level hypotheses, and they have not yet yielded definitive answers to the main questions they address: How are place fields formed? How are they stabilized? and What are they used for, and how? Nevertheless, computational models remain an extremely powerful way to explore large-scale theories of hippocampal function.

CONCLUSIONS

Hippocampal neurons exhibit robust changes in firing that are highly correlated with an animal's location in the environment. O'Keefe & Nadel (1978) suggested that these neurons, which they labeled place cells, form the basic units of Tolman's cognitive maps, with each cell representing a particular region of space and a collection of cells representing an entire environment. Presumably, as an animal moves through the world, the changing activity of hippocampal place cells provides information to the rest of the brain regarding the animal's current location in allocentric space. Despite the compelling nature of place-field activity and the effects of hippocampal lesions on spatial navigation, there is considerable evidence from lesion and recording studies that the concept of a cognitive map might not totally capture the function of the hippocampus and that place cells might represent far more than locations in space.

Hippocampal lesions have been found to cause disruption in some explicitly nonspatial tasks, for which a cognitive-mapping strategy appears to be irrelevant. Such lesions disrupt Pavlovian trace conditioning and performance on certain nonspatial "working-memory" tasks (Rawlins et al 1993) and tasks that require nonspatial configural or contextual processing (Wiener et al 1989, Sutherland & Rudy 1989). In humans, hippocampal lesions disrupt formation of new episodic memories (Squire 1992). Other studies have shown a close relationship between hippocampal pyramidal-cell activity and factors other than the rat's location in space (Eichenbaum et al 1987, Hampson et al 1993). For instance, hippocampal neurons show well-defined reliable conditioned responses in Pavlovian conditioning paradigms (Berger et al 1980, McEchron & Disterhoft 1999). Such evidence suggests that perhaps the hippocampus is involved in a more general fundamental process, of which cognitive mapping is a specific example.

Although the exact nature of the contribution made by the hippocampus to cognition is still uncertain, the importance of the extraordinary discovery of place-field activity and of the development of the comprehensive cognitive-mapping theory is undeniable. The discovery of place cells has dramatically impacted the study of the role of the hippocampus in behavior and has produced a significant paradigm shift (Kuhn 1970) in the study of brain mechanisms of behavior. The data gathered over the past 30 years of place-cell research have yielded profound insight into the nature of the neural mechanisms underlying the "where" of experience.

ACKNOWLEDGMENTS

This work was supported by the National Science Foundation (IBN-9816612 to P.J.B. and IBN-9808664 to A.A.M.). We thank Simona Doboli for her helpful contributions.

Visit the Annual Reviews home page at www.AnnualReviews.org

LITERATURE CITED

Arleo A, Gerstner W. 2000. Spatial cognition and neuro-mimetic navigation: a model of hippocampal place cell activity. *Biol. Cybern.* In press

Barnes CA, McNaughton BL, Mizumori SJY, Leonard BW, Lin LH. 1990. Comparison of spatial and temporal characteristics of neuronal activity in sequential stages of hippocampal processing. *Prog. Brain Res.* 83:287–300

Barnes CA, Suster MS, Shen J, McNaughton BL. 1997. Multistability of cognitive maps in the hippocampus of old rats. *Nature* 388: 272–75

Battaglia FP, Treves A. 1998. Attractor neural networks storing multiple space representations: a model for hippocampal place fields. *Phys. Rev. E* 58:7738–53

Berger TW, Laham RI, Thompson RF. 1980. Hippocampal unit-behavior correlations during classical conditioning. *Brain Res.* 193: 229–48

Best PJ, Ranck JB. 1975. Reliability of the relationship between hippocampal unit activity and behavior in the rat. *Soc. Neurosci. Abstr.* 1:837

Best PJ, Ranck JB. 1982. The reliability of the relationship between hippocampal unit activity and sensory-behavioral events in the rat. *Exp. Neurol.* 75:655–64

Bingman VP, Yates G. 1992. Hippocampal lesions impair navigational learning in experienced homing pigeons. *Behav. Neurosci.* 106:229–32

Blair HT, Sharp PE. 1995. Anticipatory head direction signals in anterior thalamus: evidence for a thalamocortical circuit that integrates angular head motion to compute head direction. *J. Neurosci.* 15:6260–70

Blum KI, Abbott LF. 1996. A model of spatial map formation in the hippocampus of the rat. *Neural Comput.* 8:85–93

Bostock E, Muller RU, Kubie JL. 1991. Experience-dependent modifications of hippocampal place cell firing. *Hippocampus* 1:193–206

Breese CR, Hampson RE, Deadwyler SA. 1989. Hippocampal place cells: stereotypy and plasticity. *J. Neurosci.* 9:1097–111

Brown EN, Frank LM, Tang T, Quirk MC, Wilson MA. 1998. A statistical paradigm for neural spike-train decoding applied to position prediction from ensemble firing patterns of rat hippocampal place cells. *J. Neurosci.* 18:7411–25

Brown MA, Sharp PE. 1995. Simulation of spatial learning in the Morris water maze by a neural network model of the hippocampal formation and the nucleus accumbens. *Hippocampus* 5:171–88

Brunel N, Trullier O. 1998. Plasticity of directional place fields in a model of rodent CA3. *Hippocampus* 8:651–65

Burgess N, O'Keefe J. 1996. Neuronal computations underlying the firing of place cells and their role in navigation. *Hippocampus* 7: 749–62

Burgess N, O'Keefe J, Recce M. 1993. Using hippocampal place cells for navigation, exploiting phase coding. In *Advances in Neural Information Processing Systems 5*, ed. SJ Hanson, SJ Cowan, CL Giles, pp. 929–36. San Mateo, CA: Morgan Kaufmann

Burgess N, Recce M, O'Keefe J. 1994. A model of hippocampal function. *Neural Netw.* 7:1065–81

Cressant A, Muller RU, Poucet B. 1997. Failure of centrally placed objects to control the firing fields of hippocampal place cell. *J. Neurosci.* 17:2531–42

Cressant A, Muller RU, Poucet B. 1999. Further study of the control of place cell firing by intra-apparatus objects. *Hippocampus* 4:423–31

Doboli S, Minai AA, Best PJ. 2000. Latent attractors: a model for context-dependent place representations in the hippocampus. *Neural Comput.* 12:1003–37

Eichenbaum H, Kuperstein M, Fagan A, Nagode J. 1987. Cue sampling and goal-approach correlates of hippocampal unit activity in rats performing an odor-discrimination task. *J. Neurosci.* 7:716–32

Foster DJ, Morris RGM, Dayan P. 2000. A model of hippocampally dependent navigation, using the temporal difference learning rule. *Hippocampus* 10:1–16

Fox SE, Ranck JB Jr. 1975. Localization and anatomical identification of theta and complex spike cells in dorsal hippocampal formation of rats. *Exp. Neurol.* 49:299–313

Fox SE, Ranck JB Jr. 1981. Electrophysiological characteristics of hippocampal complex-spike and theta cells. *Exp. Brain Res.* 41:399–410

Georgopoulos AP, Schwartz A, Kettner RE. 1986. Neuronal population coding of movement direction. *Science* 233:1416–19

Gerstner W, Abbott LF. 1996. Learning navigation maps through potentiation and modulation of hippocampal cells. *J. Comput. Neurosci.* 4:79–94

Gothard KM, Skaggs WE, Moore KE, McNaughton BL. 1996a. Binding of hippocampal CA1 neural activity to multiple reference frames in a landmark-based navigation task. *J. Neurosci.* 16:823–35

Gothard KL, Skaggs WE, McNaughton BL. 1996b. Dynamics of mismatch correction in the hippocampal ensemble code for space: interaction between path integration and environmental cues. *J. Neurosci.* 16:8027–40

Gustafsson B, Wigstrom H, Abraham WC, Huang Y-Y. 1987. Long-term potentiation in the hippocampus using depolarizing current pulses as the conditional stimulus for single volley synaptic potentials. *J. Neurosci.* 7:774–80

Hampson RE, Heyser CJ, Deadwyler SA. 1993. Hippocampal cell firing correlates of delayed-match-to-sample performance in the rat. *Behav. Neurosci.* 107:715–39

Hebb DO. 1961. *The Organization of Behavior, a Neuropsychological Theory.* New York: Sci. Ed.

Hetherington PA, Shapiro ML. 1993. A simple network model simulates hippocampal place fields. 2. Computing goal-directed trajectories and memory fields. *Behav. Neurosci.* 107:434–43

Hetherington PA, Shapiro ML. 1997. Hippocampal place fields are altered by the removal of single visual cues in a distance dependent manner. *Behav. Neurosci.* 111: 20–34

Hill AJ, Best PJ. 1980. Effects of deafness and blindness on the spatial correlates of hippocampal unit activity in the rat. *Exp. Neurol.* 74:204–17

Jeffery KJ, O'Keefe JM. 1999. Learned interaction of visual and idiothetic cues in the control of place field orientation. *Exp. Brain Res.* 127:151–61

Jones RSG. 1993. Entorhinal-hippocampal connections: a speculative view of their function. *Trends Neursosci.* 16:58–64

Jung MW, McNaughton BL. 1993. Spatial selectivity of unit activity in the hippocampal granular layer. *Hippocampus* 3:165–82

Jung MW, Wiener SI, McNaughton BL. 1994. Comparison of spatial firing characteristics of units in dorsal and ventral hippocampus of the rat. *J. Neurosci.* 14:7347–56

Kentros C, Hargreaves E, Hawkins RD, Kandel ER, Shapiro M, Muller RV. 1998. Abolition of long-term stability of new hippocampal place cell maps by NMDA receptor blockade. *Science* 280:2121–26

Knierim JJ, Kudrimoti HS, McNaughton BL. 1995. Place cells, head direction cells, and the learning of landmark stability. *J. Neurosci.* 15:1648–59

Knierim JJ, Kudrimoti HS, McNaughton BL. 1998. Interactions between idiothetic cues and external landmarks in the control of place cells and head direction cells. *J. Neurophysiol.* 80:425–46

Knierim JJ, McNaughton BL, Poe GR. 2000. Three-dimensional spatial selectivity of hippocampal neurons during space flight. *Nat. Neurosci.* 3:211–12

Kubie JL, Muller RU. 1991. Multiple representations in the hippocampus. *Hippocampus* 1:240–42

Kubie JL, Muller RU, Bostock EM. 1990. Spatial firing properties of hippocampal theta cells. *J. Neurosci.* 10:1110–23

Kudrimoti HS, Barnes CA, McNaughton BL. 1999. Reactivation of hippocampal cell assemblies: effects of behavioral state, experience, and EEG dynamics. *J. Neurosci.* 19:4090–101

Kuhn T. 1970. *The Structure of Scientific Revolutions*, 2nd ed. Chicago: Univ. Chicago Press

Leutgeb S, Mizumori SJY. 1999. Excitotoxic septal lesions result in spatial memory deficits and altered flexibility of hippocampal single-unit representations. *J. Neurosci.* 19:6661–72

Levy WB, Steward O. 1983. Temporal contiguity requirements for long-term associative potentiation/depression in the hippocampus. *Neuroscience* 8:791–97

Lopes da Silva FH, Witter MP, Boeijinga PH, Lohman AHM. 1990. Anatomic organization and physiology of the limbic cortex. *Physiol. Rev.* 70:453–511

Marighetto A, Yee BK, Rawlins JN. 1998. The effects of cytotoxic entorhinal lesions and electrolytic medial septal lesions on the acquisition and retention of a spatial working memory task. *Exp. Brain Res.* 119:517–28

Markus EJ, Qin YL, Leonard B, Skaggs WE, McNaughton BL, Barnes CA. 1995. Interactions between location and task affect the spatial and directional firing of hippocampal neurons. *J. Neurosci.* 15:7079–94

McEchron MD, Disterhoft JF. 1999. Hippocampal encoding of non-spatial trace conditioning. *Hippocampus* 9:385–96

McHugh TJ, Blum KI, Tsien JZ, Tonegawa S, Wilson MA. 1996. Impaired hippocampal representation of space in CA1-specific NMDAR1 knockout mice. *Cell* 87:1339–49

McNaughton BL, Barnes CA, Gerrard JL,

Gothard K, Jung MW, et al. 1996. Deciphering the hippocampal polyglot: the hippocampus as a path integration system. *J. Exp. Biol.* 199:173–85

McNaughton BL, Barnes CA, Meltzer J, Sutherland RJ. 1989. Hippocampal granule cells are necessary for normal spatial learning but not for spatially selective pyramidal cell discharge *Exp. Brain Res.* 76:485–96

McNaughton BL, Barnes CA, O'Keefe J. 1983. The contributions of position, direction, and velocity to single unit activity in the hippocampus of freely moving rats. *Brain Res.* 52:41–49

McNaughton BL, Nadel L. 1990. Hebb-Marr networks and the neurobiological representation of action in space. In *Neuroscience and Connectionist Theory*, ed. MA Gluck, DE Rumelhart, pp. 1–63. Hillsdale, NJ: Erlbaum

McNaughton BL, O'Keefe J, Barnes CA. 1983. The stereotrode: a new technique for simultaneous isolation of several single units in the central nervous system from multiple unit records. *J. Neurosci. Methods* 8:391–97

McNaughton BL, Leonard B, Chen L. 1989. Cortical-hippocampal interactions and cognitive mapping: a hypothesis based on reintegration of the parietal and inferotemporal pathways for visual processing. *Psychobiology* 17:230–35

Mehta M, Barnes CA, McNaughton BL. 1997. Experience-dependent, asymmetric expansion of hippocampal place fields. *Proc. Natl. Acad. Sci. USA* 94:8918–21

Miller VM, Best PJ. 1980. Spatial correlates of hippocampal unit activity are altered by lesions of the fornix and entorhinal cortex. *Brain Res.* 194:311–23

Mittelstaedt ML, Mittelstaedt H. 1980. Homing by path integration in a mammal. *Naturwissenschaften* 67:566–67

Mizumori SJ, Barnes CA, McNaughton BL. 1989. Reversible inactivation of the medial septum: selective effects on the spontaneous unit activity of different hippocampal cell types. *Brain Res.* 500:99–106

Morris RGM, Garrud P, Rawlins JNP, O'Keefe J. 1982. Place navigation impaired in rats with hippocampal lesions. *Nature* 297:681–83

Muller RU, Kubie JL. 1987. The effects of changes in the environment on the spatial firing properties of hippocampal place cells. *J. Neurosci.* 7:1951–68

Muller RU, Kubie JL, Ranck JB Jr. 1987. Spatial firing patterns of hippocampal complex-spike cells in a fixed environment. *J. Neurosci.* 7:1935–50

Muller RU, Stead M. 1996. Hippocampal place cells connected by Hebbian synapses can solve spatial problems. *Hippocampus* 6:709–19

O'Keefe J. 1976. Place units in the hippocampus of the freely moving rat. *Exp. Neurol.* 51:78–109

O'Keefe J. 1991. An allocentric spatial model for the hippocampal cognitive map. *Hippocampus* 1:230–35

O'Keefe J, Burgess N. 1996. Geometric determinants of the place fields of hippocampal neurons. *Nature* 381:425–28

O'Keefe J, Conway DH. 1978. Hippocampal place units in the freely moving rat: why they fire where they fire. *Exp. Brain Res.* 31:573–90

O'Keefe J, Dostrovsky J. 1971. The hippocampus as a spatial map: preliminary evidence from unit activity in the freely moving rat. *Brain Res.* 34:171–75

O'Keefe J, Nadel L. 1978. *The Hippocampus as a Cognitive Map.* Oxford, UK: Clarendon

O'Keefe J, Recce M. 1993. Phase relationship between hippocampal place units and the EEG theta rhythm. *Hippocampus* 3:317–30

O'Keefe J, Speakman A. 1987. Single unit activity in the rat hippocampus during a spatial memory task. *Exp. Brain Res.* 68:1–27

Olton DS, Branch M, Best PJ. 1978. Spatial correlates of hippocampal unit activity. *Exp. Neurol.* 58:387–409

Partlo LA, Sainsbury RS. 1996. Influence of medial septal and entorhinal cortex lesions on theta activity recorded from the hippocampus and median raphe nucleus. *Physiol. Behav.* 59:887–95

Phillips RG, Eichenbaum H. 1998. Comparison of ventral subicular and hippocampal neuron spatial firing patterns in complex and simplified environments. *Behav. Neurosci.* 112:707–13

Poucet B. 1997. Searching for spatial unit firing in the prelimbic area of the rat medial prefrontal cortex. *Behav. Brain Res.* 84:151–59

Poucet B, Thinus-Blanc C, Muller RU. 1994. Place cells in the ventral hippocampus of rats. *NeuroReport* 5:2045–48

Quirk GJ, Muller RU, Kubie JL. 1990. The firing of hippocampal place cells in the dark depends on the rat's recent experience. *J. Neurosci.* 10:2008–17

Quirk GJ, Muller RU, Kubie JL, Ranck JB Jr. 1992. The positional firing properties of medial entorhinal neurons: description and comparison with hippocampal place cells. *J. Neurosci.* 12:1945–63

Ranck JB Jr. 1973. Studies on single neurons in dorsal hippocampal formation and septum in unrestrained rats. *Exp. Neurol.* 41:461–531

Ranck JB Jr. 1985. Head direction cells in the deep cell layer of dorsal presubiculum in freely moving rats. In *Electrical Activity of Archicortex*, ed. G Buzsaki, CH Vanderwolf, pp. 217–20. Budapest: Akademai Kiado

Rawlins JNP, Lyford GL, Seferiades A, Deacon RMJ, Cassaday HJ. 1993. Critical determinants of nonspatial working memory deficits in rats with conventional lesions of the hippocampus or fornix. *Behav. Neurosci.* 3:420–33

Recce M, Harris KD. 1996. Memory for places: a navigational model in support of Marr's theory of hippocampal function. *Hippocampus* 6:735–48

Redish AD. 1999. *Beyond the Cognitive Map: From Place Cells to Episodic Memory.* Cambridge, MA: MIT Press

Redish AD, Touretzky DS. 1996. Modeling interactions of the rat's place and head direction systems. In *Advances in Neural Information Processing 8*, ed. DS Touretzky, M Mozer, ME Hasselmo, pp. 61–71. Cambridge, MA: MIT Press

Redish AD, Touretzky DS. 1997a. Cognitive maps beyond the hippocampus. *Hippocampus* 7:15–35

Redish AD, Touretzky DS. 1997b. Separating hippocampal maps. In *Spatial Functions of the Hippocampal Formation and the Parietal Cortex*, ed. N Burgess, K Jeffery, J O'Keefe. Oxford, UK: Oxford Univ. Press

Redish AD, Touretzky DS. 1998. The role of the hippocampus in solving the Morris water maze. *Neural Comput.* 10:73–111

Restle F. 1957. Discrimination of cues in mazes: a resolution of the 'place-vs-response' question. *Psychol. Rev.* 64:217–28

Rotenberg A, Muller RU. 1997. Variable place-cell coupling to a continuously viewed stimulus: evidence that the hippocampus acts as a perceptual system. *Philos. Trans. R. Soc. London Ser. B* 352:1505–13

Samsonovich A, McNaughton BL. 1997. Path integration and cognitive mapping in a continuous attractor neural network model. *J. Neurosci.* 17:5900–20

Save E, Cressant A, Thinus-Blanc C, Poucet B. 1998. Spatial firing of hippocampal place cells in blind rats. *J. Neurosci.* 18:1818–26

Shapiro ML, Eichenbaum H. 1999. Hippocampus as a memory map: synaptic plasticity and memory encoding by hippocampal neurons. *Hippocampus* 4:365–84

Shapiro ML, Hetherington PA. 1993. A simple network model simulates hippocampal place fields. 1. Parametric analysis and psychological predictions. *Behav. Neurosci.* 107:34–50

Shapiro ML, Simon DK, Olton DS, Gage FH III, Nilsson OG, Bjorklund A. 1989. Intrahippocampal grafts of fetal basal forebrain tissue alter place fields in the hippocampus of rats with fimbria-fornix lesions. *Neuroscience* 32:1–18

Shapiro ML, Tanila H, Eichenbaum H. 1997. Cues that hippocampal place cells encode: dynamic and hierarchical representation of local distal stimuli. *Hippocampus* 7:624–42

Sharp PE. 1991. Computer simulation of hippocampal place cells. *Psychobiology* 19:103–15

Sharp PE. 1997. Subicular cells generate similar spatial firing patterns in two geometrically and visually distinctive environments: comparison with hippocampal place-cells. *Behav. Brain Res.* 85:71–92

Sharp PE, Blair HT, Brown M. 1996. Neural network modeling of the hippocampal formation spatial signals and their possible role in navigation: a modular approach. *Hippocampus* 6:735–48

Sharp PE, Green C. 1994. Spatial correlates of firing patterns of single cells in the subiculum of the freely-moving rat. *J. Neurosci.* 14:2339–56

Skaggs WE, McNaughton BL. 1996. Replay of neuronal firing sequences in rat hippocampus during sleep following spatial experience. *Science* 271:1870–73

Skaggs WE, McNaughton BL. 1998. Spatial firing properties of hippocampal CA1 populations in an environment containing two visually identical regions. *J. Neurosci.* 18:8455–66

Skaggs WE, McNaughton BL, Gothard KM, Markus EJ. 1993. An information-theoretic approach to deciphering the hippocampal place code. In *Advances in Neural Information Processing Systems 5*, ed. SJ Hanson, JD Cowan, CL Giles, pp. 1030–37. Cambridge, MA: MIT Press

Skaggs WE, McNaughton BL, Wilson MA, Barnes CA. 1996. Theta phase precession in hippocampal neuronal populations and the compression of temporal sequences. *Hippocampus* 6:149–72

Squire LR. 1992. Memory and the hippocampus: a synthesis from the findings with rats, monkeys, and humans. *Psychol. Rev.* 99:195–231

Sutherland RJ, Rudy JW. 1989. Configural

association theory: the role of the hippocampal formation in learning, memory, and amnesia. *Psychobiology* 17:129–44

Sutton RS. 1988. Learning to predict by the method of temporal difference learning. *Mach. Learn.* 3:9–44

Tanila H. 1999. Hippocampal place cells can develop distinct representations of two visually identical environments. *Hippocampus* 9:235–46

Tanila H, Shapiro ML, Eichenbaum H. 1997. Discordance of spatial representations in ensembles of hippocampal place cells. *Hippocampus* 7:613–23

Tanila H, Shapiro M, Gallagher M, Eichenbaum H. 1997a. Brain aging: changes in the nature of information coding in the hippocampus. *J. Neurosci.* 17:5155–66

Tanila H, Sipila P, Shapiro M, Eichenbaum H. 1997b. Brain aging: impaired coding of novel environmental cues. *J. Neurosci.* 17:5167–74

Taube JS. 1995a. Head direction cells recorded in the anterior thalamic nuclei of freely moving rats. *J. Neurosci.* 15:70–86

Taube JS. 1995b. Place cells recorded in the parasubiculum of freely moving rats. *Hippocampus* 5:569–83

Taube JS, Muller RU. 1998. Comparisons of head direction cell activity in the postsubiculum and anterior thalamus of freely moving rats. *Hippocampus* 8:87–108

Taube JS, Muller RU, Ranck JB Jr. 1990a. Head direction cells recorded from the postsubiculum in freely moving rats. I. Description and quantitative analysis. *J. Neurosci.* 10:420–35

Taube JS, Muller RU, Ranck JB Jr. 1990b. Head direction cells recorded from the postsubiculum in freely moving rats. II. Effects of environmental manipulations. *J. Neurosci.* 10:436–47

Thompson LT, Best PJ. 1989. Place cells and silent cells in the hippocampus of freely-behaving rats. *J. Neurosci.* 9:2382–90

Tolman EC. 1948. Cognitive maps in rats and men. *Psychol. Rev.* 55:189–208

Touretzky DS, Redish AD. 1996. Theory of rodent navigation based on interacting representations of space. *Hippocampus* 6:247–70

Tsien JZ, Huerta PT, Tonegawa S. 1996. The essential role of hippocampal CA1 NMDA receptor-dependent synaptic plasticity in spatial memory. *Cell* 87:1327–38

Tsodyks MV. 1999. Attractor neural network models of spatial maps in hippocampus. *Hippocampus* 9:481–89

Tsodyks MV, Skaggs WE, Sejnowski TJ, McNaughton BL. 1996. Population dynamics and theta rhythm phase precession of hippocampal place cell firing: a spiking neuron model. *Hippocampus* 6:271–80

Wallenstein GV, Hasselmo ME. 1997. Gabaergic modulation of hippocampal population activity: sequence learning, place field development, and the phase precession effect. *J. Neurophys.* 78:393–408

Wan HS, Redish AD, Touretzky DS. 1994a. Towards a computational theory of rat navigation. *Proc. 1993 Connect. Models Summer Sch.*, ed. M Mozer, P Smolensky, D Touretzky, J Elman, A Weigend, pp. 11–19. Hillsdale, NJ: Erlbaum

Wan HS, Redish AD, Touretzky DS. 1994b. Computing goal locations from place codes. In *Proc. Annu. Conf. Cogn. Sci. Soc., 16th,* pp. 922–27. Hillsdale, NJ: Erlbaum

Wiener SI, Paul CA, Eichenbaum H. 1989. Spatial and behavioral correlates of hippocampal neuronal activity. *J. Neurosci.* 9:2737–63

Wilson MA, McNaughton BL. 1993. Dynamics of the hippocampal ensemble code for space. *Science* 261:1055–58

Wilson MA, McNaughton BL. 1994. Reactivation of hippocampal ensemble memories during sleep. *Science* 265:676–79

Wilson MA, Tonegawa S. 1997. Synaptic plasticity, place cells and spatial memory: study with second generation knockouts. *Trends Neurosci.* 20:102–6

Ylinen A, Soltesz I, Bragin A, Penttonen M, Sik A, Buzsaki G. 1995. Intracellular correlates of hippocampal theta rhythm in identified pyramidal cells, granule cells, and basket cells. *Hippocampus* 5:78–90

Zhang K, Ginzburg I, McNaughton BL, Sejnowski TJ. 1998. Interpreting neuronal population activity by reconstruction: unified framework with application to hippocampal place cells. *J. Neurophys.* 79:1017–44

Zipser D. 1985. A computational model of hippocampal place fields. *Behav. Neurosci.* 99:1006–18

Zipser D. 1986. Biologically plausible models of place recognition and goal location. In *Parallel Distributed Processing: Explorations in the Microstructure of Cognition*, Vol. 2, ed. JL McClelland, DE Rumelhart, pp. 423–70. Cambridge, MA: MIT Press

Annu. Rev. Neurosci. 2001. 24:487–517

THE VANILLOID RECEPTOR: A Molecular Gateway to the Pain Pathway

Michael J Caterina[1] and David Julius[2]

[1]*Department of Biological Chemistry, Johns Hopkins University School of Medicine, Baltimore, Maryland 21205; e-mail: caterina@jhmi.edu*
[2]*Department of Cellular and Molecular Pharmacology, University of California at San Francisco, San Francisco, California 94143; e mail: julius@socrates.ucsf.edu*

Key Words capsaicin, pain, nociceptor, ion channel, heat sensation, hyperalgesia

■ **Abstract** The detection of painful stimuli occurs primarily at the peripheral terminals of specialized sensory neurons called nociceptors. These small-diameter neurons transduce signals of a chemical, mechanical, or thermal nature into action potentials and transmit this information to the central nervous system, ultimately eliciting a perception of pain or discomfort. Little is known about the proteins that detect noxious stimuli, especially those of a physical nature. Here we review recent advances in the molecular characterization of the capsaicin (vanilloid) receptor, an excitatory ion channel expressed by nociceptors, which contributes to the detection and integration of pain-producing chemical and thermal stimuli. The analysis of vanilloid receptor gene knockout mice confirms the involvement of this channel in pain sensation, as well as in hypersensitivity to noxious stimuli following tissue injury. At the same time, these studies demonstrate the existence of redundant mechanisms for the sensation of heat-evoked pain.

INTRODUCTION

A great deal is now known about the molecular basis of sensory perception, especially in the visual, olfactory, and gustatory systems for which cell surface proteins that detect changes in our physical or chemical environment have been identified. These molecules endow sensory neurons with specialized receptive properties, thereby determining how we perceive the world around us. Pain is also a sensory modality in which specialized primary afferent neurons, called nociceptors, detect noxious stimuli. Nociceptors are remarkable and unusual sensory cells because they respond to a broad range of physical (e.g. heat, cold, and pressure) and chemical (e.g. acid, irritants, and inflammatory mediators) stimuli, but do so only at stimulus intensities capable of causing tissue damage. However, in contrast to other sensory systems, little is known about the molecules that account for the

0147-006X/01/0621-0487$14.00

unique properties of the nociceptor. Thus, a fundamental goal in pain biology is to provide a molecular understanding of how physical and chemical stimuli are detected by the nociceptor, how intensity thresholds are specified, and how these thresholds are reset in the setting of tissue injury or disease.

Our understanding of somatosensory mechanisms that detect changes in pressure, touch, or temperature is limited by the fact that receptors for these physical stimuli have not been clearly identified. Why has progress in this area lagged behind that for other sensory systems? One factor may relate to the intrinsic difficulty of carrying out biochemical studies in a system in which sensory nerve fibers and their receptors are diffusely spread throughout the body. This is in sharp contrast to the visual system, for example, in which the primary signal transduction apparatus and associated proteins are highly concentrated in one easily obtainable tissue. In the case of olfaction and taste, the sensory apparatus is also localized to a single organ, but biochemical studies are not feasible owing to the relatively small number of sensory neurons in these tissues. Nevertheless, electrophysiological and pharmacological studies have shown that signal transduction in these systems occurs through a well-characterized paradigm involving the activation of G-protein-coupled receptors, opening the way to molecular studies through homology-based screening strategies (Adler et al 2000, Buck & Axel 1991, Jones & Reed 1989, McLaughlin et al 1992). In the case of the "pain pathway," significantly less information is available regarding the types of molecules or signal transduction mechanisms involved in the detection of noxious stimuli, particularly those of a physical nature, limiting the utility of database searches and polymerase chain reaction-based screening strategies.

Despite these limitations, pharmacological methods offer alternative strategies for initiating molecular studies, especially when a drug or agent that has robust and selective effects on the cellular or physiological system of interest is identified. Natural products from microbes, sea creatures, and plants have a spectacular track record in this regard because evolution works to maximize their capacity for seizing control of endogenous signaling pathways. Understanding how such compounds elicit their effects can yield tremendous insight into the molecular mechanisms underlying complex physiological processes. In fact, the pain pathway has provided some of the best examples of this approach, as illustrated by morphine and aspirin, two natural plant products whose analgesic actions led to the discovery of opiate receptors and cyclooxygenases, respectively (Farah & Rosenberg 1980, Snyder 1977). Similar goals have propelled research on capsaicin and related vanilloid compounds, fueling the search for a bona fide receptor and the endogenous stimuli that these compounds mimic.

HISTORICAL PERSPECTIVE

"Hot" peppers have been cultivated in South America for over 7000 years and in the rest of the world since the 16th century. Today, nearly one-fourth of the world's population consumes hot peppers or related foods on a daily basis (Szallasi & Blumberg 1999). Why do these peppers "burn" when we bite into them? The answer comes from the combined efforts of chemists and physiologists, who have worked on this problem for over 100 years. In the mid-19th century, Thresh isolated the principal pungent component of peppers of the genus *Capsicum* and named it capsaicin (Thresh 1846). Several decades later, Hogyes proposed that *Capsicum* extracts act selectively on sensory neurons to promote a sensation of pain and trigger heat loss through sweating (Hogyes 1878). In 1919, Nelson reported the structure of capsaicin as being an acylamide derivative of homovanillic acid, 8-methyl-*N*-vanillyl-6-noneamide (Nelson 1919; Figure 1). Little more was learned about the actions of capsaicin until Jancso demonstrated in the 1950s and 1960s that this compound not only activates sensory neurons but also renders animals resistant to painful stimuli (Jancso et al 1967). Since this discovery, capsaicin sensitivity has proven to be an extremely useful functional marker for a subset of neurons that are specialized to detect unpleasant or painful (noxious) stimuli. Studies of capsaicin action have provided insights into the activation of primary afferent nociceptors and have revealed the ability of some nociceptors to act in an efferent capacity by stimulating inflammation, smooth muscle contraction, or secretion in target tissues. Moreover, an appreciation of the mechanisms by which capsaicin desensitizes neurons has provided a rational basis for the use of capsaicin and related compounds in the treatment of painful disorders ranging from diabetic neuropathy to arthritis.

A number of authoritative and scholarly reviews have been written on the subjects of capsaicin pharmacology and capsaicin-sensitive neurons (Bevan & Szolcsanyi 1990, Holzer 1991, Szallasi & Blumberg 1999, Szolcsanyi et al 1994, Wood 1993). In the present review, we provide only a brief description of primary afferent neurons of the pain pathway and highlight some of the major pharmacological and physiological findings of the past 30 years that relate to capsaicin action, including evidence for the existence of specific "vanilloid" receptors. We then describe the complementary DNA (cDNA) cloning and functional characterization of a vanilloid receptor (VR1) and a homologous molecule, vanilloid receptor-like protein 1 (VRL-1). Finally, we outline the phenotypic properties of mice lacking VR1 and discuss implications for the role of VR1 in nociception.

PRIMARY AFFERENT NEURONS OF THE PAIN PATHWAY

The cell bodies of nociceptors, like those of most other primary afferent neurons, reside in one of three locations: (*a*) dorsal root ganglia, which innervate the trunk, limbs, and viscera and project centrally to the spinal cord dorsal horn; (*b*) trigeminal ganglia, which innervate the head, oral cavity, and neck and project centrally to the brain stem trigeminal nucleus; and (*c*) nodose ganglia, whose peripheral terminals innervate visceral tissues and whose central terminals project to the floor of the fourth ventricle [for general reviews of nociceptor anatomy and physiology, see Fields (1987), Millan (1999), Raja et al (1999), and Snider & MacMahon (1998)].

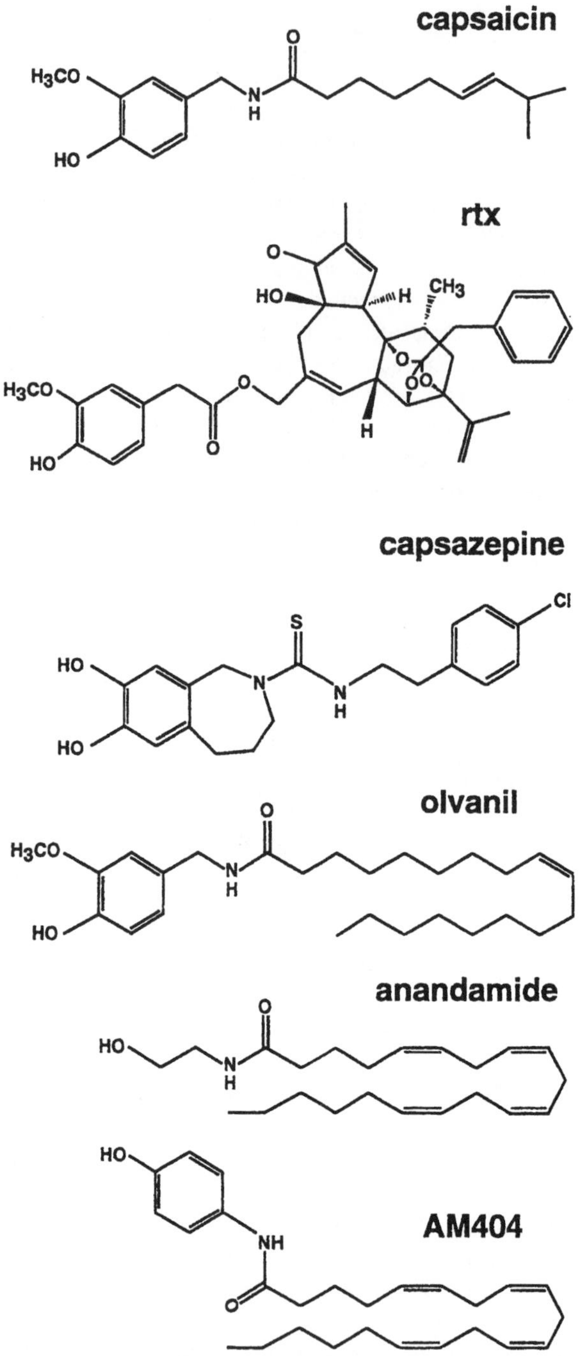

Figure 1   Structural comparison of several molecules capable of activating or inhibiting vanilloid receptors.

Nociceptors can be found among two categories of primary afferent neurons, C fibers and Aδ fibers. C fiber nociceptors have small-diameter, unmyelinated, slowly conducting axons and small (<30 μm)-diameter cell bodies. Aδ fiber nociceptors, in contrast, have medium-diameter, lightly myelinated peripheral axons whose conduction velocities are intermediate between those of C fibers and the rapidly conducting, large-diameter Aβ fibers. Despite considerable overlap, Aδ fiber cell bodies are, on average, larger than those of C fibers (Harper & Lawson 1985). These two neuronal populations can also be distinguished by the ability of A fibers to bind monoclonal antibodies directed against neurofilament proteins (Lawson & Waddell 1985). C and Aδ nociceptors can be further subclassified on functional and anatomical grounds. Members of one group of C fibers, known as polymodal nociceptors, respond to all three pain-producing modalities (i.e. mechanical, chemical, and thermal), while others respond only to subsets of these modalities (Bessou & Perl 1969, Kumazawa & Mizumura 1977, Lang et al 1990, Lynn & Carpenter 1982). Anatomically, most C fibers fall into one of two categories (Snider & MacMahon 1998). One population contains pro-inflammatory peptides such as substance P and calcitonin gene-related peptide and is regulated by nerve growth factor. The other population is nonpeptidergic but can be identified histologically by the presence of specific enzymes (e.g. thiamine monophosphate or fluoride-resistant acid phosphatase) or binding sites for the isolectin B4 (IB4). While the latter neurons are nerve growth factor-dependent during embryogenesis, their neurotrophin dependence switches during early postnatal life such that they instead require glial cell line-derived neurotrophic factor. Peptidergic and nonpeptidergic C fiber nociceptors also exhibit distinct projection patterns to the spinal cord dorsal horn, with nonpeptidergic fibers terminating at a slightly deeper layer than peptidergic fibers. It does not appear that these two anatomical populations differ from one another with regard to the stimulus modalities to which they respond. Rather, they appear to contribute differentially to the enhanced pain responsiveness that follows nerve injury as opposed to target tissue injury.

Aδ nociceptors can be functionally subdivided into two categories (Dubner et al 1977, Leem et al 1993, Meyer & Campbell 1981, Treede et al 1995). Type I Aδ nociceptors can be activated by intense mechanical stimuli or by noxious heat at temperatures higher than 52°C. Type II Aδ nociceptors are also sensitive to both mechanical and heat stimuli but exhibit a lower temperature threshold of 43°C, similar to that of nociceptive C fibers. Many Aδ nociceptors can also be directly activated or sensitized by irritant chemical stimuli such as protons, prostaglandins, or bradykinin (Lang et al 1990, Martin et al 1987, Steen et al 1992).

Capsaicin-Sensitive Primary Afferent Neurons

Capsaicin sensitivity is considered to be a principal pharmacological trait of a major subpopulation of nociceptive sensory neurons. Most capsaicin-sensitive nociceptors are C fibers (Jancso et al 1977, Szolcsanyi 1977), but another, less numerous population consists of Aδ fibers (Lawson & Nickels 1980, Nagy et al 1983). It will be of no surprise to anyone who has handled chili peppers that the skin (Foster & Ramage 1981, Kenins 1982, Szolcsanyi 1977), cornea (Belmonte et al 1991, Szolcsanyi & Jancso-Gabor 1975), and mucous membranes of the mouth (Szolcsanyi 1977, Szolcsanyi & Jancso-Gabor 1975) are rich in capsaicin-sensitive neurons. In addition, capsaicin-sensitive afferent fibers innervate the muscles (Kaufman et al 1982), joints (He et al 1988), and a host of visceral organs in the cardiovascular, respiratory, and genitourinary systems (Coleridge & Coleridge 1977, 1984; Maggi et al 1986; Szolcsanyi 1993). Visceral capsaicin-sensitive afferent neurons are involved in both reflex autonomic responses to visceral stimuli (e.g. changes in blood flow, heart rate, or respiratory rate) and the conscious perception of visceral discomfort (Ness & Gebhart 1990). Some neurons in the preoptic hypothalamus have also been reported to exhibit capsaicin sensitivity (Hori 1984, Jancso-Gabor et al 1970). It has been proposed that these cells represent "warm" receptors involved in the regulation of core body temperature.

CAPSAICIN EFFECTS ON SENSORY NEURONS

Capsaicin-Evoked Excitation

Early in vivo and in vitro physiological recordings revealed that capsaicin depolarizes sensory neurons by promoting the influx of sodium and calcium ions (Baccaglini & Hogan 1983, Heyman & Rang 1985, Marsh et al 1987, Taylor et al 1984, Williams & Zieglgansberger 1982). Whole-cell voltage clamp experiments on cultured sensory neurons (Baccaglini & Hogan 1983, Bevan & Forbes 1988, Bevan & Docherty 1993, Liu & Simon 1994) have revealed that capsaicin stimulates a cationic current response that exhibits a relative preference for divalent cations and little distinction among monovalent ions. Robust permeability to cations has also been demonstrated biochemically by measuring $^{45}Ca^{2+}$ influx and [^{3}H]guanidinium efflux (Wood et al 1988) and microscopically by visualizing Co^{2+} or Ca^{2+} influx with histochemical stains (Winter 1987) and calcium-sensitive fluorescent dyes (Bleakman et al 1990). Electrophysiological studies of membrane patches excised from sensory neurons have further demonstrated that capsaicin triggers the membrane-delimited opening of discrete ion channels (Bevan & Docherty 1993, Forbes & Bevan 1988, Oh et al 1996). Like whole-cell currents, these single-channel responses exhibit cation selectivity and an outwardly rectifying current-voltage relationship such that at a given absolute transmembrane potential, outward conductance is greater than inward conductance. As discussed below, substantial outward rectification is also characteristic of responses mediated by the cloned capsaicin receptor (Caterina et al 1997). This property may serve to enhance calcium influx during the depolarizing phase of the action potential spike, thereby altering action potential kinetics, calcium-dependent regulation of ion channel activity, or the release of neurotransmitters and neuromodulators from the nociceptor [for further discussion see Clapham (1997), Gunthorpe et al (2000)].

Efferent Functions of Capsaicin-Sensitive Neurons

Peripheral activation of capsaicin-sensitive afferent neurons triggers the vesicular release of glutamate and neuromodulatory peptides from their central terminals in the spinal cord dorsal horn, thereby eliciting an acute pain response. A peculiar feature of capsaicin-sensitive neurons, however, is that stimulation by capsaicin or other noxious stimuli can also trigger the release of neuropeptides from their *peripheral* terminals [for reviews, see Holzer (1991, 1993), Maggi (1993), and Maggi & Meli (1988)]. Chief among these neuropeptides are substance P (which triggers plasma extravasation) and calcitonin gene-related peptide (which triggers vasodilatation). The peripheral release of neuropeptides can be triggered by several routes: (*a*) direct activation of a peripheral capsaicin-sensitive terminal, (*b*) activation of a collateral terminal of the same nociceptor (accounting for the flare response commonly observed some distance from the site of tissue injury), and (*c*) retrograde activation of a capsaicin-sensitive afferent neuron by an antidromic electrical stimulus. While activation by the second and third routes can be blocked by local anesthetics that target voltage-gated sodium channels, peptide release occurring after the direct activation of a terminal is insensitive to these agents. Likewise, the direct stimulation of neurogenic inflammation by capsaicin does not depend on the activity of voltage-gated calcium channels, suggesting that the local calcium influx produced by vanilloid receptor activation is a sufficient stimulus for this process. What is the physiological importance of this efferent function? One obvious role is the recruitment of serum factors and inflammatory cells to promote healing at the site of injury and to ward off infection. The neurogenic release of peptides and other vesicular contents can also trigger noninflammatory responses such as secretion and smooth muscle contraction. It has therefore been suggested that tonic baseline efferent activity may serve roles unrelated to injury, such as the trophic maintenance of target tissue or the regulation of visceral tone and blood flow. However, under pathological circumstances, excessive neurogenic peptide release by capsaicin-sensitive afferent neurons is thought to contribute to the maladaptive inflammation associated with such conditions as asthma, inflammatory bowel disease, interstitial cystitis, and arthritis (Campbell 1993, Lundberg 1993). Thus, drugs that antagonize capsaicin receptor-evoked responses might be useful in the treatment of these disorders.

Capsaicin-Evoked Desensitization

Exposure to capsaicin leads initially to nociceptor firing and a period of enhanced sensitivity to painful thermal and mechanical stimuli. This phase is typically followed by a refractory period during which the individual is relatively resistant to capsaicin and certain other painful stimuli (Jancso 1992, Jancso et al 1967). Depending on the capsaicin dose, duration of treatment, route of administration, and subject age and species, this refractory state may last anywhere from hours to the lifetime of the subject. Functional changes are often accompanied by morphological changes that range from mild swelling of axon terminals and mitochondria to complete degeneration of the neuron. In the most extreme situation, neonatal rats or mice treated systemically with 50 mg of capsaicin/kg of body weight exhibit a selective degeneration of C fiber (and some A fiber) axons and an irreversible loss of >80% of small-diameter sensory neuron cell bodies (Jancso 1984, Jancso et al 1977, Lawson & Nickels 1980, Nagy et al 1983, Scadding 1980). As adults, these animals are unresponsive to noxious chemical stimuli such as capsaicin, mustard oil, and xylene and exhibit no neurogenic inflammation in response to these compounds or to antidromic electrical stimulation of cutaneous nerves. Many investigators have reported that rats or mice treated neonatally with capsaicin also exhibit reduced responsiveness to noxious thermal and/or mechanical stimuli (Cervero & McRitchie 1981, Doucette et al 1987, Hayes et al 1981, Holzer et al 1979, Jancso 1984). However, these latter effects vary widely in the literature, possibly owing to the use of different assays of thermal and mechanical nociception and to different extents of C fiber degeneration.

Humans, too, exhibit a sequence of sensory changes with local administration of capsaicin. Intradermal or topical application of this agent results in an initial burning sensation and hyperalgesia to mechanical and thermal stimuli. This is followed by a period of decreased sensitivity to painful chemical, mechanical, or thermal stimuli, as well as mild thermal stimuli (Simone et al 1987, 1998; Simone & Ochoa 1991). Histological analysis of the injection site reveals a reduction in the number of epidermal and subepidermal nerve fibers (Simone et al 1998). These effects underlie the clinical use of cutaneous capsaicin for treatment of burning pain associated with diabetic neuropathy or human immunodeficiency virus-related neuropathy (Robbins 2000), as well as the intravesical infusion of vanilloids for the treatment of hyperactive bladder conditions associated with spinal cord injury (Chancellor & de Groat 1999).

Capsaicin-evoked desensitization and degeneration of sensory neurons can also be observed in vitro (Dray et al 1990a, Marsh et al 1987, Williams & Zieglgansberger 1982, Wood et al 1988, Yeats et al 1992). Prolonged exposure of explanted or cultured, dissociated sensory neurons to capsaicin produces an electrophysiological response that reaches a peak and then subsides despite the continued presence of agonist. Likewise, repeated application of capsaicin at short interstimulus intervals produces a series of responses that decrease in magnitude, especially between the first and second applications. In culture, the removal of extracellular calcium from the bath solution greatly reduces the extent of both electrophysiological desensitization and neuronal degeneration. Chelation of intracellular calcium with 1,2-bis(2-Aminophenoxy)-ethane-N,N,N′,N′-tetraacetic acid (BAPTA) or addition of antagonists of the calcium-dependent phosphatase calcineurin also inhibits capsaicin-evoked desensitization (Cholewinski et al 1993, Docherty et al 1996, Koplas et al 1997).

One possible explanation for these results is that capsaicin-gated ion channels are more active when phosphorylated and calcium-dependent dephosphorylation therefore results in decreased channel activity. In addition, calcium-dependent processes almost certainly contribute to capsaicin-evoked desensitization and neuronal degeneration in vivo. Indeed, neurodegeneration in capsaicin-treated animals is often accompanied by the accumulation of calcium in mitochondria (Jancso et al 1978, 1984). Still, other processes, such as the depletion of substance P from neuronal vesicles (Yaksh et al 1979) and nonspecific blockade of voltage-gated channels (Docherty et al 1991, Petersen et al 1987), are likely to contribute to capsaicin-evoked functional desensitization of nociceptors both in vitro and in vivo.

SPECIFIC VANILLOID RECEPTORS

The exquisite selectivity of capsaicin suggested early on that certain sensory neurons express a specific receptor for this compound. Initial support for the existence of a specific capsaicin "site" was provided by the observation that capsaicin analogues exhibit structure-activity relationships (Szolcsanyi & Jancso-Gabor 1975, 1976; Walpole & Wrigglesworth 1993). Further support came from the study of the phorbol ester derivative resiniferatoxin (Figure 1). This compound, produced by the cactuslike plant *Euphorbia resinifera*, is a far more potent irritant than other phorbol esters, an observation which led Blumberg & colleagues to recognize that resiniferatoxin and capsaicin are structurally related by virtue of a common vanillyl moiety (deVries & Blumberg 1989, Szallasi & Blumberg 1989). Experimentally, they and others (Winter et al 1990) found that resiniferatoxin exhibits excitatory and desensitizing properties similar to those exhibited by capsaicin but at 1000-fold-lower doses. This relatively high specific activity makes [^{3}H]resiniferatoxin a viable radioligand for identifying saturable, capsaicin-displaceable binding sites on membranes derived from dorsal root ganglia (Szallasi & Blumberg 1990, Winter et al 1993). Taken together, these observations support the notion that capsaicin and resiniferatoxin activate sensory neurons by binding to specific vanilloid receptors. The identification of [^{3}H]resiniferatoxin binding sites on the peripheral terminals, cell bodies, and central terminals of sensory neurons (Acs et al 1994, Szallasi et al 1995, Winter et al 1993) corroborated functional studies showing vanilloid sensitivity at all of these subcellular locations (Heyman & Rang 1985, Such & Jancso 1986, Yaksh et al 1979). Final pharmacological support for the existence of a bona fide vanilloid receptor came from the development of capsazepine (Figure 1), a competitive antagonist of both vanilloid binding and vanilloid-evoked responses (Bevan et al 1992).

Molecular Cloning of the Vanilloid Receptor VR1

Whole-cell and excised-membrane patch clamp recordings from sensory neurons suggested that vanilloid receptors either have intrinsic ion channel activity or promote the opening of cationic channels through a membrane-delimited signaling mechanism. However, these studies provided no specific information about the molecular structure of vanilloid receptors or associated ion channels. A screening strategy based on function was therefore chosen as a relatively unbiased and potentially straightforward approach for obtaining a functional cDNA encoding this receptor (Caterina et al 1997). Two assumptions were essential for the success of this screen: (*a*) activation of vanilloid receptors in nonneuronal cells would lead to an increase in cytoplasmic calcium levels, and (*b*) the product of a single cDNA would be sufficient to confer functional sensitivity to capsaicin in such a heterologous cellular environment. With these caveats in mind, cDNA pools consisting of several thousand independent clones from a rodent dorsal root ganglion library were transiently expressed in human embryonic kidney-derived HEK293 cells. Transfected cells were loaded with a calcium-sensitive dye (Fura-2) and microscopically examined for capsaicin-evoked increases in intracellular free calcium by the use of a standard ratiometric imaging system. Once identified, a positive cDNA pool was iteratively subdivided and rescreened by this facile and sensitive assay until a single clone (VR1) that rendered HEK293 cells responsive to capsaicin or resiniferatoxin was obtained.

For VR1 to fulfill the identity of a bona fide vanilloid receptor, several functional and anatomical criteria had to be met. The cloning strategy ensured that the most basic element—sensitivity to vanilloid agonists—was satisfied at the outset. This was validated by direct electrophysiological measurements in which VR1-expressing HEK293 cells or *Xenopus* oocytes showed large membrane current responses to bath-applied capsaicin or resiniferatoxin with electrophysiological properties resembling those of native vanilloid receptors, including a nonselective cation flux with relatively high membrane permeability to calcium, an outwardly rectifying current-voltage relationship at both the whole-cell and single-channel levels, and calcium-dependent desensitization (Caterina et al 1997). Moreover, these responses were attenuated by capsazepine or by ruthenium red, a presumptive pore blocker that noncompetitively antagonizes capsaicin-evoked responses in neurons (Dray et al 1990b). Relative potencies and Hill coefficients for these pharmacological agents at the cloned receptor are similar to values reported for native vanilloid receptors on rat sensory neurons (Caterina et al 1997, Tominaga et al 1998, Jerman et al 2000). The same is true for apparent binding affinities as determined by displacement of [3H]resiniferatoxin from VR1-transfected HEK293 or Chinese hamster ovary cell membranes (Szallasi et al 1999). Indeed, even vanilloid-evoked excitotoxicity can be conferred by VR1, as evidenced by the finding that HEK293 cells expressing this protein are rapidly and efficiently killed after exposure to capsaicin (Caterina et al 1997).

As mentioned above, vanilloid sensitivity is a hallmark of many small-to medium-diameter neurons in dorsal root, trigeminal, and nodose sensory ganglia, and exposure to capsaicin in vitro or in vivo can lead to excitotoxic death of these cells. Thus, one would expect VR1-specific nucleic acid or antibody probes to label a significant fraction of these cells. Double-labeling studies have shown that VR1 is indeed expressed by the majority of unmyelinated peptidergic and nonpeptidergic neurons within these ganglia and can therefore account for the vanilloid sensitivity of these cells (Tominaga et al 1998, Guo et al 1999, Michael & Priestly 1999). It is interesting that VR1 immunoreactivity is not localized to any one region of the primary afferent neuron but is found throughout the cell (including the soma and both the peripheral and central terminals), consistent with functional studies mentioned above.

Within the spinal cord, VR1 staining is prominent in superficial regions of the dorsal horn (laminae I and II), consistent with known central projection patterns of most C fibers and binding profiles observed with [^{3}H]resiniferatoxin. At the same time, VR1 antibodies have revealed previously undetected heterogeneity among sensory neurons, particularly within the IB4-positive class. These cells send their axons to the inner zone of lamina II (IIi), but VR1 antibodies label only the subset of IB4-positive fibers that terminate in the medial part of IIi, not those that terminate more laterally (Tominaga et al 1998). Because medial and lateral regions of the lumbar dorsal horn receive input from distal and proximal regions of the hindlimb, respectively, these observations suggest that neurochemically and functionally heterogeneous subpopulations of primary afferent neurons innervate topographically distinct regions of the body. Additional signs of anatomical segregation have been reported by Elde & colleagues, who found that VR1-positive fibers in the spinal cord or skin showed surprisingly little costaining with substance P and calcitonin gene-related peptide antibodies (Guo et al 1999). It is currently unclear how this observation can be reconciled with the extensive colocalization seen in cell bodies or with the ability of capsaicin to promote neuropeptide release from central and peripheral C fiber terminals. On a similar note, vanilloid receptors have recently been reported to exist throughout the brain, as determined by polymerase chain reaction, in situ hybridization, and immunohistochemical staining methods (Mezey et al 2000, Sasamura et al 1998, Schumacher et al 2000). However, this result is controversial because [^{3}H] resiniferatoxin binding sites are not found on central nervous system neurons (at least not to an extent that would match proposed levels of VR1 expression) and there are no reports of vanilloid-evoked electrophysiological responses in the brain. It is conceivable that closely related subtypes were detected in this analysis, a possibility that can now be addressed by using VR1-deficient mice for similar expression studies.

VR1 Defines a New Subfamily of TRP-Like Ion Channels

The predicted amino acid sequence of VR1 reveals that it is a member of a growing family of ion channels first identified in the fly visual pathway. In 1969, Benzer & colleague described a mutation in *Drosophila melanogaster* that renders fruit flies blind. Electroretinograms showed that whereas normal fly eyes produced a sustained depolarizing response to a flash of light, these mutants exhibited only transient responses to light, earning them the name *trp* for *transient receptor potential* (Hotta & Benzer 1969). Montell & Rubin later cloned the *TRP* locus and found that it encodes a protein with six putative transmembrane domains plus an additional short hydrophobic region connecting transmembrane segments 5 and 6 (Montell & Rubin 1989). This polytopic arrangement resembles the proposed core structure of voltage-gated potassium channels of the *Shaker* family or cyclic-nucleotide-gated ion channels from olfactory neurons and vertebrate photoreceptor cells. TRP channels also contain multiple ankyrin repeats within their N-terminal cytoplasmic region, a signature feature that is shared by VR1 and many (but not all) members of this growing family of cation channels [for a recent review of TRP channel structure and function, see Harteneck et al (2000)]. Three homologous TRP channel subunits have been found in the *Drosophila* eye, where they are thought to form several different homotetrameric and heterotetrameric channels. TRP homologues have now been identified in many organisms, ranging from worms to humans, and in cell types ranging from sensory neurons to lung cells. On the basis of its predicted primary amino acid sequence, VR1 defines its own subfamily of mammalian TRP-like channels, which is now known to include at least two additional members (see below). The closest invertebrate homologue appears to be the product of the *osm-9* gene from the worm *Caenorhabditis elegans* (Colbert et al 1997). Behavioral phenotypes associated with mutations in the *osm-9* gene have implicated this channel in signaling pathways that detect odorants, mechanical stimuli, or changes in osmotic strength.

How are TRP channels activated? In the insect eye, photostimulation of rhodopsin results in the G-protein-dependent activation of phospholipase C. This enzyme, which cleaves phosphatidylinositol bisphosphate into inositol trisphosphate (IP_3) and diacylglycerol, somehow triggers the gating of TRP channels, allowing sodium and calcium to enter and depolarize the photoreceptor cell (Ranganathan et al 1995). Heterologous-expression studies have demonstrated that vertebrate members of the TRP channel family can also be gated by occupancy of G-protein-coupled receptors (Zhu et al 1996) or neurotrophin receptors (Li et al 1999) that stimulate phospholipase C. The precise mechanism(s) by which these channels are gated, however, has been the subject of much debate. Some have suggested that certain TRP channels act as "store-operated" calcium channels that are gated by the phospholipase C-mediated depletion of IP_3-sensitive intracellular calcium stores (Berridge 1995, Friel 1996). It has been further suggested that this gating mechanism involves a direct or indirect conformational coupling between TRP channels and the IP_3 receptor, analogous to that between dihydropyridine receptors and ryanodine receptors in muscle (Berridge 1995, Kiselyov et al 1998). Other studies, however, suggest a role for the diacylglycerol limb of the phospholipase C signaling pathway in TRP channel activation. For example, genetic studies in fruit flies have shown that TRP channels can be activated in the absence of IP_3 receptors (Acharya et al 1997, Scott & Zucker 1998). Moreover, several groups have now shown that fruit fly and mammalian TRP channels can be directly activated in excised membrane patches by diacylglycerol or its derivatives, such as arachidonic or linoleic acid (Chyb et al 1999, Hofmann et al 1999). These findings suggest that polyunsaturated fatty acids or other structurally related lipids may serve as endogenous ligands for TRP channels in the fruit fly eye or elsewhere, an issue that we discuss in greater detail below.

As mentioned above, VR1 is the founding member of a growing subfamily of mammalian TRP channels. With the aid of database searches, we identified a cDNA sharing ~50% sequence identity with VR1 (Caterina et al 1999). HEK293 cells and *Xenopus* oocytes expressing this clone are insensitive to capsaicin, protons, and moderate heat but do show nonselective cationic currents when challenged with a high-temperature stimulus exceeding a threshold of ~52°C. This novel channel was dubbed vanilloid receptor-like protein 1 (VRL-1) to reflect its structural similarity to VR1 but its insensitivity to capsaicin. Within sensory ganglia, VRL-1 is expressed primarily by medium- to large-diameter neurons, making it a candidate transducer of high-threshold heat responses by these cells, which probably constitute the lightly myelinated Aδ nociceptors of the type I class. VRL-1 transcripts are also found in a variety of other tissues, including spinal cord, spleen, brain, and lung, suggesting that this channel is activated by physiological stimuli other than heat. In fact, a mouse orthologue of VRL-1 [termed growth-factor-regulated channel (GRC)] was subsequently identified in a polymerase chain reaction-based screen for novel growth factor-regulated channels (Kanzaki et al 1999). Chinese hamster ovary cells transfected with this clone show increased cell surface expression of GRC and enhanced permeability to calcium after exposure to insulin-like growth factor. Thus, the VRL-1/GRC channel may be regulated by growth factors through mechanisms involving increased translocation to the plasma membrane, second-messenger-mediated changes in channel activity, or both. A similar concept may apply to another, more distantly related member of this TRP subfamily, referred to as an epithelial calcium channel (ECaC), which shares <30% identity with VR1 (Hoenderop et al 1999). ECaC was cloned from a rabbit kidney library by screening for cDNAs that enhanced calcium uptake in *Xenopus* oocytes. The pharmacological and physiological properties of ECaC suggest that it constitutes a vitamin D_3-regulated apical calcium channel activity characterized in kidney epithelium.

VRL-1 and ECaC show structural similarity to VR1, but they are clearly encoded by separate genes. In addition to these homologues, a number of cDNAs that share large regions of nucleotide identity with VR1, and probably represent alternatively spliced transcripts derived from the VR1 gene, have been described. One of these variants is called SIC, a putative stretch-inhibitable channel that is missing a significant portion of the VR1 N terminus (the coding region begins at amino acid position 308 of VR1) and shows divergence within the C-terminal cytoplasmic domain (Suzuki et al 1999). Another N-terminally truncated variant, VR.5'sv, is predicted to begin at the same position as SIC but is otherwise identical to VR1 (Schumacher et al 2000). Heterologous cells transfected with VR.5'sv cDNA do not respond to capsaicin, protons, or heat. Thus, it is not yet clear whether these putative splice variants yield functional channels or complex with VR1 or related molecules to produce channels having novel pharmacological or electrophysiological properties in vitro or in vivo.

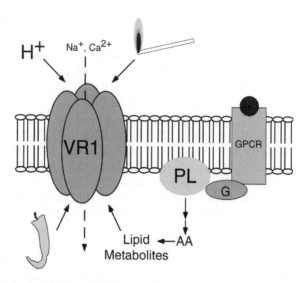

Figure 2   Cationic flux through VR1 can be regulated by the convergent actions of multiple pain-producing stimuli. Membrane-permeant second messengers, such as lipid metabolites, may modulate nociceptor activity in an autocrine fashion (as diagrammed here), or in a paracrine manner if produced by neighboring neural or nonneural cell types. Abbreviations: AA, arachidonic acid; G, heterotrimeric G protein; GPCR, G protein-coupled receptor; PL, phospholipase. Capsaicin and noxious heat are represented by pepper and flame, respectively.

VR1 IS A POLYMODAL DETECTOR OF NOXIOUS PHYSICAL AND CHEMICAL STIMULI

A major impetus to identifying the capsaicin receptor was to determine whether known pain-producing stimuli might exert their effects by acting at this same site. With a functional vanilloid receptor cDNA in hand, this fascinating question could finally be addressed by challenging VR1-expressing cells with a variety of chemical and physical stimuli capable of activating primary afferent neurons in vitro or in vivo. As summarized below, these studies have produced a number of interesting candidates for physiologically relevant vanilloid receptor agonists (Figure 2).

Heat

Many proteins, including receptors and channels, show alterations in their structure or activity as a function of temperature, but VR1 has thermal response characteristics that make it especially interesting in the context of nociception. Most notably, VR1 is gated by heat, but only when ambient temperatures exceed $\sim 43°C$ (Tominaga et al 1998), a threshold matching that of heat-evoked pain responses in humans and animals or heat-evoked electrophysiological responses in primary afferent nerve fibers or cultured sensory neurons (Cesare & McNaughton 1996, LaMotte & Campbell 1978, Raja et al 1999). Evidence for a direct relationship between VR1 expression and heat sensitivity is supported by several observations (Tominaga et al 1998), the most basic of which is that sensitivity to capsaicin and heat sensitivity are significantly correlated (in both frequency and magnitude) among VR1-transfected HEK293 cells. Responses to these stimuli also show cross-desensitization, and both capsaicin- and heat-evoked currents are attenuated by the vanilloid receptor antagonists capsazepine and ruthenium red. In addition, both responses are characterized by outwardly rectifying current-voltage relations and relatively high permeability to calcium ions. Some distinctions do exist, such as quantitative differences in cation permeability ratios or requirements for extracellular calcium in desensitization, suggesting that vanilloids and heat activate VR1 through overlapping but distinct mechanisms. In VR1-transfected HEK293 cells (as in sensory neurons), capsaicin or heat evokes single-channel currents in excised membrane patches, demonstrating that channel activation occurs via a membrane-delimited mechanism that does not require the action of diffusible cytoplasmic second messengers (Tominaga et al 1998). Site-specific mutations in VR1 can significantly alter capsaicin potency or thermal activation thresholds (Jordt et al 2000), providing further evidence that VR1 itself transduces responses to these stimuli when expressed in heterologous cells.

These observations of heterologous nonneuronal systems have led us to propose that VR1 functions as a molecular transducer of noxious thermal stimuli in vivo. Consistent with this hypothesis, sensitivity to capsaicin and sensitivity to noxious heat are also well correlated among small-diameter sensory neurons in culture (Kirschstein et al 1997; Nagy & Rang 1999a,b). Moreover, VR1 and native heat-evoked currents have a number of properties in common, including similar current-voltage relationships (Cesare & McNaughton 1996, Nagy & Rang 1999b, Reichling & Levine 1997), selective permeability to cations (Cesare & McNaughton 1996, Nagy & Rang 1999b, Reichling & Levine 1997), and, in some studies, sensitivity to vanilloid receptor antagonists (Kirschstein et al 1999). At the same time, numerous discrepancies between native heat- and vanilloid-evoked responses have been reported, including differences in relative permeability to calcium and sodium ions (Cesare & McNaughton 1996, Nagy & Rang 1999b) and sensitivity to vanilloid receptor antagonists (Nagy & Rang 1999b, Reichling & Levine 1997). Moreover, Nagy & Rang have recently shown that the amplitudes of the capsaicin- and heat-evoked responses among individual sensory neurons are not tightly correlated, contrary to the scenario expected if the same channel responds to both stimuli. Most significantly, capsaicin- and heat-evoked responses showed poor cosegregation at the single-channel level in membrane patches excised from cultured rat sensory neurons (Nagy & Rang 1999b). That is, most patches were sensitive to capsaicin or heat, but only a few patches responded to both stimuli (although the frequency of dually responsive patches was significantly higher than one would predict for random inclusion of two independent channels in the same patch).

From these findings, Nagy & Rang proposed that distinct ion channels respond to capsaicin and heat. These channels could consist of entirely different molecules or different functional isoforms of VR1 generated through alternative RNA splicing, post-translational modification, or association with other cellular proteins. Unfortunately, similar patch clamp analyses have not been carried out with VR1-expressing HEK293 cells; such experiments might help to address these issues.

At least some of the apparent discrepancies listed above may be accounted for by the fact that sensory neurons express multiple forms of heat-activated channels that differ in their biophysical or pharmacological properties. Indeed, as described above a recently identified VR1 homologue (VRL-1) is insensitive to capsaicin or protons but does respond to high-threshold heat stimuli ($>50°C$) when expressed in nonneuronal cells (Caterina et al 1999). This threshold is similar to that reported for a subset of medium- to large-diameter sensory neurons in culture (Nagy & Rang 1999a) and for some thin myelinated (Aδ) nociceptors in vivo (Raja et al 1999, Treede et al 1995). In fact, VRL-1 expression within rat and mouse sensory ganglia is confined primarily to neurons with these same anatomical properties (Caterina et al 1999). We have therefore proposed that VRL-1 accounts for the "high-threshold" thermal sensitivity of this subset of nociceptors while VR1 detects moderate-intensity heat stimuli in small-diameter, unmyelinated (C fiber) nociceptors. However, this model is based largely on correlative evidence, owing to the paucity of selective and potent pharmacological agents with which to manipulate vanilloid receptors in vivo.

Protons

Tissue damage, such as that associated with infection, inflammation, or ischemia, produces an array of chemical mediators that activate or sensitize nociceptor terminals to elicit pain and promote tenderness at the site of injury (Handwerker & Reeh 1991, Levine & Taiwo 1994). Protons constitute one important component of this pro-algesic response, reducing the extracellular pH to levels below the physiological norm of $\sim$7.6. Extracellular protons elicit both transient and sustained excitatory responses in cultured sensory neurons, the latter of which is believed to account for persistent pain associated with local tissue acidosis (Bevan & Geppetti 1994). Protons are capable of modulating the activity of a number of receptors and ion channels expressed by primary afferent nociceptors, including acid-sensitive channels of the degenerin family (Chen et al 1998; Lingueglia et al 1997; Waldmann et al 1997a,b), ATP-gated channels (Li et al 1997, Stoop et al 1997), and vanilloid receptors (Caterina et al 1997, Jordt et al 2000, Kress et al 1996, Martenson et al 1994, Petersen & LaMotte 1993, Tominaga et al 1998). Which, if any, of these entities contributes to acid-evoked pain is presently unclear, but electrophysiological and genetic studies of native and cloned vanilloid receptors suggest that they play a significant role in mediating sustained proton responses in vivo.

Because protons have been proposed to act as modulators of native vanilloid receptors, there is significant interest in understanding the relationship between sensitivity to capsaicin and proton sensitivity at the cellular and molecular levels. We have shown, using both VR1-expressing mammalian cells and *Xenopus* oocytes, that moderately acidic bath conditions augment capsaicin-evoked responses by increasing agonist potency (50% effective concentration = 90 nM at pH 7.4 versus 36 nM at pH 6.4) without altering efficacy (Caterina et al 1997, Jordt et al 2000, Tominaga et al 1998). Importantly, extracellular protons also potentiate heat-activated currents (Tominaga et al 1998). Temperature response curves generated in VR1-expressing oocytes or HEK293 cells show that a reduction in extracellular pH produces markedly larger responses at temperatures that are noxious to mammals (>43°C). Moreover, a reduction in pH dramatically lowers the threshold for channel activation, such that at pH 6.3, substantial currents can be seen at temperatures as low as 35°C, conditions under which the channel is normally closed (at pH 7.6). This augmentation of VR1 thermal responsiveness by protons closely resembles the increase in nociceptor thermal sensitivity associated with inflammation (Handwerker & Reeh 1991). In both cases, there is a significant decrease in the threshold for heat-evoked responses and an increase in response magnitudes at temperatures above the initial pain threshold. Importantly, VR1 shows especially dynamic modulation of heat-evoked currents between pH 8 and 6 (Jordt et al 2000), a sensitivity range that matches the extent of local acidosis attained during most forms of tissue injury. Below pH 6, sustained membrane currents can be observed in VR1-expressing HEK293 cells at room temperature (22°C), with a half-maximal effective pH of 5.4 (Tominaga et al 1998). Whether these proton-evoked responses result simply from a decrease in the channel's thermal response threshold or involve additional steps is not entirely clear, but recent structure-function studies suggest that proton-evoked channel activation and proton-mediated potentiation can be functionally uncoupled (Jordt et al 2000; see below).

How do protons modulate VR1 activity? Electrophysiological studies of native and cloned receptors suggest that extracellular protons act primarily to increase the probability of channel opening (Tominaga et al 1998, Baumann and Martenson 2000) rather than by increasing unitary conductance or interacting directly with a vanilloid binding site [which may be intracellular (Jung et al 1999)]. In fact, acidic bath solutions evoke ionic currents when applied to outside-out, but not inside-out, membrane patches excised from VR1-expressing HEK293 cells, suggesting that protons interact with an extracellular site(s) on the channel complex (Tominaga et al 1998). Candidate sites for such interactions include several acidic amino acids located within putative extracellular loops of VR1. Site-directed mutational analysis has pinpointed two glutamate residues of particular interest, one at position 600 and another at position 648 (Jordt et al 2000). The first of these residues is located between the putative fifth transmembrane domain and pore loop segment. Introduction of neutral or positive residues at this site increases the responses of VR1-expressing cells to capsaicin or heat. For example, E600Q mutant channels show a >10-fold increase in sensitivity to capsaicin (no change in efficacy) and E600K mutants show a dramatic decrease in thermal activation threshold

(30–32°C). Conversely, introduction of a more acidic residue at this position (E600D) decreases sensitivity to these stimuli. Thus, E600 appears to play a critical role in determining the pH sensitivity range of channel activation by noxious stimuli. Interestingly, mutants bearing nontitratable amino acids at position 600 (e.g. E600Q) can still be activated by low-pH solutions, suggesting that proton-evoked channel activation and proton-mediated potentiation involve titration of different sites. Further support for this idea comes from mutational analysis of another putative extracellular glutamate residue. Oocytes expressing E648A mutant channels are essentially insensitive to activation by low-pH (4.0) solutions but retain sensitivity to capsaicin and heat and show normal proton-mediated potentiation of these stimuli. The selective nature of this phenotype also suggests that protons, vanilloids, and heat promote channel opening through distinct pathways having one or more stimulus-specific steps. These site-directed mutagenesis studies provide additional evidence that protons interact with specific amino acids on the extracellular surface of VR1 to allosterically modulate channel activity. However, they do not by themselves pinpoint exact sites of titration by extracellular protons.

Lipids

Capsaicin bears structural similarity to a number of lipid-derived second messengers (e.g. arachidonic acid), suggesting that VR1 may be activated by an endogenous ligand of this sort. Indeed, capsazepine and other competitive vanilloid antagonists were developed with the idea that they might block receptor activation under circumstances (e.g. inflammation) in which endogenous capsaicin-like agonists would be produced (Bevan et al 1992). A number of recent observations, both from within and outside the vanilloid receptor field, provide compelling arguments for the existence of endogenous capsaicin-like ligands that may activate VR1 or alter its sensitivity to other stimuli.

One line of evidence comes from the analysis of phototransduction in the fruit fly eye. As mentioned above, *Drosophila* TRP can be activated in vitro by polyunsaturated fatty acids, such as arachidonic or linoleic acid (Chyb et al 1999). Moreover, heterologously expressed mammalian TRPC3 and TRPC6 channels can be activated by diacylglycerol (Hofmann et al 1999), further implicating lipids as potential in vivo regulators of TRP channel function. Another line of evidence comes from structural and functional connections between cannabinoid and vanilloid receptor pharmacology. For example, synthetic vanilloid receptor ligands, such as olvanil or other long-chain N-acyl-vanillyl amides, bear structural similarity to the endogenous cannabinoid receptor agonist anandamide (arachidonylethanolamide; Figure 1). Olvanil also resembles AM404 (Figure 1), a synthetic anandamide transport inhibitor, and both block reuptake of anandamide into cells (Melck et al 1999, Beltramo and Piomelli 1999). It therefore follows that some cannabinoid receptor ligands might interact with vanilloid receptors. Indeed, this turns out to be the case for both anandamide and AM404 (Zygmunt et al 1999, 2000a). These compounds are significantly less potent than capsaicin, and they elicit responses with somewhat slower kinetics, but both evoke outwardly rectifying, nonselective cationic currents in HEK293 cells or *Xenopus* oocytes expressing VR1. These responses appear to be specific because they are inhibited by capsazepine, but not by cannabinoid receptor antagonists, and a number of other synthetic or endogenous cannabinoid receptor agonists have little or no effect on VR1 function.

In each case in which bioactive lipids have been shown to modulate TRP-like channels in vitro, the estimated 50% effective concentration falls in the range of 10^{-7} to 10^{-5} M (the 50% effective concentration for anandamide activation of VR1 is ∼5 μM). This raises questions as to whether such concentrations are attained under normal or pathophysiological conditions. Of course, in vitro electrophysiological estimates of agonist potency are determined by perfusing aqueous suspensions of these hydrophobic agents over a cell surface or membrane bilayer, but how much actually reaches the appropriate site on the channel protein is difficult, if not impossible, to estimate. Polyunsaturated fatty acids or structurally related metabolites are produced in vivo through enzymatic cleavage of membrane lipids, and their access to receptors within the same or adjacent cells may occur with significantly higher efficiency, particularly if much of the action is confined to the hydrophobic environment of the bilayer. Moreover, under inflammatory conditions, macrophages and endothelial cells may release substantial amounts of anandamide and other lipid messengers into a confined intercellular space, such that the local concentrations of these agents may approach the micromolar range.

Anandamide produces analgesia with potency in the nanomolar range and inhibits release of calcitonin gene-related peptide in the skin through its actions at cannabinoid receptors. These findings have led some to argue that anandamide exerts its actions on sensory neurons via cannabinoid receptors alone, without the involvement of vanilloid receptors (Szolcsanyi 2000). However, modulation of sensory neuron function by these receptor systems need not be viewed as a mutually exclusive possibility since anandamide may interact with VR1 to produce physiological effects that are distinct from those mediated via cannabinoid receptors [see Smart & Jerman (2000) and Zygmunt et al (2000b) for more-detailed discussions]. The principal action of anandamide on vanilloid receptors may be to potentiate responses to other stimuli, most notably heat or protons. In fact, under conditions of elevated temperatures or decreased pH, the potency of anandamide as a VR1 agonist may be significantly increased. Thus pro-algesic agents such as ATP, bradykinin, serotonin, leukotrienes, and prostanoids may sensitize the primary afferent neuron, in part, by stimulating the production of lipid-derived second messengers that sensitize VR1 and consequently enhance nociceptor excitability. Additionally, anandamide produced by nonneuronal cells may synergize with these agents by modulating the activity of VR1 channels on neighboring sensory nerve terminals. Finally, anandamide may be just one of several as-yet-unidentified lipids capable of activating VR1. VR1-expressing cell lines provide the necessary tools to characterize novel, and possibly more potent, lipid agonists by means of calcium imaging, electrophysiological, or radioligand binding assays. In fact, a high-throughput screen of >1000 bioactive substances reidentified anandamide as an activator of VR1 (Smart et al 2000), and a directed analysis of arachidonic acid metabolites identified lipooxygenase products [e.g. 12- or 15-(S)-Hydroperoxyeicosa-5Z, 8Z, 11Z, 13E-tetraenoic acid] as agonists of this channel (Hwang et al 2000). Maximal efficacies for these compounds differ by as much as fivefold, and all of the substances exhibit potencies in the $1–10 \, \mu M$ range at room temperature.

Where on the VR1 protein do vanilloids and cannabinoids bind, and how do they promote channel opening? Sites of interaction have not yet been mapped, and little is known about the mechanism(s) of channel gating. Capsaicin can activate VR1 when it is applied to either side of an excised membrane patch (Caterina et al 1997), consistent with the idea that vanilloids can permeate or cross the lipid bilayer to mediate their effects. Recent electrophysiological studies using hydrophilic capsaicin derivatives suggest that vanilloids interact with an intracellular site on VR1 (Jung et al 1999), but specific ligand binding domains have not been mapped. Although capsazepine blocks the actions of both capsaicin and anandamide, it is not clear whether these compounds compete for binding to the same site. While capsaicin potency is clearly enhanced under moderately acidic (pH 6.4) conditions (Caterina et al 1997, Tominaga et al 1998), this appears not to be the case for anandamide (Smart et al 2000), suggesting that these agonists interact with VR1 in nonidentical ways. Further molecular and biochemical studies will be required to resolve these interesting questions.

GENETIC ANALYSIS OF CAPSAICIN RECEPTOR FUNCTION IN VIVO: VR1 Knockout in Mice

The studies described above present a strong circumstantial case for the involvement of VR1 not only in the actions of vanilloid compounds but also in sensory responses to noxious heat, acid stimuli, and perhaps endogenous lipids. One approach to addressing this issue directly has been to generate "knockout" mice in which the *VR1* gene is disrupted (Caterina et al 2000, Davis et al 2000). These mice are viable and fertile and exhibit a normal appearance as well as normal gross behavior. Despite the absence of VR1, sensory ganglion development (as assessed by the presence of several histological markers) is apparently unaltered in these animals. Functionally, however, disruption of the *VR1* gene produces an array of specific defects related to nociception, as described below.

Deficits in Cellular Physiology and Acute Nociception

Neurons derived from the dorsal root ganglia of *VR1*-null mice exhibit no vanilloid-evoked electrophysiological responses either in culture or in sensory nerve fibers innervating an excised patch of skin (the excised-skin nerve preparation). In the intact mouse, these deficits are manifest as drastic reductions in paw licking and neurogenic inflammation evoked by intraplantar injection of either capsaicin or resiniferatoxin. Sensory neurons of the trigeminal system also depend on VR1 for vanilloid responsiveness: whereas wild-type mice avoid the consumption of capsaicin-containing water, littermates lacking VR1 exhibit no such aversion. Moreover, *VR1*-null mice fail to exhibit the profound hypothermia observed in wild-type mice after subcutaneous injection of capsaicin. Taken together, these data demonstrate that VR1 is essential for transducing the nociceptive, inflammatory, and hypothermic effects of vanilloid compounds.

The involvement of VR1 in proton-evoked nociceptive responses is supported by data from multiple in vitro assays. Whereas 30% of cultured wild-type dorsal root ganglion neurons exhibit large, sustained current responses following exposure to pH 5 medium, <7% of neurons from *VR1*-null mice exhibit such responses. These results are mirrored in the skin nerve preparation, in which the absence of VR1 results in an ~90% reduction in the proportion of acid-sensitive C fiber nociceptors. Most likely, the residual proton-evoked responses in *VR1*-null-mouse C fibers are mediated by members of the acid-sensing ion channel (ASIC) family (Chen et al 1998; Lingueglia et al 1997; Waldmann et al 1997a, b). These proteins are part of a superfamily of channels that also includes the amiloride-sensitive epithelial sodium channels, peptide-gated ion channels from snails, and the so-called degenerins, putative mechanosensory channels first identified in *C. elegans*. At least five different ASIC subtypes have been identified, and many are expressed in sensory afferent neurons. Indeed, the expression of ASIC3 is largely restricted to these cells.

To what extent do VR1 and ASICs, respectively, contribute to proton-evoked pain or other in vivo responses associated with tissue acidification? Ischemic, inflammatory, or infectious events commonly reduce the local tissue pH to <7, with reductions sometimes to <6. VR1 can be strongly activated by a pH of 5.9 at 22°C or by a pH of 6.4 at 37°C, making it well poised to detect these tissue insults. Some ASIC family members, either alone or in combination, can also respond to protons in this concentration range, although many are activated only at lower pH values. Retrograde labeling of cardiac afferent neurons, followed by electrophysiological recording of labeled cells, has revealed that many afferent fibers innervating myocardium are capsaicin-insensitive neurons whose acid-evoked current responses most closely resemble those of heterologously expressed ASIC3 (Benson et al 1999). Thus, in this physiological setting, ASICs may play a predominant proton-sensing role. Still, the profoundly reduced prevalence of proton-evoked responses in neurons derived from *VR1*-null mice strongly suggests that these animals will exhibit deficits in acid-evoked sensory excitation in other visceral, muscular, or cutaneous locations.

Thermal nociception in *VR1*-null mice has also been examined, using a collection of in vitro and in vivo assays. Voltage-clamped, cultured sensory neurons derived from wild-type rats or mice exhibit one of three response patterns during exposure to a brief heat ramp to 60°C (Caterina et al 2000, Nagy & Rang 1999a). Approximately half of these neurons do not respond to the heat stimulus.

Another 40% exhibit large, inward currents once the temperature exceeds ~45°C. These latter cells are capsaicin sensitive, consistent with the notion that responses to both stimuli are mediated by VR1. The final 10% of neurons respond to heat only at temperatures >52°C. This high threshold, together with the capsaicin insensitivity of these cells, suggests that VRL-1 may account for this final class of responses. Consistent with this (perhaps oversimplified) interpretation, sensory neurons derived from *VR1*-null mice exhibit a normal prevalence of the high-threshold heat-evoked current responses but none of the moderate-threshold responses.

Deficient heat-evoked responses among the sensory neurons of *VR1*-null mice are also observed in single-unit recordings from the skin nerve preparation. Here, however, the picture is more complex. Among C fiber nociceptors, those of *VR1*-null mice exhibit a reduction in the proportion of heat-responsive units, dropping from the wild-type incidence of 13 in 24 to 4 in 24. The remaining heat-responsive C fibers derived from *VR1*-null mice exhibit a normal threshold but a reduced firing rate at higher noxious temperatures. In contrast to these deficits, mechanical sensitivity in this preparation is unaffected by the absence of VR1 (Caterina et al 2000).

VR1-null mice also exhibit a selective reduction in thermal nociceptive input to the spinal cord dorsal horn (evoked by heating of the hindpaw), without concomitant decreases in mechanically evoked input. Curiously, thermonociceptive input to the more deeply situated wide-dynamic-range neurons of the spinal cord dorsal horn appears to be more drastically affected by the absence of VR1 than is input to projection neurons in the more superficial layers of the dorsal horn (Caterina et al 2000). This finding may reflect distinct integration by these neurons of information from VR1-positive versus VR1-negative primary afferent neurons.

In one of the knockout studies (Caterina et al 2000), it was found that *VR1*-null mice exhibited selective deficits in several behavioral assays of acute heat-evoked nociception, including the tail immersion, hot plate, and radiant paw heating assays. In each case, heat-evoked withdrawal at relatively low noxious temperatures was comparable with that of wild-type littermates, while at higher temperatures examined, *VR1*-null mice exhibited a longer withdrawal latency, consistent with impaired thermal nociception. Indeed, the absence of statistically significant differences in heat-evoked responses between genotypes in the study by Davis & colleagues (Davis et al 2000) most likely stems from the relatively low noxious temperatures used in that study. An analysis of this knockout strain at higher temperatures and in a homogeneous background might therefore resolve this apparent discrepancy. The selective impairment of heat-evoked nociception at high temperatures in *VR1*-null mice (and preservation at lower noxious temperatures) is consistent with the relatively greater deficit in C fiber heat coding at higher temperatures but is apparently at odds with the relatively moderate in vitro threshold for VR1 activation and the absence of moderate-threshold heat-evoked responses in cultured sensory neurons derived from *VR1*-null mice. Several conclusions arise from these findings. First, not all functional properties of nociceptors are faithfully recapitulated once they are dissociated and placed in culture. Second, it appears that one cannot necessarily predict the quantitative range of a sensory deficit in a knockout-mouse model from the response properties exhibited by a single receptor molecule. At present, the significance of this observation is unclear, particularly since we do not yet know the identities or properties of all receptor molecules involved in thermosensation. These findings do, however, underscore the complexity of behavioral "readouts" of nociception, as well as the value of approaching the analysis of knockout mice from multiple physiological levels.

In any case, the findings of both *VR1* knockout studies make it clear that there exist VR1-independent mechanisms for the detection of noxious heat. VRL-1 is one candidate mediator of these responses (Caterina et al 1999). The profile of heat-evoked currents observed in cultured sensory neurons derived from *VR1*-null mice is consistent with this explanation. However, the temperature threshold exhibited by residual heat-sensitive C fibers in *VR1*-null mice is significantly below that reported for recombinant VRL-1. In addition, immunofluorescence studies of rat dorsal root ganglia have suggested that VRL-1 is not expressed at detectable levels in C fibers. Thus, species differences in the pattern of VRL-1 expression and an altered VRL-1 activation threshold must be invoked to account completely for the residual heat-evoked responses observed in the *VR1*-null mice, unless other molecules act as receptors for noxious heat. The generation and analysis of *VR1 VRL-1* doubly null mice, as well as the identification and characterization of additional vanilloid receptor homologues, might clarify this situation.

Deficits in Tissue Injury-Induced Thermal Hyperalgesia

As described above, electrophysiological studies of the cloned vanilloid receptor have revealed that VR1 is responsive to multiple stimuli, including heat, protons, and lipid metabolites. Injury brings on many changes that affect the activity of the nociceptor, including local tissue acidosis and the production of pro-algesic agents, such as bradykinin, ATP, monoamines, and arachidonic acid metabolites. The net result is one in which response thresholds to noxious stimuli are decreased, thereby contributing to the development of thermal and mechanical hypersensitivity. The capacity of VR1 to detect and integrate information from diverse physical and chemical inputs makes this channel potentially well suited for assessing the physiological environment of the primary afferent nerve terminal and for altering nociceptor excitability in the setting of tissue injury. We have therefore hypothesized that VR1 contributes to peripheral mechanisms underlying thermal hypersensitization.

This prediction is clearly borne out by the analyses of VR1-deficient mice. In these studies, tissue injury was elicited by treatment of the hindpaw with an inflammatory agent or nerve injury was produced by partial ligation of the sciatic nerve. Wild-type mice exhibit increased sensitivity to both thermal and mechanical stimuli after such treatments. In contrast, *VR1*-null mice show hypersensitivity to mechanical stimuli, as well as normal thermal hypersensitivity following nerve injury, but they do not show increased sensitivity to thermal stimuli following tissue injury. Thus, VR1 appears to be essential for the development of thermal hypersensitivity associated with tissue inflammation, but not that associated with nerve injury. This selective phenotype represents what may be the most significant practical outcome of the genetic studies because it highlights physiological settings in which vanilloid receptor antagonists have the potential to serve as effective analgesic agents.

Clearly, inflammation-induced thermal hypersensitivity may result from the convergent actions on VR1 of heat, low pH, and other inflammatory mediators. However, other mechanisms, such as up-regulation of VR1 expression or sensitization of VR1 by post-translational modification may also come into play. Moreover some of the effects of inflammation might also be manifest downstream of VR1, at the level of general nociceptor excitability. Further mechanistic studies will be required to dissect these possibilities.

FUTURE PROSPECTS

The study of vanilloid receptors has provided valuable insights into the molecular mechanisms by which nociceptors evaluate their physical and chemical environments. Still, it is clear from the data presented above that much more remains to be learned about these processes. For instance, what are the molecules that account for the residual heat-evoked nociception observed in VR1-null mice? What are the relative contributions of different heat sensors to the subjective experiences of warmth and heat-evoked pain? Do protons or endogenous lipids regulate vanilloid receptor activity in vivo? How are heat, protons, and lipids mechanistically integrated by VR1? What molecules account for the perception of mechanically evoked pain? Finally, how are the levels and/or activities of these molecules changed in pathological pain states? The continued integration of molecular, anatomical, behavioral, and physiological approaches in the study of pain should allow these questions to be answered.

Visit the Annual Reviews home page at www.AnnualReviews.org

LITERATURE CITED

Acharya JK, Jalink K, Hardy RW, Hartenstein V, Zuker CS. 1997. InsP3 receptor is essential for growth and differentiation but not for vision in *Drosophila*. *Neuron* 18:881–87

Acs G, Palkovits M, Blumberg PM. 1994. Comparison of [³H]resiniferatoxin binding by the vanilloid (capsaicin) receptor in dorsal root ganglia, spinal cord, dorsal vagal complex, sciatic and vagal nerve and urinary bladder of the rat. *Life Sci.* 55:1017–26

Adler E, Hoon MA, Mueller KL, Chandrashekar J, Ryba NJ, Zuker CS. 2000. A novel family of mammalian taste receptors. *Cell* 100:693–702

Baccaglini PI, Hogan PG. 1983. Some rat sensory neurons in culture express characteristics of differentiated pain sensory cells. *Proc. Natl. Acad. Sci. USA* 80:594–98

Baumann TK, Martenson ME. 2000. Extracellular protons both increase the activity and reduce the conductance of capsaicin-gated channels. *J. Neurosci.* 20:RC80.

Belmonte C, Gallar J, Pozo MA, Rebollo I. 1991. Excitation by irritant chemical substances of sensory afferent units in the cat's cornea. *J. Physiol.* 437:709–25

Beltramo M, Piomelli D. 1999. Anandamide transport inhibition by the vanilloid agonist olvanil. *Eur. J. Pharmacol.* 364:75–78.

Benson CJ, Eckert SP, McCleskey EW. 1999. Acid-evoked currents in cardiac sensory neurons: a possible mediator of myocardial ischemic sensation. *Circ. Res.* 84:921–28

Berridge MJ. 1995. Capacitative calcium entry. *Biochem. J.* 312:1–11

Bessou P, Perl ER. 1969. Response of cutaneous sensory units with unmyelinated fibers to noxious stimuli. *J. Neurophysiol.* 32:1025–43

Bevan S, Forbes CA. 1988. Membrane effects of capsaicin on rat dorsal root ganglion neurones in cell culture. *J. Physiol.* 398:28P

Bevan S, Geppetti P. 1994. Protons: small stimulants of capsaicin-sensitive sensory nerves. *Trends Neurosci.* 17:509–12

Bevan S, Hothi S, Hughes G, James IF, Rang HP, et al. 1992. Capsazepine: a competitive antagonist of the sensory neuron excitant capsaicin. *Br. J. Pharmacol.* 107:544–52

Bevan S, Szolcsanyi J. 1990. Sensory neuron-specific actions of capsaicin: mechanisms and applications. *Trends Pharmacol. Sci.* 11:330–33

Bevan SJ, Docherty RJ. 1993. Cellular mechanisms of the action of capsaicin. See Wood 1993, pp. 27–44

Bleakman D, Brorson JR, Miller RJ. 1990. The effect of capsaicin on voltage-gated calcium currents and calcium signals in cultured dorsal root ganglion cells. *Br. J. Pharmacol.* 101:423–31

Buck L, Axel R. 1991. A novel multigene family may encode odorant receptors: a molecular basis for odor recognition. *Cell* 65:175–87

Campbell E. 1993. Clinical applications of capsaicin and its analogues. In *Capsaicin and the Study of Pain,* ed. J Wood, pp. 255–72. London: Academic

Caterina MJ, Leffler A, Malmberg AB, Martin WJ, Trafton J, et al. 2000. Impaired nociception and pain sensation in mice lacking the capsaicin receptor. *Science* 288:306–13

Caterina MJ, Rosen TA, Tominaga M, Brake AJ, Julius D. 1999. A capsaicin receptor homologue with a high threshold for noxious heat. *Nature* 398:436–41

Caterina MJ, Schumacher MA, Tominaga M, Rosen TA, Levine JD, Julius D. 1997. The capsaicin receptor: a heat-activated ion channel in the pain pathway. *Nature* 389:816–24

Cervero F, McRitchie HA. 1981. Neonatal capsaicin and thermal nociception: a paradox. *Brain Res.* 215:414–18

Cesare P, McNaughton P. 1996. A novel heat-activated current in nociceptive neurons and its sensitization by bradykinin. *Proc. Natl. Acad. Sci. USA* 93:15435–39

Chancellor MB, de Groat WC. 1999. Intravesical capsaicin and resiniferatoxin therapy: spicing up the ways to treat the overactive bladder. *J. Urol.* 162:3–11

Chen CC, England S, Akopian AN, Wood JN. 1998. A sensory neuron-specific, proton gated ion channel. *Proc. Natl. Acad. Sci. USA* 95:10240–45

Cholewinski A, Burgess GM, Bevan S. 1993. The role of calcium in capsaicin-induced desensitization in rat cultured dorsal root ganglion neurons. *Neuroscience* 55:1015–23

Chyb S, Raghu P, Hardie RC. 1999. Polyunsaturated fatty acids activate the *Drosophila* light-sensitive channels TRP and TRPL. *Nature* 397:255–59

Clapham DE. 1997. Some like it hot: spicing up ion channels. *Nature* 389:783–84

Colbert HA, Smith TL, Bargmann CI. 1997. Osm9, a novel protein with structural similarity to ion channels, is required for olfaction, mechanosensation and olfactory adaptation in *Caenorhabditis elegans. J. Neurosci.* 17:8259–69

Coleridge JC, Coleridge HM. 1977. Afferent

C-fibers and cardiorespiratory chemoreflexes. *Am. Rev. Respir. Dis.* 115:251–60

Coleridge JC, Coleridge HM. 1984. Afferent vagal C fibre innervation of the lungs and airways and its functional significance. *Rev. Physiol. Biochem. Pharmacol.* 99:1–10

Davis JB, Gray J, Gunthorpe MJ, Hatcher JP, Davey PT, et al. 2000. Vanilloid receptor-1 is essential for inflammatory thermal hyperalgesia. *Nature* 405:183–87

deVries DJ, Blumberg PM. 1989. Thermoregulatory effects of resiniferatoxin in the mouse: comparison with capsaicin. *Life Sci.* 44:711–15

Docherty RJ, Robertson B, Bevan S. 1991. Capsaicin causes prolonged inhibition of voltage-activated calcium currents in adult rat dorsal root ganglion neurons in culture. *Neuroscience* 40:513–21

Docherty RJ, Yeats JC, Bevan S, Boddeke HW. 1996. Inhibition of calcineurin inhibits the desensitization of capsaicin-evoked currents in cultured dorsal root ganglion neurones from adult rats. *Pflugers Arch.* 431:828–37

Doucette R, Theriault E, Diamond J. 1987. Regionally selective elimination of cutaneous thermal nociception in rats by neonatal capsaicin. *J. Comp. Neurol.* 261:583–91

Dray A, Bettaney J, Forster P. 1990a. Actions of capsaicin on peripheral nociceptors of the neonatal rat spinal cord-tail in vitro: dependence of extracellular ions and independence of second messengers. *Br. J. Pharmacol.* 101:727–33

Dray A, Forbes CA, Burgess GM. 1990b. Ruthenium red blocks the capsaicin-induced increase in intracellular calcium and activation of membrane currents in sensory neurones as well as the activation of peripheral nociceptors in vitro. *Neurosci. Lett.* 110:52–59

Dubner R, Price DD, Beitel RE, Hu JW. 1977. Peripheral neural correlates of behavior in monkey and human related to sensory-discriminative aspects of pain. In *Pain in the Trigeminal Region,* ed. DJ Anderson, B Matthews, pp. 57–66. Amsterdam: Elsevier

Farah AE, Rosenberg F. 1980. Potential therapeutic applications of aspirin and other cyclooxygenase inhibitors. *Br. J. Clin. Pharmacol.* 10(Suppl. 2):261S–78S

Fields HL. 1987. *Pain.* New York: McGraw-Hill. 354 pp.

Forbes CA, Bevan S. 1988. Single channels activated by capsaicin in patches of membrane from adult rat sensory neurones in culture. *Neurosci. Lett. Suppl.* 32:S3

Foster RW, Ramage AG. 1981. The action of some chemical irritants on somatosensory receptors of the cat. *Neuropharmacology* 20:191–98

Friel DD. 1996. TRP: its role in phototransduction and store-operated Ca^{2+} entry. *Cell* 85:617–19

Gunthorpe MJ, Harries MH, Prinjha RK, Davis JB, Randall A. 2000. Voltage- and time-dependent properties of the recombinant rat vanilloid receptor (rVR1). *J. Physiol.* 525:747–59

Guo A, Vulchanova L, Wang J, Li X, Elde R. 1999. Immunocytochemical localization of vanilloid receptor 1 (VR1): relationship to neuropeptides, the P2X3 purinoceptor and IB4 binding sites. *Eur. J. Neurosci.* 11:946–58

Handwerker HO, Reeh PW. 1991. Pain and inflammation. In *Proceedings of the 6th World Congress on Pain,* ed. MR Bond, JE Charlton, CJ Woolf, pp. 59–70. Amsterdam: Elsevier

Harper AA, Lawson SN. 1985. Conduction velocity is related to morphological cell type in rat dorsal root ganglion neurones. *J. Physiol.* 359:31–46

Harteneck C, Plant TD, Schultz G. 2000. From worm to man: three subfamilies of TRP channels. *Trends Neurosci.* 23:159–66

Hayes AG, Scadding JW, Skingle M, Tyers MB. 1981. Effects of neonatal administration of capsaicin on nociceptive thresholds in the mouse and rat. *J. Pharm. Pharmacol.* 33:183–85

He X, Schmidt RF, Schmittner H. 1988. Effects of capsaicin on articular afferents of the cat's knee joint. *Agents Actions* 25:222–24

Heyman I, Rang HP. 1985. Depolarizing responses to capsaicin in a subpopulation of rat dorsal root ganglion cells. *Neurosci. Lett.* 56:69–75

Hoenderop JG, van der Kemp AW, Hartog A, van de Graaf SF, van Os CH, et al. 1999. Molecular identification of the apical Ca^{2+} channel in 1,25-dihydroxyvitamin D_3-responsive epithelia. *J. Biol. Chem.* 274: 8375–78

Hofmann T, Obukhov AG, Schaefer M, Harteneck C, Gudermann T, Schultz G. 1999. Direct activation of human TRPC6 and TRPC3 channels by diacylglycerol. *Nature* 397:259–63

Hogyes E. 1878. Beitrage zur physiologischen Wirkung der Bestandteile des Capsicum annum. *Arch. Exp. Pathol. Pharmakol.* 9:117–30

Holzer P. 1991. Capsaicin: cellular targets, mechanisms of action, and selectivity for thin sensory neurons. *Pharmacol. Rev.* 43:143–201

Holzer P. 1993. Capsaicin-sensitive nerves in the control of vascular effector mechanisms. See Wood 1993, pp. 191–218

Holzer P, Jurna I, Gamse R, Lembeck F. 1979. Nociceptive threshold after neonatal capsaicin treatment. *Eur. J. Pharmacol.* 58:511–14

Hori T. 1984. Capsaicin and central control of thermoregulation. *Pharmacol. Ther.* 26:389–416

Hotta Y, Benzer S. 1969. Abnormal electroretinograms in visual mutants of *Drosophila*. *Nature* 222:354–56

Hwang SW, Cho H, Kwak J, Lee SY, Kang CJ, et al. 2000. Direct activation of capsaicin receptors by products of lipoxygenases: endogenous capsaicin-like substances. *Proc. Natl. Acad. Sci. USA* 97:6155–60

Jancso G. 1984. Sensory nerves as modulators of inflammatory reactions. In *Antidromic Vasodilatation and Neurogenic Inflammation*, ed. L A Chahl, J Szolcsanyi, F Lembeck, pp. 207–22. Budapest: Akademiai Kiado

Jancso G. 1992. Pathobiological reactions of C-fibre primary sensory neurones to peripheral nerve injury. *Exp. Physiol.* 77:405–31

Jancso G, Karcsu S, Kiraly E, Szebeni A, Toth L, et al. 1984. Neurotoxin induced nerve cell degeneration: possible involvement of calcium. *Brain Res.* 295:211–16

Jancso G, Kiraly E, Jancso-Gabor A. 1977. Pharmacologically induced selective degeneration of chemosensitive primary sensory neurons. *Nature* 270:741–43

Jancso G, Savay G, Kiraly E. 1978. Appearance of histochemically detectable ionic calcium in degenerating primary sensory neurons. *Acta Histochem.* 62:165–69

Jancso N, Jancso-Gabor A, Szolcsanyi J. 1967. Direct evidence for neurogenic inflammation and its prevention by denervation and by pretreatment with capsaicin. *Br. J. Pharmacol.* 31:138–51

Jancso-Gabor A, Szolcsanyi J, Jancso N. 1970. Stimulation and desensitization of the hypothalamic heat-sensitive structures by capsaicin in rats. *J. Physiol.* 208:449–59

Jerman JC, Brough SJ, Prinjha R, Harries MH, Davis JB, Smart D. 2000. Characterization using FLIPR of rat vanilloid receptor (rVR1) pharmacology. *Br. J. Pharmacol.* 130:916–22

Jones DT, Reed RR. 1989. Golf: an olfactory neuron-specific G protein involved in odorant signal transduction. *Science* 244:790–95

Jordt SE, Tominaga M, Julius D. 2000. Acid potentiation of the capsaicin receptor determined by a key extracellular site. *Proc. Natl. Acad. Sci. USA* 97:8134–35

Jung J, Hwang SW, Kwak J, Lee SY, Kang CJ, et al. 1999. Capsaicin binds to the intracellular domain of the capsaicin-activated ion channel. *J. Neurosci.* 19:529–38

Kanzaki M, Zhang YQ, Mashima H, Li L, Shibata H, Kojima I. 1999. Translocation of a calcium-permeable cation channel induced by insulin-like growth factor-I. *Nat. Cell Biol.* 1:165–70

Kaufman MP, Iwamoto GA, Longhurst JC, Mitchell JH. 1982. Effects of capsaicin and bradykinin on afferent fibers with endings in skeletal muscle. *Circ. Res.* 50:133–39

Kenins P. 1982. Responses of single nerve fibres to capsaicin applied to the skin. *Neurosci. Lett.* 29:83–88

Kirschstein T, Busselberg D, Treede RD. 1997. Coexpression of heat-evoked and capsaicin-evoked inward currents in acutely dissociated rat dorsal root ganglion neurons. *Neurosci. Lett.* 231:33–36

Kirschstein T, Greffrath W, Busselberg D, Treede RD. 1999. Inhibition of rapid heat responses in nociceptive primary sensory neurons of rats by vanilloid receptor antagonists. *J. Neurophysiol.* 82:2853–60

Kiselyov K, Xu X, Mozhayeva G, Kuo T, Pessah I, et al. 1998. Functional interaction between InsP3 receptors and store-operated Htrp3 channels. *Nature* 396:478–82

Koplas PA, Rosenberg RL, Oxford GS. 1997. The role of calcium in the desensitization of capsaicin responses in rat dorsal root ganglion neurons. *J. Neurosci.* 17:3525–37

Kress M, Fetzer S, Reeh PW, Vyklicky L. 1996. Low pH facilitates capsaicin responses in isolated sensory neurons of the rat. *Neurosci. Lett.* 211:5–8

Kumazawa T, Mizumura K. 1977. The polymodal receptors in the testis of dog. *Brain Res.* 136:553–58

LaMotte RH, Campbell JN. 1978. Comparison of response of warm and nociceptive c-fiber afferents in monkey with human judgments of thermal pain. *J. Neurophysiol.* 41:509–28

Lang E, Novak A, Reeh PW, Handwerker HO. 1990. Chemosensitivity of fine afferents from rat skin in vitro. *J. Neurophysiol.* 63:887–901

Lawson SN, Nickels SM. 1980. The use of morphometric techniques to analyse the effect of neonatal capsaicin treatment on rat dorsal root ganglia and dorsal roots. *J. Physiol.* 303:12P

Lawson SN, Waddell PJ. 1985. The anti-body RT97 distinguishes between sensory cell bodies with myelinated and unmyelinated peripheral processes in the rat. *J. Physiol.* 371:59P

Leem JW, Willis WD, Chung JM. 1993. Cutaneous sensory receptors in the rat foot. *J. Neurophysiol.* 69:1684–99

Levine J, Taiwo Y. 1994. Inflammatory pain. In *Textbook of Pain*, ed. PD Wall, R Melzack, pp. 45–56. Edinburgh: Churchill Livingstone

Li C, Peoples RW, Weight FF. 1997. Enhancement of ATP-activated current by protons in dorsal root ganglion neurons. *Eur. J. Physiol.* 433:446–54

Li HS, Xu XZ, Montell C. 1999. Activation of a TRPC3-dependent cation current through the neurotrophin BDNF. *Neuron* 24:261–73

Linguegllia E, de Weille JR, Bassilana F, Heurteaux C, Sakai H, et al. 1997. A modulatory subunit of acid sensing ion channels in brain and dorsal root ganglion cells. *J. Biol. Chem.* 272:29778–83

Liu L, Simon SA. 1994. A rapid capsaicin-activated current in rat trigeminal ganglion neurons. *Proc. Natl. Acad. Sci. USA* 91:738–41

Lundberg JM. 1993. Capsaicin-sensitive sensory nerves in the airways: implications for protective reflexes and disease. See Wood 1993, pp. 220–37

Lynn B, Carpenter SE. 1982. Primary afferent units from the hairy skin of the rat hind limb. *Brain Res.* 238:29–43

Maggi CA. 1993. The pharmacological modulation of neurotransmitter release. See Wood 1993, pp. 161–89

Maggi CA, Meli A. 1988. The sensory-efferent function of capsaicin-sensitive sensory neurons. *Gen. Pharmacol.* 19:1–43

Maggi CA, Santicioli P, Giuliani S, Furio M, Meli A. 1986. The capsaicin-sensitive innervation of the rat urinary bladder: further studies on mechanisms regulating micturition threshold. *J. Urol.* 136:696–700

Marsh SJ, Stansfield CE, Brown DA, Davey R, McCarthy D. 1987. The mechanism of action of capsaicin on sensory C-type neurons and their axons in vitro. *Neuroscience* 23:275–89

Martenson ME, Ingram SL, Baumann TK. 1994. Potentiation of rabbit trigeminal responses to capsaicin in a low pH environment. *Brain Res.* 651:143–47

Martin HA, Basbaum AI, Kwiat GC Goetzl EJ, Levine JD. 1987. Leukotriene and prostaglandin sensitization of cutaneous high-threshold C- and A-delta mechanonociceptors in the hairy skin of rat hindlimbs. *Neuroscience* 22:651–59

McLaughlin SK, McKinnon PJ, Margolskee RF. 1992. Gustducin is a taste cell-specific G-protein closely related to the transducins. *Nature* 357:563–69

Melck D, Bisogno T, De Petrocellis L, Chuang H, Julius D, et al. 1999. Unsaturated long-chain N-acyl-vanillyl-amides (N-AVAMs): vanilloid receptor ligands that inhibit anandamide-facilitated transport and bind to CB1 cannabinoid receptors. *Biochem. Biophys. Res. Commun.* 262:275–84

Meyer RD, Campbell JN. 1981. Myelinated nociceptor afferents account for the hyperalgesia that follows a burn to the hand. *Science* 213:1527–29

Mezey E, Toth ZE, Cortright DN, Arzubi MK, Krause JE, et al. 2000. Distribution of mRNA for vanilloid receptor subtype 1 (VR1), and VR1-like immunoreactivity, in the central nervous system of the rat and human. *Proc. Natl. Acad. Sci. USA* 97:3655–60

Michael GJ, Priestly JV. 1999. Differential expression of the mRNA for the vanilloid receptor subtype 1 in cells of the adult rat dorsal root and nodose ganglia and its downregulation by axotomy. *J. Neurosci.* 19:1844–54

Millan MJ. 1999. The induction of pain: an integrative review. *Prog. Neurobiol.* 57:1–164

Montell C, Rubin GM. 1989. Molecular characterization of the *Drosophila trp* locus: a putative integral membrane protein required for phototransduction. *Neuron* 2:1313–23

Nagy I, Rang H. 1999a. Noxious heat activates all capsaicin-sensitive and also a subpopulation of capsaicin-insensitive dorsal

root ganglion neurons. *Neuroscience* 88: 995–97

Nagy I, Rang HP. 1999b. Similarities and differences between the responses of rat sensory neurons to noxious heat and capsaicin. *J. Neurosci.* 19:10647–55

Nagy JI, Iversen LL, Goedert M, Chapman D, Hunt SP. 1983. Dose-dependent effects of capsaicin on primary sensory neurons in the neonatal rat. *J. Neurosci.* 3:399–406

Nelson EK. 1919. The constitution of capsaicin, the pungent principal of capsicum. *J. Am. Chem. Soc.* 41:1115–21

Ness TJ, Gebhart GF. 1990. Visceral pain: a review of experimental studies. *Pain* 41:167–234

Oh U, Hwang SW, Kim D. 1996. Capsaicin activates a nonselective cation channel in cultured neonatal rat dorsal root ganglion neurons. *J. Neurosci.* 16:1659–67

Petersen M, LaMotte RH. 1993. Effect of protons on the inward current evoked by capsaicin in isolated dorsal root ganglion cells. *Pain* 54:37–42

Petersen M, Pierau FK, Weyrich M. 1987. The influence of capsaicin on membrane currents in dorsal root ganglion neurones of guinea-pig and chicken. *Pflugers Arch.* 409:403–10

Raja SN, Meyer RA, Ringkamp M, Campbell JN. 1999. Peripheral neural mechanisms of nociception. In *Textbook of Pain*, ed. PD Wall, R Melzack, pp. 11–57. Edinburgh: Churchill Livingstone

Ranganathan R, Malicki DM, Zuker CS. 1995. Signal transduction in *Drosophila* photoreceptors. *Annu. Rev. Neurosci.* 18:283–317

Reichling DB, Levine JD. 1997. Heat transduction in rat sensory neurons by calcium-dependent activation of a cation channel. *Proc. Natl. Acad. Sci. USA* 94:7006–11

Robbins W. 2000. Clinical applications of capsaicinoids. *Clin. J. Pain* 16:S86–89

Sasamura T, Sasaki M, Tohda C, Kuraishi Y. 1998. Existence of capsaicin-sensitive glutamatergic terminals in rat hypothalamus. *Neuroreport* 9:2045–48

Scadding JW. 1980. The permanent anatomical effects of neonatal capsaicin on somatosensory nerves. *J. Anat.* 131:471–82

Schumacher MA, Moff I, Sudanagunta SP, Levine JD. 2000. Molecular cloning of an N-terminal splice variant of the capsaicin receptor: Loss of N-terminal domain suggests functional divergence among capsaicin receptor subtypes. *J. Biol. Chem.* 275:2756–62

Scott K, Zucker C. 1998. TRP, TRPL and trouble in photoreceptor cells. *Curr. Opin. Neurobiol.* 8:383–88

Simone DA, Ngeow JY, Putterman GJ, LaMotte RH. 1987. Hyperalgesia to heat after intradermal injection of capsaicin. *Brain Res.* 418:201–3

Simone DA, Nolano M, Johnson T, Wendelschafer-Crabb G, Kennedy WR. 1998. Intradermal injection of capsaicin in humans produces degeneration and subsequent reinnervation of epidermal nerve fibers: correlation with sensory function. *J. Neurosci.* 18:8947–59

Simone DA, Ochoa J. 1991. Early and late effects of prolonged topical capsaicin on cutaneous sensibility and neurogenic vasodilatation in humans. *Pain* 47:285–94

Smart D, Gunthorpe MJ, Jerman JC, Nasir S, Gray J, et al. 2000. The endogenous lipid anandamide is a full agonist at the human vanilloid receptor (hVR1). *Br. J. Pharmacol.* 129:227–30

Smart D, Jerman JC. 2000. Anandamide: an endogenous activator of the vanilloid receptor. *Trends Pharmacol. Sci.* 21:134

Snider WD, MacMahon SB. 1998. Tackling pain at the source: new ideas about nociceptors. *Neuron* 20:629–32

Snyder SH. 1977. Opiate receptors and internal opiates. *Sci. Am.* 236:44–56

Steen KH, Reeh PW, Anton F, Handwerker HO. 1992. Protons selectively induce lasting excitation and sensitization to mechanical stimulation of nociceptors in rat skin, in vitro. *J. Neurosci.* 12:86–95

Stoop R, Surprenant A, North A. 1997. Different sensitivity to pH of ATP-induced currents at four cloned P2X receptors. *J. Neurophysiol.* 78:1837–40

Such G, Jancso G. 1986. Axonal effects of capsaicin: an electrophysiological study. *Acta Physiol. Hung.* 67:53–63

Suzuki M, Sato J, Kutsuwada K, Ooki G, Imai M. 1999. Cloning of a stretch-inhibitable nonselective cation channel. *J. Biol. Chem.* 274:6330–35

Szallasi A, Blumberg PM. 1989. Resiniferatoxin, a phorbol-related diterpene, acts as an ultrapotent analog of capsaicin, the irritant constituent in red pepper. *Neuroscience* 30:515–20

Szallasi A, Blumberg PM. 1990. Specific binding of resiniferatoxin, an ultrapotent capsaicin analog, by dorsal root ganglion membranes. *Brain Res.* 524:106–11

Szallasi A, Blumberg PM. 1999. Vanilloid (capsaicin) receptors and mechanisms. *Pharmacol. Rev.* 51:159–212

Szallasi A, Blumberg PM, Annicelli LL, Krause JE, Cortright DN. 1999. The cloned rat vanilloid receptor VR1 mediates both R-type binding and C-type calcium response in dorsal root ganglion neurons. *Mol. Pharmacol.* 56:581–87

Szallasi A, Nilsson S, Farkas-Szallasi T, Blumberg PM, Hokfelt T, Lundberg JM. 1995. Vanilloid (capsaicin) receptors in the rat: distribution in the brain, regional differences in the spinal cord, axonal transport to the periphery and depletion following systemic vanilloid treatment. *Brain Res.* 703:175–83

Szolcsanyi J. 1977. A pharmacological approach to elucidation of the role of different nerve fibres and receptor endings in mediation of pain. *J. Physiol.* 73:251–59

Szolcsanyi J. 1993. Actions of capsaicin on sensory neurons. See Wood 1993, pp. 1–26

Szolcsanyi J. 2000. Anandamide and the question of its functional role for activation of capsaicin receptors. *Trends Pharmacol. Sci.* 21:203–4

Szolcsanyi J, Jancso-Gabor A. 1975. Sensory effects of capsaicin congeners. I. Relationship between chemical structure and pain-producing potency of pungent agents. *Drug Res.* 25:1877–81

Szolcsanyi J, Jancso-Gabor A. 1976. Sensory effects of capsaicin congeners. II. Importance of chemical structure and pungency in desensitizing activity of capsaicin-like compounds. *Drug Res.* 26:33–37

Szolcsanyi J, Porszasz R, Petho G. 1994. Capsaicin and pharmacology of nociceptors. In *Peripheral Neurons in Nociception: Physiopharmacological Aspects*, ed. JM Besson, pp. 109–24. Paris: John Libbey

Taylor DCM, Pierau F-K, Szolcsanyi J, Krishtal O, Petersen M. 1984. Effect of capsaicin on rat sensory neurones. In *Thermal Physiology*, ed. JRS Hales, pp. 23–27. New York: Raven

Thresh LT. 1846. Isolation of capsaicin. *Pharm. J.* 6:941

Tominaga M, Caterina MJ, Malmberg AB, Rosen TA, Gilbert H, et al. 1998. The cloned capsaicin receptor integrates multiple pain-producing stimuli. *Neuron* 21:1–20

Treede R, Meyer RA, Srinivasa RN, Campbell JN. 1995. Evidence for two different heat transduction mechanisms in nociceptive primary afferents innervating monkey skin. *J. Physiol.* 483:747–58

Waldmann R, Bassilana F, Weille J, Champigny G, Heurteaux C, Lazdunski M. 1997a. Molecular cloning of a non-inactivating proton-gated Na^+ channel specific for sensory neurons. *J. Biol. Chem.* 272:20975–78

Waldmann R, Champigny G, Bassilana F, Heurteaux C, Lazdunski M. 1997b. A proton-gated channel involved in acid sensing. *Nature* 386:173–77

Walpole CSJ, Wrigglesworth R. 1993. Structural requirements for capsaicin agonists and antagonists. See Wood 1993, pp. 63–81

Williams JT, Zieglgansberger W. 1982. The acute effects of capsaicin on rat primary afferents and spinal neurons. *Brain Res.* 253:125–31

Winter J. 1987. Characterization of capsaicin-sensitive neurones in adult rat dorsal root ganglion cultures. *Neurosci. Lett.* 80:134–40

Winter J, Dray A, Wood JN, Yeats JC, Bevan S. 1990. Cellular mechanism of action of resiniferatoxin: a potent sensory neuron excitotoxin. *Brain Res.* 520:131–40

Winter J, Walpole CS, Bevan S, James IF. 1993. Characterization of resiniferatoxin binding and capsaicin sensitivity in adult rat dorsal root ganglia. *Neuroscience* 57:747–57

Wood JN, ed. 1993. *Capsaicin in the Study of Pain.* San Diego: Academic. 286 pp.

Wood JN, Winter J, James IF, Rang HP, Yeats J, Bevan S. 1988. Capsaicin-induced ion fluxes in dorsal root ganglion cells in culture. *J. Neurosci.* 8:3208–20

Yaksh TL, Farb DH, Leeman SE, Jessell TM. 1979. Intrathecal capsaicin depletes substance P in the rat spinal cord and produces prolonged thermal analgesia. *Science* 206:481–83

Yeats JC, Docherty RJ, Bevan S. 1992. Calcium-dependent and -independent desensitization of capsaicin-evoked responses in voltage-clamped adult rat dorsal root ganglion (DRG) neurones in culture. *J. Physiol.* 446:390P

Zhu X, Jiang M, Peyton M, Boulay G, Hurst R, et al. 1996. *trp*, a novel mammalian gene family essential for agonist-activated capacitative Ca^{2+} entry. *Cell* 85:661–71

Zygmunt PM, Chuang H, Movahed P, Julius D, Hogestatt ED. 2000a. The anandamide transport inhibitor AM404 activates vanilloid receptors. *Eur. J. Pharmacol.* 396:39–42

Zygmunt PM, Julius I, Di Marzo I, Hogestatt ED. 2000b. Anandamide: the other side of the coin. *Trends Pharmacol. Sci.* 21:43–44

Zygmunt PM, Petersson J, Andersson DA, Chuang H, Sorgard M, et al. 1999. Vanilloid receptors on sensory nerves mediate the vasodilator action of anandamide. *Nature* 400:452–57

Annu. Rev. Neurosci. 2001. 24:519–50

PRION DISEASES OF HUMANS AND ANIMALS:
Their Causes and Molecular Basis

John Collinge

MRC Prion Unit and Department of Neurogenetics, Imperial College School of Medicine at St. Mary's, London, United Kingdom; e-mail: J.Collinge@ic.ac.uk

Key Words prion protein, Creutzfeldt-Jakob disease, spongiform encephalopathy, scrapie

■ **Abstract** Prion diseases are transmissible neurodegenerative conditions that include Creutzfeldt-Jakob disease (CJD) in humans and bovine spongiform encephalopathy (BSE) and scrapie in animals. Prions appear to be composed principally or entirely of abnormal isoforms of a host-encoded glycoprotein, prion protein. Prion propagation involves recruitment of host cellular prion protein, composed primarily of α-helical structure, into a disease specific isoform rich in β-sheet structure. The existence of multiple prion strains has been difficult to explain in terms of a protein-only infectious agent, but recent studies suggest that strain specific phenotypes can be encoded by different prion protein conformations and glycosylation patterns. The ability of a protein to encode phenotypic information has important biological implications. The appearance of a novel human prion disease, variant CJD, and the clear experimental evidence that it is caused by exposure to BSE has highlighted the need to understand the molecular basis of prion propagation, pathogenesis, and the barriers limiting intermammalian transmission. It is unclear if a large epidemic of variant CJD will occur in the years ahead.

INTRODUCTION

Historical Background

The prion diseases are a closely related group of neurodegenerative conditions that affect both humans and animals. They have previously been described as subacute spongiform encephalopathies, slow virus diseases, and transmissible dementias. The prototypic disease is scrapie, a naturally occurring disease affecting sheep and goats. Scrapie has been recognized in Europe for over 200 years (McGowan 1922) and is present in many countries worldwide. More recently recognized animal diseases include transmissible mink encephalopathy (Marsh 1992), chronic wasting disease of mule deer and elk (Williams & Young 1980), and bovine spongiform encephalopathy (BSE) (Wells et al 1987). The more recently described feline

0147-006X/01/0621-0519$14.00

spongiform encephalopathy of domestic cats (Wyatt et al 1991) and spongiform encephalopathies of a number of zoo animals (Jeffrey & Wells 1988, Kirkwood et al 1990) are also recognized as animal prion diseases.

Traditionally, human prion diseases have been classified into Creutzfeldt-Jakob disease (CJD), Gerstmann-Sträussler syndrome (GSS) (also known as Gerstmann-Sträussler-Scheinker disease), and kuru. Although rare neurodegenerative disorders, affecting per annum about one person per million worldwide, these diseases have had remarkable attention focused on them recently. This is because of the unique biology of the transmissible agent or prion, and also because of fears that through dietary exposure to infected tissues, an epidemic of a newly recognized bovine prion disease, (BSE,) could pose a threat to public health.

In 1936, scrapie was demonstrated to be transmissible by inoculation between sheep (and goats) following prolonged incubation periods (Cuillé & Chelle 1936). It was assumed that some type of virus must be the causative agent, and in 1954 Sigurdsson coined the term slow virus infection. There was considerable interest in the 1950s in an epidemic, among the Fore linguistic group of the Eastern Highlands of Papua New Guinea, of a neurodegenerative disease, kuru, characterized principally by a progressive ataxia. Subsequent field work by a number of investigators suggested that kuru was transmitted during cannibalistic feasts. In 1959, Hadlow drew attention to the similarities between kuru and scrapie at the neuropathological, clinical, and epidemiological levels, leading to the suggestion that these diseases may also be transmissible (Klatzo et al 1959, Hadlow 1959). A landmark in the field was the transmission by intracerebral inoculation with brain homogenates into chimpanzees of first kuru (Gajdusek et al 1966) and then CJD (Gibbs et al 1968). Transmission of GSS followed in 1981 (Masters et al 1981). This work led to the concept of "transmissible dementias." The term Creutzfeldt-Jakob disease (CJD) was introduced in 1922 by Spielmeyer, who drew from earlier case reports of Creutzfeldt and Jakob. In subsequent years, the term was used to describe a range of neurodegenerative conditions, many of which would not meet modern diagnostic criteria for CJD. The new criterion of transmissibility allowed the diagnostic criteria for CJD to be assessed and refined. Atypical cases could be classified as CJD on the basis of their transmissibility. Both animal and human conditions share common histopathological features. The classical triad of spongiform vacuolation (affecting any part of the cerebral grey matter), neuronal loss, and astrocytic proliferation may be accompanied by amyloid plaques (Beck & Daniel 1987).

The nature of the transmissible agent in these diseases has been a subject of heated debate for many years. The understandable initial assumption that the agent must be some form of virus was challenged, however, both by the failure to directly demonstrate such a virus (or indeed any immunological response to it) and by evidence indicating that the transmissible agent showed remarkable resistance to treatment expected to inactivate nucleic acids (such as ultraviolet radiation or treatment with nucleases). As early as 1966, such findings had led to suggestions that the transmissible agent may be devoid of nucleic acid (Alper et al 1966, 1967).

They also led Griffith (1967) to suggest that the transmissible agent might be a protein. Progressive enrichment of brain homogenates for infectivity resulted in the isolation by Bolton et al (1982) of a protease-resistant sialoglycoprotein, designated the prion protein (PrP). This protein was the major constituent of infective fractions and was found to accumulate in affected brains and sometimes to form amyloid deposits. The term prion (from the first letters of proteinaceous infectious particle) was proposed (Prusiner 1982) to distinguish the infectious pathogen from viruses or viroids. Prions were defined as "small proteinaceous infectious particles that resist inactivation by procedures which modify nucleic acids" (Prusiner 1982).

The protease-resistant PrP extracted from affected brains was of 27–30 kDa and became known as PrP^{27-30}. At the time, PrP was assumed to be encoded by a gene within the putative slow virus thought to be responsible for these diseases. However, amino acid sequencing of part of PrP^{27-30} led to the recovery of cognate cDNA clones using an isocoding mixture of oligonucleotides. PrP^{27-30} was demonstrated in 1985 to be encoded by a single-copy chromosomal gene rather than by a putative nucleic acid in fractions enriched for scrapie infectivity. PrP^{27-30} is derived from a larger molecule of 33–35 kDa, designated PrP^{Sc} (denoting the scrapie isoform of the protein) (Oesch et al 1985). The normal product of the PrP gene, however, is protease sensitive and designated PrP^{C} (denoting the cellular isoform of the protein). No differences in amino acid sequence between PrP^{Sc} and PrP^{C} have been identified. PrP^{Sc} is known to be derived from PrP^{C} by a posttranslational process (Borchelt et al 1990, Caughey & Raymond 1991).

Animal Prion Diseases

An increasing number of animal prion diseases are being recognized. Scrapie, a naturally occurring disease of sheep and goats, has been recognized in Europe for over 200 years and is present endemically in many countries. Accurate epidemiology is lacking, although scrapie may be common in some countries. Remarkably little is known about its natural routes of transmission. Transmissible mink encephalopathy (Marsh 1992) and chronic wasting disease of mule deer and elk (Williams & Young 1980) were described in captive animals from the 1940s onward, principally in the United States. It has more recently become apparent that chronic wasting disease may be a common condition in wild deer and elk in certain areas of Colorado (Spraker et al 1997). Again the routes of transmission are unclear (Miller et al 1998). Transmissible mink encephalopathy has occurred as infrequent epidemics among ranched mink and may result from foodborne prion exposure (Marsh et al 1991).

The appearance in UK cattle in 1986 of BSE, which rapidly evolved into a major epidemic (Wilesmith et al 1988, Anderson et al 1996), was widely attributed to transmission of sheep scrapie, endemic in the United Kingdom and many other countries, to cattle via contaminated feed prepared from rendered carcasses (Wilesmith et al 1988). However, an alternative hypothesis is that epidemic BSE resulted from recycling of rare sporadic BSE cases, as cattle were also rendered to produce cattle feed. Whether or not BSE originated from sheep scrapie, it became clear in 1990, with the occurrence of novel spongiform encephalopathies among domestic and captive wild cats, that its host range was different from scrapie. Many new species—including greater kudu, nyala, Arabian oryx, Scimitar horned oryx, eland, gemsbok, bison, ankole, tiger, cheetah, ocelot, puma, and domestic cats—have developed spongiform encephalopathies coincident with or following the arrival of BSE. Several of these have been confirmed to be caused by a BSE-like prion strain (Bruce et al 1994, Collinge et al 1996), and it is likely that most or all of these are BSE related. More than 180,000 BSE cases have been confirmed in cattle in the United Kingdom, although the total number of infected animals has been estimated at around one million (Anderson et al 1996). BSE has since been reported in a number of other (mainly European) countries, with significant epidemics reported in Switzerland (Doherr et al 1999), Ireland, and Portugal.

Human Prion Diseases

Human prion diseases have been traditionally classified into Creutzfeldt-Jakob disease (CJD), Gerstmann-Sträussler-Scheinker disease (GSS), and kuru, and they can be further divided into three etiological categories: sporadic, acquired, and inherited. Acquired prion diseases include iatrogenic CJD and kuru and arise from accidental exposure to human prions through medical or surgical procedures or participation in cannibalistic feasts. Epidemiological studies do not provide any evidence for an association between sheep scrapie and the occurrence of CJD in humans (Brown et al 1987). Sporadic CJD occurs in all countries with a random case distribution and an annual incidence of one per million. Around 15% of human prion disease is inherited, and all cases to date have been associated with coding mutations in the prion protein gene (*PRNP*), of which over 20 distinct types are recognized (Collinge 1997). The inherited prion diseases can be diagnosed by *PRNP* analysis, and the use of these definitive genetic diagnostic markers has allowed the recognition of a wider phenotypic spectrum of human prion disease to include a range of atypical dementias and fatal familial insomnia (Collinge et al 1990, 1992; Medori et al 1992a,b). No such pathogenic *PRNP* mutations are present in sporadic and acquired prion disease. However, a common PrP polymorphism at residue 129, where either methionine or valine can be encoded, is a key determinant of genetic susceptibility to acquired and sporadic prion diseases, the large majority of which occur in homozygous individuals (Collinge et al 1991, Palmer et al 1991, Windl et al 1996). This protective effect of *PRNP* codon 129 heterozygosity is also seen in some of the inherited prion diseases (Baker et al 1991, Hsiao et al 1992).

The appearance in the United Kingdom in 1995 of a novel human prion disease, variant CJD (vCJD), and the experimental evidence that it is caused by the same prion strain that causes BSE in cattle (see below), has raised the possibility that a major epidemic of vCJD will occur in the United Kingdom and other countries as a result of dietary or other exposure to BSE prions (Cousens et al 1997, Ghani et al 1999, Collinge 1999). These concerns, together with those of potential iatrogenic transmission of preclinical vCJD via medical and surgical procedures, have led to intensification of efforts to understand the molecular basis of prion propagation and to develop rational therapeutics.

Many of the key advances in understanding the pathogenesis of the prion diseases have come from study of the various forms of human prion disease. In particular, the recognition that the familial forms of the human diseases are autosomal dominant inherited conditions, associated with *PRNP* coding mutations (Owen et al 1989, Hsiao et al 1989), as well as being transmissible to laboratory animals by inoculation, strongly supported the contention that the transmissible agent, or prion, was composed principally of an abnormal isoform of prion protein.

Clinical Features of Human Prion Disease

With advances in our understanding of their etiology, it now seems more appropriate to divide the human prion diseases into inherited, sporadic, and acquired forms, with CJD, GSS, and kuru clinicopathological syndromes placed within a wider spectrum of disease. Classical (sporadic) CJD is a rapidly progressive, multifocal dementia, usually with myoclonus. Onset usually occurs in the 45- to 75-year age group, with peak onset between 60 and 65 years. The clinical progression is typically weeks long, progressing to akinetic mutism and death often in 2–3 months. Around 70% of those afflicted die in under 6 months. Prodromal features, present in approximately one third of the cases, include fatigue, insomnia, depression, weight loss, headaches, general malaise, and ill-defined pain sensations. In addition to mental deterioration and myoclonus, frequent additional neurological features include extrapyramidal signs, cerebellar ataxia, pyramidal signs, and cortical blindness.

Kuru reached epidemic proportions among a defined population living in the Eastern Highlands of Papua New Guinea and provides by far our largest experience of acquired human prion disease (Alpers 1987). The earliest cases are thought to date back to the early part of the century. Kuru affected the people of the Fore linguistic group and their neighbors, with whom they intermarried. Kuru predominantly affected women and children (of both sexes) (Alpers 1987) and among women in affected villages was the most common cause of death. Kuru was transmitted during cannibalistic feasts when deceased relatives were consumed by their close relatives and others in the immediate community. Women and children predominantly participated in the feasts and ate the brain and internal organs, which is thought to explain the differential age and sex incidence. The epidemic is thought to have originated when a case of sporadic CJD, known to occur at random in all populations, occurred in a member of this population and was, as were most deceased individuals, eaten. The recycling of prions within this relatively isolated population led to a substantial epidemic that became the major cause of death among children and adult women. Prior to the cessation of cannibalism in the late 1950s, feasts were a common occurrence, and the multiple exposures that individual kuru patients may have had complicated precise estimates of their incubation periods.

However, studies of later cases with well-defined exposures provided more precise estimates (Klitzman et al 1984). What is the range of incubation periods seen? Very infrequent cases of kuru were recorded in children as young as 4.5 years, indicating incubation periods of this order or less. However, although it is assumed that dietary exposure to kuru was the principal route of transmission, inoculation with brain or other tissue, either via cuts or sores, or into the conjunctiva (following eye rubbing), was also likely (Alpers 1987). Because such routes of transmission in experimental animals are known to result in shorter mean incubation periods than does oral exposure, these cases of very short incubation periods for kuru may not represent oral transmission. At the other extreme, occasional cases of kuru still occur in the Fore region in patients exposed during some of the last feasts held in their villages and are consistent with incubation periods exceeding 40 years (J Whitfield, MP Alpers, & J Collinge, et al, manuscript in preparation). Mean incubation periods have been estimated to be approximately 12 years (MP Alpers, personal communication).

Kuru affects both sexes, and onset of disease has ranged from age 5 to over 60. The mean clinical duration of illness is 12 months, with a range of 3 months to 3 years; the course tends to be shorter in children. The central clinical feature is progressive cerebellar ataxia. In sharp contrast to CJD, dementia is often absent, although in the terminal stages, the faculties of many patients are obtunded (Alpers 1987).

Although prion diseases can be transmitted to experimental animals by inoculation, they are not contagious in humans. Documented case-to-case spread has occurred only by cannibalism (kuru) or following accidental inoculation with prions. Such iatrogenic routes include the use of inadequately sterilized intracerebral electrodes, dura mater, and corneal grafting, and from the use of human cadaveric pituitary–derived growth hormone or gonadotrophin. It is interesting to note that cases arising from intracerebral or optic inoculation manifest clinically as classical CJD, with a rapidly progressive dementia, whereas those resulting from peripheral inoculation frequently initially present with a progressive cerebellar ataxia, reminiscent of kuru. It is not surprising that the incubation period in intracerebral cases is short (19–46 months for dura mater grafts) compared with peripheral cases (mean estimated at around 15 years).

vCJD has a clinical presentation in which behavioral and psychiatric disturbances predominate, and in some cases there are marked sensory phenomena (notably dysaesthesiae or pain in the limbs or face) (Zeidler et al 1997, Hill et al 1999b). Initial referral is often to a psychiatrist, and the most prominent feature is depression, but anxiety, withdrawal, and behavioral changes are also frequent. Other features include delusions, emotional lability, aggression, insomnia, and auditory and visual hallucinations. In most patients, a progressive cerebellar syndrome develops, with gait and limb ataxia. Dementia usually develops later in the clinical course. Myoclonus is seen in most patients, in some cases preceded by chorea. The age at onset ranges from 16 to 51 years (mean 29 years), and the clinical course is unusually prolonged (9–35 months, median 14 months). All cases to date are homozygous for methionine at *PRNP* codon 129 (Collinge et al 1996, Hill et al 1999b). vCJD can be diagnosed by detection of characteristic PrP immuno-staining and PrP^{Sc} on tonsil biopsy (Collinge et al 1997, Hill et al 1999b). It is important that PrP^{Sc} is only detectable in tonsil and other lymphoreticular tissues in vCJD and not in other forms of human prion disease, indicating that it has a distinctive pathogenesis. The PrP^{Sc} type (see below) detected on Western blot in vCJD tonsil has a characteristic pattern designated type 4t. The neuropathological appearances of vCJD are striking and consistent (Will et al 1996). Although there is widespread spongiform change, gliosis and neuronal loss, most severe in the basal ganglia and thalamus, the most remarkable feature was the abundant PrP amy-loid plaques in cerebral and cerebellar cortex. These consist of kuru-like, "florid" (surrounded by spongiform vacuoles), and multicentric plaque types. The florid plaques, seen previously only in scrapie, are a particularly unusual but highly con-sistent feature. There is also abundant pericellular PrP deposition in the cerebral and cerebellar cortex and PrP deposition in the molecular layer of the cerebellum. Some of the features of vCJD are reminiscent of kuru (Alpers 1987), in which behavioral changes and progressive ataxia predominate. In addition, peripheral sensory disturbances are well recognized in the kuru prodrome. Kuru plaques are seen in approximately 70% of cases and are especially abundant in younger kuru patients. The observation that iatrogenic prion disease related to peripheral expo-sure to human prions has a more kuru-like than CJD-like clinical picture may well be relevant and would be consistent with a peripheral prion exposure in vCJD also. The relatively stereotyped clinical presentation and neuropathology of vCJD con-trasts sharply with sporadic CJD. This may be because vCJD is caused by a single prion strain, and it may also suggest that a relatively homogeneous, genetically susceptible subgroup of the population with short incubation periods to BSE has been selected to date.

GSS is an autosomal dominant disorder that presents classically as a chronic cerebellar ataxia with pyramidal features, with dementia occurring later in a much more prolonged clinical course than in CJD. The mean duration is approximately 5 years, with onset usually in either the third or fourth decade. Histologically, the hallmark is the presence of multicentric PrP-amyloid plaques. Although first asso-ciated with the P102L PRNP mutation (Hsiao et al 1989), GSS is now recognized as a pathological syndrome associated with several different PRNP mutations and forms a part of the phenotypic spectrum of inherited prion disease.

The identification of one of the pathogenic *PRNP* mutations in a case with neurodegenerative disease allows diagnosis of an inherited prion disease and sub-classification according to mutation (Collinge et al 1989). Over 20 pathogenic mutations have been described in two groups: (*a*) point mutations resulting in amino acid substitutions in PrP, or in one case production of a stop codon re-sulting in expression of a truncated PrP; and (*b*) insertions encoding additional integral copies of an octapeptide repeat present in a tandem array of five copies in the normal protein (see Figure 1; see color insert). All are autosomal dominantly inherited conditions. Kindreds with inherited prion disease have been described with phenotypes of classical CJD and GSS, and also with a range of other neurodegenerative syndromes. Some families show remarkable phenotypic variability that can encompass both CJD- and GSS-like cases as well as other cases that do not conform to either CJD or GSS phenotypes (Collinge et al 1992). Such atypical prion diseases may lack the classical histological features of a spongiform encephalopathy entirely, although PrP immunohistochemistry is usually positive (Collinge et al 1990). Progressive dementia, cerebellar ataxia, pyramidal signs, chorea, myoclonus, extrapyramidal features, pseudobulbar signs, seizures, and amyotrophic features are seen in variable combinations.

PRION PROTEINS AND PRION PROPAGATION

The Structural Properties of Prion Proteins

A wide body of data now supports the idea that prions consist principally or entirely of an abnormal isoform of a host-encoded protein, the prion protein (PrP), designated PrP^{Sc} ([for a review, see Prusiner (1991)].) PrP^{Sc} is derived from PrP^C by a posttranslational mechanism (Borchelt et al 1990, Caughey & Raymond 1991). Neither amino acid sequencing nor systematic study of known covalent posttranslational modifications have shown any consistent differences between PrP^C and PrP^{Sc} (Stahl et al 1993). It is proposed that PrP^{Sc} acts as a template that promotes the conversion of PrP^C to PrP^{Sc} and that this conversion involves only conformational change. It is clear that a full understanding of prion propagation will require knowledge both of the structure of PrP^C and PrP^{Sc} and of the mechanism of conversion between them.

The Conformation and Stability of PrP^C

The conformation of the cellular isoform was first established by nuclear magnetic resonance (NMR) measurements made on recombinant mouse protein (Riek et al 1996). Since then, NMR measurements on recombinant hamster (James et al 1997) and human PrP (Hosszu et al 1999) show that they have essentially the same conformation; however, despite strenuous efforts, no group has yet determined the three-dimensional structure of PrP^C by crystallographic methods.

Following cleavage of an N-terminal signal peptide and removal of a C-terminal peptide on addition of a glycosylphosphatidylinositol (GPI) anchor, the mature PrP^C species consists of an N-terminal region of about 100 amino acids, which is unstructured in the isolated molecule in solution, and a C-terminal segment, also approximately 100 amino acids in length. The C-terminal domain is folded into a largely α-helical conformation (three α-helices and a short antiparallel β-sheet) and stabilized by a single disulphide bond linking helices 2 and 3 (Riek et al 1996). There are two asparagine-linked glycosylation sites (see Figure 2; see color insert).

The N-terminal region contains a segment of five repeats of an eight–amino acid sequence (the octapeptide-repeat region), expansion of which by insertional

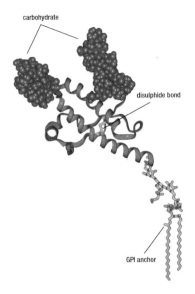

Figure 2 Model of glycosylated human prion protein indicating positions of N-linked glycans (in blue), the single disulphide bond joining helixes 2 and 3, and the glycosylphosphatidylinositol anchor to the outer surface of cell membrane.

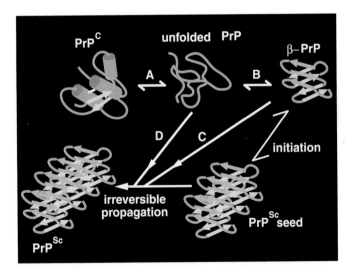

Figure 3 Possible mechanism for prion propagation. Largely α-helical cellular isoforms of the prion protein (PrP^C) proceed via an unfolded state (*A*) to refold into a largely β-sheet form, β-PrP (*B*). β-PrP is prone to aggregation in physiological salt concentrations. Prion replication may require a critical "seed" size. Further recruitment of β-PrP monomers (*C*) or unfolded PrP (*D*) then occurs as an essentially irreversible process.

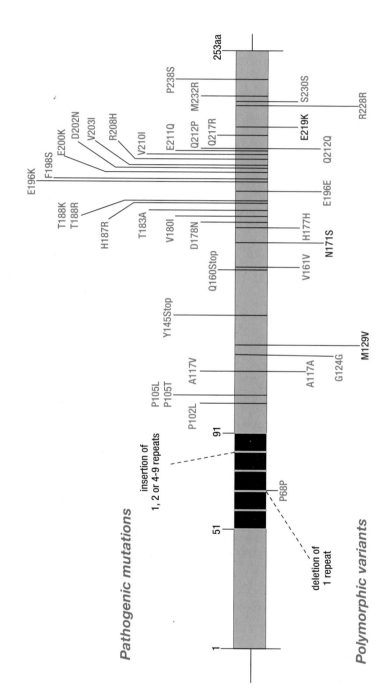

Figure 1 Pathogenic mutations and polymorphic variants of the human prion protein.

mutation leads to inherited prion disease. Although unstructured in the isolated molecule, this region contains a tight binding site for a single Cu^{2+} ion with a dissociation constant (K_d) of 10^{-14} M. A second tight copper site $(K_d = 10^{-13}$ M) is present upstream of the octa-repeat region but before the structured C-domain (GS Jackson, IA Murray, LLP Hosszu, N Gibbs, JP Waltho, AR Clarke, & J Collinge, submitted for publication). These values for copper-binding affinity are some eight or nine orders of magnitude tighter than previously reported (Brown et al 1997a; Hornshaw et al 1995a,b; Stockel et al 1998). Clearly, it is possible that the unstructured N-terminal region may acquire structure following copper binding. A role for PrP in copper metabolism or transport seems likely, and disturbance of this function by the conformational transitions between isoforms of PrP could be involved in prion-related neurotoxicity.

The structured C-domain folds and unfolds reversibly in response to chaotropic denaturants, and recent work on the folding kinetics of mouse PrP^C (Wildegger et al 1999) demonstrates that there are no populated intermediates in the folding reaction and that the protein displays unusually rapid rates of folding and unfolding. These findings have been reinforced by hydrogen/deuterium exchange measurements on the human protein, which show that the overall equilibrium constant describing the distribution of folded and unfolded states is the same as the protection factor (Hosszu et al 1999). This shows that no partially unfolded forms or intermediates have a population greater than the unfolded state. The data suggest that PrP^{Sc} is unlikely to be formed from a kinetic folding intermediate, as has been hypothesized in the case of amyloid formation in other systems. In fact, on the basis of population it would be more likely that PrP^{Sc} were formed from the unfolded state of the molecule (see Figure 3; see color insert).

Inherited prion diseases may produce disease by destabilizing PrP^C, which would predispose the molecule to aggregate. Alternatively, a mutation could facilitate the interaction between PrP^C and PrP^{Sc} or affect the binding of a ligand or coprotein. In order to relate the folding stability of PrP^C to its propensity for forming PrP^{Sc}, several of the human mutations have been copied into recombinant mouse protein (Liemann & Glockshuber 1999). Although this work broadly concluded that there is no absolute correlation between stability and disease, all the fully penetrant pathogenic mutations show significant destabilization, whereas nonpathogenic polymorphisms have little effect.

Structural Studies of PrP^{Sc}

PrP^{Sc} is extracted from affected brains as highly aggregated, detergent-insoluble material that is not amenable to high-resolution structural techniques. However, Fourier transform infrared spectroscopic methods show that PrP^{Sc}, in sharp contrast to PrP^C, has a high β-sheet content (Pan et al 1993). PrP^{Sc} is covalently indistinguishable from PrP^C (Stahl et al 1993, Pan et al 1993).

During infection, the underlying molecular events that lead to the conversion of PrP^C to the scrapie agent remain ill defined. The most coherent and general model thus far proposed is that the protein, PrP, fluctuates between a dominant native state, PrP^C, and a series of minor conformations, one or a set of which can self-associate in an ordered manner to produce a stable supramolecular structure, PrP^{Sc}, composed of misfolded PrP monomers. Once a stable "seed" structure is formed, PrP can then be recruited, leading to an explosive, autocatalytic formation of PrP^{Sc}. Such a system would be extremely sensitive to three factors: (*a*) overall PrP^C concentration; (*b*) the equilibrium distribution between the native conformation and the self-associating conformation; and (*c*) complementarity between surfaces that come together in the aggregation step. All three of these predictions from this minimal model are manifest in the etiology of prion disease: an inversely proportional relationship between PrP^C expression and prion incubation period in transgenic mice (Prusiner et al 1990, Telling et al 1995, Bueler et al 1993, Collinge et al 1995b); predisposition by relatively subtle mutations in the protein sequence (Collinge 1997); and a requirement for molecular homogeneity for efficient prion propagation (Prusiner et al 1990, Palmer et al 1991).

Little is known for certain about the molecular state of the protein that constitutes the self-propagating, infectious particle itself. There are examples of infectivity in the absence of detectable PrP^{Sc} (Collinge et al 1995a, Wille et al 1996, Lasmezas et al 1997, Shaked et al 1999), and different strains of prions (see below) are known to differ in their degree of protease resistance. A single infectious unit corresponds to approximately 10^5 PrP molecules (Bolton et al 1982). It is unclear whether this indicates that a large aggregate is necessary for infectivity or, at the other extreme, whether only a single one of these PrP^{Sc} molecules is actually infectious. This relationship of PrP^{Sc} molecules to infectivity could simply, however, relate to the rapid clearance of prions from the brain known to occur on intracerebral challenge.

In Vitro Production of Disease-like PrP Isoforms

Direct in vitro mixing experiments (Kocisko et al 1994, 1995; Bessen et al 1995) have been performed in an attempt to produce PrP^{Sc}. In such experiments, an excess of PrP^{Sc} is used as a seed to convert recombinant PrP^C to a protease-resistant form (designated PrP^{RES}). However, the relative inefficiency of these reactions has precluded determining whether new infectivity has been generated. An artificial species barrier has, however, been exploited to address this issue, and such conversion products, expected to have a different host specificity (and thus which can be bioassayed in the presence of an excess of starting material), have not shown any detectable infectivity (Hill et al 1999a). These results argue that acquisition of protease resistance by PrP^C is not sufficient for the propagation of infectivity. Despite the obvious limitations of such experiments, they may represent an initial step in the generation of the infectious isoform of PrP, which requires additional, as-yet-unknown cofactors for the acquisition of infectivity.

The difficulty in performing structural studies on native PrP^{Sc} has led to attempts to produce soluble β-sheet–rich forms of PrP that may be amenable to NMR or crystallographic structure determination.

It is now recognized that the adage "one sequence, one conformation" is not strictly true. Depending on solvent conditions, probably any protein chain can adopt a variety of conformations in which there is a degree of periodic order (that is, extensive regions of secondary structure). For instance, a recent systematic study of the conformations adopted by the glycolytic enzyme, phosphoglycerate kinase, shows that in different media, the chain can adopt five distinct states (Damaschun et al 1999). However, such alternative states do not have precisely and tightly packed side chains, which are the hallmark of the native state of orthodox globular proteins.

Although at physiological pH recombinant fragments of PrP unfold via a two-state mechanism, an alternative folding pathway is observed at acidic pH. Studies on recombinant human PrP residues 90–231 (Swietnicki et al 1997) and mouse PrP encompassing residues 121–231 identified a distinct, detergent-stabilized equilibrium folding intermediate at pH 4.0. Circular dichroism spectroscopy indicated this intermediate was structured with predominantly β-sheet topology (Hornemann & Glockshuber 1998), and it has been proposed that this may be an intermediate on the pathway to PrP^{Sc} formation. Recent studies on a large fragment of the human prion protein (PrP^{91-231}) have shown that at acidic pH, PrP can fold to a soluble monomer comprised almost entirely of β-sheet in the absence of denaturants (Jackson et al 1999). Reduction of the native disulphide bond was a prerequisite for β-sheet formation, and these observations of alternative folding pathways dependent on solvent pH and redox potential could have important implications for the mechanism of conversion to PrP^{Sc}. Indeed, this monomeric β-sheet state was prone to aggregation into fibrils with partial resistance to proteinase K digestion, characteristic markers of PrP^{Sc}. Although unusual for a protein with a predominantly helical fold, the majority of residues in PrP^{91-231} have a preference for β-conformation (55% of non–glycine/proline residues). In view of this property, it is possible that the PrP molecule is delicately balanced between radically different folds with a high-energy barrier between them: one dictated by local structural propensity (the β-conformation), and one requiring the precise docking of side chains (the native α-conformation). Such a balance would be influenced by mutations causing inherited human prion diseases (Collinge 1997). It is also worthy of note that individuals homozygous for valine at polymorphic residue 129 of human PrP (where either methionine or valine can be encoded) are more susceptible to iatrogenic CJD (Collinge et al 1991), and valine has a much higher β-propensity than does methionine.

The precise subcellular localization of PrP^{Sc} propagation remains controversial. However, there is considerable evidence implicating either late-endosome–like organelles or lysosomes (Arnold et al 1995, Mayer et al 1992, Taraboulos et al 1992, Laszlo et al 1992). The environments of these organelles are evolved to facilitate protein unfolding at low pH prior to degradation by acid-activated proteases. It is possible that the α-PrP-to-β-PrP conversion, caused by reduction and mild acidification, is relevant to the conditions that PrP^{C} would encounter within the cell, following its internalization during recycling (Shyng et al 1993). Such a mechanism could underlie prion propagation and account for the transmitted, sporadic, and inherited etiologies of prion disease (Figure 3, see color insert). Initiation of a pathogenic self-propagating conversion reaction, with accumulation of aggregated β-PrP, may be induced by exposure to a "seed" of aggregated β-PrP following prion inoculation, or as a rare stochastic conformational change or an inevitable consequence of expression of a pathogenic PrP^C mutant that is predisposed to form β-PrP. It remains to be demonstrated whether such alternative conformational states of the protein are sufficient to cause prion disease in an experimental host or whether other cellular cofactors are also required.

Normal Cellular Function of PrP

PrP is highly conserved among mammals, has been identified in marsupials (Windl et al 1995) and birds (Harris et al 1993), and may be present in all vertebrates. It is expressed during early embryogenesis and is found in most tissues in adults (Manson et al 1992). However, highest levels of expression are seen in the central nervous system, in particular in association with synaptic membranes. PrP is also widely expressed in cells of the immune system (Dodelet & Cashman 1998). As a GPI-anchored cell-surface glycoprotein, it has been speculated that it may have a role in cell adhesion or signaling processes, but its precise cellular function has remained obscure. Mice lacking PrP as a result of gene knockout ($Prnp^{o/o}$) showed no gross phenotype (Bueler et al 1992), although they were completely resistant to prion disease following inoculation and did not replicate prions (Bueler et al 1993). However, these mice were then shown to have abnormalities in synaptic physiology (Collinge et al 1994) and in circadian rhythms and sleep (Tobler et al 1996). In particular, these mice had abnormalities of inhibitory synaptic transmission in hippocampal slices with a reduced amplitude of the maximal inhibitory postsynaptic current (IPSC) that could be isolated under standardized conditions, and a depolarizing shift in the reversal potential of the IPSC (Collinge et al 1994). This phenotype was rescued by a human PrP transgene (Whittington et al 1995). Two other groups, using different methods, notably much smaller IPSCs (minimal stimulus-evoked or spontaneous miniature IPSCs) and subphysiological temperatures, failed to replicate this result in hippocampal (Lledo et al 1996) and cerebellar (Herms et al 1995) slices. However, the latter group recently found that IPSCs can be modulated by free radicals differently in $Prnp^{o/o}$ mice (Herms et al 1999), and they proposed that this could explain the previous discrepancies due to the systematic difference in temperature between the original studies at physiological temperatures (Collinge et al 1994) and room temperature (Lledo et al 1996).

A second physiological phenotype for $Prnp^{o/o}$ mice was reported, a reduction of slow after-hyperpolarizations evoked by trains of action potentials (Colling et al 1996). This phenotype has subsequently been confirmed by another group in cerebellar slices (Herms et al 1998). It is interesting that hamsters infected with scrapie also have depressed slow after-hyperpolarizations (Barrow et al 1999). The most parsimonious mechanism for the various observations on the $Prnp^{o/o}$ mice is that responses of intracellular Ca^{2+} to depolarization are lower than normal, presumably as a result of weakened influx and/or abnormal homeostasis.

Although none of these observations defines a molecular role for PrP^C, it has been argued that PrP may act as a receptor for an as-yet-unidentified extracellular ligand. Newly synthesized PrP^C is transported to the cell surface and then cycles rapidly via a clathrin-mediated mechanism, with a transit time of approximately an hour, between the surface and early endosomes (Shyng et al 1994). This type of behavior is associated with other cell-surface receptors, for instance those for transferrin and low-density lipoproteins. Potential partner proteins identified by two-hybrid screening include the laminin receptor (Rieger et al 1997). However, because of the methodology used for identification, these are of questionable relevance. PrP does not fold in the cytoplasm of cells, and thus, candidates so far identified are unlikely to interact specifically with PrP.

Although *Prnp$^{o/o}$* mice are completely resistant to prion infection, reconstitution of such mice with either mouse or hamster PrP transgenes restores susceptibility in a species-specific manner (Bueler et al 1993). This then allows a reverse genetics approach in transgenic mice to study structure-function relationships in PrP by expressing truncated or mutated PrP in *Prnp$^{o/o}$* mice. Expression of N-terminal deletion mutants to residue 106 was tolerated and allowed prion propagation. However, deletion beyond this led to severe ataxia and neuronal loss in the granular cell layer of the cerebellum (Shmerling et al 1998), symptoms and pathology that were completely absent in the original *Prnp$^{o/o}$* mice. This has led to the hypothesis that PrP and a structural homologue compete for the same receptor or ligand. An additional gene (*Prnd*) has recently been discovered downstream of the *Prnp* locus. It encodes a 179-residue protein, with between 20% and 24% identity to PrP (depending on the species), that is a plausible candidate for the putative competitor (Moore et al 1999).

PrP in its entirety is unnecessary for prion propagation. Not only can the unstructured N-terminal 90 amino acids be deleted, so can the first α-helix, the second β-strand, and part of helix 2. In transgenic animals, a 106–amino acid fragment of the protein comprising PrPΔ23-88Δ141-176 was all that was required to confer susceptibility to and propagation of prions (Muramoto et al 1996, Supattapone et al 1999).

A number of lines of evidence argue that PrP may be a metalloprotein in vivo. It has been demonstrated that two different PrP^{Sc} types, characteristic of clinically distinct subtypes of sporadic CJD, can be interconverted in vitro by altering the metal ion occupancy (Wadsworth et al 1999). Also copper chelators can induce spongiform change in experimental animals (Pattison & Jebbett 1971), and it has been claimed that the levels of copper in the brains of PrP-null mice are lower than in wild-type mice, although this finding has not been replicated by other workers (Brown et al 1997a, Waggoner et al 2000). Moreover, it has been reported that recombinant PrP possesses superoxide dismutase activity when refolded in the presence of high concentrations of copper chloride (5 mM) (Brown et al 1999). However, binding of copper ions was found to occur only if added to the denatured protein before refolding. PrP has also been proposed to function as a copper transport protein for internalization of copper (II) ions (Pauly & Harris 1998).

With regard to physical measurements of metal interactions, it has been shown that synthetic peptides corresponding to the octapeptide repeat region of PrP bind copper (II) ions (Hornshaw et al 1995a,b). The authors concluded that copper ions bound specifically to a peptide encompassing residues 60–91 with a K_d of 6.7 μM and a 4:1 stoichiometry. A similar binding affinity (K_d 5.9 μM) was reported for a longer fragment encompassing residues 23–98 (Brown et al 1997a). The stoichiometry was reported to be 5.6 coppers per peptide. Binding of metal ions to full-length recombinant hamster protein (SHaPrP$^{29-231}$) has also been studied, and here the authors concluded that binding was specific for copper ions (Stockel et al 1998). Saturation of binding was reached at approximately 1.8 copper ions per PrP molecule, with an average K_d of 14 μM. The puzzling aspect of these studies is the extremely weak binding affinity for copper (II). Binding constants in the micromolar range would lead to the conclusion that such interactions are physiologically irrelevant. Also, given the nature of the octa-peptide repeat region alone, with its five histidine side chains, one would expect copper (II) to bind with a far greater affinity. For instance, for the square planar coordination of Cu (II) by four independent imidazole groups, the effective binding affinity is 3×10^{-13} M, and even for simple organic oxo-acids such as malonate, the affinity is 10^{-8} M (Anonymous 1986).

However, two high-affinity binding sites for divalent transition metals within the human prion protein have now been characterized, consistent with affinities seen with authentic metal binding proteins (GS Jackson, IA Murray, LLP Hosszu, N Gibbs, JP Waltho, AR Clarke, & J Collinge, submitted for publication). One is in the N-terminal octapeptide-repeat segment and has a K_d for copper (II) of 10^{-14} M. Other metals (Ni^{2+}, Zn^{2+}, and Mn^{2+}) bind three or more orders of magnitude more weakly, with relative affinities consistent with histidine coordination. NMR and fluorescence data reveal a second site around histidines 96 and 111, a region of the molecule known to be important for prion propagation, and there are less-marked resonance shifts in the globular, structured region along helices 2 and 3. The K_d for copper (II) at this site is 4×10^{-14} M whereas nickel (II), zinc (II), and manganese (II) bind 6, 7, and 10 orders of magnitude more weakly, respectively, regardless of whether the protein is in its oxidized α-helical (α-PrP) or reduced β-sheet (β-PrP) conformation. A role for PrP in copper metabolism or transport now seems likely, and disturbance of this function by the conformational transition from α- to β-isoforms of PrP may be involved in prion-related neurotoxicity.

PRION STRAINS

The Conundrum of Multiple Prion Strains

A major problem for the "protein-only" hypothesis of prion propagation has been how to explain the existence of multiple isolates, or strains, of prions. Multiple distinct strains of naturally occurring sheep scrapie were isolated in mice. Such strains are distinguished by their biological properties: They produce distinct incubation periods and patterns of neuropathological targeting (so-called lesion profiles) in defined inbred mouse lines [for a review, see Bruce et al (1992).] As they can be serially propagated in inbred mice with the same *Prnp* genotype, they cannot be encoded by differences in PrP primary structure. Furthermore, strains can be reisolated in mice after passage in intermediate species with different PrP primary structures (Bruce et al 1994). Conventionally, distinct strains of conventional pathogen are explained by differences in their nucleic acid genome. However, in the absence of such a scrapie genome, alternative possibilities must be considered. Weissmann (1991) proposed a "unified hypothesis" where, though the protein alone was argued to be sufficient to account for infectivity, it was suggested that strain characteristics could be encoded by a small cellular nucleic acid, or "coprion." Although this hypothesis leads to the testable prediction that strain characteristics, unlike infectivity, would be sensitive to ultraviolet irradiation, no such test has been reported. At the other extreme, the protein-only hypothesis (Griffith 1967) would have to explain how a single polypeptide chain could encode multiple disease phenotypes. Clearly, understanding how a protein-only infectious agent could encode such phenotypic information is of considerable biological interest.

The Molecular Basis of Prion Strain Diversity

Support for the idea that strain specificity may be encoded by PrP itself was provided by study of two distinct strains of transmissible mink encephalopathy prions that can be serially propagated in hamsters, designated hyper (HY) and drowsy (DY). These strains can be distinguished by differing physiochemical properties of the accumulated PrP^{Sc} in the brains of affected hamsters (Bessen & Marsh 1992). Following limited proteolysis, strain-specific migration patterns of PrP^{Sc} on polyacrylamide gels were seen that related to different N-terminal ends of HY and DY PrP^{Sc} following protease treatment and implying differing conformations of HY and DY PrP^{Sc} (Bessen & Marsh 1994).

Recently, several human PrP^{Sc} types have been identified that are associated with different phenotypes of CJD (Parchi et al 1996, Collinge et al 1996). The different fragment sizes seen on Western blots following treatment with proteinase K suggests that there are several different human PrP^{Sc} conformations. However, although such biochemical modifications of PrP are clearly candidates for the molecular substrate of prion strain diversity, it is necessary to be able to demonstrate that these properties fulfill the biological properties of strains, in particular that they are transmissible to the PrP in a host of both the same and different species. This has been demonstrated in studies with CJD isolates, with both PrP^{Sc} fragment sizes and the ratios of the three PrP glycoforms (diglycosylated, monoglycosylated, and unglycosylated PrP) maintained on passage in transgenic mice expressing human PrP (Collinge et al 1996). Furthermore, transmission of human prions and bovine prions to wild-type mice results in murine PrP^{Sc} with fragment sizes and glycoform ratios that correspond to the original inoculum (Collinge et al 1996). Variant CJD is associated with PrP^{Sc} glycoform ratios that are distinct from those seen in classical CJD. Similar ratios are seen in BSE in cattle and BSE when transmitted to several other species (Collinge et al 1996). These data strongly support the protein-only hypothesis of infectivity and suggest that strain variation is encoded by a combination of PrP conformation and glycosylation. Furthermore, polymorphism in PrP sequence can influence the generation of particular PrP^{Sc} conformers (Collinge et al 1996). Transmission of PrP^{Sc} fragment sizes from two different subtypes of inherited prion disease to transgenic mice expressing a chimeric human mouse PrP has also been reported (Telling et al 1996). As PrP glycosylation occurs before conversion to PrP^{Sc}, the different glycoform ratios may represent selection of particular PrP^{C} glycoforms by PrP^{Sc} of different conformations. According to such a hypothesis, PrP conformation would be the primary determinant of strain type, with glycosylation being involved as a secondary process. However, because it is known that different cell types may glycosylate proteins differently, PrP^{Sc} glycosylation patterns may provide a substrate for the neuropathological targeting that distinguishes different prion strains (Collinge et al 1996). Particular PrP^{Sc} glycoforms may replicate most favorably in neuronal populations, with a similar PrP glycoform expressed on the cell surface. Such targeting could also explain the different incubation periods that also discriminate strains, since targeting of more critical brain regions, or regions with higher levels of PrP expression, might be expected to produce shorter incubation periods. Further supportive evidence for the involvement of PrP glycosylation in prion strain propagation has come from the study of transgenic mice expressing PrP with mutations interfering with N-linked glycosylation (DeArmond et al 1997).

Recent work has shown strain-specific protein conformation to be influenced by metal binding to PrP^{Sc} (Wadsworth et al 1999). Two different human PrP^{Sc} types, seen in clinically distinct subtypes of classical CJD, can be interconverted in vitro by altering the metal-ion occupancy. The dependence of PrP^{Sc} conformation on the binding of copper and zinc represents a novel mechanism for posttranslational modification of PrP, and for the generation of multiple prion strains. This finding may also explain differences in molecular classification of classical CJD. Collinge et al (Collinge et al 1996) described three PrP^{Sc} types among cases of sporadic and iatrogenic CJD and a distinctive type 4 pattern in all cases of variant CJD. An earlier study (Parchi et al 1996) of PrP^{Sc} types in classical CJD had described only two types of PrP^{Sc}, and these authors have argued that the types 1 and 2 of Collinge et al (1996) correspond to their type 1, whereas the type 3 pattern of Collinge at al corresponds to their type 2 (Parchi et al 1997). However, these authors concede a degree of heterogeneity in their type 1 cases (Parchi et al 1996). In a large-scale study of PrP^{Sc} types in CJD in conjunction with the UK National CJD Surveillance Unit, we demonstrated that patients classified as type 1 and type 2 using our criteria have distinct disease phenotypes, confirming the validity of our molecular classification. Type 1 human CJD is a distinct human prion disease with an aggressive clinical course and remarkably short clinical duration (Wadsworth et al 1999). In the presence of the metal chelators, human PrP^{Sc} types 1 and 2 produce a similar-sized fragment after proteinase K digestion, designated type 2-.

It is possible, therefore, that the discrepancy between these two different molecular classification systems may be explained by differing methodologies.

Molecular strain typing of prion isolates can now be applied to molecular diagnosis of vCJD (Collinge et al 1996, Hill et al 1997) and to produce a new classification of human prion diseases, with implications for epidemiological studies investigating the etiology of sporadic CJD. Such methods allow strain typing to be performed in days rather than the 1–2 years required for classical biological strain typing. This technique may also be applicable to determining whether BSE has transmitted to other species (Collinge et al 1996), for instance to sheep (Hill et al 1998, Kuczius et al 1998, Hope et al 1999), and thereby poses a threat to human health.

Such ability of a single polypeptide chain to encode information specifying distinct phenotypes of disease raises intriguing evolutionary questions. Do other proteins behave in this way? The novel pathogenic mechanisms involved in prion propagation may be of far wider significance and may be relevant to other neurological and nonneurological illnesses; indeed, other prion-like mechanisms have now been described (Milner & Medcalf 1991), and the field of yeast and fungal prions has emerged (Wickner & Masison 1996, Wickner 1997).

PRION TRANSMISSION BARRIERS

The "Species Barrier"

Transmission of prion diseases between different mammalian species is restricted by a "species barrier" (Pattison 1965). On primary passage of prions from species A to species B, usually not all inoculated animals of species B develop disease. Those that do have much longer and more variable incubation periods than those that are seen with transmission of prions within the same species, where typically all inoculated animals would succumb within a relatively short, and remarkably consistent, incubation period. On second passage of infectivity to further animals of species B, transmission parameters resemble within-species transmissions, with most, if not all, animals developing the disease with short and consistent incubation periods. Species barriers can therefore be quantitated by measuring the fall in mean incubation period on primary and second passage or, perhaps more rigorously, by a comparative titration study. The latter involves inoculating serial dilutions of an inoculum in both the donor and host species and comparing the mean lethal doses (LD_{50}) obtained. The effect of a very substantial species barrier (for instance that between hamsters and mice) is that few, if any, animals succumb to disease on primary passage, and only then at incubation periods approaching the natural lifespan of the species concerned.

Early studies of the molecular basis of the species barrier argued that it resided principally in differences in PrP primary structure between the species from which the inoculum was derived and the inoculated host. Transgenic mice expressing hamster PrP were, unlike wild-type mice, highly susceptible to infection with Sc237 hamster prions (Prusiner et al 1990). That most sporadic and acquired CJD occurred in individuals homozygous at *PRNP* polymorphic codon 129 supported the view that prion propagation proceeded most efficiently when the interacting PrP^{Sc} and PrP^{C} were of identical primary structure (Collinge et al 1991, Palmer et al 1991). However, it has been long recognized that prion strain type affects ease of transmission to another species. It is interesting that with BSE prions, the strain component to the barrier seems to predominate, with BSE not only transmitting efficiently to a range of species but also maintaining its transmission characteristics even when passaged through an intermediate species with a distinct PrP gene (Bruce et al 1994). For instance, transmission of CJD prions to conventional mice is difficult, with few if any inoculated mice succumbing after prolonged incubation periods, consistent with a substantial species barrier (Collinge et al 1995b, Hill et al 1997). In sharp contrast, transgenic mice expressing only human PrP are highly susceptible to CJD prions, with 100% attack rate and consistent short incubation periods that are unaltered by second passage, consistent with a complete lack of species barrier (Collinge et al 1995b). However, vCJD prions (again comprising human PrP of identical primary structure) transmit much more readily to wild-type mice than do classical CJD prions, whereas transmission to transgenic mice is relatively less efficient than with classical CJD (Hill et al 1997). The term species barrier does not seem appropriate to describe such effects and "species-strain barrier" or simply "transmission barrier" may be preferable (Collinge 1999). Both PrP amino acid sequence and strain type affect the three-dimensional structure of glycosylated PrP, which will presumably, in turn, affect the efficiency of the protein-protein interactions thought to determine prion propagation.

Prion Transmission Barrier: Molecular Basis

Mammalian PrP genes are highly conserved. Presumably only a restricted number of different PrP^{Sc} conformations (that are highly stable and can therefore be serially propagated) will be permissible thermodynamically and will constitute the range of prion strains seen. PrP glycosylation may be important in stabilizing particular PrP^{Sc} conformations. Although a significant number of different such PrP^{Sc} conformations may be possible among the range of mammalian PrPs, only a subset of these would be allowable for a given single mammalian PrP. Substantial overlap between the favored conformations for PrP^{Sc} derived from species A and species B might therefore result in relatively easy transmission of prion diseases between these two species, while two species with no preferred PrP^{Sc} conformations in common would have a large barrier to transmission (and indeed transmission would necessitate a change of strain type). According to such a model of a prion transmission barrier, BSE may represent a thermodynamically highly favored PrP^{Sc} conformation that is permissive for PrP expressed in a wide range of different species, accounting for the remarkable promiscuity of this strain in mammals. Contribution of other components to the species barrier are possible and may involve interacting cofactors that mediate the efficiency of prion propagation, although no such factors have yet been identified.

Recent data have further challenged our understanding of transmission barriers (Hill et al 2000). The assessment of species barriers has relied on the development of a clinical disease in inoculated animals. On this basis there is a highly efficient barrier limiting transmission of hamster Sc237 prions to mice. Indeed, the hamster scrapie strain Sc237 (Scott et al 1989) [which is similar to the strain classified as 263K (Kimberlin & Walker 1977, 1978)] is regarded as nonpathogenic for mice [with no clinical disease in mice observed for up to 735 days postinoculation (Kimberlin & Walker 1978)] and was used in studies of species barriers in transgenic mice (Kimberlin & Walker 1979, Scott et al 1989, Prusiner et al 1990). It was demonstrated that transgenic mice expressing hamster PrP (in addition to endogenous mouse PrP), in sharp contrast to conventional mice, were highly susceptible to Sc237 hamster prions, with consistent, short incubation periods that were inversely correlated to hamster PrP expression levels (Scott et al 1989, Prusiner et al 1990). It is important, however, that these studies defined transmission using clinical criteria and did not report PrP^{Sc} levels and types, or prion titers, in the brains of clinically unaffected animals. However, although not developing a clinical disease, and indeed living as long as mock-inoculated mice, Sc237-inoculated mice may accumulate high levels of prions in their brains (Hill et al 2000). Previous studies on the species barrier between hamsters and mice (using the Sc237 or 263K strain) did not report whether PrP^{Sc} and/or infectivity were present in clinically unaffected animals (Scott et al 1989, Prusiner et al 1990) or have attempted passage from mice only up to 280 days postinoculation (Kimberlin & Walker 1978). The barrier to primary passage appears in this case to be to the development of rapid neurodegeneration and the resulting clinical syndrome rather than a barrier to prion propagation itself.

BOVINE SPONGIFORM ENCEPALOPATHY AND RISKS TO PUBLIC HEATH

Variant CJD

A novel form of human prion disease, variant CJD (vCJD), was recognized in the United Kingdom in 1996 (Will et al 1996) and implied the arrival of a new risk factor for CJD (Collinge & Rossor 1996). These epidemiological studies argued for a link with BSE, and this was strongly supported by molecular strain typing studies (Collinge et al 1996). All cases of vCJD are associated with type 4 PrP^{Sc}. Type 4 PrP^{Sc} has a high proportion of the diglycosylated form of PrP^{Sc} and is distinct from the PrP^{Sc} types seen in classical CJD (types 1–3), with differing fragment sizes following proteinase K digestion. Also, types 1–3 are associated with a high proportion of monoglycosylated PrP^{Sc}. The glycoform ratios of proteinase K–digested PrP^{Sc} in vCJD were closely similar, however, to those seen in BSE passaged in a number of mammalian species. Furthermore, when prions isolated from either bovine brain or human brain are transmitted to experimental mice, PrP^{Sc} isolated from the infected hosts is indistinguishable, either by site of proteinase K cleavage or by glycoform ratio (Collinge et al 1996). In addition, vCJD and BSE show closely similar transmission properties in both transgenic and conventional mice, with indistinguishable neuropathology in both transgenic mice and a variety of inbred strains of mice (Hill et al 1997, Bruce et al 1997). That vCJD is human BSE is therefore supported by compelling experimental data. Moreover, it was recently found that vCJD can be further distinguished by the detection of PrP^{Sc} in the lymphoreticular system (Hill et al 1999b), a tissue distribution specific to vCJD.

Modeling the Transmission Barrier Between Cattle and Humans

The species barrier between cattle BSE and humans cannot be directly measured, but it can be modeled in transgenic mice expressing human PrP^C, which produce human PrP^{Sc} when challenged with human prions (Collinge et al 1995b). When such mice, expressing both human PrP valine 129 (at high levels) and mouse PrP, are challenged with BSE, three possibilities could be envisaged: These mice could produce human prions, murine prions, or both. In fact, only mouse prion replication could be detected. Although there are caveats with respect to this model, particularly that human prion propagation in mouse cells may be less efficient than that of mouse prions, this result would be consistent with the bovine-to-human barrier being higher than the bovine-to-mouse barrier for this *PRNP* genotype. In the second phase of these experiments, mice expressing only human PrP were challenged with BSE. Although CJD isolates transmit efficiently to such mice at approximately 200 days, only infrequent transmissions at over 500 days were seen with BSE, consistent with a substantial species barrier for this human *PRNP* genotype (Hill et al 1997). The *PRNP* valine 129 genotype was studied initially in attempts to produce an animal model of human prion disease, as this genotype was over-represented among early cases of iatrogenic CJD (Collinge et al 1991), which suggests increased susceptibility or shorter incubation periods in this genotype. However, it is important to repeat these studies in mice expressing only human PrP methionine 129 and in heterozygotes. So far, BSE appears to have transmitted only to humans of *PRNP* codon 129 methionine homozygous genotype (Collinge et al 1996; Hill et al 1999b).

Predictions of Epidemic Size

BSE can be readily transmitted to mice with most, if not all, inoculated animals succumbing to disease on primary passage (a high "attack rate"). This relatively modest species barrier has been formally measured by comparative titration studies of the same BSE isolate by intracerebral inoculation into cattle and mice. It indicates an approximately 1000-fold barrier (i.e. it takes 1000 times more BSE to kill a mouse than a cow) (Wells et al 1998). The effect of this barrier on incubation periods is to increase mean incubation periods by approximately threefold and to dramatically increase the range of incubation periods seen.

Such experiments are usually performed using the most efficient, intracerebral, route of transmission. A formal titration of BSE in mice to determine an oral LD_{50} has not been reported. However, oral challenge with approximately 10 g of BSE-affected cow brain killed the majority of exposed mice (Barlow & Middleton 1990). If the bovine-to-human species barrier were similar to that for mice, it would suggest an oral LD_{50} in humans of an order of magnitude also similar to that for mice (approximately 10 g). Clearly, it is hoped that the species barrier limiting transmission of BSE to humans will be of a far higher order. However, if we assume a pessimistic scenario, that the barrier is similar (and it remains possible it could be lower), extrapolation with the known incubation periods in the acquired human prion diseases, such as growth hormone–related iatrogenic CJD or kuru, where transmission does not involve a species barrier (and mean incubation periods are approximately 10–15 years), would suggest mean incubation periods of BSE in humans of perhaps 30 years or more, and a range extending from 10 years to, or exceeding, a normal human lifespan. Such estimations (Collinge 1999), based on extensive experience of transmission studies across species from many research groups over several decades, suggest—only 4 years after recognition of vCJD— the need for caution with respect to optimistic assessments of likely human BSE epidemic size.

Iatrogenic Transmission of vCJD Prions

Considerable concern has been expressed that blood and blood products from asymptomatic donors incubating vCJD may pose a risk for the iatrogenic transmission of vCJD. Reports of infectivity of blood from patients with classical CJD are infrequent and have been questioned (Brown 1995). Infectivity of blood from patients in the clinical phase of vCJD is unknown. However, PrP^{Sc} is consistently found in the lymphoreticular system in vCJD (Hill et al 1999b), lymphocytes express significant levels of PrP^{C} (Cashman et al 1990), and, in mice, B lymphocytes (although not necessarily expressing PrP^{C}) are required for prion neuroinvasion following peripheral inoculation (Klein et al 1997, 1998; Collinge & Hawke 1998). UK policy is now to leucodeplete all whole blood, a practice already in use (for other health reasons) in some countries, and to acquire plasma for plasma products from outside the United Kingdom.

A further possible route of transmission of vCJD is via contaminated surgical instruments. Iatrogenic transmission of classical CJD via neurosurgical instruments has been reported (Bernoulli et al 1977), and normal hospital sterilization procedures are not likely to completely inactive prions. Recent evidence suggests that classical CJD may also be transmitted by other surgical procedures (Collins et al 1999). Although in the United Kingdom all surgical instruments used on patients with suspected CJD are quarantined and not reused unless an alternate nonprion diagnosis is unequivocally confirmed, the extensive lymphoreticular involvement in vCJD, which is likely to be present from a relatively early preclinical stage, raises the possibility that instruments could be contaminated in particular during those procedures that involve contact with lymphoreticular tissues. This includes the common procedures of tonsillectomy, appendicectomy, and lymph node and gastrointestinal biopsy. Recent studies have demonstrated that prions can adhere easily to metal surfaces, and prion-contaminated metal wires are an efficient vehicle for experimental transmission of prion disease (Zobeley et al 1999).

PRION NEURODEGENERATION AND POTENTIAL THERAPEUTIC APPROACHES

Cell Death and Prion Disease

The precise molecular nature of the infectious agent and the cause of neuronal cell death remain unclear. The current working hypothesis is that an abnormal isoform of PrP is the infectious agent, and to date, the most highly enriched preparations contain 1 infectious unit per 10^5 PrP monomers (Bolton et al 1982). Various hypotheses have been proposed to explain the mechanism of spongiform change and neuronal cell loss. These have included direct neurotoxic effects from a region of the prion protein encompassing residues 106–126 (Forloni et al 1993, Tagliavini et al 1993, Brown et al 1994) to increased oxidative stress in neurones as a result of PrP^C depletion, which has been proposed to function as an antioxidant molecule (Brown et al 1997b). Neurotoxicity of PrP 106–126 is, however, controversial (Kunz et al 1999). It has also been suggested that PrP^C plays a role in regulating apoptosis, with disturbance of normal cellular levels of PrP during infection leading to cell death (Kurschner & Morgan 1995, 1996). Certainly there have been numerous recent reports of apoptotic cells being identified in the neuronal tissue of prion disease brains (Williams et al 1997). Although PrP^C expression is required for susceptibility to the disease, a number of observations argue that PrP^{Sc}, and indeed prions (whether or not they are identical), may not themselves be highly neurotoxic. Prion diseases in which PrP^{Sc} is barely or not detectable have been described (Medori et al 1992a, Collinge et al 1995a, Hsiao et al 1990, Lasmezas et al 1997). Mice with reduced levels of PrP^C expression have extremely high levels of PrP^{Sc} and prions in the brain and yet remain well for several months after their wild-type counterparts succumb (Bueler et al 1994). Conversely, Tg20 mice, with high levels of PrP^C, have short incubation periods and yet produce low levels of PrP^{Sc} after inoculation with mouse prions (Fischer et al 1996). In addition, brain grafts producing high levels of PrP^{Sc} do not damage adjacent tissue in PrP knockout ($Prnp^{o/o}$) mice (Brandner et al 1996). The cause of neurodegeneration in prion diseases remains unclear. It remains possible that prion neurodegeneration is related, at least in part, to loss of function of PrP^C. That $Prnp^{o/o}$ mice [other than those associated with overexpression of the $Prnp$-like gene Dpl (Moore et al 1999)] do not develop neurodegeneration could be due to compensatory adaptations during neurodevelopment. Complete or near-complete ablation of PrP expression in adult mice using conditional gene expression methods has not yet been achieved. A recent study has demonstrated that mice inoculated with Sc237 hamster prions replicate prions to high levels in their brains but do not develop clinical signs of prion disease during their normal lifespan, arguing that PrP^{Sc} and indeed prions (whether or not they are identical) may not themselves be highly neurotoxic (Hill et al 2000). An alternative hypothesis for prion-related neurodegeneration is that a toxic, possibly infectious, intermediate is produced in the process of conversion of PrP^C to PrP^{Sc}, with PrP^{Sc}, present as highly aggregated material, being a relatively inert end product. The steady state level of such a toxic monomeric or oligomeric PrP intermediate could then determine rate of neurodegeneration. One possibility is that Sc237-inoculated mice propagate prions very slowly and that such a toxic intermediate is generated at extremely low levels that are tolerated by mice (Hill et al 2000).

Approaches to Therapeutics

The prion diseases are now among the best understood of the degenerative brain diseases, and the development of rational treatments is appearing realistic. Various compounds, some known to bind PrP^{Sc}, including Congo red (Ingrosso et al 1995), polyene antibiotics (Pocchiari et al 1987), anthracycline (Tagliavini et al 1997), dextran sulphate, pentosan polysulphate and other polyanions (Ehlers & Diringer 1984, Farquhar & Dickinson 1986, Kimberlin & Walker 1986), and β-sheet breaker peptides (Soto et al 2000), have been shown to have limited effects in animal models of prion disease. Unfortunately, most show a significant effect only if administered long before clinical onset (in some cases with the inoculum) and/or are impractical treatments because of toxicity or bioavailability.

The precise molecular events that bring about the conversion of PrP^C to PrP^{Sc} and the molecular nature of the neurotoxic species remain ill defined, a fact that might seem to preclude screening for compounds that inhibit the process. However, any ligand that selectively stabilizes the PrP^C state will prevent its rearrangement and might reasonably be expected to block prion replication (and presumably production of any putative toxic intermediate forms of PrP on the pathway to PrP^{Sc} formation). Such an approach has recently been applied to block p53 conformational rearrangements, which are involved in tumorigenesis (Foster et al 1999). Advances in therapeutics will have to be matched by advances in early diagnosis of prion disease to provide effective intervention before extensive neuronal loss has occurred.

CONCLUDING REMARKS

Prion diseases appear to be diseases of protein conformation, and elucidating their precise molecular mechanisms may, in addition to allowing us to progress with tackling key public health issues posed by vCJD, be of far wider significance in pathobiology. It is of considerable interest that many of the more common neurodegenerative diseases, such as Alzheimer's and Parkinson's diseases, and the polyglutamine repeat disorders (such as Huntington's disease) are also associated with abnormal protein aggregates. In addition, the apparent ability of a single polypeptide chain to encode information and specify distinctive phenotypes is unprecedented. It seems likely that evolution will have used this mechanism in many other ways.

Although the protein-only hypothesis of prion propagation is supported by compelling experimental data and now appears also able to encompass the phenomenon of prion strain diversity, the goal of the production of prions in vitro remains. PrP^{Sc}-like forms of PrP have recently been produced from purified recombinant material, but as yet none have been shown in experimental animals to be capable of producing disease that can be serially propagated. Success in such an endeavor would not only prove the protein-only hypothesis, it would also serve as the essential model by which the mechanism of prion propagation can be understood in molecular detail.

ACKNOWLEDGMENTS

I thank Jonathan Beck for producing Figure 1, Dr. Richard Sessions for producing the molecular model of human PrP, Professor Tony Clarke for Figure 3, and Ray Young for preparation of these illustrations.

Visit the Annual Reviews home page at www.AnnualReviews.org

LITERATURE CITED

Anonymous. 1986. Stabilty constants of metal complexes. In *Data* for *Biochemical Research*, ed. RMC Dawson, D Elliot, W Elliot, KM Jones, pp. 399–415. Oxford, UK: Clarendon. 3rd ed.

Alper T, Cramp WA, Haig DA, Clarke MC. 1967. Does the agent of scrapie replicate without nucleic acid? *Nature* 214:764–66

Alper T, Haig DA, Clarke MC. 1966. The exceptionally small size of the scrapie agent. *Biochem. Biophys. Res. Commun.* 22:278–84

Alpers MP. 1987. Epidemiology and clinical aspects of kuru. See Prusiner & McKinley 1987, pp. 451–65

Anderson RM, Donnelly CA, Ferguson NM, Woolhouse MEJ, Watt CJ, et al. 1996. Transmission dynamics and epidemiology of BSE in British cattle. *Nature* 382:779–88

Arnold JE, Tipler C, Laszlo L, Hope J, Landon M, Mayer RJ. 1995. The abnormal isoform of the prion protein accumulates in late-endosome-like organelles in scrapie-infected mouse brain. *J. Pathol.* 176:403–11

Baker HE, Poulter M, Crow TJ, Frith CD, Lofthouse R, et al. 1991. Aminoacid polymorphism in human prion protein and age at death in inherited prion disease. *Lancet* 337:1286

Barlow RM, Middleton DJ. 1990. Dietary transmission of bovine spongiform encephalopathy to mice. *Vet. Rec.* 126:111–12

Barrow PA, Holmgren CD, Tapper AJ,

Jeffreys JGR. 1999. Intrinsic physiological and morphological properties of principal cells of the hippocampus and neocortex in hamsters infected with scrapie. *Neurobiol. Disease* 6:406–23

Beck E, Daniel PM. 1987. Neuropathology of transmissible Spongiform Encephalopathies. See Prusiner & McKinley 1987, pp. 331–85

Bernoulli C, Siegfried J, Baumgartner G, Regli F, Rabinowicz T, et al. 1977. Danger of accidental person-to-person transmission of Creutzfeldt-Jakob disease by surgery. *Lancet* 1:478–79

Bessen RA, Kocisko DA, Raymond GJ, Nandan S, Lansbury PT, Caughey B. 1995. Non-genetic propagation of strain-specific properties of scrapie prion protein. *Nature* 375:698–700

Bessen RA, Marsh RF. 1992. Biochemical and physical properties of the prion protein from two strains of the transmissible mink encephalopathy agent. *J. Virol.* 66:2096–101

Bessen RA, Marsh RF. 1994. Distinct PrP properties suggest the molecular basis of strain variation in transmissible mink encephalopathy. *J. Virol.* 68:7859–68

Bolton DC, McKinley MP, Prusiner SB. 1982. Identification of a protein that purifies with the scrapie prion. *Science* 218:1309–11

Borchelt DR, Scott M, Taraboulos A, Stahl N, Prusiner SB. 1990. Scrapie and cellular prion proteins differ in their kinetics of synthesis and topology in cultured cells. *J. Cell Biol.* 110:743–52

Brandner S, Isenmann S, Raeber A, Fischer M, Sailer A, et al. 1996. Normal host prion protein necessary for scrapie-induced neurotoxicity. *Nature* 379:339–43

Brown DR, Herms J, Kretzschmar HA. 1994. Mouse cortical cells lacking cellular PrP survive in culture with a neurotoxic PrP fragment. *NeuroReport* 5:2057–60

Brown DR, Qin K, Herms JW, Madlung A, Manson J, et al. 1997a. The cellular prion protein binds copper in vivo. *Nature* 390:684–87

Brown DR, Schulz-Schaeffer WJ, Schmidt B,

Kretzschmar HA. 1997b. Prion protein-deficient cells show altered response to oxidative stress due to decreased SOD-1 activity. *Exp. Neurol.* 146:104–12

Brown DR, Wong BS, Hafiz F, Clive C, Haswell SJ, Jones IM. 1999. Normal prion protein has an activity like that of superoxide dismutase. *Biochem. J.* 344:1–5

Brown P. 1995. Can Creutzfeldt-Jakob disease be transmitted by transfusion? *Curr. Opin. Hematol.* 2:472–77

Brown P, Cathala F, Raubertas RF, Gajdusek DC, Castaigne P. 1987. The epidemiology of Creutzfeldt-Jakob disease: conclusion of a 15-year investigation in France and review of the world literature. *Neurology* 37:895–904

Bruce M, Chree A, McConnell I, Foster J, Pearson G, Fraser H. 1994. Transmission of bovine spongiform encephalopathy and scrapie to mice: strain variation and the species barrier. *Philos. Trans. R. Soc. London Ser. B* 343:405–11

Bruce ME, Fraser H, McBride PA, Scott JR, Dickinson AG. 1992. The basis of strain variation in scrapie. See Prusiner et al 1992, pp. 497–508

Bruce ME, Will RG, Ironside JW, McConnell I, Drummond D, et al. 1997. Transmissions to mice indicate that "new variant" CJD is caused by the BSE agent. *Nature* 389:498–501

Bueler H, Aguzzi A, Sailer A, Greiner RA, Autenried P, et al. 1993. Mice devoid of PrP are resistant to scrapie. *Cell* 73:1339–47

Bueler H, Fischer M, Lang Y, Bluethmann H, Lipp H-P, et al. 1992. Normal development and behaviour of mice lacking the neuronal cell-surface PrP protein. *Nature* 356:577–82

Bueler H, Raeber A, Sailer A, Fischer M, Aguzzi A, Weissmann C. 1994. High prion and PrPSc levels but delayed onset of disease in scrapie-inoculated mice heterozygous for a disrupted PrP gene. *Mol. Med.* 1:19–30

Cashman NR, Loertscher R, Nalbantoglu J, Shaw I, Kascsak RJ, et al. 1990. Cellular isoform of the scrapie agent protein participates in lymphocyte activation. *Cell* 61:185–92

Caughey B, Raymond GJ. 1991. The scrapie-associated form of PrP is made from a cell surface precursor that is both protease- and phospholipase-sensitive. *J. Biol. Chem.* 266:18217–23

Colling SB, Collinge J, Jefferys JGR. 1996. Hippocampal slices from prion protein null mice: disrupted Ca^{2+}-activated K^+ currents. *Neurosci. Lett.* 209:49–52

Collinge J. 1997. Human prion diseases and bovine spongiform encephalopathy (BSE). *Hum. Mol. Genet.* 6:1699–705

Collinge J. 1999. Variant Creutzfeldt-Jakob disease. *Lancet* 354:317–23

Collinge J, Beck J, Campbell T, Estibeiro K, Will RG. 1996. Prion protein gene analysis in new variant cases of Creutzfeldt-Jakob disease. *Lancet* 348:56–56

Collinge J, Brown J, Hardy J, Mullan M, Rossor MN, et al. 1992. Inherited prion disease with 144 base pair gene insertion. II: Clinical and pathological features. *Brain* 115:687–710

Collinge J, Harding AE, Owen F, Poulter M, Lofthouse R, et al. 1989. Diagnosis of Gerstmann-Straussler syndrome in familial dementia with prion protein gene analysis. *Lancet* 2:15–17

Collinge J, Hawke S. 1998. B lymphocytes in prion neuroinvasion: central or peripheral players. *Nat. Med.* 4:1369–70

Collinge J, Hill AF, Ironside J, Zeidler M. 1997. Diagnosis of new variant Creutzfeldt-Jakob disease by tonsil biopsy—authors' reply to Arya and Evans. *Lancet* 349:1322–23

Collinge J, Owen F, Poulter M, Leach M, Crow TJ, et al. 1990. Prion dementia without characteristic pathology. *Lancet* 336:7–9

Collinge J, Palmer MS, Dryden AJ. 1991. Genetic predisposition to iatrogenic Creutzfeldt-Jakob disease. *Lancet* 337:1441–42

Collinge J, Palmer MS, Sidle KCL, Gowland I, Medori R, et al. 1995a. Transmission of fatal familial insomnia to laboratory animals. *Lancet* 346:569–70

Collinge J, Palmer MS, Sidle KCL, Hill AF,

Gowland I, et al. 1995b. Unaltered susceptibility to BSE in transgenic mice expressing human prion protein. *Nature* 378:779–83

Collinge J, Rossor M. 1996. A new variant of prion disease. *Lancet* 347:916–17

Collinge J, Sidle KCL, Meads J, Ironside J, Hill AF. 1996. Molecular analysis of prion strain variation and the aetiology of "new variant" CJD. *Nature* 383:685–90

Collinge J, Whittington MA, Sidle KCL, Smith CJ, Palmer MS, et al. 1994. Prion protein is necessary for normal synaptic function. *Nature* 370:295–97

Collins S, Law MG, Fletcher A, Boyd A, Kaldor J, Masters CL. 1999. Surgical treatment and risk of sporadic Creutzfeldt-jakob disease: a case-control study. *Lancet* 353:693–97

Cousens SN, Vynnycky E, Zeidler M, Will RG, Smith PG. 1997. Predicting the CJD epidemic in humans. *Nature* 385:197–98

Cuillé J, Chelle PL. 1936. La maladie dite tremblante du mouton est- elle inocuable? *C. R. Acad. Sci.* 203:1552–54

Damaschun G, Damaschun H, Gast K, Zirwer D. 1999. Proteins can adopt totally different folded conformations. *J. Mol. Biol.* 291:715–25

DeArmond SJ, Sánchez H, Yehiely F, Qiu Y, Ninchak-Casey A, et al. 1997. Selective neuronal targeting in prion disease. *Neuron* 19:1337–48

Dodelet VC, Cashman NR. 1998. Prion protein expression in human leukocyte differentiation. *Blood* 91:1556–61

Doherr MG, Heim D, Vandevelde M, Fatzer R. 1999. Modelling the expected numbers of preclinical and clinical cases of bovine spongiform encephalopathy in Switzerland. *Vet. Rec.* 145:155–60

Ehlers B, Diringer H. 1984. Dextran sulphate 500 delays and prevents mouse scrapie by impairment of agent replication in spleen. *J Gen. Virol.* 65:1325–30

Farquhar CF, Dickinson AG. 1986. Prolongation of scrapie incubation period by an injection of dextran sulphate 500 within the month before or after infection. *J. Gen. Virol.* 67:463–73

Fischer M, Rulicke T, Raber A, Sailer A, Oesch B, et al. 1996. Prion protein (PrP) with amino terminal deletions restoring susceptibility of PrP knockout mice to scrapie. *EMBO J.* 15:1255–64

Forloni G, Angeretti N, Chiesa R, Monzani E, Salmona M, et al. 1993. Neurotoxicity of a prion protein fragment. *Nature* 362:543–46

Foster BA, Coffey HA, Morin MJ, Rastinejad F. 1999. Pharmacological rescue of mutant p53 conformation and function. *Science* 286:2507–10

Gajdusek DC, Gibbs CJ Jr, Alpers MP. 1966. Experimental transmission of a kuru-like syndrome to chimpanzees. *Nature* 209:794–96

Ghani AC, Ferguson NM, Donnelly CA, Hagenaars TJ, Anderson RM. 1999. Epidemiological determinants of the pattern and magnitude of the vCJD epidemic in Great Britain. *Proc. R. Soc. London Ser. B* 265:2443–52

Gibbs CJ Jr, Gajdusek DC, Asher DM, Alpers MP, Beck E, et al. 1968. Creutzfeldt-Jakob Disease (spongiform encephalopathy): transmission to the chimpanzee. *Science* 161:388–89

Griffith JS. 1967. Self replication and scrapie. *Nature* 215:1043–44

Hadlow WJ. 1959. Scrapie and kuru. *Lancet* 2:289–90

Harris DA, Lele P, Snider WD. 1993. Localization of the mRNA for a chicken prion protein by in situ hybridization. *Proc. Natl. Acad. Sci. USA* 90:4309–13

Herms J, Tings T, Gall S, Madlung A, Giese A, et al. 1999. Evidence of presynaptic location and function of the prion protein. *J. Neurosci.* 19:8866–75

Herms JW, Kretzschmar HA, Titz S, Keller BU. 1995. Patch-clamp analysis of synaptic transmission to cerebellar Purkinje cells of prion protein knockout mice. *Eur. J. Neurosci.* 7:2508–12

Herms JW, Tings T, Dunker S, Kretzsch-

mar HA. 1998. Prion protein in modulates Ca^{2+}-activated K^+-currents in cerebellar Purkinje cells. *Soc. Neurosci. Abstr.* 24:516

Hill A, Antoniou M, Collinge J. 1999a. Protease-resistant prion protein produced in vitro lacks detectable infectivity. *J. Gen. Virol.* 80:11–14

Hill AF, Butterworth RJ, Joiner S, Jackson GS, Rossor MN, et al. 1999b. Investigation of variant Creutzfeldt Jakob disease and other human prion disease with tonsil biopsy samples. *Lancet* 353:183–89

Hill AF, Desbruslais M, Joiner S, Sidle KCL, Gowland I, Collinge J. 1997. The same prion strain causes vCJD and BSE. *Nature* 389:448–50

Hill AF, Joiner S, Linehan J, Desbruslais M, Lantos PL, Collinge J. 2000. Species barrier independent prion replication in apparently resistant species. *Proc. Natl. Acad. Sci. USA* 97:10248–53

Hill AF, Sidle KCL, Joiner S, Keyes P, Martin TC, et al. 1998. Molecular screening of sheep for bovine spongiform encephalopathy. *Neurosci. Lett.* 255:159–62

Hope J, Wood SCER, Birkett CR, Chong A, Bruce ME, et al. 1999. Molecular analysis of ovine prion protein identifies similarities between BSE and an experimental isolate of natural scrapie, CH1641. *J. Gen. Virol.* 80:1–4

Hornemann S, Glockshuber R. 1998. A scrapie-like unfolding intermediate of the prion protein domain PrP(121–231) induced by acidic pH. *Proc. Natl. Acad. Sci. USA* 95:6010–14

Hornshaw MP, McDermott JR, Candy JM. 1995a. Copper binding to the N- terminal tandem repeat regions of mammalian and avian prion protein. *Biochem. Biophys. Res. Commun.* 207:621–29

Hornshaw MP, McDermott JR, Candy JM, Lakey JH. 1995b. Copper binding to the N-terminal tandem repeat region of mammalian and avian prion protein structural studies using synthetic peptides. *Biochem. Biophys. Res. Commun.* 214:993–99

Hosszu LLP, Baxter NJ, Jackson GS, Power A,

Clarke AR, et al. 1999. Structural mobility of the human prion protein probed by backbone hydrogen exchange. *Nat. Struct. Biol.* 6:740–43

Hsiao K, Baker HF, Crow TJ, Poulter M, Owen F, et al. 1989. Linkage of a prion protein missense variant to Gerstmann-Straussler syndrome. *Nature* 338:342–45

Hsiao K, Dlouhy SR, Farlow MR, Cass C, DaCosta M, et al. 1992. Mutant prion proteins in Gerstmann-Sträussler-Sheinker disease with neurofibrillary tangles. *Nat. Genet.* 1:68–71

Hsiao KK, Scott M, Foster D, Groth DF, DeArmond SJ, Prusiner SB. 1990. Spontaneous neurodegeneration in transgenic mice with mutant prion protein. *Science* 250:1587–90

Ingrosso L, Ladogana A, Pocchiari M. 1995. Congo red prolongs the incubation period in scrapie-infected hamsters. *J. Virol.* 69:506–8

Jackson GS, Hosszu LLP, Power A, Hill AF, Kenney J, et al. 1999. Reversible conversion of monomeric human prion protein between native and fibrilogenic conformations. *Science* 283:1935–37

James TL, Liu H, Ulyanov NB, Farr-Jones S, Zhang H, et al. 1997. Solution structure of a 142-residue recombinant prion protein corresponding to the infectious fragment of the scrapie isoform. *Proc. Natl. Acad. Sci. USA* 94:10086–91

Jeffrey M, Wells GA. 1988. Spongiform encephalopathy in a nyala (*Tragelaphus angasi*). *Vet. Pathol.* 25:398–99

Kimberlin RH, Walker CA. 1977. Characteristics of a short incubation model of scrapie in the golden hamster. *J. Gen. Virol.* 34:295–304

Kimberlin RH, Walker CA. 1978. Evidence that the transmission of one source of scrapie agent to hamsters involves separation of agent strains from a mixture. *J. Gen. Virol.* 39:487–96

Kimberlin RH, Walker CA. 1979. Pathogenesis of scrapie: agent multiplication in brain at the first and second passage of hamster scrapie in mice. *J. Gen. Virol.* 42:107–17

Kimberlin RH, Walker CA. 1986. Suppression of scrapie infection in mice by heteropolyanion 23, dextran sulfate, and some other polyanions. *Antimicrob. Agents Chemother.* 30:409–13

Kirkwood JK, Wells GA, Wilesmith JW, Cunningham AA, Jackson SI. 1990. Spongiform encephalopathy in an arabian oryx (*Oryx leucoryx*) and a greater kudu (*Tragelaphus strepsiceros*). *Vet. Rec.* 127:418–20

Klatzo I, Gajdusek DC, Zigas V. 1959. Pathology of kuru. *Lab. Invest.* 8:799–847

Klein M, Frigg R, Raeber A, Flechsig E, Hegyi I, et al. 1998. PrP expression in B-lymphocytes is not required for prion neuroinvasion. *Nat. Med.* 4:1429–33

Klein MA, Frigg R, Flechsig E, Raeber AJ, Kalinke U, et al. 1997. A crucial role for B cells in neuroinvasive scrapie. *Nature* 390:687–90

Klitzman RL, Alpers MP, Gajdusek DC. 1984. The natural incubation period of kuru and the episodes of transmission in three clusters of patients. *Neuroepidemiology* 3:3–20

Kocisko DA, Come JH, Priola SA, Chesebro B, Raymond GJ, et al. 1994. Cell-free formation of protease-resistant prion protein. *Nature* 370:471–74

Kocisko DA, Priola SA, Raymond GJ, Chesebro B, Lansbury PT Jr, Caughey B. 1995. Species specificity in the cell-free conversion of prion protein to protease-resistant forms: a model for the scrapie species barrier. *Proc. Natl. Acad. Sci. USA* 92:3923–27

Kuczius T, Haist I, Groschup MH. 1998. Molecular analysis of bovine spongiform encephalopathy and scrapie strain variation. *J. Infect. Dis.* 178:693–99

Kunz B, Sandmeier E, Christen P. 1999. Neurotoxicity of prion peptide 106–126 not confirmed. *FEBS Lett.* 458:65–68

Kurschner C, Morgan JI. 1995. The cellular prion protein (PrP) selectively binds to Bcl-2 in the yeast two-hybrid system. *Brain Res. Mol. Brain Res.* 30:165–68

Kurschner C, Morgan JI. 1996. Analysis of interaction sites in homo- and heteromeric complexes containing Bcl-2 family members and the cellular prion protein. *Brain Res. Mol. Brain Res.* 37:249–58

Lasmezas CI, Deslys JP, Robain O, Jaegly A, Beringue V, et al. 1997. Transmission of the BSE agent to mice in the absence of detectable abnormal prion protein. *Science* 275:402–5

Laszlo L, Lowe J, Self T, Kenward N, Landon M, et al. 1992. Lysosomes as key organelles in the pathogenesis of prion encephalopathies. *J. Pathol.* 166:333–41

Liemann S, Glockshuber R. 1999. Influence of amino acid substitutions related to inherited human prion diseases on the thermodynamic stability of the cellular prion protein. *Biochemistry* 38:3258–67

Lledo P-M, Tremblay P, DeArmond SJ, Prusiner SB, Nicoll RA. 1996. Mice deficient for prion protein exhibit normal neuronal excitability and synaptic transmission in the hippocampus. *Neurobiology* 93:2403–7

Manson J, West JD, Thomson V, McBride P, Kaufman MH, Hope J. 1992. The prion protein gene: a role in mouse embryogenesis? *Development* 115:117–22

Marsh RF. 1992. Transmissible mink encephalopathy. See Prusiner et al 1992, pp. 300–307

Marsh RF, Bessen RA, Lehmann S, Hartsough GR. 1991. Epidemiological and experimental studies on a new incident of transmissible mink encephalopathy. *J. Gen. Virol.* 72:589–94

Masters CL, Gajdusek DC, Gibbs CJ Jr. 1981. Creutzfeldt-Jakob disease virus isolations from the Gerstmann-Straussler syndrome with an analysis of the various forms of amyloid plaque deposition in the virus-induced spongiform encephalopathies. *Brain* 104:559–88

Mayer RJ, Landon M, Laszlo L, Lennox G, Lowe J. 1992. Protein processing in lysosomes: the new therapeutic target in neurodegenerative disease. *Lancet* 340:156–59

McGowan JP. 1922. Scrapie in sheep. *Scott. J. Agric.* 5:365–75

Medori R, Montagna P, Tritschler HJ, LeBlanc A, Cortelli P, et al. 1992a. Fatal familial insomnia: a second kindred with mutation of prion protein gene at codon 178. *Neurology* 42:669–70

Medori R, Tritschler HJ, LeBlanc A, Villare F, Manetto V, et al. 1992b. Fatal familial insomnia, a prion disease with a mutation at codon 178 of the prion protein gene. *N. Engl. J. Med.* 326:444–49

Miller MW, Wild MA, Williams ES. 1998. Epidemiology of chronic wasting disease in captive Rocky Mountain elk. *J Wildl. Dis.* 34:532–38

Milner J, Medcalf EA. 1991. Cotranslation of activated mutant p53 with wild type drives the wild-type p53 protein into the mutant conformation. *Cell* 65:765–74

Moore RC, Lee IY, Silverman GL, Harrison PM, Strome R, et al. 1999. Ataxia in prion protein (PrP)-deficient mice is associated with upregulation of the novel PrP-like protein Doppel. *J. Mol. Biol.* 292:797–817

Muramoto T, Scott M, Cohen FE, Prusiner SB. 1996. Recombinant scrapie-like prion protein of 106 amino acids is soluble. *Proc. Natl. Acad. Sci. USA* 93:15457–62

Oesch B, Westaway D, Walchli M, McKinley MP, Kent SB, et al. 1985. A cellular gene encodes scrapie PrP 27–30 protein. *Cell* 40:735–46

Owen F, Poulter M, Lofthouse R, Collinge J, Crow TJ, et al. 1989. Insertion in prion protein gene in familial Creutzfeldt-Jakob disease. *Lancet* 1:51–52

Palmer MS, Dryden AJ, Hughes JT, Collinge J. 1991. Homozygous prion protein genotype predisposes to sporadic Creutzfeldt-Jakob disease. *Nature* 352:340–42

Pan K-M, Baldwin MA, Nguyen J, Gasset M, Serban A, et al. 1993. Conversion of α-helices into β-sheets features in the formation of the scrapie prion proteins. *Proc. Natl. Acad. Sci. USA* 90:10962–66

Parchi P, Capellari S, Chen SG, Petersen RB, Gambetti P, et al. 1997. Typing prion isoforms. *Nature* 386:232–33

Parchi P, Castellani R, Capellari S, Ghetti B, Young K, et al. 1996. Molecular basis of phenotypic variability in sporadic Creutzfeldt-Jakob disease. *Ann. Neurol.* 39:669–80

Pattison IH. 1965. Experiments with scrapie with special reference to the nature of the agent and the pathology of the disease. In *Slow, Latent and Temperate Virus Infections, NINDB Monogr.*, ed. DC Gajdusek, CJ Gibbs, MP Alpers, 2:249–57. Washington DC: US Gov. Print. Off.

Pattison IH, Jebbett JN. 1971. Histopathological similarities between scrapie and cuprizone toxicity in mice. *Nature* 230:115

Pauly PC, Harris DA. 1998. Copper stimulates endocytosis of the prion protein. *J. Biol. Chem.* 273:33107–10

Pocchiari M, Schmittinger S, Masullo C. 1987. Amphotericin B delays the incubation period of scrapie in intracerebrally inoculated hamsters. *J. Gen. Virol.* 68:219–23

Prusiner SB. 1982. Novel proteinaceous infectious particles cause scrapie. *Science* 216:136–44

Prusiner SB. 1991. Molecular biology of prion diseases. *Science* 252:1515–22

Prusiner SB, Collinge J, Powell J, Anderton B, ed. 1992. *Prion Diseases in Human and Animals.* London: Ellis Horwood

Prusiner SB, McKinley MP, eds. 1987. *Prions: Novel Infectious Pathogens Causing Scrapie and Creutzfeldt-Jakob Disease.* San Diego: Academic

Prusiner SB, Scott M, Foster D, Pan KM, Groth D, et al. 1990. Transgenetic studies implicate interactions between homologous PrP isoforms in scrapie prion replication. *Cell* 63:673–86

Rieger R, Edenhofer F, Lasmézas CI, Weiss S. 1997. The human 37-kDa laminin receptor precursor interacts with the prion protein in eukaryotic cells. *Nat. Med.* 3:1383–88

Riek R, Hornemann S, Wider G, Billeter M, Glockshuber R, Wuthrich K. 1996. NMR structure of the mouse prion protein domain PrP (121-231). *Nature* 382:180–82

Scott M, Foster D, Mirenda C, Serban D, Coufal F, et al. 1989. Transgenic mice expressing hamster prion protein produce species-specific scrapie infectivity and amyloid plaques. *Cell* 59:847–57

Shaked GM, Fridlander G, Meiner Z, Tarabulos A, Gabizon R. 1999. Protease-resistant and detergent-insoluble prion protein is not necessarily associated with prion infectivity. *J. Biol. Chem.* 274:17981–86

Shmerling D, Hegyi I, Fischer M, Blättler T, Brandner S, et al. 1998. Expression of amino-terminally truncated PrP in the mouse leading to ataxia and specific cerebellar lesions. *Cell* 93:203–14

Shyng S-L, Heuser JE, Harris DA. 1994. A glycolipid-anchored prion protein is endocytosed via clathrin-coated pits. *J. Cell Biol.* 125:1239–50

Shyng S-L, Huber MT, Harris DA. 1993. A prion protein cycles between the cell surface and an endocytic compartment in cultured neuroblastoma cells. *J Biol. Chem.* 268(21):15922–28

Soto C, Kascsack RJ, Saborío GP, Aucouturier P, Wisniewski T, et al. 2000. Reversion of prion protein conformational changes by synthetic beta-sheet breaker peptides. *Lancet* 355:192–97

Spielmeyer W. 1922. Die histopathologische Forschung in der Psychiatrie. *Klin. Wochenschrift* 2:1817–19

Spraker TR, Miller MW, Williams ES, Getzy DM, Adrian WJ, et al. 1997. Spongiform encephalopathy in free-ranging mule deer (*Odocoileus hemionus*), white-tailed deer (*Odocoileus virginianus*) and Rocky Mountain elk (*Cervus elaphus nelson*) in northcentral Colorado. *J Wildl. Dis.* 33:1–6

Stahl N, Baldwin MA, Teplow DB, Hood L, Gibson BW, et al. 1993. Structural studies of the scrapie prion protein using mass spectrometry and amino acid sequencing. *Biochemistry* 32:1991–2002

Stockel J, Safar J, Wallace AC, Cohen FE, Prusiner SB. 1998. Prion protein selectively binds copper(II) ions. *Biochemistry* 37:7185–93

Supattapone S, Bosque P, Muramoto T, Wille H, Aagaard C, et al. 1999. Prion protein of 106 residues creates an artificial transmission barrier for prion replication in transgenic mice. *Cell* 96:869–78

Swietnicki W, Petersen R, Gambetti P, Surewicz WK. 1997. pH-dependent stability and conformation of the recombinant human prion protein PrP(90–231). *J. Biol. Chem.* 272:27517–20

Tagliavini F, McArthur RA, Canciani B, Giaccone G, Porro M, et al. 1997. Effectiveness of anthracycline against experimental prion disease in syrian hamsters. *Science* 276:1119–22

Tagliavini F, Prelli F, Verga L, Giaccone G, Sarma R, et al. 1993. Synthetic peptides homologous to prion protein residues 106–147 form amyloid-like fibrils in vitro. *Proc. Natl. Acad. Sci. USA* 90:9678–82

Taraboulos A, Raeber A, Borchelt DR, Serban D, Prusiner SB. 1992. Synthesis and trafficking of prion proteins in cultured cells. *Mol. Biol. Cell* 3:851–63

Telling GC, Parchi P, DeArmond SJ, Cortelli P, Montagna P, et al. 1996. Evidence for the conformation of the pathologic isoform of the prion protein enciphering and propagating prion diversity. *Science* 274:2079–82

Telling GC, Scott M, Mastrianni J, Gabizon R, Torchia M, et al. 1995. Prion propagation in mice expressing human and chimeric PrP transgenes implicates the interaction of cellular PrP with another protein. *Cell* 83:79–90

Tobler I, Gaus SE, Deboer T, Achermann P, Fischer M, et al. 1996. Altered circadian activity rhythms and sleep in mice devoid of prion protein. *Nature* 380:639–42

Wadsworth JDF, Hill AF, Joiner S, Jackson GS, Clarke AR, Collinge J. 1999. Strain-specific prion-protein conformation determined by metal ions. *Nat. Cell Biol.* 1:55–59

Waggoner DJ, Drisaldi B, Bartnikas TB, Casareno RLB, Prohaska JR, et al. 2000. Brain copper content and cuproenzyme activity do not vary with prion protein expression level. *J. Biol. Chem.* 275:7455–58

Weissmann C. 1991. A "unified theory" of prion propagation. *Nature* 352:679–83

Wells GAH, Hawkins SAC, Green RB, Austin AR, Dexter I, et al. 1998. Preliminary observations on the pathogenesis of experimental bovine spongiform encephalopathy (BSE): an update. *Vet. Rec.* 142:103–6

Wells GAH, Scott AC, Johnson CT, Gunning RF, Hancock RD, et al. 1987. A novel progressive spongiform encephalopathy in cattle. *Vet. Rec.* 31:419–20

Whittington MA, Sidle KCL, Gowland I, Meads J, Hill AF, et al. 1995. Rescue of neurophysiological phenotype seen in PrP null mice by transgene encoding human prion protein. *Nat. Genet.* 9:197–201

Wickner RB. 1997. A new prion controls fungal cell fusion incompatibility. *Proc. Natl. Acad. Sci. USA* 94:10012–14

Wickner RB, Masison DC. 1996. Evidence for two prions in yeast: [URE3] and [PSI]. *Curr. Top. Microbiol. Immunol.* 207:147–60

Wildegger G, Liemann S, Glockshuber R. 1999. Extremely rapid folding of the C-terminal domain of the prion protein without kinetic intermediates. *Nat. Struct. Biol.* 6:550–53

Wilesmith JW, Wells GA, Cranwell MP, Ryan JB. 1988. Bovine spongiform encephalopathy: epidemiological studies. *Vet. Rec.* 123:638–44

Will RG, Ironside JW, Zeidler M, Cousens SN, Estibeiro K, et al. 1996. A new variant of Creutzfeldt-Jakob disease in the UK. *Lancet* 347:921–25

Wille H, Zhang GF, Baldwin MA, Cohen FE, Prusiner SB. 1996. Separation of scrapie prion infectivity from PrP amyloid polymers. *J. Mol. Biol.* 259:608–21

Williams A, Lucassen PJ, Ritchie D, Bruce M. 1997. PrP deposition, microglial activation, and neuronal apoptosis in murine scrapie. *Exp. Neurol.* 144:433–38

Williams ES, Young S. 1980. Chronic wasting disease of captive mule deer: a spongiform encephalopathy. *J. Wildl. Dis.* 16:89–98

Windl O, Dempster M, Estibeiro P, Lathe R. 1995. A candidate marsupial *PrP* gene reveals two domains conserved in mammalian PrP proteins. *Gene* 159:181–86

Windl O, Dempster M, Estibeiro JP, Lathe R, De Silva R, et al. 1996. Genetic basis of Creutzfeldt-Jakob disease in the United Kingdom: a systematic analysis of predisposing mutations and allelic variation in the *PRNP gene. Hum. Genet.* 98:259–64

Wyatt JM, Pearson GR, Smerdon TN, Gruffydd-Jones TJ, Wells GAH, Wilesmith JW. 1991. Naturally occurring scrapie-like spongiform encephalopathy in five domestic cats. *Vet. Rec.* 129:233–36

Zeidler M, Stewart GE, Barraclough CR, Bateman DE, Bates D, et al. 1997. New variant Creutzfeldt-Jakob disease: neurological features and diagnostic tests. *Lancet* 350:903–7

Zobeley E, Flechsig E, Cozzio A, Masato E, Weissmann C. 1999. Infectivity of scrapie prions bound to a stainless steel surface. *Mol. Med.* 5:240–43

Annu. Rev. Neurosci. 2001. 24:551–600

VIKTOR HAMBURGER AND RITA LEVI-MONTALCINI: The Path to the Discovery of Nerve Growth Factor

W. Maxwell Cowan

Howard Hughes Medical Institute, Chevy Chase, Maryland 20815

Key Words   neuronal development, motoneurons, sensory ganglion cells, chick embryos, cell death

■ **Abstract**   The announcement in October 1986 that the Nobel Prize for physiology or medicine was to be awarded to Rita Levi-Montalcini and Stanley Cohen for the discoveries of NGF and EGF, respectively, caused many to wonder why Viktor Hamburger (in whose laboratory the initial work was done) had not been included in the award. Now that the dust has settled, the time seems opportune to reconsider the antecedent studies on the relation of the developing nervous system to the peripheral structures it innervates. The studies undertaken primarily to investigate this issue culminated in the late 1950s in the discovery that certain tissues produce a nerve growth–promoting factor that is essential for the survival and maintenance of spinal (sensory) ganglion cells and sympathetic neurons. In this review, the many contributions that Viktor and Rita made to this problem, both independently and jointly, are reexamined by considering chronologically each of the relevant research publications together with some of the retrospective memoirs they have published in the years since the discovery of NGF was first reported.

> This review is dedicated to Viktor Hamburger on the occasion of his 100th birthday on July 9, 2000, and to Rita Levi-Montalcini to mark her 91st birthday on April 22, 2000, with admiration and affection.

INTRODUCTION

The announcement in October 1986 that the Nobel Prize for physiology or medicine was to be awarded jointly to Rita Levi-Montalcini and Stanley Cohen for their discoveries of nerve growth factor (NGF) and epidermal growth factor (EGF), respectively, was greeted by most developmental biologists with enthusiasm but also with some misgiving. On the one hand there was a general sense of elation that developmental biology had again been recognized in this prestigious and very public way. On the other hand, many developmental neurobiologists felt that by not including Viktor Hamburger, the Nobel committee had failed to appreciate the significance of his earlier contributions that had paved the way to the discovery

0147-006X/01/0621-0551$14.00   **551**

of NGF. As several of those familiar with the NGF saga pointed out, his direct involvement in the discovery appeared to have been overlooked: Not only was the work carried out in his department at Washington University, but it was through his efforts that Rita Levi-Montalcini, and later Stanley Cohen, were brought to his department. Moreover, his name had appeared as a coauthor on the first reports of a diffusible substance that promoted the growth of sensory and sympathetic neurons (subsequently identified as NGF). The comments of Dale Purves and Josh Sanes, two of Viktor's colleagues at Washington University, were typical of those who felt this way. "Many neuroscientists are puzzled by the omission of Viktor Hamburger from the prize," they wrote, "because his exclusion tends to obscure a line of research that now spans more than 50 years" (Purves & Sanes 1987). This feeling extended well beyond St. Louis: As reported in the *New York Times* of October 14, 1986, Dr. Jean Lauder of the University of North Carolina, at the time President of the International Society of Developmental Neuroscience, said that she and others were considering writing to *Science* and *Nature* to express their view that Hamburger should have shared the Nobel Prize.

The fact that prior to the announcement of the award, Viktor (as Hamburger is known to virtually everyone in developmental biology and neuroscience) and Rita (as Levi-Montalcini is generally referred to in the United States) had always been closely associated and mutually supportive made the Nobel committee's decision especially puzzling to their friends and colleagues in this country and abroad.[1] Before 1986, Viktor and Rita had written a number of reviews and personal memoirs in which they had each acknowledged their close scientific association and the importance of their independent and joint contributions to the work that preceded the discovery of NGF. For example, in a volume of essays published on the occasion of Viktor's 80th birthday, Rita wrote at length about her discovery during World War II of Viktor's seminal paper on the effects of early limb removal on the development of the motor columns in chicks (Hamburger 1934) and how this discovery had served as the impetus for much of her own work with her mentor Giuseppe Levi. She went on to describe how excited she had been when Viktor invited her to spend a year in his department and how graciously she had been treated on her arrival in St. Louis. She especially commented on how helpful Viktor had been in assisting her in writing her first papers in English. Touchingly, she ended her essay with the following token: "Viktor, we can look at this work, at our friendship, at the past we so much enjoyed, and at the future, that may or not materialize, in a *sub-specie-aeternitas* frame of mind, in a crystal-clear atmosphere uncorrupted by the turbulence of human passions and sorrow" (Levi-Montalcini 1981). The same good feeling is evident in Rita's dedication of her review (Levi-Montalcini 1982) entitled "Natural History of Nerve Growth Factor" that appeared in the 1982 volume of the *Annual Review of Neuroscience* "to Viktor Hamburger."

---

[1] That Stan Cohen was specifically recognized for the discovery of EGF was universally judged to be appropriate—as was linking his name with Rita's since he had also played a critical role in the isolation of NGF.

Viktor was equally complimentary in his remarks about Rita, acknowledging the combination of neuroanatomical expertise, experimental skill and intuitive insight that she brought to their joint work, and the energy and drive with which she pursued the isolation and testing of NGF.

Unfortunately, the omission of Viktor from the Nobel Prize resulted in a sea change in their relationship. This was marked by the appearance in the popular press and elsewhere of a number of unnecessarily critical and insensitive remarks by both Viktor and Rita. The resulting rift in their once harmonious and mutually supportive relationship has caused some of their colleagues to feel compelled to "take sides." This, in turn, has had the effect of clouding the actual history of the events that led to the discovery of NGF.

It is not my purpose to dwell on the rift between Viktor and Rita that developed following the awarding of the Nobel Prize. Rather, the purpose of this review is first to reexamine Viktor's contribution to the central problem of the relationship between the developing central nervous system and the peripheral structures it innervates, and second to consider the impact that Rita had on this work from the time she joined his laboratory in 1947 and culminating in the discovery of NGF and the elucidation of its role in the development of the sympathetic nervous system and the spinal (sensory) ganglia. To place their separate and joint work in its historic context, I have considered at some length each of the relevant studies and have included brief biographical accounts, which are largely based on their own published memoirs.

VIKTOR HAMBURGER: A Brief Biographical Background

Viktor was born in the small Silesian town of Landeshut on July 9, 1900. After graduating from the Gymnasium, he spent two summers at the University of Heidelberg, where his aunt was a senior assistant in the Zoological Institute. It was here that his interest in developmental biology was stimulated by taking an advanced course in experimental biology taught by Curt Herbst. This, in turn, led him to Freiburg (where, as he has remarked, the skiing and climbing were much better than at Heidelberg) and to the zoology department of the University of Freiburg. Here Hans Spemann, the head of the department, had assembled an outstanding group of students and junior faculty, including Otto Mangold, Hilde Proescholdt (soon to be Hilde Mangold), and Johannes Holtfreter. Not surprisingly, much of the interest in the department at the time was centered on the mechanism of primary induction and the role of the dorsal lip of the blastopore as the "organizer" (Spemann 1938).

Spemann discouraged Viktor from working on this problem, remarking that "there are already too many people hanging from the lip of blastopore" (Cowan 1981). Instead Viktor was assigned to reexamine an earlier report by Dürken to the effect that early eye removal in frog larvae could lead to a variety of limb abnormalities, which were thought to result from a cascading series of effects consequent upon the denervation of the optic tectum (Dürken 1913). Though today this project may seem an unlikely beginning to a career, in the 1920s the possibility that eye removal could affect the development of the limbs seemed no less improbable than that the transplantation of a small piece of tissue from the blastopore could lead to the formation of a second embryo. In the event, Viktor showed that the limb abnormalities Dürken had observed were almost certainly due to the poor conditions under which the larvae were raised (Hamburger 1925). However, this experience later led Viktor to explore more fully the relationship between the developing nervous system and the limbs, and to the study of limb development following removal of portions of the spinal cord.

After completing his PhD in 1925, Viktor spent a year in Göttingen in the laboratory of Professor Alfred Kühn, who worked on pigmentary patterns such as the eye spots in butterflies and moths. Kühn suggested to Viktor that he examine color vision in fish with his senior assistant, Karl Henke This was to be one of Viktor's few departures from the study of limb innervation prior to 1960. It was during this time also that he became reacquainted with Marthe Ficke, whom he had first met in Freiburg and to whom he was married in 1928.

In 1926, Viktor was offered and accepted an assistantship in Otto Mangold's department of experimental embryology at the Kaiser Wilhelm Institute for Biology in Berlin-Dahlem. During his brief stay in Mangold's department, he carried out an extensive series of experiments on the development of the hind limbs in frogs (*Rana temporaria*) from which the relevant segments of the neural tube had been resected on one or both sides. Despite the problem of cord regeneration in the hemiextirpation experiments, and the problem of the swimming impairment in the animals with bilateral lesions, enough cases survived through metamorphosis to allow him to make a detailed analysis of the fate of the skeletal and muscular components of the nerveless limbs. In every case the skeleton, and (initially) the muscles, appeared perfectly normal. In those cases in which the spinal cord had partially regenerated, the peripheral nerves (to the extent that they were formed) seemed to follow the same basic pattern as in control preparations (Hamburger 1928, 1929). As Viktor was to return to this problem later, when he studied aneurogenic limbs in chicks, I defer further comment on these early studies except to note two things. First, the experiments were conducted with the same care and the results analyzed with the same thoroughness that were to be the hallmark of his later work. Second, these studies provided the definitive proof that Dürken's hypothesis—that the development of the limbs was in some way under the direct control of the central nervous system—was wrong.

In 1927, Spemann offered Viktor what was in effect an instructorship in his Institute, and it was with a sense of relief that he returned to Freiburg, even though it lacked the cultural amenities he had enjoyed in Berlin. His principal responsibilities were to teach introductory and advanced laboratory courses in experimental embryology. However, with the rather limited time he had available for research, he continued working on aspects of developmental genetics that he had begun in

Berlin, under the influence of Goldschmidt and his colleagues. It was his hope that it might be possible to apply the approaches of contemporary genetics to some of the problems of amphibian embryology, although he recognized that this was a formidable task: Amphibians were not exactly optimal for genetic studies since no mutants were available and the experimental generation of mutants was virtually impossible. He, therefore, turned his attention to making hybrids between different species of salamander—*Triturus cristatus* and *T. taeniatus*—whose forelimbs and digits grew at different rates. After slogging through several breeding seasons and constructing growth curves of the parental species and reciprocal hybrids, he finally abandoned the project and began the transplantation experiments he had earlier planned to carry out.

The autumn of 1932 marked a major turning point in Viktor's personal life and professional career. For some time he had hoped to visit the United States, and in particular to spend time at the University of Chicago, where Professor Frank R. Lillie, a long-term friend and admirer of Spemann, had established the best known and most successful experimental group studying chick development. Lillie's 1908 book *The Development of the Chick* (Lillie 1908) had set out many of the advantages that chicks offer for developmental studies. For example, he had shown that it was possible to destroy comparatively large portions of early embryos without adversely impacting their overall development, and furthermore that such experiments were not likely to be confounded by either regeneration or regulation. As early as 1909, one of his students, MC Shorey, had published an important paper on the effects of wing bud ablation on the development of the spinal cord and the neighboring sensory ganglia, which is discussed later (Shorey 1909).

In 1932, Viktor was awarded a Rockefeller Fellowship to work in Lillie's laboratory, but by the time he arrived in October, Lillie was no longer active in research, having taken on the responsibilities of Dean of Biological and Medical Sciences. He had been replaced as Professor of Embryology in the zoology department by Dr. Benjamin Willier [with whom Viktor and Paul Weiss were later to edit the influential volume *Analysis of Development* (Willier et al 1955)]. Knowing of Viktor's work on limb development in amphibians, Lillie reminded him of Shorey's study and suggested that Viktor repeat her experiments using the refined microsurgical techniques (using glass needles and hair loops) that Viktor had perfected in Spemann's laboratory, rather than the electrocauterization method Shorey had used. Viktor was of course aware of the experiments of Ross Harrison's student, Samuel Detwiler, who had analyzed the effects of limb removals, limb translocations, and other experimental manipulations on the development of the spinal cord and the sensory ganglia in *Ambystoma* (Detwiler 1920, 1924, 1927, 1936). And he was particularly aware of the discrepancy between Detwiler's findings and those of Shorey with respect to the motor columns of the cord. So repeating Shorey's experiments was not simply a repetitive and possibly futile exercise. Rather it held the prospect of clarifying a significant issue regarding the relationship between peripheral structures and the central nervous system. In a sense this was the reverse of Viktor's earlier studies on the effect of the nervous system on the development of the limbs, a topic that he would again take up once he had transformed himself into a "chick developmental neurobiologist."

With the help of one of Willier's research associates, Dr. Mary Rawles, Viktor soon mastered the techniques of experimenting on chick embryos *in ovo* and before long he had succeeded in selectively ablating wing buds and in transplanting supernumerary wings to the flanks of host embryos. Thus was his career in the United States launched, and his fascination with center-periphery relations was to engage his interest for almost 50 years, as the following account documents. But before he could get fully started, the tranquility of his life was disrupted in April 1933 by a letter he received from the Dean of the Faculty at Freiburg informing him that under the recently promulgated law "for the cleansing of the professions," he had been dismissed from his assistantship in the zoology department. At about the same time he received a letter from Spemann pointing out that as the universities were state controlled, there was nothing he could do to circumvent this decision, and that he should try to find a position in the United States. Fortunately the Rockefeller Foundation responded quickly to the new Nazi policy and created an emergency fund to support displaced German scholars. Through this fund Viktor was assured of a further 2 years of support, which enabled him to complete his initial study in Chicago as an assistant in the zoology department. In 1934, he was offered an assistant professorship in zoology at Washington University, which he took up in the fall of 1935. He was to remain at Washington University for the rest of his career, becoming an associate professor in 1939 and, following the departure of FO Schmitt for MIT, a professor and chairman of the department in 1941. In 1969, Viktor was named Mallinckrodt Distinguished Professor Emeritus. Among his many awards and honors, I need mention only his election to the US National Academy of Sciences in 1953, and his receipt of the Wakeman Award in 1978, the National Medal of Science in 1989, the Ralph W. Girard Prize of the Society for Neuroscience in 1985, and the Karl Lashley Award of the American Philosophical Society in 1990. In 1950 and 1951, he served as President of the Society for Growth and Development (now the Society for Developmental Biology) and in 1955 as President of the American Society of Zoologists.

THE EFFECTS OF WING BUD EXTIRPATION ON THE DEVELOPMENT OF THE CENTRAL NERVOUS SYSTEM IN CHICK EMBRYOS (1934)

Viktor's first published study after moving to the United States (Hamburger 1934) was to set the pattern for much of his future work, and through its impact on Rita Levi-Montalcini, it was to have a profound influence on the future of developmental neurobiology (see Levi-Montalcini 1981). As indicated above, the study was undertaken at Lillie's suggestion, but for the way in which the experiments were carried out and the careful way in which the results were analyzed, one need only look at Viktor's earlier work in Berlin (Hamburger 1925, 1928, 1929).

For this study, the wing buds of embryos, 68–72 hours after the onset of incubation, were removed using fine glass needles. At this stage, the core of the wing bud consists of an undifferentiated mass of mesoderm, and in the brachial segments of the spinal cord the earliest motoneurons are visible. Some motoneurons have sent out axons into what will later be recognizable as the ventral root. As Tello (1922) had pointed out, although some fibers have reached the base of the limb bud at this stage, none has yet entered it. Most of the embryos were allowed to develop for a further 4–6 days, some were fixed after 2–3 days, and two were allowed to survive for 9–10 days after the operation. In the best cases, the limb was completely missing at the time the embryo was fixed for histology; in a few cases a much reduced wing had formed, and in others, the wing bud and also part of the body wall were missing.

The analysis of the resulting changes was limited to the spinal cord and the brachial nerves. In a typical case killed 5 days after the operation, the most striking changes were seen in the anterior horn of the cord, where on the operated side, the large motoneurons in the lateral motor column (LMC) were reduced in number by about 60%. The medial motor column (that innervates the trunk muscles) was unaffected, but the volume and number of cells in the posterior horn were reduced by just over 20%. Only the volumes of the related spinal ganglia were measured: Ganglion 13 was reduced in volume by 18%, ganglion 14 by 47%, ganglion 15 by 39% and ganglion 16 by 35%. Sample cell counts for two ganglia (14 and 15) showed a reduction in number of about 28%. This was less than the observed reduction in volume, which Viktor attributed to the marked loss of neuropil in the ganglia. The ganglia at this age consist of two distinct groups of cells: a population of small, dorsomedially located neurons, and a surrounding population of distinctly larger cells. On the operated side the ratio of small to large cells was 1:1.35; on the control side 1:1.2. The initial paths taken by the peripheral nerves on the operated side were normal, but their distribution beyond the brachial plexus was grossly abnormal.

The changes in the other cases examined in comparable detail were qualitatively similar but differed in the degree of cellular "hypoplasia" seen in the LMC, depending on the amount of muscle tissue in the surviving portion of the operated limb. From reconstructions of the limb musculature, the loss of muscle tissue was found to range between 31% and 96%; the corresponding reduction in cell number in the LMC ranged from 22% to 60%, whereas in the posterior horn it varied from 14% to 21%. The average hypoplasia in the relevant spinal ganglia was more nearly constant, ranging from 37% to 54%. In light of later work it is important to note that it is specifically stated that "no degenerated neurones were found in the area affected." This critical (but, as we shall see later, mistaken) observation had a significant influence on Viktor's interpretation of his experimental findings.

The extensive and very detailed discussion of this 1934 paper is important in several respects, first and most significantly because of its conclusions that:

> The different peripheral structures while growing, are in some direct connection with their appropriate centers in the nervous system. Thus, they are enabled not only to control the growth of their own centers in general but even to regulate this growth in quantitative adaptation to their own progressing increase in size.

He goes on to say that

> [e]very structure within the growing limb, muscle as well as sensory organs, send[s] stimuli to the central nervous system. Each part of the peripheral field controls directly its own nervous center, i.e., the limb muscles affect the lateral motor centers, the sensory fields control the ganglia.

Second, the discussion gives appropriate credit to Shorey's earlier study (1909), in that it explicitly acknowledges that the central finding of a "hypoplasia" of the LMC was clearly anticipated in her work (p. 479). And because it specifically cites two cases in her study in which the spinal ganglia had been inadvertently damaged, she is also credited with having demonstrated that the loss of motoneurons is not attributable to some impairment of sensory input or to a mechanism that involves some form of reflex arc (p. 474). I stress this point because in most later reports, Shorey's work (although often cited) is rarely afforded adequate credit. The fact is she was the first to demonstrate that removal of a limb bud in chicks leads to a marked reduction in the LMC of the related brachial segments of the spinal cord, although as Viktor pointed out, the definitive observations in her paper derived from only three critical cases: "Miss Shorey's results are corroborated [by his findings] in every detail." Furthermore, he noted that she had established "by counting the anterior horn cells and measuring their sizes that the [hypoplasia of the LMC] is due to reduction in cell number and not in size of the single cell." He also agreed with her conclusion that because no degenerating cells were seen, the effect of the wing ablations was "a typical hypoplasia." He adds that "the same conclusion was reached in our studies." In this context it is important to note that Viktor was not uncritical of Shorey's speculations about the possible mechanism underlying the cell loss—including the notion that the products of muscular metabolism somehow filter into the lymph system and are carried to the spinal cord, where they act as a stimulus for motoneuron growth and maintenance.

The discussion section of Viktor's paper is important for a third reason, namely that it includes the first statement of Viktor's own view of the mechanism underlying the hypoplasia of the LMC. This is the notion that the first motoneuron axons to reach the periphery (which he called pathfinders) in some way sense the extent of the field to be innervated and signal this to the spinal cord:

> We must charge the end organs [i.e. the growth cones] of these first pathfinders with the double task of locating the peripheral field, and in some

way 'reporting' back centripetally to the central organ [i.e. the developing motor column or sensory ganglion] the approximate size of the field to be innervated. The fibers would communicate the result of their exploration to their own cell bodies which would thus become the first relay station for the stimulus to be transmitted. . . . By such a kind of mechanism, or by transmission of true nervous excitations or of substances, stimuli must be transferred to the growing nerve centers. These centers on their part would be put in a state of corresponding physiological activity and *that condition would enable them to induce presumptive neuroblasts to join their group* [emphasis added].

Viktor would later allude to this discussion as anticipating the concept of the retrograde transport of growth factors (like NGF), which later became generally accepted. But more correctly it anticipated his subsequent, more fully developed notion that the periphery serves to induce the differentiation of motoneurons from a population of undifferentiated precursor cells (see below).

Lastly I must comment on the thoroughness and fairness of the Discussion. He mentions essentially every previous study on the effect of peripheral (and, in a few cases, intracentral) ablations on the development of the spinal cord and the sensory ganglia, and in several instances he comments at length on the critical findings. Moreover, he does not limit his discussion to similar work in birds but includes a discussion of many studies in anurans, urodeles, mammals, and even human cases with developmental limb anomalies. It is an unfortunate commentary on contemporary publications that such thorough reviews of the relevant literature appear to be anathema to modern editors.

THE EFFECTS OF LIMB TRANSPLANTATION ON THE SPINAL CORD AND SENSORY GANGLIA (1939)

Following the publication of his 1934 paper on wing bud ablations, Viktor returned to his earlier interest in limb development and published two interesting papers on this topic. The first (Hamburger 1938) concerned the morphological and axial differentiation of transplanted limbs—a topic that had been considered by his second embryological hero, Ross Harrison, some years earlier (Harrison 1921). The second (Hamburger 1939a) followed his own earlier work on the pattern of peripheral innervation of transplanted limbs in amphibia. I do not consider these studies here (important as they are for a more general account of Viktor's scientific contributions), but rather proceed to a consideration of another study he published in 1939, in which he examined the effects of transplanted limbs on the development of the spinal cord and sensory ganglia (Hamburger 1939b).

This study, which was the logical extension of his 1934 paper on wing extirpations, took advantage of the extensive number of wing and hind-limb bud transplantations that he had prepared earlier for the studies on limb axis determination and the patterns of peripheral limb innervation. The supernumerary limb buds were transplanted into host embryos 60–70 h after incubation by making a small slit in the flank, near the somites and immediately adjacent to the host's fore– or hind–limb. The LMC of the spinal cord and the sensory ganglia were examined 8 or 9 days after the operation. In all the successful cases the transplanted limbs were innervated by nerves from the adjoining spinal segments that often formed plexuses comparable to those innervating normal limbs, but the numbers of nerve fibers entering the transplants were always appreciably smaller.

The enlargement of the sensory ganglia contributing to the transplants was often conspicuous, but in most cases comparatively few fibers entered the limbs, and the brachial and lumbosacral enlargements on the operated and control sides of the spinal cord were not noticeably different, even when the transplant was located very close to the host limb. When counts were made of the numbers of motoneurons in the two sides, the differences were striking in only a few cases. Thus, in the brachial enlargement, the numbers on the two sides differed only by 1.5%, 5.5%, and 8% in the three cases analyzed (the experimental sides always being the larger). In the lumbosacral region the differences ranged from −1.7% to 16.5% in the eight cases that were analyzed quantitatively. However, when counts were made over the length of every segment that contributed fibers to the normal and transplanted limbs, it was evident that the numbers of motoneurons in the segments related to the transplanted limbs were increased. For example, in one case of a wing bud transplant, the numbers of motoneurons in segments 15 and 16 exceeded those on the control side by 13% and 26.5%, respectively. And in the most successful case where a hind-limb bud had been transplanted just rostral to the host limb, the numbers of motoneurons in segments 23 and 24 were increased by 88% and 22%, respectively, compared with the control side. It is interesting that although the spinal ganglia were clearly enlarged when the transplants were in the trunk region, there was no detectable hyperplasia in the motor columns over the corresponding thoracic segments.

No cell counts were done to determine the degree of hyperplasia in the "over-loaded" sensory ganglia, but an estimate of their respective volumes was made using a technique introduced for this purpose by Detwiler. This involved tracing the outline of sections through the ganglia onto paper, cutting out the tracings for each ganglion on the experimental and control sides, and then weighing them. This provided a reasonable measure of the relative size of the enlarged ganglia on the side of the transplant, compared with those on the control side. Although the degree of enlargement of the ganglia innervating the transplants varied considerably from case to case (the range was 15% to over 200%), in essentially every case the affected ganglia were increased in volume, including those that innervated the transplants inserted into the trunk region.

There had been a number of previous studies of this kind in anurans and urodeles, all of which had demonstrated an enlargement of the overloaded sensory ganglia. Detwiler, whose work on *Ambystoma* was perhaps best known, had claimed that there was no comparable hyperplasia in the spinal cord, including the LMC of the anterior horn. However, Dürken (1911) and May (1930, 1933) (both of whom had used frogs) had observed a degree of enlargement of the anterior horn. Viktor's work was distinguished from these earlier studies by his careful attempts to quantify the effects of peripheral overloading, and in this he was greatly aided by the fact that the LMC are much better defined in chicks than in amphibians. His observation that the limb transplants did not result in a generalized enlargement of the LMC, but rather in a focal increase in cell numbers in those segments that contributed to the innervation of the transplant, was the first clear indication that the relationship between the limb musculature and the growth of the motor columns was highly specific. In keeping with his earlier work, he concluded that the simplest explanation for the findings of a motor hyperplasia was "to assume a growth-controlling agent travelling in centripetal direction from the periphery along the first motor fibers to the growing motor centers" (Hamburger 1939b:281).

THE EFFECTS OF PERIPHERAL FACTORS ON PROLIFERATION AND DIFFERENTIATION IN THE CHICK SPINAL CORD (1944)

Viktor's next contribution to the center/periphery relationship appeared in 1944, in a study with his student Eugene L. Keefe of the numbers of mitoses and estimates of overall cell numbers, in the brachial region of the cord in experimental and control embryos (Hamburger & Keefe 1944). In his 1934 paper, Viktor had concluded (in the absence of observable cell death in the LMC) that the hypoplasia that occurs after early limb ablation could be attributed to (*a*) an effect on the proliferation of motoneuron progenitors, (*b*) the failure of the young neurons to migrate into the motor column, or (*c*) the failure of the first motor cells whose axons reach the periphery to induce the differentiation of other motoneurons from a preexisting pool of undifferentiated cells. A test of the first of these possibilities, he argued, would be to determine if wing bud removals resulted in an observable reduction in the numbers of mitotic figures in the neuroepithelium lining the central canal of the cord at brachial levels (Hamburger 1934).

To establish the time course of proliferation in the brachial cord, a number of control preparations were made and an enormous number of mitoses counted. But for our purposes, the key observation was that none of the animals in which one wingbud had been removed showed a consistent difference in the numbers of mitoses on the operated and control sides. In the aggregate the numbers were 20381 and 20950, respectively. In 9 of the 11 animals examined on days 5 and 6 (when cell proliferation in the cord is at its peak), the numbers of mitoses on the operated side were lower than on the control side, but the differences were fairly slight, amounting at most to ~10%. Since comparable differences were observed on the two sides of the animals in the control group, the significance of this finding is difficult to assess. In light of this finding, a series of estimates was made of the total numbers of cells on the control and operated sides in a number of cases (some of which had been used in his 1934 and 1939 studies, and two of which had been prepared by Bueker for his 1943 study—see below) in which there was a marked loss of neurons in the LMC. Similar counts were done at the level of nerve 16 in one of the chicks used in Viktor's 1939 study that had shown an appreciable hyperplasia at this level, following the transplantation of a supernumerary wing. To reduce the amount of cell counting required, the estimates of cell number were limited to the ventral halves of the spinal cord.

Again, the results seemed clear. The total numbers of cells—including motoneurons and nonmotor cells—on the two sides were essentially the same in the animals in which there was a significant reduction in the motor cell column and in those in which the LMC was said to be "hyperplastic." To cite just one example, in an animal in which there were 1388 fewer motoneurons on the operated side (5 days after the removal of a wing) compared with the control side, this loss was almost exactly matched by an excess of 1403 nonmotor cells on the affected side.

Viktor was to return to the patterning of cell proliferation in the chick cord in 1948, but based on the findings of this study, he reaffirmed his earlier position that the hypoplasia seen in the motor columns was not due to an effect on cell proliferation but to a reduction in the inductive influence of the periphery on an undifferentiated pool of cells in the cord. And conversely, that the "hyperplasia" seen after supernumerary limb transplantation was due to a sensing of the expanded periphery by the axons of the first differentiated motoneurons. As he expressed it:

> We arrive then at the concept of a "histogenetic gradient field" which has its center in the small cluster of pioneer motor neurons and which spreads over adjacent undifferentiated cells, inducing them to differentiate. The newly recruited neurons are added to the lateral motor column and increase the strength of the field, that is the radius of its expansion. This process of augmentation is not a self-perpetuating mechanism, however, but under the "remote control" of conditions prevailing at the periphery.

While not entirely embracing the most elaborate hypothesis put forward in the previous year by Barron (1943), based on studies of motoneuron differentiation in sheep, Viktor was obviously much influenced in his thinking by Barron's ideas and especially by his suggestion that motoneuron dendrites play a role in the induction and recruitment of other cells to the motor column (Barron 1943). Summarizing his own views, based on his earlier studies and his findings with Keefe, Viktor stated:

> The entire mechanism of peripheral control has clearly three different components which we are trying to analyze separately: The setting up of a stimulus by the peripheral fields to be innervated; the changes occurring in the primary neurons as a result of this stimulus; and the inductive effect of the primary neurons on indifferent cells. Of the many detailed questions that remain unsolved, that of the primary stimulus is one of the most puzzling.

FURTHER ANALYSES OF MITOTIC PATTERNS IN THE CHICK SPINAL CORD (1948)

Viktor followed his study with Keefe with a long and detailed analysis of the patterns of distribution of mitotic figures in the chick spinal cord over much of its rostro-caudal length and at each of several stages from day 3 through the day 9 of incubation. In conception this study followed an earlier but essentially similar study by Coghill (1933) on cell proliferation in the spinal cord of *Ambystoma*, which had been done to determine whether there were dorso-ventral differences in proliferation over the extent of the hind-limb innervating segments that could be related to the individuation of specific reflexes. Since Viktor's (Hamburger 1948) study of cell proliferation in the chick cord does not bear significantly on the center-periphery issue, I need mention only a few of the more important findings, although this hardly does justice to the enormous amount of effort that went into the study.

The most relevant findings from the present perspective are the following: Although no clear rostro-caudal pattern in the numbers of mitoses was detectable (and, most interesting, the numbers in the brachial segments were not greatly different from those in the pre- and post-brachial regions), there was a very striking ventral-to-dorsal gradient. Thus, whereas the numbers of mitoses in the basal plate reach their peak on day 3, the peak period of proliferation in the alar plate occurs on day 6. In terms of the aggregate numbers of mitotic figures observed, the numbers in the alar plate exceed those in the basal plate by more than a factor of two. Some of the developmental factors that might contribute to the different proliferative patterns in the basal and alar plates are discussed at some length. However, the only points of interest in the present context are the references to Viktor's earlier experiments, in which the brachial segments had been isolated from the pre- and post-brachial regions by tantalum foil inserts, and to Visintini & Levi-Montalcini's (1939b) experiments, in which all long descending pathways to the limb segments had been interrupted. In neither instance were changes observed in the proliferative patterns in the cord. Of greater interest is the reference to a paper by Levi-Montalcini and Hamburger (at the time unpublished) with the following statement: "On the other hand, the same operation (i.e. early wing bud ablation) has a marked and permanent effect on the mitotic activity in the spinal ganglia."

RITA LEVI-MONTALCINI AND THE EVENTS THAT LED TO HER JOINING VIKTOR'S LABORATORY IN 1947

In October 1947, Rita Levi-Montalcini joined Viktor's laboratory from her native Italy. She had come at Viktor's invitation and expected to stay for just a few months or a year at most. In the event, she remained associated with Washington University for 30 years, and no one could have predicted just how momentous her arrival was to be for the future of developmental neurobiology.

Before considering their first joint research endeavor, I should say something about Rita, about her remarkable life in fascist Italy in the years leading up to and during World War II, and about her early ventures into neuroembryology. Rita herself has poignantly described this phase in her life in her autobiography *In Praise of Imperfection* (Levi-Montalcini 1988) and in a number of shorter articles (Levi-Montalcini 1975, 1981, 1982), so this account can be brief.

Born into an intellectual Jewish family in Turin in 1909, Rita, like most Italian women of her generation, had virtually no exposure to science until she and her cousin Eugenia decided that they wanted to study medicine. With the help of a private tutor, who put them through a crash course in mathematics and science, she gained admission to the University of Turin's medical school in 1930. Here she had the good fortune to get to know two other medical students, Salvador Luria and Renato Dulbecco (both future Nobel laureates) and, most important, to fall under the influence of the leading Italian neurohistologist of his generation, Professor Giuseppe Levi. Despite his notoriously ferocious manner, Rita found in Levi a brilliant and challenging mentor who shared with her the humiliation and personal abuse heaped upon Jewish academics by Mussolini's black-shirted followers. After graduating from medical school in 1936, Rita stayed on for further training in neurology and psychiatry until 1938, when Mussolini issued his *Manifesto for the Defense of the Race*, which prohibited Jews from studying and teaching in state schools and universities. This caused her to leave Italy in 1939, and for a short period she worked in a research institute in Brussels. Shortly after the outbreak of World War II she returned to Turin. As she tells it, she

> first met Viktor in a cattle car in northern Italy . . . on a day in . . . that fateful June of 1940 when Mussolini declared war on France . . . I was sitting on the floor of one of these railway cars . . . reading a reprint lent to me by Giuseppe Levi on the effects of wing bud extirpation on the development of the central nervous system of chick embryos. The article was dated 1934, and as Levi had informed me, it had been written by a pupil of Hans Spemann.
>
> (Levi-Montalcini 1981)

Determined to repeat this experiment, Rita set up a simple laboratory in her bedroom, obtained fertile eggs from a local farmer, kept them in a make-shift incubator, forged her own microsurgical instruments, and on completing the experiments removed and fixed the embryos for histological study, and then proceeded to eat the rest of the eggs! With Levi's help and encouragement, she did more than just repeat Hamburger's 1934 experiments; she extended them in three significant ways. First, she analyzed the effects of the limb bud removals (she chose to ablate the hind-limb bud rather than the wing buds), focusing as much on the spinal (sensory) ganglia as on the motor columns of the spinal cord. Second, she examined embryos over a wider range of survival periods (from days 4 through 20 of incubation). Third, in addition to conventional Nissl staining for cell bodies, she prepared many of the animals for staining by De Castro's modification of Cajal's silver method, which she had learned from Levi. Facing continuous risk of discovery, and working under the most trying circumstances, she managed over the course of the next 2 years to complete this and another series of experiments, as well as to get her work published in the *Belgian Archives of Biology* (Levi-Montalcini & Levi 1942, 1943).

Her experiments on limb bud removals confirmed Viktor's observation that depriving the relevant spinal cord segments of their peripheral field results in a marked reduction in the lateral motor cell column. But more important, she observed that what Viktor had interpreted as a hypoplasia—and attributed to an inductive failure in the recruitment of motor cells—was in fact the direct consequence of the death of previously differentiated motoneurons. Even more striking was the observable degeneration of cells in the spinal ganglia, where cell counts showed that by day 12 of incubation and later, as many as 60%–70% of the neurons in ganglion 25 were lost on the operated side. And since in the silver-stained preparations it was possible to distinguish between fully differentiated neurons and undifferentiated cells, it was clear that the principal effect of early limb ablations was the degeneration of differentiated neurons. Rita's interpretation of her findings (no doubt aided by Levi's considerable experience) was thus quite different from Viktor's. The effect of removing the peripheral innervation fields of both motoneurons and sensory ganglion cells, she concluded, was not on their proliferation or differentiation but rather on their survival. This little-known study was to have a major influence on all future studies of center/periphery relationships, but it was not until after the war, when Viktor came across it, that its importance was recognized. His response on reading the paper was characteristic. As he later wrote:

> Of course, I accepted her version, but I felt that the analysis of the effect of limb extirpation could be carried further . . . I wrote to Dr. Levi and asked whether Dr. Levi-Montalcini would be interested in working in my laboratory for a year. She consented and arrived in St. Louis in the fall of 1947.

> (Hamburger 1996)

THE DEVELOPMENT OF SPINAL GANGLIA UNDER NORMAL AND EXPERIMENTAL CONDITIONS (1949)

Rita's arrival in St. Louis was a major landmark in the history of developmental neurobiology at Washington University. Although her earliest publications that appeared during the war were not widely known outside of Italy, it is clear on reading them that she had become an excellent neuroanatomist, had mastered the often capricious silver methods, especially De Castro's en-bloc staining technique, and had a sure grasp of most of the important issues in neuroembryology. In addition to her seminal paper on the effects of limb bud extirpation, she had published an excellent account of the early development of the accessory abducens nucleus in the chick (Levi-Montalcini & Levi 1942). And before that, she had published two papers with a neurophysiologist in Turin's Clinic for Nervous and Mental Diseases, Dr. Fabio Visintini (Visintini & Levi-Montalcini 1939a,b). The first of these papers describes various aspects of the early morphology of the chick spinal cord and lower brainstem. The second illustrates the normal development of the cochlear and vestibular nuclei and the effects of surgical and cauterizing lesions at rostral brainstem and diencephalic levels on the development of the motor columns of the cord. She extended the analysis of the cochleo-vestibular complex in a paper published from St. Louis, but these studies with Visintini are noteworthy for their extensive use of silver-stained preparations and, in the case of the second paper, for its inclusion of a lengthy section describing various physiological and behavioral observations on the experimentally manipulated chicks (Visintini & Levi-Montalcini 1939a,b).

By both background and inclination, Viktor was at heart an experimental embryologist in the tradition of Roux, Spemann, and Harrison, so the appearance in his laboratory of an experienced neurologist and neuroanatomist opened up new possibilities for work on the central and peripheral nervous systems. Commenting on this, Hamburger once remarked:

> [Rita and I] came from entirely different backgrounds. I came from experimental and analytical embryology, of which Rita hadn't the foggiest idea. . . . Rita was a neurologist from medical school and knew the nervous system, of which I had only the foggiest idea. And she brought to St. Louis a most important tool, the silver staining method. [see McGrayne 1996].

Viktor was especially interested in reexamining with Rita the effects of limb ablations, and this formed the basis of their first joint study (Hamburger & Levi-Montalcini 1949). In retrospect it is not clear why they chose to focus on the sensory ganglia rather than on the motor columns of the cord since it was Viktor's observations on the motor system and his interpretation of the observed changes in the motor columns that had been called into question by Rita's study with Levi. Writing about their early interactions, Viktor recently stated:

> [We] agreed to repeat the limb bud extirpation experiment once more, and on the first step, to pay special attention to the finest details in the response of the sensory ganglia. *Fortunately we chose her preference; if my preference of the motor columns, which are more homogeneous then the ganglia, had prevailed, NGF would not have been discovered in my laboratory.* (Hamburger 1996, emphasis added)

As we have not found any comparable (or contradictory) statement from Rita, we may assume that Viktor's recollection of what happened is correct; it is certainly consonant with the fact that the spinal ganglia featured more prominently in her paper with Levi than did the motor columns, and it is perhaps for this reason also that Viktor later failed to recall Rita's having ever worked on the motor system (V Hamburger, personal communication).

Their 1949 paper on the development of the spinal ganglia is unquestionably one of the most important in all of neuroembryology. Its style suggests it was written by Viktor, but by his own admission, the experimental work and the initial analysis of the data were done by Rita. Again to cite Viktor's autobiographical memoir:

> The experiments and observations on the slides were done by
> Dr. Levi-Montalcini. [But he adds:] I followed her work and discoveries
> with intense interest, and we were in close communication all the time.

And as Rita looking back on this time wrote:

> What I liked most was the clarity of [Viktor's] thinking and his superb
> control of the English language. Writing a scientific paper was a new
> experience for me, and I concentrated on the effort of learning how to do it.

The paper itself included a detailed morphological description of the appearance of the lumbo-sacral sensory ganglia from days 3 through 20 of incubation, a careful analysis of a large number of cases in which either a wing or hind-limb bud had been extirpated at $2\frac{1}{2}$–3 days of incubation, and a further group of 25 cases of wing or leg transplantations that had been carried out at the same developmental stage and allowed to survive generally to days 5 or 8 of incubation, but in some cases as late as days 9–17. In addition to staining with Heidenhain's hematoxylin, many of the preparations were stained according to a variant of De Castro's method.

The account of the normal development of the sensory ganglia followed closely that given in the paper by Levi-Montalcini & Levi (1943) in which they had recognized three developmental phases, the first beginning with the migration of the neural crest precursors (on day 2) through day 8 of incubation. This phase is marked mainly by the proliferation of ganglion cell precursors, the differentiation of the large sensory neurons in the ventrolateral part of the ganglia, and the appearance of substantial neuronal degeneration in the non–limb-related ganglia. By day 5, the central and peripheral processes of the originally bipolar neurons reach the spinal cord and dermis, respectively. The second phase continues through day 12 and is mainly distinguished by the appearance of the smaller dorsomedially located ganglion cells. The third phase extends beyond day 15, by which time it is difficult to distinguish the two populations of neurons. The first reflex responses to peripheral stimulation were observed on day 11, when the smaller cells were clearly differentiated and Tello (1922) had observed the innervation of muscle spindles.

Differences in the numbers of mitoses on the two sides could be seen in the experimental material as early as day 4 and were striking by day 5. In the case of the wing bud extirpations, the differences in ganglia 14–16 ranged from 11.8% to 37%; in the case of the supernumerary wing transplantations, the changes were less marked, ranging from +2.0% to +26.0% in ganglia 16–18. In some individual cases the observed differences were not statistically significant, but when all the experimental data were pooled the findings were clear, leading them to conclude that

> the peripheral field controls the mitotic activity of the spinal ganglia; its reduction decreases the numbers of mitotic figures in ganglia which participate in its innervation, and its enlargement increases it.
>
> (Hamburger & Levi-Montalcini 1949:474)

In the cases of the wing transplants, these findings were indirectly confirmed by the observation that all the ganglia that contributed fibers to the transplants showed a distinct numerical hyperplasia on days 9–9.5, shortly after proliferation had ceased.

Their findings on the influence of the periphery on cell proliferation in the ganglia were in general agreement with the earlier studies of Detwiler (1920, 1936) and Carpenter (1933) in urodeles. But whereas the earlier studies had been based on counts of the numbers of surviving neurons in the ganglia, the influence of enlarging or reducing the periphery on cell proliferation was inferred rather than directly measured. The study by Hamburger & Levi-Montalcini (1949) provided the first convincing evidence, based on mitotic counts, for a direct effect of the periphery on cell proliferation in the sensory ganglia.

Since the effects of peripheral manipulations on cell proliferation were found to occur before the limbs are innervated, it was difficult to account for them on the basis of the mechanism that Hamburger (1934) and Hamburger & Keefe (1944) had suggested for the changes seen in the motor cell columns. Although this issue was not discussed at length, it was proposed that it might be due to some kind of "field effect," but the nature of the "field" and how it might be influenced by the periphery was left open (and presumably to the reader's imagination).

By contrast, the degenerative changes seen in the ganglia were considered at length and clearly represent the principal focus of the study. The key observations were twofold. First, relatively few degenerating neurons could be observed in the limb-related ganglia during normal development, whereas they were abundant in the upper cervical and thoracic ganglia. And second, following wing or hind-limb bud extirpations, large numbers of degenerating cells could be seen in the brachial and lumbo-sacral ganglia, where they were found to involve almost exclusively the large, differentiated neurons in ventrolateral parts of the ganglia. A particularly important observation was that the experimentally induced degeneration in the limb-related ganglia occurs at the same stages in which degenerating neurons were observed in the normal cervical and thoracic ganglia (days 5–7), as evidenced in both hematoxylin-stained preparations and ganglia treated with supravital trypan blue (which is especially useful for revealing degenerating cells). In addition to the marked degeneration affecting the large ventrolateral ganglion cells, the smaller, dorsomedially located cells were also affected. However, in the latter the effect of limb extirpation took the form of a slow progressive atrophy rather than an acute, sharply defined phase of cell death.

While there is much else of interest in this paper (including a lengthy discussion of the significance of the terms hyper- and hypoplasia, which were widely, and often confusingly, used in the earlier embryological literature on this topic), the major findings, and those that were to prove to be of greatest importance for all later work on center/periphery relations, were the discovery of naturally occurring degeneration in the non–limb-related ganglia and the substantial degeneration of ganglion cells following limb bud extirpations. To account for the rapid degeneration of the large ventrolateral cells, the authors proposed two possible mechanisms: a breakdown in some essential metabolic or growth-related process [they mention as a possibility "axon flow," a process that had recently been described by Weiss & Hiscoe (1948)]; or alternatively, a disturbance in "a metabolic exchange between the growing neurite and the substrate on which it grows." In light of their future work, it is interesting to recall that they elaborated on this second possibility by stating that

> [s]ubstances necessary for neurite and neuroblast growth and maintenance would not be provided in adequate quantities when the limb bud is removed.

Later they added the following:

> Not the functional but the physical or chemical conditions at the periphery are ultimately responsible for the 'peripheral' effects on the development of nerve centers.

What is particularly significant about this study is not so much its recognition of neuronal degeneration as an important factor in neurogenesis, since this had been previously reported by Levi-Montalcini & Levi (1942, 1944). It had also been observed in the chick ciliary ganglion after early eye removal and in the trochlear nucleus [in a then unpublished study by Dunnebacke, one of Viktor's students (see Dunnebacke 1953)]. Rather, its significance lies in the extraordinary care and clarity with which the experimental findings were described and in the thoroughness with which they were discussed and evaluated in relation to virtually all the relevant literature. In these regards the paper was to serve as a model for all later studies on this topic. It also laid a sound foundation for much of the work that Viktor and Rita subsequently carried out. But before considering their later work, it is necessary to digress briefly to consider a surprising series of experiments by another of Viktor's former students.

ELMER BUEKER AND THE USE OF MOUSE TUMORS TO PROBE THE DEVELOPMENT OF THE NERVOUS SYSTEM (1943, 1945, 1948)

While working in Viktor's laboratory, Elmer Bueker had examined the effects of radical limb ablations and of supernumerary limb transplantations at different positions along the body wall. And, in a further series of experiments, he had grafted lengths of spinal cord about five segments long into the region adjoining the spinal cord, of host embryos (Bueker 1943, 1945). The radical limb extirpations served to establish that virtually no neurons could be found in the related LMC by day 9 of incubation, but apart from this they added little to the findings of Hamburger and Keefe. Perhaps not surprisingly, his interpretation of these findings closely followed Viktor's. The experiments with transplanted cord segments were more difficult to interpret, in part because of the inevitable distortion of the tissue, the interference with the normal pattern of innervation from the host spinal cord, and the variable outgrowth of axons from the spinal cord graft into the host limb. However, these experiments are of historical interest because they suggested to Bueker an alternative approach to the problem of center/periphery relations that was to have lasting consequences. The reasoning behind Bueker's next series of experiments followed directly from his limb transplantation experiments: If providing a second, expanded area for innervation by the LMC increased the number of motoneurons found in the LMC, it might be possible, he reasoned, to achieve a similar effect by substituting some other rapidly growing tissue for the growing limb (Bueker 1948). The tissues he chose to explore for this purpose were a mouse mammary adenocarcinoma, a mouse sarcoma (referred to as sarcoma 180), and the Rous fowl sarcoma. All were known from previous studies to grow rapidly in chicks, and to maintain their histogenetic characteristics.

It is not clear what led Bueker to pursue this course, but the generally held view that this idea was suggested to him by Viktor is probably without foundation. By the time Bueker conducted these experiments, he had been away from St. Louis for some years and had held positions at the Medical College of South Carolina and at Georgetown Medical School. And as we shall see below, Viktor seems to have been unaware of the experiments until the work was published.

Most of Bueker's experiments involved removing the hind-limb buds of chicks or making a slit in the somatopleure lateral to somites 24–30, on day 3 of incubation, and then transplanting a small piece of the tumor into the exposed region; the tumors were allowed to grow in situ for periods ranging from 1 to 5 or 6 days. Most of the chicks bearing the Rous sarcoma died from extensive hemorrhages before reaching day 8, and most of the mammary adenocarcinomas failed to grow and were resorbed by day 8. In the few cases that survived to day 9, there was no evidence that nerve fibers had innervated the tumors, but in every case the lateral motor columns of the cord and the related spinal ganglia were markedly reduced in size. The findings in the animals bearing the sarcoma 180 transplants, however, were strikingly different. In nearly every case, some nerve bundles could be seen growing into the tumor mass (which had infiltrated the surrounding region), and as judged by the weight of cut-out tracings, the volumes of the spinal ganglia were increased by about 33%. Conversely, the numbers of cells in the related LMC were reduced, on average, by 35%. The findings that the spinal ganglia were increased in size in the animals with sarcoma 180 transplants (due, it was thought, to both a hypertrophy of individual ganglion cells and an actual increase in their number) was of particular interest. It suggested that the tumor could provide an effective growth-supporting periphery, at least for sensory neurons. Conversely, the loss of motor cells in the cord was attributable to the extirpation of the developing limb and the inability of the growing sarcoma to provide "an alternative periphery" for motoneurons.

In retrospect it seems that Bueker did not fully appreciate the significance of the finding that sarcoma 180 was capable of selectively promoting the growth of sensory ganglion cells. At least there is no indication that he intended to follow up this study. Indeed, he published nothing further on this topic for several years, and nothing further might have come of the finding had Viktor not brought Bueker's paper to Rita's attention some months after its appearance.

FUTHER EVIDENCE FOR THE IMPORTANCE OF NEURONAL DEGENERATION DURING DEVELOPMENT (1949, 1950)

While carrying out the study of the effects of limb extirpation on the development of the spinal ganglia, Rita continued to work on the acoustic and vestibular centers of the brainstem that had interested her during her association with Visintini, and in 1949 she published a detailed analysis of the effects of early extirpation of the otocyst (Levi-Montalcini 1949). Although this study is somewhat tangential to the main thrust of the present review, it is worth recalling because it remains one of the best-documented analyses of the effects of depriving developing neurons of their afferent input. In brief, Rita was able to show that whereas the initial development of the acoustic centers is unaffected by the removal of the otocyst, in the period following the normal arrival of their afferent fibers, one of the deprived acoustic nuclei (the *nucleus angularis*) undergoes a very profound cell loss and a marked atrophy of the remaining neurons. A second center (the *nucleus magnocellularis*), while obviously affected, shows an appreciably less severe hypoplasia; the third center, the *nucleus laminaris* (which does not receive a direct input from the cochlear nerve but is in receipt of fibers from the nuclei angularis and magnocellularis), is not affected by removal of the otocyst (at least not until day 17 of incubation, which is as far as the study was continued).

We need not discuss this paper further except to note two things. First, we still do not know what factors afferent fibers provide for the trophic maintenance of their target neurons. And, second, Rita thanks Viktor "for his constructive criticism and help in editing the manuscript." Of more immediate relevance is another paper that Rita published the following year (Levi-Montalcini 1950), which provided the most striking evidence that substantial death occurs during the normal development of the spinal cord, as it does in the spinal ganglia.

This study involved a reexamination of the development of the nucleus of the origin of the preganglionic sympathetic fibers, usually referred to by chick embryologists as the *column of Terni*. Terni (1924) had claimed that from their first appearance the preganglionic sympathetic neurons occupied their position just lateral to the central canal of the cord, roughly midway between the dorsal and ventral horns. Rita, on the other hand, by following the development of the neurons in a closely spaced series of embryos stained by the De Castro method, had concluded that the cells reach their definitive location by secondarily migrating in a dorsomedial direction from the region of the lateral motor column. There can be no question about the correctness of her conclusion. But what is of particular interest in the present context is that in studying the origin of Terni's column, Rita was obliged to reexamine the development of the entire rostro-caudal extent of the spinal cord. This brought to light the unexpected finding that substantial neuronal degeneration occurs at certain well-defined levels of the cord during normal development. As her findings made clear, at early stages there is a distinct visceral system (comparable to the nucleus of Terni and the associated rami communicantes) in the cervical cord. But between 4.5 and 5 days of incubation, the cervical visceral system undergoes complete disintegration such that "during this period the number of degenerating cells is so large as to obscure the presence of intact cells" (Levi-Montalcini 1950:266).

No secondary cell migrations nor degeneration was seen in the brachial or lumbosacral segments, but there was evidence for the appearance of a small sacral preganglionic (parasympathetic) column that developed in much the same way as the nucleus of Terni. That the mechanism underlying this developmental pattern is intrinsic to the spinal cord was evident from a single experiment in which the thoraco-lumbar region of stage 25 embryo had been transplanted between the brachial cord and the wing bud of a host embryo; in this case the segregation and migration of the visceral outflow followed the same pattern as in normal embryos.

In addition to clarifying the origin of the preganglionic sympathetic and the sacral parasympathetic outflow from a common viscero-somatic motor column, this largely neglected paper was important in providing the first clear evidence that large-scale neuronal degeneration occurs during the normal development of the CNS as it does in the sensory ganglia. It was important also in that Rita drew attention to earlier studies of what later came to be known as "naturally occurring cell death" during development. In particular Rita recalled the work of Collin (1906), who seems to have been the first investigator to report the presence of degenerating neurons in the spinal cord, and also the findings of Ernst (1926), who had concluded that cell degeneration was a general feature in the development of all organs.

Stimulated by Rita's findings on the nucleus of Terni, Paul Shieh, one of Viktor's graduate students, analyzed the development of the Terni's nucleus in segments of the cervical spinal cord transplanted to thoracic levels. Because this work is peripheral to my primary purpose, I shall not elaborate on it, except to note that Shieh observed the same pattern of cell degeneration in the transplanted cervical cord segments as Rita had described earlier. However, in the most caudal portions of the transplants there was evidence for a presumptive preganglionic sympathetic outflow, although not all the neurons involved followed the characteristic migratory pattern that Rita had described. The mechanism responsible for this transformation in the lower cord was left undetermined: It could (it was argued) be due to a specific inductive influence operating at thoracic levels, or it might result from the removal of an inhibitory agent that normally prevents the appearance of a preganglionic system at cervical levels.

LEVI-MONTALCINI AND HAMBURGER REPEAT BUEKER'S EXPERIMENTS (1951)

Shortly after they had completed their study of the development of the sensory ganglia, Viktor and Rita carried out a series of experiments in which they had hoped to see if the transplantation of a more homogeneous mass of tissue than an entire limb could affect the development of the ganglia, the motor columns, and the associated peripheral nerves. The results of these experiments were never published, but they were referred to in the introduction of their next joint paper (Levi-Montalcini & Hamburger 1951). The transplanted tissues included portions of muscle, brain, skin, and liver that were introduced in place of a limb. As they stated:

> In this way, we hope to create specifically favorable conditions for the growth of one component. *Our preliminary results are not conclusive* [emphasis added].

Shortly thereafter, Rita later recalled, some time "in the fall of 1948, one year after my arrival in St. Louis ... Viktor showed me a short article ... which was to change entirely the direction of my research" (Levi-Montalcini 1975). The article in question was Elmer Bueker's paper describing the results of his experiments with various transplanted tumors. His success with sarcoma 180 was of special interest since not only were the tumor masses invaded by nerve fibers, but the nearby sensory ganglia were clearly enlarged. After writing to Bueker to request his permission for them to repeat (and expand) on his study, Viktor obtained mice bearing several different tumors from the Jackson Laboratories in Bar Harbor, Maine. In addition to sarcoma 180, they received a second mouse sarcoma (sarcoma 37), and two different mammary gland adenocarcinomas.

In repeating Bueker's experiments, Rita inserted small pieces of each tumor into a slit at the base of the limb bud (most transplants were at the level of the hind limb) in 3-day-old chick embryos that were allowed to survive for periods ranging from day 4 or 5 to day 17. In a parallel series of experiments, she transplanted portions of placenta from 15-day-old mouse fetuses. As before, the chick embryos were prepared for staining either with a variant of Heidenhain's hematoxylin or with De Castro's method. As Bueker had found, the transplanted adenocarcinomas were quickly resorbed (even though they had been found to grow well on the chorioallantoic membrane) and only the transplanted mouse sarcomas yielded useful results. For our purposes it is sufficient to summarize the data obtained from the 73 successful sarcoma transplants as follows.

Prior to the ingrowth of nerve fibers into the tumors, there was no evidence that they had affected either cell proliferation, early differentiation, or the initial outgrowth of sensory fibers from the ganglia, and no changes could be seen in the spinal cord. In some cases, the growth of peripheral nerves appeared to have been blocked by the tumor mass and the lateral motor columns, and the sensory ganglia were hypoplastic. Appreciable neuronal degeneration was observed among the large ventrolateral (VL) ganglion cells. From day 7 onward, however, large numbers of nerve fibers began to invade the tumors and this continued through day 17. In a typical case the LMC was markedly hypoplastic (due to the obvious obstruction to the growth of the limb nerves), but the adjoining spinal ganglia were greatly enlarged, as were the nearby paravertebral sympathetic ganglia. The ingrowth of sensory and sympathetic fibers into the tumors was especially clear in the silver-stained preparations, which often showed small bundles of fibers surrounding clusters of tumor cells.

The enlargement of the sympathetic ganglia was especially striking; in fact it often led to the apparent fusion of adjoining ganglia, some of which were six times the volume of the corresponding ganglia on the contralateral side. Cell counts at 11.5, 13, and 17 days showed striking increases in ganglion cell number compared with the controls (1.7x, 2.1x, and 3.07x, respectively). The fact that these numbers did not match the increase in the overall volumes of the ganglia was (as the silver preparations showed) due to the marked increase in the size of individual ganglion cells. Interestingly, the nucleus of Terni was unaffected in any of the experiments, but in cases in which the tumor had grown close to the suprarenal glands, large groups of sympathetic neurons were seen where normally only an occasional sympathetic ganglion cell was to be found. And in some cases sympathetic fibers from the contralateral side had grown across the midline to reach the tumor. In the spinal ganglia, the large VL neurons were severely depleted in number (presumably as a result of the blockage of access to the limb by the tumor mass), but the smaller dorsomedially located (DM) neurons were obviously increased in number and this was borne out by an observable increase in mitotic activity in the dorsomedial parts of the affected ganglia at day 7.

The discussion section of this paper is characteristically thoughtful and detailed, but only the following points merit comment. (a) The absence of any growth-stimulating effect from the transplanted fragments of E15 mouse placentas was taken as evidence that the action of the two sarcomas is quite specific and not simply a generalized growth-promoting mechanism. (b) This conclusion is strengthened by the findings that the VL sensory neurons and the motor cells in the LMC were not positively affected (as noted, their observed hypoplasia was attributed to the physical obstruction of their processes by the tumor masses). (c) The presence of the tumors strongly promotes the proliferation, differentiation, and growth of DM sensory neurons and sympathetic ganglion cells in the paravertebral ganglia. (d) The growth-stimulating effect of the tumor does not seem to be limited to neurons whose processes invade the tumors, but can also affect adjoining cells of the appropriate kind. (e) The effects do not depend on the establishment of synaptic-type junctions or sensory receptor-type endings upon the tumor cells. In conclusion the authors stated that

> all [the] available data indicate that the sarcomas 180 and 37 produce specific growth promoting agents which stimulate selectively the growth of some types of nerve fibers but not of others . . . the effects are mediated by the nerve fibers to their respective centers.

EVIDENCE THAT THE NEURAL GROWTH-PROMOTING EFFECTS OF SARCOMAS 180 AND 37 ARE DUE TO A DIFFUSIBLE FACTOR (1952, 1953)

One of the more astute observations that Rita made during the work on sarcomas 180 and 37 was that sympathetic ganglia that had not sent fibers into the tumor mass, and other collections of sympathetic neurons that had no direct connection with the transplanted tumors were appreciably enlarged. This immediately suggested to both Viktor and Rita the possibility that the causative agent was a diffusible factor released by the cells of the tumor either into the surrounding tissue fluid or into the vascular system.

It was still their contention that the "growth-promoting factor" was taken up by the processes of the affected neurons and transported back to their cell bodies, but their earlier view that it was released only at the focal sites of interaction between the terminals of sensory and sympathetic fibers and the targets they normally innervated obviously needed to be revised if the factor released by the sarcomas could diffuse for some appreciable distance from its site of production.

Rita wrote about this finding (Levi-Montalcini 1975, original emphasis).

> It was a Spring day in 1951 when the block [to her acceptance that the findings of the tumor transplantation experiments could not be fitted into their previously held views] was suddenly removed, and it dawned on me that the tumor effect was *different* from that of normal embryonic tissue in that the tumor acted by *releasing* a growth factor of unknown nature rather than by making available to the nerve fibers a larger-than-usual field of innervation.

The fact that the sympathetic fibers had extensively invaded some of the adjoining viscera long before the onset of their normal innervation was for her proof positive that a diffusible factor must be involved. As it happened, her Italian mentor, Giussepe Levi, visited St. Louis at this time and when shown slides from some of the tumor transplant bearing chicks, "shook his powerful leonine head . . . and said 'How can you say such nonsense? Don't you see that these are collagenous and not nerve fibers?'" Rita was greatly relieved when shortly afterward Viktor reassured her that she was correct, and as she noted, he "immediately grasped the far reaching significance of these findings" (Levi-Montalcini 1975).

To establish that the factor involved was indeed diffusible, Rita then carried out a very extensive series of experiments in which fragments of the two sarcomas were implanted, not near the base of the limb bud but at three remote sites: into the coelomic cavity, onto the yolk sac (from which they became incorporated into the umbilical cord), and onto the allantoic vesicle. Many grafts from the latter two sites failed "to take," but those that survived were later found on the chorioallantoic membrane. Despite an unusually high mortality rate, the results from the successful experiments were unequivocal. Large numbers of sympathetic fibers grew into some of the developing viscera (the mesonephros was especially heavily innervated) and into veins; the sympathetic ganglia, including the superior cervical ganglion and Remak's ganglion in the lumbo-sacral region, which were well removed from the growing tumor mass, were greatly enlarged. In addition several unusual sympathetic ganglion-like masses were found behind the aorta and embedded in the adrenal gland. These effects did not extend to the parasympathetic ciliary ganglion or to the enteric plexuses.

From the enlarged sympathetic ganglia, large bundles of nerve fibers could be traced in silver preparations to various viscera that normally receive only modest innervation or, in the case of the mesonephros, no innervation at all. Among the organs affected by this hyperneurotization were the ovaries, spleen, thyroid, parathyroid, metanephros, and, to a lesser extent, the liver, thymus, bone marrow, and gut. However, in a few cases the nerve outgrowth was directed almost exclusively to the implanted tumor mass. One wholly unexpected finding was the invasion of small and medium-sized veins, which in some instances was so great as to completely occlude the vessel.

Since the extraembryonic transplants were far removed from the sensory and sympathetic ganglia and were connected with them only by way of the vascular system, the ineluctable conclusion to be drawn from these experiments was that the growth-promoting agent must be diffusible. And, furthermore, since its effects on its neuronal targets were so much greater than had ever been seen, even in the most successful supernumerary limb transplants, it must be extremely potent. Its effects, however, were not only quantitatively different from those seen when an enlarged "natural target" tissue was provided, they were also qualitatively different, since it resulted in the neoformation of ganglionic masses, the hyperneurotization of viscera, the invasion of blood vessels, and the rampant and uncontrolled growth of the sympathetic system.

These exciting findings were first presented by Rita at a meeting on *The Chick Embryo in Biological Research* held at the New York Academy of Sciences in the summer of 1951 and subsequently published in the Annals of the Academy (Levi-Montalcini 1952). Although this report included all the essential observations mentioned above, a more lengthy account, with Viktor as coauthor, appeared elsewhere (Levi-Montalcini & Hamburger 1953). The fact that Rita was the sole author on the initial report is understandable since by the time the work on the murine sarcomas began, Rita had taken responsibility for essentially all the experimental work and was responsible for most of the observations.

When, some years later, Rita was asked what part Viktor had played in the work, she pointed out that when the transplant experiments were carried out, Viktor was in Cambridge, MA, having previously committed himself to spending a semester at the Massachusetts Institute of Technology (MIT) to assist his former colleague FO Schmitt (who had left Washington University to become chairman of the department of biology at MIT) in the development of a new biology curriculum (Levi-Montalcini 1981, 1988). However, there is no suggestion that Viktor was deliberately excluded from the work. In fact, Rita specifically remarked that "I kept Viktor informed weekly of the progress of my studies and of my growing interest in this extraordinary effect." And further, "Upon his return to St. Louis in the Spring of 1950, Viktor shared my enthusiasm and my belief that the growth response elicited by the tumor differed in many respects from those called forth by supernumerary limbs."

It is also evident from the style and form of the 1951 publication on the murine sarcoma transplants that Viktor—although appearing as second author—was largely responsible for writing the paper. This is true also of their second (1953) paper on this topic, which clearly bears the stamp of Viktor's hand. It is long, detailed, and carefully argued. But much of the material it contains is just as clearly due to Rita; this is especially evident in the lengthy sections on the "neuronal development of the sympathetic system" and "the response of sympathetic ganglia to tumors." There is a hint, however, that Viktor was somewhat uneasy about Rita's independent report at the meeting of the NY Academy of Sciences, which is referred to in a footnote on the second page of their 1953 paper: "A preliminary report of this work has appeared in the *Ann. N.Y. Acad. Sci.* 55, 1952." Viktor's recollections of these exciting days was that he "actively participated in the early phases of this work [that led to the discovery of the nerve growth factor] and in the preparation of the first two publications [i.e. Levi-Montalcini & Hamburger 1951, 1953] but withdrew from the project in 1953 to pursue other interests" (Hamburger 1989).

In fact his name was to appear on two further papers on the subject in 1954, but it is evident that his role in these later studies was much less direct.

THE "GOLDEN HALO": A Bioassay for the Nerve Growth-Promoting Factor (1952–1954)

The initial excitement over the discovery that murine sarcomas 180 and 37 produce a diffusible factor that has a profound, but selective, effect on neurons in sensory ganglia and the sympathetic nervous system was soon tempered by the realization that if the discovery was to be taken further, two difficult problems would have to be confronted. The first and most obvious problem concerned the nature of the growth-promoting factor, and the second was its mode of action on the responsive neurons. Viktor and Rita's immediate reaction was to see if simple chemical extracts of the two tumors injected into embryos at the appropriate stages could replicate the effects seen after tumor transplants. It is not clear who suggested this approach, how the extracts were made, or how many experiments of this type were carried out. There was no formal mention of this work in the papers published over the next few years, but it is mentioned in passing in Rita's autobiography.

> Two weeks later [after the NY Academy meeting], I was back in the lab attempting to reproduce the tumors' effects by injecting their extracts into embryos at early stages of development. Persistently negative results led me to resort to other techniques. (Levi-Montalcini 1988:152)

The obvious next approach was to see if the critical finding could be demonstrated in vitro, and in the fall of 1952, Rita set out to do just that.

While working in Turin, Rita had come to know Hertha Meyer, who had set up and maintained a tissue culture facility for Professor Levi's studies of axonal growth. Hertha (as Rita affectionately refers to her) had been trained and had worked in Germany, but when the Nazis seized power she moved to Italy. Later, when the Italian fascists began to flex their muscles she moved again, this time to Brazil, where she joined the Institute of Biophysics of the University of Rio de Janeiro, headed by Professor Carlos Chagas. Rita had kept in contact with her from time to time, and so when the next step in Rita's work called for a culture approach, it was natural that she should turn to her friend for help. Fortunately, Viktor was able to persuade the Rockefeller Foundation to provide Rita with a travel grant to enable her to visit Rio for 3 months. After returning to Italy for a brief visit to see her family, Rita traveled to Rio in late September 1952, accompanied by two white mice (each bearing a transplanted sarcoma) concealed either in her coat or in her purse (this small point varies in Rita's later accounts of her visit).

It was during the first 2 months of her visit that Rita discovered that when small fragments of the sarcoma were placed within 1–2 mm of explanted sensory ganglia from 6– to 7-day-old chick embryos, there was a striking outgrowth of neuronal processes, giving the ganglia a characteristic halo-like appearance. Throughout her stay in Rio, Rita conscientiously kept Viktor informed of her initial disappointments and later successes. She also sent him a series of pen-and-ink drawings she had made of the appearance of the stimulated ganglia, which he returned to her many years later when she was preparing to leave St. Louis for a new position in Rome. After returning to St. Louis in January 1953, Rita set up her own tissue culture facility and carried out an extensive series of experiments involving not only the two mouse sarcomas, but also adenocarcinoma and neuroblastoma cell lines. These and the initial series of experiments in Rio formed the subject of a full length paper that appeared the following year (Levi-Montalcini et al 1954).

This paper begins by setting out the rationale for the in vitro approach. Two specific reasons are given: (*a*) that the culture method obviates the possibility that the in vivo effects of the tumor on the nervous system are secondary to some generalized metabolic influence on the embryo; and (*b*) that this method might provide a useful bioassay for screening the action of the tumors (and, later, the active factor itself). The decision to focus on the sensory ganglia (rather than sympathetic ganglia, even though they had responded more vigorously to the transplanted sarcomas) is said to have been based on the fact that in 1949 they had given a very detailed account of the development of the sensory ganglia, which could serve as a control. It may also have been influenced by the fact that Hertha Meyer had used sensory ganglia in a study she had done with Levi in Turin (Levi & Meyer 1941), although it should be mentioned that Levi & Delorenzi (1935) had earlier grown sympathetic ganglia in vitro. Probably the decisive reason was that it is considerably more difficult to dissect out sympathetic ganglia from 6- or 7-day chick embryos than the larger sensory ganglia. In any event, just over 100 sympathetic ganglia were cultured (42 with fragments of the mouse sarcomas) and a total of 668 spinal ganglia were cultured. In Rita's earliest experiments in Rio, the explanted tumor tissue was taken directly from the carrier mice; however, she soon found that such tissue fragments were considerably less effective than she had found before when the tumors were transplanted into chicks. Rita therefore tried passaging the tumor tissues through chick embryos before using them for her in vitro experiments; this proved to be very effective and so for all the later experiments such chick-passaged tumor tissue was used.

The results from the two sarcomas (180 and 37) were essentially the same. When small fragments were cultured within 2 mm of the explanted ganglia, they had a profound effect on the outgrowth of processes from the sensory neurons. As early as 16 h after coculture, large numbers of fibers had grown out of the ganglia, whereas in control preparations few or no fibers grew out at this time. The fiber outgrowth in the cocultures was always more conspicuous on the side facing the tumor, but by 24 h the entire ganglia were surrounded by haloes of nerve fibers. Conversely, the outgrowth of spindle-shaped cells (presumably fibroblasts or satellite cells) was suppressed in the presence of the tumors compared with the control preparations. This appearance persisted through 48 h of culture, by which time the haloes were very dense. Essentially the same pattern was seen when the sarcoma tissue was grown close to sympathetic ganglia (from 8– to 13-day–old embryos), the only differences being that the fibers were generally finer; in these experiments the halo, if anything, was more dense.

Additional experiments involving cocultures of the tumors with fragments of chick heart and spinal cord explants were uninformative, and to the extent they were analyzed, it seemed that the tumors had no effect on these tissues; it is particularly noteworthy that they did not increase the limited outgrowth of nerve fibers seen in control spinal cord explants. Three other tumor types were used: sarcoma 1, which when cocultured with spinal ganglia resulted in enhanced fiber outgrowth—but neither as consistently nor as markedly as with sarcomas 180 and 37; mammary adenocarcinoma DBRB, which did not provoke fiber outgrowth from the ganglia; and neuroblastoma C1300, which seemed to actually inhibit the outgrowth of both cells and fibers from the ganglia. The most important additional experiments reported involved coculturing chick spinal ganglia with fragments of heart tissue from embryonic, fetal, and newborn mice. Unlike chick heart explants that had earlier shown no effect on the ganglia, the mouse tissues stimulated fiber outgrowth from the ganglia within 24 h, and this was even more marked by the end of the second day. The appearance of the ganglia in these experiments was different from that seen with the sarcomas, but the essential finding that the mouse tissues promoted fiber outgrowth was unquestionable and later proved to be of considerable interest (see below).

The conclusions to be drawn from these in vitro experiments were clear-cut. They confirmed that sarcomas 180 and 37 had a distinct, and evidently selective, influence on the outgrowth of nerve fibers from the ganglia and that this effect was mediated by a diffusible factor, as evidenced by the fact that it did not require the tumor and ganglion explants to be in contact. Furthermore, as the distance between the two explants was progressively increased, the neurite growth-promoting effect was proportionately reduced. The experiments served also to resolve an unanswered question from the prior in vivo studies, namely the possibility that the growth-promoting agent acted to "break down" some barrier that normally limited the degree to which organs and tissues can be innervated. Rather the in vitro experiments established that the product of the tumors acts directly on the ganglion cells to promote outgrowth of their processes. The coculture experiments using normal mouse tissues and chick sensory ganglia were also significant in suggesting that the growth-promoting factor was not the abnormal product of transformed tissues but might be produced and released by a variety of normal mouse tissues. But most important, these in vitro experiments raised the possibility, for the first time, that the growth-promoting factor might be isolated, and its presence assayed, at least semiquantitatively, by the use of the "halo effect." That this was a real possibility was soon to be demonstrated.

STANLEY COHEN AND THE ISOLATION OF THE NERVE GROWTH-PROMOTING FACTOR FROM MURINE SARCOMAS 180 AND 37

At some time during Rita's visit to Rio, both she and Viktor seem to have realized that if further progress were to be made and, in particular, if they were going to be able to isolate and characterize the nerve growth-promoting factor, they would need the assistance of a trained biochemist. Fortunately, exactly the right person was available: Stanley Cohen who was just completing his postdoctoral training and looking for a position and for a new challenge.

Stan (as he is generally known) was born in Brooklyn in 1922 of Russian immigrant parents. After high school he entered Brooklyn College, where he majored in biology and chemistry, and then went on to do a master's degree at Oberlin College. Transferring from Oberlin, he completed his PhD in biochemistry at the University of Michigan in 1948. This was followed by a few years as an instructor in the department of pediatrics and biochemistry, where he was engaged in metabolic studies of premature infants under the direction of Professor Harvey Gordon. In 1952, he moved to Washington University in St. Louis on an American Cancer Society fellowship, to work with Martin Kamen in the department of radiology. Here he came into contact with an intellectually stimulating and supportive group of scientists associated with Carl and Gerti Cory in the department of biochemistry, and with Arthur Kornberg and the remarkable group of colleagues he had attracted to the department of microbiology. Over the years Viktor and Rita had been drawn into this circle, and it was through this association that Viktor learned that Stan Cohen might be available for the planned assault on the nerve growth-promoting factor. As Rita recalled, on hearing from Viktor that he had invited Stan to join them and had obtained funds from the Rockefeller Foundation to support his work, she wrote to Viktor from Rio: "From the way you describe him he seems the right person to tackle the difficult problem of identifying the factor released by mouse sarcomas." Later, when she had worked with Stan for a while, she was to say:

> I have often asked myself what lucky star caused our paths to cross. . . If I, in fact, knew nothing of biochemistry, Stan when he joined us had but vague notions of the nervous system. . . 'Rita,' Stan said one day, 'you and I are good, but together we are wonderful (Levi-Montalcini 1988)

Their immediate task was to prepare sufficient tissue from the two mouse sarcomas (after passage through chick embryos) and to set up the in vitro bioassay that Rita had developed in Rio. The work went surprisingly well, and by June 1954, Viktor was able to submit a paper to the Proceedings of the National Academy of Sciences describing the isolation of the growth-promoting factor (Cohen et al 1954). This was to be the fourth, and last, paper on this topic that bore Viktor's name. As he wrote in his autobiographical statement: "In the mid-1950's I withdrew from the project. I could no longer contribute to it because of its biochemical nature; but, of course, I followed its progress with keen interest" (Hamburger 1996).

The paper reporting the isolation of the nerve growth-promoting factor (it was not yet called nerve growth factor or NGF, for short) was brief, focused, and wholly convincing. "We have. . . found," it stated, "that cell free homogenates of the tumors [S180 and S37] can duplicate in culture, the effect of the actively growing tissue." From the initial tissue fractionations it was evident that biological activity was limited to the microsomal fraction that contained about 16% of the dry weight of the tumor. Further fractionation of the microsomal preparation using streptomycin to precipitate the highly polymerized nucleic acids and nucleoproteins yielded a fraction that possessed essentially all the activity of the whole homogenate. After treatment, a solution was obtained that showed a typical nucleoprotein absorption curve with a peak at 260 nm. The active material in this solution was heat labile and nondialyzable, and in the best preparation represented 3% of the dry weight of the tumor. It consisted of 66% protein, 26% RNA, and less than 0.3% DNA.

While convincing, the paper showed some signs of having been hurriedly written and without the usual attention to detail seen in most papers bearing Viktor's name. For example, each culture used for the in vitro assay was said to contain "a sympathetic ganglion isolated from a 10-day chick embryo," but the second group of four photomicrographs show only "silver impregnated sensory ganglia." Despite this caveat, from a historical point of view, this paper marked a critical turning point and paved the way for the next surprising discovery and, ultimately, for the isolation of NGF from an unexpected source.

SNAKE VENOM AND MOUSE SALIVARY GLANDS (1956, 1960)

What followed was one of the most remarkably serendipitous events in the history of neuroscience. Since this aspect of the NGF saga has been recounted on many occasions I need only deal with it briefly. It began with a conversation between Stan Cohen and Arthur Kornberg, at that time head of microbiology at Washington University and already distinguished for his contributions to DNA replication. Stan was concerned to know whether the factor he had isolated from the two mouse sarcomas was simply a protein or a "nucleoprotein" (i.e. a protein bound to RNA or DNA). Kornberg suggested that he treat the preparation with an available snake venom that was known to be a good source of the enzyme phosphodiesterase, which would degrade whatever nucleic acids were present. If this treatment resulted in the loss of biological activity, it would strongly suggest that the active ingredient was in the nucleic acid fraction; on the other hand, if the preparation retained its activity, one could conclude that the active material was a protein (or a mixture of proteins). Stan promptly carried out the necessary experiment and gave the treated and control material to Rita to assay in her hanging-drop cultures. Within several hours Rita found that the preparation that contained the snake venom had produced an extraordinary halo radiating out from the ganglion. Since this preparation also contained the extract from sarcoma 180, it was not clear whether the observed result was due to the direct action of the snake venom on the ganglion cells, or whether some component of the venom caused the removal of a hitherto undetected inhibitory factor in the sarcoma extract. This issue was quickly resolved. The addition of a small quantity of snake venom by itself to a ganglion culture resulted in an equally dramatic outgrowth of nerve fibers. The conclusion was as unequivocal as it was surprising: The venom must contain a nerve growth-promoting factor either the same as or very similar to that in the original murine sarcomas.

These findings were reported in 1956 in a brief paper in the *Proceedings of the National Academy of Sciences* that was communicated by Viktor (Cohen & Levi-Montalcini 1956). The paper documents the methods and materials used (commercially available venom from two different species of snakes—the moccasin, *A. piscivorus*, and the rattlesnake, *Crotalus adementeus*) and summarizes some of the properties of the active factor in the venom, including the fact that it was heat labile and nondialyzable. But the most important conclusion was that in each case the factor had a specific activity (on a protein basis) of at least 1000 times that of their best purified tumor fractions.

It is not clear from the published reports who first raised the possibility that it would be worth examining the salivary glands of mice (the mammalian homologues of the venom producing glands in snakes) to see if they too contained a nerve growth-promoting activity. As was mentioned earlier, during Rita's stay in Rio she had carried out some in vitro experiments using both normal chick and mouse tissues and had found that whereas the chick tissues were ineffective, several of the mouse tissues examined caused a demonstrable outgrowth from the cocultured ganglia (Levi-Montalcini et al 1954). At the time this seemed to be just a curious, even uncomfortable anomaly (since it was then thought that the growth-promoting factor was most likely a feature of neoplastic tissues). As she later wrote:

> The mouse effect was a message I was not really capable of receiving, since I could not help thinking that it diminished—to the extent of annuling—the significance of the induction of the fibrillar halo by S180 and S37.
>
> (Levi-Montalcini 1988)

In a letter she had written to Viktor from Rio, she indicated that she was going to put aside for the time being the "mouse effect," describing it as "an unpleasant and complicated finding" (see Levi-Montalcini 1988). Regardless of whose idea it was, the decision to explore the issue was made, and in another publication, Stan reported the results of an extensive series of experiments aimed at isolating a growth-promoting factor from mouse salivary glands (Cohen 1960). In two companion papers published in the same volume, Rita and one of her graduate students, Barbara Booker, described the effects of the factor Stan had isolated and of an antiserum that he had raised against the protein (Levi-Montalcini & Booker 1960a,b).

Again it is unnecessary to describe the methods used to isolate the active fraction from the submaxillary glands of mice and the way it was assayed using sensory ganglia from 8- to 9-day-old chick embryos. However, several specific findings are worth noting. The first is that there was essentially no activity detectable in the fractions from the glands of young mice between birth and 17 days of age; thereafter the activity became increasingly evident and appeared to reach its maximum at about 50 days. It is interesting also that the specific activity of the factor isolated from the submaxillary glands of male mice was, on average, about fivefold higher than that from females. Comparable fractions from the submaxillary glands of hamsters and rats were also active, but at a level about a thousandth that found in adult male mice. The sublingual gland yielded a fraction with about a hundredth the potency of the submaxillary gland, and there seemed to be no detectable activity in the parotid glands. The activity in male submaxillary glands exceeded that in several other mouse tissues (heart, striated muscle, thymus, kidney, and serum) by a factor of approximately 5000.

Different modes of preparation of the submaxillary gland tissue yielded fractions of markedly different potency, but two in particular, identified as CM^2 and CM^3, yielded a substantial (3^+) response in the "halo" assay, at concentrations as low as 0.045 and 0.015 μg/ml, respectively. Like the factor isolated from snake venom, the submaxillary factor was heat labile, nondialyzable, and essentially removed by treatment with pepsin and chymotrypsin. Injections of the CM^2 and CM^3 factors into newborn mice resulted in a marked (~sixfold) increase in the protein concentration of the superior cervical ganglia, and a two- to threefold increase in RNA and DNA, without affecting overall body weight.

Stan raised a polyclonal antiserum to the growth-promoting factor in rabbits. When introduced into their mouse bioassay system, the antiserum had the effect of completely blocking the biological activity of the mouse submaxillary factor; it also reduced the activity of the factor in snake venom, indicating that there must be some degree of cross-reactivity. (Conversely a commercially available antivenom did not affect the activity of the mouse factor.) Subcutaneous injections of the antiserum resulted in the rapid and near-total destruction of nerve cells in the sympathetic ganglia—a finding described more fully in the second paper by Rita & Booker (1960b).

This work was carried out at Washington University, but by the time the paper was published, Stan had taken up a position in the department of biochemistry at Vanderbilt University. According to Viktor, budgetary constraints made it impossible for him to offer Stan a faculty position in the zoology department. Stan and Rita were told of this decision in December 1958, and in the summer of 1959 Stan took up his new position. Viktor did, however, communicate Stan's paper to the National Academy, and on a personal level relations between them remained warm and supportive. In a sense, this marked the end of Stan's active participation in the work on what was becoming known as NGF, but he followed its further exploration with interest, albeit at a distance. Before he left St. Louis, Stan made another wholly unanticipated discovery. This was the finding that mice injected with a partially purified preparation of the mouse submaxillary factor showed premature opening of the eyelids (as early as 7 days rather than 12–14 days, which is normal) and precocious eruption of the incisor teeth (at 6–7 days instead of 8–10 days). A less astute observer, and especially one focused only on the changes in the nervous system, would have missed these findings and, in the process, missed the discovery of a second and in many respects equally interesting factor—epidermal growth factor (EGF). Stan described the isolation of this new factor in a paper in the *Journal of Biological Chemistry* in May 1962 (Cohen 1962). In 1986, when he shared the Nobel Prize with Rita, he was specifically cited for the independent discovery of EGF and its further development.

THE EFFECTS OF THE NERVE GROWTH-PROMOTING FACTOR ISOLATED FROM MOUSE SUBMAXILLARY GLANDS AND THE ACTIONS OF AN ANTISERUM DIRECTED AGAINST THE FACTOR ON THE SYMPATHETIC NERVOUS SYSTEM (1960)

The effects of injecting the submaxillary nerve growth-promoting factor on the mouse sympathetic nervous system that Stan had mentioned in his last paper on this topic, and the consequences of injecting the rabbit antiserum he had raised, were only alluded to in a brief paragraph in his paper, but they were described more fully in the two papers by Rita & Barbara Booker (1960a,b). Rita also described them in a review she wrote in 1958 (Levi-Montalcini 1958).

The experimental section of the first paper begins with an account of the treatment of sensory and sympathetic ganglia isolated from four human fetuses (at 2.5 and 3.5 months of gestation). Using their usual hanging drop culture preparation, these ganglia were exposed to the mouse tumor extract, snake venom, and the purified submaxillary gland factor. They were found to respond in the same way as chick and mouse ganglion explants, producing a dense halo of outgrowing nerve fibers during the first 24 h in culture. Over the next 48 h the human sympathetic ganglion cultures underwent considerable liquefaction and were therefore discontinued.

Of greater interest were the experiments on newborn and adult mice injected with different submaxillary gland preparations. For these experiments large numbers of mice (10–50 in each treated group) were injected with differing concentrations of two of the fractions that Stan had isolated (fractions CM^1 and CM^3). The sympathetic chains were dissected out and usually stained as whole amounts; in some cases the superior cervical ganglia were sectioned and stained for histological examination and for counts of the numbers of mitotic figures. From a further group of 150 adult and 30 weanling mice, serum was isolated and tested (using 8-day-old chick sensory ganglion cultures) for the presence of the nerve growth-promoting factor.

It is hardly necessary to review all the experimental data analyzed in this paper. Suffice it to say that in all the mice injected with the submaxillary gland factors, the sympathetic ganglia were enlarged—up to six times in some cases. The degree of enlargement varied with the age of the animals at the time of injection (it was maximal in newborns), the amount of material injected, and the purity of the fraction used (the CM^3 fraction was consistently the most potent). The enlargement of the sympathetic ganglia was due to both the hypertrophy of individual ganglion cells and to an increase in mitotic activity (which was maximal at 5 days postnatally). Sympathetic neurons in male mice were, on average, larger than those in females; this appeared to be correlated with the appreciably higher concentration of the growth factor in the serum of adult male animals.

Although largely confirmatory of the results reported in brief by Stan, this study involved a considerable amount of work, and the documentation of the data is extremely detailed and compelling. It also marked a major departure for Rita from her previous work that had been almost exclusively focused on chick embryos and isolated chick sensory and sympathetic ganglia. From this time on, most of her work was carried out on mice (and to a lesser extent on other mammals, especially rats and hamsters). For many scientists who were trained in mammalian neurobiology, the documentation that the nerve growth-promoting material acted on mammalian neurons (as opposed to those of chicks) was considered especially important.

The second paper by Rita & Booker (1960b) dealt with the dramatic effects of injecting the antiserum that Stan had raised on the development and maintenance of the peripheral sympathetic system. Most of these experiments were carried out on newborn mice, but a few similar experiments were performed on newborn rats, rabbits, a pair of kittens, and one 7-day-old squirrel monkey.

Again, the most important findings in this study can be briefly summarized. Mice that were injected with the antiserum each day from birth to 25 days of age developed normally and, on superficial inspection, were indistinguishable either from control animals injected with normal rabbit serum or from their untreated littermates (No attempt, however, seems to have been made to test the animals under conditions that would normally have stressed the sympathetic system.) On examination, the sympathetic chain and its associated ganglia were markedly reduced in size. Counts of the numbers of neurons in the superior cervical ganglion at 20 and 25 days showed that they were reduced to between 0.6% and 1.7% of the number seen in control mice. As early as day 4, the volume of the ganglia was no more than one sixth that in normal animals. Counts of the numbers of mitotic figures showed them to be clearly reduced in the youngest animals analyzed (just 1 day after beginning the antiserum injections) and by 2 and 3 days they reached a very low level; also at this time appreciable numbers of degenerating cells could be seen throughout the ganglia. Since small numbers of neurons were still present after 25 days, a few animals, injected for 8–20 days, were allowed to survive for periods ranging from 90 days to 4 months. In these mice the percentage of neurons that survived varied between 0.84 and 2.56, which suggested that beyond the first several days, no further neuronal loss occurred. No attempt was made to determine whether injections of even larger amounts of antiserum would completely eliminate all neurons from the ganglia.

Fewer experiments were attempted in other mammals, but the results all pointed in the same direction: After as few as seven daily injections (adjusted for body weight) the ganglia in the treated animals were reduced in volume by 90%–99% and the percentage of surviving cells was reduced to between 7% and 16%.

The interpretation of these findings remained open; as the authors discuss, the effect on the sympathetic ganglia could be due to the neutralization of a circulating growth factor (and the presence of the factor in the serum of male mice was considered consonant with this view) or to a direct cytotoxic action of the antiserum. However, the findings proved to be of considerable interest to neuroscientists. They raised a number of questions that would only be resolved several years later, such as the natural source of the nerve growth-promoting material: Was it produced in only a few select organs (like the submaxillary gland) or by most tissues innervated by the sympathetic system? Were the relatively rare cases of dysautonomia reported in the medical literature due to a comparable autoimmune mechanism or to some other selective, developmental disorder? But for the short-term future, the discovery of a practical method for immunosympathectomy provided developmental biologists with yet another useful tool.

Since by 1960 essentially all the ground work on the NGF saga had been done, this is a convenient point to bring this section to a close. The accompanying review by Dr. Eric Shooter continues where this account leaves off and documents the ensuing decades of work on the chemistry and molecular biology of NGF and the many later discoveries bearing on its biological role.

VIKTOR'S FURTHER CONTRIBUTIONS TO CENTER/PERIPHERY RELATIONS AND HIS BELATED RETURN TO NGF

Although Viktor did not participate in the work on the growth-promoting factor once it had moved into its "biochemical phase," he continued for a while to be interested in the center-periphery issue, especially as it bore on the development of the LMC of the spinal cord. For a number of years he was principally engaged in studies of the ontogeny of behavior in chicks and rats (see Cowan 1981). Also while it is widely believed that Bueker had essentially lost interest in the problem after his initial observations on the effects of the murine sarcoma 180, he published half a dozen papers of interest after Viktor and Rita had taken up the subject (Bueker & Hilderman 1953; Bueker et al 1960; Bueker & Schenkein 1964; Schenkein & Bueker 1962, 1964). These papers have rarely been cited—indeed it is only in the extensive review by Rita and Pietro Angeletti that they are nearly all listed, but even here only the paper by Schenkein & Bueker (1964), which suggested that the active material might consist of two related components, is discussed (Levi-Montalcini & Angeletti 1968).

Before Viktor turned to the problems of early behavior, he made one further contribution of note to the periphery's influence on the development of the motor system (Hamburger 1958). This was essentially a follow-up of his work with Rita on the development of the sensory ganglia, but it focused specifically on the normal development of the LMC and the effects of early limb bud extirpations. The study confirmed that during normal development, cell proliferation in the ventral part of the cord is essentially over by day 4 of incubation and that by day 5.5 the LMC is fully assembled. The temporal separation of cell proliferation and migration and the subsequent outgrowth of motor fibers in some respects make the development of the LMC easier to analyze than the spinal ganglia. This point became especially clear when it was recognized that in normal development there is a considerable degree of cell death in the chick LMC between days 6 and 8, and when it was found that this degeneration is markedly accentuated following early limb extirpation. These observations finally settled the issue of the effect of the periphery on the LMC: Like sensory ganglion cells, motoneurons are dependent for their survival on the periphery; by contrast, the periphery has no effect on their earlier differentiation and migration. To this extent the study both confirmed and amplified Levi-Montalcini and Levi's earlier work and served to bring the motor system into line with the work on the sensory ganglia. While the essential findings are indeed confirmatory, Viktor's paper bears all the hallmarks of his other studies: It is carefully reasoned, the methods and the findings are described in detail, and the general conclusions drawn are both clear-cut and convincing.

In the mid-1970s, some 17 years after this last study, and after publishing 10 research papers and a number of influential reviews on the ontogeny of behavior in chicks and rats, Viktor returned to the problem of cell death in the LMC (Hamburger 1975). Taking advantage of the fact that the motor cells in the columns are large and easily distinguished from the time the column is first recognizable, and since they are not so numerous as to make estimates of their number difficult, he undertook a systematic analysis of the numbers of motoneurons in the lumbar cord from day 5.5 of incubation to just after hatching. The importance of such systematic cell counts to determine the time course of what came to be known as "naturally occurring neuronal loss" had been pointed out in a number of previous papers and reviews (see, for example, Hughes 1961, Cowan & Wenger 1967, Prestige 1970, Cowan 1970, Rogers & Cowan 1973). In the absence of firm evidence about how long it takes for a neuron to die and for the resulting cellular debris to be removed, it was difficult to determine the real magnitude of the cell loss from counts of degenerating neurons. Therefore Viktor thought it important to document the scale and time course of the naturally occurring loss of motoneurons by serial cell counts at 5.5, 6, 7, 8, 9, 12, and 18 days of incubation, and on day 5 after hatching.

The principal observation in this study is that in normal chick embryos, the lumbar lateral motor column, when first fully assembled at day 5.5, contains approximately 20,000 motoneurons; this number persists at day 6, but by day 7 it is reduced to an average of about 18,400 cells and by day 8 to just over 16,500. By day 9 it is further reduced to about 13,000 and to just over 1200 by day 12. From then on the number remains constant until after hatching. The rapidity of the cell loss over just a 3–day period, its magnitude ($\sim$40%), and its timing (corresponding to the period between the arrival of motor axons at the periphery and the establishment of the initial innervation of the limb muscles, as was documented from an examination of silver-stained preparations) are all striking, and most easily interpreted in the following terms: There is an initial overproduction of neurons followed by the subsequent degeneration of roughly half the initial number of cells. The fact that this naturally occurring cell loss begins at the time the axons of the cells first reach their target field and ends about the time the innervation of the target field is complete suggests that the axons compete within their target field for a limited supply of an essential maintenance factor (which in the case of spinal sensory and sympathetic neurons would be NGF). The cells that are unsuccessful in this competition die while those that are successful survive throughout the life of the organism. Removing the target field (e.g. by early extirpation of the developing limb bud) leads to an accentuation of this cell loss that in radical cases may be total (for review, see Cowan 1970, Oppenheimer 1981).

While this general hypothesis was consonant with virtually all the available data at the time Viktor published his study, several key elements remained to be determined. Chief among these was the following question: Given that most muscle fibers are initially innervated by several axons (although generally only one persists), why cannot all the axons compete equally for the available maintenance factor? And of course, the nature of the putative maintenance factors remained unresolved. By 1975 only NGF had been identified, but its selective action on spinal sensory and sympathetic neurons suggested that there might be a number of comparable trophic factors essential for the long-term survival of other classes of neurons (including motoneurons). There was also the question as to what maintains the neurons in the interval between the time they first differentiate and the time their axons reach their target fields. Prestige (1970) had postulated that neuroblasts are supplied ab initio with a supply of a maintenance factor that supports them until they innervate their targets; this may be so, but alternatively, it is conceivable that the cells draw on some other trophic support from their local environment or from the successive environments traversed by their axons. Lastly, there was the question of whether axons compete for specific contact sites, rather than for trophic substances. The example of NGF argued strongly for the latter view, but it was not clear whether this paradigm would hold for all classes of neurons. As Viktor stated:

> The available data do not permit a decision between the different alternatives. Once the analysis has been carried to the molecular level, the difference between a competition for contact sites and a competition for 'trophic' agents might disappear. (Hamburger 1975)

In the 1970s Viktor had a succession of postdoctoral fellows working in his laboratory, who, under his guidance, revisited a number of the issues that he and Rita had jointly or individually examined earlier. Among these fellows were Margaret (Peggy) Hollyday, Judy Brunso-Bechtold, and JW Yip, who between 1976 and 1981 published five important papers bearing on our present theme. While the fellows were generally responsible for the experimental work and the preparation of the material, Viktor actively participated both in the collection of the data (including often counting many thousands of neurons) and, most important, in its analysis.

The first of these studies (Hollyday & Hamburger 1976) was prompted by the suggestion, first clearly articulated in a review of the role of cell death in the regulation of neuronal number, that the so-called hyperplasia observed in the LMC and sensory ganglia after supernumerary limb transplants might be due, not to an increase in cell proliferation or neuronal differentiation, but to a reduction in naturally occurring cell death (Cowan 1970). To test this possibility, Peggy Hollyday repeated Viktor's earlier experiments with limb transplants (Hamburger 1939) but added an important dimension to the earlier work by systematically counting the numbers of motoneurons in the LMC in chicks with supernumerary hind-limb transplants before and after the period of naturally occurring cell loss, as defined in Viktor's 1975 paper.

The limbs were transplanted between stages 17 and 18 of the Hamburger & Hamilton (1951) series, and only those that looked morphologically normal at 6 days (stage 28) or exhibited normal patterns of motility at 11–12 days (stages

37–38) and again at 18 days were used for the analyses of cell numbers. The transplanted limbs were innervated from thoracic segment 22 and from lumbar segments 23–25. Corrected cell counts in the animals killed at 6 days (before the normal onset of cell death) showed no difference in the numbers of motoneurons in the LMC on the experimental (transplant) and control sides. In the 12-day-old animals, by contrast, the number of motoneurons was consistently higher on the side bearing the transplants, with the percentage increase ranging from 11% to 27.5%. In the two cases examined at 18 days, the numbers on the two sides were comparable to those seen at 12 days, with the transplant side having 11%–12% more motoneurons then the control side.[2] It is interesting that although one might have expected most of the increased number of cells to be limited to the rostral levels of the LMC (since it is from these levels that the transplanted limbs received their innervation), the data indicate that the increased motoneuron survival on the experimental sides was spread out over most of the rostro-caudal extent of the lumbar column.

The general interpretation of these findings was straightforward. Since the cell proliferation that gives rise to the lateral motor column extends from stage 17 through stage 24, it might have been argued that limbs transplanted at stage 17.5 could have influenced the genesis of the relevant motoneuron precursor pool. However, this interpretation is precluded by the finding that at day 6 (stage 28), when cell proliferation has ceased, the numbers of motoneurons on the control and experimental sides are the same. This leaves open only one plausible view, namely that the presence of the supernumerary limbs enables more motoneurons to survive than would occur normally. In other words, by expanding the "target field" of the motor column, the number of naturally occurring cell deaths is reduced.

As the authors point out, in light of this finding it is misleading to use the term hypoplasia for the change seen in the LMC after limb transplants (since it implies an increase either in cell proliferation or differentiation). Instead they suggested the terms neurothanasia for the process of naturally occurring cell death and hypothanasia for the reduction in cell death seen when the target field is expanded. To their disappointment, neither of these terms has come into general use. Their findings did not bear on the question of whether the survival of an increased number of motoneurons was due to an increase in the number of available innervation sites or to the increased availability of the postulated trophic or maintenance factor. In the absence of data about the actual location of the motoneurons that innervated specific muscles in the transplanted limbs, they could also throw no light on the unexpected finding that the hypothanasia extended over the entire length of the motor column.

The second study addressed the exact period during which motoneurons are generated in the brachial and lumbar segments of the chick spinal cord (Hollyday

---

[2]In his 1989 autobiographical review, Viktor noted that his colleague, Josh Sanes, had pointed out that if the comparison is made between the numbers of motoneurons before the onset of the naturally occurring degeneration at day 6 and after its termination at day 12, the actual increased cell survival in these experiments would be about 30%.

& Hamburger 1977). Previous evidence bearing on this point was equivocal. For the most part it was based on mitotic counts in the relevant regions of the cord, or on reports of the stage at which the first cytologically identifiable motoneurons could be recognized. By the 1970s the use of [^{3}H]thymidine autoradiography to determine the time of origin (or birth dates) of neurons had been well established (see Angevine 1965), and a variant of this approach had been used to study the birth dates of neurons in the spinal cord (Fujita 1964) and the multilayered optic tectum of chicks (LaVail & Cowan 1971). Since labeled thymidine introduced into the egg remains available for incorporation into DNA for a considerable period of time, the method used relies on the appearance of unlabeled neurons in the population (rather than labeled cells, as in pulse labeling studies as generally done in mammals).

The main finding of the study was that at least 95% of the motoneurons in the brachial cord are generated between stages 15 (2.5 days) and 23 (4 days) and in the lumbar cord between stages 17 and 23. There was also a clear medial to lateral (or inside-out) gradient in the time of appearance of the motoneurons in the LMC. And the paper also helped to clarify the origin and permanent location of early formed large cells in the alar region of the cord, which others had suggested might later migrate into the motor column.

Peggy Hollyday's third paper (with Viktor and Juanita Farris) examined the cells of origin of the fibers that innervate one specified muscle (the gastrocnemius) in the transplanted and normal control limbs, using as a marker the retrograde transport of horseradish peroxidase (HRP) injected into the muscle. This showed that in normal limbs the muscle is innervated by a central dorsal cluster of motoneurons in segments 26–29, whereas the gastrocnemius in the supernumerary limb consistently received its innervation from a medial cluster of neurons in segments 23–25. Although no attempt was made to determine which muscles are normally innervated by motoneurons in the latter region, the obvious conclusion was that the muscles in transplanted limbs are innervated by different cells than their normal counterparts, a finding that is of some interest for the question of neuronal specificity during development (Hollyday et al 1977).

As we have seen, when Viktor first studied the effects of early limb extirpations on the development of the LMC, he concluded that the first motor axons to grow out sensed in some way the overall extent of the field to be innervated and signaled this "estimate" back to the emerging motor pool to regulate the induction of more (or fewer) motoneurons from a population of as-yet-undifferentiated cells. Later, when it became clear from Rita's work with Levi (Levi-Montalcini & Levi 1942a,b; 1943) and his own studies with Rita (Hamburger & Levi-Montalcini 1949), that the hypoplasia that occurs in such experiments is due, not to a reduction in the inductive influence of the periphery, but rather to the death of previously differentiated cells, he accepted the view that the causative mechanism involved the availability of a trophic or maintenance factor in the target region. This led him some years later (Hamburger 1989) to re-interpret his 1934 hypothesis and to imply that the inductive signal he had originally postulated actually corresponded to the retrograde transport of a trophic agent. In reality the evidence for the retrograde transport of such an agent from the periphery was largely indirect. For the most part it was derived from such observations as the quantitative relationship between the magnitude of the observed changes in the motor column and the extent to which the peripheral field was either reduced or enlarged. Interestingly, most of the work that had been done on NGF, in vivo, did not focus on this issue since it usually involved either tumor implants or the systemic injection of the growth-promoting factor. In 1978 Viktor revisited this issue with Judy Brunso-Bechtold, taking advantage of the availability of ^{125}I-labeled NGF prepared in the laboratory of Dr. Ralph Bradshaw.

Judy inserted small pellets of polyacrymalide gel that were impregnated with ^{125}I-labeled NGF, into the knee region of chicks at about stage 36 (day 10); and 8 h later sections of the lumbar region (including the spinal cord and spinal ganglia) were prepared for autoradiography. In the four successful cases, the ipsilateral lumbar spinal ganglia, especially ganglion 23, were intensely labeled, as were the peripheral nerves leading from the site of the ^{125}I-labeled NGF pellets. Labeling over the contralateral ganglia, the sympathetic ganglia, and the motor columns and alar region of the spinal cord never exceeded background levels.

In retrospect this simple experiment was the first clear demonstration that NGF could be selectively taken up by sensory nerves and retrogradely transported to their cell bodies in the spinal ganglia. To this extent it was valuable in providing one of the missing elements in the overall NGF saga.

Of the five papers from this period, the last, which it is worth noting was dedicated to Rita, is in some respects the most important. It is also noteworthy that Viktor appeared as the first author on this paper, and to those familiar with his work, the paper clearly bears the stamp of his mind and style (Hamburger et al 1981). The purpose of the study was to see if an exogenous source of NGF could prevent, or at least limit, the amount of naturally occurring cell death in the spinal ganglia. But the paper went well beyond this and, among other things, served to correct an error in what by this time was usually referred to as "the classic paper" by Viktor and Rita on the development of the spinal ganglia.

The design of the study was straightforward. Daily injections of NGF were made into the yolk sac between stages 21 (3.5 days) and 38 (day 12), and at appropriate intervals careful counts were done of the numbers of degenerating neurons in thoracic ganglion 18 and brachial ganglion 15. These counts were then compared with similar counts in the same ganglia from normal (untreated) animals. An important part of the experimental design was to separately examine the scope of the neuronal degeneration in the two subdivisions of the ganglia: the large-celled ventrolateral (VL) division and the smaller celled dorsomedial (DM) population.

The paper begins with a description of thoracic ganglion 18 as it appears at several stages [between day 4.5 (stage 24) and day 8.5 (stage 35)] and then proceeds to document the numbers of degenerating neurons seen in its VL and DM divisions over the period from day 4.5 to 12. As Viktor and Rita had reported previously, cell death in the VL division reaches its peak at about day 5 and then declines to a fairly low level by day 6.5. In the DM division, the numbers of degenerating cells show a rather dramatic peak at day 8 and the numbers remain quite high through day 10. Following the administration of endogenous NGF, there is an appreciable reduction in the numbers of degenerating cells in both divisions of the ganglia. It is most marked in the DM division, where the number of such cells remains low throughout the entire period from days 5.5 to 12. In the VL division, the peak number of degenerating cells in the treated embryos is less than half that seen in the control preparations at day 5.5, but the number rises above the control level around day 8.

In interpreting the findings, Viktor pointed out that throughout the period studied, there is a level of what he terms sporadic cell deaths. These had been reported previously, and it was generally assumed that such deaths are caused either by errors in DNA replication or the later phases of the mitotic cycle, or by some intrinsic metabolic process. But the most important new finding was that there are distinct periods of cell degeneration in the two divisions of the ganglion with only minimal overlap between them. Whereas it had been assumed that no degeneration occurred in the brachial ganglia (Hamburger & Levi-Montalcini 1949), the new findings on brachial ganglion 15 showed unequivocally that here too there are distinct phases of cell death in the DM and VL divisions, although the levels are appreciably lower than those in thoracic ganglion 18.

Of equal importance is the clear evidence that exogenous NGF can supplement that normally produced within the target fields of the sensory neurons and can effectively eliminate most of the naturally occurring cell deaths in the ganglia. That this is true of the cells in the VL division (as well as the DM population) also disposed of the earlier notion that VL neurons are unresponsive to NGF. This view had been based on the early work with the implanted murine sarcomas (Levi-Montalcini & Hamburger 1951), but as Viktor points out in this last study, since the processes of cells from the VL division did not invade the tumor mass until day 7 (i.e. after the period of maximal responsiveness to NGF), they were not in a position to respond to the growth-promoting effects of the sarcomas. The same explanation could also account for the failure of the separated VL division to respond by forming a halo of outgrowing fibers in Rita's (Levi-Montalcini 1962) study that involved sensory ganglia from day-9 embryos (again, by which time the VL cells had lost their responsiveness to NGF).

The paper ends with the following concluding remarks, which may serve as a fitting summary of Viktor's career-long interest in this problem, for which he deserves the last word:

> We have demonstrated that NGF can rescue sensory neurons in the embryo at exactly the time they would have died without NGF supplementation. ...These findings strengthen the notion that NGF is indeed the naturally produced trophic agent for sensory ganglia.

SUMMARY AND CONCLUSIONS

This review began with some comments about the reaction of the biological community to the announcement of the award of the 1986 Nobel Prize to Rita and Stan Cohen, and so it is perhaps appropriate to end by summarizing the antecedent history that may have led to the Nobel committee's decision to omit Viktor from the award. However, as was pointed out, the purpose of this review is not to challenge or call into question the committee's decision, but to indicate the contributions that Viktor and Rita made—both individually and jointly—to the problem of the relationship between the nervous system and the periphery that culminated, after more than 30 years, in the discovery of NGF.

What unquestionably emerges from this reexamination of the history is that Viktor's work on early limb development, and especially his studies of aneurogenic limbs, set the stage for much of what was to follow. In particular these early works paved the way for Viktor's study (Hamburger 1934) of the effects of early limb bud extirpation on the development of the motor columns of the spinal cord, which proved not only to be the impetus for Rita's first work on the center/periphery issue but also to set the standard by which all later studies of this issue have been judged. That Viktor's interpretation of the findings in his study was subsequently shown to be incorrect does not diminish its importance. In the context of its time, and especially given Viktor's background as "one of Spemann's students" (to use Rita's term), it is entirely understandable that he would consider the observed reduction in the number of motor neurons after limb removal in terms of the failure of an inductive interaction. But what is even more impressive is his insightful conclusion that whatever its ultimate cause, the effect must result from a signal detected at the periphery by the first outgrowing motor axons, which is then retrogradely transmitted to their cells of origin in the motor column. It would be more than 20 years before the nature of the "signal" was discovered (at least for the neurons in the neighboring sensory ganglia), but the basic idea was clearly articulated in Viktor's seminal paper (Hamburger 1934).

Viktor's erroneous conclusion, that the "hypoplasia" seen in the motor columns was due to failure of the first motoneurons to induce the differentiation of other such cells from a pool of uncommitted precursors, stemmed in large part from his examination of the LMC at only a few selected stages in development. Had he examined them at more closely spaced time intervals (as Rita and her mentor Giuseppe Levi did some 8 years later), he would have discovered that the LMC is fully assembled prior to the onset of the "hypoplasia" and that the role of the periphery is to maintain the survival of motoneurons, not to induce their differentiation.

Rita made this important discovery while repeating Viktor's study and working under the most appalling circumstances (Levi-Montalcini & Levi 1942). By showing that the periphery acts this way on both the motor columns of the cord and the sensory ganglia, she established that the regulation of neuronal growth and maintenance involves mechanisms operating within their projection fields, and suggested that this is probably a general phenomenon in neural development.

It is impossible to know what might have become of Rita's finding had Viktor not come across her papers in 1946. Since the paper had been published during the war and in a relatively obscure journal, it is understandable that it seems to have had little or no immediate impact. But when Viktor saw that her findings called in question his interpretation of the effects of limb extirpations, his immediate reaction was to invite her to join him in St. Louis to reexamine the issue. Because he made her visit possible (with the help of the Rockefeller Foundation), it is perhaps understandable that some of his colleagues believed that among Viktor's greatest discoveries was his "discovery of Rita." But to say this is to do Rita an injustice. As I have pointed out, when she joined his laboratory in 1947, she was not, as some have portrayed her, a naive, postdoctoral fellow: She was a well-trained neurologist, technically proficient in handling chick embryos, knowledgeable about the anatomy of the central nervous system, skilled in the use of the best available neurohistological methods, and already the author of half a dozen important (if little known) papers.

Her technical skill and neurological expertise were evident in the first papers she published after moving to St. Louis and most important, in her study with Viktor on the effect of limb bud extirpation on the spinal ganglia (Hamburger & Levi-Montalcini 1949). This paper is not only one of the classics of developmental neurobiology, it is also a landmark in the field. Among other things it established beyond question that cell death is a normal (and probably widespread) feature of neural development, and that although the periphery may have an effect on cell proliferation in some systems, its principal role is to regulate the numbers of neurons that survive. Inherent in this last conclusion is the notion that for their survival and maintenance, neurons are dependent on the availability of some form of trophic factor within their target field.

It is no exaggeration to say that this study set the agenda for much of neuroembryology for the next two or three decades, and that it set Rita and Viktor on the course that finally led to the discovery of NGF. That the discovery of NGF depended on a number of fortuitous events (and, at one point, a completely serendipitous finding) is too well known to be repeated here. Suffice it to say, it began with Viktor's receiving a reprint of Bueker's paper on the effects of the murine sarcoma 180 on the sensory ganglia, was followed by the more detailed analysis of the actions of this and a second sarcoma by Rita and Viktor, and led to Rita's discovery that the tumors had an even more profound effect on the sympathetic system— including sympathetic ganglia that were not in direct contact with the tumor mass. The striking demonstration that the implanted tumors release a diffusible factor that could act on the sympathetic system through the vascular system set the stage for an all-out effort to identify the factor involved.

Exciting though their in vivo experiments were, it was clear that if further progress was to be made, and especially if the active principle released from the tumors was to be isolated, a more manageable assay system would have to be developed. Furthermore, since neither Viktor nor Rita had the necessary biochemical expertise, they would need to recruit a well-trained biochemist. The first development was met by Rita's visit to Rio de Janeiro, where with the help of her friend Hertha Meyer, she developed the in vitro coculturing system in which the outgrowth of fibers from sensory and sympathetic ganglia provided a semiquantitative assay for the presence of the nerve growth-promoting factor produced by the two murine sarcomas and certain other tissues. Although Rita was careful to keep Viktor informed about the progress of her work in Rio [and despite the fact that his name appeared on the paper reporting the use of the in vitro system (Levi-Montalcini et al 1954)], Viktor's role in this work was minimal. However, he played the key role in obtaining the necessary funds and in recruiting Stan Cohen to join the laboratory, which proved decisive as the work moved to the next important stage.

Again, although Viktor's name appeared as a coauthor on the paper that first reported the isolation of the growth-promoting factor from the murine sarcomas (Cohen et al 1954), by his own admission as the work became increasingly biochemical, he left the field to Rita and Stan and for the next decade or more devoted his efforts to the study of the ontogeny of behavior (see Cowan 1981). The most surprising discovery that snake venom (which had been used to remove nucleic acids from the partially purified preparations of the tumor-derived factor) possessed nerve growth-promoting activity soon led to the discovery that the same (or at least a very similar) factor was present in substantial quantities in the salivary glands of male mice. The generation of an antiserum against the factor (which by this time was referred to as nerve growth factor, or NGF) provided the first proof that the growth and survival of sympathetic neurons (and, by inference, sensory ganglion cells) was critically dependent during early development on the availability of NGF. As if that were not enough, while studying the effects of the partially purified factor derived from salivary glands, Stan made the equally exciting discovery of a second factor, soon to be known as epidermal growth factor (EGF), that caused precocious opening of the eyelids and eruption of the incisor teeth in mouse pups.

While Viktor was not directly involved in any of these last studies, he followed the work closely, as evidenced by his communicating a number of the papers to the Proceedings of the National Academy of Sciences. Finally, after a hiatus of almost 15 years, in the 1970s Viktor turned his attention once again to some of the as-yet-unresolved issues in the center/periphery problem and provided the first unequivocal demonstration that the administration of exogenous NGF to developing chicks can effectively eliminate naturally occurring cell loss in the spinal ganglia (Hamburger et al 1981).

By focusing narrowly on the discovery of NGF and EGF, one could reasonably conclude that the Nobel committee was correct in its selection of Rita and Stan for the 1986 prize. But viewed from a wider historical perspective, Viktor's contributions both prior to and following these discoveries, were both numerous and substantial. Fortunately, now that the "dust has settled" we can perhaps better recognize that the work which led to the isolation of NGF and EGF (and all the work that followed) is what really matters. In this conclusion Viktor and Rita would surely concur.

LITERATURE CITED

Angevine JB. 1965. Time of neuron origin in the hippocampal region. An autoradiographic study in the mouse. *Exp. Neurol. Suppl.* 2:1–70

Barron DH. 1943. The early development of the motor cells and columns in the spinal cord of the sheep. *J. Comp. Neurol.* 78:1–27

Brunso-Bechtold JK, Hamburger V. 1979. Retrograde transport of nerve growth factor in chicken embryo. *Proc. Natl. Acad. Sci. USA* 76:1494–96

Bueker ED. 1943. Intracentral and peripheral factors in the differentiation of motor neurons in transplanted lumbosacral spinal cords of chick embryos. *J. Exp. Zool.* 93:99–129

Bueker ED. 1945. The influence of a growing limb on the differentiation of somatic motor neurons in transplanted avian spinal cord segments. *J. Comp. Neurol.* 82:335–61

Bueker ED. 1948. Implantation of tumors in the hind limb field of the embryonic chick and the developmental response of the lumbosacral nervous system. *Anat. Rec.* 102:369–90

Bueker ED, Hilderman HL. 1953. Growth stimulating effects of mouse sarcomas 1, 37 and 180 on spinal and sympathetic ganglia of chick embryos as contrasted with effects of other tumors. *Cancer* 6:397–415

Bueker ED, Schenkein I. 1964. Effects of daily subcutaneous injections of nerve growth stimulating protein fractions on mice during postnatal to adult stage. *Ann. NY Acad. Sci.* 118:147–232

Bueker ED, Schenkein I, Bane JL. 1960. The problem of nerve growth factor specific for spinal and sympathetic ganglia. *Cancer Res.* 20:1220–28

Carpenter R. 1933. Spinal ganglion responses to the transplantation of limbs after metamorphosis in *Amblystoma punctatum. J. Exp. Zool.* 64:287–301

Coghill GE. 1933. Correlated anatomical and physiological studies of the growth of the nervous system of Amphibia. XI. The proliferation of cells in the spinal cord as a factor in the individuation of reflexes of the hind leg of *Amblystoma punctatum. J. Comp. Neurol.* 47:327–60

Cohen S. 1960. Purification of a nerve-growth promoting protein from the mouse salivary gland and its neurotoxic antiserum. *Proc. Natl. Acad. Sci. USA* 46:302–11

Cohen S. 1962. Isolation of a mouse submaxillary gland protein accelerating incisor eruption and eyelid opening in the newborn animal. *J. Biol. Chem.* 237:1535–62

Cohen S, Levi-Montalcini R. 1956. A nerve growth-stimulating factor isolated from snake venom. *Proc. Natl. Acad. Sci. USA* 42:571–74

Cohen S, Levi-Montalcini R, Hamburger V. 1954. A nerve growth-stimulating factor isolated from sarcoma 37 and 180. *Proc. Natl. Acad. Sci. USA* 40:1014–18

Collin R. 1906. Recherches cytologiques sur le developpement de la cellule nerveuse. *Le Nevraxe* 8:185–309

Cowan WM. 1970. Neuronal death as a regulative mechanism in the control of cell number in the nervous system. In *Development and Aging in the Nervous System*, pp. 19–41. New York: Academic

Cowan WM 1979. Selection and control in neurogenesis. In *The Neurosciences: Fourth Study Program*, ed. FO Schmitt. Boston: MIT Press

Cowan WM. 1981. Viktor Hamburger's contribution to developmental neurobiology: an appreciation. In *Studies in Developmental Neurobiology: Essays in Honor of Viktor Hamburger*, ed. WM Cowan, pp. 3–21. New York: Oxford Univ. Press

Cowan WM, Wenger E. 1967. Cell loss in the trochlear nucleus of the chick during normal development and after radical extirpation of the optic vesicle. *J. Exp. Zool.* 164:267–80

Detwiler SR. 1920. On the hyperplasia of nerve centers resulting from excessive peripheral loading. *Proc. Natl. Acad. Sci. USA* 6:96–101

Detwiler SR. 1924. The effects of bilateral extirpation of the anterior limb rudiments in *Amblystoma embryos. J. Comp. Neurol.* 37:1–14

Detwiler SR. 1927. The effects of extensive muscle loss upon development of spinal ganglia in *Ablystoma. J. Exp. Zool.* 48:1–26

Detwiler SR. 1936. *Neuroembryology: An Experimental Study.* New York: MacMillan

Dunnebacke TH. 1953. The effects of the extirpation of the superior oblique muscle on the trochlear nucleus in the chick embryo. *J. Comp. Neurol.* 98:155–77

Dürken B. 1911. Über frühzeitige Extirpation von Extremitätenanlagen beim Frosch. *Z. Wiss. Zool.* 99:189–355

Dürken B. 1913. Über einseitige Augenexstirpation bei jungen Froschlarven. *Z. Wiss. Zool.* 105:192–242

Ernst M. 1926. Über Untergang von Zellen während der normalen Entwicklung bei Wirbeltieren. *Z. Anat. Entwickl.* 79:228–62

Fujita S. 1964. Analysis of neuron differentiation in the central nervous system by tritiated thymidine autoradiography. *J. Comp. Neurol.* 122:311–28

Hamburger V. 1925. Über den Einfluss des Nervensystems auf die Entwicklung der Extremitäten von Rana fusca. *Roux Arch. Entwickl.* 105:149–201

Hamburger V. 1928. Die Entwicklung experimenteller erzeugter nervenloser und schwach innervierter Extremitäten von Anuren. *Roux Arch.* 114:272–362

Hamburger V. 1929. Experimentelle Beiträge zur Entwicklungsphysiologie der Nervenbahnen in der Frosch Extremität. *Roux Arch.* 119:47–99

Hamburger V. 1934. The effects of wing bud extirpation on the development of the central nervous system in chick embryos. *J. Exp. Zool.* 68:449–94

Hamburger V. 1938. Morphogenetic and axial self-differentiation of transplanted limb primordia of 2-day chick embryos. *J. Exp. Zool.* 77:379–99

Hamburger V. 1939a. The development and innervation of transplanted limb primordia of chick embryos. *J. Exp. Zool.* 80:347–89

Hamburger V. 1939b. Motor and sensory hyperplasia following limb-bud transplantations in chick embryos. *Physiol. Zool.* 12:268–84

Hamburger V. 1948. The mitotic patterns in the spinal cord of the chick embryo and their relation to histogenetic processes. *J. Comp. Neurol.* 88:221–83

Hamburger V. 1958. Regression versus peripheral control of differentiation in motor hypoplasia. *Am. J. Anat.* 102:365–410

Hamburger V. 1975. Cell death in the development of the lateral motor column of the chick embryo. *J. Comp. Neurol.* 160:535–46

Hamburger V. 1989. The journey of a neuroembryologist. *Annu. Rev. Neurosci.* 12:1–12

Hamburger V. 1996. Viktor Hamburger. In *The History of Neuroscience in Autobiography*, ed. LR Squire, pp. 1:222–50. Washington, DC: Soc. Neurosci.

Hamburger V, Brunso-Bechtold JK, Yip JW. 1981. Neuronal death in the spinal ganglia of the chick embryo and its reduction by nerve growth factor. *J. Neurosci.* 1:60–71

Hamburger V, Hamilton H. 1951. A series of normal stages in the development of the chick embryo. *J. Morphol.* 88:49–92

Hamburger V, Keefe EL. 1944. The effects of peripheral factors on the proliferation and differentiation in the spinal cord of chick embryos. *J. Exp. Zool.* 96:223–42

Hamburger V, Levi-Montalcini R. 1949. Proliferation, differentiation and degeneration in the spinal ganglia of the chick embryo under normal and experimental conditions. *J. Exp. Zool.* 111:457–502

Harrison RG. 1921. On relations of symmetry in transplanted limbs. *J. Exp. Zool.* 32:1–136

Hollyday M, Hamburger V. 1976. Reduction of the naturally occurring motor neuron loss by enlargement of the periphery. *J. Comp. Neurol.* 170:311–20

Hollyday M, Hamburger V. 1977. An autoradiographic study of the formation of the lateral motor column in the chick embryo. *Brain Res.* 132:197–208

Hollyday M, Hamburger V, Farris JMG. 1977. Localization of motor neuron pools supplying identified muscles in normal and supernumerary legs of chick embryos. *Proc. Natl. Acad. Sci. USA* 74:3582–86

Hughes A. 1961. Cell degeneration in the larval ventral horn of *Xenopus laevis* (Daudin) *J. Embryol. Exp. Morphol.* 9:269–84

LaVail JH, Cowan WM. 1971. The development of the chick optic tectum. II. Autoradiographic studies. *Brain Res.* 28:421–41

Levi G, Delorenzi E. 1935. Trasformazione degli elementi dei gangli spinali e sympatici coltivati *in vitro*. *Arch. Ital. Anat.* 33:443–517

Levi G, Meyer H. 1941. Nouvelles recherches sur le tissu nerveux cultivé *in vitro*: morphologie, croissance et relations réciproques des neurons. *Arch. Biol.* 52:133–278

Levi-Montalcini R. 1949. The development of the acoustico-vestibular centers in the chick embryo in the absence of the afferent root fibers and of descending fiber tracts. *J. Comp. Neurol.* 91:209–41

Levi-Montalcini R. 1950. The origin and development of the visceral system in the spinal cord of the chick embryo. *J. Morphol.* 86:253–83

Levi-Montalcini R. 1952. The effect of mouse tumor transplantation on the nervous system. *Ann. NY Acad. Sci.* 55:330–43

Levi-Montalcini R. 1958. Chemical stimulation of nerve growth. In *The Chemical Basis of Development*, ed. WD McElroy, B Glass, pp. 646–54. Baltimore: Johns Hopkins Univ. Press

Levi-Montalcini R. 1975. NGF: an uncharted route. In *The Neurosciences, Paths of Discovery*, ed. FG Worden, JP Swazey, G Adelman. Cambridge, MA: MIT Press

Levi-Montalcini R. 1981. One of Hans Spemann's students. See Cowan 1981, pp. 22–43.

Levi-Montalcini R. 1982. Developmental neurobiology and the natural history of nerve growth factor. *Annu. Rev. Neurosci.* 5:341–62

Levi-Montalcini R. 1988. *In Praise of Imperfection*. Transl. L Attardi. New York: Basic Books (From Italian)

Levi-Montalcini R, Angeletti PU. 1968. Nerve growth factor. *Physiol. Rev.* 48:534

Levi-Montalcini R, Booker B. 1960a. Excessive growth of the sympathetic ganglia evoked by a protein isolated from mouse salivary glands. *Proc. Natl. Acad. Sci. USA* 46:373–84

Levi-Montalcini R, Booker B. 1960b. Destruction of the sympathetic ganglia in mammals by an antiserum to a nerve growth protein. *Proc. Natl. Acad. Sci. USA* 46:384–91

Levi-Montalcini R, Hamburger V. 1951. Selective growth stimulating effects of mouse sarcoma on the sensory and sympathetic nervous system of the chick embryo. *J. Exp. Zool.* 116:321–61

Levi-Montalcini R, Hamburger V. 1953. A diffusible agent of mouse sarcoma producing hyperplasia of sympathetic ganglia and hyperneurotization of viscera in the chick embryo. *J. Exp. Zool.* 123:233–87

Levi-Montalcini R, Levi G. 1942a. Origine ed evoluzione del nucleo accessorio del nervo abducente nell' embrione di pollo. *Acta Acad. Pont. Sci.* 6:335–46

Levi-Montalcini R, Levi G. 1942b. Les conséquences de la destruction d'un territoire d'innervation périphérique sur le développement des centres nerveux correspondants dans l'embryon de poulet. *Arch. Biol. Liège* 53:53745

Levi-Montalcini R, Levi G. 1943. Recherches quantitatives sur la marche du processus de différenciation des neurons dans les ganglions spinaux de l'embryon de poulet. *Arch. Biol. Liège* 54:189–200

Levi-Montalcini R, Levi G. 1944. Correleziani nello svillugo tra varie parti del sistema nervoso. I. Consequenze della demolizione delle abbozzo di un arts sui centri nervosi nell' embrione di pollo. Comment. *Pontif. Acad. Sci.* 8:527–68

Levi-Montalcini R, Meyer H, Hamburger V. 1954. *In vitro* experiments on the effects of mouse sarcomas 180 and 37 on the spinal sympathetic ganglia of the chick embryo. *Cancer Res.* 14:49–57

Lillie FR. 1908. *The Development of the Chick.* New York: Holt

May RM. 1930. Répercussions de la greffe de moelle sur le systéme nerveux chez l'embryon de l'anoure. *Biol. Bull. Fr. Belg.* 64:355–87

May RM. 1933. Réactions neurongéniques de lal moelle á la greffe en surnombre, ou á l'ablation d'une ébauche de patte posterieure chez l'embryon de l'anoure, *Discoglossus pictus. Biol. Bull. Fr. Belg.* 67:327–40

May RM, Detwiler SR. 1925. The relation of transplanted eyes to developing nerve centers. *J. Exp. Zool.* 43:88–103

McGrayne SB. 1996. Rita Levi-Montalcini. In *Nobel Prize Women in Science*, pp. 201–23. New York: Birch Lane

Oppenheimer RW. 1981. Neuronal cell death and some related regressive phenomena during neurogenesis: a selective historical review and progress report. See Cowan 1981, pp. 74–133

Prestige MC. 1967. The control of cell number in the lumbar ventral horns during the development of *Xenopus laevis* tadpoles. *J. Embryol. Exp. Morphol.* 18:359–87

Prestige MC. 1970. Differentiation, degeneration and the role of the periphery: quantitative consideration. In *The Neurosciences: Second Study Program*, ed. FO Schmitt, pp. 73–82. New York: Rockefeller Univ. Press

Purves D, Sanes JR. 1987. The 1986 Nobel Prize in physiology or medicine. *Trends Neurosci.* 10:231–35

Rogers LA, Cowan WM. 1973. The development of the mesencephalic nucleus of the trigeminal nerve in the chick. *J. Comp. Neurol.* 147:291–320

Schenkein I, Bueker ED. 1962. Dialyzable cofactor in nerve growth promoting protein from mouse salivary glands. *Science* 137:433–34

Schenkein I, Bueker ED. 1964. The nerve growth factor as two essential components. *Ann. NY Acad. Sci.* 118:171–82

Shorey MC. 1909. The effect of destruction of peripheral areas on the differentiation of the neuroblasts. *J. Exp. Zool.* 7:25–63

Spemann H. 1938. *Embryonic Development and Induction.* New Haven, CT: Yale Univ. Press

Stone LS. 1930. Heteroplastic transplantation of eyes between larvae of two species of *Amblystoma. J. Exp. Zool.* 55:193–261

Tello JF. 1922. Die Entstehung der motorischen und sensiblen Nervendigungen in dem lokomotorischen System der höheren Wirbeltiere. *Z. Anat. Entwickl.* 64:348–440

Terni T. 1924. Recherche anatomiche sul sistema nervoso autonomo degli ucceli. *Arch. Ital. Anat. Embryol.* 20:433–510

Visintini F, Levi-Montalcini R. 1939a,b. Relazione tra differenziazione strutturale e funzionale dei centri e delle vie nervose nell' embrione di pollo. *Arch. Suisses Neurol. Psychiatr.* 44:119–50, and pp. 381–93

Weiss PA, Hiscoe HB. 1948. Experiments on the mechanism of nerve growth. *J. Exp. Zool.* 107:315–96

Willier BH, Weiss P, Hamburger V, eds. 1955. *Analysis of Development*. Philadelphia: Saunders

Annu. Rev. Neurosci. 2001. 24:601–29

Early Days of the Nerve Growth Factor Proteins

Eric M. Shooter

Department of Neurobiology, Stanford University School of Medicine, Stanford, California 94305-5125; e-mail: eshooter@cmgm.stanford.edu

Key Words   NGF, 7S NGF, protein subunits, precursors, kallikrein

■ **Abstract**   Adult male mouse submaxillary glands served as the preferred starting material for the isolation of the nerve growth factor (NGF) proteins in most of the isolation studies done. Two types of NGF proteins were isolated from extracts of the gland, a high-molecular-weight 7S NGF complex and a low-molecular-weight protein variously called NGF, βNGF, or 2.5S NGF. The latter, which mediated all known biological functions of NGF, were closely related forms of a basic NGF dimer in which the N and C termini of two monomers (chains) were modified by proteolytic enzymes to different extents with no effect on biological activity. The βNGF dimer showed a novel protein structure in which the two chains interacted non-covalently over a wide surface. Correspondingly, the βNGF dimer was found to be unusually stable and the form through which NGFs actions were mediated at physiological concentrations. The βNGF dimer was one of three subunits in 7S NGF; the other two were the γ subunit, an arginine esteropeptidase or kallikrein, and the α subunit, an inactive kallikrein. Two zinc ions were also present in the complex and contributed greatly to its stability. There was much debate about whether 7S NGF was a specific protein complex of interacting subunits and, if so, what functions it might play in the biology of NGF. Observations of the inhibition of the enzyme activity of the γ subunit and of the biological activity of βNGF in 7S NGF were important in determining that 7S NGF was a naturally occurring complex and the sole source of NGF in the gland extract or in saliva. Specific interactions between the active site of the γ subunit and the C-terminal arginine residues of the NGF chains, confirmed in the three-dimensional structure of 7S NGF, suggested a role for the γ subunit in pro-NGF processing during the assembly of 7S NGF. In spite of the detailed knowledge of 7S NGF structure, no information on the role of this complex in the neurobiology of NGF has emerged. With the exception of the submaxillary gland of an African rodent, no other source of NGF has been convincingly shown to synthesize the α and γ subunits, and they may well be irrelevant to NGFs actions.

601

INTRODUCTION

In the preceding chapter in this volume, Cowan (2001) describes the events that led up to the discovery of nerve growth factor (NGF) and its role in controlling cell death during development in specific neuronal populations. That particular phase of the NGF story ended in the early 1960s. It was known by then that NGF was a protein. This chapter carries the story further, outlining the research that led to a detailed description of the NGF proteins.

This review, a rather personal account, is not an exhaustive review of the biochemical and biological properties of the NGF proteins. Rather, it is a description of the interplay of biologists and biochemists in this new field as each tried to assess the contributions of the other. The final outcome was a consensus from several laboratories about the nature of the NGF proteins and the creation of a body of new information that allowed the field to expand at a prodigious rate. The discovery of NGF (Cowan 2001) was one of the great scientific discoveries of the twentieth century, and what is discussed here is only one small facet of the whole story. Many excellent reviews are available that cover the whole field in detail (Levi-Montalcini 1987, Thoenen 1995, Davies 1994, McAllister et al 1999).

NGF IN MOUSE SUBMAXILLARY GLANDS

The finding that NGF was present in adult male mouse submaxillary glands was the key to the ultimate isolation of the NGF protein. The submaxillary gland has two morphologically and functionally distinct domains: the acini, where mucopolysaccharides and amylases are concentrated; and the convoluted tubules, where proteolytic and hydrolytic enzymes and NGF are found (Levi-Montalcini & Angeletti 1964, Burdman & Goldstein 1965). NGF is synthesized in the gland (Burdman & Goldstein 1965, Levi-Montalcini & Angeletti 1968, Ishii & Shooter 1975), with the amount present markedly increasing during development. In keeping with the known sexually dimorphic character of the female submaxillary gland, concentrations are much higher in the male gland.

As noted by Cowan (2001), NGF was initially isolated from both snake venom (Cohen 1959) and mouse submaxillary glands (Cohen 1960). Although the two preparations were similar in their biological properties, they differed in specific activities: behavior on ion exchange chromatography and size, as determined by sedimentation coefficient. The latter, 2.2S and 4.3S for the snake venom and submaxillary gland NGF, respectively, suggested apparent respective molecular weights of 22,000 and 44,000, respectively. It was not possible at this stage to determine whether they were actually two different NGF proteins or whether the differences stemmed from different methods of isolation. Because most of the subsequent studies of NGF purification used mouse submaxillary glands as starting material, the major sections of this review focus on the data obtained with mouse NGF. The original Cohen (1960) procedure included (*a*) precipitation of the gland extract by streptomycin sulfate, alcohol, and ammonium sulfate and (*b*) three ion-exchange chromatographic steps. In retrospect, the behavior of the biologically active fractions on the three ion-exchange columns should have provided a clue to the nature of NGF in the extract. Although the elution characteristics on the first two ion-exchange columns suggested that the biological activity was associated with a neutral or acidic protein (specifically one with an isoelectric point of less than 7.4), the requirement of high salt concentrations to elute the active fraction from the third carboxymethyl (CM)-cellulose column at acid pH suggested a basic protein. In short, the characteristics of the active NGF protein changed during purification.

Other research at this time suggested that NGF activity required the association of two disparate proteins. Schenkein & Bueker (1964) analyzed the active fraction from the second diethyaminoethyl (DEAE)-chromatography column of Cohen's method by paper electrophoresis and resolved three inactive proteins. The combination of two of the fractions, A and C, an acidic protein of molecular weight 3500, produced NGF activity in the bioassay. Fraction A showed two components with sedimentation coefficients of 2.4S and 4.4S, and the authors concluded that the association of the small protein C with one or both proteins in fraction A was necessary for NGF activity. An extension of this work that purported to recover a NGF protein of incredibly high activity is discussed below.

It is worth emphasizing that the methods for protein purification in the late 1950s and early 1960s were limited compared with the present, and this was also true of the analytical methods for assaying protein purification. Commercial sources of ion exchange or gel filtration materials were just becoming available (they were not available for Cohen's work), and the principal analytical methods were sedimentation velocity and starch gel electrophoresis, both of which required relatively high concentrations of protein. Moreover, the NGF biological assay developed by Levi-Montalcini et al (1954) was an essential tool in any of the purification procedures. This is a highly sensitive assay, but none of the reagents was available in kit form. The assay measures the fiber outgrowth response of 8-day-old chick embryonic sensory ganglia suspended in a rooster plasma clot. The NGF solution being assayed is mixed in with the plasma before clotting and the clot suspended in a Maximow slide. Because fiber outgrowth response is bell shaped with respect to increasing concentrations of NGF, a series of dilutions of the unknown are required to determine the concentration for optimal response. The results are expressed as biological units per milliliter of sample and 1 BU/ml is defined as the NGF concentration in an assay system showing the optimal morphological response. The activity of a given sample is, therefore, calculated as the reciprocal of the dilution factor applied to achieve the maximum response. Given the protein concentration of this sample, it is then possible to express its specific NGF activity as nanograms of protein per biological unit. Purified preparations of NGF have typical specific activities of 10–20 ng/BU. Put another way, optimal NGF activity is achieved at concentrations of 10–20 ng/ml, or approximately 0.5–1.0 nM. The need to bioassay multiple fractions during a purification procedure and to do each fraction in triplicate made progress slow.

THE ISOLATION OF 7S NGF AND THE INITIAL CHARACTERIZATION OF ITS SUBUNITS

Some of the confusion over the nature of mouse NGF protein was solved when it became clear, during the work that led to the ultimate purification of NGF from the mouse submaxillary gland, that it existed in two forms of differing molecular weight, and that the low-molecular-weight form was reversibly released from the high-molecular-weight form by mild acid or alkaline pH (Varon et al 1967a,b). The starting point for the isolation procedure was the observation (S Varon, unpublished data; see Varon et al 1967a) that NGF activity in the submaxillary extract eluted from a Sephadex G-100 column at an apparent molecular weight of approximately 100,000 or higher. The elution of NGF in this high-molecular-weight fraction removed inhibitors of NGF activity, resulting in an increased total activity recovered in the fraction, as well as the separation of NGF from other known growth factors, such as epidermal growth factor (Cohen 1964) and mesenchymal growth factor (Gandini Attardi et al 1965). Two further fractionation steps were required to isolate the high-molecular-weight form of NGF. These included direct absorption of the active fraction from the G-100 Sephadex column onto DEAE-cellulose at neutral pH and a stepwise elution with increasing salt concentrations, and the further removal of lower-molecular-weight contaminants from the DEAE-fraction by gel filtration on G-150 Sephadex. The final material was recovered in about 95% purified form, as judged by analytical sedimentation velocity, sedimentation in sucrose gradients, or electrophoresis in acrylamide or starch gel, with a yield of about 80% of the original biological activity in the extract and with a specific activity of 13 ng/BU (Varon et al 1967a). This preparation was subsequently named 7S NGF, in line with the sedimentation coefficient, 7.1S, of the protein, or more precisely of the protein complex. An apparent molecular weight of 140,000 was assigned to this complex on the basis of its sedimentation behavior.

In exploring why 7S NGF differed in size from the earlier NGF preparations, 7S NGF was subjected to chromatography on CM-cellulose initially at acid pH, one of the steps in the original Cohen (1960) method. NGF activity was recovered from this column in a basic protein fraction whose sedimentation coefficient was only 2.6S. This observation was one of several that suggested 7S NGF was a protein complex that could be dissociated under appropriate conditions. Another was the observation that NGF activity decreased when the protein concentration of a fraction fell below a certain level. The reversible dissociation of 7S NGF at acid or alkaline pH was readily demonstrated by simple dialysis experiments and analyses by sedimentation in sucrose gradients (Varon et al 1967b). Recovery of the initial biological activity after exposure to either acid or alkaline pH was high (approximately 80%), although the activity measured in the sample at the changed pH was much lower. The analysis of 7S NGF on CM-cellulose equilibrated at pH 4.4 produced three different protein fractions: an acidic protein, a more neutral protein, and a basic protein, the only one with biological activity. All three proteins had the same sedimentation coefficient of 2.6S and could not, therefore, be distinguished on the basis of size. When the three fractions, but not any two fractions, were mixed together with a neutral pH in amounts proportional to their recoveries from the CM-cellulose column, a significant amount of 7S NGF, a protein that had the same sedimentation and electrophoretic properties as the starting 7S NGF material and the same specific biological activity, was reconstituted (Varon et al 1967b). The reversible dissociation of 7S NGF and the characterization of the proteins in the 7S NGF complex were further explored in detail (Varon et al 1968). The three proteins were labeled the α-, γ-, and β-subunits of 7S NGF, and the new technique of acrylamide electrophoresis was used to determine their net charges and to show that each subunit displayed hetergeneity. The NGF protein, known as the β-subunit or βNGF, was very basic and its isoelectric point was later determined to be 9.2–9.3 (Greene et al 1971). βNGF was optimally active in the same concentration range of 10–20 ng/ml as 7S NGF. Because of its high positive charge, βNGF was also readily isolated in pure form by CM-cellulose chromatography from the penultimate DEAE-fraction without subjecting the latter to the final purification step on G-150 Sephadex (Varon et al 1968). Given these observations, it seemed likely that the first two steps of the Cohen (1960) procedure were dealing with 7S NGF and that the last step isolated βNGF in association with another protein, possibly the α-subunit.

Both α- and γ-subunits were heterogeneous on electrophoresis, with the α-subunit showing three major and one minor component and the γ-subunit three components. All the individual α- or γ-subunits showed the same ability to recombine with βNGF to form a 7S NGF molecule, as did the parent α- and γ-subunits used in the initial recombination. These results suggested that the 7S NGF contained multiple forms of the 7S species, all with the same general subunit composition but differing in the types of subunits they contained (Smith et al 1968). A preparation of 7S NGF from one lobe of the submaxillary gland showed the same distribution of α- and γ-subunits.

It is interesting to look back and ask why it took between 2 and 3 years to complete the isolation and characterization of 7S NGF and βNGF. After all, 7S NGF was an abundant protein, accounting for 2% of the total protein in mouse submaxillary gland extract, and the procedure recovered 80% of the initial biological activity. One answer, as noted above, was the need to use the NGF biological assay to screen all the fractions from the various isolation steps. The dissection of chicken embryo sensory ganglia and their suspension in rooster plasma clots was very labor intensive and time consuming. The other major reason lay in the nature of the NGF protein as part of a larger complex, 7S NGF, that was readily dissociated outside the pH range of 5 to 9 or by dilution, properties that were obviously unknown at the start. A secondary consideration was the initial lack of suitable electrophoretic techniques to quickly follow the isolation procedure. The two initially available procedures, starch gel electrophoresis and paper electrophoresis, were procedures in which the initially applied protein sample was not concentrated at the start of electrophoresis. Protein fractions whose concentrations were less than 1 g%, as typically occurred in the 7S NGF isolation procedure, were therefore unsuitable for analysis. The introduction of acrylamide gel electrophoresis was a major breakthrough (e.g. Davies 1964/1965), although initially it introduced its own problems. The original method employed a short stacking gel in a cylindrical glass tube that permitted relatively dilute protein solutions loaded on top of it to be concentrated into a thin layer in the stacking gel prior to entering the resolving gel. This technique resulted in excellent separation of even closely related proteins. What was not realized at first was that the pH in the stacking gel reached above 10.0 when current was applied, resulting in the dissociation of subunit-containing proteins like 7S NGF prior to the electrophoretic separation. Even though the isolation procedure being developed was indeed producing fractions with increasing purity of 7S NGF, this was obscured by multiple bands produced by its dissociation. When this finally became apparent, acrylamide gels made and run in regular, neutral pH buffers without the use of a stacking gel provided more than adequate resolution of 7S NGF or its subunits (Varon et al 1967a). During this time, Jovin (1973) worked out the theory of discontinuous gel electrophoresis, and this enabled him and his colleagues to devise a whole new range of buffer systems to take advantage of the exquisite separating power of the acrylamide gel system at any specified pH (Jovin et al 1970, Smith et al 1968). The introduction by LKB, a Swedish company, of the ampholine system that allowed proteins to be separated in acrylamide on the basis of their isoelectric points was another major technical advance in the late 1960s.

Development of Other Methods for Isolating NGF from the Submaxillary Gland

The key advances from this research were the realization that βNGF was present in the 7S NGF complex in the mouse submaxillary gland extract and that it could be released from this complex at mild extremes of pH and readily purified because of its own relatively extreme basicity. These discoveries met with some skepticism, the basis of which was, in part, the different approaches that at least two groups took. The Stanford group consisted of Silvio Varon, trained as an engineer and as a medical doctor and who had just arrived from Rita Levi-Montalcini's laboratory at Washington University; Junichi Nomura, a psychiatrist from Osaka University who was learning biochemistry; and myself, a biochemist who had spent at least part of the past decade studying the subunit interactions in the hemoglobins. The discovery of the 7S complex seemed to fit in well with my view of nature, for were not all interesting proteins part of larger protein complexes? The St. Louis and Rome groups certainly had more than competent biochemists, and of course, they were led by Rita Levi-Montalcini. The realization of the significance of her discoveries was expanding rapidly in the late 1950s and 1960s, as neuroscience itself expanded, and she was now intent on seeking molecular explanations for the unique biological functions of NGF. Having isolated a low-molecular-weight NGF protein (2.5S NGF, see later) that possessed all the known biological activities of NGF (Zanini et al 1968), she was puzzled why the Stanford group was paying so much attention to the non-NGF proteins in 7SNGF, particularly as they did not appear to influence the activities of NGF. The debate continued for some time, as described here, but ultimately, Rita Levi-Montalcini's view was proven to be correct. The 7S NGF complex is, as far as we know, found only in the submaxillary gland, where its functions are unknown but probably not directed to the neurobiology of NGF. The low-molecular-weight NGF proteins (2.5S NGF and βNGF) are the sole entities that have NGF activity. If 7S NGF has any importance in neurobiology, it is because it led to the isolation of these latter proteins.

The skepticism that was generated was aimed at both βNGF and 7S NGF. With respect to βNGF, some neurochemists persisted in believing that NGF should have an activity more potent than that of βNGF or of Cohen's original preparation. Indeed, Schenkein et al (1968) published a method, based in part on Cohen's method, that produced two proteins, NGFA and NGFB, with the latter being a relatively acidic protein optimally active at 10^{-6} ng/ml, at least 10^{7} times more active than βNGF. Banks et al (1968), with the venom of *Vipera russelli* as starting material, also obtained a basic protein with a molecular weight of approximately 40,000 and active (unstably) at 10^{-5} ng/ml. For a time, the idea persisted. That these superactive proteins must be the authentic NGF because of their high activity was debated at meetings, and βNGF was envisaged simply as a carrier of these more active forms. More-detailed characterization of βNGF and, in particular, its sequencing was required to refute these notions.

The existence of the 7S NGF complex was also challenged because, as noted above and discussed in detail below, the non-NGF subunits in the complex appeared to have no influence on NGF activity. Some of the doubts about 7S NGF surfaced as a result of the development of two other methods to isolate the NGF protein. An active NGF protein comparable in size to βNGF was isolated from the mouse submaxillary gland by Zanini et al (1968). The method followed very closely that of Varon et al (1967a), using virtually identical gel filtration and DEAE-cellulose chromatography steps. The authors noted that the NGF activity during gel filtration was of higher molecular weight than that of the final product. The final fractionation step and chromatography on CM-cellulose at pH 5.0 occurred under conditions similar to those employed by Varon et al (1968) for the separation of the 7S NGF subunits. Not suprisingly, NGF activity was eluted last with a basic protein with physical properties similar to those of βNGF, e.g. a sedimentation coefficient of 2.3S. Bocchini & Angeletti (1969) went one step further, or more accurately backward. They subjected the active fraction from the G-100 Sephadex chromatography (the first step in the isolation of 7S NGF) to separation on the CM-cellulose column, as noted above, and again found the NGF activity associated with the last eluted, basic protein. The sedimentation coefficient of this protein was 2.43S, a value that remained constant over a range of concentrations, and the molecular weight measured from sedimentation and diffusion data or by meniscus depletion was estimated at 32,000 and 30,000, respectively. On immunodiffusion, a single precipitin band was obtained either with the original NGF antiserum or with antisera against this new protein. A different definition for the biological unit of NGF was used in these two studies. In this newer method, 1 BU was defined as the amount of NGF present in the assay that produced optimal fiber outgrowth. To compare this unit with the original definition of biological units, the specific activity in nanograms per biological unit by the new method has to be multiplied by 20 (Bocchini & Angeletti 1969). When this was done, the specific NGF activities of these NGF proteins were similar to that of βNGF.

The proteins isolated by these two procedures were simply called NGF. Once more it was noted that the NGF protein changed its size during isolation from an estimated 80,000–90,000 apparent molecular weight on the G-100 Sephadex column to the 32,000-dalton value found for the basic protein. Given the similarities in the physical and biological properties of the NGF proteins isolated by these two procedures and βNGF, it was clear that they must be closely related. However, to preserve a distinction between the methods of isolation until further information was available, the products of the Bocchini & Angeletti (1969) and Zanini et al (1968) procedures were named 2.5S NGF when it was chosen for sequencing (Angeletti & Bradshaw 1971).

The molecular heterogeneity observed during the Zanini et al (1968) procedure was ascribed to different aggregation states of the final NGF (2.5S NGF) protein by itself or with other proteins. It was noted that the inactive proteins did not contribute to the biological activity of NGF protein. The basis of this statement was not that an attempt was made to recombine the NGF protein with the inactive proteins, an experiment that probably could have revealed little because of the heterogeneity of the latter, compared with the purified α- and γ-subunits, but rather that the recovery of NGF activity in the isolated NGF protein was high, averaging 45% of the original activity in the gland extract and reaching as high as 75% on occasion (Bocchini & Angeletti 1969). This suggested that the aggregation of other gland extract proteins with the NGF protein was nonspecific and did not affect the activity of the latter. On the other hand, given the size of the higher-molecular-weight NGF protein observed in these two procedures, it seemed reasonably clear that it was 7S NGF. The latter eluted from G-100 Sephadex at an apparent molecular weight of 100,000 (Varon et al 1967a), and because the resolution of the gel matrix is at its limit in this size range, the value of 80,000–90,000 observed by Bocchini & Angeletti (1969) was indistinguishable from that of 7S NGF. Varon et al (1967a,b) noted that the α- and γ-subunits apparently did affect the activity of βNGF. The specific biological activities of 7S NGF and βNGF were similar. Because βNGF accounted for only 20% of the total 7S NGF protein, the βNGF subunit in 7S NGF had, apparently, an approximately fivefold higher activity in the complex than when free. Whether this represented regulatory interactions of the α- and γ-subunits with βNGF or whether it reflected the different stabilities or accessibilities of 7S NGF and βNGF in the complex milieu of the bioassay was debated but not answered (Varon et al 1968). In hindsight, it would be an extremely difficult question to answer without a different bioassay with a much shorter timescale. For these reasons, the apparent lack of an effect of the non-NGF gland proteins on the activity of NGF (Zanini et al 1968) was not sufficient by itself to exclude specific protein interactions.

Subunit Interactions in 7S NGF

That the subunits of 7S NGF could interact to regulate the activity of one subunit quickly became apparent with the discovery that the γ-subunit was an enzyme. This came about as Silvio Varon left Stanford to join the faculty of the University of California at San Diego and in turn was joined by a new graduate student, Lloyd Greene, later to achieve fame as, among other things, the developer of the PC12 pheochromocytoma cell line. The collaboration between the original members of the Stanford group continued, but now at a distance. It was hard to imagine two more idyllic settings than La Jolla and Stanford for such a collaboration. In the 1960s, it was not clear exactly what activities growth factors might display. Given the reports that esteropeptidase activity was associated with the mesenchymal growth factor (Gandini Attardi et al 1965), that esteropeptidases affected the growth of cultured rat hepatoma cells (Grossman et al 1969), and that the esterase thrombin had NGF-like activity (Hoffman & McDougall 1968), 7S NGF was assayed for proteolytic and esterase activities. Although 7S NGF showed little, if any, trypsin-like activity, it had a potent arginine esteropeptidase activity on α,N-benzoyl-L-arginine ethylester (BAEE) (Greene et al 1968). The hydrolysis of BAEE by 7S NGF exhibited a long phase before a maximal linear rate was established. Instead of finding this activity associated with βNGF, as had been theorized, it turned out to be a property of the γ-subunits. Unlike 7S NGF, the γ-subunits hydrolyzed BAEE at a constant rate from the start. All the different γ-subunits displayed the same specific enzymatic activity. The differences between the kinetics of hydrolysis of BAEE with 7S NGF and the γ-subunit were tentatively explained on the basis of an association-dissociation equilibrium between the two proteins, with the enzyme activity expressed by the free γ-subunit being inhibited in the 7S NGF complex (Greene et al 1969). For example, high pH or high ionic strength in dilute solution promoted dissociation of 7S NGF and released fully active γ-subunit. Conversely, the addition of excess free α-subunit and βNGF favored aggregation of the γ-subunit into 7S NGF and decreased the measured activity of the mixture. It was not possible to decide whether 7S NGF had a little or no activity, but it was clear that down-regulation of γ-subunit activity occurred through its interaction with α-subunit and βNGF.

As new techniques emerged, it was later possible to return to the question of whether the biological activity of βNGF was regulated in the 7S NGF complex. As noted earlier, the conditions in the in vivo bioassay were too complex and the timescale of the assay too long to deal with a situation involving two potentially active NGF species in equilibrium. The first new method employed chemical crosslinking with dimethylsuberimidate, employing a pH jump from pH 7.0, where 7S NGF was stable, to pH 9.5, where 7S NGF was less stable, but the cross-linking reaction was most efficient. A single cross-linked complex of molecular weight

132,000 was isolated (by the same sedimentation equilibrium technique, 7S NGF gave a molecular weight of 125,000) that contained all three subunits of 7S NGF (Stach & Shooter 1980). It was, at least, 300-fold less active than βNGF or 7S NGF. Either the cross-linking reaction itself eliminated the NGF biological activity, or it effectively prevented dissociation of the 7S NGF complex, in which case the activity of βNGF was inhibited by its interactions with the α- and γ-subunits. In favor of the latter interpretation was the previous finding (Stach & Shooter 1974) that βNGF cross-linked in the same way was still biologically active. A more direct approach measured the binding of 7S NGF to the two specific NGF receptors that had just been identified on embryonic chicken sensory ganglia (Sutter et al 1979). A 7S NGF complex labeled in the βNGF subunit with ^{125}I and stabilized in the presence of zinc ions (see later) by excess α- and γ-subunits failed to bind to either receptor (Harris-Warrick et al 1980). Clearly, βNGF was biologically inactive in 7S NGF. These findings led to a trivial explanation for the apparent difference in the specific biological activity of βNGF in the complex and when free. It seemed highly likely that the slow release of βNGF from 7S NGF in the bioassay was delivering βNGF to the sensory ganglion in an optimal manner. In contrast, when βNGF alone was added to the plasma clot, significant amounts of the protein were lost by absorption (Almon & Varon 1978) and its apparent activity decreased. Nevertheless, it was clear that there were very specific interactions between the subunits in 7S NGF.

No activity was immediately ascribed to the α-subunits. A unique feature of the α-subunits was the finding (Smith et al 1969) that they were in a mobile equilibrium with the parent 7S NGF complex. The half-life of the α-subunit exchange was approximately 300 min at 27°C, and this and other data suggested that the rate-limiting step in the exchange was the slow rate at which 7S NGF dissociated. This was latter confirmed by quantitative measurements of 7S NGF stability (Bothwell & Shooter 1978).

7S NGF as the Major NGF Protein in Mouse Submaxillary Glands

What the publication of the two shortcut methods for isolating NGF (Zanini et al 1968, Bocchini & Angeletti 1969) did accomplish was to stimulate the discoverers of 7S NGF to spend time proving that 7S NGF was the naturally occurring protein complex in mouse submaxillary gland. The methods for demonstrating that 7S NGF was a real entity were pretty standard for the time (Perez-Polo et al 1972). They could also demonstrate that the same 7S complex with the same specific NGF and enzyme activities could be purified from the submaxillary glands of a wide range of different mouse species (Bamburg et al 1971), or that the same complex was obtained by an entirely different isolation procedure (Perez-Polo & Shooter 1974, Stach et al 1977). In the selective filtration procedure (Perez-Polo & Shooter 1974), (*a*) low-molecular-weight proteins were separated from the gland extract by filtration through an appropriate membrane (XM-100) at neutral pH, (*b*) the unfiltered higher-molecular-weight concentrate was brought to pH 9.6

to dissociate 7S NGF and allow its subunits now to pass through the XM-100 filter, (c) the pH of the filtrate was adjusted back to neutrality, and (d) the recombined 7S NGF was eluted as the only high-molecular-weight protein on Sephadex G-150. This method relied on the fact that the affinities of the α- and γ-subunits and βNGF were so high and specific that recombination to 7S NGF would occur even in the presence of other proteins. These properties of the subunits were also apparent from the observation that over 60% of the 7S NGF in the gland extract was recovered as such after the pH of the gland extract was gradually raised to 10.3, the 7S NGF dissociated, and the pH lowered back to 6.8. Stach et al (1977) devised a rapid and efficient method for isolating 7S NGF using only two steps of gel filtration. Yet another method for the isolation of 7S NGF was based on the observations that NGF was secreted into saliva during either pilocarpine (Angeletti et al 1967) or α-adrenergic stimulation (Wallace & Partlow 1976). The major form of NGF in saliva was found to be 7S NGF, and a single gel filtration step on Sephadex G-100 was sufficient to isolate it in pure form (Burton et al 1978). The physical, biological, and enzymatic properties of the 7S NGF were also constant from preparation to preparation in the same or different procedures, as were the proportions of the α- and γ-subunits and βNGF in each 7S NGF preparation.

All these data led to the conclusion that 7S NGF was a major form of NGF in extracts of mouse submaxillary glands, but it did not answer the question of whether significant amounts of βNGF were also present. Mobley et al (1976), in devising yet another simplified procedure for the isolation of βNGF, found that the basic proteins in the gland extract were absorbed to CM-cellulose at pH 6.8. Very little NGF activity (about 0.3% of the NGF activity loaded) was recovered when this material was eluted from the column showing that essentially all the basic βNGF protein was complexed in 7S NGF in the gland extract. Parallel work on the protein chemistry of the epidermal growth factor (EGF) complex in mouse submaxillary glands led to a similar conclusion. EGF was shown to exist in the gland as a high-molecular-weight complex (HMW-EGF) that contained two EGF peptide chains and two copies of another arginine esteropeptidase, the EGF-binding protein (Taylor et al 1970, 1974). These additional findings in an entirely different growth factor system lent support to the idea that the two complexes, 7S NGF and HMW-EGF, might well have significant roles in the biological activities of the two factors.

THE CHARACTERIZATION OF THE LOW-MOLECULAR-WEIGHT FORMS, βNGF, AND 2.5S NGF

Two studies in the early 1970s led to the conclusion that βNGF and 2.5S NGF were dimers of (almost) identical peptide chains. In high concentrations of urea or guanidine hydrochloride, a single component with a molecular weight of approximately 12,000, half the size of βNGF, was observed on gel chromatography of βNGF (Greene et al 1971). The same-sized species appeared also on electrophoresis of reduced and carboxymethylated βNGF in the presence of sodium dodecyl sulfate. The dissociation of the βNGF dimer was followed after reduction in the same buffer system. The intact protein with a molecular weight of approximately 24,000 in this system was slowly replaced by the 12,000–molecular weight monomer with a $t_{1/2}$ at 37°C of 2 h. Because dissociation occurred in the absence of a reducing agent and no free groups were detected in βNGF, it appeared that the half-cystine residues were involved in intra- rather than inter-chain disulfide links. Angeletti et al (1971) reached the same conclusions by chemical methods. Digestion of S-carboxymethyl 2.5S NGF with trypsin produced 15 tryptic peptides that contained five S-carboxymethylcysteine residues. A sixth such residue was found in a thermolysin-derived peptide, which suggests the presence of two peptide chains, each with three intra-chain disulfide links. The two peptide chains in βNGF/2.5S NGF were clearly held together by noncovalent forces. Sedimentation data agreed with this conclusion, and significant dissociation of the dimer into monomer was only observed at high pH but not at pH 4.0 (Pignatti et al 1975).

It was the complete chemical sequencing of 2.5S NGF that finally and definitively settled all the questions about the nature of the lower-molecular-weight NGF protein (Angeletti & Bradshaw 1971, Angeletti et al 1973a,b). This was the first of a series of papers from Ralph Bradshaw's laboratory, continuing to the present, that had a major and lasting impact on the field. For its time, this sequencing of 2.5S NGF was a tour de force not only for its completion in a relatively short time, but also for its accuracy. The cloning of the 2.5S NGF/βNGF gene some years later (Scott et al 1983) showed that the chemical sequence was exact even down to the placement of the amide residues on aspartic and glutamic acid residues. Two peptide chains were identified and sequenced in 2.5S NGF. One, the A-chain, contained 118 amino acid residues; the second, the B-chain, was identical to the A-chain but lacked the first eight residues at the N terminus. The N-terminal residue of the A-chain was serine and the B-chain methionine. Clearly, during its isolation, the A-chains of 2.5S NGF underwent an unusual proteolytic cleavage at the eighth histidine-methionine peptide bond to create the B-chains. The B-chains were also referred to as des-(1-8)- or des-octaNGF chains. The order of the disulfide linking of the half-cystine residues was found to be overlapping, 1-4, 2-5, and 3-6 giving the first clues as to why native 2.5S NGF was so stable on denaturation or proteolytic cleavage. The molecular weight of 2A-chains in an intact NGF dimer was calculated to be 25,518.

Bisdes(1-8)-βNGF was isolated from βNGF after incubation with acidified submaxillary gland extract (Mobley et al 1976). The preparation retained the full activity of βNGF (see also Drinkwater et al 1993), remained a dimer, and was able to recombine with α- and γ-subunits to reform a 7S NGF complex. The βNGF endopeptidase responsible for the cleavage of the octapeptide was subsequently purified from the submaxillary gland (Wilson & Shooter 1979) and shown to be a kallikrein (Bothwell et al 1979, Fahnestock et al 1991). Taiwo et al (1991) found that the N-terminal octapeptide itself produced hyperalgesia in the hairy skin of rats but, in contrast to bradykinin, only after tissue injury.

A remaining question was the relationship of βNGF to 2.5S NGF. The physical properties of the two proteins had already been shown to be closely related, and this was born out by their almost identical amino acid composition (Bocchini 1970, Varon & Shooter 1970, Angeletti & Bradshaw 1971). The completion of the sequencing of βNGF (Mobley 1974) confirmed that βNGF and 2.5S NGF were essentially identical proteins, but it highlighted one of the differences between the two. Very little of the B-chain was found in the sequencing of βNGF (Perez-Polo et al 1972), an indication that the N terminus of βNGF was largely protected from proteolysis in 7S NGF. The other difference between the two proteins was at the C terminus. The sequence at the C terminus of 2.5S NGF was -arg.alaCOOH, (Angeletti & Bradshaw 1971, Angeletti et al 1973a) and an identical sequence was found in βNGF (Moore et al 1974). The C-terminal arginine residue was susceptible to cleavage by carboxypeptidase B-like enzymes in the gland extract to the extent that a typical 2.5S NGF preparation contained 60% of the protein (β^1NGF) with 2 C-terminal arginine residues, 30% of a protein (β^2NGF) with 1 C-terminal arginine residue, and 10% (β^3NGF) with no C-terminal residues.

In contrast, it was found that the C-terminal arginine residues in βNGF were protected from cleavage in 7S NGF and that βNGF comprised 90% of β^1NGF and only 10% of β^2 NGF. Removal of one or both C-terminal arginine residues from βNGF had no effect on its specific NGF activity (Moore et al 1974, Perez-Polo & Shooter 1975). The finding that β^3NGF, containing only chains lacking C-terminal arginine residues, failed to recombine with α- and γ-subunits implied a critical role for the C-terminal residues in the subunit interactions in 7S NGF. Given that the γ enzyme had specificity for cleavage after arginine residues (Greene et al 1971), it was hypothesized that the γ-subunit might cleave an extrapeptide sequence beyond the C terminal of the chains present in a precursor (Angeletti & Bradshaw 1971). A similar proposal was made by Taylor et al (1970, 1974) for the role of the single esteropeptidase subunit in the high-molecular-weight complex of EGF. The differences between 2.5S NGF and βNGF, therefore, resided solely in the extent of specific proteolytic cleavage at both N and C termini. Partially dissociating 7S NGF in the presence of submaxillary gland enzymes, as occurred when the intermediate G-100 or DEAE-pools were acidified prior to chromatography, allowed greater access of the enzymes to the two ends of the peptide chains in the A-chain dimer. Both termini were effectively protected in the 7S NGF complex.

A method using discontinuous isoelectric focusing in the presence of 8 M urea was developed to distinguish, in a single system, intact A-chains from the three chains that result from cleavage at either or both N and C termini (Mobley et al 1976). Because neither cleavage had any effect on the specific NGF activity, the βNGF and 2.5S NGF preparations were indistinguishable from each other and from a completely intact NGF dimer in this regard. With this knowledge, two rapid procedures for the isolation of the NGF protein were devised (Jeng et al 1979, Mobley et al 1976). In the former procedure, almost complete cleavage of the N-terminal octapeptide sequence occurred, whereas the latter produced a protein similar to 2.5S NGF in its N- and C-terminal modifications. For most purposes it is sufficient to refer to either preparation as the NGF protein or NGF dimer, especially now that the recombinant NGF protein lacking N- or C-termini modification is readily available.

The Stability of the NGF Dimer

Further studies on the NGF protein showed that it could be chemically cross-linked into a covalent dimer without loss of biological activity or ability to reform 7S NGF (Stach & Shooter 1974, Pulliam et al 1975). Some attention was also given to determine whether the monomer chains of NGF were biologically active. In an ingenious approach, Frazier et al (1973) prepared a Sepharose-bound derivative of the monomer of NGF by carrying out the linking reaction in the presence of 6 M guanidine hydrochloride and found it to be active in the sensory ganglion bioassay. A number of controls were made to eliminate the possibility that the activity was due not to the unsolubilized NGF but to soluble NGF leaking from the Sepharose derivatives. No activity, for example, could be detected in solutions kept in contact with an excess of NGF-Sepharose beads for the time of the bioassay. These experiments provided evidence that the monomer chains of NGF were active.

In alternative approaches to this problem, different methods were used to measure the stability of the NGF dimer. One was based on the ability to form hybrid β^2NGF dimers comprising one NGF chain with a C-terminal arginine and one without this residue by reversibly exposing a mixture of β^1NGF and β^3NGF to 8 M urea. The formation of the hybrid dimer, β^2NGF, was followed by isoelectric focusing in acrylamide gels, which easily resolved all three species, β^1NGF, β^2NGF, and β^3NGF (Moore & Shooter 1975). The amount of the three dimers formed in this experiment was close to that expected from random association of monomers. With this as background, it was possible to explore the formation of β^2NGF under more physiological conditions and, from the $t_{1/2}$ of this reaction, with certain reasonable assumptions, to calculate the equilibrium constant of the monomer $\leftrightarrow$ dimer equilibrium. At pH 4.0 and 4°C, where the formation of β^2NGF was most pronounced, an equilibrium constant of approximately 3×10^{-10} M was found. At pH 6.0 and above pH 9.0, the $t_{1/2}$ for the formation of β^2NGF was significantly increased, which suggested that the equilibrium constant could be orders of magnitude less than the value at pH 4.0. The low solubility of βNGF in low-ionic-strength buffers between pH 6.0 and 9.0 prevented similar experiments being carried out over this pH range. Bothwell & Shooter (1977) continued these studies using sedimentation equilibrium, sucrose gradient sedimentation, and gel filtration chromatography of both βNGF and its iodinated derivative, ^{125}I-βNGF. No dissociation of the dimer was detected by any of these techniques with NGF down to 5×10^{-9} M or ^{125}I-βNGF to 5×10^{-12} M or even less at pH 7.0. As a control to make sure that monomer would have been resolved, for example, during chromatography on Bio-Gel P60, a heavily succinylated ^{125}I-βNGF that retained the dimer form at micromolar concentrations was chromatographed on

Bio-Gel P60 at a series of decreasing concentrations. At initial loading concentrations of 1.5×10^{-11} M and 7×10^{-12} M, and after prolonged incubation times, this derivative showed up to 50% of a monomer with a molecular weight of approximately 13,000. The data indicated that the equilibrium constant for ^{125}I-βNGF was 10^{-13} M or lower. Removal of neither all the C-terminal arginine residues nor most of the N-terminal octapeptide sequences had any effect on the result. At neutral pH, βNGF was therefore a remarkably stable dimer. Although equilibria take a long time to establish for equilibrium constants in this range, it was still clear that βNGF was exclusively a dimer at concentrations (10 ng/ml or approximately 4×10^{-10} M) where it is optimally active in the in vitro assay and possibly also in vivo.

Although these results were not at variance with the equilibrium constants determined by Moore & Shooter (1975), they differed significantly from that reported by Young et al (1976), 10^{-7} M. Although the two groups used the two different NGF preparations, βNGF and 2.5S NGF, respectively, Bothwell & Shooter (1977) had already shown that heterogeneity at the N and/or C termini had no effect on the equilibrium constant. The differences arose in the use of chromatography on Sephadex G-75 using buffers containing 1 ng of serum albumin per ml (Young et al 1976), conditions where NGF interacted with the gel in such a way that its elution was retarded and incorrectly low values of molecular weight were obtained. In the sedimentation velocity techniques used in the same study, only a very limited amount of dissociation was observed in line with the data obtained by Bothwell & Shooter (1977).

The Three-Dimensional Structure of NGF

The first successful crystallization of NGF was achieved in 1975 (Wlodawer et al 1975). The crystals were hexagonal bipyramids, and the more-perfect crystals were obtained with preparations with less, rather than more, modification at the C-terminal ends of the chains. The preliminary X-ray analysis of the crystals showed that the unit cell consisted of six molecules of the NGF dimer in 12 assymetrical units. It took another 16 years before the three-dimensional structure was solved (McDonald et al 1991). The structure was novel and defined the first member of a superfamily with predominantly β-strand secondary structure and a unique cystine-knot domain in the protein monomer. In the NGF dimer, two of these monomers (chains) assembled about their flat faces and with their long axes roughly parallel. The hydrophic nature of the dimer interface readily accounted for the extremely tight association of the two NGF chains observed by Bothwell & Shooter (1977) and again emphasized that the NGF dimer and not the NGF monomer chain, as suggested in other studies, was the active NGF entity under physiological conditions. Many of the variable residues between the NGF proteins of different species or between NGF and the other members of the neurotrophin family were located in three β-hairpin loops and in a single reverse turn in the NGF structure. How these structural differences relate to the interaction of NGF with its two classes of receptors and to the functions of the different neurotrophins has been the subject of many reviews (e.g. Bothwell 1995, Ip & Yancopoulos 1996, Segal & Greenberg 1996, Lewin & Barde 1996)

THE INFLUENCE OF ZINC IONS ON THE STABILITY OF 7S NGF

An important addition to the chemistry of 7S NGF was the finding that the complex contained zinc ions as an integral part of its structure (Pattison & Dunn 1975) and that the enzymatic activity of the γ-subunit was inhibited by the zinc ion. Two zinc ions were estimated to be present in one 7S NGF molecule (Pattison & Dunn 1975, Au & Dunn 1977). That 7S NGF contains two copies of the γ-subunit was demonstrated by the finding that the Kunitz trypsin inhibitor formed a 2:1 complex with dissociated 7S NGF, but only a 1:1 complex with free γ-subunits (Au & Dunn 1977). The dissociation constant of the zinc ion from the 7S NGF-Zn complex was measured as 10^{-11} M and of the zinc ion from the γ-subunit as 10^{-6} M (Pattison & Dunn 1975). Removal of zinc by chelation with EDTA led to the activation of the esteropeptidase activity of 7S NGF, and two models were proposed to account for this phenomenon. The first of these was that removal of the zinc ion was sufficient in itself to give full enzymatic activity without causing dissociation of 7S NGF (Pattison & Dunn 1976, Au & Dunn 1977). The second model held that the 7S NGF was stabilized by the zinc ion and that removal promoted dissociation of the complex and release of the free and enzymatically active γ-subunit (Greene et al 1971, Bothwell & Shooter 1978). Using gel filtration chromatography or inhibition of enzymatic activity in the 7S NGF complex as a function of increasing concentration, all in zinc-free conditions, Bothwell & Shooter (1978) found that the equilibrium constant for the dissociation of the γ-subunit of 7S NGF was 1×10^{-7} M. Both techniques demonstrated that 7S NGF was stabilized by the zinc ion, and using data from Pattison & Dunn (1975 1976), the value of the dissociation constant of the γ-subunit from 7S NGF in the presence of 1 μM zinc ion was calculated as 2×12^{-12} M. In other words, 1 μM zinc ion should cause dissociation of the γ-subunit from 7S NGF to be negligible, and this is what was found experimentally. As noted earlier (Greene et al 1971), the enzymatic activity of 7S NGF in the presence of excess α-subunits and βNGF was very low and, after the addition of 1 μM zinc ion measurements, showed that 7S NGF had negligible enzymatic activity. Further experiments showed that the enzymatic activity of 7S NGF at any zinc concentration could be accounted for entirely in terms of 7S NGF dissociation. Moreover, the competitive inhibition of γ-subunit activity by βNGF or βNGF with the α-subunit could only be explained by the initially exclusive binding of βNGF or the substrate used to measure activity. The suggestion that zinc-free 7S NGF has significant γ-subunit activity was incorrect. These quantitative measurements of 7S NGF stability supported the idea that the active site of the γ-subunit interacted with the C-terminal arginine residue of the NGF chain and that this interaction was critically important in stabilizing the interactions of these two subunits.

Alternative Proposals for the Forms of NGF in the Gland and in Saliva

The 7S NGF model faced one last challenge. Michael Young's group at Harvard was drawn into the field partly through a study (Pantazis et al 1977) of the stability of purified 7S NGF in aqueous solution and of 7S NGF in the submaxillary gland extracts. Their data suggested that 7S NGF was a relatively unstable complex even at neutral pH. From this they drew the conclusion that the 7S NGF complex did not exist in the submaxillary gland but arose as a nonspecific complex during purification of βNGF, and that it was unlikely that the α- and γ-subunits could affect the biological activity of βNGF. However, once the role of zinc ions in effecting 7S NGF stability was known, the data of Pantazis et al (1977) could be reinterpreted. Using essentially zinc-free solvents, these authors had noted significant dissociation of purified 7S NGF at concentrations of 50 μg/ml (4×10^{-7} M) and 10 μg/ml (8×10^{-6} M). Given that the equilibrium dissociation constant of zinc-free 7S NGF was later determined to be 10^{-7} M (Bothwell & Shooter 1978), the dissociation of 7S NGF, as noted above, is exactly what would be anticipated. Furthermore, their additional observation that 7S NGF in a dilution of the original submaxillary gland extract and at a concentration of 10^{-9} M (as 7S NGF) was not dissociated but still of high molecular weight was readily explained by the high zinc-ion concentration in gland extracts (Burton et al 1978) and the added significant effect of zinc ions on 7S NGF stability (Bothwell & Shooter 1978). The same arguments could also be used to reinterpret the data obtained with the forms of NGF in mouse saliva (Murphy et al 1977). It was noted in this work that the properties of NGF in mouse saliva were identical to those in homogenates of the submaxillary gland (Pantazis et al 1977, Murphy et al 1977). It follows that their high-molecular-weight form in saliva was zinc-stabilized 7S NGF. It was likely that the low-molecular-weight form observed in saliva was the NGF dimer lacking one (β^2NGF) or both (β^3NGF) C-terminal arginine residues. These proteolytic modifications of the NGF dimer significantly decreased its affinity for the other subunits (Moore et al 1974, Perez-Polo & Shooter 1975). As noted earlier, 7S NGF was sufficiently stabilized by the zinc ion concentrations present in saliva for it to be isolated as the only NGF form in saliva in a signal gel filtration step (Burton et al 1978). Moreover, using specific antibody staining, it was possible to show that all three subunits in 7S NGF were present in the same tubule cells in the submaxillary gland (LE Burton, unpublished data). Isackson et al (1987), using specific oligonucleotide probes, also demonstrated that all three subunits were coordinately experienced in the gland and that their synthesis was much higher in male than female glands. In follow-up studies, Young et al (1978) again emphasized that 7S NGF was not the form in the gland but rather that there were up to six different species, five of which were unstable and gave rise to NGF on dilution. The sixth (NGF$_I$), although isolated in low yield compared with 7S NGF, was presumed to be the original form in the gland. It had a molecular weight of 116,000 and an equilibrium constant of 10^{-9} M. However, when the partial specific volume used in this work was changed to the value (0.72 ml/g) determined from the amino acid composition of 7S NGF, the molecular weight increased to 133,000, close to that of 7S NGF. The unstable species were most likely the binary complexes formed by interaction of either the α- or γ-subunits with βNGF (Server & Shooter 1977, Silverman & Bradshaw 1982).

In spite of the apparent similarities between the NGF$_I$ isolated by Young et al (1978) and 7S NGF, an entirely different model was proposed for the former. They agreed that NGF$_I$ could display arginine esteropeptidase (γ-subunit) activity at low concentrations and also noted that it had a protease activity capable of converting plasminogen to plasmin (Orenstein et al 1978). Based on kinetic analyses, they concluded that the appearance of esteropeptidase activity on dilution of NGF$_I$ was not due to dissociation of the complex, as suggested by Greene et al (1969), but rather to its autocatalytic activation by NGF (Young 1979). In this sense, the complex acted as an inactive NGF-zymogen until diluted and, as such, was unlikely to be responsible for converting a pro-NGF to NGF. Further analysis indicated that NGF$_I$ contained one zinc ion per molecule, and that zinc was responsible for inhibiting the single active site in NGF$_I$ (Young & Koroly 1980). NGF$_I$ apparently contained a βNGF protein, but this was released from the complex autocatalytically (see also Guerina et al 1986) and not by dissociation. The situation remained at an impasse until the characterization of the mRNA for the NGF precursor precisely defined the size of the initial gene product (Scott et al 1983) and showed that it was much smaller than that predicted by NGF$_I$. The Stanford and La Jolla groups were somewhat relieved to have this controversy behind them and were particularly cheered up when Richard Murphy and his colleagues published a detailed comparison of NGF$_I$ and 7S NGF (Guerina et al 1986) showing that NGF$_I$ was identical in molecular weight and subunit composition to 7S NGF. Richard Murphy has since made many major contributions to the NGF field, only a few of which are acknowledged in this review.

The Microheterogeneity in the γ-Subunits

A combination of physical and chemical sequencing methods were used to characterize the microheterogeneity of the γ-subunits (Stach et al 1976, Burton & Shooter 1981, Thomas et al 1981a,b). The two species, γ^1 and γ^2, identified by isoelectric focusing were found to contain both two-chain and three-chain forms of the esteropeptidase, the conversion between the two forms being due to yet another gland endopeptidase acting on an internal peptide bond. The removal of the C-terminal lysine residue created by this conversion and of the original C-terminal arginine on the shorter of the two original chains by a carboxypeptidase B-like enzyme created one extra form in the γ^2 species and the γ^3 species itself, for a total of six forms. The only difference that was noted between them was a preference for the three-chain γ^3 species over that of γ^1 and γ^2 to recombine with α-subunit and βNGF to form 7S NGF. In detailed studies (Nichols & Shooter 1983, Blaber et al 1989), this difference was quantified, and the most proteolytically modified form, γ^3, found to have an affinity for its natural ligand βNGF and for several synthetic arginine subtrates, was an order of magnitude greater than γ^1, with γ^2 being intermediate. This observation helped resolve some of the differences between the 7S NGF and NGF_I (Young 1979). The complete amino acid sequence of the chains in the γ-subunits (Thomas et al 1981a,b) emphasized its position as a member of the closely related arginine esteropeptidase, or kallikreins of the submaxillary gland. The cDNA for the γ-subunit was characterized (Ullrich et al 1984). It confirmed the amino acid sequence and also provided more detail on the posttranslational modifications.

The Microheterogeneity of the α-Subunits

Isackson & Bradshaw (1984) determined most of the amino acid sequence of the α-subunit and found it to be highly homologous to the γ-subunit. It differed in one important way. It lacked esteropeptidase activity, failing to cleave synthetic arginine substrates or peptides or to be labeled by [³H]disopropyl fluorophosphate. Subsequently, the cDNA coding for the α-subunit precursor was cloned (Isackson et al 1984) and several amino acid changes from the γ-subunit were noted that probably accounted for the lack of enzyme activity in the α-subunit. Among these changes was a substitution near the cleavage site of the activation peptide that would, almost certainly, prevent release of this peptide leaving the α-subunit in the zymogen form. It seemed likely that the microheterogeneity of the α-subunit, like that of the γ-subunit, was generated by cleavage at the lysine residues 135 or 137 and subsequent removal of these residues by endopeptidase (Isackson et al 1987). Attempts to define the precursor of the α-subunit in cell free systems (Isackson et al 1987) or in the submaxillary gland (Darling et al 1983) were complicated by the cross-reactivity of the α-subunit antibodies with the homologous γ-subunit and other kallikreins. Both studies identified an apparent precursor(s) with a molecular weight of approximately 32,000.

THE NGF PRECURSOR

As noted above, the specific interaction of the C-terminal arginine residues of βNGF with the γ-subunit led to the idea that the latter was involved in cleaving a propeptide that extended beyond the C-terminal arginine residue (Angeletti & Bradshaw 1971). A similar suggestion was made for a precursor for EGF (Taylor et al 1970, 1974). In this instance, the esteropeptidase EGF-binding protein was hypothesized to release propeptide material extending beyond the C-terminal arginine residue of EGF, resulting in the formation of the HMW-EGF complex analogous to the formation of 7S NGF. Evidence in favor of these ideas, besides the inhibition of the activities of both the γ-subunit and the EGF-binding protein in their respective complexes (Greene et al 1969, Bothwell & Shooter 1978, Server et al 1976a, Nichols & Shooter 1985) and the failure to form these complexes when the C-terminal arginine residues were removed from NGF and EGF, respectively (Moore et al 1974, Server et al 1976a), was the inability of EGF binding protein to replace the homologous γ-subunit in 7S NGF, indicative of a high degree of specificity of the cleaving enzyme (Server & Shooter 1976). Direct evidence for a NGF precursor came from in vivo and in vitro labeling experiments. These showed that adult male mouse submaxillary glands incorporated L-(^{35}S)cysteine into βNGF (Berger & Shooter 1978). When the labeling period was reduced, another major radioactive species precipitated with anti-NGF antiserum had a molecular weight of 22,000 and contained all the cysteine-containing peptides of NGF (Berger & Shooter 1977). The kinetics of labeling of the 22,000-dalton species and of βNGF identified the former as a precursor of βNGF (Berger & Shooter 1977). Conversion of the precursor to NGF was achieved with the γ-subunit, but the finding that other proteolytic enzymes, including EGF-binding protein, carried out the same conversion raised questions about the specificity of the reaction and the possible involvement of the α-subunit in conferring this specificity. It is interesting to note that in the corresponding EGF system, a 9000-dalton precursor of EGF was identified and only the EGF-binding protein, but not the γ-subunits, converted this precursor to EGF (Frey et al 1979), as would be anticipated for a highly specific cleavage reaction.

The interpretation of these NGF precursor studies became clear when the cDNAs for mouse and human NGF mRNAs were characterized (Scott et al 1983, Ullrich et al 1983). The initial gene product prepro-NGF had a calculated molecular weight of 33,800, and after removal of the signal sequence, the calculated molecular weight of pro-NGF was 30,800. Because the pro-sequence contained both a pair of arginine residues and a group of four contiguous basic amino acid residues, cleavage at these sites would produce potential intermediates with molecular weight of 22,000 and 18,500. The 22,000-dalton precursor found in the initial studies (Berger & Shooter 1977) was one of these intermediates. Subsequent work using NGF antibodies with different specificities identified all the predicted intermediates in the conversion of prepro-NGF to NGF (Darling et al 1983). Edwards et al (1986) identified a shorter transcript initiated from an internal methionine residue encoding a protein of 241 amino acid residues. The same intermediates would be produced from this sequence. Although the γ-subunit was able to convert the biosynthetic precursors of NGF to NGF after expression in cells in culture (Edwards et al 1988) and in a cell-free system, the need for greater-than-stoichiometric ratios of γ-subunit to precursor again raised doubts about the physiological significance of the γ-subunit cleavages (Jongstra-Bilen et al 1989). Recent evidence, indeed, has suggested that the conversion of pro-NGF to NGF at its N terminus used the convertase, furin (Seidah et al 1996). This enzyme was able to cleave pro-NGF in cells containing regulatory and/or constitutive secretion pathways. N-linked glycosylation of the pro-sequence and trimming of the oligosaccharide chains were required for transport of pro-NGF from the endoplasmic reticulum to the *trans* Golgi, where the conversion occurred. The observation that furin and NGF were colocalized in the tubule cells of the submaxillary gland from early development supported this view (Farhadi et al 1997).

The big surprise from the pro-NGF sequence was the finding that the larger propeptide was located at the N and not the C terminus of the NGF reading frame. As predicted, there was a pro-sequence at the C terminus, but it was only the dipeptide arg.glyCOOH. Whether this sequence was removed as a dipeptide or sequentially as amino acid residues was not determined, but the enzyme responsible was clearly the γ-subunit because this was the enzyme found uniquely in the 7S NGF complex.

The Three-Dimensional Structure of 7S NGF

Knowledge of the molecular weights of all the subunits in 7S NGF from sedimentation and sequence data, as well as the molecular weight of 7S NGF, suggested that the composition of 7S NGF was $\alpha_2\beta\gamma_2$:2Zn, where β is the NGF dimer. The final proof of the validity of the 7S NGF model and of its structure came with the determination of its three-dimensional structure (Bax et al 1997). This structure confirmed that two α-subunits, two γ-subunits, one NGF dimer, and two zinc ions were present in the complex. Within the complex, the two γ-subunits had multiple interactions with each other. The resolution was sufficient to show that the C-terminal residues of the NGF chains were in the active site of the γ-subunits. The two α-subunits interacted extensively with the NGF dimer, and the structure showed that they were in the inactive zymogen conformation. The two zinc ions were located in the interfaces between the α- and γ-subunits. The NGF dimer was buried within the 7S NGF complex in a way that explained its lack of activity in the complex. All these details added up to a most satisfying confirmation of the many biochemical and biological studies that had reached the same conclusions.

THE FORM OF NGF FROM OTHER SOURCES

The important question raised by the finding of 7S NGF as the sole NGF-containing protein of mouse submaxillary gland was whether this was always the form in which NGF is produced. It has been shown that the physiological sources of NGF, besides the submaxillary gland, are the target tissues of sensory, sympathetic, and central nervous system cholinergic neurons that are NGF responsive (Thoenen & Barde 1980). Determining the potential expression of the α- and γ-subunits in these same tissues would be a potentially daunting task because of the extreme homology between these subunits and the rest of the kallikrein family (Isackson et al 1984, Evans & Richards 1985). This problem has been approached in several different ways. In at least two examples of cells in culture, mouse fibroblasts and rat irides, the cells synthesized and secreted NGF but not the α- or γ-subunits (Pantazis 1983, Murphy et al 1986). Another approach was to isolate and characterize an NGF complex from another source rich in NGF. Aloe et al (1981) noted that the submaxillary gland of the African rat *Mastomys natalensis* was probably the richest source of NGF known, with both males and females having enlarged NGF-containing glands. Using a purification scheme that was a modification of the original one for 7S NGF (Varon et al 1967a), Burcham et al (1991) obtained a purified *M. natalensis* NGF complex that was about 30,000 less in molecular weight than 7S NGF and was composed of subunits similar to those found in 7S NGF. The NGF amino acid sequence derived from its cDNA was homologous to mouse NGF (see also Fahnestock & Bell 1988) and had the same bioactivity. The presence of the characteristic α-subunit bands on isoelectric focusing and a cross-reactivity with a mouse α-subunit antiserum was reasonably convincing proof for the presence of α-subunits. Similarly, a γ-subunit band and an esteropeptidase activity suggested a γ-subunit in the complex. However, the enzyme activity of the complex behaved like a free γ-subunit activity and was not entirely inhibited, like γ-subunit, with aprotinin. It is interesting that the presence in *M. natalensis* NGF of a proline residue before the characteristic arg.arg.gly C terminus of mouse pro-NGF could prevent the *M. natalensis* γ-subunit from remaining bound to the C-terminal arginine residue after removal of the dipeptide arg.gly. If this happened, the enzyme activity of the subunit would not, as observed, be inhibited in the complex. In spite of the fact that the structure of the *M. natalensis* NGF complex was of great interest, the expense of keeping these beautiful but particularly nasty little animals under the reasonable restrictions of the US Fish and Game Service became too much and the work was discontinued.

In contrast, NGF complexes from other rich sources did not show the presence of α- and γ-subunits. High levels of NGF have been reported in guinea pig prostate gland (Harper et al 1979) and bovine seminal fluid (Harper et al 1982). The partial amino acid sequence (Rubin & Bradshaw 1981) and the sequencing of cDNA clone for guinea pig NGF (Schwarz et al 1989) revealed an 86% identity to the mouse NGF amino acid sequence. In spite of the fact that guinea pig NGF precursor has the same C-terminal arg.arg.gly sequence as mouse NGF, no γ-subunits (or α-subunits) were detected in the prostate gland (Rubin & Bradshaw 1981, Isackson et al 1985, Dunbar & Bradshaw 1987). The bovine NGF contains C-terminal gln.arg.ala that probably would not be cleaved by a putative γ-subunit (Meier et al 1986), and this is true of other NGFs, like chicken NGF. Guinea pig NGF has been found to form a complex with other as-yet-unknown proteins (Rubin & Bradshaw 1981).

Finally, some mention of snake venom NGF is appropriate. The NGF from the Indian cobra *Naja naja* has been studied in great detail (Angeletti 1970, Hogue-Angeletti et al 1976), and partial amino acid sequence and cloning (Selby et al 1987) showed significant homology with mouse NGF. In spite of this, the *N. naja* NGF did not interact with the α- and γ-subunits of mouse 7S NGF (Server et al 1976b).

CONCLUSION

It seems unlikely with the information we have at the moment that 7S NGF or a similar complex exists anywhere other than in mouse or rat submaxillary gland. The formation of 7S NGF appears to have been driven evolutionarily by the large numbers and quantities of kallikrein enzymes in the gland for purposes that are unique to the gland but yet unknown. In other words, the properties and functions of the α- and γ-subunits are probably irrelevant to the functions of NGF. For all the scientists who have spent time exploring and characterizing the 7S NGF complex, this is, indeed, a conclusion full of irony, if only because a lot of good protein chemistry went into this work! In one of the lighter moments of the debate around 7S NGF, Rita Levi-Montalcini (1975) referred to its subunits as "the mystic Trinity." With the mysticism taken out of them, the α- and γ-subunits now emerge for what they are, regular proteins that were lucky enough to attach themselves to NGF. They are not the exception. I doubt if anyone who has worked in this field does not consider him or herself, as I do, extremely lucky to have been a participant, however small, in one of the most fascinating arenas in neurobiology.

ACKNOWLEDGMENTS

This review is dedicated to Drs. Viktor Hamburger and Rita Levi-Montalcini in honor of their recent birthdays and their extraordinary contributions to developmental neurobiology. It also acknowledges the excitement generated in the early debates on the NGF proteins with Rita Levi-Montalcini and the long standing friendship that grew out of them.

Visit the Annual Reviews home page at www.AnnualReviews.org

LITERATURE CITED

Almon RR, Varon S. 1978. Associations of beta nerve growth factor with bovine serum albumin as well as with the alpha and gamma subunits of the 7S macromolecule. *J. Neurochem.* 30:1459–67

Aloe L, Cozzarti C, Levi-Montalcini R. 1981. The submaxillary glands of the African rodent (*Praomys mastomys*) natalensis as the richest source of the nerve growth factor. *Exp. Cell Res.* 133:475–80

Angeletti P, Calissano P, Chen JS, Levi-Montalcini R. 1967. Multiple molecular forms of the nerve growth factor. *Biochim. Biophys. Acta* 147:180–82

Angeletti RH. 1970. Nerve growth factor from cobra venom. *Proc. Natl. Acad. Sci. USA* 65:668–74

Angeletti RH, Bradshaw RA. 1971. Nerve growth factor from mouse submaxillary gland: amino acid sequence. *Proc. Natl. Acad. Sci. USA* 68:2417–20

Angeletti RH, Bradshaw RA, Wade RD. 1971. Subunit structure and amino acid composition of mouse submaxillary gland nerve growth factor. *Biochemistry* 10:463–69

Angeletti RH, Hermodson MA, Bradshaw RA. 1973a. Amino acid sequences of mouse 2.5S nerve growth factor. II. Isolation and characterization of the thermolytic and peptic peptides and the complete covalent structure. *Biochemistry* 12(1):100–15

Angeletti RH, Mercanti D, Bradshaw RA. 1973b. Amino acid sequences of mouse 2.5S nerve growth factor. I. Isolation and characterization of the soluble tryptic and chymotryptic peptides. *Biochemistry* 12(1):90–100

Au AM, Dunn MF. 1977. Reaction of the basic trypsin inhibitor from bovine pancreas with the chelator-activated 7S nerve growth factor esteropeptidase. *Biochemistry* 16:3958–66

Bamburg JR, Derby MA, Shooter EM. 1971. Isolation and characterization of the 7S nerve growth factor protein from inbred strains of mice. *Neurobiology* 1:115–20

Banks BE, Banthorpe DV, Berry AR, Davies HS, Doonan S, et al. 1968. The preparation of nerve growth factors from snake venom. *Biochem. J.* 108:157–58

Bax B, Blundell TL, Murray-Rust J, McDonald NQ. 1997. Structure of mouse 7S NGF: a complex of nerve growth factor with four binding proteins. *Structure* 5:1275–85

Berger EA, Shooter EM. 1977. Evidence for pro-beta-nerve growth factor, a biosynthetic precursor to beta-nerve growth factor. *Proc. Natl. Acad. Sci. USA* 74:3647–51

Berger EA, Shooter EM. 1978. Biosynthesis of beta nerve growth factor in mouse submaxillary glands. *J. Biol. Chem.* 253:804–10

Blaber M, Isackson PJ, Marsters JC Jr, Burnier JP, Bradshaw RA. 1989. Substrate specificities of growth factor associated kallikreins of the mouse submandibular gland. *Biochemistry* 28:7813–19

Bocchini V. 1970. The nerve growth factor. Amino acid composition and physicochemical properties. *Eur. J. Biochem.* 15:127–31

Bocchini V, Angeletti PU. 1969. The nerve growth factor: purification as a 30,000-molecular-weight protein. *Proc. Natl. Acad. Sci. USA* 64:787–94

Bothwell M. 1995. Functional interactions of neurotrophins and neurotrophin receptors. *Annu. Rev. Neurosci.* 18:223–53

Bothwell MA, Shooter EM. 1977. Dissociation equilibrium constant of beta nerve growth factor. *J. Biol. Chem.* 252:8532–36

Bothwell MA, Shooter EM. 1978. Thermodynamics of interaction of the subunits of 7S nerve growth factor. The mechanism of activation of the esteropeptidase activity by chelators. *J. Biol. Chem.* 253:8458–64

Bothwell MA, Wilson WH, Shooter EM. 1979. The relationship between glandular kallikrein and growth factor-processing proteases of mouse submaxillary gland. *J. Biol. Chem.* 254:7287–94

Burcham TS, Sim I, Bolin LM, Shooter EM. 1991. The NGF complex from the African rat *Mastomys natalensis*. *Neurochem. Res.* 16:603–12

Burdman JA, Goldstein MN. 1965. Synthesis and storage of a nerve growth protein in mouse submandibular glands. *J. Exp. Zool.* 160:183–88

Burton LE, Shooter EM. 1981. The molecular basis of the heterogeneity of the gamma subunit of 7S nerve growth factor. *J. Biol. Chem.* 256:11011–17

Burton LE, Wilson WH, Shooter EM. 1978. Nerve growth factor in mouse saliva. Rapid isolation procedures for and characterization of 7S nerve growth factor. *J. Biol. Chem.* 253:7807–12

Cohen S. 1959. Purification of metabolic effects of a nerve growth-promoting protein from snake venom. *J. Biol. Chem.* 234:1129–37

Cohen S. 1960. Purification of a nerve growth promoting protein from the mouse salivary gland and its neurotoxic antiserum. *Proc. Natl. Acad. Sci. USA* 46:302

Cohen S. 1964. Isolation and biological effects of an epidermal growth-stimulating protein. *Natl. Cancer Inst. Monogr.* 13:13–38

Cowan WM. 2001. Viktor Hamburger and Rita Levi-Montalcini: the path of discovery of nerve growth factor. *Annu. Rev. Neurosci.* 24:551–600

Darling TL, Petrides PE, Beguin P, Frey P, Shooter EM, et al. 1983. The biosynthesis and processing of proteins in the mouse 7S nerve growth factor complex. *Cold Spring Harbor Symp. Quant. Biol.* 48:427–34

Davies AM. 1994. The role of neurotrophins in the developing nervous system. *J. Neurobiol.* 25:1334–48

Davis BJ. 1964/1965. Disc electrophoresis. II. Method & application to human serum proteins. *Ann. NY Acad. Sci.* 121:404–27

Drinkwater CC, Barker PA, Suter U, Shooter EM. 1993. The carboxyl terminus of nerve growth factor is required for biological activity. *J. Biol. Chem.* 268:23202–7

Dunbar JC, Bradshaw RA. 1987. Amino acid sequence of guinea pig prostate kallikrein. *Biochemistry* 26:3471–78

Edwards RH, Selby MJ, Mobley WC, Weinrich SL, Hruby DE, Rutter WJ. 1988. Processing and secretion of nerve growth factor: expression in mammalian cells with a vaccinia virus vector. *Mol. Cell. Biol.* 8:2456–64

Edwards RH, Selby MJ, Rutter WJ. 1986. Differential RNA splicing predicts two distinct nerve growth factor precursors. *Nature* 319:784–87

Evans BA, Richards RI. 1985. Genes for the alpha and gamma subunits of mouse nerve growth factor are contiguous. *EMBO J.* 4:133–38

Fahnestock M, Bell RA. 1988. Molecular cloning of a cDNA encoding the nerve growth factor precursor from *Mastomys natalensis. Gene* 69:257–64

Fahnestock M, Woo JE, Lopez GA, Snow J, Walz DA, et al. 1991. β-NGF-endopeptidase: structure and activity of a kallikrein encoded by the gene mGK-22. *Biochemistry* 30:3443–50

Farhadi H, Pareek S, Day R, Dong W, Chretien M, et al. 1997. Prohormone convertases in mouse submandibular gland: co-localization of furin and nerve growth factor. *J. Histochem. Cytochem.* 45:795–804

Frazier WA, Hogue-Angeletti RA, Sherman R, Bradshaw RA. 1973. Topography of mouse 2.5S nerve growth factor. Reactivity of tyrosine and tryptophan. *Biochemistry* 12:3281–93

Frey P, Forand R, Maciag T, Shooter EM. 1979. The biosynthetic precursor of epidermal growth factor and the mechanism of its processing. *Proc. Natl. Acad. Sci. USA* 76:6294–98

Gandini Attardi D, Levi-Montalcini R, Wenger BS, Angeletti PU. 1965. Submaxillary gland of mouse: effects of a fraction on tissues of mesodermal origin in vitro. *Science* 150:1307–9

Greene LA, Shooter EM, Varon S. 1968. Enzymatic activities of mouse nerve growth factor and its subunits. *Proc. Natl. Acad. Sci. USA* 60:1383–88

Greene LA, Shooter EM, Varon S. 1969. Subunit interaction and enzymatic activity of mouse 7S nerve growth factor. *Biochemistry* 8:3735–41

Greene LA, Varon S, Piltch A, Shooter EM. 1971. Substructure of the β subunit of the mouse 7S nerve growth factor. *Neurobiology* 1:37–48

Grossman A, Lele KP, Sheldon J, Schenkein I, Levy M. 1969. The effect of esteroproteases from mouse submaxillary gland on growth of rat hepatoma cells in tissue culture. *Exp. Cell Res.* 54:260–63

Guerina NG, Pantazis NJ, Siminoski K, Anderson JK, McCarthy M, et al. 1986. Comparison of 7S nerve growth factor and nerve growth factor I from mouse submandibular glands. *Biochemistry* 25:754–60

Harper GP, Barde YA, Burnstock G, Carstairs JR, Dennison ME, et al. 1979. Guinea pig prostate is a rich source of nerve growth factor. *Nature* 279:160–62

Harper GP, Glanville RW, Thoenen H. 1982. The purification of nerve growth factor from bovine seminal plasma. Biochemical characterization and partial amino acid sequence. *J. Biol. Chem.* 257:8541–48

Harris-Warrick RM, Bothwell MA, Shooter EM. 1980. Subunit interactions inhibit the binding of beta nerve growth factor to receptors on embryonic chick sensory neurons. *J. Biol. Chem.* 255:11284–89

Hogue-Angeletti RA, Frazier WA, Jacobs JW,

Niall HD, Bradshaw RA. 1976. Purification, characterization, and partial amino acid sequence of nerve growth factor from cobra venom. *Biochemistry* 15:26–34

Hoffman H, McDougall. 1968. Some biological properties of proteins of the mouse submaxillary gland as revealed by growth of tissues on electrophoretic acrylamide gels. *Exp. Cell Res.* 51:485–503

Ip NY, Yancopoulos GD. 1996. The neurotrophins and CNTF: two families of collaborative neurotrophic factors. *Annu. Rev. Neurosci.* 19:491–515

Isackson PJ, Bradshaw RA. 1984. The alphasubunit of mouse 7S nerve growth factor is an inactive serine protease. *J. Biol. Chem.* 259:5380–83

Isackson PJ, Dunbar JC, Bradshaw RA, Ullrich A. 1985. The structure of murine 7S nerve growth factor: implications for biosynthesis. *Int. J. Neurosci.* 26:95–108

Isackson PJ, Nisco SJ, Bradshaw RA. 1987. Expression of the alpha subunit of 7S nerve growth factor in the mouse submandibular gland. *Neurochem. Res.* 12:959–66

Isackson PJ, Ullrich A, Bradshaw RA. 1984. Mouse 7S nerve growth factor: complete sequence of a cDNA coding for the alphasubunit precursor and its relationship to serine proteases. *Biochemistry* 23:5997–6002

Ishii DN, Shooter EM. 1975. Regulation of nerve growth factor synthesis in mouse submaxillary glands by testosterone. *J. Neurochem.* 25:843–51

Jeng I, Andres RY, Bradshaw RA. 1979. Mouse nerve growth factor: a rapid isolation procedure for the alpha and gamma subunits. *Anal. Biochem.* 92:482–88

Jongstra-Bilen J, Coblentz L, Shooter EM. 1989. The in vitro processing of the NGF precursors by the gamma-subunit of the 7S NGF complex. *Brain Res. Mol. Brain Res.* 5:159–69

Jovin TM. 1973. Multiphasic zone electrophoresis. I. Steady-state moving-boundary systems formed by different electrolyte combinations. *Biochemistry* 12:871–79

Jovin TM, Dante ML, Chrambach A. 1970. *Multiphasic Buffer Systems Output. PB 196085–196092 and 203016.* Springfield, VA: Natl. Tech. Inf. Serv.

Levi-Montalcini R. 1975. NGF: an unchartered route. In *Neurosciences: Paths of Discovery*, ed. FG Wooden, JP Swazey, G Adelman, pp. 243–65. Cambridge, MA: MIT. 622 pp.

Levi-Montalcini R. 1987. The nerve growth factor 35 years later. *Science* 237:1154–62

Levi-Montalcini R, Angeletti PU. 1964. Hormonal control of the NGF content in the submaxillary glands of mice. In *Salivary Glands and Their Secretions*, ed. LM Sreebny & J Meyer, pp. 12–41. Oxford, UK: Pergamon

Levi-Montalcini R, Angeletti PU. 1968. Nerve growth factor. *Physiol. Rev.* 48:534–69

Levi-Montalcini R, Meyer H, Hamburger V. 1954. *In vitro* experiments on the effects of mouse surcomes 180 and 37 on the spinal and sympathetic ganglia of the chick embryo. *Cancer Res.* 14:49–57

Lewin GR, Barde YA. 1996. Physiology of the neurotrophins. *Annu. Rev. Neurosci.* 19:289–317

McAllister AK, Katz LC, Lo DC. 1999. Neurotrophins and synaptic plasticity. *Annu. Rev. Neurosci.* 22:295–318

McDonald NQ, Lapatto R, Murray-Rust J, Gunning J, Wlodawer A, Blundell TL. 1991. New protein fold revealed by a 2.3-A resolution crystal structure of nerve growth factor. *Nature* 354:411–14

Meier R, Becker-Andre M, Gotz R, Heumann R, Shaw A, Thoenen H. 1986. Molecular cloning of bovine and chick nerve growth factor (NGF): delineation of conserved and unconserved domains and their relationship to the biological activity and antigenicity of NGF. *EMBO J.* 5:1489–93

Mobley WC. 1974. *Structural studies on nerve growth factor.* PhD thesis. Stanford, CA:Stanford Univ. 160 pp.

Mobley WC, Schenker A, Shooter EM. 1976. Characterization and isolation of proteolytically modified nerve growth factor. *Biochemistry* 15:5543–52

Moore JB Jr, Mobley WC, Shooter EM. 1974. Proteolytic modification of the beta nerve growth factor protein. *Biochemistry* 13:833–40

Moore JB Jr, Shooter EM. 1975. The use of hybrid molecules in a study of the equilibrium between nerve growth factor monomers and dimers. *Neurobiology* 5:369–81

Murphy RA, Landis SC, Bernanke J, Siminoski K. 1986. Absence of the alpha and gamma subunits of 7S nerve growth factor in denervated rodent iris: immunocytochemical studies. *Dev. Biol.* 114:369–80

Murphy RA, Saide JD, Blanchard MH, Young M. 1977. Molecular properties of the nerve growth factor secreted in mouse saliva. *Proc. Natl. Acad. Sci. USA* 74:2672–76

Nichols RA, Shooter EM. 1983. Characterization of the differential interaction of the microheterogeneous forms of the gamma subunit of 7S nerve growth factor with natural and synthetic ligands. *J. Biol. Chem.* 258:10296–303

Nichols RA, Shooter EM. 1985. Subunit interactions of the nerve and epidermal growth factor complexes: protection of the biological subunit from proteolytic modification. *Dev. Neurosci.* 7:216–29

Orenstein NS, Dvorak HF, Blanchard MH, Young M. 1978. Nerve growth factor: a protease that can activate plasminogen. *Proc. Natl. Acad. Sci. USA* 75:5497–500

Pantazis NJ. 1983. Nerve growth factor synthesized by mouse fibroblast cells in culture: absence of alpha and gamma subunits. *Biochemistry* 22:4264–71

Pantazis NJ, Murphy RA, Saide JD, Blanchard MH, Young M. 1977. Dissociation of the 7S-nerve growth factor complex in solution. *Biochemistry* 16:1525–30

Pattison SE, Dunn MF. 1975. On the relationship of zinc ion to the structure and function of the 7S nerve growth factor protein. *Biochemistry* 14:2733–39

Pattison SE, Dunn MF. 1976. On the mechanism of divalent metal ion chelator induced activation of the 7S nerve growth factor

esteropeptidase. Thermodynamics and kinetics of activation. *Biochemistry* 15:3696–703

Perez-Polo JR, De Jong WW, Straus D, Shooter EM. 1972. The physical and biological properties of 7S and beta-NGF from the mouse submaxillary gland. *Adv. Exp. Med. Biol.* 32:91–97

Perez-Polo JR, Shooter EM. 1974. Subunit affinities in the 7S nerve growth factor complex. *Neurobiology* 4:197–209

Perez-Polo JR, Shooter EM. 1975. The preparation and properties of nerve growth factor protein at alkaline pH. *Neurobiology* 5:329–38

Pignatti PF, Baker ME, Shooter EM. 1975. Solution properties of beta nerve growth factor protein and some of its derivatives. *J. Neurochem.* 25:155–59

Pulliam MW, Boyd LF, Baglan NC, Bradshaw RA. 1975. Specific binding of covalently cross-linked mouse nerve growth factor to responsive peripheral neurons. *Biochem. Biophys. Res. Commun.* 67:1281–89

Rubin JS, Bradshaw RA. 1981. Isolation and partial amino acid sequence analysis of nerve growth factor from the guinea pig prostate. *J. Neurosci. Res.* 6:451–64

Schenkein I, Bueker ED. 1964. The nerve growth factor as two essential components. *Ann. NY Acad. Sci.* 118:147–232

Schenkein I, Levy M, Bueker ED, Tokarsky E. 1968. Nerve growth factor of very high yield and specific activity. *Science* 159:640–43

Schwarz MA, Fisher D, Bradshaw RA, Isackson PJ. 1989. Isolation and sequence of a cDNA clone of beta-nerve growth factor from the guinea pig prostate gland. *J. Neurochem.* 52:1203–9

Scott J, Selby M, Urdea M, Quiroga M, Bell GI, Rutter WJ. 1983. Isolation and nucleotide sequence of a cDNA encoding the precursor of mouse nerve growth factor. *Nature* 302:538–40

Segal RA, Greenberg ME. 1996. Intracellular signaling pathways activated by neurotrophic factors. *Annu. Rev. Neurosci.* 19:463–89

Seidah NG, Benjannet S, Pareek S, Savaria

D, Hamelin J, et al. 1996. Cellular processing of the nerve growth factor precursor by the mammalian pro-protein convertases. *Biochem. J.* 314:951–60

Selby MJ, Edwards RH, Rutter WJ. 1987. Cobra nerve growth factor: structure and evolutionary comparison. *J. Neurosci. Res.* 18:293–98

Server AC, Herrup K, Shooter EM, Hogue-Angeletti RA, Frazier WA, Bradshaw RA. 1976a. Comparison of the nerve growth factor proteins from cobra venom (*Naja naja*) and mouse submaxillary gland. *Biochemistry* 15:35–39

Server AC, Shooter EM. 1976. Comparison of the arginine esteropeptidases associated with the nerve and epidermal growth factor. *J. Biol. Chem.* 251:165–73

Server AC, Sutter A, Shooter EM. 1976b. Modification of the epidermal growth factor affecting the stability of its high molecular weight complex. *J. Biol. Chem.* 251:1188–96

Server AC, Shooter EM. 1977. Nerve growth factor. *Adv. Prot. Chem.* 31:339–409

Silverman RE, Bradshaw RA. 1982. Nerve growth factor: subunit interactions in the mouse submaxillary gland 7S complex. *J. Neurosci. Res.* 8:127–36

Smith AP, Greene LA, Fish HR, Varon S, Shooter EM. 1969. Subunit equilibria of the 7S nerve growth factor protein. *Biochemistry* 8:4918–26

Smith AP, Varon S, Shooter EM. 1968. Multiple forms of the nerve growth factor protein and its subunits. *Biochemistry* 7:3259–68

Stach RW, Server AC, Pignatti PF, Piltch A, Shooter EM. 1976. Characterization of the gamma subunits of the 7S nerve growth factor complex. *Biochemistry* 15:1455–61

Stach RW, Shooter EM. 1974. The biological activity of cross-linked beta nerve growth factor protein. *J. Biol. Chem.* 249:6668–74

Stach RW, Shooter EM. 1980. Cross-linked 7S nerve growth factor is biologically inactive. *J. Neurochem.* 34:1499–505

Stach RW, Wagner BJ, Stach BM. 1977. A more rapid method for the isolation of the 7S nerve growth factor complex. *Anal. Biochem.* 83:26–32

Sutter A, Riopelle RJ, Harris-Warrick RM, Shooter EM. 1979. Nerve growth factor receptors. Characterization of two distinct classes of binding sites on chick embryo sensory ganglia cells. *J. Biol. Chem.* 254:5972–82

Taiwo YO, Levine JD, Burch RM, Woo JE, Mobley WC. 1991. Hyperalgesia induced in the rat by the amino-terminal octapeptide of nerve growth factor. *Proc. Natl. Acad. Sci. USA* 88:5144–48

Taylor JM, Cohen S, Mitchell WM. 1970. Epidermal growth factor: high and low molecular weight forms. *Proc. Natl. Acad. Sci. USA* 67:164–71

Taylor JM, Mitchell WM, Cohen S. 1974. Characterization of the high molecular weight form of epidermal growth factor. *J. Biol. Chem.* 249:3198–203

Thoenen H. 1995. Neurotrophins and neuronal plasticity. *Science* 270:593–98

Thoenen H, Barde YA. 1980. Physiology of nerve growth factor. *Physiol. Rev.* 60:1284–335

Thomas KA, Baglan NC, Bradshaw RA. 1981a. The amino acid sequence of the gamma-subunit of mouse submaxillary gland 7S nerve growth factor. *J. Biol. Chem.* 256:9156–66

Thomas KA, Silverman RE, Jeng I, Baglan NC, Bradshaw RA. 1981b. Electrophoretic heterogeneity and polypeptide chain structure of the gamma-subunit of mouse submaxillary 7S nerve growth factor. *J. Biol. Chem.* 256:9147–55

Ullrich A, Gray A, Berman C, Dull TJ. 1983. Human beta-nerve growth factor gene sequence highly homologous to that of mouse. *Nature* 303:821–25

Ullrich A, Gray A, Wood WI, Hayflick J, Seeburg PH. 1984. Isolation of a cDNA clone coding for the gamma-subunit of mouse nerve growth factor using a high-stringency selection procedure. *DNA* 3:387–92

Varon S, Nomura J, Shooter EM. 1967a. The

isolation of the mouse nerve growth factor protein in a high molecular weight form. *Biochemistry* 6:2202–9

Varon S, Nomura J, Shooter EM. 1967b. Subunit structure of a high-molecular-weight form of the nerve growth factor from mouse submaxillary gland. *Proc. Natl. Acad. Sci. USA* 57:1782–89

Varon S, Normura J, Shooter EM. 1968. Reversible dissociation of the mouse nerve growth factor protein into different subunits. *Biochemistry* 7:1296–303

Varon S, Shooter EM. 1970. The nerve growth factor proteins of the mouse submaxillary gland. In *Biochemistry of Brain and Behavior*, ed. R Bowman, SP Data, pp. 812–27. New York: Plenum

Wallace LJ, Partlow LM. 1976. β-Adrenergic regulation of secretion of mouse saliva rich in nerve growth factor. *Proc. Natl. Acad. Sci. USA* 73:4210–14

Wilson WH, Shooter EM. 1979. Structural modification of the NH 2 terminus of nerve growth factor. Purification and characterization of beta-nerve growth factor endopeptidase. *J. Biol. Chem.* 254:6002–9

Wlodawer A, Hodgson KO, Shooter EM. 1975. Crystallization of nerve growth factor from mouse submaxillary glands. *Proc. Natl. Acad. Sci. USA* 72:777–79

Young M. 1979. Proteolytic activity of nerve growth factor: a case of autocatalytic activation. *Biochemistry* 18:3050–55

Young M, Koroly MJ. 1980. Nerve growth factor zymogen. Stoichiometry of the active-site serine and role of zinc(II) in controlling autocatalytic self-activation. *Biochemistry* 19:5316–21

Young M, Saide JD, Murphy RA, Arnason BG. 1976. Molecular size of nerve growth factor in dilute solution. *J. Biol. Chem.* 251:459–64

Young M, Saide JD, Murphy RA, Blanchard. 1978. Nerve growth factor: multiple dissociation products in homogenates of the mouse submandibular gland. Purification and molecular properties of the intact undissociated form of the protein. *Biochemistry* 17:1490–98

Zanini A, Angeletti P, Levi-Montalcini R. 1968. Immunochemical properties of the nerve growth factor. *Proc. Natl. Acad. Sci. USA* 61:835–42

Annu. Rev. Neurosci. 2001. 24:631–51

SEQUENTIAL ORGANIZATION OF MULTIPLE MOVEMENTS: Involvement of Cortical Motor Areas

Jun Tanji

Department of Physiology, Tohoku University School of Medicine, Sendai 980, Japan, and The Core Research for Evolutional Science and Technology Program, Kawaguchi 332-0012, Japan; e-mail: tanjij@mail.cc.tohoku.ac.jp

Key Words temporal sequence, voluntary movement, primates, supplementary motor area, neuronal activity

■ **Abstract** Much of our normal behavior depends on the sequential execution of multiphased movements, or the execution of multiple movements arranged in a correct temporal order. This article deals with the issue of motor selection to arrange multiple movements in an appropriate temporal order, rather than the issue of constructing spatio-temporal structures in a single action. Planning, generating, and controlling the sequential motor behavior involves multiple cortical and subcortical neural structures. Studies on human subjects and nonhuman primates, however, have revealed that the medial motor areas in the frontal cortex and the basal ganglia play particularly important roles in the temporal sequencing of multiple movements. Cellular activity observed in the supplementary and presupplementary motor areas while performing specifically designed motor tasks suggests the way in which these areas take part in constructing the time structure for the sequential execution of multiple movements.

INTRODUCTION

In performing motor tasks, in order to attain our objective we usually need to execute more than one movement. Multiple single movements must be linked in a variety of spatial and temporal configurations for purposeful motor behavior. Thus, the generation and control of a sequence of movements is of crucial importance in many aspects of our life (Rosenbaum 1991). Lashley (1951) called the problem of coordinating constituent actions into organized sequential temporal patterns the "action syntax" problem. This article reviews current knowledge of how cortical structures are involved in arranging the voluntary limb movements of primates in a variety of temporal structures; it does not deal with more automated motor

0147-006X/01/0621-0631$14.00

631

behavior, such as occurs in locomotion, respiration, or mastication. Needless to say, the performance of complex motor actions, such as throwing a ball or swinging a tennis racket, requires sequential motor control. Although the issues of motor coordination and combining movement segments into a unified single action are of great interest (Soechting & Flanders 1992, Mussa-Ivaldi 1999), I do not go into that aspect of neural mechanisms. Rather, in this paper, I focus on the issue of selecting appropriate movements sequentially in a purposeful manner. At the end of this review, I propose a hypothesis to explain the cortical mechanism for sequencing multiple movements in temporal order.

CNS STRUCTURES INVOLVED

Overview

Although multiple cortical and subcortical structures are doubtlessly involved in planning, generating, and controlling sequential motor behavior, each area appears to be involved selectively in different aspects of sequential motor control. The primary motor cortex, known to be active during the execution (Evarts 1981) and preparation of movements (Tanji & Evarts 1976), appears to increase its activity when subjects perform sequential movements (Shibasaki et al 1993, Karni et al 1995, Catalan et al 1998). This increase in activity can be explained largely in terms of enhanced demand for signals specifying activation and suppression of motor apparatus to determine multiple motor parameters (Kettner et al 1996a), or additional requirements for preparatory processes (Kettner et al 1996b). A putative role of the primary motor cortex in some cognitive aspects may, in part, influence this activity (Georgopoulos 2000).

The cerebellum, which is essential for motor coordination as well as adaptive and predictive motor control (Ito 1984), is inevitably involved in programming and executing sequential movements. It is not yet known, however, whether the cerebellar activation observed during performance of sequential motor tasks (Seitz & Roland 1992, Doyon et al 1998) indicates its extensive involvement in coordinating or stabilizing movements (Thach et al 1992) or whether it suggests possible cerebellar involvement in more cognitive aspects of motor control (Middleton & Strick 1998, Imamizu et al 2000). The involvement of the basal ganglia in sequential motor tasks should also be considered (Middleton & Strick 2000) and is discussed in later sections.

On the other hand, posterior parietal areas, which have been implicated in visuospatial guidance of movements (Sakata et al 1995, Jeannerod et al 1995, Batista et al 1999), are active during sequential motor tasks (Catalan et al 1998, Grafton et al 1998). Because sequential motor tasks call for particularly elaborate spatial control of limb movements or spatial patterning of actions, it is understandable that enhanced parietal activation would follow. The premotor cortex is also thought to be involved in spatial guidance of movements (di Pellegrino & Wise

1993, Kurata 1994, Rizzolatti et al 1998, Binkofski et al 1999b), and this is likely to be a dominant factor causing activation of the premotor cortex while subjects perform sequential movements (Catalan et al 1998, Harrington et al 2000). The role of premotor cortex in temporal ordering of multiple movements has not been studied precisely under behavioral conditions minimizing spatial factors, although an interesting observation of premotor neurons during preparation for sequential movements was made (Kettner et al 1996b). Which areas are of particular importance in temporal, rather than spatial, sequencing? The following sections deal with this issue.

Involvement of Cortical Areas in Temporal Organization of Movements

Clinical Reports   By the 1930s, it was known that lesions in the frontal "premotor cortex" could impair the serial organization of movements without causing problems in performing single movements or producing defects in spatial motor control. Luria (cf Luria 1966) explicitly described disturbances in organizing movements in correct temporal sequences in patients with lesions in the "parasagittal division of the premotor cortex." A decade later, Laplane et al (1977) reported that patients with circumscribed ablations of the supplementary motor area (SMA), as defined by Penfield & Welch (1951), had a deficiency in performing alternating serial movements with both hands. Subsequently, a number of clinical reports have hinted at the involvement of the SMA in temporal sequencing of limb movements and speech (for review, see Goldberg 1985). In a systematic functional analysis of a patient with an SMA lesion, Dick et al (1986) reported a deficit in programming sequential limb movements. More recently, Halsband et al (1993) reported that patients with lesions in the SMA (and adjacent areas) had poor temporal control of finger movements. For oculomotor controlling regions, Pierrot-Deseilligny et al (1995) reported that patients with lesions in the supplementary eye field (located anterolateral to the SMA) were poor at performing sequential saccades, particularly in the absence of visual guidance. These reports all point to the importance of medial motor areas in the temporal sequencing of motor behavior.

There are few reports of involvement of premotor areas in the lateral hemisphere in temporal sequencing of motor behavior. This may be because of a paucity of clinical cases with circumscribed lesions limited to the lateral premotor cortex, or it may suggest that the lateral premotor cortex is not crucial to temporal organization.

The involvement of the prefrontal cortex in motor sequencing is well known (Kolb & Milner 1981, Kimura 1982). In temporal organization, however, the involvement of the prefrontal cortex is primarily in the aspect of structuring general patterns of behavior, making use of purposeful processing of perception and memory (Goldman-Rakic 1996, Shallice & Burgess 1996, Fuster 1997). The timekeeping function of the prefrontal cortex covers a whole range of behavioral aspects and is beyond the scope of this review.

Lesion Studies in Subhuman Primates   Only a small number of lesion studies have specifically looked for the involvement of particular areas of the brain in sequential motor performance. Brinkman (1984) made the first detailed report of the effects of lesions in the SMA on the motor sequence. Preoperatively, monkeys removed bait from slots in a Perspex plate in a systematic manner, proceeding from one slot to the next. Postoperatively, the systematic sequence for bait retrieval was lost. Passingham and his coworkers compared the effects of lesioning the lateral premotor cortex (PM) and the SMA. They showed that PM lesions impaired the monkey's ability to associate visual information with particular movements (Halsband & Passingham 1982). This finding is in line with a proposal that the PM plays a crucial role in the visual guidance of movements (Wise 1985, di Pellegrino & Wise 1993, Kurata 1994, Rizzolatti et al 1996, 1998). However, the ability of monkeys with PM lesions to perform different manipulations to an object in a particular sequence was not impaired (Halsband & Passingham 1982, Passingham 1985). Subsequently, they found that the ability of monkeys with SMA lesions to perform a sequence of movements to an object without visual cues was impaired (Halsband & Passingham 1987, Chen et al 1995).

Transient Chemical Inactivation   Two recent studies attempted to inactivate the portions of the medial frontal cortex now defined as the presupplementary motor area (pre-SMA) (see Matsuzaka et al 1992; cf Luppino et al 1991) and the newly defined SMA (Tanji 1996) transiently. Shima & Tanji (1998) locally injected muscimol into these areas while monkeys were performing a variety of motor tasks. Monkeys were not impaired in performing a simple target-reaching task with the arm after bilateral injection of muscimol into either the SMA or pre-SMA. The reaction time and movement time were not altered. However, when monkeys were required to perform three different arm movements separately in a remembered order (the task is outlined below), they were unable to perform the task. Nevertheless, at that stage of inactivation, the monkeys were still able to select and perform the three movements if instructed with a visual signal. Nakamura et al (1999) injected muscimol while monkeys performed a sequential button-pressing task. Injection into either the SMA or pre-SMA barely affected the performance of a well-learned, sequential button-pressing task, although the reaction time was slightly prolonged. The seemingly discrepant results from the two laboratories can be explained as follows. The sequential button-pressing task required continuous execution of limb manipulation, following a variety of trajectories in space, where the spatial factor was the critical control element. In contrast, the task of Shima & Tanji required the temporal arrangement of three discrete movements that were performed separately at variable intervals. Moreover, spatial factors were minimized and temporal sequencing was critical. Combining the results of the two studies indicates that the SMA and pre-SMA are not essential for performing movements involving simply reaching out to a target or a very well-learned, continuously performed motor task, even if it has multiple phases. In contrast, both areas are critically involved in sequencing multiple, separately performed movements over time. Furthermore, Nakamura et al showed that after the pre-SMA injection, there were more errors in performing a novel sequence of button presses. This may suggest a role of the pre-SMA in acquiring a novel motor sequence (Shima et al 1996) or its role in motor learning (Hikosaka et al 1999).

Brain Imaging   Numerous studies measuring regional changes in cerebral blood flow or metabolism have found active foci in the cerebral cortex associated with the performance of a variety of sequential movements by human subjects. Early studies that attempted to measure regional cerebral blood flow changes with ^{123}Xe injections reported focal activation of the medial frontal cortex, presumably covering the traditionally defined SMA (Orgogozo & Larsen 1979, Roland et al 1980). Since then, a large number of studies using positron emissions tomography techniques have looked at activation foci when subjects perform sequential movements. The literature on these studies is too large to overview, but I mention some typical findings. Excluding the primary motor and somatosensory areas, active foci related to sequential motor performance are found primarily in four cortical regions: the medial frontal cortex, including the SMA and cingulate cortex; the lateral premotor cortex; the lateral prefrontal cortex, including areas 9 and 46; and the parietal cortex (Deiber et al 1991, Grafton et al 1998). Selection or guidance of movements in space seems to be the most prominent factor calling for activation of the parietal cortex and lateral premotor cortex (Hazeltine et al 1997, Binkofski et al 1999b) whereas a high demand for cognitive behavioral control seems to require prefrontal activity. On the other hand, self determination of a movement sequence or complexity in the temporal structure leads to activation of the medial motor areas (Deiber et al 1991, Sergent et al 1992, Anderson et al 1994, Jenkins et al 1994, Picard & Strick 1996). This view is supported by reports of event-related cortical potentials (Lang et al 1992) and by a report using functional magnetic resonance imaging (Deiber et al 1999). Some studies have reported increased regional cerebral blood flow in the primary motor cortex, as well as in the SMA, while performing sequential hand movements (Shibasaki et al 1993). It needs to be clarified whether this increase merely reflects the elevated demand for sending excitatory or inhibitory output to produce appropriate muscular activity, or whether it is associated with planning and generating temporal structure for sequential motor output.

Magnetic Stimulation   A technique of transcranial magnetic stimulation that produces temporary functional lesions has been applied to human subjects for 15 years (Pascual-Leone et al 2000). Gerloff et al (1997) performed a systematic study in which multiple regions of the cerebral cortex were stimulated while human subjects performed sequential finger movements of differing complexity. They found that stimulation over the mesial frontal cortical areas (including the

SMA) induced errors only in the complex sequence whereas stimulation over the primary motor cortex induced errors in both the complex and the simple sequences. Stimulation over the lateral frontal or parietal areas did not interfere with the sequential performance at all. These findings indicate the critical role of the medial frontal cortex in organizing movements in complex motor sequences. This view is also supported by an earlier report (Muri et al 1995) that transcranial magnetic stimulation targeted at the cortical region, including the supplementary eye field, caused frequent errors in performing memory-guided sequential saccades.

Involvement of Subcortical Areas

Basal Ganglia   Basal ganglia exert their control functions on motor behavior through (a) the thalamocortical networks and (b) the brainstem networks (DeLong & Georgopoulos 1981). The modern concept of basal ganglia functions has been greatly advanced by the discovery of two principles. First, the main circuit of the basal ganglia, composed of serial inhibitory connections, links discrete sections of the basal ganglia with separate cortical areas (Parent & Hazrati 1995). Second, basal ganglia neurons exhibit behaviorally context-dependent activity during the execution of learned motor and cognitive tasks, as well as during learning of novel motor behavior (Hikosaka et al 2000). It is within this conceptual framework that the participation of the basal ganglia in organizing sequential motor actions is to be understood.

It is well established that each part of the basal ganglia is intimately connected with multiple areas in the cerebral cortex, forming cortico-basal ganglia loops (Alexander et al 1986, Middleton & Strick 2000). Given that cortical motor areas are involved in the performance of sequential motor behavior, it is reasonable to postulate that the basal ganglia that are intimately connected to them (Takada et al 1998) are also involved. Indeed, Benecke et al (1987) found a disturbance of sequential movements in patients with Parkinson's disease, using a task in which two hand/arm movements were performed sequentially in rapid succession. They found that not only were the individual movements slower, the interval between movements was also prolonged, and sometimes the switch from the first to the second movement could not be made. Later studies (Harrington & Haaland 1991, Agostino et al 1992, Martin et al 1994) also found impairments of sequential motor tasks in Parkinsonian patients and in patients with Huntington's disease (Thompson et al 1988).

These clinical studies on human subjects are supported by animal lesion studies. Berridge (Berridge & Whishaw 1992, Cromwell & Berridge 1996) reported that typical rule-governed sequential behavior seen in rat grooming behavior was impaired by the ablation of the striatum, but not by ablation of the neocortex or cerebellum. In primates, carbachol and atropine injection in the caudate nucleus affected the performance of sequential behavior (Van den Bercken & Cools 1982).

Miyachi et al (1997) studied the role of the basal ganglia in learning and executing sequential movements by training monkeys to perform a sequential button-press task (a 2 × 5 task) and examining how it was affected by muscimol injection. They found that injections in the anterior caudate and putamen affected the learning of new sequences, whereas injections in the middle-posterior putamen disrupted the execution of well-learned sequences. Recently, Matsumoto et al (1999) injected a neurotoxin, MPTP, into the caudo-putamen of monkeys and concluded that the nigrostriatal dopamine system was necessary for both learning and executing a sequential motor task.

Brain-imaging studies in human subjects have confirmed the involvement of the basal ganglia in sequential motor behavior. For instance, Boecker et al (Petersen et al 1985) found (with H_2 [15]OPET) an active focus in the anterior globus pallidus when subjects performed sequential finger movements, and the increase in rCBF was correlated with increasing sequence complexity. Jueptner & Weiller (1998) reported that the anterior striatum and globus pallidus were active during learning of new motor sequences whereas the posterior putamen was more active during performance of prelearned sequences. Stimulus properties and attentional constraints influenced activation of the basal ganglia and cortical areas (Hazeltine et al 1997).

Cerebellum   Thach (1998) proposed recently that the role of the cerebellum is to provide a substrate for the combination of single-response elements into larger groupings, so that the occurrence of a sensory or experiential context automatically triggers a combined response, reviewing the key contribution of parallel fibers to this function. According to this view, the cerebellum is likely to be involved in sequencing of multiple movements, in addition to its role in coordinating and combining movements into single actions. Computational frameworks that describe how the cerebellum executes its role in such operations have also been provided (Wolpert & Kawato 1998). However, despite numerous neuroimaging studies of the cerebellum in task-performing subjects (Desmond & Fiez 1998), its involvement in temporal sequencing of actions (and not in spatial patterning) has not been unequivocally demonstrated with this technique. On the other hand, the cerebellum has been implicated in learning sequential movements (Sanes et al 1990, Molinari et al 1997, Lidow et al 1989, Jenkins et al 1994, Jueptner & Weiller 1998). Unfortunately, results of an experimental study in monkeys did not accord with the brain-imaging data. Muscimol injection into any of the three cerebellar nuclei of monkeys failed to affect learning of new motor sequences (Lu et al 1998). Instead, injections into the dorsal and central part of the dentate nucleus increased the number of errors in performing learned motor sequences, and movements were slower. Mushiake & Strick (1993) analyzed neuronal activity in the dentate nucleus of monkeys performing a sequential button-pressing task, under remembered-sequence and visual-tracking condition. Neurons in the ventrolateral dentate were preferentially more active during the tracking condition, whereas other dentate neurons did not show a preference. These studies indicate the involvement of cerebellar structures in generating spatial trajectories or in developing the motor skill necessary to exert visuospatial control. More work is necessary to examine the role of cerebellum in performing, or learning to perform, multiple movements in a correct temporal order.

LEARNING

The learning of sequential processes is a very important aspect that must be studied to understand the neural mechanisms for generating and controlling the sequential performance of motor behavior. This subject was reviewed and discussed recently in excellent review articles (Hikosaka et al 1998, 1999). Mechanisms both for procedural learning to acquire skills in performing continuous execution of multiphased actions and for learning to perform visuospatial sequences have been studied. On the other hand, neural mechanisms for learning to organize multiple actions in temporal sequences have scarcely been studied.

CELLULAR ACTIVITY DURING SEQUENTIAL TASK PERFORMANCE

Barone & Joseph (1989) trained monkeys to perform a delayed spatial sequencing task, in which they had to move their eyes and an arm in complex sequences. A visual target appeared in various spatial sequences, and after a delay the animal followed the sequence with saccades and then arm reaching. They found two types of cellular activity related to the task sequence in the prefrontal cortex. First, visual-tonic type cells were active selectively depending on the spatial sequence of illumination of the visual target. Second, context-dependent type cells were active during visual fixation of a given spatial target, but only selectively in a context that depended on which other targets had been or were going to be presented. The first type of activity appeared to be useful for memorizing spatial sequences of events. The authors interpreted the second type of context-dependent activity as the neural trace of a representation of the state of the sequence constructed from cognitive information. That report did not describe sequence-selective saccade-related or reaching-related activity.

In contrast to activities reflecting sequences of sensory and cognitive information, cellular activity selective for the spatial sequence of planned movements was reported in the traditionally defined supplementary motor area (Mushiake et al 1990). For instance, Mushiake et al found cellular activity before initiating a motor task that involved pressing buttons in the sequence top-bottom-right. The activity was not observed when the motor sequence was top-right-bottom or top-left-right. Subsequently, the same authors compared cellular activity in the supplementary, premotor, and primary motor areas during visually guided and internally determined sequential movements (Mushiake et al 1991). Monkeys were trained to push three buttons sequentially under two conditions. In one condition, three button presses were individually guided and triggered with illumination of the buttons. In the other, three sequential presses were guided by memorized information. More than half of the SMA cells were preferentially active during the sequential task based on memory. In contrast, more than half of the cells in the lateral premotor cortex were more active during the visually guided sequential task. Although a simple dichotomy was ruled out, the sequence-selective SMA activity relied more on memory-based information whereas the premotor activity relied more on visual information. In the primary motor cortex, cellular activity occurred during both motor preparation and motor execution. Most of the activity in this area was not affected by whether the motor sequence was memory based or visually guided.

Early studies on neuronal activity in the basal ganglia have hinted at its participation in sequential motor behavior (Kimura 1990, Brotchie et al 1991, Kimura et al 1992). In 1995, studies of neuronal activity in the caudate nucleus (Kermadi & Joseph 1995) and the pallidum (Mushiake & Strick 1995) revealed two interesting aspects of activity related to sequential motor performance. One was activity selective to a particular spatial sequence of target-reaching arm movements or saccades (sequence-selective activity). The other was movement-related activity selective to a particular serial position in the sequence (rank-order-selective or phase-selective activity). In addition, a variety of context-dependent activities were found among caudate neurons.

PLANNING AND EXECUTING MULTIPLE MOVEMENTS IN A SEQUENCE

In the single-cell studies described above, animal subjects performed sequential reaching movements where the spatial sequence was crucial. In these studies, selection of spatial trajectories in successive reaching was inevitably critical to control. On the other hand, selection of multiple movements performed separately, but in succession, is a separate issue from trajectory determination. When we intend to drink beer, we place a glass on a table, use a bottle opener, and pour beer into the glass. When we select the three movements in the correct sequence, we plan the time sequence of discrete movements performed separately, rather than the trajectories of the movements.

To study how neurons in the medial motor areas participate in performing sequential multiple movements that are individually separated in time, we analyzed neuronal activity in two areas (Tanji & Shima 1994, Shima & Tanji 2000): the pre-SMA and the newly defined SMA (Matsuzaka & Tanji 1996). Monkeys were trained to perform three different movements, separated by a waiting time, in six different orders. Initially, each series of movements was learned during five trials guided by visual signals that indicated the correct movements. The monkeys subsequently executed the three movements in the memorized order without the visual signals. Three types of neuronal activity were of particular interest and appeared to be crucially involved in sequencing the multiple motor tasks in different orders. First, we found activity changes that were selective for a particular sequence of three movements that monkeys were prepared to perform (Tanji & Shima 1994). Figure 1 shows a typical example of this sequence-selective activity. The

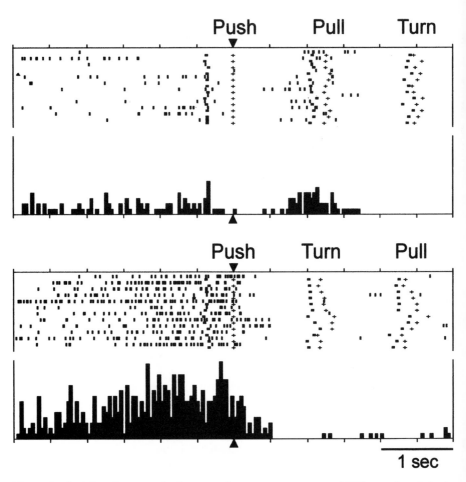

Figure 1 Activity of a neuron in the presupplementary motor area exhibiting preferential relation to a specific order of three movements performed without sensory guidance. This neuron is active during a waiting period before initiating the first movement, but only if the sequence of upcoming movements is in the order of PUSH, TURN, PULL (bottom). (Top) PUSH-PULL-TURN sequence. In raster displays, each row represents a trial, and dots represent individual discharges of this neuron. Small crosses denote the time of occurrence of the movement onset. In histograms, discharges over 12 trials are summated. Bin width for the display purpose is 40 ms. Triangles at the top of each raster indicate the start of the first movement.

neuronal activity recorded in the SMA increased before the animal initiated the first movement, PUSH, only when the second movement was TURN, and the third movement was PULL. The selective activity was not observed when the sequence was PUSH-PULL-TURN or any four of others. The sequence-selective activity ceased when the monkeys initiated the first movement. Second, we found interval-selective activity that appeared in the interval between one particular movement and another particular movement. In the SMA neuron shown in Figure 2, the activity was most prominent during the interval after the execution of PULL and before the execution of PUSH. Third, we found neuronal activity representing the rank-order of three movements arranged chronologically; that is, the activity differed selectively in the process of preparing the first, second, or third movements in individual trials. An example of rank-order–selective preparatory activity is shown in Figure 3. The pre-SMA neuron was active only during the third preparatory period, irrespective of the sequence of the three movements. Interval-selective activity was more prevalent in the SMA, whereas rank-order–selective activity was more frequently recorded in the pre-SMA. These results suggest how the neurons in both the SMA and pre-SMA are involved in sequencing multiple movements over time.

Our previous study inactivating the SMA and pre-SMA (Shima & Tanji 1998) indicated that the neuronal activities described above play a number of crucial

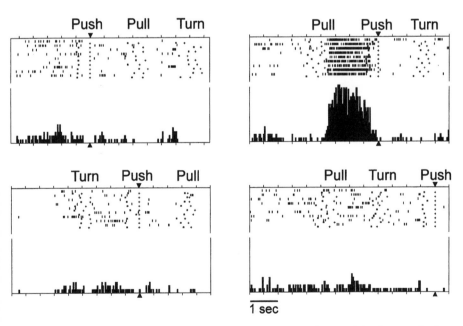

Figure 2 Discharges of a supplementary motor area neuron whose activity increased selectively during the interval after completion of a particular movement, PULL, and before the initiation of another particular movement, PUSH. Display format is the same as in Figure 1.

Figure 3 Discharges of a presupplementary motor area neuron whose activity increased while the monkey was preparing to initiate the third movement, irrespective of the sequence of three movements.

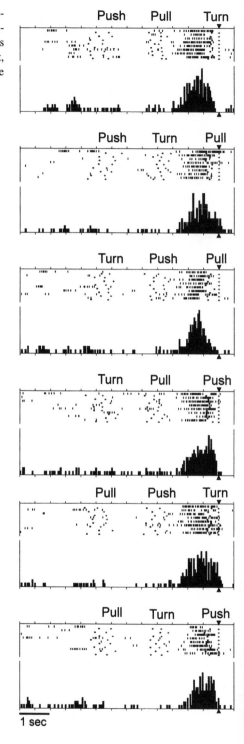

roles in organizing multiple movements in the correct temporal order. Therefore, in the next section, we hypothesize how each of these types of neuronal activity can be used to accomplish the sequencing task.

How Neuronal Activity Arranges Multiple Movements in the Correct Temporal Order—A Hypothesis

Let us consider how the neuronal activities in the SMA and pre-SMA described above might work together to provide the signals necessary to perform movements *A*, *B*, and *C*, in that order (Figure 4 shows the processes and actions necessary to perform the sequence). What kinds of neural components, with what properties, are required to provide the necessary signals? First, we need a component that retrieves and carries the information that the currently required sequence is *A-B-C*. The sequence-selective, preparatory activity found in the SMA and pre-SMA serves this function. We also found plenty of preparatory activity selective for the upcoming movement. This neuronal activity would make the preparations to start the first movement, *A*. The addition of a trigger signal following this preparatory process automatically starts the first movement, *A*. Once movement *A* is accomplished, the sequence information carried in the sequence-selective, preparatory neurons ceases, as seen in the neuronal activity exemplified in Figure 1. What mechanism (*a*) indicates that the next movement is movement *B* and (*b*) holds that information during the waiting period until receiving the signal to start the second movement? This requires an element that carries information to connect the occurrence of movement *B* after the occurrence of movement *A*. We propose that the interval-selective neuronal activity we observed serves as the element linking the occurrence of *B* after *A*, because this activity starts after movement *A* has occurred and ends before the initiation of movement *B*. This tonic signal is useful for feeding information to the neural component involved in the preparatory process before initiating movement *B*. Alternatively, the tonic signal may continuously activate a neural component that triggers a command to initiate movement *B* after receiving an external signal. When movement *B* is executed, the next step is to connect the occurrence of movement *C* after movement *B*. This can be done by a linking element that is activated after movement *B* and lasts until the initiation of movement *C* (see Figure 2). This linking element feeds tonic input to the next element, which produces the output to initiate movement *C*. When the third movement is complete, neuronal activity after the third movement might be an "end" signal, reporting completion of the sequence. We observed such activity predominantly in the pre-SMA.

If this hypothesis is correct, then it follows that the linking elements described should be involved in the process as follows. To perform the sequence *A-B-C*, elements linking *A* → *B* and *B* → *C* should be activated. The element linking *A* → *B* should come into play after the end of the first movement in the sequence and long before the initiation of the second movement. The element linking *B* → *C* should come into play after completion of the second movement in the sequence and long

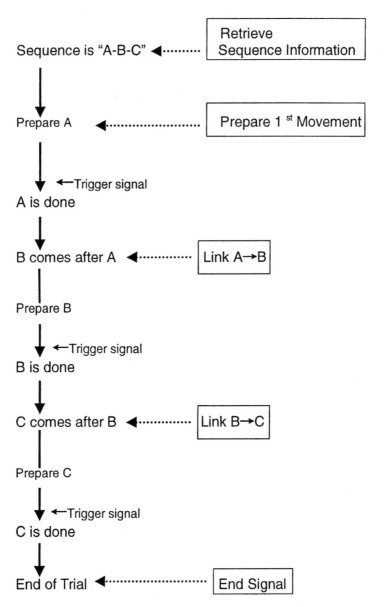

Figure 4 Processes and necessary actions for the orderly performance of three movements with time intervals when the sequence is in the order of A, B, and C. The information for the correct sequence should lead to a chain of events that gives rise to orderly delivery of output signals appropriate for commanding three movements in a correct sequence. The timing of initiation of each movement is externally determined with a trigger signal. [Adapted from Shima & Tanji (2000).]

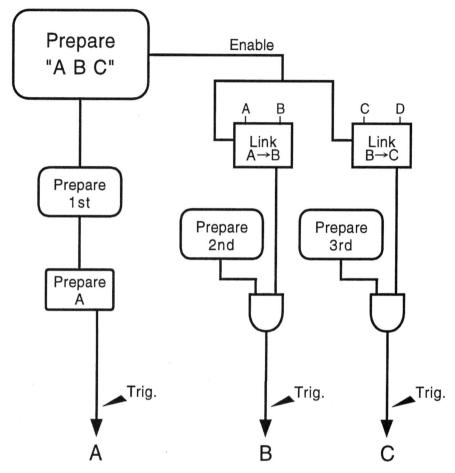

Figure 5   A diagram illustrating a hypothetical network connecting the neural elements found in medial cortical motor areas. If the preparatory elements delivering signals with four different properties feed outputs to linking elements and output elements in the way drawn, then for automatic delivery of output signals they will contribute in the order of *A*, *B*, and *C*. [Reproduced with permission from Shima & Tanji (2000).]

before initiation of the third movement. These steps require information about which of the three movements in the sequence is next. This information can be provided by the many rank-order–selective neurons that we found. If the neural system has a way of telling whether to prepare the second or third movement, the elements linking $A \rightarrow B$ and $B \rightarrow C$ should be incorporated at the appropriate times. In other words, the rank-order–selective elements serve to regulate the linking-element information. In our studies (Tanji & Shima 1994, Shima & Tanji 2000), we observed three types of interval-selective activity: activity that appeared (*a*) nonselectively before the second and third movement, (*b*) selectively before the second movement, or (*c*) selectively before the third movement. These neurons are candidates for the linking elements that appear before and after regulatory actions by the rank-order elements. Our hypothesis on the use of neural elements found in the SMA and pre-SMA is summarized in Figure 5. The diagram illustrates the basic actions of each element, which could be hardwired to generate the signals appropriate to performing the three movements in the order *A*, *B*, and *C*. We hasten to add, however, that much study remains to determine whether the neural structures in the SMA and pre-SMA actually operate in a way similar to that proposed. Moreover, the illustration may be too simple to account for the actual operation of neuronal elements. Neural structures for the sequential control of motor behavior may include components that are not readily comprehensible intuitively. Neural components may interact in a complex manner and may undergo use-dependent alteration.

REMAINING QUESTIONS

We are beginning to understand how neural structures generate and control sequential motor behavior, but many questions remain unanswered. This article is aimed at mechanisms for temporal, rather than spatial, aspects of behavioral sequence, discussing extensively the participation of medial motor areas. We need to study more the participation of other cortical and subcortical areas. The process of learning or acquiring sequential motor acts remains a central issue. We need to study whether the neural components found to participate in a particular sequential motor behavior are used to control other sequential behaviors in a general-purpose fashion (i.e. for the sequence *X-Y-X* as well as for *A-B-C*). Moreover, it will be interesting to study the mechanisms for the control of actions with long sequences, such as the playing of musical instruments.

Visit the Annual Reviews home page at www.AnnualReviews.org

LITERATURE CITED

Agostino R, Berardelli A, Formica A, Accornero N, Manfredi M. 1992. Sequential arm movements in patients with Parkinson's disease, Huntington's disease and dystonia. *Brain* 115 (Part 5):1481–95

Alexander GE, DeLong MR, Strick PL. 1986. Parallel organization of functionally segregated circuits linking basal ganglia and cortex. *Annu. Rev. Neurosci.* 9:357–81

Anderson TJ, Jenkins IH, Brooks DJ, Hawken MB, Frackowiak RS, Kennard C. 1994. Cortical control of saccades and fixation in man. A PET study. *Brain* 117(Part 5):1073–84

Barone P, Joseph JP. 1989. Prefrontal cortex and spatial sequencing in macaque monkey. *Exp. Brain Res.* 78:447–64

Batista AP, Buneo CA, Sbyder LH, Andersen RA. 1999. Reach plans in eye-centered coordinates. *Science* 285:257–260

Benecke R, Rothwell JC, Dick JP, Day BL,

Marsden CD. 1987. Disturbance of sequential movements in patients with Parkinson's disease. *Brain* 110:361–79

Berridge KC, Whishaw IQ. 1992. Cortex, striatum and cerebellum: control of serial order in a grooming sequence. *Exp. Brain Res.* 90(2):275–90

Binkofski F, Buccino G, Posse S, Seitz RJ, Rizzolatti G, Freund H. 1999a. A fronto-parietal circuit for object manipulation in man: evidence from an fMRI-study. *Eur. J. Neurosci.* 11(9):3276–86

Binkofski F, Buccino G, Stephan KM, Rizzolatti G, Seitz RJ, Freund HJ. 1999b. A parieto-premotor network for object manipulation: evidence from neuroimaging. *Exp. Brain Res.* 128(1–2):210–13

Brinkman C. 1984. Supplementary motor area of the monkey's cerebral cortex: short- and long-term deficits after unilateral ablation and the effects of subsequent callosal section. *J. Neurosci.* 4:918–29

Brotchie P, Iansek R, Horne MK. 1991. Motor function of the monkey globus pallidus. 2. Cognitive aspects of movement and phasic neuronal activity. *Brain* 114:1685–702

Catalan MJ, Honda M, Weeks RA, Cohen LG, Hallet M. 1998. The functional neuroanatomy of simple and complex sequential finger movements: a PET study. *Brain* 121:253–64

Chen YC, Thaler D, Nixon PD, Stern CE, Passingham RE. 1995. The functions of the medial premotor cortex. II. The timing and selection of learned movements. *Exp. Brain Res.* 102(3):461–73

Cromwell HC, Berridge KC. 1996. Implementation of action sequences by a neostriatal site: a lesion mapping study of grooming syntax. *J. Neurosci.* 16(10):3444–58

Deiber MP, Honda M, Ibanez V, Sadato N, Hallett M. 1999. Mesial motor areas in self-initiated versus externally triggered movements examined with fMRI: effect of movement type and rate. *J. Neurophysiol.* 81(6):3065–77

Deiber MP, Passingham RE, Colebatch JG,

Friston KJ, Nixon PD, Frackowiak RS. 1991. Cortical areas and the selection of movement: a study with positron emission tomography. *Exp. Brain Res.* 84:393–402

DeLong MR, Georgopoulos AP. 1981. Motor functions of the basal ganglia. In *Handbook of Physiology, The Nervous System II, Part 2,* ed. VB Brooks, 23:1017–61. Bethesda, MD: Am. Physiol. Soc.

Desmond JE, Fiez JA. 1998. Neuroimaging studies of the cerebellum: language, learning and memory. *Trends Cogn. Sci.* 2:355–62

Dick JPR, Benecke R, Rothwell JC, Day BL, Marsden CD. 1986. Simple and complex movements in a patient with infarction of the right supplementary motor area. *Mov. Disord.* 1(4):255–66

di Pellegrino G, Wise SP. 1993. Visuospatial versus visuomotor activity in the premotor and prefrontal cortex of a primate. *J. Neurosci.* 13:1227–43

Doyon J, Laforce R, Bouchard G, Gaudreau D, Roy J, et al. 1998. Role of the striatum, cerebellum and frontal lobes in the automatization of a repeated visuomotor sequence of movements. *Neuropsychologia* 36:625–41

Evarts EV. 1981. Role of motor cortex in voluntary movements in primates. See DeLong & Georgopoulos 1981, pp. 1083–1120

Fuster JM. 1997. *The Prefrontal Cortex: Anatomy, Physiology, and Neuropsychology of the Frontal Lobe.* Philadelphia: Lippincott-Raven

Georgopoulos AP. 2000. Neural aspects of cognitive motor control. *Curr. Opin. Neurobiol.* 10:238–41

Gerloff C, Corwell B, Chen R, Hallett M, Cohen LG. 1997. Stimulation over the human supplementary motor area interferes with the organization of future elements in complex motor sequences. *Brain* 120(Part 9):1587–602

Goldberg G. 1985. Supplementary motor area structure and function: review and hypotheses. *Behav. Brain Res.* 8:567–615

Goldman-Rakic PS. 1996. The prefrontal landscape: implications of functional architecture for understanding human mentation and the central executive. *Philos. Trans. R. Soc. London Ser.* B 351:1445–53

Grafton ST, Hazeltine E, Ivry RB. 1998. Abstract and effector-specific representations of motor sequences identified with PET. *J. Neurosci.* 18(22):9420–28

Halsband U, Ito N, Tanji J, Freund HJ. 1993. The role of premotor cortex and the supplementary motor area in the temporal control of movement in man. *Brain* 116(Part 1):243–66

Halsband U, Passingham RE. 1982. The role of premotor and parietal cortex in the direction of action. *Brain Res.* 240:368–72

Halsband U, Passingham RE. 1987. Higher distrubances of movement in monkeys (*Macaca mulatta*). In *Motor Control*, ed. GN Gantchev, B Dimitev, PC Gatev, pp. 79–85. New York: Plenum

Harrington DL, Haaland KY. 1991. Sequencing in Parkinson's disease. Abnormalities in programming and controlling movement. *Brain* 114:99–115

Harrington DL, Rao SM, Haaland KY. 2000. Specialized neural systems underlying representations of sequential movements. *J. Cogn. Neurosci.* 12:56–77

Hazeltine E, Grafton ST, Ivry R. 1997. Attention and stimulus characteristics determine the locus of motor-sequence encoding. A PET study. *Brain* 120(Part 1):123–40

Hikosaka O, Miyashita K, Miyachi S, Sakai K, Lu X. 1998. Differential roles of the frontal cortex, basal ganglia, and cerebellum in visuomotor sequence learning. *Neurobiol. Learn. Mem.* 70:137–49

Hikosaka O, Nakahara H, Rand MK, Sakai K, Lu X, et al. 1999. Parallel neural networks for learning sequential procedures. *Trends Neurosci.* 22(10):464–71. Erratum 1999. *Trends Neurosci.* 22(12):569

Hikosaka O, Takikawa Y, Kawagoe R. 2000. Role of the basal ganglia in the control of purposive saccadic eye movements. *Physiol. Rev.* 80(3):953–78

Imamizu H, Miyauchi S, Tamada T, Sasaki Y, Takino R, et al. 2000. Human cerebellar activity reflecting an acquired internal model of a new tool. *Nature* 403:192–95

Ito M. 1984. *The Cerebellum and Neural Control*. New York: Raven

Jeannerod M, Arbib MA, Rizzolatti G, Sakata H. 1995. Grasping objects: the cortical mechanisms of visuomotor transformation. *Trends Neurosci.* 18:314–20

Jenkins IH, Brooks DJ, Nixon PD, Frackowiak RS, Passingham RE. 1994. Motor sequence learning: a study with positron emission tomography. *J. Neurosci.* 14:3775–90

Jueptner M, Weiller C. 1998. A review of differences between basal ganglia and cerebellar control of movements as revealed by functional imaging studies. *Brain* 121 (Part 8):1437–49

Karni A, Meyer G, Jezzard P, Adams MM, Turner R, Ungerleider LG. 1995. Functional MRI evidence for adult motor cortex plasticity during motor skill learning. *Nature* 377:155–58

Kermadi I, Joseph JP. 1995. Activity in the caudate nucleus of monkey during spatial sequencing. *J. Neurophysiol.* 74(3):911–33

Kettner RE, Marcario JK, Clark-Phelps MC. 1996a. Control of remembered reaching sequences in monkey. I. Activity during movement in motor and premotor cortex. *Exp. Brain Res.* 112(3):335–46

Kettner RE, Marcario JK, Port NL. 1996b. Control of remembered reaching sequences in monkey. II. Storage and preparation before movement in motor and premotor cortex. *Exp. Brain Res.* 112(3):347–58

Kimura D. 1982. Left-hemisphere control of oral and brachial movements and their relation to communication. *Philos. Trans. R. Soc. London Ser.* B 298(1089):135–49

Kimura M. 1990. Behaviorally contingent property of movement-related activity of the primate putamen. *J. Neurophysiol.* 63:1277–96

Kimura M, Aosaki T, Hu Y, Ishida A, Watanabe K. 1992. Activity of primate putamen neurons is selective to the mode of voluntary movement: visually guided, self-initiated or memory-guided. *Exp. Brain Res.* 89:473–77

Kolb B, Milner B. 1981. Performance of complex arm and facial movements after focal brain lesions. *Neuropsychologia* 19(4):491–503

Kurata K. 1994. Information processing for motor control in primate premotor cortex. *Behav. Brain Res.* 61(2):135–42

Lang W, Beisteiner R, Lindinger G, Deecke L. 1992. Changes of cortical activity when executing learned motor sequences. *Exp. Brain Res.* 89(2):435–40

Laplane D, Talairach J, Meininger V, Bancaud J, Orgogozo JM. 1977. Clinical consequences of corticectomies involving the supplementary motor area in man. *J. Neurol. Sci.* 34:301–14

Lashley KS. 1951. The problem of serial order in behavior. In *Cerebral Mechanisms in Behavior*, ed. LA Jeffress, pp. 112–36. New York: Wiley

Lidow MS, Goldman-Rakic PS, Gallager DW, Rakic P. 1989. Quantitative autoradiographic mapping of serotonin 5-HT1 and 5-HT2 receptors and uptake sites in the neocortex of the rhesus monkey. *J. Comp. Neurol.* 280:27–42

Lu X, Hikosaka O, Miyachi S. 1998. Role of monkey cerebellar nuclei in skill for sequential movement. *J. Neurophysiol.* 79(5):2245–54

Luppino G, Matelli M, Camarda RM, Gallese V, Rizzolatti G. 1991. Multiple representations of body movements in mesial area 6 and the adjacent cingulate cortex: an intracortical microstimulation study in the macaque monkey. *J. Comp Neurol.* 311(4):463–82

Luria AR. 1966. *Higher Cortical Functions in Man.* London: Tavistokc

Martin KE, Phillips JG, Iansek R, Bradshaw JL. 1994. Inaccuracy and instability of sequential movements in Parkinson's disease. *Exp. Brain Res.* 102(1):131–40

Matsumoto N, Hanakawa T, Maki S, Graybiel AM, Kimura M. 1999. Role of [corrected] nigrostriatal dopamine system in learning to perform sequential motor—tasks in a predic-tive manner. *J. Neurophysiol.* 82(2):978–98. Erratum 1999. *J. Neurophysiol.* 82(5):13

Matsuzaka Y, Aizawa H, Tanji J. 1992. A motor area rostral to the supplementary motor area (presupplementary motor area) in the monkey: neuronal activity during a learned motor task. *J. Neurophysiol.* 68(3):653–62

Matsuzaka Y, Tanji J. 1996. Changing directions of forthcoming arm movements: neuronal activity in the presupplementary and supplementary motor area of monkey cerebral cortex. *J. Neurophysiol.* 76(4):2327–42

Middleton FA, Strick PL. 1998. The cerebellum: an overview. *Trends Neurosci.* 21(9):367–69

Middleton FA, Strick PL. 2000. Basal ganglia and cerebellar loops: motor and cognitive circuits. *Brain Res. Rev.* 31:236–50

Miyachi S, Hikosaka O, Miyashita K, Karadi Z, Rand MK. 1997. Differential roles of monkey striatum in learning of sequential hand movement. *Exp. Brain Res.* 115(1):1–5

Molinari M, Leggio MG, Solida A, Ciorra R, Misciagna S, et al. 1997. Cerebellum and procedural learning: evidence from focal cerebellar lesions. *Brain* 120(Part 10):1753–62

Muri RM, Rivaud S, Vermersch AI, Leger JM, Pierrot-Deseilligny C. 1995. Effects of transcranial magnetic stimulation over the region of the supplementary motor area during sequences of memory-guided saccades. *Exp. Brain Res.* 104(1):163–66

Mushiake H, Inase M, Tanji J. 1990. Selective coding of motor sequence in the supplementary motor area of the monkey cerebral cortex. *Exp. Brain Res.* 82(1):208–10

Mushiake H, Inase M, Tanji J. 1991. Neuronal activity in the primate premotor, supplementary, and precentral motor cortex during visually guided and internally determined sequential movements. *J. Neurophysiol.* 66(3):705–18

Mushiake H, Strick PL. 1993. Preferential activity of dentate neurons during limb movements guided by vision. *J. Neurophysiol.* 70:2660–64

Mushiake H, Strick PL. 1995. Pallidal neuron activity during sequential arm movements. *J. Neurophysiol.* 74(6):2754–58

Mussa-Ivaldi FA. 1999. Modular features of motor control and learning. *Curr. Opin. Neurobiol.* 9:713–17

Nakamura K, Sakai K, Hikosaka O. 1999. Effects of local inactivation of monkey medial frontal cortex in learning of sequential procedures. *J. Neurophysiol.* 82(2):106368

Orgogozo JM, Larsen B. 1979. Activation of the supplementary motor area during voluntary movement in man suggests it works as a supramotor area. *Science* 206:847–50

Parent A, Hazrati L-N. 1995. Functional anatomy of the basal ganglia. I. The cortico-basal ganglia-thalamo-cortical loop. *Brain Res. Rev.* 20:91–127

Pascual-Leone A, Walsh V, Rothwell J. 2000. Transcranial magnetic stimulation in cognitive neuroscience. *Curr. Opin. Neurobiol.* 10(2):232–37

Passingham RE. 1985. Premotor cortex: sensory cues and movement. *Behav. Brain Res.* 18(2):175–85

Penfield W, Welch K. 1951. The supplementary motor area of the cerebral cortex. *Arch. Neurol. Psychiatr.* 66:289–317

Petersen SE, Robinson DL, Keys W. 1985. Pulvinar nuclei of the behaving rhesus monkey: visual responses and their modulation. *J. Neurophysiol.* 54:867–86

Picard N, Strick PL. 1996. Motor areas of the medial wall: a review of their location and functional activation. *Cereb. Cortex* 6(3):342–53

Pierrot-Deseilligny C, Rivaud S, Gaymard B, Muri R, Vermersch A-I. 1995. Cortical control of saccades. *Ann. Neurol* 37:557–67

Rizzolatti G, Fadiga L, Gallese V, Fogassi L. 1996. Premotor cortex and the recognition of motor actions. *Cogn. Brain Res.* 3(2):131–41

Rizzolatti G, Luppino G, Matelli M. 1998. The organization of the cortical motor system: new concepts. *Electroencephalogr. Clin. Neurophysiol.* 106(4):283–96

Roland PE, Larsen B, Lassen NA, Skinhoj E.

1980. Supplementary motor area and other cortical areas in organization of voluntary movements in man. *J. Neurophysiol.* 43:118–36

Rosenbaum DA. 1991. *Human Motor Control.* San Diego, CA: Academic

Sakata H, Taira M, Murata A, Mine S. 1995. Neural mechanisms of visual guidance of hand action in the parietal cortex of the monkey. *Cereb. Cortex* 5:429–38

Sanes JN, Dimitrov B, Hallett M. 1990. Motor learning in patients with cerebellar dysfunction. *Brain* 113(Part 1):103–20

Seitz R, Roland PE. 1992. Learning of sequential finger movements in man: A combined kinematic and positron emission tomograpy (PET) study. *Eur. J. Neurosci.* 4:154–65

Sergent J, Zuck E, Terriah S, MacDonald B. 1992. Distributed neural network underlying musical sight-reading and keyboard performance. *Science* 257(5066):106–9

Shallice T, Burgess P. 1996. The domain of supervisory processes and temporal organization of behaviour. *Philos. Trans. R. Soc. London Ser. B* 351(1346):1405–11

Shibasaki H, Sadato N, Lyshkow H, Yonekura Y, Honda M, et al. 1993. Both primary motor cortex and supplementary motor area play an important role in complex finger movement. *Brain* 116(Part 6):1387–98

Shima K, Mushiake H, Saito N, Tanji J. 1996. Role for cells in the presupplementary motor area in updating motor plans. *Proc. Natl. Acad. Sci. USA* 93(16):8694–98

Shima K, Tanji J. 1998. Both supplementary and presupplementary motor areas are crucial for the temporal organization of multiple movements. *J. Neurophysiol.* 80(6):3247–60

Shima K, Tanji J. 2000. Neuronal activity in the supplementary and presupplementary motor areas for temporal organization of multiple movements. *J. Neurophysiol.* 84:2148–60

Soechting JF, Flanders M. 1992. Organization of sequential typing movements. *J. Neurophysiol.* 67:1275–90

Takada M, Tokuno H, Nambu A, Inase M.

1998. Corticostriatal projections from the somatic motor areas of the frontal cortex in the macaque monkey: segregation versus overlap of input zones from the primary motor cortex, the supplementary motor area, and the premotor cortex. *Exp. Brain Res.* 120(1):114–28

Tanji J. 1996. New concepts of the supplementary motor area. *Curr. Opin. Neurobiol.* 6(6):782–87

Tanji J, Evarts EV. 1976. Anticipatory activity of motor cortex neurons in relation to direction of an intended movement. *J. Neurophysiol.* 39(5):1062–68

Tanji J, Shima K. 1994. Role for supplementary motor area cells in planning several movements ahead. *Nature* 371(6496):413–16

Thach WT. 1998. A role for the cerebellum in learning movement coordination. *Neurobiol. Learn. Mem.* 70:177–88

Thach WT, Goodkin HP, Keating JG. 1992. The cerebellum and the adaptive coordination of movement. *Annu. Rev. Neurosci.* 15:403–42

Thompson PD, Berardelli A, Rothwell JC, Day BL, Dick JP, et al. 1988. The coexistence of bradykinesia and chorea in Huntington's disease and its implications for theories of basal ganglia control of movement. *Brain* 111 (Part 2):223–44

Van den Bercken JH, Cools AR. 1982. Evidence for a role of the caudate nucleus in the sequential organization of behavior. *Behav. Brain Res.* 4(4):319–27

Wise SP. 1985. The primate premotor cortex: past, present, and preparatory. *Annu. Rev. Neurosci.* 8:1–19

Wolpert DM, Kawato M. 1998. Internal models in the cerebellum. *Trends Cogn. Sci.* 2:338–47

Annu. Rev. Neurosci. 2001. 24:653–75

INFLUENCE OF DENDRITIC CONDUCTANCES ON THE INPUT-OUTPUT PROPERTIES OF NEURONS

Alex Reyes

Center for Neural Science, New York University, New York, New York 10003; e-mail: reyes@cns.nyu.edu

Key Words dendrite, channels, integration, firing, conductance

■ **Abstract** A fundamental problem in neuroscience is understanding how a neuron transduces synaptic input into action potentials. The dendrites form the substrate for consolidating thousands of synaptic inputs and are the first stage for signal processing in the neuron. Traditionally, dendrites are viewed as passive structures whose main function is to funnel synaptic input into the soma. However, dendrites contain a wide variety of voltage- and time-dependent ion channels. When activated, the currents through these channels can alter the amplitude and time course of the synaptic input and under certain conditions even evoke all-or-none regenerative potentials. The synaptic input that ultimately reaches the soma is likely to be a highly transformed version of the original signal. Thus, a key step in understanding the relationship between synaptic input and neuronal firing is to elucidate the signal processing that occurs in the dendrites.

INTRODUCTION

To appreciate fully the complexities in the input/output properties introduced by dendritic conductances, it is useful to consider first a neuron with passive dendrites and an active soma. For this simple model, dendritic processing of excitatory postsynaptic potentials (EPSPs) is minimal. The major transformations that an EPSP undergoes as it propagates to the soma are a decrease in amplitude and an increase in width resulting from transmembrane leak and filtering (Rall 1964, Stuart & Spruston 1998). Integration of EPSPs in the dendrites is governed by a few simple rules (Rall 1964). EPSPs originating from sites that are electrotonically distant from each other sum linearly; otherwise, EPSPs sum sublinearly because of a decrease in the driving force of the synaptic current and an increase in shunting. If the composite EPSP that finally reaches the soma is sufficiently large, an action potential initiates at the axon hillock (Coombs et al 1957a,b). The action potential propagates back to the soma and forward to the axon, where it is transmitted to other parts of the central nervous system. Sustained stimulus causes the neuron to discharge repetitively at a rate proportional to the magnitude of the synaptic

0147-006X/01/0621-0653$14.00

current reaching the soma (Granit et al 1966, Powers et al 1992, Schwindt & Crill 1996).

Unfortunately, very few of the above rules are obeyed if the dendrites are active. A quick survey of recent experiments reveals a wide range of often contradictory results. Depending on which dendritic conductances are activated, EPSPs can be boosted or attenuated prior to reaching the soma. EPSPs can sum linearly, sublinearly, or even supralinearly. Action potentials and other regenerative events can be initiated at the dendrites resulting in a variety of spiking patterns ranging from repetitive firing to bursts. Hence, defining general rules that describe accurately the input/output properties of neurons is likely to be very difficult. An important first step in elucidating the process is to examine closely the biophysical properties of channels in the dendrites.

Scope of This Review

Excellent recent reviews have discussed the active properties of dendrites (Magee et al 1998, Hausser et al 2000). This review focuses specifically on the effects that dendritic conductances have on the input/output properties of neurons. Instead of cataloging the channels described in the many cell types within the past few years, the discussion is confined primarily to neocortical pyramidal neurons, hippocampal CA1 pyramidal neurons, and spinal motoneurons. These neurons contain many of the dendritic conductances and exhibit firing patterns that are found in many cell types. The first two sections summarize the active properties of dendrites and the biophysical properties of voltage-gated channels. The subsequent sections examine how these channels affect the manner in which synaptic input is transformed, integrated, and transduced into action potentials. The last section discusses the regulation of channels by neuromodulators.

OVERVIEW OF ACTIVE PROPERTIES OF DENDRITES

Dendrites can support at least three broad classes of regenerative events. These potentials can be triggered in an all-or-none fashion by synaptic inputs and, as discussed below, can have a tremendous impact on the way neurons process signals.

Na^+-Mediated Action Potentials in the Dendrites

The dendrites of several classes of neurons can support Na^+-mediated action potentials. Among these neurons are neocortical pyramidal cells (Kim & Connors 1993; Stuart & Sakmann 1994; Larkum et al 1999a,b; Zhu 2000), hippocampal CA1 and CA3 pyramidal cells (Turner et al 1991, Wong & Stewart 1992, Andreasen & Lambert 1995, Golding & Spruston 1998, Golding et al 1999) and interneurons (Martina et al 2000), retinal ganglion cells (Velte & Masland 1999), subicular pyramidal neurons (Colbert & Johnston 1996), substantia nigra neurons (Hausser et al 1995), thalamic neurons (Williams & Stuart 2000b), mitral cells (Chen et al 1997) and motoneurons (Larkum et al 1996).

Simultaneous whole-cell recordings from the dendrites, soma, and/or the axon hillock of neurons demonstrated that action potentials can be triggered at different compartments. Because the threshold is lowest in the axon hillock or axon, depolarization of the dendrites with injection of current pulses or with stimulus-evoked EPSPs usually triggers an action potential first at the axon hillock and then at the soma and dendrites (Stuart & Sakmann 1994, Colbert & Johnston 1996, Larkum et al 1996, Schiller et al 1997, Stuart et al 1997). The dendritic action potentials are typically broader and shorter than the somatic action potential and are not accompanied by prominent afterhyperpolarizations (AHP). Action potentials can also be triggered first at the dendrites, particularly at higher stimulus intensities (Stuart & Sakmann 1994, Chen et al 1997, Golding & Spruston 1998, Schwindt & Crill 1999, Martina et al 2000). These dendritically-initiated action potentials actively propagate forward to the soma, albeit with varying reliability.

Ca^{2+}-Mediated Regenerative Potentials

Dendrites can also support Ca^{2+}-mediated regenerative potentials. These regenerative events are considerably broader than the dendritic action potentials and are blocked by divalent cations but not by the Na^+ channel blocker tetrodotoxin (TTX) (Kim & Connors 1993, Andreasen & Lambert 1995, Schiller et al 1997, Stuart et al 1997, Schwindt & Crill 1999, Golding et al 1999, Zhu 2000). The Ca^{2+}-mediated regenerative potentials initiate in the dendrites and generally do not propagate actively to the soma (Schiller et al 1997, Schwindt & Crill 1997a, Stuart et al 1997). In neocortical pyramidal neurons, active spread of the regenerative potentials is limited to a relatively restricted region of the dendrite by activation of K^+ conductances. Schwindt & Crill (1997a) recorded intracellulary at the soma and evoked Ca^{2+}-mediated regenerative potentials by iontophoresing glutamate on the apical dendrite. Moving the iontophoretic electrode away from the soma decreased the amplitude of the regenerative potential. This indicated that the regenerative potentials are initiated near the iontophoretic electrode and propagate passively to the soma. When the K^+ channels were blocked, the amplitude of the regenerative potential measured at the soma increased and did not vary with the location of the iontophoretic electrode.

Ca^{2+}-Mediated Plateau Potentials

The third type of all-or-none event, the plateau potential, has been documented in the dendrites of motoneurons (Hounsgaard & Kiehn 1993, Lee & Heckman 1996, Bennett et al 1998, Delgado-Lezama et al 1999; for a review, see Kiehn & Eken 1998) and neocortical pyramidal neurons (Schwindt & Crill 1999). In motoneurons, plateau potentials are typically observed in the presence of neuromodulators (see below). A hallmark of plateau potentials is that a transient stimuli will shift the membrane potential to a higher, stable level until a brief hyperpolarization (such as an inhibitory postsynaptic potential) shifts the membrane potential back to rest. Plateau potentials are mediated by Ca^{2+}: They are blocked with divalent cations in pyramidal neurons (Schwindt & Crill 1999) and additionally by the L-type calcium channel blocker, nifedipine, in motoneurons (Hounsgaard & Kiehn 1989, 1993; Hsiao et al 1998; Carlin et al 2000). In pyramidal cells, plateau potentials and Ca^{2+}-mediated regenerative potentials may be mediated by the same set of Ca^{2+} channels (Schwindt & Crill 1999).

Plateau potentials arise when the relationship between net ionic current (I) through voltage-gated channels and membrane voltage (V) is "N-shaped" (Bennett et al 1970; for a review, see Kiehn & Eken 1998). Two stable membrane potentials (one at resting potential and another at a more depolarized level) emerge when the voltage dependence of persistent inward and outward currents are such that the steady-state I-V curve crosses the zero current line three times. Plateau potentials can be modulated by shifting the balance between the inward and outward current (Hounsgaard & Kiehn 1993, Booth et al 1997). Thus, blockade of K^+ currents enhances plateau potentials (Reuveni et al 1993, Hoffman et al 1997) whereas blockade of Ca^{2+} currents has the opposite effect (Hounsgaard & Kiehn 1989, 1993; Carlin et al 2000). Several neuromodulators facilitate plateau potentials by upregulating Ca^{2+} channels or alternatively by downregulating K^+ channels (see below).

Several lines of evidence indicate that plateau potentials can be initiated in the dendrites (Hounsgaard & Kiehn 1993, Lee & Heckman 1996, Carlin et al 2000). In motoneurons, voltage clamping the soma (to prevent action potentials) while stimulating afferents revealed all-or-none, sustained inward currents that outlasted the stimulus (Lee & Heckman 1996). The fact that the evoked currents were not under voltage clamp control indicated that they originated at some distal and poorly space-clamped site in the dendrites. Similar techniques were used to show that plateau potentials in pyramidal neurons are also triggered in the dendrites (Schwindt & Crill 1999). Finally, plateau potentials were evoked in motoneurons in vitro by applying an electrical field across the slice that depolarized the dendrite but hyperpolarized the soma (Hounsgaard & Kiehn 1993, Delgado-Lezama et al 1999).

OVERVIEW OF DENDRITIC CONDUCTANCES

Activation Maps

The dendrites contain a wide variety of ion channels that exhibit different voltage-dependence and kinetics. An important consequence is that depending on the amplitude and time course of the synaptic input, different sets of conductances will be activated (see below). To assess which channels are likely to be activated, it is useful to construct "activation maps," which show the range of membrane potentials and times that each channel can be activated by a depolarizing stimulus. In Figure 1 (see color insert), the color-coded regions are the activation maps for some of the dendritic channels discussed below. The ordinate is the membrane potential of the neuron and the abscissa is time. The origin is aligned with the onset of a stimulus (lower traces in Figures 1B,C) on the abscissa and with the cell's resting potential (-70 mV, dashed line) on the ordinate.

For each activation map, the lower bound on the ordinate is set by the activation threshold of the channel. The arrows point to the direction of increasing activation. The leftmost boundary is set by the channel's time constant of activation and represents the delay from the time the stimulus is applied to the time the channel starts to activate. For example, a depolarizing step to -40 mV applied at time zero will activate I_{DR} (red) after a delay of 30 ms. Increasing the step to -30 mV will increase activation and decrease the delay slightly.

The length of time that each channel is active (rightmost part of each map) is delimited by either the time constant of inactivation, if the channel inactivates (like I_A), or by the time constant of deactivation, if the channel deactivates with depolarization (like I_h). Arrows to the right signify that the channel is persistent or inactivates slowly. For example, a depolarizing step to -60 mV will activate I_A (purple) after a delay of 1 ms ($=$ time constant of activation) for about 45 ms ($=$ time constant of inactivation) before it starts to inactivate. At higher potentials (-40 mV), the time constant for inactivation decreases and hence the length of time that I_A is maximally active decreases.

In Figures 1B, and C, a transient and a sustained depolarizing input are superimposed on the activation maps. The transient input is equivalent to a single synchronous volley of EPSPs, whereas the tonic input is equivalent to a continuous, asynchronous volley. The activation states of channels whose maps are intersected by each input's waveforms (black) will change during the stimulus. Thus the transient input (Figure 1B) will activate I_{NaP} and I_A. The tonic input (Figure 1C) will activate initially the same conductances as transient EPSPs. However, with time, I_A will inactivate and I_h will deactivate. By 50 ms, only I_{NaP} remains.

It should be stressed that the activation maps do not reveal how much current is actually generated by the conductances. The regions bounded by the maps depict the voltage and time ranges where the channel is optimally activated. Outside the boundaries, the ionic currents are not necessarily zero but are probably less than the currents generated within the boundaries. Also, because biophysical properties of most of the channels were measured at room temperature, the maps are likely to change at physiological temperatures.

Na$^+$ Channels

I_{NaT}, I_{NaP} Current through dendritic Na$^+$ channels, like that through somatic Na$^+$ channels, consists of a transient (I_{NaT}) and a persistent (I_{NaP}) component (Magee & Johnston 1995, Colbert et al 1997, Jung et al 1997, Mittmann et al 1997, Schwindt & Crill 1997b, Mickus et al 1999). Na$^+$ channels activate rapidly starting at -50 mV and inactivate starting at -90 mV. The time course of inactivation is described by two exponentials with time constants of 5 and 50 ms (at -40 mV). The fast time constant probably contributes to I_{NaT} whereas the slow time constant contributes to I_{NaP}. Whether the two components are mediated by the same or different channels is yet unclear (for a review, see Crill 1996). Functionally, the activation threshold of I_{NaP} is approximately 10 mV below that of I_{NaT}: The inactivation associated with I_{NaT} is postulated to counter activation such that only a few channels can open with small depolarization (Crill 1996). The activation maps for I_{NaT} (Figure 1, dark blue) and I_{NaP} (Figure 1, light blue) are plotted separately. Brief (Figure 1*B*) and tonic (Figure 1*C*) inputs, if sufficiently large, can activate both I_{NaT} and I_{NaP}. However, at the end of the tonic input, only I_{NaP} remains active. I_{NaP} boosts the amplitude of EPSPs in the subthreshold range (see below) whereas I_{NaT} underlies the upstroke of the action potential in the suprathreshold range.

In neocortical and hippocampal CA1 pyramidal neurons, Na^+ channels are uniformly distributed along the somatodendritic axis (Stuart & Sakmann 1994, Magee & Johnston 1995, Colbert & Johnston 1996). However, there are systematic differences in the kinetics of Na^+ channels. Na^+ channels in hippocampal CA1 neurons can enter a slow inactivation state that requires seconds for recovery. The recovery from slow inactivation is much slower in the dendrites than in the soma (Colbert et al 1997, Mickus et al 1999). These regional differences are reflected in the behavior of action potentials in the two compartments. During a stimulus train, the amplitudes of dendritic but not somatic action potentials decrease systematically (Callaway & Ross 1995, Spruston et al 1995, Tsubokawa & Ross 1997).

K^+ Channels

I_{DR} The dendrites of neocortical (Bekkers 2000, Korngreen & Sakmann 2000) and hippocampal CA1 (Hoffman et al 1997) pyramidal neurons have K^+ channels with properties similar to those of delayed rectifier K^+ channels. The I_{DR} channels activate with a time constant of ~30 ms starting at -40 mV and exhibit either very slow or no inactivation. As its activation map shows, I_{DR} (Figure 1, red) is active primarily in the suprathreshold range. Its main role is to repolarize dendritic action potentials (Bekkers 2000).

I_A Dendrites also contain transient A-type (I_A) K^+ channels. I_A channels activate rapidly near resting potential with a time constant of ~1 ms (Hoffman et al 1997, Bekkers 2000, Johnston et al 2000, Korngreen & Sakmann 2000). The time constant for inactivation is short for steps above -20 mV (~5–8 ms) but is substantially longer (~50 ms) near the resting potential (for neocortical pyramidal neurons) (Korngreen & Sakmann 2000). The activation map (Figure 1, purple) shows that the optimal stimulus for activation I_A is a brief transient input (Figure 1*B*). Tonic input (Figure 1*C*) will inactivate I_A eventually.

I_A channels have several roles in the dendrites, all of which reduce neuronal excitability (Andreasen & Lambert 1995, Hoffman et al 1997, Bekkers 2000). First, I_A attenuates the amplitude of transient EPSPs by increasing the overall input conductance and by countering the boosting effects of Na^+ and Ca^{2+} channels. Second, I_A limits the amplitudes of back-propagating action potentials and increases the threshold for action potential initiation in the dendrites. Third, I_A aids in the repolarization of dendritic action potentials. Finally, I_A raises the threshold for activating Ca^{2+}-mediated bursts.

The effects of I_A are amplified in the dendrites of hippocampal CA1 neurons. In these neurons, I_A increases over fivefold along the somatodendritic axis (Hoffman et al 1997). In addition, the activation curve for I_A in the dendrites is shifted by -12 mV compared to that in the soma. Such gradients in I_A do not exist in neocortical pyramidal neurons (Bekkers 2000).

I_D A putative D-type K^+ current (I_D) has been documented in the dendrites of hippocampal CA1 pyramidal neurons (Golding et al 1999). I_D, when activated by a somatic action potential, raises the threshold for calcium spikes in the soma. This may, in part, account for the fact that dendritic current injection is more likely to evoke bursts of action potentials than somatic current injection (see below).

$I_{K(Ca)}$ Ca^{2+}-dependent K^+ ($I_{K(Ca)}$) channels are found in the dendrites of several cell types (Hounsgaard & Kiehn 1993, Kang et al 1996, Poolos & Johnston 1999, Golding et al 1999). Relatively little is known about the biophysical properties of dendritic $I_{K(Ca)}$ channels although patch recordings from the proximal dendrites and somata of neocortical pyramidal neurons revealed no gross differences in the voltage or calcium dependence of $I_{K(Ca)}$ (the large conductance BK channel) in the two compartments (Kang et al 1996). In both neocortical and hippocampal pyramidal neurons, $I_{K(Ca)}$ channels seem to be less dense in the apical dendrite (Kang et al 1996, Poolos & Johnston 1999, Johnston et al 2000).

$I_{K(Ca)}$ may serve several roles in the dendrites with the general effect of reducing neuronal excitability. In hippocampal CA1 pyramidal neurons, $I_{K(Ca)}$ limits the duration of the Ca^{2+}-mediated action potentials (Golding et al 1999). Blockade of $I_{K(Ca)}$ channels lengthens the burst duration without affecting the threshold for the Ca^{2+}-mediated spike. Interestingly, $I_{K(Ca)}$ aids in the repolarization of somatic but not dendritic action potentials (Poolos & Johnston 1999, Johnston et al 2000). In neocortical pyramidal neurons, $I_{K(Ca)}$ attenuates the net synaptic current reaching the soma from the dendrites (Schwindt & Crill 1997b). Finally, in motoneurons, $I_{K(Ca)}$ may prevent plateau potentials (Booth et al 1997).

Ca^{2+} Channels

Low-Voltage–Activated Ca^{2+} Channels Low-threshold T-type (I_{CaT}) channels have been characterized in the dendrites of several cell types (Karst et al 1993, Magee & Johnston 1995, Mouginot et al 1997). I_{CaT} activates relatively quickly (time-to-peak: $\sim$10–30 ms) starting at ~-70 mV and inactivates moderately fast (time constant: $\sim$30–50 ms). In hippocampal CA1 neurons, I_{CaT} is uniformly distributed along the somatodendritic axis (Magee & Johnston 1995). As its activation map shows (Figure 1, gold), I_{CaT} is activated transiently at subthreshold membrane potentials. As discussed below, I_{CaT} actively boosts EPSPs and may underlie the supralinear summation of EPSPs.

High-Voltage–Activated Ca^{2+} Channels Ca^{2+} channels with properties similar to those of the L-type Ca^{2+} (I_{CaL}) channels have been documented in the dendrites of hippocampal CA1 pyramidal neurons (Magee & Johnston 1995) and motoneurons (Carlin et al 2000). In CA1 neurons, I_{CaL} channels activate at ~-25 mV and do not inactivate. Activation is relatively fast; I_{CaL} reaches peak value within 10 ms (Magee & Johnston 1995). The density of I_{CaL} channels seems to be greater at the soma and proximal dendrites than at the distal dendrites. I_{CaL} contributes to plateau potentials in motoneurons (Hounsgaard & Kiehn 1993, Booth et al 1997, Carlin et al 2000) and burst firing in hippocampal CA1 pyramidal neurons (Andreasen & Lambert 1995).

Hippocampal CA1 pyramidal cell dendrites also have Ca^{2+} channels with properties similar to those of R-type and N-type Ca^{2+} channels (Magee & Johnston 1995). These channels activate moderately fast (time-to-peak of ensemble current: $\sim$10 ms) starting at $\sim$−40 mV and both seem to be uniformly distributed along the somatodendritic axis.

G Protein–Activated Inward Rectifying K^+ Channels   Immunocytochemical (Drake et al 1997) and electrophysiological (Takigawa & Alzheimer 1999) studies show that G protein–activated inward rectifying K^+ channels (GIRKs) are present in the dendrites. GIRKs are activated with hyperpolarization and are enhanced by the neuromodulators adenosine and serotonin and by the GABA-B receptor agonist baclofen. All three agonists evoke a greater response in dendritic segments than in somata, indicating a difference in the density of GIRK channels and/or in the effectiveness of metabotropic receptors along the somatodendritic axis. In the dendrites, GIRKs may provide a means for neuromodulators to regulate the excitability of neurons by shifting the resting potential and by decreasing the amplitude of EPSPs via shunting.

Hyperpolarization-Activated Cation Current

The hyperpolarization-activated cation current (I_h) in the dendrites of neocortical L5 and hippocampal pyramidal neurons is partially activated near resting potential and is activated further with increasing hyperpolarization (Magee 1998, Williams & Stuart 2000a). With depolarization, I_h deactivates with a time constant of approximately 20–40 ms (at −70 mV). The activation map for I_h (green) is shown in Figure 1.

The overall effect of I_h is to reduce the overall neuronal excitability. I_h contributes to the cell's resting potential and reduces the cell's input resistance (Schwindt & Crill 1997b; Magee 1998; Williams & Stuart 2000a). The decreased resistance results in EPSPs with smaller amplitudes. In hippocampal and neocortical pyramidal neurons, there is a huge systematic increase in I_h along the somatodendritic axis, suggesting that the effects of I_h will be more prominent in the dendrites than at the soma.

TRANSFORMATION OF SYNAPTIC INPUTS IN THE DENDRITES

Many of the channels in the dendrites can be activated by subthreshold EPSPs. Calcium imaging (Markram & Sakmann 1994) and cell-attached patch recordings (Magee & Johnston 1995) demonstrate that stimulus-evoked compound EPSPs generate sufficient depolarization in the dendrites to open Na^+ and low-voltage–activated Ca^{2+} channels. A unitary EPSP evoked when a single presynaptic cell fires may also be able to alter the activation states of some of the channels: Passive models of pyramidal neurons suggest that a 1 mV unitary EPSP measured at the soma could easily be 5–20 mV at the site of the synaptic contact in the dendrite (Stuart & Spruston 1998).

As can be seen from the activation maps (Figure 1), there are several channels that can theoretically be affected by subthreshold EPSPs. These include I_{NaP}, I_{CaT}, I_A, and I_h. When activated, the current through channels will change the shapes of the EPSPs. Activation of channels that generate inward currents, such as I_{NaP} or I_{CaT}, will boost the amplitudes of EPSPs whereas activation of channels that generate outward currents, such as I_A, will attenuate EPSPs. Deactivation of channels that are already active at rest, such as I_h, will increase the neuron's input resistance and will increase the EPSP amplitude. In actuality, an EPSP is likely to change the activation states of several channels simultaneously. The net effect on the EPSP's shape will depend on the balance between channels that generate inward and those that generate outward currents.

Amplification of EPSPs

Dendritic I_{NaP} and I_{CaT} were shown to boost EPSPs evoked in hippocampal CA1 pyramidal cells (Lipowski et al 1996, Gillessen & Alzheimer 1997). In these experiments, whole-cell recordings were made at the soma, and EPSPs were evoked by stimulating afferents on the distal apical dendrite. Focal application of blockers to the proximal segment of the apical dendrite, in between the terminals and the recording electrode, reduced the EPSP amplitude, indicating that dendritic I_{Na} and I_{CaT} actively boosted the EPSP.

I_{NaT} and the high voltage-activated Ca^{2+} channels can also amplify EPSPs. In fact, the most dramatic examples of boosting are when stimulus-evoked EPSPs trigger Na^+-mediated action potentials, Ca^{2+}-mediated regenerative potentials, or plateau potentials in the dendrites (Schiller et al 1997; Golding & Spruston 1998, Schwindt & Crill 1999).

The importance of the EPSP amplitude and duration in determining the degree of boosting is underscored in two sets of experiments with layer 5 pyramidal cells. In one study, simultaneous whole-cell recordings were performed at the soma and the apical dendrite of neocortical pyramidal neurons (Stuart & Sakmann 1995). Time-varying currents were injected through the dendritic electrode to evoke voltage deflections that resembled stimulus-evoked EPSPs (termed simulated EPSPs). Because of transmembrane leak and filtering in the dendrite, the simulated EPSPs that propagated to the soma were considerably smaller and broader. Focal application of TTX at the soma decreased substantially the amplitude of the simulated EPSP whereas focal application at the dendrite had relatively little effect. The conclusion was that the majority of the boosting occurred at the soma, not at the dendrites (a similar conclusion was reached in CA3 pyramidal cells by Urban et al

1998). However, the difference in boosting may have arisen partly because of the differences in the shape and size of the EPSPs. Because some time is needed for I_{NaP} to reach steady state activation, the broader somatic EPSP may have simply enabled I_{NaP} to become activated to a greater degree than the briefer dendritic EPSP. Moreover, the much larger dendritic EPSP may have activated I_A, which would also counter the effects of I_{NaP} (see below). Finally, the boosting effects of dendritic I_{NaP} may have been partly masked by I_h, which reduces the EPSP amplitude. I_h is much larger in the dendrite (Williams & Stuart 2000a) and so would have a considerably greater effect on dendritic EPSPs.

In another study, Schwindt & Crill (1995) voltage-clamped the somata of neocortical neurons and iontophoretically applied glutamate onto the apical dendrite for 1–2 s. This stimulus tonically depolarized the distal portions of the dendrites, which were too far to be adequately voltage-clamped. Only the voltage-gated channels at the distal dendrites could be activated. The transaxial current that was measured at the soma during the stimulus was therefore a sum of the current evoked by glutamate and the current through voltage-gated channels. Because the transaxial current decreased when Na^+ channels were blocked, they concluded that dendritic I_{NaP} boosted synaptic input. As can be seen from the activation maps (Figure 1C), tonic depolarization would accentuate the boosting effects of I_{NaP}. Tonic depolarization would permit I_{NaP} to reach steady state activation and would eventually deactivate I_h and inactivate I_A, both of which attenuate EPSPs (see below).

Attenuation of EPSPs

In the experiments discussed above, I_A was most likely co-activated with I_{NaP} and I_{CaT}: All three are activated in the same voltage range, as indicated by the overlap in their activation maps (Figure 1). Therefore, boosting by the inward currents may have been, to some degree, reduced by I_A. This effect of I_A was demonstrated in hippocampal CA1 pyramidal neurons (Hoffman et al 1997). In these experiments, the amplitudes of stimulus-evoked and simulated EPSPs increased when I_A was blocked. When TTX was subsequently added to the bath, the amplitude of the simulated EPSP decreased.

$I_{K(Ca)}$ was also shown to counteract Na^+ or Ca^{2+}-mediated boosting of synaptic input (Schwindt & Crill 1997b). The transaxial current evoked with iontophoresis of glutamate in the dendrite increased when $I_{K(Ca)}$ was reduced.

I_h is tonically active at resting potential (Figure 1) and so in effect reduces the cell's resistance. The EPSPs evoked from rest are therefore smaller while I_h is active. As expected, when I_h is blocked, the amplitudes and decay times of simulated EPSPs increase (Schwindt & Crill, 1997b, Magee 1998, Williams & Stuart 2000a).

INTEGRATION OF SYNAPTIC INPUTS

An important step in the transduction of synaptic input into neuronal firing is the integration of EPSPs in the subthreshold range. Because the unitary EPSPs that reach the soma are small (Markram et al 1997, Reyes & Sakmann 1999), many

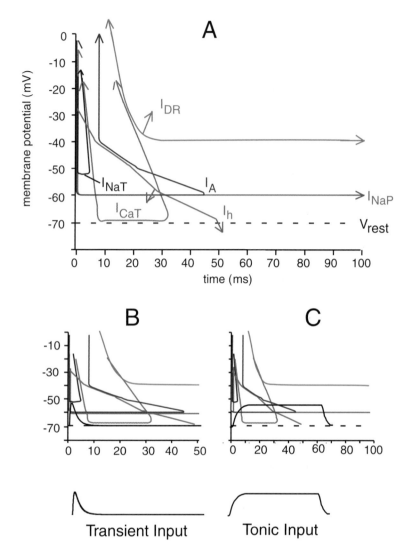

Figure 1 (*A*) Activation map showing the range of membrane potentials (ordinate) and times (abscissa) that different ion channels can be activated by a depolarizing stimulus. (*B,C*) Transient (*B*) and tonic (*C*) synaptic inputs are superimposed on the activation map to show which channels they are likely to activate.

inputs are needed in order to initiate an action potential. The requisite number of inputs depends on the manner in which the individual EPSPs sum.

The rules for EPSP summation are difficult to assess in neurons with active dendrites. Because the ion channels are voltage- and time-dependent, EPSPs are expected to sum nonlinearly: Channels not activated by individual EPSPs may however be activated by several EPSPs arriving simultaneously. There are several conductances in the subthreshold range that can be affected by EPSPs. Whether summation is linear, sublinear, or supralinear depends on which sets of conductances are activated predominantly.

Summation of EPSPs Originating from Different Dendritic Branches

Passive models predict that EPSPs from electrotonically distant parts of the dendrite will sum linearly at the soma (Rall 1964). Experiments generally support this prediction. In one experiment, simultaneous whole-cell recordings were performed from three interconnected pyramidal cells (Reyes & Sakmann 1996). Two of these cells were presynaptic to the third. The main finding was that the composite EPSP evoked when the two presynaptic cells were stimulated simultaneously was equal to the linear sum of the unitary EPSPs evoked when each cell was stimulated separately. Subsequent morphological reconstruction of the pre- and postsynaptic neurons confirmed that the presynaptic terminals were on different branches of the postsynaptic cell's basal dendrites and hence were probably electrotonically distant from each other. In other studies, two EPSP-like depolarizations (termed iontophoretically-evoked EPSPs) were evoked in a neuron by brief iontophoresis of glutamate through two pipettes (in lieu of afferent stimulation). When the two pipettes were placed on different dendritic branches, the iontophoretically-evoked EPSPs summed linearly (Skydsgaard & Hounsgaard 1994, Cash & Yuste 1999).

Summation of EPSPs Converging onto the Same Dendritic Branch

Passive models of dendrites predict that EPSPs evoked from electrotonically close inputs will sum sublinearly (Rall 1964). In some experiments, sublinear summation of EPSPs was indeed observed. In CA3 pyramidal neurons, the composite EPSP evoked by concurrent stimulation of two separate afferent systems was less than that predicted by a linear sum of the individual EPSPs evoked by stimulating each afferent separately. However, sublinear summation occurred not because of a decrease in the driving force for the synaptic current (Rall 1964) but rather because I_A was activated. When I_A was blocked, summation became linear (Urban & Barrionuevo 1998). A similar observation was made in hippocampal CA1 pyramidal neurons where sublinear summation of iontophoretically-evoked EPSPs in the apical dendrite became linear when I_A was blocked (Cash & Yuste 1999). The study showed further that the degree of sublinearity increased as the iontophoretic electrodes were moved further away from the soma. This increase paralleled the increase in I_A along the somatodendritic axis (Hoffman et al 1997).

In other experiments, activation of I_{Ca} and I_{Na} caused EPSPs to summate supralinearly. In neocortical pyramidal neurons, the composite EPSP evoked by concurrent stimulation of afferents from L1 and L3-5 were, on average, greater than the predicted linear sum of individual EPSPs (Nettleton & Spain 2000). Summation became linear when I_{Na} and I_{Ca} were blocked. In general supralinear summation of EPSPs becomes inevitable whenever enough afferents are activated to trigger Ca^{2+}-mediated regenerative potentials (Schiller et al 1997, Stuart et al 1997), dendritic action potentials (Golding et al 1999), or plateau potentials (Schwindt & Crill 1999).

Finally, in some experiments, summation was linear. In young hippocampal pyramidal neurons, two iontophoretically-evoked EPSPs summed linearly when the pulses were delivered to adjacent sites (Cash & Yuste 1998). In this case, linear summation occurred because the EPSPs activated the right balance of conductances that generate inward and outward currents.

Summation Is Time-Dependent

The fact that dendritic conductances are time-dependent means that the summation properties of EPSPs will be dynamic. As discussed above, the conductances active during the start of a tonic stimulus may differ from those active at the end (Figure 1C, see color insert).

In CA1 pyramidal cells, two trains of iontophoretically-evoked EPSPs (delivered at 25 Hz in the apical dendrites) summed sublinearly at the onset and then linearly near the end of the stimulus train (Cash & Yuste 1999). The likely explanation is that the prolonged depolarization caused by the temporal fusing of the EPSPs inactivated I_A.

In motoneurons, transient EPSPs evoked by brief stimulation of different afferents summed linearly (Burke 1967). However, sustained inputs evoked by high frequency stimulation (1 s duration) of afferents summed sublinearly (Powers & Binder 2000). This occurred in part because steady depolarization spread more evenly throughout the dendritic tree (Larkum et al 1998) and hence decreased the driving force for synaptic current. However, voltage-dependent conductances also contributed, as evidenced by a switch to supralinear summation when K^+ channels were blocked.

Summation Is Voltage-dependent

Because the dendritic conductances are voltage-dependent, the manner in which EPSPs summate will vary with the level of depolarization as different sets of conductances become activated. In CA3 pyramidal neurons, summation of EPSPs switched from being sublinear to being linear when the postsynaptic cell was depolarized (Urban & Barrionuevo 1998). As above, the likely explanation is that the prolonged depolarization inactivated I_A. In neocortical pyramidal cells, summation of Epsps became more supralinear when the neurons were hyperpolarized (Nettleton & Spain 2000). In this case, hyperpolarization removed inactivation of the T-type Ca^{2+} channel. Finally, in motoneurons, sublinear summation of tonic synaptic inputs became linear when the neurons were depolarized to near threshold (Powers & Binder 2000).

TRANSDUCTION OF SYNAPTIC POTENTIALS INTO NEURONAL FIRING

As mentioned in the Introduction, a simple model that describes the input/output properties of a neuron consists of passive dendrites attached to an active soma. For this model, the cell's firing output is determined by the average synaptic current that reaches the soma. Early experiments in motoneurons demonstrated that synaptic input and current injection at the soma has equivalent effects on the firing rate (Granit et al 1966, Kernell et al 1969, Schwindt & Calvin 1973). Thus, the relationship between synaptic input and firing output can be determined experimentally by injecting a series of suprathreshold current steps (I) through a recording electrode, measuring the neuron's firing rate (F), and plotting F vs I (F-I relation). For a relatively large portion of the F-I relation, F varies linearly with I. Therefore, the slope of the F-I multiplied by the synaptic input gives the firing output, at least within the range that the firing rate is not saturated.

In some cases, the slope of the F-I relation could indeed be used to predict the change in firing rate caused by synaptic input. Powers et al (1992) and Powers & Binder (1995) recorded intracellularly from a motoneuron in an intact cat and determined the F-I relation with current injection. Subsequently, they stimulated afferents repetitively at high frequencies and measured the resultant changes in the neuron's firing rates. They found that the firing rate was equal to the slope of the F-I relation multiplied by the magnitude of the net synaptic current reaching the soma (measured using a modified voltage clamp technique). A similar result was obtained in neocortical pyramidal cells when glutamate was iontophoresed onto the dendrites, but only under low-stimulus conditions (Schwindt & Crill 1995).

The simple model fails when regenerative potentials are triggered in the dendrites. As discussed below, the neuron can exhibit several firing patterns under different conditions that cannot be accounted for by the model.

Dendritic Input Evokes Several Firing Patterns

Tonic stimuli delivered to the dendrites can evoke at least three types of firing patterns at the soma. The first type, repetitive action potentials (Figure 2, bottom trace), can be evoked in neocortical pyramidal neurons (Schwindt & Crill 1999), hippocampal CA1 pyramidal neurons (Andreasen & Lambert 1995), and motoneurons (Powers & Binder 2000). The frequency of action potentials increases with the magnitude of the stimulus, within a certain range.

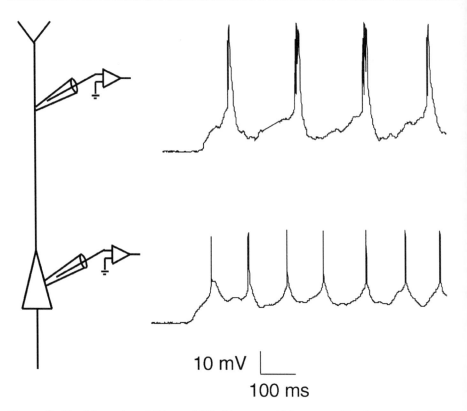

Figure 2 Possible modes of firing exhibited by pyramidal neurons. Simultaneous whole-cell recordings were performed at the soma and dendrite. Tonic current evoked bursts (top trace) when injected at the dendrite and evoked repetitive firing (bottom trace) when injected at the soma. Both voltage traces were recorded with the somatic electrode. Adapted from Oviedo & Reyes (2000).

The second type of firing, repetitive bursts (Figure 2, top trace), has been documented in cortical (Schwindt & Crill 1999, Williams & Stuart 1999, Zhu 2000) and hippocampal (Andreasen & Lambert 1995, Golding & Spruston 1998) neurons. At the soma, bursts are characterized by a rapid succession of 2–4 fast action potentials followed by an afterhyperpolarization. At the dendrite, burst waveforms are more complex and less stereotyped. The waveforms are extremely broad and in some cases the peaks fuse together for the duration of the stimulus (Andreasen & Lambert 1995, Williams & Stuart 1999, Zhu 2000). Bursting has been observed in mature motoneurons although it is yet unclear whether the bursts are initiated in the dendrites (Del Negro et al 1999).

The third firing pattern, plateau-mediated repetitive firing, can be evoked with dendritic stimulation in neocortical pyramidal neurons (Schwindt & Crill 1999) and motoneurons (Hounsgaard & Kiehn 1993, Lee & Heckman 1996, Bennett et al

1998, Delgado-Lezama et al 1999). Dendritic plateau-potentials causes repetitive single action potentials at the soma. This differs from the first type of firing in that the underlying drive to the soma is mediated by current through voltage-gated channels rather than by synaptic current. This difference has important implications for the input/output properties of neurons, which will be discussed later. Whether or not hippocampal CA1 pyramidal neurons can exhibit plateau-mediated firing is yet unclear, although these neurons seem to have the channels necessary for supporting plateau potentials.

At least one cell type, neocortical layer 5 pyramidal neurons, can generate all three firing patterns. Schwindt & Crill (1999) recorded from the somata of neocortical pyramidal neurons and iontophoresed glutamate at the apical dendrites. At low stimulus levels, neurons fired repetitive action potentials. With increasing iontophoretic current, the neuron switched to repetitive bursts. With a further increase, the firing switched back to repetitive action potentials. This differed from the firing evoked with low iontophoretic current in that the firing rate was relatively independent of stimulus intensity. Increasing the iontophoretic current increased the duration of the response but not the firing rate. They showed subsequently that this behavior occurred because plateau potentials were triggered in the dendrites.

The Evoked Firing Depends on the Location of the Synapses

Most neurons receive inputs from several classes of presynaptic cells, some of which come from different parts of the brain. The axon terminals from each class of presynaptic cells are not scattered randomly throughout the different cellular compartments but are instead segregated on specific parts of the dendritic tree. In hippocampus, for example, the inputs from the stratum oriens layer are on the basal dendrites while those from the lacunosum-moleculare layer are on the distal dendrites. The simple model does not distinguish between the different inputs: The only variable that is important for firing is the magnitude of the net current that reaches the soma. Recent experiments, however, suggest strongly that inputs to different compartments of the neuron will generate different firing patterns.

In neocortical and hippocampal pyramidal neurons, the firing pattern depends substantially on whether the stimulus is delivered at the dendrite or at the soma. Stimuli delivered at the dendrite are more likely to evoke bursts of action potentials (Figure 2, top trace) whereas stimuli delivered at the soma are more likely to evoke repetitive action potentials (Figure 2, bottom trace) (Wong & Stewart 1992, Kim & Connors 1993, Andreasen & Lambert 1995, Schwindt & Crill 1999, Golding et al 1999, Williams & Stuart 1999, Oviedo & Reyes 2000, Zhu 2000).

Burst firing involves a synergistic interaction between back-propagating action potentials and dendritic Ca^{2+}-mediated regenerative potentials (Schwindt & Crill 1999, Williams & Stuart 1999, Larkum et al 1999a,b). The sequence of events leading to burst firing seems to be as follows. An action potential initiated at the axon hillock propagates backwards to the distal dendrites and triggers a Ca^{2+}-mediated regenerative potential. The regenerative potential then propagates passively to the soma, crosses threshold, and triggers another action potential, thereby repeating the process. The process is terminated when K^+ channels such as I_A (Hoffman et al 1997) or $I_{k(Ca)}$ (Andreasen & Lambert 1995, Golding et al 1999) activate.

Dendritic input preferentially triggers bursts probably because the large dendritic EPSP summates with the back-propagating action potential to bring the membrane potential closer to the threshold for the Ca^{2+}-mediated regenerative potential. Somatic inputs are unlikely to generate comparably large depolarizations at the dendrite because of the long electrotonic distance.

An intriguing implication of this location-dependent bursting is that different populations of presynaptic cells may evoke different firing patterns, depending on where their synaptic contacts are located on the postsynaptic cell. Hence, information about the identity of the active presynaptic cells may be encoded in the discharge patterns of a neuron.

Dendritic Regenerative Events Cause Nonlinear Summation of Firing Rates

An important issue is whether the principles of synaptic input summation in the subthreshold range can be extended in the firing regime. For example, consider two separate sets of inputs into a neuron that generate synaptic currents I_{syn1} and I_{syn2}. According to the simple model, their individual effects on the firing rate (F_1 and F_2), will be given by $C(I_{syn1})$ and $C(I_{syn2})$, where C is the slope of the F-I relation. The total firing rate (F_{total}) evoked when both inputs are activated concurrently is therefore given by: $F_{total} = F_1 + F_2 = C(I_{syn1}) + C(I_{syn2}) = C(I_{syn1} + I_{syn2})$. As long as C is constant, the individual firing rates should sum linearly. Generally, summation of firing rates should parallel summation of subthreshold synaptic inputs. For example, if I_{syn1} and I_{syn2} sum nonlinearly, then so should F_1 and F_2. This prediction was upheld in some motoneurons where synaptic currents from two different afferent systems and their individual effects on firing rate both summed sublinearly (Powers & Binder 2000).

The rules for summation of firing rates are not likely to be obeyed when neurons enter different firing regimes. It is yet unclear what the neuron's overall firing rate will be if inputs that generate bursts are activated concurrently with inputs that generate repetitive action potentials. Moreover, two inputs that each generate the same firing pattern may combine to generate a completely different pattern. In some pyramidal cells, for example, somatic current injection and iontophoresis of glutamate into the dendrite causes repetitive single-action potentials when each is delivered separately. However, when both stimuli are applied concurrently, the neuron fires bursts (Schwindt & Crill 1999).

Activation of plateau potentials causes supralinear summation of firing rates. In motoneurons, injecting a slowly rising ramp of current in the soma causes an associated ramp increase in firing rate (Bennett et al 1998). At some point, however, plateau potentials are triggered and the firing rate jumps to a higher level. This suggests that when a requisite number of inputs is recruited, summation of firing rates will inevitably become supralinear.

In the extreme, plateau potentials uncouple the relationship between firing rate and synaptic input. In motoneurons, brief stimulation of afferents into the dendrites causes repetitive firing that persists long after the stimulus has ended (Hounsgaard & Kiehn 1993, Lee & Heckman 1996). In pyramidal cells, increasing the iontophoretic current (see above) produces no additional increase in firing rate when plateau potentials are triggered (Schwindt & Crill 1999). In effect, the evoked firing becomes independent of synaptic input.

REGULATION BY NEUROMODULATORS

Nearly all the dendritic ionic conductances can be modulated by neurotransmitters. Many of these neurotransmitters can change the kinetics of several channels simultaneously (Hille 1994, Nicoll et al 1990), resulting in a complex shift in the balance between the conductances that generate inward and those that generate outward currents.

Several of the channels that play a crucial role in the transformation and integration of EPSPs can be modulated. I_A is downregulated by activation of β-adrenergic and muscarinic acetylcholine receptors, by activators for protein kinases A and C (Hoffman & Johnston 1998, 1999), and by arachidonic acid (Colbert & Pan 1999). Arachindonic acid in addition enhances sustained outward currents. Downregulation of I_A will effectively unmask the boosting effects of I_{NaP} and I_{CaT} and may linearize the summation of EPSPs. Inhibition of I_{NaP} by agonists for muscarinic receptors (Mittmann & Alzheimer 1998) may lead to a decrease in boosting. Enhancement of GIRK-mediated current by agonists for metabotropic receptors, including cis-ACPD, serotonin, adenosine, baclofen, and somatostatin (e.g. Sodickson & Bean 1998, Takigawa & Alzheimer 1999), may also lead to a decrease in the EPSP amplitude.

The frequency-dependent decrease in the amplitudes of dendritic action potentials (Callaway & Ross 1995, Spruston et al 1995, Tsubokawa & Ross 1997) is reduced by protein kinase C activators (Colbert & Johnston 1998), agonists for muscarinic receptors (Tsubokawa & Ross 1997), and by Ca^{2+} acting as a second messenger (Tsubokawa et al 2000). Minimizing attenuation of backpropagating action potentials will in effect increase the interaction between Ca^{2+} regenerative sites in the dendrites and the action potential initiation region in the axon hillock.

Several neurotransmitters enhance plateau potentials in spinal and neocortical pyramidal neurons. These include serotonin, glutamate, muscarine, and substance P (Greene et al 1992, Svirskis & Hounsgaard 1998, Delgado-Lezama et al 1997, Russo et al 1997). Serotonin has a number of effects, including inhibition of N- and P-type Ca^{2+} currents, some K^+ currents, and Ca^{2+}-dependent K^+ currents (Nicoll et al 1990, Bayliss et al 1995).

CONCLUSION

The presence of dendritic conductances gives the neuron much more flexibility in processing signals. The transformation and integration of synaptic inputs in the dendrites are not hardwired to the geometry of the dendritic tree but rather are dependent on which conductances are active at a particular membrane potential and time. Different inputs are not processed equally: Transient and tonic stimuli will activate a different set of conductances as will more complex waveforms often evoked in vivo with natural stimuli. The neuron is therefore capable of filtering out specific inputs while accentuating others (Reyes et al 1996).

The presence of regenerative sites in the dendrites expands the neuron's firing repertoire and may enhance its ability to encode information. Repetitive, burst, or plateau-mediated firing can be evoked under different conditions. Qualitative differences in the firing pattern may contain information about the identity of the active presynaptic cells.

Finally, the fact that the channels can be regulated by neuromodulators means that the input/output properties of neurons are highly dynamic. Subtle changes in the properties of a channel can shift the balance between inward and outward currents and may alter radically how neurons process incoming signals.

Future Directions

It is becoming increasingly clear that somatocentric models do not adequately describe the input/output properties of neurons. By the same token, highly sophisticated dendritic models incorporating Hodgkin and Huxley kinetics are impractical for simulating a large network of neurons. The challenge therefore is to develop realistic input/output functions that incorporate all the complexities introduced by active dendrites. At the moment, the main stumbling block is that there seem to be very few, if any, generalizable rules that govern signal processing in the dendrites.

Visit the Annual Reviews home page at www.AnnualReviews.org

LITERATURE CITED

Andreasen M, Lambert JDC. 1995. Regenerative properties of pyramidal cell dendrites in area CA1 of the rat hippocampus. *J. Physiol.* 483:421–41

Bayliss DA, Umemiya M, Berger AJ. 1995. Inhibition of N- and P-type calcium currents and the after hyperpolarization in rat motoneurones by serotonin. *J. Physiol.* 485(3):635–47

Bekkers JM. 2000. Distribution and activation of voltage-gated potassium channels in cell-attached and outside-out patches from large layer 5 cortical pyramidal neurons of the rat. *J. Physiol.* 525(3):611–20

Bennett DJ, Hultborn H, Fedirchuk B, Gorassini M. 1998. Synaptic activation of plateaus in hindlimb motoneurons of decerebrate cats. *J. Neurophysiol.* 80:2023–37

Bennett MVI, Hille B, Obara S. 1970. Voltage threshold in excitable cells depends on stimulus waveform. *J. Neurophysiol.* 33:585–94

Booth V, Rinzel J, Kiehn O. 1997. Compartmental model of vertebrate motoneurons for Ca^{2+}-dependent spiking and plateau potentials under pharmacological treatment. *J. Neurophysiol.* 78:3371–85

Callaway JC, Ross WN. 1995. Frequency-dependent propagation of sodium action potentials in dendrites of hippocampal CA1 pyramidal neurons. *J. Neurophysiol.* 74:1395–403

Carlin KP, Jones KE, Jiang Z, Jordan LM, Brownstone RM. 2000. Dendritic L-type calcium currents in mouse spinal motoneurons: implications for bistability. *Eur. J. Neurosci.* 12:1635–46

Cash S, Yuste R. 1998. Input summation by cultured pyramidal neurons is linear and position-independent. *J. Neurosci.* 18:10–15

Cash S, Yuste R. 1999. Linear summation of excitatory inputs by CA1 pyramidal neurons. *Neuron* 22:383–94

Chen WR, Midtgaard J, Shepherd GM. 1997. Forward and backward propagation of dendritic impulses and their synaptic control in mitral cells. *Science* 278:463–67

Colbert CM, Johnston D. 1998. Protein kinase C activation decreases activity-dependent attenuation of dendritic Na^+ current in hippocampal CA1 pyramidal neurons. *J. Neurophysiol.* 79:491–95

Colbert CM, Magee JC, Hoffman DA, Johnston D. 1997. Slow recovery from inactivation of Na^+ channels underlies the activity-dependent attenuation of dendritic action potentials in hippocampal CA1 pyramidal neurons. *J. Neurosci.* 17:6512–21

Colbert CM, Pan E. 1999. Arachidonic acid reciprocally alters the availability of transient and sustained dendritic K^+ channels in hippocampal CA1 pyramidal neurons. *J. Neurosci.* 19:8163–71

Colbert M, Johnston D. 1996. Axonal action-potential initiation and Na^+ channel densities in the soma and axon initial segment of subicular pyramidal neurons. *J. Neurosci.* 16:6676–86

Coombs JS, Curtis DR, Eccles JC. 1957a. The interpretation of spike potentials of motoneurones. *J. Physiol.* 139:198–231

Coombs JS, Curtis DR, Eccles JC. 1957b. The generation of impulses in motoneurones. *J. Physiol.* 139:232–49

Crill WE. 1996. Persistent sodium current in mammalian central neurons. *Annu. Rev. Physiol.* 58:349–62

Del Negro CA, Hsiao C, Chandler SH. 1999. Outward currents influencing bursting dynamics in guinea pig trigeminal motoneurons. *J. Neurophysiol.* 81:1478–85

Delgado-Lezama R, Perrier JF, Hounsgaard J. 1999. Local facilitation of plateau potentials in dendrites of turtle motoneurons by synaptic activation of metabotropic receptors. *J. Physiol.* 515.1:203–7

Delgado-Lezama R, Perrier JF, Nedergaard S, Svirskis G, Hounsgaard J. 1997. Metabotropic synaptic regulation of intrinsic response properties of turtle spinal motoneurons. *J. Physiol.* 504(1):97–102

Drake CT, Bausch SB, Milner TA, Chavkin C. 1997. GIRK1 immunoreactivity is present predominantly in dendrites, dendritic spines, and somata in the CA1 region of hippocampus. *Proc. Natl. Acad. Sci. USA* 94:1007–12

Gillessen T, Alzheimer C. 1997. Amplification of EPSPs by low Ni^{2+}- and amiloride-sensitive Ca^{2+} channels in apical dendrites of rat CA1 pyramidal neurons. *J. Neurophysiol.* 77:1639–43

Golding NL, Jung H, Mickus T, Spruston N. 1999. Dendritic calcium spike initiation and repolarization are controlled by distinct potassium channel subtypes in CA1 pyramidal neurons. *J. Neurosci.* 19:8789–98

Golding NL, Spruston N. 1998. Dendritic sodium spikes are variable triggers of axonal action potentials in hippocampal CA1 pyramidal neurons. *Neuron* 21:1189–200

Granit R, Kernell D, Lamarre Y. 1966. Algebraical summation in synaptic activation of motoneurones firing within the 'primary range' to injected currents. *J. Physiol.* 187:379–99

Greene CC, Schwindt PC, Crill WE. 1992.

Metabotropic receptor-mediated afterdepolarization in neocortical neurons. *Eur. J. Pharmacol.* 226:279–80

Hausser M, Spruston N, Stuart GJ. 2000. Diversity and dynamics of dendritic signaling. *Science* 290:739–44

Hausser M, Stuart G, Racca C, Sakmann B. 1995. Axonal initiation and active dendritic propagation of action potentials in substantia nigra neurons. *Neuron* 15:637–47

Hille B. 1994. Modulation of ion-channel function by G-protein-coupled receptors. *Trends Neurosci.* 17:531–36

Hoffman DA, Johnston D. 1998. Downregulation of transient K$^+$ channels in dendrites of hippocampal CA1 pyramidal neurons by activation of PKA and PKC. *J. Neurosci.* 18:3521–28

Hoffman DA, Johnston D. 1999. Neuromodulation of dendritic action potential. *J. Neurophysiol.* 81:408–11

Hoffman DA, Magee JC, Colbert CM, Johnston D. 1997. K$^+$ channel regulation of signal propagation in dendrites of hippocampal pyramidal neurons. *Nature* 387:869–75

Hounsgaard J, Kiehn O. 1989. Serotonin-induced bistability of turtle motoneurones caused by a nifedipine-sensitive calcium plateau potential. *J. Physiol.* 414:265–82

Hounsgaard J, Kiehn O. 1993. Calcium spikes and calcium plateaux evoked by differential polarization in dendrites of turtle motoneurones in vitro. *J. Physiol.* 468:245–59

Hsiao C, Del Negro CA, Trueblood PR, Chandler SH. 1998. Ionic basis for serotonin-induced bistable membrane properties in guinea pig trigeminal motoneurons. *J. Neurophysiol.* 79:2847–56

Johnston D, Hoffman DA, Magee JC, Poolos NP, Watanabe S, et al. 2000. Dendritic potassium channels in hippocampal pyramidal neurons. *J. Physiol.* 525.1:75–81

Jung HY, Mickus T, Spruston N. 1997. Prolonged sodium channel inactivation contributes to dendritic action potential attenuation in hippocampal pyramidal neurons. *J. Neurosci.* 17:6639–46

Kang J, Huguenard JR, Prince DA. 1996. Development of BK channels in neocortical pyramidal neurons. *J. Neurophysiol.* 76:188–98

Karst H, Joels M, Wadman WJ. 1993. Low threshold calcium current in dendrites of the adult rat hippocampus. *Neurosci. Lett.* 164:154–58

Kernell D. 1969. Synaptic conductance changes and the repetitive impulse discharge of spinal motoneurones. *Brain Res.* 15:291–94

Kiehn O, Eken T. 1998. Functional role of plateau potentials in vertebrate motor neurons. *Curr. Opin. Neurobiol.* 8:746–52

Kim HG, Connors BW. 1993. Apical dendrites of the neocortex: correlation between sodium- and calcium-dependent spiking and pyramidal cell morphology. *J. Neurosci.* 13:5301–11

Korngreen A, Sakmann B. 2000. Voltage-gated K$^+$ channels in layer 5 neocortical pyramidal neurones from young rats: subtypes and gradients. *J. Physiol.* 525(3):621–39

Larkum ME, Kaiser KMM, Sakmann B. 1999a. Calcium electrogenesis in distal apical dendrites of layer 5 pyramidal cells at a critical frequency of back-propagating action potentials. *Proc. Natl. Acad. Sci. USA* 96:14600–4

Larkum ME, Launey T, Dityatev A, Luscher HR. 1998. Integration of excitatory postsynaptic potentials in dendrites of motoneurons of rat spinal cord slice cultures. *J. Neurophysiol.* 80:924–35

Larkum ME, Rioult MG, Luscher HR. 1996. Propagation of action potentials in the dendrites of neurons from rat spinal cord slice cultures. *J. Neurophysiol.* 1996 75:154–70

Larkum ME, Zhu JJ, Sakmann B. 1999b. A new cellular mechanism for coupling inputs arriving at different cortical layers. *Nature* 398:338–41

Lee RH, Heckman CJ. 1996. Influence of voltage-sensitive dendritic conductances on bistable firing and effective synaptic current in cat spinal motoneurons in vivo. *J. Neurophysiol.* 76:2107–10

Lipowski R, Gillessen T, Alzheimer C. 1996. Dendritic Na$^+$ channels amplify EPSPs in hippocampal CA1 pyramidal cells. *J. Neurophysiol.* 76:2181–91

Magee J, Hoffman D, Colbert C, Johnston D. 1998. Electrical and calcium signaling in dendrites of hippocampal pyramidal neurons. *Annu. Rev. Physiol.* 60:327–46

Magee JC. 1998. Dendritic hyperpolarization-activated currents modify the integrative properties of hippocampal CA1 pyramidal neurons. *J. Neurosci.* 18:7613–24

Magee JC, Johnston D. 1995. Synaptic activation of voltage-gated channels in the dendrites of hippocampal pyramidal neurons. *Science* 268:301–4

Markram H, Sakmann B. 1994. Calcium transients in dendrites of neocortical neurons evoked by single subthreshold excitatory postsynaptic potentials via low-voltage activated calcium channels. *Proc. Natl. Acad. Sci. USA* 91:5207–11

Markram H, Lubke J, Frotscher M, Roth A, Sakmann B. 1997. Physiology and anatomy of synaptic connections between thick tufted pyramidal neurones in the developing rat neocortex. *J. Physiol.* 500:409–40.

Martina M, Vida I, Jonas P. 2000. Distal initiation and active propagation of action potentials in interneuron dendrites. *Science* 287:295–300

Mickus T, Jung HY, Spruston N. 1999. Properties of slow, cumulative sodium channel inactivation in rat hippocampal CA1 pyramidal neurons. *Biophys. J.* 76:846–60

Mittmann T, Alzheimer C. 1998. Muscarinic inhibition of persistent Na^+ current in rat neocortical pyramidal neurons. *J. Neurophysiol.* 79:1579–82

Mittmann T, Linton SM, Schwindt P, Crill W. 1997. Evidence for persistent Na^+ current in apical dendrites of rat neocortical neurons from imaging of Na^+ sensitive dye. *J. Neurophysiol.* 78:1188–92

Mouginot D, Bossu J, Gahwiler BH. 1997. Low-threshold Ca^{2+} currents in dendritic recordings from purkinje cells in rat cerebellar slice cultures. *J. Neurosci.* 17:160–170

Nettleton JS, Spain WJ. 2000. Linear to supra-

linear summation of AMPA-mediated EPSPs in neocortical pyramidal neurons. *J. Neurophysiol.* 83:3310–22

Nicoll RA, Malenka RC, Kauer JA. 1990. Functional comparison of neurotransmitter receptor subtypes in mammalian central nervous system. *Physiol. Rev.* 70:513–65

Oviedo HV, Reyes AD. 2000. Dendritic and somatic injection of computer-generated synaptic inputs into L5 pyramidal neurons evoke different discharge patterns. *Soc. Neurosci. Abstr.* 26:1122

Poolos NP, Johnston D. 1999. Calcium-activated potassium conductances contribute to action potential repolarization at the soma but not the dendrites of hippocampal CA1 pyramidal neurons. *J. Neurosci.* 19:5205–12

Powers RK, Binder MD. 1995. Effective synaptic current and motoneuron firing rate modulation. *J. Neurophysiol.* 74:793–801

Powers RK, Binder MD. 2000. Summation of effective synaptic currents and firing rate modulation in cat spinal motoneurons. *J. Neurophysiol.* 83:483–500

Powers RK, Robinson FR, Konodi MA, Binder MD. 1992. Effective synaptic current can be estimated from measurements of neuronal discharge. *J. Neurophysiol.* 68:964–68

Rall W. 1964. Theoretical significance of dendritic trees for neuronal input-output relations. In *Neural Theory and Modeling*, ed. RF Reiss, pp. 73–97. Palo Alto, CA: Stanford Univ. Press

Reuveni I, Friedman A, Amitai Y, Gutnick MJ. 1993. Stepwise repolarization from Ca^{2+} plateaus in neocortical pyramidal cells: evidence for nonhomogeneous distribution of HVA Ca^{2+} channels in dendrites. *J. Neurosci.* 13:4609–21

Reyes AD, Rubel EW, Spain WJ. 1996. In vitro analysis of optimal stimuli for phase-locking and time-delayed modulation of firing in avian nucleus laminaris neurons. *J. Neurosci.* 16:993–1007

Reyes AD, Sakmann B. 1996. Summation of synaptic potentials in layer V pyramidal neurons. *Soc. Neurosci. Abstr.* 22:792

Reyes AD, Sakmann B. 1999. Developmental switch in the short-term modification of unitary EPSPs evoked in layer 2/3 and layer 5 pyramidal neurons of rat neocortex. *J. Neurosci.* 19:3827–35

Russo RE, Nagy F, Hounsgaard J. 1997. Modulation of plateau properties in dorsal horn neurones in a slice preparation of the turtle spinal cord. *J. Physiol.* 499:459–74

Schiller J, Schiller Y, Stuart G, Sakmann B. 1997. Calcium action potentials restricted to distal apical dendrites of rat neocortical pyramidal neurons. *J. Physiol.* 505(3):605–16

Schwindt P, Crill W. 1995. Amplification of synaptic current by persistent sodium conductance in apical dendrite of neocortical neurons. *J. Neurophysiol.* 74:2220–24

Schwindt P, Crill W. 1996. Equivalence of amplified current flowing from dendrite to soma measured by alteration of repetitive firing and by voltage clamp in layer 5 pyramidal neurons. *J. Neurophysiol.* 76:3731–39

Schwindt P, Crill W. 1997a. Local and propagated dendritic action potentials evoked by glutamate iontophoresis on rat neocortical pyramidal neurons. *J. Neurophysiol.* 77:2466–83

Schwindt P, Crill W. 1997b. Modification of current transmitted from apical dendrite to soma by blockade of voltage- and Ca^{2+} dependent conductances in rat neocortical pyramidal neurons. *J. Neurophysiol.* 78:187–98

Schwindt P, Crill W. 1999. Mechanisms underlying burst and regular spiking evoked by dendritic depolarization in layer 5 cortical pyramidal neurons. *J. Neurophysiol.* 81:1341–54

Schwindt PC, Calvin WH. 1973. Equivalence of synaptic and injected current in determining the membrane potential trajectory during motoneuron rhythmic firing. *Brain Res.* 59:389–94

Skydsgaard M, Hounsgaard J. 1994. Spatial integration of local transmitter responses in motoneurones of the turtle spinal cord in vitro. *J. Physiol.* 479.2:223–46

Sodickson DL, Bean BP. 1998. Neurotransmitter activation of inwardly rectifying potassium current in dissociated hippocampal CA3 neurons: interactions among multiple receptors. *J. Neurosci.* 18:8153–62

Spruston N, Schiller Y, Stuart G, Sakmann B. 1995. Activity-dependent action potential invasion and calcium influx into hippocampal CA1 dendrites. *Science* 268:297–300

Stuart G, Sakmann B. 1994. Active propagation of somatic action potentials into neocortical pyramidal cell dendrites. *Nature* 367:69–72

Stuart G, Sakmann B. 1995. Amplification of EPSPs by axosomatic sodium channels in neocortical pyramidal neurons. *Neuron* 15:1065–76

Stuart G, Schiller J, Sakmann B. 1997. Action potential initiation and propagation in rat neocortical pyramidal neurons. *J. Physiol.* 505(3):617–32

Stuart G, Spruston N. 1998. Determinants of voltage attenuation in neocortical pyramidal neuron dendrites. *J. Neurosci.* 18:3501–10

Svirskis G, Hounsgaard J. 1998. Transmitter regulation of plateau properties in turtle motoneurons. *J. Neurophysiol.* 79:45–50

Takigawa T, Alzheimer C. 1999. G protein-activated inwardly rectifying K^+ (GIRK) currents in dendrites of rat neocortical pyramidal cells. *J. Physiol.* 517.2:385–90

Tsubokawa H, Offermanns S, Simon M, Kano M. 2000. Calcium-dependent persistent facilitation of spike backpropagation in the CA1 pyramidal neurons. *J. Neurosci.* 20:4878–84

Tsubokawa H, Ross WN. 1997. Muscarinic modulation of spike backpropagation in the apical dendrites of hippocampal CA1 pyramidal neurons. *J. Neurosci.* 17:5782–91

Turner RW, Meyers DER, Richardson TL, Barker JL. 1991. The site for initiation of action potential discharge over the somatodendritic axis of rat hippocampal CA1 pyramidal neurons. *J. Neurosci.* 11:2270–80

Urban NN, Barrionuevo G. 1998. Active summation of excitatory postsynaptic potentials in hippocampal CA4 pyramidal neurons. *Proc. Natl. Acad. Sci. USA* 95:11450–55

Urban NN, Henze DA, Barrionuevo G. 1998. Amplification of perforant-path EPSPs in CA3 pyramidal cells by LVA calcium and sodium channels. *J. Neurophysiol.* 80:1558–61

Velte TB, Masland RH. 1999. Action potentials in the dendrites of retinal ganglion cells. *J. Neurophysiol.* 81:1412–17

Williams SR, Stuart GJ. 1999. Mechanisms and consequences of action potential burst firing in rat neocortical pyramidal neurons. *J. Physiol.* 521.2:467–82

Williams SR, Stuart GJ. 2000a. Site independence of EPSP time course is mediated by dendritic Ih in neocortical pyramidal neurons. *J. Neurophysiol.* 83:3177–82

Williams SR, Stuart GJ. 2000b. Action potential backpropagation and somato-dendritic distribution of ion channels in thalamocortical neurons *J. Neurosci.* 20:1307–17

Wong RKS, Stewart M. 1992. Different firing patterns generated in dendrites and somata of CA1 pyramidal neurones in guinea-pig hippocampus. *J. Physiol.* 457:675–87

Zhu JJ. 2000. Maturation of layer 5 neocortical pyramidal neurons: amplifying salient layer 1 and layer 4 inputs by Ca^{2+} action potentials in adult rat tuft dendrites. *J. Physiol.* 526:571–87

Annu. Rev. Neurosci. 2001. 24:677–736

NEUROTROPHINS: Roles in Neuronal Development and Function*

Eric J Huang[1] and Louis F Reichardt[2]

[1]*Department of Pathology, University of California, San Francisco, California 94143;*
e-mail: ejhuang@itsa.ucsf.edu
[2]*Department of Physiology, University of California, San Francisco, California 94143,*
and Howard Hughes Medical Institute, San Francisco, California 94143;
e-mail: lfr@cgl.ucsf.edu

Key Words   Trk receptor, nerve growth factor, apoptosis, plasticity, synapse, signaling, survival, differentiation

■ **Abstract**   Neurotrophins regulate development, maintenance, and function of vertebrate nervous systems. Neurotrophins activate two different classes of receptors, the Trk family of receptor tyrosine kinases and p75NTR, a member of the TNF receptor superfamily. Through these, neurotrophins activate many signaling pathways, including those mediated by ras and members of the cdc-42/ras/rho G protein families, and the MAP kinase, PI-3 kinase, and Jun kinase cascades. During development, limiting amounts of neurotrophins function as survival factors to ensure a match between the number of surviving neurons and the requirement for appropriate target innervation. They also regulate cell fate decisions, axon growth, dendrite pruning, the patterning of innervation and the expression of proteins crucial for normal neuronal function, such as neurotransmitters and ion channels. These proteins also regulate many aspects of neural function. In the mature nervous system, they control synaptic function and synaptic plasticity, while continuing to modulate neuronal survival.

INTRODUCTION

Neurotrophins are important regulators of neural survival, development, function, and plasticity (for reviews, see Korsching 1993, Eide et al 1993, Segal & Greenberg 1996, Lewin & Barde 1996, Reichardt & Fariñas 1997, McAllister et al 1999, Sofroniew et al 2001). As the central concept of the neurotrophic factor hypothesis, targets of innervation were postulated to secrete limiting amounts of survival factors that function to ensure a balance between the size of a target organ and the number of innervating neurons (reviewed in Purves 1988). Nerve growth factor (NGF), the first such factor to be characterized, was discovered during a

---

search for such survival factors (reviewed in Levi-Montalcini 1987). There are four neurotrophins characterized in mammals. NGF, brain-derived neurotrophic factor (BDNF), neurotrophin-3 (NT-3), and neurotrophin-4 (NT-4) are derived from a common ancestral gene, are similar in sequence and structure, and are therefore collectively named neurotrophins (e.g. Hallbook 1999). Although members of other families of proteins, most notably the glial cell–derived neurotrophic factor (GDNF) family and the neuropoietic cytokines, have been shown to also regulate survival, development, and function in the nervous system, this review focuses on the neurotrophins, examining mechanisms by which they signal and control development and function of the nervous system. A companion review by others describes the roles of these fascinating proteins in supporting the injured and aging nervous systems (Sofroniew et al 2001).

As the first neurotrophic factors to be discovered, the neurotrophins have had an unusually important influence on biology. The experiments leading to the discovery of NGF revealed the essential role of cellular interactions in development. Now almost all cells are believed to depend on their neighbors for survival (see Raff et al 1993). Almost a decade before endocytosis and transport were studied seriously in nonneural cells, NGF was shown to be internalized by receptor-dependent mechanisms and to be transported for vast distances along axons in small membrane vesicles by an energy and microtubule-dependent mechanism with eventual degradation of NGF in lysosomes. Now almost all cells are known to utilize similar mechanisms for trafficking of receptors and their ligands. Finally, neurotrophins have been shown to activate receptor tyrosine kinases. Within neural precursors and neurons, the pathways regulated by tyrosine kinases include proliferation and survival, axonal and dendritic growth and remodeling, assembly of the cytoskeleton, membrane trafficking and fusion, and synapse formation and function. Recent studies on the neurotrophins have shown that they regulated each of these functions and have increased our understanding of the molecular mechanisms underlying each. Thus, studies on these factors continue to provide insights of widespread interest to modern biologists.

Sources of Neurotrophins

NGF was purified as a factor able to support survival of sympathetic and sensory spinal neurons in culture (Levi-Montalcini 1987). Anti-NGF injections demonstrated that this factor is important in maintaining survival of sympathetic neurons in vivo as well as in vitro. Development of a two-site ELISA assay and of an NGF mRNA assay, using as probe the cloned NGF gene, made it possible to demonstrate that NGF is synthesized and secreted by sympathetic and sensory target organs (reviewed in Korsching 1993). From these sources, it is captured in nerve terminals by receptor-mediated endocytosis and is transported through axons to neuronal cell bodies where it acts to promote neuronal survival and differentiation. Within the target organs, synthesis of NGF and of other neurotrophins is associated with end organs, such as hair follicles, which become innervated by the axons of these neurons.

Subsequent work has demonstrated that there are other sources of neurotrophins. First, after peripheral nerve injury, macrophages infiltrate the nerve as part of an inflammatory response and release cytokines, which induce the synthesis of NGF in Schwann cells and fibroblasts within the injured nerve (reviewed by Korsching 1993). NGF is also synthesized in mast cells and is released following mast cell activation (reviewed in Levi-Montalcini et al 1996). NGF and other neurotrophic factors synthesized in damaged nerve are believed to be essential for survival and regeneration of injured neurons. Second, during development, neurotrophins are expressed in regions being invaded by sensory axons en route to their final targets, so they may provide trophic support to neurons that have not yet contacted their final targets (e.g. Fariñas et al 1996, 1998; Huang et al 1999a; Ringstedt et al 1999). Third, many neurons also synthesize neurotrophins. For example, several populations of sensory neurons have been shown to synthesize BDNF (e.g. Mannion et al 1999, Brady et al 1999). Although some evidence has been presented suggesting that BDNF may act in an autocrine or paracrine fashion to support dorsal root ganglion (DRG) sensory neurons (Acheson et al 1995, Robinson et al 1996), in other instances it may be transported anterogradely and act *trans*-synaptically on targets of the central afferents of these neurons within the brain (Brady et al 1999; see also Altar et al 1997, Fawcett et al 1998, von Bartheld et al 1996). Finally, when overexpressed in skin, sufficient target-derived NGF is released from the somata of trigeminal sensory neurons to support aberrant innervation by NGF-dependent sympathetic fibers (Davis et al 1998, Walsh et al 1999b). Thus, in some circumstances, a neurotrophin provided by one cell not only is effective at supporting neurons whose axons are in its vicinity, it also can provide support to more distant neurons via transcellular transport.

Neurotrophins and Their Receptors

Currently, six neurotrophins have been isolated: NGF, BDNF, NT-3, NT-4 (also known as NT-5), NT-6, and NT-7. There is substantial evidence that they all arose through successive duplications of the genome of an ancestral chordate (Hallbook 1999). The NT-6 and NT-7 genes have been identified only in fish and probably do not have mammalian or avian orthologues (Gotz et al 1994, Nilsson et al 1998). NT-4 has not been detected in avian species. Neurotrophins generally function as noncovalently associated homodimers, but at least some neurotrophin subunits are able to form heterodimers with other neurotrophin subunits. NGF, NT-6, and NT-7 appear to act on very similar and perhaps identical populations of neurons. BDNF and NT-4 have also very similar targets (e.g. Ip et al 1993). Thus the neurotrophins can be divided into three classes based upon target neuron populations, and all vertebrate species are likely to have at least one neurotrophin in each class. The structures of NGF, NT-3, and NT-4 and of NT-3/BDNF and NT-4/BDNF dimers have been solved and novel features of their structures—a tertiary fold and cystine knot—are present in several other growth factors, including platelet-derived growth factor and transforming growth factor β (McDonald et al 1991; Fandl et al 1994; Robinson et al 1995, 1999; Butte et al 1998; reviewed in McDonald & Chao 1995).

Initial efforts to identify NGF receptors resulted in discovery of a receptor now named p75NTR. For many years this was believed to be a low-affinity receptor specific for NGF. More recently, it has been shown to bind to all of the neurotrophins with a very similar affinity (Rodriguez-Tebar et al 1991). p75NTR is a distant member of the tumor necrosis factor receptor family (Chao 1994, Bothwell 1995). The cytoplasmic domain of this receptor contains a "death" domain structurally similar to those in other members of this receptor family (Liepinsch et al 1997). For many years after its discovery, it was not certain whether this receptor transmitted any signals or whether it functioned simply as a binding protein. Work during the past few years has shown, however, that this protein transmits signals important for determining which neurons survive during development. Signaling by this receptor is discussed at length below.

In a dramatic advance, the three members of the Trk (tropomyosin-related kinase) receptor tyrosine kinase family were shown to be a second class of neurotrophin receptors (reviewed in Bothwell 1995). The neurotrophins have been shown to directly bind and dimerize these receptors, which results in activation of the tyrosine kinases present in their cytoplasmic domains. NGF is specific for TrkA. BDNF and NT-4 are specific for TrkB. NT-3 activates TrkC and is also able to activate less efficiently each of the other Trk receptors. The most important site at which Trk receptors interact with neurotrophins has been localized to the most proximal immunoglobulin (Ig) domain of each receptor. The three-dimensional structures of each of these Ig domains has been solved (Ultsch et al 1999), and the structure of NGF bound to the TrkA membrane proximal Ig domain has also been determined (Wiesmann et al 1999). This exciting structural information has provided detailed information about interactions that regulate the strength and specificity of binding between neurotrophins and Trk receptors (e.g. Urfer et al 1998).

The unique actions of the neurotrophins made it seem likely that they would prove to have receptors and signal transduction pathways completely different from those of the mitogenic growth factors, such as platelet-derived growth factor or epidermal growth factor, whose receptors were known to be receptor tyrosine kinases. Thus, it was surprising when Trk receptors were identified as functional, survival-promoting receptors for neurotrophins. During the past few years, however, members of other neurotrophic factor families have also been shown to activate tyrosine kinases. These include GDNF and its relatives and ciliary neurotrophic factor (CNTF) and other neuropoietic cytokines (reviewed in Reichardt & Fariñas 1997). These tyrosine kinases activate many of the same intracellular signaling pathways regulated by the receptors for mitogens. Appreciation of this shared mechanism of action has been a major conceptual advance of the past decade.

Control of Neurotrophin Responsiveness by Trk Receptors

Tyrosine kinase–mediated signaling by endogenous Trk receptors appears to promote survival and/or differentiation in all neuronal populations examined to date. With a few exceptions, ectopic expression of a Trk receptor is sufficient to confer a neurotrophin-dependent survival and differentiation response (e.g. Allsopp et al 1994, Barrett & Bartlett 1994). Usually, endogenous expression of a Trk receptor confers responsiveness to the neurotrophins with which it binds, but this generalization is oversimplified for several reasons. First, differential splicing of the TrkA, TrkB, and TrkC mRNAs results in expression of proteins with differences in their extracellular domains that affect ligand interactions (Meakin et al 1992, Clary & Reichardt 1994, Shelton et al 1995, Garner et al 1996, Strohmaier et al 1996). The presence or absence of short amino acid sequences in the juxtamembrane domains of each receptor has been shown to affect the ability of some neurotrophins to activate these receptors. Although BDNF, NT-4, and NT-3 are capable of activating the TrkB isoform containing these amino acids, the TrkB isoform lacking them can only be activated by BDNF (Strohmaier et al 1996). These isoforms of TrkB have been shown to be expressed in nonoverlapping populations of avian sensory neurons, so splicing of this receptor almost certainly has important functional consequences (Boeshore et al 1999). Similarly, an isoform of TrkA containing a short juxtamembrane sequence is activated by both NGF and NT-3, whereas the isoform lacking these amino acids is much more specifically activated by NGF (Clary & Reichardt 1994). Although this short polypeptide sequence was not localized in the three-dimensional structure of the NGF-TrkA ligand binding domain complex, the organization of the interface between the two proteins is compatible with the possibility that these residues may directly participate in binding (Wiesmann et al 1999). The abilities of NT-3 to activate TrkA and of NT-3 and NT-4 to activate TrkB are also negatively regulated by high levels of the pan-neurotrophin receptor p75NTR (Bennedetti et al 1993, Lee et al 1994b, Clary & Reichardt 1994, Bibel et al 1999). Thus, factors that regulate differential splicing of extracellular exons in Trk receptor genes and signaling pathways that control expression of p75NTR affect the specificity of neuronal responsiveness to neurotrophins.

Important also has been the discovery of differential splicing of exons encoding portions of the Trk receptor cytoplasmic domains. Not all isoforms of TrkB and TrkC contain tyrosine kinase domains (reviewed in Reichardt & Fariñas 1997). Differential splicing generates isoforms of both TrkB and TrkC, which lack these domains. The functions of nonkinase-containing isoforms of TrkB and TrkC in nonneuronal cells may include presentation of neurotrophins to neurons. Within neurons, these same receptors are likely to inhibit productive dimerization and activation of full-length receptors, thereby attenuating responses to neurotrophins (e.g. Eide et al 1996). There is also evidence suggesting that ligand binding to truncated isoforms of TrkB and TrkC can modulate intracellular signaling pathways more directly (Baxter et al 1997, Hapner et al 1998). Differential splicing has also been shown to result in expression of an isoform of TrkC, which contains an amino acid insert within the tyrosine kinase domain. This insert does not eliminate the kinase activity of TrkC but does appear to modify its substrate specificity (e.g. Guiton et al 1995, Tsoulfas et al 1996, Meakin et al 1997).

Finally, in some central nervous system (CNS) projection neurons, Trk receptors appear to be largely sequestered in intracellular vesicles (Meyer-Franke et al

1998). Only in the presence of a second signal, such as cAMP or Ca^{2+}, are the receptors inserted efficiently into the plasmalemma. In these neurons, expression of a kinase-containing isoform of a Trk receptor may not be sufficient to confer responsiveness to a neurotrophin if the neurons are not incorporated into a signaling network that results in production of these second messengers. Thus, neurotrophin responsiveness is controlled by many factors in addition to regulators of Trk receptor gene expression.

Control of Neurotrophin Responsiveness by the Pan-Neurotrophin Receptor p75NTR

Each neurotrophin also binds to the low-affinity neurotrophin receptor p75NTR, which is a member of the tumor necrosis factor receptor superfamily (see Frade & Barde 1998). In vitro studies on p75NTR have documented that it can potentiate activation of TrkA by subsaturating concentrations of NGF (e.g. Mahadeo et al 1994, Verdi et al 1994). What is surprising is that it does not appear to potentiate activation of the other Trk receptors by their ligands in vitro, even though these also bind to p75NTR. A role for p75NTR in potentiating actions of neurotrophins in vivo, however, provides one possible explanation of the deficits in multiple classes of sensory neurons observed in the p75NTR mutant (Stucky & Koltzenburg 1997, Bergmann et al 1997, Kinkelin et al 1999). As discussed above, studies in cell culture also indicate that p75NTR reduces responsiveness of Trk receptors to noncognate ligands (Benedetti et al 1993, Clary & Reichardt 1994, Lee et al 1994b). Recently, NT-3 has also been shown to maintain survival of TrkA-expressing sympathetic neurons in vivo more effectively in the absence than in the presence of p75NTR (Brennan et al 1999). The presence of p75NTR has also been shown to promote retrograde transport of several neurotrophins (e.g. Curtis et al 1995, Ryden et al 1995, Harrison et al 2000). Most intriguing, both in vitro and in vivo evidence now indicates that ligand engagement of p75NTR can directly induce neuronal death via apoptosis (reviewed in Frade & Barde 1998; see also Friedman, 2000). Analysis of the p75NTR mutant phenotype has demonstrated that regulation of apoptosis by ligand engagement of p75NTR is important during peripheral nervous system as well as CNS development in vivo (e.g. Bamji et al 1998, Casademunt et al 1999). Finally, absence of p75NTR signaling perturbs axon growth in vitro and both axon growth and target innervation in vivo (e.g. Lee et al 1994a, Yamashita et al 1999b, Bentley & Lee 2000, Walsh et al 1999a,b).

REGULATION OF SIGNALING BY NEUROTROPHINS

Trk Receptor-Mediated Signaling Mechanisms

Ligand engagement of Trk receptors has been shown to result in phosphorylation of cytoplasmic tyrosine residues on the cytoplasmic domains of these receptors

(Figure 1; see color insert). Trk receptors contain 10 evolutionarily conserved tyrosines in their cytoplasmic domains, of which three–Y670, Y674, and Y675 (human TrkA sequence nomenclature)–are present in the autoregulatory loop of the kinase domain that controls tyrosine kinase activity (e.g. Stephens et al 1994, Inagaki et al 1995). Phosphorylation of these residues further activates the receptor. Phosphorylation of the other tyrosine residues promotes signaling by creating docking sites for adapter proteins containing phosphotyrosine-binding (PTB) or src-homology-2 (SH-2) motifs (reviewed in Pawson & Nash 2000). These adapter proteins couple Trk receptors to intracellular signaling cascades, which include the Ras/ERK (extracellular signal–regulated kinase) protein kinase pathway, the phosphatidylinositol-3-kinase (PI-3 kinase)/Akt kinase pathway, and phospholipase C (PLC)-γ1 (see Reichardt & Fariñas 1997, Kaplan & Miller 2000). Two tyrosines not in the kinase activation domain (Y490 and Y785) are major sites of endogenous phosphorylation, and most research has focused on interactions mediated by these sites with Shc and PLC-γ1, respectively (Stephens et al 1994). Five of the remaining seven conserved tyrosines also contribute to NGF-induced neurite outgrowth, however, so interactions mediated by Y490 and Y785 can mediate only a subset of Trk receptor interactions important in neurotrophin-activated signaling (Inagaki et al 1995). Recent work has resulted in identification of additional adapter proteins that interact with Trk receptors at different sites and has demonstrated that transfer of Trk receptors to various membrane compartments controls the efficiency with which these receptors can associate with and activate adapter proteins and intracellular signaling pathways (e.g. Qian et al 1998; Saragovi et al 1998; York et al 2000; C Wu, C-F Lai, WC Mobley, unpublished observations).

PLC-γ1 Signaling

Phosphorylation of Y785 on TrkA has been shown to recruit PLC-γ1 directly, which is activated by phosphorylation and then acts to hydrolyse phosphatidyl inosities to generate inositol tris-phosphate and diacylglycerol (DAG) (Vetter et al 1991). Inositol tris-phosphate induces release of Ca^{2+} stores, increasing levels of cytoplasmic Ca^{2+}. This results in activation of various enzymes regulated by cytoplasmic Ca^{2+}, including Ca^{2+}-calmodulin–regulated protein kinases and phosphatases and Ca^{2+}-regulated isoforms of protein kinase C. Formation of DAG stimulates the activity of DAG-regulated protein kinase C isoforms. In PC12 cells, protein kinase C (PKC)δ, a DAG-regulated PKC, is activated by NGF and is required for neurite outgrowth and for activation of the ERK cascade (Corbit et al 1999). Inhibition of PKCδ has been shown to inhibit activation of MEK [mitogen-activated protein kinase kinase (MAPKK)/ERK kinase)] but not of c-raf, so PKCδ appears to act between Raf and MEK in the ERK kinase cascade.

RAS-ERK Signaling

Activation of Ras is essential for normal differentiation of PC12 cells and neurons. In many cells, Ras activation also promotes survival of neurons, either by activation of PI-3 kinase or through activation of the ERK family of MAP kinases. Transient vs prolonged activation of the MAP kinase pathway has been closely associated, respectively, with a proliferation-inducing vs a differentiation-promoting response to neurotrophin application (e.g. Grewal et al 1999).

The pathways leading to activation of Ras are surprisingly complex. In the first pathway to be characterized, phosphorylation on Y490 was shown to result in recruitment and phosphorylation of the adapter protein Shc, with binding mediated by the Shc PTB domain (Stephens et al 1994; reviewed in Kaplan & Miller 2000). Shc is then phosphorylated by Trk, resulting in recruitment of a complex of the adapter protein Grb-2 and the Ras exchange factor SOS. Activation of Ras by SOS has many downstream consequences, including stimulation of PI-3 kinase, activation of the c-raf/ERK pathway, and stimulation of the p38 MAP kinase/MAP kinase-activated protein kinase 2 pathway (e.g. Xing et al 1996). Downstream targets of the ERK kinases include the RSK kinases (ribosomal S6 kinase). Both RSK and MAP kinase-activated protein kinase 2 phosphorylate CREB (cAMP-regulated enhancer binding protein) and other transcription factors (Xing et al 1998). These transcription factors in turn control expression of many genes known to be regulated by NGF and other neurotrophins. Among these, CREB regulates genes whose products are essential for prolonged neurotrophin-dependent survival of neurons (Bonni et al 1999, Riccio et al 1999).

Neurotrophin signaling through Shc/Grb-2/SOS mediates transient, but not prolonged, activation of ERK signaling pathways (e.g Grewal et al 1999). Prolonged ERK activation has been shown to depend on a distinct signaling pathway involving the adapter protein Crk, the exchange factor C3G, the small G protein rap1, and the serine-threonine kinase B-raf (York et al 1998). Neurotrophins activate this signaling pathway by utilization of a distinct adapter named FRS-2 (fibroblast growth factor receptor substrate-2) or SNT(suc-associated neurotrophic factor-induced tyrosine-phosphorylated target), which competes with Shc for phosphorylated Y490 on TrkA (Meakin et al 1999). FRS-2 is phosphorylated by Trk activation and has been shown to have binding sites for several additional proteins, including the adapter proteins Grb-2 and Crk, the cytoplasmic tyrosine kinase Src, the cyclin-dependent kinase substrate p13^{suc1}, and the protein phosphatase SH-PTP-2 (e.g. Meakin et al 1999). Crk associates with phosphorylated FRS-2 and then binds and activates the exchange factor C3G (e.g. Nosaka et al 1999). Activation by C3G of the small G protein Rap1 results in stimulation of B-raf, which activates the ERK kinase cascade. As predicted by this model, overexpression of FRS-2 or Crk results in differentiation of pheochromocytoma (PC)-12 cells (Tanaka et al 1993, Matsuda et al 1994, Hempstead et al 1994, Meakin et al 1999). In addition to providing a crucial link to a pathway that appears to be essential for prolonged MAP kinase activation, FRS-2 provides a mechanism not dependent on Shc for activation of the Grb-2/SOS/Ras pathway. This adapter protein also provides a link to the Src family tyrosine kinases, which have been implicated in receptor endocytosis and other cellular responses (e.g. Wilde et al 1999, Beattie et al 2000). Finally, binding to FRS-2 of the protein phosphatase SH-PTP-2 also facilitates activation of the ERK pathway, probably by inactivation of an inhibitor, such as Ras-GAP or MAPK phosphatase (Wright et al 1997).

PI-3 Kinase Signaling

Activation of phosphatidylinositol-3-kinase (PI-3 kinase) is essential for survival of many populations of neurons. In collaboration with the phosphatidylinositide-dependent kinases, phosphatidyl inositides generated by PI-3 kinase activate the protein kinase Akt/protein kinase B. Akt then phosphorylates and controls the biological functions of several proteins important in modulating cell survival (reviewed in Datta et al 1999, Yuan & Yankner 2000). Among the substrates of Akt are BAD, a Bcl-2 family member that promotes apoptosis by binding to Bcl-xL, which in the absence of binding would inhibit the proapoptotic activity of Bax. Phosphorylation of BAD results in its association with 14-3-3 proteins and prevents it from promoting apoptosis (Datta et al 1997). BAD is also a substrate for MAP kinases, which similarly inactivate its apoptosis-promoting function (Bonni et al 1999). Another demonstrated target of Akt is IκB (reviewed in Datta et al 1999). Phosphorylation of IκB results in its degradation and activation of NFκB, which is normally sequestered by IκB in the cytoplasm. Transcription activated by nuclear NFκB has been shown to promote neuronal survival (e.g. Middleton et al 2000). A third Akt substrate of potential relevance for neuronal survival is the forkhead transcription factor FKHRL1, which controls expression of apoptosis-promoting gene products, such as FasL (Brunet et al 1999). Another Akt substrate is human but not mouse caspase-9 (Brunet et al 1999). Glycogen synthase kinase 3-β (GSK3β) is also stimulated by trophic factor withdrawal and negatively regulated by Akt phosphorylation (Hetman et al 1999). In cultured cortical neurons, elevated GSK3β promotes apoptosis (Hetman et al 1999). Many additional proteins in the cell death cascade, including Bcl-2, Apaf-1, caspase inhibitors, and caspases, have a consensus site for Akt phosphorylation but have not been shown to be phosphorylated by this kinase (Datta et al 1999). Analyses of mouse mutants have documented the importance of many, but not all, of these proteins (reviewed in Yuan & Yankner 2000). Mutants lacking caspase-9 or Bax have reductions in neuronal apoptosis, whereas a mutant lacking Bcl-x-L has an increase in neuronal apoptosis during development (e.g. Deckwerth et al 1996, Shindler et al 1998). Absence of BAD, however, does not detectably alter neuronal apoptosis during CNS development, which suggests that Akt-mediated phosphorylation of this protein is not an essential link in the PI-3 kinase–dependent survival cascade in vivo (Shindler et al 1998). It is important to note that not all substrates of Akt are involved in cell survival. S6 kinase, for example, is important for promoting translation of a subset of mRNAs, including certain cyclins essential for cell cycle progression.

PI-3 kinase is activated by Ras. In many but not all neurons, Ras-dependent activation of PI-3 kinase is the major pathway by which neurotrophins convey survival-promoting signals (e.g. Vaillant et al 1999). PI-3 kinase and signaling pathways dependent on PI-3 kinase function can also be activated through Shc and Grb-2 by a Ras-independent mechanism. Recruitment by phosphorylated Grb-2 of the adaptor protein Gab-1 results in subsequent binding to this complex of PI-3 kinase, which is then activated (Holgado-Madruga et al 1997, reviewed by Kaplan & Miller 2000). In some cells, but not in PC-12 cells, insulin receptor substrate (IRS)-1 has been shown to be phosphorylated in response to neurotrophins and in turn to recruit and activate PI-3 kinase (Yamada et al 1997).

In addition to providing an adapter that facilitates activation of PI-3 kinase, Gab-1 has also been shown to function as an adapter that nucleates formation of a complex that includes the protein tyrosine phosphatase Shp-2 (Shi et al 2000). Shp-2 has been shown to enhance activation of the RAS-RAF-MEK-ERK pathway by a mechanism that is not clear, but that appears to involve dephosphorylation of a 90-kDa protein that is also associated with the Gab-1 complex.

Control of the Actin Cytoskeleton

The neurotrophins induce rapid ruffling and cytoskeletal rearrangements similar to those induced by other growth factors (e.g. Connolly et al 1979). These have been shown by many laboratories to involve small G proteins of the Cdc-42/Rac/Rho family, which regulate the polymerization and turnover of F-actin (reviewed in Kjoller & Hall 1999, Bishop & Hall 2000). Several exchange factors for this family of G proteins are known to be expressed in neurons and to be regulated by tyrosine phosphorylation and/or phosphatidyl inositides generated by PI-3 kinase activity (e.g. Liu & Burridge 2000). Many of these are undoubtedly regulated by Trk receptor signaling. SOS also has a latent activity as an exchange factor for rac in addition to its activity as an exchange factor for ras (Nimnual et al 1998). Activated ras has been shown to activate the exchange factor activity of SOS for rac through a mechanism dependent on PI-3 kinase. Thus SOS provides a mechanism for the coordination of ras and rac activities.

Control of Trk Signaling by Membrane Trafficking

Recent work has added complexity to the scheme presented above by providing evidence that the ability of Trk receptors to activate specific signaling pathways is regulated by endocytosis and membrane sorting. It has long been appreciated that communication of survival signals from nerve terminals to neuronal cell bodies requires retrograde transport (e.g. Thoenen & Barde 1980). Several groups have demonstrated during the past few years that NGF and activated Trk receptors are transported together in endocytotic vesicles (e.g. Grimes et al 1997, Riccio et al 1997, Tsui-Pierchala & Ginty 1999; CL Howe, E Beattie, JS Valletta, WC Mobley, unpublished observations). More recently, evidence has accumulated indicating that membrane sorting determines which pathways are activated by Trk receptors. In one set of experiments, cells were exposed to a complex of NGF and a monoclonal antibody (mAb) that does not interfere with receptor binding but induces unusually rapid internalization of the mAb-NGF-TrkA complex (Saragovi et al 1998). This NGF-mAb complex was shown to promote transient MAP kinase activation, Shc phosphorylation, and PC12 cell survival. In contrast, FRS-2 was not phosphorylated and the cells did not differentiate normally. The results suggest that recruitment of FRS-2 by ligand receptor complexes occurs on the cell surface with comparatively slow kinetics. Perhaps FRS-2, which is myristoylated, is segregated into a compartment that is not immediately accessible to the TrkA receptor.

In another set of experiments, a thermosensitive dynamin that functions as a dominant negative protein at high temperature was used to reversibly inhibit ligand-receptor internalization (Zhang et al 2000). Inhibition of internalization did not inhibit survival of PC12 cells but did strongly inhibit their differentiation. This observation suggests that ligand-receptor complexes must be internalized to activate efficiently pathways essential for differentiation. As previous work described above has strongly suggested that FRS-2 signaling through Crk is essential for prolonged MAP kinase activation and normal differentiation, the data suggest that activation of this pathway requires internalization of the NGF-TrkA signaling complex.

Consistent with the involvement of PI-3 kinase products in endocytosis (Wendland et al 1998), inhibitors of PI-3 kinase have been shown to reduce retrograde transport and to affect the activation of NGF-dependent intracellular signaling pathways (Kuruvilla et al 2000, York et al 2000). Activation of Ras has been shown to occur in the absence of TrkA internalization and absence of PI-3 kinase activity (York et al 2000). In contrast, activation of Rap-1 and B-raf and sustained ERK activation require internalization and PI-3 kinase activity (York et al 2000). To activate B-raf, TrkA must be transported to a brefeldin-A–sensitive population of endosomes (C Wu, C-F Lai, WC Mobley, unpublished observations). Examination of the distributions of Ras and Rap-1 provide a possible explanation. Although there is prominent expression of Ras on the cell surface, expression of Rap-1 appears to be restricted to small intracellular vesicles. Thus, the data suggest that for TrkA to activate Rap-1, which in turn activates B-raf and the ERK kinase cascade, it must be internalized into membrane vesicles that fuse with vesicles containing Rap-1 (York et al 2000, C Wu, C-F Lai, WC Mobley, unpublished observations). Thus, sustained activation of the ERK pathway, which is essential for normal differentiation, is regulated by both the kinetics and specificity of membrane transport and sorting. Because there are so many mechanisms for regulating membrane transport and sorting, these recent papers suggest many interesting directions for future research.

Control of Trk Signaling by Other Adapters

Results described above predict that both survival and differentiation pathways will depend on interactions of Shc or Frs-2 with the phosphorylated Y490 site. Despite this, mice homozygous for a targeted Y to F mutation of this site in TrkB are viable and have a much milder phenotype than is observed in mice lacking the TrkB kinase domain (Minichiello et al 1998). Clearly, other sites in

TrkB must be capable of activating intracellular signaling pathways important for neuronal survival and differentiation. Recent results of particular interest have suggested that the phosphorylated tyrosines in the activation loop of the Trk tyrosine kinase domain have dual functions. In addition to controlling activity of the kinase, they appear to function as docking sites for adapter proteins. Grb-2 has been shown to interact directly with a phosphorylated tyrosine residue in 2-hybrid assays and by coimmunoprecipitation (MacDonald et al 2000). Grb-2 also interacts with other sites, including the PLC-γ1 site Y785, but it is not certain these interactions are direct. Two additional adapters, rAPS and SH2-B, are similar proteins that contain a PH domain, an SH2 domain, and tyrosines phosphorylated in response to Trk activation. Both have been shown to interact with phosphorylated tyrosines in the activation loops of all three Trk receptors (Qian et al 1998). Both of these adapter proteins may also interact with other sites in the Trk receptor cytoplasmic domains. These two proteins form homodimers and also associate with each other. Both also bind to Grb-2, providing a potential link to the PI-3 kinase and Ras signaling cascades. Antibody perturbation and transfections using dominant negative constructs implicate rAPS in NGF-dependent survival, MAP kinase activation, and neurite outgrowth in neonatal sympathetic neurons (Qian et al 1998). Taken together, these results indicate that initial models of Trk receptor signaling pathways were far simpler than Trk receptor signaling is in reality.

Our current understanding of Trk receptor signaling is incomplete. First, not all functionally important interactions with Trk receptors may depend on phosphotyrosine-dependent associations. Recent work suggests that the c-abl tyrosine kinase interacts with the juxtamembrane domain of TrkA, whether or not the tyrosines in this region are phosphorylated (Yano et al 2000). A deletion in this region has been shown to block differentiation of PC12 cells without preventing mitotic responses or phosphorylation of SHC or FRS-2 (Meakin & MacDonald 1998). As c-abl is involved in many aspects of neuronal differentiation (e.g. see Hu & Reichardt 1999), it will be interesting to determine whether it has a role in Trk-mediated signaling that is perturbed by this juxtamembrane deletion.

To provide a few more examples of proteins whose roles in signaling pathways are poorly understood, in cultured cortical neurons, the insulin receptor substrates (IRS)-1 and -2 are phosphorylated in response to BDNF, which promotes sustained association with and activation of PI-3 kinase (Yamada et al 1997, 1999). This is not seen after ligand engagement of Trk receptors in PC12 cells, the most popular cellular model for neurotrophin signaling studies, which suggests that a critical adapter protein is missing from these cells. As another example, CHK, a cytoplasmic protein kinase that is a homologue of CSK (control of src kinase), has been shown to interact with TrkA in PC12 cells and to enhance ERK pathway-dependent responses, including neurite outgrowth (Yamashita et al 1999a). The pathway by which CHK affects ERK activation is not understood. Finally, a transmembrane protein with three extracellular immunoglobulin and four cytoplasmic tyrosine motifs has been shown to provide a docking site for recruitment of the protein phosphatase Shp-2 and to enhance BDNF-dependent activation of the PI-3 kinase pathway by mechanisms not prevented by mutation of the four tyrosine residues (Araki et al 2000a,b). Again, mechanisms are not understood.

In summary, although there has been rapid progress in understanding many pathways controlled by Trk receptor signaling, there are still many loose ends. This discussion has proceeded as if all signaling molecules were present in all cells, but this is certainly not so. Differences in their concentrations within different neuronal populations undoubtedly contribute to the diversity of responses seen in different neuronal populations. From the discussion above, it would be obvious to assume that TrkA, TrkB, and TrkC each activate very similar signaling pathways because of the very high similarities between them in their cytoplasmic domains. Although probably true, examples are described later in this review where it is clear that signaling through different Trk receptors has quite different actions on the same cell, as assessed by nonredundant effects on survival, differentiation, or axon guidance (e.g. Carroll et al 1998, Ming et al 1999). In some tumor cells, neurotrophin-activated Trk receptor signaling has even been shown to induce apoptosis (e.g. Kim et al 1999). Clearly, many of the most interesting details of signaling by these receptors remain to be discovered.

p75NTR Receptor-Mediated Signaling Mechanisms: NFκB Activation

As mentioned previously, p75NTR binds with approximately equal affinity to each of the neurotrophins. Ligand engagement of p75NTR has been shown to promote survival of some cells and apoptosis of others (e.g. Barrett & Bartlett 1994). p75NTR-mediated signaling also affects axonal outgrowth both in vivo and in vitro (e.g. Bentley & Lee 2000, Yamashita et al 1999b; Walsh et al 1999a,b).

Several signaling pathways are activated by p75NTR and in some cases the pathways are known in detail (see Figure 2; see color insert). An important pathway promoting cell survival of many cell populations involves activation of NFκB. For example, cytokines promote neuronal survival by activation of the NFκB signaling pathway (Middleton et al 2000). In both embryonic sensory and sympathetic neurons, neurotrophins have been shown to promote p75NTR-dependent activation of NFκB and NFκB–dependent neuronal survival (Maggirwar et al 1998, Hamanoue et al 1999). All neurotrophins have been shown to promote association of p75NTR with the adapter protein TRAF-6 (Khursigara et al 1999). In other systems, TRAF-6 has been shown to activate the protein kinase NIK (NFκB-interacting kinase), which phosphorylates IKK (inhibitor of IκB kinase), which in turn phosphorylates IκB, resulting in release and nuclear translocation of NFκB (reviewed in Arch et al 1998). It is interesting that although all neurotrophins bind to p75NTR, in rat Schwann cells, only NGF is able to induce NFκB nuclear translocation (Carter et al 1996).

p75NTR Receptor-Mediated Signaling Mechanisms: Jun Kinase Activation

The Jun kinase signaling cascade is activated following NGF withdrawal and by binding of neurotrophins to p75NTR (Xia et al 1995, Eilers et al 1998, Aloyz et al 1998). Apoptosis mediated by p75NTR requires activation of p53 through the Jun kinase–mediated signaling pathway (Aloyz et al 1998). P53 controls cell survival in many cells besides neurons (e.g. Agarwal et al 1998). Among its targets, the activation of the Jun kinase cascade has been shown to induce expression of Fas ligand in neuronal cells, which promotes apoptosis by binding to the Fas receptor (Le-Niculescu et al 1999). p53 has many gene targets, including the proapoptic gene Bax. In both PC12 cells and sympathetic neurons, activation of the Jun kinase cascade and apoptosis following trophic withdrawal involve Cdc-42 because apoptosis is strongly inhibited by a dominant negative Cdc-42 (Bazenet et al 1998). The MAP kinase kinase kinase named apoptosis signal-regulating kinase-1 (ASK1) is in the pathway controlled by Cdc-42 because overexpression of a kinase-inactive mutant of ASK-1 strongly inhibits cell death promoted by either NGF withdrawal or expression of a constitutively active Cdc-42 (Kanamoto et al 2000). The kinase providing a link between ASK-1 and Jun kinase has not been identified but may be the Jun kinase kinase named MKK7 in sympathetic neurons (Kanamoto et al 2000). Embryos lacking both JNK1 and JNK2 show aberrant, region-specific perturbations of neuronal cell apoptosis in early brain development (Kuan et al 1999), whereas the neurons of animals lacking JNK3 are resistant to excitotoxicity-induced apoptosis (Yang et al 1997). Thus, the Jun kinase cascade is important in regulating apoptosis of neurons in vivo.

p75NTR Receptor-Mediated Stimulation of Sphingolipid Turnover

Ligand engagement of p75NTR has also been shown to activate acidic sphingomyelinase, which results in generation of ceramide (Dobrowsky et al 1995). Ceramide has been shown to promote apoptosis and mitogenic responses in different cell types through control of many signaling pathways, including the ERK and Jun kinase cascades and NFκB. For example, ceramide binds to Raf and may induce formation of inactive Ras-Raf complexes, effectively inhibiting the ERK signaling cascade (Muller et al 1998). Many groups have shown that ceramide also inhibits signaling mediated through PI-3 kinase (e.g. Zhou et al 1998). Recent experiments suggest that ceramide inhibits the activity of PI-3 kinase in cells by modifying the association of receptor tyrosine kinases and PI-3 kinase with caveolin-1 in lipid rafts (Zundel et al 2000). In fibroblasts, the sensitivity of growth factor–stimulated PI-3 kinase activity to ceramide inhibition was increased and decreased by overexpression and reduced expression of caveolin-1, respectively. Ceramide may also inhibit directly PI-3 kinase activity (Zhou et al 1998). Thus, ceramide inhibits at least two of the survival and differentiation-promoting pathways activated by Trk receptor signaling.

Adapter Proteins That Bind to p75NTR

In addition to TRAF-6, several additional proteins that interact with p75NTR have been identified, each of which is a candidate to mediate Jun kinase activation or sphingolipid turnover. NRIF (neurotrophin receptor interacting factor) is a widely expressed Zn-finger–containing protein that interacts with both the juxtamembrane and death domains of p75NTR (Casademunt et al 1999). Overexpression of NRIF has been shown to kill cells in culture, and in mice lacking NRIF, there are reductions in developmentally regulated cell death among neuronal populations that are very similar to the reductions observed in mice lacking p75NTR. At this time, it is not known whether NRIF's activity or association with p75NTR is regulated by neurotrophins. It is also unclear which downstream signaling pathways are activated by this interesting protein. Another protein named NRAGE [neurotrophin receptor–interacting MAGE (melanoma-associated antigen) homologue] has recently been shown to associate with p75NTR and to be recruited to the plasma membrane when NGF is bound to p75NTR (Salehi et al 2000). NRAGE prevents the association of p75NTR with TrkA, and overexpression of NRAGE promotes NGF-stimulated, p75NTR-dependent cell cycle arrest and death of MAH (v-*myc*-infected, adrenal-derived, HNK-1-positive) cells. SC-1 (Schwann cell-1) is a distinct Zn-finger–containing protein, which has been shown to associate with p75NTR and to redistribute from the cytoplasm to the nucleus after treatment of p75-expressing cos cells with NGF (Chittka & Chao 1999). Nuclear expression of SC-1 correlates with cell cycle arrest, which suggests that nuclear localization of this protein may be involved causally in growth arrest. Thus, both NRAGE and SC-1 appear to be interesting proteins involved in the signaling events promoted by p75NTR. Finally, when overexpressed in 293 cells, several (TNF receptor-associated factor) proteins in addition to TRAF-2 can associate with either monomeric or dimeric p75NTR, and some of these promote apoptosis of these cells (Ye et al 1999). Although the interactions of these adapter proteins with p75NTR are interesting, much work remains to be done to characterize their expression patterns and signaling mechanisms in neurons.

Control of the Cytoskeleton by p75NTR

In addition to regulating neuronal cell survival, ligand engagement of p75NTR has been reported to directly enhance neurite outgrowth by ciliary neurons in culture (e.g. Yamashita et al 1999b). In contrast, it inhibits neurite outgrowth by sympathetic neurons in culture (Kohn et al 1999). Sensory and motor neurons extend axons more slowly toward their peripheral targets in mouse embryos lacking p75NTR (Yamashita et al 1999b, Bentley & Lee 2000). In adult animals, perturbations of target innervation patterns are also seen in these mice with some, but not all, targets lacking normal innervation (e.g. Lee et al 1994a, Peterson et al 1999,

Kohn et al 1999). Recent work has demonstrated an interaction between p75NTR and RhoA (Yamashita et al 1999b). p75NTR was observed to activate RhoA. Neurotrophin binding to p75NTR eliminated activation of RhoA by p75NTR. Pharmacological inactivation of RhoA and ligand engagement of p75NTR have similar stimulatory effects on neurite outgrowth by ciliary ganglion neurons, a neuronal population that does not express Trk receptors. Results from this interesting paper suggest that unliganded p75NTR tonically activates RhoA, which in turn is known to reduce growth cone motility. This observation does not provide an immediately obvious explanation for the reduced axon outgrowth by sensory and motor neurons observed in embryonic p75NTR−/− animals. Perhaps the presence of ligand-engaged p75 effectively sequesters RhoA in its inactive form. Alternatively, the presence of p75NTR has been shown to promote retrograde transport of NGF, BDNF, and NT-4 (Curtis et al 1995, Harrison et al 2000). Reductions in retrograde transport may result in reduced axon growth and neuronal survival. As a third possibility, Schwann cell migration has been shown to depend on p75NTR-mediated signaling and is clearly deficient in this mutant (Anton et al 1994, Bentley & Lee 2000). Perhaps deficits in Schwann cell migration indirectly reduce the rate of axonal outgrowth.

Reciprocal Regulation of Signaling by Trk Receptors and p75NTR

Activation of Trk receptors has profound effects on p75NTR-dependent signaling. Neurotrophins are much more effective at inducing apoptosis through p75NTR in the absence than in the presence of Trk receptor activation (e.g. Davey & Davies 1998, Yoon et al 1998). In the initial experiments demonstrating that NGF induced sphingomyelin hydrolysis and ceramide production, activation of a Trk receptor was shown to completely suppress this response (Dobrowsky et al 1995). Trk receptor activation also suppresses activation of the Jun kinase cascade (Yoon et al 1998). Activation of Ras in sympathetic neurons has been shown to suppress the Jun kinase cascade (Mazzoni et al 1999). In these neurons, activation by Ras of PI-3 kinase is essential for efficient suppression of this cascade. In recent studies utilizing nonneural cells, c-raf has been shown to bind, phosphorylate, and inactivate ASK-1 (Chen & Fu 2000). If this pathway functions efficiently in neurons, it provides a mechanism by which activation of Trk receptors may suppress p75NTR-mediated signaling through the Jun kinase cascade. It is notable that although Trk receptor kinase-mediated signaling suppresses proapoptotic responses mediated by p75NTR, Trk signaling does not inhibit induction by p75NTR of the NFκB cascade (Yoon et al 1998). Thus, in the presence of Trk signaling, activation of the NFκB cascade makes a synergistic contribution to survival (Maggirwar et al 1998, Hamanoue et al 1999).

Although kinase activity of Trk receptors suppresses signaling pathways mediated by p75NTR, Trk signaling is not invariably completely efficient at suppressing p75NTR-mediated apoptosis. NGF is able to increase apoptosis of cultured motoneurons from wild-type but not from p75NTR−/− embryos (Wiese et al 1999). In PC12 cells, BDNF binding to p75NTR has been reported to reduce NGF-dependent autophosphorylation of TrkA, possibly by promoting phosphorylation of a serine residue in the TrkA cytoplasmic domain (MacPhee & Barker 1997).

The overall picture that emerges from these studies is that the proapoptotic signals of p75NTR are largely suppressed by activation of Ras and PI-3 kinase by neurotrophins. Thus, p75NTR appears to refine the ligand-specificity of Trk receptors and may promote elimination of neurons not exposed to an appropriate neurotrophic factor environment. Consistent with this possibility, reductions in apoptosis have been observed in the retina and spinal cord in p75NTR−/− embryos (Frade & Barde 1999). It is less obvious why there are fewer sensory neurons in p75NTR mutant animals (Stucky & Koltzenburg 1997). The reductions in retrograde transport of many neurotrophins observed in the p75NTR mutant may increase apoptosis. In addition, the reduced rate of sensory axon outgrowth observed in these animals may create situations where axons are not exposed to adequate levels of neurotrophins. As neurotrophin expression is regulated during development by local tissue interactions not dependent on innervation (Patapoutian et al 1999), delays in innervation could have catastrophic consequences.

REGULATION OF NEURONAL SURVIVAL BY NEUROTROPHINS

The important experiments described above of Hamburger, Levi-Montalcini, and later workers demonstrated that NGF has essential roles in maintaining the viability of nociceptive sensory and sympathetic neurons in vivo (see Levi-Montalcini 1987, Purves 1988). These results suggested that all neurons may depend on trophic support derived from their targets for continued survival, not only during development but also in the adult nervous system. The observations also raised the possibility that neural precursors and developing neurons whose axons have not contacted their ultimate targets may also require trophic support.

Major extensions of this work have been made possible by development of gene targeting technology. With this technology, mice with deletions in genes encoding each of the neurotrophins and their receptors have been generated. Mice with deletions in almost all of the genes encoding GDNF, GDNF family members, and their receptors are also available and the few exceptions will almost certainly become available shortly. A summary of neuronal losses in the various knockouts of the neurotrophic factors and their receptors is presented in Table 1 and Table 2. In addition, Figure 3 provides a schematic overview of these losses in the peripheral nervous system. In many instances, a particular ganglion has not been examined in an individual mutant, frequently because the known expression of the receptor or responsiveness of the neurons in cell culture made a phenotype appear very unlikely. In other instances, the initial report focused on the most obvious phenotype,

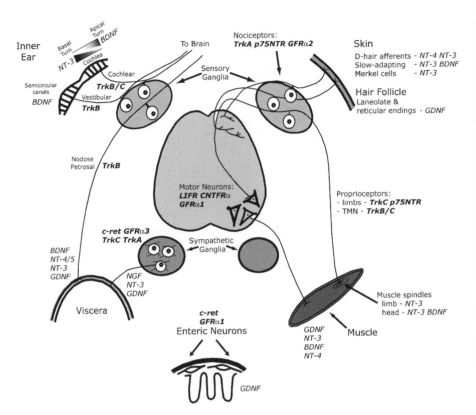

Figure 3 Summary of survival functions of neurotrophins in the peripheral nervous system. This diagram illustrates various components in the peripheral nervous system, including sensory ganglia, sympathetic ganglia, and enteric neurons. Only ligands or receptors with definitive loss of function phenotype are indicated in this figure. Ligands are indicated in italics and receptors are indicated in bold italics. (See text for abbreviations.)

and examination of other possible phenotypes was deferred, often indefinitely. To summarize briefly, cranial ganglia neurons that transmit different modalities of sensory information tend to be segregated into different ganglia, so neurons in specific ganglia are often dramatically affected by loss of an individual signaling pathway. Within DRG sensory ganglia, modalities are mixed and mutants typically have severe effects only on functionally distinct subpopulations of cells.

Sensory Ganglia Survival

Trigeminal and Dorsal Root Ganglia   In DRG and trigeminal ganglia, neurons conveying different modalities of sensory information are present in the same ganglion, and substantial progress has been made in demonstrating differential neurotrophic factor dependencies of neurons with different sensory modalities.

TABLE 1 Neuronal losses in neurotrophin and Trk-deficient mice[a]

Determinant	TrkA	NGF	TrkB	BDNF[b]	NT-4/5[c]	TrkC[d]	NT-3	TrkB/TrkC[d]	NT-3/BDNF	NT4/BDNF[e]	NT-3/BDNF/NT-4/5[f]	p75NTR[g]
Sensory ganglia												
Trigeminal	70%	75%	60%	30%	NS	21%	60%	ND	74%	9%	88%	ND
N-P	ND	ND	90%	45%	40%	14%	30%	ND	62%	90%	96%	ND
Vestibular	NS	ND	60%	85%	NS	15%	20%	100%	100%	89%	100%	ND
Cochlear	NS	NS	15%	7%	ND	50%	85%	65%	100%	ND	ND	ND
Dorsal root	70–90%	70%	30%	35%	NS	20%	60%	41%	83%	NS	92%	Small
Geniculate	ND	ND	ND	ND	ND	11%	35%	ND	ND	ND	100%	ND
TMN[h]	ND	ND	38%	41%	8%[k]	45%	57%[m]	ND	88%	46%	95%	—[n]
Comments	—[i]	—[j]	—[j]	—[j]	—[k]	—[l]	—[m]					
Sympathetic ganglia												
Superior cervical	>95%	>95%	ND	ND	NS	NS	50%		ND	ND	47%	NS
Motor												
Facial	ND	ND	ND	NS	ND	ND	ND	ND	ND	NS	22%	ND
Spinal cord	ND	ND	NS	NS	NS	ND	—[q]	—[r]	ND	NS	20%	ND
CNS	—[o]	ND	ND	—[p]	ND	ND						—[s]
Viability	P	P	VP	PM	G	M	VP	VP	VP	ND	VP	G

[a]Note: Neuronal losses are expressed as the percentage of neurons lost in the mutants compared with the wild-type controls. This table is updated from a similar table in Reichardt & Fariñas (1997), in which original references for older papers is provided. N-P, Nodose-petrosal; TMN, Trigeminal mesencephalic nucleus neurons; ND, not done; NS, not significant; P, poor; VP, very poor; PM, poor to moderate; M, moderate; G, good.

[b]Brady et al 1999.

[c]Stucky et al 1998.

[d]Minichiello et al 1996.

[e]Liebl et al 2000.

[f]Lin & Jaenisch 2000.

[g]Fundin et al 1997.

[h]Fan et al 2000.

[i]Small CGRP (nociceptive) and BS1 (thermoceptive) positive neurons missing.

[j]Myelinated and nonmyelinated axon lost. Ia afferents completely lost.

[k]D-hair afferents completely lost.

[l]Proprioceptive neurons missing.

[m]Proprioceptive and cutaneous mechano-receptors missing. Partial losses of nociceptors. Partial losses of D-hair and SA fibers.

[n]Partial deficits in all neurons.

[o]Cholinergic basal forebrain neurons present. Reduced hippocampal innervation.

[p]Deficits in NPY, calbindin, and parvalbumin expression. Cerebellar foliation defect.

[q]Proprioceptive neurons missing.

[r]No clear deficits.

[s]Increase in the number of forebrain cholinergic neurons.

TABLE 2 Neuronal losses in other neurotrophic factor and receptor-deficient mice[a]

Determinant	GDNF[g]	Neurturin[b,g]	GFRα1[c,g]	GFRα2[d]	GFRα3[e]	c-ret[f,g]	CNTFRα	LIFR
Sensory ganglia								
Trigeminal	NS	70% reduction of GFRα2(+) neurons	NS	NS	NS		NS	
N-P	40%	NS	15%	NS				
Vestibular	NS		NS					
Cochlear	ND							
Dorsal root	23%	45% reduction of pfGFRα2(+) neurons	NS	NS	NS		NS	
Comments	GFRα1 neurons lost in trigeminal ganglion	68% and 45% loss of GFRα2-expressing neurons in trigeminal and DRG		Only 10% of trigeminal neurons express GFRα2				
Sympathetic ganglia								
SC	35%	NS	NS	ND	ML	ML	NS	
Parasympathetic ganglia								
Ciliary	40%	48%	ND	NS	NS	48%	NS	
Submandibular	36%	45%	33%	81%		30%		
Otic	86%	NS (reductions in neuronal size)	NS			99%		

Enteric nervous system								
Stomach	ML 100%		ML 100%					
Intestine/colon	NS	Reduced VIP+ & SP+ fibers; reduced neuronal size	ML 100%	Reduced fiber density			NS	NS
Motor								
Facial	NS		NS		40%		35%	
Trigeminal	19%	22%	22%		35%		40%	
Spinal cord	22%	NS	NS	24%			40%	
CNS	No deficit in TH+ neurons		No deficit in TH+ neurons					
Viability	VP	G	VP	PM	G	G	VP	VP

[a]Note: Similar to Table 1, neuronal losses are expressed as the percentage of neurons lost in the mutants compared with the wild-type controls. See Reichardt & Fariñas (1997) for references of older papers. NS, Not significant; ND, not done; N-P, nodose-petrosal; SC, superior cervical; DRG, dorsal root ganglia; ML, most lost; VP, very poor; G, good; PM, poor to moderate.

[b]Heuckeroth et al 1999.

[c]Cacalano et al 1998, Enomoto et al 1998.

[d]Rossi et al 1999.

[e]Nishino et al 1999.

[f]Durbec et al 1996, Taraviras et al 1999.

[g]Enomoto et al 2000.

Thus, almost all nociceptive neurons express TrkA at some time during their development, and essentially all of these neurons are lost in the *TrkA* and *NGF* mutants (Crowley et al 1994, Smeyne et al 1994). A second major population of DRG neurons expresses TrkC from the time of initial neurogenesis. Most of these neurons differentiate into proprioceptive neurons conveying information from end organs, such as muscle spindles and tendon organs. At spinal cord levels, these neurons are completely lost in *NT-3* and *TrkC* mutants together with end organs, such as muscle spindles, whose morphogenesis requires the presence of sensory axons. NT-3 expression is observed in both muscle spindles and the ventral spinal cord, both targets of proprioceptive Ia afferents, consistent with NT-3 functioning as a target-derived trophic factor. These neurons are lost almost immediately after neurogenesis, however, which suggests that they depend on NT-3 provided initially by intermediate targets (e.g. Fariñas et al 1996).

The expression patterns of Trk receptors in sensory neurons of mouse DRG and trigeminal ganglia have been characterized and correlate with the neuronal deficits in these ganglia. It has become clear that most neurons in both ganglia in mice, with few exceptions, express one Trk receptor during neurogenesis (Fariñas et al 1998, Huang et al 1999a). Although expression of Trk receptors remains unchanged in most sensory neurons, a small fraction of neurons show dynamic changes in switching of neurotrophin receptors and neurotrophin dependence (see, e.g. Enokido et al 1999).

Several transcription factors have been shown to regulate expression of Trk receptors in sensory ganglia. For instance, targeted deletion of the POU domain transcription factor Pou4f1 (Brn-3a/Brn-3.0) prevents initiation of TrkC expression in the trigeminal ganglion and results in downregulation of TrkA and TrkB in this ganglion at later times, resulting in apoptosis of neurotrophin-dependent neurons (McEvilly et al 1996, Huang et al 1999b). Absence of Pou4f1 also prevents normal expression of TrkC in the spiral ganglion (W Liu, EJ Huang, B Fritzsch, LF Reichardt, M Xiang, unpublished observations). The basic helix-loop-helix (bHLH) factor NeuroD also controls regulation of Trk receptor expression. Its absence results in severely reduced expression of TrkB and of TrkC in the embryonic vestibulocochlear ganglion (Kim et al 2001). Expression of TrkB and TrkC appears to be relatively normal in the trigeminal ganglion in this mutant, so the effects of this mutation on Trk receptor expression are surprisingly specific. Consistent with this, a recent analysis of *cis*-elements in the *TrkA* enhancer has identified sites required for global expression and sites that are specifically required for expression within sympathetic, DRG, or trigeminal neurons (Ma et al 2000). Thus, the transcriptional machinery that specifies TrkA expression is not the same in each of these neuronal populations. The specificity in the phenotypes of the mutants described above implies that transcriptional control of *TrkB* and *TrkC* must be equally complex. In summary, these results show that accurate control of the transcription of neurotrophin receptor genes is essential for ensuring normal neuronal survival and differentiation. In order to extend this work, it will be necessary to characterize the transcriptional machinery in each population of neurotrophin-responsive neurons.

Cutaneous Sensory Receptors and Innervation   Within both DRG and trigeminal ganglia, a well-defined subpopulation of the nociceptive neurons initiates expression of c-ret plus one or more of the GFR (GDNF family receptor)-α adapter subunits, eventually losing expression of TrkA (Molliver et al 1997, Huang et al 1999b). These neurons appear to be affected in mutants lacking constituents of these signaling pathways. In some instances, mutations have been shown to result in almost complete deficits in innervation of peripheral targets. For example, there are specific deficits in transverse lanceolate endings and reticular endings within the hair follicle in *GDNF* mutant heterozygotes, whereas other endings are not detectably affected (Fundin et al 1999).

Analyses of neurotrophin mutants also provided important information regarding the roles of neurotrophins in the development of cutaneous receptors. For example, in addition to the deficit in proprioceptors, NT-3 deficiency has been shown to result in deficits in specific cutaneous [D-hair (Down hair receptor) and SA (slow-adapting)] innervation and the development of Merkel cells (Airaksinen et al 1996), which are manifested only postnatally. The time course of development of these deficits indicates that NT-3 functions as a target-derived trophic factor for these neurons. NT-4 also functions as an essential survival factor for D-hair afferents (Stucky et al 1998). The dependence of these neurons on NT-4 appears to follow their dependence on NT-3. Unlike NT-4, BDNF is required for the survival of mechanical functions of SA fibers but plays no role in the survival of D-hair receptors (Carroll et al 1998). Together, these data indicate that BDNF and NT-4 have distinct, nonoverlapping roles in cutaneous innervation and that the expression of these neurotrophins may show spatial or temporal differences during the development of cutaneous receptors.

The presence of NT-4 has also been shown recently to be required for normal survival of TrkB-expressing DRG sensory neurons at the time of DRG formation (Liebl et al 2000). Since D-hair afferents are present during the first few postnatal weeks in the *NT-4* mutant (Stucky et al 1998), they cannot be derived from this embryonic population of TrkB-expressing neurons.

Trigeminal Mesencephalic Nucleus   The trigeminal mesencephalic neurons are a group of neural crest–derived sensory neurons that reside in the brainstem at the pontomedullary junction. These neurons morphologically resemble sensory neurons in the peripheral ganglia and convey proprioceptive information from the head region. However, unlike proprioceptive neurons at the trunk level, which are completely dependent on NT-3, trigeminal mesencephalic neurons are only partially lost in the absence of NT-3 or TrkC, and partial neuronal deficits are also generated by the loss of BDNF or TrkB (Fan et al 2000, Matsuo et al 2000). It is interesting that the majority of these neurons are lost in double mutants lacking

NT-3 and BDNF, and all these neurons are lost in triple mutants lacking NT-3, BDNF, and NT-4 (Fan et al 2000). Expression of a $BDNF^{lacZ}$ reporter has been detected in a subset of muscle spindles in one target of these neurons, the masseter muscle. It seems likely that each of the neurotrophins will prove to be expressed in subsets of these spindles, and that this explains why each neurotrophin supports the survival of a subset of these neurons.

Vestibular and Cochlear (Spiral) Ganglia   During development, these two ganglia first emerge as one single ganglion, which subsequently separates into two distinct ganglia that innervate the semicircular canals and cochlea. Because of their well-documented axon projection patterns, the vestibular and cochlear ganglia are an ideal system in which to investigate the effect of neurotrophins. The initial analyses of neurotrophin mutants showed that almost all neurons in the vestibular ganglion depend on BDNF for survival, whereas the vast majority of neurons in the cochlear (spiral) ganglion require NT-3 (Ernfors et al 1994 a,b; Jones et al 1994; Fariñas et al 1994). In a double *NT-3/BDNF* mutant, essentially all neurons in both ganglia are lost (Ernfors et al 1995).

The apparent dependence of separate populations of cochlear neurons on these two neurotrophins has stimulated investigations to understand the differences in the neurons or cochlea that explain the phenotypes. The first publications suggested that cell type–specific expression of NT-3 in inner hair cells and BDNF in outer hair cells controls the survival of the type I and type II neurons, which are responsible, respectively, for innervating each of these two hair cell populations (Ernfors et al 1995). Absence of TrkC was reported to result in preferential loss of innervation of inner hair cells, whereas deficiency in TrkB appeared to cause loss of innervation to outer hair cells (Schimmang et al 1995). Later investigations challenged these data, however, because neuronal losses and innervation deficits in these two sets of mutants were shown not to be distributed uniformly throughout the cochlea but to be distributed in gradients along the cochlear turns (Fritzsch et al 1997, 1998). Neurons in the basal turn of the cochlea were completely missing whereas those in the apical turn were much more mildly affected in the *NT-3* and *TrkC* mutants. In the *BDNF* and *TrkB* mutants, the only obvious deficits were observed among neurons in the apical turn (Bianchi et al 1996, Fritzsch et al 1997, 1998).

Recent observations demonstrate that all neurons in the cochlear (spiral) ganglion express both TrkB and TrkC, indicating that they can be supported by either neurotrophin (I Fariñas, KR Jones, L Tessarollo, AJ Vigers, E Huang, M Kirstein, DC De Caprona, V Coppola, C Backus, LF Reichardt, B Fritzsch, unpublished observations). The phenotype of the *NT-3* mutant can be explained by a spatial apical-to-basal gradient of BDNF expression, which in the absence of NT-3 causes a complete absence of trophic support for these neurons in the basal turn during a brief, but crucial, period of development. Using a β-galactosidase (LacZ) reporter integrated into either the *NT-3* (*NT-3^{lacZ}*) or *BDNF* (*BDNF^{lacZ}*) locus to monitor gene expression, rapid changes in the expression patterns of both neurotrophins were seen as development proceeded. Approximately one day before the loss of neurons in the *NT-3* mutant, however, expression of BDNF was barely detectable in the cochlea, with only weak expression in the apical turn and no detectable expression in the developing middle and basal turns. This suggested that in the *NT-3* mutant, neurons were lost in the basal turn because BDNF was not present there to compensate for its absence, whereas neurons were partially spared in the apical turn because of the presence of low levels of BDNF. This model predicts that expression of BDNF under control of the *NT-3* gene promoter and regulatory elements will rescue neuronal losses in the *NT-3* mutant. This mouse has been generated and homozygotes are completely deficient in NT-3 (V Coppola, J Kucera, ME Palko, J Martinez-De Velasco, WE Lyons, B Fritzsch, L Tessarollo, unpublished observations). As predicted, neurons innervate normally all regions of the cochlea, including the basal turn, at E13.5 and P0, and there is almost complete rescue of basal turn spiral neurons (I Fariñas, KR Jones, L Tessarollo, AJ Vigers, E Huang, M Kirstein, DC De Caprona, V Coppola, C Backus, LF Reichardt, B Fritzsch, unpublished observations; V Coppola, J Kucera, ME Palko, J Martinez-De Velasco, WE Lyons, B Fritzsch, L Tessarollo, unpublished observations). Thus, there is convincing evidence that a spatial-temporal gradient of neurotrophin expression controls survival of these cells.

Nodose-Petrosal Ganglion   The nodose-petrosal ganglion contains neurons that are responsible for visceral sensory innervation. In this ganglion, all neurons express TrkB (Huang et al 1999a) and are lost in the *BDNF–NT-4* double mutant or in the *TrkB* mutant (Conover et al 1995). Approximately half of the nodose-petrosal neurons are lost in the absence of either BDNF or NT-4 alone. The dopaminergic neurons responsible for innervation of the carotid body and other sensors of blood pH and pressure are completely dependent on BDNF, which is synthesized by these target organs during the initial period of innervation (Erickson et al 1996, Brady et al 1999). Thus, mice lacking BDNF show lack of innervation to chemo- and baroreceptors, resulting in deficits in control of breathing. In this ganglion, a neuron appears to depend on either BDNF or NT-4 alone, depending on which neurotrophin is expressed in the target that it innervates.

Sympathetic Ganglia Survival

Other populations of peripheral neurons also show strong dependencies on particular signaling pathways. As expected from the work of Levi-Montalcini, sympathetic neurons are almost completely lost in the absence of NGF to TrkA signaling (Crowley et al 1994, Smeyne et al 1994). Consistent with the expression patterns of TrkA in the sympathetic neurons, extensive cell death occurs perinatally in the sympathetic ganglion of *TrkA* mutants. In fact, a significant deficit is already present at E17.5 and develops progressively after birth (Fagan et al 1996). Unlike the prominent effects of NGF/TrkA on sympathetic neurons, the view on how NT-3 affects the development of sympathetic neurons has undergone a major revision in recent years. A number of experiments demonstrate that NT-3 is able to support the survival of early sympathetic neuroblasts in vitro (Birren et al 1993, Verdi & Anderson 1994, Verdi et al 1996). Consistent with these results, TrkC mRNA has been detected in early sympathetic ganglia. It is interesting that the level of TrkC mRNA decreases in sympathetic ganglion during late embryogenesis as the expression of TrkA increases. These studies suggested that NT-3 signaling through TrkC followed by NGF signaling through TrkA provides sequential support for cells in sympathetic ganglia during early and late stages of development, respectively. Therefore, it was not surprising when an initial analysis of the *NT-3* mutant reported an increase of cell death among proliferating sympathetic precursor cells during initial formation of this ganglion (ElShamy et al 1996). However, no deficit was detected in a *TrkC* mutant (Fagan et al 1996), and later workers examining the *NT-3* mutant have not detected an early phenotype in either precursors or neurons (Wyatt et al 1997, Francis et al 1999). Instead, the deficits in the *NT-3* and *NGF* mutants appeared after E15.5 and developed during very similar stages of development (Wyatt et al 1997, Francis et al 1999). In agreement with these findings, *NT-3^{lacZ}* is expressed in the target tissues of sympathetic innervation but is not present in or around the ganglion before E15.5 (Francis et al 1999). Thus, these data indicate that NT-3 and NGF are both required for the survival of sympathetic neurons, not neuroblasts. Because no deficits are seen in the TrkC mutant, the survival-promoting effects of NT-3 must be transmitted through TrkA. The ability of TrkA to mediate NT-3 signaling during development is promoted by the absence of p75NTR (Brennan et al 1999).

Although recent results do not support the concept that NT-3 and NGF act sequentially during development of sympathetic ganglia, members of the GDNF family of ligands do play essential early roles in development of one sympathetic ganglion, the superior cervical ganglion (SCG). The requirement for GDNF family members clearly precedes the dependence of SCG neurons on NT-3, NGF, and TrkA. Initially, mice lacking c-ret, the tyrosine kinase activated by GDNF, were shown to lack all neurons in the SCG without any obvious phenotype in the sympathetic chain at the trunk level (Durbec et al 1996). The absence of c-ret was shown to result in loss of a common set of neural crest–derived precursors for both the SCG and the enteric nervous system. In the absence of c-ret, these cells failed to survive and migrate to the future site of the SCG from the vicinity of the postotic hindbrain. More recently, mice lacking the GFRα3 binding subunit have been shown to have a somewhat similar, but significantly less severe, deficit attributable in part to an effect on early precursors of the SCG (Nishino et al 1999). GFRα3 mediates activation of c-ret by Artemin, a protein closely related to GDNF. In mice lacking GFRα3, the initial ventral migration to the aorta of SCG precursors is not obviously perturbed, but a later rostral migration to the future site of the SCG does not occur. Although a majority of the precursors exit the cell cycle and differentiate into immature neurons, these neurons do not mature, are not successful in contacting their normal targets, and undergo progressive, largely postnatal apoptosis. Artemin is expressed in the vicinity of the SCG precursors at E12.5, so a deficit in Artemin-mediated signaling is very likely to account for the early deficit in migration and may explain the later phenotype also. Because the deficit in the *GFRα3* mutant is clearly less severe than that in the *c-ret* mutant, signaling through other adapter subunits must contribute to early development of SCG precursors. No deficit within the SCG has been reported in mice lacking the GFRα1 or GFRα2 subunits (Cacalano et al 1998, Rossi et al 1999). The phenotype of mice lacking the GFRα4 subunit has not been reported. It will be interesting to see whether double or triple mutants in *GFRα* subunit genes develop a phenotype in the SCG that is as severe as that observed in the *c-ret* mutant. Mice lacking GDNF have approximately 35% fewer SCG neurons at birth than do control animals (Moore et al 1996). Although the embryonic development of this phenotype has not been studied, it seems likely that a deficit in GDNF-mediated signaling contributes to the very early phenotype seen in the *c-ret* mutant.

Parasympathetic Ganglia and Enteric Neuron Survival

Enteric neurons in the intestine are almost completely lost in the absence of GDNF, GFRα1, or c-ret, and mice lacking neurturin or GFRα2 have neurons of reduced size with less fiber density (Table 2) (Sanchez et al 1996, Pichel et al 1996, Moore et al 1996, Durbec et al 1996, Cacalano et al 1998, Enomoto et al 1998). Absence of c-ret appears to affect not only early survival and migration of precursors, but also later differentiation of the precursors and the neurons arising from them (Pachnis et al 1998, Taraviras et al 1999, Natarajan et al 1999). C-ret is also required for normal development of the human enteric nervous system. Patients with Hirschsprung's disease (or congenital megacolon) have been shown to have mutations in *c-ret*, which results in impaired migration of enteric neurons into the colon (Romeo et al 1994, Edery et al 1994).

The trophic factor dependencies of parasympathetic neurons are less completely defined. The GFRα2 subunit and c-ret are expressed in parasympathetic neurons (e.g. Nishino et al 1999), and major deficits in several parasympathetic ganglia have been detected in mice lacking Neurturin or GFRα2 (Rossi et al 1999, Heuckeroth et al 1999). No deficits have been reported in mice lacking GFRα1, GFRα3, or GDNF. However, parasympathetic ganglia may have been examined closely only in the *GFRα3* mutants (Nishino et al 1999). These neurons are not affected by mutations in any of the neurotrophin or Trk receptor genes.

Motor Neuron Survival

With the exception of NGF, each of the neurotrophins is able to promote survival of purified motor neurons in vitro (reviewed in Reichardt & Fariñas 1997). GDNF, CNTF, and other CNTF-related cytokines are also potent survival factors for these neurons in vitro. Despite this, the vast majority of motor neurons are spared in mice lacking any single one of these factors in vivo. For example, no single neurotrophin or Trk mutant has a significant reduction in total number of motor neurons. Even a triple mutant lacking BDNF, NT-3, and NT-4 has only a 20% deficit in facial and spinal motor neurons (Liu & Jaenisch 2000). Deficits in the *GDNF* and *GFRα1* mutants are small, approximately 20% in the spinal motor neuron population. The most severe deficits were seen in mice lacking either the CNTF receptor-α or the leukemia inhibitory factor receptor-β subunit, where deficits of approximately 40% were reported in the facial motor nucleus (see Reichardt & Fariñas 1997). Until recently, these results were believed to reflect functional redundancy, i.e. the concept that any motor neuron has access to multiple trophic factors in vivo. However, more recent data suggest strongly that different motor neuron pools are dependent on different trophic factors and that the deficits observed in mutants are small because there is so much diversity in the trophic factor dependence of the different pools. Analysis of the *NT-3* mutant argued several years ago that continued survival of γ-motor neurons requires the presence of this neurotrophin (Kucera et al 1995). These efferents innervate muscle spindles, which express NT-3. These spindles are also innervated by NT-3–dependent sensory neurons, so motor neurons and sensory neurons that innervate the same end organ appear to require the same neurotrophin. In contrast, survival of α-motor neurons that innervate skeletal myotubes is not detectably reduced in the *NT-3* mutant (Kucera et al 1995). During the past several years, different pools of motor neurons have been shown to express different combinations of transcription factors (e.g. Tanabe & Jessell 1996) and receptors for neurotrophic substances (e.g. Oppenheim et al 2000, Garces et al 2000). Recently, analyses of the *GDNF* and *GFRα1* mutants have shown that specific pools and subpopulations of motor neurons are severely affected in each mutant, whereas other pools and subpopulations are not detectably affected (e.g. Oppenheim et al 2000, Garces et al 2000). When these studies are extended to other mutants, it now seems likely that mice lacking individual trophic factors or their receptors will prove to have severe deficits in individual populations of motor neurons.

CNS Neuron Survival

Perhaps most striking of all is the paucity of survival deficits observed in populations of CNS neurons, many of which are responsive to these same factors in cell culture. For example, CNS neurons responsive to NGF in vitro include basal forebrain and striatal cholinergic neurons. Although differentiation of these neurons is affected (Smeyne et al 1994), survival of these populations is not detectably affected perinatally in the *NGF* or *TrkA* knockout animals. Postnatal atrophy of NGF-dependent populations of cholinergic forebrain neurons has been observed in adult NGF mutant heterozygotes, however; thus, these neurons appear to retain some dependence on this neurotrophin (Chen et al 1997).

BDNF and NT-4–responsive neurons include cerebellar granule cells, mesencephalic dopaminergic neurons, and retinal ganglion cells. Although a modest increase in postnatal apoptosis of hippocampal and cerebellar granule cells is observed in *TrkB* and *TrkB/TrkC* mutants (Minichiello & Klein 1996, Alcantara et al 1997), it is not comparable to the dramatic losses observed in these same mutant animals in sensory neurons. To examine the roles of TrkB in the maintenance and function of adult nervous system, mice with cell type–specific deletions of TrkB have been generated using the Cre/loxP recombination system (Minichiello et al 1999, Xu et al 2000b). Selective deletion of *TrkB* in the pyramidal neurons of the neocortex leads to altered dendritic arborization and compression in cortical layers II/III and V (Xu et al 2000b). At later times, loss of TrkB also results in progressive elimination of SCIP (suppressed, cAMP-inducible POU)-expressing neurons in the somatosensory and visual cortices, whereas the Otx-1–expressing neurons are not lost (Xu et al 2000b). Taken together, the data described above indicate that neurotrophin-mediated signaling through TrkB is important for maintaining survival of neuronal populations in the CNS. Because weeks, not hours, are required to observe significant losses of neurons, it is uncertain whether interruption of the same signaling pathways important for survival of peripheral neurons accounts for these phenotypes.

Multiple Neurotrophic Factor Dependence

In summarizing dependencies of sensory neurons on neurotrophins, it is worth noting that, in general, the neurotrophin and Trk receptor mutant phenotypes are consistent with in vitro observations of ligand and receptor specificities. Analyses of the mutant phenotypes have provided valuable information for reconstructing at least four models that demonstrate unique receptor-ligand interactions in individual sensory ganglia. In the first model, a single neurotrophin appears to interact with a sole receptor. As a result, absence of either the ligand or receptor results in a similar phenotype. As one example, both the *NGF* and *TrkA* mutants appear to lack the same sensory neuron populations in the DRG and cranial ganglia (Table 1). In the second model, deficits in neuron numbers are larger in receptor-deficient than in neurotrophin-deficient mice. In the nodose-petrosal ganglion, the great majority of neurons express TrkB, and targeted deletion of *TrkB* leads to an almost complete absence of these neurons. However, BDNF and NT-4 appear to be expressed in different target fields, so each neurotrophin supports a separate subpopulation of neurons within this ganglion. In the third model, a single ligand interacts with multiple receptors. Several reports have documented the promiscuous role in vitro of NT-3 in activating TrkA and TrkB, in addition to activating TrkC (e.g. Ip et al 1993, Clary & Reichardt 1994, Davies et al 1995). In the DRG and trigeminal ganglia in an *NT-3* mutant, many TrkA- and TrkB-expressing neurons undergo apoptotic cell death shortly after they are born (Fariñas et al 1998, Huang et al 1999a). These neurons are not killed in the *TrkC* mutant, so NT-3 must be directly activating the other Trk receptors. Thus, there must be high enough local concentrations of NT-3 in vivo to support neurons expressing TrkA or TrkB. Indeed, expression of NT-3 has been detected in the immediate vicinity of these ganglia during the period of neurogenesis (Fariñas et al 1998, Huang et al 1999a). By examining the expression of both *NT-3^{lacZ}* and *BDNF^{lacZ}*, we have identified distinct regions in the embryonic branchial arches where only the *NT-3* gene is expressed between E11.5 and E12.5 (EJ Huang, unpublished data). Presumably, TrkB-expressing neurons that project to these regions will require NT-3 for their survival because BDNF is not available. This model predicts that replacement of NT-3 by BDNF should rescue some TrkB-expressing neurons. Consistent with this prediction, the phenotype in the DRG of mice in which *BDNF* is inserted into the *NT-3* gene has recently been reported (V Coppola, J Kucera, ME Palko, J Martinez-De Velasco, WE Lyons, B Fritzsch, L Tessarollo, unpublished observations). In the complete absence of NT-3, the expression of BDNF in the same pattern as NT-3 can indeed delay and partially rescue neuronal deficits in the DRG. Finally, in the fourth model, survival of neurons that express more than one Trk receptor will be controlled by the patterns of neurotrophin expression. As described above, in the cochlear ganglion, neuronal survival in the *NT-3* mutant is determined by an apical-to-basal gradient of BDNF. In the trigeminal mesencephalic ganglion, survival depends on whether NT-3 or BDNF is present in a particular muscle spindle (Fan et al 2000).

The data in Table 1 and Table 2, although incomplete, indicate that there is excellent correspondence between the specificities of ligand interactions with receptors as defined in vitro and the phenotypes of mutants lacking these proteins in vivo. This is evident from the analyses of mice lacking neurotrophins or Trk receptors. There are also striking similarities in the phenotypes of the *GDNF* and *GFR-α1* and *Neurturin* and *GFR-α2* mutants, which also agrees with the majority of biochemical studies characterizing interactions of these ligands with these two receptor subunits.

Switching of Neurotrophic Factor Dependence

The data in Table 1 also make it clear that many neurons require more than one neurotrophic factor-receptor signaling pathway to survive. In the vast majority of instances, this appears to reflect requirements at different developmental stages, made necessary by changes in ligand accessibility or in receptor expression. Examination in more detail of the development of the phenotypes of these mutants has suggested that ligand activation of tyrosine kinases is important for survival of neural precursors and immature neurons, as well as of mature neurons in contact with their final targets. In some instances, the same factor is essential at many stages of development. In others, different factors become important as development proceeds. For example, the SCG shows sequential dependency on c-ret–mediated followed by TrkA-mediated signaling during embryogenesis. Thus, the SCG is completely absent in the *c-ret*, *NGF*, and *TrkA* mutants, and less-severe deficiencies are seen in mice lacking GDNF, GFRα3, or NT-3. As discussed above, the requirement for c-ret clearly precedes that for the neurotrophins. A similar transient dependence on NT-3 of TrkA- and TrkB-expressing trigeminal neurons also appears to reflect the presence of NT-3, but not other neurotrophins, in the mesenchyme invaded by the axons of these neurons immediately after neurogenesis (Huang et al 1999a). Expression of NT-3 in the mesenchyme has been shown to be induced by epithelial-mesenchymal interactions, which are likely to be mediated by Wnt proteins secreted by the epithelium (Patapoutian et al 1999, O'Connor & Tessier-Lavigne 1999).

Although much attention has been focused on switches in neurotrophin responsiveness during early development of the trigeminal ganglion, more recent work suggests that there is comparatively little switching in expression of Trk receptors (Huang et al 1999a; but see also Enokido et al 1999). Instead, similar to the DRG (Ma et al 1999), neurons expressing different Trk receptors are born in waves that peak at different times. Much later in development, however, defined subsets of TrkA-expressing DRG and trigeminal ganglion neurons become responsive to GDNF family members, acquire expression of c-ret and one or more GFR adapter subunits, and lose responsiveness to NGF (e.g. Molliver et al 1997, Huang et al 1999b, Fundin et al 1999). Specific expression patterns of GDNF family ligands in targets and GFR-α adapter subunits in both targets and neurons indicate that these changes reflect the sorting out of subsets of sensory neurons specialized for innervation of specific targets and transmission of specific modalities of sensory information (Snider & McMahon 1998, Fundin et al 1999). There are many other examples, too numerous to review here, in which changes in ligand or receptor expression have been shown to explain the requirement for more than one neurotrophic factor by a single population of neurons.

Precursor Cell Survival and Proliferation

A large number of studies characterizing effects of neurotrophins on CNS neuroepithelial precursors, neural crest cells, or precursors of the enteric nervous system have demonstrated effects on both proliferation and differentiation of these cells in vitro and, in some instances, in vivo. Some of these have been described above, most notably the effect of c-ret–mediated signaling on SCG and enteric precursors. Only a few examples are described here, but they are representative of the interesting actions of these proteins on immature cells of the nervous system (for more extended descriptions, see Reichardt & Fariñas 1997).

In cell culture, many populations of CNS precursors are regulated by neurotrophins. Nestin-positive cells from the rat striatum can be induced to proliferate with NGF (Cattaneo & McKay 1990). The proliferation and survival of oligodendrocyte precursors (O2A progenitors) have been shown to be promoted by NT-3 in vitro and in vivo (Barres et al 1994). In other instances, neurotrophin application has been shown to induce differentiation of precursors. For example, NT-3, but not other neurotrophins, promotes differentiation of rodent cortical precursors (Ghosh & Greenberg 1995). The differentiation of hippocampal neuron precursors is promoted by BDNF, NT-3, and NT-4 (Vicario-Abejón et al 1995). Although these observations suggest that neurotrophins regulate the size of precursor pools and neurogenesis in vivo, analyses of the various neurotrophin and Trk receptor knockouts has provided no evidence that development of these cell populations within the CNS is perturbed during embryogenesis.

NT-3 has been shown to promote proliferation of cultured neural crest cells (e.g. Sieber-Blum 1991, 1999). In vivo, however, initial formation of sensory and sympathetic ganglia appears to occur normally in mouse embryos (Fariñas et al 1996, Wilkinson et al 1996). Only after initiation of neurogenesis is elevated apoptosis observed in the *NT-3* mutant, and apoptosis appears to be restricted to cells that have left the cell cycle and express neuronal markers (Wilkinson et al 1996; Fariñas et al 1996, 1998; Huang et al 1999a). This is consistent with evidence in murine sensory ganglia that expression of Trk receptors follows and does not precede withdrawal from the cell cycle (Fariñas et al 1998, Huang et al 1999a). Although direct effects of neurotrophin deficiency appear to be restricted to neurons, evidence for an indirect effect on precursors has been obtained in the DRG sensory ganglia of both *NT-3* and *NGF* mutants (Fariñas et al 1996; I Fariñas, unpublished data). In these ganglia, the loss of neurons through apoptosis stimulates the differentiation of precursors into neurons without affecting their survival or rate of proliferation. As a result, the precursor pool is depleted during late stages of neurogenesis, and many fewer neurons than normal are born. Increased differentiation of precursors in these mutants could potentially be caused by a reduction in lateral inhibition mediated by the Delta-to-Notch signaling pathway.

The failure to detect Trk receptor expression in murine sensory ganglia precursors does not mean that these receptors are not expressed in any proliferating neural crest cell derivatives in mouse embryos. The mesenchymal cells derived from cardiac neural crest have been shown to be important in cardiac development in avian embryos, and a similar population has recently been identified in murine embryos (Jiang et al 2000). *NT-3* and *TrkC* mutant embryos have severe cardiovascular abnormalities, including atrial and ventricular septal defects and pulmonary stenosis, which seemed likely to be caused by deficits in a neural crest–derived population of cells (Donovan et al 1996, Tessarollo et al 1997). More recently, however, expression of both NT-3 and TrkC have been detected in embryonic cardiac myocytes, and TrkC receptor-dependent signaling has been shown to promote proliferation of these myocytes during early development (Lin et al 2000). This raises the possibility that abnormalities in valve and outflow tracts observed in *NT-3* and *TrkC* mutants are not caused by a signaling deficit in neural crest cell progeny. Effects could potentially be indirect. It is interesting that BDNF deficiency also results in abnormalities in the heart (Donovan et al 2000). BDNF activation of TrkB receptors expressed in the cardiac vasculature is essential for maintaining the survival of cardiac endothelial cells and integrity of the cardiac vascular bed in early postnatal animals.

ESSENTIAL ROLES IN DIFFERENTIATION AND FUNCTION

Sensory and Sympathetic Neuron Development

In a few instances, neurotrophins have been shown to regulate the pathways of differentiation selected by neural precursors and to regulate the differentiation process, helping to determine the levels of expression of proteins essential for the normal physiological functions of differentiated neurons, such as neurotransmitters, ion channels, and receptors. For example, in vitro and in vivo, NGF promotes the differentiation of sympathoadrenal precursors into sympathetic neurons as opposed to adrenal chromaffin cells (Levi-Montalcini 1987, Anderson 1993). In contrast, glucocorticoids have been shown to suppress these responses to NGF, inhibiting differentiation into neurons and promoting differentiation into mature adrenal chromaffin cells. As glucocorticoids are present in high concentrations in the adrenal gland, they seem likely to regulate differentiation of sympathoadrenal precursors similarly in vivo. In vitro and in vivo, the actions of NGF and glucocorticoids on these precursors are largely irreversible and dramatic. The sympathetic neurons formed are permanently dependent on NGF for survival, whereas the chromaffin cells are not. The two differentiated cell types differ in their predominant transmitter (norepinephrine vs epinephrine). Morphologically they are distinct, and this clearly reflects differences in many molecular constituents, which are regulated either directly or indirectly by NGF and the glucocorticoids. In PC12 cells, a very brief exposure to NGF has been shown to result in long-term induction of a sodium channel gene (Toledo-Aral et al 1995). This may serve as a model system for investigating how neurotrophins can cause irreversible fate changes within neurons.

As described above, precursors of murine sensory neurons do not appear to express Trk receptors in vivo or be affected directly by deficiencies in neurotrophins (Fariñas et al 1998, 2000; Huang et al 1999a). In mouse DRG, at least, initial generation of TrkB- and TrkC-expressing versus TrkA-expressing neurons appears to occur in two waves, dependent on sequential expression of neurogenins 2 and 1 (Ma et al 1999). The neurotrophins function to maintain the viability of these neurons. They do not appear to guide their initial determination. There is some evidence indicating that the situation may be more complicated in the chicken embryo. There, evidence has been obtained that suggests a subpopulation of DRG precursors expresses TrkC, and available evidence suggests there is more dynamic regulation of Trk receptor expression after neurogenesis than is observed in mouse sensory ganglia (e.g. Rifkin et al 2000). It seems unlikely, however, that the role of neurotrophins is very different there than in murine sensory ganglia.

Neurotrophins are important in regulating aspects of later sensory neuron development that, in some instances, control important aspects of neuronal phenotype. For example, NGF-responsive sensory neurons primarily convey nociceptive information and extend either unmyelinated C-fibers or thinly myelinated Aδ fibers. They express small peptide transmitters, such as CGRP and substance P, specific receptors such as the capsaicin receptor, and distinct Na^+ channels isoforms (e.g. Amaya et al 2000, reviewed by Mendell et al 1999). Expression levels of most of these proteins are regulated by NGF (e.g. Fjell et al 1999; reviewed in Lewin & Barde 1996, Mendell 1999; Mendell et al 1999). Although the absence of NGF during embryogenesis results in loss of almost all nociceptive neurons, at later ages withdrawal of NGF no longer kills these neurons. Instead, perturbation of NGF levels results in phenotypic changes. When NGF is sequestered during the early postnatal period, the properties of Aδ fibers are dramatically changed (Ritter et al 1991). Normally, many of these fibers are responsive to high threshold mechanical stimulation and are classified as high-threshold mechanoreceptors (HTMRs). Postnatal deficiency in NGF results in almost complete loss of HTMRs with a proportional increase in D-hair fibers, which respond to light touch. As this change is not associated with elevated apoptosis, the results suggest that the neurons of origin for the HTMR fibers are not lost but instead undergo a change in phenotype, becoming D-hair fibers that function as touch, but not pain, receptors. The central projections of D-hair and HTMR fibers have different termination zones in the substantia gelatinosa, and it is interesting that the phenotypic conversion induced by withdrawal of NGF does not result in inappropriate innervation by D-hair fibers of zones normally innervated by HTMR fibers (Lewin & Mendell 1996). Central projections appear to be regulated to maintain appropriate modality connectivity.

Overexpression of NGF in skin using the keratin-14 promoter results in increased survival of both the C and Aδ classes of nociceptive neurons and, in addition, affects the functional properties of these neurons (Stucky et al 1999, Mendell et al 1999, Stucky & Lewin 1999). The percentage of Aδ fibers responsive to nociceptive stimuli increases from 65% to 97%, which may reflect selective survival of these neurons. Even more notably, the percentage of C-fibers responsive to heat increases from 42% to 96%, an effect too large to be accounted for by their selective survival. In addition, the chronic presence of NGF in skin also affects the functional properties of heat-sensitive C-fibers, increasing their thermal responsiveness and lowering their mechanical responsiveness. It seems likely that regulation of VR-1, the capsaicin receptor, may be involved in these phenotypic changes.

NGF is not the only neurotrophic factor to regulate the phenotype of nociceptive neurons. Although all nociceptors are believed initially to express TrkA, a proportion of these begin subsequently to express c-ret together with one or more of the GFR adapter subunits (e.g. Snider & McMahon 1998). In addition to expressing c-ret, these neurons are distinguished by expression of a binding site for the plant lectin isolectin B4. As development proceeds, their survival becomes dependent on GDNF family members (Molliver et al 1997). Targeted disruption of the GFR-α2 gene does not cause loss of cells expressing the isolectin B4 ligand but does result in a three-fold reduction in the percentage of isolectin B4 ligand-expressing neurons sensitive to heat (CL Stucky, J Rossi, MS Airaksinen, GR Lewin, unpublished observations). The results indicate that signaling mediated by a GDNF family member, most likely neurturin, is necessary for these neurons to manifest a nociceptive phenotype.

Continued presence of D-hair afferents has been shown to depend on NT-3 in early postnatal development and on NT-4 at later times (Airaksinen et al 1996, Stucky et al 1998). It is not certain whether these phenotypes are caused by neuronal loss or changes in neuronal phenotype.

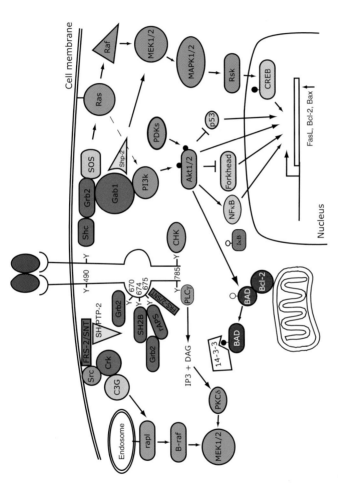

Figure 1 Schematic diagram of Trk receptor-mediated signal transduction pathways. Binding of neurotrophins to Trk receptors leads to the recruitment of proteins that interact with specific phosphotyrosine residues in the cytoplasmic domains of Trk receptors. These interactions lead to the activation of signaling pathways, such as the Ras, phosphatidylinositol-3-kinase (PI3k), and phospholipase C (PLC)-γ pathways, and ultimately result in activation of gene expression, neuronal survival, and neurite outgrowth (see text for detailed discussions and abbreviations). The nomenclature for tyrosine residues in the cytoplasmic domains of Trk receptors are based on the human sequence for TrkA. In this diagram, adaptor proteins are red, kinase green, small G proteins blue, and transcription factors brown.

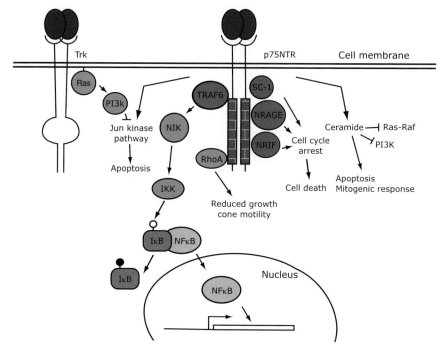

Figure 2   Schematic diagram of p75NTR-mediated signal transduction pathways. P75NTR interacts with proteins, including TRAF6, RhoA, NRAGE (neurotrophin receptor— interacting MAGE homologue), SC-1, and NRIF, and regulates gene expression, the cell cycle, apoptosis, mitogenic responses, and growth cone motility. Binding of neurotrophins to p75NTR has also been shown to activate the Jun kinase pathway, which can be inhibited by activation of the Ras-phosphatidylinositol-3-kinase (PI3K) pathway by Trk receptors. Similar to Figure 1, adaptor proteins are red, kinase green, small G proteins blue, and transcription factors brown. (See text for abbreviations.)

Control of Target Innervation

Each of the neurotrophins has been shown to promote neurite outgrowth by responsive neurons in vitro. Elegant experiments have demonstrated that local NGF regulates the advance of sympathetic neuron growth cones (Campenot 1977). The presence of NGF within a compartment was shown to be essential for axons to grow into that compartment, even when neurons received adequate trophic support. When neurons were seeded between two chambers, one with NGF and one with no neurotrophin, axons invaded only the chamber containing NGF. If at a later time NGF was withdrawn from a chamber, the axons stopped growing and slowly retracted. In addition to promoting growth, gradients of neurotrophins are able to steer growth cones in vitro (Gundersen & Barrett 1979). It is intriguing that in these assays, whether a neurotrophin acts as a chemoattractant or a chemorepellent depends on cyclic nucleotide levels within neurons (Song et al 1997, Song & Poo 1999). The chemoattractive activities of NGF and BDNF, acting through TrkA and TrkB, respectively, are converted to chemorepellent activities by inhibitors of the cAMP signaling cascade. Effects of a PI-3 kinase inhibitor and of NGF signaling through a TrkA mutant lacking a putative PI-3 kinase docking site suggest that activation of PI-3 kinase is required for the chemoattractive response (Ming et al 1999). It is intriguing that although the different Trk receptors are believed to function through similar signal transduction pathways, the chemoattractive activity of NT-3, acting through TrkC, is not affected by agents that affect cAMP-mediated signaling. Instead, inhibitors of cGMP signaling convert this chemotrophic response from attractive to repulsive (Song & Poo 1999). These observations argue persuasively that there are fundamental differences in the signaling mediated by different Trk receptors.

Since the discovery of NGF, it has been appreciated that systematically applied neurotrophins affect innervation patterns in vivo (e.g. Levi-Montalcini 1987). NGF was shown to increase innervation of tissues that receive sympathetic or sensory innervation normally and to induce aberrant innervation of tissues that normally are not innervated. In adults, neurotrophins are generally concentrated in targets of sensory and sympathetic targets (e.g. see Reichardt & Fariñas 1997). Analyses of transgenic animals either lacking or expressing ectopically neurotrophins have provided many examples where disruptions of normal expression patterns of a neurotrophin results in perturbations of innervation, including aberrant routing of axons and interference with innervation of specific targets. For example, elevation of NGF in pancreatic islets using the insulin promoter induces dense sympathetic innervation of cells within the pancreatic islets, which normally are not innervated (Edwards et al 1989). Elevation of NGF in the epidermis using the keratin-14 promoter induces similarly dense sympathetic innervation of the epidermis (Guidry et al 1998). In this case also, the pattern of innervation is perturbed. Sympathetic innervation of the footpad vasculature and sweat glands is strongly inhibited. Instead, the sympathetic fibers are found in a plexus together with sensory fibers in the dermis. Sympathetic innervation is also distributed aberrantly in the vicinity of the mystacial pads (Davis et al 1997). Overexpression of NGF under control of the keratin-14 promoter also increases greatly the density of sensory innervation, selectively promoting innervation by NGF-dependent nociceptors (Stucky et al 1999). In contrast to sympathetic fibers, aberrant targeting of these endings was not detected. As a final example, in mice that overexpress BDNF under control of the nestin promoter, sensory fibers dependent on this neurotrophin appear to stall at sites of ectopic BDNF expression at the base of the tongue and fail to reach the gustatory papillae (Ringstedt et al 1999). Fibers that do not traverse these sites of ectopic BDNF expression are able to reach and innervate their targets normally. Taken together, the results of these studies indicate that elevated expression of a neurotrophin in a region usually results in an increased density of innervation by axons from neurons that normally innervate that region. Overexpression in regions that are normally not innervated often, but not always, results in aberrant innervation by neurons responsive to that neurotrophin. This suggests that guidance and targeting clues are either masked or overridden by the presence of high levels of a neurotrophin. Consistent with this concept, recent studies in tissue culture have shown that uniform exposure of TrkA-expressing neurons to NGF results in a desensitization of chemotactic responses to gradients of netrin, BDNF, or myelin-associated glycoprotein, which suggests that these factors share common cytosolic signaling pathways (Ming et al 1999).

In some instances, deficits or aberrancies in innervation observed in transgenic animals may reflect competitive phenomena. In analyses of innervation by sensory neuron fibers of mystacial pads, elevated innervation by TrkA-dependent sympathetic fibers was observed in mutants lacking BDNF or TrkB (Rice et al 1998). Excessive innervation by TrkA-dependent sensory endings was seen in mice lacking BDNF, NT-4, or TrkB. When innervation of different classes of endings that detect mechanosensation was examined in these mutants, each ligand and Trk receptor was shown to support innervation of at least one type of mechanoreceptor (Fundin et al 1997). Innervation of some endings is dependent on more than one neurotrophin or Trk receptor. For example, NT-3 is important for formation of all types of endings, but it may signal through different Trk receptors as development proceeds. In addition, the results suggested that BDNF signaling through TrkB may suppress Merkel innervation whereas NT-3 signaling through TrkC suppresses Ruffini innervation. There is not a single compelling explanation for these observations. Some sprouting may be attributable to loss of competition for a neurotrophin. Absence of TrkC-dependent endings, for example, may result in less NT-3 being transported out of the region. The elevated level of NT-3 remaining in the region may mediate sprouting of endings otherwise supported by NGF alone. Not all observations appear compatible with this model, however. Instead, the results suggest that neurotrophin-mediated signaling regulates, both positively and negatively, responses to other factors involved in sensory fiber targeting and differentiation.

Sensory Neuron Function

Neurotrophins have multiple interesting effects on the functional properties of sensory neurons extending beyond regulation of their survival. Essentially all modalities of sensory information are modulated in different ways by changes in the levels of these factors.

During early postnatal rodent development, both NT-3 and BDNF have been shown to regulate the development of the synapses formed between Ia afferents and motor neurons (Seebach et al 1999). Chronic NT-3 results in larger monosynaptic excitatory postsynaptic potentials (Epsps) and reduced polysynaptic components, whereas BDNF actually reduces the size of the monosynaptic Epsps and increases the contribution of polysynaptic signaling. Infusion with TrkB-Ig also results in the appearance of larger monosynaptic Epsps, which suggests that endogenous BDNF is an important modulator of development of these synapses. Infusion with TrkC-Ig had little effect. These data argue that levels of endogenous BDNF within the spinal cord control the comparative efficiencies of monosynaptic and polysynaptic signaling between Ia afferents and motor neurons. The role of endogenous NT-3 is less certain.

During the first postnatal week, but not subsequently, direct application of NT-3 has been shown to acutely potentiate the α-amino-3-hydroxy-5-methyl-4-isoxazolepropionic acid (AMPA)/kainate receptor-mediated monosynaptic EPSP at the synapses formed by Ia afferents on motor neurons (Arvanov et al 2000). Potentiation is long lasting and is prevented by an inhibitor of Trk receptor kinase activity. Initiation, but not maintenance, of potentiation requires N-methyl-D-aspartate (NMDA) receptors and postsynaptic Ca^{2+}. The dependence on NMDA receptor function and Ca^{2+} are similar to the requirements for generation of long-term potentiaion (LTP) in the hippocampus (see Malenka & Nicoll 1999). It is intriguing that EPSPs with similar properties evoked by stimulation of a different innervation pathway are not potentiated by NT-3, so the effects of NT-3 are synapse specific. NT-3 is effective at potentiating the strength of the Ia afferent-motor neuron synapse only during the first week, very likely because NMDA receptors are downregulated at later times.

Effects of peripheral nerve transection on monosynaptic Ia EPSPs, however, suggest that NT-3 from peripheral sources is also important in regulating the efficiency of synaptic function in adult animals. Peripheral transection results in long-term declines in conduction velocity and monosynaptic EPSP amplitude (Mendell et al 1999). These declines can be prevented by infusion of NT-3 in the vicinity of the cut nerve ending, which suggests that interruption of NT-3 transport is the cause of the synaptic deficiencies observed after transection.

Although overexpression of NGF induces the changes described in the previous section that are likely to involve gene expression, acute NGF also has striking effects on nociceptors. Application of NGF to the undersurface of a patch of skin acutely sensitizes nociceptive C and Aδ fibers to heat within 10 min (e.g. Shu & Mendell 1999a). Sensitization is not seen in the skin of animals depleted of mast cells, so a major pathway mediating this response is believed to involve activation of mast cells by NGF, resulting in secretion from these cells of serotonin, histamine, and other agents, including NGF (e.g. see Levi-Montalcini et al 1996). Acute application of NGF also sensitized the subsequent response of sensory neurons to capsaicin (Shu & Mendell 1999b). As sensitization is seen in dissociated sensory neuron cultures, NGF must be acting directly on the sensory neurons.

One of the proteins known to be upregulated by NGF in sensory neurons is the neurotrophin BDNF (Michael et al 1997). There is evidence that BDNF is transported to both peripheral and central terminals of nociceptive sensory neurons. In the periphery, BDNF and NT-4 have been shown to acutely sensitize nociceptive fibers by a pathway that requires the presence of mast cells (Shu et al 1999, Rueff & Mendell 1996). Sensory neuron–derived BDNF also appears to act centrally (Mannion et al 1999, Woolf & Costigan 1999). Perfusion of the spinal cord with TrkB-IgG has been shown to prevent the progressive hypersensitivity elicited by low-intensity tactile stimulation of inflamed tissues.

Cortical Circuitry and Function

Several neurotrophins are expressed in the neocortex and hippocampus during development, and their expression continues in adult animals, which suggests that they have functions extending beyond initial development. NGF, for example, is widely expressed in both the developing and adult neocortex (e.g. Large et al 1986). Projections from the cholinergic basal forebrain extend throughout the neocortex and hippocampus (e.g. Mesulam et al 1983). The fibers of these projections express TrkA (e.g. Sobreviela et al 1994), and expression in these neurons of proteins associated with cholinergic function, such as choline-o-acetyl transferase, is increased by infusion of NGF (e.g. Hefti et al 1989). NGF infusion has been shown to attenuate the behavioral deficits associated with cholinergic atrophy (e.g. Fischer et al 1987). Maintenance of normal function of these neurons in adult animals is sensitive to small perturbations of NGF levels. For example, animals heterozygous for a mutation in the *NGF* gene express approximately half the normal level of NGF mRNA and protein and have significant deficits in memory acquisition and retention, which can be corrected by prolonged infusion of NGF (Chen et al 1997). Mice lacking TrkA have also been shown to have deficits in cholinergic projections from the basal forebrain (Smeyne et al 1994).

Both BDNF and TrkB are widely expressed in the developing and adult hippocampus and neocortex (e.g. Cellerino et al 1996). BDNF mRNA is present in excitatory pyramidal neurons, but not in GABAergic inhibitory interneurons. TrkB is expressed by both classes of neurons, although its expression is higher in inhibitory interneurons. In addition, expression of BDNF is regulated by both sensory input and electrical activity. For example, induction of seizures in the hippocampus strongly induces BDNF expression (e.g. Kornblum et al 1997). In the visual and somatosensory cortices, expression has been shown to be regulated by sensory inputs, with deprivation reducing expression of this neurotrophin (e.g. Castren et al 1992, Rocamora et al 1996, Singh et al 1997). Several promoters control expression of BDNF mRNA, and one of these is regulated by Ca^{2+} acting through Ca^{2+}-calmodulin–dependent protein kinase IV to phosphorylate and activate the transcription factor CREB (Tao et al 1998, Shieh et al 1998). BDNF is sorted into a regulated secretory pathway in hippocampal neurons (e.g. Farhadi et al 2000), so increases in neuronal activity should both activate transcription of the *BDNF* gene and increase secretion of the BDNF protein.

Expression of TrkB has also been shown to be modestly increased by activity (e.g. Castren et al 1992). Equally important, surface expression of TrkB is also regulated by activity (Meyer-Franke et al 1998). In the absence of activity, this protein appears to be largely sequestered into cytoplasmic vesicles. This result suggests that neurons become more responsive to BDNF as a result of activity. At the subcellular level, regulation of TrkB distribution may provide a mechanism by which active and inactive synapses differ in their responsiveness to BDNF, thereby regulating actin dynamics, glutamate receptor activity, and other functions important for adjusting synaptic function. In the absence of a direct demonstration of activity-regulated TrkB trafficking in vivo, however, this should be considered only an intriguing possibility.

In the neocortex, BDNF signaling through TrkB has been implicated in both development and maintainance of cortical circuitry. BDNF expression in excitatory neurons is promoted by activity, whereas increased release of BDNF can be expected to enhance the effectiveness of inhibitory interneurons. This has raised the possibility that BDNF-to-TrkB signaling modulates an autoregulatory circuit between excitatory pyramidal cells and inhibitory interneurons. In mixed cultures of postnatal rat cortical neurons, activity blockage has been shown to reduce reversibly GABA expression in interneurons and to reduce GABA-mediated inhibition on pyramidal cells (Rutherford et al 1997). In these cultures, the rates of firing are stabilized by scaling of the amplitude of AMPA receptor-mediated synaptic inputs (Rutherford et al 1998). These effects appear to be modulated by endogenous BDNF, as effects of activity blockade can be prevented by exogenous BDNF and effects of activity blockade are mimicked by a BDNF scavenger (Desai et al 1999). It is attractive to imagine that this autoregulatory circuit functions in vivo.

Formation of Ocular Dominance Columns

The density of innervation of layer IV by afferents from the thalamus is increased by exogenous BDNF and reduced by a scavenger of endogenous BDNF, TrkB-IgG (Cabelli et al 1997). Both agents appear to interfere with sorting of these afferents into ocular dominance columns, raising the possibility that competition for limiting amounts of BDNF by these afferents is involved in some manner in the sorting mechanism. In addition, infusion of NT-4 into the visual cortex during the critical period has been shown to prevent many of the consequences of monocular deprivation (Gillespie et al 2000). In the presence of NT-4, neurons remain responsive to stimuli from the deprived eye. Even after responses to the deprived eye are lost, infusion of NT-4 is able to restore them. These observations suggest that TrkB activation during the critical period promotes connectivity independent of correlated activity.

The function of inhibitory interneurons is essential for formation of ocular dominance columns (e.g. Hensch et al 1998), and maturation of these neurons is regulated by BDNF in vitro and in vivo. In initial analysis of the *BDNF* mutant, several deficits in interneuron maturation were detected, including expression of Ca^{2+}-binding proteins and peptide neurotransmitters (Jones et al 1994). Recently, BDNF has been overexpressed in excitatory pyramidal neurons by use of the Ca^{2+}-calmodulin–dependent protein kinase II promoter (Huang et al 1999c). In these animals, maturation of interneurons is accelerated, as assessed by expression and synaptic localization of glutamate decarboxylase, expression of parvalbumin, and the strength of inhibitory postsynaptic potentials. In addition, the critical period of ocular dominance plasticity begins and terminates precociously, and the acuity of vision increases on an accelerated time course (Huang et al 1999c, Hanover et al 1999). As visual stimulation also increased expression of BDNF within pyramidal neurons, the results suggest that early sensory stimulation acts to promote maturation of interneurons through BDNF-to-TrkB signaling. Interneurons in turn promote the refinements in synaptic connectivity needed for maturation of the cortex. The results suggest that refinement of cortical circuitry is driven by intracortical mechanisms, which then drive the sorting of thalamic afferents. Recent work supports this model (e.g. Trachtenberg et al 2000).

Another mechanism by which neurotrophins may control development and changes in cortical circuitry is through control of dendritic and axonal arbors. Neurotrophins affect neuronal morphologies at many levels in the visual pathway. Local application of BDNF or of a BDNF scavenger, for example, has been shown to decrease and increase, respectively, the complexity of Xenopus retinal ganglion cell dendritic arbors (Lom & Cohen-Cory 1999). Local application to the optic tectum of these same agents has different effects on axonal branching patterns, with BDNF increasing the complexity of retinal ganglion cell–derived axonal arbors and a BDNF scavenger having the opposite effect. NT-4 has been shown to prevent the atrophy of lateral geniculate neurons seen after monocular visual deprivation (Riddle et al 1995). Applications of BDNF, NT-3, or NT-4 to slices of neonatal neocortex have been shown to regulate the dendritic morphologies of pyramidal cells over comparatively short time spans (e.g. Horch et al 1999; reviewed in McAllister et al 1999). The effects are distinct, cell specific, and layer specific. For example, BDNF was observed to promote dendritic arborization of neurons in layers IV and V, but BDNF actually inhibited arborization by neurons in layer VI (McAllister et al 1997; see also Castellani & Boltz 1999). In layer IV, NT-3 was shown to oppose the stimulation of dendritic arborization promoted by BDNF. In layer VI, BDNF inhibited the stimulation of arborization induced by NT-3. Apical and basal dendrites of the same neurons responded differently to the same neurotrophin (e.g. McAllister et al 1995). Many of the observed effects were prevented by blocking electrical activity (e.g. McAllister et al 1996). Specific deletion of the *TrkB* gene in pyramidal neurons also results in striking changes in these cells during postnatal development, with significant retraction of dendrites observed at 6 weeks and loss of many neurons seen at 10 weeks of age (Xu et al 2000b). Although the changes in dendritic morphology and subsequent loss of neurons were cell autonomous, requiring deletion of the *TrkB* gene within the affected neurons, changes in gene expression were also seen that were not cell autonomous, i.e. they were observed in neurons that continued to express TrkB. It seems likely that circuit perturbation as a result of alterations of dendritic morphologies accounts for these changes.

Taken together, the results described above indicate that neurotrophins are important in regulating establishment and function of cortical circuits. Within the visual system, they regulate development of retinal ganglion cell axonal and dendritic arbors, thalamic afferents, cortical pyramidal cells, and cortical interneurons, with profound effects on cortical function. Although mechanisms by which neurotrophins influence axonal and dendritic morphologies in vivo have not be examined, they almost certainly involve regulation of the Cdc-42/Rac/Rho family of small GTPases. Aberrant expression of constitutively active and dominant negative mutants of these proteins have been shown to have dramatic effects on dendritic branch patterns and spine density (e.g. Li et al 2000, Nakayama et al 2000).

Synaptic Strength and Plasticity

Mechanisms underlying establishment of LTP between afferents from CA3 pyramidal cells and postsynaptic CA1 pyramidal neurons in the hippocampus have been of intense interest, as these mechanisms are believed to provide a paradigm for regulation of synaptic strength and plasticity (reviewed in Malenka & Nicoll 1999). BDNF is expressed in CA3 and CA1 pyramidal neurons within the hippocampus, and TrkB is expressed by almost all hippocampal neurons, including dentate granule cells, CA3 and CA1 pyramidal cells, and inhibitory interneurons. It is interesting that LTP is greatly reduced in *BDNF* mutants, both in homozygotes and in heterozygotes (e.g. Korte et al 1995, Patterson et al 1996). Long-lasting, protein synthesis-dependent LTP is not seen in these mutants (Korte et al 1998). There are also deficits in long-lasting LTP and in memory consolidation in the hippocampus in mutant mice lacking NT-4 (Xie et al 2000). CA1 LTP is also reduced in TrkB mutant heterozygotes and in a mouse mutant that expresses reduced levels of TrkB (Minichiello et al 1999, Xu et al 2000a). Signaling through p75NTR does not appear to be important, because there is very little expression of this receptor within the hippocampus and because functional antibodies to p75NTR do not affect LTP (Xu et al 2000a). Loss of TrkB from excitatory pyramidal neurons in the hippocampus and forebrain interferes with memory acquisition and consolidation in many learning paradigms (Minichiello et al 1999).

In these mutants, the observed reductions in synaptic plasticity probably reflect functional, not developmental, deficits. First, the hippocampi of these animals appear to be morphologically normal. Second, a very similar inhibition of LTP can be seen following acute application of the BDNF/NT-4 scavenger TrkB-IgG to hippocampal slices (e.g. Chen et al 1999). Finally, the deficits observed in BDNF mutant heterozygotes and homozygotes can be rescued by exposure of hippocampal slices to BDNF (Korte et al 1995, Patterson et al 1996). The enhanced efficiency of synaptic transmission observed after induction of LTP is largely mediated by an NMDA receptor-dependent increase in AMPA receptor function (reviewed in Malenka & Nicoll 1999). Most evidence suggests, however, that BDNF-to-TrkB signaling is not directly involved in the biochemical changes underlying LTP within the postsynaptic cells, but instead modulates the competence of presynaptic nerve terminals to generate the repetitive exocytotic events needed to modify the responses of these postsynaptic neurons. In one set of experiments, LTP was generated normally in a TrkB hypomorph by a low-frequency–paired depolarization protocol that specifically assesses properties of postsynaptic cells and puts minimal demands on presynaptic terminal function (Xu et al 2000a). In addition, AMPA and NMDA receptor functions appeared to be normal in the postsynaptic neurons. In contrast, the ability of presynaptic nerve terminals to respond to repetitive pulses of stimulation was clearly impaired. Consistent with a presynaptic deficit, BDNF has been shown to enhance synaptic vesicle release in response to tetanic stimulation, possibly by promoting docking of synaptic vesicles to the presynaptic membrane (e.g. Gottschalk et al 1998, Pozzo-Miller et al 1999). Also consistent with a presynaptic deficit, LTP is further reduced in a *TrkB* mutant heterozygote by elimination of the remaining functional *TrkB* gene from excitatory pyramidal cells in both the CA3 and CA1 regions, eliminating TrkB from both presynaptic and postsynaptic neurons (Minichiello et al 1999). LTP is not further reduced, however, by deletion of the *TrkB* gene solely within the postsynaptic CA1 pyramidal neurons in the mouse mutant that expressed reduced TrkB levels (Xu et al 2000a). All these data argue that BDNF-to-TrkB signaling is crucial in presynaptic nerve terminals in CA1 that are derived from neurons in CA3. In addition, BDNF has been shown to decrease inhibitory postsynaptic currents on CA1 pyramidal cells (Tanaka et al 1997, Frerking et al 1998), so it is possible that part of the LTP deficit reflects an increase in inhibitory signaling by GABAergic interneurons in the absence of normal TrkB function within these cells. BDNF has been shown to affect NMDA function in hippocampal neurons in culture by increasing the open probability of their channels (Levine et al 1998). The experiments described above suggest that TrkB did not modify the function of these channels during tetanic stimulation.

At many developing and mature synapses, application of a neurotrophin acutely stimulates neurotransmitter release. This has been most intensely studied at the CA1 synapse in the hippocampus and in developing Xenopus neuromuscular synapses (e.g. Kang & Schuman 1996, Wang & Poo 1997). In the developing neuromuscular cultures, the neurotrophins have both pre- and postsynaptic effects, increasing spontaneous and evoked release of synaptic vesicles and changing the kinetics of opening of the acetylcholine receptor (e.g. Wang & Poo 1997; Schinder et al 2000). Low levels of neurotrophin act synergistically with synaptic terminal depolarization (e.g. Boulanger & Poo 1999b). To be effective, the cell must be depolarized during the period of neurotrophin exposure. A cAMP agonist also synergizes with BDNF to potentiate spontaneous and action potential–evoked neurotransmitter release (Boulanger & Poo 1999a). The synergistic effect of depolarization appears to be caused by an increase in the level of cAMP. What is surprising is that a very similar presynaptic potentiation induced by NT-3 application is not affected by agents that inhibit cAMP-mediated signaling.

In hippocampal cultures, BDNF also potentiates release from presynaptic nerve terminals, but potentiation depends on inositol tris-phosphate gated Ca^{2+} stores in presynaptic nerve terminals (Li et al 1998a,b). LTP of synaptic transmission at the CA1 synapse in hippocampal slices has been seen by some, but not all, workers (e.g. Kang & Schuman 1995, Figurov et al 1996). The difference appears to be caused by differences in slice culture and conditions of application of BDNF. BDNF must be applied rapidly to slices to observe potentiation (Kang et al 1996). In conditions where it is observed, LTP also depends on inositol tris-phosphate gated Ca^{2+} stores, but in addition it requires local protein synthesis (Kang & Schuman 1996, 2000). In studies using slices of postnatal rat visual cortex, potentiation of synaptic transmission from layer IV cells to cells in layers II/III was seen only with very high concentrations of BDNF (Akaneya et al 1997). At lower concentrations, BDNF enhanced the magnitude of LTP without potentiating basal synaptic transmission. In both the hippocampus and postnatal visual cortex, BDNF enhances LTP in conditions where it does not potentiate synaptic transmission. The acute changes in BDNF concentration needed to potentiate synaptic transmission probably occur only rarely in vivo. BDNF also potentiates transmitter release from brain synaptosomes (Jovanovic et al 2000). In this case, it has been elegantly demonstrated that a MAP kinase phosphorylation of synapsin I mediates this response. The response is not seen in synaptosomes isolated from a *synapsin I* mutant and is prevented by inhibitors of the MAP kinase cascade. Phosphorylation of synapsins by MAP kinase has been shown to regulate their interactions with the actin cytoskeleton (Jovanovic et al 1996), so the MAP kinase cascade may potentiate synaptic transmission by releasing synaptic vesicles from the cytoskeleton, facilitating their entry into a exocytosis-competent pool. It will be interesting to determine whether BDNF-to-TrkB signaling regulates generation of LTP in mice lacking synapsin I.

CONCLUSION

The discovery of NGF and the neurotrophins was made possible by the observation that many populations of neurons depend on cell-cell interactions, specifically neuron-target interactions, for survival during embryonic development. The concept that limiting amounts of survival factors ensured a match between the number of neurons and the requirement for their functions was attractive. This model also provided a potential means for eliminating mistakes through death of neurons with aberrantly projecting axons. Thus, this was visualized as a mechanism for constructing the complex nervous systems of vertebrates. Subsequent studies have shown that these proteins are involved in many more aspects of neural development and function. Cell fate decisions, axon growth, dendrite pruning, synaptic function, and plasticity are all regulated by the neurotrophins. Studies of this small group of proteins have provided illuminating insights into most areas of contemporary neuroscience research and will almost certainly continue to do so in the future.

ACKNOWLEDGMENTS

We wish to thank our many colleagues who shared papers and preprints with us. Without their help, this review could not have been written. We apologize to many of these same colleagues for our inability to discuss all work in this field because of space and time constraints. We thank Ardem Patapoutian and Michael Stryker for comments on the manuscript. Work from our laboratories has been supported by the National Institutes of Health and by the Howard Hughes Medical Institute.

Visit the Annual Reviews home page at www.AnnualReviews.org

LITERATURE CITED

Acheson A, Conover JC, Fandl JP, DeChiara TM, Russell M, et al. 1995. A BDNF autocrine loop in adult sensory neurons prevents cell death. *Nature* 374:450–53

Agarwal ML, Taylor WR, Chernov MV, Chernova OB, Stark GR. 1998. The p53 network. *J. Biol. Chem.* 273:1–4

Airaksinen MS, Koltzenburg M, Lewin GR, Masu Y, Helbig C, et al. 1996. Specific subtypes of cutaneous mechanoreceptors require neurotrophin-3 following peripheral target innervation. *Neuron* 16:287–95

Akaneya Y, Tsumoto T, Kinoshinta S, Hatanaka H. 1997. Brain-derived neurotrophic factor enhances long-term potentiation in rat visual cortex. *J. Neurosci.* 17:6707–16

Alcantara S, Frisen J, del Rio JA, Soriano E, Barbacic M, Silos-Santiago I. 1997. TrkB signaling is required for postnatal survival of CNS neurons and protects hippocampal and motor neurons from axotomy-induced cell death. *J. Neurosci.* 17:3623–33

Allsopp TE, Robinson M, Wyatt S, Davies AM. 1994. TrkA mediates an NGF survival response in NGF-independent sensory neurons but not in parasympathetic neurons. *Gene Ther.* 1(Suppl. 1):S59

Aloyz RS, Bamji SX, Pozniak CD, Toma JG, Atwal J, et al. 1998. P53 is essential for developmental neuron death as regulated by the TrkA and p75 neurotrophin receptors. *J. Cell Biol.* 143:1691–703

Altar CA, Cai N, Bliven T, Juhasz M, Conner JM, et al. 1997. Anterograde transport of brain-derived neurotrophic factor and its role in the brain. *Nature* 389:856–60

Amaya F, Decosterd I, Samad TA, Plumpton C, Tate S, et al. 2000. Diversity of expression of the sensory neuron-specific TTX-resistant voltage-gated sodium ion channels SNS and SNS2. *Mol. Cell. Neurosci.* 16:331–42

Anderson DJ. 1993. Cell fate determination in the peripheral nervous system: the sympathoadrenal progenitor. *J. Neurobiol.b* 24:185–98

Anton ES, Weskamp G, Reichardt LF, Matthew WD. 1994. Nerve growth factor and its low-affinity receptor promote Schwann cell migration. *Proc. Natl. Acad. Sci. USA* 91: 2795–99

Araki T, Yamada M, Ohnishi H, Sano SI, Hatanaka H. 2000a. BIT/SHPD-1 enhances brain-derived neurotrophic factor-promoted neuronal survival in cultured cerebral cortical neurons. *J. Neurochem.* 75:1502–10

Araki T, Yamada M, Ohnishi H, Sano SI, Uetsuki T, Hatanaka H. 2000b. Shp-2 specifically regulates several tyrosine-phosphorylated proteins in brain-derived neurotrophic factor signaling in cultured cerebral cortical neurons. *J. Neurochem.* 74:659–68

Arch RH, Gedrich RW, Thompson CB. 1998.

Tumor necrosis factor receptor-associated factors (TRAFs)-a family of adapter proteins that regulates life and death. *Genes Dev.* 12:2821–30

Arvanov VL, Seebach BS, Mendell LM. 2000. NT-3 evokes an LTP-like facilitation of AMPA/kainate receptor-mediated synaptic transmission in the neonatal rat spinal cord. *J. Neurophysiol.* 84:752–58

Bamji SX, Majdan M, Pozniak CD, Belliveau DJ, Aloyz R, et al. 1998. The p75 neurotrophin receptor mediates neuronal apoptosis and is essential for naturally occurring sympathetic neuron death. *J. Cell Biol.* 140:911–23

Barres BA, Raff MC, Gaese F, Bartke I, Dechant G, Barde YA. 1994. A crucial role for neurotrophin-3 in oligodendrocyte development. *Nature* 367:371–75

Barrett GL, Bartlett PF. 1994. The p75 nerve growth factor receptor mediates survival or death depending on the stage of sensory neuron development. *Proc. Natl. Acad. Sci. USA* 91:6501–5

Baxter GT, Radeke MJ, Kuo RC, Makrides V, Hinkle B, et al. 1997. Signal transduction mediated by the truncated TrkB receptor isoforms, TrkB.T1 and TrkB.T2. *J. Neurosci.* 17:2683–90

Bazenet CE, Mota MA, Rubin LL. 1998. The small GTP-binding protein Cdc42 is required for nerve growth factor withdrawal-induced neuronal death. *Proc. Natl. Acad. Sci. USA* 95:3984–89

Beattie EC, Howe CL, Wilde A, Brodsky FM, Mobley WC. 2000. NGF signals through TrkA to increase clathrin at the plasma membrane and enhance clathrin-mediated membrane trafficking. *J. Neurosci.* 20:7325–33

Benedetti M, Levi A, Chao MV. 1993. Differential expression of nerve growth factor receptors leads to altered binding affinity and neurotrophin responsiveness. *Proc. Natl. Acad. Sci. USA* 90:7859–63

Bentley CA, Lee K-F. 2000. p75 is important for axon growth and Schwann cell migration

during development. *J. Neurosci.* 20:7706–15

Bergmann I, Priestley JV, McMahon SB, Bröcker EB, Toyka KV, Koltzenburg M. 1997. Analysis of cutaneous sensory neurons in transgenic mice lacking the low affinity neurotrophin receptor p75. *Eur. J. Neurosci.* 9:18–28

Bianchi LM, Conover JC, Fritzsch B, DeChiara T, Lindsay RM, Yancopoulos GD. 1996. Degeneration of vestibular neurons in late embryogenesis of both heterozygous and homozygous BDNF null mutant mice. *Development* 122:1965–73

Bibel M, Hoppe E, Barde YA. 1999. Biochemical and functional interactions between the neurotrophin receptors Trk and p75NTR. *EMBO J.* 18:616–22

Birren SJ, Lo L, Anderson DJ. 1993. Sympathetic neuroblasts undergo a developmental switch in trophic dependence. *Development* 119:597–610

Bishop AL, Hall A. 2000. Rho GTPases and their effector proteins. *Biochem. J.* 348 (Part 2):241–55

Boeshore KL, Luckey CN, Zigmond RE, Large TH. 1999. TrkB isoforms with distinct neurotrophin specificities are expressed in predominantly nonoverlapping populations of avian dorsal root ganglion neurons. *J. Neurosci.* 19:4739–47

Bonni A, Brunet A, West AE, Datta SR, Takasu MA, Greenberg ME. 1999. Cell survival promoted by the Ras-MAPK signaling pathway by transcription-dependent and -independent mechanisms. *Science* 286:1358–62

Bothwell M. 1995. Functional interactions of neurotrophins and neurotrophin receptors. *Annu. Rev. Neurosci.* 18:223–53

Boulanger L, Poo MM. 1999a. Gating of BDNF-induced synaptic potentiation by cAMP. *Science* 284:1982–84

Boulanger L, Poo MM. 1999b. Presynaptic depolarization facilitates neurotrophin-induced synaptic potentiation. *Nat. Neurosci.* 2:346–51

Brady R, Zaidi SI, Mayer C, Katz DM. 1999.

BDNF is a target-derived survival factor for arterial baroreceptor and chemoafferent primary sensory neurons. *J. Neurosci.* 19:2131–42

Brennan C, Rivas-Plata K, Landis SC. 1999. The p75 neurotrophin receptor influences NT-3 responsiveness of sympathetic neurons in vivo. *Nat. Neurosci.* 2:699–705

Brunet A, Bonni A, Zigmond MJ, Lin MZ, Juo P, et al. 1999. Akt promotes cell survival by phosphorylating and inhibiting a forkhead transcription factor. *Cell* 96:857–68

Butte MJ, Hwang PK, Mobley WC, Fletterick RJ. 1998. Crystal structure of neurotrophin-3 homodimer shows distinct regions are used to bind its receptors. *Biochemistry* 37:16846–52

Cabelli RJ, Shelton DL, Segal RA, Shatz CJ. 1997. Blockade of endogenous ligands of TrkB inhibits formation of ocular dominance columns. *Neuron* 19:63–76

Cacalano G, Fariñas I, Wang LC, Hagler K, Forgie A, et al. 1998. GFRalpha1 is an essential receptor component for GDNF in the developing nervous system and kidney. *Neuron* 21:53–62

Campenot RB. 1977. Local control of neurite development by nerve growth factor. *Proc. Natl. Acad. Sci. USA* 74:4516–19

Carroll P, Lewin GR, Koltzenberg M, Toyka KV, Thoenen H. 1998. A role for BDNF in mechanosensation. *Nat. Neurosci.* 1:42–46

Carter BD, Kalschmidt C, Kaltschmidt B, Offenhauser N, Bohm-Matthaei R, et al. 1996. Selective activation of NF-kB by nerve growth factor through neurotrophin receptor p75. *Science* 272:542–45

Casademunt E, Carter BD, Benzel I, Frade JM, Dechant G, Barde YA. 1999. The zinc finger protein NRIF interacts with the neurotrophin receptor p75NTR and participates in programmed cell death. *EMBO J.* 18:6050–61

Castellani V, Bolz J. 1999. Opposing roles for neurotrophin-3 in targeting and collateral formation of distinct sets of developing cortical neurons. *Development* 126:3335–45

Castren E, Zafra F, Thoenen H, Lindholm D.

1992. Light regulates expression of brain-derived neurotrophic factor mRNA in rat visual cortex. *Proc. Natl. Acad. Sci. USA* 89:9444–48

Cattaneo E, McKay R. 1990. Proliferation and differentiation of neuronal stem cells regulated by nerve growth factor. *Nature* 347:762–65

Cellerino A, Maffei L, Domenici L. 1996. The distribution of brain-derived neurotrophic factor and its receptor TrkB in parvalbumin-containing neurons of the rat visual cortex. *Eur. J. Neurosci.* 8:1190–97

Chao MV. 1994. The p75 neurotrophin receptor. *J. Neurobiol.* 25:1373–85

Chen J, Fu H. 2000. Regulation of apoptosis signal-regulating kinase (ASK)1-induced cell death by Raf-1. *Molec. Biol. Cell* 11(S):1807 (Abstr.)

Chen G, Kolbeck R, Barde YA, Bonhoeffer T, Kossel A. 1999. Relative contribution of endogenous neurotrophins in hippocampal long-term potentiation. *J. Neurosci.* 19:7983–90

Chen KS, Nishimura MC, Armanini MP, Crowley C, Spencer SD, Phillips HS. 1997. Disruption of a single allele of the nerve growth factor gene results in atrophy of basal forebrain cholinergic neurons and memory deficits. *J. Neurosci.* 17:7288–96

Chittka A, Chao MV. 1999. Identification of a zinc finger protein whose subcellular distribution is regulated by serum and nerve growth factor. *Proc. Natl. Acad. Sci. USA* 96:10705–10

Clary DO, Reichardt LF. 1994. An alternatively spliced form of the nerve growth factor receptor TrkA confers an enhanced response to neurotrophin 3. *Proc. Natl. Acad. Sci. USA* 91:11133–37

Connolly JL, Greene LA, Viscarello RR, Riley WD. 1979. Rapid, sequential changes in surface morphology of PC12 pheochromocytoma cells in response to nerve growth factor. *J. Cell Biol.* 82:820–27

Conover JC, Erickson JT, Katz DM, Bianchi LM, Poueymirou WT, et al. 1995. Neuronal deficits, not involving motor neurons, in mice lacking BDNF and/or NT4. *Nature* 375:235–38

Corbit KC, Foster DA, Rosner MR. 1999. Protein kinase Cδ mediates neurogenic but not mitogenic activation of mitogen-activated protein kinase in neuronal cells. *Mol. Cell. Biol.* 19:4209–18

Crowley C, Spencer SD, Nishimura MC, Chen KS, Pitts-Meek S, et al. 1994. Mice lacking nerve growth factor display perinatal loss of sensory and sympathetic neurons yet develop basal forebrain cholinergic neurons. *Cell* 76:1001–11

Curtis R, Adryan KM, Stark JL, Park JS, Compton DL, et al. 1995. Differential role of the low affinity neurotrophin receptor (p75) in retrograde axonal transport of the neurotrophins. *Neuron* 14:1201–11

Datta SR, Brunet A, Greenberg ME. 1999. Cellular survival: a play in three Akts. *Genes Dev.* 13:2905–27

Datta SR, Dudek H, Tao X, Masters S, Fu H, et al. 1997. Akt phosphorylation of BAD couples survival signals to the cell-intrinsic death machinery. *Cell* 91:1–20

Davey F, Davies AM. 1998. TrkB signalling inhibits p75-mediated apoptosis induced by nerve growth factor in embryonic proprioceptive neurons. *Curr. Biol.* 8:915–18

Davies AM, Minichiello L, Klein R. 1995. Developmental changes in NT3 signalling via TrkA and TrkB in embryonic neurons. *EMBO J.* 14:4482–89

Davis BM, Fundin BT, Albers KM, Goodness TP, Cronk KM, Rice FL. 1997. Overexpression of nerve growth factor in skin causes preferential increases among innervation to specific sensory targets. *J. Comp. Neurol.* 387:489–506

Davis BM, Goodness TP, Soria A, Albers KM. 1998. Over-expression of NGF in skin causes formation of novel sympathetic projections to TrkA-positive sensory neurons. *NeuroReport* 9:1103–7

Deckwerth TL, Elliott JL, Knudson CM, Johnson EM Jr, Snider WD, Korsmeyer SJ. 1996.

BAX is required for neuronal death after trophic factor deprivation and during development. *Neuron* 17:401–11

Desai NS, Rutherford LC, Turrigiano GG. 1999. BDNF regulates the intrinsic excitability of cortical neurons. *Learn. Mem.* 6:284–91

Dobrowsky RT, Jenkins GM, Hannun YA. 1995. Neurotrophins induce sphingomyelin hydrolysis. Modulation by co-expression of p75^{NTR} with Trk receptors. *J. Biol. Chem.* 270:22135–42

Donovan MJ, Hahn R, Tessarollo L, Hempstead BL. 1996. Identification of an essential nonneuronal function of neurotrophin 3 in mammalian cardiac development. *Nat. Genet.* 14:210–13

Donovan MJ, Lin MI, Wiege P, Ringstedt T, Kraemer R, et al. 2000. Brain derived neurotrophic factor is an endothelial cell survival factor required for intramyocardial vessel stabilization. *Development.* 127:4531–40

Durbec PL, Larsson-Blomberg LB, Schuchardt A, Costantini F, Pachnis V. 1996. Common origin and developmental dependence on c-ret of subsets of enteric and sympathetic neuroblasts. *Development* 122:349–58

Edery P, Lyonnet S, Mulligan LM, Pelet A, Dow E, et al. 1994. Mutations of the RET proto-oncogene in Hirschsprung's disease. *Nature* 367:378–80

Edwards RH, Rutter WJ, Hanahan D. 1989. Directed expression of NGF to pancreatic beta cells in transgenic mice to selective hyperinnervation of the islets. *Cell* 58:161–70

Eide FF, Lowenstein DH, Reichardt LF. 1993. Neurotrophins and their receptors: Current concepts and implications for neurologic disease. *Exp. Neurol.* 121:200–14

Eide FF, Vining ER, Eide BL, Zang K, Wang XY, Reichardt LF. 1996. Naturally occurring truncated TrkB receptors have dominant inhibitory effects on brain-derived neurotrophic factor signaling. *J. Neurosci.* 16:3123–29

Eilers A, Whitfield J, Babji C, Rubin LL, Ham J. 1998. Role of the jun kinase pathway in the regulation of c-jun expression and apoptosis in sympathetic neurons. *J. Neurosci.* 18:1713–24

ElShamy WM, Linnarsson S, Lee K-F, Jaenisch R, Ernfors P. 1996. Prenatal and postnatal requirements of NT-3 for sympathetic neuroblast survival and innervation of specific targets. *Development* 122:491–500

Enokido Y, Wyatt S, Davies AM. 1999. Developmental changes in the response of trigeminal neurons to neurotrophins: influence of birth date and the ganglion environment. *Development* 126:4365–73

Enomoto H, Araki T, Jackman A, Heuckeroth RO, Snider WD, et al. 1998. GFR alpha1-deficient mice have deficits in the enteric nervous system and kidneys. *Neuron* 21:317–24

Enomoto H, Heuckeroth RO, Golden JP, Johnson Jr. EM, Milbrandt J. 2000. Development of cranial parasympathetic ganglia requires sequential actions of GDNF and neurturin. *Development* 127:4877–89

Erickson JT, Conover JC, Borday V, Champagnat J, Barbacid M, et al. 1996. Mice lacking brain-derived neurotrophic factor exhibit visceral sensory neuron losses distinct from mice lacking NT4 and display a severe developmental deficit in control of breathing. *J. Neurosci.* 16:5361–71

Ernfors P, Lee KF, Jaenisch R. 1994a. Mice lacking brain-derived neurotrophic factor develop with sensory deficits. *Nature* 368:147–50

Ernfors P, Lee KF, Kucera J, Jaenisch R. 1994b. Lack of neurotrophin-3 leads to deficiencies in the peripheral nervous system and loss of limb proprioceptive afferents. *Cell* 77:503–12

Ernfors P, Van De Water T, Loring J, Jaenisch R. 1995. Complementary roles of BDNF and NT-3 in vestibular and auditory development. *Neuron* 14:1153–64. Erratum. 1995. *Neuron* 15(3):739

Fagan AM, Zhang H, Landis SC, Smeyne RJ, Silos-Santiago I, Barbacid M. 1996. TrkA, but not TrkC, receptors are essential for survival of sympathetic neurons in vivo. *J. Neurosci.* 16:6208–18

Fan G, Copray S, Huang EJ, Jones K, Yan Q, et al. 2000. Formation of a full complement of cranial proprioceptors requires multiple neurotrophins. *Dev. Dyn.* 218(2):359–70

Fandl JP, Tobkes NJ, McDonald NQ, Hendrickson WA, Ryan TE, et al. 1994. Characterization and crystallization of recombinant human neurotrophin-4. *J. Biol. Chem.* 269: 755–59

Farhadi HF, Mowla SJ, Petrecca K, Morris SJ, Seidah NG, Murphy RA. 2000. Neurotrophin-3 sorts to the constitutive secretory pathway of hippocampal neurons and is diverted to the regulated secretory pathway by coexpression with brain-derived neurotrophic factor. *J. Neurosci.* 20:4059–68

Fariñas I, Jones KR, Backus C, Wang XY, Reichardt LF. 1994. Severe sensory and sympathetic deficits in mice lacking neurotrophin-3. *Nature* 369:658–61

Fariñas I, Wilkinson GA, Backus C, Reichardt LF, Patapoutian A. 1998. Characterization of neurotrophin and Trk receptor functions in developing sensory ganglia: direct NT-3 activation of TrkB neurons in vivo. *Neuron* 21:325–34

Fariñas I, Yoshida CK, Backus C, Reichardt LF. 1996. Lack of neurotrophin-3 results in death of spinal sensory neurons and premature differentiation of their precursors. *Neuron* 17:1065–78

Fawcett JP, Bamji SX, Causing CG, Aloyz R, Ase AR, et al. 1998. Functional evidence that BDNF is an anterograde neuronal trophic factor in the CNS. *J. Neurosci.* 18:2808–21

Figurov A, Pozzo-Miller LD, Olafsson P, Want T, Lu B. 1996. Regulation of synaptic resposnes to high-frequency stimulation and LTP by neurotrophins in the hippocampus. *Nature* 318:706–9

Fischer W, Wictorin K, Bjorklund A, Williams LR, Varon S, Gage FH. 1987. Amelioration of cholinergic neurons atrophy and spatial memory impairment in aged rats by nerve growth factor. *Nature* 329:65–68

Fjell J, Cummins TR, Davis BM, Albers KM, Fried K, et al. 1999. Sodium channel expression in NGF-overexpressing transgenic mice. *J. Neurosci. Res.* 57:39–47

Frade JM, Barde YA. 1998. Nerve growth factor: two receptors, multiple functions. *BioEssays* 20:137–45

Frade JM, Barde YA. 1999. Genetic evidence for cell death mediated by nerve-growth factor and the neurotrophin receptor p75 in the developing mouse retina and spinal cord. *Development* 126:683–90

Francis N, Fariñas I, Rivas-Plata K, Backus C, Reichardt LF, Landis S. 1999. NT-3, like NGF, is required for survival of sympathetic neurons, but not their precursors. *J. Dev. Biol.* 210:411–27

Frerking M, Malenka RC, Nicoll RA. 1998. Brain-derived neurotrophic factor (BDNF) modulates inhibitory, but not excitatory, transmission in the CA1 region of the hippocampus. *J. Neurophysiol.* 80:3383–86

Friedman W. 2000. Neurotrophins induce death of hippocampal neurons via the p75 receptor. *J. Neurosci.* 20:6340–46

Fritzsch B, Barbacid M, Silos-Santiago I. 1998. The combined effects of TrkB and TrkC mutations on the innervation of the inner ear. *Int. J. Dev. Neurosci.* 16:493–505

Fritzsch B, Fariñas I, Reichardt LF. 1997. Lack of neurotrophin 3 causes losses of both classes of spiral ganglion neurons in the cochlea in a region-specific fashion. *J. Neurosci.* 17:6213–25

Fundin BT, Mikaels A, Westphal H, Ernfors P. 1999. A rapid and dynamic regulation of GDNF-family ligands and receptors correlate with the developmental dependency of cutaneous sensory innervation. *Development* 126:2597–610

Fundin BT, Silos-Santiago I, Ernfors P, Fagan AM, Aldskogius H, et al. 1997. Differential dependency of cutaneous mechanoreceptors on neurotrophins, Trk receptors, and P75 LNGFR. *Dev. Biol.* 190:94–116

Garces A, Haase G, Airaksiner MS, Livet J, Filippi P, de Lypeyriere O. 2000. GFRα1 is required for differentiation of distinct subpopulations of motoneuron. *J. Neurosci.* 20:4992–5000

Garner AS, Menegay HJ, Boeshore KL, Xie XY, Voci JM, et al. 1996. Expression of TrkB receptor isoforms in the developing avian visual system. *J. Neurosci.* 16:1740–52

Ghosh A, Greenberg ME. 1995. Distinct roles for bFGF and NT-3 in the regulation of cortical neurogenesis. *Neuron* 15:89–103

Gillespie DC, Crair MC, Stryker MP. 2000. Neurotrophin-4/5 alters responses and blocks the effect of monocular deprivation in cat visual cortex during the critical period. *J. Neurosci.* 20:9174–86

Gottschalk W, Pozzo-Miller LD, Figurov A, Lu B. 1998. Presynaptic modulation of synaptic transmission and plasticity by brain-derived neurotrophic factor in the developing hippocampus. *J. Neurosci.* 18:6830–39

Gotz R, Koster R, Winkler C, Raulf F, Lottspeich F, et al. 1994. Neurotrophin-6 is a new member of the nerve growth factor family. *Nature* 372:266–69

Grewal SS, York R, Stork PJS. 1999. Extracellular signal-regulated kinase signalling in neurons. *Curr. Opin. Neurobiol.* 9:544–53

Grimes ML, Beattie E, Mobley WC. 1997. A signaling organelle containing the nerve growth factor-activated receptor tyrosine kinase, TrkA. *Proc. Natl. Acad. Sci. USA* 94:9909–14

Guidry G, Landis SC, Davis BM, Albers KM. 1998. Overexpression of nerve growth factor in epidermis disrupts the distribution and properties of sympathetic innervation in footpads. *J. Comp. Neurol.* 393:231–43

Guiton M, Gunn-Moore FJ, Glass DJ, Geis DR, Yancopoulos GD, Tavare JM. 1995. Naturally occurring tyrosine kinase inserts block high affinity binding of phospholipase C gamma and Shc to TrkC and neurotrophin-3 signaling. *J. Biol. Chem.* 270:20384–90

Gundersen RW, Barrett JN. 1979. Neuronal hemotaxis: chick dorsal root ganglion axons turn toward high concentrations of nerve growth factor. *Science* 206:1079–80

Hallbook F. 1999. Evolution of the vertebrate

neurotrophin and Trk receptor gene families. *Curr. Opin. Neurobiol.* 9:616–21

Hamanoue M, Middleton G, Wyatt S, Jaffray E, Hay RT, Davies AM. 1999. p75-mediated NF-kB activation enhances the survival response of developing sensory neurons to nerve growth factor. *Mol. Cell. Neurosci.* 14:28–40

Hanover JL, Huang ZJ, Tonegawa S, Stryker MP. 1999. Brain-derived neurotrophic factor overexpression induces precocious critical period in mouse visual cortex. *J. Neurosci.* 19:U12–16 (Online)

Hapner SJ, Boeshore KL, Large TH, Lefcort F. 1998. Neural differentiation promoted by truncated TrkC receptors in collaboration with p75(NTR). *Dev. Biol.* 201:90–100

Harrison SMW, Jones ME, Uecker S, Albers KM, Kudrycki KE, Davis BM. 2000. Levels of nerve growth factor and neurotrophin-3 are affected differentially by the presence of p75 in sympathetic neurons in vivo. *J. Comp. Neurol.* 424:99–110

Hefti F, Hartikka J, Knussel B. 1989. Function of neruotrophic factors in the adult and aging brain and their possible use in the treatment of neurodegenerative diseases. *Neurobiol. Aging* 10:515–33

Hempstead BL, Birge RB, Fajardo JE, Glassman R, Mahadeo D, et al. 1994. Expression of the v-crk oncogene product in PC12 cells results in rapid differentiation by both nerve growth factor- and epidermal growth factor-dependent pathways. *Mol. Cell. Biol.* 14:1964–71

Hensch T, Fagiolini M, Mataga N, Stryker MP, Baekkeskov S, Kash SF. 1998. Local GABA circuit control of expreience-dependent plasticity in developing visual cortex. *Science* 282:1504–8

Hetman M, Cavanaugh JE, Kimelman D, Xia Z. 1999. Role of glycogen synthase kinase-3β in neuronal apoptosis induced by tophic withdrawal. *J. Neurosci.* 20:2567–74

Heuckeroth RO, Enomoto H, Grider JR, Golden JP, Hanke JA, et al. 1999. Gene targeting reveals a critical role for neurturin in the de-

velopment and maintenance of enteric, sensory, and parasympathetic neurons. *Neuron* 22:253–63

Holgado-Madruga M, Moscatello DK, Emlet DR, Dieterich R, Wong AJ. 1997. Grb2-associated binder-1 mediates phosphatidylinositol-3-OH kinase activation and the promotion of cell survival by nerve growth factor. *Proc. Natl. Acad. Sci. USA* 94:12419–24

Horch HW, Kruttgen A, Portbury SD, Katz LC. 1999. Destabilization of cortical dendrites and spines by BDNF. *Neuron* 23:353–64

Hu S, Reichardt LF. 1999. From membrane to cytoskeleton: enabling a connection. *Neuron* 22:419–22

Huang EJ, Wilkinson GA, Fariñas I, Backus C, Zang K, et al. 1999a. Expression of Trk receptors in the developing mouse trigeminal ganglion: In vivo evidence for NT-3 activation of TrkA and TrkB in addition to TrkC. *Development* 126:2191–2203

Huang EJ, Zang K, Schmidt A, Saulys A, Xiang M, Reichardt LF. 1999b. POU domain factor Brn-3a controls the differentiation and survival of trigeminal neurons by regulating Trk receptor expression. *Development* 126:2869–82

Huang ZJ, Kirkwood A, Pizzorusso T, Porciatti V, Morales B, et al. 1999c. BDNF regulates the maturation of inhibition and the critical period of plasticity in mouse visual cortex. *Cell* 98:739–55

Inagaki N, Thoenen H, Lindholm D. 1995. TrkA tyrosine residues involved in NGF-induced neurite outgrowth of PC12 cells. *Eur. J. Neurosci.* 7:1125–33

Ip NY, Stitt TN, Tapley P, Klein R, Glass DJ, et al. 1993. Similarities and differences in the way neurotrophins interact with the Trk receptors in neuronal and nonneuronal cells. *Neuron* 10:137–49

Jiang X, Rowitch DH, Soriano P, McMahon AP, Sucov HM. 2000. Fate of the mammalian cardiac neural crest. *Development* 127:1607–16

Jones KR, Fariñas I, Backus C, Reichardt LF. 1994. Targeted disruption of the BDNF gene perturbs brain and sensory neuron development but not motor neuron development. *Cell* 76:989–99

Jovanovic JN, Benfenati F, Siow YL, Sihra TS, Sanghera JS, et al. 1996. Neurotrophins stimulate phosphorylation of synapsin I by MAP kinase and regulate synapsin I-actin interactions. *Proc. Natl. Acad. Sci. USA* 93:3679–83

Jovanovic JN, Czernik AJ, Fienberg AA, Greengard P, Sihra TS. 2000. Synapsins as mediators of BDNF-enhanced neurotransmitter release. *Nat. Neurosci.* 3:323–29

Kanamoto T, Mota MA, Takeda K, Rubin LL, Miyazopo K, et al. 2000. Role of apoptosis signal-regulating kinase in regulation of the c-Jun N-terminal kinase pathway and apoptosis in sympathetic neurons. *Mol. Cell. Biol.* 20:196–204

Kang H, Jia LZ, Suh KY, Tang L, Schuman EM. 1996. Determinants of BDNF-induced hippocampal synaptic plasticity: Role of the TrkB receptor and the kinetics of neurotrophin delivery. *Learn. Mem.* 3:188–96

Kang H, Schuman EM. 1995. Long-lasting neurotrophin-induced enhancement of synaptic transmission in the adult hippocampus. *Science* 267:1658–62

Kang H, Schuman EM. 1996. A requirement for local protein synthesis in neurotrophin-induced hippocampal synaptic plasticity. *Science* 273:1402–6

Kang H, Schuman EM. 2000. Intracellular Ca(2+) signaling is required for neurotrophin-induced potentiation in the adult rat hippocampus. *Neurosci. Lett.* 282:141–44

Kaplan DR, Miller FD. 2000. Neurotrophin signal transduction in the nervous system. *Curr. Opin. Neurobiol.* 10:381–91

Khursigara G, Orlinick JR, Chao MV. 1999. Association of the p75 neurotrophin receptor with TRAF6. *J. Biol. Chem.* 274:2597–600

Kim W-Y, Fritzsch B, Serls A, Bakel LA, Huang EJ, et al. 2001. NeuroD-null mice are deaf due to a severe loss of the inner ear sensory neurons during development. *Development* 128:417–26

Kim JYH, Sutton ME, Lu DT, Cho TA, Goumnerova LC, et al. 1999. Activation of neurotrophin-3 receptor TrkC induces apoptosis in medulloblastomas. *Cancer Res.* 59:711–19

Kinkelin I, Stucky CL, Koltzenburg M. 1999. Postnatal loss of Merkel cells, but not of slowly adapting mechanoreceptors in mice lacking the neurotrophin receptor p75. *Eur. J. Neurosci.* 11:3963–69

Kjoller L, Hall A. 1999. Signaling to Rho GTPases. *Exp. Cell Res.* 253:166–79

Kohn J, Aloyz RS, Toma JG, Haak-Frendsho M, Miller FD. 1999. Functionally antagonistic interactions between the TrkA and p75 neurotrophin receptors regulate sympathetic neuron growth and target innervation. *J. Neurosci.* 19:5393–408

Kornblum HI, Sankar R, Shin DH, Wasterlain CG, Gall CM. 1997. Induction of brain derived neurotrophic factor mRNA by seizures in neonatal and juvenile rat brain. *Brain Res. Mol. Brain Res.* 44:219–28

Korsching S. 1993. The neurotrophic factor concept: a reexamination. *J. Neurosci.* 13:2739–48

Korte M, Carroll P, Wolf E, Brem G, Thoenen H, Bonhoeffer T. 1995. Hippocampal long-term potentiation is impaired in mice lacking brain-derived neurotrophic factor. *Proc. Natl. Acad. Sci. USA* 92:8856–60

Korte M, Kang H, Bonhoeffer T, Schuman E. 1998. A role for BDNF in the late-phase of hippocampal long-term potentiation. *Neuropharmacology* 37:553–59

Kuan CY, Yang DD, Roy DRS, Davis RJ, Rakic P, Flavell RA. 1999. The Jnk1 and Jnk2 protein kinases are required for regional specific apoptosis during early brain development. *Neuron* 22:667–76

Kucera J, Ernfors P, Walro J, Jaenisch R. 1995. Reduction in the number of spinal motor neurons in neurotrophin-3-deficient mice. *Neuroscience* 69:321–30

Kuruvilla R, Ye H, Ginty DD. 2000. Spatially and functionally distinct roles of the P13-K effector pathway during NGF signaling in sympathetic neurons. *Neuron.* 27:499–512

Large TH, Bodary SC, Clegg DO, Weskamp G,

Otten U, Reichardt LF. 1986. Nerve growth factor gene expression in the developing rat brain. *Science* 234:352–55

Lee KF, Bachman K, Landis S, Jaenisch R. 1994a. Dependence on p75 for innervation of some sympathetic targets. *Science* 263:1447–49

Lee KF, Davies AM, Jaenisch R. 1994b. p75-deficient embryonic dorsal root sensory and neonatal sympathetic neurons display a decreased sensitivity to NGF. *Development* 120:1027–33

Le-Niculescu H, Bonfoco E, Kasuya Y, Claret FX, Green D, Karin M. 1999. Withdrawal of survival factors results in activation of the JNK pathway in neuronal cells leading to Fas ligand induction and cell death. *Mol. Cell. Biol.* 19:751–63

Levi-Montalcini R. 1987. The nerve growth factor 35 years later. *Science* 237:1154–62

Levi-Montalcini R, Skaper SD, Dal Toso R, Petrelli L, Leon A. 1996. Nerve growth factor: from neurotrophin to neurokine. *Trends Neurosci.* 19:514–20

Levine ES, Crozier R, Black IB, Plummer MR. 1998. Brain-derived neurotrophic factor modulates hippocampal synaptic transmission by increasing N-methyl-D-aspartic acid receptor activity. *Proc. Natl. Acad. Sci. USA* 95:10235–9

Lewin GR, Barde YA. 1996. Physiology of the neurotrophins. *Annu. Rev. Neurosci.* 19:289–317

Lewin GR, Mendell LM. 1996. Maintenance of modality-specific connections in the spinal cord after neonatal nerve growth factor deprivation. *Eur. J. Neurosci.* 8:1677–84

Li YX, Xu Y, Ju D, Lester HA, Davidson N, Schuman EM. 1998a. Expression of a dominant negative TrkB receptor, T1, reveals a requirement for presynaptic signaling in BDNF-induced synaptic potentiation in cultured hippocampal neurons. *Proc. Natl. Acad. Sci. USA* 95:10884–89

Li YX, Zhang Y, Lester HA, Schuman EM, Davidson N. 1998b. Enhancement of neurotransmitter release induced by brain-derived neurotrophic factor in cultured hippocampal neurons. *J. Neurosci.* 18:10231–40

Li Z, Van Aelst L, Cline HT. 2000. Rho GTPases regulate distinct aspects of dendritic arbor growth in Xenopus central neurons in vivo. *Nat. Neurosci.* 3:217–25

Liebl DJ, Klesse LJ, Tessarollo L, Wohlman T, Parada LF. 2000. Loss of brain-derived neurotrophic factor-dependent neural crest-derived sensory neurons in neurotrophin-4 mutant mice. *Proc. Natl. Acad. Sci. USA* 97(5):2297–302

Liepinsh E, Ilag LL, Otting G, Ibanez CF. 1997. NMR structure of the death domain of the p75 neurotrophin receptor. *EMBO J.* 16:4999–5005

Lin MI, Das I, Schwartz GM, Tsoulfas P, Mikawa T, Hempstead BL. 2000. Trk C receptor signaling regulates cardiac myocyte proliferation during early heart development in vivo. *Dev. Biol.* 226:180–91

Liu BP, Burridge K. 2000. Vav2 activates rac1, cdc42, and RhoA downstream from growth factor receptors but not beta1 integrins. *Mol. Cell. Biol.* 20:7160–69

Liu X, Jaenisch R. 2000. Severe peripheral sensory neuron loss and modest motor neuron reduction in mice with combined deficiency of brain-derived neurotrophic factor, neurotrophin 3 and neurotrophin 4/5. *Dev. Dyn.* 218:94–101

Lom B, Cohen-Cory S. 1999. Brain-derived neurotrophic factor differentially regulates retinal ganglion cell dendritic and axonal arborization in vivo. *J. Neurosci.* 19:9928–38

Ma Q, Fode C, Guillemot F, Anderson DJ. 1999. Neurogenin-1 and neurogenin-2 control two distinct waves of neurogenesis in developing dorsal root ganglia. *Genes Dev.* 13:1717–28

Ma L, Merenmies J, Parada LF. 2000. Molecular characterization of the TrkA/NGF receptor minimal enhancer reveals regulation by multiple cis elements to drive embryonic neuron expression. *Development* 127:3777–88.

MacDonald JI, Gryz EA, Kubu CJ, Verdi JM, Meakin SO. 2000. Direct binding of the signaling adapter protein Grb2 to the activation loop tyrosines on the nerve growth factor receptor tyrosine kinase, TrkA. *J. Biol. Chem.* 275:18225–33

MacPhee IJ, Barker PA. 1997. Brain-derived neurotrophic factor binding to the p75 neurotrophin receptor reduces TrkA signaling whereas increasing serine phosphorylation in the TrkA intracellular domain. *J. Biol. Chem.* 272:23547–51

Maggirwar SDB, Sarmiere PD, Dewhurst S, Freeman RS. 1998. Nerve growth factor-dependent activation of NF-kb contributes to survival of sympathetic neurons. *J. Neurosci.* 18:10356–65

Mahadeo D, Kaplan L, Chao MV, Hempstead BL. 1994. High affinity nerve growth factor binding displays a faster rate of association than p140Trk binding. Implications for multi-subunit polypeptide receptors. *J. Biol. Chem.* 269:6884–91

Malenka RC, Nicoll RA. 1999. Long-term potentiation–a decade of progress? *Science* 285:1870–74

Mannion RJ, Costigan M, Decosterd I, Amaya F, Ma QP, et al. 1999. Neurotrophins: peripherally and centrally acting modulators of tactile stimulus-induced inflammatory pain hypersensitivity. *Proc. Natl. Acad. Sci. USA* 96:9385–90

Matsuda M, Hashimoto Y, Muroya K, Hasegawa H, Kurata T, et al. 1994. CRK protein binds to two guanine nucleotide-releasing proteins for the Ras family and modulates nerve growth factor-induced activation of Ras in PC12 cells. *Mol. Cell. Biol.* 14:5495–500

Matsuo S, Ichikawa H, Silos-Santiago I, Arends JJ, Henderson TA, et al. 2000. Proprioceptive afferents survive in the masseter muscle of TrkC knockout mice. *Neuroscience* 95:209–16

Mazzoni IE, Said FA, Aloyz R, Miller FD, Kaplan D. 1999. Ras regulates sympathetic neuron survival by suppressing the p53-mediated cell death pathway. *J. Neurosci.* 19:9716–27

McAllister AK, Katz LC, Lo DC. 1996. Neu-

rotrophin regulation of cortical dendritic growth requires activity. *Neuron* 17:1057–64

McAllister AK, Katz LC, Lo DC. 1997. Opposing roles for endogenous BDNF and NT-3 in regulating cortical dendritic growth. *Neuron* 18:767–78

McAllister AK, Katz LC, Lo DC. 1999. Neurotrophins and synaptic plasticity. *Annu. Rev. Neurosci.* 22:295–318

McAllister AK, Lo DC, Katz LC. 1995. Neurotrophins regulate dendritic growth in developing visual cortex. *Neuron* 15:791–803

McDonald NQ, Chao MV. 1995. Structural determinants of neurotrophin action. *J. Biol. Chem.* 270:19669–72

McDonald NQ, Lapatto R, Murray-Rust J, Gunning J, Wlodawer A, Blundell TL. 1991. New protein fold revealed by a 2.3-A resolution crystal structure of nerve growth factor. *Nature* 354:411–14

McEvilly RJ, Erkman L, Luo L, Sawchenko PE, Ryan AF, Rosenfeld MG. 1996. Requirement for Brn-3.0 in differentiation and survival of sensory and motor neurons. *Nature* 384:574–77

Meakin SO, MacDonald JI. 1998. A novel juxtamembrane deletion in rat TrkA blocks differentiative but not mitogenic cell signaling in response to nerve growth factor. *J. Neurochem.* 71:1875–88

Meakin SO, Gryz EA, MacDonald JI. 1997. A kinase insert isoform of rat TrkA supports nerve growth factor-dependent cell survival but not neurite outgrowth. *J. Neurochem.* 69:954–67

Meakin SO, MacDonald JIS, Gryz EA, Kubu CJ, Verdi JM. 1999. The signaling adapter FRS-2 competes with Shc for binding to the nerve growth factor receptor TrkA. *J. Biol. Chem.* 274:9861–70

Meakin SO, Suter U, Drinkwater CC, Welcher AA, Shooter EM. 1992. The rat Trk protooncogene product exhibits properties characteristic of the slow nerve growth factor receptor. *Proc. Natl. Acad. Sci. USA* 89:2374–78

Mendell LM. 1999. Neurotrophin action on

sensory neurons in adults: an extension of the neurotrophic hypothesis. *Pain* (Suppl. 6):S127–32

Mendell LM, Albers KM, Davis BM. 1999. Neurotrophins, nociceptors, and pain. *Microsc. Res. Tech.* 45:252–61

Mesulam MM, Mufson EJ, Wainer BH, Levey AI. 1983. Central cholinergic pathways in the rat: an overview based on an alternative nomenclature (Ch10Ch6). *Neuroscience* 10:1185–201

Meyer-Franke A, Wilkinson GA, Kruttgen A, Hu M, Munro E, et al. 1998. Depolarization and cAMP elevation rapidly recruit TrkB to the plasma membrane of CNS neurons. *Neuron* 21:681–93

Michael GJ, Averill S, Nitkunan A, Rattray M, Bennett DL, et al. 1997. Nerve growth factor treatment increases brain-derived neurotrophic factor selectively in TrkA-expressing dorsal root ganglion cells and in their central terminations within the spinal cord. *J. Neurosci.* 17:8476–90

Middleton G, Hamanoue M, Enokido Y, Wyatt S, Pennica D, et al. 2000. Cytokine-induced nuclear factor kappa B activation promotes the survival of developing neurons. *J. Cell Biol.* 148:325–32

Ming G, Song H, Berninger B, Inagaki N, Tessier-Lavigne M, Poo MM. 1999. Phospholipase C-gamma and phosphoinositide 3-kinase mediate cytoplasmic signaling in nerve growth cone guidance. *Neuron* 23:139–48

Minichiello L, Casagranda F, Tatche RS, Stucky CL, Postigo A, et al. 1998. Point mutation in TrkB causes loss of NT4-dependent neurons without major effects on diverse BDNF responses. *Neuron* 21:335–45

Minichiello L, Klein R. 1996. TrkB and TrkC neurotrophin receptors cooperate in promoting survival of hippocampal and cerebellar granule neurons. *Genes Dev.* 10:2849–58

Minichiello L, Korte M, Wolfer D, Huhn R, Unsicker K, et al. 1999. Essential role for TrkB receptors in hippocampus-mediated learning. *Neuron* 24:401–14

Minichiello L, Piehl F, Vazquez E, Schimmang T, Hokfelt T, et al. 1995. Differential effects of combined Trk receptor mutations on dorsal root ganglion and inner ear sensory neurons. *Development* 121:4067–75

Molliver DC, Wright DE, Leitner M, Parsadanian AS, Doster K, et al. 1997. IB4-binding DRG neurons switch from NGF to GDNF dependence in early postnatal life. *Neuron* 19:849–61

Moore MW, Klein RD, Fariñas I, Sauer H, Armanini M, et al. 1996. Renal and neuronal abnormalities in mice lacking GDNF. *Nature* 382:76–79

Muller G, Storkz P, Bourtelle S, Doppler H, Pfizenmaier K, et al. 1998. Regulation of Raf-1 kinase by TNF via its second messenger ceramide and crosstalk with mitogenic signalling. *EMBO J.* 17:732–42

Nakayama AY, Harms MB, Luo L. 2000. Smal GTPases Rac and Rho in the maintenance of dendritic spines and branches in hippocampal pyramidal neurons. *J. Neurosci.* 20:5329–38

Natarajan D, Grigoriou M, Marcos-Gutierrez CV, Atkins C, Pachnis V. 1999. Multipotential progenitors of the mammalian enteric nervous system capable of colonising aganglionic bowel in organ culture. *Development* 126:157–68

Nilsson AS, Fainzilber M, Falck P, Ibanez CF. 1998. Neurotrophin-7: a novel member of the neurotrophin family from the zebrafish. *FEBS Lett.* 424:285–90

Nimnual AS, Yatsula BA, Bar-Sagi D. 1998. Coupling of Ras and Rac guanosine triphosphatases through the Ras exchanger Sos. *Science* 279:560–63

Nishino J, Mochida K, Ohfuji Y, Shimazaki T, Meno C, et al. 1999. GFR-alpha-3, a component of the artemin receptor, is required for migration and survival of the superior cervical ganglion. *Neuron* 23:725–36

Nosaka Y, Arai A, Miyasaka N, Miura O. 1999. CrkL mediates ras-dependent activation of the Raf/ERK pathway through the guanine nucleotide exchange factor C3G in hematopoietic cells stimulated with erythropoietin or interleukin-3. *J. Biol. Chem.* 274:30154–62

O'Connor R, Tessier-Lavigne M. 1999. Identification of maxillary factor, a maxillary process-derived chemoattractant for developing trigeminal sensory axons. *Neuron* 24:165–78

Oppenheim RW, Houenou LJ, Parsadanian AS, Prevette D, Snider WD, Shen L. 2000. Glial-cell line-derived neurotrophic factor and developing mammalian motoneurons: regulation of programmed cell death among motoneuron subtypes. *J. Neurosci.* 20:5001–11

Pachnis V, Durbec P, Taraviras S, Grigoriou M, Natarajan D. 1998. III. Role of the RET signal transduction pathway in development of the mammalian enteric nervous system. *Am. J. Physiol. Gastrointest. Liver Physiol.* 275:G183–86

Patapoutian A, Backus C, Kispert A, Reichardt LF. 1999. Regulation of neurotrophin-3 expression by epithelial-mesenchymal interactions: The role of wnt factors. *Science* 283:1180–83

Patterson SL, Abel T, Deuel TAS, Martin KC, Rose JC, Kandel ER. 1996. Recombinant BDNF rescues deficits in basal synaptic transmission and hippocampal LTP in BDNF knockout mice. *Neuron* 16:1137–45

Pawson T, Nash P. 2000. Protein-protein interactions define specificity in signal transduction. *Genes Dev.* 14:1027–47

Peterson DA, Dickinson-Anson HA, Leppert JT, Lee KF, Gage FH. 1999. Ceptral neuronal loss and behavioural impairment in mice lacking neurotrophin receptor p75. *J. Comp. Neurol.* 404:1–20

Pichel JG, Shen L, Sheng HZ, Granholm AC, Drago J, et al. 1996. Defects in enteric innervation and kidney development in mice lacking GDNF. *Nature* 382:73–76

Pozzo-Miller LD, Gottschalk W, Zhang L, Mc-Dermott K, Du J, et al. 1999. Impairments in high frequency transmission, synaptic vesicle docking and synaptic protein distribution

in the hippocampus of BDNF knockout mice. *J. Neurosci.* 19:4972–83

Purves D. 1988. *Body and Brain. A Trophic Theory of Neural Connections.* Cambridge, MA: Harvard Univ. Press

Qian X, Riccio A, Zhang Y, Ginty DD. 1998. Identification and characterization of novel substrates of Trk receptors in developing neurons. *Neuron* 21:1017–29

Raff MC, Barres BA, Burne JF, Coles HS, Ishizaki Y, Jacobson MD. 1993. Programmed cell death and the control of cell survival: lessons from the nervous system. *Science* 262:695–700

Reichardt LF, Fariñas I. 1997. Neurotrophic factors and their receptors. Roles in neuronal development and function. In *Molecular and Cellular Approaches to Neural Development,* ed. WM Cowan, TM Jessell, SL Zipursky, pp. 220–63. New York: Oxford Univ. Press

Riccio A, Ahn S, Davenport CM, Blendy JA, Ginty DD. 1999. Mediation by a CREB family transcription factor of NGF-dependent survival of sympathetic neurons. *Science* 286:2358–61

Riccio A, Pierchala BA, Ciarallo CL, Ginty DD. 1997. An NGF-TrkA-mediated retrograde signal to transcription factor CREB in sympathetic neurons. *Science* 277:1097–1100

Rice FL, Albers KM, Davis BM, Silos-Santiago I, Wilkinson GA, et al. 1998. Differential dependency of unmyelinated and A delta epidermal and upper dermal innervation on neurotrophins, Trk receptors, and p75LNGFR. *Dev. Biol.* 198:57–81

Riddle DR, Lo DC, Katz LC. 1995. NT-4 mediated rescue of lateral geniculate neurons from effects of monocular deprivation. *Nature* 378:189–91

Rifkin JT, Todd VJ, Anderson LW, Lefcort F. 2000. Dynamic expression of neurotrophin receptors during sensory neuron genesis and differentiation. *Dev. Biol.* 227:465–80

Ringstedt T, Ibáñez CF, Nosrat CA. 1999. Role of brain-derived neurotrophic factor in target invasion in the gustatory system. *J. Neurosci.* 19:3507–18

Ritter AM, Lewin GR, Kremer NE, Mendell LM. 1991. Requirement for nerve growth factor in the development of myelinated nociceptors in vivo. *Nature* 350:500–2

Robinson M, Buj-Bello A, Davies AM. 1996. Paracrine interactions of BDNF involving NGF-dependent embryonic sensory neurons. *Mol. Cell. Neurosci.* 7:143–51

Robinson RC, Radziejewski C, Spraggon G, Greenwald J, Kostura MR, et al. 1999. The structures of the neurotrophin 4 homodimer and the brain-derived neurotrophic factor/neurotrophin 4 heterodimer reveal a common Trk-binding site. *Protein Sci.* 8:2589–97

Robinson RC, Radziejewski C, Stuart DI, Jones EY. 1995. Structure of the brain-derived neurotrophic factor/neurotrophin 3 heterodimer. *Biochemistry* 34:4139–46

Rocamora N, Welker E, Pascual M, Soriano E. 1996. Upregulation of BDNF mRNA expression in the barrel cortex of adult mice after sensory stimulation. *J. Neurosci.* 16:4411–19

Rodriguez-Tebar A, Dechant G, Barde YA. 1991. Neurotrophins: structural relatedness and receptor interactions. *Philos. Trans. R. Soc. London Ser. B* 331:255–8

Romeo G, Ronchetto P, Luo Y, Barone V, Seri M, et al. 1994. Point mutations affecting the tyrosine kinase domain of the RET protooncogene in Hirschsprung's disease. *Nature* 367:377–8

Rossi J, Luukko K, Poteryaev D, Laurikainen A, Sun YF, et al. 1999. Retarded growth and deficits in the enteric and parasympathetic nervous system in mice lacking GFRalpha-2, a functional neurturin receptor. *Neuron* 22:243–52

Rueff A, Mendell LM. 1996. Nerve growth factor NT-5 induce increased thermal sensitivity of cutaneous nociceptors in vitro. *J. Neurophysiol.* 76:3593–96

Rutherford LC, DeWan A, Lauer HM, Turrigiano GG. 1997. Brain-derived neurotrophic factor mediates the activity-dependent regulation of inhibition in neocortical cultures. *J. Neurosci.* 17:4527–35

Rutherford LC, Nelson SB, Turrigiano GG.

1998. BDNF has opposite effects on the quantal amplitude of pyramidal neuron and interneuron excitatory synapses. *Neuron* 21:521–30

Ryden M, Murray-Rust J, Glass D, Ilag LL, Trupp M, et al. 1995. Functional analysis of mutant neurotrophins deficient in low-affinity binding reveals a role for p75LNGFR in NT-4 signalling. *EMBO J.* 14:1979–90

Salehi AH, Roux PP, Kubu CJ, Zeindleer C, Bhakar A, et al. 2000. NRAGE, a novel MAGE protein, interacts with the p75 neurotrophin receptor and facilitates nerve growth factor-dependent apoptosis. *Neuron* 27:279–88

Sanchez MP, Silos-Santiago I, Frisen J, He B, Lira SA, Barbacid M. 1996. Renal agenesis and the absence of enteric neurons in mice lacking GDNF. *Nature* 382:70–73

Saragovi HU, Zheng W, Maliartchouk S, DiGugliemo GM, Mawal YR, et al. 1998. A TrkA-selective, fast internalizing nerve growth factor-antibody complex induces trophic but not neuritogenic signals. *J. Biol. Chem.* 273:34933–40

Schimmang T, Minichiello L, Vazquez E, San Jose I, Giraldez F, et al. 1995. Developing inner ear sensory neurons require TrkB and TrkC receptors for innervation of their peripheral targets. *Development* 121:3381–91

Schinder AF, Berninger B, Poo MM. 2000. Postsynaptic target specificity of neurotrophin-induced presynaptic potentiation. *Neuron* 25:151–63

Seebach BS, Arvanov V, Mendell LM. 1999. Effects of BDNF and NT-3 on development of Ia/motoneuron functional connectivity in neonatal rats. *J. Neurophysiol.* 81:2398–2405

Segal RA, Greenberg ME. 1996. Intracellular signaling pathways activated by neurotrophic factors. *Annu. Res. Neurosci.* 19:463–89

Shelton DL, Sutherland J, Gripp J, Camerato T, Armanini MP, et al. 1995. Human Trks: molecular cloning, tissue distribution, and expression of extracellular domain immunoadhesins. *J. Neurosci.* 15:477–91

Shi ZQ, Yu DH, Park M, Marshall M, Feng GS. 2000. Molecular mechanism for the Shp-2 tyrosine phosphatase function in promoting growth factor stimulation of Erk activity. *Mol. Cell. Biol.* 20:1526–36

Shieh PB, Hu SC, Bobb K, Timmusk T, Ghosh A. 1998. Identification of a signaling pathway involved in calcium regulation of BDNF expression. *Neuron* 20:727–40

Shindler KS, Yunker AM, Cahn R, Zha J, Korsmeyer SJ, Roth KA. 1998. Trophic support promotes survival of bcl-x-deficient telencephalic cells in vitro. *Cell Death Differ.* 5:901–10

Shu XQ, Llinas A, Mendell LM. 1999. Effects of TrkB and TrkC neurotrophin receptor agonists on thermal nociception: a behavioral and electrophysiological study. *Pain* 80:463–70

Shu XQ, Mendell LM. 1999a. Neurotrophins and hyperalgesia. *Proc. Natl. Acad. Sci. USA* 96:7693–96

Shu XQ, Mendell LM. 1999b. Nerve growth factor acutely sensitizes the response of adult rat sensory neurons to capsaicin. *Neurosci. Lett.* 274:159–62

Sieber-Blum M. 1991. Role of the neurotrophic factors BDNF and NGF in the commitment of pluripotent neural crest cells. *Neuron* 6:949–55

Sieber-Blum M. 1999. The neural crest colony assay: assessing molecular influences in development in culture. In *The Neuron in Tissue Culture*, ed. LW Haynes, pp. 5–22. New York: Wiley

Singh TD, Mizuno K, Kohno T, Nakamura S. 1997. BDNF and TrkB mRNA expression in neurons of the neonatal mouse barrel field cortex: normal development and plasticity after cauterizing facial vibrissae. *Neurochem. Res.* 22:791–97

Smeyne RJ, Klein R, Schnapp A, Long LK, Bryant S, et al. 1994. Severe sensory and sympathetic neuropathies in mice carrying a disrupted Trk/NGF receptor gene. *Nature* 368:246–409

Snider WD, McMahon SB. 1998. Tackling pain at the source: new ideas about nociceptors. *Neuron* 20:629–32

Sobreviela T, Clary DO, Reichardt LF, Brandabur MM, Kordower JH, Mufson EJ. 1994. TrkA-immunoreactive profiles in the central nervous system: colocalization with neurons containing p75 nerve growth factor receptor, choline acetyltransferase, and serotonin. *J. Comp. Neurol.* 350:587–611

Sofroniew MV, Howe CL, Mobley WC. 2001. Nerve growth factor signaling, neuroprotection and neural repair. *Annu. Rev. Neurosci.* 24:1217–81

Song HJ, Ming GL, Poo MM. 1997. cAMP-induced switching in turning direction of nerve growth cones. *Nature* 388:275–79

Song HJ, Poo MM. 1999. Signal transduction underlying growth cone guidance by diffusible factors. *Curr. Opin. Neurobiol.* 9:355–63

Stephens RM, Loeb DM, Copeland TD, Pawson T, Greene LA, Kaplan DR. 1994. Trk receptors use redundant signal transduction pathways involving SHC and PLC-gamma 1 to mediate NGF responses. *Neuron* 12:691–705

Strohmaier C, Carter B, Urfer R, Barde YA, Dechant G. 1996. A splice variant of the neurotrophin receptor TrkB with increased specificity for brain-derived neurotrophic factor. *EMBO J.* 15:3332–37

Stucky CL, Koltzenburg M. 1997. The low-affinity neurotrophin receptor p75 regulates the function but not the selective survival of specific subpopulations of sensory neurons. *J. Neurosci.* 17:4398–4405

Stucky CL, Koltzenburg M, DeChiara T, Lindsay RM, Yancopoulos GD. 1998. Neurotrophin 4 is required for the survival of a subclass of hair follicle receptors. *J. Neurosci.* 18:7040–46

Stucky CL, Koltzenburg M, Schneider M, Engle MG, Albers KM, Davis BM. 1999. Overexpression of nerve growth factor in skin selectively affects the survival and functional properties of nociceptors. *J. Neurosci.* 19:8509–16

Stucky CL, Lewin GR. 1999. Isolectin B$_4$-positive and -negative nociceptors are functionally distinct. *J. Neurosci.* 19:6497–6505

Tanabe Y, Jessell TM. 1996. Diversity and pattern in the developing spinal cord. *Science* 274:1115–23

Tanaka S, Hattori S, Kurata T, Nagashima K, Fukui Y, et al. 1993. Both the SH2 and SH3 domains of human CRK protein are required for neuronal differentiation of PC12 cells. *Mol. Cell. Biol.* 13:4409–15

Tanaka T, Saito H, Matsuki N. 1997. Inhibition of GABAa synaptic responses by brain-derived neurotrophic factor (BDNF) in rat hippocampus. *J. Neurosci.* 17:2959–66

Tao X, Finkbeiner S, Arnold DB, Shaywitz AJ, Greenberg ME. 1998. Ca2+ influx regulates BDNF transcription by a CREB family transcription factor-dependent mechanism. *Neuron* 20:709–26. Erratum. 1998. *Neuron* 20(6):1297

Taraviras S, Marcos-Gutierrez CV, Durbec P, Jani H, Grigoriou M, et al. 1999. Signalling by the RET receptor tyrosine kinase and its role in the development of the mammalian enteric nervous system. *Development* 126:2785–97

Tessarollo L, Tsoulfas P, Donovan MJ, Palko ME, Blair-Flynn J, et al. 1997. Targeted deletion of all isoforms of the TrkC gene suggests the use of alternate receptors by its ligand neurotrophin-3 in neuronal development and implicates TrkC in normal cardiogenesis. *Proc. Natl. Acad. Sci. USA* 94:14776–81

Thoenen H, Barde YA. 1980. Physiology of nerve growth factor. *Physiol. Rev.* 60:1284–335

Toledo-Aral JJ, Brehm P, Halegoua S, Mandel G. 1995. A single pulse of nerve growth factor triggers long-term neuronal excitability through sodium channel gene induction. *Neuron* 14:607–11

Trachtenberg JT, Trepel C, Stryker MP. 2000. Rapid extragranular plasticity in the absence of thalamocortical plasticity in the developing primary visual cortex. *Science* 287:2029–32

Tsoulfas P, Stephens RM, Kaplan DR, Parada LF. 1996. TrkC isoforms with inserts in the kinase domain show impaired signaling responses. *J. Biol. Chem.* 271:5691–97

Tsui-Pierchala BA, Ginty DD. 1999. Characterization of an NGF-P-TrkA retrograde-signaling complex and age-dependent regulation of TrkA phosphorylation in sympathetic neurons. *J. Neurosci.* 19:8207–18

Ultsch MH, Wiesmann C, Simmons LC, Henrich J, Yang M, et al. 1999. Crystal structures of the neurotrophin-binding domain of TrkA, TrkB and TrkC. *J. Mol. Biol.* 290:149–59

Urfer R, Tsoulfas P, O'Connell L, Hongo JA, Zhao W, Presta LG. 1998. High resolution mapping of the binding site of TrkA for nerve growth factor and TrkC for neurotrophin-3 on the second immunoglobulin-like domain of the Trk receptors. *J. Biol. Chem.* 273:5829–40

Vaillant AR, Mazzoni I, Tudan C, Boudreau M, Kaplan DR, Miller FD. 1999. Depolarization and neurotrophins converge on the phosphatidylinositol 3-kinase-Akt pathway to synergistically regulate neuronal survival. *J. Cell Biol.* 146:955–66

Verdi JM, Anderson DJ. 1994. Neurotrophins regulate sequential changes in neurotrophin receptor expression by sympathetic neuroblasts. *Neuron* 13:1359–72

Verdi JM, Birren SJ, Ibanez CF, Persson H, Kaplan DR, et al. 1994. p75LNGFR regulates Trk signal transduction and NGF-induced neuronal differentiation in MAH cells. *Neuron* 12:733–45

Verdi JM, Groves AK, Fariñas I, Jones K, Marchionni MA, et al. 1996. A reciprocal cell-cell interaction mediated by NT-3 and neuregulins controls the early survival and development of sympathetic neuroblasts. *Neuron* 16:515–27

Vetter ML, Martin-Zanca D, Parada LF, Bishop JM, Kaplan DR. 1991. Nerve growth factor rapidly stimulates tyrosine phosphorylation of phospholipase C-gamma 1 by a kinase activity associated with the product of the Trk protooncogene. *Proc. Natl. Acad. Sci. USA* 88:5650–54

Vicario-Abejón C, Johe KK, Hazel TG, Collazo D, McKay RDG. 1995. Functions of basic fibroblast growth factor and neurotrophins in the differentiation of hippocampal neurons. *Neuron* 15:105–14

von Bartheld CS, Byers MR, Williams R, Bothwell M. 1996. Anterograde transport of neurotrophins and axodendritic transfer in the developing visual system. *Nature* 379:830–33

Walsh GS, Krol KM, Crutcher KA, Kawaja MD. 1999a. Enhanced neurotrophin-induced axon growth in myelinated portions of the CNS in mice lacking the p75 neurotrophin receptor. *J. Neurosci.* 19:4155–68

Walsh GS, Krol KM, Kawaja MD. 1999b. Absence of the p75 neurotrophin receptor alters the pattern of sympathosensory sprouting in the trigeminal ganglia of mice overexpressing nerve growth factor. *J. Neurosci.* 19:258–73

Wang XH, Poo MM. 1997. Potentiation of developing synapses by postsynaptic release of neurotrophin-4. *Neuron* 19:825–35

Wendland B, Emr SD, Riezman H. 1998. Protein traffic in the yeast endocytic and vacuolar protein sorting pathways. *Curr. Opin. Cell Biol.* 10:513–22

Wiese S, Metzger F, Holtmann B, Sendtner M. 1999. The role of p75NTR in modulating neurotrophin survival effects in developing motoneurons. *Eur. J. Neurosci.* 11:1668–76

Wiesmann C, Ultsch MH, Bass SH, de Vos AM. 1999. Crystal structure of nerve growth factor in complex with the ligand-binding domain of the TrkA receptor. *Nature* 401:184–88

Wilde A, Beattie EC, Lem L, Riethof DA, Liu SH, et al. 1999. EGF receptor signaling stimulates SRC kinase phosphorylation of clathrin, influencing clathrin redistribution and EGF uptake. *Cell* 96:677–87

Wilkinson GA, Fariñas I, Backus C, Yoshida CK, Reichardt LF. 1996. Neurotrophin-3 is a survival factor in vivo for early mouse trigeminal neurons. *J. Neurosci.* 16:7661–69

Woolf CJ, Costigan M. 1999. Transcriptional and posttranslational plasticity and the generation of inflammatory pain. *Proc. Natl. Acad. Sci. USA* 96:7723–30

Wright JH, Drueckes P, Bartoe J, Zhao Z, Shen SH, Krebs EG. 1997. A role for the SHP-2 tyrosine phosphatase in nerve growth-induced PC12 cell differentiation. *Mol. Biol. Cell* 8:1575–85

Wyatt S, Pinon LGP, Ernfors P, Davies AM. 1997. Sympathetic neuron survival and TrkA expression in NT3-deficient mouse embryos. *EMBO J.* 16:3115–23

Xia Z, Dickens M, Raingeaud J, Davis RJ, Greenberg ME. 1995. Opposing effects of ERK and JNK-p38 MAP kinases on apoptosis. *Science* 270:1326–31

Xie CW, Sayah D, Chen QS, Wei WZ, Smith D, Liu X. 2000. Deficient long-term memory and long-lasting long-term potentiation in mice with a targeted deletion of neurotrophin-4 gene. *Proc. Natl. Acad. Sci. USA* 97:8116–21

Xing J, Ginty DD, Greenberg ME. 1996. Coupling of the RAS-MAPK pathway to gene activation of RSK2, a growth factor-regulated CREB kinase. *Science* 273:959–63

Xing J, Kornhauser JM, Xia Z, Thiele EA, Greenberg ME. 1998. Nerve growth factor activates extracellular signal-regulated kinase and p38 mitogen-activated protein kinase pathways to stimulate CREB serine 133 phosphorylation. *Mol. Cell. Biol.* 18:1946–55

Xu B, Gottschalk W, Chow A, Wilson R, Schnell E, et al. 2000a. The role of BDNF receptors in the mature hippocampus: Modulation of long term potentiation through a presynaptic mechanism. *J. Neurosci.* 20:6888–97

Xu B, Zang K, Ruff NL, Zhang YA, McConnell SK, et al. 2000b. Cortical degeneration in the absence of neurotrophin signaling: Dendritic retraction and neuronal loss after removal of the receptor TrkB. *Neuron* 26:233–45

Yamada M, Enokido Y, Ikeuchi T, Hatanaka H. 1995. Epidermal growth factor prevents oxygen-triggered apoptosis and induces sustained signalling in cultured rat cerebral cortical neurons. *Eur. J. Neurosci.* 7:2130–38

Yamada M, Ohnishi H, Sano SI, Araki T, Nakatami A, et al. 1999. Brain-derived neurotrophic factor stimulates interactions of Shp2 with phosphatidylinositol 3-kinase and Grb2 in cultured cerebral cortical neurons. *J. Neurochem.* 73:41–49

Yamada M, Ohnishi H, Sano SI, Nakatami A, Ikeuchi T, Hatanaka H. 1997. Insulin receptor substrate (IRS)-2 and IRS-2 are tyrosine-phosphorylated and associated with phosphatidylinositol 3-kinase in response to brain-derived neurotrophic factor in cultured cerebral cortical neurons. *J. Biol. Chem.* 48:30334–39

Yamashita H, Avraham S, Jiang S, Dikic I, Avraham H. 1999a. The Csk homologous kinase associates with TrkA receptors and is involved in neurite outgrowth of PC12 cells. *J. Biol. Chem.* 274:15059–65

Yamashita T, Tucker KL, Barde YA. 1999b. Neurotrophin binding to the p75 receptor modulates Rho activity and axonal outgrowth. *Neuron* 24:585–93

Yang DD, Kuan C, Whitmarsh HJ, Rincon M, Zheng TS, et al. 1997. Absence of excitotoxicity-reduced apoptosis in the hippocampus of mice lacking the Jnk3 gene. *Nature* 389:865–70

Yano H, Cong F, Birge RB, Goff SP, Chao MW. 2000. Association of the Abl tyrosine kinase with the Trk nerve growth factor receptor. *J. Neurosci. Res.* 59:356–64

Ye X, Mehlen P, Rabizadeh S, VanArsdale T,

Zhang H, et al. 1999. TRAF family proteins interact with the common neurotrophin receptor and modulate apoptosis induction. *J. Biol. Chem.* 274:30202–8

Yoon SO, Carter BD, Cassaccia-Bonnefil P, Chao MV. 1998. Competitive signaling between TrkA and p75 nerve growth factor receptors determines cell survival. *J. Neurosci.* 18:3273–81

York RD, Mollivar DC, Grewal SS, Steinberg PE, McCleskey EW, Stork PJS. 2000. Role of phosphoinositide 3-kinase and endocytosis in NGF induced extracellular signal-regulated kinase activation via Ras and Rap1. *Mol. Cell. Biol.* 20:18069–83

York RD, Yao H, Dillon T, Ellig CL, Eckert SP, et al. 1998. Rap1 mediates sustained MAP kinase activation induced by nerve growth factor. *Nature* 393:622–26

Yuan J, Yankner BA. 2000. Apoptosis in the nervous system. *Nature.* 407:802–9

Zhang Y, Moheban DB, Conway BR, Bhattacharyya A, Segal RA. 2000. Cell surface Trk receptors mediate NGF-induced survival while internalized receptors regulate NGF-induced differentiation. *J. Neurosci.* 20:5671–78

Zhou H, Summers S, Birnbaum MJ, Pittman RN. 1998. Inhibition of Akt kinase by cell-permeable ceramide and its implications for ceramide-induced apoptosis. *J. Biol. Chem.* 273:16568–75

Zundel W, Swiersz LM, Giaccia A. 2000. Caveolin 1-mediated regulation of receptor tyrosine kinase-associated phosphatidylinositol 3-kinase activity by ceramide. *Mol. Cell. Biol.* 20:1507–14

NOTE ADDED IN PROOF

In a recent study of the roles of GDNF and neurturin in the development of parasympathetic ganglia, GDNF has been shown to be required for precursor development, while neurturin regulates neuronal survival in the sphenopalatine and otic ganglia (Enomoto H et al 2000). A change in expression of GFRα receptors provides the most likely explanation for this switch.

Annu. Rev. Neurosci. 2001. 24:737–77

Contributions of the Medullary Raphe and Ventromedial Reticular Region to Pain Modulation and Other Homeostatic Functions

Peggy Mason

Department of Neurobiology, Pharmacology, and Physiology and the Committee on Neurobiology, University of Chicago, Chicago, Illinois 60637;
e-mail: p-mason@uchicago.edu

Key Words   serotonin, raphe magnus, vasomotion, thermoregulation, sexual function

■ **Abstract**   The raphe magnus is part of an interrelated region of medullary raphe and ventromedial reticular nuclei that project to all areas of the spinal gray. Activation of raphe and reticular neurons evokes modulatory effects in sensory, autonomic, and motor spinal processes. Two physiological types of nonserotonergic cells are observed in the medullary raphe and are thought to modulate spinal pain processing in opposing directions. Recent evidence suggests that these cells may modulate stimulus-evoked arousal or alerting rather than pain-evoked withdrawals. Nonserotonergic cells are also likely to modulate spinal autonomic and motor circuits involved in thermoregulation and sexual function. Medullary serotonergic cells have state-dependent discharge and are likely to contribute to the modulation of pain processing, thermoregulation, and sexual function in the spinal cord. The medullary raphe and ventromedial reticular region may set sensory, autonomic, and motor spinal circuits into configurations that are appropriate to the current behavioral state.

INTRODUCTION

More than 20 years ago two reviews advanced a model for endogenous pain modulation that centered on descending projections from the brainstem to the spinal dorsal horn (Basbaum & Fields 1978, Fields & Basbaum 1978). This model proposed that a discrete set of interconnected brain regions in the forebrain and midbrain converges upon medullary raphe magnus (RM) neurons that project to the spinal dorsal horn, where they specifically mediate analgesia. Since its inception, this model has served as a valuable impetus for motivating basic investigations into pain modulation and medullary raphe function. However, recent evidence suggests that the selective association between RM and pain modulation may not only be overstated but may also obscure a true understanding of medullary raphe function. RM is part of an interrelated group of raphe and reticular nuclei that has been implicated in thermoregulation, vasomotor control, sleep/wake cycles, motor control, and the control of sexual function, as well as in pain modulation. This review considers the contributions of medullary raphe and ventromedial reticular neurons to the modulation of thermoregulation and sexual function, as well as pain, and advances the idea that these neurons modulate multiple homeostatic functions in coordination with an animal's state.

THE NATURE OF PAIN AND PAIN MODULATION

Pain has traditionally been defined as an unpleasant sensory and emotional experience elicited by actual or potential tissue damage. It is commonly held that pain can only be studied in verbal humans and that nociception is the term of choice where animals are concerned. However, in recognition of the biological continuity between verbal, nonverbal, and preverbal humans and their phylogenetic relationship to nonhuman primates and other animals, pain is likely to occur outside of verbal humans. Therefore, I will use the term pain and nociception interchangeably.

Pain is not a unitary sensory modality. The sharp, stabbing pain associated with $A\delta$ nociceptor activation differs from the dull, aching pain associated with C fiber activation in terms of both perception and neural pathways involved (Ochoa & Torebjork 1989, Torebjork et al 1984). Persistent pain, caused by either inflammation or nerve injury, involves plastic changes in neural pathways that overlap with those involved in signaling acute pain. Because of the neuroanatomical separation between different sensory qualities (sharp, dull, acute versus persistent), it is possible that descending systems target different aspects of each type of pain. This review concerns pain modulation in the intact rat that encounters acute painful stimuli in the course of living its "normal" daily life and not the injured animal that faces chronic pain.

In the words of Sherrington, painful stimuli evoke reactions in a "curiously imperative" manner (quoted in Blessing 1997, p. 269). The presence of pain is overwhelming—it overtakes or influences virtually every neural process. In turn, the reaction to pain is influenced by a myriad of factors such that an organism's perceptual or motor reaction to a known stimulus cannot be predicted. The extreme variability in the pain stimulus–response relationship is due to the presence of an endogenous pain modulatory system and shows both this system's potential magnitude and pervasiveness.

In the majority of studies directed at understanding descending pain modulation in the rodent, the latency of tail withdrawal from noxious heat ("tail flick latency") is used as the primary, and sometimes sole, measure of nociceptive responsiveness. The main advantage of this paradigm is that an almost perfect correspondence exists between the ability of a test narcotic to block the tail flick and the ability of that drug to suppress pain in humans (Grumbach 1966). Yet many experimental and pharmacological manipulations that block the tail flick have little or no effect on more organized pain behaviors such as the jumping or paw licking elicited by being placed on a heated plate ("the hot plate test") (e.g. Drower & Hammond 1988, Thorat & Hammond 1997). Furthermore, the tail flick can be pharmacologically dissociated from the noxious heat-evoked paw withdrawal (Ackley et al 1999; Drower & Hammond 1988; Fang & Proudfit 1996, 1998), suggesting that the tail flick is not even representative of simple withdrawal movements. These results highlight the danger inherent to viewing nociceptive modulation as a generalized system, the outcome of which can be represented by a single measure. It is more likely that differential modulation occurs with respect to (*a*) the modality of the noxious stimulus; (*b*) the body location where the stimulus is applied (Benedetti et al 1999); and (*c*) response type (somatomotor, vasomotor, cardiovascular, endocrine, affective, etc) elicited. Understanding how brainstem modulatory circuits impact pains of different modalities or locations or different responses to painful stimuli is a challenge for the future.

In 1969 Reynolds demonstrated that upon microstimulation of sites near the midbrain periaqueductal gray (PAG) rats tolerate an abdominal incision and skin retraction without anesthesia. Although the stimulated rats did not react adversely to the surgery, they startled in response to unexpected visual or auditory stimuli. This remarkable report identified the PAG as a powerful site for the production of analgesia. Bolstered by reports that PAG and periventricular stimulation produced analgesia in patients with intractable pain (Hosobuchi et al 1977, Richardson & Akil 1977, Young et al 1985), a fascinating field of research was born. Work over the ensuing 15 years led to a heuristic model for how PAG stimulation produces analgesia (Basbaum & Fields 1978, 1984). Briefly, this model holds that cells in PAG project primarily to RM cells that in turn send their axons to terminate in the spinal dorsal horn. When activated, RM neurons exert a net inhibitory effect on nociceptive sensory processing, presumably by inhibiting superficial and deep dorsal horn cells that receive primary afferent nociceptor input. This same system could be activated by exogenous administration or endogenous release of opioids.

THE MEDULLARY RAPHE AND VENTROMEDIAL RETICULAR NUCLEI

The medullary raphe and ventromedial reticular region includes three raphe and three reticular nuclei that provide a major projection to the spinal cord (Figure 1). The nuclei of this region are indistinctly delineated on cytoarchitechtonic grounds (Taber et al 1960). Further, the dendritic arbors of cells in these nuclei extend beyond the boundaries of their parent nuclei (Gao & Mason 1997, Mason et al 1990, Newman 1985, Potrebic & Mason 1993). Projections into the medullary raphe and ventromedial reticular region arise principally from the hypothalamus, notably the medial preoptic region, amygdala, midbrain PAG, and pontine parabrachial nuclei

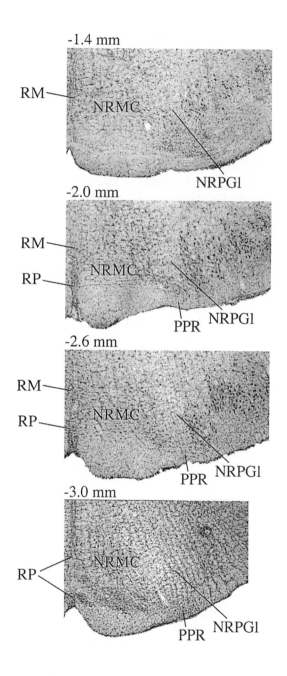

(Abols & Basbaum 1981, Gang et al 1990, Hermann et al 1997, Holstege 1987, 1988, Holstege et al 1985, Murphy et al 1999, Van Bockstaele et al 1991). These afferents reach the cytoarchitechtonic domains of several of the medullary raphe and ventromedial reticular nuclei and extend into the dendritic domains of most neurons in the regional nuclei. In each of the medullary raphe and ventromedial reticular nuclei, both serotonergic and nonserotonergic neurons are found; cells of both populations contain a variety of neuropeptide cotransmitters such as substance P and thyrotropin-releasing hormone. Whereas cells in any one nucleus may provide the major input to one spinal region, there are at least some cells in the other nuclei that target that region as well. For instance, although the predominant projection to the thoracic intermediolateral cell column from this region arises from cells in nucleus paragigantocellularis pars lateralis (NRPGl), raphe pallidus (RP), and raphe obscurus (RO), many neurons in raphe magnus (RM) and nucleus reticularis magnocellularis (NRMC) also project strongly to the intermediate horn (Helke et al 1989, Skagerberg & Bjorklund 1985).

Just as the neurochemistry and anatomy of neurons in the medullary raphe and ventromedial reticular nuclei overlap, these cells' functions may also be indistinct. In support of the idea that the neurons throughout the medullary raphe and ventromedial reticular nuclei share common, overlapping functions, the electrophysiological characteristics of RM, RP, and NRMC cells are often indistinguishable. Furthermore, inactivation of more than one of these medullary nuclei is needed to block descending pain-modulatory effects (Sandkühler & Gebhart 1984, Urban & Smith 1994). These data suggest that neurons in different regional nuclei contribute to a common spinal modulatory function, albeit with differing emphases (Taber et al 1960). To understand the function of cells in any one of these nuclei, it is thus critical to consider those neurons in the context of the cells in neighboring nuclei. Therefore, although this review will concentrate on RM and NRMC cells, information regarding RP, RO, NRPGl, and parapyramidal raphe (PPR) cells will also be considered.

Figure 1 Photomicrographs of the ventromedial medulla at four rostrocaudal levels (level relative to interaural zero is listed above each section). The medullary raphe consists of the rostrally located raphe magnus (RM), the ventrally located raphe pallidus (RP), and the dorsally located raphe obscurus (RO; not illustrated) (Kruger et al 1995, Newman 1985, Taber et al 1960). At the most anterior level illustrated (-1.4 mm), the rostral RM remains, but RP is no longer present. Caudally, at the level of the inferior olives RP is present but RM is not. The reticular region overlying the pyramids, lateral to RM and RP, is the nucleus reticularis magnocellularis [NRMC; referred to as nucleus reticularis gigantocellularis pars alpha by Paxinos & Watson (1986), but defined as NRMC by Kruger et al (1995), Newman (1985), and Taber et al (1960)]. Lateral to the pyramids the most ventral region is termed the parapyramidal region (PPR) (Helke et al 1989, Sasek & Helke 1989). Dorsal to the PPR the nucleus paragigantocellularis pars lateralis (NRPGl) extends from the middle level of the inferior olive to the level of the trapezoid body (Andrezik et al 1981).

The anatomy and the physiological effects evoked by activation of medullary raphe and ventromedial reticular neurons have implicated these cells in two major types of functions. First, activation of neurons in RM and NRMC modulates dorsal horn nociceptive sensory processing. Second, neurons in RP, RO, NRPGl, and the PPR have been implicated in efferent autonomic control and modulation of both sympathetic and parasympathetic targets. For example, projections from raphe cells to the thoracolumbar intermediolateral cell column are implicated in the sympathetic control of physiological functions such as brown adipose metabolism and cutaneous vasoconstriction (see medullary raphe and temperature regulation section below). Projections from reticular cells to the sacral intermediate and ventral horns are implicated in the modulation of autonomic and somatomotor components of sexual climax. This chapter reviews evidence for the participation of RM and NRMC in nociceptive modulation and in selected forms of autonomic regulation in the rat. A possible scenario that merges the at-first-glance, disparate functions of the medullary raphe is then considered.

Compared to the body of anatomical, microstimulation, and pharmacological studies on medullary raphe and ventromedial reticular neurons, electrophysiological studies are far less numerous and principally include recordings from sites in and immediately around RM and RP. Most electrophysiological experiments have focused on the relationship between cell discharge and a single physiological variable, thereby coercing researchers' interpretations of cell function towards dedicated, and away from integrative, roles. Furthermore, most nonpharmacological manipulations employed in electrophysiological studies challenge homeostasis and thus elicit a panoply of behavioral effects in addition to the one or two being measured. For instance, most electrophysiological studies that use noxious stimulation use the latency of the evoked motor withdrawal as the behavioral measure, although changes in vasomotion, blood pressure, heart rate, hormone secretion, and cortical activity may occur concurrently. Because painful and homeostasis-challenging stimuli unavoidably elicit pleiotropic effects, it is especially important to consider that cell discharge may be related directly to variables that are not being measured and only secondarily to the measured variable.

DESCENDING PROJECTIONS OF THE MEDULLARY RAPHE AND VENTROMEDIAL RETICULAR REGION

Understanding the function of a neuron or group of neurons in the mammalian central nervous system is best accomplished by integrating information about the efferent trajectory of the cells and the physiological effects evoked by their activation. Afferent input and a cell's pattern of discharge are extremely useful pieces of information, but they are more indicative of the conditions under which the cell becomes engaged or disengaged than of its output function. Understanding the function of neurons in the medullary raphe and ventromedial reticular region is further complicated by the anatomical, neurochemical, and physiological heterogeneity of these cells. The work of discerning the specific projection patterns of and effects evoked by functionally distinct, but anatomically intermingled, neuronal subpopulations presents a major challenge.

The primary efferent target of cells in all nuclei of the medullary raphe and reticular region is the spinal cord (Basbaum et al 1978, Brodal et al 1960, Holstege & Kuypers 1982, Watkins et al 1980). Few neurons project rostrally, and those that do are concentrated dorsally and rostral to the facial nucleus. Because relatively less is known about the projections from medullary raphe and reticular neurons to lower brainstem regions, projections to the spinal cord at all segmental levels are the focus of this section. Surprisingly, the percentage of all medullary raphe and ventromedial reticular neurons that project to the spinal cord is unknown. There is a general topographical organization to the spinal projections of medullary raphe and ventromedial reticular neurons (Basbaum & Fields 1979, Skagerberg & Bjorklund 1985). Rostrally located RM, NRMC, and NRPGl cells tend to project through the dorsolateral funiculus and target the dorsal horn (Kwiat & Basbaum 1992). Cells in rostral RO and RP project through the middle of the lateral funiculus and target the intermediate horn, including the intermediolateral cell column. Finally, cells in caudal RP and RO project through the ventrolateral and ventral funiculi to target the ventral horn.

Early axonal-tract tracing studies in both cat and rat emphasized RM and NRMC projections to the dorsal horn, via the dorsolateral funiculus, but reported at least sparse terminals in all other regions of the spinal gray (Basbaum & Fields 1979, Holstege & Kuypers 1982). In addition to the superficial dorsal horn, the deep dorsal horn, the intermediolateral cell column, and the central canal region also receive dense innervation from RM neurons (Antal et al 1996, Bullitt & Light 1989, Jones & Light 1990). Retrograde transport studies confirm that neurons in RM and NRMC project to the thoracic intermediolateral cell column, the central canal region, and even to the ventral horn (Du 1989, Masson et al 1991, Sasek et al 1990). Unfortunately, these studies cannot distinguish between projections to different cell types present in the same region, such as ventral horn motoneurons and neighboring interneurons.

Results obtained using transneuronal retrograde transport of viruses, principally pseudorabies, have provided new information regarding the specific cell types targeted by medullary raphe and reticular neurons. In rat pseudorabies virus is preferentially taken up by sympathetic and parasympathetic terminals and more weakly by somatic motoneuron endings (Rotto-Percelay et al 1992, Strack et al 1989a,b). The virus is then retrogradely transported and exocytosed at sites of synaptic contact, where it is endocytosed by the presynaptic terminal and so on (Card et al 1993). Because virus-labeled cells are rarely observed in the dorsal root ganglia (Strack et al 1989b) and the pattern of labeling in the dorsal horn is unaffected by dorsal rhizotomies (Rotto-Percelay et al 1992), pseudorabies virus does not appear to be well transported by sensory afferents in the rat. Thus, dorsal horn cells are most likely to acquire virus retrogradely from their postsynaptic targets. In sum, the pseudorabies virus technique is a powerful way to retrogradely trace oligosynaptic connections to specific peripheral targets.

RM, RP, NRMC, and NRPGl cells are consistently labeled when virus is injected into most autonomic target tissues including the bladder, heart, and penis (Marson 1997, Marson et al 1993, Ter Horst et al 1996). Both sympathetic and parasympathetic pathways from medullary cells to target tissues have been demonstrated by injecting virus into selectively innervated structures such as the kidney, adrenal medulla, or lung (Hadziefendic & Haxhiu 1999, Haxhiu et al 1993, Huang & Weiss 1999, Schramm et al 1993, Strack et al 1989a,b). The specific pathway involved has been confirmed by combining virus injections with lesions of either parasympathetic or sympathetic nerves or tracts (Gerendai et al 1998, Hadziefendic & Haxhiu 1999). The timing of the appearance of virus within medullary cells suggests that whereas some medullary cells directly innervate the second order preganglionic neuron, others innervate either spinal interneurons (pre-preganglionic neurons) or raphe-spinal cells and are therefore fourth order (Cano et al 2000).

RM neurons are labeled after virus injection into the kidney, bladder, uterus, pancreas, colon, spleen, lung, ovaries, penis, interscapular brown adipose tissue, ventral tail artery, adrenal medulla, or pterygopalatine ganglion (Bamshad et al 1999, Ding et al 1993, Gerendai et al 1998, Hadziefendic & Haxhiu 1999, Marson 1997, Papka et al 1998, Sly et al 1999, Smith et al 1998, Strack et al 1989b, Valentino et al 2000, Vizzard et al 1995). The strikingly consistent finding that RM cells project to most autonomic targets raises the question of whether RM cells diverge to innervate multiple autonomic targets. To address this issue, Loewy and colleagues injected two different virus strains into the stellate ganglion and the adrenal medulla (Jansen et al 1995). They observed both double- and single-labeled cells in the medullary raphe and reticular region. The existence of double-labeled cells is very interesting because it raises the possibility that single medullary cells modulate multiple different neuronal populations. If such divergence were common among spinally projecting neurons, a minority of raphe and reticular cells could account for the bulk of the efferent connections. Alternatively, several different medullary cell populations may exist, each innervating a restricted subset of spinal targets (see further discussion of this issue in anatomical divergence section below). At present, there are no data to distinguish between these two possibilities.

The technique of transneuronal retrograde virus transport has also been used to study the brainstem control of somatic motoneurons (Billig et al 1999, Daniels et al 1999, Fay & Norgren 1997a–c, Marson & McKenna 1996, Rotto-Percelay et al 1992). At early survival times, virus labeled cells are present in known pre-motor areas. With longer survival times, during which virus traverses an unknown number of synapses (likely ≥2), many cells, including medullary raphe and ventromedial reticular neurons, are labeled, regardless of the specific muscle injected. It is possible that medullary raphe and reticular neurons are labeled because the virus is taken up by sympathetic terminals present in skeletal muscle and then transported retrogradely through sympathetic control pathways (Rotto-Percelay et al 1992).

However, this is unlikely because similar patterns of labeling are observed after injection of virus into the striated muscle of either sympathectomized or control animals (Fay & Norgren 1997a–c).

Although a monosynaptic connection between RM and dorsal horn cells has been emphasized, the results presented above strongly suggest that medullary raphe and reticular cells, including RM and NRMC neurons, project to preganglionic autonomic neurons via monosynaptic and disynaptic routes (Bacon et al 1990). RM and NRMC neurons may also influence somatic motoneurons, probably via projections to central pattern generator neurons or other interneurons. The relative preponderance of RM and NRMC cell projections to the dorsal, intermediate, and ventral horns is unclear and is a difficult question to address experimentally. It is notable, in this regard, that no RM or NRMC cells were labeled after very small tracer injections into any portion of the dorsal horn, although cells were labeled after similarly sized injections into the central canal region, an area that contains pre-preganglionic as well as premotoneuronal interneurons (Du 1989).

SEROTONERGIC AND NONSEROTONERGIC COMPONENTS OF THE RAPHE-SPINAL PROJECTION

Controversy exists regarding the proportions of serotonergic and nonserotonergic medullary raphe and ventromedial reticular cells that project to the spinal cord. Bowker and colleagues reported that up to 85% of the raphe-spinal neurons are serotonergic and up to 90% of the serotonergic raphe cells project to the spinal cord (Bowker & Abbott 1990). Other reports suggest that well under half of the spinal projecting neurons are serotonergic and that 30–60% of medullary serotonergic raphe cells project to the spinal cord (Jones & Light 1992, Skagerberg & Bjorklund 1985). The discrepancies are likely due to differences in the anatomical regions studied and in the transport efficiency of the different retrograde tracers used. Regardless of the precise numbers, there is now no doubt that the descending projection from the medullary raphe and ventromedial reticular nuclei includes serotonergic and nonserotonergic neurons, both of which are likely to be functionally important. This is true not only of the spinally projecting neurons as a whole but also of the dorsal horn-projecting neurons that travel through the dorsolateral funiculus. This latter projection is primarily nonserotonergic (Johannessen et al 1984, Kwiat & Basbaum 1992) but contains most of the serotonergic axons that innervate the dorsal horn (Bullitt & Light 1989). Pharmacological and physiological studies implicate both serotonergic and nonserotonergic mechanisms in the efferent effects evoked by medullary stimulation (reviewed in Mason & Gao 1998).

Virtually all of the spinal serotonin derives from neurons in RM, RP, RO, ventral NRMC, PPR, and NRPGl, with mesencephalic cells making only a very small contribution to cord serotonin (Oliveras et al 1977, Skagerberg & Bjorklund 1985). Like undifferentiated medullary raphe and ventromedial reticular cell terminals, serotonergic axons within the spinal cord are found in all spinal laminae and at all segmental levels. Serotonergic terminals are present on spinothalamic tract cells, preganglionic sympathetic and parasympathetic neurons, and somatic motoneurons (Appel et al 1987, Helke et al 1986, Wu et al 1993, Wu & Wessendorf 1992). Both serotonergic and nonserotonergic medullary cells are labeled after pseudorabies injections into autonomic and motor targets (Haxhiu et al 1993; Huang & Weiss 1999; Jansen et al 1995; Loewy et al 1994; Marson et al 1993; Smith et al 1998; Spencer et al 1990; Strack et al 1989a,b). A lesion study in two cats demonstrated that partial lesions of RM decrease serotonin levels in the superficial dorsal horn, the central canal region, and the ventral horn of the lumbar cord (Oliveras et al 1977). Yet the relative targeting by serotonergic cells in different medullary locations of different terminal fields in the spinal cord remains unclear. Unfortunately, the most straightforward approach to understanding which spinal targets are specific to which serotonergic cells—immunostaining anterogradely labeled axons—has failed for technical reasons (Jones & Light 1990), leaving this issue unresolved.

RAPHE MAGNUS AS A PAIN MODULATORY CENTER

Early studies leading to the idea that raphe magnus (RM) and nucleus reticularis magnocellularis (NRMC) are critical to descending pain modulation have been reviewed (Basbaum & Fields 1978, 1984; Sandkühler 1996; Willis 1988) and are only summarized here. RM was initially identified as critical to periaqueductal gray (PAG) stimulation–evoked pain modulation because (*a*) the effects of PAG stimulation are observed within the dorsal horn, (*b*) few PAG or adjacent dorsal raphe cells project to the cervical cord and virtually none to the lumbar cord, (*c*) PAG cells project strongly to RM, (*d*) RM cells project strongly to the dorsal horn via the dorsolateral funiculus, (*e*) RM stimulation produces similar effects on nociceptive withdrawals and on nociceptive dorsal horn neurons as does PAG stimulation, and (*f*) RM or dorsolateral funiculus lesions attenuate PAG-evoked antinociceptive effects. Early studies identified NRMC as an equally likely relay for descending antinociceptive pathways as was RM. The presence of opioid peptides and receptors in RM and NRMC led to the idea that RM and NRMC neurons are as critical to opioid-mediated forms of nociceptive modulation as they are to PAG stimulation–evoked nociceptive modulation.

In the past decade it has became clear that endogenous pathways can either inhibit or facilitate pain. RM and NRMC stimulation can either suppress or facilitate nociceptive dorsal horn cells and nociceptive reflexes (Zhuo & Gebhart 1990, 1992), whereas lesioning or inactivation of the RM and NRMC attenuates both the suppression and the facilitation of nociceptive transmission (Behbehani & Fields 1979, Gebhart et al 1983, Kaplan & Fields 1991, Sandkühler & Gebhart 1984). These results led to the idea that RM contains two populations of neurons that, when activated, have opposing efferent effects (Fields et al 1991). One population of cells was hypothesized to mediate nociceptive inhibition such as that evoked by morphine, whereas activation of a distinct neuronal population was thought to mediate nociceptive facilitation such as that evoked during naloxone-precipitated opioid withdrawal (Deakin & Dostrovsky 1978, Kaplan & Fields 1991).

In 1983 Howard Fields identified two physiological classes of putative nociceptive modulatory cells in RM and NRMC (Fields et al 1983a). Subsequent work, reviewed below, led to the idea that these two cell types—ON and OFF cells— mediate nociceptive facilitation and inhibition, respectively. Some skepticism has greeted the attribution of fundamental and far-reaching functions to ON and OFF cell types that are defined by a very simple physiological criterion (see original description section below). Yet as described below, the ON and OFF cell classification system is robust in the sense that several physiological properties, examined after the original description of ON and OFF cells, differ according to class membership— being one way for OFF cells and another way for ON cells. Both ON and OFF cells are nonserotonergic (Gao & Mason 2000, Mason 1997, Potrebic et al 1994). Serotonergic neurons within RM and NRMC comprise a distinct physiological and functional class of neurons (Gao & Mason 2000, Mason 1997). In the sections below, the physiology and possible function of ON, OFF, and serotonergic cells in pain modulation are described.

ON AND OFF CELLS AND NOCICEPTIVE MODULATION

Original Description

RM OFF cells were originally defined in the lightly anesthetized rat by the pause in their discharge just prior to the initiation of a tail flick withdrawal (Fields et al 1983a). The tail heat-evoked inhibition was reported to be "steeper" when aligned to the flick than when aligned to a specific tail temperature. These response properties were interpreted as evidence that OFF cells gate motor withdrawals and that a cessation of their discharge is a requirement for the occurrence of nocifensive withdrawals. Subsequent studies have shown that OFF cells discharge continuously, rather than in bursts, after analgesic doses of μ opioid receptor agonists are administered systemically, by microinjection into the PAG, RM, or intrathecally (Barbaro et al 1986, Cheng et al 1986, Fields et al 1983b, Heinricher & Drasner 1991, Heinricher et al 1992, 1994). This increase in OFF cell discharge is associated with tail flick suppression and has been interpreted as evidence that OFF cells are the RM and NRMC cell type whose activation leads to an inhibitory effect on nociceptive transmission.

From the outset, the role of RM ON cells in modulating nociceptive transmission has been harder to discern than that of OFF cells. In particular, it has been difficult to distinguish between a nociceptive-permissive and a nociceptive-facilitatory function for ON cell discharge. ON cells are excited by noxious stimulation and inhibited by μ opioid receptor agonists (Barbaro et al 1986, Cheng et al 1986, Fields et al 1983a, Heinricher et al 1992). Evidence for a facilitatory role comes from a study of RM cell discharge during the tail flick hyperalgesia (i.e. reduction in tail flick latency; see Kaplan & Fields 1991) that accompanies naloxone-precipitated morphine withdrawal (Bederson et al 1990). The absolute activity averaged over 3 s for 7 ON cells was inversely correlated to the tail flick latency recorded concurrently. Because OFF cells were mostly inactive during both baseline and hyperalgesic periods, the decrease in tail flick latency was attributed to the correlated increase in ON cell discharge (Bederson et al 1990).

The idea that ON cells actively facilitate nociceptive responses has recently been challenged. Microinjection of kynurenate, a nonselective excitatory amino acid receptor antagonist, into RM blocks the heat-evoked excitation of ON cells, has no effect on the heat-evoked OFF cell inhibition, and does not change tail flick latency (Heinricher & McGaraughty 1998). In contrast, kynurenate microinjection suppresses systemic morphine's excitation of OFF cells, has no effect on morphine's inhibition of ON cells, and attenuates morphine's suppression of the noxious heat-evoked tail flick response (Heinricher et al 1999). Thus, modification of the tail flick withdrawal is associated with kynurenate-mediated changes in OFF, but not ON, cell discharge. These results suggest that changes in ON cell discharge alone do not alter the tail flick latency.

Do OFF Cells Suppress Nociceptive Withdrawals?

To understand the function of OFF cells, it is worthwhile to consider nociceptive responses during "spontaneous" periods of OFF cell discharge. This issue has been examined in both the anesthetized and unanesthetized rat. In the anesthetized rat ON and OFF cells discharge in bursts that are reciprocally timed such that OFF cells are active when ON cells are silent and inactive when ON cells discharge (Barbaro et al 1989). The tail flick occurs during both ON and OFF cell bursts (Heinricher et al 1989). In the unanesthetized rat, despite OFF cell discharge during slow wave sleep, animals continue to withdraw from noxious stimulation (Leung & Mason 1999). These situations, in which nociceptive withdrawals continue to occur despite OFF cell activity, imply at a minimum, that OFF cell discharge is not sufficient to completely suppress nociceptive withdrawals. Additionally, these results raise the possibility that OFF cell discharge does not cause the suppression of nociceptive transmission but rather reflects it. According to this possibility, OFF cells, which are inhibited by somatic stimulation, would be disinhibited when spinal nociceptive transmission is suppressed by PAG stimulation, opioid administration, or any other means. Just as OFF cell activity may result from rather than cause changes in nociception, ON cell discharge may be directly related to activity in ascending somatic pathways. In support of this idea, decreasing ascending input from spinal sensory pathways, by administration of intrathecal lidocaine, activates OFF cells and inhibits ON cells (Heinricher & Drasner 1991).

Although it is unlikely that coordinated OFF cell discharge alone can completely suppress nociceptive withdrawals, it may contribute to the inhibition of the tail flick. In support of this idea, the tail flick occurs at a slightly longer latency

(0.4–0.7 s) when tail heat is applied at the peak of an OFF cell burst or during ON cell silence than during an OFF cell silence or ON cell burst (Heinricher et al 1989). Yet a causal relationship between ON and OFF cell discharge and changes in tail flick latency remains unproven, and other explanations of these data are possible. For instance, both ON and OFF cell discharge and nociceptive sensitivity change in concert with blood pressure. ON cells burst during spontaneous decreases in blood pressure, and OFF cells burst during spontaneous increases (Leung & Mason 1996). In normotensive rats and humans nociceptive sensitivity decreases during periods of elevated blood pressure (Ghione 1996). Therefore, it is possible that hypertension inhibits ascending nociceptive transmission, via effects on peripheral or spinal neurons, which in turn leads to an increase in OFF cell activity. Alternatively, because the mechanism of hypertension-associated hypoalgesia is unclear and may even involve OFF cells, the possibility remains that OFF cell activity both contributes to and reflects changes in nociceptive transmission.

ON and OFF Cells: Function in Stimulus-Evoked Arousal?

If ON and OFF cells do not have a major impact on nociceptive withdrawals, then what other effects could they have? One clue comes from the dynamics of ON and OFF cell responses to noxious stimulation. In anesthetized rats, ON and OFF cell responses can begin after the initiation of the motor withdrawal but always continue after withdrawal completion—typically for an additional minute or more (Leung & Mason 1998). Thus, the motor withdrawal is complete within 10 s, whereas the noxious-evoked responses of ON and OFF cells have a time course of tens of seconds or minutes (Figure 2). In both anesthetized and unanesthetized, sleeping animals, noxious stimulation elicits changes in blood pressure, heart rate, and vasomotion and a decrease in synchronized electroencephalographic (EEG) delta band activity that last for tens of seconds to minutes (Blessing & Nalivaiko 2000, Grahn & Heller 1989, Mason et al 2001). In unanesthetized, awake rats noxious stimulation elicits an increase in cardiovascular measures and has no obvious effect on the already desynchronized EEG, but evokes an increase in exploration that lasts for minutes. These data demonstrate that the time course of ON and OFF cell responses to noxious stimulation is similar to that of cardiovascular and EEG state changes evoked by noxious stimulation.

Another clue to the functions of ON and OFF cells comes from an analysis of how these cells are activated under natural conditions. From the limited number of recordings of ON and OFF cells in unanesthetized rats, two features stand out. First, ON and OFF cell activity in the unanesthetized rat is state dependent. ON cells are most active during paradoxical sleep, active during waking, and silent during slow wave sleep, whereas OFF cells are most active during slow wave sleep, only sporadically active during waking, and silent during paradoxical sleep (Leung & Mason 1999). Because paradoxical sleep and waking are both marked by cortical arousal but have opposite levels of motor activity, cortical state, rather than motor activity, appears to be the critical factor in determining ON and OFF cell

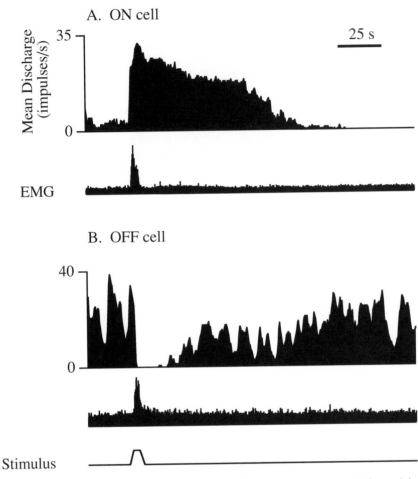

Figure 2 ON and OFF cell responses to noxious tail heat (bottom trace) persist beyond the initiation and completion of the evoked motor withdrawal (EMG traces). The top trace in each panel is the mean discharge rate calculated from three to five trials.

spontaneous discharge. These results are consistent with observations in lightly anesthetized rats that RM cell discharge changes only when the cortical "state" changes as occurs in response to noxious stimulation or body warming (Grahn & Heller 1989).

The state-dependent discharge of ON and OFF cells makes it likely that the target of ON and OFF cell modulation changes in a state-dependent manner. In this light, it is interesting to consider how behavioral responses to noxious stimuli are modulated across sleep/wake cycles. In humans thermal stimuli that are judged moderately painful during waking fail to awaken subjects from slow wave sleep, and the occurrence of these stimuli is not remembered upon awakening in the morning (Drewes et al 1997, Lavigne et al 2000). Although such thermal stimulation often evokes brief (<30 s) EEG changes, these changes do not represent full arousals (Lavigne et al 2000). Similarly, in rats noxious heat applied during slow wave sleep elicits short periods of EEG desynchronization and occasionally a postural shift, but rats soon return to the synchronized EEG pattern of slow wave sleep (Leung & Mason 1999, Mason et al 2001). In contrast to the brief and limited response during slow wave sleep, rats respond to noxious heat applied during waking by exploring the scene of the insult for several minutes (Mason et al 2001). The muted arousal evoked by a single painful stimulus applied during slow wave sleep may be an important homeostatic mechanism for ensuring that sufficient time is spent in the slow wave sleep state. In contrast, the exploration evoked by noxious stimulation applied during waking may help an animal identify a nearby attack threat and learn the cues associated with danger in order to avoid potentially injurious situations in the future. It should be noted that ethical considerations do not permit the application of intensely noxious or frankly damaging stimuli to either humans or animals. Yet one can imagine that such stimuli fully arouse an animal, regardless of whether the animal is asleep or awake.

The second notable feature of ON and OFF cell activity in the unanesthetized rat is that ON cells are excited by and OFF cells inhibited by innocuous auditory and somatic stimuli (Leung & Mason 1999; Oliveras et al 1989, 1990). ON and OFF cells respond to innocuous stimuli at least as strongly as to noxious stimuli—even though innocuous stimuli do not evoke a motor withdrawal. These results reinforce the idea that ON and OFF cell discharge modulates something other than the motor withdrawal from noxious stimulation. Like noxious stimulation, unexpected innocuous stimuli evoke brief microarousals when presented during slow wave sleep and an alerting response when presented during waking. Therefore, the state-dependent discharge of ON and OFF cells may modulate the arousal evoked by unexpected external stimuli, noxious or innocuous. External events–sound, touch, cool, warmth, sudden movement across the visual field—can represent signs of impending attack or threat to the body's homeostasis just as frankly noxious stimuli do. The responses of ON and OFF cells to innocuous but unexpected stimuli in the unanesthetized rat may then be interpreted as evidence that ON and OFF cell function is fundamentally concerned with "safety" rather than pain per se.

If stimulus-evoked arousals are a primary target of ON and OFF cells, then one would predict that OFF cell discharge suppresses stimulus-evoked arousals, whereas ON cell discharge facilitates stimulus-evoked alerting. Consistent with this idea, bicuculline microinjection into RM, which disinhibits OFF cells (Heinricher & Tortorici 1994), blocks the EEG desynchronization evoked by noxious paw heat in anesthetized rats (Nason & Mason 2000). Yet medullary raphe and ventromedial reticular neurons, caudal to the level of the facial genu, do not project to the cerebral cortex or basal forebrain (Hermann et al 1996, Martin et al 1985, Vertes 1988). Therefore, if RM and NRMC cells influence cortical state they are most likely to do so by modulating the sensory input to forebrain circuits that determine behavioral state. Interneurons projecting to tail motoneurons are a separate population of neurons from lumbar neurons that project to the ventrobasal thalamus, supporting the idea that rostrally projecting spinal neurons can be targeted independently of spinal neurons controlling motor withdrawal (Jasmin et al 1997). Furthermore, stimulation of flexor reflex muscle afferents during slow wave sleep elicits a flexion reflex but not cortical arousal (Giaquinto et al 1964), physiological evidence for a distinction between motor- and cortical-alerting spinal neurons.

In summary, the primary target of ON and OFF cell modulation is unlikely to be noxious stimulus–evoked motor withdrawals. Instead, recent evidence favors the idea that one modulatory target of ON and OFF cells is the pathway from the spinal dorsal horn to forebrain areas involved in controlling state. Because ON and OFF cells respond to innocuous stimuli in the same way as they do to noxious stimuli, it is possible that the arousal evoked by any sensory input, not only nociceptive input, is modulated by the medullary raphe. Indeed, bicuculline microinjection into RM shortens the microarousal evoked by a brief air puff applied during slow wave sleep (Foo & Mason 2000). By modulating ascending sensory neurons, ON and OFF cells may influence the salience that external stimuli achieve and thereby the resulting state of the animal.

SEROTONERGIC CELLS IN THE MEDULLARY RAPHE AND VENTROMEDIAL RETICULAR NUCLEI

The caudal raphe group includes medullary serotonergic neurons, whereas the rostral raphe group includes dorsal and median raphe cells in the pons and midbrain (Taber et al 1960). Serotonergic neurons in the caudal and rostral raphe groups discharge slowly and steadily in a state-dependent manner and share common pharmacological profiles (Jacobs & Azmitia 1992). Although these two groups of serotonergic cells are similar in these ways, they have different efferent projections. Serotonergic cells in the caudal group project caudally to the lower brainstem and spinal cord, whereas those in the rostral group project rostrally to the diencephalon and telencephalon. There are also physiological differences between serotonergic cells located in different nuclei. For instance, cells in raphe pallidus (RP), but not in dorsal raphe, are activated when the animal is placed in a cold environment (Martin-Cora et al 2000). Therefore, data gathered from study of rostral serotonergic cells should be considered, but may not always apply, when trying to understand the function of caudal serotonergic cells.

Physiological Identification of Serotonergic Cells

In evaluating electrophysiological studies on the function of medullary serotonergic cells, it is important to appreciate the methods by which serotonergic cells have been identified physiologically. Aghajanian and colleagues have accumulated a wealth of evidence that serotonergic cells in the rat dorsal raphe have long-duration action potentials, a slow and regular discharge pattern, and are sensitive to 5HT-1A receptor agonists (Aghajanian et al 1968, Aghajanian & Vandermaelen 1982, Vandermaelen & Aghajanian 1983, Wang & Aghajanian 1982). Subsequent investigators working in other raphe regions in the cat have then proposed, by analogy to rat serotonergic dorsal raphe neurons, that cells with long-duration action potentials and a slow and regular discharge pattern are serotonergic (Auerbach et al 1985, Fornal et al 1985, Martin-Cora et al 2000, Trulson & Jacobs 1979). Others have considered neurons whose conduction velocities are within a specified low range to be serotonergic (Chiang & Gao 1986, Wessendorf & Anderson 1983, Wessendorf et al 1981).

More recently, medullary serotonergic neurons have been positively identified by combining intracellular labeling with immunocytochemistry for either serotonin or tryptophan hydroxylase, the synthesizing enzyme for serotonin (Bayliss et al 1997; Gao et al 1997, 1998; Gao & Mason 2000; Li & Bayliss 1998a,b; Mason 1997). The conclusion from these studies is that serotonergic neurons in RM, RP, and raphe obscurus (RO), both in vivo and in vitro, discharge slowly and steadily but not necessarily regularly (Figure 3). Thus, most cells that are qualitatively identified as serotonergic cells by virtue of their slow and regular discharge are likely to contain serotonin. However, there may be a population of slowly and irregularly firing serotonergic cells that are mistakenly presumed to be nonserotonergic and therefore are not studied.

Serotonergic Cells and Antinociception

Although serotonergic neurons comprise a minority of the cells in the rat medullary raphe and ventromedial reticular nuclei (Dahlstrom & Fuxe 1964, Moore 1981, Potrebic et al 1994, Steinbusch 1981), they are the source of serotonin in the spinal cord (Dahlstrom & Fuxe 1964, Oliveras et al 1977). Early experiments that led to the idea that serotonin is an important mediator of descending antinociception have been summarized previously (LeBars 1988, Mason & Gao 1998, Sawynok 1989). Briefly, these experiments provide evidence that (*a*) serotonergic cells in the medullary raphe project to the superficial dorsal horn, (*b*) iontophoretic application of serotonin inhibits the responses of dorsal horn cells to noxious stimulation (Belcher et al 1978, Jordan et al 1978, Randic & Yu 1976), (*c*) intrathecal administration of serotonin attenuates nociceptive withdrawals (Yaksh & Wilson 1979), and (*d*) intrathecal administration of serotonin receptor antagonists attenuates the antinociception evoked by RM activation (Barbaro et al 1985, Hammond & Yaksh 1984, Jensen & Yaksh 1984). Because the decrease in tail flick latency evoked by nucleus reticularis magnocellularis (NRMC) stimulation is also attenuated by intrathecal methysergide (Zhuo & Gebhart 1991), medullary serotonergic cells may contribute to nociceptive modulation in the facilitatory as well as inhibitory direction.

Physiological studies of both presumed (by indirect means) and identified (by immunocytochemistry) serotonergic cells do not provide support for the

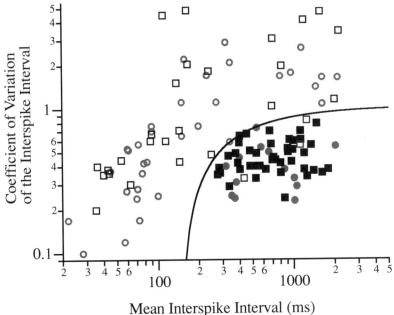

Original Study:
● serotonin immunoreactive
○ not serotonin immunoreactive
Subsequent Test Cells:
■ serotonin immunoreactive
□ not serotonin immunoreactive

Figure 3   Rate and regularity of discharge are used to distinguish serotonergic from non-serotonergic RM and NRMC cells. Rate is defined as the mean interspike interval and regularity is defined as the coefficient of variation of the interspike interval. These discharge characteristics are shown for cells containing serotonin-immunoreactivity (filled symbols) and for those that lack serotonin-immunoreactivity cells (open symbols). The original study of 46 cells was used to calculate a discriminant function that distinguishes serotonergic from nonserotonergic cells (Mason 1997). Cells that have been labeled subsequently comprise a cohort of "test" cells. Cells with discharge rates <0.4 impulses/s are not included in this graph. A line representing the discriminant function defines the optimal linear boundary between serotonergic and nonserotonergic cells and is illustrated on this same graph.

participation of serotonergic cells in descending antinociception. Morphine can suppress the tail flick in the absence of any change in serotonergic cell discharge (Auerbach et al 1985, Chiang & Pan 1985, Gao et al 1998) or release of serotonin (Chiang & Xiang 1987, Matos et al 1992). Similarly, PAG stimulation suppresses tail flick withdrawal without exciting serotonergic RM cells (Gao et al 1997).

Finally, stimulation in RM produces antinociception without coincident release of serotonin (Sorkin et al 1993). Although these data demonstrate that an increase in the discharge of serotonergic RM cells is not necessary for stimulation-evoked or opioid antinociception, the sensitivity of descending antinociception to serotonin receptor antagonists suggests that tonically released serotonin contributes to antinociception.

Serotonergic Cells and State-Related Pain Modulation

As is true for noradrenergic and histaminergic cells, serotonergic cells have state-dependent discharge. The specific pattern of discharge across sleep/wake cycles varies for serotonergic neurons located in different medullary nuclei. Serotonergic RM cells discharge at their highest rates during alert waking, at progressively decreasing rates during quiet waking and slow wave sleep, and are often nearly silent during paradoxical sleep (Fornal et al 1985). Serotonergic RP and RO cells discharge at slightly decrementing but nearly constant rates across waking and slow wave sleep states and at significantly lower rates during paradoxical sleep (Heym et al 1982; C Fornal, personal communication). Extracellular serotonin levels largely follow reported serotonergic cell discharge rates, being highest during waking and lowest during paradoxical sleep (Wilkinson et al 1991). However, serotonin levels do not go to zero during paradoxical sleep, perhaps because of the brief duration and intermittent nature of paradoxical sleep episodes (Iwakiri et al 1993).

Serotonergic RM cells fail to respond to most stimuli unless the stimuli evoke a change in the sleep/wake state. For instance, when dilute formalin is injected into the paw, animals become extremely agitated and shake and lick the injected paw. During this reaction the discharge rate of serotonergic RM cells does not exceed the typical active waking discharge rate (Auerbach et al 1985). This strong dependence of discharge upon behavioral state suggests that serotonergic RM cells may function to modulate target processes in a state-related manner. Because medullary serotonergic cells do not discharge in relation to either sleep spindles or ponto-geniculate-occipital waves and do not project to the forebrain (Fornal et al 1985, Heym et al 1982, Martin et al 1985), they are most likely to exert their modulatory effects on spinal neurons. The primary effect of serotonin on nociceptive dorsal horn neurons and nocifensive movements is inhibitory (Yaksh & Wilson 1979). Therefore, during waking, when serotonergic cell discharge is highest, high levels of serotonin in the dorsal horn would be expected to suppress nociceptive transmission. During slow wave sleep, when serotonergic cell discharge is greatly reduced, lowered levels of serotonin in the dorsal horn would be expected to result in a relative decrease in the suppression of nociceptive transmission that would appear as a disinhibition.

What aspect of nociceptive transmission is modulated by medullary serotonergic cells? As explained above, we expect the target of serotonergic cell modulation to be relatively disinhibited during slow wave sleep compared with during waking.

This pattern of response is shown by motor withdrawals from noxious stimulation. In rats, the paw withdrawal from noxious heat occurs more briskly during slow wave sleep than during waking (Mason et al 2001; cf. Kshatri et al 1998). In cats, the amplitude of the jaw-opening reflex (a trigeminal withdrawal reflex) is greater during slow wave sleep than during waking (Chase 1970). It is therefore possible that one target of serotonergic RM cell modulation is motor responsiveness to noxious stimulation.

MEDULLARY RAPHE AND NONPAIN MODULATORY FUNCTIONS

The electrophysiological data presented above provide correlational evidence that ON, OFF, and serotonergic cells could modulate various aspects of pain processing. Yet these data do not exclude the possibility that medullary raphe and ventromedial reticular neurons also play a role in autonomic modulation. Results from anatomical and microstimulation experiments also support the idea that the medullary raphe region contributes to autonomic control as well as to pain modulation. In lieu of evaluating the evidence for all of raphe's possible functions, I have chosen two exemplars—temperature regulation and the control of sexual climax—to discuss below.

The Medullary Raphe and Temperature Regulation

Activation of caudal raphe neurons stimulates thermoeffectors involved in heat production and skin vasomotion in anesthetized animals (Blessing et al 1999, Morrison 1999, Morrison et al 1999, Rathner & McAllen 1999). In rats cutaneous vasoconstriction is a primary mode of heat conservation. In addition to conserving body temperature, cutaneous vasoconstriction serves nonthermoregulatory functions such as distributing blood volume away from the body surface in response to a painful stimulus or during illness (Blessing 1997). Nonshivering heat production is largely accomplished by increasing the metabolic activity of brown adipose tissue, about a quarter of which is concentrated interscapularly (BAT-is). Direct excitation of RM neurons by glutamate microinjection elicits an increase in the activity of the vasoconstrictor nerves to the tail skin in most rats (Rathner & McAllen 1999). Disinhibition of raphe neurons, by bicuculline microinjection, evokes an increase in the activity of sympathetic nerves to the BAT-is (Morrison 1999, Morrison et al 1999). Although multiple sympathetically mediated effects, such as tachycardia and hypertension, accompany activation of BAT-is and tail vasoconstriction, renal and splanchnic nerve activity remain unchanged by raphe activation (Cao & Morrison 2000, Morrison 1999, Morrison et al 1999, Rathner & McAllen 1999), evidence for a partially generalized sympathetic activation.

Morrison reports that RP is the most effective bicuculline microinjection site for activating BAT-is activity (Morrison 1999, Morrison et al 1999). Yet effective sites are located in both RM and RP, which as described above, are arbitrarily delineated (Morrison 1999, Morrison et al 1999, Nason & Mason 2000). Furthermore, the efficacy of microinjections, particularly small ones such as those used in these studies, may be directly related to cell density. In this regard, the indistinct demarcation between RM and RP is best recognized by the change from a high cell density in RP to a low density in RM (Newman 1985, Taber et al 1960). The sites where stimulation evokes heat production are not even restricted to the medulla because electrical stimulation within the midbrain dorsal raphe evokes a 2°C increase in BAT-is temperature relative to core temperature (Dib et al 1994).

Modulation of Thermal Afference   The medullary raphe may affect temperature regulation through modulating thermal afferent information that reaches the hypothalamus. A small group of RM cells, concentrated in the dorsal and rostral edges of the nucleus, projects to the medial preoptic area, the hypothalamic center for temperature regulation (Hermann et al 1996, Leanza et al 1991, Murphy et al 1999, Vertes 1988). In rat small lesions of RM block the responses of hypothalamic neurons to changes in scrotal temperature, whereas lesions of the neighboring medial lemniscus have no such effect (Taylor 1982). Because the lesions are electrolytic, it is possible that the observed suppression is actually a result of activating cells or axons by passing large currents in the RM. Nonetheless, the simplest interpretation of these data is that thermal afferents to the hypothalamus either pass through or synapse in RM. Consistent with the latter possibility, both serotonergic and nonserotonergic RM cells respond to innocuous changes of cutaneous temperature in the rat (Dickenson 1977, Hellon & Taylor 1982, McAllen et al 2000, Young & Dawson 1987). Because lesions of the rostral raphe nuclei, both dorsal and central, also block the responses of hypothalamic neurons to peripheral temperature changes (Werner & Bienek 1985, 1990), neurons in rostral RM may comprise an accessory pathway to that arising from the rostral raphe nuclei. Finally, RM stimulation suppresses the responses of dorsal horn cells to innocuous changes in scrotal temperature (Sato 1993; cf. Dawson et al 1981), raising the possibility of a descending route for modulation of thermoafferent input.

Modulation of Thermoregulatory Output   Recent evidence has emphasized a role for the medullary raphe in the modulation of efferent thermoregulatory pathways, more so than in the modulation of thermal afferent input to the hypothalamus. Muscimol microinjection into the medullary raphe blocks the increase in BAT-is nerve activity elicited by cooling the hypothalamus in anesthetized rats (Morrison 2000). Similarly, RM inactivation, by lidocaine microinjection, compromises the thermoregulatory response to cooling the hypothalamus in awake rats (Berner et al 1999). These data implicate the mono- and disynaptic projections from caudal raphe cells to preganglionic sympathetic neurons in the intermediolateral cell column in modulating the efferent limb of temperature regulation. Pseudorabies virus experiments demonstrate that RM, RP, and NRMC neurons innervate sympathetic pathways that target the BAT-is and the ventral tail artery

(Bamshad et al 1999, Smith et al 1998). The compound action potential recorded in BAT-is nerves after electrical stimulation of raphe is compelling physiological support for an efferent pathway from medullary raphe to the BAT-is via neurons in the intermediolateral cell column and sympathetic ganglia (Morrison et al 1999). Thus, medullary raphe neurons may serve as specific "premotoneurons" to relay cold information from the hypothalamus to heat-producing (BAT-is activation) and -conserving (tail vasoconstriction) effectors (Morrison et al 1999).

The preganglionic sympathetic neurons that innervate BAT-is and tail artery are centered in T2 and L1 segments, respectively. However, putative premotoneurons in a descending temperature-regulatory pathway are unlikely to project selectively to the intermediate horns of T2 and L1. Brown adipose tissue is present in areas other than interscapularly, most notably in the retroperitoneum, which receives its sympathetic innervation from lower thoracic spinal segments. Further, changes in cutaneous vasomotion involve correlated vasomotor activity in multiple regions that lack fur, regions that are innervated by preganglionic sympathetic neurons located in a number of spinal segments (Bini et al 1980, Nalivaiko & Blessing 1999, Smith & Gilbey 2000). Therefore, medullary raphe neurons involved in the descending modulation of thermoregulation, like those involved in nociceptive modulation, would be expected to project to multiple spinal segments.

If the caudal raphe participates in cold defense by serving as an intermediary between the hypothalamus and peripheral sympathetic neurons, then raphe stimulation would be expected to evoke heat production and conservation, and raphe inactivation would be expected to attenuate autonomic responses that defend against the cold. Remarkably, large injections (1 μl) of lidocaine into the rostral pole of the rat RM elicit a rapid decline of 2°C in body and hypothalamic temperature and a decrease in oxygen consumption and muscle activity (Berner et al 1999). Similarly, inhibition of medullary raphe neurons by muscimol microinjection elicits a small increase in resting cutaneous blood flow, which would increase heat dissipation leading to a drop in core temperature (Blessing & Nalivaiko 2000). Yet electrical stimulation of RM also decreases body temperature, lowers metabolism, and inhibits shivering in hypothermic animals (Hinckel et al 1983). Furthermore, after electrolytic RM lesions, both heat and cold challenges are more efficiently countered and therefore produce smaller changes in core temperature (Szelenyi & Hinckel 1987). The confusing picture that emerges from these studies is likely due to the manipulation of a heterogeneous region such as the medullary raphe with gross stimulation and inactivation techniques that ignore physiological, anatomical, and pharmacological differences between cells. Perhaps the only conclusion to draw from these studies is that the medullary raphe can influence circuits that maintain temperature homeostasis.

Thermoregulatory homeostasis is maintained by making appropriate adjustments whenever the actual temperature deviates from a particular temperature, commonly referred to as the set-point. The set point is not a fixed temperature but is determined empirically as the temperature at which activity in thermal effectors is minimal. The range of temperatures that evokes a minimum of thermoregulatory responses has been termed the thermoneutral zone (Satinoff 1978). If the lower limit of the thermoneutral zone is shifted toward a warmer temperature, cold defense effectors will be activated—assuming that ambient temperature is below the lower limit of the thermoneutral zone, as is almost always the case. It is possible that raphe cells modulate the limits of the spinal thermoneutral zone by modifying the input to or the excitability of preganglionic sympathetic neurons. Such a modification of the thermoneutral zone could result in the preparation's temperature, previously interpreted as neutral, becoming suprathreshold for activation of BAT-is nerve and tail vasoconstrictor nerve activity. Current data fail to distinguish between a direct effect of raphe on preganglionic sympathetic neurons and an indirect effect of raphe on neurons or terminals presynaptic to preganglionic sympathetic neurons. Resolving this issue presents an important challenge.

Cell Types Involved   There is evidence implicating both nonserotonergic and serotonergic raphe neurons in temperature regulation. The projection from RM to the hypothalamus is largely nonserotonergic (Martin et al 1985). Similarly, most RP and RM neurons transneuronally labeled after pseudorabies virus injection into the tail artery are nonserotonergic (Smith et al 1998). A sizable minority of nonserotonergic (45%) RM cells respond to innocuous changes in skin temperature in the anesthetized rat (Young & Dawson 1987). On the other hand, 40% of serotonergic RM cells also respond to innocuous changes in skin temperature (Dickenson 1977). It is interesting to note that these responses persist after decerebration, suggesting that serotonergic RM cells receive thermoafferent input from extra-hypothalamic sources, probably in addition to hypothalamic sources. The conduction velocity of the compound action potential elicited by RM stimulation in BAT-is nerves is just under 1 m/s, consistent with an unmyelinated bulbospinal projection from raphe to the intermediolateral cell column (Morrison et al 1999). This is consistent with a serotonergic projection because all serotonergic axons in the dorsolateral funiculus are unmyelinated in the rat (Basbaum et al 1988). However, this evidence is not definitive, because nonserotonergic raphe cells may also send unmyelinated axons into the spinal cord.

Finally, it is possible that the influence of the caudal raphe on temperature regulation is part of a more general modulation of spinal sympathetic circuits. As stated above, tachycardia and an increase in blood pressure accompany BAT-is activation evoked by raphe disinhibition. Further, just as raphe inhibition blocks the increase in BAT-is nerve activity elicited by cooling the hypothalamus (Morrison 2000), it also abolishes the cutaneous vasoconstriction response evoked by noxious pinch (Blessing & Nalivaiko 2000). These results support the idea that medullary raphe cells do not simply relay thermoregulatory information as pre-motoneurons but instead influence a variety of types of afferent input to preganglionic sympathetic neurons.

The Medullary Raphe and Sexual Function

In anesthetized and spinalized male and female rats stimulation of the urethra elicits synchronous bursting in the smooth and striated musculature of the genitalia

(McKenna et al 1991). This urethrogenital reflex results from the activation of a lumbosacral central pattern generator for sexual climax in males and females, including ejaculation in males. Ejaculation is facilitated by sensory input from the penis (Sachs & Meisel 1988), but is tonically inhibited by rostral medullary sites (Marson & McKenna 1990). Bilateral lesions of nucleus paragigantocellularis pars lateralis (NRPGl), at the level of the facial nucleus, partially release the urethrogenital reflex from tonic inhibition by the brainstem in anesthetized rats (Marson & McKenna 1990) and decrease the latency to ejaculation and the number of intromissions required for ejaculation in awake, freely behaving rats (Yells et al 1992). Because NRPGl lesions do not completely disinhibit either the urethrogenital reflex or ejaculation, neurons outside of NRPGl also contribute to the inhibitory regulation of sexual climax.

Consistent with their role in controlling sexual climax motor patterns, NRPGl neurons are labeled after pseudorabies virus injections into the corpus cavernosus of the penis, the clitoris, or the striated muscles of the penis (Marson 1995, Marson & McKenna 1996, Marson et al 1993). Because parapyramidal region (PPR), raphe pallidus (RP), and raphe magnus (RM) neurons are consistently labeled after such virus injections, these medullary raphe and reticular neurons, in addition to neurons in NRPGl, likely contribute to the control of sexual climax. The latency to virus transport to medullary somata is consistent with medullary cells innervating interneurons as well as somatic motoneurons and preganglionic sympathetic neurons. Therefore, the anatomical evidence cannot distinguish whether neurons in the medullary raphe and ventral reticular region directly modulate somatic and autonomic efferents or whether they indirectly modulate the effect of afferent stimulation upon the urethrogenital central pattern generator. According to the latter idea, medullary cells would tonically suppress afferent input into the urethrogenital central pattern generator and thereby increase the amount of stimulation, achieved by mounting and intromission, necessary for ejaculation. This view would predict that lesioning the involved medullary cells would lead to a decrease in the amount of afferent stimulation, attained through intromissions, necessary for ejaculation. As predicted, NRPGl lesions have precisely this effect (Yells et al 1992).

Serotonergic and Nonserotonergic Cell Involvement   Serotonin has an inhibitory effect on sexual behavior that is at least partially mediated by an effect on spinal circuitry (Labbate et al 1998, Yells et al 1995, Zajecka et al 1997). Intrathecal administration of a serotonin neurotoxin reduces spinal serotonin levels by 90% and attenuates the tonic inhibition of the urethrogenital reflex in intact rats (Marson & McKenna 1994). Conversely, in spinalized and anesthetized rats, intrathecal serotonin increases the threshold needed to evoke an urethrogenital reflex (Marson & McKenna 1992). In support of roles for both serotonergic and nonserotonergic medullary cells, midline neurons labeled after pseudorabies virus injections into the penis are mostly serotonergic, whereas no more than half of the virus-labeled cells in the PPR and NRPGl are serotonergic (Marson & McKenna

1996, Marson et al 1993). Thus, medullary cells, located in both raphe and more lateral reticular regions, are likely to contribute to the modulation of sexual climax motor patterns.

CELLULAR INTEGRATION OF MULTIPLE OUTPUTS

Evidence for the participation of the medullary raphe in nociceptive modulation, temperature regulation, and sexual function has been reviewed. Although evidence that raphe contributes to other physiological functions, such as gastric acid secretion (see for example Yang et al 2000), is similarly compelling, full consideration of such data is precluded by space.

There are two principal ways in which the medullary raphe can contribute to the modulation or control of multiple physiological functions. First, individual neuronal populations could be dedicated to the modulation of each separate function. Alternatively, individual cells could contribute to the modulation or control of many different neural circuits. In the latter case, cells could either contribute to multiple modulatory functions simultaneously via divergent projections or could serve different functions at different times by some type of network reconfiguration. Of course, there are many intermediate possibilities; individual cell populations may contribute to a restricted subset of functions such that the number of populations is a fraction of the number of functions affected. This section discusses evidence in support of each of these possibilities.

Anatomical Divergence

One way to distinguish between dedicated and multifunctional cell populations is to determine whether single cells project to one or more different cell types. Unfortunately, there is only scant information available on the question of whether individual neurons project to multiple targets. After injections into two levels of the spinal cord or into the spinal cord and caudal medulla, both single- and double-labeled RM and NRMC neurons are observed (Allen & Cechetto 1994, Huisman et al 1981). However, even if a cell projects to multiple spinal segments, the targeted neurons could be of a single type. As mentioned above, Loewy and colleagues used the pseudorabies virus method to demonstrate that individual neurons, both serotonergic and nonserotonergic, project to pathways that innervate the stellate ganglion and the adrenal medulla (Jansen et al 1995). Although this is clear evidence for divergence, both the adrenal medulla, which is responsible for releasing catecholamines into the blood, and the stellate ganglion, which provides sympathetic innervation to the heart, promote cardiovascular arousal. It remains unclear whether single medullary cells diverge to targets with less closely related, or even antagonistic, physiological functions.

The spinal terminations of single myelinated axons descending from RM have been examined with anatomical and electrophysiological techniques (Fields et al 1995, Light 1985). In the cat intracellular labeling of dorsolateral funicular axons that were directly activated by medullary raphe stimulation revealed two termination patterns (Light 1985). One group of axons terminated in laminae I, II, V, and X, whereas a second group of axons terminated in laminae V, VII, and X. An antidromic mapping study revealed that 32 out of 48 ON and OFF cell axons traveled within the dorsolateral funiculus, and 17 of these 32 terminated within lamina I, II, and V of the dorsal horn (Fields et al 1995). The obverse of these findings is that up to a third of the recorded ON and OFF cells did not send axons into the dorsolateral funiculus and that almost half of the ON and OFF cells traveling within the dorsolateral funiculus did not terminate within the adjacent cervical dorsal horn. Whether these axons terminated within the dorsal horn at other spinal levels is unknown.

Unfortunately, the evidence on medullary cell projections does not distinguish between many cell populations with dedicated functions and a few cell populations with divergent functions. Evidence to date leaves open to interpretation the extent to which single medullary raphe cells contact different types of spinal neurons. For instance, a cell that projects only to laminae I, II, and V in a single segment could contact nociceptive, thermoreceptive, low threshold somatosensory, and pre-preganglionic interneurons. Conversely, a cell that projects to laminae V, VII, and X in multiple spinal segments could selectively target interneurons to preganglionic sympathetic neurons that selectively innervate brown adipose tissue. In summary, current data are insufficient to determine how many different types of spinal neurons are contacted by each medullary raphe cell class.

Physiological Heterogeneity

The physiological characteristics of cells in the medullary raphe and ventromedial reticular nuclei are sufficiently varied that a large number of functional subpopulations may exist. In addition to ON and OFF cells, RM and NRMC contain a third major type of nonserotonergic cell—the NEUTRAL cell. NEUTRAL cells are unaffected by noxious tail heat or opioid administration in the anesthetized rat (Barbaro et al 1986, Cheng et al 1986, Fields et al 1983a), and the role, if any, of these cells in nociceptive modulation has never been clear. Although NEUTRAL cells are defined by their lack of response to noxious tail heat in the anesthetized condition, most respond to noxious tail clamp and colorectal distension in the anesthetized state (Brink & Mason 2000, Leung & Mason 1998) and half respond to noxious heat in the absence of anesthesia (Leung & Mason 1999). Furthermore, in unanesthetized rats NEUTRAL cells exhibit state-dependent discharge patterns like those of ON and OFF cells (Leung & Mason 1999). Although NEUTRAL cells are unlikely to be truly unresponsive to somatic stimulation, it remains possible that a neutral response to noxious tail heat in the anesthetized preparation accurately predicts opioid sensitivity and perhaps efferent function.

The existence of four major physiological cell classes—ON, OFF, NEUTRAL, and serotonergic—within RM and NRMC belies an even greater potential for heterogeneity. The spontaneous discharge rates of ON, OFF, and NEUTRAL cells vary by greater than two orders of magnitude, and a subset of the neurons in each of these classes has a regular discharge pattern that lacks the frank bursts and pauses present in the discharge of most nonserotonergic cells (Gao & Mason 2000, Leung & Mason 1998). The noxious heat–evoked responses of ON and OFF cells also vary by more than 100-fold. The response to opioids of a large and diverse selection of ON, OFF, and NEUTRAL cells has never been tested and may be less homogeneous than is currently thought. In support of this idea, several nonserotonergic cells that were activated by noxious pinch or heat were also activated by morphine administration in the awake cat (Auerbach et al 1985). Serotonergic cells are also physiologically heterogeneous. Some serotonergic cells respond to noxious heat, retroperitoneal stimulation, and baroreceptor activation, whereas others do not (Gao & Mason 2000, Genzen et al 1997; K Gao & P Mason, unpublished observations). There are subsets of serotonergic cells that discharge in relation, either negative or positive, to blood pressure, whereas the discharge of other serotonergic cells appears unrelated to blood pressure. The responses of serotonergic cells to natural stimuli in the unanesthetized cat are also heterogeneous. For instance, most but not all, serotonergic RM neurons respond when the cat is held by the scruff of the neck (Auerbach et al 1985). As mentioned above, both serotonergic and nonserotonergic raphe cells contain various combinations of neuropeptides and amino acid neurotransmitters. Therefore, medullary raphe and ventromedial reticular neurons are diverse in terms of axonal projections, neurochemical content, and physiological characteristics. Unfortunately, the critical question of how this extreme cellular diversity corresponds to a heterogeneity in output function remains unanswered.

Convergent Afferent Input

Many medullary raphe and ventromedial reticular neurons receive convergent inputs related to multiple spinal systems, suggesting that they may serve compound functions. For instance, nonserotonergic RM cells that are excited by cooling the skin are also excited by noxious pinch, whereas cells that are excited by warming the skin are inhibited by noxious pinch (Young & Dawson 1987; cf. McAllen et al 2000). Similarly, many nonserotonergic neurons in NRMC and nucleus reticularis gigantocellularis (NRGC) that respond to low-intensity mechanical stimulation of the penis also respond to noxious pinch of the paws or ears (Hubscher & Johnson 1996)[1]. One interpretation of convergent inputs to medullary neurons is that the recipient neurons modulate the multiple systems represented by the effective inputs. An alternative possibility is that raphe cells only modulate one spinal system, such as nociception or thermoregulation, but do so in response to convergent input from multiple systems.

---

[1] Although physiological work by Rose has often been cited as evidence that medullary cells in the region of the NRPGl respond to genital stimulation, Rose's medullary recordings were actually in the nucleus ambiguus and the lateral reticular nucleus of the cat (Rose 1975, Rose & Sutin 1973).

BEHAVIORAL STATE AND THE MEDULLARY RAPHE

Compelling anatomical, physiological, and electrophysiological evidence suggests that the medullary raphe and ventromedial reticular nuclei are involved in multiple efferent functions. Is there any way to understand these multiple functions in a unified and coherent manner? One possibility is that medullary raphe and ventromedial reticular neurons ensure that somatomotor and autonomic spinal circuits are set appropriately for the behavioral state or context of the animal.

As discussed above, the spontaneous discharge of serotonergic and nonserotonergic RM cells is state dependent. Most stimuli that elicit responses in serotonergic RM neurons concurrently change the state of arousal, whereas stimuli that fail to change the state of arousal typically fail to elicit serotonergic cell responses. The responses of nonserotonergic RM cells to external stimuli may also be related to state changes. In anesthetized rats skin temperature alters the discharge of nonserotonergic neurons only when cortical and electromyographic status changes occur simultaneously (Grahn & Heller 1989). Because noxious stimulation elicits cortical desynchronization (Grahn & Heller 1989) and genital stimulation alters the state of arousal, most or all RM cell responses may be intertwined with state-changing effects.

The association between RM cell responses and state has been interpreted as evidence that the cell responses are secondary to state changes and therefore are "nonspecific" by-products of arousal. However, the nonserotonergic RM cellular response to noxious stimulation precedes the evoked cortical desynchronization (Leung & Mason 1999) and is therefore unlikely to be a secondary response. Furthermore, although state-related effects may be considered nonspecific in some systems, this is unlikely to be the case for medullary raphe and ventromedial reticular neurons. Instead, the physiological inputs that evoke responses in serotonergic and nonserotonergic RM cells are diverse in terms of modality but similar in terms of their specific ability to alter the animal's state. In this light, it is understandable why the leading ideas regarding function of the medullary raphe have focused on pain modulation and thermoregulatory control and recently on sexual function. Pain and thermal challenges are two principal threats to homeostasis and they elicit state changes. During sleep such stimuli elicit a microarousal (Lavigne et al 2000). During waking they change the behavior and motivation of an animal. Whereas stimuli in other modalities also elicit state changes, they may only do so when presented at more extreme intensities than are necessary for pain or temperature. Because pain and temperature, and sexual stimulation, stand out in their potential for influencing an animal's state changes, it should not be surprising that they also stand out in the physiology of medullary raphe cells.

Beyond the correlation between RM cell discharge and behavioral state, physiological experiments also implicate the medullary raphe and adjacent reticular nuclei in state control. As mentioned above, microinjection of bicuculline into RM, which increases neuronal activity by disinhibition, shortens the microarousal evoked by an air puff stimulus (Foo & Mason 2000). Conversely, inactivation of

RM neurons, by large (1 μl) injections of lidocaine, awakens rats from slow wave sleep (Berner et al 1999). These experiments are consistent with the idea that one effect of RM activation is to suppress physiological arousal evoked by external events. However, such a function is unlikely to be the only output of the physiologically and anatomically heterogeneous medullary raphe and ventromedial reticular nuclei. Indeed, in cats electrical stimulation of rostral RM desynchronizes cortical EEG and elicits exploratory movements (Mori et al 1989). This report is in superficial conflict with the idea that RM stimulation promotes sleep maintenance, perhaps because the stimulation sites were at the rostral pole of RM, where numerous rostrally projecting neurons are found. Alternatively, the conflicting results may arise from the use of gross manipulations on heterogeneous but intermingled cells. The most worthwhile conclusion from such studies is likely to be that neurons in the medullary raphe and ventromedial reticular nuclei can modulate the expression of behavioral state.

A connection between behavioral state and the medullary raphe provides a unifying perspective on the divergent functions in which the medullary raphe has been implicated. Consider an animal in slow wave sleep. Remarkably, an animal remains in this quiescent and vulnerable state despite occasional external disturbances. When a sound, touch, or cool wind occurs, a microarousal, but not cortical arousal and awakening, results. Dampening the autonomic effects evoked by an external stimulus aids in preventing a full awakening. At the same time, in order to remain safe, simple motor behaviors, such as withdrawals, and the autonomic arousal necessary to support those withdrawals, continue.

The physiological diversity of medullary raphe cell types is greater than the diversity of behavioral states, of which there are only three major types—waking, slow wave sleep, and paradoxical sleep. This leads to an apparent imbalance between the large combinatorial diversity achievable by multiple cell types with state-related discharge and the small number of behavioral states. However, substates have been described for both slow wave sleep and paradoxical sleep. Furthermore, waking is likely to be an even more diversified state with emotional substates that are as varied physiologically as the "shapes of rocks on a New Hampshire farm" (William James as quoted by Lang 1994). Afferents to the medullary raphe and ventromedial reticular region arise from regions that are important in setting the emotional, autonomic, and behavioral state of the animal (see above). Thus, combined input from the amygdala, hypothalamus, PAG, and parabrachial nuclei may inform medullary raphe neurons of the desired state.

Medullary raphe and ventromedial reticular cells may aid in the production of a coherent behavioral state by specifically reconfiguring the sensitivity of spinal circuits to external and internal stimuli. Raphe modulation is likely to target a number of different spinal circuits, each linked to specific behaviors. For example, the sensitivity of supraspinally projecting dorsal horn cells could be modulated to affect the probability that a somatosensory stimulus evokes a change in cortical synchronization. Modulation of the activity of nociceptive neurons that contact ventral horn interneurons and motoneurons could modify the motor responses to painful stimuli. Modulation of the sensitivity of the interneurons and sympathetic efferents that control BAT-is activity could alter the range of thermoneutral temperatures that an animal tolerates. Modulation of the sensitivity of spinal vasomotor circuits could alter the range of inputs—visceral, nociceptive, or thermal—that elicit cutaneous vasoconstriction. Likewise, the sensitivity of the urethrogenital central pattern generator could be modulated to influence an animal's receptivity to sexual advances. The output of all of the modulated circuits define a state. Failure to properly configure the spinal cord could result in counterproductive or paradoxical behavioral combinations such as walking while sleeping, cutaneous vasodilation during a fight, or sexual climax during exposure to severe cold. When operating properly the medullary raphe and ventromedial reticular region would promote favored combinations and prohibit disadvantageous combinations of sensory, autonomic, and motor function in accordance with the behavioral context of the animal.

ACKNOWLEDGMENTS

This research was supported by NINDS, NIMH, and the Brain Research Foundation. Thanks to my colleagues, Drs. WW Blessing, RW Hurley, and H Foo, and to the members of my laboratory, TS Brink and MW Nason, for helpful conversations and comments on the manuscript. This paper is dedicated to Dr. Jay Goldberg with gratitude.

Visit the Annual Reviews home page at www.AnnualReviews.org

LITERATURE CITED

Abols IA, Basbaum AI. 1981. Afferent connections of the rostral medulla of the cat: a neural substrate for midbrain-medullary interactions in the modulation of pain. *J. Comp. Neurol.* 201:285–97

Ackley MA, Hurley RW, Hammond DL. 1999. *Kappa* opioid receptor agonists in the rostral ventromedial medulla modulate synaptic transmission *in vitro* and are antinociceptive *in vivo. Soc. Neurosci.* 25:1437 (Abstr.)

Aghajanian GK, Foote WE, Sheard MH. 1968. Lysergic acid diethylamide: sensitive neuronal units in the midbrain raphe. *Science* 161:706–8

Aghajanian GK, Vandermaelen CP. 1982. Intracellular recordings from serotonergic dorsal raphe neurons: pacemaker potentials and the effect of LSD. *Brain Res.* 238:463–69

Allen GV, Cechetto DF. 1994. Serotoniner-

gic and nonserotoninergic neurons in the medullary raphe system have axon collateral projections to autonomic and somatic cell groups in the medulla and spinal cord. *J. Comp. Neurol.* 350:357–66

Andrezik JA, Chan PV, Palay SL. 1981. The nucleus paragigantocellularis lateralis in the rat. Conformation and cytology. *Anat. Embryol.* 161:355–71

Antal M, Petko M, Polgar E, Heizmann CW, Storm-Mathisen J. 1996. Direct evidence of an extensive GABAergic innervation of the spinal dorsal horn by fibres descending from the rostral ventromedial medulla. *Neuroscience* 73:509–18

Appel NM, Wessendorf MW, Elde RP. 1987. Thyrotropin-releasing hormone in spinal cord: coexistence with serotonin and with substance P in fibers and terminals apposing identified preganglionic sympathetic neurons. *Brain Res.* 415:137–43

Auerbach S, Fornal C, Jacobs BL. 1985. Response of serotonin-containing neurons in nucleus raphe magnus to morphine, noxious stimuli, and periaqueductal gray stimulation in freely moving cats. *Exp. Neurol.* 88:609–28

Bacon SJ, Zagon A, Smith AD. 1990. Electron microscopic evidence of a monosynaptic pathway between cells in the caudal raphe nuclei and sympathetic preganglionic neurons in the rat spinal cord. *Exp. Brain Res.* 79:589–602

Bamshad M, Song CK, Bartness TJ. 1999. CNS origins of the sympathetic nervous system outflow to brown adipose tissue. *Am. J. Physiol. Regul. Integr. Comp. Physiol.* 276:R1569–78

Barbaro NM, Hammond DL, Fields HL. 1985. Effects of intrathecally administered methysergide and yohimbine on microstimulation-produced antinociception in the rat. *Brain Res.* 343:223–29

Barbaro NM, Heinricher MM, Fields HL. 1986. Putative pain modulating neurons in the rostral ventral medulla: reflex-related activity predicts effects of morphine. *Brain Res.* 366:203–10

Barbaro NM, Heinricher MM, Fields HL. 1989. Putative nociceptive modulatory neurons in the rostral ventromedial medulla of the rat display highly correlated firing patterns. *Somatosens. Mot. Res.* 6:413–25

Basbaum AI, Clanton CH, Fields HL. 1978. Three bulbospinal pathways from the rostral medulla of the cat: an autoradiographic study of pain modulating systems. *J. Comp. Neurol.* 178:209–24

Basbaum AI, Fields HL. 1978. Endogenous pain control mechanisms: review and hypothesis. *Ann. Neurol.* 4:451–62

Basbaum AI, Fields HL. 1979. The origin of descending pathways in the dorsolateral funiculus of the spinal cord of the cat and rat: further studies on the anatomy of pain modulation. *J. Comp. Neurol.* 187:513–31

Basbaum AI, Fields HL. 1984. Endogenous pain control systems: brainstem spinal pathways and endorphin circuitry. *Annu. Rev. Neurosci.* 7:309–38

Basbaum AI, Zahs K, Lord B, Lakos S. 1988. The fiber caliber of 5-HT immunoreactive axons in the dorsolateral funiculus of the spinal cord of the rat and cat. *Somatosens. Mot. Res.* 5:177–85

Bayliss DA, Li YW, Talley EM. 1997. Effects of serotonin on caudal raphe neurons: activation of an inwardly rectifying potassium conductance. *J. Neurophysiol.* 77:1349–61

Bederson JB, Fields HL, Barbaro NM. 1990. Hyperalgesia during naloxone-precipitated withdrawal from morphine is associated with increased on-cell activity in the rostral ventromedial medulla. *Somatosens. Mot. Res.* 7:185–203

Behbehani MM, Fields HL. 1979. Evidence that an excitatory connection between the periaqueductal gray and nucleus raphe magnus mediates stimulation produced analgesia. *Brain Res.* 170:85–93

Belcher G, Ryall RW, Schaffner R. 1978. The differential effects of 5-hydroxytryptamine, noradrenaline and raphe stimulation on nociceptive and non-nociceptive dorsal horn interneurones in the cat. *Brain Res.* 151:307–21

Benedetti F, Arduino C, Amanzio M. 1999. Somatotopic activation of opioid systems by target-directed expectations of analgesia. *J. Neurosci.* 19:3639–48

Berner NJ, Grahn DA, Heller HC. 1999. 8-OH-DPAT-sensitive neurons in the nucleus raphe magnus modulate thermoregulatory output in rats. *Brain Res.* 831:155–64

Billig I, Foris JM, Card JP, Yates BJ. 1999. Transneuronal tracing of neural pathways controlling an abdominal muscle, rectus abdominis, in the ferret. *Brain Res.* 820:31–44

Bini G, Hagbarth KE, Hynninen P, Wallin BG. 1980. Regional similarities and differences in thermoregulatory vaso- and sudomotor tone. *J. Physiol.* 306:553–65

Blessing WW. 1997. *The Lower Brainstem*

and Bodily Homeostasis. New York: Oxford Univ. Press. 575 pp.

Blessing WW, Nalivaiko E. 2000. Regional blood flow and nociceptive stimuli in rabbits: patterning by medullary raphe, not ventrolateral medulla. *J. Physiol.* 524(Part 1):279–92

Blessing WW, Yu YH, Nalivaiko E. 1999. Raphe pallidus and parapyramidal neurons regulate ear pinna vascular conductance in the rabbit. *Neurosci. Lett.* 270:33–36

Bowker RM, Abbott LC. 1990. Quantitative re-evaluation of descending serotonergic and non-serotonergic projections from the medulla of the rodent: evidence for extensive co-existence of serotonin and peptides in the same spinally projecting neurons, but not from the nucleus raphe magnus. *Brain Res.* 512:15–25

Brink TS, Mason P. 2000. Raphe magnus neurons respond to noxious visceral stimulation. *Soc. Neurosci.* 26:654 (Abstr.)

Brodal A, Taber E, Walberg F. 1960. The raphe nuclei of the brainstem of the cat. II. Efferent connections. *J. Comp. Neurol.* 114:239–60

Bullitt E, Light AR. 1989. Intraspinal course of descending serotoninergic pathways innervating the rodent dorsal horn and lamina X. *J. Comp. Neurol.* 286:231–42

Cano G, Card JP, Rinaman L, Sved AF. 2000. Connections of Barrington's nucleus to the sympathetic nervous system in rats. *J. Auton. Nerv. Syst.* 79:117–28

Cao WH, Morrison SF. 2000. Cardiovascular effects following microinjection of a $GABA_A$ receptor antagonist into rostral raphe pallidus (RPA) of the anesthetized rats. *Soc. Neurosci.* 26:1186 (Abstr.)

Card JP, Rinaman L, Lynn RB, Lee BH, Meade RP, et al. 1993. Pseudorabies virus infection of the rat central nervous system: ultrastructural characterization of viral replication, transport, and pathogenesis. *J. Neurosci.* 13:2515–39

Chase MH. 1970. The digastric reflex in the kitten and adult cat: paradoxical amplitude fluctuations during sleep and wakefulness. *Arch. Ital. Biol.* 108:403–22

Cheng ZF, Fields HL, Heinricher MM. 1986. Morphine microinjected into the periaqueductal gray has differential effects on 3 classes of medullary neurons. *Brain Res.* 375:57–65

Chiang C-Y, Gao B. 1986. The modification by systemic morphine of the responses of serotonergic and non-serotonergic neurons in nucleus raphe magnus to heating the tail. *Pain* 26:245–57

Chiang C-Y, Pan ZZ. 1985. Differential responses of serotonergic and non-serotonergic neurons in nucleus raphe magnus to systemic morphine in rats. *Brain Res.* 337:146–50

Chiang C-Y, Xiang XK. 1987. Does morphine enhance the release of 5-hydroxytryptamine in the rat spinal cord? An in vivo differential pulse voltammetry study. *Brain Res.* 411:259–66

Dahlstrom A, Fuxe K. 1964. Evidence for the existence of monoamine-containing neurons in the central nervous system I. Demonstration of monoamine in the cell bodies of brain stem neurons. *Acta Physiol. Scand.* 232:1–36

Daniels D, Miselis RR, Flanagan-Cato LM. 1999. Central neuronal circuit innervating the lordosis-producing muscles defined by transneuronal transport of pseudorabies virus. *J. Neurosci.* 19:2823–33

Dawson NJ, Dickenson AH, Hellon RF, Woolf CJ. 1981. Inhibitory controls on thermal neurones in the spinal trigeminal nucleus of cats and rats. *Brain Res.* 209:440–45

Deakin JF, Dostrovsky JO. 1978. Involvement of the periaqueductal grey matter and spinal 5-hydroxytryptaminergic pathways in morphine analgesia: effcts of lesions and 5-hydroxytryptamine depletion. *Br. J. Pharmacol.* 63:159–65

Dib B, Rompre PP, Amir S, Shizgal P. 1994. Thermogenesis in brown adipose tissue is activated by electrical stimulation of the rat dorsal raphe nucleus. *Brain Res.* 650:149–52

Dickenson AH. 1977. Specific responses of rat raphe neurones to skin temperature. *J. Physiol.* 273:227–93

Ding ZQ, Li YW, Wesselingh SL, Blessing

WW. 1993. Transneuronal labelling of neurons in rabbit brain after injection of herpes simplex virus type 1 into the renal nerve. *J. Auton. Nerv. Syst.* 42:23–31

Drewes AM, Nielsen KD, Arendt NL, Birket SL, Hansen LM. 1997. The effect of cutaneous and deep pain on the electroencephalogram during sleep-an experimental study. *Sleep* 20:632–40

Drower EJ, Hammond DL. 1988. GABAergic modulation of nociceptive threshold: effects of THIP and bicuculline microinjected in the ventral medulla of the rat. *Brain Res.* 450:316–24

Du HJ. 1989. Medullary neurons with projections to lamina X of the rat as demonstrated by retrograde labeling after HRP microelectrophoresis. *Brain Res.* 505:135–40

Fang F, Proudfit HK. 1996. Spinal cholinergic and monoamine receptors mediate the antinociceptive effect of morphine microinjected in the periaqueductal gray on the rat tail, but not the feet. *Brain Res.* 722:95–108

Fang F, Proudfit HK. 1998. Antinociception produced by microinjection of morphine in the rat periaqueductal gray is enhanced in the foot, but not the tail, by intrathecal injection of alpha-1-adrenoceptor antagonists. *Brain Res.* 790:14–24

Fay RA, Norgren R. 1997a. Identification of rat brainstem multisynaptic connections to the oral motor nuclei in the rat using pseudorabies virus. II. Facial muscle motor systems. *Brain Res. Rev.* 25:276–90

Fay RA, Norgren R. 1997b. Identification of rat brainstem multisynaptic connections to the oral motor nuclei using pseudorabies virus. I. Masticatory muscle motor systems. *Brain Res. Rev.* 25:255–75

Fay RA, Norgren R. 1997c. Identification of rat brainstem multisynaptic connections to the oral motor nuclei using pseudorabies virus. III. Lingual muscle motor systems. *Brain Res. Rev.* 25:291–311

Fields HL, Basbaum AI. 1978. Brainstem control of spinal pain-transmission neurons. *Annu. Rev. Physiol.* 40:217–48

Fields HL, Bry J, Hentall I, Zorman G. 1983a. The activity of neurons in the rostral medulla of the rat during withdrawal from noxious heat. *J. Neurosci.* 3:2545–52

Fields HL, Heinricher MM, Mason P. 1991. Neurotransmitters in nociceptive modulatory circuits. *Annu. Rev. Neurosci.* 14:219–45

Fields HL, Malick A, Burstein R. 1995. Dorsal horn projection targets of ON and OFF cells in the rostral ventromedial medulla. *J. Neurophysiol.* 74:1742–59

Fields HL, Vanegas H, Hentall ID, Zorman G. 1983b. Evidence that disinhibition of brain stem neurones contributes to morphine analgesia. *Nature* 306:684–86

Foo H, Mason P. 2000. Bicuculline microinjection into the raphe magnus suppresses air puff-evoked awakenings in behaving rats. *Soc. Neurosci.* 26:660 (Abstr.)

Fornal C, Auerbach S, Jacobs BL. 1985. Activity of serotonin-containing neurons in nucleus raphe magnus in freely moving cats. *Exp. Neurol.* 88:590–608

Gang S, Mizuguchi A, Kobayashi N, Aoki M. 1990. Descending axonal projections from the medial parabrachial and Kolliker-Fuse nuclear complex to the nucleus raphe magnus in cats. *Neurosci. Lett.* 118:273–75

Gao K, Chen DO, Genzen JR, Mason P. 1998. Activation of serotonergic neurons in the raphe magnus is not necessary for morphine antinociception. *J. Neurosci.* 18:1860–68

Gao K, Kim YH, Mason P. 1997. Serotonergic pontomedullary neurons are not activated by antinociceptive stimulation in the periaqueductal gray. *J. Neurosci.* 17:3285–92

Gao K, Mason P. 1997. Somatodendritic morphology and axonal anatomy of intracellularly labeled serotonergic neurons in the rat medulla. *J. Comp. Neurol.* 389:309–28

Gao K, Mason P. 2000. Serotonergic raphe magnus cells that respond to noxious tail heat are not ON or OFF cells. *J. Neurophysiol.* 84:1719–25

Gebhart GF, Sandkühler J, Thalhammer JG, Zimmermann M. 1983. Inhibition of spinal nociceptive information by stimulation in midbrain of the cat is blocked by lidocaine microinjected in nucleus raphe magnus and medullary reticular formation. *J. Neurophysiol.* 50:1446–59

Genzen JR, Gao K, Chen DO, Mason P. 1997. Participation of pontomedullary serotonergic neurons in cardiovascular modulation. *Soc. Neurosci.* 23:157 (Abstr.)

Gerendai I, Toth IE, Boldogkoi Z, Medveczky I, Halasz B. 1998. Neuronal labeling in the rat brain and spinal cord from the ovary using viral transneuronal tracing technique. *Neuroendocrinology* 68:244–56

Ghione S. 1996. Hypertension-associated hypalgesia. Evidence in experimental animals and humans, pathophysiological mechanisms, and potential clinical consequences. *Hypertension* 28:494–504

Giaquinto S, Pompeiano O, Somogyi I. 1964. Supraspinal modulation of heteronymous monosynaptic and of poysynaptic reflexes during natural sleep and wakefulness. *Arch. Ital. Biol.* 102:245–81

Grahn DA, Heller HC. 1989. Activity of most rostral ventromedial medulla neurons reflect EEG/EMG pattern changes. *Am. J. Physiol. Regul. Integr. Comp. Physiol.* 257:R1496–505

Grumbach L. 1966. The prediction of analgesic activity in man by animal testing. In *Pain*, ed. RS Knighton, PR Dumke, pp. 163–82. Boston: Little, Brown & Co.

Hadziefendic S, Haxhiu MA. 1999. CNS innervation of vagal preganglionic neurons controlling peripheral airways: a transneuronal labeling study using pseudorabies virus. *J. Auton. Nerv. Syst.* 76:135–45

Hammond DL, Yaksh TL. 1984. Antagonism of stimulation-produced antinociception by intrathecal administration of methysergide or phentolamine. *Brain Res.* 298:329–37

Haxhiu MA, Jansen AS, Cherniack NS, Loewy AD. 1993. CNS innervation of airway-related parasympathetic preganglionic neurons: a transneuronal labeling study using pseudorabies virus. *Brain Res.* 618:115–34

Heinricher MM, Barbaro NM, Fields HL. 1989.

Putative nociceptive modulating neurons in the rostral ventromedial medulla of the rat: firing of on- and off-cells is related to nociceptive responsiveness. *Somatosens. Mot. Res.* 6:427–39

Heinricher MM, Drasner K. 1991. Lumbar intrathecal morphine alters activity of putative nociceptive modulatory neurons in rostral ventromedial medulla. *Brain Res.* 549:338–41

Heinricher MM, McGaraughty S. 1998. Analysis of excitatory amino acid transmission within the rostral ventromedial medulla: implications for circuitry. *Pain* 75:247–55

Heinricher MM, McGaraughty S, Farr DA. 1999. The role of excitatory amino acid transmission within the rostral ventromedial medulla in the antinociceptive actions of systemically administered morphine. *Pain* 81:57–65

Heinricher MM, Morgan MM, Fields HL. 1992. Direct and indirect actions of morphine on medullary neurons that modulate nociception. *Neuroscience* 48:533–43

Heinricher MM, Morgan MM, Tortorici V, Fields HL. 1994. Disinhibition of off-cells and antinociception produced by an opioid action within the rostral ventromedial medulla. *Neuroscience* 63:279–88

Heinricher MM, Tortorici V. 1994. Interference with GABA transmission in the rostral ventromedial medulla: disinhibition of off-cells as a central mechanism in nociceptive modulation. *Neuroscience* 63:533–46

Helke CJ, Sayson SC, Keeler JR, Charlton CG. 1986. Thyrotropin-releasing hormone-immunoreactive neurons project from the ventral medulla to the intermediolateral cell column: partial coexistence with serotonin. *Brain Res.* 381:1–7

Helke CJ, Thor KB, Sasek CA. 1989. Chemical neuroanatomy of the parapyramidal region of the ventral medulla in the rat. *Prog. Brain Res.* 81:17–28

Hellon RF, Taylor DC. 1982. An analysis of a thermal afferent pathway in the rat. *J. Physiol.* 326:319–28

Hermann DM, Luppi PH, Peyron C, Hinckel P, Jouvet M. 1996. Forebrain projections of the rostral nucleus raphe magnus shown by iontophoretic application of choleratoxin b in rats. *Neurosci. Lett.* 216:151–54

Hermann DM, Luppi PH, Peyron C, Hinckel P, Jouvet M. 1997. Afferent projections to the rat nuclei raphe magnus, raphe pallidus and reticularis gigantocellularis pars alpha demonstrated by iontophoretic application of choleratoxin (subunit b). *J. Chem. Neuroanat.* 13:1–21

Heym J, Steinfels GF, Jacobs BL. 1982. Activity of serotonin-containing neurons in the nucleus raphe pallidus of freely moving cats. *Brain Res.* 251:259–76

Hinckel P, Cristante L, Bruck K. 1983. Inhibitory effects of the lower brain stem on shivering. *J. Therm. Biol.* 8:129–31

Holstege G. 1987. Some anatomical observations on the projections from the hypothalamus to brainstem and spinal cord: an HRP and autoradiographic tracing study in the cat. *J. Comp. Neurol.* 260:98–126

Holstege G. 1988. Anatomical evidence for a strong ventral parabrachial projection to nucleus raphe magnus and adjacent tegmental field. *Brain Res.* 447:154–58

Holstege G, Kuypers HG. 1982. The anatomy of brain stem pathways to the spinal cord in cat. A labeled amino acid tracing study. *Prog. Brain Res.* 57:145–75

Holstege G, Meiners L, Tan K. 1985. Projections of the bed nucleus of the stria terminalis to the mesencephalon, pons, and medulla oblongata in the cat. *Exp. Brain Res.* 58:379–91

Hosobuchi Y, Adams JE, Linchitz R. 1977. Pain relief by electrical stimulation of the central gray matter in humans and its reversal by naloxone. *Science* 197:183–86

Huang J, Weiss ML. 1999. Characterization of the central cell groups regulating the kidney in the rat. *Brain Res.* 845:77–91

Hubscher CH, Johnson RD. 1996. Responses of medullary reticular formation neurons to input from the male genitalia. *J. Neurophysiol.* 76:2474–82

Huisman AM, Kuypers HG, Verburgh CA. 1981. Quantitative differences in collateralization of the descending spinal pathways from red nucleus and other brain stem cell groups in rat as demonstrated with the multiple fluorescent retrograde tracer technique. *Brain Res.* 209:271–86

Iwakiri H, Matsuyama K, Mori S. 1993. Extracellular levels of serotonin in the medial pontine reticular formation in relation to sleep-wake cycle in cats: a microdialysis study. *Neurosci. Res.* 18:157–70

Jacobs BL, Azmitia EC. 1992. Structure and function of the brain serotonin system. *Physiol. Rev.* 72:165–229

Jansen AS, Nguyen XV, Karpitskiy V, Mettenleiter TC, Loewy AD. 1995. Central command neurons of the sympathetic nervous system: basis of the fight-or-flight response. *Science* 270:644–46

Jasmin L, Carstens E, Basbaum AI. 1997. Interneurons presynaptic to rat tail-flick motoneurons as mapped by transneuronal transport of pseudorabies virus: few have long ascending collaterals. *Neuroscience* 76:859–76

Jensen TS, Yaksh TL. 1984. Spinal monoamine and opiate systems partly mediate the antinociceptive effects produced by glutamate at brainstem sites. *Brain Res.* 321:287–97

Johannessen JN, Watkins LR, Mayer DJ. 1984. Non-serotonergic origins of the dorsolateral funiculus in the rat ventral medulla. *J. Neurosci.* 4:757–66

Jones SL, Light AR. 1990. Termination patterns of serotoninergic medullary raphespinal fibers in the rat lumbar spinal cord: an anterograde immunohistochemical study. *J. Comp. Neurol.* 297:267–82

Jones SL, Light AR. 1992. Serotoninergic medullary raphespinal projection to the lumbar spinal cord in the rat: a retrograde immunohistochemical study. *J. Comp. Neurol.* 322:599–610

Jordan LM, Kenshalo DR Jr, Martin RF, Haber LH, Willis WD. 1978. Depression of primate spinothalamic tract neurons by iontophoretic application of 5-hydroxytryptamine. *Pain* 5:135–42

Kaplan H, Fields HL. 1991. Hyperalgesia during acute opioid abstinence: evidence for a nociceptive facilitating function of the rostral ventromedial medulla. *J. Neurosci.* 11:1433–39

Kruger L, Saporta S, Swanson LW. 1995. *Photographic Atlas of the Rat Brain*. Cambridge, UK: Cambridge Univ. Press. 299 pp.

Kshatri AM, Baghdoyan HA, Lydic R. 1998. Cholinomimetics, but not morphine, increase antinociceptive behavior from pontine reticular regions regulating rapid-eye-movement sleep. *Sleep* 21:677–85

Kwiat GC, Basbaum AI. 1992. The origin of brainstem noradrenergic and serotonergic projections to the spinal cord dorsal horn in the rat. *Somatosens. Mot. Res.* 9:157–73

Labbate LA, Grimes JB, Arana GW. 1998. Serotonin reuptake antidepressant effects on sexual function in patients with anxiety disorders. *Biol. Psychiatry* 43:904–7

Lang PJ. 1994. The varieties of emotional experience: a meditation on James-Lange theory. *Psychol. Rev.* 101:211–21

Lavigne G, Zucconi M, Castronovo C, Manzini C, Marchettini P, Smirne S. 2000. Sleep arousal response to experimental thermal stimulation during sleep in human subjects free of pain and sleep problems. *Pain* 84:283–90

Leanza G, Pellitteri R, Russo A, Stanzani S. 1991. Neurons in raphe nuclei pontis and magnus have branching axons that project to medial preoptic area and cervical spinal cord. A fluorescent retrograde double labeling study in the rat. *Neurosci. Lett.* 123:195–99

LeBars D. 1988. Serotonin and pain. In *Neuronal Serotonin*, ed. NN Osborne, M Hamon, pp. 171–226. New York: Wiley

Leung CG, Mason P. 1996. Spectral analysis of arterial blood pressure and raphe magnus neuronal activity in anesthetized rats. *Am. J. Physiol. Regul. Integr. Comp. Physiol.* 271:R483–89

Leung CG, Mason P. 1998. A physiological survey of medullary raphe and magnocellular reticular neurons in the anesthetized rat. *J. Neurophysiol.* 80:1630–46

Leung CG, Mason P. 1999. Physiological properties of medullary raphe neurons during sleep and waking. *J. Neurophysiol.* 81:584–95

Li YW, Bayliss DA. 1998a. Electrophysical properties, synaptic transmission and neuromodulation in serotonergic caudal raphe neurons. *Clin. Exp. Pharmacol. Physiol.* 25:468–73

Li YW, Bayliss DA. 1998b. Presynaptic inhibition by 5-HT1B receptors of glutamatergic synaptic inputs onto serotonergic caudal raphe neurones in rat. *J. Physiol.* 510:121–34

Light AR. 1985. The spinal terminations of single, physiologically characterized axons originating in the pontomedullary raphe of the cat. *J. Comp. Neurol.* 234:536–48

Loewy AD, Franklin MF, Haxhiu MA. 1994. CNS monoamine cell groups projecting to pancreatic vagal motor neurons: a transneuronal labeling study using pseudorabies virus. *Brain Res.* 638:248–60

Marson L. 1995. Central nervous system neurons identified after injection of pseudorabies virus into the rat clitoris. *Neurosci. Lett.* 190:41–44

Marson L. 1997. Identification of central nervous system neurons that innervate the bladder body, bladder base, or external urethral sphincter of female rats: a transneuronal tracing study using pseudorabies virus. *J. Comp. Neurol.* 389:584–602

Marson L, McKenna KE. 1990. The identification of a brainstem site controlling spinal sexual reflexes in male rats. *Brain Res.* 515:303–8

Marson L, McKenna KE. 1992. A role for 5-hydroxytryptamine in descending inhibition of spinal sexual reflexes. *Exp. Brain Res.* 88:313–20

Marson L, McKenna KE. 1994. Serotonergic neurotoxic lesions facilitate male sexual reflexes. *Pharmacol. Biochem. Behav.* 47: 883–88

Marson L, McKenna KE. 1996. CNS cell groups involved in the control of the ischiocavernosus and bulbospongiosus muscles: a transneuronal tracing study using pseudorabies virus. *J. Comp. Neurol.* 374:161–79

Marson L, Platt KB, McKenna KE. 1993. Central nervous system innervation of the penis as revealed by the transneuronal transport of pseudorabies virus. *Neuroscience* 55:263–80

Martin GF, DeLorenzo G, Ho RH, Humbertson AO Jr, Waltzer R. 1985. Serotonergic innervation of the forebrain in the North American opossum. *Brain Behav. Evol.* 26:196–228

Martin-Cora FJ, Fornal CA, Metzler CW, Jacobs BL. 2000. Single-unit responses of serotonergic medullary and pontine raphe neurons to environmental cooling in freely moving cats. *Neuroscience* 98:301–9

Mason P. 1997. Physiological identification of pontomedullary serotonergic neurons in the rat. *J. Neurophysiol.* 77:1087–98

Mason P, Escobedo I, Burgin C, Bergan J, Lee JH, et al. 2001. Nociceptive responsiveness during slow wave sleep and waking in the rat. *Sleep* 24:13–17

Mason P, Floeter MK, Fields HL. 1990. Somatodendritic morphology of on- and off-cells in the rostral ventromedial medulla. *J. Comp. Neurol.* 301:23–43

Mason P, Gao K. 1998. Raphe magnus serotonergic neurons tonically modulate nociceptive transmission. *Pain Forum* 7:143–50

Masson RL Jr, Sparkes ML, Ritz LA. 1991. Descending projections to the rat sacrocaudal spinal cord. *J. Comp. Neurol.* 307:120–30

Matos FF, Rollema H, Brown JL, Basbaum AI. 1992. Do opioids evoke the release of serotonin in the spinal cord? An in vivo microdialysis study of the regulation of extracellular serotonin in the rat. *Pain* 48:439–47

McAllen RM, Rathner JA, Owens NC, Trevaks D. 2000. Medullary raphe neurons and autonomic responses to cold. *Proc. Aust. Neurosci. Soc.* 11:12 (Abstr.)

McKenna KE, Chung SK, McVary KT. 1991. A model for the study of sexual function in anesthetized male and female rats. *Am. J. Physiol. Regul. Integr. Comp. Physiol.* 261:R1276–85

Moore RY. 1981. The anatomy of central serotonin neuron systems in the rat brain. In *Serotonin Neurotransmission and Behavior*, ed. BL Jacobs, A Gelperin, pp. 35–71. Cambridge, MA: MIT Press

Mori S, Sakamoto T, Ohta Y, Takakusaki K, Matsuyama K. 1989. Site-specific postural and locomotor changes evoked in awake, freely moving intact cats by stimulating the brainstem. *Brain Res.* 505:66–74

Morrison S. 2000. Differential regulation of sympathetic outflows to vasoconstrictor and thermoregulatory effectors. *Physiologist* 43:271 (Abstr.)

Morrison SF. 1999. RVLM and raphe differentially regulate sympathetic outflows to splanchnic and brown adipose tissue. *Am. J. Physiol. Regul. Integr. Comp. Physiol.* 276:R962–73

Morrison SF, Sved AF, Passerin AM. 1999. GABA-mediated inhibition of raphe pallidus neurons regulates sympathetic outflow to brown adipose tissue. *Am. J. Physiol. Regul. Integr. Comp. Physiol.* 276:R290–97

Murphy AZ, Rizvi TA, Ennis M, Shipley MT. 1999. The organization of preopticmedullary circuits in the male rat: evidence for interconnectivity of neural structures involved in reproductive behavior, antinociception and cardiovascular regulation. *Neuroscience* 91:1103–16

Nalivaiko E, Blessing WW. 1999. Synchronous changes in ear and tail blood flow following salient and noxious stimuli in rabbits. *Brain Res.* 847:343–46

Nason MW, Mason P. 2000. Bicuculline microinjection into raphe magnus elicits sympathetic responses and facilitates the heat-evoked paw withdrawal. *Soc. Neurosci.* 26:561 (Abstr.)

Newman DB. 1985. Distinguishing rat brainstem reticulospinal nuclei by their neuronal morphology. I. Medullary nuclei. *J. Hirnforsch.* 26:187–226

Ochoa J, Torebjork E. 1989. Sensations evoked by intraneural microstimulation of C nociceptor fibres in human skin nerves. *J. Physiol.* 415:583–99

Oliveras JL, Bourgoin S, Hery F, Besson JM, Hamon M. 1977. The topographical distribution of serotoninergic terminals in the spinal cord of the cat: biochemical mapping by the combined use of microdissection and microassay procedures. *Brain Res.* 138:393–406

Oliveras JL, Martin G, Montagne J, Vos B. 1990. Single unit activity at ventromedial medulla level in the awake, freely moving rat: effects of noxious heat and light tactile stimuli onto convergent neurons. *Brain Res.* 506:19–30

Oliveras JL, Vos B, Martin G, Montagne J. 1989. Electrophysiological properties of ventromedial medulla neurons in response to noxious and non-noxious stimuli in the awake, freely moving rat: a single-unit study. *Brain Res.* 486:1–14

Papka RE, Williams S, Miller KE, Copelin T, Puri P. 1998. CNS location of uterine-related neurons revealed by trans-synaptic tracing with pseudorabies virus and their relation to estrogen receptor-immunoreactive neurons. *Neuroscience* 84:935–52

Paxinos G, Watson C. 1986. *The Rat Brain in Stereotaxic Coordinates.* San Diego: Academic

Potrebic SB, Fields HL, Mason P. 1994. Serotonin immunoreactivity is contained in one physiological cell class in the rat rostral ventromedial medulla. *J. Neurosci.* 14:1655–65

Potrebic SB, Mason P. 1993. Three-dimensional analysis of the dendritic domains of on- and off-cells in the rostral ventromedial medulla. *J. Comp. Neurol.* 337:83–93

Randic M, Yu HH. 1976. Effects of 5-hydroxytryptamine and bradykinin in cat dorsal horn neurones activated by noxious stimuli. *Brain Res.* 111:197–203

Rathner JA, McAllen RM. 1999. Differential control of sympathetic drive to the rat tail artery and kidney by medullary premotor cell groups. *Brain Res.* 834:196–99

Reynolds DV. 1969. Surgery in the rat during electrical analgesia induced by focal brain stimulation. *Science* 164:444–45

Richardson DE, Akil H. 1977. Pain reduction by electrical brain stimulation in man. Part 1: Acute administration in periaqueductal and periventricular sites. *J. Neurosurg.* 47:178–83

Rose JD. 1975. Response properties and anatomical organization of pontine and medullary units responsive to vaginal stimulation in the cat. *Brain Res.* 97:79–93

Rose JD, Sutin J. 1973. Responses of single units in the medulla to genital stimulation in estrous and anestrous cats. *Brain Res.* 50:87–99

Rotto-Percelay DM, Wheeler JG, Osorio FA, Platt KB, Loewy AD. 1992. Transneuronal labeling of spinal interneurons and sympathetic preganglionic neurons after pseudorabies virus injections in the rat medial gastrocnemius muscle. *Brain Res.* 574:291–306

Sachs BD, Meisel RL. 1988. The physiology of male sexual behavior. In *The Physiology of Reproduction,* ed. E Knobil, J Neill, pp. 1393–485. New York: Raven

Sandkühler J. 1996. The organization and function of endogenous antinociceptive systems. *Prog. Neurobiol.* 50:49–81

Sandkühler J, Gebhart GF. 1984. Relative contributions of the nucleus raphe magnus and adjacent medullary reticular formation to the inhibition by stimulation in the periaqueductal gray of a spinal nociceptive reflex in the pentobarbital-anesthetized rat. *Brain Res.* 305:77–87

Sasek CA, Helke CJ. 1989. Enkephalin-immunoreactive neuronal projections from the medulla oblongata to the intermediolateral cell column: relationship to substance P-immunoreactive neurons. *J. Comp. Neurol.* 287:484–94

Sasek CA, Wessendorf MW, Helke CJ. 1990.

Evidence for co-existence of thyrotropin-releasing hormone, substance P and serotonin in ventral medullary neurons that project to the intermediolateral cell column in the rat. *Neuroscience* 35:105–19

Satinoff E. 1978. Neural organization and evolution of thermal regulation in mammals. *Science* 201:16–22

Sato H. 1993. Raphe-spinal and subcoeruleospinal modulation of temperature signal transmission in rats. *J. Therm. Biol.* 18:211–21

Sawynok J. 1989. The 1988 Merck Frosst Award. The role of ascending and descending noradrenergic and serotonergic pathways in opioid and non-opioid antinociception as revealed by lesion studies. *Can. J. Physiol. Pharmacol.* 67:975–88

Schramm LP, Strack AM, Platt KB, Loewy AD. 1993. Peripheral and central pathways regulating the kidney: a study using pseudorabies virus. *Brain Res.* 616:251–62

Skagerberg G, Bjorklund A. 1985. Topographic principles in the spinal projections of serotonergic and non-serotonergic brainstem neurons in the rat. *Neuroscience* 15:445–80

Sly DJ, Colvill L, McKinley MJ, Oldfield BJ. 1999. Identification of neural projections from the forebrain to the kidney, using the virus pseudorabies. *J. Auton. Nerv. Syst.* 77:73–82

Smith JE, Gilbey MP. 2000. Coherent rhythmic discharges in sympathetic nerves supplying thermoregulatory circulations in the rat. *J. Physiol.* 523(Part 2):449–57

Smith JE, Jansen AS, Gilbey MP, Loewy AD. 1998. CNS cell groups projecting to sympathetic outflow of tail artery: neural circuits involved in heat loss in the rat. *Brain Res.* 786:153–64

Sorkin LS, McAdoo DJ, Willis WD. 1993. Raphe magnus stimulation-induced antinociception in the cat is associated with release of amino acids as well as serotonin in the lumbar dorsal horn. *Brain Res.* 618:95–108

Spencer SE, Sawyer WB, Wada H, Platt KB, Loewy AD. 1990. CNS projections to the pterygopalatine parasympathetic preganglionic neurons in the rat: a retrograde transneuronal viral cell body labeling study. *Brain Res.* 534:149–69

Steinbusch HW. 1981. Distribution of serotonin-immunoreactivity in the central nervous system of the rat-cell bodies and terminals. *Neuroscience* 6:557–618

Strack AM, Sawyer WB, Hughes JH, Platt KB, Loewy AD. 1989a. A general pattern of CNS innervation of the sympathetic outflow demonstrated by transneuronal pseudorabies viral infections. *Brain Res.* 491:156–62

Strack AM, Sawyer WB, Platt KB, Loewy AD. 1989b. CNS cell groups regulating the sympathetic outflow to adrenal gland as revealed by transneuronal cell body labeling with pseudorabies virus. *Brain Res.* 491:274–96

Szelenyi Z, Hinckel P. 1987. Changes in cold- and heat-defence following electrolytic lesions of raphe nuclei in the guinea-pig. *Pflugers Arch. Eur. J. Physiol.* 409:175–81

Taber E, Brodal A, Walberg F. 1960. The raphe nuclei of the brain stem in the cat. I. Normal topography and cytoarchitechture and general discussion. *J. Comp. Neurol.* 114:161–87

Taylor DC. 1982. The effects of nucleus raphe magnus lesions on an ascending thermal pathway in the rat. *J. Physiol.* 326:309–18

Ter Horst GJ, Hautvast RW, De Jongste MJ, Korf J. 1996. Neuroanatomy of cardiac activity-regulating circuitry: a transneuronal retrograde viral labelling study in the rat. *Eur. J. Neurosci.* 8:2029–41

Thorat SN, Hammond DL. 1997. Modulation of nociception by microinjection of delta-1 and delta-2 opioid receptor ligands in the ventromedial medulla of the rat. *J. Pharmacol. Exp. Ther.* 283:1185–92

Torebjork HE, Schady W, Ochoa J. 1984. Sensory correlates of somatic afferent fibre activation. *Hum. Neurobiol.* 3:15–20

Trulson ME, Jacobs BL. 1979. Raphe unit activity in freely moving cats: correlation with level of behavioral arousal. *Brain Res.* 163:135–50

Urban MO, Smith DJ. 1994. Nuclei within the rostral ventromedial medulla mediating morphine antinociception from the periaqueductal gray. *Brain Res.* 652:9–16

Valentino RJ, Kosboth M, Colflesh M, Miselis RR. 2000. Transneuronal labeling from the rat distal colon: anatomic evidence for regulation of distal colon function by a pontine corticotropin-releasing factor system. *J. Comp. Neurol.* 417:399–414

Van Bockstaele EJ, Aston-Jones G, Pieribone VA, Ennis M, Shipley MT. 1991. Subregions of the periaqueductal gray topographically innervate the rostral ventral medulla in the rat. *J. Comp. Neurol.* 309:305–27

Vandermaelen CP, Aghajanian GK. 1983. Electrophysiological and pharmacological characterization of serotonergic dorsal raphe neurons recorded extracellularly and intracellularly in rat brain slices. *Brain Res.* 289:109–19

Vertes RP. 1988. Brainstem afferents to the basal forebrain in the rat. *Neuroscience* 24:907–35

Vizzard MA, Erickson VL, Card JP, Roppolo JR, de Groat WC. 1995. Transneuronal labeling of neurons in the adult rat brainstem and spinal cord after injection of pseudorabies virus into the urethra. *J. Comp. Neurol.* 355:629–40

Wang RY, Aghajanian GK. 1982. Correlative firing patterns of serotonergic neurons in rat dorsal raphe nucleus. *J. Neurosci.* 2:11–16

Watkins LR, Griffin G, Leichnetz GR, Mayer DJ. 1980. The somatotopic organization of the nucleus raphe magnus and surrounding brain stem structures as revealed by HRP slow-release gels. *Brain Res.* 181:1–15

Werner J, Bienek A. 1985. The significance of nucleus raphe dorsalis and centralis for thermoafferent signal transmission to the preoptic area of the rat. *Exp. Brain Res.* 59:543–47

Werner J, Bienek A. 1990. Loss and restoration of preoptic thermoreactiveness after lesions of the rostral raphe nuclei. *Exp. Brain Res.* 80:429–35

Wessendorf MW, Anderson EG. 1983. Single unit studies of identified bulbospinal serotonergic units. *Brain Res.* 279:93–103

Wessendorf MW, Proudfit HK, Anderson EG. 1981. The identification of serotonergic neurons in the nucleus raphe magnus by conduction velocity. *Brain Res.* 214:168–73

Wilkinson LO, Auerbach SB, Jacobs BL. 1991. Extracellular serotonin levels change with behavioral state but not with pyrogen-induced hyperthermia. *J. Neurosci.* 11:2732–41

Willis WD Jr. 1988. Anatomy and physiology of descending control of nociceptive responses of dorsal horn neurons: comprehensive review. *Prog. Brain Res.* 77:1–29

Wu W, Elde R, Wessendorf MW. 1993. Organization of the serotonergic innervation of spinal neurons in rats–III. Differential serotonergic innervation of somatic and parasympathetic preganglionic motoneurons as determined by patterns of co-existing peptides. *Neuroscience* 55:223–33

Wu W, Wessendorf MW. 1992. Organization of the serotonergic innervation of spinal neurons in rats–I. Neuropeptide coexistence in varicosities innervating some spinothalamic tract neurons but not in those innervating postsynaptic dorsal column neurons. *Neuroscience* 50:885–98

Yaksh TL, Wilson PR. 1979. Spinal serotonin terminal system mediates antinociception. *J. Pharmacol. Exp. Ther.* 208:446–53

Yang H, Yuan PQ, Wang L, Tache Y. 2000. Activation of the parapyramidal region in the ventral medulla stimulates gastric acid secretion through vagal pathways in rats. *Neuroscience* 95:773–79

Yells DP, Hendricks SE, Prendergast MA. 1992. Lesions of the nucleus paragigantocellularis: effects on mating behavior in male rats. *Brain Res.* 596:73–79

Yells DP, Prendergast MA, Hendricks SE,

Miller ME. 1995. Monoaminergic influences on temporal patterning of sexual behavior in male rats. *Physiol. Behav.* 58:847–52

Young AA, Dawson NJ. 1987. Static and dynamic response characteristics, receptive fields, and interaction with noxious input of midline medullary thermoresponsive neurons in the rat. *J. Neurophysiol.* 57:1925–36

Young RF, Kroening R, Fulton W, Feldman RA, Chambi I. 1985. Electrical stimulation of the brain in treatment of chronic pain. Experience over 5 years. *J. Neurosurg.* 62:389–96

Zajecka J, Mitchell S, Fawcett J. 1997. Treatment-emergent changes in sexual function with selective serotonin reuptake inhibitors as measured with the Rush Sexual Inventory. *Psychopharmacol. Bull.* 33:755–60

Zhuo M, Gebhart GF. 1990. Characterization of descending inhibition and facilitation from the nuclei reticularis gigantocellularis and gigantocellularis pars alpha in the rat. *Pain* 42:337–50

Zhuo M, Gebhart GF. 1991. Spinal serotonin receptors mediate descending facilitation of a nociceptive reflex from the nuclei reticularis gigantocellularis and gigantocellularis pars alpha in the rat. *Brain Res.* 550:35–48

Zhuo M, Gebhart GF. 1992. Characterization of descending facilitation and inhibition of spinal nociceptive transmission from the nuclei reticularis gigantocellularis and gigantocellularis pars alpha in the rat. *J. Neurophysiol.* 67:1599–614

Annu. Rev. Neurosci. 2001. 24:779–805

ACTIVATION, DEACTIVATION, AND ADAPTATION IN VERTEBRATE PHOTORECEPTOR CELLS

Marie E Burns and Denis A Baylor

Department of Neurobiology, Stanford University Medical Center, Stanford, California 94305; e-mail: mburns@stanford.edu, dbaylor@stanford.edu

Key Words rods, cones, vision, G protein, rhodopsin

■ **Abstract** Visual transduction captures widespread interest because its G-protein signaling motif recurs throughout nature yet is uniquely accessible for study in the photoreceptor cells. The light-activated currents generated at the photoreceptor outer segment provide an easily observed real-time measure of the output of the signaling cascade, and the ease of obtaining pure samples of outer segments in reasonable quantity facilitates biochemical experiments. A quiet revolution in the study of the mechanism has occurred during the past decade with the advent of gene-targeting techniques. These have made it possible to observe how transduction is perturbed by the deletion, overexpression, or mutation of specific components of the transduction apparatus.

INTRODUCTION

The purpose of this review is to assess current progress in our understanding of visual transduction. This step initiates the visual process and thus directly impacts our perception of the outside world. Whereas notable advances have been made in understanding transduction, much remains to be done, particularly in understanding the kinetics and structural basis of the underlying molecular interactions as well as their relation to pathological changes in the photoreceptor cells.

The absolute necessity of transduction for vision has important consequences. Failure of transduction, resulting from degeneration of the photoreceptor cells, occurs in the blinding diseases macular degeneration and retinitis pigmentosa, which leave the visual system without input signals to analyze. Given the intricate nature of transduction and the large amount of a photoreceptor cell's total protein content devoted to it, it is perhaps not surprising that mutations in the genes that encode transduction proteins are now known to underlie an appreciable fraction of cases of these dread diseases (reviewed in Shastry 1997, Phelan & Bok 2000).

The manner in which transduction is accomplished sets important limits on visual perception. Thus, only photons that are transduced can be seen, and the wavelength dependence of vision as well as the trichromacy of color vision result

0147-006X/01/0621-0779$14.00

directly from the wavelength dependence of absorption by the visual pigments. The absolute sensitivity of vision seems to be determined by the requirement that the number of transduced photons exceed the number of photon-like noise signals generated within the photoreceptor cells. Similarly, the range of light intensities in which rod and cone vision operate is set by the range over which the photoreceptor cells can generate output signals in response to incident photons. The dynamics of the light-induced signals in the retinal photoreceptors set the persistence time of visual sensations as well as our visual sensitivity to flickering lights of different frequencies. Finally, recovery of visual sensitivity after exposure to bright light is limited by the recovery of sensitivity in the transduction apparatus.

Our review is topical rather than comprehensive; interested readers may consult several other recent reviews (see e.g. *Methods in Enzymology*, Vols 315–316, *Novartis Foundation Symposium*, Vol. 224, *Handbook of Biological Physics*, Vol. 3; see also Rieke & Baylor 1998b, Sakmar 1998, Pugh et al 1999). Somewhat arbitrarily, we have divided our presentation into sections treating activation, deactivation, and adaptation (sensitivity control). Much of our review focuses on rods, which are better understood than cones.

ACTIVATION

Elucidation of the activation steps in rod phototransduction is a great success story of modern biochemistry (see Stryer 1995). In the activation process light excites rhodopsin, which activates the G protein transducin, which in turn activates the cGMP phosphodiesterase (PDE), lowering the intracellular concentration of cGMP and hyperpolarizing the cell.

Rhodopsin Activation by Light

Activation of rhodopsin begins when the absorption of a photon isomerizes rhodopsin's 11-cis retinal chromophore to the all-trans configuration (Wald 1968). The ensuing changes in the protein portion of the molecule are still not well understood. How is isomerization of the deeply buried chromophore communicated to the surrounding transmembrane helices and thence to the cytoplasmic surface of the molecule? What changes at the cytoplasmic surface confer catalytic activity?

Whereas most G protein–coupled receptors detect the presence of extracellular ligands, in the visual pigments the light-absorbing chromophore, the analog of a diffusible activating ligand, is covalently attached to the protein moiety at Lys296. The covalent attachment is advantageous because it abolishes temporal constraints imposed by ligand diffusion. Fast spectroscopic measurements indicate that photoisomerization of the chromophore occurs on a subpicosecond time scale (Hayward et al 1981, Schoenlein et al 1991). Isomerization of the chromophore then leads to series of changes in the configuration of the protein, and the form (Metarhodopsin II) competent to activate the G protein appears within a few milliseconds (Baumann 1976, Dickopf et al 1998).

The chromophore is attached to the amino group of Lys296 by a protonated Schiff base linkage. The apparent pK of this linkage is unusually high (pK > 16) (Steinberg et al 1993, Deng et al 1994) as a result of stabilization by the negatively charged Glu113 counterion (Sakmar et al 1989, Zhukovsky & Oprian 1989, Nathans 1990) and a tightly packed water molecule (Deng et al 1994, Nagata et al 1997, Creemers et al 1999, Eilers et al 1999). Biophysical studies have indicated that one of the early steps leading to the formation of Metarhodopsin II is deprotonation of the Schiff base linkage (Matthews et al 1963, Doukas et al 1978, Longstaff et al 1986).

It has been proposed that deprotonation of the Schiff base linkage leads to activation by permitting rearrangement of the helix bundle. Indeed, mutations that deconstrain the helices, particularly mutations in helix 3 and helix 6, result in constituitive activation of the mutant receptor (Han et al 1998). Constraining the helices via disulfide linkages reduces rhodopsin's ability to activate transducin (Farrens et al 1996, Sheikh et al 1996, Struthers et al 2000). Definitive understanding of the structural basis for rhodopsin activation will require crystallization and diffraction studies. A giant step toward this goal was recently achieved with the X-ray determination of bovine rhodopsin's three-dimensional structure at 2.8 Å resolution (Palczewski et al 2000). The new structure (PDB ID number 1f88) can be viewed at the website of the RCSB Protein Data Bank, www.rcsb.org/pdb. The crystal structure confirmed and extended the helical arrangement previously determined by cryo-EM on frog rhodopsin (Unger et al 1997), revealing the residues that surround the chromophore and redshift its absorption spectrum. In addition, the configuration of the chromophore itself was found to be 6s-cis, 11-cis, 12s-trans, anti C=N, settling a source of some controversy (see Grobner et al 2000). Next on the "wish list" is the structure of the catalytically active intermediate Metarhodopsin II.

Spontaneous Activation of Visual Pigments

In darkness rod cells display spontaneous noise fluctuations, one component of which resembles responses to single absorbed photons and arises from thermal activation of rhodopsin (Baylor et al 1980). In a toad rod near 20°C the rate constant for thermal activation of rhodopsin was estimated as 10^{-11} s^{-1} per molecule, corresponding to an average wait to isomerization of 2000 years. The great thermal stability of rhodopsin is highly advantageous, for it allows many molecules to be packed into a single rod cell, giving high light-catching ability, while keeping noise events confusable with single photons very rare, allowing dim light to be reliably detected. The basis for rhodopsin's thermal stability, as well as the reaction mechanism that leads to thermal activation, are still not well understood.

The temperature dependence of thermal activation of rhodopsin in toad rods gave an apparent activation energy of about 22 kcal/mole (Baylor et al 1980). This is similar to the activation energy for thermal isomerization of 11-cis retinal in solution (Hubbard 1966), suggesting that thermal activation resulted from spontaneous isomerization of the retinal chromophore. However, the activation energy of both processes is much smaller than the energy stored in an early intermediate (bathorhodopsin) of photoexcited rhodopsin (36 kcal/mole, Cooper 1979), suggesting that photoexcitation of rhodopsin proceeds over a much larger energy barrier than thermal isomerization. Furthermore, the apparent rate constant for thermal activation of a rhodopsin molecule is two to three orders of magnitude lower than that for thermal isomerization of 11-cis retinal in solution (Hubbard 1966, Baylor et al 1980). Assuming that thermal activation of rhodopsin is triggered by cis-trans isomerization of the retinal chromophore, the discrepancies might be explained by assuming that thermal activation of rhodopsin proceeds by a different mechanism than photoexcitation, or that thermal activation only takes place in rhodopsin molecules that have a very improbable structural state.

A specific hypothesis along the latter lines has been put forward by Birge & Barlow (1995), who propose that only chromophores with unprotonated Schiff base linkages to lysine 296 will undergo thermal isomerization at an appreciable rate. Because of the high pK of the protonated Schiff base (see above), few rhodopsins would be capable of undergoing isomerization in any brief interval. This model is based upon theoretical calculations as well as experimental observations on the pH dependence of the thermal event rate in Limulus photoreceptors (Barlow et al 1993). The behavior of the E113Q rhodopsin mutant may be consistent with the Birge-Barlow model. In this mutant, glutamate 113, the counterion to the protonated Schiff base, is changed to a neutral glutamine, reducing the pK of the Schiff base nitrogen to about 7 (Sakmar et al 1991, Lin et al 1992). This mutation causes E113Q rhodopsin to activate the G protein without light exposure (Robinson et al 1992). It is not yet clear whether the mutant rhodopin's constitutive catalytic activity represents sudden impulsive bursts resulting from thermal isomerization of rhodopsins with unprotonated Schiff base linkages, or instead a continuous low level activity.

One problem with the Birge-Barlow model is that changes in external pH altered the relative rate of thermal events in Limulus photoreceptors considerably more than the expected change in the relative number of unprotonated rhodopsins (Barlow et al 1993). Furthermore, changes in internal pH changed the rate of thermal events in the opposite direction to changes in external pH (Corson & Fein 1980); the direction of the change ought to be the same if internal and external pH affect only the protonation state of the Schiff base linkage. Thus, it will be important to test the Birge-Barlow model further and extend the tests to vertebrate photoreceptors.

Rhodopin's dark stability may also depend on interaction of the 11-cis retinal chromophore with the amino acid residues that spectrally tune the pigment. A possible connection between a pigment's wavelength of maximal absorption and its rate of spontaneous thermal activation was suggested many years ago by Barlow (1957), who proposed that the energy of the photons to which the pigment is most sensitive was the same as the energy barrier for thermal activation. Recently, it was found that the visual pigment in long wavelength-sensitive cones of the salamander retina undergoes thermal activation at a high rate (about 600 s^{-1} per cone at room temperature). This pigment, which is maximally excited by relatively low energy photons (λmax 600 nm), is at least 300 times less stable than the pigment of short wavelength (λmax 430 nm)–sensitive cones (Rieke & Baylor 2000). However, the apparent barrier to spontaneous activation of a long wavelength pigment molecule is far less than the energy of photons at the optimal wavelength (Baylor et al 1980, Koskelainen et al 2000, Rieke & Baylor 2000). Thus, it seems that the activation energy may be related to, but certainly does not equal, the energy of the photon that the pigment maximally absorbs. The specific relation between spectral tuning and thermal stability of pigments remains to be quantitatively defined and its physical basis determined.

Amplification in the Excitation Cascade

Once activated by light, photoexcited rhodopsin (MII) generates a macroscopic electrical signal in which each photoisomerization interrupts the flow of 10^5 or more cations into the cell. This particle amplification arises primarily from three processes: (*a*) activation of many copies of the G protein by a single photoexcited rhodopsin; (*b*) hydrolysis of a large number of cGMP molecules by an activated subunit of PDE; and (*c*) the response of multiple channels to the drop in cGMP concentration, each channel allowing cations to enter at a high rate in the open state. One goal of work on the transduction mechanism is to understand in quantitative terms the contribution of each process to the observed amplification.

Because particle amplification is achieved during the relatively brief duration of the elementary response, a photoexcited rhodopsin molecule must increase PDE activity at a very high rate. Pugh & Lamb (1993) provided an elegant quantitative picture of this process. Using the known cubic dependence of the rod membrane current on cGMP concentration, they converted the rate of rise of the flash response into the rate of appearance of PDE activity. Assuming that the ratio kcat/Km (in which kcat is the maximal catalytic rate at saturating substrate concentration and Km is the substrate concentration that gives half-maximal reaction rate) for an active subunit of PDE was 7×10^6 M^{-1} s^{-1}, they deduced that the rate of activation of PDE subunits was about 5000 s^{-1}.

This estimate can be compared with rates obtained by physical and biochemical techniques. Measurements of the rate of transducin activation have been made by light scattering studies (e.g. Kuhn et al 1981, Vuong et al 1984, Bruckert et al 1988). This method (reviewed in Uhl et al 1990) detects a change in the scattering of infrared light following photoactivation of rhodopsin. Scattering measurements have suggested that the rate of activation of transducin by photoexcited rhodopsin could proceed as rapidly as 700–1000 s^{-1} (Vuong et al 1984, Kahlert & Hofmann 1991, Bruckert et al 1992) under conditions of saturating guanosine triphosphate (GTP).

In contrast, estimates of the rate of transducin activation based on biochemical methods, such as measurement of GTPγS binding to nitrocellulose filters, have repeatedly yielded much lower values (summarized in Pugh & Lamb 1993; see also Dumke et al 1994, Leskov et al 2000). Initially, the lower rates of activation were thought to be due to disruption of the rod outer segment structure and dilution of cellular components. However, in a recent study, the degree of outer segment disruption did not affect the concentration of membrane-associated transducin or the observed rate of transducin activation by rhodopsin (120 s^{-1}) (Leskov et al 2000). The large discrepancy between the rates of activation determined by light scattering and biochemical methods remains puzzling.

Additional new evidence supports the idea that transducin activation may indeed be slower than previously thought. As mentioned above, Pugh & Lamb's (1993) estimate of the rate of PDE activation depended on the catalytic power (kcat/Km) of a single activated PDE subunit. Estimated Km's for PDE have varied widely because the Km observed in vitro depends both on the amount of light activation and the extent of the remaining rod outer segment structure (reviewed in Pugh & Lamb 1993, Dumke et al 1994). Because it is nearly impossible to completely disrupt all disc structure in vitro, Leskov and colleagues (2000) measured the effective Km of PDE under conditions of minimal PDE activation using very low concentrations of GTPγS. They found that the Km of PDE systematically decreased with the extent of PDE activation and finally reached a minimum at about 10 μM. This Km, with a kcat per subunit of 2200 s^{-1}, gave a kcat/Km of 2×10^8 M^{-1} s^{-1} per subunit. The implication is that an activated subunit of PDE is roughly 30-fold more adept at hydrolyzing cGMP than previously thought. Therefore, the rate of PDE activation needed to account for the rising phase of the electrical response is 30-fold lower, or about 150 s^{-1}. These biochemical results paint a coherent new picture of the light-induced rise in PDE activity, namely a somewhat slower rate of activation and a higher catalytic activity per subunit than previously supposed.

Generation of the Electrical Signal at the cGMP-Activated Channel

The final step in activation is closure of cGMP-sensitive cation channels at the plasma membrane. In excised patches of rod outer segment membrane or internally dialyzed rod outer segments there is a cubic dependence of the membrane conductance on cGMP (e.g. Yau & Nakatani 1985; Haynes et al 1986; Zimmerman & Baylor 1986; Rieke & Baylor 1996; Ruiz & Karpen 1997, 1999). Because the value of $K_{1/2}$, the concentration of cGMP necessary to half-maximally activate the channel, is higher (ca 10–40 μM) (Fesenko et al 1985, Yau & Nakatani 1985, Nakatani & Yau 1988c, Karpen & Brown 1996) than the free intracellular cGMP concentration (perhaps 3–10 μM) (Nakatani & Yau 1988c, Pugh & Lamb 1993), a small relative change in the cGMP concentration produces a threefold larger relative change in the current through the channels under physiological conditions. Because currents through rod channels have little voltage dependence (Baylor & Nunn 1986), light-dependent changes in the membrane current can easily be converted to changes in cGMP concentration.

Technological advances in recent years have improved our understanding of the structure and function of the rod channel. Expression of tandomly linked channel subunit dimers suggests that native channels possess a symmetrical diagonal arrangement ($\alpha\beta\alpha\beta$) of subunits (He et al 2000b). Mutagenesis and chimera studies have enriched understanding of the conformational changes that occur during channel activation by cGMP (Matulef et al 1999, Paoletti et al 1999, Sunderman & Zagotta 1999). Modification of tryptophan residues by UV light has also provided new insights into the energetics of channel gating (Middendorf & Aldrich 2000, Middendorf et al 2000).

The sensitivity of the rod channel for cGMP varies widely in different preparations. For example, homomeric expressed channels show a lower sensitivity for cGMP (Karpen & Brown 1996, Ruiz & Karpen 1997) than native channels. Furthermore, there also appears to be significant heterogeneity even in native channels (e.g. Haynes et al 1986, Nakatani & Yau 1988c). These variations may also explain the low apparent cooperativities in some studies (Ruiz et al 1999). The sensitivity of the channel may be modulated by kinases/phosphatases endogenous to the rod (Gordon et al 1992; Molokanova et al 1997, 1999) or the oocyte expression system (Brown et al 2000).

It has recently been discovered that the alpha subunit of the cGMP-gated channel binds to the Na^+/Ca^{2+}-K^+ exchanger in the plasma membrane, forming a stable complex (Schwarzer et al 2000). Does this represent a mechanism for inserting channels and exchangers into the membrane in equal numbers, or does it ensure specific spatial interactions between the ions being processed? The N-terminus of the channel's beta subunit contains a glutamic acid–rich protein-like (GARP) domain that has likewise been found in other GARP proteins in rod photoreceptors (GARP-1 and GARP-2) (Korschen et al 1999). GARP-2 has been shown to bind to and inhibit light-activated PDE, but the physiological significance of this action is unknown. Furthermore, GARP-1 can bind guanylate cyclase. The landscape of vertebrate phototransduction is beginning to resemble that of *Drosophila* phototransduction, with its "transducisome" microdomains of signaling molecules (Tsunoda et al 1997). Similar microdomains of intracellular signaling have been implicated in a number of other systems (reviewed in Pawson & Scott 1997, Garner et al 2000, Sheng & Pak 2000). Future work should help to define how such domains shape the specificity and temporal characteristics of intracellular signaling.

The cone channel has a higher $K_{1/2}$ for cGMP than the rod channel (Haynes & Stotz 1997, Rebrik & Korenbrot 1998). The $K_{1/2}$ of the cone channel is also more strongly dependent on calcium (Ca) concentration in the physiological range (see adaptation section, below). Recent cloning of the channel from cones has confirmed that it is indeed a different protein than the rod channel (Bonigk et al 1996, Gerstner et al 2000).

Spontaneous Activation of PDE

Spontaneous activation of rhodopsin generates about half of the total noise variance of a rod's dark current (Baylor et al 1980). The other half of the noise variance, termed the continuous noise, can be attributed to the spontaneous activation of individual catalytic subunits of PDE (Rieke & Baylor 1996). Analysis of the continuous noise fluctuations indicated that during the 0.5 s average lifetime of an activated PDE subunit, the hydrolytic activity was 1.6×10^{-5} s^{-1}. This is an order of magnitude lower than the transducin-mediated PDE activity measured biochemically (Leskov et al 2000). Although each figure is an estimate, the difference suggests that spontaneous and G protein–mediated PDE activation may involve different mechanisms. In an intact rod the spontaneous activation of PDE has the functional effect of increasing the rate of cGMP turnover, thereby shortening the duration of the dim flash response (Rieke & Baylor 1996). As was recently shown by Nikonov et al (2000), background light further shortens the response to an incremental flash by increasing the steady level of PDE activity.

Despite the fact that there is nearly an order of magnitude more transducin than PDE in the outer segment, no detectable component of the continuous noise variance could be attributed to the spontaneous activation of transducin (Rieke & Baylor 1996). This fits with the low in vitro estimates for the spontaneous rate of nucleotide exchange in the absence of photoexcited rhodopsin, namely 1×10^{-4} s^{-1} (Ramdas et al 1991). This latter figure indicates that the average wait to spontaneous activation of a transducin molecule is about 8 days. Whereas this stability pales in comparison with that of rhodopsin, it is still much higher than that of PDE. Likewise, although guanylate cyclase is active in the dark, spontaneous bursts of guanylate cyclase activity, which could also produce impulsive changes in cGMP concentration, do not appear to contribute significantly to the dark noise variance.

Rate-Limiting Step of Activation

The rod channel responds within a few milliseconds to changes in cGMP concentration (Cobbs & Pugh 1987, Karpen et al 1988). Therefore, the rise of the single photon response is not limited by the response time of the channel, but by the rate of fall of the cGMP concentration. What is the rate-limiting step for the rise in PDE activity after photon absorption? In biochemical experiments, the rates of transducin and PDE activation were very similar ($\sim$120 s^{-1}) (Leskov et al 2000), suggesting that G protein–activation by rhodopsin rate-limits the rise in PDE activity. The rate of G protein–activation depends upon the rate of its binding to photoexcited rhodopsin as well as the rate of exchange of GTP for GDP (guanosine diphosphate) on the rhodopsin-transducin complex. A clue about which step is rate-limiting in transducin activation has recently been provided by physiological recordings from mouse rods expressing half the normal level of rhodopsin (Rh +/− rods). Photoresponses from these rods rose at twice the normal rate (PD Calvert, VI Govardovskii, N Krasnoperova, RE Anderson, J Lem, CL Makino, submitted). The authors suggest that a lower rhodopsin concentration reduces protein crowding on the disc membrane, thereby increasing rhodopsin's diffusion coefficient and its rate of encounter of transducin. This finding points to the diffusional encounter of transducin by photoexcited rhodopsin as the rate-limiting step in PDE activation.

DEACTIVATION

Just as amplification is essential for the transduction of light into a macroscopic response, timely deactivation of the response is important for good temporal resolution. Furthermore, deactivation should be reproducible if the timing of photon absorption is to be accurately encoded. In recent years good progress has been made in understanding deactivation mechanisms in rods (see Figure 1), but important problems remain.

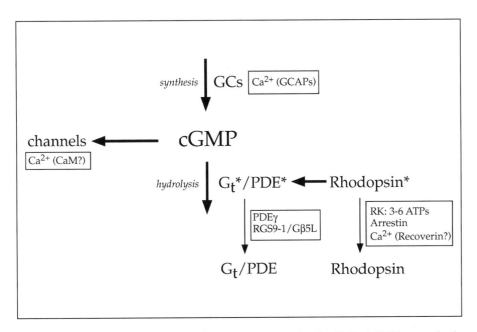

Figure 1 Scheme for regulation of visual transduction. Cyclic GMP (cGMP) controls the ion channels that generate the electrical response. Channel activation by cGMP is regulated by calcium-sensitive proteins related to calmodulin (CaM?). Synthesis of cGMP is performed by guanylate cyclases (GCs), whose activity is regulated by the calcium-dependent proteins GCAP1 and GCAP2. Hydrolysis of cGMP is accomplished by catalytically active cGMP phosphodiesterase (PDE*), which is complexed with the activated subunit of the G protein transducin (G_t^*). This complex deactivates by hydrolysis of the GTP on transducin, forming inactive G_t/PDE. This deactivation requires the formation of a complex between the gamma subunit of PDE, the accelerator protein RGS9-1, and the protein Gβ5L. Photoexcited rhodopsin, which catalytically activates the G protein and PDE, is deactivated by phosphorylation catalyzed by rhodopsin kinase, followed by the binding of arrestin. The calcium-dependent protein recoverin may control the time course of rhodopsin deactivation by regulating the activity of rhodopsin kinase.

Rhodopsin Deactivation and the Reproducibility
of the Single Photon Response

The amplitude and duration of the rod cell's single photon response are highly reproducible from trial to trial (Baylor et al 1979, Rieke & Baylor 1998a). This behavior is unexpected because the response is triggered by a single photoexcited rhodopsin molecule; typically, the activity of a single molecule varies widely from trial to trial. Statistical fluctuations in the duration of rhodopsin's catalytic activity ought to give rise to corresponding fluctuations in the single photon responses. Stages downstream from rhodopsin do not make the electrical response reproducible in the face of wide fluctuations in rhodopsin's activity, because rhodopsin continues to generate an amplified electrical response as long as it remains catalytically active. This is demonstrated by the elementary responses of rods in which rhodopsin deactivation was perturbed by removal (Chen et al 1995b) or substitution (Mendez et al 2000) of the phosphorylation sites, removal of adenosine triphosphate (ATP) (Nakatani & Yau 1988c), or deletion of rhodopsin kinase (Chen et al 1999). In these modified rods, abnormally large single photon responses could persist for up to 10 s. The behavior of the anomalous responses shows that reproducibility does not arise from a saturation (depletion of channels available to respond or depletion of an essential intermediate such as transducin, PDE, or cGMP) or from a nonlinear feedback loop that selectively attenuates responses produced by unusually large or long-lasting PDE activity. Thus, in the normal single photon response, rhodopsin's catalytic activity is apparently extinguished along a similar time course in every trial.

What makes rhodopsin's catalytic activity reproducible? Two candidate mechanisms have been suggested (Rieke & Baylor 1998a, Whitlock & Lamb 1999). One idea is that rhodopsin is shut off by a series of steps, each individually stochastic. If the catalytic activity in each trial depends on the cumulative behavior of these steps, reproducibility of any degree can theoretically be achieved (Figure 2). Another idea is that rhodopsin's catalytic activity may be terminated by a nonlinear feedback signal such as the fall in intracellular Ca concentration that occurs during the single photon response. Rieke & Baylor (1998a) argued that Ca feedback could not explain the reproducibility because reproducibility was substantially preserved when the intracellular Ca was clamped. In contrast, Whitlock & Lamb (1999) suggested that Ca might play an important role because incorporation of Ca buffers impaired reproducibility. However, Matthews (1995) has shown that in salamander rods the critical period for sensitivity to Ca concentration lasts only about 0.5 s after a flash. Furthermore, the change in Ca concentration caused by a bright flash caused only about a 30% change in the light-activated PDE activity (Matthews 1995). During a single photon response, the Ca concentration will hardly have changed within 0.5 s, and any possible effect on rhodopsin activity would be small.

If reproducibility depends on multiple steps in rhodopsin deactivation, how many steps are required? From an analysis of the variations in the single photon

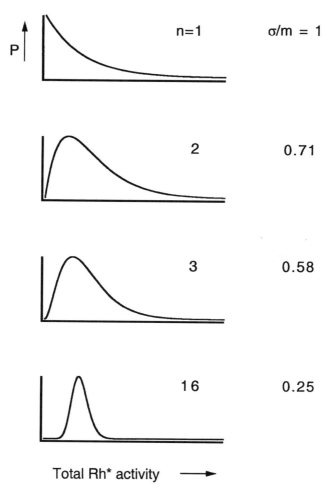

Figure 2   Multi-step mechanism for reproducibility of rhodopsin's catalytic activity. Theoretical plots of the probability (P, ordinate scale) that a molecule's total catalytic activity in a given trial will have the magnitude plotted on the abscissa. For a single step deactivation (n = 1) governed by first-order kinetics, the probability curve is exponential and the total activity has poor reproducibility across trials, with a ratio of standard deviation to mean (σ/m) of 1. As the number of steps in deactivation increases, the distributions become narrower and reproducibility improves. With n = 16, the ratio σ/m becomes 0.25, which is similar to that observed for the time integral of the single photon response (F Rieke & DA Baylor, unpublished observations). In this example it is assumed that the steps are independent, make identical contributions to lowering the catalytic activity, and have the same microscopic rate constants; the distributions have been scaled to have the same mean.

response, Rieke & Baylor (1998a) suggested that the number was 10–20. Whitlock & Lamb (1999) reported slightly more variation in the elementary response and, on the assumption that reproducibility depended on multiple steps of deactivation rather than Ca feedback, suggested that about 6 steps would suffice. For the reasons outlined, it appears that multiple-step deactivation is the more likely mechanism, and the challenge now is to identify these steps.

Biochemical studies over the years have suggested a two-step process for rhodopsin deactivation, namely phosphorylation of Metarhodopsin II followed by arrestin binding (Kuhn & Wilden 1987). Within rhodopsin's C-terminal domain there are multiple potential phosphorylation sites, six in mouse and seven in bovine. The essential role of phosphorylation in deactivation has been confirmed by the physiological experiments mentioned above (Nakatani & Yau 1988c, Chen et al 1995b, Mendez et al 2000). Mass spectrometry of purified C-terminal peptides of rhodopsin has indicated that monophosphorylated rhodopsin predominates after a light exposure (Ohguro et al 1995). However, the number and sequence of phosphorylation events that normally quench rhodopsin's catalytic activity during the single photon response have remained unclear.

Recent physiological studies on transgenic mouse rods suggest that deactivation of photoexcited rhodopsin requires multiple phosphorylation of rhodopsin (Mendez et al 2000). Single-cell recordings were made from rods expressing mutant rhodopsins whose phosphorylation sites were manipulated by site-directed mutagenesis. Mutant rhodopsins with fewer than three phosphorylation sites gave greatly prolonged single photon responses with poor reproducibility. Mutant rhodopsins with three or more phosphorylation sites gave responses that were reproducible, but all six phosphorylation sites were required for normal recovery kinetics. These results suggest that photoexcited rhodopsin may be phosphorylated at all 6 sites during normal deactivation. Whether multiple phosphorylation contributes directly to the reproducibility of rhodopsin shutoff or is merely a prerequisite for it remains to be determined.

Following phosphorylation of rhodopsin, the protein arrestin binds to rhodopsin, quenching its remaining catalytic activity (Kuhn & Wilden 1987). The essential role of arrestin binding has been demonstrated by electrical recordings from mouse rods lacking the arrestin gene; in these cells recovery of the single photon response was grossly abnormal (Xu et al 1997). Two splice-variant products of the arrestin gene, full-length arrestin (p48) and its truncated form, p44, are present in rod outer segments (Smith et al 1994). Although p44 is less abundant than full-length arrestin, p44 binds with somewhat higher affinity to phosphorylated rhodopsin (Pulvermuller et al 1997), and in vitro studies have suggested that p44 may be more efficient at turning off photoexcited rhodopsin (Langlois et al 1996). Which form of arrestin mediates rhodopsin deactivation in intact cells is not yet known.

Deactivation of Transducin and Phosphodiesterase

Although rhodopsin deactivation is essential for timely and reproducible deactivation of the light response, the downstream players, transducin and PDE, must also turn off for prompt recovery of the light response. As in other G-protein cascades, the rate limiting step for deactivation of transducin and its effector is the rate of GTP hydrolysis by transducin. As long as transducin is bound to GTP it continues to bind the gamma subunit of PDE, thus relieving gamma's inhibitory constraint on the catalytic subunits of PDE (Hurley & Stryer 1982, Wensel & Stryer 1986). The hydrolysis of GTP to GDP causes transducin to dissociate from the gamma subunit of the PDE, allowing gamma to re-inhibit PDE's catalytic subunits.

What sets the rate of GTP hydrolysis by transducin? In rod photoreceptors normal transducin deactivation requires the gamma subunit of PDE (Arshavsky & Bounds 1992, Antonny et al 1993, Tsang et al 1998) as well as the GTPase accelerator protein, RGS9-1 (Angleson & Wensel 1993, He et al 1998, Chen et al 2000). RGS9-1 is a member of the family of RGS proteins (Regulators of G-protein Signaling), which accelerate GTP hydrolysis by heterotrimeric G proteins (for review see Koelle 1997). Acceleration of transducin's GTPase activity is enhanced by PDEγ (Angleson & Wensel 1994, He et al 1998), but PDEγ exerts no effect on transducin's GTPase activity in the absence of RGS9-1 (Chen et al 2000). Furthermore, a point mutation in PDEγ that impairs the binding of transducin to PDE also reduces the rate of GTP hydrolysis and the rate of recovery of the flash response (Tsang et al 1998). Together, these results suggest that RGS9-1 stimulates GTP hydrolysis by transducin only when transducin is bound to PDE. This provides an elegant mechanism for ensuring that the activation signal from photoexcited rhodopsin is received by PDE before deactivation occurs. Deactivation of a few other G proteins also appears to be regulated by the target (Berstein et al 1992, Cook et al 2000), suggesting that effector-dependent deactivation may be a general mechanism for providing efficient activation.

Recently, another player in the deactivation of transducin/PDE has emerged. In rod outer segments RGS9-1 is tightly associated (Makino et al 1999) with an unusual isoform of the beta subunit of the G protein termed Gβ5L (Watson et al 1996). RGS9-1 and Gβ5L appear to be coexpressed, and strictly depend upon one another for proper structure and function (He et al 2000a). Binding of transducin to PDEγ increases the affinity of the RGS9-1/Gβ5L complex for transducin at least 15-fold and accelerates GTP hydrolysis by transducin to rates of 100 s^{-1} (Skiba et al 2000). The structural aspects of this association are beginning to be understood (He et al 2000a).

Restoration of cGMP Concentration by Guanylate Cyclase

Recovery of the light response requires restoration of the cGMP concentration to the dark level. In photoreceptors this is accomplished by guanylate cyclase enzymes, GC-1 and GC-2 (or GC-E and GC-F in mice), which synthesize cGMP from GTP (for review see Pugh et al 1997). Retinal GCs belong to a family of transmembrane cyclases, many of which are regulated by extracellular peptides (Yang et al 1995). Although retinal GCs possess an extracellular domain that resembles that of their cousin enzymes, no ligand is known to regulate GC activity in photoreceptor cells. If such a ligand existed it might provide a pathway over which higher-order neurons could influence photoreceptor signaling.

The activity of GCs is controlled by guanylate cyclase activating proteins, or GCAPs (reviewed in Dizhoor & Hurley 1999). There are at least two GCAPs in photoreceptors, GCAP-1 and GCAP-2, both of which appear to present in rods (Kachi et al 1999). Recently, a third member of the GCAP family, GCAP3, was also discovered (Haeseleer et al 1999). GCAPs activate GCs when Ca falls during the light response; they inhibit GCs when Ca is high. This calcium feedback regulation of GCs is one of several Ca-dependent mechanisms that oppose the light response (see below). Slowing or abolishing changes in intracellular calcium increases the amplitude and duration of the dark-adapted dim flash response (Korenbrot & Miller 1986, Lamb et al 1986; Matthews et al 1988). This indicates that one or more Ca feedback mechanisms attenuate the flash response. How much of the attenuation can be attributed to increased GC activity per se remains to be determined. However, biochemical experiments have shown that the dependence of cyclase activity on calcium is cooperative, and that GCs are half-maximally activated at a calcium concentration ($K_{1/2}$) of about 200 nM (reviewed in Dizhoor & Hurley 1999). This $K_{1/2}$ is situated at about the midpoint of the rod's physiological range of calcium levels. Thus, one would expect the rod to amplify small fluctuations in calcium into larger changes in GC activity throughout the physiological range of calcium concentrations.

The role of the multiple GCAPs is unclear. Are they redundant or do they serve distinct functions? In vitro, GCAP1 activates primarily GC1, whereas GCAP2 and GCAP3 activate both GC1 and GC2 with similar potencies (Haeseleer et al 1999). Selective expression of one or more GCAPs in a GCAP knockout background is required to determine the specific functions of each GCAP.

What Rate Limits the Recovery of the Light Response?

The "dominant time constant" of recovery in rods is about 2 s in amphibian rods (Pepperberg et al 1992, Lyubarsky et al 1996) and about 0.2 s in mouse rods (Lyubarsky & Pugh 1996, Chen et al 2000). The underlying rate-limiting process is not cGMP resynthesis, because the dominant time constant is invariant for a range of flash strengths over which GC activity varies about 10-fold (Hodgkin & Nunn 1988, Koutalos et al 1995a). Instead, the rate-limiting step is almost certainly the deactivation of rhodopsin or transducin/PDE.

Experiments in truncated toad rods suggest that rhodopsin deactivation rate-limits response recovery (Rieke & Baylor 1998a). In these experiments rhodopsin's catalytic activity was assessed at various times after a flash by observing the increase in the light response produced by a sudden increase in the internal GTP concentration. These measurements showed that the catalytic activity of rhodopsin decayed with a time constant of 2 s, identical to the dominant time constant in intact toad rods.

Experiments on truncated salamander rods have been interpreted to indicate instead that rhodopsin deactivation is fast (Sagoo & Lagnado 1997). Slowing transducin deactivation by substituting GTPγS for GTP slowed the falling phase of the flash response but did not prolong the rising phase or increase the response amplitude, indicating that significant GTP hydrolysis did not occur during the rising phase of the response. The authors concluded that transducin deactivation is slow and that rhodopsin's catalytic activity has largely ceased by the peak of the response.

Recent studies indicate that whatever the identity of the rate-limiting step, it is governed by the rate of diffusion of one or more proteins on the disc membrane. Rods with half the normal rhodopsin content in the disc membranes generated flash responses that recovered twofold faster than normal (PD Calvert, VI Govardovskii, N Krasnoperova, RE Anderson, J Lem, CL Makino, submitted). Rods that over-express membrane-associated proteins such as RGS9-1/Gβ5L and/or rhodopsin kinase may help to identify the crucial diffusing species.

Much less is known about the deactivation mechanisms of cones. Although cones use a G-protein cascade similar to that of rods, their elementary responses are much smaller and faster than those of rods. Several mechanisms may contribute to this difference. First, cones may possess unique deactivation proteins tailored for speed of operation. For example, cone photoreceptors of the ground squirrel express a specific opsin kinase (Weiss et al 1998).

A second possibility is that cones simply express higher concentrations of the proteins that mediate deactivation. Indeed, cones express more RGS9-1 than rod cells (Cowan et al 1998). Therefore, if GTP hydrolysis rate-limits recovery of the photoresponse, briefer responses may have been achieved by speeding the rate of G-protein shutoff.

A third possibility is suggested by the unique membrane topology of cones. Because the transduction machinery of cones is housed in deep invaginations of the plasma membrane rather than intracellular discs, cones have a higher ratio of membrane surface area to volume. This may permit faster changes in Ca concentration during the light response (Korenbrot 1995). Faster changes in intracellular Ca would allow Ca feedback to attenuate the elementary response earlier (e.g. Miller & Korenbrot 1994).

Finally, the faster responses of cones may reflect the behavior of an effectively light-adapted cell. The high dark noise of cone photoreceptors (Lamb & Simon 1977), particularly that of long wavelength–sensitive cones (Rieke & Baylor 2000), mimics the effect of a steady background light of low intensity. This "equivalent background" will increase the cGMP turnover time and shorten the time required to restore the cGMP to its initial level (see below).

MECHANISMS OF BACKGROUND ADAPTATION

Changing light levels confront photoreceptor cells with a formidable problem. The transduction cascade possesses high sensitivity at low light levels so that every photoisomerization produces a detectable output response. However, incident photons must continue to generate useful signals when the mean light level increases by many orders of magnitude. If a cell's steady-state sensitivity (ratio of response amplitude to light intensity) remained constant, the response amplitude would reach the maximal level at relatively modest light levels. Nature's solution is a nonlinear relation between response amplitude and steady light intensity: The slope is steep near the origin but progressively decreases with increasing intensity, so the response increases approximately with the log of intensity. In the steady state, small changes around the background level of the input produce small changes around the steady level of response up to very high background light levels. These incremental responses vary linearly with the incremental stimuli that produce them, and indeed under natural conditions transduction continues to operate linearly in this sense (Vu et al 1997).

The steady-state relation between response amplitude and light intensity may be viewed as the resultant of two factors: (a) a compressive nonlinearity that effectively operates instantaneously, conferring a dependence of response amplitude on light intensity that takes the form of a saturating exponential (Lamb et al 1981) and (b) time-dependent mechanisms that act over periods of seconds to minutes to selectively reduce the response to brighter lights, flattening the relation into its final form (Figure 3). In rods the time-dependent mechanisms involve the light-induced fall in intracellular Ca that results from the reduction in Ca influx through the cGMP-activated channels and continued efflux through the Na/Ca-K exchanger (Nakatani & Yau 1988b, Lagnado et al 1992).

Similar mechanisms probably also operate in cones, but in addition, pigment bleaching plays an important role in bright light. Bleaching makes fewer cone pigment molecules available to absorb photons, thus producing an effect comparable to reducing the incident light intensity (Rushton 1977). Unlike the rods, which become overloaded by the presence of bleached pigment (Lamb 1980), the cones seem designed to ignore it. They continue to transduce successfully when only a small fraction of their photopigment remains unbleached (Burkhardt 1994).

As background light reduces the amplitude of the incremental flash response, it shortens the effective duration, improving the time resolution of transduction (e.g. Baylor & Hodgkin 1974). Over a considerable range, the incremental flash sensitivity varies inversely with background intensity, as described by the Weber-Fechner relation (see Pugh et al 1999). Background adaptation is a necessary prerequisite for good function of the visual system, because if the output of the photoreceptors saturates, no amount of postreceptoral processing can restore useful signaling.

Ca-Dependent Mechanisms of Adaptation

The fall in intracellular Ca concentration that accompanies the light response helps to orchestrate light adaptation. Thus, when Ca is prevented from falling the time-dependent component of background adaptation largely disappears (Matthews et al 1988, Nakatani & Yau 1988a). Lowered intracellular Ca acts upon the cascade to

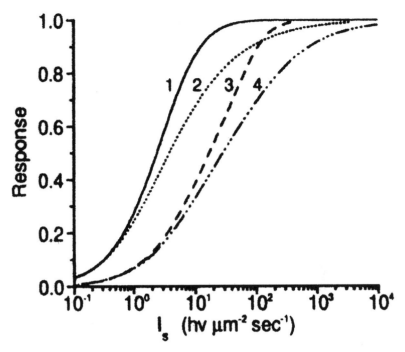

Figure 3   Ca-dependent contributions to background adaptation in a salamander rod. Calculated relations between steady-state response amplitude (ordinate) and log steady light intensity (abscissa) in a rod with intracellular Ca clamped (curve 1), with only Ca-dependent control of phosphodiesterase activity (curve 2), with only Ca-dependent control of guanylate cyclase activity (curve 3), or with both mechanisms (curve 4). The Ca-dependent reduction in phosphodiesterase activity accounts for most of the adaptation in bright light, whereas activation of cyclase activity accounts for nearly all of the sensitivity adjustment in dim light. [From Koutalos et al (1995b) by copyright permission of The Rockefeller University Press.]

oppose the effect of light, selectively reducing steady-state responses at higher stimulus intensities. Three known Ca-dependent mechanisms are thought to contribute to background adaptation: (*a*) regulation of light-dependent PDE activity, probably via regulation of the catalytic activity of rhodopsin, (*b*) regulation of guanylate cyclase activity, and (*c*) regulation of the cGMP-activated channel's sensitivity to cGMP.

Regulation of PDE Activity   Initial indications that PDE activity might be controlled by Ca were obtained by Kawamura and colleagues in biochemical experiments on frog rod outer segment membranes. They isolated from rods a factor termed S-modulin (now known as recoverin), which prolonged the lifetime of light-activated PDE in the presence of high Ca (Kawamura & Murakami 1991).

The effect required the presence of ATP (Kawamura et al 1993), which suggests that it depended on inhibition of rhodopsin phosphorylation. Indeed, subsequent work has shown that Ca-recoverin can inhibit rhodopsin kinase (Gorodovikova et al 1994, Chen et al 1995a, Klenchin et al 1995, Sanada et al 1996, Sato & Kawamura 1997, Senin et al 1997). The binding of recoverin to rhodopsin kinase appears to be inhibited when the kinase is autophosphorylated (Satpaev et al 1998).

Physiological experiments have further tested recoverin's function in rods. Incorporation of recoverin into transducing Gecko rods prolonged the time to peak and duration of the flash response (Gray-Keller et al 1993). Similarly, in truncated salamander rods recombinant myristoylated recoverin prolonged the flash response (Erickson et al 1998). Prolongation was more pronounced after bright flashes (Erickson et al 1998, Gray-Keller et al 1993). In experiments on intact rods Matthews (1997) used an ingenious strategy to separate effects of Ca on "early" components of the cascade (light-dependent PDE activity) from "late" components (guanylate cyclase activity). His work showed that the effect of Ca on the light-dependent PDE activity depended on the Ca concentration only near the time of the flash. Furthermore, experiments in truncated salamander rods have shown that the total PDE activity is reduced when the Ca concentration is low (Sagoo & Lagnado 1997). Together, these results show that Ca-recoverin may effectively prolong the flash response in the dark-adapted state. Lagnado & Baylor (1994) observed a related effect of Ca in truncated rods and suggested that Ca might control the formation of catalytically active rhodopsin. Several lines of evidence suggest that rhodopsin's initial catalytic activity is not Ca dependent (summarized in Pugh et al 1999), and therefore it seems that Ca almost certainly controls the deactivation of rhodopsin or PDE rather than the formation of the active species.

Ca regulation of PDE activity appears to be sensitive to the prevailing Ca concentration only near the time of the flash (see above). Thus, a photon absorbed during steady background adaptation would therefore elicit a smaller, briefer response. Ca-dependent PDE adaptation makes little contribution to overall light adaptation in dim backgrounds but is particularly important in bright light (Koutalos et al 1995b) (see Figure 3 above). This mechanism is also likely to be responsible for the shortened duration of responses to saturating flashes in the presence of strong background light (Matthews 1995). To summarize, it seems that the Ca-dependent regulation of PDE, which may be mediated by recoverin, requires fairly large changes in calcium and is only sensitive to the steady-state Ca concentration at the time of the flash.

Regulation of Guanylate Cyclase Activity   Activation of GC activity by the light-evoked fall in Ca increases the rate of synthesis of cGMP, opposing the light-induced drop in cGMP. Ca regulation of GC is essential for normal light adaptation (e.g. Koutalos et al 1995b). The activation of GC by steady light increases the steady state cGMP concentration and therefore reduces the steady-state response.

In bright light this fall in the steady state response rescues the cell from response saturation. Ca feedback to GCs can account for about half the total range extension (Figure 3) (Koutalos et al 1995b), accounting for roughly an order of magnitude.

Do both GCAP1 and GCAP2 contribute to GC activation during adaptation? Because both GC1 and GC2 appear to be present in rods (Yang & Garbers 1997), GCAP2 could be more important for maximal GC activation, because it can activate both GCs. Further experiments are needed to clarify the specific role of each GCAP in adaptation.

Regulation of the Channel Sensitivity   Biochemical experiments on bovine ROS by Hsu & Molday (1993) demonstrated that Ca regulates the nucleotide sensitivity of the cGMP-gated channel. Low physiological Ca concentrations were found to increase the channels' affinity for cGMP, an effect that would reduce the response to bright steady light. The Ca dependence appeared to be mediated by calmodulin (CaM) (Hsu & Molday 1993, Gordon et al 1995, Bauer 1996, Haynes & Stotz 1997), which bound to the beta subunit of the channel (Chen et al 1994, Grunwald et al 1998, Weitz et al 1998) at high Ca. The association of Ca-CaM with the rod channel increases the channel's $K_{1/2}$ for cGMP from about 30 to 40 μM without affecting the Hill coefficient (Hsu & Molday 1994). Thus, when Ca falls during the steady-state response to light CaM releases its Ca and unbinds from the channel, increasing the channel's sensitivity for cGMP, opening more channels, and extending the rod's operating range. More recent experiments suggest that an additional protein may be involved in mediating the Ca dependence of channel activation by cGMP (Gordon et al 1995, Sagoo & Lagnado 1996). The cooperativity of the Ca dependence is 1.5–2 (Nakatani et al 1995, Bauer 1996, Sagoo & Lagnado 1996), and the $K_{1/2}$ of the effect is about 50 nM Ca (Hsu & Molday 1993, Nakatani et al 1995, Sagoo & Lagnado 1996). Thus, in rods it seems that the increase of the channel's sensitivity by lowered Ca will occur primarily at relatively high background intensities.

Experiments on truncated salamander rods suggested initially that the contribution of channel modulation to light adaptation in rods is smaller than that of the other Ca-dependent mechanisms (Koutalos et al 1995a). More recently, however, Sagoo & Lagnado (1996) reported that the effect could be sizable—changing the inward current by up to an order of magnitude. The magnitude and speed of the channel effect, its specific role in adaptation, and the molecular identity of the protein(s) that mediate the Ca dependence deserve further study.

In cones a more potent effect of Ca on the channel's sensitivity for cGMP has been observed (Rebrik & Korenbrot 1998, Rebrik et al 2000). The Ca modulation of the cone channels' sensitivity occurs throughout the physiological range of Ca concentrations (Rebrik et al 2000). This effect does not seem to be mediated by CaM (Hackos & Korenbrot 1997, Haynes & Stotz 1997) but by an as yet unidentified soluble factor (Hackos & Korenbrot 1997, Rebrik & Korenbrot 1998).

ACKNOWLEDGMENTS

Work from this laboratory was supported by NIH grants EY01543 and EY05750.

Visit the Annual Reviews home page at www.annualReviews.org

LITERATURE CITED

Angleson JK, Wensel TG. 1993. A GTPase-accelerating factor for transducin, distinct from its effector cGMP phosphodiesterase, in rod outer segment membranes. *Neuron* 11:939–49

Angleson JK, Wensel TG. 1994. Enhancement of rod outer segment GTPase accelerating protein activity by the inhibitory subunit of cGMP phosphodiesterase. *J. Biol. Chem.* 269:16290–96

Antonny B, Otto-Bruc A, Chabre M, Vuong TM. 1993. GTP hydrolysis by purified alpha-subunit of transducin and its complex with the cyclic GMP phosphodiesterase inhibitor. *Biochemistry* 32:8646–53

Arshavsky V, Bownds MD. 1992. Regulation of deactivation of photoreceptor G protein by its target enzyme and cGMP. *Nature* 357:416–17

Barlow HB. 1957. Purkinje shift and retinal noise. *Nature* 179:255–56

Barlow RB, Birge RR, Kaplan E, Tallent JR. 1993. On the molecular origin of photoreceptor noise. *Nature* 366:64–66

Bauer PJ. 1996. Cyclic GMP-gated channels of bovine rod photoreceptors: affinity, density and stoichiometry of Ca(2+)-calmodulin binding sites. *J. Physiol.* 494:675–85

Baumann C. 1976. The formation of metarhodopsin380 in the retinal rods of the frog. *J. Physiol.* 259:357–66

Baylor DA, Hodgkin AL. 1974. Changes in time scale and sensitivity in turtle photoreceptors. *J. Physiol.* 242:729–58

Baylor DA, Lamb TD, Yau KW. 1979. Responses of retinal rods to single photons. *J. Physiol.* 288:613–34

Baylor DA, Matthews G, Yau KW. 1980. Two components of electrical dark noise in toad retinal rod outer segments. *J. Physiol.* 309:591–621

Baylor DA, Nunn BJ. 1986. Electrical properties of the light-sensitive conductance of rods of the salamander Ambystoma tigrinum. *J. Physiol.* 371:115–45

Berstein G, Blank JL, Jhon DY, Exton JH, Rhee SG, Ross EM. 1992. Phospholipase C-beta 1 is a GTPase-activating protein for Gq/11, its physiologic regulator. *Cell* 70:411–18

Birge RR, Barlow RB. 1995. On the molecular origins of thermal noise in vertebrate and invertebrate photoreceptors. *Biophys. Chem.* 55:115–26

Bonigk W, Muller F, Middendorff R, Weyand I, Kaupp UB. 1996. Two alternatively spliced forms of the cGMP-gated channel alpha-subunit from cone photoreceptor are expressed in the chick pineal organ. *J. Neurosci.* 16:7458–68

Brown RL, Haley TL, Snow SD. 2000. Irreversible activation of cyclic nucleotide-gated ion channels by sulfhydryl-reactive derivatives of cyclic GMP. *Biochemistry* 39:432–41

Bruckert F, Chabre M, Vuong TM. 1992. Kinetic analysis of the activation of transducin by photoexcited rhodopsin. Influence of the lateral diffusion of transducin and competition of guanosine diphosphate and guanosine triphosphate for the nucleotide site. *Biophys. J.* 63:616–29

Bruckert F, Vuong TM, Chabre M. 1988. Light and GTP dependence of transducin solubility in retinal rods. Further analysis by near infrared light scattering. *Eur. Biophys. J.* 16:207–18

Burkhardt DA. 1994. Light adaptation and photopigment bleaching in cone photoreceptors in situ in the retina of the turtle. *J. Neurosci.* 14:1091–105

Chen CK, Burns ME, He W, Wensel TG, Baylor DA, Simon MI. 2000. Slowed recovery of rod photoresponse in mice lacking the GTPase accelerating protein RGS9-1. *Nature* 403:557–60

Chen CK, Inglese J, Lefkowitz RJ, Hurley JB. 1995. Ca(2+)-dependent interaction of recoverin with rhodopsin kinase. *J. Biol. Chem.* 270:18060–66

Chen CK, Burns ME, Spencer M, Niemi GA, Chen J, et al. 1999. Abnormal photoresponses and light-induced apoptosis in rods lacking rhodopsin kinase. *Proc. Natl. Acad. Sci. USA* 96:3718–22

Chen J, Makino CL, Peachey NS, Baylor DA, Simon MI. 1995. Mechanisms of rhodopsin inactivation in vivo as revealed by a COOH-terminal truncation mutant. *Science* 267:374–77

Chen TY, Illing M, Molday LL, Hsu YT, Yau KW, Molday RS. 1994. Subunit 2 (or beta) of retinal rod cGMP-gated cation channel is a component of the 240-kDa channel-associated protein and mediates Ca(2+)-calmodulin modulation. *Proc. Natl. Acad. Sci. USA* 91:11757–61

Cobbs WH, Pugh EN Jr. 1987. Kinetics and components of the flash photocurrent of isolated retinal rods of the larval salamander, Ambystoma tigrinum. *J. Physiol.* 394:529–72

Cook B, Bar-Yaacov M, Cohen Ben-Ami H, Goldstein RE, Paroush Z, et al. 2000. Phospholipase C and termination of G-protein-mediated signalling in vivo. *Nat. Cell Biol.* 2:296–301

Cooper A. 1979. Energy uptake in the first step of visual excitation. *Nature* 282:531–33

Corson DW, Fein A. 1980. The pH dependence of discrete wave frequency in Limulus ventral photoreceptors. *Brain Res.* 193:558–61

Cowan CW, Fariss RN, Sokal I, Palczewski K, Wensel TG. 1998. High expression levels in cones of RGS9, the predominant GTPase accelerating protein of rods. *Proc. Natl. Acad. Sci. USA* 95:5351–56

Creemers AF, Klaassen CH, Bovee-Geurts PH, Kelle R, Kragl U, et al. 1999. Solid state 15N NMR evidence for a complex Schiff base counterion in the visual G-protein-coupled receptor rhodopsin. *Biochemistry* 38:7195–99

Deng H, Huang L, Callender R, Ebrey T. 1994. Evidence for a bound water molecule next to the retinal Schiff base in bacteriorhodopsin and rhodopsin: a resonance Raman study of the Schiff base hydrogen/deuterium exchange. *Biophys. J.* 66:1129–36

Dickopf S, Mielke T, Heyn MP. 1998. Kinetics of the light-induced proton translocation associated with the pH-dependent formation of the metarhodopsin I/II equilibrium of bovine rhodopsin. *Biochemistry* 37:16888–97

Dizhoor AM, Hurley JB. 1999. Regulation of photoreceptor membrane guanylyl cyclases by guanylyl cyclase activator proteins. *Methods* 19:521–31

Doukas AG, Aton B, Callender RH, Ebrey TG. 1978. Resonance Raman studies of bovine metarhodopsin I and metarhodopsin II. *Biochemistry* 17:2430–35

Dumke CL, Arshavsky VY, Calvert PD, Bownds MD, Pugh EN Jr. 1994. Rod outer segment structure influences the apparent kinetic parameters of cyclic GMP phosphodiesterase. *J. Gen. Physiol.* 103:1071–98

Eilers M, Reeves PJ, Ying W, Khorana HG, Smith SO. 1999. Magic angle spinning NMR of the protonated retinylidene Schiff base nitrogen in rhodopsin:expression of 15N-lysine- and 13C-glycine-labeled opsin in a stable cell line. *Proc. Natl. Acad. Sci. USA* 96:487–92

Erickson MA, Lagnado L, Zozulya S, Neubert TA, Stryer L, Baylor DA. 1998. The effect of recombinant recoverin on the photoresponse of truncated rod photoreceptors. *Proc. Natl. Acad. Sci. USA* 95:6474–79

Farrens DL, Altenbach C, Yang K, Hubbell WL, Khorana HG. 1996. Requirement of rigid-body motion of transmembrane helices for light activation of rhodopsin. *Science* 274:768–70

Fesenko EE, Kolesnikov SS, Lyubarsky AL. 1985. Induction by cyclic GMP of cationic conductance in plasma membrane of retinal rod outer segment. *Nature* 313:310–13

Garner CC, Nash J, Huganir RL. 2000. PDZ domains in synapse assembly and signalling. *Trends Cell Biol.* 10:274–80

Gerstner A, Zong X, Hofmann F, Biel M. 2000. Molecular cloning and functional characterization of a new modulatory cyclic nucleotide-gated channel subunit from mouse retina. *J. Neurosci.* 20:1324–32

Gordon SE, Brautigan DL, Zimmerman AL. 1992. Protein phosphatases modulate the apparent agonist affinity of the light-regulated ion channel in retinal rods. *Neuron* 9:739–48

Gordon SE, Downing-Park J, Zimmerman AL. 1995. Modulation of the cGMP-gated ion channel in frog rods by calmodulin and an endogenous inhibitory factor. *J. Physiol.* 486:533–46

Gorodovikova EN, Gimelbrant AA, Senin II, Philippov PP. 1994. Recoverin mediates the calcium effect upon rhodopsin phosphorylation and cGMP hydrolysis in bovine retina rod cells. *FEBS Lett.* 349:187–90

Gray-Keller MP, Polans AS, Palczewski K, Detwiler PB. 1993. The effect of recoverin-like calcium-binding proteins on the photoresponse of retinal rods. *Neuron* 10:523–31

Grobner G, Burnett IJ, Glaubitz C, Choi G, Mason AJ, Watts A. 2000. Observations of light-induced structural changes of retinal within rhodopsin. *Nature* 405:810–13

Grunwald ME, Yu WP, Yu HH, Yau KW. 1998. Identification of a domain on the beta-subunit of the rod cGMP-gated cation channel that mediates inhibition by calcium-calmodulin. *J. Biol. Chem.* 273:9148–57

Hackos DH, Korenbrot JI. 1997. Calcium modulation of ligand affinity in the cyclic GMP-gated ion channels of cone photoreceptors. *J. Gen. Physiol.* 110:515–28

Haeseleer F, Sokal I, Li N, Pettenati M, Rao N, et al. 1999. Molecular characterization of a third member of the guanylyl cyclase-activating protein subfamily. *J. Biol. Chem.* 274:6526–35

Han M, Smith SO, Sakmar TP. 1998. Constitutive activation of opsin by mutation of methionine 257 on transmembrane helix 6. *Biochemistry* 37:8253–61

Haynes LW, Kay AR, Yau KW. 1986. Single cyclic GMP-activated channel activity in excised patches of rod outer segment membrane. *Nature* 321:66–70

Haynes LW, Stotz SC. 1997. Modulation of rod, but not cone, cGMP-gated photoreceptor channels by calcium-calmodulin. *Vis. Neurosci.* 14:233–39

Hayward G, Carlsen W, Siegman A, Stryer L. 1981. Retinal chromophore of rhodopsin photoisomerizes within picoseconds. *Science* 211:942–44

He W, Cowan CW, Wensel TG. 1998. RGS9, a GTPase accelerator for phototransduction. *Neuron* 20:95–102

He W, Lu L, Zhang X, El-Hodiri HM, Chen CK, et al. 2000a. Modules in the photoreceptor RGS9-1/Gbeta5L GAP complex control effector coupling, GTPase acceleration, protein folding, and stability. *J. Biol. Chem.* 275:37093–100

He Y, Ruiz M, Karpen JW. 2000b. Constraining the subunit order of rod cyclic nucleotide-gated channels reveals a diagonal arrangement of like subunits. *Proc. Natl. Acad. Sci. USA* 97:895–900

Hodgkin AL, Nunn BJ. 1988. Control of light-sensitive current in salamander rods. *J. Physiol.* 403:439–71

Hsu YT, Molday RS. 1993. Modulation of the cGMP–gated channel of rod photoreceptor cells by calmodulin. *Nature* 361:76–79

Hsu YT, Molday RS. 1994. Interaction of calmodulin with the cyclic GMP-gated channel of rod photoreceptor cells. Modulation of activity, affinity purification, and localization. *J. Biol. Chem.* 269:29765–70

Hubbard R. 1966. The stereoisomerization of 11-cis-retinal. *J. Biol. Chem.* 241:1814–18

Hurley JB, Stryer L. 1982. Purification and

characterization of the gamma regulatory subunit of the cyclic GMP phosphodiesterase from retinal rod outer segments. *J. Biol. Chem.* 257:11094–99

Kachi S, Nishizawa Y, Olshevskaya E, Yamazaki A, Miyake Y, et al. 1999. Detailed localization of photoreceptor guanylate cyclase activating protein-1 and -2 in mammalian retinas using light and electron microscopy. *Exp. Eye Res.* 68:465–73

Kahlert M, Hofmann KP. 1991. Reaction rate and collisional efficiency of the rhodopsin-transducin system in intact retinal rods. *Biophys. J.* 59:375–86

Karpen JW, Brown RL. 1996. Covalent activation of retinal rod cGMP-gated channels reveals a functional heterogeneity in the ligand binding sites. *J. Gen. Physiol.* 107:169–81

Karpen JW, Zimmerman AL, Stryer L, Baylor DA. 1988. Gating kinetics of the cyclic-GMP-activated channel of retinal rods: flash photolysis and voltage-jump studies. *Proc. Natl. Acad. Sci. USA* 85:1287–91

Kawamura S, Hisatomi O, Kayada S, Tokunaga F, Kuo CH. 1993. Recoverin has S-modulin activity in frog rods. *J. Biol. Chem.* 268:14579–82

Kawamura S, Murakami M. 1991. Calcium-dependent regulation of cyclic GMP phosphodiesterase by a protein from frog retinal rods. *Nature* 349:420–23

Klenchin VA, Calvert PD, Bownds MD. 1995. Inhibition of rhodopsin kinase by recoverin. Further evidence for a negative feedback system in phototransduction. *J. Biol. Chem.* 270:16147–52

Koelle MR. 1997. A new family of G-protein regulators—the RGS proteins. *Curr. Opin. Cell Biol.* 9:143–47

Korenbrot JI. 1995. Ca2+ flux in retinal rod and cone outer segments: differences in Ca2+ selectivity of the cGMP-gated ion channels and Ca2+ clearance rates. *Cell Calcium* 18:285–300

Korenbrot JI, Miller DL. 1986. Calcium ions act as modulators of intracellular information

flow in retinal rod phototransduction. *Neurosci. Res. Suppl.* 4:S11–34

Korschen HG, Beyermann M, Muller F, Heck M, Vantler M, et al. 1999. Interaction of glutamic-acid-rich proteins with the cGMP signalling pathway in rod photoreceptors. *Nature* 400:761–66

Koskelainen A, Ala-Laurila P, Fyhrquist N, Donner K. 2000. Measurement of thermal contribution to photoreceptor sensitivity. *Nature* 403:220–23

Koutalos Y, Nakatani K, Tamura T, Yau KW. 1995a. Characterization of guanylate cyclase activity in single retinal rod outer segments. *J. Gen. Physiol.* 106:863–90

Koutalos Y, Nakatani K, Yau KW. 1995b. The cGMP-phosphodiesterase and its contribution to sensitivity regulation in retinal rods. *J. Gen. Physiol.* 106:891–921

Kuhn H, Bennett N, Michel-Villaz M, Chabre M. 1981. Interactions between photoexcited rhodopsin and GTP-binding protein: kinetic and stoichiometric analyses from light-scattering changes. *Proc. Natl. Acad. Sci. USA* 78:6873–77

Kuhn H, Wilden U. 1987. Deactivation of photoactivated rhodopsin by rhodopsin-kinase and arrestin. *J. Recept. Res.* 7:283–98

Lagnado L, Baylor DA. 1994. Calcium controls light-triggered formation of catalytically active rhodopsin. *Nature* 367:273–77

Lagnado L, Cervetto L, McNaughton PA. 1992. Calcium homeostasis in the outer segments of retinal rods from the tiger salamander. *J. Physiol.* 455:111–42

Lamb TD. 1980. Spontaneous quantal events induced in toad rods by pigment bleaching. *Nature* 287:349–51

Lamb TD, Matthews HR, Torre V. 1986. Incorporation of calcium buffers into salamander retinal rods: a rejection of the calcium hypothesis of phototransduction. *J. Physiol.* 372:315–49

Lamb TD, McNaughton PA, Yau KW. 1981. Spatial spread of activation and background desensitization in toad rod outer segments. *J. Physiol.* 319:463–96

Lamb TD, Simon EJ. 1977. Analysis of electrical noise in turtle cones. *J. Physiol.* 272:435–68

Langlois G, Chen CK, Palczewski K, Hurley JB, Vuong TM. 1996. Responses of the phototransduction cascade to dim light. *Proc. Natl. Acad. Sci. USA* 93:4677–82

Leskov IB, Klenchin VA, Handy JW, Whitlock GG, Govardovskii VI, et al. 2000. The gain of rod phototransduction: reconciliation of biochemical and electrophysiological measurements. *Neuron* 27:525–37

Lin SW, Sakmar TP, Franke RR, Khorana HG, Mathies RA. 1992. Resonance Raman microprobe spectroscopy of rhodopsin mutants: effect of substitutions in the third transmembrane helix. *Biochemistry* 31:5105–11

Longstaff C, Calhoon RD, Rando RR. 1986. Deprotonation of the Schiff base of rhodopsin is obligate in the activation of the G protein. *Proc. Natl. Acad. Sci. USA* 83:4209–13

Lyubarsky A, Nikonov S, Pugh EN Jr. 1996. The kinetics of inactivation of the rod phototransduction cascade with constant Ca2+i. *J. Gen. Physiol.* 107:19–34

Lyubarsky AL, Pugh EN Jr. 1996. Recovery phase of the murine rod photoresponse reconstructed from electroretinographic recordings. *J. Neurosci.* 16:563–71

Makino ER, Handy JW, Li T, Arshavsky VY. 1999. The GTPase activating factor for transducin in rod photoreceptors is the complex between RGS9 and type 5 G protein beta subunit. *Proc. Natl. Acad. Sci. USA* 96:1947–52

Matthews HR. 1995. Effects of lowered cytoplasmic calcium concentration and light on the responses of salamander rod photoreceptors. *J. Physiol.* 484:267–86

Matthews HR. 1997. Actions of Ca2+ on an early stage in phototransduction revealed by the dynamic fall in Ca2+ concentration during the bright flash response. *J. Gen. Physiol.* 109:141–46

Matthews HR, Murphy RL, Fain GL, Lamb TD. 1988. Photoreceptor light adaptation is mediated by cytoplasmic calcium concentration. *Nature* 334:67–69

Matthews RG, Hubbard R, Brown PK, Wald G. 1963. Tautomeric forms of metarhodopsin. *J. Gen. Physiol.* 47:215–40

Matulef K, Flynn GE, Zagotta WN. 1999. Molecular rearrangements in the ligand-binding domain of cyclic nucleotide-gated channels. *Neuron* 24:443–52

Mendez A, Burns ME, Roca A, Lem J, Wu L-W, et al. 2000. Rapid and reproducible deactivation of rhodopsin requires multiple phosphorylation sites. *Neuron* 28:153–64

Middendorf TR, Aldrich RW. 2000. Effects of ultraviolet modification on the gating energetics of cyclic nucleotide-gated channels. *J. Gen. Physiol.* 116:253–82

Middendorf TR, Aldrich RW, Baylor DA. 2000. Modification of cyclic nucleotide-gated ion channels by ultraviolet light. *J. Gen. Physiol.* 116:227–52

Miller JL, Korenbrot JI. 1994. Differences in calcium homeostasis between retinal rod and cone photoreceptors revealed by the effects of voltage on the cGMP-gated conductance in intact cells. *J. Gen. Physiol.* 104:909–40

Molokanova E, Maddox F, Luetje CW, Kramer RH. 1999. Activity-dependent modulation of rod photoreceptor cyclic nucleotide-gated channels mediated by phosphorylation of a specific tyrosine residue. *J. Neurosci.* 19:4786–95

Molokanova E, Trivedi B, Savchenko A, Kramer RH. 1997. Modulation of rod photoreceptor cyclic nucleotide-gated channels by tyrosine phosphorylation. *J. Neurosci.* 17:9068–76

Nagata T, Terakita A, Kandori H, Kojima D, Shichida Y, Maeda A. 1997. Water and peptide backbone structure in the active center of bovine rhodopsin. *Biochemistry* 36:6164–70

Nakatani K, Koutalos Y, Yau KW. 1995. Ca2+ modulation of the cGMP-gated channel of bullfrog retinal rod photoreceptors. *J. Physiol.* 484:69–76

Nakatani K, Yau KW. 1988a. Calcium and light adaptation in retinal rods and cones. *Nature* 334:69–71

Nakatani K, Yau KW. 1988b. Calcium and

magnesium fluxes across the plasma membrane of the toad rod outer segment. *J. Physiol.* 395:695–729

Nakatani K, Yau KW. 1988c. Guanosine 3′,5′-cyclic monophosphate-activated conductance studied in a truncated rod outer segment of the toad. *J. Physiol.* 395:731–53

Nathans J. 1990. Determinants of visual pigment absorbance: identification of the retinylidene Schiff's base counterion in bovine rhodopsin. *Biochemistry* 29:9746–52

Nikonov S, Lamb TD, Pugh EN Jr. 2000. The role of steady phosphodiesterase activity in the kinetics and sensitivity of the light-adapted salamander rod photoresponse. *J. Gen. Physiol.* 116:795–824

Ohguro H, Van Hooser JP, Milam AH, Palczewski K. 1995. Rhodopsin phosphorylation and dephosphorylation in vivo. *J. Biol. Chem.* 270:14259–62

Palczewski K, Kumasaka T, Hori T, Behnke CA, Motoshima H, et al. 2000. Crystal structure of rhodopsin: a G protein-coupled receptor. *Science* 289:739–45

Paoletti P, Young EC, Siegelbaum SA. 1999. C-linker of cyclic nucleotide-gated channels controls coupling of ligand binding to channel gating. *J. Gen. Physiol.* 113:17–34

Pawson T, Scott JD. 1997. Signaling through scaffold, anchoring, and adaptor proteins. *Science* 278:2075–80

Pepperberg DR, Cornwall MC, Kahlert M, Hofmann KP, Jin J, et al. 1992. Light-dependent delay in the falling phase of the retinal rod photoresponse. *Vis. Neurosci.* 8:9–18

Phelan JK, Bok D. 2000. A brief review of retinitis pigmentosa and the identified retinitis pigmentosa genes. *Mol. Vis.* 6:116–24

Pugh EN Jr, Duda T, Sitaramayya A, Sharma RK. 1997. Photoreceptor guanylate cyclases: a review. *Biosci. Rep.* 17:429–73

Pugh EN Jr, Lamb TD. 1993. Amplification and kinetics of the activation steps in phototransduction. *Biochim. Biophys. Acta* 1141:111–49

Pugh EN Jr, Nikonov S, Lamb TD. 1999. Molecular mechanisms of vertebrate pho-

toreceptor light adaptation. *Curr. Opin. Neurobiol.* 9:410–18

Pulvermuller A, Maretzki D, Rudnicka-Nawrot M, Smith WC, Palczewski K, Hofmann KP. 1997. Functional differences in the interaction of arrestin and its splice variant, p44, with rhodopsin. *Biochemistry* 36:9253–60

Ramdas L, Disher RM, Wensel TG. 1991. Nucleotide exchange and cGMP phosphodiesterase activation by pertussis toxin inactivated transducin. *Biochemistry* 30:11637–45

Rebrik TI, Korenbrot JI. 1998. In intact cone photoreceptors, a Ca2+-dependent, diffusible factor modulates the cGMP-gated ion channels differently than in rods. *J. Gen. Physiol.* 112:537–48

Rebrik TI, Kotelnikova EA, Korenbrot JI. 2000. Time course and Ca(2+) dependence of sensitivity modulation in cyclic GMP-gated currents of intact cone photoreceptors. *J. Gen. Physiol.* 116:521–34

Rieke F, Baylor DA. 1996. Molecular origin of continuous dark noise in rod photoreceptors. *Biophys. J.* 71:2553–72

Rieke F, Baylor DA. 1998a. Origin of reproducibility in the responses of retinal rods to single photons. *Biophys. J.* 75:1836–57

Rieke F, Baylor DA. 1998b. Single-photon detection by rod cells of the retina. *Rev. Mod. Phys.* 70:1027–36

Rieke F, Baylor DA. 2000. Origin and functional impact of dark noise in retinal cones. *Neuron* 26:181–86

Robinson PR, Cohen GB, Zhukovsky EA, Oprian DD. 1992. Constitutively active mutants of rhodopsin. *Neuron* 9:719–25

Ruiz M, Brown RL, He Y, Haley TL, Karpen JW. 1999. The single-channel dose-response relation is consistently steep for rod cyclic nucleotide-gated channels: implications for the interpretation of macroscopic dose-response relations. *Biochemistry* 8:10642–48

Ruiz M, Karpen JW. 1999. Opening mechanism of a cyclic nucleotide-gated channel based on analysis of single channels locked in each liganded state. *J. Gen. Physiol.* 113:873–95

Ruiz ML, Karpen JW. 1997. Single cyclic

nucleotide-gated channels locked in different ligand-bound states. *Nature* 389:389–92

Rushton WA. 1977. Visual adaptation. *Biophys. Struct. Mech.* 3:159–62

Sagoo MS, Lagnado L. 1996. The action of cytoplasmic calcium on the cGMP-activated channel in salamander rod photoreceptors. *J. Physiol.* 497:309–19

Sagoo MS, Lagnado L. 1997. G-protein deactivation is rate-limiting for shut-off of the phototransduction cascade. *Nature* 389:392–95

Sakmar TP. 1998. Rhodopsin: a prototypical G protein-coupled receptor. *Prog. Nucleic Acid Res. Mol. Biol.* 59:1–34

Sakmar TP, Franke RR, Khorana HG. 1989. Glutamic acid-113 serves as the retinylidene Schiff base counterion in bovine rhodopsin. *Proc. Natl. Acad. Sci. USA* 86:8309–13

Sakmar TP, Franke RR, Khorana HG. 1991. The role of the retinylidene Schiff base counterion in rhodopsin in determining wavelength absorbance and Schiff base pKa. *Proc. Natl. Acad. Sci. USA* 88:3079–83

Sanada K, Shimizu F, Kameyama K, Haga K, Haga T, Fukada Y. 1996. Calcium-bound recoverin targets rhodopsin kinase to membranes to inhibit rhodopsin phosphorylation. *FEBS Lett.* 384:227–30

Sato N, Kawamura S. 1997. Molecular mechanism of S-modulin action: binding target and effect of ATP. *J. Biochem.* 122:1139–45

Satpaev DK, Chen CK, Scotti A, Simon MI, Hurley JB, Slepak VZ. 1998. Autophosphorylation and ADP regulate the Ca2+-dependent interaction of recoverin with rhodopsin kinase. *Biochemistry* 37:10256–62

Schoenlein RW, Peteanu LA, Mathies RA, Shank CV. 1991. The first step in vision: femtosecond isomerization of rhodopsin. *Science* 254:412–15

Schwarzer A, Schauf H, Bauer PJ. 2000. Binding of the cGMP-gated channel to the Na/Ca-K exchanger in rod photoreceptors. *J. Biol. Chem.* 275:13448–54

Senin II, Zargarov AA, Akhtar M, Philippov PP.

1997. Rhodopsin phosphorylation in bovine rod outer segments is more sensitive to the inhibitory action of recoverin at the low rhodopsin bleaching than it is at the high bleaching. *FEBS Lett.* 408:251–54

Shastry BS. 1997. Signal transduction in the retina and inherited retinopathies. *Cell. Mol. Life Sci.* 53:419–29

Sheikh SP, Zvyaga TA, Lichtarge O, Sakmar TP, Bourne HR. 1996. Rhodopsin activation blocked by metal-ion-binding sites linking transmembrane helices C and F. *Nature* 383:347–50

Sheng M, Pak DT. 2000. Ligand-gated ion channel interactions with cytoskeletal and signaling proteins. *Annu. Rev. Physiol.* 62:755–78

Skiba NP, Hopp JA, Arshavsky VY. 2000. The effector enzyme regulates the duration of G protein signaling in vertebrate photoreceptors by increasing the affinity between transducin and RGS protein. *J. Biol. Chem.* 275:32716–20

Smith WC, Milam AH, Dugger D, Arendt A, Hargrave PA, Palczewski K. 1994. A splice variant of arrestin. Molecular cloning and localization in bovine retina. *J. Biol. Chem.* 269:15407–10

Steinberg G, Ottolenghi M, Sheves M. 1993. pKa of the protonated Schiff base of bovine rhodopsin. A study with artificial pigments. *Biophys. J.* 64:1499–502

Struthers M, Yu H, Oprian DD. 2000. G protein-coupled receptor activation: analysis of a highly constrained, "straitjacketed" rhodopsin. *Biochemistry* 39:7938–42

Stryer L. 1995. *Biochemistry.* New York: Freeman

Sunderman ER, Zagotta WN. 1999. Sequence of events underlying the allosteric transition of rod cyclic nucleotide-gated channels. *J. Gen. Physiol.* 113:621–40

Tsang SH, Burns ME, Calvert PD, Gouras P, Baylor DA, et al. 1998. Role for the target enzyme in deactivation of photoreceptor G protein in vivo. *Science* 282:117–21

Tsunoda S, Sierralta J, Sun Y, Bodner R, Suzuki

E, et al. 1997. A multivalent PDZ-domain protein assembles signalling complexes in a G-protein-coupled cascade. *Nature* 388:243–49

Uhl R, Wagner R, Ryba N. 1990. Watching G proteins at work. *Trends Neurosci.* 13:64–70

Unger VM, Hargrave PA, Baldwin JM, Schertler GF. 1997. Arrangement of rhodopsin transmembrane alpha-helices. *Nature* 389:203–6

Vu TQ, McCarthy ST, Owen WG. 1997. Linear transduction of natural stimuli by dark-adapted and light-adapted rods of the salamander, Ambystoma tigrinum. *J. Physiol.* 505:193–204

Vuong TM, Chabre M, Stryer L. 1984. Millisecond activation of transducin in the cyclic nucleotide cascade of vision. *Nature* 311:659–61

Wald G. 1968. Molecular basis of visual excitation. *Science* 162:230–39

Watson AJ, Aragay AM, Slepak VZ, Simon MI. 1996. A novel form of the G protein beta subunit Gbeta5 is specifically expressed in the vertebrate retina. *J. Biol. Chem.* 271:28154–60

Weiss ER, Raman D, Shirakawa S, Ducceschi MH, Bertram PT, et al. 1998. The cloning of GRK7, a candidate cone opsin kinase, from cone- and rod-dominant mammalian retinas. *Mol. Vis.* 4:27

Weitz D, Zoche M, Muller F, Beyermann M, Korschen HG, et al. 1998. Calmodulin controls the rod photoreceptor CNG channel through an unconventional binding site in the N-terminus of the beta-subunit. *EMBO J.* 17:2273–84

Wensel TG, Stryer L. 1986. Reciprocal control of retinal rod cyclic GMP phosphodiesterase by its gamma subunit and transducin. *Proteins* 1:90–99

Whitlock GG, Lamb TD. 1999. Variability in the time course of single photon responses from toad rods: termination of rhodopsin's activity. *Neuron* 23:337–51

Xu J, Dodd RL, Makino CL, Simon MI, Baylor DA, Chen J. 1997. Prolonged photoresponses in transgenic mouse rods lacking arrestin. *Nature* 389:505–9

Yang RB, Foster DC, Garbers DL, Fulle HJ. 1995. Two membrane forms of guanylyl cyclase found in the eye. *Proc. Natl. Acad. Sci. USA* 92:602–6

Yang RB, Garbers DL. 1997. Two eye guanylyl cyclases are expressed in the same photoreceptor cells and form homomers in preference to heteromers. *J. Biol. Chem.* 272:13738–42

Yau KW, Nakatani K. 1985. Light-suppressible, cyclic GMP-sensitive conductance in the plasma membrane of a truncated rod outer segment. *Nature* 317:252–55

Zhukovsky EA, Oprian DD. 1989. Effect of carboxylic acid side chains on the absorption maximum of visual pigments. *Science* 246:928–30

Zimmerman AL, Baylor DA. 1986. Cyclic GMP-sensitive conductance of retinal rods consists of aqueous pores. *Nature* 321:70–72

Annu. Rev. Neurosci. 2001. 24:807–43

ACTIVITY-DEPENDENT SPINAL CORD PLASTICITY IN HEALTH AND DISEASE

Jonathan R Wolpaw and Ann M Tennissen

Laboratory of Nervous System Disorders, Wadsworth Center, New York State Department of Health and State University of New York, Albany, New York 12201-0509; e-mail: wolpaw@wadsworth.org, tenniss@wadsworth.org

Key Words spinal cord injury, rehabilitation, memory, learning, conditioning, behavior

■ **Abstract** Activity-dependent plasticity occurs in the spinal cord throughout life. Driven by input from the periphery and the brain, this plasticity plays an important role in the acquisition and maintenance of motor skills and in the effects of spinal cord injury and other central nervous system disorders. The responses of the isolated spinal cord to sensory input display sensitization, long-term potentiation, and related phenomena that contribute to chronic pain syndromes; they can also be modified by both classical and operant conditioning protocols. In animals with transected spinal cords and in humans with spinal cord injuries, treadmill training gradually modifies the spinal cord so as to improve performance. These adaptations by the isolated spinal cord are specific to the training regimen and underlie new approaches to restoring function after spinal cord injury. Descending inputs from the brain that occur during normal development, as a result of supraspinal trauma, and during skill acquisition change the spinal cord. The early development of adult spinal cord reflex patterns is driven by descending activity; disorders that disrupt descending activity later in life gradually change spinal cord reflexes. Athletic training, such as that undertaken by ballet dancers, is associated with gradual alterations in spinal reflexes that appear to contribute to skill acquisition. Operant conditioning protocols in animals and humans can produce comparable reflex changes and are associated with functional and structural plasticity in the spinal cord, including changes in motoneuron firing threshold and axonal conduction velocity, and in synaptic terminals on motoneurons. The corticospinal tract has a key role in producing this plasticity. Behavioral changes produced by practice or injury reflect the combination of plasticity at multiple spinal cord and supraspinal sites. Plasticity at multiple sites is both necessary—to insure continued performance of previously acquired behaviors—and inevitable—due to the ubiquity of the capacity for activity-dependent plasticity in the central nervous system. Appropriate induction and guidance of activity-dependent plasticity in the spinal cord is an essential component of new therapeutic approaches aimed at maximizing function after spinal cord injury or restoring function to a newly regenerated spinal cord. Because plasticity in the spinal cord contributes to skill acquisition and because the spinal cord is relatively simple and accessible, this plasticity is a logical and practical starting point for studying the acquisition and maintenance of skilled behaviors.

0147-006X/01/0621-0807$14.00

INTRODUCTION

The Cinderella of the Central Nervous System

As the title of this section implies, the spinal cord has always occupied a humble place in neuroscience (Aminoff 1993). Considered part of the central nervous system (CNS) by virtue of its meningeal coverings and histology, it is commonly thought to be merely a waystation between the brain and the periphery, the home of a few hardwired reflexes and nothing else. This traditional view goes back 1800 years to Galen, who wrote that the spinal cord was essentially a protected bundle of nerves that connect the brain to the body (see Liddell 1960, Neuburger 1981, Clarke & Jacyna 1987, Clarke & O'Malley 1996 for historical review). Even its reflex functions went unrecognized until well after the Renaissance. Reflexes were believed to be entirely peripheral, produced by activity that passed from sensory receptors to muscles through anastomoses between peripheral nerves, without the participation of the brain or spinal cord. Pliny the Elder, who predated Galen, knew that this was not true, and so did Leonardo da Vinci and Stephen Hales later on, but the idea persisted until the eighteenth century, when Robert Whytt, studying the headless frog, observed that "the strong convulsive motions excited by irritation in the legs and trunk. . . cease as soon as [the spinal cord] is destroyed."

While Whytt laid to rest the Galenic belief that the spinal cord was only a big nerve, subsequent developments led to a distinction between the spinal cord and the brain that is still widely accepted and maintains the spinal cord's inferior status. In the early nineteenth century Marshall Hall distinguished between "The Cerebral, or the Sentient and Voluntary" part of the nervous system and the "True Spinal, or the Excito-motory" part. The first was the cerebral hemispheres and cerebellum and produced conscious, or volitional, behavior; the second was the spinal cord and medulla and produced reflexes. Hall thought these two systems were totally separate and even had separate sets of afferent and efferent nerves. His reasons were both scientific and religious: The separation protected the immortal soul, which resided in the brain, from domination by the external world. He coined the term "reflex arc" to describe the pathways underlying the fixed behaviors ascribed to the spinal cord.

Hall's theory of the spinal cord as a reflex center underlies current concepts. While we now know that spinal reflex pathways do not have a separate set of peripheral nerves, most neuroscientists retain something like Hall's distinction between the voluntary behaviors produced by the brain and the involuntary, or reflex, behaviors produced by the spinal cord. The spinal cord is often conceived to be a hardwired system that simply responds quickly and in a stereotyped, or "knee-jerk" fashion to afferent inputs from sensory receptors in the periphery or to descending commands from the brain. For example, sudden muscle stretch produces rapid contraction of the stretched muscle and ascending activity in dorsal column pathways, a pinprick to the foot causes rapid limb withdrawal nearly as quickly, and descending activity from motor cortex rapidly excites spinal motoneurons and produces muscle contraction. Such short-latency input-output connections are traditionally believed to be the full extent of the spinal cord's capabilities. Activity-dependent plasticity—defined as persistent CNS modification that results from past experience and affects future behavior—is considered a supraspinal capacity. Complex motor performances—standard skills such as walking and writing and special skills such as dancing and piano playing—that are acquired through practice, and thus reflect the persistent effects of activity, are generally thought to result from supraspinal plasticity that simply uses the fixed capacities of the unchanging spinal cord.

Activity-Dependent Plasticity in the Spinal Cord

This traditional view ignores a large body of evidence that has accumulated over the past century and grown rapidly in recent years. Both clinical and experimental observations indicate that the spinal cord, like other parts of the CNS, shows activity-dependent plasticity—that inputs from the periphery or from the brain can cause lasting changes in the spinal cord that affect its output far into the future. The spinal cord possesses capacities for neuronal and synaptic plasticity comparable to those found elsewhere in the CNS.

In spite of the substantial evidence for activity-dependent plasticity in the spinal cord, its role in the acquisition and maintenance of behaviors in normal states and in the aftermath of CNS injury has only recently begun to be properly recognized and explored. This new interest and effort derive largely from two factors. The first is the remarkable excitement and energy now focused on developing new treatments for spinal cord injury (Bregman et al 1997, Fawcett 1998, Amar & Levy 1999, Tuszynski & Kordower 1999). The newly recognized possibilities for CNS regeneration inevitably raise the issue of how regenerated neuronal tissue is to become useful, of how it will come to provide normal, or at least acceptable, function. A normally functioning adult spinal cord is a product of appropriate activity-dependent plasticity during early development and throughout subsequent life. Thus, a newly regenerated spinal cord will probably not be properly or even acceptably configured for effective use (Muir & Steeves 1997). It is likely to display diffuse infantile reflexes or other disordered and dysfunctional outputs. As methods for inducing spinal cord regeneration develop, methods for redeveloping spinal cord function—for re-educating the newly regenerated spinal cord—will become essential. This anticipated need compels attention to activity-dependent spinal cord plasticity, to the processes by which spinal cord neurons and synapses are shaped to serve important functions as diverse as locomotion, urination, and playing a musical instrument. Furthermore, recent appreciation of the latent capacities for plasticity of the injured unregenerated spinal cord provides additional incentive for exploring activity-dependent plasticity in the spinal cord. Understanding this plasticity is essential for understanding both the changes that occur after injury and the processes that can be accessed and guided to restore useful function.

The second factor is the growing recognition, driven by data from a variety of laboratory models, that the acquisition and maintenance of both normal motor performances and the abnormal behaviors associated with disease involve activity-dependent plasticity at multiple sites throughout the CNS, including the spinal cord. The peripheral and descending inputs that occur during practice or as a result of trauma or disease change the spinal cord, and these changes combine with changes elsewhere in the CNS to change behavior. Thus, knowledge of the mechanisms of spinal cord plasticity and its interactions with activity-dependent plasticity elsewhere in the CNS is important for understanding normal behaviors, as well as for understanding the complex disabilities produced by disorders such as spinal cord injury.

The Present Review

The focus of this review is activity-dependent spinal cord plasticity (that is, lasting change in spinal cord function produced by peripheral and/or descending inputs) and its impact on motor function in health and disease. Other kinds of spinal cord plasticity—such as the long-term effects of neurotrophins or the processes triggered by axotomy (e.g. Mendell 1988, 1999; Wilson & Kitchener 1996; Chen & Frank 1999)—are not addressed. The primary intent is to review the range of clinical and laboratory phenomena that reflect or elucidate activity-dependent spinal cord plasticity, with particular attention to its role in normal and abnormal behaviors.

The first section addresses plasticity induced by sensory input in spinal cord that has been isolated from descending influence. Activity-dependent plasticity in this setting provides insight into neuronal and synaptic mechanisms and has great relevance for the treatment of spinal cord injury. The following section addresses spinal cord plasticity that occurs in the intact CNS and often involves both sensory input from the periphery and descending input from the brain. This is the setting in which spinal cord plasticity usually occurs and is most relevant to motor function in health and disease. The final section addresses the relationships between activity-dependent plasticity and behavioral change. Activity-dependent plasticity at multiple spinal and supraspinal sites underlies the acquisition and maintenance of both normal and abnormal behaviors, and the relative simplicity and accessibility of the spinal cord facilitates recognition and exploration of this complex plasticity. Furthermore, the necessity for and inevitability of this complex plasticity suggest principles that should guide development of new therapeutic methods for promoting recovery of function after spinal cord injury.

PLASTICITY PRODUCED BY SENSORY INPUT IN THE ISOLATED SPINAL CORD

The isolated spinal cord, deprived of descending influence from the brain, has been a popular and productive experimental model for more than a century, providing, by virtue of its accessibility and relative simplicity, a large part of present-day understanding of CNS neuronal and synaptic function. Many studies have explored the persistent effects of sensory input. While most have studied the relatively short-term effects (i.e. minutes to hours) of single inputs or simple combinations of inputs, more recent studies have investigated the long-term effects (i.e. weeks to months) of complex sequences of sensory input and are directly related to treatment of spinal cord injury.

In this work the primary measures of spinal cord function have been the two most prominent classes of short-latency spinal cord responses to peripheral inputs: flexion withdrawal reflexes and proprioceptive reflexes (Matthews 1972, Baldissera et al 1981, Burke 1998, Kandel et al 2000). Flexion withdrawal reflexes are mediated by oligosynaptic pathways to spinal motoneurons from unmyelinated C nociceptive fibers and from small myelinated A-delta fibers that are also activated by noxious stimuli. They are normally limited to those muscles that withdraw the body from the painful stimuli. After spinal cord injury or with other disorders, they may become diffuse and hyperactive and thereby contribute to spasms and postural abnormalities. Proprioceptive reflexes are mediated by mono- and oligosynaptic pathways to spinal motoneurons from larger afferents that innervate muscle spindles, Golgi tendon organs, and other receptors that reflect muscle length and tension and limb position. The simplest proprioceptive reflex is the spinal stretch reflex (SSR), or tendon jerk, which is mediated largely by a monosynaptic pathway made up of the Ia afferent fiber from the muscle spindle, its synapse on the motoneuron, and the motoneuron. It produces contraction in response to sudden muscle stretch, and descending activity affects its gain through several mechanisms. Other proprioceptive reflexes, both excitatory and inhibitory, reflect more complex segmental responses to muscle length and tension and limb position, and are also regulated by descending activity.

Simple Sensory Inputs

Flexion withdrawal reflexes in the isolated spinal cord display habituation and sensitization with a variety of protocols (Mendell 1984). Most research has been oriented toward clarifying pain mechanisms and has focused on sensitization of spinal cord responses to C-fiber input. This work is the subject of numerous recent reviews (Pockett 1995, Randić 1996, Baranauskas & Nistri 1998, Alvares & Fitzgerald 1999, Woolf & Costigan 1999, Yaksh et al 1999, Dubner & Gold 1999, Herrero et al 2000) and is only briefly summarized here.

In the most thoroughly studied phenomenon, usually referred to as wind-up, (Mendell & Wall 1965), repetitive C-fiber stimulation at rates >0.3 Hz leads to progressive increase in the neuronal excitation produced by each stimulus. Short series of stimuli (e.g. 30 in 30 sec) produce sensitization lasting minutes, and longer series can produce sensitization lasting much longer. Wind-up is not limited to the dorsal horn neurons directly contacted by C fibers; other spinal cord neurons, including motoneurons in ventral horn, may also exhibit it. The underlying mechanisms vary across species, neuronal populations, and protocols, and multiple processes probably contribute in individual instances. Initially, release of the excitatory amino acids glutamate and aspartate and tachykinins such as substance P activates NMDA receptors leading to calcium entry and activates second messenger systems that affect proteins controlling membrane excitability. Eventually, activation of c-fos and other early immediate genes may take place, and expansion of the receptive fields of dorsal horn neurons and phenotypic conversion of A-beta fibers to C-fiber-like behavior can occur. As illustrated in Figure 1A, high-frequency stimulation can produce persistent increases in neuronal excitability comparable to long-term potentiation elsewhere in the CNS. The spinal cord plasticity induced by C-fiber input contributes to the clinical syndromes of spontaneous pain, abnormal sensitivity to noxious stimuli, or even innocuous stimuli, and referred pain that often follow injury to peripheral tissues.

While most evidence for spinal cord plasticity produced by afferent input relates to C-fiber stimulation, activity in other afferent populations has been linked to change in the spinal cord. Brief periods of high-frequency stimulation of the largest sensory fibers, Ia afferents from muscle spindles, produce posttetanic potentiation lasting seconds to minutes, and longer periods of stimulation can produce potentiation lasting hours (Lloyd 1949, Kandel 1977).

Over the past 70 years, associative conditioning phenomena, both classical and operant, have been repeatedly demonstrated in the spinal cords of cats, dogs, and rats (reviewed in Patterson 1976, Kandel 1977, Thompson 2001). These models normally use stimuli that generate C-fiber input. In a well-characterized classical conditioning protocol in the isolated lumbosacral spinal cord of decerebrate cats, Durkovic and colleagues used superficial peroneal nerve stimulation as the unconditioned stimulus (US), saphenous nerve stimulation as the conditioned stimulus (CS), and tibialis anterior muscle contraction as the response (Durkovic 1985, 1986). As illustrated in Figure 1B, paired presentation of the stimuli, with the CS preceding the US by one second, quickly produced a substantial increase in the response to the CS alone, while unpaired presentation led only to a decrease in the response to the CS alone. The phenomenon exhibited the features of classical conditioning, including characteristic dependence on the delay between CS and US. The spinal cord plasticity underlying this conditioning appears to involve interneurons conveying the sensory input, rather than the sensory afferents or the motoneurons themselves.

Training Regimens

Locomotion   The rhythmical and sequential activation of muscles that underlies locomotion is a characteristic function of the vertebrate CNS, and one in which the spinal cord plays a central role (Rossignol 1996, Kiehn et al 1998, Orlovsky et al 1999). Locomotion, whether swimming, flying, or walking, is produced by interconnected spinal neurons that together constitute a locomotor pattern generator (LPG). The spinal LPG is activated by supraspinal influences that descend from locomotor regions in the brainstem and thalamus, and its operation is

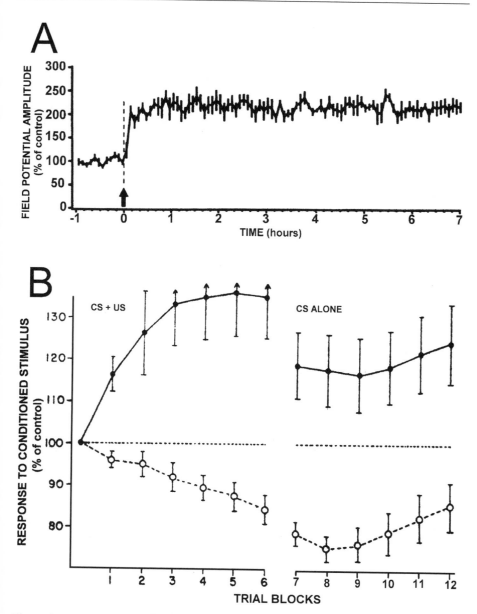

Figure 1   Activity-dependent plasticity induced by sensory input in the isolated spinal cord. (*A*) Long-term potentiation of field potentials evoked by stimulation (*arrow*) of C fibers in rat sciatic nerve. Modified from Liu & Sandkühler (1997). (*B*) Classical conditioning in the cat. Paired presentation (*solid line*) of the conditioned stimulus (CS) and unconditioned stimulus (US) for six blocks of five trials each increases the response to the CS from its control level, while unpaired presentation (*dashed line*) produces only habituation to the CS. The increased response to the CS is still evident with subsequent presentation of the CS alone. Modified from Durkovic (1985).

continually adjusted and refined by proprioceptive and cutaneous inputs from the periphery. The existence and importance of the spinal LPG is most apparent in lower vertebrates such as the lamprey or chick, in which robust locomotion is evident after removal of all supraspinal influence. Its existence is also evident in higher vertebrates such as the cat, in which the lumbosacral spinal cord is quite capable of producing well-coordinated treadmill locomotion after transection of the thoracic spinal cord has removed supraspinal control. Furthermore, studies in people with spinal cord injuries suggest the existence of an LPG in the human lumbrosacral spinal cord (Holmes 1915, Kuhn 1950, Bussel et al 1988, Calancie et al 1994, Dietz et al 1995, Dobkin et al 1995, Dimitrijevic et al 1998, Rossignol 2000). If the operation of a human LPG could be encouraged and guided—by training programs, pharmacological agents, sensory stimulation, surviving descending pathways, or other means—useful locomotion might be restored. In the work motivated by this possibility, impressive new evidence for activity-dependent plasticity in the spinal cord has emerged.

Fifty years ago, Shurrager & Dykman (1951) reported that treadmill walking in spinalized cats (i.e. cats with transected spinal cords) improved with training. Over the past 15 years several energetic research groups have confirmed this phenomenon in spinalized animals and in humans with spinal cord injuries, described its major features, and begun to reveal its mechanisms and define its potential role in therapy (Lovely et al 1986, Barbeau & Rossignol 1987, Barbeau et al 1999). In the typical protocol, cats were subjected to complete spinal cord transection at a thoracic level and then began a regimen in which they were trained for 30–60 min per day to walk on a treadmill with their hindlimbs while their forelimbs stood on a platform. Early in training, coordinated locomotion occurred only when the experimenter provided weight support for the hindquarters and a strong nonspecific sensory input such as stimulation of the perineum or tail. The primary observation was that coordinated locomotion developed and improved over days and weeks. Animals gradually walked faster, with longer steps, and for longer periods, and eventually required little or no weight support or sensory stimulation to do so. Careful electromyographic and kinematic analyses described this locomotion in detail (Bélanger et al 1996). While subtle differences from normal were found, spinal locomotion after training was in major respects comparable to that found prior to injury.

Locomotion was much better in cats exposed to treadmill training than in cats that received only normal nursing care after injury (de Leon et al 1998a). Although untrained animals regained some locomotor ability, they were clearly inferior to trained animals. Figure 2A illustrates the marked difference. The noradrenergic alpha-2 agonist clonidine facilitated early training, reducing the need for sensory stimulation and speeding improvement (Chau et al 1998). The improved locomotion produced by training persisted when training stopped, declining very little over 6 weeks, and only showing significant loss at 12 weeks (de Leon et al 1999a).

Functional improvement was found to depend quite specifically on the nature of the training regimen. A regimen in which cats were trained to stand rather than to walk clearly improved standing (de Leon et al 1998b). If stand training was confined to one leg, improvement was focused in that leg. Furthermore, animals trained to stand did not walk well on the treadmill, and animals trained to walk did not stand well. Nevertheless, animals trained to perform one task could then be trained to perform the other. In addition, locomotor deficits produced by muscle denervation abated with continued training (Bouyer et al 2001). Furthermore, studies of motor unit properties after spinal cord transection and with or without training, indicated that the training-induced improvements in walking and standing were not attributable to peripheral changes in muscle strength or other motor unit properties (Roy & Acosta 1986; Roy et al 1991, 1998). They appeared to be largely or wholly due to activity-dependent spinal cord plasticity.

Comparable treadmill training is being evaluated in humans with complete or incomplete spinal cord injuries (Barbeau & Fung 1992; Harkema et al 1997; Wernig et al 1995, 1998; Dietz et al 1995; Dobkin 1998; Field-Fote 2000). In people who retain some control of leg muscles, treadmill training assisted by weight-support (with a harness or other device) can improve locomotion, producing greater speed, strength, coordination, and endurance, and reduce need for assistive devices (e.g. Figure 2*B*). Comparison of spinal cord–injured patients who underwent treadmill training with others who simply received conventional rehabilitation indicates that treadmill training increases walking ability. Improvement has been reported both in patients recently injured and in those injured a long time ago, and can persist beyond the cessation of formal training. The encouraging initial studies await confirmation by randomized clinical trials (Dobkin 1999).

The spinal cord plasticity that underlies training-induced functional improvement clearly depends on the pattern of afferent, efferent, and interneuronal activity that occurs during training. Recent work provides some insight into crucial aspects of that activity. A certain minimum level of appropriately timed sensory input appears to be essential (Rossignol & Bouyer 2001). The importance of sensory input is evident in chicks with spinal cord hemisection, in which training to walk is much more effective than training to swim, even though the behaviors are very similar (Muir & Steeves 1995, Muir 1999). The difference is attributable to the greater phasic sensory input that occurs during walking—the input produced by foot contact and the excitation of cutaneous and proprioceptive receptors associated with it. When comparable sensory input was provided during swim training, performance improved markedly, as illustrated in Figure 2*C*. Functional electrical stimulation, both sensory and motor, can also contribute to spinal cord plasticity and peripheral change, and can thereby improve performance (Peckham & Creasey 1992, Muir & Steeves 1997, Stein 1999).

The performance improvements produced by treadmill training in cats appear to be associated with change in glycinergic inhibition in the spinal cord. The effects of the glycinergic inhibitor, strychnine, were evaluated in cats that had been trained over 12 weeks to either walk or stand (de Leon et al 1999b). In cats trained to walk locomotion proceeded as before. In contrast, in cats trained to stand locomotion was much improved when strychnine reduced glycinergic inhibition. Furthermore,

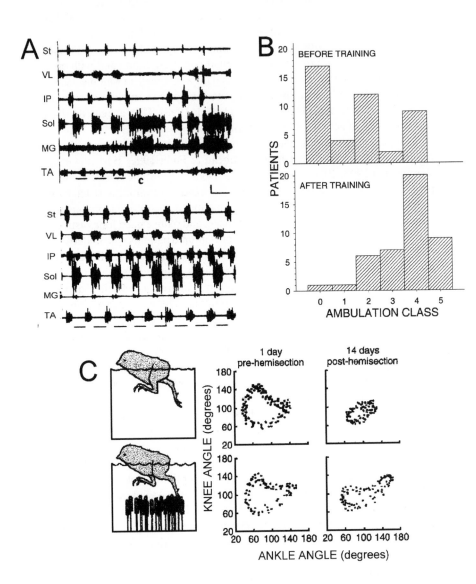

when cats trained first to walk were subsequently trained to stand, so that walking performance deteriorated, strychnine restored walking. Conversely, if cats first trained to stand were subsequently trained to walk, strychnine then had no effect on their walking. These results suggest that locomotor training reduces the level of glycinergic inhibition in the spinal cord.

Urination and Other Essential Functions   Although the effects of spinal cord injury on locomotion and other voluntary limb movements usually receive the most attention and have been the focus of most research, its devastating effects on urinary tract, bowel, and sexual function, and on blood pressure and body temperature control, are of equal or greater concern to those affected (as well as to their families and caregivers) and can have greater impact on their lives (Ronthal 1998, Biering-Sørensen et al 1999, McKinley et al 1999, Teasell et al 2000, Chen & Nussbaum 2000, Linsenmeyer 2000, Monga et al 1999, Stiens et al 1997, Weld & Dmochowski 2000). Satisfactory urinary tract function is particularly critical to the survival and well-being of people with spinal cord injuries. Urinary tract infections and eventual kidney failure contributed greatly to the dismal prognosis of spinal cord injury prior to the early twentieth century, and appropriate bladder management is an essential feature of current rehabilitation and long-term care protocols.

Normal urinary tract function depends on appropriate reciprocal relationships between activation of bladder muscle and activation of the urethral sphincter muscles (de Groat et al 1997, Yoshimura 1999). During urine accumulation, bladder muscle is relatively inactive and sphincter muscles are tonically active. Urination is triggered by stretch receptors in the bladder wall and executed by spinal and supraspinal reflex pathways that excite bladder muscles and relax sphincter muscles. After transection of the cervical or thoracic spinal cord, and after the

Figure 2   Effects of locomotor training of the isolated spinal cord. (*A*) Electromyographic activity on the treadmill in spinalized cats that have (*bottom traces*) or have not (*top traces*) undergone treadmill training. Consistent rhythmical muscle activation and consistent walking (shown by the *horizontal lines* indicating each stance phase) are present only in the trained cat. Rhythmical muscle activation and walking are imperfect and sporadic in the untrained cat. (C, collapse; St, semitendinosus; VL, vastus lateralis; IP, iliopsoas; Sol, soleus; MG, medial gastrocnemius; TA, tibialis anterior.) Scale: 1 sec and 1.0 mV (2.0 for Sol). Modified from de Leon et al (1998a). (*B*) Ambulation performance before and after treadmill training in humans with spinal cord injuries. Those in classes 0–2 are wheelchair-bound, those in class 5 can walk >5 steps without any assistive device. Ambulation is greatly improved after training. Modified from Wernig et al (1995). (*C*) Knee and ankle excursions during swimming in normal chicks and in the same chicks after spinal cord hemisection followed by 14 days of swim training with (*bottom*) or without (*top*) plantar stimulation during the phase of movement equivalent to the stance phase of walking. After training movement is closer to normal in hemisected chicks that have received planter stimulation. Modified from Muir (1999).

acute period of spinal shock and reflex depression, urination is more frequent and less effective, owing to changes in sensory afferents from the bladder and in spinal cord reflex pathways that receive this afferent input. C fibers innervating the bladder wall, which normally convey nociceptive information and have relatively high firing thresholds, undergo histological and chemical changes that lower their thresholds and make them the primary trigger for urination. Their change in threshold is thought to result from change in sodium and potassium channels, due at least in part to increases in trophic factors such as nerve growth factor (NGF) caused by chronic bladder distension. The contribution of altered afferent activity itself to spinal cord plasticity remains unclear, but the demonstrated effects of altered C-fiber input in other situations (see above) suggest that it also has a role here. Chronic spinal cord injury also produces loss of the coordinated sphincter relaxation that normally accompanies bladder contraction and urination. Spasmodic sphincter contractions interfere with urine flow and lead to incomplete emptying and bladder distension. The abnormalities in spinal pathway function that account for this loss of coordination remain undefined, and the extent to which they reflect plasticity produced by the known changes in afferent input and afferent neurons and by the loss of supraspinal reflex pathways remains to be determined.

Activity-dependent spinal cord plasticity probably contributes to the disordered urinary tract function after spinal cord injury and seems likely to also contribute to the abnormalities in other functions, such as bowel control and blood pressure regulation. Functional electrical stimulation to elicit voiding in those with spinal cord injuries is in clinical use (Brindley 1995, Rijkhoff et al 1997, Van Kerrebroeck 1998), and a preliminary effort to classically condition urination after spinal cord injury showed temporary success (Ince et al 1978). Concerted efforts to induce and guide activity-dependent plasticity to improve these functions after spinal cord injury, efforts comparable to those now focused on restoring locomotion, are likely to yield important new therapeutic advances.

PLASTICITY PRODUCED BY DESCENDING INPUT AND ASSOCIATED PERIPHERAL INPUT

Descending inputs to the spinal cord are a continual barrage of activity in a variety of pathways. The immediate short-term products of this activity—voluntary movements, responses to peripheral disturbances, activation of the spinal cord circuitry underlying locomotion, respiration, and urination, task-specific adjustments in spinal reflex pathways—are readily apparent and have received much attention. The immediate effects of loss or distortion of descending activity due to supraspinal or spinal trauma are similarly apparent. In contrast, the long-term effects of descending input on the spinal cord are not so obvious and have received less attention. Nevertheless, while the rapid effects of descending input are easy to see and convenient to study, the gradual effects are also important. They help establish and maintain spinal cord function in a state most amenable to effective performance. Activity-dependent plasticity, driven by descending input and associated peripheral input, shapes spinal cord function during development and continues to modify it throughout life.

Abnormal Descending Input

The immediate effects of spinal cord contusion or transection—loss of purposeful movement and sensation and profound suppression of spinal reflexes below the level of the lesion—have been known throughout history and are described in the oldest medical records (Ronthal 1998). Prior to the twentieth century, however, the long-term effects were largely unknown, because most victims succumbed rapidly to respiratory or urinary tract infections or to other consequences of immobility and autonomic dysfunction. The wars of the past century spurred development of comprehensive programs for acute and chronic management of the numerous problems attendant on spinal cord injury. As a result, most people with severe spinal cord injuries can now have nearly normal life spans, and the long-term effects of these injuries are extremely significant.

Destruction or severe impairment of spinal cord pathways initiates a sequence of changes in spinal cord function below the lesion that develops over days, weeks, and months (Riddoch 1917, Kuhn 1950, Mountcastle 1980, Ronthal 1998, Hiersemenzel et al 2000). At first, function is severely depressed and reflexes are difficult or impossible to elicit. This period of spinal shock abates over several weeks in humans. Then, over a longer period, the reflexes become abnormally strong. This is commonly manifested by increased resistance to passive muscle stretch, particularly in antigravity muscles (i.e. leg extensors and arm flexors), hyperactive tendon jerks, and increased flexion withdrawal responses to painful stimulation. These signs comprise the syndrome of spasticity and are accompanied by changes in a variety of electrophysiological measurements. The gradual development of these effects reflects spinal cord plasticity caused by the destruction of supraspinal connections, by the accompanying loss of descending input, and by associated changes in peripheral input.

Physiological studies give some insight into the mechanisms underlying this long-term plasticity. After cord transection in cats, Ia afferent excitatory postsynaptic potentials (EPSPs) in triceps surae motoneurons are larger than normal and have faster rise times, the motoneurons have more positive firing thresholds and shorter afterhyperpolarizations, and changes occur in motor unit properties and type distributions (Nelson & Mendell 1979, Cope et al 1986, Munson et al 1986, Hochman & McCrea 1994a–c). These effects differ across muscles. In both humans and animals spinal interneuronal pathways also change after spinal cord transection or injury (Boorman et al 1991, Thompson et al 1992, Shefner et al 1992). For example, both recurrent inhibition, mediated by the Renshaw cell, and reciprocal inhibition, mediated by the Ia inhibitory interneuron, appear to be increased. Electron microscopic analysis of motoneuron synaptic coverage after spinal cord injury reveals changes in specific terminal populations (Tai & Goshgarian 1996, Tai et al 1997).

These clinical and laboratory data demonstrate that spinal cord injury changes the distal spinal cord. However, they do not distinguish plasticity due to the change in descending input from plasticity due to pathway destruction and the accompanying retrograde and anterograde effects, or from plasticity due to accompanying peripheral changes. Spinal cord plasticity clearly attributable to a more limited and specific change in descending input was first demonstrated in the 1920s, when Anna DiGiorgio altered descending input to the spinal cord by lesioning one side of the cerebellum in anesthetized dogs, rabbits, and guinea pigs (DiGiorgio 1929, 1942). The immediate response was an asymmetric hindlimb posture: One leg was flexed and the other was extended. After a variable delay she removed the descending input responsible for this asymmetry by cutting the thoracic spinal cord. When the delay was short transection eliminated the asymmetric posture. However, when the delay was longer the asymmetric posture survived transection. It persisted even though all descending input was gone. Subsequent experiments in a variety of species confirmed her results and showed that they were not due to peripheral mediation, such as change in sensory receptor function (Manni 1950, Gerard 1961). In the 1960s Ralph Gerard and his colleagues defined in rats the time course of the development of asymmetry that survived transection (Chamberlain et al 1963). Persistent asymmetry did not occur if the delay between cerebellar lesion and spinal cord transection was only 30 min, and rose rapidly to its maximum value as the delay increased to one hour. The authors concluded that a period of 45 min was needed to establish significant persistent asymmetry. This phenomenon was considered by Gerard and others to be a good model for the fixation or consolidation of memory believed to occur at higher levels of the CNS, and was therefore called spinal fixation. Comparable phenomena occur in the spinal cord with a variety of supraspinal lesions and can also follow manipulation of labyrinthian sensory inputs (Giulio 1952, Straka & Dieringer 1995). Clearly, altered descending input that lasts for sufficient time produces spinal cord plasticity that persists after the input stops.

Descending Input in Normal Life

Spinal cord plasticity produced by long-term changes in descending activity is not limited to abnormal or pathological situations. It appears to be a feature of normal life, both during development and during later life as well. The acquisition of motor skills, both standard skills such as walking and writing and specialized skills such as dancing and piano playing, are acquired only through prolonged practice, which involves specific patterns of activity in descending pathways and associated patterns of activity in sensory afferents. The continued maintenance of these skills, in spite of peripheral and central changes associated with growth and aging, is also likely to involve prolonged adjustments in descending inputs. Diverse evidence indicates that spinal cord plasticity caused by descending activity contributes to motor development in childhood and to the learning of motor skills later in life.

Shaping of Spinal Cord Reflexes During Development   Descending input during the first years of life gradually modifies spinal reflexes and helps produce the normal adult reflex pattern. Both flexion withdrawal reflexes and muscle stretch reflexes are shaped by this influence. Schouenborg and colleagues have demonstrated the importance of descending influence in producing adult flexion withdrawal reflexes (Levinsson et al 1999). In the normal neonatal rat focal nociceptive stimulation produces diffuse and often inappropriate muscle contractions and limb movements. In contrast, in the normal adult such stimulation excites the appropriate muscles, that is, the muscles that withdraw the limb from the painful stimulus. As Figure 3A illustrates, neonatal spinal cord transection prevents development of the adult pattern, so that nonspecific and inappropriate flexion withdrawal reflexes remain in the adult.

The importance of descending input during early life in shaping the spinal cord circuitry that produces muscle stretch reflexes is shown by the effects of the perinatal supraspinal lesions associated with cerebral palsy. In normal infants muscle stretch produces very short-latency spinally mediated stretch reflexes in both the stretched muscles and their antagonists (Myklebust et al 1986, O'Sullivan et al 1991). Normally, the antagonist stretch reflexes gradually disappear during childhood, leaving the adult with standard, so called knee-jerk, reflexes, limited to the stretched muscles. However, in infants in whom supraspinal damage distorts activity in descending pathways, this normal evolution often fails to occur, so that antagonist stretch reflexes last into adulthood and contribute to motor disability. Figure 3B shows agonist and antagonist stretch reflexes from normal infants and normal adults, and from adults with cerebral palsy. In affected individuals, the original damage is supraspinal. Thus, the likely explanation for the abnormal persistence of infantile spinal reflexes into adulthood is the absence or distortion of the long-term descending input that normally eliminates or suppresses these reflexes over the course of development. The development of adult reflex patterns presumably reflects activity-dependent plasticity produced by appropriate descending input and by the appropriate patterns of peripheral input that the descending input produces by influencing muscle activation and limb movement.

Acquisition and Maintenance of Motor Skills   The acquisition of motor skills later in life is associated with less dramatic changes in spinal cord circuitry. In adults spinal reflexes correlate with the nature, intensity, and duration of motor training. The strengths of spinal reflexes depend on past physical activity and training (Rochcongar et al 1979, Goode & Van Hoven 1982, Casabona et al 1990, Koceja et al 1991, Nielsen et al 1993). The most frequently studied reflexes have been the spinal stretch reflex (SSR) (produced mainly by a monosynaptic pathway consisting of the Ia afferent from the muscle spindle, its synapse on the motoneuron, and the motoneuron itself), and its electrical analog, the H-reflex, which is elicited by direct electrical stimulation of the Ia afferents (Magladery et al 1951, Matthews 1972, Henneman & Mendell 1981, Brown 1984).

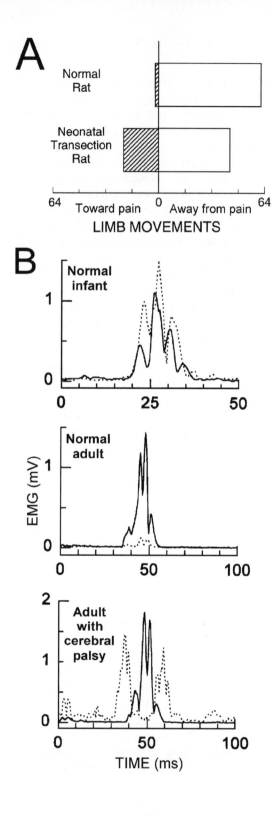

These reflexes differ between athletes and nonathletes and between different groups of athletes. A particularly valuable study reported H-reflexes in soleus muscles of people who were either sedentary, moderately active, or extremely active, or were professional ballet dancers (Nielsen et al 1993). Both the H-reflex and disynaptic reciprocal inhibition were larger in moderately active subjects than in sedentary subjects, and even larger in extremely active subjects. Because the human soleus muscle consists almost entirely of slow (i.e. type I) fibers, exercise-induced change in motor unit properties cannot readily account for the reflex increase seen with activity. Furthermore, the most remarkable finding, illustrated in Figure 4A, was that both the H-reflex and disynaptic reciprocal inhibition were lowest in the dancers, even though they were much more active than any other group. Their values were lower than those of sedentary subjects and much lower than those of active subjects. Noting that muscle cocontraction is accompanied by increased presynaptic inhibition and decreased reciprocal inhibition, the authors speculated that the prolonged cocontractions required by the classical ballet postures lead to persistent decreases in synaptic transmission at the Ia synapses, and thereby account for the reductions in H-reflexes and reciprocal inhibition. Viewed from the perspective of performance, the decreased direct peripheral influence on motoneurons indicated by the smaller reflexes may increase cortical control and allow more precise movement.

Laboratory evidence for training-induced spinal cord plasticity comes from a study in which monkeys were trained to make smooth repetitive flexion and extension movements at the elbow, and random brief perturbations were superimposed (Meyer-Lohmann et al 1986). Over months and years, the SSR elicited by the perturbation gradually increased so that it took over the task of responding to the torque pulse, while later reflex responses gradually disappeared. As illustrated in Figure 4B, the larger SSR was adaptive: it was associated with more rapid and effective correction of the change in position caused by the perturbation. The

---

←

Figure 3 Activity-dependent plasticity produced by descending input during development. (A) Direction of limb movement produced by flexion withdrawal responses to a nociceptive stimulus in normal adult rats and in adult rats subjected to spinal cord transection just after birth. Direction is almost always appropriate, i.e. away from the stimulus in normal adults, but is often inappropriate in transected adults. Neonatal transection prevents normal shaping of flexion withdrawal reflexes by descending input. Modified from Levinsson et al (1999). (B) Short-latency electromyographic responses of soleus (*solid*) and tibialis anterior (*dotted*) muscles to sudden foot dorsiflexion, which stretches the soleus and shortens the tibialis anterior, in a normal infant, a normal adult, and an adult with cerebral palsy. In the normal infant spinal stretch reflexes occur in both muscles. In a normal adult a reflex occurs only in the stretched muscle, i.e. the soleus. Little or no response occurs in the tibialis anterior. In contrast, in an adult with cerebral palsy, in whom perinatal supraspinal injury has impaired the descending input responsible for development of normal adult reflexes, the infantile pattern persists: Reflexes occur in both muscles. From B Myklebust, unpublished data (see Myklebust et al 1982, 1986 for comparable data).

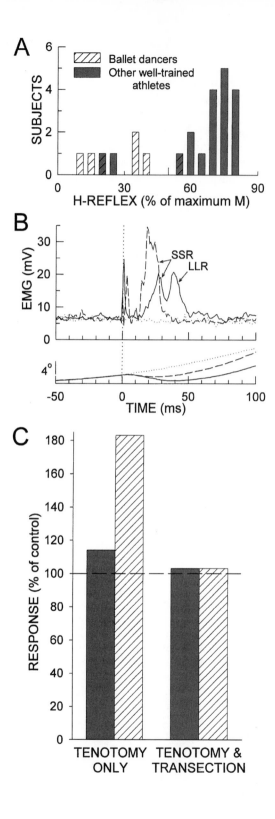

investigators concluded that the results "demonstrate a long-term functional plasticity of the sensorimotor system of adult animals and suggest a growing role for fast segmental mechanisms in the reaction to external disturbances as motor learning progresses" (p. 398).

Additional evidence for adaptive spinal cord plasticity during life and in response to specific demands comes from studies of reflex changes in humans associated with aging, space flight, and specialized training paradigms, and studies of reflex changes in animals associated with chronic alterations in peripheral input produced by tenotomy (Figure 4C) or by application of tetrodotoxin to the peripheral nerve (Beránek & Hnik 1959, Kozak & Westerman 1961, Robbins & Nelson 1970, Goldfarb & Muller 1971, Gallego et al 1979, Sabbahi & Sedgwick 1982, DeVries et al 1985, Reschke et al 1986, Webb & Cope 1992, Trimble & Koceja 1994, Angulo-Kinzler et al 1998, Yamanaka et al 1999, Zheng et al 2000).

All these data suggest that gradual activity-dependent changes in spinal cord function contribute to the acquisition and maintenance of motor skills throughout life. Their prolonged time course and dependence on repetition probably account in part for the lengthy periods and intensive practice required for acquisition and maintenance of athletic skills and other motor skills such as playing a musical instrument. At the same time, while experiments of the DiGiorgio type demonstrate that abnormal descending input can modify the spinal cord, the spinal reflex changes described in this section might conceivably be imposed by concurrent descending activity and/or might reflect peripheral changes: The spinal cord itself might have undergone no intrinsic or enduring change. Thus, these data strongly suggest, but do not demonstrate, that normal descending inputs produce spinal cord

←
_____

Figure 4   Activity-dependent spinal cord plasticity produced by descending input occurring during skill acquisition or in response to a peripheral lesion. (*A*) Soleus H reflexes are much smaller in professional ballet dancers than in other well-trained athletes (e.g. runners, swimmers, cyclists). (H-reflexes of sedentary subjects fall in between.) Modified from Nielsen et al (1993). (*B*) Working for reward, monkeys performed an elbow flexion-extension task on which brief perturbations were randomly superimposed. Biceps electromyographic activity and elbow angle (flexion is upward) for an unperturbed trial (dotted), a perturbed trial early in training (solid), and a perturbed trial late in training (dashed) are shown. Early in training perturbation produces both a spinal stretch reflex (SSR) and a long-latency polysynaptic response (LLR). After intermittent training over several years the SSR is much larger and the LLR has disappeared. The SSR has gradually taken over the role of opposing the perturbation. This improves performance: The perturbation of the smooth course of elbow flexion is smaller and briefer. Modified from Meyer-Lohmann et al (1986). (*C*) Tenotomy in cats increases the monosynaptic ventral root response to stimulation of the nerve from a tenotomized muscle (*hatched*) after 1–4 weeks. The response to stimulation of the nerve from a nontenotomized muscle (*solid*) does not change. The increase does not occur in cats in which the spinal cord was transected just prior to tenotomy. Thus, descending input in response to tenotomy appears to induce the spinal cord plasticity. Modified from Kozak & Westerman (1961).

plasticity. A simple laboratory model provides more direct evidence of activity-dependent spinal cord plasticity with skill acquisition.

Skill Acquisition in a Laboratory Model   Most of the evidence suggesting that descending and peripheral inputs during development and later in life change the spinal cord consists of changes in spinal cord reflexes, principally SSRs, H-reflexes, and flexion withdrawal reflexes. While these reflexes normally function as parts of complex behaviors, they are in themselves simple behaviors, the simplest behaviors of which the mammalian nervous system is capable, and adaptive changes in them are essentially simple skills that can be used as laboratory models for the plasticity underlying skill acquisition. Operant conditioning of the SSR, or its electrical analog the H-reflex, which has been demonstrated in monkeys, rats, and humans, has furnished clear evidence of activity-dependent spinal cord plasticity and is providing insight into its mechanisms (Wolpaw et al 1983, Evatt et al 1989, Wolpaw 1997).

In the standard protocol, used in monkeys, rats, and humans, SSR or H-reflex amplitude is measured as electromyographic activity, and reward occurs when amplitude is above (for up-training) or below (for down-training) a criterion value. The primary observation is that imposition of the reward criterion changes reflex amplitude appropriately over days and weeks. This adaptive change appears to occur in two phases, a small rapid phase 1 in the first few hours or days and a much slower phase 2 that continues for weeks (Figure 5*A*) (Wolpaw & O'Keefe

————————————————————————————→

Figure 5   Course of training of the spinal stretch reflex (SSR) pathway and associated spinal cord plasticity. (*A*) Two-phase course of SSR up-or down-training in monkeys. Rapid phase 1 change, reflecting appropriate change in descending influence over the reflex arc, occurs within 6 h. Gradual phase 2 change, reflecting spinal cord plasticity produced by the continuation of the altered descending input, develops over at least 40 days. Modified from Wolpaw & O'Keefe (1984). (*B*) Probable sites of spinal cord plasticity and altered descending input with SSR or H-reflex training. "MN" is the motoneuron, and each "IN" is one or more spinal interneuron types. Open synaptic terminals are excitatory, solid ones are inhibitory, half-open ones could be either, and the subdivided one is a cluster of C terminals. Dashed pathways imply the possibility of intervening spinal interneurons, and the dotted pathway is uncertain. The monosynaptic and possibly disynaptic H-reflex pathway from Ia and Ib afferents to the motorneuron is shown. The hypothesized sites of plasticity are circled with solid lines. Starting at the left and moving clockwise, these are: the motoneuron membrane (i.e. firing threshold and axonal conduction velocity), C terminals on the motoneuron, the Ia afferent synaptic connection, and terminals conveying disynaptic group I inhibition or excitation to the motoneuron. The corticospinal tract (CST) is shown, and the probable sites of action of the descending input responsible for the plasticity in the H-reflex pathway are circled with dashed lines. These are: connections on interneurons mediating presynaptic inhibition of the Ia synapse, supplying C terminals to the motoneuron, and/or conveying disynaptic Group I inhibition or excitation to the motoneuron. From Wolpaw (1997).

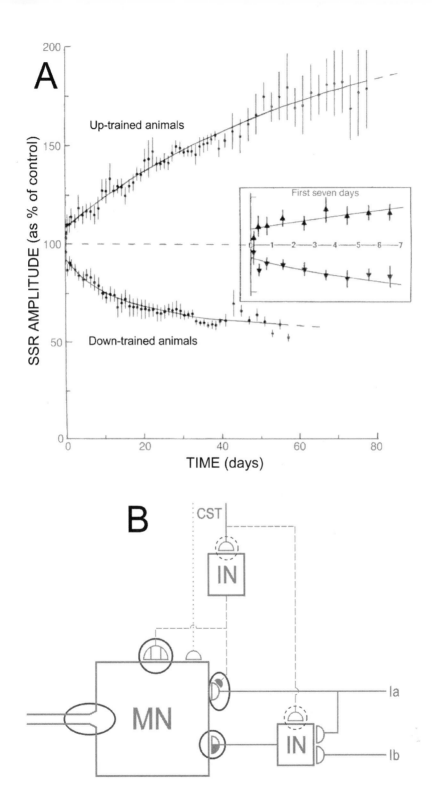

1984). Phase 1 appears to reflect rapid mode–appropriate change in descending influence over the spinal arc of the reflex, while phase 2 appears to reflect gradual spinal cord plasticity produced by the chronic continuation of the descending input responsible for phase 1. This descending input is conveyed by the corticospinal tract (Chen & Wolpaw 1997, Chen et al 2000). Training is possible in humans with partial spinal cord injuries, but does not seem to occur in those with strokes involving sensorimotor cortex (Segal & Wolf 1994, Segal 1997).

Once established, the reflex asymmetry created by this training survives removal of descending input: It persists after the spinal cord is isolated from the brain (Wolpaw & Lee 1989). Thus, the training changes the spinal cord. This spinal cord plasticity includes changes in motoneuron properties (Carp & Wolpaw 1994, Halter et al 1995, Carp et al 2001). Down-training is accompanied by a positive shift in motoneuron firing threshold and a reduction in axonal conduction velocity. Both changes suggest a positive shift in sodium channel activation voltage, and the change in threshold could account in large part for the smaller reflex. While activity-dependent synaptic plasticity has traditionally received the most attention as the probable basis of learning, the possibility that learning can also involve changes in neuronal voltage-gated ion channels has recently drawn interest (Spitzer 1999). The shift in motoneuron threshold produced by down-training appears to be an example of such neuronally based learning. Additional physiological and anatomical studies suggest that SSR or H-reflex training also affects the Ia afferent–motoneuron synapse, other synaptic terminals on the motoneuron, and interneurons that convey oligosynaptic group 1 input to the motoneuron (Carp & Wolpaw 1995, Feng-Chen & Wolpaw 1996). Down-training and up-training are not mirror images of each other, but rather have different mechanisms. Figure 5B summarizes current knowledge of the multi-site spinal cord plasticity produced by SSR or H-reflex training.

SPINAL CORD PLASTICITY AND THE PRODUCTION OF BEHAVIOR

As the preceding sections indicate, clinical and laboratory evidence indicates that activity-dependent plasticity occurs in the spinal cord throughout normal life as well as in response to trauma and disease. The sites of plasticity include synaptic connections made by incoming fibers, interneuronal populations interposed between these inputs and motoneurons, synaptic connections on motoneurons, and the motoneurons themselves. Because spinal motoneurons are, in Sherrington's phrase, "the final common path" for all movements, understanding spinal cord plasticity is central to understanding both simple and complex behaviors (Clarke & O'Malley 1996).

Furthermore, the same attributes responsible for the traditionally inferior status of the spinal cord—separation from the rest of the CNS, simpler structure, obvious role as a connector, and responsibility for simple reflex behaviors— facilitate studies of its plasticity, provide access to supraspinal plasticity, and allow exploration of how spinal and supraspinal plasticity interact to support acquisition and maintenance of motor skills. The technical accessibility of the spinal cord and its inputs and outputs permit localization and definition of plasticity occurring within it. Moreover, the well-defined pathways that connect it to the rest of the CNS and can be interrupted in the laboratory allow studies to focus on the impact of supraspinal plasticity and the manner in which it generates and interacts with spinal cord plasticity. These advantageous features of the spinal cord clarify fundamental principles of skill acquisition and maintenance and provide guidance for the design and implementation of new methods for restoring function after injury.

A Change in Behavior Involves Plasticity at Multiple Spinal and/or Supraspinal Sites

The fact that activity-dependent plasticity is ubiquitous in the CNS suggests that persistent changes in peripheral or descending input to the spinal cord—whether changes associated with development, skill acquisition (i.e. practice), trauma, or disease—will cause plasticity at multiple sites, both spinal and supraspinal. By taking advantage of the anatomical separation of the spinal cord from the rest of the CNS, several recent studies provide clear examples of behaviors that reflect multi-site activity-dependent plasticity.

A recent study of treadmill locomotion in cats compared the effects on treadmill locomotion of spinalization (i.e. spinal cord transection) followed several weeks later by unilateral denervation of ankle flexor muscles with the effects of denervation followed by spinalization (Carrier et al 1997). Spinalization followed by denervation (or denervation alone) had minimal persistent effects on the pattern or bilateral symmetry of leg movement during locomotion. Increases in hip and knee flexion soon compensated for the decrease in ankle flexion, so that locomotion was only slightly disturbed. In contrast, denervation followed by spinalization produced markedly abnormal and asymmetrical locomotion. After spinalization, the increases in hip and knee flexion that had followed denervation alone were magnified and accompanied by other marked abnormalities in amplitude and timing of muscle activity, so that locomotion was greatly disturbed and did not recover with practice. Figure 6A illustrates the difference in muscle activity during locomotion between a cat in which denervation followed spinalization and one in which spinalization followed denervation. The difference implies that the plasticity that occurred after denervation alone and was responsible for the recovery of nearly normal locomotion included modifications at both spinal and supraspinal levels. Spinal and supraspinal plasticity combined to compensate for the effects of denervation. After spinalization removed the influence of the supraspinal plasticity, the spinal cord plasticity functioned by itself, and the result was grossly abnormal spinal locomotion.

Another recent study described a related phenomenon and provided some insight into its mechanism (Whelan & Pearson 1997). During the stance phase of walking in the cat, stimulation of group I afferents in the nerves innervating ankle

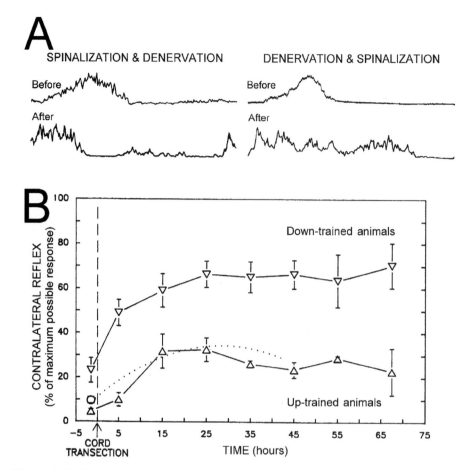

Figure 6 Evidence for multi-site activity-dependent plasticity. (*A*) Timing of muscle activation (i.e. ipsilateral sartorius) during one step cycle of treadmill locomotion in cats before and after denervation followed by spinalization (i.e. spinal cord transection) or spinalization followed by denervation. Timing of activation is nearly normal when denervation follows transection and profoundly abnormal when transection follows denervation. The difference implies that denervation results in both spinal and supraspinal plasticity, which together compensate for the deficit caused by denervation. When spinalization removes the contribution of the supraspinal plasticity so that the spinal cord plasticity functions in isolation, locomotion becomes grossly abnormal. Modified from Carrier et al (1997). (*B*) Contralateral monosynaptic reflex responses (+/− SEM) under anesthesia and with cord transection in monkeys in which the ipsilateral reflex had been increased or decreased by training over 50 days. The contralateral reflexes are much larger in down-trained animals than in up-trained animals [or in naive animals (*open circle* and *dotted line*)]. This finding was unexpected, because in the awake behaving animals the contralateral reflexes had changed little over the course of training. It indicated that training produced plasticity in the contralateral spinal cord that affected behavior only after anesthesia and transection removed descending input and/or suppressed tonic activity in the spinal cord and thereby eliminated the compensatory effect of supraspinal and/or other spinal cord plasticity. (The reflex increase in the first 15 hours is a nonspecific effect of anesthesia and surgery.) Modified from Wolpaw & Lee (1989).

extensor muscles, the LGS nerve to the lateral gastrocnemius and soleus, and the MG nerve to the medial gastrocnemius excites leg extensor muscles and delays the transition to the swing phase. Normally, LGS stimulation is more effective than MG stimulation. However, when the LGS nerve was cut and the cat continued to practice walking on the treadmill, the ability of LGS nerve stimulation to prolong the stance phase decreased to nearly zero over one month, while the effectiveness of MG nerve stimulation increased markedly in 5 days and remained high. Most important, when the cat was spinalized, the decreased effect of LGS nerve stimulation persisted in all cats and the increased effect of MG nerve stimulation persisted in some, indicating that, depending on the cat, spinal cord plasticity was wholly or partly responsible for the changes in the effectiveness of nerve stimulation. Spinal cord plasticity comparable to that revealed here, and/or other denervation-induced spinal cord plasticity, presumably accounted for the profoundly abnormal locomotion that occurred when spinalization followed ankle flexor denervation (Carrier et al 1997) (Figure 6A).

As Figure 5B summarizes, spinal stretch reflex (SSR) or H-reflex training is associated with plasticity at multiple sites in the ipsilateral spinal cord. Additional evidence that the behavioral effects of this training depend on multi-site plasticity came from measurement of the contralateral reflex in trained monkeys. Over the 50-day course of up-training or down-training of the ipsilateral H-reflex, the contralateral H-reflex remained close to its initial, or control, size (Wolpaw et al 1993). Thus, the behavioral effect of conditioning was focused on the ipsilateral H-reflex. However, when trained monkeys were anesthetized and the spinal cord was transected the contralateral reflex of down-trained animals was at least twice as large as the contralateral reflex of up-trained animals or the reflexes of naive animals (Wolpaw & Lee 1989). As shown in Figure 6B, anesthesia and cord transection uncovered a hidden effect of the training: It changed the contralateral side of the spinal cord. In the awake behaving monkey, this plasticity was not apparent, presumably because its effect on the contralateral H-reflex was cancelled out by plasticity at another site, which could be spinal or supraspinal. Anesthesia and spinalization, by removing descending influence and/or by quieting tonic activity in the spinal cord, eliminated the cancelling effect of this additional plasticity and revealed the presence of plasticity that changed the size of the contralateral reflex. What this contralateral plasticity might be and how it might relate to the H-reflex on the trained side are questions as yet unanswered. A similarly puzzling effect occurs in humans who suffer strokes affecting sensorimotor cortex of one hemisphere (Thilmann et al 1990). In addition to the well-known increased reflexes in the contralateral arm, these individuals also display decreased reflexes in the ipsilateral arm. Furthermore, unilateral hindlimb denervation in cats appears to affect contralateral as well as ipsilateral reflexes (Gossard et al 1999).

These clear demonstrations of multi-site plasticity were made possible by the anatomical separation of the spinal cord and by its capacity to produce simple behaviors in reduced preparations and in the presence of deep anesthesia. The unique central position of the spinal cord in the production of almost all behavior provides insight into the origins of multi-site plasticity.

Multi-Site Plasticity is Necessary and Inevitable

That behavioral changes as apparently simple as a larger or smaller H-reflex are associated with plasticity at multiple sites was initially surprising: The common expectation was that a simple change in behavior would be associated with a simple change in the CNS. However, the occurrence of multi-site plasticity appears to be a general principle that applies to even the simplest learning. It has been found as well in other ostensibly simple learning in vertebrate and invertebrate models (Lieb & Frost 1997, Thompson et al 1997, Cohen et al 1997, Lisberger 1998, Garcia et al 1999, Pearson 2000). As discussed below, multi-site plasticity would seem to be both necessary and inevitable, particularly for the spinal cord.

Together with its homologous brainstem nuclei, the spinal cord is the final assembly point for all neuromuscular behaviors, both simple and complex. For example, the motoneurons, interneurons, and synapses in the lumbosacral spinal cord execute all the different forms of locomotion and postural maintenance, produce a variety of specialized movements, withdraw the legs from painful stimuli, participate appropriately in actions involving all four limbs, etc. The fact that the normal spinal cord is able to support these many behaviors satisfactorily, as well as to incorporate new behaviors throughout life, suggests that its neuronal and synaptic function is appropriately adjusted and continually readjusted to accommodate the current behavioral repertoire. That such adjustments occur on a short-term basis as the CNS shifts from one behavior to another or cycles through the different phases of a single behavior is known from studies such as those showing the differences in presynaptic inhibition across standing, walking, and running or the changes that occur in the responses to group I afferent input during the step cycle (Capaday & Stein 1987, Stein 1995, Rossignol 1996, Faist et al 1996, Pearson & Ramirez 1997). The data reviewed in the preceding sections show that long-term adjustments also occur. Activity-dependent plasticity, driven by descending and peripheral inputs, is presumably responsible for maintaining spinal cord circuitry in a functional state appropriate for the execution of its current roster of behaviors. This long-term effect, a consensus produced by the different patterns of activity associated with these different behaviors, serves essentially as a coarse adjustment, establishing ranges over which the fine adjustments specific to each behavior are made. At any point in time, for example, the possible strength of Ia input to soleus motoneurons has a range that includes values appropriate to standing, walking, and running.

In this setting the neural activity that adds a new behavior to the repertoire (whether the activity is produced by daily practice and the behavior is an athletic skill, or the activity results from a peripheral or central lesion and the behavior represents or compensates for a functional deficit) is likely to cause plasticity that accommodates the new behavior as well as plasticity that maintains the old behaviors. For example, the stronger motoneuron response to Ia afferent input that underlies a new behavior (e.g. Figure 4B) is likely to affect the many other behaviors that involve primary afferent input to the motoneuron. These effects are likely to trigger additional activity-dependent plasticity that restores these other behaviors. Furthermore, simply because activity-dependent plasticity can occur at numerous sites in the spinal cord, the changes in activity caused by plasticity that supports the new behavior or maintains old behaviors are likely to trigger additional plasticity at other sites. For example, the larger reflex contralateral to an H-reflex that has been down-trained, evident only with anesthesia and spinal cord transection (Wolpaw & Lee 1989), or the smaller stretch reflexes found in the apparently normal arm contralateral to an arm paralyzed by a hemispheric stroke (Thilmann et al 1990), may represent reactive plasticity caused by change in activity in segmental pathways connecting the right and left sides of the spinal cord. The additional plasticity that maintains a nearly normal contralateral H-reflex in the awake behaving monkey might be compensatory, restoring normal function.

Thus, acquisition of any new behavior, whether it is a skill developed through prolonged practice or an abnormal response associated with supraspinal disease, is likely to involve three categories of plasticity: primary plasticity responsible for the new behavior, compensatory plasticity that maintains previous behaviors despite the impact of the primary plasticity, and reactive plasticity caused by the changes in activity resulting from primary and compensatory plasticity. This etiological categorization helps explain the multi-site plasticity associated with even the simplest change in behavior by indicating that multi-site plasticity is both necessary—to maintain the full repertoire of behaviors—and inevitable—due to the widespread capacity for activity-dependent plasticity. It also helps explain why some examples of plasticity (such as the contralateral spinal cord plasticity with H-reflex training) may bear no apparent relationship to the behavioral change with which they are associated. Furthermore, recognition of these different etiological categories of plasticity helps define factors controlling the effectiveness of therapeutic methods and thus helps guide therapeutic research.

Engagement of Spinal Cord Plasticity in Restoration of Behavior

The ultimate goal of treatment for spinal cord trauma or disease is a spinal cord that once again has normal structure and produces normal behavior. In practice, this single goal separates into two different and not necessarily compatible goals: restoration of normal structure (that is, normal neurons and glia, and intrinsic and extrinsic synaptic connections that display normal strengths and elicit normal neuronal responses) and restoration of normal behavior (that is, the normal repertoire of motor performances). Restoration of structure will guarantee restoration of behavior only if the restoration of structure is complete, is accompanied by restoration of normal peripheral and descending inputs, and is coupled with a comprehensive re-education program that eliminates the plasticity induced by injury or disease and induces the activity-dependent plasticity that occurs during normal development. Partial restoration of structure (for example, normal corticospinal tract connections to motoneurons and interneurons without normal interneuron connections to motoneurons, or normal interneuron connections without normal distribution of peripheral inputs) or restoration of connections without appropriate activity-dependent adjustment of their strengths is likely to have complex and not necessarily beneficial effects on behavior. Nevertheless, it seems clear that partial restoration of structure and imperfect re-education is all that will be possible in the near future.

At the same time, it is equally clear that restoration of behavior does not necessarily require restoration of normal neuronal and synaptic function. Nearly normal locomotion returns after selective muscle denervation even though the underlying CNS activity is different, and operant conditioning of one H-reflex does not affect the contralateral H-reflex even though the contralateral spinal cord is changed (Carrier et al 1997, Wolpaw & Lee 1989). Restoration of behavior might be achieved without restoration of normal structure. Restoration of useful locomotion or of acceptable bladder, bowel, or sexual function might be achieved more quickly by focusing directly on restoring them rather than focusing on restoring the neuronal and synaptic structure that produced them prior to injury.

In this context, the spinal cord's capacities for activity-dependent plasticity are both a challenge and an opportunity. On the one hand they contribute to the disabilities that follow spinal cord injury and will certainly affect the outcomes of new therapeutic methods that promote regeneration. On the other hand they offer the opportunity to guide restoration of neuronal and synaptic function and should allow imperfect regeneration to support substantial behavioral improvements. For both these reasons, the productive engagement of activity-dependent plasticity in the spinal cord is likely to be a key component of new therapeutic programs for spinal cord injury and other chronic neuromuscular disorders. As the preceding sections indicate, the induction and guidance of activity-dependent spinal cord plasticity requires training protocols that induce appropriate patterns of peripheral and descending inputs to the spinal cord. These protocols may benefit from incorporation of pharmacologic agents and/or artificial or exaggerated sensory inputs and from attention to injury-associated changes in spinal cord elements such as specific receptor populations (e.g. Chau et al 1998, Giroux et al 1999, Muir 1999).

The laboratory development and clinical application of activity-dependent spinal cord plasticity is still at the earliest stage. Explorations of the mechanisms of this plasticity and development of its clinical uses have just begun for locomotion and are still less developed for other important behaviors, such as urination. The methods and results of this work will be greatly affected by what is perhaps the most distinctive feature of activity-dependent spinal cord plasticity as it functions in normal life and in the presence of disease: the slow rate of its effect on behavior. Despite the rapidity of processes such as long-term potentiation, the changes in behavior that result from activity-dependent spinal cord plasticity occur gradually. In the spinal cat the improvements in treadmill locomotion produced by training develop over weeks of daily exposure. Operant conditioning of the SSR or H-reflex in rats, monkeys, or humans occurs gradually over days and weeks. The reflex changes that occur during normal development and those associated with skills such as ballet develop over months and years. This characteristic is, of course, fortunate—rapid large changes in spinal cord function would wreak havoc with movement control and require prodigious supraspinal compensation.

The characteristically gradual effect of activity-dependent spinal cord plasticity on behavior has practical implications. First and most obviously, laboratory and clinical manipulations and observations of this plasticity need to extend over sufficient time periods. Second, because of the ubiquity of activity-dependent plasticity and the inevitable interaction between primary, compensatory, and reactive types, the concordance between short-term and long-term effects of any intervention (e.g. Figure 5A), cannot be assumed in every situation. Short-term gains will not necessarily evolve into long-term improvements. Third, while spinal cord plasticity may support restoration of walking or standing (de Leon 1998b), it may not support the concurrent restoration of both behaviors. The capacity to switch rapidly and appropriately from one behavior to another may require supraspinal participation.

CNS Plasticity and Behavior

The substantial capacity for activity-dependent plasticity in the spinal cord has wide implications. First, as already noted, it suggests that most or all motor skills that are acquired gradually through prolonged practice involve spinal cord plasticity. Thus, these skills cannot be understood or explained simply by studying the changes that occur in cortex, cerebellum, or other supraspinal areas. The changes in the spinal cord need to be defined as well. Second, the fact that these motor skills depend on activity-dependent plasticity throughout the CNS suggests that gradually acquired intellectual skills, such as language mastery or mathematical facility, may also depend on widely distributed plasticity that develops slowly. The rapid changes in behavior that have traditionally engaged most research attention, such as the one-trial acquisition of a new word, may reflect minor adjustments in patterns of plasticity gradually acquired through prolonged practice, adjustments analogous to the change in presynaptic inhibition that accompanies the transition from standing to running, or the alteration in descending influence responsible for phase 1 change in the spinal stretch reflex (Figure 5A). Understanding of most skilled behaviors may require exploration of gradually acquired activity-dependent plasticity comparable to the plasticity most readily recognized and studied in the spinal cord.

CONCLUSIONS

The traditional concept of the spinal cord as a hardwired structure that simply provides rapid stereotyped responses to sensory inputs and to commands from the brain is not correct. Ample and diverse evidence indicates that activity-dependent plasticity occurs in the spinal cord during development, with skill acquisition and maintenance later in life, and in response to trauma and disease. In the isolated spinal cord, appropriate peripheral stimuli can produce a variety of persistent effects including habituation, sensitization, several forms of long-term potentiation, and both classical and operant conditioning, and treadmill training regimens can greatly improve locomotion. Abnormal descending input produced by spinal cord injury or supraspinal disorders gradually changes the spinal cord. Normal descending influence guides development of spinal cord reflexes early in life and throughout later life produces spinal cord plasticity that contributes to skill acquisition and maintenance. The spinal cord plasticity produced by peripheral and descending inputs affects input connections, interneuronal pathways, and motoneurons. Both synaptic and neuronal properties can change, and even simple behavioral changes are associated with changes at multiple spinal and supraspinal sites. This complex plasticity serves to support new behaviors and preserve old behaviors, and also reflects the ubiquity of the capacity for activity-dependent plasticity in the CNS. Engagement of activity-dependent spinal cord plasticity is a key component of new therapeutic approaches to restoring function after spinal cord injury. Appropriate guidance of this plasticity can maximize residual function and will be essential for re-educating a newly regenerated spinal cord. In these practical efforts restoration of useful behaviors might be achieved without full restoration of normal neuronal and synaptic structure, and the typically gradual development of the behavioral effects of activity-dependent spinal cord plasticity will be an important factor. Finally, activity-dependent spinal cord plasticity may help elucidate mechanisms of learning throughout the CNS.

ACKNOWLEDGMENTS

We thank Drs. Elizabeth Winter Wolpaw, Linda S Sorkin, Serge Rossignol, Keir G Pearson, Dennis J McFarland, Bruce H Dobkin, Xiang Yang Chen, Jonathan S Carp, and Laurent JG Bouyer for their valuable comments on the manuscript and Mr. Hesham Sheikh for excellent technical assistance. Work in the authors' laboratory has been supported by the National Institutes of Health, the Paralyzed Veterans of America Spinal Cord Research Foundation, the United Cerebral Palsy Research and Education Foundation, the American Paralysis Association, and the International Spinal Research Trust.

Visit the Annual Reviews home page at www.AnnualReviews.org

LITERATURE CITED

Alvares D, Fitzgerald M. 1999. Building blocks of pain: the regulation of key molecules in spinal sensory neurones during development and following peripheral axotomy. *Pain Suppl.* 6:S71–S85

Amar AP, Levy ML. 1999. Pathogenesis and pharmacological strategies for mitigating secondary damage in acute spinal cord injury. *Neurosurgery* 44(5):1027–39

Aminoff MJ. 1993. *Brown-Séquard. A Vision-*

ary of Science, p. 113. New York: Raven. 211 pp.

Angulo-Kinzler RM, Mynark RG, Koceja DM. 1998. Soleus H-reflex gain in elderly and young adults: modulation due to body position. *J. Gerontol. Ser. A* 53(2):M120–25

Baldissera F, Hultborn H, Illert M. 1981. Integration in spinal neuronal systems. In *Handbook of Physiology. Section I: The Nervous System. Vol. II: Motor Control, Part I*, ed.

VB Brooks, pp. 509–95. Baltimore, MD: Williams & Wilkins

Baranauskas G, Nistri A. 1998. Sensitization of pain pathways in the spinalcord: cellular mechanisms. *Prog. Neurobiol.* 54:349–65

Barbeau H, Fung J. 1992. New experimental approaches in the treatment of spastic gait disorders. *Med. Sports Sci.* 36:234–46

Barbeau H, McCrea DA, O'Donovan MJ, Rossignol S, Grill WM, Lemay MA. 1999. Tapping into spinal circuits to restore motor function. *Brain Res. Rev.* 30:27–51

Barbeau H, Rossignol S. 1987. Recovery of locomotion after chronic spinalization in the adult cat. *Brain Res.* 412:84–95

Bélanger M, Drew T, Provencher J, Rossignol S. 1996. A comparison of treadmill locomotion in adult cats before and after spinal transection. *J. Neurophysiol.* 76(1):471–91

Beránek R, Hnik P. 1959. Long-term effects of tenotomy on spinal monosynaptic response in the cat. *Science* 130:981–82

Biering-Sørensen F, Nielans H-M, Dørflinger T, Sørensen B. 1999. Urological situation five years after spinal cord injury. *Scand. J. Urol. Nephrol.* 33:157–61

Boorman G, Hulliger M, Lee RG, Tako K, Tanaka R. 1991. Reciprocal Ia inhibition in patients with spinal spasticity. *Neurosci. Lett.* 127:57–60

Bouyer LJG, Whalen PJ, Pearson KG, Rossignol S. 2001. Adaptive locomoter plasticity in chronic spinal cats following ankle extensor neurectomy. *J. Neurosci.* In press

Bregman BS, Diener PS, McAtee M, Dai HN, James C. 1997. Intervention strategies to enhance anatomical plasticity and recovery of function after spinal cord injury. *Adv. Neurol.* 72:257–75

Brindley GS. 1995. The sacral anterior root stimulator as a means of managing the bladder in patients with spinal cord lesions. *Bailliere's Clin. Neurol.* 4(1):1–13

Brown WF. 1984. *The Physiological and Technical Basis of Electromyography.* Boston, MA: Butterworths

Burke RE. 1998. Spinal cord: Ventral Horn. In *The Synaptic Organization of the Brain*, ed. GM Shepherd, 3:77–120. New York: Oxford Univ. Press

Bussel B, Roby-Brami A, Azouvi P, Biraben A, Yakovleff A, Held JP. 1988. Myoclonus in a patient with spinal cord transection: possible involvement of the spinal stepping generator. *Brain* 111:1235–45

Calancie B, Needham-Shropshire B, Jacobs P, Willer K, Zych G, Green BA. 1994. Involuntary stepping after chronic spinal cord injury: evidence for a central rhythm generator for locomotion in man. *Brain* 117:1143–59

Capaday C, Stein RB. 1987. A method for stimulating the reflex output of a motoneuron pool. *J. Neurosci. Methods* 21:91–104

Carp JS, Wolpaw JR. 1994. Motoneuron plasticity underlying operantly conditioned decrease in primate H-reflex. *J. Neurophysiol.* 72:431–42

Carp JS, Wolpaw JR. 1995. Motoneuron properties after operantly conditioned increase in primate H-reflex. *J. Neurophysiol.* 73:1365–73

Carp JS, Chen XY, Sheikh H, Wolpaw JR. 2001. Operant conditioning of rat H-reflexes affects motoneuron axonal conduction velocity. *Exp. Br. Res.* 136:369–73

Carrier L, Brustein E, Rossignol S. 1997. Locomotion of the hindlimbs after neurectomy of ankle flexors in intact and spinal cats: model for the study of locomotor plasticity. *J. Neurophysiol.* 77:1979–93

Casabona A, Polizzi MC, Perciavalle V. 1990. Differences in H-reflex between athletes trained for explosive contraction and non-trained subjects. *Eur. J. Appl. Physiol.* 61 (1-2):26–32

Chamberlain T, Halick P, Gerard RW. 1963. Fixation of experience in the rat spinal cord. *J. Neurophysiol.* 22:662–73

Chau C, Barbeau H, Rossignol S. 1998. Early locomotor training with clonidine in spinal cats. *J. Neurophysiol.* 79:392–409

Chen D, Nussbaum SB. 2000. The gastrointestinal system and bowel management following spinal cord injury. *Phys. Med. Rehabil. Clin. N. Am.* 11(1):45–56

Chen HH, Frank E. 1999. Development and specification of muscle sensory neurons. *Curr. Opin. Neurobiol.* 9(4):405–9

Chen XY, Chen L, Wolpaw JR. 2000. The corticospinal tract in development and maintenance of H-reflex operant conditioning in rats. *J. Soc. Neurosci. Abstr.* 26:2206

Chen XY, Wolpaw JR. 1997. Dorsal column but not lateral column transection prevents down conditioning of H-reflex in rats. *J. Neurophysiol.* 78(3):1730–34

Clarke E, Jacyna LS. 1987. *Nineteenth-Century Origins of Neuroscientific Concepts.* Berkeley: Univ. Calif. Press

Clarke E, O'Malley CD. 1996. *The Human Brain and Spinal Cord.* San Francisco: Norman

Cohen TE, Kaplan SW, Kandel ER, Hawkins RD. 1997. A simplified preparation for relating cellular events to behavior: mechanisms contributing to habituation, dishabituation, and sensitization of the Aplysia gill-withdrawal reflex. *J. Neurosci.* 17(8):2886–99

Cope TC, Bodine SC, Fournier M, Edgerton VR. 1986. Soleus motor units in chronic spinal transected cats: physiological and morphological alterations. *J. Neurophysiol.* 55:1202–20

de Groat WC, Kurse MN, Vizzard MA, Cheng C-L, Araki I, Yoshimura N. 1997. Modification of urinary bladder function after spinal cord injury. *Adv. Neurol.* 72:347–64

de Leon RD, Hodgson JA, Roy RR, Edgerton VR. 1999a. Retention of hindlimb stepping ability in adult spinal cats after the cessation of step training. *J. Neurophysiol.* 81:85–94

de Leon RD, Hodgson JA, Roy RR, Edgerton VR. 1998a. Locomotor capacity attributable to step training versus spontaneous recovery after spinalization in adult cats. *J. Neurophysiol.* 79:1329–40

de Leon RD, Hodgson JA, Roy RR, Edgerton VR. 1998b. Full weight-bearing hindlimb standing following stand training in the adult spinal cat. *J. Neurophysiol.* 80:83–91

de Leon RD, Tamaki H, Hodgson JA, Roy RR, Edgerton VR. 1999b Hindlimb locomotor and postural training modulates glycinergic inhibition in the spinal cord of the adult cat. *J. Neurophysiol.* 82:359–69

DeVries HA, Wiswell RA, Romero GT, Heckathorne E. 1985. Changes with age in monosynaptic reflexes elicited by mechanical and electrical stimulation. *Am. J. Phys. Med.* 64:71–81

Dietz V, Colombo G, Jensen L, Baumgartner L. 1995. Locomotor capacity of spinal cord in paraplegic patients. *Ann. Neurol.* 37:574–86

DiGiorgio AM. 1929. Persistenza nell'animale spinale, di asymmetrie posturali e motorie di origine cerebellare: I, II, III. *Arch. Fisiol.* 27:518–80

DiGiorgio AM. 1942. Azione del cervelletto-neocerebellum-sul tono posturale degli arti e localizzazioni cerebellari dell'animale rombencefalico. *Arch. Fisiol.* 42:25–79

Dimitrijevic MR, Gerasimenko Y, Pinter MM. 1998. Evidence for a spinal central pattern generator in humans. *Ann. NY Acad. Sci.* 860:360–76

Dobkin B. 1998. Activity-dependent learning contributes to motor recovery. *Ann. Neurol.* 44(2):158–60

Dobkin B. 1999. Overview of treadmill locomotor training with partial body weight support: a neurophysiologically sound approach whose time has come for randomized clinical trials. *Neurorehab. Neural Repair* 13:157–65

Dobkin B, Harkema S, Requejo PS, Edgerton VR. 1995. Modulation of locomoter-like EMG activity in subjects with complete and incomplete spinal cord injury. *J. Neurol. Rehabil.* 9:183–90

Dubner R, Gold M. 1999. The neurobiology of pain. *Proc. Natl. Acad. Sci. USA* 96:7627–30

Durkovic RG. 1985. Retention of a classically conditioned reflex response in spinal cat. *Behav. Neural Biol.* 43:12–20

Durkovic RG. 1986. The spinal cord: a simplified system for the study of neural mechanisms of mammalian learning and memory. In *Development and Plasticity of the Mammalian Spinal Cord*, ed. ME Goldberger, A Gorio, M Murray, pp. 149–62. Padova: Liviana

Evatt ML, Wolf SL, Segal RL. 1989. Modification of human spinal stretch reflexes: preliminary studies. *Neurosci. Lett.* 105:350–55

Faist M, Dietz V, Pierrot-Deseilligny E. 1996. Modulation, probably presynaptic in origin, of monosynaptic Ia excitation during human gait. *Exp. Brain Res.* 109:441–49

Fawcett JW. 1998. Spinal cord repair: from experimental models to human application. *Spinal Cord* 36(12):811–17

Feng-Chen KC, Wolpaw JR. 1996. Operant conditioning of H-reflex changes synaptic terminals on primate motoneurons. *Proc. Natl. Acad. Sci. USA* 93:9206–11

Field-Fote EC. 2000. Spinal cord control of movement: implications for locomotor rehabilitation following spinal cord injury. *Phys. Ther.* 80(5):477–84

Gallego R, Kuno M, Nunez R, Snider WD. 1979. Disuse enhances synaptic efficacy in spinal mononeurones. *J. Physiol.* 291:191–205

Garcia KS, Steele PM, Mauk MD. 1999. Cerebellar cortex lesions prevent acquisition of conditioned eyelid responses. *J. Neurosci.* 19(24):10940–47

Gerard RW. 1961. The fixation of experience. In *Brain Mechanisms and Learning*, ed. RW Gerard, J Konorski, pp. 21–32. Oxford: Blackwell

Giroux N, Rossignol S, Reader TA. 1999. Autoradiographic study of a1-, a2-noradrenergic and serotonin 1a receptors in the spinal cord of normal and chronically transected cats. *J. Comp. Neurol.* 406:402–14

Giulio L. 1952. Sulla funzione del midollo spinale. Persistenza di asimmetrie da eccitamento labirintico. *Boll. Soc. Ital. Biol. Sper.* 28:1651–52

Goldfarb J, Muller RU. 1971. Occurrence of

heteronymous monosynaptic reflexes following tenotomy. *Brain Res.* 28:553–55

Goode DJ, Van Hoven J. 1982. Loss of patellar and Achilles tendon reflex in classical ballet dancers. *Arch. Neurol.* 39:323

Gossard JP, Bouyer LJG, Ménard A, Leblond H. 1999. Plastic changes in the transmission of spinal pathways to pretibial flexors during fictive locomotion following a partial cutaneous denervation in cats. *Neuroscience* 25:1152 (Abstr.)

Halter JA, Carp JS, Wolpaw JR. 1995. Operantly conditioned motoneuron plasticity: possible role of sodium channels. *J. Neurophysiol.* 74:867–71

Harkema S, Hurley SL, Patel UK, Requejo PS, Dobkin B, Edgerton VR. 1997. Human lumbosacral spinal cord interprets loading during stepping. *J. Neurophysiol.* 77:797–811

Henneman E, Mendell LM. 1981. Functional organization of motoneuron pool and inputs. In *Handbook of Physiology. Sect. I. The Nervous System. Vol. II. Motor Control, Part I*, ed. VB Brooks, pp. 423–507. Baltimore, MD: Williams & Wilkins

Herrero JF, Laird JM, Lopez-Garcia JA. 2000. Wind-up of spinal cord neurones and pain sensation: much ado about something? *Prog. Neurobiol.* 61:169–203

Hiersemenzel LP, Curt A, Dietz V. 2000. From spinal shock to spasticity: neuronal adaptations to a spinal cord injury. *Neurology* 54(8):1574–82

Hochman S, McCrea DA. 1994a. Effects of chronic spinalization on ankle extensor motoneurons I. Composite monosynaptic Ia EPSPs in four motoneuron pools. *J. Neurophysiol.* 71(4):1452–67

Hochman S, McCrea DA. 1994b. Effects of chronic spinalization on ankle extensor motoneurons II. Motoneuron electrical properties. *J. Neurophysiol.* 71(4):1468–79

Hochman S, McCrea D. 1994c. Effects of chronic spinalization on ankle extensor motoneurons III. Composite Ia EPSPs in motoneurons separated into motor unit types. *J. Neurophysiol.* 71(4):1480–90

Holmes G. 1915. Spinal injuries of warfare. *Br. Med. J.* 2:815–21

Ince LP, Brucker BS, Alba A. 1978. Conditioned responding of the neurogenic bladder. *Psychosom. Med.* 40(1):14–24

Kandel ER. 1977. Neuronal plasticity and the modification of behavior. In *Handbook of Physiology, Section. I. The Nervous System, Vol. I. Cellular Biology of Neurons*, ed. JM Brookhart, VB Mountcastle, pp. 1137–82. Bethesda, MD: Williams & Wilkins

Kandel ER, Schwartz JH, Jessell TM. 2000. *Principles of Neural Science*. New York: McGraw-Hill

Kiehn O, Harris-Warrick RM, Jordan LM, Hultborn H, Kudo N, eds. 1998. *Neuronal Mechanisms for Generating Locomotor Activity*. New York: NY Acad. Sci.

Koceja DM, Burke JR, Kamen G. 1991. Organization of segmental reflexes in trained dancers. *Int. J. Sports. Med.* 12:285–89

Kozak W, Westerman RA. 1961. Plastic changes of spinal monosynaptic responses from tenotomized muscles in cats. *Nature* 189:753–55

Kuhn RA. 1950. Functional capacity of the isolated human spinal cord. *Brain* 1:1–51

Levinsson A, Luo XL, Holmberg H, Schouenborg J. 1999. Developmental tuning in a spinal nociceptive system: effects of neonatal spinalization. *J. Neurosci.* 19(23):10397–403

Liddell EGT. 1960. *The Discovery of Reflexes*. Oxford: Clarendon

Lieb JR, Frost WN. 1997. Realistic simulation of the Aplysia siphon-withdrawal reflex circuit: roles of circuit elements in producing motor output. *J. Neurophysiol.* 77(3):1249–68

Linsenmeyer TD. 2000. Sexual function and infertility following spinal cord injury. *Phys. Med. Rehabil. Clin. N. Am.* 11(1):141–56

Lisberger SG. 1998. Physiologic basis for motor learning in the vestibulo-ocular reflex. *Otolaryngol.-Head Neck Surg.* 119(1):43–48

Liu X, Sandkühler J. 1997. Characterization of long-term potentiation of C-fiber-evoked potentials in spinal dorsal horn of adult rat: essential role of NK1 and NK2 receptors. *J. Neurophysiol.* 78(4):1973–82

Lloyd DPC. 1949. Post-tetanic potentiation of response in monosynaptic reflex pathways of the spinal cord. *J. Gen. Physiol.* 33:147–70

Lovely RG, Gregor RJ, Roy RR, Edgerton VR. 1986. Effects of training on the recovery of full-weight-bearing stepping in the adult spinal cat. *Exp. Neurol.* 92:421–35

Magladery JW, Porter WE, Park AM, Teasdall RD. 1951. Electrophysiological studies of nerve and reflex activity in normal man. IV. The two-neuron reflex and identification of certain action potentials from spinal roots and cord. *Bull. John Hopkins Hosp.* 88:499–519

Manni E. 1950. Localizzazoni cerebellari corticali nella cavia. Nota 1: Il "corpus cerebelli". *Arch. Fisiol.* 49:213–37

Matthews PBC. 1972. *Mammalian Muscle Receptors and Their Central Actions*, pp. 319–409. Baltimore, MD: Williams & Wilkins

McKinley WO, Jackson AB, Cardenas DD, DeVivo MJ. 1999. Long-term medical complications after traumatic spinal cord injury: a regional model systems analysis. *Arch. Phys. Med. Rehabil.* 80(11):1402–10

Mendell LM. 1984. Modifiability of spinal synapses. *Phys. Rev.* 64:260–324

Mendell LM. 1988. Physiological aspects of synaptic plasticity: the Ia/motoneuron connection as a model. *Adv. Neurol.* 47:337–60

Mendell LM. 1999. Neurotrophin action on sensory neurons in adults: an extension of the neurotrophic hypothesis. *Pain* Suppl. 6:S127–32

Mendell LM, Wall PD. 1965. Response of single dorsal cells to peripheral cutaneous unmyelinated fibers. *Nature* 206:97–99

Meyer-Lohmann J, Christakos CN, Wolf H. 1986. Dominance of the short-latency component in perturbation induced electromyographic responses of long-trained monkeys. *Exp. Brain Res.* 64:393–99

Monga M, Bernie J, Rajasekaran M. 1999. Male infertility and erectile dysfunction in spinal cord injury: A review. *Arch. Phys. Med. Rehabil.* 80:1331–39

Mountcastle VB. 1980. Effects of spinal cord transection. In *Medical Physiology*, ed. VB Mountcastle, 1:781–86. St. Louis: Mosby

Muir GD. 1999. Locomotor plasticity after spinal injury in the chick. *J. Neurotrauma* 16(8):705–10

Muir GD, Steeves JD. 1995. Phasic cutaneous input facilitates locomotor recovery after spinal cord injury in the chick. *J. Neurophysiol.* 74:358–68

Muir GD, Steeves JD. 1997. Sensorimotor stimulation to improve locomotor recovery after spinal cord injury. *Trends Neurosci.* 20(2):72–77

Munson JB, Foehring RC, Lofton SA, Zengel JE, Sypert GW. 1986. Plasticity of medial gastrocnemius motor units following cordotomy in the cat. *J. Neurophysiol.* 55:619–34

Myklebust BM, Gottlieb GL, Agarwal GC. 1986. Stretch reflexes of the normal human infant. *Dev. Med. Child Neurol.* 28:440–49

Myklebust BM, Gottlieb GL, Penn RL, Agarwal GC. 1982. Reciprocal excitation of antagonistic muscles as a differentiating feature in spasticity. *Ann. Neurol.* 12:367–74

Nelson SG, Mendell LM. 1979. Enhancement in Ia-motoneuron synaptic transmission caudal to chronic spinal cord transection. *J. Neurophysiol.* 42:642–54

Neuburger M. 1981. Experiments on the reflex mechanism. In *The Historical Development of Experimental Brain and Spinal Cord Physiology before Flourens*, ed. M Neuberger, E Clarke, pp. 237–46. Baltimore: The Johns Hopkins Univ. Press

Nielsen J, Crone C, Hultborn H. 1993. H-reflexes are smaller in dancers from the Royal Danish Ballet than in well-trained athletes. *Eur. J. Appl. Physiol.* 66:116–21

Orlovsky GN, Deliagina TG, Grillner S. 1999. *Neuronal Control of Locomotion from Mollusc to Man.* New York: Oxford Univ. Press

O'Sullivan MC, Eyre JA, Miller S. 1991. Radiation of phasic stretch reflex in biceps brachii to muscles of the arm in man and its restriction during development. *J. Physiol.* 439:529–43

Patterson MM. 1976. Mechanisms of classical conditioning and fixation in spinal mammals. In *Advances in Psychobiology*, ed. LM Aanonsen, RF Thompson, 10:381–436. New York: Wiley

Pearson KG. 2000. Plasticity of neuronal networks in the spinal cord: modifications in response to altered sensory input. *Prog. Br. Res.* 128:61–70

Pearson KG, Ramirez JM. 1997. Sensory modulation of pattern-generating circuits. In *Neurons, Networks and Motor Behavior*, ed. PSG Stein et al, 225–35. Cambridge Mass: MIT Press. 305 pp.

Peckham PH, Creasey GH. 1992. Neural prostheses: clinical application of functional electrical stimulation in spinal cord injury. *Paraplegia* 30:96–101

Pockett S. 1995. Spinal cord synaptic plasticity and chronic pain. *Anest. Analg.* 80:173–79

Randić M. 1996. Plasticity of excitatory synaptic transmission in the spinal cord dorsal horn. *Prog. Brain Res.* 113:463–506

Reschke MF, Anderson DJ, Homick JL. 1986. Vestibulo-spinal response modification as determined with the H-reflex during the Spacelab-1 flight. *Exp. Brain Res.* 64:367–79

Riddoch G. 1917. The reflex functions of the completely divided spinal cord in man, compared with those associated with less severe lesions. *Brain* 40:264–402

Rijkhoff NJM, Wijkstra H, Van Kerrebroeck PEV, Debruyne FMJ. 1997. Urinary bladder control by electrical stimulation techniques in spinal cord injury. *Neurourol. Urodyn.* 16:39–53

Robbins A, Nelson PG. 1970. Tenotomy and the spinal monosynaptic reflex. *Exp. Neurol.* 27:66–75

Rochcongar P, Dassonville J, Le Bars R. 1979. Modifications du reflexe de Hoffmann en fonction de l'entrainement chez le sportif. *Eur. J. Appl. Physiol.* 40:165–70

Ronthal M. 1998. Spinal cord injury. In *Clinical*

Neurology, ed. RJ Joynt, RC Griggs, 47:1–28. Hagerstown, MD: Lippincott Williams & Wilkins

Rossignol S. 1996. Neural control of stereotypic limb movements. In *Handbook of Physiology*, ed. LB Rowell, JT Sheperd, pp. 173–216. New York: Oxford Univ. Press

Rossignol S. 2000. Locomotion and its recovery after spinal injury. *Curr. Opin. Neurobiol.* 10:708–16

Rossignol S, Bouyer LJG. 2001. Locomoter compensation to peripheral nerve lesions in the cat. In *Spinal Cord Plasticity: Alterations in Reflex Function*, ed. MM Patterson, JW Grau. Boston: Kluwer Acad.

Roy RR, Acosta L. 1986. Fiber type and fiber size changes in selected thigh muscles six months after low thoracic spinal cord transection in adult cats: exercise effects. *Exp. Neurol.* 92:675–85

Roy RR, Baldwin KM, Edgerton VR. 1991. The plasticity of skeletal muscle: effects of neuromuscular activity. *Exerc. Sport Sci. Rev.* 19:269–312

Roy RR, Talmadge RJ, Hodgson JA, Zhong H, Baldwin KM, Edgerton VR. 1998. Training effects on soleus of cats spinal cord transected (T12-13) as adults. *Muscle Nerve* 21:63–71

Sabbahi MA, Sedgwick EM. 1982. Age-related changes in monosynaptic reflex excitability. *J. Gerontol.* 37(1):24–32

Segal RL. 1997. Plasticity in the central nervous system: operant conditioning of the spinal stretch reflex. *Top. Stroke Rehabil.* 3(4):76–87

Segal RL, Wolf SL. 1994. Operant conditioning of spinal stretch reflex in patients with spinal cord injuries. *Exp. Neurol.* 130:202–13

Shefner JM, Berman SA, Sarkarati M, Young RR. 1992. Recurrent inhibition is increased in patients with spinal cord injury. *Neurology* 42:2162–68

Shurrager PS, Dykman RA. 1951. Walking spinal carnivores. *J. Comp. Physiol. Psychol.* 44:252–62

Spitzer NC. 1999. New dimensions of neuronal plasticity. *Nat. Neurosci.* 2(6):489–91

Stein RB. 1995. Presynaptic inhibition in humans. *Prog. Neurobiol.* 47:533–44

Stein RB. 1999. Functional electrical stimulation after spinal cord injury. *J. Neurotrauma* 16(8):713–17

Stiens SA, Bergman SB, Goetz LL. 1997. Neurogenic bowel dysfunction after spinal cord injury: clinical evaluation and rehabilitative management. *Arch. Phys. Med. Rehabil.* 78:S86–102

Straka H, Dieringer N. 1995. Spinal plasticity after hemilabyrinthectomy and its relation to postural recovery in the frog. *J. Neurophysiol.* 73:1617–31

Tai Q, Goshgarian HG. 1996. Ultrastructural quantitative analysis of glutamatergic and GABAergic synaptic terminals in the phrenic nucleus after spinal cord injury. *J. Comp. Neurol.* 372(3):343–55

Tai Q, Palazzolo KL, Goshgarian HG. 1997. Synaptic plasticity of 5-hydroxytryptamine-immunoreactive terminals in the phrenic nucleus following spinal cord injury: a quantitative electron microscopic analysis. *J. Comp. Neurol.* 386(4):613–24

Teasell RW, Arnold JM, Krassioukov A, Delaney GA. 2000. Cardiovascular consequences of loss of supraspinal control of the sympathetic nervous system after spinal cord injury. *Arch. Phys. Med. Rehabil.* 81(4):506–16

Thilmann A, Fellows S, Garms E. 1990. Pathological stretch reflexes on the "good" side of hemiparetic patients. *J. Neurol. Neurosurg. Psychiatry* 53(3):208–14

Thompson FJ, Reier PJ, Lucas CC, Parmer R. 1992. Altered patterns of reflex excitability subsequent to contusion injury of the rat spinal cord. *J. Neurophysiol.* 68:1473–86

Thompson RF. 2001. Spinal plasticity. In *Spinal Cord Plasticity: Alterations in Reflex Function*, ed. MM Patterson, JW Grau. Boston: Kluwer Acad.

Thompson RF, Bao S, Chen L, Cipriano BD,

Grethe JS, et al. 1997. Associative learning. *Int. Rev. Neurobiol.* 41:151–89

Trimble MH, Koceja DM. 1994. Modulation of the triceps surae H-reflex with training. *Int. J. Neurosci.* 76:293–303

Tuszynski MH, Kordower JE. 1999. *CNS Regeneration-Basic Science and Clinical Advances.* San Diego: Academic

Van Kerrebroeck PEV. 1998. The role of electrical stimulation in voiding dysfunction. *Eur. J. Urol.* 34(Suppl. 1):27–30

Webb CB, Cope TC. 1992. Modulation of Ia EPSP amplitude: the effects of chronic synaptic inactivity. *J. Neurosci.* 12:338–44

Weld KJ, Dmochowski RR. 2000. Effect of bladder management on urological complications in spinal cord injured patients. *J. Urol.* 163:768–72

Wernig A, Muller S, Nanassy A, Cagol E. 1995. Laufband therapy based on "rules of spinal locomotion" is effective in spinal cord injured persons. *Eur. J. Neurosci.* 7:823–29

Wernig A, Nanassy A, Muller S. 1998. Maintenance of locomotor abilities following Laufband (treadmill) therapy in para- and tetraplegic persons: follow-up studies. *Spinal Cord* 36(11):744–49

Whelan PJ, Pearson KG. 1997. Plasticity in reflex pathways controlling stepping in the cat. *J. Neurophysiol.* 78(3):1643–50

Wilson P, Kitchener PD. 1996. Plasticity of cutaneous primary afferent projections to the spinal dorsal horn. *Prog. Neurobiol.* 48:105–29

Wolpaw JR. 1997. The complex structure of a simple memory. *Trends Neurosci.* 20:588–94

Wolpaw JR, Braitman DJ, Seegal RF. 1983.

Adaptive plasticity in the primate spinal stretch reflex: initial development. *J. Neurophysiol.* 50:1296–311

Wolpaw JR, Herchenroder PA, Carp JS. 1993. Operant conditioning of the primate H-reflex: factors affecting the magnitude of change. *Exp. Brain Res.* 97:31–39

Wolpaw JR, Lee CL. 1989. Memory traces in primate spinal cord produced by operant conditioning of H-reflex. *J. Neurophysiol.* 61:563–72

Wolpaw JR, O'Keefe JA. 1984. Adaptive plasticity in the primate spinal stretch reflex: evidence for a two-phase process. *J. Neurosci.* 4:2718–24

Woolf CJ, Costigan M. 1999. Transcriptional and posttranslational plasticity and the generation of inflammatory pain. *Proc. Natl. Acad. Sci. USA* 96:7723–30

Yaksh TL, Hua X-Y, Kalcheva I, Nozaki-Taguchi N, Marsala M. 1999. The spinal biology in humans and animals of pain states generated by persistent small afferent input. *Proc. Natl. Acad. Sci. USA* 96:7680–86

Yamanaka K, Yamamoto S, Nakazawa K, Yano H, Suzuki Y, Fukunaga T. 1999. The effects of long-term bed rest on H-reflex and motor evoked potential in the human soleus muscle during standing. *Neurosci. Lett.* 266:101–4

Yoshimura N. 1999. Bladder afferent pathway and spinal cord injury: possible mechanisms inducing hyperreflexia of the urinary bladder. *Prog. Neurobiol.* 57:583–606

Zheng Z, Gibson SJ, Khalil Z, Helme R, McMeeken JM. 2000. Age-related differences in the time course of capsaicin-induced hyperalgesia. *Pain* 85:51–58

Annu. Rev. Neurosci. 2001. 24:845–67

QUANTITATIVE GENETICS AND MOUSE BEHAVIOR

Jeanne M Wehner[1], Richard A Radcliffe[2], and Barbara J Bowers[3]

[1]Institute for Behavioral Genetics and Department of Psychology, University of Colorado, Boulder, Colorado 80309; e-mail: Jeanne.Wehner@Colorado.edu
[2]Institute for Behavioral Genetics and Department of Pharmacology, University of Colorado, Denver, Colorado 80262; e-mail: Richard.Radcliffe@Colorado.edu
[3]Institute for Behavioral Genetics, University of Colorado, Boulder, Colorado 80309; e-mail: bbowers@Colorado.edu

Key Words   inbred strains, learning and memory, anxiety, seizures, drug abuse

■ **Abstract**   Quantitative differences are observed for most complex behavioral and pharmacological traits within any population. Both environmental and genetic influences regulate such individual differences. The mouse has proven to be a superb model in which to investigate the genetic basis for quantitative differences in complex behaviors. Genetically defined populations of mice, including inbred strains, heterogeneous stocks, and selected lines, have been used effectively to document these genetic differences. Recently, quantitative trait loci methods have been applied to map the chromosomal regions that regulate variation with the goal of eventually identifying the gene polymorphisms that reside in these regions.

INTRODUCTION

Geneticists have long recognized the tremendous resource that the mouse provides for studying mammalian biology. In the neurosciences, the mouse is becoming increasingly valuable due to a wealth of information that is being generated through the application of transgenic methodologies, mutagenesis techniques, and gene mapping strategies. Although single gene technologies provide a way to uncover the essential proteins in pathways leading to behavioral and pharmacological responses, it is well accepted that individual differences in complex behaviors are polygenically regulated in both humans and animals. The scope of this review is limited to quantitative traits that can best be studied in multiple genotypes. For this reason, this review is restricted to literature using more than two mouse strains for any comparison.

What is a Quantitative Trait?

Most phenotypes or traits that show continuous variation in a population are quantitative traits. Such phenotypes are distinguished from qualitative traits in that the description of the phenotype of individuals within the group is measured numerically. Multiple genes, each of which may have a small effect, contribute to the population variation for a quantitative trait. The genetic basis for individual differences is the existence of multiple alleles for the same gene in a population. In any individual, each gene has two alleles, one of maternal and the other of paternal derivation. However, in an entire species there are many different alleles for every gene. Differing alleles are said to be polymorphic because they contain differences in the DNA sequence. The molecular basis of a polymorphism is a nonlethal mutation that is preserved in the population because it is neutral in effect, has only a subtle effect, or theoretically confers some advantage to individuals carrying the mutation. Polymorphisms are the sources of genetically based individual differences in a population and exist in relatively high frequencies. In contrast, rare mutations occur at frequencies of less than 1% in a population. Examples of polymorphisms with neutral effects include those in noncoding regions of the genome and those that do not alter the amino acid composition of proteins. Polymorphisms having subtle effects may alter the enzymatic activity or structural properties of proteins. Quantitative traits result from polymorphisms in multiple genes and include both normal and abnormal physiological and behavioral traits. Often the transmission of complex traits is termed non-Mendelian because a clear pattern of inheritance for a single gene is not observed.

THE MOUSE AS A GENETIC MODEL

Isogenic Stocks and Terminology

It is well established that the mouse genome contains many, if not most, of the same genes that are found in humans. Moreover, there is remarkable synteny between the two species, such that the order of genes over large portions of the genome is identical. A huge number of differing genetic stocks of the mouse are available with known genealogies, and using defined stocks allows replication across laboratories (http:www.informatics.jax.org). Several excellent books and numerous Websites on mouse genetics are available that provide fundamental definitions, histories of the derivation of particular mouse strains, and nomenclature, as well as breeding and mapping techniques (Festing 1979, Silver 1995; http:www.informatics.jax.org). Different types of genetic stocks can be used in neurobiological research, including isogenic stocks of mice, heterogeneous stocks, selected lines, recombinant inbred strains, segregating crosses of mice, and congenic strains of mice. Unlike randomly bred populations, all of these genetic stocks are the results of systematic breeding efforts. Isogenic stocks are those in which each mouse of the population is genetically identical (with the exception of sex chromosomes). Both inbred strains and the first filial generation (F1) hybrids of two inbred strains are isogenic. Inbred mouse strains by definition are generated by systematic mating of brothers and sisters for 20 generations. This limits the number of alleles in the population and leads to genetic fixation, such that homozygosity (two copies of the same allele) is produced at virtually all gene loci. Those most commonly used inbred strains belong to the *Mus musculus* group (Silver 1995). Estimates of the genetic diversity across strains were made using dense maps of DNA polymorphisms distributed throughout the genome (www.informatics.jax.org). Many commonly used strains are estimated to be about 40–50% different based on mapping of over 300 DNA polymorphisms (Dietrich et al 1992).

Sometimes the process of inbreeding and genetic fixation results in reduced fitness (inbreeding depression). For this reason, some pharmacological and behavioral studies are performed in F1 hybrids that are generated by crossing two inbred mice. F1 mice are isogenic, heterozygous at all loci that differ between the two parental strains. Regardless of the exact isogenic population, the only source of phenotypic variation is the environment.

The impact of the environment on behavioral and physiological traits should not be dismissed. Considerable environmental variation has been observed in behavioral traits but also in body weight, even with rigorous control of testing environments (Crabbe et al 1999). It is possible that not all genotypes respond uniformly to the same environmental stimuli. Thus, gene by environment interactions must be considered. It is not acceptable to study one mouse or just a few mice in any inbred strain and assume that the phenotypic score is indicative of that for the whole strain. This is true whether the phenotype is a behavior or the amounts of RNA or protein expressed in a particular brain region.

Genetic studies of complex phenotypes, i.e. quantitative traits, require multiple genotypes, and the large repertoire of inbred strains provides a marvelous resource for maximizing genetic variation. There are several goals for quantitative genetic studies, including elucidating genes that regulate individual differences, determining whether two, or more, processes are regulated by common genes, and determining whether there are gene interactions (epistasis). Most quantitative genetic studies are initiated by screening different inbred strains of mice for a phenotype of interest. Subsequent studies may then focus on two particular strains that differ maximally and serve as the parental strains for genetic crosses. However, for polygenic traits, genetic analyses can be pursued even if the two parental strains do not differ on a particular measure. There are few, if any, complex traits that do not demonstrate quantitative variation. The fact that strain variation is observed for a particular phenotype when these strains are tested under similar environmental conditions is indicative of some degree of genetic regulation of a trait.

Other Genetically Defined Populations

Other genetically defined populations that can be studied to maximize genetic variation include heterogeneous stocks of mice and selected lines of mice. Heterogeneous stocks (HS) are created by systematic breeding of several inbred strains

(McClearn et al 1970, Demarest et al 2001). HS mice show wide variation on pharmacological and behavioral traits but are genetically defined such that the differing alleles at any particular gene locus can be traced back to the strain of origin. They have proven to be a very good tool for fine mapping of the loci regulating some behavioral traits (Talbot et al 1999, Demarest et al 2001). Unlike isogenic stocks, however, each individual is different and the source of variation in the phenotype is due both to genetic and environmental factors.

Capitalizing further on the high degree of genetic variation in the mouse can be accomplished using genetic selection to create lines of mice that differ dramatically in particular physiological and behavioral traits.

Selected lines are a way to find novel genes regulating a trait without knowing anything about those genes. To select genetically, the founding population must show genetic variation. Therefore, nonisogenic founding populations for a selection study in mice are usually from one of two different sources. The F2 and F3 generations derived from two inbred strains can serve as the founding stock for a selection study. A good example of a successful selection study using this strategy is found in the bidirectional selection of high and low open-field activity lines by DeFries et al (1978). HS also serve as an excellent founding population. There are numerous examples of their application to genetic selection (Erwin & Deitrich 1996, Marley et al 1998). The mathematical considerations for selection are discussed in Falconer (1989), and guidelines for the best selection design can be found in DeFries (1981).

The Use of Phenotypic and Genetic Correlations and Heritability to Study Complex Traits

In many instances, an interesting behavioral or physiological difference may be observed between two inbred mouse strains, and it is desirable to explore relationships between behavior and biochemical or physiological parameters. Genetic correlational analyses of F2 populations, recombinant inbred strains, inbred strain surveys, and selected lines are often used to expand the analysis.

An F2 population is derived from a cross of F1 hybrids. Because of the fundamental processes of segregation and recombination in which new arrangements of genes occur during meiosis, every individual in an F2 population is unique. Testing each individual in an F2 population for multiple phenotypes can be useful to determine whether there is an association between the various measures. However, there are limitations to the use of F2 populations for this type of study. The first is that only a phenotypic correlation can be obtained for parameters of interest, i.e. correlations may be due to both genetic and environmental influences. Second, because F2 individuals are unique, stable genetic stocks are not generated, and a single animal must be tested repeatedly.

These limitations are overcome by the use of inbred strains and recombinant inbred (RI) strains. RI strains are generated by inbreeding F2 families for 20 generations (Bailey 1971). Multiple panels of RI strains have been generated from crosses of commonly used inbred strains (see www.informatics.jax.org). As with other inbred strains, multiple measures can be obtained for a RI strain without testing the same individual mouse more than once. Moreover, because each RI strain is isogenic, a genetic correlation can be calculated using strain means. Genetic correlations provide information on whether two parameters are regulated by common genes (Hegman & Possidente 1981, Crabbe et al 1990).

Similar to the RI analysis, estimates of genetic correlations can be obtained by surveying a large number of inbred mouse strains. This provides a better estimate of whether common genes mediate several processes in a more generalized mouse sample than with either the F2 or RI approaches. There are numerous examples in neurobiology that demonstrate the utility of this approach for understanding whether two pharmacological agents or two behavioral responses to the same drug are mediated by common genes (mechanisms) (Crabbe 1983, Belknap et al 1998).

Genetic correlations can also be assessed easily in selected lines, though to be meaningful, replicate lines must also be tested (Crabbe et al 1990). A phenotypic measure that differs between the high lines and low lines and that was not used as the selected trait is considered to be a correlated trait. A finding such as this supports the hypothesis that common genes regulate variation in both the selection trait and the trait of interest.

Although most recent studies focus on elucidating the genes regulating a phenotype, numerical estimates of the degree to which a phenotype is regulated by genetic factors are very useful. When studying inbred mouse strains, this is done by performing a classical Mendelian genetic analysis and obtaining an estimate of the heritability of the trait. Heritabilities are also calculated in any genetic selection study. Using the variances in each of the populations tested, a heritability estimate (h^2), defined as the proportion of phenotypic variation due to genetic factors or $h^2 = Vg/Vp$, can be calculated (Falconer 1989). Most estimates of heritability for complex traits are modest, such that about 30–50% of the variation in the phenotype is regulated by genetic factors. It is important to note that heritability is a population-specific parameter. Heritability estimates for the same phenotype may vary from study to study depending on the genotypes in the study, the genetic design, and the environment in which testing is performed. Heritability is a useful parameter because it can guide future decisions relating to the feasibility of mapping the genes regulating variation in a phenotype.

Quantitative Trait Loci Analyses

Recently, geneticists have attempted to map quantitative traits to specific chromosomal regions. Whereas the ultimate goal of such studies is to identify the polymorphisms underlying genetic differences for a quantitative trait, the initial stage of the analysis is to map variation, not genes. Lander & Botstein (1989) developed a quantitative trait loci analysis that allows for the simultaneous mapping of multiple loci regulating a portion of the variation in a particular trait. A quantitative trait locus (QTL) is a chromosomal region that contains a gene, or genes, that regulates a portion of the genetic variation for a particular phenotype. The theoretical basis for QTL analysis is segregation and recombination, which are responsible for introducing variability into a population. Because recombination occurs by physically breaking and resealing stretches of DNA during crossing over, genes that are located far apart from each other will frequently be separated (i.e. will recombine). Genes that are located very closely are less likely to undergo recombination and are said to be "linked." QTL analysis is particularly useful for complex disorders and behaviors because it allows the localization of genes without any a priori knowledge of the gene itself.

QTL analysis requires analyzing a trait in a large number of individuals in a segregating population (F2 or backcross populations, RI strains) and then mapping these individual differences utilizing a series of DNA marker sequences that are polymorphic between two parental populations. Such polymorphic DNA markers do not need to be in a particular gene; they only need to be linked to the QTL for a QTL to be mapped (Dietrich et al 1992). After deriving phenotypic scores in a segregating population and the pattern of polymorphisms in individuals exhibiting high or low scores for the phenotype of interest, data are analyzed by a variety of statistical methods including maximum likelihood techniques for interval mapping, regression analyses, or chi-square analyses. Basically, all analyses are testing the probability that variation in the phenotype (high or low score) is associated with a particular genetically mapped polymorphism. Usually, the association of a QTL with a phenotype is reported using log of the odds scores or standard alpha levels (p-values). Discussion of the various statistical methods and acceptable significance levels are beyond the scope of this review, but the reader is referred to Lander & Kruglyak (1995).

Following the initial identification of QTLs in segregating populations, further experimentation is required to confirm them and move beyond a statistical association to a causal relationship. Depending on the number of recombinant inbred strains in a RI panel, confirmational studies are usually performed in an F2 population (Johnson et al 1992). Initially, QTLs are localized to very large chromosomal intervals that may contain hundreds of genes. Although, throughout this review, cases of overlapping QTLs across different phenotypes will be discussed, it must be acknowledged that overlap does not necessarily mean the same gene underlies the QTL across phenotypes.

It is desirable to narrow the QTL region to a size that is amenable to positional cloning or reasonable identification of candidate genes. There is no firm consensus as to the strategy that should be invoked in this phase, but a number of strategies have been proposed (Darvasi 1998). The use of congenic strains is popular because ultimately a stable strain can have immense advantages. In the construction of a congenic strain (Silver 1995), a particular QTL region from one parental strain is introgressed into the other parental strain using DNA marker-assisted breeding strategies thereby isolating the QTL of interest from all others. Typically, the introgressed region is fairly large (as much as one fourth of a full chromosome), but by creating a series of congenics with overlapping intervals, the QTL can be narrowed to a much smaller interval (Darvasi 1998). The number of strains required for this strategy is dependent on the size of the original introgressed region and the desired reduction in size of the QTL. The limitation of the congenic approach is that if epistasis between QTLs is robust, the QTL effect may not be observed in a congenic strain.

The final step in QTL analysis requires identifying the gene(s) that contains functional polymorphisms responsible for the phenotypic differences in the parental lines. A polymorphism may exist in an exon of a gene that leads to an amino acid difference in a protein. Such polymorphisms could lead to altered ligand binding, enzymatic activity, or a structural change in a protein of interest. Another type of polymorphism that is more difficult to detect could be found in regulatory regions such as enhancers or promoters. This type of polymorphism could produce quantitative differences in gene expression.

Although the criteria for proof that a polymorphism leads to a phenotypic alteration in behavior is not well established, it is likely that creation of a "knock-in" mouse containing the site-specific polymorphisms using transgenic mouse methods would provide convincing evidence that a particular polymorphism does indeed result in a shift in the behavioral phenotype. Although knock-in strategies continue to improve, the limited availability of embryonic stem cells from many inbred mouse strains (for example, DBA/2J, A/J, and BALB/c) that are used most often for studying complex traits poses a serious limitation for the study of strain specific polymorphisms using gene targeted methodologies.

QUANTITATIVE BEHAVIORAL TRAITS

The examples provided here do not represent a complete survey of quantitative behavioral genetics in the mouse, but have been selected to represent some of the most active research areas in the field.

Learning and Memory

Individual differences in specific cognitive function are documented in human behavioral genetic studies (DeFries et al 1979). Clearly, the mouse cannot serve as a model for all facets of human cognitive function, but some forms of learning are amenable to study. Many recent studies have focused on forms of hippocampal-dependent learning and fear conditioning.

Spatial Learning   The integral role of the hippocampus in regulating spatial learning (O'Keefe & Nadel 1978) prompted investigations into the genetic regulation of spatial learning as measured by the Morris water task (Morris 1981). In one version the platform is clearly visible, and in another version the platform is hidden and the mouse must find its location using distal spatial cues. The ability to locate the hidden platform is sensitive to hippocampal lesions in the mouse (Logue et al 1997b). Although a strain survey indicated a continuous distribution in latencies to find the platform during training, and in probe trial measures obtained by tracking behavior after the platform has been removed after training, there are several important concerns relating to genetic background when analyzing spatial learning in this paradigm (Upchurch & Wehner 1988, Owen et al 1997b, Rogers et al 1999). Of prominent concern is poor visual acuity in some strains that carry the retinal degeneration gene (Upchurch & Wehner 1988, Fox et al 1999, Owen et al 1997b, Rogers et al 1999). The most commonly used inbred strain, C57BL/6J (B6), is a relatively good spatial learner compared with other inbred strains, but many other inbreds including DBA/2J (D2) show poor performance on the Morris task (Upchurch & Wehner 1988). Some of the inadequacies of inbred mouse strains are a result of inbreeding depression, because significant heterosis (hybrid vigor) is exhibited in F1 hybrids created between almost any two inbred mouse strains (Owen et al 1997b). Consistent with heterosis, a classical genetic analysis of Morris water task performance between B6 and D2 strains demonstrated that most of the variability between these two strains is under the regulation of dominant genes (Upchurch & Wehner 1989). Analyses of 11 BXD RI strains generated from B6 and D2 demonstrated that variation in the Morris water task is under polygenic regulation and that BXD strain differences in spatial selectivity are significantly correlated to the amount of hippocampal, but not cortical, protein kinase C activity (Wehner et al 1990). Further analyses failed to detect significant QTLs, probably owing in part to the lack of power for QTL detection with small numbers of RI strains (JM Wehner, unpublished data).

Contextual and Cued Fear Conditioning   Contextual and cued fear conditioning is a highly conserved form of Pavlovian learning (LeDoux 2000). The ability to learn to associate the presentation of a cue (conditioning stimulus) with a mild footshock (unconditioned stimulus) is dependent on the amygdala (LeDoux 2000); additionally, learning to associate the context in which conditioning took place is dependent on the hippocampus (Anagnostaras et al 1999, Logue et al 1997b).

Both inbred strain surveys (Owen et al 1997b) and a BXD recombinant inbred strain study (Owen et al 1997a) demonstrate polygenic regulation of both contextual and cued fear conditioning. Genetic correlational analyses suggest there is some overlap between genes regulating learning of the auditory cue and the context (Owen et al 1997a). Whereas there is quantitative variation in the degree to which the various strains showed memory of the auditory cue, all strains showed fear in the form of freezing when the auditory cue was presented 24 hr after conditioning, but some strains were poor contextual learners (Owen et al 1997a,b).

QTLs for contextual fear conditioning, fear conditioning in an altered context, and auditory cued fear conditioning were mapped in an F2 intercross population generated from B6 and D2 (Wehner et al 1997) and in a backcross population generated between C3H/HeJ and B6 (Caldarone et al 1997). In both genetic crosses QTLs on the distal portion of chromosome 1 and on chromosome 3 were detected for contextual fear conditioning, suggesting that some QTLs generalize across inbred strains. Other QTLs were also detected that were unique to the particular cross. Another QTL study performed in an F2 intercross between B6 and BALB/cJ only detected a gender-specific QTL on chromosome 8 in males (Valentinuzzi et al 1998). This lack of detection of other QTLs is probably due to a small sample size.

Confirmational studies on the QTLs for contextual fear conditioning were pursued in the B6XD2 cross. In a short-term phenotypic selection, contextual fear conditioning was the selection measure (Radcliffe et al 2000b). Confirmation of QTLs on chromosome 2, 3, and 16 were obtained. Although the evidence was weaker for the QTL on chromosomes 1, this may be due to the fact that the selection was for contextual, not cued, fear conditioning and is consistent with some evidence supporting a greater role of the QTL on chromosome 1 for regulating fear or anxiety but not contextual learning (Flint et al 1995, Gershenfeld et al 1997, Wehner et al 1997). Recently, examination of congenic strains has confirmed the chromosome 1 QTL for cued fear conditioning (RA Radcliffe & JM Wehner, unpublished results).

Additional studies are needed to understand the quantitative genetic regulation of learning and memory. A rigorous examination of quantitative differences in spatial learning in a variety of other tasks has not been done, although strain differences have been demonstrated in eight-way radial arm maze-learning (Roullet & Lassalle 1995) and in a spatial recognition task (Dellu et al 2000). Although the impact of sensory deficits has been considered in the analysis of inbred strains, only recently has notice been given to the role of variation in attentional processes across inbred strains and the potential impact that attention has on learning performance (Gould & Wehner 1999).

Open-Field Activity and Anxiety

Anxiety disorders are some of the most prevalent forms of psychiatric illness in humans. Twin, adoption, and family studies indicate that genetic factors significantly contribute to the expression of pathological anxiety (Torgersen 1990) and that the heterogeneity of anxiety disorders reflects polygenic regulation. In several testing paradigms, anxiety-related behaviors can be measured in the mouse and are operationally defined as approach-avoidance behavior elicited by conflict situations. For example, many paradigms pair an aversive stimulus such as open space, bright light, or height with a rewarding stimulus such as a dark, confined space that would be perceived as nonthreatening by the animal. Anxious rodents choose the nonthreatening environment over their natural exploratory drive to approach novel environments. Three tests have been used extensively to study the polygenic nature of anxiety in the mouse: the open field, the elevated plus-maze, and the light/dark transitions test. The open field was one of the first tests used to measure emotionality in rodents and is still used extensively (Hall 1934). The open field is a large arena that serves as an anxiety-producing environment for mice. Many measures can be derived in the open field, but total activity for a particular time period may be used as a measure of locomotion, whereas combining measures of activity with defecation or deriving measures of thigmotaxis and habituation may tap into more complex phenotypes. It is assumed that an emotionally reactive rodent will exhibit decreased activity and increased defecation in the open field owing to activation of the autonomic nervous system (Hall 1934). High levels of illumination in the test are thought to increase anxiety in the animal. Strain surveys indicated that quantitative differences are observed for open field activity tested under dim (Crabbe 1986, Bolivar et al 2000) and brightly lit conditions (Logue et al 1997a), as well as for habituation and thigmotaxis (Bolivar et al 2000, Koyner et al 2000).

A classic study that provides evidence of polygenic regulation of emotionality in mice is the selection study by DeFries et al (1978), in which mice generated from an F3 cross between C57BL/6J (high activity strain) and BALB/cJ (low activity strain) were selectively bred for high and low activity in the open field and a negatively correlated response to selection was observed for defecation scores. This correlation indicates that some of the same genes regulate both activity and defecation in these lines. These selected lines have been used for QTL analyses (see below).

Both the elevated plus-maze (Lister 1987) and the light/dark transitions test are based on approach-avoidance conflicts (Crawley & Goodwin 1980, Costall et al 1989). The plus-maze pairs an open, elevated environment with a protected, enclosed (albeit elevated) space. The light/dark maze pairs a brightly lit arena with a small, dark space.

Trullas & Skolnick (1993) compared the behaviors of 16 inbred strains in the elevated plus-maze and the open field and showed that behavior in the plus-maze was substantially regulated by genetic factors: 69–78% of the variance between strains in percent time and entrances into the open arms of the maze was due to polygenic factors. Similarly, 75% of the variation in open-field activity under bright light was due to genetic factors; however, variation due to genetic influences was decreased to approximately 44% for activity under low illumination. The rank order of the strains was also different in the elevated plus-maze compared with the open field. A similar difference in rank order among seven strains was also observed by Griebel et al (2000; also see Rogers et al 1999). These data suggest that although common genes regulate a general aspect of emotionality, a unique set of genes regulate variability in specific measures of anxiety. Moreover, it is clear that additional genes, different from those regulating baseline anxiety, may regulate the anxiolytic effects of benzodiazepines (Griebel et al 2000).

QTL analyses for anxiety-related behaviors measured in the open field, light/dark transitions test, and the elevated plus-maze have been performed in populations derived from the C57BL/6J and A/J strains (Gershenfeld et al 1997, Gershenfeld & Paul 1997, Mathis et al 1995) and in the DeFries high- and low-activity selected lines that were discussed above. Several putative QTLs were identified in C57BL/6 and A/J (BXA and AXB) RI strains (Mathis et al 1995). In studies using more powerful F2 intercross populations derived from these strains, two highly significant QTLs on chromosomes 1 (101 cM) and 10 (74 cM), as well as several provisional QTLs on other chromosomes, were identified (Gershenfeld et al 1997, Gershenfeld & Paul 1997). Multiple QTLs regulating emotionality were identified in an F2 cross of the DeFries open-field lines (Flint et al 1995). In addition, Flint et al (1995) scored open-arm activity in the elevated plus-maze and activity in a Y-maze. One of the most significant findings from this study is that three common QTLs located on chromosomes 1, 12, and 15 were identified for anxiety-related behaviors from all three mazes. Turri et al (1999) confirmed and narrowed the QTL intervals for emotionality on chromosomes 1 and 15 identified by Flint et al (1995) using a recombinant inbred segregation test (Darvasi 1998). Talbot et al (1999) mapped QTLs for emotionality to a 0.8 cM region on chromosome 1 and another to chromosome 12 (see Flint et al 1995) in HS mice. Whereas the chromosome 1 QTL does not appear to be identical to that mapped in the DeFries lines, the application of HS mice to fine mapping is a considerable advancement, and this strategy is now being applied to other behaviors (Demarest et al 2001).

A QTL on chromosome 1 has been detected with striking consistency in multiple genotypes, across anxiety mazes, in both the dimly lit and brightly lit open field arena, and in fear conditioning (Flint et al 1995, Gershenfeld & Paul 1997, Koyner et al 2000). It remains to be seen whether multiple genes in the QTL region explain this overlap, or whether QTLs in this region regulate variation in response to novelty. Although it is possible that the chromosome 1 QTL could simply regulate motor activity because all of the tests require an animal to move, this seems unlikely because QTL analysis of spontaneous activity under normal lighting and housing conditions did not reveal a QTL on chromosome 1 (Toth & Williams 1999).

Seizures

With the exception of stroke, epilepsy is the most common neurological disorder in humans, with a prevalence of nearly 1% in the population (Smith et al 1998). The complexity and heterogeneity of epilepsy have made identification of the gene(s) difficult. Knowledge of underlying genetic factors would significantly advance our understanding of the neurological basis of seizure disorders. Because of neurobiological similarities in seizure pathways between humans and mice, quantitative genetic analyses of mouse populations can provide valuable information (Seyfried & Todorova 1999).

Seizures can be elicited in mice by chemical convulsants, electroshock, sound, rhythmic tossing, and withdrawal from drugs such as ethanol or pentobarbital. The results of inbred strain surveys, RI strain studies, and phenotypic distributions of F2 populations indicate that seizure susceptibility is polygenically regulated (Miner & Collins 1989; Kosobud & Crabbe 1990; Neumann & Seyfried 1990; Rise et al 1991; Ferraro et al 1997, 1999; Buck et al 1997, 1999; Gershenfeld et al 1999; Hain et al 2000; TN Ferraro personal communication). Subsequent QTL analyses and other gene mapping strategies have been performed in B6 and D2 populations because of their well-documented differences in sensitivity in a variety of seizure-induction models (Kosobud & Crabbe 1990, Hall 1947; TN Ferraro, personal communication). Comparisons across gene mapping studies could reveal shared map locations for QTLs indicating common neurobiological mechanism(s) for seizure susceptibility independent of the source of seizure induction. Recently, a series of quantitative genetic studies using three induction mechanisms, kainic acid, pentylenetetrazol, and electroshock, have identified a major seizure susceptibility locus at the distal region of chromosome 1 (80–100 cM) (Ferraro et al 1997, 1999; TN Ferraro personal communication). More precise mapping of this region using a congenic strain (B6.D2-Mtv7a/Ty) has narrowed the chromosome 1 QTL to a 3 cM interval (TN Ferraro, personal communication). In addition, a second QTL on chromosome 15 overlaps among the three studies (Ferraro et al 1997, 1999; TN Ferraro personal communication).

Further support for the existence of a major seizure-regulating gene within the chromosome 1 region comes from QTL analyses of ethanol and pentobarbital withdrawal phenotypes in B6 and D2 populations (Buck et al 1997, 1999) (see "Models of Drug Abuse," below). The biological significance of finding this overlapping QTL is strengthened by the fact that these seizure induction methods are distinctly different mechanistically and methodologically, but it remains to be determined whether a single gene or multiple genes underlie this QTL.

Although the conclusion is that the QTL on chromosome 1 contains a gene of major effect for neuronal excitability, this region has not been identified in mapping studies of other seizure models, including audiogenic seizure susceptibility (AGS) in young AGS-resistant B6 and AGS-sensitive D2 mice. Three QTLs accounting for most of the genetic variation in these strains have been located on chromosomes 12 (*asp*-1), 4 (*asp*-2), and 7 (*asp*-3) (Neumann & Seyfried 1990, Neumann & Collins 1991). QTLs specific to defects in the EL mouse, a model of human temporal lobe epilepsy (Frankel et al 1995, Rise et al 1991, Legare et al 2000) have also been identified. Analyses of other chemically induced seizures have resulted in additional sets of QTLs specific to each investigation with some overlap with other studies. The $GABA_A$ receptor inverse agonist, methyl-β-carboline-3-carboxylate (β-CMM), produces generalized seizures in mice. QTLs for β-CMM seizure thresholds in F2 and backcross populations derived from A/J and C57BL/6J mice were found on distal chromosome 4 (Gershenfeld et al 1999), which has some overlap with the withdrawal seizure locus (Buck et al 1997, 1999); two other QTLs did not correspond to any loci discussed above. A QTL study of cocaine-induced seizures in BXD RIs and F2 populations from a B6XD2 intercross found a QTL on chromosome 9 that appears to overlap with the position of the *El1* gene identified in the EL mouse strain (Rise et al 1991).

The lack of overlap in many chromosomal locations in the different seizure models points to the complexities in regulation of CNS excitability, but also to the specificity of the particular pathways that are manipulated to produce seizures. However, the consensus of the chromosome 1 QTL is encouraging, because it indicates that this region contains a gene(s) important for a common regulatory mechanism in seizure sensitivity. To date, positional cloning has not been done to identify polymorphisms in specific genes. However, candidate genes in some of these QTL regions include serotonin receptor subunits, $GABA_A$ receptor subunits, and proteins involved in transmembrane ion flux, such as subunits for Na, K-ATPase and a potassium channel gene (Ferraro et al 1997, 1999; Buck et al 1997, 1999).

MODELS OF DRUG ABUSE

The genetic regulation of phenomena related to drug abuse including initial sensitivity, tolerance, and sensitization has been actively pursued using a variety of behavioral measures with the goal of providing information about mechanisms regulating individual differences in drug-seeking behavior. The reinforcing effects of drugs of abuse have been tested in mice more directly using several paradigms, including (*a*) the preference test (measures drug consumption when given a choice between plain drinking water and drug-containing water), (*b*) conditioned place preference (subject is administered drug in one compartment of a chamber consisting of two distinct compartments; after training, a drug is considered to be rewarding when the subject spends more time in the drug-paired chamber during the test trial), and (*c*) operant learning procedures associated with drug self-administration (subjects learn to lever-press or nose-poke to receive a drug reward).

Opiates

Opiates and opiate derivatives are potent analgesics, and many of these drugs also have high abuse potential owing to their euphoric and anxiolytic properties. Opiate drugs interact with specific G-protein-coupled receptors in brain and in the periphery. This section concentrates on work related specifically to opiate abuse, but for a recent review of the quantitative genetics of pain and analgesia, see Mogil et al (2000).

Strain surveys indicated that locomotor activity and body temperature responses to morphine in the BXD RI strains were genetically correlated (Belknap & Crabbe 1992). However, across a wide range of inbred strains (Belknap et al 1998, Brase et al 1977) the two responses were not correlated. These results support the conclusion that there are some common and some unique genetic mechanisms regulating physiological responses to morphine.

Genetic variation for morphine reinforcement as measured by preference, conditioned place preference, and IV self-administration was observed in inbred strains (Semenova et al 1995, Belknap et al 1993, Horowitz et al 1977), and provisional morphine-preference QTLs were mapped in the BXD RI strains (Belknap & Crabbe 1992). Three QTLs accounting for 85–90% of the genetic variance for morphine preference were mapped in a B6 X D2 F2 intercross, and two of these were in common with two of the BXD RI QTLs (Berrettini et al 1994). The μ opiate receptor gene (*Oprm*) resides on proximal chromosome 10 where a major QTL was mapped in both studies, and it is tempting to speculate that a polymorphism in *Oprm* may account for a portion of the genetic variance for opiate responses, but none has been reported.

Nicotine

Tobacco products are used widely, but the basis of dependence is not understood. Genetic control of sensitivity to nicotine, the psychoactive component of tobacco, was demonstrated for locomotor behavior and regulation of body temperature (Marks et al 1983). It was demonstrated subsequently that mice could be selectively bred for acute nicotine-induced motor activity, further indicating that variation in acute nicotine responses was influenced by genetic factors (Smolen & Marks 1991). It is interesting that the hypothermic response to nicotine cosegregated with locomotor activity, suggesting a common genetic mechanism (Smolen & Marks 1991). A more extensive survey of 19 inbred strains confirmed that genetic factors influence variation in nicotine-induced changes in respiratory rate, heart rate, body temperature, activity measures in the Y-maze, and acoustic startle (Marks et al 1989b), as well as nicotine-induced seizures (Miner & Collins 1989). These acute responses may relate to nicotine self-administration, at least indirectly, in that nicotine preference was significantly correlated to nicotine-induced seizures among six inbred strains (Robinson et al 1996).

Nicotine binds at multimeric ligand-gated ion channels in brain, and receptor binding studies of inbred strains indicated that the number of [^{3}H]nicotine and α-[^{125}I]bungarotoxin-binding sites was strain dependent (Marks et al 1989a, Miner & Collins 1989). Moreover, [^{125}I]bungarotoxin binding density was positively correlated to seizures, and [^{3}H]nicotine binding was correlated to nicotine-induced Y-maze activity and hypothermia (Marks et al 1989a, Miner & Collins 1989). Thus, it is possible that genetic regulation of nicotinic receptors is partly the basis of strain differences in acute nicotine responses. Stitzel et al (2000a,b) reported that restriction fragment length polymorphisms for the α-4 and α-7 nicotinic subunits were significantly associated with nicotine-induced seizures in the LSXSS RI strains and in a C3H X D2 F2 population, respectively. The specific nature of these restriction fragment length polymorphisms is as yet unknown, but coding sequence variants for the α-4 subunit have been identified (JA Stitzel & AC Collins, personal communication). One of the polymorphisms is a single base substitution that results in a predicted alanine to threonine substitution, and this appears to confer differences in maximal ion conductance in α-4-containing channels (AC Collins, personal communication).

Psychostimulants

Psychostimulants such as cocaine and amphetamine produce euphoria and heightened awareness. These effects are thought to be mediated by the blockade of monoamine neurotransmitter reuptake. Research into psychostimulant responses has focused primarily on this effect on dopamine reuptake and dopamine-related responses because of dopamine's putative role in drug reinforcement.

In various inbred strains, the acute locomotor response to cocaine ranged from hyperactivity in some strains to depressed activity in others, and the effect was dose and strain dependent as indicated by significant strain-by-dose interactions (Elmer et al 1996, Marley et al 1998, Shuster et al 1977). The successful bidirectional selection of cocaine-influenced locomotor activity also indicated that variation in this response was heritable (Marley et al 1998, Smolen & Marks 1991). Following repeated administrations of cocaine, a cocaine challenge elicited increased loco- motor activity (sensitization) in several inbred strains and some of the BXD RI strains, whereas other strains showed no change (Elmer et al 1996, Phillips et al 1998b, Shuster et al 1977, Tolliver et al 1994). Cocaine sensitization includes a learning component that is also strain dependent, i.e. some strains will not show a sensitized response unless the cocaine is paired to the testing apparatus (Elmer et al 1996, Phillips et al 1998b).

Inbred strain results formed the basis for QTL mapping studies of locomotor activity and other measures following acute or repeated administrations of psycho- stimulants in the BXD RI strains (Alexander et al 1996, Belknap & Crabbe 1992, Grisel et al 1997, Jones et al 1999, Phillips et al 1998b, Tolliver et al 1994). Many provisional QTLs were detected, but none have yet been confirmed. Dopamine D1, D2, and transporter receptor densities were also mapped in naive subjects from the BXD RI strains (Jones et al 1999). Several QTLs were found in common between receptor densities and cocaine-induced activity measures, suggesting that genetic variation in expression of dopamine receptors may account for some portion of the variation in behavioral responses to cocaine.

Alcohol

Alcohol elicits a range of behavioral responses including activation, ataxia, and sedation, and its abuse potential is well documented. There is a rich literature on the genetics of ethanol-related behaviors in the mouse dating back nearly 40 years (McClearn 1962), but more recently geneticists have been exploring the molecular basis of the effects of ethanol.

Numerous studies of inbred strains, RI strains, and selected lines indicate that genetic factors contribute to variation in a wide variety of acute responses to ethanol, including loss of righting reflex (DeFries et al 1989, McClearn & Kakihana 1973), locomotor activity (Crabbe et al 1980, Phillips et al 1996), withdrawal (Crabbe et al 1983, Goldstein & Kakihana 1975), hypothermia (Crabbe 1983, Phillips et al 1990), ataxia (Dudek & Phillips 1990), preference (Fuller & Collins 1972), and conditioned place preference (Cunningham 1995). Genetic effects have also been noted for the acquisition of acute or chronic tolerance and sensitization to ethanol's effects on some of these measures (Erwin & Deitrich 1996, Phillips et al 1996). Genetic associations have been made between ethanol withdrawal susceptibility and ethanol consumption (Metten et al 1998), between measures of learning and acute ethanol tolerance (Radcliffe et al 1998), and between a measure of impulsivity and ethanol preference (Logue et al 1998). Correlations also have been observed between ethanol-induced expression of neuronal c-Fos and locomotor activity (Demarest et al 1999) and between neurotensin receptor density and loss of righting reflex (Erwin & Jones 1993).

Over 30 QTL mapping experiments have been conducted for ethanol-related behaviors primarily in the BXD and LSXSS RI strains. Using F2 and/or congenic strategies, QTLs have been confirmed for acute ethanol withdrawal (Buck et al 1997), ethanol-influenced locomotor activity (Demarest et al 2001), ethanol preference (Phillips et al 1998a), and initial sensitivity (Bennett & Johnson 1998, Markel et al 1997, Radcliffe et al 2000a). One of the most consistent observations in the ethanol QTL literature is the identification of a QTL on chromosome 2 for ethanol preference (*Alcp1*) (Belknap et al 1997, Gehle & Erwin 1998, Gill et al 1996, Melo et al 1996, Phillips et al 1998a, Rodriguez et al 1995, Tarantino et al 1998, Whatley et al 1999); however, this QTL still encompasses a large region.

Demarest et al (2001) mapped two ethanol-induced locomotor activity QTLs with high resolution in an HS population based on a QTL that was originally mapped in a B6 X D2 F2. Taking a different approach, Buck et al (1997) mapped acute ethanol withdrawal QTLs using several different B6 X D2 segregating crosses and tested the possibility that the gene for the $\gamma 2$ subunit of the $GABA_A$ chloride channel (*Gabrg2*) was the basis of a QTL on chromosome 11. It was shown subsequently that the B6 and D2 have polymorphic forms of *Gabrg2* that cause a predicted threonine to alanine substitution in position 11 of the mature peptide (Buck & Hood 1998). The distribution of the *Gabrg2* allelic variants in the BXD RI strains correlates well with acute ethanol withdrawal and also with ethanol-conditioned taste aversion, ethanol-induced motor incoordination, and ethanol-induced hypothermia (Hood & Buck 2000). However, it remains to be seen whether this polymorphism, alone or in concert with other linked genes, is the molecular source of the QTL.

SUMMARY AND CONCLUSIONS

By their very nature, polygenic systems are difficult to study, but there is notable progress in elucidating the genes underlying regulation of quantitative behavioral and pharmacological traits at several levels. The detection of common QTLs on mouse chromosome 1 associated with anxiety, locomotor activity, and fear is striking. However, no obvious candidate genes have been discovered, and this may be a QTL for which gene discovery will depend on positional cloning strategies. Notable progress is being made via sequencing of candidate genes for several QTLs. The discovery of polymorphisms in the brain nicotinic receptor genes and GABAergic receptor genes are important, especially as they relate to mapping of traits associated with alcohol and drug abuse. The overlap in QTLs regulating susceptibility to various seizure-inducing agents is also noteworthy.

This review focuses on behavioral traits, but there is strain variation in many fundamental properties of the nervous system that are also regulated by polygenic systems (Lipp et al 1989, Williams et al 1998). Increasing our understanding of developmental and neuroanatomical differences across inbred mouse strains will have important implications for any of the behaviors described here.

ACKNOWLEDGMENTS

The authors are indebted to many investigators who supplied preprints of their work, including V Bolivar, K Buck, A Collins, K Demarest, T Ferraro, L Flaherty, R Hitzemann, and J Stitzel. Owing to page limitations, we apologize to the many fine researchers whose work could not be included in this review. We thank Ms. Dawn Caillouet for her expert assistance in preparing the review. We gratefully acknowledge the support of NIH grants including: MH-53668, AA-03527, AA-11275, DA-10156, and an RCA to JMW (AA-00141).

Visit the Annual Reviews home page at www.annualReviews.org

LITERATURE CITED

Alexander RC, Wright R, Freed W. 1996. Quantitative trait loci contributing to phencyclidine-induced and amphetamine-induced locomotor behavior in inbred mice. *Neuropsychopharmacology* 15:484–90

Anagnostaras SG, Maren S, Fanselow MS. 1999. Temporally graded retrograde amnesia of contextual fear after hippocampal damage in rats: within-subjects examination. *J. Neurosci.* 19:1106–14

Bailey DW. 1971. Recombinant inbred strains. An aid to finding identity linkage and function of histocompatibility and other genes. *Transplantation* 11:325

Belknap JK, Crabbe JC. 1992. Chromosome mapping of gene loci affecting morphine and amphetamine responses in BXD recombinant inbred mice. *Ann. NY Acad. Sci.* 654:311–23

Belknap JK, Crabbe JC, Riggan J, O'Toole LA. 1993. Voluntary consumption of morphine in 15 inbred mouse strains. *Psychopharmacology* 112:352–58

Belknap JK, Richards SP, O'Toole LA, Helms ML, Phillips TJ. 1997. Short-term selective breeding as a tool for QTL mapping: ethanol preference drinking in mice. *Behav. Genet.* 27:55–66

Belknap JK, Riggan J, Cross S, Young ER, Gallaher EJ, et al. 1998. Genetic determinants of morphine activity and thermal responses in 15 inbred mouse strains. *Pharmacol. Biochem. Behav.* 59:353–60

Bennett B, Johnson TE. 1998. Development of congenics for hypnotic sensitivity to ethanol by QTL-marker-assisted counter selection. *Mamm. Genome* 9:969–74

Berrettini WH, Ferraro TN, Alexander RC, Buchberg AM, Vogel WH. 1994. Quantitative trait loci mapping of three loci controlling morphine preference using inbred mouse strains. *Nat. Genet.* 7:54–58

Bolivar V, Caldarone BJ, Reilly AA, Flaherty L. 2000. Habituation of activity in an open field: A survey of inbred strains and F1 hybrids. *Behav. Genet.* 30:285–293

Brase DA, Loh HH, Way EL. 1977. Comparison of the effects of morphine on locomotor activity, analgesia and primary and protracted physical dependence in six mouse strains. *J. Pharmacol. Exp. Ther.* 201:368–74

Buck KJ, Hood HM. 1998. Genetic association of a GABA$_A$ receptor gamma2 subunit variant with severity of acute physiological dependence on alcohol. *Mamm. Genome* 9:975–78

Buck KJ, Metten P, Belknap JK, Crabbe JC. 1997. Quantitative trait loci involved in genetic predisposition to acute alcohol withdrawal in mice. *J. Neurosci.* 17:3946–55

Buck KJ, Metten P, Belknap JK, Crabbe JC. 1999. Quantitative trait loci affecting risk for pentobarbital withdrawal map near alcohol withdrawal loci on mouse Chromosomes 1, 4, and 11. *Mamm. Genome* 10:431–37

Caldarone B, Saavedra C, Tartaglia K, Wehner

JM, Dudek BC, Flaherty L. 1997. Quantitative trait loci analysis affecting contextual conditioning in mice. *Nat. Genet.* 17:335–37

Costall B, Jones BJ, Kelly ME, Naylor RJ, Tomkins DM. 1989. Exploration of mice in a black and white test box: Validation as a model of anxiety. *Pharmacol. Biochem. Behav.* 32:777–85

Crabbe JC. 1983. Sensitivity to ethanol in inbred mice: genotypic correlations among several behavioral responses. *Behav. Neurosci.* 97:280–89

Crabbe JC. 1986. Genetic differences in locomotor activation in mice. *Pharmacol. Biochem. Behav.* 25:289–92

Crabbe JC, Janowsky JS, Young ER, Rigter H. 1980. Strain-specific effects of ethanol on open field activity in inbred mice. *Subst. Alcohol Actions Misuse* 1:537–43

Crabbe JC, Kosobud A, Young ER. 1983. Genetic selection for ethanol withdrawal severity: differences in replicate mouse lines. *Life Sci.* 33:955–62

Crabbe JC, Phillips TJ, Kosobud A, Belknap JK. 1990. Estimation of genetic correlation: Interpretation of experiments using selectively bred and inbred animals. *Alcohol. Clin. Exp. Res.* 14:141–51

Crabbe JC, Wahlsten D, Dudek BC. 1999. Genetics of mouse behavior: interactions with laboratory environment. *Science* 284:1670–72

Crawley J, Goodwin FK. 1980. Preliminary report of a simple animal behavior model for the anxiolytic effects of benzodiazepines. *Pharmacol. Biochem. Behav.* 13:167–70

Cunningham CL. 1995. Localization of genes influencing ethanol-induced conditioned place preference and locomotor activity in BXD recombinant inbred mice. *Psychopharmacology* 120:28–41

Darvasi A. 1998. Experimental strategies for the genetic dissection of complex traits in animal models. *Nat. Genet.* 18:19–24

DeFries JC. 1981. Selective breeding for behavioral and pharmacological responses in laboratory mice. In *Genetic Research Strategies for Psychobiology and Psychiatry*, ed. ES Gershon, S Matthysse, XO Breakefield, RD Ciaranello, 15:199–213. Pacific Grove, CA: Boxwood

DeFries JC, Gervais MC, Thomas EA. 1978. Response to 30 generations of selection for open-field activity in laboratory mice. *Behav. Genet.* 8:3–13

DeFries JC, Johnson RC, Kuse AR, McClearn GE, Polovina J, et al. 1979. Familial resemblance for specific cognitive abilities. *Behav. Genet.* 9:23–43

DeFries JC, Wilson JR, Erwin VG, Petersen DR. 1989. LS X SS recombinant inbred strains of mice: initial characterization. *Alcohol. Clin. Exp. Res.* 13:196–200

Dellu F, Contarino A, Simon H, Koob GF, Gold LH. 2000. Genetic differences in response to novelty and spatial memory using a two-trial recognition task in mice. *Neurobiol. Learn. Mem.* 73:31–48

Demarest K, Hitzemann B, Phillips T, Hitzemann R. 1999. Ethanol-induced expression of c-Fos differentiates the FAST and SLOW selected lines of mice. *Alcohol. Clin. Exp. Res.* 23:87–95

Demarest K, Koyner J, McCaughran J, Cipp L, Hitzemann R. 2001. Further characterization and high resolution mapping of quantitative trait loci for ethanol-induced locomotor activity. *Behav. Genet.* In press

Dietrich W, Katz H, Lincoln SE, Shin H-S, Friedman J, et al. 1992. A genetic map of the mouse suitable for typing intraspecific crosses. *Genetics* 131:423–47

Dudek BC, Phillips TJ. 1990. Distinctions among sedative, disinhibitory, and ataxic properties of ethanol in inbred and selectively bred mice. *Psychopharmacology* 101:93–99

Elmer GI, Gorelick DA, Goldberg SR, Rothman RB. 1996. Acute sensitivity vs. context-specific sensitization to cocaine as a function of genotype. *Pharmacol. Biochem. Behav.* 53:623–28

Erwin VG, Deitrich RA. 1996. Genetic selection and characterization of mouse lines for acute functional tolerance to ethanol. *J. Pharmacol. Exp. Ther.* 279:1310–17

Erwin VG, Jones BC. 1993. Genetic correlations among ethanol-related behaviors and neurotensin receptors in long sleep (LS) x short sleep (SS) recombinant inbred strains of mice. *Behav. Genet.* 23:191–96

Falconer DS. 1989. *Introduction to Quantitative Genetics.* New York: Wiley. 438 pp.

Ferraro TN, Golden GT, Smith GG, Schork NJ, St. Jean P, et al. 1997. Mapping murine loci for seizure response to kainic acid. *Mamm. Genome* 8:200–8

Ferraro TN, Golden GT, Smith GG, St. Jean P, Schork NJ, et al. 1999. Mapping loci for pentylenetetrazol-induced seizure susceptibility in mice. *J. Neurosci.* 19:6733–39

Festing MFW. 1979. *Inbred Strains in Biomedical Research.* New York: Oxford Univ. Press. 483 pp.

Flint J, Corley R, DeFries JC, Fulker DW, Gray JA, et al. 1995. A simple genetic basis for a complex psychological trait in laboratory mice. *Science* 269:1432–35

Fox GB, LeVasseur RA, Faden AI. 1999. Behavioral responses of C57BL/6, FVB/N, and 129/SvEMS mouse strains to traumatic brain injury: Implications for gene targeting approaches to neurotrauma. *J. Neurotrauma* 16:377–89

Frankel WN, Valenzuela A, Lutz CM, Johnson EW, Dietrich WF, et al. 1995. New seizure frequency QTL and the complex genetics of epilepsy in EL mice. *Mamm. Genome* 6:830–38

Fuller JL, Collins RL. 1972. Ethanol consumption and preference in mice: a genetic analysis. *Ann. NY Acad. Sci.* 197:42–48

Gehle VM, Erwin VG. 1998. Common quantitative trait loci for alcohol-related behaviors and CNS neurotensin measures: voluntary ethanol consumption. *Alcohol. Clin. Exp. Res.* 22:401–8

Gershenfeld HK, Neumann PE, Li X, St. Jean PL, Paul SM. 1999. Mapping quantitative trait loci for seizure response to a $GABA_A$

receptor inverse agonist in mice. *J. Neurosci.* 19:3731–38

Gershenfeld HK, Neumann PE, Mathis C, Crawley JN, Li X, et al. 1997. Mapping quantitative trait loci for open-field behavior in mice. *Behav. Genet.* 27:201–10

Gershenfeld HK, Paul SM. 1997. Mapping quantitative trait loci for fear-like behaviors in mice. *Genomics* 46:1–8

Gill K, Liu Y, Deitrich RA. 1996. Voluntary alcohol consumption in BXD recombinant inbred mice: relationship to alcohol metabolism. *Alcohol. Clin. Exp. Res.* 20:185–90

Goldstein DB, Kakihana R. 1975. Alcohol withdrawal convulsions in genetically different populations of mice. *Adv. Exp. Med. Biol.* 59:343–52

Gould TJ, Wehner JM. 1999. Genetic influences on latent inhibition. *Behav. Neurosci.* 113:1291–96

Griebel G, Belzung C, Perrault G, Sanger DJ. 2000. Differences in anxiety-related behaviours and in sensitivity to diazepam in inbred and outbred strains of mice. *Psychopharmacology* 148:164–70

Grisel JE, Belknap JK, O'Toole LA, Helms ML, Wenger CD, et al. 1997. Quantitative trait loci affecting methamphetamine responses in BXD recombinant inbred mouse strains. *J. Neurosci.* 17:745–54

Hain HS, Crabbe JC, Bergeson SE, Belknap JK. 2000. Cocaine-induced seizure thresholds: Quantitative trait loci detection and mapping in two populations derived from the C57BL/6 and DBA/2 mouse strains. *J. Pharmacol. Exp. Ther.* 293:180–87

Hall CS. 1934. Emotional behaviour in the rat. I. Defaecation and urination as measures of individual differences in emotionality. *J. Comp. Psychol.* 18:385–403

Hall CS. 1947. Genetic differences in fatal audiogenic seizures between two inbred strains of house mice. *J. Hered.* 38:2–6

Hegman JP, Possidente B. 1981. Estimating genetic correlations from inbred strains. *Behav. Genet.* 11:103–14

Hood HM, Buck KJ. 2000. Allelic variation in the $GABA_A$ receptor $\gamma 2$ subunit is associated with genetic susceptibility to ethanol-induced motor incoordination and hypothermia, conditioned taste aversion, and withdrawal in BXD/Ty recombinant inbred mice. *Alcohol. Clin. Exp. Res.* 24:1327–34

Horowitz GP, Whitney G, Smith JC, Stephan FK. 1977. Morphine ingestion: genetic control in mice. *Psychopharmacology* 52:119–22

Johnson TE, DeFries JC, Markel PD. 1992. Mapping quantitative trait loci for behavioral traits in the mouse. *Behav. Genet.* 22:635–53

Jones BC, Tarantino LM, Rodriguez LA, Reed CL, McClearn GE, et al. 1999. Quantitative-trait loci analysis of cocaine-related behaviours and neurochemistry. *Pharmacogenetics* 9:607–17

Kosobud AE, Crabbe JC. 1990. Genetic correlations among inbred strain sensitivities to convulsions induced by 9 convulsant drugs. *Brain Res.* 526:8–16

Koyner J, Demarest K, McCaughran J Jr, Cipp L, Hitzemann R. 2000. Identification and time dependence of quantitative trait loci for basal locomotor activity in the BXD recombinant inbred series and a B6D2 F_2 intercross. *Behav. Genet.* 30:159–170

Lander E, Kruglyak L. 1995. Genetic dissection of complex traits: guidelines for interpreting and reporting linkage results. *Nat. Genet.* 11:241–47

Lander ES, Botstein D. 1989. Mapping Mendelian factors underlying quantitative traits using RFLP linkage maps. *Genetics* 121:185–99

LeDoux JE. 2000. Emotion circuits in the brain. *Annu. Rev. Neurosci.* 23:155–84

Legare ME, Bartlett FS II, Frankel WN. 2000. A major effect QTL determined by multiple genes in epileptic EL mice. *Genome Res.* 10:42–48

Lipp H-P, Schwegler H, Crusio WE, Wolfer DP, Leisinger-Trigona M-C, et al. 1989. Using genetically-defined rodent strains for the identification of hippocampal traits relevant for two-way avoidance behavior: a noninvasive approach. *Experientia* 45:845–59

Lister RG. 1987. The use of a plus-maze to measure anxiety in the mouse. *Psychopharmacology* 92:180–85

Logue SF, Owen EH, Rasmussen DL, Wehner JM. 1997a. Assessment of locomotor activity, acoustic and tactile startle, and prepulse inhibition of startle in inbred mouse strains and F_1 hybrids: Implications of genetic background for single gene and quantitative trait loci analyses. *Neuroscience* 80:1074–86

Logue SF, Paylor R, Wehner JM. 1997b. Hippocampal lesions cause learning deficits in inbred mice in the Morris water maze and conditioned-fear task. *Behav. Neurosci.* 111:104–13

Logue SF, Swartz RJ, Wehner JM. 1998. Genetic correlation between performance on an appetitive-signaled nosepoke task and voluntary ethanol consumption. *Alcohol. Clin. Exp. Res.* 22:1912–20

Markel PD, Bennett B, Beeson MA, Gordon LG, Johnson TE. 1997. Confirmation of quantitative trait loci for ethanol sensitivity in long-sleep and short-sleep mice. *Genome Res.* 7:92–99

Marks MJ, Burch JB, Collins AC. 1983. Genetics of nicotine response in four inbred strains of mice. *J. Pharmacol. Exp. Ther.* 226:291–302

Marks MJ, Romm E, Campbell SM, Collins AC. 1989a. Variation of nicotinic binding sites among inbred strains. *Pharmacol. Biochem. Behav.* 33:679–89

Marks MJ, Stitzel JA, Collins AC. 1989b. Genetic influences on nicotine responses. *Pharmacol. Biochem. Behav.* 33:667–78

Marley RJ, Arros DM, Henricks KK, Marley ME, Miner LL. 1998. Sensitivity to cocaine and amphetamine among mice selectively bred for differential cocaine sensitivity. *Psychopharmacology* 140:42–51

Mathis C, Neumann PE, Gershenfeld H, Paul SM, Crawley JN. 1995. Genetic analysis of anxiety-related behaviors and responses to benzodiazepine-related drugs in AXB and

BXA recombinant inbred mouse strains. *Behav. Genet.* 25:557–68

McClearn GE. 1962. Genetic differences in the effects of alcohol upon behavior of mice. *Proc. 3rd Int. Conf. Alcohol Road Traffic, London,* Sept. (Abstr.)

McClearn GE, Kakihana R. 1973. Selective breeding for ethanol sensitivity in mice. *Behav. Genet.* 3:409–10

McClearn GE, Wilson JR, Meredith W. 1970. The use of isogenic and heterogenic mouse stocks in behavioral research. In *Contributions to Behavior-Genetic Analysis: The Mouse as a Prototype,* ed. G Lindzey, D Thiessen, pp. 3–22. New York: Appleton-Century-Crofts

Melo JA, Shendure J, Pociask K, Silver LM. 1996. Identification of sex-specific quantitative trait loci controlling alcohol preference in C57BL/ 6 mice. *Nat. Genet.* 13:147–53

Metten P, Phillips TJ, Crabbe JC, Tarantino LM, McClearn GE, et al. 1998. High genetic susceptibility to ethanol withdrawal predicts low ethanol consumption. *Mamm. Genome* 9:983–90

Miner LL, Collins AC. 1989. Strain comparison of nicotine-induced seizure sensitivity and nicotinic receptors. *Pharmacol. Biochem. Behav.* 33:469–75

Mogil JS, Yu L, Basbaum AI. 2000. Pain genes? Natural variation and transgenic mutants. *Annu. Rev. Neurosci.* 23:777–811

Morris RGM. 1981. Spatial localisation does not depend on the presence of local cues. *Learn. Motiv.* 12:239–60

Neumann PE, Collins RL. 1991. Genetic dissection of susceptibility to audiogenic seizures in inbred mice. *Proc. Natl. Acad. Sci. USA* 88:5408–12

Neumann PE, Seyfried TN. 1990. Mapping of two genes that influence susceptibility to audiogenic seizures in crosses of C57BL/6J and DBA/2J mice. *Behav. Genet.* 20:307–23

O'Keefe J, Nadel L. 1978. *The Hippocampus as a Cognitive Map.* Oxford: Clarendon

Owen EH, Christensen SC, Paylor R, Wehner JM. 1997a. Identification of quantitative trait loci involved in contextual and auditory-cued fear conditioning in BXD recombinant inbred strains. *Behav. Neurosci.* 111:292–300

Owen EH, Logue SF, Rasmussen DL, Wehner JM. 1997b. Assessment of learning by the Morris water task and fear conditioning in inbred mouse strains and F1 hybrids: Implications of genetic background for single gene mutations and quantitative trait loci analyses. *Neuroscience* 80:1087–99

Phillips TJ, Belknap JK, Buck KJ, Cunningham CL. 1998a. Genes on mouse chromosomes 2 and 9 determine variation in ethanol consumption. *Mamm. Genome* 9:936–41

Phillips TJ, Huson MG, McKinnon CS. 1998b. Localization of genes mediating acute and sensitized locomotor responses to cocaine in BXD/Ty recombinant inbred mice. *J. Neurosci.* 18:3023–3034

Phillips TJ, Lessov CN, Harland RD, Mitchell SR. 1996. Evaluation of potential genetic associations between ethanol tolerance and sensitization in BXD/Ty recombinant inbred mice. *J. Pharmacol. Exp. Ther.* 277:613–23

Phillips TJ, Terdal ES, Crabbe JC. 1990. Response to selection for sensitivity to ethanol hypothermia: genetic analyses. *Behav. Genet.* 20:473–80

Radcliffe RA, Bohl ML, Lowe MV, Cycowski CS, Wehner JM. 2000a. Mapping of quantitative trait loci for hypnotic sensitivity to ethanol in crosses derived from the C57BL/6 and DBA/2 mouse strains. *Alcohol. Clin. Exp. Res.* 24:1335–42

Radcliffe RA, Erwin VG, Wehner JM. 1998. Acute functional tolerance to ethanol and fear conditioning are genetically correlated in mice. *Alcohol. Clin. Exp. Res.* 22:1673–79

Radcliffe RA, Lowe MV, Wehner JM. 2000b. Confirmation of contextual fear conditioning QTLs by short-term selection. *Behav. Genet.* 30:183–191

Rise ML, Frankel WN, Coffin JM, Seyfried

TN. 1991. Genes for epilepsy mapped in the mouse. *Science* 253:669–73

Robinson SF, Marks MJ, Collins AC. 1996. Inbred mouse strains vary in oral self-selection of nicotine. *Psychopharmacology* 124:332–39

Rodriguez LA, Plomin R, Blizard DA, Jones BC, McClearn GE. 1995. Alcohol acceptance, preference, and sensitivity in mice. II. Quantitative trait loci mapping analysis using BXD recombinant inbred strains. *Alcohol. Clin. Exp. Res.* 19:367–73

Rogers DC, Jones DN, Nelson PR, Jones CM, Quilter CA, et al. 1999. Use of SHIRPA and discriminant analysis to characterise marked differences in the behavioural phenotype of six inbred mouse strains. *Behav. Brain Res.* 105:207–17

Roullet P, Lassalle JM. 1995. Radial maze learning using exclusively distant visual cues reveals learners and nonlearners among inbred mouse strains. *Physiol. Behav.* 58:1189–95

Semenova S, Kuzmin A, Zvartau E. 1995. Strain differences in the analgesic and reinforcing action of morphine in mice. *Pharmacol. Biochem. Behav.* 50:17–21

Seyfried TN, Todorova M. 1999. Experimental models of epilepsy. In *The Epilepsies: Etiologies and Prevention*, ed. P Kotagal, HO Lüders, 65:527–42. San Diego, CA: Academic

Shuster L, Yu G, Bates A. 1977. Sensitization to cocaine stimulation in mice. *Psychopharmacology* 52:185–90

Silver LM. 1995. *Mouse Genetics: Concepts and Applications.* New York/Oxford: Oxford Univ. Press. 362 pp.

Smith DF, Appleton RE, MacKenzie JM, Chadwick DW. 1998. *An Atlas of Epilepsy*, p. 14. New York/Carnforth, UK: Parthenon

Smolen A, Marks MJ. 1991. Genetic selections for nicotine and cocaine sensitivity in mice. *J. Addict. Disord.* 10:7–28

Stitzel JA, Jimenez M, Marks MJ, Tritto T, Collins AC. 2000a. Potential role of the $\alpha4$ and $\alpha6$ nicotinic receptor subunits in regulating nicotine-induced seizures. *J. Pharmacol. Exp. Ther.* 293:67–74

Stitzel JA, Lu Y, Jimenez M, Collins AC. 2000b. Genetic and pharmacological strategies identify a behavioral function of neuronal nicotinic receptors. *Behav. Brain Res.* 293:67–74

Talbot CJ, Nicod A, Cherny SS, Fulker DW, Collins AC. 1999. High-resolution mapping of quantitative trait loci in outbred mice. *Nat. Genet.* 21:305–8

Tarantino LM, McClearn GE, Rodriguez LA, Plomin R. 1998. Confirmation of quantitative trait loci for alcohol preference in mice. *Alcohol. Clin. Exp. Res.* 22:1099–1105

Tolliver BK, Belknap JK, Woods WE, Carney JM. 1994. Genetic analysis of sensitization and tolerance to cocaine. *J. Pharmacol. Exp. Ther.* 270:1230–38

Torgersen S. 1990. Genetics of anxiety and its clinical implications. In *Handbook of Anxiety.* Vol. 3: *The Neurobiology of Anxiety*, ed. M Roth, R Noyes Jr, GD Burrows, 14:381–406. New York: Elsevier Sci.

Toth LA, Williams RW. 1999. A quantitative genetic analysis of locomotor activity in CXB recombinant inbred mice. *Behav. Genet.* 29:319–28

Trullas R, Skolnick P. 1993. Differences in fear motivated behaviors among inbred mouse strains. *Psychopharmacology* 111:323–31

Turri MG, Talbot CJ, Radcliffe RA, Wehner JM, Flint J. 1999. High-resolution mapping of quantitative trait loci for emotionality in selected strains of mice. *Mamm. Genome* 10:1098–1101

Upchurch M, Wehner JM. 1988. Differences between inbred strains of mice in Morris water maze performance. *Behav. Genet.* 18:55–68

Upchurch M, Wehner JM. 1989. Inheritance of spatial learning ability in inbred mice: A classical genetic analysis. *Behav. Neurosci.* 103:1251–58

Valentinuzzi VS, Kolker DE, Vitaterna MH, Shimomura K, Whiteley A, et al. 1998. Automated measurement of mouse freezing behavior and its use for quantitative trait locus analysis of contextual fear conditioning in (BALB/cJ x C57BL/6J)F$_2$ mice. *Learn. Mem.* 5:391–403

Wehner JM, Radcliffe RA, Rosmann ST, Christensen SC, Rasmussen DL, et al. 1997. Quantitative trait locus analysis of contextual fear conditioning in mice. *Nat. Genet.* 17:331–34

Wehner JM, Sleight S, Upchurch M. 1990. Hippocampal protein kinase C activity is reduced in poor spatial learners. *Brain Res.* 523:181–87

Whatley VJ, Johnson TE, Erwin VG. 1999. Identification and confirmation of quantitative trait loci regulating alcohol consumption in congenic strains of mice. *Alcohol. Clin. Exp. Res.* 23:1262–71

Williams RW, Strom RC, Goldowitz D. 1998. Natural variation in neuron number in mice is linked to a major quantitative trait locus on Chr 11. *J. Neurosci.* 18:138–46

Annu. Rev. Neurosci. 2001. 24:869–96

EARLY ANTERIOR/POSTERIOR PATTERNING OF THE MIDBRAIN AND CEREBELLUM

Aimin Liu[1] and Alexandra L Joyner[1,2]

Howard Hughes Medical Institute and Developmental Genetics Program, Skirball Institute of Biomolecular Medicine, Departments of [1]Cell Biology and [2]Physiology & Neuroscience, New York University School of Medicine, New York, NY10016; e-mail: liua02@saturn.med.nyu.edu, joyner@saturn.med.nyu.edu

Key Words midbrain/hindbrain organizer, Fibroblast growth factor, *Otx2*, *Gbx2*, *Engrailed*

■ **Abstract** Transplantation studies performed in chicken embryos indicated that early anterior/posterior patterning of the vertebrate midbrain and cerebellum might be regulated by an organizing center at the junction between the midbrain and hindbrain. More than a decade of molecular and genetic studies have shown that such an organizer is indeed central to development of the midbrain and anterior hindbrain. Furthermore, a complicated molecular network that includes multiple positive and negative feedback loops underlies the establishment and refinement of a mid/hindbrain organizer, as well as the subsequent function of the organizer. In this review, we first introduce the expression patterns of the genes known to be involved in this patterning process and the quail-chick transplantation experiments that have provided the foundation for understanding the genetic pathways regulating mid/hindbrain patterning. Subsequently, we discuss the molecular genetic studies that have revealed the roles for many genes in normal early patterning of this region. Finally, some of the remaining questions and future directions are discussed.

INTRODUCTION

Regionalization of the central nervous system (CNS) is a critical early event in vertebrate neural development. By embryonic day 9.5 (E9.5), the main regions of mouse CNS can be clearly distinguished morphologically along the anterior/posterior (A/P) axis (Figure 1, see color insert). The forebrain consists of the telencephalon situated at the most rostral end of the neural tube and the more posterior diencephalon, which abuts the midbrain (mesencephalon). The forebrain is divided into several prosomeres, with prosomere 1 being the most caudal. The hindbrain is separated from the midbrain by a constriction that is called the mid/hindbrain junction or isthmus and is divided into the rostral metencephalon and the caudal myelencephalon. The hindbrain, also called rhombencephalon, is

0147-006X/01/0621-0869$14.00

further divided into eight rhombomeres. The metencephalon consists of rhombomere 1 (r1), which gives rise to the cerebellum and pons (Millet et al 1996, Wingate & Hatten 1999), and r2. In this review we refer to the region that gives rise to the midbrain and cerebellum and that includes the mesencephalon and anterior metencephalon (r1) as the mes/met region.

Quail-chick transplantation experiments showed that development of the mes/met region could be regulated by a group of organizer cells located at the mid/hindbrain junction (for review, see Alvarado-Mallart 1993). Subsequently, several secreted growth factor genes were found to be expressed in the isthmus, and a number of genes encoding transcription factors were found to be expressed more broadly in the mes/met (Wassef & Joyner 1997, Joyner et al 2000) (Figure 1, see color insert). Gain- and loss-of-function studies over the past 10 years have demonstrated that these genes are indeed critical regulators of vertebrate mes/met development. Significantly, Fibroblast growth factor-8 (FGF8), which is expressed in the isthmus, is both necessary for mes/met development and can induce ectopic midbrain or cerebellum development. Furthermore, the junction between the expression domains of two transcription factors, *Otx2* and *Gbx2*, defines the position of the cells with organizer function. Finally, a complicated set of genetic interactions involving these and other growth factors and transcription factors regulate both establishment of the mes/met and subsequent A/P patterning of the midbrain and cerebellum.

EXPRESSION PATTERNS OF MANY MES/MET GENES ARE CONSISTENT WITH THE MID/HINDBRAIN JUNCTION CONTAINING AN ORGANIZER

An organizer, such as the well-known Spemann's organizer, is a group of cells that produce signals that can "organize" neighboring cells into well-patterned and functional structures (for a review, see Nieto 1999). One important characteristic shared by all known organizers is that they can direct competent tissues to take on new fates that are normally taken only by their native neighbors. Meinhardt (1983) studied the formation of several well-known organizers and found that they were all formed at points of apposition of two or more differentially specified tissues. Therefore, he proposed a model in which determinants from two (or more) adjoining compartments acting on the same border cells lead to the induction of expression in the border cells of a patterning signal(s), which in turn, function(s) to sharpen the border and guide the patterning of the adjacent compartments.

The expression patterns of a number of mes/met genes during early development are consistent with patterning of the region being regulated in a manner similar to that predicted by the model proposed by Meinhardt (1983). Two differentially specified regions of the neural plate, anteriorly located *Otx2*-expressing tissue and posteriorly located *Gbx2*-expressing tissue, are found to confront each other in mice as early as E7.5 (Wassarman et al 1997). Expression of another transcription factor gene *Pax2* is then initiated at the presomite stage (E7.5) surrounding the site of confrontation, followed by similar expression of several transcription factors (Figure 1, see color insert) (for a review, see Wassef & Joyner 1997, Joyner et al 2000). Most interesting, the expression patterns of the two secreted molecules, FGF8 and WNT1, become highly restricted to adjacent narrow transverse bands at the *Otx2/Gbx2* border (Wilkinson et al 1987, Crossley et al 1995). A relatively fuzzy boundary between *Gbx2* and *Otx2* expression domains also becomes sharp at the 4- to 6-somite stage, and this boundary colocalizes with the mid/hindbrain junction that in mice is morphologically clear by E9.5. Genes such as *En1*, *En2*, and *Pax5* continue to be expressed in both the mesencephalon and the metencephalon after E9.5 (Figure 1, see color insert). This scenario of gene expression correlates well with the "organizer model" proposed by Meinhardt (1983). Indeed, numerous transplantation experiments with chicks have shown that the mid/hindbrain junction has an organizing activity (for a review, see Alvarado-Mallart 1993, Joyner et al 2000). We review in more detail the expression patterns of various mes/met genes mainly in mice in this section and then discuss the experiments that revealed the organizing function of the isthmus.

Otx2 and *Gbx2*

The transcription factors OTX1 and -2 are homologues of the Drosophila orthodenticle protein (Simeone et al 1992). Prior to gastrulation, *Otx2* is expressed throughout the epiblast and in the anterior visceral endoderm (Simeone et al 1992, 1993). As the primitive streak forms, *Otx2* expression becomes progressively restricted to an anterior region of the mouse embryo in all three germ layers and in the visceral endoderm. *Otx1* expression begins at the 1- to 3-somite stage in the anterior neuroectoderm. Subsequently, the caudal boundaries of *Otx1* and -2 expression are located in the isthmic constriction in mouse embryos. Fate mapping experiments with chicks have provided evidence that the caudal limit of *Otx2* expression as early as Hamburger & Hamilton (HH) stage 10 (10 somites) (Hamburger & Hamilton 1992) marks the boundary between the midbrain and hindbrain (Millet et al 1996).

At E7.5, the homeobox gene *Gbx2* is expressed in all three germ layers in the posterior region of a mouse embryo (Bouillet et al 1995, Wassarman et al 1997). Subsequently, its expression in the CNS becomes restricted to the anterior hindbrain by E8.5. At E9.5, *Gbx2* expression is restricted to the anterior metencephalon and four longitudinal lateral stripes through the entire length of the hindbrain and spinal cord. Starting at E10.0, *Gbx2* expression in the CNS is also found in precursors of the thalamus in the dorsal diencephalon.

Fgf8, *Fgf17*, and *Fgf18*

Fgf8 is expressed in many locations of a developing mouse embryo at different stages, which suggests that it has multiple roles in regulating cell proliferation and differentiation in diverse processes (Heikinheimo et al 1994, Crossley & Martin 1995). In mouse mes/met development, *Fgf8* is initially expressed in a broad region in the prospective anterior hindbrain starting at the 3-somite stage. This expression quickly becomes more intense and more restricted to a narrow ring just caudal to the mid/hindbrain junction and persists until E12.5.

More recently, it has been reported that two additional mouse *Fgf* genes, *Fgf17* and *Fgf18*, which share very high sequence homology with *Fgf8* (63.7% and 56.8%, respectively, at the amino acid level), are also expressed in the anterior hindbrain region at E8.5 (Maruoka et al 1998, Xu et al 1999). However, it seems that their expression levels at early somite stages are weaker than that of *Fgf8*. Furthermore, *Fgf17* expression is broader than that of *Fgf8*, it occurs in both the mesencephalon and the metencephalon, and it persists until at least E14.5, a stage when *Fgf8* is no longer expressed. The fact that *Fgf8* expression seems stronger at early somite stages and the fact of the severe mes/met phenotype of *Fgf8* loss-of-function mutants (see below) suggest that FGF8 is the major FGF molecule that is essential for early patterning of this region.

Wnt1, Lmx1b, En1/2, and *Pax2/5*

Wnt1 is a homologue of the Drosophila segmentation gene *wingless* and encodes a secreted signaling molecule (for a review, see Nusse & Varmus 1992). *En1* and *En2* are homologues of the Drosophila segmentation gene *engrailed* and encode homeodomain-containing transcription factors (Joyner & Martin 1987). *Pax2* and *Pax5* are the homologues of the Drosophila pair-rule gene *paired* and encode paired-domain-containing transcription factors (Gruss & Walther 1992). *Lmx1b* is a gene that encodes a LIM homeodomain-containing transcription factor. *Lmx1b* expression has been described in detail only from chick brain (Yuan & Schoenwolf 1999, Adams et al 2000). It is first expressed broadly in the caudal forebrain, midbrain, and hindbrain. By HH stage 10, the expression is restricted to the dorsal and ventral midline of the midbrain, overlapping with *Wnt1* expression. *Lmx1b* is also expressed in the isthmus at this stage, and its expression overlaps with both the *Wnt1* and *Fgf8* domains.

Pax2 expression in mice begins at the presomite stage in cells surrounding the *Otx2/Gbx2* boundary (Rowitch & McMahon 1995). *Wnt1* and *En1* expression then initiates within the *Pax2* domain at the 1-somite stage, with *Wnt1* expression restricted to *Otx2*-positive midbrain cells, and *En1* expression in the entire mes/met (Davis & Joyner 1988, Rowitch & McMahon 1995, Liu & Joyner 2001). *En2* and *Pax5* expression initiates around the 5-somite stage across the *Otx2/Gbx2* boundary (Davis & Joyner 1988, Adams et al 1992, Song et al 1996). The expression patterns of all these genes are dynamic in the mes/met region. For example, *Wnt1* is expressed in the entire mesencephalon at early somite stages and soon becomes restricted to a narrow ring rostral to the mid/hindbrain junction, as well as being expressed along the dorsal midline of the midbrain and caudal diencephalon, caudal hindbrain, and spinal cord. Transient expression of *Wnt1* in the ventral midline of the midbrain and diencephalon is also seen at E9.5–E10.5. *Pax2* expression also becomes restricted to the isthmus by E9, just caudal to the *Wnt1* expression domain. In contrast, the expression of *En1*, *En2*, and *Pax5* remains relatively broad on both sides of the mid/hindbrain junction (Figure 1, see color insert)

TRANSPLANTATION STUDIES IN CHICKS REVEAL THAT THE ISTHMUS HAS ORGANIZER ACTIVITY

The Isthmus Serves as an Organizing Center

Many transplantation experiments between quail and chicken embryos at the 11- to 14-somite stage have shown that a mid/hindbrain organizer is present at the mid/hindbrain junction (for a review, see Alvarado-Mallart 1993, Wassef & Joyner 1997). These experiments can be summarized into two groups. The first group is a series of inversions. When quail rostral mesencephalic tissue is inverted and transplanted into the corresponding region of a chick host, the transplant adjusts its polarity according to the new environment. However, if the transplanted tissue includes tissue from the caudal part of the mesencephalic vesicle [precursors of isthmic nuclei and cerebellum (Millet et al 1996)], it not only maintains its own polarity in most cases, it also can induce the adjacent, more-anterior host diencephalic tissue to form caudal mesencephalic structures (Figure 2*A*). In the second group of transplantation experiments, isthmus tissue is transplanted into either the diencephalic or myelencephalic region. In the diencephalic region, the transplants induce caudal mesencephalic development, whereas in the hindbrain they can induce ectopic cerebellar structures (Figure 2*B*). The reverse has also been done, and in these cases diencephalic or hindbrain tissue transplanted into the region of the isthmus is transformed into mes/met tissue.

In all the experiments where ectopic mes/met structures were induced by isthmic tissue, expression of *En2* was shown to be induced before the ectopic mes/met structures developed. The expression of *Wnt1* was also altered in the diencephalon, such that it extended from the dorsal midline where endogenous *Wnt1* expression resides to the vicinity of the graft. Based on these results, it was suggested that *En2* and *Wnt1* could be involved in the ectopic development of the midbrain and cerebellar tissues (for a review, see Wassef & Joyner 1997).

The Isthmic Organizer Can be Generated by the Apposition of Midbrain and Anterior Hindbrain Tissue

Consistent with the general proposal made by Meinhardt (1983), Irving & Mason (1999) have shown both in vitro and in vivo that when chick midbrain tissue is placed adjacent to posterior r1 tissue, expression of *Fgf8* is induced at the new boundary between the tissues (Figure 2*C*). Ectopic *Wnt1* expression and altered tectal polarity was also observed after grafting posterior r1 tissue into the midbrain.

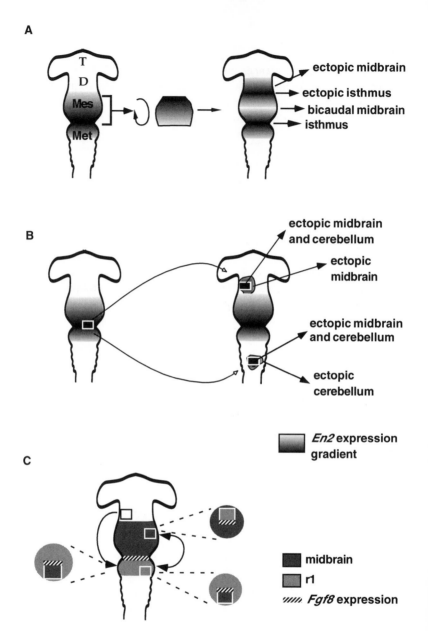

Figure 2 The isthmic region can act like a midbrain/cerebellum organizer. (*A*) When the entire mesencephalic vesicle is inverted in the chick at HH stage 10, a bicaudal midbrain is formed, and the adjacent caudal forebrain is transformed into an ectopic midbrain. (*B*) Transplantation of isthmic tissue can induce ectopic expression of mes/met genes, such as *En2* and ectopic mes/met development. The strongest endogenous *En2* expression is in the mid/hindbrain junction, and it decreased on both sides of the junction. (*C*) The apposition of caudal rhombomere 1 (r1) tissue and midbrain or caudal forebrain tissue leads to induction of *Fgf8* expression. T, telencephalon; D, diencephalon; Mes, mesencephalon; Met, metencephalon; My, myelencephalon.

These results suggest that formation of an isthmic organizer might result from an interaction between midbrain and anterior hindbrain tissues. Furthermore, *Otx2* and *Gbx2* might be involved in this process, given the correlation between the isthmic organizer and the *Otx2/Gbx2* border. Nevertheless, Irving & Mason (1999) suggested that there must be other genes involved in *Fgf8* induction because when r2 tissue, which also expresses *Gbx2*, is placed adjacent to midbrain tissue, no *Fgf8* is induced at the border. However, compared with expression in r1, *Gbx2* expression in r2 appears weaker and restricted dorsally and, thus, may not be strong enough to induce *Fgf8* expression.

Hidalgo-Sanchez et al (1999) also addressed whether juxtaposition of *Otx2*-positive and *Gbx2*-positive cells could induce an organizer using transplants. They showed that when *Otx2*-positive diencephalic tissue is transplanted into *Gbx2*-positive r1, *Gbx2* is induced and *Otx2* is repressed in the grafted cells that contact the *Gbx2*-positive host r1 tissue. *Fgf8*, *En2*, *Wnt1*, and *Pax2* are also induced in the grafts at the new *Gbx2/Otx2* border. The only complication with these studies, as mentioned by the authors, is that transient contact between the graft and *Fgf8*-expressing host cells in the isthmus is hard to avoid. Therefore, it is difficult to distinguish whether the mes/met gene expression that is induced in grafted cells is solely due to the initial confrontation between *Gbx2*- and *Otx2*-expressing cells and/or to the inductive function of *Fgf8* in the host.

FIBROBLAST GROWTH FACTORS ARE CRITICAL FOR MIDBRAIN AND CEREBELLAR DEVELOPMENT

General Properties of Fibroblast Growth Factors and Their Receptors

The mouse fibroblast growth factor (FGF) family currently consists of at least 19 members, named FGF1–19 (for a review, see Basilico & Moscatelli 1992, Coulier et al 1997). Most FGFs are secreted proteins and are expressed not only in transformed tumor cells but also in many signaling centers during normal vertebrate embryonic development. It has been shown in Xenopus that FGFs are involved in inducing mesoderm formation and posterior neural fate (for a review, see Slack et al 1996). The importance of the *Fgf* genes in mouse development has been demonstrated by the analysis of null mutants for several *Fgf* genes (Mansour et al 1993, Hebert et al 1994, Feldman et al 1995, Min et al 1998, Sekine et al 1999, Sun et al 1999).

The best-characterized FGF receptors are high-affinity, signal-transducing tyrosine kinase coupled receptors, which include at least four members, called FGFR1–4. Many FGFR isoforms are generated by alternative splicing of the mRNAs. The dimerization of the FGFRs on ligand binding triggers the activation of the receptors and downstream signaling events. Each kind of FGF molecule is a preferential ligand for a specific FGFR, and each kind of FGFR can bind several different FGF molecules (Basilico & Moscatelli 1992). It has been suggested that heparan-sulfate proteoglycans may help to stabilize the FGF-FGFR complex, and structural analysis has provided evidence for such a function (Plotnikov et al 1999).

Comparative expression studies have been done in medaka fish and in chicks to investigate the expression patterns of different *Fgfr* genes as a means of gaining information as to which gene(s) could be involved in mes/met development. In medaka fish, *Fgfr2, 3*, and *4* are expressed in the midbrain and anterior hindbrain; however, *Fgfr2* is the only gene that is expressed in cells adjacent to, and on both sides of, the mid/hindbrain junction at early somite stages (Carl & Wittbrodt 1999). In contrast, chick *Fgfr1, 2*, and *3* are all expressed throughout the presumptive neural plate during early neural induction (~HH stage 5) and elevated in the anterior neural plate before neuromeres form (~HH stage 8) (Walshe & Mason 2000). However, when neuromeres form (HH stage 9–11), only *Fgfr1* is expressed in the mes/met region. Detailed comparative study of the expression patterns of *Fgfrs* in mouse brain at early stages has not been reported, although it has been shown that *Fgfr1* and -2, but not -4, are expressed in embryonic brain (Stark et al 1991; Yamaguchi et al 1992; Orr-Urtreger et al 1993; Peters et al 1992, 1993). Our preliminary studies, however, show that in mice *Fgfr1* is weakly expressed throughout the CNS and *Fgfr2* and -3 are largely excluded from the posterior midbrain and r1 at E8.5–E9.5 (A Liu, AL Joyner, unpublished data). Taken together, the expression patterns of the FGF receptors are dynamic, and at least FGFR1 and 2 could be involved in mes/met development. All four *Fgfr* genes have been mutated in mice by gene targeting, but mes/met patterning defects have not been reported for any *Fgfr* mutant, either in mutant embryos or in chimeras containing mutant cells in the mes/met. It is possible that the FGFRs have overlapping functions (Deng et al 1994, 1996; Yamaguchi et al 1994; Colvin et al 1996; Ciruna et al 1997; Arman et al 1998, 1999; Weinstein et al 1998; Xu et al 1998).

Fgf8 was first identified as an oncogene responsible for androgen-dependent growth of mammary gland carcinoma cells and was initially called androgen-induced growth factor (Tanaka et al 1992). FGF8 has seven isoforms (Crossley & Martin 1995, MacArthur et al 1995b), and biochemical and cell transformation assays have demonstrated that most of the FGF8 isoforms activate the c isoforms of FGFR2, -3, and -4 and can transform NIH3T3 cells in culture. Among the FGF8 isoforms, FGF8b has the strongest affinity for the three receptors and has the strongest ability to transform NIH3T3 cells, whereas FGF8a shows little affinity to the receptors and very weak transforming activity (MacArthur et al 1995a,b; Blunt et al 1997).

Fgf8 is Required for Normal Development of the Vertebrate Mes/Met Region

A series of mouse *Fgf8* mutant alleles have been generated by gene targeting in ES cells (Meyers et al 1998). A null mutation in which the second and third exons are deleted [*Fgf8*$^{\Delta2, 3/\Delta2, 3}$] leads to a gastrulation defect (Meyers et al 1998, Sun et al 1999). Mice homozygous for a hypomorphic allele, *Fgf8*^{neo}, which contains an insertion of a *neo* cassette into the first intron, or trans-heterozygous for the hypomorphic and null alleles, lack most of the midbrain and cerebellum, indicating *Fgf8* is required for mes/met development (Meyers et al 1998). However, the loss of mes/met tissue either could be due to the *Fgf8* expression in the metencephalon or could be secondary to an early mild gastrulation defect. This will be addressed by analyzing conditional mutants in which the *Fgf8* gene is mutated only in mes/met precursor cells.

Zebrafish *ace* (*acerebellar*) mutants lack the isthmus and cerebellum and have a point mutation in a zebrafish homologue of *Fgf8* that leads to production of a truncated form of protein (Brand et al 1996, Reifers et al 1998). The midbrain in *ace* mutants is expanded caudally and has impaired polarity along both the A/P and dorsal/ventral axes (Picker et al 1999). Analysis of marker gene expression showed that mes/met expression of *eng*, *wnt1*, and *pax2.1* was initiated properly at the end of gastrulation but was then progressively lost. The apparent milder phenotype in *ace* mutants compared with that of *Fgf8* hypomorphic mouse mutants could be because either the *ace* mutation is a hypomorphic *Fgf8* allele or another *Fgf* family member can partially compensate for the loss of *Fgf8* in fish.

FGF8b-Soaked Beads Can Induce Mes/Met Genes and Ectopic Midbrain and Cerebellar Structures in Regions of Chick Brain

A striking finding is that FGF8b-soaked beads inserted into either the posterior diencephalon (prosomere 1 and 2) or anterior mesencephalon of stage 9–12 chicken embryos induce *En1*, *En2*, *Wnt1*, and *Pax2* expression within 24–48 h (Crossley et al 1996, Shamim et al 1999). An interesting recent finding is that *Sprouty2*, which encodes a putative FGF antagonist, is also induced by FGF8-soaked beads, but more rapidly and within an hour (Minowada et al 1999, Chambers et al 2000). Normally, *Sprouty1* and *-2* are expressed around the isthmus, and thus they could normally play a role in modulating FGF signaling from r1 (Minowada et al 1999, Chambers et al 2000, Chambers & Mason 2000).

The long-term consequence of insertion of FGF8-soaked beads into certain brain regions is induction of ectopic mes/met-derived structures. Both the rostral mesencephalon and caudal diencephalon can be transformed into caudal mesencephalic structures, with an ectopic isthmus close to the FGF8 bead (Figure 3A). FGF8b-soaked beads placed in prosomere 1 or the anterior midbrain were found to induce *Gbx2* expression and repress *Otx2* in cells around the beads (Irving & Mason 1999, Martinez et al 1999). Furthermore, in some cases the *Otx2*-negative cells in contact with the beads can form an outgrowth that protrudes from the neural tube (Martinez et al 1999). The outgrowth later develops proximally into isthmic nuclei and distally into cerebellum-like structures, which suggests that one of the normal functions for FGF8 might be to repress *Otx2* expression in the anterior hindbrain and allow, or induce, cerebellum development (Figure 3A). Recently, it was shown that although FGF8-soaked beads can induce *En2* and *Pax2* in r2, it does not lead to formation of ectopic structures (Irving & Mason 2000) (Figure 3A).

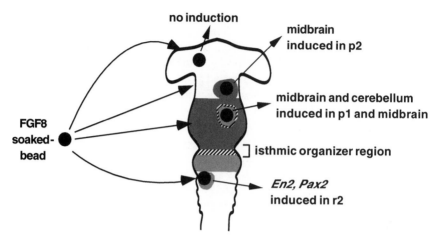

no induction

midbrain
induced in p2

midbrain and cerebellum
induced in p1 and midbrain

FGF8
soaked-
bead

isthmic organizer region

En2, Pax2
induced in r2

Figure 3   FGF8 can induce ectopic mid/hindbrain development in chicks and mice. In the diencephalon and midbrain of both chicken embryos and mouse brain explants, FGF8-soaked beads can induce mid/hindbrain gene expression and, in chicks, ectopic development of midbrain and cerebellar structures. FGF8-soaked beads inserted into the anterior forebrain do not induce mes/met gene expression or mes/met structures. In the anterior hindbrain, FGF8 can induce the expression of *En2*, *Pax2*, and *Wnt1* but no induction of mes/met structures. p, prosomere; r, rhombomere.

It is interesting that Shamim et al (1999) found that both FGF4- and FGF8-soaked beads when implanted into the midbrain can induce *Fgf8* expression, with FGF4 inducing expression in a broader region. In contrast to what was described by Crossley et al (1996), endogenous *Fgf8* was not found to be induced by FGF4 or -8 in the posterior forebrain. However, weak expression of *Fgf8* was induced in the anterior midbrain when FGF4-, but not FGF8-, soaked beads were inserted into the caudal forebrain. Taken together, these chick studies using FGF-soaked beads show that FGF8 is sufficient to induce midbrain, isthmic, and cerebellar structures. However, not all brain regions are equally competent to respond to FGF8.

Wnt1-Fgf8b Transgenic Mouse Embryos Show an Early Transformation of the Midbrain and Posterior Forebrain into an Anterior Hindbrain Fate

To study the role of *Fgf8* in mouse mes/met patterning, transgenic embryos (*Wnt1-Fgf8*) have been produced in which *Fgf8* is expressed under the control of a *Wnt1* regulatory element such that *Fgf8* expression is extended into the entire mesencephalon at early somite stages and along the dorsal midline into the caudal diencephalon after E9.0 (Lee et al 1997, Liu et al 1999). *Fgf8b*-expressing embryos show severe exencephaly and die shortly after E15.5 (Liu et al 1999). Detailed

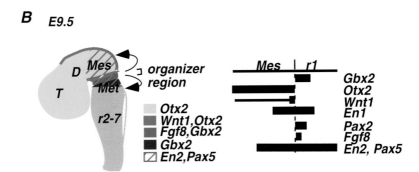

Figure 1 Expression patterns of mes/met genes in the mouse at E8.5 and E9.5. (*A*) At the 3-5 somite stage (E8.5), *Otx2* and *Wnt1* are expressed broadly in the midbrain with *Otx2* also in the forebrain. *Fgf8* and *Gbx2* are expressed broadly in the anterior hindbrain, and *En1*, *En2*, *Pax2* and *Pax5* are expressed in the entire mes/met region. (*B*) At E9.5, *Wnt1* expression is restricted to a narrow ring anterior to the mid/hindbrain junction. *Pax2*, *Fgf8* and *Gbx2* are expressed in narrow rings caudal to the mid/hindbrain junction. *En1*, *En2* and *Pax5* are expressed in regions of the midbrain and anterior hindbrain. The thick lines indicate that expression is along the entire D-V axis and the thin lines indicate that expression is only in the dorsal or ventral midline of the neural tube.

marker gene expression studies at E9.5 indicate that the midbrain and posterior forebrain do not form in such embryos, and the hindbrain is found adjacent to the remaining anterior forebrain. This phenotype appears to result from an early transformation of the midbrain and posterior forebrain into an anterior hindbrain fate, as indicated by an expansion of *Gbx2* expression, repression of *Otx2* expression, and a rostral shift of *Fgf8* expression by the 5- to 7-somite stage, soon after the transgene is first expressed. In contrast, ectopic expression of *Fgf8a* mainly causes overproliferation of the midbrain and caudal diencephalon and up-regulation of *En2*, as well as *EphrinA2*, a prospective downstream target of *En2* (Logan et al 1996) in the anterior dorsal midbrain (Lee et al 1997, Liu et al 1999). This difference between the phenotypes is consistent with results obtained in tissue culture systems demonstrating that FGF8b is a stronger isoform than FGF8a.

The patterning function of FGF8 in mouse mid/hindbrain development has also been investigated in brain explant culture systems (Shimamura & Rubenstein 1997, Liu et al 1999, Liu & Joyner 2001). Similar to the in vivo experiments in mice, FGF8 can induce *En1*, *En2*, *Pax5*, and *Gbx2*, but not *Fgf8*, in both midbrain and forebrain explants. *Pax6*, a diencephalon gene (Walther & Gruss 1991), is repressed by FGF8 in caudal forebrain explants but *Otx2* is repressed only in midbrain explants. It is interesting that both *Lmx1b* and *Wnt1* are initially induced broadly by FGF8 in midbrain explants, but subsequently *Wnt1* is excluded from cells adjacent to the FGF8 source. These explant assays provide a simple in vitro model system to study the early responses of mes/met genes to FGF8 in normal and mutant brain tissue.

Fgf17 Collaborates with *Fgf8* in Patterning the Anterior Cerebellum

As discussed above, following initiation of *Fgf8* expression, *Fgf17* and *Fgf18* are also expressed in the mid/hindbrain region. Mice lacking *Fgf17* have only a mild cerebellar defect, with a decrease in precursor cell proliferation in the medial part of the cerebellum after E11.5 (Xu et al 2000). $Fgf8^{\Delta 2, 3/+}$; $Fgf17^{-/-}$ mutant embryos have a more severe phenotype than do $Fgf17^{-/-}$ mutants, indicating that FGF17 and FGF8 have partially overlapping functions in development of the anterior hindbrain.

OTX1, OTX2, AND GBX2 ARE REQUIRED TO POSITION THE ISTHMUS ORGANIZER AND FOR LATER MIDBRAIN OR CEREBELLUM DEVELOPMENT, RESPECTIVELY

Multiple Essential Roles for *Otx2* in Anterior Patterning

Otx2 null mutant embryos have severe gastrulation defects resulting in deletion of the rostral part of the neural tube anterior to r3 (Acampora et al 1995, Matsuo et al

1995, Ang et al 1996, Rhinn et al 1998) (Figure 4*B*). The expression patterns of marker genes in *Otx2* mutants revealed that the forebrain and midbrain are absent as early as E7.75 and the metencephalon is deleted by E8.5. Chimeric embryos were made to distinguish the functions of OTX2 in the visceral endoderm and epiblast (Rhinn et al 1998). Chimeras composed mostly of wild-type cells in the epiblast and *Otx2*$^{-/-}$ cells in the visceral endoderm have a phenotype similar to

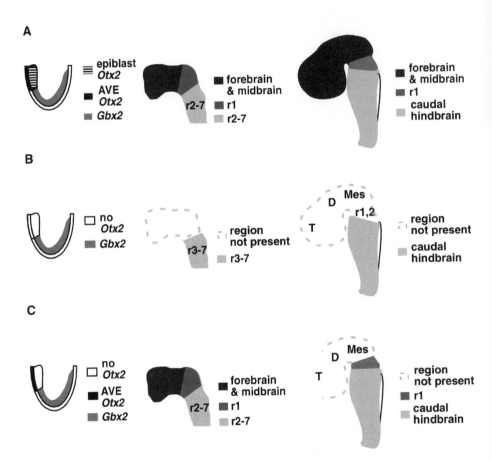

Figure 4 *Otx2* is required in the visceral endoderm for normal gastrulation and initiation of anterior brain development and in the epiblast for the maintenance of anterior brain structures. (*A*) *Otx2* is expressed in both the epiblast and visceral endoderm of wild-type embryos, and it guides normal development of anterior structures. (*B*) *Otx2*$^{-/-}$ embryos that lack *Otx2* expression in both the epiblast and visceral endoderm fail to gastrulate normally and lose structures anterior to rhombomere (r)3 at an early stage. (*C*) In chimeric embryos in which *Otx2* is only expressed in the visceral endoderm, or in embryos in which hOTX1 replaces *Otx2* and is present only in the visceral endoderm, forebrain and midbrain structures initially form at early stages, but are lost subsequently.

that of *Otx2* null mutants. In contrast, chimeras in which the epiblast is mostly composed of $Otx2^{-/-}$ cells and the visceral endoderm wild-type cells undergo nearly normal gastrulation but fail to maintain the forebrain and midbrain after E8 (Figure 4*C*). Taken together, these studies show that the first requirement for *Otx2* is in the extraembryonic endoderm and that *Otx2* plays a cell nonautonomous role in the anterior visceral endoderm in inducing the formation of anterior head structures. This role of *Otx2* likely reflects a function in generating an organizing center in the anterior visceral endoderm (for a review, see Beddington & Robertson 1999). *Otx2* also has a second role in the epiblast, where it is required to maintain the survival and/or identity of rostral brain structures.

Distinct functions for *Otx2* in the visceral endoderm and the epiblast were further demonstrated by analyzing a mouse mutant in which a human *OTX1* cDNA was inserted in place of the endogenous *Otx2* gene, such that *OTX1* mRNA is only translated in the visceral endoderm and not the epiblast. Embryos homozygous mutant for this *Otx2* allele undergo nearly normal gastrulation and initial anterior brain development but subsequently lose anterior brain structures, including the forebrain and midbrain by E8.5 (Acampora et al 1998; Y Li, AL Joyner, unpublished data) (Figure 4*C*). Nevertheless, *Fgf8* and other mes/met genes are expressed even at late stages, but in overlapping domains at the rostral tip of the mutant embryo.

$Otx1^{+/-}$; $Otx2^{+/-}$ (Suda et al 1997) or $Otx1^{-/-}$; $Otx2^{+/-}$ (Acampora et al 1997) mouse embryos have the striking phenotype that the midbrain and caudal forebrain are replaced with cerebellar tissue. Gene expression analysis in such mutants indicates that at early somite stages, expression of all the mes/met genes is initiated normally. However, *Fgf8* expression soon fails to be restricted to the rostral hindbrain region and expands into more rostral brain regions (Figure 5*B*). This expansion is followed by a retraction of *Otx2* expression and a rostral shift in the *En1/2*, *Wnt1*, and *Gbx2* domains. Therefore, the midbrain and caudal forebrain are transformed into an anterior hindbrain fate during early neural patterning, when the dose of OTX proteins is not sufficient to maintain their normal fate. These studies indicate that the caudal limit of *Otx2* could be critical for determining the position of the isthmic organizer, or the *Fgf8* expression domain, and a high level of *Otx2* is required for midbrain development.

The above *Otx1/2* compound mutant phenotype showed that both *Otx* genes have overlapping functions late in development of the midbrain and caudal forebrain. In contrast, when mouse *Otx1* is inserted into the *Otx2* locus such that it is expressed like *Otx2* in the embryo, *Otx1* can rescue the gastrulation defect of *Otx2* mutant embryos but not the deletion of most of the rostral brain region at early somite stages (Suda et al 1999). Thus, the two OTX proteins appear to have different functions during early specification of forebrain and midbrain precursors.

Finally, recent gain-of-function studies in which *Otx2* was ectopically expressed in the dorsal metencephalon from early somite stages provided further evidence that *Otx2* is not only required to position the organizer and direct midbrain development, it also is sufficient to direct both these processes. When *Otx2* was expressed in the

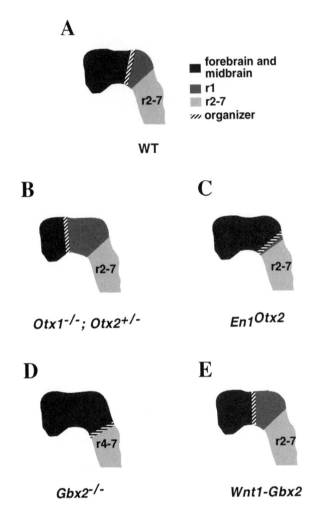

Figure 5 *Otx2* and *Gbx2* regulate mes/met development differently. (*A*) Schematic representation of a wild-type E9.5 mouse brain. (*B*) In *Otx1⁻/⁻*; *Otx2⁺/⁻* embryos, the midbrain and posterior forebrain are transformed into an anterior hindbrain. (*C*) Ectopic expression of *Otx2* in the dorsal metencephalon from the *En1* locus results in a caudal expansion of the midbrain and partial deletion of the metencephalon. (*D*) In the absence of *Gbx2*, rhombomere (r)1–r3 do not form and the midbrain expands caudally, such that the midbrain directly abuts r4. (*E*) Ectopic expression of *Gbx2* in the posterior midbrain of transgenics from E8.5 leads to a smaller midbrain and an enlarged metencephalon at E9.5.

dorsal anterior hindbrain of mouse embryos, the caudal midbrain was extended and the cerebellum partially deleted (Broccoli et al 1999) (Figure 5C). These morphological changes were accompanied by a repression of *Fgf8* and *Gbx2* in the dorsal anterior most hindbrain from early stages. Similar results were obtained in chicken embryos by electroporating anterior hindbrain tissue with an *Otx2* expression construct (Katahira et al 2000).

Gbx2 is Required for Normal Isthmic Organizer Function and Position as well as Anterior Hindbrain Development

Mouse *Gbx2* null mutants die at birth with loss of anterior hindbrain (r1-r3) derivatives, which suggests that *Gbx2* might be involved in maintaining the identity and/or survival of the anterior hindbrain (Wassarman et al 1997). Gene expression analysis between E9.5 and E12.5 showed that the caudal *Otx2* expression border was adjacent to that of the r4 marker *Hoxb1*, showing an early deletion of the anterior hindbrain. Furthermore, unlike in wild-type embryos, where *Fgf8* expression does not overlap with that of *Otx2* and *Wnt1* in the isthmus, the expression domains of the three genes overlap in *Gbx2* mutant embryos. More interesting, the morphology of the posterior midbrain is not normal, which suggests that *Gbx2* also contributes to midbrain patterning, possibly indirectly through a function in regulating the organizer.

More recently, it has been shown that in *Gbx2*$^{-/-}$ embryos, the *Otx2* domain is greatly expanded caudally by the 4- to 6-somite stage, and its caudal boundary is never sharp (Millet et al 1999) (Figure 5D). The *Wnt1* and *Fgf8* expression domains also shift caudally and *Fgf8* overlaps *Wnt1* and *Otx2*. These expression studies demonstrated that an early phenotype in *Gbx2* mutants is a transformation of presumptive r1–r3 cells into a midbrain fate. In a complementary manner, misexpression of *Gbx2* in the midbrain from a *Wnt1* promoter (*Wnt1-Gbx2* transgenics) at early somite stages results in a repression of *Otx2* expression and a rostral shift of isthmic *Wnt1* and *Fgf8* expression (Millet et al 1999) (Figure 5E). Endogenous *Gbx2* expression is also induced by the transgene. The consequence of the early changes in gene expression is a reduction of the midbrain and expansion of the hindbrain at E9.5. Taken together, these two types of mutant studies suggest GBX2 functions to antagonize *Otx2* expression and midbrain development, in addition to being involved in positioning the isthmic organizer.

It is striking that midbrain development begins to recover soon after the transgene expression is lost at E9.5, and by E12.5 the brains of *Wnt1-Gbx2* embryos look indistinguishable from those of wild types. Thus, the midbrain is still plastic at E9.5 in mice and can recover from early transient *Gbx2* misexpression. In contrast, transient ectopic expression of *Gbx2* in the chick midbrain at HH stage 10 leads to repression of *Otx2* expression only transiently in rostral midbrain regions but permanently in caudal regions. Furthermore, caudal midbrain misexpression of *Gbx2* results in a rostral shift of the mid/hindbrain junction and a smaller midbrain at late stages (Katahira et al 2000).

Taken together, the gain-of-function and loss-of-function studies of both *Gbx2* and *Otx2* mutants provide strong evidence that these genes act antagonistically, and that they are also required for anterior hindbrain and midbrain development, respectively. Furthermore, the two genes are involved in positioning and normal functioning of the isthmic organizer. However, the two genes are not required to initiate expression of *Fgf8* and other mes/met genes because in embryos lacking both *Otx2* and *Gbx2*, all the genes are expressed in the rostral CNS (Y Li, AL Joyner, manuscript in preparation).

WNT1, EN1/2, AND PAX2/5 ARE ESSENTIAL FOR NORMAL MES/MET DEVELOPMENT

Wnt1 is Required for Mes/Met Development and Positioning of the Caudal *Otx2* Border

Loss of *Wnt1* function leads to death at birth with an early large-scale deletion first of the mesencephalon and then of r1 (McMahon & Bradley 1990, Thomas & Capecchi 1990, Mastick et al 1996). Furthermore, *En1* expression is initiated normally in *Wnt1* mutants but is lost by the 27-somite stage, before the loss of the r1 tissue (McMahon et al 1992). One possible interpretation of these studies is that one function of *Wnt1* is to maintain *En1* expression. Consistent with this, a transgene with a *Wnt1* enhancer driving *En1* expression was found to rescue most of the *Wnt1* mutant phenotype (Danielian & McMahon 1996). However, in *Wnt1* mutants, *Fgf8* expression in the metencephalon is lost by the 14-somite stage, which suggests that WNT1 more directly regulates *Fgf8* expression in this region and that loss of *En1* could be secondary to the loss of *Fgf8* (Lee et al 1997). Because *En1* misexpression in chicks leads to induction of *Fgf8* in both midbrain and diencephalon (see below), transgene driven *En1* expression in *Wnt1* mutants might lead to maintenance of *Fgf8* expression and, thus, rescue of the phenotype.

Analysis of a *Wnt1* mutant allele, *Swaying*, that appears to be a hypomorph has revealed a function for *Wnt1* in maintaining a stable mid/hindbrain junction (Thomas et al 1991, Bally-Cuif et al 1995). *Swaying* contains a frame-shift mutation that causes premature termination of WNT1 protein, such that a large portion of the C terminus of WNT1 is deleted. *Wnt1$^{sw/sw}$* mutant mice have a less-severe brain phenotype compared with targeted *Wnt1* null mutants, with the midbrain and hindbrain regions being only partially reduced. It is striking that analysis of the relatively late (E9.0 and onward) midbrain and hindbrain phenotype of *Swaying* mutants showed that formation of the straight *Otx2* and *Wnt1* caudal expression borders was perturbed, and small ectopic islands of *Otx2*-positive (mes) or *Otx2*-negative (met) cells were located inappropriately within the hindbrain or midbrain, respectively. Moreover, *Wnt1* expression was induced at the border of the ectopic mes and met islands, consistent with the idea that interactions between mes (*Otx2* positive) and met (*Otx2* negative) cells positively regulate organizer gene expression, as reflected by *Wnt1* expression at the boundary.

Wnt1 gain-of-function studies have failed to demonstrate that *Wnt1* is sufficient to alter mes/met patterning. In chicks, *Wnt1* misexpression in the midbrain at HH stage 9–12 does not result in any phenotype, although in the telencephalon it causes overproliferation (Adams et al 2000). Furthermore, ectopic expression of *Wnt1* in the ventricular zone of the neural tube posterior to the r6/r7 boundary between E8.5 and E10.5 results only in expansion of the ventricular zone and has no obvious changes in patterning (Dickinson et al 1994).

A recent study showed that ectopic *Lmx1b* expression leads to up-regulation of *Wnt1* expression in chick midbrain (Adams et al 2000). Furthermore, *Lmx1b* is induced in chick caudal forebrain by FGF8-soaked beads. This study suggests that *Lmx1b* might be in the same pathway as *Wnt1* downstream of *Fgf8* and could play a role in early mes/met patterning. The requirement for *Lmx1b* in mes/met development has not been reported, although *Lmx1b* mouse mutants have been generated (Chen et al 1998).

En Genes are Required for Mes/Met Development and Likely Repress Forebrain Development

En1 mutant mice have a deletion including most of the midbrain and cerebellum, a phenotype similar to, but milder than, that of *Wnt1* null mutants (Wurst et al 1994). Unlike *En1* mutants, *En2* mutants are viable and show only subtle defects in cerebellar development, which include an early reduction in size of the cerebellar anlage and later abnormal foliation (Joyner et al 1991; Millen et al 1994, 1995). This suggests at least one role for *En2* is in patterning the cerebellum. Two experiments have demonstrated that the two EN proteins can carry out similar functions in the brain, given correct temporal and spatial expression of the genes. First, when the *En1* coding sequences are replaced with those of an *En2* cDNA using a gene-targeting approach, the mes/met phenotype of *En1* mutants is rescued (Hanks et al 1995). It is interesting that *En2* cannot fully rescue all the limb defects seen in *En1* mutants, and it is significant that Drosophila *en* can rescue the *En1* brain defects but none of the limb defects (Hanks et al 1998). Second, double mutants of the two *En* genes have a more severe early phenotype than do either of the single mutants (Liu & Joyner 2001). In such double mutants, *Wnt1*, *Fgf8*, and *Pax5* expression is initiated by the 5-somite stage but fails to be maintained in the mes/met by the 11-somite stage. Taken together, these studies show that an EN protein must be present during early somite stages when *En1*, but not *En2*, is normally expressed. Without early expression of an EN protein, further mes/met development is greatly compromised.

Based on a correlation between the decreasing EN protein gradient from posterior to anterior midbrain and the A/P polarity of the midbrain before and after 180° rotation of the anterior midbrain in chicks, it was suggested that the *En* genes could be involved in setting up the topographic axon projection map of the midbrain (Itasaki et al 1991). This suggestion was addressed by expressing *En* genes ectopically in the anterior midbrain in both chicken and mouse embryos. In chicks, ectopic expression of *En1* or *En2* in the rostral midbrain induces *EphrinA2* and *EphrinA5*, two genes involved in repulsing retinal temporal axons in the posterior midbrain and ectopic projections of retinal nasal axons in the rostral tectal tissue (Friedman & O'Leary 1996, Itasaki & Nakamura 1996, Logan et al 1996). In mice, ectopic expression of *En1* driven by a *Wnt1* enhancer causes ectopic *EphrinA2* expression in the dorsal midline of the rostral midbrain (Lee et al 1997). Thus, the expression level of *En* genes can directly, or indirectly, determine the A/P positional cues of midbrain cells.

In chicks, ectopic expression of *En2* in the diencephalic region results in expansion of tectal tissue at the expense of caudal diencephalic tissue (Araki & Nakamura 1999). It is striking that this phenotype is accompanied by repression of the diencephalon gene *Pax6* prior to induction of the mes/met genes *Pax5*, *Wnt1*, and *Fgf8*. Because it has been shown that the EN family of proteins are transcriptional repressors (Jaynes & O'Farrell 1991, Smith & Jaynes 1996, Hanks et al 1998), it is possible that EN is normally involved in setting up the mid/forebrain boundary by repressing forebrain genes. Consistent with this hypothesis, a mutant form of EN2 in which a single amino acid substitution disrupts its interaction with the corepressor, GROUCHO, cannot transform the forebrain or induce mid/hindbrain gene expression. Furthermore, when a chimeric EN protein in which the repression domain of EN2 was replaced by the transactivation domain of VP16 was expressed in chick brain, the midbrain was greatly reduced and *Pax6* expression expanded into the midbrain region (Araki & Nakamura 1999).

It has also been shown in medaka fish and frog embryos that ubiquitous expression of the fish *En2* gene, *Ol-eng2*, can induce ectopic midbrain development and repress formation of diencephalic tissue, whereas expression of an *Ol-eng2* gene with a mutation in the homeodomain does not produce such an effect (Ristoratore et al 1999). It is interesting to note that despite widespread expression of injected *Ol-eng2*, ectopic expression of *fgf8*, *pax2*, and endogenous *eng2* is restricted only to regions of the posterior forebrain. Furthermore, the ectopic midbrain that is induced always has a reversed A/P polarity. Induction of an ectopic isthmic organizer by *Ol-eng2* therefore seems to require cofactors that are expressed locally in the forebrain.

The *Pax2/5* Genes have Overlapping Functions in Development of the Mes/Met

The *Pax2* and *Pax5* genes have been shown to be required for mouse mes/met patterning using loss-of-function mutants. Several mouse *Pax2* mutant alleles have been generated and surprisingly they have different phenotypes. One targeted deletion allele that removes most of the coding sequences, referred to as *Pax2$^{ko/ko}$*, appears to show no mes/met A/P patterning defects, just a varied degree of exencephaly, depending on the genetic background (Torres et al 1996, Schwarz et al 1997). In contrast, homozygotes for the spontaneous $Pax2^{1Neu}$ allele, which contains a frame-shift mutation that truncates the protein, have a brain deletion that includes most of the mes/met region (Favor et al 1996). More recently, another mouse $Pax2$ allele was generated by gene targeting, which is thought to be a null allele, and homozygous embryos for this allele have an early deletion of the mes/met, similar to $Pax2^{1Neu}$ mutants (Bouchard et al 2000). $Pax5^{-/-}$ mutant mice, in contrast, have only a partial deletion of the inferior colliculi (posterior midbrain) and a slightly enlarged third lobe of the cerebellum (Urbanek et al 1994).

Overlapping functions of $Pax2$ and $Pax5$ have been demonstrated using double mutants. For example, $Pax2^{kol+}$; $Pax5^{-/-}$ and $Pax2^{kolko}$; $Pax5^{-/-}$ mutants lack most of the midbrain and cerebellum (Schwarz et al 1997). Although heterozygous Krd mutants, which contain a 7-cM deletion that includes the $Pax2$ gene (Keller et al 1994), and $Pax5^{+/-}$ mice are phenotypically normal, compound mutants $Krd^{+/-}$; $Pax5^{+/-}$ and $Krd^{+/-}$; $Pax5^{-/-}$ have deletions of the midbrain and cerebellum, similar to $Pax2$ null mutants (Urbanek et al 1997). Furthermore, $Pax5$ is able to rescue the brain phenotype of the $Pax2$ null allele by expressing it from the $Pax2$ locus. Therefore, like the EN proteins, a PAX2/5 protein must be expressed at early somite stages to sustain mes/met development. In addition, $Pax6$ expression, as well as the posterior commisure, a morphological landmark of the posterior diencephalon, are expanded posteriorly in $Pax2/5$ double mutants, which suggests that $Pax2/5$ might be involved in maintaining mes/met development by repressing forebrain development (Schwarz et al 1999).

The function of the PAX2/5/8 subfamily of Pax genes in brain patterning has also been studied in zebrafish. There are four genes in the $Pax2/5/8$ family in zebrafish, $pax2.1$, $pax2.2$, $pax5$, and $pax8$ (Lun & Brand 1998). Injection of a neutralizing antibody against PAX2.1 into zebrafish early embryos resulted in a deletion of most parts of the midbrain and cerebellum (Krauss et al 1992). Noi (no $isthmus$) mutants in which the zebrafish $pax2.1$ gene is mutated have a similar deletion of the caudal midbrain and cerebellum (Brand et al 1996). Expression of $pax2.2$, $pax5$, and $pax8$ is never detected in noi mutants, indicating that PAX2.1 regulates the other pax genes and that the noi mutant is actually equivalent to a $pax2.1/2.2/5/8$ quadruple mutant (Lun & Brand 1998). Similarly, PAX2 in mice has been shown to be involved in regulating $Pax5$ mes/met expression (Pfeffer et al 2000).

In noi mutants, $eng2$ and $eng3$ expression is not initiated normally, whereas $fgf8$ and $wnt1$ expression initiates normally but is lost before the mes/met tissue undergoes apoptosis (Pfeffer et al 1998). These changes in mes/met gene expression indicate that $pax2.1$ is necessary for the normal pattern of mes/met gene expression, as well as for the survival of mes/met tissue. The Pax genes also have been implicated in regulating expression of $En2$ in mice. PAX2/5/8 proteins were shown to bind to two DNA sequences in an $En2$ DNA enhancer fragment that is sufficient to direct expression of a $lacZ$ reporter gene in the isthmus (Logan et al 1993, Song et al 1996). Furthermore, mutation of the PAX2/5/8 binding sites abolishes the reporter gene expression. Deletion of the two PAX2/5/8-binding sites in the endogenous *En2* gene by gene targeting, however, showed that the binding sites are required only for normal initiation, not maintenance, of endogenous *En2* expression (Song & Joyner 2000).

Gain-of-function studies in chicks have shown that the *Pax* genes, like *En1/2* and *Fgf8*, can induce midbrain development. Ectopic expression of either *Pax2* or *Pax5* in the diencephalon is sufficient to induce ectopic expression of mes/met genes, such as *Fgf8* and *En*, and to transform diencephalic tissue into midbrain structures (Funahashi et al 1999, Okafuji et al 1999). Furthermore, *Pax5*, but not *Pax2*, misexpression in the anterior midbrain can induce *Fgf8* and *En2* expression, indicating that the two PAX proteins have developed some different functions during evolution.

CROSS-REGULATION BETWEEN THE MES/MET GENES UNDERLIES ORGANIZER FUNCTION AND MES/MET DEVELOPMENT

The expression studies and functional analyses of mes/met genes have begun to elucidate a set of complex cross-regulatory interactions between these genes during midbrain and cerebellum development. Some of these same gene families are also involved in regulatory networks in Drosophila. For example, the *Pax* homologue, *prd*, is required for initiation of *en* and *wg* expression in ectoderm segments; then *en* expression and *wg* expression become dependent on each other, and finally *en* is autoregulated (for a review, see Hooper & Scott 1992, Perrimon 1994). Some similar, but not identical, regulatory interactions are found during vertebrate mes/met development. As mentioned above, *Pax2* expression precedes and overlaps with *En1* and *Wnt1* in mice, and genetic studies implicate *Pax2/5* in regulating *En1/2* expression in zebrafish and mice. In contrast to the situation in Drosophila, aspects of *Wnt1* expression do not seem to require PAX2 function (Lun & Brand 1998, Schwarz et al 1999). There is also no evidence for autoregulation of the *En* genes (Logan 1993, Liu & Joyner 2001).

Several lines of evidence indicate that *En* and *Wnt1* are involved in regulating each other's expression, although it is not clear whether this is direct. As described above, in *Wnt1* null mutant mice, *En1* expression is initiated normally but lost, and in *En1/2* double mutant embryos, *Wnt1* expression is rapidly lost. Furthermore, when *En1* expression is maintained in *Wnt1* mutants, a nearly complete rescue of the *Wnt1* brain phenotype is observed. In addition, ectopic expression of *En* can induce ectopic *Wnt1* expression (Araki & Nakamura 1999, Ristoratore et al 1999), although the reverse does not seem to be true (Adams et al 2000).

Many gain- and loss-of-function studies have shown that *Otx2* and *Gbx2* are involved in a negative feedback loop (see above). We recently explored whether *Gbx2* acts directly or indirectly to repress *Otx2* using a midbrain explant system and FGF8-soaked beads (Liu & Joyner 2001). It is surprising that although *Gbx2* was required for repressing *Wnt1* expression, it was not required for repressing *Otx2*.

Therefore, FGF8 can regulate *Otx2* expression through a *Gbx2*-independent pathway, and it is possible that *Gbx2* can repress *Otx2* through up-regulation of *Fgf8*. Finally, as discussed above, FGF8 has been shown to be able to induce *Pax2*, *Pax5*, *En1*, *En2*, *Wnt1*, *Lmx1b*, and *Gbx2* in midbrain and caudal forebrain tissue, as well as to repress *Otx2*. Nevertheless, the normal timing of gene expression shows that *Fgf8* expressed in the metencephalon is unlikely to be responsible for inducing initial expression of these mes/met genes. Whether another FGF, possibly one expressed outside the neural tube, initiates mes/met gene expression is not clear. *Fgf8* is, however, likely to be critical for modulating and maintaining expression of many mes/met genes after the 5-somite stage in mice.

PROSPECTS AND FUTURE DIRECTION

The many genetic studies we have described have revealed that multiple genetic pathways regulate early patterning of the vertebrate midbrain and cerebellum. An early step in patterning the mes/met seems to be a genetic interaction at the border of *Otx2*- and *Gbx2*-expressing cells, which determines where the midbrain and cerebellum will form. It is also clear that by the 10-somite stage, a self-sustaining genetic network is set up within the mes/met that is controlled centrally by an isthmic organizer. Furthermore, FGF8-like molecules alone can appropriately regulate the expression of most the other mes/met genes and induce midbrain and cerebellum development. Thus, FGF8 and related factors are likely central to this self-sustaining network. It must, however, be more complicated because loss of *En*, *Wnt1*, or *Pax2/5* function alone leads to a collapse of mes/met development.

A number of key questions nevertheless remain to be addressed. For example, none of the loss-of-function mutants described seems to lack initial expression of all other genes in the mes/met region, raising the question of what signal triggers the mes/met molecular pathway before the early somite stages. Also, the epistatic relationships among the known genes need to be established using new approaches because they have been hard to explore in null mutants owing to simultaneous early loss of expression of multiple genes. Another important and long-standing question is how the signals from the mid/hindbrain organizer are transmitted across the entire length of the midbrain and r1. Does a relay mechanism exist or are signals transmitted from the isthmus directly to cells at a distance? Finally, it has been shown that isthmic tissue cannot induce ectopic mes/met development in either the anterior forebrain (Martinez et al 1991) or the spinal cord (Grapin-Botton et al 1999), raising the question of what regulates the competence of cells to respond to isthmic organizing signals.

ACKNOWLEDGMENTS

We thank Sandrine Millet and Yuanhao Li for insightful discussions and helpful comments on the manuscript. The work described from our laboratory was supported by grants from the NINDS to ALJ. ALJ is an investigator of the Howard Hughes Medical Institute.

Visit the Annual Reviews home page at www.annualreviews.org

LITERATURE CITED

Acampora D, Avantaggiato V, Tuorto F, Briata P, Corte G, Simeone A. 1998. Visceral endoderm-restricted translation of Otx1 mediates recovery of Otx2 requirements for specification of anterior neural plate and normal gastrulation. *Development* 125:5091–104

Acampora D, Avantaggiato V, Tuorto F, Simeone A. 1997. Genetic control of brain morphogenesis through Otx gene dosage requirement. *Development* 124:3639–50

Acampora D, Mazan S, Lallemand Y, Avantaggiato V, Maury M, et al. 1995. Forebrain and midbrain regions are deleted in Otx2−/− mutants due to a defective anterior neuroectoderm specification during gastrulation. *Development* 121:3279–90

Adams B, Dorfler P, Aguzzi A, Kozmik Z, Urbanek P, et al. 1992. Pax-5 encodes the transcription factor BSAP and is expressed in B lymphocytes, the developing CNS, and adult testis. *Genes Dev.* 6:1589–607

Adams KA, Maida JM, Golden JA, Riddle RD. 2000. The transcription factor Lmx1b maintains Wnt1 expression within the isthmic organizer. *Development* 127:1857–67

Alvarado-Mallart RM. 1993. Fate and potentialities of the avian mesencephalic/metencephalic neuroepithelium. *J. Neurobiol.* 24:1341–55

Ang SL, Jin O, Rhinn M, Daigle N, Stevenson L, Rossant J. 1996. A targeted mouse Otx2 mutation leads to severe defects in gastrulation and formation of axial mesoderm and to deletion of rostral brain. *Development* 122:243–52

Araki I, Nakamura H. 1999. Engrailed defines the position of dorsal di-mesencephalic boundary by repressing diencephalic fate. *Development* 126:5127–35

Arman E, Haffner-Krausz R, Chen Y, Heath JK, Lonai P. 1998. Targeted disruption of fibroblast growth factor (FGF) receptor 2 suggests a role for FGF signaling in pregastrulation mammalian development. *Proc. Natl. Acad. Sci. USA* 95:5082–87

Arman E, Haffner-Krausz R, Gorivodsky M, Lonai P. 1999. Fgfr2 is required for limb outgrowth and lung-branching morphogenesis. *Proc. Natl. Acad. Sci. USA* 96:11895–99

Bally-Cuif L, Cholley B, Wassef M. 1995. Involvement of Wnt-1 in the formation of the mes/metencephalic boundary. *Mech. Dev.* 53:23–34

Basilico C, Moscatelli D. 1992. The FGF family of growth factors and oncogenes. *Adv. Cancer Res.* 59:115–65

Beddington SP, Robertson EJ. 1999. Axis development and early asymmetry in mammals. *Cell* 96:195–209

Blunt AG, Lawshe A, Cunningham ML, Seto ML, Ornitz DM, MacArthur CA. 1997. Overlapping expression and redundant activation of mesenchymal fibroblast growth factor (FGF) receptors by alternatively spliced FGF-8 ligands. *J. Biol. Chem.* 272:3733–38

Bouchard M, Pfeffer P, Busslinger M. 2000. Functional equivalence of the transcription factors Pax2 and Pax5 in mouse development. *Development* 127:3703–13

Bouillet P, Chazaud C, Oulad-Abdelghani M, Dolle P, Chambon P. 1995. Sequence and expression pattern of the Stra7 (Gbx-2) homeobox-containing gene induced by retinoic acid in P19 embryonal carcinoma cells. *Dev. Dyn.* 204:372–82

Brand M, Heisenberg CP, Jiang YJ, Beuchle D, Lun K, et al. 1996. Mutations in zebrafish genes affecting the formation of the boundary between midbrain and hindbrain. *Development* 123:179–90

Broccoli V, Boncinelli E, Wurst W. 1999. The caudal limit of Otx2 expression positions the isthmic organizer. *Nature* 401:164–68

Carl M, Wittbrodt J. 1999. Graded interference with FGF signalling reveals its dorsoventral asymmetry at the mid-hindbrain boundary. *Development* 126:5659–67

Chambers D, Mason I. 2000. Expression of sprouty2 during early development of the chick embryo is coincident with known sites of FGF signalling. *Mech. Dev.* 91:361–64

Chambers D, Medhurst AD, Walsh FS, Price J, Mason I. 2000. Differential display of genes expressed at the midbrain-hindbrain junction identifies sprouty2: an FGF8-inducible member of a family of intracellular FGF antagonists. *Mol. Cell. Neurosci.* 15:22–35

Chen H, Lun Y, Ovchinnikov D, Kokubo H, Oberg KC, et al. 1998. Limb and kidney defects in Lmx1b mutant mice suggest an involvement of LMX1B in human nail patella syndrome. *Nat. Genet.* 19:51–55

Ciruna BG, Schwartz L, Harpal K, Yamaguchi TP, Rossant J. 1997. Chimeric analysis of fibroblast growth factor receptor-1 (Fgfr1) function: a role for FGFR1 in morphogenetic movement through the primitive streak. *Development* 124:2829–41

Colvin JS, Bohne BA, Harding GW, McEwen DG, Ornitz DM. 1996. Skeletal overgrowth and deafness in mice lacking fibroblast growth factor receptor 3. *Nat. Genet.* 12:390–97

Coulier F, Pontarotti P, Roubin R, Hartung H, Goldfarb M, Birnbaum D. 1997. Of worms and men: an evolutionary perspective on the fibroblast growth factor (FGF) and FGF receptor families. *J. Mol. Evol.* 44:43–56

Crossley PH, Martin GR. 1995. The mouse Fgf8 gene encodes a family of polypeptides and is expressed in regions that direct outgrowth and patterning in the developing embryo. *Development* 121:439–51

Crossley PH, Martinez S, Martin GR. 1996.

Midbrain development induced by FGF8 in the chick embryo. *Nature* 380:66–68

Danielian PS, McMahon AP. 1996. Engrailed-1 as a target of the Wnt-1 signalling pathway in vertebrate midbrain development. *Nature* 383:332–34

Davis CA, Joyner AL. 1988. Expression patterns of the homeo box-containing genes En-1 and En-2 and the proto-oncogene int-1 diverge during mouse development. *Genes Dev.* 2:1736–44

Deng CX, Wynshaw-Boris A, Shen MM, Daugherty C, Ornitz DM, Leder P. 1994. Murine FGFR-1 is required for early postimplantation growth and axial organization. *Genes Dev.* 8:3045–57

Deng CX, Wynshaw-Boris A, Zhou F, Kuo A, Leder P. 1996. Fibroblast growth factor receptor 3 is a negative regulator of bone growth. *Cell* 84:911–21

Dickinson ME, Krumlauf R, McMahon AP. 1994. Evidence for a mitogenic effect of Wnt-1 in the developing mammalian central nervous system. *Development* 120:1453–71

Favor J, Sandulache R, Neuhauser-Klaus A, Pretsch W, Chatterjee B, et al. 1996. The mouse Pax2(1Neu) mutation is identical to a human PAX2 mutation in a family with renal-coloboma syndrome and results in developmental defects of the brain, ear, eye, and kidney. *Proc. Natl. Acad. Sci. USA* 93:13870–75

Feldman B, Poueymirou W, Papaioannou VE, DeChiara TM, Goldfarb M. 1995. Requirement of FGF-4 for postimplantation mouse development. *Science* 267:246–49

Friedman GC, O'Leary DD. 1996. Retroviral misexpression of engrailed genes in the chick optic tectum perturbs the topographic targeting of retinal axons. *J. Neurosci.* 16:5498–509

Funahashi J, Okafuji T, Ohuchi H, Noji S, Tanaka H, Nakamura H. 1999. Role of Pax-5 in the regulation of a mid-hindbrain organizer's activity. *Dev. Growth Differ.* 41:59–72

Grapin-Botton A, Cambronero F, Weiner HL, Bonnin MA, Puelles L, Le Douarin NM. 1999. Patterning signals acting in the spinal cord override the organizing activity of the isthmus. *Mech. Dev.* 84:41–53

Gruss P, Walther C. 1992. Pax in development. *Cell* 69:719–22

Hamburger V, Hamilton HL. 1992. A series of normal stages in the development of the chick embryo. *Dev. Dyn.* 195:231–72

Hanks M, Wurst W, Anson-Cartwright L, Auerbach AB, Joyner AL. 1995. Rescue of the En-1 mutant phenotype by replacement of En-1 with En-2. *Science* 269:679–82

Hanks MC, Loomis CA, Harris E, Tong CX, Anson-Cartwright L, et al. 1998. Drosophila engrailed can substitute for mouse Engrailed1 function in mid-hindbrain, but not limb development. *Development* 125:4521–30

Hebert JM, Rosenquist T, Gotz J, Martin GR. 1994. FGF5 as a regulator of the hair growth cycle: evidence from targeted and spontaneous mutations. *Cell* 78:1017–25

Heikinheimo M, Lawshe A, Shackleford GM, Wilson DB, MacArthur CA. 1994. Fgf-8 expression in the post-gastrulation mouse suggests roles in the development of the face, limbs and central nervous system. *Mech. Dev.* 48:129–38

Hidalgo-Sanchez M, Simeone A, Alvarado-Mallart RM. 1999. Fgf8 and Gbx2 induction concomitant with Otx2 repression is correlated with midbrain-hindbrain fate of caudal prosencephalon. *Development* 126:3191–203

Hooper JE, Scott MP. 1992. The molecular genetic basis of positional information in insect segments. *Results Probl. Cell Differ.* 18:1–48

Irving C, Mason I. 1999. Regeneration of isthmic tissue is the result of a specific and direct interaction between rhombomere 1 and midbrain. *Development* 126:3981–89

Irving C, Mason I. 2000. Signalling by FGF8 from the isthmus patterns anterior hindbrain and establishes the anterior limit of Hox gene expression. *Development* 127:177–86

Itasaki N, Ichijo H, Hama C, Matsuno T, Nakamura H. 1991. Establishment of rostrocaudal polarity in tectal primordium: engrailed expression and subsequent tectal polarity. *Development* 113:1133–44

Itasaki N, Nakamura H. 1996. A role for gradient en expression in positional specification on the optic tectum. *Neuron* 16:55–62

Jaynes JB, O'Farrell PH. 1991. Active repression of transcription by the engrailed homeodomain protein. *EMBO J.* 10:1427–33

Joyner AL, Herrup K, Auerbach BA, Davis CA, Rossant J. 1991. Subtle cerebellar phenotype in mice homozygous for a targeted deletion of the En-2 homeobox. *Science* 251:1239–43

Joyner AL, Liu A, Millet S. 2000. Otx2, Gbx2 and Fgf8 interact to position and maintain a mid-hindbrain organizer. *Curr. Opin. Cell. Biol.* 12:736–41

Joyner AL, Martin GR. 1987. En-1 and En-2, two mouse genes with sequence homology to the Drosophila engrailed gene: expression during embryogenesis. *Genes Dev.* 1:29–38

Katahira T, Sato T, Sugiyama S, Okafuji T, Araki I, et al. 2000. Interaction between Otx2 and Gbx2 defines the organizing center for the optic tectum. *Mech. Dev.* 91:43–52

Keller SA, Jones JM, Boyle A, Barrow LL, Killen PD, et al. 1994. Kidney and retinal defects (Krd), a transgene-induced mutation with a deletion of mouse chromosome 19 that includes the Pax2 locus. *Genomics* 23:309–20

Krauss S, Maden M, Holder N, Wilson SW. 1992. Zebrafish pax[b] is involved in the formation of the midbrain-hindbrain boundary. *Nature* 360:87–89

Lee SM, Danielian PS, Fritzsch B, McMahon AP. 1997. Evidence that FGF8 signalling from the midbrain-hindbrain junction regulates growth and polarity in the developing midbrain. *Development* 124:959–69

Liu A, Joyner AL. 2001. EN and GBX2 play essential roles downstream of FGF8 in patterning the mouse mid/hindbrain region. *Development* 128:181–91

Liu A, Losos K, Joyner AL. 1999. FGF8 can activate Gbx2 and transform regions of the rostral mouse brain into a hindbrain fate. *Development* 126:4827–38

Logan C, Khoo WK, Cado D, Joyner AL. 1993. Two enhancer regions in the mouse En-2 locus direct expression to the mid/hindbrain region and mandibular myoblasts. *Development* 117:905–16

Logan C, Wizenmann A, Drescher U, Monschau B, Bonhoeffer F, Lumsden A. 1996. Rostral optic tectum acquires caudal characteristics following ectopic engrailed expression. *Curr. Biol.* 6:1006–14

Lun K, Brand M. 1998. A series of no isthmus (noi) alleles of the zebrafish pax2.1 gene reveals multiple signaling events in development of the midbrain-hindbrain boundary. *Development* 125:3049–62

MacArthur CA, Lawshe A, Shankar DB, Heikinheimo M, Shackleford GM. 1995a. FGF-8 isoforms differ in NIH3T3 cell transforming potential. *Cell Growth Differ.* 6:817–25

MacArthur CA, Lawshe A, Xu J, Santos-Ocampo S, Heikinheimo M, et al. 1995b. FGF-8 isoforms activate receptor splice forms that are expressed in mesenchymal regions of mouse development. *Development* 121:3603–13

Mansour SL, Goddard JM, Capecchi MR. 1993. Mice homozygous for a targeted disruption of the proto-oncogene int-2 have developmental defects in the tail and inner ear. *Development* 117:13–28

Martinez S, Crossley PH, Cobos I, Rubenstein JL, Martin GR. 1999. FGF8 induces formation of an ectopic isthmic organizer and isthmocerebellar development via a repressive effect on Otx2 expression. *Development* 126:1189–200

Martinez S, Wassef M, Alvarado-Mallart RM. 1991. Induction of a mesencephalic phenotype in the 2-day-old chick prosencephalon is preceded by the early expression of the homeobox gene en. *Neuron* 6:971–81

Maruoka Y, Ohbayashi N, Hoshikawa M, Itoh N, Hogan BLM, Furuta Y. 1998. Comparison of the expression of three highly related genes, Fgf8, Fgf17 and Fgf18, in the mouse embryo. *Mech. Dev.* 74:175–77

Mastick GS, Fan CM, Tessier-Lavigne M,

Serbedzija GN, McMahon AP, Easter SS Jr. 1996. Early deletion of neuromeres in Wnt-1−/− mutant mice: evaluation by morphological and molecular markers. *J. Comp. Neurol.* 374:246–58

Matsuo I, Kuratani S, Kimura C, Takeda N, Aizawa S. 1995. Mouse Otx2 functions in the formation and patterning of rostral head. *Genes Dev.* 9:2646–58

McMahon AP, Bradley A. 1990. The Wnt-1 (int-1) proto-oncogene is required for development of a large region of the mouse brain. *Cell* 62:1073–85

McMahon AP, Joyner AL, Bradley A, McMahon JA. 1992. The midbrain-hindbrain phenotype of Wnt-1-/Wnt-1-mice results from stepwise deletion of engrailed-expressing cells by 9.5 days postcoitum. *Cell* 69:581–95

Meinhardt H. 1983. Cell determination boundaries as organizing regions for secondary embryonic fields. *Dev. Biol.* 96:375–85

Meyers EN, Lewandoski M, Martin GR. 1998. An Fgf8 mutant allelic series generated by Cre- and Flp-mediated recombination. *Nat. Genet.* 18:136–41

Millen KJ, Hui CC, Joyner AL. 1995. A role for En-2 and other murine homologues of Drosophila segment polarity genes in regulating positional information in the developing cerebellum. *Development* 121:3935–45

Millen KJ, Wurst W, Herrup K, Joyner AL. 1994. Abnormal embryonic cerebellar development and patterning of postnatal foliation in two mouse Engrailed-2 mutants. *Development* 120:695–706

Millet S, Bloch-Gallego E, Simeone A, Alvarado-Mallart RM. 1996. The caudal limit of Otx2 gene expression as a marker of the midbrain/hindbrain boundary: a study using in situ hybridisation and chick/quail homotopic grafts. *Development* 122:3785–97

Millet S, Campbell K, Epstein DJ, Losos K, Harris E, Joyner AL. 1999. A role for Gbx2 in repression of Otx2 and positioning the mid/hindbrain organizer. *Nature* 401:161–64

Min H, Danilenko DM, Scully SA, Bolon B,

Ring BD, et al. 1998. Fgf-10 is required for both limb and lung development and exhibits striking functional similarity to Drosophila branchless. *Genes Dev.* 12:3156–61

Minowada G, Jarvis LA, Chi CL, Neubuser A, Sun X, et al. 1999. Vertebrate Sprouty genes are induced by FGF signaling and can cause chondrodysplasia when overexpressed. *Development* 126:4465–75

Nieto MA. 1999. Reorganizing the organizer 75 years on. *Cell* 98:417–25

Nusse R, Varmus HE. 1992. Wnt genes. *Cell* 69:1073–87

Okafuji T, Funahashi J, Nakamura H. 1999. Roles of Pax-2 in initiation of the chick tectal development. *Brain Res. Dev. Brain Res.* 116:41–49

Orr-Urtreger A, Bedford MT, Burakova T, Arman E, Zimmer Y, et al. 1993. Developmental localization of the splicing alternatives of fibroblast growth factor receptor-2 (FGFR2). *Dev. Biol.* 158:475–86

Perrimon N. 1994. The genetic basis of patterned baldness in Drosophila. *Cell* 76:781–84

Peters K, Ornitz D, Werner S, Williams L. 1993. Unique expression pattern of the FGF receptor 3 gene during mouse organogenesis. *Dev. Biol.* 155:423–30

Peters KG, Werner S, Chen G, Williams LT. 1992. Two FGF receptor genes are differentially expressed in epithelial and mesenchymal tissues during limb formation and organogenesis in the mouse. *Development* 114:233–43

Pfeffer PL, Bouchard M, Busslinger M. 2000. Pax2 and homeodomain proteins cooperatively regulate a 435 bp enhancer of the mouse Pax5 gene at the midbrain-hindbrain boundary. *Development* 127:1017–28

Pfeffer PL, Gerster T, Lun K, Brand M, Busslinger M. 1998. Characterization of three novel members of the zebrafish Pax2/5/8 family: dependency of Pax5 and Pax8 expression on the Pax2.1 (noi) function. *Development* 125:3063–74

Picker A, Brennan C, Reifers F, Clarke JD,

Holder N, Brand M. 1999. Requirement for the zebrafish mid-hindbrain boundary in midbrain polarisation, mapping and confinement of the retinotectal projection. *Development* 126:2967–78

Plotnikov AN, Schlessinger J, Hubbard SR, Mohammadi M. 1999. Structural basis for FGF receptor dimerization and activation. *Cell* 98:641–50

Reifers F, Bohli H, Walsh EC, Crossley PH, Stainier DY, Brand M. 1998. Fgf8 is mutated in zebrafish acerebellar (ace) mutants and is required for maintenance of midbrain-hindbrain boundary development and somitogenesis. *Development* 125:2381–95

Rhinn M, Dierich A, Shawlot W, Behringer RR, Le Meur M, Ang SL. 1998. Sequential roles for Otx2 in visceral endoderm and neuroectoderm for forebrain and midbrain induction and specification. *Development* 125:845–56

Ristoratore F, Carl M, Deschet K, Richard-Parpaillon L, Boujard D, et al. 1999. The midbrain-hindbrain boundary genetic cascade is activated ectopically in the diencephalon in response to the widespread expression of one of its components, the medaka gene Ol-eng2 *Development* 126:3769–79

Rowitch DH, McMahon AP. 1995. Pax-2 expression in the murine neural plate precedes and encompasses the expression domains of Wnt-1 and En-1. *Mech. Dev.* 52:3–8

Schwarz M, Alvarez-Bolado G, Dressler G, Urbanek P, Busslinger M, Gruss P. 1999. Pax2/5 and Pax6 subdivide the early neural tube into three domains. *Mech. Dev.* 82:29–39

Schwarz M, Alvarez-Bolado G, Urbanek P, Busslinger M, Gruss P. 1997. Conserved biological function between Pax-2 and Pax-5 in midbrain and cerebellum development: evidence from targeted mutations. *Proc. Natl. Acad. Sci. USA* 94:14518–23

Sekine K, Ohuchi H, Fujiwara M, Yamasaki M, Yoshizawa T, et al. 1999. Fgf10 is essential for limb and lung formation. *Nat. Genet.* 21:138–41

Shamim H, Mahmood R, Logan C, Doherty P,

Lumsden A, Mason I. 1999. Sequential roles for Fgf4, En1 and Fgf8 in specification and regionalisation of the midbrain. *Development* 126:945–59

Shimamura K, Rubenstein JL. 1997. Inductive interactions direct early regionalization of the mouse forebrain. *Development* 124:2709–18

Simeone A, Acampora D, Gulisano M, Stornaiuolo A, Boncinelli E. 1992. Nested expression domains of four homeobox genes in developing rostral brain. *Nature* 358:687–90

Simeone A, Acampora D, Mallamaci A, Stornaiuolo A, D'Apice MR, et al. 1993. A vertebrate gene related to orthodenticle contains a homeodomain of the bicoid class and demarcates anterior neuroectoderm in the gastrulating mouse embryo. *EMBO J.* 12:2735–47

Slack JM, Isaacs HV, Song J, Durbin L, Pownall ME. 1996. The role of fibroblast growth factors in early Xenopus development. *Biochem. Soc. Symp.* 62:1–12

Smith ST, Jaynes JB. 1996. A conserved region of engrailed, shared among all en-, gsc-, Nk1-, Nk2- and msh-class homeoproteins, mediates active transcriptional repression in vivo. *Development* 122:3141–50

Song DL, Chalepakis G, Gruss P, Joyner AL. 1996. Two Pax-binding sites are required for early embryonic brain expression of an Engrailed-2 transgene. *Development* 122:627–35

Song DL, Joyner AL. 2000. Two Pax2/5/8-binding sites in Engrailed-2 are required for proper initiation of endogenous midhindbrain expression. *Mech. Dev.* 90:155–65

Stark KL, McMahon JA, McMahon AP. 1991. FGFR-4, a new member of the fibroblast growth factor receptor family, expressed in the definitive endoderm and skeletal muscle lineages of the mouse. *Development* 113:641–51

Suda Y, Matsuo I, Aizawa S. 1997. Cooperation between Otx1 and Otx2 genes in developmental patterning of rostral brain. *Mech. Dev.* 69:125–41

Suda Y, Nakabayashi J, Matsuo I, Aizawa S. 1999. Functional equivalency between Otx2 and Otx1 in development of the rostral head. *Development* 126:743–57

Sun X, Meyers EN, Lewandoski M, Martin GR. 1999. Targeted disruption of Fgf8 causes failure of cell migration in the gastrulating mouse embryo. *Genes Dev.* 13:1834–46

Tanaka A, Miyamoto K, Minamino N, Takeda M, Sato B, et al. 1992. Cloning and characterization of an androgen-induced growth factor essential for the androgen-dependent growth of mouse mammary carcinoma cells. *Proc. Natl. Acad. Sci. USA* 89:8928–32

Thomas KR, Capecchi MR. 1990. Targeted disruption of the murine int-1 proto-oncogene resulting in severe abnormalities in midbrain and cerebellar development. *Nature* 346:847–50

Thomas KR, Musci TS, Neumann PE, Capecchi MR. 1991. Swaying is a mutant allele of the proto-oncogene Wnt-1. *Cell* 67:969–76

Torres M, Gomez-Pardo E, Gruss P. 1996. Pax2 contributes to inner ear patterning and optic nerve trajectory. *Development* 122:3381–91

Urbanek P, Fetka I, Meisler MH, Busslinger M. 1997. Cooperation of Pax2 and Pax5 in midbrain and cerebellum development. *Proc. Natl. Acad. Sci. USA* 94:5703–8

Urbanek P, Wang ZQ, Fetka I, Wagner EF, Busslinger M. 1994. Complete block of early B cell differentiation and altered patterning of the posterior midbrain in mice lacking Pax5/BSAP. *Cell* 79:901–12

Walshe J, Mason I. 2000. Expression of FGFR1, FGFR2 and FGFR3 during early neural development in the chick embryo. *Mech. Dev.* 90:103–110

Walther C, Gruss P. 1991. Pax-6, a murine paired box gene, is expressed in the developing CNS. *Development* 113:1435–49

Wassarman KM, Lewandoski M, Campbell K, Joyner AL, Rubenstein JL, et al. 1997. Specification of the anterior hindbrain and establishment of a normal mid/hindbrain organizer is dependent on Gbx2 gene function. *Development* 124:2923–34

Wassef M, Joyner AL. 1997. Early mesencephalon/metencephalon patterning and development of the cerebellum. *Perspect. Dev. Neurobiol.* 5:3–16

Weinstein M, Xu X, Ohyama K, Deng CX. 1998. FGFR-3 and FGFR-4 function cooperatively to direct alveogenesis in the murine lung. *Development* 125:3615–23

Wilkinson DG, Bailes JA, McMahon AP. 1987. Expression of the proto-oncogene int-1 is restricted to specific neural cells in the developing mouse embryo. *Cell* 50:79–88

Wingate RJ, Hatten ME. 1999. The role of the rhombic lip in avian cerebellum development. *Development* 126:4395–404

Wurst W, Auerbach AB, Joyner AL. 1994. Multiple developmental defects in Engrailed-1 mutant mice: an early mid-hindbrain deletion and patterning defects in forelimbs and sternum. *Development* 120:2065–75

Xu J, Lawshe A, MacArthur CA, Ornitz DM. 1999. Genomic structure, mapping, activity and expression of fibroblast growth factor 17. *Mech. Dev.* 83:165–78

Xu J, Liu Z, Ornitz DM. 2000. Temporal and spatial gradients of Fgf8 and Fgf17 regulate proliferation and differentiation of midline cerebellar structures. *Development* 127:1833–43

Xu X, Weinstein M, Li C, Naski M, Cohen RI, et al. 1998. Fibroblast growth factor receptor 2 (FGFR2)-mediated reciprocal regulation loop between FGF8 and FGF10 is essential for limb induction. *Development* 125:753–65

Yamaguchi TP, Conlon RA, Rossant J. 1992. Expression of the fibroblast growth factor receptor FGFR-1/flg during gastrulation and segmentation in the mouse embryo. *Dev. Biol.* 152:75–88

Yamaguchi TP, Harpal K, Henkemeyer M, Rossant J. 1994. fgfr-1 is required for embryonic growth and mesodermal patterning during mouse gastrulation. *Genes Dev.* 8:3032–44

Yuan S, Schoenwolf GC. 1999. The spatial and temporal pattern of C-Lmx1 expression in the neuroectoderm during chick neurulation. *Mech. Dev.* 88:243–47

Annu. Rev. Neurosci. 2001. 24:897–931

NEUROBIOLOGY OF PAVLOVIAN FEAR CONDITIONING

Stephen Maren

*Department of Psychology and Neuroscience Program, University of Michigan,
Ann Arbor, Michigan 48109-1109; e-mail: maren@umich.edu*

Key Words   learning, memory, amygdala, hippocampus, synaptic plasticity

■ **Abstract**   Learning the relationships between aversive events and the environmental stimuli that predict such events is essential to the survival of organisms throughout the animal kingdom. Pavlovian fear conditioning is an exemplar of this form of learning that is exhibited by both rats and humans. Recent years have seen an incredible surge in interest in the neurobiology of fear conditioning. Neural circuits underlying fear conditioning have been mapped, synaptic plasticity in these circuits has been identified, and biochemical and genetic manipulations are beginning to unravel the molecular machinery responsible for the storage of fear memories. These advances represent an important step in understanding the neural substrates of a rapidly acquired and adaptive form of associative learning and memory in mammals.

INTRODUCTION

> To sum up, we may legitimately claim the study of the formation and
> properties of conditioned reflexes as a special department of physiology.
>
> I. P. Pavlov, 1927

Seventy-five years ago, Ivan Petrovich Pavlov advocated the physiological analysis of the simple form of associative learning that carries his name: Pavlovian or classical conditioning (Pavlov 1927). If Pavlov were alive today, he would most certainly be impressed with the amazing progress we have made in delineating the brain circuits and neuronal mechanisms underlying Pavlovian conditioning in a variety of behavioral systems and species (Holland & Gallagher 1999, Krasne & Glanzman 1995, Thompson & Krupa 1994). One form of Pavlovian conditioning that has received considerable attention in the last 10 years is *fear conditioning* (Davis 1992, Fendt & Fanselow 1999, LeDoux 2000, Maren 1996). Simply stated, Pavlovian fear conditioning involves learning that certain environmental stimuli predict aversive events—it is the mechanism whereby we learn to fear people, places, objects, and animals. Evolution has crafted this form of learning to promote survival in the face of present and future threats, and it is an essential component of many mammalian defensive behavior systems (Fanselow 1994). Fear conditioning has attracted such great interest in recent years because it is squarely seated at the interface of memory and emotion (LeDoux 2000). Moreover, disturbances in fear conditioning may contribute to disorders of fear and anxiety in humans, such as panic disorder and specific phobias (Rosen & Schulkin 1998, Wolpe 1981).

John Watson and Rosalie Rayner's famous experiment with the infant, Albert B, is an instructive example of the Pavlovian fear conditioning procedure (Watson & Rayner 1920). In this experiment Watson and Rayner set out to condition fear to a white rat by sounding a loud and aversive noise after presenting the rat to "Little Albert." Before pairing the white rat with noise, the rat did not evoke fear in Albert (Figure 1A). Not surprisingly, the loud noise, which Watson generated by striking a hammer on a suspended steel bar, produced a robust fear response in Albert. Upon hearing the noise, Albert "startled violently" and "broke into a sudden crying fit." After several pairings of the rat and noise, Albert came to fear the rat. When Watson presented the rat to Albert after conditioning, Albert fell over, cried, and attempted to crawl away from the animal (Figure 1B). In Watson and Rayner's words, "This was as convincing a case of a completely conditioned fear response as could have been theoretically pictured." Although this experiment nicely illustrates fear conditioning, it is important to note that this type of experiment would not be acceptable by current ethical standards.

Appealing to the semantics of Pavlovian conditioning, Little Albert had learned that an innocuous conditional stimulus (CS; the white rat) predicted the occurrence of a noxious unconditional stimulus (US; the loud noise). Learning was manifest as a conditional response (crying) that, in this case, took the form of the unconditional response that was elicited by the loud noise prior to conditioning. Watson and Rayner's experiment with Albert exemplifies the traditional view of Pavlovian conditioning that one stimulus comes to evoke the response of another—the so-called conditioned reflex. However, as Rescorla has powerfully argued, current thinking holds that Pavlovian conditioning involves learning the hierarchical relationships among events (Rescorla 1988). Indeed, Pavlovian conditioning enables organisms to form neural representations of their worlds. Hence, the representation of the relations between aversive or traumatic events and the stimuli that predict them is at the core of Pavlovian fear conditioning.

The aim of this review is to describe recent developments in our understanding of the neurobiological basis of Pavlovian fear conditioning in mammals, including humans. The review focuses on work in rodent models, although data from other mammals and humans is included as necessary. It begins with a brief history of the brain and fear, proceeds with an outline of the neuroanatomical circuitry required for fear conditioning, and concludes with a discussion of the cellular and synaptic mechanisms within that circuitry that are responsible for the formation, storage, and expression of fear memories.

Figure 1   Still frames captured from a film of Watson and Rayner's famous experiment with the infant, Albert B ("Little Albert") (Watson 1920). (*A1-A3*) Watson presents Albert with a novel white rat. Albert responds with curiosity and reaches out to touch the animal. Although not shown, Watson subsequently paired presentations of the white rat with a loud and aversive noise. (*B1-B3*) Watson again presents Albert with the white rat after the rat had been paired with the loud noise. Unlike his initial reaction to the rat shown in *A*, Albert now responds to the rat with fear. He moves away from the rat and cries. (Still images courtesy of The Archives of the History of American Psychology, Akron, OH.)

THE BRAIN AND FEAR: HISTORICAL PERSPECTIVES

Our modern appreciation of the brain circuits involved in fear conditioning emerged from early observations of the effects of brain damage on emotional behavior in animals. In 1888, Brown & Schäfer (1888) described profound alterations in emotional reactivity following temporal lobe injuries in monkeys. Klüver & Bucy (1937) elaborated this effect in 1937. Both groups found that temporal lobe resections, which damaged both cortical and subcortical tissue, produced marked behavioral changes, including hyperorality, hypersexuality, visual agnosia, and notably, a loss of fear. For example, resected monkeys readily consumed novel and normally avoided foods, such as meat, and they would mouth inedible objects. Moreover, monkeys that once cowered in the presence of humans readily approached and contacted their caretakers after surgery. This work heralded the study of the neural substrates of emotion and focused intense interest on the role of the temporal lobes in the mediation of fear.

Subsequent work by Weiskrantz and others (Weiskrantz 1956, Zola Morgan et al 1991) demonstrated that the loss of fear in monkeys with temporal lobe lesions results from damage to the amygdala, a heterogeneous group of nuclei buried deep within the temporal lobes. Indeed, a recent study confirms that selective excitotoxic damage to amygdala neurons results in a fear reduction similar to that observed by Klüver & Bucy (Meunier et al 1999). Numerous other studies have demonstrated reduced fear ("taming") after amygdala damage in several mammalian species including rats, cats, rabbits, dogs, and humans (Goddard 1964). Moreover, both electrical stimulation of the amygdala and amygdaloid seizures are associated with autonomic and behavioral changes characteristic of fear (Davis 1992, Gloor 1960). Hence, consensus has emerged from these studies that the amygdaloid complex has an indispensable role in the regulation of fear.

Soon after the discovery of the amygdala's role in fear, several investigators set out to further quantify this function by employing learning and memory tasks. The earliest studies to investigate the involvement of the amygdala in fear-motivated learning used instrumental avoidance tasks, in which animals could avoid an aversive stimulus by making the appropriate behavioral response (Fonberg 1965, Horvath 1963, King 1958, Robinson 1963). For example, Brady and colleagues trained cats in a footshock-motivated shuttle avoidance task and found that large amygdala aspirations impaired the acquisition, but not retention, of the avoidance response (Brady et al 1954). In addition to instrumental learning, Pavlovian fear conditioning has been used to assess the involvement of the amygdala in emotional behavior. For example, Kellicutt & Schwartzbaum (1963) demonstrated a critical role for the amygdala in the acquisition of a conditioned emotional response, which they indexed by measuring bar-press suppression to a CS previously paired with shock. The Blanchards extended this work by demonstrating a direct role for the amygdala in the acquisition of contextual fear conditioning (Blanchard & Blanchard 1972), in which animals learn that the situational or contextual cues associated with conditioning predict the occurrence of footshock. Amygdala lesions completely eliminate shock-elicited freezing (somatomotor immobility), as well as unconditional freezing elicited by a predator (a cat) (Blanchard & Blanchard 1972). These studies established that forms of learning and memory that are motivated by fear require the amygdala.

In recent years several investigators have revealed an important role for the human amygdala in fear conditioning (Davidson & Irwin 1999). For example, a patient with bilateral amygdala pathology associated with the rare genetic disorder, Urbach-Wiethe disease, does not exhibit Pavlovian fear conditioning to either visual or auditory cues paired with loud noise (Bechara et al 1995). Patients who have received a unilateral amygdalectomy for the treatment of epilepsy also have deficits in auditory fear conditioning (LaBar et al 1995), and patients with amygdala damage fail to recognize fear in facial expressions (Adolphs et al 1995, 1999; Young et al 1995). Functional neuroimaging has extended these lesion studies by revealing amygdala activation to visual or vocal expressions of fear (Morris et al 1996, Phillips et al 1997, Whalen et al 1998) and during Pavlovian fear conditioning (Buchel et al 1999, LaBar et al 1998). Thus, the neural mechanisms of fear conditioning appear to exhibit homology across several mammalian species.

NEURAL SYSTEMS FOR FEAR CONDITIONING

In the light of this work, the analysis of the neural circuitry of fear conditioning has largely concentrated on describing the intrinsic anatomy of the amygdala and mapping the sensory afferents and motor efferents of the amygdala. In the rat the amygdala consists of several anatomically and functionally distinct nuclei, including (but not limited to) the lateral (LA), basolateral (BL), basomedial, and central (CE) amygdaloid nuclei (Brodal 1947, Krettek & Price 1978). Anatomical and behavioral evidence indicates that these nuclei are components of two distinct subsystems within the amygdala that are important for fear conditioning (LeDoux 1995, Maren & Fanselow 1996). The first subsystem of the amygdala is comprised of LA, BL, and basomedial. Collectively referred to as the basolateral complex (BLA), these nuclei form the primary sensory interface of the amygdala. Thus, selective lesions of the BLA produce severe deficits in both the acquisition and expression of Pavlovian fear conditioning independent of the stimulus modality used to train fear responses (Campeau & Davis 1995b, Cousens & Otto 1998, LeDoux et al 1990, Maren et al 1996a). Within the BLA, the LA appears to be essential for fear conditioning (Amorapanth et al 2000). Selective lesions of the BL do not impair fear conditioning but do attenuate the acquisition of instrumental avoidance behavior (Amorapanth et al 2000). Killcross and colleagues have reported a similar effect, although their lesions encompassed both LA and BL (Killcross et al 1997).

The second subsystem of the amygdala consists of the CE and it constitutes the amygdala's interface to fear response systems. For example, electrical stimulation of CE produces behavioral responses similar to those evoked by stimuli paired with shock (Iwata et al 1987, Kapp et al 1982). Lesions of the CE also produce profound deficits in both the acquisition and expression of conditional fear (Hitchcock & Davis 1986, Iwata et al 1986, Kim & Davis 1993, Roozendaal et al 1991, Young & Leaton 1996), and pharmacological studies suggest that this is due to a deficit in the performance of conditional fear responses, rather than an associative deficit (Fanselow & Kim 1994, Goosens et al 2000). Moreover, lesions placed in structures efferent to the CE, such as the lateral hypothalamus or periaqueductal grey, produce selective deficits in either cardiovascular or somatic conditional fear responses, respectively (Amorapanth et al 1999, De Oca et al 1998, LeDoux et al 1988). This suggests that the CE is the final common pathway for the generation of learned fear responses.

The major afferent and efferent projection systems of the amygdala that are relevant to fear conditioning are illustrated in Figure 2. For simplicity, this figure shows only unidirectional connections and ignores interconnections between extra-amygdaloid structures. It is readily appreciated that the BLA is a locus of convergence of afferents from both subcortical and cortical sensory regions (McDonald 1998, Pitkanen et al 1997, Swanson & Petrovich 1998). As such, projections from either the auditory thalamus or the auditory cortex to the BLA are essential for conditioning to auditory CSs (Campeau & Davis 1995a, LeDoux et al 1986, McCabe et al 1993, Romanski & LeDoux 1992), projections from the hippocampal formation to the BLA underlie conditioning to contextual CSs (Kim & Fanselow 1992, Maren 1999c, Maren & Fanselow 1995, Phillips & LeDoux 1992), and projections from the perirhinal cortex transmit visual CS information to the BLA (Campeau & Davis 1995a, Rosen et al 1992). Information about the aversive footshock US might reach the BLA via parallel thalamic and cortical pathways (Shi & Davis 1999). Consistent with this anatomy, single neurons in the BLA respond to auditory, visual, and somatic (shock) stimuli (Romanski et al 1993), which indicates that the amygdala is a locus of convergence for information about CSs and USs. Thus, the BLA is anatomically situated to integrate information from a variety of sensory domains.

Several models posit a role for the BLA in CS-US association during fear conditioning (Davis 1992, Fanselow & LeDoux 1999, Maren 1999a). In these models it is assumed that direct projections from the BLA to the CE enable associations established in the BLA to elicit fear responses via the CE. Indeed, the CE projects to several diverse brain areas involved in the generation of various fear responses. Hence, amygdala lesions block fear conditioning to contextual (Antoniadis & McDonald 2000, Maren 1998), auditory (Campeau & Davis 1995b, LeDoux et al 1990, Maren et al 1996a), olfactory (Cousens & Otto 1998), and visual CSs (Sananes & Davis 1992), and these deficits are manifest when one measures freezing (Cousens & Otto 1998, Maren et al 1996a), defecation (Antoniadis & McDonald 2000), hypoalgesia (Helmstetter 1992, Watkins et al 1993), potentiated acoustic startle (Campeau & Davis 1995b), increased heart rate (Antoniadis & McDonald 2000, LeDoux et al 1990), or corticosterone secretion (Goldstein et al 1996).

Association Formation and the Amygdala

An important goal of the neurobiological analysis of fear conditioning is to identify the essential substrate for the encoding and storage of fear memories (i.e. CS-US associations). There is now strong evidence that the amygdala, and the BLA in particular, is a locus for the formation and storage of CS-US associations during Pavlovian fear conditioning. This evidence has been obtained from studies employing permanent and reversible lesions of the amygdala and neurophysiological recordings of amygdala spike firing during learning.

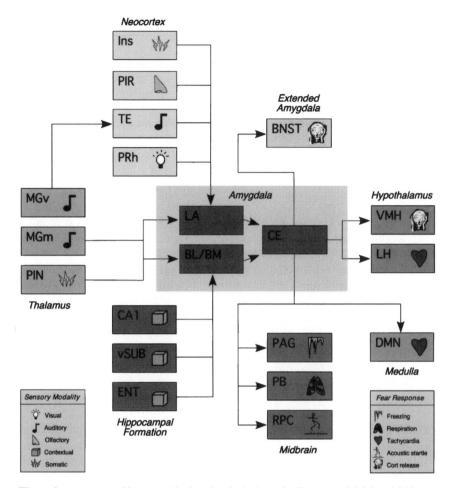

Figure 2 Anatomy of fear conditioning circuits in the brain. The amygdaloid nuclei (shown in the center) can be roughly divided into two subsystems. These include the lateral (LA), basolateral (BL), and basomedial (BM) nuclei, which together form the basolateral complex (BLA) and the central nucleus (CE). The BLA receives and integrates sensory information from a variety of sources. These include the medial and ventral divisions of the thalamic medial geniculate nucleus (MGm and MGv, auditory), the perirhinal cortex (PRh, visual), primary auditory cortex (TE), the insular cortex (INS, gustatory and somatosensory), the thalamic posterior intralaminar nucleus (PIN, somatosensory), the hippocampal formation (spatial and contextual) including area CA1, the ventral subiculum (vSUB), the entorhinal cortex (ENT), and the piriform cortex (PIR, olfactory). Thus, the BLA is a locus of sensory convergence and a plausible site for CS-US association within the amygdala. Intra-amygdaloid circuitry conveys the CS-US association to the CE, where divergent projections to the hypothalamus and brainstem mediate fear responses such as freezing (periaqueductal gray, PAG), potentiated acoustic startle (nucleus reticularis ponits caudalis, RPC), increased heart rate and blood pressure (lateral hypothalamus, LH; dorsal motor nucleus of the vagus, DMN), increased respiration (parabrachial nucleus, PB), and glucocorticoid release (paraventricular nucleus of the hypothalamus, PVN; bed nucleus of the stria terminalis, BNST). For simplicity, all projections are drawn as unidirectional connections, although in many cases these connections are reciprocal.

Permanent or Temporary Lesions of the Amygdala Disrupt the Acquisition and Expression of Conditional Fear   Selective neurotoxic lesions of the BLA severely attenuate the acquisition of fear conditioning to both contextual and discrete CSs when made before training (Cousens & Otto 1998, Maren et al 1996a, Sananes & Davis 1992). Moreover, neurotoxic BLA lesions completely abolish the expression of conditional fear responses when made either shortly after training (Campeau & Davis 1995b; Cousens & Otto 1998; Maren 1998, 1999b; Maren et al 1996a), two weeks following training (Cousens & Otto 1998, Maren et al 1996a), or even up to one month following training (Lee et al 1996, Maren et al 1996a). Posttraining neurotoxic BLA lesions also abolish conditional fear after extensive overtraining (Maren 1998, 1999b). It is important to note that neurotoxic BLA lesions do not affect footshock sensitivity nor do they alter baseline locomotor activity (Campeau & Davis 1995b, Maren 1998). Neurotoxic lesions of the CE also attenuate the acquisition and expression of fear conditioning (Campeau & Davis 1995b, Helmstetter 1992), but there is reason to believe that the CE is primarily involved in expressing, as opposed to encoding, CS-US associations (Fanselow & Kim 1994, Goosens et al 2000).

As mentioned earlier, it is well documented that amygdala damage disrupts not only learned fear, but also innate fear under some conditions. For example, rats with amygdala lesions do not exhibit freezing or analgesia in the presence of a cat (Blanchard & Blanchard 1972, Fox & Sorenson 1994), they show attenuated unconditional analgesia and heart rate responses to loud noises (Bellgowan & Helmstetter 1996, Young & Leaton 1996), and they exhibit reduced taste neophobia (Nachman & Ashe 1974). Amygdala damage does not disrupt all unconditional fear responses, however. Amygdala lesions do not affect either open arm avoidance in an elevated plus maze (Treit & Menard 1997, Treit et al 1993) or unconditional analgesia to shock (Watkins et al 1993). Nonetheless, the impact of amygdala lesions on unconditional fear raises questions regarding the nature of the deficits observed in associative tasks (Cahill et al 1999, Vazdarjanova 2000). It has been argued that deficits in conditional freezing in rats with neurotoxic BLA lesions, for example, may represent a deficit in performing the freezing response, as opposed to a deficit in learning and memory per se (Vazdarjanova & McGaugh 1998).

We have addressed this issue by submitting rats to an extensive overtraining procedure in which they receive more than 10 times the number of footshocks needed to produce asymptotic levels of freezing in controls (Maren 1998, 1999b). Under these conditions, rats with BLA lesions acquire the conditional freezing response and perform the response at the same high level as control subjects. It is noteworthy that the same overtraining procedure does not eliminate the severe deficits that are induced by posttraining BLA lesions, nor does it facilitate reacquisition of conditional fear during subsequent training (i.e. overtraining does not promote savings of the fear memory) relative to a naive group of animals (see Figure 3). These and other data reveal that fear conditioning deficits in rats with BLA lesions are not simply due to performance deficits. In contrast, they imply a role for the BLA in associative processes underlying fear conditioning (Maren 2000).

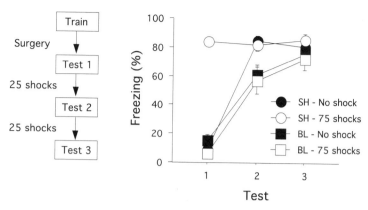

Figure 3 Overtraining does not mitigate the effects of posttraining basolateral amygdala (BLA) lesions. The left panel illustrates the phases of behavioral training and testing. All rats received fear conditioning, consisting of either 0 or 75 unsignaled footshocks in a novel chamber. For each training condition, the rats were further divided into those that received posttraining BLA lesions (BL–no shock, BL–75 shocks) or sham surgery (SH–no shock, SH–75 shocks). All rats then received three 4-min extinction tests, with 25 additional trials following both the first and second test. The right panel illustrates the mean (±) SEM percentage of freezing in these groups of rats. Posttraining BLA lesions completely eliminated conditional freezing measured during Test 1, despite presurgical overtraining. However, rats with BLA lesions were able to acquire conditional fear with additional training. BLA rats with presurgical overtraining (BL–75 shocks) acquired conditional freezing at the same rate as conditioning-naive BLA rats (BL–no shock). Overtraining did not yield savings of the fear memory in rats with BLA lesions. (Adapted from Maren 1999b.)

One experimental strategy that oversteps the problems associated with permanent brain lesions employs pharmacological agents to temporarily inactivate brain regions. This technique has now yielded important information concerning the role of the amygdala in the acquisition and expression of conditional fear. For example, inactivation of BLA neurons with muscimol, a GABA$_A$ receptor agonist, prevents both the acquisition and expression of fear conditioning (Helmstetter & Bellgowan 1994, Muller et al 1997, Wilensky et al 1999). In addition, muscimol only blocks conditioning when it is infused prior to training—immediate posttraining infusions of muscimol do not affect the acquisition of fear conditioning (Wilensky et al 1999). A similar outcome is obtained with intra-amygdaloid infusions of the N-methyl-D-aspartate (NMDA) receptor antagonist, APV (Maren et al 1996b). APV blocks the acquisition of conditional freezing when infused into the BLA before, but not immediately after, training. These results suggest that activity in BLA neurons is required when CS-US association occurs. Posttraining inactivation of the amygdala with either lidocaine or tetrodotoxin, which inhibits both cellular and axonal excitability, does in fact impair fear conditioning (Sacchetti et al 1999, Vazdarjanova & McGaugh 1999). This may indicate a role for amygdala neurons in consolidating fear memories (see below), although the effects of these drugs on *en passant* axons may be responsible for the observed deficits.

Fear Conditioning–Related Spike Firing in the Amygdala   Electrophysiological recordings of amygdaloid neuronal activity support a role for the amygdala in encoding and storing fear associations. In a series of elegant single-unit recording studies, LeDoux and colleagues have discovered that auditory fear conditioning induces short-latency plasticity in LA neurons (Quirk et al 1995, 1997). This plasticity takes the form of enhanced spike firing elicited by acoustic CSs. The short latency of learning-related changes in spike firing is consistent with plasticity in thalamo-amygdala projections, specifically, projections from the medial division of the medial geniculate nucleus (MGm). Amygdala neurons exhibit plasticity earlier in training than auditory cortical neurons, further suggesting that direct thalamo-amygdala projections, rather than cortico-amygdala projections, mediate neuronal plasticity in the LA (Quirk et al 1997). As shown in Figure 4, we have recently demonstrated that conditioning-related increases in CS-elicited spike firing in LA neurons are also evident in overtrained rats (Maren 2000). Again, the latency of peak conditional activity is consistent with plasticity in thalamo-amygdala projections. It is noteworthy that the amygdala is not essential for short-latency plasticity in the auditory cortex (Armony et al 1998), although the behavioral relevance of cortical plasticity in rats with amygdala damage is unclear.

In addition to enhancing CS-elicited spike firing in the amygdala, fear conditioning also increases anticipatory, pre-CS firing (Pare & Collins 2000). Lateral

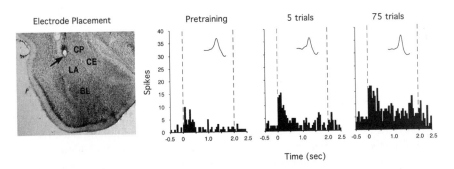

Figure 4   Conditioning-related plasticity in lateral amygdala neurons after overtraining. (*Leftmost panel*) Electrode placement in the dorsal division of the lateral amygdaloid nucleus. Spike firing rate from a single LAd neuron (*inset*, spike waveform) during three phases of training is shown in the three unit *histograms*. The *histograms* display the summed spike activity obtained during 10 auditory continual stimulus (CS) presentations (50-ms bins). Auditory CSs elicited spike firing in lateral amygdala neurons prior to fear conditioning (pretraining) 50–100 ms following CS onset (*dashed lines*, CS onset and offset). After five conditioning trials, significant increases in CS-elicited spike firing were observed in several post-CS bins, most notably the short-latency (0–50 ms) bin. Extensive overtraining (75 trials) did not mitigate the enhancement in short-latency, CS-elicited spike firing. (Adapted from Maren 2001a.)

amygdala neurons have also been found to exhibit discriminative plasticity in a differential fear-conditioning paradigm in cats (Collins & Pare 2000). In this case, LA neurons actually show decreases in spike firing elicited by the CS− over the course of training. The cellular mechanism for increases in firing to the CS+ on the one hand and decreases in firing to the CS− on the other are not known but may involve synaptic plasticity mechanisms such as long-term potentiation (LTP) and long-term depression (LTD) (Maren 1999a).

The critical role for the amygdala in both the acquisition and expression of Pavlovian fear conditioning implies that conditioning-related plasticity in amygdala neurons is due to local synaptic plasticity rather than passive transmission of plasticity from afferent brain areas. In the case of auditory fear conditioning, however, cellular plasticity develops in both the thalamic medial geniculate nucleus (medial division, MGm), which is the primary auditory afferent of the amygdala (Edeline & Weinberger 1992, McEchron et al 1995, Supple & Kapp 1989), and the auditory cortex (Weinberger 1995) after auditory fear conditioning. The latency of CS-elicited plasticity in LA is not consistent with transmission of plasticity from the cortex (Maren 2001a, Quirk et al 1997). However, transmission of plasticity from the MGm cannot be ruled out (Cahill et al 1999, Weinberger 1993). Indeed, MGm neurons are capable of LTP (Gerren & Weinberger 1983), and synaptic plasticity has been demonstrated to occur in the MGm during fear conditioning (McEchron et al 1996). Therefore, further studies are required to determine whether learning-induced changes in amygdala spike firing arise from local or remote synaptic plasticity.

Is the Amygdala Only Involved in Pavlovian Association Formation? The role of the amygdala in aversive conditioning is not limited to encoding and storing Pavlovian CS-US associations. There is substantial evidence that the amygdala is involved in consolidating memories for aversive experiences outside of the amygdaloid circuitry (Cahill & McGaugh 1998, McGaugh 2000). Amygdaloid involvement in memory consolidation is particularly robust for fear-motivated instrumental learning tasks (Liang et al 1982, Tomaz et al 1991), and the BL, in particular, has an important role in instrumental avoidance learning (Amorapanth et al 2000, Killcross et al 1997, Maren et al 1991, Poremba & Gabriel 1999). The role for the amygdala in Pavlovian association formation and memory consolidation is dissociable. For example, posttraining inactivation of the amygdala with muscimol produces deficits in the retention of inhibitory avoidance conditioning but not Pavlovian fear conditioning (Wilensky et al 2000). Therefore, the nature of the amygdala's involvement in aversive learning, whether it be local memory storage or remote memory consolidation, depends importantly on the associative structure of the conditioning situation (Kapp et al 1978).

Of course, it is also important to stress that the amygdala does not encode every aspect of an aversive learning experience. For example, humans with amygdala damage exhibit intact declarative memory for a fear conditioning experience, despite failing to exhibit conditional fear responses to stimuli paired with loud noise

(Bechara et al 1995). Similarly, rats with amygdala lesions avoid a compartment in which they have received footshock, despite failing to exhibit conditional freezing to the contextual cues associated with shock (Vazdarjanova & McGaugh 1998). These results indicate that multiple memory systems are engaged during relatively simple learning and memory tasks.

Context Processing and the Hippocampus

In a typical fear conditioning experiment rats acquire fear to not only the CS paired with the US, but also to the contextual cues associated with US delivery. As mentioned above, fear conditioning to both contextual and discrete CSs requires neurons in the amygdala. However, the neural pathways involved in processing these types of stimuli before they come into association with shock are quite different. Whereas information regarding discrete CSs appears to reach the amygdala via direct projections from primary sensory areas in both the thalamus and the cortex, information concerning contextual CSs is transmitted to the amygdala via multisensory brain areas. In fact, recent work has elucidated neural circuitry in the hippocampal formation that is responsible for assembling contextual representations and transmitting these representations to the amygdala for association with USs.

Contextual Encoding The first clues to the neural pathways involved in contextual fear conditioning came from a series of studies indicating that electrolytic lesions of the dorsal hippocampus (DH) prevented both the acquisition and expression of contextual fear conditioning (Kim & Fanselow 1992, Phillips & LeDoux 1992, Selden et al 1991). The impairment of contextual fear conditioning exhibits a temporal dependence typical of that found in human amnesia (Squire & Zola Morgan 1991). Hippocampal lesions only impaired the memory for contextual fear conditioning when made within one month of training (Kim & Fanselow 1992). This pattern of results has now been demonstrated using an elegant within-subjects design (Anagnostaras et al 1999a) and neurotoxic hippocampal lesions (Maren et al 1997). DH lesions tend to spare fear conditioning to auditory CSs (Anagnostaras et al 1999a; Kim & Fanselow 1992; Phillips & LeDoux 1992, 1994), although larger neurotoxic lesions that include the subiculum appear to produce deficits in auditory fear conditioning in many cases (Maren 1999c, Maren et al 1997, Richmond et al 1999). Nonetheless, there is considerable evidence that indicates that contextual and auditory fear conditioning are mediated, at least in part, by dissociable neural systems (Pugh et al 1997, Rudy 1993, Rudy et al 1999).

The fact that auditory conditioning is largely spared in rats with DH damage suggests that the DH does not play a direct role in CS-US association per se. What is the role, then, of the hippocampus in processing contextual information? One possibility is that the hippocampus is involved in mediating context-US associations. There is some evidence for this possibility (Frohardt et al 2000, Wilson et al 1995). However, there are many cases in which context-US associations are

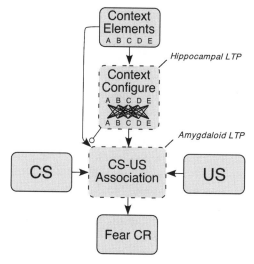

Figure 5   A model illustrating the roles for hippocampal and amygdaloid long-term potentiation (LTP) in Pavlovian fear conditioning. Contextual stimuli (elements) are assembled into configural representations in the hippocampus, and hippocampal LTP is posited as a mechanism underlying this process. Elemental or configural representations of context can come into association with footshock in the amygdala, although configural representations do so at the expense of elemental representations in intact animals (indicated by the inhibitory link, open circle). Discrete conditional stimuli (CS) and unconditional stimuli (US) converge upon amygdala neurons, and amygdaloid LTP is posited to play a role in CS-US association formation.

formed in animals with hippocampal damage (see below). Another possibility is that the hippocampus is required for assembling the elemental cues within a particular training context into a configural representation (Fanselow 1990) that then comes into association with footshock in the amygdala (see Figure 5). Support for this view comes from the finding that hippocampal lesion–induced deficits in contextual fear conditioning can be eliminated if preexposure to the context occurs one month prior to conditioning (Young et al 1994). Presumably, contextual conditioning (and the formation of a context-US association) proceeds normally in this case because the contextual representation was encoded and consolidated prior to the hippocampal damage.

Although initial reports found that electrolytic DH lesions produced impairments of contextual fear conditioning (Kim et al 1993a; Maren & Fanselow 1997; Phillips & LeDoux 1992, 1994), more recent reports indicate that axon-sparing neurotoxic lesions of the DH do not yield contextual conditioning deficits when made prior to training (Cho et al 1999, Frankland et al 1998, Gisquet-Verrier et al 1999, Maren et al 1997). However, posttraining neurotoxic DH lesions produce massive deficits in contextual fear conditioning (Maren et al 1997). The differential effects of pre- and posttraining neurotoxic DH lesions suggest the existence of alternate strategies for acquiring contextual fear (Maren et al 1997, 1998). We have proposed that intact rats use a hippocampal-dependent configural strategy in which a unified representation of the situational cues associated with training is assembled and associated with footshock. Indeed, recent work by Rudy and colleagues indicates that intact rats do use a configural strategy to acquire contextual fear conditioning (Rudy & O'Reilly 1999). Once acquired in this manner, the contextual fear memory is sensitive to posttraining hippocampal lesions. In contrast, rats with pretraining hippocampal lesions do not use a configural strategy and they default to a hippocampus-independent elemental strategy in which individual cues in the context come into association with footshock. Both strategies can be used to successfully acquire contextual fear, although only the configural strategy requires hippocampal involvement and it is presumed to operate at the expense of the elemental strategy in intact rats (Figure 5) (Anagnostaras et al 2001, Fanselow 2000, Maren et al 1997, 1998).

Why then do rats with pretraining electrolytic DH lesions have impairments in the acquisition of contextual fear conditioning? Deficits in the acquisition of contextual fear conditioning in rats with electrolytic DH lesions appear to be the result of a disruption of connections between the ventral subiculum and the nucleus accumbens. We have found that electrolytic or neurotoxic lesions of the ventral subiculum, a major afferent of the nucleus accumbens, produce contextual fear conditioning deficits (Maren 1999c). Lesions of the fornix, the tract through which subiculo-accumbens fibers travel, also impair contextual fear conditioning (Maren & Fanselow 1997, Phillips & LeDoux 1995). Moreover, Westbrook and colleagues have found that pharmacological inactivation of the accumbens produces selective deficits in the acquisition of contextual fear conditioning (Haralambous & Westbrook 1999, Westbrook et al 1997; see also Riedel et al 1997). Disruption of subiculo-accumbens projections appears to disregulate exploratory behavior and may interfere with the process by which rats sample contextual cues in their environment (Fanselow 2000, Maren 1999c).

Needless to say, the pattern of deficits following damage to the hippocampal system is complicated. Some have argued for a more parsimonious account of these data that centers around the influence of hippocampal damage on the performance of behavioral responses commonly used to assess fear, such as freezing (Gewirtz et al 2000). For example, Davis and colleagues have found that hippocampal lesions do not affect either contextual fear conditioning (McNish et al 1997) or contextual blocking (McNish et al 2000) assessed by measuring fear-potentiated startle. To account for this pattern of results, they and others have argued that deficits in contextual conditioning are due to a disruption of freezing behavior by the locomotor hyperactivity that typically accompanies hippocampal damage (Gewirtz et al 2000, Good & Honey 1997). However, it is clear that freezing-performance deficits are not sufficient to account for the full range of deficits associated with hippocampal damage (Anagnostaras et al 2001, Maren et al 1998). For example, rats with DH lesions exhibit a robust impairment for freezing when tested in a context in which they received shock 1 day before the lesion, but exhibit normal and high levels of freezing when tested in a context in which they received shock 50 days before the lesion (Anagnostaras et al 1999a). Because the same rats are freezing at high levels in one context and at low levels in another context, one cannot explain their deficits in expression of the recent memory in terms of a freezing-performance problem. Although hippocampal damage surely interacts with freezing behavior, there is strong evidence that such damage is also characterized by associative deficits in contextual fear conditioning.

Contextual Memory Retrieval   In addition to its role in encoding contextual representations, we have recently discovered an important role for the hippocampus in the regulation of memory retrieval by context (Holt & Maren 1999, Maren & Holt 2000). In many Pavlovian conditioning paradigms a CS can acquire more than one meaning. For example, in latent inhibition a phase of CS-alone presentations (CS preexposure) precedes the conditioning phase, in which the CS is paired with the US. In this paradigm the CS comes to have two meanings: It first comes to predict nothing (e.g. CS–no event) and subsequently comes to predict the US (e.g. CS–shock). Latent inhibition is characterized by a reduction in conditional responding to the CS that is produced by interference between these two conflicting memories (Bouton 1993). Contextual cues can be used to disambiguate these competing memories. Hence, if preexposure and conditioning occur in different contexts, latent inhibition is greatly reduced. Moreover, if an animal is returned to the preexposure context after the training phase latent inhibition is renewed. That is, conditional responding to the CS is once again reduced if testing occurs in the context of preexposure. It is apparent that animals can use contextual cues to retrieve the meaning of the CS appropriate to that context.

Early theories of hippocampal function posited a role for the hippocampus in this type of contextual memory retrieval (Hirsh 1974). However, recent investigations of retrieval phenomena using permanent hippocampal lesions have yielded mixed results (Frohardt et al 2000, Honey & Good 1993, Wilson et al 1995). To further investigate the role of the hippocampus in memory retrieval, we used muscimol to reversibly inactivate hippocampal neurons during a latent inhibition retrieval test (Holt & Maren 1999). The use of a reversible lesion technique for these experiments was critical because it allowed us to selectively target the hippocampus during retrieval testing. As previously reported, we found that control rats exhibited robust contextual retrieval. That is, they showed attenuated conditional responding when exposed to the CS in the context of preexposure and high levels of responding when the CS was tested in a context different from that of preexposure. It is important that hippocampal inactivation eliminated the contextual regulation of conditional responding—rats receiving intrahippocampal muscimol infusions exhibited low levels of conditional responding in both test contexts. They were unable to use contextual cues to regulate performance of the different CS memories and in fact, performed purely according to the sum of their experiences with the CS. Because they had 30 times as many CS–no event trials as CS-US trials, they performed according to the former association and exhibited little conditional responding (Maren & Holt 2000). A similar role for the hippocampus in contextual retrieval has been demonstrated in studies of human memory (Chun & Phelps 1999). Together, these studies are beginning to open up a new realm of hippocampal function in the processing of contextual cues, conditional relations, and high-order learning phenomenon such as occasion setting (Holland & Bouton 1999, Honey & Good 2000).

Fear Inhibition and the Prefrontal Cortex

In addition to understanding the processes by which fear memories are established and expressed, there is considerable interest in the mechanisms by which fear memories are inhibited. Understanding fear reduction has important clinical implications for treating disorders of fear and anxiety, such as posttraumatic stress disorder and panic disorder. There is considerable evidence implicating the prefrontal cortex (PFC) in emotional processes (Davidson & Irwin 1999), and there is an emerging, but complicated, literature suggesting a role for the PFC in the inhibition of conditional fear. For example, PFC lesions have been reported to attenuate extinction of fear under some conditions (Morgan & LeDoux 1995, Morgan et al 1993). However, others have not found an effect of PFC lesions on extinction (Gewirtz et al 1997). Recently, Quirk and colleagues reported that PFC lesions do not affect the acquisition or extinction of conditional fear per se, but either impair consolidation or affect the contextual modulation of the extinction memory (Quirk et al 2000).

One paradigm that has been adopted to study fear reduction is conditioned inhibition. A number of brain structures known to be important for excitatory fear conditioning, including the amygdala, perirhinal cortex, and medial geniculate nucleus, do not appear to be involved in conditioned inhibition of fear (Falls et al 1997, Falls & Davis 1995, Heldt & Falls 1998). Moreover, the PFC does not appear to be required for conditioned inhibition (Gewirtz et al 1997, Vouimba et al 2000), although recordings of prefrontal cortical activity reveal that there is an amygdala-dependent reduction of spike firing to a conditioned inhibitor (Garcia et al 1999). Thus, the neural substrates underlying the inhibition of fear remain elusive.

SYNAPTIC AND MOLECULAR MECHANISMS OF FEAR CONDITIONING

Considerable progress in mapping the neural circuitry underlying fear conditioning has opened the door to analyses of the synaptic and molecular mechanisms underlying the formation and storage of fear memories. In general, the focus of these studies has been to describe the properties of synaptic plasticity in the amygdala and hippocampus, to examine whether fear conditioning is accompanied by synaptic plasticity in these brain structures, and to investigate the influence on fear conditioning of manipulations that perturb synaptic plasticity in fear conditioning circuits. The evidence supports the view that synaptic plasticity in the amygdala is involved in CS-US association, whereas synaptic plasticity in the hippocampus is involved in contextual encoding.

Long-term potentiation (LTP) is the prototypical form of enduring synaptic plasticity in the mammalian brain (Bliss & Collingridge 1993, Maren & Baudry 1995, Martin et al 2000). It was first discovered in the hippocampus (Bliss & Lomo 1973), and has now been demonstrated to occur at synapses in the amygdala (Chapman et al 1990, Clugnet & LeDoux 1990, Maren & Fanselow 1995). Several properties of LTP, such as its rapid induction and associativity, make it an ideal candidate for encoding Pavlovian fear memories (Fanselow 1993, Maren 1999a, Rogan & LeDoux 1996). Although there is considerable debate concerning the role of LTP in learning and memory (Izquierdo & Medina 1995, Martin et al 2000, Shors & Matzel 1997), we have argued that Pavlovian fear conditioning is the ideal model system for examining the LTP-learning connection (Maren 2001b).

Glutamate Receptors and Fear Conditioning

The first series of studies to implicate LTP in Pavlovian fear conditioning used antagonists of the NMDA subclass of glutamate receptors. NMDA receptors are required for the induction of some forms of LTP in both the hippocampus (Bliss & Collingridge 1993, Maren & Baudry 1995) and the amygdala (Huang & Kandel 1998, Maren & Fanselow 1995). In a groundbreaking study, Davis and colleagues demonstrated that infusion of the NMDA receptor antagonist, APV, into the BLA prevents the acquisition of conditional fear to a visual CS in a fear-potentiated startle paradigm (Miserendino et al 1990). The attenuation of fear conditioning by APV was dose-dependent and was not due to an APV-induced decrease in footshock sensitivity. It is important to note that APV infusion into the BLA before testing did not affect the performance of a fear conditional response acquired in an earlier phase of training. Furthermore, APV infusion into the cerebellar interpositus nucleus, a brain structure that is not required for fear conditioning, did not affect acquisition of fear-potentiated startle. Subsequent work demonstrated that intra-amygdala APV also blocks the acquisition, but not the expression, of fear-potentiated startle to acoustic CSs (Campeau et al 1992). The deleterious effect of APV on fear-potentiated startle acquisition has also been demonstrated for second-order conditioning, suggesting that APV impairs fear conditioning by attenuating an associative mechanism, rather than affecting CS or US processing per se (Gewirtz & Davis 1997).

The effects of intra-amygdala APV have also been examined in the conditional-freezing paradigm (Maren et al 1996b). We have reported that infusions of APV into the BLA before fear conditioning produce a robust impairment in the acquisition of conditional freezing measured either immediately after footshock or 24 hours following conditioning. APV only blocked conditioning when it was infused into the BLA before training; immediate posttraining infusions of APV did not affect the acquisition of conditional freezing (Maren et al 1996b). However, unlike the results obtained from the fear-potentiated startle paradigm, we found that the effects of APV were not specific to acquisition; the expression of a previously acquired fear conditional response was also impaired by APV (Maren et al 1996b). This pattern of results has recently been replicated (Lee & Kim 1998) and may be due to the influence of NMDA receptor antagonists on evoked potentials in the amygdala (Li et al 1995, Maren & Fanselow 1995). Thus, it appears that amygdaloid NMDA receptor activation has a general role in the acquisition of fear CRs and a selective role in the expression of the conditional freezing (see Lee & Kim 1998 for a discussion of this issue). Intracerebroventricular administration of APV (Fanselow et al 1994; Kim et al 1991, 1992) and intrahippocampal APV infusions (Young et al 1994) indicate that hippocampal NMDA receptors appear to have a more selective role in contextual fear conditioning.

Recent work indicates that, as in the hippocampus (Grover & Teyler 1990), there are forms of amygdaloid LTP that do not depend on NMDA receptor activation (Chapman & Bellavance 1992, Weisskopf et al 1999) but do require the activation of voltage-gated calcium channels (Weisskopf et al 1999). Calcium channel–dependent plasticity may play an important role in fear conditioning, although this possibility has yet to be explored. Similarly, non-NMDA (i.e. AMPA) receptors also play an important role in fear conditioning. For example, Davis and colleagues have shown that intra-amygdala infusion of AMPA receptor antagonists impair both the acquisition and expression of fear-potentiated startle (Kim et al 1993b, Walker & Davis 1997). Additionally, LeDoux and colleagues have shown that AMPA receptor agonists infused into the amygdala prior to training enhance the acquisition of conditional freezing (Rogan et al 1997a). Recent studies also suggest that both cholinergic (Anagnostaras et al 1999b, Rudy 1996) and dopaminergic (Guarraci et al 1999, Nader & LeDoux 1999) neurotransmission play a role in the acquisition and expression of conditional fear. It is therefore likely that several interacting neurochemical systems regulate the synaptic plasticity in the amygdala that is critical for fear conditioning.

Synaptic Plasticity in Fear Conditioning Circuits

The foregoing studies indicate that both hippocampal and amygdaloid NMDA receptors are involved in the acquisition of Pavlovian fear conditioning in rats. These results implicate NMDA receptor–dependent LTP in these brain areas in the acquisition of conditional fear. A number of studies have used a correlational approach to examine the role of hippocampal LTP in contextual fear conditioning. Moreover, several studies have directly assessed amygdaloid synaptic transmission during, or shortly after, fear conditioning. These studies suggest that hippocampal LTP is involved in encoding contextual representations, whereas amygdaloid LTP is involved in the formation and storage of CS-US associations (Figure 5).

Correlations Between Hippocampal LTP and Contextual Fear Conditioning
To explore the relationship between hippocampal LTP and Pavlovian fear conditioning, we examined the influence of behavioral manipulations that enhance

learning rate on both the induction of hippocampal LTP and the acquisition of contextual fear conditioning. We submitted rats to acute water deprivation and found that deprivation reliably enhanced the magnitude of hippocampal LTP induced by high-frequency stimulation and augmented the rate of contextual fear conditioning; deprivation did not augment auditory fear conditioning (Maren et al 1994a,b; Maren & Fanselow 1998). Similar correlations between hippocampal LTP induction and contextual fear conditioning have emerged from studies of sex differences in LTP and fear conditioning (Anagnostaras et al 1998, Maren et al 1994c). Moreover, synaptic plasticity in hippocampo-septal projections has also been found to play a role in contextual fear conditioning (Garcia & Jaffard 1992, Vouimba et al 1998). These correlations are consistent with hypotheses that invoke hippocampal LTP as a mechanism for contextual fear conditioning (Fanselow 1997, Maren 1997). More specifically, these results suggest that hippocampal LTP has an important role in processing contextual CSs and may be involved in establishing configural representations of contextual stimuli (Maren 2001b).

Fear Conditioning Induces LTP in the Amygdala   As indicated above, the blockade of fear conditioning by NMDA receptor antagonists in the amygdala suggests that amygdaloid LTP mediates fear conditioning (Maren 1999a). This possibility has received support from a series of experiments performed by LeDoux and colleagues. Rogan & LeDoux (1995) found that induction of LTP at thalamo-amygdaloid synapses in vivo potentiates auditory evoked potentials in the amygdala that use this pathway. Auditory evoked potentials in the thalamo-amygdaloid pathway were also augmented during the acquisition of auditory fear conditioning (Rogan et al 1997b). The similar increase in auditory evoked potentials in the amygdala following both tetanic LTP induction and fear conditioning suggests that LTP-like increases in thalamo-amygdaloid synaptic transmission contribute to the acquisition of auditory fear conditioning.

McKernan & Shinnick-Gallagher (1997) have shown that fear conditioning enhances the amplitude of synaptic currents in amygdaloid neurons in vitro. Rats receiving paired CS-US trials, but not those receiving unpaired trials, exhibited a marked increase in synaptic currents evoked in amygdaloid neurons by stimulation of thalamic afferents. This increase in synaptic transmission was due to an elevation of presynaptic neurotransmitter release. Synaptic transmission in the endopyriform nucleus, which is not believed to play a role in fear conditioning, was not altered by the conditioning procedures. Insofar as tetanus-induced amygdaloid LTP is associated with both increased evoked responses and enhanced neurotransmitter release (Huang & Kandel 1998, Maren & Fanselow 1995), it would appear that fear conditioning induces a form of "behavioral" LTP. Further studies are required to determine whether these forms of plasticity share common cellular mechanisms.

Long-Term Depression   In addition to LTP, both the amygdala and hippocampus exhibit use-dependent decreases in synaptic efficacy under some conditions (Bramham & Srebro 1987, Heinbockel & Pape 2000, Li et al 1998, Wang & Gean

1999). The precise role for long-term depression (LTD) in fear conditioning is not known, although one can certainly imagine the necessity for bidirectional synaptic plasticity in fear conditioning circuits. In the amygdala LTD may be responsible for limiting synaptic transmission in CS pathways that are uncorrelated or anticorrelated with US occurrence. For example, amygdala neurons decrease their firing to a CS that is not explicitly paired with a US (Collins & Pare 2000). In general, amygdaloid LTD may be a mechanism whereby stimuli acquire inhibitory properties, which is consistent with a role for the amygdala in some forms of inhibitory learning, such as extinction (Falls et al 1992).

Molecular Cascades for Fear Memories

The elaboration of LTP as a mechanism for fear conditioning has been fostered not only by pharmacological and electrophysiological experiments but also by a new breed of experimentation that is driven by our expanding knowledge of intracellular signal transduction pathways and the molecular genetics of these pathways. These studies have taken two approaches. The first is a standard pharmacological approach, in which various components of the signal transduction cascade associated with LTP, for example, are targeted with drugs in behaving animals. The second approach takes advantage of powerful new molecular techniques to disable, eliminate, or even enhance key proteins associated with synaptic plasticity.

Protein Kinase Inhibitors   It is well documented that NMDA receptor activation is only the first step in a biochemical cascade that ultimately leads to synaptic modification. Activation of intracellular protein kinases, which are stimulated by NMDA receptor activation, is essential for the induction of LTP in both the hippocampus and amygdala (Huang & Kandel 1998, Huang et al 2000). Examinations of the role for protein kinases in fear conditioning are in their infancy, but there is already evidence that various kinases are required for establishing long-term fear memories. For instance, Kandel and colleagues have shown that posttraining intracerebroventricular (ICV) administration of protein kinase A (PKA) inhibitors impairs memory consolidation for contextual fear conditioning (Bourtchouladze et al 1998). Likewise, LeDoux and colleagues have found that posttraining ICV administration of PKA and mitogen-activated protein kinase (MAPK) inhibitors disrupts memory for contextual and auditory fear conditioning (Schafe et al 1999).

We have recently examined the influence of intra-amygdala infusions of H7, an inhibitor of protein kinase C (PKC) and PKA. This procedure allowed us to address the question of whether the attenuation of fear conditioning observed after ICV administration of kinase inhibitors was due to an effect on amygdaloid kinase activity. Consistent with the ICV data, we found that intra-amygdala infusions of H7 selectively inhibited the formation of long-term fear memories (Figure 6B)—short-term fear memories were spared (Figure 6A). Moreover, we found that H7 only affected long-term memory formation when infused into the BLA; CE infusion of H7 did not attenuate fear conditioning (Figure 6C). The

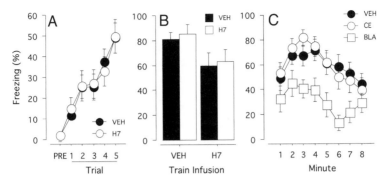

Figure 6 Intra-amygdala infusion of the protein kinase inhibitor H7 into the basolateral amygdala (BLA) selectively attenuates the acquisition of long-term, but not short-term, conditional fear memories. All panels display mean ($\pm$SEM) percentage of freezing. (*A*) Immediate postshock freezing during the conditioning session is not affected by intra-BLA infusions of H7. (*B*) Contextual freezing expressed 24 hours after fear conditioning is attenuated by pretraining, but not pretesting, (VEH, dark bar; H7, open bar) infusions of H7. This effect was not state dependent. (*C*) Infusions of H7 into the central nucleus (CE) did not affect the acquisition of contextual freezing. (Adapted from Goosens et al 2000.)

effect of H7 in the BLA on fear conditioning was not modality specific insofar as both auditory and contextual fear memories were attenuated. Also, it should be noted that H7 did not affect the expression of already learned fear memories (Figure 6*B*). These data suggest that protein kinase activation in the BLA is required for consolidating long-term fear memories. Ultimately, the consolidation (and reconsolidation) of fear memories requires de novo protein synthesis, insofar as intra-amygdala infusions of protein synthesis inhibitors, such as anisomycin (Nader et al 2000), and mRNA synthesis inhibitors, such as actinomycin-D (Bailey et al 1999), impair long-term memory formation. The induction of immediate early genes, such as *c-fos* and *zif268*, in the amygdala after fear conditioning may be a key component of the molecular cascade that leads to protein synthesis–dependent memory consolidation (Beck & Fibiger 1995, Campeau et al 1991, Malkani & Rosen 2000, Rosen et al 1998).

Genetically Modified Mice   As indicated above, it is clear that the NMDA receptor plays an important role in Pavlovian fear conditioning. Recently, transgenic techniques have been used to manipulate NMDA receptors in the hippocampus (Tsien et al 1996). Elimination of key NMDA receptor subunits in mice has been found to attenuate the acquisition of contextual fear conditioning (Kiyama et al 1998). Recent work indicates that trace fear conditioning, which is dependent on the hippocampus (McEchron et al 1998), is also impaired in mice that lack hippocampal NMDA receptors (Huerta et al 2000). Contextual fear conditioning deficits in hippocampal NMDA receptor–knockout mice are overcome by environmental enrichment (Rampon et al 2000). Perhaps one of the more interesting results to emerge in recent years is the finding that mice that overexpress the NMDA receptor 2B subunit, which prolongs the activation of NMDA receptors, actually exhibit enhanced hippocampal LTP and contextual fear conditioning (Tang et al 1999). Collectively, these data confirm pharmacological data that indicate an important role for NMDA receptors in synaptic plasticity and fear conditioning. Unfortunately, amygdala-specific NMDA receptor knockouts have not been developed. However, as discussed above, pharmacological and electrophysiological data clearly support a role for amygdaloid NMDA receptors and LTP in Pavlovian fear conditioning.

Studies using genetically modified mice also implicate various protein kinases that are linked to NMDA receptor activation in both LTP and learning and memory. Mice that lack PKCγ exhibit mild deficits in contextual, but not auditory, fear conditioning (Abeliovich et al 1993b). They also exhibit normal immediate postshock freezing, suggesting that only their long-term memory for contextual fear is impaired. The selective deficit in contextual conditioning is interesting insofar as these mice also exhibit impairments in hippocampal LTP (Abeliovich et al 1993a). Mice that lack the β isoform of PKC mice exhibit normal hippocampal LTP, but they exhibit robust impairments in both auditory and contextual fear conditioning (Weeber et al 2000). A similar pattern of behavioral results has been observed in mice that either express an inhibitory form of the regulatory subunit for PKA or overexpress Ca^{2+}-calmodulin-dependent protein kinase II , which also play a role in the induction of LTP (Abel et al 1997, Mayford et al 1996). These transgenic mouse strains exhibit long-term, but not short-term, impairments in both contextual and auditory fear conditioning, and both transgenic strains exhibit deficits in hippocampal LTP induction. Unfortunately, amygdaloid LTP was not examined in any of these studies. Deficits in amygdaloid LTP might account for the global fear conditioning impairments in PKCβ, PKA, and Ca^{2+}-calmodulin-dependent protein kinase II mice. Nonetheless, these findings are consistent with the involvement of protein kinases in both synaptic plasticity and fear conditioning. Indeed, our pharmacological data indicate that PKC and PKA activity in the amygdala is critical for both auditory and contextual fear conditioning (Goosens et al 2000).

Another recent series of studies has examined the influence of a targeted mutation of the cAMP-responsive, element-binding (CREB) protein, which is a transcription factor thought to play an important role in establishing long-term memories, on both fear conditioning and LTP. Mice with a disruption of the α and δ isoforms of CREB exhibit robust impairments in both contextual and auditory fear conditioning (Bourtchuladze et al 1994). These impairments are time dependent, insofar as freezing to both contexts and tones is intact when measured within 30 or 60 minutes of training, respectively. However, conditional freezing is nearly absent at long (24 hour) retention intervals. Thus, CREB mutants are capable of normal freezing under some conditions, and the time-dependent loss of conditional freezing over long retention intervals indicates that CREB is essential for consolidating long-term fear memories. In parallel with the time course of fear conditioning deficits, mice that lacked CREB also exhibited impairments in a late phase of hippocampal LTP; short-lasting, posttetanic potentiation is not impaired in these mice (Bourtchuladze et al 1994). The time period over which LTP decayed appears to parallel the time period over which fear memories are lost in CREB mutants. Not all CREB mutants exhibit impaired learning and synaptic plasticity, however. A recent study has found that mice expressing a dominant-negative form of CREB exhibit normal amygdaloid LTP and only minimal fear conditioning deficits in one of three transgenic lines (Rammes et al 2000). Nonetheless, Impey et al (1998) recently reported that both contextual and auditory fear conditioning rapidly induce CREB in the hippocampus and amygdala.

A more direct demonstration of a specific role for amygdaloid LTP in fear conditioning is revealed by studies of mice that lack Ras-GRF, a neuron-specific guanine nucleotide–releasing factor that is activated by both Ca^{2+} and G-protein-coupled messengers. Electrophysiological recordings from brain slices obtained from mice lacking Ras-GRF indicate a pronounced deficit in the induction of LTP in the BL (Brambilla et al 1997). These mice also exhibit impairments in consolidating long-term fear memories for both contextual and acoustic stimuli. These deficits in LTP and learning are selective for the amygdala and Pavlovian fear conditioning insofar as both hippocampal LTP and spatial learning in Ras-GRF knockouts are normal (Brambilla et al 1997). Ras-GRF modulates CREB activity through the MAPK pathway, and a role for MAPK in fear conditioning has recently been demonstrated (Atkins et al 1998, Schafe et al 1999). Together, these results provide strong support for the view that synaptic LTP in the amygdala is required for the establishment and maintenance of emotional memories. Further studies are required, however, to more precisely specify the role for amygdaloid and hippocampal synaptic plasticity in Pavlovian fear conditioning.

CONCLUSIONS

Pavlovian fear conditioning has undergone an extensive neurobiological analysis in recent years. This analysis has revealed that the amygdala and hippocampus are critical components of the neural circuitry underlying association formation and contextual processing, respectively, during fear conditioning. Moreover, synaptic plasticity mechanisms, such as LTP, in the hippocampus and amygdala play distinct and critical roles in these processes (see Figure 5). Most recently, molecular techniques are beginning to unravel the intracellular cascades that underlie the formation and storage of fear memories. Collectively, these advances yield great promise for understanding the neurobiology of learning and memory, in general, and in understanding the neurobiological basis of disorders of fear and anxiety in humans.

ACKNOWLEDGMENTS

This work was supported by the National Institute of Mental Health (R29MH-57865). I thank Ki Goosens for commenting on an earlier draft of the manuscript.

LITERATURE CITED

Abel T, Nguyen PV, Barad M, Deuel TA, Kandel ER, et al. 1997. Genetic demonstration of a role for PKA in the late phase of LTP and in hippocampus-based long-term memory. *Cell* 88:615–26

Abeliovich A, Chen C, Goda Y, Silva AJ, Stevens CF, et al. 1993a. Modified hippocampal long-term potentiation in PKC gamma-mutant mice. *Cell* 75:1253–62

Abeliovich A, Paylor R, Chen C, Kim JJ, Wehner JM, et al. 1993b. PKC gamma mutant mice exhibit mild deficits in spatial and contextual learning. *Cell* 75:1263–71

Adolphs R, Tranel D, Damasio H, Damasio AR. 1995. Fear and the human amygdala. *J. Neurosci.* 15:5879–91

Adolphs R, Tranel D, Hamann S, Young AW, Calder AJ, et al. 1999. Recognition of facial emotion in nine individuals with bilateral amygdala damage. *Neuropsychologia* 37:1111–17

Amorapanth P, LeDoux JE, Nader K. 2000. Different lateral amygdala outputs mediate reactions and actions elicited by a fear-arousing stimulus. *Nat. Neurosci.* 3:74–79

Amorapanth P, Nader K, LeDoux JE. 1999. Lesions of periaqueductal gray dissociate-conditioned freezing from conditioned suppression behavior in rats. *Learn. Mem.* 6:491–99

Anagnostaras SG, Gale GD, Fanselow MS. 2001. The hippocampus and contextual fear conditioning: recent controversies and advances. *Hippocampus.* In press

Anagnostaras SG, Maren S, DeCola JP, Lane NI, Gale GD, et al. 1998. Testicular hormones do not regulate sexually dimorphic Pavlovian fear conditioning or perforant-path long-term potentiation in adult male rats. *Behav. Brain Res.* 92:1–9

Anagnostaras SG, Maren S, Fanselow MS. 1999a. Temporally graded retrograde amnesia of contextual fear after hippocampal damage in rats: within-subjects examination. *J. Neurosci.* 19:1106–14

Anagnostaras SG, Maren S, Sage JR, Goodrich S, Fanselow MS. 1999b. Scopolamine and Pavlovian fear conditioning in rats: dose-effect analysis. *Neuropsychopharmacology* 21:731–44

Antoniadis EA, McDonald RJ. 2000. Amygdala, hippocampus and discriminative fear conditioning to context. *Behav. Brain Res.* 108:1–19

Armony JL, Quirk GJ, LeDoux JE. 1998. Differential effects of amygdala lesions on early and late plastic components of auditory cortex spike trains during fear conditioning. *J. Neurosci.* 18:2592–601

Atkins CM, Selcher JC, Petraitis JJ, Trzaskos JM, Sweatt JD. 1998. The MAPK cascade is required for mammalian associative learning. *Nat. Neurosci.* 1:602–9

Bailey DJ, Kim JJ, Sun W, Thompson RF, Helmstetter FJ. 1999. Acquisition of fear conditioning in rats requires the synthesis of mRNA in the amygdala. *Behav. Neurosci.* 113:276–82

Bechara A, Tranel D, Damasio H, Adolphs R, Rockland C, et al. 1995. Double dissociation of conditioning and declarative knowledge relative to the amygdala and hippocampus in humans. *Science* 269:1115–18

Beck CH, Fibiger HC. 1995. Conditioned fear-induced changes in behavior and in the expression of the immediate early gene c-fos: with and without diazepam pretreatment. *J. Neurosci.* 15:709–20

Bellgowan PS, Helmstetter FJ. 1996. Neural systems for the expression of hypoalgesia during nonassociative fear. *Behav. Neurosci.* 110:727–36

Blanchard DC, Blanchard RJ. 1972. Innate and conditioned reactions to threat in rats with amygdaloid lesions. *J. Comp. Physiol. Psychol.* 81:281–90

Bliss TV, Collingridge GL. 1993. A synaptic model of memory: Long-term potentiation in the hippocampus. *Nature* 361:31–39

Bliss TV, Lomo T. 1973. Long-lasting potentiation of synaptic transmission in the dentate area of the anaesthetized rabbit following stimulation of the perforant path. *J. Physiol.* 232:331–56

Bourtchouladze R, Abel T, Berman N, Gordon R, Lapidus K, Kandel ER. 1998. Different training procedures recruit either one or two critical periods for contextual memory consolidation, each of which requires protein synthesis and PKA. *Learn. Mem.* 5:365–74

Bourtchuladze R, Frenguelli B, Blendy J, Cioffi D, Schutz G, et al. 1994. Deficient long-term memory in mice with a targeted mutation of the cAMP-responsive element-binding protein. *Cell* 79:59–68

Bouton ME. 1993. Context, time, and memory retrieval in the interference paradigms of Pavlovian learning. *Psychol. Bull.* 114:80–99

Brady JV, Schreiner L, Geller I, Kling A. 1954. Subcortical mechanisms in emotional behavior: The effect of rhinencephalic injury upon the acquisition and retention of a conditioned avoidance response in cats. *J. Comp. Physiol. Psychol.* 47:179–86

Brambilla R, Gnesutta N, Minichiello L, White G, Roylance AJ, et al. 1997. A role for the Ras signalling pathway in synaptic transmission and long-term memory. *Nature* 390:281–86

Bramham CR, Srebro B. 1987. Induction of long-term depression and potentiation by low- and high-frequency stimulation in the dentate area of the anesthetized rat: magnitude, time course and EEG. *Brain Res.* 405:100–7

Brodal A. 1947. The amygdaloid nucleus in the rat. *J. Comp. Neurol.* 87:1–16

Brown S, Schäfer A. 1888. An investigation into the functions of the occipital and temporal lobes of the monkey's brain. *Philos. Trans. R. Soc. London Ser. B* 179:303–27

Buchel C, Dolan RJ, Armony JL, Friston KJ. 1999. Amygdala-hippocampal involvement in human aversive trace conditioning revealed through event-related functional magnetic resonance imaging. *J. Neurosci.* 19:10869–76

Cahill L, McGaugh JL. 1998. Mechanisms of emotional arousal and lasting declarative memory. *Trends Neurosci.* 21:294–99

Chill L, Weinberger NM, Roozendaal B, McGaugh JL. 1999. Is the amygdala a locus of "conditioned fear"? Some questions and caveats. *Neuron* 23:227–28

Campeau S, Davis M. 1995a. Involvement of subcortical and cortical afferents to the lateral nucleus of the amygdala in fear conditioning measured with fear-potentiated startle in rats trained concurrently with auditory and visual conditioned stimuli. *J. Neurosci.* 15:2312–27

Campeau S, Davis M. 1995b. Involvement of the central nucleus and basolateral complex of the amygdala in fear conditioning measured with fear-potentiated startle in rats trained concurrently with auditory and visual conditioned stimuli. *J. Neurosci.* 15:2301–11

Campeau S, Hayward MD, Hope BT, Rosen JB, Nestler EJ, et al. 1991. Induction of the c-fos proto-oncogene in rat amygdala during unconditioned and conditioned fear. *Brain Res.* 565:349–52

Campeau S, Miserendino MJ, Davis M. 1992. Intra-amygdala infusion of the N-methyl-D-aspartate receptor antagonist AP5 blocks acquisition but not expression of fear-potentiated startle to an auditory conditioned stimulus. *Behav. Neurosci.* 106:569–74

Chapman PF, Bellavance LL. 1992. Induction of long-term potentiation in the basolateral amygdala does not depend on NMDA receptor activation. *Synapse* 11:310–18

Chapman PF, Kairiss EW, Keenan CL, Brown TH. 1990. Long-term synaptic potentiation in the amygdala. *Synapse* 6:271–78

Cho YH, Friedman E, Silva AJ. 1999. Ibotenate lesions of the hippocampus impair spatial learning but not contextual fear conditioning in mice. *Behav. Brain Res.* 98:77–87

Chun MM, Phelps EA. 1999. Memory deficits

for implicit contextual information in amnesic subjects with hippocampal damage. *Nat. Neurosci.* 2:844–47

Clugnet MC, LeDoux JE. 1990. Synaptic plasticity in fear conditioning circuits: induction of LTP in the lateral nucleus of the amygdala by stimulation of the medial geniculate body. *J. Neurosci.* 10:2818–24

Collins DR, Pare D. 2000. Differential fear conditioning induces reciprocal changes in the sensory responses of lateral amygdala neurons to the CS(+) and CS(−). *Learn. Mem.* 7:97–103

Cousens G, Otto T. 1998. Both pre- and post-training excitotoxic lesions of the basolateral amygdala abolish the expression of olfactory and contextual fear conditioning. *Behav. Neurosci.* 112:1092–103

Davidson RJ, Irwin W. 1999. The functional neuroanatomy of emotion and affective style. *Trends Cogn. Sci.* 3:11–21

Davis M. 1992. The role of the amygdala in fear and anxiety. *Annu. Rev. Neurosci.* 15:353–75

De Oca BM, DeCola JP, Maren S, Fanselow MS. 1998. Distinct regions of the periaqueductal gray are involved in the acquisition and expression of defensive responses. *J. Neurosci.* 18:3426–32

Edeline JM, Weinberger NM. 1992. Associative retuning in the thalamic source of input to the amygdala and auditory cortex: receptive field plasticity in the medial division of the medial geniculate body. *Behav. Neurosci.* 106:81–105

Falls WA, Bakken KT, Heldt SA. 1997. Lesions of the perirhinal cortex interfere with conditioned excitation but not with conditioned inhibition of fear. *Behav. Neurosci.* 111:476–86

Falls WA, Davis M. 1995. Lesions of the central nucleus of the amygdala block conditioned excitation, but not conditioned inhibition of fear as measured with the fear-potentiated startle effect. *Behav. Neurosci.* 109:379–87

Falls WA, Miserendino MJ, Davis M. 1992. Extinction of fear-potentiated startle: blockade by infusion of an NMDA antagonist into the amygdala. *J. Neurosci.* 12:854–63

Fanselow MS. 1990. Factors governing one-trial contextual conditioning. *Anim. Learn. Behav.* 18:264–70

Fanselow MS. 1993. Associations and memories: The role of NMDA receptors and long-term potentiation. *Curr. Dir. Psychol. Sci.* 2:152–56

Fanselow MS. 1994. Neural organization of the defensive behavior system responsible for fear. *Psychon. Bull. Rev.* 1:429–38

Fanselow MS. 1997. Without LTP the learning circuit is broken. *Behav. Brain Sci.* 20:616

Fanselow MS. 2000. Contextual fear, gestalt memories, and the hippocampus. *Behav. Brain Res.* 110:73–81

Fanselow MS, Kim JJ. 1994. Acquisition of contextual Pavlovian fear conditioning is blocked by application of an NMDA receptor antagonist D,L-2-amino-5-phosphonovaleric acid to the basolateral amygdala. *Behav. Neurosci.* 108:210–12

Fanselow MS, Kim JJ, Yipp J, De Oca B. 1994. Differential effects of the N-methyl-D-aspartate antagonist DL-2-amino-5-phosphonovalerate on acquisition of fear of auditory and contextual cues. *Behav. Neurosci.* 108:235–40

Fanselow MS, LeDoux JE. 1999. Why we think plasticity underlying Pavlovian fear conditioning occurs in the basolateral amygdala. *Neuron* 23:229–32

Fendt M, Fanselow MS. 1999. The neuroanatomical and neurochemical basis of conditioned fear. *Neurosci. Biobehav. Rev.* 23:743–60

Fonberg E. 1965. Effect of partial destruction of the amygdaloid complex on the emotional-defensive behaviour of dogs. *Bull. Acad. Pol. Sci. Biol.* 13:429–32

Fox RJ, Sorenson CA. 1994. Bilateral lesions of the amygdala attenuate analgesia induced by diverse environmental challenges. *Brain Res.* 648:215–21

Frankland PW, Cestari V, Filipkowski RK, McDonald RJ, Silva AJ. 1998. The dorsal hippocampus is essential for context discrimination but not for contextual conditioning. *Behav. Neurosci.* 112:863–74

Frohardt RJ, Guarraci FA, Bouton ME. 2000. The effects of neurotoxic hippocampal lesions on two effects of context after fear extinction. *Behav. Neurosci.* 114:227–40

Garcia R, Jaffard R. 1992. The hippocamposeptal projection in mice: long-term potentiation in the lateral septum. *NeuroReport* 3:193–96

Garcia R, Vouimba RM, Baudry M, Thompson RF. 1999. The amygdala modulates prefrontal cortex activity relative to conditioned fear. *Nature* 402:294–96

Gerren RA, Weinberger NM. 1983. Long term potentiation in the magnocellular medial geniculate nucleus of the anesthetized cat. *Brain Res.* 265:138–42

Gewirtz JC, Davis M. 1997. Second-order fear conditioning prevented by blocking NMDA receptors in amygdala. *Nature* 388:471–74

Gewirtz JC, Falls WA, Davis M. 1997. Normal conditioned inhibition and extinction of freezing and fear-potentiated startle following electrolytic lesions of medial prefrontal cortex in rats. *Behav. Neurosci.* 111:712–26

Gewirtz JC, McNish KA, Davis M. 2000. Is the hippocampus necessary for contextual fear conditioning? *Behav. Brain Res.* 110:83–95

Gisquet-Verrier P, Dutrieux G, Richer P, Doyere V. 1999. Effects of lesions to the hippocampus on contextual fear: evidence for a disruption of freezing and avoidance behavior but not context conditioning. *Behav. Neurosci.* 113:507–22

Gloor P. 1960. Amygdala. In *Handbook of Physiology. Section 1, Neurophysiology*, ed. J Field, HW Magoun, WE Hall, 2:1395–420. Washington, DC: Am. Physiol. Soc.

Goddard GV. 1964. Functions of the amygdala. *Psychol. Bull.* 62:89–109

Goldstein LE, Rasmusson AM, Bunney BS, Roth RH. 1996. Role of the amygdala in the coordination of behavioral, neuroendocrine, and prefrontal cortical monoamine responses to psychological stress in the rat. *J. Neurosci.* 16:4787–98

Good M, Honey RC. 1997. Dissociable effects of selective lesions to hippocampal subsystems on exploratory behavior, contextual learning, and spatial learning. *Behav. Neurosci.* 111:487–93

Goosens KA, Holt W, Maren S. 2000. A role for amygdaloid PKA and PKC in the acquisition of long-term conditional fear memories in rats. *Behav. Brain Res.* 114:145–52

Grover LM, Teyler TJ. 1990. Two components of long-term potentiation induced by different patterns of afferent activation. *Nature* 347:477–79

Guarraci FA, Frohardt RJ, Kapp BS. 1999. Amygdaloid D1 dopamine receptor involvement in Pavlovian fear conditioning. *Brain Res.* 827:28–40

Haralambous T, Westbrook RF. 1999. An infusion of bupivacaine into the nucleus accumbens disrupts the acquisition but not the expression of contextual fear conditioning. *Behav. Neurosci.* 113:925–40

Heinbockel T, Pape HC. 2000. Input-specific long-term depression in the lateral amygdala evoked by theta frequency stimulation. *J. Neurosci.* 20:RC68

Heldt SA, Falls WA. 1998. Destruction of the auditory thalamus disrupts the production of fear but not the inhibition of fear conditioned to an auditory stimulus. *Brain Res.* 813:274–82

Helmstetter FJ. 1992. The amygdala is essential for the expression of conditional hypoalgesia. *Behav. Neurosci.* 106:518–28

Helmstetter FJ, Bellgowan PS. 1994. Effects of muscimol applied to the basolateral amygdala on acquisition and expression of contextual fear conditioning in rats. *Behav. Neurosci.* 108:1005–9

Hirsh R. 1974. The hippocampus and contextual retrieval of information from memory: a theory. *Behav. Biol.* 12:421–44

Hitchcock J, Davis M. 1986. Lesions of the

amygdala, but not of the cerebellum or red nucleus, block conditioned fear as measured with the potentiated startle paradigm. *Behav. Neurosci.* 100:11–22

Holland PC, Bouton ME. 1999. Hippocampus and context in classical conditioning. *Curr. Opin. Neurobiol.* 9:195–202

Holland PC, Gallagher M. 1999. Amygdala circuitry in attentional and representational processes. *Trends Cogn. Sci.* 3:65–73

Holt W, Maren S. 1999. Muscimol inactivation of the dorsal hippocampus impairs contextual retrieval of fear memory. *J. Neurosci.* 19:9054–62

Honey RC, Good M. 1993. Selective hippocampal lesions abolish the contextual specificity of latent inhibition and conditioning. *Behav. Neurosci.* 107:23–33

Honey RC, Good M. 2000. Associative components of recognition memory. *Curr. Opin. Neurobiol.* 10:200–4

Horvath FE. 1963. Effects of basolateral amygdalectomy on three types of avoidance behavior in cats. *J. Comp. Physiol. Psychol.* 56:380–89

Huang YY, Kandel ER. 1998. Postsynaptic induction and PKA-dependent expression of LTP in the lateral amygdala. *Neuron* 21:169–78

Huang YY, Martin KC, Kandel ER. 2000. Both protein kinase A and mitogen-activated protein kinase are required in the amygdala for the macromolecular synthesis-dependent late phase of long-term potentiation. *J. Neurosci.* 20:6317–25

Huerta PT, Sun LD, Wilson MA, Tonegawa S. 2000. Formation of temporal memory requires NMDA receptors within CA1 pyramidal neurons. *Neuron* 25:473–80

Impey S, Smith DM, Obrietan K, Donahue R, Wade C, et al. 1998. Stimulation of cAMP response element (CRE)-mediated transcription during contextual learning. *Nat. Neurosci.* 1:595–601

Iwata J, Chida K, LeDoux JE. 1987. Cardiovascular responses elicited by stimulation of neurons in the central amygdaloid nucleus in awake but not anesthetized rats resemble conditioned emotional responses. *Brain Res.* 418:183–88

Iwata J, LeDoux JE, Meeley MP, Arneric S, Reis DJ. 1986. Intrinsic neurons in the amygdaloid field projected to by the medial geniculate body mediate emotional responses conditioned to acoustic stimuli. *Brain Res.* 383:195–214

Izquierdo I, Medina JH. 1995. Correlation between the pharmacology of long-term potentiation and the pharmacology of memory. *Neurobiol. Learn. Mem.* 63:19–32

Kapp BS, Gallagher M, Holmquist BK, Theall CL. 1978. Retrograde amnesia and hippocampal stimulation: dependence upon the nature of associations formed during conditioning. *Behav. Biol.* 24:1–23

Kapp BS, Gallagher M, Underwood MD, McNall CL, Whitehorn D. 1982. Cardiovascular responses elicited by electrical stimulation of the amygdala central nucleus in the rabbit. *Brain Res.* 234:251–62

Kellicutt MH, Schwartzbaum JS. 1963. Formation of a conditioned emotional response (CER) following lesions of the amygdaloid complex in rats. *Psychol. Rep.* 12:351–58

Killcross S, Robbins TW, Everitt BJ. 1997. Different types of fear-conditioned behaviour mediated by separate nuclei within amygdala. *Nature* 388:377–80

Kim JJ, DeCola JP, Landeira-Fernandez J, Fanselow MS. 1991. N-methyl-D-aspartate receptor antagonist APV blocks acquisition but not expression of fear conditioning. *Behav. Neurosci.* 105:126–33

Kim JJ, Fanselow MS. 1992. Modality-specific retrograde amnesia of fear. *Science* 256:675–77

Kim JJ, Fanselow MS, DeCola JP, Landeira-Fernandez J. 1992. Selective impairment of long-term but not short-term conditional fear by the N-methyl-D-aspartate antagonist APV. *Behav. Neurosci.* 106:591–96

Kim JJ, Rison RA, Fanselow MS. 1993a. Effects of amygdala, hippocampus, and periaqueductal gray lesions on short- and long-term contextual fear. *Behav. Neurosci.* 107:1093–98

Kim M, Campeau S, Falls WA, Davis M. 1993b. Infusion of the non-NMDA receptor antagonist CNQX into the amygdala blocks the expression of fear-potentiated startle. *Behav. Neural Biol.* 59:5–8

Kim M, Davis M. 1993. Lack of a temporal gradient of retrograde amnesia in rats with amygdala lesions assessed with the fear-potentiated startle paradigm. *Behav. Neurosci.* 107:1088–92

King FA. 1958. Effect of septal and amygdaloid lesions on emotional behaviour and conditioned avoidance responses in the rat. *J. Nerv. Ment. Dis.* 126:57–63

Kiyama Y, Manabe T, Sakimura K, Kawakami F, Mori H, et al. 1998. Increased thresholds for long-term potentiation and contextual learning in mice lacking the NMDA-type glutamate receptor epsilon1 subunit. *J. Neurosci.* 18:6704–12

Klüver H, Bucy PC. 1937. "Psychic blindness" and other symptoms following bilateral temporal lobectomy in rhesus monkeys. *Am. J. Physiol.* 119:352–53

Krasne FB, Glanzman DL. 1995. What we can learn from invertebrate learning. *Annu. Rev. Psychol.* 46:585–624

Krettek JE, Price JL. 1978. A description of the amygdaloid complex in the rat and cat with observations on intra-amygdaloid axonal connections. *J. Comp. Neurol.* 178:255–79

LaBar KS, Gatenby JC, Gore JC, LeDoux JE, Phelps EA. 1998. Human amygdala activation during conditioned fear acquisition and extinction: a mixed-trial fMRI study. *Neuron* 20:937–45

LaBar KS, LeDoux JE, Spencer DD, Phelps EA. 1995. Impaired fear conditioning following unilateral temporal lobectomy in humans. *J. Neurosci.* 15:6846–55

LeDoux JE. 1995. Emotion: clues from the brain. *Annu. Rev. Psychol.* 46:209–35

LeDoux JE. 2000. Emotion circuits in the brain. *Annu. Rev. Neurosci.* 23:155–84

LeDoux JE, Cicchetti P, Xagoraris A, Romanski LM. 1990. The lateral amygdaloid nucleus: sensory interface of the amygdala in fear conditioning. *J. Neurosci.* 10:1062–69

LeDoux JE, Iwata J, Cicchetti P, Reis DJ. 1988. Different projections of the central amygdaloid nucleus mediate autonomic and behavioral correlates of conditioned fear. *J. Neurosci.* 8:2517–29

LeDoux JE, Sakaguchi A, Iwata J, Reis DJ. 1986. Interruption of projections from the medial geniculate body to an archineostriatal field disrupts the classical conditioning of emotional responses to acoustic stimuli. *Neuroscience* 17:615–27

Lee H, Kim JJ. 1998. Amygdalar NMDA receptors are critical for new fear learning in previously fear-conditioned rats. *J. Neurosci.* 18:8444–54

Lee Y, Walker D, Davis M. 1996. Lack of a temporal gradient of retrograde amnesia following NMDA-induced lesions of the basolateral amygdala assessed with the fear-potentiated startle paradigm. *Behav. Neurosci.* 110:836–39

Li H, Weiss SR, Chuang DM, Post RM, Rogawski MA. 1998. Bidirectional synaptic plasticity in the rat basolateral amygdala: characterization of an activity-dependent switch sensitive to the presynaptic metabotropic glutamate receptor antagonist 2S-alpha-ethylglutamic acid. *J. Neurosci.* 18:1662–70

Li XF, Phillips R, LeDoux JE. 1995. NMDA and non-NMDA receptors contribute to synaptic transmission between the medial geniculate body and the lateral nucleus of the amygdala. *Exp. Brain Res.* 105:87–100

Liang KC, McGaugh JL, Martinez J Jr, Jensen RA, Vasquez BJ, et al. 1982. Post-training amygdaloid lesions impair retention of an inhibitory avoidance response. *Behav. Brain Res.* 4:237–49

Malkani S, Rosen JB. 2000. Specific induction of early growth response gene 1 in the lateral nucleus of the amygdala following contextual fear conditioning in rats. *Neuroscience* 97:693–702

Maren S. 1996. Synaptic transmission and plasticity in the amygdala. An emerging physiology of fear conditioning circuits. *Mol. Neurobiol.* 13:1–22

Maren S. 1997. Arousing the LTP and learning debate. *Behav. Brain Sci.* 20:622–23

Maren S. 1998. Overtraining does not mitigate contextual fear conditioning deficits produced by neurotoxic lesions of the basolateral amygdala. *J. Neurosci.* 18:3088–97

Maren S. 1999a. Long-term potentiation in the amygdala: a mechanism for emotional learning and memory. *Trends Neurosci.* 22:561–67

Maren S. 1999b. Neurotoxic basolateral amygdala lesions impair learning and memory but not the performance of conditional fear in rats. *J. Neurosci.* 19:8696–703

Maren S. 1999c. Neurotoxic or electrolytic lesions of the ventral subiculum produce deficits in the acquisition and expression of Pavlovian fear conditioning in rats. *Behav. Neurosci.* 113:283–90

Maren S. 2000. Reply to Vazdarjanova. *Trends Neurosci.* 23:345–46

Maren S. 2001a. Auditory fear conditioning increases CS-elicited spike firing in lateral amygdala neurons even after extensive overtraining. *Eur. J. Neurosci.* In press

Maren S. 2001b. Multiple roles for synaptic plasticity in Pavlovian fear conditioning. In *Neuronal Mechanisms of Memory Formation*, ed. C Holscher, pp. 77–99. Cambridge: Cambridge Univ. Press

Maren S, Aharonov G, Fanselow MS. 1996a. Retrograde abolition of conditional fear after excitotoxic lesions in the basolateral amygdala of rats: absence of a temporal gradient. *Behav. Neurosci.* 110:718–26

Maren S, Aharonov G, Fanselow MS. 1997. Neurotoxic lesions of the dorsal hippocampus and Pavlovian fear conditioning in rats. *Behav. Brain Res.* 88:261–74

Maren S, Aharonov G, Stote DL, Fanselow MS. 1996b. N-methyl-D-aspartate recep-

tors in the basolateral amygdala are required for both acquisition and expression of conditional fear in rats. *Behav. Neurosci.* 110:1365–74

Maren S, Anagnostaras SG, Fanselow MS. 1998. The startled seahorse: Is the hippocampus necessary for contextual fear conditioning? *Trends Cogn. Sci.* 2:39–42

Maren S, Baudry M. 1995. Properties and mechanisms of long-term synaptic plasticity in the mammalian brain: Relationships to learning and memory. *Neurobiol. Learn. Mem.* 63:1–18

Maren S, DeCola JP, Fanselow MS. 1994a. Water deprivation enhances fear conditioning to contextual, but not discrete, conditional stimuli in rats. *Behav. Neurosci.* 108:645–49

Maren S, DeCola JP, Swain RA, Fanselow MS, Thompson RF. 1994b. Parallel augmentation of hippocampal long-term potentiation, theta rhythm, and contextual fear conditioning in water-deprived rats. *Behav. Neurosci.* 108:44–56

Maren S, De Oca B, Fanselow MS. 1994c. Sex differences in hippocampal long-term potentiation (LTP) and Pavlovian fear conditioning in rats: positive correlation between LTP and contextual learning. *Brain Res.* 661:25–34

Maren S, Fanselow MS. 1995. Synaptic plasticity in the basolateral amygdala induced by hippocampal formation stimulation in vivo. *J. Neurosci.* 15:7548–64

Maren S, Fanselow MS. 1996. The amygdala and fear conditioning: Has the nut been cracked? *Neuron* 16:237–40

Maren S, Fanselow MS. 1997. Electrolytic lesions of the fimbria/fornix, dorsal hippocampus, or entorhinal cortex produce anterograde deficits in contextual fear conditioning in rats. *Neurobiol. Learn. Mem.* 67:142–49

Maren S, Fanselow MS. 1998. Appetitive motivational states differ in their ability to augment aversive fear conditioning in rats (Rattus norvegicus). *J. Exp. Psychol.: Anim. Behav. Process.* 24:369–73

Maren S, Holt W. 2000. The hippocampus

and contextual memory retrieval in Pavlovian conditioning. *Behav. Brain Res.* 110:97–108

Maren S, Poremba A, Gabriel M. 1991. Basolateral amygdaloid multi-unit neuronal correlates of discriminative avoidance learning in rabbits. *Brain Res.* 549:311–16

Martin SJ, Grimwood PD, Morris RG. 2000. Synaptic plasticity and memory: an evaluation of the hypothesis. *Annu. Rev. Neurosci.* 23:649–711

Mayford M, Bach ME, Huang YY, Wang L, Hawkins RD, et al. 1996. Control of memory formation through regulated expression of a CaMKII transgene. *Science* 274:1678–83

McCabe PM, McEchron MD, Green EJ, Schneiderman N. 1993. Electrolytic and ibotenic acid lesions of the medial subnucleus of the medial geniculate prevent the acquisition of classically conditioned heart rate to a single acoustic stimulus in rabbits. *Brain Res.* 619:291–98

McDonald AJ. 1998. Cortical pathways to the mammalian amygdala. *Prog. Neurobiol.* 55:257–332

McEchron MD, Bouwmeester H, Tseng W, Weiss C, Disterhoft JF. 1998. Hippocampectomy disrupts auditory trace fear conditioning and contextual fear conditioning in the rat. *Hippocampus* 8:638–46

McEchron MD, Green EJ, Winters RW, Nolen TG, Schneiderman N, et al. 1996. Changes of synaptic efficacy in the medial geniculate nucleus as a result of auditory classical conditioning. *J. Neurosci.* 16:1273–83

McEchron MD, McCabe PM, Green EJ, Llabre MM, Schneiderman N. 1995. Simultaneous single unit recording in the medial nucleus of the medial geniculate nucleus and amygdaloid central nucleus throughout habituation, acquisition, and extinction of the rabbit's classically conditioned heart rate. *Brain Res.* 682:157–66

McGaugh JL. 2000. Memory–a century of consolidation. *Science* 287:248–51

McNish KA, Gewirtz JC, Davis M. 1997. Evidence of contextual fear after lesions of the hippocampus: a disruption of freezing

but not fear-potentiated startle. *J. Neurosci.* 17:9353–60

McNish KA, Gewirtz JC, Davis M. 2000. Disruption of contextual freezing, but not contextual blocking of fear-potentiated startle, after lesions of the dorsal hippocampus. *Behav. Neurosci.* 114:64–76

McKernan MG, Shinnick-Gallagher P. 1997. Fear conditioning induces a lasting potentiation of synaptic currents in vitro. *Nature* 390:607–11

Meunier M, Bachevalier J, Murray EA, Malkova L, Mishkin M. 1999. Effects of aspiration versus neurotoxic lesions of the amygdala on emotional responses in monkeys. *Eur. J. Neurosci.* 11:4403–18

Miserendino MJ, Sananes CB, Melia KR, Davis M. 1990. Blocking of acquisition but not expression of conditioned fear-potentiated startle by NMDA antagonists in the amygdala. *Nature* 345:716–18

Morgan MA, LeDoux JE. 1995. Differential contribution of dorsal and ventral medial prefrontal cortex to the acquisition and extinction of conditioned fear in rats. *Behav. Neurosci.* 109:681–88

Morgan MA, Romanski LM, LeDoux JE. 1993. Extinction of emotional learning: contribution of medial prefrontal cortex. *Neurosci. Lett.* 163:109–13

Morris JS, Frith CD, Perrett DI, Rowland D, Young AW, et al. 1996. A differential neural response in the human amygdala to fearful and happy facial expressions. *Nature* 383:812–15

Muller J, Corodimas KP, Fridel Z, LeDoux JE. 1997. Functional inactivation of the lateral and basal nuclei of the amygdala by muscimol infusion prevents fear conditioning to an explicit conditioned stimulus and to contextual stimuli. *Behav. Neurosci.* 111:683–91

Nachman M, Ashe JH. 1974. Effects of basolateral amygdala lesions on neophobia, learned taste aversions, and sodium appetite in rats. *J. Comp. Physiol. Psychol.* 87:622–43

Nader K, LeDoux JE. 1999. Inhibition of the

mesoamygdala dopaminergic pathway impairs the retrieval of conditioned fear associations. *Behav. Neurosci.* 113:891–901

Nader K, Schafe GE, LeDoux JE. 2000. Fear memories require protein synthesis in the amygdala for reconsolidation after retrieval. *Nature* 406:722–26

Pare D, Collins DR. 2000. Neuronal correlates of fear in the lateral amygdala: multiple extracellular recordings in conscious cats. *J. Neurosci.* 20:2701–10

Pavlov IP. 1927. *Conditioned Reflexes: An Investigation of the Physiological Activity of the Cerebral Cortex.* London: Oxford Univ. Press

Phillips ML, Young AW, Senior C, Brammer M, Andrew C, et al. 1997. A specific neural substrate for perceiving facial expressions of disgust. *Nature* 389:495–98

Phillips RG, LeDoux JE. 1992. Differential contribution of amygdala and hippocampus to cued and contextual fear conditioning. *Behav. Neurosci.* 106:274–85

Phillips RG, LeDoux JE. 1994. Lesions of the dorsal hippocampal formation interfere with background by not foreground contextual fear conditioning. *Learn. Mem.* 1:34–44

Phillips RG, LeDoux JE. 1995. Lesions of the fornix but not the entorhinal or perirhinal cortex interfere with contextual fear conditioning. *J. Neurosci.* 15:5308–15

Pitkanen A, Savander V, LeDoux JE. 1997. Organization of intra-amygdaloid circuitries in the rat: an emerging framework for understanding functions of the amygdala. *Trends Neurosci.* 20:517–23

Poremba A, Gabriel M. 1999. Amygdala neurons mediate acquisition but not maintenance of instrumental avoidance behavior in rabbits. *J. Neurosci.* 19:9635–41

Pugh CR, Tremblay D, Fleshner M, Rudy JW. 1997. A selective role for corticosterone in contextual-fear conditioning. *Behav. Neurosci.* 111:503–11

Quirk GJ, Armony JL, LeDoux JE. 1997. Fear conditioning enhances different temporal components of tone-evoked spike trains in auditory cortex and lateral amygdala. *Neuron* 19:613–24

Quirk GJ, Repa C, LeDoux JE. 1995. Fear conditioning enhances short-latency auditory responses of lateral amygdala neurons: parallel recordings in the freely behaving rat. *Neuron* 15:1029–39

Quirk GJ, Russo GK, Barron JL, Lebron K. 2000. The role of ventromedial prefrontal cortex in the recovery of extinguished fear. *J. Neurosci.* 20:6225–31

Rammes G, Steckler T, Kresse A, Schutz G, Zieglgansberger W, et al. 2000. Synaptic plasticity in the basolateral amygdala in transgenic mice expressing dominant-negative cAMP response element-binding protein (CREB) in forebrain. *Eur. J. Neurosci.* 12:2534–46

Rampon C, Tang YP, Goodhouse J, Shimizu E, Kyin M, et al. 2000. Enrichment induces structural changes and recovery from non-spatial memory deficits in CA1 NMDAR1-knockout mice. *Nat. Neurosci.* 3:238–44

Rescorla RA. 1988. Pavlovian conditioning. It's not what you think it is. *Am. Psychol.* 43:151–60

Richmond MA, Yee BK, Pouzet B, Veenman L, Rawlins JN, et al. 1999. Dissociating context and space within the hippocampus: effects of complete, dorsal, and ventral excitotoxic hippocampal lesions on conditioned freezing and spatial learning. *Behav. Neurosci.* 113:1189–203

Riedel G, Harrington NR, Hall G, Macphail EM. 1997. Nucleus accumbens lesions impair context, but not cue, conditioning in rats. *NeuroReport* 8:2477–81

Robinson E. 1963. The effect of amygdalectomy on fear-motivated behavior in rats. *J. Comp. Physiol. Psychol.* 56:814–20

Rogan MT, LeDoux JE. 1995. LTP is accompanied by commensurate enhancement of auditory-evoked responses in a fear conditioning circuit. *Neuron* 15:127–36

Rogan MT, LeDoux JE. 1996. Emotion: systems, cells, synaptic plasticity. *Cell* 85:469–75

Rogan MT, Staubli UV, LeDoux JE. 1997a. AMPA receptor facilitation accelerates fear learning without altering the level of conditioned fear acquired. *J. Neurosci.* 17:5928–35

Rogan MT, Staubli UV, LeDoux JE. 1997b. Fear conditioning induces associative long-term potentiation in the amygdala. *Nature* 390:604–7

Romanski LM, Clugnet MC, Bordi F, LeDoux JE. 1993. Somatosensory and auditory convergence in the lateral nucleus of the amygdala. *Behav. Neurosci.* 107:444–50

Romanski LM, LeDoux JE. 1992. Equipotentiality of thalamo-amygdala and thalamo-cortico-amygdala circuits in auditory fear conditioning. *J. Neurosci.* 12:4501–9

Roozendaal B, Koolhaas JM, Bohus B. 1991. Central amygdala lesions affect behavioral and autonomic balance during stress in rats. *Physiol. Behav.* 50:777–81

Rosen JB, Fanselow MS, Young SL, Sitcoske M, Maren S. 1998. Immediate-early gene expression in the amygdala following footshock stress and contextual fear conditioning. *Brain Res.* 796:132–42

Rosen JB, Hitchcock JM, Miserendino MJ, Falls WA, Campeau S, et al. 1992. Lesions of the perirhinal cortex but not of the frontal, medial prefrontal, visual, or insular cortex block fear-potentiated startle using a visual conditioned stimulus. *J. Neurosci.* 12:4624–33

Rosen JB, Schulkin J. 1998. From normal fear to pathological anxiety. *Psychol. Rev.* 105:325–50

Rudy JW. 1993. Contextual conditioning and auditory cue conditioning dissociate during development. *Behav. Neurosci.* 107:887–91

Rudy JW. 1996. Scopolamine administered before and after training impairs both contextual and auditory-cue fear conditioning. *Neurobiol. Learn. Mem.* 65:73–81

Rudy JW, Kuwagama K, Pugh CR. 1999. Isolation reduces contextual but not auditory-cue fear conditioning: a role for endogenous opioids. *Behav. Neurosci.* 113:316–23

Rudy JW, O'Reilly RC. 1999. Contextual fear conditioning, conjunctive representations, pattern completion, and the hippocampus. *Behav. Neurosci.* 113:867–80

Sacchetti B, Lorenzini CA, Baldi E, Tassoni G, Bucherelli C. 1999. Auditory thalamus, dorsal hippocampus, basolateral amygdala, and perirhinal cortex role in the consolidation of conditioned freezing to context and to acoustic conditioned stimulus in the rat. *J. Neurosci.* 19:9570–78

Sananes CB, Davis M. 1992. N-methyl-D-aspartate lesions of the lateral and basolateral nuclei of the amygdala block fear-potentiated startle and shock sensitization of startle. *Behav. Neurosci.* 106:72–80

Schafe GE, Nadel NV, Sullivan GM, Harris A, LeDoux JE. 1999. Memory consolidation for contextual and auditory fear conditioning is dependent on protein synthesis, PKA, and MAP kinase. *Learn. Mem.* 6:97–110

Selden NR, Everitt BJ, Jarrard LE, Robbins TW. 1991. Complementary roles for the amygdala and hippocampus in aversive conditioning to explicit and contextual cues. *Neuroscience* 42:335–50

Shi C, Davis M. 1999. Pain pathways involved in fear conditioning measured with fear-potentiated startle: lesion studies. *J. Neurosci.* 19:420–30

Shors TJ, Matzel LD. 1997. Long-term potentiation: What's learning got to do with it? *Behav. Brain Sci.* 20:597–634

Squire LR, Zola Morgan S. 1991. The medial temporal lobe memory system. *Science* 253:1380–86

Supple W Jr, Kapp BS. 1989. Response characteristics of neurons in the medial component of the medial geniculate nucleus during Pavlovian differential fear conditioning in rabbits. *Behav. Neurosci.* 103:1276–86

Swanson LW, Petrovich GD. 1998. What is the amygdala? *Trends Neurosci.* 21:323–31

Tang YP, Shimizu E, Dube GR, Rampon C, Kerchner GA, et al. 1999. Genetic enhancement of learning and memory in mice. *Nature* 401:63–69

Thompson RF, Krupa DJ. 1994. Organization of memory traces in the mammalian brain. *Annu. Rev. Neurosci.* 17:519–49

Tomaz C, Dickinson-Anson H, McGaugh JL. 1991. Amygdala lesions block the amnestic effects of diazepam. *Brain Res.* 568:85–91

Treit D, Menard J. 1997. Dissociations among the anxiolytic effects of septal, hippocampal, and amygdaloid lesions. *Behav. Neurosci.* 111:653–58

Treit D, Pesold C, Rotzinger S. 1993. Dissociating the anti-fear effects of septal and amygdaloid lesions using two pharmacologically validated models of rat anxiety. *Behav. Neurosci.* 107:770–85

Tsien JZ, Huerta PT, Tonegawa S. 1996. The essential role of hippocampal CA1 NMDA receptor-dependent synaptic plasticity in spatial memory. *Cell* 87:1327–38

Vazdarjanova A. 2000. Does the basolateral amygdala store memories for emotional events? *Trends Neurosci.* 23:345

Vazdarjanova A, McGaugh JL. 1998. Basolateral amygdala is not critical for cognitive memory of contextual fear conditioning. *Proc. Natl. Acad. Sci. USA* 95:15003–7

Vazdarjanova A, McGaugh JL. 1999. Basolateral amygdala is involved in modulating consolidation of memory for classical fear conditioning. *J. Neurosci.* 19:6615–22

Vouimba RM, Garcia R, Baudry M, Thompson RF. 2000. Potentiation of conditioned freezing following dorsomedial prefrontal cortex lesions does not interfere with fear reduction in mice. *Behav. Neurosci.* 114:720–24

Vouimba RM, Garcia R, Jaffard R. 1998. Opposite effects of lateral septal LTP and lateral septal lesions on contextual fear conditioning in mice. *Behav. Neurosci.* 112:875–84

Walker DL, Davis M. 1997. Double dissociation between the involvement of the bed nucleus of the stria terminalis and the central nucleus of the amygdala in startle increases produced by conditioned versus unconditioned fear. *J. Neurosci.* 17:9375–83

Wang SJ, Gean PW. 1999. Long-term depression of excitatory synaptic transmission in the rat amygdala. *J. Neurosci.* 19:10656–63

Watkins LR, Wiertelak EP, Maier SF. 1993. The amygdala is necessary for the expression of conditioned but not unconditioned analgesia. *Behav. Neurosci.* 107:402–5

Watson JB. 1920. *Experimental Investigation of Babies.* Chicago: Stoelting

Watson JB, Rayner R. 1920. Conditioned emotional reactions. *J. Exp. Psychol.* 3:1–14

Weeber EJ, Atkins CM, Selcher JC, Varga AW, Mirnikjoo B, et al. 2000. A role for the beta isoform of protein kinase C in fear conditioning. *J. Neurosci.* 20:5906–14

Weinberger NM. 1993. Learning-induced changes of auditory receptive fields. *Curr. Opin. Neurobiol.* 3:570–77

Weinberger NM. 1995. Dynamic regulation of receptive fields and maps in the adult sensory cortex. *Annu. Rev. Neurosci.* 18:129–58

Weiskrantz L. 1956. Behavioral changes associated with ablation of the amygdaloid complex in monkeys. *J. Comp. Physiol. Psychol.* 49:381–91

Weisskopf MG, Bauer EP, LeDoux JE. 1999. L-type voltage-gated calcium channels mediate NMDA-independent associative long-term potentiation at thalamic input synapses to the amygdala. *J. Neurosci.* 19:10512–19

Westbrook RF, Good AJ, Kiernan MJ. 1997. Microinjection of morphine into the nucleus accumbens impairs contextual learning in rats. *Behav. Neurosci.* 111:996–1013

Whalen PJ, Rauch SL, Etcoff NL, McInerney SC, Lee MB, et al. 1998. Masked presentations of emotional facial expressions modulate amygdala activity without explicit knowledge. *J. Neurosci.* 18:411–18

Wilensky AE, Schafe GE, LeDoux JE. 1999. Functional inactivation of the amygdala before but not after auditory fear conditioning prevents memory formation. *J. Neurosci.* 19:RC48

Wilensky AE, Schafe GE, LeDoux JE. 2000. The amygdala modulates memory consolidation of fear-motivated inhibitory avoidance learning but not classical fear conditioning. *J. Neurosci.* 20:7059–7066

Wilson A, Brooks DC, Bouton ME. 1995. The role of the rat hippocampal system in several effects of context in extinction. *Behav. Neurosci.* 109:828–36

Wolpe J. 1981. The dichotomy between classical conditioned and cognitively learned anxiety. *J. Behav. Ther. Exp. Psychiatry* 12:35–42

Young AW, Aggleton JP, Hellawell DJ, Johnson M, Broks P, et al. 1995. Face processing impairments after amygdalotomy. *Brain* 118:15–24

Young BJ, Leaton RN. 1996. Amygdala central nucleus lesions attenuate acoustic startle stimulus-evoked heart rate changes in rats. *Behav. Neurosci.* 110:228–37

Young SL, Bohenek DL, Fanselow MS. 1994. NMDA processes mediate anterograde amnesia of contextual fear conditioning induced by hippocampal damage: immunization against amnesia by context preexposure. *Behav. Neurosci.* 108:19–29

Zola Morgan S, Squire LR, Alvarez Royo P, Clower RP. 1991. Independence of memory functions and emotional behavior: separate contributions of the hippocampal formation and the amygdala. *Hippocampus* 1:207–20

Annu. Rev. Neurosci. 2001. 24:933–62

α-LATROTOXIN AND ITS RECEPTORS: Neurexins and CIRL/Latrophilins

Thomas C Südhof

Howard Hughes Medical Institute, Center for Basic Neuroscience, and the Department of Molecular Genetics, The University of Texas Southwestern Medical Center at Dallas, Texas 75390-9111, e-mail: Thomas.Sudhof@UTSouthwestern.edu

Key Words neurotransmitter release, synaptic vesicles, exocytosis, membrane fusion, synaptic cell adhesion

■ **Abstract** α-Latrotoxin, a potent neurotoxin from black widow spider venom, triggers synaptic vesicle exocytosis from presynaptic nerve terminals. α-Latrotoxin is a large protein toxin (120 kDa) that contains 22 ankyrin repeats. In stimulating exocytosis, α-latrotoxin binds to two distinct families of neuronal cell-surface receptors, neurexins and CLs (Cirl/latrophilins), which probably have a physiological function in synaptic cell adhesion. Binding of α-latrotoxin to these receptors does not in itself trigger exocytosis but serves to recruit the toxin to the synapse. Receptor-bound α-latrotoxin then inserts into the presynaptic plasma membrane to stimulate exocytosis by two distinct transmitter-specific mechanisms. Exocytosis of classical neurotransmitters (glutamate, GABA, acetylcholine) is induced in a calcium-independent manner by a direct intracellular action of α-latrotoxin, while exocytosis of catecholamines requires extracellular calcium. Elucidation of precisely how α-latrotoxin works is likely to provide major insight into how synaptic vesicle exocytosis is regulated, and how the release machineries of classical and catecholaminergic neurotransmitters differ.

INTRODUCTION

Black and brown widow spiders of the genus *Latrodectus* produce a potent venom that contains a mixture of toxins referred to as latrotoxins. Most latrotoxins target insects or crustaceans, which are the natural prey of black widow spiders. However, one of the spider toxins, α-latrotoxin, is specific for vertebrates. Clinically, black widow spider bites are not a significant health problem for humans because they rarely cause serious disease. A black widow spider is occasionally referred to as "the elegant lady who inflicts pain" (Simmons 1991). Only in the most severe cases do black widow spider bites cause latrodectism, a syndrome consisting of generalized muscle pain, abdominal cramps, profuse sweating, raised blood pressure, and tachycardia (Muller 1993, Zukowski 1993). Scientifically, α-latrotoxin was revealed to be extremely useful. α-Latrotoxin has become one of the prime tools

0147-006X/01/0621-0933$14.00

for studying synaptic transmission in neurobiology because it acts selectively on presynaptic nerve terminals to stimulate synaptic vesicle exocytosis by what turns out to be a very interesting mechanism. Examination of this mechanism has given important insights into the release process. Among others, studies on α-latrotoxin led to the discovery of two families of synaptic receptors, whose functions are now being elucidated, and to a better understanding of the various stages of synaptic vesicle exocytosis. In this review, I provide an overview of how α-latrotoxin works and what its receptors do in the framework of our current understanding of synaptic transmission.

STRUCTURE OF α-LATROTOXIN

α-Latrotoxin is a 120-kDa protein, an unusually large size for a toxin (Frontali & Grasso 1964, Frontali et al 1976). Its cDNA sequence predicts an even larger protein (approximately 160 kDa), which suggests that α-latrotoxin is synthesized as a protein precursor in the venom gland and is cleaved by endoproteases to generate the mature toxin (Kiyatkin et al 1990). Besides α-latrotoxin, three additional latrotoxins that act on invertebrates have been purified and cloned from black widow spiders: insect-specific α- and δ-latroinsectotoxins, and crustacean-specific α-latrocrustotoxin (Kiyatkin et al 1993, Dulubova et al 1996, Danilevich et al 1999). These three invertebrate latrotoxins are also proteins of more than 100 kDa that are synthesized as larger precursors similar to α-latrotoxin. All three invertebrate latrotoxins are homologous to each other and to α-latrotoxin over their entire length. The venom of black widow spiders probably contains additional toxins that act on insects, mollusks, and crustaceans and that remain to be characterized but that are likely to be also homologous. Thus far, all latrotoxins trigger neurotransmitter release in the various organisms in which they are active (Fritz et al 1980, Knipper et al 1986, Magazanik et al 1992, Elrick & Charlton 1999). Thus, black widow spider venom consists of a rich mixture of toxins that are homologous to each other and similarly trigger neurotransmitter release but that differ in their species specificity.

α-Latrotoxin and other latrotoxins are composed of four domains that are schematically illustrated for α-latrotoxin in Figure 1 (Kiyatkin et al 1990, 1993; Dulubova et al 1996; Danilevich et al 1999). Domain I is a cleaved signal peptide. Although the published cDNA sequence of α-latrotoxin lacks a signal peptide (Kiyatkin et al 1990), other latrotoxins contain a signal peptide, and the secretion of α-latrotoxin suggests that it requires a signal peptide. Domain II is a conserved N-terminal domain of 431 residues. This domain contains three invariant cysteine residues and two hydrophobic sequences of 20–26 amino acids, which could potentially span a membrane. However, the secreted toxin is highly soluble, which suggests that the hydrophobic sequences are not exposed on the surface of the soluble toxin but rather folded into the interior core of the toxin. Domain III is a central domain composed of 22 imperfect, ankyrin-like repeats covering 745 residues.

Domains:

Figure 1   Domain structure of α-latrotoxin. The four domains of the α-latrotoxin precursor are depicted schematically, with an N-terminal cleaved signal peptide (SP) (domain I), an N-terminal conserved domain with three critical cysteine residues (domain II), a central domain composed of 22 ankyrin-like repeats (domain III), and a C-terminal domain that is proteolytically removed during maturation of the toxin (domain IV). The conserved cysteine residues in domain II that were shown by mutagenesis to be essential for α-latrotoxin activity are marked by asterisks. The boundary between the N-terminal domain II and the ankyrin repeats (domain III) at which four amino acids were inserted in the mutant Ltx^{N4C} is identified with an arrow labeled N4C. The boundary between the central ankyrin-like repeats and the C-terminal domain IV where proteolytic cleavage occurs during the maturation of the toxin is also marked by an arrow. [Modified from Ichtchenko et al (1998).]

This domain is highly homologous among different latrotoxins, although the number of ankyrin repeats varies between toxins. The presence of 22 ankyrin repeats in α-latrotoxin is remarkable because such repeats are usually only found in intracellular proteins. With 22 repeats, α-latrotoxin contains the largest number of such repeats after ankyrin among currently known proteins. Domain IV is a C-terminal domain of 206 residues that is less conserved between latrotoxins and is presumably cleaved during the maturation of α-latrotoxin.

In addition to the proteolytic cleavage during the maturation of the toxin, α-latrotoxin also appears to be modified by intramolecular disulfide bonds because sulfhydryl reduction abolishes the toxicity of the toxin. No other posttranslational modifications of α-latrotoxin are known; in particular, α-latrotoxin does not appear to be glycosylated. When purified from venom glands, α-latrotoxin is isolated in a complex with a low-molecular-weight peptide of 70 residues (Kiyatkin et al 1992). This small, secreted protein contains a cleaved signal peptide and six cysteine residues that form three disulfide bonds. The low-molecular-weight component exhibits distant homology to members of the family of crustacean hyperglycemic hormones, which suggests that it is evolutionarily derived from an invertebrate neuropeptide (Gasparini et al 1994). Purified α-latrotoxin from black widow spider venom (including the small peptide component) forms a high-molecular-weight complex in solution, as judged by sucrose gradient centrifugation (Petrenko et al 1993). This complex was shown by cryoelectronmicroscopy to consist of a dimer in the absence of divalent cations, or a tetramer in the presence of divalent cations (Orlova et al 2000).

Insight into the relationship between the structure and function of α-latrotoxin was gained in studies of recombinant α-latrotoxin (Ichtchenko et al 1998, Volynski et al 1999). When the full-length α-latrotoxin precursor was synthesized in insect cells using baculovirus vectors, no active toxin was obtained. However, recombinant toxin that was truncated at the predicted C-terminal proteolytic cleavage site after the last ankyrin repeat was fully active and as potent in neurotransmitter release as native toxin (Ichtchenko et al 1998). Toxicity of the recombinant toxin did not require the small peptide that copurifies with native toxin. Thus, neither the C-terminal region (domain IV) of the α-latrotoxin precursor nor the small peptide copurifying with α-latrotoxin are required for the synthesis, folding, or activity of α-latrotoxin.

Substitution of each of the three conserved cysteine residues in domain II of α-latrotoxin for serine residues abolished toxin activity, which testifies to the importance of disulfide bonds in the conserved N-terminal domain. In addition, a mutant that carries an insertion of four amino acids between the N-terminal cysteine-rich domain and the central ankyrin repeats (referred to as Ltx^{N4C}) was unable to trigger neurotransmitter release, although it still bound to the two α-latrotoxin receptors (Ichtchenko et al 1998). This suggests that the exact distance between N-terminal and central domains in the toxin are of critical importance in activity, an idea that receives additional support from the high degree of conservation between different latrotoxins in the connecting region between the N-terminal domain and ankyrin repeats.

STIMULATION OF NEUROTRANSMITTER RELEASE BY α-LATROTOXIN: A Dual Mode of Action

α-Latrotoxin causes massive neurotransmitter release by synaptic vesicle exocytosis when applied to nerve terminals. This appears to occur by a two-stage process whereby the toxin first binds to highly specific receptors on the terminals and then triggers synaptic vesicle exocytosis. The actions of α-latrotoxin in exocytosis were first studied in detail in the frog neuromuscular junction (Longenecker et al 1970, Clark et al 1970), and later in synaptosomes (Grasso & Senni 1979, Nicholls et al 1982) and cultured hippocampal neurons (Capogna et al 1996, Augustin et al 1999). Furthermore, the toxin stimulates exocytosis of peptidergic vesicles from neuroendocrine cells such as PC12 cells (Grasso et al 1980, Meldolesi et al 1983), and studies on these cells have provided additional insight into its mechanism of action.

Applications of increasing concentrations of α-latrotoxin to presynaptic nerve terminals under controlled conditions leads to discrete dose-dependent effects on synaptic transmission. All these effects, independent of the toxin concentration, set in after a lag period of approximately 1 min following addition of the toxin. At low concentrations (subnanomolar), α-latrotoxin stimulates continuous exocytosis of synaptic vesicles. This becomes electrophysiologically manifest by an increase in the rate of spontaneous miniature postsynaptic currents ("minis"), which correspond to individual vesicle fusion events (Capogna et al 1996). The enhanced mini rate is sustained for more than 30 min even when the toxin is applied for only a few seconds and subsequently washed out. Morphologically, low doses of α-latrotoxin have no discernable effect on the nerve terminal, and the typical clusters of synaptic vesicles that normally fill the cytoplasm of a resting terminal are unchanged. Addition of higher toxin concentrations (nanomolar) causes a burst of neurotransmitter release followed by continuous steady state release that is also maintained for a considerable time (Ceccarelli et al 1979). At these concentrations, α-latrotoxin not only stimulates fusion of docked synaptic vesicles at the active zone (the "readily-releasable pool"), it also mobilizes vesicles in the reserve and resting pools in the backfield of the synapse. This effect does not depend on Ca^{2+} (see discussion below) and is quite different from the effects of hypertonic sucrose [a secretagog that also induces Ca^{2+}-independent neurotransmitter release (Rosenmund & Stevens 1996)] or even very strong electrical stimulation. In addition, stimulation by nanomolar concentrations of α-latrotoxin results in discrete morphological changes. The vesicle cluster shrinks, synaptic vesicles are depleted from the terminal, endocytosis lags behind, and the plasma membrane expands (Pumplin & Reese 1977, Gorio et al 1978, Ceccarelli et al 1979, Duchen et al 1981, Matteoli et al 1988). The total mobilization of vesicles by α-latrotoxin suggests that the toxin activates extensive signal transduction cascades. Application of high concentrations of α-latrotoxin (>10 nM) overwhelms the metabolic and secretory apparatus of the nerve terminal. High toxin concentrations not only induce massive synaptic vesicle exocytosis, they also lead to a dramatic drop in ATP levels of the nerve terminal that causes disintegration of the plasma membrane with release of cytoplasmic markers (McMahon et al 1990). Under these conditions, α-latrotoxin action loses specificity and behaves almost like a detergent. Morphologically, all vesicles are lost from nerve terminals, mitochondria swell, and the plasma membrane becomes grossly extended, especially when stimulations are carried out in the presence of Ca^{2+} (Pumplin & Reese 1977, Henkel & Betz 1995). Overall, the dose dependence of the α-latrotoxin effects reminds us that α-latrotoxin is a toxin that will provide only interpretable effects in studying neurotransmitter release at submaximal doses.

A striking feature of α-latrotoxin is that it stimulates neurotransmitter release in the absence of extracellular Ca^{2+}, at least in the case of the majority of neurotransmitters. This was observed in the initial studies of α-latrotoxin-triggered acetylcholine release at the frog neuromuscular junction (Clark et al 1970, Longenecker et al 1970, Ceccarelli et al 1979). No inhibition of acetylcholine secretion after application of α-latrotoxin was observed when Ca^{2+} was removed, whereas secretion stimulated by action potentials under the same Ca^{2+}-free conditions was abolished (Gorio et al 1978). In fact, Ca^{2+} removal was protective in that nerve terminal integrity was better after large doses of toxin, and mitochondrial swelling was not observed (Gorio et al 1978, Pumplin & Reese 1977, Henkel & Betz 1995). It is interesting that endocytosis was inhibited when α-latrotoxin was applied to neuromuscular nerve terminals in the absence of Ca^{2+}, which suggests a role for Ca^{2+} in regulating endocytosis, a concept that has since received much experimental support (Ceccarelli & Hurlbut 1980, Robinson et al 1993, Henkel &

Betz 1995, Brodin et al 2000). These initial experiments suggested that α-latrotoxin may be able to trigger exocytosis by bypassing the exocytotic Ca^{2+} sensor. In neuroendocrine PC12 cells, however, efficient stimulation of dopamine or norepinephrine secretion by α-latrotoxin clearly requires Ca^{2+} (Grasso et al 1980, Meldolesi et al 1983). Similar observations were made with synaptosomes, casting doubt on the ability of α-latrotoxin to work in the absence of Ca^{2+} (Davletov et al 1997). In hippocampal cultures, conversely, stimulation of glutamate release (similar to acetylcholine release in neuromuscular synapses) did not require extracellular Ca^{2+} (Capogna et al 1996).

The reason for these apparent contradictions probably is that there are differences in the Ca^{2+} dependence of α-latrotoxin action between various neurotransmitters. This is illustrated in a comparison of the release of different neurotransmitters from synaptosomes triggered by α-latrotoxin in the absence and presence of Ca^{2+} (Figure 2). Various transmitters respond dramatically differently to α-latrotoxin as a function of Ca^{2+} (Figure 2) (Khvotchev et al 2000). Glutamate and GABA are released equally well by α-latrotoxin in the presence or absence of Ca^{2+} (Figure 2). These neurotransmitters are used by the majority of central synapses in morphologically typical synaptic connections, with clusters of vesicles adjacent to an active zone. In contrast, vesicles containing transmitters that are not exocytosed at traditional synapses (e.g. neuropeptides and monoamines) require Ca^{2+} for α-latrotoxin-stimulated release (Figure 2) (Khvotchev & Südhof 2000a). This result suggests that α-latrotoxin has a dual, transmitter-specific mode of action: Ca^{2+} appears to be necessary for α-latrotoxin to trigger exocytosis of dense-core vesicles containing catecholamines and/or neuropeptides, but it appears to be dispensable for stimulation of exocytosis of small clear vesicles containing acetylcholine, glutamate, and GABA (Khvotchev et al 2000). In a beautiful example of this phenomenon, stimulation of neuromuscular junctions by high doses of α-latrotoxin in the absence of Ca^{2+} leads to a complete depletion of synaptic vesicles from the motor nerve terminals while neuropeptide-containing dense-core vesicles are left untouched (Matteoli et al 1988).

It is interesting that the requirement for Ca^{2+} in neurotransmitter release triggered by α-latrotoxin precisely parallels the necessity of phosphatidylinositolphosphate biosynthesis (Khvotchev & Südhof 1998). When phosphatidylinositolphosphate synthesis is inhibited with phenylarsine oxide, Ca^{2+}-dependent release of norepinephrine is inhibited, whereas glutamate and GABA are secreted normally. This suggests that the organization of the synapse may be the decisive factor in allowing α-latrotoxin to trigger release in the absence of Ca^{2+}. According to this idea, classical synapses containing characteristic vesicle clusters and active zones (e.g. glutamatergic, GABAergic, and cholinergic synapses) do not require Ca^{2+} for α-latrotoxin action, whereas nonclassical synapses without predocked vesicles require Ca^{2+}. This Ca^{2+} requirement may be due to the same mechanism that dictates the need of these synapses for phosphatidylinositolphosphate synthesis in exocytosis. One possible explanation is that Ca^{2+} and phosphatidylinositolphosphates are involved in the efficient transport and docking of the vesicles in a manner

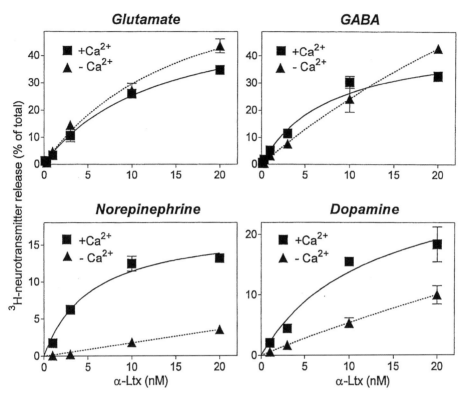

Figure 2 Ca^{2+} requirement for α-latrotoxin (α-Ltx) action in release of various neurotransmitters from synaptosomes. Dose/response curves show the amount of release of different neurotransmitters stimulated by the indicated α-latrotoxin concentrations. [Modified from Khvotchev et al (2000).] Release triggered from synaptosomes by α-latrotoxin was measured in the presence and absence of Ca^{2+}, demonstrating that there is no difference in the amount of GABA and glutamate release triggered by α-latrotoxin under these two conditions, whereas major inhibition of release is observed for norepinephrine and dopamine in the absence of Ca^{2+}. The norepinephrine release observed in the absence of Ca^{2+} is probably a lytic artifact due to high α-latrotoxin concentrations, whereas dopamine release likely has a Ca^{2+}-independent component, possibly because dopaminergic vesicles are observed both in varicosities outside of classical synapses and in classical synapses.

similar to what has been observed in PC12 cells (Martin et al 1997), and that the transport and docking of the vesicles occur by a different mechanism in classical synapses containing clouds of vesicles at the active zone.

The precise point at which Ca^{2+} is required for α-latrotoxin to trigger norepinephrine release was examined in synaptosomes (Khvotchev et al 2000). When α-latrotoxin is applied in the absence of Ca^{2+} and the synaptosomes are then washed in Ca^{2+}-free buffer, subsequent addition of Ca^{2+} potently triggers norepinephrine release. Thus, α-latrotoxin reacts with synaptosomes in the absence of

Ca^{2+} but requires Ca^{2+} for stimulus-secretion coupling in norepinephrine release. This is an important finding because it shows that the Ca^{2+} dependence operates not at the level of receptor binding of α-latrotoxin but at the level of stimulation of release. The finding agrees well with the idea that differences in the mechanism of exocytosis, such as the requirement for phosphatidylinositolphosphates, determine the differential Ca^{2+} sensitivity of norepinephrine release vs GABA and glutamate release induced by α-latrotoxin.

CHANNEL FORMATION BY α-LATROTOXIN

A very influential observation on how α-latrotoxin might stimulate secretion was made by Finkelstein et al (1976). These authors showed that α-latrotoxin added to the solution on one side of black lipid membranes spontaneously inserts into the membranes and forms a stable cation channel. Extensive characterization of the channel revealed that it requires incorporation of α-latrotoxin into the lipid bilayer, is nonselective with highest conductance for the divalent cations Ca^{2+} and Mg^{2+}, and is inhibited by such transition metals as Cd^{2+} and Ni^{2+} (Mironov et al 1986, Robello et al 1987, Lishko et al 1990). However, α-latrotoxin does not readily insert into the membranes of cultured cells when added to the medium. It appears to require high-affinity receptors on the cell surface in order to insert into the membrane, and α-latrotoxin-induced channels were probably only detectable in black lipid membranes because of the exquisite sensitivity of this detection technique. For example, in neuroendocrine cells containing α-latrotoxin receptors, the toxin causes opening of cation channels (Wanke et al 1986, Hurlbut et al 1994, Hlubek et al 2000). At least the channels observed in the latter two studies were similar to those recorded with black lipid membranes, which suggests that the toxin spontaneously inserts into membranes to form a channel in vivo. However, in Xenopus oocytes that do not express α-latrotoxin receptors, application of α-latrotoxin by itself does not cause Ca^{2+} channels to open. Only after brain proteins have been expressed by injection of total brain mRNA are α-latrotoxin-induced Ca^{2+} channels observed in the plasma membranes, presumably because α-latrotoxin receptors were synthesized from the brain mRNA (Filipov et al 1990, Umbach et al 1990). Based on these studies, it seems likely that binding of α-latrotoxin to receptors is necessary for the toxin to insert into biological membranes, where one of its actions then is to form a nonselective cation channel (Khvotchev & Südhof 2000). This mechanism, schematically depicted in Figure 3, would also explain the approximately 1-min lag period that is required for α-latrotoxin to act on a synapse.

The finding that α-latrotoxin induces formation of Ca^{2+} channels provides a ready explanation for how α-latrotoxin triggers exocytosis from chromaffin cells, PC12 cells, and noradrenergic nerve terminals in a Ca^{2+}-dependent manner (Meldolesi et al 1983, Khvotchev et al 2000). For this type of release, α-latrotoxin appears to act primarily as a Ca^{2+} channel, although even here, α-latrotoxin may have an additional direct effect on the secretory apparatus (Liu &

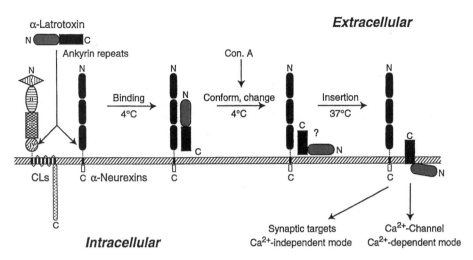

Figure 3 Model for the sequential interactions of α-latrotoxin with receptors followed by membrane insertion, channel formation, and intracellular activation of exocytosis. The model proposes that α-latrotoxin first binds to one or both of its receptors [CIRL/latrophilins (CLs) and neurexins] followed by a conformational change with subsequent membrane insertion. Receptor binding and the conformational change occur at 4°C, but membrane insertion requires a higher temperature. The lectin concanavalin A (Con A) does not inhibit receptor binding but interferes with the conformational change. Membrane insertion results in the formation of Ca^{2+} channels and the translocation of the N-terminal domain of α-latrotoxin into the presynaptic intracellular space. Exocytosis can subsequently be triggered by two mechanisms that apply differentially to different neurotransmitters (see text): Ca^{2+} influx via α-latrotoxin channels stimulates norepinephrine, dopamine, and neuropeptide release, whereas a direct Ca^{2+}-independent action of the N-terminal domain induces secretion of glutamate, GABA, and acetylcholine. The N and C termini of the various proteins are indicated by N and C, respectively. Note that only a single subunit of the α-latrotoxin multimer is shown. [Modified from Khvotchev & Südhof (2000).]

Misler 1998, Bittner et al 1998). Curiously, it has been suggested that Ca^{2+} influx is not responsible for triggering norepinephrine release by α-latrotoxin. Instead, it was proposed that Ca^{2+} has to be mobilized by α-latrotoxin from internal stores in order for release to occur (Davletov et al 1997, Rahman et al 1999). This conclusion was based entirely on the effect of high doses of thapsigargin, an inhibitor of the endoplasmic reticulum Ca^{2+}-ATPase that empties internal Ca^{2+} stores. However, thapsigargin is unable to trigger norepinephrine release by itself and does not inhibit α-latrotoxin action at more moderate doses that completely block Ca^{2+}-ATPase (Khvotchev et al 2000). This suggests that internal Ca^{2+} stores are probably not involved in α-latrotoxin action (Khvotchev et al 2000). Furthermore, the need for Ca^{2+} release from internal stores is difficult to understand under conditions that involve large Ca^{2+} influx from the external medium via α-latrotoxin-created channels. In addition, most nerve terminals lack extensive endoplasmic reticulum and mitochondria (Shepherd & Harris 1998) and do not have large internal Ca^{2+} stores, which argues against an essential role for internal Ca^{2+} stores in neurotransmitter release. Therefore, in the case of catecholamines and neuropeptides, α-latrotoxin probably acts as a Ca^{2+} channel (Rosenthal et al 1990, Khvotchev et al 2000).

The discovery of channel formation by α-latrotoxin raised the question of whether there really is a dual mode of action of α-latrotoxin whereby at least some neurotransmitters are released in a Ca^{2+}-independent fashion. Is it possible that the channels could explain the action of α-latrotoxin in triggering classical neurotransmitter release in the absence of Ca^{2+}? It was suggested that release of acetylcholine, glutamate, and GABA in the absence of extracellular Ca^{2+} could be due to leaky membranes, membrane depolarization, and mobilization of Ca^{2+} from internal stores, all mediated directly or indirectly by α-latrotoxin channels (Finkelstein et al 1976, Pumplin & Reese 1977). For example, in the absence of extracellular Ca^{2+}, Mg^{2+} and/or Na+ could flow through the α-latrotoxin channels to trigger release by depolarizing the terminal and/or mobilizing Ca^{2+} from binding sites and internal stores. A number of findings strongly argue against this hypothesis and suggest that α-latrotoxin truly does have a direct mode of action in the terminal. First, if the channel hypothesis was true, it is unclear why norepinephrine release requires Ca^{2+}—why can it not be stimulated by the same mechanism as glutamate and GABA? Second, application of Ca^{2+} ionophores, such as A23187, that are not specific for Ca^{2+} but also conduct Mg^{2+} does not mimic the effect of α-latrotoxin (Khvotchev & Südhof 1998). If these ionophores are applied in the absence of Ca^{2+} (but with Mg^{2+}), no stimulation of release is observed. Even in the presence of Ca^{2+}, these ionophores cause less neurotransmitter release than does α-latrotoxin, indicating that α-latrotoxin is better able to mobilize vesicles than are ionophores. Third, the requirement for Mg^{2+} in Ca^{2+}-free solutions for α-latrotoxin action can be partly abolished by increasing the molarity of the solution (Misler & Hurlbut 1979), which suggests that divalent cations are not absolutely required. Fourth, monoclonal antibodies to α-latrotoxin have been described that inhibit α-latrotoxin-stimulated Ca^{2+} fluxes completely but have only a partial effect on GABA release, thereby uncoupling the two processes (Pashkov et al 1993). Fifth, Cd^{2+}, which is known to inhibit the conductance of the α-latrotoxin channel (Lishko et al 1990), enhances α-latrotoxin action (Ichtchenko et al 1998). Finally, α-latrotoxin is still capable of triggering quantitatively unchanged glutamate release in two mouse mutants in which normal Ca^{2+} triggered release is severely inhibited (Geppert et al 1994, Augustin et al 1999). In these mutants, the synaptic proteins synaptotagmin I or munc13-1 have been deleted, leading to a block in the normal ability of Ca^{2+} to stimulate release. Because these mouse mutants are Ca^{2+} unresponsive but α-latrotoxin responsive, α-latrotoxin must act at a point upstream of Ca^{2+}. Together these findings establish that although α-latrotoxin forms a channel in the membrane, it also directly stimulates the secretory apparatus of a synapse, presumably by direct action of sequences that have translocated into the cytosol (Figure 3).

RECEPTOR-BASED INSERTION OF α-LATROTOXIN INTO BIOLOGICAL MEMBRANES

In order to stimulate exocytosis, α-latrotoxin needs to bind to specific neuronal high-affinity receptors (Tzeng & Siekevitz 1979). These receptors appear to be localized close to the active zone in the presynaptic plasma membrane (Valtorta et al 1984). Two classes of α-latrotoxin receptors were found in brain that differ in their Ca^{2+} dependence. Both classes exhibit a similar abundance and bind α-latrotoxin with the same nanomolar affinity. However, one receptor class is Ca^{2+} dependent, whereas the second class is Ca^{2+} independent (Rosenthal et al 1990, Geppert et al 1998). As discussed below, the Ca^{2+}-dependent class of α-latrotoxin receptors is constituted by neurexins, a large family of neuron-specific cell-adhesion molecules that were discovered in the search for α-latrotoxin receptors, whereas the Ca^{2+}-independent receptor class is composed of proteins called CLs (for CIRL/latrophilins). It should be noted that there is no connection between the Ca^{2+}-dependent and -independent classes of α-latrotoxin receptors and the Ca^{2+}-dependent and -independent classes of neurotransmitters whose release is triggered by α-latrotoxin. At first glance, there appears to be a striking similarity between the two classes of α-latrotoxin receptors and the dual mode of neurotransmitter release triggered by α-latrotoxin because both are defined in terms of their Ca^{2+} dependence. However, the requirement for Ca^{2+} in norepinephrine release triggered by α-latrotoxin occurs after receptor binding, and Ca^{2+}-independent receptors can still mediate Ca^{2+}-dependent neurotransmitter release (e.g. see Sugita et al 1998); thus, the two phenomena are unrelated.

As discussed above, formation of ion channels by α-latrotoxin depends on the presence of receptors and occurs only with a delay after α-latrotoxin application to nerve terminals. This suggests that α-latrotoxin first binds to receptors and then inserts into the presynaptic plasma membrane (Figure 3). This notion was recently confirmed biochemically when it was shown that α-latrotoxin becomes partially resistant to proteases after it has been incubated with synaptosomes (Khvotchev & Südhof 2000). Two conformational changes were detected on the basis of protease resistance patterns. The changes were distinguished because they required different incubation temperatures. At $4°C$, the toxin binds to receptors and undergoes a conformational change that alters its protease digestion pattern without causing true protease resistance. This limited conformational change was interpreted as a receptor-induced change at the membrane, possibly an exposure of the hydrophobic sequences in the N-terminal domain of α-latrotoxin (Figure 1), in preparation for membrane insertion. However, receptor binding alone was not sufficient for the initial conformational change because the change could be inhibited by the lectin concanavalin A, which had no effect on receptor binding (Khvotchev & Südhof 2000).

When synaptosomes with receptor-bound α-latrotoxin were warmed to $37°C$, the pattern of protease resistance was altered dramatically, with nearly complete protection of part of the toxin from digestion. This observation was taken as evidence for membrane insertion because protease protection was independent of the type of protease used, could be reversed on lysis of the synaptosomes, and was associated with the conversion of the toxin from a peripheral membrane protein to a detergent-insoluble form (Khvotchev & Südhof 2000). Because only parts and not the whole α-latrotoxin was protected, the toxin is not endocytosed in toto but remains partly surface accessible. Mapping of the regions of the toxins that become protease resistant, and thus presumably translocated into the cytosol, resulted in a surprise (Khvotchev & Südhof 2000). The entire N-terminal domain was found to be protected during prolonged exposure to proteases, whereas only the N-terminal few ankyrin repeats were not proteolysed (see Figure 3). This was an unexpected result because ankyrin-like repeats are generally found only in intracellular sequences, which suggests that these repeats would most likely be intracellularly active. The protected N-terminal domain that translocates into the synaptosomal membrane includes two hydrophobic sequences (Dulubova et al 1996), indicating that this domain may form part of the channel in addition to being translocated into the cytosol (Figure 3).

In solution, α-latrotoxin forms a multimer that is dependent on divalent cations (Petrenko et al 1993, Orlova et al 2000). Although the oligomerization state of α-latrotoxin in the membrane is unknown, it seems likely that it will form a multimer primarily composed of the N-terminal domain, with the hydrophobic segments assemblying into a channel (Figure 3). Here, the requirement for divalent cations, such as Mg^{2+}, for α-latrotoxin-triggered release of classical neurotransmitters (which does not require Ca^{2+}) may reflect a need for tetramerization that is mediated by these divalent cations (Orlova et al 2000). In this respect it is interesting that the divalent cation Cd^{2+} enhances α-latrotoxin-induced neurotransmitter release (Ichtchenko et al 1998) but blocks the cation channels formed by α-latrotoxin (Mironov et al 1986, Robello et al 1987, Lishko et al 1990). This suggests the possibility that divalent cations are generally required for α-latrotoxin action because they multimerize the toxin in the membrane. According to this hypothesis, Cd^{2+} and other divalent cations enter the mouth of the α-latrotoxin channels to perform their action in triggering release but do not actually have to pass through the channel.

PROPERTIES OF NEUREXINS

The first insight into the nature of α-latrotoxin receptors was gained in affinity chromatography experiments by the purification of binding proteins on immobilized α-latrotoxin (Scheer & Meldolesi 1985, Scheer et al 1986, Petrenko et al 1990). It is confusing, however, that a large number of different proteins were purified, many of which turned out to be presumably irrelevant intracellular components, for example a mitochondrial protein (Petrenko et al 1993). Nevertheless, one particular class of proteins with the properties of α-latrotoxin receptors was discovered (Ushkaryov et al 1992). These proteins, called neurexins, constitute a highly polymorphic family of neuron-specific cell-surface proteins. Vertebrates contain at least three neurexin genes (neurexins 1, 2, and 3). Each gene has two independent promoters that direct transcription of the larger α-neurexins and the shorter β-neurexins (Ushkaryov et al 1992, 1994; Ushkaryov & Südhof 1993). Structurally, neurexins resemble cell-surface receptors with extended extracellular sequences that include O-linked sugar attachment sites, a single transmembrane region, and a short intracellular sequence (Figure 4). α- and β-neurexins differ only in their extracellular N-terminal domains and contain identical transmembrane regions and intracellular C-terminal domains. The extracellular sequences of α-neurexins are composed of three overall repeats, each of which consists of a central epidermal growth factor (EGF)-like domain flanked by distantly related LNS domains (Figure 4). The LNS domains were named after laminin A, neurexins, and sex-hormone binding globulin because this domain was first identified as a repeat in the G-domain of laminin, was first shown to constitute an independently folding functional domain in neurexins, and was independently identified as a sequence motif in sex-hormone binding globulin (for a review, see Missler & Südhof 1998a). LNS domains are found in a large number of extracellular sequences, including agrin, slit, protein S, and perlecan (Ushkaryov et al 1992). They constitute protein-protein interaction domains that in neurexins, among others, bind α-latrotoxin (see below). The three-dimensional structure of the sixth LNS domain of neurexin 1α, simultaneously the only LNS domain of neurexin 1β, has been characterized by crystallography (Rudenko et al 1999). The structure revealed that the LNS domain is composed of a 14-stranded β-sandwich with a concave/convex overall shape. Although not identical to any known previously determined structure, the LNS domain structure is similar to that of pentraxin and lectins, which suggests even a possible evolutionary relationship (G Rudenko, E Hohenester, YA Muller, submitted for publication).

The primary transcripts of neurexins are subject to extensive alternative splicing, resulting in potentially thousands of isoforms (Ushkaryov et al 1992, Ullrich et al 1995). The highly variable nature of neurexins is interesting because expression of distinct neurexins on the neuronal cell surface could provide a possible mechanism by which different neurons are identified. To be physiologically meaningful, however, the alternative splicing of neurexins would have to fullfill the following criteria: (*a*) Alternative splicing should be similar for different neurexin genes; (*b*) it should be evolutionarily conserved; (*c*) it should be regulated and not random (i.e. different neurons should express distinct combinations of splice variants); (*d*) alternative splicing should determine the interactions of neurexins with other cell-surface proteins; and (*e*) it should be instrumental for the in vivo functions of neurexins. All these criteria except for the last have been confirmed at least partially for neurexins. The last criterion, a demonstration that alternative splicing regulates the in vivo functions of neurexins, has not been demonstrated because the in vivo functions of neurexins remain to be clarified.

The transcripts of all three neurexin genes are alternatively spliced, confirming the first criterion (numbered arrows in Figure 4) (Ullrich et al 1995). For

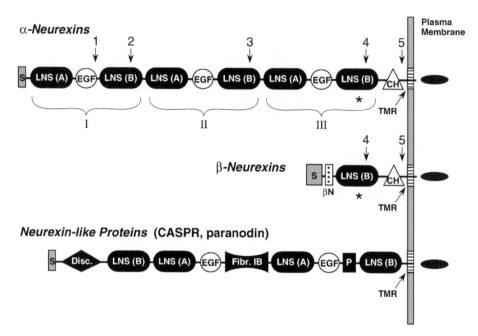

Figure 4 Domain structures of neurexins. The domain structures of α- and β-neurexins are illustrated and compared with those of neurexin-like proteins. Neurexins 1α–3α and 1β–3β are characterized by distinct extracellular N-terminal domain structures but share the same C-terminal domains, including the carbohydrate attachment sequence (CH), transmembrane region (TMR), and cytoplasmic tail. α-Neurexins are composed of three overall repeats (labeled I, II, and III) containing a central epidermal growth factor (EGF)-like sequence that is flanked by LNS domains. Although significant homology exists between all LNS domains in the repeats, the LNS-A domains N-terminal to the EGF-like sequence form a separate class from the LNS-B domains C-terminal to the EGF-like sequence. β-Neurexins lack most of the N-terminal domains found in α-neurexins and are composed only of a β-neurexin-specific sequence at the very N terminus (βN) followed by a single LNS-B domain that corresponds to the sixth overall LNS domain in α-neurexins. α- and β-neurexins are subject to extensive alternative splicing at five canonical positions for α-neurexins (splices sites numbers 1–5), the last two of which are also found in β-neurexins, as indicated by the labeled arrows. Neurexin-like proteins share some of the LNS- and EGF-like sequences found in α-neurexins but in addition contain an N-terminal discoidin-like domain (Disc.), a central fibrinogen β/γ-related sequence (Fibr.), and a PGY-motif (P). [Modified from Missler et al (1998a).]

some splice sites, the alternatively spliced inserts consist of more than 10 distinct variants. The positions of alternative splicing and at least some of the splice variants are evolutionarily conserved, as predicted by the second criterion (Ullrich et al 1995, Patzke & Ernsberger 2000). In situ hybridizations demonstrated that neurexins are expressed in overlapping differential patterns, with regional regulation of at least some of the alternative splicing, satisfying the third criterion (Ullrich et al 1995). This was particularly striking for splice site 4, which is expressed in only two variants, which are differentially expressed in various types of neurons (Ichtchenko et al 1995). The different sites of alternative splicing are used independently of each other, resulting in probably thousands of isoforms that are present in different brain regions in a characteristic combination of multiple variants. With this pattern, neurexins have the potential to represent a combinatorial code. As regards the fourth criterion, only two classes of endogenous ligands for neurexins are known, neuroligins and neurexophilins (Ichtchenko et al 1995, 1996; Petrenko et al 1996; Missler & Südhof 1998b; Missler et al 1998b). Of these, neuroligins are neuronal cell-surface proteins that only bind to β-neurexins when these lack an insert in splice site 4. Thus, the interaction of neuroligins with neurexins is tightly regulated by alternative splicing, satisfying the fourth criterion (Ichtchenko et al 1995, 1996). Neurexin binding to neuroligins forms an intercellular junction, which suggests that these proteins function as cell adhesion molecules (Nguyen & Südhof 1997). Neuroligins, when expressed in heterologous cells, can induce the generation of presynaptic specializations in neurons, indicating a function for neuroligins in synapse formation (Scheiffele et al 2000). In contrast to neuroligins, neurexophilins are secreted proteins, only bind to α-neurexins, and interact with neurexins independently of alternative splicing (Missler & Südhof 1998a,b; Missler et al 1998a,b).

Although in situ hybridizations indicate that neurexins are expressed only in neurons, no precise ultrastructural localization is available. Indirect evidence suggests that neurexins are concentrated in synapses. For example, subcellular fractionations revealed a high degree of enrichment of neurexins with active zones (Butz et al 1999), and the identification of neurexins as functional presynaptic α-latrotoxin receptors suggests a relation to synapses (see below) (Sugita et al 1999). Also, immunogold localization of CASK and neuroligin I, interacting partners for neurexins, to synaptic complexes supports a localization to synapses (Song et al 1999, Hsueh et al 1998). However, these indirect clues do not supplant the need for a direct localization of neurexins. Indeed, immunofluorescence studies in Torpedo have suggested that some neurexins may be axonal instead of synaptic (Russell & Carlson 1997), although the resolution of these studies was so low that it was not possible to determine the precise localization of neurexins.

Neurexins are evolutionarily conserved in vertebrates and in invertebrates, as revealed by the *Caenorhabditis elegans* and Drosophila genome sequences (Rubin et al 2000). However, these classical neurexins should not be confused with neurexin-like proteins (Figure 4). In Drosophila, a neurexin-like protein that is required for the formation and maintenance of septate junctions was called "neurexin IV" (Baumgartner et al 1996). Neurexin IV has caused confusion in the literature because its naming erroneously suggests that it belongs to the same family as vertebrate neurexins, whereas in fact it is the Drosophila homolog of vertebrate CASPR/paranodin proteins that exhibit a distant resemblance to neurexins (Peles et al 1997a,b; Menegoz et al 1997; Poliak et al 1999). CASPR/neurexin IV and vertebrate neurexins are similar cell surface receptor-like proteins containing, among others, LNS and EGF-like domains (Figure 3) (reviewed in Missler & Südhof 1998a). However, there are considerable differences between these proteins. Vertebrate neurexins are neuron specific, whereas Drosophila neurexin IV is expressed primarily outside of neurons; the domain structures of neurexins and CASPR/neurexin IV differ, and CASPR/neurexin IV are not polymorphic and not subject to alternative splicing. Thus, CASPR and neurexin IV should be more accurately thought of as neurexin-like proteins (Figure 4).

NEUREXINS AS α-LATROTOXIN RECEPTORS

The initial discovery of neurexins as α-latrotoxin-binding proteins during affinity chromatography suggested that they may constitute cell surface receptors for the toxin (Ushkaryov et al 1992). This was confirmed in studies demonstrating that the extracellular sequences of recombinant neurexins bind to α-latrotoxin with nanomolar affinity (Davletov et al 1995). Both α- and β-neurexins bind to α-latrotoxin directly, although only α-neurexins were initially identified because their binding is much more resistant to salt washes (Sugita et al 1999). Binding of α-latrotoxin requires micromolar concentrations of Ca^{2+} and is tightly regulated by alternative splicing of neurexins at splice site 4 in the middle of the last LNS domain (Figure 4) (Sugita et al 1999), the same splice site that also regulates binding of β-neurexins to neuroligins (Ichtchenko et al 1995, Nguyen & Südhof 1997). These data suggest that α-latrotoxin binds to the last LNS domain of neurexin close to the membrane. Because alternative splicing of this LNS domain is differentially regulated, this binding may help explain the regional differences in α-latrotoxin binding sites in brain (Malgaroli et al 1989).

The fact that neurexins are high-affinity cell surface binding proteins for α-latrotoxin, however, does not necessarily establish their function as α-latrotoxin receptors. Because α-latrotoxin stimulates the release of classical neurotransmitters without extracellular Ca^{2+} and neurexins require extracellular Ca^{2+} for binding to α-latrotoxin, the candidacy of neurexins as physiological α-latrotoxin receptors was dubious for some time. However, a series of experiments established that neurexins are "true" α-latrotoxin receptors. Analysis of neurexin 1α-knockout mice showed that α-latrotoxin binding to brain membranes was reduced approximately twofold in neurexin 1α-deficient mice in the presence of Ca^{2+} but was unchanged in the absence of Ca^{2+} (Geppert et al 1998). In synaptosomes from neurexin 1α-deficient mice, α-latrotoxin-stimulated glutamate release was normal in the absence of Ca^{2+} but severely depressed in the presence of Ca^{2+} (Geppert et al 1998). These results showed that in vivo, neurexin 1α is responsible for approximately half of the α-latrotoxin binding sites in the presence of Ca^{2+} and is required for α-latrotoxin to elicit neurotransmitter release. The function of neurexins as α-latrotoxin receptors was also studied in PC12 cells in which various neurexins were cotransfected with human growth hormone as a reporter molecule (Sugita et al 1999). In "naive" PC12 cells that have not been transfected with a neurexin,

α-latrotoxin triggers exocytosis only at relatively high toxin concentrations, probably because the levels of endogenous α-latrotoxin receptors in PC12 cells are very low. Transfection of PC12 cells with neurexin 1α or 1β makes the PC12 cells 10- to 100-fold more sensitive to α-latrotoxin (Sugita et al 1999). Neurexin 1α was as potent in this assay as the Ca^{2+}-independent receptor CL1 (discussed below), which suggests that both receptors can equally mediate α-latrotoxin action in transfected PC12 cells.

As α-latrotoxin receptors, neurexins are presumably close to the point of exocytosis in the presynaptic nerve terminal. The discovery that the putative synaptic vesicle Ca^{2+}-sensor synaptotagmin 1 and the synaptic adaptor protein CASK bind to the cytoplasmic tail of neurexins initially suggested that neurexins transduce an exocytotic signal of α-latrotoxin to the intracellular space (Hata et al 1993a, 1996). As mentioned above, however, α-latrotoxin is fully active in synaptotagmin 1 knockout mice even though in the same mice Ca^{2+}-stimulated glutamate release is severely impaired (Geppert et al 1994). This result demonstrates that synaptotagmin is not essential for α-latrotoxin action. Furthermore, deletion of the entire cytoplasmic tail of neurexin 1α has no effect on its function as an α-latrotoxin receptor in transfected PC12 cells (Sugita et al 1999). These results argue against the possibility of a direct coupling of receptor binding to activation of exocytosis and suggest that after neurexin binding, α-latrotoxin performs a second independent action in triggering neurotransmitter release.

CIRL/LATROPHILINS AS α-LATROTOXIN RECEPTORS

The purification and cloning of a Ca^{2+}-independent receptor for α-latrotoxin was simultaneously described by two groups (Davletov et al 1996; Krasnoperov et al 1996, 1997; Lelianova et al 1997). In a long-standing tradition of work on synaptic proteins, the two groups gave the receptor different names, CIRL and latrophilin. Further cloning revealed that at least three closely related forms of CIRL/latrophilin are expressed in vertebrates (Sugita et al 1998, Ichtchenko et al 1999, Matsushita et al 1999). To avoid confusion between the names of these receptors (CIRL and latrophilin), we follow the nomenclature of Sugita et al (1998) and abbreviate the two designations into a single name, CL. The three isoforms are therefore referred to as CL1, CL2, and CL3.

All CLs represent unusual G-protein-linked receptors with very large extra- and intracellular domains and are composed of the same domain structure (Sugita et al 1998): At the N terminus, a cleaved signal peptide is followed by a series of individual domains that cover approximately 500 amino acids (Figure 5). These domains include a lectin-like sequence; a region homologous to olfactomedins and myocilin; a long, previously undefined homology region that CLs share with a protein family of unknown function, referred to as BAI1-3 (brain-specific angiogenesis inhibitors) (Shiratsuchi et al 1997); and a short, cysteine-rich sequence. The short, cysteine-rich sequence outside of the first transmembrane region in CLs

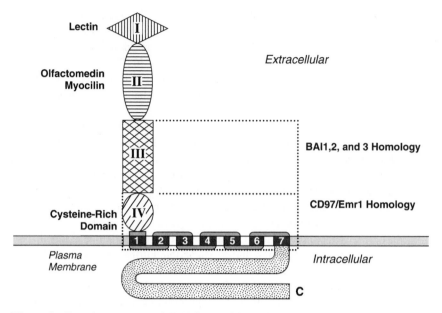

Figure 5   Domain structures of CIRL/latrophilins (CLs). CLs are G-protein-coupled receptors with unusually large extra- and intracellular sequences. The extracellular sequences are composed of five domains, shown above the plasma membrane, whereas the intracellular sequences have no discernable domain structure. The transmembrane regions of the CLs are similar to those of the calcitonin/secretin receptor family; in addition, the extracellular domains share homology with those of the BAI1-3 and the EMR1 family of G-protein-coupled receptors. [Modified from Sugita et al (1998).]

is highly homologous to sequences found in several G-protein-linked receptors (Sugita et al 1998). In CD97 and in CL1, the cysteine-rich domain probably directs the cleavage of the receptors after the last cysteine residue just N-terminal to the first transmembrane region (Gray et al 1996, Krasnoperov et al 1997). This sequence may represent a proteolytic cleavage signal during the maturation of G-protein-linked receptor and has been termed GPS domain for G-protein-coupled receptor proteolytic site domain (Gray et al 1996, Krasnoperov et al 1997, Ichtchenko et al 1999). However, copies of this domain are also present in other membrane proteins, such as polycystin-1, which suggests that it may be a general motif present outside of G-protein-linked receptors (Ponting et al 1999). In CL1, the cleaved N-terminal extracellular domain largely remains attached to the transmembrane portion of the receptor by an unknown mechanism, although some of the cleaved protein may also be shed into the medium. The sequences of the seven transmembrane regions and connecting linkers of CLs are significantly related to those of the secretin family of G-protein-coupled receptors, placing CLs into this general class of G-protein-coupled receptors (Figure 5). Following the seven transmembrane regions, CLs feature an unusually long cytoplasmic tail of approximately

500 residues that is subject to extensive alternative splicing (Sugita et al 1998, Matsushita et al 1999).

CL1, 2, and 3 are differentially expressed in a pattern that suggests a widespread function not directly related to neurons. Although CL1 was initially thought to be brain specific (Krasnoperov et al 1997, Lelianova et al 1997), more-sensitive RNA blots revealed low levels of CL1 in all tissues (Sugita et al 1998). CL2 is produced primarily outside of brain, with low levels in brain, whereas CL3 is dectectable only in brain (Sugita et al 1998, Ichtchenko et al 1999, Matsushita et al 1999). Both the initially discovered CL1 and CL2 bind α-latrotoxin with nanomolar affinity, whereas CL3 is unable to bind (Ichtchenko et al 1999). This suggests that CL1 and CL2 can function as α-latrotoxin receptors, a surprising result considering the ubiquitous distribution of CL2. This result does agree, however, with the observation that α-latrotoxin is capable of stimulating glutamate release from such nonneuronal cells as astrocytes (Parpura et al 1995). Based on their structure, it seems likely that CLs have a physiological function as cell-adhesion molecules coupled to signal transduction. This hypothesis is supported by the binding of PDZ-domain proteins called SHANKs to the C terminus of all CLs (Tobaben et al 2000). However, the in vivo roles of CLs or their natural ligands are unknown. Clearly, in view of their tissue distributions, CLs do not perform a synapse-specific, let alone neuron-specific, function.

Recent studies employing a CL1 knockout mouse demonstrated that CL1 is not essential for normal mouse development, survival, or cage life but confirms its function as an α-latrotoxin receptor (S Tobaben, TC Südhof, B Stahl, submitted for publication). Challenges of synaptosomes lacking CL1 with α-latrotoxin revealed that the majority of α-latrotoxin-evoked glutamate release was abolished, whereas depolarization-stimulated release was normal. α-Latrotoxin-induced release was impaired both in the presence and absence of Ca^{2+}, in contrast to the neurexin 1α-deficient synaptosomes, in which release was only decreased in the presence of Ca^{2+} (Geppert et al 1998). These findings confirm that CL1 is the major Ca^{2+}-independent α-latrotoxin receptor in vivo.

How do CLs function as receptors for α-latrotoxin in triggering release? After the discovery of CL1 as a G-protein-linked receptor, the initial expectation had been that α-latrotoxin activates an intracellular G-protein signal that triggers exocytosis (Krasnoperov et al 1997, Lelianova et al 1997). This expectation was enhanced by the observation of the copurification of syntaxin, a SNARE protein involved in fusion, with the receptor (Krasnoperov et al 1997). However, transfection experiments with PC12 cells and chromaffin cells revealed that CL1 lacking all cytoplasmic sequences functioned as an excellent α-latrotoxin receptor, demonstrating that intracellular signal transduction is not required (Sugita et al 1998, Hlubek et al 2000). This result parallels data obtained for neurexins (Sugita et al 1999), which suggest that at least in transfected PC12 cells, both receptors presumably serve to recruit α-latrotoxin to the membrane at the appropriate point (i.e. close to the active zone at the presynaptic plasma membrane) without in fact transducing the intracellular effect of the toxin.

THE TWO CLASSES OF α-LATROTOXIN
RECEPTORS MAY COOPERATE

α-Latrotoxin is unique among toxins because it binds to two families of high-affinity receptors with no sequence similarity or functional resemblance. Why did two receptors evolve for α-latrotoxin? Transfections of either neurexins or CLs into PC12 cells demonstrated that they can function as independent α-latrotoxin receptors in this system, and cotransfection of the two receptor families does not enhance the α-latrotoxin response over that of the singly transfected cells (Sugita et al 1998, 1999). These results show that the two receptors do not functionally co-operate in PC12 cells. However, as discussed above, α-latrotoxin has a dual mode of action, and the action of the receptors in PC12 cell exocytosis, a typical example of dense-core vesicle secretion that requires Ca^{2+}, may not be representative of receptor action in central synapses. Indeed, several observations suggest that the CL and neurexin receptor families collaborate in α-latrotoxin-triggered neurotransmitter release in central synapses, although the nature of their interaction is unclear.

One line of evidence for a collaboration between receptors was obtained from comparisons of Ca^{2+}-dependent and Ca^{2+}-independent binding of α-latrotoxin to brain membranes and of Ca^{2+}-dependent and Ca^{2+}-independent glutamate and GABA release stimulated by α-latrotoxin in synaptosomes. The number of α-latrotoxin binding sites is approximately two times higher in the presence of Ca^{2+} than in the absence of Ca^{2+}, as would be expected if the total binding was the sum of Ca^{2+}-dependent and Ca^{2+}-independent binding (Rosenthal et al 1990, Geppert et al 1998). However, glutamate and GABA release is approximately the same in the presence or absence of Ca^{2+}, independent of the α-latrotoxin dose (see Figure 2) (Khvotchev et al 2000). If the α-latrotoxin receptors acted independently, release observed in the presence of Ca^{2+} should be the sum of Ca^{2+}-dependent and Ca^{2+}-independent release, similar to binding, which is clearly not the case.

Results from studies with the neurexin 1α knockout mice (Geppert et al 1998) provide further support for the idea that the two classes of receptors functionally cooperate. In these mice, an expected nearly 50% decrease of total α-latrotoxin-binding sites in brain was observed, with a majority of the Ca^{2+}-dependent binding sites abolished. Furthermore, as expected, Ca^{2+}-independent neurotransmitter release triggered by α-latrotoxin was unchanged. Unexpected was the finding that in the same mice, α-latrotoxin triggered neurotransmitter release was reduced in the presence of Ca^{2+}. This result is puzzling because the portion of release that corresponds to the Ca^{2+}-independent mechanism should be selectively preserved in the neurexin 1α knockouts. These observations raise the possibility that Ca^{2+} inhibits α-latrotoxin action via CLs, and that this inhibition is abolished on Ca^{2+} withdrawal. Although this hypothesis is currently difficult to understand mechanistically, it should be recalled that Ca^{2+} is normally always present in the extracellular fluid, and the release in the presence of Ca^{2+} represents the more physiological situation.

Additional evidence for a cooperation between Cls and neurexins as α-latrotoxin receptors was provided by the analysis of the CL1 single and the CL1/neurexin 1α double knockout mice (S Tobaben, TC Südhof, B Stahl, submitted for publication 2001). As described above, the CL1 knockout had the expected phenotype, in that Ca^{2+}-independent glutamate release triggered by α-latrotoxin was largely abolished, but unexpectedly, Ca^{2+}-dependent glutamate release was also largely abolished although neurexin Iα was still present. Furthermore, Ca^{2+}-dependent α-latrotoxin release in the double knockout mouse was no more impaired than in either of the two single knockout mice whereas, surprisingly, Ca^{2+}-independent glutamate release in the double knockout mouse was even slightly enhanced. It is easy to understand why some α-latrotoxin-triggered release was retained in the double knockout mice because the remaining isoforms of neurexins and CLs can also serve as α-latrotoxin receptors (Sugita et al 1998, 1999; Ichtchenko et al 1999). However, the relative impairments in α-latrotoxin-triggered release by the two knockouts suggest that they do not function as independent receptors but must somehow interact. Because each receptor forms an independent high-affinity α-latrotoxin binding site in brain, this interaction must occur downstream of receptor binding at a currently unidentified process.

Based on the overall evidence, it is likely that the receptors cooperate in triggering release by α-latrotoxin from central synapses. However, this interaction is probably not due simply to receptor activation. In the α-latrotoxin mutant Ltx[N4C] described above, four amino acids were inserted between the N-terminal conserved domain and the ankyrin repeats (Figure 1). The Ltx[N4C] mutant binds to α-latrotoxin receptors with the same affinity as wild-type α-latrotoxin but causes no stimulation of neurotransmitter release (Ichtchenko et al 1998). α-Latrotoxin binding to its receptors on neurons and synaptosomes produces a number of intracellular effects, some of which are likely caused by intracellular Ca^{2+} influx. For example, α-latrotoxin stimulates massive hydrolysis of phosphatidylinositolphosphates, presumably by activation of phospholipase C (Vicentini & Meldolesi 1984, Ichtchenko et al 1998). The fact that the Ltx[N4C] mutant still stimulates hydrolysis of phosphatidylinositolphosphates suggests that it is still active as a ligand although unable to trigger release. This mutant therefore uncouples receptor activation from the stimulation of neurotransmitter release, which suggests that the two processes are mechanistically distinct.

HOW DOES α-LATROTOXIN TRIGGER RELEASE AT A CLASSICAL SYNAPSE?

As described above, considerable progress has been made in the understanding of α-latrotoxin action. The toxin has been cloned and functionally expressed, allowing analysis of mutants. The receptors for α-latrotoxin on neuronal membranes have now been identified, leading to the discovery of two new families of cell surface proteins, neurexins and CLs. The toxin has been shown to insert into the plasma membrane after receptor binding, and the N-terminal parts of the toxin that are involved in the intracellular translocation have been identified. It seems likely that the toxin multimerizes in the membrane in a reaction that requires divalent cations, although not necessarily Ca^{2+}, and that the multimerization is essential for all actions of the toxin. The intracellular signaling initiated by the toxin, such as Ca^{2+} influx and hydrolysis of phosphatidylinositols, has been uncoupled from its action in stimulated exocytosis of synaptic vesicles containing classical neurotransmitters, such as GABA and glutamate, that do not require Ca^{2+} for release triggered by α-latrotoxin. The overall picture of α-latrotoxin action that emerges from these studies is similar to that of many other toxins that bind to cell surface receptors, insert into the membrane, and then transduce intracellular signals.

α-Latrotoxin, however, also has several unusual properties that we currently do not understand. It acts via two classes of dissimilar high-affinity receptors. The rationale for having two receptors is unclear, although recent data suggest that the two receptors may in fact cooperate in triggering exocytosis. The most unusual feature of α-latrotoxin may be its dual mode of action. In what could be termed its simple mode, α-latrotoxin forms a channel in the membrane that conducts Ca^{2+}, with the resulting Ca^{2+} influx triggering a number of intracellular events, including exocytosis of dense-core vesicles. The second mode of action of α-latrotoxin, which could be called the complex mode and is probably more important, is its ability to directly stimulate synaptic vesicle exocytosis without participation of Ca^{2+}. Insertion of α-latrotoxin into the membrane with translocation of the N-terminal domain is required for this mode of action but is not sufficient because the LtxN4C mutant, which is unable to trigger release, still inserts into the membrane, although with lowered efficiency (Khvotchev & Südhof 2000). In the complex mode, α-latrotoxin does not function as a Ca^{2+} channel or channel for any other ion because this mode is not inhibited by Cd^{2+}, which does inhibit the α-latrotoxin channel (Ichtchenko et al 1998). Thus, this mode involves a direct stimulation of the secretory apparatus at the synapse.

Recent work, primarily based on studies of knockout mice and results obtained with botulinum and tetanus toxins, have allowed a description of key proteins that function in identified steps in vesicle fusion (Figure 6) (Fernandez-Chacon & Südhof 1999). According to this work, vesicles that are docked at the active zone undergo a priming/prefusion reaction before they are ready to be stimulated for release by Ca^{2+}. The priming/prefusion reaction requires initially the SM protein munc18-1 and the SNARE proteins synaptobrevin, syntaxin, and SNAP-25, which are inhibited by the botulinum and tetanus toxins, and then the active zone protein munc13-1 (Jahn & Südhof 1999). The Ca^{2+}-triggered fusion reaction then is regulated by synaptotagmin, rab3, and the rab3-effector RIM (Figure 6). Studies with botulinum and tetanus toxins, which interfere with the functions of the SNARE proteins, showed that the SNARE proteins are essential for α-latrotoxin action (Dreyer et al 1987, Capogna et al 1997; for a review, see Südhof et al 1993). The fact that these toxins can block α-latrotoxin action demonstrates that

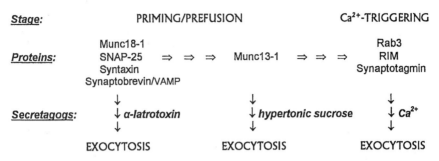

Figure 6 Point of action of α-latrotoxin in neurotransmitter release in relation to synaptic proteins. The final part of synaptic vesicle exocytosis physiologically triggered by Ca^{2+} influx during an action potential can be divided into a priming/prefusion reaction that makes the vesicles competent for a Ca^{2+} signal, and the actual Ca^{2+}-triggering reaction (top of diagram) (for a review, see Südhof 1995). The point in the reaction where each of the indicated synaptic proteins are required is shown in the middle. Based primarily on knockout studies, the priming/prefusion reaction can be subdivided into an early stage that requires the SNAREs synaptobrevin, SNAP-25, and syntaxin, as well as the SM-protein munc18-1 (Jahn & Südhof 1999), and a later stage that requires munc13-1 (Augustin et al 1999). The action of these proteins is then followed by the need for synaptotagmin I, rab3A, and the rab3-effector RIM in late stages of exocytosis during the Ca^{2+}-triggering reaction. Exocytosis can be triggered by α-latrotoxin at a stage that precedes munc13-1 action because α-latrotoxin stimulates release normally in munc13-1-deficient mice, whereas hypertonic sucrose, another secretagog that triggers release in the absence of Ca^{2+} (Rosenmund & Stevens 1996), requires munc13-1 but does not need rab3 or synaptotagmin I.

α-latrotoxin utilizes the same SNARE-dependent mechanism of secretion as does Ca^{2+}-triggered neurotransmitter release. In addition, mice lacking munc18-1 are probably unable to respond to α-latrotoxin (Verhage et al 2000). However, for the munc18-1 mutant, the situation is not totally clear because there appears to be a general lack of synaptic transmission, making it difficult to determine whether there is a true deficiency of α-latrotoxin response (Verhage et al 2000). Together these findings suggest that the core fusion proteins, synaptobrevin, syntaxin, SNAP-25, and munc18-1, are required for α-latrotoxin action (Figure 6). It is interesting, however, that mutants in munc13-1 are not impaired in the α-latrotoxin response (Augustin et al 1999). In the same mutant, most of the Ca^{2+}-triggered release of glutamate and Ca^{2+}-independent release stimulated by hypertonic sucrose are severely impaired. These results place the action of α-latrotoxin downstream of the munc18/SNARE complex but upstream of munc13-1.

The question now arises as to how α-latrotoxin acts on the core complex and munc18-1, and what its precise intracellular mode of action is. After many efforts, which may be characterized as scientific dances—two steps forward, one step back—the promise of α-latrotoxin is now beginning to be realized. We now know that (*a*) the toxin can indeed directly stimulate the secretory apparatus, (*b*) it has been demonstrated that this only occurs at classical synapses, and (*c*) the point at which this stimulation occurs has been identified. This provides a unique opportunity to identify the molecular switch or handle that α-latrotoxin uses to trigger release. The most plausible hypothesis is that the toxin acts by releasing a molecular brake that normally blocks vesicle fusion from occurring in classical synapses in the absence of Ca^{2+}. Clearly, definition of the mechanism by which α-latrotoxin operates will provide a major step forward in understanding what keeps the vesicles from automatically progressing through the fusion reaction and thus will clarify a key question of how a synapse, as opposed to a nonsecretory system, functions.

ACKNOWLEDGMENTS

I would like to thank my colleagues and collaborators Drs. B Stahl and S Tobaben (Göttingen) and Drs. M Khvotchev and S Sugita (Dallas) for helpful comments. This work was supported by NIH grant R37-MH52804-06.

Visit the Annual Reviews home page at www.AnnualReviews.org

LITERATURE CITED

Augustin I, Rosenmund C, Südhof TC, Brose N. 1999. Munc-13 is essential for fusion competence of glutamatergic synaptic vesicles. *Nature* 400:457–61

Baumgartner S, Littleton JT, Broadie K, Bhat MA, Harbecke R, et al. 1996. A Drosophila neurexin is required for septate junction and blood-nerve barrier formation and function. *Cell* 87:1059–68

Bittner MA, Krasnoperov VG, Stuenkel EL, Petrenko AG, Holz RW. 1998. A Ca2+-independent receptor for α-latrotoxin, CIRL, mediates effects on secretion via multiple mechanisms. *J. Neurosci.* 18:2914–22

Brodin L, Low P, Shupliakov O. 2000. Sequential steps in clathrin-mediated synaptic vesicle endocytosis. *Curr. Opin. Neurobiol.* 10:312–20

Butz S, Fernandez-Chacon R, Schmitz F, Jahn R, Südhof TC. 1999. The subcellular localizations of atypical synaptotagmins: synaptotagmin III is enriched in synapses and synaptic plasma membranes but not in synaptic vesicles. *J. Biol. Chem.* 274:18290–96

Capogna M, Gahwiler BH, Thompson SM. 1996. Calcium-independent actions of α-latrotoxin on spontaneous and evoked synaptic transmission in the hippocampus. *J. Neurophysiol.* 76:149–58

Capogna M, McKinney RA, O'Connor V, Gahwiler BH, Thompson SM. 1997. Ca^{2+} or Sr^{2+} partially rescues synaptic transmission in hippocampal cultures treated with botulinum toxin A and C, but not tetanus toxin. *J. Neurosci.* 17:7190–202

Ceccarelli B, Grohovaz F, Hurlbut WP. 1979. Freeze-fracture studies of frog neuromuscular junctions during intense release of neurotransmitter. I. Effects of black widow spider venom and Ca2+-free solutions on the structure of the active zone. *J. Cell Biol.* 81:163–77

Ceccarelli B, Hurlbut WP. 1980. Ca^{2+}-dependent recycling of synaptic vesicles at the frog neuromuscular junction. *J. Cell Biol.* 87:297–303

Clark AW, Mauro A, Longenecker HE Jr, Hurlbut WP. 1970. Effects of black widow spider venom on the frog neuromuscular junction. Effects on the fine structure of the frog neuromuscular junction. *Nature* 225:703–5

Danilevich VN, Luk'ianov SA, Grishin EV. 1999. Cloning and structure of gene encoded alpha-latrocrustoxin from the black widow spider venom. *Bioorg. Khim.* 25:537–47

Davletov BA, Krasnoperov V, Hata Y, Petrenko AG, Südhof TC. 1995. High affinity binding of α-latrotoxin to recombinant neurexin Iα. *J. Biol. Chem.* 270:23903–5

Davletov BA, Meunier FA, Ashton AC, Matsushita H, Hirst WD, et al. 1998. Vesicle exocytosis stimulated by α-latrotoxin is mediated by latrophilin and requires both external and stored Ca^{2+}. *EMBO J.* 17:3909–20

Davletov BA, Shamotienko OG, Lelianova VG, Grishin EV, Ushkaryov YA. 1996. Isolation and biochemical characterization of a Ca^{2+}-dependent α-latrotoxin-binding protein. *J. Biol. Chem.* 271:23239–45

Dreyer F, Rosenberg F, Becker C, Bigalke H, Penner R. 1987. Differential effects of various secretagogues on quantal transmitter release from mouse motor nerve terminals treated with botulinum A and tetanus toxin. *Naunyn Schmiedebergs Arch. Pharmacol.* 335:1–7

Duchen LW, Gomez S, Queiroz LS. 1981. The neuromuscular junction of the mouse after black widow spider venom. *J. Physiol.* 316:279–91

Dulubova IE, Krasnoperov VG, Khvotchev MV, Pluzhnikov KA, Volkova TM, et al. 1996. Cloning and structure of δ-latroinsectotoxin, a novel insect-specific member of the latrotoxin family. *J. Biol. Chem.* 271:7535–43

Dulubova IE, Sugita S, Hill S, Hosaka M, Fernandez I, et al. 1999. A conformational switch in syntaxin during exocytosis. *EMBO J.* 18:4372–82

Elrick DB, Charlton MP. 1999. α-Latrocrustatoxin increases neurotransmitter release by activating a calcium influx pathway at crayfish neuromuscular junction. *J. Neurophysiol.* 82:3550–62

Fernandez-Chacon R, Südhof TC. 1999. Genetics of synaptic vesicle function: towards the complete functional anatomy of an organelle. *Annu. Rev. Physiol.* 61:753–76

Filipov AK, Kobrinsky EM, Tsurupa GP,

Pashkov VN, Grishin EV. 1990. Expression of receptor for α-latrotoxin in Xenopus oocytes after injection of mRNA from rat brain. *Neuroscience* 39:809–14

Finkelstein A, Rubin LL, Tzeng MC. 1976. Black widow spider venom: effect of purified toxin on lipid bilayer membranes. *Science* 193:1009–11

Fritz LC, Tzeng C, Mauro A. 1980. Different components of black widow spider venom mediate transmitter release at vertebrate and lobster neuromuscular junctions. *Nature* 283:486–88

Frontali N, Ceccarelli B, Gorio A, Mauro A, Siekevitz P, et al. 1976. Purification from black widow spider venom of a protein factor causing the depletion of synaptic vesicles at neuromuscular junctions. *J. Cell Biol.* 68:462–79

Frontali N, Grasso A. 1964. Separation of three toxicologically different protein components from the venom of the spider *Latrodectus tredecimguttatus*. *Arch. Biochem. Biophys.* 106:213–18

Gasparini S, Kiyatkin N, Drevet P, Boulain JC, Tacnet F, et al. 1994. The low-molecular-weight protein which copurifies with alpha-latrotoxin is structurally related to crustacean hyperglycemic hormones. *J. Biol. Chem.* 269:19803–9

Geppert M, Goda Y, Hammer RE, Li C, Rosahl TW, et al. 1994. Synaptotagmin I: a major Ca^{2+} sensor for transmitter release at a central synapse. *Cell* 79:717–27

Geppert M, Khvotchev M, Krasnoperov V, Goda Y, Missler M, et al. 1998. Neurexin Iα is a major α-latrotoxin receptor that cooperates in a α-latrotoxin action. *J. Biol. Chem.* 273:1705–10

Gorio A, Rubin LL, Mauro A. 1978. Double mode of action of black widow spider venom on frog neuromuscular junction. *J. Neurocytol.* 7:193–202

Grasso A, Alema S, Rufini S, Senni MI. 1980. Black widow spider toxin-induced calcium fluxes and transmitter release in a neurosecretory cell line. *Nature* 283:774–76

Grasso A, Senni MI. 1979. A toxin purified from the venom of black widow spider affects the uptake and release of radioactive gamma-amino butyrate and norepinephrine from rat brain synaptosomes. *Eur. J. Biochem.* 102:337–44

Gray JX, Haino M, Roth MJ, Maguire JE, Jensen PN, et al. 1996. CD97 is a processed, seven-transmembrane, heterodimeric receptor associated with inflammation. *J. Immunol.* 157:5438–34

Hata Y, Butz S, Südhof TC. 1996. CASK: a novel dlg/PSD95 homologue with an N-terminal CaM kinase domain identified by interaction with neurexins. *J. Neurosci.* 16:2488–94

Hata Y, Davletov B, Petrenko AG, Jahn R, Südhof TC. 1993a. Interaction of synaptotagmin with the cytoplasmic domains of neurexins. *Neuron* 10:307–15

Henkel AW, Betz WJ. 1995. Monitoring of black widow spider venom (BWSV) induced exo- and endocytosis in living frog motor nerve terminals with FM1-43. *Neuropharmacology* 34:1397–406

Hlubek MD, Stuenkel EL, Krasnoperov VG, Petrenko AG, Holz RW. 2000. Calcium-independent receptor for α-latrotoxin and neurexin 1α facilitate toxin-induced channel formation: evidence that channel formation results from tethering of toxin to membrane. *Mol. Pharmacol.* 57:519–28

Hsueh YP, Yang FC, Kharazia V, Naisbitt S, Cohen AR, et al. 1998. Direct interaction of CASK/LIN-2 and syndecan heparan sulfate proteoglycan and their overlapping distribution in neuronal synapses. *J. Cell Biol.* 142:139–51

Hurlbut WP, Chieregatgti E, Valtorta F, Haimann C. 1994. α-Latrotoxin channels in neuroblastoma cells. *J. Membr. Biol.* 138:91–102

Ichtchenko K, Bittner MA, Krasnoperov V, Little AR, Chepurny O, et al. 1999. A novel ubiquitously expressed α-latrotoxin receptor is a member of the CIRL family of G-protein-coupled receptors. *J. Biol. Chem.* 274:5491–98

Ichtchenko K, Hata Y, Nguyen T, Ullrich B, Missler M, et al. 1995. Neuroligin 1: a splice-site specific ligand for β-neurexins. *Cell* 81:435–43

Ichtchenko K, Khvotchev M, Kiyatkin N, Simpson L, Sugita S, Südhof TC. 1998. α-Latrotoxin action probed with recombinant toxin: receptors recruit α-latrotoxin but do not transduce an exocytoctic signal. *EMBO J.* 17:6188–99

Ichtchenko K, Nguyen T, Südhof TC. 1996. Structures, alternative splicing, and neurexin binding of multiple neuroligins. *J. Biol. Chem.* 271:2676–82

Jahn R, Südhof TC. 1999. Membrane fusion and exocytosis. *Annu. Rev. Biochem.* 68:863–911

Khvotchev M, Lonart G, Südhof TC. 2000. Role of calcium in neurotransmitter release evoked by α-latrotoxin or hypertonic sucrose. *Neuroscience* 101:793–802

Khvotchev M, Südhof TC. 1998. Newly synthesized phosphatidylinositol phosphates are required for synaptic norepinephrine but not glutamate or GABA release. *J. Biol. Chem.* 273:21451–54

Khvotchev M, Südhof TC. 2000. α-Latrotoxin triggers transmitter release via direct insertion into the presynaptic plasma membrane. *EMBO J.* 19:3250–62

Kiyatkin N, Dulubova I, Chekhovskaya I, Lipkin A, Grishin E. 1992. Structure of the low-molecular-weight protein copurified with α-latrotoxin. *Toxicon* 7:771–74

Kiyatkin N, Dulubova I, Grishin E. 1993. Cloning and structural analysis of α-latroinsectotoxin cDNA. Abundance of ankyrin-like repeats. *Eur. J. Biochem.* 213:121–27

Kiyatkin N, Dulubova IE, Chekhovskaya IA, Grishin E. 1990. Cloning and structure of cDNA encoding α-latrotoxin from black widow spider venom. *FEBS Lett.* 270:127–31

Knipper M, Madeddu L, Breer H, Meldolesi J. 1986. Black widow spider venom-induced release of neurotransmitters: mammalian synaptosomes are stimulated by a unique venom component (α-latrotoxin), insect synaptosomes by multiple components. *Neuroscience* 19:55–62

Krasnoperov V, Beavis R, Chepurny OG, Little AR, Plotnikov AN, Petrenko AG. 1996. The calcium-independent receptor of α-latrotoxin is not a neurexin. *Biochem. Biophys. Res. Commun.* 227:868–75

Krasnoperov V, Bittner MA, Beavis R, Kuang Y, Salnikow KV, et al. 1997. α-Latrotoxin stimulates exocytosis by the interaction with a neuronal G-protein-coupled receptor. *Neuron* 18:925–37

Krasnoperov V, Bittner MA, Holz RW, Chepurny O, Petrenko AG. 1999. Structural requirements for α-latrotoxin binding and α-latrotoxin-stimulated secretion. A study with calcium-independent receptor of α-latrotoxin (CIRL) deletion mutants. *J. Biol. Chem.* 274:3590–96

Lelianova VG, Davletov BA, Sterling A, Rahman MA, Grishin EV, et al. 1997. α-Latrotoxin receptor, latrophilin, is a novel member of the secretin family of G protein-coupled receptors. *J. Biol. Chem.* 272:21504–8

Lishko VK, Sichenko EA, Storchak LG, Gimmerl'reikh NG. 1990. α-Lathrotoxin channels: permeability for divalent cations. *Biokimiia* 55:1578–83

Liu J, Misler S. 1998. Latrotoxin alters spontaneous and depolarization-evoked quantal release from rat adrenal chromaffin cells: evidence for multiple modes of action. *J. Neurosci.* 18:6113–25

Longenecker HE Jr, Hurlbut WP, Mauro A, Clark AW. 1970. Effects of black widow spider venom on the frog neuromuscular junction. Effects on end-plate potential, miniature end-plate potential and nerve terminal spike. *Nature* 225:701–3

Magazanik LG, Fedorova IM, Kovalevskaya GI, Pashkov VN, Bulgakow OV, Grishin EV. 1992. Selective presynaptic insectotoxin (α-latroinsectotoxin) isolated from black widow spider venom. *Neuroscience* 46:181–88

Malgaroli A, DeCamilli P, Meldolesi J. 1989. Distribution of alpha latrotoxin receptor in

the rat brain by quantitative autoradiography: comparison with the nerve terminal protein, synapsin I. *Neuroscience* 32:393–404

Martin TF, Loyet KM, Barry VA, Kowalchyk JA. 1997. The role of PtdIns(4,5)P2 in exocytotic membrane fusion. *Biochem. Soc. Trans.* 25:1137–41

Matsushita H, Lelianova VG, Ushkaryov YA. 1999. The latrophilin family: multiply spliced G protein-coupled receptors with differential tissue distribution. *FEBS Lett.* 443:348–52

Matteoli M, Haimann C, Torri-Tarelli F, Polak JM, Ceccarelli B, De Camilli P. 1988. Differential effect of α-latrotoxin on exocytosis from small synaptic vesicles and from large dense-core vesicles containing calcitonin gene-related peptide at the frog neuromuscular junction. *Proc. Natl. Acad. Sci. USA* 85:7366–70

McMahon HT, Rosenthal L, Meldolesi J, Nicholls DG. 1990. α-Latrotoxin releases both vesicular and cytoplasmic glutamate from isolated nerve terminals. *J. Neurochem.* 55:2039–47

Meldolesi J, Madeddu L, Torda M, Gatti G, Niutta E. 1983. The effect of α-latrotoxin on the neurosecretory PC12 cell line: studies on toxin binding and stimulation of transmitter release. *Neuroscience* 10:997–1009

Menegoz M, Gaspar P, Le Bert M, Galvez T, Burgaya F, et al. 1997. Paranodin, a glycoprotein of neuronal paranodal membranes. *Neuron* 19:319–31

Mironov SL, Sokolov YuV, Chanturiya AN, Lishko VK. 1986. Channels produced by spider venoms in bilayer lipid membrane: mechanisms of ion transport and toxic action. *Biochim. Biophys. Acta* 862:185–98

Misler S, Hurlbut WP. 1979. Action of black widow spider venom on quantized release of acetylcholine at the frog neuromuscular junction: dependence upon external Mg^{2+}. *Proc. Natl. Acad. Sci. USA* 76:991–95

Missler M, Fernandez-Chacon R, Südhof TC. 1998a. The making of neurexins. *J. Neurochem.* 71:1339–47

Missler M, Hammer RE, Südhof TC. 1998b. Neurexophilin binding to alpha-neurexins. A single LNS domain functions as an independently folding ligand-binding unit. *J. Biol. Chem.* 273:34716–23

Missler M, Südhof TC. 1998a. Neurexins: three genes and 1001 products. *Trends Genet.* 14:20–25

Missler M, Südhof TC. 1998b. Neurexophilins form a conserved family of neuropeptide-like glycoproteins. *J. Neurosci.* 18:3630–38

Muller GJ. 1993. Black and brown widow spider bites in South Africa. A series of 45 cases. *S. Afr. Med. J.* 83:399–405

Nguyen T, Südhof TC. 1997. Binding properties of neuroligin 1 and neurexin 1β reveal function as heterophilic cell adhesion molecules. *J. Biol. Chem.* 272:26032–39

Nicholls DG, Rugolo M, Scott IG, Meldolesi J. 1982. α-Latrotoxin of black widow spider venom depolarizes the plasma membrane, induces massive calcium influx, and stimulates transmitter release in guinea pig brain synaptosomes. *Proc. Natl. Acad. Sci. USA* 79:7924–28

Orlova EV, Rahman MA, Gowen B, Volynski KE, Ashton AC, et al. 2000. Structure of α-latrotoxin oligomers reveals that divalent cation-dependent tetramers form membrane pores. *Nat. Struct. Biol.* 7:48–53

Parpura V, Liu F, Brethorst S, Jeftinija K, Jeftinija S, Haydon PG. 1995. α-Latrotoxin stimulates glutamate release from cortical astrocytes in cell culture. *FEBS Lett.* 360:266–70

Pashkov V, Grico N, Tsurupa G, Storchak L, Shatursky O, et al. 1993. Monoclonal antibodies can uncouple the main α-latrotoxin effects: toxin-induced Ca^{2+} influx and stimulated neurotransmitter release. *Neuroscience* 56:695–701

Patzke H, Ernsberger U. 2000. Expression of neurexin Ialpha splice variants in sympathetic neurons: selective changes during differentiation and in response to neurotrophins. *Mol. Cell Neurosci.* 15:561–72

Peles E, Joho K, Plowman GD, Schlessinger J.

1997a. Close similarity between Drosophila neurexin IV and mammalian Caspr protein suggests a conserved mechanism for cellular interactions. *Cell* 88:745–46

Peles E, Nativ M, Lustig M, Grumet M, Schilling J, et al. 1997b. Identification of a novel contactin-associated transmembrane receptor with multiple domain implicated in protein-protein interactions. *EMBO J.* 16:978–88

Petrenko AG, Kovalenko VA, Shamotienko OG, Surkova IN, Tarasyuk TA, et al. 1990. Isolation and properties of the alpha-latrotoxin receptor. *EMBO J.* 9:2023–27

Petrenko AG, Lazareva VD, Geppert M, Tarasyuk TA, Moomaw C, et al. 1993. Polypeptide composition of the α-latrotoxin receptor. *J. Biol. Chem.* 268:1860–67

Petrenko AG, Ullrich B, Missler M, Krasnoperov V, Rosahl TW, Südhof TC. 1996. Structure and evolution of neurexophilin *J. Neurosci.* 16:4360–69

Pevsner J, Hsu SC, Scheller RH. 1994. n-Sec1: a neural-specific syntaxin-binding protein. *Proc. Natl. Acad. Sci. USA.* 91:1445–49

Poliak S, Gollan L, Martinez R, Custer A, Einheber S, et al. 1999. Caspr2, a new member of the neurexin superfamily, is localized at the juxtaparanodes of myelinated axons and associates with K+channels. *Neuron* 24:1037–47

Ponting CP, Hofmann K, Bork P. 1999. A latrophilin/CL-1-like GPS domain in polycystin-1. *Curr. Biol.* 20:R585–88

Pumplin DW, Reese TS. 1977. Action of brown widow spider venom and botulinum toxin on the frog neuromuscular junction examined with the freeze-fracture technique. *J. Physiol.* 273:443–57

Rahman MA, Ashton AC, Meunier FA, Davletov BA, Dolly JO, Ushkaryov YA. 1999. Norepinephrine exocytosis stimulated by α-latrotoxin requires both external and stored Ca^{2+} and is mediated by latrophilin, G proteins and phospholipase C. *Philos. Trans. R. Soc. London Ser. B* 354:379–86

Robello M, Fresia M, Maga L, Grasso A,

Ciani S. 1987. Permeation of divalent cations through alpha-latrotoxin channels in lipid bilayers: steady-state current-voltage relationships. *J. Membr. Biol.* 95:55–62

Robinson PJ, Sontag JM, Liu JP, Fykse EM, Slaughter C, et al. 1993. Dynamin GTPase regulated by protein kinase C phosphorylation in nerve terminals. *Nature* 365:163–66

Rosenmund C, Stevens CF. 1996. Definition of the readily releasable pool of vesicles at hippocampal synapses. *Neuron* 16:1197–1207

Rosenthal L, Zacchetti D, Madededu L, Meldolesi J. 1990. Mode of action of alpalatrotoxin: role of divalent cations in Ca^{2+}-dependent and Ca^{2+}-independent effects mediated by the toxin. *Mol. Pharmacol.* 38:917–23

Rubin GM, Yandell MD, Wortman JR, Gabor Miklos GL, Nelson CR, et al. 2000. Comparative genomics of the eukaryotes. *Science* 287:2204–15

Rudenko G, Nguyen T, Chelliah Y, Südhof TC, Deisenhofer J. 1999. The structure of the ligand-binding domain of neurexin 1β: regulation of LNS domain function by alternative splicing. *Cell* 99:93–101

Russell AB, Carlson SS. 1997. Neurexin is expressed on nerves, but not at nerve terminals, in the electric organ. *J. Neurosci.* 17:4734–43

Scheer H, Meldolesi J. 1985. Purification of the putative alpha-latrotoxin receptor from bovine synaptosomal membranes in an active binding form. *EMBO J.* 4:323–27

Scheer H, Prestipino G, Meldolesi J. 1986. Reconstitution of the purified alpha-latrotoxin receptor in liposomes and planar lipid membranes. Clues to the mechanism of toxinaction. *EMBO J.* 5:2643–48

Scheiffele P, Fan J, Choih J, Fetter R, Serafini T. 2000. Neuroligin expressed in nonneuronal cells triggers presynaptic development in contacting axons. *Cell* 101:657–69

Shepherd GM, Harris KM. 1998. Three-dimensional structure and composition of CA3→CA1 axons in rat hippocampal slices: implications for presynaptic connectivity and compartmentalization. *J. Neurosci.* 18:8300–10

Shiratsuchi T, Nishimori H, Ichise H, Nakamura Y, Tokino T. 1997. Cloning and characterization of BAI2 and BAI3, novel genes homologous to brain-specific angiogenesis inhibitor 1 (BAI1). *Cytogenet. Cell Genet.* 79:103–8

Simmons CR. 1991. The hunt is on for an elegant lady who inflicts pain. *Wall Str. J.* Aug. 15, p. 1

Song JY, Ichtchenko K, Südhof TC, Brose N. 1999. Neuroligin 1 is a postsynaptic cell-adhesion molecule of excitatory synapses. *Proc. Natl. Acad. Sci. USA* 96:1100–25

Südhof TC. 1995. The synaptic vesicle cycle: a cascade of protein-protein interactions. *Nature* 375:645–53

Südhof TC, DeCamilli P, Niemann H, Jahn R. 1993. Membrane fusion machinery: insights from synaptic proteins. *Cell* 75:1–4

Sugita S, Ichtchenko K, Khvotchev M, Südhof TC. 1998. α-Latrotoxin receptor CIRL/latrophilin 1 (CL1) defines an unusual family of ubiquitous G-protein-linked receptors. G-protein coupling not required for triggering exocytosis. *J. Biol. Chem.* 273:32715–24

Sugita S, Khvochtev M, Südhof TC. 1999. Neurexins are functional α-latrotoxin receptors. *Neuron* 22:489–96

Tobaben S, Südhof TC, Stahl B. 2000. The G protein-coupled receptor CL1 interacts directly with proteins of the Shank family. *J. Biol. Chem.* 275:36204–10

Tzeng MC, Siekevitz P. 1979. The binding interaction between α-latrotoxin from black widow spider venom and dog cerebral cortex synaptosomal membrane preparations. *J. Neurochem.* 33:263–74

Ullrich B, Ushkaryov YA, Südhof TC. 1995. Cartography of neurexins: more than 1000 isoforms generated by alternative splicing and expressed in distinct subsets of neurons. *Neuron* 14:497–507

Umbach JA, Grasso A, Gundersen CB. 1990. α-Latrotoxin triggers an increase of ionized calcium in Xenopus oocytes injected with rat brain mRNA. *Brain Res. Mol. Brain Res.* 8:31–36

Ushkaryov YA, Hata Y, Ichtchenko K, Moomaw C, Afendis S, et al. 1994. Conserved domain structure of β-neurexins. *J. Biol. Chem.* 269:11987–92

Ushkaryov YA, Petrenko AG, Geppert M, Südhof TC. 1992. Neurexins: synaptic cell surface proteins related to the α-latrotoxin receptor and laminin. *Science* 257:50–56

Ushkaryov YA, Südhof TC. 1993. Neurexin IIα: extensive alternative splicing generates membrane-bound and soluble forms in a novel neurexin. *Proc. Natl. Acad. Sci. USA* 90:6410–14

Valtorta F, Maddedu L, Meldolesi J, Ceccarelli B. 1984. Specific localization of the α-latrotoxin receptor in the nerve terminal plasma membrane. *J. Cell Biol.* 99:124–44

Verhage M, Maia AS, Plomp JJ, Brussaard AB, Heeroma JH, et al. 2000. Synaptic assembly of the brain in the absence of neurotransmitter secretion. *Science* 287:864–69

Vicentini LM, Meldolesi J. 1984. α-Latrotoxin of black widow spider venom binds to a specific receptor coupled to phosphoinositide breakdown in PC12 cells. *Biochem. Biophys. Res. Commun.* 121:538–44

Volynski KE, Nosyreva ED, Ushkaryov YA, Grishin EV. 1999. Functional expression of alpha-latrotoxin in baculovirus system. *FEBS Lett.* 442:25–28

Wanke E, Ferroni A, Gattanini P, Meldolesi J. 1986. α-Latrotoxin of the black widow spider venom opens a small, nonclosing cation channel. *Biochem. Biophys. Res. Commun.* 134:320–25

Zukowski CW. 1993. Black widow spider bite. *J. Am. Board Fam. Pract.* 6:279–81

Annu. Rev. Neurosci. 2001. 24:963–79

IMAGING AND CODING
IN THE OLFACTORY SYSTEM

John S Kauer and Joel White

Department of Neuroscience, Tufts University School of Medicine, 136 Harrison Avenue, Boston, Massachusetts 02111; e-mail: john.kauer@tufts.edu, joel.white@tufts.edu

Key Words   smell, odor coding, physiology, fluorescent dyes, functional imaging

■ **Abstract**   Functional imaging methods permit analysis of neuronal systems in which activity is broadly distributed in time and space. In the olfactory system the dimensions that describe odorant stimuli in "odorant space" are still poorly defined. One way of trying to characterize the attributes of this space is to examine the ways in which its dimensions are encoded by the neurons and circuits making up the system and to compare these responses with physical-chemical attributes of the stimuli and with the output behavior of the animal. For documenting distributed events as they occur, imaging methods are among the few tools available. We are still in the early stages of this analysis; however, a number of recent studies have contributed new information to our understanding of the odorant coding problem. This paper describes imaging results in the context of other data that have contributed to our understanding of how odors are encoded by the peripheral olfactory pathway.

INTRODUCTION

Functional imaging has played an important role in developing an understanding of olfactory function because of its ability to reveal events distributed in space and time. The distributed nature of olfactory information processing was first observed 50 years ago by Adrian (1953). Subsequent studies using serial single electrode recordings (Leveteau & MacLeod 1966) and, in some cases, multi-electrode recording (Moulton 1976) provided additional support for the idea that odor information is represented by activity occurring in parallel in different locations within the olfactory pathway. Until the advent of functional imaging methods, it was not possible to observe these events with spatial resolution that gives global views of distributed processes. These early electrophysiological studies stimulated exploration of distributed neuronal events that might not have been further scrutinized had there been more recognizable relationships between odorant structure and the single unit recordings.

The present discussion reviews the contributions of functional imaging. We focus on methods that permit observation of spatially distributed events related to

0147-006X/01/0621-0963$14.00

activity generated by odorants. Imaging methods that solely record spatial properties of structure, such as histological or immunohistochemical staining, are reviewed elsewhere. This structure/function dichotomy is somewhat blurred by recently developed vital staining and immunohistochemical methods and genetically generated molecular marking techniques that reveal functionally important topographical or biochemical features (see for example LaMantia et al 1992, Mori et al 1985, Mombaerts et al 1996, Dynes & Ngai 1998, Siegel & Isacoff 1997). While the emphasis here is on methods that permit observation of relatively rapid odor-generated events within the time frames of short sniff odorant applications, data from more static imaging methods that have contributed to ideas about odor coding are also included.

In the course of assembling a view of the coding process, we examine several levels of the olfactory pathway extending from the inhalation of odorants, to events in the olfactory sensory neurons (OSNs) in the nose, to the circuits of the olfactory bulb (OB). Most data are from vertebrate species, but invertebrate studies have contributed to a number of seminal observations that have influenced much thinking about basic olfactory system functioning (see Hildebrand & Shepherd 1997, Krieger & Breer 1999 for review and Ai et al 1998, Gervais et al 1996, Joerges et al 1997, Kimura et al 1998 for examples of specific experiments). Although technical details of the various methods are briefly discussed, this is not intended to be a comprehensive analysis of either neuronal imaging methods (e.g. see Ebner & Chen 1995, Lieke et al 1989, Kobal & Kettenmann 2000 for reviews) or olfactory coding (e.g. see Buck 1996, Christensen & White 2000, Hildebrand & Shepherd 1997, Laurent 1999, Mori et al 1999 for reviews). This discussion emphasizes functional imaging methods that resolve the spatial (and, in fortuitous cases, temporal) aspects of many neuronal events occurring simultaneously. Space and time are not immediately apparent as stimulus characteristics related to odor quality, but appear to be important neural dimensions representing aspects of odorant quality that are accessible to analysis by these methods.

IMAGING METHODS AND THE OLFACTORY SYSTEM

Imaging methods potentially provide advantages for examining a number of important functional attributes of nervous systems. These include the study of (a) spatially distributed events that occur during development, (b) changes in real time activity in response to perturbations by stimulation, (c) plastic responses to long-term external or internal environmental changes, and (d) changes due to injury, senescence, and normal (e.g. apoptotic) and abnormal (disease-associated) degeneration. To characterize each of these processes completely, one would need to obtain information with (a) spatial resolution that ranges from nanometers (molecular size) to meters (organism size); (b) temporal resolution extending from microseconds (for molecular events such as enzymatic reactions or isomerization, as in rhodopsin) to years (for developmental, plastic, and degenerative changes); (*c*) the ability to observe in three, as well as two, dimensions (using tools like confocal or multiple photon microscopy); (*d*) the ability to access the signals in tissue without invasive surgery; and (*e*) the ability to assess function without perturbation by the observation process. Methods presently available can approach some of these goals and, as with all analytical methods, one has to chose the appropriate method for the question being asked.

Figure 1 (see color insert) is an illustration of the application of voltage-sensitive dye recording to the analysis of potentials distributed through the dendritic architecture of a rat somatosensory pyramidal neuron (Antic et al 1999). These data are not taken from the olfactory system, but are presented here to illustrate several attributes of methods discussed in this paper. For most of these methods one must have optical access to the structure to be observed (Figure 1*A*). The intrinsic pixelation of imaging methods permits analysis of both a single point (pixel) on the structure over time (Figure 1*B,C*) as well as analysis of spatially distributed events over time (Figure 1*D*) by use of time-series pseudo color representations. This figure shows explicitly how a single imaging approach can provide high-resolution spatial characterization of where and when rapid electrical changes occur.

Imaging systems, in general, include use of markers for neuronal activity [radiolabels (2-deoxyglucose, *c-fos*), antibodies (*c-fos*), fluorescent dyes (voltage-sensitive or Ca^{++} reporters), or intrinsic optical properties of the tissue], optical detectors with good spatial and temporal resolution such as video cameras or photodiode arrays, and various computational devices for acquiring, digitizing, and storing the data and for manipulating, filtering, and presenting the images. These components, along with unimpeded optical access to the tissue, are required for all the imaging methods discussed here.

Imaging methods that have been applied to analysis of olfactory function include (*a*) observation of inhaled and exhaled air flows by schlieren photography (DA Kester, GS Settles & LJ Dodson-Dreibelbis, manuscript in preparation); (*b*) examination of neuronal activity by measuring glucose uptake with radiolabeled 2-deoxyglucose (2DG) (Kennedy et al 1975); (*c*) observation of changes in activity-related gene expression, such as *c-fos* (Sagar et al 1988); (*d*) measurement of intracellular calcium concentrations using fluorescent markers such as fura-2 (Grynkiewicz et al 1985) or calcium green (Friedrich & Korsching 1997); (*e*) measurement of transmembrane voltage using fluorescent voltage-sensitive dyes (Cohen & Lesher 1986, Blasdel & Salama 1986); and (*f*) measurement of activity-dependent changes in light-scattering and/or oxy- or deoxy-hemoglobin absorption (Grinvald et al 1986). Earlier results using some of these approaches are presented by Cinelli & Kauer (1992). Descriptions of other methods not applied to olfactory analysis can be found in some of the references mentioned above as well as in Lieke et al 1989 and Zochowski et al 2000.

AN EMERGING VIEW OF OLFACTORY CODING

Analysis of the results of imaging methods in the context of much new electrophysiological, biophysical, biochemical, and molecular biological data on the transduction process (for reviews see Ache & Zhainazarov 1995, Mombaerts 1999, Schild & Restrepo 1998) begins to describe some of the events that participate in the encoding of odorant information. We emphasize the peripheral pathway here because of the paucity of information on odor representation at higher levels (for a review of responses seen with electrical stimulation in animals see Litaudon et al 1997; for results using noninvasive methods with odorant stimulation in humans, see Levy et al 1999, Sobel et al 1998, Yousem et al 1999; reviews in Kobal & Kettenmann 2000, Zald & Pardo 2000).

Odorant Stimulus Intake and Distribution in the Nasal Cavity

The first step in the detection and discrimination of odors in terrestrial vertebrates is inhalation of the vapor phase substance into the nose and delivery of the stimulus to the OSNs in the olfactory epithelium (OE) lining the nasal cavity, a relatively simple structure in amphibians and reptiles, but highly convoluted in macrosmatic mammals. Distribution of odorant molecules to the OE is accomplished via the intrinsic characteristics of the inhaled air flow as well as via the aerodynamic properties of the cavity. Odor access to OSNs is also likely influenced by chromatographic interactions with the mucus lining (Mozell et al 1987). Although there are a number of studies that have calculated and measured the airflow in models (Hahn et al 1993, Keyhani et al 1997), there is surprisingly little work using direct methods to image flows either into or within the nasal cavity. Initial work has begun, however, on the nature of the access of the vapor phase stimulus to the nostrils.

Figure 2 (see color insert) shows the use of a schlieren imaging system (DA Kester, GS Settles & LJ Dodson-Dreibelbis, manuscript in preparation) to visualize air flow during the inhalation and exhalation cycle in the dog, based on temperature-dependent differences in refractive index. Dogs are macrosmats, capable of detecting concentrations of certain nitroaromatic compounds as low as 500 parts per trillion (Williams et al 1998). It has been consistently found that the inhalation and exhalation processes in these animals are complex, are not symmetrical, and are modified by the behavioral task that the dog is performing. In this side view (Figure 2, left) one can see how inhalation leads to the constrained flow of high-velocity air being drawn into the nose from the front. In a different animal (Figure 2, right) air is exhaled backward, away from the source. This redirection of flow during different parts of the sniffing cycle is governed by a muscle activated flap on the external nares.

These studies are in their early stages, but it is clear that sniffing behavior and the aerodynamics of air flow within the nasal cavity are important components of high performance detection behavior, although the precise effects on olfaction are still poorly characterized. It is likely that inhalation and exhalation processes mold the structure of the vapor pulse that is finally presented to the sensory elements and therefore do not simply present to the sensory surface replicas of the spatial and temporal properties of the original stimulus outside the nose. All coding events downstream of odor delivery depend upon this process. A deficiency in many experimental paradigms, often dictated by the constraints of the imaging method, is lack of data from normally sniffing, behaviorally attentive, animals. Another aspect of this process has been evaluated by measuring the influence of delivering odorant stimuli under different flow conditions that, although artificially manipulated, are at least related to flow changes during sniffing behavior (Kent et al 1995, 1996; Youngentob & Kent 1995) (see Figure 4, color insert).

Access to Receptor Cell Membranes

Once the odorant stimulus, which is almost always a complex mixture of compounds, is delivered to the sensory region of the nasal cavity, a number of physical events occur before transduction from chemical to neural information takes place. The stimulus molecules must diffuse through the mucus to reach the cilia of the receptor cells. This diffusion event is still poorly defined and has not been observed directly. It is possible that micro schlieren analysis in OE slices might provide information on this event. Although the mucus can be as much as 30 μm thick, it is likely that many OSN cilia float near the surface, and thus the distances over which diffusion occurs may be short.

Interactions of Odorants with Olfactory Receptor Neurons

Once through the mucus, the small odorant molecules likely bind to proteinaceous receptors (Buck & Axel 1991) in the plasma membrane of the cilia, triggering one or more transduction cascades that generally lead (in vertebrates) to depolarization of the olfactory cilia and knob. There is much new information on many aspects of these transduction events (see Ache & Zhainazarov 1995, Buck 1996, Krieger & Breer 1999, Schild & Restrepo 1998 for reviews), but details about adaptation, the intrinsic sensitivity of individual receptors, and relationships between biochemical cascades and OSN firing patterns remain undefined.

Spatial and temporal consequences of these activation events within the cilia and olfactory knob are seen by the calcium imaging experiments shown in Figure 3a (see color insert) (Leinders-Zufall et al 1998). The top row of images (A–E) shows the progression of increases in intracellular Ca^{++} taken before, 2, 4, and 16 s after stimulation with cineole in a dissociated salamander OSN. This sequence explicitly demonstrates that odorant stimulation leads first to uniform Ca^{++} changes within all the cilia. An increase in Ca^{++} then spreads to the knob, dendrite, and soma, decaying in the different compartments with different time courses. As seen in the plots (F–H), the changes in intracellular calcium concentration last longer than responses recorded using electrophysiological methods and longer than the time course of a normal sniff (about one second in amphibia, less in mammals). This imaging method extends patch clamp recordings from single ciliary membranes (Nakamura & Gold 1987) by clearly showing the degree of involvement and the dynamic spread of the initial response.

Distribution of Receptor Types in the Olfactory Epithelium

There are perhaps as many as 1000 genes in the mammal that encode the 7 transmembrane G-protein-linked receptor molecules found in the cilia. It is not yet known how many of the potential population of 1000 receptor types are expressed at any one time in a single animal. It appears that each OSN may express only one or a few of each receptor type. Based on in situ hybridization imaging OSNs expressing any one receptor are found in regional areas or zones of the nasal cavity in among other OSNs expressing other receptors. In rodents these zones have complex topographies, but are generally oriented in the anterior-posterior direction as stripes along different regions of the turbinates (see Buck 1996 for review). The task of correlating spectra of odorant responses from dissociated, individual OSNs and from OSNs in situ in the intact epithelium with the distribution of molecular receptor types has just begun (see below).

Responses of Individual OSNs

Based on many years of in vivo electrical recording, it is well established that single OSNs, in virtually all species studied, typically respond with temporally patterned bursts of action potentials to more than one odorant (e.g. see Gesteland et al 1965, Duchamp-Viret & Duchamp 1997 for review). Different OSNs have different response spectra and different sensitivities within the range of compounds to which they respond. Recent calcium imaging studies in dissociated OSNs in which the expressed receptor has been identified also show different responses to groups of test compounds, sometimes responding to relatively few chemically related stimuli, sometimes responding to broadly divergent chemical structures (Malnic et al 1999, Krautwurst et al 1998, Touhara et al 1999). These kinds of studies have begun to lay the foundation for explicit comparisons between receptor types and the structural features of odorants that bind to them. Other studies have exploited Ca^{++} imaging methods for examining details of the transduction process (Noe & Breer 1998, Schild et al 1995, Gomez et al 2000).

Calcium imaging of OSNs identified by retrograde labeling from different olfactory bulb glomerular regions has shown that cells projecting to the dorso-medial in contrast to the dorso-lateral bulb not only respond to somewhat different compounds, they also have different breadths of response (see Figure 3b and OB imaging described below) (Bozza & Kauer 1998). A remaining challenge for imaging studies is to assess how groups of simultaneously imaged, individual OSNs respond to odorants in real time when in their normal positions in the OE where the mucus, their relationships with surrounding supporting cells, and their axons are still intact. There are no observations yet available from OSNs that project to bulbar regions other than the optically accessible dorsal surface. A possible solution to this problem, if the resolution of the technique is improved, may be to use imaging methods such as functional magnetic resonance imaging (fMRI) that are not limited to surface structures and do not require surgical intervention (Yang et al 1998). Differential binding of odor ligands to different receptors is presumably the initial event in odorant discrimination, carrying the first information in the neural encoding process. As noted above, because single OSNs likely express one or few molecular receptor types and because each cell can respond to a number of odorants, the first order neurons likely do not recognize overall odorant structure, rather they recognize some chemical feature shared by the effective ligands. This hypothesis is consistent with observations and data gathered from the olfactory bulb and its essential tenet was suggested many years ago based on chemical analyses and human behavioral responses (Beets 1970, Polak 1973).

All the studies at each level of the olfactory pathway that have sought to characterize the odorant responsiveness of individual cells have been confounded by the problem of how to adequately test the universe of all odors ("odor space") to which a particular experimental animal is sensitive. Because the dimensionality of this space must be defined by behavioral measures, animals such as insects offer an opportunity for choosing odorant stimuli with behavioral relevance (Stopfer et al 1997). Whereas insects offer the possibility of observing behavior and physiology simultaneously, a number of other studies in vertebrates have also focused on correlating behavioral relevance with imaged activity (Coopersmith & Leon 1984, Galizia et al 1999, Kent et al 1995, Johnson & Leon 2000, Woo et al 1996, Youngentob & Kent 1995).

Distribution of Responses across the Olfactory Epithelium

Early electrophysiological studies in frog showed that responses to single compounds were widely distributed across the mucosa of the surgically exposed OE (Mustaparta 1971). Receptive field mapping in the salamander, also in surgically exposed preparations, showed that single mitral/tufted cells could be activated by punctate stimulation over large OE regions (Kauer & Moulton 1974; see summary in Kauer 1987). This study also provided evidence for widespread sensitivity to an odorant across the OE and showed that axons from distributed OSNs responsive to the same odorant converge onto single bulbar output cells. Other studies in which odorant responses in the amphibian OE were observed directly by voltage-sensitive dye (VSD) imaging (Kent & Mozell 1992) also showed distributed activity over wide areas that varied with odor.

None of these experiments, however, examined patterning of OE activity with respect to the aerodynamics of how the odorants were delivered to the mucosal surface. More recent imaging experiments using VSD imaging have explicitly tested the effect of odorant delivery direction and examined the relative contributions of inherent differences in OSN receptor sensitivities compared with differences imposed by air flow direction and with chromatographic effects due to odorant retention in the mucus. Figure 4 (see color insert) (Kent et al 1996) shows the dramatic differences in the patterns of VSD responses between strongly sorbed odorants (carvone and ethyl acetoacetate) and a weakly sorbed compound (propyl acetate) when the odorants were delivered either tangential (from the right) or normal (from the top) to the OE surface. The top and bottom of the figure show response profiles at high (top) and low (bottom) flow rates. The effect is best seen at the lower flow rate at the bottom. Strongly sorbed odorants delivered tangentially (left column) generate large response gradients, with the larger response close to the odorant input (right). The weakly sorbed odorant, propyl acetate, shows much less of a response gradient. The same odorants delivered from above directly to the OE (middle column), rather than along it, show more uniformly distributed response patterns. The imposed flow rate responses (right column) are generated by subtracting the middle from the left column. These data indicate that there are both response pattern differences generated by intrinsic OSN sensitivities as well as pattern differences imposed by the odorant delivery process itself.

These studies suggest that the aerodynamic events, some of which are related to sniffing behavior, and some to the physical properties of the nasal cavity, can be important for understanding how responses occur within populations of distributed OSNs. Activation of sensory receptor cell populations is clearly sensitive to delivery conditions and is a widely distributed event showing complex responses in space across the convolutions of the turbinate structure and in time across regions having differential access to the odorant stream. One could imagine that evolutionary pressures have influenced the distribution of OSNs expressing particular molecular odorant receptors so that they are appropriately located to take advantage of the distribution of odorant due to their chemical properties and to the aerodynamics of the cavity (Mozell et al 1987).

Projections of Olfactory Sensory Neurons to the Glomeruli of the Olfactory Bulb

Evidence from several experiments now shows that the neighbor relations between OSNs in the OE are not preserved in their projection to the OB (see Mombaerts et al 1996) and are, therefore, unlike the connections between the periphery and central targets in the visual, auditory, and somatosensory systems. OSNs expressing a particular molecular receptor in the left or right nasal cavity, distributed within one of the OE receptor distribution zones, project their axons in a convergent fashion onto (usually) two glomeruli symmetrically situated on either side of the ipsilateral olfactory bulb [see Figure 7 (color insert) and review by Buck 1996]. It is not known what determines which subset of OSNs from this population projects to one or the other of the two glomeruli.

The presence of paired glomeruli is interesting because there are anatomical pathways connecting homologous lateral and medial OB regions (Schoenfeld et al 1985) that might use the information coming to the two glomerular targets to enhance detection or discrimination. It is generally thought that all the OSNs projecting to the two glomeruli share the same response profile, because they express the same receptor. However, at least one study (Malnic et al 1999) has shown that dissociated OSNs ostensibly expressing the same receptor can sometimes show different responses to a particular odorant set. Much additional work is needed to relate the response properties of OSN populations expressing particular receptors to the response properties of their glomerular targets (Bozza & Kauer 1998).

Imaged Responses in the Olfactory Bulb

From the early 2DG experiments to the most recent intrinsic-signal imaging studies, the consistent finding has been that even single odorant compounds generate activity that is widely distributed across the glomerular layer of the OB. Different odorants generate different response patterns, and the patterns appear to relate in size and position to the modularity of glomeruli characteristic of this layer where the OSN axons terminate. In studies achieving spatial resolution adequate to see glomerular-sized structures it appears that different numbers of glomeruli are activated to different degrees by different odorant compounds and probably at different times (see below). It has also been shown that compounds with structural similarities to one another, as in homologous series, activate nearby structures. While here we focus on the application of imaging methods to odor coding in vivo, there are a number of studies carried out in vitro in which other attributes of OB circuit physiology have been examined using imaging approaches (Keller et al 1998; Lam et al 2000; Senseman 1996; Wachowiak & Cohen 1999; Wellis & Kauer 1993, 1994).

Imaging studies using 2DG uptake provided one of the first opportunities to examine patterns of distributed, glomerular, activity generated by odor stimulation (Stewart et al 1979, Coopersmith & Leon 1984). Results from this method have helped guide the development of odorant coding hypotheses and have specifically directed experiments that have been pursued using other methods. For example, Bell and colleagues (1987) demonstrated that propionic acid activates a particularly prominent 2DG uptake region in the dorso-medial OB. This finding was then used by Mori and colleagues (Imamura et al 1992, Mori et al 1992) to select a location at which to examine odorant response profiles of single-unit output mitral cells. These single-unit studies were among the first to characterize odorant response spectra of mitral cells in defined OB locations, giving rise to the concept of mitral cell–molecular receptive ranges for odorants with structural similarities.

The naturally discrete anatomical features of glomerular structures and the localization of odor-generated activity within the glomerular layer have long been thought to represent modules that could provide a basis for an odorant coding scheme (Shepherd 1981). Observations with VSDs showed that odorants stimulate modular-like structures across the layers of the in vivo salamander OB (Kauer & Cinelli 1993). Modules extending through the layers of the rat OB were also suggested by studies using c-fos expression (Guthrie et al 1993). Figure 5a (see color insert) shows increased c-fos expression in modular-like groups of glomeruli and regions of underlying OB layers after exposure of the animal to peppermint (*A*), isoamyl acetate (*B*), and air (*C*) for 30 min. These *c-fos* expression patterns were similar to 2DG glomerular patterns generated by the same odors, but in addition, showed broad, flask-shaped regions in the deeper layers related to the glomerular foci, similar to the voltage-sensitive dye patterns in the salamander.

The modular nature of OB activity has been further investigated by Johnson & Leon (2000) using quantitative 2DG methods in the rat. Examples of activity patterns generated by four concentrations of two related odorants are shown in Figure 5*b* (see color insert). The pseudo-colored images depict the glomerular layer unfolded on its ventral meridian as shown at the top of the figure. These studies also indicated that compounds that are perceived by humans to differ qualitatively with intensity show changes in their activity pattern as concentration is increased (see white arrows in pentanal series). A second odorant, methyl pentanoate, that is not perceived differently at different concentrations did not show qualitative activity pattern differences.

Activity patterns also have been examined by imaging fluorescent dyes or intrinsic signals in a number of species in vivo: bees (Joerges et al 1997), zebrafish (Friedrich & Korsching 1998, 1997), salamander (Cinelli et al 1995a,b; Cinelli & Kauer 1995a), and rat (Rubin & Katz 1999, Uchida et al 2000). Two examples are shown in Figure 6 (see color insert).

Figure 6*a* shows presynaptic activity in OSN terminals of zebrafish OB glomeruli labeled by anterograde transport of a Ca^{++} indicator dye delivered to the OE (Friedrich & Korsching 1997). Stimulation with various amino acids clearly generates different patterns of activity that are shared by certain glomeruli for a number of the compounds. For example, the region indicated by the white arrow (1) in the upper right "Trp" panel showed activity to all neutral amino acids; the white arrow (2) in the central "Ile" panel shows a region with activity to Val, Ile, Leu, and Met; and the white arrow (3) in the lower left "Lys" panel shows a region with activity to basic amino acids. Consistent with 2DG and *c-fos* studies, activity for all compounds was found in a number of sites with some overlap.

Figure 6*b* shows data from a study, which used intrinsic signal imaging, asking similar questions about the molecular receptive ranges of glomeruli in the rat OB (Rubin & Katz 1999). Stimulation consisted of a homologous series of n-aliphatic aldehydes with different carbon chain lengths. In (*A*), the histogram bars represent the glomerular signal intensities at positions shown by the filled circle in the inset. The different aldehyde chain lengths are shown above. Shaded bars are responses to the aldehydes that were >50% of the highest response; open bars are responses <50%. Note that a particular glomerulus responded best to aldehydes with chain lengths similar to the compound that gave the largest response. Furthermore, a number of the glomeruli within the extent of the inset showed similar response ranges. There were, of course, glomeruli in this field that showed no response to these stimuli, and one does not know what other compounds might have generated activity in these glomeruli had they been tested. Figure 6*b* (*B*) shows the ranges of aldehyde carbon chain length to which all the glomeruli observable in this experiment responded.

Another recent study using intrinsic signal imaging in the rat demonstrated two response domains in the dorsal OB that have activity to sets of odorants having different functional groups (Uchida et al 2000). This experiment showed responses similar to those described above for a series of aldehydes. Additional data were also presented that suggested there may be subdivisions within the observed OB domains in which responses not only relate to carbon chain length but also to functional groups and molecular branching patterns. Again, not all glomeruli within the areas observed responded to the compounds tested.

Taken together, these findings indicate that odorant representation in the OB has the following characteristics:

1. Monomolecular odorant stimulation leads to patterns of activity that are distributed nonhomogeneously over the glomerular sheet and underlying layers.

2. Higher concentration stimuli give larger patterns of response that may change pattern shape.

3. Activity in the deeper OB layers is often located beneath foci of activity in the glomerular layer and may reflect modules extending through the bulbar layers.

4. Spatially distributed activity patterns probably relate to certain structural features of the stimulus. Such differential patterns undoubtedly exist, and their elucidation has been a major advance in beginning to describe the coding process.

5. Some classes of compounds generate activity in identifiable glomerular regions, but within these domains, not all glomeruli respond to the tested compounds.

6. There are many other regions of the bulb that have not yet been observed because of optical inaccessibility.

The larger picture of what actually happens during the second or less that it takes an animal to make a behavioral odor discrimination is, however, likely to be more complex than these patterns suggest. There are several reasons for this. The first is the question of how short-term, temporally patterned, behaviorally relevant responses contribute to the spatial patterns observed by methods that require extended stimulation such as 2DG, *c-fos*, and imaging of intrinsic signals. The second is the question of whether the patterns are really stationary if observed in real time (Cinelli et al 1995a). For example, does activity actually rise and fall within different areas that are represented by static observation methods that average activity over time? Third, it is difficult to reconcile exclusively spatial encoding of odorant information with studies in which large lesions of the bulb that substantially disrupt regions involved in the spatial responses show relatively little effect on behavioral discrimination (Hudson & Distel 1987, Lu & Slotnick 1998, Slotnick et al 1997).

Relatively few imaging studies have used methods that have adequate temporal resolution to observe neuronal events in real time (see Cinelli et al 1995a–c). In vertebrates many electrophysiological studies have consistently shown that activity patterns in the OB measured by both electroencephalogram and single unit methods change over time (e.g. Eeckman & Freeman 1990, Freeman & Di Prisco 1986, Harrison & Scott 1986, Wellis et al 1989, Hamilton & Kauer 1989, Kauer 1974, Meredith & Moulton 1978). In invertebrates attempts to understand temporal patterning in olfactory lobe circuits have been taken a step further by Laurent and his colleagues by comparing single-cell firing patterns with field potential patterns and with behavior (see Stopfer et al 1997, Laurent 1999 for review). Such complex temporal properties likely also occur in vertebrates and probably relate to the relationships among single neuron firing patterns and synchronous firing among populations. fMRI methods may be useful for such studies in the future.

In addition to caveats relating to temporal and spatial issues, it should be noted that the system by which odorants are encoded and classified by the olfactory system need not coincide with conventional chemical classification related to, for example, functional groups and carbon chain length. Based on the response ranges of individual OSNs and their concentration sensitivities (Duchamp-Viret & Duchamp 1997), it would seem likely that olfactory molecular receptors do not recognize attributes of their cognate ligands via high affinity binding. Individual OSNs appear to be broadly responsive, rather than quite specific. Therefore, it may be unreasonable to expect to be able to define strict pharmacological relationships among ligand candidates and particular molecular receptor types. Concerns about specificity also, of course, always need to be considered in the context of how many odorants are tested, what concentrations were examined, and how long the stimuli were applied. This latter concern is especially important for the sense of smell because the olfactory system adapts so readily.

Figure 7 (see color insert) summarizes some of the attributes of the peripheral system as described above. This diagram is based on data from the salamander, but the general principles appear to hold for other vertebrates. OSNs expressing molecular receptors for recognition sites on odorant molecules are distributed within bounded regions across the OE. Axonal projections from OSNs expressing one receptor type converge on glomeruli in the OB. The groups of mitral/tufted, periglomerular, and granule cells related to the activated glomerular regions may form functional modules, a number of which participate in the encoding of odorant molecular structure. This is, of course, a highly simplified diagram, but a number of its features have been revealed by the imaging methods discussed above.

CONCLUSIONS

The use of methods for imaging neuronal activity has enhanced our understanding of the odor coding process. The results are beginning to generate complementary views that emphasize different aspects of coding in space and time. There is still no single approach that provides all of the ideal properties of noninvasive, high spatial

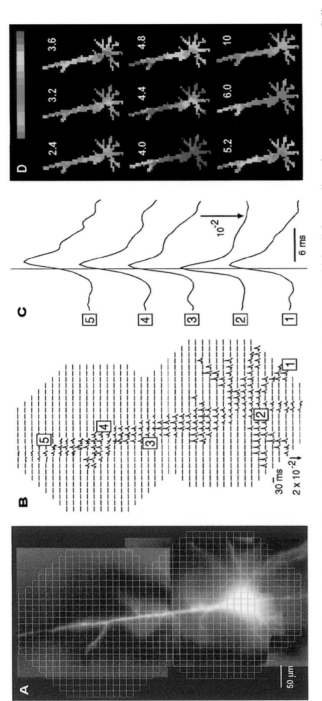

Figure 1 Multisite monitoring of a spike evoked by shock to the white matter in a pyramidal cell from a slice of rat somatosensory cortex. Cell was filled with the voltage sensitive dye JPW3028. The entire extent of the cell is covered by positioning the photodiode detector array over two positions (see two grids). (*A*) CCD camera fluorescence image of the cell. (*B*) Distributions of action potentials taken at the two diode array positions. Fluorescence recordings are normalized to resting light level. (*C*) Expanded time base action potential signals, scaled to the same height, from diodes at different locations shown in *B*. (*D*) Color coded representation of data in *B* provides information about both spatial and temporal dimensions of the electrical events (color range from blue to red indicates peak of action potential in red; numbers are times in ms after the shock). From Antic et al 1999 with permission.

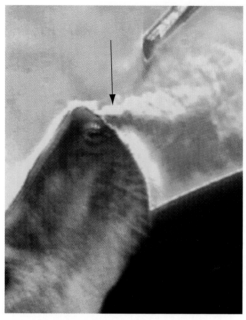

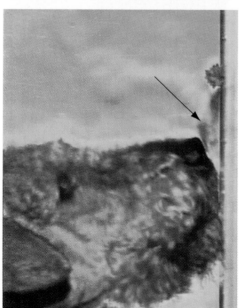

EXHALATION

INHALATION

Figure 2 Schlieren images of air flow in two different dogs during inhalation (*left*) and exhalation (*right*). Dog at the left is sniffing a small food pellet. Notice the air stream entering the nose from the front (*arrow*). Dog at the right is exhaling after sniffing from source (*probe at right*). Notice the air leaving the nose in a backwards direction (*arrow*) as a result of the movement of a muscular flap at the naris. From Settles 2000 with permission.

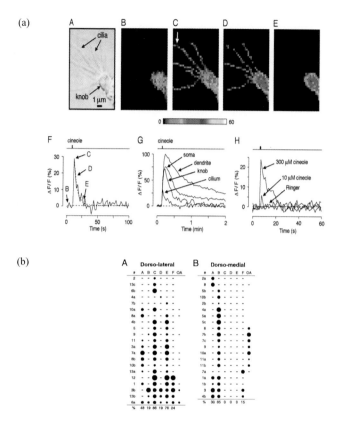

Figure 3 (*a*) Confocal measurements of transient increases in intracellular Ca^{++} concentration using fluo-3 from cilia and knob of a salamander olfactory sensory neuron (OSN) after a 1 sec pulse of cineole, 300 mM. (*A*) Phase-contrast image showing several cilia emanating from the knob. (*B—E*) Pseudocolored series (0.5 Hz) of fluorescence images taken before the odor (*B*), near peak fluorescence at 2 sec (*C*), 4 sec (*D*), and 16 sec (*E*) after the odor pulse. In *E* the ciliary signal has returned to baseline, but Ca^{++} remains high in the knob. (*F*) Time courses of ciliary Ca^{++} transient from cilium labeled by the white arrow in *C*. The decay time constant (dotted line exponential fit) was 5.3 sec. Letters indicate when frames above were taken. (*G*) Time courses of odor-induced Ca^{++} changes in various OSN compartments. Data are from the neuron shown in *A—E*. (*H*) Time course and peak amplitude of the ciliary signal depends on odorant concentration. Odor-free Ringer's gives no signal. From Leinders-Zufall et al 1998 with permission.
(*b*) Example of odorant response profiles of individual OSNs projecting to different bulbar regions. Different OSNs (row numbers) responding to various odorant mixtures [column letters A, B, C, D, E, F, OA (organic acids)]. (*A*) Dorso-lateral projecting OSNs (n = 21). (*B*) Dorso-medial projecting neurons (n = 20). Blank spaces indicate a stimulus was not tested; dashes indicate no response. The percentage of cells responding to each mixture is below each plot. Cells with the same number are from the same animal and are ordered by increasing breadth of response. Size of filled circles indicates relative response magnitudes. From Bozza & Kauer 1998 with permission.

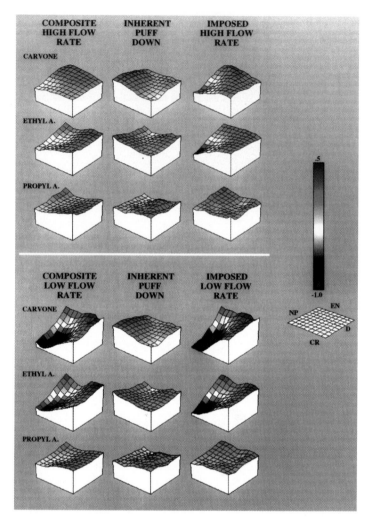

Figure 4 Voltage-sensitive dye activity patterns from an in vitro rat olfactory epithelium (OE). Color-enhanced surface plots for composite, inherent, and imposed activity patterns on the medial surface of the turbinates in response to carvone, ethyl acetoacetate, and propyl acetate when drawn across the surface at either 440 cc/min (*top*) or 100 cc/min (*bottom*). "Composite" patterns are observed when the odorant is delivered from the external naris (EN) side of the recording array. "Inherent" patterns are observed by stimulating directly down onto the OE from above so there is no tangential flow component. "Imposed" patterns are generated by subtracting the inherent from the composite responses. The z-axis shows relative response magnitudes, measured as natural logarithms, normalized to the response to amyl acetate puffed in the downward direction. Log values vary from -1.0 to 0.5 as indicated by the color bar to the right. EN, external naris; NP, nasopharynx; D, dorsal; CR, cribriform plate. From Kent et al 1996 with permission.

(a)

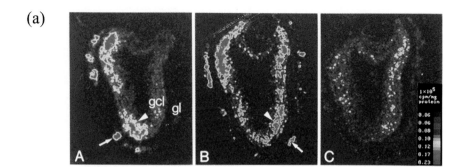

(b)

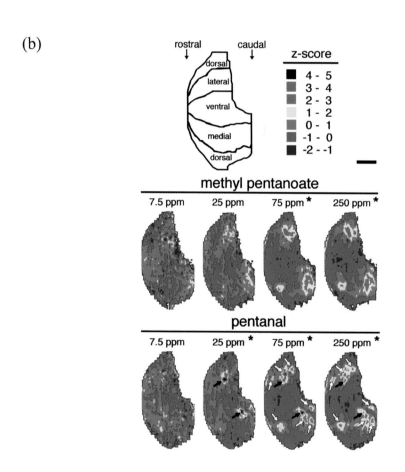

See text page C-6

See figure page C-5

Figure 5 (a) (A—C) Pseudocolored autoradiograms of rat olfactory bulb (OB) sections hybridized with ^{35}S-labeled c-fos cRNA of animals exposed to peppermint odor (A), isoamyl acetate (B), or clean air (C) for 30 min. Distinct hybridization patterns are seen in glomerular (gl, *solid arrows*) and granule cell layers (gcl, *arrowheads*) with odor (A and B), but much less with air (C). From Guthrie et al 1993 with permission. (b) Averaged patterns of 2DG uptake across the glomerular layer exposed to different concentrations of two odorants. Each image is data averaged from 3—5 rats. The orientation of the unfolded glomerular layer is shown in the upper left panel. Z-score values are relative degrees of 2DG uptake. Asterisks indicate concentrations that gave patterns different from air. For pentanal, black arrows denote patterns that were consistent across all concentrations above threshold. White arrows indicate patterns that emerged at higher concentrations. Scale bar = 2 mm. From Johnson & Leon 2000 with permission.

See figure page C-7

Figure 6 (a) Glomerular activity patterns in the zebrafish olfactory bulb. Activity was induced by 18 amino acids (aa's) (10 uM) in the aa-responsive subregion as measured by changes in fluorescence of calcium green dextran. Glomeruli were loaded by anterograde transport of the dye from the olfactory epithelium. Note each stimulus induces a unique pattern but with some shared active glomeruli. Arrows show glomerular modules responding to all neutral aa's (1); to Val, Ile, Leu, and Met (2); and to basic aa's (3). From Friedrich & Korsching 1997 with permission. (b) Individual rat glomeruli appear to have restricted receptive ranges when tested with a limited set of odorants varying along a single presumed molecular dimension. (A) For four glomeruli in the inset at right, the magnitude of the optically imaged response of each glomerulus is plotted as a percentage of the maximal response. Odorants eliciting >50% of the maximal response were considered effective in activating a glomerulus (filled bars; open bars are glomeruli with <50%response). These four glomeruli were activated by between one and three members of the homologous series. (B) Summary of molecular receptive ranges from 40 glomeruli (n = 5 bulbs in 4 animals) to the homologous series. From Rubin & Katz 1999 with permission.

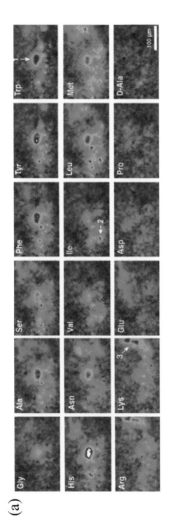

(a)

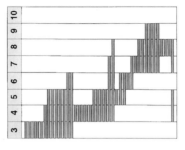

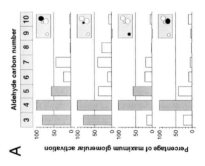

(b)

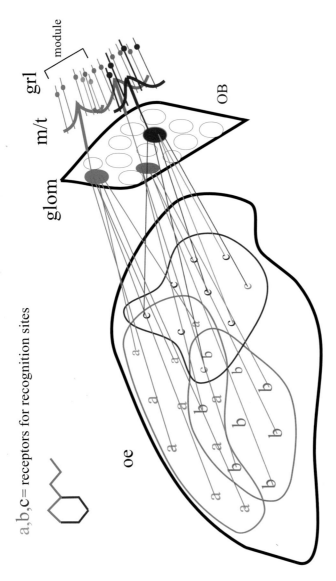

Figure 7 Stylized schematic diagram of the vertebrate (salamander) peripheral olfactory circuit. Adapted from Kauer 1987 and Kauer & Cinelli 1993. Olfactory sensory neurons (OSNs) expressing different molecular receptors (lower case letters on olfactory epithelium) project convergent axon connections to pairs of glomeruli in the olfactory bulb. Only one of the pair of target glomeruli is shown. Schematic molecular structure is shown with sites (denoted by different colors) recognized by OSNs expressing the appropriate receptors.

and temporal observation. By gathering data using approaches that complement one another and that mitigate each others' deficiencies, we begin to assemble a view of olfactory events that can be related to identifiable properties of the odorant stimulus world and to the extraordinary behavioral abilities of animals using this sensory modality. Among the many important research problems that still need to be pursued are (*a*) elucidation of the detailed molecular interactions between defined olfactory receptor molecules and odorants structure, (*b*) identification of which anatomical and physiological details in olfactory circuits are required for odor detection and discrimination in real world behaviors, and (*c*) characterization of the modes by which olfactory information is encoded in higher olfactory centers extending from prepyriform to neocortex. Given the progress so far and the history of the contributions of imaging methods, it is likely that imaging studies will also contribute significantly to the solutions of these problems in the future.

ACKNOWLEDGMENTS

This work was carried out with support from The National Institutes of Health–NIDCD, the Office of Naval Research, and the Defense Advanced Research Projects Agency. We thank Barbara Talamo for critically reading the manuscript.

Visit the Annual Reviews home page at www.AnnualReviews.org

LITERATURE CITED

Ache BW, Zhainazarov A. 1995. Dual second-messenger pathways in olfactory transduction. *Curr. Opin. Neurobiol.* 5:461–66

Adrian ED. 1953. Sensory messages and sensation. The response of the olfactory organ to different smells. *Acta Physiol. Scand.* 29:5–14

Ai H, Okada K, Hill ES, Kanzaki R. 1998. Spatio-temporal activities in the antennal lobe analyzed by an optical recording method in the male silkworm moth Bombyx mori. *Neurosci. Lett.* 258:135–38

Antic S, Major G, Zecevic D. 1999. Fast optical recordings of membrane potential changes from dendrites of pyramidal neurons. *J. Neurophysiol.* 82:1615–21

Beets MGJ. 1970. The molecular parameters of olfactory response. *Pharmacol. Rev.* 22:1–34

Bell GA, Laing DG, Panhuber H. 1987. Odour mixture suppression: evidence for a peripheral mechanism in human and rat. *Brain Res.* 426:8–18

Blasdel GG, Salama G. 1986. Voltage-sensitive dyes reveal a modular organization in monkey striate cortex. *Nature* 321:579–85

Bozza TC, Kauer JS. 1998. Odorant response properties of convergent olfactory receptor neurons. *J. Neurosci.* 18:4560–69

Buck LB. 1996. Information coding in the vertebrate olfactory system. *Annu. Rev. Neurosci.* 19:517–44

Buck LB, Axel R. 1991. A novel multigene family may encode odorant receptors: a molecular basis for odor recognition. *Cell* 65:175–87

Christensen TA, White J. 2000. Representation of olfactory information in the brain. In *The Neurobiology of Taste and Smell*, ed. TE Finger, WL Silver, D Restrepo, pp. 197–228. New York: Wiley-Liss

Cinelli AR, Hamilton KA, Kauer JS. 1995a. Salamander olfactory bulb neuronal activity observed by video-rate voltage-sensitive dye imaging. III. Spatio-temporal properties of responses evoked by odorant stimulation. *J. Neurophysiol.* 73:2053–71

Cinelli AR, Kauer JS. 1992. Voltage-sensitive dyes and functional activity in the olfactory pathway. *Annu. Rev. Neurosci.* 15:321–51

Cinelli AR, Kauer JS. 1995b. Salamander olfactory bulb neuronal activity observed by video-rate voltage-sensitive dye imaging. II. Spatio-temporal properties of responses evoked by electrical stimulation. *J. Neurophysiol.* 73:2033–52

Cinelli AR, Neff SR, Kauer JS. 1995c. Salamander olfactory bulb neuronal activity observed by video-rate voltage-sensitive dye imaging. I. Characterization of the recording system. *J. Neurophysiol.* 73:2017–2032

Cohen LB, Lesher S. 1986. Optical monitoring of membrane potential: methods of multisite optical measurement. *Soc. Gen. Physiol. Ser. B* 40:71–99

Coopersmith R, Leon M. 1984. Enhanced neural response to familiar olfactory cues. *Science* 225:849–51

Duchamp-Viret P, Duchamp A. 1997. Odor processing in the frog olfactory system. *Prog. Neurobiol.* 53:561–602

Dynes JL, Ngai J. 1998. Pathfinding of olfactory neuron axons to stereotyped glomerular targets revealed by dynamic imaging in living zebrafish embryos. *Neuron* 20:1081–1091

Ebner TJ, Chen G. 1995. Use of voltage-sensitive dyes and optical recordings in the central nervous system. *Prog. Neurobiol.* 46:463–506

Eeckman FH, Freeman WJ. 1990. Correlations between unit firing and EEG in the rat olfactory system. *Brain Res.* 528:238–44

Freeman WJ, Di Prisco GV. 1986. Relation of olfactory EEG to behavior: time series analysis. *Behav. Neurosci.* 100:753–63

Friedrich RW, Korsching SI. 1997. Combinatorial and chemotopic odorant coding in the zebrafish olfactory bulb visualized by optical imaging. *Neuron* 18:737–52

Friedrich RW, Korsching SI. 1998. Chemotopic, combinatorial, and noncombinatorial odorant representations in the olfactory bulb

revealed using a voltage-sensitive transducer. *J. Neurosci.* 18:9977–88

Galizia CG, Sachse S, Rappert A, Menzel R. 1999. The glomerular code for odor representation is species specific in the honeybee Apis mellifera. *Nat. Neurosci.* 2:473–78

Gervais R, Kleinfeld D, Delaney KR, Gelperin A. 1996. Central and reflex neuronal responses elicited by odor in a terrestrial mollusk. *J. Neurophysiol.* 76:1327–39

Gesteland RC, Lettvin JY, Pitts WH. 1965. Chemical transmission in the nose of the frog. *J. Physiol.* 181:525–59

Gomez G, Rawson NE, Cowart B, Lowry LD, Pribitkin EA, Restrepo D. 2000. Modulation of odor-induced increases in $[Ca(2+)](i)$ by inhibitors of protein kinases A and C in rat and human olfactory receptor neurons. *Neuroscience* 98:181–89

Grinvald A, Lieke E, Frostig RD, Gilbert CD, Wiesel TN. 1986. Functional architecture of cortex revealed by optical imaging of intrinsic signals. *Nature* 324:361–64

Grynkiewicz G, Poenie M, Tsien RY. 1985. A new generation of Ca^{++} indicators with greatly improved fluorescence properties. *J. Biol. Chem.* 260:3440–50

Guthrie KM, Anderson AJ, Leon M, Gall C. 1993. Odor-induced increases in *c-fos* mRNA expression reveal an antomical "unit" for odor processing in olfactory bulb. *Proc. Natl. Acad. Sci. USA* 90:3329–33

Hahn I, Scherer PW, Mozell MM. 1993. Velocity profiles measured for airflow through a large-scale model of the human nasal cavity. *J. Appl. Physiol.* 75:2273–87

Hamilton KA, Kauer JS. 1989. Patterns of intracellular potentials in salamander mitral/tufted cells in response to odor stimulation. *J. Neurophysiol.* 62:609–25

Harrison TA, Scott JW. 1986. Olfactory bulb responses to odor stimulation: analysis of response pattern and intensity relationships. *J. Neurophysiol.* 56:1571–89

Hildebrand JG, Shepherd GM. 1997. Mechanisms of olfactory discrimination: converging evidence for common principles across phyla. *Annu. Rev. Neurosci.* 20:595–631

Hudson R, Distel H. 1987. Regional autonomy in the peripheral processing of odor signals in newborn rabbits. *Brain Res.* 421:85–94

Imamura K, Mataga N, Mori K. 1992. Coding of odor molecules by mitral/tufted cells in rabbit olfactory bulb. I. Aliphatic compounds. *J. Neurophysiol.* 68:1986–2002

Joerges J, Kuttner A, Galizia CG, Menzel R. 1997. Representations of odours and odour mixtures visualized in the honeybee brain. *Nature* 387:285–88

Johnson BA, Leon M. 2000. Modular representations of odorants in the glomerular layer of the rat olfactory bulb and the effects of stimulus concentration. *J. Comp. Neurol.* 422:496–509

Kauer JS. 1974. Response patterns of amphibian olfactory bulb neurones to odour stimulation. *J. Physiol.* 243:695–716

Kauer JS. 1987. Coding in the olfactory system. In *The Neurobiology of Taste and Smell*, ed. TE Finger, WL Silver, pp. 205–31. New York: Wiley

Kauer JS, Cinelli AR. 1993. Are there structural and functional modules in the vertebrate olfactory bulb? *Microsc. Res. Tech.* 24:157–67

Kauer JS, Moulton DG. 1974. Responses of olfactory bulb neurones to odour stimulation of small nasal areas in the salamander. *J. Physiol.* 243:717–37

Keller A, Yagodin S, Aroniadou-Anderjaska V, Zimmer LA, Ennis M, et al. 1998. Functional organization of rat olfactory bulb glomeruli revealed by optical imaging. *J. Neurosci.* 18:2602–12

Kennedy C, Des Rosiers MH, Jehle JW, Reivich M, Sharp FR, Sokoloff L. 1975. Mapping of functional neural pathways by autoradiographic survey of local metabolic rate with 14C-deoxyglucose. *Science* 187:850–53

Kent PF, Mozell MM. 1992. The recording of odorant-induced mucosal activity patterns with a voltage-sensitive dye. *Journal of Neurophysiology* 68:1804–19

Kent PF, Mozell MM, Murphy SJ, Hornung DE. 1996. The interaction of imposed and inherent olfactory mucosal activity patterns and their composite representation in a mammalian species using voltage-sensitive dyes. *J. Neurosci.* 16:345–53

Kent PF, Youngentob SL, Sheehe PR. 1995. Odorant-specific spatial patterns in mucosal activity predict perceptual differences among odorants. *Journal of Neurophysiology* 74:1777–81

Keyhani K, Scherer PW, Mozell MM. 1997. A numerical model of nasal odorant transport for the analysis of human olfaction. *J. Theoret. Biol.* 186:279–301

Kimura T, Toda S, Sekiguchi T, Kawahara S, Kirino Y. 1998. Optical recording analysis of olfactory response of the procerebral lobe in the slug brain. *Learning & Memory* 4:389–400

Kobal G, Kettenmann B. 2000. Olfactory functional imaging and physiology. *Int. J. Psychophysiol.* 36:157–63

Krautwurst D, Yau KW, Reed RR. 1998. Identification of ligands for olfactory receptors by functional expression of a receptor library. *Cell* 95:917–26

Krieger J, Breer H. 1999. Olfactory reception in invertebrates. *Science* 286:720–23

Lam YW, Cohen LB, Wachowiak M, Zochowski MR. 2000. Odors elicit three different oscillations in the turtle olfactory bulb. *J. Neurosci.* 20:749–62

LaMantia AS, Pomeroy SL, Purves D. 1992. Vital imaging of glomeruli in the mouse olfactory bulb. *J. Neurosci.* 12:976–88

Laurent G. 1999. A systems perspective on early olfactory coding. *Science* 286:723–28

Leinders-Zufall T, Greer CA, Shepherd GM, Zufall F. 1998. Imaging odor-induced calcium transients in single olfactory cilia: specificity of activation and role in transduction. *J. Neurosci.* 18:5630–39

Leveteau J, MacLeod P. 1966. Olfactory discrimination in the rabbit olfactory glomerulus. *Science* 153:175–76

Levy LM, Henkin RI, Lin CS, Finley A. 1999. Rapid imaging of olfaction by functional

MRI (fMRI): identification of presence and type of hyposmia. *J. Comput. Assist. Tomogr.* 23:767–75

Lieke EE, Frostig RD, Arieli A, Tso DY, Hildesheim R, Grinvald A. 1989. Optical imaging of cortical activity: real-time imaging using extrinsic dye-signals and high resolution imaging based on slow intrinsic-signals. *Annu. Rev. Physiol.* 51:543–59

Litaudon P, Datiche F, Cattarelli M. 1997. Optical recording of the rat piriform cortex activity. *Prog. Neurobiol.* 52:485–510

Lu XC, Slotnick BM. 1998. Olfaction in rats with extensive lesions of the olfactory bulbs: implications for odor coding. *Neuroscience* 84:849–66

Malnic B, Hirono J, Sato T, Buck LB. 1999. Combinatorial receptor codes for odors. *Cell* 96:713–23

Meredith M, Moulton DG. 1978. Patterned response to odor in single neurones of goldfish olfactory bulb: influence of odor quality and other stimulus parameters. *J. Gen. Physiol.* 71:615–43

Mombaerts P. 1999. Seven-transmembrane proteins as odorant and chemosensory receptors. *Science* 286:707–11

Mombaerts P, Wang F, Dulac C, Chao SK, Names A, et al. 1996. Visualizing an olfactory sensory map. *Cell* 87:675–86

Mori K, Fujita SC, Imamura K, Obata K. 1985. Immunohistochemical study of subclasses of olfactory nerve fibers and their projections to the olfactory bulb in the rabbit. *J. Comp. Neurol.* 242:214–29

Mori K, Mataga N, Imamura K. 1992. Differential specificities of single mitral cells in rabbit olfactory bulb for a homologous series of fatty acid odor molecules. *J. Neurophysiol.* 67:786–89

Mori K, Nagao H, Yoshihara Y. 1999. The olfactory bulb: coding and processing of odor molecule information. *Science* 286:711–15

Moulton DG. 1976. Spatial patterning of response to odors in the peripheral olfactory system. *Physiol. Rev.* 56:578–93

Mozell MM, Sheehe PR, Hornung DE, Kent

PF, Youngentob SL, Murphy SJ. 1987. "Imposed" and "inherent" mucosal activity patterns. Their composite representation of olfactory stimuli. *J. Gen. Physiol.* 90:625–50

Mustaparta H. 1971. Spatial distribution of receptor responses to stimulation with different odours. *Acta Physiol. Scand.* 82:154–66

Nakamura T, Gold GH. 1987. A cyclic nucleotide-gated conductance in olfactory receptor cilia. *Nature* 325:442–44

Noe J, Breer H. 1998. Functional and molecular characterization of individual olfactory neurons. *J. Neurochem.* 71:2286–93

Polak EH. 1973. Multiple profile-multiple receptor site model for vertebrate olfaction. *J. Theor. Biol.* 40:469–84

Rubin BD, Katz LC. 1999. Optical imaging of odorant representations in the mammalian olfactory bulb. *Neuron* 23:499–511

Sagar SM, Sharp FR, Curran T. 1988. Expression of c-fos protein in brain: metabolic mapping at the cellular level. *Science* 240:1328–31

Schild D, Geiling H, Bischofberger J. 1995. Imaging of L-type Ca2+ channels in olfactory bulb neurones using fluorescent dihydropyridine and a styryl dye. *J. Neurosci. Methods* 59:183–90

Schild D, Restrepo D. 1998. Transduction mechanisms in vertebrate olfactory receptor cells. *Physiol. Rev.* 78:429–66

Schoenfeld TA, Marchand JE, Macrides F. 1985. Topographic organization of tufted cell axonal projections in the hamster main olfactory bulb: an intrabulbar associational system. *J. Comp. Neurol.* 235:503–18

Senseman DM. 1996. High-speed optical imaging of afferent flow through rat olfactory bulb slices: voltage-sensitive dye signals reveal periglomerular cell activity. *J. Neurosci.* 16:313–24

Shepherd GM. 1981. The olfactory glomerulus; its significance for sensory processing. In *Brain Mechanisms of Sensation*, ed. Y Katsuki, R Norgren, M Sto, pp. 209–23. New York: Wiley

Siegel MS, Isacoff EY. 1997. A genetically

encoded optical probe of membrane voltage. *Neuron* 19:735–41

Slotnick BM, Bell GA, Panhuber H, Laing DG. 1997. Detection and discrimination of propionic acid after removal of its 2-DG identified major focus in the olfactory bulb: a psychophysical analysis. *Brain Res.* 762:89–96

Sobel N, Prabhakaran V, Desmond JE, Glover GH, Goode RL, et al. 1998. Sniffing and smelling: separate subsystems in the human olfactory cortex. *Nature* 392:282–86

Stewart WB, Kauer JS, Shepherd GM. 1979. Functional organization of rat olfactory bulb analysed by the 2-deoxyglucose method. *J. Comp. Neurol.* 185:715–34

Stopfer M, Bhagavan S, Smith BH, Laurent G. 1997. Impaired odour discrimination on desynchronization of odour-encoding neural assemblies. *Nature* 390:70–74

Touhara K, Sengoku S, Inaki K, Tsuboi A, Hirono J, et al. 1999. Functional identification and reconstitution of an odorant receptor in single olfactory neurons. *Proc. Natl. Acad. Sci. USA* 96:4040–45. Erratum. 2000. *Proc. Natl. Acad. Sci. USA* 97(7):3782

Uchida N, Takahashi YK, Tanifuji M, Mori K. 2000. Odor maps in the mammalian olfactory bulb: domain organization and odorant structural features. *Nat. Neurosci.* 3:1035–1043

Wachowiak M, Cohen LB. 1999. Presynaptic inhibition of primary olfactory afferents mediated by different mechanisms in lobster and turtle. *J. Neurosci.* 19:8808–17

Wellis DP, Kauer JS. 1993. GABAa and glutamate receptor involvement in dendrodendritic synaptic interactions from salamander olfactory bulb. *J. Physiol.* 469:315–39

Wellis DP, Kauer JS. 1994. GABAergic and glutamatergic synaptic input to identified granule cells in salamander olfactory bulb. *J. Physiol.* 475:419–30

Wellis DP, Scott JW, Harrison TA. 1989. Discrimination among odorants by single neurons of the rat olfactory bulb. *J. Neurophysiol.* 61:1161–77

Williams M, Johnston JM, Cicoria M, Paletz E, Waggoner LP, et al. 1998. Canine detection odor signatures for explosives. In *Proc. 2nd Annu. Conf. Enforc. Secur. Technol.*, ed. AT DePersia, JJ Pennella. Boston, MA: Int. Soc. Optic. Eng.

Woo CC, Oshita MH, Leon M. 1996. A learned odor decreases the number of Fos-immunopositive granule cells in the olfactory bulb of young rats. *Brain Res.* 716:149–56

Yang X, Renken R, Hyder F, Siddeek M, Greer CA, et al. 1998. Dynamic mapping at the laminar level of odor-elicited responses in rat olfactory bulb by functional MRI. *Proc. Natl. Acad. Sci. USA* 95:7715–20

Youngentob SL, Kent PF. 1995. Enhancement of odorant-induced mucosal activity patterns in rats trained on an odorant identification task. *Brain Res.* 670:82–88

Yousem DM, Maldjian JA, Siddiqi F, Hummel T, Alsop DC, et al. 1999. Gender effects on odor-stimulated functional magnetic resonance imaging. *Brain Res.* 818:480–87

Zald DH, Pardo JV. 2000. Functional neuroimaging of the olfactory system in humans. *Int. J. Psychophysiol.* 36:165–81

Zochowski M, Wachowiak M, Falk CX, Cohen LB, Lam YW, et al. 2000. Imaging membrane potential with voltage-sensitive dyes. *Biol. Bull.* 198:1–21

Annu. Rev. Neurosci. 2001. 24:981–1004

THE ROLE OF THE CEREBELLUM IN VOLUNTARY EYE MOVEMENTS

Farrel R Robinson[1] and Albert F Fuchs[2]

[1]Department of Biological Structure and [2]Department of Physiology and Biophysics and Regional Primate Research Center, University of Washington, Seattle, Washington 98195-7420, e-mail: robinsn@u.washington.edu, fuchs@u.washington.edu

Key Words   saccades, smooth pursuit, adaptation, cerebellar nuclei, flocculus

■ **Abstract**   In general the cerebellum is crucial for the control but not the initiation of movement. Voluntary eye movements are particularly useful for investigating the specific mechanisms underlying cerebellar control because they are precise and their brain-stem circuitry is already well understood. Here we describe single-unit and inactivation data showing that the posterior vermis and the caudal fastigial nucleus, to which it projects, provide a signal during horizontal saccades to make them fast, accurate, and consistent. The caudal fastigial nucleus also is necessary for the recovery of saccadic accuracy after actual or simulated neural or muscular damage causes horizontal saccades to be dysmetric. Saccade-related activity in the interpositus nucleus is related to vertical saccades. Both the caudal fastigial nucleus and the flocculus/paraflocculus are necessary for the normal smooth eye movements that pursue a small moving spot. By using eye movements, we have begun to uncover basic principles that give us insight into how the cerebellum may control movement in general.

INTRODUCTION

Cerebellar lesions do not abolish movements, but they make them slow, inaccurate, rough, and variable. Previous findings have suggested that the cerebellum compensates for different loads and muscle lengths, improves movement accuracy and smoothness, and increases movement speed and consistency. These functions affect movement as it is occurring, so we consider them part of the cerebellum's "moment-to-moment" or short-term role. In addition to its short-term influence on movements, the cerebellum also plays a longer-term role, adapting motor commands gradually over many movements to compensate when motor commands produce consistently inaccurate movements. Motor commands may cause consistently inaccurate movements as we age or when trauma or physical disability, such as weakened muscles, changes the consequences of a command.

In this review, we describe the short- and long-term role of the cerebellum in the control of movement in primates by considering its influence on voluntary eye movements, whose premotor structures and descending commands are the best understood of any movement system.

CEREBELLAR CONTROL OF SACCADES

Saccades are the voluntary rapid eye movements that move our eyes from one visual target to another. During saccades, the eyes rotate very quickly (up to $>500°$ per s) and the movements are very brief (often <50 ms). This brevity maximizes the number of targets viewable and minimizes the saccadic transit time during which vision is impaired. Because saccades are so brief, there is not enough time for visual feedback to guide them to their targets. Thus, the brain must specify the command for a saccade before it starts. As we show below, the cerebellum helps modify this command and thus is critical for making saccades that are fast, accurate, and consistent from moment-to-moment and for maintaining their accuracy in the long term.

Saccade-Related Posterior Medial Cerebellum

Location

Most of what we know about the cerebellum's role in saccades comes from study of the saccade-related part of the posterior lobe vermis, i.e. the oculomotor vermis (Noda & Fujikado 1987), and the part of the fastigial nucleus to which it projects. The oculomotor vermis is the region within which Purkinje cells (P-cells) and background activity discharge a burst of spikes during saccades and from which one can elicit saccades by stimulating with current trains of <10 μA (Noda & Fujikado 1987). This area includes vermal lobule VII and, in some monkeys, the most posterior folium or two of lobule VI. Axons of P-cells in the oculomotor vermis terminate densely in a small oval region (1.5–2 mm A-P and $\sim$1 mm D-V) in the caudal part of the ipsilateral fastigial nucleus (CFN) and less densely in the rostral fastigial nucleus (Yamada & Noda 1987). Neurons in the rostral fastigial nucleus do not respond with saccades and are not considered further here (Büttner et al 1991).

Role in the Moment-to-Moment Control of Saccades

Effects of Lesions   When the caudal fastigial nucleus is disabled saccades are inaccurate, slow, and abnormally variable in size and speed (Vilis & Hore 1981, Robinson et al 1993). Normal saccades have gains, the size of the saccade divided by the distance to its target, of nearly 1.0. After the CFN on one side of the cerebellum is inactivated by an injection of the GABAa agonist muscimol, ipsiversive saccades are too large (gain, $\sim$1.2–1.9) and contraversive saccades are too small (gain, $\sim$0.6–0.8). In addition, saccade gains are more variable than normal (e.g. standard deviations for saccades to $10°$ horizontal targets are 1.2–4.8 times normal). Saccades to vertical targets curve strongly, ending $2°$–$9°$ to the left of their targets.

In addition to being dysmetric, both ipsilateral and contralateral saccades are slower than normal saccades of the same size. For example, after unilateral CFN inactivation, a 15° ipsiversive saccade reaches only ~84% of the velocity of a normal 15° saccade. Contraversive 15° saccades achieve velocities that average ~73% of normal for 15° saccades (Robinson et al 1993). After unilateral CFN inactivation, saccade velocities not only are abnormally slow, they are also much more variable than normal. Correlation coefficients of the relation of velocity and size typically drop from 0.95 to 0.39 (Figure 8A in Robinson et al 1993).

The dysmetrias produced by unilateral CFN inactivation suggest that each saccade is missing a contraversive component and that the net affect of saccade-related CFN activity on one side is to drive the eyes toward the contraversive side. Consistent with this interpretation, electrical stimulation of the CFN elicits saccades with large contraversive components (Noda et al 1988).

Humans with damage to one fastigial nucleus, or a lateral medullary stroke thought to reduce ipsilateral CFN activity (Wallenberg's syndrome) (Waespe & Wichmann 1990), exhibit saccade abnormalities like those of monkeys with unilateral CFN inactivation. For example, in five Wallenberg's patients, ipsiversive saccades to the lesioned side overshot by 33% and contraversive saccades undershot by 21.4% (Waespe & Baumgartner 1992).

Bilateral CFN inactivation in monkeys makes saccades in all directions too large (mean gain of saccades to 10° horizontal and vertical targets is 1.33 and 1.25, respectively) (Robinson et al 1993). Although their endpoints show abnormal scatter, saccades to vertical targets do not curve horizontally. Again, horizontal saccade velocities are slower and more variable than normal.

Effects of Vermal Lesions   Lesions of the oculomotor vermis also impair saccades. After such lesions, both leftward and rightward saccades were between 20% and 30% hypometric (Barash et al 1999). Also, postlesion gains were at least twice as variable as normal (Barash et al 1999, Takagi et al 1998). Average saccade gain returned to normal within 3–12 months after the lesion but remained as variable as it was immediately after the lesion (Barash et al 1999). Evidently, a mechanism outside the medial cerebellum slowly restores mean saccade gain to normal, but no mechanism outside the medial cerebellum can restore saccade consistency. After vermal lesions, horizontal saccades in some monkeys were ~30% slower than normal. Humans with infarctions that include the posterior vermis exhibit hypometria of both leftward and rightward saccades (Vahedi et al 1995).

In summary, without the CFN, the remaining saccade machinery produces dysmetric, slow saccades that lack their normal stereotypy. The hypometria and low velocity of saccades contraversive to a unilateral lesion suggest that CFN activity helps accelerate contraversive saccades. The hypermetria of ipsiversive saccades suggests that CFN activity helps decelerate ipsiversive saccades. We now see that the activity of CFN neurons is consistent with this suggestion.

Saccade-Related Unit Activity in the CFN   Unlike burst neurons elsewhere in the saccadic system, the vast majority of CFN neurons discharge a burst of action potentials for nearly every saccade, whatever its direction or size (Ohtsuka & Noda

1991, Fuchs et al 1993). However, for saccades of the same size and direction, there is a considerable variability in burst lead, firing frequency, and burst duration. Bursts in CFN neurons begin an average of ∼8 ms before small saccades in any direction. For contraversive saccades, there is little change in this timing as saccade size increases. For ipsiversive saccades, however, the burst occurs later for larger saccades, so that for a 20° saccade, the burst begins after saccade onset but before saccade end. For saccades ∼10° or larger, CFN neurons usually burst earlier for contralateral than for ipsilateral saccades. This pattern of burst timing suggests that the bursts are associated with the beginning of contraversive saccades and the end of ipsiversive ones (Ohtsuka & Noda 1990, 1991; Fuchs et al 1993).

CFN neurons not only are active during targeting saccades, they also respond during the fast phases of optokinetic nystagmus (Helmchen et al 1994) and during spontaneous saccades in the light (Fuchs et al 1993, Helmchen et al 1994). There is disagreement as to whether they do (Helmchen et al 1994) or do not (Ohtsuka & Noda 1992) respond during spontaneous saccades in the dark. Also, the role of the CFN might be clarified by characterizing its activity during a richer variety of voluntary gaze shifts, such as express, memory-guided, and self-paced saccades. Finally, because CFN inactivation produces dysmetria of both the head and eye components of free-head gaze shifts in cats (Goffart & Pélisson 1994), primate CFN neurons should also be examined during gaze shifts with the head unrestrained.

Although there is general agreement on the dependence of CFN burst timing on saccade direction, there is disagreement about how strongly other burst attributes are related to saccade metrics. For example, Ohtsuka & Noda (1991) report that burst and saccade durations are very well correlated ($R = 0.85$–0.97). In contrast, our lab finds a much lower correlation ($R = \sim0.6$) for our entire population of CFN neurons (Fuchs et al 1993). Because some of our neurons exhibit correlation coefficients above 0.8, perhaps Ohtsuka & Noda saw higher correlations because they limited their analysis to 19 well-behaved neurons in their sample of 96. In general we find that the properties of saccade-related bursts in CFN neurons are loosely related to saccade properties. In particular, the same saccade may be accompanied by a brisk burst on one trial and little, if any, discharge on the next.

In summary, the dependence of burst timing on saccade direction is consistent with the CFN inactivation data. For contraversive saccades, the burst occurs early in the saccade, providing a contraversive drive to help accelerate it. Without this drive, the saccade falls short. Later in the saccade, neurons in the other CFN, i.e. ipsiversive to the direction of the saccade, discharge a burst that delivers a drive opposite to the direction of the saccade to slow the movement. Without this late burst, the saccade does not decelerate as quickly as normal and overshoots. The CFN discharge somehow also makes the saccades more consistent or repeatable. How this is accomplished is currently a mystery.

Saccade-Related Unit Activity in the Posterior Vermis   Most (71%) oculomotor vermis P-cells exhibit saccade-related bursts for either ipsiversive or contraversive saccades or both (Ohtsuka & Noda 1995). For ipsiversive saccades, P-cell bursts begin before saccade onset and fall off abruptly in the last half of the saccade.

This fall-off of inhibitory P-cell activity could cause a rebound depolarization in cerebellar nuclear neurons (Aizenman & Linden 1999), perhaps promoting the onset of late bursts in CFN neurons for ipsiversive saccades.

For contraversive saccades, P-cell bursts begin before or early in the movement, peak near its middle, and continue past saccade end (Ohtsuka & Noda 1995). This pattern could help produce the postburst pauses that occur in ~19% of CFN cells. About 18% of P-cells pause for contraversive saccades. Pause timing is variable from saccade to saccade, with the average mean lead time of 17.5 ms (Ohtsuka & Noda 1995). These pauses begin sharply and so could, if synchronized across many P-cells, trigger the early bursts in CFN neurons for contraversive saccades. In general, however, we cannot account entirely for the saccade-related responses of CFN neurons by using only P-cell activity. Specifically, it is not yet clear how the brain produces the bursts that CFN neurons discharge before contraversive saccades (but see the next section). Signals from the frontal eye field (FEF) and superior colliculus (SC) on one side would have to reach the ipsilateral CFN. These anatomical connections exist. The FEF (Stanton et al 1988) and the SC (Harting 1977) project to subdivisions of the ipsilateral pontine nuclei that, in turn, project to the ipsilateral, as well as contralateral, CFN (Noda et al 1990). We schematically represent a connection between the right SC and the right CFN in Figure 1.

How Does the CFN Influence the Saccade Machinery?   CFN cells project to brain-stem regions containing saccade-related premotor neurons. A schematic of the brain-stem generator for horizontal saccades is shown in Figure 1 (Fuchs et al 1985). The burst-tonic firing pattern of abducens motor neurons for an ipsiversive saccade and its subsequent fixation position is produced by a velocity signal from excitatory burst neurons (EBNs) in the ipsilateral paramedian pontine reticular formation and a position signal from the nucleus prepositus hypoglossi in the medulla (not shown). The pause in activity associated with contraversive saccades is caused by inhibitory burst neurons (IBNs) in the contralateral medullary reticular formation. Both IBNs and EBNs are inhibited during fixation by steadily firing omnipause neurons (OPNs) in the pontine raphe, which must be inhibited to trigger a saccade to occur. Both a desired change in gaze command required by the EBNs and the trigger signal to the OPNs is thought to be provided by the SC.

Anatomical (Noda et al 1990) and recording (Scudder et al 2000) evidence indicates that the CFN projects contralaterally to EBNs, IBNs, and OPNs. Because IBN inputs to motor neurons are crossed but EBN inputs are not, CFN activity could help accelerate the eyes contralaterally via either projection (for details, see Fuchs et al 1993). If it does, there will be late off-direction bursts in IBNs, EBNs, and abducens motor neurons. Indeed, most EBNs exhibit a weak late burst for off-direction saccades (e.g. Strassman et al 1986), and ~30% of IBNs do (Scudder et al 1988) but abducens motoneurons do not (Fuchs et al 1988). CFN neurons also send a large projection to the ventrolateral region of the contralateral thalamus (Noda et al 1990), a pathway whose role has yet to be investigated.

The CFN apparently receives information with timing appropriate to produce late bursts for ipsiversive saccades and early bursts for contraversive saccades. The

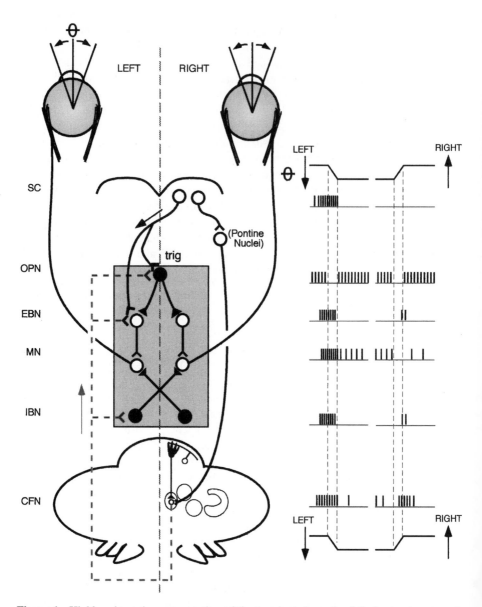

Figure 1 Highly schematic representation of the transient elements of the brain-stem saccade burst generator (shaded box) and its inputs from the superior colliculus (SC) and the cerebellum (dashed lines). Cerebellar output via the caudal fastigial nucleus (CFN) influences the excitatory burst neuron (EBN), inhibitory burst neuron (IBN), and omnipause neuron (OPN). Filled synaptic terminals indicate inhibitory synapses; open terminal symbols mark excitatory synapses. Other abbreviations: MN, motor neuron.

early bursts may originate in the SC because SC neurons burst before contraversive saccades. In addition to the SC's projection to the contralateral burst generator, it also projects to a subdivision of the ipsilateral pontine nuclei (Harting 1977) that projects bilaterally to the CFN (Noda et al 1990). If the early bursts in CFN neurons in fact originate in the SC, signals from SC neurons on one side must reach the CFN on the same side. Perhaps the pathway from the SC to the ipsilateral CFN, as shown in Figure 1, carries these signals.

CFN Models   Current models have attempted to incorporate the CFN into the saccade circuitry. Most past circuits of the saccade burst generator have used a classical control systems model with a local feedback loop that compares actual with commanded saccade amplitude at the EBN. Employing such a model and adding the CFN influence at the EBN, Dean (1995) was able to produce saccades whose size, velocity, acceleration, and deceleration were quantitatively similar to those observed after CFN inactivation.

In a recent model touted to have more predictive power (Lefévre et al 1998, Quaia et al 1999), the oculomotor vermis generates a motor error signal, shaped by feedback from the brain-stem saccade generator and signals from the SC. This error signal, in turn, activates a specific topographic locus in the CFN corresponding to the saccade to be generated. As the saccade progresses, activity in the CFN contraversive to the saccade direction moves from the activated site medially. Near the end of the saccade, the activity crosses the midline into the other CFN, and the bursts produced there activate IBNs to inhibit activity in motoneurons and end the saccade. This model also accurately simulates the abnormalities in saccade velocity, size, and direction after bilateral CFN inactivation. Unfortunately, the predictions of the model (Quaia et al 1999) that CFN neurons are topographically organized by desired saccade displacement and that some CFN cells should burst only for saccades greater than a certain amplitude are not substantiated by single-unit recordings (Ohtsuka & Noda 1991, Fuchs et al 1993). Furthermore, demonstrating a topographic organization of any sort in the CFN would be impossible because of its compact size.

CFN Role in Saccade Adaptation

Effects of Lesions   Normally primates can adapt the size of their saccades. For example, when damage to the medial longitudinal fasciculus in humans (Doslak et al 1980) or the eye muscles in monkeys (Optican & Robinson 1980) causes hypometric saccades, both primates gradually increase saccade gain toward normal. We can simulate saccade dysmetria to elicit adaptation by rapidly moving a target spot during each saccade so that the saccades seem to miss their target (McLaughlin 1967). This technique causes adaptive changes in saccade size in both humans and monkeys. Adaptation is substantially complete within $\sim$100 saccades in humans (Miller et al 1981, Deubel et al 1986) and $\sim$1000 saccades in monkeys (Straube et al 1997a).

Lesions that include the posterior medial cerebellum seem to abolish the ability to adapt saccades. After removal of the medial cerebellar cortex and nuclei, monkeys no longer could eliminate the dysmetria caused by weakened eye muscles (Optican & Robinson 1980). After a large electrolytic lesion of the medial and interpositus nuclei, a monkey could not adapt to backward-moving targets (Goldberg et al 1993). After ablations of large parts of the oculomotor vermis, monkeys were unable to adapt either to backward- (Takagi et al 1998) or forward-moving targets (Barash et al 1999). Also, humans with Wallenberg's syndrome were unable to adapt the size of their saccades to backward-moving targets (Waespe & Baumgartner 1992). Finally, we demonstrated that bilateral inactivation of the CFN alone impaired a monkey's ability to reduce its saccade gain (Robinson et al 2000).

Despite this last result, it would be premature to conclude that the CFN was necessary for saccadic gain adaptation. Perhaps adaptation occurred upstream from the inactivated CFN but could not reach the motor neurons through the anesthetized CFN output. To test this possibility, we inactivated the CFN bilaterally in two monkeys and confirmed that there was no reduction in saccade gain after >1000 overshooting saccades in each direction. The animal was then placed in the dark for 10 h so that it would have no visual targets until the muscimol dissipated. After the muscimol dissipated, saccades were hypometric, which suggests that adaptation had occurred during CFN inactivation and that adapted signals could influence saccades once the CFN was able to relay them to the brain stem (Robinson et al 2000).

CFN Changes During Saccade Adaptation   To alter saccade amplitude, CFN neurons possibly could increase or decrease their burst rates or alter the timing of the burst relative to saccade onset or end. Adaptation appears to use both mechanisms. Some neurons in the CFN increase their burst frequency and duration during adaptations that decrease the size of ipsiversive saccades (FR Robinson, CT Noto, AF Fuchs, unpublished data). Another study did not observe changes in the firing rate of CFN neurons during saccade adaptation but did observe changes in burst timing (Scudder 1998).

The Site of Saccade Adaptation   Our preliminary data indicate that the site of saccade adaptation is upstream from the CFN. Sites that have been considered include the SC, the FEF, and the oculomotor vermis. After saccade gain was reduced with intrasaccadic target movments, electrical stimulation at sites in either the FEF (Edelman & Goldberg 1995) or the SC (Fitzgibbon et al 1986) elicited the same saccades as before adaptation, which suggests that adaptation occurs upstream of both structures. On the other hand, adaptation produces no shift in the topographic map of saccade amplitude in SC neurons, which suggests that each site in the SC continues to command the same saccade after adaptation as before (Frens & van Opstal 1997). This finding indicates that a signal must be subtracted from the SC saccadic command signal at a downstream site. Potential sites include the oculomotor vermis, which receives a heavy SC input via the dorsolateral pontine nucleus and the nucleus reticularis tegmenti pontis, and also the brain-stem saccade generator (Figure 1). Because the cerebellar cortex has been implicated in other kinds of oculomotor learning, such as vestibuloocular reflex (VOR) adaptation (e.g. Ito & Kano 1982), a parsimonious suggestion is that it also is involved in saccade adaptation.

Unresolved Issues

Why is There Little Evidence of Eye Position Signals in CFN Units When Lesion Data Suggest a Position Sensitivity in Saccade Deficits?

Cerebellar lesions that include the oculomotor vermis (Ritchie 1976) or the CFN (Vilis & Hore 1981) cause saccade dysmetrias whose size depends on initial eye position. Centrifugal saccades are smaller than centripetal saccades. This is also true in human patients with infarcts that include the posterior vermis (Vahedi et al 1995). These observations led to the proposal that the output of the posterior medial cerebellum normally compensates for initial eye position so that centrifugal and centripetal saccades are nearly the same size. After bilateral inactivation of the CFN, we also found that centrifugal saccades to 20° targets were 79% of the size of centripetal saccades. Others, however, found no difference in the size of centrifugal and centripetal saccades (Ohtsuka et al 1994). Furthermore, the discharge of CFN neurons has either only a weak (Fuchs et al 1993) or no (Ohtsuka & Noda 1991, Ohtsuka et al 1994) relationship with initial eye position.

What is the CFN's Influence on Gaze Position?

Unilateral CFN inactivation in monkeys with their heads restrained causes the eyes to aim ∼1°–2° to the side of the target in the direction of the inactivated CFN (Robinson et al 1993). Monkeys consistently move their eyes away from the target by a distance equal to this offset when the target spot moves to fall on the fovea. This offset is too small to account for the ipsiversive saccade hypermetria after unilateral CFN inactivation. For example, after unilateral CFN inactivation, two monkeys overshot 10° targets by an average of 5.6° but their mean ipsiversive offset was only 1.2° (Robinson et al 1993).

In contrast to a head-fixed monkey, a cat with its head unrestrained exhibits a 10°–20° ipsiversive offset of gaze after unilateral CFN inactivation. Ipsiversive gaze shifts of any size overshoot their targets by the size of the offset, i.e. the offset accounts completely for ipsiversive gaze hypermetria. If a target appears directly ahead, the cat will shift its gaze to look 10°–20° ipsiversive of the target (Goffart & Pélisson 1998). Contraversive gaze movements are different, undershooting by an amount that depends on the size of the saccade, so that a gain of 0.25–0.68 (depending on the experiment) accounts very well for this hypometria.

We currently do not understand why CFN inactivation causes an ipsilateral offset, why animals work to maintain it, or why this offset is so much larger when the head is not restrained. Goffart & Pélisson (1998) propose that the output of the CFN on each side helps to determine perceived heading direction in the ipsiversive field.

How does Variable Activity in the CFN Make Saccades More Consistent?

Evidently CFN bursts make saccades more consistent, although the bursts themselves are quite inconsistent in their onset time, burst duration, peak frequency, and number of spikes for nearly identical saccades (Fuchs et al 1993). Saccade-related CFN neurons project directly to burst neurons in the premotor saccade-generating network (Scudder et al 2000), whose activity is machine-like in its consistent relationship to saccade metrics.

There are two general proposals about how the variable activity in the CFN could reduce saccade variability. The first asserts that the saccadic system outside the cerebellum, e.g. the SC, provides the burst generator with variable commands but that CFN output to the burst generator varies to complement this variability exactly, thereby producing consistent saccades (Robinson 1995). This interpretation leads immediately to the question of how the cerebellum can produce the appropriate complementary signal. One possibility is that the cerebellum receives an efference copy of each saccade that tailors CFN output to keep saccades on the right trajectory and to end them on target (Lefévre et al 1998).

The second proposal asserts that although the output of individual cerebellar cells is variable, their combined activity reliably specifies some feature of each saccade. Indeed, Thier et al (2000) demonstrated that although the activity of any single Purkinje cell in the oculomotor vermis poorly represents saccade timing, the average activity of many such cells accurately specifies the end of a saccade.

Saccade-Related Posterior Interpositus Nucleus and Dorsal Paraflocculus

Neurons in the ventrolateral corner of the posterior interpositus nucleus (VPIN) and the adjacent limb of the lateral nucleus discharge a burst of action potentials for nearly every saccade (Robinson et al 1996). For most VPIN neurons, the burst occurs during acceleration of downward saccades and deceleration of upward saccades. Bilateral inactivation of the VPIN deprives all saccades of a downward component, i.e. all saccades end above their targets. However, inactivation also reduces upward acceleration and downward deceleration, indicating that VPIN activity also provides an upward drive. Therefore, current data indicate that some VPIN neurons drive the eyes downward during saccades and others drive them upward, but the downward drive predominates (Robinson 2000).

Despite the apparent similarities between the CFN's influence on the horizontal component of saccades and the VPIN's influence on the vertical component, there are two noteworthy differences. First, CFN inactivation causes vertical saccades to become hypermetric. In contrast, VPIN inactivation has relatively little effect on the size of horizontal saccades. Second, after unilateral CFN inactivation, saccades to vertical targets curve dramatically toward the side of the injection. After bilateral VPIN inactivation, however, saccades to horizontal targets travel a straight trajectory to a point above their target. Therefore, the CFN and VPIN do not simply provide the same signals for saccades with orthogonal vectors.

Other Saccade-Related Regions of the Cerebellum

Basal Interstitial Nucleus   The basal interstitial nucleus (BIN) of the cerebellum is a broad flat collection of neurons on the roof of the ventricle, ventral to the lateral and interpositus cerebellar nuclei (Langer 1985). BIN neurons exhibit a burst of action potentials for saccades in every direction (Takikawa et al 1998). These bursts begin an average of 16 ms before the start of visually guided saccades and end 33 ms before the saccade ends. BIN activity is correlated with the duration of the saccade but not with the saccade's size or speed. Takikawa et al (1998) suggest that BIN neurons may contribute to the onset of the pause in OPNs but could not help sustain that pause until saccade end because their bursts end well before.

Lateral Nucleus   Although there has not yet been a systematic exploration of the dentate nucleus, it may be involved with saccade generation because its caudal portion projects, via a relay in the thalamus, to the saccade-related part of the FEF (Lynch et al 1994).

CEREBELLAR CONTROL OF SMOOTH PURSUIT

General Properties of Smooth Pursuit

The objective of smooth-pursuit eye movements is to reduce the slip of a visual image over the fovea to velocities that are slow enough to allow clear vision. Smooth pursuit is most highly developed in primates, where it has been studied by requiring them to track a small target spot, usually moving with one of two patterns. Sinusoidal motion tests smooth pursuit in the steady state but allows considerable prediction (Leung et al 2000). To determine how the smooth-pursuit system responds to unpredictable target motion, investigators use a target that steps in one direction before moving at constant velocity in the opposite direction, i.e. step-ramp motion (Rashbass 1961). If the target crosses the initial fixation point at approximately the primate's saccadic reaction time ($\sim$200 ms), the eye will accelerate smoothly to acquire the target. The initial $\sim$40 ms of acceleration is relatively unaffected by target speed and serves only to start the eye in the correct direction. Acceleration between 40 and $\sim$100 ms depends strongly on the target's velocity, position, movement direction, and contrast (Lisberger et al 1987). Thereafter, visual feedback can help guide pursuit. After the eye acquires the target, it moves at nearly target velocity, so retinal image motion is quite low. If retinal slip is eliminated completely by stabilizing the target on the retina, smooth pursuit continues, which suggests that once begun, smooth pursuit is driven by a "velocity memory" and not a visual signal (Morris & Lisberger 1987).

Eye velocity often oscillates during ramp tracking. These oscillations disappear when the target is stabilized on the retina (Lisberger et al 1987), and their period can be altered by altering the time between eye and target movement (Goldreich et al 1992). Therefore, the oscillations are under visual control (and not just the result of instabilities of an internal feedback loop with a ~100-ms visual delay). In summary, the smooth-pursuit system requires predominantly visual signals to drive eye acceleration and oculomotor signals to control maintained pursuit.

Smooth-Pursuit Adaptation

The cerebellum has been implicated in adaptation of the gains of both the VOR (e.g. Lisberger 1988) and saccades (see above). Both these movements are so fast that visual feedback can not guide them; thus, adaptation is necessary to reduce persistent visual slip (VOR) or saccadic dysmetria. In contrast, smooth pursuit is slow enough to employ visual feedback to compensate for movement deficits due to neuronal or muscular failure. Therefore, it would seem unnecessary for the smooth-pursuit system to have the capability to adapt. Nevertheless, it does.

In three of four patients with unilateral oculomotor palsies, one week of viewing with the impaired eye caused the acceleration of the normal eye to increase (Optican et al 1985). Presumably this is a consequence of increased neural drive, which goes to both eyes, to compensate for the sluggish movements of the paretic eye. This adaptive increase in acceleration occurred between about 40 and 100 ms of pursuit onset. During maintained tracking, adapted patients often displayed substantial oscillations, which suggests an increase in the effect of image slip on eye motion.

As with saccade adaptation, pursuit adaptation can be elicited in normal subjects by manipulating target motion. Subjects track a ramp target that starts at one speed and, after a brief period of smooth pursuit, assumes either a higher or lower speed. After 20–30 min of such double-speed tracking, two thirds of normal subjects showed significant changes in eye velocity starting 100–200 ms after pursuit onset (Fukushima et al 1996). When this double-speed method was used on monkeys, the largest pursuit adaptation occurred in the first 50–80 ms after the onset of smooth pursuit after about 200 trials (Kahlon & Lisberger 1996). In summary, smooth-pursuit adaptation in humans and monkeys develops after the initial direction-only component of pursuit, i.e. after ~40 ms. However, adaptation in humans appears to occur somewhat later in pursuit acceleration than it does in monkeys. Whether this reflects a difference in experimental technique or in the species is currently unclear.

Cerebellar Lesions Implicate Two Separate Areas in Smooth Pursuit

In 1973, Westheimer & Blair (1973) made the anecdotal observation that animals with complete cerebellectomies tracked a sinusoidally moving horizontal target with only saccadic eye movements. However, because there was no quantification of the deficit, it is unclear whether their monkeys retained any smooth-pursuit capability. Subsequent smaller lesions revealed that the profound smooth-pursuit deficit that Westheimer and Blair described was the result of damage to at least two separate cerebellar subregions: the flocculus/ventral paraflocculus and the posterior medial cerebellum.

Bilateral ablation of the flocculus and the ventral paraflocculus caused a 33% reduction in horizontal, steady state, smooth-pursuit gain (eye velocity/target velocity). During steady state tracking, any deficit is ameliorated by visual feedback, so this deficit is equivalent to ∼90% reduction in open loop gain (Zee et al 1981). In addition to their pursuit deficit, monkeys could not hold their eyes at an eccentric location, and there was a centripetal, exponential drift with a time constant of ∼1.6 s in the first few weeks after surgery. The ablation had no effect on saccade metrics. Finally, there was no consistent effect on the gain of the VOR (Zee et al 1981). These data indicate that the flocculus participates in smooth pursuit and also helps regulate the gaze-holding integrator.

The second cerebellar area involved with smooth pursuit is the posterior medial cerebellum, including the cerebellar vermis (lobule VIc and all of VII) (Noda & Fujikado 1987) and the fastigial nuclei to which these vermal P-cells project. In addition to the saccadic dysmetria discussed above, muscimol inactivation of the monkey CFN produces ∼30% reduction in gain of both contraversive and downward sinusoidal pursuit, with little effect on ipsiversive or upward pursuit (Robinson et al 1997). When both fastigial nuclei are inactivated, the deficits in contraversive and downward pursuit are larger and upward, and ipsiversive pursuit becomes slower [but not apparently in humans (Büttner et al 1994)]. In contrast to the effect of floccular lesions, there is no deficit in gaze holding. As after flocculus lesions, the VOR appears normal (Kurzan et al 1993) and VOR suppression, believed by some to be a smooth-pursuit capability, is impaired. Surgical ablation of the vermis, including lobules IV–VI and parts of VII and VIII, caused a 30%–40% reduction of simian sinusoidal smooth-pursuit gain (Keller 1988). Damage to the cerebellar vermis of humans also causes smooth-pursuit deficits (Furman et al 1986, Pierrot-Deseilligny et al 1990, Vahedi et al 1995).

More lateral cerebellar lesions also affect smooth pursuit (Straube et al 1997b). Indeed, smooth-pursuit–related activity has been found in monkey interpositus nucleus, to which the laterally placed dorsal paraflocculus projects (Robinson & Brettler 1998). Information about these possible lateral cerebellar smooth-pursuit areas is currently sparse and is not considered further.

In the following sections we consider which aspects of the smooth-pursuit response the CFN and the flocculus/paraflocculus might control. Currently, there is substantial evidence on their roles in the moment-by-moment control of pursuit but little concerning their roles in pursuit adaptation.

Flocculus/Ventral Paraflocculus

Sinusoidal Target Tracking

Different investigators have described firing patterns during different tracking tasks, so cerebellar neurons have not been tested under a set of standard conditions. However, most studies agree that the simple spike activity of the vast majority of floccular P-cells is modulated during smooth pursuit of a small target (Miles et al

1980). Most of these cells respond best to either horizontal (ipsiversive) or vertical (mostly down but with a small contraversive component) pursuit directions (Lisberger & Fuchs 1978; Krauzlis & Lisberger 1994, 1996; Stone & Lisberger 1990a; Fukushima et al 1999; Leung et al 2000). During pursuit of a sinusoidally moving target in a unit's best direction, firing rate modulation is roughly sinusoidal (Stone & Lisberger 1990a) and is in phase with and increases monotonically with eye velocity. Many of the cells with smooth-pursuit sensitivity also respond to whole body oscillations even when eye movements are suppressed, increasing their activity for head rotation in the same direction that elicits the best smooth-pursuit response. The sensitivities to head and eye movements often are quite similar, so when the eyes move opposite to the head during the VOR or when a rotating animal fixates a target stationary in space, these units show little or no modulation in their activity. In this paradigm, eye position relative to space, i.e. gaze, remains fixed, so neurons with this behavior have been designated gaze-velocity P-cells. Most units with horizontal best directions apparently have gaze-velocity characteristics (but see Belton & McCrea 1999).

In contrast, most of the P-cells with vertical-preferred pursuit directions have unequal sensitivities to head and eye velocity and may even have different vestibular- and pursuit-preferred directions. Therefore, they continue to exhibit modulation when the oscillating monkey tracks a target that remains stationary in space. During pursuit, most tend to fire in phase with eye velocity, but some fire in phase with eye position (Fukushima et al 1999). In the vertical system, therefore, there appears to be a continuum of P-cell types reflecting a range of vestibular sensitivities. P-cells with horizontal- and vertical-preferred directions are intermixed within the flocculus/ventral paraflocculus (Krauzlis & Lisberger 1996).

P-cells also discharge complex spikes, which are elicited by the climbing fiber input from the inferior olive. Complex spike rates also are modulated during sinusoidal pursuit. For those ipsiversive and downward gaze-velocity P-cells with robust complex-spike modulation, complex-spike firing increased during contraversive and upward pursuit (Stone & Lisberger 1990b). When averages of eye and target velocity were triggered on the occurrence of complex spikes, a transient impulse of visual image motion was detected ∼100 ms before the complex spike. Consistent with this data, an increase in climbing fiber spikes also occurred 100 ms after initiation of contraversive or upward ramp target motion for horizontal and vertical gaze-velocity P-cells, respectively. Because the complex spike causes a brief decrease in simple spike activity, its occurrence would facilitate the decrease of simple spike activity that is required to drive contraversive or downward tracking in response to image motion due to imperfect pursuit. Indeed, the visual climbing fiber responses seem more likely to occur during active pursuit than as a result of passive background motion (Stone & Lisberger 1990b). However, any climbing fiber influences are infrequent because of their very low modulation rate; thus, it seems unlikely that the complex spike is involved with immediate on-line control of smooth pursuit. Perhaps, as has been suggested for the VOR, the climbing fiber signal serves as an error signal to drive smooth-pursuit adaptation (see below).

Step-Ramp Target Tracking

To parse out the signals that effect P-cell firing, most investigators have used a step-ramp target motion, which allows separation of smooth pursuit and the accompanying unit activity into an initial directional component (first 40 ms), an open loop component that responds to target acceleration (40–100 ms), and a maintained component when eye velocity is similar to target velocity (>100 ms). Step-ramp stimuli have only been used to evaluate gaze-velocity neurons.

In the preferred smooth-pursuit direction, most P-cells exhibited a burst of simple spike firing during eye acceleration and a sustained rate during maintained pursuit. In general, the burst begins after smooth-pursuit onset and reaches its maximum near the end of acceleration (Stone & Lisberger 1990a). Discharge frequency then settles down to a lower steady rate associated with constant-velocity smooth pursuit (Stone & Lisberger 1990a, Krauzlis & Lisberger 1994). As expected from the sinusoidal tracking data, the steady firing rate increases with velocity increases in maintained pursuit. A reduction in firing rate often occurs during smooth-pursuit acceleration in the nonpreferred direction.

Because of the intimate relation of smooth pursuit with the target movement that generates it (Lisberger et al 1987), it is difficult to sort out what the simple spike discharge of P-cells is related to. After using different step-ramp combinations to dissociate eye and image motion, Krauzlis & Lisberger (1991) successfully simulated the associated discharge patterns by using visual signals to drive pursuit acceleration to change eye velocity and an eye movement signal to sustain it. Although several arguments are consistent with the contention that the burst accompanying pursuit acceleration is not solely oculomotor, no single experiment has conclusively demonstrated its visual origin (Stone & Lisberger 1990a). Frank visual responses to movement of either a small spot or the background while a monkey is fixating occurred in a small percentage of gaze-velocity P-cells (Noda & Warabi 1987, Stone & Lisberger 1990a). Stabilization of the target on the retina during maintained tracking did not change the firing of most P-cells, which supports the existence of an underlying eye-velocity signal during the maintained phase of smooth pursuit (Stone & Lisberger 1990a).

The simple spike discharge of P-cells, in turn, has the requisite characteristics to match nicely the dynamic load presented by the oculomotor plant (Krauzlis & Lisberger 1994). Therefore, the flocculus is thought to transform visual inputs to provide the dynamic but not the static (Shidara et al 1993) signals required to generate smooth pursuit. This conclusion seems at odds with the observation that the burst of most P-cells begins after smooth-pursuit onset. On the other hand, stimulation of the flocculus elicits smooth pursuit at latencies ranging from ~10 to 20 ms (Belknap & Noda 1987, Shidara & Kawano 1993, Lisberger & Pavelko 1988), and modeling of flocculus signals driving the brain-stem smooth-pursuit apparatus requires a burst lead of 9 ms (Krauzlis & Lisberger 1994). Perhaps the small minority of P-cells with early bursts are sufficient for pursuit initiation.

In conclusion, destruction of gaze-velocity P-cells probably accounts for the smooth-pursuit deficits produced by flocculus removal.

Sensitivities of Other Purkinje Cells

In addition to those P-cells that discharge with eye velocity, a minority have saccade and/or position-related simple spike activity. For some, their pause or burst is well timed with saccade onset and offset (Noda & Suzuki 1979b). However, none of the saccade-related activity appears to be important because ablation of the flocculi and paraflocculi has no effect on saccade metrics or accuracy (Zee et al 1981). The steady discharge of some P-cells is tightly related to eye position (Noda & Suzuki 1979a). A small number of P-cells with eye position sensitivity have also been reported by others (Miles et al 1980). It is possible that the loss of these cells could account for the gaze-holding deficit produced by flocculectomy (Zee et al 1981).

Neural Correlates of Pursuit Adaptation

In the double-speed paradigm, ~25% of flocculus P-cells showed changes in simple spike firing appropriate for the resulting smooth-pursuit adaptation (Kahlon & Lisberger 2001). Whether climbing fiber discharge could guide smooth-pursuit learning was unclear.

Posteriomedial Cerebellum

Oculomotor Vermis

Lying among the P-cells in the oculomotor vermis with simple spike activity related to saccades are others that discharge simple spikes during sinusoidal smooth pursuit (Sato & Noda 1992, Suzuki et al 1981, Suzuki & Keller 1988). At least half of those also discharge in phase with head velocity and the velocity of a small spot that moves through a second spot on which the monkey is fixating (Suzuki & Keller 1988). Because almost half of these P-cells responsive to head, eye, and image motion preferred the same directions for all three motions and roughly, across the population, had the same velocity sensitivities, it has been suggested that such vermal P-cells provide signals related to target velocity. No study has yet described the vectorial directional preferences of these P-cells. Unlike flocculus P-cells, the discharge of vermal P-cells waxes and wanes considerably from cycle to cycle under identical tracking and target motion conditions. Like flocculus P-cells, many of the vermal P-cells with smooth-pursuit sensitivity also show bursts or pauses in activity with saccades.

Because of the paucity of pursuit-related units in their survey of the oculomotor vermis, Sato & Noda (1992) opined that the vermis is primarily concerned with saccades. In their hands, stimulation of the vermis with low currents produced only saccades (Noda & Fujikado 1987). However, more recently, Krauzlis & Miles (1998) have shown that electrical stimulation of the vermis can produce either saccades or smooth-pursuit eye movements, depending on stimulation frequency and whether the eyes are fixating or pursuing a moving target. The evoked pursuit movements are ipsiversive and occur with a latency of 10–20 ms.

Unfortunately, there are no recordings during step-ramp tracking, so the timing of P-cell discharge to response initiation is not known. Nor do we know if changes in firing occur preferentially during the acceleration or sustained velocity phases of pursuit. Finally, there are no studies concerning climbing fiber responses in the oculomotor vermis during smooth pursuit.

Caudal Fastigial Nucleus

Some CFN neurons exhibit a modulation in firing rate when a monkey makes smooth eye movements in pursuit of a moving target (Büttner et al 1991, Fuchs et al 1994). For 72% of the neurons, the pursuit direction that elicits the most vigorous modulation is contraversive and/or downward; most of the remainder prefer ipsiversive and/or upward motion. Based on the phase of the modulation during sinusoidal pursuit, 80% of cells with contraversive/downward–preferred directions discharge most during smooth-pursuit acceleration, whereas those with the opposite-preferred directions discharge most during smooth-pursuit deceleration. These data suggest that the CFN helps accelerate the eyes during contraversive pursuit and decelerate them during ipsiversive pursuit. This interpretation is consistent with the deficits reported after CFN inactivations (Robinson et al 1997). Firing patterns during step-ramp pursuit reveal that many CFN neurons could initiate contraversive eye acceleration because their burst-sustained discharge precedes pursuit onset by >25 ms (Fuchs et al 1994). Finally, almost every CFN neuron with pursuit sensitivity also responds during yaw or pitch rotation. Most have unequal sensitivities to eye and head velocity, with some more sensitive to eye and others more sensitive to head velocity. Therefore, the apparent gaze- or target-velocity sensitivity of vermal P-cells is not reflected in the discharge of individual CFN neurons.

Like vermal P-cells, the discharge patterns of CFN cells display considerable trial-to-trial variation. Some CFN cells (~30%) with smooth-pursuit sensitivity also discharged a burst of spikes with saccades.

Summary

Based on current evidence, it is difficult to assign the flocculus or the posteriomedial cerebellum unique roles in the control of smooth pursuit. Both have neurons with a variety of burst-sustained discharge patterns during step-ramp pursuit. However, the bursts of a majority of CFN neurons lead pursuit onset, whereas the bursts of most floccular P-cells occur after pursuit is underway. Therefore, the CFN could be more important in pursuit initiation and the flocculus could serve to maintain it. The fact that floccular neurons respond to velocities >100° per s and CFN neurons to velocities of only 20°–60° per s seems consistent with this suggestion. However, the observation that bilateral CFN inactivation does not consistently delay pursuit onset (Robinson et al 1997) may argue against it.

All CFN cells and many floccular P-cells have vestibular as well as pursuit sensitivities. However, it is unclear, even in the flocculus, where many cells have equal head- and eye-velocity sensitivities, that gaze velocity is the controlled variable, especially when the head is free to rotate (Belton & McCrea 1999). Many flocculus P-cells and CFN cells respond for both smooth pursuit and saccades. Because smooth-pursuit cells in the CFN are in the immediate vicinity of saccade cells that discharge during eye acceleration, it has been suggested that the CFN provides signals to help accelerate both saccades and smooth-pursuit movements (Fuchs et al 1994). However, a similar role seems unlikely for the flocculus, where bilateral removal does not affect saccade metrics, which suggests that those saccade-related bursts have no impact downstream. With regard to the likelihood that these two structures could have a significant impact on smooth pursuit, floccular efferents impinge on neurons that are only one or two synapses from motor neurons. On the other hand, CFN efferents have no apparent smooth-pursuit target in the brain stem (Noda et al 1990). Furthermore, the activity of CFN cells waxes and wanes from trial to trial, a potentially problematic attribute for precision control. Floccular P-cell activity is considerably more reliable. In conclusion, deficits after lesions leave little doubt that the flocculus and posteriomedial cerebellum are required for accurate smooth pursuit, but we need more information to determine whether each structure indeed has a specific role in the generation of smooth pursuit.

Unresolved Issues

Do the Flocculus and the Posterior Vermis/CFN Account for all of the Cerebellum's Influence of Smooth Pursuit?

We can answer this question directly by testing a monkey's ability to produce pursuit after simultaneous inactivations of the flocculus and the CFN.

How does Pursuit-Related P-Cell Activity in the Oculomotor Vermis Shape the Discharge of CFN Neurons?

To answer this question the timing of P-cell discharges should be examined with step-ramp target motion to allow comparison with activity recorded under similar conditions in the CFN. Again, it would be informative to design an experiment that would separate the direct mossy fiber inputs to the CFN from those routed through the vermis.

Does the CFN also have a Role in Smooth-Pursuit Adaptation?

To examine this issue, the CFN first could be inactivated pharmacologically to determine whether adaptation to double-speed targets indeed is affected. If so, CFN smooth-pursuit units could be recorded while adaptation was occurring.

CEREBELLAR ROLE IN VERGENCE

In addition to their role in saccades and smooth pursuit, the CFN and VPIN also influence vergence movements. Some CFN neurons increase their firing rates during the near response, i.e. convergence and increasing accommodation (Zhang & Gamlin 1996). Inactivating the CFN with muscimol reduces the speed and size of convergence (Gamlin & Zhang 1996). Of the CFN neurons that respond with convergence, 63% also discharge during saccades (Zhang & Gamlin 1996). We think these saccade-related neurons are the same ones we have characterized because the saccade-related part of the CFN is so small. The fact that many CFN neurons respond to both saccades and convergence raises the possibility that these neurons play an important role in coordinating saccade and vergence movements, which occur together when targets are at different distances from the viewer.

Some VPIN neurons, in contrast, increase their firing rates during the far response, i.e. divergence and decreasing accommodation (Zhang & Gamlin 1998). These cells are in the same area as those with saccade-related responses, but none discharges with saccades. CFN and VPIN output may mediate vergence movement in humans as well because a patient with damage to his superior cerebellar peduncle, which carries cerebellar output, exhibited an almost complete absence of vergence eye movements (Ohtsuka et al 1993).

Clearly, we need much more information before our understanding about the cerebellum's role in vergence movements approaches that of its role in saccades and smooth pursuit.

CONCLUSION

Studies on the behaving monkey reveal that the cerebellum is necessary for the production of both accurate saccades and smooth-pursuit eye movements. Evidence is also accumulating for a cerebellar role in vergence eye movements. The neurons underlying the modification of these various eye movements have been localized and their behavior has been documented during the short-term control of individual movements. With this information, we can now evaluate the role of the cerebellum in long-term motor adaptation or learning. We can consider such fundamental issues as how neuronal discharge reflects behavioral adaptation, what might be the role of climbing fibers, and where are the synapses that undergo change during learning. Using the oculomotor system in this way, we may reveal mechanisms that are fundamental to the cerebellar control of all voluntary movements.

ACKNOWLEDGMENTS

AFF is very grateful for informative discussions concerning smooth pursuit with several colleagues, including Kikuro Fukushima, Rich Krauzlis, Leo Ling, Stephen Lisberger and Fred Miles.

Visit the Annual Reviews home page at www.AnnualReviews.org

LITERATURE CITED

Aizenman CD, Linden DJ. 1999. Regulation of the rebound depolarization and spontaneous firing patterns of deep nuclear neurons in slices of rat cerebellum. *J. Neurophysiol.* 82:1697–709

Barash S, Melikyan A, Sivakov A, Zhang M, Glickstein M, Thier P. 1999. Saccadic dysmetria and adaptation after lesions of the cerebellar cortex. *J. Neurosci.* 19:10931–39

Belknap DB, Noda H. 1987. Eye movements

evoked by microstimulation in the flocculus of the alert macaque. *Exp. Brain Res.* 67:352–62

Belton T, McCrea RA. 1999. Contribution of the cerebellar flocculus to gaze control during active head movements. *J. Neurophysiol.* 81:3105–9

Büttner U, Fuchs AF, Markert-Schwab G, Buckmaster P. 1991. Fastigial nucleus activity in the alert monkey during slow eye and head movements. *J. Neurophysiol.* 65:1360–71

Büttner U, Straube A, Spuler A. 1994. Saccadic dysmetria and intact smooth pursuit eye movements after bilateral deep cerebellar nuclei lesions. *J. Neurol. Neurosurg. Psychiatr.* 57:832–34

Dean P. 1995. Modelling the role of the cerebellar fastigial nuclei in producing accurate saccades: the importance of burst timing. *Neuroscience* 68:1059–77

Deubel H, Wolf W, Hauske G. 1986. Adaptive gain control of saccadic eye movements. *Hum. Neurobiol.* 5:245–53

Doslak MJ, Kline LB, Dell'Osso LF, Daroff RB. 1980. Internuclear ophthalmoplegia: recovery and plasticity. *Invest. Ophthalmol.* 19:1506–11

Edelman JA, Goldberg ME. 1995. Metrics of saccades evoked by electrical stimulation in the frontal eye fields are not affected by short-term saccadic adaptation. *Soc. Neurosci. Abstr.* 21:1195

Fitzgibbon EJ, Goldberg ME, Segraves MA. 1986. Short term saccadic adaptation in the monkey. In *Adaptive Processes in Visual and Oculomotor Systems*, ed. EL Keller, DS Zee, pp. 329–33. Oxford, UK: Pergamon

Frens MA, van Opstal AJ. 1997. Superior colliculus activity during short-term saccadic adaptation. *Brain Res. Bull.* 43:473–83

Fuchs AF, Kaneko CRS, Scudder CA. 1985. Brainstem control of saccadic eye movements. *Annu. Rev. Neurosci.* 8:307–27

Fuchs AF, Robinson FR, Straube A. 1993. Role of the caudal fastigial nucleus in saccade generation. I. Neuronal discharge patterns. *J. Neurophysiol.* 70:1723–40

Fuchs AF, Robinson FR, Straube A. 1994. Participation of the caudal fastigial nucleus in pursuit eye movements. I. Neuronal activity. *J. Neurophysiol.* 72:2714–28

Fuchs AF, Scudder CA, Kaneko CR. 1988. Discharge patterns and recruitment order of identified motoneurons and internuclear neurons in the monkey abducens nucleus. *J. Neurophysiol.* 60:1874–95

Fukushima K, Fukushima J, Kaneko CRS, Fuchs AF. 1999. Vertical Purkinje cells in the monkey floccular lobe: simple-spike activity during pursuit and passive whole body rotation. *J. Neurophysiol.* 82:787–803

Fukushima K, Tanaka M, Suzuki Y, Fukushima J, Yoshida T. 1996. Adaptive changes in human smooth pursuit eye movement. *Neurosci. Res.* 25:391–98

Furman JMR, Baloh RW, Yee RD. 1986. Eye movement abnormalities in a family with cerebellar vermian atrophy. *Acta Otolaryngol.* 101:371–77

Gamlin PDR, Zhang H. 1996. Effects of muscimol blockade of the posterior fastigial nucleus on vergence and ocular accommodation in the primate. *Soc. Neurosci. Abstr.* 22:2034

Goffart L, Pélisson D. 1994. Orienting gaze shifts during muscimol inactivation of caudal fastigial nucleus in the cat. I. Gaze dysmetria. *J. Neurophysiol.* 79:1942–58

Goffart L, Pélisson D. 1998. Cerebellar contribution to the spatial encoding of orienting gaze shifts in head-free cat. *J. Neurophysiol.* 72:2547–50

Goldberg ME, Musil S, Smith MK, Olson CR. 1993. The role of the cerebellum in the control of saccadic eye movements. In *Cerebellum and Basal Ganglia in the Control of Movement*, ed. EN Mano, pp. 203–11. Amsterdam: Elsevier

Goldreich D, Krauzlis RJ, Lisberger SG. 1992. Effect of changing feedback delay on spontaneous oscillations in smooth pursuit eye movements of monkeys. *J. Neurophysiol.* 67(3):625–38

Harting JK. 1977. Descending pathways from the superior colliculus: an autoradiographic analysis in rhesus monkey (*Macaca mulatta*). *J. Comp. Neurol.* 173:583–612

Helmchen C, Straube A, Büttner U. 1994. Saccade-related activity in the fastigial oculomotor region of the macaque monkey during spontaneous eye movements in light and darkness. *Exp. Brain Res.* 98:474–82

Ito M, Kano M. 1982. Long-lasting depression of parallel fiber-Purkinje cell transmission induced by conjunctive stimulation of parallel fibers and climbing fibers in the cerebellar cortex. *Neurosci. Lett.* 33:253–58

Kahlon M, Lisberger SG. 1996. Coordinate system for learning in the smooth pursuit eye movements of monkeys. *J. Neurosci.* 16:7270–83

Kahlon M, Lisberger SG. 2001. Changes in the responses of Purkinje cells in the floccular complex of monkeys after motor learning in smooth pursuit eye movements. *J. Neurophysiol.* 84:2945–60

Keller EL. 1988. Cerebellar involvement in smooth pursuit eye movement generation: flocculus and vermis. In *Physiological Aspects of Clinical Neuro-Ophthalmology*, ed. C Kennard, FC Rose, pp. 341–54. London: Chapman & Hall

Krauzlis R, Miles FA. 1998. Role of the oculomotor vermis in generating pursuit and saccades: effects of microstimulation. *J. Neurophysiol.* 80:2046–62

Krauzlis RJ, Lisberger SG. 1991. Visual motion commands for pursuit eye movements in the cerebellum. *Science* 253:568–71

Krauzlis RJ, Lisberger SG. 1994. Simple spike responses of gaze velocity Purkinje cells in the floccular lobe of the monkey during the onset and offset of pursuit eye movements. *J. Neurophysiol.* 72:2045–50

Krauzlis RJ, Lisberger SG. 1996. Directional organization of eye movement and visual signals in the floccular lobe of the monkey cerebellum. *Exp. Brain Res.* 109:289–302

Kurzan R, Straube A, Büttner U. 1993. The effect of muscimol micro-injections into the fastigial nucleus on the optokinetic response and the vestibulo-ocular reflex in the alert monkey. *Exp. Brain Res.* 94:252–60

Langer T, Fuchs AF, Chubb MC, Scudder CA, Lisberger SG. 1985. Floccular efferents in the rhesus macaque as revealed by autoradiography and horseradish peroxidase. *J. Comp Neurol.* 235:26–37

Lefévre P, Quaia C, Optican LM. 1998. Distributed model of control of saccades by superior colliculus and cerebellum. *Neural Netw.* 11:1175–90

Leung H-C, Suh M, Kettner RE. 2000. Cerebellar flocculus and paraflocculus Purkinje cell activity during circular pursuit in monkey. *J. Neurophysiol.* 83:13–30

Lisberger SG. 1988. The neural basis for learning of simple motor skills. *Science* 242:728–35

Lisberger SG, Fuchs AF. 1978. Role of primate flocculus during rapid behavioral modification of vestibuloocular reflex. I. Purkinje cell activity during visually guided horizontal smooth-pursuit eye movements and passive head rotation. *J. Neurophysiol.* 41:733–63

Lisberger SG, Morris EJ, Tychsen L. 1987. Visual motion processing and sensory-motor integration for smooth pursuit eye movements *Annu. Rev. Neurosci.* 10:97–129

Lisberger SG, Pavelko TA. 1988. Brain stem neurons in modified pathways for motor learning in the primate vestibulo-ocular reflex. *Science* 242:771–73

Lynch JC, Hoover JE, Strick PL. 1994. Input to the primate frontal eye field from the substantia nigra, superior colliculus, and dentate nucleus demonstrated by transneuronal transport. *Exp. Brain Res.* 100:181–86

McLaughlin SC. 1967. Parametric adjustment in saccadic eye movements. *Percept. Psychophysiol.* 2:359–62

Miles FA, Fuller JH, Braitman DJ, Dow BM. 1980. Long-term adaptive changes in primate vestibuloocular reflex. III. Electrophysiological observations in flocculus of normal monkeys. *J. Neurophysiol.* 43:1437–76

Miller JM, Anstis A, Templeton WB. 1981. Saccadic plasticity: parametric adaptive control of retinal feedback. *J. Exp. Psychol.* 7:356–66

Morris EJ, Lisberger SG. 1987. Different responses to small visual errors during initiation and maintenance of smooth-pursuit eye movements in monkeys. *J. Neurophysiol.* 58:1351–69

Noda H, Fujikado T. 1987. Topography of the oculomotor area of the cerebellar vermis in macaques as determined by microstimulation. *J. Neurophysiol.* 58:359–78

Noda H, Murakami S, Yamada J, Tamaki T, Aso T. 1988. Saccadic eye movements evoked by microstimulation of the fastigial nucleus of macaque monkeys. *J. Neurophysiol.* 60:1036–52

Noda H, Sugita S, Ikeda Y. 1990. Afferent and efferent connections of the oculomotor region of the fastigial nucleus in the macaque monkey. *J. Comp. Neurol.* 302:330–48

Noda H, Suzuki DA. 1979a. The role of the flocculus of the monkey in fixation and smooth pursuit eye movements. *J. Physiol.* 294:335–48

Noda H, Suzuki DA. 1979b. The role of the flocculus of the monkey in saccadic eye movements. *J. Physiol.* 294:317–34

Noda H, Warabi T. 1987. Responses of Purkinje cells and mossy fibers in the flocculus of the monkey during sinusoidal movements of a visual pattern. *J. Physiol.* 387:611–28

Ohtsuka K, Maekawa H, Sawa M. 1993. Convergence paralysis after lesions of the cerebellar peduncles. *Ophthalmologica* 206:143–48

Ohtsuka K, Noda H. 1990. Direction selective saccadic-burst neurons in the fastigial oculomotor region of the macaque. *Exp. Brain Res.* 81:659–62

Ohtsuka K, Noda H. 1991. Saccadic burst neurons in the oculomotor region of the fastigial nucleus of macaque monkeys. *J. Neurophysiol.* 65:1422–34

Ohtsuka K, Noda H. 1992. Burst discharges of fastigial neurons in macaque monkeys are driven by vision- and memory-guided saccades but not by spontaneous saccades. *Neurosci. Res.* 15:224–28

Ohtsuka K, Noda H. 1995. Discharge properties of Purkinje cells in oculomotor vermis during visually guided saccades in macaque monkey. *J. Neurophysiol.* 74:1828–40

Ohtsuka K, Sato H, Noda H. 1994. Saccadic burst neurons in the fastigial nucleus are not involved in compensating for orbital nonlinearities. *J. Neurophysiol.* 71:1976–80

Optican LM, Robinson DA. 1980. Cerebellar-dependent adaptive control of primate saccadic system. *J. Neurophysiol.* 44:1058–76

Optican LM, Zee DS, Chu FC. 1985. Adaptive response to ocular muscle weakness in human pursuit and saccadic eye movements. *J. Neurophysiol.* 54:110–22

Pierrot-Deseilligny C, Amarenco P, Roullet E, Marteau R. 1990. Vermal infarct with pursuit eye movement disorders. *J. Neurol. Neurosurg. Psychiatr.* 53:519–21

Quaia C, Lefévre P, Optican LM. 1999. Model of the control of saccades by superior colliculus and cerebellum. *J. Neurophysiol.* 82:999–1018

Rashbass C. 1961. The relationship between saccadic and smooth tracking eye movements. *J. Physiol.* 159:326–38

Ritchie L. 1976. Effects of cerebellar lesions on saccadic eye movements. *J. Neurophysiol.* 39:1246–56

Robinson FR. 1995. Role of the cerebellum in movement control and adaptation. *Curr. Opin. Neurobiol.* 5:755–62

Robinson FR. 2000. The role of the cerebellar posterior interpositus nucleus in saccades. I. Effect of temporary lesions. *J. Neurophysiol.* 84:1289–303

Robinson FR, Brettler SC. 1998. Smooth pursuit properties of neurons in the ventrolateral posterior interpositus nucleus of the monkey cerebellum. *Soc. Neurosci. Abstr.* 24:1405

Robinson FR, Fuchs AF, Noto CT. 2000. Role of the cerebellar caudal fastigial nucleus in saccade adaptation in monkey. *Soc. Neurosci. Abstr.* 26:293

Robinson FR, Fuchs AF, Straube A, Watanabe S. 1996. The role of the interpositus nucleus in saccades is different from the role of the fastigial nucleus. *Soc. Neurosci. Abstr.* 22:1200

Robinson FR, Straube A, Fuchs AF. 1993. Role of the caudal fastigial nucleus in saccade generation. II. Effects of muscimol inactivation. *J. Neurophysiol.* 70:1741–58

Robinson FR, Straube A, Fuchs AF. 1997. Participation of caudal fastigial nucleus in smooth pursuit eye movements. II. Effects of muscimol inactivation. *J. Neurophysiol.* 78:848–59

Sato H, Noda H. 1992. Posterior vermal Purkinje cells in macaques responding during saccades, smooth pursuit, chair rotation and/or optokinetic stimulation. *Neurosci. Res.* 12:583–95

Scudder CA. 1998. Discharge of fastigial nucleus neurons is altered during adaptive modification of saccade size. *Soc. Neurosci. Abstr.* 24:147

Scudder CA, Fuchs AF, Langer TP. 1988. Characteristics and functional identification of inhibitory burst neurons in the trained monkey. *J. Neurophysiol.* 59:1430–54

Scudder CA, McGee DM, Balaban CD. 2000. Connections of monkey saccade-related fastigial nucleus neurons revealed by anatomical and physiological methods. *Soc. Neurosci. Abstr.* 26:971

Shidara M, Kawano K. 1993. Role of Purkinje cells in the ventral paraflocculus in short-latency ocular following responses. *Exp. Brain Res.* 93:185–95

Shidara M, Kawano K, Gomi H, Kawato M. 1993. Inverse-dynamics model eye movement control by Purkinje cells in the cerebellum. *Nature* 365:50–52

Stanton GB, Goldberg ME, Bruce CJ. 1988. Frontal eye field efferents in the macaque monkey. II. Topography of terminal fields in midbrain and pons. *J. Comp. Neurol.* 271:493–506

Stone LS, Lisberger SG. 1990a. Visual responses of Purkinje cells in the cerebellar flocculus during smooth-pursuit eye movements in monkeys. I. Simple spikes. *J. Neurophysiol.* 63:1241–61

Stone LS, Lisberger SG. 1990b. Visual responses of Purkinje cells in the cerebellar flocculus during smooth-pursuit eye movements in monkeys. II. Complex spikes. *J. Neurophysiol.* 63:1262–75

Strassman A, Highstein SM, McCrea RA. 1986. Anatomy and physiology of saccadic burst neurons in the alert squirrel monkey. I. Excitatory burst neurons. *J. Comp. Neurol.* 249:337–57

Straube A, Fuchs AF, Usher S, Robinson FR. 1997a. Characteristics of saccadic gain adaptation in rhesus macaques. *J. Neurophysiol.* 77:874–95

Straube A, Scheuerer W, Eggert T. 1997b. Unilateral cerebellar lesions affect initiation of ipsilateral smooth pursuit eye movements in humans. *Ann. Neurol.* 42:891–98

Suzuki DA, Keller EL. 1988. The role of the posterior vermis of monkey cerebellum in smooth-pursuit eye movement control. II. Target velocity-related Purkinje cell activity. *J. Neurophysiol.* 59:19–40

Suzuki DA, Noda H, Kase M. 1981. Visual and pursuit eye movement-related activity in posterior vermis of monkey cerebellum. *J. Neurophysiol.* 46:1120–39

Takagi M, Zee DS, Tamara RJ. 1998. Effects of lesions of the oculomotor vermis on eye movements in primate: saccades. *J. Neurophysiol.* 80:1911–31

Takikawa Y, Kawagoe R, Miyashita N, Hikosaka O. 1998. Presaccadic omnidirectional burst activity in the basal interstitial nucleus in the monkey cerebellum. *Exp. Brain Res.* 121:442–50

Thier P, Dicke PW, Haas R, Barash S. 2000. Encoding of movement time by populations of cerebellar Purkinje cells. *Nature* 405:72–76

Vahedi K, Rivauud S, Amarenco P, Pierrot-Deseilligny C. 1995. Horizontal eye movement disorders after posterior vermis infarctions. *J. Neurol. Neurosurg. Psychiatr.* 58:91–94

Vilis T, Hore J. 1981. Characteristics of saccadic dysmetria in monkeys during reversible lesions of medial cerebellar nuclei. *J. Neurophysiol.* 46:828–38

Waespe W, Baumgartner R. 1992. Enduring dysmetria and impaired gain adaptivity of saccadic eye movements in Wallenberg's lateral medullary syndrome. *Brain* 115:1125–46

Waespe W, Wichmann W. 1990. Oculomotor disturbances during visual vestibular interaction in Wallenberg's lateral medullary syndrome. *Brain* 113:821–46

Westheimer G, Blair S. 1973. Oculomotor defects in cerebellectomized monkeys. *Invest. Ophthalmol.* 12:618–21

Yamada J, Noda H. 1987. Afferent and efferent connections of the oculomotor cerebellar vermis in the macaque monkey. *J. Comp. Neurol.* 1265:224–41

Zee D, Yamazaki A, Butler PH, Gücer G. 1981. Effects of ablation of flocculus and paraflocculus on eye movements in primate *J. Neurophysiol.* 46:878–99

Zhang H, Gamlin PD. 1996. Single unit activity within the posterior fastigial nucleus during vergence and accommodation in the alert primate. *Soc. Neurosci. Abstr.* 22: 2034

Zhang H, Gamlin PD. 1998. Neurons in the posterior interposed nucleus of the cerebellum related to vergence and accommodation. I. Steady-state characteristics. *J. Neurophysiol.* 79:1255–69

Annu. Rev. Neurosci. 2001. 24:1005–039

ROLE OF THE REELIN SIGNALING PATHWAY IN CENTRAL NERVOUS SYSTEM DEVELOPMENT

Dennis S Rice and Tom Curran

Department of Developmental Neurobiology, St. Jude Children's Research Hospital, Memphis, Tennessee 38105; e-mail: fos1@aol.com, dennis.rice@stjude.org

Key Words migration, reeler, brain, Dab1, tyrosine kinase

■ **Abstract** The neurological mutant mouse *reeler* has played a critical role in the evolution of our understanding of normal brain development. From the earliest neuroanatomic studies of *reeler*, it was anticipated that the characterization of the gene responsible would elucidate important molecular and cellular principles governing cell positioning and the formation of synaptic circuits in the developing brain. Indeed, the identification of *reelin* has challenged many of our previous notions and has led to a new vision of the events involved in the migration of neurons. Several neuronal populations throughout the brain secrete Reelin, which binds to transmembrane receptors located on adjacent cells triggering a tyrosine kinase cascade. This allows neurons to complete migration and adopt their ultimate positions in laminar structures in the central nervous system. Recent studies have also suggested a role for the Reelin pathway in axonal branching, synaptogenesis, and pathology underlying neurodegeneration.

INTRODUCTION

One of the fundamental issues facing the field of developmental neurobiology is to understand the mechanisms that control the deployment of neurons with similar properties into specific layers. This is a daunting question given the fact that there are so many different types of neurons that must be precisely connected for proper brain function. Neuronal positioning is critical for the formation of cytoarchitecturely distinct brain regions such as the cerebral cortex, hippocampus, and cerebellum. During development of laminated brain structures a series of coordinated migrations takes neurons from their site of origin to their final destinations, where they adopt definitive morphological phenotypes by elaborating dendritic and axonal processes (Ramón y Cajal 1955).

The analysis of neurological mutant mice has led to dramatic progress in the identification and characterization of genes important for neuronal migration in the developing brain (Rice & Curran 1999). Over a relatively short period of time, genetic and biochemical studies have identified a new signaling pathway that controls neuronal cell positioning. The gene disrupted in *reeler* mice encodes a

0147-006X/01/0621-1005$14.00

TABLE 1   Mouse genes implicated, genetically or biochemically, in the Reelin signaling pathway

Description	Gene	Name	Reference
Extracellular	*ApoE*	ApolipoproteinE	D'Arcangelo et al 1999
	Bdnf	Brain-derived neurotrophic factor	Ringstedt et al 1998
	Reln	Reelin	D'Arcangelo et al 1995
Transmembrane	*ApoER2*	Apolipoprotein E receptor 2	Trommsdorff et al 1999
	CNR	Cadherin-related neuronal receptor	Senzaki et al 1999
	itga3	α3 integrin	Dulabon et al 2000
	PS-1	Presenilin-1	Hartmann et al 1999
	Vldlr	Very low–density lipoprotein receptor	Trommsdorff et al 1999
Cytoplasmic	*Abl*	Abelson proto-oncogene	Howell et al 1997a
	CASK	Membrane-associated guanylate kinase	Hsueh et al 2000
	Cdk5	Cyclin-dependent kinase 5	Ohshima et al 1996
	Cdk5r	p35	Chae et al 1997
	Dab-1	Disabled 1	Sheldon et al 1997
	Src	Src proto-oncogene	Howell et al 1997a
Nuclear	*Emx2*	Related to *empty spiracles*	Mallamaci et al 2000
	P73	Transformation-related protein 73	Yang et al 2000
	Tbr-1	T-box brain-1	Bulfone et al 1995

large extracellular protein, Reelin, that is produced by discrete populations of cells in the brain (D'Arcangelo et al 1995). Reelin binds to transmembrane receptors, the very low–density lipoprotein receptor (Vldlr), and the apolipoprotein E receptor 2 (ApoER2) present on migrating neurons (D'Arcangelo et al 1999, Hiesberger et al 1999). The cytoplasmic domains of these receptors bind to disabled-1 (Dab1), an intracellular adapter protein (Trommsdorff et al 1999). Binding of Reelin to lipoprotein receptors induces tyrosine phosphorylation of Dab1 (Howell et al 1999a), triggering an intracellular signaling cascade that instructs neurons to occupy their proper locations in the developing central nervous system (CNS). Other genes have been implicated in the Reelin signaling pathway, although their role is not fully understood (Table 1). Collectively, these genes provide a molecular starting point for biological, biochemical, and cellular studies that are providing important and exciting insights into the molecular mechanisms that control brain development and, potentially, the pathogenesis of neurodegenerative disorders.

HISTORICAL PERSPECTIVE

The study of heritable mutations in mice can be traced back to 2000 BC in the Middle East and the Orient. While most of the world viewed mice as pests and carriers of disease, China and Japan embraced the mouse as a symbol of prosperity

(Keeler 1931). So-called white mice were known to breed true and other coat-color variations, and behavioral mutant mice were collected or "fancied" as long ago as 80 BC. Many of these fancy mice made their way from Japan to Europe through the hands of British traders. Breeding studies on the propagation of colorful traits, such as white English sables and creamy buffs, were performed on mice before the rediscovery of Mendel's Law of Heredity (Davenport 1900). Therefore, modern mouse genetics in the United States was built on a foundation of pet mice through the studies of William E Castle and his students (Morse 1981). The practice of inbreeding mouse strains to study cancer was implemented in the early twentieth century. Shortly thereafter, the first scientific report of an inheritable defect in the CNS of mice was published in the *Proceeding of the National Acadamy of Sciences* (Keeler 1924). The *rodless* phenotype, which is now known to result from a mutation in the gene encoding the β-subunit of cGMP phosphodiesterase, was identified by analysis of histological sections of mouse retinas (Pittler et al 1993). Inbreeding of mouse stocks propagated this mutation in many of the standard lines of mice used today. Inbreeding also increased the likelihood of recessive characteristics appearing in offspring, some of which produced noticeable behavioral phenotypes. Over time, more "deviant" mice were reported, and the first database of mutations affecting the development of the CNS, produced in 1965, contained over 100 records (Sidman et al 1965). *Reeler* mice represented one of the founding members of the behavioral mutants reported in the first record, and they were recognized to exhibit widespread neuroanatomic defects in the cerebral cortex, cerebellum, and hippocampus. Additional anatomical abnormalities are also evident in other important brain structures, such as the thalamus, midbrain, brain stem, and spinal cord (Yip et al 2000, Caviness et al 1988). The neuroanatomical disruptions in the *reeler* brain suggested that the mutation affected a gene that is critical for controlling cell positioning in the developing CNS.

THE NEUROLOGICAL MUTANT MOUSE *reeler*

The *reeler* mutation arose spontaneously in a stock of "snowy-bellied" mice at the Institute of Animal Genetics in Edinburgh, Scotland. Falconer (1951) described the locomoter abnormality in *reeler* and in another neurological mutant named *trembler*. In the first description of the *reeler* cerebellum Hamburgh (1960:460) reported, "The typical appearance of the folia is missing. The arrangement of Purkinje cells which normally surround the granular layer is severely disturbed. The granular layer is much reduced and the area of white matter contains large numbers of cells, which resemble Purkinje cells." There are several known alleles of *reeler*, all of which exhibit identical phenotypes (D'Arcangelo & Curran 1998). Recently, mutations in *reelin* were also reported in humans with autosomal recessive lissencephaly with cerebellar hypoplasia (Hong et al 2000). Lissencephaly is a condition of severe neuronal migration disorders that result in a smooth appearance of the brain surface in humans. Although mice are normally lissencephalic, migration disorders result in similar misplacements of neurons in laminated brain regions in both mice and humans.

THE *reeler* MUTATION DISRUPTS CELL POSITIONING IN THE BRAIN

In mammals the organization of the cerebral cortex follows a stereotypic plan during development. Neurons with similar morphologies and connections are positioned in the same layer. This cellular organization reflects the patterning of neuronal processing that occurs in the cerebral cortex, where certain layers receive input and connect to other layers, allowing multiple levels of information integration, modulation, and extraction. One of the most revealing techniques used to characterize the *reeler* brain was neuronal "birthdating" analysis. This approach employed autoradiography of brain slices after administration of tritiated thymidine to mice, which is permanently incorporated into DNA during synthesis. Cells that undergo their final division and adopt a postmitotic phenotype are strongly marked by radioactivity, whereas those that re-enter the cell cycle further dilute the label through successive rounds of cell division. Therefore, the temporal and spatial assembly of cells can be revealed by autoradiographic analysis of brains that received tritiated thymidine at specific stages during development. The relative position of neurons with respect to their birthdate in the adult cortex is shown schematically in Figure 1. Using this approach, it was found that cortical neurons in the mouse are generated between embryonic days (E) 10 to 18 and that the 6 layers of the cerebral cortex assemble in an inside-out sequence (Angevine & Sidman 1961). These studies demonstrated that, in addition to morphological and physiological characteristics, neurons positioned within specific cortical layers also share a common birthdate.

The *reeler* cerebral cortex violates this fundamental plan of cortical assembly (Caviness & Sidman 1973a). Layer I is not discernible in the mutant, and the position of cells comprising other layers is relatively inverted (Figure 1). Importantly, all major morphological cell classes are present in the *reeler* cortex, and cohorts of neurons comprising specific layers are generated on schedule (Caviness & Sidman 1973b). The neuronal classes in *reeler* cortex fail to align in an inside-out fashion like their counterparts in the normal cortex. In *reeler*, neurons born relatively late during corticogenesis reside in deep layers beneath the older neurons (Caviness & Sidman 1973b). These comparative studies demonstrated that the *reeler* mutation affects the positioning of neurons within specific layers. Thus, the molecular mechanisms controlling cell positioning are distinct from those that determine the number and proportion of cell types constituting various layers of the cerebral cortex (McConnell & Kaznowski 1991, Herrup 1987).

Disorganization of fiber patterns representing the major afferent and efferent systems accompanies cellular ectopia in the mutant cerebral cortex (Caviness et al 1988). Therefore, *reeler* mice provide a model system in which to address the

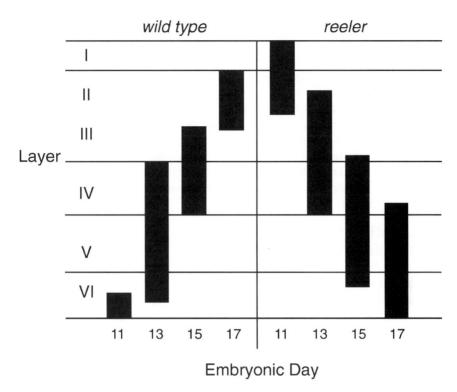

Figure 1   The position of cells born at specific developmental times is relatively inverted in the adult *reeler* cerebral cortex. In this schematic the pial surface is at the top and the white matter is at the bottom. Injections of [^{3}H] thymidine were administered at 48-hour intervals throughout corticogenesis. Cells generated on E11 are located in Layer VI in the wild-type cortex, whereas cells with similar birthdates and morphologies are located near the surface of the *reeler* brain. Other neuronal classes are generated on schedule in *reeler*, but they obtain an incorrect location in the adult cerebral cortex. A similar situation is observed in the *scrambler* cortex. This data was taken from Caviness 1982.

relationship between cell positioning and synaptic circuit formation during development of the nervous system. Careful studies of *reeler* provided conclusive evidence that the molecular mechanisms controlling cell positioning are distinct from those that control axonal guidance, which is a key first step in the formation of synaptic circuitry (Caviness & Rakic 1978). Afferent projections to the visual, olfactory, somatosensory, and motor cortices rearrange their trajectories and actually find their correct target cells even though they are located in ectopic positions (Terashima et al 1983, Simmons et al 1982, Caviness 1976). Some of the axonal projections take circuitous routes to their postsynaptic targets in the *reeler* cortex. Thalamocortical projections that are normally restricted to deep cortical layers, namely Layer IV, ascend to the superficial aspect of the *reeler* cerebral cortex before "looping" down to their correct target cells (Caviness 1976, Steindler & Colwell 1976). Studies of *reeler* demonstrated that the overall organization of the major systems and the physiological responses of individual neurons are comparable to those in normal brain (Silva et al 1991, Lemmon & Pearlman 1981, Dräger 1981). The conclusion from these studies is that the general wiring pattern of the brain is independent of cell position, implying that other genes control axon guidance. However, changes in synaptic density, distribution, and topology are present in a number of brain structures in *reeler* including the cerebellum, hippocampus, and piriform cortex (Borrell et al 1999a, Caviness 1977, Mariani et al 1977). Thus, while the overall mechanisms of axon guidance appear unaffected in *reeler*, other attributes of synaptic circuit formation are abnormal.

DEVELOPMENT OF CELL LAMINATION IN THE CEREBRAL CORTEX

The mammalian brain is assembled through a choreographed series of far-ranging migrations that results in the segregation of neurons with similar properties into discrete layers. The initial alignment of neurons in the CNS relates to the time at which they exit the cell cycle and migrate to their ultimate locations where they adopt a definitive neuronal phenotype (McConnell & Kaznowski 1991, Luskin & Shatz 1985, Rakic 1974). Two modes of migration, tangential and radial, are recognized in the developing brain. Radial migration is the most common form of migration in the developing cortex and it relies on a specialized scaffold of cells that span the cerebral wall (Hatten 1999, Rakic 1972). This scaffold is comprised of bipolar cells, known as radial glia, that were first described using the Golgi technique, but they can also be visualized with immunohistochemical markers (Bentivoglio & Mazzarello 1999). The descriptive name accurately depicts the morphology of radial glia, which send a short process from the cell soma to the ventricular surface and a long process that spans the width of the neural axis. The ascending process splits into several branches near the pial surface. Migrating cortical neurons physically associate with radial glia during their ascent to the top of the cortex, and their migratory paths are likely guided by molecules expressed on the plasma membrane (Anton et al 1996, Cameron & Rakic 1994). Radial glia are present only during neuronal development. Subsequently, they transform into astrocytes in the cerebral cortex and Bergmann fibers in the cerebellum (Hatten 1999).

Neocortical development begins with the appearance of the preplate or primordial plexiform layer above the ventricular zone (Figure 2A). The preplate is composed of subcortical afferents and two populations of postmitotic neurons, the Cajal-Retzius cells and the subplate neurons (Marin-Padilla 1998, Super et al 1998). Cajal-Retzius cells are among the earliest neurons to be generated in the mammalian neocortex, where they occupy positions near the pial surface on the superficial aspect of the brain (Meyer & Goffinet 1998, König et al 1977).

Subplate neurons are born slightly later than Cajal-Retzius cells and they serve as transient synaptic targets for thalamocortical projections (Allendoerfer & Shatz 1994). The next phase of development occurs when the cortical plate neurons exit the cell cycle near the ventricular surface and invade the preplate. Migrating neurons move past the subplate, displacing this layer away from the Cajal-Retzius cells, which remain adjacent to the pial surface in a cell-sparse area known as the marginal zone (Figure 2A). As new cortical plate neurons arrive on the radial glial, they migrate past the older subplate and cortical plate neurons before inserting directly beneath Cajal-Retzius cells. The systematic migration of younger neurons past their predecessors results in the "inside-out" pattern of development, in which the cortical plate (future Layers II-VI) develops between the marginal zone (future Layer I) and the subplate (Marin-Padilla 1998, Angevine & Sidman 1961).

Histological and ultrastructural studies on the embryonic *reeler* cortex demonstrated that cell position defects arise during the formation of the cortical plate. The preplate develops normally and migration of cortical plate neurons commences on schedule. Shortly thereafter, disorganization of the cortical plate becomes apparent when the first cohort of migrating neurons fails to invade the preplate (Figure 2B). As a result, the preplate is not split into the marginal zone and the subplate (Sheppard & Pearlman 1997, Hoffarth et al 1995, Ogawa et al 1995, Caviness et al 1988, Caviness 1982, Pinto Lord & Caviness 1979, Goffinet 1979). As additional cohorts of neurons arrive on the radial glia guides, they are unable to bypass their predecessors. This leads to formation of a disorganized cortical plate underneath the superplate. Figure 2A shows the histotypical appearance of a normal cerebral cortex compared to a *reeler* cerebral cortex at E16.5. The marginal zone in normal mice contains fibers and Cajal-Retzius cells. However, the marginal zone in *reeler* is replaced by a relatively cell-dense region known as the superplate, which contains Cajal-Retzius cells, subplate neurons and a few cortical plate neurons (Derer 1985, Caviness 1982). Cells in the *reeler* cortical plate are frequently displaced by fibers that run obliquely through the cerebral wall (Goffinet 1979). This gives the appearance of cell-free rifts in the *reeler* cortex on histological examination (Figure 2A). It is important to note that the cortical plate does form in *reeler* and that migrating cells reach this level. Ultrastructure analysis of the *reeler* cerebral cortex at E17 revealed that several postmigratory neurons remain closely associated with the radial glia (Pinto-Lord et al 1982). Leading processes of migrating cells in *reeler* are apparently unable to breach these inappropriate contacts and their ascent to the marginal zone is impeded. Moreover, the characteristic appearance of the radial glia scaffold is disturbed during later stages of corticogenesis in the *reeler* cortex, and radial fibers are deployed at oblique angles (Hunter-Schaedle 1997, Mikoshiba et al 1983). These studies led to the suggestion that the *reeler* mutation affects molecular interactions between migrating neurons and the glial guides.

An alternative proposal suggested that the *reeler* mutation affects interactions among cortical neurons destined for particular layers. This idea is based on studies conducted in vitro by De Long & Sidman (1970), in which neurons isolated from *reeler* cortex fail to show the typical patterns of cell aggregation. This observation was confirmed and extended by Hoffarth et al (1995) using a similar in vitro assay. Analysis of *reeler* neurons generated during the earliest stages of corticogenesis (ca E11–12) revealed that they clump together more frequently than their normal counterparts. Moreover, retrograde labeling of different neuronal populations revealed that the *reeler* mutation preferentially targeted the earliest born cortical neurons. These studies suggested that the mutant gene affected neuron-neuron interactions. In addition, chimeric mouse studies indicated that the gene functioned extrinsically (Terashima et al 1986, Mullen 1984). The two prevailing models at the time proposed disruptions of cellular interactions either between migrating neurons and radial fibers or among migrating neurons themselves. Whereas these models could explain certain aspects of the *reeler* phenotype, it was hoped that identification of the *reeler* gene would resolve the cellular target and clarify the molecular events underlying the neuroanatomic disruptions.

———————————————————————————————→

Figure 2   A simplified view of corticogenesis and the histological appearance of the developing cerebral wall. (*A*) Corticogenesis in mammals begins with the appearance of the preplate (pp), which is located directly above the proliferating cells in the ventricular zone (vz). The preplate forms around embryonic days (E) 11–12 in mice and it contains Cajal-Retzius neurons (CR) and subplate (SP) neurons, among others. Cells destined for the cortical plate are generated over the next week. The first cohort of cortical plate cells (gray) migrates past the subplate and stops beneath the Cajal-Retzius cells in the marginal zone (mz). Successively generated waves (gray to white) of cells migrate past their predecessors and stop beneath the Cajal-Retzius cells. Therefore, by E16.5 in the mouse young cortical plate neurons are positioned above the older neurons, with the exception of the Cajal-Retzius cells. Cortical plate neurons mature into adult neurons in the cerebral cortex by the elaboration of dendritic and axonal fibers. A disruption in cortical development in *reeler* is obvious at E16. The marginal zone in the wild-type is relatively cell-free, except for Cajal-Retzius and several other neurons. The cortical plate (cp) contains tightly packed cells with a radial alignment. The subplate is beneath the cortical plate. Many neurons destined for superficial layers migrate along radial glia in the intermediate zone (iz). Cells continue to divide in the ventricular zone (vz). Although the overall divisions of the cerebral wall are obvious in *reeler*, the cortical plate is disorganized and cells are displaced by fibers that run obliquely in the cerebral wall. Many subplate cells are located with the Cajal-Retzius cells in a relatively cell-dense region known as the superplate (spp). (*B*) The defect in the *reeler* cortex is apparent at the onset of cortical plate formation. In wild-type cortex the first cohort of cortical plate neurons (gray) positions itself between the Cajal-Retzius cells (black) that produce Reelin (*stipples*) and the subplate layer. In *reeler* the first cohort of cortical plate neurons fails to migrate past the subplate and instead accumulates in a disorganized fashion in the superplate. Subsequently, many cortical neurons lose their radial alignment and deploy dendritic and axonal projections at oblique angles.

REELIN IS A SECRETED PROTEIN

The identification and characterization of the *reelin* gene (*Reln*) have been reviewed recently (D'Arcangelo & Curran 1998). Briefly, a transgene insertion into the *reeler* locus led directly to the isolation of a large mRNA (>12 kb) that contains an open reading frame of 10,383 bases encoding a protein of approximately 385 kDa (D'Arcangelo et al 1995). Several isoforms of Reelin are present in brain extracts and in the supernatant of primary neuronal cultures. These isoforms arise via cleavage of full-length Reelin into two smaller proteins of approximately 250 and 180 kDa (D'Arcangelo et al 1999, de Rouvroit et al 1999). The N-terminus of Reelin contains a cleavable signal peptide and a region of similarity with

A

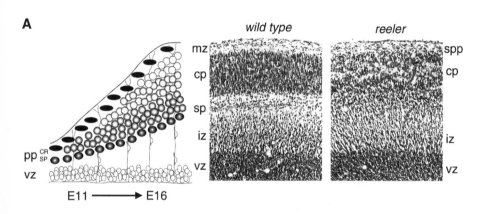

B

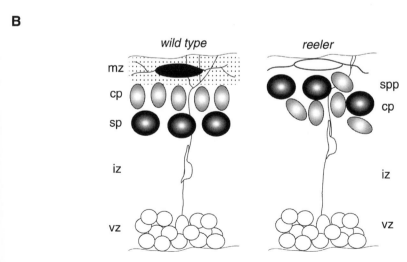

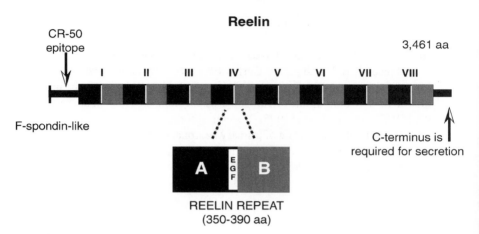

Figure 3   The *reelin* open reading frame predicts a novel protein of 3461 amino acids (aa) at a molecular mass of approximately 385 kDa. Reelin contains a cleavable signal peptide at the N-terminus, followed by a region of similarity to F-spondin, a secreted protein produced by floor plate cells that controls cell migration and neurite outgrowth. The most striking feature of Reelin is the presence of a series of eight internal repeats comprising 350–390 aa. These so-called Reelin repeats contain two related subdomains, A and B, separated by a stretch of 30 aa harboring an epidermal growth factor–like motif. A region rich in arginine residues at the C-terminus of Reelin is required for secretion. A monoclonal antibody named CR-50 recognizes an epitope defined to aa 230–346, which is C-terminal to the F-spondin region but before the first Reelin repeat. This antibody blocks Reelin function in vivo and in vitro.

F-spondin (Figure 3), a protein secreted by floor plate cells that directs neural crest cell migration and neurite outgrowth (Klar et al 1992). The main body of Reelin consists of a series of eight internal repeats (Reelin repeats) of 350–390 amino acids. Each Reelin repeat contains two related subdomains flanking a pattern of conserved cysteine residues related to epidermal growth factor–like motifs. These motifs are related to those in other extracellular proteins such as tenascin and restrictin and in the integrin family of receptors. Deletion analyses demonstrated that a short region of positively charged amino acids near the C-terminus is required for secretion of Reelin (D'Arcangelo et al 1997). In the *reeler* Orleans mutant (rl^{orl}) Reelin is made but not secreted due to the insertion of a L1 transposable element that alters the amino acid sequence of the C-terminus (de Bergeyck et al 1997).

One of the most surprising findings uncovered by the identification of *reelin* was the elucidation of its expression pattern in the developing brain. By using in situ hybridization, *reelin* was found to be present in a relatively small population of cells located in the marginal zone of the cerebral cortex (Figure 2B). These cells were identified as Cajal-Retzius neurons by their position and morphology

(D'Arcangelo et al 1995). At the same time *reelin* was described, Ogawa et al (1995) reported the generation of a monoclonal antibody by immunizing *reeler* mice with brain extracts from normal mice. Immunohistochemical studies in wild-type mice revealed that this antibody labeled Cajal-Retzius cells, which also express the calcium-binding protein calretinin. Importantly, the cerebral cortex of *reeler* mice did not react with this antibody, although Cajal-Retzius neurons are present (Ogawa et al 1995, Derer 1985). Subsequent studies demonstrated that the CR-50 antibody recognizes Reelin (D'Arcangelo et al 1997). Importantly, several experimental approaches revealed that CR-50 functions as a blocking antibody both in vivo and in vitro (Borrell et al 1999a, Miyata et al 1997, Nakajima et al 1997, Ogawa et al 1995). In an elegant series of experiments, Ogawa et al (1995) employed a cortical cell aggregation assay to address the physiological effects of the CR-50 antibody. Neurons dissociated from normal cortices distributed into a radial pattern as viewed with antibodies specific for microtubule-associated protein-2. In parallel cultures dissociated neurons obtained from the *reeler* cortex formed large clumps of microtubule-associated protein-2–positive neurons that lacked radial organization. When the CR-50 antibody was added to cultures of normal neurons at the time of cultivation the resulting cellular organization of the aggregates appeared very similar to those obtained with *reeler* neurons. The histotypic conversion of wild-type neurons to *reeler* neurons was abolished when CR-50 was preabsorbed against normal cortices but not *reeler* cortices. Recently, Reelin was shown to form multimeric complexes through the CR-50 epitope region (Utsunomiya-Tate et al 2000). These complexes were disrupted in the presence of CR-50, suggesting that multimerization of Reelin is required for its function.

Several hypotheses were proposed to explain the *reeler* phenotype and the function of Reelin in the developing cerebral cortex. The spatial and temporal expression pattern of Reelin suggested that it was important during the final phases of migration. Reelin is expressed by Cajal-Retzius before the first wave of cortical plate neurons reaches the preplate and it is secreted precisely at the location where these neurons stop migrating and detach from radial glia (Figure 2*B*). Therefore, Reelin was suggested to provide positional information to migrating neurons that instructs them to stop and detach from their guides. Reelin expression persists in the marginal zone throughout corticogenesis, implying that all migrating neurons encounter Reelin after their ascent to the top of the radial glia (Alcantara et al 1998). Reelin may act as an attractant molecule that enables each wave of cortical neurons to bypass their predecessors. Alternatively, Reelin could repel subplate neurons, thereby facilitating the invasion of the cortical plate (D'Arcangelo & Curran 1998). Although these ideas have merit in explaining the cortical defect in *reeler*, they rely on the assumption that Reelin binds cortical neurons directly. After identification of Reelin, the major question facing the field was, "How is the extracellular signal received and interpreted by migrating neurons?"

DISRUPTION OF THE *Disabled-1* GENE CAUSES *reeler*-LIKE PHENOTYPES

Davisson and colleagues at the Jackson Laboratory reported *reeler*-like phenotypes in a new neurological mutant named *scrambler* (Sweet et al 1996). *Scrambler* arose spontaneously and the mutants were ataxic by 2 weeks of age. This behavior was attributed to the abnormal cerebellum, which was small and lacked foliation. Closer examination revealed a reduction in the number of granule cells and Purkinje cell ectopia that were identical to that in *reeler* (Goldowitz et al 1997). Similarities between *reeler* and *scrambler* extended to the hippocampus and the cerebral cortex. Detailed studies of the histology and sequence of corticogenesis in *scrambler* demonstrated that Layer I was absent and the normal "inside-out" patterning of neurons was relatively inverted (Gonzalez et al 1997). These neuroanatomical defects were associated with an autosomal recessive mutation, and genetic mapping studies proved that the gene responsible was distinct from *reelin* (Sweet et al 1996). The striking similarities between *scrambler* and *reeler* suggested that the *scrambler* gene controls Reelin expression, either directly or indirectly by affecting cells that normally produce Reelin. However, immunohistochemical studies failed to reveal alterations in Reelin expression, and primary neuronal cultures obtained from *scrambler* mice produced and secreted Reelin (Gonzalez et al 1997, Goldowitz et al 1997). This genetic evidence suggested that *scrambler* functions downstream of Reelin in a common signaling pathway.

The *scrambler* locus was mapped near the *disabled-1* (*Dab1*) gene on chromosome 4, and mice with a targeted disruption of *Dab1* were found to exhibit ataxia and neuroanatomical defects indistinguishable from those in *scrambler* and *reeler* (Howell et al 1997b). Immunoblotting studies with Dab1-specific antibodies revealed a substantial decrease in Dab1 in *scrambler* brain (Sheldon et al 1997). Northern blot analysis demonstrated that the levels of the normal 5.5 kb Dab1 mRNA in *scrambler* were dramatically decreased. Furthermore, an additional mRNA species of approximately 7 kb was present that arose as a consequence of splicing defects in *Dab1* (Sheldon et al 1997, Ware et al 1997). Splicing defects at the same nucleotide in *Dab1* are also responsible for *reeler*-like traits observed in *yotari* mice, which appeared in a colony of mice carrying a disruption of the gene encoding the receptor for inositol-1,4,5-trisphosphate (Sheldon et al 1997, Yoneshima et al 1997). These studies provided convincing evidence that disruption of *Dab1*, either by spontaneous or targeted mutation, results in behavioral and anatomical abnormalities that are indistinguishable from those described in *reeler*.

Dab1 encodes a cytoplasmic protein containing a motif known as a protein interaction/phosphotyrosine binding (PI/PTB) domain (Figure 4). This domain was originally identified in the adapter protein Shc as a region required for binding to the epidermal growth factor receptor, the insulin receptor, and other tyrosine-phosphorylated proteins (Margolis 1996). Mammalian Dab1 was originally identified as a Src-binding protein in a yeast two-hybrid screen (Howell et al 1997a). It is expressed at high levels in the developing CNS and it is phosphorylated on

Disabled-1

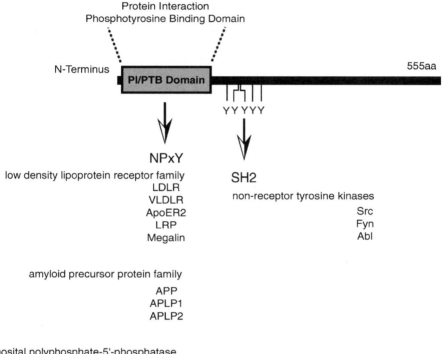

Figure 4 The *disabled-1* open reading frame predicts a protein of 555 amino acids that exhibits properties of adapter proteins. Near the N-terminus of Disabled-1 (Dab1) is a region of approximately 150 amino acids known as a protein interaction/phosphotyrosine binding domain (PI/PTB). The PI/PTB domain of Dab1 is most closely related to that in p96 (Dab2) and *Drosophila* Disabled. The PI/PTB domain binds to NPxY-containing motifs present in several proteins such as members of the low-density lipoprotein receptor family and the amyloid precursor protein family. Dab1 also binds to the NPxY motif in SHIP, an inositol polyphosphate-5′-phosphatase. A cluster of tyrosine residues (Y) downstream of the PI/PTB domain serve as docking sites for SH2 domain–containing proteins, such as the nonreceptor tyrosine kinases Src, Fyn, and Abl. LDLR, low-density lipoprotein receptor; VLDLR, very low–density lipoprotein receptor; ApoER2, Apolipoprotein E receptor 2; LRP, low-density lipoprotein receptor–related protein; APP, amyloid precursor protein; APLP1 and APLP2, amyloid precursor–like protein 1 and 2, respectively.

tyrosine residues during brain development. Tyrosine phosphorylation of Dab1 promotes an interaction with several nonreceptor tyrosine kinases, including Src, Fyn, and Abl through their SH2 domains, implying Dab1 functions in kinase signaling cascades during development (Howell et al 1997a).

Dab1-related genes in other species are involved in kinase signaling pathways that function during brain development. The *Dab* gene in *Drosophila* was first discovered as a genetic modifier of the Abl tyrosine kinase (Gertler et al 1989). Flies deficient in Abl die as adults and their eyes appear rough owing to irregular spacing of retinal cells (Henkemeyer et al 1987). Haploinsufficiency of *Dab* in an Abl-deficient fly results in axonal pathway defects, and flies that lack both Dab and Abl exhibit frequent breaks in axonal tracts in the CNS (Gertler et al 1993). Dab also physically associates with the sevenless (Sev) receptor tyrosine kinase and contributes to *Drosophila* eye development (Le & Simon 1998). Flies in which the *Dab* gene is inactivated have disorganized eye structures and a frequent loss of R7 photoreceptors. These results demonstrate that Dab is an important adapter protein in both receptor and nonreceptor tyrosine kinase signaling pathways that control formation of the nervous system in flies.

One important binding site for the PTB domain of Dab1 is the peptide sequence Asn-Pro-x-Tyr (NPxY). The NPxY motif was first recognized in members of the low-density lipoprotein receptor (LDLR) family and it is present in several other transmembrane proteins where it functions in clathrin-mediated endocytosis (Chen et al 1990). Interactions between the PTB domains in several adapter proteins and the NPxY motif are dependent on tyrosine phosphorylation (Margolis 1996). However, the PTB domain of Dab1 can bind to NPxY motifs in a phosphotyrosine-independent manner. Biochemical protein interaction assays identified F/YxNPxY motifs in the cytoplasmic domains of the LDLR family and the amyloid precursor protein (APP) family as binding partners of Dab1 (Homayouni et al 1999, Howell et al 1999b, Trommsdorff et al 1998). The PI/PTB domain of Dab1 has also been shown to bind to phosphoinositides present in phospholipid bilayers. The association between Dab1 and phosphoinositides does not interfere with its ability to bind to peptides, suggesting that these two ligands do not compete for the PI/PTB domain (Howell et al 1999b). Therefore, Dab1 could interact with the plasma membrane while simultaneously docking another protein involved in signal transduction events.

The similar neuroanatomical phenotypes observed in mice deficient in *reelin* or *Dab1* imply that these genes function in a signaling pathway that controls cell positioning in the CNS. The spatial and temporal patterns of Reelin and Dab1 expression in several brain regions support this notion (Figure 5). For example, in the cerebellum, Reelin is expressed as early as E13.5 by cells in the nuclear transitory zone (ntz) and the external germinal layer (egl), which contains granule cell precursors (Schiffmann et al 1997, Miyata et al 1996, D'Arcangelo et al 1995). Purkinje cells arise between E11 and E13 from the proliferative zone lining the fourth ventricle and they migrate along glial fibers in the direction of the cells producing Reelin (Goldowitz et al 1997). In situ hybridization and

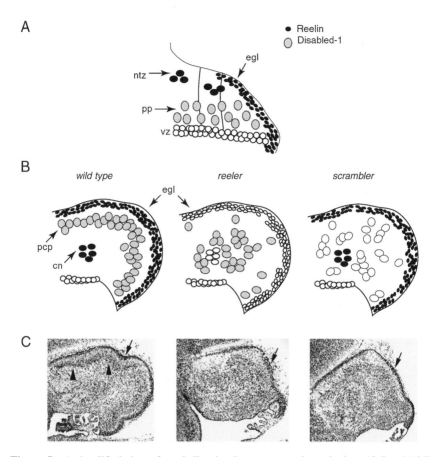

Figure 5   A simplified view of cerebellar development at embryonic days 13.5 and 16.5 and the histological appearance of the cerebellum at embryonic day 16.5 in wild-type, *reeler*, and *scrambler* mice. The IV ventricle is at the bottom of each panel. (*A*) Schematic of a sagittal view of the developing cerebellum at approximately embryonic day 13.5. Reelin (black) is present in the nuclear transitory zone (ntz) and in the external germinal layer (egl) during the initial phase of cerebellar development. The Purkinje cell precursors (pp), which express high levels of Dab1, arise from the ventricular zone (vz) located beneath these Reelin-rich areas and move towards the surface of the cerebellum. (*B*) Several days later in the wild-type, Purkinje cells form a rudimentary Purkinje cell plate (pcp) that is located beneath the egl. Reelin is produced by the granule cell precursors in the egl, which continues to increase in thickness owing to the proliferation of precursors. The cerebellar nuclei (cn), which are likely comprised of cells that moved through the ntz, are located beneath the pcp. These cells eventually lose expression of Reelin. In both *reeler* and *scrambler* Purkinje cells migrate away from the vz, but many fail to form the Purkinje cell plate beneath the egl. Instead, clusters of Purkinje cells are present in the cerebellum. (*C*) Sagittal view of the cerebellum stained with cresyl violet from wild-type, *reeler*, and *scrambler* mice. Rostral is to the left in these micrographs. The Purkinje cell plate (arrowheads) is obvious in the wild-type beneath the egl (arrow). In both *reeler* and *scrambler* the Purkinje cell plate does not form beneath the egl (arrows).

immunohistochemical studies revealed that Purkinje cells express high levels of Dab1 mRNA and protein during the migratory phase of Purkinje cell development (Rice et al 1998, Sheldon et al 1997, Howell et al 1997b). In the normal cerebellum at E16.5 Purkinje cells have already migrated away from their site of origin to form a structure known as the Purkinje cell plate located directly beneath cells that produce Reelin (Miyata et al 1996). However, the Purkinje cell plate does not form in mice lacking *reelin* or *Dab1* (Gallagher et al 1998, Goffinet 1984, Mariani et al 1977).

Two sources of Reelin are present near the Purkinje cell precursors that express Dab1 (Figure 5). Reelin is produced by cells in the ntz and egl, implying that one or both sources could provide a signal to Purkinje cells. Recent genetic evidence suggests that both sources of Reelin are required for the proper positioning of Purkinje cells. Mice with a targeted disruption of *Math-1* fail to form an egl, and consequently the cerebellum lacks granule cells (Ben-Arie et al 1997). Nevertheless, the majority of Purkinje cells migrate to the surface of the cerebellum, implying that transient expression of Reelin by cells in the ntz is sufficient for the migration of many Purkinje cells (Jensen et al 2000). The fact that some Purkinje cells fail to migrate implies that Reelin produced by granule cells in the egl is also required for proper migration. Multiple sources of Reelin are present in other structures affected by the *reeler* mutation. For example, Reelin is expressed initially in the marginal zone of developing cortex, followed by a population of deep cortical plate cells around E17 (Alcantara et al 1998). This additional source of Reelin may contribute to the proper migration of the last cohort of cortical plate neurons.

The close proximity of Reelin-producing cells to the Purkinje cells that express Dab1 is critical for the subsequent organization of the cerebellar cortex. In *reelin*- or *Dab1*-deficient mice the cerebellum lacks characteristic foliation, and the majority of Purkinje cells are found in several clusters beneath the cortex (Figure 5). The lack of cerebellar foliation in the mutants arises as a consequence of a decrease in the number of granule cells. Purkinje cells are known to produce the mitogenic factor sonic hedgehog that acts on granule cell precursors. Presumably, the overall levels of sonic hedgehog that reach the egl are reduced in the mutants owing to the misplacement of Purkinje cells (Wechsler-Reya & Scott 1999). This does not affect the remaining granule cells, which are located in an internal granule cell layer (Goldowitz et al 1997, Mariani et al 1977). Therefore, Reelin and Dab1 are required for the proper placement of Purkinje cells, but not granule cells, in the developing cerebellum.

The expression patterns of *reelin* and *Dab1* in the developing cerebral cortex are consistent with the notion that these genes function during the onset of corticogenesis. *Dab1* is expressed in the forebrain at relatively low and uniform levels throughout the ventricular zone as early as E11.5, whereas *reelin* is present in the Cajal-Retzius cells in the preplate (Rice et al 1998). A dramatic increase in *Dab1* occurs at the onset of cortical plate formation. Migrating neurons destined to form the cortical plate express *Dab1* when they invade the preplate. The first cohort of cortical plate neurons is unable to invade the preplate in *Dab1*-deficient mice, even though Reelin is present in the marginal zone. This disturbance is similar to that in *reeler* mice, in which the preplate fails to split into the marginal zone and subplate (Figure 2). Immunohistochemical studies using Dab1-specific antibodies revealed more intense staining in the *reeler* cerebral cortex at E16.5 compared with normal, suggesting that there is an increase in Dab1 levels. Immunoblotting analysis confirmed that Dab1 accumulates in *reeler* brain to a level approximately 5 to 10-fold greater than that in normal brain (Rice et al 1998). The peak time of over-expression of Dab1 corresponds to the period in which neuronal migration is underway and when Reelin is required for normal positioning of neurons in the CNS.

Recent biochemical and genetic data suggest that intracellular transduction of the Reelin signal involves activation of a tyrosine kinase cascade involving Dab1 (Howell et al 1999a). Dab1 is phosphorylated in both wild-type and *reeler* neurons, as revealed by immunoprecipitation with Dab1-specific antibodies followed by immunoblotting with antiphosphotyrosine antibodies. In the *reeler* brain the amount of phosphotyrosine in Dab1 is lower than that in normal brain, despite the presence of elevated levels of Dab1. Primary neurons exposed to Reelin-conditioned media display an increased level of phosphorylated Dab1 (Howell et al 1999a). This response is dependent on divalent cations and is not mimicked by a phosphatase inhibitor, suggesting that the Reelin action is mediated through specific receptors expressed on target neurons. The biochemical data imply that Reelin binding to receptors stimulates a kinase that phosphorylates Dab1, leading to the activation of an intracellular signaling cascade.

There are five potential sites for Reelin-induced tyrosine phosphorylation of Dab1 immediately C-terminal to the PI/PTB domain (Figure 4). Analysis of the predicted amino acid sequence indicated that this region is a good substrate for both receptor and nonreceptor tyrosine kinases, such as Src and Abl (Howell et al 2000). To determine if these sites are important during Reelin-induced signaling cascades, mice were generated that express a mutant form of Dab1, in which the five tyrosines were substituted with phenylalanine. Remarkably, these mutant mice display ataxia and disruptions of cell positioning in the cerebral cortex, cerebellum, and hippocampus that closely resemble those in *reelin*- and *Dab1*-deficient mice (Howell et al 2000). Immunoblotting studies revealed that Dab1 is present and that it lacks detectable tyrosine-phosphorylated residues. Moreover, Dab1 levels are slightly elevated in the phosphotyrosine Dab1–deficient brain, implying that Reelin signaling affects turnover of Dab1. Alterations in Dab1 levels likely represent a secondary consequence of Reelin-induced signaling and they are not responsible for the cell positioning defects observed in *reeler, scrambler,* or *yotari* mice. Dab1 is phosphorylated on tyrosine residues in both *reeler* and wild-type brain, suggesting that phosphorylation of Dab1 at basal levels is independent of Reelin. It will be interesting to determine the kinase(s) involved and the specific residues that are phosphorylated in response to Reelin. These sites of Reelin-induced phosphorylation are likely to represent docking areas for protein-protein interactions that are involved in downstream signaling events.

GENETIC EVIDENCE IMPLICATES LIPOPROTEIN
RECEPTORS IN REELIN SIGNALING

The PI/PTB in Dab1 binds to the cytoplasmic tails of the five members of the LDLR family and the three members of the APP family of proteins (Homayouni et al 1999; Trommsdorff et al 1998, 1999; Howell et al 1999b). The first indication that lipoprotein receptors function in the Reelin pathway came from a gene disruption study (Trommsdorff et al 1999). Mice deficient in both the very low–density lipoprotein receptor (*Vldlr*) and the apolipoprotein E receptor-2 (*ApoER2*) exhibit behavioral and neuratomical defects that are identical to those in *reeler*. The double mutants are smaller and they exhibit ataxia at two weeks after birth. The cerebellum is decreased in size and lacks foliation. In situ hybridization with a riboprobe recognizing *calbindin*, a gene expressed specifically in cerebellar Purkinje neurons, revealed large clusters of cells beneath the cerebellar cortex. In the hippocampus pyramidal cells are loosely arranged in multiple layers, and granule cells fail to form the histotypical layers in the dentate gyrus. Lamination defects in the cerebral cortex of *ApoER2* and *Vldlr* double mutants are also highly reminiscent of those seen in *reelin* or *Dab1* mutants, in which Layer I is not discernible, and there is an apparent increase in cell density in this area. This observation suggests that both receptors are capable of transmitting the Reelin signal to Dab1 and proper splitting of the preplate. Moreover, Dab1 levels are elevated in mice that lack both *Vldlr* and *ApoER2*, implying that signaling through Reelin and the lipoprotein receptors results in increased turnover of Dab1.

In contrast, mice deficient in only *Vldlr* or *ApoER2* exhibit subtle neuroanatomical defects that do not resemble *reeler*. For example, Layer I in the cerebral cortex appears normal in either *Vldlr* or *ApoER2* knock-out mice, implying that the early events of corticogenesis occur properly. However, cell positioning defects are obvious in other cortical layers in mice lacking *ApoER2*, as revealed by birthdating experiments (Trommsdorff et al 1999). In mice that lack *vldlr*, a small subset of Purkinje cells is inappropriately positioned in the cerebellum, which appears slightly reduced in size. The CA1 region of the hippocampus proper is thinner, and several pyramidal cells fail to align in the appropriate layer in mice lacking *ApoER2*. Abnormalities in the hippocampus are less dramatic in the *Vldlr*-deficient mice. These comparative studies suggest that each receptor is important during neuronal migration, but that they may exhibit both overlapping and distinct functions in transmission of a signal from Reelin in different brain regions.

The intracellular domains of LDLR family members bind, at least in vitro, to a variety of proteins implicated in kinase signaling, cell adhesion, cytoskelatal organization, vesicle transport, and synaptic transmission (Gotthardt et al 2000). ApoER2 contains a domain in its cytoplasmic tail that associates with the JNK family of interacting proteins, JIP-1 and JIP-2, which are highly expressed in the developing CNS (Stockinger et al 2000). JIPs are kinase scaffolding proteins important in the JNK kinase signaling pathway, and JIP-1 associates with the RhoA

GTPase exchange factor, rhoGEF (Meyer et al 1999). Therefore, extracellular guidance cues could bind to lipoprotein receptors and initiate different responses depending on the adapter molecules and signaling proteins associated with the receptor complex. This may explain the different phenotypes of mice deficient in *Vldlr* compared to those that lack *ApoER2*.

REELIN IS A LIGAND FOR LIPOPROTEIN RECEPTORS

Both *Vldlr* and *ApoER2* are expressed in the target cells of Reelin in several different brain regions during the critical period in which Reelin directs cell positioning (Trommsdorff et al 1999). For example, Vldlr and ApoER2 are expressed in the developing cortical plate and the intermediate zone. *ApoER2* is more widely expressed compared with *Vldlr*, and both receptors are present in similar cell populations that contain *Dab1* (Trommsdorff et al 1999). Although genetic evidence suggested that lipoprotein receptors functioned in the Reelin pathway, it was necessary to obtain direct biochemical evidence to support their role as Reelin receptors. Two different experimental strategies demonstrated that Reelin binds to both Vldlr and ApoER2 with an apparent affinity in the 0.5 nM range (D'Arcangelo et al 1999, Hiesberger et al 1999). The binding was inhibited by the CR-50 antibody, which binds to the N-terminus of Reelin and blocks its function in vivo and in vitro (D'Arcangelo et al 1999, Nakajima et al 1997, Ogawa et al 1995). CR-50 also decreased Reelin-induced phosphorylation of Dab1 in primary neurons, suggesting that Reelin binding to these receptors is critical for activation of a downstream kinase (Senzaki et al 1999). The association of Reelin with the lipoprotein receptors is also blocked by the receptor-associated protein, which acts as a molecular chaperon for the LDLR family (Willnow 1998). The presence of receptor-associated protein or fragments of the VLDLR receptor also blocked Reelin-induced tyrosine phosphorylation of Dab1 (Hiesberger et al 1999). These biochemical studies demonstrate that Reelin binds to Vldlr and ApoER2, resulting in internalization of Reelin and activation of a tyrosine kinase signaling cascade that results in phosphorylation of Dab1 (D'Arcangelo et al 1999). Thus, both genetic and biochemical evidence suggest that Reelin is a ligand for lipoprotein receptors.

Dab1 binds to the cytoplasmic domains of VLDLR, ApoER2, and other LDL-receptor family members, including LDLR, LRP, and Megalin (Howell et al 1999b, Trommsdorff et al 1998). The LDLR family of receptors does not possess intrinsic tyrosine kinase activity, and they have not been shown to physically associate with tyrosine kinases or phosphatases. This raises a question regarding the mechanism whereby Reelin binding to the receptors activates a tyrosine kinase that phosphorylates Dab1. The PI/PTB domain in Dab1 interacts with the receptor tails via an association with the NPxY motif, which is important in clathrin-mediated endocytosis and receptor cycling (Chen et al 1990). Reelin is internalized after binding lipoprotein receptors, suggesting that endocytosis may recruit a tyrosine kinase to Dab1 in vesicles and further activate additional signaling molecules (D'Arcangelo et al 1999). Alternatively, Reelin may also bind to coreceptors expressed on migrating neurons that have an associated kinase activity.

REELIN BINDS TO OTHER TRANSMEMBRANE PROTEINS EXPRESSED ON NEURONS

Recently, it was reported that Reelin binds to a novel class of proteins (Senzaki et al 1999). This family of transmembrane proteins, known as cadherin-related neuronal receptors (CNRs), contains six ectodomains that are similar to those found in cadherins. CNRs diverge from cadherins in the sequence of their cytoplasmic domains (Kohmura et al 1998). CNR proteins were discovered through their ability to interact specifically with the nonreceptor tyrosine kinase Fyn in yeast two-hybrid assays. Expression studies demonstrated that CNRs are present in synapses throughout the adult brain (Kohmura et al 1998). The function of CNRs is unclear at present, although the similarity of the extracellular domains to cadherins implies that CNRs are involved in homophilic or heterophilic cell-cell interactions.

To investigate the possibility that Reelin binds to CNRs, alkaline phosphatase fusion proteins containing a partial Reelin N-terminal region (amino acids 28–911) were coprecipitated with an Fc fusion protein corresponding to the extracellular domain of CNR1 (Senzaki et al 1999). Using these truncated proteins, it was found that the B domain in the first Reln repeat (Figure 3) bound to CNR1 in vitro. The binding region was mapped to a domain in CNR1 that contains an RGD motif, which functions in ligand receptor interactions (Pierschbacher & Ruoslahti 1984). Mutation of the arginine residue in this motif abolished Reelin binding to CNR1. Rotation cultures, in which cortical neurons were dissociated and allowed to self-aggregate in the presence of an antibody recognizing the RGD region of CNR1, were used to examine the function of CNRs in corticogenesis. In normal aggregates, microtubule-associated protein-2–positive neurons were located on the surface of the aggregate. However, in the presence of the CNR1 antibody many neurons were clustered in the central region, and the aggregate was smaller than controls (Senzaki et al 1999). As noted by the investigators, the histotypic appearance of aggregates obtained following treatment with CNR1 antibodies was quite distinct from that obtained in the presence of CR-50 antibody or in aggregates prepared using *reeler* neurons (Ogawa et al 1995, De Long & Sidman 1970). It is possible that CNRs participate in neurophilic interactions based on the aggregation results. The fact that the cytoplasmic domains of CNRs bind to a kinase that phosphorylates Dab1 in vitro raises the possibility that CNRs participate in a complex with lipoprotein receptors on the surface of migrating neurons. Unlike the situation with the lipoprotein receptors, genetic evidence has not been presented that supports a role for CNR genes in Reelin signaling.

Other transmembrane receptor classes function to direct cell migration in the brain. Gene disruption analysis has shown that $\alpha 3$ integrin plays a role in cell migration during corticogenesis. The migratory patterns of neurons in the cortex lacking α3 integrin are somewhat distinct from those of *reeler*. For example, neuronal heterotopias are present in the intermediate zone, implying that neurons failed to migrate properly. Neurons born on E13.5 are normally located in the deep layers of the cortical plate, but in α3 integrin–deficient mice these cells are scattered throughout the cortical plate and the intermediate zone (Anton et al 1999). The α3 integrin protein is highly expressed in ventricular zone, migrating neurons in the intermediate zone, and at low levels in the cortical plate (Anton et al 1999). Studies carried out using in vitro migration assays suggested that the primary function of α3β1 integrin is to promote association between migrating neurons and their radial glia guides. In the absence of α3 integrin, neurons alter their preference for radial glia and exhibit an increased propensity to associate with other neurons (Anton et al 1999). Thus, ectopic cells in the cortex and neuronal heterotopias in mice lacking the α3 integrin may arise because of a failure in cell-cell interactions between migrating neurons and the radial glia.

Integrin receptors bind a variety of ligands associated with the extracellular matrix and they mediate cell-cell interactions in many tissues. Functional receptors consist of heterodimers between two subunits encoded by different genes. There are over 20 integrin receptors comprising different combinations of α and β subunits, with differing ligand-binding specificity and associated intracellular signal transduction cascades (Hynes 1992). Recently, Reelin was suggested to bind α3β1 integrin, resulting in an alteration in neuronal migration in vitro (Dulabon et al 2000). Incubation with Reelin-conditioned medium reduced the rate of migration and stimulated neurons to detach from the radial glia in cortical imprint assays. Importantly, these effects were abolished when Reelin was inhibited by treatment with the CR-50 antibody. Moreover, neurons obtained from mice lacking the α3 integrin did not respond to Reelin-conditioned medium in the imprint assay. To determine if Reelin binding to α3 integrin is relevant in vivo, Reelin-coated beads were injected into the intermediate zone of the developing cerebral cortex. This treatment prevented the subsequent migration of BrdU-labeled neurons, whereas neurons that failed to express the α3 integrin ignored the Reelin-coated beads. These observations led to the proposal that Reelin acts as a negative regulator of neuronal migration by inducing changes in cell-cell interactions and detachment from radial glia. Protein association assays carried out using the CR-50 antibody and antibodies against β1 integrin, α3 integrin, and a myc epitope tag on Reelin suggested that Reelin bound to β1 and, to a lesser degree, α3 integrin (Dulabon et al 2000). Surprisingly, in contrast to the interaction of Reelin with lipoprotein receptors and in vivo and in vitro biological assays of Reelin function, CR-50 did not inhibit the interaction of Reelin with integrin. This implies that multimerization of Reelin is not required for this activity.

The cytoplasmic tail of β1 integrin contains two NPxY motifs, one of which has been suggested to function in cell migration by targeting integrin to focal adhesions (Vignoud et al 1994, 1997). Integrins are known to trigger activation of tyrosine kinases such as Fak and Syk, which are involved in signaling transduction pathways that affect cytoskeletal organization (Clark & Brugge 1995).

Interestingly, $\alpha3$ integrin does not appear to be required for the Reelin-induced tyrosine phosphorylation of Dab1 in primary neurons (Dulabon et al 2000). This implies that the kinase activity associated with Reelin binding to the lipoprotein receptors on neurons is independent of integrin signaling. Moreover, Dab1 levels are decreased in $\alpha3$ integrin–deficient mice. This is in stark contrast to the situation in *reeler* brains or those that lack both the *Vldlr* and *ApoER2* genes, in which Dab1 accumulates in neurons that go astray (Trommsdorff et al 1999, Rice et al 1998).

GENES THAT MODULATE THE REELIN SIGNALING PATHWAY IN BRAIN

A number of genes have now been suggested to play a role in the Reelin signaling pathway that controls neuronal positioning during development (Table 1). Although many of these have been directly implicated in the Reelin pathway, others appear to influence Reelin signaling through effects on the generation or survival of Cajal-Retzius cells. Targeted disruption of either cyclin-dependent kinase 5 (Cdk5) or its regulatory subunit p35 produce migration defects in brain regions affected in *reeler*. Cdk5 is a member of the cyclin family of serine/threonine kinases that is widely expressed in postmitotic neurons. Cdk5 phosphorylates a variety of substrates associated with the cytoskeleton (Tsai et al 1993). Cdk5 is activated by association with its regulatory subunits p35 or p39 that are expressed in postmitotic neurons (Cai et al 1997, Tsai et al 1994). *Cdk5*-null mice die shortly after birth with failed migrations in the cerebral cortex, cerebellum, and hippocampus (Ohshima et al 1996). The cerebellar phenotype in the Cdk5$^{-/-}$ mice is characterized as a complete block in Purkinje cell migration and a failure in the inward migration of some granule cells towards the internal granule cell layer (Ohshima et al 1999). Less severe, but similar defects are present in the p35-deficient cerebellum (Chae et al 1997). This contrasts with *reeler*, in which granule cell migration is apparently normal, and a small subset of Purkinje cells is capable of migrating to the cerebellar cortex (Mariani et al 1977).

During corticogenesis the preplate splits into the marginal zone and subplate in both Cdk5- and p35-deficient mice. This implies that the migration of the first cohort of cortical plate neurons is independent of Cdk5/p35 (Kwon & Tsai 1998, Gilmore et al 1998). Similar to the situation in the embryonic $\alpha3$ integrin–deficient cortex, there is an increase in the apparent cell density in the intermediate zone in mice that lack either Cdk5 or p35. Recently, Li et al (2000) demonstrated that $\alpha1\beta1$ integrin stimulated Cdk5 kinase activity. It is conceivable that $\alpha3\beta1$ integrin activates Cdk5 in the later cohorts of migrating neurons, allowing them to traverse the intermediate zone. Indeed, Cdk5 has been shown to negatively regulate N-cadherin-mediated adhesion via an association between p35 and β-catenin (Kwon et al 2000). This led to the proposal that Cdk5 kinase activity functions to silence N-cadherin-mediated adhesion in migrating neurons during corticogenesis (Kwon et al 2000). In this model integrin stimulation of Cdk5 would suppress N-cadherin-mediated adhesion among neurons migrating through the intermediate zone and cortical plate, which contains high levels of N-cadherin (Redies & Takeichi 1993). This may explain the failed migrations in later cohorts of neurons in mice that lack either α3 integrin or Cdk5 (Anton et al 1999, Kwon & Tsai 1998, Gilmore et al 1998).

It is interesting to compare the cortical defects observed in mice lacking *Cdk5* to those in *reeler*. In contrast to *reeler*, the preplate splits in Cdk5$^{-/-}$ mice, suggesting that the Reelin/lipoprotein receptor/Dab1 pathway is distinct from that affected by Cdk5. However, it remains possible that Reelin has different effects on early- and late-born cortical neurons. In the later phases of cortical neuron migration Reelin could influence Cdk5 kinase activity, resulting in abrogation of N-cadherin-mediating adhesion (Homayouni & Curran 2000). Thus, Reelin binding to lipoprotein receptors may simultaneously suppress neuronal-glia interactions by internalization of integrin receptors, which may stimulate neurophilic interactions by reducing Cdk5 kinase activity.

ADDITIONAL FUNCTIONS OF THE REELIN SIGNALING PATHWAY IN THE BRAIN

Recent evidence suggests that the Reelin pathway is involved in other aspects of neurodevelopment and in the mature CNS. Soriano and colleagues have shown that Reelin and Cajal-Retzius cells are important for the correct formation of synaptic circuits in the hippocampus. *Reeler* mice display several alterations in the entorhino-hippocampal projections, including reduced axonal branching, decreased number of synapses, and abnormal topography of synapses (Borrell et al 1999a, Del Rio et al 1997). Others have shown colocalization of Reelin and integrins in synaptic clefts, implicating a role for Reelin in synaptogenesis (Rodriguez et al 2000). Application of the Reelin-blocking antibody CR-50 to organotypic cultures obtained from normal entorhino-hippocampal slices recapitulated the fiber pathway abnormalities observed in *reeler*. Moreover, the length of axonal collaterals in the stratum lacunosum-moleculare is decreased in *reeler*, suggesting that Reelin promotes neurite outgrowth (Borrell et al 1999a). In addition, misrouting of commissural afferents is a feature of the *reeler* hippocampus (Borrell et al 1999b). This was attributed to loss of some Cajal-Retzius cells in hippocampus, particularly in the CA3 region, of *reeler* Orleans. These studies show that the overall layer specificity and targeting of afferent projections to the hippocampus are normal in *reeler*, which implies that Reelin may not function as a long-range attractant molecule. This conclusion is consistent with other observations on the major fiber pathways in the *reeler* mutant (Caviness and Rakic 1988). Rather, Reelin functions locally to promote synaptogenesis, axonal branching and collateral outgrowth in the developing hippocampus. Therefore, Reelin acts in conjunction with other short- and long-range chemoattractive and chemorepulsive cues during formation of synaptic circuitry (Chedotal et al 1998).

Several components of the Reelin signaling pathway have also been reported to be expressed in neurons after completion of cell positioning. In the neural retina *reelin* is expressed by retinal ganglion cells, amacrine cells, and cone bipolar cells (Rice & Curran 2000b). Reelin is present at high levels in the outer and inner synaptic layers. Dab1 becomes apparent during the first postnatal week in a subpopulation of cells in the inner nuclear layer (Rice & Curran 2000a). Interestingly, at postnatal day 6 Dab1 was found to accumulate in type AII amacrine cells of *reeler* mice, suggesting that the Reelin pathway is active in the retina during the period of synaptogenesis. Indeed, the synaptic layering of type AII amacrine cells and rod bipolar cells is defective in mice deficient in Reelin and Dab1. These findings suggest that the Reelin pathway modulates the formation of the scotopic synaptic circuitry in mammalian retina.

The complementary patterns of Reelin and Dab1 expression in other brain regions suggest that they may have functions in adults. For example, in the mature cerebellum, Reelin is expressed in the granule cell layer and the molecular layer, which contains the parallel fibers that synapse on Purkinje cell dendrites expressing Dab1 (Pesold et al 1998, Ikeda & Terashima 1997). Lipoprotein receptors are also present in adult neurons, where they could potentially contribute to a signal transduction pathway affecting the stability of the cytoskeleton (Trommsdorff et al 1999). Many neurons in the adult cerebral cortex that produce γ-aminobutyric acid (GABA) as their primary neurotransmitter also express *reelin* (Alcantara et al 1998). These inhibitory interneurons modulate cortical synaptic circuitry and they synapse on cortical pyramidal cells expressing Dab1 (Rodriguez et al 2000, Pesold et al 1999). Mitral cells and a subpopulation of periglomerular neurons express very high levels of *reelin* in the adult olfactory bulb (Alcantara et al 1998, Ikeda & Terashima 1997). The olfactory bulb actively remodels synaptic connections, suggesting that Reelin may participate in synaptic maintenance or plasticity. In addition, Reelin may be a component of the extracellular matrix that is important for the localization of neurotrophic factors, which contribute to the survival of neurons (Celio & Blumcke 1994). This may relate to the modest loss of certain subpopulations of neurons in the *reeler* brain (Wyss et al 1980, Stanfield & Cowan 1979, Mariani et al 1977).

IS THE REELIN SIGNALING PATHWAY INVOLVED IN NEUROPATHOLOGY?

The interaction of Reelin with lipoprotein receptors is inhibited in the presence of another ligand of the LDL receptor family, apolipoprotein E (ApoE), which is a major constituent of very low–density lipoproteins (D'Arcangelo et al 1999, Mahley

1988). Three alleles of *ApoE* are present in humans, and epidemiological studies have linked *ApoE4* with late-onset and sporadic Alzheimer's disease (Strittmatter & Roses 1996). ApoE3 and ApoE4, but not ApoE2, inhibit Reelin binding to the lipoprotein receptors in vitro (D'Arcangelo et al 1999). Furthermore, whereas the basal level of tyrosine phosphorylation of Dab1 is not altered by treatment of primary neurons with ApoE3 alone, Reelin-induced tyrosine phosphorylation of Dab1 is inhibited. Transgenic mice that express human ApoE4 in neurons display progressive motor abnormalities, wasting, and premature death (Tesseur et al 2000). These mice also exhibit elevated levels of hyperphosphorylated tau, a microtubule-associated protein important in the organization of the cytoskeleton. Hyperphosphorylation of tau, which leads to dissociation of microtubules, has been associated with Alzheimer's disease, and mutations in *tau* are linked to dementia (Hardy et al 1998, Hong et al 1998). A recent study has indicated that increased levels of tau phosphorylation can be detected in mice deficient in *reelin* or both *Vldlr* and *ApoER2* (Hiesberger et al 1999).

Several protein kinases have been implicated in tau hyperphosphorylation, including Cdk5. Recently, p25, a truncated isoform of p35, the regulatory subunit of Cdk5, was found to accumulate in Alzheimer's disease. This isoform constitutively activates Cdk5, resulting in elevated levels of tau phosphorylation (Patrick et al 1999). This may lead to disruption of the cytoskeletal network and formation of paired helical filaments, which are characteristic of Alzheimer's and other neurodegenerative diseases. Dab1 was reported to be an in vitro substrate of Cdk5, and it is tempting to speculate that Cdk5 activity may be modulated by the Reelin signaling pathway (Homayouni et al 1999). Dab1 may affect Cdk5 kinase activity directly or indirectly by interacting with its binding partners such as Abl (Zukerberg et al 2000). Alternatively, Dab1 could influence Cdk5 function by binding to its substrates such as APP (Iijima et al 2000, Homayouni et al 1999, Howell et al 1999b). Thus, it is possible that the Reelin signaling pathway, which controls cell positioning and branching of neuronal fibers during development, may also participate in communicating extracellular signals to alterations in the cytoskeleton, which affect cell shape and physiology in the adult brain. Conceivably, the transmission of a signal from Reelin through lipoprotein receptors to tyrosine phosphorylation of Dab1 may act antagonistically with a pathway involving ApoE, protein kinases, and tau phosphorylation.

Dab1 interacts with the cytoplasmic tails of all three members of the APP family (Homayouni et al 1999, Howell et al 1999b, Trommsdorff et al 1998). A proteolytic product of APP, amyloid β-protein, accumulates in amyloid plaques, and mutations in APP are associated with familial Alzheimer's disease (Selkoe 1998). Dab1 may mediate the intracellular sorting of the cytoplasmic cleavage product of APP, which has recently been associated with neuronal death (Lu et al 2000). Alternatively, Dab1 could affect the production or localization of the mature form of APP by interacting with its cytoplasmic NPxY motif. It is also plausible that Dab1 modulates the endocytotic pathway by competing with other adapter proteins involved in APP trafficking (Guenette et al 1999). Therefore, it will be interesting to determine if Reelin signaling through Dab1 affects generation of the amyloid β-protein peptide and other APP peptides associated with the pathological progression of Alzheimer's disease (Selkoe 1998).

CONCLUSIONS

According to legend, there is an ancient Chinese curse that states, "May you live in interesting times!" The Reelin field has now entered an interesting period. Over the past few years, a number of proteins have been shown to function in the Reelin signaling pathway. We are now faced with the challenging task of integrating almost 50 years of research on the anatomy and biology of the *reeler* mouse with knowledge of the biochemical properties and expression patterns of these proteins in the developing brain. Although the earliest descriptions of *reeler* defined the mutation as a disorder of cell migration, it is now clear that the genetic lesion compromises a signaling pathway whose biological role is context dependent. For example, in the cerebral cortex migration per se is not affected; rather, there is a failure in cell positioning in the final phase of cortical plate formation. In the cerebellum Purkinje cells complete the first phase of migration, but they fail to form a Purkinje cell plate. Hippocampal pyramidal neurons do not align appropriately, and granule cells of the dentate gyrus are dispersed. Furthermore, the Reelin pathway does not function exclusively in cell positioning, because entorhino-hippocampal projections and retinal synaptic layering are disrupted in the mutant mice. Thus, in discussing the function of Reelin it may not be appropriate to ascribe biological properties such as "attractant" or "stop signal" to an intercellular signal that depends on additional cues in each specific context to elicit an appropriate response.

The biochemical role of Reelin is emerging from studies of the molecular properties of components of the pathway. Reelin is an extracellular protein that is present at specific locations throughout the developing brain. It binds to lipoprotein receptors and is internalized, while activating a tyrosine kinase cascade that leads to phosphorylation of Dab1 (Figure 6). In the developing cerebral cortex the *reeler* phenotype is apparent when the first cohort of cortical plate neurons encounters the preplate (Caviness 1982, Goffinet 1979). This implies that Reelin alters the surface properties of migrating neurons and causes them to insert into the preplate. It is difficult to present a model that accounts for all of the events that have been associated with the Reelin pathway. It is possible that the interaction of Reelin with lipoprotein receptors changes the properties of leading processes on migrating cells or synaptic structures by redistributing and reorganizing cell surface molecules. This could be accomplished by downregulation of adhesion molecules such as integrin, which are thought to mediate the cell-cell contacts among migrating neurons and radial glia, or changes in molecules such as CNRs that could influence homotypic interactions among neurons. Alternatively, Reelin

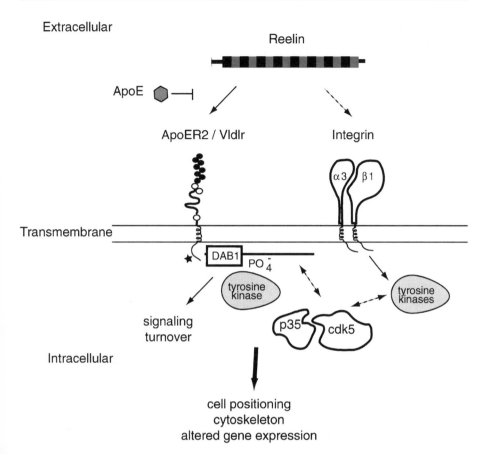

Figure 6 Model of the Reelin signaling pathway based on both genetic and biochemical data. Reelin is an extracellular protein that binds to lipoprotein receptors, Vldlr and ApoER2, expressed on neurons. Dab1 is an adapter protein that associates with an NPxY motif (star) in the cytoplasmic domain of these receptors. Reelin binding to lipoprotein receptors activates a tyrosine kinase, resulting in phosphorylation of Dab1. This effect is inhibited by ApoE. Reelin is internalized following binding to lipoprotein receptors, and Dab1 is degraded. The Reelin signaling pathway activates a complex array of biological responses. Mutations in Cdk5, p35, and integrin elicit similar responses that may be mediated through interactions with the Reelin pathway.

could provoke a more global response through effects on endocytosis, changes in the distribution of intracellular signal transduction molecules such as protein kinases, or through targeting of other proteins to the cell surface. In addition to activating second-messenger cascades, the Reelin pathway may also provoke changes in gene expression that lead to long-lasting alterations in neuronal properties. We hope clues about the molecular events initiated in the target cells of Reelin will emerge in the coming years.

ACKNOWLEDGMENTS

This work was supported in part by NIH Cancer Center Support CORE grant P30 CA21765, NIH grants F32 EY06972 (DSR), RO1 NS36558 (TC), Human Frontiers Science Program RG67/98, and the American Lebanese Syrian Associated Charities (ALSAC). Special thanks to Lakhu Keshvara, Ramin Homayouni, Susan Magdaleno, Victor Borrell, and Patricia Jensen for helpful comments.

Visit the Annual Reviews home page at www.AnnualReviews.org

LITERATURE CITED

Alcantara S, Ruiz M, D'Arcangelo G, Ezan F, de Lecea L, et al. 1998. Regional and cellular patterns of reelin mRNA expression in the forebrain of the developing and adult mouse. *J. Neurosci.* 18:7779–99

Allendoerfer KL, Shatz CJ. 1994. The subplate, a transient neocortical structure: its role in the development of connections between thalamus and cortex. *Annu. Rev. Neurosci.* 17:185–218

Angevine JB, Sidman RL. 1961. Autoradiographic study of cell migration during histogenesis of cerebral cortex in the mouse. *Nature* 192:766–68

Anton ES, Cameron RS, Rakic P. 1996. Role of neuron-glial junctional domain proteins in the maintenance and termination of neuronal migration across the embryonic cerebral wall. *J. Neurosci.* 16:2283–93

Anton ES, Kreidberg JA, Rakic P. 1999. Distinct functions of alpha3 and alpha(v) integrin receptors in neuronal migration and laminar organization of the cerebral cortex. *Neuron* 22:277–89

Ben-Arie N, Bellen HJ, Armstrong DL, McCall AE, Gordadze PR, et al. 1997. Math1 is essential for genesis of cerebellar granule neurons. *Nature* 390:169–72

Bentivoglio M, Mazzarello P. 1999. The history of radial glia. *Brain Res. Bull.* 49:305–15

Borrell V, Del Rio JA, Alcantara S, Derer M, Martinez A, et al. 1999a. Reelin regulates the development and synaptogenesis of the layer-specific entorhino-hippocampal connections. *J. Neurosci.* 19:1345–58

Borrell V, Ruiz M, Del Rio JA, Soriano E. 1999b. Development of commissural connections in the hippocampus of reeler mice: evidence of an inhibitory influence of Cajal-Retzius cells. *Exp. Neurol.* 156:268–82

Bulfone A, Smiga SM, Shimamura K, Peterson A, Puelles L, Rubenstein JL. 1995. T-brain-1: a homolog of Brachyury whose expression defines molecularly distinct domains within the cerebral cortex. *Neuron* 15:63–78

Cai XH, Tomizawa K, Tang D, Lu YF, Moriwaki A, et al. 1997. Changes in the expression of novel Cdk5 activator messenger RNA (p39nck5ai mRNA) during rat brain development. *Neurosci. Res.* 28:355–60

Cameron RS, Rakic P. 1994. Identification of membrane proteins that comprise the plasmalemmal junction between migrating neurons and radial glial cells. *J. Neurosci.* 14:3139–55

Caviness VS Jr. 1976. Patterns of cell and fiber distribution in the neocortex of the reeler mutant mouse. *J. Comp. Neurol.* 170:435–47

Caviness VS Jr. 1977. Reeler mutant mouse: a genetic experiment in developing mammalian cortex. *Neurosci. Symp.* 2:27–46

Caviness VS Jr. 1982. Neocortical histogenesis in normal and reeler mice: a developmental study based upon [^{3}H]thymidine autoradiography. *Brain Res.* 256:293–302

Caviness VS Jr, Crandall JE, Edwards MA. 1988. The Reeler Malformation: implications for neocortical histogenesis. In *Cerebral Cortex*, Vol 17, ed. A Peters, EG Jones, pp. 59–89. New York: Plenum

Caviness VS Jr, Rakic P. 1978. Mechanisms of cortical development: a view from mutations in mice. *Annu. Rev. Neurosci.* 1:297–326

Caviness VS Jr, Sidman RL. 1973a. Retrohippocampal, hippocampal and related structures of the forebrain in the reeler mutant mouse. *J. Comp. Neurol.* 147:235–54

Caviness VS Jr, Sidman RL. 1973b. Time of origin of corresponding cell classes in the cerebral cortex of normal and reeler mutant mice: an autoradiographic analysis. *J. Comp. Neurol.* 148:141–51

Celio MR, Blumcke I. 1994. Perineuronal nets–a specialized form of extracellular matrix in the adult nervous system. *Brain Res. Brain Res. Rev.* 19:128–45

Chae T, Kwon YT, Bronson R, Dikkes P, Li E, Tsai LH. 1997. Mice lacking p35, a neuronal specific activator of Cdk5, display cortical lamination defects, seizures, and adult lethality. *Neuron* 18:29–42

Chedotal A, Del Rio JA, Ruiz M, He Z, Borrell V, et al. 1998. Semaphorins III and IV repel hippocampal axons via two distinct receptors. *Development* 125:4313–23

Chen WJ, Goldstein JL, Brown MS. 1990. NPXY, a sequence often found in cytoplasmic tails, is required for coated pit-mediated internalization of the low density lipoprotein receptor. *J. Biol. Chem.* 265:3116–23

Clark EA, Brugge JS. 1995. Integrins and signal transduction pathways: the road taken. *Science* 268:233–39

D'Arcangelo G, Curran T. 1998. Reeler: new tales on an old mutant mouse. *BioEssays* 20:235–44

D'Arcangelo G, Homayouni R, Keshvara L, Rice DS, Sheldon M, Curran T. 1999. Reelin is a ligand for lipoprotein receptors. *Neuron* 24:471–79

D'Arcangelo G, Miao GG, Chen SC, Soares HD, Morgan JI, Curran T. 1995. A protein related to extracellular matrix proteins deleted in the mouse mutant reeler. *Nature* 374:719–23

D'Arcangelo G, Nakajima K, Miyata T, Ogawa M, Mikoshiba K, Curran T. 1997. Reelin is a secreted glycoprotein recognized by the CR-50 monoclonal antibody. *J. Neurosci.* 17:23–31

Davenport CB. 1900. Review of von Guaita's experiments in breeding mice. *Biol. Bull.* 2:121–28

de Bergeyck V, Nakajima K, Lambert de Rouvroit C, Naerhuyzen B, Goffinet AM, et al. 1997. A truncated Reelin protein is produced but not secreted in the 'Orleans' reeler mutation (Reln[rl-Orl]). *Brain Res. Mol. Brain Res.* 50:85–90

De Long GR, Sidman RL. 1970. Alignment deficit of reaggregating cells in culture of developing brains of reeler mutant mice. *Dev. Biol.* 22:584–600

Del Rio JA, Heimrich B, Borrell V, Forster E, Drakew A, et al. 1997. A role for Cajal-Retzius cells and reelin in the development of hippocampal connections. *Nature* 385:70–74

Derer P. 1985. Comparative localization of Cajal-Retzius cells in the neocortex of normal and reeler mutant mice fetuses. *Neurosci. Lett.* 54:1–6

de Rouvroit CL, de Bergeyck V, Cortvrindt C, Bar I, Eeckhout Y, Goffinet AM. 1999. Reelin, the extracellular matrix protein deficient in reeler mutant mice, is processed by a metalloproteinase. *Exp. Neurol.* 156:214–17

Dräger UC. 1981. Observations on the organization of the visual cortex in the reeler mouse. *J. Comp. Neurol.* 201:555–70

Dulabon L, Olson EC, Taglienti MG, Eisenhuth S, McGrath B, et al. 2000. Reelin binds alpha-3-beta-1 integrin and inhibits neuronal migration. *Neuron* 27:33–44

Falconer DS. 1951. Two new mutants trembler and reeler, with neurological actions in the house mouse. *J. Genet.* 50:192–201

Gallagher E, Howell BW, Soriano P, Cooper JA, Hawkes R. 1998. Cerebellar abnormalities in the disabled (mdab1-1) mouse. *J. Comp. Neurol.* 402:238–51

Gertler FB, Bennett RL, Clark MJ, Hoffmann

FM. 1989. Drosophila abl tyrosine kinase in embryonic CNS axons: a role in axonogenesis is revealed through dosage-sensitive interactions with disabled. *Cell* 58:103–13

Gertler FB, Hill KK, Clark MJ, Hoffmann FM. 1993. Dosage-sensitive modifiers of Drosophila abl tyrosine kinase function: prospero, a regulator of axonal outgrowth, and disabled, a novel tyrosine kinase substrate. *Genes Dev.* 7:441–53

Gilmore EC, Ohshima T, Goffinet AM, Kulkarni AB, Herrup K. 1998. Cyclin-dependent kinase 5-deficient mice demonstrate novel developmental arrest in cerebral cortex. *J. Neurosci.* 18:6370–77

Goffinet AM. 1979. An early development defect in the cerebral cortex of the reeler mouse. A morphological study leading to a hypothesis concerning the action of the mutant gene. *Anat. Embryol.* 157:205–16

Goffinet AM. 1984. Events governing organization of postmigratory neurons: studies on brain development in normal and reeler mice. *Brain Res.* 319:261–96

Goldowitz D, Cushing RC, Laywell E, D'Arcangelo G, Sheldon M, et al. 1997. Cerebellar disorganization characteristic of reeler in scrambler mutant mice despite presence of reelin. *J. Neurosci.* 17:8767–77

Gonzalez JL, Russo CJ, Goldowitz D, Sweet HO, Davisson MT, Walsh CA. 1997. Birthdate and cell marker analysis of scrambler: a novel mutation affecting cortical development with a reeler-like phenotype. *J. Neurosci.* 17:9204–11

Gotthardt M, Trommsdorff M, Nevitt MF, Shelton J, Richardson JA, et al. 2000. Interactions of the low density lipoprotein receptor gene family with cytosolic adaptor and scaffold proteins suggest diverse biological functions in cellular communication and signal transduction. *J. Biol. Chem.* 275:25616–24

Guenette SY, Chen J, Ferland A, Haass C, Capell A, Tanzi RE. 1999. hFE65L influences amyloid precursor protein maturation and secretion. *J. Neurochem.* 73:985–93

Hamburgh M. 1960. Observations on the neuropathology of "reeler", a neurological mutation in mice. *Experientia* 16:460–61

Hardy J, Duff K, Hardy KG, Perez-Tur J, Hutton M. 1998. Genetic dissection of Alzheimer's disease and related dementias: amyloid and its relationship to tau. *Nat. Neurosci.* 1:355–58

Hartmann D, De Strooper B, Saftig P. 1999. Presenilin-1 deficiency leads to loss of Cajal-Retzius neurons and cortical dysplasia similar to human type 2 lissencephaly. *Curr. Biol.* 9:719–27

Hatten ME. 1999. Central nervous system neuronal migration. *Annu. Rev. Neurosci.* 22:511–39

Henkemeyer MJ, Gertler FB, Goodman W, Hoffmann FM. 1987. The Drosophila Abelson proto-oncogene homolog: identification of mutant alleles that have pleiotropic effects late in development. *Cell* 51:821–28

Herrup K. 1987. Roles of cell lineage in the developing mammalian brain. *Curr. Top. Dev. Biol.* 21:65–97

Hiesberger T, Trommsdorff M, Howell BW, Goffinet A, Mumby MC, et al. 1999. Direct binding of Reelin to VLDL receptor and ApoE receptor 2 induces tyrosine phosphorylation of disabled-1 and modulates tau phosphorylation. *Neuron* 24:481–89

Hoffarth RM, Johnston JG, Krushel LA, van der Kooy D. 1995. The mouse mutation reeler causes increased adhesion within a subpopulation of early postmitotic cortical neurons. *J. Neurosci.* 15:4838–50

Homayouni R, Curran T. 2000. Cortical development: Cdk5 gets into sticky situations. *Curr. Biol.* 10:R331–34

Homayouni R, Rice DS, Sheldon M, Curran T. 1999. Disabled-1 binds to the cytoplasmic domain of amyloid precursor-like protein 1. *J. Neurosci.* 19:7507–15

Hong M, Zhukareva V, Vogelsberg-Ragaglia V, Wszolek Z, Reed L, et al. 1998. Mutation-specific functional impairments in distinct tau isoforms of hereditary FTDP-17. *Science* 282:1914–17

Hong SE, Shugart YY, Huang DT, Shahwan

SA, Grant PE, et al. 2000. Autosomal recessive lissencephaly with cerebellar hypoplasia is associated with human RELN mutations. *Nat. Genet.* 26:93–96

Howell BW, Gertler FB, Cooper JA. 1997a. Mouse disabled (mDab1): a Src binding protein implicated in neuronal development. *EMBO J.* 16:121–32

Howell BW, Hawkes R, Soriano P, Cooper JA. 1997b. Neuronal position in the developing brain is regulated by mouse disabled-1. *Nature* 389:733–37

Howell BW, Herrick TM, Cooper JA. 1999a. Reelin-induced tryosine phosphorylation of disabled 1 during neuronal positioning. *Genes Dev.* 13:643–48

Howell BW, Herrick TM, Hildebrand JD, Zhang YN, Cooper JA. 2000. Dab1 tyrosine phosphorylation sites relay positional signals during mouse brain development. *Curr. Biol.* 10:877–85

Howell BW, Lanier LM, Frank R, Gertler FB, Cooper JA. 1999b. The disabled 1 phosphotyrosine-binding domain binds to the internalization signals of transmembrane glycoproteins and to phospholipids. *Mol. Cell. Biol.* 19:5179–88

Hsueh YP, Wang TF, Yang FC, Sheng M. 2000. Nuclear translocation and transcription regulation by the membrane-associated guanylate kinase CASK/LIN-2. *Nature* 404:298–302

Hunter-Schaedle KE. 1997. Radial glial cell development and transformation are disturbed in reeler forebrain. *J. Neurobiol.* 33:459–72

Hynes RO. 1992. Integrins: versatility, modulation, and signaling in cell adhesion. *Cell* 69:11–25

Iijima K, Ando K, Takeda S, Satoh Y, Seki T, et al. 2000. Neuron-specific phosphorylation of Alzheimer's beta-amyloid precursor protein by cyclin-dependent kinase 5. *J. Neurochem.* 75:1085–91

Ikeda Y, Terashima T. 1997. Expression of reelin, the gene responsible for the reeler mutation, in embryonic development and adulthood in the mouse. *Dev. Dyn.* 210:157–72

Jensen P, Zoghbi HY, Goldowitz D. 2000. Dis-

section of the cellular and molecular events that direct cerebellar development. *Soc. Neurosci.* 26 (1):58 (Abstr.)

Keeler CE. 1924. The inheritance of a retinal abnormality in white mice. *Proc. Natl. Acad. Sci. USA* 10:329–33

Keeler CE. 1931. *The Laboratory Mouse: Its Origin, Heredity, and Culture.* Cambridge: Harvard Univ. Press

Klar A, Baldassare M, Jessell TM. 1992. F-spondin: a gene expressed at high levels in the floor plate encodes a secreted protein that promotes neural cell adhesion and neurite extension. *Cell* 69:95–110

Kohmura N, Senzaki K, Hamada S, Kai N, Yasuda R, et al. 1998. Diversity revealed by a novel family of cadherins expressed in neurons at a synaptic complex. *Neuron* 20:1137–51

König N, Valet J, Fulcrand J, Marty R. 1977. The time of origin of cajal-retzius cells in the rat temporal cortex, an autoradiographic study. *Neurosci. Lett.* 4:21–26

Kwon YT, Gupta A, Zhou Y, Nikolic M, Tsai LH. 2000. Regulation of N-cadherin-mediated adhesion by the p35-Cdk5 kinase. *Curr. Biol.* 10:363–72

Kwon YT, Tsai LH. 1998. A novel disruption of cortical development in p35(-/-) mice distinct from reeler. *J. Comp. Neurol.* 395:510–22

Le N, Simon MA. 1998. Disabled is a putative adaptor protein that functions during signaling by the sevenless receptor tyrosine kinase. *Mol. Cell. Biol.* 18:4844–54

Lemmon V, Pearlman AL. 1981. Does laminar position determine the receptive field properties of cortical neurons? A study of corticotectal cells in area 17 of the normal mouse and the reeler mutant. *J. Neurosci.* 1:83–93

Li BS, Zhang L, Gu JG, Amin ND, Pant HC. 2000. Integrin alpha(1)beta(1)-mediated activation of cyclin-dependent kinase 5 activity is involved in neurite outgrowth and human neurofilament protein H Lys-Ser-Pro tail domain phosphorylation. *J. Neurosci.* 20:6055–62

Lu DC, Rabizadeh S, Chandra S, Shayya RF,

Ellerby LM, et al. 2000. A second cytotoxic proteolytic peptide derived from amyloid b-protein precursor. *Nat. Med.* 6:397–404

Luskin MB, Shatz CJ. 1985. Neurogenesis of the cat's primary visual cortex. *J. Comp. Neurol.* 242:611–31

Mahley RW. 1988. Apolipoprotein E: cholesterol transport protein with expanding role in cell biology. *Science* 240:622–30

Mallamaci A, Mercurio S, Muzio L, Cecchi C, Pardini CL, et al. 2000. The lack of Emx2 causes impairment of Reelin signaling and defects of neuronal migration in the developing cerebral cortex. *J. Neurosci.* 20:1109–18

Margolis B. 1996. The PI/PTB domain: a new protein interaction domain involved in growth factor receptor signaling. *J. Lab. Clin. Med.* 128:235–41

Mariani J, Crepel F, Mikoshiba K, Changeux JP, Sotelo C. 1977. Anatomical, physiological and biochemical studies of the cerebellum from Reeler mutant mouse. *Philos. Trans. R. Soc. London* 281:1–28

Marin-Padilla M. 1998. Cajal-Retzius cells and the development of the neocortex. *Trends Neurosci.* 21:64–71

McConnell SK, Kaznowski CE. 1991. Cell cycle dependence of laminar determination in developing neocortex. *Science* 254:282–85

Meyer D, Liu A, Margolis B. 1999. Interaction of c-Jun amino-terminal kinase interacting protein-1 with p190 rhoGEF and its localization in differentiated neurons. *J. Biol. Chem.* 274:35113–18

Meyer G, Goffinet AM. 1998. Prenatal development of reelin-immunoreactive neurons in the human neocortex. *J. Comp. Neurol.* 397:29–40

Mikoshiba K, Nishimura Y, Tsukada Y. 1983. Absence of bundle structure in the neocortex of the reeler mouse at the embryonic stage. Studies by scanning electron microscopic fractography. *Dev. Neurosci.* 6:18–25

Miyata T, Nakajima K, Aruga J, Takahashi S, Ikenaka K, et al. 1996. Distribution of a reeler gene-related antigen in the developing cerebellum: an immunohistochemical study with an allogeneic antibody CR-50 on normal and reeler mice. *J. Comp. Neurol.* 372:215–28

Miyata T, Nakajima K, Mikoshiba K, Ogawa M. 1997. Regulation of Purkinje cell alignment by reelin as revealed with CR-50 antibody. *J. Neurosci.* 17:3599–609

Morse HC III. 1981. The Laboratory Mouse-a historical perspective. In *The Mouse in Biomedical Research*, ed. HL Foster, D Small, JG Fox, pp. 1–16. New York: Academic

Mullen RJ. 1984. Ontogeny and genetics of the mammalian nervous system. In *Chimeras in Developmental Biology*, ed. N LeDouarin, A McLaren, pp. 353–68. London: Academic

Nakajima K, Mikoshiba K, Miyata T, Kudo C, Ogawa M. 1997. Disruption of hippocampal development in vivo by CR-50 mAb against reelin. *Proc. Natl. Acad. Sci. USA* 94:8196–201

Ogawa M, Miyata T, Nakajima K, Yagyu K, Seike M, et al. 1995. The reeler gene-associated antigen on Cajal-Retzius neurons is a crucial molecule for laminar organization of cortical neurons. *Neuron* 14:899–912

Ohshima T, Gilmore EC, Longenecker G, Jacobowitz DM, Brady RO, et al. 1999. Migration defects of cdk5$^{(-/-)}$ neurons in the developing cerebellum is cell autonomous. *J. Neurosci.* 19:6017–26

Ohshima T, Ward JM, Huh CG, Longenecker G, Veeranna, et al. 1996. Targeted disruption of the cyclin-dependent kinase 5 gene results in abnormal corticogenesis, neuronal pathology and perinatal death. *Proc. Natl. Acad. Sci. USA* 93:11173–78

Patrick GN, Zukerberg L, Nikolic M, de la Monte S, Dikkes P, Tsai LH. 1999. Conversion of p35 to p25 deregulates Cdk5 activity and promotes neurodegeneration. *Nature* 402:615–22

Pesold C, Impagnatiello F, Pisu MG, Uzunov DP, Costa E, et al. 1998. Reelin is preferentially expressed in neurons synthesizing gamma- aminobutyric acid in cortex and hippocampus of adult rats. *Proc. Natl. Acad. Sci. USA* 95:3221–26

Pesold C, Liu WS, Guidotti A, Costa E, Caruncho HJ. 1999. Cortical bitufted, horizontal, and Martinotti cells preferentially express and secrete reelin into perineuronal nets, nonsynaptically modulating gene expression. *Proc. Natl. Acad. Sci. USA* 96:3217–22

Pierschbacher MD, Ruoslahti E. 1984. Cell attachment activity of fibronectin can be duplicated by small synthetic fragments of the molecule. *Nature* 309:30–33

Pinto-Lord MC, Caviness VS Jr. 1979. Determinants of cell shape and orientation: a comparative Golgi analysis of cell-axon interrelationships in the developing neocortex of normal and reeler mice. *J. Comp. Neurol.* 187:49–69

Pinto-Lord MC, Evrard P, Caviness VS Jr. 1982. Obstructed neuronal migration along radial glial fibers in the neocortex of the reeler mouse: a Golgi-EM analysis. *Brain Res.* 256:379–93

Pittler SJ, Keeler CE, Sidman RL, Baehr W. 1993. PCR analysis of DNA from 70-year-old sections of rodless retina demonstrates identity with the mouse rd defect. *Proc. Natl. Acad. Sci. USA* 90:9616–19

Rakic P. 1972. Mode of cell migration to the superficial layers of fetal monkey neocortex. *J. Comp. Neurol.* 145:61–83

Rakic P. 1974. Neurons in rhesus monkey visual cortex: systematic relation between time of origin and eventual disposition. *Science* 183:425–27

Ramón y Cajal S. 1955 [1911]. *Histology of the Nervous System of Man and Vertebrates.* Transl. L Swanson, N Swanson. Oxford, UK: Oxford Univ. Press (From Spanish)

Redies C, Takeichi M. 1993. Expression of N-cadherin mRNA during development of the mouse brain. *Dev. Dyn.* 197:26–39

Rice DS, Curran T. 1999. Mutant mice with scrambled brains: understanding the signaling pathways that control cell positioning in the CNS. *Genes Dev.* 13:2758–73

Rice DS, Curran T. 2000a. Disabled-1 is expressed in type AII amacrine cells in the mouse retina. *J. Comp. Neurol.* 424:327–38

Rice DS, Curran T. 2000b. Role of reelin and disabled-1 in the development of the retina. *Investig. Ophthalmol. Vis. Sci.* 41(Suppl):S7

Rice DS, Sheldon M, D'Arcangelo G, Nakajima K, Goldowitz D, Curran T. 1998. Disabled-1 acts downstream of Reelin in a signaling pathway that controls laminar organization in the mammalian brain. *Development* 125:3719–29

Ringstedt T, Linnarsson S, Wagner J, Lendahl U, Kokaia Z, et al. 1998. BDNF regulates reelin expression and Cajal-Retzius cell development in the cerebral cortex. *Neuron* 21:305–15

Rodriguez MA, Pesold C, Liu WS, Kriho V, Guidotti A, et al. 2000. Colocalization of integrin receptors and reelin in dendritic spine postsynaptic densities of adult nonhuman primate cortex. *Proc. Natl. Acad. Sci. USA* 97:3550–55

Schiffmann SN, Bernier B, Goffinet AM. 1997. Reelin mRNA expression during mouse brain development. *Eur. J. Neurosci.* 9:1055–71

Selkoe DJ. 1998. The cell biology of beta-amyloid precursor protein and presenilin in Alzheimer's disease. *Trends Cell Biol.* 8:447–53

Senzaki K, Ogawa M, Yagi T. 1999. Proteins of the CNR family are multiple receptors for reelin. *Cell* 99:635–47

Sheldon M, Rice DS, D'Arcangelo G, Yoneshima H, Nakajima K, et al. 1997. Scrambler and yotari disrupt the disabled gene and produce a reeler-like phenotype in mice. *Nature* 389:730–33

Sheppard AM, Pearlman AL. 1997. Abnormal reorganization of preplate neurons and their associated extracellular matrix: an early manifestation of altered neocortical development in the reeler mutant mouse. *J. Comp. Neurol.* 378:173–79

Sidman RL, Green MC, Appel SH. 1965. *Catalog of the Neurological Mutants of the Mouse.* Boston: Harvard Univ. Press

Silva LR, Gutnick MJ, Connors BW. 1991. Laminar distribution of neuronal membrane properties in neocortex of normal and reeler mouse. *J. Neurophysiol.* 66:2034–40

Simmons PA, Lemmon V, Pearlman AL. 1982. Afferent and efferent connections of the striate and extrastriate visual cortex of the normal and reeler mouse. *J. Comp. Neurol.* 211:295–308

Stanfield BB, Cowan WM. 1979. The morphology of the hippocampus and dentate gyrus in normal and reeler mice. *J. Comp. Neurol.* 185:393–422

Steindler DA, Colwell SA. 1976. Reeler mutant mouse: maintenance of appropriate and reciprocal connections in the cerebral cortex and thalamus. *Brain Res.* 113:386–93

Stockinger W, Brandes C, Fasching D, Hermann M, Gotthardt M, et al. 2000. The reelin receptor ApoER2 recruits JNK-interacting proteins-1 and-2. *J. Biol. Chem.* 275:25625–32

Strittmatter WJ, Roses AD. 1996. Apolipoprotein E and Alzheimer's disease. *Annu. Rev. Neurosci.* 19:53–77

Super H, Soriano E, Uylings HBM. 1998. The functions of the preplate in development and evolution of the neocortex and hippocampus. *Brain Res. Rev.* 27:40–64

Sweet HO, Bronson RT, Johnson KR, Cook SA, Davisson MT. 1996. Scrambler, a new neurological mutation of the mouse with abnormalities of neuronal migration. *Mamm. Genome* 7:798–802

Terashima T, Inoue K, Inoue Y, Mikoshiba K, Tsukada Y. 1983. Distribution and morphology of corticospinal tract neurons in reeler mouse cortex by the retrograde HRP method. *J. Comp. Neurol.* 218:314–26

Terashima T, Inoue K, Inoue Y, Yokoyama M, Mikoshiba K. 1986. Observations on the cerebellum of normal-reeler mutant mouse chimera. *J. Comp. Neurol.* 252:264–78

Tesseur I, Van Dorpe J, Spittaels K, Van den Haute C, Moechars D, Van Leuven F. 2000. Expression of human apolipoprotein E4 in neurons causes hyperphosphorylation of protein tau in the brains of transgenic mice. *Am. J. Pathol.* 156:951–64

Trommsdorff M, Borg JP, Margolis B, Herz J. 1998. Interaction of cytosolic adaptor proteins with neuronal apolipoprotein E receptors and the amyloid precursor protein. *J. Biol. Chem.* 273:33556–60

Trommsdorff M, Gotthardt M, Hiesberger T, Shelton J, Stockinger W, et al. 1999. Reeler/Disabled-like disruption of neuronal migration in knockout mice lacking the VLDL receptor and ApoE receptor 2. *Cell* 97:689–701

Tsai LH, Delalle I, Caviness VS Jr, Chae T, Harlow E. 1994. p35 is a neural-specific regulatory subunit of cyclin-dependent kinase 5. *Nature* 371:419–23

Tsai LH, Takahashi T, Caviness VS Jr, Harlow E. 1993. Activity and expression pattern of cyclin-dependent kinase 5 in the embryonic mouse nervous system. *Development* 119:1029–40

Utsunomiya-Tate N, Kubo K, Tate S, Kainosho M, Katayama E, et al. 2000. Reelin molecules assemble together to form a large protein complex, which is inhibited by the function-blocking CR-50 antibody. *Proc. Natl. Acad. Sci. USA* 97:9729–34

Vignoud L, Albiges-Rizo C, Frachet P, Block MR. 1997. NPXY motifs control the recruitment of the alpha-5-beta-1 integrin in focal adhesions independently of the association of talin with the beta-1 chain. *J. Cell Sci.* 110:1421–30

Vignoud L, Usson Y, Balzac F, Tarone G, Block MR. 1994. Internalization of the alpha-5-beta-1 integrin does not depend on "NPXY" signals. *Biochem. Biophys. Res. Commun.* 199:603–11

Ware ML, Fox JW, Gonzalez JL, Davis NM, de Rouvroit CL, et al. 1997. Aberrant splicing of a mouse disabled homolog, mdab1, in the scrambler mouse. *Neuron* 19:239–49

Wechsler-Reya RJ, Scott MP. 1999. Control of neuronal precursor proliferation in the cerebellum by Sonic Hedgehog. *Neuron* 22:103–14

Willnow TE. 1998. Receptor-associated protein (RAP): A specialized chaperone for endocytic receptors. *Biol. Chem.* 379:1025–31

Wyss JM, Stanfield BB, Cowan WM. 1980. Structural abnormalities in the olfactory bulb of the Reeler mouse. *Brain Res.* 188:566–71

Yang A, Walker N, Bronson R, Kaghad M, Oosterwegel M, et al. 2000. p73-deficient mice have neurological, pheromonal and inflammatory defects but lack spontaneous tumours. *Nature* 404:99–103

Yip JW, Yip YPL, Nakajima K, Capriotti C. 2000. Reelin controls position of autonomic neurons in the spinal cord. *Proc. Natl. Acad. Sci. USA* 97:8612–16

Yoneshima H, Nagata E, Matsumoto M, Yamada M, Nakajima K, et al. 1997. A novel neurological mutant mouse, yotari, which exhibits reeler-like phenotype but expresses CR-50 antigen/reelin. *Neurosci. Res.* 29:217–23

Zukerberg LR, Patrick GN, Nikolic M, Humbert S, Wu CL, et al. 2000. Cables links Cdk5 and c-Abl and facilitates Cdk5 tyrosine phosphorylation, kinase upregulation, and neurite outgrowth. *Neuron* 26:633–46

Annu. Rev. Neurosci. 2001. 24:1041–70

HUMAN BRAIN MALFORMATIONS AND THEIR LESSONS FOR NEURONAL MIGRATION

M Elizabeth Ross[1] and Christopher A Walsh[2]

[1]Departments of Neurology, Neuroscience and Genetics Cell Biology and Development, University of Minnesota, Minneapolis, Minnesota 55455; e-mail: rossx001@tc.umn.edu
[2]Department of Neurology, Beth Israel Deaconess Medical Center and Harvard Medical School, Boston, Massachusetts 02115; e-mail: cwalsh@caregroup.harvard.edu

Key Words   cortical dysplasia, lissencephaly, neurogenetics, epilepsy, mental retardation

■ **Abstract**   The developmental steps required to build a brain have been recognized as a distinctive sequence since the turn of the twentieth century. As marking tools for experimental embryology emerged, the cellular events of cortical histogenesis have been intensively scrutinized. On this rich backdrop, molecular genetics provides the opportunity to play out the molecular programs that orchestrate these cellular events. Genetic studies of human brain malformation have proven a surprising source for finding the molecules that regulate CNS neuronal migration. These studies also serve to relate the significance of genes first identified in murine species to the more complex human brain. The known genetic repertoire that is special to neuronal migration in brain has rapidly expanded over the past five years, making this an appropriate time to take stock of the emerging picture. We do this from the perspective of human brain malformation syndromes, noting both what is now known of their genetic bases and what remains to be discovered.

CORTICAL MALFORMATION: Disorders of Neuronal Position

Clinical syndromes involving cerebral cortical malformation are recognized after the fact by the abnormal position of neural cells. This static picture can only suggest the mechanism by which the disorganized brain developed. The movement of cells from their origins in the ventricular zone may be impaired in a number of ways. Primary neurogenesis or cell number may be altered by disturbed cell proliferation, fate determination, and programmed cell death. The failure of particular cells to differentiate or the improper timing of the birth of a neuron or glial cell may alter the fate and positional information of other cells in the region. The migration of cells from the ventricular zone may be curtailed by interference with the mechanical motors and cytoskeletal dynamics of the cells. Alternatively, the molecular signals that initiate movement, guide the cell in its migration, and inform it that the final position has been reached may be altered. Once the position of the cell body is established, assembly and consolidation of neurite/axonal projections and synapses are further established and refined, in part accounting for the survival of neurons and their associated glia.

Much recent interest is trained on the molecules affecting neuronal migration, but their actions may not be purely confined to cell movement. Thus, the events that organize brain structure include neurogenesis, early mid and late migration, as well as axon projection and guidance, all of which are anticipated to overlap considerably. Therefore, the classification of a malformation as a neuronal migration syndrome should be viewed with caution until the function of the responsible gene product is established.

Neuronal Migration Syndromes

The molecular events in neuronal migration are multifaceted and undoubtedly share many features with other migrating cells (for general discussions see Mitchison & Cramer 1996, Condic & Letourneau 1997). However, not all the molecular players in cell motility are equivalent, because there are both human and animal disorders in which migration impairment is virtually restricted to brain. Therefore, superimposed on these basic steps there must be regulatory aspects that are unique to brain. At least four requirements must be fulfilled to move and organize cells in the developing cortex (Figure 1). First, cells must receive signals to "go." Second, adhesive and contractile elements must be coordinately regulated to produce strong adhesion at the leading edge of the cell and weaker adhesion at the rear, ultimately translating into a net movement of the cell when cytoskeletal elements contract. Third, signals must determine the direction or vector of movement. Finally, "stop" signals must inform cells when they have reached their final destination. There are numerous opportunities for genetic mutation to interfere with neuronal migration. Early indications from the reeler and Kallmann syndrome mutations suggested that the unique aspects of neuronal migration in brain would derive from extracellular matrix molecules having selective expression in brain. The astonishing insight from human mutations is that the intracellular cytoskeletal machinery of migrating neurons has distinctive character as well.

To date, over a dozen molecules that are peculiar to neuronal migration in brain have been reported (Table 1). At least half were identified first from clinical human studies. The rest were discovered either in mice or in *Drosophila* (the relevance of the latter for mammalian development subsequently being confirmed in mouse). The cellular events and many of the molecular components of neuronal migration have recently been extensively reviewed in this series and are not repeated here (Hatten 1999). Instead, brain formation is discussed in view of human clinical syndromes recently shown to arise from impairment of neuronal migration. Additional syndromes thought to represent migration abnormalities, but for which genetic identification is currently lacking, are pointed out. These provide

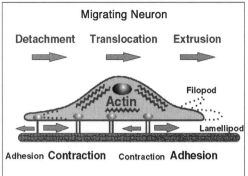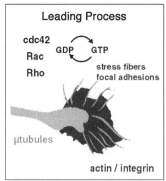

Figure 1   Molecular events in neuronal migration. The left panel summarizes the subcellular components of migration. The right panel further illustrates events at the leading process.

opportunities to uncover new pieces to the puzzle. By far the largest class of neuronal migration syndromes are the lissencephalies (from "lissos," smooth + "encephaly," brain) and are dealt with first.

The Lissencephalies

Cortical malformation often has severe consequences in humans, including mental retardation, epilepsy, paralysis and blindness. Less severe deficits occur when only selected regions of cortex or only a portion of the cortical cell population are affected. High resolution imaging has brought the realization that displaced neurons, or heterotopia, are more prevalent than once thought and that up to 25% of childhood epilepsy is associated with heterotopia (Kuzniecky et al 1993). Neuronal heterotopia are often thought to arise from interference in neuronal migration mechanisms during brain development. Among malformations with heterotopia, the lissencephalies have been grouped because of their probable shared mechanism of incomplete neuronal migration from the ventricular neuroepithelium (Barkovich et al 1996, Dobyns in press). Although these malformations have distinctive features on magnetic resonance imaging, they share a loss of normal sulcation and gyral formation, as well as thickening of the cortical gray matter.

Classical Lissencephaly   This term is used to encompass a spectrum of simplified cortex ranging from total absence of cortical convolutions (agyria) to broadened gyri (pachygyria) with abnormally thick cortex (typically 10–20 mm compared with 2.5–4 mm in the normal cortex) (Dobyns & Truwit 1995, Dobyns et al 1996a). This spectrum intersects with double cortex (DC, aka subcortical band heterotopia or SBH), in which neurons are partially hung up in their migrations, finally to reside as a poorly organized band of neurons in the white matter beneath a relatively normal cortex. Associated abnormalities can include enlarged lateral

TABLE 1 Summary of genes affecting neuronal migration in the central nervous system

Migration Molecule[a]	Gene	Mutation phenotype		References
		Murine	Human	
Platelet activating factor Acetylhydrolase 1b1 (Pafah1b1) (signaling/tubulin cytoskeletal dynamics)	LIS1	Lis1 −/+ animals with neuronal migration delay	Miller Dieker syndrome (MDS) Lissencephaly (ILS)[b] (class LISa1-4) LIS with cerebellar hypoplasia (LCH) (class LCHa)	Reiner et al 1993 LoNigro et al 1997 Hirotsune et al 1998 Ross et al 2001
Doublecortin (Dbcn) (microtubule associated protein-cytoskeletal dynamics)	DCX/XLIS		Double cortex (DC) or (SBH) subcortical band heterotopia X-linked LIS (XLIS) (class LISb1-4) LCH (class LCHa)	Gleeson et al 1998 Des Portes et al 1998 Pilz et al 1998 Ross et al 2001
Reelin [extracellular matrix (ECM) molecule]	RELN	Reeler: inverted cortex with layer 5 neurons superficial, failure of preplate to split	LCH (class LCHb)	Caviness 1977 D'Arcangelo et al 1995 Sheppard et al 1997 Hong et al 2000
Disabled (docking protein for cAbl tyrosine kinase, intracellular signaling)	DAB	Scrambler/yotari (phenotype identical to reeler)	LCH? (predicted, based on similarity of reeler and scrambler mice)	Sheldon et al 1997 Ware et al 1997
VLDLR, ApoER2 (lipoprotein receptors that also bind reelin)	VLDLR ApoER2	Knockouts with reeler-like cortex		Heisberger et al 1999 Trommsdorff et al 1999 D'Arcangelo et al 1999

Protein	Gene	Human syndrome	Mouse phenotype	Reference
Cdk5 (neurofilament phosphorylation)	*CDK5*		Knockout with reeler-like cortex	Gilmore et al 1998
p35 (activating subunit of cdk5)	*p35*		Knockout similar but less severe than *Cdk5*−/− nulls	Chae et al 1997; Kwon & Tsai 1998
Peroxisomal proteins	*PEX2* *Pxr1*	Zellweger syndrome	Deficient mice display heterotopia similar to Zellweger patients	Faust & Hatten 1997; Baes et al 1997
Filamin-1 (actin crosslinking phospoprotein)	*FLN1*	Bilateral periventricular nodular heterotopia (BPNH), neurons fail to leave the VZ		Fox et al 1998
Fukutin (putative ECM protein)	*FCMD*	Fukuyama congenital muscular dystrophy (FCMD)-cobblestone complex type II LIS & myopathy		Kobayashi et al 1998
Anosmin-1 (ECM protein)	*KAL1*	Kallmann syndrome		Hardelin et al 1992; Soussi-Vanicostas et al 1998
Astrotactin 1 (neuronal surface molecule)	*ASTN1*		Nulls with neuronal migration delay in cerebellum and cortex	Adams et al 2001

[a]Suspected role appears in parentheses.
[b]ILS = isolated lissencephaly sequence. Refers to classical LIS without the facial features of MDS.

ventricles, hypoplasia of the corpus callosum, and hypoplasia of the cerebellum, typically in the midline. Two genes, *LIS1* and *DCX*, account for the majority of classical *lissencephaly* (LIS) (Pilz et al 1998). Genotype-phenotype analyses have revealed a gradient of LIS severity that characterizes *LIS1* (posterior worse than anterior, P>A) or *DCX* (A>P) mutations, from which a detailed grading system has been developed (Dobyns et al 1999b, Dobyns in press). Such classification has proven useful in clinical diagnosis and helpful in recognizing causes of LIS and DC/SBH. In addition, despite the striking similarity of LIS caused by *LIS1* or *DCX*, the distinct gradient suggests that these genes may participate in separate though related molecular pathways (Pilz et al 1998).

Chromosome 17-Linked Lissencephaly (LIS1 Gene)   The first LIS syndrome delineated is Miller-Dieker syndrome, which is manifested by severe LIS and characteristic facial abnormalities (Dobyns et al 1993). Chromosomal analysis shows visible deletions of 17p13.3 in over 90% of Miller-Dieker syndrome patients, suggesting that this is a contiguous gene deletion syndrome in which the characteristic facial features are caused by involvement of loci neighboring *LIS1*. Isolated lissencephaly sequence (ILS) consists of classical LIS with relatively normal facial appearance. Fluorescence in situ hybridization (FISH) studies show deletions of 17p13.3 in about 40% of ILS children (Dobyns et al 1994). Extensive genotype-phenotype correlation reveals that mutations in the *LIS1* gene together with *DCX* (see below) account for 76% of classical LIS (Pilz et al 1998). Therefore, as much as 24% of ILS cases could involve another gene. The *LIS1* gene has been cloned and confirmed by identification of *LIS1* point mutations and intragenic deletion in patients with ILS (Lo Nigro et al 1997, Reiner et al 1993).

Gene dosage of *Lis1* has been examined in mice (Hirotsune et al 1998). The null state produces early embryonic lethality in mouse, which may explain why recessive inheritance of *Lis1* mutation has not been observed in humans. Mice heterozygous for *Lis1* survive and reveal abnormalities primarily in cerebral cortex, but also hippocampus and cerebellum, owing to the impaired, slower migration of neurons (Hirotsune et al 1998, Clark et al submitted). Interestingly, though these mice reveal migration defects, the malformation is distinct from the prototype neuronal migration mouse model, *reeler* (see below), in that *Lis1*+/− cortex is not inverted and the hippocampus and cerebellum are far less involved.

LIS1 encodes a noncatalytic subunit of platelet activating factor-acetylhydrolase, or Pafah1b1, and is part of a G-protein-like $(\alpha1/\alpha2)\beta$ trimer (Ho et al 1997). Hereafter, this subunit is referred to as Lis1 protein. Pafah is known to regulate platelet activating factor (PAF), a potent lipid first messenger that is involved in processes ranging from general cell activation to inflammatory and allergic reactions to carcinogenesis and apoptosis. It is not yet proven whether Lis1 exerts its influence on migration through regulation of PAF. However, in vitro studies of neuronal cell motility by Clark and colleagues have shown that exposure to excess PAF inhibits unidirectional cerebellar granule cell movement in culture along neighboring axons (Bix & Clark 1998). Thus, at least one of the roles of Lis1 protein in neuronal migration may be through regulation of PAF. Interestingly, this factor may be involved in another neuronal migration disorder, Zellweger syndrome, in which mutant peroxisomal proteins are expected to result in elevated PAF levels (see below; Faust & Hatten 1997).

Lis1 is a soluble protein with seven WD40 repeats forming a seven-bladed propeller-like structure involved in protein-protein interactions (Garcia-Higuera et al 1996). Lis1 binds tubulin and reduces microtubule catastrophe in vitro, suggesting that it may stabilize the microtubule cytoskeleton (Sapir et al 1997, 1999). Significantly, Lis1 is a highly conserved homologue of the NUDF protein in Aspergillus, in which it is required for nuclear translocation through interaction with a dynein motor (Xiang et al 1995). In fungus, NUDF interacts with several proteins including NUDC and NUDE to regulate this translocation (Chiu et al 1997, Efimov & Morris 2000). Recently, Lis1, overexpressed in mammalian cells and neurons, has been shown to bind cytoplasmic dynein and to affect microtubule organization (Feng et al 2000, Sasaki et al 2000, Smith et al 2000). In this model Lis1 promotes the peripheral, plus end–directed movement of microtubule segments by dynein motors that are attached to stable microtubules, oriented with minus ends anchored at the centrosome and plus ends at the peripheral membrane skeleton. In conditions of low Lis1 levels dynein motor activity is reduced and microtubule segments accumulate near the nucleus. One of the challenges to understanding Lis1 function is that Pafah is a ubiquitously expressed enzyme, yet loss of a single *Lis1* allele produces defects largely confined to brain. Based on Western analysis, Tsai and colleagues hypothesized that the neuronal specificity of *Lis1* haploinsufficiency relates to the unusually high levels of Lis1 expressed in wild type neurons (Smith et al 2000).

X-Linked Lissencephaly (DCX/XLIS Gene)   XLIS refers to the syndrome of classical LIS in hemizygous males and DC/SBH in heterozygous females. The clinical features and responsible gene were delineated only recently (Dobyns et al 1996a, Ross et al 1997, Gleeson et al 1998, des Portes et al 1998). The clinical manifestations of XLIS in males are very similar to those of chromosome 17 associated ILS. DC/SBH is characterized by a symmetric and circumferential band of gray matter, located just beneath the cortex and separated from it by a thin band of white matter. The phenotype arises because females, though they have two X-chromosomes, inactivate one early in embryogenesis through the process of random X-inactivation. This process ensures that on average, half of the cortical neurons in an affected female will inactivate the mutant *DCX* allele and express normal protein, whereas the other half will inactivate the normal allele and become impaired in their migration, to form the band heterotopia. Skewing of X-inactivation can lead to the most severe form of lissencephaly in a *DCX* female (Ross et al 1997). DC/SBH individuals are far less affected than classical LIS patients and may manifest seizures with cognitive impairment of varying severity. Indeed, $\sim$25% of DC/SBH patients possess normal or near normal intelligence (Dobyns et al 1996a). The degree of neurologic impairment roughly coincides with the thickness and extent of the subcortical band, which is presumed to arise from the incomplete migration of affected cells to the forming cortical gray matter (cortical plate). *DCX* accounts for ~85% of sporadic DC/SBH and 100% (11 of 11) of X-linked, DC/SBH pedigrees (Gleeson et al 1999b, Matsumoto et al 2000). Both LIS and DC/SBH due to mutations in *DCX* reveal an anterior (A>P) gradient that has been helpful in recognizing mutational heterogeneity in the gene that produces milder forms of the disorder (Gleeson et al 1999b, Matsumoto et al 2000).

The XLIS gene, *DCX*, encodes a 40 kD soluble protein named doublecortin (Dbcn) (Gleeson et al 1998, des Portes et al 1998). Unlike the ubiquitously expressed Lis1 protein, Dbcn has been detected only in neurons of the central and peripheral nervous system (Gleeson et al 1999a). The protein bears homology to the noncatalytic portion of the CAM-kinase member, DCAMKL1. Dbcn displays several structural features of interest, including a predicted motif for phosphorylation by members of the cAbl, nonreceptor tyrosine kinase family (Gleeson et al 1998). This is likely to be significant, because another neuronal migration disorder in the scrambler mouse is associated with mutation in *disabled* (*mDab*), whose gene product in *Drosophila* is a regulatory protein for cAbl (Ware et al 1997, Sheldon et al 1997). Extensive genotype-phenotype analysis of patient populations has revealed two internal repeat regions in Dbcn that are essential for its function (Gleeson et al 1999b, Matsumoto et al 2000). Mutation analysis in vitro and in situ indicates that these repeats bind to tubulin to promote precipitation and stabilization of microtubules (Gleeson et al 1999a, Taylor et al 2000, Horesh et al 1999).

LIS1 and DCX Gene Functions in Neuronal Migration   A rough schematic of the emerging relationships among these two genes and other molecules identified in mouse is provided in Figure 2. This is by no means intended to be complete, and several of the assignments are tentative. Nevertheless, the diagram provides a framework in which to consider possible mechanistic relationships. Molecules are viewed at the interface of the actin-based and microtubule-based cytoskeleton. Those depicted have been identified or implicated in neuronal migration (Lis1, Dbcn, cdk5, mDab, reelin, integrins, VLDL and ApoE receptors, Dynein motors, Rac, Rho, cdc42). Included are molecules either shown to interact with these components, or that are closely related and are demonstrated participants in axon extension [enabled (Ena), profilin, cAbl, abl related gene (Arg), Pak1].

To date, the Lis1 and Dbcn proteins have been implicated in regulation of microtubule dynamics by virtue of binding tubulin and reducing microtubule catastrophe (Sapir et al 1997, 1999; Gleeson et al 1999b; Taylor et al 2000). In addition, Clark and colleagues have demonstrated in a yeast two-hybrid system that point mutations found in ILS patients interfere with the association of Lis1 for the 29 and 30 kD subunits of Pafah (Sweeney et al 2000). The fact that all patient mutations in Lis1 examined in this system interfere with binding to Lis1 suggests that the heterotrimeric Pafah complex may be involved in regulating migration. Further

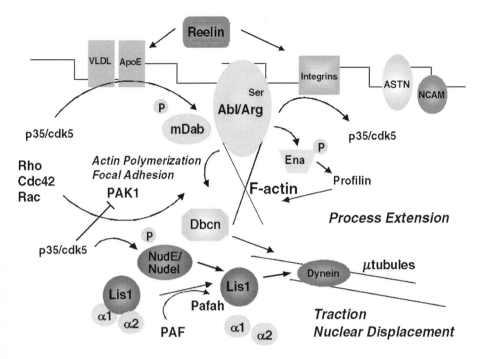

Figure 2 Interrelationships among proteins governing neuronal migration and process extension. Although several assignments are tentative, the diagram provides a framework in which to consider mechanistic interactions. Proteins depicted have been identified or implicated in neuronal migration (Lis1, Dbcn, cdk5, mDab, reelin, integrins, VLDL and ApoE receptors, dynein moters, Rac, Rho, cdc42, ASTN, NCAM). Included are proteins either shown to interact with these components (NudE, Nudel) or that are closely related participants in axon extension [enabled (Ena), profilin, cAbl, abl related gene (Arg), Pak1].

evidence indicates that PAF interferes with binding of Lis1 to the 30 kD subunit and that PAF rescues the migration phenotype of Lis1 haploinsufficient neurons. This suggests a model in which the role of PAF is to mobilize the Lis1 protein from Pafah to exert a downstream effect on migration. In such a model, PAF could rescue neurons expressing low levels of Lis1 by displacing more of Lis1 protein from the complex (GD Clark, GJ Bix, A Shinoya, S Hirotsune, ME Ross et al 2001, submitted for publication).

The downstream, non-Pafah-dependent effects of Lis1 involve additional protein-protein interactions. For example, several recent reports have described a Lis1 interaction with dynein heavy chain that recapitulates the interaction of the homologous proteins demonstrated in Aspergillus (Faukner et al 2000, Liu et al 2000, Smith et al 2000). In addition, interactions have been identified in yeast two-hybrid screens between Lis1 and two homologues of the Aspergillus *NUDE* gene, *mNudE* and *Nudel* (Efimov & Morris 2000, Feng et al 2000, Niethammer et al 2000, Sasaki et al 2000). In young mammalian neurons, mNudE and Nudel are both co-localized with Lis1 at the microtubule organizing center (centrosome). Not only does mNudE localize to the centrosome, but it simultaneously binds multiple components there that appear to determine the organization of the centrosome (Feng et al 2000). Evidence that this interaction with Lis1 is functionally important comes from the findings that (*a*) missense *Lis1* mutations block mNudE/Lis1 binding, and (*b*) disrupting the interaction of mNudE with Lis1 in *Xenopus* disrupts the lamination of the anterior nervous system (Feng et al 2000). As the cell matures, Nudel/Lis1 distribute into the axon (Sasaki et al 2000). In addition, Nudel is a substrate for phosphorylation by cdk5, an important molecule for process extension and axon fasciculation (Niethammer et al 2000, Sasaki et al 2000). Together, the mammalian data suggest a mechanism in which Lis1 participates in the organization of the microtubule-based cytoskeleton needed for neurite extension and nuclear translocation.

The relationship(s) between Lis1 or Dbcn and other migration molecules in the neuronal migration scheme are tenuous at best. Dbcn has a putative link to the cAbl family of nonreceptor tyrosine kinases through a motif that, based on sequence, should be phosphorylated by an Abl-like kinase (Gleeson et al 1998). The connection between cAbl and neuronal migration is further implicated by observations that (*a*) cAbl mutants in *Drosophila* have an axonal projection and fasciculation defect (Gertler et al 1989, Bashaw et al 2000) and (*b*) mutation in the mouse disabled (*mDab*) gene, which encodes a docking protein for Abl in the fly, produces a reeler-like phenotype (Howell et al 1997, Ware et al 1997). Moreover, binding of the reelin protein to the VLDL or ApoE2 receptor leads to phosphorylation of mDab (D'Arcangelo et al 1999, Hiesberger et al 1999, Trommsdorff et al 1999). Biochemical evidence points to JNK-interacting proteins 1 and 2 as intracellular effector molecules downstream of reelin-ApoER2 interactions (Stockinger et al 2000). This may occur via the nonreceptor tyrosine kinase Fyn, a member of the cAbl family that is known to bind to the intracellular domain of ApoER2.

The functional involvement of the Abl family kinases in neuronal movement is further supported by the demonstration that p35/cdk5, whose loss of function leads to cortical neuronal migration defects, is phosphorylated by cAbl (Chae et al 1997, Kwon & Tsai 1998, Gilmore et al 1998, Zukerberg et al 2000). cAbl in the cytoplasm has itself been demonstrated to bind filamentous actin (McWhirter & Wang 1993), and one of its binding proteins, enabled (Mena in mammals), is known to interact with profilin, a modulator of actin polymerization (Lanier et al 1999). Phosphorylation by p35/cdk5 is further implicated in actin dynamics because the p35 subunit has been demonstrated to interact directly with the small GTPase Rac, and p35/cdk5 hyperphosphorylation of Pak1 inhibits the Pak1 kinase, a modulator of actin polymerization (Nikolic et al 1998).

Additional neuronal migration signaling cascades from the membrane surface that involve adhesion molecules may indirectly impact cAbl-family kinases. For example, another twist to the story of reelin action is the identification of its binding to $\alpha3\beta1$ integrin to provide an inhibitory signal to migrating neurons

(Dulabon et al 2000). The downstream events following integrin binding have yet to be clarified, but mice lacking $\alpha 3\beta 1$ integrin express reduced levels of mDab, implicating transcriptional regulation (Dulabon et al 2000). Coming full-circle then, the demonstrated ability of Dbcn to interact with microtubules and its structural motif that suggests phosphorylation by cAbl, a kinase involved in actin cytoskeletal dynamics, position Dbcn protein as a candidate molecule for crosstalk between the actin- and microtubule-based cytoskeleton. Similarly, the multiple substrates emerging for cdk5 implicate this kinase in such a coordinating role.

Lissencephaly with Cerebellar Hypoplasia

This group is only now evolving a nomenclature and comprises six broad classes, lissencephaly with cerebellar hypoplasia (LCH)a–f, that are grouped according to distinguishing features (Ross et al 2001). Within these broader groups, three causative genes have been identified and another locus described. Syndromes of LCH have in common a lissencephaly spectrum of agyria-pachygyria, with mildly (5–10 mm) or markedly (10–20 mm) thickened gray matter. Cerebellar involvement detected by magnetic resonance imaging may be mild, with predominantly midline hypoplasia seen with *LIS1* or *DCX* mutations (LCHa) to severe cerebellar defects with hypoplasia of the cerebellar hemispheres and abnormal or absent foliation (LCHb or d) (al Shahwan et al 1995, Kato et al 1999, Ross et al 2001). Another form of LCH (LCHe) is characterized by an abrupt transition from agyria frontally to gyral simplification of parietal-occipital cortex, associated with moderate cerebellar hypoplasia. A striking phenotypic manifestation (LCHf) is total absence of the corpus callosum. The rarity and lack of families with multiple affected individuals suggests that many of the LCH phenotypes are likely to arise from autosomal recessive mutations.

Recently, mutation has been identified in the human *reelin* (*RELN*) gene that produces LCHb. This distinctive pattern possesses a moderately thickened cortex and pachygyria, markedly abnormal hippocampal formation, and severe cerebellar hypoplasia with absent folia (Hourihane et al 1993, Hong et al 2000). This is of particular interest because the reeler mouse was first proposed as an animal model of lissencephaly in the 1970s, though direct evidence has been lacking (Caviness 1977). *RELN* encodes a large 388 kD protein containing 8 EGF repeats and is secreted by the Cajal-Retzius cells of the embryonic preplate, the marginal zone, the cerebellar EGL, and pioneering cells of the hippocampus (D'Arcangelo et al 1995, Schiffmann et al 1997, Sheppard & Pearlman 1997). One of the receptor types for the Reln protein is the lipoprotein receptors VLDLR and ApoER2 (D'Arcangelo et al 1999). Reln has been recently demonstrated to direct phosphorylation of the mDab1 protein via action of ApoE receptor 2 (Hiesberger et al 1999). The fact that mutation in *Dab1* produces the scrambler mouse, with a brain phenotype nearly identical to the reeler mouse, further implicates mDab1 as an effector of neuronal migration that functions downstream of Reln (Ware et al 1997, Sheldon et al 1997,

Rice et al 1998). However, whether human *DAB1* mutations cause clinical LCH is not yet known.

Additional candidate genes for LCH include segmentation genes, growth and trophic factors identified as important for dorsal-ventral patterning of hindbrain (for summary see Millen et al 1999). One appealing LCH candidate is the neuronal migration gene, astrotactin (*ASTN1*), which was originally identified through investigation of migration-blocking antibodies generated from cerebellar tissues (Edmondson et al 1988). A mouse model lacking *Astn1* reveals reduced volumes of cerebral cortex and cerebellum and subtle disorganization of these structures (Adams N, Dietz G, Kon N, Hatten ME 2001, submitted for publication). Astn1 is a transmembrane protein with fibronectin type III repeats in the extracellular domain (Zheng et al 1996). Expressed on cerebellar granule neurons, Astn1 is required for glia-guided migration of granule cells in culture (Fishell & Hatten 1991, Zheng et al 1996). Astn1 mRNA is also expressed in cerebral cortex and hippocampus. The previous observation of much milder phenotype in *Lis1* knockout mice compared with the human presentation suggests that the clinical syndrome associated with *ASTN1* mutation would likely be more severe than in the mouse.

Cobblestone Complex (Lissencephaly)

Pathologically, cobblestone complex consists of migration of heterotopic young neurons beyond the marginal zone—future layer I—into the leptomeninges through gaps in the external basement membrane. This could result from alteration in the extracellular matrix, a hypothesis that may be supported by the identification of fukutin in *Fukuyama congenital muscular dystrophy* (see below). The movement of neurons into the leptomeninges can obliterate the subarachnoid space and give rise to communicating hydrocephalus or ventricular enlargement owing to impaired reabsorption of cerebrospinal fluid (Gelot et al 1995, Bornemann et al 1997). On first glance, the magnetic resonance imaging features of cobblestone lissencephaly may be difficult to distinguish from those of polymicrogyria. There is a thickened cortical gray matter with a knobby cobblestone surface and few or absent true sulci. Some areas appear as pachygyria with smooth surface. Bands of "white matter" that are composed of glial-fibrous tissue interrupt the gray matter. There are often small cerebellar cysts and white matter neuronal heterotopia. Hydrocephalus may be present and when severe, produces marked thinning of the cortex.

Fukuyama Congenital Muscular Dystrophy (FCMD)   This is the second most common form of inherited congenital muscular dystrophy in the Japanese. There is progressive facial and limb weakness with delayed motor development, congenital and progressive joint contractures and elevated serum muscle enzymes (CK-creatine kinase) reaching 10–50 times normal (Fukuyama et al 1981). Brain involvement includes cobblestone complex that is less severe than Walker Warburg syndrome or muscle-eye-brain disease, with minor or no eye abnormalities. There may be mild cerebellar foliation defects, but there is not usually hypoplasia (Aida et al 1996). This autosomal recessive mutation is the first human syndrome documented to arise from a retrotransposal insertion that interrupts the 3' end of the transcript on chromosome 9q31-32 (Kobayashi et al 1998). There is a strong founder effect, because 87% of FCMD has been found to carry this mutation and it is rarely encountered outside of Japan or Korea (Toda et al 2000). The gene, *FCMD*, encodes a putative extracellular matrix molecule, called fukutin (Kobayashi et al 1998).

Walker-Warburg Syndrome (WWS)   Characteristic features of WWS include severe cobblestone complex and retinal and other eye abnormalities, as well as congenital muscular dystrophy (Dobyns et al 1989, Chadani et al 2000). This is an autosomal recessive disorder that is genetically distinct from FCMD, though severely affected FCMD children may resemble WWS (Chadani et al 2000). The cortex is thickened (7–10 mm), except when it is thinned owing to increased pressure within the cranium from hydrocehpalus, with a mixture of cobblestone complex with diffuse agyria. There are diffuse white matter changes, and irregular laminar heterotopia may appear beneath the cortex. The brainstem is hypoplastic, and there is significant midline>hemisphere cerebellar hypoplasia. Many WWS patients have a retrocerebellar cyst (Dandy-Walker malformation). In some this cyst may protrude through a small defect in the skull, producing an occipital cephalocoele. Eye involvement includes retinal nonattachment or detachment or more mild retinal and optic nerve hypoplasia (Gerding et al 1993). Some have microphthalmia, buphthalmos, colobomas, or high myopia.

Muscle-Eye-Brain Disease (MEB)   The MEB gene has been mapped in a Finnish population to chromosome 1p34-p32 (Cormand et al 1999). In the same population children with WWS have been genetically excluded from the 1p34-32 locus of *MEB* or the 9q31-q33 position of *FCMD* (Ranta et al 1995, Cormand et al 2001). MEB consists of cobblestone complex, retinal and other eye abnormalities including a glial preretinal membrane, and myopathy (Haltia et al 1997). Retinal physiological abnormalities may be detected after 1 year of age on electroretinograms or visual evoked potentials. The cortex shows frontal pachygyria and less severe gyral abnormalities in the occipital region. The cerebellum is less involved than in WWS, and the midline may be hypoplastic, but hemispheres are usually normal. Dandy-Walker malformation and cephaloceles are not seen in MEB.

Cobblestone Complex Only   Patients with the appearance of cobblestone complex but without significant retinal or muscle pathology have been identified (Dobyns et al 1996b). These three patients from two consanguinous families have been excluded by linkage analysis from the 9q31 locus of *FCMD* and the 1p34 locus of *MEB*. The brain malformation may be diffuse and resemble polymicrogyria, or similar to that in MEB with an A>P gradient. The brainstem and cerebellar vermis in *cobblestone complex only* are mildly reduced in volume, whereas the cerebellar hemispheres appear normal on magnetic resonance imaging. As in other cobblestone syndromes, cerebellar microcysts may be present.

Other Disorders of Neuronal Migration

Several other disorders, though not associated with a "smooth brain," are nevertheless associated with altered migration of selected subpopulations of neurons.

Bilateral Periventricular Nodular Heterotopia (BPNH or PH)   This is characterized by neuronal heterotopia lining the lateral ventricles. Several BPNH families have demonstrated affected women with periventricular nodules, seizures, and normal intelligence (Kamuro & Tenokuchi 1993, Huttenlocher et al 1994). The familial syndrome shows a female predominance and is usually lethal in males, consistent with an X-linked inheritance, though a few severely affected males have been known to survive (Fink et al 1997). On tissue examination, the periventricular nodules contain well-differentiated cortical neurons in the subependymal zone (Eksioglu et al 1996). Like the situation in SBH/DC, the milder phenotype in females is attributed to X-inactivation, with the heterotopic neurons representing cells in which the normal allele is silent. The gene defect in BPNH has been identified as a mutation in *filamin 1* (*FLN1*), encoding a 280 kD actin-crosslinking phosphoprotein (ABP-280) (Fox et al 1998). Fln1 was first identified in white blood cells, where it is required for migration, formation of filopodia, chemotaxis, cell morphology, and platelet aggregation. Fln1 is thought to transduce ligand-receptor binding into actin remodeling necessary for motility. The identification of sporadic cases of periventricular heterotopia affecting females and males equally, suggests that additional autosomal genes, and/or X-linked genes other than *FLN1*, must exist whose function is required to initiate migration from the ventricular region (Raymond et al 1994, Guerrini & Dobyns 1998; ME Ross & CA Walsh, personal observations).

Zellweger Syndrome   This syndrome is recognized by a cortical dysplasia resembling polymicrogyria of cerebral and cerebellar cortex, sometimes with pachygyria around the sylvian fissure, and focal subcortical and subependymal heterotopia. There may be subependymal cysts and hypoplasia of the inferior olivary nucleus (Takashima et al 1992). In postmortem cortex there is an increase in cholesterol-ester very-long-chain fatty acids and decreased plasmalogens (Powers & Moser 1998). Thought to arise from errors of peroxisomal metabolic function, the primary abnormality proposed in Zellweger syndrome is a defect in the mitochondrial desaturation pathway–metabolizing Docosahexenoic (22:6n-3) acids (Infante & Huszagh 1997). Several animal models of Zellweger syndrome have been produced. Inactivation of *Pxr1*, encoding the transport receptor for peroxisomal matrix proteins, produces multiple neuronal heterotopia (Baes et al 1997). Similarly, loss-of-function mutation of Pex2, a peroxisome assembly protein, disrupts neuronal migration in mouse brain (Faust & Hatten 1997). Interestingly, such peroxisomal disorders associated with this dysplastic syndrome would be expected to alter platelet activating factor (PAF) levels in brain, perhaps hinting at a role for this potent first messenger in both classical lissencephaly (*LIS1* mutation) and Zellweger neuronal migration defects.

Kallmann Syndrome   The constellation of anosmia due to lack of olfactory bulb development, mental retardation, and hypogonadism comprises Kallmann syndrome. This was recognized early on as a failure of migration of particular neurons of the olfactory cortex and GnRH-secreting cells of the hypothalamus. Autosomal dominant, recessive, and X-linked inheritance patterns have all been described. The X-linked form is associated with mutation in *KAL1*, encoding anosmin-1, a secreted extracellular matrix protein with fibronectin type III–like domains (Hardelin et al 1992, del Castillo et al 1992, Soussi-Yanicostas et al 1998). Adhesion to anosmin-1 depends upon the presence of chondroitin sulfate and glycosaminoglycans on the cell surface. Anosmin-1 has been shown in vitro to modulate neurite outgrowth and fasciculation in a cell type–specific manner (Soussi-Yanicostas et al 1998, Hardelin et al 1999).

Disorders of Axonal Projection and Assembly

These disorders, though rarely identified in a pure form, are closely related to the neuronal migration genes noted above (Table 2). In fact, several migration genes are listed in both Tables 1 and 2 because their roles in neuronal migration and neurite extension have been examined in some detail.

Agenesis of the Corpus Callosum (ACC)   Perhaps the most striking examples of defective neurite outgrowth in humans are associated with failure of cortical axons to cross the midline, thereby producing absence or agenesis of the corpus callosum. ACC is found in an impressive array of cortical dysplasia, making it likely to arise from a number of gene defects. Yet when present, ACC speaks to the importance of the involved gene in neurite outgrowth and thus is a valuable clue to the selection of candidate genes. For example, although there may be thinning or partial agenesis of the corpus callosum in ILS/Miller-Deiker or XLIS, total ACC has not been seen in patients with the associated known mutations. ACC is a prominent feature of the CRASH (callosal agenesis, retardation, abducted thumbs, spastic paraparesis, and hydrocephalus) syndrome that has been associated with mutations in the L1 (L1CAM) neuronal adhesion molecule (Sztriha et al 2000). Moreover, a probable X-linked syndrome has recently been identified in five children with LIS associated with ACC and ambiguous genitalia (XLAG), without mutation in either *LIS1* or *DCX* genes (Dobyns et al 1999a). There are numerous examples of human brain maldevelopment associated with ACC, and the reader is referred to several excellent discussions (Marszal et al 2000, Dobyns 1996).

Candidate Genes for Defects in Neuritogenesis and Axon Pathfinding   A number of genes implicated in neuronal migration disorders have been observed to

TABLE 2 Summary of genes affecting neurite outgrowth in central nervous system

Nutrition Outgrowth[a]	Gene	Mutation phenotype Murine	Human	References
Mammalian enabled (Mena) (actin reorganization/adhesion)	*MENA*	Marked axonal projection defects in −/− Exacerbated by loss of profilin 1		Lanier et al 1999
Ableson (phosphorylates cdk5)	*cABL*	Nulls are early postnatal lethals		Zukerberg et al 2000
Abl Related Gene (tyrosine kinase regulates actin cytoskeleton)	*ARG*	Enriched in CNS, colocalizes with cAbl/F actin Neural tube defect in double Abl/Arg nulls		Koleske et al 1998
Profilin (binds Mena and regulates actin polymerization)	*Profilin I* *Profilin II*	[b]*Chickadee* mutation in *Drosophila* identical to growth cone arrest of cAbl nulls		Suetsugu et al 1998 Wills et al 1999
Slit (ligand of Robo, repels midline axons)	*slit 1* *slit 2*	Axon outgrowth from retinal explants inhibited by *Slit 2* transfected cells		Rothberg et al 1990 Ringstedt et al 2000

Roundabout (repulsive axon guidance receptor)	*Robo*	[b]Slit/robo interactions first identified in flies; *Drosophila*. Slit repels spinal motor axons in culture	Brose et al 1999 Bashaw et al 2000
Cdk5 (modulates actin dynamics via Rac GTPase and Pak1)	*CDK5*	p35/cdk5 localized to growth cones; mutant forms inhibit neurites	Nikolic et al 1996 Nikolic et al 1998
p35 (activating subunit of cdk5)	*p35*	p35 nulls reveal callosal axon guidance defects	Kwon et al 1999
Reelin (ECM protein)	*RELN*	Neurite outgrowth and neuronal synaptogenesis defects	Borrell et al 1999a,b Hong et al 2000
L1 (ECM adhesion protein)	*L1CAM*	LCH with pachygyria, marked cerebellar hypoplasia, and foliation defect. CRASH syndrome with callosal defect, retardation, adducted thumbs, spasticity, and hydrocephaly	Sztriha et al 2000

[a]Suspected role appears in parentheses.

[b]Primary evidence in *Drosophila* with relevance to mammalian systems demonstrated.

produce defects in neurite extension as well as axon pathfinding (compare Tables 1 and 2, Figure 2). For example, *reelin* mutation produces not only neuronal migration defects, but also abnormalities in neurite outgrowth (Borrell et al 1999a,b). In *Drosophila*, the cAbl tyrosine kinase produces defects of axonal fasciculation, and in mammals the kinase is capable of phosphorylating cdk5 (Zukerberg et al 2000). In turn, cdk5 and its activating subunit, p35, have both been localized to growth cones in the mouse, and mutations can interfere with neurite outgrowth, whereas p35 nulls reveal defects in callosal axon guidance (Kwon et al 1999, Nikolic et al 1998, 1996).

In addition to the tantalizing association of cAbl function with the mDab1 protein in the migration defects revealed by *scrambler* and *yotari* mice, another cAbl modulator, enabled, has been implicated in axonogenesis. For example, mice deficient in the Mena (mammalian enabled) protein display defects in neural tube closure and axonal commissure formation (Lanier et al 1999). Mena has been shown to bind directly to profilin, an actin binding protein that modulates actin polymerization (Lanier et al 1999, Suetsugu et al 1999). In *Drosophila*, both cAbl and profilin are required for axon extension (Wills et al 1999). In mammals the abl related gene (*Arg*) is a tyrosine kinase that is enriched in brain and colocalizes with cAbl and actin microfilaments (Koleske et al 1998). That double *Abl/Arg* nulls display embryonic lethality with severe neurulation defects indicates that the two kinases may compensate for loss of one another in neuronal migration defects.

Recently, a crucial ligand-receptor system for axonal pathfinding and midline decussation (crossing) has been identified in *Drosophila*, for which homologous elements are being found in mammals. The ligand, Slit, is an extracellular matrix protein required for midline glial and axon commissural pathfinding (Rothberg et al 1990). There are several Slit proteins that bind a family of receptors, Robo (round-about), that control the axonal "decision" to cross midline structures (Bashaw et al 2000, Brose et al 1999, Kidd et al 1999). In flies the Robo receptor proteins appear to function upstream of opposing actions of cAbl and enabled (Bashaw et al 2000). In mammals Slits 1 and 2 and Robo 1 and 2 are expressed in forebrain, and human Slit-2 can collapse growth cones and repel hippocampal and olfactory axons, as well as modulate retinal axon pathfinding (Nguyen Ba-Charvet et al 1999, Ringstedt et al 2000).

Because the proposed actions of many genes involved in neurite extension entail remodeling of the actin cytoskeleton, it is likely that at least some of these influence neuronal migration as well. This will undoubtedly include actions of leading process extension and contractility elements that generate the net movement forward (Figure 1). Regulation of microtubule dynamics will clearly participate in both neurite extension and cell migration, though the impact on neurites of genes like *LIS1* and *DCX* has yet to be examined in any detail. A crucial aspect of future investigations will be characterization of the interface between regulation of the actin-based and microtubule-based cytoskeleton in response to signals presented to the cell.

SYNDROMES OF CORTICAL DISORGANIZATION

Although not necessarily a primary affector of neuronal migration, a number of important processes may result in heterotopic placement of neurons if disturbed. These syndromes warrant brief discussion, as they may ultimately bear on the fine-tuning of neuronal migration and certainly pertain to the establishment of cortical structure. A partial list appears in Table 3, where they are mentioned because either clinical observation or animal models have revealed neuronal heterotopia and cortical dysplasia phenotypes.

Dysplasia Associated with Altered Neurogenesis and Cell Survival

Cortical dysplasia may arise from disturbances in the primary genesis of neural elements, both in cell number and fate. A few candidate genes that could disrupt

TABLE 3   Summary of genes affecting neurogenesis and brain organization

		Mutation phenotype		
Proliferation/Apoptosis[a]	Gene	Murine	Human	Reference
TISH (unknown, recessive gene in seizure-prone rat)	*Tish*	DC/SBH-like rat mutant due to ectopic neural cell proliferation		Lee et al 1997 Lee et al 1998
EMX2 (transcription factor homologue of *Drosophila* segmentation gene empty spiracles)	*EMX2*	−/−: disorganized cortex with abnormal subplate and marginal zone development affecting radial glia	A rare cause of open lipped schizencephaly	Granata et al 1997 Mallamaci et al 2000
Rb (oncogene regulating G1 progression)	*pRB*	−/−: ectopic mitoses and apoptosis in brain		Lee et al 1992
Caspase-3 (enzyme required for apoptosis)	*CPP32*	−/−: shows increased cell numbers, displaced neurons		Kuida et al 1996
Cyclin D2 [G1 active subunit regulating cyclin-dependent kinase 4 (cdk4)]	*Ccnd2*	−/−: thinned cortex, small cerebellum, reduced neural proliferation, and increased apoptosis		Huard et al 1999
Cyclin D1 (G1-active subunit regulating cdk4)	*Ccnd1*	−/−: runted animal, thinned cortex, motor abnormalities suggesting spasticity		Sicinski et al 1995

[a]Suspected role appears in parentheses.

cortical histogenesis in this way have been examined with regard to their effects on brain organization. Other clinically recognized cortical malformations have been associated with mutation in genes recognized for their influence on neural fate determination.

Ectopic Neurogenesis   At least two animal models have demonstrated that neural precursor proliferation can produce striking disruptions in brain histogenesis. The spontaneous appearance of an autosomal recessively transmitted mutation in the *tish* rat displays a phenotype that bears similarity to human DC/SBH (Lee et al 1997). *Tish* brain reveals a band of heterotopic neurons just beneath a relatively normal-appearing cortical gray matter, in fronto-parietal cortex. The *tish* phenotype suggests altered migration of a subpopulation of neurons toward the cortex. Surprisingly, however, pulse labeling with BrdU demonstrated that *tish* heterotopia arise from ectopic cell proliferation rather than impaired migration of young neurons just out of the cell cycle (Lee et al 1998). Band heterotopia in humans and *tish* represents an important example of similar phenotypes that arise from different mechanisms, and underscores the caveat that initial classification of a clinical syndrome as a neuronal migration or other defect must be tentative pending functional investigation once the causative gene is identified.

A second example of ectopic neurogenesis is found in mouse models that inactivate the retinoblastoma protein, Rb (Lee et al 1992). This nuclear phosphoprotein regulates passage of neural precursor cells through the G1 phase of the cell cycle. Complete inactivation of pRb is lethal by embryonic day 16, and brain is profoundly affected by multiple foci of ectopic mitoses as well as massive cell death (Lee et al 1992). Together, the *tish* and *pRb*$^{-/-}$ models indicate that ectopic cell proliferation is an important consideration when designing investigations of gene function associated with cortical malformation.

Altered Cell Survival   Several genes involved in programmed cell death have been demonstrated in mouse models to result in neuronal displacements. A dramatic example is found in the *caspase-3* (*CPP32*) null mice (Kuida et al 1996). These animals are reduced in overall size, and brain is markedly abnormal, with areas of hyperplasia, reduced apoptosis, and heterotopia. Caspace-3 is a member of the ICE-protease family of enzymes that are required for the DNA fragmentation and chromatin condensation characteristic of apoptosis.

Microcephaly   Defined as head size less than −2 standard deviations below normative curves for age, microcephaly comprises a very heterogeneous group of disorders. This is true even when considering only congenital microcephaly, defined as head circumference less than −2 standard deviations from the mean at birth. Recent reports of magnetic resonance imaging in congenital microcephaly have recognized multiple phenotypic patterns of small brains, including those with the following: normal gyral pattern, simplified gyral pattern, agenesis of the corpus callosum, thickened cortical gray matter (a.k.a. microlissencephaly), and other cortical dysplasia including polymicrogyria (Barchovich et al 1998, Sztriha et al 1998, 1999, Peiffer et al 1999). Among affected patients, the severity of cognitive and motor impairment and epilepsy correlate with both head size and the pattern of cortical malformation. Inherited genetic causes of microcephaly are indicated by the occurrence in siblings reported for many of these subtypes (Sztriha et al 1998, 1999; Peiffer et al 1999). The observation of dysplastic forms with thickened cortical gray matter (microlissencephaly) and neuronal heterotopia suggests that neuronal migration abnormalities contribute to the pathogenesis of some microcephalic children (Barkovich et al 1998).

At least 5 loci for congenital microcephaly have been mapped, including 8p22, 19q13, 9q34, 1q31, and 15q (Jackson et al 1998, Jamieson et al 1999, Roberts et al 1999, Moynihan et al 2000, Pattison et al 2000). Candidate genes for microcephaly include those that decrease proliferation in the germinal neuroepithelium or that increase apoptotic cell loss. Examples of cell cycle regulatory genes that are candidates are found in the cyclins D1 and D2. These are activating noncatalytic subunits of cyclin-dependent kinases, principally cdk4, that regulate the mid-G1 phase transition (Xiong et al 1992). Passage of cells through this restriction point commits the cell to a round of division. Null mice lacking the cyclin D1 have small brains, are small in stature and manifest motor abnormalities, suggesting central nervous system developmental defects (Sicinski et al 1995). Interestingly, although cell cycle regulatory genes would be expected to produce small but otherwise normal appearing brains, animal models suggest these genes may also affect brain organization. For example, cyclin D2 nulls are of normal body size but display reduced cerebellar and cerebral cortical volumes with altered cytoarchitecture owing to a combination of decreased cell proliferation and increased apoptosis (Huard et al 1999).

Malformations Not Yet Classified

These represent disorders in which neurons are observed to be out of place, but lack of gene identification or adequate genetic models prevents definitive classification of the primary defect.

Polymicrogyria (PMG)   In this disorder, neurons move out to the cortical surface but organize abnormally to produce multiple small gyri. Radiographically, PMG is characterized by multiple small gyri with shallow sulci and white-matter interdigitation giving the cortex a roughened appearance (Barkovich et al 1999). Thick magnetic resonance sections can give the appearance of pachygyria because signal intensities are averaged and microgyri are missed, so that images constructed at <4 mm intervals may be necessary to detect PMG. The surface of the brain may be variably smooth, when the outer molecular layer fuses over the microsulci, or may be irregular.

Pathologically, two patterns of PMG are recognized: (*a*) "four-layered" PMG, in which cortex comprises a molecular layer, an organized outer layer, a cell sparse layer, and a disorganized inner layer and (*b*) a completely disorganized "unlayered" microgyrus. PMG can arise from several intrauterine insults including hypoxemia with cortical laminar necrosis. This hypothesis of intrauterine insult is in part derived from experimental models in the rat produced by cortical lesions such as focal freezing injury (Jacobs et al 1999). However, the detection of PMG in several affected family members indicates inherited causes as well. Lesion experiments suggest mechanisms that interfere with radial glial function and secondary disturbances of neuronal migration, though this is far from established as a general rule. That involvement of radial glia may not always be required is suggested by interesting in vitro experiments that produce heterotopia beneath the marginal zone of cortical slices exposed to the trophic factor NT4 (Brunstrom et al 1997). This malformation appears to arise from aberrant migration of interneurons from the ganglionic eminence (Brunstrom & Pearlman 2000, Eisenstat et al 1999, Anderson et al 1999).

Focal patterns of PMG and transmission within families suggest several causative genes (Barkovich et al 1999). Diffuse PMG involves widespread areas of cortex and is most likely to have epigenetic causes, such as intrauterine infection or toxic exposure producing encephalopathy, though some causes may be genetic. Thirteen patients with PMG confined to frontal cortex have recently been reported. All 13 were sporadic, though there was parental consanguinity in 2 of the families, raising the possibility of autosomal recessive inheritance (Sztriha & Nork 2000, Guerrini et al 2000). Studies in 12 kindreds with perisylvian PMG indicate genetic hetrogeneity with both X-linked and autosomal (dominant with incomplete penetrance or recessive) patterns (Guerreiro et al 2000, Borgatti et al 1999). PMG involving parieto-occipital cortex has been reported in nine individuals; however, an inheritance pattern has not been established (Guerrini et al 1997).

Schizencephaly   The hallmark of schizencephaly ("split brain") is a cleft in the cerebral cortex, which may be unilateral or bilateral, that extends through the cortex to the lateral ventricular surface. This may appear as a wide cleft (open lipped) or a narrow groove (closed lipped) that is typically lined by gray matter polymicrogyria. The transcription factor, *EMX2*, accounts for a minority (likely rare cases) of open-lipped schizencephaly (Faiella et al 1997, Granata et al 1997). *EMX2* is a mammalian homologue of the *Drosophila* homeotic gene *empty spiracles*, involved in cell fate determination (Simeone et al 1992). It is expressed in the cortical ventricular zone of mammals. Mice lacking *EMX2* expression initially produce reelin at the time of cortical preplate formation but later lose this protein with adverse consequences for radial glial development, neuronal migration, and cortical plate formation (Mallamaci et al 2000a). In addition, lack of *EMX2* produces shifts in cerebral cortical–area development, such that anterior-lateral cortical areas predominate over medial-caudal identities (Bishop et al 2000, Mallamaci et al 2000b). This may in part account for the regional localization of the most severely affected cortex in Scizencephaly. *EMX2* demonstrates that gene action determining the fate of cells can have secondary effects on migration away from the germinal neuroepithelium and subsequent cortical organization.

SUMMARY

The number and nature of gene mutations that can disrupt neuronal migration is large and growing. This is not surprising, given the impressive array of molecular steps that must be taken to move a cell and position it within a system as complex as the brain. Through investigations stimulated by human clinical syndromes and animal models of brain malformation, the pieces of this intricate puzzle are coming into view and are beginning to fall into a fascinating pattern. Clearly, these genes are subserving multiple functions, and their actions will be found to span artificial boundaries between primary neurogenesis, cell survival, migration, neurite extension, axon pathfinding, and synaptogenesis. This fundamental knowledge will assist formulation of strategies to promote repair and regeneration of brain damaged by trauma, stroke, maldevelopment, mechanisms of immunity, or aging.

Visit the Annual Reviews home page at www.AnnualReviews.org

LITERATURE CITED

Aida N, Tamagawa K, Takada K, Yagishita A, Kobayashi N, et al. 1996. Brain MR in Fukuyama congenital muscular dystrophy. *AJNR Am. J. Neuroradiol.* 17:605–13

Anderson S, Mione M, Yun K, Rubenstein JL. 1999. Differential origins of neocortical projection and local circuit neurons: role of Dlx genes in neocortical interneuronogenesis. *Cereb. Cortex* 9:646–54

Baes M, Gressens P, Baumgart E, Carmeliet P, Casteels M, et al. 1997. A mouse model for Zellweger syndrome. *Nat. Genet.* 17:49–57

Barkovich AJ, Ferriero DM, Barr RM, Gressens P, Dobyns WB, et al. 1998. Microlissencephaly: a heterogeneous malformation of cortical development. *Neuropediatrics* 29:113–19

Barkovich AJ, Hevner R, Guerrini R. 1999. Syndromes of bilateral symmetrical polymicrogyria. *AJNR Am. J. Neuroradiol.* 20:1814–21

Barkovich AJ, Kuzniecky R, Dobyns WB, Jackson G, Becker LE, Evrard P. 1996. Malformations of cortical development. *Neuropediatrics* 27:59–63

Bashaw GJ, Kidd T, Murray D, Pawson T, Goodman CS. 2000. Repulsive axon guidance: Abelson and Enabled play opposing roles downstream of the roundabout receptor. *Cell* 101:703–15

Bishop KM, Goudreau G, O'Leary DD. 2000. Regulation of area identity in the mammalian neocortex by Emx2 and Pax6. *Science* 288:344–49

Bix GJ, Clark GD. 1998. Platelet-activating factor receptor stimulation disrupts neuronal migration in vitro. *J. Neurosci.* 18:307–18

Borgatti R, Triulzi F, Zucca C, Piccinelli P, Balottin U, et al. 1999. Bilateral perisylvian polymicrogyria in three generations. *Neurology* 52:1910–13

Bornemann A, Aigner T, Kirchner T. 1997. Spatial and temporal development of the gliovascular tissue in type II lissencephaly. *Acta Neuropathol.* 93:173–77

Borrell V, Del Rio JA, Alcantara S, Derer M, Martinez A, et al. 1999a. Reelin regulates the development and synaptogenesis of the layer-specific entorhino-hippocampal connections. *J. Neurosci.* 19:1345–58

Borrell V, Ruiz M, Del Rio JA, Soriano E. 1999b. Development of commissural connections in the hippocampus of reeler mice: evidence of an inhibitory influence of Cajal-Retzius cells. *Exp. Neurol.* 156:268–82

Brose K, Bland KS, Wang KH, Arnott D, Henzel W, et al. 1999. Slit proteins bind Robo receptors and have an evolutionarily conserved role in repulsive axon guidance. *Cell* 96:795–806

Brunstrom JE, Gray-Swain PA, Osborne PA, Pearlman AL. 1997. Neuronal heterotopias in the developing cerebral cortex produced by neurotrophin-4. *Neuron* 18:505–17

Brunstrom JE, Pearlman AL. 2000. Growth factor influences on the production and migration of cortical neurons. *Results Probl. Cell Differ.* 30:189–215

Caviness VS Jr. 1977. Reeler mutant mouse: a genetic experiment in developing mammalian cortex. In *Approaches to the Cell Biology of Neurons*, ed. WM Cowan et al, pp. 27–46. Bethesda, MD: Soc. Neurosci.

Chadani Y, Kondoh T, Kamimura N, Matsumoto T, Matsuzaka T, et al. 2000. Walker-Warburg syndrome is genetically distinct from Fukuyama type congenital muscular dystrophy. *J. Neurol. Sci.* 177:150–53

Chae T, Kwon YT, Bronson R, Dikkes P, Li E, Tsai L-H. 1997. Mice lacking p35, a neuronal specific activator of Cdk5, display cortical lamination defects, seizures, and adult lethality. *Neuron* 18:29–42

Chiu YH, Xiang X, Dawe AL, Morris NR. 1997. Deletion of nudC, a nuclear migration gene of Aspergillus nidulans, causes morphological and cell wall abnormalities and is lethal. *Mol. Biol. Cell* 8:1735–49

Condic ML, Letourneau PC. 1997. Ligand-induced changes in integrin expression regulate neuronal adhesion and neurite outgrowth. *Nature* 389:852–56

Cormand B, Avela K, Pihko H, Santavuori P, Talim B, et al. 1999. Assignment of the muscle-eye-brain disease gene to 1p32-p34 by linkage analysis and homozygosity mapping. *Am. J. Hum. Genet.* 64:126–35

Cormand B, Pihko H, Bayes M, Valanne L, Santavuori P, et al. 2001. Clinical and genetic distinction between Walker-Warburg syndrome and muscle-eye-brain disease. *Ann. Neurol.* In press

D'Arcangelo G, Homayouni R, Keshvara L, Rice DS, Sheldon M, Curran T. 1999. Reelin is a ligand for lipoprotein receptors. *Neuron* 24:471–79

D'Arcangelo G, Miao GG, Chen SC, Soares HD, Morgan JI, Curran T. 1995. A protein related to extracellular matrix proteins deleted in the mouse mutant reeler. *Nature* 374:719–23

del Castillo I, Cohen-Salmon M, Blanchard S, Lutfalla G, Petit C. 1992. Structure of the X-linked Kallmann syndrome gene and its homologous pseudogene on the Y chromosome. *Nat. Genet.* 2:305–10

des Portes V, Pinard JM, Billuart P, Vinet MC, Koulakoff A, et al. 1998. A novel CNS gene required for neuronal migration and involved in X-linked subcortical laminar heterotopia and lissencephaly syndrome. *Cell* 92:51–61

Dobyns WB. 1996. Absence makes the search grow longer. *Am. J. Hum. Genet.* 58:7–16

Dobyns WB. 2001. Lissencephaly: the clinical and molecular genetic basis of diffuse malformations of neuronal migration. In *Neuronal Migration Disorders*, ed. PG Barth. London: McKeith. In press

Dobyns WB, Andermann E, Anderman F, Czapansky-Beilman D, Dubeau F, et al. 1996a. X-linked malformations of neuronal migration. *Neurology* 47:331–39

Dobyns WB, Berry-Kravis E, Havernick NJ, Holden KR, Viskochil D. 1999a. X-linked lissencephaly with absent corpus callosum and ambiguous genitalia. *Am. J. Med. Genet.* 86:331–37

Dobyns WB, Carrozzo R, Ledbetter DH. 1994. Frequent deletions of the LIS1 gene in classic lissencephaly. *Ann. Neurol.* 36:489–90

Dobyns WB, Pagon RA, Armstrong D, Curry CJ, Greenberg F, et al. 1989. Diagnostic criteria for Walker-Warburg syndrome. *Am. J. Med. Genet.* 32:195–210

Dobyns WB, Patton MA, Stratton RF, Mastrobattista JM, Blanton SH, Northrup H. 1996b. Cobblestone lissencephaly with normal eyes and muscle. *Neuropediatrics* 27:70–75

Dobyns WB, Reiner O, Carrozzo R, Ledbetter DH. 1993. Lissencephaly: a human brain malformation associated with deletion of the LIS1 gene located at chromosome 17p13. *JAMA* 270:2838–42

Dobyns WB, Truwit CL. 1995. Lissencephaly and other malformations of cortical development: 1995 update. *Neuropediatrics* 26:132–47

Dobyns WB, Truwit CL, Ross ME, Matsumoto N, Pilz DT, et al. 1999b. Differences in the gyral pattern distinguish chromosome 17-linked and X-linked lissencephaly. *Neurology* 53:270–77

Dulabon L, Olson EC, Taglienti MG, Eisenhuth S, McGrath B, et al. 2000. Reelin binds alpha-3-beta-1 integrin and inhibits neuronal migration. *Neuron* 27:33–44

Edmondson JC, Liem RKH, Kuster JE, Hatten ME. 1988. Astrotactin: a novel neuronal cell surface antigen that mediates neuron-astroglial interactions in cerebellar microcultures. *J. Cell Biol.* 106:505–17

Efimov VP, Morris NR. 2000. The LIS1-related NUDF protein of Aspergillus nidulans interacts with the coiled-coil domain of the NUDE/RO11 protein. *J. Cell Biol.* 150:681–88

Eisenstat DD, Liu JK, Mione M, Zhong W, Yu G, et al. 1999. DLX-1, DLX-2, and DLX-5 expression define distinct stages of basal forebrain differentiation. *J. Comp. Neurol.* 414:217–37

Eksioglu YZ, Scheffer IE, Cardeness P, Knoll J, DiMario F, et al. 1996. Periventricular heterotopias: an X-linked dominant epilepsy locus causing aberrant cerebral cortical development. *Neuron* 16:77–87

Faiella A, Brunelli S, Granata T, D'Incerti L, Cardini R, et al. 1997. A number of schizencephaly patients including 2 brothers are heterozygous for germline mutations in the homeobox gene EMX2. *Eur. J. Hum. Genet.* 5:186–90

Faulkner NE, Dujardin DL, Tai CY, Vaughan KT, O'Connell CB, et al. 2000. A role for the lissencephaly gene LIS1 in mitosis and cytoplasmic dynein function. *Nat. Cell Biol.* 2:784–91

Faust PL, Hatten ME. 1997. Targeted deletion of the PEX2 peroxisome assembly gene in mice provides a model for Zellweger syndrome, a human neuronal migration disorder. *J. Cell Biol.* 139:1293–1305

Feng G, Olson EC, Stukenbuerg PT, Flanagan LA, Kirschner MW, Walsh CA. 2000. LIS1 regulates CNS lamination by interacting with mNudE, a central component of the centrosome. *Neuron* 28:653–64

Fink JM, Dobyns WB, Guerrini R, Hirsch BA. 1997. Identification of a duplication of Xq28 associated with bilateral periventricular nodular heterotopia. *Am. J. Hum. Genet.* 61:379–87

Fishell G, Hatten ME. 1991. Astrotactin provides a receptor system for CNS neuronal migration. *Development* 113:755–65

Fox JW, Lamperti ED, Eksioglu YZ, Hong SE, Feng Y, et al. 1998. Mutations in filamin 1 prevent migration of cerebral cortical neurons in human periventricular heterotopia. *Neuron* 21:1315–25

Fukuyama Y, Osawa M, Suzuki H. 1981. Congenital progressive muscular dystrophy of the Fukuyama type—clinical, genetic and pathological considerations. *Brain Dev.* 3:1–29

Garcia-Higuera I, Fenoglio J, Li Y, Lewis C, Panchenko MP, et al. 1996. Folding of proteins with WD-repeats: comparison of six members of the WD-repeat superfamily to the G protein beta subunit. *Biochemistry* 35:13985–94

Gelot A, de Villemeur TB, Bordarier C, Ruchoux MM, Moraine C, Ponsot G. 1995. Developmental aspects of type II lissencephaly. Comparative study of dysplastic lesions in fetal and post-natal brains. *Acta Neuropathol.* 89:72–84

Gerding H, Gullotta F, Kuchelmeister K, Busse H. 1993. Ocular findings in Walker-Warburg syndrome. *Childs Nerv. Syst.* 9:418–20

Gertler FB, Bennett RL, Clark MJ, Hoffmann FM. 1989. *Drosophila abl* tyrosine kinase in embryonic CNS axons: a role in axogenesis is revealed through dosage-sensitive interactions with disabled. *Cell* 58:103–13

Gilmore EC, Ohshima T, Goffinet AM, Kulkarni AB, Herrup K. 1998. Cyclin-dependent kinase 5-deficient mice demonstrate novel developmental arrest in cerebral cortex. *J. Neurosci.* 18:6370–77

Gleeson JG, Allen KM, Fox JW, Lamperti ED, Berkovic S, et al. 1998. *Doublecortin*, a brain-specific gene mutated in human X-linked lissencephaly and double cortex syndrome, encodes a putative signaling protein. *Cell* 92:63–72

Gleeson JG, Lin PT, Flanagan LA, Walsh CA. 1999a. Doublecortin is a microtubule-associated protein and is expressed widely by migrating neurons. *Neuron* 23:257–71

Gleeson JG, Minnerath S, Allen KM, Fox JW, Hong S, et al. 1999b. Characterization of mutations in the gene *doublecortin* in patients with double cortex syndrome. *Ann. Neurol.* 45:146–53

Granata T, Farina L, Faiella A, Cardini R, D'Incerti L, et al. 1997. Familial schizencephaly associated with EMX2 mutation. *Neurology* 48:1403–6

Guerreiro MM, Andermann E, Guerrini R, Dobyns WB, Kuzniecky R, et al. 2000. Familial perisylvian polymicrogyria: a new familial syndrome of cortical maldevelopment. *Ann. Neurol.* 48:39–48

Guerrini R, Barkovich AJ, Sztriha L, Dobyns WB. 2000. Bilateral frontal polymicrogyria: a newly recognized brain malformation syndrome. *Neurology* 54:909–13

Guerrini R, Dobyns WB. 1998. Bilateral periventricular nodular heterotopia with mental retardation and frontonasal malformation. *Neurology* 51:499–503

Guerrini R, Dubeau F, Dulac O, Barkovich AJ, Kuzniecky R, et al. 1997. Bilateral parasagittal parietooccipital polymicrogyria and epilepsy. *Ann. Neurol.* 41:65–73

Haltia M, Leivo I, Somer H, Pihko H, Paetau A, et al. 1997. Muscle-eye-brain disease: a neuropathological study. *Ann. Neurol.* 41:173–80

Hardelin JP, Julliard AK, Moniot B, Soussi-Yanicostas N, Verney C, et al. 1999. Anosmin-1 is a regionally restricted component of basement membranes and interstitial matrices during organogenesis: implications for the developmental anomalies of X chromosome-linked Kallmann syndrome. *Dev. Dyn.* 215:26–44

Hardelin JP, Levilliers J, del Castillo I, Cohen-Salmon M, Legouis R, et al. 1992. X chromosome-linked Kallmann syndrome: stop mutations validate the candidate gene. *Proc. Natl. Acad. Sci. USA* 89:8190–94

Hatten ME. 1999. Central nervous system neuronal migration. *Annu. Rev. Neurosci.* 22:511–39

Hiesberger T, Trommsdorff M, Howell BW, Goffinet A, Mumby MC, et al. 1999. Direct binding of Reelin to VLDL receptor and ApoE receptor 2 induces tyrosine phosphorylation of disabled-1 and modulates tau phosphorylation. *Neuron* 24:481–89

Hirotsune S, Fleck MW, Gambello MJ, Bix GJ, Chen A, et al. 1998. Graded reduction of *Pafah1b1* (*Lis1*) gene activity results in neuronal cell autonomous migration defects and early embryonic lethality. *Nat. Genet.* 19:333–39

Ho YS, Swenson L, Derewenda U, Serre L, Wei Y, et al. 1997. Brain Acetylhydrolase that inactivates platelet-activating factor is a G-protein-like trimer. *Nature* 385:89–93

Hong SE, Shugart YY, Huang DT, Shahwan SA, Grant PE, et al. 2000. Autosomal recessive lissencephaly with cerebellar hypoplasia is associated with human RELN mutations. *Nat. Genet.* 26:93–96

Horesh D, Sapir T, Francis F, Wolf SG, Caspi M, et al. 1999. Doublecortin, a stabilizer of microtubules. *Hum. Mol. Genet.* 8:1599–610

Hourihane JO, Bennett CP, Chaudhuri R, Robb SA, Martin ND. 1993. A sibship with a neuronal migration defect, cerebellar hypoplasia and congenital lymphedema. *Neuropediatrics* 24:43–46

Howell BW, Gertler FB, Cooper JA. 1997. Mouse disabled (mDab1): a Src binding protein implicated in neuronal development. *EMBO J.* 16:121–32

Huard J, Forster C, Carter ML, Sicinski P, Ross ME. 1999. Cerebellar histogenesis is disturbed in mice lacking cyclin D2. *Development* 126:1927–35

Huttenlocher PR, Taravath S, Mojtahedi S. 1994. Periventricular heterotopia and epilepsy. *Neurology* 44:51–55

Infante JP, Huszagh VA. 1997. On the molecular etiology of decreased arachidonic (20:4n-6), docosapentaenoic (22:5n-6) and docosahexaenoic (22:6n-3) acids in Zellweger syndrome and other peroxisomal disorders. *Mol. Cell. Biochem.* 168:101–15

Jackson AP, McHale DP, Campbell DA, Jafri H, Rashid Y, et al. 1998. Primary autosomal recessive microcephaly (MCPH1) maps to chromosome 8p22-pter. *Am. J. Hum. Genet.* 63:541–46

Jacobs KM, Hwang BJ, Prince DA. 1999. Focal epileptogenesis in a rat model of polymicrogyria. *J. Neurophysiol.* 81:159–73

Jamieson CR, Govaerts C, Abramowicz MJ. 1999. Primary autosomal recessive microcephaly: homozygosity mapping of MCPH4 to chromosome 15. *Am. J. Hum. Genet.* 65:1465–69

Kamuro K, Tenokuchi Y. 1993. Familial periventricular nodular heterotopia. *Brain Dev.* 15:237–41

Kato M, Takizawa N, Yamada S, Ito A, Honma T, et al. 1999. Diffuse pachygyria with cerebellar hypoplasia: a milder form of microlissencephaly or a new genetic syndrome? *Ann. Neurol.* 46:660–63

Kidd T, Bland KS, Goodman CS. 1999. Slit is the midline repellent for the robo receptor in Drosophila. *Cell* 96:785–94

Kobayashi K, Nakahori Y, Miyake M, Matsumura K, Kondo-Iida E, et al. 1998. An ancient retrotransposal insertion causes Fukuyama-type congenital muscular dystrophy. *Nature* 394:388–92

Koleske AJ, Gifford AM, Scott ML, Nee M,

Bronson RT, et al. 1998. Essential roles for the Abl and Arg tyrosine kinases in neurulation. *Neuron* 21:1259–72

Kuida K, Zheng TS, Na S, Kuan C, Yang D, et al. 1996. Decreased apoptosis in the brain and premature lethality in CPP32-deficient mice. *Nature* 384:368–72

Kuzniecky R, Murro A, King D, Morawetz R, Smith J, et al. 1993. Magnetic resonance imaging in childhood intractable partial epilepsy: pathologic correlations. *Neurology* 43:681–87

Kwon YT, Tsai LH. 1998. A novel disruption of cortical development in p35($-/-$) mice distinct from reeler. *J. Comp. Neurol.* 395:510–22

Kwon YT, Tsai LH, Crandall JE. 1999. Callosal axon guidance defects in p35($-/-$) mice. *J. Comp. Neurol.* 415:218–29

Lanier LM, Gates MA, Witke W, Menzies AS, Wehman AM, et al. 1999. Mena is required for neurulation and commissure formation. *Neuron* 22:313–25

Lee EY, Chang CY, Hu N, Wang YC, Lai CC, et al. 1992. Mice deficient for Rb are nonviable and show defects in neurogenesis and haematopoiesis. *Nature* 359:288–94

Lee KS, Collins JL, Anzivino MJ, Frankel EA, Schottler F. 1998. Heterotopic neurogenesis in a rat with cortical heterotopia. *J. Neurosci.* 18:9365–75

Lee KS, Schottler F, Collins JL, Lanzino G, Couture D, et al. 1997. A genetic animal model of human neocortical heterotopia associated with seizures. *J. Neurosci.* 17:6236–42

Liu Z, Steward R, Luo L. 2000. Drosophila Lis1 is required for neuroblast proliferation, dendritic elaboration and axonal transport. *Nat. Cell Biol.* 2:776–83

Lo Nigro EJ, Chong CS, Smith AC, Dobyns WB, Carrozzo R, Ledbetter DH. 1997. Point mutations and an intragenic deletion in LIS1, the lissencephaly causative gene in isolated lissencephaly sequence and Miller-Dieker syndrome. *Hum. Mol. Genet.* 6:157–64

Mallamaci A, Mercurio S, Muzio L, Cecchi C,

Pardini CL, et al. 2000a. The lack of Emx2 causes impairment of Reelin signaling and defects of neuronal migration in the developing cerebral cortex. *J. Neurosci.* 20:1109–18

Mallamaci A, Muzio L, Chan CH, Parnavelas J, Boncinelli E. 2000b. Area identity shifts in the early cerebral cortex of Emx2−/− mutant mice. *Nat. Neurosci.* 3:679–86

Marszal E, Jamroz E, Pilch J, Kluczewska E, Jablecka-Deja H, Krawczyk R. 2000. Agenesis of corpus callosum: clinical description and etiology. *J. Child Neurol.* 15:401–5

Matsumoto N, Leventer RJ, Kuc JA, Mewborn SK, Dudlicek LL, et al. 2001. Mutation analysis of the *DCX* gene and genotype/phenotype correlation in subcortical band heterotopia. *Eur. J. Hum. Genet.* In press

McWhirter JR, Wang JY. 1993. An actin-binding function contributes to transformation by the Bcr-Abl oncoprotein of Philadelphia chromosome-positive human leukemias. *EMBO J.* 12:1533–46

Millen KJ, Millonig JH, Wingate RJ, Alder J, Hatten ME. 1999. Neurogenetics of the cerebellar system. *J. Child Neurol.* 14:574–81

Mitchison TJ, Cramer LP. 1996. Actin-based cell motility and cell locomotion. *Cell* 84:371–79

Moynihan L, Jackson AP, Roberts E, Karbani G, Lewis I, et al. 2000. A third novel locus for primary autosomal recessive microcephaly maps to chromosome 9q34. *Am. J. Hum. Genet.* 66:724–27

Nguyen Ba-Charvet KT, Brose K, Marillat V, Kidd T, Goodman CS, et al. 1999. Slit2-mediated chemorepulsion and collapse of developing forebrain axons. *Neuron* 22:463–73

Niethammer M, Smith DS, Ayala R, Peng J, Ko J, et al. 2000. NUDEL is a novel cdk5 substrate that associates with LIS1 and cytoplasmic dynein. *Neuron* 28:697–711

Nikolic M, Chou MM, Lu W, Mayer BJ, Tsai LH. 1998. The p35/Cdk5 kinase is a neuron-specific Rac effector that inhibits Pak1 activity. *Nature* 395:194–98

Nikolic M, Dudek H, Kwon YT, Ramos YF,

Tsai LH. 1996. The cdk5/p35 kinase is essential for neurite outgrowth during neuronal differentiation. *Genes Dev.* 10:816–25

Pattison L, Crow YJ, Deeble VJ, Jackson AP, Jafri H, et al. 2000. A fifth locus for primary autosomal recessive microcephaly maps to chromosome 1q31. *Am. J. Hum. Genet.* 67:1578–80

Peiffer A, Singh N, Leppert M, Dobyns WB, Carey JC. 1999. Microcephaly with simplified gyral pattern in six related children. *Am. J. Med. Genet.* 84:137–44

Pilz DT, Matsumoto N, Minnerath S, Mills P, Gleeson JG, et al. 1998. *LIS1* and *XLIS/doublecortin* mutations cause most human classical lissencephaly, but different patterns of malformation. *Hum. Mol. Genet.* 7:2029–37

Powers JM, Moser HW. 1998. Peroxisomal disorders: genotype, phenotype, major neuropathologic lesions, and pathogenesis. *Brain Pathol.* 8:101–20

Ranta S, Pihko H, Santavuori P, Tahvanainen E, de la Chapelle A. 1995. Muscle-eye-brain disease and Fukuyama type congenital muscular dystrophy are not allelic. *Neuromuscul. Disord.* 5:221–25

Raymond AA, Fish DR, Stevens JM, Sisodiya SM, Alsanjari N, Shorvon SD. 1994. Subependymal heterotopia: a distinct neuronal migration disorder associated with epilepsy. *J. Neurol. Neurosurg. Psychiatry* 57:1195–202

Reiner O, Carrozzo R, Shen Y, Whenert M, Faustinella F, et al. 1993. Isolation of a Miller-Dieker lissencephaly gene containing G protein ß-subunit-like repeats. *Nature* 364:717–21

Rice DS, Sheldon M, D'Arcangelo G, Nakajima K, Goldowitz D, Curran T. 1998. Disabled-1 acts downstream of Reelin in a signaling pathway that controls laminar organization in the mammalian brain. *Development* 125:3719–29

Ringstedt T, Braisted JE, Brose K, Kidd T, Goodman C, et al. 2000. Slit inhibition of retinal axon growth and its role in retinal axon pathfinding and innervation patterns in the diencephalon. *J. Neurosci.* 20:4983–91

Roberts E, Jackson AP, Carradice AC, Deeble VJ, Mannan J, et al. 1999. The second locus for autosomal recessive primary microcephaly (MCPH2) maps to chromosome 19q13.1-13.2. *Eur. J. Hum. Genet.* 7:815–20

Ross ME, Allen KM, Srivastava AK, Featherstone T, Gleeson JG, et al. 1997. Linkage and physical mapping of X-linked lissencephaly/SBH (XLIS): A gene causing neuronal migration defects in human brain. *Hum. Mol. Genet.* 6:555–62

Ross ME, Swanson K, Dobyns WB. 2001. Lissencephaly with cerebellar hypoplasia (LCH): a heterogeneous group of cortical malformations. *Neuropediatrics* In press

Rothberg JM, Jacobs RJ, Goodman CS, Artavanis-Tsakonas S. 1990. *Slit*: an extracellular protein necessary for development of midline glia and commissural axon pathways contains both EGF and LRR domains. *Genes Dev.* 4:2169–87

Sapir T, Cahana A, Seger R, Nekhai S, Reiner O. 1999. LIS1 is a microtubule-associated phosphoprotein. *Eur. J. Biochem.* 265:181–88

Sapir T, Elbaum M, Reiner O. 1997. Reduction of microtubule catastrophe events by LIS1, platelet-activating factor acetylhydrolase subunit. *EMBO J.* 16:6977–84

Sasaki S, Shionoya A, Ishida M, Gambello MJ, Yingling J, et al. 2000. A LIS1/NUDEL/cytoplasmic dynein heavy chain complex in the developing and adult nervous system. *Neuron* 28:681–96

Schiffmann SN, Bernier B, Goffinet AM. 1997. Reelin mRNA expression during mouse brain development. *Eur. J. Neurosci.* 9:1055–71

al Shahwan SA, Bruyn GW, al Deeb SM. 1995. Non-progressive familial congenital cerebellar hypoplasia. *J. Neurol. Sci.* 128:71–77

Sheldon M, Rice DS, D'Arcangelo G, Yoneshima H, Nakajima K, et al. 1997. *Scrambler* and *yotari* disrupt the *disabled* gene and produce a *reeler-like* phenotype in mice. *Nature* 389:730–36

Sheppard AM, Pearlman AL. 1997. Abnormal reorganization of preplate neurons and their associated extracellular matrix: an early manifestation of altered neocortical development in the reeler mutant mouse. *J. Comp. Neurol.* 378:173–79

Sicinski P, Donaher JL, Parker SB, Li TS, Gardner H, et al. 1995. Cyclin D1 provides a link between development and oncogenesis in the retina and breast. *Cell* 82:621–30

Simeone A, Gulisano M, Acampora D, Stornaiuolo A, Rambaldi M, Boncinelli E. 1992. Two vertebrate homeobox genes related to the *Drosophila empty spiracles* gene are expressed in the embryonic cerebral cortex. *EMBO J.* 11:2541–50

Smith DS, Niethammer M, Zhou Y, Gambello MJ, Wynshaw-Boris A, Tsai L-H. 2000. Regulation of cytoplasmic dynein behavior and microtubule organization by mammalian Lis1. *Nat. Cell Biol.* 2:767–75

Soussi-Yanicostas N, Faivre-Sarrailh C, Hardelin JP, Levilliers J, Rougon G, Petit C. 1998. Anosmin-1 underlying the X chromosome-linked Kallmann syndrome is an adhesion molecule that can modulate neurite growth in a cell-type specific manner. *J. Cell Sci.* 111:2953–65

Stockinger W, Brandes C, Fasching D, Hermann M, Gotthardt M, et al. 2000. The reelin receptor ApoER2 recruits JNK-interacting proteins-1 and 2. *J. Biol. Chem.* 275:25625–32

Suetsugu S, Miki H, Takenawa T. 1999. Distinct roles of profilin in cell morphological changes: microspikes, membrane ruffles, stress fibers, and cytokinesis. *FEBS Lett.* 457:470–74

Sweeney KJ, Clark GD, Prokscha A, Dobyns WB, Eichele G. 2000. Lissencephaly associated mutations suggest a requirement for the PAFAH1B heterotrimeric complex in brain development. *Mech. Dev.* 92:263–71

Sztriha L, Al-Gazali L, Varady E, Nork M, Varughese M. 1998. Microlissencephaly. *Pediatr. Neurol.* 18:362–65

Sztriha L, Al-Gazali LI, Varady E, Goebel HH,

Nork M. 1999. Autosomal recessive micrencephaly with simplified gyral pattern, abnormal myelination and arthrogryposis. *Neuropediatrics* 30:141–45

Sztriha L, Frossard P, Hofstra RM, Verlind E, Nork M. 2000. Novel missense mutation in the L1 gene in a child with corpus callosum agenesis, retardation, adducted thumbs, spastic paraparesis, and hydrocephalus. *J. Child Neurol.* 15:239–43

Sztriha L, Nork M. 2000. Bilateral frontoparietal polymicrogyria and epilepsy. *Pediatr. Neurol.* 22:240–43

Takashima S, Houdou S, Kamei J, Hasegawa M, Mito T, et al. 1992. Neuropathology of peroxisomal disorders; Zellweger syndrome and neonatal adrenoleukodystrophy. *No To Hattatsu* 24:186–93

Taylor KR, Holzer AK, Bazan JF, Walsh CA, Gleeson JG. 2000. Patient mutations in Doublecortin define a repeated tubulin-binding domain. *J. Biol. Chem.* 275:34442–50

Toda T, Kobayashi K, Kondo-Iida E, Sasaki J, Nakamura Y. 2000. The Fukuyama congenital muscular dystrophy story. *Neuromuscul. Disord.* 10:153–59

Trommsdorff M, Gotthardt M, Hiesberger T, Shelton J, Stockinger W, et al. 1999. Reeler/disabled-like disruption of neuronal migration in knockout mice lacking the VLDL receptor and ApoE receptor 2. *Cell* 97:689–701

Ware ML, Fox JW, Gonzalez JL, Davis NM, Lambert de Rouvroit C, et al. 1997. Aberrant splicing of a mouse disabled homolog, mdabl, in the scrambler mouse. *Neuron* 19:1–20

Wills Z, Marr L, Zinn K, Goodman CS, Van Vactor D. 1999. Profilin and the Abl tyrosine kinase are required for motor axon outgrowth in the Drosophila embryo. *Neuron* 22:291–99

Xiang X, Osmani AH, Osmani SA, Xin M, Morris R. 1995. *NudF*, a nuclear migration gene in *Aspergillus nidulans*, is similar to the human *LIS-1* gene required for neuronal migration. *Mol. Biol. Cell* 6:297–310

Xiong Y, Zhang H, Beach D. 1992. D type cyclins associate with multiple protein kinases and the DNA replication and repair factor PCNA. *Cell* 71:504–14

Zheng C, Heintz N, Hatten ME. 1996. CNS gene encoding astrotactin, which supports neuronal migration along glial fibers. *Science* 272:417–19

Zukerberg LR, Patrick GN, Nikolic M, Humbert S, Wu CL, et al. 2000. Cables links Cdk5 and cAbl and facilitates Cdk5 tyrosine phosphorylation, kinase upregulation, and neurite outgrowth. *Neuron* 26:633–46

Annu. Rev. Neurosci. 2001. 24:1071–89

MORPHOLOGICAL CHANGES IN DENDRITIC SPINES ASSOCIATED WITH LONG-TERM SYNAPTIC PLASTICITY

Rafael Yuste[1] and Tobias Bonhoeffer[2]

[1]Department of Biological Sciences, Columbia University, New York, NY 10027;
e-mail: rmy5@columbia.edu
[2]Max Planck Institut für Neurobiologie, 82152 München-Martinsried, Germany;
e-mail: tobias.bonhoeffer@neuro.mpg.de

Key Words   LTP, hippocampus, learning, imaging, memory

■ **Abstract**   Dendritic spines are morphological specializations that receive synaptic inputs and compartmentalize calcium. In spite of a long history of research, the specific function of spines is still not well understood. Here we review the current status of the relation between morphological changes in spines and synaptic plasticity. Since Cajal and Tanzi proposed that changes in the structure of the brain might occur as a consequence of experience, the search for the morphological correlates of learning has constituted one of the central questions in neuroscience. Although there are scores of studies that encompass this wide field in many species, in this review we focus on experimental work that has analyzed the morphological consequences of hippocampal long-term potentiation (LTP) in rodents. Over the past two decades many studies have demonstrated changes in the morphology of spines after LTP, such as enlargements of the spine head and shortenings of the spine neck. Biophysically, these changes translate into an increase in the synaptic current injected at the spine, as well as shortening of the time constant for calcium compartmentalization. In addition, recent online studies using time-lapse imaging have reported increased spinogenesis. The currently available data show a strong correlation between synaptic plasticity and morphological changes in spines, although at the same time, there is no evidence that these morphological changes are necessary or sufficient for the induction or maintenance of LTP. Still, they highlight once more how form and function go hand in hand in the central nervous system.

INTRODUCTION

The search for the mechanisms underlying learning in the brain spans more than a century. In the earliest form of this debate Cajal speculated that learning required novel neuronal growth (Ramón y Cajal 1893). At the same time, recapitulating an earlier suggestion from Spencer (1862), Tanzi (1893) argued that changes in existing connections might underlie information storage in the brain. Later, Hebb incorporated both ideas into a postulate that suggested that alteration in synaptic strength, as well as formation of novel synapses, were responsible for memory storage (Hebb 1949, pp. 335).

In 1973 it was discovered that brief tetanic stimulation produced a long lasting form of synaptic plasticity, long-term potentiation (LTP), that can last for hours or days in the mammalian hippocampus (Bliss & Lømo 1973). Just before that the involvement of the hippocampal formation in memory was established by clinical data indicating that lesions of this structure in humans produce anterograde amnesia (Milner 1966). Since then, many laboratories have been studying LTP as a cellular model for information storage in the brain. Although the relation of LTP to learning is not universally accepted (e.g. see Mayford et al 1996 in favor and Zamanillo et al 1999 against it), LTP is a widely used and helpful paradigm for long-term synaptic plasticity in a central synapse. LTP furthermore nicely relates to neural network theories of brain function because it implements a local learning rule, an essential element for associative neuronal networks (Hopfield 1982), and one that ensures many of the computational features that make neural networks so attractive.

Although there have been reviews of the role of morphological changes of spines in LTP (e.g. Calverley & Jones 1990, Wallace et al 1991), a recent flurry of work, much of it using novel imaging techniques and time-lapse recordings, has added important information. In addition, there is now increasing biophysical evidence of the relation between morphological and functional parameters of the spine. We now know that the volume of the spine-head is directly proportional to the number of postsynaptic receptors (Nusser et al 1998) and to the presynaptic number of docked vesicles (Schikorski & Stevens 1999). Also, the small size of the spine head determines fast diffusional equilibration for calcium, whereas the length of the spine neck controls the time constant of calcium extrusion in spines (Majewska et al 2000a, Yuste et al 2000). This implies that the morphology of a spine directly reflects its function, and this makes it particularly relevant to investigate morphological changes of spines during synaptic plasticity.

MORPHOLOGICAL PLASTICITY OF DENDRITIC SPINES

Besides LTP, there are many different experimental or behavioral conditions that have been associated with changes in spine morphology that we will not be able to cover here in detail. Nevertheless, in the following we briefly mention some of these studies to help the reader approach this large body of literature. Many, but not all, of these studies indicate that increases of neural activity produce more spines. For example, light deprivation in mice causes a reversible reduction in the number of spines (Globus & Scheibel 1967; Valverde 1967, 1971). Similarly, increases in spine density occur after visual stimulation (Parnavelas et al 1973). Other environmental manipulations, such as rearing animals in complex environments, also alter spine morphology (Greenough & Volkmar 1973), so do social isolation (Connor &

Diamond 1982) and reportedly even space flight (Belichenko & Krasnov 1991). A reduction in the size of the spine has also been observed after the first orientation flight in honeybees (Brandon & Coss 1982). In birds, spine morphological plasticity is observed during postnatal development (Rausch & Scheich 1982), imprinting with light (Bradley & Horn 1979), and in learning tasks involving pecking (Patel et al 1988). Finally, in a fascinating study, squirrels have been documented to lose 40% of their spines during hibernation and to recover them in a few hours after arousal from hibernation (Popov & Bocharova 1992, Popov et al 1992).

Changes in spine form and number have also been observed in vitro. In dissociated cultures (Boyer et al 1998, Papa et al 1995) as well as in brain slices (Kirov et al 1999), pyramidal neurons have increased spine densities compared to those found in vivo. Pharmacological manipulations also influence spine morphology and number. Stimulation of alpha-amino-3-hydroxy-5-methyl-4-isoxazolepropionic acid (AMPA) receptors is needed for the maintenance of spines, whereas blocking AMPA receptors reduces the number of spines (McKinney et al 1999). Alternatively, synaptic blockade with high Mg^{2+} and low Ca^{2+} increases both spine number and size (Kirov & Harris 1999). This is also found in cultured neurons disinhibited with bicuculline (Papa & Segal 1996) and after stimulation of internal calcium release (Korkotian & Segal 1999).

Early on it was noticed that spines are initially overproduced and later reduced in number during normal development and aging (Ramón y Cajal 1904). More recently it turned out that even during the estrous cycle of some mammals, large numbers of spines are produced in the hippocampus and later eliminated in substantial numbers (Woolley et al 1990). Finally, many diseases, such as dementia (Mehraein et al 1975), mental retardation (Purpura 1974), Down syndrome (Marín-Padilla 1972), irradiation (Brizzee et al 1980), malnutrition (Salas 1980), fragile X syndrome (Wisniewski et al 1991), and epilepsy (Multani et al 1994), can produce abnormalities in spine morphologies.

RAPID SPINE MOTILITY

Adding to the rich field of work that has documented spine morphological plasticity in different systems, it has been recently discovered that spines show considerable motility (Fischer et al 1998). This motility had been proposed by Blomberg et al (1977) and Crick (1982) and has now been documented in dissociated cultures (Fischer et al 1998), in brain slices (Dunaevsky et al 1999), and in vivo (Lendvai et al 2000). Rapid spine motility can produce changes in the morphological classification of spines (Parnass et al 2000), it is developmentally regulated (Dunaevsky et al 1999), and it is more pronounced during the critical period (Dunaevsky et al 1999, Lendvai et al 2000). The mechanism responsible for this rapid motility is actin-dependent (Fischer et al 1998, Dunaevsky et al 1999), is interfered with by volatile anesthetics (Kaech et al 1999), and could involve the Rho family of small GTPases (Tashiro et al 2000). Clearly, spine motility occurs throughout the central nervous system (Dunaevsky et al 1999), but at present its function is still a mystery (but see Majewska et al 2000b).

EFFECTS OF LONG-TERM POTENTIATION ON SPINE MORPHOLOGY

As indicated above, in this review we focus only on LTP or synaptic stimulation experiments from mammalian hippocampus. We do not intend to provide a comprehensive overview but instead concentrate only on a few studies.

Long-Term Potentiation in the Dentate Gyrus

Fivkova Studies: Swelling of Granule Cell Spines After Tetanization Although long-lasting potentiation of the myotatic reflex in the partially deaffarented mammalian spinal cord was reported by Eccles & McIntyre (1953) it was Lømo and Bliss's description of LTP in the monosynaptic connections between the perforant pathway and the granule cells in the hippocampus that triggered enormous interest, up to present times, in the investigation of the morphological consequences of long-term synaptic enhancement in the central nervous system (Lømo 1970, Bliss & Lømo 1973).

Some of the earliest studies of the effect of LTP on the morphology of dendritic spines were carried out by Van Harrefeld & Fifkova (1975). They used an experimental protocol similar to that pioneered by Bliss and Lømo. The authors set out to test the hypothesis that LTP involved swelling of dendritic spines and reduction in neck length to produce increased synaptic activation by the input. By using a rapid freeze-substitution electron microscopy protocol that largely preserves spine morphology, the authors compared granule cell spines in the distal region of the dentate gyrus (the region that receives perforant pathway inputs) with those in the proximal region, devoid of such inputs, in both stimulated and control mice. By comparing distal and proximal spines from the same dentate gyrus, the authors controlled for potential artifacts and variability introduced by the experimental techniques. Taking samples from a narrow (10–15 μm) layer of freeze-substitute material, the authors used paper cutouts to measure the mean projected area of 100 spines. They found that in control animals distal spines had on average 13% larger areas than proximal spines, but that difference increased to 53% in stimulated animals. This effect was present as early as 2 minutes and lasted at least 60 minutes after stimulation, indicating that, similar to the synaptic enhancement produced by LTP, the increase in volume was immediately produced by the tetanization and was long lasting. The authors speculated that the increase in volume was produced by swelling due to the uptake of electrolytes, a mechanism previously proposed to underlie memory formation (Diamond et al 1970, Rall 1970). To further explore the relation between synaptic stimulation and spine swelling, in a second study the same authors carried out a temporal characterization of the effect and extended their observation time to up to 23 hours (Fifkova & Van Harrefeld 1977). Again, they observed an increase in the volume of the spines from stimulated animals compared with controls. This increase was significant at all times tested (2 minutes to 23 hours), although it was largest (39%) 10–60 minutes after the stimulation. In the same experiments no systematic differences were observed in the proximal spines between stimulated and control animals.

The effect of tetanic stimulation on the size of the spine neck was specifically examined by Fivkova & Anderson (1981). Again, this study found an increase in the volume of the spine head in distal spines after tetanic stimulation. In addition, they found major increases (up to 42%) in the width of the spine neck, as well as shortenings (as much as 31%) of its length. These changes were again specific to distal spines and were present as late as 90 minutes after the stimulation.

Taken together, these studies paint a consistent picture of the effects of tetanizing the perforant pathway: The spine heads in the distal region of the dentate gyrus become larger, while the spine necks become shorter and wider. Although these studies did not confirm electrophysiologically the LTP presumably produced by the stimulation protocols and whether the exact spines analyzed were the ones that had been tetanized, the anatomical changes were long lasting and restricted to distal spines receiving perforant path inputs, just as one would predict from the properties of LTP. Fivkova later proposed that the enlargement of the spine head was in fact a result of an increase in the size of the synapse and that its ultimate stabilization might be mediated by the mobilization of actin caused by local increases in $[Ca^{2+}]_i$ (Fifkova 1985).

Desmond & Levy Studies: Further Evidence for Spine Modifications After Long-Term Potentiation   Conclusions similar to the Fifkova studies were drawn by Desmond and Levy from a series of combined electrophysiological/morphological experiments (Desmond & Levy 1983; 1986a,b; 1988). They used the rat perforant pathway/dentate gyrus as the model system and combined field potential recordings and quantitative electron microscopy (EM) to search for changes in size and number of synapses after LTP. In these blind studies, two stimulation protocols were used: either a short burst of stimuli (massed conditioning) or a more distributed protocol with the same number of electrical shocks but spread out over time (spaced conditioning). The criterion for the inclusions of the animals in the study was a successful change of at least 50% in the initial excitatory postsynaptic potential (EPSP) slope, a relatively strict cutoff confirming the effectiveness of the physiological manipulation.

Desmond & Levy found anatomical changes following the two forms of conditioning, with up to 48% increases in the density in concave spines and up to 23% decreases in the number of nonconcave (simple and ellipsoid) spine profiles (1983, 1986a). Concave spine profiles, in their nomenclature, correspond to the population of large spines that have spinules or U-shaped profiles. These changes were more pronounced in the middle third of the molecular layers, the dentate gyrus region that was stimulated by their more medial electrode placement in the perforant pathway. Although Desmond & Levy initially observed increases in the shaft synapses (1983), in their later studies they failed to detect significant changes in the total density of shaft or spine synapses (1986a).

These observations were further extended in a quantitative study of the size of the postsynaptic densities (PSD), in which the total PSD surface area per unit volume in concave spines increased significantly, whereas the PSD area of the nonconcave spines decreased (Desmond & Levy 1986b). In addition, the mean PSD length increased significantly across all spine profiles in stimulated animals and persisted for at least 60 minutes. In a later study the authors found concomitant changes in the presynaptic terminals (Desmond & Levy 1988).

These data showed increases in spine size after LTP, but at the same time other groups provided evidence for increases in spine number (see below). This prompted Desmond & Levy to consider the possibility of an increase in the number of synapses in the dentate gyrus (1990). The authors used two morphological markers of synaptogenesis, the presence of polyribosomes and multiple synaptic contacts, and found statistically significant decreases in their incidence after LTP, arguing for a modification of existing synapses, rather than de novo synaptogenesis. This result was extended to somatic ribosomes in a later publication from the same group, indicating that LTP is associated with increased protein synthesis (Wenzel et al 1993).

In conclusion, the Desmond & Levy studies—like the ones from Fifkova—present a scenario where there is an interconversion of spine shapes during LTP, without addition of new synapses. Smaller spines become larger spines, and this is mirrored by increases in the size of the synaptic surface area. In agreement with Tanzi's early suggestion, these data argue for no new synapses but for modification of existing ones.

CA1 LTP

Effects of Long-Term Potentiation on Synapse Localization   Fifkova et al did not establish whether the stimulation protocols used indeed produced long-lasting synaptic plasticity. Therefore, Lynch's laboratory early on undertook several studies combining electrophysiological recordings with EM, examining the morphological effects of synaptic plasticity (Lee et al 1979a,b; Lee et al 1980). These experiments were carried out stimulating the Schaffer collateral pathway in vivo (Lee et al 1979b, 1980) or in vitro (Lee et al 1979a) and used blind analysis of ultrathin electron microscopic sections of the CA1 region of the hippocampus to compare the effects of low-frequency stimulation, which did not produce any long-lasting potentiation, with high-frequency stimulation, which produced LTP. After in vivo as well as in vitro stimulation, the authors found 33–50% more synaptic contacts on dendritic shafts in high-frequency stimulated animals. No significant differences were found in the number of synapses on spines, length of the spine PSD, area of spines, width of spine necks, or length of the PSD on the dendritic shafts. Nevertheless, in the potentiated group the authors found a reduction in the variance of several morphological features of the spines. These studies did not confirm the swelling of spines or shortening of spine necks described by Fifkova and coworkers (Van Harrefeld & Fifkova 1975, Fifkova & Van Harrefeld 1977; Fivkova & Anderson 1981).

The authors speculated that LTP might be associated with increased innervation of dendritic shafts. This increase in shaft synapses could have been produced by retraction of spines or by new formation of shaft synapses, although they recognized that most shaft synapses were located on dendrites from relatively aspiny cells, i.e. inhibitory interneurons. Therefore, although these studies have the strength of electrophysiological monitoring of the effects of the stimulation, the conclusion drawn from their EM data are difficult to interpret.

A similar study was carried out by Chang & Greenough (1984), who also used hippocampal brain slices and ultrastructural analysis to compare the effects of different protocols of stimulation, including high-frequency stimulation that produced LTP, high-frequency stimulation that did not produce LTP, and low-frequency stimulation and synaptic inactivation by high-Mg^{2+} and low Ca^{2+} ACSF. After LTP the authors found significant increases in the number of sessile (stubby) and shaft synapses, whereas no detectable differences were found in the total number of spines or perforated synapses. Also, statistical reductions were observed in average spine perimeter, contact length, and the percentage of "cup"-shaped synapses, although no significant changes were encountered in spine head area, bouton area, PSD length, spine neck width, or length and number of presynaptic vesicles. These authors again concluded that LTP produced an increase in the synaptic innervation of interneurons, although they raised the possibility that it also induced the transition from shaft to spine synapses in CA1 pyramidal cells. These two effects would explain the increased number of synapses on dendritic shafts and stubby spines. They also proposed that LTP produced a change in spine shape, in which typical cup spines are transformed into a flatter spine profile with reduced spine perimeter and contact length.

INCREASED SPINOGENESIS AFTER LONG-TERM POTENTIATION: Andersen et al Studies

In all the studies discussed so far the authors reported rather subtle changes in spine shape or the geometry of the postsynaptic density. Andersen and coworkers were the first to report unexpectedly large increases in the number of spines as well as rather dramatic changes in spine shape. The initial studies (Andersen et al 1987a,b; Trommald et al 1990) were all done by inducing LTP in the dentate gyrus. Serial reconstructions of electron-microscopic material showed an increase in spine number of up to 50% as well as changes in the diameter of the spine neck and an increase in so-called bifurcated spines. These are structures hypothesized to result from a splitting of a single synapse, first into a larger and supposedly more effective "perforated synapse" (Peters & Kaiserman-Abramof 1969) and from there into two synapses on different spine heads residing on a single spine trunk. Because it had been proposed that perforated synapses would be particularly effective synapses, the observation of increased numbers of such structures and bifurcated spines naturally explained an increased synaptic efficacy.

A later study by the same authors (Trommald et al 1996) elaborated on these initial findings and found that the strong effect of an increased spine density and more bifurcated spines indeed held up. The overall density of spines increased by ~30% while the number of bifurcated spines rose by a factor of more than three. Whereas this part of the earlier studies was corroborated, the changes in spine dimensions could not be confirmed. The authors reported that no statistically significant changes in the dimensions of the spines could be observed (Trommald et al 1996). This is not entirely unexpected, considering the enormous variation of shape and size in the dendritic spines that the authors report. Moreover, this result does not mean that such changes do not occur; it merely indicates that this (statistical) approach does not yield a significant result. The authors indicated that a method allowing one to observe single spines in living tissue over time would be much better suited to bring out potential morphological changes of this sort. They also noted that their results on the large increase in spine number is at odds with earlier studies that reported no increase in spines density (Desmond & Levy 1983) or even a decrease (Geinisman et al 1991). However, they have no obvious explanation for these differences other than variations in the experimental and stereological detail of the studies.

The same authors later used confocal microscopy instead of electron microscopy to further investigate morphological correlates of synaptic plasticity (Moser et al 1994). The reason for using confocal microscopy was presumably that it is much less tedious than serial EM. This in turn allows one to analyze many more spines and therefore allows for better statistical sampling to bring out small changes. In these new studies plastic changes were not induced by electrical stimulation but by an altered sensory environment, and the morphological changes were assessed in the CA1 region of the hippocampus. Rats were housed either alone or in pairs in plastic cages, and the experimental group was allowed to explore an environment with multiple platforms containing interesting items such as wooden blocks, branches, paper bags, and leaves. The animals that had the exploratory experience performed better in behavioral tasks like the Morris watermaze, but they also showed a small but significant increase in spine density, whereas all other measured parameters were unaffected. Interestingly, the enhanced spine density only occurred on basal dendrites (Moser et al 1997) and not on apical dendrites, which were also evaluated.

These studies are different from all the others discussed in this review in that the procedure applied is not necessarily expected to cause "storage" of new information; it is rather a skill that is acquired. Because the induction of LTP is often thought to electrically mimic information storage in the brain, the aforementioned studies only indirectly relate to morphological changes observed with LTP. They suggest, rather, that acquiring the capability to learn better (and not the storage of information per se) can result in morphological changes in the form of newly emerging spines.

UNALTERED SPINE SHAPES AND NUMBERS: Harris et al Studies

A number of years later Harris and co-workers (Sorra & Harris 1998) tested whether tetanus-induced LTP would also result in an enhanced number of synapses in the CA1 region. Again they used an unbiased volume sampling procedure to look for possible changes in spine density or in the geometrical parameters of preexisting spines. In line with other investigators (e.g. Lee et al 1980, Chang & Greenough 1984), they found no changes in absolute spine number. This contrasts with the findings by Andersen et al (Andersen et al 1987a,b; Trommald et al 1990), who observed that at least in the dentate gyrus in vivo such changes seem to occur. Two major differences between these studies could contribute to this difference: (*a*) The brain area investigated was different in the two studies (dentate gyrus versus CA1 region) and (*b*) the Sorra & Harris study was performed in slices, whereas the others were performed in vivo. In this respect a follow-up study (Kirov et al 1999) is interesting and worrisome at the same time: These authors found that hippocampal slices showed, one hour after the slicing procedure, roughly two times more dendritic spines than "native" perfusion-fixed hippocampus. A similar result had been reported in somatic spines from granule cells from the dentate gyrus after slicing (Wenzel et al 1994). Therefore, a possible explanation for the lack of changes in the CA1 region of slices is that so much spinogenesis had occurred after the initial preparation that additional changes were either not possible anymore or "buried" within the noise.

Interestingly, Andersen's group (Andersen & Soleng 1998) verified that the enriched environment causes spinogenesis, and they also replicated Sorra & Harris (1998) findings showing that after 4 hours of LTP no net increase in spine number is observed in CA1 of hippocampal slices.

From all these studies it is clear (and in many of them it was explicitly noted) that online observation of potential morphological modifications would be a major technical improvement in the ability to detect potential changes. It would obviate the requirement to do statistics on large samples of experimental and control tissue and at the same time use temporal correlation of observed morphological changes with the experimental manipulations as a sensitive measure to detect causality. In fact, it is clear that certain kinds of morphological changes are only detectable by online observation: For example, statistical methods have no way of excluding that an overall null change of spine density is actually caused by some new spines forming as a consequence of LTP and that this net gain of spines in the potentiated region is compensated for by the loss of spines elsewhere.

THE FIRST LONGITUDINAL EXPERIMENTS:
The Hosokawa, Rusakov, Bliss, and Fine Study

Another important drawback of most studies reported so far is that they all relied on comparison of different populations of cells from different animals in fixed tissue, thereby creating a number of potential artifacts. Therefore, time-lapse imaging of living tissue was clearly an important step forward in investigating morphological changes after LTP.

Hosokawa et al (1995) were the first to move into this experimentally challenging direction. They used confocal microscopy of hippocampal slices in which individual CA1 pyramidal cells were stained by a specially developed "DiI-microdrop technique." Synaptic potentiation was induced by "chemical LTP," produced by the application of a superfusion solution containing elevated Ca^{2+}, reduced Mg^{2+}, and tetraethylammonium. The reason for choosing this kind of LTP induction was to cause potentiation in as many synapses as possible. Using this experimental approach, the authors observed that a subpopulation of (small) spines extended, and they further reported that there was an increased range of angular displacement of spines in the potentiated tissue. All other parameters showed no significant changes. In particular, the appearance of completely new spines was a rare (and statistically insignificant) event.

In principle, the resolution of (one-photon) confocal microscopy should be able to resolve these events. However, this technique is fraught with the problem of bleaching and photodynamic damage so that in practice, especially in longitudinal studies in living tissue, the attainable resolution is limited and just barely sufficient to detect and quantify spines under physiologically acceptable conditions. Another problem is that confocal microscopy can, due to the limited penetration of visible light, only detect superficial structures. Therefore, all synapses observed had to be within 75 μm (sometimes as close as 25 μm) of the surface, a depth of the slice at which there might be damage from the preparation of the tissue.

IMAGING SPINOGENESIS "ONLINE": The 1999 Studies

The problems of photodynamic damage and depth of penetration were largely overcome by the development of two-photon microscopy (Denk et al 1990). This method provides an ideal tool to further investigate morphological changes of spines that might be associated with LTP. In 1999 two independent two-photon studies on this question were published. In the first, Maletic-Savatic et al (1999) used organotypic slices and local stimulation to address the question of spinogenesis in this preparation. Individual cells were visualized by infection of the slices with sindbisvirus-eGFP constructs. These authors observed that a strong tetanus, a stimulus that would normally induce LTP, led to local outgrowth of dendritic processes. The newly formed protrusions, however, were more akin to filopodia in that they were often >4 μm long. In some instances, however, these filopodia later turned into spine-like structures. Importantly, the emergence of new protrusions could be blocked by agents, such as APV, that interfere with LTP.

Shortly afterwards another study showed that new spines can be formed in hippocampal tissue even with much more moderate stimulation than the rather strong tetanus. This study (Engert & Bonhoeffer 1999) used a different strategy to pinpoint the locations where potential morphological changes could occur: In a relatively thin, different type of organotypic slice culture (Gähwiler 1981), all synaptic transmission was blocked by applying a medium with a relatively high concentration of Cd^{2+} ions and low Ca^{2+}. This transmission blockade was then relieved only very locally (in an area of $\sim$30 μm diameter) with a local superfusion system applying regular recording medium. This enabled the experimenters to know where the observed changes in synaptic strength must have taken place and therefore allowed them to scrutinize this area for morphological changes with a two-photon microscope while simultaneously monitoring the effect of the synaptic stimulation electrophysiologically. In every instance of successful synaptic enhancement, new spines were generated, whereas practically no new spines appeared when the enhancement was not successful (or blocked). This study therefore proved a strong correlation between the functional enhancement of synapses and the generation of new spines. These observations do not, of course, preclude that other more subtle changes in the shape of spines might also have occurred, but they clearly show that spinogenesis is a correlate of successful enhancement.

Later the same year an EM study was published that once more addressed the question of spinogenesis after LTP, but from a different angle. This study (Toni et al 1999) was an extension of earlier work from the same laboratory (Buchs & Muller 1996) using a technique to select for the stimulated synapses and therefore the location in which morphological changes were expected to occur. The assumption they used is that synapses that have just been subjected to a strong stimulus show, in electron-microscopic images, an accumulation of a calcium precipitate in the postsynaptic spine. This precipitate is visible postsynaptically and therefore "earmarks" spines activated by the electrical stimulation. Toni et al (1999) then scrutinized these spines for morphological changes and found that in many cases after stimulation there were pairs or triplets of (calcium precipitate– marked) spines making contact with the same presynaptic terminal. This then was taken as evidence that under conditions of LTP-inducing stimulation new spines were generated.

Although these studies have now established that there is a clear-cut correlation between the formation of new spines and successful enhancement of synapses, these results have to be interpreted with caution in that they show a correlation, but they do not prove causality in the sense that they do not prove that the newly observed spines with their presumptive synapses actually contribute to the strength of the measured connection . The study by Toni et al (1999) shows that in most cases the new (second) spine contacts the stimulated axons and not other unstimulated ones. Therefore, the new spines could at least potentially contribute to enhanced transmission of the stimulated pathways. Yet, the experiments of Maletic-Savatic et al (1999) and Engert & Bonhoeffer (1999) actually prove that the emergence of new spines cannot be causal, at least for the early phase of the enhancement because the enhancement occurs within minutes, whereas new spines only appear after roughly 30 minutes. However, recent studies have made the interesting observation that interfering with the actin cytoskeleton, which among many other effects could presumably also inhibit the generation of new spines, can block LTP and even its early components (Kim & Lisman 1999, Krucker et al 2000).

Therefore, after two decades of searching for morphological correlates of LTP, it seems clear that such correlates exist. The role of these morphological changes, however, is still enigmatic. Direct contributions to synaptic strength as well as other functional roles, such as the generation of new potential contact sites or better calcium compartmentalization, are just a few of many possibilities.

CONCLUSION: A Time-Line for Morphological Changes Following Long-Term Potentiation

The changes in spine morphology after LTP reported in the studies reviewed here are varied and often even contradictory. In fact, basically every possible effect, be it upregulation or downregulation of a certain parameter, has by now been reported: After dentate gyrus LTP, larger spines with shorter necks were found in some studies (Fifkova & Anderson 1981), whereas others report that these parameters are unaffected (Trommald et al 1996). These studies in turn describe increases in spine number (Trommald et al 1996), but other studies are inconsistent with this (Desmond & Levy 1990). In CA1 LTP, similar discrepancies emerge: Some studies report no (or very small) changes in spine morphology or number (Hosokawa et al 1995), whereas others show considerable numbers of new spines (Engert & Bonhoeffer 1999, Toni et al 1999) or filopodia (Maletic-Savatic et al 1999). Moreover, a true assessment of the situation is further complicated by the fact that negative findings are normally not published. Therefore, studies failing to show a relation between LTP and changes in spine morphology are most likely missing from the literature.

There are many reasons for this confusing state of affairs. One is that the changes that occur after conventional, tetanus-induced LTP can only be expected in a subset of spines, and therefore it becomes critical how the potential loci for changes are identified (Engert & Bonhoeffer 1999, Toni et al 1999). Another option is to activate many more synapses by inducing LTP chemically (Hosokawa et al 1995) or even by environmental manipulations (Andersen & Soleng 1998), procedures so different that it is perhaps not surprising that different results have been obtained. Furthermore, the electron-microscopic studies, although best suited to study the minute changes occurring on the level of spines and synapses, have— more than the other studies—the problem that they heavily rely on statistics. In this respect the advent of two-photon microscopic techniques has brought a substantial improvement in that morphological changes can be observed online, making them not only much more vivid for the human observer but also introducing the element of temporal correlation, which adds considerably to the confidence of a result. It is our feeling then that for this reason some of the recent time-lapse imaging experiments have provided more widely accepted evidence that morphological changes do occur after LTP. The functional consequences of these changes remain, however, the subject of speculation.

We therefore propose a possible sequence of morphological events that accompany LTP. Our proposal, although it tries to integrate as many previous studies as possible, for the reasons stated above, emerges largely from the consideration of more recent data. In order to reflect on the functional consequences of these data we also try to incorporate the biophysical consequences of the observed morphological changes.

Potentially, then, the sequence of events might be as follows (see Figure 1):

1. Initially, within minutes after potentiation, functional changes that are not (or at least not easily) detectable morphologically occur at the potentiated synapse. Such changes include modifications in postsynaptic receptor composition as well as alterations in synaptic release properties as postulated for the early phase of LTP (e.g. Malinow & Tsien 1990, Stevens & Wang 1994, Malenka & Nicoll 1999).

2. Approximately 30 minutes after induction, the first morphological changes become detectable with light or electron microscopy. Spine heads from stimulated spines become larger (Fifkova & Van Harrefeld 1977,

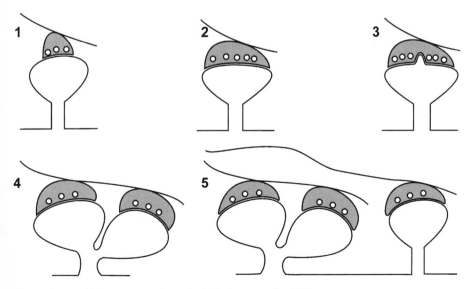

Figure 1 Possible sequence of morphological events after LTP.

Desmond & Levy 1986b) and there is a concomitant increase in synaptic area (Desmond & Levy 1988). Because of a direct relationship between spine volume and synaptic strength, reflected in an increase in the number of postsynaptic receptor and presynaptic docked vesicles (Schikorski & Stevens 1999), this increase could therefore be—at least in part—responsible for the late phase of LTP, as it would provide an enduring structural modification tied to a change in synaptic strength. Besides, enlargement of the spine heads and the widening and shortening of spine necks (Fifkova & Anderson 1981) would have immediate functional consequences bringing about major changes in calcium compartmentalization, thereby influencing the rules for synaptic plasticity (Majewska et al 2000a,b).

3. Synapses that were already large break apart, forming perforated synapses (Peters & Kaiserman-Abramof 1969, Toni et al 1999).

4. Bifurcating spines (Trommald et al 1990) and ultimately even closely associated pairs of spines emanating from one dendrite and touching the same presynaptic element occur (Toni et al 1999). These morphologies could implement the increase in independent synaptic release sites found in some studies of late-phase LTP (Bolshakov & Siegelbaum 1995, Bolshakov et al 1997).

5. Finally, new spines can also form in the vicinity of the activated spines (Engert & Bonhoeffer 1999, Toni et al 1999).

We therefore propose that there is a continuum of morphological events that can occur after the induction of LTP. They range from slight enlargements of synaptic surfaces or spines to the generation of completely new structures. Clearly, in many cases not all steps of the process will be taken, but morphological changes of lesser extent are probably the end-result of synaptic potentiation. In a way, Cajal and Tanzi were both right because we can envision LTP-related changes that require novel neuronal growth as well as those that only involve functional changes in existing connections.

It also is worthwhile to emphasize that although the correlation between LTP and some morphological events such as spinogenesis is well documented (Engert & Bonhoeffer 1999, Toni et al 1999), it is important to take it as such: correlation not necessarily proving causality. Whereas LTP can produce morphological re-arrangements, those changes may or may not contribute to the potentiation. It might seem the simplest scenario that new spines contribute to the enhanced efficacy of the observed synapses, yet other possible roles for the new spines are viable, and perhaps even more likely. It is, for instance, possible that the new spines are created as dendritic sites for future plasticity. One might hypothesize that they only contain silent, i.e. purely NMDA, synapses (Isaac et al 1995, Liao et al 1995).

Thus, over the past few years considerable progress has been made in pinpointing morphological changes that correlate with changes in synaptic strength, but there is still ample room for discovery to understand the true function of these morphological changes.

ACKNOWLEDGMENTS

RY is funded by the National Eye Institute (EY 111787 and 13237) and the NINDS (NSHD 40726). TB is funded by the Max-Planck Gesellschaft. We thank Tim Bliss, Nancy Desmond, Alan Fine, Chip Levy, and Steve Siegelbaum for comments and Chuck Stevens for discussions.

Visit the Annual Reviews home page at www.AnnualReviews.org

LITERATURE CITED

Andersen P, Blackstad T, Hulleberg G, Trommald M, Vaaland JL. 1987a. Dimensions of dendritic spines of rat dentate granule cells during long-term potentiation. *J. Physiol.* 390:P264

Andersen P, Blackstad T, Hulleberg G, et al. 1987b. Changes in spine morphology associated with LTP in rat dentate granule cells. *Proc. Physiol. Soc. PC* 50:P288

Andersen P, Soleng A. 1998. Long-term potentiation and spatial training are both associated with the generation of new excitatory synapses. *Brain Res. Rev.* 26:353–59

Belichenko PV, Krasnov I. 1991. The dendritic spines of the pyramidal neurons in layer V of the rat sensorimotor cortex following a 14-day space flight. *Biull. Eksp. Biol. Med.* 112:541–42. (In Russian)

Bliss TVP, Lømo T. 1973. Long-lasting potentiation of synaptic transmission in the dentate area of the anaesthetized rabbit following stimulation of the perforant path. *J. Physiol.* 232:331–56

Blomberg F, Cohen R, Siekevitz P. 1977. The structure of postsynaptic densities isolated from dog cerebral cortex. II. Characterization and arrangement of some of the major proteins within the structure. *J. Cell Biol.* 86:831–45

Bolshakov VY, Golan H, Kandel ER, Siegelbaum SA. 1997. Recruitment of new sites of synaptic transmission during the cAMP-

dependent late phase of LTP at CA3-CA1 synapses in the hippocampus. *Neuron* 19:635–51

Bolshakov VY, Siegelbaum SA. 1995. Regulation of hippocampal transmitter release during development and Long-Term Potentiation. *Science* 269:1730–34

Boyer C, Schikorski T, Stevens CF. 1998. Comparison of hippocampal dendritic spines in culture and in brain. *J. Neurosci.* 18:5294–300

Bradley P, Horn G. 1979. Neuronal plasticity in the chick brain: morphological effects of visual experience on neurones in hyperstriatum accessorium. *Brain Res.* 162:148–53

Brandon JG, Coss RG. 1982. Rapid dendritic spine stem shortening during one-trial learning: the honeybee's first orientation flight. *Brain Res.* 252:51–61

Brizzee KR, Ordy JM, Kaack MB, Beavers T. 1980. Effect of prenatal ionizing radiation on the visual cortex and hippocampus of newborn squirrel monkeys. *J. Neuropathol. Exp. Neurol.* 39:523–40

Buchs PA, Muller D. 1996. Induction of long-term potentiation is associated with major ultrastructural changes of activated synapses. *Proc. Natl. Acad. Sci. USA* 93:8040–45

Calverley RKS, Jones DG. 1990. Contributions of dendritic spines and perforated synapses to synaptic plasticity. *Brain Res. Rev.* 15:215–49

Chang FLF, Greenough WT. 1984. Transient

and enduring morphological correlates of synaptic activity and efficacy change in the rat hippocampal slice. *Brain Res.* 309:35–46

Connor JR, Diamond MC. 1982. A comparison of dendritic spine number and type on pyramidal neurons of the visual cortex of old adult rats from social or isolated environments. *J. Comp. Neurol.* 210:99–106

Crick F. 1982. Do spines twitch? *Trends Neurosci.* 5:44–46

Denk W, Strickler JH, Webb WW. 1990. Two-photon laser scanning fluorescence microscopy. *Science* 248:73–76

Desmond NL, Levy WB. 1983. Synaptic associative potentiation/depression: an ultrastructural study in the hippocampus. *Brain Res.* 265:21–30

Desmond NL, Levy WB. 1986a. Changes in the numerical density of synaptic contacts with long-term potentiation in the hippocampal dentate gyrus. *J. Comp. Neurol.* 253:466–75

Desmond NL, Levy WB. 1986b. Changes in the postsynaptic density with long-term potentiation in the dentate gyrus. *J. Comp. Neurol.* 253:476–82

Desmond NL, Levy WB. 1988. Synaptic interface surface area increases with long-term potentiation in the hippocampal dentate gyrus. *Brain Res.* 453:308–14

Desmond NL, Levy WB. 1990. Morphological correlates of long-term potentiation imply the modification of existing synapses, not synpatogenesis, in the hippocampal dentate gyrus. *Synapse* 5:139–43

Diamond J, Gray EG, Yasargil GM. 1970. The function of dendritic spines: a hypothesis See Andersen & Jansen 1970, pp. 213–22

Dunaevsky A, Tashiro A, Majewska A, Mason CA, Yuste R. 1999. Developmental regulation of spine motility in mammalian CNS. *Proc. Natl. Acad. Sci. USA* 96:13438–43

Eccles JC, McIntyre AK. 1953. The effects of disuse and of activity on mammalian spinal reflexes. *J. Physiol.* 121:492–516

Engert F, Bonhoeffer T. 1999. Dendritic spine changes associated with hippocampal long-term synaptic plasticity. *Nature* 399:66–70

Fifkova E. 1985. A possible mechanism of morphometric change in dendritic spines induced by stimulation. *Cell. Mol. Neurobiol.* 5:47–63

Fifkova E, Anderson CL. 1981. Stimulation-induced changes in dimensions of stalks of dendritic spines in the dentate molecular layer. *Exp. Neurol.* 74:621–27

Fifkova E, Van Harrefeld A. 1977. Long-lasting morphological changes in dendritic spines of dentate granular cells following stimualtion of the entorhinal area. *J. Neurocytol.* 6:211–30

Fischer M, Kaech S, Knutti D, Matus A. 1998. Rapid actin-based plasticity in dendritic spine. *Neuron* 20:847–54

Gähwiler B. 1981. Organotypic monolayer cultures of nervous tissue. *J. Neurosci. Methods* 4:329–42

Geinisman Y, deToledo-Morrell L, Morrell F. 1991. Induction of long-term potentiation is associated with an increase in the number of axospinous synapses with segmented postsynaptic densities. *Brain Res.* 566:77–88

Globus A, Scheibel A. 1967. The effect of visual deprivation on cortical neurons: a Golgi study. *Exp. Neurol.* 19:331–45

Greenough WT, Volkmar FR. 1973. Pattern of dendritic branching in occipital cortex of rats reared in complex environments. *Exp. Neurol.* 40:491–504

Hebb DO. 1949. *The Organization of Behaviour.* New York: Wiley

Hopfield JJ. 1982. Neural networks and physical systems with emergent collective computational abilities. *Proc. Natl. Acad. Sci. USA* 79:2554–58

Hosokawa T, Rusakov DA, Bliss TVP, Fine A. 1995. Repeated confocal imaging of individual dendritic spines in the living hippocampal slice: evidence for changes in length and orientation associated with chemically induced LTP. *J. Neurosci.* 15:5560–73

Isaac JTR, Nicoll RA, Malenka RC. 1995. Evidence for silent synapses: implications for the expression of LTP. *Neuron* 15:427–34

Kaech S, Brinkhas H, Matus A. 1999. Volatile anesthetics block actin-based motility in dendritic spines. *Proc. Natl. Acad. Sci. USA* 96:10433–37

Kim CH, Lisman JE. 1999. A role of actin filament in synaptic transmission and long-term potentiation. *J. Neurosci.* 19:4314–24

Kirov SA, Harris KM. 1999. Dendrites are more spiny on mature hippocampal neurons when synapses are inactivated. *Nat. Neurosci.* 2:878–83

Kirov SA, Sorra KE, Harris KM. 1999. Slices have more synapses than perfusion-fixed hippocampus from both young and mature rats. *J. Neurosci.* 19:2876–86

Korkotian E, Segal M. 1999. Release of calcium from stores alters the morphology of dendritic spines in cultured hippocampal neurons. *Proc. Natl. Acad. Sci. USA* 96:12068–72

Krucker T, Siggins G, Halpain S. 2000. Dynamic actin filaments are required for stable long-term potentiation (LTP) in area CA1 of the hippocampus. *Proc. Natl. Acad. Sci. USA* 97:6856–61

Lee K, Oliver M, Schottler F, Creager R, Lynch G. 1979a. Ultrastructural effetcs of repetitive synaptic stimulation in the hippocampal slice preparation: a preliminary report. *Exp. Neurol.* 65:478–80

Lee KS, Schottler F, Oliver M, Lynch G. 1979b. Synaptic change associated with the induction of long-term potentiation. *Anat. Rec.* 193:601–2

Lee KS, Schottler F, Oliver M, Lynch G. 1980. Brief bursts of high-frequency stimulation produce two types of structural change in rat hippocampus. *J. Neurophysiol.* 44:247–58

Lendvai B, Stern E, Chen B, Svoboda K. 2000. Experience-dependent plasticity of dendritic spines in the developing rat barrel cortex *in vivo*. *Nature* 404:876–81

Liao D, Hessler N, Malinow R. 1995. Activation of postsynaptically silent synapses during pairing-induced LTP in CA1 region of hippocampal slice. *Nature* 375:400–4

Lømo T. 1970. Some properties of a cortical excitatory synapse. See Andersen & Jansen 1970, pp. 207–11

Majewska A, Brown E, Ross J, Yuste R. 2000a. Mechanisms of calcium decay kinetics in hippocampal spines: role of spine calcium pumps and calcium diffusion through the spine neck in biochemical compartmentalization. *J. Neurosci.* 20:1722–34

Majewska A, Tashiro A, Yuste R. 2000b. Regulation of spine calcium dynamics by rapid spine motility. *J. Neurosci.* 20:8262–68.

Malenka RC, Nicoll RA. 1999. Long-term potentiation—a decade of progress? *Science* 285:1870–74

Maletic-Savatic M, Malinow R, Svoboda K. 1999. Rapid dendritic morphogenesis in CA1 hippocampal dendrites induced by synaptic activity. *Science* 283:1923–27

Malinow R, Tsien RW. 1990. Presynaptic enhancement shown by whole-cell recordings of long-term potentiation in hippocampal slices. *Nature* 346:177–80

Marín-Padilla M. 1972. Structural abnormalities of the cerebral cortex in human chromosomal aberrations. *Brain Res.* 44:625–29

Mayford M, Bach ME, Huang YY, Wang L, Hawkins RD, Kandel ER. 1996. Control of memory formation through regulated expression of a CaMKII transgene. *Science* 274:1678–83

McKinney RA, Capogna M, Durr R, Gahwiler BH, Thompson SM. 1999. Miniature synaptic events maintain dendritic spines via AMPA receptor activation. *Nat. Neurosci.* 2:44–49

Mehraein P, Yamada M, Tarnowska-Dziduszko E. 1975. Quantitative study on dendrites and dendritic spines in Alzheimer's disease and senile dementia. *Adv. Neurol.* 12:453–58

Milner B. 1966. Amnesia following operation on the temporal lobes. In *Amnesia: Clinical, Psychological and Medicolegal Aspects*, ed. CWM Whitty, OL Zangwill, pp. 109–133. London: Butterworths

Moser MB, Trommald M, Andersen P. 1994. An increase in dendritic spine density on hippocampal CA1 pyramidal cells following spatial learning in adult rats suggests the formation of new synapses. *Proc. Natl. Acad. Sci. USA* 91:12673–75

Moser MB, Trommald M, Egeland T, Andersen P. 1997. Spatial training in a complex environment and isolation alter the spine distribution differently in rat CA1 pyramidal cells. *J. Comp. Neurol.* 380:373–81

Multani P, Myers RH, Blume HW, Schomer DL, Sotrel A. 1994. Neocortical dendritic pathology in human partial epilepsy: a quantitative Golgi study. *Epilepsia* 35:728–36

Nusser Z, Lujan R, Laube G, Roberts JDB, Molnar E, Somogyi P. 1998. Cell type and pathway dependence of synaptic AMPA receptor number and variability in the hippocampus. *Neuron* 21:545–59

Papa M, Bundman MC, Greenberger V, Segal M. 1995. Morphological analysis of dendritic spine development in primary cultures of hippocampal neurons. *J. Neurosci.* 15:1–11

Papa M, Segal M. 1996. Morphological plasticity in dendritic spines of cultured hippocampal neurons. *Neuroscience* 71:1005–11

Parnass Z, Tashiro A, Yuste R. 2000. Analysis of spine morphological plasticity in developing hippocampal pyramidal neurons. *Hippocampus* 10:561–68

Parnavelas J, Globus A, Kaups P. 1973. Continuous illumination from birth affects spine density of neurons in the visual cortex of the rat. *Exp. Neurol.* 40:742–47

Patel S, Rose S, Stewart M. 1988. Training induced dendritic spine density changes are specifically related to memory formation processing the chick, Gallus domesticus. *Brain Res.* 463:168–73

Peters A, Kaiserman-Abramof IR. 1969. The small pyramidal neuron of the rat cerebral cortex. The synapses upon dendritic spines. *Z. Zellforsch. Mikrosk. Anat.* 100:487–506

Popov V, Bocharova L. 1992. Hibernation-induced structural changes in synaptic contacts between mossy fibres and hippocampal pyramidal neurons. *Neuroscience* 48:53–62

Popov V, Bocharova L, Bragin A. 1992. Repeated changes of dendritic morphology in the hippocampus of ground squirrels in the course of hibernation. *Neuroscience* 48:45–51

Purpura D. 1974. Dendritic spine "dysgenesis" and mental retardation. *Science* 186:1126–28

Rall W. 1970. Cable properties of dendrites and effects of synaptic location. See Andersen & Jansen 1970, pp. 175–87

Ramón y Cajal S. 1893. Neue Darstellung vom histologischen Bau des Zentralnervensystem. *Arch. Anat. Entwick.* 319–428

Ramón y Cajal S. 1904. *La Textura del Sistema Nerviosa del Hombre y los Vertebrados.* Madrid: Moya

Rausch G, Scheich H. 1982. Dendritic spine loss and enlargement during maturation of the speech control system in the mynah bird. *Neurosci. Lett.* 29:129–33

Salas M. 1980. Effects of early undernutrition on dendritic spines of cortical pyramidal cells in the rat. *Dev. Neurosci.* 3:109–17

Schikorski T, Stevens C. 1999. Quantitative fine-structural analysis of olfactory cortical synapses. *Proc. Natl. Acad. Sci. USA* 96:4107–12

Sorra K, Harris K. 1998. Stability in synapse number and size at 2 hr after long-term potentiation in hippocampal area CA1. *J. Neurosci.* 18:658–71

Spencer W. 1862. *First Principles.* London: Williams Norgate

Stevens CF, Wang YY. 1994. Changes in reliability of synaptic function as a mechanism for plasticity. *Nature* 371(6499):704–7

Tanzi G. 1893. I fatti i le indizioni nell'odierna istologi del sistema nervoso. *Riv. Sper. Freniatr.* 19:419–72

Tashiro A, Minden A, Yuste R. 2000. Regulation of dendritic spine morphology by the Rho family of small GTPases: antagonistic roles of Rac and Rho. *Cereb. Cortex* 10(10):927–38

Toni N, Buchs PA, Nikonenko I, Bron CR, Muller D. 1999. LTP promotes formation of multiple spine synapses between a single axon terminal and a dendrite. *Nature* 402:421–25

Trommald M, Hulleberg G, Andersen P. 1996. Long-term potentiation is associated with new excitatory spine synapses on rat dentate granule cells. *Learn. Mem.* 3:218–28

Trommald M, Vaaland JL, Blackstad TW, Andersen P. 1990. Dendritic spine changes in rat dentate granule cells associated with long-term potentiation. In *Neurotoxicity of Excitatory Amino Acids*, ed. A Guidotti, pp. 163–74. New York: Raven

Valverde F. 1967. Apical dendritic spines of the visual cortex and light deprivation in the mouse. *Exp. Brain Res.* 3:337–52

Valverde F. 1971. Rate and extent of recovery from dark rearing in the visual cortex of the mouse. *Brain Res.* 33:1–11

Van Harrefeld A, Fifkova E. 1975. Swelling of dendritic spines in the fascia dentata after simulation of the preforant fibers as a mechanisms of post-tetanic potentiation. *Exp. Neurol.* 49:736–49

Wallace CS, Hawrylak N, Greenough WT. 1991. Studies of synaptic structural modifications after long-term potentiation and kindling: context for a molecular morphology. In *Long-Term Potentiation: A Debate of Current Issues*, ed. M Baudry, JL Davis, pp. 189–232. Cambridge: MIT Press

Wenzel J, Desmond NL, Levy WB. 1993. Somatic ribosomal changes induced by long-term potentiation of the perforant path-hippocampal CA1 synapses. *Brain Res.* 619:331–33

Wenzel J, Otani S, Desmond NL, Levy WB. 1994. Rapid development of somatic spines in stratum granulosum of the adult hippocampus in vitro. *Brain Res.* 656:127–34

Wisniewski K, Segan S, Miezejeski C, Sersen E, Rudelli R. 1991. The Fra(X) syndrome: neurological, electrophysiological, and neuropathological abnormalities. *Am. J. Med. Genet.* 38:476–80

Woolley CS, Gould E, Frankfurt M, McEwen BS. 1990. Naturally occurring fluctuation in dendritic spine density on adult hippocampal pyramidal neurons. *J. Neurosci.* 10:4035–39

Yuste R, Majewska A, Holthoff K. 2000. From form to function: calcium compartmentalization in dendritic spines. *Nat. Neurosci.* 3:653–59

Zamanillo D, Sprengel R, Hvalby O, Jensen V, Burnashev N, et al. 1999. Importance of AMPA receptors for hippocampal synaptic plasticity but not for spatial learning. *Science* 284:1805–11

Annu. Rev. Neurosci. 2001. 24:1091–119

STOPPING TIME: The Genetics of Fly and Mouse Circadian Clocks

Ravi Allada[2], Patrick Emery[1,3], Joseph S. Takahashi[1,2], and Michael Rosbash[1,3]

[1]*Howard Hughes Medical Institute*
[2]*Northwestern University, Department of Neurobiology and Physiology, Evanston, Illinois 60208*
[3]*Brandeis University, Department of Biology, Waltham, Massachusetts 02454;
e-mail: r-allada@northwestern.edu, emery@brandeis.edu,
j-takahashi@northwestern.edu, rosbash@brandeis.edu*

Key Words   circadian, bHLH-PAS, period, cryptochrome, casein kinase I epsilon

■ **Abstract**   Forward genetic analyses in flies and mice have uncovered conserved transcriptional feedback loops at the heart of circadian pacemakers. Conserved mechanisms of posttranslational regulation, most notably phosphorylation, appear to be important for timing feedback. Transcript analyses have indicated that circadian clocks are not restricted to neurons but are found in several tissues. Comparisons between flies and mice highlight important differences in molecular circuitry and circadian organization. Future studies of pacemaker mechanisms and their control of physiology and behavior will likely continue to rely on forward genetics.

INTRODUCTION

With the completion of the human genome sequence, the era of simple gene cloning and identification is nearing an end. With the advent of high throughput technologies such as DNA microarrays, the temporal and spatial expression patterns of thousands of genes will be known. The challenge for neuroscience will be to functionally link genes, including their complex expression patterns, to the output of the nervous system: behavior. Genetics will play a crucial role in navigating through this genomic jungle.

Forward genetics, the process of identifying mutant phenotypes to isolate genes, is best applied to a problem lacking molecular description. Forward genetics used to be "genetics," but the advent of recombinant DNA and reverse genetics necessitates a more useful term. Genetic screens require no prior hypothesis about the mechanism by which a system functions. The only requirements are an efficient means of mutagenesis and phenotypic screening. If there is an abundance of correlative molecular or cellular data, genetic analysis can also be helpful in

0147-006X/01/0621-1091$14.00   **1091**

determining which variables are critical. Distinguishing between observations that are simply correlated with an output and those that are causal can be difficult. By randomly mutagenizing and assaying the consequence on a particular behavioral or physiologic parameter, one can identify key regulatory components.

Nowhere in neuroscience has the power of forward genetics been felt more acutely than in circadian rhythms. Genetics has propelled rhythms research to appear twice recently among *Science* magazine's breakthroughs of the year (1997, 1998). In this review, we discuss the key discoveries and how they were made possible by the confluence of genetics and genomics. We compare the molecular mechanisms and functions of different animal circadian systems, focusing on the fruit fly, *Drosophila melanogaster*, and the mouse. For the sake of brevity and to avoid redundancy with other recent reviews (Dunlap 1999, Ishida et al 1999, Edery 2000, Hall 2000, King & Takahashi 2000, Scully & Kay 2000, Young 2000a,b), we focus our efforts on circadian pacemakers and their systemic organization and outputs.

BACKGROUND

Animals temporally organize their behavior and metabolism to adapt to and anticipate the progression through the 24-hour solar cycle. The earth's 24-hour rotation has apparently dictated that all manner of life evolve to maximize fitness in this environment. Endogenous clocks serve to anticipate daily changes in their environment. The overt manifestations of these clocks are circadian rhythms. They are defined by periodicity in the absence of exogenous cues. Circadian rhythms are therefore not simply driven by the environment but arise from internal biological clocks. Circadian clocks only closely approximate but are not exactly 24 hours. Importantly, the timing of the clock is sensitive to light, among other environmental cues. The daily cycle of light and darkness resets internal clocks to maintain synchrony between the external solar day and internal biological clocks. Across a wide range of physiologic temperatures, the periodicity of rhythms is stable. This temperature compensation mechanism is particularly relevant in poikilotherms to prevent perturbations of the clock due to changes in the ambient temperature. However, even neural pacemaker tissues such as the mammalian suprachiasmatic nucleus (SCN) when maintained in vitro also defend against changes in temperature (Ruby et al 1999). Temperature compensation is thought to reflect an underlying feature of the clock mechanism. Clocks have been typically associated with regions of the nervous system, such as the SCN (Moore & Eichler 1972, Stephan & Zucker 1972). However, they are found in cells outside the nervous system and in organisms that do not have nervous systems. Even unicellular prokaryotes (e.g. photosynthetic cyanobacteria) exhibit robust circadian rhythms (Johnson & Golden 1999). Determining the molecular nature of these clocks awaited a forward genetic approach, first in *Drosophila* and other organisms (Konopka & Benzer 1971, Bruce 1972, Feldman & Hoyle 1973) and later in the mouse (Vitaterna et al 1994).

BEHAVIOR, GENETICS, AND THE FRUIT FLY

In the modern era the study of the genetic basis of behavior was pioneered by Seymour Benzer and colleagues, using the fruit fly, *Drosophila melanogaster.* Benzer's laboratory initiated genetic studies of many aspects of fly behavior, including circadian rhythms (Benzer 1971). The fruit fly had previously been a workhorse of geneticists, aiding in the discovery of many of the core principles of inheritance. *Drosophila* has a generation time of only 10 days, and its long history in the field of genetics has led to the development of many useful tools. Benzer's laboratory hoped to apply the genetic techniques optimized in this model system to questions related to behavior. Even then, much was known about circadian rhythms in *Drosophila* at the behavioral level. Fruit flies proceed through a series of characteristic developmental stages, beginning with the embryonic through a series of three larval stages to a pupal stage from which adults emerge or eclose. The fruit fly prefers to eclose at a particular time of day, reflecting gating by a circadian clock (Pittendrigh 1954). In fact, the word *Drosophila* means dew-loving, referring to its tendency to eclose in the morning. Ron Konopka, a graduate student in Benzer's laboratory, designed a screen of mutagenized populations, looking for mutant flies that eclosed with a different circadian phase (Konopka & Benzer 1971). In now classic studies, Konopka identified three such strains. When studying the individual strains under constant conditions, he saw that one had a long period rhythm of 29 hours, the second had a short period rhythm of 19 hours, and the third had no detectable rhythm. Remarkably, all three mutant phenotypes mapped to the X-chromosome to what appeared to be a single locus, which they called *period* (*per*).

The identification of these behavioral mutants suggested that one could study genes as a means to understanding behavior in general and circadian rhythms in particular. In part, the success of this screen reflected the unusual precision of this particular behavioral assay. Few behavioral assays can detect small differences on the order of 20% or less that distinguish their short and long period mutants. The ability to identify adult circadian rhythm mutants suggested that the mutated genes may not be vital to functions in development. If genes involved in circadian rhythms were also required for vital functions, then null mutations of such genes would never live to eclosion. The ability to identify mutants suggested that the function of the *period* gene might be relatively circumscribed. As a result, we now speak of "circadian rhythm genes" or "circadian clock genes."

TRANSCRIPTION AS NEURAL CODE: *Drosophila Period*

The work of Konopka and Benzer raised the possibility that forward genetics might more generally illuminate the circadian clockworks. However, the mutants and their intriguing phenotypes shed little light on the actual clock mechanism. This level of insight awaited the cloning of the *period* gene and more precisely, a careful analysis of its temporal expression pattern. The cloning of PER was initially misleading, assigned to a group of cell surface proteins known as proteoglycans (Jackson et al 1986, Reddy et al 1986). This observation fit intercellular models of circadian clocks. However, more careful analysis of *per* sequence, expression, and function changed this view dramatically. Both PER protein and RNA are rhythmically expressed (Siwicki et al 1988, Hardin et al 1990, Zerr et al 1990). Importantly, the timing of the RNA fluctuations was sensitive to point mutations in the *period* protein (Hardin et al 1990). This set of observations placed these molecular fluctuations at the heart of the circadian pacemaker and led to the formulation of a model in which PER feeds back on its own transcription.

A clue to the mechanism of PER feedback came from sequence homologies with the basic helix-loop-helix (bHLH) transcription factors, *single-minded (sim)*, and the aryl hydrocarbon receptor nuclear translocator (ARNT) (Crews et al 1988). However, *per* does not contain a canonical bHLH or other DNA-binding domain. The conserved domain, termed PAS (for *per-ARNT-sim*), mediates dimeric PAS-PAS interactions (Huang et al 1993). In fact, it was shown that PER could inhibit transcriptional activation by ARNT and its bHLH-PAS partner, the aryl hydrocarbon receptor (Ahr) (Lindebro et al 1995). It was proposed that PER might feed back by inhibiting the activity of such bHLH-PAS transcription factor(s). Genetics would be crucial in determining which of these factors is the relevant in vivo target (see below).

For a decade, *period* was the only known circadian gene in animals. The intensive study of *period* led to the elucidation of several principles concerning clock genes. First, the gold standard for defining a clock gene is genetic: If one disrupts a gene, does it affect circadian rhythms? Second, genes that satisfy this first criterion are often (but not always) rhythmically transcribed. Third, clock proteins feed back and regulate their own transcription. Finally, there must be delays between activation and inhibition in order to generate free-running oscillations. In the absence of such delays, the system will damp to steady state.

TRANSCRIPTIONAL FEEDBACK

Fly Transcriptional Activators: CLOCK and CYCLE

Genes on the positive arm of the cycle appear to be well conserved. In *Drosophila* a pair of bHLH-PAS–containing transcription factors, CLOCK (CLK) and CYCLE (CYC), play roles as activators of clock genes. Arrhythmic mutant alleles of each gene abolish the rhythm of *period* and *timeless (tim)* transcription and RNA abundance and peg their levels near the trough of a dynamic cycle (Allada et al 1998, Rutila et al 1998). Interestingly, *Clk*, but not *cyc*, RNA and protein levels cycle over a 24-hour period (Bae et al 1998, 2000; Darlington et al 1998; Lee et al 1998). A circadian enhancer from the *per* promoter has been identified that is necessary and sufficient to confer cycling to a reporter gene (Hao et al 1997). Of note, a known binding site for bHLH-PAS transcription factors, an E-box

(CACGTG), is required for this activity (Hao et al 1997). Similar sequences have been identified in the *timeless* promoter (Darlington et al 1998). Transfection of CLK into a CYC-expressing *Drosophila* cell line results in activation from cotransfected *per* and *timeless* enhancers in an E-box-dependent fashion (Darlington et al 1998). CLK and CYC coimmunoprecipitate from fly head extracts and interact in yeast two-hybrid assays and as in vitro translated proteins (Darlington et al 1998, Lee et al 1999, Bae et al 2000). Thus, CLK and CYC appear to work together as a canonical heterodimeric transcription factor complex.

Mouse Transcriptional Activators: mCLOCK and BMAL1

For years following the cloning of *per*, many interpreted the failure to identify mammalian homologs as a sign that the two systems are not well conserved, at least at the molecular level. The recent expansion in both fly genes and their mammalian orthologs proved these observers wrong. In the mouse, the activators work similarly but perhaps on targets somewhat distinct from those in *Drosophila*. Genetic evidence is currently available from the mouse *Clock* mutant (*mClock*). *mClock* was originally identified in a behavioral forward genetic screen for mutants with circadian rhythm phenotypes (Vitaterna et al 1994). Homozygous mutants exhibit long period (28 hr) rhythms, which damp to arrhythmicity (Vitaterna et al 1994, Antoch et al 1997, King et al 1997). In these *Clock* mutant mice, several cycling genes appear to be downregulated, including *mouse Period1-3(mPer1-3)*, *mouse Cryptochrome1-2(mCry1-2)*, and *Bmal1*, implying a possible master regulatory role as a transcriptional activator (Zylka et al 1998b, Kume et al 1999, Shearman et al 2000b). Unlike *Drosophila Clock*, *mClock* does not cycle at the RNA level in the SCN (Shearman et al 2000b). However, *Bmal1* RNA levels appear to cycle in the mouse and rat suprachiasmatic nucleus, though with a low amplitude (Abe et al 1998, Oishi et al 1998a, Shearman et al 2000b). CLOCK and BMAL1 interact in two-hybrid assays and cooperatively activate from E-box elements in the *mPer1* promoter in transfection experiments (Gekakis et al 1998, Hogenesch et al 1998). It is tempting to equate mouse CLOCK and BMAL1 with fly CLOCK and CYCLE. Recent genetic evidence supports a role for BMAL1 (Bunger et al 2000). Although the 5′ promoter region of *mPer1* has been shown to confer cycling in vivo, it is not known if the E-boxes located there are necessary or sufficient for this cycling (Kuhlman et al 2000, Yamaguchi et al 2000, Yamazaki et al 2000). Complicating the picture is the presence of other homologous bHLH-PAS transcription factors, such as MOP9. MOP9 appears to exceed BMAL1's ability to activate with mCLOCK in transfection experiments (Hogenesch et al 2000). A MOP9 knockout will be required to distinguish the role of this gene.

Fly Transcriptional Inhibitors: PERIOD and TIMELESS

The inhibitory complex in *Drosophila* seems to consist of PER and TIM. Null mutants of both *per* and *tim* are completely arrhythmic in constant darkness

(Konopka & Benzer 1971, Sehgal et al 1994). The levels of *per* and *tim* RNAs both oscillate daily (Hardin et al 1990, Sehgal et al 1995). In both arrhythmic mutants the levels of their own RNAs is middle to high in comparison with the dynamic range of a daily cycle (Hardin et al 1990, Sehgal et al 1994). These measurements of RNA levels have been largely confirmed by transcription run-on assays and in vivo measurements of promoter activity (Hardin et al 1992b, Brandes et al 1996, So & Rosbash 1997). Although TIM does not have a PAS domain, it coimmuno-precipitates with PER from fly head extracts (Zeng et al 1996). Furthermore, cotransfection of PER and TIM together represses CLK-mediated transcription from *per* and *tim* E-boxes (Darlington et al 1998, Rothenfluh et al 2000). PER and TIM coimmunoprecipitate with CLK and CYC from fly head extracts (Lee et al 1998, Bae et al 2000). This association appears to be specific to times of falling *per* and *tim* transcription. These four proteins have also been shown to associate as in vitro translated proteins (Lee et al 1999). Moreover, in vitro PER and TIM modestly reduce CLK-CYC binding to its target E-box (Lee et al 1999). Thus, there is strong evidence that PER and TIM behave as direct biochemical inhibitors of CLK/CYC-mediated transcriptional activation.

Is there functional specialization within this heterodimeric complex? TIM levels, but not PER levels, are suppressed within minutes of exposure of the organism to light (Hunter-Ensor et al 1996, Lee et al 1996, Myers et al 1996, Zeng et al 1996). TIM may therefore function to link transcriptional repression to external temporal cues. Furthermore, PER protein levels, but not *per* RNA levels, are low in tim^0 mutants, suggesting a positive role for TIM in PER stabilization (Price et al 1995). In addition to the presence of a PAS domain in PER, experimental evidence more strongly links PER than TIM to repression. Disappearance of PER but not TIM seems to correlate well with the turn-on of *per* and *tim* transcription (Marrus et al 1996, So & Rosbash 1997). Light pulses that degrade TIM, in a *tim* mutant (tim^{UL}), lead to rapid decreases in *per* RNA levels, perhaps by freeing PER monomer (Rothenfluh et al 2000). Furthermore, transfection of *per* mutants, which are constitutively nuclear in *Drosophila* S2 cells independently repress CLK/CYC transcriptional activity without TIM (Rothenfluh et al 2000). Therefore, PER may play the role of primary repressor, and TIM may transport and/or stabilize PER (Figure 1).

Mouse Transcriptional Inhibitors: mCRY and mPER

In the mouse the molecular nature of the circadian inhibitor appears to be distinct from that of *Drosophila*. The strongest candidates for components of an inhibitory complex are the cryptochromes. Cryptochromes are members of a blue-light sensitive family of proteins, which also includes UV-dependent DNA repair enzymes (photolyases) (Cashmore et al 1999). In flies strong evidence supports a role for cryptochrome in circadian photoreception (Emery et al 1998, 2000; Stanewsky et al 1998; Egan et al 1999; Ishikawa et al 1999). In mice genetic inactivation

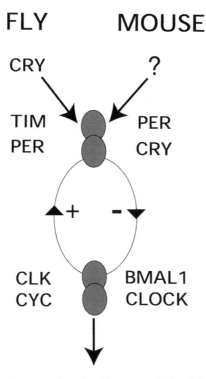

Figure 1 Schematic diagram of a circadian transcriptional feedback loop. CLK in Drosophila and BMAL1 in mammals are transcriptional activators that cycle in gene expression. CYC in Drosophila and mCLOCK in mice are transcriptional activators and heterodimeric partners of the above that do not cycle in gene expression. These complexes increase transcription of per and tim in Drosophila and mPer and mCry in mice. They are also thought to connect to output genes (arrow down). PER and mCRY play the role of primary transcriptional repressors. TIM and mPER transmit light information to the feedback loop and associate with these repressors. CRY behaves as a photoreceptor in flies. The mouse photoreceptor(s) is unknown.

of *mCry1* results in short period rhythms, whereas knockout of *mCry2* results in long period rhythms (Thresher et al 1998, van der Horst et al 1999, Vitaterna et al 1999). The double mutant displays no rhythm whatsoever, indicating a role for these genes in the central pacemaker (van der Horst et al 1999, Vitaterna et al 1999). *mCry1* and *mCry2* cycle at the RNA and protein levels in the SCN (Miyamoto & Sancar 1998, Kume et al 1999, Okamura et al 1999). In the double knockout of *mCry1* or *mCry2*, two of the CLOCK-BMAL1 target genes, *mPer1* and *mPer2*, are at high levels, consistent with a transcriptional suppressor role for the cryptochromes (Okamura et al 1999, Vitaterna et al 1999). These data are supported by transfection experiments, demonstrating potent transcriptional inhibition by both *mCry1* or *mCry2* expression (Kume et al 1999). Moreover, mCRY1/2 interact with mCLOCK and BMAL1 in yeast two-hybrid assays (Griffin et al 1999). These data support a biochemically direct inhibitory role for the mCRYs, parallel to the role of PER-TIM or perhaps PER in the fly pacemaker mechanism (Figure 1).

One important question is whether mCRY1 and mCRY2 act alone or in concert with mammalian *Per* and/or *Tim*. Each CRY represses mCLOCK-BMAL1 activation in heterologous *Drosophila* S2 cells, which do not endogenously express dPER or dTIM (Saez & Young 1996, Shearman et al 2000b). The simplest model therefore is one in which mCRY1 and/or mCRY2 repress mCLOCK-BMAL1-mediated activation. A mystery remains as to why the *mCry1* and *mCry2* single knockouts have opposing period phenotypes, one shorter and one longer than 24 hours. Clearly, the transcriptional inhibitory assay of mCRY function in tissue culture cells, which does not significantly distinguish mCRY1 from mCRY2, does not explain the whole story.

If mCLOCK and BMAL1 are the activators and mCRY1/2 are repressors, what is the function of mPER? First, at least one of the mPERs satisfies the genetic gold standard. A deletion of the PAS domain in mPER2 results in shortened periods, which grade into arrhythmicity (Zheng et al 1999). All three mouse *period* genes cycle at the RNA and protein levels (Shearman et al 1997, Sun et al 1997, Tei et al 1997, Takumi et al 1998, Zylka et al 1998b, Hastings et al 1999, Field et al 2000). In these homozygous mutants, both *mPer2* and *mPer1* transcript levels are substantially reduced (Zheng et al 1999). This is in contrast to flies in which *per* transcript levels are middle to high in *per^0* mutants (Hardin et al 1990). However, the nature of the mPER2 mutation complicates the interpretation of these results: A mutant protein might be produced and play a dominant-negative role. Indeed, transfection experiments show that this mutant allele is expressed and antagonizes wild-type mPER2 cellular localization (Shearman et al 2000b). Therefore, a true null mutation would be useful to determine the precise role of *mPer2* in the clock. Inactivation of the *mPer3* gene also has a slightly short period phenotype (Shearman et al 2000a). Despite being the otherwise most well studied of the *Period* genes, *mPer1*, when knocked out, has little effect on rhythms (Shearman et al 2000b). Functional redundancy of the *mPer* genes may be an obstacle to genetically deciphering their roles. Nonetheless, these data support a positive role for *mPer2* in promoting *mPer1* and *mPer2* RNA levels.

By analogy to *Drosophila*, one might expect mPER to be a transcriptional repressor of mCLOCK and BMAL1. The low RNA levels of *mPer1* and *mPer2* in the *mPer2* mutant animals are difficult to explain in this context. In fact, initial studies of mPER1 showed that it modestly repressed mCLOCK-BMAL1 activation (Sangoram et al 1998). However, the modest inhibition by mPER1 is dwarfed by that shown by either mCRY1 or mCRY2 alone (Kume et al 1999). In the absence of contradictory evidence, the simplest model is that mammalian *Period* genes do not behave like their repressing counterparts in flies.

How then does *mPer*(s) participate in the feedback loop, specifically *mPer2*, for which there is the strongest genetic evidence? Recently, it was proposed that *mPer2* participates positively in *Bmal1* RNA expression (Shearman et al 2000b). However,the peak values of *Bmal1* RNA are not dramatically affected in *mPer2* mutant animals (Shearman et al 2000b). Thus, there is no compelling evidence in favor of a specific function of mPER2. Interestingly, the mCRY and mPER proteins have been shown to engage in protein-protein contacts from SCN extracts (Field et al 2000). Furthermore, mPER2 protein levels are reduced in *mCry1-mCry2* double knockout mice (Shearman et al 2000b). It is therefore possible that in vivo the mPERs modulate the repressing activity of the mCRYs by direct physical association.

The mammalian *period* genes may also be involved in the response of the pacemaker to light. *mPer1* and *mPer2* RNAs as well as proteins are rapidly induced by light in the SCN (Albrecht et al 1997, Shearman et al 1997, Shigeyoshi et al 1997, Field et al 2000). Antisense oligonucleotides against *mPer1* reduce phase resetting, implicating mPER1 in this process (Akiyama et al 1999). However, the effect of *Period* gene knockout on light-induced *mPer* expression has yet to be determined. The in vivo role of the mammalian *Period* genes awaits true knockout of each of these genes as well as double and triple mutant combinations. The interactions with mCRY as well as light responsiveness implicate mPER in transducing light information, leading to a modulation of mCRY feedback, perhaps similar to the role of TIM in flies (Figure 1).

A mammalian homolog of *Drosophila timeless* has also been identified. Inactivation of *mTimeless* (*mTim*) function in vivo results in early embryonic lethality (Gotter et al 2000). The role of *mTim* has been confused by divergent results. Some reports have shown no cycling of *mTim* RNA; others show induction by light (Koike et al 1998, Sangoram et al 1998, Zylka et al 1998a, Tischkau et al 1999). In *Drosophila*, *timeless* RNA and protein cycle in constant darkness conditions (Sehgal et al 1995, Hunter-Ensor et al 1996, Lee et al 1996, Myers et al 1996, Zeng et al 1996). Moreover, *Drosophila* TIM is degraded by light (Hunter-Ensor et al 1996, Lee et al 1996, Myers et al 1996, Zeng et al 1996). The only reports of mTIM protein show no cycling and no light responsiveness (Hastings et al 1999, Field et al 2000). In addition, a new *Drosophila* gene, *timeout*, has been identified, which more closely resembles mTIM (Benna et al 2000, Gotter et al 2000). It is unclear if *timeout* has any role in fly rhythms. These observations have led to the proposal that the mammalian *Timeless* may not be involved in circadian rhythms (Gotter et al 2000). However, mTIM coimmunoprecipitates from SCN with mCRY1 and mCRY2 (Field et al 2000). Because the *mTim* knockout is lethal in mice, circadian rhythms cannot be assayed in these animals. Therefore, the jury will remain out on the in vivo role of *mTim* until conditional knockouts are reported. The *dbt* gene in *Drosophila* is just one example of a gene with developmental lethality but a clear function in the pacemaker (Price et al 1998).

In summary (see Figure 1), the *Drosophila* bHLH-PAS DNA-binding proteins, CLK and CYC, activate transcription of the *period* and *timeless* genes. PER and TIM feed back and inhibit the activity of the CLK-CYC heterodimer. In the mouse,

Clock and *cycle* orthologs (mCLOCK and BMAL1) probably activate the *mPer* and *mCry* genes. However, both CRYPTOCHROMES (mCRY1 and mCRY2) feed back and inhibit the mCLOCK-BMAL1 heterodimer. mPERs may transduce light information to mCRYs. The role of mTIM remains unclear.

CODEPENDENT LOOPS?

One of the latest waves of circadian research deals with the role of positive and negative interdependent feedback loops. The principle was first laid out in examinations of *Drosophila Clock* RNA but has infected other organisms, including the mouse. Like *per* and *tim*, *Clock* is rhythmically expressed (Bae et al 1998, Darlington et al 1998). However, *Clock* RNA cycles antiphase to *per* and *tim*, peaking in the late night and early morning [Zeitgeber Time (ZT) 23, ZT 5; ZT 0 = lights-on ("sunrise"); ZT 12 = lights-off ("sunset")]. Consistent with this altered phase of cycling, *Clock* transcript levels respond differently in arrhythmic mutants and are low in the per^0 and tim^0 strains (Bae et al 1998); *per* and *tim* transcripts are at middle to peak levels in these mutants (Hardin et al 1990, Sehgal et al 1995). In cyc^0 and Clk^{Jrk} mutants, *Clk* transcripts are high, whereas *per* and *tim* transcripts are low (Glossop et al 1999). Finally, in per^0; cyc^0 and per^0; Clk^{Jrk}, levels are similar to single cyc^0 and Clk^{Jrk} mutants (Glossop et al 1999). These data led to a model in which there are two interconnected feedback loops: the classic one regulates *per* and *tim* transcription, with CLK/CYC acting positively and PER/TIM acting negatively; the other regulates *Clk* transcription, with PER/TIM acting positively and CLK/CYC acting negatively. Epistasis analysis in double mutants indicates that the effects of *per* and *tim* operate on *Clk* RNA through CLK/CYC. Analysis of various mouse knockout strains of circadian rhythm genes came to similar conclusions, although the names have changed. mPER2 is proposed to act as a positive regulator of *Bmal1* transcripts (Shearman et al 2000b), although these effects are modest (see above).On the other hand, mCRY1 and mCRY2 are thought to play the major negative role on *mPer* and *mCry* transcription (see above).

Interdependent loops even appear to be involved in the circadian rhythms of the fungus, *Neurospora crassa* (Lee et al 2000). In *Neurospora*, *frequency* (*frq*) was the first clock component identified (reviewed in Dunlap 1999). Like *per*, it is rhythmically expressed and feeds back negatively on its own transcription. *White collar-1* (*WC-1*) encodes a transcriptional activator that promotes expression of the *frq* gene. Unlike *frq*, *WC-1* RNA is not rhythmically expressed. However, its protein product (WC-1) does oscillate, consistent with circadian posttranscriptional regulation (Lee et al 2000). As expected, *frq* overexpression downregulates expression of the endogenous *frq* gene. On the other hand, this excess *frq* positively regulates WC-1 expression.

One problem with the interdependent feedback models, particularly from flies and mice, is the reliance on arrhythmic null mutants. Genetic studies of putative rhythm genes are the gold standard in determining whether or not they are clock genes. However, analysis of mutant phenotypes can be difficult to interpret and even misleading about the specific role of the gene. Like setting off a row of dominoes, the absence of the gene(s) throughout development may result in consequences that may obscure true gene function in the adult. A second problem is the absence of corroborating biochemical data to support interdependent loops. As all rhythms we measure are abolished by per^0 and Clk^{Jrk}, it is not surprising that per rhythms (or any rhythm) depend on Clk or that Clk rhythms depend on per. Therefore, these analyses may lend little mechanistic insight into the exact nature of this regulation.

POSTRANSLATIONAL REGULATION

Gated Nuclear Entry?

Transcriptional feedback loops are ubiquitous in nature. A distinctive feature of circadian clocks is their timing. It has been proposed that it is the delay between synthesis and feedback that is necessary for circadian oscillations to occur. Furthermore, the magnitude of this delay may dictate the daily oscillatory frequency. Evidence is accumulating that posttranslational regulation of certain circadian rhythm proteins is crucial to setting up these delays and therefore to timing feedback.

Several features of circadian rhythm gene expression appear to be subject to posttranslational regulation. In the case of *Drosophila per*, the transport of PER from cytoplasm to nucleus seems to be circadianly gated, at least in the pacemaker lateral neurons (LNs) (Curtin et al 1995). As PER protein accumulates (ZT 8–17), immunohistochemical staining is detectable predominantly in the cytoplasm, appearing in a so-called doughnut pattern. After ZT 17, PER staining rapidly (over the next few hours) moves from predominantly cytoplasmic to predominantly nuclear in pacemaker neurons. These observations have been independently reproduced (Lee et al 1996, Matsumoto et al 1999). The importance of the gating of nuclear entry of PER is supported by genetic evidence. In per^L mutants the timing of nuclear entry is also delayed (Curtin et al 1995). These observations support a model of regulated nuclear entry of PER.

PER association with its heterodimeric partner TIME has been implicated in the temporal control of PER nuclear entry. TIME strongly associates with PER both in vitro and in vivo. The *time* gene was cloned in part by its ability to interact with PER in a yeast two-hybrid assay (Gekakis et al 1995). These in vitro findings have been confirmed by in vivo observations as well. In coimmunoprecipitation experiments with fly head extracts, PER and TIM are strongly associated (Zeng et al 1996). In tim^0 mutants, PER proteins levels are suppressed, even though RNA levels are relatively high, suggesting a role for TIM in PER stability (Price et al 1995). Furthermore, in tim^0 mutants, PER staining is constitutively cytoplasmic (Vosshall et al 1994). In cultured *Drosophila* S2 cells, nuclear localization of transfected PER is completely dependent upon cotransfection of TIM (Saez & Young 1996). In per^0 mutants, TIM protein is predominantly cytoplasmic (Hunter-Ensor et al 1996, Myers et al 1996). These observations led to a model in which association of PER with TIM is a prerequisite for nuclear co-entry of this heterodimer.

Though this model has been largely accepted, stray bits of evidence may ultimately undermine this nuclear entry model. Beyond the pacemaker lateral neurons, PER cycling is evident in several parts of the fly nervous system and in many different tissues throughout the fly. If gating is a necessary step to generate circadian macromolecular oscillations, then "doughnuts" should be observed in many other PER-expressing cells. In stark contrast to the current model, no cytoplasmic accumulation is observed prior to nuclear entry, despite robust PER cycling, as noted by Curtin et al 1995. PER has been previously shown to be a transcriptional repressor, and its entry to the nucleus has been proposed to be required for its activity. Gating of entry would be an elegant means of controlling the timing of repression and therefore the clock. However, the timing of nuclear entry does not coincide with the downturn of *per* transcription and in fact occurs several hours too late at about the time transcription has nearly reached trough levels (Curtin et al 1995, So & Rosbash 1997). Finally, gated nuclear entry has not yet been described for circadian systems of other organisms. If TIM is not gating PER nuclear entry, TIM's predominant function in *Drosophila* may therefore be to regulate PER stability. More careful studies in the ever-growing list of circadian model systems will be required to determine if gated entry is a ubiquitous feature of pacemakers.

Phosphorylation

Phosphorylation may also impose a delay on PER feedback. The phosphorylation state and hence the mobility of PER protein varies systematically over a 24-hour time course (Edery et al 1994). PER accumulates during the late day/early night (ZT 8–16). As night progresses, PER migrates more slowly, i.e. at larger apparent molecular weights. As PER disappears in the early morning, PER phosphorylation peaks, as measured by mobility. Immunoprecipitation of PER and in vitro phosphatase treatment returns PER to its baseline mobility, indicating a crucial role of phosphorylation in these mobility changes. Studies of TIM protein show similar progressive changes in mobility due to phosphorylation (Zeng et al 1996). In fact, TIM phosphorylation seems to occur in parallel with PER, implying similar regulation. The coincidence of the peak in phosphorylation with protein disappearance implicates phosphorylation as a key signal for protein degradation.

A major step forward in the understanding of the mechanistic basis of circadian phosphorylation awaited the identification of *doubletime* (*dbt*), a kinase involved in circadian rhythms. *dbt* was originally identified in genetic screens for mutants with altered circadian locomotor activity rhythms (Price et al 1998). Three *dbt* alleles were identified: one with short period rhythms (hence the mutant name; dbt^S), one with long period rhythms (27 h; dbt^L), and one homozygous lethal allele (dbt^P). As predicted, the long and short period alleles alter in parallel the metabolism of PER and TIM expression, lending little mechanistic insight to the function of this gene (Price et al 1998). In this regard, the lethal allele turned out to be extremely important in establishing a model for *doubletime* function.

It has been believed that a clock operates in *Drosophila* from early in larval development (Brett 1955, Sehgal et al 1992). Subsequent studies have shown that PER and TIM are rhythmically expressed in the larval central nervous system, including the precursors of the adult pacemaker neurons (Kaneko et al 1997). These 16–20 neurons, called the ventral lateral neurons (LNVS), are located deep inside each hemisphere of the *Drosophila* brain and control circadian locomotor activity (see below). As the lethal *dbt* allele, dbt^P, did not induce lethality until after the larval stage, PER and TIM expression could be examined in homozygotes (Price et al 1998). In these homozygotes, PER and TIM cycling is abolished. PER specifically accumulates to very high levels in a hypophosphorylated form. The cloning of *doubletime* revealed that it encodes a kinase that is homologous to human casein kinase I epsilon (Kloss et al 1998). DBT coimmunoprecipitates with PER in cotransfection experiments (Kloss et al 1998). This model posits that DBT phosphorylation of PER is a key signal for PER degradation. However, no study has yet shown that DBT directly phosphorylates PER in *Drosophila*, although this is the likely working hypothesis.

Remarkably, genetic data also support a role for the mammalian homolog of DBT in rhythms. These data involved a spontaneously mutant golden hamster named *tau* (Ralph & Menaker 1988). *tau* is a semidominant mutant with shortened circadian periods. The *tau* locus was molecularly cloned and found to encode the hamster homolog of *Drosophila doubletime* or casein kinase I epsilon (Lowrey et al 2000). The shortened period of the *tau* hamster was attributed to the disrupted biochemical activity of the mutant TAU kinase (Lowrey et al 2000). It was found that this enzyme has a markedly reduced maximal velocity (V_{max}). The mutant protein is still able to associate with PER, although its ability to phosphorylate PER in vitro is reduced. Purified human casein kinase I epsilon (hCKI epsilon) also phosphorylates human PER1 in vitro (Keesler et al 2000). Moreover, in co-transfection experiments, hCKI epsilon associates with and significantly shifts the mobility of hPER1 owing to phosphorylation (Keesler et al 2000). Consistent with genetic evidence from *Drosophila*, hCKI epsilon also destabilizes hPER1 (Keesler et al 2000). Similar findings were made with mouse CKI (Vielhaber et al 2000). As in flies, casein kinase I epsilon appears to play a crucial role in the circadian rhythms in mammals, and the enzymes appear to play similar roles in both systems. Nonetheless, it remains unclear for both systems whether PER is the true target of DBT. In fact, preliminary mammalian studies do not find circadian regulation of electrophoretic mobility of mPERs in vivo as in *Drosophila* (Field et al 2000). Additional antibodies will help determine if mPERs undergo cyclic phosphorylation changes.

How might *doubletime* contribute to circadian timekeeping? Current models of DBT/CKI epsilon function focus on PER phosphorylation. One possibility is that PER, through its association with DBT, imposes circadian regulation on this kinase activity. Perhaps PER prevents DBT from autophosphorylation, which has been shown to inactivate the enzyme (Rivers et al 1998, Gietzen & Virshup 1999). Thus, DBT activity rises with increasing PER levels, leading to an increase in PER phosphorylation. Phosphorylated PER may be a more avid transcriptional repressor, promoting or speeding up transcriptional feedback until PER is degraded and the cycle restarts. In mammals, CKI epsilon also increases PER phosphorylation, but this may lead to a reduction or slowing in mCRY negative feedback (Figure 1). These subtle differences may explain why reductions in kinase activity in mutants lead to opposing period changes: long in flies and short in hamsters (Lowrey et al 2000, Suri et al 2000).

CIRCADIAN ORGANIZATION

The Mammalian Pacemaker: The Suprachiasmatic Nucleus

It has been clear to physiologists that several parameters of biological function are subject to circadian influence. Moreover, studies of the suprachiasmatic nucleus indicated that its destruction resulted in a dramatic loss of much of this rhythmicity (Moore & Eichler 1972, Stephan & Zucker 1972). Moreover, SCN transplantation from another animal could rescue rhythmicity (Drucker-Colin et al 1984, Sawaki et al 1984, Lehman et al 1987). These studies led to an SCN-centric view of circadian systems. The SCN is a symmetrical paired group of approximately 20,000 neurons located just above the optic chiasm. Measurements of metabolic activity using 2-deoxyglucose uptake indicated that its metabolic activity was circadianly regulated (Schwartz & Gainer 1977). Furthermore, dissociated individual neurons showed clear circadian rhythms of spontaneous activity, indicating that rhythmicity was likely generated intracellularly (Welsh et al 1995). However, intercellular mechanisms cannot be excluded as important modifiers of such a rhythm. Key experiments to demonstrate the pivotal role of SCN as a pacemaker involved the short period *tau* hamster described above (Ralph & Menaker 1988). Lesion/transplantation experiments with *tau* animals demonstrated that the period of lesioned/transplanted animals was invariably controlled by the donor SCN (Ralph et al 1990).

Fly Circadian Organization

One of the unexpected dividends of the molecular genetic approach has been a deeper understanding of how circadian systems are organized. In *Drosophila*, the expression of *per* and other clock genes is not restricted to a handful of brain neurons but is widespread in diverse tissue types and organ systems (Liu et al 1988). These clock genes are not just widely expressed. In most cases, they continue to cycle, implying that their expression is relevant to various circadian functions (Hardin 1994).

Previous studies demonstrated that a diffusible signal from the *Drosophila* head confers rhythmicity. Transplantation of the brains of *per^S* animals into the abdomens of arrhythmic per^0 animals rescued circadian activity rhythms (Handler & Konopka 1979). Based on these data, it might be expected that separating peripheral oscillators from the central pacemaker neurons in the *Drosophila* head would completely abolish cycling of gene expression. Consistent with this view, developmental mutants that significantly disrupt *per*-expressing LNVS result in largely behaviorally arrhythmic flies. For example, the *disconnected* (*disco*) mutation eliminates connections between photoreceptor cells in the *Drosophila* eye and targets in the optic lobe (Steller et al 1987). Furthermore, PER-positive lateral neurons are not detectable (Zerr et al 1990). As a result of this disruption in brain architecture, the flies are largely arrhythmic (Dushay et al 1989). However, peripheral cycling is robust in light-dark cycles and persists with some damping over time in constant darkness (Hardin et al 1992a). Consistent with these neuroanatomic mutants, decapitation does not abolish peripheral oscillations (Hege et al 1997). Impressively, when the Malpighian tubules (a *Drosophila* kidney analog) are transplanted from one fly to another, these tubules retain the circadian molecular cycling of the donor, even if out-of-phase with the recipient (Giebultowicz et al 2000). Therefore, although there are hormonal signals in flies, they do not strongly entrain peripheral molecular oscillators. Thus, the lateral neurons do not appear to be essential for peripheral cycling.

Surprisingly, these peripheral clocks respond and entrain to light. Transgenic flies containing the *per* promoter fused to firefly luciferase (*per-luc*) express this reporter gene in the spatial and temporal distribution of the *per* gene (Brandes et al 1996). Luciferase has been used previously as a method of continuous monitoring of gene expression in live organisms (Kay 1993). Because of the relatively rapid turnover of the luciferase enzyme, gene transcription changes are readily reflected in enzyme activity changes. Thus, transgenic *per-luc* flies exhibit cycling bioluminescence when fed on the substrate luciferin. Although clocks are light-sensitive, the degree of bioluminescence is not sufficient to affect behavioral rhythms. Such noninvasive monitoring systems have made the repeated sampling of RNA or protein values unnecessary and have allowed an "on-line" view of gene expression in individual animals.

To test whether these peripheral oscillators respond to light and cycle independently, investigators systematically removed various parts of the *Drosophila* body from a *per-luc* transgenic animal (Plautz et al 1997). They then placed legs, wings, antennae, and other parts into culture media containing luciferin. Once the parts were separated from the central pacemaker, investigators observed that rhythmic gene expression in these persisted but appeared to damp over time in constant darkness. The luminescence technique could not determine if the loss of rhythmicity was due to asynchronous oscillators between different cells or a loss of intrinsic oscillator amplitude within each cell. These separated tissues also demonstrated photoreceptive properties (Plautz et al 1997). Robust rhythmicity, lost after several days in constant darkness, could be reinitiated by reexposure to light cycles. Taken together, these studies reveal how molecular genetic studies can illuminate the organization of circadian systems.

Mammalian Circadian Organization

Studies of the organization of the *Drosophila* circadian system differ markedly with that found in mammals. As mentioned above, lesioning of the SCN abolishes activity rhythms, and transplants restore them to SCN-ablated animals (Ralph et al 1990). Transplantation of SCNs in semipermeable capsules also restores rhythmicity in hamsters (Silver et al 1996). Therefore, similar to *Drosophila* (Handler & Konopka 1979), diffusion of humoral molecules is sufficient to restore some rhythmicity in mammals. As in the case of fly rhythm genes, mammalian rhythm genes are not restricted to the suprachiasmatic nucleus. For example, *mClock* transcripts are found in several neural and nonneural tissues, such as the heart and lungs (King et al 1997). Examination of the temporal profile of *Per* expression in peripheral tissues (e.g. liver and skeletal muscle) finds robust molecular cycling as well as an interesting temporal lag when compared with the oscillations in the SCN (Balsalobre et al 1998, Zylka et al 1998b). Even lymphocytes in peripheral blood exhibit circadian gene cycling (Oishi et al 1998b).

Nevertheless, there are marked differences between flies and mammals in the control of circadian rhythmicity in peripheral oscillators. Unlike the case of the fly, lesioning of the central pacemaker, the SCN, abolishes most peripheral rhythmicity, at least as assayed at the tissue level (Sakamoto et al 1998). Again, loss of rhythmicity at the tissue level may be due either to asynchronous oscillations between different cells or to a loss of intrinsic oscillation amplitude within each cell. Although these observations, in general, differ substantially from *Drosophila*, autonomous light-sensitive peripheral oscillators also exist in mammals: Hamster retinal cells show robust and light-entrainable oscillation in culture (Tosini & Menaker 1996).

Studies using *per*-driven luciferase reporters also emphasize the heavy reliance of peripheral oscillators on central oscillators in mammals. Transgenic rats were constructed using the mouse *mPer1* promoter fused to a luciferase reporter (Yamazaki et al 2000). Various tissues and organs were removed from these animals and assayed for rhythmic bioluminescence in vitro. In culture, the SCNs from these animals exhibited robust rhythms for up to 32 days. Peripheral oscillators such as the liver, lung, and skeletal muscle were also rhythmic, but damped after a few cycles. This damping of peripheral oscillators is reminiscent of *Drosophila* peripheral oscillators. The authors then examined the response to rapid changes, i.e. advances or delays, in the environmental light cycle as might occur to someone experiencing jet lag. They observed that the SCN rhythm shifted rapidly to the new environmental light regime. However, the bioluminescence rhythm in the peripheral tissues was either lost for several days or lagged for many days before shifting to the new light cycle. The asynchrony between environmental light cycles and the SCN on the one hand and peripheral oscillators on the other hand may explain the symptomatology of jet lag.

Immortalized cell lines derived from rat SCN neurons exhibit circadian rhythms of metabolism and gene expression (Earnest et al 1999a,b). Remarkably, transplantation of these SCN cell lines to the third ventricle restored circadian activity rhythm to SCN-lesioned animals (Earnest et al 1999b). Thus, these SCN cell lines may release appropriate phase-setting signals. Transformed cell lines, such as Rat-1 fibroblasts, exhibit circadian patterns of gene expression after a brief exposure of high concentrations of serum (Balsalobre et al 1998). It has been proposed that serum contains important circadian phase-setting factors. Cycling of genes in rat lymphocytes suggests that similar humoral factors may drive circadian clocks in these cells (Oishi et al 1998b).

Prominent among circadianly regulated hormones in humans is melatonin. However, peripheral rhythms persist in melatonin-deficient mice (Zylka et al 1998b). Another important class of circadian-controlled hormones is the glucocorticoids (Tronche et al 1998). Consistent with a possible role of glucocorticoids as output mediators, dexamethasone, a glucocorticoid analog, induces circadian rhythmicity in cultured Rat-1 fibroblasts (Balsalobre et al 2000). Moreover, dexamethasone transiently phase-shifts the rhythm of peripheral organs such as the liver. Finally, dexamethasone has no effect on the SCN rhythm, consistent with an absence of glucocorticoid receptor expression there. However, liver-specific knockout of the glucocorticoid receptor did not abolish circadian rhythmicity in the liver (Balsalobre et al 2000). Thus, glucocorticoids cannot be the sole entraining signals.

In zebrafish, rhythms of gene expression persist in peripheral organs such as the kidney and heart in vitro, i.e. separated from the brain (Whitmore et al 1998). Furthermore, as in the case of *Drosophila*, these peripheral oscillators appear to be directly light sensitive (Whitmore et al 1998). In fact, a zebrafish cell line has been shown to exhibit rhythmic gene expression in vitro (Whitmore et al 2000). Moreover, these rhythms are also entrainable by light. These data are more consistent with the more autonomous fly circadian organization.

Why these differences in organization? In small and partially transparent animals like *Drosophila* and zebrafish, light can easily penetrate into the body of the animal. Therefore, the presence of a photoreceptor (CRY in flies) allows a very simple way to synchronize the peripheral oscillators with the environment. On the other hand, larger and more opaque animals like mammals may need neurohormonal control of circadian oscillations in internal organs. Strikingly, the mammalian retina, which obviously receives light, has maintained light-sensitive cell-autonomous oscillation. The nature of the photoreceptor in this tissue or any mammalian circadian photoreceptor is still mysterious. In zebrafish, though, *zCRY4*, which is in sequence much closer to *Drosophila* CRY than to the mammalian CRYs, looks like an interesting candidate for study (Kobayashi et al 2000).

CIRCADIAN OUTPUT

Direct Output Targets

Although much progress has been made in identifying clock genes, little is known about how these genes connect to outputs. In many organisms, various fractions of whole genomes appear to be under circadian control, from the vast majority in photosynthetic cyanobacteria to 1–5% in *Drosophila* (Liu et al 1995, Van Gelder et al 1995). Given the size of the *Drosophila* genome, hundreds of genes should cycle. These data implicate transcription in the control of various output genes and therefore in the control of behavioral and physiologic outputs. Most of these clock-controlled genes are probably indirectly controlled; many fewer are likely to be direct targets of clock genes. In our opinion, a direct interaction with a clock component has not yet been conclusively demonstrated for any output gene.

What then should be the criteria for a direct output target? As identification of cycling RNAs seems to be a straightforward strategy, it is likely that a large fraction of these will be controlled at the level of transcription. In this case, there should be direct binding of a circadian transcription factor, such as *Clock* to the promoter of the regulated gene. There are two main ways to show this convincingly. First, the construction of altered specificity mutants, i.e. mutants that alter DNA-binding specificity need to be coupled to mutants with correspondingly altered binding sites to prove direct binding in vivo. Second, chromatin immunoprecipitations could show that a factor is physically bound to its putative target in vivo. However, as these strategies are relatively difficult, studies of output gene regulation have heavily relied on transient transfection into tissue culture cells. Typically, DNAs expressing the relevant clock genes as well as the target promoters driving a reporter are transfected and allowed to express for 24–48 hours. The long temporal delay between expression and the downstream target gene allows for several intermediate steps to occur. Thus, the observed effects could easily be indirect. The relevance of the putative target site of the transcription factor needs to be assessed in vivo. Promoter elements need to confer cycling on reporter genes in transgenic animals. Mutation of the appropriate binding sites should abolish or reduce this cycling.

Most studies of output genes have tried to identify the targets of the CLK and CYC transcription complex, CACGTG. By simple randomness, this sequence will occur approximately once every four kilobases. The random probability that one will find such a sequence in the vicinity of a cycling gene is therefore relatively high. Moreover, expression is typically far higher in transfection experiments than what would be observed in vivo. The transfected target promoter DNA does not have the typical repressing chromatin structure of genomic DNA. Thus, under these highly artificial conditions, overexpressed transcription factors may find these vulnerable E-boxes, when they would not be accessible in vivo. Therefore, results based only on these types of experiments need to be interpreted very cautiously.

The neuropeptide gene, *arginine vasopressin*, is one putative output target in mammals. Besides many other functions, such as the physiologic control of water balance, this gene is also rhythmically expressed in the suprachiasmatic nucleus (Uhl & Reppert 1986). In *Clock* mutants the levels of *vasopressin* RNA are dramatically reduced and rhythmicity is abolished in the SCN (Jin et al 1999). In cell culture experiments, mCLOCK and its partner BMAL1 activate from E-boxes in the *vasopressin* promoter, providing a plausible explanation for the in vivo results and a model for how output might be generated (Jin et al 1999). However, it remains unclear whether the E-box sequence found in the *vasopressin* promoter is required for its rhythmic expression in vivo. Moreover both the genetic evidence and the cell culture data leave open the possibility that the effect of mCLOCK-BMAL1 on *vasopressin* is indirect.

A more likely direct target output gene is the D-box binding element protein (DBP). DBP is rhythmically expressed in the liver and SCN among other tissues (Fonjallaz et al 1996). DBP activates the promoters of many enzymes involved in hepatic processes, including cholesterol metabolism (Lavery et al 1999). Mice lacking DBP exhibit a subtle circadian behavioral phenotype, implying that DBP may have a primary function in mediating output (Lopez-Molina et al 1997). Deletion of intragenic DBP regions substantially reduce promoter-driven expression (Ripperger et al 2000). Studies of in vivo promoter occupancy identified DNase I hypersensitive sites, suggesting that an exclusive focus on E-boxes is overly simplistic. Protein binding to DNA often renders the local DNA sensitive to DNase I. In the *Dbp* genomic region, as many as five different loci, some of which are intragenic, undergo circadian regulation of DNase I hypersensitivity. Control of output genes may therefore be much more complex than regulation by one or two factors. E-box motifs were identified and shown to bind mCLOCK in vitro. However, this binding activity does not cycle, raising questions about whether mCLOCK is the factor cyclically bound in vivo. Nonetheless, DBP regulates several enzymes circadianly expressed in the liver and is therefore an important link between the circadian pacemaker and the final outputs.

In combination with genetic approaches, recent work in *Drosophila* has focused on molecular methods to find clock-controlled genes. In a search for genes that are differentially expressed between wild-type and *per[0]* animals, the zinc finger transcription factor *vrille* (*vri*) was identified (Blau & Young 1999). Previous studies of the *vri* gene demonstrated that it was required for embryonic development (George & Terracol 1997). Like *per*, *vri* is rhythmically expressed in circadian pacemaker neurons (Blau & Young 1999). Furthermore, as in the case of *vasopressin*, the *Drosophila* orthologs of mCLOCK and BMAL1, CLOCK and CYCLE, activate a reporter gene fused to the *vri* promoter, which contains E-box target sites (Blau & Young 1999). Constitutive overexpression of *vri* lengthened behavioral rhythms and reduced expression of the *period* and *timeless* RNAs, indicating a potential role of this gene in the central pacemaker. This expression also downregulates the output molecule, pigment dispersing factor (PDF; see below). However, phenotypes due to misexpression or overexpression can be misleading, as they may induce functions that normal expression does not. Heterozygous deletions of the *vri* locus yield very subtle circadian phenotypes. Conditional rescue of the embryonic lethality and subsequent analysis of homozygous null mutant adults will be required to more fully specify the role of *vri* in the pacemaker.

Using a similar subtractive hybridization approach, another clock-regulated gene was identified called *takeout* (*to*) (So et al 2000). *takeout* was originally identified on the basis of its low expression in the *cyc[0]* mutant. Studies of a *takeout* mutant suggest that TAKEOUT protein is involved in the response to starvation (Sarov-Blat et al 2000). *takeout* expression is reduced in all circadian rhythm mutants tested, which distinguishes it from any other studied transcript. Although *takeout* is also rhythmically expressed at the RNA level, the phase of this oscillation is distinct from other cycling RNAs from *Clk, per*, and *tim*. As in the case of *vri* and *vasopressin*, a search was undertaken to identify E-box target sites. Not only was an E-box identified, but there was also remarkable similarity in promoter regions flanking the E-box, suggesting that this region was functional. Surprisingly, this E-box was not sufficient to drive cycling when tested in vivo (which was not done for *vri, vasopressin*, or *Dbp* promoters). Alternatively, the lack of cycling could have been due to an artifact of the transgene. Nonetheless, the *takeout* mRNA cycling is not likely due to a direct effect of CLK and CYC. These studies highlight the importance of testing enhancer elements in vivo for circadian function.

Circadian Behavioral Output Molecules

Though the discovery of several new output genes and phenomena illustrates how pacemakers may generate output, they did not reveal any key mediator of behavioral rhythms. PDF is a neuropeptide that has been well studied in invertebrates and has been shown to phase-shift circadian clocks when injected into cockroaches (Petri & Stengl 1997). Initial studies with *Drosophila pdf* indicated that its expression was largely restricted to the head and absent in *disco* mutants (Park 1998). These mutants are largely arrhythmic and lack pacemaker lateral neurons (Dushay et al 1989, Zerr et al 1990). *pdf* mRNA and protein are specifically expressed in a ventral subset of these lateral neurons (LNvs) (Helfrich-Forster 1995, Park et al 2000). Furthermore, PDF is rhythmically expressed in the termini of these neurons (Park et al 2000). The role of *pdf* in these neurons was more clearly elucidated by the discovery of a *pdf* mutant, *pdf^{01}* (Renn et al 1999). Like *disco* mutants, these mutants display only weak or no rhythms. Because *disco* mutants exhibit far more neuroanatomical defects beyond just an absence of pacemaker lateral neurons, the *pdf* promoter was used to direct expression of the proapoptotic genes, *head-involution defective* (*hid*) and *reaper*, to the ventral lateral neurons. This expression resulted in the complete ablation of these neurons. The behavioral consequence was a circadian phenotype virtually identical to that of *pdf^{01}*, indicating that the prinicipal mediator of the circadian signal from these key pacemaker neurons is *pdf* (Renn et al 1999).

Is PDF synaptically released? To address the role of chemical synaptic transmission, the tetanus-toxin light chain (TeTxLC) was expressed in all pacemaker-containing cells (Kaneko et al 2000). TeTxLC blocks synaptic transmission by cleaving the synaptic protein, synaptobrevin. Although clock gene cycling was unaffected in these animals, rhythmic behavior was substantially reduced, indicating a role for synaptic transmission. Surprisingly, targeted expression of TeTxLC exclusively to the LN_{vs} did not affect circadian behavior, indicating that *pdf* release may operate through nonsynaptic mechanisms (Kaneko et al 2000). Furthermore, other cells may also be relevant to behavioral rhythms. In contrast to the SCN, which relies on the eyes for synchronizing its activity with the environment, the LNVS also contain a circadian photoreceptor and are therefore directly light sensitive (Emery et al 2000). They are circadianly self-sufficient because they contain an input pathway, a molecular pacemaker, and an output. There does not appear to be a mammalian ortholog of PDF. It will be interesting to see what the mammalian version(s) of this output signal will be.

Sleep in Flies?

One of the most controversial notions regarding circadian output in flies is that the behavioral cycles of flies mimic the mammalian sleep-wake cycle. Close observations of fly behavior during their 24 hour cycle demonstrate that flies can be immobile for long periods of time (over 2 hours in some cases) (Hendricks et al 2000, Shaw et al 2000). During this immobile state, flies exhibit an increased threshold to arousing, sensory stimuli. Importantly, this state is homeostatically regulated, like sleep. Rest-deprived flies will increase their rest subsequently. Similar stimulation during wake periods has no subsequent effect on rest behavior. Drugs that increase or decrease sleep in mammals, such as antihistamines and caffeine, increase and decrease rest in flies. As in the case of circadian rhythms, molecular and genetic studies of sleep (or this sleep-like state if you prefer) in flies may prove to be very influential in our understanding of the neurobiology of sleep.

CONCLUSION

The problem of circadian rhythms has been reduced to the molecular realm: a problem of understanding transcriptional feedback loops and the posttranslational regulation of their loop components. The past few years have seen a remarkable increase in the number of identified clock genes. There remains much to learn about how these genes regulate each other, especially in mammals, where we are truly at the dawn of the molecular era. In the next few years an important focus will be on posttranslational regulation and feedback. The identification of additional kinases and phosphatases as well as their regulatory features will be central to a better understanding of timing. In addition to understanding how the clock couples to output, the identification of key output molecules in mammals may have medical applications. Many of these discoveries will undoubtedly continue to capitalize on the power of forward genetics.

Visit the Annual Reviews home page at www.AnnualReviews.org

LITERATURE CITED

Abe H, Honma S, Namihira M, Tanahashi Y, Ikeda M, Honma K. 1998. Circadian rhythm and light responsiveness of BMAL1 expression, a partner of mammalian clock gene

Clock, in the suprachiasmatic nucleus of rats. *Neurosci. Lett.* 258:93–96

Akiyama M, Kouzu Y, Takahashi S, Wakamatsu H, Moriya T, et al. 1999. Inhibition of light- or glutamate-induced mPer1 expression represses the phase shifts into the mouse circadian locomotor and suprachiasmatic firing rhythms. *J. Neurosci.* 19:1115–21

Albrecht U, Sun ZS, Eichele G, Lee CC. 1997. A differential response of two putative mammalian circadian regulators, mper1 and mper2, to light. *Cell* 91:1055–64

Allada R, White NE, So WV, Hall JC, Rosbash M. 1998. A mutant Drosophila homolog of mammalian Clock disrupts circadian rhythms and transcription of period and timeless. *Cell* 93:791–804

Antoch MP, Song EJ, Chang AM, Vitaterna MH, Zhao Y, et al. 1997. Functional identification of the mouse circadian Clock gene by transgenic BAC rescue *Cell* 89:655–67

Bae K, Lee C, Hardin PE, Edery I. 2000. dCLOCK is present in limiting amounts and likely mediates daily interactions between the dCLOCK–CYC transcription factor and the PER-TIM complex. *J. Neurosci.* 20:1746–53

Bae K, Lee C, Sidote D, Chuang KY, Edery I. 1998. Circadian regulation of a Drosophila homolog of the mammalian Clock gene: PER and TIM function as positive regulators *Mol. Cell. Biol.* 18:6142–51

Balsalobre A, Brown SA, Marcacci L, Tronche F, Kellendonk C, et al. 2000. Resetting of circadian time in peripheral tissues by glucocorticoid signaling. *Science* 289:2344–47

Balsalobre A, Damiola F, Schibler U. 1998. A serum shock induces circadian gene expression in mammalian tissue culture cells. *Cell* 93:929–37

Benna C, Scannapieco P, Piccin A, Sandrelli F, Zordan M, et al. 2000. A second timeless gene in Drosophila shares greater sequence similarity with mammalian tim. *Curr. Biol.* 10:R512–13

Benzer S. 1971. From the gene to behavior. *JAMA* 218:1015–22

Blau J, Young MW. 1999. Cycling vrille expression is required for a functional Drosophila clock. *Cell* 99:661–712

Brandes C, Plautz JD, Stanewsky R, Jamison CF, Straume M, et al. 1996. Novel features of Drosophila period transcription revealed by real-time luciferase reporting. *Neuron* 16:687–92

Brett W. 1955. Persistent diurnal rhythmicity in Drosophila emergence. *Ann. Entomol. Soc. Am.* 48:119–31

1997. Breakthrough of the year. The runners-up. *Science* 278:2039–42

1998. Breakthrough of the year. The runners-up. *Science* 282:2157–61

Bruce VG. 1972. Mutants of the biological clock in Chlamydomonas reinhardi. *Genetics* 70:537–48

Bunger MK, Wilsbacher LD, Moran SM, Clendenin C, Radcliffe LA, et al. 2000. Mop3 is an essential component of the master circadian pacemaker in mammals. *Cell* 103:1009–17

Cashmore AR, Jarillo JA, Wu YJ, Liu D. 1999. Cryptochromes: blue light receptors for plants and animals. *Science* 284:760–65

Crews ST, Thomas JB, Goodman CS. 1988. The Drosophila single-minded gene encodes a nuclear protein with sequence similarity to the per gene product. *Cell* 52:143–51

Curtin KD, Huang ZJ, Rosbash M. 1995. Temporally regulated nuclear entry of the Drosophila period protein contributes to the circadian clock. *Neuron* 14:365–72

Darlington TK, Wager-Smith K, Ceriani MF, Staknis D, Gekakis N, et al. 1998. Closing the circadian loop: CLOCK–induced transcription of its own inhibitors per and tim. *Science* 280:1599–603

Drucker-Colin R, Aguilar-Roblero R, Garcia-Hernandez F, Fernandez-Cancino F, Bermudez Rattoni F. 1984. Fetal suprachiasmatic nucleus transplants: diurnal rhythm recovery of lesioned rats. *Brain Res.* 311:353–57

Dunlap JC. 1999. Molecular bases for circadian clocks. *Cell* 96:271–90

Dushay MS, Rosbash M, Hall JC. 1989. The disconnected visual system mutations in *Drosophila melanogaster* drastically disrupt circadian rhythms. *J. Biol. Rhythms* 4:1–27

Earnest DJ, Liang FQ, DiGiorgio S, Gallagher

M, Harvey B, et al. 1999a. Establishment and characterization of adenoviral E1A immortalized cell lines derived from the rat suprachiasmatic nucleus. *J. Neurobiol.* 39:1–13

Earnest DJ, Liang FQ, Ratcliff M, Cassone VM. 1999b. Immortal time: circadian clock properties of rat suprachiasmatic cell lines. *Science* 283:693–95

Edery I. 2000. Circadian rhythms in a nutshell. *Phys. Genomics* 3:59–74

Edery I, Zwiebel LJ, Dembinska ME, Rosbash M. 1994. Temporal phosphorylation of the Drosophila period protein. *Proc. Natl. Acad. Sci. USA* 91:2260–64

Egan ES, Franklin TM, Hilderbrand-Chae MJ, McNeil GP, Roberts MA, et al. 1999. An extraretinally expressed insect cryptochrome with similarity to the blue light photoreceptors of mammals and plants. *J. Neurosci.* 19:3665–73

Emery P, So WV, Kaneko M, Hall JC, Rosbash M. 1998. CRY, a Drosophila clock and light-regulated cryptochrome, is a major contributor to circadian rhythm resetting and photosensitivity. *Cell* 95:669–79

Emery P, Stanewsky R, Helfrich-Forster C, Emery-Le M, Hall JC, Rosbash M. 2000. Drosophila CRY is a deep brain circadian photoreceptor. *Neuron* 26:493–504

Feldman JF, Hoyle MN. 1973. Isolation of circadian clock mutants of Neurospora crassa. *Genetics* 75:605–13

Field MD, Maywood ES, O'Brien JA, Weaver DR, Reppert SM, Hastings MH. 2000. Analysis of clock proteins in mouse SCN demonstrates phylogenetic divergence of the circadian clockwork and resetting mechanisms. *Neuron* 25:437–47

Fonjallaz P, Ossipow V, Wanner G, Schibler U. 1996. The two PAR leucine zipper proteins, TEF and DBP, display similar circadian and tissue-specific expression, but have different target promoter preferences. *EMBO J.* 15:351–62

Gekakis N, Saez L, Delahaye-Brown AM, Myers MP, Sehgal A, et al. 1995. Isolation of timeless by PER protein interaction: defective interaction between timeless protein and long-period mutant PERL. *Science* 270:811–15

Gekakis N, Staknis D, Nguyen HB, Davis FC, Wilsbacher LD, et al. 1998. Role of the CLOCK protein in the mammalian circadian mechanism. *Science* 280:1564–69

George H, Terracol R. 1997. The vrille gene of Drosophila is a maternal enhancer of decapentaplegic and encodes a new member of the bZIP family of transcription factors. *Genetics* 146:1345–63

Giebultowicz JM, Stanewsky R, Hall JC, Hege DM. 2000. Transplanted Drosophila excretory tubules maintain circadian clock cycling out of phase with the host. *Curr. Biol.* 10:107–10

Gietzen KF, Virshup DM. 1999. Identification of inhibitory autophosphorylation sites in casein kinase I epsilon *J. Biol. Chem.* 274:32063–70

Glossop NR, Lyons LC, Hardin PE. 1999. Interlocked feedback loops within the Drosophila circadian oscillator. *Science* 286:766–68

Gotter AL, Manganaro T, Weaver DR, Kolakowski LF Jr, Possidente B, et al. 2000. A time-less function for mouse timeless. *Nat. Neurosci.* 3:755–56

Griffin EA Jr, Staknis D, Weitz CJ. 1999. Light–independent role of CRY1 and CRY2 in the mammalian circadian clock. *Science* 286:768–71

Hall JC. 2000. Cryptochromes: sensory reception, transduction, and clock functions subserving circadian systems. *Curr. Opin. Neurobiol.* 10:456–66

Handler AM, Konopka RJ. 1979. Transplantation of a circadian pacemaker in Drosophila. *Nature* 279:236–38

Hao H, Allen DL, Hardin PE. 1997. A circadian enhancer mediates PER–dependent mRNA cycling in *Drosophila melanogaster. Mol. Cell. Biol.* 17:3687–93

Hardin PE. 1994. Analysis of period mRNA cycling in Drosophila head and body tissues indicates that body oscillators behave differently from head oscillators. *Mol. Cell. Biol.* 7211–18

Hardin PE, Hall JC, Rosbash M. 1990. Feedback of the Drosophila period gene product on circadian cycling of its messenger RNA levels. *Nature* 343:536–40

Hardin PE, Hall JC, Rosbash M. 1992a. Behavioral and molecular analyses suggest that circadian output is disrupted by disconnected mutants in *D. melanogaster. EMBO J.* 11:1–6

Hardin PE, Hall JC, Rosbash M. 1992b. Circadian oscillations in period gene mRNA levels are transcriptionally regulated. *Proc. Natl. Acad. Sci. USA* 89:11711–15

Hastings MH, Field MD, Maywood ES, Weaver DR, Reppert SM. 1999. Differential regulation of mPER1 and mTIM proteins in the mouse suprachiasmatic nuclei: new insights into a core clock mechanism. *J. Neurosci.* 19:RC11

Hege DM, Stanewsky R, Hall JC, Giebultowicz JM. 1997. Rhythmic expression of a PER-reporter in the Malpighian tubules of decapitated Drosophila: evidence for a brain-independent circadian clock. *J. Biol. Rhythms* 12:300–8

Helfrich-Forster C. 1995. The period clock gene is expressed in central nervous system neurons which also produce a neuropeptide that reveals the projections of circadian pacemaker cells within the brain of *Drosophila melanogaster. Proc. Natl. Acad. Sci. USA* 92:612–16

Hendricks JC, Finn SM, Panckeri KA, Chavkin J, Williams JA, et al. 2000. Rest in Drosophila is a sleep-like state. *Neuron* 25:129–38

Hogenesch JB, Gu YZ, Jain S, Bradfield CA. 1998. The basic-helix-loop-helix-PAS orphan MOP3 forms transcriptionally active complexes with circadian and hypoxia factors. *Proc. Natl. Acad. Sci. USA* 95:5474–79

Hogenesch JB, Gu YZ, Moran SM, Shimomura K, Radcliffe LA, et al. 2000. The basic helix-loop-helix-PAS protein MOP9 is a brain-specific heterodimeric partner of circadian and hypoxia factors. *J. Neurosci.* 20:RC83

Huang ZJ, Edery I, Rosbash M. 1993. PAS is a dimerization domain common to Drosophila period and several transcription factors. *Nature* 364:259–62

Hunter-Ensor M, Ousley A, Sehgal A. 1996. Regulation of the Drosophila protein timeless suggests a mechanism for resetting the circadian clock by light *Cell* 84:677–85

Ishida N, Kaneko M, Allada R. 1999. Biological clocks. *Proc. Natl. Acad. Sci. USA* 96:8819–20. Erratum. 2000. *Proc. Natl. Acad. Sci. USA* 97(16):9347

Ishikawa T, Matsumoto A, Kato T Jr, Togashi S, Ryo H, et al. 1999. DCRY is a Drosophila photoreceptor protein implicated in light entrainment of circadian rhythm. *Genes Cells* 4:57–65

Jackson FR, Bargiello TA, Yun SH, Young MW. 1986. Product of per locus of Drosophila shares homology with proteoglycans. *Nature* 320:185–88

Jin X, Shearman LP, Weaver DR, Zylka MJ, de Vries GJ, Reppert SM. 1999. A molecular mechanism regulating rhythmic output from the suprachiasmatic circadian clock. *Cell* 96:57–68

Johnson CH, Golden SS. 1999. Circadian programs in cyanobacteria adaptiveness and mechanism *Annu. Rev. Microbiol.* 53:389–409

Kaneko M, Helfrich-Forster C, Hall JC. 1997. Spatial and temporal expression of the period and timeless genes in the developing nervous system of Drosophila: newly identified pacemaker candidates and novel features of clock gene product cycling. *J. Neurosci.* 17:6745–60

Kaneko M, Park JH, Cheng Y, Hardin PE, Hall JC. 2000. Disruption of synaptic transmission or clock-gene-product oscillations in circadian pacemaker cells of Drosophila cause abnormal behavioral rhythms. *J. Neurobiol.* 43:207–33

Kay SA. 1993. Shedding light on clock controlled cab gene transcription in higher plants. *Semin. Cell Biol.* 4:81–86

Keesler GA, Camacho F, Guo Y, Virshup D,

Mondadori C, Yao Z. 2000. Phosphorylation and destabilization of human period I clock protein by human casein kinase I epsilon. *NeuroReport* 11:951–55

King DP, Takahashi JS. 2000. Molecular genetics of circadian rhythms in mammals. *Annu. Rev. Neurosci.* 23:713–42

King DP, Zhao Y, Sangoram AM, Wilsbacher LD, Tanaka M, et al. 1997. Positional cloning of the mouse circadian clock gene. *Cell* 89:641–53

Kloss B, Price JL, Saez L, Blau J, Rothenfluh A, et al. 1998. The Drosophila clock gene double-time encodes a protein closely-related to human casein kinase-I-epsilon. *Cell* 94:97–107

Kobayashi Y, Ishikawa T, Hirayama J, Daiyasu H, Kanai S, et al. 2000. Molecular analysis of zebrafish photolyase/cryptochrome family: two types of cryptochromes present in zebrafish. *Genes Cells* 5:725–38

Koike N, Hida A, Numano R, Hirose M, Sakaki Y, Tei H. 1998. Identification of the mammalian homologues of the Drosophila timeless gene, Timeless1. *FEBS Lett.* 441:427–31

Konopka RJ, Benzer S. 1971. Clock mutants of *Drosophila melanogaster. Proc. Natl. Acad. Sci. USA* 68:2112–16

Kuhlman SJ, Quintero JE, McMahon DG. 2000. GFP fluorescence reports Period 1 circadian gene regulation in the mammalian biological clock. *NeuroReport* 11:1479–82

Kume K, Zylka MJ, Sriram S, Shearman LP, Weaver DR, et al. 1999. mCRY1 and mCRY2 are essential components of the negative limb of the circadian clock feedback loop. *Cell* 98:193–205

Lavery DJ, Lopez-Molina L, Margueron R, Fleury-Olela F, Conquet F, et al. 1999. Circadian expression of the steroid 15 alpha-hydroxylase (Cyp2a4) and coumarin 7-hydroxylase (Cyp2a5) genes in mouse liver is regulated by the PAR leucine zipper transcription factor DBP. *Mol. Cell. Biol.* 19:6488–99

Lee C, Bae K, Edery I. 1998. The Drosophila CLOCK protein undergoes daily rhythms

in abundance, phosphorylation, and interactions with the PER–TIM complex. *Neuron* 21:857–67

Lee C, Bae K, Edery I. 1999. PER and TIM inhibit the DNA binding activity of a Drosophila CLOCK-CYC/dBMAL1 heterodimer without disrupting formation of the heterodimer: a basis for circadian transcription. *Mol. Cell. Biol.* 19:5316–25

Lee C, Parikh V, Itsukaichi T, Bae K, Edery I. 1996. Resetting the Drosophila clock by photic regulation of PER and a PER-TIM complex. *Science* 271:1740–44

Lee K, Loros JJ, Dunlap JC. 2000. Interconnected feedback loops in the Neurospora circadian system. *Science* 289:107–10

Lehman MN, Silver R, Gladstone WR, Kahn RM, Gibson M, Bittman EL. 1987. Circadian rhythmicity restored by neural transplant. Immunocytochemical characterization of the graft and its integration with the host brain. *J. Neurosci.* 7:1626–38

Lindebro MC, Poellinger L, Whitelaw ML. 1995. Protein-protein interaction via PAS domains: role of the PAS domain in positive and negative regulation of the bHLH/PAS dioxin receptor-Arnt transcription factor complex. *EMBO J.* 14:3528–39

Liu X, Lorenz L, Yu QN, Hall JC, Rosbash M. 1988. Spatial and temporal expression of the period gene in *Drosophila melanogaster. Genes Dev.* 2:228–38

Liu Y, Tsinoremas NF, Johnson CH, Lebedeva NV, Golden SS, et al. 1995. Circadian orchestration of gene expression in cyanobacteria. *Genes Dev.* 9:1469–78

Lopez-Molina L, Conquet F, Dubois-Dauphin M, Schibler U. 1997. The DBP gene is expressed according to a circadian rhythm in the suprachiasmatic nucleus and influences circadian behavior. *EMBO J.* 16:6762–71

Lowrey PL, Shimomura K, Antoch MP, Yamazaki S, Zemenides PD, et al. 2000. Positional syntenic cloning and functional characterization of the mammalian circadian mutation tau. *Science* 288:483–92

Marrus SB, Zeng H, Rosbash M. 1996. Effect

of constant light and circadian entrainment of pers flies: evidence for light-mediated delay of the negative feedback loop in Drosophila. *EMBO J.* 15:6877–86

Matsumoto A, Tomioka K, Chiba Y, Tanimura T. 1999. timrit Lengthens circadian period in a temperature-dependent manner through suppression of PERIOD protein cycling and nuclear localization. *Mol. Cell. Biol.* 19:4343–54

Miyamoto Y, Sancar A. 1998. Vitamin B2-based blue-light photoreceptors in the retinohypothalamic tract as the photoactive pigments for setting the circadian clock in mammals. *Proc. Natl. Acad. Sci. USA* 95:6097–102

Moore RY, Eichler VB. 1972. Loss of a circadian adrenal corticosterone rhythm following suprachiasmatic lesions in the rat. *Brain Res.* 42:201–6

Myers MP, Wager-Smith K, Rothenfluh-Hilfiker A, Young MW. 1996. Light-induced degradation of TIMELESS and entrainment of the Drosophila circadian clock. *Science* 271:1736–40

Oishi K, Sakamoto K, Okada T, Nagase T, Ishida N. 1998a. Antiphase circadian expression between BMAL1 and period homologue mRNA in the suprachiasmatic nucleus and peripheral tissues of rats. *Biochem. Biophys. Res. Commun.* 253:199–203

Oishi K, Sakamoto K, Okada T, Nagase T, Ishida N. 1998b. Humoral signals mediate the circadian expression of rat period homologue (rPer2) mRNA in peripheral tissues. *Neurosci. Lett.* 256:117–19

Okamura H, Miyake S, Sumi Y, Yamaguchi S, Yasui A, et al. 1999. Photic induction of mPer1 and mPer2 in cry-deficient mice lacking a biological clock. *Science* 286:2531–34

Park JH, Hall JC. 1998. Isolation and chronobiological analysis of a neuropeptide pigment-dispersing factor gene in Drosophila melanogaster. *J. Biol. Rhythms* 13:219–28

Park JH, Helfrich-Forster C, Lee G, Liu L, Rosbash M, Hall JC. 2000. Differential regula-

tion of circadian pacemaker output by sepa rate clock genes in Drosophila. *Proc. Natl Acad. Sci. USA* 97:3608–13

Petri B, Stengl M. 1997. Pigment-dispersing hormone shifts the phase of the circadian pacemaker of the cockroach Leucophaea maderae. *J. Neurosci.* 17:4087–93

Pittendrigh C. 1954. On temperature independence in the clock-system controlling emer gence time in Drosophila. *Proc. Natl. Acad Sci. USA* 40:1018–29

Plautz JD, Kaneko M, Hall JC, Kay SA. 1997 Independent photoreceptive circadian clocks throughout Drosophila. *Science* 278:1632–35

Price JL, Blau J, Rothenfluh A, Abodeely M, Kloss B, Young MW. 1998. double-time i a novel Drosophila clock gene that regulates PERIOD protein accumulation. *Cell* 94:83–95

Price JL, Dembinska ME, Young MW, Rosbash M. 1995. Suppression of PERIOD protein abundance and circadian cycling by the Drosophila clock mutation timeless. *EMBO J.* 14:4044–49

Ralph MR, Foster RG, Davis FC, Menaker M. 1990. Transplanted suprachiasmatic nucleus determines circadian period. *Science* 247:975–78

Ralph MR, Menaker M. 1988. A mutation of the circadian system in golden hamsters. *Science* 241:1225–27

Reddy P, Jacquier AC, Abovich N, Petersen G, Rosbash M. 1986. The period clock locus o D. melanogaster codes for a proteoglycan *Cell* 46:53–61

Renn SC, Park JH, Rosbash M, Hall JC, Taghert PH. 1999. A pdf neuropeptide gene mutation and ablation of PDF neurons each cause severe abnormalities of behavioral circadian rhythms in Drosophila. *Cell* 99:791–802. Erratum. 2000. *Cell* 101(1): following 113

Ripperger JA, Shearman LP, Reppert SM, Schibler U. 2000. CLOCK, an essential pace maker component, controls expression of the circadian transcription factor DBP. *Gene Dev.* 14:679–89.

Rivers A, Gietzen KF, Vielhaber E, Virshup DM. 1998. Regulation of casein kinase I epsilon and casein kinase I delta by an in vivo futile phosphorylation cycle. *J. Biol. Chem.* 273:15980–84.

Rothenfluh A, Young MW, Saez L. 2000. A TIMELESS-independent function for PERIOD proteins in the Drosophila clock. *Neuron* 26:505–14

Ruby NF, Burns DE, Heller HC. 1999. Circadian rhythms in the suprachiasmatic nucleus are temperature-compensated and phase-shifted by heat pulses in vitro. *J. Neurosci.* 19:8630–36

Rutila JE, Suri V, Le M, So WV, Rosbash M, Hall JC. 1998. CYCLE is a second bHLH-PAS clock protein essential for circadian rhythmicity and transcription of Drosophila period and timeless. *Cell* 93:805–14

Saez L, Young MW. 1996. Regulation of nuclear entry of the Drosophila clock proteins period and timeless. *Neuron* 17:911–20

Sakamoto K, Nagase T, Fukui H, Horikawa K, Okada T, et al. 1998. Multitissue circadian expression of rat period homolog (rPer2) mRNA is governed by the mammalian circadian clock, the suprachiasmatic nucleus in the brain. *J. Biol. Chem.* 273:27039–42

Sangoram AM, Saez L, Antoch MP, Gekakis N, Staknis D, et al. 1998. Mammalian circadian autoregulatory loop: a timeless ortholog and mPer1 interact and negatively regulate CLOCK-BMAL1-induced transcription. *Neuron* 21:1101–13

Sarov-Blat L, So WV, Liu L, Rosbash M. 2000. The Drosophila takeout gene is a novel molecular link between circadian rhythms and feeding behavior. *Cell* 101:647–56

Sawaki Y, Nihonmatsu I, Kawamura H. 1984. Transplantation of the neonatal suprachiasmatic nuclei into rats with complete bilateral suprachiasmatic lesions. *Neurosci. Res.* 1:67–72

Schwartz WJ, Gainer H. 1977. Suprachiasmatic nucleus: use of 14C-labeled deoxyglucose uptake as a functional marker. *Science* 197:1089–91

Scully AL, Kay SA. 2000. Time flies for Drosophila. *Cell* 100:297–300

Sehgal A, Price J, Young MW. 1992. Ontogeny of a biological clock in *Drosophila melanogaster*. *Proc. Natl. Acad. Sci. USA* 89:1423–27

Sehgal A, Price JL, Man B, Young MW. 1994. Loss of circadian behavioral rhythms and per RNA oscillations in the Drosophila mutant timeless. *Science* 263:1603–6

Sehgal A, Rothenfluh-Hilfiker A, Hunter-Ensor M, Chen Y, Myers MP, Young MW. 1995. Rhythmic expression of timeless: a basis for promoting circadian cycles in period gene autoregulation. *Science* 270:808–10

Shaw PJ, Cirelli C, Greenspan RJ, Tononi G. 2000. Correlates of sleep and waking in *Drosophila melanogaster*. *Science* 287:1834–37

Shearman LP, Jin X, Lee C, Reppert SM, Weaver DR. 2000a. Targeted disruption of the mPer3 gene: subtle effects on circadian clock function. *Mol. Cell. Biol.* 20:6269–75

Shearman LP, Sriram S, Weaver DR, Maywood ES, Chaves I, et al. 2000b. Interacting molecular loops in the mammalian circadian clock. *Science* 288:1013–19

Shearman LP, Zylka MJ, Weaver DR, Kolakowski LF Jr, Reppert SM. 1997. Two period homologs: circadian expression and photic regulation in the suprachiasmatic nuclei. *Neuron* 19:1261–69

Shigeyoshi Y, Taguchi K, Yamamoto S, Takekida S, Yan L, et al. 1997. Light-induced resetting of a mammalian circadian clock is associated with rapid induction of the mPer1 transcript. *Cell* 91:1043–53

Silver R, LeSauter J, Tresco PA, Lehman MN. 1996. A diffusible coupling signal from the transplanted suprachiasmatic nucleus controlling circadian locomotor rhythms. *Nature* 382:810–13

Siwicki KK, Eastman C, Petersen G, Rosbash M, Hall JC. 1988. Antibodies to the period gene product of Drosophila reveal diverse tissue distribution and rhythmic changes in the visual system. *Neuron* 1:141–50

So WV, Rosbash M. 1997. Post-transcriptional regulation contributes to Drosophila clock gene mRNA cycling. *EMBO J.* 16:7146–55

So WV, Sarov-Blat L, Kotarski CK, McDonald MJ, Allada R, Rosbash M. 2000. Takeout, a novel Drosophila gene under circadian clock transcriptional regulation. *Mol. Cell. Biol.* 20:6935–44

Stanewsky R, Kaneko M, Emery P, Beretta B, Wager-Smith K, et al. 1998. The cryb mutation identifies cryptochrome as a circadian photoreceptor in Drosophila. *Cell* 95:681–92

Steller H, Fischbach KF, Rubin GM. 1987. Disconnected: a locus required for neuronal pathway formation in the visual system of Drosophila. *Cell* 50:1139–53

Stephan FK, Zucker I. 1972. Circadian rhythms in drinking behavior and locomotor activity of rats are eliminated by hypothalamic lesions. *Proc. Natl. Acad. Sci. USA* 69:1583–86

Sun ZS, Albrecht U, Zhuchenko O, Bailey J, Eichele G, Lee CC. 1997. RIGUI, a putative mammalian ortholog of the Drosophila period gene. *Cell* 90:1003–11

Suri V, Hall JC, Rosbash M. 2000. Two novel doubletime mutants alter circadian properties and eliminate the delay between RNA and protein in Drosophila. *J. Neurosci.* 20:7547–55

Takumi T, Taguchi K, Miyake S, Sakakida Y, Takashima N, et al. 1998. A light-independent oscillatory gene mPer3 in mouse SCN and OVLT. *EMBO J.* 17:4753–59

Tei H, Okamura H, Shigeyoshi Y, Fukuhara C, Ozawa R, et al. 1997. Circadian oscillation of a mammalian homologue of the Drosophila period gene. *Nature* 389:512–16

Thresher RJ, Vitaterna MH, Miyamoto Y, Kazantsev A, Hsu DS, et al. 1998. Role of mouse cryptochrome blue-light photoreceptor in circadian photoresponses. *Science* 282:1490–94

Tischkau Y, Barnes JA, Lin FJ, Myers EM, Barnes JW, et al. 1999. Oscillation and light

induction of timeless mRNA in the mammalian circadian clock. *J. Neurosci.* 19:RC15

Tosini G, Menaker M. 1996. Circadian rhythm in cultured mammalian retina. *Science* 272:419–21

Tronche F, Kellendonk C, Reichardt HM, Schutz G. 1998. Genetic dissection of glucocorticoid receptor function in mice. *Curr. Opin. Genet. Dev.* 8:532–38

Uhl GR, Reppert SM. 1986. Suprachiasmatic nucleus vasopressin messenger RNA: circadian variation in normal and Brattleboro rats. *Science* 232:390–93

van der Horst GT, Muijtjens M, Kobayashi K, Takano R, Kanno S, et al. 1999. Mammalian Cry1 and Cry2 are essential for maintenance of circadian rhythms. *Nature* 398:627–30

Van Gelder RN, Bae H, Palazzolo MJ, Krasnow MA. 1995. Extent and character of circadian gene expression in *Drosophila melanogaster*: identification of twenty oscillating mRNAs in the fly head. *Curr. Biol.* 5:1424–36

Vielhaber E, Eide E, Rivers A, Gao ZH, Virshup DM. 2000. Nuclear entry of the circadian regulator mPER1 is controlled by mammalian casein kinase I epsilon. *Mol. Cell Biol.* 20:4888–99

Vitaterna MH, King DP, Chang AM, Kornhauser JM, Lowrey PL, et al. 1994. Mutagenesis and mapping of a mouse gene Clock, essential for circadian behavior. *Science* 264:719–25

Vitaterna MH, Selby CP, Todo T, Niwa H, Thompson C, et al. 1999. Differential regulation of mammalian period genes and circadian rhythmicity by cryptochromes 1 and 2. *Proc. Natl. Acad. Sci. USA* 96:12114–19

Vosshall LB, Price JL, Sehgal A, Saez L, Young MW. 1994. Block in nuclear localization of period protein by a second clock mutation timeless. *Science* 263:1606–9

Welsh DK, Logothetis DE, Meister M, Reppert SM. 1995. Individual neurons dissociated from rat suprachiasmatic nucleus express independently phased circadian firing rhythms. *Neuron* 14:697–706

Whitmore D, Foulkes NS, Sassone-Corsi P. 2000. Light acts directly on organs and cells in culture to set the vertebrate circadian clock. *Nature* 404:87–91

Whitmore D, Foulkes NS, Strahle U, Sassone-Corsi P. 1998. Zebrafish Clock rhythmic expression reveals independent peripheral circadian oscillators. *Nat. Neurosci.* 1:701–7

Yamaguchi S, Mitsui S, Miyake S, Yan L, Onishi H, et al. 2000. The 5′ upstream region of mPer1 gene contains two promoters and is responsible for circadian oscillation. *Curr. Biol.* 10:873–76

Yamazaki S, Numano R, Abe M, Hida A, Takahashi R, et al. 2000. Resetting central and peripheral circadian oscillators in transgenic rats. *Science* 288:682–85

Young MW. 2000a. Circadian rhythms. Marking time for a kingdom. *Science* 288:451–53

Young MW. 2000b. The tick-tock of the biological clock. *Sci. Am.* 282:64–71

Zeng H, Qian Z, Myers MP, Rosbash M.

1996. A light-entrainment mechanism for the Drosophila circadian clock. *Nature* 380:129–35

Zerr DM, Hall JC, Rosbash M, Siwicki KK. 1990. Circadian fluctuations of period protein immunoreactivity in the CNS and the visual system of Drosophila. *J. Neurosci.* 10:2749–62

Zheng B, Larkin DW, Albrecht U, Sun ZS, Sage M, et al. 1999. The mPer2 gene encodes a functional component of the mammalian circadian clock. *Nature* 400:169–73

Zylka MJ, Shearman LP, Levine JD, Jin X, Weaver DR, Reppert SM. 1998a. Molecular analysis of mammalian timeless. *Neuron* 21:1115–22

Zylka MJ, Shearman LP, Weaver DR, Reppert SM. 1998b. Three period homologs in mammals: differential light responses in the suprachiasmatic circadian clock and oscillating transcripts outside of brain. *Neuron* 20:1103–10

Annu. Rev. Neurosci. 2001. 24:1121–159

NEURODEGENERATIVE TAUOPATHIES

Virginia M-Y Lee,[1] Michel Goedert,[2] and John Q Trojanowski[1]

[1]Center for Neurodegenerative Disease Research, Department of Pathology and
Laboratory Medicine, University of Pennsylvania School of Medicine, Philadelphia,
Pennsylvania 19104; e-mail: vmylee@mail.med.upenn.edu,
trojanow@mail.med.upenn.edu
[2]Medical Research Council, Laboratory of Molecular Biology, Hills Road, Cambridge,
CB2 2QH, United Kingdom; e-mail: mg@mrc-lmb.cam.ac.uk

Key Words Alzheimer's disease, frontotemporal dementia, mutation,
neurodegenerative disease, filamentous deposits, pathology, tau protein

■ **Abstract** The defining neuropathological characteristics of Alzheimer's disease
are abundant filamentous tau lesions and deposits of fibrillar amyloid β peptides.
Prominent filamentous tau inclusions and brain degeneration in the absence of
β-amyloid deposits are also hallmarks of neurodegenerative tauopathies exemplified by
sporadic corticobasal degeneration, progressive supranuclear palsy, and Pick's disease, as well as by hereditary frontotemporal dementia and parkinsonism linked to
chromosome 17 (FTDP-17). Because multiple *tau* gene mutations are pathogenic
for FTDP-17 and tau polymorphisms appear to be genetic risk factors for sporadic
progressive supranuclear palsy and corticobasal degeneration, tau abnormalities are
linked directly to the etiology and pathogenesis of neurodegenerative disease. Indeed,
emerging data support the hypothesis that different *tau* gene mutations are pathogenic
because they impair tau functions, promote tau fibrillization, or perturb *tau* gene splicing, thereby leading to formation of biochemically and structurally distinct aggregates of tau. Nonetheless, different members of the same kindred often exhibit diverse
FTDP-17 syndromes, which suggests that additional genetic or epigenetic factors influence the phenotypic manifestations of neurodegenerative tauopathies. Although these
and other hypothetical mechanisms of neurodegenerative tauopathies remain to be
tested and validated, transgenic models are increasingly available for this purpose,
and they will accelerate discovery of more effective therapies for neurodegenerative
tauopathies and related disorders, including Alzheimer's disease.

INTRODUCTION

The study of sporadic and familial neurodegenerative diseases over the past decade
has led to the realization that many of these disorders are characterized by distinct
hallmark brain lesions that have in common the formation of filamentous deposits of abnormal brain proteins. Thus, a group of heterogeneous dementias and

movement disorders that are characterized neuropathologically by prominent intracellular accumulations of abnormal filaments formed by the microtubule-associated protein tau appears to share common mechanisms of disease. They are collectively known as neurodegenerative tauopathies (Table 1). Despite their diverse phenotypic manifestations, brain dysfunction and degeneration in tauopathies is linked to the progressive accumulation of filamentous tau inclusions, and this, together with the absence of other disease-specific neuropathological abnormalities, provided circumstantial evidence implicating abnormal tau in disease onset and/or progression. However, this view remained unproven and highly controversial until 1998, when multiple *tau* gene mutations were discovered in frontotemporal dementia and parkinsonism linked to chromosome 17 (FTDP-17), thereby providing unequivocal evidence that tau abnormalities alone are sufficient to cause neurodegenerative disease (Foster et al 1997, Hutton et al 1998, Poorkaj et al 1998, Spillantini et al 1998c). This seminal finding opened up new avenues for

TABLE 1 Diseases with tau-based neurofibrillary pathology

Alzheimer's disease

Amyotrophic lateral sclerosis/parkinsonism–dementia complex[a]

Argyrophilic grain dementia[a]

Corticobasal degeneration[a]

Creutzfeldt-Jakob disease

Dementia pugilistica[a]

Diffuse neurofibrillary tangles with calcification[a]

Down's syndrome

Frontotemporal dementia with parkinsonism linked to chromosome 17[a]

Gerstmann-Sträussler-Scheinker disease

Hallervorden-Spatz disease

Myotonic dystrophy

Niemann-Pick disease, type C

Non-Guamanian motor neuron disease with neurofibrillary tangles

Pick's disease[a]

Postencephalitic parkinsonism

Prion protein cerebral amyloid angiopathy

Progressive subcortical gliosis[a]

Progressive supranuclear palsy[a]

Subacute sclerosing panencephalitis

Tangle only dementia[a]

[a]Diseases in which tau-positive neurofibrillary pathology are the most predominant neuropathologic feature.

investigating the role of tau abnormalities in mechanisms of brain dysfunction and degeneration.

This review is designed to integrate and interpret the remarkable recent advances that have led to new insights into the mechanistic role that tau abnormalities play in neurodegenerative disease. It begins with a brief summary of our current understanding of the human *tau* gene and the functions of the six alternatively spliced tau isoforms that are expressed in the normal adult human brain. What is known about the role of tau abnormalities in Alzheimer's disease (AD) is considered next, and this is followed by an overview of several prototypical, as well as some novel sporadic, tauopathies. This information provides the context and perspective in which to present an update on the increasing number of pathogenic *tau* gene mutations, as well as a summary of the neuropathological and biochemical profiles of the tau pathologies in FTDP-17. Finally, the review concludes with an update of the current status of efforts to develop transgenic (TG) and other animal models of human neurodegenerative tauopathies.

STRUCTURE, FUNCTION, AND MOLECULAR GENETICS OF TAU

Tau proteins are low M_r microtubule-associated proteins that are abundant in the central nervous system (CNS), where they are expressed predominantly in axons. They are also expressed in axons of peripheral nervous system neurons but are barely detectable in CNS astrocytes and oligodendrocytes (Cleveland et al 1977b, Binder et al 1985, Shin et al 1991, Couchie et al 1992, LoPresti et al 1995). Human tau proteins are encoded by a single gene consisting of 16 exons on chromosome 17q21, and the CNS isoforms are generated by alternative mRNA splicing of 11 of these exons (Neve et al 1986, Goedert et al 1988, Andreadis et al 1992) (Figure 1). In adult human brain, alternative splicing of exons (E)2, E3, and E10 generates six tau isoforms ranging from 352 to 441 amino acids in length (Figure 1), which differ by the presence of either three (3R-tau) or four (4R-tau) carboxy-terminal tandem repeat sequences of 31 or 32 amino acids each that are encoded by E9, E10, E11, and E12 (Goedert et al 1989a,b; Andreadis et al 1992). Additionally, the triplets of 3R-tau and 4R-tau isoforms differ as a result of alternative splicing of E2 and E3 to generate tau isoforms without (0N) or with either 29 (1N) or 58 (2N) amino acid inserts of unknown functions (Goedert et al 1989b). In adult human brain, the ratio of 3R-tau to 4R-tau isoforms is ~1, but the 1N, 0N, and 2N tau isoforms comprise about 54%, 37%, and 9%, respectively, of total tau (Goedert & Jakes 1990, Hong et al 1998). In addition, the alternative splicing of tau is developmentally regulated such that only the shortest tau isoform (3R/0N) is expressed in fetal brain, whereas all six isoforms appear in the postnatal period of the human brain (Goedert et al 1989a). In the peripheral nervous system, inclusion of E4a in the amino-terminal half results in the expression of higher M_r proteins termed big tau (Georgieff et al 1991, Couchie et al 1992, Goedert et al 1992c).

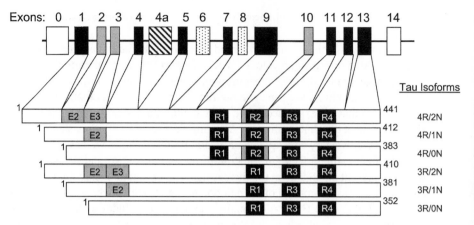

Figure 1   Schematic representation of the human tau gene and the six central nervous system (CNS) tau isoforms generated by alternative mRNA splicing. The human tau gene contains 16 exons, including exon (E)0, which is part of the promoter. Alternative splicing of E2, E3, and E10 (gray boxes) produces the six tau isoforms. E6 and E8 (stippled boxes) are not transcribed in the human CNS. E4a (striped box), which is also not transcribed in the human CNS, is expressed in the peripheral nervous system, leading to the larger tau isoforms, termed big tau (see text). The black bars depict the 18–amino acid microtubule binding repeats and are designated R1 to R4. The relative sizes of the exons and introns are not drawn to scale.

Since its discovery >25 years ago, a number of well-defined functions of tau protein have been discovered and extensively characterized (for review, see Buée et al 2000). Most notably, tau binds to and stabilizes microtubules (MTs), in addition to promoting MT polymerization (Weingarten et al 1975, Cleveland et al 1977b). The MT binding domains of tau are localized to the carboxy-terminal half of the molecule within the four MT binding motifs. These motifs are composed of highly conserved 18–amino acid long binding elements separated by flexible, but less conserved, interrepeat sequences that are 13–14 amino acids long (Himmler et al 1989, Lee et al 1989, Butner & Kirschner 1991). The binding of tau to MTs is a complex process mediated in part by a flexible array of weak MT binding sites that are distributed throughout the MT binding domain delineated by these repeats and their interrepeat sequences (Lee et al 1989, Butner & Kirschner 1991). In addition, sequences flanking the repeats contribute to microtubule binding (Gustke et al 1994). 4R-tau isoforms are more efficient at promoting MT assembly and have a greater MT binding affinity than do 3R-tau isoforms (Goedert & Jakes 1990, Butner & Kirschner 1991). It is interesting to note that the interrepeat sequence between the first and second MT binding repeats has more than twice the binding affinity of any individual MT binding repeat (Goode & Feinstein 1994). This region is unique to 4R-tau and is believed to be responsible for the higher MT binding affinity of 4R-tau isoforms compared with 3R-tau isoforms (Goedert & Jakes 1990). Notably, because other microtubule-associated proteins probably perform similar functions, it is possible they can compensate for a deficiency or loss of tau, which may account for the report that tau knockout mice did not have any overt phenotype (Harada et al 1994). However, a more recent study has reported some subtle behavioral impairments in older knockout mice (Ikegami et al 2000).

TAU PHOSPHORYLATION AND REGULATION OF TAU FUNCTIONS

There are 79 potential serine (Ser) and threonine (Thr) phosphate acceptor residues in the longest tau isoform, and phosphorylation at $\sim$30 of these sites has been reported in normal tau proteins (reviewed in Billingsley & Kincaid 1997, Buée et al 2000). Tau phosphorylation is developmentally regulated such that fetal tau is more highly phosphorylated in embryonic compared with adult CNS (Kanemaru et al 1992, Bramblett et al 1993, Goedert et al 1993, Watanabe et al 1993), and the degree of phosphorylation of all six tau isoforms decreases with age, probably because of the activation of phosphatases (Mawal-Dewan et al 1994). The tau phosphorylation sites are clustered in regions flanking the MT binding repeats, and it is well established that increasing tau phosphorylation negatively regulates MT binding (Drechsel et al 1992, Bramblett et al 1993, Yoshida & Ihara 1993, Biernat et al 1993). The importance of individual sites in tau in regulating MT binding has been debated for some time. For example, phosphorylation of Ser262, which lies within the first MT binding domain, is thought to play a dominant role in reducing the binding of tau to MTs (Biernat et al 1993). A similar role may be considered for phosphorylation of Ser396, which is located adjacent to the carboxy-terminal end of the fourth MT binding domain (Bramblett et al 1993), but other data argue that phosphorylation of neither of these sites is sufficient to eliminate the binding of tau to MTs (Seubert et al 1995). It is interesting that both sites are phosphorylated in fetal tau and that they are hyperphosphorylated in all six tau isoforms that form abnormal paired helical filaments (PHFs) in the neurofibrillary tangles (NFTs) of the AD brain (Seubert et al 1995). The regulation of the binding of tau to MTs may be also be regulated by sites outside the MT binding repeats. For example, although the evidence is fragmentary, it has been suggested that a heptapeptide sequence ($_{224}$KKVAVVR$_{230}$) located within a proline-rich domain, amino-terminal to the MT binding domains, promotes MT binding in combination with the repeat regions (Goode et al 1997). Thus, it is likely that phosphorylation at multiple sites, especially those flanking the MT binding repeats, but also additional intra- and intermolecular interactions, regulate the MT binding function of tau. Only little is known about the regulation of tau phosphorylation. One study has reported increased phosphorylation of tau at Ser202/Thr205 in mice that lack Reelin or both the very-low-density lipoprotein receptor and the apolipoprotein E receptor 2 (Hiesberger et al 1999). It suggests that tyrosine phsophorylation of the disabled-1 adaptor protein may play a role in regulating the level of tau phosphorylation. It remains to be seen whether this finding can be extended to additional phosphorylation sites in tau and to mice mutant for other components of this developmental pathway.

For the reasons discussed above, considerable research has focused on the protein kinases and protein phosphatases that regulate tau phosphorylation, and a large number of Ser/Thr protein kinases have been implicated or have been suggested as playing a role in regulating tau functions in vivo (reviewed in Billingsley & Kincaid 1997, Buée et al 2000, Hong et al 2000). These kinases include mitogen-activated protein kinase (Drewes et al 1992, Drechsel et al 1992, Goedert et al 1992a), glycogen synthase kinase 3β (GSK-3β) (Hanger et al 1992, Mandelkow et al 1992), cyclin-dependent kinase 2 (cdk2) (Baumann et al 1993), cdk5 (Baumann et al 1993, Kobayashi et al 1993), cAMP-dependent protein kinase (Litersky & Johnson 1992), Ca^{2+}/calmodulin-dependent protein kinase II (Baudier & Cole 1987), and MT-affinity regulating kinase (Drewes et al 1997). In addition, several members of the family of stress-activated protein kinases also phosphorylate tau at multiple sites (Goedert et al 1997; Reynolds et al 1997a,b). Nonetheless, it is important to emphasize that many of these studies provide only in vitro evidence to implicate specific kinases and that it remains unclear what role they play in the in vivo phosphorylation of tau.

Recent data have implicated two protein kinases, GSK-3β and cdk5, in the in vivo regulation of tau phosphorylation. GSK-3β is a Ser/Thr kinase that is abundant in brain and associates with MTs (Mandelkow et al 1992, Ishiguro et al 1994, Singh et al 1995, Takahashi et al 1995, Cohen 1999). Cotransfection of nonneuronal cells with human tau and GSK-3β induces hyperphosphorylation of tau associated with a loss of MT binding (Lovestone et al 1996). In cultured neuronal cells, GSK-3β-mediated phosphorylation of tau is inhibited by insulin and IGF-1 via a phosphatidylinositol 3-kinase and protein kinase B–dependent signaling pathway (Hong & Lee 1997). In addition, direct inhibition of GSK-3β by lithium salts or ATP competitive inhibitors reduces tau phosphorylation and affects MT stability (Hong et al 1997, Munoz-Montano et al 1997, Lovestone et al 1999, Takahashi et al 1999, Leost et al 2000). Cdk5 is a Ser/Thr protein kinase highly enriched in neurons that colocalizes to the cytoskeleton and contributes to the phosphorylation of tau (Baumann et al 1993, Kobayashi et al 1993, Lew & Wang 1994). It is activated by interaction with regulatory subunits, the best characterized of which is p35 (Ishiguro et al 1994, Lew et al 1994, Tsai et al 1994). Recently, Sobue et al (2000) demonstrated that cdk5 complexes with tau in a manner that depends on the phosphorylation of tau, and that tau anchors cdk5 to MTs. Moreover, cdk5-mediated tau phosphorylation stimulates further phosphorylation of tau by GSK-3β (Yamaguchi et al 1996, Sengupta et al 1997). However, further work is needed to determine the relative contributions of individual kinases to tau phosphorylation in vivo.

Because protein phosphatases are required for counterbalancing the effects of tau protein kinases, they have also been an intense focus of research. A number of studies have implicated several phosphatases in regulating tau phosphorylation, including PP1, PP2A, PP2B, and PP2C (reviewed in Billingsley & Kincaid 1997,

Buée et al 2000). They all dephosphorylate tau in vitro with overlapping specificities; however, their role in vivo is unclear. Both PP2A and PP2B are present in
human brain tissue, and they dephosphorylate tau in a site-specific manner. Both
enzymes dephosphorylate Ser396 (Matsuo et al 1994), whereas PP2A also dephosphorylates tau at multiple additional sites (Goedert et al 1992a, Drewes et al 1993,
Billingsley & Kincaid 1997). Of the phosphatase activity in rat brain, PP2A is the
major activity toward tau phosphorylated by a number of protein kinases (Goedert
et al 1992a, 1995a). PP1 and PP2A bind to tau, and this interaction is believed
to mediate an association with MTs (Sontag et al 1995, Liao et al 1998). PP2A
has also been demonstrated to bind directly to MTs, an interaction that regulates
its activity in vitro (Sontag et al 1999). Lastly, inhibition of PP1 and PP2A by
okadaic acid in cultured human NT2N neurons results in increased tau phosphorylation, accompanied by decreased tau binding to MTs, selective destruction of
stable MTs, and rapid degeneration of axons (Merrick et al 1997). As with the
kinases implicated in the phosphorylation of tau, further studies are necessary to
define the specific role of individual phosphatases in the in vivo regulation of the
phosphorylation state of tau.

AD NEUROFIBRILLARY PATHOLOGY IS MADE OF ABNORMALLY PHOSPHORYLATED TAU

Filamentous neuronal or neuronal and glial tau inclusions associated with the degeneration of affected brain regions are the defining neuropathological features of
tauopathies. In AD, NFTs and neuropil thread pathology are found in conjunction
with the deposition of β-amyloid (Aβ) fibrils in the extracellular space. By light
microscopy, the neurofibrillary lesions of AD are stained with anti-tau antibodies (Brion et al 1985, Grundke-Iqbal et al 1986). Ultrastructurally, the dominant
components of neurofibrillary lesions in AD are paired helical filaments (PHFs)
and straight filaments (Kidd 1963). PHFs are composed of two strands of filament
twisted around one another with a periodicity of 80 nm and a width varying from
8 to 20 nm (Crowther & Wischik 1985) whereas straight filaments lack this helical
periodicity (Crowther 1991). Both PHFs and straight filaments are composed predominantly of abnormally hyperphosphorylated tau proteins (Goedert et al 1988,
Kondo et al 1988, Kosik et al 1988, Wischik et al 1988, Lee et al 1991). Analysis
of PHFs purified from AD brains by sodium dodecyl sulphate–polyacrylamide
gel electrophoresis has revealed three major bands of approximately 68, 64, and
60 kDa, as well as a minor band of approximately 72 kDa (Greenberg & Davies
1990, Lee et al 1991) (Figure 2). Upon dephosphorylation, six bands are resolved
that correspond to the six isoforms of tau found in adult human brain (Lee et al
1991, Greenberg et al 1992, Goedert et al 1992b). The relative proportions of
the tau isoforms observed in AD PHFs are similar to those that are characteristic of the six soluble tau isoforms observed in normal adult human brain (see
Trojanowski & Lee 1994, Morishima-Kawashima et al 1995, Goedert et al 1995b,
Hong et al 1998). Although many phosphorylation sites identified in PHFtau

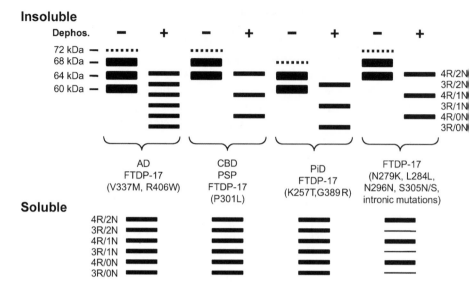

Figure 2 Schematic representation of Western blot banding patterns of insoluble and soluble tau from different tauopathies. The cartoon depicts the typical banding pattern of nondephosphorylated and dephosphorylated insoluble (filamentous) tau (top panels) as well as soluble tau (bottom panels) from brains of patients with the diseases indicated following resolution by sodium dodecyl sulphate–polyacrylamide gel electrophoresis and immunoblotting with phosphorylation-independent anti-tau antibodies. Nondephosphorylated insoluble tau from the brain of patients with Alzheimer's disease (AD) and some frontotemporal dementia and parkinsonism linked to chromosome 17 (FTDP-17), mutations that do not affect splicing (V337M and R406W), runs as three major bands of 68, 64, and 60 kDa and as a minor, variable band of 72 kDa. When dephosphorylated, it resolves into six bands that correspond to soluble tau. In corticobasal degeneration (CBD) and progressive supranuclear palsy (PSP), as well as FTDP-17 with the P301L mutation, the two prominent 68- and 64-kDa insoluble tau bands are detected (the 72-kDa minor band is variably detected) and align with four tandem repeat sequences (4R-tau) following dephosphorylation. The soluble fraction shows all six isoforms, indicating that there is selective aggregation of 4R-tau. In contrast, in FTDP-17 mutations that affect mRNA splicing, there is expression of predominantly soluble 4R-tau throughout the entire brain. Only 4R-tau is deposited in a filamentous form. In Pick's disease (PiD) and some FTDP-17 mutations (K257T, G389R) that do not affect splicing, the lower two 64- and 60-kDa insoluble tau bands predominate. Following dephosphorylation, a predominance of 3R-tau is observed. All six tau isoforms are expressed in the soluble fraction in PiD. Major tau proteins are depicted by solid bars and the thickness of the bars correlates with the relative abundance of the specific tau isoform. A dashed bar is used to depict the minor, and more variable, 72- and 68-kDa tau isoforms.

have also been found to be phosphorylated to some extent in tau proteins isolated from biopsies of normal human brain (Matsuo et al 1994), it is clear that PHFtau is hyperphosphorylated and abnormally phosphorylated (Morishima-Kawashima et al 1995, Hasegawa et al 1996, Hoffmann et al 1997, Zheng-Fischhofer et al 1998).

Numerous protein kinases and protein phosphatases have been implicated in the dysregulation of tau phosphorylation in the AD brain (for detailed reviews, see Billingsley & Kincaid 1997, Buée et al 2000, Hong et al 2000). A recent study has suggested that cdk5 may play a specific role in this process (Patrick et al 1999). It showed that p25, a truncated form of p35, accumulates in neurons in the brains of AD patients. The calcium-dependent cysteine protease calpain is believed to generate p25 from p35 (Kusakawa et al 2000, Lee et al 2000, Nath et al 2000). The accumulation of p25 correlated with increased cdk5 kinase activity and the binding of p25 to cdk5 constitutively activated cdk5. Finally, the expression of the cdk5/p25 complex in various cell lines increased tau phosphorylation and disrupted the cytoskeletal network. Thus, it is possible that the cdk5/p25 complex may play a mechanistic role in the conversion of normal tau into PHFtau in AD; however, it will be important to confirm and extend these results in additional in vitro and in vivo studies.

The mechanisms underlying PHF formation in neurons are still unclear, but it is possible that hyperphosphorylation disengages tau from MTs, thereby increasing the pool of unbound tau. Unbound tau may be more resistant to degradation and more prone to aggregate than MT-bound tau. This suggests that an increased ability of pathological tau to intereact with MTs may be beneficial. The organic osmolytes trimethylamine N-oxide and betaine have been shown to increase tau-promoted assembly of MTs (Tseng & Graves 1998). These compounds were also shown to restore the ability of tau phosphorylated by cAMP-dependent protein kinase or GSK-3β to promote MT assembly (Tseng et al 1999). Tau is a highly flexible, extended molecule with little secondary structure (Schweers et al 1994, Goedert et al 1999a, Barghorn et al 2000). By circular dichroism spectroscopy, it appears as a random coil (Cleveland et al 1977a), even when carrying FTDP-17 mutations (Goedert et al 1999a, Barghorn et al 2000) or in the presence of trimethylamine N-oxide (Tseng & Graves 1998). Binding of tau to MTs generates some ordered structures, indicating that MTs can induce conformational changes in tau (Woody et al 1983). Trimethylamine N-oxide and betaine probably induce a tubulin and/or tau conformational change that favors assembly.

Along related lines, Lu et al (1999) demonstrated that the prolyl isomerase Pin1 binds to a single phosphothreonine residue (Thr231) in tau and that this restored the ability of tau phosphorylated by cdc2 kinase to intereact with MTs. Pin1 was reported to copurify with PHFs, resulting in a depletion of soluble Pin1 in AD brain. Because depletion of Pin1 is believed to induce mitotic arrest and apoptotic cell death, its sequestration in NFTs could contribute to neurodegeneration.

SYNTHETIC TAU FILAMENTS

Hyperphosphorylation is believed to be an early event in the pathway that leads from soluble to insoluble and filamentous tau protein (Braak et al 1994). However, it is unclear whether it is sufficient for assembly into filaments. Currently, there is no experimental evidence that links hyperphosphorylation of tau to filament assembly. The availability of large quantities of recombinant human tau isoforms and the ease with which tau fragments can be expressed has facilitated studies aimed at producing synthetic tau filaments. Early experiments had shown that PHF-like filaments can be assembled in vitro from bacterially expressed, non-phosphorylated 3R fragments of tau (Crowther et al 1992, Wille et al 1992). The formation of these filaments lent strong support to the view that the repeat region of tau is the only component necessary for the morphological appearance of the PHF. However, these studies failed to provide any insight into filament formation in vivo because the tau filaments were obtained only with truncated tau under non-physiological conditions. This contrasts with PHFs from AD brain, which consist of full-length tau protein (Goedert et al 1992b).

More recently, experiments using sulphated glycosaminoglycans (GAGs) to stimulate phosphorylation of tau by a number of protein kinases have led to the observation that sulphated GAGs induce the assembly of full-length tau into filaments (Goedert et al 1996, Perez et al 1996). Assembly of individual 3R-tau isoforms gave filaments with a typical paired-helical-like morphology, when incubated with heparin or heparan sulphate, whereas assembly of individual 4R-tau isoform gave filaments with a straight appearance. By immunoelectron microscopy, the paired-helical-like filaments were decorated by antibodies directed against the amino and carboxy termini of tau, but not by an antibody against the MT-binding region (Goedert et al 1996). A short amino acid sequence (VQIVYK) in the third MT-binding repeat of tau has recently been shown to be essential for heparin-induced filament assembly (von Bergen et al 2000).

Assembly of tau into filaments in the presence of sulphated GAGs occurs after a lag period and is heavily concentration dependent, consistent with a nucleation-dependent process (Goedert et al 1996, Perez et al 1996, Arrasate et al 1997, Hasegawa et al 1997, Friedhoff et al 1998). Phosphorylation of tau at Ser/Thr-Pro sites does not significantly influence heparin-induced assembly (Goedert et al 1996). However, it has been reported that phosphorylation at other sites, such as Ser214 and Ser262, is strongly inhibitory toward assembly (Schneider et al 1999). Subsequent to this work, RNA (Kampers et al 1996, Hasegawa et al 1997) and arachidonic acid (Wilson & Binder 1997) were also shown to induce the bulk assembly of full-length recombinant tau into filaments. This work has provided robust methods for the assembly of full-length tau into filaments. Pathological colocalization of sulphated GAGs (Snow et al 1990, Goedert et al 1996, Verbeek et al 1999) and RNA (Ginsberg et al 1998) with hyperphosphorylated tau protein suggests that these findings may also be relevant for the assembly of tau in AD.

SPORADIC TAUOPATHIES

The amyloid cascade hypothesis for the pathogenesis of AD proposes that the deposition and fibrillization of Aβ peptides to form extracellular senile plaques is the central event that causes formation of NFTs and neuronal loss (Hardy & Allsop

1991). Compelling evidence in support of a causative pathologic role of tau protein in neurodegeneration is provided by recent studies of tauopathies other than AD, which share an abundant filamentous tau pathology and brain degeneration in the absence of extracellular amyloid deposits (Table 1). Progressive supranuclear palsy (PSP), corticobasal degeneration (CBD), and Pick's disease (PiD) are three such disorders. A recent consensus conference has classified them as disorders belonging to a group of diseases known as frontotemporal dementia (FTD) (Pick's Conference 2001). Clinically, PSP is characterized by supranuclear gaze palsy as well as by prominent postural instability (Steele et al 1964). Neuropathologically, PSP is characterized by atrophy of the basal ganglia, subthalamus, and brainstem, with corresponding neuronal loss and gliosis. Within these brain regions, there is a high density of fibrillary tau pathology, including neuropil threads, and NFTs that are typically round or globose (Pollock et al 1986, Hauw et al 1994, Litvan et al 1996). Glial fibrillary tangles in both astrocytes (tufted astrocytes) and oligodendrocytes (coiled bodies) are also often present (Hauw et al 1990, Yamada et al 1992, Komori 1999). In contrast to AD, ultrastructural analysis of these neurofibrillary lesions has revealed 15- to 18-nm straight filaments, and filaments with a long periodicity have also been observed (Tellez-Nagel & Wisniewski 1973, Roy et al 1974).

The filamentous tau pathology of PSP correlates with the biochemical identification of insoluble, hyperphosphorylated tau in affected brain regions. However, in contrast to the three major bands identified in AD, only the two high M_r bands (68 and 64 kDa) are present (the minor 72-kDa band is variably detected) (Flament et al 1991, Vermersch et al 1994) (Figure 2). These bands are made of hyperphosphorylated tau isoforms with four MT repeats (Spillantini et al 1997, Sergeant et al 1999). They exhibit the same profile of phosphorylation-dependent tau epitopes as those detected in PHFtau from AD brains (Schmidt et al 1996). Furthermore, in PSP, the relative abundance of tau mRNA containing E10 has been reported to be increased in the brainstem but not in the cortex, which is consistent with the distribution of the neurofibrillary pathology (Chambers et al 1999).

Polymorphisms in the *tau* gene may contribute to the risk of developing PSP because a polymorphic dinucleotide repeat in the intron between E9 and E10 of the *tau* gene has been linked to PSP (Conrad et al 1997). Subjects with the homozygous tau allele A0, characterized by 11 TG repeats, were found to be overrepresented in PSP patients (95.5%) compared with controls (57.4%) and AD (49.7%) patients. Subsequent studies have confirmed this correlation in the Caucasian (but not Asian) population (Bennett et al 1998, Higgins et al 1998, Hoenicka et al 1999, Morris et al 1999a). Moreover, two extended *tau* gene haplotypes consisting of eight common single-nucleotide polymorphisms in addition to the dinucleotide repeat polymorphism have been described (Baker et al 1999). The haplotypes are in complete linkage dysequilibrium and span the entire human *tau* gene. The more common haplotype, H1, is significantly overrepresented in Caucasians with PSP. In addition, two missense mutations in E4a are associated with the H1 haplotype and have been linked to PSP, and a 238-bp deletion in the intron flanking E10 of the *tau* gene is inherited as part of the less-common H2 extended haplotype and thus shows a negative assciation with PSP (Higgins et al 1998, Baker et al 1999, Bonifati et al 1999, Ezquerra et al 1999). The relationships of the H1 haplotype and the A0 allele to the pathogenesis of PSP are unknown, but it is possible that the 238-bp deletion flanking E10 in the H1 haplotype might affect E10 splicing, thereby increasing the relative proportion of 4R-tau.

CBD is an adult-onset progressive neurodegenerative disorder involving the cerebral cortex, deep cerebellar nuclei, and substantia nigra, in association with prominent neuronal achromasia (Rebeiz et al 1967, 1968). Neuropathological examination shows depigmentation of the substantia nigra, as well as an asymmetric frontoparietal atrophy that is often most severe in the pre- and postcentral regions. In affected regions, there is neuronal loss with spongiosis, gliosis, and prominent glial and neuronal intracytoplasmic filamentous tau pathology (Iwatsubo et al 1994, Mori et al 1994). The glial tau pathology in CBD consists of characteristic astrocytic plaques (Feany & Dickson 1995), as well as numerous tau-immunoreactive inclusions in the white matter in both astrocytes and oligodendrocytes (coiled bodies) (Komori et al 1998, Komori 1999). A striking feature of CBD is the extensive accumulation of tau-immunoreactive neuropil threads throughout gray and white matter (Feany & Dickson 1995, Feany et al 1996). The tau filaments in CBD include both PHF-like filaments and straight tubules (Ksiezak-Reding et al 1994, Komori 1999).

The biochemical profile of insoluble tau in CBD is similar to that of PSP in that it consists of two major bands of 64 and 68 kDa and a variable, minor band of 72 kDa (Ksiezak-Reding et al 1994, Buée-Scherrer et al 1996) (Figure 2). However, the isoforms present in the tau pathology of CBD may differ from those found in PSP. Antibodies specific for the insert encoded by E3 did not detect the fibrillary tau pathology in CBD either biochemically or immunohistochemically in studies from one group (Ksiezak-Reding et al 1994, Feany et al 1995). However, another report failed to confirm this finding (Sergeant et al 1999). The latter study demonstrated that the fibrillary inclusions in CBD are composed predominantly of 4R-tau isoforms that also contain the inserts encoded by E2 and to E3. Another similarity between PSP and CBD was recently described by Di Maria et al (2000), who showed that CBD is associated with the A0 allele of the *tau* gene, as well as the H1 haplotype. Thus, the current biochemical and genetic data strongly suggest that there is a substantial overlap between PSP and CBD. This is also apparent with respect to the clinical (Hauw et al 1994) and pathological (Feany et al 1996) features. Rather than representing two separate and distinct disorders, PSP and CBD may be different phenotypic manifestations of the same underlying disease process.

PiD, a variant of FTD, is defined neuropathologically by the presence of tau-immunoreactive Pick bodies (Constantinidis et al 1974, Pollock et al 1986, Feany et al 1996, Pick's Conference 2001). Neuropathologically, it is characterized by a frontotemporal lobar and limbic atrophy associated with marked neuronal loss, spongiosis, and gliosis, with ballooned neurons and Pick bodies (Lund &

Manchester Groups 1994, Dickson 1998). Pick bodies are detected by antibodies to hyperphosphorylated tau and are most numerous in layers II and VI of the neocortex and in the dentate granule neurons of the hippocampus (Iwatsubo et al 1994, Probst et al 1996). Ultrastructurally, Pick bodies are composed of a mixture of wide, straight filaments and wide, long-period twisted filaments (Munoz-Garcia & Ludwin 1984, Murayama et al 1990, Dickson 1998).

Western blot analyses have revealed that the insoluble tau in PiD is distinct from that in AD, CBD, and PSP in that it comprises two major bands of 60 and 64 kDa and a variable, minor band of 68 kDa (Buée-Scherrer et al 1996, Delacourte et al 1996, Lieberman et al 1998) (Figure 2). Because the two major PiD tau bands appear to lack the MT binding repeat encoded by E10, they are believed to be composed exclusively of 3R-tau (Sergeant et al 1997, Mailliot et al 1998). Ser262 has been shown not to be phosphorylated, in contrast to AD, CBD, or PSP (Probst et al 1996, Delacourte et al 1998). However, a separate study has detected a signal in Pick bodies and PiD tau using an antibody specific for phosphorylated Ser262 (Lieberman et al 1998). This may reflect heterogeneity of phosphorylation at Ser262 in PiD.

In contrast to PiD, the majority of patients with FTD show frontotemporal neuron loss, gliosis, and microvacuolar (spongiform) change but no disease-specific diagnostic lesions. This neuropathological entity is referred to by several names, including frontotemporal lobar degeneration (FTLD) and dementia lacking distinctive histology (DLDH), but there is no agreement on the most appropriate nomenclature for this form of FTD (Mann 1998). However, this may change soon because one study has defined this neuropathology more precisely using quantitative morphometric methods (Arnold et al 2000). Moreover, recent evidence suggests that FTLD (DLDH) may be a novel tauopathy that is caused by a selective reduction or complete loss of all six brain tau isoforms in affected and unaffected brain regions (Zhukareva et al 2001). The explanation for this is enigmatic because there was no concomitant loss of tau mRNA compared with control and AD brains. Although the majority of FTD patients with FTLD (DLDH) neuropathology showed a dramatic loss of tau protein in brain regions with and without neuronal degeneration, others showed less-substantial but still statistically significant reductions in brain tau levels. Although the pathogenic mechanism underlying this marked reduction in all six brain tau proteins in FTLD (DLDH) is not known, the consequence of this loss may be similar to the losses of tau function resulting from some of the *tau* gene mutations in FTDP-17.

FAMILIAL TAUOPATHIES—FTDP-17 SYNDROMES

The group of syndromes known as FTDP-17 consists of autosomal-dominantly inherited neurodegenerative diseases with diverse, but overlapping, clinical and neuropathological features (Foster et al 1997). Neuropathologically, they all show the presence of an abundant filamentous tau pathology in nerve cells, and for some in glial cells (reviewed in Spillantini et al 1998a, Crowther & Goedert 2000, Hong et al 2000). The first such disorder was linked to chromosome 17 in 1994, when Wilhelmsen et al (1994) described a familial disease they called "disinhibition-dementia-parkinsonism-amyotrophy complex" and demonstrated genetic linkage of this disease to chromosome 17q21-22. Subsequently, a number of related neurodegenerative disorders were linked to the same region on chromosome 17 (Wijker et al 1996, Bird et al 1997, Foster et al 1997, Heutink et al 1997, Murrell et al 1997, Lendon et al 1998). Clinically, they are characterized primarily by FTD and parkinsonism (Foster et al 1997), but the different FTDP-17 syndromes appear to reflect the burden of tau pathology and degeneration in brain regions known to subserve specific cognitive, executive, or motor functions. Despite this phenotypic heterogeneity, the neuropathology of FTDP-17 is characterized by marked neuronal loss in affected brain regions, with extensive neuronal or neuronal and glial fibrillary pathology composed of hyperphosphorylated tau protein, but without evidence of Aβ deposits or other disease-specific brain lesions in the majority of the cases (Murrell et al 1999, Lippa et al 2000, Rizzini et al 2000, Spillantini et al 2000).

Because the *tau* gene had been localized to chromosome 17q21-22, it was an obvious candidate for the disease locus. In 1998, several groups identified pathogenic mutations in the *tau* gene that segregated with FTDP-17 (Clark et al 1998, Dumanchin et al 1998, Hutton et al 1998, Poorkaj et al 1998, Spillantini et al 1998c). To date, >20 distinct pathogenic mutations in the *tau* gene have been identified in a large number of families with FTDP-17 (Table 2, Figure 3). Eleven missense mutations in coding regions of the *tau* gene are known, including mutations in E9 [K257T (Pickering-Brown et al 2000, Rizzini et al 2000), I260V (M Hutton, personal communication), and G272V (Hutton et al 1998)], E10 [N279K (Clark et al 1998, Delisle et al 1999, Yasuda et al 1999, Arima et al 2000), P301L (Clark et al 1998, Dumanchin et al 1998, Hutton et al 1998, Bird et al 1999, Houlden et al 1999, Mirra et al 1999, Kodama et al 2000), P301S (Bugiani et al 1999, Sperfeld et al 1999), and S305N (Iijima et al 1999)], in E12 [V337M (Poorkaj et al 1998) and E342V (Lippa et al 2000)], and in E13 [G389R (Murrell et al 1999, Pickering-Brown et al 2000)] and R406W (Hutton et al 1998, Van Swieten et al 1999)]. Three silent mutations in E10 [L284L (D'Souza et al 1999), N296N (Spillantini et al 2000), and S305S (Stanford et al 2000)] as well as a deletion mutation [ΔK280 (Rizzu et al 1999)] have also been identified. In addition, five substitutions in six different positions of the intron following E10 have been identified at positions +3 (Spillantini et al 1998c, Tolnay et al 2000), +12 (Yasuda et al 2000), +13 (Hutton et al 1998), +14 (Clark et al 1998, Hutton et al 1998), and +16 (Hutton et al 1998, Hulette et al 1999, Goedert et al 1999b, Morris et al 1999b). Besides mutations in the intron following E10, additional pathogenic mutations may be present in other introns of the *tau* gene. Thus, a mutation in the intron following E9 has been described in a patient with familial FTD (Rizzu et al 1999). It disrupts one of the several (A/T)GGG repeats that may play a role in the regulation of the alternative splicing of E10.

TABLE 2   Tau mutations identified in FTDP-17[a]

Mutation	Location	E10 Splicing	MT Assembly	Phenotype	Reference
K257T	E9, R1	No change	Reduced	PiD-like	Pickering-Brown et al (2000), Rizzini et al (2000)
I260V	E9, R1	ND	ND	NA	M Hutton, personal communication
G272V	E9, R1	No change	Reduced	FTDP-17	Hutton et al (1998)
N279K	E10, IR1-2	Increased	No effect	PSP-like	Clark et al (1998)
Δ280K	E10, IR1-2	Decreased	Reduced	FTDP-17	Rizzu et al (1999)
L284L	E10, IR1-2	Increased	No effect	AD-like	D'Souza et al (1999)
N296N	E10, R2	Increased	No effect	CBD-like	Spillantini et al (2000)
P301L	E10, R2	No change	Reduced	FTDP-17, CBD-like, PSP-like	Hutton et al (1998)
P301S	E10, R2	No change	Reduced	FTDP-17, CBD-like	Bugiani et al (1999), Sperfeld et al (1999)
S305N	E10, IR2-3	Increased	No effect	CBD-like	D'Souza et al (1999), Hasegawa et al (1999), Iijima et al (1999)
S305S	E10, IR2-3	Increased	No effect	PSP-like	Stanford et al (2000)
E10+3	I10	Increased	No effect	FTDP-17	Spillantini et al (1998c)
E10+12	I10	Increased	No effect	FTDP-17	Yasuda et al (2000)
E10+13	I10	Increased	No effect	NA	Hutton et al (1998)
E10+14	I10	Increased	No effect	FTDP-17, PSP-like	Hutton et al (1998)
E10+16	I10	Increased	No effect	FTDP-17, PSP-like CBD-like	Hutton et al (1998)
E9+33	I9	ND	ND	NA	Rizzu et al (1999)
V337M	E12, IR3-4	No change	Reduced	FTDP-17	Poorkaj et al (1998)
E342V	E12, IR3-4	ND	ND	FTDP-17	Lippa et al (2000)
G389R	E13	No change	Reduced	PiD-like	Murrell et al (1999)
R406W	E13	No change	Reduced	PSP-like	Hutton et al (1998)

[a]FTDP-17, frontotemporal demential and parkinsonism linked to chromosome 17; E, exon; I, intron; R, microtubule (MT) binding repeat; IR, interrepeat regions; ND, not determined; NA, not available; PiD, Pick's disease; PSP, progressive supranuclear palsy; AD, Alzheimer's disease; CBD, corticobasal degeneration; increased, enhanced E10 utilization; decreased, reduced E10 utilization.

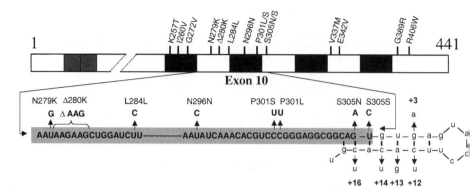

Figure 3   Schematic representation of mutations in the tau gene identified in frontotemporal dementia and parkinsonism linked to chromosome 17. The longest human brain tau isoform is shown with known coding region mutations indicated above. The gray boxes near the amino terminus represent the alternatively spliced inserts encoded by exons (E)2 and E3, whereas the black boxes represent each of the four microtubule (MT) binding repeats (not drawn to scale). The second MT binding repeat is encoded by E10. Part of the mRNA sequence encoding E10 and the intron following E10 is shown. Mutations in E10 and the downstream intron are indicated. Intronic nucleotides that are part of intron 10 are shown in lower case.

Data emerging from several laboratories continue to add increasing support in favor of the hypothesis that FTDP-17 mutations lead to tau dysfunction and disease by one or more of three distinct mechanisms. Intronic and some exonic mutations affect the alternative splicing of E10 and consequently alter the relative proportion of 4R-tau and 3R-tau. The other exonic mutations impair the ability of tau to bind MTs and to promote MT assembly. Some of these mutations also promote the assembly of tau into filaments. Moreover, additional mechanisms may play a role in the case of some coding region mutations (Yen et al 1999, Goedert et al 2000). The intronic mutations clustered around the 5′ splice site of E10, as well as several mutations within E10 (N279K, L284L, N296N, S305N, and S305S), increase the ratio of 4R-tau to 3R-tau by altering the splicing of E10 (Hong et al 1998; Hutton et al 1998; Spillantini et al 1998c; D'Souza et al 1999; Delisle et al 1999; Grover et al 1999; Hasegawa et al 1999; Varani et al 1999; Yasuda et al 1999, 2000; Spillantini et al 2000; Stanford et al 2000). As a result of these mutations, there is a relative increase in E10-containing tau mRNAs, and this probably reflects increased utilization of the E10 5′ splice site, as demonstrated in exon trapping experiments. Biochemical analysis of insoluble tau extracted from autopsied FTDP-17 brain tissue of patients with these mutations reveals exclusively 4R-tau isoforms (Spillantini et al 1997, 1998c; Clark et al 1998; Hong et al 1998; Reed et al 1998; Hulette et al 1999; Goedert et al 1999b; Yasuda et al 2000). Furthermore, 4R-tau protein levels are increased in both affected and unaffected regions of FTDP-17 brains (Hong et al 1998, Spillantini et al 1998c, Goedert et al 1999b, Yasuda et al 2000).

The regulation of splicing of E10 in the *tau* gene appears to be complex and may involve multiple *cis*-acting regulatory elements that either enhance or inhibit the utilization of the E10 5′ splice site, many of which are affected by the mutations identified in the *tau* gene (D'Souza et al 1999, Grover et al 1999, D'Souza & Schellenberg 2000, Gao et al 2000, Jiang et al 2000). Splicing regulatory elements within E10 appear to include an exon-splicing enhancer (ESE) and an exon splicing silencer (ESS) (D'Souza et al 1999, D'Souza & Schellenberg 2000, Gao et al 2000). The ESE consists of three domains, a potential SC35 binding element, a purine-rich sequence, and an AC-rich sequence (D'Souza & Schellenberg 2000). Immediately downstream of the ESE within E10 is a purine-rich ESS. The flanking exons of the *tau* gene also appear to affect E10 splicing (Gao et al 2000). For example, it appears that E9 and E11 exert opposite effects, i.e. E9 may promote E10 splicing, whereas E11 may suppress it. Lastly, intronic sequences immediately downstream of E10 inhibit its splicing (D'Souza et al 1999, Grover et al 1999, D'Souza & Schellenberg 2000, Gao et al 2000, Jiang et al 2000). This inhibition may be secondary to the formation of a stem-loop structure that sequesters the E10 5′ splice site from the splicing machinery, including the U1- and U6-snRNPs (Grover et al 1999; Varani et al 1999, 2000; Jiang et al 2000) (Figure 3). The determination of the three-dimensional structure of a 25-nucleotide-long RNA from the E10 5′-intron junction by nuclear magnetic resonance spectroscopy has shown that this sequence forms a stable, folded stem-loop structure (Varani et al 1999, 2000). The stem consists of a single G-C base pair that is separated from a double helix of 6 bp by an unpaired adenine (Figure 3). As is often the case with single-nucleotide purine bulges, the unpaired adenine at position −2 does not extrude into solution but intercalates into the double helix. The apical loop consists of six nucleotides that adopt multiple conformation in rapid exchange. Known intronic mutations and the mutations in codon 305 are located in the upper part of the stem and reduce the thermodynamic stability of the stem loop (Varani et al 1999, 2000; Yasuda et al 2000). Moreover, the relative proportions of 3R-tau and 4R-tau isoforms from nonhuman species correlate with the predicted stability of this stem-loop structure (Grover et al 1999). However, another study has concluded that this ESS may function as a linear sequence that is independent of a stem-loop structure (D'Souza & Schellenberg 2000).

Pathogenic FTDP-17 mutations in the *tau* gene may alter E10 splicing by affecting several of the regulatory elements described above. Thus, the intronic mutations, as well as the exonic mutations at codon 305 (S305N and S305S), may destabilize the inhibitory stem-loop structure (Grover et al 1999, Varani et al 1999) (Figure 3). The S305N mutation and the +3 intronic mutation may also enhance E10 splicing by increasing the strength of the 5′ splice site (Spillantini et al 1998c, Iijima et al 1999). However, the finding that the S305S mutation that weakens the E10 5′ splice site also leads to a predominance of 4R-tau argues against this effect of these mutations (Stanford et al 2000). The N279K mutation may improve the function of the ESE by lengthening the purine-rich sequence within this regulatory element (TAAGAA to GAAGAA), thus enhancing E10 splicing

(D'Souza & Schellenberg 2000). Moreover, the thymidine nucleotide present in the wild-type (WT) sequence may function as an inhibitor of splicing (Tanaka et al 1994). The observation that the ΔK280 mutation, which deletes the three adjacent purine residues (AAG), abolishes E10 splicing supports this hypothesis (D'Souza et al 1999). The silent L284L mutation that enhances E10 splicing may do so by disrupting a potential ESS (UUAG to UCAG) (Si et al 1998, D'Souza et al 1999). However, because mutation of this consensus sequence does not increase E10 splicing (D'Souza & Schellenberg 2000), a second possibility is that the mutation lengthens the AC-rich element within the ESE. Thus, the L284L mutation may affect either an enhancing or an inhibiting regulatory splicing element. The effect of the N296N mutation on splicing of E10 is probably due to disruption of an ESS (D'Souza & Schellenberg 2000, Spillantini et al 2000).

The mechanisms by which these changes in the ratio of 3R-tau to 4R-tau (3R/4R-tau) lead to neuronal and glial dysfunction and cell death remain unclear. 4R-tau and 3R-tau may bind to distinct sites on MTs (Goode & Feinstein 1994, Goode et al 1997), and it is possible that a specific ratio of tau isoforms is necessary for normal MT function. Thus, the altered ratio of 3R/4R-tau may directly affect MT function. In addition, overproduction of 4R-tau may lead to an excess of free tau in the cytoplasm, leading to its hyperphosphorylation and assembly into filaments.

In contrast to the FTDP-17 mutations discussed above, other mutations alter the ability of tau to interact with MTs. Specifically, mutations K257T, G272V, ΔK280, P301L, P301S, V337M, G389R, and R406W reduce the binding of tau to MTs and decrease its ability to promote MT assembly in in vitro assays (Hasegawa et al 1998, Hong et al 1998, Bugiani et al 1999, D'Souza et al 1999, Murrell et al 1999, Rizzu et al 1999, Barghorn et al 2000, Pickering-Brown et al 2000, Rizzini et al 2000). These effects are not observed with the tau missense mutations that affect E10 splicing (Hong et al 1998, D'Souza et al 1999, Hasegawa et al 1999). Similar effects on MT function are observed when tau is expressed in a variety of cell lines, including SHSY5Y neuroblastoma cells (Dayanandan et al 1999), Chinese hamster ovary (CHO) cells (Dayanandan et al 1999, Matsumura et al 1999, Vogelsberg-Ragaglia et al 2000), monkey kidney (COS) cells (Arawaka et al 1999, Sahara et al 2000), and Sf9 insect cells (Frappier et al 1999). Expression of a variety of tau missense mutations including G272V, Δ280K, P301L, V337M, and R406W in these cells caused varying degrees of reduced MT binding, disorganized MT morphology, and defects in MT assembly and MT instability. However, in two studies, many mutations had only a modest or no effect on MT binding and/or function, both in in vitro assays and in transfected cell lines (DeTure et al 2000, Sahara et al 2000). The discrepancies between these and other studies are most likely due to the differences in the levels of expression, the methods used for the quantification of tau levels, and the binding of tau to MTs. Nevertheless, even if these missense mutations cause only a modest reduction in MT binding affinity, this could have large cumulative effects on affected neurons over the human life span. Furthermore, increased cytosolic concentrations of unbound mutant tau proteins may facilitate aggregation of these abnormal proteins, with or without their WT counterparts, into filamentous inclusions.

A subset of missense *tau* gene mutations may cause FTDP-17, at least in part, by promoting tau aggregation. Several studies have demonstrated that some of these mutations, including K257T, G272V, Δ280K, P301L, P301S, V337M, and R406W, promote heparin- or arachidonic acid–induced tau filament formation in vitro relative to WT tau (Arrasate et al 1999, Nacharaju et al 1999, Goedert et al 1999a, Barghorn et al 2000, Gamblin et al 2000, Rizzini et al 2000). Furthermore, aggregation of mutant tau proteins in intact cells also has been demonstrated. Thus, CHO cells expressing tau with the ΔK280 mutation, but not other mutations (V337M, P301L, and R406W), formed insoluble amorphous and fibrillar tau aggregates (Vogelsberg-Ragaglia et al 2000). In addition, expression of the ΔK280, and R406W mutants in CHO and other cells led to reduced levels of tau phosphorylation relative to other mutant constructs and WT tau (Dayanandan et al 1999, Matsumura et al 1999, Perez et al 2000, Sahara et al 2000, Vogelsberg-Ragaglia et al 2000).

All known mutations in the *tau* gene lead to the formation of filaments made of hyperphosphorylated tau protein (Crowther & Goedert 2000). However, the fibrillary lesions observed with mutations in E10 or the intron following E10 are biochemically and ultrastructurally distinct from the lesions caused by mutations that are located outside E10. Coding region mutations located outside E10 affect all six isoforms of tau. Thus, as one might predict, tau fibrillary lesions are composed of all six tau isoforms (Hong et al 1998, Murrell et al 1999, Van Swieten et al 1999) (Figure 2). For some mutations (V337M and R406W), the morphologies and biochemical characteristics of tau filaments are indistinguishable from those of AD (Spillantini et al 1996, Hong et al 1998, Van Swieten et al 1999). Other coding region mutations located outside E10 (K252T, G272V, E342V, and G389R) give rise to a tau pathology that closely resembles that of PiD (Spillantini et al 1998b, Murrell et al 1999, Lippa et al 2000, Rizzini et al 2000). In contrast, mutations located within E10 itself or the intron following E10 lead to aggregation of predominantly 4R-tau (Clark et al 1998; Hong et al 1998; Hutton et al 1998; Reed et al 1998; Spillantini et al 1998b,c; Hulette et al 1999; Mirra et al 1999; Nasreddine et al 1999; Goedert et al 1999b; Yasuda et al 2000). Ultrastructurally, these lesions are composed of twisted ribbons that are similar to the filaments observed in 4R-tau disorders, particularly CBD (Reed et al 1998, Spillantini et al 1998b, Bird et al 1999, Bugiani et al 1999, Delisle et al 1999, Hulette et al 1999, Iijima et al 1999, Mirra et al 1999, Goedert et al 1999b, Yasuda et al 2000).

Finally, a family with a syndrome known as hereditary dysphasic disinhibition dementia 2 (HDDD2), which appears similar to some of the syndromes seen in FTDP-17 kindreds, has been reported to show linkage to 17q21-22 with a lod (logarithm of odds) score of 3.68; however, no *tau* gene mutation or any other genetic abnormality has been identified in this family (Lendon et al 1998). It is surprising, however, that recent studies of three brains from affected members of the HDDD2 kindred revealed that HDDD2 shares significant neuropathological and biochemical abnormalities with sporadic FTD patients classified as FTLD (DLDH) (Lendon et al 1998, Zhukareva et al 2001). As discussed above regarding FTLD (DLDH), the loss of tau proteins in several of the HDDD2 brains, together with the preservation of tau mRNA, suggests that the abundance of tau protein may be controlled posttranscriptionally, either at the level of tau mRNA translation or through mechanisms that regulate mRNA stability. Thus, the HDDD2 kindred appears to be the familial counterpart of sporadic FTLD (DLDH), both of which may define a novel and distinct tauopathy caused by a reduction of brain tau.

Although the biochemical and structural characteristics of the tau aggregates in FTDP-17 appear to be predictable, based on our understanding of the functions of tau protein and tau gene splicing, the basis for the clinical phenotypes and topographical distributions of pathology is more enigmatic. For example, it is not clear why the clinical and neuropathologic phenotype of individuals with FTDP-17 mutations ranges from FTD (including subtypes thereof, such as PiD, CBD, and PSP) to multisystem neurodegeneration. However, some *tau* gene mutations cause a similar phenotype in different families or in different members of the same family. For instance, the N279K mutation typically causes a phenotype reminiscent of PSP with superimposed dementia (Reed et al 1998, Delisle et al 1999, Yasuda et al 1999). In contrast, there are several clinical and pathologic descriptions of families with the P301L mutation that demonstrate a highly variable phenotype ranging from PSP to CBD to PiD (Spillantini et al 1998b, Bird et al 1999, Mirra et al 1999, Nasreddine et al 1999). Even more perplexing is the report of two brothers from one P301L family (Bird et al 1999), with frontal lobe degeneration in one individual and PSP-like pathology in the other. Similarly, in a family with the P301S mutation, one individual presented with FTD whereas his son presented clinically with CBD (Bugiani et al 1999). Although only a few reports of this kind have been published, they suggest that there is extensive overlap between the various tau-related disorders and that the clinical and pathologic distinctions between them may be due to other genetic and/or epigenetic factors that modify the effects of the primary mutation. Currently, the specific genetic and/or environmental modifiers that might determine the phenotype of a specific individual remain unknown, but these are fields of active investigation, and the generation of animal models of tau-mediated neurodegeneration may facilitate this research (Figure 4).

EXPERIMENTAL ANIMAL MODELS OF TAUOPATHIES

Experimental and TG animal models of tauopathies will serve as informative systems for elucidating the role of abnormalities in tau in the onset and progression of a variety of neurodegenerative disorders. In addition, they may be useful models for the development and testing of novel therapies. Early efforts to produce animal models with tau pathology were largely based on the hypothesis that the development of extracellular $A\beta$ pathology in TG mice would induce intraneuronal tau pathology. However, although various TG mouse lines accumulate $A\beta$ plaques, none has developed AD-like tau pathology (Games et al 1995, Hsiao et al 1996, Sturchler-Pierrat et al 1997). More recently, several animal models of tau pathology were produced by overexpressing human tau proteins (Table 3).

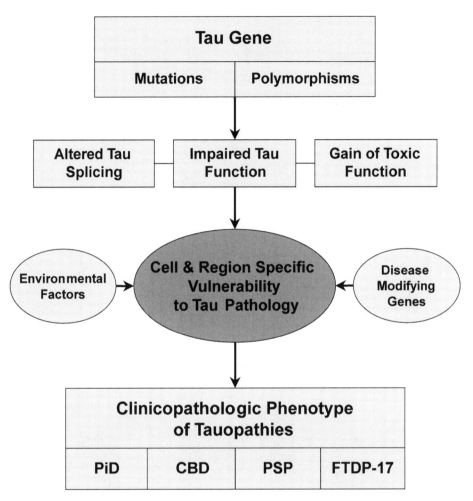

Figure 4 Model of disease pathways in tauopathies. Mutations and/or polymorphisms in the tau gene in conjunction with environmental and additional genetic factors initiate pathogenic processes that cause regional and cell type–specific tau pathology and neurodegeneration, thus leading to specific clinicopathologic phenotypes. PiD, Pick's disease; CBD, corticobasal degeneration; PSP, progressive supranuclear palsy; FTDP-17, frontotemporal dementia and parkinsonism linked to chromosome 17.

Initial reports described TG mouse lines expressing 4R2N or 3R0N human tau utilizing cDNA constructs with either the Thy1 or the 3-hydroxy-3-methylglutaryl coenzyme A reductase promoters (Götz et al 1995, Brion et al 1999). Both lines developed somatodendritic expression of tau, suggestive of "pretangle" pathology, but no filamentous tau inclusions were observed and the animals were phenotypically normal. The lack of filament formation may have been due to the relatively modest expression levels of human tau, and this notion is supported by the

TABLE 3 Transgenic models of tauopathies

Gene	Promoter	Tau Pathology	Phenotype	Reference
Tau, 4R/2N	Thy1	Somatodendritic tau expression	Normal	Götz et al (1995)
Tau, 3R/0N	HMG-CoA reductase	Somatodendritic tau expression		Brion et al (1999)
Tau, 3R/0N	Prion protein	Somatodendritic tau expression; tau immunoreactive spheroids in brain and spinal cord; neurofibrillary tangles in brain at 18 months or older	Axonopathy with muscle weakness	Ishihara et al (1999, 2001)
Tau, 4R/2N	Thy1	Somatodendritic tau expression; tau immunoreactive spheroids in brain and spinal cord	Axonopathy with sensorimotor dysfunction	Spittaels et al (1999)
Tau, 4R/2N	Thy1	Somatodendritic tau expression; tau immunoreactive spheroids in brain and spinal cord	Axonopathy with muscle weakness	Probst et al (2000)
Tau, 4R/0N, P301L	Prion protein	Neurofibrillary tangles in brain and spinal cord; somatodendritic tau expression	Motor and behavioral deficits; amyotrophy	Lewis et al (2000)
Tau, 4R/2N, P301L	Thy1	Neurofibrillary tangles in brain and spinal cord; somatodendritic tau expression		Götz et al (2001)
Tau, genomic	Endogenous	Axonal expression	Normal	Duff et al (2000)
ApoE4	Multiple	Phosphorylated tau expression in neocortex, hippocampus, and amygdala	Motor dysfunction and amyotrophy	Tesseur et al (2000)
p25	Neuron specific enolase	Phosphorylated tau expression in cortex, amygdala, and thalamus	Increased locomotor activity	Ahlijanian et al (2000)
Anti-NGF IgH/Igκ	Cytomegalovirus early region	Phosphorylated tau expression in cortex and hippocampus with associated neuron loss	Spatial memory and object recognition impairment	Capsoni et al (2000)

finding that massive overexpression of 4R/2N human tau in lamprey reticulospinal neurons led to the formation of PHF-like tau inclusions with degeneration of a subset of these neurons (Hall et al 1997, 2000). Subsequently, a series of papers has described lines of TG mice expressing high levels of 3R/0N or 4R/2N human tau utilizing cDNA constructs with either the Thy1 or the prion protein promoter (Ishihara et al 1999, Spittaels et al 1999, Duff et al 2000, Probst et al 2000). These mice developed numerous abnormal tau-immunoreactive nerve cell bodies and dendrites and large numbers of pathologically enlarged axons containing tau-immunoreactive spheroids. These changes were most prominent in spinal cord but were also seen in brain. They were accompanied by histological and behavioral signs of amytrophy. Mice doubly transgenic for 4R2N tau and GSK-3β showed increased levels of tau phosphorylation and a marked reduction in the number of spheroids and associated histological and behavioral changes at 3–4 months of age (Spittaels et al 2000). In this system, therefore, hyperphosphorylation of tau correlates inversely with pathology. Although the tau pathology most closely resembles that observed in the amyotrophic lateral sclerosis/parkinsonism–dementia complex (Matsumoto et al 1990), it differs in several respects from that found in human diseases. Thus, besides tau, the spheroids in these TG mice also contain neurofilament proteins and tubulin. They are not detected by Congo red and Thioflavin S and do not bind Gallyas silver. However, a recent report has shown that as these tau TG mice aged to over 18 months, modest numbers of filamentous tau tangles with similar properties to those found in AD could be detected in hippocampus and entorhinal cortex (Ishihara et al 2001). Moreover, like NFTs in AD, the tau tangles in these aged mice were ubiquitinated, did not contain neurofilaments or tubulin, and were detected by Congo red, Thioflavin S, and Gallyas silver staining methods. It is interesting that mice transgenic for the entire human *tau* gene expressed all six tau isoforms but failed to develop significant pathology (Duff et al 2000).

The discovery of mutations in the tau gene in FTDP-17 is leading to the production of TG mouse lines expressing mutant human tau in neurons and glial cells. Lewis et al (2000) developed several lines expressing modest levels of 4R/0N tau with and without the P301L mutation. In contrast to mice expressing WT tau, mice expressing tau with the P301L mutation exhibited an age- and gene dose-dependent accumulation of tau tangles in both brain and spinal cord, with associated nerve cell loss and reactive gliosis. Similar results have been reported by Götz et al (2001) in TG lines expressing 4R2N tau with the P301L mutation. The tau tangles found in these mice appeared to comprise only the mutant human tau, which suggests that the P301L mutation probably causes neurodegeneration by promoting the aggregation of the mutant tau. This is supported by recent biochemical studies using antibodies specific for P301L tau that demonstrated recovery of mutant but not WT tau from the insoluble fraction isolated from brain tissue of individuals with the P301L *tau* gene mutation (Rizzu et al 2000).

Other approaches to the development of tau pathology in TG mice have made use of molecules known to interact with tau. For example, TG mice overexpressing human p25, an activator of cdk5, develop disturbances in cytoskeletal architecture and behavioral alterations (Ahlijanian et al 2000). However, there was no biochemical evidence of an accumulation of insoluble, hyperphosphorylated tau in these mice. TG mice expressing apolipoprotein E4, an allelic risk factor for sporadic AD (Corder et al 1993), showed an age-dependent increase in tau phosphorylation that correlated with the level of apolipoprotein E4 expression (Tesseur et al 2000). Although these mice showed somatodendritic expression of phosphorylated tau, there was no evidence of fibrillary pathology. Finally, TG mice expressing antibodies to nerve growth factor inside nerve cells developed a prominent age-dependent neurodegenerative pathology, including neuronal loss and hyperphosphorylated, insoluble tau in cortex and hippocampus (Capsoni et al 2000). However, as for the apolipoprotein E4 TG mice described above, no evidence of fibrillary tau pathology was presented. In summary, although several TG mouse models show features of various tau-related disorders, they still fall short of demonstrating the entire constellation of the most characteristic features of human tauopathies.

CONCLUSION

The accumulation of filamentous tau inclusions is a common feature of a wide variety of neurodegenerative disorders, many of which are distinguished by the distinct topographic and cell type–specific distributions of inclusions. The biochemical and ultrastructural characteristics of the *tau* abnormalities, which are frequently related to the inclusion or exclusion of E10, also reveal a significant phenotypic overlap. The discovery of multiple mutations in the *tau* gene that lead to the abnormal aggregation of tau and cause FTDP-17 demonstrates that tau dysfunction is sufficient to produce neurodegenerative disease. The mutations lead to specific cellular alterations, including altered expression, function, and biochemistry of tau protein. The finding that specific polymorphisms and mutations lead to diverse phenotypes raises the possibility that the clinical and pathological expression of these disorders may be influenced by other genetic and epigenetic factors.

All these disorders have as a common theme accumulation of hyperphosphorylated tau protein in a filamentous form, which almost certainly perturbs the function of MTs and interferes with axonal transport. It remains to be established whether a protein kinase/phosphatase imbalance is an early mechanistic step leading to the generation of filamentous tau in some tauopathies. Genetic and/or environmental factors could initiate a cascade of events that leads to the abnormal phosphorylation of tau through incompletely defined pathways (Figure 4). It also remains to be seen whether the mere presence of tau filaments inside brain cells is sufficient to cause them to die. Similarly, the precise mechanisms by which tau protein assembles into filaments in human brain remain to be discovered. Further investigation into the mechanisms of tau dysfunction, as well as the identification of potential disease-modifying factors, will provide additional insight into novel strategies for disease treatment and prevention. The development of additional animal models of tauopathies that more closely recapitulate human diseases will facilitate this undertaking.

The aggregation of tau in AD and various tauopathies is but one example of abnormal protein-protein interactions that result in the intracellular accumulation of filamentous proteins. Abnormal protein aggregation is observed in a large number of neurodegenerative disorders (Prusiner 1998, Goedert et al 1998, Trojanowski et al 1998). Thus, besides tau pathology, AD is characterized by the extracellular accumulation of Aβ fibrils in the form of amyloid plaques; Lewy body disorders contain intracytoplasmic filamentous aggregates of α-synuclein; trinucleotide repeat disorders have intranuclear inclusions composed of fibrous polyglutamines; and spongiform encephalopathies demonstrate aggregates of proteinase-resistant prion protein. Aggregation of proteins in the brain is a common theme in a diverse group of disorders, and insight into the pathogenesis of any one of these disorders may have implications for our understanding of the mechanisms that underlie all these diseases.

ACKNOWLEDGMENTS

VM-YL is the John H Ware third Chair of Alzheimer's disease research at the University of Pennsylvania. Work done in our laboratories is supported by grants from the National Institute of Aging of the National Institutes of Health, the Dana Foundation, the US Alzheimer's Association, the UK Medical Research Council, and the UK Alzheimer's Research Trust. We thank Drs. MS Forman and V van Deerlin for their input.

Visit the Annual Reviews home page at www.AnnualReviews.org

LITERATURE CITED

Ahlijanian MK, Barrezueta NX, Williams RD, Jakowski A, Kowsz KP, et al. 2000. Hyperphosphorylated tau and neurofilament and cytoskeletal disruptions in mice overexpressing human p25, an activator of cdk5. *Proc. Natl. Acad. Sci. USA* 97(6):2910–15

Andreadis A, Brown WM, Kosik KS. 1992. Structure and novel exons of the human tau gene. *Biochemistry* 31(43):10626–33

Arawaka S, Usami M, Sahara N, Schellenberg GD, Lee G, Mori H. 1999. The tau mutation (val337met) disrupts cytoskeletal networks of microtubules. *NeuroReport* 10(5):993–97

Arima K, Kowalska A, Hasegawa M, Mukoyama M, Watanabe R, et al. 2000. Two brothers with frontotemporal dementia and

parkinsonism with an N279K mutation of the tau gene. *Neurology* 54(9):1787–95

Arnold SE, Han LY, Clark CM, Grossman M, Trojanowski JQ. 2000. Quantitative neurohistological features of frontotemporal degeneration. *Neurobiol. Aging* 21(6):913–919

Arrasate M, Perez M, Armas-Portela R, Avila J. 1999. Polymerization of tau peptides into fibrillar structures. The effect of FTDP-17 mutations. *FEBS Lett.* 446(1):199–202

Arrasate M, Perez M, Valpuesta JM, Avila J. 1997. Role of glycosaminoglycans in determining the helicity of paired helical filaments. *Am. J. Pathol.* 151(4):1115–22

Baker M, Litvan I, Houlden H, Adamson J, Dickson D, et al. 1999. Association of an extended haplotype in the tau gene with progressive supranuclear palsy. *Hum. Mol. Genet.* 8(4):711–15

Barghorn S, Zheng-Fischhöfer Q, Ackmann M, Biernat J, von Bergen M, et al. 2000. Structure, microtubule interactions, and paired helical filament aggregation by tau mutants of frontotemporal dementias. *Biochemistry* 39(38):11714–21

Baudier J, Cole RD. 1987. Phosphorylation of tau proteins to a state like that in Alzheimer's brain is catalyzed by a calcium/calmodulin-dependent kinase and modulated by phospholipids. *J. Biol. Chem.* 262(36):17577–83

Baumann K, Mandelkow EM, Biernat J, Piwnica-Worms H, Mandelkow E. 1993. Abnormal Alzheimer-like phosphorylation of tau-protein by cyclin-dependent kinases cdk2 and cdk5. *FEBS Lett.* 336(3):417–24

Bennett P, Bonifati V, Bonuccelli U, Colosimo C, De Mari M, et al. 1998. Direct genetic evidence for involvement of tau in progressive supranuclear palsy. European Study Group on Atypical Parkinsonism Consortium. *Neurology* 51(4):982–85

Biernat J, Gustke N, Drewes G, Mandelkow EM, Mandelkow E. 1993. Phosphorylation of Ser262 strongly reduces binding of tau to microtubules: distinction between PHF-like immunoreactivity and microtubule binding. *Neuron* 11(1):153–63

Billingsley ML, Kincaid RL. 1997. Regulated phosphorylation and dephosphorylation of tau protein: effects on microtubule interaction, intracellular trafficking and neurodegeneration. *Biochem. J.* 323(3):577–91

Binder LI, Frankfurter A, Rebhun LI. 1985. The distribution of tau in the mammalian central nervous system. *J. Cell Biol.* 101(4):1371–78

Bird TD, Nochlin D, Poorkaj P, Cherrier M, Kaye J, et al. 1999. A clinical pathological comparison of three families with frontotemporal dementia and identical mutations in the tau gene (P301L). *Brain* 122(4):741–56

Bird TD, Wijsman EM, Nochlin D, Leehey M, Sumi SM, et al. 1997. Chromosome 17 and hereditary dementia: linkage studies in three nonAlzheimer families and kindreds with late-onset FAD. *Neurology* 48(4):949–54

Bonifati V, Joosse M, Nicholl DJ, Vanacore N, Bennett P, et al. 1999. The tau gene in progressive supranuclear palsy: exclusion of mutations in coding exons and exon 10 splice sites, and identification of a new intronic variant of the disease-associated H1 haplotype in Italian cases. *Neurosci. Lett.* 274(1):61–65

Braak E, Braak H, Mandelkow EM. 1994. A sequence of cytoskeleton changes related to the formation of neurofibrillary tangles and neuropil threads. *Acta Neuropathol.* 87(6):554–67

Bramblett GT, Goedert M, Jakes R, Merrick SE, Trojanowski JQ, Lee VM-Y. 1993. Abnormal tau phosphorylation at Ser396 in Alzheimer's disease recapitulates development and contributes to reduced microtubule binding. *Neuron* 10(6):1089–99

Brion JP, Passareiro H, Nunez J, Flament-Durand J. 1985. Mise en évidence immunologique de la protéine tau au niveau des lésions de dégénérescence neurofibrillaire de la maladie d'Alzheimer. *Arch. Biol.* 95:229–35

Brion JP, Tremp G, Octave JN. 1999. Transgenic expression of the shortest human tau affects its compartmentalization and its phosphorylation as in the pretangle stage of Alzheimer's disease. *Am. J. Pathol.* 154(1):255–70

Buée L, Bussière T, Buée-Scherrer V, Delacourte A, Hof PR. 2000. Tau protein isoforms, phosphorylation and role in neurodegenerative disorders. *Brain Res. Rev.* 33:(1):95–130

Buée-Scherrer V, Hof PR, Buée L, Leveugle B, Vermersch P, et al. 1996. Hyperphosphorylated tau proteins differentiate corticobasal degeneration and Pick's disease. *Acta Neuropathol.* 91(4):351–59

Bugiani O, Murrell JR, Giaccone G, Hasegawa M, Ghigo G, et al. 1999. Frontotemporal dementia and corticobasal degeneration in a family with a P301S mutation in tau. *J. Neuropathol. Exp. Neurol.* 58(6):667–77

Butner KA, Kirschner MW. 1991. Tau protein binds to microtubules through a flexible array of distributed weak sites. *J. Cell Biol.* 115(3):717–30

Capsoni S, Ugolini G, Comparini A, Ruberti F, Berardi N, Cattaneo A. 2000. Alzheimer-like neurodegeneration in aged antinerve growth factor transgenic mice. *Proc. Natl. Acad. Sci. USA* 97(12):6826–31

Chambers CB, Lee JM, Troncoso JC, Reich S, Muma NA. 1999. Overexpression of 4R-tau mRNA isoforms in progressive supranuclear palsy but not in Alzheimer's disease. *Ann. Neurol.* 46(3):325–32

Clark LN, Poorkaj P, Wszolek Z, Geschwind DH, Nasreddine ZS, et al. 1998. Pathogenic implications of mutations in the tau gene in pallido-ponto-nigral degeneration and related neurodegenerative disorders linked to chromosome 17. *Proc. Natl. Acad. Sci. USA* 95(22):13103–7

Cleveland DW, Hwo SY, Kirschner MW. 1977a. Physical and chemical properties of purified tau factor and the role of tau in microtubule assembly. *J. Mol. Biol.* 116(2):227–47

Cleveland DW, Hwo SY, Kirschner MW. 1977b. Purification of tau, a microtubule-associated protein that induces assembly of microtubules from purified tubulin. *J. Mol. Biol.* 116(2):207–25

Cohen P. 1999. The Croonian Lecture 1998. Identification of a protein kinase cascade of major importance in insulin signal transduction. *Philos. Trans. R. Soc. London Ser. B* 354:485–95

Conrad C, Andreadis A, Trojanowski JQ, Dickson DW, Kang D, et al. 1997. Genetic evidence for the involvement of tau in progressive supranuclear palsy. *Ann. Neurol.* 41(2):277–81

Constantinidis J, Richard J, Tissot R. 1974. Pick's disease. Histological and clinical correlations. *Eur. Neurol.* 11(4):208–17

Corder EH, Saunders AM, Strittmatter WJ, Schmechel DE, Gaskell PC, et al. 1993. Gene dose of apolipoprotein E type 4 allele and the risk of Alzheimer's disease in late onset families. *Science* 261(5123):921–23

Couchie D, Mavilia C, Georgieff IS, Liem RK, Shelanski ML, Nunez J. 1992. Primary structure of high molecular weight tau present in the peripheral nervous system. *Proc. Natl. Acad. Sci. USA* 89(10):4378–81

Crowther RA. 1991. Straight and paired helical filaments in Alzheimer disease have a common structural unit. *Proc. Natl. Acad. Sci. USA* 88(6):2288–92

Crowther RA, Goedert M. 2000. Abnormal tau-containing filaments in neurodegenerative diseases. *J. Struct. Biol.* 130(2–3):271–79

Crowther RA, Olesen OF, Jakes R, Goedert M. 1992. The microtubule binding repeats of tau protein assemble into filaments like those found in Alzheimer's disease. *FEBS Lett.* 309(2):199–202

Crowther RA, Wischik CM. 1985. Image reconstruction of the Alzheimer paired helical filament. *EMBO J.* 4(13B):3661–65

Dayanandan R, van Slegtenhorst M, Mack TG, Ko L, Yen SH, et al. 1999. Mutations in tau reduce its microtubule binding properties in intact cells and affect its phosphorylation. *FEBS Lett.* 446(2–3):228–32

Delacourte A, Robitaille Y, Sergeant N, Buée L, Hof PR, et al. 1996. Specific pathological Tau protein variants characterize Pick's disease. *J. Neuropathol. Exp. Neurol.* 55(2):159–68

Delacourte A, Sergeant N, Wattez A, Gauvreau D, Robitaille Y. 1998. Vulnerable neuronal subsets in Alzheimer's and Pick's disease are distinguished by their tau isoform distribution and phosphorylation. *Ann. Neurol.* 43(2):193–204

Delisle MB, Murrell JR, Richardson R, Trofatter JA, Rascol O, et al. 1999. A mutation at codon 279 (N279K) in exon 10 of the Tau gene causes a tauopathy with dementia and supranuclear palsy. *Acta Neuropathol.* 98(1):62–77

DeTure M, Ko LW, Yen S, Nacharaju P, Easson C, et al. 2000. Missense tau mutations identified in FTDP-17 have a small effect on tau-microtubule interactions. *Brain Res.* 853(1):5–14

Dickson DW. 1998. Pick's disease: a modern approach. *Brain Pathol.* 8(2):339–54

Di Maria E, Tabaton M, Vigo T, Abbruzzese G, Bellone E, et al. 2000. Corticobasal degeneration shares a common genetic background with progressive supranuclear palsy. *Ann. Neurol.* 47(3):374–77

Drechsel DN, Hyman AA, Cobb MH, Kirschner MW. 1992. Modulation of the dynamic instability of tubulin assembly by the microtubule-associated protein tau. *Mol. Biol. Cell* 3(10):1141–54

Drewes G, Ebneth A, Preuss U, Mandelkow EM, Mandelkow E. 1997. MARK, a novel family of protein kinases that phosphorylate microtubule-associated proteins and trigger microtubule disruption. *Cell* 89(2):297–308

Drewes G, Lichtenberg-Kraag B, Döring F, Mandelkow EM, Biernat J, et al. 1992. Mitogen activated protein (MAP) kinase transforms tau protein into an Alzheimer-like state. *EMBO J.* 11(6):2131–38

Drewes G, Mandelkow EM, Baumann K, Goris J, Merlevede W, Mandelkow E. 1993. Dephosphorylation of tau protein and Alzheimer paired helical filaments by calcineurin and phosphatase-2A. *FEBS Lett.* 336(3):425–32

D'Souza I, Poorkaj P, Hong M, Nochlin D, Lee VM-Y, et al. 1999. Missense and silent tau gene mutations cause frontotemporal dementia with parkinsonism-chromosome 17 type, by affecting multiple alternative RNA splicing regulatory elements. *Proc. Natl. Acad. Sci. USA* 96(10):5598–603

D'Souza I, Schellenberg D. 2000. Determinants of 4 repeat tau expression: coordination between enhancing and inhibitory splicing sequences for exon 10 inclusion. *J. Biol. Chem.* 275(23):17700–9

Duff K, Knight H, Refolo LM, Sanders S, Yu X, et al. 2000. Characterization of pathology in transgenic mice over-expressing human genomic and cDNA tau transgenes. *Neurobiol. Dis.* 7(2):87–98

Dumanchin C, Camuzat A, Campion D, Verpillat P, Hannequin D, et al. 1998. Segregation of a missense mutation in the microtubule-associated protein tau gene with familial frontotemporal dementia and parkinsonism. *Hum. Mol. Genet.* 7(11):1825–29

Ezquerra M, Pastor P, Valldeoriola F, Molinuevo JL, Blesa R, et al. 1999. Identification of a novel polymorphism in the promoter region of the tau gene highly associated to progressive supranuclear palsy in humans. *Neurosci. Lett.* 275(3):183–86

Feany MB, Dickson DW. 1995. Widespread cytoskeletal pathology characterizes corticobasal degeneration. *Am. J. Pathol.* 146(6):1388–96

Feany MB, Ksiezak-Reding H, Liu WK, Vincent I, Yen SH, Dickson DW. 1995. Epitope expression and hyperphosphorylation of tau protein in corticobasal degeneration: differentiation from progressive supranuclear palsy. *Acta Neuropathol.* 90(1):37–43

Feany MB, Mattiace LA, Dickson DW. 1996. Neuropathologic overlap of progressive supranuclear palsy, Pick's disease and corticobasal degeneration. *J. Neuropathol. Exp. Neurol.* 55(1):53–67

Flament S, Delacourte A, Verny M, Hauw JJ, Javoy-Agid F. 1991. Abnormal tau proteins in progressive supranuclear palsy. Similarities and differences with the neurofibrillary degeneration of the Alzheimer type. *Acta Neuropathol.* 81(6):591–96

Foster NL, Wilhelmsen K, Sima AA, Jones MZ, D'Amato CJ, et al. 1997. Frontotemporal dementia and parkinsonism linked to chromosome 17: a consensus conference. *Ann. Neurol.* 41(6):706–15

Frappier T, Liang NS, Brown K, Leung CL, Lynch T, et al. 1999. Abnormal microtubule packing in processes of SF9 cells expressing the FTDP-17 V337M tau mutation. *FEBS Lett.* 455(3):262–66

Friedhoff P, von Bergen M, Mandelkow EM, Davies P, Mandelkow E. 1998. A nucleated assembly mechanism of Alzheimer paired helical filaments. *Proc. Natl. Acad. Sci. USA* 95(26):15712–17

Gamblin TC, King ME, Dawson H, Vitek MP, Kuret J, et al. 2000. In vitro polymerization of tau protein monitored by laser light scattering: method and application to the study of FTDP-17 mutants. *Biochemistry* 39(20):6136–44

Games D, Adams D, Alessandrini R, Barbour R, Berthelette P, et al. 1995. Alzheimer-type neuropathology in transgenic mice overexpressing V717F beta-amyloid precursor protein. *Nature* 373(6514):523–27

Gao QS, Memmott J, Lafyatis R, Stamm S, Sreaton G, Andreadis A. 2000. Complex regulation of tau exon 10, whose missplicing causes frontotemporal dementia. *J. Neurochem.* 74(2):490–500

Georgieff IS, Liem RK, Mellado W, Nunez J, Shelanski ML. 1991. High molecular weight tau: preferential localization in the peripheral nervous system. *J. Cell Sci.* 100(1):55–60

Ginsberg SD, Galvin JE, Chiu TS, Lee VM-Y, Masliah E, Trojanowski JQ. 1998. RNA sequestration to pathological lesions of neurodegenerative diseases. *Acta Neuropathol.* 96(5):487–94

Goedert M, Cohen ES, Jakes R, Cohen P. 1992a. p42 MAP kinase phosphorylation sites in microtubule-associated protein tau are dephosphorylated by protein phosphatase 2A₁. Implications for Alzheimer's disease. *FEBS Lett.* 312(1):95–99

Goedert M, Hasegawa M, Jakes R, Lawler S, Cuenda A, Cohen P. 1997. Phosphorylation of microtubule-associated protein tau by stress-activated protein kinases. *FEBS Lett.* 409(1):57–62

Goedert M, Jakes R. 1990. Expression of separate isoforms of human tau protein: correlation with the tau pattern in brain and effects on tubulin polymerization. *EMBO J.* 9(13):4225–30

Goedert M, Jakes R, Crowther RA. 1999a. Effects of frontotemporal dementia FTDP-17

mutations on heparin-induced assembly of tau filaments. *FEBS Lett.* 450(3):306–11

Goedert M, Jakes R, Crowther RA, Six J, Lübke U, et al. 1993. The abnormal phosphorylation of tau protein at Ser202 in Alzheimer disease recapitulates phosphorylation during development. *Proc. Natl. Acad. Sci. USA* 90(11):5066–70

Goedert M, Jakes R, Qi Z, Wang JH, Cohen P. 1995a. Protein phosphatase 2A is the major enzyme in brain that dephosphorylates tau protein phosphorylated by proline-directed protein kinases or cyclic AMP-dependent protein kinase. *J. Neurochem.* 65(6):2804–7

Goedert M, Jakes R, Spillantini MG, Hasegawa M, Smith MJ, Crowther RA. 1996. Assembly of microtubule-associated protein tau into Alzheimer-like filaments induced by sulphated glycosaminoglycans. *Nature* 383(6600):550–53

Goedert M, Satumtira S, Jakes R, Smith MJ, Kamibayashi C, et al. 2000. Reduced binding of protein phosphatase 2A to tau protein with frontotemporal dementia and parkinsonism linked to chromosome 17 mutations. *J. Neurochem.* 75(5):2155–62

Goedert M, Spillantini MG, Cairns NJ, Crowther RA. 1992b. Tau proteins of Alzheimer paired helical filaments: abnormal phosphorylation of all six brain isoforms. *Neuron* 8(1):159–68

Goedert M, Spillantini MG, Crowther RA. 1992c. Cloning of a big tau microtubule-associated protein characteristic of the peripheral nervous system. *Proc. Natl. Acad. Sci. USA* 89(5):1983–87

Goedert M, Spillantini MG, Crowther RA, Chen SG, Parchi P, et al. 1999b. Tau gene mutation in familial progressive subcortical gliosis. *Nat. Med.* 5(4):454–57

Goedert M, Spillantini MG, Davies SW. 1998. Filamentous nerve cell inclusions in neurodegenerative diseases. *Curr. Opin. Neurobiol.* 8(5):619–32

Goedert M, Spillantini MG, Jakes R, Crowther RA, Vanmechelen E, et al. 1995b. Molecular dissection of the paired helical filament. *Neurobiol. Aging* 16(3):325–34

Goedert M, Spillantini MG, Jakes R, Rutherford D, Crowther RA. 1989a. Multiple isoforms of human microtubule-associated protein tau: sequences and localization in neurofibrillary tangles of Alzheimer's disease. *Neuron* 3(4):519–26

Goedert M, Spillantini MG, Potier MC, Ulrich J, Crowther RA. 1989b. Cloning and sequencing of the cDNA encoding an isoform of microtubule-associated protein tau containing four tandem repeats: differential expression of tau protein mRNAs in human brain. *EMBO J.* 8(2):393–99

Goedert M, Wischik CM, Crowther RA, Walker JE, Klug A. 1988. Cloning and sequencing of the cDNA encoding a core protein of the paired helical filament of Alzheimer disease: identification as the microtubule-associated protein tau. *Proc. Natl. Acad. Sci. USA* 85(11):4051–55

Goode BL, Denis PE, Panda D, Radeke MJ, Miller HP, et al. 1997. Functional interactions between the proline-rich and repeat regions of tau enhance microtubule binding and assembly. *Mol. Biol. Cell* 8(2):353–65

Goode BL, Feinstein SC. 1994. Identification of a novel microtubule binding and assembly domain in the developmentally regulated interrepeat region of tau. *J. Cell Biol.* 124(5):769–82

Götz J, Chen F, Barmettler R, Nitsch RM. 2001. Tau filament formation in transgenic mice expressing P301L tau. *J. Biol. Chem.* 276(1):529–34

Götz J, Probst A, Spillantini MG, Schafer T, Jakes R, et al. 1995. Somatodendritic localization and hyperphosphorylation of tau protein in transgenic mice expressing the longest human brain tau isoform. *EMBO J.* 14(7):1304–13

Greenberg SG, Davies P. 1990. A preparation of Alzheimer paired helical filaments that displays distinct tau proteins by polyacrylamide gel electrophoresis. *Proc. Natl. Acad. Sci. USA* 87(15):5827–31

Greenberg SG, Davies P, Schein JD, Binder LI. 1992. Hydrofluoric acid-treated tau PHF proteins display the same biochemical properties as normal tau. *J. Biol. Chem.* 267(1):564–69

Grover A, Houlden H, Baker M, Adamson J, Lewis J, et al. 1999. 5′ splice site mutations in tau associated with the inherited dementia FTDP-17 affect a stem-loop structure that regulates alternative splicing of exon 10. *J. Biol. Chem.* 274(21):15134–43

Grundke-Iqbal I, Iqbal K, Tung YC, Quinlan M, Wisniewski HM, Binder LI. 1986. Abnormal phosphorylation of the microtubule-associated protein tau (tau) in Alzheimer cytoskeletal pathology. *Proc. Natl. Acad. Sci. USA* 83(13):4913–17

Gustke N, Trinczek B, Biernat J, Mandelkow EM, Mandelkow E. 1994. Domains of tau protein and interactions with microtubules. *Biochemistry* 33(32):9511–22

Hall GF, Chu B, Lee G, Yao J. 2000. Human tau filaments induce microtubule and synapse loss in an in vivo model of neurofibrillary degenerative disease. *J. Cell Sci.* 113(8):1373–87

Hall GF, Yao J, Lee G. 1997. Human tau becomes phosphorylated and forms filamentous deposits when overexpressed in lamprey central neurons in situ. *Proc. Natl. Acad. Sci. USA* 94(9):4733–38

Hanger DP, Hughes K, Woodgett JR, Brion JP, Anderton BH. 1992. Glycogen synthase kinase-3 induces Alzheimer's disease-like phosphorylation of tau: generation of paired helical filament epitopes and neuronal localisation of the kinase. *Neurosci. Lett.* 147(1):58–62

Harada A, Oguchi K, Okabe S, Kuno J, Terada S, et al. 1994. Altered microtubule organization in small-calibre axons of mice lacking tau protein. *Nature* 369(6480):488–91

Hardy J, Allsop D. 1991. Amyloid deposition as the central event in the aetiology of Alzheimer's disease. *Trends Pharmacol. Sci.* 12(10):383–88

Hasegawa M, Crowther RA, Jakes R, Goedert M. 1997. Alzheimer-like changes in microtubule-associated protein tau induced by sulfated glycosaminoglycans. Inhibition of microtubule binding, stimulation of phosphorylation, and filament assembly depend on the degree of sulfation. *J. Biol. Chem.* 272(52):33118–24

Hasegawa M, Jakes R, Crowther RA, Lee VM-Y, Ihara Y, Goedert M. 1996. Characterization of mAb AP422, a novel phosphorylation-dependent monoclonal antibody against tau protein. *FEBS Lett.* 384(1):25–30

Hasegawa M, Smith MJ, Goedert M. 1998. Tau proteins with FTDP-17 mutations have a reduced ability to promote microtubule assembly. *FEBS Lett.* 437(3):207–10

Hasegawa M, Smith MJ, Iijima M, Tabira T, Goedert M. 1999. FTDP-17 mutations N279K and S305N in tau produce increased splicing of exon 10. *FEBS Lett.* 443(2):93–96

Hauw JJ, Daniel SE, Dickson D, Horoupian DS, Jellinger K, et al. 1994. Preliminary NINDS neuropathologic criteria for Steele-Richardson-Olszewski syndrome (progressive supranuclear palsy). *Neurology* 44(11):2015–19

Hauw JJ, Verny M, Delaere P, Cervera P, He Y, Duyckaerts C. 1990. Constant neurofibrillary changes in the neocortex in progressive supranuclear palsy. Basic differences with Alzheimer's disease and aging. *Neurosci. Lett.* 119(2):182–86

Heutink P, Stevens M, Rizzu P, Bakker E, Kros JM, et al. 1997. Hereditary frontotemporal dementia is linked to chromosome 17q21-q22: a genetic and clinicopathological study of three Dutch families. *Ann. Neurol.* 41(2):150–59

Hiesberger T, Trommsdorff M, Howell BW, Goffinet A, Mumby MC, et al. 1999. Direct binding of Reelin to VLDL receptor and ApoE receptor 2 induces tyrosine phosphorylation of disabled-1 and modulates tau phosphorylation. *Neuron* 24(2):481–89

Higgins JJ, Litvan I, Pho LT, Li W, Nee LE. 1998. Progressive supranuclear gaze palsy is in linkage disequilibrium with the tau and not the alpha-synuclein gene. *Neurology* 50(1):270–73

Himmler A, Drechsel D, Kirschner MW, Martin DW Jr. 1989. Tau consists of a set of proteins with repeated C-terminal microtubule-binding domains and variable N-terminal domains. *Mol. Cell Biol.* 9(4):1381–88

Hoenicka J, Perez M, Perez-Tur J, Barabash A, Godoy M, et al. 1999. The tau gene A0 allele and progressive supranuclear palsy. *Neurology* 53(6):1219–25

Hoffmann R, Lee VM-Y, Leight S, Varga I, Otvos L Jr. 1997. Unique Alzheimer's disease paired helical filament specific epitopes involve double phosphorylation at specific sites. *Biochemistry* 36(26):8114–24

Hong M, Chen DC, Klein PS, Lee VM-Y. 1997. Lithium reduces tau phosphorylation by inhibition of glycogen synthase kinase-3. *J. Biol. Chem.* 272(40):25326–32

Hong M, Lee VM-Y. 1997. Insulin and insulin-like growth factor-1 regulate tau phosphorylation in cultured human neurons. *J. Biol. Chem.* 272(31):19547–53

Hong M, Trojanowski JQ, Lee VM-Y. 2000. Tau-based neurofibrillary lesions. In *Neurodegenerative Dementias*, ed. CM Clark, JQ Trojanowski, pp. 161–75. New York: McGraw Hill

Hong M, Zhukareva V, Vogelsberg-Ragaglia V, Wszolek Z, Reed L, et al. 1998. Mutation-specific functional impairments in distinct tau isoforms of hereditary FTDP-17. *Science* 282(5395):1914–17

Houlden H, Baker M, Adamson J, Grover A, Waring S, et al. 1999. Frequency of tau mutations in three series of non-Alzheimer's degenerative dementia. *Ann. Neurol.* 46(2):243–48

Hsiao K, Chapman P, Nilsen S, Eckman C, Harigaya Y, et al. 1996. Correlative memory deficits, Aβ elevation, and amyloid plaques in transgenic mice. *Science* 274(5284):99–102

Hulette CM, Pericak-Vance MA, Roses AD, Schmechel DE, Yamaoka LH, et al. 1999.

Neuropathological features of frontotemporal dementia and parkinsonism linked to chromosome 17q21–22 (FTDP-17): Duke Family 1684. *J. Neuropathol. Exp. Neurol.* 58(8):859–66

Hutton M, Lendon CL, Rizzu P, Baker M, Froelich S, et al. 1998. Association of missense and 5′-splice-site mutations in tau with the inherited dementia FTDP-17. *Nature* 393(6686):702–5

Iijima M, Tabira T, Poorkaj P, Schellenberg GD, Trojanowski JQ, et al. 1999. A distinct familial presenile dementia with a novel missense mutation in the tau gene. *NeuroReport* 10(3):497–501

Ikegami S, Harada A, Hirokawa N. 2000. Muscle weakness, hyperactivity, and impairment in fear conditioning in tau-deficient mice. *Neurosci. Lett.* 279(3):129–32

Ishiguro K, Kobayashi S, Omori A, Takamatsu M, Yonekura S, et al. 1994. Identification of the 23 kDa subunit of tau protein kinase II as a putative activator of cdk5 in bovine brain. *FEBS Lett.* 342(2):203–8

Ishihara T, Hong M, Zhang B, Nakagawa Y, Lee MK, et al. 1999. Age-dependent emergence and progression of a tauopathy in transgenic mice overexpressing the shortest human tau isoform. *Neuron* 24(3):751–62

Ishihara T, Zhang B, Higuchi M, Yashiyama Y, Trojanowski JQ, Lee VM-Y. 2001. Age dependent induction of congophilic neurofibrillary inclusions in tau transgenic mice. *Am. J. Pathol.* 158(2):555–62

Iwatsubo T, Hasegawa M, Ihara Y. 1994. Neuronal and glial tau-positive inclusions in diverse neurologic diseases share common phosphorylation characteristics. *Acta Neuropathol.* 88(2):129–36

Jiang Z, Cote J, Kwon JM, Goate AM, Wu JY. 2000. Aberrant splicing of tau pre-mRNA caused by intronic mutations associated with the inherited dementia frontotemporal dementia with parkinsonism linked to chromosome 17. *Mol. Cell Biol.* 20(11):4036–48

Kampers T, Friedhoff P, Biernat J, Mandelkow EM, Mandelkow E. 1996. RNA stimulates aggregation of microtubule-associated protein tau into Alzheimer-like paired helical filaments. *FEBS Lett.* 399(3):344–49

Kanemaru K, Takio K, Miura R, Titani K, Ihara Y. 1992. Fetal-type phosphorylation of the tau in paired helical filaments. *J. Neurochem.* 58(5):1667–75

Kidd M. 1963. Paired helical filaments in electron microscopy of Alzheimer's disease. *Nature* 197:192–94

Kobayashi S, Ishiguro K, Omori A, Takamatsu M, Arioka M, et al. 1993. A cdc2-related kinase PSSALRE/cdk5 is homologous with the 30 kDa subunit of tau protein kinase II, a proline-directed protein kinase associated with microtubule. *FEBS Lett.* 335(2):171–75

Kodama K, Okada S, Iseki E, Kowalska A, Tabira T, et al. 2000. Familial frontotemporal dementia with a P301L tau mutation in Japan. *J. Neurol. Sci.* 176(1):57–64

Komori T. 1999. Tau-positive glial inclusions in progressive supranuclear palsy, corticobasal degeneration and Pick's disease. *Brain Pathol.* 9(4):663–79

Komori T, Arai N, Oda M, Nakayama H, Mori H, et al. 1998. Astrocytic plaques and tufts of abnormal fibers do not coexist in corticobasal degeneration and progressive supranuclear palsy. *Acta Neuropathol.* 96(4):401–8

Kondo J, Honda T, Mori H, Hamada Y, Miura R, et al. 1988. The carboxyl third of tau is tightly bound to paired helical filaments. *Neuron* 1(9):827–34

Kosik KS, Orecchio LD, Binder L, Trojanowski JQ, Lee VM-Y, Lee G. 1988. Epitopes that span the tau molecule are shared with paired helical filaments. *Neuron* 1(9):817–25

Ksiezak-Reding H, Morgan K, Mattiace LA, Davies P, Liu WK, et al. 1994. Ultrastructure and biochemical composition of paired helical filaments in corticobasal degeneration. *Am. J. Pathol.* 145(6):1496–508

Kusakawa G, Saito T, Onuki R, Ishiguro K, Kishimoto T, Hisanaga S. 2000. Calpain-dependent proteolytic cleavage of the p35 cyclin-dependent kinase 5 activator to p25. *J. Biol. Chem.* 275(22):17166–72

Lee G, Neve RL, Kosik KS. 1989. The micro-tubule binding domain of tau protein. *Neuron* 2(6):1615–24

Lee MS, Kwon YT, Li M, Peng J, Friedlander RM, Tsai LH. 2000. Neurotoxicity induces cleavage of p35 to p25 by calpain. *Nature* 405(6784):360–64

Lee VM-Y, Balin BJ, Otvos L Jr, Trojanowski JQ. 1991. A68: a major subunit of paired helical filaments and derivatized forms of normal tau. *Science* 251(4994):675–78

Lendon CL, Lynch T, Norton J, McKeel DW Jr, Busfield F, et al. 1998. Hereditary dysphasic disinhibition dementia: a frontotemporal dementia linked to 17q21–22. *Neurology* 50(6):1546–55

Leost M, Schultz C, Link A, Wu YZ, Biernat J, et al. 2000. Paullones are potent inhibitors of glycogen synthase kinase-3β and cyclin-dependent kinase 5/p25. *Eur. J. Biochem.* 267(19):5983–94

Lew J, Huang QQ, Qi Z, Winkfein RJ, Aebersold R, et al. 1994. A brain-specific activator of cyclin-dependent kinase 5. *Nature* 371(6496):423–26

Lew J, Wang JH. 1994. Neuronal cdc2-like kinase. *Trends Biochem. Sci.* 20:33–37

Lewis J, McGowan E, Rockwood J, Melrose H, Nacharaju P, et al. 2000. Neurofibrillary tangles, amyotrophy and progressive motor disturbance in mice expressing mutant (P301L) tau protein. *Nat. Genet.* 25(4):402–5

Liao H, Li Y, Brautigan DL, Gundersen GG. 1998. Protein phosphatase 1 is targeted to microtubules by the microtubule-associated protein tau. *J. Biol. Chem.* 273(34):21901–8

Lieberman AP, Trojanowski JQ, Lee VM-Y, Balin BJ, Ding XS, et al. 1998. Cognitive, neuroimaging, and pathological studies in a patient with Pick's disease. *Ann. Neurol.* 43(2):259–65

Lippa CF, Zhukareva V, Kawarai T, Uryu H, Shafiq M, et al. 2000. Frontotemporal dementia with novel tau pathology and a Glu342Val *tau* gene mutation. *Ann. Neurol.* 48(6):850–58

Litersky JM, Johnson GV. 1992. Phosphoryla-tion by cAMP-dependent protein kinase inhibits the degradation of tau by calpain. *J. Biol. Chem.* 267(3):1563–68

Litvan I, Agid Y, Calne D, Campbell G, Dubois B, et al. 1996. Clinical research criteria for the diagnosis of progressive supranuclear palsy (Steele-Richardson-Olszewski syndrome): report of the NINDS-SPSP international workshop. *Neurology* 47(1):1–9

LoPresti P, Szuchet S, Papasozomenos SC, Zinkowski RP, Binder LI. 1995. Functional implications for the microtubule-associated protein tau: localization in oligodendrocytes. *Proc. Natl. Acad. Sci. USA* 92(22):10369–73

Lovestone S, Davis DR, Webster MT, Kaech S, Brion JP, et al. 1999. Lithium reduces tau phosphorylation: effects in living cells and in neurons at therapeutic concentrations. *Biol. Psychiatry* 45(8):995–1003

Lovestone S, Hartley CL, Pearce J, Anderton BH. 1996. Phosphorylation of tau by glycogen synthase kinase-3 beta in intact mammalian cells: the effects on the organization and stability of microtubules. *Neuroscience* 73(4):1145–57

Lu PJ, Wulf G, Zhou XZ, Davies P, Lu KP. 1999. The prolyl isomerase Pin1 restores the function of Alzheimer-associated phosphorylated tau protein. *Nature* 399(6738):784–88

Lund and Manchester Groups. 1994. Clinical and neuropathological criteria for frontotemporal dementia. The Lund and Manchester Groups. *J. Neurol. Neurosurg. Psychiatry* 57(4):416–18

Mailliot C, Sergeant N, Bussière T, Caillet-Boudin ML, Delacourte A, Buée L. 1998. Phosphorylation of specific sets of tau isoforms reflects different neurofibrillary degeneration processes. *FEBS Lett.* 433(3):201–4

Mandelkow EM, Drewes G, Biernat J, Gustke N, Van Lint J, et al. 1992. Glycogen synthase kinase-3 and the Alzheimer-like state of microtubule-associated protein tau. *FEBS Lett.* 314(3):315–21

Mann DM. 1998. Dementia of frontal type and dementias with subcortical gliosis. *Brain Pathol.* 8(2):325–38

Matsumoto S, Hirano A, Goto S. 1990. Spinal cord neurofibrillary tangles of Guamanian amyotrophic lateral sclerosis and parkinsonism-dementia complex: an immunohistochemical study. *Neurology* 40(6):975–79

Matsumura N, Yamazaki T, Ihara Y. 1999. Stable expression in Chinese hamster ovary cells of mutated tau genes causing frontotemporal dementia and parkinsonism linked to chromosome 17 (FTDP-17). *Am. J. Pathol.* 154(6):1649–56

Matsuo ES, Shin RW, Billingsley ML, Van deVoorde A, O'Connor M, et al. 1994. Biopsy-derived adult human brain tau is phosphorylated at many of the same sites as Alzheimer's disease paired helical filament tau. *Neuron* 13(4):989–1002

Mawal-Dewan M, Henley J, Van de Voorde A, Trojanowski JQ, Lee VM-Y. 1994. The phosphorylation state of tau in the developing rat brain is regulated by phosphoprotein phosphatases. *J. Biol. Chem.* 269(49):30981–87

Merrick SE, Trojanowski JQ, Lee VM-Y. 1997. Selective destruction of stable microtubules and axons by inhibitors of protein serine/threonine phosphatases in cultured human neurons. *J. Neurosci.* 17(15):5726–37

Mirra SS, Murrell JR, Gearing M, Spillantini MG, Goedert M, et al. 1999. Tau pathology in a family with dementia and a P301L mutation in tau. *J. Neuropathol. Exp. Neurol.* 58(4):335–45

Mori H, Nishimura M, Namba Y, Oda M. 1994. Corticobasal degeneration: a disease with widespread appearance of abnormal tau and neurofibrillary tangles, and its relation to progressive supranuclear palsy. *Acta Neuropathol.* 88(2):113–21

Morishima-Kawashima M, Hasegawa M, Takio K, Suzuki M, Yoshida H, et al. 1995. Hyperphosphorylation of tau in PHF. *Neurobiol. Aging* 16(3):365–71

Morris HR, Janssen JC, Bandmann O, Daniel SE, Rossor MN, et al. 1999a. The tau gene A0 polymorphism in progressive supranuclear palsy and related neurodegenerative diseases. *J. Neurol. Neurosurg. Psychiatry* 66(5):665–67

Morris HR, Perez-Tur J, Janssen JC, Brown J, Lees AJ, et al. 1999b. Mutation in the tau exon 10 splice site region in familial frontotemporal dementia. *Ann. Neurol.* 45(2):270–71

Munoz-Garcia D, Ludwin SK. 1984. Classic and generalized variants of Pick's disease: a clinicopathological, ultrastructural, and immunocytochemical comparative study. *Ann. Neurol.* 16(4):467–80

Munoz-Montano JR, Moreno FJ, Avila J, Diaz-Nido J. 1997. Lithium inhibits Alzheimer's disease-like tau protein phosphorylation in neurons. *FEBS Lett.* 411(2–3):183–88

Murayama S, Mori H, Ihara Y, Tomonaga M. 1990. Immunocytochemical and ultrastructural studies of Pick's disease. *Ann. Neurol.* 27(4):394–405

Murrell JR, Koller D, Foroud T, Goedert M, Spillantini MG, et al. 1997. Familial multiple-system tauopathy with presenile dementia is localized to chromosome 17. *Am. J. Hum. Genet.* 61(5):1131–38

Murrell JR, Spillantini MG, Zolo P, Guazzelli M, Smith MJ, et al. 1999. Tau gene mutation G389R causes a tauopathy with abundant Pick body-like inclusions and axonal deposits. *J. Neuropathol. Exp. Neurol.* 58(12):1207–26

Nacharaju P, Lewis J, Easson C, Yen S, Hackett J, et al. 1999. Accelerated filament formation from tau protein with specific FTDP-17 missense mutations. *FEBS Lett.* 447(2–3):195–99

Nasreddine ZS, Loginov M, Clark LN, Lamarche J, Miller BL, et al. 1999. From genotype to phenotype: a clinical pathological, and biochemical investigation of frontotemporal dementia and parkinsonism (FTDP-17) caused by the P301L tau mutation. *Ann. Neurol.* 45(6):704–15

Nath R, Davis M, Probert AW, Kupina NC, Ren X, et al. 2000. Processing of cdk5 activator p35 to its truncated form (p25) by calpain in acutely injured neuronal cells. *Biochem. Biophys. Res. Commun.* 274(1):16–21

Neve RL, Harris P, Kosik KS, Kurnit DM, Donlon TA. 1986. Identification of cDNA clones for the human microtubule-associated protein tau and chromosomal localization of the genes for tau and microtubule-associated protein 2. *Brain Res.* 387(3):271–80

Patrick GN, Zukerberg L, Nikolic M, de la Monte S, Dikkes P, Tsai LH. 1999. Conversion of p35 to p25 deregulates Cdk5 activity and promotes neurodegeneration. *Nature* 402(6762):615–22

Perez M, Lim F, Arrasate M, Avila J. 2000. The FTDP-17-linked mutation R406W abolishes the interaction of phosphorylated tau with microtubules. *J. Neurochem.* 74(6):2583–89

Perez M, Valpuesta JM, Medina M, Montejo G, Avila J. 1996. Polymerization of tau into filaments in the presence of heparin: the minimal sequence required for tau-tau interaction. *J. Neurochem.* 67(3):1183–90

Pickering-Brown S, Baker M, Yen SH, Liu WK, Hasegawa M, et al. 2000. Pick's disease is associated with mutations in the *tau* gene. *Ann. Neurol.* 48(5):806–8

Pollock NJ, Mirra SS, Binder LI, Hansen LA, Wood JG. 1986. Filamentous aggregates in Pick's disease, progressive supranuclear palsy, and Alzheimer's disease share antigenic determinants with microtubule-associated protein, tau. *Lancet* 2(8517): 1211

Poorkaj P, Bird TD, Wijsman E, Nemens E, Garruto RM, et al. 1998. Tau is a candidate gene for chromosome 17 frontotemporal dementia. *Ann. Neurol.* 43(6):815–25

Probst A, Götz J, Wiederhold KH, Tolnay M, Mistl C, et al. 2000. Axonopathy and amyotrophy in mice transgenic for human four-repeat tau protein. *Acta Neuropathol.* 99(5):469–81

Probst A, Tolnay M, Langui D, Goedert M, Spillantini MG. 1996. Pick's disease: hyperphosphorylated tau protein segregates to the somatoaxonal compartment. *Acta Neuropathol.* 92(6):588–96

Prusiner SB. 1998. Prions. *Proc. Natl. Acad. Sci. USA* 95(23):13363–83

Rebeiz J, Kolodny EH, Richardson EP Jr. 1967. Corticodentatonigral degeneration with neuronal achromasia: a progressive disorder in late adult life. *Trans. Am. Neurol. Assoc.* 92:23–26

Rebeiz JJ, Kolodny EH, Richardson EP Jr. 1968. Corticodentatonigral degeneration with neuronal achromasia 2. *Arch. Neurol.* 18(1):20–33

Reed LA, Schmidt ML, Wszolek ZK, Balin BJ, Soontornniyomkij V, et al. 1998. The neuropathology of a chromosome 17-linked autosomal dominant parkinsonism and dementia ("pallido-ponto-nigral degeneration"). *J. Neuropathol. Exp. Neurol.* 57(6):588–601

Rep. Work Group Frontotemporal Dementia Pick's Disease. 2001. Clinical and neuropathological diagnosis of frontotemporal dementias. *Neurobiol. Aging.* In press

Reynolds CH, Nebreda AR, Gibb GM, Utton MA, Anderton BH. 1997a. Reactivating kinase/p38 phosphorylates tau protein in vitro. *J. Neurochem.* 69(1):191–98

Reynolds CH, Utton MA, Gibb GM, Yates A, Anderton BH. 1997b. Stress-activated protein kinase/c-jun N-terminal kinase phosphorylates tau protein. *J. Neurochem.* 68(4):1736–44

Rizzini C, Goedert M, Hodges JR, Smith MJ, Jakes R, et al. 2000. Tau gene mutation K257T causes a tauopathy similar to Pick's disease. *J. Neuropathol. Exp. Neurol.* 59(11):990–1001

Rizzu P, Joose M, Ravid R, Hoogeveen A, Kamphorst W, et al. 2000. Mutation-dependent aggregation of tau protein and its selective depletion from the soluble fraction in brain of P301L FTDP-17 patients. *Hum. Mol. Genet.* 9(20):3075–82

Rizzu P, Van Swieten JC, Joosse M, Hasegawa M, Stevens M, et al. 1999. High prevalence of mutations in the microtubule-associated protein tau in a population study of frontotemporal dementia in the Netherlands. *Am. J. Hum. Genet.* 64(2):414–21

Roy S, Datta CK, Hirano A, Ghatak NR, Zimmerman HM. 1974. Electron microscopic study of neurofibrillary tangles in Steele-Richardson-Olszewski syndrome. *Acta Neuropathol.* 29(2):175–79

Sahara N, Tomiyama T, Mori H. 2000. Missense point mutations of tau to segregate with FTDP-17 exhibit site-specific effects on microtubule structure in COS cells: a novel action of R406W mutation. *J. Neurosci. Res.* 60(3):380–87

Schmidt ML, Huang R, Martin JA, Henley J, Mawal-Dewan M, et al. 1996. Neurofibrillary tangles in progressive supranuclear palsy contain the same tau epitopes identified in Alzheimer's disease PHFtau. *J. Neuropathol. Exp. Neurol.* 55(5):534–39

Schneider A, Biernat J, von Bergen M, Mandelkow E, Mandelkow EM. 1999. Phosphorylation that detaches tau protein from microtubules (Ser262, Ser214) also protects it against aggregation into Alzheimer paired helical filaments. *Biochemistry* 38(12):3549–58

Schweers O, Schönbrunn-Hanebeck E, Marx A, Mandelkow E. 1994. Structural studies of tau protein and Alzheimer paired helical filaments show no evidence for beta-structure. *J. Biol. Chem.* 269(39):24290–97

Sengupta A, Wu Q, Grundke-Iqbal I, Iqbal K, Singh TJ. 1997. Potentiation of GSK-3-catalyzed Alzheimer-like phosphorylation of human tau by cdk5. *Mol. Cell Biochem.* 167(1–2):99–105

Sergeant N, David JP, Lefranc D, Vermersch P, Wattez A, Delacourte A. 1997. Different distribution of phosphorylated tau protein isoforms in Alzheimer's and Pick's diseases. *FEBS Lett.* 412(3):578–82

Sergeant N, Wattez A, Delacourte A. 1999. Neurofibrillary degeneration in progressive supranuclear palsy and corticobasal degeneration: tau pathologies with exclusively "exon 10" isoforms. *J. Neurochem.* 72(3):1243–49

Seubert P, Mawal-Dewan M, Barbour R, Jakes R, Goedert M, et al. 1995. Detection of phosphorylated Ser262 in fetal tau, adult tau, and

paired helical filament tau. *J. Biol. Chem.* 270(32):18917–22

Shin RW, Iwaki T, Kitamoto T, Tateishi J. 1991. Hydrated autoclave pretreatment enhances tau immunoreactivity in formalin-fixed normal and Alzheimer's disease brain tissues. *Lab. Invest.* 64(5):693–702

Si ZH, Rauch D, Stoltzfus CM. 1998. The exon splicing silencer in human immunodeficiency virus type 1 Tat exon 3 is bipartite and acts early in spliceosome assembly. *Mol. Cell Biol.* 18(9):5404–13

Singh TJ, Zaidi T, Grundke-Iqbal I, Iqbal K. 1995. Modulation of GSK-3-catalyzed phosphorylation of microtubule-associated protein tau by nonproline-dependent protein kinases. *FEBS Lett.* 358(1):4–8

Snow AD, Mar H, Nochlin D, Sekiguchi RT, Kimata K, et al. 1990. Early accumulation of heparan sulfate in neurons and in the beta-amyloid protein-containing lesions of Alzheimer's disease and Down's syndrome. *Am. J. Pathol.* 137(5):1253–70

Sobue K, Agarwal-Mawal A, Wei L, Sun W, Miura Y, Paudel HK. 2000. Interaction of neuronal cdc2 like protein kinase with microtubule associated protein tau. *J. Biol. Chem.* 275(22):16673–80

Sontag E, Nunbhakdi-Craig V, Bloom GS, Mumby MC. 1995. A novel pool of protein phosphatase 2A is associated with microtubules and is regulated during the cell cycle. *J. Cell Biol.* 128(6):1131–44

Sontag E, Nunbhakdi-Craig V, Lee G, Brandt R, Kamibayashi C, et al. 1999. Molecular interactions among protein phosphatase 2A, tau, and microtubules. Implications for the regulation of tau phosphorylation and the development of tauopathies. *J. Biol. Chem.* 274(36):25490–98

Sperfeld AD, Collatz MB, Baier H, Palmbach M, Storch A, et al. 1999. FTDP-17: an early-onset phenotype with parkinsonism and epileptic seizures caused by a novel mutation. *Ann. Neurol.* 46(5):708–15

Spillantini MG, Bird TD, Ghetti B. 1998a. Frontotemporal dementia and parkinsonism linked to chromosome 17: a new group of tauopathies. *Brain Pathol.* 8(2):387–402

Spillantini MG, Crowther RA, Goedert M. 1996. Comparison of the neurofibrillary pathology in Alzheimer's disease and familial presenile dementia with tangles. *Acta Neuropathol.* 92(1):42–48

Spillantini MG, Crowther RA, Kamphorst W, Heutink P, Van Swieten JC. 1998b. Tau pathology in two Dutch families with mutations in the microtubule-binding region of tau. *Am. J. Pathol.* 153(5):1359–63

Spillantini MG, Goedert M, Crowther RA, Murrell JR, Farlow MR, Ghetti B. 1997. Familial multiple system tauopathy with presenile dementia: a disease with abundant neuronal and glial tau filaments. *Proc. Natl. Acad. Sci. USA* 94(8):4113–18

Spillantini MG, Murrell JR, Goedert M, Farlow MR, Klug A, Ghetti B. 1998c. Mutation in the tau gene in familial multiple system tauopathy with presenile dementia. *Proc. Natl. Acad. Sci. USA* 95(13):7737–41

Spillantini MG, Yoshida H, Rizzini C, Lantos PL, Khan N, et al. 2000. A novel tau mutation (N296N) in familial dementia with swollen achromatic neurons and corticobasal inclusion bodies. *Ann. Neurol.* 48(6):939–43

Spittaels K, Van den Haute C, Van Dorpe J, Bruynseels K, Vandezande K, et al. 1999. Prominent axonopathy in the brain and spinal cord of transgenic mice overexpressing four-repeat human tau protein. *Am. J. Pathol.* 155(6):2153–65

Spittaels K, Van den Haute C, Van Dorpe J, Geerts H, Mercken M, et al. 2000. Glycogen synthase kinase-3beta phosphorylates protein tau and rescues the axonopathy in the central nervous system of human four-repeat tau transgenic mice. *J. Biol. Chem.* 275(52):41340–49

Stanford PM, Halliday GM, Brooks WS, Kwok JB, Storey CE, et al. 2000. Progressive supranuclear palsy pathology caused by a novel silent mutation in exon 10 of the tau gene: expansion of the disease phenotype caused by tau gene mutations. *Brain* 123(5):880–93

Steele JC, Richardson JC, Olszewski J. 1964. Progressive supranuclear palsy. *Arch. Neurol.* 10:333–59

Sturchler-Pierrat C, Abramowski D, Duke M, Wiederhold KH, Mistl C, et al. 1997. Two amyloid precursor protein transgenic mouse models with Alzheimer disease-like pathology. *Proc. Natl. Acad. Sci. USA* 94(24):13287–92

Takahashi M, Tomizawa K, Ishiguro K, Takamatsu M, Fujita SC, Imahori K. 1995. Involvement of tau protein kinase I in paired helical filament-like phosphorylation of the juvenile tau in rat brain. *J. Neurochem.* 64(4):1759–68

Takahashi M, Yasutake K, Tomizawa K. 1999. Lithium inhibits neurite growth and tau protein kinase I/glycogen synthase kinase-3β-dependent phosphorylation of juvenile tau in cultured hippocampal neurons. *J. Neurochem.* 73(5):2073–83

Tanaka K, Watakabe A, Shimura Y. 1994. Polypurine sequences within a downstream exon function as a splicing enhancer. *Mol. Cell Biol.* 14(2):1347–54

Tellez-Nagel I, Wisniewski HM. 1973. Ultrastructure of neurofibrillary tangles in Steele-Richardson-Olszewski syndrome. *Arch. Neurol.* 29(5):324–27

Tesseur I, Van Dorpe J, Spittaels K, Van den Haute C, Moechars D, Van Leuven F. 2000. Expression of human apolipoprotein E4 in neurons causes hyperphosphorylation of protein tau in the brains of transgenic mice. *Am. J. Pathol.* 156(3):951–64

Tolnay M, Spillantini MG, Rizzini C, Eccles D, Lowe J, Ellison D. 2000. A new case of frontotemporal dementia and parkinsonism resulting from an intron 10 +3-splice site mutation in the tau gene: clinical and pathological features. *Neuropathol. Appl. Neurobiol.* 26(4):368–78

Trojanowski JQ, Goedert M, Iwatsubo T, Lee VM-Y. 1998. Fatal attraction: abnormal protein aggregation and neuron death in

Parkinson's disease and Lewy body dementia. *Cell Death Differ.* 5:832–37

Trojanowski JQ, Lee VM-Y. 1994. Paired helical filament tau in Alzheimer's disease. The kinase connection. *Am. J. Pathol.* 144(3):449–53

Tsai LH, Delalle I, Caviness VS Jr, Chae T, Harlow E. 1994. p35 is a neural-specific regulatory subunit of cyclin-dependent kinase 5. *Nature* 371(6496):419–23

Tseng HC, Graves DJ. 1998. Natural methylamine osmolytes, trimethylamine N-oxide and betaine, increase tau-induced polymerization of microtubules. *Biochem. Biophys. Res. Commun.* 250(3):726–30

Tseng HC, Lu Q, Henderson E, Graves DJ. 1999. Phosphorylated tau can promote tubulin assembly. *Proc. Natl. Acad. Sci. USA* 96(17):9503–8

Van Swieten JC, Stevens M, Rosso SM, Rizzu P, Joosse M, et al. 1999. Phenotypic variation in hereditary frontotemporal dementia with tau mutations. *Ann. Neurol.* 46(4):617–26

Varani L, Hasegawa M, Spillantini MG, Smith MJ, Murrell JR, et al. 1999. Structure of tau exon 10 splicing regulatory element RNA and destabilization by mutations of frontotemporal dementia and parkinsonism linked to chromosome 17. *Proc. Natl. Acad. Sci. USA* 96(14):8229–34

Varani L, Spillantini MG, Goedert M, Varani G. 2000. Structural basis for recognition of the RNA major groove in the tau exon 10 splicing regulatory element by aminoglycoside antibiotics. *Nucleic Acids Res.* 28(3):710–19

Verbeek MM, Otte-Holler I, van den Born J, van den Heuvel LP, David G, et al. 1999. Agrin is a major heparan sulfate proteoglycan accumulating in Alzheimer's disease brain. *Am. J. Pathol.* 155(6):2115–25

Vermersch P, Robitaille Y, Bernier L, Wattez A, Gauvreau D, Delacourte A. 1994. Biochemical mapping of neurofibrillary degeneration in a case of progressive supranuclear palsy: evidence for general cortical involvement. *Acta Neuropathol.* 87(6):572–77

Vogelsberg-Ragaglia V, Bruce J, Richter-

Lansberg C, Zhang B, Hong M, et al. 2000. Distinct FTDP-17 missense mutations in tau produce tau aggregates and other phathological phenotypes in transfected CHO cells. *Mol. Biol. Cell.* 11(12):4093–104

von Bergen M, Friedhoff P, Biernat J, Heberle J, Mandelkow EM, Mandelkow E. 2000. Assembly of tau protein into Alzheimer paired helical filaments depends on a local sequence motif ((306)VQIVYK(311)) forming beta structure. *Proc. Natl. Acad. Sci. USA* 97(10):5129–34

Watanabe A, Hasegawa M, Suzuki M, Takio K, Morishima-Kawashima M, et al. 1993. In vivo phosphorylation sites in fetal and adult rat tau. *J. Biol. Chem.* 268(34):25712–17

Weingarten MD, Lockwood AH, Hwo SY, Kirschner MW. 1975. A protein factor essential for microtubule assembly. *Proc. Natl. Acad. Sci. USA* 72(5):1858–62

Wijker M, Wszolek ZK, Wolters EC, Rooimans MA, Pals G, et al. 1996. Localization of the gene for rapidly progressive autosomal dominant parkinsonism and dementia with pallido-ponto-nigral degeneration to chromosome 17q21. *Hum. Mol. Genet.* 5(1):151–54

Wilhelmsen KC, Lynch T, Pavlou E, Higgins M, Nygaard TG. 1994. Localization of disinhibition-dementia-parkinsonism-amyotrophy complex to 17q21–22. *Am. J. Hum. Genet.* 55(6):1159–65

Wille H, Drewes G, Biernat J, Mandelkow EM, Mandelkow E. 1992. Alzheimer-like paired helical filaments and antiparallel dimers formed from microtubule-associated protein tau in vitro. *J. Cell Biol.* 118(3):573–84

Wilson DM, Binder LI. 1997. Free fatty acids stimulate the polymerization of tau and amyloid beta peptides. In vitro evidence for a common effector of pathogenesis in Alzheimer's disease. *Am. J. Pathol.* 150(6):2181–95

Wischik CM, Novak M, Thogersen HC, Edwards PC, Runswick MJ, et al. 1988. Isolation of a fragment of tau derived from the core of the paired helical filament of

Alzheimer disease. *Proc. Natl. Acad. Sci. USA* 85(12):4506–10

Woody RW, Clark DC, Roberts GC, Martin SR, Bayley PM. 1983. Molecular flexibility in microtubule proteins: proton nuclear magnetic resonance characterization. *Biochemistry* 22(9):2186–92

Yamada T, McGeer PL, McGeer EG. 1992. Appearance of paired nucleated, tau-positive glia in patients with progressive supranuclear palsy brain tissue. *Neurosci. Lett.* 135(1):99–102

Yamaguchi H, Ishiguro K, Uchida T, Takashima A, Lemere CA, Imahori K. 1996. Preferential labeling of Alzheimer neurofibrillary tangles with antisera for tau protein kinase (TPK) I/glycogen synthase kinase-3 beta and cyclin-dependent kinase 5, a component of TPK II. *Acta Neuropathol.* 92(3):232–41

Yasuda M, Kawamata T, Komure O, Kuno S, D'Souza I, et al. 1999. A mutation in the microtubule-associated protein tau in pallido-nigro-luysian degeneration. *Neurology* 53(4):864–68

Yasuda M, Takamatsu J, D'Souza I, Crowther RA, Kawamata T, et al. 2000. A novel mutation at posititon +12 in the intron following exon 10 of the tau gene in familial frontotemporal dementia (FTD-Kumamoto). *Ann. Neurol.* 47(4):422–29

Yen SH, Hutton M, DeTure M, Ko LW, Nacharaju P. 1999. Fibrillogenesis of tau: insights from tau missense mutations in FTDP-17. *Brain Pathol.* 9(4):695–705

Yoshida H, Ihara Y. 1993. Tau in paired helical filaments is functionally distinct from fetal tau: assembly incompetence of paired helical filament-tau. *J. Neurochem.* 61(3):1183–86

Zheng-Fischhöfer Q, Biernat J, Mandelkow EM, Illenberger S, Godemann R, Mandelkow E. 1998. Sequential phosphorylation of tau by glycogen synthase kinase-3β and protein kinase A at Thr212 and Ser214 generates the Alzheimer-specific epitope of antibody AT100 and requires a paired-helical-filament-like conformation. *Eur. J. Biochem.* 252(3):542–52

Zhukareva V, Vogelsberg-Ragaglia V, Van Deerlin VMD, Bruce J, Schuck T, et al. 2001. Loss of brain tau defines novel sporadic and familial tauopathies with frontotemporal dementia. *Ann. Neurol.* 49(1):165–75

Annu. Rev. Neurosci. 2001. 24:1161–192

MATERNAL CARE, GENE EXPRESSION, AND THE TRANSMISSION OF INDIVIDUAL DIFFERENCES IN STRESS REACTIVITY ACROSS GENERATIONS

Michael J Meaney

Developmental Neuroendocrinology Laboratory, Douglas Hospital Research Centre, Department of Psychiatry and McGill Centre for the Study of Behavior, Genes and Environment, McGill University, Montréal, Canada; e-mail: mdmm@muscia.mcgill.ca

Key Words corticotropin-releasing factor (hormone), gene expression, hippocampal development, oxytocin receptors

■ **Abstract** Naturally occurring variations in maternal care alter the expression of genes that regulate behavioral and endocrine responses to stress, as well as hippocampal synaptic development. These effects form the basis for the development of stable, individual differences in stress reactivity and certain forms of cognition. Maternal care also influences the maternal behavior of female offspring, an effect that appears to be related to oxytocin receptor gene expression, and which forms the basis for the intergenerational transmission of individual differences in stress reactivity. Patterns of maternal care that increase stress reactivity in offspring are enhanced by stressors imposed on the mother. These findings provide evidence for the importance of parental care as a mediator of the effects of environmental adversity on neural development.

PARENTAL CARE AND THE HEALTH OF OFFSPRING

The quality of family life influences the development of individual differences in vulnerability throughout life to illness. As adults, victims of childhood physical or sexual abuse are at considerably greater risk for mental illness, as well as for obesity, diabetes, and heart disease (e.g. Bifulco et al 1991, Brown & Anderson 1993, McCauley et al 1997, Felitti et al 1998). Children need not be beaten to be compromised. Persistent emotional neglect, family conflict, and conditions of harsh, inconsistent discipline all serve to compromise growth (e.g. Montgomery et al 1997) and intellectual development (Ammerman et al 1986, Trickett & McBride-Chang 1995) and to increase the risk for adult obesity (Lissau & Sorensen 1994), depression, and anxiety disorders (Holmes & Robbins 1987, 1988; Gottman 1998) to a level comparable to that for abuse.

0147-006X/01/0621-1161$14.00 **1161**

More subtle relationships also exist. Low scores on measures of parental bonding, reflecting cold, distant parent-child relationships are associated with a significantly increased risk of depression and anxiety in later life (e.g. Canetti et al 1997, Parker 1981). And again, the risk is not unique to mental health. Russak & Schwartz (1997) found that by midlife, those individuals who, as undergraduate students, rated their relationships with parents as cold and detached had a fourfold greater risk of chronic illness, including depression and alcoholism as well as heart disease and diabetes. The sword cuts both ways. Family life can also serve as a source of resilience in the face of chronic stress (Rutter 1979). Thus, warm, nurturing families tend to promote resistance to stress and to diminish vulnerability to stress-induced illness (Smith & Prior 1995).

Parental factors also serve to mediate the effects of environmental adversity on development. For example, the effects of poverty on emotional and cognitive development are mediated by parental factors to the extent that if such factors are controlled, there is no discernible effect of poverty on child development (Eisenberg & Earls 1975, Conger et al 1994, McLloyd 1998). Moreover, treatment outcomes associated with early intervention programs are routinely correlated with changes in parental behavior: In cases where parental behavior proves resistant to change, treatment outcomes for the children are seriously limited.

A critical question concerns the mechanisms that mediate these enduring parental influences on the health of offspring. The relationship between early life events and health in adulthood appears to be, in part, mediated by parental influences on the development of neural systems that underlie the expression of behavioral and endocrine responses to stress (Seckl & Meaney 1994, Nemeroff 1996, Sroufe 1997, Francis & Meaney 1999, Francis et al 1999a, Heim et al 2001). Physical and sexual abuse in early life, for example, increases endocrine and autonomic responses to stress in adulthood (DeBellis et al 1994, Heim et al 2001). There are two critical assumptions here: First, that prolonged activation of neural and hormonal responses to stress can promote illness; and second, that early environmental events influence the development of these responses. There is strong evidence in favor of both ideas.

RESPONSES TO STRESS

Stress is a risk factor for a variety of diseases, ranging from autoimmune disorders to mental illness. It is ironic that the pathways by which stressful events promote the development of such divergent forms of illness involve the same hormones that ensure survival during a period of stress (Chrousos & Gold 1992, McEwen & Stellar 1993, McEwen 1998). These effects can, to some extent, be understood in terms of the normal set of adaptive responses elicited by stressors (Dallman et al 1987, 1995; Chrousos & Gold 1992; McEwen & Stellar 1993; De Kloet et al 1998). The increased sympathoadrenal release of catecholamines, primarily adrenaline and noradrenaline, as well as the adrenal glucocorticoids, orchestrate a move to catabolism, mobilizing lipid and glucose reserves, and to insulin antagonism

(Munck et al 1984, Baxter & Tyrrell 1987, Brindley & Rolland 1989, Dallman et al 1995). The increase in circulating levels of catecholamines and glucocorticoids also promotes increased cardiovascular tone. These actions serve to increase the availability and distribution of energy substrates. Although these responses serve to meet the metabolic demands posed by the stressor, prolonged exposure to elevated levels of these "stress hormones" can promote insulin resistance, hypertension, hyperlipidemia, hypercholesterolemia, abdominal fat deposition, and an increased risk of arterial damage, all of which are associated with an increased risk for heart disease (Brindley & Rolland 1989, Rosmond et al 1995).

There are also cognitive responses to stressors that include systems that mediate attentional processes, as well as learning and memory (Arnsten 1998). During stress, individuals become hypervigilant; the level of attention directed to the surrounding environment is increased at the expense of effortful concentration on tasks that are not essential for survival. As a result of these changes in attentional processes, as well as of the effects of glucocorticoids on such relevant brain structures as the hippocampus, episodic memory capacity is diminished during periods of stress (Landfield & Pitler 1984, Joels & De Kloet 1989, Diamond et al 1992, Starkman et al 1992, Sapolsky 1992, Lupien et al 1998, Lupien & Meaney 1998, De Kloet et al 1998, McEwen et al 1999). At the same time, glucocorticoids act on such areas of the brain as the amygdala to enhance learning and memory for emotionally salient events (e.g. Davis et al 1997, Quirarte et al 1997, Pitkanen et al 1998, Cahill & McGaugh 1998). Stress also provokes altered emotional states: Feelings of apprehension and fear predominate during a stressful experience. Although these responses are highly adaptive, chronic activation of these systems can promote the emergence of specific forms of cognitive impairments, states of anxiety and dysphoria, sleep disorders, etc (Koob et al 1994, Nemeroff 1996, Rosen & Schulkin 1998).

Herein lies the dilemma: The same responses that permit survival during stress can ultimately promote disease. Consequently, individual differences in endocrine and sympathetic responses to stress can serve as a source of vulnerability (or resistance) to pathology over a life span (McEwen & Stellar 1993, Chrousos & Gold 1992). In human and nonhuman populations, individuals who show exaggerated hypothalamic-pituitary-adrenal (HPA) and sympathetic responses to stress are at increased risk for a variety of disorders, including heart disease, diabetes, anxiety, depression, and drug addiction.

Corticotropin-Releasing Factor Systems

Central corticotropin-releasing factor (CRF) systems furnish the critical signal for the activation of behavioral, emotional, autonomic, and endocrine responses to stressors. There are two major CRF pathways regulating the expression of these stress responses. First, a CRF pathway from the parvocellular regions of the paraventricular nucleus of the hypothalamus (PVNh) to the hypophysial-portal system of the anterior pituitary serves as the principal mechanism for the transduction of a neural signal into a pituitary-adrenal response (Rivier & Plotsky 1986, Plotsky 1991, Antoni 1993, Whitnall 1993, Pacak et al 1995). In responses to stressors, CRF, as well as such cosecretagogues as arginine vasopressin, are released from PVNh neurons into the portal blood supply of the anterior pituitary, where it stimulates the synthesis and release of adrenocorticotropin hormone (ACTH). Pituitary ACTH, in turn, causes the release of glucocorticoids from the adrenal gland.

CRF neurons in the central nucleus of the amygdala project directly to the locus coeruleus and increase the firing rate of locus coeruleus neurons, resulting in increased noradrenaline release in the vast terminal fields of this ascending noradrenergic system. Thus, intracerebroventricular infusion of CRF increases extracellular noradrenaline levels (Lavicky & Dunn 1993, Emoto et al 1993, Page & Valentino 1994, Valentino et al 1998). The amygdaloid CRF projection to the locus coeruleus (Moga & Gray 1989, Koegler-Muly et al 1993, van Bockstaele et al 1996, Gray & Bingaman 1996, Valentino et al 1998) is also critical for the expression of behavioral responses to stress: Microinjections of CRF receptor antagonists into the locus coeruleus attenuate fear-related behaviors (Butler et al 1990, Liang et al 1992, Sweirgel et al 1993, Koob et al 1994, Schulkin et al 1994, Bakshi et al 2000) and are emerging as a major target for drug development in the treatment of affective disorders. In contrast, CRF overproduction is associated with increased fearfulness (e.g. Stenzel-Poore et al 1994). Hence, the CRF neurons in the PVNh and the central n. of the amygdala serve as important mediators of both behavioral and endocrine responses to stress. It is not surprising that chronically increased CRF levels have been associated with serious mood disorders (Chrousos & Gold 1992, Nemeroff 1996).

These findings provide an understanding of how stress can influence health. Yet the influence of stress can only be fully appreciated when we factor into the equation some appreciation of the individual's response to stress. After all, not all individuals fall sick under conditions of stress, and questions concerning the basis for such individual differences are central to understanding the etiology of chronic disease. The hypothesis that guides research on the development of psychopathology focuses on the role of early life events in determining individual differences in vulnerability to stress. This hypothesis rests on the assumption that chronic activation of central and endocrine stress responses can promote illness (see references cited above). Thus, early life events that increase stress reactivity result in a greater vulnerability to stress-induced illness over a life span.

ENVIRONMENTAL REGULATION OF HPA AND BEHAVIORAL RESPONSES TO STRESS

Postnatal Handling Studies

Perhaps the strongest evidence for the environmental regulation of the development of responses to stress is from postnatal handling research with rodents. Handling involves a brief (i.e. 3–15 min), daily period of separation of the pup from the mother for the first few weeks of life and results in decreased stress reactivity in adulthood (Levine 1957, 1962; Levine et al 1967; Ader & Grota 1969; Hess et al 1969; Zarrow et al 1972; Meaney et al 1989; Viau et al 1993; Bhatnagar et al 1995). As adults, rats handled neonatally show decreased fearfulness and more modest HPA responses to stress; such effects are apparent in animals tested as late as 26 months of age (Meaney et al 1988, 1992).

The handling effects on the development of HPA responses to stress have important consequences for health. Glucocorticoid levels often rise with age in rats and are associated with hippocampal degeneration and the emergence of learning and memory deficits (Landfield et al 1981, Landfield & Pitler 1984, Sapolsky et al 1984, Issa et al 1990). Such age-related increases in basal and stress-induced pituitary-adrenal activity is significantly less apparent in the handled animals, and thus these animals show little evidence of hippocampal aging (Meaney et al 1988, 1991). Likewise, handled animals also show more modest stress-induced suppression of immune function compared with nonhandled rats (Bhatnagar et al 1996).

Such findings have tempted researchers to believe that the handled animals are in some ways hardier than nonhandled animals. However, this misses the point. Handled animals are not better adapted than nonhandled animals, they are simply different. The environmental context then serves to determine the adaptive value of increased or decreased stress reactivity. In the examples cited above, it would appear that the handled animals were at some advantage. But this is not universal. Laban et al (1995) found that nonhandled animals were more resistant to the induction of experimental allergic encephalomyelitis (EAE) than were handled animals. Normally, glucocorticoids are protective against the development of EAE, which can be fatal (Mason 1991). Adrenalectomized animals, for example, rarely survive EAE. Hence, the increased HPA responsivity of nonhandled animals appears to have rendered these animals some advantage. The cost of such resistance is increased vulnerability to glucocorticoid-induced illness, but it is not difficult to imagine a scenario whereby such a cost is an acceptable trade-off. The point here is that these animals differ in stress reactivity and such differences are derived from early life experience. The critical question concerns the mechanisms that mediated these differences in stress reactivity.

Considering the importance of the CRF systems to both behavioral and HPA responses to stress, it is probably not surprising that these systems are critical targets for the handling effect on stress reactivity. Compared with nonhandled rats, adult animals exposed to postnatal handling show decreased CRF mRNA expression in the PVNh and the central n. of the amygdala (Plotsky & Meaney 1993; Viau et al 1993; PM Plotsky, C Caldji, S Sharma S, MJ Meaney, submitted for publication), decreased CRF content in the locus coeruleus, and decreased CRF receptor levels in the locus coeruleus. Together, these findings suggest that in the handled animals, there would be decreased CRF-induced activation of the locus coeruleus during stress. At least two recent findings are consistent with this idea. By comparison to nonhandled rats, acute stress in handled animals produces (*a*) a smaller stress-induced increase in cFOS immunoreactivity (ir) neurons in the locus coeruleus (Pearson et al 1997) and (*b*) more modest increases in extracellular noradrenaline levels in the PVNh (Liu et al 2000). We propose that postnatal handling can decrease the expression of behavioral responses to stress by altering the development of the central n. of the amygdala–locus coeruleus CRF system.

Postnatal handling also affects the development of neural systems that regulate CRF gene expression. Levels of CRF mRNA and protein in PVNh neurons are subject to inhibitory regulation via glucocorticoid negative feedback (Dallman et al 1987, 1993; De Kloet 1991). Handled rats show increased negative feedback sensitivity to glucocorticoids (Meaney et al 1989, Viau et al 1993). This effect, in turn, is related to the increased glucocorticoid receptor expression in the hippocampus and frontal cortex (Meaney et al 1985, 1989; Sarrieau et al 1988; Viau et al 1993; O'Donnell et al 1994), regions known to mediate the inhibitory effects of glucocorticoids over CRF synthesis in PVNh neurons (see de Kloet 1991, Jacobson & Sapolsky 1991, Diorio et al 1993, De Kloet et al 1998). The alterations in glucocorticoid receptor expression are a critical feature for the effect of the early environment on negative feedback sensitivity and HPA responses to stress; reversing the differences in hippocampal glucocorticoid receptor levels eliminates the differences in HPA responses to stress between handled and nonhandled animals (Meaney et al 1989).

CRF activity within the amygdaloid–locus coeruleus pathway is subject to GABAergic inhibition (Owens et al 1991, deBoer et al 1992). It is interesting that handled rats also show increased GABAA and central benzodiazepine (CBZ) receptor levels in the noradrenergic cell body regions of the locus coeruleus and the n. tractus solitarius, as well as in the basolateral and central n. of the amygdala. These effects were associated with increased expression of the mRNA for the $\gamma2$ subunit of the GABAA receptor, which encodes for the CBZ site. Handled animals also showed increased levels of mRNA encoding for the $\alpha1$ subunit of the GABAA receptor. These findings suggest that the composition of the GABAA receptor complex in brain regions that regulate stress reactivity is influenced by early life events. Handling increases $\alpha1$ and $\gamma2$ subunit expression (Caldji et al 1999, 2000a,b), and more importantly, this profile is associated with increased GABA binding (see Wilson 1996, Mehta & Ticku 1999). It is interesting that in humans, individual differences in CBZ receptor sensitivity are associated with vulnerability for anxiety disorders (e.g. Glue et al 1995).

Together, the effects of handling on glucocorticoid and GABAA/CBZ receptor gene expression could serve to dampen CRF synthesis and release, and to decrease the effect of CRF at critical target sites, such as the locus coeruleus. This model provides a reasonable working hypothesis for the mechanisms underlying the handling effect on endocrine and behavioral responses to stress.

Maternal Separation and Responses to Stress

The handling procedure involves briefly separating the pup from the mother for a period of ~15 min. In the course of normal mother-pup interactions, the dam is regularly away from the nest, and the pups, for periods of 20–30 min (Jans &

Woodside 1990, Rosenblatt 1994). Thus, the handling manipulation does not result in an abnormal period of separation or loss of maternal care. But what about longer periods of separation, where there is a clear privation of maternal care? Plotsky & Meaney (1993) studied adult animals that were separated from their mothers once per day for 180 min from days 2–14 of life. This manipulation was based on the observation of Calhoun (1962) that in seminaturalistic conditions, subordinate females were often obligated to locate nests at some distance from food and water sources, resulting in periods of separation from their pups that extended for as long as 2–3 h.

The effects of maternal separation on stress reactivity were precisely the opposite of those associated with postnatal handling. As adults, animals exposed to repeated periods of maternal separation showed significantly increased pituitary-adrenal responses to acute stress (Plotsky & Meaney 1993, Liu et al 2000). Maternal separation also resulted in decreased glucocorticoid receptor binding in the hippocampus, hypothalamus, and frontal cortex and resulted in blunted negative feedback sensitivity. As expected, adult animals exposed to maternal separation as neonates also exhibited a marked increase in hypothalamic CRF mRNA and CRF peptide content.

Maternal separation was also associated with (a) a twofold increase in CRF mRNA levels in the central n. of the amygdala, (b) increased CRF-like immunoreactivity at the level of the amygdala, the locus coeruleus, and the neighboring parabrachial nucleus, and (c) increased CRF receptor levels in the locus coeruleus and the raphé nucleus (Ladd et al 1996; PM Plotsky, C Caldji, S Sharma, MJ Meaney, submitted for publication). These findings suggest that maternal separation-induced changes in CRF systems might regulate both noradrenergic and serotonergic responses to stress [for comparable findings in nonhuman primates, see Kraemer et al (1989), Higley et al (1991), Suomi (1997)]. Indeed, we found that PVNh levels of noradrenaline during stress were elevated in maternal separation animals (Liu et al 2000). Predictably, the maternal separation animals were highly fearful in behavioral tests of novelty (see Caldji et al 2000b). These effects involved reduced exploration or feeding in a novel environment, and an increased acoustic startle responsivity, all of which are mediated, in part at least, by CRF effects on noradrenergic release (Kalin 1985, Dunn & Berridge 1990, Koob et al 1994, Nemeroff 1996, Valentino et al 1998).

WHAT ARE THE CRITICAL FEATURES OF THESE ENVIRONMENTAL MANIPULATIONS?

The decreased mother-pup contact resulting from extended periods of maternal separation is likely to be critical for the effects of this procedure on behavioral and HPA responses to stress. But does this imply that under normal conditions maternal care actively contributes to the development of neural systems that mediate stress responses, or simply that the absence of the mother is so disruptive to pup physiology that it affects the development of these systems? If maternal care is indeed critical under normal conditions, then what are the relevant features of mother-pup interactions, and how do they influence neural development?

We examined this question by attempting to define naturally occurring variations in maternal behavior over the first 8 days after birth through the simple albeit time-consuming observation of mother-pup interactions in normally reared animals. There was considerable variation in two forms of maternal behavior—licking/grooming of pups and arched-back nursing (Stern 1997). Licking/grooming included both body as well as anogenital licking. Arched-back nursing, also referred to as "crouching," is characterized by a dam nursing her pups with her back conspicuously arched and legs splayed outward. Although common, it is not the only posture from which dams nurse. A blanket posture, where the mother is almost lying on the suckling pups, represents a more relaxed version of the arched-back position. As you can imagine, it provides substantially less opportunity for such movements as nipple switching. Dams also nurse from their sides and often will move from one posture to another over the course of a nursing bout. It is interesting that the frequency of licking/grooming and arched-back nursing was highly correlated ($r = +0.91$) across animals, and thus we were able to define mothers according to both behaviors: high or low licking/grooming (LG)–arched-back nursing (ABN) mothers. For the sake of most of the studies described here, high and low LG-ABN mothers were identified as females whose scores on both measures were ±1 SD above (high) or below (low) the mean for their cohort. It is important that high and low LG-ABN mothers do not differ in the amount of contact time with pups; differences in the frequency of LG or ABN do not occur simply as a function of time in contact with pups. The results of three independent studies have failed to reveal any relationship between the frequency of LG or ABN and either litter size or gender composition (for all, $r < 0.10$). The latter is an important consideration because it has been reported (e.g. Moore 1995) that male pups are licked more frequently than females. However, this refers only to anogenital licking. Our studies measure both anogenital and body licking and focus on the first week of life. The gender differences in anogenital LG reported by Moore & Power (1986) appear only later in development, toward the second week of life. It is important that both groups raise a comparable number of pups to weaning, and there are no differences in the weaning weights of the pups, which suggests an adequate level of maternal care across the groups. These findings also suggest that we are examining the consequences of variations in maternal care that occur within a normal range.

The differences in maternal behavior in the high and low LG-ABN mothers were not unique to the first litter (Francis, Champagne & Meaney, unpublished). Across dams there was a correlation of $+0.84$ between the licking/grooming of the first and second litters and a correlation of $+0.72$ between the licking/grooming scores for the first and third litters Thus, the individual differences in maternal behavior are rather stable. These findings are comparable to those of primate studies in which individual differences in maternal behavior remained consistent across infants (e.g. Fairbanks 1996).

The critical question, of course, concerns the potential consequences of these differences in maternal behavior for the development of behavioral and neuroendocrine responses to stress. As adults, the offspring of high LG-ABN mothers showed reduced plasma ACTH and corticosterone responses to acute stress in comparison to adult offspring of low LG-ABN mothers. The high LG-ABN offspring also showed significantly increased hippocampal glucocorticoid receptor mRNA expression, enhanced glucocorticoid negative feedback sensitivity, and decreased hypothalamic CRH mRNA levels. Moreover, the magnitude of the corticosterone response to acute stress was significantly correlated with the frequency of both maternal LG ($r = -0.61$) and ABN ($r = -0.64$) during the first week of life, as was the level of hippocampal glucocorticoid receptor mRNA and hypothalamic CRH mRNA expression (for all, $r > 0.70$) (Liu et al 1997).

The offspring of the high and low LG-ABN mothers also differed in behavioral responses to novelty (Caldji et al 1998). As adults, offspring of high LG-ABN mothers showed decreased startle responses, increased open-field exploration, and shorter latencies to eat food provided in a novel environment. The offspring of high LG-ABN mothers also showed decreased CRF receptor levels in the locus coeruleus, increased GABAA and CBZ receptor levels in the basolateral and central n. of the amygdala, as well as in the locus coeruleus (Caldji et al 1998), and decreased CRF mRNA expression in the central nucleus of the amygdala (DD Francis, D Diorio & MJ Meaney, unpublished data). Predictably, stress-induced increases in PVNh levels of noradrenaline were significantly higher in the offspring of low LG-ABN mothers (Caldji et al 1998).

The adult offspring of high LG-ABN mothers also showed significantly higher levels of GABAA and CBZ receptor binding in the basolateral and central n. of the amygdala, as well as in the locus coeruleus. These findings provide a mechanism for increased GABAergic inhibition of amygdala–locus coeruleus activity. A series of in situ hybridization studies have illustrated the molecular mechanism for these differences in receptor binding and suggest that variations in maternal care might actually permanently alter the subunit composition of the GABAA receptor complex in the offspring. The offspring of high LG-ABN mothers show increased levels of the mRNAs for the $\gamma 1$ and $\gamma 2$ subunits, which contribute to the formation of a functional CBZ binding site. Such differences are not unique to the γ subunits. Levels of mRNA for the $\alpha 1$ subunit of the GABAA/CBZ receptor complex are significantly higher in the amygdala and locus coeruleus of high compared with low LG-ABN offspring. The $\alpha 1$ subunit appears to confer higher affinity for GABA, providing the most efficient form of the GABAA receptor complex, through increased receptor affinity for GABA. The adult offspring of low LG-ABN mothers actually show increased expression of the mRNAs for the $\alpha 3$ and $\alpha 4$ subunits in the amygdala and the locus coeruleus. It is interesting that GABAA/CBZ receptors composed of the $\alpha 3$ and $\alpha 4$ subunits show a reduced affinity for GABA compared with the $\alpha 1$ subunit. Moreover, the $\alpha 4$ subunit does not contribute to the formation of a CBZ receptor site.

These differences in GABAA receptor subunit expression are also reflected in the CBZ receptor binding. Although the $\alpha 3$ subunit contributes to the formation of a BZ receptor binding site, those sites are of the type II rather than type I variety (Hadingham et al 1993). [³H]zolpidem can be used to distinguish type I and type II BZ receptor sites because this radioligand has little affinity for the type II receptor (e.g. Arbilla et al 1986). The previously reported differences in BZ receptor binding capacity between the adult offspring of high compared with low LG-ABN mothers lie in the density of type I sites: Differences in [³H]zolpidem binding map onto, and in fact exceed, those observed using [³H]flunitrazepam, which labels both type I and type II receptor sites. It appears as though the $\alpha 1$ subunit alone produces a type I BZ receptor site. GABAA receptor complexes containing $\alpha 3$ subunit display type II BZ receptor pharmacology. The $\alpha 4$ subunit, as mentioned above, does not produce a BZ receptor type of either variety (e.g. Khan et al 1996).

Thus, variations in the subunit profiles in both groups actively contribute to the differences in type I CBZ and GABAA receptor binding observed in the offspring of high and low LG-ABN mothers. These differences in subunit expression are tissue specific; no such differences are apparent in the hippocampus, hypothalamus, or cortex. Thus, differences in GABAA/CBZ receptor binding are due not simply to a deficit in subunit expression in the offspring of low LG-ABN mothers, but also to an apparently active attempt to maintain a specific GABAA/CBZ receptor profile in selected brain regions. Maternal care during the first week of life permanently alters subunit expression and, thus, GABAA/CBZ receptor composition in adulthood in brain regions that regulate stress reactivity.

Summary

It is interesting that postnatal handling increases maternal LG and ABN, whereas maternal separation has precisely the opposite effect (Liu et al 1997; DD Francis & MJ Meaney, unpublished data). These findings support the long-held belief (Levine 1975, Smotherman & Bell 1980) that the effects of such early environmental manipulations are in fact mediated by alterations in maternal behavior. Together, the results of these studies suggest that the behavior of a mother toward her offspring can "program" behavioral and neuroendocrine responses to stress in adulthood. These effects are associated with sustained changes in the expression of genes in brain regions that mediate responses to stress and form the basis for stable differences between individuals in stress reactivity. These findings provide a potential mechanism for the influence of parental care on vulnerability/resistance to stress-induced illness over a life span.

THE INTERGENERATIONAL TRANSMISSION OF INDIVIDUAL DIFFERENCES IN MATERNAL CARE TO THE OFFSPRING

Individual differences in behavioral and neuroendocrine responses to stress in rats are, in part, derived from naturally occurring variations in maternal care. Such effects might serve as a possible mechanism by which selected traits are transmitted from one generation to another. Indeed, low LG-ABN mothers are more fearful and show increased HPA responses to stress compared with high LG-ABN dams (Francis et al 2000). Individual differences in stress reactivity are apparently transmitted across generations: Fearful mothers beget more stress-reactive offspring. Likewise, as adults, the female offspring of high LG-ABN mothers show significantly more LG and ABN nursing than did female offspring of low LG-ABN mothers (Francis et al 1999b). Hence, the differences in maternal behavior and stress reactivity are transmitted from one generation to the next.

The obvious question is whether the transmission of these traits occurs only as a function of genomic-based inheritance. If this is the case, then the differences in maternal behavior may be simply an epiphenomena and not causally related to the development of individual differences in behavioral and neuroendocrine responses to stress or to maternal behavior. The issue here is not one of inheritance, that much seems clear. The question concerns the mode of inheritance.

The results of recent studies provide evidence for a nongenomic transmission of individual differences in stress reactivity and maternal behavior (Francis et al 1999b). One study involved a reciprocal cross-fostering of the offspring of low and high LG-ABN mothers. The primary concern here was that the wholesale fostering of litters between mothers is known to affect maternal behavior (Maccari et al 1995). In order to avert this problem and maintain the original character of the host litter, no more than 2 of 12 pups were fostered into or from any one litter (McCarty & Lee 1996). The critical groups of interest are the biological offspring of low LG-ABN mothers fostered onto high LG-ABN dams, and vice versa. The limited cross-fostering design did not result in any effect on group differences in maternal behavior. Hence, the frequency of pup LG and ABN across all groups of high LG-ABN mothers was significantly higher than that for any of the low LG-ABN dams, regardless of litter composition.

The results of the behavioral studies are consistent with the idea that variations in maternal care are causally related to individual differences in the behavior of the offspring. The biological offspring of low LG-ABN dams reared by high LG-ABN mothers were significantly less fearful under conditions of novelty than were the offspring reared by low LG-ABN mothers, including the biological offspring of high LG-ABN mothers. A separate group of female offspring were then mated, allowed to give birth, and observed for differences in maternal behavior. The effect on maternal behavior followed the same pattern as that for differences in fearfulness. As adults, the female offspring of low LG-ABN dams reared by high LG-ABN mothers did not differ from normal, high LG/ABN offspring in the frequency of pup LG or ABN. The frequency of LG and ABN in animals reared by high LG-ABN mothers was significantly higher than in any of the low LG-ABN groups, and again this included female pups originally born to high LG-ABN mothers but reared by low LG-ABN dams. Individual differences in fearfulness or maternal behavior mapped onto those of the rearing mother rather than the biological mother.

Francis et al (1999b) also addressed this question using a variation of the study described above in which the female offspring of high and low LG-ABN mothers were mated. The pups of female offspring were then either handled or nonhandled during the first 2 weeks of life. Again, the offspring of high LG-ABN mothers showed significantly more LG and ABN of pups than did the offspring of low LG-ABN mothers. Handling the pups of these mothers, as expected (see above), increased maternal LG and ABN. It is interesting that handling affected the maternal behavior only of the female offspring of low LG-ABN mothers. Low LG -ABN mothers with handled pups showed significantly more LG and ABN of pups than did the low LG/ABN-derived mothers of nonhandled pups. This finding was expected based on earlier studies (Lee & Williams 1974, 1975; Liu et al 1997).

As adults, the animals showed the predictable differences in behavioral and HPA responses to stress. On measures of plasma corticosterone responses to stress or behavioral fearfulness under conditions of novelty, the handled offspring of low LG-ABN mothers did not differ from either the handled or nonhandled offspring of high LG-ABN mothers. They were, after all, handled pups. Predictably, the nonhandled offspring of low LG-ABN mothers showed significantly increased HPA responses to stress and increased fearfulness in responses to novelty.

The critical part of the study concerns the maternal behavior of these animals. If the differences in maternal behavior are transmitted only through genetic inheritance, then the prediction is that the offspring of low LG-ABN mothers should also be low LG-ABN mothers regardless of whether or not they were handled in early life. A behavioral mode of transmission would suggest that the maternal behavior of the handled offspring of low LG-ABN mothers should resemble that of high LG-ABN mothers, which is in character with the maternal behavior if not with the pedigree of their mothers. The answer was clear: The handled offspring of low LG-ABN mothers did not differ from the offspring of high LG-ABN mothers in their frequency of LG or ABN. As would be expected, the nonhandled offspring of low LG-ABN mothers were themselves low LG-ABN mothers. These findings provide further evidence for a nongenomic mechanism of inheritance.

The same was true of the effects on fearfulness. As adults, the offspring of the handled LG-ABN mothers, recipients of high levels of maternal LG, resembled the offspring of either handled or nonhandled, high LG-ABN mothers on measures of fearfulness. The offspring of nonhandled LG-ABN mothers, as would be expected, showed greater fearfulness in novel surroundings. Hence, the handling experience was transmitted to the next generation via the alteration in maternal behavior. The same pattern was observed for measures of CRF or glucocorticoid receptor gene expression: For hypothalamic CRF mRNA or hippocampal glucocorticoid receptor mRNA measures, adult offspring of handled LG-ABN mothers resembled the offspring of either handled or nonhandled, high LG-ABN mothers. Thus, it appears that individual differences in maternal behavior can be transmitted from one generation to the next through a behavioral mode of transmission. Moreover, the effects of an environmental event occurring in early life can also be transmitted into the next generation, and this effect is mediated by alterations in maternal behavior.

These findings suggest that environmental events can alter the trajectory of development not only in the affected offspring but also into the next generation. Almost 40 years ago, Denenberg (1964) provided evidence for such nongenomic transmission. These researchers compared the offspring of handled-handled matings with those of nonhandled-nonhandled matings and found that, as adults, the offspring of handled parents were significantly less fearful in response to novelty than were the offspring of nonhandled parents, thus providing evidence for a transgenerational effect. For reasons I have never understood, despite being published in *Nature*, the results of this remarkable study have remained almost ignored. The contribution of our laboratory (Francis et al 1999b) to this story is to have identified maternal behavior as a potential mediator for such transgenerational effects.

These findings are also consistent with those of recent studies on the potential effects of maternal behavior on the development of behavior and endocrine responses to stress in BALBc mice. BALBc mice are normally extremely fearful and show elevated HPA responses to stress. However, those cross-fostered to C57 mothers are significantly less fearful, with lower HPA responses to stress (Zaharia et al 1996). It is important that C57 mothers lick/groom their pups about twice as frequently as do BALBc mothers (Anisman et al 1998). Comparable findings have emerged with rat strains. Typically, Fisher 344 rats are more responsive to novelty and have increased HPA responses to acute stress compared with Long-Evans rats (Dhabhar et al 1993). Predictably, Long-Evans dams lick/groom their offspring significantly more often than do Fisher 344 mothers (Moore & Lux 1998). These findings are consistent with a behavioral transmission hypothesis. The nexus of this hypothesis is not to underestimate the importance of genetic-based inheritance but to underscore the potential for traits to move from one generation to another via a behavioral mode of transmission that involves variations in maternal behavior.

Under normal circumstances, of course, BALBc mice are reared by BALBc mothers. The genetic and environmental factors conspire to produce excessively fearful animals. This is usually the reality of gene-environment interactions. The child of the depressed mother inherits not only the genetic vulnerability but also the depressed parent (e.g. Field 1998). This is also the reason why many epidemiological studies based on linear regression models often find that the epigenetic factors, such as parental care, do not add predictive value above that of genetic inheritance. The environment the parent provides commonly works in the same direction as the genetic influences; they are redundant forces. Knowledge of an animal's BALBc pedigree is sufficient to predict a high level of timidity in adulthood. Additional information on maternal care would add statistically little to the predictability—the two factors work in the same direction. But this is clearly different from concluding that maternal care is not relevant, and the results of the cross-fostering studies attest to the importance of such epigenetic influences. The misunderstanding on this point illustrates the degree to which the inappropriate use and interpretation of linear regression models to resolve the futile nature-nurture debate has served as a serious obstruction in developmental sciences.

MATERNAL CARE AND HIPPOCAMPAL DEVELOPMENT

In addition to the long-term effects described above, maternal care has immediate impact on endocrine function in infant rats. Tactile stimulation derived from mothers serves to dampen HPA activity in neonates, protecting the animals against the highly catabolic effects of adrenal glucocorticoids during a period of rapid development (see Levine 1994). Likewise, tactile stimulation from mothers stimulates the release of growth hormone (Schanberg et al 1984). Pups exposed to prolonged periods of maternal separation show increased levels of glucocorticoids and decreased levels of growth hormone. These effects can be reversed with "stroking" with a brush, a manipulation that mimics the tactile stimulation derived from maternal LG.

The results of these studies suggest that maternal LG can serve to promote an endocrine state that fosters growth and development. Variations in maternal care also appear to be related to individual differences in the synaptic development of selected neural systems that mediate cognitive development. As adults, the offspring of high LG-ABN mothers show enhanced spatial learning/memory in the Morris water maze (see Liu et al 2000) as well as in object recognition (Bredy et al 2000). The performance in both tasks is dependent on hippocampal function (e.g. Morris et al 1982, Squire 1992, Whishaw 1998, Wood et al 1999) and maternal care–altered hippocampal synaptogenesis. At either day 18 or day 90, there were significantly increased levels of neural cell adhesion molecule (N-CAM) or synaptophysin-like immunoreactivity on Western blot analyses in hippocampal samples from the high LG-ABN offspring, which suggests increased synapse formation/survival.

The influence of the hippocampus in spatial learning is thought to involve, in part at least, cholinergic innervation emerging from the medial septum (e.g. Quirion et al 1995). It is interesting that in microdialysis studies of adult offspring of high LG-ABN mothers, increased hippocampal choline acetyltransferase (ChAT) activity, acetylcholinesterase staining, and hippocampal basal and K+-stimulated acetylcholine release was found (Liu et al 2000). These findings suggest increased cholinergic synaptic number in the hippocampus of high LG-ABN offspring. There was also increased hippocampal levels of brain-derived neurotrophic factor (BDNF) mRNA in high LG-ABN offspring on day 8 of life (Liu et al 2000). BDNF is associated with the survival of cholinergic synapses in rat forebrain (Alderson et al 1990, Thoenen 1995, Friedman et al 1995). It is interesting that BDNF expression is enhanced by tactile stimulation in early development and decreased by maternal deprivation (Zhang et al 1997) and by elevated glucocorticoid levels (Chao et al 1998).

The expression of BDNF is regulated by N-methyl-D-aspartate (NMDA) receptor activation, and tactile stimulation has been shown to increase NMDA receptor expression in the barrel cells of mice (Jablonska et al 1996). There is increased mRNA expression of both the NR2A and NR2B subunits of the NMDA receptor in the offspring of high compared with low LG-ABN mothers at day 8 of age

(Liu et al 2000). These effects are associated with increased NMDA receptor binding. The results of a recent DNA array study (Diorio et al 2000) revealed the major class of effects on gene expression: (*a*) genes related to cellular metabolic activity (glucose transporter, cFOS, cytochrome oxydase, low-density-lipoprotein receptor, etc); (*b*) genes related to glutamate receptor function, including effects on the glycine receptor as well as those mentioned for the NMDA receptor subunits; and (*c*) genes encoding for growth factors, including BDNF, basic fibroblast growth factor (bFGF), and β-nerve growth factor. In each case, expression was greater than threefold higher in hippocampal samples from day-8 offspring of high LG-ABN mothers.

Naturally occurring variations in maternal LG and ABN were associated with the development of cholinergic innervation to the hippocampus, as well as with differences in the expression of NMDA receptor subunit mRNAs. In adults, there was increased hippocampal NR1 mRNA expression. These findings provide a mechanism for the differences observed in spatial learning and memory in adult animals. In adult rats, spatial learning and memory is dependent on hippocampal integrity; lesions of the hippocampus result in profound spatial learning impairments (e.g. Morris et al 1982, Squire 1992, Whishaw 1998, Wood et al 1999). Moreover, spatial learning is regulated by both cholinergic or NMDA receptor activation (e.g. Gage & Bjorkland 1986, Morris et al 1986, Quirion et al 1995) or NR1 subunit knockout animals (McHugh et al 1996). Likewise, hippocampal long-term potentiation, often considered a potential neural model for learning and memory (Bliss & Collingridge 1993, Bailey et al 1996), is enhanced by treatments that increase acetylcholine release (Calbresi et al 1998) or overexpression of NMDA receptor subunits at the level of the hippocampus (Tang et al 1999). Taken together, these findings suggest that maternal care increases hippocampal NMDA receptor levels, resulting in elevated BDNF expression, increased hippocampal synaptogenesis, and, thus, enhanced spatial learning in adulthood. These results are also consistent with the idea that maternal behavior actively stimulates hippocampal synaptogenesis in offspring through systems known to mediate experience-dependent neural development (e.g. Schatz 1990, Kirkwood et al 1993).

INDIVIDUAL DIFFERENCES IN MATERNAL BEHAVIOR

Variations in maternal behavior appear to have profound impact on neural development in rats. For these rodents, this may not be surprising. The first weeks of life do not hold a great deal of stimulus diversity for rat pups. Stability is the theme of the burrow, and the social environment in the first days of life is defined by the mother and the littermates. The mother, then, serves as a direct link between the environment and the developing animal. Thus, it seems reasonable that variations in mother-pup interaction would serve to carry so much importance for development. However, the parental mediation of environmental effects is not unique to the isolated confines of the rodent burrow. In humans, parental factors also serve to mediate the effects of environmental adversity on development. For example, the effects of poverty on emotional and cognitive development are mediated by parental factors, to the extent that if such factors are controlled, there is no discernible effect of poverty on child development (Eisenberg & Earls 1975, Conger et al 1994, McLloyd 1998). These findings suggest that environmental adversity alters the behavior of the parents, which in turn affects the development of the child.

Human clinical research suggests that the social, emotional, and economic contexts are overriding determinants of the quality of the relationship between parent and child (Eisenberg 1990). Human parental care is disturbed under conditions of chronic stress. Conditions that most commonly characterize abusive and neglectful homes involve economic hardship, martial strife, and a lack of social and emotional support (Eisenberg 1990). Such homes, in turn, breed neglectful parents. Perhaps the best predictor of child abuse and neglect is the parents own history of childhood trauma. More subtle variations in parental care also show continuity across generations. Scores on the Parental Bonding Index, a measure of parent-child attachment, are highly correlated across generations of mothers and daughters (Miller et al 1997). In nonhuman primates, there is also strong evidence for the transmission of stable individual differences in maternal behavior (Berman 1990, Fairbanks 1996, Maestripieri 1999). But what are the neural mechanisms underlying such variations in maternal care, and how might they be regulated by environmental adversity?

The stress reactivity of offspring mirrors that of their mothers. Low LG-ABN mothers are more fearful than are high LG-ABN dams (Francis et al 2000). Likewise, their offspring are more fearful than are those of high LG-ABN mothers. The differences in fearfulness may, in fact, be a crucial point in understanding the neural basis for the individual differences in maternal behavior. Maternal behavior in rats emerges as a resolution of an interesting conflict (Rosenblatt 1994, Stern 1997, Fleming 1999). Unless they are in late pregnancy or lactating, female rats generally show an aversion to pups. Typical of the generally neophobic adult rat, it is the novelty of the pups that is the source of the aversion. Habituation to the novelty results in an altered set of responses. Thus, continuous exposure to the novel pups renders virgin females more likely to exhibit maternal behavior (i.e. pup sensitization) (Rosenblatt 1994; Bridges 1994, 1996). For maternally responsive females, the positive cues associated with pups are tactile, gustatory, and auditory (Stern 1997). Thus, pup stimuli can either be aversive, eliciting withdrawal, or positive, eliciting approach. The onset of maternal behavior clearly depends on decreasing the negative-withdrawal tendency associated with neophobia, and increasing the positive-approach responses. Amygdaloid lesions, which dampen fearful reactions to novelty, increase maternal responsivity in nulliparous females (Fleming et al 1980). These finding suggest that the less-fearful female offspring of high LG-ABN mothers might be less averse to pups than the offspring of low LG-ABN mothers. This appears to be the case. Meaney & Champagne (2000) used the pup sensitization paradigm and found that the virgin female offspring of high LG-ABN mothers became fully maternal in less than half the time that was required to induce a comparable effect in the offspring of low LG-ABN mothers. Moreover, simply screening adult female rats obtained from the breeder using the pup-sensitization paradigm showed that females that became maternal more readily (i.e. required a shorter period of exposure to pups) were subsequently high LG-ABN mothers.

Of course, even a primiparous female rat is fully maternal immediately after and even during parturition. For lactating females, pups are an intense source of attraction (e.g. Fleming 1999). The differences in the responses to pups of virgin vs lactating females are associated with the endocrine events of late pregnancy. Thus, a hormonal regimen that mimics the endocrine changes occurring in late pregnancy facilitates the expression of maternal behavior in virgin female rats (Bridges 1994, 1996; Rosenblatt 1994). It is interesting that the same treatment also reduces the animals' fear of novelty (Fleming et al 1989). One of the key components of these endocrine events is an estrogen surge that serves to induce oxytocin receptors in multiple reproductive tissues as well as in brain regions (Insel 1990, Pedersen 1995) that are known to mediate the expression of maternal behavior (Numan & Sheehan 1997). The increased sensitivity to oxytocin is associated with both a reduced level of fearfulness (McCarthy et al 1996, Windle et al 1997, Uvnas-Moberg 1997, Neumann et al 2000) and an increased maternal responsivity (Pedersen 1995). Among nonlactating females, there are no differences in oxytocin receptor levels except in the central n. of the amygdala, where receptor levels were higher in high LG-ABN mothers (Francis et al 2000). It is interesting that oxytocin appears to act at the central n. of the amygdala to reduce fearfulness (I Neumann, personal communication). Among lactating females, there were significantly higher levels of oxytocin receptors in the medial preoptic area, the bed n. of the stria terminalis, and the lateral septum in all animals (for a review, see Pederson 1995); however, the lactation-induced increase in receptors levels was substantially greater in high LG-ABN mothers (Francis et al 2000). Each of these brain regions has been implicated in the expression of maternal behavior in rats (Numan & Sheehan 1997). The results of a recent study (F Champagne, J Diorio, MJ Meaney, unpublished data) reflect the functional importance of such differences in oxytocin receptor levels. Thus, intracerebroventricular infusion of an oxytocin receptor antagonist on day 3 postpartum completely eliminated the differences in pup licking/grooming between high and low LG-ABN mothers.

Throughout most of pregnancy, progesterone levels are high and accompanied by moderate levels of estrogen. Then, prior to parturition, progesterone levels fall and there occurs a surge in estrogen levels. Both events are obligatory for the onset of maternal behavior. Estrogen appears to act at the level of the medial preoptic area to enhance the expression of maternal behavior (see Rosenblatt 1994). The influence of ovarian hormones on the onset of maternal behavior in rats appears to be mediated, in part, by effects on central oxytocinergic systems (Pederson 1995). Estrogen increases oxytocin receptor gene expression and receptor binding (e.g. de Kloet et al 1986, Johnson et al 1989, Bale et al 1995, Young et al 1997).

Intracerebroventricular administration of oxytocin rapidly stimulates maternal behavior in virgin rats (Pederson et al 1979, Fahrbach et al 1985), and the medial
preoptic area appears to be a critical site. The effect of oxytocin is abolished
by ovariectomy and reinstated with estrogen treatment. Moreover, treatment with
oxytocin-antisera or receptor antagonists blocks the effects of ovarian steroid treatments on maternal behavior (Pederson et al 1985, Fahrbach et al 1985). The mechanism underlying the differential effects of lactation on the induction of oxytocin
receptors in high and low LG-ABN mothers appears to involve differences in
estrogen sensitivity. Among ovariectomized females given estrogen replacement,
there was a significantly greater estrogen effect on oxytocin receptor levels in the
medial preoptic area in high compared with low LG-ABN animals (Champagne &
Meaney 2000). The effect was apparent across a wide range of doses, and indeed
there was little evidence for any effect of estrogen on oxytocin receptor levels in
the medial preoptic area of low LG-ABN females. The fact that such differences
occurred even in the nonlactating, ovariectomized state suggests the existence of
stable differences in estrogen sensitivity in these animals. Although the mechanism for such differences in estrogen sensitivity is not yet clear, it is possible that
these findings represent an active process of "feminization" such that the behavior
of the mother toward her female offspring sensitizes selected brain regions to the
effects of estrogen in adulthood and, thus, forms the basis for the transmission of
individual differences in maternal behavior.

ENVIRONMENTAL REGULATION
OF MATERNAL BEHAVIOR

Under natural conditions, and the sanctity of the burrow, rat pups have little direct
experience with the environment. Instead, conditions such as the scarcity of food,
social instability, low dominance status, etc, directly affect the emotional state of
the mother and, thus, of maternal care. The effects of these environmental challenges on the development of the pups are then mediated by alterations in maternal
care (see Figure 1). Variations in maternal care can thus serve to transduce an environmental signal to the pups. The environmentally driven alterations in maternal
care then influence the development of neural systems that mediate behavioral and
HPA responses to stress (see Figure 1). Animals that are more fearful and anxious,
such as low LG-ABN mothers, are more neophobic and lower in maternal responsivity to pups than are the less-fearful animals. These effects could then serve as
the basis for comparable patterns of maternal behavior in offspring (F1) and for the
transmission of these traits to the subsequent generation (F2) (see Figure 1). These
individual differences are transmitted to the offspring through effects on the development of neural systems mediating the expression of fearfulness. Perhaps the
pivotal finding is that maternal care in infancy regulates the development of central
CRF systems that serve to activate behavioral, endocrine, and autonomic responses
to stress. Variations in maternal care in infant rats also influence the development

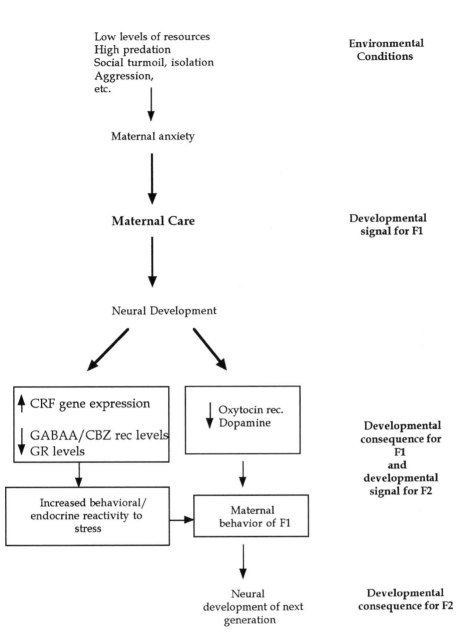

Figure 1 A schema representing the potential outcomes of the proposed relationship between environmental adversity and infant care. The key feature of this formulation is the hypothesized relationship between fearfulness and maternal behavior (for a review, see Fleming 1998). Thus, variations in maternal care affect the development of neural systems that mediate stress reactivity, which may then serve to influence maternal behavior. These effects then serve to influence the development of the subsequent generation and thus provide a basis for the transmission of individual differences in stress reactivity from one generation to the next. CRF, corticotropin-releasing factor; CBZ, central benzodiazepine; GR, glucocorticoid receptor.

of neural systems, such as glucocorticoid and GABAA receptor systems, which provide an inhibitory tone over CRF synthesis and release.

Perhaps the most compelling evidence for this process emerges from the studies of Rosenblum, Coplan, and their colleagues (e.g. Coplan et al 1996, 1998). Bonnet macaque mother-infant dyads were maintained under one of three foraging conditions: low foraging demand, where food was readily available; high foraging demand, where ample food was available, but required long periods of searching; and variable foraging demand (VFD), a mixture of the two conditions on a schedule that did not allow for predictability. At the time these conditions were imposed, there were no differences in the nature of the mother-infant interactions. However, following a number of months of these conditions, there were highly significant differences. The VFD condition was clearly the most disruptive (Rosenblum & Andrews 1994). Mother-infant conflict increased in the VFD condition. Infants of mothers housed under these conditions were significantly more timid and fearful. Remarkably, these infants showed the signs of depression commonly observed in maternally separated macaque infants, even though the infants were in contact with their mothers. As adolescents, the infants reared in the VFD conditions were more fearful and submissive and showed less social play behavior.

More recent studies have demonstrated the effects of these conditions on the development of neural systems that mediate behavioral and endocrine responses to stress. As adults, monkeys reared under VFD conditions showed increased cerebrospinal fluid levels of CRF (Coplan et al 1996, 1998). Increased central CRF drive would suggest altered noradrenergic and serotonergic responses to stress, and this is exactly what was seen in adolescent VFD-reared animals. It will be fascinating to see if these traits are transmitted to the next generation.

The critical issue here is the effect of environmental adversity on maternal behavior. In rats, females exposed to stress during pregnancy showed increased retrieval latencies (Fride et al 1985, Moore & Power 1986, Kinsley & Bridges 1988), a finding that would seem to reflect an effect of stress on maternal responsivity. In a recent study, Champagne & Meaney (2000) examined the effect of such gestational stress on maternal behavior in high and low LG-ABN mothers. Females previously defined as high or low LG-ABN mothers with their first litter were exposed either to restraint stress during the last half of gestation or to control conditions. We found that gestational stress decreased the frequency of maternal LG and ABN in high but not low LG-ABN mothers. Thus, a stressful environmental signal during gestation was sufficient to reverse completely the pattern of maternal behavior in high LG-ABN mothers. These findings led to us to question whether such effects of gestational stress would be apparent with a subsequent litter, even in the absence of any further stress. Indeed, the effects of gestational stress were fully evident with the third litter. The maternal behavior of high LG-ABN mothers exposed to gestational stress during an earlier pregnancy was indistinguishable from that of low LG-ABN mothers. The results raise what, in our minds, is a fascinating possibility: Chronic stress during gestation results in a sustained alteration in one or more neural systems that mediate the expression of maternal behavior, resulting in long-term changes in maternal behavior.

Summary

Taken together, these findings suggest that environmental adversity alters the emotional well-being of the mother: Chronic stress increases anxiety and fearfulness and, thus, decreases maternal responsivity, which in turn influences the development of stress reactivity in the offspring. For humans, these are not isolated conditions: One in five teens and one in six adult women experience abuse during pregnancy (Newberger et al 1992, Parker et al 1994). Also, in humans, Fleming (1988) reported that many factors contribute to the quality of the mother's attitude toward her newborn, but none were correlated more highly than the women's level of anxiety. Mothers who felt depressed and anxious were, not surprisingly, less positive toward their babies (also Field 1998). Moreover, there is evidence for the behavioral transmission of anxiety. Highly anxious mothers are more likely to have children who are shy and timid, and the behavior of the mother predicts the level of such behavioral inhibition in the child (Hershfeld et al 1997a,b). It was recently found that scores on parental bonding measures were correlated with autonomic, HPA and mesolimbic dopamine responses to stress (Preussner et al 2000): Young adults who described cold, distant relationships with their parents showed increased glucocorticoid and cardiovascular responses to stress, as well as evidence for increased dopamine release in the ventral striatum. More extreme variations in parental care have predictable results. Heim et al (2000) recently reported that, as adults, victims of abuse in early life show increased endocrine and autonomic responses to stress.

CONCLUSIONS

These patterns of transmission likely reflect very adaptive patterns of development. Children inherit not only genes from their parents but also an environment (West & King 1987): Englishmen inherit England, as Francis Galton remarked. We believe that the findings on intergenerational transmission via maternal behavior represent an adaptive approach to development. Under conditions of increased environmental demand, it is commonly in an animal's interest to enhance its behavioral (e.g. vigilance, fearfulness) and endocrine (HPA and metabolic/cardiovascular) responsivity to stress (see above). These responses promote detection of potential threat, avoidance learning, and metabolic/cardiovascular responses that are essential under the increased demands of the stressor. Because offspring usually inhabit niches similar to those of their parents, the transmission of these traits from parent to offspring could serve to be adaptive. A metaphor for this argument exists in the physiology of the thrifty phenotype in rodents (Neel 1962, Hales & Barker 1992). In response to the deprivation of energy substrates in fetal life, rodents show a pattern of development that favors energy conservation and an increased capacity for both gluconeogenesis and lipolysis in adulthood. Both effects appear to reflect "anticipatory" patterns of development that would be adaptive under repeated periods of food shortages. It is interesting that these effects are mediated by sustained changes in the expression of genes in hepatic tissues that mediate glucose and fat metabolism (Desai et al 1995, Bauer et al 1998, Phillips 1998, Seckl et al 1999). I believe that the effects of maternal care of the expression of genes involved in the regulation of behavioral and endocrine responses to stress reflect a comparable effect.

The key issue here is that of the potential adaptive advantage of the increased level of stress reactivity apparent in the offspring of low LG-ABN mothers. This point was addressed earlier in relation to the handling studies, and the potential advantages of increased HPA responsivity to stress in the nonhandled animals. In the current context, the research of Farrington et al (1988) and Haapasap & Tremblay (1994), for example, on young males growing-up in low socio-economic status (SES) and high crime environments provides an excellent illustration of the potential advantages of increased stress reactivity. In this environment, the males who were most successful in avoiding the pitfalls associated with such a "criminogenic" environment were those who were shy and somewhat timid. Under such conditions, a parental rearing style that favored the development of a greater level of stress reactivity to threat would be adaptive. It is thus perhaps understandable that parents occupying a highly demanding environment would transmit to their young an enhanced level of stress reactivity in "anticipation" of a high level of environmental adversity. Such a pessimistic developmental profile (see Figure 1) would be characterized (*a*) by an increased level of hypothalamic and amygdaloid CRF gene expression and (*b*) in patterns of gene expression that dampen the capacity of inhibitory systems, such as the GABAA/CBZ receptor complex and the hippocampal glucocorticoid receptor system. In addition, a pessimistic developmental profile provides less investment in metabolically expensive synaptic systems, such as the hippocampal circuits (see also Sapolsky 1992). This might also be adaptive because elevated glucocorticoid levels associated with environmental adversity and increased stress reactivity would serve to damage hippocampal systems (see Meaney et al 1988, Sapolsky 1992). In contrast, more favorable environments would encourage an optimistic pattern of development, characterized by more modest levels of stress reactivity and increased hippocampal synaptogenesis. The quality of the environment influences the behavior of the parent, which in turn is the critical factor in determining whether development proceeds along an optimistic or a pessimistic pattern of development. The obvious conclusion is that there is no single ideal form of parenting: Various levels of environmental demand require different traits in offspring.

A final issue here concerns the cost of such increased stress reactivity. A shy and timid child in an urban slum may be at an advantage with respect to the demands of a menacing environment. The question is whether such traits would later also confer an increased risk for stress-induced illness. I would argue that it does, and that this risk reflects the cost of adaptation to a high level of environmental demand, such as a low socioeconomic environment, in early life, a process mediated by effects of adversity on parental care.

Visit the Annual Reviews home page at www.AnnualReviews.org

LITERATURE CITED

Ader R, Grota LJ. 1969. Effects of early experience on adrenocortical reactivity. *Physiol. Behav.* 4:303–5

Alderson RF, Alterman AL, Barde Y-A, Lindsay RM. 1990. Brain-derived neurotrophic factor increases survival and differentiated functions of rat spetal cholinergic neurons in culture. *Neuron* 5:297–306

Ammerman RT, Cassisi JE, Hersen M, van Hasselt VB. 1986. Consequences of physical abuse and neglect in children. *Clin. Psychol. Rev.* 6:291–310

Anisman H, Zaharia MD, Meaney MJ, Merali Z. 1998. Do early life events permanently alter behavioral and hormonal responses to stressors? *Int. J. Dev. Neurosci.* 16:149–64

Antoni FA. 1993. Vasopressinergic control of pituitary adrenocorticotropin secretion comes of age. *Front. Neuroendocrinol.* 14:76

Arbilla S, Allen J, Wick A, Langer SZ. 1986. High affinity [^{3}H]Zolpidem binding in the rat brain: an inidazopyridine with agonist properties at central benzodiazepine receptors. *Eur. J. Pharmacol.* 130:257–63

Bailey CH, Bartsch D, Kandel ER. 1996. Toward a molecular definition of long-term memory storage. *Proc. Natl. Acad. Sci. USA* 96:13445–52

Bakshi VP, Shelton SE, Kalin NH. 2000. Neurobiological correlates of defensive behaviors. *Prog. Brain Res.* 122:105–15

Bale TL, Pedersen CA, Dorsa DM. 1995. CNS oxytocin receptor mRNA expression and regulation by gonadal steroids. *Adv. Exp. Med. Biol.* 395:269–80

Bauer MK, Harding JE, Bassett NS, Brier BH, Oliver MH, et al. 1998. Fetal growth and placental function. *Mol. Cell. Endocrinol.* 140:115–20

Baxter JD, Tyrrell JB. 1987. The adrenal cortex. In *Endocrinology and Metabolism*, ed. P Felig, JD Baxter, AE Broadus, L Frohman, pp. 385–511. New York: McGraw-Hill

Berman CM. 1990. Intergenerational transmission of maternal rejection rates among free-ranging rhesus monkeys on Cayo Santiago. *Anim. Behav.* 44:247–58

Bhatnagar S, Shanks N, Meaney MJ. 1995. Hypothalamic-pituitary-adrenal function in handled and nonhandled rats in response to chronic stress. *J. Neuroendocrinol.* 7:107–19

Bhatnagar S, Shanks N, Meaney MJ. 1996. Plaque-forming cell responses and antibody titers following injection of sheep-red blood cells in nonstressed, acute and/or chronically stressed handled and nonhandled animals. *Dev. Psychobiol.* 29:171–81

Bifulco A, Brown GW, Adler Z. 1991. Early sexual abuse and clinical depression in adult life. *Br. J. Psychiatr.* 159:115–22

Bliss TVP, Collingridge GL. 1993. A synaptic model of memory: long-term potentiation in the hippocampus. *Nature* 361:31–39

Brady LS, Gold PW, Herkenham M, Lynn AB, Whitfield HW. 1992. The antidepressants fluoxetine, idazoxan and phenelzine alter corticotropin-releasing hormone and tyrosine hydroxylase mRNA levels in rat brain: therapeutic implications. *Brain Res.* 57:117–25

Bredy T, Cain DP, Meaney MJ. 2000. Peripubertal environmental enrichment reverses the functional effects of maternal care on hippocampal development. *Soc. Neurosci.* (Abstr.) 26:1132

Bridges RS. 1994 . The role of lactogenic hormones in maternal behavior in female rats. *Acta Paediatr. Suppl.* 397:33–39

Bridges RS. 1996. Biochemical basis of parental behavior in the rat. *Adv. Study Behav.* 25:215–42

Brindley DN, Rolland Y. 1989. Possible connections between stress, diabetes, obesity, hypertension and altered lipoprotein metabolism that may result in atherosclerosis. *Clin. Sci.* 77:453–61

Brown GR, Anderson B. 1993. Psychiatric morbidity in adult inpatients with childhood histories of sexual and physical abuse. *Am. J. Psychiatry* 148:55–61

Butler PD, Weiss JM, Stout JC, Nemeroff CB. 1990. Corticotropin-releasing factor produces fear-enhancing and behavioural activating effects following infusion into the locus coeruleus. *J. Neurosci.* 10:176–83

Cahill L, McGaugh JL. 1998. Mechanisms of emotional arousal and lasting declarative memory. *Trends Neurosci.* 21:294–99

Calbresi P, Centonze D, Gubellini P, Pisani A, Bernadi G. 1998. Blockade of M2-like muscarinic receptors enahnces long-term potentiation at corticostriatal synapses. *Eur. J. Neurosci.* 1:3020–23

Caldji C, Diorio J, Meaney MJ. 2000a. Maternal behavior in infancy regulates the development of GABAA receptor levels, its subunit composition and GAD. *Soc. Neurosci. Abstr.* 26:489

Caldji C, Francis D, Sharma S, Plotsky PM, Meaney MJ. 2000b. The effects of early rearing environment on the development of GABAA and central benzodiazepine receptor levels and novelty-induced fearfulness in the rat. *Neuropsychopharmacology* 22:219–29

Caldji C, Tannenbaum B, Sharma S, Francis D, Plotsky PM, Meaney MJ. 1998. Maternal care during infancy regulates the development of neural systems mediating the expression of behavioral fearfulness in adulthood in the rat. *Proc. Natl. Acad. Sci. USA* 95:5335–40

Calhoun JB. 1962. *The Ecology and Sociology of the Norway Rat.* Bethesda, MD: Health, Educ., Welfare, Public Health Serv.

Canetti L, Bachar E, Galili-Weisstub E, De-Nour AK, Shalev AY. 1997. Parental bonding and mental health in adolescence. *Adolescence* 32:381–94

Champagne F, Meaney MJ. 2000. Gestational stress effects on maternal behavior. *Soc. Neurosci. Abstr.* 26:2035

Chao HM, Sakai RR, Ma LY, McEwen BS. 1998. Adrenal steroid regulation of neurotrophic factor expression in the rat hippocampus. *Endocrinology* 139:3112–18

Chrousos GP, Gold PW. 1992. The concepts of stress and stress system disorders. *JAMA* 267:1244–52

Conger R, Ge X, Elder G, Lorenz F, Simons R. 1994. Economic stress, coercive family process and developmental problems of adolescents. *Child Dev.* 65:541–61

Coplan JD, Andrews MW, Rosenblum LA, Owens MJ, Friedman S, et al. 1996. Persistent elevations of cerebrospinal fluid concentrations of corticotropin-releasing factor in adult nonhuman primates exposed to early-life stressors: implications for the pathophysiology of mood and anxiety disorders. *Proc. Natl. Acad. Sci. USA* 93:1619–23

Coplan JD, Trost RC, Owens MJ, Cooper TB, Gorman JM, et al. 1998. Cerebrospinal fluid concentrations of somatostatin and biogenic amines in grown primates reared by mothers exposed to manipulated foraging conditions. *Arch. Gen. Psychiatry* 55:473–77

Dallman M, Akana S, Caren C, Darlington D, Jacobson L, Levin N. 1987. Regulation of ACTH secretion variations on a theme of B. *Rec. Prog. Horm. Res.* 43:113–73

Dallman MF, Akana SF, Scribner KA, Bradbury MJ, Walker C-D, et al. 1993. Stress, feedback and facilitation in the hypothalamo-pituitary-adrenal axis. *J. Neuroendocrinol.* 4:517–26

Dallman MF, Akana SF, Strack AM, Hanson ES, Sebastian RJ. 1995. The neural network that regulates energy balance is responsive to glucocorticoids and insulin and also regulates HPA axis responsivity at a site proximal to CRF neurons. *Ann. NY Acad. Sci.* 771:730–42

Davis M, Walker DL, Lee Y. 1997. Amygdala and bed nucleus of the stria terminalis: different roles in fear and anxiety measured with the acoustic startle reflex. *Philos. Trans. R. Soc. London Ser. B* 352:1675–87

DeBellis MD, Chrousos GP, Dom LD, Burke L, Helmers K, et al. 1994. Hypothalamic pituitary adrenal dysregulation in sexually abused girls. *J. Clin. Endocrinol. Metab.* 78:249–55

deBoer SF, Katz JL, Valentino RJ. 1992. Common mechanism underlying the proconflict effects of corticotropin-releasing factor, a benzodiazepine inverse agonist and electric footshock. *J. Pharmacol. Exp. Ther.* 262:335–42

De Kloet ER. 1991. Brain corticosteroid receptor balance and homeostatic control. *Front. Neuroendocrinol.* 12:95–164

De Kloet ER, Voorhuis TAM, Elands J. 1986. Estradiol induces oxytocin binding sites in rat hypothalamic ventromedial nucleus. *Eur. J. Pharmacol.* 118:185–86

De Kloet ER, Vregdenhil E, Oitzl MS, Joels M. 1998. Brain corticosteroid receptor balance in health and disease. *Endocrinol. Rev.* 19:269–301

Denenberg VH. 1964. Critical periods, stimulus input, and emotional reactivity: a theory of infantile stimulation. *Psychol. Rev.* 71:335–51

Desai M, Crowther NJ, Ozanne SE, Lucas A, Hales CN. 1995. Adult glucose and lipd metabolism may be programmed during early life. *Biochem. Soc. Trans.* 23:331–35

Dhabhar FS, McEwen BS, Spencer RL. 1993. Stress response, adrenal steroid receptor levels and corticoteroid binding-globulin—a comparison of Sprague Dawley, Fisher 344 and Lewis rats. *Brain Res.* 616:89–98

Diamond DM, Bennett MC, Fleshner M, Rose GM. 1992. Inverted-U relationship between the level of peripheral corticosterone and the magnitude of hippocampal primed burst potentiation. *Hippocampus* 2:421–28

Diorio D, Viau V, Meaney MJ. 1993. The role of the medial prefrontal cortex (cingulate gyrus) in the regulation of hypothalamic-pituitary-adrenal responses to stress. *J. Neurosci.* 13:3839–49

Dorio J, Weaver ICG, Meaney MJ. 2000. A DNA array study of hippocampal gene expression regulated by maternal behavior in infancy. *Soc. Neurosci. Abstr.* 26:1366

Dunn AJ, Berridge CW. 1990. Physiological and behavioral responses to corticotropin-releasing factor administration: Is CRF a mediator of anxiety or stress responses? *Brain Res. Rev.* 15:71–100

Eisenberg L. 1990. The biosocial context of parenting in human families. In *Mammalian Parenting: Biochemical, Neurobiological, and Behavioral Determinants.* ed. NA Krasnegor, RS Bridges pp. 9–24. New York: Oxford Univ. Press

Eisenberg L, Earls FJ. 1975. Poverty, social depreciation and child development. In *American Handbook of Psychiatry*, ed. DA Hamburg, 6:275–91. New York: Basic Books

Emoto H, Yokoo H, Yoshida M, Tanaka M. 1993. Corticotropin-releasing factor enhances noradrenaline release in the rat hypothalamus assessed by intracerebral microdialysis. *Brain Res.* 601:286–88

Fahrbach SE, Morrell JI, Pfaff DW. 1985. Possible role for endogenous oxytocin in estrogen-facilitated maternal behavior in rats. *Neuroendocrinology* 40:526–32

Fairbanks LM. 1996. Individual differences in maternal style. *Adv. Study Behav.* 25:579–611

Farrington DA, Gallagher B, Morley L, St Ledger RJ, West DJ. 1988. Are there any successful men from criminogenic backgrounds? *Psychiatry* 51:116–30

Felitti VJ, Anda RF, Nordenberg D, Williamson DF, Spitz AM, et al. 1998. Relationship of childhood abuse and household dysfunction to many of the leading causes of death in adults. *Am. J. Prev. Med.* 14:245–58

Field T. 1998. Maternal depression effects on infants and early interventions. *Prev. Med.* 27:200–3

Fleming AS. 1988. Factors influencing maternal responsiveness in humans: usefulness of an animal model. *Psychoneuroendocrinology* 13:189–212

Fleming AS. 1999. The neurobiology of mother-infant interactions: experience and central nervous system plasticity across development and generations. *Neursci. Biobehav. Rev.* 23:673–85

Fleming AS, Cheung U, Myhal N, Kessler Z. 1989. Effects of maternal hormones on "timidity" and attraction to pup-related odors in female rats. *Physiol. Behav.* 46:449–53

Fleming AS, Vaccarino F, Leubke C. 1980. Amygdaloid inhibition of maternal behavior in the nulliparous female rat. *Physiol. Behav.* 25:731–43

Francis D, Meaney MJ. 1999. Maternal care and the development of stress responses. *Curr. Opin. Neurobiol.* 9:128–34

Francis DD, Caldji C, Champagne F, Plotsky PM, Meaney MJ. 1999a. The role of corticotropin-releasing factor—norepinepherine systems in mediating the effects of early experience on the development of behavioral and endocrine responses to stress. *Biol. Psychiatry* 46:1153–66

Francis DD, Champagne F, Meaney MJ. 2000. Variations in maternal behaviour are associated with differences in oxytocin receptor levels in the rat. *J. Neuroendocrinol.* 12:1145–48

Francis DD, Diorio J, Liu D, Meaney MJ. 1999b. Nongenomic transmission across generations in maternal behavior and stress responses in the rat. *Science* 286:1155–58

Fride E, Dan Y, Gavish M, Weinstock M. 1985. Prenatal stress impairs maternal behavior in a conflict situation and reduces hippocampal benzodiazepine receptors. *Life Sci.* 36:2103–9

Friedman B, Klienfeld DP, Ip NY, Verge VM, Moulton R, et al. 1995. BDNF and NT-4/5 exert neurotrophic influences on injured spinal motor neurons. *J. Neurosci.* 15:1044–56

Gage FH, Bjorklund A. 1986. Cholinergic speptal grafts into hippocampal formation improve spatial learning and memory in aged rat by an atropine-sensitive mechanism. *J. Neurosci.* 6:2837–47

Glue P, Wilson S, Coupland N, Ball D, Nutt D. 1995. The relationship between benzodiazepine receptor sensitivity and neuroticism. *J. Anxiety Dis.* 9:33–45

Gottman JM. 1998. Psychology and the study of marital processes. *Annu. Rev. Psychol.* 49:169–97

Gray TS, Bingaman EW. 1996. The amygdala: corticotropin-releasing factor, steroids, and stress. *Crit. Rev. Neurobiol.* 10:155–68

Haapasap J, Tremblay RE. 1994. Physically aggressive boys from ages 6 to 12: family background, parenting behavior, and prediction of delinquency. *J. Consult. Clin. Psychol.* 62:1044–52

Hadingham KL, Wingrove P, Le Bourdelles B, Palmer KJ, Ragan CI, Whitiing PJ. 1993. Cloning of cDNA sequences encoding human a2 and $\alpha3$ g-aminobutyric acidA receptor subunits and characterization of the benzodiazepine pharmacology of recombinant $\alpha1$-, $\alpha2$-, $\alpha3$- and $\alpha5$-containing human g-aminobutyric acidA receptors. *Mol. Pharmacol.* 43:970–75

Hales CN, Barker DJP. 1992. Type 2 (noninsulin-dependent diabetes) mellitus: the thrify phenotype hypothesis. *Diabetologia* 35:595–601

Heim C, Owens MJ, Plotsky PM, Nemeroll CH. 1997. The role of early adverse life events in the etiology of depression and posttraumatic stress disorder: focus on corticotropin-releasing factor. *Ann. NY Acad. Sci.* 821:194–207

Hess JL, Denenberg VH, Zarrow MX, Pfeifer WD. 1969. Modification of the corticosterone response curve as a function of handling in infancy. *Physiol. Behav.* 4:109–12

Higley JD, Haser MF, Suomi SJ, Linnoila M. 1991. Nonhuman primate model of alcohol abuse: effects of early experience, personality and stress on alcohol consumption. *Proc. Natl. Acad. Sci. USA* 88:7261–65

Hirschfeld DR, Biederman J, Brody L, Faraone

SV, Rosenbaum JF. 1997a. Associations between expressed emotion and child behavioral inhibition and psychopathology: a pilot study. *J. Am. Acad. Child Adolesc. Psychiatry* 36:205–13

Hirschfeld DR, Biederman J, Brody L, Faraone SV, Rosenbaum JF. 1997b. Expressed emotion towards children with behavioral inhibition: association with maternal anxiety disorder. *J. Am. Acad. Child Adolesc. Psychiatry* 36:910–17

Holmes SJ, Robins LN. 1987. The influence of childhood disciplinary experience on the development of alcoholism and depression. *J. Child. Psychol. Psychiatr. Allied Discipl.* 28:399–15

Holmes SJ, Robins LN. 1988. The role of parental disciplinary practices in the development of depression and alcoholism. *Psychiatry* 51:24–36

Insel TR. 1990. Regional changes in brain oxytoin receptors post-partum: time course and relationship to maternal behaviour. *J. Neuroendocrinol.* 2:539–45

Issa A, Gauthier S, Meaney MJ. 1990. Hypothalamic-pituitary-adrenal activity in aged cognitively impaired and cognitively unimpaired aged rats. *J. Neurosci.* 10:3247–57

Jablonska B, Kossut M, Skangiel-Kramska J. 1996. Transient increase of AMPA and NMDA receptor binding in the barrel cortex of mice after tactile stimulation. *Neurobiol. Learn. Mem.* 66:36–43

Jacobson L, Sapolsky RM. 1991. The role of the hippocampus in feedback regulation of the hypothalamic-pituitary-adrenal axis. *Endocrinol. Rev.* 12:118–34

Jans J, Woodside BC. 1990. Nest temperature: effects on maternal behavior, pup development, and interactions with handling. *Dev. Psychobiol.* 23:519–34

Joels M, De Kloet ER. 1989. Effect of glucocorticoids and norepinephrine on excitability in the hippocampus. *Science* 245:1502

Johnson AE, Coirini H, Ball GF, McEwen BS. 1989. Anatomical localization of the effects of 17-estradiol on oxytocin receptor binding in the ventromedial hypothalamic nucleus. *Endocrinology* 124:207–11

Kalin NH. 1985. Behavioral effects of ovine corticotropin-releasing factor administered to rhesus monkeys. *Fed. Proc.* 44:249–58

Khan ZU, Gutierrez A, De Blas AL. 1996. The $\alpha 1$ and $\alpha 6$ subunits can coexist in the same cerebellar GABAA receptor maintaining their individual benzodiazepine-binding specificities. *J. Neurochem.* 66:685–91

Kinsley CH, Mann PE, Bridges RS. 1988. Prenatal stress alters morphine- and stress-induced analgesia in male and female rats. *Pharmacol. Biochem. Behav.* 50:413–19

Kirkwood A, Dudek SM, Gold JT, Aizenman CD, Bear MF. 1993. Common forms of synaptic plasticity in the hippocampus and neocortex in vitro. *Science* 260:1518–21

Koegler-Muly SM, Owens MJ, Ervin GN, Kilts CD, Nemeroff CB. 1993. Potential corticotropin-releasing factor pathways in the rat brain as determined by bilateral electrolytic lesions of the central amygdaloid nucleus and paraventricular nucleus of the hypothalamus. *J. Neuroendocrinol.* 5:95–98

Koob GF, Heinrichs SC, Menzaghi F, Pich EM, Britton KT. 1994. Corticotropin-releasing factor, stress and behavior. *Semin. Neurosci.* 6:221–29

Kraemer GW, Ebert MH, Schmidt DE, McKinney WT. 1989. A longitudinal study of the effect of different social rearing conditions on cerebrospinal fluid norepinephrine and biogenic amine metabolites in rhesus monkeys. *Neuropsychopharmacology* 2:175–89

Laban O, Dimitrijevic M, van Hoersten S, Markovic BM, Jancovic BD. 1995. Experimental allergic encephalomyelitis in the rat. *Brain Behav. Immun.* 9:9–19

Ladd CO, Owens MJ, Nemeroff CB. 1996. Persistent changes in corticotropin-releasing factor neuronal systems induced by maternal depriviation. *Endocrinology* 137:1212–18

Landfield P, Baskin RK, Pitler TA. 1981. Brain-aging correlates: retardation by hormonal-pharmacological treatments. *Science* 214:581–85

Landfield PW, Pitler TA. 1984. Prolonged Ca-dependent afterhyperpolarizations in hippocampal neurons of the aged rat. *Science* 226:1089–93

Lavicky J, Dunn AJ. 1993. Corticotropin-releasing factor stimulates catecholamine release in hypothalamus and prefrontal cortex in freely moving rats as assessed by microdialysis. *J. Neurochem.* 60:602–12

Lee MHS, Williams DI. 1974. Changes in licking behaviour of rat mother following handling of young. *Anim. Behav.* 22:679–81

Lee MHS, Williams DI. 1975. Long term changes in nest condition and pup grouping following handling of rat litters. *Dev. Psychobiol.* 8:91–95

Levine S. 1957. Infantile experience and resistance to physiological stress. *Science* 126:405–6

Levine S. 1962. Plasma-free corticosteroid response to electric shock in rats stimulated in infancy. *Science* 135:795–96

Levine S. 1975. Psychosocial factors in growth and development. In *Society, Stress and Disease*, ed. L Levi, pp. 43–50. New York: Oxford Univ. Press

Levine S. 1994. Maternal behavior as a mediator of pup adrenocortical function. *Ann. NY Acad. Sci.* 746:260–75

Levine S, Haltmeyer GC, Karas GG, Denenberg VH. 1967. Physiological and behavioral effects of infantile stimulation. *Physiol. Behav.* 2:55–63

Liang KC, Melia KR. Miserendino MJ, Fails WA, Campeau S, Davis M. 1992. Corticotropin releasing factor: long-lasting facilitation of the acoustic startle retlex. *J. Neurosci.* 12:2303–12

Lissau I, Sorensen TIA. 1994. Parental neglect during childhood and increased risk of obesity in young adulthood. *Lancet* 343:324–27

Liu D, Caldji C, Sharma S, Plotsky PM, Meaney MJ. 2000. The effects of early life events on in vivo release of norepinephrine in the paraventricular nucleus of the hypothalamus and hypothalamic-pituitary-adrenal responses during stress. *J. Neuroendocrinol.* 12:5–12

Liu D, Tannenbaum B, Caldji C, Francis D, Freedman A, et al. 1997. Maternal care, hippocampal glucocorticoid receptor gene expression and hypothalamic-pituitary-adrenal responses to stress. *Science* 277:1659–62

Lupien S, Meaney MJ. 1998. Stress, glucocorticoids, and hippocampus aging in rat and human. In *Handbook of Human Aging*, ed. E Wang, S Snyder, pp. 19–50. New York: Academic

Lupien S, Sharma S, Nair NPV, Hauger R, McEwen BS, et al. 1998. Glucocorticoids and human brain aging. *Nat. Neurosci.* 1:69–73

Maccari S, Piazza PV, Kabbaj M, Barbazanges A, Simon H, Le Moal M. 1995. Adoption reverses the long-term impairments in glucocorticoid feedback induced by prenatal stress. *J. Neurosci.* 15:110–16

Maestripieri D. 1999. The biology of human parenting: insights from nonhuman primates. *Neurosci. Biobehav. Rev.* 23:411–22

Mason D. 1991. Genetic variation in the stress response: susceptibility to experimental allergic encephalomyelitis and implications for human inflammatory disease. *Immunol. Today* 12:57–60

McCarthy MM, McDonald CH, Brooks PJ, Goldman D. 1996. An anxiolytic action of oxytocin is enhanced by estrogen in the mouse. *Physiol. Behav.* 60:1209–15

McCarty R, Lee JH. 1996. Maternal influences on adult blood pressure of SHRs: a single pup cross-fostering study. *Physiol. Behav.* 5:71–75

McCauley J, Kern DE, Kolodner K, Dill L, Schroeder AF, et al. 1997. Clinical characteristics of women with a history of childhood abuse: unhealed wounds. *JAMA* 277:1362–68

McEwen BS. 1998. Protective and amaging effects of stress mediators. *New Engl. J. Med.* 338:171–79

McEwen BS, de Leon MJ, Lupien S, Meaney MJ. 1999. Corticosteroids, the damaging brain and cognition. *Trends Endocrinol. Metabol.* 10:92–97

McEwen BS, Steller E. 1993. Stress and the individual: mechanisms leading to disease. *Arch. Intern. Med.* 153:2093–101

McHugh TJ, Blum KI, Tsien JZ, Tonegawa S, Wilson MA. 1996. Impaired hippocampal representation of space in CA1-specific NMDAR1 knockout mice. *Cell* 87:1339–49

McLloyd VC. 1998. Socioeconomic disadvantage and child development. *Am. Psychol.* 53:185–204

Meaney MJ, Aitken DH, Bhatnagar S, Van Berkel Ch, Sapolsky RM. 1988. Postnatal handling attenuates neuroendocrine, anatomical, and cognitive impairments related to the aged hippocampus. *Science* 238:766–68

Meaney MJ, Aitken DH, Bodnoff SR, Iny LJ, Tatarewicz JE, Sapolsky RM. 1985. Early, postnatal handling alters glucocorticoid receptor concentrations in selected brain regions. *Behav. Neurosci.* 99:760–65

Meaney MJ, Aitken DH, Sapolsky RM. 1991. Environmental regulation of the adrenocortical stress response in female rats and its implications for individual differences in aging. *Neurobiol. Aging* 21:323–31

Meaney MJ, Aitken D, Sharma S, Viau V. 1992. Basal ACTH, corticosterone, and corticosterone-binding globulin levels over the diurnal cycle, and hippocampal type I and type II corticosteroid receptors in young and old, handled and nonhandled rats. *Neuroendocrinology* 55:204

Meaney MJ, Aitken DH, Sharma S, Viau V, Sarrieau A. 1989. Postnatal handling increases hippocampal type II, glucocorticoid receptors and enhances adrencocortical negative-feedback efficacy in the rat. *Neuroendocrinology* 51:597–604

Meaney MJ, Champagne F. 2000. Latency to maternal behavior in high and low LG-ABN mothers/offspring. *Soc. Neurosci. Abstr.* 26:2035

Mehta AK, Ticku MJ. 1999. An update on GABAA receptors. *Brain Res. Rev.* 29:196–217

Miller L, Kramer R, Warner V, Wickramaratne P, Weissman M. 1997. Intergenerational transmission of parental bonding among women. *J. Am. Acad. Child Adolesc. Psychiatry* 36:1134–39

Moga MM, Gray TS. 1989. Evidence for corticotropin-releasing factor, neurotensin, and somatostatin in the neural pathway from the central nucleus of the amygdala to the parabrachial nucleus. *J. Comp. Neurol.* 241:275–84

Montgomery SM, Bartley MJ, Wilkinson RG. 1997. Family conflict and slow growth. *Arch. Dis. Child.* 77:326–30

Moore CL. 1995. Maternal contributions to mammalian reproductive development and the divergence of males and females. *Adv. Study Behav.* 24:47–118

Moore CL, Lux BA. 1998. Effects of lactation on sodium intake in fischer-344 and Long-Evans rats. *Dev. Psychobiol.* 32:51–56

Moore CL, Power KL. 1986. Parental stress affects monther-infant interaction in Norway rats. *Dev. Psychobiol.* 19:235–45

Morris RG, Anderson E, Lynch GS, Baudry M. 1986. Selective impairment of learning and blockade of long-term potentiation by an N-methyl-D-aspartate receptor antagonist, AP5. *Nature* 319:774–76

Morris RGM, Garrard P, Rawlins JNP, O'Keefe J. 1982. Place navigation is impaired in rats with hippocampal lesions. *Nature* 297:681–83

Munck A, Guyre PM, Holbrook NJ. 1984. Physiological functions of glucocorticoids in stress and their relations to pharmacological actions. *Endocrinol. Rev.* 5:25–44

Neel JV. 1962. Diabetes mellitus: "a thrifty gentotype" rendered determinal by "progress"? *Am. J. Hum. Genet.* 14:353–62

Nemeroff CB. 1996. The corticotropin-releasing factor (CRF) hypothesis of depression: new findings and new directions. *Mol. Psychiatry* 1:336–42

Neumann ID, Torner L, Wigger A. 2000. Brain oxytocin: differential inhibition of neuroendocrine stress responses and anxiety-related behaviour in virgin, pregnant and lactating rats. *Neuroscience* 95:567–75

Newberger EH, Barkan SE, Lieberman ES, McCormick MC, Yllo K, et al. 1992. Abuse of pregnant women and adverse birth outcomes. Current knowledge and implications for practice. *JAMA* 267:2370–72

Numan M, Sheehan TP. 1997. Neuroanatomical circuitry for mammalian maternal behavior. *Ann. NY Acad. Sci.* 807:101–25

O'Donnell D, Larocque S, Seckl JR, Meaney MJ. 1994. Postnatal handling alters glucocorticoid, but not mineralocorticoid mRNA expression in adult rats. *Mol. Brain Res.* 26:242–48

Owens MJ, Vargas MA, Knight DL, Nemeroff CB. 1991. The effects of alprazoplam on corticotropin-releasing factor neurons in the rat brain: acute time course, chronic treatment and abrupt withdrawal. *J. Pharmacol. Exp. Ther.* 258:349–56

Pacak K, Palkovits M, Kopin I, Goldstein DS. 1995. Stress-induced norepinepherine release in hypothalamic paraventricular nucleus and pituitary-adrenrocortical and sympathoadrenal activity: in vivo microdialysis studies. *Front. Neuroendocrinol.* 16:89–150

Page ME, Valentino RJ. 1994. Locus coeruleus activation by physiological challenges. *Brain Res. Bull.* 35:557–60

Parker B, McFarlane J, Soeken K. 1994. Abuse during pregnancy: effects on maternal complications and birth weight in adult and teenage women. *Obstet. Gynecol.* 84:323–28

Parker G. 1981. Parental representations of patients with anxiety neurosis. *Acta Psychiatr. Scand.* 63:33–36

Pearson D, Sharma S, Plotsky PM, Pfaus JG, Meaney MJ. 1997. The effect of postnatal environment on stress-induced changes in hippocampal FOS-like immunoreactivity in adult rats. *Soc. Neursci. Abstr.* 23:1849

Pedersen CA. 1995. Oxytocin control of maternal behavior. Regulation by sex steroids and offspring stimuli. *Ann. NY Acad. Sci.* 126–45

Pedersen CA, Prange AJ Jr. 1979. Induction of maternal behavior in virgin rats after intracerebroventricular administration of oxytocin. *Proc. Natl. Acad. Sci. USA* 76:6661–65

Phillips DJW. 1998. Birth weight and the future development of diabetes. *Diabet. Care* 21:B150–55

Pitkanen A, Savander V, LeDoux JE. 1998. Organization of intra-amygdaloid circuitries in the rat: an emerging framework for understanding functions of the amygdala. *Trends Neurosci.* 20:517–23

Plotsky PM. 1991. Pathways to the secretion of adrenocorticotropin: a view from the portal. *J. Neuroendocrinol.* 3:1–9

Plotsky PM, Meaney MJ. 1993. Early, postnatal experience alters hypothalamic corticotropin-releasing factor (CRF) mRNA, median eminence CRF content and stress-induced release in adult rats. *Mol. Brain Res.* 18:195–200

Quirarte GL, Roozendaal B, McGaugh JL. 1997. Glucocorticoid enhancement of memory storage involves noradrenergic activation in the basolateral amygdala. *Proc. Natl. Acad. Sci. USA* 94:14048–53

Quirion R, Wilson A, Rowe WB, Doods HN, White N, Meaney MJ. 1995. Facilitation of acetylcholine release and cognitive performance by an M2 muscarinic receptor antagonist in aged memory-impaired rats. *J. Neurosci.* 15:1455–62

Rivier C, Plotsky PM. 1986. Mediation by corticotropin-releasing factor of adenohypophysial hormone secretion. *Annu. Rev. Physiol.* 48:475–89

Rosen JB, Schulkin J. 1998. From normal fear to pathological anxiety. *Psychol. Rev.* 105:325–50

Rosenblatt JS. 1994. Psychobiology of maternal behavior: contribution to the clinical understanding of maternal behavior among humans. *Acta Paediatr. Suppl.* 397:3–8

Rosenblum LA, Andrews MW. 1994. Influences of environmental demand on maternal behavior and infant development. *Acta Paediatr. Suppl.* 397:57–63

Rosenblum LA, Coplan JD, Friedman S, Bassoff T, Gorman JM, Andrews MW. 2001.

Adverse early experiences affect noradrenergic and serotonergic functioning in adult primates. *Biol. Psychiatr.* In press

Rosmond R, Dallman MF, Bjorntorp P. 1995. Stress-related cortisol secretion in men: relationship with abdominal obesity and endocrine, emtabolic and hemodynamic abnormalities. *J. Clin. Endocrinol. Metabol.* 83:1853–59

Russak LG, Schwartz GE. 1997. Feelings of parental care predict health status in midlife: a 35 year follow-up of the Harvard Mastery of Stress Study. *J. Behav. Med.* 20:1–11

Rutter M. 1979. Protective factors in children's responses to stress and disadvantage. *Prim. Prev. Psychopathol.* 3:49–74

Sapolsky RM. 1992. *Stress, the Aging Brain, and the Mechanisms of Neuron Death.* Cambridge, MA: MIT Press. 423 pp.

Sarrieau A, Sharma S, Meaney MJ. 1988. Postnatal development and environmental regulation of hippocampal glucocorticoid and mineralocorticoid receptors in the rat. *Dev. Brain Res.* 43:158–62

Schanberg SM, Evoniuk G, Kuhn CM. 1984. Tactile and nutritional aspects of maternal care: specific regulators of neuroendocrine function and cellular development. *Proc. Soc. Exp. Biol. Med.* 175:135–46

Schatz CJ. 1990. Impulse activity and the patterning of connections during CNS development. *Neuron* 5:745–56

Schulkin J, McEwen BS, Gold PW. 1994. Allostasis, the amygdala and anticipatory angst. *Neurosci. Biobehav. Rev.* 18:1–12

Seckl JR, Meaney MJ. 1994. Early life events and later development of ischaemic heart disease. *Lancet* 342:1236

Seckl JR, Nyirenda MJ, Walker BR, Chapman KE. 1999. Glucocorticoids and fetal programming. *Biochem. Soc. Trans.* 27:74–78

Smith J, Prior M. 1995. Temperament and stress resilience in school-age children: a within-families study. *J. Am. Acad. Child Adolesc. Psychiatry* 34:168–79

Squire LR. 1992. Memory and the hippocampus: a synthesis from findings with rats, monkeys, and humans. *Psychol. Rev.* 9:195–215

Sroufe LA. 1997. Psychopathology as an outcome of development. *Dev. Psychopathol.* 9:251–68

Starkman MN, Gebarski SS, Berent S, Schteingart DE. 1992. Hippocampal formation volume, memory dysfunction, and cortisol levels in patients with Cushing's syndrome. *Biol. Psychiatry* 32:756–66

Stenzel-Poore NIP, Heinrichs SC, Rivest S, Knob OF, Vale WW. 1994. Overproduction of corticotropin-releasing factor in transgenic mice: a genetic model of anxiogenic behavior. *J. Neurosci.* 14:2579–84

Stern JM. 1997. Offspring-induced nurturance: animal-human parallels. *Dev. Psychobiol.* 31:19–37

Suomi SJ. 1997. Early determinants of behaviour: evidence from primate studies. *Br. Med. Bull.* 53:170–84

Sweirgel AH, Takahashi LK, Kahn NH. 1993. Attenuation of stress-induced by antagonism of corticotropin-releasing factor receptors in the central amygdala in the rat. *Brain Res.* 623:229–34

Tang YP, Shimizu E, Dube GR, Rampon C, Kerchner GA, et al. 1999. Genetic enhancement of learning and memory in mice. *Nature* 401:63–69

Thoenen H. 1995. Neurotrophins and neuronal plasticity. *Science* 270:593–98

Trickett PK, McBride-Chang C. 1995. The developmental impact of different forms of child abuse and neglect. *Dev. Rev.* 15:311–37

Uvnas-Moberg K. 1997. The physiological and endocrine effects of social contact. *Ann. NY Acad. Sci.* 807:146–63

Valentino RJ, Curtis AL, Page ME, Pavcovich LA, Florin-Lechner SM. 1998. Activation of the locus cerulus brain noradrenergic system during stress: circuitry, consequences, and regultion. *Adv. Pharmacol.* 42:781–84

van Bockstaele EJ, Colago EEO, Valentino RJ. 1996. Corticotropin-releasing factor-containing axon terminals synapse onto catecholamine dendrites and may presynaptically modulate other afferents in the rostral pole of the nucleus locus coeruleus in the rat brain. *J. Comp. Neurol.* 364:523–34

Viau V, Sharma S, Plotsky PM, Meaney MJ. 1993. The hypothalamic-pituitary-adrenal response to stress in handled and nonhandled rats: differences in stress-induced plasma secretion are not dependent upon increased corticosterone levels. *J. Neurosci.* 13:1097–105

West MJ, King AP. 1987. Settling nature and nurture into an ontogenetic niche. *Dev. Psychol.* 20:549–62

Whishaw IQ. 1998. Place learning in hippocampal rats and the path integration hypothesis. *Neurosci. Biobehav. Rev.* 22:209–20

Whitnall MH. 1993. Regulation of the hypothalamic corticotropin-releasing hormone neurosecretory system. *Prog. Neurobiol.* 40:573–629

Wilson MA. 1996. GABA physiology: modulation by benzodiazepines and hormones. *Crit. Rev. Neurobiol.* 10:1–37

Windle RJ, Shanks N, Lightman SL, Ingram CD. 1997. Central oxytocin administration reduces stress-induced corticosterone release and anxiety behavior in rats. *Endocrinology* 138:2829–34

Wood ER, Dudchenko PA, Eichenbaum H. 1999. The global record of memory in hippocampal neuronal activity. *Nature* 39:561–63

Young LJ, Muns S, Wang Z, Insel TR. 1997. Changes in oxytocin receptor mRNA in rat brain during pregnancy and the effects of estrogen and interleukin-6. *J. Neuroendocrinol.* 9:859–65

Zaharia MD, Shanks N, Meaney MJ, Anisman H. 1996. The effects of postnatal handling on Morris water maze acquisition in different strains of mice. *Psychopharmacology* 128:227–39

Zarrow MX, Campbell PS, Denenberg VH. 1972. Handling in infancy: increased levels of the hypothalamic corticotropin releasing factor (CRF) following exposure to a novel situation. *Proc. Soc. Exp. Biol. Med.* 356:141–43

Zhang LX, Xing GO, Levine S, Post RM, Smith MA. 1997. Maternal deprivation induces neuronal death. *Soc. Neurosci. Abstr.* 23:1113

Annu. Rev. Neurosci. 2001. 24:1193–216

NATURAL IMAGE STATISTICS AND NEURAL REPRESENTATION

Eero P Simoncelli

Howard Hughes Medical Institute, Center for Neural Science, and Courant Institute of Mathematical Sciences, New York University, New York, NY 10003; e-mail: eero.simoncelli@nyu.edu

Bruno A Olshausen

Center for Neuroscience, and Department of Psychology, University of California, Davis, Davis, California 95616; e-mail: baolshausen@ucdavis.edu

Key Words efficient coding, redundancy reduction, independence, visual cortex

■ **Abstract** It has long been assumed that sensory neurons are adapted, through both evolutionary and developmental processes, to the statistical properties of the signals to which they are exposed. Attneave (1954) and Barlow (1961) proposed that information theory could provide a link between environmental statistics and neural responses through the concept of coding efficiency. Recent developments in statistical modeling, along with powerful computational tools, have enabled researchers to study more sophisticated statistical models for visual images, to validate these models empirically against large sets of data, and to begin experimentally testing the efficient coding hypothesis for both individual neurons and populations of neurons.

INTRODUCTION

Understanding the function of neurons and neural systems is a primary goal of systems neuroscience. The evolution and development of such systems is driven by three fundamental components: (*a*) the tasks that the organism must perform, (*b*) the computational capabilities and limitations of neurons (this would include metabolic and wiring constraints), and (*c*) the environment in which the organism lives. Theoretical studies and models of neural processing have been most heavily influenced by the first two. But the recent development of more powerful models of natural environments has led to increased interest in the role of the environment in determining the structure of neural computations.

The use of such ecological constraints is most clearly evident in sensory systems, where it has long been assumed that neurons are adapted, at evolutionary, developmental, and behavioral timescales, to the signals to which they are exposed.

Because not all signals are equally likely, it is natural to assume that perceptual systems should be able to best process those signals that occur most frequently. Thus, it is the statistical properties of the environment that are relevant for sensory processing. Such concepts are fundamental in engineering disciplines: Source coding, estimation, and decision theories all rely heavily on a statistical "prior" model of the environment.

The establishment of a precise quantitative relationship between environmental statistics and neural processing is important for a number of reasons. In addition to providing a framework for understanding the functional properties of neurons, such a relationship can lead to the derivation of new computational models based on environmental statistics. It can also be used in the design of new forms of stochastic experimental protocols and stimuli for probing biological systems. Finally, it can lead to fundamental improvements in the design of devices that interact with human beings.

Despite widespread agreement that neural processing must be influenced by environmental statistics, it has been surprisingly difficult to make the link quantitatively precise. More than 40 years ago, motivated by developments in information theory, Attneave (1954) suggested that the goal of visual perception is to produce an efficient representation of the incoming signal. In a neurobiological context, Barlow (1961) hypothesized that the role of early sensory neurons is to remove statistical redundancy in the sensory input. Variants of this "efficient coding" hypothesis have been formulated by numerous other authors (e.g. Laughlin 1981, Atick 1992, van Hateren 1992, Field 1994, Riecke et al 1995).

But even given such a link, the hypothesis is not fully specified. One needs also to state which environment shapes the system. Quantitatively, this means specification of a probability distribution over the space of input signals. Because this is a difficult problem in its own right, many authors base their studies on empirical statistics computed from a large set of example images that are representative of the relevant environment. In addition, one must specify a timescale over which the environment should shape the system. Finally, one needs to state which neurons are meant to satisfy the efficiency criterion, and how their responses are to be interpreted.

There are two basic methodologies for testing and refining such hypotheses of sensory processing. The more direct approach is to examine the statistical properties of neural responses under natural stimulation conditions (e.g. Laughlin 1981, Rieke et al 1995, Dan et al 1996, Baddeley et al 1998, Vinje & Gallant 2000). An alternative approach is to "derive" a model for early sensory processing (e.g. Sanger 1989, Foldiak 1990, Atick 1992, Olshausen & Field 1996, Bell & Sejnowski 1997, van Hateren & van der Schaaf 1998, Simoncelli & Schwartz 1999). In such an approach, one examines the statistical properties of environmental signals and shows that a transformation derived according to some statistical optimization criterion provides a good description of the response properties of a set of sensory neurons. In the following sections, we review the basic conceptual framework for linking environmental statistics to neural processing, and we discuss a series of examples in which authors have used one of the two approaches described above to provide evidence for such links.

BASIC CONCEPTS

The theory of information was a fundamental development of the twentieth century. Shannon (1948) developed the theory in order to quantify and solve problems in the transmission signals over communication channels. But his formulation of a quantitative measurement of information transcended any specific application, device, or algorithm and has become the foundation for an incredible wealth of scientific knowledge and engineering developments in acquisition, transmission, manipulation, and storage of information. Indeed, it has essentially become a theory for computing with signals.

As such, the theory of information plays a fundamental role in modeling and understanding neural systems. Researchers in neuroscience had been perplexed by the apparent combinatorial explosion in the number of neurons one would need to uniquely represent each visual (or other sensory) pattern that might be encountered. Barlow (1961) recognized the importance of information theory in this context and proposed that an important constraint on neural processing was informational (or coding) efficiency. That is, a group of neurons should encode as much information as possible in order to most effectively utilize the available computing resources. We will make this more precise shortly, but several points are worth mentioning at the outset.

1. The efficiency of the neural code depends both on the transformation that maps the input to the neural responses and on the statistics of the input. In particular, optimal efficiency of the neural responses for one input ensemble does not imply optimality over other input ensembles!

2. The efficient coding principle should not be confused with optimal compression (i.e. rate-distortion theory) or optimal estimation. In particular, it makes no mention of the accuracy with which the signals are represented and does not require that the transformation from input to neural responses be invertible. This may be viewed as either an advantage (because one does not need to incorporate any assumption regarding the form of representation, or the cost of misrepresenting the input) or a limitation (because such costs are clearly relevant for real organisms).

3. The simplistic efficient coding criterion given above makes no mention of noise that may contaminate the input stimulus. Nor does it mention uncertainty or variability in the neural responses to identical stimuli. That is, it assumes that the neural responses are deterministically related to the input signal. If these sources of external and internal noise are small compared with the stimulus and neural response, respectively, then the criterion described is approximately optimal. But a more complete solution should take noise into account, by maximizing the information that the responses provide about the stimulus (technically, the mutual information between stimulus and response). This quantity is generally difficult to measure, but Bialek et al (1991) and Rieke et al (1995) have recently developed approximate techniques for estimating it.

If the efficient coding hypothesis is correct, what behaviors should we expect to see in the response properties of neurons? The answer to this question may be neatly separated into two relevant pieces: the shape of the distributions of individual neural responses and the statistical dependencies between neurons.

Efficient Coding in Single Neurons

Consider the distribution of activity of a single neuron in response to some natural environment.[1] In order to determine whether the information conveyed by this neuron is maximal, we need to impose a constraint on the response values (if they can take on any real value, then the amount of information that can be encoded is unbounded). Suppose, for example, that we assume that the responses are limited to some maximal value, R_{max}. It is fairly straightforward to show that the distribution of responses that conveys maximal information is uniform over the interval $[0, R_{max}]$. That is, an efficient neuron should make equal use of all of its available response levels. The optimal distribution depends critically on the neural response constraint. If one chooses, for example, an alternative constraint in which the variance is fixed, the information-maximizing response distribution is a Gaussian. Similarly, if the mean of the response is fixed, the information-maximizing response distribution is an exponential.[2]

Efficient Coding in Multiple Neurons

If a set of neurons is jointly encoding information about a stimulus, then the efficient coding hypothesis requires that the responses of each individual neuron be optimal, as described above. In addition, the code cannot be efficient if the effort of encoding any particular piece of information is duplicated in more than one neuron. Analogous to the intuition behind the single-response case, the joint responses should make equal use of all possible combinations of response levels. Mathematically, this means that the neural responses must be statistically independent. Such a code is often called a factorial code, because the joint probability distribution of neural responses may be factored into the product of the individual response probability distributions. Independence of a set of neural responses also means that one cannot learn anything about the response of any one neuron by observing the responses of others in the set. In other words, the conditional probability distribution of the response of one neuron given the responses of other neurons should be a fixed distribution (i.e. should not depend on the

---

[1]For the time being, we consider the response to be an instantaneous scalar value. For example, this could be a membrane potential, or an instantaneous firing rate.

[2]More generally, consider a constraint of the form $\varepsilon[\phi(x)] = c$, where x is the response, ϕ is a constraint function, ε indicates the expected or average value over the responses to a given input ensemble, and c is a constant. The maximally informative response distribution [also known as the maximum entropy distribution (Jaynes 1978)] is $\mathcal{P}(x) \propto e^{-\lambda\phi(x)}$, where λ is a constant.

a.   **b.**   **c.**

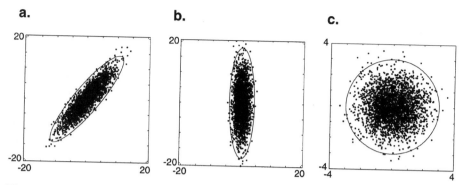

Figure 1: Illustration of principal component analysis on Gaussian-distributed data in two dimensions. (*a*) Original data. Each point corresponds to a sample of data drawn from the source distribution (i.e. a two-pixel image). The ellipse is three standard deviations from the mean in each direction. (*b*) Data rotated to principal component coordinate system. Note that the ellipse is now aligned with the axes of the space. (*c*) Whitened data. When the measurements are represented in this new coordinate system, their components are distributed as uncorrelated (and thus independent) univariate Gaussians.

response levels of the other neurons). The beauty of the independence property is that unlike the result for single neurons, it does not require any auxilliary constraints.

Now consider the problem faced by a "designer" of an optimal sensory system. One wants to decompose input signals into a set of independent responses. The general problem is extremely difficult, because characterizing the joint histogram of the input grows exponentially with the number of dimensions, and thus one typically must restrict the problem by simplifying the description of the input statistics and/or by constraining the form of the decomposition. The most well-known restriction is to consider only linear decompositions, and to consider only the second-order (i.e. covariance or, equivalently, correlation) properties of the input signal. The solution of this problem may be found using an elegant and well-understood technique known as principal components analysis (PCA)[3]. The principal components are a set of orthogonal axes along which the components are decorrelated. Such a set of axes always exists, although it need not be unique. If the data are distributed according to a multi-dimensional Gaussian,[4] then the components of the data as represented in these axes are statistically independent. This is illustrated for a two-dimensional source (e.g. a two-pixel image) in Figure 1.

---

[3]The axes may be computed using standard linear algebraic techniques: They correspond to the eigenvectors of the data covariance matrix.

[4]A multidimensional Gaussian density is simply the extension of the scalar Gaussian density to a vector. Specifically, the density is of the form $\mathcal{P}(\vec{x}) \propto \exp[-\vec{x}^T \Lambda^{-1} \vec{x}/2]$, where Λ is the covariance matrix. All marginal and conditional densities of this density are also Gaussian.

After transforming a data set to the principal component coordinate system, one typically rescales the axes of the space to equalize the variance of each of the components (typically, they are set to one). This rescaling procedure is commonly referred to as "whitening," and is illustrated in Figure 1.

When applying PCA to signals such as images, it is commonly assumed that the statistical properties of the image are translation invariant (also known as stationary). Specifically, one assumes that the correlation of the intensity at two locations in the image depends only on the displacement between the locations, and not on their absolute locations. In this case, the sinusoidal basis functions of the Fourier transform are guaranteed to be a valid set of principal component axes (although, as before, this set need not be unique). The variance along each of these axes is simply the Fourier power spectrum. Whitening may be achieved by computing the Fourier transform, dividing each frequency component by the square root of its variance, and (optionally) computing the inverse Fourier transform. This is further discussed below.

Although PCA can be used to recover a set of statistically independent axes for representing Gaussian data, the technique often fails when the data are non-Gaussian. As a simple illustration, consider data that are drawn from a source that is a linear mixture of two independent non-Gaussian sources (Figure 2). The non-Gaussianity is visually evident in the long tails of data that extend along two oblique axes. Figure 2 also shows the rotation to principal component axes and the whitened data. Note that the axes of the whitened data are not aligned with those of the space. In particular, in the case when the data are a linear mixture of non-Gaussian sources, it can be proven that one needs an additional rotation of the coordinate system to recover the original independent axes.[5] But the appropriate rotation can only be estimated by looking at statistical properties of the data beyond covariance (i.e. of order higher than two).

Over the past decade, a number of researchers have developed techniques for estimating this final rotation matrix (e.g. Cardoso 1989, Jutten & Herault 1991, Comon 1994). Rather than directly optimize the independence of the axis components, these algorithms typically maximize higher-order moments (e.g. the kurtosis, or fourth moment divided by the squared second moment). Such decompositions are typically referred to as independent component analysis (ICA), although this is a bit of a misnomer, as there is no guarantee that the resulting components are independent unless the original source actually was a linear mixture of sources with large higher-order moments (e.g. heavy tails). Nevertheless, one can often use such techniques to recover the linear axes along which the data are most independent.[6] Fortuitously, this approach turns out to be quite successful in the case of images (see below).

---

[5]Linear algebraically, the three operations (rotate-scale-rotate) correspond directly to the singular value decomposition of the mixing matrix.
[6]The problem of blind recovery of independent sources from data remains an active area of research (e.g. Hyvarinen & Oja 1997, Attias 1998, Penev et al 2000).

a.   b.

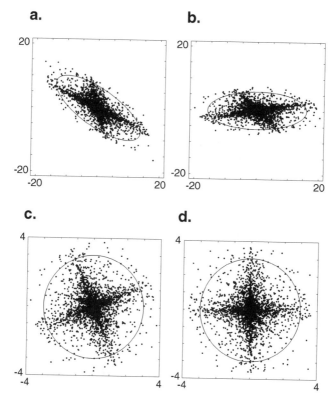

c.   d.

Figure 2   Illustration of principal component analysis and independent component analysis on non-Gaussian data in two dimensions. (*a*) Original data, a linear mixture of two non-Gaussian sources. As in Figure 1, each point corresponds to a sample of data drawn from the source distribution, and the ellipse indicates three standard variations of the data in each direction. (*b*) Data rotated to principal component coordinate system. Note that the ellipse is now aligned with the axes of the space. (*c*) Whitened data. Note that the data are not aligned with the coordinate system. But the covariance ellipse is now a circle, indicating that the second-order statistics can give no further information about preferred axes of the data set. (*d*): Data after final rotation to independent component axes.

IMAGE STATISTICS: CASE STUDIES

Natural images are statistically redundant. Many authors have pointed out that of all the visual images possible, we see only a very small fraction (e.g. Attneave 1954, Field 1987, Daugman 1989, Ruderman & Bialek 1994). Kersten (1987) demonstrated this redundancy perceptually by asking human subjects to replace missing pixels in a four-bit digital image. He then used the percentage of correct guesses to estimate that the perceptual information content of a pixel was approximately 1.4 bits [a similar technique was used by Shannon (1948) to estimate the redundancy of written English]. Modern technology exploits such redundancies every day in order to transmit and store digitized images in compressed formats. In the following sections, we describe a variety of statistical properties of images and their relationship to visual processing.

Intensity Statistics

The simplest statistical image description is the distribution of light intensities in a visual scene. As explained in the previous section, the efficient coding hypothesis predicts that individual neurons should maximize information transmission. In a nice confirmation of this idea, Laughlin (1981) found that the contrast-response function of the large monopolar cell in the fly visual system approximately satisfies the optimal coding criterion. Specifically, he measured the probability distribution of contrasts found in the environment of the fly, and showed that this distribution is approximately transformed to a uniform distribution by the function relating contrast to the membrane potential of the neuron. Baddeley et al (1998) showed that the instantaneous firing rates of spiking neurons in primary and inferior temporal visual cortices of cats and monkeys are exponentially distributed (when visually stimulated with natural scenes), consistent with optimal coding with a constraint on the mean firing rate.

Color Statistics

In addition to its intensity, the light falling on an image at a given location has a spectral (wavelength) distribution. The cones of the human visual system represent this distribution as a three-dimensional quantity. Buchsbaum & Gottshalk (1984) hypothesized that the wavelength spectra experienced in the natural world are well approximated by a three-dimensional subspace that is spanned by cone spectral sensitivities. Maloney (1986) examined the empirical distribution of reflectance functions in the natural world, and showed not only that it was well-represented by a low-dimensional space, but that the problem of surface reflectance estimation was actually aided by filtering with the spectral sensitivities of the cones.

An alternative approach is to assume the cone spectral sensitivities constitute a fixed front-end decomposition of wavelength, and to ask what processing should be performed on their responses. Ruderman et al (1998), building on previous work by Buchsbaum & Gottschalk (1983), examined the statistical properties of log cone responses to a large set of hyperspectral photographic images of foliage. The use of the logarithm was loosely motivated by psychophysical principles (the Weber-Fechner law) and as a symmetrizing operation for the distributions. They found that the principal component axes of the data set lay along directions corresponding to $\{L+M+S, L+M-2S, L-M\}$, where $\{L,M,S\}$ correspond to the log responses of the long, middle, and short wavelength cones. Although the similarity of these axes to the perceptually and physiologically measured "opponent" mechanisms is intriguing, the precise form of the mechanisms depends on the experiment used to measure them (see Lennie & D'Zmura 1988).

a.

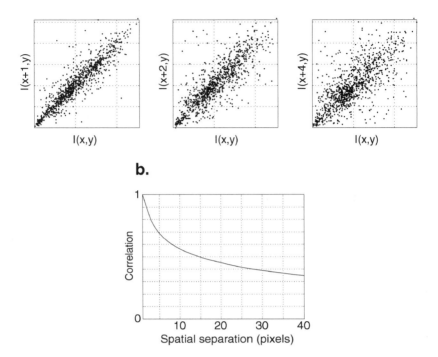

b.

Figure 3 (*a*) Joint distributions of image pixel intensities separated by three different distances. (*b*) Autocorrelation function.

Spatial Correlations

Even from a casual inspection of natural images, one can see that neighboring spatial locations are strongly correlated in intensity. This is demonstrated in Figure 3, which shows scatterplots of pairs of intensity values, separated by three different distances, and averaged over absolute position of several different natural images. The standard measurement for summarizing these dependencies is the autocorrelation function, $C(\Delta x, \Delta y)$, which gives the correlation (average of the product) of the intensity at two locations as a function of relative position. From the examples in Figure 3, one can see that the strength of the correlation falls with distance.[7]

By computing the correlation as a function of relative separation, we are assuming that the spatial statistics in images are translation invariant. As described above,

---

[7]Reinagel & Zador (1999) recorded eye positions of human observers viewing natural images and found that correlation strength falls faster near these positions than generic positions.

the assumption of translation invariance implies that images may be decorrelated by transforming to the frequency (Fourier) domain. The two-dimensional power spectrum can then be reduced to a one-dimensional function of spatial frequency by performing a rotational average within the two-dimensional Fourier plane. Empirically, many authors have found that the spectral power of natural images falls with frequency, f, according to a power law, $1/f^p$, with estimated values for p typically near 2 [see Tolhurst (1992) or Ruderman & Bialek (1994) for reviews]. An example is shown Figure 4.

The environmental causes of this power law behavior have been the subject of considerable speculation and debate. One of the most commonly held beliefs is that it is due to scale invariance of the visual world. Scale invariance means that the statistical properties of images should not change if one changes the scale at which observations are made. In particular, the power spectrum should not change shape under such rescaling. Spatially rescaling the coordinates of an image by a factor of α leads to a rescaling of the corresponding Fourier domain axes by a factor of $1/\alpha$. Only a Fourier spectrum that falls as a power law will retain its shape under this transformation. Another commonly proposed theory is that the $1/f^2$ power spectrum is due to the presence of edges in images, because edges themselves

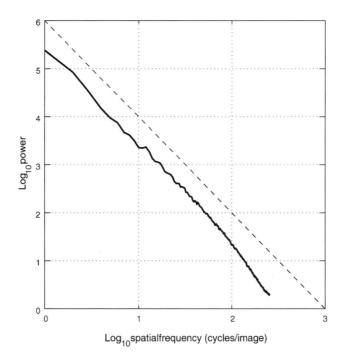

Figure 4   Power spectrum of a natural image (solid line) averaged over all orientations, compared with $1/f^2$ (dashed line).

have a $1/f^2$ power spectrum. Ruderman (1997) and Lee & Mumford (1999) have argued, however, that it is the particular distribution of the sizes and distances of objects in natural images that governs the spectral falloff.

Does the visual system take advantage of the correlational structure of natural images? This issue was first examined quantitatively by Srinivasan et al (1982). They measured the autocorrelation function of natural scenes and then computed the amount of subtractive inhibition that would be required from neighboring photoreceptors in order to effectively cancel out these correlations. They then compared the predicted inhibitory surround fields to those actually measured from first-order interneurons in the compound eye of the fly. The correspondence was surprisingly good and provided the first quantitative evidence for decorrelation in early spatial visual processing.

This type of analysis was carried a step further by Atick & Redlich (1991, 1992), who considered the problem of whitening the power spectrum of natural images (equivalent to decorrelation) in the presence of white photoreceptor noise. They showed that both single-cell physiology and the psychophysically measured contrast sensitivity functions are consistent with the product of a whitening filter and an optimal lowpass filter for noise removal (known as the Wiener filter). Similar predictions and physiological comparisons were made by van Hateren (1992) for the fly visual system. The inclusion of the Wiener filter allows the behavior of the system to change with mean luminance level. Specifically, at lower luminance levels (and thus lower signal-to-noise ratios), the filter becomes more low-pass (intuitively, averaging over larger spatial regions in order to recover the weaker signal). An interesting alternative model for retinal horizontal cells has been proposed by Balboa & Grzywacz (2000). They assume a divisive form of retinal surround inhibition, and show that the changes in effective receptive field size are optimal for representation of intensity edges in the presence of photon-absorption noise.

Higher-Order Statistics

The agreement between the efficient coding hypothesis and neural processing in the retina is encouraging, but what does the efficient coding hypothesis have to say about cortical processing? A number of researchers (e.g. Sanger 1989, Hancock et al 1992, Shonual et al 1997) have used the covariance properties of natural images to derive linear basis functions that are similar to receptive fields found physiologically in primary visual cortex (i.e. oriented band-pass filters). But these required additional constraints, such as spatial locality and/or symmetry, in order to achieve functions approximating cortical receptive fields.

As explained in the introduction, PCA is based only on second-order (covariance) statistics and can fail if the source distribution is non-Gaussian. There are a number of ways to see that the distribution of natural images is non-Gaussian. First, we should be able to draw samples from the distribution of images by generating a set of independent Gaussian Fourier coefficients (i.e. Gaussian white noise), unwhitening these (multiplying by $1/f^2$) and then inverting the Fourier transform.

a.   **b.**

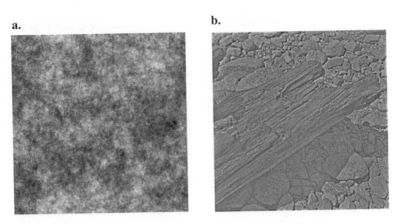

Figure 5   (*a*) Sample of 1/f Gaussian noise; (*b*) whitened natural image.

Such an image is shown in Figure 5*a*. Note that it is devoid of any edges, contours, or many other structures we would expect to find in a natural scene. Second, if it were Gaussian (and translation invariant), then the Fourier transform should decorrelate the distribution, and whitening should yield independent Gaussian coefficients (see Figure 5). But a whitened natural image still contains obvious structures (i.e. lines, edges, contours, etc), as illustrated in Figure 5*b*. Thus, even if correlations have been eliminated by whitening in the retina and lateral geniculate nucleus, there is much work still to be done in efficiently coding natural images.

Field (1987) and Daugman (1989) provided additional direct evidence of the non-Gaussianity of natural images. They noted that the response distributions of oriented bandpass filters (e.g. Gabor filters) had sharp peaks at zero, and much longer tails than a Gaussian density (see Figure 6). Because the density along any axis of a multidimensional Gaussian must also be Gaussian, this constitutes direct

Figure 6 Histogram of responses of a Gabor filter for a natural image, compared with a Gaussian distribution of the same variance.

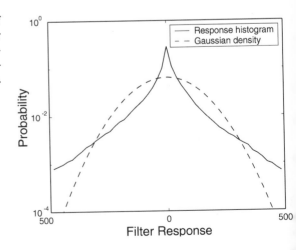

evidence that the overall density cannot be Gaussian. Field (1987) argued that the representation corresponding to these densities, in which most neurons had small amplitude responses, had an important neural coding property, which he termed sparseness. By performing an optimization over the parameters of a Gabor function (spatial-frequency bandwidth and aspect ratio), he showed that the parameters that yield the smallest fraction of significant coefficients are well matched to the range of response properties found among cortical simple cells (i.e. bandwidth of 0.5–1.5 octaves, aspect ratio of 1–2).

Olshausen & Field (1996; 1997) reexamined the relationship between simple-cell receptive fields and sparse coding without imposing a particular functional form on the receptive fields. They created a model of images based on a linear superposition of basis functions and adapted these functions so as to maximize the sparsity of the representation (number of basis functions whose coefficients are zero) while preserving information in the images (by maintaining a bound on the mean squared reconstruction error). The set of functions that emerges after training on hundreds of thousands of image patches randomly extracted from natural scenes, starting from completely random initial conditions, strongly resemble the spatial receptive field properties of simple cells—i.e. they are spatially localized, oriented, and band-pass in different spatial frequency bands (Figure 7). This method may also be recast as a probabilistic model that seeks to explain images in terms of

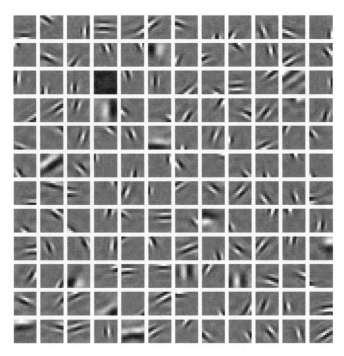

Figure 7   Example basis functions derived using sparseness criterion (see Olshausen & Field 1996).

components that are both sparse and statistically independent (Olshausen & Field 1997) and thus is a member of the broader class of ICA algorithms (see above). Similar results have been obtained using other forms of ICA (Bell & Sejnowski 1997, van Hateren & van der Schaaf 1998, Lewicki & Olshausen 1999), and Hyvärinen & Hoyer (2000) have derived complex cell properties by extending ICA to operate on subspaces. Physiologically Vinje & Gallant (2000) showed that responses of neurons in primary visual cortex were more sparse during presentation of natural scene stimuli.

It should be noted that although these techniques seek statistical independence, the resulting responses are never actually completely independent. The reason is that these models are limited to describing images in terms of linear superposition, but images are not formed as sums of independent components. Consider, for example, the fact that the light coming from different objects is often combined according to the rules of occlusion (rather than addition) in the image formation process. Analysis of the form of these statistical relationships reveals nonlinear dependencies across space as well as across scale and orientation (Wegmann & Zetzche 1990, Simoncelli 1997, Simoncelli & Schwartz 1999).

Consider the joint histograms formed from the responses of two nonoverlapping linear receptive fields, as shown in Figure 8a. The histogram clearly indicates that the data are aligned with the axes, as in the independent components decomposition described above. But one cannot determine from this picture whether the responses are independent. Consider instead the conditional histogram of Figure 8b. Each column gives the probability distribution of the ordinate variable r_2, assuming the corresponding value for the abscissa variable, r_1. That is, the data are the

a.   **b.**

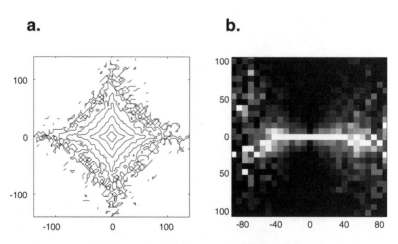

Figure 8   (a) Joint histogram of responses of two nonoverlapping receptive fields, depicted as a contour plot. (b) Conditional histogram of the same data. Brightness corresponds to probability, except that each column has been independently rescaled to fill the full range of display intensities (see Buccigrossi & Simoncelli 1999, Simoncelli & Schwartz 1999).

same as those in Figure 8*a*, except that each column has been independently normalized. The conditional histogram illustrates several important aspects of the relationship between the two responses. First, they are (approximately) decorrelated: The best-fitting regression line through the data is a zero-slope line through the origin. But they are clearly not independent, because the variance of r_2 exhibits a strong dependence on the value of r_1. Thus, although r_2 and r_1 are uncorrelated, they are still statistically dependent. Furthermore, this dependency cannot be eliminated through further linear transformation.

Simoncelli & Schwartz (1999) showed that these dependencies may be eliminated using a nonlinear form of processing, in which the linear response of each basis function is rectified (and typically squared) and then divided by a weighted sum of the rectified responses of neighboring neurons. Similar "divisive normalization" models have been used by a number of authors to account for nonlinear behaviors in neurons (Reichhardt & Poggio 1973, Bonds 1989, Geisler & Albrecht 1992, Heeger 1992, Carandini et al 1997). Thus, the type of nonlinearity found in cortical processing is well matched to the non-Gaussian statistics of natural images. Furthermore, the weights used in the computation of the normalization signal may be chosen to maximize the independence of the normalized responses. The resulting model is surprisingly good at accounting for a variety of neurophysiological observations in which responses are suppressed by the presence of nonoptimal stimuli, both within and outside of the classical receptive field (Simoncelli & Schwartz 1999, Wainwright et al 2001). The statistical dependency between oriented filter responses is at least partly due to the prevalence of extended contours in natural images. Geisler et al (2001) examined empirical distributions of the dominant orientations at nearby locations and used them to predict psychophysical performance on a contour detection task. Sigman et al (2001) showed that these distributions are consistent with cocircular oriented elements and related this result to the connectivity of neurons in primary visual cortex.

Space-Time Statistics

A full consideration of image statistics and their relation to coding in the visual system must certainly include time. Images falling on the retina have important temporal structure arising from self-motion of the observer, as well as from the motion of objects in the world. In addition, neurons have important temporal response characteristics, and in many cases it is not clear that these can be cleanly separated from their spatial characteristics. The measurement of spatiotemporal statistics in natural images is much more difficult than for spatial statistics, though, because obtaining realistic time-varying retinal images requires the tracking of eye, head, and body movements while an animal interacts with the world. Nevertheless, a few reasonable approximations allow one to arrive at useful insights.

As with static images, a good starting point for characterizing joint space-time statistics is the autocorrelation function. In this case, the spatio-temporal autocorrelation function $C(\Delta x, \Delta y, \Delta t)$ characterizes the pairwise correlations of image pixels as a function of their relative spatial separation $(\Delta x, \Delta y)$ and temporal separation Δt. Again, assuming spatio-temporal translation invariance, we find that this function is most conveniently characterized in the frequency domain.

The problem of characterizing the spatio-temporal power spectrum was first studied indirectly by van Hateren (1992), who assumed a certain image velocity distribution and a $1/f^2$ spatial power spectrum and inferred from this the joint spatio-temporal spectrum, assuming a $1/f^2$ spatial power spectrum. Based on this inferred power spectrum, van Hateren then computed the optimal neural filter for making the most effective use of the postreceptoral neurons' limited channel capacity (similar to Atick's whitening filter). He showed from this analysis that the optimal neural filter matches remarkably well the temporal response properties of large monopolar cells in different spatial frequency bands. He was also able to extend this analysis to human vision to account for the spatio-temporal contrast sensitivity function (van Hateren 1993).

Dong & Atick (1995a) estimated the spatio-temporal power spectrum of natural images directly by computing the three-dimensional Fourier transform on many short movie segments (each approximately 2–4 seconds in length) and averaging together their power spectra. This was done for an ensemble of commercial films as well as videos made by the authors. Their results, illustrated in Figure 9, show an interesting dependence between spatial and temporal frequency. The slope of the spatial-frequency power spectrum becomes shallower at higher temporal

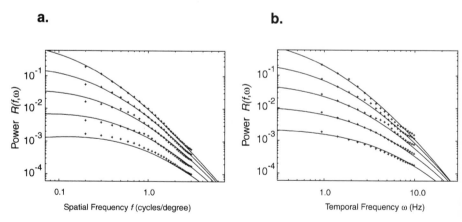

Figure 9 Spatiotemporal power spectrum of natural movies. (*a*) Joint spatiotemporal power spectrum shown as a function of spatial-frequency for different temporal frequencies (1.4, 2.3, 3.8, 6, and 10 Hz, from top to bottom). (*b*) Same data, replotted as a function of temporal frequency for different spatial frequencies (0.3, 0.5, 0.8, 1.3, and 2.1 cy/deg., from top to bottom). Solid lines indicate model fits according to a power-law distribution of object velocities (from Dong & Atick 1995b).

frequencies. The same is true for the temporal-frequency spectrum—i.e. the slope becomes shallower at higher spatial frequencies. Dong & Atick (1995a) showed that this interdependence between spatial and temporal frequency could be explained by assuming a particular distribution of object motions (i.e. a power law distribution), similar in form to van Hateren's assumptions. By again applying the principle of whitening, Dong & Atick (1995b) computed the optimal temporal filter for removing correlations across time and showed that it is closely matched (at low spatial frequencies) to the frequency response functions measured from lateral geniculate neurons in the cat.

The analysis of space-time structure in natural images may also be extended to higher-order statistics (beyond the autocorrelation function), as was previously described for static images. Such an analysis was recently performed by van Hateren & Ruderman (1998) who applied an ICA algorithm to an ensemble of many local image blocks (12×12 pixels by 12 frames in time) extracted from movies. They showed that the components that emerge from this analysis resemble the direction-selective receptive fields of V1 neurons—i.e. they are localized in space and time (within the $12 \times 12 \times 12$ window), spatially oriented, and directionally selective (see Figure 10). In addition, the output signals that result from filtering images with the learned receptive fields have positive kurtosis, which suggests that time-varying natural images may also be efficiently described in terms of a sparse code in which relatively few neurons are active across both space and time. Lewick & Sejnowski (1999) and Olshausen (2001) have shown that these output signals may be highly sparsified so as to produce brief, punctate events similar to neural spike trains.

Although the match between theory and experiment in the above examples is encouraging, it still does not answer the question of whether or not visual neurons perform as expected when processing natural images. This question was addressed directly by Dan et al (1996) who measured the temporal frequency spectrum of LGN neuron activity in an anaesthetized cat in response to natural movies. Consistent with the concept of whitening, the output power of the cells in response to the movie is fairly flat, as a function of temporal frequency. Conversely, if one plays a movie of Gaussian white noise, in which the input spectrum is flat, the output spectrum from the LGN cells increases linearly with frequency, corresponding to the temporal-frequency response characteristic of the neurons. Thus, LGN neurons do not generically whiten any stimulus, only those exhibiting the same correlational structure as natural images.

DISCUSSION

Although the efficient coding hypothesis was first proposed more than forty years ago, it has only recently been explored quantitatively. On the theoretical front, image models are just beginning to have enough power to make interesting predictions. On the experimental front, technologies for stimulus generation and neural

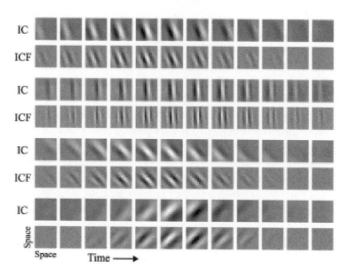

Figure 10 Independent components of natural movies. Shown are four space-time basis functions (rows labeled "IC") with the corresponding analysis functions (rows labeled "ICF"), which would be convolved with a movie to compute a neuron's output (from van Hateren & Ruderman 1998).

recording (especially multiunit recording) have advanced to the point where it is both feasible and practical to test theoretical predictions. Below, we discuss some of the weaknesses and drawbacks of the ideas presented in this review, as well as several exciting new opportunities that arise from our growing knowledge of image statistics.

The most serious weakness of the efficient coding hypothesis is that it ignores the two other primary constraints on the visual system: the implementation and the task. Some authors have successfully blended implementation constraints with environmental constraints (e.g. Baddeley et al 1998). Such constraints are often difficult to specify, but clearly they play important roles throughout the brain. The tasks faced by the organism are likely to be an even more important constraint. In particular, the hypothesis states only that information must be represented efficiently; it does not say anything about what information should be represented. Many authors assume that at the earliest stages of processing (e.g. retina and V1), it is desirable for the system to provide a generic image representation that preserves as much information as possible about the incoming signal. Indeed, the success of efficient coding principles in accounting for response properties of neurons in the retina, LGN, and V1 may be seen as verification of this assumption. Ultimately, however, a richer theoretical framework is required. A commonly proposed example of such a framework is Bayesian decision/estimation theory, which includes both a prior statistical model for the environment and also a loss or reward function that specifies the cost of different errors, or the desirability of different behaviors.

Such concepts have been widely used in perception (e.g. Knill & Richards 1996) and have also been considered for neural representation (e.g. Oram et al 1998).

Another important issue for the efficient coding hypothesis is the timescale over which environmental statistics influence a sensory system. This can range from millenia (evolution), to months (neural development), to minutes or seconds (short-term adaptation). Most of the research discussed in this review assumes the system is fixed, but it seems intuitively sensible that the computations should be matched to various statistical properties on the time scale at which they are relevant. For example, the $1/f^2$ power spectral property is stable and, thus, warrants a solution that is hardwired over evolutionary time scales. On the other hand, several recent results indicate that individual neurons adapt to changes in contrast and spatial scale (Smirnakis et al 1997), orientation (Muller et al 1999), and variance (Brenner et al 2000) on very short time scales. In terms of joint response properties, Barlow & Foldiak (1989) have proposed that short-term adaptation acts to reduce dependencies between neurons, and evidence for this hypothesis has recently been found both psychophysically (e.g. Atick et al 1993, Dong 1995, Webster 1996, Wainwright 1999) and physiologically (e.g. Carandini et al 1998, Dragoi et al 2000, Wainwright et al 2001).

A potential application for efficient coding models, beyond predicting response properties of neurons, lies in generating visual stimuli that adhere to natural image statistics. Historically, visual neurons have been characterized using fairly simple test stimuli (e.g. bars, gratings, or spots) that are simple to parameterize and control, and that are capable of eliciting vigorous responses. But there is no guarantee that the responses measured using such simple test stimuli may be used to predict neural responses to a natural scene. On the other hand, truly naturalistic stimuli are much more difficult to control. An interesting possibility lies in statistical texture modeling, which has been used as a tool for understanding human vision (e.g. Julesz 1962, Bergen & Adelson 1986). Knill et al (1990) and Parraga et al (1999) have shown that human performance on a particular discrimination task is best for textures with natural second-order (i.e. $1/f^2$) statistics, and degraded for images that are less natural. Some recent models for natural texture statistics offer the possibility of generating artificial images that share some of the higher-order statistical structure of natural images (e.g. Heeger & Bergen 1995, Zhu et al 1998, Portilla & Simoncelli 2000).

Most of the models we have discussed in this review can be described in terms of a single-stage neural network. For example, whitening could be implemented by a set of connections between a set of inputs (photoreceptors) and outputs (retinal ganglion cells). Similarly, the sparse coding and ICA models could be implemented by connections between the LGN and cortex. But what comes next? Could we attempt to model the function of neurons in visual areas V2, V4, MT, or MST using multiple stages of efficient coding? In particular, the architecture of visual cortex suggests a hierarchical organization in which neurons become selective to progressively more complex aspects of image structure. In principle, this can allow for the explicit representation of structures, such as curvature, surfaces, or even entire objects (e.g. Dayan et al 1995, Rao & Ballard 1997), thus providing a principled basis for exploring the response properties of neurons in extra-striate cortex.

Although this review has been largely dedicated to findings in the visual domain, other sensory signals are amenable to statistical analysis. For example, Attias & Schreiner (1997) have shown that many natural sounds obey some degree of self-similarity in their power spectra, similar to natural images. In addition, M S Lewicki (personal communication) finds that the independent components of natural sound are similar to the "Gammatone" filters commonly used to model responses of neurons in the auditory nerve. Schwartz & Simoncelli (2001) have shown that divisive normalization of responses of such filters can serve as a nonlinear whitening operation for natural sounds, analogous to the case for vision. In using natural sounds as experimental stimuli, Rieke et al (1995) have shown that neurons at early stages of the frog auditory system are adapted specifically to encode the structure in the natural vocalizations of the animal. Attias & Schreiner (1998) demonstrated that the rate of information transmission in cat auditory midbrain neurons is higher for naturalistic stimuli.

Overall, we feel that recent progress on exploring and testing the relationship between environmental statistics and sensation is encouraging. Results to date have served primarily as post-hoc explanations of neural function, rather than predicting aspects of sensory processing that have not yet been observed. But it is our belief that this line of research will eventually lead to new insights and will serve to guide our thinking in the exploration of higher-level visual areas.

ACKNOWLEDGMENTS

The authors wish to thank Horace Barlow and Matteo Carandini for helpful comments. EPS was supported by an Alfred P. Sloan Research Fellowship, NSF CAREER grant MIP-9796040, the Sloan Center for Theoretical Neurobiology at NYU and the Howard Hughes Medical Institute. BAO was supported by NIMH R29-MH57921.

Visit the Annual Reviews home page at www.AnnualReviews.org

LITERATURE CITED

Atick JJ. 1992. Could information theory provide an ecological theory of sensory processing? *Netw. Comput. Neural Syst.* 3:213–51

Atick JJ, Li Z, Redlich AN. 1993. What does post-adaptation color appearance reveal about cortical color representation? *Vis. Res.* 33(1):123–29

Atick JJ, Redlich AN. 1991. *What does the retina know about natural scenes?* Tech.

Rep. IASSNS-HEP-91/40, Inst. Adv. Study, Princeton, NJ

Atick JJ, Redlich AN. 1992. What does the retina know about natural scenes? *Neural Comput.* 4:196–210

Attias H. 1998. Independent factor analysis. *Neural Comput.* 11:803–51

Attias H, Schreiner CE. 1997. Temporal low-order statistics of natural sounds. In

Advances in Neural Information Processing Systems, ed. MC Mozer, M Jordan, M Kearns, S Solla, 9:27–33. Cambridge, MA: MIT Press

Attias H, Schreiner CE. 1998. Coding of naturalistic stimuli by auditory midbrain neurons. In *Advances in Neural Information Processing Systems*, ed. M Jordan, M Kearns, S Solla, 10:103–9. Cambridge, MA: MIT Press.

Attneave F. 1954. Some informational aspects of visual perception. *Psychol. Rev.* 61:183–93

Baddeley R, Abbott LF, Booth MC, Sengpiel F, Freeman T, et al. 1998. Respones of neurons in primary and inferior temporal visual cortices to natural scenes. *Proc. R. Soc. London Ser. B* 264:1775–83

Balboa RM, Grzywacz NM. 2000. The role of early lateral inhibition: more than maximizing luminance information. *Vis. Res.* 17:77–89

Barlow HB. 1961. Possible principles underlying the transformation of sensory messages. In *Sensory Communication*, ed. WA Rosenblith, pp. 217–34. Cambridge, MA: MIT Press

Barlow HB, Foldiak P. 1989. Adaptation and decorrelation in the cortex. In *The Computing Neuron*, ed. R Durbin, C Miall, G Mitchinson, 4:54–72. New York: Addison-Wellesley

Bell AJ, Sejnowski TJ. 1997. The "independent components" of natural scenes are edge filters. *Vis. Res.* 37(23):3327–38

Bergen JR, Adelson EH. 1986. Visual texture segmentation based on energy measures. *J. Opt. Soc. Am. A* 3:99

Bialek W, Rieke F, de Ruyter van Steveninck RR, Warland D. 1991. Reading a neural code. *Science* 252:1854–57

Bonds AB. 1989. Role of inhibition in the specification of orientation selectivity of cells in the cat striate cortex. *Vis. Neurosci.* 2:41–55

Brenner N, Bialek W, de Ruyter van Steveninck RR. 2000. Adaptive rescaling maximizes information transmission. *Neuron* 26:695–702

Buccigrossi RW, Simoncelli EP. 1999. Image compression via joint statistical characterization in the wavelet domain. *IEEE Trans. Image Proc.* 8(12):1688–701

Buchsbaum G, Gottschalk A. 1983. Trichromacy, opponent color coding, and optimum colour information transmission in the retina. *Proc. R. Soc. London Ser. B* 220:89–113

Buchsbaum G, Gottschalk A. 1984. Chromaticity coordinates of frequency-limited functions. *J. Opt. Soc. Am. A* 1(8):885–87

Carandini M, Heeger DJ, Movshon JA. 1997. Linearity and normalization in simple cells of the macaque primary visual cortex. *J. Neurosci.* 17:8621–44

Carandini M, Movshon JA, Ferster D. 1998. Pattern adaptation and cross-orientation interactions in the primary visual cortex. *Neuropharmacology* 37:501–11

Cardoso JF. 1989. Source separation using higer order moments. In *Int. Conf. Acoustics Speech Signal Proc.*, pp. 2109–12. IEEE Signal Process. Soc.

Common P. 1994. Independent component analysis, a new concept? *Signal Process* 36:387–14

Dan Y, Atick JJ, Reid RC. 1996. Efficient coding of natural scenes in the lateral geniculate nucleus: experimental test of a computational theory. *J. Neurosci.* 16:3351–62

Daugman JG. 1989. Entropy reduction and decorrelation in visual coding by oriented neural receptive fields. *IEEE Trans. Biomed. Eng.* 36(1):107–14

Dayan P, Hinton GE, Neal RM, Zemel RS. 1995. The Helmholtz machine. *Neural Comput.* 7:889–904

Dong DW. 1995. Associative decorrelation dynamics: a theory of self-organization and optimization in feedback networks. In *Advances in Neural Information Processing Systems*, ed. G Tesauro, D Touretzky, T Leen. 7:925–32

Dong DW, Atick JJ. 1995a. Statistics of natural time-varying images. *Netw. Comput. Neural Syst.* 6:345–58

Dong DW, Atick JJ. 1995b. Temporal decorrelation: a theory of lagged and nonlagged responses in the lateral geniculate nucleus. *Netw. Comput. Neural Syst.* 6:159–78

Dragoi V, Sharma J, Sur M. 2000. Adaptation-induced plasticity of orientation tuning in adult visual cortex. *Neuron* 28:287–88

Field DJ. 1987. Relations between the statistics of natural images and the response properties of cortical cells. *J. Opt. Soc. Am. A* 4(12):2379–94

Field DJ. 1994. What is the goal of sensory coding? *Neural Comput.* 6:559–601

Foldiak P. 1990. Forming sparse representations by local anti-hebbian learning. *Biol. Cybernet.* 64:165–70

Geisler WS, Albrecht DG. 1992. Cortical neurons: isolation of contrast gain control. *Vis. Res.* 8:1409–10

Geisler WS, Perry JS, Super BJ, Gallogly DP. 2001. Edge co-occurance in natural images predicts contour grouping performance. *Vis. Res.* 41:711–24

Hancock PJB, Baddeley RJ, Smith LS. 1992. The principal components of natural images. *Network* 3:61–72

Heeger D, Bergen J. 1995. Pyramid-based texture analysis/synthesis. In *Proc. Assoc. Comput. Mach. Special Interest Groups Graph*, pp. 229–38

Heeger DJ. 1992. Normalization of cell responses in cat striate cortex. *Vis. Neurosci.* 9:181–98

Hyvärinen A, Hoyer P. 2000. Emergence of topography and complex cell properties from natural images using extensions of ica. In *Advances in Neural Information Processing Systems*, ed. SA Solla, TK Leen, K-R Müller, 12:827–33, Cambridge, MA: MIT Press

Hyvärinen A, Oja E. 1997. A fast fixed-point algorithm for independent component analysis. *Neural Comput.* 9:1483–92

Jaynes ET. 1978. Where do we stand on maximum entropy? In *The Maximal Entropy Formalism*, ed. RD Levine, M Tribus, pp. 620–30. Cambridge, MA: MIT Press

Julesz B. 1962. Visual pattern discrimination. *IRE Trans. Inf. Theory*, IT-8

Jutten C, Herault J. 1991. Blind separation of sources. Part I: An adaptive algorithm based on neuromimetic architecture. *Signal Process* 24(1):1–10

Kersten D. 1987. Predictability and redundancy of natural images. *J. Opt. Soc. Am. A* 4(12):2395–400

Knill DC, Field D, Kersten D. 1990. Human discrimination of fractal images. *J. Opt. Soc. Am. A* 7:1113–23

Knill DC, Richards W, eds. 1996. *Perception as Bayesian Inference.* Cambridge, UK: Cambridge Univ. Press

Laughlin SB. 1981. A simple coding procedure enhances a neuron's information capacity. *Z. Naturforsch.* 36C:910–12

Lee AB, Mumford D. 1999. An occlusion model generating scale-invariant images. In *IEEE Workshop on Statistical and Computational Theories of Vision*, Fort Collins, CO. Also at http://www.cis.ohiostate.edu/~szhu/SCTV99.html

Lennie P, D'Zmura M. 1988. Mechanisms of color vision. *CRC Crit. Rev. Neurobiol.* 3:333–400

Lewicki MS, Olshausen BA. 1999. Probabilistic framework for the adaptation and comparison of image codes. *J. Opt. Soc. Am. A* 16(7):1587–601

Lewicki M, Sejnowski T. 1999. Coding time-varying signals using sparse, shift-invariant representations. In *Advances in Neural Information Processing Systems*, ed. MS Kearns, SA Solla, DA Cohn, 11:815–21. Cambridge, MA: MIT Press

Maloney LT. 1986. Evaluation of linear models of surface spectral reflectance with small numbers of parameters. *J. Opt. Soc. Am. A* 3(10):1673–83

Müller JR, Metha AB, Krauskopf J, Lennie P. 1999. Rapid adaptation in visual cortex to the structure of images. *Science* 285:1405–8

Olshausen BA. 2001. Sparse codes and spikes. In *Statistical Theories of the Brain*, ed. R Rao, B Olshausen, M Lewicki. Cambridge, MA: MIT Press. In press

Olshausen BA, Field DJ. 1996. Emergence of simple-cell receptive field properties by learning a sparse code for natural images. *Nature* 381:607–9

Olshausen BA, Field DJ. 1997. Sparse coding with an overcomplete basis set: a strategy employed by V1? *Vis. Res.* 37:3311–25

Oram MW, Foldiak P, Perrett DI, Sengpiel F. 1998. The "ideal homunculus": decoding neural population signals. *Trends Neurosci.* 21(6):259–65

Parraga CA, Troscianko T, Tolhurst DJ. 2000. The human visual system is optimised for processing the spatial information in natural visual images. *Curr. Biol.* 10:35–38

Penev P, Gegiu M, Kaplan E. 2000. Fast convergent factorial learning of the low-dimensional independent manifolds in optical imaging data. In *Proc. 2nd Int. Workshop Indep. Comp. Anal. Signal Separation*, pp. 133–38. Helsinki, Finland

Portilla J, Simoncelli EP. 2000. A parametric texture model based on joint statistics of complex wavelet coefficients. *Int. J. Comput. Vis.* 40(1):49–71

Rao RPN, Ballard DH. 1997. Dynamic model of visual recognition predicts neural response properties in the visual cortex. *Neural Comput.* 9:721–63

Reichhardt W, Poggio T. 1979. Figure-ground discrimination by relative movement in the visual system of the fly. *Biol. Cybernet.* 35:81–100

Reinagel P, Zador AM. 1999. Natural scene statistics at the centre of gaze. *Netw. Comput. Neural Syst.* 10:341–50

Rieke F, Bodnar DA, Bialek W. 1995. Naturalistic stimuli increase the rate and efficiency of information transmission by primary auditory afferents. *Proc. R. Soc. London B* 262:259–65

Ruderman DL. 1997. Origins of scaling in natural images. *Vis. Res.* 37:3385–98

Ruderman DL, Bialek W. 1994. Statistics of natural images: scaling in the woods. *Phys. Rev. Lett.* 73(6):814–17

Ruderman DL, Cronin TW, Chiao CC. 1998. Statistics of cone responses to natural images: implications for visual coding. *J. Opt. Soc. Am. A* 15(8):2036–45

Sanger TD. 1989. Optimal unsupervised learning in a single-layer network. *Neural Netw.* 2:459–73

Schwartz O, Simoncelli E. 2001. Natural sound statistics and divisive normalization in the auditory system. In *Advances in Neural Information Processing Systems*, ed. TK Leen, TG Dietterich, V Tresp, Vol. 13. Cambridge, MA: MIT Press. In Press

Shannon C. 1948. The mathematical theory of communication. *Bell Syst. Tech. J.* 27:379–423

Shouval H, Intrator N, Cooper LN. 1997. BCM Network develops orientation selectivity and ocular dominance in natural scene environment. *Vis. Res.* 37(23):3339–42

Sigman M, Cecchi GA, Gilbert CD, Magnasco MO. 2001. On a common circle: natural scenes and gestalt rules. *Proc. Natl. Acad. Sci.* 98(4):1935–40

Simoncelli EP. 1997. *Statistical Models for Images: Compression, Restoration and Synthesis.* Asilomar Conf. Signals, Systems, Comput. 673–78. Los Alamitos, CA: IEEE Comput. Soc. http://www.cns.nyu.edu/~eero/publications.html

Simoncelli EP, Schwartz O. 1999. Image statistics and cortical normalization models. In *Advances in Neural Information Processing Systems*, ed. MS Kearns, SA Solla, DA Cohn. 11:153–59

Smirnakis SM, Berry MJ, Warland DK, Bialek W, Meister M. 1997. Adaptation of retinal processing to image contrast and spatial scale. *Nature* 386:69–73

Srinivasan MV, Laughlin SB, Dubs A. 1982. Predictive coding: A fresh view of inhibition in the retina. *J. R. Soc. London Ser. B* 216:427–59

van Hateren JH. 1992. A theory of maximizing sensory information. *Biol. Cybern.* 68:23–29

van Hateren JH. 1993. Spatiotemporal contrast sensitivity of early vision. *Vis. Res.* 33:257–67 van Hateren JH, van der Schaaf A. 1998. Independent component filters of natural images compared with simple cells in primary visual cortex. *Proc. R. Soc. London Ser. B* 265:359–66

Vinje WE, Gallant JL. 2000. Sparse coding and decorrelation in primary visual cortex during natural vision. *Science* 287:1273–76

Wainwright MJ. 1999. Visual adaptation as optimal information transmission. *Vis. Res.* 39:3960–74

Wainwright MJ, Schwartz O, Simoncelli EP. 2001. Natural image statistics and divisive normalization: modeling nonlinearity and adaptation in cortical neurons. In *Statistical Theories of the Brain*, ed. R Rao, B Olshausen, M Lewicki. Cambridge, MA: MIT Press. In press

Webster MA. 1996. Human colour perception and its adaptation. *Netw. Comput. Neural Syst.* 7:587–634

Wegmann B, Zetzsche C. 1990. Statistical dependence between orientation filter outputs used in an human vision based image code. In *Proc. SPIE Vis. Commun. Image Processing*, 1360:909–22. Lausanne, Switzerland: Soc. Photo-Opt. Instrum. Eng.

Zhu SC, Wu YN, Mumford D. 1998. FRAME: Filters, random fields and maximum entropy—towards a unified theory for texture modeling. *Int. J. Comp. Vis.* 27(2):1–20

Annu. Rev. Neurosci. 2001. 24:1217–281

Nerve Growth Factor Signaling, Neuroprotection, and Neural Repair

Michael V Sofroniew

Department of Neurobiology and Brain Research Institute, University of California Los Angeles, Los Angeles, California 90095-1763; e-mail: sofroniew@mednet.ucla.edu

Charles L Howe

Department of Neurology and Neurological Sciences, Stanford University, Stanford, California 94305-5489; e-mail: c.howe@stanford.edu

William C Mobley

Department of Neurology and Neurological Sciences, Stanford University, Stanford, California 94305; e-mail: ngfv1@leland.stanford.edu

Key Words neurotrophins, NGF, TrkA, $p75^{NTR}$, neurodegeneration, neuroregeneration, excitotoxicity, tyrosine kinase

■ **Abstract** Nerve growth factor (NGF) was discovered 50 years ago as a molecule that promoted the survival and differentiation of sensory and sympathetic neurons. Its roles in neural development have been characterized extensively, but recent findings point to an unexpected diversity of NGF actions and indicate that developmental effects are only one aspect of the biology of NGF. This article considers expanded roles for NGF that are associated with the dynamically regulated production of NGF and its receptors that begins in development, extends throughout adult life and aging, and involves a surprising variety of neurons, glia, and nonneural cells. Particular attention is given to a growing body of evidence that suggests that among other roles, endogenous NGF signaling subserves neuroprotective and repair functions. The analysis points to many interesting unanswered questions and to the potential for continuing research on NGF to substantially enhance our understanding of the mechanisms and treatment of neurological disorders.

INTRODUCTION

In mammals and other vertebrates, soluble peptide growth factors play essential roles in intercellular communication. They exert their effects by signaling through surface membrane receptors that interact with diverse types of intracellular second-messenger systems. In a sometimes surprising manner, many growth factors have been found to subserve a wide variety of functions by acting on many cell types at different stages of development or in adult life.

0147-006X/01/0621-1217$14.00

Nerve growth factor (NGF) was discovered 50 years ago as a molecule that regulates the survival and maturation of developing neurons in the peripheral nervous system (PNS) (Levi-Montalcini & Hamburger 1951, 1953), and ideas about the biological role of NGF have been dominated by concepts that arose from studies on the differentiation and survival of young neurons. Until recently, the expectation was that the biology of NGF would center around the classical target-derived neurotrophic factor paradigm in which NGF released by postsynaptic targets acts on presynaptic neurons to build or maintain functional contacts and enhance the function of well-defined neural circuits. Although this paradigm undoubtedly plays a critical role in both the PNS and central nervous system (CNS), it does not appear to be the sole role for NGF actions. With the availability of tools that allow sensitive and specific measurements of mRNA and protein levels for NGF and its receptors, it has become apparent that NGF actions extend beyond the developmental period, beyond nerve cells, and even beyond the nervous system. Indeed, NGF and its receptors are produced throughout adult life and during aging by many different cell types. The dynamically regulated expression of NGF and its receptors throughout adult life suggests multiple functions for NGF signaling, many of which are poorly understood. NGF and NGF receptor expression can be upregulated during the response to injury in both the PNS and CNS, and a growing body of evidence suggests that among other roles, endogenous NGF signaling through both neurons and nonneuronal cells subserves neuroprotective functions and facilitates neural repair.

One of the major advances of molecular neuroscience in the past 25 years has been to recognize that much of the cellular damage resulting from such CNS insults as stroke, trauma, and degenerative disease may be caused by a limited number of endogenously generated molecules with neurotoxic activities. Less well developed is the idea that endogenous mechanisms exist to provide neuroprotection, and that endogenous molecules may be produced specifically to subserve neuroprotective signaling functions (Mattson 1997). For NGF to be viewed as a specifically expressed, neuroprotective molecule with widespread activity in the CNS, several criteria must be fulfilled: (*a*) NGF and NGF receptor expression must occur in cellular compartments where it could influence the neural response to injury; (*b*) NGF signaling should be able to influence cellular events involved in the response to insults and injury; (*c*) NGF should exert protective effects; and (*d*) failure of NGF signaling should be associated with increased degeneration and vulnerability to injury. In this review, we consider evidence supporting these criteria and conclude that NGF does play a role in endogenous neuroprotection.

STRUCTURE, EXPRESSION, AND REGULATION OF NGF AND ITS RECEPTORS

The NGF gene is located on human chromosome 1 and is expressed as two major splice variants (Edwards et al 1986, 1988). The mature, fully processed form of biologically active NGF appears to be similar in all tissues and consists of a dimer of

13-kDa polypeptide chains, each of which has three intrachain disulfide bridges. The crystal structure of NGF has been resolved (McDonald et al 1991). The NGF dimer has an elongated shape with a core, or "waist," that is formed by twisted beta sheets; the molecule also features a cysteine-knot motif, a reverse turn at one end (loop 3) and three beta-hairpin loops at the other (loops 1, 2, and 3). The amino terminus of NGF is not defined in the crystal structure. An octapeptide derived from the NGF amino terminus has potent bradykinin-like activity (Taiwo et al 1991) and is normally produced in the mouse submandibular gland in response to stress, but whether it is found under physiological conditions in other tissues is unknown (Fahnestock et al 1991). NGF is part of the neurotrophin family of molecules, which share a high degree of structural homology and includes brain-derived neurotrophic factors (BDNF), neurotrophin-3 (NT-3), and neurotrophin-4 (NT-4) (Butte et al 1998; Ibanez 1994; Robinson et al 1995, 1999). Neurotrophins are found in both mammals and lower vertebrates, and the neurotrophin homologues NT-6 and NT-7 were recently cloned in fish (Gotz et al 1994, Lai et al 1998).

NGF has two known receptors, TrkA and $p75^{NTR}$ (Bothwell 1995, Kaplan & Miller 1997). TrkA is a single-pass transmembrane protein that serves as a receptor tyrosine kinase (RTK) for NGF. NGF signaling through TrkA elicits many of the classical neurotrophic actions ascribed to NGF (Loeb et al 1991). TrkA is a member of the Trk gene family, which includes TrkB, the receptor for BDNF and NT-4, and TrkC, the receptor for NT-3 (Kaplan & Miller 1997). NGF activates only TrkA; NT-3 activates TrkA but only does so at much higher concentrations than does NGF. Two isoforms for TrkA exist that differ in their extracellular domain through the inclusion of six additional amino acids near the transmembrane domain of one of the variants ($TrkA_{II}$). Inclusion of the insert appears to relax the specificity of TrkA activation; NT-3 mediated signaling is markedly enhanced through this receptor isoform (Clary & Reichardt 1994). $p75^{NTR}$ is a transmembrane glycoprotein that binds all members of the neurotrophin family with approximately equal nanomolar affinity. $p75^{NTR}$ regulates signaling through TrkA; in addition, as discussed below, NGF binding to $p75^{NTR}$ activates signaling pathways that are characteristic for this receptor (Casaccia-Bonnefil et al 1999; Dobrowsky et al 1994, 1995; Friedman & Greene 1999).

Recent findings for the three-dimensional structure of NGF bound to its TrkA receptor provide a structural explanation for many of the results provided by mutagenesis studies (Wiesmann et al 1999). They show that NGF engages the TrkA second immunoglobulin (Ig)-like domain through two distinct patches (Wiesmann et al 1999). The first patch involves the four beta sheets that form the "waist" of the NGF molecule together with the first loop (residues 29–33); it includes NGF domains that show considerable homology with the other neurotrophins (Wiesmann et al 1999). It is likely that NGF and its neurotrophin family members engage each of their Trk receptors through this patch. The second patch is formed by the amino terminus of NGF, which in the NGF-TrkA structure is well defined (Wiesmann et al 1999). The lack of homology of the NGF amino terminus with that of other neurotrophins suggests that the second patch serves to specify NGF binding to TrkA. As yet there is no three-dimensional structure for NGF binding to $p75^{NTR}$.

Mutagenesis studies for NGF binding to p75^{NTR} point to the importance of mostly different domains (i.e. the first, third, and fourth loops and the carboxy-terminus) (Ibanez et al 1992, Ryden & Ibanez 1997, Urfer et al 1994) than those identified for binding to TrkA. The findings suggest that NGF could bind to both TrkA and p75^{NTR} simultaneously (Wiesmann et al 1999).

Both NGF and its receptors are produced during development, adult life, and aging by many cell types in the CNS and PNS, immune and inflammatory system, and various tissues. Given the wide range of neuronal and nonneuronal cells that have the potential to produce and/or respond to NGF, clues to the different functions that might be played by NGF signaling have been obtained by examining the expression of NGF and its receptors. During development, expression of NGF by target cells is compatible with its role as a survival and maturation factor for afferent neurons. In addition, as discussed in this section, a large body of evidence demonstrates that in response to numerous stimuli there is dynamic regulation of NGF and NGF receptor expression. It is interesting that NGF and/or its receptors are markedly upregulated by many cell types after tissue injury or insult. Documenting the patterns for NGF and NGF receptor gene expression in specific cells and tissues is required for documenting the plurality of NGF actions and for interpreting their physiological significance.

Peripheral Nervous System and Peripheral Tissues

NGF Receptor Expressing Cells   Sympathetic neurons and small diameter peripheral sensory neurons that mediate nociception, the first identified NGF-responsive neurons, express both TrkA and p75^{NTR} during development and in the adult (Ruit et al 1990, Verge et al 1989). Most, if not all, α-motor neurons, whose cell bodies reside in the CNS and send projections through peripheral nerve to muscle targets, transiently express p75^{NTR} during the phase of axon elongation that occurs in development; expression is downregulated to undetectable levels in adults but returns after peripheral nerve injury (Ernfors et al 1989, Wood et al 1990). Among PNS glial cells, Schwann cells in peripheral nerve express p75^{NTR} during development. In the normal adult, p75^{NTR} expression is reduced to levels that are only one percent of those seen during development (Heumann et al 1987b). Schwann cells markedly upregulate p75^{NTR} in response to the loss of contact with axons that follows axotomy, to local tissue injury, or if stimulated with inflammatory cytokines (Heumann et al 1987b, Lemke & Chao 1988, Mirsky & Jessen 1999, Taniuchi et al 1988). Expression patterns for NGF receptors in the PNS suggest that distinct functions are carried out during development, normal adult life, and following injury.

NGF-Producing Cells   Nonneuronal target cells of sympathetic and sensory neurons throughout the body produce NGF during development. These include targets in the skin (e.g. keratinocytes and melanocytes), vascular and other smooth muscle cells, and various endocrine tissues, such as the testis and ovary, pituitary, thyroid and parathyroid, and exocrine salivary (e.g. submandibular) glands. Most of these cells continue to produce NGF during adult life and modulate NGF production in response to stimuli (reviewed by Levi-Montalcini et al 1995, 1996). In some tissues, including skin and viscera such as the bladder, experimental evidence suggests that NGF production is markedly upregulated after injury or in response to tissue inflammation or injury, but the NGF-producing cell types have not yet been characterized (Dmitrieva et al 1997, McMahon et al 1995, Mendell et al 1999). Among PNS glia, immature Schwann cells and satellite cells produce NGF during development (Mirsky & Jessen 1999). In adults, mature myelinating Schwann cells downregulate NGF expression to undetectable levels, but after nerve injury, reactive and dedifferentiated Schwann cells markedly upregulate NGF production in vivo; in vitro, NGF expression by Schwann cells is upregulated by cytokines and other inflammatory mediators (Lindholm et al 1987, Mirsky & Jessen 1999). As for its receptors, the patterns for NGF expression suggest roles that extend beyond development and beyond its classical role as a target-derived neurotrophic factor.

Central Nervous System

NGF Receptor Expressing Cells $p75^{NTR}$ gene expression in the CNS is widespread, especially during development. In addition to both major populations of forebrain cholinergic neurons, $p75^{NTR}$ mRNA and protein are found in a number of developing neuronal populations in both the brain and brainstem (Longo et al 1993). $p75^{NTR}$ expression is more restricted in the adult, and several populations, including cholinergic neurons of the caudate-putamen and cranial nerve nuclei of the brainstem, show markedly reduced or no expression (Koh & Higgins 1991). Cerebellar Purkinje neurons, hippocampal pyramidal neurons, and retinal ganglion neurons also downregulate expression to undetectable levels in adults but reexpress $p75^{NTR}$ after injury (Brann et al 1999, Eckenstein 1988, Martínez-Murillo et al 1998, Yamashita et al 1999b). The majority of $p75^{NTR}$-expressing neurons do not also express TrkA, but developing horizontal cells and amacrine cells of the retina express TrkA and potentially $p75^{NTR}$ (Karlsson et al 1998), whereas cholinergic neurons of the septal-basal forebrain complex express both TrkA and $p75^{NTR}$ during development and throughout adult life (Holtzman et al 1992). It is interesting that expression of TrkA, but not of $p75^{NTR}$, in these neurons is significantly decreased in aged animals (Cooper et al 1994, Hasenöhrl et al 1997) and is particularly reduced in aged patients with Alzheimer's disease (Mufson et al 1997). Expression of both TrkA and $p75^{NTR}$ in forebrain neurons is upregulated by NGF (Gage et al 1989, Holtzman et al 1992). Adult cholinergic neurons of the extended striatal complex (caudate, putamen, accumbens, etc) express only TrkA; however, $p75^{NTR}$ is upregulated to detectable levels, and TrkA expression is increased by local tissue injury or NGF infusions (Gage et al 1989, Holtzman et al 1995). Adult neurons that express TrkA, but not $p75^{NTR}$, are found in the thalamic paraventricular nuclei, rostral and intermediate subnuclei of the interpeduncular nucleus, and various other brain regions (Holtzman et al 1995, Venero et al 1994), and also in the spinal cord in regions associated with regulation of the autonomic outflow (Michael et al 1997). TrkA mRNA has been detected in CNS regions where its cellular localization is yet to be established. Some hippocampal pyramidal neurons may also express very low levels of TrkA (Cellerino 1995), and a recent immunocytochemical study points to the presence of TrkA and p75^{NTR} proteins in pyramidal cells of the somatosensory cortex of the mature rat (Pitts & Miller 2000). If confirmed, these results would contribute significantly to our understanding of NGF production and actions in the CNS. As detection methods increase in sensitivity, it is likely that other NGF receptor-expressing neurons will be identified in the CNS.

Among glial cells, light microscopic studies show that CNS astrocytes in vivo rarely stain for p75^{NTR} (P Belichenko & WC Mobley, unpublished observations). However, as many as one fifth of astrocytes in the dentate gyrus were immunoreactive for p75^{NTR} in a recent immuno-EM study (Dougherty & Milner 1999). This result suggests that very low levels of p75^{NTR} are present in many mature astrocytes. p75^{NTR} and, more controversially, TrkA are also expressed by astrocytes in vitro, particularly after exposure to NGF or inflammatory cytokines (Hutton et al 1992, Hutton & Perez-Polo 1995, Kumar et al 1993, Semkova & Krieglstein 1999). A detailed analysis of NGF receptor expression by reactive astrocytes after CNS injury would provide information for detailing the actions of neurotrophins in the CNS. Astrocytes are not alone in expressing NGF receptors. Oligodendrocytes express p75^{NTR} (Casaccia-Bonnefil et al 1996, Kumar et al 1993). Microglia have the capacity to express p75^{NTR} and TrkA, and expression levels are modulated by inflammatory stimuli, such as cytokines and bacterial lipopolysaccharide (Elkabes et al 1998). The diversity of NGF receptor expression in the CNS is at least as great as that in the PNS and suggests that NGF signaling mediates many different functions.

NGF-Producing Cells   NGF is produced in the CNS during development and throughout adult life. NGF-producing cells are present in the cortical target regions of basal forebrain cholinergic neurons. Most such cells are neurons, including pyramidal neurons, though glial cells are occasionally found to contain NGF (Pitts & Miller 2000). In the hippocampal formation, pyramidal and dentate granule neurons express NGF, as do subpopulations of GABAergic interneurons (French et al 1999, Gall & Isackson 1989, Pascual et al 1998). These neurons also serve as targets of cholinergic innervation. In striatum, NGF is produced by a subpopulation of small interneurons (Bizon et al 1999). NGF expression in hippocampus is regulated by neuronal activity; increases are caused by glutamatergic and cholinergic neurotransmission, and decreases are caused by GABAergic neurotransmission (Berzaghi et al 1993, Knipper et al 1994, French et al 1999). Neuronal NGF expression in vivo is markedly upregulated by seizures, forebrain ischemia, marked hypoglycemia, and tissue injury (Gall & Isackson 1989, Lindvall et al 1994, Zafra et al 1991). Studies in vivo and in vitro indicate that cerebral insults influence NGF gene expression via excitatory amino acid neurotransmission as well as through other pathways (Lindvall et al 1994).

Among glial cells, NGF is produced throughout the CNS by astrocytes and microglia, and NGF expression in both cell types is markedly upregulated by local tissue injury, inflammation, cytokines, and bacterial lipopolysaccharide (both in vivo and in vitro) (Arendt et al 1995, Elkabes et al 1996, Heese et al 1998, Micera et al 1998, Yoshida & Gage 1992). In astrocytes, NGF expression is also upregulated by fibroblast growth factor, interleukin-1, glutamate agonists, reactive oxygen species, high potassium, ischemia, and traumatic brain injury (Abiru et al 1998; Friedman et al 1996; Goss et al 1998; Gottlieb & Matute 1999; Pechan et al 1992, 1993; Strauss et al 1968; Yoshida & Gage 1991). The data for NGF expression in the uninjured brain are largely consistent with a role for NGF in target-derived trophic support. Increased NGF levels in the injured CNS suggest that astrocytes and microglial cells could serve as local sources of NGF for injured neurons and other NGF responsive cell types.

Immune and Inflammatory System

In recent years, a great deal of interest has focused on NGF and NGF receptor gene expression in cells of the immune and inflammatory system. Several types of bone marrow-derived leukocytes have the capacity to express TrkA, including mast cells, CD4+ T lymphocytes, B lymphocytes, monocytes, and macrophages; follicular dendritic cells and B lymphocytes express p75[NTR] (Labouyrie et al 1997, Levi-Montalcini et al 1996, Torcia et al 1996). Many of the same types of leukocytes also have the capacity to express NGF. These include mast cells, monocytes and macrophages, T lymphocytes (CD3+ and CD4+ T cells), and B lymphocytes (Lambiase et al 1997, Leon et al 1994, Levi-Montalcini et al 1996, Mizuma et al 1999, Torcia et al 1996). Both NGF and NGF receptor expression are dynamically regulated in leukocytes such that expression is increased by inflammatory and other stimuli as well as in activated cells (Barouch et al 2000, Lambiase et al 1997, Levi-Montalcini et al 1996, Mizuma et al 1999, Torcia et al 1996). A previously unexpected role for NGF in immune and inflammatory functions is suggested by these findings.

NGF SIGNALING MECHANISMS

Cellular responses to NGF are elicited through binding and activation of its receptors, TrkA and p75[NTR] (Bothwell 1995). NGF signaling is now recognized as being broad based, dynamically regulated, and context dependent. Numerous intracellular signaling cascades are triggered by NGF receptor activation, and there is evidence for convergence of, and direct interactions between, NGF signaling and signaling triggered by other molecules. Studies on the intracellular signaling cascades triggered by NGF have relied heavily on in vitro models using primary cell cultures or cell lines, in particular the rat pheochromocytoma cell line PC12. In fact, many of the signaling cascades discussed in the following section have only been delineated in PC12 cells. However, the insights gained from analysis of such cell culture models are useful in the context of an instructive role for further investigation of NGF signaling within neurons and other neural cells. Likewise, studies of NGF signaling have largely focused on developmental processes, such as neuronal differentiation and neurite outgrowth, but information about NGF signaling mechanisms in other contexts, such as degeneration, death, and neuroprotection is increasingly available.

NGF Signaling Through TrkA

TrkA Activation   TrkA, a single transmembrane-spanning polypeptide chain member of the receptor tyrosine kinase (RTK) superfamily, was initially discovered as an oncogenic fusion protein isolated from human colon carcinoma (Martin-Zanca et al 1986a,b). Genetic analysis revealed that in normal cells the proto-oncogene encoded a 140-kDa glycosylated protein containing an extracellular region comprised of several immunoglobulin-like binding domains, a short, single transmembrane domain, and an intracellular domain encoding a tyrosine kinase (Martin-Zanca et al 1989). Following its initial discovery in 1986, the receptor remained an "orphan receptor" until 1991, when it was discovered that NGF evoked a rapid tyrosine phosphorylation of endogenous TrkA in PC12 cells and of exogenous TrkA in transfected fibroblasts (Kaplan et al 1991a,b; Klein et al 1991). Furthermore, TrkA was found to elicit signaling cascades necessary for the biological responses of PC12 cells and neurons to NGF. Upon binding of NGF to TrkA, the receptor is subjected to a series of events that characterize RTK signaling. These include receptor dimerization and transphosphorylation of activation loop tyrosines leading to activation of kinase activity, followed by autophosphorylation of tyrosines outside of the activation loop (Cunningham et al 1997). These autophosphorylation sites serve as binding sites for specific signaling proteins and adaptors such as PLCγ and Shc. Subsequent phosphorylation and activation of accessory proteins lead to the generation of a cascade of receptor-independent signaling pathways (Greene & Kaplan 1995).

Ras Pathway   Tyrosines 490 and 785 are two autophosphorylation targets that are transphosphorylated following TrkA kinase activation (Loeb et al 1994, Middlemas et al 1994, Stephens et al 1994). Shc, an adaptor protein that is critical to activation of the Ras signaling cascade (Figure 1) binds to phosphorylated tyrosine 490 (Basu et al 1994, Obermeier et al 1994). Following binding and phosphorylation of Shc, the Grb2-Sos complex binds to phospho-Shc via an SH2 interaction (Rozakis-Adcock et al 1992), thereby bringing Sos into proximity to membrane-associated Ras and activating the MAP kinase signaling cascade. Sos is a Ras GTP exchange factor that promotes the transition from inactive Ras-GDP to active Ras-GTP (McCormick 1994). Ras is targeted to the plasma membrane via farnesylation (Casey 1995) and resides at the plasma membrane in an inactive, GDP-bound state. Upon recruitment of Sos to the membrane, Ras is activated by exchange of GDP for GTP (McCormick 1994). Ras then recruits the serine-threonine kinase C-Raf to the plasma membrane (Marshall 1994, Van Aelst et al 1993, Wood et al 1992). In PC12 cells, Raf family members (Jaiswal et al 1994, Oshima et al 1991, Traverse & Cohen 1994) mediate NGF signaling by phosphorylating and thereby activating the dual-specificity MAP kinase kinase MEK1 at serine 217 and serine 221 (Jaiswal et al 1994, Lange-Carter & Johnson 1994, Vaillancourt et al 1994). MEK1 activation leads to the phosphorylation of two members of the MAP kinase family, extracellular signal-related kinases 1 and 2 (Erk 1/2) (Crews et al 1992, Crews & Erikson 1992). Erk1/2 are phosphorylated on threonine 202 and tyrosine 204 by MEK1 (Payne et al 1991), leading to activation and translocation of Erk1/2 into the nucleus (Chen et al 1992). Erk1/2 are proline-directed serine-threonine kinases that phosphorylate several substrates, including Elk-1 (Miranti et al 1995). Phosphorylation of Elk-1 at serine 383 and serine 389 stimulates its interaction with the transcription factor serum response factor (SRF) and with the CAGGAT binding site of the serum response element (SRE) within the *c-fos* gene (Gille et al 1995, Hill et al 1993, Mueller & Nordheim 1991, Treisman 1992). *c-fos* is an immediate early gene that is rapidly transcribed in response to many extracellular stimuli, including NGF, and is an early component of a series of transcriptional events necessary for initiation and maintenance of differentiation (Ginty et al 1994, Greenberg et al 1986, Sheng & Greenberg 1990).

Additional transcription factors contribute to the regulation of c-fos transcription in response to NGF signaling. The cAMP regulatory element binding protein (CREB) is a transcription factor that binds to a site called the CRE, or cAMP response element, within the *c-fos* promoter (Berkowitz et al 1989). NGF signaling leads to the phosphorylation of CREB at serine 133 via a Ras-dependent mechanism (Ginty et al 1994). This allows CREB to interact with SRF and Elk-1 (Bonni et al 1995, Ramirez et al 1997), possibly via the transcriptional coactivator protein CREB binding protein (CBP), which binds to phosphorylated serine 133 in CREB (Chrivia et al 1993). CBP also binds to SRF (Ramirez et al 1997) and Elk-1 family members (Janknecht et al 1993). CREB may also play an important role in transcriptional regulation of several NGF-specific delayed response genes, including the *VGF* gene. Mutation of the CREB binding site within the *VGF* gene significantly reduced NGF-induced *VGF* transcription (Hawley et al 1992). It is interesting that *VGF* transcription may require the cooperation of CREB with an as yet unidentified transcription factor product of an immediate early gene. CREB is persistently phosphorylated at serine 133 for several hours after an initial NGF stimulus, and this may permit accumulated immediate early gene proteins to interact with activated CREB. In contrast, EGF stimulation, which does not lead to *VGF* transcription, only transiently phosphorylates CREB, such that by the time sufficient immediate early gene product is present, activated CREB may no longer be available to cooperatively stimulate *VGF* transcription (Bonni et al 1995). This may be one mechanism by which NGF and EGF activate different transcriptional programs leading to either differentiation or proliferation (Marshall 1994).

Map Kinase Temporal Dynamics   The difference in temporal control of CREB phosphorylation induced by NGF or EGF is a specific example of a more general temporal difference elicited in the MAP kinase pathway by these two growth factors. In PC12 cells treated with NGF, there is a sustained activation of the MAP kinase pathway that persists for several hours. In contrast, EGF stimulation only transiently activates the MAP kinase pathway (Muroya et al 1992, Qui & Green

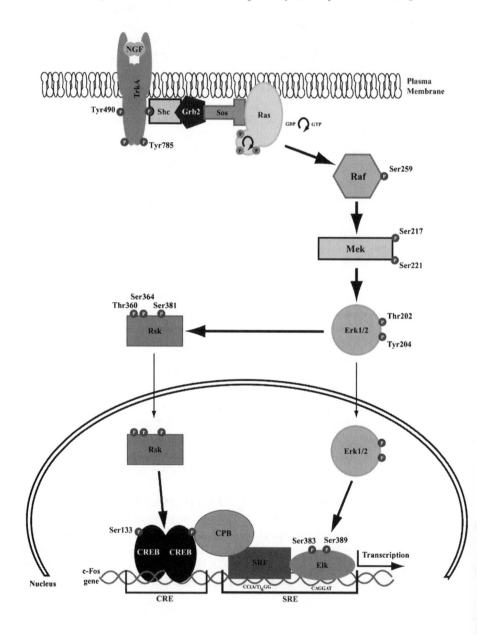

1992, Traverse et al 1992), suggesting that the temporal dynamics of Erk1/2 activation may account for a differentiative versus proliferative signaling outcome. One explanation for how two RTKs linked to very similar signaling pathways might induce such very different MAP kinase activation kinetics requires a better understanding of the specific isoforms of certain adaptor proteins utilized in these cascades. For example, while both NGF and EGF appear to utilize the classic Shc/Grb2/Sos/Ras/C-Raf/MEK pathway to activate Erk, NGF also utilizes an accessory route to Erk activation that utilizes Gab2/CrkL/C3G/Rap1/B-Raf/MEK (Figure 1). This second pathway, which may be unique to NGF signaling, promotes sustained activation of Erk1/2 (CB Wu, CF Lai, WC Mobley, submitted for publication; York et al 1998). The persistant Erk activation that follows NGF stimulation of the Rap1 pathway may induce expression of immediate early gene proteins that interact with activated CREB, induce transcription of novel delayed response genes, or both. Rap1 signaling through MAP kinase does not regulate all aspects of differentiation, nor can one exclude a role for Ras. Expression of a mutant Rap that blocks sustained Erk activation in response to NGF does not block neurite outgrowth in PC12 cells (York et al 1998). On the other hand, complete inhibition of Erk activation, either by pharmacological inhibition of MEK or transfection with a dominant-interfering MEK mutant, does block NGF-induced neurite outgrowth (Cowley et al 1994, Pang et al 1995), and inhibition of Ras activity by microinjection of a Ras-neutralizing antibody also blocks differentiation (Hagag et al 1986). Thus, Ras-dependent signaling is apparently important for NGF-induced differentiation. It is likely that some early event triggered by a Ras- and C-Raf-mediated activation of the Erk pathway is necessary for priming the cell to respond to the later and sustained activation of Erk by the Rap1 and B-Raf pathway.

Figure 1 The Ras-MAP kinase cascade downstream from TrkA. Following phosphorylation of tyrosine 490 within TrkA, Shc is recruited to the receptor via either an SH2- or phosphotyrosine-binding domain-based interaction. Consequently, Shc is bound by the Grb2-Sos complex. Recruitment of Sos to the membrane brings it into proximity of Ras, where it functions as a GTP-exchange factor, activating Ras. Activated Ras recruits and activates Raf. Raf is a serine-threonine kinase that phosphorylates the MAP kinase kinase MEK on 2 serines. This phosphorylation event initiates activity of the dual-specificity kinase, leading to activation of the MAP kinases Erk1/2 via phosphorylation of threonine 202 and tyrosine 204. Phosphorylated Erk1/2 then participate in at least two cascades. Erk1/2 may translocate into the nucleus, where they phosphorylate the transcription factor Elk-1, or they may phosphorylate the kinase Rsk. Phosphorylation of Elk-1 allows it to interact with the accessory transcription factor SRF, after which it binds to the serum response element (SRE) within the *c-fos* promoter region and contributes to initiation of transcription. Phosphorylation of Rsk leads to its nuclear translocation and consequent phosphorylation of CREB on serine 133. Phosphorylated CREB is bound by the transcriptional coactivator protein CPB, which also binds to the SRF-Elk complex, creating an extended transcriptional factor complex that leads to *c-fos* transcription.

Rsk Pathway   A further level of control of NGF-induced immediate early gene transcription and translation comes from parallel activation of the Rsk pathway downstream from Ras. The Rsk serine-threonine kinase was originally isolated as a 90-kDa cell-cycle regulated kinase that phosphorylated the S6 protein of the 40S ribosomal subunit (Erikson & Maller 1991, Erikson et al 1991). This p90 kinase (ribosomal S6 kinase, hence Rsk) was itself found to be regulated by serine-threonine phosphorylation, and Erk1/2 were subsequently identified as the kinases responsible for this regulatory phosphorylation (Sturgill et al 1988, Zhao et al 1996). The Rsk family is comprised of Rsk1, Rsk2, and Rsk3, each showing unique patterns of tissue expression (Moller et al 1994, Zhao et al 1995). Rsk2 was identified as a Ras-dependent protein kinase that phosphorylates CREB on serine 133 (Ginty et al 1994, Xing et al 1996), thereby regulating its transcriptional activation. Rsk family members are also involved in phosphorylation of the estrogen receptor-α, IκBα/NFκB, and c-fos (Ghoda et al 1997, Joel et al 1998, Schouten et al 1997, Xing et al 1996). Rsks also bind to the transcriptional coactivator CBP (Nakajima et al 1996) and phosphorylate several members of the ribosomal complex (Angenstein et al 1998). Sos, a substrate for Rsk, appears to be negatively regulated by Rsk kinase activity, suggesting that Rsk activation downstream from activation of Erk1/2 may feed back to truncate Ras signaling (Douville & Downward 1997). Recently, all three members of the Rsk family were found to be activated by NGF in PC12 cells, and all were able to phosphorylate CREB at serine 133 (Xing et al 1998). Hence, the Ras pathway is able to regulate c-fos induction by using a parallel and cooperative pathway in which Erk phosphorylation of Elk-1 converges upon Rsk phosphorylation of CREB (Xing et al 1996). Thus, the Erk pathway is marked by both divergent and convergent signaling, in which an early divergence at the level of Shc versus Gab2 can control the temporal dynamics of Erk activation, and convergence at the level of Elk-1 and CREB regulation of c-fos can control gene transcription and protein translation.

Src and PKC Pathways   Convergence of control over the MAP kinase pathway may also occur between Ras, PKC, and Src. Src is a member of a large family of nonreceptor protein tyrosine kinases that share significant sequence homology. This family includes Fyn, Yes, Yrk, Blk, Fgr, Hck, Lck, Lyn, Frk/Rak, and Iyk/Bsk (Brown & Cooper 1996, Cance et al 1994, Lee et al 1994a, Thomas & Brugge 1997, Thuveson et al 1995, Welch & Maridonneau-Parini 1997). Src kinases regulate a wide range of cellular events, ranging from cell proliferation, cytoskeletal alterations, and differentiation, to survival, adhesion, and migration. RTKs interact with Src kinases and use them to transduce several signaling pathways (Erpel & Courtneidge 1995). Involvement of Src or an Src family member in NGF-mediated differentiative signaling was first proposed when it was discovered that infection of PC12 cells with the oncogenic form of Src recapitulated the neurite outgrowth induced by NGF (Alema et al 1985). Further analysis showed that neutralization of Ras by microinjection of anti-Ras antibodies blocked the neuritogenic effects of both Src and NGF (Hagag et al 1986, Kremer et al 1991). In contrast, neutralization of Src activity by antibody microinjection did not block neurite outgrowth induced by infection with oncogenic Ras (Bar-Sagi & Feramisco 1985, Kremer et al 1991, Noda et al 1985) but did inhibit NGF-induced neuritogenesis. It also caused retraction of established neurites induced by NGF or FGF treatment (Kremer et al 1991). Finally, both oncogenic Src and oncogenic Ras are able to "prime" PC12 cells, such that subsequent NGF treatment elicits a more rapid and robust neuritogenesis than NGF treatment of unprimed cells (Thomas et al 1991). It is interesting that oncogenic Src activated the N-terminal c-jun kinase (JNK), a member of the MAP kinase family, without activating Erk1/2 (Kuo et al 1997). Hence, one possible explanation for the role that both Src and Ras play in differentiation is that they control the activity of a common MEK family member that is upstream of both Erk1/2 and JNK (Ellinger-Ziegelbauer et al 1997, Lewis et al 1998). This model is compatible with data showing that pharmacological inhibition of MEK in PC12 cells abrogated neurite outgrowth in response to NGF (Pang et al 1995). MEK activity is also regulated by several PKC isoforms (Berra et al 1993, 1995; Schonwasser et al 1998, van Dijk et al 1997), and overexpression of either PKCι or PKCζ resulted in enhanced NGF-induced neurite outgrowth and enhanced NGF-induced JNK activation (Wooten et al 1999), while inhibition of atypical PKC isoforms blocked NGF-induced activation of JNK (Wooten et al 1999). PI3 kinase is also implicated in signaling to JNK, as NGF-induced JNK activation was impaired by either wortmannin or LY294002, and overexpression of PI3 kinase resulted in neurite outgrowth and JNK activation in the absence of Erk activation (Kobayashi et al 1997). Thus, a signaling cascade including Src, PI3 kinase, PKC, and JNK appears to be involved in neurite outgrowth and differentiative signaling and may either complement or parallel the Ras-Raf-MEK-Erk1/2 cascade.

Signaling through Src, PI3 kinase, PKC, and JNK may also play a role in cell survival signaling. Overexpression of either Src or PKCι enhanced PC12 cell survival in serum-free conditions, and both increased the activation of the transcription factor NFκB (Wooten et al 2000, 1999), apparently via JNK signaling. Moreover, inhibition of Src or atypical PKC isoforms promoted cell death (Seibenhener et al 1999, Wooten et al 2000). Likewise, inhibition of PI3 kinase activity blocked cell survival and reduced NGF-induced NFκB activation (Wooten et al 2000). These findings are compatible with data showing that activation of NFκB promotes cell survival and resistance to apoptosis, and that NGF induction of NFκB is primarily dependent on signaling through the JNK pathway (Wooten et al 2000). Thus, both differentiative and survival signaling may be controlled in part by a signaling unit that includes Src, PI3 kinase, and PKC.

PI3 Kinase Pathway   PI3 kinase and Src are also implicated in survival signaling via the common substrate Akt, a serine-threonine kinase also known as protein kinase B (PKB), or related to A and C protein kinase (RAC-PK). Akt is regulated by growth factor and serum factor signaling through PI3 kinase (Alessi et al 1996; Andjelkovic et al 1996; Burgering & Coffer 1995; Franke et al 1995, 1997; Klippel et al 1997). PI3 kinase is a heterodimer composed of an 85-kDa regulatory subunit and a 110-kDa catalytic subunit. Activation of the kinase involves binding of the regulatory subunit either directly or via adaptors to activated RTKs. This interaction with the cytoplasmic domain of an RTK results in recruitment of the 110-kDa catalytic subunit to the plasma membrane, where it can interact with and phosphorylate membrane phosphoinositides. Such phosphorylation results in the production of PI-3,4-P$_2$ and PI-3,4,5-P$_3$. Akt interacts with PI-3,4-P$_2$ or PI-3,4,5-P$_3$, and with the 3-phosphoinositide-dependent kinase (PDK1). PDK1 contains a pleckstrin homology domain that binds PI-3,4-P$_2$ or PI-3,4,5-P$_3$, and this binding is necessary to permit PDK1 to phosphorylate and activate Akt (Alessi et al 1997a,b; Cohen et al 1997; Stephens et al 1998; Stokoe et al 1997). Hence, TrkA signaling via PI3 kinase presumably signals to generate 3-phosphoinositides that bind PDK1 and induce the activation of Akt. PDK1 phosphorylates the activation loop of several other serine-threonine kinases, including certain isoforms of PKC (Chou et al 1998, Le Good et al 1998), suggesting that PI3 kinase-mediated generation of 3-phosphoinositides may also control differentiative or survival signaling via PKC activation.

Mediation of TrkA survival signaling by PI3 kinase is indicated by the results of experiments showing that two inhibitors of PI3 kinase activity, wortmannin and LY294002, induce apoptosis in PC12 cells and sympathetic neurons supported by NGF (Crowder & Freeman 1998, Yao & Cooper 1995). The role of Akt in regulation of cell survival downstream from PI3 kinase is suggested by the fact that overexpression of Akt in primary cultures of cerebellar neurons or sympathetic neurons provides protection against death induced by serum withdrawal or inhibition of PI3 kinase, while expression of dominant-interfering forms of Akt blocked NGF-mediated survival (Crowder & Freeman 1998, Dudek et al 1997). The mechanism by which Akt mediates survival is unclear, though Akt has been reported to bind and phosphorylate Bad, a member of the Bcl-2 family of proteins (Figure 2) (Datta et al 1997, del Peso et al 1997). Phosphorylation of Bad prevents it from binding the anti-apoptotic Bcl-2 family members Bcl-2 and Bcl-X$_L$ (Zha et al 1996), shifting the cell to contain more Bcl-2 homodimers than Bcl-2/Bax heterodimers. The Bcl family is composed of two groups of proteins, one that promotes cell survival and includes Bcl-2 and Bcl-X$_L$, and the other that promotes cell death and includes Bad and Bax (Boise et al 1995, Kroemer 1997, Steller 1995). The members of the Bcl family form homo- and heterodimers, and the balance of each dimer within the cell is considered to regulate the maintenance of survival or the induction of death. In the absence of phosphorylation of Bad on serine 112 and serine 136, Bad signals to promote cell death, apparently by forming heterodimers with Bcl-X$_L$. Formation of these heterodimers leads to the generation of Bax homodimers. Homodimerization of Bax induces its translocation into mitochondria and insertion into the mitochondrial membrane (Gross et al 1998). There it leads to altered mitochondrial membrane potential via ion channel formation and to generation of cytotoxic reactive oxygen species (Xiang et al 1996). In contrast, the phosphorylation of Bad promotes cell survival by inducing an interaction between Bad and the 14-3-3 protein. This interaction effectively

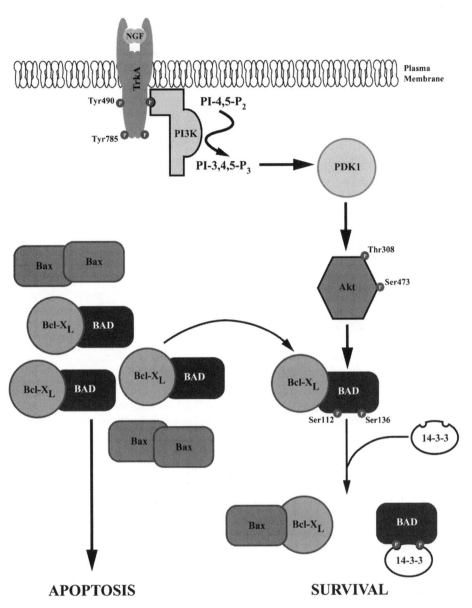

Figure 2   TrkA survival signaling TrkA phosphorylation leads to the activation of PI3 kinase. PI3 kinase catalyzes the production of 3-phosphoinositides, including PI-3,4,5-P$_3$, which bind to and activate PDK1. PDK1 associates with and phosphorylates the serine-threonine kinase Akt. Akt then phosphorylates Bad, inducing its association with the 14-3-3 protein and sequestering it from heterodimerization with Bcl-X$_L$. As a result of Bad sequesteration, Bcl-X$_L$ is able to heterodimerize with Bax, preventing Bax homodimerization. Homodimerized Bax is a key element in apoptotic signaling, via its role in altering mitochondrial membrane potential, and the balance of Bax:Bax homodimers versus Bax:Bcl-X$_L$ heterodimers may determine whether the cell lives or dies.

sequesters Bad from any interaction with Bcl-X_L, keeping the balance of Bcl-X_L/Bax heterodimers high and preventing Bax homodimerization (Zha et al 1996). Hence, TrkA survival signaling involves PI3 kinase-mediated activation of Akt and the consequent maintenance of Bcl-X_L/Bax heterodimers. Src is also implicated in the activation of Akt via a mechanism that involves PI3 kinase and SHP-2 (Datta et al 1996, Hakak et al 2000). This interaction may explain the finding, presented above, that inhibition of Src promotes cell death, and it suggests that additional complexity may exist in the mechanism by which TrkA signaling induces cell survival.

TrkA activation may be linked to the phosphotidylinositol 3-kinase (PI3 kinase) pathway via binding of Grb2 and the Grb2-associated binder-1 (Gab1) protein to tyrosine 490. Gab1 was initially identified as a Grb2-associated protein in a human glial tumor expression library and was also identified in a yeast 2-hybrid screen using the Met RTK as bait (Holgado-Madruga et al 1996, Weidner et al 1996). Gab1 is a member of a family of adaptor proteins that includes Gab2, IRS-1, IRS-2, and Dos, all of which exhibit sequence homology, and all of which link plasma membrane RTKs to intracellular signaling cascades (Bausenwein et al 2000, Gu et al 1998). Gab1 contains several SH2 and SH3 binding domains that recognize PI3 kinase and SHP-2, as well as Grb2, Nck, and Crk (Holgado-Madruga et al 1996, Weidner et al 1996). Gab1 is tyrosine phosphorylated in response to signaling downstream from TrkA (Holgado-Madruga et al 1997), and it is also induced to associate with PI3 kinase, recruiting the p85 subunit to the plasma membrane and eliciting activation. Furthermore, overexpression of Gab1 reduced the concentration of NGF necessary for mediating cell survival in serum-free conditions, while expression of a mutant Gab1 lacking the PI3 kinase binding sites enhanced apoptosis (Holgado-Madruga et al 1997). These data suggest that anti-apoptotic TrkA signaling to PI3 kinase and the Akt pathway is mediated by Gab1. This is supported by the finding that adenovirus-mediated expression of Gab1 in PC12 cells is sufficient to support enhanced survival, even in the absence of NGF signaling, and that this enhancement is correlated with increased PI3 kinase signaling (Korhonen et al 1999). However, Gab1 appears to utilize both the PI3 kinase pathway and the MAP kinase pathway to mediate its effect on cell survival, as pharmacological inhibition of both MEK and PI3 kinase was required to fully suppress Gab1-mediated cell survival (Korhonen et al 1999). Finally, adenovirus-expressed Gab1 enhanced neurite outgrowth in response to NGF via a mechanism that was sensitive to either MEK inhibition or PI3 kinase inhibition (Korhonen et al 1999). These results suggest that Gab1 plays a role as an adaptor protein for both the PI3 kinase pathway and the MAP kinase pathway downstream from TrkA signaling. However, another member of the Gab family, Gab2, was recently identified as a substrate for tyrosine phosphorylation downstream of TrkA, and Gab2 was found in complex with CrkL, C3G, and SHP-2 following NGF treatment of PC12 cells (CB Wu, CF Lai, WC Mobley, submitted for publication). This finding suggests that Gab2 may adapt TrkA to the Rap1/B-Raf pathway by inducing NGF-dependent activation of C3G, a Rap GTP exchange factor. In that activation of the Rap1 pathway leads to MEK activation in parallel with the Ras pathway, as described above, it is possible that overexpressed Gab1 subsumes the role of endogenous Gab2 in mediation of neurite outgrowth.

FRS-2 In addition to binding Shc and Gab, tyrosine 490 also appears to mediate the interaction of TrkA with FRS-2, a novel membrane-anchored adaptor protein that is tyrosine phosphorylated in response to NGF (Kouhara et al 1997, Ong et al 2000). Phosphorylated FRS-2 binds to the Grb2-Sos signaling unit, forming a multi-protein complex that includes Crk and the protein tyrosine phosphatase SHP-2 (Hadari et al 1998, Kouhara et al 1997, Meakin et al 1999). Formation of this complex is necessary for FRS-2 activation of the MAP kinase pathway. FRS-2 competes with Shc for binding to tyrosine 490 on TrkA, adding an interesting layer of complexity to the signaling cascades elicited by NGF treatment (Meakin et al 1999). FRS-2 may or may not be identical to SNT (Friedman & Greene 1999, Kouhara et al 1997), a protein that may be a candidate for the factor that controls the decision between cell-cycle progression and cell-cycle arrest, a critical component of differentiative signaling. The ability of SNT to bind the cyclin-dependent kinase substrate p13^{suc1}, and the fact that it is rapidly tyrosine phosphorylated in response to NGF (Rabin et al 1993) suggests that SNT may be the mediator of this key decision. While the relationship between SNT and FRS-2 is still unresolved, recent evidence indicates that human FRS-2 does bind p13^{suc1} in a constituitive manner (Meakin et al 1999), strengthening the possibility that FRS-2 is an SNT.

It is interesting to note that mutations in tyrosine 490 of TrkA do not abolish NGF induction of the MAP kinase signaling pathway. However, cells expressing TrkA with a double mutation at tyrosine 490 and tyrosine 785 do not exhibit MAP kinase activation or neurite outgrowth in response to NGF (Stephens et al 1994). This finding suggests that there is an as yet undiscovered complexity or redundancy to the interaction of adaptor proteins with tyrosines 490 and 785. One possible component in this additional complexity is the recent finding that Grb2 binds directly to activated TrkA at both tyrosine 785 and the kinase activation loop tyrosines (MacDonald et al 2000). This additional route to the Ras pathway may circumvent loss of either tyrosine 490 or tyrosine 785, but not both.

PLCγ and PKC Pathways Tyrosine 785, near the C terminus of TrkA, is within a consensus site for the binding of the SH2 domain of phospholipase C-γ (PLCγ). This tyrosine is required for NGF-dependent recruitment of PLCγ to TrkA and for the phosphorylation and activation of PLCγ (Vetter et al 1991). Following binding to tyrosine 785 of TrkA, PLCγ is activated and induced to hydrolyze phosphatidylinositol 4,5-bisphosphate (PI 4,5-P$_2$). PLCγ-mediated hydrolysis of PI 4,5-P$_2$ yields two products that each function as intracellular second messengers: inositol 1,4,5-P$_3$ (IP$_3$), which interacts with its specific receptor on the endoplasmic reticulum to induce the release of intracellular calcium, and diacylglycerol (DAG), which is a potent activator of protein kinase C (PKC) isoforms (Lee & Rhee 1995). IP$_3$-mediated release from intracellular calcium stores leads to the activation of calcium-dependent proteins within the cell and to the generation of further IP derivatives such as IP_4, IP_5, and IP_6, which are able to interact with other intracellular proteins (Menniti et al 1993). DAG is an activator of several isoforms of the serine-threonine calcium-dependent kinase PKC. These include several classical, novel, and atypical PKC isoforms (Bell & Burns 1991; Nishizuka 1988; Liyanage et al 1992; Ono et al 1988; Osada et al 1990, 1992; Marais et al 1998). DAG cooperates with calcium, phosphatidylserine, cis-unsaturated fatty acids, and lysophosphatidylcholine to activate the classical PKC isoforms, and it cooperates with phosphatidylserine and cis-unsaturated fatty acids to activate the δ and ε isoforms of novel PKC. PLCγ activation is often accompanied by phospholipase A2–mediated hydrolysis of phosphatidylcholine, directly generating cis-unsaturated fatty acid and lysophosphatidylcholine (Asaoka et al 1992, Nishizuka 1992). These factors, in combination with DAG, serve to tune PKC activation to signaling downstream from TrkA, leading to phosphorylation of several proteins critical to survival and differentiation (Coleman & Wooten 1994; Wooten et al 1994, 1997, 1999). One such substrate of PKC is Raf, which is directly activated by PKC-mediated phosphorylation (Carroll & May 1994, Kolch et al 1993, Schonwasser et al 1998, Sozeri et al 1992, van Dijk et al 1997). The association of PKC with Raf appears to be mediated by binding of the scaffolding protein 14-3-3 (Freed et al 1994, Fu et al 1994, Irie et al 1994, van der Hoeven et al 2000). A PKC-(14-3-3)-Raf complex may also contribute to PKCθ- and PKCμ-mediated regulation of the MAP kinase cascade (Hausser et al 1999, Meller et al 1996) and may account for PKCε-mediated activation of Raf (Cacace et al 1996, Ueffing et al 1997). PKC might also mediate activation of the MAP kinase cascade by directly activating Ras, leading to the formation of a (Ras-GTP)-Raf complex (Marais et al 1998). This finding is consistent with evidence that PKC-mediated activation of Raf is blocked by mutation in the Ras-binding domain of Raf (Luo et al 1997). Finally, PKC can directly phosphorylate the c-jun protein product, which is also under the control of phosphorylation by Erk and which is able to bind to the c-fos protein product to form the transcriptional regulatory complex AP-1 (Oberwetter et al 1993).

Abl Pathway   The juxtamembrane region of TrkA, a unique region in the cytoplasmic domain of the receptor, has also been implicated in carrying out several specific signaling functions downstream from NGF binding. This region apparently mediates the association of activated TrkA with Abl, a non–receptor tyrosine kinase that is involved in the regulation of adhesion-dependent signaling and cytoskeletal remodeling that occurs during neuronal differentiation (Yano et al 2000). The association of Abl with TrkA may lead to its activation, and consequently to the phosphorylation of paxillin (Matsuda et al 1994, Ribon & Saltiel 1996, Teng et al 1995, Torres & Bogenmann 1996). It is interesting to note that tyrosine phosphorylation of paxillin is critical to the increased cell adhesion necessary for neurite outgrowth, and that Abl is involved in this pathway in Drosophila (Gertler et al 1989, 1993; Wills et al 1999).

rAPS- and SH2-B-Mediated Pathways   Two other adaptor proteins that do not appear to interact with either tyrosine 490 or tyrosine 785 are rAPS and SH2-B, which were recently identified as TrkA substrates in developing cortical and sympathetic neurons (Qian et al 1998). Both rAPS and SH2-B were found in complex with Grb2, and either adaptor was able to mediate NGF induction of MAP kinase activation. In nnr5 PC12 cells that express extremely low levels of TrkA, cotransfection with rAPS and a TrkA mutant lacking all tyrosines except those in the kinase activation loop, or with SH2-B and this TrkA mutant, led to robust neurite outgrowth (Qian et al 1998). Moreover, while the interaction between rAPS and Grb2 is at least partially dependent on tyrosine phosphorylation of rAPS, Grb2 appears to bind to SH2-B constituitively via an SH3 interaction. Finally, antibodies to SH2-B inhibited NGF-dependent survival of cultured neonatal sympathetic neurons, and transfection with a dominant-interfering mutant of SH2-B completely blocked the elaboration of axons by cultured sympathetic neurons. This suggests that SH2-B and rAPS are critical elements in the TrkA signaling pathway necessary for both neurite outgrowth and survival, but that their interaction with TrkA may utilize a novel association mechanism.

NGF Signaling Through p75^{NTR}

p75^{NTR} was the first identified NGF receptor and for many years was believed to be the only such receptor. However, following the discovery of a receptor tyrosine kinase for NGF that exhibited readily identifiable signaling properties, p75^{NTR} was largely relegated to the role of modulating and modifying TrkA signaling. While such a role continues to be an important area of investigation, it has become increasingly clear that p75^{NTR} is a signaling receptor in its own right. In fact, the signals initiated by p75^{NTR} are likely to be as complex as those for TrkA and to be critically influenced by the cells in which such signaling arises (Friedman & Greene 1999, Kaplan & Miller 1997). The function of NGF signaling via p75^{NTR} in the context of cell death and regeneration may be important for understanding NGF actions in controlling the processes of neural repair and neuroprotection.

p75^{NTR} is the first identified member of a superfamily of receptors that includes CD27, CD30, CD40, OX40, Fas (CD95), and the tumor necrosis factor receptors (TNF-R) (Bazan 1990, Cosman et al 1990, Mallett & Barclay 1991, Smith et al 1994). These receptors share several common signaling features, including the ability to control cell viability via regulation of apoptosis. For example, in the embryonic chick retina, neural precursor cells expressing p75^{NTR} in the absence of TrkA undergo NGF-dependent apoptosis, suggesting that developmentally programmed death in these cells is mediated by p75^{NTR} (Bredesen & Rabizadeh 1997, Carter & Lewin 1997, Frade et al 1996). Furthermore, p75^{NTR} mediates NGF-induced death of cultured oligodendrocytes (Casaccia-Bonnefil et al 1996, Gu et al 1999, Yoon et al 1998) and cultured hepatic stellate cells (Trim et al 2000), and BDNF signaling via p75^{NTR} was shown to induce apoptosis of postnatal sympathetic neurons in culture (Bamji et al 1998). Moreover, an increased number of sympathetic neurons are found in BDNF-deficient mice, and there is a delay in sympathetic cell death in p75^{NTR} homozygous knockout mice (Bamji et al 1998). BDNF-dependent trigeminal neurons are killed via binding of NT-4 to p75^{NTR}, even though p75^{NTR} is necessary to the cell survival induced by BDNF (Agerman et al 1999). This indicates that p75^{NTR} signaling is not only dependent on cell context but also on neurotrophin binding specificity.

Ceramide Signaling   One signal transduction pathway ascribed to p75^{NTR} that may be involved in apoptotic signaling involves generation of the lipid second messenger ceramide via activation of sphingomyelinase. In fibroblasts expressing p75^{NTR} but not TrkA, NGF induced the production of ceramide. Furthermore, in T9 glioma cells, NGF induced the activation of sphingomyelinase and the production of ceramide, and inhibited growth and fiber formation, a process that was mimicked by incubation with membrane-permeant ceramide analogs (Dobrowsky et al 1994). Other members of the p75^{NTR} superfamily, such as TNF-RI and Fas, also signal via ceramide production (Cifone et al 1994). This signaling function appears to be mediated at least in part by a region within TNF-RI and Fas termed the death domain, a C-terminal region in the cytoplasmic domain that is necessary for apoptotic signaling downstream from these receptors (Tartaglia et al 1993, Watanabe-Fukunaga et al 1992). Analysis of the p75^{NTR} sequence shows that a homologous death domain region exists within the intracellular region of this receptor (Liepinsh et al 1997). Recent experiments suggest that the death domain serves to mediate protein:protein interactions. For example, this region mediates Fas and TNF-RI intracellular domain aggregation (Boldin et al 1995a, Song et al 1994), and a homologous region has been found within ankyrin, a protein that anchors transmembrane proteins to the cytoskeleton (Boldin et al 1995b).

Chopper   Another death signaling domain was recently discovered within the p75^{NTR} juxtamembrane region. This domain, a 29-residue sequence named chopper, is necessary and sufficient to induce cell death in several cell types, including neurons. It is interesting that a peptide corresponding to the chopper domain only signaled cell death when associated with the plasma membrane via a lipid anchor. Nonanchored chopper peptide did not mediate cell death and, in fact, acted in a dominant-negative manner to p75^{NTR}-mediated death signaling (Coulson et al 2000), suggesting that palmitoylation of p75^{NTR} is a crucial factor in mediating signaling from the receptor. This finding also suggests the possibility that proteolytic cleavage of the intracellular domain may play a role in controlling p75^{NTR} signaling.

Ligand-Independent p75^{NTR} Signaling   Another possible mechanism of p75^{NTR}-mediated cell death was suggested by the observation that overexpression of the intracellular domain of p75^{NTR} induced cell death in several neuronal populations within the central and peripheral nervous systems (Majdan et al 1997). This finding, plus the observation that immortalized neural cells overexpressing p75^{NTR}

exhibit enhanced cell death following serum withdrawal (Rabizadeh et al 1993), suggests that p75^{NTR} may signal pro-apoptotically in the absence of ligand binding. In this model, binding of NGF to p75^{NTR} induces a conformational change that blocks the production of a death signal. Further support for this idea comes from work showing that antisense-induced downregulation of p75^{NTR} in neonatal dorsal root ganglia sensory neurons enhanced survival (Barrett & Bartlett. 1994). Moreover, identification of an alternatively spliced isoform of p75^{NTR} lacking the neurotrophin-binding domain supports the model of ligand-independent signaling (Dechant & Barde 1997). The receptor produced by this alternative splice event contains the transmembrane and intracellular domains, but lacks the ability to bind neurotrophin and may therefore exhibit enhanced cell death signaling consistent with the function of the death domains described above. Finally, p75^{NTR} appears to exhibit ligand-independent signaling through the RhoA pathway. In cells transfected with p75^{NTR}, RhoA activation was generated in the absence of ligand and was abolished by addition of ligand, suggesting that p75^{NTR} can signal to reorganize the actin cytoskeleton in a manner that is negatively modulated by the presence of neurotrophin (Yamashita et al 1999b).

NFκB Pathway   Many proteins in the p75^{NTR} superfamily interact with TNF receptor-associated factors (TRAFs) that modulate signaling through the JNK and NFκB pathways. Six such factors have been identified in signaling evoked by TNF-R, CD30, CD40, and the IL-1 receptor (Arch et al 1998, Rothe et al 1995), and recently p75^{NTR} was shown to associate with TRAF-2, TRAF-4, and TRAF-6 following treatment with NGF (Khursigara et al 1999, Ye et al 1999). Interestingly, the association of TRAF-6 with p75^{NTR} is mediated by the receptor's juxtamembrane domain (Khursigara et al 1999) within a sequence that is absolutely conserved between human, rat, and chicken p75^{NTR} (Large et al 1989), suggesting that the interaction with TRAF-6 is critical to p75^{NTR} function. TRAF-6 is recruited to the IL-1 receptor via binding to IRAK, the IL-1 receptor-associated serine-threonine kinase (Cao et al 1996a,b), and TRAF-6 also signals through NIK, the NFκB inducing kinase (Malinin et al 1997), suggesting that one role of the p75^{NTR}-(TRAF-6) interaction may be to couple p75^{NTR} to several different kinase cascades. The use of adaptor proteins such as TRAF-6 potentially permits p75^{NTR}, which lacks any intrinsic kinase activity, to recruit and noncatalytically activate several cytoplasmic non–receptor kinases, thereby linking NGF binding to p75^{NTR} to NFκB activation.

In addition to apoptosis-related signaling, p75^{NTR} binding of NGF also activates the transcription factor NFκB in neuroblastoma cells (Korner et al 1994), cultured sensory and sympathetic neurons (Maggirwar et al 1998, Wood 1995), Schwann cells (Carter et al 1996, Khursigara et al 1999), and oligodendrocytes (Ladiwala et al 1998, Yoon et al 1998). The activation of NFκB downstream from most inducer proteins involves the degradation of the IκB protein, an inhibitory factor that binds heterodimers of the NFκB p50 and p65 subunits and prevents them from translocating into the nucleus (Ghosh et al 1998). IκB degradation results in NFκB nuclear translocation and in upregulated transcription of several genes, including the IκB gene. In oligodendrocytes, in which p75^{NTR} appears to signal via both NFκB and the JNK pathway, expression of TrkA abrogates NGF-induced cell death in a manner that correlates with cessation of JNK signaling, whereas the NFκB signal downstream from p75^{NTR} is unaffected (Yoon et al 1998). This suggests that p75^{NTR} may evoke two separate pathways, one pro-apoptotic, the other anti-apoptotic. The balance of these two pathways, as modulated by TrkA signaling in some cells, may control the ultimate fate of the cell. However, the exact role that NFκB plays is unresolved—in some systems it exhibits anti-apoptotic signaling (Maggirwar et al 1998, Mattson et al 1997), but in others it is associated with pro-apoptotic signaling (Schneider et al 1999, Schwaninger et al 1999). The TNF receptor, generally associated with death signaling, also activates NFκB in a pathway that appears to promote survival of lymphoid cells and fibroblasts (Liu et al 1996, Van Antwerp et al 1996, Wang et al 1996). Likewise, in hippocampal neurons that do not express TrkA, NGF signaling through p75^{NTR} protects these cells from glucose deprivation-induced apoptosis (Cheng & Mattson 1991). Furthermore, p75^{NTR} appears to play a role in protecting Schwann cells following axotomy. In the normal adult animal, Schwann cells do not express p75^{NTR}. However, following nerve injury, Schwann cells distal to the injury site dramatically upregulate p75^{NTR} expression (Heumann et al 1987b, Taniuchi et al 1986), and exhibit increased NFκB activation (Gentry et al 2000). This increase in NFκB activation is correlated with the absence of apoptosis in Schwann cells distal to the injury (Grinspan et al 1996). It is interesting that during development Schwann cells require axonal contact for trophic support, and loss of such contact results in cell death. Hence, injury induced expression of p75^{NTR} and consequent signaling through NFκB may serve in the adult to maintain Schwann cells in the absence of trophic support from the axon, thereby providing time for the axon to regrow.

Interactions Between p75^{NTR} and TrkA

Prior to defining p75^{NTR} signaling pathways, a great deal of attention was focused on a role for p75^{NTR} in regulating NGF signaling through TrkA. A wealth of data show that p75^{NTR} does impact TrkA signal transduction. It does so by enhancing binding of NGF to TrkA, by increasing the specificity of TrkA for NGF binding, and through effects on TrkA signal transduction. Moreover, TrkA signaling also impacts signaling through p75^{NTR}. These efffects may be mediated, at least in part, through the direct association of TrkA and p75^{NTR}, as revealed in studies using a variety of techniques, including photobleaching (Wolf et al 1995), copatching (Ross et al 1996), cross-linking (Ross et al 1998), and coimmunoprecipitation (Bibel et al 1999, Gargano et al 1997, Huber & Chao 1995). Recently, TrkA and p75^{NTR} were colocalized to caveolae-like domains of PC12 cells, and both TrkA (Huang et al 1999) and p75^{NTR} (Bilderback et al 1997) signal from these membranes. These findings highlight an important additional level of complexity for NGF signaling and point to the need for understanding the cell biology of receptor trafficking and signaling.

There are two classes of binding sites for NGF: Low-affinity receptors bind NGF with nanomolar affinity, whereas high-affinity receptors bind NGF with an affinity that is 100-fold greater (i.e. 10^{-11} M) (Meakin & Shooter 1992, Sutter et al 1979). The two classes are distinguished by the much slower rate of dissociation from high-affinity receptors (Landreth & Shooter 1980, Meakin & Shooter 1992, Schechter & Bothwell 1981, Woodruff & Neet 1986). High-affinity receptors are thought to play an important role in mediating NGF actions. Dissociation of NGF from TrkA is slow (Meakin et al 1992), which suggests that TrkA contributes to the formation of these receptors. TrkA is often referred to as the high-affinity receptor, a designation that suggests TrkA alone binds NGF with high affinity. However, although there is a small amount of high-affinity binding of NGF in cells expressing only TrkA, most NGF binding to such cells is of low affinity (Mahadeo et al 1994). In fact, most high-affinity binding appears to reflect the interaction of p75^{NTR} with TrkA. p75^{NTR} has been shown to increase the rate of association of NGF with TrkA, thereby increasing the number of high-affinity receptors (Mahadeo et al 1994). Moreover, p75^{NTR} enhances activation of TrkA (Barker & Shooter 1994). Of note, a recent study showed that some receptor complexes from which NGF was slowly released contained p75^{NTR} (Huang et al 1999).

p75^{NTR} may also interact with TrkA to modify binding specificity. In fibroblasts that express only TrkA, NT-3 and NT-4/5 are able to activate the receptor, whereas in PC12 cells, which express both p75^{NTR} and TrkA, only NGF is able to activate TrkA (Berkemeier et al 1991, Ip et al 1993). Likewise, mutant PC12 cells that express only very low levels of p75^{NTR} exhibit NT-3-induced TrkA activation (Benedetti et al 1993). Finally, postnatal sympathetic neurons normally exhibit very limited survival in culture in response to NT-3, but these same neurons isolated from p75^{NTR} transgenic knockout mice show a much more robust NT-3-induced survival response (Lee et al 1994c). These data suggest that p75^{NTR} may function to tune individual neurons to specific neurotrophin responsiveness, thereby controlling the ability of such neurons to compete for target-derived neurotrophic support. It is interesting to note that sympathetic neurons normally undergo a switch in trophic dependence, from an early dependence upon NT-3 to a later dependence on NGF, and that this switch is temporally correlated to the onset of p75^{NTR} expression (Birren et al 1993). Furthermore, as NGF signaling via TrkA appears to control p75^{NTR} expression in these cells (Miller et al 1991, 1994; Verdi & Anderson 1994; Verge et al 1992; Wyatt et al 1990), it is possible that first contact between the innervating sympathetic fibers and NGF available from the target field elicits the trophic dependency switch. Also, the expression of p75^{NTR} by cells that have received an NGF signal from the target may increase the sensitivity of those neurons to low levels of target-derived NGF, leading to a situation in which those neurons that express p75^{NTR} are better able to compete for synaptic space within the target. Hence, the ability of p75^{NTR} to sharpen TrkA binding specificity may play a significant role in the maturation of target innervation, and may control the competition that defines the adult pattern of innervation. Whether p75^{NTR} plays such a role in synaptic competition within the central nervous system remains to be determined.

Perhaps the most interesting facet of the interaction of TrkA and p75^NTR is evidence for reciprocal effects on signaling. Barker and colleagues provided evidence that signaling through p75^NTR inhibits signaling though TrkA. Addition of BDNF to PC12 cells markedly reduced the TrkA activation and downstream signaling events that were elicited by treatment with an NGF mutant that only binds to TrkA (MacPhee & Barker 1997). BDNF is known to activate sphingomyelinase and increase ceramide levels (Dobrowsky et al 1995). Ceramide addition produced changes similar to those seen with BDNF. It is interesting that both BDNF and ceramide treatment were shown to increase phosphoserine content on the intracellular domain of TrkA (MacPhee & Barker 1997), which suggests that BDNF acts through p75^NTR and ceramide production to influence TrkA signaling. Completing the analysis of signaling interactions, there are examples in which TrkA has a negative or restraining effect on p75^NTR signaling. Although NGF effectively induced sphingomyelin hydrolysis in cells expressing p75^NTR in the absence of TrkA, NGF did not do so in PC12 cells. This effect is apparently mediated by TrkA signaling because inhibition of TrkA signaling by the inhibitor K252a restored the ability of NGF to hydrolyze sphingomyelin (Dobrowsky et al 1995). In another example, when oligodendrocytes were transfected with TrkA, treatment with NGF induced activation of the MAPK pathway, suppressed JNK activity, and prevented cell death without influencing the activation of NFκB (Yoon et al 1998). These studies document the existence of robust, reciprocal interactions between TrkA and p75^NTR. An important goal is the elucidation of the molecular basis for these interactions and the definition of their physiological significance.

Signaling Endosomes

Internalization of the NGF-TrkA complex plays an important role in intracellular signaling, particularly in neurons, where retrograde transport of the "signal" from distant axon terminals is required to trigger signaling in the cell body. Considerable evidence suggests that this internalization involves endocytosis and the formation of "signaling endosomes," organelles in which NGF continues to be bound to its activated receptors (Grimes et al 1996, 1997; Riccio et al 1997; Tsui-Pierchala & Ginty 1999; Ure & Campenot 1997; Watson et al 1999). TrkA and p75^NTR activation appears to occur predominantly in caveolae-like membranes that contain many of the intermediates of their signaling pathways (Bilderback et al 1999, Huang et al 1999). It is possible that signaling endosomes are derived from these membranes, but it is noteworthy that at least some TrkA appears to be internalized via clathrin-coated membranes (CL Howe, JS Valletta, WC Mobley, submitted for publication). In fact, in recent studies we have shown that NGF increased the association of clathrin with membranes and induced the formation of complexes containing activated TrkA, clathrin heavy chain, and the plasma membrane specific adaptor complex, AP2. Moreover, we discovered that clathrin-coated vesicles isolated from NGF-treated cells contained NGF bound to activated TrkA, linking formation of the TrkA—clathrin complex to endocytosis via clathrin-coated membranes. It is exciting to note that Shc was recruited to these membranes, as was Ras and activated Erk1/2. Importantly, we found that NGF-induced clathrin-coated vesicles were able to signal in an in vitro kinase assay to propagate the NGF signal from Erk to Elk. Our findings indicate that NGF signals from endosomes and that clathrin-coated vesicles are one source of signaling endosomes produced in response to NGF treatment (CL Howe, JS Valletta, WC Mobley, submitted for publication). In other recent work, we have also isolated additional membranes that contain the NGF signal, suggesting the existence of a variety of signaling endosome species (CB Wu, CF Lai, WC Mobley, submitted for publication). Whether $p75^{NTR}$ signals from endosomes is an interesting possibility that requires further study, though preliminary evidence suggests that $p75^{NTR}$ does utilize the clathrin pathway for internalization, hinting at the existence of clathrin-coated vesicles that contain $p75^{NTR}$ and $p75^{NTR}$-associated signaling elements (CL Howe, AP Krüttgen, E Shooter, WC Mobley, unpublished observations). These findings are consistent with the concept that NGF signaling initiates the endocytosis of specialized membrane regions to form signaling endosomes that are enriched both in NGF receptors and their downstream signaling second-messenger target molecules. Signaling endosomes may exist as signaling complexes in the cytoplasm that can be transported from the site of their formation along neurites to the cell body.

Positive Feedback

There are several means by which positive feedback could be exerted in NGF signaling loops. First, as described above, NGF upregulates the expression of its own receptors. Second, NGF upregulates expression of such effector molecules as acetylcholine (Mobley et al 1985) or substance P and related tachykinins (Lindsay & Harmar 1989), which on their release would be expected to upregulate the expression of NGF by target tissues (Berzaghi et al 1993, French et al 1999, Woolf et al 1994). These changes may lead to reinforcement and strengthening of NGF signaling in a positive feedback manner (Sofroniew & Mobley 1993). The functional consequences of these effects are poorly understood. In nociceptive neurons, NGF upregulates the expression of neuropeptides such as substance P and calcitonin gene-related peptide, and positive feedback in this system may facilitate the induction of sensitization and hyperalgesia in response to tissue injury (Malcangio et al 1997; Mendell 1996; Verge et al 1995, 1996). In the forebrain, positive feedback may represent a means of reinforcing heavily used cholinergic connections, with significant ramifications for learning- and memory-related plasticity (Howe & Mobley 2001).

Interactions Between NGF and Other Molecular Signaling

Glutamate   Both NGF and glutamate signaling have well documented effects on the regulation of neuronal survival and neurite outgrowth during development.

Reports also exist of synergistic interactions between glutamate and NGF (Cohen-Cory et al 1991) or other neurotrophins (Morrison & Mason 1998) in sculpting neuronal survival or morphology during development. Interactions between glutamate and neurotrophin signaling that modulate neuronal excitablility persist in mature neurons (McAllister 1999). The intracellular signaling mechanisms involved in these effects are not yet known, and evidence for direct convergence of intracellular signaling mechanisms has not yet been reported, but it appears possible along several pathways, including modulation of cytoplasmic Ca^{2+} levels (Mattson 1996).

Estrogen   Estrogen enhances neuronal growth and differentiation and regulates cytoskeletal and growth-associated gene expression. There is now evidence for both colocalization of estrogen and NGF receptors in the same cells (Toran-Allerand et al 1992), and for direct convergence of NGF and estrogen signaling through the MAPK pathway (Singer et al 1999, Singh et al 1999). Such mechanisms may contribute to the many modulatory effects of estrogen on neural function. An important unresolved issue is how estrogen signals to induce MAPK activation.

Intracellular Signals   Much attention has focused on NGF signaling effects on intracellular Ca^{2+}, and there is reason to believe that Ca^{2+} plays an important role in many aspects of the biology of NGF and the other neurotrophins. In recent studies using PC12 cells, and fibroblasts transfected with TrkA or p75^{NTR}, NGF was shown to signal through TrkA and p75^{NTR} to cause acute and transient increases in intracellular Ca^{2+} as a result of increased uptake through L-type Ca^{2+} channels (Jia et al 1999, Jiang et al 1999). TrkA activation also increased intracellular Ca^{2+} mobilization (Jiang et al 1999). A number of functions can be envisioned for the increased intracellular Ca^{2+} that follows NGF treatment. The "Ca^{2+} set point hypothesis" posits that the level of cytoplasmic Ca^{2+} determines the degree of NGF signaling required to suppress cell death mechanisms during development (Johnson & Deckwerth 1993). Effects of NGF signaling on cytoplasmic Ca^{2+} levels also represent a means for NGF to influence the plasticity and vulnerability of mature neurons (Mattson et al 1995). It is tempting to suggest that through acute changes in intracellular Ca^{2+}, NGF could influence the behavior of synapses through increased release of neurotransmitters or of other neurotrophins (Berninger & Poo 1996, Krüttgen et al 1998).

EFFECTS AND FUNCTIONS OF NGF

Given the many cell types expressing NGF and NGF receptors, and the diverse intracellular signaling cascades triggered by NGF, it is not surprising that NGF signaling is implicated in many different functions during development and in adults. NGF mediates several types of intercellular communication and has been shown to act as (*a*) a retrogradely transported, target-derived factor that influences afferent neurons, (*b*) a locally released paracrine factor that affects both neurons and nonneuronal cells, (*c*) an autocrine factor acting on the same cells that produce and release it, and (*d*) an endocrine factor that acts after transport through the blood stream (Levi-Montalcini et al 1995, 1996).

NGF Roles in Development

NGF has a number of roles in the development of neuronal and nonneuronal cells. During development NGF promotes the survival and maturation of several populations of neurons that express TrkA and p75^{NTR}: in the PNS, sympathetic and sensory neurons (Conover & Yancopoulos 1997, Johnson et al 1986, Lehmann et al 1999, Snider 1994) and in the CNS, basal forebrain, and striatal cholinergic neurons (Kew et al 1996, Li et al 1995, Svendsen et al 1994). For each of these populations, NGF participates in classical target-derived neurotrophic relationships. The importance of these relationships is evident from the results of experiments in which the genes for NGF or the NGF receptors were disrupted. Knocking out the NGF gene resulted in the loss of most small nociceptive dorsal root ganglion (DRG) neurons and sympathetic neurons in the PNS (Crowley et al 1994). In the CNS of animals heterozygous for disruption of the NGF gene, there was a clear reduction in the number of basal forebrain cholinergic neurons, atrophy of these cells, and reduction in the cholinergic innervation of the hippocampus (Chen et al 1997). Further study is needed to explore the effect of NGF gene disruption on striatal cholinergic neurons. However, both basal forebrain and striatal cholinergic neurons are reduced in number and size in animals in which the gene for TrkA was disrupted (Fagan et al 1997). Also, there was marked depletion of both small DRG neurons and sympathetic neurons in TrkA knockout animals (Smeyne et al 1994).

Changes are also seen with p75^{NTR} gene disruption in both the PNS and CNS. In the PNS, there are defects in sensory innervation of skin and in sympathetic innervation of the pineal gland and sweat glands (Lee 1992, 1994b,c). Remarkably, disrupting the gene for p75^{NTR} had little if any effect on the number of sympathetic neurons in animals in which both copies of the NGF gene were present (Brennan et al 1999). However, disrupting p75^{NTR} increased the number of sympathetic neurons in animals in which the NGF gene was also disrupted (Brennan et al 1999). The mechanism for this surprising and interesting result is not yet established. However, one possible interpretation is that NGF signaling through p75^{NTR} restrains the normal survival and development of sympathetic neurons. Another is that p75^{NTR} gene disruption allows NT-3 to signal through TrkA to enhance neuron survival, a suggestion for which there is experimental evidence (Brennan et al 1999). Further studies will be needed to clarify the interaction.

Paralleling the results for sympathetic neurons, p75^{NTR} gene disruption appears to increase the number and size of basal forebrain cholinergic neurons (Greferath et al 2000, Yeo et al 1997, but see Peterson et al 1999). While it is interesting to speculate that p75^{NTR} signaling inhibits the normal development and function of these neurons, the mechanistic basis for this is yet to be determined. Though additional studies are required to detail NGF effects in the developing nervous system, gene disruption studies have documented an important role for NGF in the survival and differentiation of both PNS and CNS neurons.

Interesting additional themes have emerged that have considerably modified the original neurotrophic hypothesis. First, CNS and PNS neurons are not continuously dependent on the constitutive supply of a single target-derived factor throughout life. Rather, a variety of different molecules from different sources influence developmental survival and maturation. For example, sensory neurons are transiently dependent for survival on different neurotrophins at different time points as they progress through phases of development (Davies 1994). Second, transiently required growth factors may derive from sources other than the final target region, such as local interactions around the cell bodies (Enokido et al 1999), or intermediate targets that axons encounter and then grow past en route to final destinations (Wang & Tessier-Lavigne 1999). Third, NGF signaling can also mediate axon sprouting, as well as growth cone turning and local guidance (Campenot 1977, Gallo et al 1997, Patel et al 2000, Rice et al 1998, Tuttle & O'Leary 1998). For example, a recent study shows that NGF is critical for the elongation of the peripheral but not the central processes of sensory neurons (Patel et al 2000). Fourth, NGF can also induce the death of certain developing neurons by signaling through p75^{NTR} in the absence of TrkA, as in the retina (Frade & Barde 1998, Frade et al 1996). Regarding glia, NGF may regulate the development of oligodendrocytes, but here too, rather than promoting survival, NGF signaling via p75^{NTR} can under certain circumstances induce the death of these cells (Casaccia-Bonnefil et al 1996, Chao et al 1998, Gu et al 1999). Taken together, these findings point to the complex neurotrophic environment that guides the development of the nervous system. A role for TrkA signaling in NGF actions during development is well established and appears to be the dominant theme in signaling events that are required for survival and differentiation. Nevertheless, it is apparent that p75^{NTR} has important roles to play in modulation of TrkA signaling and may also independently regulate cell survival. The stage is now set for exploring the details of NGF signaling in the developing nervous system.

NGF Roles in Adults

Both NGF and NGF receptors continue to be expressed and dynamically regulated by many different cells types throughout adult life and aging. Information about different NGF functions in adults is emerging for both NGF-responsive neurons and nonneuronal cell types. Once mature, most neurons lose absolute dependence on target-derived growth factors for acute survival. In adults, the focus of NGF signaling shifts away from the regulation of neuronal survival to the regulation of neuronal phenotype and function. In cases well studied thus far, CNS and PNS neurons that are developmentally dependent upon NGF for survival become independent of a constitutive supply of target-derived NGF for acute survival once they have established their connections and reached maturity. In adults, sensory and CNS cholinergic neurons do not die for many months after NGF withdrawal (Johnson & Deckwerth 1993; Sofroniew et al 1990, 1993; Svendsen et al 1994). Adult sympathetic neurons also do not die acutely after immunologically induced NGF withdrawal; instead, they undergo a gradual cell death of about 25% after one month, which increases to about 40% after 3 months (Ruit et al 1990). The resistance of these neurons to NGF withdrawal may be caused by loss of the c-fos induction that normally follows such withdrawal in young neurons and by consequent interruption of a cascade that involves both c-fos and Bax (Easton et al 1997). Although these mature NGF-responsive neurons become independent of NGF for acute survival, they all undergo atrophic changes if subjected to NGF withdrawal. These changes take the form of cell shrinkage (which is often severe) and reduced transmitter-related gene expression.

It is likely that TrkA signaling mechanisms are implicated in these changes, since NGF has been shown to increase cell size in neurons that do not express $p75^{NTR}$ (Holtzman et al 1995). Because mature NGF receptor expressing neurons do not require a constitutive supply of NGF for acute survival, the possibility exists that acute fluctuations in NGF signaling dynamically regulate various types of activities in mature NGF-responsive cells. As discussed in this section, these functions include modulation of the plasticity of NGF-responsive neurons. A particularly well-documented and striking example of this is the regulation by NGF of mature nociceptive neurons (Woolf et al 1996).

The widespread production of NGF by glial cells and other nonneuronal cells is leading to new ideas about other types of NGF functions, prominent among which appear to be roles in inflammation and the response to injury in the CNS, PNS, and peripheral tissue. NGF appears likely to have other functions that are currently not well understood. For example, NGF infused into the lateral cerebral ventricles induces hypophagia and weight loss in rats (Winkler et al 2000), and NGF treatment has been reported to affect appetite in patients in clinical trials (Petty et al 1994).

Plasticity of NGF-Responsive Neurons   The adult nervous system exhibits a remarkable ability to alter both its structure and function in response to stimuli, a capacity commonly referred to as plasticity. Neurotrophins are implicated as molecular mediators of specific forms of both structural and functional plasticity. NGF has thus far been associated in particular with effects on structural plasticity and has far fewer reported direct effects on neuronal activity in comparison with other neurotrophins, such as BDNF (McAllister 1999). Nevertheless, NGF has reported effects on stimulus-dependent activity in adult somatosensory cortex (Cellerino & Maffei 1996, Gu et al 1994, Prakash et al 1996) that appear to be mediated by TrkA and facilitated by $p75^{NTR}$ (Pizzorusso et al 1999). The mechanistic basis for these effects is unclear but may involve NGF-dependent modulation of cholinergic function and subsequent modification of cortical plasticity (Howe & Mobley 2001). In the PNS, NGF regulates the cell body size, terminal sprouting, dendritic arborization, and gene expression of sympathetic neurons and small nociceptive sensory neurons (Johnson et al 1986, Ruit et al 1990). In the CNS, NGF regulates the cell body size, dendritic arborization, terminal sprouting, and gene expression of basal forebrain and striatal cholinergic neurons (Cuello 1996, Debeir et al 1999, Howe & Mobley 2001). As discussed above, NGF increases the expression of its own receptors, as well as of transmitters or transmitter-producing enzymes, and may do so to modulate cell function in a context-specific manner. It is interesting that in both the CNS and PNS, target-derived as well as locally applied NGF will exert these effects, and there are both target-derived and local sources of NGF-producing cells available to the neurons. The significance of, and interactions between, local and target-derived signaling for NGF-responsive neurons are not yet understood.

Nociception   Small nociceptive sensory neurons express both types of NGF receptor throughout life, and NGF has a variety of effects on these cells, including upregulation of TrkA and p75^{NTR}, CGRP (calcitonin gene related peptide), and tachykinin expression, as well as modulation of cell size, activity, and neuropeptide release (Malcangio et al 1997; Mendell 1996; Verge et al 1995, 1996). NGF signaling in these cells leads to hypersensitivity to nociceptive stimuli in the form of allodynia and hyperalgesia in both animals and patients given NGF in clinical trials for peripheral neuropathies (Petty et al 1994, Shu & Mendell 1999). NGF is expressed and released in many tissues in response to injury, and blockade of NGF signaling using function blocking antibodies in experimental animals with skin injury and inflammation prevents the development of hyperalgesia (McMahon et al 1995). NGF elicits both mechanical and thermal hyperalgesia. With respect to the latter, NGF acts as a peripheral sensitizing agent to alter the response of nociceptors to noxious stimuli. It may accomplish this effect in part by inducing mast cell degranulation and the consequent release of serotonin, histamine, and NGF itself (Shu & Mendell 1999). In terms of mechanical hyperalgesia, NGF may act centrally by upregulating CGRP, substance P, and BDNF (Shu & Mendell 1999). Recent findings suggest that PKC may mediate the actions of NGF on peripheral nociception (Khasar et al 1999). Thus, NGF plays an important role in the nociceptive response that follows tissue injury by inducing both peripheral and central sensitization through signaling that involves several cell types, including leukocytes and nociceptive sensory and sympathetic neurons (Mendell et al 1999, Shu & Mendell 1999, Woolf et al 1996). NGF may also be an important mediator in pain due to visceral inflammation (Dmitrieva et al 1997) and in neuropathic pain syndromes induced by peripheral nerve irritation.

Immune and Inflammatory System   NGF has numerous effects on immune and inflammatory cells that are generally directed at inducing their state of activation and effector functions (Levi-Montalcini et al 1996, Otten et al 1994, Simone et al 1999). NGF increases mast cell number, induces mast cell degranulation, and increases mast cell expression of cyclooxygenase and interleukin-6 (Marshall et al 1999, Simone et al 1999). NGF activates monocytes, macrophages, and CNS microglia by increasing their phagocytic activity and by inducing their expression of interleukin-1, Fcγ receptor, and lysosomal proteases of the cathepsin family (Liuzzo et al 1999; Susaki et al 1996, 1998). NGF is chemotactic for neutrophils (Gee et al 1983). NGF influences T and B lymphocyte proliferation, is an autocrine survival factor for B lymphocytes, and stimulates immunoglobulin production (Levi-Montalcini et al 1996, Otten et al 1994, Torcia et al 1996). Through its interactions with immune and inflammatory cells, as well as with glia and neurons, NGF has been suggested to play a role in various diseases with postulated autoimmune and inflammatory components, including arthritis and demyelinating diseases (Aloe 1998, Bonini et al 1999, Levi-Montalcini et al 1996).

NGF AND NEUROPROTECTION

Substantial evidence suggests that among other functions, NGF acts to protect neurons from endogenous toxic events generated during the response to tissue injury and that NGF signaling facilitates regrowth and repair. The signaling mechanisms engaged in neuroprotection have not been defined. However, it appears that the protective effects of NGF extend both to neurons known to express NGF receptors and to those that are not known to express such receptors.

NGF Protection of Neurons Known to Express NGF Receptors

Protection from Axotomy   Throughout the PNS and CNS, adult neurons vary considerably in their vulnerability to axotomy. For unknown reasons, some axotomized neurons survive with few obvious changes, others survive but atrophy to moderate or severe degrees, and others die either rapidly or over a prolonged time course (Sofroniew 1999). Among NGF-responsive neurons, sensory neurons and cholinergic neurons in the basal nucleus survive axotomy but exhibit moderate-to-severe atrophy (Sofroniew et al 1983, Verge et al 1996). In contrast, about 50% (but not all) of axotomized sympathetic neurons and septal cholinergic neurons rapidly die (O'Brien et al 1990, Ruit et al 1990, Tuszynski et al 1990). Neither the mechanism of axotomy-induced cell death, nor the reasons that some neurons die while others survive but atrophy, are understood. In the adult septum, axotomy-induced cell death is not due to loss of NGF signaling because target lesion and NGF-depletion studies show that these neurons are not acutely dependent on NGF for survival (Kordower et al 1993; Sofroniew et al 1990, 1993). Nevertheless, the death of these neurons can be largely prevented in both rodents and primates by NGF infusions at the time of the axotomy (Hefti 1986, Tuszynski & Gage 1995b, Williams et al 1986). In the septum, this effect appears to be mediated via TrkA signaling (Lucidi-Phillipi et al 1996), and NGF need only be given transiently for a few weeks after the axotomy and can then be discontinued without subsequent loss of neurons (Tuszynski & Gage 1995a). In addition, NGF is able to prevent axotomy-induced atrophy in cells that are not killed by axotomy, as well as to reverse atrophy that has already occurred (Cuello 1994). The intracellular pathways through which NGF signaling protects TrkA and p75^{NTR} expressing neurons from axotomy-induced death are not defined.

Protection from Glutamate Excitotoxicity   Aloe (1987) first reported a protective effect of NGF against glutamate receptor–mediated excitotoxicity, such that NGF infusions reduced the overall size of excitotoxic lesions in the striatum and prevented the death of cholinergic, NGF receptor–expressing, striatal neurons. These observations have been confirmed by others using local infusions of NGF or grafts of cells genetically modified to secrete NGF (Davies & Beardsall 1992, Frim et al 1993, Martinex-Serrano & Bjorklund 1996, Schumacher et al 1991). NGF has also been reported to protect PC12 cells from anoxia and glucose deprivation, or from nitric oxide cytotoxicity (Boniece & Wagner 1993, Wada et al 1996). To examine the potential role of endogenous NGF signaling in the protection of NGF-responsive neurons from excitotoxicity, we recently studied septal cholinergic neurons that express NGF receptors and project to hippocampus and are thus, in contrast to striatal cholinergic neurons, well separated from their target cells that produce NGF. Both local and target-derived (i.e. retrogradely transported) NGF signaling significantly attenuated glutamate receptor–mediated excitotoxic death of these neurons in young adult rats. In addition, aged rats that have a reduced capacity to retrogradely transport NGF, and young adult rats given target lesions that deplete access to retrogradely derived NGF, both exhibited (*a*) significantly increased vulnerability of cholinergic neurons to glutamate receptor–mediated toxicity, (*b*) significantly reduced protective effects of local NGF, and (*c*) significantly reduced levels of TrkA. Chronic intracerebroventricular NGF significantly restored TrkA and the protective effect of local NGF (Horner et al 1999; HHD Lam, CH Horner, J Berke, JD Cooper, RE Brown, SB Dunnett, MV Sofroniew, submitted for publication). These findings suggest that in adult forebrain, signaling through TrkA serves ongoing neuroprotective functions. They also suggest that loss of endogenous NGF signaling leads to increased vulnerability of these cells to excitotoxicity. However, the intracellular pathways through which NGF signaling protects TrkA and p75^{NTR} expressing neurons from glutamate receptor–mediated death are not yet defined.

NGF Protection of CNS Neurons that do not Appear to Express NGF Receptors

Although readily detectable expression of TrkA and p75^{NTR} is confined to relatively few populations of neurons in both the PNS and CNS, NGF is reported to protect a broad spectrum of neurons from ischemia, glutamate receptor–mediated excitotoxicity, and metabolic insults such as glucose deprivation and oxidative stress. Protective effects of NGF on neurons not known to express NGF receptors have been described and confirmed by numerous research groups using many different in vivo and in vitro models.

Protection from Glutamate Excitotoxicity and Ischemia In Vivo   The initial report by Aloe (1987) of NGF protection from glutamate receptor–mediated excitotoxicity in striatum described a reduction in lesion size greater than that which would be expected if only TrkA- and p75^{NTR}-expressing neurons had been protected. A generalized and widespread neuroprotective effect of NGF has been confirmed in several in vivo experimental models. In response to excitotoxic glutamate analogues infused into the striatum, widespread and generalized neuronal protection is achieved both by simultaneous direct infusions of NGF (Holtzman et al 1996) and by previously placed grafts of genetically modified cells (using various different cell types) that express and release NGF (Frim et al 1993, Martinex-Serrano & Bjorklund 1996, Schumacher et al 1991). In addition, NGF delivered by infusion, grafts, or transgenic expression or induced in astrocytes by β2-adrenoreceptor activation protects diverse populations of retinal, hippocampal, cortical, and other forebrain neurons that do not express detectable levels of TrkA and p75^{NTR} from ischemia or excitotoxic glutamate analogues (Culmsee et al 1999, Guegan et al 1997, Shigeno et al 1991, Siliprandi et al 1993). It is interesting that neutralizing antibodies to NGF block the protective effects of interleukin-1 against glutamate excitotoxicity, further indicating that endogenous NGF can exert widespread protection (Carlson et al 1999, Strijbos & Rothwell 1995).

Protection from Toxins and Metabolic Insults In Vitro   NGF neuroprotective effects have been described and characterized extensively in vitro for many types of neurons not known to express TrkA and p75^{NTR}. NGF protects cortical, hippocampal, striatal, retinal, and other types of neurons grown in dispersed cell tissue cultures from glutamate excitotoxicity, hypoglycemia, and oxidative stress, as first reported by Mattson and colleagues (Cheng & Mattson 1991, Mattson et al 1995) and confirmed, and extended to include ethanol toxicity, in many laboratories (Cunha et al 1998, Heaton et al 1993, Luo et al 1997, Mattson & Marck 1996, Singer et al 1999).

Potential Mechanisms of NGF-Mediated Neuroprotection of Non-NGF Receptor-Expressing Neurons   The signaling mechanisms underlying NGF-mediated protection of neurons that do not express NGF receptors are not understood. In some cases, protection has been shown to involve stabilization of intracellular Ca^{2+} levels and prevention of the surge in cytoplasmic Ca^{2+} associated with cell death, as triggered by excess glutamate signaling or oxidative stress (Mattson 1996). In a recent study, protection of cortical neurons from glutamate excitotoxicity in vitro by both NGF and estrogen were found to require activation of the MAPK pathway (Singer et al 1999). The means by which NGF signaling leads to widespread neuroprotective effects is not known, and there are several different options to consider. The first is simply that TrkA and p75^{NTR} expression is more widespread among neurons than is currently appreciated, and very low levels of TrkA or p75^{NTR} might be able to mediate the protective signaling. The second option is that a novel NGF receptor exists that is widely expressed.

It is difficult simply to dismiss this possibility. Although radiolabeled NGF binding to brain sections defines the known NGF-responsive populations, Altar et al (1991) also found significant binding in the hippocampus, in the subiculum, and in the cingulate, frontal, parietal, and occipital cortex. It is likely that much of this binding is attributable to NGF receptor expression on the axons of known NGF responsive neurons. However, the formal possibility also exists that this binding represents the expression of a novel receptor on cells in these regions. A third option is that widespread neuroprotective effects of NGF on neurons that do not appear to express TrkA and p75^{NTR} are mediated via NGF signaling through such NGF-responsive nonneuronal cells as glia and inflammatory cells. Several lines of evidence support this possibility. As described above, nonneuronal cells (*a*) express both NGF and NGF receptors, and expression is generally upregulated by injury and other CNS insults, (*b*) respond to NGF signaling, and (*c*) generate molecules that are potentially toxic to neurons, such as nitric oxide and reactive oxygen species, as part of their response to injury and inflammation. Potentially neuroprotective effects of NGF include the rapid inhibition of reactive oxygen species generation in vitro (Dugan et al 1997) and the rapid inhibition of basal and glutamate receptor–induced nitric oxide synthase in vivo (Lam et al 1998).

NGF AND NEURAL REPAIR

The molecular signals that control cellular interactions during the response to injury and attempts at repair in the PNS and CNS are beginning to be understood. Available evidence suggests a complex interplay of molecules, including numerous growth factors and cytokines. Changes in the expression of NGF and NGF receptors in both PNS and CNS are compatible with roles in repair, and their involvement warrants further investigation.

NGF and PNS Repair

Glia and Inflammatory Cells   After a peripheral nerve injury that causes axotomy, both meylinating and nonmyelinating Schwann cells distal to the injury dedifferentiate and reenter the cell cycle. Proliferating and reactive Schwann cells produce growth factors, cytokines, and growth-associated proteins, which are likely to play key roles in axon regeneration and nerve repair (Frostick et al 1998, Mirsky & Jessen 1999, Verge et al 1996). Changes in gene expression by reactive Schwann cells include a marked upregulation of both NGF and p75^{NTR} (Frostick et al 1998, Lindholm et al 1987, Mirsky & Jessen 1999). Peripheral nerve injury also leads to substantial infiltration of inflammatory cells, and among these, macrophages, mast cells, and subsets of T cells have the capacity to express NGF. The precise roles and functions of NGF signaling for different cell types during the response to peripheral nerve injury are not certain. An overall effect of exogenously administered NGF is to increase both the number and myelination of regenerating axons in experimental entubation repair in which transected peripheral nerve stumps regrow through bridging tubes containing artificial matrices (Derby et al 1993, Rich et al 1989). This may be due to effects of NGF signaling both on regenerating nerve fibers, as discussed below, and on Schwann and inflammatory cells. Exogenous NGF promotes Schwann cell migration (Anton et al 1994), and Schwann cell migration is thought to precede and promote axon elongation into entubation repair sites (Madison & Archibald 1994). Denervation is the trigger for Schwann cell production of NGF and NGF receptors, and the proximity of regenerating axons suppresses this expression, probably through diffusible molecules (Taniuchi et al 1988). Downregulation of p75^{NTR} on Schwann cells induced by contact with axons is likely to lead to Schwann cell accumulation around regenerating axons and may promote eventual remyelination (Madison & Archibald 1994).

Neurons   NGF receptor expression in peripheral nerve cells changes after axotomy; TrkA and p75^{NTR} expression by sensory neurons decreases (Verge et al 1996), whereas motor neurons begin to reexpress detectable levels of p75^{NTR} (Ernfors et al 1988, Wood et al 1990). Neither the reasons for, nor the consequences of, the decline in TrkA expression by axotomized sensory neurons are understood. Nevertheless, sensory neurons regenerate robustly and, after regeneration, reexpress TrkA at preinjury levels. As described above, motor neurons express high levels of p75^{NTR} during the period of axon outgrowth during development and downregulate expression to undetectable levels after contact with target structures. After injury in adult PNS, motor neurons that are allowed to regenerate recapitulate this developmental event and reexpress p75^{NTR} during the period of axon regeneration. Reexpression does not occur in neurons that are not allowed to regenerate and requires retrograde transport from axons growing through injured peripheral nerve tissue (Bussmann & Sofroniew 1999, Wood et al 1990). It is not yet clear whether p75^{NTR} reexpression plays an essential role in the regeneration of either sensory or motor neurons; however, recent findings show that p75^{NTR} signaling through Rho and ceramide pathways may be involved in promoting axon elongation (Brann et al 1999, Lehmann et al 1999, Yamashita et al 1999b).

NGF and CNS Repair

In the CNS, as in the PNS, injury or insults such as trauma, ischemia, or degenerative disease trigger rapid and substantial upregulation in the expression of NGF and NGF receptors by cell types involved in the repair process, including (*a*) local astrocytes and microglia, (*b*) invading inflammatory cells, including macrophages, mast cells, and subsets of T cells, and (*c*) certain neurons. The purpose underlying such changes in expression levels is not yet known. Several lines of evidence suggest that NGF signaling may in some of these instances facilitate the repair or reorganization of neural connections.

Axon Sprouting and Growth   Forebrain cholinergic neurons that express TrkA and p75[NTR] exhibit neurite outgrowth in response to NGF in adults, particularly after injury (Heisenberg et al 1994, Kawaja & Gage 1991, Kordower et al 1994), suggesting that NGF may stimulate regrowth and reorganziation of connectivity of receptor bearing neurons after injury. Several other populations of CNS neurons transiently express p75[NTR] during development and, in a manner similar to primary motor neurons, downregulate this expression after maturity. These populations include cerebellar Purkinje, hippocampal pyramidal neurons, and retinal neurons. Whether these neurons reexpress p75[NTR] after injury, as do regenerating motor neurons, and whether NGF signaling might increase their axon regeneration, as in the PNS, is a point for further study. In support of this possibility, NGF induces neurite outgrowth in developing hippocampal pyramidal neurons in vitro, via a p75[NTR]-ceramide–mediated signaling pathway (Brann et al 1999), and in retinal neurons, p75[NTR] signaling activates Rho pathways, which have been associated with axon growth (Lehmann et al 1999, Yamashita et al 1999b).

Other intriguing recent observations suggest that stimulation with NGF and other growth factors may regulate the intrinsic capacity of neurons to regenerate transected axons through the hostile extracellular environment of the injured adult CNS. In adult spinal cord, NGF and other growth factor infusions stimulate the regrowth of fibers of receptor-expressing sensory neurons across the PNS-CNS border in a manner that does not occur in the absence of growth factor, which suggests that the growth factors (including NGF) enabled these axons to overcome environmental cues that inhibit axon elongation in the CNS (Ramer et al 2000). The signaling mechanism for this effect is not established, but evidence from other studies suggests it may involve cAMP. It has been known for some time that if the peripheral nerves containing peripheral branches of DRG sensory neurons receive two lesions within a short period of time, axon regeneration occurs more rapidly after the second injury. Moreover, a conditioning lesion made to the peripheral branches of DRG neurons a number of days prior to a lesion of the central branches results in increased regeneration of these central branches into PNS-grafts (Richardson & Issa 1984). Neumann & Woolf (1999) recently extended these observations by showing that a peripheral conditioning lesion also enables considerable regeneration of the central branches of DRG neurons in the absence of any other intervention after a second, delayed injury in the adult CNS. The findings of Cai et al (1999) suggest a possible mechanism underlying these observations by showing that NGF and other growth factors can mimic the effects of peripheral conditioning lesions by increasing the capacity of DRG neurons to grow on otherwise inhibitory CNS myelin substrates in vitro. The intracellular signaling pathway for this effect involves elevation of neuronal cAMP levels and requires PKA activity. NGF and other growth factors are generated within the injured peripheral nerve tissue and are transported back to the injured neurons within the time frame of the observed enhancements of axon regeneration by the conditioning lesion. Such findings point toward the potential to exploit signaling mechanisms of NGF and other growth factors to facilitate repair in the CNS.

Inappropriate Connectivity   It is important to consider that increases in NGF and other growth factors after injury may stimulate formation of inappropriate connections and concomitant unwanted side effects. For example, after experimental spinal cord injury, endogenous NGF-induced sprouting of small-diameter afferent fibers has been implicated as a primary cause of autonomic dysreflexia (Krenz et al 1999), an important clinical problem of patients with spinal cord injury. Thus, not all fiber regeneration and connectivity triggered by NGF and other growth factors after injury may be beneficial, and aberrant and inappropriate pathways may also be formed. Such possibilities must be studied and understood before treatment strategies are considered.

FAILURE OF NGF SIGNALING AND NEURODEGENERATION

Early ideas about the functions of NGF and other members of the neurotrophin family naturally focused on their roles as neuronal survival factors (Barde 1989). Strict interpretation of the neurotrophic hypothesis led to the suggestion that neurons might be dependent on a continuous, constitutive supply of a single target-derived neurotrophin not only during development but throughout adult life and aging. An obvious extension of this idea is that an interruption in this continuously required supply of target-derived neurotrophic support might be a direct cause of neuronal death in aging and degenerative disease. Evidence accumulated to date has not substantiated this idea. As discussed above, developmentally NGF-dependent neurons for the most part become independent of a constitutive supply of target-derived NGF for acute survival once they have established their connections and reached maturity. In addition, there is no evidence for a decline in NGF production in aging or degenerative disease; neither NGF mRNA nor protein are reduced in the cerebral cortex of aged animals (Alberch et al 1991, Crutcher & Weingartner 1991) or patients with Alzheimer's disease, where NGF protein levels may even be increased (Crutcher et al 1993, Goedert et al 1986, Scott et al 1995). These observations do not, however, preclude other effects precipitated by an aging- or disease-related failure of NGF signaling, and there is evidence that failure of NGF signaling contributes to certain neurodegenerative processes.

NGF and the Age-Related Atrophy and Vulnerability of Forebrain Cholinergic Neurons

TrkA and p75^{NTR} expressing basal forebrain cholinergic neurons undergo moderate degenerative changes in aging, including cell atrophy, downregulation of transmitter-synthesizing enzyme, and mild cell loss in both animals and humans, and these degenerative changes are markedly exacerbated in Alzheimer's disease (de Lacalle et al 1991a,b; Finch 1993; Fischer et al 1987, 1991; Pearson et al 1983; Whitehouse et al 1982). With the exception of cell loss, such atrophic changes can also be subacutely induced in young adult animals by disrupting access of these neurons to their target cells, which produce NGF and other neurotrophins (Kordower et al 1993; Sofroniew et al 1990, 1993). Furthermore, atrophic changes can be reversed in aged animals by infusions of exogenous NGF (Fischer et al 1987). These findings suggest that failure of neurotrophin signaling may underlie or contribute to the atrophic neuronal changes observed in aging or degenerative disease. However, as described above, most studies agree that NGF levels are not substantially reduced in aging or Alzheimer's disease. Alternatively, available evidence does suggest that intrinsic neuronal changes might compromise the ability of aged basal forebrain cholinergic neurons to derive neurotrophic support. In aged rats, forebrain cholinergic neurons exhibit a reduced capacity for generalized retrograde transport and a pronounced reduction specifically in the retrograde transport of NGF; cholinergic neurons that do not transport NGF are severely shrunken and downregulate TrkA expression (Cooper et al 1994, de Lacalle et al 1996). Even though NGF protein levels are not reduced in the cortex, they are substantially reduced in the basal forebrain of aged animals and patients with Alzheimer's disease (Alberch et al 1991, Scott et al 1995), as well as in mice with Down's syndrome (JC Cooper, A Salehi, P Belichenko, J-D Delcroix, J Chua-Couzens, J Kilbridge, CL Howe, WC Mobley, submitted for publication). These findings all suggest a disturbance in retrograde transport and NGF signaling mechanisms in these neurons in aging and Alzheimer's disease. Age-related declines of axonal transport have been reported in other neuronal systems, such as sciatic nerve (McMartin & O'Connor 1979), and disturbances in neurofilaments and axonal transport have been proposed as a possible mechanism contributing to neurodegenerative changes associated with aging and such age-related CNS diseases as Alzheimer's disease (Gadjusek 1985, Saper et al 1987). Abnormal phosphorylation of the microtubule-associated protein Tau is the basis for the formation of tangles in aged neurons (Goedert 1996), and forebrain cholinergic neurons exhibit tangle formation in aging and Alzheimer's disease, with likely disruption of microtubule-dependent axonal transport.

Taken together, these observations suggest that an intrinsic, age-related reduction in the capacity of basal forebrain cholinergic neurons to sustain retrograde NGF signaling may contribute to the degenerative changes seen in basal forebrain cholinergic neurons in aging that is markedly exacerbated in Alzheimer's disease. Basal forebrain cholinergic neurons may be exquisitely sensitive to interruptions in NGF signaling as a by-product of the positive feedback in NGF signaling discussed above, such that an initially small failure in NGF signaling may be rapidly exacerbated by downregulation of NGF receptors and other changes (Sofroniew & Mobley 1993). The consequences of this failure in NGF signaling may be both direct, by inducing cellular atrophy and changes in gene expression, and indirect, by increasing the vulnerability of the atrophic neurons to other insults, such as glutamate signaling, which may become excitotoxic in the absence of NGF signaling. As discussed above, recent studies have shown NGF significantly attenuates glutamate receptor–mediated death of forebrain cholinergic neurons, and that aged rats with reduced capacity to retrogradely transport NGF exhibit significantly increased vulnerability of cholinergic neurons to glutamate toxicity. These findings suggest that endogenous NGF serves neuroprotective functions, and that loss of target-derived NGF signaling, by increasing vulnerability to excitotoxicity, may contribute to the gradual loss of basal forebrain cholinergic neurons in aging and Alzheimer's disease. The potential for failure of neuroprotective mechanisms provided by endogenous growth factor signaling to cause or exacerbate neurodegeneration is not widely considered in the context of mechanisms that may be operant in neurodegeneration. A better understanding of such mechanisms may facilitate the development of protective interventions.

NGF SIGNALING AND THERAPEUTIC INTERVENTION

The many potentially beneficial effects of NGF signaling in neural protection and repair raise possibilities of therapeutic intervention using either NGF protein or small-molecule analogues (LeSauteur et al 1996, Longo & Mobley 1996, Yuen & Mobley 1996). Nevertheless, the diversity of cells that respond to NGF signaling may predispose to unwanted side effects when NGF or NGF agonists/antagonists are delivered systemically. For example, in a randomized and placebo-controlled study on the safety of intravenous or subcutaneous administration of NGF, patients experienced sustained hyperalgesia, which varied in a dose-dependent manner (Petty et al 1994). Infusions of NGF into the cerebroventricular system also led to a variety of undesirable and unwanted effects in patients and animals (Eriksdotter-Jönhagen et al 1998, Winkler et al 2000). Thus, ways of achieving site-specific delivery or identifying means of achieving cell-type selective activation of NGF receptors, or of NGF signalling pathways, may be needed to realize the therapeutic potential that NGF signaling appears to hold.

The diversity of NGF-responsive cells also raises the possibility that NGF signaling may, under certain circumstances, spill over from one cell compartment to another, causing unwanted and undesirable effects, and may thus represent part of a pathophysiological mechanism. For example, NGF produced and released for signaling intended for regulation of neurons or inflammatory cells may affect glia and present a potential means by which oligodendrocytes could be killed by p75^{NTR}-signaling. Such potential effects will need to be understood before there are attempts to exploit NGF signaling for therapeutic interventions in the CNS. Therapeutic trials must also be preceded by careful studies in realistic animal models of disease or injury. Even when these steps have been completed and have defined a role for NGF treatment, there is no guarantee of success in humans. The recent phase III trial of NGF in diabetic neuropathy failed to show an effect of NGF. This may well have resulted from the failure to deliver NGF in adequate amounts to sensory neurons, a consequence of the fact that pain limited the NGF dose that could be given. These and other factors will need to be addressed before NGF or NGF analogues can be successfully used to intervene in neurodegenerative diseases.

CONCLUSIONS

Growth factors play important roles as intercellular signaling molecules throughout development, adult life, and aging. Many growth factors influence a wide range of cell types and take part in numerous functions. NGF is no exception, and ideas about its functions have steadily expanded. In addition to its long-recognized developmental effects on the survival and maturation of a few restricted neuronal populations in the PNS and CNS, NGF has effects on numerous types of neurons and nonneuronal cells throughout adult life and during aging. Among its different functions, NGF signaling appears to play important roles in the response to injury or disease that subserve neuroprotection and neural repair.

ACKNOWLEDGMENTS

The authors wish to acknowledge the support of the McGowan Charitable Trust, the Adler Foundation, the Christopher Reeve Paralysis Foundation, and NIH Grant NS24054. CL Howe is supported by a Howard Hughes Medical Institute Predoctoral Fellowship and the Adler Foundation. We also thank numerous colleagues for discussions, information, and criticism, including J-D Delcroix, C-B Wu, P Belichenko, A Salehi, J Valletta, S Lai, A Langer-Gould, and A Krüttgen.

Visit the Annual Reviews home page at www.AnnualReviews.org

LITERATURE CITED

Abiru Y, Katoh-Semba R, Nishio C, Hatanaka H. 1998. High potassium enhances secretion of neurotrophic factors from cultured astrocytes. *Brain Res.* 809:115–26

Agerman K, Canlon B, Duan M, Ernfors P. 1999. Neurotrophins, NMDA receptors, and nitric oxide in development and protection of the auditory system. *Ann. NY Acad. Sci.* 884:131–42

Alberch J, Perez-Navarro E, Arenas E, Marsal J. 1991. Involvement of nerve growth factor and its receptor in the regulation of the cholinergic function in aged rats. *J. Neurochem.* 57:1483–87

Alema S, Casalbore P, Agostini E, Tato F. 1985. Differentiation of PC12 phaeochromocytoma cells induced by v-src oncogene. *Nature* 316:557–9

Alessi DR, Caudwell FB, Andjelkovic M, Hemmings BA, Cohen P. 1996. Molecular basis for the substrate specificity of protein kinase B; comparison with MAPKAP kinase-1 and p70 S6 kinase. *FEBS Lett.* 399:333–38

Alessi DR, Deak M, Casamayor A, Caudwell FB, Morrice N, et al. 1997a. 3-Phosphoinositide-dependent protein kinase-1 (PDK1): structural and functional homology with the Drosophila DSTPK61 kinase. *Curr. Biol.* 7:776–89

Alessi DR, James SR, Downes CP, Holmes AB, Gaffney PR, et al. 1997b. Characterization of a 3-phosphoinositide-dependent protein kinase which phosphorylates and activates protein kinase Balpha. *Curr. Biol.* 7:261–69

Aloe L. 1987. Intracerebral pretreatment with nerve growth factor prevents irreversible brain lesions in neonatal rats injected with ibotenic acid. *Biotechnology* 5:1085–86

Aloe L. 1998. Nerve growth factor and autoimmune diseases: role of tumor necrosis factor-alpha. *Adv. Pharmacol.* 42:591–94

Altar CA, Dugich-Djordjevic M, Armanini M, Bakhit C. 1991. Medial-to-lateral gradient of neostriatal NGF receptors: relationship to cholinergic neurons and NGF-like immunoreactivity. *J. Neurosci.* 11:828–36

Andjelkovic M, Jakubowicz T, Cron P, Ming XF, Han JW, Hemmings BA. 1996. Activation and phosphorylation of a pleckstrin homology domain containing protein kinase (RAC-PK/PKB) promoted by serum and protein phosphatase inhibitors. *Proc. Natl. Acad. Sci. USA* 93:5699–704

Angenstein F, Greenough WT, Weiler IJ. 1998. Metabotropic glutamate receptor-initiated translocation of protein kinase p90rsk to polyribosomes: a possible factor regulating synaptic protein synthesis. *Proc. Natl. Acad. Sci. USA* 95:15078–83

Anton ES, Weskamp G, Reichardt LF, Matthew WD. 1994. Nerve growth factor and its low-affinity receptor promote Schwann cell migration. *Proc. Natl. Acad. Sci. USA* 91:2795–99

Arch RH, Gedrich RW, Thompson CB. 1998. Tumor necrosis factor receptor-associated factors (TRAFs)—a family of adapter proteins that regulates life and death. *Genes Dev.* 12:2821–30

Arendt T, Brückner MK, Krell T, Pagliusi S, Kruska L, Heumann R. 1995. Degeneration of rat cholinergic basal forebrain neurons and reactive changes in nerve growth factor expression after chronic neurotoxic injury. II. Reactive expression of the nerve growth factor gene in astrocytes. *Neuroscience* 65:647–59

Asaoka Y, Oka M, Yoshida K, Sasaki Y, Nishizuka Y. 1992. Role of lysophosphatidylcholine in T-lymphocyte activation: involvement of phospholipase A2 in signal transduction through protein kinase C. *Proc. Natl. Acad. Sci. USA* 89:6447–51

Bamji SX, Majdan M, Pozniak CD, Belliveau DJ, Aloyz R, et al. 1998. The p75 neurotrophin receptor mediates neuronal apoptosis and is essential for naturally occurring sympathetic neuron death. *J. Cell Biol.* 140:911–23

Bar-Sagi D, Feramisco JR. 1985. Microinjection of the ras oncogene protein into PC12 cells induces morphological differentiation. *Cell* 42:841–48

Barde YA. 1989. Trophic factors and neuronal survival. *Neuron* 2:1525–34

Barker PA, Shooter EM. 1994. Disruption of NGF binding to the low affinity neurotrophin receptor p75LNTR reduces NGF binding to TrkA on PC12 cells. *Neuron* 13:203–15

Barouch R, Appel E, Kazimirsky G, Braun A, Renz H, Brodie C. 2000. Differential regulation of neurotrophin expression by mitogens and neurotransmitters in mouse lymphocytes. *J. Neuroimmunol.* 103:112–21

Barrett GL, Bartlett PF. 1994. The p75 nerve growth factor receptor mediates survival or death depending on the stage of sensory neuron development. *Proc. Natl. Acad. Sci. USA* 91:6501–5

BartLett SE, Reynolds AJ, Weible M, Heydon K, Hendry IA. 1997. In sympathetic but not sensory neurones, phosphoinositide-3 kinase is important for NGF-dependent survival and the retrograde transport of 125I-βNGF. *Brain Res.* 761:257–62

Basu T, Warne PH, Downward J. 1994. Role of Shc in the activation of Ras in response to epidermal growth factor and nerve growth factor. *Oncogene* 9:3483–91

Bausenwein BS, Schmidt M, Mielke B, Raabe T. 2000. In vivo functional analysis of the daughter of sevenless protein in receptor tyrosine kinase signaling. *Mech. Dev.* 90:205–15

Bazan JF. 1990. Structural design and molecular evolution of a cytokine receptor superfamily. *Proc. Natl. Acad. Sci. USA* 87:6934–8

Beattie EC, Howe CL, Wilder A, Brodsky FM, Mobley WC. 2001. NGF signals through TrkA to increase clathrin at the plasma membrane and enhance clathrin-mediated membrane trafficking. *J. Neurosci.* In press

Bell RM, Burns DJ. 1991. Lipid activation of protein kinase C. *J. Biol. Chem.* 266:4661–64

Benedetti M, Levi A, Chao MV. 1993. Differential expression of nerve growth factor receptors leads to altered binding affinity and neurotrophin responsiveness. *Proc. Natl. Acad. Sci. USA* 90:7859–63

Berg-von der Emde K, Dees WL, Hiney JK, Hill D, Dissen GA, et al. 1995. Neurotrophins and the neuroendocrine brain: different neurotrophins sustain anatomically and functionally segregated subsets of hypothalamic dopaminergic neurons. *J. Neurosci.* 15:4223–37

Berkemeier LR, Winslow JW, Kaplan DR, Nikolics K, Goeddel DV, Rosenthal A. 1991. Neurotrophin-5: a novel neurotrophic factor that activates trk and trkB. *Neuron* 7:857–66

Berkowitz LA, Riabowol KT, Gilman MZ. 1989. Multiple sequence elements of a single functional class are required for cyclic AMP responsiveness of the mouse c-fos promoter. *Mol. Cell. Biol.* 9:4272–81

Berninger B, Poo M. 1996. Fast actions of neurotrophic factors. *Curr. Opin. Neurobiol.* 6:324–30

Berra E, Diaz-Meco MT, Dominguez I, Municio MM, Sanz L, et al. 1993. Protein kinase C zeta isoform is critical for mitogenic signal transduction. *Cell* 74:555–63

Berra E, Diaz-Meco MT, Lozano J, Frutos S, Municio MM, et al. 1995. Evidence for a role of MEK and MAPK during signal transduction by protein kinase C zeta. *EMBO J.* 14:6157–63

Berzaghi MP, Cooper JD, Castren E, Zafra F, Sofroniew MV, et al. 1993. Cholinergic regulation of brain-derived neurotrophic factor (BDNF) and nerve growth factor (NGF) but not neurotrophin-3 (NT-3) mRNA levels in the developing rat hippocampus. *J. Neurosci.* 13:3818–26

Bibel M, Hoppe E, Barde YA. 1999. Biochemical and functional interactions between the neurotrophin receptors trk and p75NTR. *EMBO J.* 18:616–22

Bilderback TR, Gazula VR, Lisanti MP, Dobrowsky RT. 1999. Caveolin interacts with Trk A and p75(NTR) and regulates neu-

rotrophin signaling pathways. *J. Biol. Chem.* 274:257–63

Bilderback TR, Grigsby RJ, Dobrowsky RT. 1997. Association of p75(NTR) with caveolin and localization of neurotrophin-induced sphingomyelin hydrolysis to caveolae. *J. Biol. Chem.* 272:10922–27

Birren SJ, Lo L, Anderson DJ. 1993. Sympathetic neuroblasts undergo a developmental switch in trophic dependence. *Development* 119:597–610

Bizon JL, Lauterborn JC, Gall CM. 1999. Subpopulations of striatal interneurons can be distinguished on the basis of neurotrophic factor expression. *J. Comp. Neurol.* 408:283–98

Boise LH, Gottschalk AR, Quintans J, Thompson CB. 1995. Bcl-2 and Bcl-2-related proteins in apoptosis regulation. *Curr. Top. Microbiol. Immunol.* 200:107–21

Boldin MP, Mett IL, Varfolomeev EE, Chumakov I, Shemer-Avni Y, et al. 1995a. Self-association of the "death domains" of the p55 tumor necrosis factor (TNF) receptor and Fas/APO1 prompts signaling for TNF and Fas/APO1 effects. *J. Biol. Chem.* 270:387–91

Boldin MP, Varfolomeev EE, Pancer Z, Mett IL, Camonis JH, Wallach D. 1995b. A novel protein that interacts with the death domain of Fas/APO1 contains a sequence motif related to the death domain. *J. Biol. Chem.* 270:7795–98

Boniece I, Wagner J. 1993. Growth factors protect PC12 cells against ischemia by a mechanism that is independent of PKA, PKC, and protein synthesis. *J. Neurosci.* 13:4220

Bonini S, Lambiase A, Bonini S, Levi-Schaffer F, Aloe L. 1999. Nerve growth factor: an important molecule in allergic inflammation and tissue remodelling. *Int. Arch. Allergy Immunol.* 118:159–62

Bonni A, Ginty DD, Dudek H, Greenberg ME. 1995. Serine 133-phosphorylated CREB induces transcription via a cooperative mechanism that may confer specificity to neurotrophin signals. *Mol. Cell. Neurosci.* 6:168–83

Bothwell M. 1995. Functional interactions of neurotrophins and neurotrophin receptors. *Annu. Rev. Neurosci.* 18:223–53

Brann AB, Scott R, Neuberger Y, Abulafia D, Boldin S, et al. 1999. Ceramide signaling downstream of the p75 neurotrophin receptor mediates the effects of nerve growth factor on outgrowth of cultured hippocampal neurons. *J. Neurosci.* 19:8199–206

Bredesen DE, Rabizadeh S. 1997. p75NTR and apoptosis: Trk-dependent and Trk-independent effects. *Trends Neurosci.* 20: 287–90

Brennan C, Rivas-Plata K, Landis SC. 1999. The p75 neurotrophin receptor influences NT-3 responsiveness of sympathetic neurons in vivo. *Nat. Neurosci.* 2:699–705

Brown MT, Cooper JA. 1996. Regulation, substrates and functions of src. *Biochim. Biophys. Acta* 1287:121–49

Burgering BM, Coffer PJ. 1995. Protein kinase B (c-Akt) in phosphatidylinositol-3-OH kinase signal transduction. *Nature* 376:599–602

Bush TG, Puvanachandra N, Horner CH, Polito A, Ostenfeld T, et al. 1999. Leukocyte infiltration, neuronal degeneration and neurite outgrowth after ablation of scar-forming, reactive astrocytes in adult transgenic mice. *Neuron* 23:297–308

Butte MJ, Hwang PK, Mobley WC, Fletterick RJ. 1998. Crystal structure of neurotrophin-3 homodimer shows distinct regions are used to bind its receptors. *Biochemistry* 37:16846–52

Cacace AM, Ueffing M, Philipp A, Han EK, Kolch W, Weinstein IB. 1996. PKC epsilon functions as an oncogene by enhancing activation of the Raf kinase. *Oncogene* 13:2517–26

Cai D, Shen Y, De Bellard ME, Tang S, Filbin MT. 1999. Prior exposure to neurotrophins blocks the inhibition of axonal regeneration by MAG and myelin via a cAMP-dependent mechanism. *Neuron* 22:89–101

Campenot RB. 1977. Local control of neurite development by nerve growth factor. *Proc. Natl. Acad. Sci. USA* 74:4516–19

Cance WG, Craven RJ, Bergman M, Xu L, Alitalo K, Liu ET. 1994. Rak, a novel nuclear tyrosine kinase expressed in epithelial cells. *Cell Growth Differ.* 5:1347–55

Cao Z, Henzel WJ, Gao X. 1996a. IRAK: a kinase associated with the interleukin-1 receptor. *Science* 271:1128–31

Cao Z, Xiong J, Takeuchi M, Kurama T, Goeddel DV. 1996b. TRAF6 is a signal transducer for interleukin-1. *Nature* 383:443–46

Carlson N, Wieggel W, Chen J, Bacchi A, Rogers S, Gahring L. 1999. Inflammatory cytokines IL-1 alpha, IL-1 beta, IL-6, and TNF-alpha impart neuroprotection to an excitotoxin through distinct pathways. *J. Immunol.* 163:3963–68

Carroll MP, May WS. 1994. Protein kinase C-mediated serine phosphorylation directly activates Raf-1 in murine hematopoietic cells. *J. Biol. Chem.* 269:1249–56

Carter BD, Kaltschmidt C, Kaltschmidt B, Offenhauser N, Bohm-Matthaei R, et al. 1996. Selective activation of NF-kappa B by nerve growth factor through the neurotrophin receptor p75. *Science* 272:542–45

Carter BD, Lewin GR. 1997. Neurotrophins live or let die: does p75NTR decide? *Neuron* 18:187–90

Casaccia-Bonnefil P, Carter BD, Dobrowsky RT, Chao MV. 1996. Death of oligodendrocytes mediated by the interaction of nerve growth factor with its receptor p75. *Nature* 383:716–19

Casademunt E, Carter BD, Benzel I, Frade JM, Dechant G, Barde YA. 1999. The zinc finger protein NRIF interacts with the neurotrophin receptor p75(NTR) and participates in programmed cell death. *EMBO J.* 18:6050–61

Casey PJ. 1995. Mechanisms of protein prenylation and role in G protein function. *Biochem. Soc. Trans.* 23:161–66

Cellerino A. 1995. Expression of messenger RNA coding for the nerve growth factor receptor trkA in the hippocampus of the adult rat. *Neuroscience* 70:613–16

Cellerino A, Maffei L. 1996. The action of neurotrophins in the development and plasticity of the visual cortex. *Prog. Neurobiol.* 49:53–71

Chao M, Casaccia-Bonnefil P, Carter B, Chittka A, Kong H, Yoon SO. 1998. Neurotrophin receptors: mediators of life and death. *Brain Res. Rev.* 26:295–301

Chen KS, Nishimura MC, Armanini MP, Crowley C, Spencer SD, Phillips HS. 1997. Disruption of a single allele of the nerve growth factor gene results in atrophy of basal forebrain cholinergic neurons and memory deficits. *J. Neurosci.* 17:7288–96

Chen RH, Sarnecki C, Blenis J. 1992. Nuclear localization and regulation of erk- and rsk-encoded protein kinases. *Mol. Cell. Biol.* 12:915–27

Cheng B, Mattson MP. 1991. NGF and bFGF protect rat hippocampal and human cortical neurons against hypoglycemic damage by stabilizing calcium homeostasis. *Neuron* 7:1031–41

Chou MM, Hou W, Johnson J, Graham LK, Lee MH, et al. 1998. Regulation of protein kinase C zeta by PI 3-kinase and PDK-1. *Curr. Biol.* 8:1069–77

Chrivia JC, Kwok RP, Lamb N, Hagiwara M, Montminy MR, Goodman RH. 1993. Phosphorylated CREB binds specifically to the nuclear protein CBP. *Nature* 365:855–59

Cifone MG, De Maria R, Roncaioli P, Rippo MR, Azuma M, et al. 1994. Apoptotic signaling through CD95 (Fas/Apo-1) activates an acidic sphingomyelinase. *J. Exp. Med.* 180:1547–52

Clary DO, Reichardt LF. 1994. An alternatively spliced form of the nerve growth factor receptor TrkA confers an enhanced response to neurotrophin 3. *Proc. Natl. Acad. Sci. USA* 91:11133–37

Cohen P, Alessi DR, Cross DA. 1997. PDK1, one of the missing links in insulin signal transduction? *FEBS Lett.* 410:3–10

Cohen-Cory S, Dreyfus CF, Black IB. 1991.

NGF and excitatory neurotransmitters regulate survival and morphogenesis of cultured cerebellar Purkinje cells. *J. Neurosci.* 11:462–71

Coleman ES, Wooten MW. 1994. Nerve growth factor-induced differentiation of PC12 cells employs the PMA-insensitive protein kinase C-zeta isoform. *J. Mol. Neurosci.* 5:39–57

Conover JC, Yancopoulos GD. 1997. Neurotrophin regulation of the developing nervous system: analyses of knockout mice. *Rev. Neurosci.* 8:13–27

Cooper JD, Lindholm D, Sofroniew MV. 1994. Reduced transport of 125I-NGF by cholinergic neurons and downregulated TrkA expression in the medial septum of aged rats. *Neuroscience* 62:625–29

Cosman D, Lyman SD, Idzerda RL, Beckmann MP, Park LS, et al. 1990. A new cytokine receptor superfamily. *Trends Biochem. Sci.* 15:265–70

Coulson EJ, Reid K, Baca M, Shipham KA, Hulett SM, et al. 2001. Chopper, a new death domain of the p75 neurotrophin receptor that mediates rapid neuronal cell death. *J. Biol. Chem.* In press

Cowley S, Paterson H, Kemp P, Marshall CJ. 1994. Activation of MAP kinase kinase is necessary and sufficient for PC12 differentiation and for transformation of NIH 3T3 cells. *Cell* 77:841–52

Crews CM, Alessandrini A, Erikson RL. 1992. Erks: their fifteen minutes has arrived. *Cell Growth Differ.* 3:135–42

Crews CM, Erikson RL. 1992. Purification of a murine protein-tyrosine/threonine kinase that phosphorylates and activates the Erk-1 gene product: relationship to the fission yeast byr1 gene product. *Proc. Natl. Acad. Sci. USA* 89:8205–9

Crowder RJ, Freeman RS. 1998. Phosphatidylinositol 3-kinase and Akt protein kinase are necessary and sufficient for the survival of nerve growth factor-dependent sympathetic neurons. *J. Neurosci.* 18:2933–43

Crowley C, Spencer SD, Nishimura MC, Chen KS, Pitts-Meek S, et al. 1994. Mice lacking nerve growth factor display perinatal loss of sensory and sympathetic neurons yet develop basal forebrain cholinergic neurons. *Cell* 76:1001–11

Crutcher KA, Scott SA, Liang S, Everson WV, Weingartner J. 1993. Detection of NGF-like activity in human brain tissue: increased levels in Alzheimer's disease. *J. Neurosci.* 13:2540–50

Crutcher KA, Weingartner J. 1991. Hippocampal NGF levels are not reduced in the aged fischer 344 rat. *Neurobiol. Aging* 12:449–54

Cuello AC. 1994. Trophic factor therapy in the adult CNS: remodelling of injured basalocortical neurons. *Prog. Brain Res.* 100:213–21

Cuello AC. 1996. Effects of trophic factors on the CNS cholinergic phenotype. *Prog. Brain Res.* 109:347–58

Culmsee C, Semkova I, Krieglstein J. 1999. NGF mediates the neuroprotective effect of the β2-adrenoceptor agonist clenbuterol in vitro and in vivo: evidence from an NGF-antisense study. *Neurochem. Int.* 35:47–57

Cunha GMA, Moraes RA, Moraes GA, Franca MC Jr, Moraes MO, Viana GSB. 1998. Nerve growth factor, ganglioside and vitamin E reverse glutamate cytotoxicity in hippocampal cells. *Eur. J. Pharmacol.* 367:107–12

Cunningham ME, Stephens RM, Kaplan DR, Greene LA. 1997. Autophosphorylation of activation loop tyrosines regulates signaling by the TRK nerve growth factor receptor. *J. Biol. Chem.* 272:10957–67

Datta K, Bellacosa A, Chan TO, Tsichlis PN. 1996. Akt is a direct target of the phosphatidylinositol 3-kinase. Activation by growth factors, v-src and v-Ha-ras, in Sf9 and mammalian cells. *J. Biol. Chem.* 271:30835–39

Datta SR, Brunet A, Greenberg ME. 1999. Cellular survival: a play in three Akts. *Genes Dev.* 13:2905–27

Datta SR, Dudek H, Tao X, Masters S, Fu H, et al. 1997. Akt phosphorylation of BAD couples survival signals to the cell-intrinsic death machinery. *Cell* 91:231–41

Davies AM. 1994. The role of neurotrophins in the developing nervous system. *J. Neurobiol.* 25:1334–48

Davies SW, Beardsall K. 1992. Nerve growth factor selectively prevents excitotoxin induced degeneration of striatal cholinergic neurons. *Neurosci. Lett.* 140:161–64

Debeir T, Saragovi HU, Cuello AC. 1999. A nerve growth factor mimetic TrkA antagonist causes withdrawal of cortical cholinergic boutons in the adult rat. *Proc. Natl. Acad. Sci. USA* 96:4067–72

Dechant G, Barde YA. 1997. Signalling through the neurotrophin receptor p75NTR. *Curr. Opin. Neurobiol.* 7:413–18

de Lacalle S, Cooper JD, Svendsen CN, Dunnett SB, Sofroniew MV. 1991a. Degenerative changes in basal forebrain cholinergic neurons in aged rats. A quantitative analysis of cells stained for ChAT or NGFr, or retrogradely labelled with fluorescent tracer. *Soc. Neurosci. Abstr.* 17:1508

de Lacalle S, Cooper JD, Svendsen CN, Dunnett SB, Sofroniew MV. 1996. Reduced retrograde labeling with fluorescent tracer accompanies neuronal atrophy of basal forebrain cholinergic neurons in aged rats. *Neuroscience* 75:19–27

de Lacalle S, Iraizoz I, Gonzalo LM. 1991b. Differential changes in cell size and number in topographic subdivisions of human basal nucleus in normal aging. *Neuroscience* 43:445–56

del Peso L, Gonzalez-Garcia M, Page C, Herrera R, Nunez G. 1997. Interleukin-3-induced phosphorylation of BAD through the protein kinase Akt. *Science* 278:687–89

Derby A, Engleman VW, Frierdich GE, Neises G, Rapp SR, Roufa DG. 1993. Nerve growth factor facilitates regeneration across nerve gaps: morphological and behavioral studies in rat sciatic nerve. *Exp. Neurol.* 119:176–91

Dmitrieva N, Shelton D, Rice A, McMahon S. 1997. The role of nerve growth factor in a model of visceral inflammation. *Neurosci.* 78:449–59

Dobrowsky RT, Jenkins GM, Hannun YA. 1995. Neurotrophins induce sphingomyelin hydrolysis. Modulation by co-expression of p75NTR with Trk receptors. *J. Biol. Chem.* 270:22135–42

Dobrowsky RT, Werner MH, Castellino AM, Chao MV, Hannun YA. 1994. Activation of the sphingomyelin cycle through the low-affinity neurotrophin receptor. *Science* 265:1596–99

Dougherty K, Milner T. 1999. p75NTR immunoreactivity in the rat dentate gyrus is mostly within presynaptic profiles but is also found in some astrocytic and postsynaptic profiles. *J. Comp. Neurol.* 407:77–91

Douville E, Downward J. 1997. EGF induced SOS phosphorylation in PC12 cells involves P90 RSK-2. *Oncogene* 15:373–83

Dudek H, Datta SR, Franke TF, Birnbaum MJ, Yao R, et al. 1997. Regulation of neuronal survival by the serine-threonine protein kinase Akt. *Science* 275:661–65

Dugan LL, Creedon DJ, Johnson EM, Holtzman DM. 1997. Rapid suppression of free radical formation by nerve growth factor involves the mitogen-activated protein kinase pathway. *Proc. Natl. Acad. Sci. USA* 94:4086–91

Easton RM, Deckwerth TL, Parsadanian AS, Johnson EM Jr. 1997. Analysis of the mechanism of loss of trophic factor dependence associated with neuronal maturation: a phenotype indistinguishable from bax deletion. *J. Neurosci.* 17:9656–66

Eckenstein F. 1988. Transient expression of NGF-receptor-like immunoreactivity in postnatal rat brain and spinal cord. *Brain Res.* 446:149–54

Edwards RH, Selby MJ, Mobley WC, Weinrich SL, Hruby DE, Rutter WJ. 1988. Processing and secretion of nerve growth factor: expression in mammalian cells with a vaccinia virus vector. *Mol. Cell. Biol.* 8:2456–64

Edwards RH, Selby MJ, Rutter WJ. 1986. Differential RNA splicing predicts two distinct nerve growth factor precursors. *Nature* 319:784–87

Elkabes S, DiCicco-Bloom E, Black I. 1996. Brain microglia/macrophages express neurotrophins that selectively regulate microglial proliferation and function. *J. Neurosci.* 16:2508–21

Elkabes S, Peng L, Black IB. 1998. Lipopolysaccharide differentially regulates microglial trk receptor and neurotrophin expression. *J. Neurosci. Res.* 54:117–22

Ellinger-Ziegelbauer H, Brown K, Kelly K, Siebenlist U. 1997. Direct activation of the stress-activated protein kinase (SAPK) and extracellular signal-regulated protein kinase (ERK) pathways by an inducible mitogen-activated protein kinase/ERK kinase kinase 3 (MEKK) derivative. *J. Biol. Chem.* 272:2668–74

Enokido Y, Wyatt S, Davies AM. 1999. Developmental changes in the response of trigeminal neurons to neurotrophins: influence of birthdate and the ganglion environment. *Development* 126:4365–73

Eriksdotter-Jönhagen M, Nordberg A, Amberla K, Backman L, Ebendal T, et al. 1998. Intracerebroventricular infusion of nerve growth factor in three patients with Alzheimer's disease. *Dement. Geriatr. Cogn. Disord.* 9:246–57

Erikson E, Maller JL. 1991. Purification and characterization of ribosomal protein S6 kinase I from Xenopus eggs. *J. Biol. Chem.* 266:5249–55

Erikson E, Maller JL, Erikson RL. 1991. Xenopus ribosomal protein S6 kinase II. *Methods Enzymol.* 200:252–68

Ernfors P, Hallbook F, Ebendal T, Shooter EM, Radeke MJ, et al. 1988. Developmental and regional expression of beta-nerve growth factor receptor mRNA in the chick and rat. *Neuron* 1:983–96

Ernfors P, Henschen A, Olson L, Persson H. 1989. Expression of nerve growth factor receptor mRNA is developmentally regulated and increased after axotomy in rat spinal cord motoneurons. *Neuron* 2:1605–13

Erpel T, Courtneidge SA. 1995. Src family protein tyrosine kinases and cellular signal transduction pathways. *Curr. Opin. Cell Biol.* 7:176–82

Facchinetti F, Dawson VL, Dawson TM. 1998. Free radicals as mediators of neuronal injury. *Cell. Mol. Neurobiol.* 18:609–20

Fagan AM, Garber M, Barbacid M, Silos-Santiago I, Holtzman DM. 1997. A role for TrkA during maturation of striatal and basal forebrain cholinergic neurons in vivo. *J. Neurosci.* 17:7644–54

Fahnestock M, Woo JE, Lopez GA, Snow J, Walz DA, et al. 1991. beta-NGF-endopeptidase: structure and activity of a kallikrein encoded by the gene mGK-22. *Biochemistry* 30:3443–50

Finch CE. 1993. Neuron atrophy during aging: programmed or sporadic. *Trends Neurosci.* 16:104–10

Fischer W, Chen KS, Gage FH, Björklund A. 1991. Progressive decline in spatial learning and integrity of forebrain cholinergic neurons in rats during aging. *Neurobiol. Aging* 13:9–23

Fischer W, Wictorin K, Bjorklund A, Williams LR, Varon S, Gage FH. 1987. Amelioration of cholinergic neuronal atrophy and spatial memory impairment in aged rats by nerve growth factor. *Nature* 329:65–68

Frade J, Barde Y. 1998. Microglia-derived nerve growth factor causes cell death in the developing retina. *Neuron* 20:35–41

Frade JM, Rodriguez-Tebar A, Barde Y. 1996. Induction of cell death by endogenous nerve growth factor through its p75 receptor. *Nature* 383:166–68

Franke TF, Kaplan DR, Cantley LC. 1997. PI3K: downstream AKTion blocks apoptosis. *Cell* 88:435–37

Franke TF, Yang SI, Chan TO, Datta K, Kazlauskas A, et al. 1995. The protein kinase encoded by the Akt proto-oncogene is a target of the PDGF-activated phosphatidylinositol 3-kinase. *Cell* 81:727–36

Freed E, Symons M, Macdonald SG, McCormick F, Ruggieri R. 1994. Binding of 14-3-3 proteins to the protein kinase Raf and effects on its activation. *Science* 265:1713–16

French SJ, Humby T, Horner CH, Sofroniew MV, Rattray M. 1999. Hippocampal neurotrophin and trk receptor mRNA levels are altered by local administration of nicotine, carbachol and pilocarpine. *Mol. Brain Res.* 67:124–36

Friedman WJ, Greene LA. 1999. Neurotrophin signaling via Trks and p75. *Exp. Cell Res.* 253:131–42

Friedman WJ, Thakur S, Seidman L, Rabson AB. 1996. Regulation of nerve growth factor mRNA by interleukin-1 in rat hippocampal astrocytes is mediated by NFκB. *J. Biol. Chem.* 271:31115–20

Frim DM, Uhler TA, Short MP, Ezzedine ZD, Klagsbrun M, et al. 1993. Effects of biologically delivered NGF, BDNF and bFGF on striatal excitotoxic lesions. *NeuroReport* 4:367–70

Frostick SP, Yin Q, Kemp GJ. 1998. Schwann cells, neurotrophic factors, and peripheral nerve regeneration. *Microsurgery* 18:397–405

Fu H, Xia K, Pallas DC, Cui C, Conroy K, et al. 1994. Interaction of the protein kinase Raf-1 with 14-3-3 proteins. *Science* 266:126–29

Gadjusek DC. 1985. Hypothesis: interference with axonal transport of neurofilament as a common pathogenic mechanism in certain diseases of the central nervous system. *N. Engl. J. Med.* 312:714–19

Gage FH, Batchelor P, Chen KS, Chin D, Higgins GA, et al. 1989. NGF receptor reexpression and NGF-mediated cholinergic neuronal hypertrophy in the damaged adult neostriatum. *Neuron* 2:1177–84

Gall CM, Isackson PJ. 1989. Limbic seizures increase neuronal production of messenger RNA for nerve growth factor. *Science* 245:758–61

Gallo G, Lefcort FB, Letourneau PC. 1997. The trkA receptor mediates growth cone turning toward a localized source of nerve growth factor. *J. Neurosci.* 17:5445–54

Gargano N, Levi A, Alema S. 1997. Modulation of nerve growth factor internalization by direct interaction between p75 and TrkA receptors. *J. Neurosci. Res.* 50:1–12

Gee AP, Boyle MD, Munger KL, Lawman MJ, Young M. 1983. Nerve growth factor: stimulation of polymorphonuclear leukocyte chemotaxis in vitro. *Proc. Natl. Acad. Sci. USA* 80:7215–18

Gentry JJ, Casaccia-Bonnefil P, Carter BD. 2000. Nerve growth factor activation of nuclear factor κB through its p75 receptor is an anti-apoptotic signal in RN22 Schwannoma cells. *J. Biol. Chem.* 275:7558–65

Gertler FB, Bennett RL, Clark MJ, Hoffmann FM. 1989. Drosophila abl tyrosine kinase in embryonic CNS axons: a role in axonogenesis is revealed through dosage-sensitive interactions with disabled. *Cell* 58:103–13

Gertler FB, Hill KK, Clark MJ, Hoffmann FM. 1993. Dosage-sensitive modifiers of Drosophila abl tyrosine kinase function: prospero, a regulator of axonal outgrowth, and disabled, a novel tyrosine kinase substrate. *Genes Dev.* 7:441–53

Ghoda L, Lin X, Greene WC. 1997. The 90-kDa ribosomal S6 kinase (pp90rsk) phosphorylates the N-terminal regulatory domain of IkappaBalpha and stimulates its degradation in vitro. *J. Biol. Chem.* 272:21281–88

Ghosh S, May MJ, Kopp EB. 1998. NF-kappa B and Rel proteins: evolutionarily conserved mediators of immune responses. *Annu. Rev. Immunol.* 16:225–60

Gille H, Kortenjann M, Thomae O, Moomaw C, Slaughter C, et al. 1995. ERK phosphorylation potentiates Elk-1-mediated ternary complex formation and transactivation. *EMBO J.* 14:951–62

Ginty DD, Bonni A, Greenberg ME. 1994. Nerve growth factor activates a Ras-dependent protein kinase that stimulates c-fos transcription via phosphorylation of CREB. *Cell* 77:713–25

Goedert M, Fine A, Hunt SP, Ullrich A. 1986. Nerve growth factor mRNA in peripheral and central rat tissues and in the human central nervous system: lesion effects in the rat brain

and levels in Alzheimer's disease. *Mol. Brain Res.* 1:85–92

Goedert M. 1996. Tau protein and the neurofibrillary pathology of Alzheimer's disease. *Ann. NY Acad. Sci.* 777:121–31

Goss JR, O'Malley ME, Zou L, Styren SD, Kochanek PM, DeKosky ST. 1998. Astrocytes are the major souce of nerve growth factor upregulation following traumatic brain injury in the rat. *Exp. Neurol.* 149:301–9

Gottlieb M, Matute C. 1999. Expression of nerve growth factor in astrocytes of the hippocampal CA1 area following transient forebrain ischemia. *Neuroscience* 91:1027–34

Gotz R, Koster R, Winkler C, Raulf F, Lottspeich F, et al. 1994. Neurotrophin-6 is a new member of the nerve growth factor family. *Nature* 372:266–69

Greene LA, Kaplan DR. 1995. Early events in neurotrophin signalling via Trk and p75 receptors. *Curr. Opin. Neurobiol.* 5:579–87

Greenberg ME, Hermanowski AL, Ziff EB. 1986. Effect of protein synthesis inhibitors on growth factor activation of c-fos, c-myc, and actin gene transcription. *Mol. Cell. Biol.* 6:1050–57

Greferath U, Bennie A, Kourakis A, BartLett PF, Murphy M, Barrett GL. 2000. Enlarged cholinergic forebrain neurons and improved spatial learning in p75 knockout mice. *Eur J. Neurosci.* 12:885–93

Grimes ML, Beattie E, Mobley WC. 1997. A signaling organelle containing the nerve growth factor-activated receptor tyrosine kinase, TrkA. *Proc. Natl. Acad. Sci. USA* 94:9909–14

Grimes ML, Zhou J, Beattie EC, Yuen EC, Hall DE, et al. 1996. Endocytosis of activated TrkA: evidence that nerve growth factor induces formation of signaling endosomes. *J. Neurosci.* 16:7950–64

Grinspan JB, Marchionni MA, Reeves M, Coulaloglou M, Scherer SS. 1996. Axonal interactions regulate Schwann cell apoptosis in developing peripheral nerve: neuregulin receptors and the role of neuregulins. *J. Neurosci.* 16:6107–18

Gross A, Jockel J, Wei MC, Korsmeyer SJ. 1998. Enforced dimerization of BAX results in its translocation, mitochondrial dysfunction and apoptosis. *EMBO J.* 17:3878–85

Gu C, Casaccia-Bonnefil P, Srinivasan A, Chao MV. 1999. Oligodendrocyte apoptosis mediated by caspase activation. *J. Neurosci.* 19:3043–49

Gu H, Pratt JC, Burakoff SJ, Neel BG. 1998. Cloning of p97/Gab2, the major SHP2-binding protein in hematopoietic cells, reveals a novel pathway for cytokine-induced gene activation. *Mol. Cell* 2:729–40

Gu Q, Liu Y, Cynader MS. 1994. Nerve growth factor-induced ocular dominance plasticity in adult cat visual cortex. *Proc. Natl. Acad. Sci. USA* 91:8408–12

Guegan C, Onteniente B, Makiura Y, Merad-Boudia M, Ceballos-Picot I, Sola B. 1997. Reduction of cortical infarction and impairment of apoptosis in NGF-transgenic mice subjected to permanent focal ischemia. *Mol. Brain Res.* 55:133–40

Hadari YR, Kouhara H, Lax I, Schlessinger J. 1998. Binding of Shp2 tyrosine phosphatase to FRS2 is essential for fibroblast growth factor-induced PC12 cell differentiation. *Mol. Cell. Biol.* 18:3966–73

Hagag N, Halegoua S, Viola M. 1986. Inhibition of growth factor-induced differentiation of PC12 cells by microinjection of antibody to ras p21. *Nature* 319:680–82

Hakak Y, Hsu YS, Martin GS. 2000. Shp-2 mediates v-Src-induced morphological changes and activation of the anti-apoptotic protein kinase Akt. *Oncogene* 19:3164–71

Hasenöhrl RU, Söderström S, Mohammed AH, Ebendal T, Huston JP. 1997. Reciprocal changes in expression of mRNA for nerve growth factor and its receptors TrkA and LNGFR in brain of aged rats in relation to maze learning deficits. *Exp. Brain Res.* 114:205–13

Hausser A, Storz P, Link G, Stoll H, Liu YC, et al. 1999. Protein kinase C mu is negatively regulated by 14-3-3 signal transduction proteins. *J. Biol. Chem.* 274:9258–64

Hawley RJ, Scheibe RJ, Wagner JA. 1992. NGF induces the expression of the VGF gene through a cAMP response element. *J. Neurosci.* 12:2573–81

Heaton MB, Paiva M, Swanson DJ, Walker DW. 1993. Modulation of ethanol neurotoxicity by nerve growth factor. *Brain Res.* 620:78–85

Heese K, Fiebich BL, Bauer J, Otten U. 1998. NF-κB modulates lipopolysaccharide-induced microglial nerve growth factor expression. *Glia* 22:401–7

Hefti F. 1986. Nerve growth factor promotes survival of septal cholinergic neurons after fimbrial transections. *J. Neurosci.* 6:2155–62

Heisenberg C-P, Cooper JD, Berke J, Sofroniew MV. 1994. NMDA potentiates NGF-induced sprouting of septal cholinergic neurons. *NeuroReport* 5:413–16

Heumann R, Korsching S, Bandtlow C, Thoenen H. 1987a. Changes of nerve growth factor synthesis in nonneuronal cells in response to sciatic nerve transection. *J. Cell Biol.* 104:1623–31

Heumann R, Lindholm D, Bandtlow C, Meyer M, Radeke MJ, et al. 1987b. Differential regulation of mRNA encoding nerve growth factor and its receptor in rat sciatic nerve during development, degeneration, and regeneration: role of macrophages. *Proc. Natl. Acad. Sci. USA* 84:8735–39

Hill CS, Marais R, John S, Wynne J, Dalton S, Treisman R. 1993. Functional analysis of a growth factor-responsive transcription factor complex. *Cell* 73:395–406

Holgado-Madruga M, Emlet DR, Moscatello DK, Godwin AK, Wong AJ. 1996. A Grb2-associated docking protein in EGF- and insulin-receptor signalling. *Nature* 379:560–64

Holgado-Madruga M, Moscatello DK, Emlet DR, Dieterich R, Wong AJ. 1997. Grb2-associated binder-1 mediates phosphatidylinositol 3-kinase activation and the promotion of cell survival by nerve growth factor. *Proc. Natl. Acad. Sci. USA* 94:12419–24

Holtzman DM, Kilbridge J, Li Y, Cunningham ET, Lenn NJ, et al. 1995. TrkA expression in the CNS: evidence for the existence of several novel NGF-responsive CNS neurons. *J. Neurosci.* 15:1567–76

Holtzman DM, Li Y, Parada LF, Kinsman S, Chen CK, et al. 1992. p140trk mRNA marks NGF-responsive forebrain neurons: evidence that *trk* gene expression is induced by NGF. *Neuron* 9:465–78

Holtzman DM, Sheldon RA, Jaffe W, Cheng Y, Ferriero DM. 1996. Nerve growth factor protects the neonatal brain against hypoxic-ischemic injury. *Ann. Neurol.* 39:114–22

Horner CH, Lam HHD, Berke J, Cooper JD, Sofroniew MV. 1999. Local and target-derived nerve growth factor protect basal forebrain cholinergic neurons from glutamate receptor-mediated toxicity. *Soc. Neurosci. Abstr.* 25:770

Horvath CM, Wolven A, Machadeo D, Huber J, Boter L, et al. 1993. Analysis of the trk NGF receptor tyrosine kinase using recombinant fusion proteins. *J. Cell Sci.* 17:223–28

Howe CL, Mobley WC. 2001. Nerve growth factor effects on cholinergic modulation of hippocampal and cortical plasticity. In *Neurobiology of the Neurotrophins*, ed. I Mocchetti, FP Graham, pp. 255–307

Huang CS, Zhou J, Feng AK, Lynch CC, Klumperman J, et al. 1999. Nerve growth factor signaling in caveolae-like domains at the plasma membrane. *J. Biol. Chem.* 274:36707–14

Huber LJ, Chao MV. 1995. A potential interaction of p75 and trkA NGF receptors revealed by affinity crosslinking and immunoprecipitation. *J. Neurosci. Res.* 40:557–63

Hutton L, deVellis J, Perez-Polo J. 1992. Expression of p75NGFR TrkA, and TrkB mRNA in Rat C6 glioma and type U astrocyte cultures. *J. Neurosci. Res.* 32:375–83

Hutton L, Perez-Polo J. 1995. In vitro glial responses to nerve growth factor. *J. Neurosci. Res.* 41:185–96

Ibanez CF. 1994. Structure function relationships in the neurotrophin family. *J. Neurobiol.* 25:1349–61

Ibanez CF, Ebendal T, Barbany G, Murray-Rust J, Blundell TL, Persson H. 1992. Disruption of the low affinity receptor-binding site in NGF allows neuronal survival and differentiation by binding to the trk gene product. *Cell* 69:329–41

Ip NY, Stitt TN, Tapley P, Klein R, Glass DJ, et al. 1993. Similarities and differences in the way neurotrophins interact with the Trk receptors in neuronal and nonneuronal cells. *Neuron* 10:137–49

Irie K, Gotoh Y, Yashar BM, Errede B, Nishida E, Matsumoto K. 1994. Stimulatory effects of yeast and mammalian 14-3-3 proteins on the Raf protein kinase. *Science* 265:1716–19

Jaiswal RK, Moodie SA, Wolfman A, Landreth GE. 1994. The mitogen-activated protein kinase cascade is activated by B-Raf in response to nerve growth factor through interaction with p21ras. *Mol. Cell. Biol.* 14:6944–53

Janknecht R, Ernst WH, Pingoud V, Nordheim A. 1993. Activation of ternary complex factor Elk-1 by MAP kinases. *EMBO J.* 12:5097–104

Jia M, Li M, Liu XW, Jiang H, Nelson PG, Guroff G. 1999. Voltage-sensitive calcium currents are acutely increased by nerve growth factor in PC12 cells. *J. Neurophysiol.* 82:2847–52

Jiang H, Takeda K, Lazarovici P, Katagiri Y, Yu ZX, et al. 1999. Nerve growth factor (NGF)-induced calcium influx and intracellular calcium mobilization in 3T3 cells expressing NGF receptors. *J. Biol. Chem.* 274:26209–16

Joel PB, Smith J, Sturgill TW, Fisher TL, Blenis J, Lannigan DA. 1998. pp90rsk1 regulates estrogen receptor-mediated transcription through phosphorylation of Ser-167. *Mol. Cell. Biol.* 18:1978–84

Johnson EM, Deckwerth TL Jr. 1993. Molecular mechanisms of developmental neuronal death. *Annu. Rev. Neurosci.* 16:31–46

Johnson EM, Rich KM, Yip HK. 1986. The role

of NGF in sensory neurons in vivo. *Trends Neurosci.* 9:33–37

Kaplan DR, Hempstead BL, Martin-Zanca D, Chao MV, Parada LF. 1991a. The trk proto-oncogene product: a signal transducing receptor for nerve growth factor. *Science* 252:554–58

Kaplan DR, Martin-Zanca D, Parada LF. 1991b. Tyrosine phosphorylation and tyrosine kinase activity of the trk proto-oncogene product induced by NGF. *Nature* 350:158–60

Kaplan DR, Miller FD. 1997. Signal transduction by the neurotrophin receptors. *Curr. Opin. Cell. Biol.* 9:213–21

Karlsson M, Clary DO, Lefcort FB, Reichardt LF, Karten HJ, Hallbook F. 1998. Nerve growth factor receptor TrkA is expressed by horizontal and amacrine cells during chicken retinal development. *J. Comp. Neurol.* 400:408–16

Kawaja MD, Gage FH. 1991. Reactive astrocytes are substrates for the growth of adult CNS axons in the presence of elevated levels of nerve growth factor. *Neuron* 7:1019–30

Kew JNC, Smith DW, Sofroniew MV. 1996. Nerve growth factor withdrawal induces the apoptotic death of developing septal cholinergic neurons *in vitro*: protection by cAMP analogue and high K$^+$. *Neuroscience* 70:329–39

Khasar SG, Lin YH, Martin A, Dadgar J, McMahon T, et al. 1999. A novel nociceptor signaling pathway revealed in protein kinase C epsilon mutant mice. *Neuron* 24:253–60

Khursigara G, Orlinick JR, Chao MV. 1999. Association of the p75 neurotrophin receptor with TRAF6. *J. Biol. Chem.* 274:2597–600

Klein R, Jing SQ, Nanduri V, O'Rourke E, Barbacid M. 1991. The trk proto-oncogene encodes a receptor for nerve growth factor. *Cell* 65:189–97

Klesse LJ, Parada LF. 1998. p21 ras and phosphatidylinositol-3 kinase are required for survival of wild-type and NF1 mutant sensory neurons. *J. Neurosci.* 8:10420–28

Klippel A, Kavanaugh WM, Pot D, Williams LT. 1997. A specific product of phos-

phatidylinositol 3-kinase directly activates the protein kinase Akt through its pleckstrin homology domain. *Mol. Cell. Biol.* 17:338–44

Knipper M, Berzaghi MP, Blochl A, Breer H, Thoenen H, Lindholm D. 1994. Positive feedback between acetylcholine and the neurotrophins nerve growth factor and brain-derived growth factor in the rat hippocampus. *Eur. J. Neurosci.* 6:668–71

Kobayashi M, Nagata S, Kita Y, Nakatsu N, Ihara S, et al. 1997. Expression of a constitutively active phosphatidylinositol 3-kinase induces process formation in rat PC12 cells. Use of Cre/loxP recombination system. *J. Biol. Chem.* 272:16089–92

Koh S, Higgins GA. 1991. Differential regulation of the low-affinity nerve growth factor receptor during postnatal development of the rat brain. *J. Comp. Neurol.* 313:494–508

Kolch W, Heidecker G, Kochs G, Hummel R, Vahidi H, et al. 1993. Protein kinase C alpha activates RAF-1 by direct phosphorylation. *Nature* 364:249–52

Kordower JH, Burke-Watson M, Roback JD, Wainer BH. 1993. Stability of septohippocampal neurons following excitotoxic lesions of the rat hippocampus. *Exp. Neurol.* 117:1–16

Kordower JH, Winn SR, Liu Y, Mufson EJ, Sladek JR, et al. 1994. The aged monkey basal forebrain: rescue and sprouting of axotomized basal forebrain neurons after grafts of encapsulated cells secreting human nerve growth factor. *Proc. Natl. Acad. Sci. USA* 91:10898–902

Korhonen JM, Said FA, Wong AJ, Kaplan DR. 1999. Gab1 mediates neurite outgrowth, DNA synthesis, and survival in PC12 cells. *J. Biol. Chem.* 274:37307–14

Korner M, Tarantino N, Pleskoff O, Lee LM, Debre P. 1994. Activation of nuclear factor kappa B in human neuroblastoma cell lines. *J. Neurochem.* 62:1716–26

Kouhara H, Hadari YR, Spivak-Kroizman T, Schilling J, Bar-Sagi D, et al. 1997. A lipid-anchored Grb2-binding protein that links

FGF-receptor activation to the Ras/MAPK signaling pathway. *Cell* 89:693–702

Kremer NE, D'Arcangelo G, Thomas SM, De-Marco M, Brugge JS, Halegoua S. 1991. Signal transduction by nerve growth factor and fibroblast growth factor in PC12 cells requires a sequence of src and ras actions. *J. Cell Biol.* 115:809–19

Krenz NR, Meakin SO, Krassioukov AV, Weaver LC. 1999. Neutralizing intraspinal nerve growth factor blocks autonomic dysreflexia caused by spinal cord injury. *J. Neurosci.* 19:7405

Kroemer G. 1997. The proto-oncogene Bcl-2 and its role in regulating apoptosis. *Nat. Med.* 3:614–20

Krüttgen A, Möller J, Heymach JV, Shooter EM. 1998. Neurotrophins induce release of neurotrophins by the regulated secretory pathway. *Proc. Natl. Acad. Sci. USA* 95:9614–19

Kumar S, Pena L, de Vellis J. 1993. CNS glial cells express neurotrophin receptors whose levels are regulated by NGF. *Brain Res. Mol. Brain Res.* 17:163–68

Kuo WL, Chung KC, Rosner MR. 1997. Differentiation of central nervous system neuronal cells by fibroblast-derived growth factor requires at least two signaling pathways: roles for Ras and Src. *Mol. Cell. Biol.* 17:4633–43

Labouyrie E, Parrens M, de Mascarel A, Bloch B, Merlio JP. 1997. Distribution of NGF receptors in normal and pathologic human lymphoid tissues. *J. Neuroimmunol.* 77:161–73

Ladiwala U, Lachance C, Simoneau SJ, Bhakar A, Barker PA, Antel JP. 1998. p75 neurotrophin receptor expression on adult human oligodendrocytes: signaling without cell death in response to NGF. *J. Neurosci.* 18:1297–304

Lai KO, Fu WY, Ip FCF, Ip NY. 1998. Cloning and expression of a novel neurotrophin, NT-7, from carp. *Mol. Cell. Neurosci.* 11:64–76

Lam HHD, Bhardwaj A, O'Connell MT, Hanley DF, Traystman RJ, Sofroniew MV. 1998. Nerve growth factor rapidly suppresses basal, NMDA-evoked, and AMPA-evoked nitric oxide synthase activity in rat hippocampus *in vivo. Proc. Natl. Acad. Sci. USA* 95:10926–31

Lambiase A, Bracci-Laudiero L, Bonini S, Bonini S, Starace G, et al. 1997. Human CD4+ T cell clones produce and release nerve growth factor and express high-affinity nerve growth factor receptors. *J. Allergy Clin. Immunol.* 100:408–14

Landreth GE, Shooter EM. 1980. Nerve growth factor receptors on PC12 cells: ligand-induced conversion from low- to high-affinity states. *Proc. Natl. Acad. Sci. USA* 77:4751–55

Lange-Carter CA, Johnson GL. 1994. Ras-dependent growth factor regulation of MEK kinase in PC12 cells. *Science* 265:1458–61

Large TH, Weskamp G, Helder JC, Radeke MJ, Misko TP, et al. 1989. Structure and developmental expression of the nerve growth factor receptor in the chicken central nervous system. *Neuron* 2:1123–34

Le Good JA, Ziegler WH, Parekh DB, Alessi DR, Cohen P, Parker PJ. 1998. Protein kinase C isotypes controlled by phosphoinositide 3-kinase through the protein kinase PDK1. *Science* 281:2042–45

Lee J, Wang Z, Luoh SM, Wood WI, Scadden DT. 1994a. Cloning of FRK, a novel human intracellular SRC-like tyrosine kinase-encoding gene. *Gene* 138:247–51

Lee KF, Bachman K, Landis S, Jaenisch R. 1994b. Dependence on p75 for innervation of some sympathetic targets. *Science* 263:1447–49

Lee KF, Davies AM, Jaenisch R. 1994c. p75-deficient embryonic dorsal root sensory and neonatal sympathetic neurons display a decreased sensitivity to NGF. *Development* 120:1027–33

Lee KF, Li E, Huber LJ, Landis SC, Sharpe AH, et al. 1992. Targeted mutation of the gene encoding the low affinity NGF receptor p75 leads to deficits in the peripheral sensory nervous system. *Cell* 69:737–49

Lee SB, Rhee SG. 1995. Significance of PIP2

hydrolysis and regulation of phospholipase C isozymes. *Curr. Opin. Cell Biol.* 7:183–89

Lehmann M, Fourneir A, Selles-Navarro I, Dergham P, Sebok A, et al. 1999. Inactivation of rho signaling pathway promotes CNS axon regeneration. *J. Neurosci.* 19:7537–47

Lemke G, Chao M. 1988. Axons regulate Schwann cell expression of the major myelin and NGF receptor genes. *Development* 102:499–504

Leon A, Buriani A, Dal Toso R, Fabris M, Romanello S, et al. 1994. Mast cells synthesize, store, and release nerve growth factor. *Proc. Natl. Acad. Sci. USA* 91:3739–43

LeSauteur L, Cheung NV, Lisbona R, Saragovi HU. 1996. Small molecule nerve growth factor analogues image receptors in vivo. *Nat. Biotechnol.* 14:1120–22

Levi-Montalcini R, Hamburger V. 1951. Selective growth stimulating effects of mouse sarcoma on the sensory and sympathetic nervous system of chick embryo. *J. Exp. Zool.* 116:321–61

Levi-Montalcini R, Hamburger V. 1953. A diffusible agent of mouse sarcoma, producing hyperplasia of sympathetic ganglia and hyperneurotization of viscera in the chick embryo. *J. Exp. Zool.* 123:233–87

Levi-Montalcini R, Skaper SD, Toso RD, Petrelli L, Leon A. 1996. Nerve growth factor: from neurotrophin to neurokine. *Trends Neurosci.* 19:514–20

Levi-Montalcini R, Toso RD, Valle FD, Skaper SD, Leon A. 1995. Update of the NGF saga. *J. Neurol. Sci.* 130:119–27

Lewis TS, Shapiro PS, Ahn NG. 1998. Signal transduction through MAP kinase cascades. *Adv. Cancer Res.* 74:49–139

Li Y, Holtzman DM, Kromer LF, Kaplan DR, Chua-Couzens J, et al. 1995. Regulation of TrkA and ChAT expression in developing rat basal forebrain: evidence that both exogenous and endogenous NGF regulate differentiation of cholinergic neurons. *J. Neurosci.* 15:2888–905

Liepinsh E, Ilag LL, Otting G, Ibanez CF. 1997. NMR structure of the death domain of the p75

neurotrophin receptor. *EMBO J.* 16:4999–5005

Lindholm D, Heumann R, Meyer M, Thoenen H. 1987. Interleukin-1 regulates synthesis of nerve growth factor in non-neuronal cells of rat sciatic nerve. *Nature* 330:658–59

Lindsay RM, Harmar AJ. 1989. Nerve growth factor regulates expression of neuropeptide genes in adult sensory neurons. *Nature* 337:362–64

Lindvall O, Kokaia Z, Bengzon J, Elmer E, Kokaia M. 1994. Neurotrophins and brain insults. *Trend. Neurosci.* 17:490–96

Liu YZ, Chrivia JC, Latchman DS. 1998. Nerve growth factor upregulates the transcriptional activity of CBP through activation of the p42/p44(MAPK) cascade. *J. Biol. Chem.* 273:32400–407

Liu ZG, Hsu H, Goeddel DV, Karin M. 1996. Dissection of TNF receptor 1 effector functions: JNK activation is not linked to apoptosis while NF-kappaB activation prevents cell death. *Cell* 87:565–76

Liuzzo J, Petanceska S, Devi L. 1999. Neurotrophic factors regulate cathepsin S in macrophages and microglia: a role in the degradation of myelin basic protein and amyloid beta peptide. *Mol. Med.* 5:334–43

Liyanage M, Frith D, Livneh E, Stabel S. 1992. Protein kinase C group B members PKC-delta, -epsilon, -zeta and PKC-L(eta). Comparison of properties of recombinant proteins in vitro and in vivo. *Biochem. J.* 283:781–87

Loeb DM, Maragos J, Martin-Zanca D, Chao MV, Parada LF, Greene LA. 1991. The trk proto-oncogene rescues NGF responsiveness in mutant NGF-nonresponsive PC12 cell lines. *Cell* 66:961–66

Loeb DM, Stephens RM, Copeland T, Kaplan DR, Greene LA. 1994. A Trk nerve growth factor (NGF) receptor point mutation affecting interaction with phospholipase C-gamma 1 abolishes NGF-promoted peripherin induction but not neurite outgrowth. *J. Biol. Chem.* 269:8901–10

Longo FM, Holtzman DM, Grimes ML, Mobley WC. 1993. Nerve growth factor: actions in the peripheral and central nervous systems. In *Neurotrophic Factors*, ed. SE Loughlin, JH Fallon. San Deigo, CA: Academic. 209–56

Longo FM, Mobley WC. 1996. Minimized hormones grow in stature. *Nat. Biotechnol.* 14:1092

Lucidi-Phillipi CA, Clary DO, Reichardt LF, Gage FH. 1996. TrkA activation is sufficient to rescue axotomized cholinergic neurons. *Neuron* 16:653–63

Luo J, West JR, Pantazis NJ. 1997. Nerve growth factor and basic fibroblast growth factor protect rat cerebellar granule cells in culture against ethanol-induced cell death. *Alcohol Clin. Exp. Res.* 21:1108–20

Luo Z, Diaz B, Marshall MS, Avruch J. 1997. An intact Raf zinc finger is required for optimal binding to processed Ras and for ras-dependent Raf activation in situ. *Mol. Cell. Biol.* 17:46–53

MacDonald JI, Gryz EA, Kubu CJ, Verdi JM, Meakin SO. 2000. Direct binding of the signaling adapter protein Grb2 to the activation loop tyrosines on the nerve growth factor receptor tyrosine kinase, TrkA. *J. Biol. Chem.* 275:18225–33

MacPhee IJ, Barker PA. 1997. Brain-derived neurotrophic factor binding to the p75 neurotrophin receptor reduces TrkA signaling while increasing serine phosphorylation in the TrkA intracellular domain. *J. Biol. Chem.* 272:23547–51

Madison RD, Archibald SJ. 1994. Point sources of Schwann cells result in growth into a nerve entubulation repair site in the absence of axons: effects of freeze-thawing. *Exp. Neurol.* 128:266–75

Mahadeo D, Kaplan L, Chao MV, Hempstead BL. 1994. High affinity nerve growth factor binding displays a faster rate of association than p140trk binding. Implications for multi-subunit polypeptide receptors. *J. Biol. Chem.* 269:6884–91

Maggirwar SB, Sarmiere PD, Dewhurst S, Freeman RS. 1998. Nerve growth factor-dependent activation of NF-kappaB con-

tributes to survival of sympathetic neurons. *J. Neurosci.* 18:10356–65

Majdan M, Lachance C, Gloster A, Aloyz R, Zeindler C, et al. 1997. Transgenic mice expressing the intracellular domain of the p75 neurotrophin receptor undergo neuronal apoptosis. *J. Neurosci.* 17:6988–98

Malcangio M, Garrett NE, Tomlinson DR. 1997. Nerve growth factor treatment increases stimulus-evoked release of sensory neuropeptides in the rat spinal cord. *Eur. J. Neurosci.* 9:1101–4

Malinin NL, Boldin MP, Kovalenko AV, Wallach D. 1997. MAP3K-related kinase involved in NF-kappaB induction by TNF, CD95 and IL-1. *Nature* 385:540–44

Mallett S, Barclay AN. 1991. A new super-family of cell surface proteins related to the nerve growth factor receptor. *Immunol. Today* 12:220–23

Marais R, Light Y, Mason C, Paterson H, Olson MF, Marshall CJ. 1998. Requirement of Ras-GTP-Raf complexes for activation of Raf-1 by protein kinase C. *Science* 280:109–12

Marshall CJ. 1994. MAP kinase kinase kinase, MAP kinase kinase and MAP kinase. *Curr. Opin. Genet. Dev.* 4:82–89

Marshall JS, Gomi K, Blennerhassett MG, Bienenstock J. 1999. Nerve growth factor modifies the expression of inflammatory cytokines by mast cells via a prostanoid-dependent mechanism. *J. Immunol.* 162:4271–76

Martin-Zanca D, Hughes SH, Barbacid M. 1986a. A human oncogene formed by the fusion of truncated tropomyosin and protein tyrosine kinase sequences. *Nature* 319:743–48

Martin-Zanca D, Mitra G, Long LK, Barbacid M. 1986b. Molecular characterization of the human trk oncogene. *Cold Spring Harb. Symp. Quant. Biol.* 51:983–92

Martin-Zanca D, Oskam R, Mitra G, Copeland T, Barbacid M. 1989. Molecular and biochemical characterization of the human trk proto-oncogene. *Mol. Cell. Biol.* 9:24–33

Martinex-Serrano A, Bjorklund A. 1996. Protection of the neostriatum against excitotoxic damage by neurotrophin-producing, genetically modified neural stem cells. *J. Neurosci.* 16:4604

Martínez-Murillo R, Fernández AP, Bentura ML, Rodrigo J. 1998. Subcellular localization of low-affinity nerve growth factor receptor-immunoreactive protein in adult rat purkinje cells following traumatic injury. *Exp. Brain Res.* 119:47–57

Matsuda M, Hashimoto Y, Muroya K, Hasegawa H, Kurata T, et al. 1994. CRK protein binds to two guanine nucleotide-releasing proteins for the Ras family and modulates nerve growth factor-induced activation of Ras in PC12 cells. *Mol. Cell. Biol.* 14:5495–500

Mattson M. 1996. Calcium and free radicals: mediators of neurotrophic factor and excitatory transmitter-regulated developmental plasticity and cell death. *Perspect. Dev. Neurobiol.* 3:79–91

Mattson M, Marck R. 1996. Excitotoxicity and excitoprotection in vitro. *Adv. Neurol.* 71:1–30

Mattson MP. 1997. Neuroprotective signal transduction: relevance to stroke. *Neurosci. Biobehav. Rev.* 21:193–206

Mattson MP, Goodman Y, Luo H, Fu W, Furukawa K. 1997. Activation of NF-kappaB protects hippocampal neurons against oxidative stress-induced apoptosis: evidence for induction of manganese superoxide dismutase and suppression of peroxynitrite production and protein tyrosine nitration. *J. Neurosci. Res.* 49:681–97

Mattson MP, Lovell MA, Furukawa K, Markesbery WR. 1995. Neurotrophic factors attenuate glutamate-induced accumulation of peroxides, elevation of intracellular Ca^{2+} concentration, and neurotoxicity and increase antioxidant enzyme activities in hippocampal neurons. *J. Neurochem.* 65:1740–51

McAllister AK. 1999. Neurotrophins and synaptic plasticity. *Annu. Rev. Neurosci.* 22:295–318

McCormick F. 1994. Activators and effectors of ras p21 proteins. *Curr. Opin. Genet. Dev.* 4:71–76

McDonald NQ, Lapatto R, Murray-Rust J, Gunning J, Wlodawer A, Blundell TL. 1991. New protein fold revealed by a 2.3-A resolution crystal structure of nerve growth factor. *Nature* 354:411–14

McMahon SB, Bennete DLH, Priestley JV, Shelton DL. 1995. The biological effects of endogenous nerve growth factor on adult sensory neurons revealed by a trkA-IgG fusion molecule. *Nat. Med.* 1:774

McMartin DN, O'Connor JA. 1979. Effect of age on axoplasmic transport of cholinesterase in rat sciatic nerves. *Mech. Ageing Dev.* 10:241–48

Meakin SO, MacDonald JI, Gryz EA, Kubu CJ, Verdi JM. 1999. The signaling adapter FRS-2 competes with Shc for binding to the nerve growth factor receptor TrkA. A model for discriminating proliferation and differentiation. *J. Biol. Chem.* 274:9861–70

Meakin SO, Shooter EM. 1992. The nerve growth factor family of receptors. *Trends Neurosci.* 15:323–31

Meakin SO, Suter U, Drinkwater CC, Welcher AA, Shooter EM. 1992. The rat trk protooncogene product exhibits properties characteristic of the slow nerve growth factor receptor. *Proc. Natl. Acad. Sci. USA* 89:2374–78

Meller N, Liu YC, Collins TL, Bonnefoy-Berard N, Baier G, et al. 1996. Direct interaction between protein kinase C theta (PKC theta) and 14-3-3 tau in T cells: 14-3-3 overexpression results in inhibition of PKC theta translocation and function. *Mol. Cell. Biol.* 16:5782–91

Mendell LM. 1996. Neurotrophins and sensory neurons: role in development, maintenance and injury. A thematic summary. *Philos. Trans. R. Soc. London Ser. B* 351:463–67

Mendell LM, Albers KM, Davis BM. 1999. Neurotrophins, nociceptors, and pain. *Microsc. Res. Technol.* 45:252–61

Menniti FS, Oliver KG, Putney JW Jr, Shears SB. 1993. Inositol phosphates and cell signaling: new views of InsP5 and InsP6. *Trends Biochem. Sci.* 18:53–56

Merry DE, Korsmeyer SJ. 1997. cl-2 gene family in the nervous system. *Annu. Rev. Neurosci.* 20:245–67

Micera A, Vigneti E, Aloe L. 1998. Changes of NGF presence in nonneuronal cells in response to experimental allergic encephalomyelitis in lewis rats. *Exp. Neurol.* 154:41–46

Michael GJ, Kaya E, Averill S, Rattry M, Clay DO, Priestley JV. 1997. TrkA immunoreactive neurones in the rat spinal cord. *J. Comp. Neurol.* 385:441–55

Middlemas DS, Meisenhelder J, Hunter T. 1994. Identification of TrkB autophosphorylation sites and evidence that phospholipase C-gamma 1 is a substrate of the TrkB receptor. *J. Biol. Chem.* 269:5458–66

Miller FD, Mathew TC, Toma JG. 1991. Regulation of nerve growth factor receptor gene expression by nerve growth factor in the developing peripheral nervous system. *J. Cell Biol.* 112:303–12

Miller FD, Speelman A, Mathew TC, Fabian J, Chang E, et al. 1994. Nerve growth factor derived from terminals selectively increases the ratio of p75 to trkA NGF receptors on mature sympathetic neurons. *Dev. Biol.* 161:206–17

Miranti CK, Ginty DD, Huang G, Chatila T, Greenberg ME. 1995. Calcium activates serum response factor-dependent transcription by a Ras- and Elk-1-independent mechanism that involves a Ca^{2+}/calmodulin-dependent kinase. *Mol. Cell. Biol.* 15:3672–84

Mirsky R, Jessen KR. 1999. The neurobiology of Schwann cells. *Brain Pathol.* 9:293–311

Mizuma H, Takagi K, Miyake K, Takagi N, Ishida K, et al. 1999. Microsphere embolism-induced elevation of nerve growth factor level and appearance of nerve growth factor immunoreativity in activated T-lymphocytes in the rat brain. *J. Neurosci. Res.* 55:749–61

Mobley WC, Rutkowski JL, Tennekoon GI, Buchanan K, Johnston M. 1985. Choline

acetyltransferase activity in striatum of neonatal rats increased by nerve growth factor. *Science* 229:284–87

Moller DE, Xia CH, Tang W, Zhu AX, Jakubowski M. 1994. Human rsk isoforms: cloning and characterization of tissue-specific expression. *Am. J. Physiol.* 266: C351–59

Morrison ME, Mason CA. 1998. Granule neuron regulation of purkinje cell development: striking a balance between neurotrophin and glutamate signaling. *J. Neurosci.* 18:3563–73

Mueller CG, Nordheim A. 1991. A protein domain conserved between yeast MCM1 and human SRF directs ternary complex formation. *EMBO J.* 10:4219–29

Mufson EJ, Lavine N, Jaffar S, Kordower JH, Quirion R, Saragovi HU. 1997. Reduction in p140-TrkA receptor protein within the nucleus basalis and cortex in Alzheimer's disease. *Exp. Neurol.* 146:91–103

Muragaki Y, Chou TT, Kaplan DR, Trojanowski JQ, Lee VM. 1997. Nerve growth factor induces apoptosis in human medulloblastoma cell lines that express TrkA receptors. *J. Neurosci.* 17:530–42

Muroya K, Hattori S, Nakamura S. 1992. Nerve growth factor induces rapid accumulation of the GTP-bound form of p21ras in rat pheochromocytoma PC12 cells. *Oncogene* 7:277–81

Nakajima T, Fukamizu A, Takahashi J, Gage FH, Fisher T, et al. 1996. The signal-dependent coactivator CBP is a nuclear target for pp90RSK. *Cell* 86:465–74

Neumann S, Woolf CJ. 1999. Regeneration of dorsal column fibers into and beyond the lesion site following adult spinal cord injury. *Neuron* 23:83–91

Nishizuka Y. 1988. The molecular heterogeneity of protein kinase C and its implications for cellular regulation. *Nature* 334:661–65

Nishizuka Y. 1992. Intracellular signaling by hydrolysis of phospholipids and activation of protein kinase C. *Science* 258:607–14

Noda M, Ko M, Ogura A, Liu DG, Amano

T, et al. 1985. Sarcoma viruses carrying ras oncogenes induce differentiation- associated properties in a neuronal cell line. *Nature* 318:73–75

Obermeier A, Bradshaw RA, Seedorf K, Choidas A, Schlessinger J, Ullrich A. 1994. Neuronal differentiation signals are controlled by nerve growth factor receptor/Trk binding sites for SHC and PLC gamma. *EMBO J.* 13:1585–90

Oberwetter JM, Conrad KE, Gutierrez-Hartmann A. 1993. The Ras and protein kinase C signaling pathways are functionally antagonistic in GH4 neuroendocrine cells. *Mol. Endocrinol.* 7:915–23

O'Brien TS, Svendsen CN, Isacson O, Sofroniew MV. 1990. Loss of True blue labelling from the medial septum following transection of the fimbria-fornix: evidence for the death of cholinergic and noncholinergic neurons. *Brain Res.* 508:249–56

Ong SH, Guy GR, Hadari YR, Laks S, Gotoh N, et al. 2000. FRS2 proteins recruit intracellular signaling pathways by binding to diverse targets on fibroblast growth factor and nerve growth factor receptors. *Mol. Cell. Biol.* 20:979–89

Ono Y, Fujii T, Ogita K, Kikkawa U, Igarashi K, Nishizuka Y. 1988. The structure, expression, and properties of additional members of the protein kinase C family. *J. Biol. Chem.* 263:6927–32

Osada S, Mizuno K, Saido TC, Akita Y, Suzuki K, et al. 1990. A phorbol ester receptor/protein kinase, nPKC eta, a new member of the protein kinase C family predominantly expressed in lung and skin. *J. Biol. Chem.* 265:22434–40

Osada S, Mizuno K, Saido TC, Suzuki K, Kuroki T, Ohno S. 1992. A new member of the protein kinase C family, nPKC theta, predominantly expressed in skeletal muscle. *Mol. Cell. Biol.* 12:3930–38

Oshima M, Sithanandam G, Rapp UR, Guroff G. 1991. The phosphorylation and activation of B-raf in PC12 cells stimulated by nerve growth factor. *J. Biol. Chem.* 266:23753–60

Otten U, Scully JL, Ehrhard PB, Gadient RA. 1994. Neurotrophins: signals between the nervous and immune systems. *Prog. Brain Res.* 103:293–305

Pang L, Sawada T, Decker SJ, Saltiel AR. 1995. Inhibition of MAP kinase kinase blocks the differentiation of PC-12 cells induced by nerve growth factor. *J. Biol. Chem.* 270:13585–88

Pascual M, Rocamora N, Acsády L, Freund TF, Soriano E. 1998. Expression of nerve growth factor and neurotrophin-3 mRNAs in hippocampal interneurons: morphological characterization, levels of expression, and colocalization of nerve growth factor and neurotrophin-3. *J. Comp. Neurol.* 395:73–90

Patel TD, Jackman A, Rice FL, Kucera J, Snider WD. 2000. Development of sensory neurons in the absence of NGF/TrkA signaling in vivo [see comments]. *Neuron* 25:345–57

Payne DM, Rossomando AJ, Martino P, Erickson AK, Her JH, et al. 1991. Identification of the regulatory phosphorylation sites in pp42/mitogen-activated protein kinase (MAP kinase). *EMBO J.* 10:885–92

Pearson RCA, Sofroniew MV, Cuello AC, Powell TPS, Eckenstein F, et al. 1983. Persistance of cholinergic neurons in the basal nucleus in a brain with senile dementia of the Alzheimer's type demonstrated by immunohistochemical staining for choline acetyltransferase. *Brain Res.* 289:375–79

Pechan PA, Chowdhury K, Gerdes W, Seifert W. 1993. Glutamate induces the growth factors NGF and bFGF, the receptors FGF-R1 and c-fos mRNA expression in rat astrocyte culture. *Neurosci. Lett.* 16:111–14

Pechan PA, Chowdhury K, Seifert W. 1992. Free radicals induce gene expression of NGF and bFGF in rat astrocyte culture. *NeuroReport* 3:469–72

Peterson DA, Dickinson-Anson HA, Leppert JT, Lee KF, Gage FH. 1999. Central neuronal loss and behavioral impairment in mice lacking neurotrophin receptor p75. *J. Comp. Neurol.* 404:1–20

Petty B, Cornblath DR, Adornato BT, Chaudry

V, Flexner C, et al. 1994. The effect of sytemically administered recombinant human nerve growth factor in healthy human subjects. *Ann. Neurol.* 36:244–46

Pitts AF, Miller MW. 2000. Expression of nerve growth factor, brain-derived neurotrophic factor, and neurotrophin-3 in the somatosensory cortex of the mature rat: coexpression with high-affinity neurotrophin receptors. *J. Comp. Neurol.* 418:241–54

Pizzorusso T, Berardi N, Rossi FM, Viegi A, Venstrom K, et al. 1999. TrkA activation in the rat visual cortex by antirat trkA IgG prevents the effect of monocular deprivation. *Eur J. Neurosci.* 11:204–12

Prakash N, Cohen-Cory S, Frostig RD. 1996. Rapid and opposite effects of BDNF and NGF on the functional organization of the adult cortex in vivo. *Nature* 381:702–6

Qian X, Riccio A, Zhang Y, Ginty DD. 1998. Identification and characterization of novel substrates of Trk receptors in developing neurons. *Neuron* 21:1017–29

Qui MS, Green SH. 1992. PC12 cell neuronal differentiation is associated with prolonged p21ras activity and consequent prolonged ERK activity. *Neuron* 9:705–17

Rabin SJ, Cleghon V, Kaplan DR. 1993. SNT, a differentiation-specific target of neurotrophic factor-induced tyrosine kinase activity in neurons and PC12 cells. *Mol. Cell. Biol.* 13:2203–13

Rabizadeh S, Oh J, Zhong LT, Yang J, Bitler CM, et al. 1993. Induction of apoptosis by the low-affinity NGF receptor. *Science* 261:345–48

Ramer MS, Priestley JV, McMahon SB. 2000. Functional regeneration of sensory axons into the adult spinal cord. *Nature* 403:312–16

Ramirez S, Ait-Si-Ali S, Robin P, Trouche D, Harel-Bellan A. 1997. The CREB-binding protein (CBP) cooperates with the serum response factor for transactivation of the c-fos serum response element. *J. Biol. Chem.* 272:31016–21

Ribon V, Saltiel AR. 1996. Nerve growth factor stimulates the tyrosine phosphorylation of endogenous Crk-II and augments its association with p130Cas in PC-12 cells. *J. Biol. Chem.* 271:7375–80

Riccio A, Ahn S, Davenport CM, Blendy JA, Ginty DD. 1999. Mediation by a CREB family transcription factor of NGF-dependent survival of sympathetic neurons. *Science* 286:2358–61

Riccio A, Pierchala BA, Ciarello CL, Ginty DD. 1997. An NGF-TrkA-mediated retrograde signal to transcription factor CREB in sympathetic neurons. *Science* 277:1097–100

Rice FL, Albers KM, Davis BM, Silos-Santiago I, Wilkinson GA, et al. 1998. Differential dependency of unmyelinated and A delta epidermal and upper dermal innervation on neurotrophins, trk receptors, and p75LNGFR. *Dev. Biol.* 198:57–81

Rich KM, Alexander TD, Pryor JC, Hollowell JP. 1989. Nerve growth factor enhances regeneration through silicone chambers. *Exp. Neurol.* 105:162–70

Richardson PM, Issa VM. 1984. Peripheral injury enhances entral regeneration of primary sensory neurones. *Nature* 309:791–93

Robinson RC, Radziejewski C, Spraggon G, Greenwald J, Kostura MR, et al. 1999. The structures of the neurotrophin 4 homodimer and the brain-drived neurotrophic factor/neurotrophin 4 heterodimer reveal a common Trk-binding site. *Protein Sci.* 8:2589–97

Robinson RC, Radziejewski C, Stuart DI, Jones EY. 1995. Structure of the brain-derived neurotrophic factor/neurotrophin 3 heterodimer. *Biochemistry* 34:4139–46

Ross AH, Daou MC, McKinnon CA, Condon PJ, Lachyankar MB, et al. 1996. The neurotrophin receptor, gp75, forms a complex with the receptor tyrosine kinase TrkA. *J. Cell Biol.* 132:945–53

Ross GM, Shamovsky IL, Lawrance G, Solc M, Dostaler SM, et al. 1998. Reciprocal modulation of TrkA and p75NTR affinity states is mediated by direct receptor interactions. *Eur. J. Neurosci.* 10:890–98

Rothe M, Sarma V, Dixit VM, Goeddel DV.

1995. TRAF2-mediated activation of NF-kappa B by TNF receptor 2 and CD40. *Science* 269:1424–27

Rothstein JD, Dykes-Hoberg M, Pardo CA, Bristol LA, Jin L, et al. 1996. Knockout of glutamate transporters reveals a major role for astroglial transport in excitotoxicity and clearance of glutamate. *Neuron* 16:675–86

Rozakis-Adcock M, McGlade J, Mbamalu G, Pelicci G, Daly R, et al. 1992. Association of the Shc and Grb2/Sem5 SH2-containing proteins is implicated in activation of the Ras pathway by tyrosine kinases. *Nature* 360:689–92

Ruit KG, Osborne PA, Schmidt RE, Johnson EM Jr, Snider WD. 1990. Nerve growth factor regulates sympathetic ganglion cell morphology and survival in the adult mouse. *Neuroscience* 10:2412–19

Ryden M, Ibanez CF. 1997. A second determinant of binding to the p75 neurotrophin receptor revealed by alanine-scanning mutagenesis of a conserved loop in nerve growth factor. *J. Biol. Chem.* 272:33085–91

Saper CB, Wainer BH, German DC. 1987. Axonal and transneuronal transport in the transmission of neurological disease: potential role in systems degeneration. *Neuroscience* 23:389–97

Schechter AL, Bothwell MA. 1981. Nerve growth factor receptors on PC12 cells: evidence for two receptor classes with differing cytoskeletal association. *Cell* 24:867–74

Schneider A, Martin-Villalba A, Weih F, Vogel J, Wirth T, Schwaninger M. 1999. NF-kappaB is activated and promotes cell death in focal cerebral ischemia. *Nat. Med.* 5:554–59

Schonwasser DC, Marais RM, Marshall CJ, Parker PJ. 1998. Activation of the mitogen-activated protein kinase/extracellular signal-regulated kinase pathway by conventional, novel, and atypical protein kinase C isotypes. *Mol. Cell. Biol.* 18:790–98

Schouten GJ, Vertegaal AC, Whiteside ST, Israel A, Toebes M, et al. 1997. IkappaB alpha is a target for the mitogen-activated 90

kDa ribosomal S6 kinase. *EMBO J.* 16:3133–44

Schumacher JM, Short MP, Hyman BT, Breakfield XO, Isacson O. 1991. Intracerebral implantation of nerve growth factor-producing fibroblasts protects striatum against neurotoxic levels of excitatory amino acids. *Neuroscience* 43:567–70

Schwaninger M, Sallmann S, Petersen N, Schneider A, Prinz S, et al. 1999. Bradykinin induces interleukin-6 expression in astrocytes through activation of nuclear factor-kappaB. *J. Neurochem.* 73:1461–66

Scott SA, Mufson EJ, Weingartner JA, Skau KA, Crutcher KA. 1995. Nerve growth factor in Alzheimer's disease: increased levels throughout the brain coupled with declines in nucleus basalis. *J. Neurosci.* 15:6213–21

Seibenhener ML, Roehm J, White WO, Neidigh KB, Vandenplas ML, Wooten MW. 1999. Identification of Src as a novel atypical protein kinase C-interacting protein. *Mol. Cell. Biol. Res. Commun.* 2:28–31

Semkova I, Krieglstein J. 1999. Cilliary neurotrophic factor enhances the expression of NGF and p75 low-affinity NGF receptor in astrocytes. *Brain Res.* 838:184–92

Sheng M, Greenberg ME. 1990. The regulation and function of c-fos and other immediate early genes in the nervous system. *Neuron* 4:477–85

Shigeno T, Mima T, Takakura K, Graham DI, Kato G, et al. 1991. Amelioration of delayed neuronal death in the hippocampus by nerve growth factor. *J. Neurosci.* 11:2914–19

Shu XQ, Mendell LM. 1999. Neurotrophins and hyperalgesia. *Proc. Natl. Acad. Sci. USA* 96:7693–96

Siliprandi R, Canella R, Carmignoto G. 1993. Nerve growth factor promotes functional recovery of retinal ganglion cells after ischemia. *Invest. Ophthalmol. Vis. Sci.* 34:3232–45

Simone M, De Santis S, Vigneti E, Papa G, Amadori S, Aloe L. 1999. Nerve growth factor: a survey of activity on immune and hematopoietic cells. *Hematol. Oncol.* 17:1–10

Singer CA, Figueroa-Masot XA, Batchelor RH, Dorsa DM. 1999. The mitogen-activated protein kinase pathway mediates estrogen neuroprotection after glutamate toxicity in primary cortical neurons. *J. Neurosci.* 19:2455

Singh M, Sétáló GJ, Guan X, Warren M, Toran-Allerand CD. 1999. Estrogen-induced activation of mitogen-activated protein kinase in cerebral cortical explants: convergence of estrogen and neurotrophin signaling pathways. *J. Neurosci.* 19:1179–88

Smeyne RJ, Klein R, Schnapp A, Long LK, Bryant S, et al. 1994. Severe sensory and sympathetic neuropathies in mice carrying a disrupted Trk/NGF receptor gene [see comments]. *Nature* 368:246–49

Smith CA, Farrah T, Goodwin RG. 1994. The TNF receptor superfamily of cellular and viral proteins: activation, costimulation, and death. *Cell* 76:959–62

Snider WD. 1994. Functions of the neurotrophins during nervous system development: what the knockouts are teaching us. *Cell* 77:627–38

Sofroniew MV. 1999. Neuronal responses to axotomy. In *CNS Regeneration: Basic Science and Clinical Advances*, ed. M Tuszynski, J Kordower, pp. 3–26. San Diego, CA: Academic

Sofroniew MV, Cooper JD, Svendsen CN, Crossman P, Ip NY, et al. 1993. Atrophy but not death of adult septal cholinergic neurons after ablation of target capacity to produce mRNAs for NGF, BDNF and NT-3. *J. Neurosci.* 13:5263–76

Sofroniew MV, Galletly NP, Isacson O, Svendsen CN. 1990. Survival of adult basal forebrain cholinergic neurons after loss of target neurons. *Science* 247:338–42

Sofroniew MV, Mobley WC. 1993. On the possibility of positive feedback in trophic interactions between afferent and target neurons. *Semin. Neurosci.* 5:309–12

Sofroniew MV, Pearson RCA, Eckenstein F,

Cuello AC, Powell TPS. 1983. Retrograde changes in the basal forebrain of the rat following cortical damage. *Brain Res.* 289:370–74

Song HY, Dunbar JD, Donner DB. 1994. Aggregation of the intracellular domain of the type 1 tumor necrosis factor receptor defined by the two-hybrid system. *J. Biol. Chem.* 269:22492–95

Sozeri O, Vollmer K, Liyanage M, Frith D, Kour G, et al. 1992. Activation of the c-Raf protein kinase by protein kinase C phosphorylation. *Oncogene* 7:2259–62

Steller H. 1995. Mechanisms and genes of cellular suicide. *Science* 267:1445–49

Stephens L, Anderson K, Stokoe D, Erdjument-Bromage H, Painter GF, et al. 1998. Protein kinase B kinases that mediate phosphatidylinositol 3,4,5-trisphosphate-dependent activation of protein kinase B. *Science* 279:710–14

Stephens RM, Loeb DM, Copeland TD, Pawson T, Greene LA, Kaplan DR. 1994. Trk receptors use redundant signal transduction pathways involving SHC and PLC-gamma 1 to mediate NGF responses. *Neuron* 12:691–705

Stokoe D, Stephens LR, Copeland T, Gaffney PR, Reese CB, et al. 1997. Dual role of phosphatidylinositol-3,4,5-trisphosphate in the activation of protein kinase B. *Science* 277:567–70

Strauss S, Otten U, Joggerst B, Pluss K, Volk B. 1968. Increased levels of nerve growth factor (NGF) protein and mRNA and reactive gliosis following kainic acid injection into the rat striatum. *Neurosci. Lett.* 168:193–96

Strijbos P, Rothwell N. 1995. Interleukin-1 beta attenuates excitatory amino acid-induced neurodegeneration in vitro: involvement of nerve growth factor. *J. Neurosci.* 15:3468

Sturgill TW, Ray LB, Erikson E, Maller JL. 1988. Insulin-stimulated MAP-2 kinase phosphorylates and activates ribosomal protein S6 kinase II. *Nature* 334:715–18

Susaki Y, Shimizu S, Katakura K, Watanabe N, Kawamoto K, et al. 1996. Functional properties of murine macrophages promoted by nerve growth factor. *Blood* 88:4630–37

Susaki Y, Tanaka A, Honda E, Matsuda H. 1998. Nerve growth factor modulates Fc gamma receptor expression on murine macrophage J774A.1 cells. *J. Vet. Med. Sci.* 60:87–91

Sutter A, Riopelle RJ, Harris-Warrick RM, Shooter EM. 1979. Nerve growth factor receptors. Characterization of two distinct classes of binding sites on chick embryo sensory ganglia cells. *J. Biol. Chem.* 254:5972–82

Svendsen CN, Kew JNC, Staley K, Sofroniew MV. 1994. Death of developing septal cholinergic neurons following NGF withdrawal in vitro: protection by protein synthesis inhibition. *J. Neurosci.* 14:75–87

Taiwo YO, Levine JD, Burch R, Woo JE, Mobley WC. 1991. Hyperalgesia induced in the rat by the amino-terminal octapeptide of nerve growth factor. *Proc. Natl. Acad. Sci. USA* 88:5144–48

Taniuchi M, Clark HB, Johnson EM Jr. 1986. Induction of nerve growth factor receptor in Schwann cells after axotomy. *Proc. Natl. Acad. Sci. USA* 83:4094–98

Taniuchi M, Clark HB, Schweitzer JB, Johnson EM. 1988. Expression of nerve growth factor receptors by Schwann cells of axotomized peripheral nerves: ultrastructural location, suppression by axonal contact, and binding properties. *J. Neurosci.* 8:664–81

Tartaglia LA, Ayres TM, Wong GH, Goeddel DV. 1993. A novel domain within the 55 kd TNF receptor signals cell death. *Cell* 74:845–53

Teng KK, Lander H, Fajardo JE, Hanafusa H, Hempstead BL, Birge RB. 1995. v-Crk modulation of growth factor-induced PC12 cell differentiation involves the Src homology 2 domain of v-Crk and sustained activation of the Ras/mitogen-activated protein kinase pathway. *J. Biol. Chem.* 270:20677–85

Thomas SM, Brugge JS. 1997. Cellular functions regulated by Src family kinases. *Annu. Rev. Cell Dev. Biol.* 13:513–609

Thomas SM, Hayes M, D'Arcangelo G, Armstrong RC, Meyer BE, et al. 1991. Induction of neurite outgrowth by v-src mimics critical aspects of nerve growth factor-induced differentiation. *Mol. Cell. Biol.* 11:4739–50

Thuveson M, Albrecht D, Zurcher G, Andres AC, Ziemiecki A. 1995. iyk, a novel intracellular protein tyrosine kinase differentially expressed in the mouse mammary gland and intestine. *Biochem. Biophys. Res. Commun.* 209:582–89

Toran-Allerand CD, Miranda RC, Bentham WD, Sohrabji F, Brown TJ, et al. 1992. Estrogen receptors colocalize with low-affinity nerve growth factor receptors in cholinergic neurons of the basal forebrain. *Proc. Natl. Acad. Sci. USA* 89:4668–72

Torcia M, Bracci-Laudiero L, Lucibello M, Nencioni L, Labardi D, et al. 1996. Nerve growth factor is an autocrine survival factor for memory B lymphocytes. *Cell* 85:345–56

Torres M, Bogenmann E. 1996. Nerve growth factor induces a multimeric TrkA receptor complex in neuronal cells that includes Crk, SHC and PLC-gamma 1 but excludes P130CAS. *Oncogene* 12:77–86

Traverse S, Cohen P. 1994. Identification of a latent MAP kinase kinase kinase in PC12 cells as B-raf. *FEBS Lett.* 350:13–18

Traverse S, Gomez N, Paterson H, Marshall C, Cohen P. 1992. Sustained activation of the mitogen-activated protein (MAP) kinase cascade may be required for differentiation of PC12 cells. Comparison of the effects of nerve growth factor and epidermal growth factor. *Biochem. J.* 288:351–55

Treisman R. 1992. The serum response element. *Trends Biochem. Sci.* 17:423–26

Trim N, Morgan S, Evans M, Issa R, Fine D, et al. 2000. Hepatic stellate cells express the low affinity nerve growth factor receptor p75 and undergo apoptosis in response to nerve growth factor stimulation. *Am. J. Pathol.* 156:1235–43

Tsui-Pierchala BA, Ginty DD. 1999. Characterization of an NGF-P-TrkA retrograde-signaling complex and age-dependent regulation of TrkA phosphorylation in sympathetic neurons. *Neuroscience* 19:8207–18

Tuszynski MH, Armstrong DM, Gage FH. 1990. Basal forebrain cell loss following fimbria/fornix transection. *Brain Res.* 508:241–48

Tuszynski MH, Gage FH. 1995a. Bridging grafts and transient nerve growth factor infusions promote long-term central nervous system neuronal rescue and partial functional recovery. *Proc. Natl. Acad. Sci. USA* 92:4621–25

Tuszynski MH, Gage FH. 1995b. Maintaining the neuronal phenotype after injury in the adult CNS. Neurotrophic factors, axonal growth substrates, and gene therapy. *Mol. Neurobiol.* 10:151–67

Tuttle R, O'Leary DDM. 1998. Neurotrophins rapidly modulate growth cone response to the axon guidance molecule, collapsin-1. *Mol. Cell. Neurosci.* 11:1–8

Twiss JL, Shooter EM. 1995. Nerve growth factor promotes neurite regeneration in PC12 cells by translational control. *J. Neurochem.* 64:550–57

Ueffing M, Lovric J, Philipp A, Mischak H, Kolch W. 1997. Protein kinase C-epsilon associates with the Raf-1 kinase and induces the production of growth factors that stimulate Raf-1 activity. *Oncogene* 15:2921–27

Ure DR, Campenot RB. 1997. Retrograde transport and steady-state distribution of ^{125}I-nerve growth factor in rat sympathetic neurons in compartmented cultures. *J. Neurosci.* 17:1282–90

Urfer R, Tsoulfas P, Soppet D, Escandon E, Parada LF, Presta LG. 1994. The binding epitopes of neurotrophin-3 to its receptors trkC and gp75 and the design of a multifunctional human neurotrophin. *EMBO J.* 13:5896–909

Vaillancourt RR, Gardner AM, Johnson GL. 1994. B-Raf-dependent regulation of the MEK-1/mitogen-activated protein kinase pathway in PC12 cells and regulation by cyclic AMP. *Mol. Cell. Biol.* 14:6522–30

Vaillant AR, Mazzoni I, Tudan C, Boudreau M, Kaplan DR, Miller FD. 1999. Depolarization and neurotrophins converge on the phosphatidylinositol 3-kinase-Akt pathway to synergistically regulate neuronal survival. *J. Cell Biol.* 146:955–66

Van Aelst L, Barr M, Marcus S, Polverino A, Wigler M. 1993. Complex formation between RAS and RAF and other protein kinases. *Proc. Natl. Acad. Sci. USA* 90:6213–17

Van Antwerp DJ, Martin SJ, Kafri T, Green DR, Verma IM. 1996. Suppression of TNF-alpha-induced apoptosis by NF-kappaB. *Science* 274:787–89

van der Hoeven PCJ, van der Wal JCM, Ruurs P, van Dijk MCM, van Blitterswijk WJ. 2000. 14-3-3 isotypes facilitate coupling of protein kinase C-zeta to Raf-1: negative regulation by 14-3-3 phosphorylation. *Biochem. J.* 345:297–306

van Dijk MC, Hilkmann H, van Blitterswijk WJ. 1997. Platelet-derived growth factor activation of mitogen-activated protein kinase depends on the sequential activation of phosphatidylcholine-specific phospholipase C, protein kinase C-zeta and Raf-1. *Biochem. J.* 325:303–7

Venero JL, Knusel B, Beck KD, Hefti F. 1994. Expression of neurotrophin and trk receptor genes in adult rats with fimbria transections: effect of intraventricular nerve growth factor and brain-derived neurotrophic factor administration. *Neuroscience* 59:797–815

Verdi JM, Anderson DJ. 1994. Neurotrophins regulate sequential changes in neurotrophin receptor expression by sympathetic neuroblasts. *Neuron* 13:1359–72

Verge VM, Gratto KA, Karchewski LA, Richardson PM. 1996. Neurotrophins and nerve injury in the adult. *Philos. Trans. R. Soc. London Ser. B* 351:423–30

Verge VM, Merlio JP, Grondin J, Ernfors P, Persson H, et al. 1992. Colocalization of NGF binding sites, trk mRNA, and low-affinity NGF receptor mRNA in primary sensory neurons: responses to injury and infusion of NGF. *J. Neurosci.* 12:4011–22

Verge VMK, Richardson PM, Benoit R, Riopelle RJ. 1989. Histochemical characterization of sensory neurons with high-affinity receptors for nerve growth factor. *J. Neurocytol.* 18:583–91

Verge VM, Richardson PM, Wiesenfeld-Hallin Z, Hökfelt T. 1995. Differential influence of nerve growth factor on neuropeptide expression in vivo: a novel role in peptide suppression in adult sensory neurons. *J. Neurosci.* 15:2081–96

Vetter ML, Martin-Zanca D, Parada LF, Bishop JM, Kaplan DR. 1991. Nerve growth factor rapidly stimulates tyrosine phosphorylation of phospholipase C-gamma 1 by a kinase activity associated with the product of the trk protooncogene. *Proc. Natl. Acad. Sci. USA* 88:5650–54

von Bussmann KA, Sofroniew MV. 1999. Reexpression of p75^{NTR} by adult motor neurons after axotomy is triggered by retrograde transport of a positive signal from regenerating peripheral nerve. *Neuroscience* 91:273–81

Wada K, Okada N, Yamamura Y, Koizumi S. 1996. Nerve growth factor induces resistance of PC12 cells to nitric oxide cytotoxicity. *Neurochem. Int.* 29:461–67

Wang CY, Mayo MW, Baldwin AS Jr. 1996. TNF- and cancer therapy-induced apoptosis: potentiation by inhibition of NF-kappaB. *Science* 274:784–87

Wang H, Tessier-Lavigne M. 1999. En passant neurotrophic action of an intermediate axonal target in the developing mammalian CNS. *Nature* 401:765–69

Watanabe-Fukunaga R, Brannan CI, Copeland NG, Jenkins NA, Nagata S. 1992. Lymphoproliferation disorder in mice explained by defects in Fas antigen that mediates apoptosis. *Nature* 356:314–17

Watson FL, Heerssen HM, Moheban DB, Lin MZ, Sauvageot CM, et al. 1999. Rapid nuclear responses to target-derived neurotrophins require retrograde transport of ligand-receptor complex. *J. Neurosci.* 9:7889–900

Weidner KM, Di Cesare S, Sachs M, Brinkmann V, Behrens J, Birchmeier W. 1996. Interaction between Gab1 and the c-Met receptor tyrosine kinase is responsible for epithelial morphogenesis. *Nature* 384:173–76

Welch H, Maridonneau-Parini I. 1997. Lyn and Fgr are activated in distinct membrane fractions of human granulocytic cells. *Oncogene* 15:2021–29

Whitehouse PJ, Price DL, Struble RG, Clark AW, Coyle JT, DeLong MR. 1982. Alzheimer's disease and senile dementia: loss of neurons in the basal forebrain. *Science* 215:1237–39

Wiesmann C, Ultsch MH, Bass SH, de Vos AM. 1999. Crystal structure of nerve growth factor in complex with the ligand-binding domain of the TrkA receptor. *Nature* 401:184–88

Williams LR, Varon S, Peterson GM, Wictorin K, Fischer W, et al. 1986. Continuous infusion of nerve growth factor prevents basal forebrain neuronal death after fimbria fornix transection. *Proc. Natl. Acad. Sci. USA* 83:9231–35

Wills Z, Marr L, Zinn K, Goodman CS, Van Vactor D. 1999. Profilin and the Abl tyrosine kinase are required for motor axon outgrowth in the Drosophila embryo. *Neuron* 22:291–99

Wilson JX. 1997. Antioxidant defense of the brain: a role for astrocytes. *Can. J. Physiol. Pharmacol.* 75:1149–63

Winkler J, Ramirez GA, Thal LJ, Waite JJ. 2000. Nerve growth factor (NGF) augments cortical and hippocampal cholinergic functioning afer p75NGF receptor-mediated deafferentation but impairs inhibitory avoidance and induces fear-related behaviors. *J. Neurosci.* 20:834–44

Wolf DE, McKinnon CA, Daou MC, Stephens RM, Kaplan DR, Ross AH. 1995. Interaction with TrkA immobilizes gp75 in the high affinity nerve growth factor receptor complex. *J. Biol. Chem.* 270:2133–38

Wood JN. 1995. Regulation of NF-kappa B activity in rat dorsal root ganglia and PC12 cells by tumour necrosis factor and nerve growth factor. *Neurosci Lett.* 192:41–44

Wood KW, Sarnecki C, Roberts TM, Blenis J. 1992. ras mediates nerve growth factor receptor modulation of three signal-transducing protein kinases: MAP kinase, Raf-1, and RSK. *Cell* 68:1041–50

Wood SJ, Pritchard J, Sofroniew MV. 1990. Re-expression of nerve growth factor receptor after axonal injury recapitulates a developmental event in motor neurons: differential regulation when regeneration is allowed or prevented. *Eur. J. Neurosci.* 2:650–57

Woodruff NR, Neet KE. 1986. Beta nerve growth factor binding to PC12 cells. Association kinetics and cooperative interactions. *Biochemistry* 25:7956–66

Woolf C, Ma Q, Allchorne A, Poole S. 1996. Peripheral cell types contributing to the hyperalgesic action of nerve growth factor in inflammation. *J. Neurosci.* 16:2716–23

Woolf CJ, Safieh-Garabedian B, Ma QP, Crilly P, Winter J. 1994. Nerve growth factor contributes to the generation of inflammatory sensory hypersensitivity. *Neuroscience* 62:327–31

Wooten MW, Seibenhener ML, Neidigh KB, Vandenplas ML. 2000. Mapping of atypical protein kinase C within the nerve growth factor signaling cascade: relationship to differentiation and survival of PC12 cells. *Mol. Cell. Biol.* 20:4494–504

Wooten MW, Seibenhener ML, Zhou G, Vandenplas ML, Tan TH. 1999. Overexpression of atypical PKC in PC12 cells enhances NGF-responsiveness and survival through an NF-kappaB dependent pathway. *Cell Death Differ.* 6:753–64

Wooten MW, Zhou G, Seibenhener ML, Coleman ES. 1994. A role for zeta protein kinase C in nerve growth factor-induced differentiation of PC12 cells. *Cell Growth Differ.* 5:395–403

Wooten MW, Zhou G, Wooten MC, Seibenhener ML. 1997. Transport of protein kinase C isoforms to the nucleus of PC12 cells by nerve growth factor: association of atypical zeta-PKC with the nuclear matrix. *J. Neurosci. Res.* 49:393–403

Wyatt S, Shooter EM, Davies AM. 1990. Expression of the NGF receptor gene in sensory neurons and their cutaneous targets prior to and during innervation. *Neuron* 4:421–27

Xiang J, Chao DT, Korsmeyer SJ. 1996. BAX-induced cell death may not require interleukin 1 beta-converting enzyme-like proteases. *Proc. Natl. Acad. Sci. USA* 93:14559–63

Xing J, Ginty DD, Greenberg ME. 1996. Coupling of the RAS-MAPK pathway to gene activation by RSK2, a growth factor-regulated CREB kinase. *Science* 273:959–63

Xing J, Kornhauser JM, Xia Z, Thiele EA, Greenberg ME. 1998. Nerve growth factor activates extracellular signal-regulated kinase and p38 mitogen-activated protein kinase pathways to stimulate CREB serine 133 phosphorylation. *Mol. Cell. Biol.* 18:1946–55

Xue L, Murray JH, Tolkovsky AM. 2000. The Ras/phosphatidylinositol 3-kinase and Ras/ERK pathways function as independent survival modules each of which inhibits a distinct apoptotic signaling pathway in sympathetic neurons. *J. Biol. Chem.* 275:8817–24

Yamashita H, Avraham S, Jiang S, Dikie I, Avraham H. 1999a. The Csk homologous kinase associates with TrkA receptors and is involved in neurite outgrowth of PC12 cells. *J. Biol. Chem.* 274:15059–65

Yamashita T, Tucker KL, Barde Y. 1999b. Neurotrophin binding to the p75 receptor modulates rho activity and axonal outgrowth. *Neuron* 24:585–93

Yano H, Cong F, Birge RB, Goff SP, Chao MV. 2000. Association of the Abl tyrosine kinase with the Trk nerve growth factor receptor. *J. Neurosci. Res.* 59:356–64

Yao R, Cooper GM. 1995. Requirement for phosphatidylinositol-3 kinase in the prevention of apoptosis by nerve growth factor. *Science* 267:2003–6

Ye X, Mehlen P, Rabizadeh S, VanArsdale T, Zhang H, et al. 1999. TRAF family proteins interact with the common neurotrophin receptor and modulate apoptosis induction. *J. Biol. Chem.* 274:30202–8

Yeo TT, Chua-Couzens J, Butcher LL, Bredesen DE, Cooper JD, et al. 1997. Absence of p75NTR causes increased basal forebrain cholinergic neuron size, choline acetyltransferase activity, and target innervation. *J. Neurosci.* 17:7594–605

Yoon SO, Casaccia-Bonnefil P, Carter B, Chao MV. 1998. Competitive signaling between TrkA and p75 nerve growth factor receptors determines cell survival. *J. Neurosci.* 18:3273–81

York RD, Yao H, Dillon T, Ellig CL, Eckert SP, et al. 1998. Rap1 mediates sustained MAP kinase activation induced by nerve growth factor. *Nature* 392:622–26

Yoshida K, Gage FH. 1991. Fibroblast growth factors stimulate nerve growth factor synthesis and secretion by astrocytes. *Brain Res.* 538:118–26

Yoshida K, Gage FH. 1992. Cooperative regulation of nerve growth factor synthesis and secretion in fibroblasts and astrocytes by fibroblast growth factor and other cytokines. *Brain Res.* 569:14–25

Yuen EC, Mobley WC. 1996. Therapeutic potential of neurotrophic factors for neurological disorders. *Ann. Neurol.* 40:346–54

Zafra F, Castren E, Thoenen H, Lindholm D. 1991. Interplay between glutamate and gamma-aminobutyric acid transmitter systems in the physiological regulation of brain-derived neurotrophic factor and nerve growth factor in hippocampal neurons. *Proc. Natl. Acad. Sci. USA* 88:10037–41

Zha J, Harada H, Yang E, Jockel J, Korsmeyer SJ. 1996. Serine phosphorylation of death agonist BAD in response to survival factor results in binding to 14-3-3 not BCL-X(L). *Cell* 87:619–28

Zhao Y, Bjorbaek C, Moller DE. 1996. Regulation and interaction of pp90(rsk) isoforms with mitogen-activated protein kinases. *J. Biol. Chem.* 271:29773–79

Zhao Y, Bjorbaek C, Weremowicz S, Morton CC, Moller DE. 1995. RSK3 encodes a novel pp90rsk isoform with a unique N-terminal sequence: growth factor-stimulated kinase function and nuclear translocation. *Mol. Cell. Biol.* 15:4353–63

Annu. Rev. Neurosci. 2001. 24:1283–309

FLIES, GENES, AND LEARNING

Scott Waddell and William G Quinn

Department of Brain and Cognitive Sciences, Department of Biology, Center for Learning and Memory, Massachusetts Institute of Technology, Cambridge, Massachusetts 02139; e-mail: waddell@mit.edu, cquinn@mit.edu

Key Words *Drosophila*, learning, memory, drugs of abuse, genetics, mutants

■ **Abstract** Flies can learn. For the past 25 years, researchers have isolated mutants, engineered mutants with transgenes, and tested likely suspect mutants from other screens for learning ability. There have been notable surprises—conventional second messenger systems co-opted for intricate associative learning tasks, two entirely separate forms of long-term memory, a cell-adhesion molecule that is necessary for short-term memory. The most recent surprise is the mechanistic kinship revealed between learning and addictive drug response behaviors in flies. The flow of new insight is likely to quicken with the completion of the fly genome and the arrival of more selective methods of gene expression.

INTRODUCTION

When Benzer started his genetic dissection of behavior the conventional reaction was mixed, to put it politely. After a large seminar at Woods Hole an entrenched physiologist was heard to remark, "We have nothing to fear from this man." Nevertheless, other neurophysiologists, including some great ones, recognized a smart man with a genuinely new approach—one that could truly alter the field (Weiner 1999).

Thirty years later the sanguine and sourpuss viewpoints both have their merits. Forward-genetic studies of behavior—measuring a behavioral response in some genetically opportune animal, mutagenizing, selecting mutant individuals that perform the behavior aberrantly, mapping and cloning the affected genes—have provided simple, startling insights into learning and memory mechanisms. With mutants, with luck, one can leap in a single bound from a molecule to a memory process. However, the leap is perilous and insubstantial—it omits information about intermediate stages of cell biology, neurophysiology, and anatomy—and filling out the picture necessarily depends on inferences from other methods.

A principal advantage, and a concomitant disadvantage, of the forward-genetic approach is that it circumvents rational thought. Mutating, mapping, and cloning genes that affect a behavior such as learning can provide information totally foreign to current hypotheses. This can revolutionize our thinking, as the discovery of homeobox gene complexes did to development (Gehring & Hiromi 1986). Nevertheless, the genetic approach, viewed from inside, often looks like mindless stamp collecting—a fishing expedition, in the words of study section reports.

Research with invertebrate systems, including insects, is similarly attractive and repulsive at the same glance. Invertebrates have simple genomes, simple nervous systems, and anthropomorphizable behaviors, such as recollection. They also often have neuronal circuitry that is reproducible from animal to animal, so one can compile a map of synaptic connections and a catalog of individual neuronal functions by assembling information from many animals. However, until recently, invertebrate animals were often taken as models of themselves rather than as simplified versions of us. This attitude has changed lately, with the demonstrated universality of genes and mechanisms and with the evident sequence similarity among the worm, fly, mouse, and human genomes (Rubin et al 2000).

Mutation-based studies of fruit fly learning started 30 years ago (Quinn et al 1974, Dudai et al 1976). This endeavour has been reviewed by Davis (1996) and Dubnau & Tully (1998). Comparisons with other invertebrate systems, particularly the marine snail *Aplysia californica* and honeybees, *Apis mellifera*, are covered in detail by Mayford & Kandel (1999), Menzel (1999), and Carew (2000). Here we concentrate on *Drosophila*, with particular attention to the relation between learning and memory and behavioral responses to drugs.

Learning in *Drosophila*

Fruit flies can learn a lot of things. Most impressively for their trainers, they can learn to run away from specific odors that they previously experienced with electric shock (Quinn et al 1974, Tully & Quinn 1985). In contrast, hungry flies can learn to run toward odors previously associated with sugar reward (Tempel et al 1983).

Flies can also learn visual, tactile, proprioceptive, and perhaps spatial cues (Menne & Spatz 1977, Folkers 1982, Guo et al 1996, Booker & Quinn 1981, Wustmann et al 1996, Wustmann & Heisenberg 1997). Not surprisingly, after introspection, male flies learn to attenuate their courtship behavior after experiencing rejection from females (Siegel & Hall 1979, Gailey et al 1984).

Memory after olfactory training persists for different periods depending on the specific training regime. Training to sugar reward appears to elicit longer memories than equivalent training to electric shock punishment, with surprising linear additivity of behavioral performance following the two reinforcements (Tempel et al 1983). Very strikingly, repetitive training with rest intervals interspersed (spaced training) can produce memory that lasts for several days (Tully et al 1994). This apparent requirement for spaced training parallels the protocol required for very-long-term memory formation in other species, including humans.

GETTING LEARNING MUTANTS

The genetic approach to learning is straightforward: (*a*) Mutagenize flies to produce individual mutant progeny; (*b*) breed these advantageously to generate progeny populations that are all affected at one gene; (*c*) test the mutant flies for their ability to learn or to remember; (*d*) assess them for behavioral, morphological, and developmental normalcy; (*e*) genetically map the affected gene; and (*f*) clone it if possible.

1) Mutagenesis

The methods used are critical. Chemical mutagenesis (using ethyl methane sulfonate or ethyl nitroso urea) works with very high efficiency, but changes DNA subtly, usually at a single nucleotide. Chemically induced mutants are therefore relatively easy to produce but hard to clone. In contrast, mutagenesis by mobilization of transposable elements (P-elements in *Drosophila*) can greatly facilitate cloning because the identified transposon leaves a molecular tag on the interrupted gene (Cooley et al 1988). On the downside, mutagenesis with transposons is between threefold and tenfold less efficient than mutagenesis with chemicals, depending how much faith and time one wishes to invest scoring transposition events. Furthermore, transposon hops tend to disrupt genes wholesale. This is in contrast to the single-nucleotide microsurgery that was critical to the behaviorally based isolation of the dnc^1 and rut^1 mutations as specific to learning. Benzer, Quinn, and their colleagues (Dudai et al 1976, Quinn et al 1979, Livingstone et al 1984, Folkers et al 1993) mutagenized (with chemicals), bred, and behaviorally tested about 5000 lines to obtain 4–6 good learning mutants. This effort took them about four person-years. With less efficient transposon mutagenesis such an effort would plausibly have entailed between 12 and 40 person-years—without the nucleotide microsurgery. Rapid clonability comes at a price, unless researchers use smart tricks. There are two established tricks: (*a*) selecting for genes that are preferentially expressed in anatomical learning centers and (*b*) selecting for genetic suppressors of female sterility caused by the mutant *dunce*. These are discussed later.

2) Fly Breeding Methods

Fly breeding methods can dramatically affect the scope of the screen and also the number of human lifetimes expended in getting mutants. About 20% of the fly genome is on the X-chromosome—one copy in males. Most new mutations are recessive (masked by a normal copy of the gene). Thus, finding new genes is much easier if they are X-linked. Special fly stocks (with attached X chromosomes) make such screening even easier—tenfold less labor than for other (autosomal) mutants. This buys time, but eventually the rest of the genome has to be dealt with.

3) Testing for Learning

To date, all the mutants isolated have been identified in tests for deficiencies in olfactory learning. However, flies demonstrably learn to many cues (Quinn et al 1974, Tully & Quinn 1985, Tempel et al 1983, Menne & Spatz 1977, Guo et al 1996, Booker & Quinn 1981, Wustmann et al 1996, Wustmann & Heisenberg 1997). It is a matter of time and youth until new researchers isolate new mutants using other tasks.

Also to date, a disproportionate number of the informative mutants (*dunce*, *rutabaga*, *amnesiac*) have been isolated with a relatively crude, behaviorally fragile test—that of Quinn et al (1974). Some new mutants, including one really interesting one (*volado*, see below), were isolated with a behaviorally more robust test. However, it is possible that the original screen is not as benighted as it looked 10 years ago. Seemingly less effective "sensitized" screens may be the best way to get informative mutants.

In the first behavioral screen (Quinn et al 1974, 1979; Dudai et al 1976; Livingstone et al 1984), about 4000 fly stocks carrying random, chemically induced mutations in single genes were tested for their ability to learn in the olfactory conditioning test. The first olfactory learning mutant identified in this paradigm, by Byers in Benzer's lab, was *dunce* (Dudai et al 1976). *dunce* flies learn very poorly in the olfactory paradigm, although they can sense odorants and shock and they appear normal in other behaviors. Continuing this approach, several additional mutants [*rutabaga* (Livingstone et al 1984), *turnip* (Choi et al 1991), *cabbage* (Aceves-Pina & Quinn 1979), *amnesiac* (Quinn et al 1979), and *radish* (Folkers et al 1993)] were identified—most by Sziber in Quinn's lab.

A variant olfactory test (Tully & Quinn 1985) has now been used as a primary screen (more than 2000 stocks tested for 3-h memory) to isolate 3 new mutants, *latheo* (Boynton & Tully 1992), *linotte* (Dura et al 1993), and *nalyot* (DeZazzo et al 2000) and as a secondary screen to confirm 4 candidate learning mutants, *leonardo*, (Skoulakis & Davis 1996), dPKA-RI (Goodwin et al 1997), *volado* (Grotewiel et al 1998), and *NF1* (Guo et al 2000).

4a) Measurements of Sensory Acuity

With a mutant in hand, one has to measure the ability of mutant flies to perceive the cues (conditioned stimulus) and reinforcement (unconditioned stimulus) and to carry out the behavior. For olfactory learning, this means ability to smell the odors, sense the electric shock, and run with apparent coordination in an appropriate direction.

4b) General Assays for Normalcy

In evaluating a mutant, if one is honest, one has to decide whether the fly looks and behaves normally and if not, whether the abnormality is related to the lack of learning. Not all sick mutants are irrelevant. Severe *dunce* mutations confer partial lethality and nearly total female sterility. Nevertheless, the original behaviorally isolated *dunce* mutants (involving very partial loss of enzymatic function) live and behave normally and breed almost normally, but have learning deficiencies comparable to the severest *dunce* gene mutants. Homozygous *turnip* flies can barely get off the ground; however, heterozygous *turnip* flies live, walk, fly, and learn fine, but they forget very rapidly (Quinn et al 1979, Tully & Quinn 1985, Mihalek et al 1997). Deletion of the complete *volado* gene is lethal (compromising other, more subtle, behavioral assays). However, removal of either one of the two major splice forms of the gene transcript yields a healthy but learning-deficient fly (Grotewiel et al 1998).

5) Genetic Mapping

There is no rocket science here, but hard work sometimes. Traditional recombination mapping using a behavioral assay is arduous. It involves recombining the chromosome containing the behavioral mutation with a homologous chromosome containing several evenly spaced morphological marker mutations, then generations of back-crosses, crosses to generate populations, and behavioral assays of the populations to assess each recombinant chromosome (See Dudai et al 1976). When necessary, mutations can be more finely mapped by using the extensive collection of chromosomal deletions available in *Drosophila* (Byers et al 1981, Folkers et al 1993). This allows positional cloning of otherwise inaccessible genes (see below).

All these steps are circumvented if the mutation is transposon generated. In this case, one simply takes mutant larvae from the population, squashes their salivary glands, hybridizes cloned DNA from the transposon to the polytene chromosome squashes, and notes the location to which transposon DNA has hybridized (see Feany & Quinn 1995).

The availability of the entire fly genome sequence makes cloning and localizing P-element–generated mutations even simpler. Identified DNA sequence from that flanking the P-element can be used to search the fly genome database. The precise insertion location of the transposon is then apparent in minutes.

6) Cloning the Gene

This is the payoff. In some cases [*dunce* = cAMP phosphodiesterase (Byers et al 1981, Davis & Kiger 1981, Chen et al 1986) and rutabaga = adenylyl cyclase (Livingstone et al 1984, Levin et al 1992)] it consolidates lucky enzymatic guesses as to learning machinery into engraved metabolic truth. In other cases [*amnesiac* = PACAP-like neuropeptide (Quinn et al 1979, Feany & Quinn 1995) and *volado* = α-integrin (Grotewiel et al 1998)], the forward-genetic approach can provide genuinely new intellectual entrees into the learning process.

6a) Selective Expression of Cloned Genes

Constructing transgenic flies with an extra, cloned gene is straightforward nowadays (Spradling 1986). The question is where, when, and how to express it. The two promoters most frequently used to drive fly transgenes are the hsp70 promoter and the GAL4 upstream activating sequence (GAL4-UAS).

The heat shock–inducible hsp70 promoter allows temporal control of transgene expression, but promiscuously in all tissues. Transgenes inserted downstream from this promoter can be induced before, during, or after a learning experience to assess the role of a particular gene in learning or memory. Dominant-negative (blocking) transgenes can be used to interrupt learning at defined times (Drain et al 1991, Yin et al 1994). Alternatively, the wild-type gene can be reintroduced and acutely expressed to restore normal retention to a mutant fly stock (Grotewiel et al 1998, Guo et al 2000). Such restoration (called "transformation rescue" by fly people) can provide definitive evidence that the correct gene has been identified from the mutant, and it allows one to ask when and where that gene product might act. On the downside, Hsp70-driven transformation rescue does not always work. Failure to rescue with an acutely expressed, Hsp70-driven, transgene may occur because (*a*) misguided researchers have identified the wrong gene, (*b*) the gene is required developmentally rather than acutely at the time of learning, (*c*) promiscuous expression of the gene in all tissues destroys the anatomical specificity required for learning a specific task, or (*d*) the transgene happens not to be expressed at the critical level required for appropriate cell signaling.

The GAL4 system (Brand & Perrimon 1993), when used astutely, allows introduced genes to be expressed selectively in chosen tissues or subsets of cells, but usually without temporal control. The GAL4 system is binary, involving two separate transgene constructs in flies. The yeast transcription factor GAL4 lacks a functional homolog in flies. It can therefore be used to drive expression of other transgenes that are introduced in separate constructs downstream from a GAL4-responsive promoter, GAL4-UAS. Briefly, one transforms *Drosophila* with a desired transgene, downstream from a GAL4-UAS. Normally, this transgene will not be expressed. One then makes, or sends away for, another fly stock containing a second P-element transgene—this time consisting of the gene for the GAL4 transcription factor on an intrinsically weak promoter. In this case, experience shows, the introduced GAL4 gene will usually respond to the collection of enhancers in its neighborhood and will frequently be expressed in the same subset of cells or tissues as its nearest-neighbor gene. There are several ways, using stainable reporter genes, to identify the set of tissues expressing GAL4 protein in a particular fly stock.

Next one simply crosses the two fly stocks containing the two separate transgene constructs—one with the regionally expressed GAL4 transcription factor, the other with the GAL4-UAS promoter–driven, learning-related gene of choice. Double-transgenic progeny from this cross will express the learning-related protein with the regional specificity of the GAL4 line used. This technique is universally important in *Drosophila* because a large number of region-restricted GAL4 driver lines have been generated and catalogued that express GAL4 in different subregions of the adult fly brain (Armstrong et al 1995, Yang et al 1995). Consequently, given a cloned gene that might be related to learning (be it wild type, hyperactive, or engineered for dominant-negative blocking), one can, by simple crosses, arrange to express the gene in a number of brain structures of interest and generally to see what genes influence what memories in what tissues.

Tissue-specific expression of transgenes can also be used to rescue learning and memory of mutant flies. With this method one can identify those brain structures where expression is sufficient for learning (Zars et al 2000, Waddell et al 2000). In one defining instance, appropriate tissue-specific expression of a wild-type transgene has rescued learning and memory performance in a mutant when acute and developmentally sustained expression of that gene with the hsp70 promoter has worked poorly (DeZazzo et al 1999, Waddell et al 2000).

6b) Analysis of Cloned Gene Product(s)

This is often conceptually mundane, but it can yield momentous results. For example, Davis and colleagues (Davis & Kiger 1981, Chen et al 1986, Levin et al 1992, Nighorn et al 1991, Han et al 1992) cloned or helped clone the *dunce* and *rutabaga* genes, found that both genes were highly expressed in anatomical structures (the mushroom bodies) that were known to be involved in olfactory learning, and thereafter undertook a successful search for other genes that were expressed with similar regional specificity (Skoulakis & Davis 1996, Grotewiel et al 1998), with important findings.

INFORMATION WE HAVE GAINED FROM THE MUTANTS

Chemical Mutagenesis

A fragile, seemingly primitive olfactory learning paradigm was used, along with chemical mutagenisis and brute labor, to isolate the mutants *dunce*, *rutabaga*, *amnesiac*, *radish*, and *turnip*.

dunce The first *dunce* mutant (Dudai et al 1976) was isolated by its deficient olfactory learning in the assay of Quinn et al (1974). Other mutations in this gene came from a screen for female sterility mutants (Mohler 1977). The *dnc* locus is very complex, stretched over at least 148 kilobases (Davis & Davidson 1984, 1986; Chen et al 1986, 1987; Qui et al 1991). The *dnc* gene encodes at least 10 alternative RNA splice-forms and presumably many variant DNC proteins.

rutabaga The original *rut* mutation (Livingstone et al 1984) was isolated, like *dunce*, because of its learning deficiency in the assay of Quinn et al (1974). Other

P-element-induced alleles of *rut* were isolated because they suppressed the female sterility of *dnc* mutants (Bellen et al 1987, Levin et al 1992).

Molecular cloning of both the *dnc* and *rut* genes was aided, first by the directed identification in these mutant stocks of biochemical defects (Byers et al 1981, Davis & Kiger 1981, Livingstone et al 1984). The correspondence of metabolic function of the *dnc* and *rut* gene products was astonishing. The two enzymes lay in the same biochemical pathway, a pathway used throughout the animal kingdom for cellular responses to outside messengers. The *rut* gene encodes a Ca^{2+}/Calmodulin-stimulated (type I) adenylate cyclase, AC (Levin et al 1992); the *dnc* gene encodes cAMP phosphodiesterase (Chen et al 1986, Qiu et al 1991). Therefore, RUT makes cAMP and DNC degrades it.

amnesiac   The first *amnesiac* mutation was identified in a deliberate screen for mutants that affected memory (Quinn et al 1979). However, the *amnesiac* gene was cloned by a trick from a P-element-induced allele (Feany & Quinn 1995) and has been repeatedly cloned since (e.g. Moore et al 1998, Toba et al 1999). The *amnesiac* gene encodes a protein that has sequence features of a pre-pro-neuropeptide and that has limited homology to the mammalian neuropeptide/hormone pituitary adenylyl cyclase–activating peptide (PACAP) (Vaudry et al 2000). This finding makes sense because neuropeptides often act in parallel with conventional monoamine neurotransmitters to provide reinforcement. The homology to PACAP, together with supporting genetic and biochemical evidence, indicates that the AMN peptide stimulates cAMP synthesis. The piece of evidence that led to a clonable *amn* allele (Feany & Quinn 1995) is this: Severe alleles of the learning mutant *dnc* are learning defective and female sterile. A P-element insertion in the *amn* gene rescues the female fertility phenotype of mutant *dnc* females. Therefore, mutations in the *amn* gene act to counter the effect of *dunce* mutations (i.e. too much cAMP). Furthermore, *amn* mutants are hypersensitive to ethanol (Moore et al 1998) as well as being forgetful (discussed below). Feeding *amn* flies forskolin (an activator of adenylate cyclases) or increasing PKA activity reverts ethanol sensitivity to wild-type levels (Moore et al 1998).

No one knows whether the AMN peptide is a true homolog of mammalian PACAP. Nevertheless, artificial application of mammalian PACAP38 induces changes in synaptic signaling at the fly larval neuromuscular junction (NMJ) (Zhong & Pena 1995). These PACAP-induced changes in flies are mediated by the cAMP cascade—as with PACAP in mammals and, evidently, the AMN peptide(s) in flies (Feany & Quinn 1995, Moore et al 1998). The PACAP38 peptide, applied to the fly NMJ, elicits a slow inward current lasting tens of seconds, followed by an enhanced outward K^+ current. These responses to PACAP are absent in *rut* mutants (Zhong 1995), indicating that RUT adenylyl cyclase activation, and hence cAMP signaling, is required for PACAP action. This NMJ response actually requires the simultaneous activation of both the Ras and cAMP pathways; activation of either pathway alone is insufficient for synaptic change (Zhong 1995). Analysis of the PACAP response in flies mutant for the Ras GTPase-activating protein, NF1, which negatively regulates ras in mammals, indicates that the signal tranduction events downstream from PACAP are complex. The PACAP response is abolished in *NF1* mutants, but it can be restored by pharmacological stimulation of the cAMP cascade (Guo et al 1997).

radish, Anesthesia-Resistant Memory, and Long-Term Memory   Long-term memory has been historically defined as (*a*) memory that persists for a long time; (*b*) memory that is dependent on spaced training; (*c*) memory that is resistant to disruption by anesthesia, electroconvulsive shock, cooling, concussion, or other agents that interfere with patterned neural activity; (*d*) memory that is dependent on new protein synthesis; and (*e*) memory that requires the cAMP response-element binding protein (CREB) transcription factor (Yin et al 1994, Tully et al 1994). In brief, *Drosophila* mutants seem to further resolve the previous definitions of long-term memory into two separate parts.

The *radish* mutation selectively eliminates anesthesia-resistant memory (ARM) (Folkers et al 1993), leaving protein synthesis–dependent memory intact (Tully et al 1994). ARM occurs after ordinary training and lasts at least 3 days (Tully et al 1994). Therefore, ARM is a legitimate form of long-term memory.

Transgenic flies with an inhibitory form of a fly CREB transcription factor, dCREBb, have normal ARM (see above). However, they are completely devoid of protein synthesis–dependent long-term memory (Yin et al 1994, Tully et al 1994). This form of long-term memory lasts for at least 7 days and requires spaced training, protein synthesis, and the transcription factor CREB. We call it long lasting long-term memory (LLTM) to distinguish it from ARM. Additionally, startling results have been reported (Yin et al 1995a) that indicate that transgenic flies with a superabundance of active CREB have "flashbulb memory"—memory that persists for days after a (normally ephemeral) single training trial. These experiments have not been pursued as expected.

The *radish* mutant provides a unique handle on ARM (Folkers et al 1993). Twenty-four hours after spaced training, memory is composed of two experimentally separable portions: Half the 24-hour memory can be abolished by inhibiting protein synthesis, and the other half can be abolished by the *rsh* mutation (Tully et al 1994, Yin et al 1994). Flies that are fed the protein synthesis inhibitor cycloheximide, or flies that express an inhibitory CREB transgene, display only half the normal 24-hour memory. Flies with both the drug inhibitor and the transgene still have half this memory, indicating that blocking CREB transcription and blocking most of protein synthesis affect the same process to the same extent. Mutant *rsh* flies also show only half-normal 24-hour memory. However, introducing the inhibitory CREB transgene into the *rsh* mutant eliminates all 24-hour memory, indicating that *rsh* and CREB affect entirely separate memory processes. There are two interpretations of these experiments: (*a*) Protein synthesis–dependent memory is exactly equivalent to CREB-dependent memory. The CREB transcription factor lies at or near the top of the gene regulation cascade that leads to LLTM and is a bottleneck for that cascade; all transcription-dependent memory goes through it. (*b*) A second, completely separable, form of long-term memory requires a normal *rsh* gene (Folkers et al 1993). Mutant *rsh* flies are entirely lacking in consolidated ARM (Folkers et al 1993, Tully et al 1994).

The *rsh* mutation has been localized to a 180-kb region of the X-chromosome by genetic mapping. DNA spanning the interval has been cloned and sequenced in its entirety by the genome project. Several interesting candidate genes within the interval are currently being tested for a *rsh*-specific mutation.

turnip   This odd mutant implicates the protein kinase C (PKC) pathway (Choi et al 1991). Homozygous *tur* mutant flies are sluggish and have a reduced response to electric shock (Mihalek et al 1997). However, heterozygous *tur*/+ flies are healthy and behaviorally responsive. They learn normally, but memory decays rapidly (Quinn et al 1979, Tully & Quinn 1985, Choi et al 1991, Mihalek et al 1997). The *tur* mutation comaps, in a dose-dependent manner, with low PKC activity. However, the mutation does not lie in or near any of the identified fly genes encoding PKC family members (Choi et al 1991).

P-Element Screening

Tully and colleagues have performed a P-element-based behavioral screen for learning and memory mutants. The genes affected in P-element-induced mutations are readily clonable. They have reported three new mutants, *latheo*, *linotte*, and *nalyot* (Boynton & Tully 1992, Dura et al 1993, DeZazzo et al 2000).

latheo   The *latheo* gene encodes a component of the origin recognition complex (Pinto et al 1999). Complete loss-of-function *lat* mutations are lethal. Partial loss-of-function mutants learn poorly but lack mushroom bodies, a finding that bodes ill for learning specificity of the gene. Intriguingly, however, LAT protein is detectable in presynaptic boutons, and the NMJ of *lat* mutants has abnormal synaptic properties (Rohrbough et al 1999). Therefore, it is conceivable that *lat* encodes a multifunctional protein involved in both DNA replication and synaptic plasticity. This is an example of the unique strength of a forward-genetic approach that assumes nothing besides the fact that single gene mutations can impact learning performance. It has provided completely unexpected novel information.

linotte   The identity of the gene affected in the *linotte* mutant (Dura et al 1993) is disputed. The *lio* gene either encodes a novel protein (Bolwig et al 1995) or it is an allele of the *derailed* receptor tyrosine kinase (Dura et al 1995). Regardless, *lio* mutants have structural brain defects that extend to the mushroom bodies and central complex (Moreau-Fauvarque et al 1998, Simon et al 1998). Whether these defects are responsible for the retarded learning performance, however, is undetermined.

nalyot   *nalyot* is an allele of the myb-related Adf1 transcription factor (DeZazzo et al 2000). *Adf1* is an essential gene, but partial loss-of-function mutants are viable. Such a partial mutation, nal^{P1}, has a mild effect on learning and a pronounced effect on long-term memory. The nal^{P1} mutation also causes a modest reduction in the number of synaptic boutons at the larval NMJ. On the contrary, increased nal expression appears to cause a modest increase in the number of boutons. The authors propose (a) that the Adf1 transcription factor is directly involved in regulating the structural aspect of synaptic plasticity in concert with a *Drosophila* CREB that regulates functional plasticity (see below) and (b) that the lack of structural plasticity measured in the mutant NMJ may underlie the long-term memory deficit observed in *nal* mutant flies after spaced training. However, the experiments published do not directly address these interpretations. Although they restored initial learning of nal^{P1} flies to wild-type levels with *nal* transgenes, they did not report rescue of the longer-term (1 or 7 day) memory defects. Therefore, conclusive evidence that the long-term memory defect depends directly on the NAL gene product remains to be provided. By the authors' own assessment, "the level of performance at earlier memory phases is not a reliable predictor of performance at later memory stages"(p155).

Accurate *nal* expression is critical. Ectopic expression of *nal* under the control of several neural-specific or glial-specific GAL4 promoters is lethal. Furthermore, ubiquitous expression of NAL, driven at high levels by the heat-shock promoter, is actually deleterious to olfactory memory.

Using Neuroanatomy to Screen for Learning Genes

The products of the *dunce* and *rutabaga* genes are expressed at high levels in the fly mushroom bodies (MBs) (Nighorn et al 1991, Han et al 1992), structures that are central to olfactory learning (discussed below). In addition, the PKA catalytic subunit encoded by the *DC0* gene is also more abundant in MBs (Skoulakis et al 1993). However, reduced *dnc* expression in the MBs does not necessarily correlate with learning deficiency (Qiu & Davis 1993). Furthermore, MB-enhanced expression has been argued to be due to the unusually high cell density and parallel organization of the MB (Ito et al 1998). Notably though, not all genes appear to be preferentially expressed in MBs, and Davis and coworkers have identified the new learning mutants *leonardo* (Skoulakis & Davis 1996) and *volado* (Grotewiel 1998) using P-element-based enhancer trapping to visualize enriched expression in the MBs.

volado   Studies with this mutant implicate alteration of cell adhesion in the process of short-term memory (Grotewiel et al 1998). Although a mechanistic understanding is currently lacking, we believe that studies of this gene product will eventually provide key insight into the molecular events of synaptic remodeling that are believed to underlie learning and memory.

The *vol* gene was isolated because it is preferentially expressed in MBs. The gene encodes two splice variants of an α-integrin. Integrins are cell-surface receptors that mediate cell adhesion and signal transduction (Hynes 1992). Mutant *vol* flies are markedly reduced in short-term olfactory memory (Grotewiel et al 1998). However, memory in *vol* mutants can be fully rescued by heat-shock induction of the short *vol* cDNA transcript in adult flies, just prior to olfactory training. Strikingly, the ability to rescue memory decays with the same kinetics as *vol* RNA expression, a result strongly suggesting that the VOL integrin is acutely needed for memory.

Rohrbough et al (2000) have reported a synaptic role for VOL, and other *Drosophila* integrins also influence synaptic morphology and function (Beumer et al 1999). It remains to be determined whether the memory deficit of *vol* mutants is due to chronic alteration in synaptic structure (caused by changess in cell adhesion) that prevents modulation, or whether it is acute VOL signaling that is critical for memory formation. It is also not known if cAMP signaling regulates VOL-mediated cell adhesion.

leonardo   The *leonardo* gene was isolated, like *volado*, because it is preferentially expressed in MBs (Skoulakis & Davis 1996). *leo* mutant flies are defective in olfactory learning and short-term memory. The *leo* mutation affects the zeta isoform of the mundanely named protein 14-3-3.

Proteins of the 14-3-3 family are involved in several intracellular signaling pathways. They can activate and repress protein-kinase-C (PKC) activity (Aitken et al 1995, Xiao et al 1995), activate tyrosine hydroxylase and tryptophan hydroxylase, the rate-limiting enzymes in catecholamine and serotonin biosynthesis (Ichimura et al 1995), and interact with several signal-transduction cascades, including RAF-1 in the mitogen-activated protein kinase (MAPK) pathway (Fantl et al 1994, Freed et al 1994, Irie et al 1994, Li et al 1995).

The LEO protein is enriched in presynaptic termini, and *leo* mutants have reduced synaptic transmission at the larval NMJ, especially under stress conditions (Broadie et al 1997). The LEO gene product has been proposed as a candidate to mediate voltage-dependent Ca^{2+} influx and presynaptic vesicle exocytosis.

LEO does in fact demonstrably interact with the presynaptically located *Drosophila* calcium-dependent potassium (K_{Ca}) channel Slowpoke (dSlo) via the slowpoke-binding-protein Slob (Zhou et al 1999). Through this interaction LEO regulates the voltage sensitivity of the dSlo channel (DiChiara & Reinhart 1995, Cui et al 1997). Whether LEO also exerts an effect on synaptic efficacy via activation of PKA and/or the RAS/RAF mitogen-activated protein kinase cascade is unknown.

Testing Available Mutants for Learning Deficiency

A shortcut approach to studying learning is to behaviorally assess ready-made mutant stocks, in the hope that they have defects in learning or memory that are relatively specific. Sometimes there are quick payoffs.

Ddc   *Ddc* mutations lie in the structural gene for aromatic-amino-acid-decarboxylase, a necessary enzyme on the pathways to the major monoamine neurotransmitters—serotonin, dopamine, and octopamine in invertebrates. Constitutive *Ddc* mutations are lethal, but temperature-sensitive mutant stocks can be acutely blocked in adulthood and are viable. Livingstone & Tempel (1983) and Tempel et al (1984) temperature-shifted such mutants and tested their biochemistry and learning behavior. *Ddc* mutant flies have lower serotonin and dopamine levels after a few days at the restrictive temperature. In these flies learning was reduced in accordance with the reduction in DDC enzyme activity (Tempel et al 1984). Both our lab and others have had difficulty in repeating this work—in getting temperature-shifted flies that were healthy and learning-deficient at the same time.

NF1   The inferred role of NF1 (a GTPase-activating protein for *ras*), discussed above with *amnesiac*, prompted direct learning measurements of *Drosophila NF1* mutant stocks. The mutant flies are in fact defective in olfactory learning (Guo et al 2000). This defect can be rescued in adult flies by heat-shock induction of either an *NF1* transgene or a PKA transgene. Therefore, the cell-signaling pathway (identified in studies of the larval neuromuscular junction) leading from PACAP-binding to synaptic enhancement also appears to mediate olfactory learning in the adult. In humans the *NF1* gene is linked to the human disease neurofibromatosis (Shen et al 1996). Some neurofibromatosis patients have learning disabilities; however, it is not known whether the learning impairment is due to an acute role for NF1 in synaptic signaling or to brain-developmental consequences of the mutation. Results from flies provide intriguing information bearing on this issue. *NF1* flies are learning defective and small in size (Guo et al 2000, The et al 1997). Their small size can be rescued by induction either of an NF1 or a PKA transgene during development but not in adulthood (The et al 1997). In contrast, the learning deficiency of "small" *NF1* mutant flies can be rescued by induction either of an NF1 or a PKA transgene in adults (Guo et al 2000)

Genetically Engineered Alterations of Learning and Memory

Generation of transgenic flies is straightforward and quick. Genes can be made transiently inducible by cloning them downstream from the heat-shock-inducible hsp70 promoter. Obtaining direct gene knockouts in flies is now possible (Rong & Golic 2000) although it is methodologically tricky at present, but making and introducing dominant-negative gene products is easy. This approach has been useful in studies of learning and memory.

cAMP-Dependent Protein Kinase   Knowing that levels of cAMP were central to learning, Drain et al (1991) directed their attention to the obvious downstream target of the cAMP signal, cAMP-dependent protein kinase (PKA). Induction of inhibitory fragments of PKA before training blocked olfactory learning. This was the first demonstration of transgenic alteration of learning. The role of PKA in learning was later confirmed by direct studies with mutants (Skoulakis et al 1993, Li et al 1996, Goodwin et al 1997).

dCREB2   Similar experiments with the *Drosophila* gene for the cAMP-response-element-binding protein (CREB) (Yin et al 1995b) were more consequential. They showed that CREB-dependent gene expression is required for long-lasting long-term memory (LLTM) after associative learning (discussed above in connection with *radish*). Expression of an inhibitory (blocking) CREB gene abbreviates such memory (Yin et al 1994). This is consistent with work in other species and suggests that CREB-dependent transcription is critical for long-term memory formation. Expression of an activating CREB has been reported to induce hypertrophied long-term memory after one short training session (Yin et al 1995a). This result is potentially revolutionary to memory studies However, it has not been followed up in the manner one would have expected.

Analysis of the larval neuromuscular junction (NMJ) in learning mutants and in flies carrying the inhibitory and activated CREB transgenes has indicated that the cAMP cascade and CREB are directly involved in synaptic plasticity. Increased neuronal activity (elevated via *ether-a-go-go* and *Shaker* mutants) and increased cAMP concentration (elevated by the *dnc* mutant) both induce exuberant presynaptic growth and increased synaptic transmission (measured physiologically at the NMJ). The structural synaptic growth is accompanied by a reduction in levels of the neural cell adhesion molecule (N-CAM) homolog *FasII* (Schuster et al 1996). Indeed, a concomitant reduction in *FasII*-mediated cell adhesion is critical to allow synaptic growth. Mutant larvae with low levels of *FasII* show exuberant synaptic arborization at the NMJ. In contrast, larvae with higher levels of *FasII* show reduced arborization. Structure does not necessarily reflect function. In *FasII* larvae the average synaptic output is reduced, suggesting the normal aliquot of synaptic release machinery is shared among an increased number of synapses.

NMJs of *dnc* mutant larvae are different. They have increased arborization and an increased average output per bouton (Budnik et al 1990, Zhong et al 1992). These results suggest that cAMP levels affect both structural and functional synaptic change, whereas *FasII* affects only structure.

cAMP alters functional change, among other ways, via the CREB transcription factor. Expression of the inhibitory CREB transgene in *dnc* mutant larvae blocks functional but not structural plasticity (Davis et al 1996). In contrast, an activated CREB transgene increases presynaptic transmitter release. These results suggest that cAMP regulates functional synaptic plasticity via CREB and also structural plasticity via a *FasII*-dependent pathway.

Adenylate Cyclase-Stimulatory G Protein   Unregulated Gs signaling apparently blocks learning. Disruption of Gs_α adenylate cyclase-stimulatory G protein mediated signaling in the MBs (done with GAL4 drivers), but not in the central complex, absolutely abolished olfactory learning (Connolly et al 1996). These experiments confirmed the idea that signaling through the cAMP second-messenger system—in the MBs—is essential for olfactory learning. These results are surprising in that the extent of the learning defect (they do not learn at all) greatly exceeds that of flies that either have grossly disorganized or reduced MBs (Heisenberg et al 1985) or that lack MBs altogether as a result of chemical treatment that ablates them (de Belle & Heisenberg 1994).

Ca²⁺/Calmodulin-Dependent Protein Kinase Calcium is an important factor in neuronal signaling. A major intracellular respondent of Ca^{2+} is type II Ca^{2+}/calmodulin-dependent protein kinase (CAMKII). A potential role for this enzyme in associative and nonassociative behavioral plasticity was assessed by expressing an inhibitory transgene (Griffith et al 1993, Jin et al 1998) under the control of the heat-shock promoter in a similar manner to that employed for PKA by Drain et al (1991). Induction of a CAMKII inhibitory transgene inhibits associative learning measured in the courtship conditioning paradigm (Griffith et al 1993). The performance of these flies in olfactory associative conditioning has not been reported.

The *Drosophila* CAMKII gene gives rise to multiply spliced mRNAs. Eight different isoforms have been identified that differ at the junction of the regulatory and association domains of the kinase (Griffith & Greenspan 1993). This alternative splicing produces CAMKII enzymes with altered substrate specificity and differing sensitivity to Ca^{2+}/calmodulin binding (GuptaRoy et al 2000). *Drosophila* CAMKII has many reported targets, most interestingly for the purposes of this review, the eag K^+ channel subunit, the *leonardo*-associated protein Slob (Zhou et al 1999), the Adf1 transcription factor encoded by the *nalyot* gene (GuptaRoy et al 2000), and Discs large protein (a PDZ family protein), which regulates the clustering of synaptic molecules (Koh et al 1999).

ANATOMY OF *DROSOPHILA* LEARNING

Studies in many insects have indicated the importance of mushroom bodies (MBs) in learning (Strausfeld et al 1998, Zars 2000). Insightful analysis of the MBs in *Drosophila* has confirmed a role for these structures in olfactory learning and memory. Mutant flies that were identified based on their defective MB anatomy do not learn olfactory tasks (Heisenberg et al 1985). Similarly, chemical ablation of MBs abolishes olfactory learning (de Belle & Heisenberg 1994).

The MBs are only two synapses away from olfactory reception. Information from olfactory receptors on the antennae and maxillary palps travels via the antennal lobes to the MB calyces. These calyces contain dendrites of the intrinsic MB neurons, the Kenyon cells. Axons from these Kenyon cells project from the calyx down the stalk-like pedunculus. Toward the front of the central brain the pedunculus splits into five lobes (α, α', β, β', and γ) and the spur. The lobes are assumed to be the synaptic output region of the MB, although input also comes into the lobes.

The function of the MBs is not exclusively olfactory (Heisenberg 1998, Strausfeld et al 1998, Zars 2000). Although the *Drosophila* MBs are dispensable for several types of learning, including visual, tactile, and motor (Wolf et al 1998), they are believed to receive multimodal sensory information. In fact, visual deprivation reduces MB calycal volume (Barth & Heisenberg 1997). This structural plasticity is absent in the learning mutants dnc^1 and amn^1 (see below) and is therefore believed to reflect functional adaptation and long-term memory. Whether MBs are required for consolidated memory of visual stimuli or any other nonolfactory task has not been tested.

A stunning study (Liu et al 1999) has demonstrated that the MBs are required for context generalization in visual learning, a basic cognitive process. Flies can learn to associate visual patterns with the presence or absence of heat punishment (Guo et al 1996). Following learning, they fly toward the pattern predicted to avoid the heat. Wild-type flies can associate visual patterns with heat and are unaffected by changes in the illumination conditions between training and testing trials—for example, a change from monochromatic color to white light, or from intermittent to steady light (Liu et al 1999). In contrast, flies that lack MBs—as a result of chemical ablation—are unable to learn and remember the visual task if illumination conditions are changed between training and testing. Nevertheless, these MB-less flies are able to learn if the light conditions are kept constant. These results suggest the MBs are essential for the fly to be able to extract relevant information from multiply variable visual stimuli.

How the MBs actually function remains essentially mysterious (Heisenberg 1998). Their intricate anatomical organization, with lobes projecting in three orthogonal directions in the fly brain, is conceptually intriguing (Strausfeld 1976, Strausfeld et al 1998, Crittenden et al 1998). The only simple MB feature that we can see is a functional parallel with another system that is critical to learning, the hippocampal–entorhinal cortex system in mammals. Both systems show elegantly regular, only slightly scrutable anatomical organization and appear suited to deal with complex, multimodal assemblies of information.

It is now clear that even among the intrinsic cells of the MB, the Kenyon cells, there is great diversity. Multiple subpopulations of Kenyon cells with different projection patterns are distinguishable by different gene expression (Yang et al 1995, Ito et al 1997). Localization of learning-related gene products has indicated the MBs as a critical site of cAMP cascade action in olfactory learning (see Figure 1). The products of the *dnc*, *rut*, and *DC0* genes are all preferentially expressed in the MB Kenyon cells (Nighorn et al 1991, Han et al 1992, Skoulakis et al 1993). In fact, expression of *rut* in the MBs (using the GAL4 system) is sufficient for olfactory learning (Zars et al 2000).

We, working with others, have also used the GAL4 method to confirm the importance in memory of two large Dorsal Paired Medial (DPM) cells that express the *amn* gene product (Waddell et al 2000). Restoration of *amn* gene expression to these cells reestablishes normal olfactory memory. AMN neuropeptide is provided to the MB lobes by these large DPM neurons.

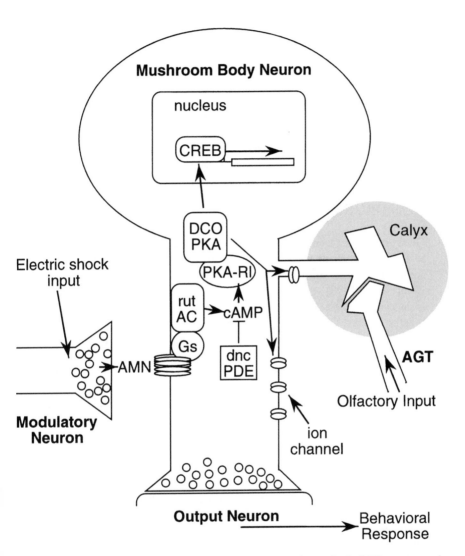

Figure 1   A model for olfactory learning in *Drosophila*. A mushroom body (MB) neuron receives convergent sensory input from olfactory presentation [via the antennoglomerular tract (AGT) interneurons that synapse in the MB calyx] and electric shock (via modulatory neurons that release the AMN neuropeptide, perhaps with a monoamine). Coincident activity of these two input paths triggers a synergistic stimulation of the RUT adenylate cyclase and subsequently, elevation of cAMP levels. Depending on the training paradigm, the cAMP elevation results either in a short-lived modification of MB neuron excitability (short-term memory) or a long-lasting functional and structural change (long-term memory). The duration of protein kinase (PKA) activation is a critical factor (Muller 2000). Persistent PKA activation supports long-term memory in part through activated cAMP response-element binding protein (CREB)-dependent transcription.

A caveat of studies that use the GAL4 method is that regional redundancy of expression cannot be easily discounted. Few GAL4 lines express only in the desirable brain region. It is conceivable that expression in the noncommon tissue or the background expression is sufficient for the rescue. Using several different GAL4 driver-lines that have overlapping expression patterns can reduce this problem.

In a screen for learning mutants it is reasonable that one will obtain flies that are grossly affected in structural brain anatomy and circuitry, as well as more subtle mutations that leave the brain largely intact. This has been the case. Both *latheo* and *linotte* have notable brain defects in the mushroom bodies (Moreau-Fauvarque et al 1998, Simon et al 1998, Pinto et al 1999).

FLIES ON DRUGS

Genes that affect learning are also involved in the response of flies and mammals to drugs of human abuse (Berke & Hyman 2000). The first findings on this in flies came from work with the inebriometer (Cohan & Hoffman 1986, Weber 1988, Moore et al 1998). In this device flies on precarious, slanted perches inside a glass cylinder are exposed to ethanol vapor in an air current. As the alcohol they breathe increases, they (understandably) turn about, stagger, fall off their perches, and finally tumble down through the inebriometer tube into a fraction collector. Wild-type flies "elute" from the column with a peak at about 20 minutes. Various sensitive or resistant mutants elute at different peak times. This is drug behavior reduced to the methods of conventional chemistry.

The first ethanol-hypersensitive mutant characterized, *cheapdate*, turned out to be a P-element allele of the memory mutant *amnesiac*. Following up on this finding, Moore et al (1998) found that other learning mutations in the cAMP pathway— *rutabaga* and *DCO*—also altered the flies' susceptibility to ethanol. Apparently, the signaling pathway first identified in memory formation in flies influences their ability to handle ethanol.

Thus the same second-messenger pathway is implicated in fly learning and fly ethanol behavior. Moreover, a handful of neurotransmitters known to stimulate this pathway—the AMN neuropeptide and a number of monamime transmitters—also appear as central in genetic studies of both behaviors.

Flies, acutely exposed to ethanol, stagger and fall over, as we do. Chronically exposed flies become less sensitive to the sedative effects of the drug—also like us (Scholz et al 2000). This functional drug tolerance is plausibly believed to reflect adaptive neuronal changes resembling learning (Cunningham et al 1983, Fadda & Rossetti 1998). Flies that have reduced octopamine levels (because of a mutation in the tyramine β-hydroxylase enzyme that synthesizes it) are impaired in their ability to develop tolerance to ethanol (Scholz et al 2000). Octopamine is plausibly involved in fly learning (Dudai et al 1987), and in the honeybee it substitutes for positive reinforcement (sugar reward) in classical conditioning of the proboscis extension response (Hammer & Menzel 1998).

Cocaine is another drug of human abuse. Volatilized cocaine, administered to flies, induces several odd behaviors (McClung & Hirsh 1998, Bainton et al 2000). The easiest such behavior to measure is this: Following mechanical agitation that shakes flies to the bottom of a tube, normal flies race to the top of the tube—negative geotaxis. Cocaine decreases this upward mobility in a quantifiable, dose-dependent manner (Bainton et al 2000).

Flies develop a behavioral sensitization to cocaine (McClung & Hirsh 1998, 1999). An enhanced behavioral response is observed following subsequent cocaine exposure even after a single dose. Sensitization is thought to be a contributing factor to addiction. As with ethanol-induced behaviors, the usual suspect molecular components shown to affect learning influence cocaine-induced behaviors: monoamines, G-protein-coupled receptor signaling, and cAMP cascade regulation.

The monoamine tyramine appears to be required for this sensitization (McClung & Hirsh 1999). *inactive* mutant flies have low tyramine levels and do not show behavioral sensitization, although their initial response to cocaine is similar to wild-type flies. Tyramine may compete with octopamine for receptor binding or it may potentiate the effects of the other monoamines, dopamine and serotonin, by inhibiting their uptake. In most cases, changes in cAMP synthesis will be central to the downstream effects. In fact, type II cAMP-dependent protein kinase regulatory subunit mutant flies have decreased sensitivity to cocaine (and ethanol) and fail to sensitize to repeated cocaine exposure (Park et al 2000).

Dopamine modulates the response of *Drosophila* to cocaine, ethanol, and nicotine (Bainton et al 2000, Li et al 2000). Therefore, although the cellular targets of these drugs are likely different, they all engage the fly dopaminergic system, and hence the cAMP cascade, perhaps in a reward-based manner.

FUTURE PROSPECTS

Lucky for us fly teachers: New molecular genetic tools arrive quickly at our door, often from fly researchers who have no interest in flies' learning. A tool that looks to be required for continued rapid progress, and one that seems imminent, is a spatially selective method of regulating transgene expression with the addition of temporal control.

The GAL4 method (Brand & Perrimon 1993) for tissue-selective transgene expression brought new meaning to our little fly lives. Nevertheless—especially for learning researchers—the method is retarded by a lack of ability to turn the critical genes on and off at appropriate times.

It is quite likely that a given GAL4 driver, with a desired expression in the adult brain, expresses in other, often unrelated tissue during development. Such idiosyncratic expression of a transgene might well be developmentally lethal or affect the behavior of the mature fly. In a time-dependent field like learning that deals with acquisition and recollection, storage and retrieval, long-term and short-term, time control is essential to dissecting processes. Fly learning researchers face a quandary. However, a solution beckons. It should be possible to temporally block expression from GAL4 lines by combining the GAL4/UAS system with the GAL80 protein that represses the GAL4 transcription factor (Lee & Luo 1999). If a temperature-sensitive GAL80 can be developed it should be possible to keep the developmentally expressed GAL4 (and any transgene under GAL4/UAS control) inactive until adulthood simply by manipulating the temperature. In this manner we would be able to add temporal control to the entire complement of brain region–specific GAL4 lines that are already available.

Genetic dissection of neuroanatomy with the GAL4 system (e.g. Armstrong et al 1995, Yang et al 1995, Ito et al 1998) is helping us map the neural networks in the fly brain and is concurrently providing us with tools to explore gene function in distinct neurons. Understanding the functionality of the circuits is critical if we hope to understand and model their properties. Despite plentiful analyses of the larval NMJ and dissociated neuronal preparations (Wu et al 1998, Lee & O'Dowd 2000), current technology is limiting physiological analyses of the intact adult *Drosophila* brain. However, circuit physiology is currently being productively studied in the MBs and connected neuropil of several larger insects—primarily cockroaches (Mizunami et al 1998), locusts (Stopfer & Laurent 1999), and honeybees (Faber et al 1999). Honeybees actually have smaller neurons on average than do flies. It is likely that the methodology can be adapted to flies, given a scientist with sufficient courage and dexterity. The combination of genetics and functional recording would change the field.

Tully and colleagues are using current DNA chip technology to search for learning-related genes. Finding the memory-relevant CREB target genes is of great interest. The most exciting use of such genes is the possibility of histological reporter systems for learning-related gene expression, or stainable antibodies for proteins that are selectively upregulated or modified following learning. Such a "tag" would allow one to identify the neurons that are modified in a functional circuit underlying a particular learning behavior. This would promote *Drosophila* from one of the anatomically least tractable animals, with respect to functional anatomy, to one of the most informative.

Interesting mutants come from all directions: anatomy screens, neurochemistry screens, and fertility screens. Most recently, scientists have been isolating new fly mutants that show an altered response to drugs (Moore et al 1998, Singh & Heberlein 2000). In view of the demonstrated commonality of the genes involved in the drug response and learning, some new molecular players will be relevant to the learning process. The supply of candidate mutants should increase dramatically.

The availability of the entire fly genome sequence has already quickened research. More interestingly, the Berkeley *Drosophila* Genome Project has generated and catalogued a large number of P-element insertion lines (Spradling et al 1999). If you are a lucky researcher, both a DNA sequence and a mutant fly stock for your favorite gene are already available by mail. This availability should stir the hearts of fly investigators to test their favorite mutants for learning defects, because the approach is very easy and potentially very interesting.

Nevertheless, in the end there will be no substitute for the head-butting forward-genetic approach. As ever, this amounts to mutagenizing flies, selecting mutants that have aberrant behavior, and cloning the affected genes. At heart this amounts to walking the beaches and peeking into bottles, with the hope that nature might have left a message inside. Some such messages have enlightened us about pattern formation in development. With luck, they will enlighten us about our capacity to remember.

ACKNOWLEDGMENTS

SW was funded by a Wellcome Prize Travelling Research Fellowship and a Merck/MIT collaborative fellowship. Work in our lab is supported by a grant from the NIH to WGQ.

Visit the Annual Reviews home page at www.AnnualReviews.org

LITERATURE CITED

Aceves-Pina EO, Quinn WG. 1979. Learning in normal and mutant *Drosophila* larvae. *Science* 206:93–96

Aitken A, Howell S, Jones D, Madrazo J, Martin H, et al. 1995. Post-translationally modified 14-3-3 isoforms and inhibition of protein kinase C. *Mol. Cell. Biochem.* 149-150:41–49

Armstrong JD, Jones C, Kaiser K. 1995. *Flytrap*. http://brainbox.gla.ac.uk/flytrap

Bainton RJ, Tsai LT, Singh CM, Moore MS, Neckameyer WS, et al. 2000. Dopamine modulates acute responses to cocaine, nicotine and ethanol in *Drosophila*. *Curr. Biol.* 10:187–94

Barth M, Heisenberg M. 1997. Vision affects mushroom bodies and central complex in *Drosophila melanogaster*. *Learn. Mem.* 4:219–29

Bellen HJ, Gregory BK, Olsson CL, Kiger JA Jr. 1987. Two *Drosophila* learning mutants, *dunce* and *rutabaga*, provide evidence of a maternal role for cAMP on embryogenesis. *Dev. Biol.* 121:432–44

Berke JD, Hyman SE. 2000. Addiction, dopamine, and the molecular mechanisms of memory. *Neuron* 25:515–32

Beumer KJ, Rohrbough J, Prokop A, Broadie K. 1999. A role for PS integrins in mor-phological growth and synaptic function at the postembryonic neuromuscular junction of *Drosophila*. *Development* 126:5833–46

Bolwig GM, Del Vecchio M, Hannon G, Tully T. 1995. Molecular cloning of *linotte* in *Drosophila*: a novel gene that functions in adults during associative learning. *Neuron* 15:829–42

Booker R, Quinn WG. 1981. Conditioning of leg position in normal and mutant *Drosophila*. *Proc. Natl. Acad. Sci. USA* 78: 3940–44

Boynton S, Tully T. 1992. *latheo*, a new gene involved in associative learning and memory in *Drosophila melanogaster*, identified from P element mutagenesis. *Genetics* 131:655–72

Brand AH, Perrimon N. 1993. Targeted gene expression as a means of altering cell fates and generating dominant phenotypes. *Development* 118:401–15

Broadie K, Rushton E, Skoulakis EM, Davis RL. 1997. Leonardo, a *Drosophila* 14-3-3 protein involved in learning, regulates presynaptic function. *Neuron* 19:391–402

Budnik V, Zhong Y, Wu CF. 1990. Morphological plasticity of motor axons in *Drosophila* mutants with altered excitability. *J. Neurosci.* 10:3754–68

Byers D, Davis RL, Kiger JA. 1981. Defect in cyclic AMP phosphodiesterase due to the *dunce* mutation of learning in *Drosophila melanogaster*. *Nature* 289:79–81

Carew TJ. 2000. Persistent activation of cAMP-dependent protein kinase and the induction of long-term memory. *Neuron* 27:7–8

Chen CN, Denome S, Davis RL. 1986. Molecular analysis of cDNA clones and the corresponding genomic coding sequences of the *Drosophila dunce+* gene, the structural gene for cAMP phosphodiesterase. *Proc. Natl. Acad. Sci. USA* 83:9313–17

Chen CN, Malone T, Beckendorf SK, Davis RL. 1987. At least two genes reside within a large intron of the *dunce* gene of *Drosophila*. *Nature* 329:721–24

Choi KW, Smith RF, Buratowski RM, Quinn WG. 1991. Deficient protein kinase C activity in *turnip*, a *Drosophila* learning mutant. *J. Biol. Chem.* 266:15999–6006

Cohan FM, Hoffman AA. 1986. Genetic divergence under uniform selection. II. Different responses to selection for knockdown resistance to ethanol among *Drosophila melanogaster* populations and their replicate lines. *Genetics* 114:145–63

Connolly JB, Roberts IJ, Armstrong JD, Kaiser K, Forte M, et al. 1996. Associative learning disrupted by impaired Gs signaling in *Drosophila* mushroom bodies. *Science* 274:2104–7

Cooley L, Kelley R, Spradling A. 1988. Insertional mutagenesis of the *Drosophila* genome with single P-elements. *Science* 239:1121–28

Crittenden JR, Skoulakis EM, Han KA, Kalderon D, Davis RL. 1998. Tripartite mushroom body architecture revealed by antigenic markers. *Learn. Mem.* 5:38–51

Cui J, Cox DH, Aldrich RW. 1997. Intrinsic voltage dependence and Ca^{2+} regulation of mslo large conductance Ca-activated K^+ channels. *J. Gen. Physiol.* 109:647–73

Cunningham CL, Crabbe JC, Rigter H. 1983. Pavlovian conditioning of drug-induced changes in body temperature. *Pharmacol Ther.* 23:365–91

Davis RL. 1996. Physiology and biochemistry of *Drosophila* learning mutants. *Physiol. Rev.* 76:299–317

Davis RL, Davidson N. 1984. Isolation of the *Drosophila melanogaster dunce* chromosomal region and recombinational mapping of *dunce* sequences with restriction site polymorphisms as genetic markers. *Mol. Cell. Biol.* 4:358–67

Davis RL, Davidson N. 1986. The memory gene *dunce+* encodes a remarkable set of RNAs with internal heterogeneity. *Mol. Cell. Biol.* 6:1464–70

Davis RL, Kiger JA Jr. 1981. *dunce* mutants of *Drosophila melanogaster*: mutants defective in the cyclic AMP phosphodiesterase enzyme system. *J. Cell Biol.* 90:101–7

Davis GW, Schuster CM, Goodman CS. 1996. Genetic dissection of structural and functional components of synaptic plasticity. III. CREB is necessary for presynaptic functional plasticity. *Neuron* 17:669–79

de Belle JS, Heisenberg M. 1994. Associative odor learning in *Drosophila* is abolished by chemical ablation of mushroom bodies. *Science* 263:692–95

DeZazzo J, Sandstrom D, de Belle S, Velinzon K, Smith P, et al. 2000. *nalyot*, a mutation of the *Drosophila* myb-related Adf1 transcription factor, disrupts synapse formation and olfactory memory. *Neuron* 27:145–58

DeZazzo J, Xia S, Christensen J, Velinzon K, Tully T. 1999. Developmental expression of an *amn(+)* transgene rescues the mutant memory defect of *amnesiac* adults. *J. Neurosci.* 19:8740–46

DiChiara TJ, Reinhart PH. 1995. Distinct effects of Ca^{2+} and voltage on the activation and deactivation of cloned Ca^{2+}-activated K^+ channels. *J. Physiol.* 489:403–18

Drain P, Folkers E, Quinn WG. 1991. cAMP-dependent protein kinase and the disruption of learning in transgenic flies. *Neuron* 6:71–82

Dubnau J, Tully T. 1998. Gene discovery in

Drosophila: new insights for learning and memory. *Annu. Rev. Neurosci.* 21:407–44

Dudai Y, Buxbaum J, Corfas G, Ofarim M. 1987. Formamidines interact with the *Drosophila* octopamine receptors, alter the flies' behavior and reduce learning ability. *J. Comp. Physiol.* 15:739–46

Dudai Y, Jan YN, Byers D, Quinn WG, Benzer S. 1976. *dunce*, a mutant of *Drosophila* deficient in learning. *Proc. Natl. Acad. Sci. USA* 73:1684–88

Dura JM, Preat T, Tully T. 1993. Identification of *linotte*, a new gene affecting learning and memory in *Drosophila melanogaster*. *J. Neurogenet.* 9:1–14

Dura JM, Taillebourg E, Preat T. 1995. The *Drosophila* learning and memory gene *linotte* encodes a putative receptor tyrosine kinase homologous to the human RYK gene product. *FEBS Lett.* 370:250–54

Faber T, Joerges J, Menzel R. 1999. Associative learning modifies neural representations of odors in the insect brain. *Nat. Neurosci.* 2:74–78

Fadda F, Rossetti ZL. 1998. Chronic ethanol consumption: from neuroadaptation to neurodegeneration. *Prog. Neurobiol.* 56:385–31

Fantl WJ, Muslin AJ, Kikuchi A, Martin JA, MacNicol AM. 1994. Activation of Raf-1 by 14-3-3 proteins. *Nature* 371:612–14

Feany MB, Quinn WG. 1995. A neuropeptide gene defined by the *Drosophila* memory mutant *amnesiac*. *Science* 68:869–73

Folkers E. 1982. Visual learning and memory of *Drosophila melanogaster* wild-type C-S and the mutants *dunce*[1], *amnesiac*, *turnip* and *rutabaga*. *J. Insect Physiol.* 28:535–39

Folkers E, Drain P, Quinn WG. 1993. *radish*, a *Drosophila* mutant deficient in consolidated memory. *Proc. Natl. Acad. Sci. USA* 90:8123–27

Freed E, Symons M, Macdonald SG, McCormick F, Ruggieri R. 1994. Binding of 14-3-3 proteins to the protein kinase Raf and effects on its activation. *Science* 265:1713–16

Gailey DA, Jackson FR, Siegel RW. 1984.

Conditioning mutations in *Drosophila melanogaster* affect an experience-dependent behavioral modification in courting males. *Genetics* 106:613–23

Gehring WJ, Hiromi Y. 1986. Homeotic genes and the homeobox. *Annu. Rev. Genet.* 20:147–73

Goodwin SF, Del Vecchio M, Velinzon K, Hogel C, Russell SR, et al. 1997. Defective learning in mutants of the *Drosophila* gene for a regulatory subunit of cAMP-dependent protein kinase. *J. Neurosci.* 17:8817–27

Griffith LC, Greenspan RJ. 1993. The diversity of calcium/calmodulin-dependent protein kinase II isoforms in *Drosophila* is generated by alternative splicing of a single gene. *J Neurochem.* 61:1534–37

Griffith LC, Verselis LM, Aitken KM, Kyriacou CP, Danho W, et al. 1993. Inhibition of calcium/calmodulin-dependent protein kinase in *Drosophila* behavioral plasticity. *Neuron* 10:501–9

Grotewiel MS, Beck CD, Wu KH, Zhu XR, Davis RL. 1998. Integrin-mediated short-term memory in *Drosophila*. *Nature* 391:455–60

Guo A, Li L, Xia SZ, Feng CH, Wolf R, Heisenberg M. 1996. Conditioned visual flight orientation in *Drosophila*: dependence on age, practice, and diet. *Learn. Mem.* 3:49–59

Guo HF, Tong J, Hannan F, Luo L, Zhong Y. 2000. A neurofibromatosis-1-regulated pathway is required for learning in *Drosophila*. *Nature* 403:895–98

Guo HF, The I, Hannan F, Bernards A, Zhong Y. 1997. Requirement of *Drosophila* NF1 for activation of adenylyl cyclase by PACAP38-like neuropeptides. *Science* 276:795–98

GuptaRoy B, Marwaha N, Pla M, Wang Z, Nelson HB, et al. 2000. Alternative splicing of *Drosophila* calcium/calmodulin-dependent protein kinase II regulates substrate specificity and activation. *Brain Res. Mol. Brain Res.* 8:26–34

Hammer M, Menzel R. 1998. Multiple sites of associative odor learning as revealed by local brain microinjections of octopamine in honeybees. *Learn. Mem.* 5:146–56

Han PL, Levin LR, Reed RR, Davis RL. 1992. Preferential expression of the *Drosophila rutabaga* gene in mushroom bodies, neural centers for learning in insects. *Neuron* 9:619–27

Heisenberg M. 1998. What do the mushroom bodies do for the insect brain? An introduction. *Learn. Mem.* 5:1–10

Heisenberg M, Borst A, Wagner S, Byers D. 1985. *Drosophila* mushroom body mutants are deficient in olfactory learning. *J. Neurogenet.* 2:1–30

Hynes RO. 1992. Integrins: versatility, modulation, and signalling in cell adhesion. *Cell* 69:11–25

Ichimura T, Uchiyama J, Kunihiro O, Ito M, Horigome T, et al. 1995. Identification of the site of interaction of the 14-3-3 protein with phosphorylated tryptophan hydroxylase. *J. Biol. Chem.* 270:28515–18

Irie K, Gotoh Y, Yashar BM, Errede B, Nishida E, et al. 1994. Stimulatory effects of yeast and mammalian 14-3-3 proteins on the Raf protein kinase. *Science* 265:1716–19

Ito K, Awano W, Suzuki K, Hiromi Y, Yamamoto D. 1997. The *Drosophila* mushroom body is a quadruple structure of clonal units each of which contains a virtually identical set of neurones and glial cells. *Development* 124:761–771

Ito K, Suzuki K, Estes P, Ramaswami M, Yamamoto D, et al. 1998. The organization of extrinsic neurons and their implications in the functional roles of the mushroom bodies in *Drosophila melanogaster* Meigen. *Learn. Mem.* 5:52–77

Jin P, Griffith LC, Murphey RK. 1998. Presynaptic calcium/calmodulin-dependent protein kinase II regulates habituation of a simple reflex in adult *Drosophila*. *J. Neurosci.* 18:8955–64

Koh YH, Popova E, Thomas U, Griffith LC, Budnik V. 1999. Regulation of DLG localization at synapses by CaMKII-dependent phosphorylation. *Cell* 98:353–63

Lee T, Luo L. 1999. Mosaic analysis with a repressible neurotechnique cell marker for studies of gene function in neuronal morphogenesis. *Neuron* 22:451–61

Lee D, O'Dowd DK. 2000. cAMP-dependent plasticity at excitatory cholinergic synapses in Drosophila neurons: alterations in the memory mutant *dunce*. *J. Neurosci.* 20:2104–11

Levin LR, Han PL, Hwang PM, Feinstein PG, Davis RL, et al. 1992. The *Drosophila* learning and memory gene *rutabaga* encodes a Ca^{2+}/calmodulin-responsive adenylyl cyclase. *Cell* 68:479–89

Li H, Chaney S, Roberts IJ, Forte M, Hirsh J. 2000. Ectopic G-protein expression in dopamine and serotonin neurons blocks cocaine sensitization in *Drosophila melanogaster*. *Curr. Biol.* 10:211–14

Li S, Janosch P, Tanji M, Rosenfeld GC, Waymire JC, et al. 1995. Regulation of Raf-1 kinase activity by the 14-3-3 family of proteins. *EMBO J.* 14:685–96

Li W, Tully T, Kalderon D. 1996. Effects of a conditional *Drosophila* PKA mutant on learning and memory. *Learn. Mem.* 2:320–33

Liu L, Wolf R, Ernst R, Heisenberg M. 1999. Context generalization in *Drosophila* visual learning requires the mushroom bodies. *Nature* 400:753–56

Livingstone MS, Sziber PP, Quinn WG. 1984. Loss of calcium/calmodulin responsiveness in adenylate cyclase of *rutabaga*, a *Drosophila* learning mutant. *Cell* 37:205–15

Livingstone MS, Tempel BL. 1983. Genetic dissection of monoamine neurotransmitter synthesis in *Drosophila*. *Nature* 303:67–70

Mayford M, Kandel ER. 1999. Genetic approaches to memory storage. *Trends Genet.* 15:463–70

McClung C, Hirsh J. 1998. Stereotypic behavioral responses to free-base cocaine and the development of behavioral sensitization in *Drosophila*. *Curr. Biol.* 8:109–12

McClung C, Hirsh J. 1999. The trace amine tyramine is essential for sensitization to cocaine in *Drosophila. Curr. Biol.* 9:853–60

Menne D, Spatz HC. 1977. Colour vision in *Drosophila melanogaster. J. Comp. Physiol.* 114:301–12

Menzel R. 1999. Memory dynamics in the honeybee. *J. Comp. Physiol.* 185:323–40

Mihalek RM, Jones CJ, Tully T. 1997. The *Drosophila* mutation *turnip* has pleiotropic behavioral effects and does not specifically affect learning. *Learn. Mem.* 3:425–44

Mizunami M, Okada R, Li Y, Strausfeld NJ. 1998. Mushroom bodies of the cockroach: activity and identities of neurons recorded in freely moving animals. *J. Comp. Neurol.* 402:501–19

Mohler JD. 1977. Developmental genetics of the *Drosophila* egg. I. Identification of 59 sex-linked cistrons with maternal effects on embryonic development. *Genetics* 85:259–72

Moore MS, DeZazzo J, Luk AY, Tully T, Singh CM, et al. 1998. Ethanol intoxication in *Drosophila*: genetic and pharmacological evidence for regulation by the cAMP signaling pathway. *Cell* 93:997–1007

Moreau-Fauvarque C, Taillebourg E, Boissoneau E, Mesnard J, Dura JM. 1998. The receptor tyrosine kinase gene *linotte* is required for neuronal pathway selection in the *Drosophila* mushroom bodies. *Mech. Dev.* 78:47–61

Muller U. 2000. Prolonged activation of cAMP-dependent protein kinase during conditioning induces long-term memory in honeybees. *Neuron* 27:159–68

Nighorn A, Healy MJ, Davis RL. 1991. The cyclic AMP phosphodiesterase encoded by the *Drosophila dunce* gene is concentrated in the mushroom body neuropil. *Neuron* 6:455–67

Park SK, Sedore SA, Cronmiller C, Hirsh J. 2000. Type II cAMP-dependent protein kinase-deficient *Drosophila* are viable but show developmental, circadian, and drug response phenotypes. *J. Biol. Chem.* 275:20588–96

Pinto S, Quintana DG, Smith P, Mihalek RM,

Hou ZH, et al. 1999. *latheo* encodes a subunit of the origin recognition complex and disrupts neuronal proliferation and adult olfactory memory when mutant. *Neuron* 23:45–54

Qiu YH, Chen CN, Malone T, Richter L, Beckendorf SK, et al. 1991. Characterization of the memory gene *dunce* of *Drosophila melanogaster. J. Mol. Biol.* 222:553–65

Qiu Y, Davis RL. 1993. Genetic dissection of the learning/memory gene *dunce* of *Drosophila melanogaster. Genes Dev.* 7:1447–58

Quinn WG, Harris WA, Benzer S. 1974. Conditioned behavior in *Drosophila melanogaster. Proc. Natl. Acad. Sci. USA.* 71:708–12

Quinn WG, Sziber PP, Booker R. 1979. The *Drosophila* memory mutant *amnesiac. Nature* 277:212–14

Rohrbough J, Grotewiel MS, Davis RL, Broadie K. 2000. Integrin-mediated regulation of synaptic morphology, transmission, and plasticity. *J. Neurosci.* 20:6868–78

Rohrbough J, Pinto S, Mihalek RM, Tully T, Broadie K. 1999. *latheo*, a *Drosophila* gene involved in learning, regulates functional synaptic plasticity. *Neuron* 23:55–70

Rong YS, Golic KG. 2000. Gene targeting by homologous recombination in *Drosophila. Science* 288:2013–18

Rubin GM, Yandell MD, Wortman JR, Gabor Miklos GL, Nelson CR, et al. 2000. Comparative genomics of the eukaryotes. *Science* 287:2204–15

Scholz H, Ramond J, Singh CM, Heberlein U. 2000. Functional ethanol tolerance in *Drosophila. Neuron* 28:261–71

Schuster CM, Davis GW, Fetter RD, Goodman CS. 1996. Genetic dissection of structural and functional components of synaptic plasticity. I. Fasciclin II controls synaptic stabilization and growth. *Neuron* 17:641–54

Shen MH, Harper PS, Upadhyaya M. 1996. Molecular genetics of neurofibromatosis type 1 (NF1). *J. Med. Genet.* 33:2–17

Siegel RW, Hall JC. 1979. Conditioned responses in courtship behavior of normal and mutant *Drosophila. Proc. Natl. Acad. Sci. USA.* 76:3430–34

Simon AF, Boquet I, Synguelakis M, Preat T. 1998. The *Drosophila* putative kinase linotte (derailed) prevents central brain axons from converging on a newly described interhemispheric ring. *Mech. Dev.* 76:45–55

Singh CM, Heberlein U. 2000. Genetic control of acute ethanol-induced behaviors in Drosophila. *Alcohol Clin. Exp. Res.* 24:1127–36

Skoulakis EM, Davis RL. 1996. Olfactory learning deficits in mutants for *leonardo*, a *Drosophila* gene encoding a 14-3-3 protein. *Neuron* 17:931–44

Skoulakis EM, Kalderon D, Davis RL. 1993. Preferential expression in mushroom bodies of the catalytic subunit of protein kinase A and its role in learning and memory. *Neuron* 11:197–208

Spradling AC. 1986. P-element mediated transformation. In *Drosophila: A Practical Approach*, ed. DB Roberts. Oxford: IRL Press. 175 pp.

Spradling AC, Stern D, Beaton A, Rhem EJ, Laverty T, et al. 1999. The Berkeley Drosophila Genome Project gene disruption project: single P-element insertions mutating 25% of vital Drosophila genes. *Genetics* 153:135–77

Stopfer M, Laurent G. 1999. Short-term memory in olfactory network dynamics. *Nature* 402:664–68

Strausfeld NJ. 1976. *Atlas of an Insect Brain.* Berlin: Springer

Strausfeld NJ, Hansen L, Li Y, Gomez RS, Ito K. 1998. Evolution, discovery, and interpretations of arthropod mushroom bodies. *Learn Mem.* 5:11–37

Tempel BL, Bonini N, Dawson DR, Quinn WG. 1983. Reward learning in normal and mutant *Drosophila. Proc. Natl. Acad. Sci. USA* 80:1482–86

Tempel BL, Livingstone MS, Quinn WG. 1984. Mutations in the dopa decarboxylase gene affect learning in *Drosophila. Proc. Natl. Acad. Sci. USA* 81:3577–81

The I, Hannigan GE, Cowley GS, Reginald S, Zhong Y, et al. 1997. Rescue of a *Drosophila* NF1 mutant phenotype by protein kinase A. *Science* 276:791–94

Toba G, Ohsako T, Miyata N, Ohtsuka T, Seong KH, Aigaki T. 1999. The gene search system. A method for efficient detection and rapid molecular identification of genes in *Drosophila melanogaster. Genetics* 151:725–37

Tully T, Preat T, Boynton SC, Del Vecchio M. 1994. Genetic dissection of consolidated memory in *Drosophila. Cell* 79:35–47

Tully T, Quinn WG. 1985. Classical conditioning and retention in normal and mutant *Drosophila melanogaster. J. Comp. Physiol. A* 157:263–77

Vaudry D, Gonzalez BJ, Basille M, Yon L, Fournier A, et al. 2000. Pituitary adenylate cyclase-activating polypeptide and its receptors: from structure to functions. *Pharmacol. Rev.* 52:269–324

Waddell S, Armstrong JD, Kitamoto T, Kaiser K, Quinn WG. 2000. The *amnesiac* gene product is expressed in two neurons in the *Drosophila* brain that are critical for memory. *Cell* 103:805–13

Weber KE. 1988. An apparatus for measurement of resistance to gas-phase reagants. *Drosoph. Inf. Serv.* 67:91–93

Weiner J. 1999. *Time, Love, Memory: A Great Biologist and His Quest for the Origins of Behavior.* New York: Knopf

Wolf R, Wittig T, Liu L, Wustmann G, Eyding D, et al. 1998. *Drosophila* mushroom bodies are dispensable for visual, tactile, and motor learning. *Learn. Mem.* 5:166–78

Wu CF, Renger JJ, Egel JE. 1998. Activity-dependent functional and developmental plasticity of Drosophila neurons. *Adv. Insect Physiol.* 27:385–440

Wustmann G, Heisenberg M. 1997. Behavioral manipulation of retrieval in a spatial memory task for *Drosophila melanogaster. Learn. Mem.* 4:328–36

Wustmann G, Rein K, Wolf R, Heisenberg M.

1996. A new paradigm for operant conditioning of *Drosophila melanogaster*. *J. Comp. Physiol. A* 179:429–36

Xiao B, Smerdon SJ, Jones DH, Dodson GG, Soneji Y, et al. 1995. Structure of a 14-3-3 protein and implications for coordination of multiple signaling pathways. *Nature* 376:188–91

Yang MY, Armstrong JD, Vilinsky I, Strausfeld NJ, Kaiser K. 1995. Subdivision of the *Drosophila* mushroom bodies by enhancer-trap expression patterns. *Neuron* 15:45–54

Yin JC, Del Vecchio M, Zhou H, Tully T. 1995a. CREB as a memory modulator: induced expression of a dCREB2 activator isoform enhances long-term memory in *Drosophila*. *Cell* 81:107–15

Yin JC, Wallach JS, Del Vecchio M, Wilder EL, Zhou H, et al. 1994. Induction of a dominant negative CREB transgene specifically blocks long-term memory in *Drosophila*. *Cell* 7:49–58

Yin JC, Wallach JS, Wilder EL, Klingensmith J, Dang D, et al. 1995b. A *Drosophila* CREB/CREM homolog encodes multiple isoforms, including a cyclic AMP-dependent protein kinase-responsive transcriptional activator and antagonist. *Mol. Cell. Biol.* 15:5123–30

Zars T. 2000. Behavioral functions of the insect mushroom bodies. *Curr. Opin. Neurobiol.* 10:790–95

Zars T, Fischer M, Schulz R, Heisenberg M. 2000. Localization of a short-term memory in *Drosophila*. *Science* 288:672–75

Zhong Y. 1995. Mediation of PACAP-like neuropeptide transmission by coactivation of Ras/Raf and cAMP signal transduction pathways in *Drosophila*. *Nature* 375:588–92

Zhong Y, Budnik V, Wu CF. 1992. Synaptic plasticity in *Drosophila* memory and hyperexcitable mutants: role of cAMP cascade. *J. Neurosci.* 12:644–51

Zhong Y, Pena LA. 1995. A novel synaptic transmission mediated by a PACAP-like neuropeptide in *Drosophila*. *Neuron* 14:527–36

Zhou Y, Schopperle WM, Murrey H, Jaramillo A, Dagan D, et al. 1999. A dynamically regulated 14-3-3, Slob, and Slowpoke potassium channel complex in *Drosophila* presynaptic nerve terminals. *Neuron* 22:809–18

SUBJECT INDEX

1311

CUMULATIVE INDEXES

CONTRIBUTING AUTHORS, VOLUMES 15–24

1331

CHAPTER TITLES, VOLUMES 15–24

1335

Circadian and Other Rhythms

Clinical Neuroscience

Glia, Schwann Cells, and Extracellular Matrix

Motor Systems